Chemistry
of
Coal Utilization

Chemistry of Coal Utilization

Supplementary Volume

H. H. Lowry, Editor

Prepared by the Committee on Chemistry of Coal,
Division of Chemistry and Chemical Technology,
National Academy of Sciences-National Research Council
A. R. Powell, Chairman

JOHN WILEY & SONS, New York • Chichester • Brisbane • Toronto

6 7 8 9 10

Library of Congress Catalog Card Number: 45:5498

Printed in the United States of America

ISBN # 471-55158-9

PREFACE

World-wide acceptance of the first two volumes of *Chemistry of Coal Utilization*, prepared by a Committee of the National Research Council and published by John Wiley and Sons in 1945, was extremely gratifying. It was reprinted in 1947, and a Russian translation was published in 1951.

Advances in coal science and in the technology of coal utilization led to many inquiries regarding the possibility of a revision of this publication. The magnitude and cost of such a revision did not seem to justify consideration of this possibility. In 1956, however, Bituminous Coal Research, Inc. (BCR) initiated discussions of a new volume. Correspondence with Dr. Clem O. Miller, Executive Secretary, Division of Chemistry and Chemical Technology, National Academy of Sciences-National Research Council (NAS-NRC) led to several preliminary meetings in 1957, which were also attended by Dr. F. D. Rossini, Chairman of the Division of Chemistry and Chemical Technology, NAS-NRC.

At a meeting held on June 13, 1957, it was decided to organize a committee on "Chemistry of Coal" within the Division of Chemistry and Chemical Technology, NAS-NRC, to explore the possibility of bringing *Chemistry of Coal Utilization* up to date. Dr. A. R. Powell of Koppers Company was appointed Chairman in October 1957. In December of the same year, BCR authorized $15,000 to finance the editorial expenses involved. The following committee was appointed and held its first meeting in Washington, D.C., in February 1958:

A. H. Brisse, United States Steel Corporation

F. B. Brown, Union Carbide Olefins Company

J. Mitchell, Eastern Gas and Fuel Associates

H. Perry, Bureau of Mines, United States Department of the Interior

E. H. Reichl, Consolidation Coal Company

H. J. Rose, Bituminous Coal Research, Inc.

C. W. Stahl, Bethlehem Steel Corporation

H. H. Storch,* American Cyanamid Corporation

P. L. Walker, Jr., Pennsylvania State University

H. H. Lowry was appointed a nonvoting member to serve as editor-secretary.

Subsequent meetings were held on March 5 and May 13, 1958. At these meetings it was decided that a revision of the first two volumes was impractical, partly because of cost and partly because of unequal technical advances in the subjects covered by the individual chapters of the first two volumes. The alternative of a supplementary volume that would be self-contained was accepted. The organization of the present volume resulted from these discussions and from correspondence on selection of the authorities to be invited to participate in its preparation.

Invitations to the individual authorities selected to prepare the twenty-three chapters were sent out in June 1958. Acceptance carried with it an appointment to the Committee on Chemistry of Coal. The response was almost unanimous and demonstrated a recognition of the need for the present volume. A contract for publication of this volume was entered into in October 1959.

In a collaborative effort such as this, unequal treatment must be anticipated. Authors were requested to submit outlines of their respective chapters by April 1959 to help the editor avoid duplication of effort. It was

* Deceased

requested that manuscripts be submitted by January 1960, and that they include critical reviews of the important literature through 1958. Complete bibliographic coverage was to be considered less important than reference to the more significant publications. Both sides of controversial subjects were to be documented. Each chapter was to be reviewed by competent workers in the field of activity covered, the reviewers being selected by consultation with members of the Committee on Chemistry of Coal. The review system was designed to avoid undue emphasis on particular subjects solely because of immediate personal interests of the authors. The serious consideration given the manuscripts by the following reviewers is gratefully acknowledged:

A. H. Brisse, United States Steel Corporation

F. B. Brown, Union Carbide Olefins Company

R. C. Corey, Bureau of Mines, United States Department of the Interior

W. L. Crentz, Bureau of Mines, United States Department of the Interior

H. G. Davis, Union Carbide Olefins Company

S. Ergun, Bureau of Mines, United States Department of the Interior

H. C. Howard,* Bureau of Mines, United States Department of the Interior

J. Mitchell, Eastern Gas and Fuel Associates

M. B. Neuworth, Consolidation Coal Company

L. L. Newman, Bureau of Mines, United States Department of the Interior

W. H. Ode, Bureau of Mines, United States Department of the Interior

A. A. Orning, Bureau of Mines, United States Department of the Interior

O. W. Rees, Illinois Geological Survey

E. H. Reichl, Consolidation Coal Company

H. J. Rose, Bituminous Coal Research, Inc.

C. C. Russell, Koppers Company

C. H. Sawyer, Eastern Gas and Fuel Associates

R. A. Sherman, Battelle Memorial Institute

H. H. Storch,* American Cyanamid Company

G. Thiessen, Koppers Company

E. W. Volkmann, Koppers Company

J. G. Walters, Bureau of Mines, United States Department of the Interior

I. Wender, Bureau of Mines, United States Department of the Interior

D. E. Wolfson, Bureau of Mines, United States Department of the Interior

* Deceased

The first manuscript was received by the editor in December 1959 and the last in September 1961. Several of the authors of the chapters received first included in their revised copy references to work published subsequent to 1958. It must be acknowledged, however, that in certain fields of research that are currently the most active, the treatment would need revision to reflect present thought.

Although each author is to be regarded as solely responsible for the conclusions drawn in his chapter, the editor must acknowledge responsibility for many changes made primarily for uniformity of style. Incomplete or incorrect references were emended by the editor. Since each author saw only galley proof of his chapter and not the edited manuscript, an indeterminate share of the responsibility for correctness of statement and citation must be acknowledged by the editor. Care was taken not to change any expression of personal opinion by an author, even though the opinion was distinctly contrary to that held by the editor. In addition to reading galley and page proof and checking the illustrations, the editor prepared both the author and subject indexes. The indexes provide principal cross referencing to investigations considered by more than one author.

To all those who have contributed so generously to this cooperative work, the Committee on Chemistry of Coal acknowledges its gratitude.

H. H. Lowry
Editor-Secretary
Committee on Chemistry of Coal
Division of Chemistry and
Chemical Technology
National Academy of Sciences-
National Research Council

Wilmerding, Pennsylvania
October, 1962

CONTENTS

*Deceased

CHAPTER 1

ORIGIN, PETROGRAPHY, AND CLASSIFICATION OF COAL

Bryan C. Parks

Bureau of Mines, U.S. Department of Interior, Pittsburgh, Pa.

ORIGIN AND FORMATION OF COAL

That coal is formed from plant substances preserved from complete decay in a favorable environment and later altered by various chemical and physical agencies is the theory now almost universally accepted. The concept of coal formation thus stated implies that there are at least two broad stages in the process of coal formation as originally proposed by White,[1] the biochemical or peat stage and the dynamochemical or metamorphic stage. These stages represent the period of accumulation and preservation of plant remains as peat deposits, and the period of conversion of these organic remains or peat deposits to coal. Although coal workers of varied scientific background are in general agreement with the accepted theory of coal origin, there is no universal agreement on what chemical and physical changes take place in the two stages of coal formation. This is quite apparent from the voluminous literature available on the subject of coal formation and the varied lines of investigation pursued in the many coal research laboratories directed toward solving the problems of the nature and origin of coal.

The accumulation of plant remains under favorable environmental conditions as peat deposits is generally considered a geological process. Thus the study of the basic accumulation problems pertaining to the origin of peat has been made by coal geologists or those scientists of allied training and interests referred to as coal paleobotanists and coal petrographers. Since the actual transformation of organic remains of peat to coal involves chemical and physical changes, this phase of coal formation has increasingly stimulated the interest of scientists trained in the fields of chemistry and physics. We do not mean to imply, however, that mutually exclusive fields of work have been established. Chemical processes take place in the peat stage, and the chemical and physical alterations occurring during the dynamochemical period of coalification are brought about at least in part by geological agencies. Geologists have ventured into the subject of coal metamorphism with suggestions of causes and effects, and chemists have investigated the processes of change that begin in the peat stage of coal formation.

The problems of peat accumulation, as the parent material of coal deposits, received the attention of geologists for all of the nineteenth century and perhaps even longer before the controversial subject was resolved in a manner acceptable to most geologists. They concluded that peat accumulated for the most part in an environment exceptionally favorable for plant growth and for preservation from complete destruction of the plant remains by biochemical agencies. They recognized that some peat originated from plant remains transported to the site of deposition by wind and water agencies, but such drift plant material was not considered to be of prime importance in the formation of peat deposits as a whole. The various controversial theories of peat accumulation have been discussed in the first volume of *Chemistry of Coal Utilization*[2] and need not be repeated here.

Formation of Different Types of Peat

According to Thiessen,[3] the origin of different types of coal depended on the environment. He considered factors such as (1) the

[1] White, D., *Econ. Geol.*, **3,** 292–318 (1908).

[2] Hendricks, T. A., *Chemistry of Coal Utilization*, John Wiley and Sons, New York, Vol. 1, pp. 1–24, 1945

[3] Thiessen, R., *U.S. Bur. Mines, Inform. Circ.* No. 7397 (1947), 53 pp.

nature of the topographic depression and the consequent size and form of the peat swamp it produced, (2) the nature of the underlying rock and soil, (3) the prevailing climate including length of seasons, precipitation, and wind, and (4) the physiography of the surrounding area including its latitude and altitude, and inlets and outlets to the swamp. All these physical factors and perhaps others influenced the types of plant associations that thrived in the peat swamp environment and furnished the source material for the accumulating peat. Another factor in the peat swamp that had important bearing on the type of peat accumulated and preserved was the nature of the swamp water under both surface and subsurface conditions—that is, its degree of acidity, oxygen content, and mineral matter both in solution and in suspension. The factor of acidity, sometimes referred to as toxicity or stagnancy of the swamp water, determined the aerobic or anaerobic conditions in the swamp water, the nature of its bacteria population, and consequently the rate and extent of biochemical decay. The rate of decay is also greatly influenced by the inherent resistivity to bacterial attack of the various plant parts, tissues, organs, and secretory products.

Thiessen also recognized that coal deposits represent a long interval of peat accumulation during which the various environmental conditions of the swamp changed as well as the plant associations. These changes within the peat swamp resulted in marked differences in the composition of the plant material from bottom to top of the deposit, and these profile changes in composition and texture of the peat material are reflected in coal deposits by the vertical variation in petrographic composition observed in normal banded coal.

On the basis of paleobotanical and petrographic study of the brown coal deposits of the Cologne basin—which revealed the kind of plant material present in the coal and the relative abundance of the various plant substances—and of a comparative study of environmental conditions and plant associations that prevail in present-day coastal plains peat swamps, Teichmüller[4] attempted graphic reconstruction of the environments and plant associations that probably existed in the Miocene peat swamps which produced the Cologne brown coals. This study of Cologne brown coals was supplemented and the conclusions were confirmed by an investigation[5] of the geological history of the lower Rhine basin in middle and late Tertiary times. The Miocene peat swamp reconstructions are novel in the study of the origin of coal[4] because they show that different types of peat, accumulated under a variety of environmental conditions, do not necessarily succeed each other in time sequence but may exist contemporaneously and in proximity to each other.

From these petrographic and paleobotanic studies of the Cologne brown coals, Teichmüller reconstructed and described four different types of swamp environments and their plant associations which were considered to be the main sources of the Cologne brown coal deposits.

1. Along the shore of a shallow and probably broad swamp lake accumulated coarse and finely divided plant remains admixed with some inorganic sediments. Much of the accumulating plant material was transported into the swamp lake by wind and water agencies; biochemical activity was probably at a maximum in the open water, and for the most part only the highly resistant plant substances survived biochemical destruction to form attrital type peat.

2. Inland, and adjacent to the swamp lake, was a rather broad low-lying swamp area taken over principally by reed grasses and floating water plants, with some herbaceous vegetation. Attrital peat also accumulated in this favorable environment.

3. Further inland, adjacent to the reed swamp, lay a rather broad wooded swamp frequently flooded and inhabited by water-thriving trees of the cypress (*taxodium*) and tupelo, gum (*nyssa*) varieties. This wet, wooded swamp was bordered by a less frequently flooded and drier swamp inhabited by a bush and brushy vegetation of the bog myrtle, bayberry, and titi varieties.

4. Adjacent to the bush swamp and on slightly higher ground existed a heavily wooded swamp inhabited by trees of the

[4] Teichmüller, M., *Fortschr. Geol. Rheinld u. Westf.*, **2**, 599–612 (1958).

[5] Teichmüller, R., *Fortschr. Geol. Rheinld u. Westf.*, **2**, 721–50 (1958).

redwood (*sequoia*) type, and in this environment a woody-type peat was produced.

Thus two broad concepts are available to account for the origin of different coal types in the peat stage.

1. The idea that varied kinds of peat deposits accumulated under greatly different environmental conditions but essentially in the same locality, the changing environments succeeding each other in time sequence during the life history of the swamp and peat deposition.
2. The concept that varied types of peat accumulated in greatly differing environments which existed contemporaneously and in spatial relationship to each other.

These two concepts are essentially similar in depicting the physical nature, botanical relationships, and types of peat produced in the varied swamp environments. The principal difference is in the time of occurrence of the variable swamp environments and peat accumulations. The first concept may offer the best explanation for the origin of the different coal types that are commonly found in vertical succession in typical banded or layered coal deposits, whereas the second concept better explains the lateral changes in petrographic types of a coal deposit.

On the basis of information from a critical analytical study of North American and European coals and peats considered against a broad background of geological information, Mott[6] deduced certain generalizations concerning the origin and composition of coals. He stated that probably the most important coal deposits of the world are of Carboniferous age laid down in coal basins subjected to frequent and perhaps rhythmic subsidence. These are mainly *in situ* deposits (autochthonous coals) which characteristically are relatively thin, are extensive laterally, and normally have low content of ash-forming inorganic constituents. In certain minor Carboniferous coal fields, coal deposits are exceptionally thick, although not extensive laterally, and are rather impure. These coals may be considered to have developed in large part by the accumulation of vegetable matter which drifted into an enclosed basin (allochthonous coals). The coals originating in geological periods later than the Carboniferous are considerably thicker on the average than Carboniferous deposits.

Origin of Durain and Similar Attrital Coals

According to Mott,[6] durain in thin and thick bands and layers is a typical feature of most British coals of comparatively low rank, under 87% C content (Parr basis), but occurs less commonly in the coals of higher rank. Examined under the microscope, durain reveals a mixed composition of finely divided plant debris, in which the bands of vitrain characteristic of bright coal are extremely rare but which is much richer in spores and plant substances of similar chemical make-up. The characteristic attrital nature of durain and the abundance of spore material have led to the conclusion that durain, as well as the cannel and boghead coals, accumulated in excessively wet swamp environments, which were perhaps frequently flooded, or even in open pools of water within the peat swamps. However, a drift origin of durain is not as likely as it is for cannel and boghead coals. The microstructure of durain, revealed in the parallel orientation of the various plant fragments, shows that it is to be regarded as a banded coal, although it is sometimes referred to as microbanded since the banding is not usually discernible without magnification. Usually, durain and bright coal are found together in deposits. On the other hand, cannel and boghead coals are nonbanded coals in which orientation of plant fragments is characteristically random, deposits usually being lenticular in outline, and local in extent at the top of a coal bed.

Examination of durain in thin section under the microscope shows varying amounts of opaque matter—the opaque attritus of Thiessen's[7] splint coal. Splint coals have a concentration of more than 30% opaque attritus.[8] Opaque attritus is called micrinite by others, and is similar to fusain in its degree of opacity to transmitted light but does not have the same microstructure as fusain. Mott[6] concluded that durain was probably formed as an

[6] Mott, R. A., *Fuel*, **21**, 129–35 (1942); **22**, 20–6 (1943).

[7] Thiessen, R., and Sprunk, G. C., *Fuel*, **15**, 304–15 (1936).

[8] Parks, B. C., and O'Donnell, H. J., *Trans. Am. Inst. Mining Met. Engrs.*, **177**, 535–55 (1948).

in situ deposit under dry conditions which promoted more extensive decay of the vegetable debris and thorough maceration to fine particle size.

Hickling[9] has suggested withholding final judgment on the origin of durain and cannel coals until more extended and systematic studies have been made. The question is whether durains and cannels should be regarded as different portions of a widely varying but substantially continuous group of attrital coals, or whether they are essentially different materials of completely different history, which do not blend together through intermediary types. Examination of these materials shows that there is every possible gradation between them, and attempts to define a typical durain and typical cannel coal have not met with general agreement.

The minutely fragmented condition of these attrital coals indicates their subjection to conditions especially favorable to decomposition and maceration of plant remains.[10] Hence they are characterized by prevalence of more resistant plant substances such as spores, cuticles, resin, and waxes, with some opaque attritus (micrinite) and fusinite, and not infrequently minutely divided vitrinite. Micrinite in considerable proportion is characteristic of many durains but not universally so, despite claims that micrinite is a distinctive mark of durain, nor is this opaque substance entirely absent from cannel and boghead coals.

Cannel coals, usually considered as being distinctly spore coals, show on occasion infusion of fossil oil algae, which are usually considered characteristic of boghead coals. Oil algae remains have also been found[11] in a type of attrital coal that is transitional toward normal banded coals.

Schopf[12] has agreed that there is a relationship between durain and cannel coals and would even extend the relationship to include the bogheads, torbanites, and oil shales. He pointed out that the distinction between durain, cannels, bogheads, and oil shales is very obscurely drawn.

[9] Hickling, G., *Fuel*, **21**, 135–6 (1942).

[10] Sprunk, G. C., Ode, W. H., Selvig, W. A., and O'Donnell, H. J., *U.S. Bur. Mines, Tech. Paper* No. 615 (1940), 59 pp.

[11] Kosanke, R. M., *Am. J. Sci.*, **249**, 444–50 (1951).

[12] Schopf, J. M., *Econ. Geol.*, **44**, 68–71 (1949).

Origin of Fusain

Marshall[13] made a microscopic and chemical examination of fusain from the Waterloo seam of England; the microscopical study revealed the presence of two dominant types, designated long-fiber and granular fusinite. The excellent preservation of physical structure of the long-fiber fusinites suggested that they had been preserved as the result of selective chemical or biochemical activity during the peat stage of the coal seam's history. These fusinites also often appeared to have close associations with the vitrifusinite (semi-translucent fusinite of vitrinitic appearance) and normal vitrinite; transitions between these substances were commonplace and the detailed structure was generally well preserved. The second main type of black, lusterless, charcoal-like fusain which disintegrated readily to granular, angular powder usually revealed only a coarse cellular structure which was often mineralized. Normally it did not show well-preserved or fine structure, and from the visual evidence could have been formed through the agency of fire. The fusinite samples exhibited a wide range of chemical variations exceeding that of vitrain samples. Of particular interest was the fact that both main types of fusinite fell on a rank curve based on analytical data, that paralleled normal rank variation at a generally lower hydrogen content than that of vitrinites.

Terres and associates[14] attacked the problem of formation of fusain on the basis of its specific heat. They worked from the premise that the specific heat of a carbonaceous material is higher the lower the temperature was in its prior history. The average specific heat of a number of fusains indicated they were formed by a thermal process presumably in forest fires, the average temperature being calculated to be between 700° and 1000°C, although other fusains were formed by other than thermal processes. Petrographic inspection was invoked to confirm these conclusions.

Microscopic study of fusain from a wide variety of coals shows some features that cannot be accepted logically as supporting evidence of forest fire origin. As an example,

[13] Marshall, C. E., *Fuel*, **33**, 134–44 (1954).

[14] Terres, E., Dähne H., Nandi, B., Scheidel, C., and Trappe, K., *Brennstoff-Chem.*, **37**, 269–77 (1956).

fusinized tissues generally reveal excellent preservation of cellular forms, sometimes in the finest microscopic details. The conversion of such delicately organized tissues to fusain at the high temperatures involved in forest fires seems improbable. Frequently a fusinized plant substance is observed that is recognized to be secretory in nature some of which originally was probably waxy-resinous material of a relatively low melting temperature that would be expected to volatilize at forest fire temperatures. Finally, the cell walls of woody tissues so frequently fusinized appear normal in morphological form with little or none of the shrinkage that would be expected if the original tissues had been subjected to devolatilization and other reactions as the result of high forest fire temperatures. On the whole, this fidelity of morphological preservation is rather striking in comparison with the great change in chemical composition represented by the conversion of various plant substances to fusain.

These arguments do not obviate, of course, the possibility of fusain being of dual origin.[13] The commonplace occurrence of fusain particles in sedimentary deposits such as shales and sandstones associated with coal beds also suggests that such fusain particles were gathered along with inorganic mineral detritus by running water from an area under erosion and deposited in sedimentary beds. Perhaps detrital fusinite was formed by forest fires, not necessarily in a peat swamp, and was disintegrated and dispersed as detrital particles with the mineral sediments by the action of running water. The deposition of detrital fusain in coal beds seems entirely possible but has not been proved.

Schopf[15] has pointed out certain significant facts pertaining to the origin and formation of fusain as shown in Fig. 1. This illustration shows that the fusinization of plant remains (phyteral alteration) has an early inception and progresses rapidly mostly in the peat and early lignitic stages, after which the metamorphic changes in fusain are almost imperceptible. Whatever theory of the origin of fusain is favored, there is general acceptance that plant substances are altered to form fusain early in the coalification sequence before heavy overburden accumulates. Thus fusain probably should be regarded as diagenetic in origin even though the nature of the process is controversial. In contrast to fusinization, vitrinization is shown as a progressive alteration through all stages of rank development of coal. In greatest contrast to fusinization, the waxy-resinous substances show little alteration in composition and morphological form in the peat and brown coal stages, slow change in later lignitic and high-volatile bituminous stages, and then a rapid change in the low-volatile bituminous and early anthracitic stages.

Another important fact about the formation of fusain is that the same types of plant substances that are capable of coalification by other processes are also susceptible to fusinization, and there seems to be no particular relationship between the botanical nature of the plant remains and susceptibility to fusinization.

Coalification

In discussing certain aspects of the chemical changes occurring in the peat and the succeeding stages of coalification, Mott[6] concluded that forest peats not rich in resinous and waxy substances but essentially woody such as those of carboniferous age tended to form completely ulmified material approximating the molecular formula $C_{65}H_4O_{30}$. This alteration occurred under acid conditions without any covering of the peat, by loss of H_2O, some CH_4, and a little CO_2. On the basis of $C + H + N + O = 100$, the limit of this stage was completely ulmified material of approximately the foregoing formula. Advent of a covering with probably a base exchange with sodium chloride to give alkaline conditions resulted in anaerobic bacterial activity and decomposition, causing carbon dioxide to evolve with deoxygenation of ulmins and an increase in hydrogen and carbon contents. This was the second or lignite stage of coalification, the end product usually being a low-rank bituminous coal approximating $C_{79}H_{5.5}O_{14.1}$, but the extent of rank development may have depended on the point at which alkaline roof conditions developed.

The third and fourth stages of coalification were conversion of low-rank bituminous coal to a medium-rank bituminous coal and then to high-rank bituminous coal. During the third stage, the hydrogen content remained about

[15] Schopf, J. M., *Econ. Geol.*, **43**, 207–25 (1948).

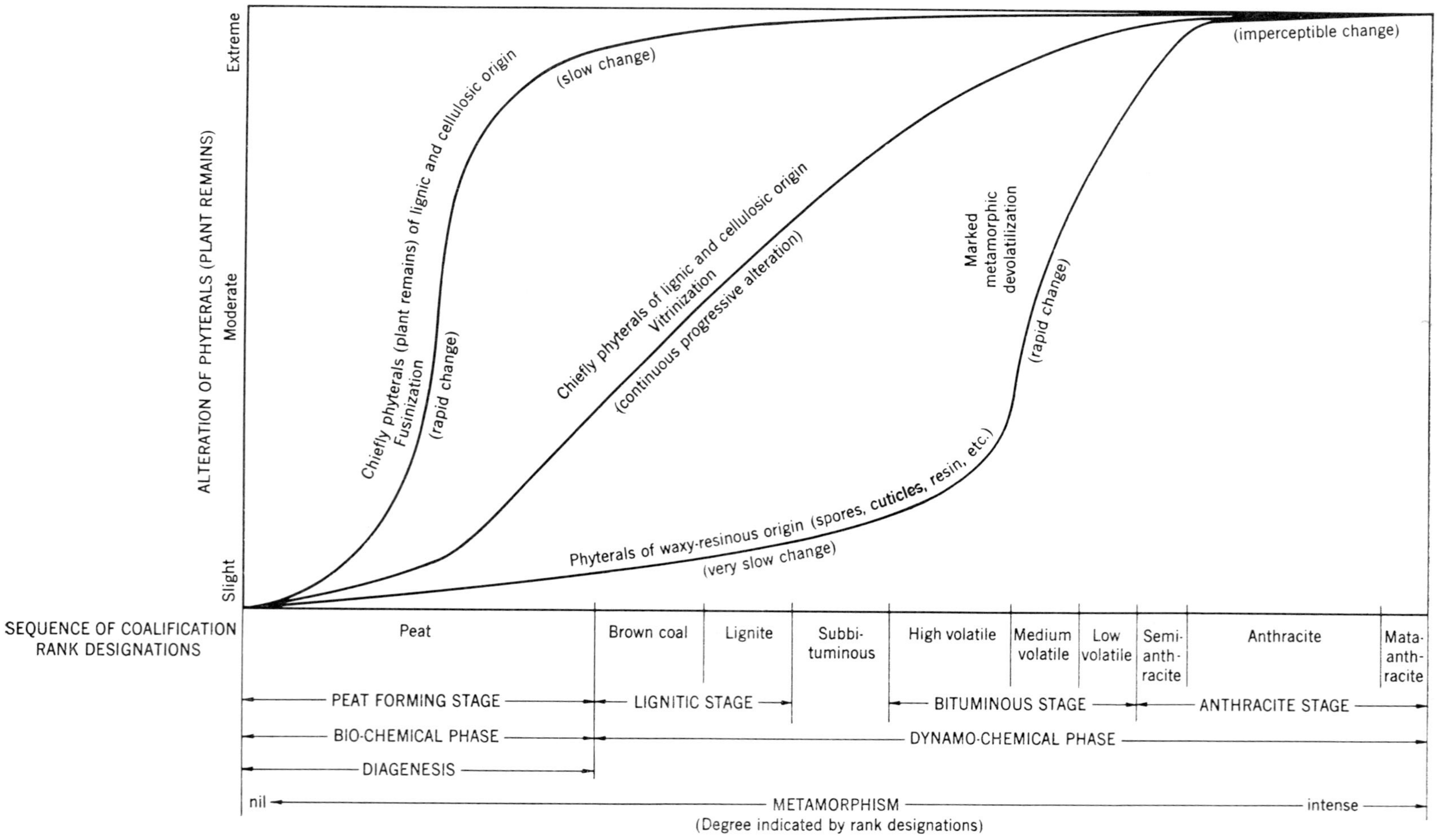

Fig. 1. Metamorphic alteration of phyteral constituents.[15]

constant and oxygen fell until only one atom of oxygen remained in the coal molecule. In the fourth stage, hydrogen fell but oxygen was reduced more slowly than in the earlier stages. During these two stages, CH_4, CO_2, and probably H_2O were the byproducts of coalification.

In the fifth stage of coalification—that of anthracitization—oxygen remained constant and hydrogen content fell more rapidly than in any of the earlier stages, the chief byproducts being CH_4, with little H_2O but no CO_2.

According to Mott[6] continuation of the first stage of coalification before a roof cover was provided gave subhydrous material which on subjection to the second and third stages of coalification became durain, and a continuation of the second stage of coalification beyond the normal end-point gave cannel coal.

Thermal Metamorphism

Roberts[16] has cited examples in Scotland and elsewhere of the transition of relatively immature coals to coals of increasingly higher stages of rank development. The coal field of the Stirling area of Scotland is a particularly striking illustration of the influence and effect of hot, intrusive, magmatic rocks on the character and properties of overlying coal seams. This coal field is underlaid by a great sheet of igneous rock of considerable thickness. This rock when molten, at a temperature estimated to have been about 1200°C, was thrust more-or-less parallel with the bedding into the coal-bearing strata, probably at a geological period not long after the coal beds were deposited. After this magmatic intrusion there followed an interval of cooling, during which a relatively large quantity of heat was transmitted through the overyling rocks and coal seams until a state of thermal equilibrium was reached. Since a number of coal seams lie directly above the igneous mass, the effect of the cooling magmatic rock in altering these coals can be studied in vertical sequence. The extent of their devolatilization and their carbon content—criteria of their stage of rank—is directly related to the distance of the individual seams above the igneous mass. As revealed in a shaft sinking, the lowest coal bed, about 200 ft above the igneous rock, is anthracite of 89% fixed carbon content. In five coals lying above this anthracite and extending to a distance of about 1170 ft above the igneous rock, the fixed carbon content progressively decreases to 57% in the uppermost bed of bituminous rank coal.

Roberts also cited an example of transition in coal rank caused by thermal metamorphism of Sumatran coals.[17] The coal field in question is of relatively recent geological age, Pliocene, and its coals range in rank from lignite to anthracite. At the close of the Pliocene epoch, following deposition of the coals, volcanic activity brought about the intrusion of a dome-like mass of igneous rock which pushed up through the sedimentary strata, arched the overlying beds, and was responsible for a vast amount of heat being transmitted into these sedimentary deposits, including the associated brown coal seams. Under the influence of heat from this laccolithic magma, brown coal deposits were converted to subbituminous coals, bituminous coals, anthracite, and natural coke. The degree of alteration of the brown coal was directly related to its distance from the intrusive laccolith. At a certain distance, the brown coal remained unchanged, but the closer it approached the igneous rock, the greater was the alteration in rank and the more mature the coal became. When the carbon content exceeded 84% to about 89%, the coking properties of the coal were strongly developed and diminished to zero in the anthracite. This led Kreulen to consider anthracite as a natural product of low-temperature carbonization under pressure.

It seems evident from the examples of thermal metamorphism in the Scotland and Sumatra coal fields described that anthracitization of low-rank coals can be regarded as a process of heat treatment under pressure.

Another discussion of thermal alteration of coking coals in Colorado has been given by Johnson[18] who reported that halos of high-rank coal occur under certain circumstances adjacent to intrusive bodies of igneous rock. This relationship has been observed frequently in western United States coal-mining districts. The factors that control such local increases

[16] Roberts, J., *Coke and Gas*, **18**, 25–8, 68–72 (1956).

[17] Kreulen, D. J. W., *Elements of Coal Chemistry*, Nijgh and van Ditmar, Rotterdam, 1948, 204 pp.

[18] Johnson, V. H., *Mining Eng.*, **4**, 391–5 (1952).

in rank are understood only in a general way to be (1) the regional rank of the coal before the emplacement of the intrusive body; (2) temperature and composition of the magma; (3) size and form of the intrusive body; (4) amount and character of the overburden; and (5) distance of coal from the igneous body. In the Paonia and adjacent districts of Colorado, Tertiary intrusive bodies are numerous and some are large. An example of such thermal metamorphism may be observed at the Richardson Mine on Carbon Creek, Gunnison County, Colorado. A laccolithic intrusive of quartz monzonite porphyry lies immediately north of the mine and cuts the coal-bearing formation at a low angle. At the mine, the main mass lies about 100 ft above the coal bed. Field study and chemical analyses reveal that, in a distance of 2000 ft beyond the flank of the intrusive body, the coal varies in rank from semianthracite to high-volatile bituminous. The former evidently is the highest rank that has been reached as result of the thermal metamorphism, and the latter is approximately the regional rank resulting from normal coalification of the coals in the Mesa Verde formation. Between the two extremes there is an irregular belt of medium- to low-volatile bituminous coal that has good coking qualities. The coal of semianthracite rank is coextensive with the igneous body where it lies but a short distance above the coal. In the same general area of Colorado, additional evidence of thermal metamorphism has been observed near Placita in Pitkin County where coking coal locally occurs in the vicinity of an intrusive body whereas the regional rank of the coal is high-volatile bituminous.

From the information of field studies and of chemical analyses of coals of various areas where they are known to occur in the proximity of igneous rocks such as the examples cited in Scotland, Sumatra, and Colorado, it is apparent that local advances in rank above the stage reached in regional rank development are the result of thermal distillation of volatile substances including oxygen, which occurred during and subsequent to the emplacement of magmatic rock masses.

Development of Rank by Orogenic Forces

Although thermal metamorphism by magmatic heat is important in changing coals to higher ranks under local conditions, this process cannot be generalized to account for all the extensive developments of high rank in coal as represented by the vast deposits of coking coals and anthracites that are known. Although it may be assumed that heat also has played an important role in development of high rank in coking coals and anthracite deposits, in the great majority of coal fields the original source of heat is not so obvious as it is in the coal fields that have been intruded by magmatic igneous rocks.

A geological relationship that invariably seems to exist with coals of relatively high rank, especially anthracites, is the fact of their association with sedimentary strata that have been subjected to considerable rock disturbance such as folding, faulting, and fracturing. These disturbances of rock strata are usually caused by orogenic or tectonic faces that accompany uplift and mountain-building activity in the crust of the earth.

On the basis of certain laboratory experimentation pertaining to the behavior of coal when subjected to heat treatment under controlled conditions, Roberts[13] determined that the temperature of formation of anthracite must be from 500 to 550°C for a typical anthracite. He reported that these and even higher temperatures are attained in rock formations and coals subjected to the lateral thrusts of mountain-building forces. The supposed effect of these forces in the development of extensive deposits of anthracite can be studied in a number of areas of the world, notably South Wales in Britain, Pennsylvania (U.S.A.), Rocky Mountains (U.S.A. and Canada), and the Donetz basin (U.S.S.R.).

The great lateral pressures of the orogenic and tectonic forces resulted in the folding, faulting, and fracturing of rock strata and associated coal beds, during which activity there was considerable differential movement along numerous planes of the sedimentary strata. The combination of pressure, movement, and frictional heat acted to drive off the volatile constituents of the coals and thus was responsible for their increase in rank.

Dryden[19] has discussed the subject of origin and formation of coal in the light of some conflicting hypotheses of a number of recent investigations. Dryden pointed out that coals

[19] Dryden, I. G. C., *Coke and Gas*, **18**, 123–8, 181–4 (1956).

appear to belong to a continuous metamorphic series starting with peat and ending with anthracite, but direct proof of conversion from one class to another is difficult to find since the process took place over hundreds of millions of years. Evidence, therefore, is mainly circumstantial, and the truth probably lies somewhere between two extreme points of view. In one point of view, every main class of coal is regarded as an intermediate product in a continuous metamorphic series, and in the other a separate origin from plant debris may be postulated for each class of material.

Chemical Aspects of Coalification

If coals do form a continuous series, it is important to demonstrate that the conversion of one class to another is chemically possible. An attempt to relate the postulated genesis with the chemical structures of various kinds of coal has been made by Huck and Karweil.[20] These two researchers divided coalification into two main phases, the biochemical and the geochemical, and accepted the theory that metamorphism in the geochemical stage is caused by increase in temperature produced by the depth at which the material was buried and the effects of pressure of overlying strata. They considered that the observed phenomena can be accounted for with temperatures not exceeding 200°C, except in the exceptional cases.

Huck and Karweil first reviewed the changes in the biochemical stage after listing the original plant materials present in order of their resistance to attack or stability as follows.

Fats, Waxes, and Resins. These substances are quite resistant to the attack of anaerobic bacteria, which organisms derive their energy by using oxygen-containing organic compounds such as cellulose instead of elementary oxygen. These substances are oxidized more readily by aerobic organisms, which acquire atmospheric oxygen directly.

Lignin. Lignin is more stable and resistant to bacterial action but is slowly oxidized by aerobic bacteria to the complex, aromatic "humic acids" that form a large part of peat and may be responsible for most of the aromatic structure in coals.

Cellulose. Aerobic activity destroys cellulose more readily into CO_2 and H_2O through simple sugars. In anaerobic fermentation cellulose is a medium for oxygen supply, but a small part is converted into simpler nonaromatic substances, but the possibility of aromatization of some portion cannot be excluded. It is possible though not established that aromatic humic acids may be formed to some extent by interaction between proteins and cellulose, or between their degradation products. Biochemical and purely chemical reactions intermingle throughout this phase in the formation of coal; among the important determining factors are climate, acidity, and availability of oxygen.

Development of Coal Molecule. Huck and Karweil assumed in accordance with their earlier conclusions[21] that the structural unit of the coal molecule contains a flat aromatic core surrounded by a methylene system (or M-system), which is relatively rich in hydrogen and oxygen and is largely made up of methylene groups containing carboxyl, hydroxyl, and carbonyl constituent groups in aliphatic and alicyclic combinations. The unit is equivalent in concept, though differs in detail, to the "cluster unit" of Dryden and Griffith.[22] To make the model more precise, the aromatic nucleus was considered on the basis of the earlier work[21] to contain about ten rings (the hydrocarbon ovalene was taken as example). The M-system was then regarded as a side chain, attached to the nucleus, of length decreasing with increasing rank of coal. They considered units of this kind to be grouped in three-dimensional molecular structures, less regular in low-rank coals but becoming more regular in coals of higher rank—the turbostratic crystallites of Riley. Huck and Karweil then considered that coal develops from brown coal to anthracite as follows. First there is a gradual splitting off of the oxygen-rich groups in the M-system from the aromatic nucleus of nearly constant size. Above about 90% carbon content when the M-system is small, methane-forming condensation reactions between units outweigh reactions involving the splitting of carbon dioxide and water from the M-system. The aromatic nuclei at this later stage thus become interlinked; in the absence of fringes they also can be readily aligned, by cohesive

[20] Huck, G., and Karweil, J., *Brennstoff-Chem.*, **36,** 1–11 (1955).

[21] Huck, G., and Karweil, J., *Brennstoff-Chem.*, **34,** 97–129 (1953).

[22] Dryden, I. G. C., and Griffith, M., *Fuel*, **32,** 199–210 (1953).

forces, one above the other in approximately parallel stacks.

The average nitrogen content of the vegetable matter that formed coal is relatively small, and was estimated by Huck and Karweil to have been between 0.6 and 1.0%. Because of the natural reactions of the coalification process, there is a gradual enrichment of nitrogen content up to about 45% volatile matter content after which it tends to decrease slowly. They suggested that the polar character of nitrogen is significant in the formation of coal, and that nitrogen, as well as oxygen in the M-system, by determining the balance of cohesive forces, influences the ability of the coal to swell and soften when heated. The sulfur content of wood and vegetable matter of peat in general is too small to account for the amount usually found in coals. Oxidation of inorganic sulfur to sulfate followed by the production of organic sulfur through the action of sulfate-reducing bacteria was suggested as a supplementary source.

The broader conclusions of Huck and Karweil on the thermodynamics of coal metamorphism are that thermodynamic criteria favor, at temperatures up to about 200°C, the formation of coal by reactions involving condensation of aromatic systems to form larger units with elimination, for example, of methane from side chains, and reactions in which carboxyl groups are lost, forming carbon dioxide.

Although the biochemical phase is of overriding importance during the formation of peat, the stage reached by the coals that today are of higher rank was determined by the rates of reaction during the later geochemical phase.

Radioactive Coalification

In a review of the geochemistry of coal, Breger[23] has noted that the intensive surveys of various sedimentary deposits for new sources of uranium led to the discovery of low-rank coals containing appreciable percentages of this radioactive element.[24] In some instances, the uranium content of the coal was sufficient to warrant consideration of such coals as possible sources of uranium rather than as fuels.

Schopf and Gray[25] have described results of microscopic studies of certain uraniferous coals, in particular the uranium-bearing lignite of Harding County, South Dakota. They concluded that no apparent correlation exists between any single coal petrographic component and uranium content, but the observation that the samples richest in uranium contain relatively large amounts of attrital humic matter and the greatest amount of yellow, resinous-waxy matter seemed significant. Such residual constituents of attrital type coal are representative of highly degraded plant matter and are characteristic of coal derived from plant debris that was subjected to intensive microbial decay. The association of high uranium content in humic attrital coal may indicate that plant material that has been most subjected to decay is also very receptive to uranium emplacement.

Association of Uranium in Coalified Wood. In addition to the association of uranium in low-rank coal deposits, as previously described, other natural occurrences of uranium in organic materials are known such as the uraniferous fossil plant debris described as coalified wood or logs of the Colorado Plateau. This coalified woody material occurs in Triassic and Jurassic sedimentary rocks of the Colorado Plateau. Coalification by natural metamorphic processes of the woody material was essentially completed by the end of Triassic and Jurassic time, and much of the coalified wood has lost all vestiges of cellular structure. The coalification seems to have been completed before introduction of the uranium probably in late Cretaceous or early Tertiary time, so that irrespective of its geological age, the coalified wood has been subjected to irradiation by the associated uranium for approximately the same duration of geological time, perhaps 60 to 90 million years.

Microscopic and autoradiographic studies of the samples have revealed that the uranium is dispersed throughout the coalified material and further suggest that the uranium was absorbed by the coalified wood after entering the material along microfractures, and was subsequently reduced from U^{+6} to U^{+4} in the form of relatively insoluble minerals of

[23] Breger, I. A., *Econ. Geol.*, **53**, 823–41 (1958).

[24] Denson, N. M., Bachman, G. O., Gill, J. R., Hail, W. J. Jr., Love, J. D., Masursky, H., Moore, C. W., Pipiringos, G. N., Vine, J. D., and Zellar, H. D., *U.S. Atomic Energy Comm., U.S. Geol. Survey*, TEM 341-A (1952), 44 pp.

[25] Schopf, J. M., and Gray, R. J., *U.S. Geol. Survey, Circ.* No. 343 (1954), 10 pp.

uraninite or coffinite and dispersed as colloidal particles throughout the coalified material.

In studies by Breger[26] of the relationship of this coalified material to uranium content, the analysis of the samples indicated that the composition and equivalent rank of the coalified wood are dependent on the environment of burial. Although the coalified wood has usually been regarded as corresponding to subbituminous rank coal, the rank of the samples as determined from Bureau of Mines analysis on an as-received basis showed the range in rank to be from subbituminous to high-volatile A bituminous, and none of the samples analyzed contained more than 84% C. This rather sharp limiting value of carbon seems to indicate that coalification took place during the relatively rapid biochemical stage, and that dynamochemical metamorphism may not have been an important factor in coalification. Results of microscopic observations of the samples revealed an apparent correlation between the opacity of the coalified woody material to transmitted light of the prepared specimens and the volatile content, indicating that the mode of preservation rather than metamorphism was principally responsible for the wide range of analytical values.

Further analysis of results of chemical studies of the samples revealed there was no apparent dependence or correlation of uranium and organic carbon content, but when organic hydrogen is plotted against uranium the hydrogen content decreases with increasing amounts of uranium for samples containing more than 0.1% uranium. On the assumption that this relationship is linear, it was calculated that hydrogen decreases 0.25% for every incremental increase of 1.0% of uranium retained in the coal. This more or less inverse relationship between uranium and hydrogen was explained as arising from radiochemical dehydrogenation of the coalified material by the irradiation of alpha particles from uranium and its daughter products, which is a rather common phenomenon when organic material is irradiated. The high-energy radiation is believed to initiate chain mechanisms that result in cross linking of molecules.

Dehydrogenation by alpha-particle bombardment was also thought to explain the decrease in volatile matter content and in Btu values of the samples with increase in uranium—relationships that follow a similar pattern to that of uranium and hydrogen contents.

A study has been made[27] of the physical and chemical properties of vitrain from the coalified woody material of certain of Breger's Colorado Plateau samples. The investigation included chemical analyses, density determinations, reflectance, and X-ray scattering measurements of nine samples of vitrain selected from the uraniferous coalified woody material collected and studied by Breger. Results of chemical analysis revealed that the carbon content of these vitrains had a rather narrow range from 81.5 to 83.6%, but a relatively wide spread of hydrogen contents from 3.4 to 7.7%. Hydrogen:carbon ratios versus oxygen:carbon ratios showed no conformity in trend with similar plots of normal vitrinites as is shown in Fig. 2. Specific volumes were determined by helium displacement, and results are plotted in the diagram of Fig. 3, which shows specific volume versus hydrogen content of normal coals of different rank and of the vitrains associated with uranium. Most of the uraniferous vitrinites had high densities and did not conform to the relation observed for normal coals. One sample had an unusually high density of 2.133 g/cc close to the density of natural graphite. Reflectance measurements indicated an increase with decrease in H:C ratio, which was interpreted to be the result of increased density caused by radiochemical dehydrogenation. For normal coals, however, increase in reflectance, mainly, is attributed to increase in aromatic carbon content. X-ray scattering intensities differed greatly from those of normal vitrinites. The samples containing high hydrogen content resembled exinites in their scattering character. With decrease in the hydrogen-carbon ratio, the intensity patterns became more diffuse, which is the reverse of normal coals. Total intensities corrected for polarization, absorption, and air scattering are plotted in Fig. 4 where 2θ is the angle of scatter and λ is the wavelength of X rays. On the figure shown, the scattering intensities of a sample of a high-volatile bituminous A vitrinite of 83.0% carbon

[26] Breger, I. A., *Proc. 3rd Conf. on Origin and Constitution of Coal*, Crystal Cliffs, Nova Scotia, June 16–19, 1956.

[27] Ergun, S., Donaldson, W. F., and Breger, I. A., *Fuel*, **39**, 71–7 (1960).

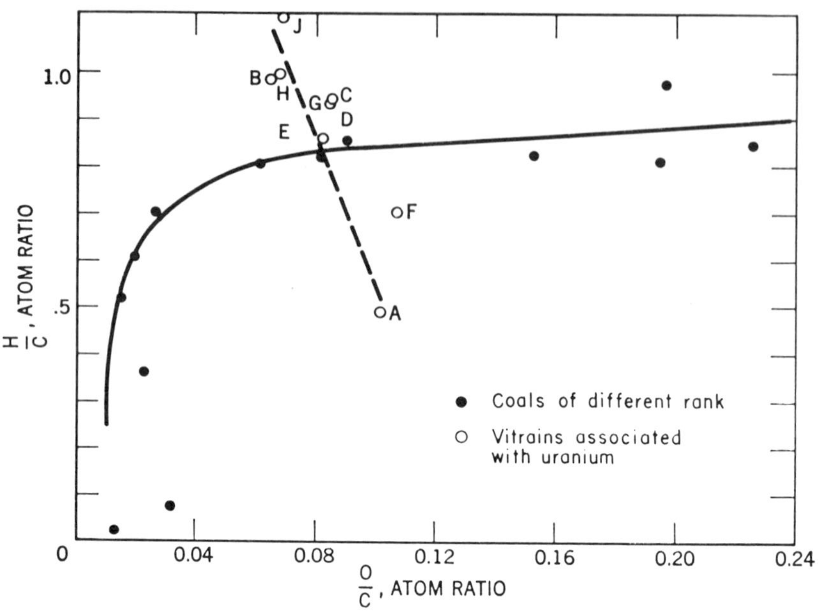

Fig. 2. Composition diagram of normal coals of different rank and vitrains associated with uranium.

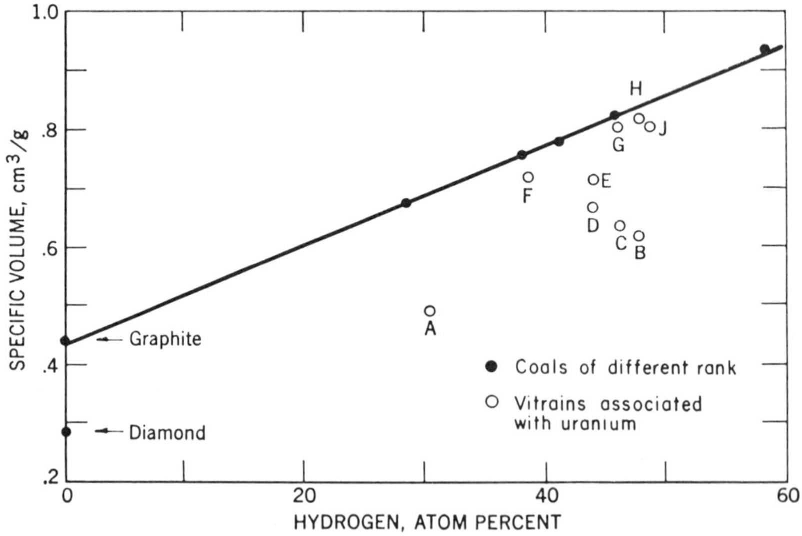

Fig. 3. Specific volume as a function of hydrogen content of normal coals of different rank and vitrains associated with uranium.

content are also included for comparison. The results of the study suggested that alpha irradiation causes cross linking of alicyclic units rather than aromatization.

Radioactive Alteration in European Coals. Stach[28] has studied the effect of irradiation on coalification in an impure coal of the Erda bed containing small zircon crystals which by virtue of their thorium content possessed radioactive properties and thereby had been converted to the mineral *malakon*. All the macerals of the coal had been altered by radioactive coalification. The degree of coalification was determined by the relative hardness and reflectance to be equivalent to

[28] Stach, E., *Brennstoff-Chem.*, **39**, 329–31 (1958).

low-volatile coal. Irradiation of the coal had broken the hydrogen and oxygen bridges and left the stronger aromatic groups intact.

Microscopic study of this association of thorium minerals in coal led the investigator to conclude that coalification can be promoted by radioactivity and that irradiation can be a factor in coalification. The microscopic observations revealed to the investigator that a high-volatile coal of 81% C was converted by the irradiation into a low-volatile coal of 91% C and to anthracite. Differential hardness of the coal produced by rank development combined with gradual compaction of the coal mass resulted in the development of concentric arched layering of the different rank coals about the radioactive malakon crystals as centers. The mechanism of the radioactive coalification was interpreted to have been initiated by bombardment of the coalified mass by alpha particles emitted from the thorium minerals. These acted upon the aromatic condensed ring system, breaking the weaker hydrogen and oxygen bridges, and as these elements were expelled, the aromatic lamellae became larger, resulting in an increase in the ratio of aromatic to nonaromatic carbon. Irradiation of the organic material

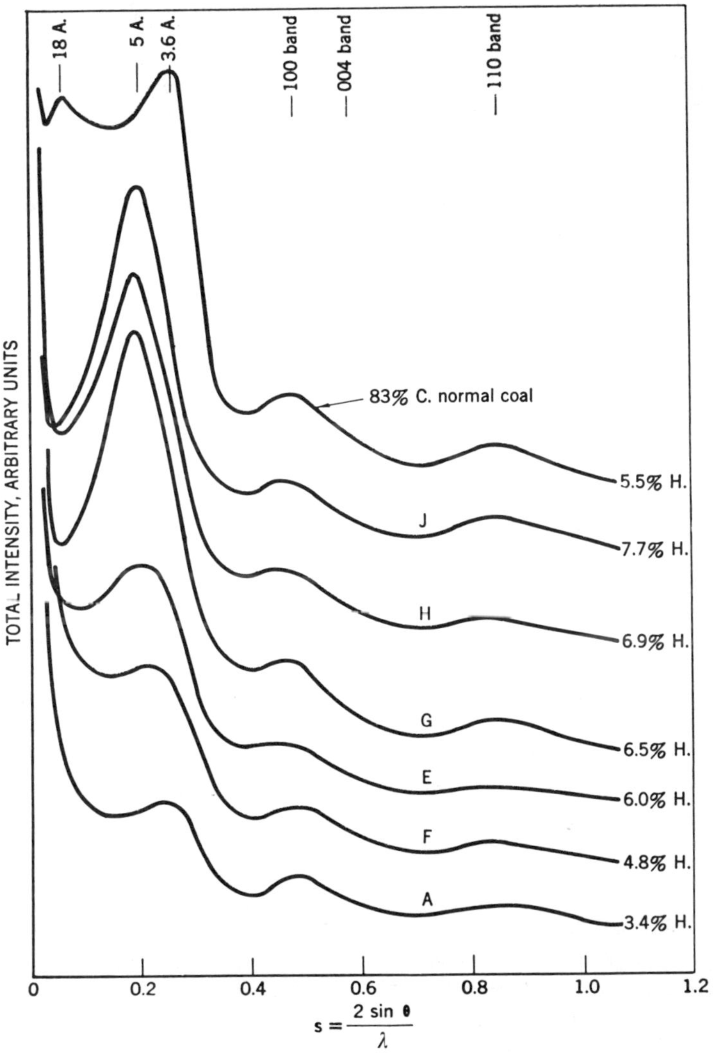

Fig. 4. X-ray scattering of some vitrains associated with uranium.

began when the radioactive mineral crystals were incorporated into the original peat mass and continued concurrently with the normal processes of coalification to the present time, which apparently was of sufficient duration to permit formation from the original peat of high-volatile, medium-volatile, and low-volatile coals, and anthracite.

In another study[29] of the effect of radioactivity on coalification, the high-volatile bituminous rank of coal of the Cambrian kolm of Sweden was attributed to radioactivity of finely disseminated uranium which seemed to be very closely associated with, perhaps bound to, the organic matter. In the course of normal coalification, only a product of brown coal rank would be expected to develop in the same geological environment. Microscopic evidence was also found of radioactive coalification affecting the organic matter incorporated into the variegated sandstone of Wrexen in Hesse. For the most part, the radioactive substance in this coal also appeared to be closely associated with the organic matter and to a lesser extent with the inorganic constituents. The coals of the Wrexen locality not affected by the radioactivity range from high-volatile to medium-volatile bituminous rank, whereas the irradiated coal is of anthracitic rank as determined by reflectance.

Summary

The gradual and progressive alteration of coal resulting from geologically long-continued application of heat and pressure of sedimentary overburden during the dynamochemical period of coal metamorphism is perhaps the most popular theory of the origin of rank development in coal and is commonly referred to as normal coalification. On the other hand, the heat and pressure arising from rock movement in crustal disturbances involved in mountain-building uplift is regarded as an extraordinary process responsible for upgrading coal to a rank above that reached in normal coalification occurring only on a regional scale.

Thermal metamorphism of coal, arising from intrusion of coal beds by hot magmatic rock in molten condition and of relatively high temperature is similarly regarded as an extraordinary process of coalification responsible for local upgrading of coal to a rank above that of normal coalification.

As brought out in the previous discussion of radioactive coalification, field and laboratory evidence has been presented by different investigators in recent years to indicate still another extraordinary process of coalification. These investigators have revealed that coalification, the promotion of rank in coal, can be effected through the bombardment of the coal by alpha particles emitted from uranium or radioactive minerals incorporated in the coal. But it is not likely that this process played an overwhelming role in the metamorphism of coal deposits except in unusual cases of high concentration of radioactive substances in coal.

To summarize rather briefly without emphasis on the mechanisms of processes, it is generally agreed that coalification has its inception in the peat stage by the action of forces categorized as biochemical in nature. At least two paths of coalification for the organic material of peat are believed to be initiated in the biochemical stage: (1) fusinization, the formation of fusain, which progresses very rapidly in the biochemical stage; and (2) vitrinization, the formation of vitrain, which seems to make much slower progress than fusinization. The waxy-resinous plant substances of peat are very resistant to microbial activity and thus seem to survive the biochemical stage with little or no alteration.

When the organic debris identified as peat, is buried beneath a sedimentary cover, dynamochemical forces take over the metamorphism of coal, the most important probably being heat and pressure developed by the overlying load of sedimentary cover. This is normal coalification and is marked during the dynamochemical phase by very gradual, almost undetectable, change in fusain, and the slow but progressive alteration of anthraxylon (vitrain) is easier to determine than that of fusain, and so overwhelmingly important is this component in most coals that the alteration of vitrain is commonly accepted as the standard criterion for determining the degree of metamorphism or rank of coal. The alteration of waxy-resinous plant substances progresses very slowly during the dynamochemical phase until acted upon by

[29] Teichmüller, M., and Teichmüller, R., *Geol. en Mijnbouw*, **20**, 41–66 (1958).

the extraordinary forces of coal metamorphism, marked by the development of very high rank in coal designated as anthracitization. During the phase of anthracitization the alteration of all coal components becomes practically undetectable as the coal approaches homogeneity in the ultimate goal of complete carbonization or graphitization.

PETROGRAPHY OF COAL

Developments in the field of coal petrography beginning about 1930, particularly in Germany and the United States, were marked by efforts to establish coal petrography as a practical and useful tool in coal technological work.

Reflected Light Technique

The study of coal by means of reflected light microscopes and polished surface specimens of coal was developed to a point of practical usefulness in the laboratory of the Prussian Geological Survey at Berlin, Germany, principally by Stach,[30] Hoffmann, Kühlwein, and Krüpe. These coal scientists working together perfected the use of the oil-immersion lens in reflected light microscopes for examination of highly polished surfaces of coal specimens. This method of coal petrography subsequently became standard procedure for coal petrographic investigations in Germany.

In 1932, a laboratory designed for practical application of the newly developed methods of coal study to investigations of Ruhr coals was established at Bochum after preliminary studies by Lehmann and Stach had clearly demonstrated the practical importance of this application.

As interest in the work at Bochum spread in the coal industry, coal petrographers trained at the Bochum center of coal petrographic studies and at the Berlin Laboratory dispersed and established laboratories at other locations mainly in the Ruhr. As an example, Mackowsky transferred to Essen in 1940 and set up a laboratory for applied coal petrography primarily to study the nature of ash-forming minerals of the Ruhr coals. This laboratory was sponsored and consistently supported by the Ruhr coal operators association, now called Steinkohlenbergbauverein, and has become the largest and most important laboratory of its kind in Germany. Scientists and technologists from coal-producing countries all over the world come to this laboratory to observe and study petrographic methods and their application.

In 1943, Mackowsky and Abramski,[31] convinced that coal petrography had acquired such importance that a critical review was needed to show its development and practical importance, jointly published a short series of articles. Although intended as a summary of German coal petrographic techniques, the scope of the subject matter was sufficiently comprehensive, in absence of other more detailed information, to serve as a laboratory guide to the German procedure. However, laboratory equipment, especially microscopic and optical apparatus recommended by the German coal petrographers in their work based primarily on reflected light microscopy, was not generally available outside Germany. Hence, close duplication of German techniques of coal petrography was not possible.

A more complete and detailed reference book covering various phases of German coal petrography published in 1949[32] describes in considerable detail various microscopes, optical equipment, and accessories developed for use in applied coal petrography by the reflected light technique. The various types of petrographic analysis such as qualitative and quantitative bed profile analysis of banded ingredients, microconstituent analysis of maceral components of the banded ingredients, coke cellularity, and porosity analysis, were described. The preparation of the coal and coke specimens for microscopic examination was also covered.

As a result of Mackowsky's and Stach's publications, the German reflected light method of coal petrography could for the first time be studied in detail and compared with other techniques. To fill the need for further detailed information on the reflected light method of coal petrography, particularly for laboratory purposes, a number of the leading German coal petrographers collaborated to prepare a valuable reference and laboratory

[30] Stach, E., *Handbuch der Mikroskopie in der Technik*, Umschau, Frankfurt am Main, 1952, vol. 2, pp. 3–63.

[31] Mackowsky, M. T., and Abramski, C., *Feuerungstechnik*, **31**, 25–33, 49–64 (1943).

[32] Stach, E., *Lehrbuch der Kohlenmikroskopie*, Glückauf, Kettwig, 1949, 285 pp.

guide for those interested in following the German method.[33] The usefulness of the atlas is greatly enhanced by the abundance of photomicrographs of coal made with oil-immersion lens and reflected light microscope, and by the fact that the explanatory captions of the many photomicrographs are printed in English as well as in German and French. The atlas for applied coal petrography was followed in 1952 by a supplemental volume[34] which also represents the collaboration of many leading German specialists in the field of coal and coke microscopy.

When these two comprehensive reference works appeared, interest in German coal petrography spread rapidly, and laboratories for the petrographic studies of coal were established in most of the coal-producing countries of the world. In addition to the Western European coal-producing countries and the U.S.S.R., coal petrographic investigations are now carried on to a limited extent in American countries, including the United States, Canada, Mexico, Columbia, Venezuela, Brazil, Argentina, and Chile; also in the Asian countries of India, China, and Japan; and in Australia and South Africa.

Seyler's Stepwise Reflectance Theory

In England during the period from about 1930 to 1945, studies of coal by reflected light technique were pursued independently of the German work, particularly by Seyler who was especially interested in making accurate measurements of the reflectance property of coal. Seyler used the Berek slit microphotometer with oil-immersion lens. Using cedar oil as immersion observational medium, he obtained measurements of reflectance which he considered to be highly accurate. His studies of the reflectance measurements of a number of coals resulted in a proposal that attracted considerable attention and interest among coal scientists and technologists. Although essentially completed in 1941, Seyler's[35] results were first published in 1943. He proposed that all coals originating from lignocellulosic tissues derived mainly from the woody and bark tissues, and other parts of plants consist of a limited number of optically recognizable components, the reflectance of which increases in rank by a series of jumps rather than by continuous variation. He reported nine steps related to each other by a geometrical pattern so that the reflectance in air (R) could be represented by the formula $R = 0.26 \times 1.363N^r$, where N^r is the number of the component in the series from 1 to 9. The component of lowest reflectance at any stage of metamorphism was the predominant component of the bright coals, vitrain and clarain, and was designated vitrinite; at the end point the component of constant reflectance was called fusinite. The seven intervening components were classed as intermediates. In general, only a few—three or four—of the optical components were found in any particular coal. The theory also postulated that, as coals advance in rank, the optical components present suddenly mutate to the next higher step of reflectance. An example of how the theory works is as follows. In a coal of relatively low rank, say a subbituminous coal, there might be a predominance of component 3, vitrinite, with subordinate amounts of higher reflectance components. In progressively higher rank coals, the predominant vitrinite might be components 4, 5, 6, 7, and so on until ultimately a coal that had reached the limit of metamorphic alteration would consist wholly of component 9 which could be called either vitrinite or fusinite.

Perhaps the most significant feature of Seyler's theory of reflectance of coal that particularly appealed to coal technologists primarily interested in practical applications of coal petrography was that it indicated for the first time the possibility of developing an objective method of determining the petrographic composition of coal. It seemed reasonable to assume that if the petrographic components of a particular coal consisted of a limited number of optically discrete substances, each having a fixed measurable reflectance value, the qualitative and quantitative determination of such petrographic composition should be readily amenable to instrumental measurement with laboratory equipment especially designed or assembled for the purpose. Such an analytical method when developed would afford a welcome replacement for the more subjective methods

[33] *Atlas für angewandte Steinkohlenpetrographie*, Glückauf, Essen, 1951, 329 pp.

[34] Freund, H., *Handbuch der Mikroskopie in der Technik*, Umschau, Frankfurt am Main, 1952, Vol. 2, 759 pp.

[35] Seyler, C. A., *J. Inst. Fuel*, **16**, 134–41 (1943).

inherent in the conventional polished-surface, reflected-light, and thin-section, transmitted-light techniques of examining coal.

Despite the considerable amount of interest aroused in coal science laboratories in the theory of the discontinuous reflectance property of coal at the time of its proposal, scarcely any laboratory work was done for some time to prove or disprove the validity of the proposal. The failure to begin immediate laboratory testing probably can be attributed to the facts that very few laboratories were equipped to duplicate Seyler's technique, and that the early papers did not give detailed laboratory procedures for making independent reflectance studies. The lack of information about methods was at least partially corrected in a later paper published by Seyler and Edwards.[36] They stated that the method depends on consideration of coal as a sedimentary rock-like deposit consisting of homogeneous optical components called macerals that are analogous to the minerals of rocks. It was pointed out that petrographic analysis by conventional methods of microscopic examination of coal merely give subjectively derived approximations of coal composition in terms of banded ingredients, which in reality are not petrological components comparable to minerals, but rather are complex rock types. Hence, such analyses do not truly characterize the coal in regard to its petrological composition. The macerals of coal as defined by Seyler were claimed to be truly homogeneous components and analogous to the mineral components of rock. Thus, according to Seyler, by quantitative determination of the maceral composition a true petrological analysis of coal is obtained. These macerals were characterized microscopically by the amount of light normally reflected from polished surfaces with an oil-immersion lens. The immersion oil had to have an index of refraction (n) of about 1.515.

Some laboratory studies of Seyler's theory. After publication of Seyler's papers on coal reflectance, a number of coal scientists visited Seyler's laboratory and seemed to obtain results that essentially confirmed Seyler's findings.

Probably the first fruitful results in attempting to experimentally duplicate Seyler's procedure in an independent laboratory were obtained by McCartney[37] of the United States Bureau of Mines, working with apparatus assembled to follow the procedure as described by Seyler in published papers and personal correspondence. The main element of McCartney's instrumental setup was a Berek slit microphotometer mounted on a Leitz Panphot microscope. Reflectance determinations were made on anthraxylon or vitrain bands in sixteen coals of varying rank. The results of these reflectance measurements strongly indicated a continuous change in reflectance that with a few exceptions showed a general increase with rank, a fact that had been discovered by previous investigators. When the averages of McCartney's measurements of reflectances of sixteen coals were plotted along a line representing the percent reflectance, it was clearly evident from distribution of the averages that there was little or no agreement with the steps of Seyler's reflectance series plotted along the same line.

Holubár and Krejcík[38] developed a new method of reflectivity measurement of coal in which an adaptor for comparison photometry was used instead of the Berek microphotometer. With this instrument the reflectivity of coals from all coal basins in Czechoslovakia were measured. Considering the accuracy achieved and the systematic selection of the investigated surfaces, it was claimed that the existence of Seyler's reflectance steps could be proved for samples from all coal basins in Czechoslovakia.

Huntjens and van Krevelen[39] made a critical study of reflectances of vitrinites in Cedar oil and found results, the averages of which tended to group about centers that approximated the values of Seyler's components.

Wege[40] made measurements of reflectance by a photocell technique of about 180 specimens of coal. It was concluded from these tests that reflectance increases continuously with coalification and that the data did not support the existence of optical components as defined by reflectance steps.

[36] Seyler, C. A., and Edwards, W. J., *Fuel*, **28**, 121–7 (1949).

[37] McCartney, J. T., *Econ. Geol.*, **47**, 202–10 (1952).

[38] Holubár, V., and Krejcík, Z., *Uhli*, **7**, 381–7 (1957).

[39] Huntjens, F. J., and van Krevelen, D. W., *Fuel*, **33**, 88–103 (1954).

[40] Wege, E., *Brennstoff-Chem.*, **35**, 33–41 (1954).

Broadbent and Shaw[41] re-examined the mathematical model for reflectance data and obtained new experimental data by using a photocell fitted to a Berek microscope. The main conclusions reached in this study were (1) that the published data are inconsistent, (2) that continuous variation in reflectance is possible, and (3) that many of the uses of reflectance are independent of the existence of steps or preferences.

As a result of critical testing in various independent laboratories, the consensus of scientific opinion seems to be that Seyler's concept of a limited number of discrete optical components of coal characterized by discontinuous variation in reflectance values with coalification is not amenable to conclusive experimental proof and hence is not acceptable as a practical theory. Furthermore, the idea of discontinuous change in the physical property reflectance is not supported by any chemical evidence from the many analyses of vitrain (anthraxylon), which have shown no discontinuous variation in any element with change in rank.

Regardless of the failure of various investigations to concur with Seyler's reflectance proposal, his experimental procedure of reflectance measurement and description of the variation in reflectance of the components of coal derived from woody tissue form a significant and valuable contribution to the petrography of coal. Reflectance measurements offer an objective basis for classifying the petrographic constituents that may become an important complement to the more subjective qualitative descriptions commonly presented. And it may yet be possible to establish arbitrary classes or limits of reflectance as a basis for quantitative analysis of coal samples.

Transmitted Light Technique

The method of studying thin, translucent sections of coal by transmitted light probably was the most successfully developed at the United States Bureau of Mines by Thiessen and associates. To describe the microscopic composition of coal as seen in these thin sections by transmitted light, Thiessen[42] in 1920 introduced a terminology in which the microcomponents of coal were designated in terms of their botanical origin. Thiessen's method involved preparing a completely representative series of thin sections from a columnar profile sample cut from a full seam thickness of coal. A critical study of the thin section method has been discussed in the paper by Parks and O'Donnell.[8] This paper described the technique of thin section analysis and discussed results of laboratory tests to ascertain the accuracy of the method.

The practical application of Thiessen's method of coal petrography began in 1930 at the Bureau of Mines in connection with a cooperative project with the American Gas Association in a survey of gas, coke, and byproduct making properties of American coals. The first report of this survey was also the first report published in which the petrographic composition of a commercial coal was included.[43]

A review of the Bureau of Mines work in coal petrography up to 1956[44] presents detailed information, mostly from a study of columnar profile samples, on the petrographic characteristics and properties of 160 coal beds mined in various parts of the United States, includes chemical analyses of the various samples, and gives results of many incidental tests and determinations pertaining to the technological properties of the coals. The report also explains the Bureau's system of describing and classifying coal by microscopic examination of thin sections as developed by Thiessen.

Although not generally adopted as a standard procedure in all American coal petrography laboratories, probably because of difficulties in mastering the technique of thin section preparation and analysis, the thin-section–transmitted-light method of coal microscopy is familiar to most active American coal petrographers and can be considered the standard reference for characterizing the petrographic composition of American coals. In consideration of the great amount of published literature on thin section investigations of American coals, the transmitted light method is frequently referred to as the American system of coal petrography in

[41] Broadbent, S. R., and Shaw, A. J., *Fuel*, **34**, 385–403 (1955).

[42] Thiessen, R., *J. Geol.*, **28**, 185–209 (1920).

[43] Fieldner, A. C., Davis, J. D., Thiessen, R., Kester, E. D., and Selvig, W. A., *U.S. Bur. Mines, Bull.*, No. 344 (1931), 107 pp.

[44] Parks, B. C., and O'Donnell, H. J., *U.S. Bur. Mines, Bull.*, No. 550 (1956), 193 pp.

contrast to equally frequent reference to the reflected-light method as the German system of coal petrography.

The independent development of two basically different approaches to the microscopic study of coal, each with its completely different scheme of coal description and classification terminology, inevitably led to considerable confusion in the field of coal petrography. As a result, very few workers completely understand both methods, and the interchange of this type of information about coal is less than satisfactory.

International Standardization of Coal Petrography Nomenclature

With the spread of coal petrographic investigative work after the close of World War II, the coal petrographers of Western Europe, where the most interest in the work had previously been centered, realized the need for some degree of standardization of laboratory techniques and especially of nomenclatural classification and description used in the petrographic analysis of coal, if any semblance of international cooperation in the dissemination of petrographic information concerning various coals was to be achieved. The problems of international exchange of information were already complicated by the fact that the use of two basically different methods of coal petrography and systems of nomenclature and description was relatively widespread.

The first serious attempt to resolve the vexing problem of nomenclatural confusion was made at the Second International Congress for the Study of the Stratigraphy of Carboniferous Rocks held at Heerlen, Netherlands, in 1935. Here a new nomenclature for reflected-light coal microscopy was proposed, discussed, and adopted.[45] The new scheme retained Stopes' original designations of the banded ingredients (also called rock types), as vitrain, clarain, durain, etc.[46] The distinction between these banded ingredients, or rock types, and the microscopic constituents (also called maceral components) was made by use of the terminations *ite* or *inite*, as vitrinite, fusinite, etc. A slight modification in the English terms to conform with accepted German usage was made by substituting the *it* or *ite* termination for *ain*, as in vitrit (vitrite), clarit (clarite), etc.

Unfortunately, no American representative was present at the 1935 Heerlen Conference to present Thiessen's system of thin-section–transmitted-light terminology for consideration in the proposed international coal nomenclature. Hence with respect to work with the reflected-light method, the action of the European group may have clarified the system of nomenclatural description and classification for reflected-light coal microscopy but left unsolved the problem of correlation of reflected-light and transmitted-light terminologies.

At the first meeting of the International Committee on Coal Petrology held at Geleen, Netherlands, in June 1953, a subcommittee on petrographic nomenclature was organized to prepare a comprehensive glossary of terms used in coal petrography, redefining them if necessary, and standardizing accepted definitions. This subcommittee met in various cities of western Europe at intervals of about 6 months. At each meeting it considered a list of previously prepared terms drawn from current usage in coal petrography. These terms were discussed at length and necessary changes were made until agreement was reached on the adoption or rejection of each term in the list. A report on the status of work of the international committee was read by Mackowsky[47] at the International Congress of Coal Petrology at Heerlen in September 1958. The report reviewed reasons for establishing the international committee.

According to Mackowsky's report, the results of work up to 1958 led to conclusions that maceral analysis of coal is possible with satisfactory accuracy and is easiest with the point counter. However, less satisfactory results are obtained with analysis of microlithotypes mainly because of the defining limitation of 5% used in distinguishing microlithotypes. The most promising method appears to be a combination of maceral analysis and analysis of microlithotypes. In any event, analysis of coals by rank offers no

[45] Jongmans, W. J., Koopmans, R. G., and Roos, G., *Fuel*, **15**, 14–5 (1936).

[46] Stopes, M. C., *Proc. Roy. Soc.*, **90B**, 470–87 (1919).

[47] Mackowsky, M. T., paper read at Intern. Congr. of Coal Petrology at Heerlen, September, 1958, translated by G. H. Cady, and N. O. Frederickson, SLA, 59–10393 (1958), 22 pp.

particular difficulty regardless of method if the coals are of distinctly different rank.

An American representative was appointed to the subcommittee, and a sincere attempt was made by American coal petrographers working primarily through the Coal Group of the Geological Society of America to cooperate with the European workers in preparation of a contemplated glossary, especially in the inclusion of American coal petrography terms. However, it has not been practical for an American representative to attend all the meetings of the subcommittee on nomenclature, and hence the American group has worked somewhat at a disadvantage by having to depend largely on correspondence for exchange of ideas and points of view. Nevertheless, despite the problems inherent in a project of this kind, progress has been made, and the first edition of the glossary appeared in 1957.[48] This glossary contains a list of coal petrography terms, each of which is briefly discussed as to its origin, description, occurrence in coal, and practical importance. Excellent photomicrographs are included to supplement the descriptive information.

At the Geleen meeting in 1953 of the International Committee on Coal Petrology, a method[49] involving a combination of the reflected-light and transmitted-light techniques was outlined with the hope that it would lead to a better understanding of the inherent differences in the two techniques and correlation of results. Teichmüller developed a technique of achieving a high degree of polish on the surface of a normal thin section of coal. By means of a universal-type microscope with built-in transmitted- and reflected-light sources, these polished thin sections can be studied alternately by both methods of microscopic examination, and in examining a petrographic constituent of somewhat doubtful identity the observer by choice of illumination can make identification reasonably certain. Thus by practical elimination of errors in identification of various microscopic components, a direct comparison of qualitative and quantitative results of petrographic analysis can be obtained.

In a joint study by Parks and Teichmüller[50] of polished thin sections of an American coal by transmitted- and reflected-light methods for comparison, results were obtained that offered a rational explanation for discrepancies in quantitative analysis of the same coal by the two methods. Conclusions from the comparative test indicated there may be an unavoidable tendency for the observer using the thin-section–transmitted-light method under certain conditions of microscopical observation to underestimate the amount of pale-colored substances called exinite such as spores, cuticles, and resin and to overestimate the quantity of opaque and semiopaque substances. In general, the examination of translucent substances such as anthraxylon (vitrain) is more satisfactorily done by transmitted light, whereas the opaque and semiopaque substances in coal are best observed by reflected light. Significant differences may be obtained in results of microscopic analysis by the two methods, making direct comparison of analytical data on the same coal virtually impossible. The terms of Thiessen's system of nomenclature based on microscopic appearances in thin sections by transmitted light are not positively interchangeable with terms of the European nomenclature system based on microscopic appearances by reflected light. Therefore, the terminologies of the two methods of coal petrography cannot be correlated precisely, but approximate correlation is both possible and useful. Table 1 shows correlation of banded components and attrital constituents of the Thiessen nomenclatural system with maceral and maceral-groups of the Stopes-Heerlen nomenclature. Table 2 shows correlation of types of coal of the Thiessen system with the microlithotypes of the Stopes-Heerlen nomenclature.

The work of the international committee for coal petrology on clarification of the problem of confusion in petrographic nomenclatures is continuing, and it is hoped that the somewhat chaotic situation of terminology, which is a detriment to progress in coal

[48] International Committee for Coal Petrology, Nomenclature Subcommittee. Decimal Classification 1957 Edition, Loose Leaf. Centre National Recherche Scientifique, Paris, France.

[49] Teichmüller, M., *Proc. Intern. Comm. Coal Petrology*, **1**, 25–32, 1954, Ernst Van Allst, Maastricht, Holland.

[50] Teichmüller, M., and Parks, B. C., *Proc. Intern. Comm. on Coal Petrology*, **2**, 21–7 1955, R. Louis, Brussels, Belgium.

TABLE 1

CORRELATION OF THE BANDED COMPONENTS AND ATTRITAL CONSTITUENTS OF THIESSEN'S-BUREAU OF MINES NOMENCLATURAL SYSTEM WITH THE MACERALS AND MACERAL GROUPS OF THE STOPES-HEERLEN NOMENCLATURAL SYSTEM

Transmitted-Light–Thin-Section Examination			Reflected-Light–Polished-Surface Examination	
Banded Components	Constituents of Attritus		Macerals	Maceral-Groups
Anthraxylon			Vitrinite with more than 14-μ band width	Vitrinite
Attritus	Translucent attritus	Translucent humic matter	Vitrinite width less than 14-μ band width	
		Spores, pollen, cuticles, algae	Sporinite, cutinite, alginite	Exinite
		Resinous and waxy substance	Resinite	
		Brown matter	Weak reflecting semifusinite, weak reflecting massive micrinite, weak reflecting sclerotinite, strong reflecting resinite	Inertinite
	Opaque attritus	Granular opaque matter	Granular micrinite	
		Amorphous (massive) opaque matter	Fusinite with less than 37-μ band width strong reflecting massive micrinite strong reflecting sclerotinite	
Fusain			Fusinite and semifusinite with more than 37-μ band width	

petrography in particular and coal science in general will be cleared up in the not too distant future.

STUDIES OF FUNDAMENTAL PROPERTIES AND BEHAVIOR OF PETROGRAPHIC COMPONENTS

Petrographic investigations of coal have led to a better understanding of the relation of petrographic composition to technological properties of coal. A more thorough knowledge of the fundamental chemical and physical properties of the maceral constituents would increase the usefulness of petrographic composition in predicting their behavior characteristics in various processing operations and in improving coals for specific uses by preparation methods, in addition to contributing to better understanding of the origin and nature of coals. To conduct such basic research, it has been necessary to obtain in reasonably pure form petrographic components from desired coals in quantities sufficient for analysis and tests.

There has been no real difficulty in obtaining required amounts of certain banded petrographic components such as anthraxylon (vitrain) and fusain. These commonly occur in bands or layers of most coals sufficiently thick so that they can be cut out of lump specimens of coal and upgraded to visual purity under a stereoscopic microscope.

Fisher and associates[51] have determined the hydrogenation properties of anthraxylon

[51] Fisher, C. H., Sprunk, G. C., Eisner, A., Clarke, L., Fein, M. L., and Storch, H. H., *Fuel*, **19**, 132–8, 162–72 (1940).

TABLE 2

CORRELATION OF THE TYPES OF COAL OF THIESSEN'S NOMENCLATURAL SYSTEM WITH THE MICROLITHOTYPES OF THE STOPES-HEERLEN NOMENCLATURE

	Transmitted-Light–Thin-Section Examination		Reflected-Light–Polished-Section Examination
	Types of Coal	Quantitative Details	Microlithotypes
Banded coals	Bright coal	More than 5% anthraxylon Less than 20% opaque attritus	Vitrite, clarite, duroclarite in part
	Semisplint coal	More than 5% anthraxylon 20 to 30% opaque attritus	Duroclarite in part, clarodurite in part, vitrinertite in part
	Splint coal	More than 5% anthraxylon More than 30% opaque attritus	Clarodurite in part, duroclarite in part, vitrinertite in part
Nonbanded coals	Cannel coal	Less than 5% anthraxylon predominantly translucent attritus (little or no algae)	Cannel coal
	Boghead coal	Less than 5% anthraxylon, predominantly translucent attritus, with algae abundant	Boghead

(vitrain) so separated from a variety of coals including peat, brown coal, lignite, subbituminous coal, bituminous coal, and anthracite. Tests of hydrogenation properties of fusain similarly obtained showed that this supposedly inert component was not totally inert to hydrogenation treatment.[52] Certain maceral constituents such as spores and resin, as well as the banded components and petrographic types of coal, were included in these hydrogenation tests.[53] By determining the hydrogenation properties of the various petrographic constituents contained in practically all coals irrespective of rank, Bureau of Mines technologists developed a useful empirical formula with which, when supplied with the petrographic analysis of an untested coal, its hydrogenation characteristics with reference to amenability to hydrogenation treatment and liquefaction yields could be predicted fairly successfully.

[52] Fisher, C. H., Sprunk, G. C., Eisner, A., Clarke, L., and Storch, H. H., *Ind. Eng. Chem.*, **31**, 190–5 (1939).

[53] Fisher, C. H., Sprunk, G. C., Eisner, A., Clarke, L., and Storch, H. H., *Fuel*, **18**, 132–41 (1939).

In another field of coal technology, Sprunk and Thiessen[54] studied the relation of petrographic composition to chemical, coking, and byproduct properties. In fact, this investigation covered a period of about 20 years and included study of a considerable number of American coals of a wide range of rank and petrographic types.

Fundamental Studies of Coal Petrographic Macerals

Whereas the banded components, vitrain and fusain, and such distinctive lithotypes as bright coal and durain present no severe problem in concentrating substantial quantities for analysis and tests, the maceral (microscopic) substances such as exinite, micrinite, and resinite, in addition to being of small particle size, are also so intimately mixed and tightly compacted in attrital coal that it is exceedingly difficult to free and concentrate them in reasonably "pure" form. Furthermore, since certain maceral substances as semifusinite, and massive micrinite often

[54] Sprunk, G. C., and Thiessen, R., *Ind. Eng. Chem.*, **27**, 446–51 (1935).

are gradational in nature, it is virtually impossible to obtain concentrations of them in petrographically "pure" form. However, despite the difficulty of obtaining concentrations of "pure" macerals, a number of research workers have attacked the problem of obtaining such samples of macerals mainly for the purpose of making fundamental studies of their chemical and physical properties and technological behavior.

Kröger[55] separated certain macerals from a number of coals in the following degrees of purity: vitrinite 99%, exinite 94 to 97%, and micrinite 91 to 95%. Kröger's procedure for separating the coal macerals was based on gravity separation in a heavy liquid medium (carbon tetrachloride and toluene) aided by a moderately high-speed centrifuge. The first step was to obtain a maximum freeing of the often closely associated and tightly compacted raw materials. Particle size reduction to bring about the optimum maceral separation was effected by an elastic impact grinder in which the fineness obtained amounted to 1 to 20 μ for vitrinite and micrinite and 1 to 100 μ for exinite. After a preliminary separation, a sharp separation was then made in a centrifuge operated at 12,000 to 25,000 rpm. Chemical investigation of the concentrated macerals included the proximate and elementary analysis and the determination of the heating value and of the hydroxyl and carbonyl oxygen. The data indicated that coalification of all the macerals proceeds continuously and is not stepwise.

In addition to the study of chemical and physical properties of the macerals, Kröger and Pohl[56] studied the calorific effects during thermal decomposition. By differential-thermal analysis of the heating process of maceral components of four Ruhr coals, five effects were found. The first was determined by the evolution of adsorbed liquid, the second and third were caused by elimination of water of formation and melting processes, the fourth was exothermic, and the fifth was a chemical transformation. These effects were smallest in micrinite.

Horton[57] used a similar technique of fine grinding and float-and-sink separation in a heavy organic liquid (carbon tetrachloride and benzene) with the aid of a centrifuge to effect the separation of maceral constituents into fractions of different densities, which were subjected to petrological and chemical studies.

A procedure of maceral separation of coal developed by van Krevelen and associates,[58] differs from that previously described in the use of an inorganic solution as the heavy liquid separation medium. The coal was crushed in ball mill for 24 hr to less than 10-μ particle size and then dispersed in a zinc chloride solution of 1.25 specific gravity and separated in a centrifuge. These researchers occasionally obtained concentrations of a particular maceral of over 90% "purity," but normally the final separations were less than this owing to the intimate mixing and compactness of maceral constituents in the coals used as raw material. The ultimate, proximate, and petrographic compositions and densities of the maceral concentrates were determined. With these data and reflectance values, it was possible to calculate the composition and structural parameters of the pure macerals and to plot their coalification tracks. Results indicated that micrinite might possibly be formed by coalification of humic acid flocculations, that is, by coalification preceded by an oxidation process.

Ladam and associates[59] investigated a number of methods of concentrating exinite from high-volatile bituminous coal with reasonably satisfactory success by two methods. One of these was heavy liquid separation based on the fact that the exinite of the coal investigated had a density of 1.1 compared to 1.3 for vitrinite of the same coal. In this technique, the coal was broken to −0.5-mm particle size and screened at 0.2 mm to eliminate a high proportion of vitrinite fines. The 0.5 by 0.2-mm-sized product was given a preliminary separation in a solution of calcium chloride of 1.2 specific gravity, and the resulting concentrate was pulverized to −50 μ and

[55] Kröger, C., *Brennstoff-Chem.*, **37,** 183–9 (1956).

[56] Kröger, C., and Pohl, A., *Brennstoff-Chem.*, **38,** 179–83 (1957).

[57] Horton, L., *Fuel*, **31,** 341–54 (1952).

[58] Dormans, H. N. M., Huntjens, F. J., and van Krevelen, D. W., *Fuel*, **36,** 321–39 (1957).

[59] Ladam, A., Iselin, P., and Alpern, B., *Symposium on Physico-Chemical Studies of Coal Macerals, Second Intern. Conf. on Coal Science*, Valkenburg, The Netherlands, May, 1957, Paper C5.

separated in an organic liquid (carbon tetrachloride plus benzene) of 1.2 specific gravity with settling of the concentrate in the heavy liquid accelerated by means of a centrifuge. By this technique, a concentrate of 90% exinite was obtained. A chemical method of selective dissolution of macerals was developed when it was discovered that α-naphthol at its boiling point of 279° dissolved the pure vitrinite of the Lorraine coal in a period of 3 to 5 hr compared to 30 hr required for exinite of the same coal. Separation of exinite from the solution was effected by dilution with pyridine and filtering. Exinite is practically insoluble in pyridine. Comparative chemical studies of the exinite and the vitrinite of the same high-volatile bituminous coal showed that the exinite has an aromatic structure less condensed than that of the vitrinite. Exinite resists the action of potassium hydroxide and sulfuric acid, but it is more sensitive to the action of hydrogen under pressure and to the action of strong oxidizing agents. The coking properties also are very different.

Ergun and associates[60] obtained concentrates of macerals of a high-volatile A bituminous coal by fine grinding and centrifugal separation in a solution of zinc chloride of adjusted specific gravity. Maceral concentrate samples of 94% vitrinite, 86% exinite, 87% micrinite, and 96% fusinite were obtained. The various chemical and physical properties of the maceral concentrates including chemical composition, helium density, surface area, indices of refraction, and X-ray scattering intensities were determined. The four maceral constituents were found to differ considerably in chemical composition and in most of the physical properties. Some of the physical properties could be related to chemical composition.

Ergun and associates[61] also studied the rate of thermal decomposition of certain of the coal macerals by means of observation of their behavior on a modified, hot-stage microscope. The time required for fusion and resolidification at constant temperatures within a restricted range were determined. Data collected on exinite permitted a systematic kinetic anaylsis. The temperature coefficients of time lapse before fusion for two exinites of 7.6 and 6.5 hydrogen contents were 86 and 61 kcal/mole, respectively. However, temperature coefficients of total reaction times were about the same, 44 kcal/mole and 50 kcal/mole, respectively. Similar observations of the vitrinite macerals were not entirely satisfactory. Reactions of micrinite and fusinite could not be studied quantitatively under the microscope.

Practical Applications of Coal Petrography

The practical application of coal petrography probably has received its greatest impetus in coal-producing regions where there is a short supply of superior coals desired for specific uses. As a case in point, in the Lorraine basin of eastern France, most of the locally produced high-volatile coal is of poor quality for making metallurgical coke when used alone without blending or special treatment. Formerly, to make acceptable coke for the steel industry of Lorraine, it was necessary to import good coking coals, mainly from Germany, to blend with the local Lorraine coals to make a coke that would meet steel mill requirements. Therefore, from the standpoint of economic and political necessity, it seemed important to the Lorraine industry to increase the use of local coals and thereby decrease the importation of foreign coals for coke-making in Lorraine.

A study of German coal petrography practices in connection with improvement of Ruhr coals for coke-making led to consideration of coal preparation from the point of view of petrographic composition of coal, as a promising approach to the problem of improvement of Lorraine coals for making coke. Burstlein[62] designed and built a preparation plant at the steel-making center of Thionville, incorporating in its operation the principle of elastic or differential breakage of coal components based on their inherent differences in hardness[63] as well as differences in coking properties.

The basis of the coal preparation method developed by Burstlein, now called the Longwy-Burstlein, or Sovaco process, is the control of

[60] Ergun, S., McCartney, J. T., and Mentser, M., *Econ. Geol.*, **54**, 1068–77 (1959).

[61] Ergun, S., O'Donnell, H. J., and Parks, B. C., *Fuel*, **38**, 205–10 (1959).

[62] Burstlein, E., *Chaleur & ind.*, **30**, 351–70 (1954); **31**, 14–28 (1955).

[63] Lehmann, K., and Hoffmann, E., *Brennstoff-Chem.*, **13**, 21–9 (1932).

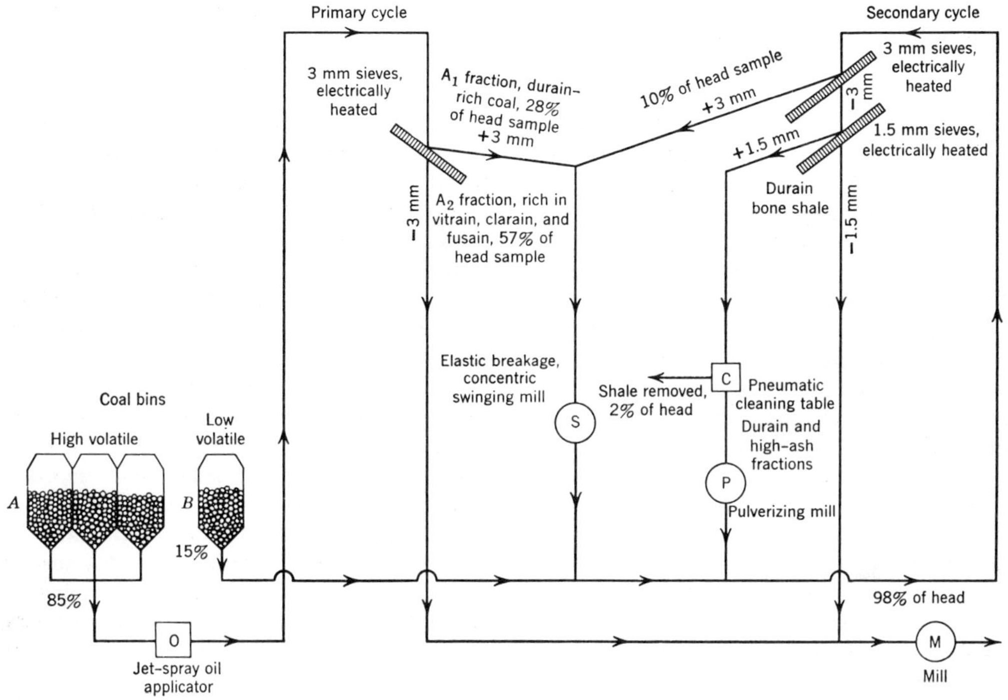

Fig. 5. Diagrammatic flow sketch showing working principle of petrographic coal preparation as used in the Sovaco process.[62]

uniformity of grain size of coal petrographic components with screens and impact breakage mills. The system takes advantage of the differences in resistance to mechanical breakage of the various petrographic components to effect the desired grain size control. The vitrain and vitrinite-rich clarain fractions possessing good coking qualities but also marked friability are kept within the optimum particle size ranging from about 3 mm to $1\frac{1}{2}$ mm. Durain, the hardest component and usually inferior to vitrain and clarain in coking quality, is broken to $-1\frac{1}{2}$-mm grain size and at the same time cleaned of associated mineral and rock impurities. Fusain, having virtually no resistance to mechanical breakage, is reduced to fine-grained, powdery particle size. This component also has virtually no coking quality, but in fine particle size and in limited quantity can be incorporated into the plastic coal mass with beneficial results as a strengthening agent. All sizes of petrographically broken coal are then brought together and thoroughly mixed into the desired blend. A flow diagram of the preparation procedure is shown in Fig. 5. It has been claimed that the objectives set forth in planning the preparation plant were fully met; that is, the amount of Lorraine coal used in coke blend was successfully increased by a considerable amount, up to about 50%.

Blending of coking coal mixtures based on petrography has been investigated in Australia.[64] It was found that the caking and swelling properties of coal blends are markedly increased by addition of vitrain and vitrinite-rich clarain fines. The fluidity was increased by vitrain enrichment. An examination of the raw mined coal revealed that vitrain and vitrinite-rich clarain tended to concentrate in the $-\frac{1}{4}$-in. fines, and the durain, which usually has less satisfactory caking qualities, accumulated in the coarser coal of the mined product. By testing varying proportions of vitrain-rich fines and the coarser sizes, optimum blend ratios were established for producing the most satisfactory coke product.

[64] Anon., *Coal Research* (Australia, CSIRO), No. 6, 2–6 (1959).

Taylor[65] of the same research group studied the behavior of the petrological components of coal on carbonization. In this investigation, the carbonization of the maceral constituents of some Ruhr and New South Wales coals of comparable rank was examined using microscopic techniques including determination of volume change. It was revealed that the vitrinite of the coals studied formed fused coke with 55% of the volume of the original coal. Semifusinite showed some distortion but no fusion or appreciable overall shrinkage; micrinite appeared to shrink slightly but did not fuse. No alteration could be detected in fusinite and sclerotinite. Exinite appeared to volatilize almost completely leaving a very small residue.

In the Soviet Union, Korobchanskiĭ and associates[66] investigated selective crushing of Donetz Basin coals with the objective of eliminating durain and concentrating vitrain-rich coal to prepare coal for coke-making. A selective centrifugal crusher was used which gave about 73% yields of vitrain-rich coal having 9.2% ash and 2.6% sulfur, and 22.4% yields of durain-rich coal having 24% ash and 5.8% sulfur. The remainder represented the dust fraction composed mostly of fusain. The recognized superiority of the vitrain-rich bright coal concentrates for coking purpose as well as its low ash content made this coal the most valuable product of the preparation procedure and suggested that the less satisfactory high-ash durain be used either for making semicoke or for generating steam.

Fenton and Smith[67] have presented a short review and discussion of some observations on the effect of coal petrology and coal rank in regard to carbonization. They concluded that coal petrology or petrography promises to find real use in the industry, but it remains to be proved whether or not it will become an indispensable discipline and fully earn its cost.

By no means has all investigative work concerned with improvement in coke-making by coal preparation and blending practices involving the petrographic composition of coal, carried on in various coal-producing countries, been cited here, but in general most of the investigation programs have followed the same principles as described and accomplished similar results. The cumulative results of many such investigations of practical application of coal petrography in coal technology seem to establish rather convincingly the fact that the various petrographic components of coal possess marked differences in coking properties as well as in other technological properties, and that the natural mixture in which they occur in a particular coal is not always the best for producing the best quality of coke or other desired results. Various experimental tests have shown that improvement in the coke-making quality of an unsatisfactory coal usually can be obtained by preparation practices which take advantage of significant differences in the inherent physical qualities of coal components such as hardness and density to effect a selective concentration of the most desirable coal for preparing coking blends that will give optimum results in quality of coke produced.[68]

Summary

Reports covering various lines of investigation have been cited to show that coal petrography has progressed from mere routine examination of coal to determine proportional amounts of certain visually recognizable banded components to a more basic study of the chemical and physical properties of what are assumed to be the ultimate optically indentifiable entities of coal—the maceral constituents. These maceral constituents have been successfully separated and concentrated from various coals and studied by a variety of chemical and physical techniques to determine their fundamental properties and characteristics in hopes that such information and data would lead to better understanding of the true nature and origin of coal. It is also hoped that this more sophisticated approach to petrographic investigation of coal may eventually result in

[65] Taylor, G. H., *Fuel*, **36,** 221–35 (1957).

[66] Korobchanskiĭ, I. E., Kutznetsov, M. D., Eĭdelman, E. Ya., Potashnikova, M. M., Korobchanskiĭ, P. I., and Sirenko, N. P., *Koks i Khim*, **1956,** No. 6, 8–13.

[67] Fenton, G. W., and Smith, A. H. V., *Gas World*, **149,** Coking Sect. 81–7 (1959).

[68] Spackman, W., *Trans. N. Y. Acad. Sci.*, **20,** No. 5, 411–23 (1958). Spackman, W., Berry, W. F., Dutcher, R. R., and Brisse, A. H., *Yearbook Am. Iron Steel Inst.*, **1960,** 403–49. Schapiro, N., and Gray, R. J., *Proc. Illinois Mining Inst.*, **1960,** 83–97. Kötter, K., *Brennstoff-Chem.*, **41,** 263–72 (1960).

the development of better objective methods of petrographic characterization of coal and reliable means of predicting and explaining the behavior of coal under various conditions of technological processing. Of especial interest to coal science and technology in general is the fact that a considerable number of laboratories located in most of the important coal-producing countries of the world are now engaged in much investigative work in which coal petrography plays a primary or integral part.

The scope of the problems that engage the coal petrographers and their scientific associates in the widely separated coal research centers is remarkably similar. The principal problems that confront these workers can be summarized as follows.

1. Improvement of the technique of petrographic classification of coal substances including classification into coal types based primarily on genetic origins and macroscopic characterization.
2. Nomenclatural classification and description of microcomponents of coal and microscopic characterization of coal types.
3. Study of the botanically derived components of coal and the nature of their conversion to coal substances by vitrinization, incorporation, fusinization, or other processes of coalification.
4. Development of methods of separating or enrichment of the microcomponents of coal in relatively pure form for fundamental studies of their chemical and physical properties and technical behavior.
5. Development of methods for determining the rank of coal not only on qualitative considerations but also on the basis of quantitative determination of such properties as reflectance, refraction, density, porosity, and others.
6. Combined chemical, physical, petrographic, and technological studies of coal and coal components to determine the quality and suitability of coal for specific uses such as coking, hydrogenation, and gasification, and the improvement of less desirable coals for special uses by methods of beneficiation and preparation.
7. Improvement of qualitative and quantitative methods, including search for strictly objective techniques, for determining the proportional relationship in coal of various coal components and coal types of different properties and technological behavior.
8. Study of mineral matter in coal including its genesis, nature, distribution, and concentration in coal, and including the so-called rare or trace elements such as germanium and uranium which may have economic importance independent of the associated coal substance.
9. Study of varied geological problems pertaining to coal such as the nature and stability of roof rock, nature and lateral continuity of rock partings and their effect on seam thicknesses, lateral uniformity or variation of coal types in seams, and the correlation of coal seams by means of petrographic study and spore analysis.

CLASSIFICATION OF COAL

The present-day classification of coal has evolved from many years of investigation of the chemical and physical properties of coal. Rose[69] reviewed the early developments of various methods of coal classification and discussed the different parameters used for defining coals in the various systems.

With regard to efforts in the field of coal classification, in 1946, the coal survey officers of the Department of Scientific and Industrial Research (DSIR) developed and proposed for use in England[70] a system of coal classification based on the parameters (1) volatile matter calculated to the dry ash-free (d.a.f.) basis, and (2) coking property as determined by the Gray-King coke type assay. This system, with a few minor revisions made in 1952 for a more exact definition of coke type, and with the specification that coal containing more than 7.5% ash should be washed before analysis, has been accepted by the National Coal Board as a satisfactory system for defining British coals. However, in view of subsequent information and increasing knowledge of coal, further revisions were made in 1956.[71] These modifications mainly were (1) calculation of volatile matter on a dry, mineral matter free (d.m.m.f.) basis and (2) extension of the

[69] Rose, H. J., *Chemistry of Coal Utilization*, John Wiley and Sons, New York, 1945, Vol. 1, pp. 25–85.

[70] *Gt. Brit. Dept. Sci. Ind. Research, Fuel Research Survey Paper* No. 58 (1946), 25 pp.

[71] Anon., *Gas Times*, **91,** 246 (1957).

TABLE 3

THE COAL CLASSIFICATION SYSTEM USED BY THE NATIONAL COAL BOARD

Group (S)	Class	Volatile Matter dry, mineral matter-free %	Gray-King Coke Type*	General Description
100		Under 9.1	A	Anthracites
	101†	Under 6.1	A	
	102†	6.1–9.0		
200		9.1–19.5	A-G8	Low-volatile steam coals
	201	9.1–13.5	A-G	Dry steam coals
	201*a*	9.1–11.5	A-B	
	201*b*	11.6–13.5	B-C	
	202	13.6–15.0	B-G	Coking steam coals
	203	15.1–17.0	B-G4	
	204	17.1–19.5	G1-G8	
	206	9.1–19.5	A-B for V. M. 9.1–15.0 A-D for V. M. 15.1–19.5	Heat-altered low-volatile steam coals
300		19.6–32.0	A-G9 and over	Medium-volatile coals
	301	19.6–32.0	G4 and over	Prime coking coals
	301*a*	19.6–27.5	G4 and over	
	301*b*	27.6–32.0		
	305	19.6–32.0	G-G3	(Mainly) heat-altered medium-volatile coals
	306	19.6–32.0	A-B	
400–900		Over 32.0	A-G9 and over	High-volatile coals
400		Over 32.0	G9 and over	Very strong caking coals
	401	32.1–36.0	G9 and over	
	402	Over 36.0		
500		Over 32.0	G5-G8	Strongly caking coals
	501	32.1.3		
	502	Over 32.1.0	G5-G8	
600		Over 32.0	G5-G4	Medium caking coals
	601	32.1–36.0		
	602	Over 36.0	G1-G4	
700		Over 32.0	B-G	Weakly caking coals
	701	32.1–36.0		
	702	Over 36.0	B-G	
800		Over 32.0	C-D	Very weak caking coals
	801	32.1–36.0	C-D	
	802	Over 36.0		
900		Over 32.0	A-B	Noncaking coals
	901	32.1–36.0	A-B	
	902	Over 36.0		

* Coals of groups 100 and 200 are classified by using the parameter of volatile matter alone. The Gray-King coke types quoted for these coals indicate the ranges found in practice and are not criteria for classification.

† To divide anthracites into two classes, it is sometimes convenient to use a hydrogen content of 3.35% (d.m.m.f.) instead of a volatile matter of 6.0% as the limiting criterion. In the original Coal Survey rank coding system the anthracites were divided into four classes then designated 101, 102, 103, and 104. Although the present division into two classes satisfies most requirements, it may sometimes be necessary to recognize four or five classes.

class of prime coking coals of medium volatile matter content. An outline of the present system of classification used by the National Coal Board is given in Table 3.

After an intensive study of coal classification based upon ultimate and proximate analyses, Mott[72] proposed a new scheme of coal classification in which coals are divided into broad groups according to volatile matter, and further divided into divisions according to calorific value. Mott's system is usually represented by a classification chart in which volatile matter and calorific value are used as the principal rectangular coordinates, whereas hydrogen and oxygen are subsidiary coordinates inclined at an angle of 40° to the main coordinates. The plot of a variety of coals of different rank on this rectangular coordinate chart shows that a majority of the coals tend to form a distribution pattern along a somewhat broad curving path that is described as the coal band. A similar classification chart featuring a coal band was proposed as early as 1899 by Seyler[73] who revised it in 1931 and in 1938 (Fig. 6). Despite the scientific excellence of both Seyler's and Mott's systems of coal classification, their rather complicated nature has precluded their widespread acceptance and usage.

International System of Coal Classification

The greatly increased volume of trade in coal among various nations following World War II, particularly among those of Western Europe, emphasized the need for an international system of coal classification. The various coal-producing countries concerned possessed their own national classification systems, but these had been developed to fit the characteristics of domestic coals and often the needs of the national coal-using industry without reference to coals of other countries. This practice of developing national coal classification systems resulted in the evolution of different terminologies for describing similar or identical coals, which naturally led to confusion in evaluating and comparing the coals shipped in international trade.

In 1949, because of this situation, the Coal Committee of the Economic Commission for Europe, Geneva, Switzerland, established a classification working party to develop an international system of coal classification. Experts from Austria, Belgium, Czechoslovakia, Denmark, the Federal Republic of Germany, France, Greece, Hungary, Italy, the Netherlands, the Union of Soviet Socialist Republics, East Germany, the United Kingdom, the United States of America, and Yugoslavia participated in various stages of the classification work.

The preliminary study by the classification working party consisted of exchanging information on the existing national systems of classification and on methods of sampling and analyzing coal. The exchange of samples of representative coals gave much valuable information about properties of coals from different countries and provided data for comparing analytical results by the standardized procedures used in each country. It was found that many national systems had features in common. Furthermore, the results of analysis and calorific value determinations made by the various national methods were close enough to be used for international classification until standard international methods were made available.

Hard Coal Classification

At its eighth session in June 1953, the classification working party agreed on a system designated "International Classification of Hard Coal by Type" and submitted it to the Coal Committee, and also recommended a simplified system of grouping coals for statistical purposes, Fig. 7. Later in 1953, the Coal Committee invited the participating countries to give the two proposed classification schemes a trial period of not longer than two years and in 1956, after noting that the trials had been successful, recommended to the various governments concerned the practical application of the two schemes. A paper describing and illustrating the two systems was published as a United Nations document in 1956.[74]

A report by Ode and Frederic[75] reviewed the system of hard coal classification by type

[72] Mott, R. A., *J. Inst. Fuel*, **22**, 2–11 (1948).

[73] Seyler, C. A., *Proc. S. Wales Inst. Engr.*, **21**, 483–526 (1899); **22**, 112–20 (1900); **53**, 254–327, 396–407 (1938). *Fuel*, **3**, 15–26, 41–9, 79–83 (1924).

[74] *United Nations*, Publ. No. 1956 II, E.4, E/ECE/247, E/ECE/Coal/110, 52 pp.

[75] Ode, W. H., and Frederic, W. H., *U.S. Bur. Mines, Rept. Invest.* No. 5435 (1958), 19 pp.

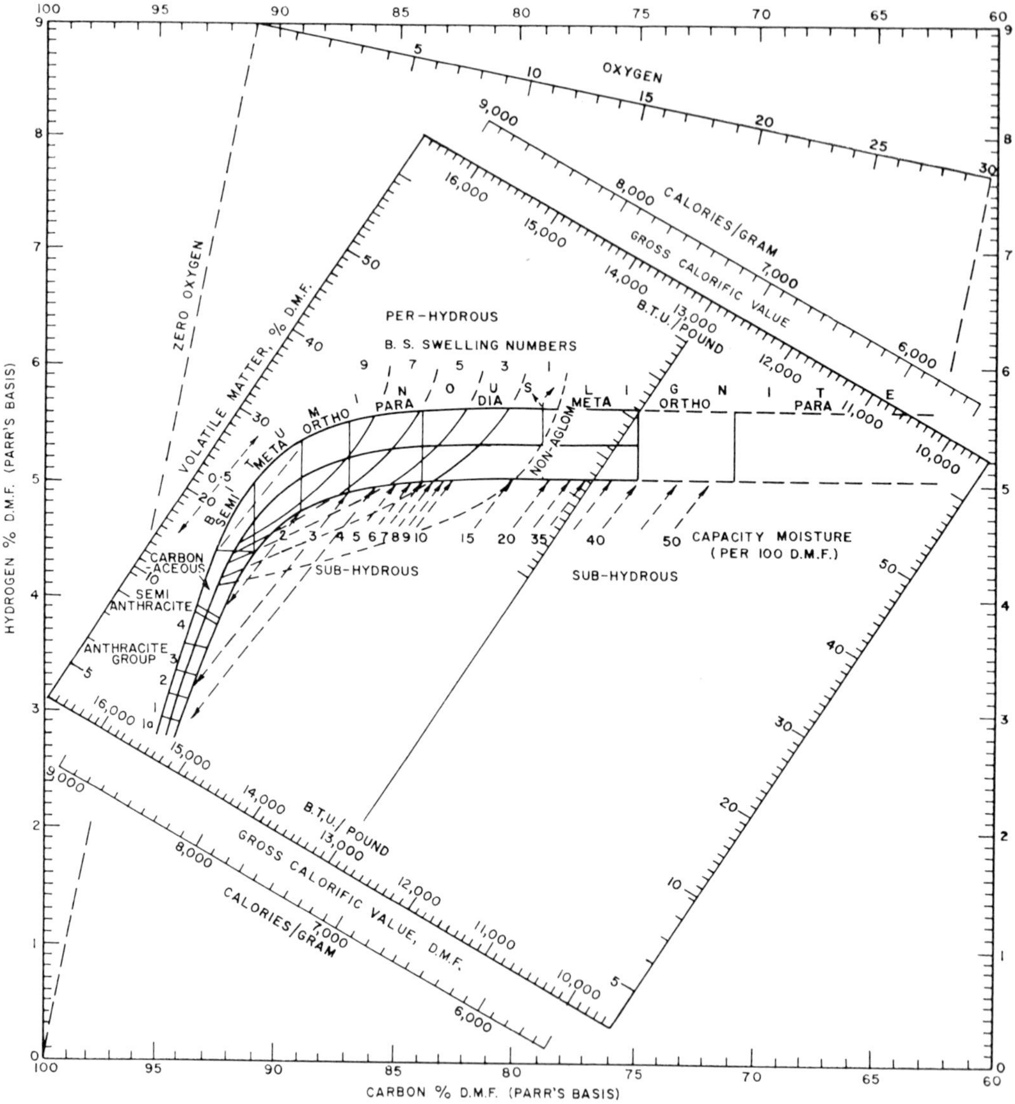

Fig. 6. Seyler's coal chart 1951. All results on dry, mineral-matter free basis (Parr's basis) except capacity moisture and B.S. swelling numbers.

with particular emphasis on its application to American coals. Figure 8 shows a grouping of American coals in the International System of Classification. The new International System of Coal Classification has also been described and reviewed from the standpoint of European coal scientists by King[76] in England, by Radmacher[77] in Germany, and by Deshalit[78] in the Soviet Union and others. King observed that it is implicit in the classification that the methods of analysis and test should be subject to international agreement, and that the International Standards Organization (I.S.O.) is making notable progress in this direction, as he had reported in earlier papers.[79]

[76] King, J. G., *Coke and Gas*, **19,** 152–60 (1957).

[77] Radmacher, W., *Brennstoff-Chem.*, **38,** 58–60 (1956).

[78] Deshalit, G. I., *Koks i Khim.*, **1957,** No. 6, 55–8.

[79] King, J. G., *Coke and Gas*, **16,** 55–60 (1954); **17,** 305–11 (1955).

GROUPS (determined by caking properties)			CODE NUMBERS										SUBGROUPS (determined by coking properties)		
GROUP NUMBER	ALTERNATIVE GROUP PARAMETERS: Free-swelling index (crucible-swelling number)	ALTERNATIVE GROUP PARAMETERS: Roga index	The first figure of the code number indicates the class of the coal, determined by volatile-matter content up to 33% V. M. and by calorific parameter above 33% V. M. The second figure indicates the group of coal, determined by caking properties. The third figure indicates the subgroup, determined by coking properties.										SUBGROUP NUMBER	ALTERNATIVE SUBGROUP PARAMETERS: Dilatometer	ALTERNATIVE SUBGROUP PARAMETERS: Gray-King
3	> 4	> 45					435	535	635				5	> 140	> G_8
						334	434	534	634				4	> 50–140	G_5–G_8
						333	433	533	633	733			3	> 0–50	G_1–G_4
						332 a / 332 b	432	532	632	732	832		2	⩽ 0	E–G
2	$2\frac{1}{2}$–4	> 20–45				323	423	523	623	723	823		3	> 0–50	G_1–G_4
						322	422	522	622	722	822		2	⩽ 0	E–G
						321	421	521	621	721	821		1	Contraction only	B–D
1	1–2	> 5–20			212	312	412	512	612	712	812		2	⩽ 0	E–G
					211	311	411	511	611	711	811		1	Contraction only	B–D
0	0–$\frac{1}{2}$	0–5		100 (A, B) I	200 II	300	400	500	600	700	800	900	0	Nonsoftening	A
CLASS NUMBER →			0	1	2	3	4	5	6	7	8	9			
CLASS PARAMETERS	Volatile matter (dry, ash-free) →		0–3	> 3–10 (A: >3–6.5; B: >6.5–10)	> 10–14	> 14–20	> 20–28	> 28–33	> 33	> 33	> 33	> 33			
	Calorific parameter [a] →		—	—	—	—	—	—	> 13,950	> 12,960–13,950	> 10,980–12,960	> 10,260–10,980			

Statistical groupings (roman numerals): I (100 A, B); II (200); III (300, 311, 400, 411); IV (312–323, 412–423); V_A (333, 334); V_B (433, 434); V_C (534, 535, 634, 635); V_D (432, 532, 533, 632, 633, 732, 733, 832); VI_A (521–823); VI_B (512–812); VII (500, 511, 600, 611, 700, 711, 800, 811, 900).

CLASSES
(Determined by volatile matter up to 33% V. M. and by calorific parameter above 33% V. M.)

As an indication, the following classes have an approximate volatile-matter content of:
Class 6 33–41% volatile matter
7 33–44% " "
8 35–50% " "
9 42–50% " "

Note: (i) Where the ash content of coal is too high to allow classification according to the present systems, it must be reduced by laboratory float-and-sink method (or any other appropriate means). The specific gravity selected for flotation should allow a maximum yield of coal with 5 to 10 percent of ash.
(ii) 332a . . . >14–16% V. M.
332b . . . >16–20% V. M.

[a] Gross calorific value on moist, ash-free basis (30° C., 96% relative humidity) B. t. u./ lb.

Fig. 7. International classification of hard coals by types with statistical groupings, shown bounded by heavier lines. The statistical groupings are designated by roman numerals.[74]

GROUPS (determined by caking properties)			CODE NUMBERS										SUBGROUPS (determined by coking properties)		
GROUP NUMBER	ALTERNATIVE GROUP PARAMETERS: Free-swelling index (crucible-swelling number)	ALTERNATIVE GROUP PARAMETERS: Roga index	The first figure of the code number indicates the class of the coal, determined by volatile-matter content up to 33% V. M. and by calorific parameter above 33% V. M. The second figure indicates the group of coal, determined by caking properties. The third figure indicates the subgroup, determined by coking properties.										SUBGROUP NUMBER	ALTERNATIVE SUBGROUP PARAMETERS: Dilatometer	ALTERNATIVE SUBGROUP PARAMETERS: Gray-King
3	>4	> 45					435	535	635				5	> 140	> G_8
						334	434	534	634	734			4	>50–140	G_5–G_8
						333	433	533	633	733			3	> 0–50	G_1–G_4
						332 a / 332 b	432	532	632	732	832		2	⩽ 0	E–G
2	$2\frac{1}{2}$–4	>20–45				323	423	523	623	723	823		3	> 0–50	G_1–G_4
						322	422	522	622	722	822		2	⩽ 0	E–G
						321	421	521	621	721	821		1	Contraction only	B–D
1	1–2	> 5–20			212	312	412	512	612	712	812		2	⩽ 0	E–G
					211	311	411	511	611	711	811		1	Contraction only	B–D
0	0–$\frac{1}{2}$	0–5	000	100 (A, B)	200	300	400	500	600	700	800	900	0	Nonsoftening	A
CLASS NUMBER →			0	1	2	3	4	5	6	7	8	9	As an indication, the following classes have an approximate volatile-matter content of: Class 6 33–41% volatile matter; 7 33–44% " "; 8 35–50% " "; 9 42–50% " "		
CLASS PARAMETERS	Volatile matter (dry, ash-free) →		0–3	>3–10 (A: >3–6.5; B: >6.5–10)	>10–14	>14–20	>20–28	>28–33	>33	>33	>33	>33			
	Calorific parameter [a] →		—	—	—	—	—	—	>13,950	>12,960–13,950	>10,980–12,960	>10,260–10,980			
CLASSES (Determined by volatile matter up to 33% V. M. and by calorific parameter above 33% V. M.)															

Note: (i) Where the ash content of coal is too high to allow classification according to the present systems, it must be reduced by laboratory float-and-sink method (or any other appropriate means). The specific gravity selected for flotation should allow a maximum yield of coal with 5 to 10 percent of ash.

(ii) 332a . . . >14–16% V. M.
332b . . . >16–20% V. M.

(iii) From information available, American coals are represented by open-faced blocks. There are probably no commercial deposits of American coals represented by ⁚⁚⁚

[a] Gross calorific value on moist, ash-free basis (30° C., 96% relative humidity) B. t. u./ lb.

Fig. 8. Grouping of American coals in international system of classification.[75]

TABLE 4

INTERNATIONAL CLASSIFICATION OF COALS WITH A GROSS CALORIFIC VALUE BELOW 5700 kcal/kg*[80] (STATISTICAL GROUPING)

Group Number	Group Parameter: Tar Yield % (dry, ash-free)	Code Number					
40	25	1040	1140	1240	1340	1440	1540
30	20–25	1030	1130	1230	1330	1430	1530
20	15–20	1020	1120	1220	1320	1420	1520
10	10–15	1010	1110	1210	1310	1410	1510
00	10 and less	1000	1100	1200	1300	1400	1500
	Class Number	10	11	12	13	14	15
Class Parameter	Total Moisture % (ash-free)	20 and less	>20 to 30	>30 to 40	>40 to 50	>50 to 60	>60 to 70

Notes: The total moisture content refers to freshly mined coal.
For internal purposes, coals with a gross calorific value over 5700 kcal/kg, considered in the country of origin as brown coals or lignites but classified as hard coals for international purposes, may be classified under this system, to ascertain, in particular, their suitability for processing.
When the total moisture content is over 30%, the gross calorific value is always below 5700 kcal/kg.
* Moist, ash-free basis (30°C and 96% relative humidity).

Brown Coal Classification

The term hard coal as used in the international system is based on European usage and is defined as coal with a gross calorific value of more than 10,260 Btu/lb (5700 kcal/kg) on the moist ash-free basis. Coals classified in the American Society for Testing Materials (ASTM) system as anthracitic, bituminous, and the higher-rank subbituminous coals are included in the international system for classification of hard coals by type. The soft or brown coals of European terminology, which in the ASTM system are lower-rank subbituminous and lignitic coals, were subjected to further study by the classification working group with the objective of creating a separate system of classification.

The significance to the European coal industry of developing an international system of brown coal classification is apparent when the increasing importance of brown coal in the European economy is considered. As a result of developments in recent years, not only has the use of brown coal for energy production been substantially increased, but with the growth of the chemical industry brown coals have become an important source of raw material for chemical processing. Brown coals occur and are worked in nineteen European countries from which production reached a total of over 400 million tons in 1955.

After a very thorough study of a large number of different coals, the *ad hoc* group of experts on brown coal of the E.C.E. coal classification working party proposed the establishment of a dividing line between hard coals and brown coals on the basis of gross calorific value set at 5700 kcal/kg with a tolerance of ±60 kcal/kg on a moist, ash-free basis. The international scheme of classification for brown coals is based on two intrinsic principal characteristics which indicate the value of brown coal as a fuel, and as a raw material for chemical purposes: (1) the total moisture on ash-free basis, and (2) the tar yield on dry, ash-free basis.

The coal classification working party submitted the proposed scheme of brown coal classification to the Coal Committee[80] in 1957, (Table 4) with request for a 1-year trial period. This proposal and request were endorsed by the Coal Committee, and after a year's trial period the proposed scheme of brown coal classification was officially adopted and recommended to the various participating governments for practical application.

[80] *United Nations*, Publ. No. 1957 II. E., Min. 20., 67 pp.

CHAPTER 2

PHYSICAL PROPERTIES OF COALS

H. TSCHAMLER and E. DE RUITER

*Union Carbide European Research Associates S. A., Brussels, Belgium**

In contrast to chemical methods, physical methods have the advantage that by their use the structure being investigated is not changed at all or only to a very slight extent. On the other hand, for both theoretical and practical reasons, physical measurements may be applicable only under rather circumscribed experimental conditions, whereas chemical studies may be applied much more widely. Thus chemical methods may be regarded as an essential adjunct to physical measurements and vice versa.

Numerous basically different physical methods can be applied to the study of molecular structures, but they are significant only when the basis for the structural sensitivity is theoretically understood or, at the very least, qualitatively worked out. The establishment of a structural influence must be achieved through the calibration of the effect with a completely known structure. But this involves interpolation, or more likely extrapolation, from known to unknown, and can only be done with certainty when the physical conditions are completely understood.

Most of the numerical data deduced from experimentally proven techniques must be regarded as reliable. If the structure is known in advance, a given characteristic can usually be uniquely measured; if the structure is not known in advance, however, the measured characteristic may be consistent with more than one structure and hence only the following limited statement is possible. Certain structures are compatible with the experimental results, but others are excluded. Hence, to arrive at a precise structural conclusion, the largest possible number of different observations must be brought to bear on the problem.

Since different physical methods place restrictions of widely varying precision on the structural variations with which they are compatible, each must be assigned a different weight in arriving at a final decision. The objective evaluation of such results, avoiding all prejudice, requires experience, critical examination, and scientific judgment. It is understood that the possibility of detailed structural conclusions decreases as the number of atoms in the molecule increases and as the molecular symmetry decreases.

A schematic presentation of the physical methods reviewed, together with an indication of the quantities about which each permits conclusions relating to coal structure, may assist in evaluating the importance of each and aid in an overall survey of the material.

Measurement of	Yields Information on:
X-ray diffraction	Size distribution of aromatic ring systems Average diameter of aromatic lamellae Mean bond length Average thickness of the packets of lamellae
Ultraviolet and visible absorption	Aromaticity Average aromatic ring size
Reflectance Refractive index (molecular refraction)	Aromatic surface area Optical anisotropy
Infrared absorption	Characteristic groups such as OH, CH_{ar}, CH_{al}, $(C{=}C)_{ar}$ The H_{ar}/H_{al} ratio
Electron spin resonance	Free radicals
Proton spin resonance	The H_{ar}/H_{al} ratio
Electrical conductivity	Changes in size of aromatic ring systems
Diamagnetic susceptibility (molar diamagnetic susceptibility)	Average number of aromatic rings
Dielectric constant	Dipole moment
Sound velocity	Aromaticity
Density (molar volume)	Aromaticity Ring condensation index

* The authors express their grateful appreciation to Dr. R. H. Gillette and Dr. C. E. Sunderlin for useful discussions and their help in translating the text.

X-RAY DIFFRACTION

(In collaboration with W. Ruland)

When X rays pass through matter a portion of the energy is elastically scattered as coherent radiation, inelastically scattered as incoherent (Compton) radiation, and emitted as fluorescent radiation. All three of these types of secondary rays arise from the interaction of the X-ray quanta with the electron shells of the atoms in the material irradiated. For the coherent radiation, the electrons acted on by the field of the incident beam reradiate X rays of the same wavelength; for Compton and fluorescent radiation there is a displacement of the wavelength. The superposition of the coherently scattered rays from the various scattering centers (atoms or ions) leads to interference effects, the pattern of which is a function of the geometric arrangement of the scattering units; Compton and fluorescence radiations, because of their incoherence, produce no interference effects and yield no information concerning atomic arrangement.

The diffraction of X rays by crystals gives sharp interference maxima of the coherent scattered radiation, an effect which is produced by a regular array of atoms over regions which are infinitely large in comparison with the wavelength of the X rays used in the study. As is well known, the conditions that must be satisfied if the interaction of the X-ray beam with ordered structures is to produce sharp interferences is given by Bragg's law.

$$2d \sin \theta = n\lambda$$

where d is an interplanar distance, θ the Bragg angle, λ the wavelength of the primary X rays, and n the order of the diffraction.

If monochromatic incident radiations were employed with large crystals, the sharp interference peaks—the locations of which are given by Bragg's law—would rise above a continuous scatter background produced by the Compton and fluorescent radiations and permit a simple separation of the coherent portion. As the crystal dimensions decrease, the interference peaks, which are initially very narrow, become broadened and, under suitable conditions, the crystal dimensions can be estimated from this broadening since the half-width of a peak is, in first approximation, inversely proportional to the crystal dimension.

An amorphous substance, in contrast to a crystal, shows no interference peaks for the coherent scattering but rather a continuous scattering curve with broad maxima and minima. The separation of fluorescence and Compton radiation from the coherent scattering is not as simple as with crystals. If fluorescent radiation cannot be avoided by suitable choice of wavelength of the incident beam, it can only be eliminated by monochromatization of the secondary beam. The Compton scattering can be evaluated from the sum of the coherent and incoherent radiation because it stands in a definite intensity ratio to the coherent scattering. The shape of the scattering curve for amorphous substances is determined by the ordering of atoms in the molecules, by the ordering of the molecules with respect to one another, and possibly by some superstructure (micelles, pores, etc.). The influence on the scattering curve of these different types of order is dependent on wavelength and scattering angle. At very small angles, the influence is essentially determined by the micelle or pore structure; in the middle angle range, the short-range order of the molecules plays the main role, and for large angles the structure of the molecules is principally responsible for the features of the curve.

Experimental Methods

In the medium- and large-angle range, the scattering curve for coal can be measured with a camera or diffractometer. As samples, specimens pressed from coal powder or blocks of coal may be used, the latter especially for texture studies. Recording of the secondary beam can be accomplished by film, when cameras are used, or by a variety of counters—Geiger, proportional or scintillation—when diffractometers are used. The technique of counting quanta has the considerable advantage that the quantitative intensities are obtained directly and it is not necessary to photometer the film. For photographic registration, X radiation is best monochromatized by reflection from a suitably chosen crystal, although the beam intensity is seriously weakened by this procedure. Therefore, this technique unfortunately cannot be used with diffractometers. The best procedure, when diffractometers are used, is monochromatization by the filter-difference method, or when proportional or scintillation counters are used, by a pulse height discriminator.

The counter diffractometer provides the possibility of eliminating fluorescent radiation. In coal diffraction diagrams, when copper radiation is used, iron fluorescence becomes a significant factor; even at low iron contents ($<1\%$) considerable fluorescent intensity is produced.[1] Since coal samples nearly always contain iron, scattering measurements which do not monochromatize the secondary beam may be seriously in error. Since the corresponding monochromatization cannot be satisfactorily applied to photographic techniques, the only alternative is extremely careful chemical purification to remove traces of iron or the use of another primary radiation, for example, cobalt, which does not excite the iron fluorescence.

A further source of error in X-ray measurements on coals is air-scattering. When counters are used this can be evaluated by a blank measurement and usually proves to be negligible. However, when film is used this effect can lead to serious error because, in film cameras, it is nearly always true that the ratio of the air volume irradiated by the primary beam to the irradiated volume of the sample is much larger than with a counter diffractometer. The usual way to eliminate this error is to work with an evacuated camera or one filled with hydrogen or helium.

For a precise measurement of scattering, the slight absorption of the sample must be considered, which can falsify the measured scattering curve by smearing out the maxima and minima, by displacing the whole curve on the θ-scale, or by an angle-dependent intensity factor. These corrections, however, depend on the geometry of the measuring equipment and must be separately evaluated for each different kind of camera or diffractometer.

Interpretation of the X-Ray Scattering Curves of Coal

The X-ray patterns of coal show diffuse peaks at the positions of the most prominent graphite bands, and the peaks become less

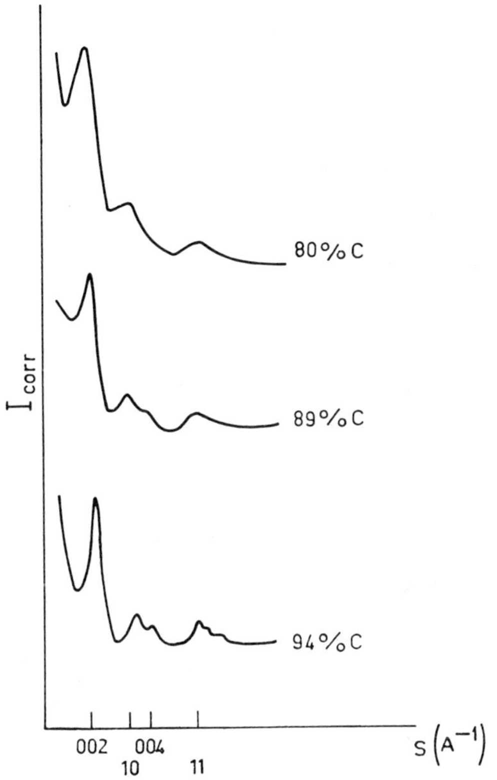

Fig. 1. X-ray scattering intensity curves for vitrains corrected for incoherent scattering.[3]

diffuse with increasing rank. A suggested classification for coking coals was based on this fact.[2] The scattering curves, indexed according to the graphite bands for three vitrains of different ranks,[3] are shown in Fig. 1. Fusains[4,5] and micrinites[6] have X-ray patterns similar to but somewhat sharper than those of vitrains and vitrinites. The X-ray pattern of the four major petrographic components exinite (86% purity), vitrinite (94% purity), micrinite (87% purity), and

[1] Ruland, W., *Acta Cryst.*, **12,** 679–83 (1959).

[2] Suzuki, K., Kitazaki, U., and Yagishita, H., *Japan. J. Geol. and Geography, Trans.*, **22,** 241–56 (1952).

[3] Hirsch, P. B., *Proc., Roy. Soc. (London)*, **A226,** 143–69 (1954).

[4] Kasatochkin, V. I., Zolotarevskaya, E. Yu., and Razumova, L. L., *Doklady Akad. Nauk S.S.S.R.*, **79,** 315–18 (1951).

[5] Siever, R., *Proc. 2nd Conf. on Origin and Constitution of Coal*, Crystal Cliffs, Nova Scotia, 1952, pp. 341–60.

[6] Cartz, L., Diamond, R., and Hirsch, P. B., *Nature*, **177,** 500–502 (1956).

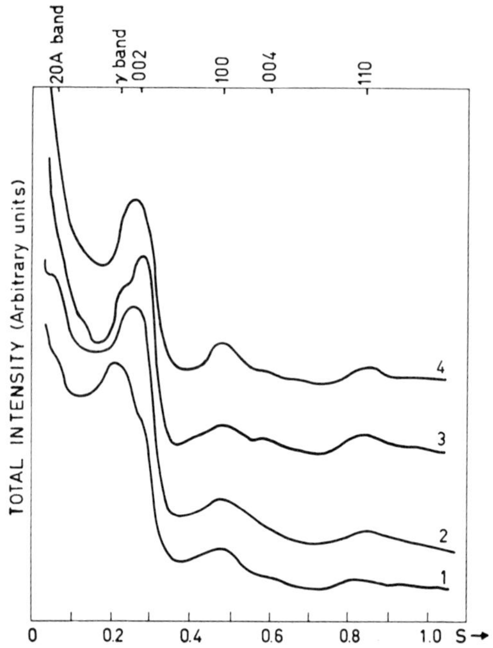

Fig. 2. X-ray patterns of an exinite (1), vitrinite, (2), micrinite (3), and a fusinite (4), each separated from a single dull coal.[7]

fusinite (96% purity), all separated from a single dull coal,[7] are shown in Fig. 2.

The diffuseness of the X-ray pattern of coal has been attributed to particles in which the arrangement of carbon atoms is that of a graphite crystal, but with extremely small size of the elemental crystallites.[8] An explanation,[9] accepted even today, is that coal possesses a "turbostratic structure,"[10,11] which means that coal contains stacked aromatic layers which are roughly parallel and equidistant, but with each layer having a completely random orientation in plane and about the layer normals. The similarity of the X-ray scattering curves for coal to those of microcrystalline graphite suggested the application of the standard methods for interpreting crystalline patterns to coal. The *d*-value associated with each maximum was calculated from Bragg's law and, from the half-width of the peaks, "crystallite" dimensions were estimated.[12] This procedure gives only qualitative results.

A number of methods of interpretation have been published based on the assumption that coal is an amorphous substance with high aromaticity. The ordering of the carbon atoms in the aromatic rings and the parallel packing of these ring systems determine the X-ray scattering curve at medium and large angles, and the aromatic systems are associated with the maxima indexed (10) and (11) whereas the parallel ordering of the aromatic layers is associated with the (002) and (004) peaks. There are two methods in general use for the interpretation of such curves: (1) use of the Fourier transform of the scattering curve, as a radial distribution function,[13,14] or as a one-dimensional Patterson projection (cosine transform);[3,15] and (2) comparison with theoretical scattering curves.[3,13,16,17]

The evaluation of the radial distribution function from the scattering curves for liquids and amorphous substances is a method frequently employed:[18] the coherent portion of the curve is calculated, normalized, and corrected with suitable angle-dependent factors. From the resulting curve, the sine transform is calculated and from it the radial distribution of atoms, giving essentially, on a time and space average, the number of atoms having a given separation from any atom of the material under investigation. The method has the advantage that it is based on the single assumption that the molecules are randomly oriented in space, but it has the disadvantage that the evaluation of the radial atomic distribution curve, especially for mixtures, is extremely difficult since the effect of the

[7] Ergun, S., McCartney, J. T., and Mentser, M., *Econ. Geol.*, **54,** 1068–77 (1959).

[8] Brusset, H., Devaux, J., and Guinier, A., *Compt. rend.*, **216,** 152–4 (1943).

[9] Blayden, H. E., Gibson, J., and Riley, H. L., *Proc. Conf. on Ultrafine Structure of Coals and Cokes*, BCURA, London, 1944, pp. 176–231.

[10] Warren, B. E., *Phys. Rev.*, **59,** 693–8 (1941).

[11] Biscoe, J., and Warren, B. E., *J. Appl. Phy.*, **13,** 364–71 (1942).

[12] Peiser, H. S., Rooksby, M. P., and Wilson, A. J. C., *X-Ray Diffraction of Polycrystalline Materials*, The Institute of Physics, London, 1955, pp. 409–29.

[13] Franklin, R. E., *Acta Cryst.*, **3,** 107–21 (1950).

[14] Nelson, J. B., *Fuel*, **33,** 381–93 (1954).

[15] Diamond, R., Ph. D. Thesis, University of Cambridge, England, 1956, 286 pp.

[16] Diamond, R., *Acta Cryst.*, **10,** 359–64 (1957).

[17] *Ibid.*, **11,** 129–38 (1958).

[18] Peiser, H. S., Rooksby, M. P., and Wilson, A. J. C., *X-Ray Diffraction of Polycrystalline Materials*, The Institute of Physics, London, 1955, pp. 438–53.

ordering of the molecules with respect to one another is difficult to distinguish from the intramolecular ordering. Nevertheless, it should be mentioned that the radial distribution functions of vitrinites (77 to 89% C) would be consistent with condensed aromatic nuclei having an average size not greater than four rings.[14]

For the one-dimensional Patterson projection (cosine transform) only a portion of the scattering curve is used, preferably the region around the (002) reflection. From this portion, after normalization and correction with appropriate angle-dependent factors, the cosine transform is evaluated and, from this, the cosine transform in a direction perpendicular to the plane of the aromatic lamellae is calculated. This last calculation gives the average number of lamellae at a given perpendicular distance from an arbitrary lamella and, from this, provides information on the packing distribution—that is, how many single lamellae are present, how many lamellae in packets of two, three, four, etc. are to be found in the coal.

From the Fourier transform of the scattering curve, information is obtained on the geometry of the structural elements in the form of a statistical distribution of the various interatomic distances, and the latter are compared with the distribution demanded by the proposed structural models. Comparison of experimental with theoretical scattering curves is equivalent to the reverse of the foregoing process; by using the atomic distributions inherent in the structural model being tested, theoretical scattering curves are evaluated and compared with the experimentally measured curves, and the results of the comparison are used to interpret the experimental curve. Clearly this method presupposes that enough is already known concerning the structure—for example, from a study of the radial distribution curve for atoms—so that only a limited number of structural models need be considered. It has the advantage that different structural parameters may be determined from different parts of the scattering curve. For example, the region of the (002) peak can, with the help of the theoretical scattering curve, provide information on the stacking of parallel aromatic lamellae[13] whereas the (10) or (11) peaks can be compared with theoretical scattering curves for hexagonal honeycomb structures of different dimensions.[10,19,20]

The method of structure determination by comparison with theoretical scattering curves has been considerably extended.[15] Assuming that the scattering curves for coals from a certain θ value onward can be expressed as a sum of theoretical scattering curves of aromatic systems of different dimensions, the theoretical curves for a series of such structures of different shapes and sizes (see Fig. 3) have been calculated.[15]

By averaging the scattering curves for four aromatic structures of the same size but of different shapes, the mean scattering curve for each of a series of different sized systems can be determined. Such theoretical scattering curves could be calculated by use of the Debye formula for the scattering of individual molecules. This formula can include more atoms, for example, α C-atoms in side groups, than are accounted for by the aromatic lamellae themselves. As a first step, the theoretical scattering curves of six different aromatic size classes (in Fig. 3 the models for $\bar{L}$ = 30 A are not included) as well as a scattering curve for the carbon atoms not included in aromatic systems have been calculated and are employed as standard scattering curves. A given experimental curve

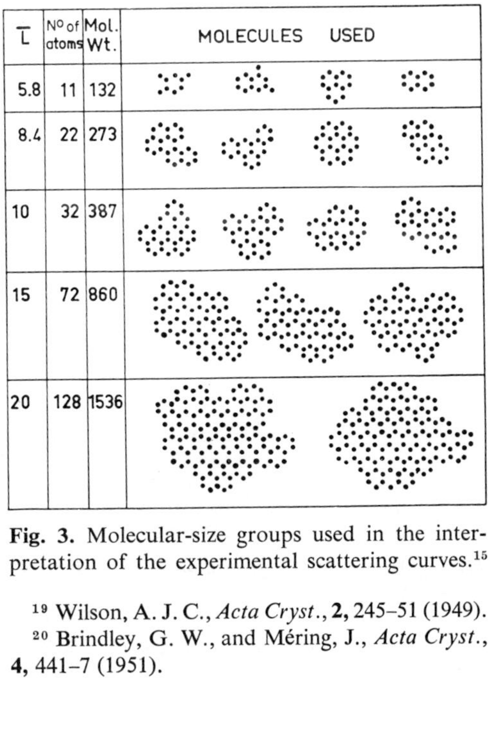

Fig. 3. Molecular-size groups used in the interpretation of the experimental scattering curves.[15]

[19] Wilson, A. J. C., *Acta Cryst.*, **2,** 245–51 (1949).

[20] Brindley, G. W., and Méring, J., *Acta Cryst.*, **4,** 441–7 (1951).

can then be approximated rather closely as a combination of these standard curves. The participation of each size group in the experimental curve is evaluated by means of a matrix calculation. From the solution of this equation, the contribution of each theoretical standard scattering curve as well as the contribution from the "amorphous carbon" can be deduced, and hence a distribution function of aromatic systems of different sizes can be determined. From this distribution the average diameter $\bar{L}$ of aromatic structures or, what amounts to almost the same thing, the average number of carbon atoms $\bar{N}$ per aromatic lamella can be very accurately calculated ($\bar{N} = 0.32\bar{L}^2$). Since the theoretical scattering curves were calculated with a single carbon-carbon bond length of 1.405 A, it is also possible to calculate the mean bond length l in the material being investigated. The method has been thoroughly tested using amorphous polymers having known sizes of aromatic clusters and showed excellent basic agreement between theory and experiment.[1] However, Fourier transforms of the Debye equation may not always make it possible to distinguish between graphite-like and diamond-like structures.[21]

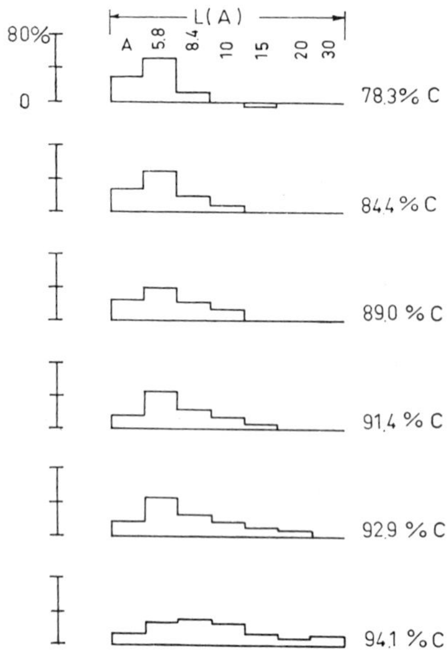

Fig. 4. Molecular-size histograms for a series of vitrains.[22]

[21] Ergun, S., and Tiensuu, V. H., *Acta Cryst.*, **12,** 1050–51 (1959).

Coals and Their Petrographic Components

Aromatic Lamellae

Size distribution. The histograms of vitrains[22] presented in Fig. 4 show that up to about 90% C, the size distribution of the aromatic clusters does not vary much, and that a relatively narrow distribution is present. Above 90% C, the size distribution becomes significantly broader. There are four essential points in interpreting these histograms:

1. To draw conclusions concerning the percentage contributions of each of the fundamental structures shown in Fig. 3 is to expect too much of the method. A self-analysis of each class[17] generally gave correct results only for the largest and smallest classes; those in between usually showed a distribution over neighboring classes. Furthermore, it was found that the histogram of an aromatic system lying between two classes does not differ from what would be obtained from a mixture of the larger and smaller systems of the two neighboring classes having the same average lamella diameter.[23]

2. The innermost atoms of edge groups are included to some extent in the layer when layer sizes are calculated[15] which must be kept in mind if conclusions are drawn from the average lamella diameter derived from the layer size distribution.[24]

3. The histograms obtained are based on the assumption that the layers are perfect. This does not seem to be so because in general the (10) band is more intense than it should be if the layers were perfect and if the size distribution was that calculated and fitted to the (11) region.[24,25] Therefore, the presence of imperfections makes the precise interpretation of the molecular sizes somewhat uncertain.[25]

4. Negative values in a histogram are produced only by the mathematical treatment of the problem.

[22] Hirsch, P. B., *Proc. Conf. on Science in the Use of Coal*, Sheffield, Inst. Fuel, London, 1958, pp. A29–33.

[23] Ruland, W., and Tschamler, H., *Brennstoff-Chem.*, **39,** 363–5 (1958).

[24] Ergun, S., *Fuel*, **37,** 365–9 (1958).

[25] Diamond, R., and Hirsch, P. B., *Proc. Conf. on Industrial Carbon and Graphite*, London, 1957, Soc. Chem. Ind., 1958, pp. 197–203.

In the histograms of Fig. 4, the "amorphous" portion is designated by A, which refers to the structural elements of the coal not associated with the hexagonal network characteristic of the aromatic clusters with the adjoining side groups. This contribution, as may be seen from the figure, decreases with increasing rank and runs completely parallel to the "nonaromatic" fraction ($1 - f_a$, where f_a = aromaticity) determined for vitrains by an entirely independent method.[26] The evaluation of the "nonaromatic" portion of the coal structure from the "amorphous" portion requires an accurate knowledge of the influence of different factors such as foreign elements and nonaromatic structures on the "amorphous" fraction.

A qualitative study of exinites, vitrinites, and micrinites led to the conclusion[6,27] that for exinites the average lamella is slightly smaller than that of the corresponding vitrintes whereas in the micrinites the aromatic clusters are distinctly larger. An analysis of the (11) bands for the distribution of "amorphous carbon" and layer size groups for exinite, vitrinite, micrinite, and fusinite, all separated from a single dull coal, shows[7] that "amorphous carbon" and smaller layer groups are characteristic of exinite and vitrinite whereas larger groups predominate in micrinite and fusinite.

Average lamella diameter ($\bar{L}$). The $\bar{L}$ values determined by application of relations between line width and lamella diameter, for the (10) or (11) peaks,[9,10] appear from a survey of the more recent literature to be too large. The $\bar{L}$ values for vitrains of different rank which are today to be regarded as most accurate are presented in Fig. 5.[22] Between 80 and 91.5% carbon $\bar{L}$ increases only from 7.0 to 7.5 A, corresponding to an increase of $\bar{N}$ from 16 to 18; above 93% these quantities no longer remain approximately constant but increase sharply. This behavior of the $\bar{L}$ values has been known for some time,[5,9] but in the earlier work the absolute values themselves were too large. An attempt has been made to calculate the aromaticity from the $\bar{N}$-values.[28]

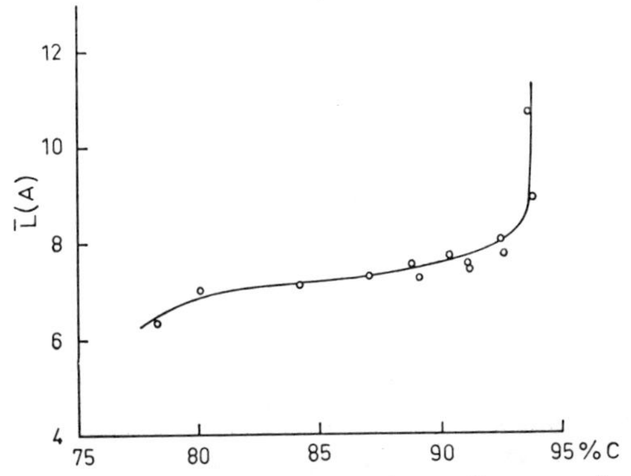

Fig. 5. Variation of average layer diameter, $\bar{L}$, with carbon content.[22]

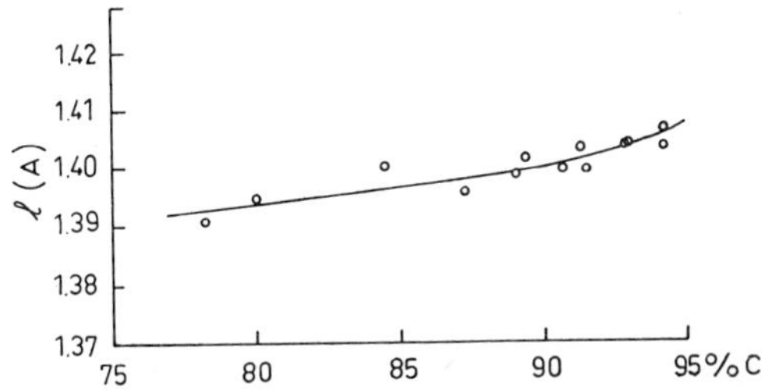

Fig. 6. Variation of the mean bond length, l, with carbon content.[22]

Mean bond length (l). In Fig. 6 the dependence of the mean bond length l on rank in a series of vitrains is presented and shows a slow nearly uniform increase. An interpretation of these l values is still extremely difficult since they contain the contributions of the nearest atoms of side chains and would be strongly influenced by deformation of the hexagonal network, for example, by the presence of five-membered rings or hydroaromatic structures.

Packing of the Aromatic Lamellae

The parallel arrangement of the aromatic lamellae produces the so-called (00l) reflections of which the most intense is (002). The average interlayer distance of the lamellae can be computed from the position of the (002) reflection, whereas the average dimension of the packets of lamellae in the direction perpendicular to their planes may be estimated from its half-width.[9] However, work on the scattering function of parallel stacked lamellae has shown conclusively that the average separation can only be calculated from the (002) reflection if this is corrected for the influence of the (00) reflection from the

[26] van Krevelen, D. W., Chermin, H. A. G., and Schuyer, J., *Fuel*, **36**, 313–20 (1957).

[27] Kröger, C., and Ruland, W., *Brennstoff-Chem.*, **39**, 1–13 (1958).

[28] Dryden, I. G. C., *Fuel*, **37**, 444–68 (1958).

lamellae;[13,29] the latter in this region is proportional to f_C^2/s^2 [f_C = atomic form factor for carbon and $s = 2(\sin \theta/\lambda)$]. Such a correction displaces the (002) reflection to larger angles by an amount which increases with the half-width of this reflection. For this reason the calculation of the separation of the lamellae is only meaningful after this correction has been applied. The use of the (002) reflection after such correction to calculate, from its half-width, the average packet thickness and hence the average number of lamellae per packet leads to the correct result only when all packets possess the same number of lamellae —a result not to be expected in coal. Application of such a half-width method where there is a distribution of packet thicknesses over-emphasizes the larger packets and hence does not lead to a true average.

In vitrinites the (002) peak position progressively changes toward higher angles with increasing rank, and at the same time the band becomes sharper.[30] Micrinites, as compared with the corresponding vitrinites, show a more perfect short-range order for the lamellae, and this leads to sharper, more intense (002) reflections.

The interlayer distance decreases (3.9 → 3.5 A) with increasing rank, a trend which has been observed generally for both coals and their petrographic components.[3,5,31–34] Taking into account the "turbostratic structure" for coals, the minimum separation between aromatic lamellae would be 3.44 A.[9,35] In vitrinites, the spacing, extrapolated linearly to 100% C, coincides with the interlayer distance of graphite.[30]

The packing distribution can be deduced from the (002) reflection by means of a cosine transform.[3,15] Two such distribution curves are presented in Fig. 7[3] where it may be seen that the parallel ordering of the aromatic lamellae is better in an 89% C vitrain than in one of 84% C.

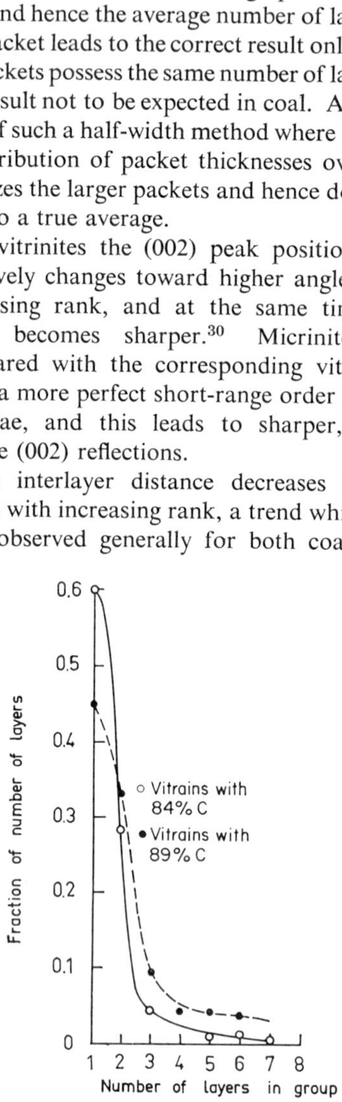

Fig. 7. Fraction of the number of layers occurring in groups of *n* layers, as a function of *n*, for vitrains with 84% C and 89% C.[3]

The Gamma Band

In addition to or even instead of the (002) reflection, some coals show a broad intense scattering peak with a *d* value of from 4 to 5 A which has been indexed as the "gamma band."[9] The origin of this peak has not yet been successfully explained. It may be due to a packing effect of aliphatic side chains,[36] to alicyclic structures with parallel stacking,[37] or to a short-range order of aromatic lamellae which has been disturbed by side groups.[6,27,33] In vitrinites (76.9 to 93.1% C) the γ band may not be observed,[30,38] but appears as one of the most prominent features of the scattering curves of exinites[6,7,14,27,30,36] (see Fig. 2). Whether it represents a unique structural element of this maceral is not yet certain.

In coals and coal components rich in hydrogen (more than 6% by weight) which all show the γ band, the (10) and (11) reflections cannot be interpreted unequivocally as only aromatic structures.[39] A theoretical calculation of the X-ray curves for an aromatic

[29] Ruston, W. R., *Fuel*, **32**, 52–61 (1953).

[30] Nelson, J. B., *Fuel*, **33**, 153–75 (1954).

[31] Dhar, J., and Niyogi, B. B., *Proc. Nat. Inst. Sci. India*, **A8**, 127–35 (1942).

[32] Mitra, G. B., *J. Sci. Ind. Research. (India)*, **12B**, 88–90 (1953).

[33] Inouye, K., Tani, H., and Abiko, M., *J. Fuel Soc. Japan*, **32**, 292–6 (1953).

[34] Kessler, M. F., and Večeřiková, V., *Uhli*, **6**, 28–30 (1956).

[35] Franklin, R. E., *Acta Cryst.*, **4**, 253–61 (1951).

[36] Kasatochkin, V. I., *Izvest. Akad. Nauk S.S.S.R., Otdel. Tekh. Nauk*, **1951**, 1321–34.

[37] Ergun, S., and Wender, I., *Fuel*, **37**, 503–506 (1958).

[38] Nelson, J. B., *Research* (London) **5**, 489–91 (1952).

[39] Ergun, S., and Tiensuu, V. H., *Nature*, **183**, 1668–70 (1959).

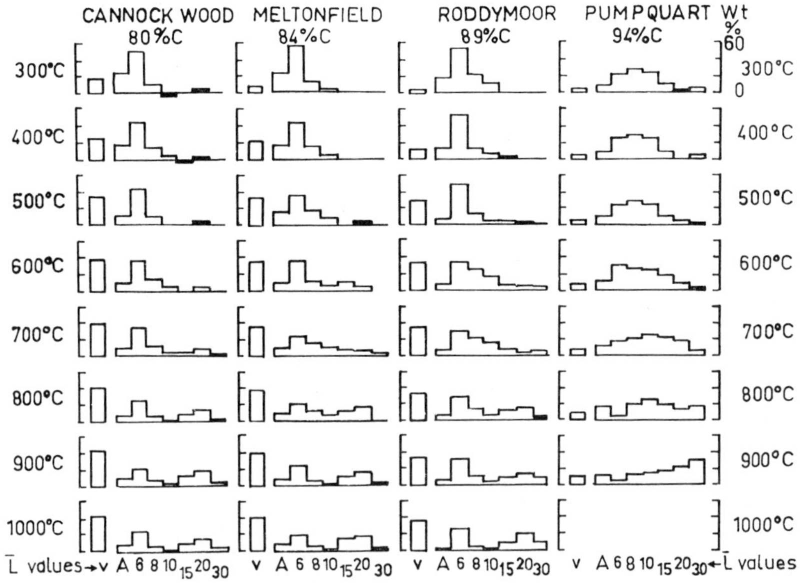

Fig. 8. Variation of molecular-size histograms with carbonization temperature for a series of vitrains.[15]

molecule and its alicyclic counterparts shows that the (10) and (11) reflections are approximately at the same positions and have the same intensities. Therefore it has been concluded that the X-ray pattern of such coals and coal components can be explained by the presence of both alicyclic and aromatic structures.[39]

Carbonization

Some general trends in changes of the X-ray pattern with carbonization temperature have been found.[9,15,40] There is an initial sharpening of the (002) band with increase in carbonization temperature up to about 500°C, which then becomes more diffuse up to 700°C but sharpens again on further increase in temperature. The (10) and (11) peaks also sharpen gradually with carbonization temperature and more rapidly beginning at about 500°C.

Aromatic Lamellae

Size distribution. The histograms for vitrains of different rank carbonized at various temperatures presented in Fig. 8[15] show that below 500°C no significant change in the aromatic clusters occurs. Above 600°C however, there is a progressive broadening of the distribution and correspondingly a decrease in the contributions of the smaller classes to the total distribution. These results can be explained by a condensation of the aromatic lamellae above 600°C to form larger clusters, but it is notable that even above 900°C there is an appreciable contribution by the smallest classes in the first three samples. This effect may be due to some small layers which have not grown during carbonization, but it could also arise from large layers containing holes of the diameter of these small size groups.[25] The "amorphous" portion (A) decreases in all coals up to 500°C while at the same time the fraction of material volatilized (v) increases. The decrease of the "amorphous" fraction on coking had previously been noted using radial distribution methods.[40]

Average lamella diameter ($\bar{L}$). Whereas it was found[9,40] that $\bar{L}$ generally increases with coking, other studies[15,41] have shown that the $\bar{L}$ values of vitrains of different rank scarcely change up to 500°C. Above this temperature a fairly steady increase in the $\bar{L}$ values is noted and is correlated with the increase in size distribution above 600°C, as shown in Fig. 9.[15] As a consequence up to 500°C the

[40] Kasatochkin, V. I., and Razumova, L. L., *Izvest. Akad. Nauk S.S.S.R., Ser. Fiz.* **20,** 751–4 (1956).

[41] Young, R. S., and Robertson, S., *Econ. Geol.,* **47,** 609–33 (1952).

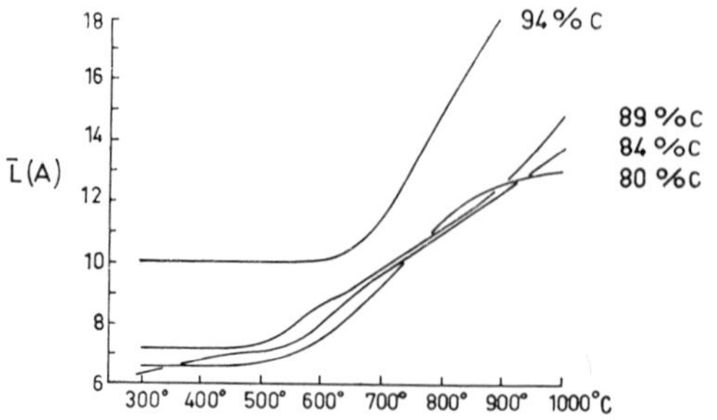

Fig. 9. Variation of average layer diameter $\tilde{L}$ with carbonization temperature.[15]

average number of atoms per aromatic cluster $\bar{N}$ remains practically constant, but then increases, for example, in the vitrains (80 to 89% C) from 16 to 60 at 1000°C.

Mean bond length (l). This quantity at first decreases with increasing coking temperature, passes through a minimum between 400 and 500°C, and then increases.[15] The increase in the l values is to be expected since it should approach the graphite value 1.42 A as the aromatic clusters become very large. The minimum diminishes with increasing rank; anthracite has only a very shallow minimum and shows a value close to 1.410 A at all temperatures.[42]

Packing of Aromatic Lamellae

The parallel ordering of the aromatic lamellae increases with coking temperature up to about 500°C where it passes through a maximum and begins to decrease; at about 700°C a minimum is reached and the parallel ordering increases again.[9,15,40,43,44] For anthracites a regular decrease to about 1100°C was found, followed by a sharp increase above this temperature.[41] For a fusain the parallel ordering of the aromatic lamellae seems to remain constant over the entire temperature range.[41] The average number of lamellae per stack, $\bar{n}$, for two vitrains of 84 and 89% C as, a function of coking temperature is shown in Fig. 10.[15]

The interlayer distance between the parallel stacked lamellae at first decreases as the carbonization temperature increases, passes through a flat minimum between 600 and 700°C, and increases again up to 1000°C,[15] which may be because of the growth of the layers leaving obstructions within the stacks.[42] On the other hand, a steady decrease in interlayer distance with increasing carbonization temperature has been found[45] for coals of 83 to 91% C.

Interpretation of the Coking Process up to 1000°C

Up to about 500°C there is a continuous splitting off of side chains and a loss of small ring systems. These degradation products are found in the "volatile matter" which is particularly strongly produced in this temperature region. As a result of the loss of aliphatic side chains, the average bond length decreases since the side chain atoms closest to the aromatic rings are at the normal aliphatic bond distance which is generally appreciably larger than the aromatic bond length. The loss of side chains also removes in some measure the factors hindering the parallel stacking of the aromatic lamellae so that, even though the size of the lamellae as well as their size distribution do not change, the packing becomes better, the average packet thickness increases, and the distance between lamellae reaches a minimum. Above 600°C, the aromatic lamellae begin to condense, the size distribution gets broader, and the average lamella diameter increases. As expected, the average bond length increases and, as a result of the broader size distribution, the packing

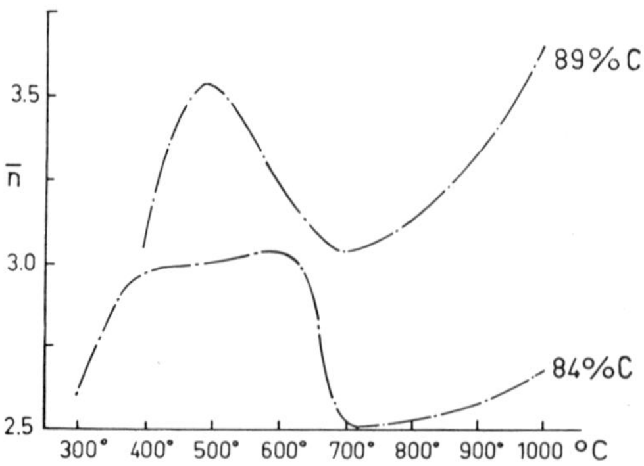

Fig. 10. Variation of the average number of layers per stack, $\bar{n}$ with carbonization temperature.[15]

[42] Diamond, R., *Proc. 3rd Conf. on Carbon*, Buffalo, 1957, Pergamon Press Ltd., London 1959; pp. 367–75.

[43] Kasatochkin, V. I., *Doklady Akad. Nauk S.S.S.R.*, **86,** 759–62 (1952).

[44] Durif, S., *Brennstoff-Chem.*, **37,** 402–404 (1956).

[45] Kessler, M. F., and Večeříková, V., *Paliva*, **36,** 192–9 (1956).

of the lamellae becomes poorer, the average packet thickness decreases, and the average lamella separation increases. Above 800°C, the aromatic clusters have grown sufficiently large to show a detectable displacement of the size distribution, and this produces a further increase in the average diameter of a lamella and in the mean bond length. The parallel ordering becomes better, and the average packet thickness again increases.

Coal Extracts

Qualitative investigations of coal extracts[9] [pyridine soluble–chloroform soluble (γ fraction), pyridine soluble–chloroform insoluble (β fraction) and pyridine insoluble (α fraction)], have shown that in particular the region of the (002) reflection was strongly influenced by extraction. The deformation of this maximum and its shift to smaller angles with continued extraction were explained by assuming the presence of a peak at about 4.6 to 4.8 A which, because of its intensity in the γ fraction, was referred to as the γ band. This peak was thought to be produced by the "bitumen molecules" of the coal.

The X-ray scattering of a vitrain with 84% C was quantitatively studied in the region of the (11) reflection and compared with that of the soluble and insoluble fractions of a series of progressive extractions (pyridine-trichlorobenzene–benzene–carbon tetrachloride).[23] It was found that neither the size distribution of aromatic clusters nor the average number of carbon atoms per cluster (17 ± 2) was markedly affected by the extraction. The mean bond length was somewhat larger in the soluble as compared to the corresponding insoluble fraction and became progressively larger as the extraction continued.[23]

Coals (80 to 89% C) have been extracted with ethylene diamine. Although the X-ray patterns at high angles were very similar to those of the starting materials, differences occurred at low angles.[6]

The effect of a preoxidation (60 hr at 120°C in air) on the structure and its change on coking was studied on the α, β, and γ fractions of a bituminous coal of 88.5% C.[46] The starting material and its extraction products showed no significant change either in the dimensions of the aromatic clusters or their dependence on coking temperature as a result of the oxidation. The thickness of the packets in the extraction products seemed always to be larger than in the corresponding oxidized samples. In coking, both the original and preoxidized samples showed the same dependence of the packet thickness on temperature; the α and β fractions showed no maximum at 500°C, whereas the γ fraction showed a definite maximum at this temperature and increased sharply after passing through a minimum at about 800°C.[46,47]

High-volatile A and B bituminous coals, as well as oxidized high-volatile B, have been extracted with pyridine, chloroform, and hexane.[48] In general, there is an indication that the extracts possess a higher degree of orientation than the starting materials.

Effects Induced by Physical Treatment of Coals

Ultrasonic Treatment

An investigation of the effect of ultrasonic treatment on the size of aromatic packets in a coal with 76% C showed that the parallel ordering of the lamellae can be influenced in this way.[49] The effect is frequency dependent and produced maximum ordering at 250 kc/sec.

Pressure Treatment

The influence of high pressures on the orientation of lamellae (packets) in coals with 82, 88, 90 and 94% C has been investigated.[50] After treatment at 20 tons/cm^2 for 10 min, the coals with 88 and 90% C showed an orientation of their lamellae perpendicular to the direction of the applied pressure; the other two samples did not show this effect. The orientation showed itself in a non-uniform intensity distribution in the (002)

[46] Blayden, H. E., Gibson, J., and Riley, H. L., *J. Chem. Soc. (London)* **1948,** 1693–700.

[47] Riley, H. L., *Bull. soc. chim. Belges*, **57,** 400–15 (1948).

[48] Pierron, E. D., Rees, O. W., and Clark, G. L., *Illinois State Geol. Survey, Circ.* No. 269 (1959), 36 pp.

[49] Berkowitz, N., *Can. J. Technol.*, **33,** 378–87 (1955).

[50] Razumova, L. L., Kasatochkin, V. I., and Volarovich, M. P., *Doklady Akad. Nauk S.S.S.R.*, **103,** 1033–4 (1955).

interference ring when the sample was irradiated perpendicular to the direction in which the pressure was applied.

In macerals, a relation has been observed between compressibility and the intensity of the γ band; the greater its intensity the greater the compressibility.[27] This led to the conclusion that the γ band belongs to a short-range ordering structure which is considerably more compressible than a structure consisting of aromatic packets with an average lamellar separation of 3.5 A which produces the (002) reflection.

Chemically Treated Coals

Oxidation

After oxidation in air at 190 to 200°C for about 200 hr, the interlayer distance of vitrains (81 to 88% C) decreased;[51] this decrease could be due to an improved parallel stacking as a direct result of oxidation (removal of side chains).

A bituminous coal with 88.5% C was heated in air for 60 hr at 120°C, and then the starting coal and this preoxidized product were both coked.[46] This treatment did not produce a detectable change either in the dimensions of the lamellae or in their dependence on the coking temperature; the same was true for the thickness of the aromatic packets. Upon coking the original coal, however, maximum packing of lamellae was found at 500°C, but the preoxidized sample showed a continuous decrease in packet thickness up to 600°C. The same behavior has been found for a Russian coal of the same rank and was thought to be due to more oxygen bridges in the oxidized coal.[52]

Acid Treatment

In some bituminous coals of different origin which were heated to 500 to 600°C and then treated 40 hr with boiling HCl, a peak at about 4.2 A appeared in the X-ray scattering curve in addition to the (002) reflection.[53] This peak disappeared upon heating to 600°C in a nitrogen atmosphere. The phenomenon has not been explained.

Reduction with Lithium in Ethylenediamine

Anthraxylons (78 to 89% C) have been reduced by lithium in ethylenediamine and the X-ray patterns of the products compared with those of the starting materials.[37] A major change was found in the (002) region where this peak was broadened and displaced as the degree of hydrogenation increased. This corresponds to an increase in the layer separation from 3.6 to 4.8 A in the case of a Pittsburgh anthraxylon, which means that the carbon atoms in the hydrogenated layers were now slightly displaced from planar positions to form buckled layers.[37] Furthermore, the line shapes as well as peak positions of the (11) reflections showed extensive changes. The line profile of the strongly hydrogenated sample could no longer be explained on the basis of planar aromatic molecules.[37] Calculation of the mean bond length l showed an increase from 1.395 to 1.47 A in the highly reduced sample. The latter value is obtained by analogy with planar molecules[15] and corresponds to the projected bond length on a plane parallel to the buckled layers.[37]

High-Vacuum Distillates of Coal

The vitrinite of a high-volatile bituminous coal (Pittsburgh bed) was heated to from 450 to 510°C under 10^{-3} mm pressure.[24] In general, the X-ray patterns of the starting material, the distillate, and the residue were not dissimilar. This indicates that some aromatic clusters were set free during distillation under these conditions; the process therefore cannot be regarded as a simple removal of peripheral groups.

An increased broadening of the (10) and (11) peaks in the residue was observed. Of the three X-ray patterns, those for distillate and residue exhibited the greatest similarity.

A quantitative analysis of the (11) peaks showed that the "amorphous carbon" fraction in vitrinite was much larger than in the distillate and residue.[24] The average lamella diameter ($\bar{L}$) in vitrinite was found to be 7.6 A (3 to 5 condensed aromatic rings), whereas $\bar{L}$ of the distillate and residue was 5.8 to 6 A (1 to 3 rings).

The fact that $\bar{L}$ in the vitrinite was larger than $\bar{L}$ in the distillate and in the residue

[51] Ali, S. Z., *Proc. Symposium on the Nature of Coal*, Central Fuel Research Inst., Jealgora, India, 1960, pp. 95–104.

[52] Kasatochkin, V. I., and Razumova, L. L., *Doklady Akad. Nauk S.S.S.R.*, **88**, 91–4 (1953).

[53] Inouye, K., *Bull. Chem. Soc. Japan*, **29**, 644–5 (1956).

might be attributed to coplanar coalescence of smaller units in the vitrinite as a result of long-term pressure-induced orientation and packing.[24] The orientation of the layers would be disturbed by thermal treatment and depolymerization and cracking would occur. Some of the structural entities would be set free, whereas others would combine, giving rise to a highly disordered carbon skeleton.[24]

Determination of Mineral Matter

The crystalline mineral matter in coal gives sharp interference maxima which rise above the much less specific scattering of the coal itself and can usually be separated from the latter without trouble. By measuring the Bragg angles and the intensities of the peaks which appear, it is possible to obtain a qualitative evaluation of the minerals present in the coal.

Comparison of the X-ray patterns with ash analyses on English coals of high mineral content established the presence of mica, kaolin, quartz, and carbonates,[54] whereas Indian coals contained clay minerals, quartz, pyrites, anatase, dolomite, and apatite.[55] Kaolinite, pyrite, quartz, calcite, and dolomite have been found in coals from Logan County (West Virginia).[41]

The hypothesis has been proposed that in coals of high mineral content the minerals and the organic coal material occur in alternate layers of molecular dimensions.[56]

An interesting discovery is the fact that there is no significant difference in the mineral content (mainly kaolinite and dolomite) of the macerals vitrinite, exinite, and micrinite from different seams in the Ruhr region.[27]

X-Ray Small-Angle Scattering

If a material shows X-ray scattering at very small angles, it possesses inhomogeneities in electron density extending over relatively large regions; these inhomogeneities may, for example, be caused by micelles, by macromolecules, or by pores. To evaluate the small-angle scattering, it is necessary to distinguish between *dilute* and *densely packed* systems.[57–59] In the former for example, macromolecules in dilute solution, the form of the scattering curve is determined by the shape of the scattering particles[57] and conclusions may be drawn as to size and shape of these particles. In densely packed systems, the scattering from the individual particles is partially overlaid by scattering due to the ordering of the particles with respect to one another and, since it is rarely possible to separate the two satisfactorily, the interpretation leads to less detailed conclusions.[58]

Experimental Methods

The choice of a camera for studying small-angle scattering depends on the resolving power desired, that is, the largest d value which can be reached with the given camera. For studies down to Bragg angles corresponding to spacings of about 200 A, a collimator consisting of three slits suffices.[30] If the scattering out to spacings of 5000 A is to be studied, a double-crystal spectrometer[60] has been used.[3]

Coals and their petrographic components. The intense small-angle scattering of coals frequently possesses a maximum at an angle corresponding to a d value of 20 A, and the intensity increases rapidly toward smaller angles. In vitrinites,[27,30,61] the 20-A maximum appears beyond 84% C and reaches its maximum intensity at about 90% C, becoming more and more diffuse toward higher carbon contents.[3] In exinites,[27] the 20-A maximum appears first at 87% C, and the intensity increases steadily with rank until exinites and vitrinites become petrographically indistinguishable. In micrinites,[27] the 20-A maximum could not be observed, but micrinites of about

[54] Nagelschmidt, G., and Hicks, D., *Proc. Conf. on Ultrafine Structure of Coals and Cokes*, BCURA, London, 1944, pp. 240–5.

[55] Mitra, G. B., *Fuel*, **33**, 316–30 (1954).

[56] Riley, D. P., *Proc. Conf. on Ultrafine Structure of Coals and Cokes*, BCURA, London, 1944, pp. 246–71.

[57] Guinier, A., and Fournet, G., *Small Angle Scattering of X-Rays*, Chapman and Hall Ltd., London, 1955, 268 pp.

[58] Porod, G., *Kolloid Z.*, **124**, 83–114 (1951); **125**, 51–7, 108–22 (1952).

[59] Luzzati, V., *Acta Cryst.*, **10**, 136–8, 643–8 (1957).

[60] Kaesberg, P., Ritland, H. N., and Beeman, W. W., *Phys. Rev.*, **74**, 71–3 (1948).

[61] Lahiri, A., and Iyengar, M. S., *J. Sci. Ind. Research (India)*, **17A**, 60–73 (1958).

87% C show a point of inflection in the scattering curves at a point corresponding to a spacing of about 30 A. In fusains, no 20-A band has been found.[61]

The interpretation of the 20-A maximum is difficult and not yet unequivocal. If it is assumed (1) that the "particles" are identical with the aromatic packets which have a significantly higher electron density—because of their close packing—than the amorphous portion of the coal structure, and (2) that there is a "liquid-like" ordering of these particles, the 20-A maximum could be interpreted as the result of the short-range order of densely packed spherical particles of 30 to 35 A diameter.[62] Coals having the 20-A maximum would then be interpreted as having a "liquid-like" structure, and those not showing this maximum would be said to have a "gas-like" particle distribution.[62]

However, the 20-A maximum could also be interpreted as arising from an average favored distance between aromatic packets with an average diameter of only 8 A when the space between the particles is filled with nonaromatic side groups and single lamellae.[3]

Still another possible interpretation of the 20-A maximum is that it arises from particles with no particular ordering but having a core with a higher electron density and a surface region with a lower electron density than that of the medium surrounding the particles.[63] The outside diameter of the surface region would have to be about 20 A and the diameter of the core about 8 A.

In view of the fact that the 20-A maximum is more intense when a block of high-rank coal is irradiated parallel to the plane of the seam than when the primary beam is incident perpendicular to this direction [for the (002) reflection the opposite is true], it was concluded that there is a certain ordering of the aromatic packets parallel to the plane of the seam.[3]

Finally, there seems to exist a correlation between the presence of the 20-A band and the absence of hydrogen-bonded OH groups and vice versa.[6]

It should be mentioned, furthermore, that during carbonization of a caking coal the 20-A maximum increases in intensity up to about 500°C. In a coking coal, very little change of the 20-A band occurs. At about 700°C in both cases the 20-A maximum disappears into the continuous background of small-angle scattering.[6,15] If the 20-A maximum is considered as an interference effect between the aromatic packets and their surroundings, better packing would lead to greater intensity and poorer packing to decreased intensity in accord with these observations.

The sharp increase in intensity of small-angle scattering in passing to very small angles has been studied in one case to a *d* value of 250 A[30] and in another case to a *d* value of 5000 A.[3] The intensity of this scattering passes through a minimum between 88 and 90% C. Since the porosity of coals also reaches its minimum at this rank,[64] it is tempting to consider that this part of the small-angle scattering curve is produced by the pore structure. On this assumption, the shape of the scattering curve and the total scattering can be explained by a system of disk-like pores with a distribution of sizes (300 to 5000 A) and a preferred orientation parallel to the plane of the seam.[3]

In a few coals, a completely different small-angle scattering curve was observed; from this, particles with a radius of gyration of 50 A could be deduced.[8]

MEASUREMENTS IN THE ULTRAVIOLET AND VISIBLE REGION

Absorption in the Ultraviolet and Visible Region

All substances absorb in the vacuum ultraviolet (1400 to 1850 A), many also in the ultraviolet (1850 to 4100 A), and some in the ultraviolet and visible region (4100 to 7500 A). Generally speaking, the energy of the rays absorbed in this region leads to the excitation of bonding electrons and of free electron pairs.[65] For single bonds, a very high excitation energy (vacuum ultraviolet) is necessary,

[62] Riley, D. P., *Proc. Conf. on Ultrafine Structure of Coals and Cokes*, BCURA, London, 1944, pp. 232–9.

[63] Brown, J. K., and Hirsch, P. B., *Nature*, **175**, 229–33 (1955).

[64] King, J. G., and Wilkins, E. T., *Proc. Conf. on Ultrafine Structure of Coals and Cokes*, BCURA, London, 1944, pp. 46–56.

[65] Reid, C., *Excited States in Chemistry and Biology*, Butterworths, London, 1957, 215 pp.

and because of the experimental difficulties this region has been relatively little investigated. If multiple bonds are present, however, absorption occurs in the ultraviolet or ultraviolet and visible region. This corresponds usually to the excitation of π electrons (π-π transitions).[66] In compounds containing N, O, S, or P which possess relatively weakly bound electron pairs ("nonbonding" or "unshared" pairs), n-π transitions[67] may occur, generally in the ultraviolet region. To distinguish between a π-π and an n-π transition it is usually sufficient to measure the absorption in a nonpolar and in a polar solvent. Whereas for a π-π transition a red shift will be noted in the polar solvent, for a n-π transition a violet shift is found.[65] All transitions obey certain selection rules which are determined by molecular symmetry.[68]

In electronic molecular spectra, bands are observed instead of lines; this observation is a result of the fact that in addition to the electronic excitation, vibrational and rotational transitions are simultaneously excited. Hence a multiplicity of lines is to be expected and can only be resolved in the gaseous state and then only partially. In solution and more particularly in the solid state, only bands are to be expected as a result of intermolecular interactions.

For structural studies, absorption spectra in the ultraviolet and in the visible have proved to be especially useful for the identification of condensed aromatic hydrocarbons[69] and certain chromophoric groups. On the other hand, a clear-cut proof of the existence of alkyl groups, OH-groups, and ether oxygen is impossible.

Optical Density (Extinction), Specific and Molar Extinction Coefficient

The transmission of a material is measured experimentally, and from the ratio of the initial intensity I_0 to the transmitted intensity I, the optical density D may be calculated:

$$D = \log \frac{I_0}{I}$$

If quantitative comparisons between different substances are to be made, the D-values must be independent of the thickness of the absorbing layer l (in centimeters) and of the concentration c (in grams/liter) or c' (in moles/liter). This leads to the specific extinction coefficient, E_{spec}, defined by

$$E_{\text{spec}} = \frac{D}{lc}$$

and to the molar extinction coefficient, E_{mol}, defined by

$$E_{\text{mol}} = \frac{D}{lc'}$$

Strictly speaking, these relations, which depend on the Lambert-Beer's law[70,71] are only valid for dilute solutions. However, for purposes of comparison, they may also be applied to solid suspensions (e.g., KBr windows) or even to the solid itself (e.g., in thin sections). If, as in coal, the molecular weight is not known, only E_{spec} can be determined.

Index of Refraction and Molar Refraction

If a ray of light falls upon the boundary surface between two media I and II in which the velocity of light is different (C_{I} and C_{II}), a portion of the ray will be transmitted into the medium. This transmitted ray, for all except perpendicular incidence, will be changed in direction by an amount given by Brewster's law[70,71]:

$$\frac{C_{\text{I}}}{C_{\text{II}}} = \frac{\sin i}{\sin r} = n$$

where i = the angle between the incident ray and the normal to the boundary surface (angle of incidence), r = the angle between the transmitted (refracted) ray and the normal to the boundary surface (angle of refraction), n = the index of refraction for the pair of media in question. Generally the medium which the light first traverses is taken to be a

[66] Klevens, H. B., and Platt, J. R., *J. Chem. Phys.*, **17**, 470–81 (1949).

[67] McMurry, H. L., and Mulliken, R. S., *Proc. Nat. Acad. Sci. U.S.*, **26**, 312–7 (1940).

[68] Sponer, H., and Teller, E., *Rev. Mod. Phys.*, **13**, 75–170 (1941).

[69] Clar, E., *Aromatische Kohlenwasserstoffe-Polycyclische Systeme*, Springer, Berlin, Göttingen, Heidelberg, 1952, 481 pp.

[70] Partington, J. R., *An Advanced Treatise on Physical Chemistry*, Vol. IV, Longmans, Green and Co., London, New York, Toronto, 1953, 678 pp.

[71] Pohl, R. W., *Einführung in die Optik*, Springer, Berlin, 1943, 320 pp.

vacuum and then n depends only on the properties of the other medium, the wavelength, and the temperature.

If the index of refraction n, the molecular weight M, and the density d of a material are known, a quantity known as the molar refraction, R_M,[70,71] can be calculated from the Lorenz-Lorentz formula:

$$R_M = \frac{n^2 - 1}{n^2 + 2}\frac{M}{d}$$

It has been known for many years that this quantity can be calculated for molecules containing only single bonds with a considerable degree of accuracy as $R'_M = \Sigma a_i r_i$ where a_i is the number of atoms of type i in the molecule and r_i are the corresponding atomic refractions. The r_i values may be taken from standard tables.[72] When unsaturation is present, R_M is always greater than R'_M and the difference between the two is known as the molar refractometric increment I_M:

$$I_M = R_M - R'_M$$

In coal, R_M cannot be calculated because the molecular weight is not known. However, from the elementary analysis it is possible to calculate a molecular weight per carbon atom, M/C, from the formula

$$\frac{M}{C} = \frac{1201}{\%C} = 12.01 + 1.008\frac{H}{C} + 16.00\frac{O}{C} + 14.008\frac{N}{C} + 32.06\frac{S}{C}$$

where H/C, O/C, etc., are the atomic ratios from the analysis.[73, 74] By using M/C, R_M/C may be calculated. Furthermore, R'_M/C may be derived from

$$\frac{R'_M}{C} = r_C + r_H\frac{H}{C} + r_O\frac{O}{C} + r_N\frac{N}{C} + r_S\frac{S}{C}$$

where r_H, r_O, r_N, r_S, are the atomic refractions for H, O, N, S. It is thus possible to calculate I_M/C:

$$\frac{I_M}{C} = \frac{R_M}{C} - \frac{R'_M}{C}$$

If all the double bonds in coal are aromatic, I_M can be associated only with the aromatic carbon atoms, so if the "aromaticity," that is, the ratio of aromatic carbon to total carbon C_{ar}/C, is known, it is possible to calculate I_M/C_{ar}.

If I_M/C_{ar} is plotted against "aromatic surface areas" for a series of known condensed aromatic hydrocarbons, it is found that I_M/C_{ar} increases with an increasing number of rings. Such a diagram[73–75] may then be used to determine the "aromatic surface area" of coal from the calculated value of I_M/C_{ar}.

Reflectance and Absorption Index

If light passing through a nonabsorbing medium I of refractive index n_1 falls perpendicularly on the boundary surface between this medium and medium II of index of refraction n_2 and absorption index K_2, a portion of the light will be reflected and, for this case, the reflectance r (ratio of reflected to incident intensity), is given by the Beer's relation[70,71]

$$r = \frac{(n_2 - n_1)^2 + n_2^2K_2^2}{(n_2 + n_1)^2 + n_2^2K_2^2}.$$

To calculate n_2 and K_2 of an unknown substance, two measurements of reflectance must be made in two different media, the refractive indices of which are known and are as different as possible, for example, cedar oil and air.

If n is known, the absorption index K can also be calculated by use of Lambert's law from the diminution in light intensity of light of wavelength λ (in vacuum) after passing through a thickness l of absorbing material:

$$K = \frac{D}{l}\frac{\lambda}{4\pi n}$$

or

$$K = 183\frac{\lambda d E_{spec}}{n} = 183\frac{\lambda d E_{mol}}{nM}$$

Optical Anisotropy

If optical properties (e.g., reflectance) are dependent on the orientation of the sample, the material is said to be optically anisotropic. In principle, all molecules which are not completely symmetric are optically anisotropic, but normally the random orientation

[72] Vogel, A. I., Crosswell, W. I., Jefferey, G. H., and Leicester, J., *J. Chem. Soc.*, **1952,** 514–49.

[73] van Krevelen, D. W., *Brennstoff-Chem.*, **34,** 167–82 (1953).

[74] Schuyer, J., and van Krevelen, D. W., *Fuel*, **33,** 176–83 (1954).

[75] Schuyer, J., Blom, L., and van Krevelen, D. W., *Trans. Faraday Soc.*, **49,** 1391–1401 (1953).

of the unsymmetric molecules in solution or in the amorphous state precludes the possibility of such observations. The greater the ordering of such molecules, however, the more distinctly the optical anisotropy reveals itself —for example, as maximum and minimum reflectance.

The Absorption Spectra of Coal, Petrographic Components, and Coal Extracts

Experimental Methods and Preparation of Samples

For measurements in the ultraviolet and visible region standard spectrophotometers may be used, and the method of diffuse reflectance may in principle also be applied.[76,77]

Thin sections.[78–81] The main advantage of using thin sections is the almost complete absence of scattering effects. To measure the ultraviolet absorption accurately, sections of about 0.1 μ thickness are required. This method suffers from the disadvantages that preparation of the thin sections is extremely time consuming, it is almost impossible to obtain them with uniform thickness, and a sufficiently accurate measurement of the actual thickness in the range 0.1 to 1 μ is extraordinarily difficult. The use of thin sections several microns in thickness is limited to measurements in the visible region.

KBr windows.[79–81] Finely ground coal pressed in a KBr window may be regarded as a suspension of coal in the KBr, and the concentration (g/liter KBr) and the layer thickness are easy to measure. On the debit side, a nonuniform distribution and a range of particle sizes are difficult if not impossible to avoid. Furthermore, scattering effects can lead to apparent absorptions which are too large, especially for substances showing low absorption, or the complete absorption of the incident light by the larger particles (in highly absorbing materials) may lead to apparent absorptions which are too low.

For quantitative measurements, a uniform distribution and a narrow range of particle sizes are necessary.[82] Also the "apparent" optical densities should be corrected to their "true" values.[83]

Films of coal.[78] In this method a film of coal is deposited on a transparent carrier from a suspension by evaporation of the dispersing agent, CCl_4.[78] Only the average thickness of such films can be measured, and surface tension effects during evaporation usually produce some conglomeration of smaller particles.

Solutions. This technique can be applied only to coal extracts[84–89] but the results are accurate and perfectly reproducible. A serious difficulty arises from the fact that all good solvents (e.g., pyridine, benzene, dimethyl formamide, etc.) possess absorption bands of their own in the ultraviolet and hence permit measurements only beyond 250 to 300 mμ. If coal extracts form partially colloidal suspensions, scattering effects and deviations from the Lambert-Beer's law are to be expected.

Coals and Their Petrographic Components

Changes of the absorption with wavelength and with rank. From a thin section, as well as from KBr disks, of a Pittsburgh vitrain of 84% C, it was possible to show that only a very unspecific absorption spectrum could be

[76] Kortüm, B. G., *Rev. universelle mines*, [9] **15,** 495–506 (1958).

[77] Kortüm, B. G., Theilacker, W., and Schreyer, G., *Z. physik. Chem. (Frankfurt)*, **11,** 182–95 (1957).

[78] Huntjens, F. J., and van Krevelen, D. W., *Fuel*, **33,** 88–103 (1954).

[79] Friedel, R. A., *Nature*, **179,** 1237–8 (1957).

[80] Friedel, R. A., and Queiser, J. A., *Fuel*, **38,** 369–80 (1959).

[81] Friedel, R. A., *Proc. 4th Conference on Carbon*, Buffalo, 1959, Pergamon Press, London, 1960, pp. 321–36.

[82] Otvos, J. W., Stone, H., and Harp, W. R., *Spectrochim. Acta*, **9,** 148–56 (1957).

[83] Brown, T. H., and Turkevich, J., *Proc. 3rd Conference on Carbon*, Buffalo, 1957, Pergamon Press, London, 1959, pp. 113–20.

[84] Roy, M. M., *Fuel*, **36,** 344–54 (1957).

[85] Barclay, L. R. C., and Layton, T. M., *Fuel*, **35,** 31–7 (1956).

[86] Orchin, M., Golumbic, C., Anderson, J. E., and Storch, H. H., *U.S. Bur. Mines, Bull.*, No. 505 (1951), 15 pp.

[87] de Ruiter, E., and Tschamler, H., *Brennstoff-Chem.*, **39,** 362–3 (1958).

[88] Wada, M., Nawata, S., and Suzuki, Y., *Bull. Research Inst. Mineral Dressing and Met., Tohoku Univ.*, **12,** 19–26 (1956); *Chem. Abs.*, **50,** 16072 (1956).

[89] Sakabe, T., Inouye, K., Ouchi, K., and Honda, H., *J. Fuel Soc. Japan*, **32,** 610–26 (1953).

observed.[79-81] The absorption increased markedly in going from the visible to the ultraviolet and displayed an obvious shoulder at 2650 A. With increasing rank this shoulder shows up more clearly.[90] This band may be assigned to aromatic nuclei[91] because many aromatic hydrocarbons of different size and type of condensation possess a band in this wavelength region.[66] The strong band at 2600 A in graphite is related to transitions of electrons from the valence band to the conduction band,[91] an interpretation which may also be valid for high-rank anthracites. Furthermore it should be mentioned that kojic acid has a strong band near 2650 A, and β diketones also have their strongest band in this region, and only small amounts of such structures would be needed to produce the shoulder in the absorption spectra of coals.[80]

The value of E_{spec} for the Pittsburgh vitrain of 84% C increases uniformly from 0.8 at 7750 A to 20.0 at 3000 A. The fact that known condensed aromatic hydrocarbons possess values of E_{spec} which are forty to fifty times larger has been interpreted as an indication that the concentration of such systems in coal would at best be rather small.[79] The validity of this conclusion is called in question by the fact that comparisons were made of the E_{spec} values at the band maxima of model substances in solution with those of coal which absorbs nonspecifically.[92,93] Furthermore, values of E_{spec} in the visible for a pitch fraction of high aromaticity and an average ring number of 5 to 7 are of the same order of magnitude as those of the Pittsburgh vitrain.[92]

A comparison has been made of the integrated areas under the absorption curves of both model substances and a coal sample, which leads again to a low aromaticity for the coal sample.[80,81] This conclusion needs further confirmation.

Investigations of the present authors by the KBr window technique of exinites, vitrinites, and micrinites of different rank in the visible region have shown that the unspecific absorption curves of all three petrographic components increase toward shorter wavelengths—exinites strongest, micrinites least. The E_{spec} values of exinites are initially less than those of vitrinites, but increase faster with increasing rank so that beyond about 90% C the two are approximately the same. Micrinites have distinctly larger E_{spec} values than the other two petrographic components throughout the range studied. If absorption in the visible is taken as a measure of the average size of aromatic lamellae, exinites would possess smaller and micrinites larger aromatic systems than the corresponding vitrinites.

The absorption in the ultraviolet and visible of a reduced anthraxylon (81.4% C) is much less than that of the original anthraxylon and may be due to a decrease both in aromatic structures and in free radicals.[94]

The color of coal. Coal is only black in thick layers. In thin layers it has a color which depends on the relative thickness.[95,96] The color observed may be explained by the presence of large condensed aromatic systems,[73,97] by the otherwise confirmed presence of stable free radicals,[79,80,98] by the mobility of electrons in the atomic coal network,[43] or by the presence of charge transfer complexes.[99,100] The black coprecipitate from an H_2SO_4 solution of pyranthrone, isoviolanthrone, and violanthrone, which are individually red or blue-violet has been cited as a model to explain the black appearance of coal.[100,101]

[90] Fujii, S., *Nenryô Kyôkaishi*, **38**, 384–7 (1959); *Chem. Abs.*, **53**, 18440 (1959).

[91] Humphreys-Owen, S. P. F., and Gilbert, L. A., *Proc. Conf. on Industrial Carbon and Graphite*, London, 1957, Soc. Chem. Ind., London, 1958, pp. 37–41.

[92] van Krevelen, D. W., Groenewege, M. P., and Rietveld, B. J., *Nature*, **181**, 640–1 (1958).

[93] Shannon, J. S., *Fuel*, **37**, 352–4 (1958).

[94] Reggel, L., Raymond, R., Friedman, S., Friedel, R. A., and Wender, I., *Fuel*, **37**, 126–8 (1958).

[95] Cannon, C. G., and George, W. H., *Proc. Conf. on Ultrafine Structure of Coals and Cokes*, BCURA, London, 1944, pp. 290–315.

[96] Mackowsky, M. T., *Brennstoff-Chem.*, **39**, S 40–7 (1958).

[97] van Krevelen, D. W., *Ann. mines Belg.* **53**, 750–7 (1954).

[98] Friedel, R. A., *J. Chem. Phys.*, **31**, 280–1 (1959).

[99] Leicester, J., *J. Appl. Chem. (London)*, **6**, 96–101 (1956).

[100] Schuyer, J., *Brennstoff-Chem.*, **37**, 74–8 (1956).

[101] Akamatu, H., and Nagamazu, K., *J. Colloid Sci.*, **2**, 593–8 (1947).

Coal Extracts

Absorption spectra. In coal extracts in solution as in coal itself, the ultraviolet and visible absorption is unspecific and increases regularly from longer to shorter wavelengths.[86–89] This is true for all coal extracts so far measured whether of American[86,87] or of Japanese[88,89] origin. If it is assumed that the ultraviolet and visible absorption is mainly caused by condensed aromatic systems, it proves possible to reconstruct the nonspecific absorption curve of coal extracts by suitably chosen mixtures of nine condensed aromatic hydrocarbons; the fit of the pyridine extract of a Pittsburgh vitrain (84% C) is shown in Fig. 11.[87]

Table 1 lists the nine model substances chosen and gives the weight percent of each needed in the mixture to best reproduce the absorption-curves of successive extracts (pyridine followed by trichlorobenzene followed by benzene followed by carbon tetrachloride; extraction in each case for 24 hr under purified nitrogen at the boiling point of the solvent). The difference between the sum of the weight percentages of the models and 100% consists of nonabsorbing material and represents the nonaromatic fraction in the different coal extracts. If benzene rings possibly present are neglected, as well as some functional groups known to cause appreciable red shifts—for example, hydroxyquinone structures—it may be concluded from the data of Table 1 that (1) the aromaticity decreases with progressive extraction from 78 to 70%, (2) an average aromatic ring number of 3 to 4 is present, (3) in every case the insoluble fraction requires for its reconstruction proportionately more large ring systems than does the soluble fraction, and (4) at each stage of the extraction the spectrum of the starting material can be reconstructed from that of the extract and residue if the extraction yield is taken into account.

If the pyridine extract of a vitrain (84% C) is dissolved in concentrated H_2SO_4 and the absorption compared with that of the same material dissolved in pyridine, a strong broad additional band is observed between 300 and 850 mμ; this band disappears on dilution of the acid solution with water.[102] Certain

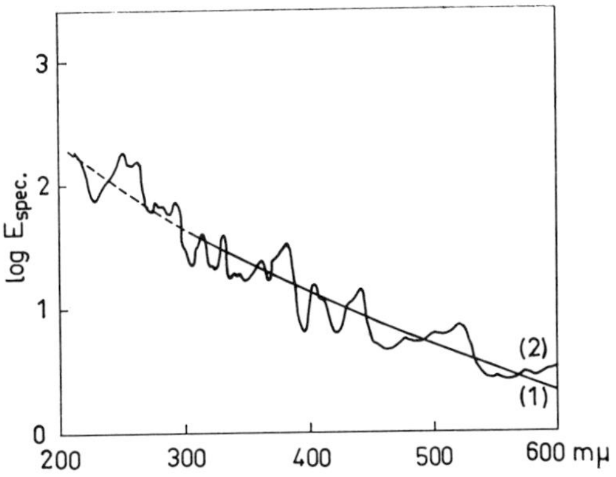

Fig. 11. Absorption spectra of a pyridine extract (1) and of a mixture of nine condensed aromatic hydrocarbons (2).[87]

condensed aromatic hydrocarbons in concentrated H_2SO_4 compared with a pyridine solution of the same hydrocarbons also show strong extra absorption bands in the long wavelength region, which are to be attributed to the formation of carbonium and/or positive ions. The additional absorption bands also disappear on dilution of the acid solution and it therefore seems evident that the same explanation applies to the coal extract.[102] It should be remarked, however, that quinones also show strong, long wavelength absorption in concentrated H_2SO_4—because of carbonium ion formation. Hence if these are present in the coal extract they will contribute to this effect.[102]

Progress in identification of constituents in coal extracts. The pyridine- and chloroform-soluble portion (γ fraction) of a Joggins (31.7% volatile matter) and of a Dominion (40.0% volatile matter) coal from which acidic and basic constituents had been separated were fractionated by chromatography on Al_2O_3.[85] The γ fraction of the Joggins coal was also fractionated by vacuum distillation. All of these separated fractions in dioxane solution have essentially nonspecific absorption which increases toward shorter wavelengths and shows recognizable peaks at 230 and 260 mμ, the peaks being attributed to naphthalenic structures. Very similar spectra were found for the β fractions (pyridine-soluble, chloroform-insoluble) of both coals, as well as the methanol-benzene extract of the Joggins coal. It was concluded that coal is relatively uniform in structure.[85]

The pyridine extract of an Assam coal (Baragolai-Colliery, 78.2% C) was separated

[102] de Ruiter, E., and Tschamler, H., *Brennstoff-Chem.*, **40**, 43–5 (1959).

TABLE 1

AROMATIC HYDROCARBONS AND WEIGHT PERCENTAGES USED IN THE RECONSTRUCTION OF THE VARIOUS COAL EXTRACTS[87]

Aromatic Hydrocarbons Used	Pyridine Soluble	Trichlorobenzene Soluble	Benzene Soluble	CCl_4 Soluble	Benzene Insoluble	CCl_4 Insoluble
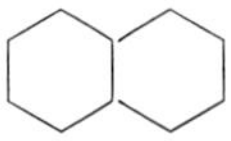	15.0	17.0	20.0	21.0	10.0	16.0
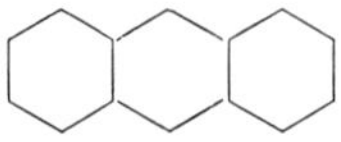	7.5	8.5	12.0	13.0	4.0	8.0
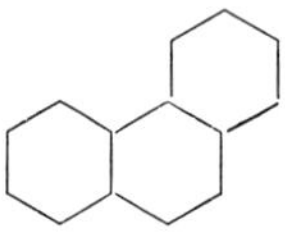	7.5	8.5	12.0	13.0	4.0	8.0
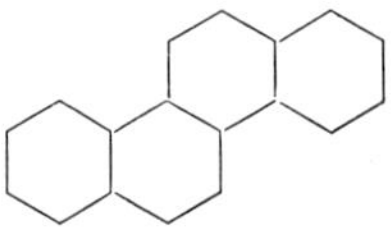	7.5	9.0	10.0	8.0	8.0	9.0
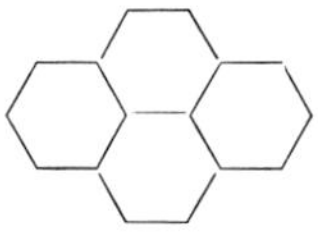	15.0	11.0	6.0	5.0	15.0	12.5
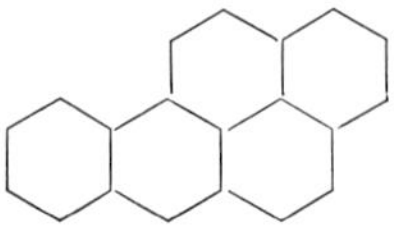	15.0	11.5	6.0	5.0	18.0	12.5
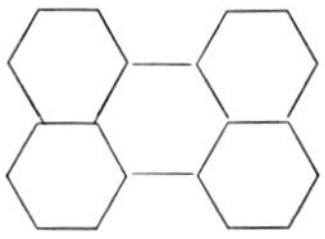	6.0	5.5	3.0	2.0	9.0	6.0
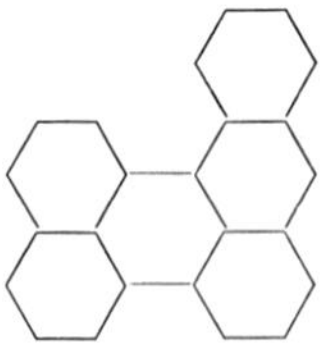	1.8	1.4	0.4	0.4	2.7	1.3
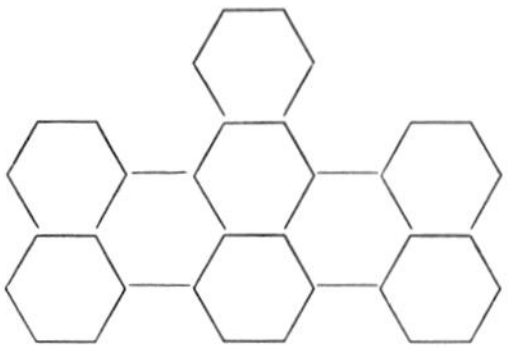	2.5	1.8	0.4	0.3	3.8	1.6

by chromatography on a silica gel column, the γ fraction on Al_2O_3 and the ethylene-diamine extract on bentonite, silica gel, and Al_2O_3.[84] Attempts were made to identify the individual fractions by elementary analysis, molecular weight determinations, and ultraviolet absorption studies. The principal constituents proved to be: 2-methylnaphthalene, 2,6 xylenol, γ-phenylpyridine, *p*-phenylphenol, diphenylether and diphenylmethane. From these results obtained on extracts it was concluded that the Assam coal contains benzene and naphthalene rings, cyclic ethers, heterocyclic nitrogen, OH groups, and some CH_2 groups.[84]

Reflectance of Coal

Measurements of reflecting power of coals are important not only for fundamental studies but also for applications since the rank, the content of the different petrographic components[103] (two factors of importance for the quality of coke[104–106]), and the carbonization temperature[107] may be derived from them. For the determination of the petrographic components, it is an open question as to whether reflectance or transmission measurements are to be preferred.[108]

Experimental Methods and Sample Preparation

Reflectance is usually measured on a polished surface with polarized light at vertical incidence by either the single- or the double-beam method.[109,110]

In the single-beam method, the brightness of the sample is compared under a microscope with that of standard coals embedded together with the sample[104] or even with that of the embedding wax itself,[111] or the comparison can be made by use of standard glasses.[107] In place of visual observations a photoelectric arrangement[112,113] may be used. The sample may also be slowly moved under the objective to render the measurement independent of the particular spot where the reading was taken.[112]

In the double-beam method, a microphotometer developed by Berek[114,115] is employed.[109,116–120] With experience, readings can be made with a relative reproducibility of ± 2 to 3%.[78,105,106,121] A variant on the Berek technique uses a grey scale with twelve gradations in the field of the sample in place of the matching of the two fields which, with increasing fatigue, tends to lead to purely subjective judgment of optimal matching.[122]

The Berek photometer has been said to possess distinct advantages over single-beam measurements with a photo cell:[120] (1) the photo cell can only measure accurately a microscopically homogeneous field having no disturbing structures, holes, or inclusions which falsify the brightness; (2) with the photo cell, the field must be completely filled with light which is frequently impossible with very small samples; and (3) especially for fusain and sclerotinite it is possible to measure an average between the dark cells

103 Stach, E., *Brennstoff-Chem.*, **33,** 361–70 (1952).

104 Ammossov, I. I., and Mussial, S. A., *Doklady Akad. Nauk S.S.S.R.*, **84,** 1223–6 (1952).

105 Mackowsky, M. T., *Brennstoff-Chem.*, **35,** 193–201, 232–5 (1954).

106 van Krevelen, D. W., Chermin, H. A. G., Dormans, H. N. M., and Huntjens, F. J., *Brennstoff-Chem.*, **39,** S 77–83 (1958).

107 Holubar, V., and Krejcik, Z., *Uhli*, **7,** 25–9 (1957).

108 Teichmüller, M., and Parks, B. C., *Proc. Intern. Comm. for Coal Petrology*, No. 2, 1956, pp. 21–27.

109 Stach, E., *Lehrbuch der Kohlenmikroskopie*, Glückauf, Kettwig, 1949, Vol. 1, pp. 1–285.

110 Rodriguez, V. H., *Bol. inf. inst. nac. Carbon*, **6,** 115–34 (1957).

111 Huntjens, F. J., Bons, J. M., and van Krevelen, D. W., *Rev. ind. minérale*, (Numéro Spécial, July 15, 1958), **40,** 138–42 (1958).

112 McCartney, J. T., and Hofer, L. J. E., *Anal. Chem.*, **27,** 1320–5 (1955).

113 Siever, R., *Illinois State Geol. Survey, Circ.* No. 241 (1957), pp. 1–11.

114 Berek, M., *Z. Krist.*, **77,** 1–22 (1931).

115 Hoffmann, E., and Jenkner, A., *Glückauf*, **68,** 81–8 (1932).

116 Dahme, A., and Mackowsky, M. T., *Brennstoff-Chem.*, **31,** 129–38 (1950).

117 McCartney, J. T., *Econ. Geol.*, **47,** 202–10 (1952).

118 Seyler, C. A., *Fuel*, **31,** 159–70 (1952).

119 Wege, E., *Brennstoff-Chem.*, **35,** 1–6, 33–41 (1954).

120 Stach, E., and Michels, H. C., *Geol. Jahrb.*, **71,** 113–43 (1955).

121 Huntjens, F. J., and van Krevelen, D. W., *Proc. Intern. Comm. for Coal Petrology*, No. 1, 1954, pp. 39–50.

122 Stach, E., *Rev. ind. minérale*, (Numéro Spécial, July 15, 1958) **40,** 159–70 (1958).

and the much lighter cell walls. The measurements are generally carried out in air or under cedar oil or another immersion medium. The oil immersion leads to lower brightness values but the differences are more clearly distinguished.[120] Still greater differences in brightness may be produced by the methylene iodide immersion technique recently introduced.[123]

The reflectance can also be measured at oblique incidence.[95] The sample is set in the center of a spectrometer table and light source and the photo cell can be rotated with respect to it. Such measurements can be made only in air. Similarly, pressed pellets of bituminous materials can be measured with a Brewster refractometer.[124] Aside from the fact that the microscopic structure cannot be observed in this manner and hence only averages are obtained, with highly absorbing materials such as coal only perpendicularly incident light should be used. Measurements of diffuse reflection on blocks of coal as well as those of total reflection of powders lead to few unambiguous results.[125] Because reflectance is somewhat wavelength dependent,[81,126] it is best to use monochromatic light.

Sample preparation. The reflecting surface can be polished on a block of coal or on a coal powder embedded in a suitable binder, usually carnauba wax.[105,109,113,116,127,128] It has been reported that the preparation of the surface with no special precautions leads to reproducible results, and that two different methods of polishing lead to concordant values.[117] There is, however, some indication that the polishing may influence the results.[120]

Coals and their Petrographic Components

Changes of reflectance with rank. The surprising discovery that reflectance varies stepwise[118,125,127,128–134] or at least not fully continuously[78,121,135,136] with increasing rank stands in sharp contradiction to other reports of a completely continuous variation.[116,117,119,137–141] The reason for the appearance of these steps may be ascribed to too few measurements and technical difficulties.

Whereas the reflectance of coals,[89,95,104,107,116,117,118,125,127,129,130,142,143] vitrains[136,137,144] exinites,[120,145,146] vitrinites,[73,78,113,119–121,127,138–141,145–147] and micrinites,[78,113,121,145,146] all increase with increasing rank when compared at the same wavelength, fusains[113,145,146] appear to show a reflectance which is independent of rank. When light of 5460 A is used, exinites up to 89% C reflect less than vitrinites[7,120,145,146] and these less than micrinites[7,145,146] up to 94% C. Fusains possess even higher reflecting power than micrinites.[7,113,145,146] Beyond 89% C, exinites can no longer be distinguished from vitrinites and beyond 94% C the latter can no longer be distinguished from micrinites and fusains.[145,146]

[123] Stach, E., *Brennstoff-Chem.*, **39,** 15–20 (1958).

[124] Palumbo, M., *Olii minerali, grassi e saponi, colori e vernici*, **32,** 191–3 (1955).

[125] Sherlock, E., *Fuel*, **30,** 31–9, 75–9 (1951).

[126] Friedel, R. A., *Science*, **130,** 220–1 (1959).

[127] Seyler, C. A., *J. Inst. Fuel*, **16,** 134–41 (1943).

[128] Seyler, C. A., and Edwards, W. J., *Fuel*, **28,** 121–7 (1949).

[129] Mukherjee, B. C., *Sci. and Culture* (*Calcutta*) **17,** 393–5 (1952).

[130] Mukherjee, B. C., *Fuel*, **31,** 153–8 (1952).

[131] Seyler, C. A., *Proc. Conf. on Ultrafine Structure of Coals and Cokes*, BCURA, London, 1944, pp. 270–89.

[132] Seyler, C. A., *Brennstoff-Chem.*, **32,** 270–4 (1951).

[133] *Ibid.*, 353–7.

[134] Seyler, C. A., *Econ. Geol.*, **47,** 211–15 (1952).

[135] Seyler, C. A., and Chandra, D., *Fuel*, **34,** 503–12 (1955).

[136] Basu, T. N., and Bagchi, D., *Proc. Symposium on the Nature of Coal*, The Central Fuel Research Inst., Jealgora, India, 1960, pp. 137–144.

[137] Dahme, A., and Mackowsky, M. T., *Brennstoff-Chem.*, **32,** 175–86 (1951).

[138] Broadbent, S. R., and Shaw, A. J., *Fuel*, **34,** 385–403 (1955).

[139] Murchison, D. G., *Brennstoff-Chem.*, **39,** S 47–50 (1958).

[140] McCartney, J. T., and Ergun, S., *Fuel*, **37,** 272–82 (1958).

[141] McCartney, J. T., and Ergun, S., *Proc. 3rd Conf. on Carbon*, Buffalo, 1957, Pergamon Press, London, 1959, pp. 223–31.

[142] Balyi, K., and Papp, F., *Földtani Közlöny*, **75–76,** 42–4 (1945/1946).

[143] Ueji, T., *J. Geol. Soc. Japan*, **50,** 137–52 (1943); *Chem. Abs.*, **41,** 7703 (1947).

[144] Kojima, K., *J. Coal Research Inst. Japan*, **4,** 38–48 (1953); *Chem. Abs.*, **49,** 7221 (1955).

[145] Dormans, H. N. M., Huntjens, F. J., and van Krevelen, D. W., *Fuel*, **36,** 321–39 (1957).

[146] Huntjens, F. J., Dormans, H. N. M., and van Krevelen, D. W., *Brennstoff-Chem.*, **39,** S 67–71 (1958).

[147] van Krevelen, D. W., *Ann. mines Belg.*, **53,** 591–8 (1954).

Although certain investigators[119,138–141] have reported maximum and minimum values for the reflectance (i.e., optical anisotropy, the degree of which is strongly dependent on geological factors[119]) of vitrains ranging from 80 to 85% C, two r values have always been found in coals of rank higher than 85% C[73,78,95,119,137–141,147,148–152] (see Fig. 12[73]).

It appears that high rank micrinites and vitrinites may be distinguished by the fact that the former are isotropic while the vitrinites are not. Above 94% C the reflectivity of micrinites is equal to the maximum reflectivity of the vitrinites.[145]

Finally, it was found that the log r_{max} values for vitrinites measured in oil plotted against H/C ratios yielded a straight line corresponding to the empirical relation:[140,141] $r_{max(vitrinite)} = r_{max(graphite)}e^{-3.6H/C}$.

Changes of reflectance with carbonization. The higher the rank of the starting material, the higher is the pyrolysis temperature (300 to 500°C) at which the reflectance begins to increase sharply.[107,118,133,143,153,154] Coals carbonized above 600°C show higher reflectance than noncarbonized coals of the same hydrogen content or of the same content of volatile matter.[153] Since these samples can show higher reflectance values than the highest rank noncarbonized coals, two new reflectance steps (α and β coke) were added to the nine steps observed in natural coals.[118,133] Although two coals of 83.6 and 84.6% C showed only one r value even after carbonization at temperatures up to 900°C, the difference between r_{max} and r_{min} of a coal of 92% C became significantly larger upon heating above 400°C.[153,154]

Changes of reflectance with chemical treatment. Two samples of coal powder—81.8 and 83.4% C—were oxidized in air from

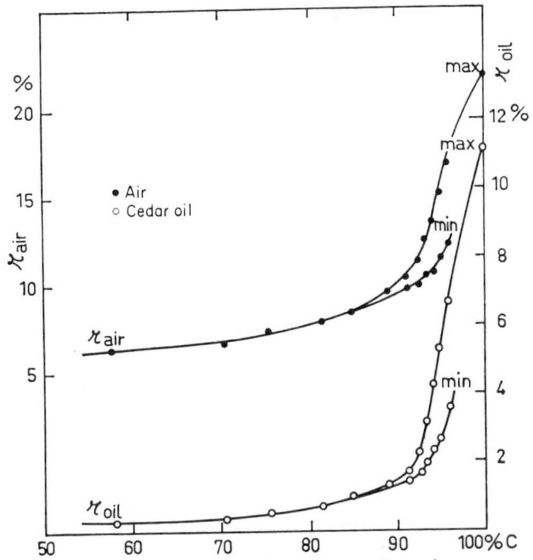

Fig. 12. Variation of reflectance with carbon content measured at 5460 A in air and in cedar oil.[73]

18 to 90 hr at 150°C and from 200 to 418 days at 110°C without showing any measurable change in reflecting power.[155] This suggests that reflectance measurements might be used to determine the original rank of weathered or oxidized coals, which is not possible by elementary analysis (carbon content) or content of volatile matter.[155,156] On the other hand, weathering, or natural oxidation, of coals has been shown to produce a difference in reflectance.[157] A coal of about 87% C oxidized at 275°C for some hours showed a distinctly higher reflectance for the oxidized edges (oxy-coal) visible under the microscope than the unoxidized matrix.[158] Similar results have been reported for Japanese coals.[159] Furthermore, as rank increases the difference in reflection between oxy-coal ($r_{oil} \sim 2\%$, $r_{air} \sim 11\%$) and the coal matrix decreases and vanishes completely with the anthracites.[158,160]

148 Duparque, A., *Ann. soc. géol. Nord* (*Lille*), **69,** 103–70 (1949).

149 Cannon, C. G., *J. Inst. Fuel*, **16,** 51–60 (1942).

150 Cannon, C. G., and George, W. H., *Nature*, **151,** 53–4 (1943).

151 Travin, A. B., *Doklady Akad. Nauk S.S.S.R.*, **110,** 655–7 (1956).

152 Chandra, D., *Nature*, **180,** 297 (1957).

153 Chandra, D., and Bond, R. L., *Proc. Intern. Comm. Coal Petrology*, No. 2, 1956, pp. 47–51.

154 Bond, R. L., Chandra, D., and Dryden, I. G. C., *Rev. ind. minérale*, (Numéro Spécial, July 15, 1958) **40,** 171–81 (1958).

155 Chandra, D., *Econ. Geol.*, **53,** 102–8 (1958).

156 Lahiri, K. C., and Basu, T. N., *J. Sci. Ind. Research* (*India*), **11B,** 486–90 (1952).

157 Nakayanagi, Y., *J. Coal Research Inst. Japan*, **7,** 225–32 (1956); *Chem. Abs.* **51,** 1581 (1957).

158 van Krevelen, D. W., and Schuyer, J., *Coal Science*, Elsevier, Amsterdam, 1957, pp. 312–35.

159 Nakagome, T., Yoshida, S., and Sugimura, H., *Nenryô Kyôkaishi*, **37,** 465–72 (1958).

160 Schroeder, H., *Brennstoff-Chem.*, **35,** 14–23 (1954).

A considerable increase in reflecting power has been observed on brown coals and bituminous coals treated with H_2SO_4 under such conditions that sulfonation has taken place.[161]

Index of Refraction

Changes of the index of refraction with rank and with carbonization. The refractive index, n, of coal is determined by measuring the reflectance in two different media, for example, air and cedar oil. As the reflectance is somewhat wavelength dependent, the same must be true for the index of refraction.[81,126] It has been found quite generally that at a given wavelength the n values of coals,[107,140,141,144,148,149,156,162] and their petrographic components[30,78,95,121,139,145–147,150] increase with increasing rank up to about 90% C.

As soon as optical anisotropy becomes important, a maximum reflectivity value is observed for polarized light vibrating parallel to the bedding plane and a minimum reflectivity value for polarized light vibrating perpendicular to the bedding plane, thus leading to two different n values. In vitrains, n_{max} increases up to 92% C and then remains constant[73,95,150] or seems to decrease again slightly.[138,140,141] On the other hand n_{min} also increases at first but above 92% C the results are in disagreement; either it remains constant[73] or there is a definite decrease.[95,138,140,141,150] The cause for the decrease may be a loosening of the crystallite packing because of the formation of more-ordered layers of larger diameter which do not pack as closely, resulting in a decrease in n perpendicular to the layer planes.[141] This idea seems to be in accordance with the change of the X-ray scattering maximum at 20 A in the anthracite range and with its interpretation.[3]

From measurements on exinites, vitrinites, and micrinites with carbon contents up to about 90%, it may be seen that micrinites show the highest n values and exinites the lowest[7,145,146] (see Table 3). The index of refraction of a fusinite of 91.5% C ($n_D = 1.91$) was found to be between the n values of vitrinite of 85.2% C ($n_D = 1.81$) and micrinite of 85.9% C ($n_D = 2.01$).[7] All three petrographic components were separated from the same dull coal.

Between 80 and 92% C the following relation seems to be valid[162,163]

$$n = \frac{313.5 - \%\text{ volatile matter}}{150}.$$

Since, among other things, n is not independent of wavelength this is obviously only a first approximation.

Finally, it has been found[154] that the refractive index of two low-rank coals (83.6% C, $n = 1.81$ and 84.6% C, $n = 1.88$) remains constant when the coals are heated at temperatures below 300°C; between this temperature and 700°C the n values increase to about 2.3 and above 700°C decrease ($n = 2.05$ at 900°C). With a high-rank coal (92% C, $n = 2.05$), no change of n_{max} with carbonization temperature up to 900°C could be observed, whereas n_{min} seems to increase between 400 and 900°C. This surprising result, however, may be only due to a higher degree of inaccuracy in the measurement of r_{min}.[154]

Structural parameters derived from the refractive index. It has already been pointed out that from a knowledge of n, d, the aromaticity and the elementary analysis, it is possible to calculate I_M/C_{ar} for coals and from this quantity a number of important structural parameters may be evaluated.[73,74,99,100,164–166] For vitrinites it was found that I_M/C_{ar} follows the theoretical curve up to 85% C, but then deviates, especially from 91% C content onwards.[73,74,75] The falling off in this region can be explained by the formation of "crystallites" which decrease the polarizability and hence the refractometric increment.[99,100] At best, the "aromatic surface area" and hence the number of aromatic carbon atoms per structural unit can only be calculated from I_M/C_{ar} values for samples of $\leq 85\%$ C.[74,99] Nevertheless, it was found empirically that a linear relation between "aromatic surface area" and (100 – % C) seems to exist for samples of less than 85% C, and this curve extrapolates rather well to

[161] Stach, E., and Teichmüller, M., *Brennstoff-Chem.*, **34,** 333–8 (1953).

[162] Krylowa, N. M., *Doklady Akad. Nauk S.S.S.R.*, **85,** 875–8 (1952).

[163] Krylowa, N. M., *Zapiski Vsesoyuz. Mineral. Obshchestva*, **83,** 37–47 (1954).

[164] Dryden, I. G. C., *Fuel*, **34,** S 48–53 (1955).

[165] van Krevelen, D. W., *Ind. chim. belge*, **20,** S 123–7 (1955).

[166] Schuyer, J., *Chem. Weekblad*, **52,** 234–40 (1956).

TABLE 2

STRUCTURAL PARAMETERS AND COMPOSITIONS OF MEAN STRUCTURAL UNITS OF VITRINITES[167]

% C	d_4^{20}	$\bar{n}$	f_a	I_M/C_{ar}	$S(A^2)$	C_{ar}	C	H	O	N	M	R
70.5	1.425	1.705	0.67	1.119	33.9	14	21	18	5	(0.3)	350	6
75.5	1.385	1.730	0.76	1.106	33.9	14	18	14	3	(0.3)	290	5
81.5	1.320	1.775	0.83	1.324	45.2	18	22	17	2	(0.4)	330	6
85.0	1.283	1.815	0.84	1.517	55.6	23	27	20	2	(0.4)	380	7
89.0	1.296	1.88	0.89	1.572	—	31	35	24	1	(0.6)	470	8
91.2	1.352	1.94	0.93	1.582	—	39	42	25	1	(0.6)	550	11
92.5	1.400	1.98	0.95	1.502	—	45	48	24	1	1	620	14
93.4	1.452	1.99	0.97	1.368	—	52	53	23	1	1	680	17
94.2	1.511	1.99	1.00	1.190	—	59	59	22	1	1	750	19
95.0	1.587	1.99	1.00	1.034	—	68	68	21	1	1	860	24
96.0	1.698	1.99	1.00	0.814	—	85	85	19	1	1	1060	34

$\bar{n}$ = average values for index of refraction.
f_a = ratio of aromatic carbon to total carbon.
I_M/C_{ar} = refractometric increment per aromatic carbon atom.
S = aromatic surface area.
C_{ar}, C, etc. = number of aromatic carbon atoms, total number of carbon atoms, etc. per mean structural unit.
M = molecular weight per mean structural unit.
R = number of rings per mean structural unit.

the value for graphite itself. Hence this empirical curve may also be used to estimate the "aromatic surface area" in higher rank coals. By taking the surface area of an aromatic carbon atom to be 2.46 A², one may then estimate the number of aromatic carbons per mean structural unit (C_{ar}):[167]

$$C_{ar} \approx \frac{340}{100 - \% C}$$

From the value of the aromaticity C_{ar}/C and the atomic H/C, O/C, and N/C ratios, the number of C, H, O, and N atoms per mean structural unit and the molecular weight may be approximated (presented in Table 2[167]). Similar results for exinites, vitrinites, and micrinites are summarized in Table 3;[167] it is seen that for a given rank exinite has the smallest and micrinite the largest C_{ar} value per mean structural unit. The I_M/C_{ar} values of carbonized coals are the same as those of natural coals of the same carbon content. This leads to the conclusion that the same structural changes seem to occur on pyrolysis as in natural coalification.[154]

[167] van Krevelen, D. W., and Schuyer, J., *Coal Science*, Elsevier, Amsterdam, 1957, pp. 151–89, 231–43.

Absorption Index

The absorption index, K, of coal can either be calculated from two reflectance measurements in two different media (e.g., air and cedar oil), or it may be derived from Lambert's Law, in which case the optical density is also needed. Both of these possibilities have their disadvantages; the r value does not depend strongly on K, especially when the absorption is small, and hence the results are not very accurate.[79] On the other hand, the determination of the optical density by means of thin sections or coal films is extremely difficult because nearly all the light is absorbed. As an example, if K is 0.08 and the film thickness of the sample is 1 μ it will be found that 99% of the available light is absorbed.[167]

The fact that a K value of 0.025 was found at 5460 A for a Pittsburgh vitrinite of 84% C as measured by transmission through a thin section[79] whereas the same material studied[167]

TABLE 3

TYPICAL DATA OF PETROGRAPHIC COMPONENTS[167]

% C of the Vitrinite	Maceral	Ultimate Analysis, Dry, Ash-Free C	H	O	N	S	$d_{(calc)}$	f_a	r_{oil} (max)	r_{air} (max)	$\bar{n}$	C_{ar}
	E	82.2	7.4	8.5	1.3	0.6	1.174	0.62	0.13	5.71	1.628	9
81.5	V	81.5	5.15	11.7	1.25	0.4	1.326	0.83	0.67	7.91	1.788	18
	M	83.6	3.95	10.5	1.35	0.6	1.380	0.90	1.27	9.70	1.910	21
	E	85.7	6.5	5.8	1.4	0.6	1.210	0.75	0.24	6.32	1.671	10
85.0	V	85.0	5.4	8.0	1.2	0.4	1.288	0.84	0.92	8.52	1.836	23
	M	87.2	4.15	6.7	1.35	0.6	1.398	0.92	1.50	10.31	1.952	27
	E	87.7	5.85	4.4	1.45	0.6	1.242	0.84	0.44	7.18	1.729	13
87.0	V	87.0	5.35	5.9	1.25	0.5	1.282	0.86	1.07	9.01	1.864	26
	M	89.1	4.2	4.7	1.4	0.6	1.384	0.93	1.66	10.70	1.982	31
	E	89.6	5.2	3.3	1.3	0.6	1.283	0.89	0.82	8.30	1.817	18
89.0	V	89.0	5.1	4.0	1.3	0.6	1.292	0.89	1.26	9.62	1.899	31
	M	90.8	4.1	3.2	1.3	0.6	1.389	0.94	1.90	11.15	2.008	37
	E	90.4	4.9	2.8	1.3	0.6	1.304	0.91	1.12	9.20	1.873	22
90.0	V	90.0	4.94	3.2	1.35	0.5	1.303	0.91	1.39	10.03	1.921	34
	M	91.5	3.95	2.6	1.35	0.6	1.403	0.95	2.08	11.40	2.017	40
	E	91.5	4.5	2.3	1.2	0.5	1.341	0.93	1.64	10.63	1.960	30
91.2	V	91.2	4.55	2.6	1.15	0.5	1.335	0.93	1.64	10.63	1.960	39
	M	92.2	3.65	2.2	1.35	0.6	1.435	0.96	2.44	11.81	2.020	43

E = Exinite
V = Vitrinite
M = Micrinite
r_{oil}, r_{air} = reflectance as measured in oil or air.
$\bar{n}$ = average values for index of refraction.

either by reflection at vertical incidence or by transmission of a coal film on teflon gave $K = 0.06$, together with the further observations that a Velsen vitrinite (84.5% C)[167] gave $K = 0.08$ by transmission and $K = 0.06$ by reflection, may be ascribed to experimental difficulties.

Reflectance measurements with perpendicularly incident polarized light show that at a given wavelength K increases slightly with increasing rank—especially strongly beyond 92% C.[73,78,121,138,140,141,147] From reflectance measurements with obliquely incident polarized light, it has also been deduced that, for a given sample, K increases with decreasing wavelengths,[95,149] but these K values are two to three times higher than with perpendicularly incident light. Results for K derived from n and E_{spec} showed the same wavelength dependence, but the absolute values are rather small.[81,126]

If coals show optical anisotropy, maximum and minimum values of K may be calculated from the maximum and minimum values of the reflectance.

From measurements made at ~5500 A it was shown that log K_{max} vs. H/C seems to follow a straight line relationship.[141]

Finally, it has been found that the absorption index increases with increasing pyrolysis temperature and follows a curve very similar to that of natural coalification.[154]

Optical Anisotropy

It has already been established that the appearance of two reflectance values in coals must be attributed to optical anisotropy. This effect is a uniaxial negative birefringence; the refractive index of the extraordinary ray is always less than that of the ordinary ray. The optical axis is perpendicular to the layering or stratification of the coal samples.[95,121,137,167] This optical anisotropy is a consequence of mutual orientation of the structural elements (see Fig. 13[137]) produced by external influences—principally pressure.[50,119,137,148,151,168,169]

Coals with less than 85% C show in transmission through a thin section a weak but noticeable anisotropy,[162,170,171] whereas by reflection (polished surface) this effect has not always been observed. A possible explanation for this discrepancy might be the fact that inner reflections could mask the anisotropy in the reflected light.[170,171]

In coals as well as vitrains and vitrinites with more than 85% C, anisotropy is observed both in reflection and in transmission. However, this anistropy becomes strong in reflection only beyond 88% C or H/C < 0.700 (see Fig. 14[141]). Micrinites are isotropic.[145]

Attempts have been made to deduce the degree of orientation from the optical anisotropy measured by reflection.[167] This would lead to the conclusion that even at 96% C the orientation of the lamellae is still far from perfect.

Initially isotropic coals show no anisotropy when pyrolyzed up to temperatures of 900°C; if the starting sample is demonstrably anisotropic, this effect will be strengthened by pyrolysis.[153,154]

INFRARED ABSORPTION

Classical electrodynamics first predicted and quantum theory later completely confirmed that an interaction between matter and infrared radiation in the form of absorption or emission can occur only if it is associated with a change of the electrical dipole moment. If this is true, rotations and vibrations, or more frequently a combination of the two, may be observed as "infrared active" bands. The fundamental rotation-vibration spectrum stretches from about 2.5 μ to about 50 μ ($1\ \mu = 10^4$ A $= 10^{-4}$ cm). The number of bands observable for a given molecule depends on the number of atoms it contains and its symmetry. The assignment of an absorption band to a given mode of vibration of the molecular system is closely tied to the *position* of the band and to some extent to its *intensity*.

The *position* of infrared bands, that is the position of their maximum intensity, is given either in wavelengths, λ, measured in microns, μ, or as wave numbers, $\bar{\nu}$, measured in cm^{-1}. The two units are connected by the relation

$$\bar{\nu} = \frac{1}{\lambda(\text{cm})} = \frac{10{,}000}{\lambda(\mu)}.$$

It is a matter of experience that there is a correlation between certain groupings of atoms which appear as structural features in molecules and the position of corresponding bands in the infrared spectrum. This correlation has led to the idea of "characteristic" or "key" bands which are determined more or less completely by the forces between atoms in a portion of the molecule and depend only slightly on the structure of the rest of the molecule. This idea can only be a rough approximation since it is known that in general *all* the atoms participate in a fundamental mode of vibration. Nevertheless, small shifts of the position in some of these "characteristic" bands give valuable clues as to the exact structures from which they arise. For example, all CH-stretching vibrations in hydrocarbons lie between 3.0 and 3.6 μ; but $CH_{\text{acetylenes}}$ lie in the range 3.0 to 3.1 μ, CH_{olefins} between 3.15 and 3.30 μ, CH_{aromatic} between 3.2 and 3.3 μ and $CH_{\text{aliphatic}}$ between 3.36 and 3.6 μ.

As molecules increase in complexity, or if a mixture of different substances is present, it becomes more difficult to interpret the spectrum because of the partial or even complete overlapping of "characteristic" bands. Nevertheless, an assignment of the strongest bands to definite atomic groupings is nearly always possible, but it is more difficult to arrive at structural conclusions without uncertainty.

The *intensity* of a given band depends on the number of molecules absorbing (or emitting) and on the change in electric moment associated with the vibration in question, that is, essentially to the change in the distribution of

[168] Williams, E., *Fuel*, **32**, 89–99 (1953).

[169] Mincher, D., *Godishnik Sofiskiya Univ., Biol.-Geol.-Geogr. Fak., Kniga 2-Geol.*, **49**, 65–71 (1954/1955).

[170] Mackowsky, M. T., *Glückauf*, **89**, 1013–8 (1953).

[171] Mackowsky, M. T., *Proc. Intern. Comm. for Coal Petrology*, No. 1, 1954, p. 49.

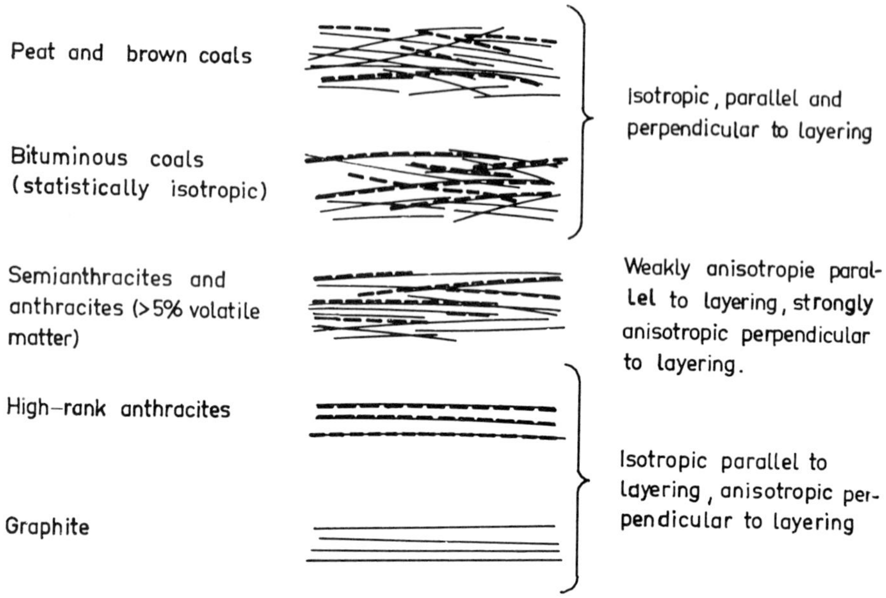

Fig. 13. Rank, degree of layering, and variation in anisotropy.[137]

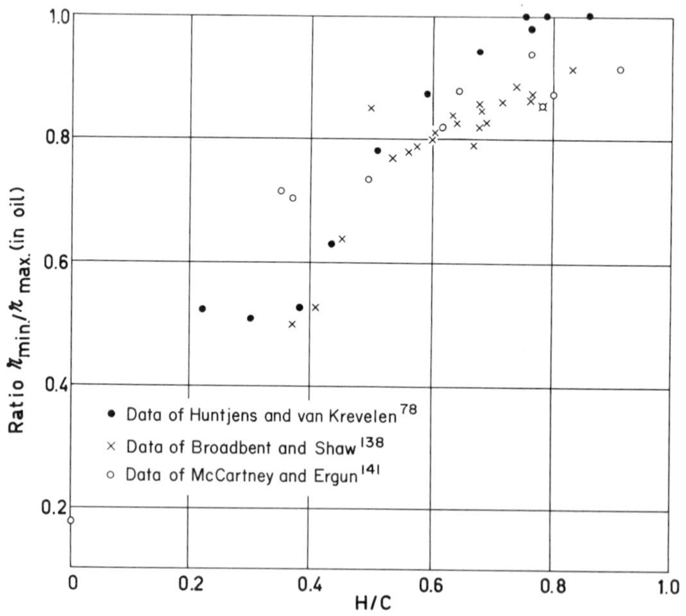

Fig. 14. Degree of anisotropy for vitrains of different rank.[141]

electron density within the structure produced by the vibration. To solve this problem completely theoretically would require detailed knowledge of the effect of the remaining atoms on the charge distribution in each bond—a degree of refinement not yet achieved. Hence it is necessary to resort to empirical correlations. Either the maximal extinction or the integral extinction may be used as a quantitative measure of band intensity. The latter quantity is far less dependent on the characteristics of the particular equipment than is the former and hence is more reproducible from laboratory to laboratory (for example, if two different laboratories operate with different effective slit widths, the area under a given band will be about the same in both cases but the laboratory operating with the wider slit will have a broader band and consequently a lower maximum extinction).

The difficulties besetting a quantitative analysis are chiefly overlapping of bands, which necessitates an uncertain correction for the wing of a neighboring band and a difficultly evaluated correction, if the background is not constant.

If it appears from what has been said that an unequivocal analysis of a complicated molecule is difficult, it must be pointed out that in one sense infrared analysis provides valuable assistance: the absence of a particular "characteristic" band in the completely unmodified material permits the definite conclusion that this group is either not present or is present only in such a small concentration that it cannot be observed. A further advantage of present day infrared techniques is that they permit measurements which are independent of the state of aggregation and, in particular, of whether or not the material is soluble.

Technique of Measurement

Measurements in the region of infrared fundamental vibrations can now be made with any of a number of commercially available single- or double-beam spectrophotometers. To study coals, a number of techniques have been developed.

Thin sections.[172–178] Thin sections of coals between 0.13 and 0.012 mm in thickness have been prepared and their infrared spectra measured. This method has the advantage that there is little scattered radiation.[178] In comparison with decidedly simpler methods of sample preparation, the spectra obtained are in no respect more specific,[176] but there is the disadvantage that, during the relatively long time necessary for preparing the thin section, oxidation may easily take place.[177]

Mulls.[173–184] In the most used method a suspension of coal in Nujol is prepared in which coal in about the same weight of oil is ground to an average particle size of 1 μ. It has been found that there is little use in grinding longer than 8 hr since the scattered light is not further reduced.[176] To obtain usable spectra, samples up to 0.05 mm thick must be used, and these require such broad slits as to reduce resolving power appreciably. Another reason for using broad slits is the very low transparency of the samples caused by scattering of the incident light by the embedded particles.[174] Attempts have been made to improve relative transparency by compensation with cloudy NaCl windows[175,178] or with Nujol mulls of spectroscopic carbon[174] and by holding the background absorption as constant as possible by suitable variation of the slit width.[176]

The advantages of this method are associated with the fact that the region of OH stretching absorption (2.7 to 3.1 μ) can be measured

[172] Sutherland, G. B. B. M., Fellgett, P. B., and Willis, H. A., *Proc. Conf. on Ultrafine Structure of Coals and Cokes*, BCURA, London, 1944, pp. 330–3.

[173] Cannon, C. G., and Sutherland, G. B. B. M., *Trans. Faraday Soc.*, **41**, 279–88 (1945).

[174] Gordon, R. R., Adams, W. N., and Jenkins, G. I., *Nature*, **170**, 317 (1952).

[175] Friedel, R. A., and Pelipetz, M. G., *J. Opt. Soc. Am.*, **43**, 1051–2 (1953).

[176] Brown, J. K., *J. Chem. Soc. (London)*, **1955**, 744–52.

[177] van Vucht, H. A., Rietveld, B. J., and van Krevelen, D. W., *Fuel*, **34**, 50–9 (1955).

[178] Friedel, R. A., and Queiser, J. A., *Anal. Chem.*, **28**, 22–30 (1956).

[179] Cannon, C. G., and Sutherland, G. B. B. M., *Nature*, **156**, 240 (1945).

[180] Gordon, R. R., Adams, W. N., Pitt, G. J., and Watson, G. H., *Nature*, **174**, 1098–9 (1954).

[181] Brooks, J. D., Durie, R. A., and Sternhell, S., *Australian J. Appl. Sci.*, **9**, 63–80 (1958).

[182] den Hertog, W., and Berkowitz, N., *Fuel*, **37**, 253–71 (1958).

[183] Kasatochkin, V. I., and Smutkina, Z. S., *Doklady Akad. Nauk S.S.S.R.*, **113**, 1314–17 (1957).

[184] Brown, J. K., *Fuel*, **38**, 55–63 (1959).

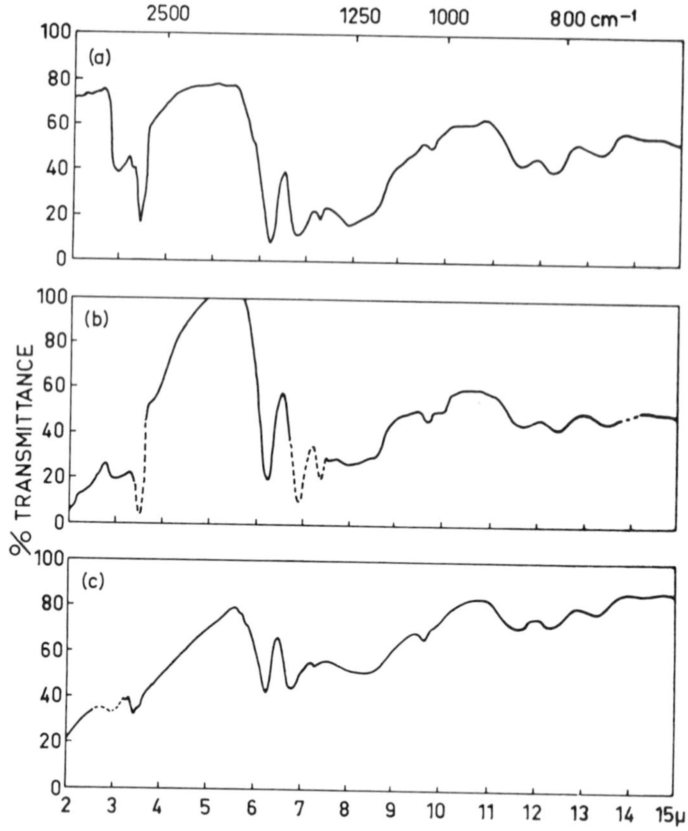

Fig. 15. The infrared spectrum of a vitrain (84% C) measured as a: (*a*) thin section (0.020 mm),[178] (*b*) Nujol mull (1:1; 0.038 mm)[178] (– – – – – uncertain intensities; Nujol absorption bands), and (*c*) KBr window (3 mg vitrain/1 g KBr) (– – – – – – uncertain intensity; OH absorption of KBr).

since Nujol has no band in this region. Among the disadvantages may be mentioned the strong bands in Nujol itself (3.44 μ, 6.86 μ, 7.26 μ) which preclude studying samples in these regions and the fact that at best only qualitative results are possible; a quantitative reproducibility is nearly impossible because of different scatter losses and the different distribution of particles.

For studies between 2.5 and 5 μ mulls made with hexachlorobutadiene, which has no bands in this region, have proved useful.[176,181,184] Advantages and disadvantages are the same as those of Nujol mulls.

Pressed windows[48,177,178,180,182,185–192] In this method a mixture of coal and KBr powder is ground in an inert atmosphere, and the uniform mixture is pressed to a window under vacuum. The coal may amount to 0.3 to 0.5 wt % of the mixture; KCl has been substituted for KBr without improving the results.[181,193] Advantages of this method are that the pressed windows are relatively easy to prepare and that in very pure KBr or KCl there are no absorption bands below 15 μ. Nevertheless, quantitative measurements and complete

[185] Bergmann, G., Huck, G., Karweil, J., and Luther, H., *Brennstoff-Chem.*, **35,** 175–6 (1954).

[186] Kojima, K., Sakashita, K., and Yoshino, T., *Nippon Kagaku Zasshi*, **77,** 1432–4 (1956).

[187] Bergmann, G., Huck, G., Karweil, J., and Luther, H., *Brennstoff-Chem.*, **39,** S 20–4 (1958).

[188] Alexanian, M. Ch., *Compt. rend.*, **246,** 1192–5 (1958).

[189] Szymanowski, W., Wisniewski, R., Golebiowski, R., and Olszewska, I., *Bull. acad. polon. sci.*, **6,** 223–31 (1958).

[190] Kinney, C. R., and Doucette, E. I., *Nature*, **182,** 785–6 (1958).

[191] Elofson, R. M., *Can. J. Chem.*, **35,** 926–31 (1957).

[192] Fujii, S., and Yokayama, F., *Nenryô Kyôkaishi*, **37,** 643–7 (1958).

[193] Brooks, J. D., and Sternhell, S., *Australian J. Appl. Sci.*, **8,** 206–21 (1957).

reproducibility are difficult to achieve because of difficulties in repeating the exact fineness of grinding,[185,194] the same distribution of the particles, the same effective thickness of the sample,[83,185] and an exact determination of the contribution from scattered light.[185] Although a band which frequently appears in commercial KBr at 7.22 μ, which results from contamination by traces of nitrates, can be easily compensated, the moisture content of KBr is such that it is nearly impossible to achieve exact compensation of the OH stretching band; consequently, exact measurements of absorption in that neighborhood are very difficult.

Potassium bromide windows have been heated to 100°C in a porcelain crucible without any detectable deterioration of the quality of the windows. Chemical changes in the coal due to KBr or oxidation by atmospheric oxygen during the heating have never been observed.[187] In fact, recent measurements have shown that even heating to 200°C produces no other spectral changes than the disappearance of the OH band in the KBr. This suggests that some of the previously mentioned difficulties of the method may be eliminated by this procedure.

Compensation by means of a wire grating has been tried in an attempt to diminish the disturbing nonspecific background.[187]

In conclusion, it may be said that the maximum information concerning the infrared spectra of coals can only be achieved by employing in each spectral region the method most useful. Nevertheless, it is clear from Fig. 15[178] that as a result of the high degree of nonspecificity of coal spectra the results are almost completely independent of the method used. It is also worth mentioning that the spectra of coals show no sharpening of the bands even when measured at −100°C.[187]

Infrared Spectra of Coals and Their Petrographic Components

The type of spectrum shown in Fig. 15 is obtained from coal samples independently of their origin[178,187,195] and their rank.[173,176,181,186,190,192,195] The bands present do not permit unequivocal conclusions concerning the degree of aromaticity or the size of the condensed aromatic ring systems in the samples, but they are not inconsistent with a high degree of aromaticity.[63,185]

Mineral constituents are usually recognizable in the region 9.6 to 11 μ whenever the ash content is above 2%.[178,185,187,196] Chemical removal of the mineral constituents does not appear to modify the spectrum of the coal itself.[187]

The Background Absorption

In all coal samples, there is a nonspecific background absorption which is usually small up to 90% C but which increases rapidly beyond 91% C.[63,176,177,197] There are basic differences of opinion concerning the origin of this absorption. Thus, this background absorption has been ascribed to intense scattering caused by insufficient grinding of the coal or to a reagglomeration of the coal particles in a mull.[180] Another view is that the effect is not produced by scattering, since a microscopic examination showed that the particle sizes in mulls of coals between 89 and 94% C did not essentially differ from one another; however, the background absorption increases very sharply in such a series.[176] Apparently the basic cause is electronic absorption,[172,176,177] because it was found that an absorption edge in cellophane chars moved from the visible into the infrared with progressive carbonization[198] and a 2 μ thick layer of a dispersion of graphite in Nujol showed a continuous unspecific absorption of 70 to 80% throughout the visible and infrared regions.[197]

The marked increase in background absorption of coals having carbon contents higher than about 89% may be explained in the following way.[176] In semiconductors, of which coal is assumed to be an example, the energy difference between filled and empty electronic levels decreases with the number of π electrons in the molecule, and this in turn is directly related to the number of condensed aromatic rings. Finally, the difference in energy for electronic transitions becomes sufficiently small to lie in the infrared. It is also possible that the optical anisotropy and the growth of

[194] Monnot, G. A., and Morel, P., *Appl. Spectroscopy*, **9**, 180–3 (1955).

[195] Kasatochkin, V. I., *Proc. Symposium on the Nature of Coal*, Central Fuel Research Inst., Jealgora, India, 1960, pp. 64–73.

[196] Urbański, T., Kucziński, W., Hofman, W., Urbanik, H., and Witanowski, M., *Bull. acad. polon. sci.*, **7**, 207–14 (1959).

[197] Cannon, C. G., *Nature*, **171**, 308 (1953).

[198] Kmetko, E. A., *Phys. Rev.*, **82**, 456–7 (1951).

crystalline regions, transition from "aromatic" to "graphite-like" structures, may have an influence on the increase in background absorption.[197]

Investigations of one of the authors (H. Tschamler) on exinites (85.5 to 89.2% C), vitrinites (83.5 to 88.9% C), and micrinites (86.8 to 89.7% C) separated from four Ruhr coals of different rank have been carried out. (The separation was carried out by Prof. C. Kröger, Technische Hochschule, Aachen, Germany.) For all twelve samples, the same instrumental and experimental conditions were used for the measurement of the nonspecific background absorption at the wavelength 10.5 μ. It was found that in the micrinites, independent of rank, the nonspecific background absorption was always the highest and almost constant. Exinites had the lowest nonspecific background absorption which increased with rank and became approximately the same as that of the vitrinites at 89% C; the background absorption of vitrinites was almost independent of rank and distinguishably lower than in micrinites. If it is assumed that the nonspecific background absorption results from electronic absorption, the foregoing observations would mean that micrinites contain larger condensed aromatic systems than vitrinites, which, up to 89% C, possess larger systems than exinites.

Assignment of the Infrared Bands of Coals to Definite Functional Groups

In Table 4, a complete survey of the positions of characteristic infrared bands and their assignments by various authors is presented. In addition the following points should be noted.

1. The broad band with maximum near 3300 cm^{-1} is undoubtedly to be ascribed to the OH groups appearing in the organic material of coal since this absorption persists after drying the coal at 105°C for 24 hr[199] or even at these conditions in high vacuum.[176] That it consists essentially of phenolic OH groups seems to be unequivocally proved by the strong diminution in the intensity of this band and by bands which appear at other characteristic positions when the OH groups are acetylated. The position of this band indicates hydrogen-bonded OH groups; this hydrogen bonding must be regarded as an essential feature of the coal structure.[197] Strong intramolecular hydrogen bonding, as it occurs, for example, in six-membered chelate rings, cannot be excluded, but, as seen from the spectra of model substances, the small intensity and extraordinary band width of the OH absorption in such structures make them generally difficult to detect.[176,178] In this respect, it is important to remark that the 3300 cm^{-1} absorption in coal is partially overlaid by a weak and very broad band covering the region from about 3000 to 2700 cm^{-1}, which shows above the nonspecific background and the correct interpretation of which may be of fundamental significance for the elucidation of coal structure.[176] Hydrogen-bonded NH groups may also contribute to the intensity of the band at 3300 cm^{-1}, but it is clear that this can be at most small since the total amount of nitrogen in coals is not large.

2. There is no doubt about the assignment of the specific bands occurring in the 3000 to 2800 cm^{-1} region. The very small intensity of the 3030 cm^{-1} band suggests, on the assumption of high aromaticity in coals, that the aromatic systems present are highly substituted or highly crosslinked or both.[63] The only exception to the generally low intensity of this band is a Baragolai coal (Assam) in which this band is very prominent.[199] The fact that the band near 2980 cm^{-1} is only detected as a shoulder led to the conclusion that the CH_3 content in coals may be small, but CH_3 groups on aromatic systems do absorb mainly at 2925 and 2860 cm^{-1}. Aliphatic CH_3, CH_2, and CH groups all contribute to the intensity of the two bands at 2925 cm^{-1} and 2860 cm^{-1}, but the dominant contribution may come from CH_2 groups. Nevertheless, it is not possible to decide whether the CH_2 groups are present in aliphatic chains, in hydroaromatic structures, or in cycloparaffins.

3. In brown coals a distinct band occurs at about 1700 cm^{-1}, which in coals of higher rank appears only as a relatively weak shoulder on the strong and broad band at 1600 cm^{-1} or may not be detectable at all.[187,192,195] In brown coals, this band is unquestionably to be assigned to C=O groups in the organic structure, but in the bituminous coals this absorption might also be the result of a slight oxidation of the coal during sampling, drying, or grinding.[187]

[199] Roy, M. M., *Fuel*, **36**, 249–50 (1957).

TABLE 4

BAND POSITIONS IN THE INFRARED SPECTRA OF COALS AND THEIR ASSIGNMENTS BY DIFFERENT AUTHORS*

Band Position cm^{-1}	Band Position μ	Assignment	References
3300	3	—OH str. / —NH str.: phenolic OH .... O	173, 176–178, 181, 185, 186, 189, 192, 196, 199
		—OH (Hydroperoxide)	178
		>NH .... N	176, 181, 189, 196
3030	3.3	ar. CH str.	173, 176–178, 185, 186, 192, 195, 197, 199
2978 sh.	3.36	CH_3 str.	186, 195
2925 2860	3.42 3.5	CH_3 str., CH_2 str., al. CH str.	173, 176–178, 181, 185, 186, 189, 192, 195, 197, 199
1700	5.9	C═O str.	178, 181, 185, 192, 193, 195, 197, 199
1600	6.25	ar. C═C str.	48, 173, 176, 181, 185–187, 189, 191, 192, 195, 199, 200
		C═O .... HO—	48, 176, 178, 186, 192, 195, 199
1500	6.65	ar. C═C str.	191
1450	6.9	CH_3 assym. def. CH_2 scissor ar. C═C str.	173, 176, 178, 181, 185, 186, 189, 191, 192, 199 176, 187
1380	7.25	CH_3 sym. def.	173, 176, 178, 181, 185, 186, 189, 191, 192, 199
1300 to 1000	7.7 to 10.0	C—O str. (phenols) OH def.	176, 181, 197, 199
		C_{ar}—O—C_{ar} str.	173, 176, 178, 195, 199
		C—O str. (alcohols)	176, 186, 189
		C_{ar}—O—C_{al} str.	176, 178, 189, 195
		C_{al}—O—C_{al} str.	176, 189, 199
900 to 700	11.1 to 14.3	"aromatic bands"	172, 185, 191, 192
860 820 750	11.6 12.2 13.3	ar. HCC rocking (single and condensed rings)	81, 176, 177, 178, 181, 186, 189, 195, 197, 199
873 816 751	11.5 12.3 13.3	subst. benzene ring with isolated H subst. benzene ring with 2 neigh. H *o*-subst. benzene ring	186
893 820 758	11.2 12.2 13.2	H atoms in meso position (anthracene) angular cond. ring systems *o*-subst. benzene rings mono-subst. benzene rings cond. ring systems	201

* ar. = aromatic; al. = aliphatic; str. = stretching; def. = deformation; sh. = shoulder; subst. = substituted.

4. Most of the discussion of coal spectra centers about the very strong, broad absorption always occurring near 1600 cm^{-1}. The problem to be solved is whether the intensity of this band is produced entirely by aromatic C═C bonds or whether a contribution comes from special oxygen-containing functional groups. The view that this band is due only to aromatic C═C bonds and hence that a high concentration of aromatic carbon is present in coal is based on the following considerations:

(*a*) A very intense band at 1600 cm^{-1} occurs also in high molecular weight aromatic petroleum fractions and even in those containing less than 1% oxygen.[187]

(*b*) The fact that the 1600 cm^{-1} band seems not to be influenced either in position or in intensity by acetylation or by oxidation of coals[187] which manifests itself in a band near 1700 cm^{-1}.

(*c*) The discovery that ether oxygen as in diphenyl ether and in

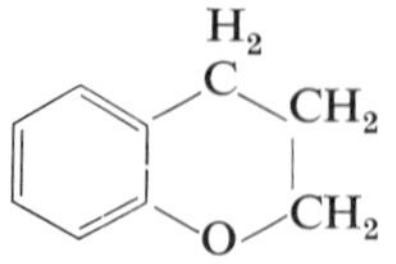

or the oxygen in phenols may produce a marked increase in the intensity of absorption of the aromatic C=C bond.[176,178]

(*d*) The following ether

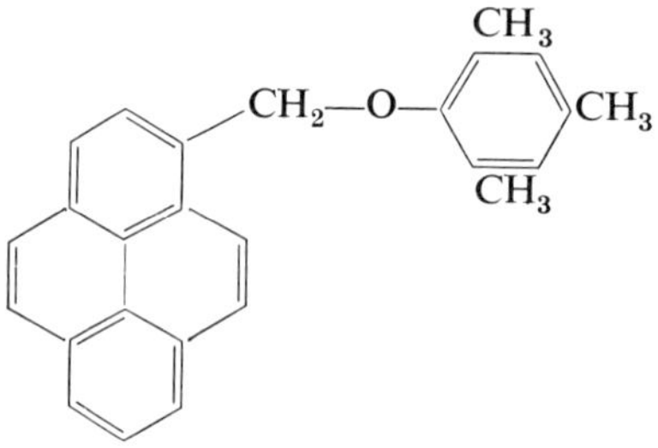

which was studied as a model material does not possess a notable absorption at 1600 cm^{-1} but the char obtained from it at 425°C, under which circumstances the splitting off of CO was established, shows a strong absorption at 1600 cm^{-1} and in this essentially oxygen-free material the band can only be associated with highly condensed aromatic systems.[200]

(*e*) The lack of suitable aromatic structures which can serve as models for the elucidation of the coal structure.[187]

(*f*) The spectrum of a coal (89% C) has been compared with that of (1) a condensed pitch, (2) a high-molecular-weight pitch fraction, (3) a pitch, (4) a mixture of twenty-five aromatic model substances and (5) each of a number of highly condensed aromatic hydrocarbons. It was found that the optical density of the band near 1600 cm^{-1}, in all of these spectra, with the single exception of the pitch, is about the same.[202]

The view that in addition to aromatic C=C bonds, significant contributions to the breadth[48] and intensity of the 1600 cm^{-1} band are made by oxygen-containing functional groups is based on the following considerations:

(*a*) All compounds with strong intramolecular hydrogen bonding in the form of six-membered chelate rings (e.g., hydroxyquinones, 2 hydroxy-acetophenone, etc.) or five-membered chelate rings (e.g., kojic acid) possess very strong absorption at 1600 cm^{-1} (diphenoquinone itself possesses a strong band in this region), and the presence of quinone oxygen in coals seems to be proved.[203]

(*b*) Char prepared from *n*-octacosane at 450°C produces most of the typical coal-like spectrum, except that the 1600 cm^{-1} band is much weaker than in coal.[81] Oxidative pyrolysis of *n*-octacosane at 300°C produces the same spectrum but with a strong 1600 cm^{-1} band, these results strongly indicating that the 1600 cm^{-1} band is a carbonyl band including aromatic C=C stretching vibrations.[81]

(*c*) Carboxylates absorb strongly in this region so that an intensity contribution to the 1600 cm^{-1} band might be expected from chemical bonds between the organic material in coal and inorganic elements to be found in the ash.[181]

5. The absorption at 1500 cm^{-1} typical of the C=C bonds in benzene rings has been clearly established in lignites and subbituminous coals,[191] but could not be observed in bituminous coals.[187,191,199] This may be explained by the fact that increased substitution on the benzene ring is known generally to weaken this band[191] and the fact that as the degree of condensation increases, this band is displaced to smaller wave numbers[190,191] and finally disappears in the strong absorption at 1450 cm^{-1}.[176,187]

6. The strong absorption at 1450 cm^{-1} is undoubtedly due in part to CH_2 groups, but contributions may originate from CH_3 groups, aromatic C=C bonds, and strongly hydrogen-bonded OH groups.

7. From the rather weak absorption at 1380 cm^{-1} which can be assigned to CH_3 groups, it was concluded that the CH_3 content in coals should be relatively small, but investigations on model compounds have shown that the intensity of this band can vary to a great extent. A contribution to this absorption may arise from CH_2 groups in cyclic structures.[192]

[200] Depp, E. A., and Neuworth, M. B., *Brennstoff-Chem.*, **39,** S 90–2 (1958).

[201] Hadzi, D., *J. phys. radium*, **15,** 194–6 (1954).

[202] Groenewege, M. P., *Brennstoff-Chem.*, **39,** Sondersausdrücke, 24 (1958).

[203] Given, P. H., and Peover, M. E., *Proc. Conf. on Science in the Use of Coal*, Sheffield, Inst. Fuel, London, 1958, pp. A33–7.

8. The broad but poorly resolved absorption between 1300 and 1000 cm^{-1} has its principal origin in the C—O stretching vibrations of various ethers but may contain also some aliphatic and aromatic HCC wagging motions.

9. The three bands between 900 and 700 cm^{-1} characteristic of coals of more than 81% C are mainly due to aromatic HCC rocking vibrations in benzene and condensed aromatic ring systems. This assignment seems to be established by the fact that a completely deuterated paraffin charred in the presence of oxygen showed no "aromatic bands" in its spectrum, these having been shifted to longer wavelength.[81] However, contributions are to be expected from the aliphatic HCC rocking motions and may occur from C=C—C in-plane deformation vibrations of the carbon skeleton. The bands between 900 and 700 cm^{-1} have also been interpreted as resulting only from individual benzene rings substituted in particular ways[178,186] as indicated in Table 4. Whether the absorption bands in this region are to be regarded as an average spectrum of polycondensed aromatics[177] or whether they represent special vibration modes permitting conclusions as to specific points of substitution on the aromatic framework cannot yet be decided. From the absence of distinct absorption bands near 700 cm^{-1} it has been concluded that no long aliphatic chains are present in coals.[178]

In conclusion, it must be noted that a thin section of a coal studied between 700 and 400 cm^{-1} did not show any infrared active bands at all.[178]

The Infrared Spectra of Petrographic Components of Coal

The spectrum of a vitrain differs only slightly from that of a bright coal. Durains also show, aside from a distinct shoulder near 1700 cm^{-1}, spectra which are quite comparable to those of vitrains.[176,185] The spectrum of a fusain (hand picked from Yallourn brown coal) has a strong broad absorption at 3400 cm^{-1}, shows no aromatic CH bands and little evidence of aliphatic hydrogen content, but has a pronounced C=O absorption near 1700 cm^{-1} beside the strong absorption at 1610 cm^{-1}.[181] A fusain of German origin showed essentially the same features with the exception of the 1700 cm^{-1} band. During the pyrolysis of wood, an absorption appears in the short wavelength region which increases in intensity with increase in pyrolysis temperature and at the same time moves toward longer wavelengths. The spectrum of fusain may therefore be due to plant material which was heated to temperatures between 400 and 700°C, for example by a forest fire.[187]

From a "charbon flambant sec de La Houve" (Lorraine) fractions very much enriched in exinite (81% C, 8.4% H, 8% O) and in vitrinite (82.1% C, 5.1% H, 12.1% O) were separated.[204] The spectrum of neither maceral showed unequivocal evidence of OH, or of aromatic and aliphatic CH stretching vibrations, but both possessed definite bands at 1730 cm^{-1} (C=O stretching; esters) and at 1690 cm^{-1} (C=O stretching; aldehydes + ketones). The relatively higher ester content of the exinite was associated with the higher spore and cuticle content,[204] both of which contain ester groups. The strong broad band at 1620 cm^{-1} was weaker in exinite than in vitrinite. Although a "triplet" near 1445 cm^{-1} (CH_2- and CH_3-groups) was considerably stronger in exinite than in vitrinite, in the broad strong absorption between 1300 and 1000 cm^{-1} the opposite was true. Furthermore, the exinite showed two clearly defined bands at 870 and 725 cm^{-1}, whereas in the vitrinite broad bands at 870, 840, 740, and 725 cm^{-1} were found.[204]

Comparison of the infrared spectra of an exinite, a vitrinite, and a micrinite separated from a Yorkshire durain showed that the exinite fraction ($d < 1.25$) had fewer aromatic CH bonds, more saturated CH bonds, and fewer aromatic CC bonds than the vitrinite. Furthermore, the micrinite fraction ($d > 1.35$) possessed, also in comparison with the vitrinite, fewer aromatic and aliphatic CH bonds but more aromatic CC bonds.[145]

It is noteworthy that a product obtained from the formaldehyde condensation of pitch possessed an infrared spectrum differing from that of a vitrinite only in having a weaker aliphatic CH stretching absorption, a stronger band at 740 cm^{-1}, and a new band at 690 cm^{-1}.[205]

[204] Monnot, G. A., and Ladam, A., *Compt. rend.*, **241**, 1939–41 (1955).

[205] van Krevelen, D. W., Schors, A., Bos, H., Groenewege, M. P., and Westrik, R., *Fuel*, **35**, 230–43 (1956).

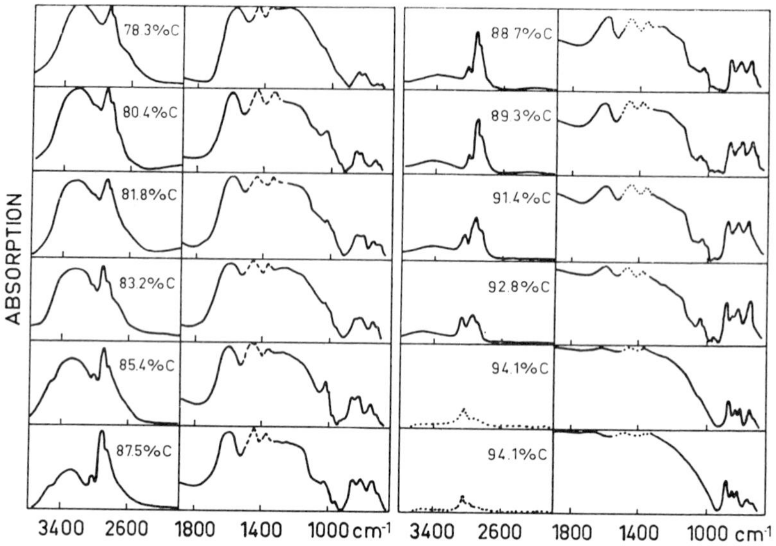

Fig. 16. Variation of infrared spectra of vitrains with carbon content.[176] Region 4000 to 2000 cm^{-1} measured as hexachlorobutadiene mulls (. . . . . uncertain intensities). Region 2000 to 650 cm^{-1} measured as Nujol mulls (– – – – – Nujol absorption bands).

Changes of the Infrared Spectra of Coals with Rank

As may be seen in Fig. 16,[176] there is a nearly continuous transition in spectra in proceeding from brown coals through bituminous coals to anthracites. That these changes are small in comparison with those observed by other methods may be attributed to the overlapping of the infrared bands of similar groups which are present in great variety in the molecular units of coals.[187] In this connection it is especially noteworthy that in the spectra of the brown coals which have high total oxygen contents, the absorption regions associated with oxygen containing functional groups are in no way outstanding as regards intensity.[181]

In the bituminous region, on the other hand, the decrease of total oxygen content makes itself felt in noticeable decrease in the absorption bands of the corresponding functional groups. Thus the OH absorption at 3300 cm^{-1} decreases with increasing rank and beyond 89% C is extremely weak.[63,176,178,188,190–192,197]

The aromatic CH stretching band increases with increasing rank whereas the aliphatic CH stretching at first decreases slowly but beyond 89% C falls off rapidly.[63,176,190,192]

The strong 1600 cm^{-1} band varies little with rank,[63,178,182] and it can be assumed that the concentration of the groups contributing to this absorption changes little with increasing coalification.[63] From this it must be concluded

TABLE 5

VARIATION OF POSITION AND INTENSITY OF THE "AROMATIC BANDS" WITH CARBON CONTENT[176]

78% C	860 cm^{-1}	825 cm^{-1}	750 cm^{-1} Significantly weaker than the other two
90% C	874 cm^{-1} Shifted	814 cm^{-1} Shifted	750 cm^{-1} Intensity now same as the others
93% C	874 cm^{-1} More intense	814 cm^{-1} More intense	750 cm^{-1} More intense
94% C	874 cm^{-1} More intense	814 cm^{-1} Less intense	750 cm^{-1} Less intense

that throughout the bituminous region there is no marked change in the number of aromatic CC bonds or in the concentration of those oxygen-containing groups which may absorb at 1600 cm^{-1}. On the assumption that, as a result of the sharp decrease in total oxygen content with increasing rank, all oxygen-containing groups uniformly decrease whereas total nitrogen content is nearly independent of rank, it has been concluded that nitrogen-containing rings (pyridines, quinolines, etc.) contribute somewhat to the intensity of the 1600 cm^{-1} band.[182]

The broad absorption between 1300 and 1000 cm^{-1} also decreases in intensity with increase in rank[176,178,188,190,192,195,197] but there remains a noticeable absorption apparently due to aromatic ethers.[176]

The "aromatic bands" between 900 and 700 cm^{-1} are also subject to certain variations in position and intensity with increasing rank.[63,176,178,192] The most significant of these changes are summarized in Table 5.[176]

If these bands are considered to be produced only by aromatic HCC rocking motions, the foregoing shifts and variations in intensity may, by analogy with model substances (substituted benzenes and substituted small condensed systems), be interpreted in the following way.[63,176] In the coal of 78% C, the degree of substitution on the aromatic groups present is very high and this becomes lower with increasing rank; between 90 and 93% C there is substantially no change in substitution; beyond 94% C there are condensed aromatic rings present, the greater part of which possess isolated CH bonds.

The exinites, vitrinites, and micrinites mentioned in this chapter have also been investigated between 1800 cm^{-1} and 1400 cm^{-1}. All twelve samples were studied under the same experimental conditions (3 mg macerals/1 g KBr). To compare the variation of the intensity of the 1600 cm^{-1} and 1450 cm^{-1} bands in the macerals, it was necessary to bring them all to as equal a background as possible, which was achieved by using KBr windows with different concentrations of graphite for compensation. In the lowest-rank exinite, the 1600 cm^{-1} band was much weaker than the 1450 cm^{-1} band; with increasing rank, the difference became smaller, and, in the exinite with 89% C, the two bands possessed approximately the same intensity. In the lowest-rank micrinite, the 1600 cm^{-1} band was much stronger than the 1450 cm^{-1} band, the difference in intensity decreasing with increasing rank. However, the highest-rank micrinite showed a somewhat stronger band at 1600 cm^{-1} than at 1450 cm^{-1}. In the vitrinites, independent of rank, the 1600 cm^{-1} band had a slightly stronger or approximately the same intensity as the 1450 cm^{-1} band.

In addition to all the aforesaid qualitative observations, attempts have also been made to derive quantitative information from the variations of the infrared absorption bands with rank. These attempts have been principally concerned with the change in OH-content and with the variation of the ratio $H_{aromatic}/H_{aliphatic}$.

Change of OH Content with Rank

Following the discovery that heating a KBr window for 24 hr at 150°C eliminated the more or less intense absorption at 3300 cm^{-1} (due to moisture), similar windows containing vitrains with carbon contents between 80 and 90% in concentrations of 3 mg coal per gram KBr were heated to 150°C (and measured at the same temperature) until no further change in the 3300 cm^{-1} absorption could be observed (see Fig. 17*a*; work done in the laboratory of the authors). The integral extinction of OH groups, corrected for the same unspecific background absorption for all samples, was calculated, remembering that the broad, weak band extending over the CH stretching region, and almost certainly due to strongly hydrogen bonded OH groups, must be included. From Fig. 17*b* it may be seen that these results lead to a good linear relation between chemically and spectroscopically determined hydroxyl contents. Thus, by use of a sufficiently refined technique (scale expander), a calibration curve could be established to permit quantitative OH determination in coal samples by infrared methods.

Variation of H_{ar}/H_{al} with Rank

The band maximum at 3030 cm^{-1} was taken as a measure of the aromatic hydrogen content whereas that at 2920 cm^{-1} was similarly used for aliphatic hydrogen.[176,186] It can be seen from Fig. 16 that, over a wide range of rank, the intensity of the aliphatic CH stretching absorption is significantly greater than that of the aromatic CH stretching. From measurements on a wide variety of model substances of mixed aliphatic-aromatic character, it was found that the absorption intensity of a single aliphatic CH bond is always greater than that of a single aromatic CH bond. Thus the number of aliphatic H atoms in coal, although larger than the number of aromatic H atoms, is by no means as much larger as would appear from the differences in absorption intensities.[186]

Using special experimental conditions to hold the background absorption as constant as possible,[176] or by use of the base line method,[186] it is possible from the experimental intensity curves to calculate the optical densities D_{ar} and D_{al}. For coals of carbon content below 84%, the overlapping of the region of interest by the OH band is extremely serious and, for coals of more than 94% C, the nonspecific background is so intense as to preclude quantitative measurements, so that accurate D-values can only be computed for coals between 84 and 94% C.[176]

In this connection, it follows that a correction for the broad OH absorption over the region of the CH stretching frequencies must be considered because this correction may influence more or less strongly the conclusions drawn. As may be seen from Fig. 18,[176] the values of D_{ar}/D_{al} increase monotonically throughout the bituminous region but especially sharply beyond about 90% C content.

For the Japanese coals Mi-ike (84% C), Yubari (84.9% C) and Yatake (89.5% C), values of D_{ar}/D_{al} of 0.06, 0.08, and 0.28 respectively were found,[186] the results being in fair agreement with the curve of Fig. 18. The fact that the Yatake coal had a D_{ar}/D_{al} ratio four times that of the other two seems to be due to the fact that a halving of the number of aliphatic CH bonds is associated with a doubling of the number of aromatic CH bonds and, hence, it was concluded that between 84 and 89% C an aromatization of aliphatic structures had occurred.[186]

The optical densities D_{ar} and D_{al} are related to the extinction coefficients ϵ_{ar} and ϵ_{al} for a single CH bond by the relations:

$$D_{ar} = n_{ar}\epsilon_{ar}$$

$$D_{al} = n_{al}\epsilon_{al}$$

where n_{ar} and n_{al} are the numbers of the respective bonds. From this,

$$D_{ar}/D_{al} = \frac{n_{ar}}{n_{al}} \frac{\epsilon_{ar}}{\epsilon_{al}}$$

Since the exact structure of coal is unknown, some means must be found for evaluating

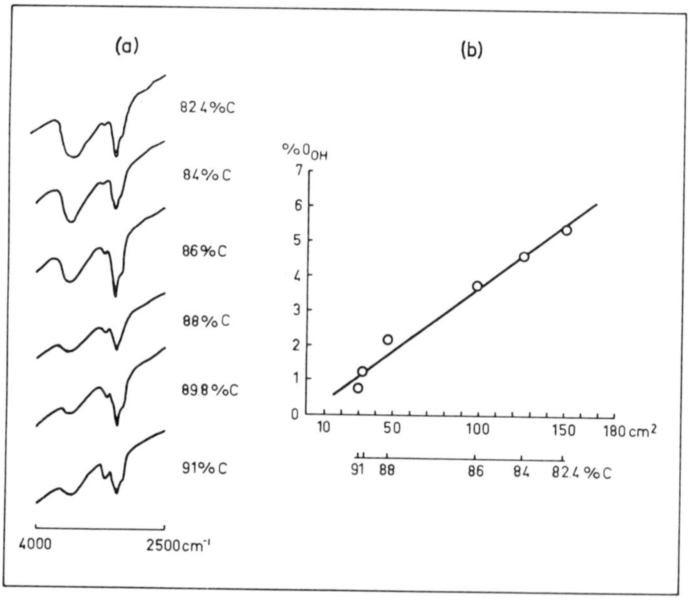

Fig. 17. (*a*) Changes in the infrared spectra of vitrains between 4000 and 2500 cm^{-1} (3 mg sample/1 g KBr heated 24 hr at 150°C and measured at this temperature). (*b*) Relation between average % O as OH, determined by different chemical methods, and the integral extinction for the OH absorption.

$\epsilon_{ar}/\epsilon_{al}$ in order to arrive at values of n_{ar}/n_{al}. The D_{ar}/D_{al} values have been determined experimentally for a large number of aromatic-aliphatic materials of the most different sorts and, since n_{ar}/n_{al} was always known, it was possible to evaluate $\epsilon_{ar}/\epsilon_{al}$. Values of the last ratio were found to lie between 0.3 and 1.0 but most frequently near 0.5.[176] Investigations by the authors on two amorphous high polymers, polyallylnaphthalene and polyallylfluorene, gave $\epsilon_{ar}/\epsilon_{al}$ values between 0.3 and 0.4.

If the value of $\epsilon_{ar}/\epsilon_{al} = 0.5$ is taken as valid for coals, the n_{ar}/n_{al} ratio can be directly evaluated from the experimentally measured D_{ar}/D_{al} ratio. It is found in this way that n_{ar}/n_{al} for a coal of 84% C is 0.25 (1 aromatic H for 4 aliphatic H's) and 1.0 for a coal of 92% C (1 aromatic H for 1 aliphatic H). By use of the atomic H/C ratio from elementary analysis and under the simplifying assumption that all oxygen in coal is present in hydroxyl groups and all aliphatic hydrogen in CH_2 groups, it is possible to calculate the aromaticity f_a.[63] For a coal of 84% C, f_a was found to be about 0.7 and, for one of 92% C, $f_a \simeq 0.9$. Even if only 60% of the oxygen is present as OH and if a small portion of the aliphatic hydrogen occurs in CH_3 groups, it can be shown that the values of f_a are little affected. If, however, $\epsilon_{ar}/\epsilon_{al}$ is taken as 0.3 or 1.0 instead of 0.5, both the n_{ar}/n_{al} and the f_a values are changed in a way which cannot be neglected.[28] It should be mentioned that a value $\epsilon_{ar}/\epsilon_{al} = 0.5$ for coals could be a good choice but D_{ar}/D_{al} may be too low in low-rank coals because it is certainly difficult to establish D_{ar} because of the overlapping of the 3030 cm^{-1} band by the long-wavelength wing of the OH absorption at 3300 cm^{-1}.

In spite of all the uncertainties outlined, this method for determination of the hydrogen distribution represents a highly valuable contribution to structural determinations of coals.

Infrared Spectra of the Distillation Products and Residues of Thermally Treated Coals

Measurements have been carried out on coals which have been carbonized under vacuum[178,206] or in nitrogen[183,207,208] and on KBr windows heated to 600°C.[187,209] Both the

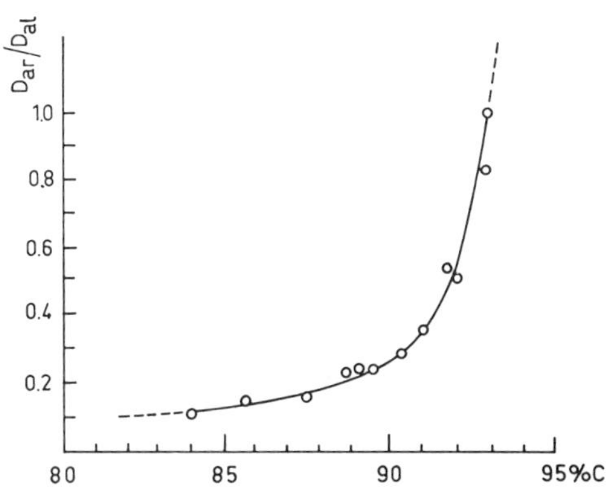

Fig. 18. Dependence of the ratio D_{ar}/D_{al} of vitrains on carbon content.[176]

volatile materials[187,209,210] and the residues[178,183,206,207,208] were investigated.

It has been found in general that the infrared spectra of the oily condensates obtained from coals up to temperatures as high as 500°C are very similar to one another and comparable with those of the coals themselves.[209,210] A strong absorption near 3300 cm^{-1} (OH-groups),[187,209,210] an unquestionable C═O stretching absorption near 1700 cm^{-1},[187,209] a strong absorption at 1600 cm^{-1},[187,209,210], a band near 1500 cm^{-1},[187] a strong composite band near 1460 cm^{-1},[187,209,210] bands between 1230 and 1180 cm^{-1} assigned to branched aliphatic groups[187] and the "aromatic bands"[209,210] have been observed.

In general it has been found that the nonspecific background absorption of the residues obtained from thermal treatment of coals of bituminous rank increases with increasing carbonization temperature and moves progressively from short wavelengths to longer ones. For samples treated at 600°C, this absorption extends to 1100 cm^{-1}.[187] Above 600°C, the samples become opaque and no longer show specific absorptions between 4000 and 650 cm^{-1}.[178,187,207,208]

Particularly impressive are the changes in spectra on carbonization of a Northumberland

[206] Adams, W. N., and Pitt, G. J., *Fuel*, **34,** 383–4 (1955).

[207] Roy, M. M., *Fuel*, **36,** 250–2 (1957).

[208] Brown, J. K., *J. Chem. Soc.* (*London*), **1955,** 752–7.

[209] Bergmann, G., Huck, G., Karweil, J., and Luther, H., *Brennstoff-Chem.*, **38,** 193–9 (1957).

[210] Orning, A. A., and Greifer, B., *Fuel*, **35,** 381–3 (1956).

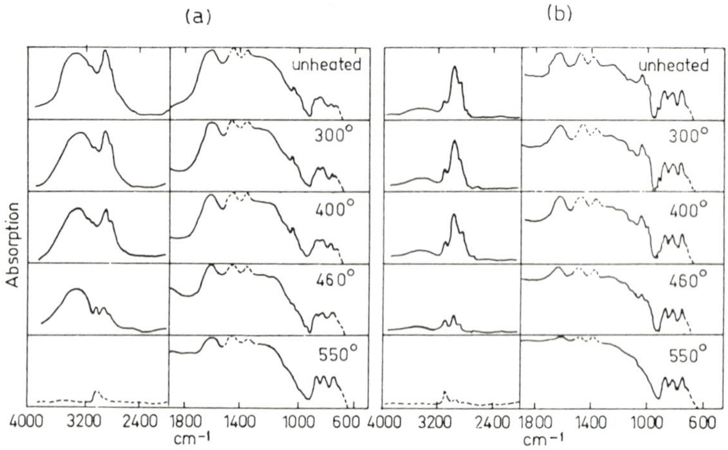

Fig. 19. Variations in infrared spectra of (*a*) a caking coal and (*b*) a coking coal with carbonization temperature.[208] Region 4000 to 2000 cm^{-1} measured as hexachlorobutadiene mulls (-------- uncertain intensities). Region 2000 to 600 cm^{-1} measured as Nujol mulls (- - - - - - Nujol absorption bands).

vitrain (Ellington High Main, 81.8% C, caking coal) and a Durham Coal (South Garsfield Victoria Seam, 88.8% C, coking coal) shown respectively in Fig. 19*a* and 19*b*.[208] From these figures together with the results of other authors,[178,187,206,207,211] several conclusions may be drawn concerning the spectra of carbonized bituminous coals. Up to 400°C, the changes in spectra compared with the starting material are extremely small, there being a small decrease in the intensity of the OH absorption at 3300 cm^{-1} and of the aliphatic CH stretching absorption.[178,187,201,206,207,211] Between 460 and 500°C, the intensity of the OH band decreases noticeably,[183,207,208,211] and in the coals of higher ranks, the band can scarcely be detected.[178,208] The aliphatic CH stretching frequencies, as well as the bands near 1440 cm^{-1} and 1370 cm^{-1},[211] also fall off markedly in intensity.[207,208,211] Bands observed in a Russian flame coal around 1670 cm^{-1} and 1560 cm^{-1} were ascribed to C=C bonds conjugated to aromatic systems and to bonds between the clusters.[183] In a carbonized Pittsburgh seam anthraxylon, in an Assam coal (Baragolai colliery), and in the flame coal of Russian origin definite decreases in the intensity of the broad absorption between 1300 and 1000 cm^{-1}, especially on the long wavelength side, were observed.[178,183,207] The three bands between 900 and 700 cm^{-1} are little changed with respect to the starting materials with a slight increase in intensity of the 750 cm^{-1} band.[178,208] At 550°C, the OH band has practically disappeared and the aliphatic CH stretching vibrations have been greatly decreased in intensity, whereas the bands between 900 and 700 cm^{-1} have generally gained in intensity.[178,207,208] At 600°C, the residue of the Assam coal contained only a trace of aliphatic CH stretching absorption and there was a further increase in intensity of the bands between 900 and 700 cm^{-1} accompanied by slight shifts in position.[207]

Especially noteworthy is the fact that the intensity of the 1600 cm^{-1} band is substantially independent of the temperature of carbonization,[178,187,208,211] in fact, in the Assam coal this band actually gains somewhat in intensity,[207] showing that this absorption must have its origin in extremely stable groups. Attempts have been made to relate these results on carbonized coals to the structure of the starting material,[187,208] but the conclusions are far from unequivocal and this kind of study needs to be carried further.

The Infrared Spectra of Coal Extracts

These spectra have been determined on films left after evaporation of a solvent,[86,173,212,213] solutions in pyridine[173,213] or CS_2,[86] Nujol[184,196,213] or hexachlorobutadiene mulls,[184] and

[211] Maher, T. P., Harris, J. M., and Yohe, G. R. *Illinois State Geol. Survey, Rep. Invest.* No. 212 (1959), 75 pp.

[212] Hadzi, D., *Sloven. akad. znanosti umetnosti, Razred mat. fiz. teh. vede*, **3**, 99–107 (1951).

[213] Kirkby, W. A., Lakey, J. R. A., and Sarjant, R. J., *Fuel*, **33**, 480–96 (1954).

KBr windows.[48] It has been found that so long as there is no reaction between the organic material of the coal and the extractant the spectra of the extracts are very similar, not only to those of the starting materials[48,63,86,173,184,212–215] but also to those of the residues not dissolved.[48,182,184,216] These results appear to establish the close relation between the chemical structure of solvent extracts and of the parent coal.[184] Polymer systems with a wide range of degrees of polymerization are said to behave similarly.[63] The spectra of the extracts also appear to be substantially independent of the method used for extraction whether in a Soxhlet extractor at the boiling point of the solvent or by stirring the sample in the solvent at room temperature under nitrogen.[182]

A systematic study[182] has shown that solvents can be divided—albeit not sharply—into two classes: those which give low extraction yields, such as benzene, toluene, xylene, acetone, chloroform, the extracts showing notably enhanced details in their spectra and evidently not representative of the starting materials; and those giving high extraction yields, such as methanol, ethanol, pyridine, quinoline, the solutions frequently being not optically clear and sometimes unstable. The extracts from the more potent solvents have a spectrum very similar to that of the starting material and may be taken as representative of the organic matter present in the original coal.

Spectra of ethylenediamine extracts[199] of coal show all the absorption regions of the starting material but also bands which are evidently associated with reaction products of the coal substance with the solvent (the nitrogen content of the extract may increase by as much as 15.4%). Dimethylformamide and ethanolamine[215] behave similarly.

A large number of bright coals (82 to 93% C) and durains (84 to 90% C) were extracted with pyridine for 20 hr at room temperature, the stirring being accomplished by ultrasonic waves and the soluble portions separated by column chromatography.[213] With increasing rank of the bright coals, the extracts showed a decrease of the bands at 1700 cm^{-1} (C=O stretching) and 1385 cm^{-1} (CH_3 groups) and an increase of the broad band between 1200 and 1100 cm^{-1} (C—O—C stretching). These intensity changes were related to the British Standard swelling number.[213] In the spectra of some extracts, bands with distinctive structure are observed between 2300 and 1850 cm^{-1}. These bands are probably due to overtones of the skeletal frequencies between 900 and 700 cm^{-1}. It was especially interesting that the extracts of the durains showed, as their rank increased, an absorption at 1700 cm^{-1} and a band of increasing intensity at 1385 cm^{-1}.[213]

An interesting experiment has been performed in which the solid pyridine extract of a coal (75.5% C) was extracted with hot water-pyridine mixtures of 20%, 40%, 60%, and 80% pyridine.[182] Study of the infrared spectra of the various extracts showed that with increasing pyridine content, the 3300 cm^{-1} band is unchanged, the aliphatic CH stretching absorption increases, the 1700 cm^{-1} band gets weaker, and the bands at 1600 cm^{-1} (aromatic CC-bonds) and at 1280 cm^{-1} (C—O stretching) are both unchanged. On the basis of an X-ray diffraction study of the same samples, and on the assumption that the pyridine extract is representative of the starting material, it has been suggested that coals may be composed of a series of clathrate compounds—the two components of these compounds being closely related to lignin and cellulose.[182]

Three brown coals ($\approx$66% C) and three bituminous coals (83, 88, and 90.5% C) have been extracted with a 30/70 mixture of ethanol and benzene.[196] A marked increase of the background absorption with increasing rank was observed. The absorption in the region of 3350 cm^{-1} was very strong in the lignites, was reduced considerably with increasing carbon content and was almost absent in the materials with 88 to 90% C.[196]

Peculiar results have been found on extracting coals (84.5 to 87.5% C) with phenol at 150°C for periods up to 240 hr in a nitrogen atmosphere.[217] Since the extraction still continues even after such a long time, it was concluded that the phenol not only dissolves but also depolymerizes the coal. If the phenol extracts are again extracted with chloroform

[214] Brown, J. K., and Wyss, W. F., *Chem. & Ind. (London)*, **1955,** 1118.

[215] Berti, V., Casu, B., Manci, C., Padovani, C., and Siniramed, C., *Riv. combustibili*, **9,** 5–34 (1955).

[216] Brown, J. K., BCURA *Bull.*, **17,** 449–64 (1954).

[217] Kirkby, W. A., and Sarjant, R. J., *Nature*, **170,** 597–9 (1952).

and separated by chromatography, the products are found to contain a large number of bands in the region 2500 to 1800 cm^{-1} where the coal itself does not absorb.[217]

Infrared Spectra of Irradiated Coals

Five coals (70 to 90.5% C) were exposed to pile irradiation.[218] The infrared spectra of the irradiated high rank coals were much more diffuse than those of the nonirradiated samples but band positions and relative intensities were unchanged. The increased diffuseness of the spectra and the observed increased hardness of the samples after irradiation indicated appreciable polymerization.[218]

Infrared Spectra of Chemically Treated Coals and Coal Extracts

Experiments have been carried out on coal powders oxidized in air[177,185,187,206,211,219] and in oxygen.[207] The principal differences in the spectra compared with those of the starting materials are a strong increase in OH absorption at 3300 cm^{-1},[206,219] a decrease of the aliphatic CH absorption,[185,207,211,219] a new band at 1700 cm^{-1} ascribed to the C=O stretching vibration of COOH groups[187,206,211,219] which increases with increasing oxidation,[185,187,206,207,211,219] an increase in intensity of the 1600 cm^{-1} band[219] in one case only, an increase in the broad absorption over the region 1300–1150 cm^{-1},[206,219] and a tendency for the bands between 900 and 700 cm^{-1} to disappear with increasing oxidation.[206,207]

It was concluded that CH_3 groups may have been oxidized to COOH groups,[207,219] and that carboxylic acids and phenols are formed at the expense of peripheral CH bonds of condensed aromatic systems.[206,207]

Three coals (74.5, 79.2, and 81.3% C) were treated with lithium in anhydrous ethylenediamine at 90 to 100°C for $2\frac{1}{2}$ hours.[94] Compared to the spectra of the starting materials, an increased aliphatic CH stretching absorption and a remarkable decrease of the broad band at 1250 cm^{-1} and of the "aromatic bands" were observed. Furthermore the absorption at 1600 cm^{-1} was no longer the most prominent band. After reduction of the pyridine extract from the coal with 79.2% C, the same spectral changes were found as for the reduced coal itself.[94]

From studies on hydrogenation products of coals[185,187,215] and the fractions obtained from them by chromatography,[207,215] it was found that the OH absorption seemed to become weaker,[185,207] that all the aliphatic CH vibrations became stronger,[185,207] that a new absorption at 4130 cm^{-1} (SH stretch?) appeared in one case,[207] that the 1600 cm^{-1} band remained unchanged[187,207] or slightly decreased,[185] that the absorption at 1250 cm^{-1} became weaker, and that the "aromatic bands" between 900 and 700 cm^{-1} became sharper and two new absorptions at 820 and 730 cm^{-1} appeared.[207]

A Pittsburgh seam coal has been hydrogenated and the reaction product extracted with benzene and hexane. This fraction gave an oil distillable at 105°C and 3 mm pressure ("light oil"). The main differences in infrared spectra between the "light oil" and the residue of the distillation ("heavy oil") were a strong increase in the OH and 1250 cm^{-1} absorptions in the "light oil." A comparison between the "heavy oil" and the "asphaltene" (benzene soluble-hexane insoluble fraction) showed a small decrease in OH, a decrease of the 1600 cm^{-1} band, an increase of the 1375 cm^{-1} band and a strong decrease of the broad absorption in the 1280 cm^{-1} to 1100 cm^{-1} region in the "heavy oil."[220]

It has been further observed that the spectrum of a completely hydrogenated coal is, in its essential features, identical with that of a completely hydrogenated phenanthrene oil.[177] Also a hydrogenated fraction of a Sulcis bituminous coal gave a spectrum which closely resembled in general structure that of 1,9-dimethylphenanthrene.[215] Finally, attempts have been made, using completely hydrogenated fractions in hexane or ether solution, to deduce the CH_3 content from the intensity of the 1380 cm^{-1} band by use of a calibration curve determined from model substances.[221]

Studies have been carried out on methylated brown coals[181] and showed that the significant spectral changes were: (1) a decrease in the OH absorption still present in the completely

[218] Friedel, R. A., and Breger, I. A., *Science*, **130**, 1762–3 (1959).

[219] Fujii, S., and Yokayama, F., *Nenryô Kyôkaishi*, **38**, 98–106 (1959); *Chem. Abs.*, **53**, 15530 (1959).

[220] Pelipetz, M. G., and Friedel, R. A., *Fuel*, **38**, 8–16 (1959).

[221] Schuhmacher, J. P., van Vucht, H. A., Groenewege, M. P., Blom, L., and van Krevelen, D. W., *Fuel*, **35**, 281–90 (1956).

methylated sample, (2) an increase in the absorption due to aliphatic CH groups (2920 cm^{-1}, 1460–1440 cm^{-1}), (3) a strong absorption at 1720 cm^{-1}, (4) a shift in the band near 1615 cm^{-1} to 1610 cm^{-1} with a notable decrease in intensity, and (5) only a slight change in the band at 1380 cm^{-1}.

The acetylation products of coal, independently of rank, as well as acetylated coal extracts show spectra different from those of the original coals in the following respects: (1) the OH absorption decreases but does not disappear,[181,214,222] (2) a marked absorption at 1760 cm^{-1} which from studies on model substances, may be assigned to the C=O stretching vibration in aliphatic ester groups attached to aromatic rings,[181,214,222] (3) the acetylated products of brown coals show a decrease in the intensity of the 1615 cm^{-1} band,[181] but this has not been unequivocally observed in acetylated bituminous coals,[222] (4) an increase of absorption at 1365 cm^{-1},[181,222] (5) a broad, strong new absorption at 1200 cm^{-1},[181,214,222] which can definitely be assigned to the C—O stretching vibration in ester groups on aromatic rings,[181,214,222] and (6) in acetylated brown coals well-defined bands occur at 1020 and 900 cm^{-1}. These observations, especially points 1, 2, and 5 permit the generally valid conclusion that practically only phenolic OH groups are present in coals and coal extracts. Making use of the base line technique for measurements of optical densities, it was found that Lambert-Beer's law is obeyed for concentrations between 0.25 to 1.0% by weight in KCl.[222] Acetylated samples of brown coals and bituminous coals showed a linear relationship between the optical densities of either the 1365 cm^{-1} or the 1755 cm^{-1} bands and the chemically estimated acetyl content.[222]

The spectra of brown coals[181] chlorinated with PCl_5 in boiling nitrobenzene compared with those of the starting materials show a small decrease in OH absorption, but this remains strong, a decrease of absorption at 2920 and 1450 cm^{-1} from which it must be concluded that brown coals possess "active" aliphatic hydrogen which can be replaced with chlorine, an unexplained nearly complete disappearance of the 1700 cm^{-1} absorption, a considerable decrease in intensity and shift of the band at 1615 to 1610 cm^{-1}, a considerable decrease in the intensity of the broad absorption between 1400 and 1000 cm^{-1}, and a new absorption at 800 cm^{-1} to be assigned to the C—Cl stretching vibration.[181]

Suspensions of coals of 81.7 and 89.7% C in carbon tetrachloride was irradiated with a Hannovia S 500 medium pressure mercury arc in a chlorine atmosphere and with vigorous stirring for periods from 14 to 72 hr.[223] After this treatment the aliphatic CH absorption was decreased, that is, the aliphatic side chains were chlorinated, and new absorption bands appeared at 785 and 759 cm^{-1} due to the C—Cl stretching vibrations.[223]

When suspensions of these two coals in water were irradiated under nitrogen or air for 72 hr using a Hannovia S 500 medium pressure mercury arc as the source, the infrared spectra of the irradiated samples were practically identical with those of the starting materials. It would appear that no significant changes in structure had occurred. It is, however, not excluded that appreciable changes may have occurred on the surface of the particles but, because these represent such a small fraction of the total sample, they do not make themselves felt in the spectra.[223]

The benzene extract of the coal of 81.7% C (Soxhlet extraction, 24 hr under nitrogen at the boiling point of the solvent) was irradiated in solution in the presence of air and, as the time of radiation increased, showed a strong bleaching of the initial color and the formation of a precipitate which, after 72 hr irradiation amounted to 30% of the starting material. Analysis of the precipitate indicated a considerable uptake of oxygen and, its infrared spectrum showed, in comparison with the starting material, weaker absorption due to aromatic and aliphatic CH bonds and a strong, broad maximum at 1720 cm^{-1} due to the C=O stretching of COOH groups. The spectrum of the material remaining in solution showed stronger aliphatic and aromatic CH absorption and a weaker C=O stretching absorption at 1720 cm^{-1} than did the precipitate. If the benzene extract of the same coal was irradiated in the absence of air, there was no change in color even after long irradiation and the precipitate formed was only one-fourth that obtained when air was present. The infrared

[222] Durie, R. A., and Sternhell, S., *Australian J. Chem.*, **12**, 205–17 (1959).

[223] Bent, R., and Brown, J. K., *Proc. Conf. on Science in the Use of Coal*, Sheffield, Inst. Fuel, London, 1958, pp. A48–51.

spectrum showed the same features as that from the sample irradiated in air but they were considerably less marked. From all these observations it was concluded that a copious photooxidation was produced by the irradiation.[223]

Treatment of brown coals in 5*N* HCl produced a strong increase in the intensity of the 1700 cm^{-1} band and a simultaneous sharpening of the 1615 cm^{-1} absorption.[181,191] Treatment of the same brown coals with 5*N* NaOH caused the 1700 cm^{-1} band to disappear completely, but when the coal was reacidified the same spectra were obtained as those produced by the initial acid treatment. It follows from these observations that the 1700 cm^{-1} band in brown coal is due to the C=O stretching of carboxyl groups and that carboxylates are present which contribute to the intensity of the 1615 cm^{-1} band.[181,193]

ELECTRON SPIN RESONANCE (PARAMAGNETIC RESONANCE)

Electron spin resonance, which has achieved considerable importance, depends on the absorption of electromagnetic radiation by unpaired electrons in a magnetic field.[224–228] Most stable chemical systems are in their lowest energy state when all their electron spins are paired. Such substances have no net electron spin angular momentum and are diamagnetic, that is, they are repelled by a magnetic field.

In special cases, however, structures may exist in which one or more electrons remain unpaired. The ions of the transition elements are an example. The oxygen molecule which contains an even number of electrons is one of the relatively rare cases where the lowest energy state is paramagnetic. Most molecules containing an odd number of electrons react with one another to complete the pair. In special cases, however, especially those where several resonance structures are available, molecules with a single unpaired electron may be so stabilized as to have comparatively long lifetimes. Nitric oxide is a well-known simple case of a molecule having fifteen electrons where the compound shows substantially no tendency to dimerize. In large aromatic systems, stabilization by resonance may also occur. Materials with one or more unpaired electrons may possess a net electron spin angular momentum and hence are paramagnetic and are attracted by magnetic fields.

The completely paired ground state is said to be a "singlet state" and a system with a single unpaired electron is in a "doublet state." Even normal diamagnetic molecules with only paired electrons can be made temporarily paramagnetic by exciting one of the bonding electrons to a higher orbital where it no longer pairs with its partner. This excitation to a "triplet state" may be accomplished thermally, by light absorption or by electron or ion impact. In the case of naphthalene, the optically excited "triplet state" has been observed to have a relatively long lifetime because transitions back to the ground state are usually forbidden in dipole radiation.[229]

When materials having unpaired electrons are placed in a uniform magnetic field, the spins orient themselves either parallel or antiparallel to the direction of the magnetic field. Because of the quantum conditions these states are sharply defined, and energy is required to force a spin with parallel orientation to assume an antiparallel orientation. The difference in energy is given by

$$\Delta E = g\beta_e H [c^2 g s^{-2}]$$

where g is the gyromagnetic splitting factor, β_e is the Bohr magneton (in units of $c^{5/2} g^{1/2} s^{-1}$) and H is the applied magnetic field (in units of $c^{-1/2} g^{1/2} s^{-1}$). By the well-known Planck relation, the frequency ν of radiation which must be applied to a system to effect a transition from one state to another is given by

$$h\nu = \Delta E = g\beta_e H$$

whence

$$\nu = \frac{g\beta_e H}{h}$$

where h is Planck's constant.

[224] Ingram, D. J. E., *Free Radicals as Studied by Electron Spin Resonance*, Butterworths, London, 1958, 274 pp.

[225] Ingram, D. J. E., *Spectroscopy at Radio- and Microwave Frequencies*, Butterworths, London, 1955, 322 pp.

[226] Abragam, A., *Cours sur les phénomènes liés à la résonance magnétique*, Centre d'études nucléaires de Saclay, 1956–1957, 193 pp.

[227] Uebersfeld, J., *Ann. phys.*, [13], **1**, 395–461 (1956).

[228] Wertz, J. E., *Chem. Rev.*, **55**, 829–955 (1955).

[229] Hutchinson, C. A., and Mangum, B. W , *J. Chem. Phys.*, **29**, 952–3 (1958).

Thus the magnitude of this frequency, in contradistinction to the usual measurements of electromagnetic spectra, is determined by the magnitude of the magnetic field and scarcely at all by the nature of the material under investigation. For technical reasons, the fields generally used range from 3200 to 13,000 gauss and this requires frequencies between 9000 and 36,000 megacycles or wavelengths from 3.1 to 0.8 cm.

In any system at approximate thermal equilibrium, the lower energy state will be more highly populated than the upper. Because of the equal probability of induced emission and absorption, the electron spin resonance phenomenon represents a net absorption of energy. However, if only radiative transitions were involved, the upper level would soon attain, through absorption of the applied resonant energy, the same population as the lower state and no net absorption could be observed. This condition is known as "saturation." It turns out, however, that the important relaxation mechanism which leads to loss of energy to the surroundings is the radiationless transition back to the ground state. The spin-lattice relaxation time T_1 characterizes the loss of energy to the medium in which the electron spin is considered to be embedded (always referred to as the "lattice"). The spin-spin relaxation time T_2 is usually used to represent the interaction of the unpaired electrons with each other.

The absorption bands found in paramagnetic substances are usually characterized by their intensity, half-width, gyromagnetic factor, and temperature dependence. The intensity, other conditions remaining constant, is a measure of the total number of unpaired electrons in the sample in question. The half-width is a measure of the interaction between the unpaired electrons and their surroundings; the wider the band, the greater the interaction. The gyromagnetic factor (g or Landé factor) depends on the spin-orbit coupling of the system, and for most free radicals is very close to the free electron value 2.0023. From the temperature dependence it should be possible to distinguish between stable radicals and excited (triplet) states. In stable radicals, the number of unpaired electrons remains constant as temperature is lowered, but the lower state is more highly populated and hence the band increases in intensity. If the unpaired electrons

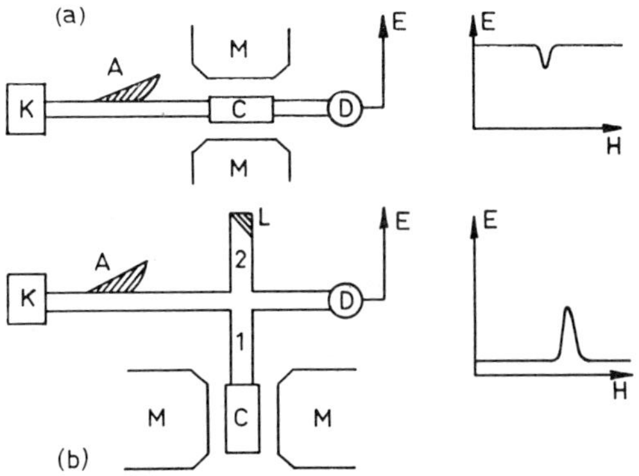

Fig. 20. Diagrams for apparatus used in measuring electron spin resonance, (*a*) by transmission and (*b*) by reflection. The plot of E versus H illustrates type of signal observed in each case. K. Klystron; A. Attenuator; M. Magnet; C. Cavity; D. crystal detector; L. absorbing load; E. energy detected; H. magnetic field.

arise from excitation, which in the absence of light can only be produced thermally, the lowering of the temperature should result in less excitation, hence fewer unpaired electrons and a consequent decrease in band intensity.

Experimental Methods

Apparatus

The electron-spin resonance phenomenon can be measured either by transmission[224, 230–233] or by reflection.[224,227,234,235]

A typical apparatus for transmission measurements is shown schematically in Fig. 20*a*. Microwaves from a tunable Klystron pass through an attenuator to the material to be measured in a resonant cavity and then on to a crystal detector. When the resonance condition is satisfied, energy will be absorbed by the system and the signal picked up by the detector will decrease. To make the absorption

[230] Bowers, K. D., Kamper, R. A., and Knight, R. B. D., *J. Sci. Instr.*, **34,** 49–53 (1957).

[231] Buckmaster, H. A., and Scovil, H. E. D., *Can. J. Phys.*, **34,** 711–21 (1956).

[232] Llewellyn, P. M., *J. Sci. Instr.*, **34,** 236–9 (1957).

[233] Sarkar, G. N., Mukherji, A., Chatterjee, R. N., and Ghosh, U. S., *Indian J. Phys.*, **33,** 117–22 (1959).

[234] Jarrett, H. S., and Sloan, G. J., *J. Chem. Phys.*, **22,** 1783–4 (1954).

[235] *Varian-V-4500 Spectrometer*, Varian Associates, 611 Hansen Way, Palo Alto, California, U.S.A., 4 pp.

band visible on a cathode ray oscilloscope, it is customary to hold the frequency constant at the resonant cavity frequency and modulate the magnetic field on the two sides of this condition sufficiently to trace out the band. In such an arrangement, the energy not absorbed by the resonating system is always large so that the nonmodulated energy reaching the detector at all times is very large in comparison to the absorbed intensity. This is no basic disadvantage but somewhat complicates the technique of detection.[236]

In the arrangement described, the resonant cavity may be compared to the absorption cell in ordinary spectroscopy. Both cylindrical and rectangular forms are used. The quality Q of a cavity, that is, the ratio of stored energy to energy losses which is proportional to the volume to surface ratio, is more favorable in cylindrical than in rectangular cavities. The cylindrical are somewhat larger than the rectangular, a fact which is useful in some cases, for example, for introducing a Dewar flask into the cavity. On the other hand, heating of the sample in situ is better accomplished in a rectangular cavity.[237]

In measurement by reflection (Fig. 20*b*), the energy coming from the Klystron is divided equally between arms 1 and 2 and, in the absence of absorption in the cavity which is located in arm 1, these two energy streams destroy one another substantially completely by destructive interference and hence almost no energy reaches the detector. When, however, the resonance condition is reached, the two arms become unbalanced and an amount of energy proportional to this unbalance reaches the detector and the signal increases. This signal is no longer a small difference between two large energy quantities and hence is easier to measure accurately. The band is made visible by the same means as in the transmission case.

It is theoretically possible to calculate the number of unpaired electrons in a sample from the absolute magnitude of all changes in Q of the cavity at the resonant condition caused by introduction of the sample. In practice, however, this is too difficult and the cavity is calibrated by comparing the band area of the unknown with that from a material having a known number of unpaired electrons, for example, diphenylpicrylhydrazyl (DPPH), which is known to exist entirely as free radicals. Another standard used is $CuSO_4 \cdot 5H_2O$.[238] Also, Linde synthetic ruby—Al_2O_3, containing about 0.5% Cr^{+3} by weight—at the correct orientation has been used as an excellent internal standard.[239]

Simple arrangements permit the detection of 10^{15} to 10^{16} free radicals, that is, unpaired electrons, when the half-width is 2 gauss. With a highly stabilized arrangement and special detection methods (super heterodyne detection, phase sensitive detection, and recording), the sensitivity can be increased to the point where 10^{11} to 10^{12} unpaired electrons can be measured.

Preparation of the Sample

The amount of coal required ranges from a few micrograms to a few milligrams depending on the sensitivity of the apparatus being used. Since the experimental results may be affected by oxygen, it is preferable to seal the sample in a 2 to 3 mm quartz tube under a vacuum of at least 10^{-2} mm Hg.

The particle size is important only insofar as an increase of the surface area increases the number of free radicals accessible to oxygen. The effect of oxygen is due to the interaction of the paramagnetic moment of the oxygen molecule with the unpaired electrons in coals, and can result either in a broadening of the band or a decrease in the spin lattice relaxation time.

Coals and Their Petrographic Components

All coals and their petrographic components studied so far show electron spin resonance absorption.[227,233,237,238,240–263] That this effect

[236] Feher, G., *Bell System Tech. J.*, **36**, 449–84 (1957).

[237] Austen, D. E. G., Ingram, D. J. E., and Tapley, J. G., *Trans. Faraday Soc.*, **54**, 400–8 (1958).

[238] Miyasita, I., and Yokozawa, Y., *Bull. Research Inst. Appl. Elec. (Japan)*, **7**, 162–71 (1955).

[239] Singer, L. S., *J. Appl. Phys.*, **30**, 1463–4 (1959).

[240] van Gerven, J., van Itterbeek, A., and De Wolf, E., *J. phys. radium*, **17**, 140–2 (1956).

[241] Austen, D. E. G., and Ingram, D. J. E., *Proc. Conf. on Science in the Use of Coal, Sheffield*, Inst. Fuel, London, 1958, pp. A62–6.

[242] Smidt, J., *Arch. sci., Geneva*, **11**, 180–4 (1958).

[243] Smidt, J., *Nature*, **181**, 176 (1958).

is produced by stable free radicals may be concluded from the fact that a g factor of 2.003 was found[244,246,260,262] and the band intensity increases with decrease in temperature as required by Curie's law.[237,240] Paramagnetic or ferromagnetic mineral components, for which g would be >2 or thermally excited triplet states for which the absorption intensity should decrease with decrease in temperature, are therefore excluded. Whether the radicals in coals involve carbon (broken σ bonds[264] or unpaired π electrons[255,263]) or oxygen (semiquinone structures[61]) or both has not been established.

Coal samples obtained from different depths of a mine (760 to 2876 ft) have the same free radical concentrations and half-width.[233] Furthermore, it has been observed that the g factor and the half-width between 90 and 295°K with frequencies of 30, 300 and 9400 Mc remain constant.[262]

Changes of the ESR-Effect with Rank

Radical concentration. It has been established,[237,244,246,247,251,255,256,259–261] as shown in Fig. 21,[237] that the radical concentration in coals of increasing carbon content (decreasing volatile matter) at first increases slowly, then rises sharply between 90 and 94% C content and soon thereafter decreases just as sharply to a value below the limits of detectability. Up to 92% C, the number of free radicals measured in air or under vacuum is in principle the same, but above 92% C the radical concentration found under vacuum is much higher than that measured in air (see Fig. 21). An extensively degassed sample (96.5% C) showed[259] a remarkably higher number of free radicals than the maximum value in Fig. 21.

It is assumed that the radicals arise through chemical reactions or degradations and are stabilized by resonance in the aromatic system.[237,244,248,255,256,263] It might be expected that such stabilization would be greater the greater the average size of the aromatic system. This view is in accord with X-ray measurements which show that the average size of the aromatic lamellae increases only slowly between 80 and 90% carbon content, but grows rapidly above 90% C. The decrease in radical concentration beyond 95% C is ascribed to the rapidly increasing electrical conductivity which permits electron pairing.[237,255,256]

In a coal of 70% C, there is one radical for every 50,000 carbon atoms, whereas for an anthracite with 94% C there is one radical for every 1000 carbon atoms.[227,237,242,246,251,255,256] Attention has been called to the fact that the accuracy with which the number of free

[244] Uebersfeld, J., Etienne, A., and Combrisson, J., *Nature*, **174**, 614 (1954).

[245] Ingram, D. J. E., and Bennett, J. E., *Phil. Mag.*, **45**, 545–7 (1954).

[246] Ingram, D. J. E., Tapley, J. G., Jackson, R., Bond, R. L., and Murnaghan, A. R., *Nature*, **174**, 797–8 (1954).

[247] Etienne, A., and Uebersfeld, J., *J. chim. phys.*, **51**, 328 (1954).

[248] Bennett, J. E., Ingram, D. J. E., and Tapley, J. G., *J. Chem. Phys.*, **23**, 215 (1955).

[249] Bennett, J. E., Ingram, D. J. E., and Tapley, J. G., *Rept. Conf. on Defects in Crystalline Solids*, Bristol, 1954, pp. 65–73.

[250] Yokozawa, Y., Miyasita, I., Kugo, M., and Higasi, K., *Bull. Chem. Soc., Japan*, **28**, 536 (1955).

[251] Ingram, D. J. E., *Discussions Faraday Soc.*, No. 19, 179–81 (1955).

[252] Uebersfeld, J., and Erb, E., *J. phys. radium*, **16**, 340 (1955).

[253] Garif'ianov, N. S., and Kozyrev, B. M., *J. Exptl. Theoret. Phys. U.S.S.R.*, **30**, 272–6 (1956); *Soviet Phys. JETP*, **3**, 255–8 (1956).

[254] Kröger, C., *Brennstoff-Chem.*, **39**, Sondersausdrücke 29 (1958).

[255] Austen, D. E. G., and Ingram, D. J. E., *Brennstoff-Chem.*, **39**, Sondersausdrücke 25–48, (1958).

[256] Ingram, D. J. E., *Proc. of 3rd Conference on Carbon*, Buffalo, 1957, Pergamon Press, London, 1959, pp. 93–102.

[257] Galkin, A. A., and Kichigin, D. A., *Khim. i Tekhnol. Topliva i Masel*, **3**, No. 7, 8–14 (1958).

[258] Garif'ianov, N. S., and Kozyrev, B. M., *Zhur. Eksptl. i Teoret Fiz.*, **30**, 1160 (1956).

[259] Smidt, J., and van Krevelen, D. W., *Fuel*, **38**, 355–68 (1959); *Chim. & ind. (Paris)*, **82**, 487–93 (1959).

[260] Losev, B. I., and Bylyna, E. A., *Doklady Akad. Nauk S.S.S.R.*, **125**, 814–6 (1959); *Zhur. Priklad. Khim.*, **32**, 2359–61 (1959).

[261] Honda, H., Chitoku, K., Yokozawa, Y., and Higasi, K., *Bull. Chem. Soc. Japan*, **31**, 890–2 (1958).

[262] Garif'ianov, N. S., and Aksel'rod, M. M., *Khim. i Tekhnol. Topliv i Masel*, **4**, 64–8 (1959).

[263] Uebersfeld, J., *31st Cong. Intern. de Chimie Industrielle, Liege*, **1958**; *Ind. chim. belge*, **1**, 413–5 (1959).

[264] Mrozowski, S., *Proc. 4th Conf. on Carbon*, Buffalo, 1959, Pergamon Press, London, 1960, pp. 271–8.

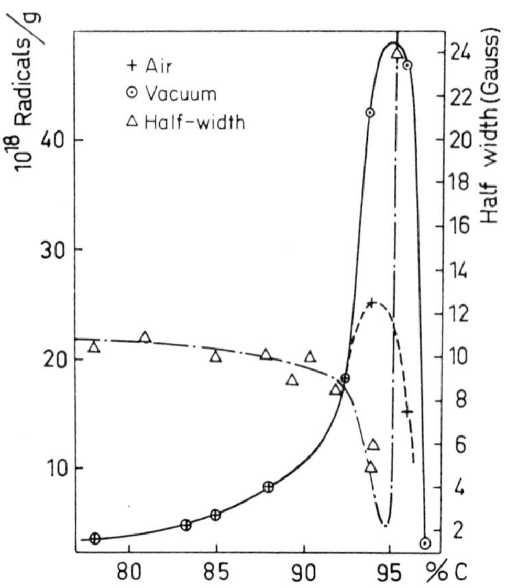

Fig. 21. Dependence of radical concentration (measured in air and under vacuum, scale on the left) and half width (scale on the right) on carbon content.[237]

radicals can be determined is so great that, with the help of a calibration curve similar to Fig. 21, it should be possible to deduce the rank of an unknown sample.[257]

Vitrains behave similarly to coal as regards change of radical concentration with rank. Exinites and vitrinites also show an increase of radical concentration with rank but, in complete agreement with petrographic results, beyond 89% C the two can no longer be distinguished[254] (see Fig. 22). Micrinites show a completely different behavior since their radical concentration seems to decrease with rank[254] (see Fig. 22).

As a result of strong saturation effects caused by the relatively long spin-lattice relaxation time, especially in brown coals and high-volatile bituminous coals, much too low radical concentrations may often be obtained. Investigations on a series of vitrains[259] in air and under vacuum show that the spin-lattice relaxation time T_1 seems to increase slowly with increasing rank (for 86 to 88% C: $T_1 \sim 9 \times 10^{-5}$ sec) whereas for anthracites[258,259] T_1 is smaller than 10^{-5} sec. With increasing rank the difference between T_1 and T_2 decreases and both relaxation times are of the same order of magnitude in anthracites.[262] Hence it is seen that there is a much more rapid exchange of energy between the spin and the lattices in high rank coals which can undoubtedly be related to a much higher ordering.[243] Mineral constituents may reduce spin-lattice relaxation time independently of rank.[243]

Half-width. The variation of the half-width with increasing rank when measured under vacuum is also shown in Fig. 21.[237] Between 70 and 90% C this value is seen to decrease slowly from 12 to 8 gauss, then very rapidly to 2 gauss for older anthracites, after which it climbs very sharply above 95% C.[237,259] The lowest measured half-width, 0.7 gauss, was found with a Donetz anthracite.[240]

Up to a carbon content of about 92%, the half-width can be interpreted as the result of spin-spin interactions between the unpaired electrons and protons.[237,242,255,256] The following sharp decrease to about 95% C is to be ascribed on the one hand to the sharp decrease in the number of protons and on the other to an exchange narrowing as a result of the greater mobility of the electrons.[237,242,255,256,262]

An attempt has been made to deduce from these facts a mechanism of coalification.[242] As the carbon content increases up to 92%, the number of edge protons remains essentially constant while band-like condensation of aromatic systems takes place, primarily in a single direction. Above 92% C, these bands further condense with the splitting out of edge protons to form large aromatic clusters.[242] The increase in half-width beyond 95% C is ascribed to interaction between the unpaired electrons and conduction electrons.[237,255,256]

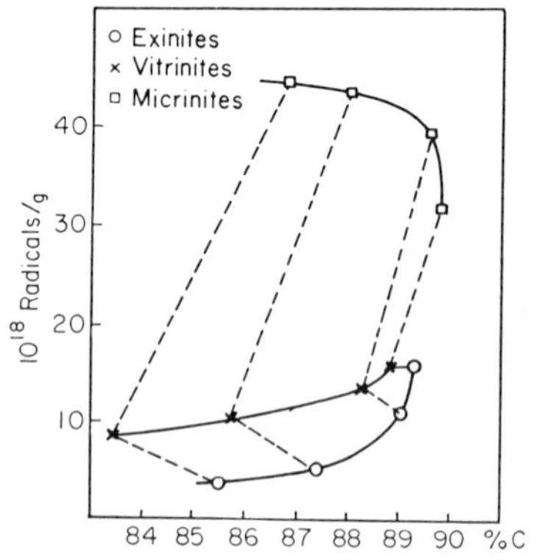

Fig. 22. Relation of radical concentration of exinites, vitrinites, and micrinites to carbon content.[254] The dotted lines connect the three macerals separated from a single coal of a given rank.

The half-width measured in air for samples containing up to 92% C is the same as that measured under vacuum, but above 92% C it increases rapidly,[259] due to interactions between the unpaired spins and atmospheric oxygen. As a consequence of this effect the half-width may also depend on particle size as observed in some anthracites[237,240,241,255] and in a fusain (Investigations of E. de Ruiter).

By using sugar charcoals, the influence of oxygen on the absorption has been carefully studied.[83,251,265–270] In sugar charcoal produced above 500°C, a strong ESR effect is observed in vacuum but this can scarcely be detected if the measurements are made in air. The process is strictly reversible. This observation is undoubtedly to be explained by spin-spin interaction between the spins of the free radicals and those of the paramagnetic oxygen of the air, which serves to broaden the band so that it can no longer be detected.[269,270] There may also be another reversible chemical effect[237,255,256,265] having a different origin than the purely physical one.

ESR Effect and Carbonization

Change of radical concentration with pyrolysis temperature. In the earliest measurements, a strong increase in radical concentration was observed in the range 300 to 550°C which then rapidly decreased until with samples treated at 600°C free radicals could no longer be detected.[227,238,244,246,271] It was shown, however, that the band broadening by atmospheric oxygen above 300°C was by no means negligible, especially above 600°C.[237,241,255] When measured under vacuum, a bituminous coal of 89% C attained a maximum radical concentration between 500 and 600°C[241,255] (see Fig. 23).[241]

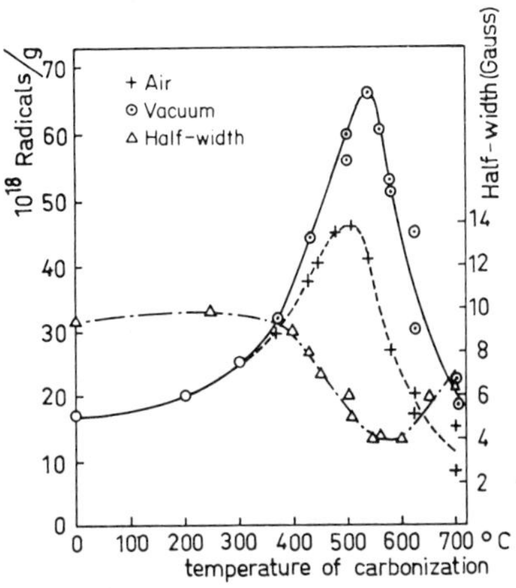

Fig. 23. Dependence of radical concentration (measured in air and under vacuum, scale on the left) and half-width (scale on the right) on carbonization temperature.[241] Rank of coal sample, 89% C.

In a systematic study on vitrinites of 71.7 to 96.5% C, those of 75.9 to 93.7% C showed an increase of the radical concentration with increasing pyrolysis temperature reaching a maximum between 450 and 600°C. The maximum number of free radicals increased up to about 87% C and seemed then to be nearly constant for vitrinites having higher carbon contents.[259] These results are in disagreement with earlier work where it was found that although the number of free radicals measured in air increased more rapidly with pyrolysis temperature in the younger coals, all samples independently of rank seem to have the same maximum radical concentration when treated at about 550°C.[227]

Above 600°C, the concentration of radicals decreased but their presence has been observed up to pyrolysis temperatures of 900°C.[272] The observed decrease of radicals has been explained by the increase in conductivity.[237,241,255] On the other hand, experiments on sugar charcoals carried out with complete exclusion of oxygen have shown that under these conditions there is only a slight decrease in radical concentration when pyrolysis occurs between 600 and 900°C.[270] From other experiments on charcoal, it follows that up to 650°C the radical concentration varies linearly with the C/H ratio which also seems to be true

[265] Ingram, D. J. E., and Tapley, J. G., *Chem. & Ind. (London)*, **1955**, 568–9.

[266] Austen, D. E. G., and Ingram, D. J. E., *Chem. & Ind. (London)*, **1956**, 981–2.

[267] Uebersfeld, J., and Erb, E., *Compt. rend.*, **242**, 3050–51 (1956).

[268] *Ibid.*, **243**, 363–5.

[269] Singer, L. S., and Spry, W. J., *Bull. Am. Phys. Soc.*, [2], **1**, 214–5 (1956).

[270] Singer, L. S., Spry, W. J., and Smith, W. H., *Proc. 3rd Conf. on Carbon*, Buffalo, 1957, Pergamon Press, London, 1959, pp. 121–8.

[271] Castle, J. G., Jr., *Phys. Rev.*, **98**, 1564 (1955).

[272] Uebersfeld, J., and Combrisson, J., *Discussions Faraday Soc.*, No. 19, 181–2 (1955).

for natural coals with a carbon content $\leq 95\%$.[273]

Coals pyrolyzed above 1500°C again begin to show strong electron spin resonance absorption.[274] It is possible that in the region between 900° and 1400°C there is such a broadening of the bands that absorption is no longer detectable and that the resonance effect observed above 1500°C is of the same kind as that below 900°C.[255] On the other hand, there is little fundamental difference between the electron spin resonance absorption in samples treated above 1500°C and that in graphite itself.[275] Newer measurements seem to indicate that the graphite absorption is due to conduction electrons.[276]

Finally, it should be mentioned that the concentration of free radicals is a function not only of the carbonization temperature but also of the heating rate.[277]

Change of the half-width with pyrolysis temperature. As seen in Fig. 23,[241] the half-width measured under vacuum of a coal with 89% C at first remains constant with increasing pyrolysis temperature,[237,241,242,255] then decreases and passes through a minimum at approximately the temperature of the maximum radical concentration, and then increases. The same behavior was found in vitrinites ranging from 80.2 to 93.7% C.[259] Comparison of Figs. 21 and 23 will show that the variation of radical concentration and half-width with increasing rank runs parallel to the variation of the same quantities with pyrolysis temperature. Thus a pyrolyzed younger coal would correspond to an unpyrolyzed older coal.

The spin-lattice relaxation time for a vitrinite of 88.9% C increases slightly with temperature up to 400°C, but then falls off very strongly.[259] With increasing carbonization temperature, T_1 and T_2 show the same trend as with increasing rank.[262]

Coal Extracts

In all coal extracts so far studied, free radicals have been found but always in lower concentrations than in the starting material.[107,223,241,250,255] There is a connection between the radical concentration in the extract and the extraction yield, so that the higher the yield the greater the radical concentration in the extract.[241,255] Measurements on a vitrain of 84% C extracted with pyridine show that the radical concentrations in the extract and residue can be combined by use of the extraction yield to give the radical concentration of the starting material.[102] Hence, no radicals are lost or created in the extraction process.

Irradiated Coals

Five coals ranging in rank from lignite to low-volatile bituminous coal were exposed to pile irradiation.[218] Although no effect of radiation was observed either in free radical concentration or in half-width up to 82% C, the high-rank coals showed a somewhat increased radical concentration after irradiation.[218]

Chemically Treated Coals

Oxidation in air up to 200°C. When a vitrain of 84% C was heated to 145°, 175°, and 200°C in air, the number of free radicals increased in a relatively short time to double that present in the starting material, the half-width remaining constant at 8 gauss.[278] After this vitrain was heated about 100 hr, the radical concentration dropped below that of the starting material, and the half-width decreased to 6 gauss. These values, especially the radical concentration, are extremely dependent on particle size.[278] When the same vitrain was acetylated the radical concentration on air oxidation at 175°C also increased, but appreciably less than in the original vitrain. After long heating in air, the value also decreased below that of the starting material.[278] During oxidation of the acetylated vitrain, there was no change of half-width which remained at 6 gauss. From these results it was concluded that the radicals are produced by the oxidation of hydroxyl groups to resonance-stabilized semiquinone structures.[278] The decrease on long heating is doubtless to be ascribed to

[273] Uebersfeld, J., and Erb, E., *Proc. 3rd Conf. on Carbon*, Buffalo, 1957, Pergamon Press, London, 1959, pp. 103–11.

[274] Castle, J. G., Jr., *Phys. Rev.* **95,** 846–7 (1954).

[275] *Ibid.*, **92,** 1063 (1953).

[276] Wagoner, G., *Bull. Am. Phys. Soc.*, [2], **4,** 22 (1959).

[277] Smidt, J., and van Krevelen, D. W., *Nature*, **184,** 719–20 (1959).

[278] de Ruiter, E., and Tschamler, H., *Brennstoff-Chem.*, **40,** 41–3 (1959).

oxidative degradation.[279] The same effect was found with sugar charcoals.[279,280]

Reduction. Upon reduction with lithium in ethylamine, coals of different ranks show a strong decrease in their free radical concentrations, the effect being greatest for a coal of about 90% C.[241]

A dimethylformamide extract showed no change in free radical concentration when reduced in a polarograph at a potential of −1.0 v.[241] When the extract was reduced at −1.6 v, however, the number of free radicals first decreased sharply (uptake of one electron), but increased again strongly when the potential was raised to −2.1 v,[281] which may be attributed to formation of singly charged aromatic anions which are paramagnetic.

Acetylation. The acetylation of coals changes the radical concentration either not at all or only slightly.[241,255,256,278] In a vitrain of 84% C, the half-width of 8 gauss was reduced to 6 gauss by acetylation.[278]

Chlorination. In the case of chlorination, only the change of the half-width has been studied,[241] and it was shown that this decreased in a manner similar to that already described for oxidation in air. The effect was thought to be the result of splitting off of protons.[241]

Treatment with concentrated sulfuric acid. When a vitrain of 84% C, its pyridine extract, or the extraction residue were treated with concentrated H_2SO_4 at room temperature, the number of free radicals increased more than tenfold and the half-width remained constant.[102] When the H_2SO_4 was removed, for example by washing with water, the radical concentration returned to approximately the value for the starting vitrain, the effect being therefore practically reversible. By visible and ultraviolet spectral studies on a concentrated H_2SO_4 solution of the pyridine soluble extract, it could be shown that this ESR effect may be explained by the formation of singly charged aromatic positive ions.[102]

Photochemical reactions. Powdered samples of an Ellington High Main coal (81.7% C) and a South Garesfield coking coal (89.7% C) as vigorously stirred suspensions in deaerated distilled water have been irradiated with a Hanovia S 500 medium-pressure mercury arc in air and under nitrogen up to 72 hr.[223] The first mentioned coal (7.9×10^{18} radicals/g) showed, after 72 hr in the absence of air, a concentration of 9.3×10^{18} and, in the presence of air, 10.2×10^{18} free radicals/g. An increase in free radical concentration is not unexpected, although the reason for the greater increase in the presence of air is not clear but may be associated with paramagnetic surface oxide complexes.[223] The South Garesfield coal, irradiated either in the presence or absence of oxygen, showed the same radical concentrations as that of the nonirradiated sample.

The chloroform extract, prepared from South Garesfield coal shock-carbonized at 420°C, contained 2.6×10^{18} free radicals/g.[223] If the chloroform solution was irradiated for 72 hr in air or in absence of air, precipitates were formed which contained 10.7×10^{18} and 9.0×10^{18} free radicals/g respectively, a marked increase compared to the starting material.

PROTON SPIN RESONANCE (NUCLEAR MAGNETIC RESONANCE)

Proton spin resonance is produced by the absorption of electromagnetic energy by nuclei in a magnet field resulting in change of the orientation of their spins with respect to the field.[226,228,282–288]

As is well known, the proton, like the electron, possesses an intrinsic angular momentum known as spin. Other nuclei may also possess spins but, as it happens, the principal isotopes

[279] Ingram, D. J. E., *Intern. Conf. Chemical Engineering in the Coal Industry*, Stoke Orchard, Cheltenham, England, June, 1956, Pergamon Press, London, pp. 23–4.

[280] Jackson, C., Harker, H., and Wynne-Jones, N. F. K., *Nature*, **182,** 1154–5 (1958).

[281] Austen, D. E. G., Given, P. H., Ingram, D. J. E., and Peover, M. E., *Fuel*, **38,** 309–14 (1959).

[282] Bloch, F., *Phys. Rev.*, **70,** 460–74 (1946).

[283] Bloembergen, N., Purcell, E. M., and Pound, R. V., *Phys. Rev.* **73,** 679–712 (1948).

[284] Jacobsohn, B. A., and Wangsness, R. K., *Phys. Rev.* **73,** 942–6 (1948).

[285] Grivet, P., *La Résonance Paramagnétique Nucléaire*, Centre National de la Recherche Scientifique, Paris, 1955, 298 pp.

[286] Andrew, E. R., *Nuclear Magnetic Resonance*, The University Press, Cambridge, 1955, 265 pp.

[287] Brown, J. K., *BCURA Bull.*, **20,** 105–18 (1956).

[288] Kopfermann, H., *Kernmomente*, Second edition, Frankfurt a. M., 1956, 462 pp. Schneider, E. E., *Nuclear moments*, Academic Press, New York, 1958, 505 pp.

of both carbon and oxygen, the main elements in coal other than hydrogen, have an even number of neutrons and protons in their nuclei and hence no net spin. Thus at the moment only proton magnetic resonance is important in coal studies.

The spin angular momentum of the proton must, by quantum rules, be oriented parallel or antiparallel to an external magnetic field. This orientation produces two energy states, the separation of which is proportional to the magnitude of the applied magnetic field. The energy difference is given by

$$\Delta E = g'\beta_p H$$

where g' is the spectroscopic splitting factor or gyromagnetic ratio, β_p is the proton spin, and H is the magnitude of the applied magnetic field. Transitions between the two levels can be induced by radiation of frequency ν given by

$$h\nu = \Delta E = g'\beta_p H,$$

whence

$$\nu = \frac{g'\beta_p H}{h}$$

where h is Planck's constant. In principle, there is no fundamental difference between proton spin resonance and electron spin resonance treated in the previous section. However, the experimental methods are quite different since the proton spin is 1836 times smaller than the electron spin. Moreover, the proton has a g' value of 5.58 so that, for example, the frequency in a field of 5000 gauss is 21.3 megacycles/sec compared to about 14,000 megacycles/sec for an electron.

There is another characteristic difference between proton and electron spin resonance. Although the electron usually ranges through the whole (often large) radical, the proton is strictly localized by its own bond. The electron may therefore be subjected to a number of different influences depending on its instantaneous position and this may produce, given sufficient resolution, considerable structure in the absorption curve. The proton in a given chemical species on the other hand is subjected to relatively constant influences, principally due to electron screening and spin-spin interactions with other nuclei having spins. The former weakens the applied field and produces the so-called "chemical shift;" the latter are frequently an indirect interaction through the molecular bonding electrons. In general, therefore, a single absorption line is expected, the position of which is characteristic for a given proton in a given group, for example,

$$-CH_3,\quad >CH_2,\quad >\!-CH,\quad =CH,\quad -OH,\ \text{etc.}$$

Because of the often rather strong interaction of proton spins and also in part for reasons associated with the apparatus the absorption is usually observed as a band.

There is, however, an important advantage in proton spin resonance (PSR) spectroscopy compared to all other spectroscopic methods. The areas of the different bands in the PSR spectrum are, under suitable experimental conditions, for example, radio frequency level low enough to avoid saturation, directly proportional to the number of hydrogen atoms involved and independent of their chemical nature.

In solution, the broadening of the single absorption line is generally small because molecular motions are usually far faster than the nuclear precession and local variations of the applied field due to spin-spin interactions may be averaged out so that band widths of 0.1 to 100 milligauss are observed.

In the solid state on the other hand, and this is a severe disadvantage of proton spin resonance, spin-spin interactions may produce extreme band broadening, the bands due to different groups tending to fuse together into a single very broad absorption of width 0.1 to 10 gauss. It is customary to take the "mean square width" or "second moment" of the absorption band as a measure of the effectiveness of spin-spin interactions.[289] It has been shown that the second moment increases with a decrease in temperature down to about 100°K, below which temperature it remains constant because molecular motions are fairly well restricted.[290] In view of the fact that the broadening measured by the second moment may be produced by inter- as well as by intramolecular interactions, they are extremely difficult to calculate even with known model substances.[291]

[289] Purcell, E. M., *Science*, **118**, 431–6 (1953).

[290] Andrew, E. R., and Eades, R. G., *Proc. Roy. Soc. (London)*, **A216**, 398–412 (1953); **A218**, 537–52 (1953).

[291] Dijkstra, H., *Brennstoff-Chem.*, **39**, Sondersausdrücke 31–2 (1958).

Since coal itself can only be investigated in the solid state, it is necessary to deal with a very broad absorption band. An interpretation of the broad line spectrum must therefore be based on reasonable assumptions concerning the variety of hydrogens present and on the contributions of such hydrogens to the second moment as found from model substances.

Calculations of the second moment have been made in the solid state at 90°K for a series of hydroaromatic, mixed aromatic-aliphatic, and aromatic hydrocarbons and the contributions to the second moment for protons in various groups evaluated:[290,292] $H_{CH_2} = 27.5$, $H_{CH_3} = 10.0$, and $H_{ar} = 9.7$, all in gauss^2. The significant difference in second moment between protons in CH_2 groups as compared to CH_3 groups may be explained by the fact that, even at 90°K, the methyl groups are still undergoing hindered rotation about their threefold symmetry axis (C_3) and hence tend to average out the spin-spin interactions. So far the only exception to this observation is the case of 1,4,5-trimethylnaphthalene for which a second moment of 22.4 gauss^2 was found for $H_{CH_3(peri)}$.[292]

If the contributions to the second moment for protons in groups of CH_{ar}, CH_2, CH_3, and CH_3(peri) are used from model substances, and if it is assumed that these values may be added to give the total second moment of the coal sample, the following relations for the experimentally observed second moment, ΔH^2, may be derived:[292]

If we define a and b by

$$\frac{H_{ar} + H_{CH_3}}{H_{CH_2} + H_{CH_3(peri)}} = \frac{a}{b}$$

the foregoing assumption of additivity leads to

$$\Delta H^2 = \frac{a}{a+b} 9.7 + \frac{b}{a+b} 27.5$$

whence $$\frac{a}{b} = \frac{27.5 - \Delta H^2}{\Delta H^2 - 9.7}$$

Experimental Methods

Apparatus

Either the applied magnetic field or the frequency may be freely chosen to determine PSR. For practical reasons, magnetic fields between 2000 and 15,000 gauss are used so that frequencies between 10 and 60 megacycles/sec —wavelengths of 30 to 5 m—are required. For measurements of proton spin resonance, either the induction method[293–295] or the radio-frequency bridge method[283,296,297] may be used. They both produce equally satisfactory results.

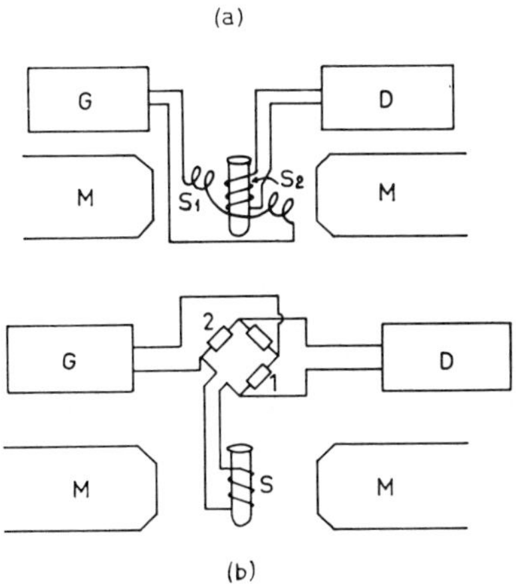

Fig. 24. Diagrams for apparatus used in measuring proton spin resonance by: (*a*) the induction method and (*b*) the radio frequency bridge method. G. Radio frequency generator; M. magnet; D. radio frequency amplifier and detector; S. induction coil.

The apparatus for the induction method is drawn schematically in Fig. 24*a*. The radio-frequency power from a suitable generator is applied to a coil S_1 in which the sample is placed. A second induction coil S_2 is located very close to the sample with its axis perpendicular to that of S_1 so that theoretically there would be no induced current from S_2. When, however, the frequency is correctly regulated in comparison to the field, a weak current will be induced in S_2 as a result of the resonance and this may be amplified and detected. Usually this is accomplished with a cathode-ray oscillograph and to this end the magnetic

[292] Bell, C. L. M., Richards, R. E., and Yorke, R. W., *Brennstoff-Chem.* **39**, Sondersausdrücke 30–3 (1958).

[293] Bloch, F., Hansen, W. W., and Packard, M. E., *Phys. Rev.*, **70**, 474–85 (1946).

[294] Proctor, W. G., *Phys. Rev.* **79**, 35–44 (1950).

[295] *Model-V*-4300, Varian Associates, Palo Alto, California.

[296] Purcell, E. M., Torrey, H. C., and Pound, R. V., *Phys. Rev.*, **69**, 37–8 (1946).

[297] Gutowsky, H. S., Meyer, L. H., and McClure, R. E., *Rev. Sci. Instr.*, **24**, 644–52 (1953).

field is generally modulated to sweep out the absorption peak.

The corresponding apparatus for the radio-frequency bridge method is shown schematically in Fig. 24*b*. The sample is placed in a coil in one of the two arms of the radio frequency bridge, and after application of the RF field the bridge is balanced. When the modulated magnetic field passes through the resonance condition, absorption of radio-frequency energy takes place, the bridge becomes unbalanced, and the unbalanced signal can be detected by means of an oscilloscope.

With both methods the ultimate sensitivity obtainable at present permits the measurement of 10^{16} to 10^{17} proton spins;[298] in favorable cases 10^{14} to 10^{15} proton spins can be detected.[295]

Preparation of Samples

A solid cylinder a few centimeters high and 0.5 to 1 cm in diameter may be cut from a block of coal.[262,292,299] To exclude the influence of water adsorbed on coals, the samples were heated to about 80°C under 0.2 mm pressure for 4 hr,[262,299] but in the case of low rank coals it seems necessary to use a temperature of 165°C under 10^{-5} mm pressure.[300]

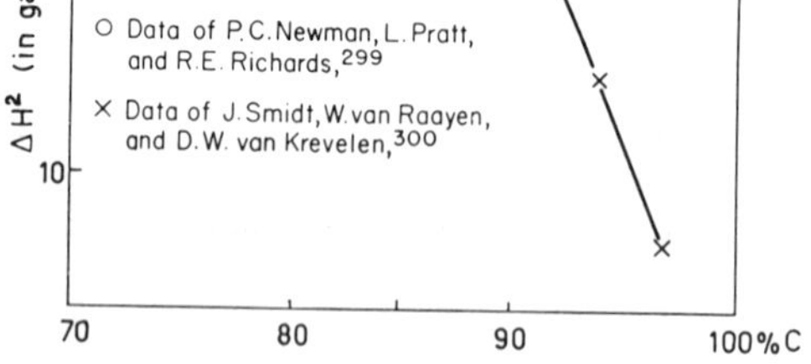

Fig. 25. Variation of the second moment ΔH^2 with carbon content of vitrains.

[298] Shoolery, J. N., *Anal. Chem.*, **26**, 1400–1403 (1954).

[299] Newman, P. C., Pratt, L., and Richards, R. E., *Nature*, **175**, 645 (1955).

[300] Smidt, J., van Raayen, W., and van Krevelen, D. W., *Fuel*, **38**, 539–40 (1959).

Coals and Their Petrographic Components

Measurements on three bright coals of 79.2, 83.1, and 92.6% C and a vitrain content >50% have been made at 90°K.[299] The

TABLE 6

SECOND MOMENTS, *a*/*b* RATIOS, AROMATICITY VALUES, AND AVERAGE AROMATIC RING SIZES FOR THREE BRIGHT COALS[28,299]

% C	ΔH^2(gauss²)	a/b	$(a/b)_{corr}$	f_a	R_{ar}
79.2	21.5	0.5	0.4	~0.7	~3
83.1	19.5	0.8	0.7	~0.8	~4
92.6	16.0	1.8	1.7	~0.9	~8

second moments, ΔH^2, and the *a*/*b* ratios calculated from them, assuming that tertiary CH groups are not present, are given in Table 6.

If the conclusion from infrared measurements is correct that there is very little aliphatic hydrogen in coals in the form of CH_3 groups,[186,195] the ratio *a*/*b* is essentially a measure of the ratio H_{ar}/H_{al}. Since it is not possible by PSR to distinguish protons in OH groups from aromatic hydrogens, a correction on *a*/*b* for OH groups in coals must be applied.[28] The results given in Table 6, column 4, as $(a/b)_{corr}$ are so corrected. From these values it may be seen that the coal of 79.2% C has approximately four aromatic per ten aliphatic hydrogens whereas the coal with 92.6% C has seventeen aromatic per ten aliphatic hydrogens. Taking into account further assumptions, average values for the aromaticity f_a and the aromatic ring sizes R_{ar} can be determined[28] (Table 6, columns 5 and 6). As can be seen from Fig. 25, in which the ΔH^2 values are connected with a broken line, the second moment decreases steadily with increasing rank.[299]

Recent measurements on vitrains, ranging from 71.7 to 96.5% C at 77.2°K, however, have shown that ΔH^2 seems to be constant up to 85% C, but then decreases strongly,[300] (see Fig. 25, full line). Whereas the discrepancy in the low-rank coals may be due to insufficient drying of the samples,[300] the differences in the anthracite range need further investigation. From the fact that the sample of 96.5% C gave $\Delta H^2 = 8.6$ gauss² and calculated values reported earlier have a minimum value of 9.7 gauss², the results concerning coal structure can at the moment be regarded only as approximations.[300] The same changes of ΔH^2 with rank have been observed whether the measurements are made at 77.2 or at 110°K.[300] Investigations on Russian coals have shown that the broad signal obtained at 90°K is superimposed at 295°K and at 410°K by a small narrower signal which increases with temperature and disappears again on cooling.[262]

Soluble Products Derived from Coals

It is evident that high resolution spectra obtained in solution offer much more information than broad line spectra obtained for the solid state, but the solvent should at the same time dissolve a sufficient quantity of the products and its own PSR spectrum should not interfere with that of the material. CS_2[81,98] and methylene chloride[301] have been used for the few samples studied up to now. Completely deuterated pyridine and benzene will doubtless have important advantages as solvents in further work in this field.

For all samples so far studied in solution, the same type of proton spin resonance spectrum as that shown in Fig. 26[98] has been found.[81,98,301] As can be seen, the peaks are still broad, an effect which may be due to the molecular complexity of the samples under investigation,[81,98,301] to the possible presence of colloidal systems,[301] and to the presence of free radicals.[81,98,301] On the other hand, in contrast to the results with solid samples, three definitive peaks are found in the solution spectra. Based on information from model substances, peak 1 in Fig. 26 has to be assigned to hydrogens in aromatic CH and phenolic OH groups;[81,98,301] peaks 2 and 3 certainly have their origin in hydrogens on saturated carbon atoms.[81,98,301] It seems probable that peak 2 includes hydrogens on saturated carbon atoms adjacent to aromatic rings (α C atoms) and peak 3 includes hydrogens on saturated carbon atoms further removed from aromatic rings.[81,98,301] A peak occasionally observed on top of peak 3 (see Fig. 26) may indicate protons on freely rotating chains or CH_3 groups.[81,98]

The very important parameters H_{ar}/H and H_{al}/H can be established quantitatively from the clear separation of the aromatic from the aliphatic protons. It is possible that not all the material in solution contributes to the high resolution spectra, however, because some protons may have given a broad line pattern which could not be detected.[301]

[301] Brown, J. K., Ladner, W. R., and Sheppard, N., *3rd. Intern. Conf. on Coal Science*, Valkenburg, The Netherlands, 1959.

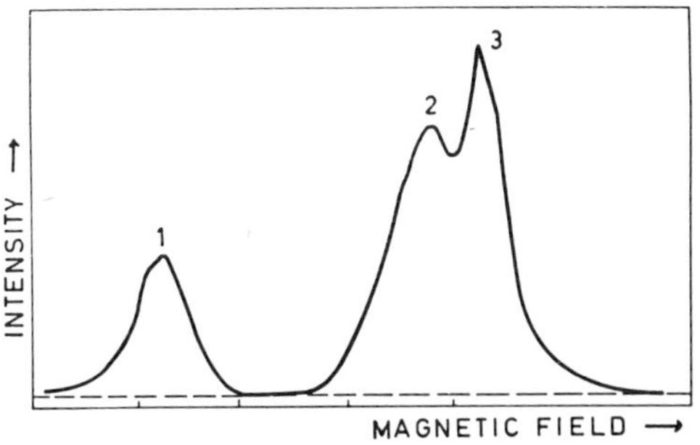

Fig. 26. PSR spectrum of "asphaltene" in CS_2 at 125°C.[98]

From the PSR spectrum of an "asphaltene"—benzene-soluble and hexane-insoluble fraction of a hydrogenated coal—it was concluded that this product consists of large molecules with many tetralin-type structures and a few polynuclear condensed aromatics.[81,98]

Vacuum distillates from coals of different rank heated to temperatures of about 400°C have been investigated and the qualitative conclusions are that with increasing rank of the coals both the proportion of aromatic to non-aromatic hydrogen and the proportion of aromatic hydrogen to that on α C atoms increases.[301] The former means that the aromaticity increases, the latter that the degree of aromatic substitution decreases.[301]

DOUBLE MAGNETIC RESONANCE

The idea was suggested several years ago that it should be possible to observe not only the influence of the surroundings on the magnetic properties of coal, for example, broadening of the ESR signal by oxygen, but also the influence of coal on the magnetic properties of the surroundings, using the "Double Magnetic Resonance" method.[273] If a system composed of one electron spin and one proton spin is placed in a magnetic field, it has four energy levels (Zeeman sublevels) defined by the four combinations of the magnetic quantum numbers. The energy levels are given by:

$$E = -H(g\beta_e m + g'\beta_p m')$$

where H is the intensity of the external magnetic field, g and g' are the gyromagnetic splitting factors for the electron and proton, β_e and β_p are the Bohr magnetons of the electron and proton spins, and m and m' the electron and proton magnetic quantum numbers ($\pm\frac{1}{2}$).

The following transitions are possible among the four energy levels:

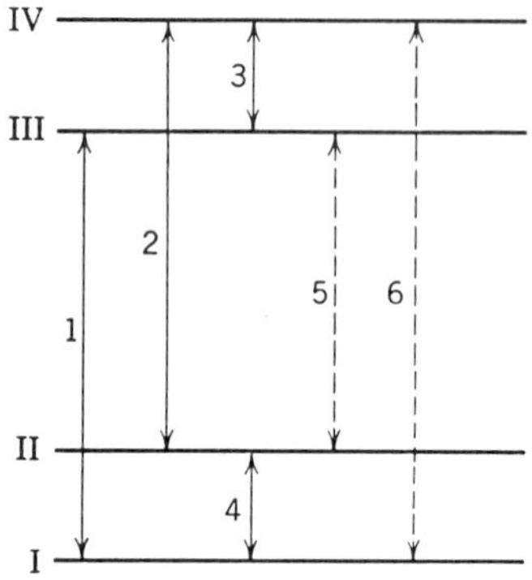

If there is no interaction between the electron spin and the proton spin, only the true ESR, transitions 1 and 2, and PSR effect, transitions 3 and 4, can be observed. If there are interactions between the electron spin and the proton spin, two cases must be considered: (1) A dynamic dipole-dipole interaction which is time dependent and the average value of which is zero. Again only the true ESR and PSR effects can be found but saturation of transition 1 or 2 changes the population at all four energy levels in such a way as to produce an increased PSR signal, which is known as the "Overhauser effect."[302,303] (2) A static dipole-dipole interaction which is time independent. In this case, transitions 5 or 6, which involve at the same time a change in orientation of both the electron spin and the proton spin, are no longer forbidden.[304] Saturation of transition 5 leads to an increased population in the energy level III and therefore to an increased PSR signal—absorption. If transition 6 is saturated, the population of energy level IV is increased and again a strong PSR signal appears—emission. These phenomena are called collectively the "double Overhauser effect."[304,305,306,307]

[302] Overhauser, A. W., *Phys. Rev.*, **89**, 689–700 (1953).

[303] Abragam, A., *Phys. Rev.* **98**, 1729–35 (1955).

[304] Abragam, A., and Proctor, W. G., *Compt. rend.*, **246**, 2253–6 (1958).

Experimental Method

From the foregoing theoretical background of the double magnetic resonance effect, it follows that the usual ESR and PSR techniques can be applied, but to achieve the highest saturation possible it is necessary to work with high microwave power. A cylindrical cavity is favored and special attention has to be given to the correct alignment of the RF coil.[308]

Experimental Results

Systems have been studied in which liquid benzene or toluene on the one hand and gaseous ammonia or hydrogen sulfide on the other hand were adsorbed on coals or charcoals and the "double Overhauser effect" observed.[307] If benzene adsorbed on charcoals prepared at different temperatures (350 to 600°C) is used, the "double Overhauser effect" decreases with increasing carbonization temperature to zero at about 600°C.[305–307] For charcoals prepared above 600°C, the "Overhauser effect" is obtained.[306,307] A tentative conclusion from these results would be that the free radicals in charcoals are at first localized but become more and more delocalized with increasing carbonization temperature.[307]

MAGNETIC SUSCEPTIBILITY

If a substance is brought into a magnetic field it is either repelled—diamagnetism—or attracted—paramagnetism—by the field. Materials, the atoms or molecules of which have only closed electron shells, are diamagnetic; those with unpaired electrons are, as a consequence of uncompensated electron spins, paramagnetic. Ferromagnetism is a special case of paramagnetism but of far greater magnitude, and for this reason traces of a ferromagnetic impurity may greatly falsify magnetic measurements on a given sample. To eliminate the effects of such impurities it is possible to saturate them by means of strong fields, for example, to extrapolate to infinite field

[305] Erb, E., Motchane, J. L., and Uebersfeld, J., *Compt. rend.*, **246**, 2121–3 (1958).

[306] *Ibid.*, 3050–2.

[307] Uebersfeld, J., and Jacubowicz, M., *Proc. 4th Conf. on Carbon*, Buffalo, 1959, Pergamon Press, London, 1960, pp. 267–70.

[308] Motchane, J. L., Erb, E., and Uebersfeld, J., *Compt. rend.*, **246**, 1833–5 (1958).

strength.[309] Nevertheless, it is far better to remove such ferromagnetic components chemically before the magnetic measurement rather than to try the field saturation method.[309,310] Diamagnetism and paramagnetism may occur simultaneously in the same material. The former depends only slightly on temperature whereas the latter, according to Curie's law, decreases with increasing temperature and hence may be eliminated by extrapolation to infinite temperature.[309]

Since the diamagnetic volume susceptibility represents the magnetization of 1 cm^3 in a field strength of 1 gauss, the specific diamagnetic susceptibility or the diamagnetic mass susceptibility χ is the magnetization of 1 g by a field of 1 gauss and is of the order -10^{-6} cgs units in organic substances—paramagnetic susceptibility has a positive sign. The molar diamagnetic susceptibility, χ_M, is the product of χ with the molecular weight for the given material and is the sum of atomic susceptibilities of the constituent atoms χ_a and increments λ for multiple bonds:[309,311,312]

$$\chi_M = \Sigma\chi_a + \Sigma\lambda.$$

The χ value may be dependent on the orientation of the sample in the magnetic field and results in diamagnetic anisotropy. A particularly significant property of aromatic ring systems is their strong diamagnetic anisotropy in the solid state—the value of χ_M perpendicular to the plane of the ring is many times larger than parallel to it and has a magnitude which increases regularly with the number of condensed rings in the system.[312,313] This property is a result of the fact that mobility of the π-electrons in the plane of the ring is many times greater than in a direction perpendicular to it. In powdered samples, χ is an average value of the susceptibilities in each of the three directions. There appears to be a relation between the average lamella diameter $\bar{L}$ and χ. Below $\bar{L}$ values of ~40 to 50 A, only a slight increase of χ with increasing $\bar{L}$ has been found, but with $\bar{L}$ changing from ~50 to ~150 A the diamagnetic mass susceptibility increases from about -1×10^{-6} to about -7×10^{-6} cgs units.[310,314–318] As in graphite, χ parallel to the layer plane is about the same as in small aromatic systems, that is, -0.5×10^{-6} cgs units, but χ perpendicular to the layer plane is -22×10^{-6} cgs units.[316] This sharp increase of χ is certainly due to the change of χ perpendicular to the layer plane.

On the basis of the relation between diamagnetic anisotropy and the number of condensed rings derived from measurements on aromatic hydrocarbons, the molar diamagnetic anisotropy[89,312] and the specific diamagnetic anisotropy[309] may be used to estimate the average number of aromatic rings in a substance.

For aromatic ring systems containing O and N as heteroatoms and without consideration of the type of condensation, the following empirical relation for the molar diamagnetic susceptibility has been derived:[319,320]

$$-10^6\chi_M = 6.00\,\mathrm{C} + 2.93\,\mathrm{H} + 5.57\,\mathrm{N} + 4.61\,\mathrm{O}_{-\mathrm{O}-} + 3.36\,\mathrm{O}_{=\mathrm{O}} - 12.5\,\mathrm{O}_{\mathrm{quinone}} + 30\,R_{\mathrm{ar}} - \mathrm{C}_{\mathrm{ar}} - \mathrm{N}_{\mathrm{ar}} + 15$$

where C, H, N, etc. = number of C, H, N ... atoms

R_{ar} = number of aromatic rings

C_{ar} = number of aromatic carbon atoms

N_{ar} = number of heterocyclic nitrogen atoms

Experimental Methods

Most measurements have been made by the Gouy balance method.[311] The force F exerted

[309] Honda, H., and Ouchi, K., *Fuel*, **36**, 159–75 (1957).

[310] Akamatu, H., Inokuchi, H., Takahashi, H., and Matsunaga, Y., *Bull. Chem. Soc. Japan*, **29**, 574–81 (1956).

[311] Selwood, P. W., *Magnetochemistry*, Interscience, New York, 1956, 435 pp.

[312] Akamatu, H., and Matsunaga, Y., *Bull. Chem. Soc. Japan*, **26**, 364–72 (1953).

[313] Selwood, P. W., *J. Research Natl. Bur. Standards*, **41**, 151–62 (1948).

[314] Kiive, P., and Mrozowski, S., *Proc. 3rd Conf. on Carbon*, Buffalo, 1957, Pergamon Press, London, 1959, pp. 165–72.

[315] Adamson, A. F., and Blayden, H. E., *Proc. 3rd Conf. on Carbon*, Buffalo, 1957, Pergamon Press, London, 1959, pp. 147–57.

[316] Wynne-Jones, W. F. K., Blayden, H. E., and Iley, R., *Brennstoff-Chem.*, **33**, 268–73 (1952).

[317] Marchand, A., *Compt. rend.*, **238**, 460–2 (1954).

[318] Pinnick, H. T., *Phys. Rev.*, **91**, 228–9 (1953).

[319] Schuyer, J., *Naturwissenschaften*, **42**, 388 (1955).

[320] van Krevelen, D. W., and Schuyer, J., *Coal Science*, Elsevier, Amsterdam, 1957, pp. 249–63.

by the field on the sample obtained by measuring the increase in apparent weight of the sample when the field is switched on is found to be proportional to the diamagnetic susceptibility, the field strength, and the field gradient. With a coal sample of about 1 g and a field strength of the order of 20,000 gauss, the value of F is in the range of 1 mg. To make the measurement independent of the exact value of the field strength and the field gradient, the unknown sample A is compared with a known standard material B, usually water. The specific diamagnetic susceptibility of A is then calculated from

$$\chi_A = \frac{F_A}{F_B}\frac{m_B}{m_A}\chi_B$$

where χ_B = specific diamagnetic susceptibility of water, -0.7218×10^{-6} cgs units,

F_A, F_B = forces exerted by the field on A and B,

m_A, m_B = masses of A and B.

As already remarked, χ must be determined by extrapolation to $T = \infty$ in order to eliminate the effect of paramagnetic contributions,[309] and measurements must be carried out over a wide range of temperatures.[321] For coal, a temperature of about 115°C must not be exceeded if structural changes or oxidation are to be avoided.[309]

Another method of measuring susceptibilities is to suspend the sample between the poles of a magnet and, by use of a microscope, measure its deflection when the field is switched on. This method is independent of the quantity of the sample.[322]

The Krishnan method,[311,313] whereby the period of the sample suspended as a pendulum between the pole faces of the magnet is measured with and without the field, may also be applied to coals.

Measurements may be made both on blocks[316,322] and on powders.[309,310,316,323–326]

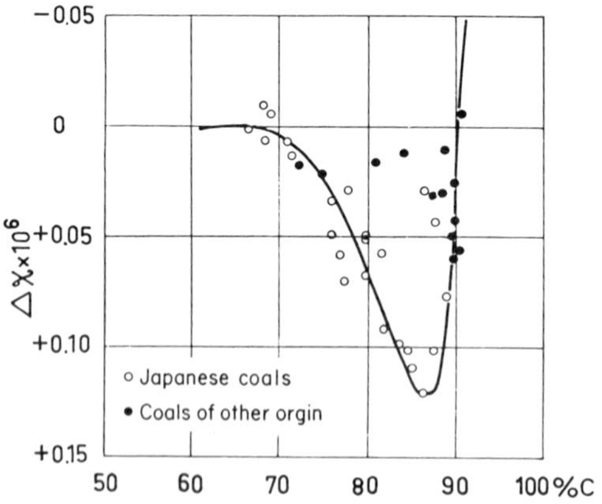

Fig. 27. Dependence of $\Delta\chi$ (cgs units) on carbon content.[309]

Coals and Their Petrographic Components

The specific diamagnetic susceptibilities of the pure organic matter in a large number of demineralized coals,[309,323,325–327] vitrains,[309,323] clarains,[323] and durains[323] were found to lie between -0.5 and -0.6×10^{-6} cgs units. The sole exception was a Kôkô anthracite with $\chi = -1.152 \times 10^{-6}$ cgs units.[309] Measurements with coals and anthracites in block form showed that the latter exhibit anisotropy effects.[322]

Dependence of χ on Rank

If the difference $\Delta\chi$ between χ_{exp} measured at room temperature and χ_{calc} from the sum of the atomic susceptibilities is plotted against carbon content for coals of various origins (Fig. 27[309]), it is seen that between 70 and 90% C the experimental values become smaller than those calculated, thus indicating the presence of a paramagnetic contribution.[309]

If this paramagnetic part is eliminated by measuring the diamagnetic susceptibility at a series of temperatures and extrapolating to $T = \infty$, the curve shown in Fig. 28[309] is obtained for the true diamagnetic mass susceptibility as a function of carbon content. It is seen that between 80 and 90% C the increase with rank is less than above and below this

[321] Honda, H., Ouchi, K., and Hirose, Y., *J. Chem. Soc. Japan*, Pure Chem. Sect., **76**, 154–8 (1955).

[322] Wooster, W. A., and Wooster, N., *Proc. Conf. on Ultrafine Structure of Coals and Cokes*, BCURA, London, 1944, pp. 322–9.

[323] Honda, H., Kimura, H., and Ouchi, K., *J. Fuel Soc. Japan*, **32**, 669–81 (1953).

[324] Honda, H., and Ouchi, K., *Sci. Repts. Research Inst. Tohoku Univ.*, **37**, 55–63 (1953).

[325] Honda, H., *J. Fuel Soc. Japan*, **32**, 486–93 (1953).

[326] Ouchi, K., *Rept. Resources Research Inst.*, No. 31 (1955), 109 pp.

[327] Honda, H., and Ouchi, K., *Rept. Resources Research Inst.* No. 1 (1952), 66 pp.; *BCURA Bull.*, **17**, 321 (1953).

range.[309] Three Japanese coals in the region between 80 and 90% C, on the other hand, lay very close to the curve extrapolated from lower carbon contents. This fact was ascribed to the presence of other sorts of side chains, bridging bonds, or naphthenic systems.[309]

If it is assumed that the only double bonds present in coals are aromatic in character, the difference $\Delta\chi'$ between the true diamagnetic mass susceptibility and the sum of the atomic susceptibilities per gram gives the contribution of the π-electrons to the diamagnetic mass susceptibility, but it should be kept in mind that the absolute values of $\Delta\chi'$ are very strongly dependent on the shape of the aromatic nuclei. As seen in Fig. 29,[309] $\Delta\chi'$ increases slowly up to 85% C but much more rapidly beyond this value.

By comparison of these results with studies on pure aromatic hydrocarbons, it was concluded that from brown coals to bituminous coals of 88% C essentially 1 to 4 aromatic rings per mean structural unit are present. Larger ring systems are probably present above 90% C. If naphthenic rings are present, the values of $\Delta\chi'$ must be corrected which would lead to somewhat larger condensed aromatic ring systems.[309] The type of condensation may play a major rôle.[309,312]

If the number of aromatic rings per mean structural unit is calculated from the empirical relation given earlier, significantly higher values are obtained between 70 and 90% C (4 to 8 rings).[320]

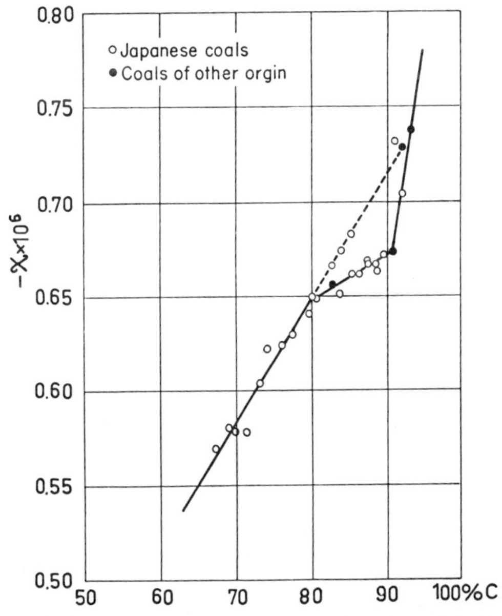

Fig. 28. Dependence of true diamagnetic mass susceptibility (cgs units) on carbon content.[309]

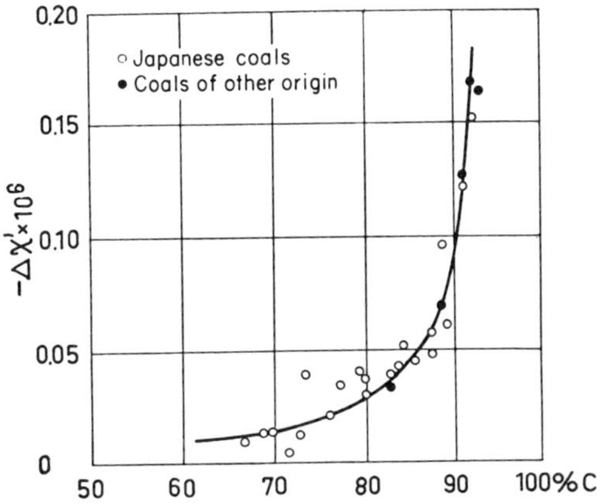

Fig. 29. Dependence of $\Delta\chi'$ (cgs units) on carbon content.[309]

Dependence of χ on Carbonization Temperature

Coal samples of 66.0 to 93.3% C, pitches and oil cokes as well as glucose and cellulose, were pyrolyzed in steps between 200 and 2000°C. The χ vs T curve shows four distinct zones[314,315,324,326,328–331] as illustrated in Fig. 30.[331]

1. Up to 700°, but especially beyond 400° the susceptibility decreases, probably as a result of free radicals produced by decomposition. This explanation alone, however, seems to be insufficient, since the minimum in diamagnetic susceptibility, which was observed during carbonization up to 1000°C of three coals of 66.0, 83.3, and 93.3% C between 600° and 800°C, occurs at a somewhat higher temperature than the maximum radical concentration.[331]

2. Between 700 and 1000°C, the susceptibility increases. This fact, along with the

[328] Honda, H., Ouchi, K., and Nagata, K., *J. Chem. Soc. Japan*, Pure Chem. Sect., **74**, 720–3 (1953).

[329] Honda, H., Ouchi, K., and Miyake, A., *J. Chem. Soc. Japan*, Pure Chem. Sect. **74**, 802–5 (1953).

[330] Honda, H., and Ouchi, K., *J. Fuel Soc. Japan*, **33**, 579–86 (1954).

[331] Honda, H., *Proc. 3rd Conf. on Carbon*, Buffalo, 1957, Pergamon Press, London, 1959, pp. 159–64.

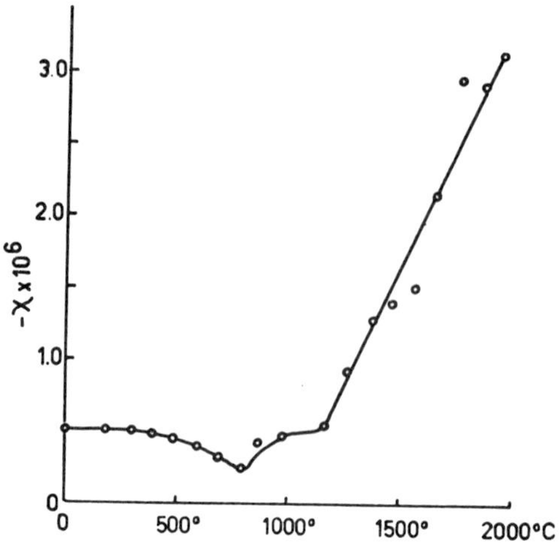

Fig. 30. Dependence of diamagnetic mass susceptibility (cgs units) measured at room temperature of a Yubari coal (83.3% C) on carbonization temperature.[331]

remarkable change in this region of other physical properties, leads to the conclusion that larger, three-dimensional systems are being formed and the number of free radicals is decreasing. The rank of a coal seems to have only slight influence on the changes in magnetic susceptibility between room temperature and 1100°C.[331] From the similarity between the shape of the susceptibility curve as a function of temperature up to 1000°C and that of the change of $\Delta\chi$ with rank (cf. Fig. 27), it has been concluded that natural coalification proceeds by a mechanism similar to that of pyrolysis.[309,326,332] For cokes of different origin, a positive temperature dependence of χ has been observed that disappears at about 1100°C.[314,315,326]

3. Between 1000 and 1300°C, the susceptibility of all samples remains nearly constant. It is possible that there is a simultaneous decomposition with the formation of new free radicals and the formation of larger crystallites resulting in a simultaneous decrease and increase of χ or that the condensed aromatic ring systems remain unchanged.

4. Between 1300 and 2000°C, a sharp steady increase of the susceptibility occurs that may be attributed to the formation of a two-dimensional graphite-like lattice,[331] because from about 1300°C or a χ value of -1.0×10^{-6} cgs units onward the appearance of a negative temperature dependence of the susceptibility has been observed.[326,331] This negative temperature dependence gets stronger with increasing carbonization temperature and may indicate the degree of graphitization.[331] Still an unequivocal theory for the sharp rise in susceptibility with temperature cannot be given.[314,315]

5. On further increase of temperature to >2000°C, the susceptibility approaches the value characteristic of graphite, -7.5×10^{-6} cgs units.[331] The relation between diamagnetic mass susceptibility of coals carbonized at 2000°C and rank of the original coals is shown in Fig. 31.[331] From the behavior of χ with rank it was concluded that the graphitization of coals up to 2000°C is easiest for coals of about 85% C.[326,331]

Coal Extracts

Benzene extracts of eight different coals ranging from 80 to 88% C obtained at 280°C under pressure all showed the same susceptibility as the starting material.[309] It was concluded from this that the extracts have essentially the same structure as the starting material.[309]

In another set of experiments a Yubari coal (83.8% C) was extracted with benzene under pressure to yield "bitumen" and the residue was further extracted after mild "hydrogenolysis" to yield "pseudobitumen."[89] Both the "bitumen" and "pseudobitumen" were separated into seven fractions by extraction with ligroin, methanol, ether, acetone, and carbon tetrachloride and the magnetic susceptibility of each fraction was measured.[89] Since the

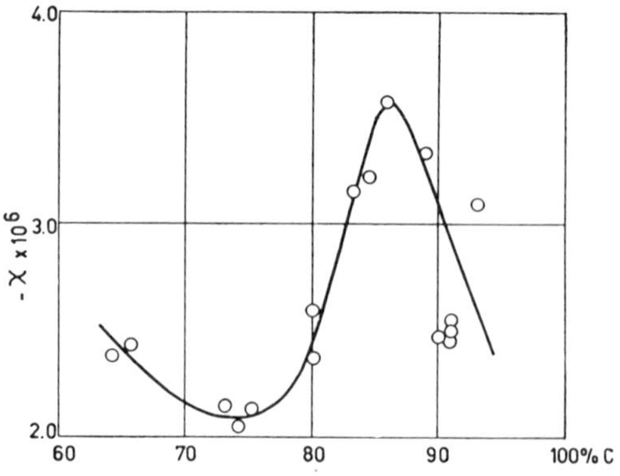

Fig. 31. Dependence of diamagnetic mass susceptibility (cgs units) of coals carbonized at 2000°C on the carbon content of the original coals.[331]

[332] Ouchi, K., *Bull. Resources Research Inst.*, **5**, 70 (1953).

average molecular weight was known, the molar diamagnetic susceptibility could be calculated and the difference between the values obtained and the corresponding sums of the atomic susceptibilities was taken to be the contribution per mole of the π electrons from the aromatic structures. From comparison with model substances[89,327,333] and after consideration of the various possibilities of substitution,[333] it was concluded that all fractions have 3 to 4 condensed aromatic rings.[89] More hydroaromatic and polyphenyl structures are to be expected in "bitumen" than in the "pseudobitumen," but the latter should contain more aliphatic side chains. If diamagnetic volume susceptibility is plotted against density,[89] characteristic lines are obtained for paraffins, hydroaromatics, condensed aromatics, and polyphenyl compounds.[333,334] The previously mentioned fourteen fractions, as well as certain pitch fractions,[335] lie between the lines for condensed aromatics and polyphenyl compounds and hence polyphenyl structures should also be taken into account.

From the different dependence of the magnetic susceptibility of "bitumen" and "pseudobitumen" on pyrolysis temperature, it was concluded that condensation reactions between the phenols originally present begin to occur between 300 and 500°C in the presence of CH_3 groups—"pseudobitumen"—but only at 700°C in the absence of CH_3 groups—"bitumen."[326]

Finally, magnetic studies on the petroleum ether soluble fraction of a hydrogenated Yubari coal of 83.8% C led to the conclusion that 1 to 4 ring systems were present.[336]

ELECTRICAL CONDUCTIVITY

Any material when subjected to an electric field conducts current to some extent. According to the magnitude of this conductivity materials are classified as insulators, semiconductors, and conductors, and the differences among these groups may be regarded as a measure of the freedom of motion of electrons. In a conductor such as a metal, some of the electrons are released from their binding to particular atoms and wander more or less freely through the structure as conduction electrons. In an insulator such as quartz, the electrons possess an extremely limited mobility. Between these two classes lie the semiconductors which show distinct but relatively weak conductivity. Where the boundaries between these classes are to be placed is a matter of practical expediency.

The quantity of electricity transported per second through a unit cross section under the action of an electric field of strength E in v/centimeter is a measure of conducting ability. If it is only desired to compare various materials in this respect, the values of E and the cross section are immaterial so long as they remain the same. In view of this fact, it is customary when comparing the conductivity of different materials to eliminate quantities not characteristic of the material under study, such as length l and cross section q of the sample, and to report a "specific conductance" K defined by

$$K = \frac{l}{qR} = \frac{1}{\rho} [\Omega^{-1}\ \mathrm{cm}^{-1}]$$

where R is the measured resistance of the sample in ohms and ρ is the "specific resistance" or "resistivity." It is clear that ρ is the resistance of a sample of 1 cm length and 1 cm^2 cross section and is measured in Ω-cm. As a very rough indication, conductors may be taken to be materials having a specific resistance less than 1 Ω-cm, whereas insulators will have corresponding values greater than 10^{15} ohm-cm.

The specific resistance of a substance depends on temperature and pressure and, if the material is hygroscopic, on the moisture content. When materials are anisotropic and contain many crystals, as in graphite, specific resistance also depends on direction of the applied electric field.

In metallic conductors, ρ increases with increasing temperature as a consequence of greater scattering of the conduction electrons by lattice waves. In semiconductors, the increase in the number of current carriers—electrons or holes—produced by the relatively

[333] Honda, H., *J. Fuel Soc. Japan*, **33**, 134–43 (1954).

[334] *Ibid.*, **32**, 447–60 (1953).

[335] Kikkawa, S., Honda, H., and Ouchi, K., *J. Fuel. Soc. Japan*, **32**, 535–40 (1953); *Chem Abs.*, **48**, 5481 (1954).

[336] Sakabe, T., Ouchi, K., Sassa, R., and Honda, H., *Nenryô Kyokaishi* **36**, 712–22 (1957); *Chem. Abs.*, **52**, 3307 (1958).

TABLE 7

COMPARISON OF ELECTRON EXCITATION ENERGIES OF AROMATIC COMPOUNDS USING THREE DIFFERENT METHODS

	Sample	R_{ar}	Energy Gap (eV) from		
			Photoconductivity	Optical Absorption	$\Delta\epsilon$
Aromatic Hydrocarbons	Benzene	1		3.7†	
	Naphthalene	2		2.6†	3.7§,**
	Anthracene	3		1.8†	1.9‖
				3.0‡,**	2.7‡,**
	Phenanthrene	3		2.7†	
	Tetracene	4			1.7‖
	Chrysene	4		2.5†	2.2‖
	1,2-benzanthracene	4		2.0†	
	Pyrene	4		2.1†	2.0‖
					2.4‖,**
	Triphenylene	4		2.9†	
	Pentacene	5			1.5–1.7‡,‖
	1,2,5,6-dibenzanthracene	5		2.3†	
	Perylene	5			1.9‖
	Anthanthrene	6			1.6–1.7‖,¶
	Coronene	7		2.4†	
	Mesonaphthodianthrene	8			1.2¶
	Pyranthrene	8	0.85*	0.8*,‡	1.1*
	Violanthrene	9	0.88*	0.8*	0.85*
	Isoviolanthrene	9	0.85*		0.8*
	Ovalene	10	1.20*	1.2*	1.1*
Quinones	Anthanthrone	6			1.7¶
	Mesonaphthodianthrone	8			1.3‡
	Pyranthrone	8	1.14*	0.85*,‡	1.1*
	Violanthrone	9	0.84*	0.75*	0.8*
	Isoviolanthrone	9	0.93*	0.75*	0.75*
Miscellaneous Polycyclics	Indanthrazine	7	0.80*	0.7*,‡	0.7*
	Indanthrone	7	0.74*	0.65*	0.6*
	Flavanthrone	8	0.76*	0.75*,‡	0.7*
	Cyananthrone	11			0.2‡
	Fluorene	2		3.0†	

* From H. Inokuchi.[346]
† From D. S. McClure.[343]
‡ From H. Akamatu and H. Inokuchi.[342]
§ From H. Pick and W. Wissman.[344]
‖ From D. C. Northrop and O. Simpson.[345]
¶ From H. Inokuchi.[340]
** Measurements on a single crystal.

low energies inherent in thermal excitation far outweighs the greater scattering caused by more vigorous lattice vibrations and therefore the net resistivity decreases with temperature.[337]

Pressures above 100 kg/cm² begin to produce deformation of molecular lattices and lead to an increase in the number of conduction electrons and consequently to a slight decrease in the specific resistance. This effect has been exhaustively studied in condensed aromatic systems.[338]

As the specific resistance of water is about 10^6 Ω-cm, the measured ρ values of samples with $\rho > 10^6$ Ω-cm may be seriously falsified

[387] Cusack, N., *The Electrical and Magnetic Properties of Solids*, Longmans, Green and Co., London, 1958, 428 pp.

[338] Inokuchi, H., *Bull. Chem. Soc. Japan*, **28**, 570–2 (1955).

by adsorbed water unless care is exercised to make certain that the water is eliminated.

Graphite may be cited as a typical example of the effect of anisotropy. Its specific resistance when measured parallel to the layer plane is very much lower—10^{-4} Ω-cm—than when measured perpendicular to it—2 Ω-cm.[339]

It has been shown that the specific resistance of condensed aromatic hydrocarbons[340,341] when measured at 80 kg/cm^2 pressure show a strong temperature dependence of specific resistance. If log ρ is plotted against $1/T$, a linear relation results. From this a temperature dependence of the Boltzmann type may be deduced:

$$\rho = \rho_\infty e^{\Delta\epsilon/2kT}$$

where ρ_∞ is the value for $1/T = 0$ and the quantity $\Delta\epsilon$, which can be evaluated from the slope of the log ρ vs $1/T$ curve, is known as the electron excitation energy or energy barrier. It is effectively a measure of the energy needed to excite a bound electron to a state in which it becomes conducting. For aromatic compounds, it has been found that $\Delta\epsilon$ derived from the temperature dependence of ρ is in good agreement with the electron excitation energy from photoconductivity and light absorption (Table 7), a fact which characterizes these compounds as molecular crystals.[342]

As can be seen from Table 7, in aromatic systems the magnitude of $\Delta\epsilon$ depends on the size of the aromatic system R_{ar} and the type of condensation. $\Delta\epsilon$ is smaller the greater the number of rings in the condensed system. On the other hand, catacondensed systems have smaller $\Delta\epsilon$ values than pericondensed systems of the same R_{ar} value. Hence, $\Delta\epsilon$ is a structural parameter of at least qualitative interest.

In solid solutions, such as tetracene in anthracene, and anthanthrene in chrysene, pyrene, or anthracene, the plot of log ρ vs $1/T$ consists of two linear segments, leading to two $\Delta\epsilon$ values.[345] $\Delta\epsilon$ derived from the lower temperature region is that of the dissolved solid and $\Delta\epsilon$ in the higher temperature region that of the solid solvent,

Since $\Delta\epsilon$ is derived only from the temperature dependence of ρ and not from ρ itself, and since this temperature dependence is influenced less by experimental conditions, for example, contact resistances,[347] than is ρ itself, it follows that $\Delta\epsilon$ is far simpler and more accurate to determine than the absolute value of ρ.

Experimental Methods

In principle, resistance measurements can be carried out on a block of coal,[100,348–352] on a coal powder under pressure,[149,349–351,353–355] or on pellets of pressed powdered coals.[352,356]

The preparation of blocks presupposes a certain mechanical strength and hence cannot be achieved with coals of all ranks. Lead plates covered with graphite powder[348] or graphite rods[349] may be applied to or pressed on to the sample as electrodes. For bituminous coals and even for anthracites, powders under pressure may be used. The apparatus for such a measurement is shown in Fig. 32.[340]

[339] Krishnan, K. S., and Ganguli, N., *Nature*, **144**, 667 (1939).

[340] Inokuchi, H., *Bull. Chem. Soc. Japan*, **24**, 222–6 (1951).

[341] Akamatu, H., and Inokuchi, H., *J. Chem. Phys.*, **18**, 810–1 (1950).

[342] Akamatu, H., and Inokuchi, H., *Proc. 3rd Conf. on Carbon*, Buffalo, 1957, Pergamon Press, London, 1959, pp. 51–62.

[343] McClure, D. S., *J. Chem. Phys.*, **17**, 905–13 (1949).

[344] Pick, H., and Wissman, W., *Z. Physik*, **138**, 436–40 (1954).

[345] Northrop, D. C., and Simpson, O., *Proc. Phys. Soc. (London)*, **67B**, 892–4 (1954); *Proc. Roy. Soc. (London)*, **A234**, 124–35 (1956).

[346] Inokuchi, H., *Bull. Chem. Soc. Japan*, **27**, 22–7 (1954).

[347] Hirabayashi, H., and Toyoda, H., *Proc. 4th Conf. on Carbon*, Buffalo, 1959, Pergamon Press, London, 1960, pp. 227–39.

[348] Schuyer, J., and van Krevelen, D. W., *Fuel*, **34**, 213–8 (1955).

[349] Aono, T., and Yamauchi, G., *J. Electrochem. Soc. Japan*, **19**, 221–4 (1950).

[350] *Ibid.*, **20**, 114–6 (1951).

[351] Krajewski, L., and Herszderfer, S., *Prace Glównego Inst. Górnictwa, Komun.* No. 78, Katowice (1951), 5 pp.

[352] Pope, M. I., and Gregg, S. J., *Brit. J. Appl. Phys.*, **10**, 507–10 (1959).

[353] Kröger, C., *Brennstoff-Chem.*, **37**, 186–9 (1956).

[354] Agroskin, A. A., *Freiberger Forschungsh.*, **B9**, 286–94 (1957).

[355] Mints, B. V., *Tsvetnye Metal.*, No. 12, 65–76 (1940).

[356] Gregg, S. J., and Pope, M. I., *3rd. Intern. Conf. on Coal Science*, Valkenburg, The Netherlands, 1959.

One electrode E_1, which is grounded, serves as a piston for applying the necessary pressure to the coal powder. The second electrode E_2, so formed that it holds the sample, is completely insulated on a pyrex or teflon block P. By means of a heating mantle H, the sample may be brought to any desired temperature. The whole apparatus may be flushed with nitrogen to avoid insofar as possible the effects of atmospheric oxygen, especially at the higher temperatures.

The moisture content of coals strongly affects the results,[351,356–358] and its disturbing effect is best avoided by starting the series of measurements at higher temperatures, above ~200°C, and then proceed to the lower temperatures, below ~100°C, under dry, oxygen-free nitrogen.[348]

There are several opinions as to the pressures needed to produce reliable results by the powder method. Measurements on condensed aromatic hydrocarbons[340,341] have shown that, at pressures of 80 to 100 kg/cm², the specific resistance is practically independent of pressure. In contradistinction to these results, it has been established, by measurements on pyrolyzed coals,[350] wood charcoal,[350,357] petroleum coke,[350] carbonized pitch,[357] and graphite,[357] that extremely high pressures are needed before the specific resistance of the pressed powder reaches that of the material measured in blocks. Even though small contact resistances between the coal particles may still persist at pressures of 100 kg/cm², these certainly play no important rôle in the determination of $\Delta\epsilon$.[347] The fact that powders of coals of different ranks measured at room temperature at pressures of 80, 100, and 150 kg/cm² showed at first a decrease in ρ with increasing pressure, whereas at the highest measured pressure ρ increased again[359] is difficult to explain.

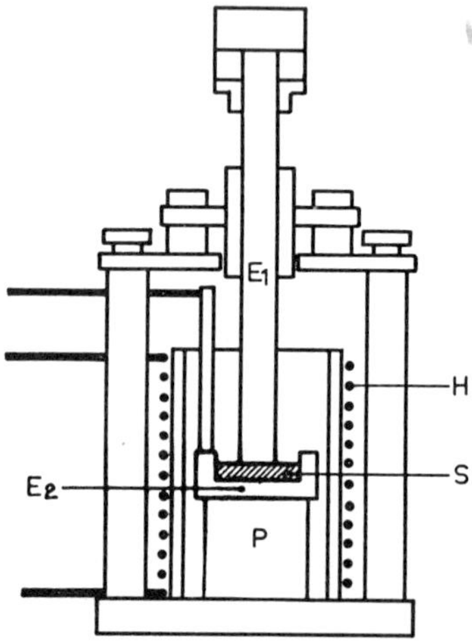

Fig. 32. Diagram for an apparatus used for measuring electrical resistance of powders under different pressures and temperatures.[340] E_1, E_2. Electrodes; H. heating mantle; P. pyrex or teflon block; S. sample.

The particle size when coal powders are measured is also important both for reproducibility and for intercomparison of results.[355,360]

In summarizing all these different effects, it must be kept in mind that the electrical conductivity of any compacted powder will depend on the conductivity of the sample itself, the contact resistance, and the adsorbed liquids and/or gases.[356]

For resistances larger than $10^9\,\Omega$, a fluxmeter—a condenser discharged through the sample—has been used.[361] Ohmmeters, such as standard Wheatstone bridge methods,[349,362] or electrometers[352,356] have also been employed. If the resistance is less than 100 Ω, ammeter-voltmeter methods have been used successfully.[349,358,363] For the sake of completeness, it may be mentioned that the same results are obtained with either d-c or a-c measurements[348,358] and that neither the magnitude of the voltage applied for the measurement[354] nor its frequency[348] affect the results significantly.

Coals and Their Petrographic Components

The electrical conductivity of bituminous coals, anthracites, and pyrolyzed coals undoubtedly permits the conclusion to be drawn

[357] Inokuchi, H., and Tooi, K., *Tanso (Carbons)*, **5**, 7–11 (1955); *Chem. Abs.*, **50**, 12571 (1956).

[358] Bondarenko, S. T., and Yakovleva, A. I., *Izvest. Akad. Nauk S.S.S.R., Otdel Tekh. Nauk*, **1957**, No. 2, 132–6.

[359] Toporetzs, S. A., *Doklady Akad. Nauk S.S.S.R.*, **120**, 629–31 (1958).

[360] Walker, P. L., and Rusinko, F., Jr., *Fuel*, **36**, 43–50 (1957).

[361] Agroskin, A. A., *Khim. i Tekhnol. Topliv i Masel*, **4**, 7–13 (1959).

[362] Mukherjee, B., *J. Sci. Research (New Delhi)*, **13B**, 53–5 (1954); *Chem. Zentr.*, **1955**, 8531.

[363] Akamatu, H., and Inokuchi, H., *J. Chem. Soc. Japan*, Pure Chem. Sec., **70**, 185–9 (1949).

that the electrons in the atomic network have a certain mobility.[43] Whether the electrons move freely in the atomic planes and only require activation energy to pass from one plane to another,[100,340,348,354] or whether, as has been deduced from measurements on mixtures of aromatic hydrocarbons, the conductivity is to be explained as essentially[345] or partly[364] due to "triplet states" or, finally, how large a rôle is played by the free radicals[352,365] known to be present, are all questions which cannot yet be answered. That the foreign elements built into the carbon lattice in coals are responsible for their conductivity[366] is improbable,[367] but the mineral constituents seem to play a certain part.[368,369]

Coals, with the exception of anthracites, were up to a few years ago generally regarded as insulators[148,362] and were thought to conduct only when they were moist,[362] or after having been pyrolyzed at temperatures above 500 to 600°C.[149,362] This conclusion was undoubtedly the result of experimental difficulties, since anthracites at room temperature have specific resistances between 1 and 10^5 Ω-cm[100,240,348,349,355] whereas bituminous coals have corresponding values between 10^{10} and 10^{14} Ω-cm[349,351,354,358,370,371] and are consequently to be classified as semiconductors. It has further been found that fusains have relatively low specific resistances of 10^2 to 10^4 Ω-cm, and vitrains, clarains, and durains have comparatively high values of $>10^8$ Ω-cm.[351] Finally, anthracites show a directional dependence, or anisotropy, in ρ.[348,349]

[364] Pullman, B., *Proc. 3rd. Conf. on Carbon*, Buffalo, 1957, Pergamon Press, London, 1959, pp. 3–20.

[365] Winslow, F. H., Baker, W. O., and Yager, W. A., *J. Am. Chem. Soc.*, **77**, 4751–6 (1955).

[366] Onusaitis, B. A., and Yur'evskaya, N. P., *Bull. acad. sci. U.R.S.S., Classe sci. tech.*, **1946**, 989–1000, 1635–47; *Chem. Abs.*, **41**, 2557, 3278 (1947).

[367] Mrozowski, S., *Phys. Rev.*, **85**, 609–20 (1952); **86**, 1056 (1952).

[368] Toporetzs, S. A., *Doklady Akad. Nauk S.S.S.R.*, **118**, 177–80 (1958).

[369] *Ibid.*, **122**, 286–8 (1958).

[370] Agroskin, A. A., and Petrenko, I. G., *Zavodskaya Lab.*, **14**, 807–12 (1948); *Chem. Abs.*, **44**, 305 (1950).

[371] Agroskin, A. A., and Petrenko, I. G., *Izvest. Akad. Nauk S.S.S.R., Otdel Tekh. Nauk*, **1950**, 89–100; *Chem. Abs.*, **45**, 9242 (1951).

Dependence of Electrical Conductivity on Rank

The specific resistance ρ decreases with increasing rank for vitrains, clarains, and durains whereas it seems to be less dependent on rank for fusains.[359,368,372]

Resistance measurements performed by the authors (E. de Ruiter, R. Leutner, and H. Tschamler) on powdered vitrains of 80 to 90% C at a pressure of 100 kg/cm² under purified nitrogen between 285°C and room temperature showed that at a given temperature ρ decreases with increasing rank and that the plot of log ρ vs $1/T$ consists of two linear segments, hence two different $\Delta\epsilon$ values can be calculated. By analogy with results obtained on solid solutions of aromatic hydrocarbons,[345] it may be assumed that the higher $\Delta\epsilon$ values at 285 to 230°C are the more representative ones for the vitrains. These $\Delta\epsilon$ values decrease from 1.6 to 1.5 ev with increasing rank. The $\Delta\epsilon$ values of anthracites of 92 to 96% C as pressed powders and as blocks,[348] with measurements of ρ between 200 and 100°C, agree well with one another at from 0.5 to 0.2 ev. On the other hand, from measurements of ρ between ~150 and 550°C for an anthracite of 94.0% C a $\Delta\epsilon$ value of 1.0 ev has been calculated.[352]

It has been found empirically, from resistance measurements on condensed aromatic hydrocarbons measured as powders under pressure[340] and on anthracites in block form[348] that a linear relation seems to exist between $\Delta\epsilon$ and the reciprocal of the "aromatic surface area."[348]

Dependence of Electrical Conductivity on Carbonization Temperature

Aside from the fact that brown coals and bituminous coals of high moisture content show an increase of specific resistance on heating to 100 to 200°C as a consequence of the loss of water,[352,354,358,361,370] it is generally true that both specific resistance[149,330,349,350,352,354,355,358,361,362,366,371–377] and electron excitation energy[330,376] decrease sharply with increase in temperature. Nevertheless, the very small

[372] Kröger, C., and Dobmaier, N., *Proc. Symposium on the Nature of Coal*, Central Fuel Research Inst., Jealgora, India, 1960, pp. 267–71.

[373] Sapozhnikov, V. Yu., *Byull. Vsesoyuz. Nauk-Issledovatel. Ugol'nogo Inst.* No. 3, 87–96 (1939); *Chem. Abs.*, **37**, 743 (1943).

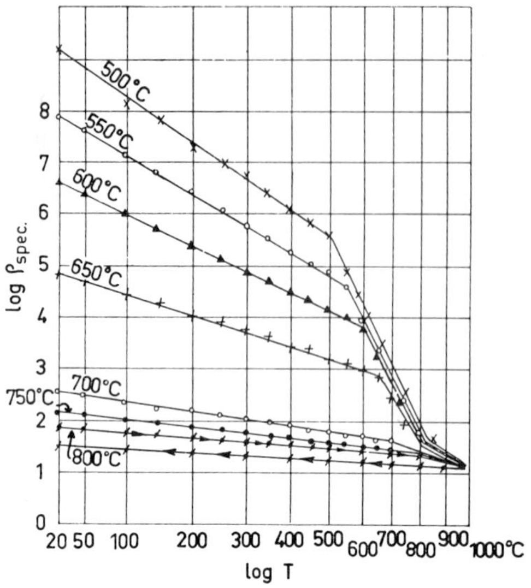

Fig. 33. Dependence of specific resistance on temperature for a series of cokes.[377] The coke samples were prepared from coal "Anna," 28.4% VM by heat treatment for 2 hr each at the temperatures indicated on the individual curves. The reversible effect is shown only for the 800°C sample.

conductivity of bituminous coals increases so much faster with pyrolysis temperature than the much higher initial conductivity of anthracites[349,358,366,371] that after treatment at sufficiently high temperatures ~1000°C, only relatively small differences in conductivity are observed[349,352,366] ($\rho = 10 - 10^{-2}$ Ω-cm).

If the specific resistance is measured at the pyrolysis temperature itself, rather than on a sample cooled to room temperature, it is necessary to distinguish between two overlapping effects: (1) the regular increase in conductivity with temperature to be expected of semiconductors which should be reversible, and (2) a very considerable increase in conductivity starting at a definite temperature depending on rank—300 to 400°C in the bituminous range, 500 to 700°C in anthracites—which is associated with structural changes in the coal.[352,354,361,366] This latter effect is irreversible,[349,354,358,361] and attempts have been made to explain it, especially in the region above 600°C, as a result of increase in the size of the aromatic systems.[352,354,361] It has also been suggested[367] that more "electron holes" appear above 600°C.

With the help of these two effects as well as the transition temperature between the two, it should be possible to deduce the highest temperature to which a coal or a coke has been subjected[377] (Fig. 33[377]). In this manner it was concluded from conductivity data that fusains were produced at temperatures near 700°C.[372]

The dominance of one or the other of these two effects in different samples explains the apparently contradictory result that, for many coals, the decrease of specific resistance with temperature (T) up to about 800°C is given by $\log \rho = a - bT$ ($a = 11.2$ to 12.4, $b = 1.29$ to 1.24×10^{-2})[354,361,370,371] whereas for some anthracites,[373] a noticeable decrease in specific resistance is observed only between 700 and 1000°C.

Recent investigations on a series of vitrains of from 78.5 to 94.0% C showed that the rate of change of ρ with temperature passes through a number of definite maxima and minima which have been related to desorption phenomena and changes in structure.[352]

There appears also to be a correlation between the increase in conductivity and the loss of "volatile matter" during pyrolysis. Thus, in measurements on "Bauer cokes" from exinites, vitrinites, and micrinites of different ranks, it was established that the specific resistance, independently of starting material, depends mainly on the volatile matter of the coke.[353] The specific resistance at room temperature of samples of different origins treated between 600 and 1000°C could be well reproduced by the formula $\rho = AB^V$, where A and B are constants and V is percentage of volatile matter.[378] Finally, it was found that the resistivity of carbon electrodes increases with increasing volatile matter[379] and that the resistivity of cokes decreases with increasing coking time.[362]

It is questionable whether there is a direct

[374] Vosanchuk, S. I., *Nauk Zapiski L'vov Politekh. Inst.*, No. 35, 17–30 (1955); *Chem. Abs.*, **52**, 20990 (1958).

[375] Zielinsky, H., *Koks, Smola, Gaz*, **1**, 128–33 (1956).

[376] Sandor, J., *Proc. Conf. on Ultrafine Structure of Coals and Cokes*, BCURA, London, 1944, pp. 342–9.

[377] Kröger, C., and Dobmaier, N., *Brennstoff-Chem.*, **40**, 1–8 (1959).

[378] Takahashi, H., *J. Chem. Soc. Japan*, Pure Chem. Sect., **71**, 29–32 (1950); *Chem. Abs.*, **45**, 4988 (1951).

[379] Mantell, C. L., *Industrial Carbon*, D. Van Nostrand, New York, 1946, 472 pp.

connection between the decrease of internal surface observed above 600°C and the increase in conductivity.[149]

The observation that the specific resistance of a gas coal fell from 10^9 Ω-cm to 1 Ω-cm upon heating to 400°C in a high-frequency field, whereas bringing it to the same temperature in an oven heated with a Bunsen burner caused a decrease of specific resistance only to 10^6 Ω-cm, is unexplained and must be ascribed to another mechanism.[380]

Coal Extracts

Pyridine extracts and residues of vitrains ranging from 80 to 90% C have been investigated by the authors (E. de Ruiter and H. Tschamler). The resistance measurements were carried out between 285 and 100°C on powders pressed at 100 kg/cm² under purified nitrogen in accordance with the thermal stability of the different samples. For the extracts and residues, the same type of temperature dependence of ρ already described for the vitrains was observed. With increasing rank of the vitrains, the $\Delta\epsilon$ values of the extracts decreased from 3.0 to 1.9 ev. For the residues, $\Delta\epsilon$ varies from 1.3 to 1.5 ev. If the molecules in the pyridine extracts and residues possess only one aromatic cluster, $\Delta\epsilon$ would be a qualitative measure for the aromatic ring size, and therefore the systems in the pyridine extracts would be considerably smaller than in the corresponding residues. From X-ray studies, however, it has been found[23] that the cluster size in both types of sample is of the same order, but the much lower $\Delta\epsilon$ in the residues may be explained by the fact that $\Delta\epsilon$ in molecules with more than one aromatic cluster mainly linked together by direct bonds is lowered.

DIELECTRIC CONSTANT AND DIPOLE MOMENT

In a given electrical condenser, an amount of electricity can be stored which depends not only on the area and separation of the plates but also on the nature of the dielectric material occupying the space between the plates. The ratio of the capacity of a condenser filled with a given dielectric material to the capacity of the same condenser with a vacuum between the plates is called the static dielectric constant ϵ of the material in question. The quantity ϵ, sometimes called the specific inductive capacity, is a measure of the electrostatic polarizability $\bar{\alpha}$ of the dielectric. The quantities ϵ and $\bar{\alpha}$ are related by:

$$\frac{4\pi}{3} N\bar{\alpha} = \frac{\epsilon - 1}{\epsilon + 2}\frac{M}{d} = P$$

which defines the molar polarization P. Here M is the molecular weight, d, the density, and N, the Avogadro number. P may also be written as the sum of terms due to electron, atom, and orientation (dipole) polarizations:

$$P = P_E + P_A + P_O.$$

The first two terms on the right are together referred to as displacement polarization. In the visible region P_E is related to the optical polarizability α and the index of refraction n by

$$P_E = \frac{4\pi}{3} N\alpha = \frac{n^2 - 1}{n^2 + 2}\frac{M}{d} = R$$

where R is the molar refraction already defined.

Since P_A is usually small in comparison to P_E, R can therefore be taken as a good approximation of the displacement polarization $(P_E + P_A)$. In organic molecules having no permanent dipole, $P_O = 0$ and $P = R$, and, assuming P_A to be small, $n^2 = \epsilon$ (the Maxwell relation). Hence, whenever $n^2 < \epsilon$, the presence of a permanent dipole is indicated.

The Debye relation between the orientation polarization P_O and the dipole moment μ is

$$P_O = \frac{4\pi N\mu^2}{9kT}$$

and hence, μ may be calculated either from

$$\mu^2 = \frac{9kT}{4\pi N}\frac{M}{d}\left(\frac{\epsilon - 1}{\epsilon + 2} - \frac{n^2 - 1}{n^2 + 2}\right)$$

or from the corresponding simplified expression

$$\mu = A\left[(\epsilon - \epsilon_\infty)\frac{M}{d}\right]^{1/2}$$

where A is a constant derived from the Debye and Onsager relations[381,382] and ϵ_∞ is the

[380] Sevast'yanov, Yu. L., Dolgopolov, N. N., Bur'yan, Yu. L., and Margolin, V. S., *Doklady Akad. Nauk S.S.S.R.*, **74,** 755–7 (1950).

[381] Partington, J. R., *An Advanced Treatise on Physical Chemistry*, Longmans Green, London, New York, Toronto, 1953, Vol. V, 552 pp.

[382] Fröhlich, H., *Theory of Dielectrics*, Clarendon Press, 1958, 190 pp.

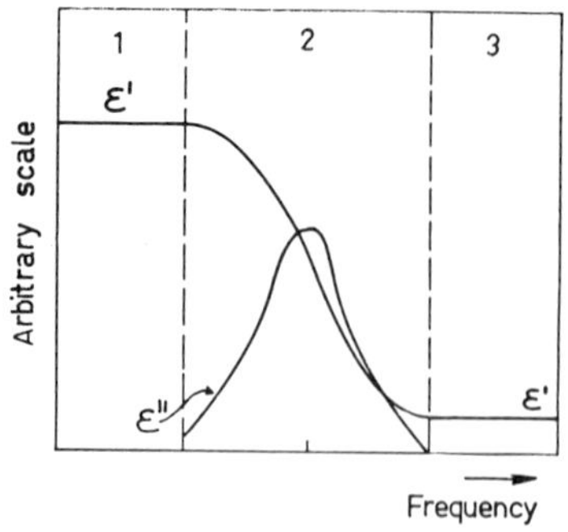

Fig. 34. Dependence on frequency of ϵ' (real part) and ϵ'' (imaginary part or loss factor) of the complex dielectric constant ϵ^*.

dielectric constant at sufficiently high frequencies—of the order 10^9 cycles/sec—that it is approximately equal to n^2.

In an alternating field, displacement and orientation polarization play the same role as in static fields. However, since orientation polarization in particular requires a certain time to adjust itself, it is quite understandable that as the frequency increases there is a progressively greater phase-angle lag behind that of the applied field. Under these circumstances the static dielectric constant ϵ must be replaced by a complex dielectric constant $\epsilon^* = \epsilon' - \epsilon''$. The real part ϵ' satisfies the same theoretical relations as does ϵ. The imaginary part ϵ'' is also called the loss factor or dielectric absorption coefficient, and measures the energy losses in the dielectric.

For organic molecules with permanent dipole moments which are free to rotate, both ϵ' and ϵ'' show a characteristic dependence on frequency which can be divided into three separate regions (see Fig. 34). In frequency region 1, which extends to perhaps 10^3 cycles/sec, both the displacement and the orientation polarization can follow the field and ϵ' is substantially identical with ϵ; ϵ'' is generally so small as to be practically unmeasurable. In frequency region 2, known as the region of dielectric dispersion and extending to perhaps 10^6 cycles/sec, the orientation of the dipoles can no longer keep up with the oscillating field and begin to show phase lags and hence energy absorption; ϵ'' begins to increase whereas ϵ' begins to decrease. As the frequency continues to rise, more and more dipoles fail to follow the field and consequently ϵ'' begins to fall; the energy losses in the dielectric again decrease. The position of the maximum of ϵ'' is generally characteristic of a given dipole, since it may be restricted in movement in a given case, and consequently, in a heterogeneous material like coal where a distribution of different dipoles is doubtless present, a maximum such as that shown in Fig. 34 is scarcely to be expected. In frequency region 3, above about 10^9 cycles/sec, the orientation polarization can no longer make itself felt and only displacement polarization is observed. In this region evidently $\epsilon' \sim n^2$ and ϵ'' is practically zero. At still higher frequencies, other dispersion effects, due, for example, to atomic displacements which can no longer follow the field, may be expected, but these are not of concern for the present discussion.

An almost identical behavior has to be found when fine particles of a semiconductor are embedded in an insulator.[383] For suspensions of coal powder in paraffin, the following equation applies as a first approximation:[384]

$$\left(\frac{\epsilon - 1}{\epsilon + 2}\right)_{\text{suspension}} = \left(\frac{\epsilon_1 - 1}{\epsilon_1 + 2}\right)\phi_1 + \left(\frac{\epsilon_2 - 1}{\epsilon_2 + 2}\right)\phi_2$$

in which ϕ_1 and ϕ_2 are the volume fractions of components 1 and 2. Thus from the experimentally determined value of ϵ for the suspension and a knowledge of the dielectric constant, ϵ_1, of the dispersing medium it is possible to calculate the dielectric constant, ϵ_2, of the suspended coal.

Experimental Methods

In measuring dielectric constants it is necessary to distinguish between the methods

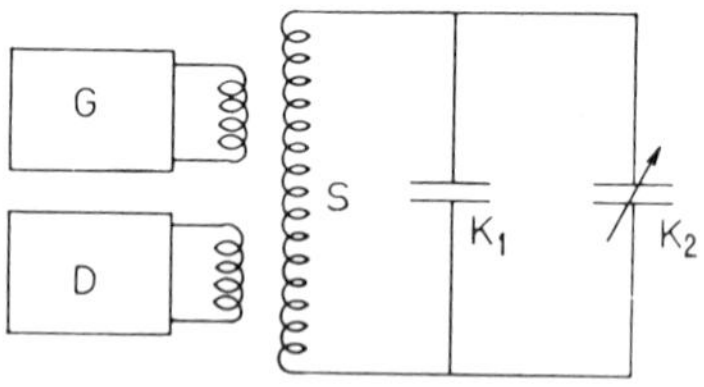

Fig. 35. Schematic diagram of high-frequency circuit for measurements of dielectric constant. G. High-frequency generator; S. induction coil; K_1, K_2. condensers; D. detector.

[383] von Hippel, A. R., *Dielectrics and Waves*, John Wiley and Sons, New York, 1954, 284 pp.

[384] Groenewege, M. P., Schuyer, J., and van Krevelen, D. W., *Fuel*, **34**, 339–44 (1955).

applicable in the high-frequency and in the low-frequency ranges.[385]

The experimental arrangement for high-frequency measurements is shown schematically in Fig. 35. The induction coil S, the parallel arrangement of the measuring condenser K_1, and the variable condenser K_2 form a parallel resonant circuit. The introduction of the sample between the plates of K_1 throws the circuit out of resonance and the amount by which K_2 must be changed to restore this resonant condition is measured. From this value, after suitable calibration, ϵ' can be calculated directly. Most commercially available dielectric constant meters operating in the region of 10^6 cycles/sec utilize this principle although frequently a second oscillator is present to produce a standard frequency.

Another technique frequently employed is based on the use of a commercially available Q-Meter, a device for measuring "Q," the ratio of reactance to resistance of the coil or condenser in a resonant circuit. The Q of the circuit at resonance is measured with and without the sample in the condenser. In this way both ϵ' and ϵ'' can be evaluated.[386]

At low frequencies, the standard methods of measuring the capacity of a condenser by means of an impedance bridge can be applied.

In both high- and low-frequency measurements, cylindrical sample cells are used. The outer surface may be one plate of the condenser and an internal cylinder the other, or one or both electrodes may be in the form of plates.

Preparation of Samples

Measurements on coals have been carried out with disk-shaped blocks of a diameter of 40 mm and a thickness of 5 mm,[386,387] with mulls of coal powders in diisopropyl ether, ethyl acetate, amyl acetate, or pentanol,[384] or the powders may be mixed with molten paraffin and cast into blocks.[388–391] Since water has an extremely large dielectric constant of about 81, the moisture content of a sample can seriously falsify the value of the dielectric constant, so great care must be taken to dry the samples before measurement—actually one method of measuring the moisture content of coals is based on dielectric constant determinations.[385,392] The drying may be carried out at room temperature in a desiccator over P_2O_5, when it takes several months,[386–389] or in a dry, oxygen-free nitrogen stream at 105°C when $1\frac{1}{2}$ hours are required.[384] Drying must always be continued until ϵ no longer decreases.

Coals and Their Petrographic Components

Dependence of Dielectric Constant on Rank

Figure 36[384] presents the values of ϵ and n^2 for a series of dry vitrains of 72 to 92% C measured at room temperature in liquid suspension at 1 megacycle/sec. Three zones may

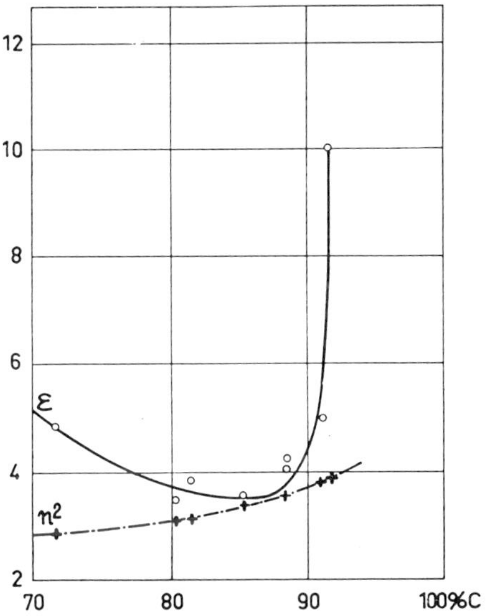

Fig. 36. Dependence of dielectric constant ϵ and the square of the index of refraction n^2 on carbon content.[384]

385 Badzioch, S., and Cornford, G. B., *BCURA Bull.*, **16,** 77–89 (1952).

386 Miyasita, I., and Higasi, K., *Bull. Chem. Soc. Japan*, **30,** 513–7 (1957).

387 Miyasita, I., Miura, Y., and Higasi, K., *Bull. Chem. Soc. Japan*, **28,** 147–8 (1955).

388 Higasi, K., Miyasita, I., and Ozawa, Y., *Bull. Chem. Soc. Japan*, **30,** 546–50 (1957).

389 Miyasita, I., Higasi, K., and Kugo, M., *Bull. Chem. Soc. Japan*, **30,** 550–5 (1957).

390 Miyasita, I., *Bull. Research Inst. Appl. Elec. Japan*, **5,** 123–5 (1953); *Chem. Abs.*, **48,** 11755 (1954).

391 Zyomoto, Y., *J. Fuel Soc. Japan*, **37,** 248–56 (1958).

392 Funasaka, W., Ando, T., and Kamino, Y., *J. Fuel Soc. Japan*, **36,** 69–74 (1957); *BCURA Bull.*, **21,** Abstract 5562 (1957).

be recognized. From 70 to 87% C, ϵ is greater than n^2. The deviation from the Maxwell relation $\epsilon = n^2$ may be explained by the presence of appreciable numbers of polar groups which, like the difference between ϵ and n^2, decrease with increasing rank. In the neighborhood of 87% C, ϵ is approximately equal to n^2. This means either that no polar groups are present or, if they are present, they must be symmetrically arranged on the ring systems and hence compensate one another. Above 87% C, once again ϵ is greater than n^2. This is to be explained by increasing electrical conductivity, a conclusion which is further strengthened by results from chemical studies on polar groups.

Measurements of the ϵ values have been made for a series of (mainly) Japanese coals of 66.6 to 91.3% C on disk-shaped blocks[386,388] and as mulls in paraffin.[388,389,391,393] The ϵ values calculated from the measurements on suspensions were practically identical with those measured on the coals themselves.[388] The values are in good agreement with those shown in Fig. 36 and the same explanation has been given[389,391,394] as that outlined earlier. The minimum of ϵ seems to occur in coals of about 85% C.[391]

Except in brown coals, only a small frequency dependence of both ϵ' and ϵ'' can be observed in the range 300 cycles/sec to 50 megacycles/sec.[386,388–390,393,395] From this, it was concluded that bituminous coals do not possess conducting particles, such as graphite-like structures, embedded in a nonconducting material, such as bitumen.[389,390,393]

In a series of Japanese, English, American, and Indian coals between 88 and 90% C, it was observed that both ϵ' and ϵ'' for the Japanese coals were practically independent of frequency whereas all the other samples showed a strong decrease of both ϵ' and ϵ'' with increasing frequency.[389] This behavior was ascribed to the semiconducting character of the samples. The ϵ' values of the Japanese coals were considerably smaller than those of the other samples. Since higher ϵ' values are probably due to the strong polarizability of the π electrons in larger condensed aromatic systems,[389,396] this would suggest that such aromatic systems are present in the Japanese coals to a distinctly smaller extent.[389]

Dependence of Dielectric Constant on Pyrolysis Temperature

Measurements on powdered cokes in paraffin yield a value of dielectric constant three times larger than that of the original coal which had 25% volatile matter.[390] ϵ'' showed a maximum at about 1 megacycle/sec. This behavior is typical for a semiconductor (coke) embedded in an insulator (paraffin).[388,390] Systematic studies on Japanese, American, and Indian coals showed that ϵ' increases slowly with pyrolysis temperature up to 600°C, but beyond this temperature the increase is rapid.[391] Pyrolyzed samples of the same content of volatile matter but different origin do not have the same dielectric constant.[391]

The pyrolysis products of a Calcutta coal containing 79.4% C, 8.67% ash, and 39.37% volatile matter, were mulled with paraffin and the change of ϵ'' at 1 megacycle/sec was studied as a function of increasing pyrolysis temperature.[397] It could be shown that the loss factor passed through a maximum at a pyrolysis temperature between 600 and 700°C, decreased to a minimum, and then increased sharply again toward still higher temperatures.

Estimation of the Dipole Moment of Coal

The dipole moments of coals have been calculated both from dielectric constant measurements[384,388,394,396,398] and as a sum of the dipole moments of functional groups.[391] Assuming a M/d value of about 300, the dipole moments for coals of 66.6 to 89.3% C were calculated from dielectric constant measurements to lie between 1.8 and 0 Debye units.[384,388,390,394,396] From the functional groups present, values several times larger were calculated.[391] On the other hand, coals of higher rank than 90% C showed higher dipole

[393] Miyasita, I., *Bull. Research Inst. Appl. Elec. Japan*, **7**, 17–26 (1955); *Chem. Abs.*, **50**, 5272 (1956).

[394] Higasi, K., Higasi, Y., and Miyasita, I., *Bull. Chem. Soc. Japan*, **30**, 556–61 (1957).

[395] Miyasita, I., *Bull. Research Inst. Appl. Elec. Japan*, **6**, 117–23 (1954); *Chem. Abs.*, **50**, 5272 (1956).

[396] Higasi, K., *Kagakû (Tokyo)*, **27**, 148, 630 (1957).

[397] Shiromoto, Y., and Higasi, K., *Kagakû (Tokyo)*, **26**, 638 (1956); *Chem. Abs.*, **51**, 5388 (1957).

[398] Chitoku, K., *Oyô Denki Kenkyûjo Ihô*, **10**, 214–25 (1958); *Chem. Abs.*, **53**, 11799 (1959).

moment values from the dielectric constant measurements than those calculated from the functional groups. The discrepancy for coals containing less than 89% C is easy to explain: the dipoles present in coals are not completely free and may, especially when symmetrically arranged partially compensate one another.[384, 391] Hence the value calculated as the sum of the moments of all polar groups present is much too high. Above 90% C, the difference between ϵ and n^2 is no longer determined by the dipoles present but by the strong polarizability of the π electrons of the aromatic clusters,[384,391] and consequently the calculation of the dipole moments from measurements of ϵ no longer has any significance.[391]

SOUND VELOCITY

For nonassociated liquids a relation between sound velocity u in m/sec and density d has been found to be valid:[399]

$$u^{1/3}\frac{M}{d} = u^{1/3}V = A$$

where M is the molecular weight, V the molar volume, and A a quantity which is independent of the temperature and characteristic for the liquid under investigation. Although A does not have the dimensions of a molar volume it is reasonable to assume that the same additivity rules would apply to A as to V. This assumption has proved to be valid since the difference in the values of A between successive members of a homologous series is constant and is independent of the type of compound measured.[399]

In the application of this equation to solid substances, such as high polymers, it was found[400] that it is only valid when values ranging from $u^{1/5}$ to $u^{1/9}$ are taken instead of $u^{1/3}$. The following general relationship was therefore developed in order to eliminate this difficulty:[400]

$$u^{1/3}\frac{M}{d}\left[\frac{1+\sigma}{3(1-\sigma)}\right]^{1/6} = A$$

where σ is the Poisson ratio, which ranges for polyethylene from 0.40 to 0.45. Since $\sigma = 0.5$ for nonassociated liquids, the second equation reduces to the first for such substances.

[399] Rao, M. R., *J. Chem. Phys.*, **9**, 682–5(1941).

[400] Schuyer, J., *Nature*, **181**, 1394–5 (1958); *J. Polymer Sci.*, **36**, 475–83 (1959).

The quantity A can either be calculated from the second equation or derived from the atomic contributions "a" and the structural increments "I":

$$A = \Sigma_i a_i n_i + \Sigma I$$

where n_i is the number of atoms of type i. From different sets of sound velocity measurements,[399,401–404] the following values for the atomic contributions and structural increments have been established:[405] $C = 8.5$; $H = 93.1$; $O_{OH} = 38.3$; $O_{-O-} = 84.7$; $O_{=O} = 181.7$; $N_{-N=} = 159.3$; double bond $_{C=C} = 120.5$. The last mentioned value is valid for both olefinic and aromatic double bonds.

If it is assumed that carbon-carbon bonds in coal are present only as single bonds and aromatic double bonds, that nitrogen is present only in heterocyclic form, and that sulfur may be neglected, the third equation then becomes

$$\begin{aligned} A = {} & 8.5C + 93.1H + 38.1O_{OH} \\ & + 84.7O_{-O-} + 181.7O_{C=O} \\ & + 159.3N_{-N=} + 120.5n_{(C=C)_{ar}}, \end{aligned}$$

where C, H, O, and N are the numbers of atoms of each and $n_{(C=C)_{ar}}$ the number of aromatic double bonds per molecule or structural unit. The quantity $n_{(C=C)_{ar}}$ can be replaced by $C_{ar}/2$ since there are two carbons for each aromatic double bond.

As the average molecular weight of coal is not known, A is determined per gram atom of carbon present in the coal:[405]

$$\begin{aligned} \frac{A}{C} = {} & 8.5 + 93.1\frac{H}{C} + 38.1\frac{O_{OH}}{C} \\ & + \cdots + 60.2\frac{C_{ar}}{C}. \end{aligned}$$

Since C_{ar}/C is by definition the aromaticity "f_a," this structural parameter can be calculated[405] from data obtained by elementary analysis, the chemical determination of oxygen distribution, the measurements of the sound

[401] Lagemann, R. T., and Corry, J. E., *J. Chem. Phys.*, **10**, 759 (1942).

[402] Lagemann, R. T., and Dunbar, W. S., *J. Phys. Chem.*, **49**, 428–36 (1945).

[403] Weissler, A., *J. Am. Chem. Soc.*, **71**, 419–21 (1949).

[404] Natta, G., and Baccaredda, M., *Makromol. Chem.*, **4**, 134–55 (1949).

[405] van Krevelen, D. W., Chermin, H. A. G., and Schuyer, J., *Fuel*, **38**, 483–8 (1959).

velocity and helium density, and a Poisson ratio, $\sigma = 0.345$.[406] The upper limit for the determination of f_a by this method is reached,[405] however, as soon as elastic anisotropy becomes important[406] at carbon contents higher than ~93%.

Several marked advantages exist for using the sound velocity method for determinations of structural parameters, since A is independent of temperature and also independent of the state of aggregation of the substance under investigation.[399,400]

Experimental Method

A pulse technique for sound generation[406,407] has been employed with a frequency of 2 megacycles/sec.[406] Water was used for calibration since both its sound velocity and the temperature dependence of the latter are well known.[408] Measurements were carried out at 20°C on coal samples in the shape of disks, each having a diameter of 2 cm and a thickness of 0.5 cm.[406]

Results on Vitrains

Vitrains of 81.5 to 93.4% C have been investigated[405] and the aromaticities obtained are summarized in Table 8.

TABLE 8

SOUND VELOCITIES u AND AROMATICITIES f_a FOR VITRAINS OF DIFFERENT RANK[405]

%C	$u(m/s)$	f_a
81.5	2318	0.85
85.0	2320	0.85
89.0	2336	0.87
91.2	2342	0.91
92.5	2391	0.95
93.4	2481	0.99

DENSITY

(In collaboration with V. Sandor)

The density of a substance is defined as the weight per unit volume and is generally given in grams per cubic centimeter or grams per milliliter. The specific gravity of a substance is its density referred to density of water at 4°C. The specific volume is by definition the volume in cubic centimeters occupied by 1 g of the material and is therefore the reciprocal of the density. It is clear from the definition of density that very accurate methods for determining both weight and volume are necessary for an accurate density determination. Since weights are relatively easy to measure with high precision, the uncertainty in density determinations is associated with the uncertainty in measuring volumes accurately. The greatest difficulty in volume measurements is encountered with solid porous materials, of which coal is an example. For such materials we must distinguish between true and apparent density.

To determine the true density of a porous substance, the pore structure must be completely filled with a fluid medium, which has no interaction at all with the porous substance, and therefore permits an accurate measurement of the true volume. Thus it is clear that a porous substance can only have one true density.

On the other hand the same porous substance may have several apparent densities, depending on particle size and porosity of the substance as well as on the molecular dimensions of the medium and its degree of interaction with the substance itself. In other words, the actual value obtained for an apparent density will depend on the ratio of total pore volume to volume filled by the fluid medium which is a function of pore diameters in the substance and molecular diameters of the fluid used; on forces of interaction between the fluid and the porous material; and possibly on penetration time of the fluid.

Experimental Methods

Determination of Helium Density

The requirements of a measuring fluid, at least for studies of coal, are very satisfactorily fulfilled by helium since the smallest pores of coal are 5 to 10 A in diameter[409] and the effective diameter of the helium atom is about 1.78 A. Furthermore, helium is not adsorbed on the coal surface, or at least not to an extent

[406] Schuyer, J., Dijkstra, H., and van Krevelen, D. W., *Fuel*, **33**, 409–18 (1954).

[407] Terry, N. B., *Fuel*, **38**, 125–46 (1959).

[408] Smith, A. H., and Lawson, A. W., *J. Chem. Phys.*, **22**, 351–9 (1954).

[409] van Krevelen, D. W., *Brennstoff-Chem.*, **35**, 257–62 (1954).

that would influence the determination of the density.[410,411] The objection has been raised that helium may penetrate into intermolecular spaces which are no longer to be characterized as true pores.[412]

Particle size has no influence on the density determinations in coal using helium since the coal structure seems to have no closed-off pores[413] which are not accessible to the helium atoms. This has been established by a series of density determinations on a vitrain of 89.5% C over a range of temperatures from 77.3 to 373°K. It was found that a linear relation existed between specific volume and the temperature of measurement.[414] Hence even at very low temperatures, and therefore independently of temperature, all pores are accessible.

There are two methods for helium density determination; by direct pressure measurement[411,413,415–419] and by indirect volume determination.[410,420,421] In direct pressure measurement the dead volume of the apparatus is first determined, and then a weighed sample and a quantity of helium, the pressure and volume of which are known, are introduced. The change in pressure at constant temperature is measured; the volume can then be calculated from the gas laws. The method of indirect volume determination differs from the first method in that the pressure decrease in the presence of the sample is not measured but compensated by introducing mercury. The quantity needed for this purpose can be very accurately determined and its volume equals that of the sample. A simpler apparatus for measuring helium density has been developed in which under constant pressure the volume of the helium displaced by the sample is measured directly.[422]

In the helium displacement method, either particles larger than 0.1 mm or pellets pressed from coal powder are used. For the volume determination, it is necessary to free the surface and pores from adsorbed water and other gases. This liberation is accomplished by heating for 15 to 18 hr at temperatures of 90 to 120°C and under a vacuum of at least 10^{-4} mm.[413,423] For cokes and active carbons, temperatures of 300°C are used.[410]

Determination of Apparent Density

For determining apparent density, the pycnometer method,[411–413,418,424–432] the Mohr-Westphal balance method,[433] or the flotation method[430,434,435] may be used. Of these, the pycnometer method has been most used.

It has already been mentioned that the apparent density depends on particle size and the measuring medium and its penetration time. Too small particle sizes are particularly undesirable in the pycnometer method since they may float on the measuring liquid.

[410] Howard, H. C., and Hulett, G. A., *J. Phys. Chem.*, **28**, 1082–95 (1924).

[411] Bond, R. L., and Spencer, D. H. T., *Conf. on Industrial Carbon and Graphite*, London, 1957, Soc. Chem. Ind., London, 1958, pp. 231–53.

[412] Kröger, C., and Bakenecker, J., *Brennstoff-Chem.*, **38**, 82–7 (1957).

[413] Franklin, R. E., *Trans. Faraday Soc.*, **45**, 274–86 (1949).

[414] Zwietering, P., and van Krevelen, D. W., *Fuel*, **33**, 331–7 (1954).

[415] Smith, R. C., Jr., and Howard, H. C., *Ind. Eng. Chem.*, **34**, 438–41 (1942).

[416] Schumb, W. C., and Rittner, E. S., *J. Am. Chem. Soc.*, **65**, 1692–5 (1943).

[417] Krieger, K. A., *Ind. Eng. Chem., Anal. Ed.*, **16**, 398–9 (1944); **17**, 532 (1945).

[418] Agrawal, P. L., *Proc. Symposium on the Nature of Coal*, Central Fuel Research Inst., Jealgora, India, 1960, pp. 121–8.

[419] Smith, J. W., *Chem. & Ind.* **1958**, 885–6.

[420] Rossman, R. P., and Smith, W. R., *Ind. Eng. Chem.*, **35**, 972–6 (1943).

[421] Waterman, H. I., and Wolfs, P. M. J., *Appl. Sci. Research*, **6A**, 372–84 (1957).

[422] Pruss, W., *Brennstoff-Chem.*, **40**, 392–3 (1958).

[423] Malherbe, P., *Fuel*, **30**, 97–109 (1951).

[424] Wandless, A. M., and Macrae, J. C., *Fuel*, **13**, 4–15 (1934).

[425] Sherlock, E., *Fuel*, **29**, 245–52 (1950).

[426] Dulhunty, J. A., and Penrose, R. E., *Fuel*, **30**, 109–13 (1951).

[427] Mertins, A. G., *Braunkohle*, **4**, 353–66 (1952).

[428] Heredy, L., Neuberger, V., and Rona, V., *Földtani Int. Evkönyve*, **45**, 259–73 (1956).

[429] Durka, M., *Przegląd Górniczy*, **14**, 288–91 (1958); *Chem. Abs.*, **53**, 2577 (1959).

[430] Whitaker, J. W., *Fuel*, **29**, 33–41 (1950).

[431] Akamatu, H., and Nakagawa, T., *Tanso (Carbons)*, **5**, 59–64 (1956); *Chem. Abs.*, **51**, 6122 (1957).

[432] Weltner, M., and Borbely, B., *Nehezvegyip. Kut. Int. Közl. Veszprem.*, **1**, 229–32 (1959).

[433] Parks, B. C., *Econ. Geol.*, **44**, 376–424 (1949).

[434] Hutchison, C. A., and Johnston, H. L., *J. Am. Chem. Soc.*, **62**, 3165–8 (1940); *J. Chem. Phys.*, **8**, 869–77 (1940).

[435] Bernal, J. D., and Crowfoot, D., *Nature*, **134**, 809–10 (1934).

Furthermore, if the particles are very small there is always the possibility that, because of the large surface area, the adsorbed air may not be completely displaced. Apparently particles of 0.1 to 0.3 mm diameter are best.[428,436,437] The measuring fluid used should have no interaction with the sample because of surface effects, swelling, etc., should have the lowest possible surface tension to insure good wetting, and, insofar as possible, should penetrate the different sized pores. Since the molecules frequently used—water, methanol, benzene, carbon tetrachloride, etc.—have widely differing sizes, they fill the pore system to different extents, and hence one and the same coal will usually show different apparent densities, depending on the measuring liquid selected. Furthermore, the time required for penetrating the pores is different with different fluids, and the penetration times required to reach a constant density value may vary widely.[411,437]

In the case of brown coals, only apparent densities can be measured because the drying conditions necessary to remove all adsorbed water may cause structural changes.[426]

Calculation of Helium Density from Methanol Density

The apparent density measured with methanol is always larger than the helium density (see Fig. 37), which has been explained as a volume contraction due to an adsorption effect in the smallest pores.[413] It has further been found that this contraction is proportional to the heat of wetting of the sample by methanol—a contraction of 0.0026 cm^3 corresponds to 1 cal of heat of wetting.[413] Hence attempts have been made to calculate the helium density from that measured in methanol by using the following equation:[413]

$$\Delta V = V_{He} - V_{CH_3OH} = \frac{1}{d_{He}} - \frac{1}{d_{CH_3OH}} = 0.0026 Q_{CH_3OH}$$

$$d_{He(calc)} = \frac{d_{CH_3OH}}{0.0026 Q_{CH_3OH}\, d_{CH_3OH} + 1}$$

where $d_{He(calc)}$ is the calculated helium density, d_{CH_3OH} is the density of the sample measured with CH_3OH, and Q_{CH_3OH} is the heat of wetting of the sample by CH_3OH, measured in cal/g coal.

[436] Thibaut, Ch. G., and Vigneron, P., *BCURA Bull.*, **24**, 64 (1960).

[437] Franklin, R. E., *Trans. Faraday Soc.*, **45**, 668–82 (1949).

Density Correction for Mineral Matter

All coals and their petrographic components contain mineral constituents which, because of their usually high density, lead to incorrect values for the density of the organic part of coal. Although it is chemically possible to remove the mineral matter quantitatively before measuring the density, there is always the chance that structural changes may have occurred.

When the coal has a high mineral content, the minerals present can frequently be identified by X-ray analysis of the coal sample itself and of the coal ash. With the known density values of the specific minerals present, an accurate correction for mineral content can be made.[438–440] When less mineral matter is present, the correction is approximated from the ash and sulfur content[441] or from ash, sulfur, carbon dioxide, and chlorine contents.[442] Attempts have been made to derive the mineral matter composition from the ash analysis.[443]

Since it has been found that the density of the mineral matter present in coal differs little from that of the ash itself, the corrected density value of a coal, d_{corr}, with ash content $A\%$ and ash density d_a may be calculated from the experimental density, d_{exp}, by the formula:[424]

$$d_{corr} = \frac{(d_a)(d_{exp})(100 - A)}{100(d_a) - (d_{exp})A}$$

Where the ash density has not been determined, average ash densities of 2.7 or 3 have been used.[167,444]

[438] Nagelschmidt, G., *Spec. Rep. Series*, Med. Res. Council, London, No. 244, Sect. D (1943).

[439] Hicks, D., *Spec. Rep. Series*, Med. Res. Council, London, No. 244, Sect. F (1943).

[440] Nelson, J. B., *BCURA Bull.*, **17**, 41–55 (1953).

[441] Parr, S. W., and Wheeler, W. F., *Univ. Illinois Eng. Exp. Sta., Bull.*, **37**, (1909), 68 pp.

[442] King, J. G., Maries, M. B., and Crossley, H. E., *J. Soc. Chem. Ind. (London)*, **55**, 277–81 (1936).

[443] Schuster, F., *Brennstoff-Chem.*, **32**, 366–8 (1951).

[444] Turner, H. G., *Trans. Am. Inst. Mining Met. Engrs.*, **108**, 330–43 (1934).

Coals and Their Petrographic Components

Change of Density with Rank

The helium density of bright coals has been measured over a wide range of rank.[413,418,445] As seen from Fig. 37,[413] the helium densities at first decrease with increasing rank, pass through a flat minimum at about 85% C, and then increase again above 90% C to reach the value of 2.25 for graphite (not shown in Fig. 37).

The methanol densities of the same bright coals are always greater than the helium densities (see Fig. 37[413]). This effect has been explained, first as a contraction of CH_3OH in the pores,[413] because the density of the first adsorbed layer of CH_3OH is 20% greater than the density of CH_3OH itself,[446] and second as an interaction between CH_3OH and coal whereby complexes are formed which take up less space than the sum of the volumes of the coal and the CH_3OH.[167]

The water densities of the bright coals up to 85% C lie generally above the helium densities[413,426] (see Fig. 37). In coals of 80 to 84% C, the high densities obtained with water are explained by a contraction similar to that observed with methanol[446,447] or as a result of a relatively high concentration of polar groups in these high-oxygen coals.[167] The fact that beyond 85% C the water density is always less than the helium density may be related to the strongly hydrophobic character of these coals.[167,413]

Finally it has been found that mercury densities of bright coals of 82.3 to 93.2% C are of the order 1.22 to 1.30 and thus the lowest of the different apparent densities.[418]

Although helium and water densities are independent of the penetration time throughout the whole range of carbon content, this is not true for the densities determined with methanol, *n*-hexane, and benzene.[411,413] With methanol, the measured density slowly increases for about 24 hr when it reaches a stationary value, which is independent of particle size. With *n*-hexane and benzene it

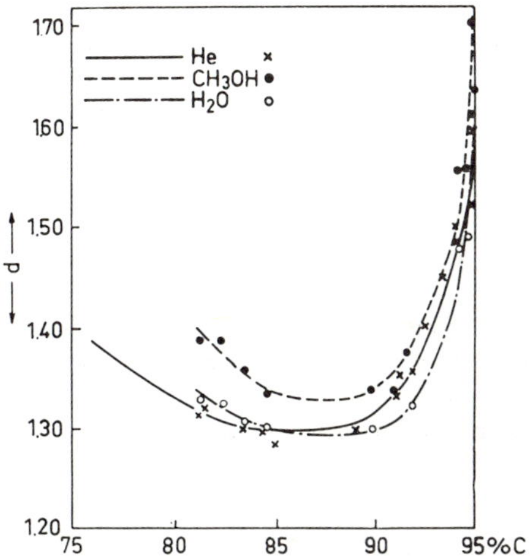

Fig. 37. Dependence of helium and apparent densities of bright coals on carbon content.[413]

could be shown that in coals of less than 89 or more than 93% C there is a strong dependence on penetration time; the increase in density measured after 24 hr amounted to as much as 2.2% relative to the value obtained immediately. With coals between 89 and 93% C, no significant effect of this sort could be found.[413] These results can be explained by the relatively poor penetrating power of these media which possess comparatively large molecular dimensions.[167,413] The minimal accessibility between 89 and 93% C has been said to be associated with the micropores which become smallest in this region.[413]

Apparent densities measured in dibutylphthalate are independent of penetration time although its molecular dimensions are larger than the pores in coal. Measurements in carbon tetrachloride have shown the longest penetration time of all fluids so far investigated.[411]

The helium densities of vitrains over the rank from 70 to 96% C show almost the same values and give a curve of the same shape as those of the bright coals.[73,145,448] The helium densities of vitrains, clarodurains, and fusains of different rank[145] show that for a given carbon content of the different petrographic components, vitrains and clarodurains have about the same densities whereas fusains lie somewhat higher.

[445] Alpern, B., *Rev. ind. minérale*, **38**, 170–81 (1956).

[446] Danforth, J. D., and De Vries, T., *J. Am. Chem. Soc.*, **61**, 873–6 (1939).

[447] Stamm, A. J., and Hansen, L. A., *J. Phys. Chem.*, **41**, 1007–16 (1937).

[448] van Krevelen, D. W., and Chermin, H. A. G., *Fuel*, **33**, 79–87 (1954).

From H_2O density measurements of durains, clarains, vitrains, and fusains of different rank, it was concluded that for a given rank black durains have the lowest density, clarains are somewhat less dense, and grey durains more dense than vitrains, whereas fusains have the highest densities.[424]

By starting with clarodurains, clarains, and durains, a series of fractions of enriched concentrations of different macerals were obtained by flotation procedures in $ZnCl_2$ solutions. The helium densities were measured, and from the known maceral concentrations as determined by petrography and hydrogen content

TABLE 9

METHANOL DENSITIES AND CALCULATED HELIUM DENSITIES OF MACERALS OF DIFFERENT RANK[412]

Macerals	% C	d_{CH_3OH}	$d_{He(calc)}$
Exinites	85.49	1.201	1.187
	87.41	1.213	1.193
	89.10	1.288	1.267
	89.29	1.347	1.325
Vitrinites	83.5	1.345	1.304
	85.74	1.334	1.304
	88.36	1.317	1.295
	88.84	1.368	1.338
Micrinites	86.77	1.463	1.435
	87.98	1.415	1.386
	89.59	1.414	1.389
	89.78	1.413	1.385

it was possible to extrapolate to the helium densities of the constituent macerals exinite, vitrinite, micrinite, and fusinite.[145,167] The results for the first three are given in Table 3. The separated fusinite with 94.2% C had a density of 1.511.[167] Helium densities of fusinites of 92 to 95% C seem to be independent of rank and are of the order of 1.51.[449]

The helium densities have been measured[445] for enriched concentrations of exinites of 78.9 to 82.2% C, vitrinites of 79.2 to 88.5% C, and micrinites of 85 to 90.1% C separated from French coals by flotation methods and showed the same general tendency.

[449] Huntjens, F. J., Dormans, H. N. M., and van Krevelen, D. W., *Brennstoff-Chem.*, **39**, S 67–70 (1958).

The helium densities[7,450] of vitrinites of 84 to 95% C and the four major petrographic components—exinite, vitrinite, micrinite, and fusinite—separated from a single dull coal are comparable with those given in Table 3.

The CH_3OH densities of exinites, vitrinites, and micrinites of different rank are presented in Table 9[412] along with the helium densities calculated after the heats of wetting were determined.[412]

The same general trends as those described earlier for helium densities of enriched concentrations of macerals extrapolated to 100% purity[167] are found for helium densities of the pure exinites, vitrinites, and micrinites calculated from their methanol densities.[412]

If the helium densities of exinites, vitrinites, and micrinites of the same rank are compared, it is found that exinites have the lowest densities up to 89% C and micrinites the highest up to 94% C;[167,412,450,451] fusinites always have higher densities than micrinites.[449–451]

Change of Density During Carbonization

Since the densities of technological cokes are often used in practice with other physical properties as a measure of quality,[452–454] the change of density with carbonization temperature has been carefully investigated.

The density values of thermally treated coals depend on the rank of the original coal, the rate of heating, the final temperature reached, and whether the coal was heated in vacuum or in air. Particularly exhaustive studies have been made of the helium and apparent densities of bright coals in the range 82.4 to 94.2% C which were heated in vacuum to temperatures between 300 and 1650°C.[437,455] The changes of density with carbonization temperature occur in steps, as can be seen for one coal in Fig. 38. If changes in helium and apparent densities occur on carbonization up

[450] Ergun, S., Mentser, M., and Howard, H. C., *Fuel*, **38**, 495–9 (1959).

[451] Ziotkowski, J., *Prace Glównego Inst. Górnictwa, Komun (Stalinogród)* **229** (1959), 12 pp.

[452] Mantel, W. A., *Brennstoff-Chem.*, **31**, 161–72 (1950).

[453] Das Gupta, N. N., Das Gupta, A. R., and Choudhury, A. K. R., *Indian Mining J.*, **4**, No. 8, 1–4 (1956); *Chem. Abs.*, **51**, 5388 (1957).

[454] Kijewska, A., *Prace Inst. Hutnic.*, **10**, 124–32 (1958); *Chem. Abs.*, **52**, 20997 (1958).

[455] Franklin, R. E., *Fuel*, **27**, 46–9 (1948).

to 500°C, they are usually small and are dependent on the rank of the original coal.[326,411,418,437] In the temperature range of 500 to 1000°C, and especially above 600°C, both the helium and apparent densities increase sharply independently of the rank of the starting material.[326,411,418,437] The sudden increase is certainly due to the fundamental changes in structure caused by thermal decomposition. In this temperature range, the penetration time is also an important factor which depends on the molecular dimensions of the medium, the carbonization temperature, and the rank of the original coal. Methanol densities often start to decrease from about 700°C onward, this temperature depending strongly on the rank and origin of the starting sample. The decrease in density is related to the decreasing accessibility of the pore structure to methanol with increasing temperature.[326,437] As can be seen in Fig. 38, even helium from 900°C onward may not be able to penetrate into the carbonized coal, an effect which is responsible for the step in the helium density curve.[418,437] The apparent densities of the solid products prepared from coal between 1000 and 1650°C are independent of the measuring medium and increase smoothly with increasing temperature. Whereas no difference between helium and apparent density could be detected in a coal of 83.5% C heated to 1400°C or beyond, in a coal of 89.7% C this effect was observed from 1100°C onward.[437]

The penetration time for six different measuring fluids—methanol, ethanol, benzene, carbon tetrachloride, *n*-hexane, and decalin—and its change with carbonization temperature up to 900°C has been extensively studied on a vitrain of 82.4% C and a durain of 92% C.[411] Although there does not seem to be a clear relation between the apparent density and the observed penetration time for both coals and coals carbonized up to 600°C, that there is any relation indicates a molecular-sieve behavior. From the results with benzene and methanol on the samples carbonized at 900°C, it was concluded that the accessibility to molecules of the measuring fluid decreases strongly up to this temperature.[411] As carbonization temperature increases further to 1100°C, a similar trend has been observed when using helium, neon, hydrogen, argon, or krypton.[411]

With a series of measuring fluids of considerably different but well-known molecular

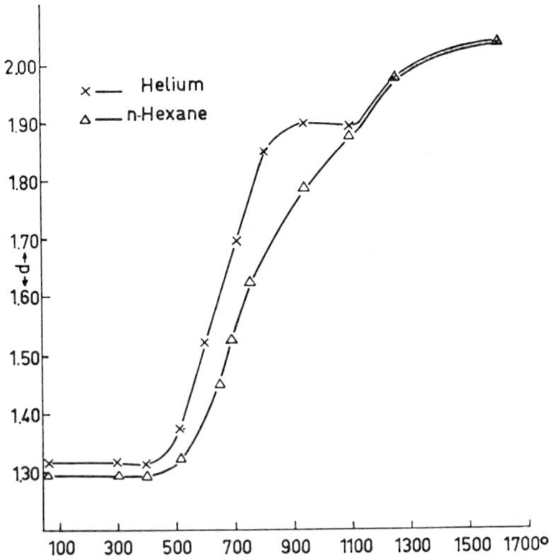

Fig. 38. Dependence of helium and *n*-hexane densities on carbonization temperature. Rank of the bright coal, 89.7% C.[437]

diameters, density measurements have been carried out on coals between 68.1 and 87.5% C and in the temperature range 400 to 900°C.[456] The densities were determined with helium (1.78 A), water (4 A), ethylene glycol (6 A), tri-*o*-cresyl phosphate (10 A), and a silicone oil (about 25 A). The results showed that the cokes from brown coals and bituminous coals have similar pore structures. In both cases, a micro- and a macropore structure were observed. In the cokes from bituminous coals, pores between 25 and 400 A were apparently not present, and in cokes from brown coals those between 40 and 250 A seemed to be absent.

To compare the morphological changes occurring in natural coalification with those during carbonization, the helium and apparent densities of a lignite and a semianthracite and of the cokes obtained from them at temperatures of 500, 600, and 900°C were measured.[457] The results showed that the internal surface of the coals suffered much greater changes during carbonization than during coalification.

Finally, an empirical relation has been established[418] to calculate the helium densities of carbonized products:

$$\frac{1}{d} = \frac{1}{d_0} + \frac{0.443 - \dfrac{1}{d_0}}{100 - C_0}(C - C_0)$$

[456] Cameron, A., and Stacy, W. O., *Australian J. App. Sci.*, **9**, 283–302 (1958).
[457] Ingles, O. G., *Fuel*, **36**, 252–4 (1957).

where d is the helium density of the carbonized sample, d_0 the helium density of the starting material, C_0 and C the percentages of carbon before and after carbonization.

Coal Extracts

The helium density of the pyridine extract of a coal of 82.4% C showed a 2.5% density decrease compared to the original coal.[455] Similarly, the pyridine extract of a vitrinite of 83.9% C showed a decrease of about 1% (1.283 to 1.269); the benzene extract of this pyridine soluble fraction had a helium density 1.138, 10% less than that of the pyridine extract (results of V. Sandor).

Apparent densities in methanol at 30°C of the chloroform soluble and insoluble portions of the pyridine extract of a bituminous coal of 87.8% C were 1.28 and 1.38.[458]

Chemically Treated Coals

Oxidation

A coal of 82.4% C heated at 105°C for 36 days in air showed a helium density 10% larger than the starting material, whereas under the same conditions a coal of 89.7% C increased only 3.5% in density.[455] A coal of 83.6% C having a helium density of 1.298 was oxidized in a fluidized bed at 150°C in air and achieved an 11% increase to 1.479 in 330 hr.[459] Coals of ranks ranging from 80.7 to 90.3% C have been oxidized in air at 170°C for several hundred hours, and it was found that the increase of the helium densities after oxidation was of the order of 25% for the younger coals and 20% for the older ones.[460] Furthermore, it has been observed that a linear relationship exists between the increase in oxygen content and the increase in density.[158]

Acetylation

The helium density at 25°C of a vitrinite of 84.4% C increased from 1.283 to 1.363 after acetylation (results of V. Sandor).

Relation Between Density or Specific Volume and Hydrogen Content

When water densities of dry ash-free vitrains, clarains, fusains, and both black and grey durains were plotted versus the hydrogen content in weight percent, a linear relationship has been observed.[424] This was confirmed for both water and helium densities of coals varying widely in rank.[430,461]

The following linear relationship has been established between the specific volume $1/d$ and the hydrogen content in weight percent:[455]

$$\frac{1}{d} = 0.54 + 0.043\text{H}$$

This equation has been found to be valid for coals of 80.6 to 95.2% C,[418,455] for oxidized coals of 73.2 and 85.4% C,[455] and even a pyridine extract.[455] From water densities of coals of 62.5 to 97.3% C, it was concluded that the validity of the foregoing equation is limited to the range 80 to 95% C.[426] Furthermore, if bright coals alone are considered, there is an indication of an inflection point near 5.2% H.[462]

A similar linear relationship has been found for the helium densities of a progressively oxidized coal, which originally contained 83.6% C.[459] Petrographic components such as exinites, vitrinites, micrinites, and fusinites also follow this relationship,[7,412,450,451] although an inflection point may exist at about 5% H.[412]

A generally accepted explanation for the remarkable linear dependence of the specific volume on the hydrogen content is that only a single common basic mean structure predominates over the wide range of rank from 80 to 95% C.[7,450,455] Extrapolation of the foregoing equation to zero hydrogen content gives a specific volume of 0.54 cm³, or density of 1.85, a value which has been assumed to be the specific volume of "amorphous" carbon.[412,455] However, different physical properties including density when plotted versus carbon content tend to the values for graphite, for example, $d_{\text{helium}} = 2.264$ or $1/d = 0.44$ cm³. This apparent discrepancy may be due to "nongraphite-like" structures in the broad

[458] Sakabe, T., *Fuel*, **35**, 310–17 (1956).

[459] Farenden, P. J., and Spencer, D. H. T., *Fuel*, **34**, 131–2 (1955).

[460] Mazumdar, B. K., Chakrabartty, S. K., Saha, M., Anand, K. S., and Lahiri, A., *Fuel*, **38**, 469–82 (1959).

[461] Sevenster, P. G., *J. S. African Chem. Inst.*, **7**, 41–59 (1954).

[462] Dryden, I. G. C., *Fuel*, **32**, 82–8 (1953).

range of the validity of the previous equation.[7,450,455,463] A proof of the correctness of this idea is the behavior of specific volume versus percent hydrogen of carbonized coals differing in rank.[418,455,463] In this case, two different linear relationships seem to exist, one for coals carbonized at temperatures up to 500 to 600°C which tends to the value of "amorphous" carbon, and another for samples carbonized at higher temperatures which tends to the value for graphite.

An interesting result is found if, for coals and their petrographic components, the specific volume is plotted against atom percent hydrogen. Here also a linear relation completely independent of rank is obtained which tends to $1/d = 0.47$ cm^3, a value much closer to that of graphite than that derived from the plot against weight percent.[7,450]

Structural Parameters Derived from Density

Statistical methods[464]—ring analysis—which have been worked out for the investigation of petroleum fractions have been applied to coals,[73,448,465,466] taking into account the necessary modifications. To use such analyses, it must be assumed that coal is an amorphous material with a structure corresponding to that of a supercooled liquid and that there are no olefinic double bonds and triple bonds in the coal structure.[73,448]

Since the molecular weight of coal cannot be determined, it is necessary to work with a "reduced molecular weight per carbon atom" calculable from the elementary analysis by the relation:[73,448]

$$M_C = 12.01 + 1.008\frac{H}{C} + 16.00\frac{O}{C} + 14.008\frac{N}{C} + 32.06\frac{S}{C} = \frac{12.01}{\%C}$$

where H/C, O/C, etc. are atomic ratios. Two important structural parameters can be derived from M_C and the density d: (1) the "aromaticity," f_a, the ratio of aromatic carbon to total carbon, $f_a = C_{ar}/C$, and (2) the "ring condensation index," $2(R - 1)/C$, where R is the total number of rings and C the number of carbon atoms per molecule. These two parameters are related by the formula,

$$f_a = 2 - \frac{H}{C} - \frac{2(R-1)}{C}$$

an equation exactly valid only for pure hydrocarbons.

Two statistical methods have been developed: the densimetric method,[73,448] in which the f_a value is determined from the R/C value, and the graphical densimetric method[465,466] by which the "ring condensation index" is deduced from the "aromaticity."

The first of these methods is based on a comparison of the experimentally determinable reduced molar volume $(M_C/d)_{exp}$ with the calculated reduced molar volume $(M_C/d)_{calc}$, the latter depending on the assumed additivity of atomic volumes.[467] The method leads to the calculation of R/C, the number of rings per carbon atom. Since the equation requires the ring condensation index $2(R - 1)/C$ rather than R/C for the calculation of f_a, it is only possible to calculate this quantity to a first approximation based on the assumption that $R \gg 1$. Because of these inherent uncertainties, this method is rarely used.

The graphical densimetric method was originally based on a M/d versus H/C diagram determined from the values of liquid benzene and graphite, for both of which $f_a = 1$, and the value of the paraffinic CH_2 group, for which $f_a = 0$. The validity of the uniform subdivision of the line connecting the points $f_a = 0$ and $f_a = 1$ has been checked on a great number of known mixed aromatic-aliphatic hydrocarbons.[465] Recent investigations on pitch fractions ($f_a \approx 1$)[26] as well as on diphenanthryl as an undercooled liquid[468] and on 1,3,5-tri-α-naphthylbenzene,[468] which forms a glass on cooling, both with $f_a = 1$, show that they do not lie on the straight line connecting liquid benzene and graphite. As a consequence the original diagram has been modified.[26] The aromaticity of any pure hydrocarbon can therefore be read from the diagram in Fig. 39

[463] Róna, V., *Kohászati Lapok*, **14,** 423–7 (1959).

[464] Vlugter, J. C., Waterman, H. I., and van Westen, H. A., *J. Inst. Petrol. Technologists*, **21,** 661–76, 701–8 (1935).

[465] van Krevelen, D. W., and Chermin, H. A. G., *Fuel*, **33,** 338–47 (1954).

[466] van Krevelen, D. W., *Brennstoff-Chem.*, **35,** 289–98 (1954).

[467] Traube, J., *Ber. deut. chem. Ges.* **28,** 2722–8 (1895).

[468] van Krevelen, D. W., Bigot, J. A., and Fortuin, J. M. N., *Fuel*, **38,** 535–6 (1959).

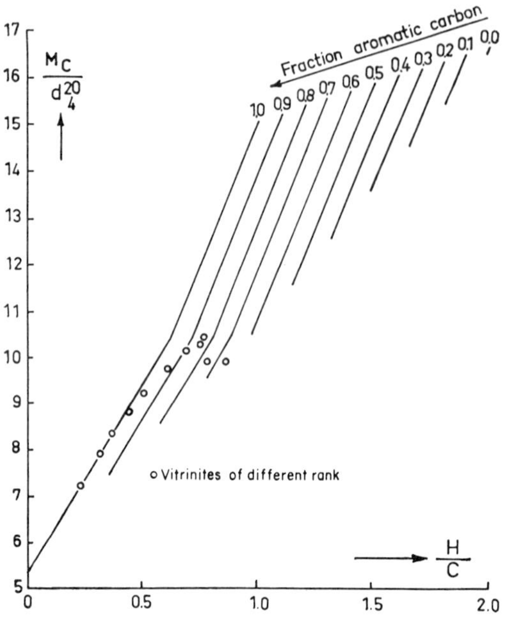

Fig. 39. Diagram for the determination of aromaticity.[26]

when elementary analysis and density are known.

Studies were made to see whether the line for $f_a = 1$ in the diagram of Fig. 39, which has been based on relatively few model substances could not be justified theoretically. If it is assumed that the transition from benzene to graphite takes place essentially by pericondensation, the following relation must exist between M_C/d and H/C:[469]

$$\frac{M_C}{d} = V_C + V_H \frac{H}{C} + \frac{I(H/C)^2}{6}$$

where V_C is the atomic volume of the aromatic carbon (6.6 cm³ in the amorphous[470] and 6.0 cm³ in the crystalline state[467,469]) V_H is the atomic volume of the hydrogen (3.45 cm³),[470,471] and I is the molar volume increment (31.2 cm³ in the amorphous[471] and 25.0 cm³ in the crystalline state[467]). If the values for amorphous carbon and the molar volume increment are used in this equation for the terms V_C and I, a continuous curve is obtained for the limiting line $f_a = 1$. The theoretical curve is in close agreement with the line for $f_a = 1$ in Fig. 39, which is based on limited experimental data.

To apply the graphical densimetric method to coals, "foreign" elements such as O, N, and S must be eliminated by suitable corrections to M_C/d and H/C, that is, the actual organic part of coal must be calculated to a pure hydrocarbon basis. As a first approximation, oxygen is assumed to be present only as OH groups or as ether-oxygen, nitrogen only in heterocyclic rings, and sulfur only in the form of —S— groups.[465,466] Under these assumptions, H/C need not be corrected and $(M_C/d)_{corr}$ is calculated from the $(M_C/d)_{exp}$ by the following formula:[465,466]

$$\left(\frac{M_C}{d}\right)_{corr} = \left(\frac{M_C}{d}\right)_{exp} - \left(8.1\frac{O}{C} + 6.4\frac{N}{C} + 12.5\frac{S}{C}\right)$$

As a result of recent investigations on coal extracts it can be concluded that coal has oxygen not only in OH groups and in "nonreactive" form but also in quinone structures.[472] Therefore, both the M_C/d and the H/C ratios must be corrected. If it is further taken into account that the value 8.1 in the equation may not be correct,[450] the aromaticity values for coals as read from the diagram may be a few percent too high or too low.

The "ring condensation index" is valid for the mean molecule in coal. If it is assumed that coal is a high polymer, R is certainly much larger than 1, so that:[73,448]

$$2 - f_a - \frac{H}{C} \approx \frac{2R}{C}$$

The "ring condensation index" for the mean structural unit in coal can also be derived if the number of links between the structural units building up the molecule are taken into account:[473]

$$2 - f_a - \frac{H_u}{C_u} = \frac{2(R_u - 1)}{C_u} + \frac{B_u}{C_u}$$

where H_u and C_u are the numbers of H and C atoms, R_u is the total number of rings, and B_u is the number of links per mean structural unit.

[469] Róna, V., *Brennstoff-Chem.*, **40,** 346–54 (1959).

[470] Heredy, L., Neuberger, V., and Róna, V., *Acta Chim. Acad. Sci. Hung.*, **12,** 35–56 (1957).

[471] Kurtz, S. S., and Sankin, A., *Ind. Eng. Chem.*, **46,** 2186–91 (1954).

[472] Delavarenne, S., Halleux, A., and Tschamler, H., *3rd Intern. Conf. on Coal Science*, Valkenburg, The Netherlands, 1959.

[473] Dryden, I. G. C., *Fuel*, **37,** 444–68 (1958).

The right-hand side of this equation can also be written[474,475] as $2(R_u - 1/m)/C_u$ where m is the number of structural units in the molecule. The quantity $1/m$ is certainly small compared to R_u and as $H_u/C_u \approx H/C$,

$$2 - f_a - \frac{H}{C} \approx \frac{2R_u}{C_u}$$

If all substituents on the aromatic skeletons in the mean structural unit are replaced by hydrogen, thus forming a pure aromatic hydrocarbon ($f_a = 1$),[473]

$$1 - \frac{H_{ar}}{C_{ar}} = \frac{2(R_{ar} - 1)}{C_{ar}}$$

An attempt has been made to derive the numerical value for the left-hand side of this equation.[473] From master diagrams plotting $2(R_{ar} - 1)/C_{ar}$ vs R_{ar}, C_{ar}, or H_{ar} for both catacondensed and pericondensed aromatic hydrocarbons, the values for R_{ar}, C_{ar}, and H_{ar} per structural unit can be estimated.[473] Results obtained for R_{ar} are summarized in Table 10.

Values of aromaticity f_a and "ring condensation index," $2(R - 1)/C$, have been determined for a series of vitrinites of 70.5 to 96% C[26,167] (see Fig. 39 and Table 2). These vitrinites are all highly aromatic, and the aromaticity increases with rank until $f_a \approx 1$ above 94% C, which can be seen also from Fig. 40.[26,167] The change of the "ring condensation index" with increasing rank is similar to the change of helium density with rank (see Fig. 37), so that the minimum in the density curve might be correlated with a minimal number of rings per carbon atom.[167] That the "ring condensation index" and the helium density both pass through a minimum while the aromaticity is progressively increasing may be explained by the opening of nonaromatic rings in low-rank coals. The sharp increase in "ring condensation index" in high-rank coals corresponds to a rapid growth of the aromatic clusters.[167] A theoretical study based on the additivity rules for the molar volume predicts a minimum value for the density at about 87% C.[462]

TABLE 10

AVERAGE NUMBER OF CONDENSED AROMATIC RINGS PER STRUCTURAL UNIT IN VITRAINS OF DIFFERENT RANK[473]

	R_{ar}	
% C	Peri	Cata
82.5	2.0	2.4
85.0	2.2	3.3
87.5	2.5	—
90.0	3.5	—
92.5	5.9	—

[474] van Krevelen, D. W., *Fuel*, **38,** 245–7 (1959).
[475] van Krevelen, D. W., Wolfs, P. M. J., and Waterman, H. I., *Brennstoff-Chem.*, **40,** 371–7 (1959).

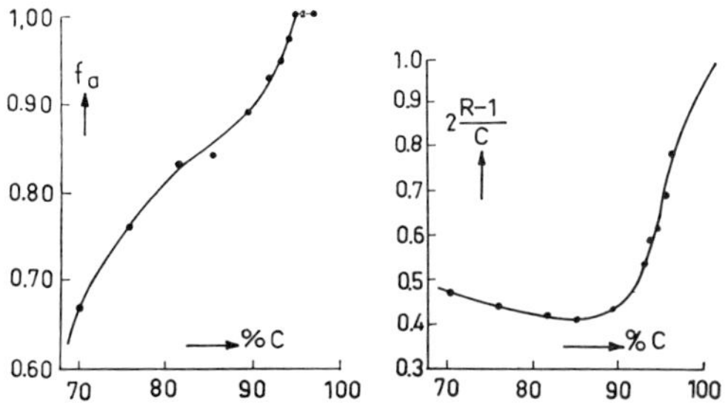

Fig. 40. Changes in aromaticity f_a and "ring condensation index $2(R - 1)/C$ of vitrinites of different rank.[26,167]

Results for exinites, vitrinites, and micrinites of different rank and for a single fusinite are presented in Table 11.[26,167] This table shows that for the petrographic components of the same rank, exinites up to 89% C have distinctly lower f_a values than vitrinites, whereas micrinites have a decidedly higher aromaticity. The fact that differences in physical properties between the different petrographic components decrease with increasing rank and disappear at 89% C for exinites and vitrinites and at 94% C for vitrinites and micrinites is confirmed also by the aromaticity values.

Investigations have been carried out in the laboratory of the authors (V. Sandor and H. Tschamler) on a vitrinite of 83.9% C which was extracted with pyridine and this extract in turn with benzene. Helium densities of the extracts and the residues together with the values of f_a and $2(R - 1)/C$ calculated with the help of the graphical densimetric method are presented in Table 12.

It should be emphasized that the helium density values of the extraction products which are given in Table 12 can be very satisfactorily recombined using the extraction yields to reproduce the density of the starting material. The disproportionation which is clearly to be seen in the data of Table 12 increases with successive extractions.

TABLE 11

VALUES FOR AROMATICITY, f_a, AND "RING CONDENSATION INDEX", $\frac{2(R-1)}{C}$, FOR PETROGRAPHIC COMPONENTS OF DIFFERENT RANK[26,167]

Maceral	% C	% H	f_a	$\frac{2(R-1)}{C}$
E	82.2	7.40	0.62	0.31
V	81.5	5.15	0.83	0.42
M	83.6	3.95	0.90	0.54
E	85.7	6.50	0.75	0.35
V	85.0	5.40	0.84	0.40
M	87.2	4.15	0.92	0.51
E	87.7	5.85	0.84	0.37
V	87.0	5.35	0.86	0.41
M	89.1	4.2	0.93	0.51
E	89.6	5.2	0.89	0.42
V	89.0	5.1	0.89	0.43
M	90.8	4.1	0.94	0.52
E	90.4	4.90	0.91	0.44
V	90.0	4.94	0.91	0.43
M	91.5	3.95	0.95	0.54
E	91.5	4.50	0.93	0.48
V	91.2	4.55	0.93	0.48
M	92.2	3.65	0.96	0.57
F	94.2	3.0	1.0	0.62

E—Exinite; V—Vitrinite; M—Micrinite; F—Fusinite.

TABLE 12

VALUES FOR HELIUM DENSITY, AROMATICITY, AND "RING CONDENSATION INDEX" FOR A VITRINITE OF 83.9% C AND CERTAIN EXTRACTION PRODUCTS

Sample	d_{He}^{25}	f_a	$\frac{2(R-1)}{C}$
Vitrinite	1.283	0.81	0.41
Pyridine-soluble	1.269	0.78	0.38
Pyridine-insoluble	1.290	0.83	0.42
Benzene-soluble	1.138	0.72	0.30
Benzene-insoluble	1.330	0.81	0.43

When a Maurits Seam coal was oxidized, the aromaticity at first increased with oxidation proving that the nonaromatic portion is first attacked and broken down. At the same time, the "ring condensation index" decreased slightly, suggesting an initial attack on the periphery of the aromatic clusters.[158] On the other hand, a slight increase in the aromaticity and a remarkable increase in the "ring condensation index" have been found for coals of 80.7 to 90.3% C oxidized in air at 170°C for several hundred hours.[460]

APPENDIX

Publications Which Have Come to the Attention of the Authors but Are Not Included as References in the Text

X-Ray Diffraction

Davydova, M. A., and Plavnik, G. M., *Izvest. Akad. Nauk S.S.S.R., Otdel. Tekh. Nauk, Met. i Toplivo*, **1959**, No. 5, 185–7. Mazankina, K. T., and Kasatochkin, V. I., *Trudy Inst. Goryuchikh Iskopaemykh, Akad. Nauk S.S.S.R.*, **8,** 104–12(1959). Kessler, M. F., and Vecerikova, V., *Paliva*, **39,** 90–93 (1959). Cartz, L., and Hirsch, P. B., *Phil. Trans. Roy. Soc. London, Ser. A*, **252,** 557–604 (1960). Diamond, R., *Phil. Trans. Roy. Soc. London, Ser. A*, **252,** 193–223 (1960). Ergun, S., Donaldson, W. F., and Breger, I. A., *Fuel*, **39,** 71–7 (1960). Ergun, S., and Wender, I., *J. Appl. Chem. (London)*, **10,** 189–92 (1960). Kessler, M. F., and Vecerikova, V., *Czechoslov. J. Phys.*, **10,** 125–35 (1960). Ali, S. Z., *Fuel*, **39,** 369–76 (1960). Davydova, M. A., and Plavnik, G. M., *Izvest. Akad. Nauk S.S.S.R., Otdel Tekh. Nauk, Met. i Toplivo*, **1960,** No. 6, 150–55. Ergun, S., Mentser, M., and O'Donnell, H. J., *Science*, **132,** 1314–16 (1960). Kröger, C., and Mues, G., *Brennstoff-Chem.*, **42,** 33–40, 77–84 (1961). Ergun, S., and McCartney, J. T., *Science*, **134,** 1620–22 (1961). Brown, H. R., and Taylor, G. H., *Fuel*, **40,** 211–24 (1961).

Measurements in the Ultraviolet and Visible Region

Holubar, V., and Barcolova, L., *Sbornik prác hornickcho ústavu Csav*, **1959,** 162–181. Ergun, S., McCartney, J. T., and Walline, R. E., *Nature*, **187,** 1014–15 (1960). Ergun, S., Donaldson, W. F., and Breger, I. A., *Fuel*, **39,** 71–7 (1960). Ergun, S., and McCartney, J. T.,

Fuel, **39,** 449–54 (1960). Gilbert, L. A., *Fuel*, **39,** 393–400 (1960); *Nature*, **186,** 143–4 (1960). Bent, R., and Ladner, W. R., *Fuel*, **39,** 479–84 (1960). Ritter, H., and Juranek, G., *Brennstoff-Chem.*, **41,** 170–76 (1960). Kötter, K., *Brennstoff-Chem.*, **41,** 263–72 (1960). Friedel, R. A., and Queiser, J. A., *ASTM, Spec. Tech. Publ.* No. 269, 218–26 (1960). de Ruiter, E., and Tschamler, H., *Brennstoff-Chem.*, **42,** 74–7 (1961). Ergun, S., McCartney, J. T., and Walline, R. E., *Fuel*, **40,** 109–17 (1961). Friedel, R. A., *Chem. & Ind. (London)*, **1961,** 462–3. Gilbert, L. A., *Fuel*, **40,** 72–3 (1961). Reggel, L., Raymond, R., Steiner, W. A., Friedel, R. A., and Wender, I., *Fuel*, **40,** 339–56 (1961). Murchison, D. G., and Boult, E. H., *Fuel*, **40,** 389–406 (1961). Brown, H. R., and Taylor, G. H., *Fuel*, **40,** 211–24 (1961).

Infrared Adsorption

Fujii, S., *J. Chem. Soc. Japan*, Ind. Chem. Sect., **62,** 1574–7 (1959). Urbanski, T., Kuczynski, W., Andrzejak, A., Hofman, W., and Witanowski, M., *Bull. acad. polon. sci.*, **8,** 19–22 (1960). Durie, R. A., Lynch, B. M., and Sternhell, S., *Australian J. Chem.*, **13,** 156–68 (1960). Lynch, B. M., and Durie, R. A., *Australian J. Chem.*, **13,** 567–81 (1960). Brown, J. K., and Ladner, W. R., *Fuel*, **39,** 87–96 (1960). Czuchajowski, L., *Fuel*, **39,** 377—85 (1960). Given, P. H., and Peover, M. E., *J. Chem. Soc.*, **1960,** 394–400. Alexanian, C., and Morel, P., *Publ. group. avance. méthodes spectrog.*, **1960,** 211–20. Larina, N. K., and Kasatochkin, V. I., *Trudy Inst. Goryuchikh Iskopaemykh*, **1960,** No. 14, 98–107. Bent, R., and Brown, J. K., *Fuel*, **40,** 47–56 (1961). Czuchajowski, L., *Fuel*, **40,** 361–74 (1961). Oth, J. F. M., de Ruiter, E., and Tschamler, H., *Brennstoff-Chem.*, **42,** 378–80 (1961). Reggel, L.. Raymond, R., Steiner, W. A., Friedel, R. A., and Wender, I., *Fuel*, **40,** 339–56 (1961).

Electron Spin Resonance (Paramagnetic Resonance)

Tikhomirova, N. N., Nikolaeva, I. V., and Voevodskiĭ, V. V., *Zhur. Strukturnoi Khim.*, **1,** 99–102 (1960). Smidt, J., Thesis, Univ. of Delft, 1960, 107 pp. Cerutti, M., Uebersfeld, J., Millett, J., and Parisot, J., *J. chim. phys.*, **57,** 907–11 (1960). Kichigin, D. A., *Ukrain. Fiz. Zhur.*, 6, 255–62 (1961). Berkowitz, N., Cavell, P. A., and Elofson, R. M., *Fuel*, **40,** 279–88 (1961). de Ruiter, E., and Tschamler, H., *Brennstoff-Chem.*, **42,** 290–2, 311–2 (1961). Reggel, L., Raymond, R., Steiner, W. A., Friedel, R. A., and Wender, I., *Fuel*, **40,** 339–56 (1961). Duchesne, J., Depireux, J., and van der Kaa, J. M., *Geochim. et Cosmochim. Acta*, **23,** 209–18 (1961).

Proton Spin Resonance (Nuclear Magnetic Resonance)

Brown, J. K., Ladner, W. R., and Sheppard, N., *Fuel.*, **39,** 79–86 (1960). Brown, J. K., and Ladner, W. R., *Fuel*, **39,** 87–96 (1960). Richards, R. E., and Yorke, R. W., *J. Chem. Soc.*, **1960,** 2489–97. Oth, J. F. M., and Tschamler, H., *Fuel*, **40,** 119–22 (1961). Ladner, W. R., and Stacey, A. E., *Fuel*, **40,** 295–305 (1961). Oth, J. F. M., de Ruiter, E., and Tschamler, H., *Brennstoff-Chem.*, **42,** 378–80 (1961).

Magnetic Susceptibility

Cavell, P. A., and Berkowitz, N., *Fuel*, **39,** 401–12 (1960). Pacault, A., Marchand, A., Bothorel, P., Zanchetta, J., Boy, F., Cherville, J., and Overlin, M., *J. chim. phys.*, **57,** 892–906 (1960). Berkowitz, N., Cavell, P. A., and Elofson, R. M., *Fuel*, **40,** 279–88 (1961).

Electrical Conductivity

Onusaitis, B. A., and Koslova, N. I., *Zavodskaya Lab.*, **24,** 181–3 (1958). Koslovtseva, Z. I., and Makarov, G. N., *Trudy Moskov. Khim. Tekhnol. Inst. im. D. I. Mendeleeva*, **1959,** No. 28, 89–95. Gregg, S. J., and Pope, M. I., *Fuel*, **39,** 301–311 (1960). Sharma, S. K., Sen Gupta, S. K., Chatterjee, N. N., Das Gupta, N. N., and Lahiri, A., *J. Sci. Ind. Res. (India)*, **20D,** 247–51 (1961). Pope, M. I., and Gregg, S. J., *Fuel*, **40,** 123–9 (1961). Brown, H. R., and Taylor, G. H., *Fuel*, **40,** 211–24 (1961). McGeer, J. P., *Proc. 1960 Anthracite Conf., Penn State Univ. Mineral Inds. Expt. Sta. Bull.* No. 75, 106–117 (1961). Toporets, S. A., *Doklady Akad. Nauk S.S.S.R.*, **140,** 451–4 (1961).

Dielectric Constant and Dipole Moment

Sen, D. K., Sen Gupta, S. K., Kini, K. A., and Lahiri, A., *J. Sci. Ind. Res. (India)*, **19B,** 184–6 (1960). Chitoku, K. and Ihigasi, K., *Bull. Chem. Soc. Japan*, **34,** 320–5 (1961).

Sound Velocity

Padmini, P. R. K. L., Rao, K. S., and Rao, B. R., *Trans Faraday Soc.*, **56,** 1404–8 (1960). Martinov, E. G., and Matveev, A. K., *Primenenie Ul'traakustiki k Issledovaniyu Veshchestva,* **1960,** No. 10, 147–56. Gooberman, G., *Nature*, **191,** 693–4 (1961).

Density

Ergun, W. F., Donaldson, W. F., and Breger I. A., *Fuel*, **39,** 71–7 (1960). Ettinger, I. L., and Zhupakhina, E. S., *Fuel*, **39,** 387–92 (1960). Mentser, M., and Ergun, S., *Fuel*, **39,** 509–10 (1960). Bose, B. N., Bondopadhye, J., Das Gupta, N. N., and Lahiri, A., *Brennstoff-Chem.*, **41,** 11–14 (1960). Polansky, T. S., Donald, H. J., and Kinney, C. R., *Nature*, **186,** 792–3 (1960). Maggs, F. A. P., Schwabe, P. H., and Williams, J. H., *Nature*, **186,** 956–8 (1960). Kijewska, A., *Koks, Smola, Gaz*, **5,** 3–9 (1960). Yamaguchi, T., *J. Chem. Soc. Japan*, Ind. Chem. Sect., **63,** 217–21 (1960). Ziólkowski, J., *Prace Glównego Inst. Górnictwa*, **1960,** No. 253, 3–14. Posyl'nyi, V. Ya., *Doklady Akad. Nauk S.S.S.R.*, **137,** 178–81 (1961). Sandor, V., *Brennstoff-Chem.*, **42,** 192–4 (1961).

CHAPTER 3

MECHANICAL PROPERTIES

R. L. BROWN and F. J. HIORNS

British Coal Utilisation Research Association, Leatherhead, Surrey, England.

In the first edition of this book, Yancey and Geer surveyed the work, up until about 1940, on the hardness, friability, and grindability of coal.[1] As is often the case when complex processes—such as involuntary degradation, crushing and grinding, and mining practice—are under investigation, the first attempts to provide a scientific foundation took the form of model tests. Thus, the greater part of the earlier work was concerned with establishing satisfactory ways of measuring friability by drop shatter and tumbler tests and grindability by a variety of methods. A great change has taken place during the past twenty years. In part, this change was made possible by the general advance of science, particularly in respect of the study of mechanical properties of nonhomogeneous materials.[2] Several lines of work have been pursued. First, a determined attack has been made on the direct measurement of the mechanical properties of coal and the relationship of these measurements to notions of physical structure and mining practice.[2] Second, coal and the breakage processes to which it is subjected have been treated as a branch of statistical mechanics, giving a statistical basis for the variation of the strength of coal with its size, and for the size distribution of the products of fragmentation. Third, there has been exploitation of the model tests described by Yancey and Geer. These three lines of investigation are beginning to meet in a most encouraging way. However, more work is needed before all these studies can be evaluated.

The full effects of the measurements of mechanical properties during the last decade on ideas about the structure of coal have yet to be realized. However, it is apparent that mechanical tests will be as important as other tests in understanding coal structure. Even simple experiments have thrown some light on the gross structure. For example, Terry, in experiments on the closure of cracks under load, was able to estimate the extent of flaws in a coal as being of the order of 2 μ/mm.[3] Measurements of internal friction have already thrown some light on the behavior of coking coals.[4] The results of these experiments are all subject to difficulties of interpretation arising from the flawed and heterogenous nature of the structure and the lack of a coherent theory of the behavior of nonmetallic materials. The flaws in coal have demanded that great care be given to the preparation of specimens for elastic and rheological measurements. Flaws must be taken into account in interpreting the results of stress-strain measurements and of elastic moduli found by dynamic methods. The latter have the particular value of determining the internal friction or damping of waves passing through the coal.

Another important group of mechanical properties centers around the strength of coal in tension and in compression. Here interest has been aroused by the strength-size relations disclosed by the experiments. This work, and that on triaxial compression, has led to an increased knowledge of the way in which ease of mining is affected by the orientation of major faults.[5] Further applications can be anticipated in the design of coal-getting machines, minimization of energy consumed in mechanical mining, and the production of desired sizes.

[1] Yancey, H. R., and Geer, M. R., *Chemistry of Coal Utilization*, John Wiley and Sons, New York, 1945, Vol. 1, pp. 145–59.

[2] Walton, W. H., Ed., *Mechanical Properties of Non-Metallic Brittle Materials*, Butterworths, London, 1958, 492 pp.

[3] Terry, N. B., *Fuel*, **38**, 125–46 (1959).

[4] Inouye, K., and Tani, H., *Bull. Chem. Soc. Japan*, **26**, 157–60, 458–63 (1953).

[5] Walton, W. H., Ed., *Op. cit.*, pp. 399–418.

Statistical mechanical studies have attempted to relate strength of piece and size distribution of broken coal to the flaws and weaknesses in the coal. Broadbent and Callcott's proposal to represent the interdependent processes of selection and resistance to degradation by means of matrices has proved valuable.[6] Berenbaum has applied this idea in the study of shatter tests, incorporating also a strength-size relation expressed as a probability of survival of pieces differing in size.[7] The work has placed the drop shatter test on a secure basis, in respect of mechanical properties, and enables the results to be reported in such a way as to permit prediction of the degradation ensuing from transport. As an example of the direct practical use of fundamental studies, this is the most important advance in the last decade.

In contrast, the present position of work on comminution is less satisfactory. A large number of investigations of the comminution of many different types of materials have not yet led to a generally accepted basis for assessing the energy required to pulverize coal. What appears to be needed is a deepening of the ideas of Epstein and Lowry,[8–10] Bennett[11] and Brown,[12,13] and Evans[14] and Pomeroy[15] on the influence of flaws in coal on its breakage and of Carey and Stairmand[16] on the energy associated with the production of natural size distributions, in the light of recent work on the kinetics of grinding.

During the period up to about 1950, much useful work of a more empirical nature than that referred to above was reported. Attention was given to the by-no-means trivial question of measuring and interpreting the sieve analyses of broken coal, leading in the United Kingdom to a system of size grading of coal leaving the colliery[17] and to the further exploitation of the size relations due to Rosin and Rammler[18] and Gaudin.[19] A good picture of the published information is given in a review on fundamental aspects of breakage and structure of coal, methods of particle size analysis, and crushing and grinding in several industries.[20] The bibliography contains several hundred papers on each of a number of topics including fundamental aspects, various methods of size reduction (including milling), and papers dealing with a variety of materials (including some two-hundred papers on coal and coke). Furthermore, there is much relevant material in a second edition of *Micromeritics* by Dallavalle.[21] The most recent of several international meetings covering scientific and practical data relevant to the subject matter of this chapter is that of the American Institution of Mining Engineers early in 1960. Also, several papers on comminution were presented to the Third International Mineral Processing Congress.[22]

ELASTICITY AND RHEOLOGY

Attempts to measure the elasticity and strength of coal,[2] have been made over a number of years, primarily in connection with mining problems. Recent refinements of experimental technique have permitted several studies of rheological properties, using both static and dynamic methods.

[6] Broadbent, S. R., and Callcott, T. G., *Phil. Trans. Roy. Soc., London*, **Ser. A., 249**, 99–123 (1956).

[7] Berenbaum, R., *J. Inst. Fuel*, **34**, 367–74 (1961); **35**, 346–51, 396–9 (1962).

[8] Epstein, B., *J. Appl. Phys.*, **19**, 140–7 (1948).

[9] Lowry, H. H., and Epstein, B., *AIME, Proc. Blast Furnace, Coke Oven and Raw Materials Comm.* **7**, 3–27 (1948).

[10] Epstein, B., *J. Franklin Inst.*, **244**, 471–7 (1947).

[11] Bennett, J. G., *J. Inst. Fuel*, **10**, 22–39 (1936).

[12] Bennett, J. G., and Brown, R. L., *J. Inst. Fuel*, **14**, 135–43 (1941).

[13] Brown, R. L., *Research*, **1**, 93–6 (1947).

[14] Evans, I., *Proc. Conf. on Science in the Use of Coal*, Inst. Fuel, London, 1958, pp. B.24–9.

[15] Walton, W. H., Ed., *Op. cit.*, pp. 5–28.

[16] Carey, W. F., and Stairmand, C. J., *Recent Developments in Mineral Dressing*, Inst. Mining and Metallurgy, London, 1953, pp. 117–36.

[17] Brown, R. L., *Statistical Report on the Size Grading of Coals*, BCORA and BCURA Joint Publication, London, 1947, 44 pp.

[18] Rosin, P., Rammler, E., and Sperling, K., *Wärme*, **56**, 783 (1933); *J. Inst. Fuel*, **7**, 29–36 (1933). Rammler, E., *Braunkohlenarch.*, **58**, 52–7 (1951). Rammler, E., and Glockner, E., *Bergakademie, Die Freiberger Forsch., Bergbau A*, No. 15, 5–21 (1953). Rammler, E., *Freiberger Forschungsh.*, **A50**, 5–47 (1956).

[19] Gaudin, A. M., *Trans. Am. Inst. Mining Met. Engrs.*, **73**, 253–313 (1926).

[20] Bickle, W. H., Ed., *Crushing and Grinding—A Bibliography*, DSIR, H.M.S.O., London, 1958, 425 pp.

[21] Dallavalle, J. M., *Micromeritics*, Second edition, Pitman, New York, 1948, 555 pp.

[22] *Third International Minerals Processing Congress*, Inst. Mining Met., London 1960, 1118 pp.

Preparation of Specimens

The measurement of mechanical properties requires the preparation of suitable specimens, and techniques have been developed which provide for the production of slabs, strips, and cylinders of a range of coals. McKee and Smith described several techniques and noted that when using, for example, thin-slitting wheels of abrasive in rubber, the grit size must be chosen to avoid clogging, while the debris should be removed by an air blast.[23] Wet-cutting or polishing are occasionally used,[24,25] but the mechanical property examined may be affected by the water content of the coal, and hysteresis effects, due to water adsorption, may be expected to operate. It is essential to make slow and shallow cuts to avoid over-heating the coal, and a degree of plastic deformation of the surface may occur. The importance of this depends on the test. For example, a theoretical analysis and comparison of measurements by bending, or by another method, suggested that both the elasticity and strength of the coal specimens used in the bending tests were significantly affected by a surface layer.[26–28]

Stress-Strain Relations

Measurements of the elastic properties of coals are closely concerned with the presence of cracks. This is apparent when specimens are deformed in different ways. Thus, although coal specimens strained by bending may give linear stress-strain relations, compression of cubes both perpendicular and parallel to the bedding plane has shown nonlinearity between stress and strain.[25,29,30] Heywood found a nonlinearity such that the increase in strain was greater over the first than over the second part of the stress range.[29] Further, a specimen compressed perpendicularly to the bedding plane showed a greater initial increase in strain and a higher failure stress than when the direction of compression was parallel to the bedding plane. Other results[31] did not confirm this finding, but a later detailed investigation[28] suggested that Heywood's observations could be explained satisfactorily by assuming that the first effect of compressive stress was to close up cracks in the coal, the initial slope of the stress-strain curve being determined by the severity of the cracks perpendicular to the direction of the compression. Terry[3] then showed conclusively that this was so.

This behavior may be interpreted as elastic anisotropy of the coal substance. There is some evidence for anisotropy due to other causes, but these are not determined, although they are certainly associated with the bedding plane.[28,32] The permeation of coal by pores and cracks affects the measurement of elastic moduli by reducing the amount of load-bearing substance.[28]

Elastic Moduli from Static Tests

Mechanical measurements of static elastic moduli and Poisson's ratio have been attempted.[25,28,29,33–35] Measurements of elastic properties of rocks *in situ* have also been made,[36] and Jones has shown that the velocity of longitudinal waves in porous rocks[37] and in coal cubes[38] is affected by the presence of static stress. Attempts to measure seam stresses by this method do not appear to have been made.

[23] McKee, J. H., and Smith, G. W., *BCURA, Inform. Circ.* No. 35 (1950), 4 pp.

[24] Gaddy, F. L., *Bull. Va. Polytech. Inst.*, **49** (1956), 27 pp.

[25] Macrae, J. C., and Mitchell, A. R., *Fuel*, **36,** 423–41 (1957).

[26] Berenbaum, R., and Brodie, J., *Brit. J. Appl. Phys.*, **10,** 281–6 (1959).

[27] Berenbaum, R., and Brodie, J., *J. Inst. Fuel*, **32,** 320–7 (1959).

[28] Morgans, W. T. A., and Terry, N. B., *Fuel*, **37,** 201–21 (1958).

[29] Heywood, J., *J. Inst. Fuel*, **9,** 94–105 (1935).

[30] Holland, C. T., *Trans. Am. Inst. Mining Met. Engrs.*, **149,** 75–93 (1942).

[31] South Staffs., and Warwick Inst. Mining Engrg., *Trans. Inst. Mining Engrs. (London)*, **105,** 458–503 (1945)–6.

[32] Schuyer, J., Dijkstra, H., and van Krevelen, D. W., *Fuel*, **33,** 409–18 (1954).

[33] Bangham, D. H., and Maggs, F. A. P., *Proc. Conf. on Ultra-fine Structure of Coals and Cokes, BCURA, London*, 1944, pp. 118–30.

[34] Inouye, K., and Tani, H., *Bull. Chem. Soc. Japan*, **26,** 458–63 (1953).

[35] Lawall, C. E., and Holland, C. T., *Trans. Am. Inst. Mining Met. Engrs.*, **101,** 100–16 (1932); *West Va. Univ. Bull.*, No. 17 (1937), 50 pp.

[36] Habib, P., *Sols et Fond*, No. 3 (1950). Brown, P. D., and Robertshaw, J., *Géotechnique*, **3,** 283–6 (1952–3). Evison, F. F., *Géotechnique*, **6,** 118–23 (1956).

[37] Jones, R., *Nature*, **165,** 39–40 (1950).

[38] Jones, R., Private Communication from the Road Research Laboratory, DSIR, (U.K.) (1953).

Measurements of Young's modulus in compression and in bending have been made for a hard coal and for an anthracite, the mean values being about 3.4 and 4.3 $\times 10^{10}$ dyne cm^{-2}, respectively, with the stress parallel to the bedding plane.[28] Bending tests (using center-point loading) on Japanese and American coals gave lower values of the modulus, which may have been due to the existence of gross cracks in the specimens.[4] Especially in a compression test, it is difficult to avoid progressive failure of grossly cracked coals under load. Some coals resist compression until failure takes place explosively, whereas others crumble under the load. It is doubtful whether static measurements of 'elasticity' of this latter type of coal have much meaning.

An alternative method of deriving a static Young's modulus was described by Bangham and Maggs.[33] Measurements of the swelling of coal resulting from the adsorption of methyl alcohol were used to calculate the modulus. The swelling depends on the amount of internal surface available, and the formula for the modulus involves the assumption of a relation between the measured surface and its effect on the modulus; namely,

$$E = 100\rho\epsilon/\lambda \qquad (1)$$

where E is the modulus, ϵ the measured internal surface, ρ the density, and λ a constant relating the observed swelling to the lowering of surface energy. The work was exploratory, but good agreement was obtained for several coals between moduli calculated in this way as well as for those from measurements made in compression. The importance of this method lies in the fact that it would appear to measure the elasticity of the coal substance, rather than that of the bulk coal. However, it involves some assumptions about the distribution of internal surface, and the mechanism of wetting and further work is needed.[33]

Attempts to correlate static moduli with rank of coal have met with little success, at least partly because of the cracked and friable nature of many coals. These attempts have been criticized on the grounds that viscoelastic behavior vitiates the measurements. Published data are few. Huck and Karweil calculated a relation between modulus of elasticity and percentage volatile matter derived from a chemical model of coal constitution and stated that measurements of static and dynamic elasticity by a number of authors for a large range of coals followed the general trend of the curve. There was, however, a very considerable scatter.[39]

Time Dependence

The argument that static measurements of moduli of elasticity are disturbed by time effects was taken up by Morgans and Terry[28] who measured a strain-time curve for a cube of Barnsley Hard coal under compression. The observed behavior (Fig. 1) was that of a viscoelastic solid in which irrecoverable strain represented only about 0.5% of the total strain. In other experiments with samples of clarain at 100°C, a figure of 1% was obtained.[25] Hence, it would not be expected that viscoelastic behavior could influence room-temperature measurements of static moduli. Experiments by Pomeroy[40] with an anthracite did not show any time-dependent strain, either in compression or bending, whereas bending experiments with the Barnsley Hard coal used by Morgans and Terry gave creep and recovery curves which appeared to confirm the existence of small amounts of recoverable and irrecoverable strain. It was also noted that, whereas explosive failure in compression, typical of hard coals, was invariably found with anthracite, very slow compression of the Barnsley coal resulted in slow failure. Pomeroy suggested that this behavior was paralleled in the incidence of bursts in mines, which are found most frequently in anthracite areas. The "viscosity" associated with the reversible creep of the Barnsley coal was estimated to be of the order of 10^{16} poises.[40] Similar results have been obtained by Griggs[41] for limestone rocks. Inouye and Tani have suggested that coals, like glasses and a number of other substances, show static fatigue, so that the breakage strength decreases as the time of load is increased.[42]

Another type of nonrecoverable deformation of coal is that first observed by Boddy.[43] Coal

[39] Huck, G., and Karweil, J., *Brennstoff-Chem.*, **34**, 97–102, 129–34 (1953).

[40] Pomeroy, C. D., *Nature*, **178**, 279–80 (1956).

[41] Griggs, D. T., *J. Geol.*, **47**, 225–51 (1939).

[42] Inouye, K., and Tani, H., *Bull. Chem. Soc. Japan*, **26**, 200–204, 359–63 (1953).

[43] Boddy, R. G. H. B., *Nature*, **151**, 54 (1943); *Fuel*, **22**, 56–63 (1943); *Proc. Conf. on Ultra-fine Structure of Coals and Cokes*, *BCURA*, London, 1944, pp. 336–41.

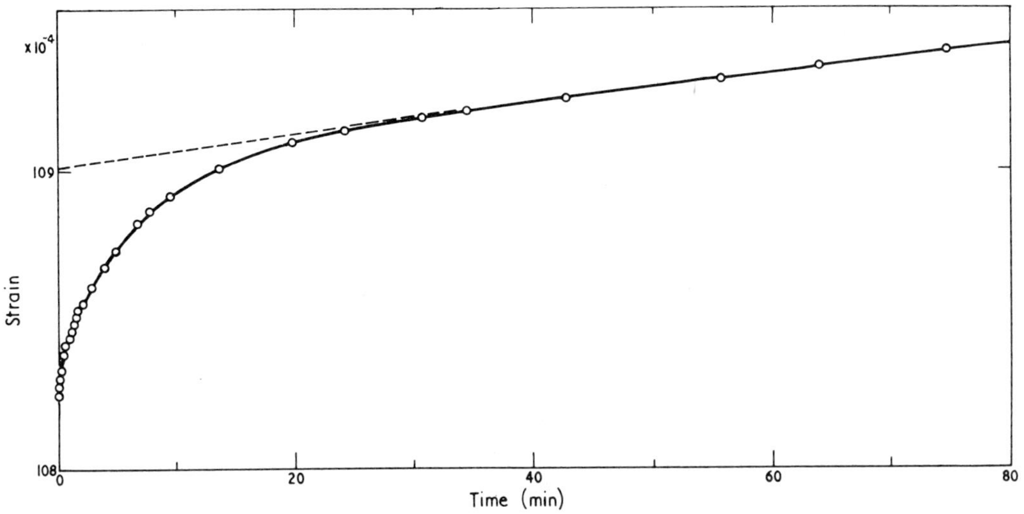

Fig. 1. Strain-time curve for Barnsley hards $1\frac{1}{2}$-in. cube. Load 1.98 ton. Compression perpendicular to the bedding plane. (After Morgans and Terry.[28])

and other particles of a diameter below about 50 μ were compressed under a cover glass while being observed through a microscope. The coal particles were observed to deform plastically to thin translucent sheets, and this behavior was contrasted with that of the so-called "brittle" materials, gypsum and limestone, although later it was shown by Bangham and Berkowitz that particles of a wide range of materials could be made to deform plastically by refinement of this experimental procedure.[44] Similar behavior has been observed in small coal particles in briquettes.[45] Bangham and Berkowitz measured a "critical deforming force" F_c, and related it to the volatile content of the coal.[44] The force F_c was found to have a complex dependence on temperature, indicating irreversible changes. In contrast, F_c for pitch varied linearly with temperature and was independent of temperature for cupric oxide particles. The complex relationships observed with coal particles as the temperature was raised appear to have been due at least in part to adsorption of vapors. Although Bangham and Berkowitz observed similar behavior in particles of a wide range of substances, they considered that the plastic deformation of coal was evidence of shear failure at the boundaries of a micellar structure. It has been argued[25,46] that the effect arises from a rise in temperature resulting from friction, but the work of Bridgman[47] and of King and Tabor[48] suggests that the experimental conditions resulted in a hydrostatic stress condition favorable to plastic flow. If this explanation is correct, the observed deformation does not provide evidence in favor of "plastic" properties of coal, but of the difficulties of obtaining stress conditions favorable to the brittle fracture of small particles.

Elastic Properties from Dynamic Tests

More sensitive measurement of rheological behavior may be made by vibration methods and the measurement of dynamic elastic constants and internal friction have been attempted by several investigators.[3,4,28,32,49–54] Whereas the static moduli are obtained by successive

[44] Bangham, D. H., and Berkowitz, N., *Bituminous Coal Research*, **2,** 139–50 (1945); *Research (London)*, **1,** 86–93 (1947).

[45] Newman, P. C., *Brit. J. Appl. Phys.*, **6,** 348–49 (1955).

[46] Hiorns, F. J., *BCURA Bull.*, **19,** 533–43 (1955).

[47] Bridgman, P. W., *The Physics of High Pressure*, G. Bell & Sons Ltd., London, 1931, 445 pp.

[48] King, R. F., and Tabor, D., *Proc. Roy. Soc. (London)*, **A223,** 225–38 (1954).

[49] Inouye, K., *Fuel*, **34,** 249–50 (1955).

[50] Hiorns, F. J., *Fuel*, **32,** 113–14 (1953).

[51] Inouye, K., *Colloid Sci.*, **6,** 190–210 (1951).

[52] Walton, W. H., Ed., *Mechanical Properties of Non-Metallic Brittle Materials*, Butterworths Scientific Publications, London, 1958, pp. 259–68.

[53] Walton, W. H., Ed., *Op. cit.*, pp. 239–58.

[54] Terry, N. B., *Fuel*, **37,** 309–22 (1958).

measurements of the strain, at a number of levels of stress, in customarily simply defined stress systems, the determination of a dynamic modulus involves either the vibration of a specimen, with measurement of the frequency, or measurement of the velocity of sound in the specimen. The effect of the cracks in the coal differs between static and dynamic tests as well as between the different types of test in the two classes. Thus, in the static tensile test, and in the dynamic measurement by lateral vibration of a strip, the specimen may be so oriented that the bedding plane or cleat cracks are under tension, and the result of the test will be strongly affected by this condition. Morgans and Terry have pointed out that, in the propagation of sound through coal, the transit time is primarily determined by the path taken through the coal substance and is artificially lengthened by the presence of the cracks.[28] Because the elastic modulus and the velocity are connected by a relation of the form:

$$(K/\rho)^{1/2} = l/t \qquad (2)$$

where K = elastic modulus, ρ = density, l = path length, and t = time. Use of a specimen length $l' < l$ instead of l, in equation 2 results in a low value of K. In the measurement of static moduli, a proportion of the strain represents opening or closing of the cracks, again giving a low value for the modulus. Hence, the static modulus is in general lower than that found in a dynamic test.

Apart from the different behavior of cracks in dynamic and static tests, measured dynamic moduli are generally higher than the corresponding static values because of the viscoelastic behavior of the coal.[28,49] The observed modulus depends on the rate of loading, that is the frequency in the vibration test, and may be expected to approach the static value at low frequencies. Measurements of dynamic elastic moduli have been made in the kilocycle region by a magnetostrictive method[28] and at lower frequencies by electrostatic excitation of a fixed-free strip.[52] Hard coals suitable for cutting into small specimens were used in both of these methods. Larger specimens of a wide range of Japanese and American coals were tested by magnetic excitation of flexural vibrations in bar specimens supported at the nodes.[51] The latter experiments were aimed at determining the effect of ash content on the modulus, and complex relationships were claimed to have been found. It was also suggested that the presence of moisture affected the elasticity of the coal samples; complex relations between Young's modulus and water content have been found for a British coal.[52]

The elastic constants of a series of pure vitrains, covering the range of carbon content from 82 to 96%, have been calculated from sound velocity measurements. The method involved the propagation of pulses of longitudinal or transverse waves at 2 megacycles/sec through small samples of the coals immersed in a water column between transducers.[32] The variation with carbon content of the measured elastic constants and the derived Young's modulus indicated that the coals containing up to 92% C were isotropic, while those of higher carbon content became increasingly transversely anisotropic with increasing carbon content. In view of the investigations of Terry,[3] this apparent anisotropy needs further study. Possible effects of water content also appear to have been ignored.[3,52] In addition to Young's modulus (which showed a maximum at 85% C and a minimum at 90% C), the bulk modulus and Poisson's ratio were also calculated.

The coefficient of restitution has been used to distinguish between coals. The variation of coefficient of restitution with impact energy was measured by Cannon and George for a number of coals, using steel balls dropped on to polished coal surfaces.[55] In general, the coefficient varied little with change in impact energy, except for low-rank bright coals, which showed a minimum in coefficient of restitution at about 30 g/cm. It was suggested that this might represent an energy absorbing process connected with the formation of weaknesses in the coal.

Internal Friction

The measurement of internal friction by the conventional method of determining the half-amplitude width of the resonance curve is much more liable to errors than the measurement of the modulus, especially when the resonance is sharp and the small difference of two large frequencies is required. However, the losses in coal are large, so that the ratio (resonance frequency/half-amplitude frequency difference) can be measured with reasonable precision

[55] Cannon, C. G., and George, W. H., *Nature*, **155**, 787–8 (1945).

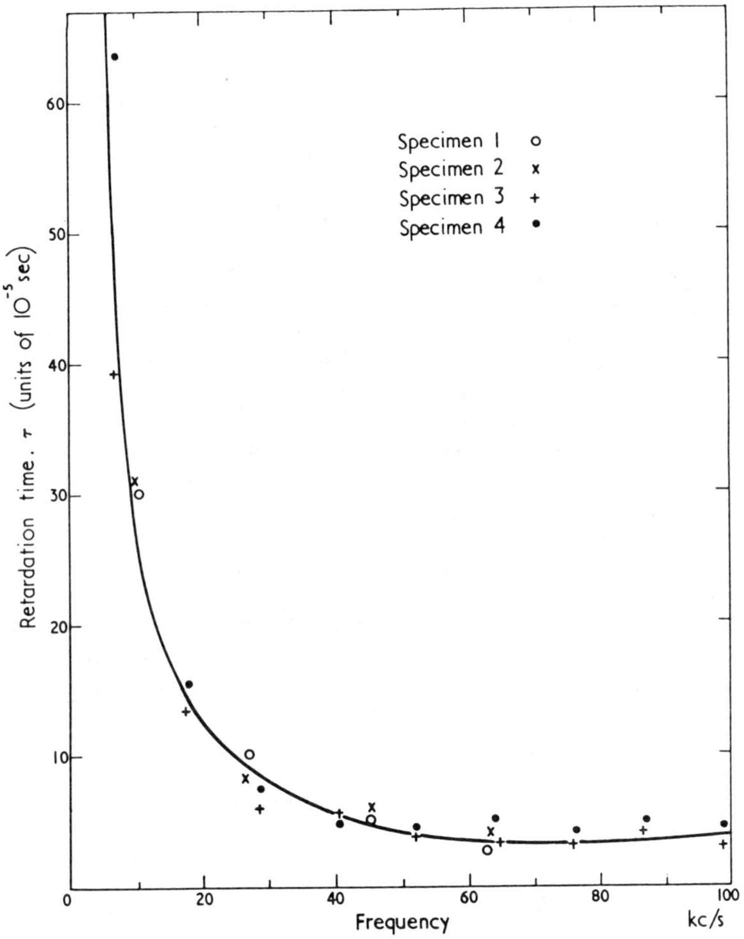

Fig. 2. Variation of retardation time with frequency for Barnsley Hards coal.[54]

over a wide range of frequencies. This ratio is proportional to Q, the mechanical quality factor of the solid. The effects of cracks and flaws are, however, appreciable even for specimens of hard coals. Terry obtained values of Q in the region 5 to 10 kc/sec.[54] For two anthracites. Q ranged from 92 to 130 and from 110 to 270; for Barnsley Hard coal of 34% volatile matter, from 50 to 61; and for a soft friable coal, from 14 to 77. It was concluded that only from specimens similar to the Barnsley coal could useful conclusions be drawn from internal friction measurements. Nevertheless, the values quoted reflect the general physical differences between the three types of coal. Similar values of Q for a hard coal were obtained by Hiorns at much lower frequencies.[52] By using a simple Burgers' model consisting of the series connection of a Maxwell element (that is, spring and dashpot in series) and a Voigt element (that is, spring and dashpot in parallel), Terry was able to calculate an attenuation constant and retardation times for a range of frequencies (obtained by harmonic excitation) for four specimens.[54] The frequency range covered was from about 6–100 kc/sec, and the calculations were based on earlier measurements of the static Young's modulus E_s for the same coal:

$$\text{Retardation time } \tau = \frac{c\left(\dfrac{\rho\omega^2}{E_s} + \dfrac{\alpha^2 - \omega^2}{c^2}\right)}{2\alpha^2} \tag{3}$$

where c = elastic wave velocity, ρ = density, α = attenuation constant, $\omega = 2\pi \times$ (frequency), and E_s is the static modulus.

The variation of retardation time with frequency is shown in Fig. 2.[54] This variation is not consistent with the simple Burgers' model, which contains a single Voigt element, and the single Voigt element was replaced by a more complex model consisting of a chain of Voigt elements in series, characterized by springs $E_1, E_2 \ldots$ and dashpot viscosities $\eta_1, \eta_2 \ldots$. Supporting evidence for this model was obtained from the creep curve referred to earlier.[28,40]

Measurements of internal friction have been made also by Inouye,[51] and Inouye and Tani.[4] In these experiments, a small iron plate was attached to one end of the specimen, which was in the form of a bar clamped at the other end. The oscillation was detected by an electrochemical method. The specimens were cut parallel to the bedding plane, thus minimizing its effect, but were about 2 × 1 cm in cross section, whereas those used by Terry[54] and Hiorns[52] were much smaller and were chosen to be free from visible flaws. Inouye measured the width of the resonance curve and calculated a viscosity η. Five coals were tested, each over a wide range of ash contents, and some evidence was obtained for a linear increase of η with ash content characteristic of each coal. These relations were extrapolated back to give a value η_o, supposed to refer to ash-free coal: η_o decreased as volatile matter increased. η was found to decrease as frequency increased over the range 400 to 1000 c/sec. The frequency was altered by changing the free length of the specimen; it has been pointed out that results using this technique must be interpreted with caution, as changing the dimensions may result in effects arising from the heterogeneity and crack distribution of the specimen.[52]

Inouye and Tani's experiments were aimed at interpretation of changes in caking properties of coals with volatile content, and further comparative experiments were carried out with specimens of the same coals as before, first on the raw coal and then after each of two periods of oxidation for 30 hr at 120°C in air.[4] The extrapolated viscosity η_o was tabulated for three of the five coals tested, before and after oxidation. Two of the coals, both strongly caking, showed increases due to oxidation of about one order in η_o, whereas a noncaking coal was almost unaffected. As a result, the value of η_o for the oxidized caking coal approached that for the raw noncaking coal, in agreement with observed effects of oxidation on caking properties.

Effect of Ambient Conditions on Elasticity and Rheology

It would be expected that ambient conditions would have some effect on the mechanical properties of coal. On the effect of water content, Inouye and Tani have, as already noted,[4] found some irregular and generally minor effects of relative humidity on the dynamic modulus. Hiorns[52] found considerable changes in the dynamic modulus with changes in temperature over the range 20 to 100°C and ascribed these changes to changes in water content. Terry and Morgans[53] stated, however, that the dynamic modulus of a similar coal changed by only 1% from the air-dry to the (desiccated) dry condition, and that small changes of temperature had little effect. The discrepancy between these results may arise from differences in experimental conditions: In the former experiment, the changes in specimen water content were probably much larger. It is known that the mechanical properties of other materials are affected by the presence of liquids and gases, and coal would be expected to show similar effects. The point deserves further study. The effect of temperature is complicated by the interrelation between temperature and water content. Inouye and Tani made their measurements of viscosity η at four temperatures in the range 30 to 70°C and applied the Andrade formula:

$$\eta = Ae^{u/RT} \tag{4}$$

where u = activation energy, T = the temperature in °K, and R = the gas constant, to obtain an activation energy of about 1 kcal/mol for a caking coal.[4] Although there have been many studies of the plastic behavior of coals at high temperatures,[56] the variation of properties over the lower range of temperature does not appear to have been studied further.

[56] Anon., *Coal Research*, **17**, No. 2, 5 (1957). Fitzgerald, D., *J. Sci. Instr.*, **32**, 359–61 (1955). Fitzgerald, D., *Nature*, **175**, 515–6 (1955). Hoffmann, H., and Hoehne, K., *Brennstoff-Chem.*, **41**, 5–11, 70–9, 142–50, 204–8, 235–43 (1960). Huntjens, F. J., Dormans, H. N. M., and van Krevelen, D. W., *Fuel*, **35**, 462–74 (1956). Seyler, C. A., *Fuel Research Board Annual Rept.*, 1938, pp. 65–8. Ritter, H., and Juranek, G., *Brennstoff-Chem.*, **41**, 170–6 (1960). Rusinko, F., Weinstein, A., and Walker, P. L., 136*th Meeting, Am. Chem. Soc.*, **7K**, Abs. 25 (1959).

Hardness

Measurements of the hardness of coals have been made largely as a means of examining change in properties with rank. Van Krevelen, for example, measured the change in Vickers microhardness with carbon content and found a minimum at 90% C which was interpreted as a minimum in binding forces.[57] Honda and Sanada[58] have measured the hardness of coals by the Vickers and Shore scleroscope methods and obtained similar changes with rank in each case. A minimum in hardness was also found at 90% C, and a maximum found at around 80%, supporting van Krevelen's results. Both static and dynamic Young's moduli and Mohs hardness were calculated from the measurements, and reasonable agreement was obtained between the calculated moduli and those measured by Inouye and Tani.[4,51] Similar hardness measurements were also made on carbonized coals.[59] Heinze has measured Vickers microhardness and macrohardness, microscratch hardness, and also the Späth vibrohardness, of a number of German coals and their constituents.[60] Similar variations with volatile matter were obtained for each measurement except the vibrohardness test, in which the polished coal surface was subjected to impulsive loading by a bit. Generally, there was a rapid fall in hardness from 5% volatile matter to a minimum at about 15% and, thereafter, a steady increase up to 40% volatile matter. This increase was generally shown by the petrographic constituents, except for the durite component.

Friction

It has been shown by Spurr[61] that there is a relation between the Mohs hardness and the coefficient of sliding friction of minerals, and it is clear that both hardness and friction are important properties in mining and pulverizing. Nevertheless, little work on coal friction appears to have been published. Brown and Pomeroy[62] reported initial measurements of the friction of nine coals on brass and steel, using loads up to 1000 psi. The friction coefficient of coal on steel decreased as the load was increased, and it was found that this behavior could be reasonably well-described by a model (due to Archard) in which asperities in the coal surface deformed elastically, and the number of asperities making contact depended on load. The friction between coal and brass was made more complex by transfer of brass to coal as well as of coal to brass. Coefficients of friction at the start of sliding were in the range 0.4 to 0.6, and a well-defined curve of dependence on rank was obtained, with a peak in friction at about 38% and a minimum at about 20% dry, ash-free volatile matter, when a load of 1000 psi was applied. Because there was heavy transfer of coal to the metal surface, the frictional forces appeared to arise from the shear of the coal asperities.

STRENGTH

The strength of coal is of direct importance in mining and pulverizing. A number of standard laboratory tests may be applied to prepared coal specimens, and such tests have been used as the basis of discussions of theories of coal strength. However, the relations between the results of laboratory strength tests and the strength of coals as evidenced in mining and pulverizing may be difficult to derive. In many pulverizing operations, for example, several different types of stress system operate, and a laboratory test may, in such cases, be less satisfactory as a guide to mill performance than an empirical model test. On the other hand, the behavior of coal in the seam, under considerable static load and lateral restraint, may be quite different from that of the freed coal in the laboratory, unless the laboratory test simulates, in addition, the conditions of the seam.

Both the strength and mode of failure are dependent on the rank and condition of the coal and on the form of applied stresses. Mined coals differ greatly in the severity of the crack systems they contain, and differences between fracture propagation in coals may arise both from the extent of cracks and from heterogeneity.

[57] van Krevelen, D. W., *Brennstoff-Chem.*, **34**, 167–82 (1953).

[58] Honda, H., and Sanada, Y., *Fuel*, **35**, 451–61 (1956); **37**, 141–50 (1958).

[59] Honda, H., and Sanada, Y., *Fuel*, **36**, 403–16 (1957).

[60] Heinze, G., *Bergbau-Arch.*, **19**, 71–93 (1958).

[61] Spurr, R. T., *Brit. J. Appl. Phys.*, **9**, 486–7 (1958).

[62] Walton, W. H., Ed., *Op. cit.*, pp. 419–31.

TABLE 1

VARIATION OF COMPRESSIVE STRENGTH OF CUBES OF DEEP DUFFRYN COAL[15]
PERCENTAGE OF CUBES FAILING IN STRESS RANGES (lb/in.^{2})

Size	600 to 1200	1200 to 1800	1800 to 2400	2400 to 3000	3000 to 3600	3600 to 4200	4200 to 4800	4800 to 5400	Over 5400	Mean Strength
$\frac{1}{4}$ in.	2	6	14	14	18	16	15	10	5	1530
$\frac{1}{2}$ in.	6	10	23	29	19	6	6	1	0	840
1 in.	6	21	37	26	2	8	0	0	0	710

As an example of the dependence of fracture on the presence of cracks, an experiment by Bond may be cited.[63] Cubes of coal, apparently free from major flaws, were found to withstand a certain number of light impacts on the uppermost face and then fail suddenly into two or three pieces along planes perpendicular to the direction of maximum tensile stress in compression. On the other hand, the strength of coal is affected by its physical properties in ways which may be illustrated, for example, by the experiments of Macrae and Mitchell.[25] The fracture of coal cubes in compression perpendicular to the bedding plane was studied at room temperature and at temperatures above 100°C. At room temperature, fracture occurred along planes parallel to the direction of compression and, at the higher temperatures, along planes of maximum shear stress.

Strength in Compression

Effect of Size

The strength of coal specimens has been most commonly studied by means of the compression test, because the results have direct application to the estimation of the load-bearing capacities of pillars in coal mines. For this reason, the variation of compressive strength with size of specimen has been studied extensively, although the results of compression tests are liable to be misleading. This is because specimens of accurate dimensions with plane parallel faces are difficult to prepare (especially when large), and frictional effects between the platens and the faces of the specimen are important and difficult to control. Crack orientation also affects the results, as it does in all measurements of coal strength and, further, particularly with severely cracked coals, it is often difficult to define the load at which failure occurs. For these reasons various strength-size relationships which have been claimed to hold must be interpreted with caution. Again, Protodyakanov[64,65] has shown, using empirical tests, that strength may vary by as much as 30% over the width of a seam; this is a reminder that adequate numbers of specimens must be tested and care taken in choice of specimens for comparative purposes.

The compressive strengths of a number of pieces of the same size cover a wide range. The mean compressive strength decreases with increasing size of specimen, as shown in Table 1, derived from histograms given by Evans and Pomeroy.[15]

Using such data, Evans and Pomeroy have discussed the concept of probability of survival of cubical specimens in the compression test, and endeavored to show by the weakest-link theory that the variable affecting strength was the length of the cube side. An expression relating the probabilities of survival of cubes of side a and b was derived:

$$P_b = P_a^{n/m} \tag{5}$$

where cubes a and b contain n and m elements, respectively, each element corresponding to a link in a chain. The ratio n/m may be written $(b/a)^\beta$ where β is determined according to the effect of the elements on breaking strength: $\beta = 1$, 2 or 3 would imply a dependence on cube side, area or volume respectively.
Hence,

$$\beta \log (b/a) = \log (\log P_b/\log P_a) \tag{6}$$

and from experimentally derived relations of the probability of survival with strength, for a range of sizes, the two sides of the equation may be plotted to give a value for β. Evans

[63] Bond, R. L., *Fuel*, **33**, 249–50 (1954).

[64] Protodyakanov, M. M., *Ugol'*, **25**, No. 9, 20–4 (1950).

[65] Protodyakanov, M. M., *Izvest. Akad. Nauk S.S.S.R.*, **1953**, No. 2, 283–98.

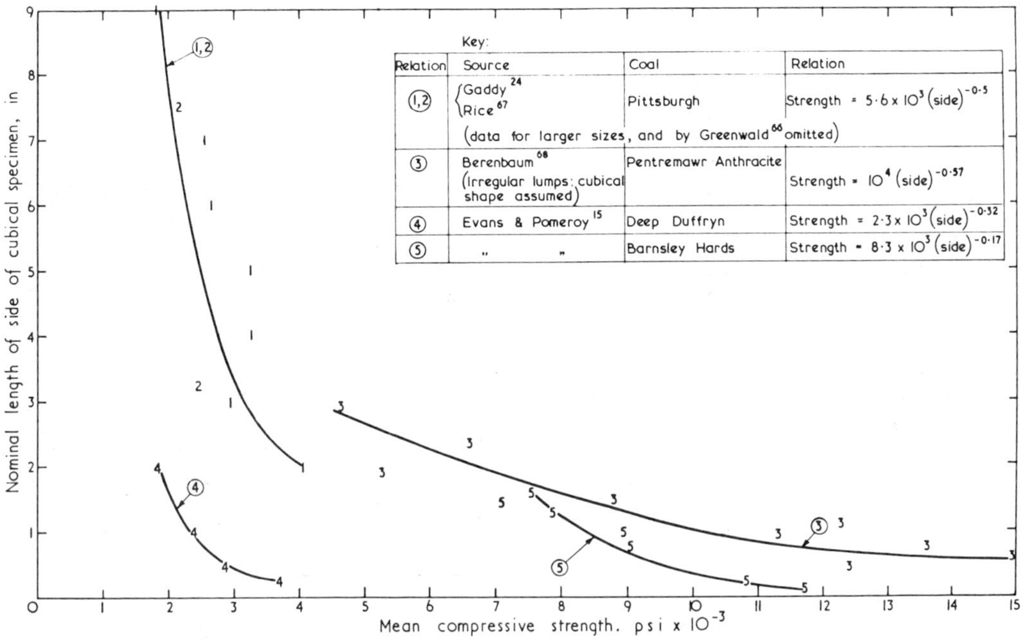

Fig. 3. Compressive strength and size.

and Pomeroy obtained $\beta = 1.00 \pm 0.15$ for the friable Deep Duffryn coal and 1.27 ± 0.12 for Barnsley Hard coal. This suggested breaking stress to be more closely related to the length of side of the cube than to the area or volume.

The variation of mean strength, expressed as breaking stress in lb/in.^{2}, with size of specimen, is illustrated in Fig. 3 over the range of sizes of most immediate practical importance. The sources of the data are given in Table 2.

It will be noted that the breaking stress decreases with increasing specimen size, at first rapidly and, then, for sizes over a certain value, slowly. Various forms have been proposed for this relation,[15,24,66,67,69] the first three of which are algebraically equivalent, namely:

$$\text{crushing load (lb wt)} \propto (\text{volume or weight})^{1/2} \quad (7)$$

$$\text{crushing load/unit volume} \propto (\text{volume or weight})^{-1/2} \quad (8)$$

$$\text{breaking stress (lb wt/in.}^2) \propto (\text{side})^{-1/2} \quad (9)$$

$$\text{breaking stress (lb wt/in.}^2) \propto (\text{side})^{-r} \quad (10)$$

where r is a variable exponent. Evans and Pomeroy,[15] using equation 10, found a variation in r by loading a coal in three different ways, obtaining values between 0.25 and 0.52. It should be noted that a variable exponent in equation 7 would then range only from 0.49 to 0.58. Clearly equations 9 and 10 are more sensitive than equations 7 and 8.

Evans and Pomeroy pointed out that the range of sizes used in their experiments was not large enough to fix the exponent r precisely and suggested that over a wider range the slope of the log-log plot might not be constant. Gaddy[24] took samples from the same part of the seam as was used by Greenwald[66] and Rice,[67] although at widely differing dates, and claimed that these data could be treated together to cover the size range of 2 to 64 in. He found good agreement with equation 9, but it should be noted that only one specimen was available for each of the larger sizes. Millard, Newman and Phillips[69] found good agreement with equation 7 for a wide range of irregular pieces weighing 5×10^{-9} to 5×10^{-6} g and 10^{-2} to 10^3 g. While this equation undoubtedly shows the trend, their data plotted on a log-log scale are not particularly well-represented by a single straight line. Further, they pointed out that some of the very small particles exhibited flow and were excluded from the analysis.

[66] Greenwald, H. P., *U.S. Bur. Mines, Rept. Invest.* No. 3575 (1941), 6 pp.

[67] Rice, G. S., *U.S. Bur. Mines, Bull.* No. 303 (1929), 44 pp.

[68] Evans, I., Pomeroy, C. D., and Berenbaum, R., *Colliery Eng.*, **38,** 75–80, 123–37, 172–8 (1961).

[69] Millard, D. J., Newman, P. C., and Phillips, J. W., *Proc. Phys. Soc. (London)*, **B68,** 723–8 (1955).

TABLE 2

RANGE OF DATA ON VARIATION OF STRENGTH WITH SIZE

Source	Coal	Size Range	Notes
Greenwald[66]	Pittsburgh	31–64 in.	*In situ* measurements
Rice[67]	Pittsburgh	2½–54 in.	Laboratory tests
Gaddy[24]	Pittsburgh	2–9 in.	Samples from site used by Greenwald
Gaddy[24]	Clintwood	2–7 in.	
	Marker	2–7 in.	
	Pocahontas	2–7 in.	
	Harlan	2–7 in.	
Evans and Pomeroy[15]	Deep Duffryn	¼–2 in.	Friable
	Barnsley	⅛–1.6 in.	Hards
Berenbaum[68]	Deep Duffryn	0.6 –2.85 in.	Irregular lumps
	Barnsley	0.4 –3.4 in.	Irregular lumps
	Pentremawr Anthracite	0.35–2.9 in.	Irregular lumps
Millard, Newman and Phillips[69]		Approximate Values	
	Betteshanger, Langwith	15–150 μ	Irregular particles
		0.05–1.7 in.	Irregular lumps
	Llandebie	15	Irregular particles
		0.25–4.5 in.	Cubes

Phillips (see ref. 2, p. 25) has reiterated that their data were obtained with a friable, extensively cracked coal, and that the half-power relation would not necessarily apply to other coals. This variation with the type of coal is probably related to the differences in exponent β in equation 6 already noted. Evans and associates have extended the work on irregular lumps by taking into account the area of contact.[68]

Both Evans and Pomeroy, and Millard, Newman and Phillips, applied the Griffith theory[70] to their results. Although the calculations show some agreement with experiment, the limitations of the simple Griffith theory are such that its application to heavily cracked solids is of doubtful value. In their discussion, Evans and Pomeroy examined the consequences of the assumption of different types of flaw density distribution on the strength-size relationship. The problem has been treated more generally by Epstein, who summarized the types of strength-size relations which would be expected to hold as a result of the assumption of various types of distribution of flaw severities.[8]

Further measurements of compressive strength have been reported by Holland[30,71] and Greenwald,[66] among others.

Compressive Strength and Rank

The variation of compressive strength with rank of British coals was given by Pomeroy and Foote, and a plot of strength against volatile matter shows the customary minimum at 20 to 25% dry, ash-free volatile matter (cf. Fig. 4) for compression both perpendicular and parallel to the bedding plane.[72] They also obtained figures for cleat frequency for the nine coals tested, and a plot of their data shows that compressive strength falls off as cleat frequency increases.

Triaxial Compression

The compressive strengths of coals, as measured in the laboratory, are frequently less than the pressures withstood in the seam, and

[70] Griffith, A. A., *Phil. Trans. Roy. Soc. London*, **A221**, 163–98 (1920–1); *Proc. 1st Intern. Congr. Appl. Mech.*, **1924**, 55–63.

[71] Holland, C. T., *Colliery Eng.*, **32,** 110–14, 145–53 (1955).

[72] Pomeroy, C. D., and Foote, P., *Colliery Eng.*, **37,** 146–54 (1960).

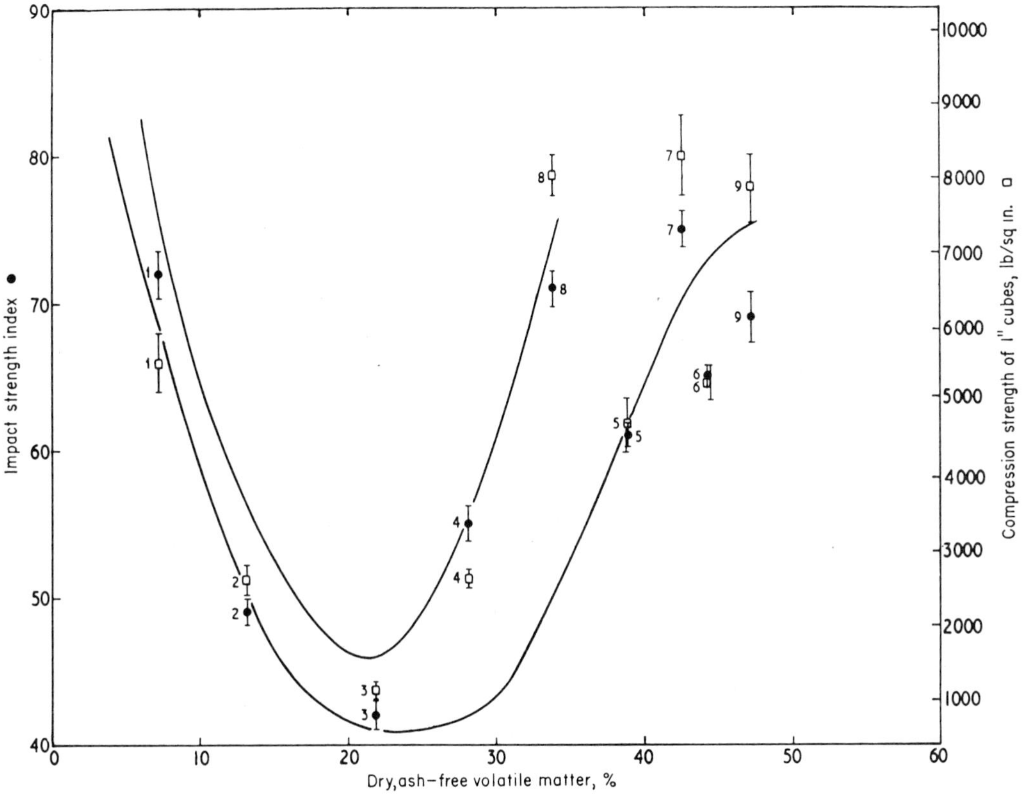

Fig. 4. Strength and rank.[72]

measurements of compressive strength under lateral compression have been made. Murrell[73] applied triaxial compression to coal cylinders, using lateral pressures in the range of 0 to 4000 psi. The work represented an extension of that of Müller,[74] in that orientation effects were taken into account. The results, which confirmed, in general, the conclusions of Müller, are summarized in Table 3. A striking feature of these results is the close similarity in strength of the two coals at the maximum confining pressure.

Failure in the triaxial compression tests was found to occur at surfaces inclined to the axes of the cylinders, and attempts were made to fit the measured stresses to the Mohr failure theory.[75] Whereas fairly good agreement was obtained between observed and calculated strengths, the directions of failure were not in good agreement with predicted values. Approximate estimates of tensile strength and fracture angle were also obtained from the Griffith[70] theory, but all the predicted values were too high. In the application of the Griffith theory, it was necessary to assume both a simplified (biaxial) stress system and that there was no interference between cracks. The author suggested that the discrepancy between observed and predicted angles of fracture might arise from the presence of three mutually perpendicular crack planes.

[73] Walton, W. H., Ed., *Op. cit.*, pp. 123–46.

[74] Müller, O., *Glückauf*, **66,** 1601–12 (1930).

[75] Sokolovski, V. V., *Statics of Soil Media*, Butterworths, London, 1960, 260 pp.

Strength in Tension

Few measurements have been made of the tensile strengths of coals. The standard laboratory tensile test has little meaning, especially with specimens containing gross cracks. Further, there seemed, at one time, to be no practical application for such measurements. However, it has been shown that the influence of cracks on the strength of coal in the seam is affected by the overburden load, and investigations in the United Kingdom into the method of mining by ploughing have led to the

TABLE 3

VARIATION OF AXIAL FRACTURE STRESS WITH LATERAL PRESSURE[73]

Orientation of Cylinder Axis	Confining Pressure, (psi)	Mean Fracture Stress, psi	
		Barnsley Hard Coal	Cwmtillery Coal
Parallel to bedding plane and major cleat plane	0	3540 ± 380	840 ± 80
	3980	18,290 ± 580	15,530 ± 320
Parallel to bedding plane, perpendicular to major cleat plane	0	6210 ± 640	1550 ± 40
	3980	17,970 ± 350	16,450 ± 260
Perpendicular to bedding plane, parallel to major cleat plane	0	4660 ± 140	1150 ± 70
	3980	19,300 ± 430	15,010 ± 690

realization that tensile strength, measured in a suitable tensile or compressive test, may be directly relevant. Of the measurements of strength in tension reported here, the method of Pomeroy and Morgans,[76] used by them in work of an exploratory nature, is generally applicable only to the more coherent coals. Four-point bending was applied to strip specimens 4 in. long, with widths between 0.25 and 1 in. and thicknesses from 0.1 to 0.375 in. Four coals were tested, with the bedding plane both parallel and perpendicular to the face of the strip. Apart from the usual dependence of the results on the orientation of major cracks in the original seam, the greatest tensile strength was obtained with an anthracite at 880 psi and the least with a friable coking coal at 70 psi. Of the four coals used, one (Barnsley Hards), which generally showed a clean fracture, was tested in a further experiment by a method of successive breakage. Coal strips of several widths were broken and then joined with a synthetic glue and rebroken, up to a maximum of ten times. A general increase in strength was found as the strips were successively broken and rejoined. For example, for strips $\frac{1}{4}$ in. wide, an initial tensile strength was obtained of 830 psi (mean of 22 measurements), and this increased to a strength of 1460 psi (mean of 11 measurements) after ten breaks. A general increase in strength up to the tenth break was found for all widths over the range 0.09 to 1.26 in., although the greatest increase in strength was for the narrow strips. This was taken as evidence of a general effect of size on strength, whereas the steady increase in strength with number of breaks indicated a wide range of severities of weakness.

Berenbaum and Brodie[26] described the measurement of tensile strength by methods which do not suffer the disadvantage, so far as friable coals are concerned, of requiring the preparation of standard strip or tensile specimens. In one method, coal slabs were compressed between indenters with plane ends. The specimens were observed to break in tension, and analysis showed that tensile breakage under the test conditions could occur only in a substance having a shear strength substantially greater than the tensile strength.

An alternative method, which is more convenient for coal, was described in the same paper[26] and consisted of diametral compression of a disk. In this test, the disks failed in tension along the direction of compressive stress—the relation between tensile stress σ, and applied load W being given by:

$$\sigma = 2W/\pi\, ds \qquad (11)$$

where d and s are the disk diameter and thickness. The advantage of this test, apart from the ease of preparation of specimens, is that the effects of orientation may be examined by taking successive disks from a cylindrical sample. In another paper, Berenbaum and Brodie[27] reported results of a series of measurements of tensile strength of coal both by indentation and disk methods, showing reasonably good agreement between the two. By slicing disks from cylinders cut in four orientations, it was possible to obtain a complete picture of the variation of tensile strength with orientation.

[76] Pomeroy, C. D., and Morgans, W. T. A., *Brit. J. Appl. Phys.*, 7, 243–6 (1956).

Among the general conclusions drawn from this work, it was found that for the two very different coals examined, one hard and compact, the other cracked and friable, the change in tensile strength was a maximum in relation to the orientation of the bedding plane, even for the heavily cleated coal. It thus appeared that the large numbers of relatively small cracks found in the bedding planes represented a greater source of weakness than the relatively few, but more severe, cleat cracks. The measurement of tensile strength in three orientations was also included in the series of experiments by Pomeroy and Foote,[72] referred to above. The variations of tensile strength and compressive strength with volatile matter are similar.

Relevance of Strength Measurements to the Mining of Coal

The relevance of measurements of the strength of coals to problems of mining by ploughing was discussed in four papers[5,77–79] presented at a conference on mechanical properties of nonmetallic brittle materials held in 1958. These papers were concerned with laboratory studies in which various aspects of ploughing were simulated. Dumbleton, O'Dogherty and Shepherd[5] investigated the effectiveness of blades of different angles at various orientations to cleat planes and found that orientation was critical with a heavily cleated coal. Under given conditions, the load required on the cutting tool was proportional to the compressive strengths of equal-sized cubes of the coals tested. This laboratory work was extended by Pomeroy,[77] who simulated mining conditions by applying a confining load to the sample. He showed that when the cut was deep enough, the cutting load first increased to a maximum, and then decreased with increase in confining pressure. The results suggested that the value of confining pressure at which the cutting load started to decrease was dependent on the compressive or tensile strength of the coal used; for the three coals tested, the peak in cutting force occurred at a higher load for Deep Duffryn (compressive strength c. 2500 psi) than for Oakdale coal (c. 1000 psi), whereas the cutting force for Barnsley Hards (compressive strength c. 6500 psi) increased steadily over the range of confining pressures used, that is up to about 500 psi.

In another experiment, Evans and Murrell measured the resistance, to penetration by a wedge, of coal samples having only one unconfined face and found that, to a first approximation, the penetration resistance and the compressive strength were equal, the relation being affected by the presence of complex frictional forces.[78] The effective coefficient of friction acting on the wedge was estimated at the low value of about 0.1, because the frictional forces, which caused the coal to fail in shear along the faces of the wedge, were limited by the low shear strength.

Pomeroy and Foote[72] extended earlier experiments on the relation between ploughability and rank by correlating the grooving force required for a model plough with impact, tensile, and compressive strengths and with cleat frequency. They concluded that ease of ploughing might most readily be assessed by a penetrometer[80] or by an *in situ* tensile test machine. One such machine, which has been described by Il'nitskaya[81] was found unsatisfactory in its present form. The possibility of assessing ploughability from volatile matter, suggested by Potts and Shuttleworth,[82] was not confirmed because of the importance of cleat as an additional factor.

Factors Affecting Strength

As already mentioned, there is a little evidence that the elastic properties of coal are dependent on moisture content and temperature. These factors do affect the pulverizing of coal, but, because they also influence the transport of coal within the mill, the results are complex. Mott[83] has compared wet and dry fine grinding of coal and of coke and found, for example, that the size limit below which agglomeration of coke particles occurred was 0.3 μ in dry grinding, but 0.03 μ in wet

[77] Walton, W. H., Ed., *Op. cit.*, pp. 469–79.

[78] *Ibid.*, pp. 432–50.

[79] *Ibid.*, pp. 451–68.

[80] Evans, I., Pomeroy, C. D., and Davies, R., *Colliery Eng.*, **6**, 234–40 (1959).

[81] Il'nitskaya, E. I., *Doklady Akad. Nauk S.S.S.R.*, **104**, 90–100 (1955).

[82] Potts, E. L. J., and Shuttleworth, P., *Trans. Inst. Mining Engrs.*, **117**, 519–53 (1957–8).

[83] Mott, R. A., *J. Soc. Chem. Ind.*, **69**, 346–9 (1950).

grinding. There is, it appears, an intermediate range of moisture contents in which grinding is, for one reason or another, unsatisfactory. Transport is much improved by the use of a liquid medium but is reasonably satisfactory for dry coals. Work on the Hardgrove index[84] and experiments with a ring-ball mill[85] suggested that coals become more difficult to grind when the temperature is raised unduly.

Little work appears to have been done on the effect of fluids other than water on the grinding of coal, but evidence for other materials[86] suggests that such a study might be worthwhile.

Adsorbed gases also influence coal strength. Nikolskiĭ[87] has studied the effect of gas on the passage of shock waves through coal, whereas Ettinger and his colleagues have investigated the strength of coals in various gases as providing information both on adsorption and on ease of mining.[88,89] Ettinger and Lamba[89] measured the strength of coals in hydrogen, methane, and carbon dioxide by means of the strength test due to Protodyakanov.[64] The quantity of fines produced under the standard test conditions was in greater proportion in CO_2 than in air and generally followed the regression line:

$$\text{Fines in } CO_2 = 1.34 \times \text{fines in air} \quad (12)$$

so that the weak, heavily cracked coals, which produced most fines, showed the greatest difference in behavior in the two media. The relative effects on strength of atmospheres of carbon dioxide, methane, and hydrogen decreased in the order given, indicating that effectiveness of adsorption was the deciding factor, and that molecular size, that is, ease of access to internal pores, was less important.

[84] Callcott, T. G., Private Communication (1956).

[85] Hiorns, F. J., and Sayles, C. P., *J. Inst. Fuel*, **32**, 464–75 (1959).

[86] Frangiskos, A. Z., and Smith, H. G., *Proc. Intern. Conf. Mineral Dressing*, **1957**, 67–84. von Szantho, E., *Erzmetall*, **2**, 353–60 (1949).

[87] Nikolskiĭ, T. A., *Doklady Akad. Nauk S.S.S.R.*, **96**, 25–8 (1954).

[88] Ettinger, I. L., Lamba, E. G., and Adamov, V. G., *Doklady Akad. Nauk S.S.S.R.*, **99**, 1057–60 (1954). Ettinger, I. L., and Lamba, E. G., *Ugol'*, **7**, 35–6 (1957).

[89] Ettinger, I. L., and Lamba, E. G., *Fuel*, **36**, 298–306 (1956).

STATISTICAL MECHANICS

To relate the fundamental physical measurements described in the first two sections to practical occurrences and to the results of model tests, it is necessary to evolve an adequate mechanics. As already noted for coal in the seam, the relation depends on the application of measurements of mechanical properties *in situ*. In the mechanics of coal broken away from the seam, account must be taken of the complexity of the breakage forces to which the coal is subjected. Added to these difficulties are the inhomogeneity and flaws in pieces of coal; it is not surprising that statistical representations have found favor.

Matrix Representation of Breakage

It will help in discussing the main statistical theories that have been proposed to describe first the general method of representing coal breakage given by Broadbent and Callcott.[6,90] A sieve analysis with five sieves is written:

$$\mathbf{f} = (f_1, f_2, f_3, f_4, f_5, f_6) \quad (13)$$

where f_1 is the percentage or fraction oversize, f_2 is percentage or fraction between the top two sieves, and so on; f_6 is found by difference. If the feed $\mathbf{f}$ is changed into a product $\mathbf{p}$ of a different size, each size fraction, say f_3, may contribute to that size and all smaller sizes in the product, namely, the percentages or fractions p_3, p_4, p_5, p_6.

Thus:

$$\left.\begin{aligned} p_1 &= a_1 f_1 \\ p_2 &= b_1 f_1 + b_2 f_2 \\ p_5 &= e_1 f_1 + e_2 f_2 + e_3 f_3 + e_4 f_4 + e_5 f_5 \end{aligned}\right\} \quad (14)$$

and the undersize p_6 is found by difference. Clearly, a fraction a_1 of the oversize coal remains oversize, the remainder being distributed among the smaller sizes, and so on. The manner of this distribution depends jointly on the way in which each size is selected for breakage and the sizes of the fragments produced in the breakage. The latter can be calculated from size distribution relations. The former depends in part on the mechanical properties of the coal (namely, variation of strength with size) and in part on factors such as the shielding of small particles by big ones when beds of coal are crushed.

[90] Broadbent, S. R., and Callcott, T. G., *J. Inst. Fuel*, **29**, 524–39 (1956); **30**, 13–25 (1957).

Equations 14 can be arranged conveniently in the matrix form:

$$\begin{pmatrix} p_1 \\ p_2 \\ p_3 \\ p_4 \\ p_5 \end{pmatrix} = \begin{pmatrix} a_1 & 0 & 0 & 0 & 0 \\ b_1 & b_2 & 0 & 0 & 0 \\ c_1 & c_2 & c_3 & 0 & 0 \\ d_1 & d_2 & d_3 & d_4 & 0 \\ e_1 & e_2 & e_3 & e_4 & e_5 \end{pmatrix} \begin{pmatrix} f_1 \\ f_2 \\ f_3 \\ f_4 \\ f_5 \end{pmatrix} \quad (15)$$

or

$$\mathbf{p} = \mathbf{Mf} \quad (16)$$

where **M** is a lower triangular matrix.

Equation 16 has the merit that it is as accurate as—and no more accurate than—the histogram of the sieve analysis. It is also amenable to straightforward numerical calculations. In contrast, replacement of the sieve analysis by an algebraic form can lead to difficult mathematical problems.

Any single process which changes the sieve analysis can be represented by a matrix. Thus, size classification is a diagonal matrix, the elements of the main diagonal representing the efficiency of classification of each sieve fraction, all other elements being zero. Then the equations of closed-circuit grinding can be solved to show the way in which the breakage in a mill and the efficiency of size classification impinge on each other and jointly determine the recirculating load.[91] Callcott has extended the method in a study of varying ash content of sieve fractions, showing that inadequate experimental data could easily lead to erroneous conclusions as to the degree of true selectivity obtained by grinding.[92]

Broadbent and Callcott separated the matrix **M** into two portions representing, respectively, the *breakage* **B** effected on each sieve fraction and the *selection* **S** of the particles for breakage. In the simplest case, a constant proportion, π, was selected from each sieve fraction, independently of particle size. A simple exponential relation was then assumed for calculation of the elements of the breakage matrix **B**. The method was applied successfully to the breakage occurring in laboratory ring-ball, ball, and beater mills[6] and to the shatter test data of Gilmore, Nicolls and Connell.[6,93] Broadbent and Callcott suggested (but see the theory of the shatter test below) that the results of shatter tests can, in many cases, be translated into a single figure, the proportion of material broken, which in turn is related to the extent of degradation occurring in coal-handling plant.[90] A more complex selection matrix was needed for the beater mill, while the ring-ball and ball mill data showed that repeated fracture took place necessitating powers (including fractional powers) of the breakage matrix **B**. Ellis, Ray, and Venables have since determined the matrix **M** experimentally, in an open circuit laboratory ring-ball mill, showing that with a feed of coal consisting of a mixture of sizes, the presence of other sizes disturbs the breakage measured for each size grade separately.[94] Similarly, using a radioactive tracer technique, Gaudin, Speddon and Kaufman found that a feed of mixed sizes of albite gave relatively fewer fines than a feed of one-sieve fraction only.[95] More work is needed on interaction.

It is evident that the matrix representation of breakage helps to disentangle breakage processes from the properties of the coal. There are not yet enough data to give rules for specifying the selection and breakage matrices **S** and **B**. The general form of these rules can be seen, however, from the several statistical theories of breakage that have been proposed. Generally, selection is of two kinds—that arising from the spread of strength values of pieces of a given size and that which is determined by the combined characteristics of the coal and the stress system, for example, the interrelation between particle shape, friction, and angle of nip in a ring-ball pulverizer.

Statistical Theories of Breakage

Two main theories have been developed. That initiated by Bennett in 1936, and developed by his colleagues in a series of papers from 1941 to 1948, was centered round the idea that breakage, occurring preferentially at preexisting flaws, amounts to a separation into pieces with size distribution determined by the spatial distribution of the weaknesses in the

[91] Brown, R. L., *Brit. Chem. Eng.*, **4,** 463–6 (1959).

[92] Callcott, T. G., *Proc. Conf. on Science in the Use of Coal*, Inst. Fuel, London, 1958, pp. B35–40.

[93] Gilmore, R. E., Nicolls, J. H. H., and Connell, G. P., *Can. Dept. Mines, Mines Branch, Invest. Fuels and Fuel Testing*, No. 762 (1935), 102 pp.

[94] Ellis, J. E., Ray, W. D., and Venables, J., *Proc. Conf. on Science in the Use of Coal*, Inst. Fuel, London, 1958, pp. B30–4.

[95] Gaudin, A. M., Speddon, H. R., and Kaufman, D. F., *Mining Eng.*, **3,** 969–70 (1951).

original piece.[11,12,96–98] This theory is concerned with the breakage process described by the matrix **B**. The other followed on the idea that the first failure of a piece subjected to stress occurs at the weakest link.[15] Epstein generalized the previous theoretical studies in 1948,[8] and Evans and Pomeroy presented the first experimental evidence for coal in a paper published in 1958.[15] This theory is concerned mainly with selection, as described by matrices **S**.

A starting point for understanding the concept of selection for breakage is Evans and Pomeroy's interpretation of data on the compressive strength of cubes of side x in terms of their probability of survival, C^x, written by Berenbaum[7] as:

$$C^x = e^{x \ln C} \tag{17}$$

where C is the probability of survival of a cube of unit size. Equation 17 is a simplified form of the general strength-size relations derived by Epstein from alternative assumptions as to the statistical distribution of the strengths of flaws in the specimen. As already noted, the experimental evidence is in accord with several of these alternatives and does not permit a choice to be made between them. If the relevant size characteristic is the side (and not the area or volume of the piece), purely statistical theories using area or volume are not applicable.[99] Equation 17 describes that part of the selection process that is determined by the inherent resistance to fracture of pieces of different size. The other part is determined by the ways in which stresses are applied to the pieces.

Consider now the size distribution resulting from breakage. Having in mind involuntary degradation during the handling of coal, Bennett postulated that the flaws were distributed at random within the pieces, and that the stresses were adequate to cause fracture simultaneously at a number of flaws.[11] Now the random disposition of points on a line gives a simple Poisson or exponential distribution for the distances between adjacent points.[100] Bennett and others[11,101,102] attempted to derive from this concept the Rosin-Rammler-Sperling size distribution relation:

$$100 - U(x) = 100\, e^{-bx^n}, \quad \text{(Rosin-Rammler relation)} \tag{18}$$

where $U(x)$ is the percentage underside and x, b, and n are constants. This relation had been found experimentally for pulverized coal.[18] None of these attempts at a basic theory was entirely satisfactory.

Fot x small enough, the Rosin-Rammler relation is approximated by the well-known equation:[19]

$$U(x) = 100\left(\frac{x}{X}\right)^n \quad \text{(Gaudin relation)} \tag{19}$$

where X is the top size.

In applying these relations, it has to be remembered that partial degradation results in a residue, similar in size to the original pieces, and a complement consisting of fragments of all sizes (up to an upper size limit) usually formed from the regions of greatest stress concentration. Bennett, Brown and Crone showed experimentally that a single application of breakage forces on large pieces, wherein breakage was severe enough to eliminate the residue, resulted in a simple exponential distribution of sizes (that is, $n = 1$ in equation 18).[96] Partial fracture of smaller pieces, a few inches in size, gave a complement following Gaudin's equation with $n = 1$: on account of the upper size limit, it was not experimentally possible to detect an exponential. It would seem, therefore, that a single fracture of coal results in finer fragments having a simple exponential distribution up to an upper size limit that would be determined by the strength of the piece and the severity of the breakage. Following this result, Brown calculated the consequences of a few repetitions of fracture in which all pieces were broken equally each time (i.e., selection of every size for breakage was perfect) and showed that the distributions were well-approximated by the Rosin-Rammler relation.[97] The exponent n proved to be determined mainly by the number of cycles of breakage. In practice, stable size distributions

[96] Bennett, J. G., Brown, R. L., and Crone, H. G., *J. Inst. Fuel*, **14,** 111–28 (1941).

[97] Brown, R. L., *J. Inst. Fuel*, **14,** 129–34 (1941).

[98] Brown, R. L., *Trans. Inst. Mining Engrs.*, **103,** 88–106 (1944).

[99] Manning, A. B., *J. Inst. Fuel*, **25,** 31–2 (1952).

[100] Lienau, C. C., *J. Franklin Inst.*, **221,** 485–94, 673–86, 769–80 (1936).

[101] Manning, A. B., *J. Inst. Fuel*, **10,** 117–18 (1936). Bennett, J. G., *Proc. Inst. Mech. Engrs. (London)*, **125,** 383–4 (1933). Stange, K., *Ingenieur-Arch.*, **21,** 368–80 (1953).

[102] Zehler, V., *Kolloid-Z.*, **160,** 16–20 (1958).

having constant n, often near 0.7, were found. This was attributed to selection not being the same for all sizes. Thus, the exponent n in the Rosin-Rammler and Gaudin relations has been found experimentally to be unity for a single fracture. The exponent in the Rosin-Rammler relation has smaller values for multiple fracture.

Evans has made the interesting suggestion that a size distribution relation can be derived from the selection function given in equation 17 on the argument that the proportions of pieces present in broken coal will be related to the probability of their survival.[14] In effect, Evans assumed Gaudin's relation and, therefore, tacitly, the effects of a few repetitions of fracture and proposed:

$$U(x) = \frac{1}{\Gamma(\mu)} \int_0^{x/\bar{x}} x^{\mu-1}\, dx \text{ (incomplete gamma relation)} \quad (20)$$

This relation is not experimentally distinguishable from the Rosin-Rammler relation. The validity of this attempt turns on whether it can be demonstrated that the sizes of the fragments produced by breakage are related to the distribution of weakest links that determines the selection function. This would require study of the way in which a fragment is actually formed and, therefore, of the "next-weakest" links, and so on.

One other relation must be mentioned. Epstein has pointed out that, for numerous breakage steps, in each of which are accidental occurrences, the distribution will tend to a logarithmico-normal form.[10,103] This form does not appear to be valid for the complement, possibly either on account of imperfect selection or because the complement is not several orders of magnitude smaller than the starting material. Recently, however, Taylor[104] and Lázàr[105] have found it to be applicable to the residue produced in some crushing and degradation processes. Possibly this is because the numerous accidental events—which are all that it is necessary to postulate for a Gaussian or logarithmico-normal relation to be obtained—can be associated with the variation in strength between pieces of the same size found with some materials.

It is worth noting here that the Gaussian relation and its variants are characteristic of conservation processes. In contrast, breakage is a destructive process in which the presence of one piece influences the proportions of pieces of different size. Hence, quite apart from the large number of parameters involved in empirical proposals[106,107] to use relations having Rosin-Rammler, Gaudin, Gaussian, and logarithmico-normal relations as special cases, important physical facts can be obscured.[108] It would seem, rather, that the size distribution of broken coal can exhibit one, two, or more modes, each having a characteristic distribution determined by the process to which the coal has been subjected.[104]

The difficulty inherent in a statistical theory, leading to both a selection function as given in equation 17 and a breakage function as given in equation 18 or 19, can be seen from Brown's attempt[13] to derive a selection function, or strength-size relation, from Bennett's postulate. In involuntary degradation, coal is subjected to a range of stresses and to a range of repetitions of stress, leading to an assembly of pieces differing in size. The size distribution of this assembly, in the absence of any process of classification, is related to the previous history of the coal, in the sense that part of the energy has gone into the formation of fresh flaws and part into the creation of fresh external surface.[12] The energy dissipated as heat and kinetic energy of fragments is ignored. Let $U(x)$ be the fraction undersize x in the assembly. Consider now the breakage in a shatter test of a number of pieces of size x taken from the assembly. Owing to the wide variations in strength that can be induced by mild treatment,[12,109] it is clear that the breakage in the shatter test will be greater or less, depending on whether the energy put into the assembly has gone more or less into the formation of fresh flaws. If all the energy had appeared as fresh external surface, the size distribution of the assembly would be that found

[103] Epstein, B., *Ind. Eng. Chem.*, **40,** 2289–91 (1948).

[104] Taylor, J., *J. Inst. Fuel*, **26,** 133–8 (1953).

[105] Lázàr, J., *Acta Tech. Acad. Sci. Hung.*, **17,** 175–217 (1957).

[106] Svensson, J., *Trans. Roy. Inst. Technol. Stockholm*, No. 88 (1955), 53 pp.

[107] Harris, C. C., *2nd Symposium on Coal Preparation*, University of Leeds, 191–210 (1957).

[108] Brown, R. L., *Metallurgia*, **39,** 277–8 (1949).

[109] Brown, R. L., *Rev. ind. minérale*, **31,** 615–30 (1950/1).

by Griffith.[110] Comparing the actual size distribution of the assembly with that due to Griffith leads to the strength-size relation:

Shatter strength of size x

$$= a\left(\frac{100 - U}{dU/dx}\right) + d \quad (21)$$

where a and d are constants characteristic of the coal. The term $\left(\frac{100 - U}{dU/dx}\right)$ has the dimensions of a length and may be regarded as an 'apparent' size of pieces having "actual" size x. Equation 21, relating prior history to future breakage, was derived following a number of assumptions[13] which need reconsideration in the light of later work, particularly that on probability of survival. Nevertheless, experimental evidence was given for four different prehistories of one coal. If this result were found to be generally valid, it would suggest that the probability of survival C, in equation 17, of pieces of unit size, can be altered by suitable pretreatment.

To sum up, two statistical theories, concerned respectively with selection and breakage functions, have been proposed. Each suggests, but does not prove from first principles, a simple exponential form for the size distribution from breakage of a single fracture. This has been verified experimentally. The Rosin-Rammler and Gaudin forms have been shown to follow from a few repetitions of breakage. A strength-size relation associated with failure under uniform stress is inherent in one theory. The other leads to a strength-size relation possibly relevant to fracture of pieces by a range of stresses. The general accord between the theories and experiment is good enough to warrant the view that their further development may serve to limit possible forms for the matrix **M**.

It is, however, emphasized that statistical theories will not aid in an ultimate physical understanding of the mechanical properties of coal. For example, behind these theories lies the notion of a predetermined fracture pattern, as first proposed by Carey.[111] The fracture pattern can be given greater or less physical content by making various models related to the physical structure of coal, such as those due to Bennett,[11] Millard, Newman and Phillips,[69] and Evans and Pomeroy.[15] In the following discussion on comminution, reference is made to Bond, who has given particular attention to the length and propagation of cracks.[112] It would seem probable that modern work on the propagation and bifurcation of cracks may provide direct experimental evidence for a physical picture of fracture that can be used profitably as a basis for one statistical theory covering both the selection and breakage functions.

Theory of Shatter Test

Berenbaum, following on Broadbent and Callcott's analysis of shatter tests by the matrix method to obtain a single figure for the probability of breakage,[6] has incorporated equation 17, relating probability of survival to particle size, into the interpretation of shatter test data.[7] This enabled the change in size grading that occurs in several successive drops of a batch of coal to be taken into account. With some coals, constant values of C were obtained. In other cases, differences were evident. In his further work,[113] Berenbaum studied the effect of height of drop, h, showing that for Deep Duffryn coal, C varied as e^{-ih} where i is a constant. Table 4 shows typical results read from a graph.[113]

TABLE 4

EFFECT OF SIZE OF PIECE AND HEIGHT OF DROP ON PROBABILITY OF SURVIVAL[113]
[DEEP DUFFRYN COAL, $i = 0.013$ (in. ft)$^{-1}$]

x in. \ h ft	5	10	20
1	0.88	0.74	0.54
3	0.68	0.40	0.16
10	0.27	0.05	0.00

It will be appreciated that a shatter test cannot be refined sufficiently to bring out small differences in the form of the selection function, and it would not be expected that a selection function more complicated than the simple exponential used by Berenbaum could be discerned.

[110] Griffith, L., *Can. J. Research*, **A21**, 57–64 (1943).

[111] Carey, W. F., *Trans. Inst. Chem. Engrs.* (*London*), **12**, 179–85 (1934).

[112] Bond, F. C., *Chem. Eng.*, **59**, 169–71 (1952); *Trans. Am. Inst. Mining Met. Engrs.*, **193**, 484–94 (1952).

[113] Berenbaum, R., *J. Inst. Fuel*, **35**, 396–9 (1962).

Comminution

The complex degradation occurring in pulverizers has led to a number of studies differing in character from the foregoing. This work has had, as its main objective, the explanation of the relationship between energy input and size reduction. A large body of laboratory and plant data has been accumulated for many materials in recent years.[114,115] Although little of the work has been done with coal, it is worth reviewing briefly the lines of investigation, because, later, these may prove to be applicable.

Energy input varies with fineness of grinding, ranging from 3 to 4 kwhr/ton for coarse crushing, through 20 to 30 kwhr/ton for fine grinding, to 100 to 1000 kwhr/ton for superfine grinding. All but a fraction of the energy is dissipated as heat. Carey and Halton have calculated that of 30 kwhr needed to grind a ton of coal in a Hardinge mill, 8 kwhr were lost to the atmosphere, 9 kwhr went into heating the air purge, 7 kwhr were used to evaporate moisture, and 6 kwhr heated (a fraction going into breakage) the coal.[116] In other words, only 1 to 10% of the energy input to a mill is employed usefully in size reduction; moreover, the energy required for fracture is only 0.1 to 1% of the theoretical energy required to separate new surfaces in a homogeneous uncracked solid.[117] Hence, there is an unknown factor, somewhere between 1 and 100, required to link practical results with some ultimate physical explanation. This is to some extent accounted for in the proposal by Carey and Stairmand to relate the energy used in practical milling to the energy associated with the natural size distribution of material produced by free crushing.[16] In their test, a number of closely sized particles, spaced on a platen so that there was no mutual interference, were crushed simultaneously: This averaged effects due to the orientation of the particles. By limiting the travel of the platens to give a reduction ratio not greater than four, the authors claimed that energy

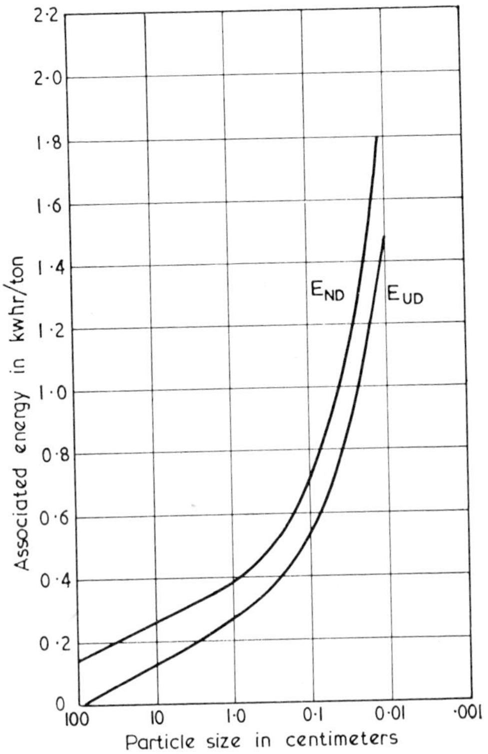

Fig. 5. Associated energy of natural and uniform distributions of coal. (South Hetton duff) (After Carey & Stairmand[16])

losses due to aggregation of the crushed material were minimized. The size distribution of the broken material was interpreted as a straight line on a Gaudin plot and called a natural size distribution. Treating the associated energy of particles of 1 m in size as zero, the energy, *ND*, required to produce natural distributions from a series of uniformly sized feeds was measured. The energy, *UD*, to produce a uniform distribution of particles all of one size, was calculated from these data. These two associated energies are plotted against particle size in Fig. 5. Given these curves, and the size distributions of the feed and product from a mill, it is possible to calculate the associated energy for the size reduction effected in the mill by splitting the product distribution into natural and uniform components. In so far as the crushing obtaining in this test is regarded as the most efficient practical crushing that can be obtained, then a practical efficiency of the mill can be calculated. The authors obtained an overall efficiency of about 5% for a tube-mill grinding coal by this method. The main assumptions

[114] Riley, R. V., *Chem. & Process Eng.*, **40,** 423–6 (1959); **41,** 23–5 (1960).

[115] Work, L. T., *Ind. Eng. Chem.*, **51,** 395–7 (1959); **52,** 178–80 (1960).

[116] Carey, W. F., and Halton, E. M., *Trans. Inst. Chem. Engrs.* (*London*), **24,** 102–8 (1946).

[117] Joffé, A. F., *Trans. Faraday Soc.*, **24,** 65–72 (1928).

appear to be that energy consumption is independent of the stress applied, that the particles are subjected to breaking forces in a manner substantially independent of their size, and that for coal of any rank, aggregation of the products of breakage of the single particles does not occur. In its present state of development, the method is somewhat laborious for extensive practical use.

The more usual approach has been to seek a fundamental relation between energy input and size reduction. Some have followed Kick's law, according to which the energy needed to effect equal reduction of pieces does not vary with the size of the starting piece. Fundamentally this law is applicable to homogeneous, elastic bodies having a yield point close to the elastic limit, so that work done in plastic deformation can be neglected.[118] Other workers have followed Rittinger's law, according to which the energy is proportional to the increase in surface area consequent upon the size reduction.[119] In the latter case, energy lost in plastic deformation and thermal vibrations is neglected, and in both cases the kinetic energy of fragments is neglected. Holmes[120] and Schuhmann[121] have suggested that many reported "proofs" of these two laws may be only apparent demonstrations of their validity, and other workers[122] have suggested conditions for their applicability.

On the basis of a large number of laboratory and plant tests,[123] a relation of energy to particle size x, intermediate between Rittinger's law and Kick's law, has been proposed, arguing that the length of cracks, which determines the energy for comminution, is inversely proportional to $x^{1/2}$. Holmes[120] and Charles[124] independently introduced a variable exponent r of particle size, namely, x^{-r} and showed that values between 0 (Kick) and 1 (Rittinger) could be obtained with different materials. Charles' analysis of data for coal obtained by Hukki[125] by breakage in a "frictionless crusher," gave $r = 0.9$; the Gaudin plot had $n = 1.0$. For all the hard brittle materials tested, values of r approximately equal to n were obtained.

Since 1950 a number of papers on the kinetics of grinding have appeared, and it may be that these will serve in due course to carry further the ideas discussed above and to lead to a better understanding of fine comminution. Bradshaw has suggested that the heat produced in grinding can weld small particles to larger ones, thus leading to a dynamic equilibrium.[126] Sedlatschek and Bass,[127] Bass,[128] and Zehler[102] have studied the rate of change of various size fractions with milling time. Theimer treated the equilibrium between grinding and agglomeration analogously to "thermal equilibrium" and derived, therefrom, equilibrium particle-size distribution[129] and rate equations.[130] Brenner and Vidmajer[131] and Patat[132] have extended this work. Simmonds has treated the formation of crack nuclei as an activated rate process and resulting subdivision as an energy distribution system.[133] It may be suggested that direct assessment of the breakage mechanism, as has been attempted by Reiners for impact comminution[134] and by

[118] Kick, F., *Dinglers Polytech. J.*, **247,** 1–5 (1883).

[119] Rittinger, P. von, *Lehrbuch der Aufbereitungskunde*, Ernst und Korn, Berlin 1867, 595 pp.

[120] Holmes, J. A., *Trans. Inst. Chem. Engrs. (London)*, **35,** 125–41 (1957).

[121] Schuhmann, R., *Mining Eng.*, **12,** 161–4 (1960).

[122] Gosset, M., *L'Équipment Mécanique des Mines, Carrières et Grandes Enterprises*, May, 1949, pp. 5–12. Walker, D. R., and Shaw, M. C., *Mining Eng.*, **6,** 313–20, 1106–8 (1954).

[123] Bond, F. C., and Wang, J.-T., *Trans. Am. Inst. Mining Met. Engrs.*, **187,** 871–8 (1950). Bond, F. C., *Mining Eng.*, **4,** 484–94 (1952).

[124] Charles, R. J., *Trans. Am. Inst. Mining Met. Engrs.*, **9,** 80–8 (1957).

[125] Hukki, R. J., Thesis, Mass. Inst. of Technology (1943).

[126] Bradshaw, B. C., *J. Chem. Phys.*, **19,** 1057–9 (1951).

[127] Sedlatschek, K., and Bass, L., *Powder Met. Bull.*, **6,** 148–53 (1953).

[128] Bass, L., *Z. angew. Math. u. Phys.*, **5,** 283–92 (1954).

[129] Theimer, O., *Kolloid Z.*, **128,** 1–6 (1952). Theimer, O., and Moser, F., *Kolloid Z.*, **128,** 68–74 (1952).

[130] Theimer, O., *Kolloid Z.*, **132,** 134–41 (1953); **133,** 44–50 (1953).

[131] Brenner, R., and Vidmajer, A., *Kolloid Z.*, **143,** 154–61 (1955).

[132] Patat, F., *Génie chim.*, **85,** 61–74 (1960). Patat, F., and Langemann, H., *Chem. Ing. Tech.*, **31,** 561–8 (1959).

[133] Simmonds, W. H. C., *Chem. Eng. News*, **33,** 2206–11 (1955); *Chem. Eng. Progr.*, **52,** 139–42 (1956).

[134] Reiners, E., *Chem. Ing. Tech.*, **32,** 136–42 (1960).

Nijman[135] for ball mills, must also be made if satisfactory explanations of comminution are to be obtained. Because comminution results in a product largely in the subsieve-size range, it is not unexpected that there is little information on the lower limit of size, except that implicit in direct measurements of specific surface, as have been made, for example, by Schellinger,[136] Piret,[137] and others. Because equilibrium theories have reference to a lower limit of size, much experimental work is, in this respect, inadequate. Meldau[138] and McCartney[139] have suggested that, for coal, the lower limit is from 200 to 500 Å, and this is small enough to call for special experimental procedures.

Moreover, many of the ways of representing data depend on an average property of the ground product, which for similar size distributions (for example, the same n in equations 18 or 19) can be replaced adequately by the percentage through a fixed mesh, such as the 200 mesh commonly used with pulverized coal. This statement does not imply that the more fundamental studies have no value; they are essential if crushing and grinding of coal are to be improved and the true interrelations between the mechanical properties of coal and the various breakage processes are to be found.

BROKEN COAL

Because of the importance of size grading in coal utilization, much practical work has gone into ways of designating coal sizes, especially those graded between two screens and into standard methods of determining sieve analyses. Where the coal covers a wide range of sizes, various ways of representing the sieve analyses have been proposed. In some cases, use has been made of the size-distribution relations already described in the preceding theoretical section.

Designation of Size and Shape

It has long been recognized that the designation of size and shape of an irregular solid such as coal requires careful attention to detail. Material larger than about 200 mesh can be analyzed into sieve fractions. Rules are available which enable representative samples to be taken and reliable analyses to be obtained.[140] These show that sieve analyses must be made and interpreted most carefully. Smaller material can be analyzed by microscope counting, sedimentation in a fluid, and other methods.[21,141,142] When it is necessary to combine data obtained by different methods, account must be taken of the shapes of the particles.[141,143] Industrial coal is commonly characterized by analysis on a standard series of sieves.[144] With pulverized fuel, a subsieve-size analysis is sometimes made as well. For some grades of coal, a full-sieve analysis is not necessary, and various abbreviated designations have been proposed.[145]

[135] Nijman, J., *Brit. Chem. Eng.*, **3**, 76–80 (1958).

[136] Schellinger, A. K., *Mining Eng.*, **3**, 518–22 (1951); **4**, 369–74 (1952).

[137] Kwong, J. M. S., Adams, J. T., Johnson, N. J. F., and Piret, E. L., *Chem. Eng. Progr.*, **45**, 508–16 (1949). Adams, J. T., Johnson, N. J. F., and Piret, E. L., *Chem. Eng. Progr.*, **45**, 655–60 (1949). Johnson, N. J. F., Axelson, J., and Piret, E. L., *Chem. Eng. Progr.*, **45**, 708–15 (1949).

[138] Meldau, R., *Verfahrenstechnik*, **4**, 103–6 (1940).

[139] McCartney, J. T., *U.S. Bur. Mines, Rept. Invest.*, No. 3827 (1945), 11 pp.

[140] Am. Soc. Testing Materials, ASTM Standards D 410–38, D 310–34, D 197–30, D 311–30. Brit. Standard No. 735 (1944), 75 pp; No. 1293 (1946), 54 pp.

[141] *Physics of Particle Size Analysis, Brit. J. Appl. Phys., Supplement* No. 3 (1954), 218 pp.

[142] Am. Soc. Testing Materials, *Spec. Tech. Publ.* No. 234 (1958), 310 pp. Rose, H. E., *The Measurement of Particle Size*, Constable, London, 1953, 127 pp.

[143] Skinner, D. G., Boas-Traube, S., Brown, R. L., and Hawksley, P. G. W., *Determination of Particle Size in the Sub-sieve Range*, BCORA-BCURA, (1944), 69 pp. Robins, W. H. M., *Physics of Particle Size Analysis, Brit. J. Appl. Physics, Supplement* No. 3, 82–5 (1954). Guruswamy, S., Roy, L. C., Dasvarma, R. L., and Srinivasan, S. R., *J. Sci. Ind. Research (India)*, **12B**, 91–101 (1953). Harris, C. C., and Howitt, D., *Nature*, **187**, 401–3 (1960).

[144] Am. Standard, E11–58T (1959). Catalogue No. 53, W. S. Tyler Co., Cleveland, Ohio (1952). Brit. Standards, No. 481 (1953), 9 pp.; No. 1669 (1950), 10 pp.; No. 1812 (1951), 17 pp. French Standards, M82–516 (1948); X11–501 (1938); X12–101 (1941). German Standards, DIN 1170 (1939); DIN 1171 (1934).

[145] Am. Soc. Testing Materials, ASTM Standard D431–38 (1942).

These proposals are based on practical experience and, as such, serve their purpose adequately. Indeed, a similar straightforward method can be adopted fruitfully in the study of research and operational data. For example, important results on the breakage of coal in pick-and-roll crushers were deduced by Davies and Wilkins[146] by studying the proportions passing a $\frac{5}{16}$ in. screen. Likewise, it is common practice to interpret data on pulverizing in terms of the percentage of the product passing 200 mesh.

The value of a severely practical approach is further illustrated by the proposal in the United Kingdom to designate the size of graded coals by the nominal sizes of the screens used at the colliery, coupled with an actual percentage undersize limitation found from a sieve analysis.[147] It was found that coals prepared at the colliery with screens differing in nominal size could have the same actual sieve analysis, and contrariwise.[17] This fact permitted the term doubles, for example, to be used for coals prepared by passing through a round hole screen in the range $1\frac{3}{4}$ to $2\frac{1}{4}$ in. and over a screen in the range 1 to $1\frac{1}{2}$ in. with such efficiency as to give a sieve analysis having 25% or less through $\frac{7}{8}$ in. square mesh and 3% or less through $\frac{1}{8}$ in. square mesh.

Interpretation of Sieve Analyses

When it is desired to examine the whole size analysis of broken coal, for example, run-of-mine, slacks, or smalls graded through a screen, or mine dust, it is customary to plot a graph of the percentage weight undersize or oversize against sieve size, or of the percentage number undersize or oversize against microscope size. Various special grids have been developed so that the graphs may approximate to straight lines.[21,148,149] Also numerous procedures have been proposed for calculating various statistical measures of the size distribution.[21,148,150] The more important of these measures are the mean size (on a weight, area, or number basis) and the dispersion. For many practical purposes, for example, combustion of pulverized fuel or flow through bunkers, the largest or top size can be important.

Differentiation of the weight undersize or oversize curve gives the weight frequency distribution. An approximation can be obtained directly from the sieve analysis by plotting a histogram in which the ordinate is the percentage between adjacent sieves divided by the difference of sieve sizes.

Examination of frequency curves or histograms is particularly important for coal broken in crushers or shatter tests. In such cases several distinct modes or humps may appear.[151] This may be due to the simultaneous presence of several stages of size reduction or to the incomplete or partial degradation of the original material. Now, in a broken coal having a residue over (say) 1 in. and a complement under (say) $1\frac{1}{2}$ in., the percentage through $\frac{1}{2}$ in. depends on both the amount of the complement and its size characteristics. In such a case, useful information is obtainable from the data by relating the percentages through $\frac{1}{2}$ in. to that through other sizes. This simple idea led Brown to propose the use of *ratios* of the weight between any two sieves divided by the weight between two fixed sieves, the fixed sieves being chosen to lie wholly in the complement or wholly in the residue.[152] The method has been applied to the fineness of cuttings in underground workings,[153] to untreated smalls,[154] and to washed smalls.[155] Regularities

[146] Davies, D. T., and Wilkins, E. T., *Trans. Inst. Mining Engrs.* (*London*), **96**, 55–81 (1938).

[147] *The Size Grading of British Coals*, BCORA-BCURA, (1946), 8 pp.

[148] Taggart, A. F., *Handbook of Mineral Dressing*, John Wiley and Sons, New York, 1950, pp. 145–50.

[149] Herdan, G., *Small Particle Statistics*, 2nd edition, Butterworths, London, 1960, pp. 90–105.

[150] Kneschke, A., *VDI*, **101**, 413–8 (1959). Langemann, H., *Chem. Ing. Tech.*, **27**, 27–32 (1955). Scott, G. S., *U.S. Bur. Mines, Rept. Invest.* No. 3732 (1943), 9 pp. Landers, W. S., and Reid, W. J., *U.S. Bur. Mines, Inform. Circ.* No. 7346 (1946), 5 pp. Hüttig, G. F., *Monatsh. Chem.*, **84**, 272–7 (1953). Bull, F. A., *J. Inst. Fuel*, **28**, 163–70 (1955).

[151] Heywood, H., *Symposium on Particle Size Analysis*, Supplement to *Trans. Inst. Chem. Engrs.* (*London*), **25**, 14–24 (1947); *Chem. & Ind. Rev.*, **56**, 149–54 (1937).

[152] Brown, R. L., *Colliery Eng.*, **25**, 295–9 (1948).

[153] Brown, R. L., *Brit. Colliery Owners Res. Assoc.*, **7**, 9–21 (1945).

[154] Brown, R. L., and Boswell, I., *Colliery Eng.*, **26**, 18–22, 58–62 (1949).

[155] Brown, R. L., *Colliery Eng.*, **25**, 420–2 (1948).

TABLE 5

COMPARISON OF SIZE DISTRIBUTION RELATIONS

Type of Coal	Method of Sizing	Gaudin Slope	Rosin-Rammler n	Rosin-Rammler $\bar{x}$	Gamma Function n	Gamma Function $\bar{x}$†
Run of mine	Sieving	0.72	0.82	5.9 cm	0.82‡	3.7 cm
$\frac{1}{2}$ in. untreated smalls	Sieving	0.79	0.88	0.38 cm	0.88‡	0.26 cm
Pulverized fuel	Sieving and microscope	0.92	1.08	100 μ	1.08‡	82 μ
Mine dust	Microscope*	2.44	2.64	36 μ	3.36	29 μ

* Converted to weight basis on assumption that weight was proportional to x^3.
† Taken as the value of the undersize at 50% on the Gaudin plot.
‡ For these coals, n was taken to be the same as for the Rosin-Rammler value.

in the sieve analyses resulting from different types of breakage have been found. For many untreated coals,

$$\left.\begin{aligned}&\text{Relative fineness}\\&=\frac{\text{percentage through } \frac{1}{8} \text{ in.}}{\text{percentage through } \frac{3}{4} \text{ in.}} \simeq \frac{1}{3}\\&\text{Relative dustiness}\\&=\frac{\text{percentage through 60 mesh}}{\text{percentage through } \frac{1}{8} \text{ in.}} \simeq \frac{1}{5}\end{aligned}\right\} \quad (22)$$

This work led to a method of designating the fines content of smalls, that is, the percentage through $\frac{1}{8}$ in., produced normally at collieries in the United Kingdom.[17,147] Although the ratios may be used without assuming or proving any size distribution relation, it may be noted that a given ratio is nearly constant for all materials covering similar, but not identical size ranges, and following the Gaudin relation with the same slope n.

Size Distribution Relations

As already mentioned, the sieve analysis of the complement produced by shattering 3 in. and similar sized lumps by dropping from $1\frac{1}{2}$ to 6 ft was approximately linear when plotted in the Gaudin form.[96] The slope n is a measure of the dispersion, which increases as n decreases from the value of unity that obtains for a single fracture.[96,156] The wide applicability of the Gaudin relation is shown, for example, by its use in studies of fragmentation of stone blocks by explosives,[157] of crushing in hammer mills, jaw and roll crushers,[158] and in analyzing the sub-sieve size debris resulting from crushing 1 in. cubes of coal.[159]

To give some idea of the applicability of the various relations found from statistical mechanical studies, data for a run-of-mine coal,[11] an untreated $\frac{1}{2}$ in. smalls,[160] a pulverized coal (unpublished data), and a North Staffordshire mine dust[161] are plotted in the form log $U(x)$ against log x in Fig. 6. In all cases, the lower left hand portion of the data is near a straight line, that is, close to Gaudin's relation. The curvature at the top right-hand corner is taken equally well into account by the Rosin-Rammler and the incomplete gamma relations, as shown by the curves of best fit in the figure. The constants are given in Table 5.

It will be noted that the slope of the straight line in the Gaudin relation is slightly less than the exponent n in the Rosin-Rammler relation. A good fit for the incomplete gamma function was obtained with the run-of-mine coal, $\frac{1}{2}$-in. untreated smalls, and pulverized fuel by taking the same exponent as that found from the Rosin-Rammler plot. For the mine dust, where n departed markedly from unity, this was not feasible, and a method of successive approximation had to be used. Slopes near 2.5 were also found by McCartney for Rosin-Rammler plots of the ultrafine fraction, 0.1 to

[156] Rammler, E., *Verfahrenstechnik*, **4**, 6–11 (1940).

[157] Walton, W. H., Ed., *Op. cit.*, pp. 380–9.

[158] Ratcliffe, A., *Proc. Inst. Mech. Engrs. (London)*, **162**, 378–91 (1950).

[159] Walton, W. H., Ed., *Op. cit.*, pp. 365–79.

[160] MacDonald, E. J., and Murray, M. V., *J. Inst. Fuel*, **25**, 308–17 (1953).

[161] Wynn, A. H. A., and Dawes, J. G., *S.M.R.E. Research Rept.* No. 28, London, H.M.S.O. (1951), 46 pp.

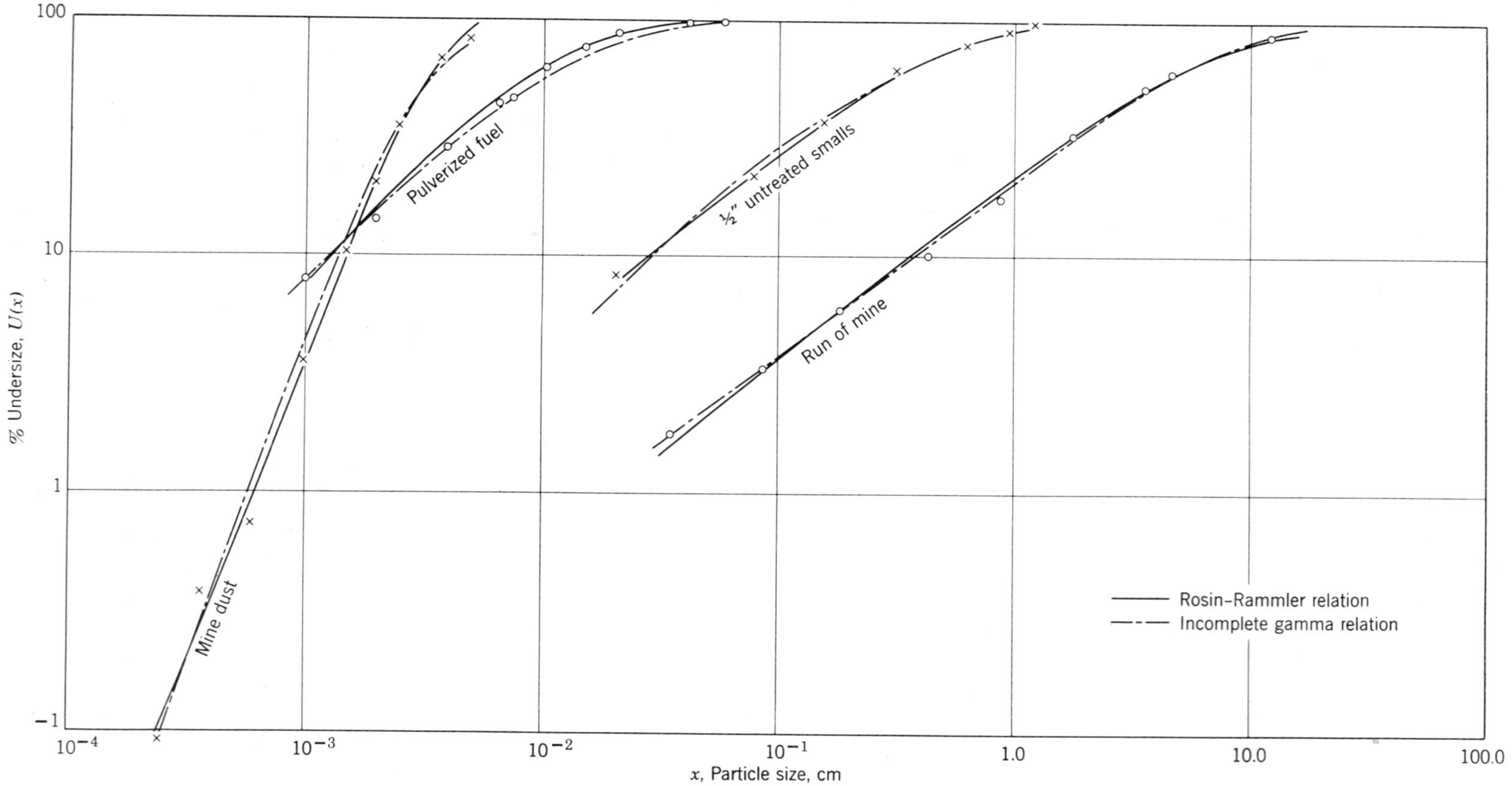

Fig. 6. Size distribution relations.

2 μ, extracted from a suspension of subsieve sizes and sized with an electron microscope.[139] Figure 6 demonstrates the empirical value of relation (18), to which Rammler has recently drawn attention.[162] Because repeated shatter gives slopes less than unity, those powders giving values exceeding unity owe their size distribution to some other mechanism.

Consideration of Fig. 6 shows that experimental data, as usually obtained, do not permit an unequivocal choice of size distribution relation. The curvature at the larger sizes is often suspect because some classification may have occurred. Moreover, in practice, more complex size distributions having two or more modes are found. Examples have already been given in the reference to partial degradation of lump coal: Bimodal distributions have also been found with some mine dusts.[161] Even with unimodal distributions, operating conditions in coal preparation plant can result in other forms, for example, normal or log normal distributions.

MODEL TESTS

Once coal has been gotten from the seam, it is subjected to a variety of forces, the nature of which is not describable precisely. Consequently, model tests, simulating practical occurrences, serve a useful purpose in assessing the degradation that coal undergoes. The more important occurrences are shattering, as simulated by the friability test, and grinding, representing a combination of attrition and crushing. The work reported is prior to that of Berenbaum and does not make use of a matrix representation.

Friability

At the time when Yancey and Geer[1] presented their comprehensive survey of drop shatter and tumbler tests for assessing the friability of coal, standard tests were not available.

The standards published soon afterwards by the ASTM epitomize the careful preparatory work referred to above. The latest standard was revised in 1949.[163] Friability is expressed as:

$$(100\text{—size stability}) \tag{23}$$

where size stability is:

$$100\left(\frac{\text{average size after test}}{\text{average size before test}}\right) \tag{24}$$

The test involves two 6 ft drops on any chosen initial size of coal. In the tumbler test 1.06 × 1.50 in. coal is used.

In a discussion of friability data and its applications, Fish noted that friability increased with rank through bituminous coals, to a maximum near 75% fixed carbon, and then decreased for the anthracites. There was a sharp decrease as the inherent moisture increased to 5% and, thereafter, a slower, but steady fall.[164] Guruswamy has proposed, as an alternative method of assessing the breakage in a shatter test, to derive a "toughness" factor from the amount of breakage and the sieve analysis of the broken product. Using this factor in a study of breakage of different sizes and specific gravity fractions, he found the toughness factor to increase as size decreased from 6 to 2 in., and to decrease with increasing specific gravity.[165]

Before passing to other work on coal, it should be noted that the thorough examination of shatter and tumbler tests on cokes, published by Lowry and Epstein in 1948, led to the conclusion that data relating to more than one stage of breakdown are needed if effective practical use is to be made of laboratory tests.[9] Gayle and Auvil have since examined the precision of the shatter test on coke.[166] Protodyakanov has described six rapid methods ranging from penetration of tips into a coal surface to a drop hammer test on lumps of coal over 0.4 in. in size.[64] Pozzetto has discussed the influence of friability of the coal on the manufacture of briquettes.[167]

[162] Rammler, E., and Glockner, E., *Freiberger Forschungsh.*, **A44**, 96–108 (1955). Rammler, E., *Forsch. u. Fortschr.*, **30**, 1–9 (1956).

[163] Am. Soc. Testing Materials, ASTM Standards D 440–49, 992–96; D 441–45.

[164] Fish, B. G., *Colliery Eng.*, **28**, 317–19, 371–3, 413–14 (1951).

[165] Guruswamy, S., and Roy, L. C., *J. Sci. Ind. Research (India)*, **11**, 482–5 (1952). Guruswamy, S., *J. Sci. Ind. Research (India)*, **12**, 270–4 (1953).

[166] Gayle, J. B., and Auvil, H. S., *U.S. Bur. Mines, Rept. Invest.* No. 4987 (1953), 13 pp.

[167] Pozzetto, L., *Rev. ind. minérale*, **42**, 40–52 (1960).

In 1948, Brown examined drop shatter and tumbler tests on coal in the light of the idea that breakage occurs at pre-existing flaws and cracks.[168] He showed that the strength of a given size of coal in a given consignment can be measured only if the sample is taken without either decreasing its strength through causing invisible breakage or increasing its strength through producing visible breakage. He drew attention to the need to relate the strength of coal to some other measurable characteristic of a consignment. The sieve analysis of the consignment was later suggested as an appropriate measure.[13]

Recently, Pomeroy[169] made a careful study of the model test for impact strength proposed earlier by Protodyakanov[64] and showed that care was needed in standardizing the conditions of the test. This is evident from Carey's earlier work[111] on the energy used in crushing beds of coal particles. Twice as much energy was needed for a size reduction in the proportion 1:0.43 as in a reduction, only a little less severe, of 1:0.50. In Pomeroy's test, a bed of 100 g of $\frac{3}{8} \times \frac{1}{8}$ in. coal was broken by a 4 lb weight dropped twenty times from a height of 12 in. The impact strength was assessed by the percentage remaining on $\frac{1}{8}$ in. Figure 4 shows that this impact strength is closely related to the compressive strength of 1 in. cubes. It should be noted that it does not show effects due to anisotropy, nor does it have relation to tensile strength. Also shown in Fig. 4 is a band, derived from the bright coal band shown for the relation, given below, of grindability and rank. The band in Fig. 4 has ordinate $(98-y/2)$, where y is the ordinate in Fig. 7. The conversion factor is necessarily arbitrary. It is evident that the variations of grindability, impact, and compressive strengths with rank are similar. It would be interesting, however, to compare data for these properties all measured on the same coal. It may be noted that the probability of survival of 1 in. lumps varies with rank in the same way as the band in Fig. 4.[68]

Grindability

The tests of grindability most commonly employed are the ball mill test and that devised by Hardgrove. Standard forms of these tests have been formulated by the ASTM.[170,171] Alternative measures of grindability are in use or have been suggested: For example, Jensen and Fryer[172] have described a method using a pair of rolls driven at different speeds with a fixed gap, while Gründer[173] has described a test in which the energy consumption is related to the percentage of fines in the product. Bond has given grindability figures for a large number of materials.[174] Brown, Ivison, and Birney proposed the use of an edge-runner mill on the grounds of convenience.[175] Zeisel has proposed criteria, relating ease of grinding to surface area and grinding conditions.[176]

The Hardgrove method is generally preferred to the ball mill test, by reason of its greater speed, and the ASTM standard for the ball mill test was discontinued in 1951. Correlations between the ball mill and Hardgrove grindability indices have been obtained by the ASTM,[171] and for British coals by Fitton, Hughes and Hurley.[177] Earlier, Heywood at the 1947 Pulverized Fuel conference had given reasons for preferring the ball mill test.[178] In its original formulation, the Hardgrove index purported to be related to the specific surface of the ground product. It has been shown recently, however, that the index is a poor measure of the increase in external surface area with grinding. Following on work by Romer,[179] Austin, Gardner, and Walker measured the

[168] Brown, R. L., *Fuel*, **27**, 82–95 (1948).

[169] Pomeroy, C. D., *J. Inst. Fuel*, **30**, 50–4 (1957).

[170] Am. Soc. Testing Materials, ASTM Standards, D409–51.

[171] Am. Soc. Testing Materials, ASTM Standards, D408–37T.

[172] Jensen E. J. and Fryer, J. F., *Research Council Alberta (Can.), Circ.* No. 26 (1958), 37 pp.

[173] Gründer, W., *Glückauf*, **74**, 641–6 (1938); *Verfahrenstechnik*, **2**, 17–23 (1938).

[174] Bond, F. C., and Maxson, W. L., *Am. Inst. Mining Met. Engrs., Tech. Publ.* No. 1579 (1943), 11 pp. Bond, F. C., *Am. Inst. Mining Met. Engrs., Tech. Publ.* No. 2180 (1947), 11 pp.; *Trans. Am. Inst. Mining Met. Engrs.*, **153**, 362–72 (1943); *Mining Technol.*, **11**, No. 4 (1947), 17 pp.

[175] Brown, J., Ivison, N. J., and Birney, J. W., *Proc. Pulverised Fuel Conf.*, Inst. Fuel, Harrogate, 1947, pp. 390–7.

[176] Zeisel, H. G., *Schriftenreihe Zement-Ind.*, **14**, 31–72 (1953).

[177] Fitton, A., Hughes, T. H., and Hurley, T. F., *J. Inst. Fuel*, **30**, 54–65 (1957).

[178] Heywood, H., *Proc. Pulverised Fuel Conf.*, Inst. Fuel, Harrogate, 1947, pp. 594–611.

[179] Romer, J. B., *ASTM, Proc.*, **41**, 1152–66 (1941).

shape factor of coal and used it to derive the specific surface from the sieve analysis.[180] It appeared that high and low rank coals break to more flaky forms than do those of medium rank, and this affects the conversion factor. It has been suggested by Callcott that the results of the test are adequately reported simply as the percentage passing the 200 mesh sieve, losses being included.[181]

Factors affecting the result of a grindability measurement by the Hardgrove test are (1) method of making the test, (2) presence of different petrographic constituents, and (3) moisture and temperature. Callcott has shown that both the profile of the bowl, and the method of preparing the sample, can affect the measurement.[181] Differences of up to 5 units could arise if the mean size of the sample approached closer to or further from the larger permitted sieve size (16 mesh U.S.S., 14 mesh B.S.). Subsequent correspondence indicated that greater attention to sample preparation would be given in the revised standard.[171] The method of sample preparation also affects the measured index through variations in the petrographic constitution of the prepared sample. Ghosal and others, in a study of Indian coals, have shown that the different petrographic constituents can exercise a considerable influence on grindability.[182] This is not unexpected because macerals can be concentrated by controlled size reduction and classification.[183]

Jutte has suggested that failure to take ash into account means that the test is not a true measure of grindability.[184] The effect of ash content has been observed by Fitton, Hughes and Hurley who measured the Hardgrove grindability indices of a number of coals, covering a range of ash contents for each coal.[177] The ash had a Hardgrove index of about 75, and it was shown that the indices of both the hard and soft coals converged towards this value as the ash content was increased.

There has been little published work on the effect of temperature on grindability, although it has been shown that oven-dried samples gave a higher index.[177] The effect of temperature is inevitably bound up in practice with the effect of moisture content, and Callcott[185] and Venables[186] have shown that these effects are complex. The effect of moisture content has been particularly studied in low-rank coals, where it presents a special problem. Ellman and Belter studied the variation in Hardgrove index of a Dakota Star lignite and found an index of 55 at 10% moisture, falling to 43 at 26% and rising again to 55 at 38%.[187] Similar variations were found with other lignites and were attributed to interparticulate cohesion at the higher moisture contents and to a weakening of the physical structure at the lower moisture contents. Similar behavior in Alberta lignites was demonstrated by Jensen and Fryer, who showed that results from their alternative test, referred to above, varied in a more regular and perhaps more useful manner.[172] It was, however, admitted that further work on this test was needed.

Correlations between grindability index and rank are of practical interest as providing a general guide to the ease of grinding of a coal of a particular rank, although it has been asserted that the rank of a coal is not a sufficient guide to grindability and a direct measurement should be made.[177] Callcott has suggested that the index is a first order measure of the mechanical characteristics.[181] Surveys of Hardgrove grindability indices have been made of American,[188] British,[177,181] and Indian coals.[182] The relation between grindability and rank is shown diagrammatically in Fig. 7. This diagram was prepared from the original data given in the surveys by setting limits for the

[180] Austin, L. G., Gardner, R. P., and Walker, P. L., *136th Mtg. Am. Chem. Soc., 6K, Abstrs.* **21, 22, 23** (1959).

[181] Callcott, T. G., *J. Inst. Fuel*, **29**, 207–17 (1956); **30**, 466–71 (1957).

[182] Ghosal, A., Roy, L. C., Bose, R. N., and Lahiri, A., *J. Inst. Fuel*, **31**, 50–5 (1958).

[183] Kröger, C., and Bade, E., *Glückauf*, **96**, 863–71 (1960). Winnicki, V., *Koks, Smola, Gaz*, **4**, 166–71 (1959). Das Gupta, N. N., Rao, V. V., and Lahiri, A., *Mines, J., Metals & Fuels*, **7**, 53–8 (1959). Winnicki, V., Olszewska, K., and Magnes, C., *Prace Glównego Inst. Górnictwa, Komm.* No. 260 (1960), 16 pp.

[184] Jutte, K., *Brennstoff-Wärme-Kraft*, **6**, 92–3 (1954).

[185] Callcott, T. G., Private Communication (1956).

[186] Venables, J., Private Communication (1958).

[187] Ellman, R. C., and Belter, J. W., *U.S. Bur. Mines, Rept. Invest.* No. 5167 (1955), 38 pp.

[188] Babcock & Wilcox Co., *Bull.* No. 3 (1938), 34 pp. Gillmore, D. W., Wright, C. C., and Walker, P. L., *Penn. State Univ., Mineral Ind. Expt. Sta., Bull.* No. 70 (1957), pp. 67–9.

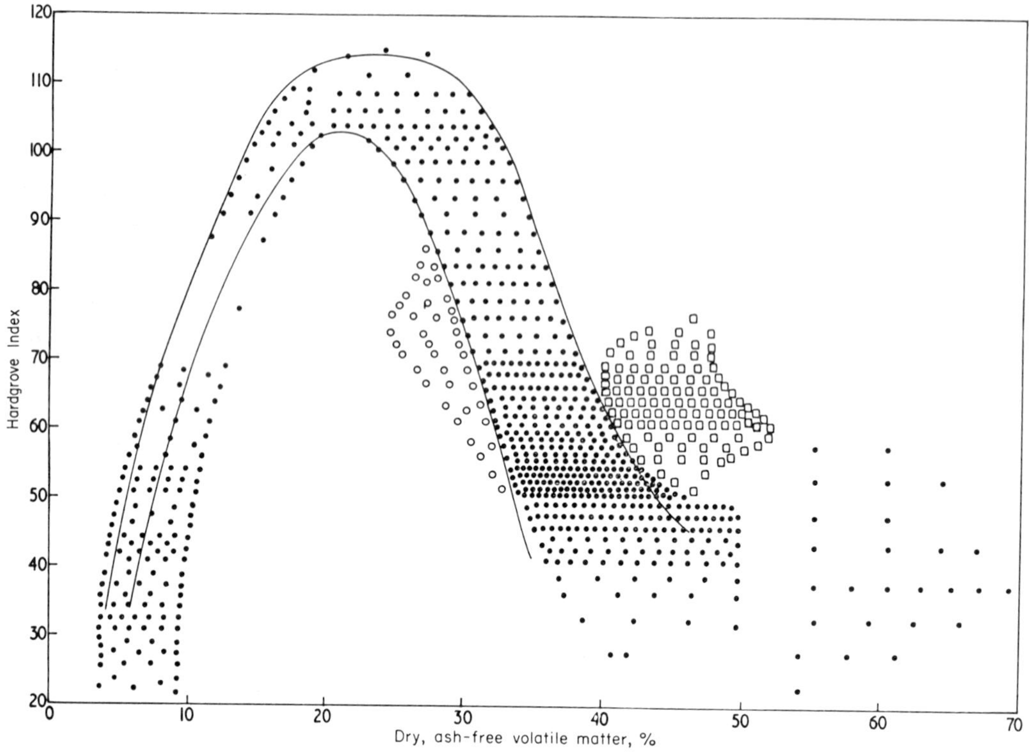

Fig. 7. Grindability and rank.

range of volatile matter within which 90% of readings lay, in groups of indices from 20 to 25, 25 to 30, and so on and plotting a dot for each coal tested. This stylized diagram, therefore, shows the general trend. The curves define the range within which British bright coals lie, mostly having medium or low ash content. Some Indian coals, mainly from the Jahria field, have rather low grindability indices. This may be due to the effect of intrusive dykes.[189] Lignitic coals, mainly from the southern American States, also lie outside the curves and have rather high grindabilities. The effect of rank has been studied in a different way by Tachikawa and Ikai,[190] using criteria based on a Rosin-Rammler plot, and this equation has also been used in a comparison of the grindability of bituminous coals and lignites.[191] Lignites were found to be more resistant to grinding than either bituminous or brown coals, and this was attributed to the "plastic" character of the lignites. Heinze, using Gründer's "structure test," has shown a considerable variation to occur over the range of volatile matter from 5 to 40%.[60]

[189] Francis, W., *Coal*, Arnold, London, 1954, pp. 464–9.

[190] Tachikawa, S., and Ikai, S., *J. Fuel Soc. Japan*, **29,** 156–61 (1950).

[191] Buimovich, D., and Bykleshanu, D., *Rev. electrotech. energét. Acad. rép. populaire Roumaine*, **2,** 291–9 (1957).

Grinding Machines

No recent work has been reported on the correlation of grindability with ease of pulverizing, apart from studies of the operation of a ring-ball mill.[85,192] The performance of the mill, expressed as a power per ton of useful output (through 200 mesh B.S.), was related to the Hardgrove indices of the coals used, which covered the range 54 to 104. The power requirement for a given output tended to increase more rapidly as the Hardgrove index of the coal was reduced below about 70.

[192] Fuel Research Board and British Coal Utilisation Research Association, *J. Inst. Fuel*, **30,** 269–76 (1957). Hiorns, F. J., and Reed, L. E., *J. Inst. Fuel*, **31,** 383–92 (1958). Brown, R. L., and Hurley, T. F., *5th World Power Conf.*, Section 6, Paper 116 (1956).

Studies of grinding machines are not examined in detail in this chapter. For accounts of progress in grinding techniques, the reader is referred to the articles by Riley[114] and Work,[115] to a survey by Behrens[193] and to a summary by Speddon[194] of the papers on comminution presented to the Third International Mineral Processing Congress.[22] The mechanism of tube mills has been studied by Rose and Sullivan,[195] and Callcott has reviewed types of ball mill and their operation.[196] Further papers of interest on crushing and grinding include work on impact comminution,[197] jaw crushers;[198] hammer mills;[199] rolls;[200] vibration mills;[201] pinned disk mills;[202] and metal abrasion.[203] There is a review of nonmechanical methods of size reduction[204] and a study of breakage in jets.[205] Many other useful papers may be found in the DSIR bibliography.[20] Papers concerned with the general principles of operation of grinding machines include those of Andreasen[206] and Heywood.[207] Taylor has examined the size distribution of the product obtained when coal and coke are broken in several types of mill.[208]

[193] Behrens, D., *Staub*, **20**, 101–8 (1960).

[194] Speddon, H. R., *Mining Eng.*, **12**, 677A–677C (1960).

[195] Rose, H. E., and Sullivan, R. M. E., *Ball, Tube and Rod Mills*, Constable, London, 1958, 258 pp.

[196] Callcott, T. G., *BCURA Monthly Bull.*, **21**, 101–20, 153–71 (1957). Frenay, E., and Masson, A., *Rev. matériaux construct. et trav. publ.*, No. 520, 1–9 (1959). Frick, C. H., *Coal Technology*, **1**, 1–8 (1946). Guerrero, P. K., and Arbiter, N., *Mining Eng.*, **12**, 488–90 (1960). Mortensen, M., and Hansen, H. J., *Tidsskr. Kjemi Bergvesen Met.*, **15**, 141–6 (1955). Perow, W. A., and Brand, W. J., *Fernmallen der Erge*, V. E. B. Verlag Technik, Berlin, 1955, 192 pp. Rose, H. E., and Blunt, G. D., *Proc. Inst. Mech. Engrs. (London)*, **170**, 793–800 (1956).

[197] Rumpf, H., *Dechema Monograph.*, **24**, 58–104 (1955); *Chem. Ing. Tech.*, **32**, 129–35, 335–42 (1960). Schranz, H., *3rd Intern. Coal Preparation Conf., Liège*, **1958**, pp. 77–86.

[198] Levenson, L. B., *Mekhanizatsiya Stroitel.*, **1954**, No. 1, 27–31. Gauldie, K., *Engineering*, **176**, 456–8, 485–6 (1953). Birebent, A., and Mamillan, M., *Ann. inst. tech. bâtiment et trav. publ.*, **7**, 965–84 (1954).

[199] Bechdoldt, H., *Mitt. Ver. Grosskesselbesitzer*, **22**, 326–8 (1953). Beleni, I., *Acta Tech. Acad. Sci. Hung.*, **10**, 355–92 (1955). Berezin, B. V., *Koks i Khim.*, **5**, 51–5 (1957). Emel'yanov, V. S., *Elek. Stantsii*, **28**, 18–9 (1957). Schiller, A., *Brennstoff-Wärme-Kraft*, **7**, 223–7 (1955). Callcott, T. G., *J. Inst. Fuel*, **33**, 529–39 (1960).

[200] Kircher, A., *Glückauf*, **92**, 794–9 (1956). Seguiti, T., *Rev. ind. minêrale*, **33**, 537–46 (1952).

[201] Chetaev, N. G., *Izvest. Akad. Nauk S.S.S.R., Otdel, Tekh. Nauk*, **1957**, No. 3, 49–56. Rose, H. E., *Trans. Inst. Chem. Engrs. (London)*, **38**, 107–24 (1960).

[202] Batel, W., *Chem. Ing. Tech.*, **32**, 448–53 (1960).

[203] Yancey, H. F., Geer, M. R., and Price, J. D., *Trans. Am. Inst. Mining Met. Engrs.*, **190**, 262–8 (1951). Sokolov, N. V., and Osokin, V. P., *Teplotekhnika*, **6**, 37–41 (1959).

[204] Murphy, E., Ross, F. F., and Sharpe, G. C. H., *BCURA Monthly Bull.*, **11**, 221–35 (1947).

[205] Batten, J. J., Thesis, London Univ., 1958, 177 pp.

[206] Andreasen, A. H. M., *Ing. Vetenskaps Akad.*, **26**, 86–93 (1955).

[207] Heywood, H., *Chemical Engineering Practice*, Butterworths, London, 1957, Vol. 3, pp. 97–108.

[208] Taylor, J., *J. Inst. Fuel*, **27**, 249–54 (1954).

CHAPTER 4

THE PLASTIC PROPERTIES OF COAL

R. Loison, A. Peytavy, A. F. Boyer, and R. Grillot

Centre d'Etudes et Recherches des Charbonnages de France, Paris

A rise in temperature affects the behavior of coal in two ways: The coal undergoes chemical transformations, giving rise to gas and condensable vapors and leaves a solid residue consisting almost exclusively of carbon; and the coal softens and becomes more or less plastic, the particles caking to form a bubbly compact mass which swells and then resolidifies, forming a solid coherent body with a porous structure, that is, coke.

Numerous types of laboratory apparatus have been used for studying the behavior of coal during heating. Tests have been devised which allow this behavior to be characterized by numerical indices that are not generally physical constants but serve as reference points. These indices are of great practical use for classifying coals and predicting their behavior during combustion, gasification, and carbonization in industrial plants. Several attempts have been made, however, to measure physical values relating to the general principles of the mechanics of solid or liquid bodies.

DIFFERENT METHODS OF STUDYING PLASTIC PROPERTIES

Microscopic Examination

Direct observation was probably the first method which gave an indication of the plastic properties of coal during heating. Observation with the naked eye gives only a rough and qualitative indication of the phenomena. Examination under the microscope gives more detailed information.

In one method, the sample is kept in the field of a microscope while being heated, and the changes are observed continuously. A current of inert gas is passed over the sample to prevent oxidation and to remove gases and vapors as they are liberated. The temperature of the sample is measured by a thermocouple. The Leitz heating microscope used by various authors[1,2] and the apparatus of Potonié and Stockfisch[3] are based on this principle. The sample examined may consist of a small individual particle of coal,[4,5,6] or a thin, transparent section observed on a heating stage. The changes can be followed visually or photographically. In the latter case, a "slow motion" movie gives a very informative representation of the phenomena.[7]

In another method, the sample, for example, a cut cube of coal,[8,9] or a pellet of briquetted coal fines,[10] is placed in a furnace where it is heated at a controlled heating rate to a specific temperature and cooled rapidly. It is then examined by ordinary microscopic methods on a polished surface or thin section. By repeating the experiment, using different temperatures at intervals between 0 and 1000°C, the changes taking place between these temperatures can be observed.

For examination of a polished surface of briquetted coal fines, the sample must be impregnated under vacuum with a hardenable

[1] Gieseler, K., *Glückauf*, **69**, 604–7 (1933).

[2] Kimura, H., *J. Fuel Soc. Japan*, **37**, 423–4 (1958).

[3] Potonié, R., and Stockfisch, J. K., *Z. deut. geol. Ges.*, **1952**, 433.

[4] Ball, A. M., and Curtis, H. A., *Ind. Eng. Chem.*, **22**, 137–40 (1930).

[5] Kattwinkel, R., *Brennstoff-Chem.*, **11**, 329–30 (1930).

[6] Kattwinkel, R., *Glückauf*, **68**, 518–22 (1932).

[7] Mackowsky, M. T., *Brennstoff-Chem.*, **39**, Sonderausgabe, 40–7 (1958).

[8] Mott, R. A., *J. Soc. Chem. Ind.* (*London*), **46**, 85–92T (1927); *Fuel*, **6**, 227–31 (1927).

[9] Thiessen, R., and Sprunk, G. C., *Fuel*, **13**, 116–25 (1934).

[10] Coppens, L., and Noel, R., *Chim. & ind.* (*Paris*), *Compt. rend. congr. intern. chim. ind., 27th Congr. Brussels*, **1954, II**, 33–41.

plastic material. The plastic material is introduced between the particles without deforming them or modifying their relative positions; after hardening it forms a firm structure which can be cut and polished like a rock.

The general pattern of the phenomena observed by this method in the case of a coking coal is as follows.[10] At a certain temperature, which may be called the initial softening point, the particles of coal—hitherto of angular shape—start to undergo a slight deformation and become rounded; deformation then becomes more intense, the particles come into contact with each other, and at the same time, small, nearly spherical bubbles appear inside of them. The particles then become joined together and large bubbles are formed between them. Bubbles with regular outlines denote a coal of high plasticity; in places, their walls, which are of different thickness, may be formed of complete particles, more or less deformed, inside of which small bubbles can be detected. The size of the bubbles increases with the temperature. At a certain temperature, that is, the resolidification point, the appearance undergoes no further change. The sample then consists of a solid porous mass, the walls of the bubbles often showing small fissures.

Examination under the microscope allows the softening point and resolidification points to be determined fairly well, but it can give only a qualitative assessment of the plastic properties. The principal importance of these methods is to characterize the behavior of the individual petrographic components, whereas the methods to be described in the subsequent sections can only give average indications. On the other hand, the disadvantage of the technique of thin sections mounted on heating stages is that the sample is only several hundredths of a millimeter thick, and the swelling phenomena observed with particles of several millimeters may be substantially modified or even eliminated.

Free Swelling

In free-swelling tests, a sample of finely powdered coal is placed in a crucible or tube without being compacted to any considerable extent; it is heated according to a fixed heating rate, until a temperature of about 800°C is attained. If the coal is infusible under these temperature conditions, it distills without undergoing a change in appearance, and the sample remains in powdered form. When the coal is fusible, the particles soften, fuse together, and the mass of coal tends to swell; the only resistance which this swelling meets is that offered by the walls of the crucible. After resolidification and cooling, the coal sample is transformed into a porous coke, the volume of which is greater than the initial volume of the sample. The degree of swelling can be assessed from the ratio between the volume of the coke obtained and the initial volume of the coal or, more frequently, by comparing the profile of the coke obtained with a series of reference profiles which enables a "swelling index" to be defined.

These tests are also sometimes called agglomerating tests.[11] In actual fact, they indicate the tendency of the particles to fuse together and form a compact mass and allow a clear limit to be drawn between caking and noncaking coals. Coals, the particles of which do not adhere together during heating, are called noncaking, although this distinction is only valid for the particular rate of heating used for the tests. The tests make it possible to classify both weakly and medium caking coals, but they are of less use for distinguishing between strongly caking coals, such as good coking coals.

These tests are quite empirical and make no claim to measure a characteristic physical value of coal. The numerical indices obtained have no value unless all details of the operational procedure are carefully standardized. If this condition is satisfied, they give very reproducible results and can be carried out rapidly, which explains their popularity in operational practice.

Two tests have been standardized and are widely employed—the BSI swelling test,[12] standardized in Great Britain (the equivalent tests are standardized in the United States[13] and France),[14] and the Gray-King assay.[15] Both have been retained as criteria for the classification of coals in several national

[11] Brewer, R. E., *Chemistry of Coal Utilization*, John Wiley and Sons, New York, 1945, Vol. 1, pp. 160–309.

[12] Brit. Standards, 1016 (1942).

[13] ASTM Standards on Coal and Coke, **1959,** 83–7.

[14] AFNOR, Paris, N.F. M 11–001 (1944).

[15] *Dept. Sci. Ind. Research* (*Brit. Gov.*), *Fuel Research Tech. Paper*, No. 44, 53–68 (1940).

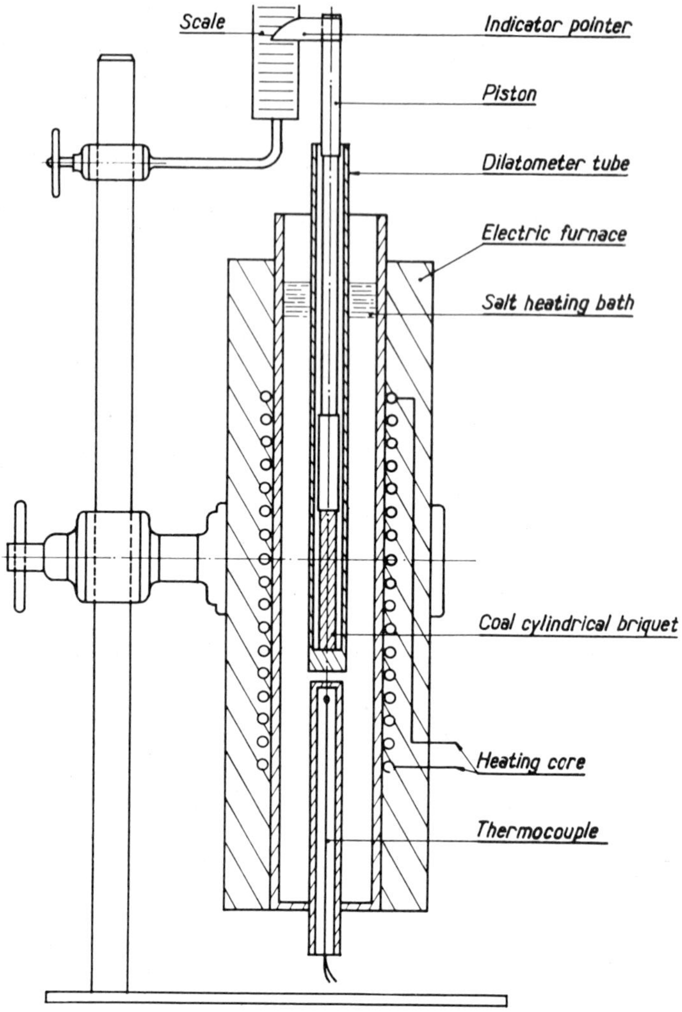

Fig. 1. Audibert-Arnu dilatometer.[16,18]

systems of classification and in the International System of Classification of Hard Coals by Type.[16] The significance of the results obtained by use of the BSI swelling test has been discussed by Kreulen.[17]

Dilatometers

Dilatometers enable the variations in the length of a coal sample heated at a definite rate to be followed; the coal sample may consist of a cut piece of coal or, more often, a pellet obtained by the compression of pulverized coal. The coal may be subjected to a weight or not.

Two categories of phenomena have been studied by means of dilatometers: (1) softening and swelling, resulting in the formation of semicoke and taking place in the temperature range 300 to 500°C; and (2) the behavior of semicoke above the resolidification point and up to 1000°C. Among the many types of apparatus used in the lower temperature range, the most typical is the Audibert-Arnu dilatometer, which is employed in various countries and has been proposed for standardization in the International system.[16,18] The latest apparatus proposed for use at the high temperatures is the Chévenard-Joumier dilatometer.[19]

[16] United Nations Publ., 1956 II.E.4, E/ECE/247, E/ECE/Coal/110 (1956), 52 pp.

[17] Kreulen, D. J. W., *Fuel*, **29,** 112–17 (1950).

[18] Anon., ISO/TC/27, Doc. 221, AFNOR, Paris (1958), 10 pp.

[19] Chévenard, P., *Bull. soc. franç. minéral. et crist.*, **76,** 165–88 (1953).

Audibert-Arnu dilatometer (Fig. 1). About 2 g of finely ground and slightly moistened coal is compressed in a mold by a special press to form a slightly tapered cylindrical "pencil" of diameter 6.5 mm and height 60 mm. The pencil is placed at the bottom of a metal tube having an inside diameter of 8 mm, closed at the lower end. Above the pencil, there is inserted a piston of diameter 7.8 mm which slides freely inside the tube and is extended by a rod equipped with a position indicator moving along a graduated scale. The weight of the piston and its rod, which rests on the coal sample, is about 150 g. The tube is placed inside an electric furnace, the temperature of which can be regulated. This furnace contains a salt bath or aluminum block[20] to insure a uniform temperature. The same block may also contain several tubes so that several tests can be carried out simultaneously. The piston rod can be connected to a device which records the height of the piston with respect to time.

The apparatus is used in two different ways: In the first, the furnace is brought to a predetermined temperature and maintained at this temperature; the sample is introduced and the movement of the piston is followed as a function of time. Figure 2 gives the shape of the curves obtained when a coking coal is heated at several temperatures.[21] It gives the displacements of the piston as a function of time, expressed as a percentage of the initial height of the coal pencil. When the temperature is below a certain limit (320° in the case of Fig. 2), no displacement of the piston is observed. At a temperature a little above this

[20] Chevallier, L., and Boyer, A. F., *Compt. rend. 4th congr. intern. chauff. ind.*, **I**, Groupe II, Sect. 24, No. 49; *Chaleur & ind.*, Paris (1952), 8 pp.

[21] Audibert, E., *Rev. ind. minérale, I, Mém.*, **6,** 115–36 (1926). *Fuel,* **5,** 229–44 (1926); **6,** 185 (1927).

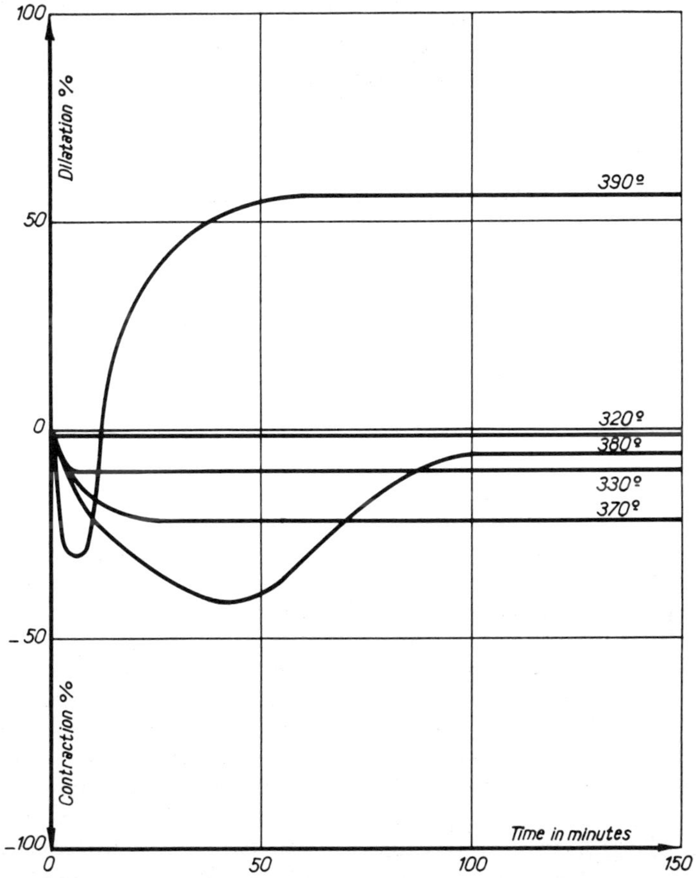

Fig. 2. Audibert-Arnu Dilatometer. Swelling of Lens coal when heated at different constant temperatures.[21]

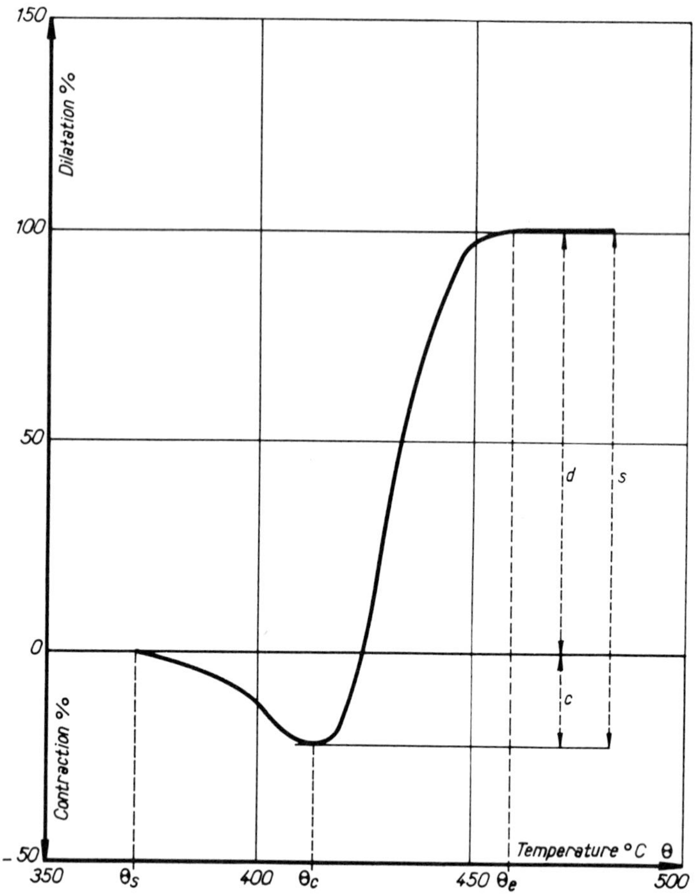

Fig. 3. Audibert-Arnu dilatometer. Heating rate 2°C/min dilatometric curve.

limit, a certain contraction, followed by stabilization, may be noted. At higher temperatures, the contraction is followed by expansion, then by stabilization. The rate of both contraction and expansion is greater the higher the temperature. The maximum expansions increases regularly with temperature. This method of operation is seldom employed when a coal is to be characterized, because the tests must be repeated; it has mainly been used for studying plasticity of coal.[21,22] Nevertheless, it enables the initial softening point and initial swelling point to be determined.

In the second, the furnace is brought to a temperature somewhat below the softening temperature of the coal to be tested; the sample is introduced, and the movement of the piston is followed as a function of time, while the temperature of the furnace is raised at a constant rate.

For a coking coal, the furnace is then regulated in such a way that the temperature of the sample increases linearly, for example, 2 or 3°C/min, when the curve has the shape indicated by Fig. 3. The following characteristics, defined fairly accurately by Fig. 3, can be deduced from the shape of the curve: softening point, θ_s; maximum contraction temperature, θ_c, or initial swelling point; final swelling temperature, θ_e; contraction, c; dilatation, d; swelling, $s = c + d$, where c, d, and s are expressed as a percentage of the initial length of the sample, plasticity index, $c/(\theta_c - \theta_s)$ which represents the mean slope of the descending branch of the curve. All these characteristics have no significance unless the dimensions of the dilatometer and all the details of the operational procedure, particularly the rate of

[22] Audibert, E., and Delmas, L., *Rev. ind. minérale*, I, Mém., **7**, 1–22 (1927). *Fuel*, **6**, 131–40, 182–9 (1927). Correction to both references: *Fuel*, **8**, 232–3 (1929).

heating, are carefully defined. The shapes of the curves recorded vary considerably with the nature of the coal. For certain coals, no contraction is obtained; for others, contraction is not followed by expansion ($s = 0$).

A comparison of the observations obtained with the microscope and those provided by the dilatometer enables the nature of the dilatometer curves to be interpreted.[10,21–24] Contraction is caused in the first place by softening of the particles which, under the effect of the weight of the piston, are deformed like a plastic mass. The particles tend to fill the interstices by becoming fused together. In addition, the pencil, the initial diameter of which is less than that of the tube, is pushed against the bottom of the tube, occupying the whole cross section. Concomitantly with the softening of the particles, bubbles of gas appear either inside the latter or between those particles which have already become fused and tend to make the mass swell. When the second phenomenon outweighs the first, the apparent contraction ceases and expansion commences.

The commencement of contraction thus corresponds to that of softening; in actual fact, the softening point is poorly defined. It decreases with increasing weight of the piston, as was demonstrated by the experiments of Griffin and Storch,[25] who employed conditions differing little from those of the Audibert-Arnu dilatometer. On the other hand, the rate of contraction, that is, the slope of the contraction curve, does not appreciably depend on the load.

If swelling does not occur and the plasticity becomes very high, the coal will collect at the bottom of the sample tube and the contraction will have the value: $c_m = (v_1 - v_0)/v_1$ where v_0 is the initial volume of the pencil of coal and v_1 is the free volume of the tube below the piston at the commencement of the test; the contraction could even exceed this value if the reduction in the gaps between the particles is taken into account. In point of fact, this value is never attained because swelling occurs fairly soon. The maximum value of the contraction does not characterize the plasticity attained by the coal, because it is, itself, the result of softening and swelling phenomena. The slope of the contraction curve appears to be a better characteristic of the plasticity.

Swelling takes place as soon as the particles are sufficiently fused together to offer appreciable resistance to the flow of the volatile matter. The degree of swelling depends both on the rate of liberation of volatile matter and the plasticity. Swelling ceases when the coal has lost its plasticity and begins to resolidify. The final swelling point is, therefore, close to the resolidification point.

The dilatation, d, and the swelling, s, defined in the preceding paragraph, have no physical significance, because they depend on the relative dimensions of the coal pencil and the dilatometer tube. Van Krevelen[26] suggested that the term "true expansion," δ, should be introduced; for this purpose, the ratio between the volume of the sample v and its initial volume v_0 was used, whereas the dilatation, d, refers to the volume v_1 of the tube under the piston.

$$\delta = \frac{v - v_0}{v_0}; \qquad d = \frac{v - v_1}{v_1}$$

hence

$$\delta = \frac{v_1}{v_0}(1 + d) - 1$$

When this equation is used, it is possible to calculate a curve which gives the true expansion as a function of temperature from the experimental curve. The importance of this method of representation is that it makes possible the prediction of the shape of the dilatometer curve of a coal mixture from the dilatometer curves of the components, assuming that the "true expansions" can be formed additively.

Chévenard-Joumier dilatometer.[19,20] The general principle is similar to that of the Audibert-Arnu dilatometer, but the operation differs essentially in that the test can be carried out up to 1000°C, and in that the equipment includes a very accurate recording device similar to the Chévenard device used in metallurgy for many years for studying the transformation of metals. It records expansion directly as a function of temperature by means

[23] Davies, R. G., and Wheeler, R. V., *Fuel*, **10**, 100–108 (1931).

[24] Mott, R. A., and Wheeler, R. V., *The Quality of Coke*, Chapman and Hall, London, 1939, 464 pp.

[25] Griffin, H. K., and Storch, H. H., *Ind. Eng. Chem., Anal. Ed.*, **9**, 280–6 (1937).

[26] van Krevelen, D. W., Chermin, H. A. G., Dormans, H. N. M., and Huntjens, F. J., *Brennstoff-Chem.*, **39**, Sonderausgabe, 77–83 (1958).

of a system of levers operated simultaneously by the stem of the piston which rests on the pencil of coal and a pyrometric standard brought to the same temperature as the pencil.

By means of this apparatus, it is possible to study softening and swelling in the same way as with a conventional dilatometer, but its main interest is that it enables a close analysis to be made of the contraction which the coal undergoes after its resolidification. Figure 4 gives an example of the curve obtained with a weakly caking coal having a high volatile matter content. From this curve, by simple graphical derivation, it is possible to deduce the contraction coefficient defined by:

$$\alpha = \frac{1}{l_0} \cdot \frac{dl}{d\theta},$$

where l_0 is the length of the sample at the moment of resolidification, and l is its length at the temperature θ. The curve of α as a function of θ has the shape shown in Fig. 5 with a very pronounced maximum in the vicinity of the resolidification point, after which it decreases rapidly. It has a much flatter maximum in the vicinity of 700°C. The maximum value of α is an important characteristic of the coals used in carbonization.

The Sheffield dilatometer. Two models, one of silica, the other of steel, have been described by Davies and Mott[27] and Mott and Spooner,[24,28] respectively. The method of operation differs from that with the Audibert-Arnu dilatometer by the fact that the coal sample is not previously formed into a pencil but is compressed at the bottom of a tube of diameter $\frac{5}{8}$ in. by allowing a weight of 130 g to

[27] Davies, R. G., and Mott, R. A., *Fuel*, **12**, 294–303 (1933).

[28] Mott, R. A., and Spooner, C. E., *Fuel*, **16**, 4–14 (1937).

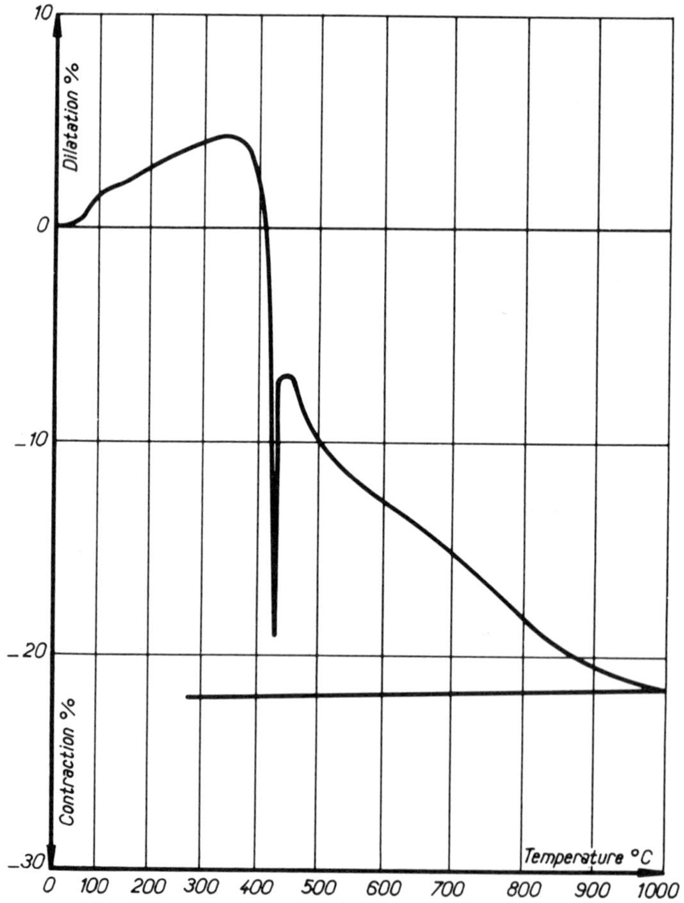

Fig. 4. Chévenard-Joumier dilatometer. Heating rate 2°C/min dilatometric curve of Ste Fontaine coal.[20]

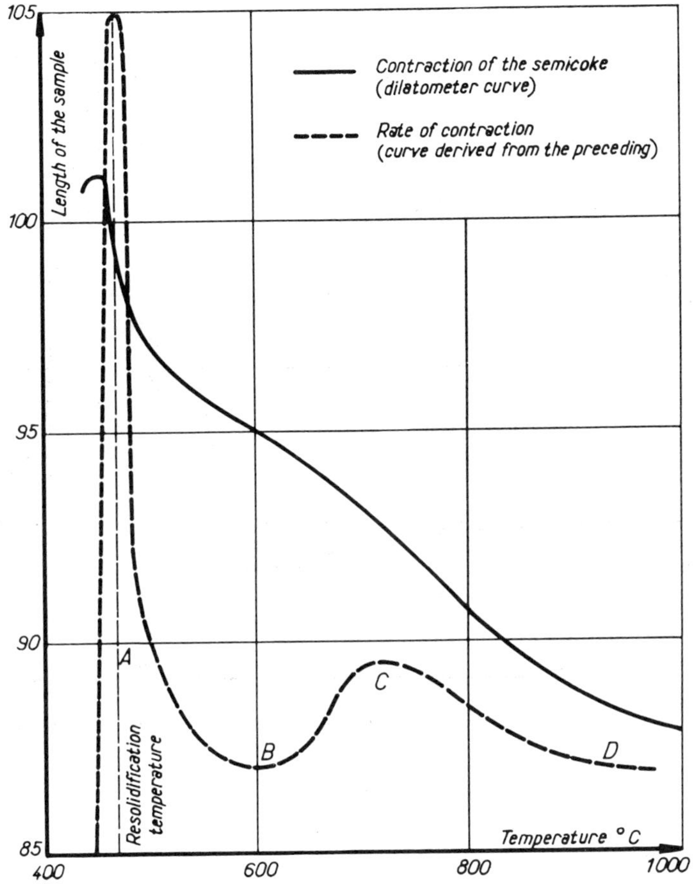

Fig. 5. Contraction and rate of contraction curves (coefficients of contraction).

fall on it. As a result, the contraction observed is less. The piston is, moreover, suspended from a pulley by a wire with a counterweight and it is, therefore, easy to change the pressure exerted by the piston. The diameter of the piston is considerably less than that of the diameter of the tube, so that the piston can penetrate the mass of coals that pass through a highly plastic state. The apparatus thus functions as a penetrometer, and the curve recorded has the shape indicated in Fig. 6.

The resolidification point can be determined by blocking the piston when maximum swelling is attained, freeing it several instants later when the temperature has reached a certain value θ, and checking to determine whether the piston has penetrated the coal or not. By repeating the test at different values of the temperature, the resolidification point can be determined as the temperature above which the piston does not penetrate. If the coal is of the strongly swelling type, the difference between the final swelling point and the resolidification point may be considerable.

The Hoffmann dilatometer.[29] This apparatus is very similar to the previous one. The coal sample is again introduced directly into the retort, but the degree of compaction is very slight. In addition, the diameter of the piston is considerably less than that of the crucible.

The curves recorded with this type of dilatometer are appreciably different from those given by the Arnu dilatometer. From their shape, Hoffmann[30] classified coals into four categories (Fig. 7). With very plastic coals, in particular, a falling off after maximum expansion is noted. This phenomenon is infrequently

[29] Hoffmann, H., *Oel u. Kohle*, **40**, 531–51 (1944).

[30] Hoffmann, H., and Hoehne, K., *Brennstoff-Chem.*, **35**, 202–11, 236–46, 269–75, 298–304 (1954). van Krevelen, D. W., Huntjens, F. J., and Wilms, A. H., *3rd Intern. Conf. on Coal Sci.*, Valkenburg, April 27–30, 1959, Commun. IV, 10 pp.

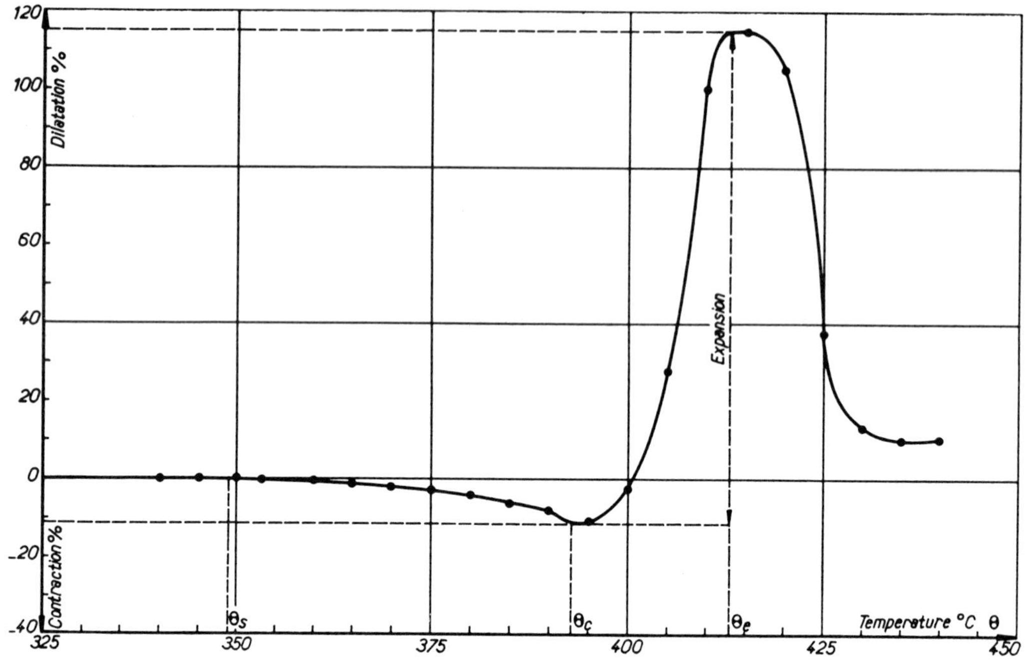

Fig. 6. Sheffield laboratory dilatometer. Dilatometer curve of a high swelling power coal (heating rate: 1°C/min).[24]

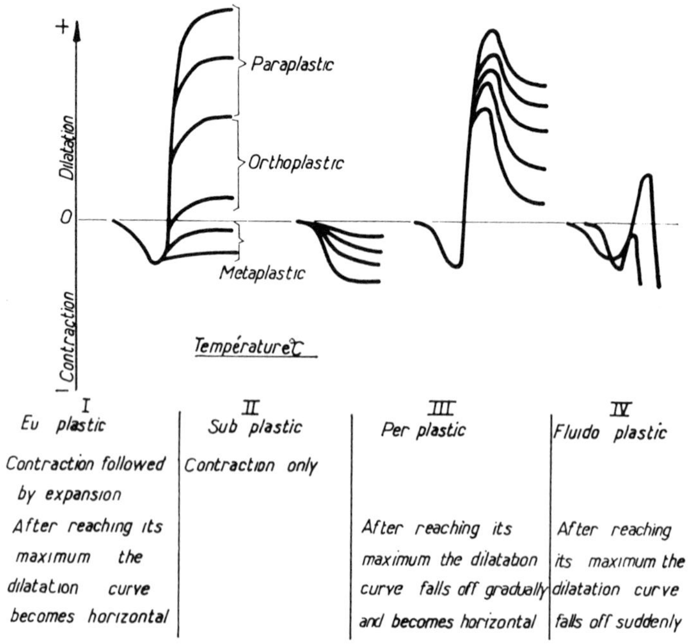

Fig. 7. Coal classification system based on the dilatometer curve.[30]

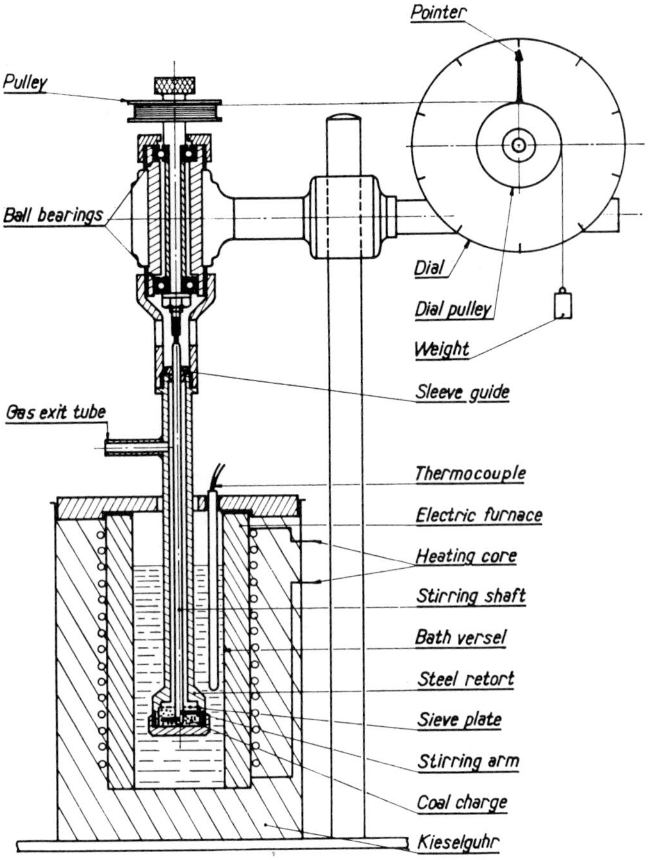

Fig. 8. Gieseler plastometer.

observed with the Arnu dilatometer and is never very conspicuous because of the very small clearance between the piston and the crucible.

The Hoffmann dilatometer is particularly suitable for the study of weakly coking coals; in addition, as a result of its fairly large diameter (23 mm) it can accommodate a coal sample which has not been very finely ground, which is an essential factor for studying the influence of the particle size of coals.

Plastometers

Plastometers are used to study the resistance offered to the rotation of a movable unit in the middle of a mass of powdered coal subjected to a particular rate of heating. They can be placed in two categories: Constant-torque plastometers in which the movable unit is subjected to a constant couple and the rate of rotation is measured; and variable-torque plastometers in which the movable unit is rotated at constant speed and the torque opposing its rotation is measured.

Constant-torque plastometers. This type of apparatus is represented by the plastometer described by Gieseler in 1934.[31–37] A variant was published by the ASTM in 1954.[38,39] The general characteristics are as follows (Fig. 8). The finely powdered coal is first compacted in

[31] Gieseler, K., *Glückauf*, **70,** 178–83 (1934).

[32] Yung, G., *Glückauf*, **71,** 1141–8 (1935).

[33] Chevallier, L., *Compt. rend., 4th congr. intern. chauff. ind.* **I.,** Group II, Sect. 24, No. 20; *Chaleur & ind.*, Paris (1952), 7 pp.

[34] Boulangier, F., *Compt. rend., congr. ind. gaz, 74th Congr., Bordeaux*, **1957,** 181–8.

[35] Macura, H., *Oel u. Kohle ver. Erdoel u. Teer*, **14,** 1097–1107 (1938); **15,** 1–12 (1939).

[36] Hoehne, K., *Glückauf*, **75,** 941–5 (1939).

[37] Brewer, R. E., and Triff, J. E., *Ind. Eng. Chem., Anal. Ed.*, **11,** 242–7 (1939).

[38] ASTM Standards on Coal and Coke, 1954, 129–33.

[39] Soth, G. C., and Russell, C. C., *ASTM, Proc.*, **43,** 1176–93 (1943).

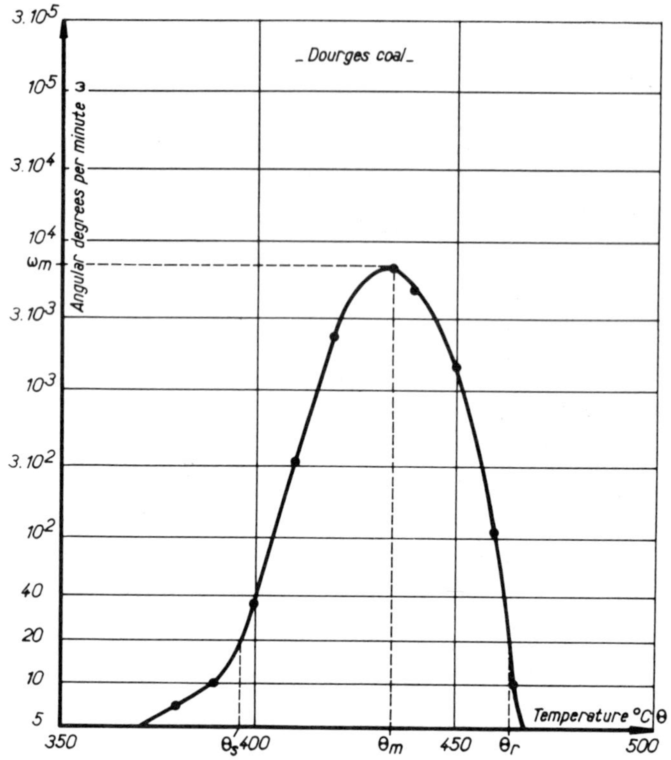

Fig. 9. Plasticity curve obtained with the Gieseler plastometer. Heating rate: 2°C/min.

a small cylindrical capsule of diameter 1 to 2 cm and height 2 to 4 cm, equipped with a stirrer having blades or arms. The axis of the stirrer is subjected to a constant torque as a result of a weight suspended from a wire wound around a pulley. This torque is always weak, of the order of 100 g/cm. The capsule is placed inside a furnace, the temperature of which can be regulated. The rotation of the stirrer is read on a dial or recorded.

The temperature of the sample is generally brought up to 300°C rapidly and then at a constant rate of 2 to 3°/min. The rate of rotation, expressed in deg/min, varies as a function of the temperature, as is shown in Fig. 9. Before the coal becomes plastic, the stirrer is blocked by the coal compacted in the capsule, remains stationary, and only begins to rotate when the coal reaches a certain degree of plasticity. The rate of rotation increases with temperature, passes through a maximum which is frequently very pronounced, and then decreases until it becomes zero when the coal resolidifies. Because the commencement and the end of the movement of the plastometer are very gradual, to determine a softening point, θ_s, and a resolidification point, θ_r, with a fair degree of accuracy, it is best to select a certain minimum speed, for example, 20°/min.[40] The temperatures determined in this way are only conventional indices. The temperature, θ_m, corresponding to the maximum rate of rotation and the value of this maximum rate, ω_m, can also serve as indices. All these values, particularly the maximum rate, are markedly influenced by the exact characteristics of the apparatus and the method of operation. It is difficult to obtain results which are reproducible from one laboratory to another but, on the other hand, the same operator using the same apparatus can often obtain fairly reproducible results. The numerical values are very sensitive to variations, however slight, of the properties of the coals, particularly those due to oxidation. This technique is, therefore, very suitable for the identification of coals and for checking the constancy of their characteristics.

The apparatus can be calibrated, and its results expressed in "viscosity" units (poises),

[40] Hoffmann, H., and Hoehne, K., *Brennstoff-Chem.*, **35,** 202–11 (1954).

by operating at different temperatures with a substance such as pitch, the viscosity of which has been previously determined by means of a conventional viscosimeter as a function of temperature.[40,41] The "viscosity" measured in this way is, however, only an apparent viscosity and has no simple physical significance, because the mobility of the stirrer depends not only on the true plasticity of the coal, but also on the character of the foam structure within the capsule of the instrument.

Instead of submitting the sample to a constant rate of heating, it can be taken up rapidly to a given temperature and kept at the latter, while the rate of rotation as a function of time is determined. This method of operation has been used principally for theoretical studies.[42,43]

Variable-torque plastometers. A variable-torque plastometer was described for the first time by Davis in 1931.[11,38,44] Two other types of variable-torque plastometers, differing in design from the Davis plastometer, have been developed more recently.

In the Davis plastometer, 18 g of −20 mesh coal is placed in a horizontal cylindrical retort which is rotated at a uniform speed of 2 rpm by means of a synchronous motor. A shaft with rabble arms, fixed to the framework of the apparatus by means of a spring, is located inside the retort. The coal is set in movement by the retort and, itself, tends to move the shaft; the deformation of the spring, which is very slight, enables the torque acting on the shaft to be assessed. The retort is placed inside of an electric furnace. With a constant rate of heating, a typical change of torque as a function of the temperature is illustrated in Fig. 10. Until the coal softens, the resistance remains low and then increases rapidly as the particles adhere together. As the plasticity of the coal increases, the torque decreases and passes through a minimum which is sometimes not much greater than the corresponding torque in cold coal. The reduction in the plasticity preceding solidification is accompanied by a very considerable increase in the torque; the semicoke formed then breaks up, and the torque falls rapidly, the second maximum of the curve corresponding to the resolidification point. The Davis plastometer thus allows the softening point, θ_s, and the resolidification point, θ_r, to be determined with a high degree of accuracy. The difference between these temperatures, that is, the plasticity range, is considered to be an important characteristic of coals; it is greater than that determined with the Gieseler plastometer because the torques produced are much higher. The maximum value of the torque attained at the moment of resolidification also gives an idea of the mechanical resistance of the semicoke.

Echterhoff[45] has modified the Brabender plastograph to adapt it for the study of coals. In contrast to the Davis plastometer, it is the retort which is fixed and the shaft with rabble arms which moves. The latter is moved at constant speed by a motor mounted on pivots. A mechanical device allows the torque, which acts on the shaft with rabble arms, to be measured and recorded. The rabble arms fixed on the movable shaft and on the walls of the retort are arranged in such a way that the whole mass of coal is stirred and set in motion. Because the retort is large, holding 10 g of coal, the coal need not be ground to less than 1 mm. The curve recorded has a shape similar to that given by the Davis plastometer.

The Cerchar plastometer[46] was specially devised to determine the resolidification point to within ±2°C. As with the Brabender apparatus, the retort is fixed and the movable shaft is rotated by a motor mounted on pivots. This manner of installation permits the torque exerted on the shaft to be measured after mechanical or optical amplification. The retort and the shaft with rabble arms are similar to those of the Gieseler plastometer, the shaft having been reinforced to support higher torques. In addition, the retort has a square section in order to prevent the whole mass of coal being carried around bodily in the vicinity of the resolidification point. This apparatus differs from the two preceding types by the fact

[41] Stephens, J. N., *Fuel Research (Australia)*, **M 130**, June (1957).

[42] Van Krevelen, D. W., Huntjens, F. J., and Dormans, H. N. M., *Fuel*, **35**, 462–75 (1956).

[43] Pitt, G. J., *Brennstoff-Chem.*, **39**, Sonderausgabe, 87–90 (1958). Oxley, G. R., and Pitt, G. J., *Fuel*, **37**, 19–21 (1958).

[44] Davis, J. D., *Ind. Eng. Chem., Anal. Ed.*, **3**, 43–5 (1931).

[45] Echterhoff, G., *Erdöl u. Kohle*, **8**, 294–8 (1955); *Glückauf*, **90**, 510–16 (1954).

[46] Boyer, A. F., and Lahouste, J., *Rev. ind. minérale*, **35**, 1107–33 (1954).

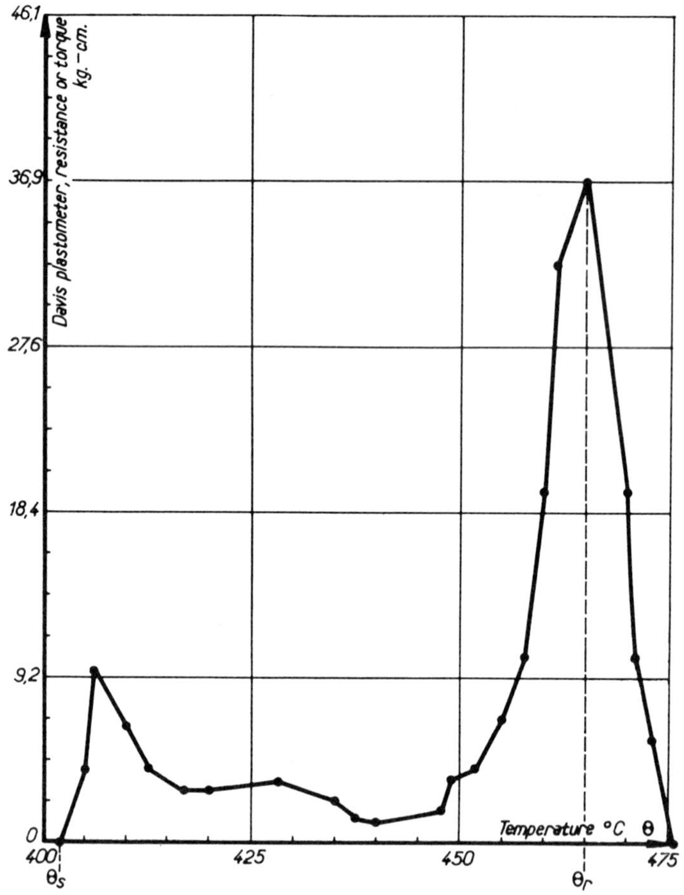

Fig. 10. Typical plasticity curve of a high-volatile A coal determined in the Davis plastometer.[37]

that it has a much slower rotation, of the order of 1 rev/hr. It is only set in motion when the coal is already in the plastic state. The total displacement of the shaft is, therefore, very low, of the order of $\frac{1}{4}$ to $\frac{1}{2}$ a turn. The shape of the curve obtained is illustrated in Fig. 11. The abscissa at *A* represents the commencement of resolidification, the abscissa at *B*, the resolidification point. As the rate of rotation increases from 1 rev/hr to 2 rev/hr, the resolidification point changes not more than 2 to 3°C. In a variant of the Cerchar plastometer,[47] the retort has a much greater volume, holding about 12 g of coal, and the coal, therefore, does not need to be finely ground. To obtain uniform heating, the retort is ring-shaped, and it is heated simultaneously from the interior and exterior.

[47] Boyer, A. F., Lahouste, J., and Malzieu, R., *Rev. ind. minérale*, **39**, 232–56 (1957).

Penetrometers

To characterize the plasticity of coal, a metal needle which, under a given load penetrates the coal, may be employed instead of a movable rotating device like the plastometer, its displacement as a function of the temperature being observed. The rate of penetration of the needle into the coal depends on the plasticity of the coal. The rate of penetration as a function of temperature is similar to that obtained with a plastometer. Sometimes a piece of cut coal is used in penetrometers, but it is more usual to employ coal pellets in these tests. They allow the softening temperature to be determined with reasonable accuracy, but are less suitable for the examination of plasticity over the whole range. If the coal is very plastic, the needle penetrates the entire sample rapidly; if, however, the coal has a low plasticity, the movement of the needle may be disturbed by the swelling and the apparatus behaves like a dilatometer. In actual fact, there is no exact demarcation between dilatometers and penetrometers. It has been pointed out that dilatometers in which the diameter of the piston is considerably less than that of the dilatometer tube behave in the case of very plastic coals like penetrometers.

Few improvements have been made in the early types of penetrometers.[11] A new type of apparatus, however, has been developed recently by Cerchar[46] for the accurate determination of the resolidification point. Four identical samples of powdered coal are heated simultaneously at constant rates of heating in four cells located in the same isothermal copper block. They are successively brought directly underneath a slender rod of diameter 2 to 3 mm loaded by a weight. The curve of the displacement of the rod with respect to time is recorded, and the resolidification point is the limiting temperature above which there is no displacement of the rod as determined by interpolation between two successive tests. It can be determined with an accuracy of approximately ±2°C.

Agglutination Tests

Agglutination tests were probably the first tests to be employed systematically in attempts to assess the coking capacity of coals. They are intended for the determination of the capacity of coal particles to agglutinate with each other during fusion and to agglutinate with an inert substance with which they are mixed, so as to give a solid and coherent residue. Agglutination tests may be divided into two categories. In the first, a mixture of given proportions of the coal to be examined and a selected inert substance is made; the mechanical strength of the coke button obtained after this mixture has been heated is determined either by a crushing test or abrasion test. In the other, the proportion of inert

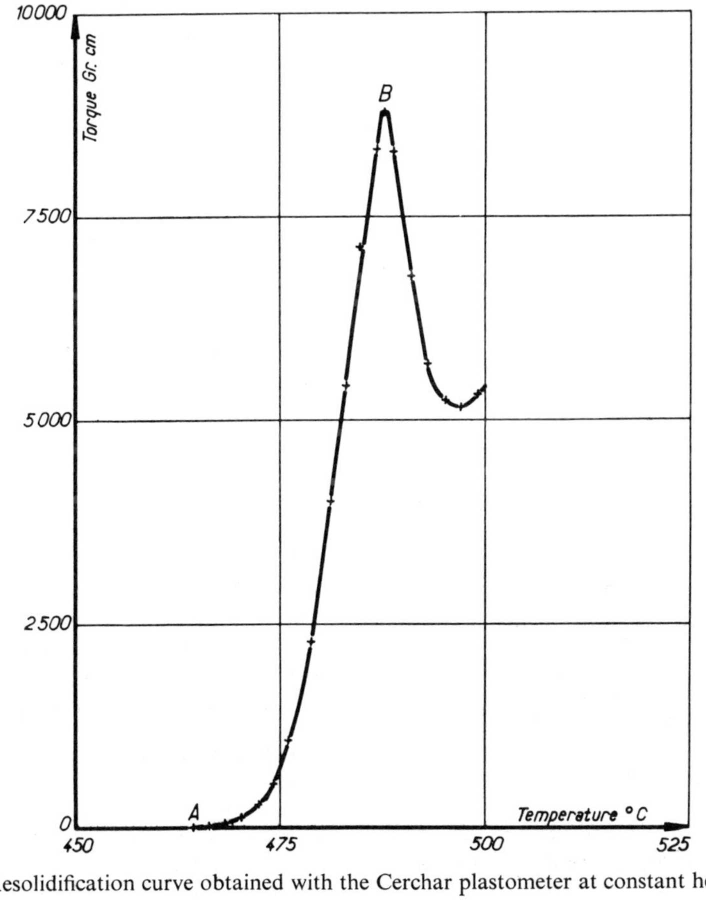

Fig. 11. Resolidification curve obtained with the Cerchar plastometer at constant heating rate.[46]

substance which must be added to the coal in order that the coke button obtained has a given mechanical strength or a given swelling is determined by successive tests. Reviews of the various agglutination tests proposed were given by Marshall and Bird[48] in 1929 and Brewer[11] in 1945.

It is difficult to associate the "agglutination coefficients" obtained with any physical property of the coal. Undoubtedly they are complex functions of the plasticity, and swelling of the coal, and the surface tension of the plastic coal, and they are related to the mechanical strength of the semicoke. Hence, a method of this type should allow an over-all evaluation of the coking capacity. In order that the test should give reproducible results, it is necessary that all the parameters concerned, in particular the bulk density of the sample and the nature of the inert substance, are carefully standardized; the latter's importance as a factor is mainly with respect to its porosity and surface properties. Sand, anthracite, electrode coke, or silicon carbide are generally used as the inert substance.

The principal tests are those of Campredon,[49] Meurice,[50] Damm,[51] and Kattwinkel[52] which are still employed in certain laboratories. Two other methods have been standardized. The Roga test has long been in use,[53] but it has been employed outside of Poland only in recent years. It represents one of the criteria of the Geneva International Classification System.[16] A mixture of 1 g of coal crushed to 0.2 mm and 5 g of anthracite sized to between 0.3 and 0.4 mm is compacted in a porcelain crucible under a load of 6 kg applied for 30 sec. After being brought to a temperature of 850° in 15 min, the coke button is weighed (Q g) and screened at 1 mm, the weight of the >1 mm fraction being designated a. This material is placed in a rotating drum of diameter 20 cm and is treated for three periods of 5 min, at the end of which the <1 mm fraction is removed by screening. If b, c, and d are the weights of the >1 mm residues at the end of these periods, the Roga agglutination coefficient defined by

$$\frac{\frac{a+d}{2}+b+c}{3} \times \frac{100}{Q}$$

ranges from 0 to 70 in practice.

In the ASTM agglutination test,[54] derived from the work of Selvig, Beattie, and Clelland,[52,55] 1.25 g of coal ground to 0.075 mm is mixed with 18.75 g of silicon carbide screened between 0.25 and 0.35 mm. The mixture is compressed under carefully determined conditions and heated to 950° in 15 sec. The crushing strength is determined with an accuracy of 0.1 kg by means of a press. The agglutination coefficient is the mean of 6 crushing strength tests, expressed in kg.

Agglomeration tests also include the Gray-King test in which the proportion of inert substance (electrode carbon), which must be added to coal to obtain a given swelling, is determined, as mentioned earlier.

Tests with Unidirectional Heating

In all of the tests described so far, the coal being investigated is heated in such a way that temperature gradients throughout the sample are kept to a minimum. This differs from conditions present in coke ovens where the heat is transmitted from the heating walls toward the center of the oven, and isothermal surfaces are approximately parallel to the heating walls with a temperature gradient of the order of 50 to 200°C/cm through the plastic range. To obtain an analogous rate of heating, various types of apparatus have been suggested in which a coal sample of cylindrical shape is placed on one of its bases in contact with a heating surface. Its lateral surface and the other base are thermally insulated, thus providing unidirectional heating (Fig. 12).[56] By

[48] Marshall, S. M., and Bird, B. M., *Trans. Am. Inst. Mining Met. Engrs.*, **88**, 546–88 (1930).

[49] Campredon, L., *Compt. rend.*, **121**, 820–22 (1895); *Rev. chim. ind.*, **7**, 8–9 (1896).

[50] Meurice, A. A., *Ann. mines Belg.*, **19**, 625 (1914); *Colliery Guardian*, **117**, 1559–60 (1919); *Fuel*, **2**, 305–8 (1923); *Chaleur & ind.*, **4**, 45–52 (1923).

[51] Damm, P., *Glückauf*, **64**, 1073–80, 1105–11 (1928).

[52] Kattwinkel, R., *Glückauf*, **62**, 972–5 (1926).

[53] Roga, B., *Z. oberschles. berg. -u. hüttenmänn. Ver., Katowice*, **70**, 565–74 (1931).

[54] ASTM Standards on Coal and Coke, 1954, 151–5.

[55] Selvig, W. A., Beattie, B. B., and Clelland, J. B., *ASTM, Proc.*, **33**, Part II, 741–60 (1933).

[56] Peytavy, A., and Lahouste, J., *Compt. rend. congr. ind. gaz*, 71*th Congr.* 415–22 (1954).

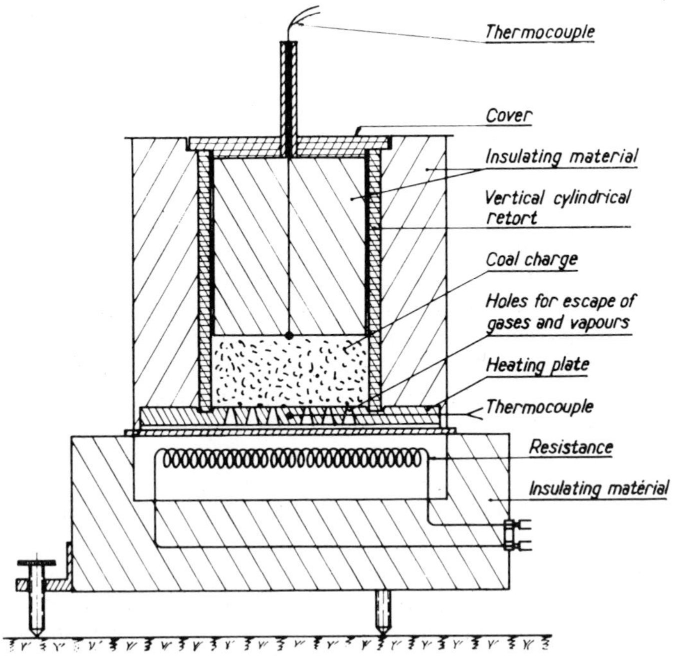

Fig. 12. Diagram of a unidirectional electrically heated oven.[56]

suitable regulation of the temperature rise of the heating surface, it is possible to produce a rate of heating in the coal sample completely similar to that present in a coke oven. A plastic zone is formed initially at the contact of the heated surface and moves through the mass of coal until it reaches the other face. In certain types of apparatus where the two bases are heated identically and the two plastic zones move together and meet in the center, the analogy with a coke oven is still closer.

Apparatus of varying sizes have been designed on this principle, ranging from laboratory ovens holding about a hundred grams of coal to large-scale ovens treating several hundred kilograms. Only the main types of apparatus and, in particular, those enabling the plastic properties to be determined will be described.

Most of the fixed-wall ovens with two-sided heating are true models on a small scale of industrial ovens. They permit 200 to 300 kg of coke to be produced under conditions very similar to those of industrial ovens.[57–59] The properties of the coke may be evaluated by standard methods: the shatter test and the tumbler test. This type of apparatus provides a most reliable means for predicting the quality of coke in relation to the type of coal and the conditions of coking: charge density, rate of heating, water content, etc. There are smaller units treating from 5 to 20 kg of coal.[60] The properties of the coke are still very closely related to those of the coke produced in industrial ovens, but it is not possible to determine them by standard tests which require a larger quantity of coke. Special tests must be used; consequently, their results cannot be applied directly to industrial practice.

Constant-volume ovens with a movable wall and two-sided heating[61–63] differ from the

[57] Reed, F. H., Jackman, H. W., Rees, O. W., Yohe, G. R., and Henline, P. W., *Illinois State Geol. Survey, Bull.* No. 71 (1947), 132 pp. Reed, F. H., Jackman, H. W., and Henline, P. W., *Ind. Eng. Chem.*, **37,** 560–6 (1945).

[58] Foch, P., and Meimarakis, G., *Compt. rend. congr. ind. gaz, 74th Congr.* 329–46 (1957).

[59] Reerink, W., Echterhoff, H., and Beck, K. G., *Glückauf,* **94,** 102–10 (1958).

[60] Durand, G., and Lahouste, J., *Compt. rend. congr. ind. gaz, 76th Congr.* (1959).

[61] Koppers, H., and Jenkner, A., *Glückauf,* **67,** 353–62 (1931); *Fuel,* **10,** 232–9, 273–7 (1931).

[62] Russell, C. C., *Trans. Am. Inst. Mining Met. Engrs.,* **139,** 313–27 (1940). ASTM Standards on Coal and Coke, 1954, 138–41.

[63] Gayle, J. B., and Auvil, H. S., *U.S. Bur. Mines, Rep. Invest.* No. 4923 (1952), 17 pp.

preceding by the fact that one of the walls is supported by a dynamometric device for measuring the pressure to which it is subjected during coking. These ovens are generally designated as movable-wall ovens, but, in reality, the displacement of the wall is extremely small, so that the volume of the oven remains practically constant. This apparatus, treating from 200 to 400 kg of coal, reproduces, in their entirety, all the phenomena which proceed in an industrial oven. It allows the pressure to which the walls of ovens are subjected to be measured. It can also be used for predicting the quality of the coke under the same condition as the ovens with fixed walls.

In constant-pressure ovens with one-sided heating, the coal layer is placed between a fixed wall—heated at a given rate—and a movable wall in the form of a piston and subjected to a constant force. The displacement of the piston during coking is recorded. This apparatus is, therefore, somewhat similar to a dilatometer, differing principally in size and mode of heating. The final dilatation expressed in percentage of the original thickness of the coal layer is often used to characterize the result of the test. It may be positive (expansion) or negative (contraction) according to the nature of the coal.

One form of the constant-pressure oven most frequently used is the "Heinrich Koppers A.G. Laboratory Apparatus"[64] treating 100 g of coal and for which various modifications have been suggested.[11,65] Several have been built on a larger scale (20 to 100 kg of coal) and are widely used in the United States, the walls being either vertical (vertical slot oven)[66] or horizontal (sole-heated oven).[13,66–68] They are mainly used to measure the pressure that may be expected to be exerted on the walls of the coke oven, but may also be used to predict the coke quality. The value of the results obtained is, however, very questionable, because unidirectional heating does not permit the heating conditions in industrial ovens to be entirely reproduced. It seems more advisable to use ovens with two-sided heating.

Sapozhnikov and Basilevich[69] developed a method for measurement of the thickness of the plastic layer. Applications of the method have been described by Brewer.[11] Its use has been developed in all countries of Eastern Europe[70,71] and Devecchi[72] has also employed it for various European and American coals. It serves as a base for the classification of caking coals in the USSR.[73]

Peytavy and Lahouste[56] have examined the plastic layer in a unidirectionally heated 100-mm diameter oven by means of X rays, and assembled successive photographs to form a "time-lapse" film. The results were correlated with those provided by a dilatometer and plastometer by means of temperatures measured at different levels in the oven. As shown in Fig. 13, the plastic zone is very evident in a good coking coal, four zones from top to bottom can be distinguished. In zone 1, the coal is still unchanged; in zone 2, corresponding to the contraction zone of the dilatometer, the particles cake and form a compact mass which is more opaque; in zone 3, which is less opaque, the coal is very fluid, and its bulk density is reduced by the presence of numerous bubbles; and in zone 4, two subzones must be distinguished, according to the indications of the penetrometer, zone 4*a* corresponding to the end of the plastic zone and zone 4*b* representing definitely resolidified semicoke. In the case of a high-volatile coal, fusible but with a rather low swelling index (B.S. Swelling Index: 6), zone 2 is not very conspicuous, and zone 3 cannot be distinguished from zone 4. With a weakly caking coal (B.S. Swelling Index of 3–4), the transformation of the coal to coke

[64] H. Koppers A. G., *Koppers Mitt.*, **12**, No. 1, 1–23 (1930).

[65] Wilson, J. E., and Naugle, B. W., *U.S. Bur. Mines, Rep. Invest.* No. 5583 (1960), 11 pp.

[66] Auvil, H. S., and Davis, J. D., *U.S. Bur. Mines, Rep. Invest.*, No. 3403 (1938), 18 pp. Auvil, H. S., Davis, J. D., and McCartney, J. T., *U.S. Bur. Mines Rep. Invest.*, No. 3451 (1939), 21 pp.

[67] Brown, W. T., *AGA, Proc.* **1938**, 640–79; *Blast Furnace and Steel Plant*, **27**, 172–4 (1939).

[68] Gayle, J. B., and Eddy, W. H., *U.S. Bur. Mines, Rep. Invest.* No. 5464 (1959), 35 pp.

[69] Sapozhnikov, L. M., and Basilevitch, L. P., *Khim. Tverdogo Topliva*, **3**, Nos. 2–3, 79–102 (1932).

[70] Kosina, A., Herda, M., and Swenby, A., *Paliva*, **30**, 920–16 (1950).

[71] Nadziakiewicz, J., *Coke and Gas*, **19**, 193–8 (1957).

[72] Devecchi, W., *Schweiz. Ver. Gas- u. Wasserfach., Monatsbull.* **32**, 117–29 (1952).

[73] Glussmann, L. D., Nikolsky, N. A., and Edelmann, J. J., *Tech. Com. de l'Etat Ukrainien*, Charkov, Kiev 1937, 432 pp.

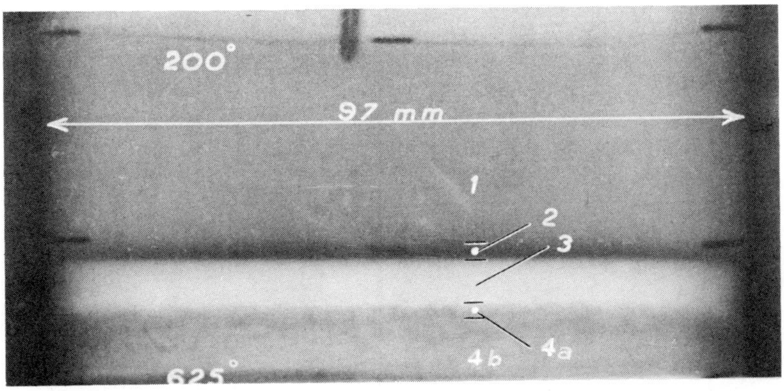

Fig. 13. Radiograph of a coking coal during its carbonization.[56]

is not indicated by any variation in the opacity. X-ray examination also makes it possible to observe the deformation of the semicoke after resolidification and the development of fissures.[74]

If a flat piece of coal, about 1 cm thick, is subjected to unidirectional heating, it is found that it curves but does not crack—the convexity being directed towards the heating surface. It is this phenomenon which accounts for the "cauliflower" shape of coke lumps located near the heating walls. It results from the variation of the contraction coefficient above the resolidification point (Fig. 5). Boyer, Ladam, and Soulè[74] have shown that the total curvature that results when a cake is heated to 900 to 1000°C is proportional to the thermal gradient and the contraction coefficient, α_m, at the resolidification point. By measuring the curvature with a spherometer it is, therefore, possible to determine the coefficient α_m. But this method is less accurate than the method employing the Chévenard-Joumier dilatometer.

Comparison between the Different Methods of Observation

The different methods of observation agree with and complement each other sufficiently well to give a qualitative description of the physical transformations which coals undergo during heating. As regards the quantitative aspect, the only measured values which appear to have physical significance are the softening point, the resolidification point, and the contraction coefficient after resolidification. Even these values are not "physical constants," and for their clear definition it is necessary that the method of their determination be carefully specified.

The softening point can be measured by a dilatometer, plastometer, penetrometer, Foxwell test, or microscopic observation. The most accurate method is that employing a dilatometer specially designed for this purpose, but other methods give results which do not differ greatly.

The resolidification point can only be measured by a variable—torque plastometer or penetrometer, the method proposed by Cerchar[47] being, in our opinion, the most accurate. On the other hand, the temperature of the completion of swelling indicated by dilatometers or the "stoppage" temperature of constant-torque plastometers are generally lower than the resolidification point.

The contraction coefficient after resolidification can be determined quite definitely by means of a sufficiently accurate dilatometer.[19]

On the other hand, it is far more difficult to characterize the plasticity; this results from the fact that the softening, agglutination, and swelling phenomena are superimposed and "interlocked." In consequence, only their resultant can be characterized by the various types of apparatus. We have seen, for example, that the indications of a plastometer can be expressed in poises, but the viscosity measured in this way is only an apparent viscosity which depends not only on the viscosity of the plastic phase but also on its foam characteristics.

The majority of the tests developed for characterizing the plasticity of coals are, moreover, not intended for the measurement of a physical

[74] Boyer, A. F., Ladam, A., and Soulè, J. L., *Rev. ind. minérale*, **34**, 592–616 (1953).

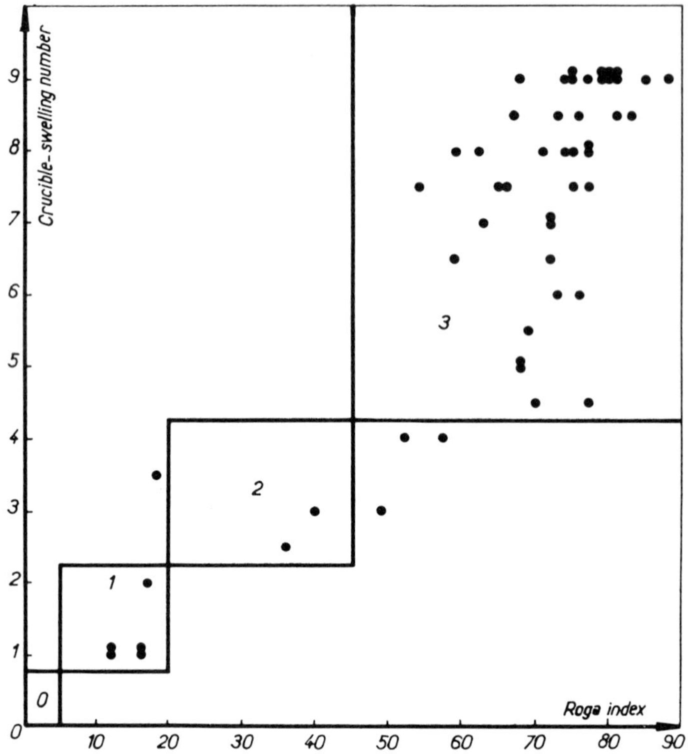

Fig. 14. Correlation between the results of the crucible-swelling test and the Roga test.[16]

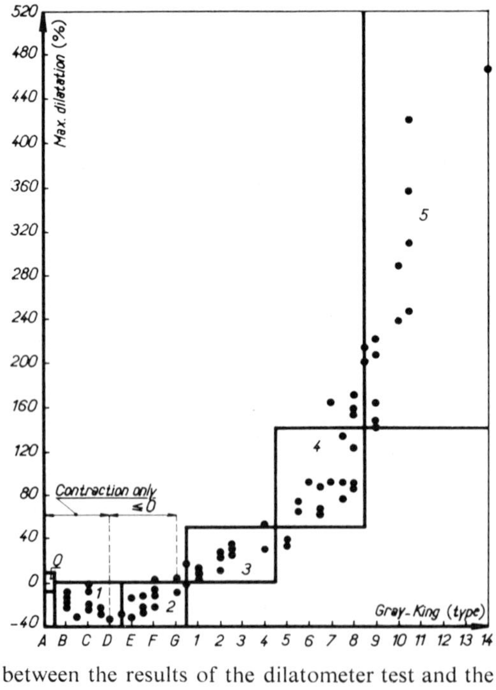

Fig. 15. Correlation between the results of the dilatometer test and the Gray-King assay.[16]

value but for the establishment of an index. These indices are only of value if all the experimental conditions are strictly standardized. The very wide range of operating parameters employed by different investigators, each of whom has his own particular manner of perfecting the test, makes it difficult to compare the results obtained in different laboratories. The standardization on the international level of 4 tests employed as classification criteria by the Geneva International Classification System[16] represents an important step in this direction. These 4 tests are: the Free Swelling Index, the Roga Index, the dilatation measured with the Audibert-Arnu dilatometer, and the Gray-King Index. Rapid heating is used for the first two and slow heating for the others. As indicated by Figs. 14 and 15, there is a certain correlation between these different indices.

FACTORS GOVERNING PLASTIC PROPERTIES

Heating Rate

Whatever the method of observation employed, an increase in the heating rate enhances the plastic properties of the coal; the fluidity and swelling are increased; and the plastic range is broadened. Coals, which show no softening when heated slowly, soften and may swell when heated more rapidly. The influence of the heating rate has been studied by numerous authors; in particular by Ferrero,[75] Kreulen,[76] Macura,[77] Brown,[78] Van Krevelen, Huntjens and Dormans,[42] Nadziakiewicz,[71] Brown and co-workers,[79] and Boyer, Ladam, and Soulè.[74]

Typical results are given in Figs. 16 and 17, the former obtained with the Audibert-Arnu dilatometer, and the latter with the Gieseler plastometer. To summarize the results, an increase in the heating rate in the 0.5 to 10°C/min range leads to several changes. In general, the softening temperature is slightly higher, but Atkinson, Brewer, and Davis[80] found that the fusion point of Powellton coal was lower by several degrees. On the other hand, the resolidification point shows a very definite increase with rate of heating, the plastic range being, therefore, considerably broadened. According to Boyer,[81] when the heating rate is increased from 2°C/min to 3°C/min the resolidification point is increased from 12 to 14°C for medium-volatile and from 15 to 20°C for high-volatile coals. Swelling measured with the dilatometer and plasticity measured with the plastometer show an increase, depending to a considerable degree on the nature of the coal. The temperature of plasticity and the swelling temperature of maximum swelling are displaced parallel to the resolidification point. The coefficient of contraction after resolidification shows only a slight increase.[74]

When coal is heated in the absence of oxygen at a temperature about 50°C below the softening point, even for prolonged periods, its plastic properties are not modified; different conclusions have been drawn by certain authors, but it is probable that there was some oxidation. When preheating is carried out closer to the softening temperature it appears that—with some coals at least—a reduction in the plastic properties takes place even in the absence of oxidation.[78,80] The characteristic temperatures are generally the same, but the maximum plasticity measured in a plastometer is reduced. If coal is kept at a temperature above its softening point, even for a few minutes, its plastic properties are considerably altered, even if the volatile matter liberated during preheating is only a small amount.

Particle Size

Most laboratory devices do not lend themselves to the study of the effect of the particle size because the coal used must generally be very finely ground in order to obtain a representative sample for small scale equipment. Comparing fine sizes of the order of several tenths of a millimeter, various authors[17,82–84]

[75] Ferrero, P., *Schweiz. Ver. Gas- u. Wasserfach., Monatsbull.*, **30,** 275–86 (1950).

[76] Kreulen-Van Selms, F. G., *Chem. en Pharm. Tech. (Dordrecht)*, **5,** 181–6 (1950).

[77] Macura, H., *Oel, u. Kohle*, **37,** 727–40 (1941).

[78] Brown, H. R., *Coke and Gas*, **18,** 390–2 (1956).

[79] Brown, J. K., Dryden, I. G. C., Dunevein, D. H., Joy, W. K., and Pankhurst, K. S., *J. Inst. Fuel*, **31,** 259–73 (1958).

[80] Atkinson, R. G., Brewer, R. E., and Davis J. D., *Ind. Eng. Chem.*, **29,** 840–4 (1937).

[81] Cerchar, *Rapport annuel*, **1955,** p. 77.

[82] Ghosh, S. R., Das Gupta, N. N., and Lahiri, A., *J. Sci. Ind. Research (India)*, **16B,** 89–95 (1957).

[83] Bürstlein, E., *Chaleur & ind.*, **26,** 215–32 (1950); **30,** 351–70 (1954); **31,** 14–28 (1955).

[84] Mönnig, H., *Glückauf*, **72,** 752–7 (1936).

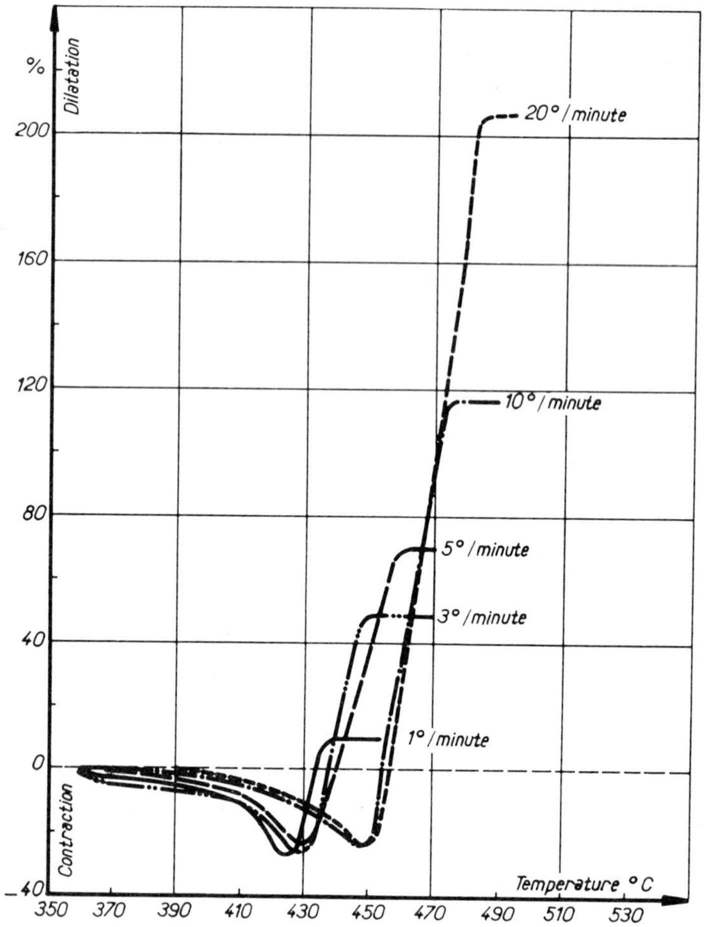

Fig. 16. Effect of heating rate on the dilatation of a coking coal as determined in the Audibert-Arnu dilatometer.[75]

have concluded that very fine grinding of coal reduces its plastic properties, that is, free swelling in the dilatometer and plasticity. The effect is particularly pronounced in the case of dull coals; it is not so appreciable for bright coals, where an increase in the fluidity has sometimes been noted.[85]

On the other hand, larger sizes—which are more closely related to the sizes used in industrial coking plants—can be studied with the Hoffmann dilatometer which accomodates itself to rather coarse coal samples. A systematic study of the Saar and Lorraine coals led Hoffmann to conclude[29] that the effect of particle size varies according to the type of coal and to the petrographic constituents.

[85] Sapoznikov, L. M., *Coke Smokeless-Fuel Age*, **1**, 91–4, 129–31 (1939).

With the vitrains of the Saar coals, contraction decreased very appreciably with particles of several tenths of a millimeter to 2 to 3 mm, becoming negligible for 1 to 2 mm particles. Swelling decreased slightly in the case of strongly swelling coals and increased for coals which, when finely ground, are slightly swelling or not swelling, this effect being more marked as the load on the piston of the dilatometer increased. With the durain of the same coals, the contraction was a maximum for a particle size of 0.6 to 1.2 mm, at least when there was a heavy load on the piston of from 150 to 200 g. The dilatation of such durains remained slight, whatever the particle size was. The particle size of the coal is of great importance in the case of a mixture of coals, but this will be considered in a later section.

Nature of the Coal

The plastic properties vary considerably from one coal to another. Numerous relations of a statistical character between the plastic properties and other characteristics of coals have been investigated and proposed.

In general, the plastic properties become evident when the volatile matter content exceeds 13 to 15%. Initially, these properties increase with the volatile matter content, and are at a maximum for coals with 25 to 30% volatile matter. Above this content they decrease and are no longer present when the volatile matter exceeds 35 to 40%. When the volatile matter content increases from 15 to 40%, the softening point initially decreases very rapidly and then very slowly from 430 to 450°C to 330 to 350°C. Above a volatile matter content of 30% the softening point does not change greatly. A better relation is found between the resolidification point and the volatile matter content. The resolidification point decreases almost linearly from 510 to 450°C. The difference between these two temperatures, the plastic range, increases initially with the volatile matter content, passes through a maximum at 25 to 30% volatile matter, and then decreases. The temperatures of maximum plasticity and maximum swelling vary almost parallel with the resolidification point.

The swelling measured in the dilatometer is a maximum for a volatile matter content of about 25 to 28%. For coals with 15 to 18% volatile matter, as for those with 35 to 40%, the dilatometer tests leads to contraction without swelling. The fluidity measured with the Gieseler plastometer reaches a maximum at a volatile matter content of about 28 to 32%, as does also the Gray-King index. The swelling index determined by the BSI Swelling Test shows a very flattened maximum because the majority of coals with a volatile matter content of between 21 and 32% have an index near the maximum value.

Whatever test is employed for characterizing the plastic properties, the correlation of the indices and volatile matter content is very approximate as is clearly shown in Fig. 18.[86] Two coals having the same volatile matter content can have very different Gray-King indices, especially in the range of 35 to 40% volatile matter. This fact has long been recognized and is the reason why the majority of systems of coal classification, particularly

[86] King, J. G., *Fuel*, **23**, 61–4 (1944).

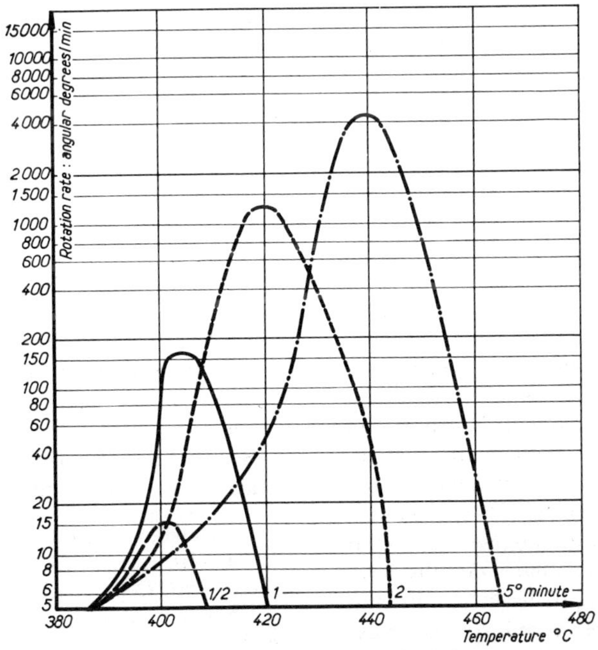

Fig. 17. Effect of heating rate on plastic properties of coal of a high-volatile coal as determined in a Gieseler plastometer.[33]

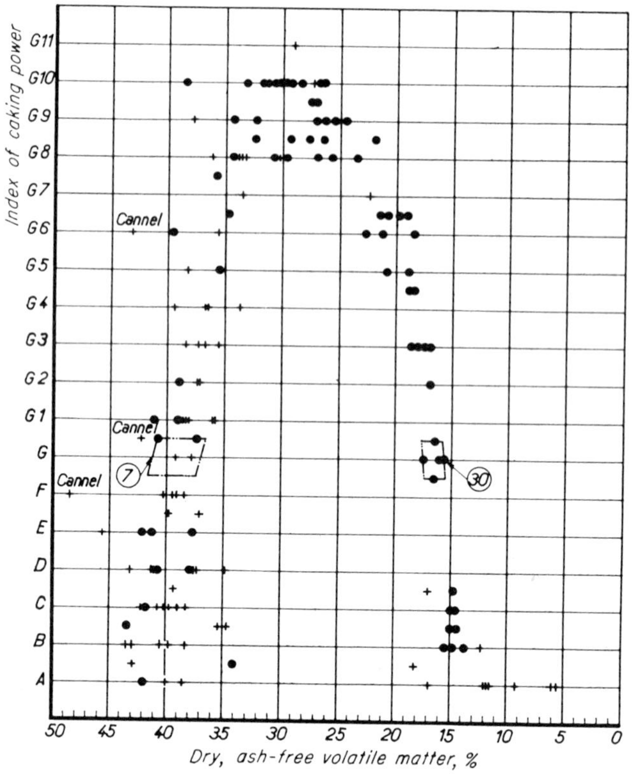

Fig. 18. Relation between Gray-King index and volatile-matter content.[86]

the international classification, have adopted a second criterion—an index characteristic of the plastic properties—in addition to the volatile matter content.

The coefficient of maximum contraction after resolidification increases regularly with the volatile matter content. It is about 4×10^{-4} for coals with 16 to 18% volatile matter and about 11 to 12 $\times 10^{-4}$ for coals with 38 to 40% volatile matter.[46]

The plastic properties are dependent on the dry, ash-free carbon content in almost the same way as they are on volatile matter content. Between 81 and 92% they are evident and reach a maximum at about 89%.[86] The correlation with the carbon content is better than the correlation with the volatile matter content for low-rank coals but is still approximate. Closer statistical relations have been obtained by taking into consideration either the oxygen content or hydrogen content. Graphs have been constructed by Spooner,[87] Jowett and collaborators,[88] and Mott and Spooner[89] expressing the Gray-King index, the BSI Swelling Number, or the swelling in the dilatometer as a function of the hydrogen and carbon content. Qualitatively, for equal carbon contents, the plastic properties increase with the hydrogen content and decrease with the oxygen content. Certain coals, such as coals from Rasa (Jugoslavia)—with an organic sulfur content of 10%—have unusual plastic properties.[90] Their plastic range begins about 100°C lower than conventional coking coals and extends to 480°C. They are highly fluid. The fluidity is little affected by rate of heating or by oxidation. In these coals, it appears that the sulfur has taken the place of the carbon rather than that of the oxygen—the (C+ S + N) content is almost the same as in coking coals.

Because the majority of the properties of coals change fairly regularly with the rank, as

[87] Spooner, C. E., *Fuel*, **30**, 193–202 (1951).

[88] Jowett, J. H., Price, P. H., and Sarjant, R. J., *J. Inst. Fuel*, **23**, 304–7 (1950).

[89] Spooner, C. E., and Mott, R. A., *Fuel*, **16**, 96–106 (1937).

[90] Lavrenčič, B., *Brennstoff-Chem.*, **39**, Sonderausgabe, 73–7 (1958).

characterized by the carbon content, it is not surprising that correlations can be found between these different properties. These correlations only denote the fact that the coals form a continuous series with a single parameter. Spooner,[87] for example, constructed graphs connecting the Gray-King index and the BSI Swelling Number with the volatile matter content and the calorific capacity. Russell and Perch[91] noted a close correlation between the fluidity observed in a Gieseler plastometer and the parameter: volatile matter content + natural moisture content. A fairly close parallel has been shown between the plastic properties and the porosity[92] with a minimum at a carbon content of about 89%, the most plastic coals being, therefore, those with lowest porosity. Heat of wetting by methanol also shows a very marked minimum for these coals.[93,94] The correlation between plastic properties and porosity or heat of wetting does not appear more exact than the correlation between plastic properties and rank. The relations between the plastic properties and the coal fraction soluble in certain solvents such as benzene, pyridine, and chloroform[95–97] have been discussed by Berkowitz[98] and do not appear to be of general significance.

Pressure

Coals which soften, swell, and cake when heated under atmospheric pressure no longer give a coherent residue when heated under vacuum; the coal particles soften and swell slightly but do not cake together. Conversely, when coal is kept under pressure during heating, the caking of the particles is promoted and the swelling increases.[99,100] A high pressure of the order of many atmospheres is required to obtain an appreciable effect. Low-density cokes with large pores and very fine walls have been obtained by Cerchar at a gas pressure of 100 kg/sq cm.[99] When a high-volatile coal is used, the coke obtained is anisotropic in polarized light, but this is not the case when heating is carried out at atmospheric pressure.

Oxidation

The plastic properties are very markedly modified by oxidation, even slight oxidation. The change in these properties is often observed before any change in the composition. More intense oxidized coals may not even soften or swell.[17,20,45,72,91,101–103] In general, oxidation causes an increase in the softening temperature and a decrease in the resolidification temperature, the change in the latter being much less than that in the former. The plastic range decreases to zero by intense oxidation. At the same time, the fluidity and the swelling decrease. An example of the effect of oxidation on the plasticity of coal as determined in the Gieseler plastometer is shown in Fig. 19. In certain cases, it has been found that slight oxidation causes an increase in swelling measured either by the BSI Swelling Test[17,102] or the dilatometer.[101] This is probably due to a reduction in the fluidity. In every case, more intense oxidation leads to a reduction in the swelling.

The effects of oxidation vary greatly from one coal to another. For some coals oxidation occurs at room temperature (20°) when the coal is stored as fines, 0 to 6 mm or 0 to 10 mm, for several days. On the other hand, other coals may be stored for months without change. As a general rule, the liability to oxidation is greater the lower the rank of the

[91] Russell, C. C., and Perch, M., *AGA*, *Proc.* **31,** 372–89 (1949).

[92] King, J. G., and Wilkins, E. T., *Proc. Conf. Ultrafine Structure of Coals and Cokes*, 1943, BCURA, London, 1944, pp. 46–56.

[93] Griffith, M., and Hirst, W., *Proc. Ultrafine Structure of Coals and Cokes*, 1943, BCURA, London, 1944, pp. 80–94.

[94] Dryden, I. G. C., *J. Inst. Fuel*, **30,** 193–222 (1957).

[95] Illingworth, S. R., *Fuel*, **1,** 213–19 (1922).

[96] Bone, W. A., Pearson, A. R., and Quarendon, R., *Proc. Roy. Soc.* (*London*), **A105,** 608–25 (1924); Bone, W. A., Horton, L., and Tei, L. J., *Proc. Roy. Soc.* (*London*), **A120,** 523–45 (1928).

[97] Funasaka, W., Yokokawa, C., Suga, S., Kito, Y., and Awaya, T., *Bull. Inst. Chem. Research, Kyoto Univ.*, **25,** 18 (1951).

[98] Berkowitz, N., *Fuel*, **28,** No. 5, 97–102 (1949).

[99] Cerchar, *Rapport annuel*, **1953,** p. 72.

[100] Nadziakiewicz, J., *Fuel*, **37,** 361–3 (1958).

[101] Boyer, A. F., *Proc. Intern. Conf. on Chem. Eng. in the Coal Ind.*, 1956 (Stoke Orchard), Pergamon Press, London, New York, Paris, 1957, 141 pp.

[102] Wildenstein, R., *Chaleur & ind.*, **29,** 233–6 (1953).

[103] Sarkar, S., Rao, V. V., Das Gupta, S. C., and Das Gupta, N. N., *J. Sci. Ind. Research* (*India*), **16B,** 539–47 (1957).

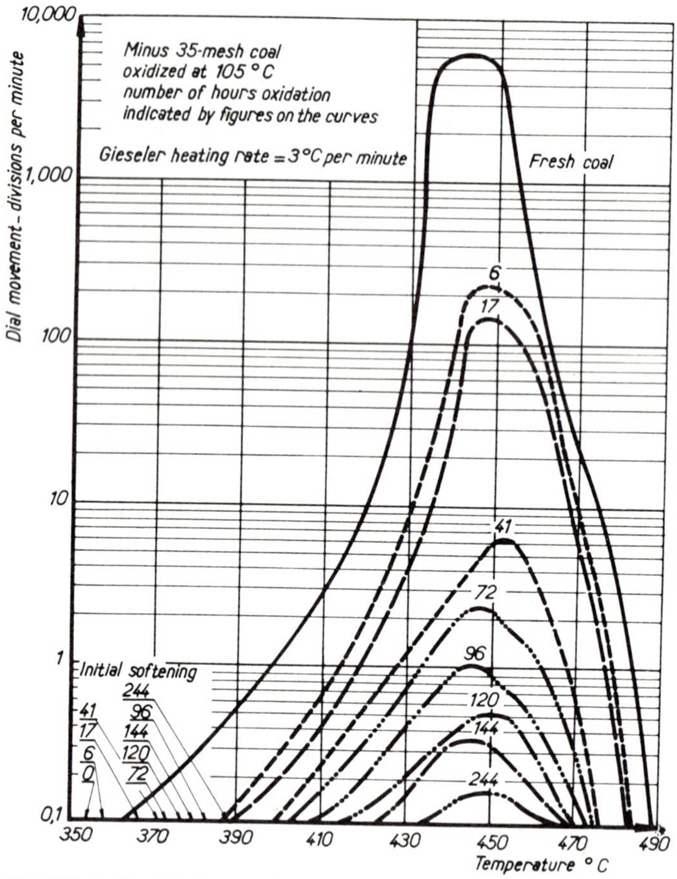

Fig. 19. Effect of oxidation on the plastic properties of an Eagle Seam coal.[91]

coal.[103–105] Georgiadis and Gaillard[106] considered that the rate of oxidation is principally a function of the oxygen content of the coal and increases with it; but this statement is not contradictory to the former because, in general, the oxygen content increases with a reduction in rank. According to Oreshko,[107] oxidation at low temperature increases with the surface area of the open pores, but the latter also varies inversely with rank for high-volatile coals. It has also been observed that coals with a high natural moisture content oxidize more rapidly than those with a low content, but this effect can again be interpreted as the influence of rank. Mandal and Orning[108] have stated that dull coals are more susceptible to oxidation than bright coals, but this does not seem to be general and the reverse effect has sometimes been noted.

Oxidation of coals in contact with air is in general a harmful phenomenon when the coal is used in a coking plant, and efforts are made to reduce it by suitable stocking conditions.[109, 110] Investigations on the use of inhibitors such as NH_3 and ammonia compounds have recently been carried out.[111] On the other hand,

[104] Brewer, R. E., Reynolds, D. A., Steiner, W. A., and Van Gilder, R. D., *Ind. Eng. Chem.*, **40,** 1243–54 (1948).

[105] Frey, W. A., *Brennstoff-Chem.*, **24,** 603 (1943).

[106] Georgiadis, C., and Gaillard, G., *Chim. & ind. (Paris)*, **70,** 383–96 (1953).

[107] Oreshko, V. F., *Doklady Akad. Nauk S.S.S.R.*, **70,** 445–8 (1950).

[108] Mandal, A. K., and Orning, A. A., *Fuel*, **31,** 33–6 (1952).

[109] Schmidt, L. D., *Chemistry of Coal Utilization*, John Wiley and Sons, New York, 1945, Vol. 1, pp. 627–76.

[110] Loison, R., *Note Tech., Charbonnages de France*, **1953,** October, 30 pp.

[111] Pruss, W., *Brennstoff-Chem.*, **39,** 348 (1958).

this phenomena can be employed for the deliberate oxidation of coal to reduce its plastic properties in certain problems, such as the distillation of briquettes, semidistillation of fines in a rotary furnace, or fluidized bed to produce "char," the carbonization of coals which are too fluid or too "swelling." By suitable regulation of the temperature and duration of oxidation as a function of the particle size of the coal it is, therefore, possible to bring the plastic properties to a desired level.

The oxidation of coal can be effected not only by keeping it in air or oxygen, but also by keeping it in an oxidizing solution. Thus, the oxidation of a coal in chloroform by perbenzoic acid has been studied by Storch and his co-workers.[112] A change in the plastic properties, similar to that caused by oxidation in air, was noted. The addition to coal of small quantities of powdered Fe_2O_3 strongly reduces its plastic properties, whereas Fe_3O_4 behaves like an inert substance.

Action of Sulfur and Various Reagents

The reaction of sulfur with coal presents a certain similarity with that of oxygen. It also leads to a considerable reduction in the plastic properties.[113–116] To observe this effect it is sufficient to mix the coal with a slight amount of sulfur and the reaction takes place at a temperature below 170°C.[116] The addition of chlorine[117,118] and selenium[119] also effects a reduction in the softening and swelling capacity. On the other hand, Bunte and Imhof[120] have stated that an increase in the swelling results from the addition of boric acid, which could be the result of a reduction in the fluidity.[121]

Hydrogenation of Coal

Mild hydrogenation of coal, without any major change in its chemical composition, causes a considerable change of the plastic properties. Generally, the swelling and, in particular, the fluidity are increased accompanied by an appreciable reduction in the softening point, a slight rise in the resolidification point, and a broadening of the plastic range.

Treatment of a fresh subbituminous coal with aluminum and lithium hydride dissolved in tetrahydrofuran, or sodium dissolved in liquid ammonia[112,122] had little effect. On the other hand, such treatment almost completely regenerated the plastic properties of a bituminous coal, the swelling index of which had been reduced from 7 to 4 by oxidation in air.

Similar effects have been found by several investigators[112,123–125] using pure hydrogen, coke-oven gas, or a mixure of CO and steam at temperatures of 300 to 400°C and pressures of about 100 to 200 kg/sq cm. It is possible to reduce the pressure by adding a suitable catalyst such as ferric oxide alone or mixed with lime,[124] molybdic acid or ammonium molybdate,[124,125] or tin or zinc + ammonium chloride.[112]

The changes in the chemical composition of the coal caused by hydrogenation depend on the nature of the coal. In general, a slight

112 Orchin, M., Golumbic, C., Anderson, J. E., and Storch, H. H., *U.S. Bur. Mines Bull.* No. 505 (1951), 15 pp.

113 Cerchar, *Rapport annuel*, **1952,** p. 147.

114 Postowsky, J. J., and Harlampovich, A. B., *Fuel*, **15,** 229–32 (1936).

115 Patterson, W. S., and Metcalfe, W. J., *J. Soc. Chem. Ind.* (*London*), **56,** 388–91T (1937).

116 Berkowitz, N., and McCulloch, A., *Fuel*, **25,** 36–41 (1946).

117 Marsh, A., McCulloch, A., and Parrish, E., *J. Soc. Chem. Ind.* (*London*), **48,** 167–74T (1929).

118 Kay, H., and McCulloch, A., *J. Soc. Chem. Ind.* (*London*), **51,** 186–90T (1932); **52,** 47–49T (1933).

119 Boll, H., *Gas u. Wasserfach*, **80,** 209 (1937).

120 Bunte, K., and Imhof, H., *Gas u. Wasserfach*, **82,** 805–12 (1939).

121 Lambris, G., *Gas u. Wasserfach*, **76,** 1–3 (1933).

122 Orchin, M., Goldbach, G. L., Wolak, M., and Storch, H. H., *U.S. Bur. Mines Rept. Invest.*, No. 4499 (1949), 10 pp.

123 Crawford, A., Williams, F. A., King, J. G., and Sinnatt, F. S., *G. Brit. Dept. Sci. Ind. Research, Fuel Research Tech. Paper* No. 29, **1931,** 37 pp. Lander, C. H., Sinnatt, F. S., King, J. G., and Crawford, A., Brit. Pat. 301. 720. Pampuch, P., Nadziakiewicz, J., *Koks, Smola, Gaz*, **1,** 45–52 (1956). Lambris, G., and Gerde, J., *Brennstoff-Chem.*, **22,** 139–42 (1941). Hovers, T., Koopmans, H., and Pieters, H. A. J., *Fuel*, **15,** 233–5 (1936). Kukharenko, T. A., and Taits, E. M., *Bull. acad. sci. U.R.S.S., Classe sci. tech.*, **1942,** Nos. 1–2, 97.

124 Ahuja, L. D., Sharma, J. N., Kini, K. A., and Lahiri, A., *J. Sci. Ind. Research* (*India*), **17A,** 27–9 (1958).

125 Cerchar, *Rapport annuel*, **1956,** p. 70.

increase in the hydrogen content and a definite decrease in the oxygen content are observed. It appears, therefore, that the first stage of hydrogenation is characterized primarily by "deoxidation," and its effect on the plastic properties is qualitatively the reverse of the effect of oxidation. The plastic properties of a hydrogenated coal are, however, not identical with those of a natural coal having the same oxygen content.

Plastic Properties of Macerals

The plastic properties of macerals can be studied by direct observation by the *microscope* of their behavior during heating of a thin section,[3,7,126] or by use of the various methods described earlier on a sample composed of only one maceral or the macerals of one group. Methods for recovery of the macerals include hand-sorting based on differences in appearance, separation by heavy media based on differences in density, crushing and screening based on differences in hardness, and solubility in a reagent. Whereas it is fairly easy to prepare almost pure samples of vitrinite, it is more difficult with inertinite and exinite because of the finely dispersed state in which these macerals are often found. One must, therefore, prepare more or less rich "concentrates." The properties of the pure macerals can only be determined by extrapolation, which is uncertain when an additive property is not concerned.[127,128]

Several conclusions may be drawn from the different studies carried out by these methods.[7,129] The plastic properties of each maceral vary considerably with the rank of the coal in which it is contained. Because the elementary compositions of the macerals of the same coal sample are different, the rank is denoted by the carbon content of the vitrinite, which is generally the most abundant maceral. The differences between these properties are less marked the higher the rank and become imperceptible when the carbon content exceeds 90 to 92%. Inertinite is almost infusible, whatever the rank, and it behaves like an inert substance. It is also distinguished from vitrinite of the same rank by a lesser hydrogen content, a lower volatile matter content and a higher density. Exinite differs from vitrinite of the same rank by a higher volatile matter content, a higher hydrogen content, and a lower density. Its softening point is lower, its plastic range is wider, its fluidity observed in a Gieseler plastometer is much greater, and its swelling measured in a dilatometer is generally higher. In high-volatile coals, the volatile matter content of exinite reaches values above 70% and carbonization leaves little residue that is very fragile.

Plastic Properties of Lithotypes

The determination of the plastic properties of lithotypes does not present serious difficulties. The plastic properties are, to a certain extent, the resultant of the properties of the macerals composing it, but may equally depend on the manner of distribution of the macerals and, in particular, on the greater or lesser dispersion of the exinite and inertinite within the vitrinite. In actual fact, this dispersion determines the degree of interaction of the different macerals during their fusion. It is, therefore, impossible to predict the behavior of a lithotype with accuracy from the data of its maceral analysis alone. Several authors[40,130] have studied mixtures of macerals in variable proportions in order to reconstitute artificial lithotypes with a well-defined and regulatable maceral composition. The results obtained must be interpreted with caution. The properties of such a mixture are not necessarily identical with those of a natural lithotype having the same maceral composition, because the manner of distribution and, therefore, the interaction of the different macerals may be different.

With the exception of fusinite, which is always infusible, the plastic properties of lithotypes vary greatly with rank. If the lithotypes

[126] Pieters, H. A. J., and Koopmans, H., *Fuel*, **11**, 447–51 (1932); Ergun, S., O'Donnell, H. J., and Parks, B. C., *Fuel*, **38**, 205–10 (1959).

[127] Alpern, B., *Ind. chim. belge*, **20**, 9–12 (1955).

[128] Alpern, B., Ladam, A., and Iselin, P., *Brennstoff-Chem.*, **39**, Sonderausgabe, 58–61 (1958); *Uhli*, **8**, 140–1 (1958).

[129] Mackowsky, M. T., *Brennstoff-Chem.*, **30**, 44–60, 141–7, 218–25 (1949); **31**, 129–38 (1950); **32**, 175–86 (1951). Alpern, B., *Rev. ind. minérale*, **38**, 170–81 (1956). Kröger, C., *Brennstoff-Chem.*, **37**, 186–9 (1956). Van Krevelen, D. W., Dormans, H. N. M., and Huntjens, F. J., *Fuel*, **38**, 165–82 (1959). Ergun, S., McCartney, J. T., and Mentser, M., *Econ. Geol.* **54**, 1068–77 (1959).

[130] Kröger, C., *Brennstoff-Chem.*, **37**, 186–9 (1956). Patteiwsky, K., *Brennstoff-Chem.*, **37**, 277–83 (1956).

of the same coal are compared, the order in which the vitrite, clarite, and durite are classified for a given test is not always the same. In general, it varies little for coals from the same basin and having fairly similar characteristics, but it may be reversed from one family to another. General results cannot, therefore, be given. From a study of coals from the Saar and Lorraine in the Hoffmann dilatometer, Hoffmann and Hoehne[40] have stated that for Saar coals of about 32 to 34% volatile matter, the swelling decreases in the order vitrite, clarite, durite. On the other hand, for the lowest rank Lorraine coals the order was found to be clarite, durite, vitrite. In the case of Ruhr coking coals, Patteiwsky[130] found, on the contrary, that the swelling measured in a dilatometer was always higher for vitrite than for clarite or durite.

INTERPRETATION OF PLASTICITY PHENOMENA

At heating rates of 1 to 3°C/min, the softening point observed with the dilatometer or plastometer differs little from the decomposition point. In order to interpret the phenomenon of "melting," it is important to know whether or not decomposition precedes softening. Because softening and decomposition are both gradual, they are not clearly defined. They depend on the experimental method employed to determine them and can only be accurately defined by adopting an arbitrary standard. Accurate methods have been proposed for determining either the decomposition point[131,132] or the softening point.[133] In a critical analysis of the problem, it was concluded[134] that "if softening is recognized by the attainment of visible fluidity" it is always preceded by a "slight thermal degradation of the solid to other solids and liquids." However, "a degree of softening may occur before active decomposition," the latter being termed the temperature at which decomposition starts to increase.[131]

Numerous investigators have studied the pyrolytic reactions of coals by determining the loss in weight at constant temperature or, more frequently, with progressive heating at constant rate.[135,136] Figure 20 gives an example of an experimental curve recorded at a heating rate of 2°C/min. The dotted curve is the derived curve representing the rate of loss in weight. First, a loss of moisture is observed a little above 100°C, which may sometimes be continued up to about 200°C. Between 200 and 350°, the loss in weight is negligible, but it becomes appreciable at a temperature θ_1, which may be called the decomposition point. The rate shows a very pronounced maximum at a temperature θ_i.

Whatever the method employed to determine it, the decomposition point θ_1 increases when the heating rate is increased.[135] In the case of coking coals, an increase of 35 to 40°C is observed when the heating rate is increased from 0.5 to 5°C/min.[134] The softening point, θ_s, on the other hand, changes little, with the result that the difference $(\theta_1 - \theta_s)$ increases with the heating rate. At higher rates of 10°C/min or more, softening may be observed before there is an appreciable loss in weight. Both the decomposition point, θ_1, and the softening point, θ_s, decrease with the rank of the coal. The difference $(\theta_1 - \theta_s)$ is often greater for coals with higher fusibility and swelling than for poorly fusible coals.[134] The temperature, θ_1, corresponding to the maximum rate of loss in weight depends on the heating rate and the rank of the coal in the same way as does the maximum fluidity temperature and the resolidification point determined in the plastometer.[137]

Several investigators[138] have determined the internal surface by the heat of wetting of

[131] Holroyd, R., and Wheeler, R. V., *J. Chem. Soc.*, **1928**, 3197–203.

[132] Hibbott, H. W., and Wheeler, R. V., *J. Chem. Soc.*, **1934**, 1084–6.

[133] Seyler, C. A., *J. Inst. Fuel*, **12**, 301–3 (1939).

[134] Dryden, I. G. C., and Griffith, M., *Bull. Brit. Coal Utilisation Research Assoc.*, **18**, 62–86 (1954).

[135] Boyer, A. F., *Compt. rend. 69th congr. ind. gaz*, **1952**, 653–71. van Krevelen, D. W., van Herden, C., and Huntjens, F. J., *Fuel*, **30**, 253–9 (1951). Fitzgerald, D., *Gas J.*, **293**, 42–4 (1958).

[136] Fitzgerald, D., and van Krevelen, D. W., *Fuel*, **38**, 17–37 (1959). Kröger, C., and Pohl, A., *Brennstoff-Chem.*, **38**, 102–7 (1957).

[137] Boyer, A. F., *Chaleur & ind.*, **30**, 324–32 (1954).

[138] Cannon, C. G., Griffith, M., and Hirst, W., *Proc. Conf. Ultrafine Structure of Coals & Cokes*, 1943, BCURA, London, 1944, pp. 131–46. Bond, R. L., BCURA, Private communication. Bond, R. L., and Spencer, H. T., *Proc. Conf. Ind. Carbon and Graphite, London*, 1957, Soc. Chem. Ind., London, 1958, pp. 231–51.

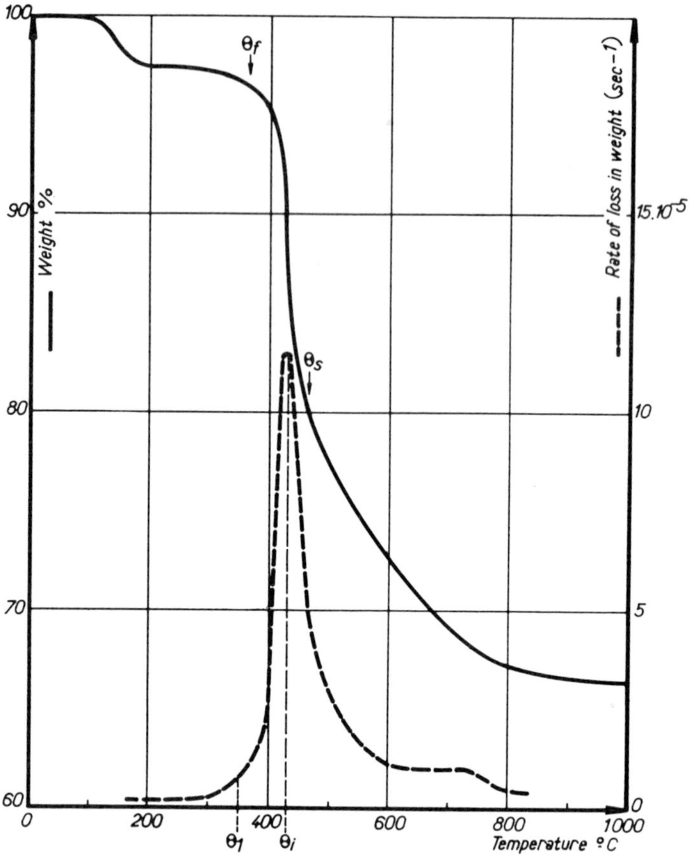

Fig. 20. Degasification of coal at a heating rate of 2°C/min.

methanol or by the adsorption of neon and the microporosity, by density in helium, of coal samples brought to increasing temperatures and cooled rapidly. The changes in surface area and microporosity are relatively slight, up to 500 to 600°C, suggesting that the fine structure of coal is little modified by fusion.[139] But this opinion is not universally held. Studies of graphitization[140] suggest that the carbon atoms begin to be arranged in a "pregraphitic" lattice during the fusion phase.

The solubility of a coking coal in chloroform is increased if it has been brought to a temperature near the softening point. This phenomenon, discovered by Illingworth,[95] has been studied in detail by Dryden and Pankhurst[141] and Pitt[43] who employed the following procedure: The coal was brought rapidly to a temperature, θ, kept there for a time, t, cooled rapidly, and then was extracted with chloroform. The rate of formation of the soluble fraction increased with temperature. Up to 400°C this fraction was stable, but above this temperature it decomposed, giving an insoluble substance and volatile matter. With a noncaking coal, the same treatment did not increase the soluble fraction. The residue after extraction of a coking coal was not fusible but, if the extract was mixed with a noncoking coal, the latter fused and acquired swelling characteristics. The reduction in the swelling power of a coking coal caused by controlled oxidation is also accompanied by a reduction in the yield of chloroform extract.[79] All these facts show that the formation of an extract soluble in chloroform during the heating of a coking coal is closely linked to the mechanism of softening.

[139] Brown, H. R., Brooks, J. D., Cameron, A., and Stacy, W., *Proc. Residential Conf. on Sci. in the Use of Coal*, Inst. Fuel, 1958, pp. E-42–7.

[140] Fránklin, R. E., *Brennstoff-Chem.*, **34**, 359–61 (1953). Durif, S., *Brennstoff-Chem.*, **37**, 402–4 (1956).

[141] Dryden, I. G. C., and Pankhurst, K. S., *Fuel*, **34**, 363–6 (1955).

The Mechanism of Softening

The mechanism of softening has been described by many authors, but there are very few theories proposing an accurate and relatively detailed system of the phenomenon. Vitrinite, the principal constituent of the majority of coals, appears to give the phenomenon its general pattern. The theories proposed to explain softening may be classed as physical or physicochemical. In the former, softening is considered as a paste-like melting resulting from the increase in thermal agitation and is similar to that which thermoplastic materials undergo. Parallel to this fusion and independently of it, pyrolysis develops and modifies the fusibility. The behavior of the coal is, therefore, the result of the superposition of the physical phenomenon of melting and the chemical phenomena of pyrolysis. It may be assumed either that the whole mass of coal melts simultaneously in a homogenous manner, or that only a fraction of the coal melts and "plasticizes" the remainder. The various physical theories differ with regard to the hypotheses put forward on the nature of the fusible fraction and the mode of action of the plasticizing fraction.

On the other hand, in the physicochemical theories, softening is a *consequence* of pyrolysis, being due either to the pressure of the gas and vapor produced forcing a path through the porous structure of the coal,[142] or to the formation of a fluid substance which "plasticizes" the mass of coal (thermobitumen theory).

Homogeneous Melting

According to a theory proposed by Audibert in 1926[143] and later adopted by various authors,[144] coal undergoes true melting throughout the entire mass but decomposes at the same time, giving rise to solid infusible products and gas or vapor. It forms, therefore, a mixture of fusible and infusible products, the composition of which depends on time and temperature. Resolidification corresponds to the completion of the transformation into infusible products.

Audibert's interpretation accounts for the influence of the rate of heating: When the latter increases, the softening point is not modified because it is a physical mechanism; the pyrolytic reactions, on the other hand, shift to higher temperatures, because at a given temperature, the quantity of nondecomposed fused coal is greater, and the plasticity increases. In addition, the temperature of maximum fluidity and the resolidification temperature should increase, which agrees well with experience. This theory also accounts for the fact that, for certain coals at least, the commencement of softening is a reversible phenomenon.[78] But it does not account for the fact that, in experiments at constant temperature, fluidity increases as a function of time (Fig. 21): It should attain its maximum immediately and then decrease. According to Audibert's theory, the behavior of a coal depends essentially on the difference between the softening point and the decomposition point. Plastic properties must be more marked, the greater the value of this difference. Noncaking coals are those for which the decomposition point is clearly lower than the temperature at which coal could melt if it did not decompose.

The theory of homogeneous melting cannot account for two important experimental results: the marked reduction of fusibility and caking power by carbonization under vacuum and the slight variation of the fine structure during melting; in fact, it seems improbable that after melting, the fissuration due to cooling, or the bubble structure caused by the liberation of gas, gives rise to a microporosity of the same order as that of the initial coal.[134] Moreover, by extrapolation of measurements at lower temperatures, the fluidity of undecomposed fused coal would be too low to account for actual values.[145]

The theory of Berkowitz,[98] based on a micellar model of coal, can be linked with homogeneous melting. When the temperature rises, the atoms or groups of superficial atoms of the micelles acquire a certain mobility which allows the micelles to change their relative positions while they themselves remain solid. These superficial atoms thus behave like a "two-dimensional liquid," playing the role of a lubricant. Whereas conventional homogeneous melting is due to a relative movement of the molecules, plasticity is due here to the relative movement of micelles, the

[142] Mott, R. A., *Fuel*, **6**, 217–31 (1927).
[143] Audibert, E., *Fuel*, **5**, 229–44 (1926).
[144] Gillet, A., *Nature*, **167**, 406–7 (1951).
[145] Van Krevelen, D. W., and Huntjens, F. J., *Chim. & ind. (Paris)*, **80**, 234–9 (1958).

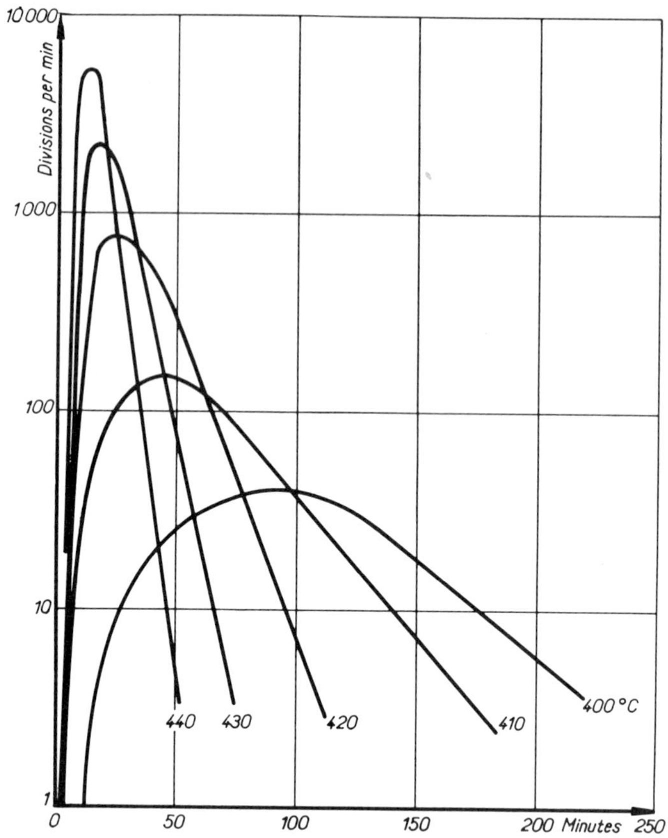

Fig. 21. Fluidity of Union coal as a function of carbonizing time at different temperatures.[43]

latter retaining their individuality and rigidity. Thus, there is, principally, a difference of scale of the mobile elements, the micelles in the Berkowitz model having diameters of several hundred Angströms. This theory presents almost the same characteristics as that of homogeneous melting but also offers an explanation as to why the fine structure of coal varies little during melting. The internal surface and the porosity are identified respectively by the surface of the micelles and the intermicellar volume and undergo little modification during softening because the micelles retain their individuality.

Melting of a Coal Component

Older theories on fusion considered that only a fraction of the coal can fuse, and that the fluidity of this fraction is sufficient to give the whole a certain plasticity. This hypothesis can explain why the fine structure undergoes little modification during softening: This structure is retained by all the unmelted fraction, which is dispersed in the fusible fraction. The fusible fraction or "coking principle" has often been identified with the fraction of coal soluble in organic liquids such as benzene and chloroform and termed "bitumen."[92,95,146] This opinion is based on several arguments, none of which is decisive. The extracts are very fusible, whereas the extraction residues have hardly any plastic properties. However, the mixture of the extract and residue does not have the exact properties of the initial coal.[147] Moreover, the addition of the extract of a coking coal to the residue of a noncaking coal does not give a fusible product.[148] Although a certain correlation has been found between the yield of soluble products and the plastic properties of

[146] Pott, A., Broche, H., Nedelmann, H., and Scheer, W., *Fuel*, **13,** 91–125, 134 (1934). Kiebler, M. W., *Ind. Eng. Chem.*, **32,** 1389–94 (1940).

[147] Dryden, I. G. C., *Fuel*, **29,** 197–207 (1950).

[148] Broche, H., and Schmitz, H., *Brennstoff-Chem.*, **13,** 81–5 (1932).

coal, this correlation does not appear general.[94,98] That controlled hydrogenation of coal[149] or an addition of oil[150] bring about a parallel increase in the yield of extracts and the plasticity is not necessarily relevant.

Various hypotheses have been advanced with regard to the form in which the fusible "component" is present, closely connected with the models proposed for the representation of coal structure. In the older theories, it was assumed that coal is composed of a true physical mixture of several components having different chemical compositions. The following were distinguished: bitumens—fusible component, and humic substances—infusible component. This representation of coal is no longer acceptable. Certain vitrinites are found to be perfectly homogeneous even under the electron microscope,[151] although they fuse well.

In the model proposed by Riley and his co-workers, based primarily on X-ray observations,[152] coal is composed of lamellae stacked more or less upon each other and forming crystallites. The diameter of these crystallites is of the order of 20 to 30 A; the number of lamellae of which each crystallite is composed varies. The smallest crystallites represent the fusible "component"; they are soluble in organic solvents and it is these which become mobile when the temperature rises and which, by acting like a lubricating agent, give the mass of still unfused coal its plastic properties.

The representation of coal by an *isogel* has been discussed by Hirst and Lahiri,[153] according to which coal is composed of an assemblage of large molecules (or micelles) and smaller molecules, all molecules having analogous compositions. At low temperatures these molecules are linked by van der Waals forces, and the mass is solid. When the temperature rises, these forces of attraction are weakened, the small molecules become mobile, "melt," and the micelles, which have remained solid, become plastic. These small molecules also form the fraction soluble in organic solvents. This theory has the advantage over its predecessors in that it agrees with the apparent homogeneity of vitrinite, but like all the physical theories, it does not account for the influence of pressure.

Thermobitumen and Metaplast

The close parallelism between the pyrolytic reactions, the development of plastic properties, and the particular fact that coals only attain a definitely plastic state when the pyrolytic reactions develop at a sufficient rate has led the majority of modern authors[42,79,136,154,155] to consider the softening of coal as a consequence of pyrolysis. Pyrolysis transforms coal into fluid products either completely or partially. It is the appearance of these fluid products which causes softening and plasticity. Their disappearance either by evaporation or thermal decomposition leads to resolidification. By analogy with the term "bitumens," introduced in previous theories, these fluid products were initially called "thermobitumens," signifying that they are *produced* by pyrolysis and do not pre-exist in coal. Following this general idea, several systems may be considered in attempts to specify the nature and mode of action of "thermobitumens." The simplest hypothesis is to consider that coal decomposes into a fluid product and a solid residue. The fluid product, or thermobitumen, forms the primary pyrolytic product, mainly primary tars and a little gas. At the temperature at which they form, the vapor pressure of these tars is low, so that they remain for a certain time, in the liquid state, dispersed in the coal, making it plastic. Subsequently, they disappear, either by simple evaporation and liberation as volatile matter or by thermal decomposition giving rise to gas or vapor—which escapes—and a solid residue which aggregates to form a

[149] Horton, L., King, J. G., and Williams, F. A., *J. Inst. Fuel*, **7**, 85–97 (1933).

[150] Onusaitis, B. A., *Fuel Abstr.*, **3**, 1642 (1948).

[151] Pregermain, S., and Guillemot, G., *4th Intern. Kongr. Electronenmikroskopie*, Springer-Verlag, Heidelberg.

[152] Blayden, H. E., Gibson, J., and Riley, H. L., *Inst. Fuel, Wartime Bull.* No. 18, 117–129 (1945); *Proc. Conf. Ultrafine Structure of Coals and Cokes*, BCURA, London, 1944, pp. 176–231.

[153] Hirst, W., *Proc. Conf. Ultrafine Structure of Coals and Cokes, BCURA, London*, 1944, pp. 71–9. Lahiri, A., *Fuel*, **30**, 241–7 (1951).

[154] Oele, A. P., *Brennstoff-Chem.*, **33**, 231–8 (1952). Kugo, M., *J. Chem. Soc. Japan*, Ind. Chem. Sect., **54**, 628 (1951).

[155] Swietoslawski, W., *Coke Formation Process and Physico-chemical Properties of Coals*, Polish Inst. Arts and Sciences in America, New York, 1942, 145 pp.

primary solid residue, semicoke. The system is, therefore, as follows:

$$\begin{cases} P \rightarrow \text{liquid } Th + R_1 \\ \text{liquid } Th \rightarrow Th \text{ gas} \\ Th \rightarrow G + R_2 \end{cases}$$

where P is the initial coal, Th the thermobitumen, G the gaseous products resulting from the decomposition of the thermobitumen, R_1 the primary solid residue, R_2 the secondary solid residue, and $R_1 + R_2$ form the semicoke.

It has been assumed[42,156,157] that the coal as a whole (or vitrinite, at least) is transformed by a depolymerization reaction into a fluid mass called "fluid coal" or "metaplast." This metaplast is decomposed, in turn, by a cracking reaction into gas or vapor, which forms the volatile matter and is eliminated, and by a solid residue which forms the semicoke. These two reactions of depolymerization and cracking overlap in time, so that, during the softening phase, undecomposed coal (solid), metaplast (fluid), and semicoke (solid) and to a secondary extent, gas—in process of liberation through the metaplast—exist simultaneously. The plasticity is greater, the higher the proportion of metaplast; its disappearance leads to resolidification. The semicoke subsequently undergoes secondary decomposition, giving rise to gas, but this reaction does not affect the development of the plastic properties.

Whatever the exact system adopted, the idea of thermobitumen or metaplast allows several facts to be explained.

1. When carbonization is carried out under vacuum, the plasticity of the coal is markedly reduced or even suppressed, possibly due to the fact that the thermobitumens are immediately carried away in the form of vapor or mist which can then no longer act as plasticizing agent.

2. When coal has been brought to a temperature within the plastic range and then cooled rapidly and reheated, its softening temperature during the second heating is lower than during the first.[156] Part of the metaplast formed during the first heating is not destroyed by the rapid cooling and, during the second heating, it softens and causes softening of the coal as soon as its viscosity is sufficiently reduced by the rise in temperature. It should be noted, however, that this phenomenon may not always be observed.[78]

3. When a small proportion of a porous substance such as alumina, activated carbon, chalk, coke breeze, or semicoke is mixed with coal, the plasticity is reduced and the solidification point is lowered.[47] This effect is attributed to the absorption of a fraction of the thermobitumens by the porous substance, which may be shown by observation of the particles of the porous substance after cooling. In general, this effect is more marked the greater the specific surface and porosity of the porous material. However, the action of semicokes prepared at 600 to 700°C is not very different from that of coke breeze, although their surfaces are greater. Neither surface nor total porosity are the only factors to be considered, because it is probable that the large molecules of thermobitumens cannot penetrate the finest pores.

4. The absence of plasticity in lean and low rank coals may be due to two different reasons. Fluid thermobitumens are not formed in lean coals, because depolymerization of the coal gives rise to a product which is still solid. In low rank coals, thermobitumens are abundant but have little thermostability, decomposing rapidly so that their concentration is always very low.

A mathematical analysis of the action of heat on coal has been made[156] assuming first-order reactions for the transformation of the coal (P) to the metaplast (M), for that of the metaplast to semicoke and primary gas (G_1) and for that of the semicoke to coke and secondary gas. The development of the system can be described by the following system of equations:

$$\frac{dP}{dt} = k_1 P$$

$$\frac{dM}{dt} = k_1 P - k_2 M$$

$$\frac{dG_1}{dt} = k_2 M$$

k_1 and k_2 being the velocity constants of the first two reactions. If coal is kept at constant temperature, k_1 and k_2 are constants. The dependence of M/P_0 (P_0 being the initial weight of the coal) on $k_1 t$ is shown on Fig. 22,

[156] Chermin, A. G., and Van Krevelen, D. W., *Fuel*, **36**, 85–104 (1957).

[157] Fitzgerald, D., *Trans. Faraday Soc.*, **52**, 362–9 (1956).

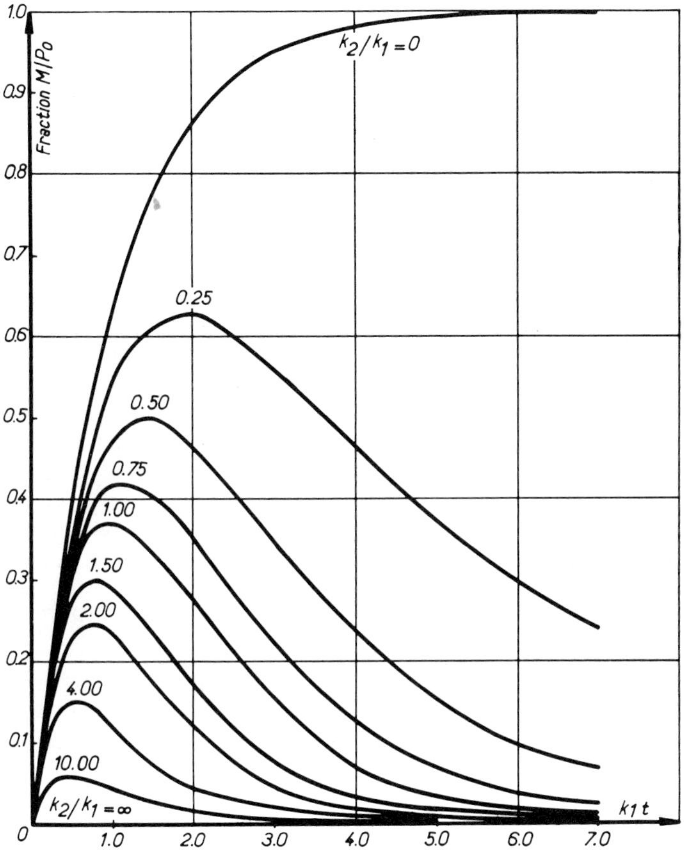

Fig. 22. Dependence of M/P_0 on k_1t for different values of k_2/k_1.[156]

the different curves corresponding to different values of k_2/k_1. If it is assumed that plasticity is a function of the concentration of metaplast (M/P_0), these curves must represent the change of plasticity as a function of time. Qualitatively they have much the same shape as the experimental curves (Fig. 21). In particular, it is found that the plasticity does not immediately attain its maximum value as required by the physical theories. The smaller the ratio k_2/k_1, the greater the plasticity, that is, the decomposition of the metaplast is slower than its formation.

If coal is subjected to *heating at constant rate*, k_1 and k_2 are functions of temperature and, therefore, of time. From the mathematical analysis, it may be concluded that if the activation energies, E_1 and E_2, are of the same order of value, the temperature, θ_i, corresponding to the maximum of the decomposition rate, and the temperature, θ_m, corresponding to the maximum metaplast concentration (and, therefore, the plasticity maximum) vary in a parallel manner when the rate of heating v increases, their difference remaining less than 20°C. The maximum rate of decomposition is proportional to v, and θ_i is related to v by $1/\theta_i = a - b \log v$, which agrees with experiment. The coefficient of proportionality, b, determined by experiment, makes it possible to evaluate the activation energy E_1 at about 50 to 60 kcal, which is also that observed for depolymerization of several polymers, and corresponds to the carbon-carbon bonding energy. The mathematical analysis requires that the maximum metaplast concentration is independent of v. It is 0.37 when $E_1 = E_2$ which disagrees with the experimental fact that the maximum plasticity measured in the plastometer increases with v. In point of fact, the plastometer does not provide a true viscosity measurement. Its indications are also influenced by the bubble structure, which depends on the rate of liberation of gas, and the latter increases with v. In addition, the viscosity of coal must be a function not only of the concentration of metaplast but also of its viscosity. This viscosity decreases when the temperature rises, and the temperature, θ_m, of the maximum plasticity increases when v increases. Although this point has not been precisely formulated, the softening point, θ_s, which corresponds to the appearance of a certain concentration of metaplast must, when v increases, be displaced towards the high temperature such as θ_i, θ_m and the decomposition point, θ_1, even though experiment shows that the difference $(\theta_1 - \theta_s)$ increases with v. To attempt to explain this discrepancy with experiment, it may be considered that, for softening to take place, it is necessary not only that the metaplast concentration be high enough, but also that the viscosity of the metaplast be sufficiently low, requiring that the temperature be sufficiently high. When the heating rate v is low, the decomposition point, θ_1, being low, softening would only occur at a temperature, θ_s, appreciably higher than θ_1.

Consideration has been given to identifying the chloroform-soluble product formed during heating with the metaplast. The development of the chloroform-soluble product does parallel that of the metaplast. When the temperature increases, the concentration of this product increases, passes through a maximum, and then decreases at the same time as the plasticity does.[43,79,141] Like the plasticity, it increases with pressure.[100] Its structure determined by its infrared spectrum is very similar to that of the initial coal.[79] It decomposes under the effect of heat, giving gas and an insoluble residue. A comparison of the plastometric curves (Fig. 21) and the yield curves of the chloroform extracts shows that the plastic properties cannot be interpreted quantitatively by the concentration of extract; in particular, the fluidity maximum occurs appreciably later than the maximum yield of extract.[43] Pitt concluded from this that the metaplast cannot be identified with the chloroform extract.

The temperature difference between the maximum fluidity and the maximum yield of extract could be explained if the fluidity depends not only on the metaplast concentration and temperature, but increases slightly with time as a result of the progressive development of the porous structure.[158] It is also true that the chloroform extract is only a fraction of the metaplast, and its composition differs from both that of the initial coal and that of the metaplast, its hydrogen content being higher.

Carbonization of Models

Several authors have attempted to devise models of known composition and structure which, like coal, have the capacity to soften, swell, and resolidify. To illustrate the theory of homogeneous melting, Audibert[143] quoted the example of potassium chlorate, the fusion of which is followed by decomposition which involves the liberation of gas and leads to swelling and resolidification. But this is far removed from reality and gives only a general representation of the proposed theory. More recently models have been proposed in which attempts have been made to represent a structure similar to that believed to be possessed by coal.[42,145,159,160] The analogy between the behavior of the model and that of coal makes it possible to test the hypotheses advanced on the structure of coal and the mechanism of its pyrolysis.

An important study in this field was made by condensing formaldehyde with aromatic hydrocarbons or phenols producing polymers of the type:[161]

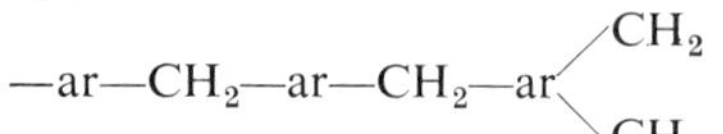

and a study was made of their pyrolysis and behavior in the dilatometer. When the "ar" radical is represented by phenanthrene or pyrene, some of these polymers have a behavior very similar to that of coking coals. If the degree of polymerization is low, melting begins before any decomposition occurs; it is, therefore, purely "physical." The softening point rises with the degree of polymerization. With higher degrees of polymerization, melting begins only when sufficient depolymerization has

[158] Finlayson, P. C., and Macrae, J. C., *Proc. Residential Conf. Sci. in the Use of Coal*, Sheffield, Inst. Fuel, London, 1958, pp. C 15–17.

[159] Neuworth, M. B., and Depp, E. A., *Brennstoff-Chem.*, **39**, Sonderausgabe, 90–2 (1958).

[160] Hadzi, D., *Ragprave Akad. Ljubljana*, **6**, 43–57 (1954).

[161] Wolfs, P. M. J., Van Krevelen, D. W., and Waterman, H. I., *Brennstoff-Chem.*, **40**, 189–94 (1959).

occurred. In all these cases, the fluidity increases as pyrolysis progresses; it passes through a maximum and then decreases up to resolidification, as a result of the formation of additional C—C bonds between the "ar" groups. If the initial degree of polymerization is still higher (very ramified structure), no further softening is observed.

If the "ar" group is a phenol, softening can take place before decomposition, and the initial degree of polymerization is fairly low. But generally, condensation reactions due to the OH groups prevail over depolymerization reactions from the outset, so that pyrolysis is immediately accompanied by the commencement of resolidification. If the degree of polymerization is higher no softening is observed.

The preceding discussion has been concerned with the softening behavior of vitrinites, which constitute the principal fraction of most coals. Much less is known about the softening phenomena associated with the other petrographic constituents present in most coals. The macerals of the inertinite group do not become plastic, and it is probable that they are not dissolved by fused vitrinite. Yet they do not behave exactly like inert substances. Fusain is finely porous and can play the same role with respect to "thermobitumens" as adsorbent substances with large specific surfaces. Exinite and resins are fusible. The mechanism of their fusion and resolidification appears to be the same as that of vitrinite, although there are important quantitative differences in low rank coals. When heated by itself, exinite passes through a state of extreme fluidity and only leaves a small residue of coke. In its natural state, it is very finely dispersed, most frequently in vitrinite. It appears, therefore, that interdissolution of the two macerals and the formation of a homogeneous mixed phase occur. In some high-volatile coals, vitrinite is completely infusible, whereas exinite is definitely fusible. The clarain of these coals becomes moderately plastic during heating and appears to give a homogeneous coke. This shows that exinite, or the products from its pyrolysis, acts as "plasticizing thermobitumens" with respect to vitrinite.

Mechanism of Swelling

The swelling of coals can appear in two forms: the swelling of an isolated coal particle and the swelling of an assembly of coal particles. Although the latter has been studied the most, the first is also easy to observe. In an assembly of coal particles, swelling can originate either from the swelling of each of these particles (intragranular swelling) or from the expansion of the gaps between the particles (intergranular swelling). The majority of authors consider swelling as a consequence of the softening of coal and the liberation of gas in matter which has been made less permeable by softening. This explanation applies both to intragranular and to intergranular swelling.

Intragranular swelling is supposed to result from the formation of a fluid phase, partially or totally replacing the initial solid coal, that can obstruct the fissures and macropores through which the gas would otherwise flow. A gas pressure results that tends to deform the plastic mass of coal by forming bubbles. Intergranular swelling results in the closing of the passages open to the circulation of gas between the particles, which soften and fuse together, and of intragranular swelling which tends to bring them closer together. This sequence of events is accentuated by closer initial contact of the particles, explaining the influence of bulk density on swelling, which is very marked for coals having slight plasticity. It also explains the fact that swelling increases with the heating rate because the yield of gas liberated increases with this rate.

The fact that swelling can be caused by a mechanism of this type has been shown[162] by heating an intimate mixture of powdered glass and calcium carbonate resulting in a porous product with a structure similar to that of coke. Carbon dioxide, originating from the decomposition of the carbonate, is liberated at a temperature at which the glass particles soften and agglutinate. Another investigation[42] of a mixture of a polymer capable of softening at about 65°, a substance causing condensation and resolidification of the polymer at about 100°C, and a substance decomposing with intense liberation of gas at about 80°C in the dilatometer and plastometer gave curves quite similar to those obtained with coal. Although these experiments show that the proposed mechanism of swelling is possible, they do not prove that swelling cannot be caused by some other mechanism.

[162] Bangham, D. H., MacKee, J. H., and Pirani, M., *J. Soc. Glass Technol.*, **31**, 267–9 (1947).

According to the proposed mechanism, the extent of swelling must be a function of the rate of liberation of gas and of the plasticity, a certain plasticity being necessary for the passages taken by the gas to be closed. Once this plasticity is attained, the swelling would be greater the lower the plasticity, or the higher the viscosity. This explains why a slight oxidation may increase swelling by reducing the plasticity.

A quantitative analysis of swelling in the dilatometer has been proposed,[156] based on the concept of coal first pyrolyzing into a metaplast with subsequent decomposition to semicoke and gas, as already discussed in consideration of the mechanism of melting. It assumes that the height of the coal pellet, h, is proportional, on the one hand, to the viscosity, μ, of the plastic phase and, on the other, to the rate of liberation of gas, dG/dt, which is itself proportional to the rate of decomposition of the metaplast, $h = K\mu(dG/dt)$. Swelling attains its maximum value, h_m, at the temperature, θ_i, corresponding to the maximum decomposition rate. The maximum decomposition rate occurs at higher temperatures as the heating rate, v, increases according to a relation of the form $1/\theta_i = a - b \log v$. Furthermore, it can be shown that h_m increases with v according to a relation of the form: $h_m = k\mu_{\theta_i}v$. If μ did not change with temperature, h_m would increase proportionally with v; in fact, μ decreases when the temperature rises and θ_i rises when v increases. The consequence is that h_m increases with v a little slower than proportionally. These results are in close agreement with experiment, for coking coals at least.

It is difficult to know the relative importance of intragranular and intergranular swelling. MacKee[163] attempted to determine this by investigating mixtures of coal and inert diluents and concluded that intragranular swelling was only of very slight importance. But, as shown by the analysis of these results by Dryden and Griffith,[134] this conclusion is debatable. It is certain, moreover, that the relative importance of the two types of swelling depends both on the particle size and on the nature of the coal. Intragranular swelling is only to be considered for very weakly caking coals. If the coals produce a coherent coke with a continuous solid phase, intergranular swelling must necessarily occur. If the coal is very plastic, the two types of swelling may not be distinguishable.

The proposed mechanism of swelling is not accepted by Berkowitz,[98] who has considered swelling as a phenomenon almost independent of softening. He attributed swelling to the pressure of the gas forcing a path through the micropores of the coal, the latter being formed by the interstices between the micelles. The resistance to the passage of gas, which generates the pressure, is, therefore, not attributed—as in previous theories—to the plastic deformation of the coal, but to the constrictions existing in the coal in its initial solid state. This mechanism permits an explanation of intragranular swelling, but not of intergranular swelling. The principal argument in favor of Berkowitz's theory rests on the correlation observed between the swelling capacity of coals of different origin and their heat of wetting or their inherent moisture content, both of which are functions of the internal surface.[164] The most strongly swelling coals are those for which the heat of wetting is lowest. Lean coals and low rank coals are also distinguished by the fact that their pores are more difficultly accessible,[165] and that they offer a higher resistance to the liberation of gas formed by pyrolysis. Because the majority of the properties of coal change regularly with rank, it is possible to find numerous correlations between two of the properties, and the fact that there is a close correlation between porosity and swelling does not imply that one is the *cause* of the other. They may both be manifestations of the same cause, such as the oxygen content. Certain "abnormal" coals, which show considerable swelling in spite of low rank and high oxygen content,[164] actually have an abnormally low moisture content, making it possible to explain their swelling in accordance with Berkowitz's theory. But it is possible that in these coals the oxygen is present in an unusual form which would account for both the low porosity and the melting and swelling capacity.[134]

Boyer[166] has observed swelling according to the mechanism postulated by Berkowitz in a

[163] MacKee, J. H., *Fuel*, **29**, 195–6 (1950).

[164] Berkowitz, N., *Fuel*, **29**, 138–43 (1950).

[165] Franklin, R. E., Ph.D. Thesis, Cambridge, 1946.

[166] Boyer, A. F., Private communication.

dilatometer test on samples of coal cut from a lump of coal. In some instances, a slight localized and rapid swelling of the order of 1 to 2% was observed *before* softening became evident by the collapse of the sample. But there are serious objections against the use of this mechanism to represent the principal swelling shown by high volatile coals at the end of their plastic state. In particular, an explanation cannot be offered for the formation of bubbles of the order of 0.1 to 1 mm which are found in the coke. Investigation with the mercury porosimeter[167] has shown that the main part of the volume of coke is accessible by passages of the order of a micron and not of tens of Angströms. Constrictions of molecular dimensions cannot, therefore, be offered as an explanation of the difficulty of evolution of the gas. Berkowitz's theory can account for only one aspect—probably of little importance—of swelling phenomena.

Mechanism of Resolidification

The mechanism of resolidification has already been partly discussed in connection with that of softening, because the majority of theories of softening include an interpretation of resolidification. All the theories, including the purely physical theories of softening, attribute the cause of resolidification to a chemical reaction. They assume that the plastic phase, whether of the coal as a whole, the "fusible component," the thermobitumens, or the metaplast, undergoes thermal decomposition, giving rise to gaseous products which escape and a residue which is difficultly fusible or solid. The exact nature of this reaction is specified to a greater or lesser extent according to the authors. The general trend is that of cracking reactions followed by condensation.[42,134] The medium-sized molecules forming the plastic phase split up. The smaller fragments resulting from this decomposition form the gas and tars and evaporate; the heavier fragments remain and are linked together by free valences, thus forming the large molecules of which semicoke is composed.

The mechanism of resolidification has not been very closely analyzed. If coal is regarded as an isogel,[168] in which plasticity results from the mobility of small molecules forming a fluid phase, "plasticizing" the micelles which have remained solid, resolidification would be due to the coagulation of the micelles promoted by the volatilization—partially, at least—of the fluid phase and increased by cross linking between the carbon atoms of the micelles. Gillet[144] considered that the decomposition of the plastic phase gives rise to a solid product which coagulates and is precipitated. Others[169] have also compared the formation of semicoke to coagulation. Plastic coal undergoing resolidification was described as a suspension of the solid decomposition product in the "fluid coal," the progressive increase in the viscosity, μ, being due to the increase in the concentration of solid, C. Assuming a relation of the form

$$\mu = \mu_0(1 + \lambda C),$$

μ_0 being the viscosity of the fluid coal, and λ a constant, it is possible to relate the change in the viscosity to the rate of the decomposition and to determine the kinetic characteristics of the latter by viscosity measurements. Measurements of the change of viscosity at constant temperature by means of the Gieseler plastometer led to the conclusion that the decomposition reaction is first order and that its activation energy is about 50 kcal/mole, which agrees well with the value found by other methods.

Fitzgerald[157] tried to show that the values found in this way are characteristic of coal and are independent of factors such as pressure, bulk density, and particle size. But he did not take into account the influence exerted by the gas bubbles formed during the plastic range on the apparent viscosity determined by the plastometer, which may make his conclusions incorrect.[156]

Contraction after Resolidification

The contraction which coal undergoes after resolidification has long been known, but a mathematical description of the phenomenon has only been given recently as a result of the development of apparatus sufficiently accurate

[167] Boyer, A. F., Durand, G., and Payen, P., *Chim. & ind. (Paris)*, **80,** 220–33 (1958).

[168] Lahiri, A., *loc. cit.* (ref. 153).

[169] Bronowski, J., Fitzgerald, D., Gillings, D. W., and Rhys Jones, D. C., *Nature*, **171,** 389 (1953).

for quantitative measurement.[74,170–172] This contraction is the resultant of the loss in weight of the coal, which continues to liberate volatile matter above its resolidification point and up to about 1000°C, and of the increase in its density, which increases approximately from 1.4 to 1.9. It is possible to calculate the contraction from the change in weight and density and the values found agree with the measured values. These two phenomena are themselves the consequence of the decomposition and condensation reactions which the semicoke undergoes and which convert it progressively into coke.

In Figs. 4 and 5, it was shown that, in the vicinity of the resolidification point, the contraction is very rapid. This rate of contraction varies markedly from one coal to another. If resolidification occurs a little before the end of the primary pyrolytic reactions, the contraction immediately afterwards is very rapid, as happens with high-volatile coals. If resolidification occurs at a slightly higher temperature, the subsequent contraction is much less rapid,[173] as observed with coking coals. The value of the contraction coefficient at the resolidification point is, therefore, essentially determined by the relative position of the resolidification point and the temperature at which the primary pyrolytic reactions come to an end. When the heating rate is increased, these two temperatures increase in almost the same manner, thus explaining why the contraction coefficient at the resolidification point is only slightly dependent on the heating rate.[74,173] Between 550 and 1000°C, the rate of contraction is less and is almost the same for all coals. It passes through a relative maximum at about 700°C, which probably corresponds to liberation of hydrogen and the structural modifications which result. The total linear contraction is about 10%.

[170] Macura, H., *Oel u. Kohle*, **40**, 227–31 (1944). Taits, E. M., *Chem. Abstr.*, **50**, 16,079 (1956). Nadziakiewicz, J., *Coke and Gas*, **21**, 46–55 (1959).

[171] Dainton, A. D., and Phillips, J. W., *Proc. Residential Conf. on Sci. in the Use of Coal*, Inst. of Fuel, London, 1958, pp. C-40-4.

[172] Phillips, J. W., Dainton, A. D., and Kaye, W. G., *3rd Intern. Conf. Coal Sci.*, Valkenburg, Comm. IV B 4, 27–30, April, 1959.

[173] Boyer, A. F., *Brennstoff-Chem.*, **37**, 226–32 (1956).

Conclusion

The mechanism by which certain coals soften, swell, and then resolidify is still imperfectly known. It is open to discussion and should be the object of future research. The following simplified representation of this process may, however, be given.

It is possible that the beginning of softening is, sometimes at least, a purely physical phenomenon independent of pyrolytic reactions. This phenomenon of "paste-like" melting is analogous to that observed for numerous macromolecular substances. If this phenomenon exists, which has not been proved, it is of little importance and becomes rapidly masked by a phenomenon of chemical origin. High plasticity can only be observed when the decomposition reactions have been definitely initiated; it is probably the result of these reactions. The latter, the initial phase of which is considered to be depolymerization, gives rise to fluid products. Whether the coal as a whole (or the vitrinite, at least) is transformed into fluid products, or whether it is only a fraction of the decomposition products which acquires a true fluidity and "plasticizes" the other components (which have remained solid, or almost solid) cannot be definitely decided; however, the latter appears more probable.

Swelling is, in the main, caused by the resistance offered to the liberation of gas through the macropores and fissures of the coal particles and through the interstices between the particles; the obstruction of these passages is, itself, caused by the plasticity and fluidity acquired by the coal. This swelling is, therefore, a consequence of the plasticity. It is possible that, in its initial phase, a small proportion of the swelling may be due to the distension of coal, still in the solid state, under the effect of the gas forcing a passage through the micropores of the coal; however, this appears to be a secondary phenomenon.

Resolidification is the consequence of the disappearance of fluid products. This disappearance is caused partly by evaporation and removal as volatile matter, and partly by decomposition, which gives rise to vapor and a solid residue.

Finally, the decomposition which the coal undergoes after resolidification leads to a loss in weight and an increase in density, causing a contraction of the solid phase.

THE PLASTIC PROPERTIES OF COAL AND THE FORMATION OF COKE

The Formation of Coke

The cold coal charge introduced into an oven is in contact with two walls heated to about 1200 to 1500°C. Heat is transmitted by conduction, convection, and radiation from the walls towards the center of the charge. The isothermal surfaces are planes parallel to the walls that move towards the interior. At a given instant, the layer of coal included between the isotherms representing the softening point, θ_s, and the resolidification point, θ_r, is in the plastic state, the two plastic layers being arranged symmetrically inside the oven.[56,78,174] Between the two plastic layers, the mass of coal has not yet undergone plastic transformation and is composed of uncaked particles. Within the plastic layers the coal particles soften, swell, and fuse to give a porous caked mass which, when solidified, forms semicoke. The phenomenon is similar to that observed in the dilatometer, but the expansion is limited. A pressure results which is transmitted, on the one hand, to the coal of the inner zone, which it tends to compress, and on the other hand, to the walls through the medium of the semicoke of the outer zone. Between the plastic layers and the walls of the oven is a mass of solid. The surface in contact with the walls takes on a curvature, the convexity being towards the walls; fissures appear, perpendicular to the walls, and divide the mass into fragments with a "cauliflower" shape of varying size. As coking progresses, the two plastic zones move toward the interior of the oven. When they meet, the pressure exerted on the walls of the oven increases, passes through a maximum—sometimes very pronounced—and then decreases rapidly and drops to zero.

If the coke is to be coherent, the coal particles must soften and swell sufficiently to fuse together and form a continuous solid phase. This agglutination capacity of the particles appears to be characterized fairly well by the swelling in the dilatometer; it is also related to the plastic range, which determines the width of the plastic zone in the oven.[175] Agglutination is favored by increased bulk density and increased heating rate.

It is difficult to ascertain the relative importance of the different phenomena participating in the formation of coke. The intrinsic strength of the walls of the bubbles probably has little influence, because the cokes with the weakest walls often have the greatest abrasion strength.[173,176] On the other hand, the agglutination of the particles plays an essential part, at least in the case of weakly caking coals or mixtures having a considerable proportion of infusible substances. In point of fact, the loss of cohesion shown by abrasion or crushing may be initiated by the separation of particles insufficiently fused together. But as soon as a certain minimum swelling is attained, permitting the particles to fuse, which happens in nearly all industrial cokes, this factor is of no importance, and the microstrength is determined mainly by the porous structure.

Fissuration Factors

The majority of authors agree that the cause of the fissuration of coke is the contraction which the charge undergoes in the oven. Some of these take into consideration the total contraction between the initial temperature of the coal and the final coking temperature, that is, the sum of the contraction before the resolidification point and the contraction afterwards.[175,177] This value characterizes the "carbonization shrinkage," on which the ease of removal of the coke charge from the oven depends. The contraction also determines the fissuration.[177,178] To reduce fissuration it would be advisable to lower the resolidification temperature in order to avoid the pore shrinkage which occurs during the final period of the plastic state.[178] Fissuration is also determined by the behavior of the coal in the plastic layer and, in particular, by the relative value of the stresses set up by the loss of volatile matter and by the viscous forces of cohesion, the fissuration being greater the

[174] Peytavy, A., and Lahouste, J., *Compt. rend.*, **234,** 934–5 (1952).

[175] Sapozhnikov, L. M., and Basilevich, L. P., *Investigation of the Coking Process*, State Publishing House of Ukraine, Kharkov, 1935, pp. 5–33. Sapozhnikov, L. M., *Coals and Metallurgical Coke*, Akad. Nauk S.S.S.R., Moscow, 1941.

[176] Alpern, B., *Compt. rend.*, **242,** 653–6 (1956).

[177] Nadziakiewicz, J., *loc. cit.* (ref. 170).

[178] Volodine, P. F., *Bull. acad. sci. U.R.S.S.*, Classe sci. tech., No. 7, **1950,** 1025–32.

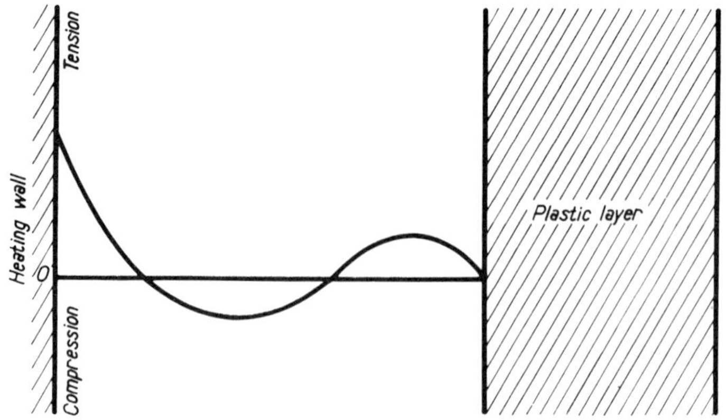

Fig. 23. Stress distribution in semicoke.[184]

lower the viscosity.[179] It is difficult to visualize how the contraction of the plastic layer can produce stresses in the solid mass of semicoke and cause its fissuration. No mechanical theory has been proposed to attempt to justify this opinion. It is more widely agreed that only when contraction occurs in an already solidified substance can it result in stress which cause fissuration.

Audibert[180] considered that fissuration is determined by contraction after resolidification, that is, between θ_r and 1000°C. In reality, the magnitude of this contraction does not vary greatly from one coal to another, whereas there are considerable differences in fissuration. It was suggested that fissuration is linked not only with the value of contraction after resolidification but also with the dependence of this contraction on temperature; the high degree of fissuration observed with gas coals is due to a rapid and irregular contraction.[181]

A detailed study of the contraction curve of semicoke has been undertaken for this purpose by various authors.[74,182] On the basis of the experimental curve determined in the Chevenard dilatometer (Fig. 4), or the derived curve (Fig. 5) giving the contraction coefficient $\alpha = 1/l_0(dl/d\theta)$ as a function of the temperature, Soulé[74,183] developed a quantitative theory of fissuration. By employing the equations for the mechanics of solid elastic bodies, he calculated the stress at each point of the mass of semicoke included at a given instant between the wall of the oven and the plastic layer. He assumed that, at the moment when the plastic coal solidifies at the boundary of the plastic layer, it is not subjected to any stress, the temperature distribution in the mass of semicoke is known and assumed to be linear, and that, above the resolidification point, the semicoke contracts as shown in the dilatometer (Fig. 4). The stress distribution in relation to the distance from the wall then has the character shown in Fig. 23;[184] some zones are in a state of tension, others are under compression. The tension is a maximum at the contact with the wall; by definition, it is zero at the contact with the plastic layer.

When the plastic zone progresses towards the center of the oven, the maximum tension increases until it reaches the breaking point. Fissures, perpendicular to the wall, are formed at the contact with the latter that lead to a reduction in the stress. The mean distance

[179] Gillings, D. W., and Lawson, W., *J. Inst. Fuel*, **30**, 446–60 (1957).

[180] Audibert, E., *Rev. ind. minérale*, **9**, Part 1, No. 192, 151–87 (1929); **10**, Part 1, No. 217, 1–24 (1930). See also: Arnu, C., *Rev. ind. minérale*, **13**, Part 1, No. 296, 155–73 (1933); *Chaleur & ind.*, **15**, 278–96 (1934); *Chim. & ind. (Paris)*, **32**, 276–92, 530–40 (1934); *Commun. 15th Congr. Chim. Ind., Bruxelles*, September, 22–28, 1935, Chim. & ind. (Paris), 7 pp. Damm, P., *Brennstoff-Chem.*, **9**, 293–5 (1928); **10**, 191–5, 217–21 (1929); *Glückauf*, **64**, 1073–80, 1105–11 (1928).

[181] Onusaitis, B. A., and Jourievskaya, N. P., *Bull. acad. sci. U.R.S.S., Classe sci. tech.*, **1947**, No. 7, 881–94.

[182] Macura, H., *Oel u. Kohle*, **39**, 113–6, 365–77 (1943). Erkin, L. J., and Gorbounova, L. J., *Zavodskaya Lab.*, **14**, 801–7 (1948).

[183] Soulé, J. L., *Fuel*, **34**, 68–77 (1955).

[184] Loison, R., *Chauffage industriel et utilisation des combustibles*, Technique et Documentation, Paris, Chap. 17, 1959, 374 pp.

between the fissures must be such that at each point the tension is reduced below the breaking point. Calculations show that this mean distance is inversely proportional to the maximum contraction coefficient α_m, which is virtually the contraction coefficient at the resolidification point, and to the temperature gradient in the vicinity of the walls in the 500 to 600°C region. This last conclusion agrees with the fact that the size of the coke and the "stability factor" decrease when the temperature of the oven is increased.[63,185] It is not assumed *a priori* that fissuration is determined by the contraction at resolidification, α_m. This relation results from calculations on the mechanical model. Other authors have, on the contrary, assigned a determining role to the contraction maximum observed in the neighborhood of 700°, but this was unsubstantiated.

Another mechanism of fissure formation is based on the components of the coal charge having different contraction rates so that stresses appear in the vicinity of the contact surfaces of the different components.[186] This occurs, in particular, when the coal charge includes infusible substances such as mineral material, coke breeze, or low-volatile coals, the contraction of which is nil or low. If the particles of these substances are fairly large, fissures are formed in their neighborhood that have no definite orientation and are generally of only slight extent. It would be infrequent that this phenomenon would play a major role in the fissuration of coke.

Relation between the Mechanical Characteristics of Coke and the Characteristics of Coal

Numerous attempts have been made to express the mechanical characteristics of coke as a function of elementary composition of the coal charged to the oven, in particular, the carbon content;[187] swelling, measured in the dilatometer;[180,188] BSI, associated with the volatile matter content;[189] extent of the plastic range ($\theta_r - \theta_s$), either alone[190] or in association with the volatile matter content[179] or the total contraction;[175] the viscosity measured in the plastometer, either alone[191] or in association with the volatile matter content;[192] the contraction coefficient at resolidification;[74,193] the calorific value of the distillation gas, in association with the swelling in the dilatometer;[194] or the losses in weight during the different phases of the transformation of coal: below θ_s, between θ_s and θ_r, and above θ_r.[195]

Because the coke quality depends not only on the nature of the coal, but also on all the parameters governing the conditions of preparation and heating of the coal charge, an exact relation with the nature of the coal can only be found by considering constant coking conditions: particle size, moisture content, oven temperature, width of the ovens, etc. Even when this condition has been satisfied, no exact and general relation applicable to all coals has been firmly established. Approximate relations may be found that are valid only for one group of coals of neighboring origin whose characteristics are not very different. The reason for this difficulty is that the coke characteristics determined by established methods are the resultant of several coke properties. A low value of the "hardness factor" may, for

[185] Foch, P., and Busso, R., *Compt. rend. congr. ind. gaz., 72nd Congr., Paris*, **1955**, 298–326; *Compt. rend. congr. ind. gaz, 76th Congr.*, **1959**, 14 pp.

[186] Macura, H., *Oel u. Kohle*, **40**, 227–31 (1944).

[187] Lowry, H. H., Landau, H. G., and Naugle, L. L., *Trans. Am. Inst. Mining Met. Engrs.*, **149**, 297–330 (1942). Evans, E. C., and Ridgion, J. M., *Iron Steel Inst. (London), Spec. Rept.* No. 17 (1937), 62 pp.

[188] Brooke, W. J., Walshaw, H. R. B., and Lee, A. W., *J. Iron Steel Inst. (London)*, **134**, 287–325P (1936); *Colliery Guardian*, **153**, 567–70, 620–23 (1936).

[189] Swartzman, E., Burrough, E. J., and Strong, R. A., *Can. Dept. Mines, Mines Branch*, No. 737–3 (1933), 24 pp.; *Invest. Fuels Fuel Testing*, No. 737, 36–57 (1934).

[190] Brewer, R. E., and Atkinson, R. G., *Ind. Eng. Chem., Anal. Ed.*, **8**, 443–9 (1936).

[191] Perch, M., and Russell, C. C., *Trans. Am. Inst. Mining Met. Engrs.*, **184**, 205–14 (1949).

[192] McKellar, H., Burns, W., and Giedroyc, V., *J. Inst. Fuel*, **31**, 528–34 (1958).

[193] Boyer, A. F., and Loison, R., *La Documentation Métallurgique*, No. spec. SK 1, Congr. Centenaire *Rev. ind. minérale*, **1955**, 3–18 (1956). Loison, R., and Foch, P., *AIME, Proc. Blast Furnace, Coke Oven & Raw Materials Comm.*, **17**, 188–96 (1958).

[194] Deringer, H., *Schweiz. Ver. Gas u. Wasserfach, Monatsbull.*, **34**, 316–8 (1954).

[195] Melzer, W., *Arch. Eisenhüttenw.*, **4**, 169–75 (1930); **6**, 89–93 (1932). Porsch, H., and Dietrich, O., *Glückauf*, **91**, 520–23 (1955).

example, be due either to insufficient agglomeration of the particles, the presence of sponge coke, an excessively fine porosity, or excessive fissuration, and each of these properties is associated with different characteristics of the coal.

From the preceding analysis of the mechanism of coke formation, it follows that the coal characteristics which play the main role in determining coke characteristics are:[26,173] swelling, determined in the dilatometer; the width of the plastic zone, that is, the difference between θ_r and θ_s; and the contraction coefficient at resolidification. The first two affect mainly the microstrength, while fissuration is mainly influenced by the third. This opinion is not shared by all authors. Of the relations of statistical origin which have been proposed for expressing the characteristics of coke, some concern other properties of coal such as maximum fluidity and BSI, and they agree with the experience obtained for a fairly wide range of coals. But it is known that all the properties of coal are more or less correlated with their rank. It is, therefore, natural that the characteristics of coke can be expressed statistically as a function of any one property of coal, even if it does not have a direct influence on the coke quality. It is possible, for example, to employ the volatile matter content instead of the contraction coefficient at resolidification because these two values are closely correlated. But some coals do not show this correlation and, in such cases, it is more accurate to employ the contraction coefficient.[46]

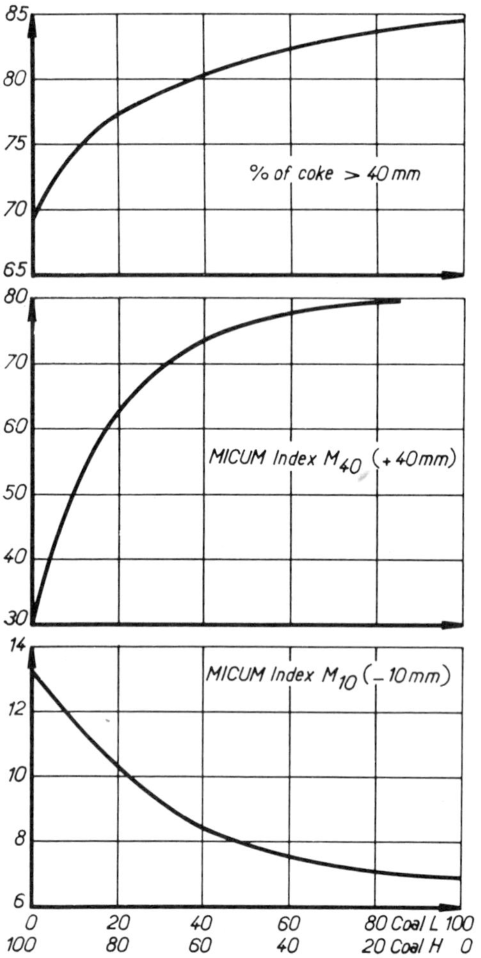

Fig. 24. Influence of the composition of a mixture of two coals on coke quality.[184]

Blending of Coals

Because the supplies of good coking coals are very limited in certain countries, it has been necessary to use mixtures. This method is increasingly used and it allows the range of coals suitable for coking plants to be considerably enlarged. Numerous investigations have been carried out,[196–200] usually at the coke plant itself, to determine, empirically, the influence of the composition of a mixture on coke quality and establish rules for mixing. As a result of the great variety of coals that are hardly comparable from one country to another, these rules can be employed accurately only for the coals used to establish them but,

[196] Oboukhovskii, J. M., *Ugol'*, **25,** No. 12, 15–19 (1950). Nelson, H. W., Ward, J. J., *AIME, Blast Furnace, Coke Oven & Raw Materials Comm.*, **12,** 182–95 (1953). Bargett, H., *J. Inst. Fuel*, **27,** 274–98 (1954). Brit. Coke Research Assoc., London, *Tech. Paper* No. 4 (1950), 23 pp. Powell, A. R., and Russell, C. C., *Contr. 4th Conf. Brit. Coke Research Assoc.*, London, October 25, 1950, 22 pp. Jenkner, A., *Glückauf*, **88,** 363–7 (1952).

[197] Kröger, C., Kaunert, P., and Kuthe, F., *Brennstoff-Chem.*, **36,** 142–51 (1955). Gillings, D. W., Lawson, W., and Willing, E. G. J., *Ind. chim. belge*, **20,** 62–7 (1955).

[198] Reed, F. H., Jackman, H. V., Rees, O. W., and Henline, P. W., *Illinois State Geol. Survey*, Circ., No. 178 (1952), 8 pp.

[199] Foch, P., and Busso, R., *Rev. ind. minérale*, **40,** 669–85 (1958).

[200] Brisse, A. H., and Harris, H. E., *Blast Furnace & Steel Plant*, **45,** 301–7 (1957).

qualitatively at least, they may have a certain general value.

Generally, a high volatile coal, "H," the carbonization of which gives a fairly well-caked, or very well-caked but excessively fissured coke, is blended with a coal "L" having a lower volatile content. When the proportion of L increases from 0 to 100% (Fig. 24), the strength and hardness of the coke increase very rapidly up to 20 to 30% and then more slowly; above 50 to 60% there is hardly any improvement. Figure 24 is taken from the work of the Marienau Experimental Station.[184,193] but similar results have been obtained by numerous authors.[197] It is established, undeniably, that the law of additivity is far from exact. It would be remarkable if it were so, because the mechanical characteristics of coke are not physical values, but only indices supplied by empirical tests.

If different L coals are added to the same H coal, the quality of the coke varies greatly with the characteristics of the L coal. It is, therefore, important to select the most suitable L coal for each type of H coal. By way of example, the results obtained at the Marienau Experimental Station on mixture of equal proportions of L coals from the Ruhr and Aachen districts and H coals from the Saar and Lorraine[193,199] are represented diagrammatically in Fig. 25. The volatile matter content of the L coals is plotted on the abscissa and the M_{40} and M_{10} Micum indices on the ordinate. Each curve corresponds to a given H coal. The characteristics of the coals are summarized subsequently.

For a given H coal, the strength and hardness of the coke pass through a maximum as a function of the rank of the L coal. The optimum rank of the L coal is lower, the lower the rank of the H coal itself. The particular

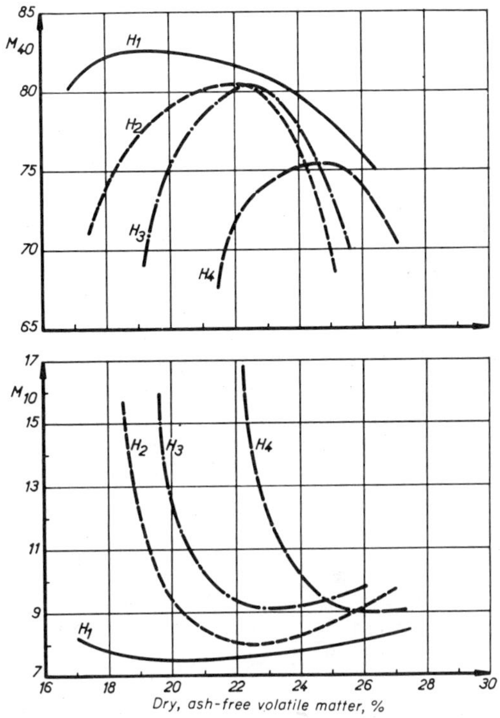

Fig. 25. Influence of the characteristics of the components on the coke quality.[199]

features of Fig. 25 and their interpretation will be discussed later. At this point, it should be only noted that the mixing law obtained from these results is exactly opposite to that based on the assumption of a constant volatile matter content of the mixture, because this requires that the volatile matter content of the L coal must increase with that of the H coal.

Other mixing laws, based not on the characteristics of each of the components, but on the overall characteristics of the mixture, have been advanced by various authors. Thus,

TABLE 1

	Volatile Matter, % Dry, Ash-Free	Carbon Content, %	British Swelling Index	Gieseler Plastometer Maximum Dial, Divisions per Minute	Dilatometer Swelling, %	Contraction Coefficient at Resolidification, α_m	Resolidification point, θ_r
L_1	18	90.5	7.5	5	0	5×10^{-4}	500°C
L_2	23	90	9	20	50	4.5×10^{-4}	495
L_3	25	89	9	2000	160	6×10^{-4}	490
H_1	35	87	8.5	3000	100	8×10^{-4}	475
H_2	38	85	6.5–7	200	30	14×10^{-4}	465
H_3	38	83.5	4–5	20	2	17×10^{-4}	455
H_4	38.5	82	2–3	0	0	20×10^{-4}	440

according to Hoffmann and Hoehne,[40] the swelling of the mixture in the dilatometer must be kept between two limits, and the breadth of the plastic zone must be as high as possible; the fluidity determined in the Gieseler plastometer should not be too high, and should be greater, the lower the rank of the *H* coal. Van Krevelen[26] considered that a good mixture should have a volatile matter content of approximately 20 to 25%, a swelling in the dilatometer of 40 to 90%, and a plasticity range of 100°C. Gillings and Lawson[179] proposed a relation of statistical origin, expressing the $+1\frac{1}{2}$ in. Shatter Test as a function of the plastic range $(\theta_r - \theta_s)$ and the volatile matter content:

$$+1\tfrac{1}{2} = 0.447\,(\theta_r - \theta_s) - 3.801 \text{ Vol. Mat} + 190.2$$

Formation of Coke from a Coal Mixture

The extent that coal particles of a coal charge become mixed together during melting, to form a more or less homogeneous phase, has been determined both by X-ray measurements of the degree of graphitization of cokes[201] and observations under the microscope in polarized light.[202] The latter are the most convincing and are based on the fact that the anisotropy of cokes increases with the rank of the coals from which they are formed. If a coke obtained from a mixture of two coals of very different rank is examined in polarized light, it is possible to identify the elements originating from one or another of the coals. It is found that the particles of coal have retained a certain individuality during fusion but, in the vicinity of the contact surface of two particles of different nature, there is a narrow zone of several hundredths of a millimeter, the optical properties of which are intermediate between those of the two components. This zone undoubtedly results from the interaction of particles intermediate between those of the two components. When the coal particles are relatively large (1 to 2 mm), this "mixed phase" occupies only a very small volume, and the particles, therefore, retain their individuality almost completely during fusion. If, however, the particles are very small (<0.05 mm) the "mixed phase" occupies the whole of the coal charge, and the particles cannot be distinguished. The result of this is that the behavior of the mixture may be influenced markedly by the particle size of the coals; in particular, it is necessary to employ great caution in the interpretation of the results of conventional laboratory tests in which the coal is very finely ground (generally less than 0.1 to 0.2 mm), whereas the particle size of industrial coal charges is far greater.

In general, the divergence from the additivity law is slight in the swelling test,[17] but may be greater in either direction in the Roga agglutination test.[155] In the Gieseler *plastometer* test, the apparent viscosity of the mixture is definitely greater than that indicated by the additivity law.[197] The curve representing the viscosity of a mixture of two *H* and *L* coals as a function of the proportion of the components has a similar shape to that of the hardness of the coke (Fig. 24). Gillings and Lawson[179] concluded that the mechanical strength of coke is determined by the viscosity of the mixture, but this conclusion has been disputed by Boyer.[203] A lower viscosity than that indicated by the additivity law has also been observed,[37,204] and it appears difficult to predict the viscosity of a mixture from the plastometric curves of its components.

The *swelling* measured in the dilatometer also does not follow the additivity law, the swelling of a mixture is generally lower than that predicted by it.[205] The divergence is greater, the greater the difference between the ranks of the components. The swelling of the mixture may be even less than the swelling of either of the components. Van Krevelen[129] has proposed an interpretation of these observations and showed that the swelling of a mixture may be predicted from the dilatometer curves of the components, provided that it is assumed that the "true expansions" follow the additivity law. From the experimental dilatometer curves of the components, he calculated the "true expansion" curves and applied the additivity law to these. From the true expansion curve calculated in this way, he plotted the dilatometer curve of the mixture and deduced the maximum swelling from this curve. The results obtained agreed well with experiment.

[201] Durif, S., *Brennstoff-Chem.*, **37,** 402–4 (1956).

[202] Alpern, B., *Brennstoff-Chem.*, **37,** 194–8 (1956).

[203] Boyer, A. F., *J. Inst. Fuel*, **31,** 465–6 (1958).

[204] Echterhoff, H., *Glückauf*, **90,** 510–16 (1954).

[205] Barking, H., and Eymann, C., *Gas u. Wasserfach*, **99,** 1037–45 (1958).

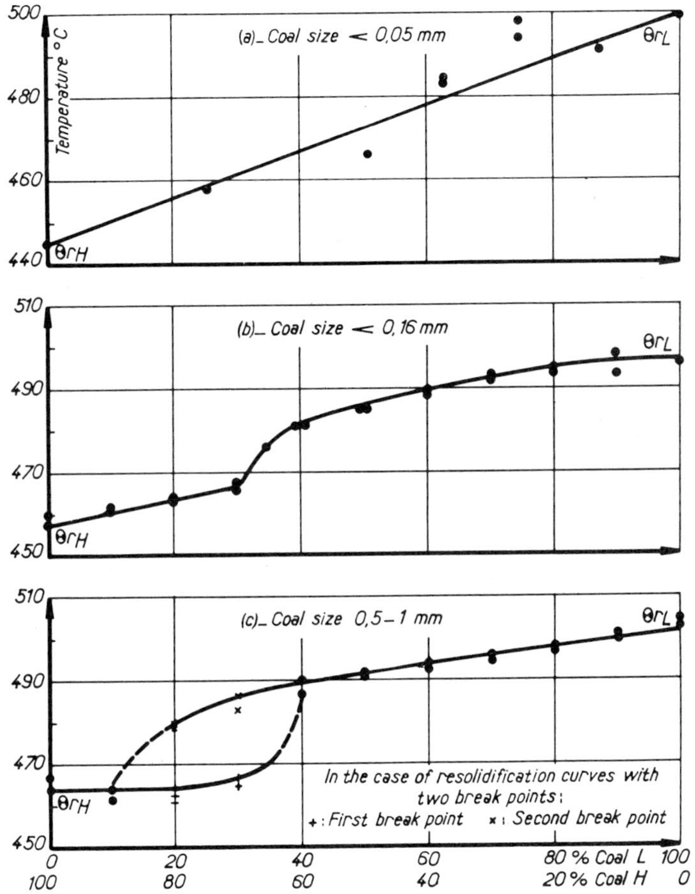

Fig. 26. Variation of the resolidification point of mixtures of two *H* and *L* coals with different particle sizes.[47]

The Cerchar plastometer[46] has been used in a study of the variation of the resolidification point of a mixture of two *H* and *L* coals as a function of particle size.[47] When the coals were ground to less than 0.05 mm, the resolidification point varied linearly with the composition of the mixture (Fig. 26*a*). It diverged appreciably from a linearity for slightly larger particles, <0.16 mm (Fig. 26*b*). When the particle size of the coal was between 0.5 and 1 mm, the phenomenon was more complex (Fig. 26*c*); if the proportion of *L* coal was less tan 10%, the resolidification point of the mixture was equal to the resolidification point of the *H* coal, namely, θ_{r_H}; if the proportion of *L* coal was between 10 and 30%, two resolidification points were observed, and the curve recorded by the plastometer has two maxima, the first always being very close to θ_{r_H} and the second being definitely higher, although less than θ_{r_L}; and if the proportion of *L* coal exceeded 30 to 40%, one resolidification point was observed close to θ_{r_L} and increased linearly with the concentration of *L* coal.

These observations are readily interpreted if account is taken of the interaction of the particles.[173,193] When the coals are very finely ground, they mix during fusion and give an almost homogeneous "mixed phase," the properties of which are intermediate between those of the components. The resolidification point naturally follows the law of mixtures. When the coal particles are fairly large, this mixed phase, limited to the surface of the particles, is of little extent, and the particles retain their individuality. If the proportion of *L* coal is greater than about 40%, the particles of *H* coal are completely enclosed at the moment of their resolidification by the *L* coal,

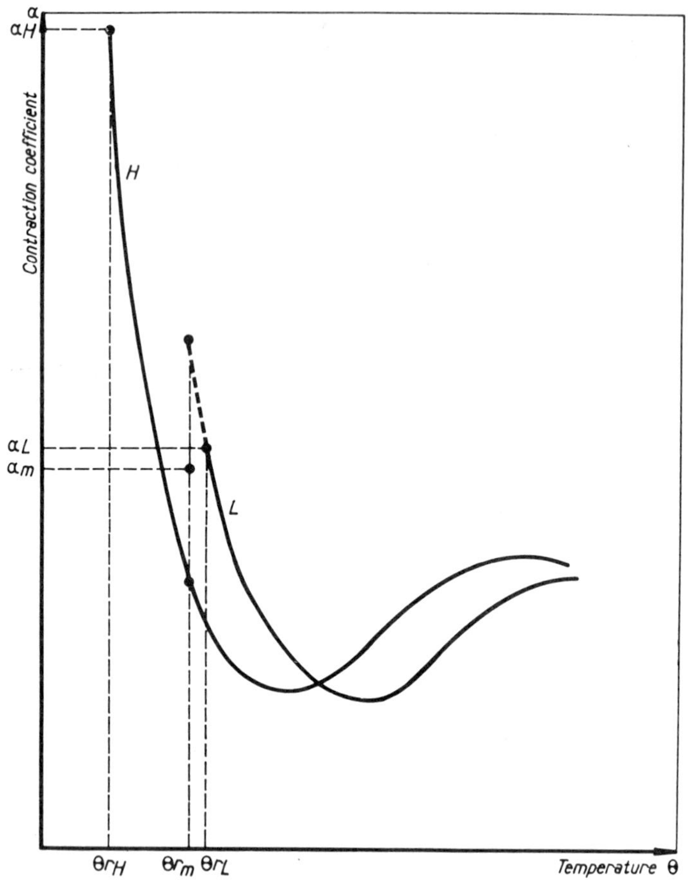

Fig. 27. Contraction coefficient of a mixture of two H and L coals at the resolidification point.[184]

which is still plastic; the mixture, therefore, remains plastic up to the resolidification point of the L coal. If no interaction of the coals took place, the resolidification point of the mixture would be exactly equal to θ_{r_L}, but it is slightly lowered as a result of the interaction of the particles. If the proportions of L coal is between 10 and 30%, it is insufficient to enclose the H coal completely, and the solidification of the latter at the temperature θ_{r_H} gives rise to a solid framework, which is fragile and is broken by the plastometer. The mixture is, therefore, replasticized to a certain extent, until the L coal resolidifies and results in the two resolidification points observed in the dilatometer. Finally, if the proportion of L coal is very low, the resolidification of the H coal gives rise to a solid mass of H coal containing only a few plastic inclusions of L coal, and only one resolidification point, θ_{r_H}, is observed.

According to Soulé's theory,[183] the mean distance between fissures in the coke is determined by the contraction coefficient at the resolidification point, and this is also true when the coal charge consists of a mixture. It is, therefore, necessary to consider the contraction coefficient of the mixture at the resolidification point of the latter. This value can be predicted from the contraction curves of the components owing to the fact, proved by experiment, that the contractions after resolidification are additive.[47,173,193]

The contraction curves of two H and L coals are shown in Fig. 27. As long as the proportion of L coal in the mixture is more than 40%, the resolidification point of the mixture is very little less than θ_{r_L}, the fineness of the coals being assumed to be similar to that in industrial coking. From the relative location of the H and L contraction curves, the contraction coefficient of the mixture α_m differs little from the contraction coefficient, α_L, of the L coal. The fissuration of the mixture must, therefore, differ little from that of the L coal alone. On the contrary, when the proportion of L coal falls below 40%, θ_{r_m} falls with increasing rapidity, tending towards θ_{r_H}, and α_m increases at an increasing rate, tending towards α_H. This interpretation explains the shape of the curve in Fig. 24, showing coke strength as a function of the proportion of the components.

In Fig. 25, coke strength indices were shown for 50:50 mixtures of several high-volatile (H) coals with low-volatile (L) coals. The results can be interpreted, qualitatively at least;[193,194] by assuming that coke quality is determined mainly by two characteristics: the swelling of the mixture in the dilatometer and the contraction coefficient of the mixture at the resolidification point. The contraction coefficient, α_L, decreases only slightly with decreasing volatile matter, and θ_{r_L} increases.

From Fig. 27, it follows that the contraction coefficient, α_m, of the mixture decreases. The higher the rank of the L coal, the lower will, therefore, be the fissuration of the coke, provided that the L coal remains plastic. If the H coal is very fluid and strongly swelling, as in the case of the coal H_1, the swelling of the mixture will always be sufficient, whatever the nature of L, to ensure adequate agglutination of the coal particles, and the coke quality is determined almost entirely by its fissuration. The best L coal is, therefore, one which, although it has a low plasticity, is characterized by a contraction without swelling in the dilatometer and possesses the highest resolidification point. If the H coal is moderately fluid and swelling, as in the case of H_2 or H_3 coal, it is also necessary to take into account the swelling in the dilatometer. When the volatile matter of the L coal decreases, the swelling of the mixture decreases, because the swelling of L decreases, and also because its plastic range diverges from that of the H coal. There is an optimum value for the rank of the L coal and, if its rank increases above this optimum, the swelling of the mixture decreases, and the microstrength of the coke falls. This leads to a reduction in the hardness and, in spite of an increasingly lesser tendency to fissuration, to a reduction in the strength. If the rank of the L coal falls below this optimum, the swelling remains important and ensures a high microstrength; the fissuration increases, leading to a considerable reduction in the strength and to a slight decrease in the hardness. Because the fluidity and swelling of H coals decrease with increasing volatile matter content, the optimum rank of the L coal is lower.

This interpretation also takes into account a rather curious fact shown by Fig. 25. If different H coals are added to a coal L_3, with a volatile matter content of about 25%, it is found that the coke strength characterized by the M_{40} index, increases in the order H_2, H_3, H_4, that is, in the reverse order to that observed when the coals H_2, H_3, and H_4 are coked alone. More generally, whatever the L coal to which they are added, the relative behavior of two H coals is often very different when they are coked alone or in a mixture with an L coal. These facts are readily explained by an examination of the contraction curves (Fig. 28). When the H coals are coked alone, the fissuration is determined by the contraction coefficients at their resolidification point, that is α_{H_1}, and α_{H_2}, but when they are coked in a mixture, it is the contraction coefficient at the resolidification point θ_{r_m} of the mixture, that is, α'_{H_1} or α'_{H_2}, which is the determining factor.

Addition of Infusible Substances

Infusible substances present in a coal charge influence both the plastic properties and the changes that occur in the semicoke after resolidification. The mineral matter of the coal and the infusible petrographic components—fusinite and micrinite—represent infusible matter naturally present. Other infusible substances are introduced intentionally to improve the coke quality. In practice, coke breeze, semicoke, lean coals, or anthracite are used and are frequently called "inerts." The addition of an inert to a fusible coal decreases its plastic properties, that is, it affects the fluidity measured in the plastometer,[206] the free swelling,[205] the swelling in the dilatometer,[207] and the caking indices.[155] The effect is generally greater than that indicated by the mixing law,[26] and at a certain concentration, which varies considerably with the nature of the coal, the plastic properties are completely

[206] Sarkar, S., Das Gupta, N. N., and Lahiri, A., *J. Sci. Ind. Research* (*India*), **17B**, 127–31 (1958). Brewer, R. E., *Ind. Eng. Chem.*, **36**, 1165–8 (1944).

[207] Georgiadis, C., and Gaillard, G., *Chaleur & ind.*, **32**, 247–58 (1956).

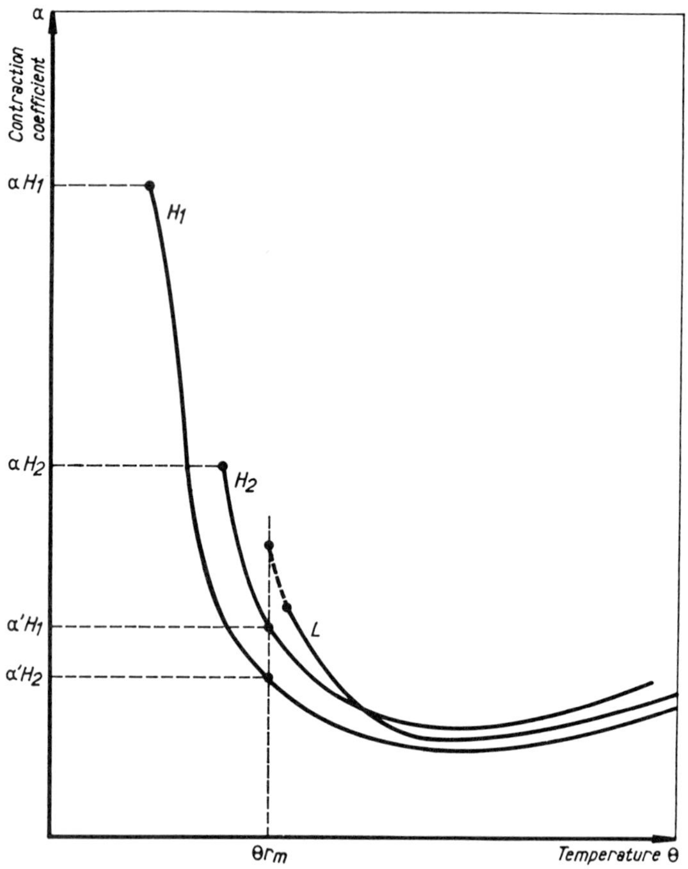

Fig. 28. Comparison of two H_1 and H_2 coals mixed with an L coal.[184]

destroyed. A low concentration sometimes has a lesser action than that corresponding to the additivity law, and it only becomes truly evident above a certain concentration.[208] The action of the infusible substances is generally greater, the finer the particle size[47,209] and the higher the porosity.[47,207] The resolidification point is not modified by a nonporous infusible substance such as quartz but is lowered by porous substances such as coke breeze, semi-coke, wood charcoal.[47,207] The effect increases with the porosity and fineness.

Several mechanisms have been put forward to explain the action of inerts during the plastic phase. Owing to the sole fact that the inert neither softens nor swells, the viscosity is increased and the swelling is decreased. If these effects appeared alone and the swelling of the coal would not be modified by the presence of the inert, the "true expansion" δ of a mixture containing $x\%$ by volume of inert would be

$$\delta = \delta_0\left(1 - \frac{x}{100}\right),$$

where δ_0 is the true expansion of the coal. The fact that the observed true expansion is less than this value[26,209] shows that dilution is inadequate to explain entirely the influence of the inert. This equation does not take into account the fineness or porosity of the inert. The inert may absorb a fraction of the tars or thermobitumens produced by pyrolysis, and this effect appears to be well-proved by experience. It is undoubtedly responsible for the lowering of the solidification point and, partly, for the reduction in the fluidity. It is higher the greater the fineness and porosity of the infusible substance. The points of contact between two inert particles form weak points which reduce the microstrength of the coke.

[208] Shotts, R. Q., *Fuel*, **31**, 448–61 (1952).

[209] Van Krevelen, D. W., Dormans, H. N. M., and Huntjens, F. J., *loc. cit.* (ref. 129).

The same is true, but to a lesser degree, of the contact points between inert particles and particles of weakly caking coals to which they are liable to be poorly fused. This effect must increase more rapidly than in proportion to the concentration of inert. It has also been suggested that the presence of the infusible substance modifies the intergranular swelling by facilitating the liberation of gas, but this effect is purely hypothetical. Certain infusible substances may react chemically with the coal and modify its plasticity; this is the case, in particular, for hematite or limonite which are reducible at low temperature. These iron oxides markedly reduce the fluidity and the swelling, which may be explained by the oxidation of coal in the plastic region.[210] On the other hand, magnetite Fe_3O_4 cannot react with coal below its resolidification point and behaves like an infusible inert such as quartz.[205,211]

The temperature coefficient of expansion of an inert is generally different from that of the coal with which it is mixed. An extreme case is that in which the inert does not undergo transformation, such as coke breeze or mineral matter which increase slightly in volume as a result of thermal expansion. It is, therefore, opposed to the contraction of the semicoke surrounding it. The contraction of the mixture is thereby reduced, and this decreases the tendency of the coke to fissuration. Calculations show that this effect is more pronounced the greater the modulus of elasticity of the inert. It is difficult to measure the modulus of elasticity of a powdery substance such as coke breeze, but the microhardness, measured by a ball test, generally parallels the modulus of elasticity. Experience definitely shows a close correlation between the microhardness of the inert and the reduction of the contraction.[47,193] When the inert is fine and porous, it lowers the resolidification point slightly, resulting in a slight increase in the contraction coefficient of the coal at the resolidification point. This effect partially compensates the preceding one. Because it is more marked, the greater the fineness and porosity of the inert, the total effect, that is, the reduction of the contraction coefficient of the mixture at the resolidification point is the less marked for a given inert, the finer the grinding of the latter.[47]

Each infusible particle is strongly compressed by the semicoke surrounding it. Thus, the semicoke is subjected to considerable stress, which may give rise to small fissures. The microstrength of the coke is decreased. This effect is greater, the larger and harder the infusible particles.

When the inert contracts above the resolidification point, as in the case of semicoke and lean coal, the two preceding effects are less marked and may not appear. To assess them, it is necessary to compare the temperature coefficient of contraction of the inert with that of the coal with which it is mixed. The reduction of the contraction coefficient of the mixture at the resolidification point is determined by the contraction of the inert in the temperature range from 500 to 600°C. The reduction of the microstrength of the coke is mainly determined by the *total* contraction of the inert between the resolidification point of the coal and the final temperature.

From the above analysis, it follows that the properties of an inert which determine its influence are its particle size, porosity, modulus of elasticity or microhardness, total contraction, and the contraction coefficient in the vicinity of the resolidification point of the coal with which it is mixed.

The effect of an addition of an inert on coke quality depends on its characteristics, the nature of the coal charge to which it is added, and the coking conditions. The addition of an inert may be advantageous, useless, or harmful.[212] This complex character undoubtedly results from the varied nature of the modes of action of the inert, the above analysis of which certainly does not cover all aspects.

The general trend of the development of coke quality as a function of the proportion of inerts is illustrated by Fig. 29.[184,193,213] It relates to the addition of coke breeze ground to 80% less than 0.2 mm in size to a very fluid high-volatile coal (H_1 coal of Table 1). An addition of from 5 to 10% leads both to a slight

[210] Titov, N. G., Fedorovskaya, N. P., Jarkova, A. V., and Sourina, N. L., *Izvest. Akad. Nauk S.S.S.R. Otdel. Tekh. Nauk*, **9**, 114–23 (1954).

[211] Barking, H., and Eymann, C., *Glückauf*, **88**, 1090–4 (1952). Jappelt, A., *Brennstoff-Chem.*, **34**, 226–32 (1953).

[212] Blayden, H. E., Riley, R. L., and Shaw, F., *Fuel*, **25**, 13–23 (1946).

[213] Foch, P., and Loison, R., *Rev. ind. minérale*, **39**, 127–58 (1957); **40**, 1–34 (1958); **41**, 277–312 (1959); **42**, 169–95 (1960).

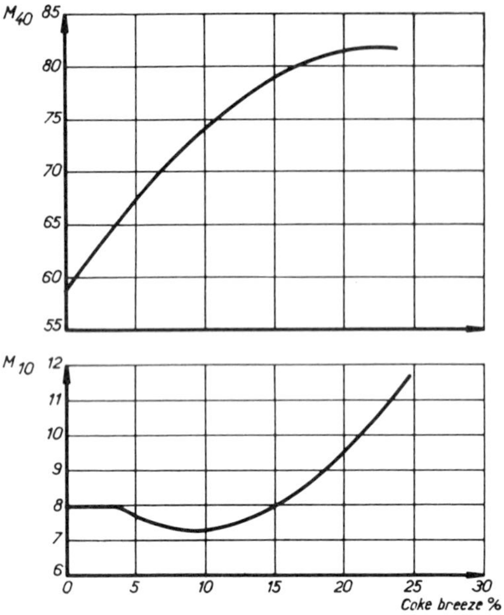

Fig. 29. Coke quality as a function of the proportion of coke breeze.[183]

improvement in the hardness (M_{10} index), resulting from the disappearance of the spongy masses, and a marked improvement in the strength (M_{40} index), resulting from the reduction of contraction after resolidification. A larger addition leads to a rapid decrease in hardness, resulting from reduction of fluidity, swelling, and caking and from the difference between the contraction of coke breeze and that of the coal after resolidification. There is an optimum value of coke breeze concentration which, in this example, would be of the order of 15 to 20%, but varies greatly with the nature of the coal. If the latter is weakly caking, the decrease in hardness becomes significant at low concentrations of the inert, and the optimum concentration is low or even zero. The particle size of the coke breeze exerts a very great influence: When it increases, the favorable effect on fissuration increases slightly (undoubtedly as a result of the influence on the resolidification point), but the unfavorable effect of the difference between the contraction of the coke breeze and that of the coal on the microstrength increases markedly. There is, therefore, an optimum particle size, but it is always of the order of 80% less than 0.2 mm, at least for blast furnace coke. The optimum concentration and the optimum particle size naturally depend on the relative value attached to the hardness and to the strength of the coke and are not the same for foundry coke and blast furnace coke.

Semicoke has a smaller influence than coke breeze on the fissuration of coke, probably because of its low hardness. On the other hand, it has a less unfavorable influence on the microstrength because its total contraction is comparable to that of the coal charge. It can, therefore, be introduced in an increasing proportion without the necessity of grinding it very fine. When added to coal charges it inhibits formation of spongy masses.

Certain low-volatile coals and anthracites have a behavior intermediate between that of semicoke and coke breeze, but generally closer to the latter.

Magnetite, Fe_3O_4, exerts an influence comparable to that of coke breeze and can be incorporated in proportions up to 50% with strongly caking high-volatile coals.[211] On the other hand, ferric oxides, Fe_2O_3, can be employed only as an addition in small amounts to high-rank coals that resist oxidation strongly.

SUMMARY

Substantial progress has been made in the last fifteen years with regard to the knowledge of the mechanism of coke formation and the influence of the various factors that determine its quality. This progress makes it possible to explain a considerable proportion of the facts observed by direct investigation at coking plants. It also provides a guide to the choice of components, the ratios of mixtures, and manufacturing conditions. Our knowledge, however, is still very incomplete; although it often allows known experimental facts to be explained, it has seldom led to the discovery of new processes and techniques. The knowledge gained provides a guide for research on conditions of coke production that is usually only qualitative in character. It is seldom useful to predict exactly and quantitatively the characteristics of a coke manufactured under given conditions from coals whose characteristics have been determined only in the laboratory.

Investigations at first were mainly characterization of the behavior of coal during coking by overall tests and comparison of the results with the characteristics of the coke also determined by an overall test. As a result of the complexity of the phenomena, the correlations between such tests were not very exact and had little general value. The only valid overall test is that which reproduces all the coke-oven conditions exactly, such as an experimental oven with a charge of 200 to 300 kg. The greatest progress has been attained by analysis of the phenomena of softening, swelling, and contraction, resulting from the heating of coal, and by measurement of true physical values such as the viscosity, dilatation, melting point, resolidification point, and contraction coefficient. At the same time, efforts have been made to analyze the different stages of coke formation in the coke oven by trying to determine the influence of the various properties of coal. This method should be followed up, because it has not yet provided all the results which can be expected from its application. It would be particularly desirable to carry out a closer analysis of the formation of the porous structure of coke in the plastic layer. It is equally necessary to characterize the mechanical strength of coke not only by one or two overall tests, as at present, but by measuring the physical properties of which these tests are the resultant.

CHAPTER 5

COAL ANALYSIS AND MINERAL MATTER

W. H. Ode

Bureau of Mines, U.S. Department of Interior, Washington, D.C.

COAL ANALYSIS

With the rapidly expanding use of coal brought about by the industrial revolution during the past century, accepted standard methods of coal analysis became necessary. Much work, particularly by national standardizing bodies, has led to the development of reliable methods. For example, the American Society for Testing Materials has had uninterrupted work in this field for more than fifty years.[1] The same type of development in standardization of methods has taken place in all of the major coal-producing countries. Everyone who is interested in methods of coal analysis should be familiar with the standardized methods adopted in his own country. However, the methods adopted are not the only suitable procedures. New methods are being developed and present methods are being modified to increase the accuracy and precision of results. In buying and selling coal, the national standard methods normally are used because they have legal status when controversy develops.

With the great increase in international trade in coal following World War II, standardization on an international basis became a natural development. The work has been done under the direction of the International Organization for Standardization (ISO) through its Technical Committee 27 on Solid Mineral Fuels. Membership of the committee is composed of national standardizing bodies representing 21 participating and 14 observer countries. Rapid progress has been made covering the broad field of sampling, analysis, and testing of coal and coke. It is expected that international methods soon will become available as ISO Recommendations.[2]

PROXIMATE ANALYSIS

The proximate analysis of coal was developed as a simple means for determining the distribution of products obtained during heating under standard conditions. It is the most used type of analysis for characterizing coals in connection with their utilization. Proximate analysis separates the compounds into four groups: (1) water or moisture; (2) volatile matter consisting of gases and vapors driven off during the pyrolysis; (3) fixed carbon, remaining as the nonvolatile fraction of the pyrolyzed coal; and (4) ash, derived from the mineral impurities. The fixed carbon is a calculated figure obtained by subtracting from 100 the sum of the percentages of moisture, volatile matter, and ash.

WATER

The role of water in coal and its quantitative determination have occupied the attention of coal-research workers for many years.[3] The difficulties in accurate assessment of water in coal are brought about by the fact that water is present in the pure coal substance in more than one form. Gauger,[4] in his Priestley

[1] Fieldner, A. C., *ASTM, Bull.* No. 204, 16–18 (1955).

[2] King, J. G., *Coke and Gas*, **14**, 363–7 (1952); **16**, 55–60 (1954); **17**, 305–11 (1955); **20**, 461–4, 509–13 (1958). Selvig, W. A., *Mag. of Standards*, **26**, 282–3 (1955).

[3] Gauger, A. W., *Chemistry of Coal Utilization*, John Wiley and Sons, New York, 1945, Vol. 1, pp. 600–615.

[4] Gauger, A. W., *Fuel*, **26**, 138–9 (1947).

lecture, reported that different ranks of coal contain varying quantities of water present in part as a mechanical and in part as a physical mixture. The mechanically admixed water on the surface and inside cracks and large capillaries has a normal vapor pressure, which he termed "free water," with the same physical properties as ordinary water. The physically-held water in the internal pore structure of the coal has a vapor pressure lower than normal, which he termed "bound water." This form of moisture is held in the smaller capillaries or pores. Agde and Schürenberg[5] determined the different forms of water by a graphic evaluation of the inflections in the logarithmic curve of vapor pressure versus water concentration. They proposed the terms lyosorption water, capillary water, and coarsely combined water. Bond, Griffith, and Maggs[6] concluded, as a result of a study of water adsorption isotherms, adsorption swelling measurements, attempts to refreeze the adsorbed water, and direct tests on relation between adsorbed and total water, that the submicroscopic pores of coal in contact with saturated vapor become filled with adsorbed water with properties greatly different from those of normal liquid water. Kreulen[7] reported that, for brown coals, a distinction must be made between strongly and weakly attached water. The latter obeys the sorption laws, and the calculated heat of sorption corresponds approximately to the molar heat of evaporation of the water; the binding of this water is apparently due to the gel structure of the coal. Strongly attached water is not removed at ordinary temperatures, which indicates that it is connected with the inner surface of the coal and not with its gel structure.

Several investigators have suggested that although sorption in the coal-water system is mainly physical in nature, part of the moisture is held in the pore structure by chemisorption. Lecky, Hall, and Anderson[8] and Malherbe and Carman[9] reported that the true surface area of coal is not measured by adsorption of methyl alcohol or water, because these polar compounds are not physically adsorbed in the usual sense. They suggested that polar compounds such as water react with coal-forming hydrogen bonds or "salt" type linkages.

Work by Indian investigators[10] on sorption studies of coal also suggests the chemisorption theory. Mukherjee, Nandi, and Lahiri proposed two mechanisms to account for the sorption characteristics of coal. In the first, hysteresis and the effect of temperature on sorption were attributed to the availability of chelated hydroxyl groups in the coal. The second postulated the formation of a coal-water complex that retains water on coal by secondary valence forces. If part of the water is held in chemical combination with the coal molecule, Storch[11] has recommended that analytical procedures for determining moisture in coal, particularly those of low rank, should be based on heat-adsorption versus temperature curves, rather than on empirically selected temperatures in thermal drying or in distillation with liquids such as xylene.

Hoeppner, Fowkes, and McMurtrie[12] pointed out that the distinction between physically and chemically combined moisture in the younger fuels, brown coals, and lignites may become obscure, because of thermally unstable organic molecules present as a residue of the original plant substance, as well as the large capillary pore system of the coal. In an attempt to differentiate and classify various types of water in lignite, they investigated the release of water and carbon dioxide from lignite by heating in inert-gas atmospheres at

[5] Agde, G., and Schürenberg, H., *Brennstoff-Chem.*, **22,** 241–4 (1941).

[6] Bond, R. L., Griffith, M., and Maggs, F. A. P., *Fuel*, **29,** 83–93 (1950); *Discussions Faraday Soc.*, **1948,** No. 3, 28–40.

[7] Kreulen, D. J. W., *Combustibles (Zaragoza)*, **14,** No. 72, 63–71 (1954); *Brennstoff-Chem.*, **36,** 266–71 (1955).

[8] Lecky, J. A., Hall, W. K., and Anderson, R. B., *Nature*, **168,** 124–5 (1951).

[9] Malherbe, P. le R., and Carman, P. C., *Fuel*, **31,** 210–19 (1952).

[10] Mukherjee, P. N., Basak, N. G., and Lahiri, A., *Fuel*, **30,** 215–16 (1951); *J. Sci. Ind. Research (India)*, **12B,** 15–24 (1953). Iyengar, M. S., and Lahiri, A., *J. Sci. Ind. Research (India)*, **15B,** 593–9 (1956). Mukherjee, P. N., Nandi, S. P., and Lahiri, A., *J. Sci. Ind. Research (India)*, **15B,** 638–45 (1956).

[11] Storch, H. H., *2nd Conf. on the Origin and Constitution of Coal*, Crystal Cliffs, Nova Scotia, 1952, pp. 369–70.

[12] Hoeppner, J. J., Fowkes, W. W., and McMurtrie, R., *U.S. Bur. Mines, Rept. Invest.* No. 5215 (1956), 25 pp. Fowkes, W. W., Hoeppner, J. J., and McMurtrie, R., *Fuel*, **36,** 469–74 (1957).

different temperature levels and time intervals. The continued release of water and carbon dioxide with heating, together with the effect of gas atmosphere on rate of release, pointed directly to control of rate by a diffusion mechanism, apparently that of diffusion from the surface of release out through the capillary or pore system of the lignite. Release in a given gas atmosphere was a continuous and progressive process with both time and temperature and with no evidence of breakpoints in rate that might correspond to changes in the mechanism of the attachment of the water.

Determination

The total moisture in coal may be determined in a one-stage process, or the as-received sample may be air-dried at about room temperature and the residual moisture determined in the laboratory sample. The latter method is usually referred to as the two-stage method. The purpose of air-drying is to reduce the moisture in the sample to approximate equilibrium with the air of the laboratory and, thereby, minimize changes in moisture content when the sample is handled during crushing and analysis. The two-stage method is preferable for preparing laboratory samples that are intended for highly accurate analysis.[13]

Methods for determining moisture in coal have been classified by Brown[14] as follows:

1. Thermal drying methods, involving:
   (a) determination of the loss of weight of coal heated to various temperatures in different ways, and in various atmospheres, or in vacuum;
   (b) determination of the gain in weight of a vessel packed with a desiccant, through which the gases and vapors evolved when coal is heated as in (a) are passed;
2. Desiccator method, involving determination of the loss in weight of coal contained in a desiccator in the presence of a desiccant, either at normal, or at a reduced pressure, but at room temperature;
3. Distillation method, involving collection and measurement, either gravimetrically or volumetrically, of the water distilled from coal when the coal is boiled in a solvent immiscible with water;
4. Extraction and solution methods, in which the coal is shaken:
   (a) with a solvent that will extract the water from the coal, followed by determination of the water content of the solvent,
   (b) with a standardized reagent that will be altered in concentration by the water in the coal;
5. Chemical methods, in which the water content is measured by direct chemical titration, or by chemical reaction with reagents which react with water to evolve gases, the volume of which can be measured;
6. Electrical method, involving the measurement of the dielectric constant of the coal.

The determination of moisture by measuring the loss in weight by heating at a temperature slightly higher than the boiling point of water has been used by various laboratories for many years. Air or nitrogen usually is used to sweep the water vapor away from the coal. Some objection has been raised to methods of this type for coals that contain absorbed gases, such as methane, that are driven off during heating. Also, if the coal is dried in air, it is liable to oxidation, and the gain in weight from fixation of the oxygen correspondingly affects the results; this source of error is especially apparent for high-moisture, low-rank coals. For example, the latest British Standard[15] permits the use of air for drying high-rank coals, and only then if the coal is coarsely crushed. For low-rank coals, oxygen-free nitrogen must be substituted or, alternatively, the coal is dried in a vacuum oven.

Standard methods for determining moisture by loss of weight generally require heating for periods of 1 hr or more. To speed up analyses for control purposes, various modifications of the basic procedure have been proposed. Several investigations have been made of the Brabender[16] semi-automatic moisture tester. It consists of a combination electrically-heated drying-oven and balance with a capacity for holding 10-g samples. Weighings are made

[13] Fieldner, A. C., and Selvig, W. A., *U.S. Bur. Mines, Bull.* **492,** 5 (1951). Brit. Standards Inst., Brit. Standard 1016, Part 1 (1957).

[14] Brown, G. M., *Natl. Gas Bull.* (*Melbourne*), **17,** No. 1, 14–21 (1953).

[15] Brit. Standards Inst., Brit. Standard 1016, Parts 1 and 3 (1957).

[16] Brabender, C. W., U.S. Patent 2,047,765 (1936).

without removing the samples from the oven. Winkler[17] found that the results by this method agreed well with those by the xylene-distillation method. Rees and Bursack[18] concluded, from tests of bituminous coal ranging in moisture content from 5 to 19%, that the tester gave moisture values at the temperatures and sizes studied, which checked closely with those by the ASTM procedure. The drying time with the Brabender equipment was 10 to 40 min as compared to 90 min by the standard procedure. Cooper, Tarpley, and Abernethy[19] found that although the instrument does not meet several requirements of the ASTM Standard, moisture values for high-rank coals, except anthracite, are within permissible tolerances of the ASTM procedure.

Other rapid methods also have been described. Belcher and Spooner[20] used a modification of the British Standard method whereby 0.5 g of coal was spread in a thin layer in a large boat and dried in a rapid stream of nitrogen at 110°C; drying of air-dried coals was accomplished in about 10 min. Methods of heating investigated include high-frequency currents,[21] infrared rays,[22] electrically heated plates,[23] and specially constructed ovens.[24]

To minimize possible sources of error such as may occur with thermal processes in which the moisture is measured by loss of weight, a direct method is becoming increasingly popular, particularly in European laboratories, in which the water evolved during heating in an inert atmosphere is weighed. Thus, errors associated with absorption of gas or loss of fixed gases, such as methane in certain coals, the possible effect of oxidation, or the reabsorption of moisture by the coal are eliminated.

Briefly, the method consists of heating the coal in a tube through which a current of nitrogen at 105 to 110°C is passed, and the moisture driven off is collected in an absorption tube.[25] The moisture is calculated from the increase in weight of the absorption tube. Belcher and Spooner[26] also recommended the method as a rapid determination; they concluded that little danger of oxidation occurs when using air, because drying is completed in 15 min.

The effect of the temperature of heating in the direct method has been studied by a number of investigators. Kreulen[7] found that an appreciable amount of water was retained in coal dried at 100°C, especially in coals of low-rank, on progressive heating to higher temperatures. Sharp breaks were obtained at some temperature above 180°C. Decomposition of the coal was measurable at this temperature, but below, decomposition reactions are not important. Crawford[27] found in tests of 5 coals covering the range of 1 to 14% moisture that there was a small but significant increase in the moisture figure obtained when the temperature was raised from 105 to 150°C. The increase was much of the same order as observed when xylene distillation was used to determine the moisture content. He also reported that air must not be used as purging gas at any of the temperatures investigated, because on changing from oxygen-free nitrogen to air, additional water was evolved which must have had its origin in oxidation reactions occurring in the coal substance. Hoeppner, Fowkes, and McMurtrie[12] showed that when lignite is heated in inert-gas atmosphere at temperatures from 105 to 175°C, water is released at measurable rates for as long as 144 hr. The inert-gas atmosphere also was found to have a significant effect, the rate of release in a helium atmosphere at 175°C exceeding that in nitrogen and argon.

Another direct method that is finding increased use is a distillation procedure whereby coal is heated in a liquid that has a

[17] Winkler, G., *Braunkohle*, **40**, 665–71 (1941).

[18] Rees, O. W., and Bursack, K. F., *Ind. Eng. Chem., Anal. Ed.*, **14**, 729–31 (1942).

[19] Cooper, H. M., Tarpley, E. C., and Abernethy, R. F., *U.S. Bur. Mines, Rept. Invest.* No. 4304 (1948), 15 pp.

[20] Belcher, R., and Spooner, C. E., *Fuel*, **23**, 46–9 (1944).

[21] Liplavk, I. L., *Zavodskaya Lab.*, **12**, 555–8 (1946).

[22] Podbel'skiĭ, G. N., and Budrina, O. V., *Zavodskaya Lab.*, **22**, 70–3 (1956).

[23] Grigorenko, N. F., and Milashevskiĭ, V. P., *Ugol'*, **29**, No. 12, 31–2 (1954).

[24] Fedorovskaya, N. P., Miesserova, L. V., and Borozdina, L. A., *Zavodskaya Lab.*, **22**, 73–5 (1956).

[25] Šimek, B. G., and Ludmila, J., *Fuel*, **26**, 135–7 (1947). Deinum, H. W., and Goedkoop, M. L., *Chem. Weekblad*, **48**, 170–3 (1952). Netherlands Standard 924 (1952). Brit. Standard 1016, Part 3, 13–7 (1957).

[26] Belcher, R., and Spooner, C. E., *Fuel*, **26**, 55–6 (1947).

[27] Crawford, A., *Fuel*, **36**, 7–25 (1957).

boiling point higher than water and is immiscible with it. The water and liquid distilling off are condensed, and the volume of water is measured. Liquids usually employed are either toluene, xylene, or a petroleum fraction of selected boiling range. The distillation method is considered to be advantageous, particularly for low-rank coals, because air is excluded from the coal by the liquid and its vapor.[28] The method has been adopted as an alternative technique in several national standards,[29] and certain improvements in design of apparatus have been reported.[30]

As a substitute for thermal methods, attempts have been made to use extraction, solution, and chemical techniques as an entirely different approach to the determination of moisture. Although interesting and novel in their concept, their use generally has been quite limited. Mielecki[31] determined moisture in peat and brown coal by shaking the coal with alcohol and then decanting off the solution and measuring its density. A similar technique is the subject of a Japanese patent.[32] Hartmann and Bachelard[33] measured the critical solution temperature (cryohydric point) of the resulting alcohol-water solution with an equal volume of kerosine. Belcher and Spooner[20] extracted the coal with methyl alcohol: An aliquot of the filtrate was treated with an equal volume of carbon tetrachloride, and the solution was titrated with water until a slight turbidity was evident. Schmidt and Seymour,[34] in an investigation of effect of bulk density on carbonizing properties of coal, determined surface moisture by quickly washing with alcohol and measuring the specific gravity of the resulting solution. Radmacher and Mohrhauer[35] used a similar principle by measuring the change in density of a $ZnCl_2$-LiCl solution.

Gray and Whelan[36] found that the maximum rise in temperature recorded when a 5-g sample of coal was mixed with concentrated sulfuric acid was linearly related to the moisture content of the coal. By this method, percentage moisture contents between 15 and 40 could be estimated with an accuracy of ± 5 in a few minutes.

The well-known Karl Fischer method for determining moisture has been applied to coal by several investigators.[37] The results generally have been found to check well with conventional methods. Other chemical methods for rapid determination of moisture include reaction with quicklime and the maximum rise in temperature measured.[38] The evolution of acetylene by reaction of calcium carbide with the moisture in coal has also been used.[39]

Two methods of measuring moisture in various materials by capacitance or resistance measurement are used by industry, particularly for moving streams. Such methods[40] have been applied to air-dried brown coals in the briquetting industry and also coals charged to coke ovens. In the capacitance method, the stream of coal is passed between two electrodes, from which a signal is transmitted to a recording device. The resistance method uses

[28] Goodman, J. B., Gomez, M., and Parry, V. F., *U.S. Bur. Mines, Rept. Invest.* No. 4969 (1953), 20 pp. Brown, G. M., *Australian J. Appl. Sci.*, **6,** 365–75 (1955).

[29] Deutscher Normenauschuss, DIN 51718–1950. Polish Standard Comm., PN–57 C–04315. Brit. Standards Inst., Brit. Standard 1016 (1957), Part 1.

[30] Simon, W., *Braunkohle*, **41,** 53 (1942). Šimek, B. G., and Ludmila, J., *Brennstoff-Chem.*, **23,** 223–7 (1942). Marskell, W. G., and Rayner, J. E., *Fuel*, **26,** 49–50 (1947). Šimek, B. G., and Ludmila, J., *Fuel*, **26,** 132–7 (1947). Bainbridge, J. R., Eisenklam, P., and Scanlan, P. G., *Fuel*, **26,** 150–3 (1947). Bainbridge, J. R., Scanlan, P. G., and Belyea, I., *Fuel*, **28,** 88–90 (1949).

[31] Mielecki, T., *Przeglad Górniczy*, **5,** No. 1, 28–31 (1949).

[32] Baba, A., and Fujiwara, M., Japan. Pat. 1100 (1954).

[33] Hartmann, H. F., and Bachelard, H. S., *Natl. Gas Bull.* (*Melbourne*), **17,** No. 4, 21–3 (1953).

[34] Schmidt, L. D., and Seymour, W., *U.S. Bur. Mines, Rept. Invest.* No. 3811 (1945), 11 pp.

[35] Radmacher, W., and Mohrhauer, P., *Glückauf*, **92,** 166–71 (1956).

[36] Gray, V. R., and Whelan, P. F., *Chem. & Ind.* (*London*), **1955,** 126–8.

[37] Salyi, G., and Siniramed, C., *Riv. combustibili*, **4,** 94–105 (1950). Bryson, A., and Pickering, W. F., *J. Inst. Fuel*, **25,** 28–30 (1952). Funasaka, W., Yokakawa, C., and Kojima, T., *Kooru Taaru*, **5,** 51–2 (1953).

[38] Schaller, R. H., *Chem. & Met. Eng.*, **52,** No. 11, 116–7 (1945).

[39] Voskuil, J., and Zwaan, P., *Chem. Weekblad*, **39,** 257–61 (1942). Lusinchi, J., CERCHAR 690 (1956), 5 pp.

[40] Badzioch, S., and Cornford, G. B., BCURA *Bull.* **16,** 77–89 (1952).

two probes inserted in the coal and a signal sent to a recorder. Electrical methods also have been reported.[41]

Bed or Equilibrium Moisture

In the ASTM Standard Specifications for Classification of Coals by Rank,[42] and the international system for hard coals,[43] high volatile coals are classified according to calorific value on the moist basis. "Moist" refers to coal containing its natural bed moisture but not including visible water on the surface of the coal. As a means of estimating the bed moisture of coal that is either wet and showing visible surface moisture or coal that may have lost some moisture, an equilibration method is used that restores the coal to essentially its bed moisture condition.[44] Briefly, the method consists of wetting the coal with water, draining off excess water, and equilibrating the sample at 30°C in a vacuum desiccator containing a saturated solution of potassium sulfate that maintains a relative humidity of 96 to 97%. Equilibration requires 48 to 72 hr depending on the rank of coal. The total moisture in the equilibrated sample is then determined. The method also may be used to estimate the surface or extraneous moisture of wet coal, such moisture being the difference between total moisture and equilibrium moisture. The method is based on the principle originally developed by Stansfield and Gilbart.[45] Kreulen[46] investigated the method and recommended that the test sample be completely wetted with water before beginning the equilibration test, as otherwise low results would be obtained for low-rank coals if they had lost an appreciable amount of their bed moisture. Selvig and Ode[47] concluded that, for practical purposes of coal classification, bituminous and subbituminous coals, even though air-dried to relatively low moisture contents, can be restored to virtually their natural bed moisture by the equilibration method; for lignites, equilibration tests are preferably made on freshly mined samples. Although the method is satisfactory for all ranks of American coals, Rammler and Knopfe[48] reported that the equilibrium moisture method is not suitable for classification of certain European low-rank coals, because the equilibrium moisture is less than total or bed moisture. Drekopf and Steiner[49] found that the period of equilibration could be reduced to 4 hr if the air film around the coal particles was first removed by means of a vacuum.

In a different approach to the determination, Dunningham[50] removed the adherent moisture from wet coal by blotting with filter paper. Dulhunty[51] determined maximum inherent or bed moisture by means of controlled evaporation of the adherent moisture at pressures just below the equilibrium pressure at a relative humidity of approximately 99%—slightly above room temperature.

VOLATILE MATTER

The volatile matter obtained during pyrolysis of coal consists mainly of the combustible gases, hydrogen, carbon monoxide, methane,

[41] Asbach, H. R., *Glückauf*, **87**, 18–22 (1951). Tomków, K., *Práce Glòwnego Inst. Gòrnictwa, Komun.* No. 126, Series B (1952), 9 pp. Klempt, W., *Brennstoff-Chem.*, **34**, 294–8 (1953). Pohlmann, —, *Braunkohle*, **6**, 448–53 (1954). Berliner, M. A., and Sokolov, V. S., *Elek. Stantsii*, **25**, 11–13 (1954); *Colliery Guardian*, **189**, 279–81 (1954). Wickfield, H., *Bergbautechnik*, **5**, 596–601 (1955). Nishio, J., Tsuzaki, K., and Maeda, M., *Nenryô Kyôkaishi*, **35**, 591–9 (1956). *Australian J. Instr. Technol.*, **13**, No. 2, 31–6 (1957). Funasaka, W., Ando, T., and Kamino, Y., *Nenryô Kyôkaishi*, **36**, 69–74 (1957). Zolotukhin, A. I., Lazovskiĭ, I. M., and Filyashin, K. Ya., *Koks i Khim.*, **1958**, No. 6, 6–10.

[42] Am. Soc. Testing Materials, 1958 ASTM Standards, Part 8, 1078–83 (1959).

[43] United Nations' Economic Commission for Europe, *International Classification of Hard Coals by Type*, E/ECE/247, E/ECE/COAL/110: Pub. Sales No. II.E. 4 (1956).

[44] Am. Soc. Testing Materials, 1958 ASTM Standards, Part 8, 1066–8 (1959).

[45] Stansfield, E., and Gilbart, K. C., *Trans. Am. Inst. Min. Met. Eng.*, Coal Division, **101**, 125–43 (1932).

[46] Kreulen, D. J. W., *Chem. en Pharm. Tech. (Dordrecht)*, **7**, 23–4 (1951).

[47] Selvig, W. A., and Ode, W. H., *U.S. Bur. Mines, Rept. Invest.* No. 4968 (1953), 10 pp.

[48] Rammler, E., and Knopfe, E., *Freiberg. Forschungsh.*, **A44**, 60–79 (1955).

[49] Drekopf, K., and Steiner, H., *Glückauf*, **92**, 461–3 (1956).

[50] Dunningham, A. C., *Proc. Conf. Ultra-Fine Structure of Coals and Cokes*, Brit. Coal Utilisation Research Assoc., 1944, pp. 57–70.

[51] Dulhunty, J. A., *J. Proc. Roy. Soc., N.S. Wales*, **81**, 60–8 (1947).

and other hydrocarbons; tar vapors; and, some incombustible gases, such as carbon dioxide and water vapor. It does not include the moisture in coal that can be removed by heating to temperatures slightly higher than the boiling point of water, but does include the water formed during thermal decomposition of the coal substance. The composition of the volatile matter varies greatly for different ranks of coal, with the proportion of incombustible gases increasing as the rank decreases.

It is one of the most important determinations in coal analysis, because volatile matter is used as a parameter in certain classification systems and for evaluating coals for combustion and carbonization. Because of the importance of the volatile matter determination in connection with smoke ordinances where the permissible volatile matter content is specified, the American Society for Testing Materials has adopted a standard method covering both the sampling and analytical procedures to be used.[52]

Determination

All methods for determining volatile matter are based on the same principle—the differences being arbitrary.[53] Essentially, they consist of heating a weighed sample of coal, usually 1 g, in a covered crucible, to a prescribed temperature. The loss in weight less the moisture is termed "volatile matter." The chief differences in the methods are variations in the size, weight, and materials of the crucibles used; the rate of temperature rise, the final temperature, and duration of heating; and modifications required for coals that decrepitate or lose particles by entrainment during extremely rapid escape of water vapor or volatile matter. As all of the variables affect the results,[54] the method is necessarily empirical, and to obtain satisfactory results, standardized procedures must be followed closely.

Because of the arbitrary nature of the test, details of various national standards will not be given here. The temperature range is from 875 to 1050°C, the duration of heating from 3 to 20 min, and crucibles are of platinum, silica, or ceramic material. The German standard specifies a temperature of 875°C to agree with industrial coking practice. On the other hand, certain standards specify temperatures of 1000 to 1050°C, as evolution of volatile matter above these temperatures is reported to be insignificant. A temperature of 950°C is specified in the ASTM Standard. Normally, most standards specify a single crucible, although in France and Belgium, what is known as the double-crucible method may be used. According to this method, the coal is heated in a crucible which is enclosed in a larger crucible, the space between them being filled with charcoal. The advantage claimed for this modification is that errors due to partial oxidation are prevented, particularly for cokes and chars.

[52] Am. Soc. Testing Materials, *D* 980–53.

[53] Shakhno, A. P., *Trudy Vsesoyuz. Konf. Anal. Khim.*, **3**, 143–68 (1944).

[54] Mott, R. A., and Spooner, C. E., *Fuel*, **24**, 106–10 (1945). Rees, O. W., Coolican, F. A., and Pierron, E. D., *Illinois State Geol. Survey, Circ.* No. 240 (1957), 8 pp.

ASH

Ash is the residue derived from the mineral matter during complete incineration of the coal. It is different in chemical composition and usually is less than the mineral matter originally present in the coal. During incineration, various weight changes take place such as loss of water of constitution of the silicate minerals, loss of carbon dioxide from carbonate minerals, oxidation of iron pyrites to iron oxide, and fixation of oxides of sulfur by bases such as calcium and magnesium. As the conditions of incineration determine the extent to which the weight changes take place, the determination is empirical. It is essential that standardized procedures be closely followed to obtain reproducible results.

Determination

The amount of ash is normally determined by burning a 1- to 2-g sample of coal in an adequately ventilated muffle furnace to temperatures of 700 to 850°C. The prescribed ashing temperature in the American Standard is 700 to 750°C for coal, and not more than 950°C for coke; Japan, 750°C; Germany, Netherlands, and the United Kingdom, 775 ± 25°C; France, 825°C; and, Belgium and Poland, 800 to 850°C. The most important factor causing variation in the amount of determined ash is the effect of retention of sulfur. If the amount of pyrite and carbonate minerals in high-rank coals is low, sulfur retention is not critical, and ashing may be done rapidly. On the other hand, for coals with considerable amounts of pyrite and calcite, the normal practice is to burn the coal at low temperatures to decompose the pyrite before the decomposition point of the carbonate minerals is reached. Thus, less sulfur is left in the coal to react with the bases formed at higher temperatures.[55] Ashing of thin layers of coal also aids in reducing sulfur retention. Another method which involves more work is to use Parr's[56] sulfated ash method as modified by Rees.[57] Briefly, the method consists of first ashing the coal as usual, then treating the ash residue with a few drops of sulfuric acid, and finally igniting to constant weight. The amount of ash is calculated back to the calcium carbonate basis by subtracting three times the equivalent of carbon present as mineral carbonate from the ash as weighed.

Composition

When coal is burned, the mineral constituents form as ash residue composed chiefly of compounds of silicon, aluminum, iron, and calcium, with smaller quantities of compounds of magnesium, titanium, sodium, and potassium. Although the constituents are reported as oxides by the analyst, they occur in the ash largely as a mixture of silicates, oxides, and sulfates, with smaller quantities of other compounds. The silicates have their origin in the shale and clay minerals. The principal source of the iron oxide is pyrite, which burns to form ferric oxide and sulfur oxides. The calcium and magnesium oxides result from decomposition of the carbonate minerals. The sulfates are formed largely from interaction among carbonate minerals, pyrite, and oxygen.

The composition of coal ash varies widely, depending on the mineral constituents associated with the coal. Typical limits of ash composition of bituminous coals are shown in Table 1. The ash composition of American subbituminous coals may be higher in CaO, MgO, and SO_3 than for bituminous coals; the trend may even be more pronounced for lignite ash.[58] Rosin and Fehling[59] also have reported the same trend for German brown coals.

TABLE 1

TYPICAL LIMITS OF ASH COMPOSITION OF BITUMINOUS COALS

Constituent	USA[60]	England[61]	Germany[59]
SiO_2	20–60	25–50	25–45
Al_2O_3	10–35	20–40	15–21
Fe_2O_3	5–35	0–30	20–45
CaO	1–20	1–10	2–4
MgO	0.3–4	0.5–5	0.5–1
TiO_2	0.5–2.5	0–3	—
$Na_2O + K_2O$	1–4	1–6	—
SO_3	0.1–12	1–12	4–10

Analysis

The analyses of coal ash often are made by modifications of classical wet-chemical methods originally developed for silicate rocks.[62] Such methods are accurate and are considered as standards, but, when comparing other types of analytical procedures, they are time-consuming, because of the numerous separations required to complete an analysis. To overcome this objection, particularly when a large volume of analyses must be make, rapid methods that are comprehensive and also reasonably accurate have been developed during the past few years. They employ some of the more modern physico-chemical techniques that include spectrophotometry, flame photometry, polarography, emission spectroscopy, and chelatometric titration.

[55] Rees, O. W., and Selvig, W. A., *Ind. Eng. Chem., Anal. Ed.*, **14,** 209–12 (1942).

[56] Parr, S. W., *Illinois State Geol. Survey, Illinois Coal Mining Invest., Bull.* No. 3, 35 (1916).

[57] Rees, O. W., *Ind. Eng. Chem., Anal. Ed.*, **9,** 307–9 (1937).

[58] Selvig, W. A., and Gibson, F. H., *U.S. Bur. Mines, Bull.* No. 567 (1956), p. 8.

[59] Rosin, P., and Fehling, R., *Ber. Reichskohlenrat*, **54/55** (1935).

[60] Selvig, W. A., and Gibson, F. H., ref. 58, p. 32.

[61] King, J. G., and Crossley, H. E., *Gr. Brit. Dept. Sci. Ind. Research, Fuel Research, Survey Paper* No. 28 (1933), 20 pp.

[62] Brit. Standards Inst., Brit. Standard 1016 (1942), 95–116. *Gr. Brit. Dept. Sci. Ind. Research, Fuel Research, Survey Paper* No. 50 (1949), 40 pp. Fieldner, A. C., and Selvig, W. A., *U.S. Bur. Mines, Bull.* No. 492 (1951), 44–9. Chavarria Domingo, P., *Combustibles (Zaragoza)*, **18,** 105–17 (1958).

Węclewska[63] determined silica, iron, aluminum, titanium, and phosphorus by spectrophotometric methods, calcium by titration, and magnesium by spectrophotometric analysis of the phosphorus combined with it.

Bannerjee and Colliss[64] developed colorimetric methods based on spectrophotometric techniques for determining silicon, aluminum, iron, titanium, manganese, magnesium, phosphorus, germanium, and vanadium. Calcium was determined by precipitation of the oxalate, and titration with permanganate.

A chromatographic method developed by Ellington and Stanley[65] at the Coal Research Establishment of the National Coal Board in Great Britain employed ion-exchange resins for the separation of iron, sodium, potassium, titanium, aluminum, magnesium, and calcium. The elements then could be determined using normal analytical techniques.

Pringle[66] applied wet oxidation directly to coal to destroy the organic matter and bring the majority of the constituents into solution. The bulk of the residue was silica, which was determined gravimetrically. Aluminum, iron, titanium, phosphorus, and manganese were determined in the solution by photometric methods. After removal of the iron and aluminum from another portion of the solution, calcium, magnesium, sodium, and potassium were isolated by precipitation and the precipitates used for the determination of these elements by photometric methods. Durie and Schafer,[67] in their investigation of the method, made several modifications that included determination of sodium, potassium, and calcium by flame photometry.

Radmacher and Schmitz[68] employed photometric methods for determining silicon, iron, aluminum, titanium, phosphorus, manganese, and vanadium. Calcium and magnesium were determined by titration with a standard solution of the sodium salt of ethylene diamine tetra-acetic acid. Polarographic methods were used for zinc, copper, nickel, and cobalt, and flame photometry for sodium and potassium.

Archer, Flint, and Jordan[69] described methods used by the British Coal Utilisation Research Association for analysis of coal ash, slag, and boiler deposits. The procedure followed the general pattern of the method developed by Shapiro and Brannock[70] for the analysis of silicate rocks, but with slight differences in detail. Colorimetric procedures were used for silicon, aluminum, iron, titanium, manganese, and phosphorus. Calcium and magnesium were determined by titration with a standard solution of sodium salt of ethylene diamine tetra-acetic acid, and sodium and potassium by flame photometry. Sulfur was determined gravimetrically. Comparison of the results obtained by the rapid methods with those by the classical wet-chemical methods indicated that the accuracy of the rapid methods was better than 0.4% for all constituents, except Al_2O_3 which was about 1%.

Investigations of the flame photometric method for determining sodium and potassium were reported by Jackson and Smith[71] and by Schuhknecht and Schinkel.[72] In these methods, aluminum was added to the sample solution to suppress the interference of calcium in the determination of sodium. Bond and Hutton[73] recommended that a small excess of sulfuric acid should be present to stabilize and reduce calcium emission. Another method recommended that calcium be removed by ion exchange.[74] On the other hand, Archer and co-workers[75] reported that calcium interference in their method could be ignored for normal samples.

[63] Węclewska, M., *Prace Głównego Inst. Górnictwa, Kowun.* No. 163, Series B. (1954), 15 pp.

[64] Bannerjee, N. N., and Colliss, B. A., *Fuel*, **34,** S71–83 (1955).

[65] Ellington, F., and Stanley, L., *Nat. Coal Board, Sci. Dept., C.R.E. Rept.* No. 1275 (1955) 19 pp.; *Analyst*, **80,** 313–14 (1955).

[66] Pringle, W. J. S., *Fuel*, **36,** 257–76 (1957).

[67] Durie, R. A., and Schafer, H. N. S., *Australia, Commonwealth Sci. Ind. Research Organization, Fuel Research*, **23,** (1957), 62 pp.

[68] Radmacher, W., and Schmitz, W., *Brennstoff-Chem.*, **38,** 225–30, 270–4, 308–12 (1957); **39,** 164–72 (1958).

[69] Archer, K., Flint, D., and Jordan, J., *Fuel*, **37,** 421–43 (1958).

[70] Shapiro, L., and Brannock, W. W., *U.S. Geol. Survey, Bull.* No. 1036C (1956), 56 pp.

[71] Jackson, P. J., and Smith, A. C., *J. Appl. Chem. (London)*, **6,** 547–59 (1956).

[72] Schuhknecht, W., and Schinkel, H., *Brennstoff-Chem.*, **38,** 275–7 (1957).

[73] Bond, R. D., and Hutton, J. T., *Analyst*, **83,** 684–6 (1958).

[74] Boiler Availability Committee (London), *J. Inst. Fuel*, **28,** 550 (1955).

[75] Archer, K., Flint, D., and Jordan, J., *Fuel*, **37,** 421–43 (1958).

Edgcombe and Hewitt[76] applied a flame photometric method for determining calcium, which gave good results for samples containing up to 10% calcium oxide. Edgcombe and Gregory[77] and Kessler and Dočkalová[78] reported good results using a polarographic method for determination of magnesium.

Emission spectrographic analysis, which is the most rapid of any of the methods used for coal ash, has been reported by a number of investigators. Headlee and Hunter[79] used a total-energy method for quantitative determination of both major and minor constituents. Waring and Annell[80] developed a semiquantitative method for use when analyses of the highest accuracy are not required. Kessler and Veleska[81] determined silicon, iron, and aluminum without prior combustion of the coal. Shubnikov and co-workers[82] described a method for major constituents in which the average deviation from the results by classical chemical methods ranged from 1 to 10%. Dixon[83] used a spectrochemical method for major constituents; analysis was made either of the solution resulting from wet-oxidation of the coal or a solution obtained by alkali fusion of the coal ash. Methods for minor or trace elements[84] and potassium[85] also have been developed.

[76] Edgcombe, L. J., and Hewitt, D. R., *Analyst*, **79,** 755–8 (1954).

[77] Edgcombe, L. J., and Gregory, G. R. E. C., *Analyst*, **80,** 236–7 (1955).

[78] Kessler, M. F., and Dočkalová, L., *Práce ústavu pro výzkum a využiti paliv*, **8,** Publ. No. 17–26, 188–213 (1957).

[79] Hunter, R. G., and Headlee, A. J. W., *Anal. Chem.*, **22,** 441–5 (1950). Headlee, A. J. W., and Hunter, R. G., *W. Va. Geol. Survey, Bull.* **13A,** 36–9 (1955).

[80] Waring, C. L., and Annell, C. S., *Anal. Chem.*, **25,** 1174–9 (1953).

[81] Kessler, M. F., and Veleska, F., *Paliva*, **37,** 158–60 (1957).

[82] Shubnikov, A. K., Soboleva, G. N., Bronovets, T. M., and Khrisanfova, A. I., *Khim. i Tekhnol. Topliva i Masel*, **1957,** No. 10, 1–5.

[83] Dixon, K., *Analyst*, **83,** 362–71 (1958).

[84] Carlsson, C. G., and Larsson, A., *Jernkontorets Ann.*, **138,** 737–43 (1954). Benko, I., and Szádeczky-Kardoss, G., *Magyar Kém. Folyóirat*, **63,** 78–84 (1957). Smith, A. C., *J. Appl. Chem. (London)*, **8,** 636–45 (1958).

[85] Rekus, A. F., *Appl. Spectroscopy*, **12,** 141–2 (1958).

ULTIMATE ANALYSIS

Ultimate analysis expresses the composition of coal in percentages of carbon, hydrogen, nitrogen, sulfur, oxygen, and ash, regardless of their origin. The carbon includes that present in the organic coal substance as well as that originally present as mineral carbonates. The hydrogen includes that of the organic coal substance and the hydrogen present in the form of moisture and the water of constitution of the silicate minerals. In American practice, the reported figure for hydrogen in the as-received and air-dried sample includes both the hydrogen in the coal substance and in the moisture. In certain other countries, the value for the determined hydrogen is corrected for the moisture in the coal, and the corrected hydrogen and moisture are reported separately. All of the nitrogen is present as a part of the organic coal substance. The sulfur normally is present in three forms—organic sulfur compounds in the coal substance, pyrite or marcasite (FeS_2), and inorganic sulfates. If suitable corrections are made for the carbon, hydrogen, and sulfur derived from the inorganic material and for conversion of ash to mineral matter, the ultimate analysis represents the elemental composition of the organic coal substance in terms of carbon, hydrogen, nitrogen, sulfur, and oxygen.

CARBON AND HYDROGEN

Determination

All methods for determining carbon and hydrogen are based on the principle whereby a weighed sample is burned in a closed system and the products of combustion fixed in an absorption train after removal of interfering substances. The most widely-used methods based on this principle are the classic Liebig procedure, adopted for a variety of organic materials, and what generally has become known as the Sheffield high-temperature combustion method.

In the Liebig method, adopted in many national standards, the coal is burned in a stream of oxygen at 800 to 900°C and the products of combustion passed over heated cupric oxide, which insures complete conversion of all of the carbon to carbon dioxide and all of the hydrogen to water. Oxides of sulfur and chlorine are removed by passing the products of combustion over heated lead chromate and silver gauze. The carbon and hydrogen are calculated from the increase in weight of the absorbants used to collect the water and carbon dioxide.

Edwards[86] and Mendoza[87] reported that during combustion of coal in oxygen at 800°C some nitric oxide is formed which may partially oxidize to nitrogen dioxide. As a result, the nitrogen dioxide is absorbed and weighed with the water and carbon dioxide. Although the small error in the hydrogen determination can be ignored, the figure for the carbon can be high about 0.2%, unless the nitrogen dioxide is first removed. Removal may be accomplished by absorption in manganese dioxide.[88] While Mott and Wilkinson[89] reported that oxides of nitrogen are negligible in amount, they claimed that the manganese dioxide absorber is valuable in that it promises a factor of safety by removing the last traces of oxides of sulfur that may have escaped the lead chromate packing after prolonged use.

Although the Liebig method on a macro-scale using 0.2 to 0.5-g samples is often employed, microanalytical techniques developed by Pregl for organic compounds are the basis of methods for determination of carbon and hydrogen in coal. The advantages claimed for such methods are that they are as accurate and more rapid than the macro-methods. Belcher and Smith[90] and Titov[91] have developed methods based on the Pregl technique using samples of about 50 mg. A similar procedure is provided as an alternative to the macro-method in the German standard.[92]

The Sheffield high-temperature combustion method was devised by Belcher and Spooner.[93] The method in its original form consisted of a gravimetric procedure for the simultaneous determination of carbon, hydrogen, sulfur, and chlorine. The coal was heated to 1350°C in an unpacked tube through which a fast stream of oxygen was passed. Combustion was completed in 10 min. The water and carbon dioxide were taken up in suitable absorbants and the sulfur and chlorine retained on a roll of silver gauze. At the end of the test, the gauze was washed with water to remove silver sulfate and with ammonia to remove silver chloride. The sulfur and chloride were calculated from the determination of the sulfate and chloride ions in the washing. Because sulfur and chlorine can be determined by simpler and more accurate methods, Mott and Wilkinson[94] modified the method for determining carbon and hydrogen by making certain improvements in design of the equipment. The important claim for the method is that it is simple in operation and much quicker than the Liebig method. The accuracy of both methods have been reported to be about the same[95] and are alternative standards used in Great Britain.[96]

NITROGEN

Determination

As van Krevelen and Schuyer[97] have pointed out, scarcely any information is known of the mode of occurrence of nitrogen in coal except that it all exists as part of the organic coal substance. The best known methods for determining nitrogen in solid fuels whereby the nitrogen is released from the organic coal substance are based on three principles—decomposition, oxidation, and reduction.[98] The decomposition principle was introduced by Kjeldahl,[99] in 1883, who reported that many nitrogen-containing compounds are attacked with boiling concentrated H_2SO_4, with liberation of nitrogen as ammonium sulfate. Dumas introduced the oxidation method in which a mixture of fuel and copper oxide are heated in an inert atmosphere to form carbon dioxide,

[86] Edwards, A. H., *Fuel*, **37**, 415–20 (1958).

[87] Mendoza, M. P., *Fuel*, **37**, 409–14 (1958).

[88] Belcher, R., and Ingram, G., *Anal. Chim. Acta*, **4**, 401–10 (1950).

[89] Mott, R. A., and Wilkinson, H. C., *Fuel*, **37**, 62–70 (1958).

[90] Belcher, R., and Smith, F., *Fuel*, **19**, 183–6 (1940).

[91] Titov, E. M., *Zavodskaya Lab.*, **13**, 308–14 (1947).

[92] Deutscher Normenauschuss, DIN 51721, 1950.

[93] Belcher, R., and Spooner, C. E., *Fuel*, **20**, 130–4 (1941).

[94] Mott, R. A., and Wilkinson, H. C., *Fuel*, **34**, 169–84 (1955).

[95] Edgcombe, L. J., *Fuel*, **34**, 185–90 (1955).

[96] Brit. Standards Inst., Brit. Standard 1016, Part 6 (1958).

[97] van Krevelen, D. W., and Schuyer, J., *Coal Science*, Elsevier, Amsterdam, 1957, p. 216.

[98] Gonzalez Sanchez, F., and Isobel Garcia, R., *Combustibles (Zaragoza)*, **14**, 249–65 (1954).

[99] Kjeldahl, J., *Z. anal. Chem.*, **22**, 366–82 (1883).

water, and nitrogen. The carbon dioxide is absorbed in a suitable reagent, the water condensed, and the nitrogen measured volumetrically. The water-gasification method consists in reducing the coal in an atmosphere of steam to form ammonia, the amount of which is determined titrimetrically.

Kjeldahl Method

The Kjeldahl method is the most widely used method for determining nitrogen in solid fuels. Wilfarth,[100] in 1885, found that various metallic compounds, especially mercuric oxide, added as catalysts, decreased the digestion period required. In 1889, Gunning[101] reported that the addition of a neutral salt to raise the boiling point is an effective means for increasing the rate of digestion. Of the various salts, potassium sulfate is one of the most effective and is almost universally used. Because of the early modifications, the method is often referred to as the Kjeldahl-Wilfarth-Gunning method.

The fact that the necessary conditions for reduction of the nitrogen to ammonia must coexist with oxidizing conditions for the remainder of the decomposed organic molecule is an indication of the narrow oxidation-reduction range in which the decomposition must be carried out. For example, at no time must the composition of the mixture of sulfuric acid and potassium sulfate approach that of potassium acid sulfate, or ammonia will be lost. Also, the use of oxidizing agents such as $KMnO_4$ to speed up the digestion must be used judiciously so that ammonia is not oxidized to nitric acid. Only when a sufficient quantity of other reducing agents are present, which will remove oxidizing agents preferentially, as compared with the removal of ammonia or other reduced nitrogen, is the use of an oxidizing agent considered safe.

The macro-Kjeldahl method is used in many laboratories[102] and is the national standard in many countries.[103] Essentially, the method consists in boiling 1 g of the pulverized sample with 30 ml of concentrated H_2SO_4 containing 7 to 10 g of K_2SO_4 and a suitable catalyst. The catalyst usually is mercury or a mercury salt, selenium or one of its compounds, or a mixture of the two. The use of selenium is considered to be advantageous in reducing the digestion time.[104] The boiling is continued about 2 hr after the solution reaches a straw-colored stage. The total time of digestion may be as much as 3 to 4 hr except for coke and anthracite, which may require as long as 12 to 16 hr. Chromic oxide or potassium permanganate is helpful in reducing the digestion period for highly carbonaceous materials. After cooling, the solution is diluted with cold water, K_2S solution is added to precipitate the mercury, if used, and finally the solution is made distinctly alkaline with strong caustic solution. The flask is then connected to a condensing apparatus and the ammonia distilled into a boric acid solution or a measured amount of a standard H_2SO_4 solution. The ammonia content is then determined titrimetrically.

During the past 20 years, the Kjeldahl semimicro-method has been introduced, particularly in European coal laboratories. The claim for the method is that the determination is speeded up significantly with no loss of accuracy. It is one of the British, French, and German alternative standards.[105] In the method as originally developed by Beet and Belcher,[106] 0.1 g of coal was digested with 3 ml of concentrated sulfuric acid containing 1 g of a catalyst mixture composed of 32 parts K_2SO_4, 5 parts $HgSO_4$, and 1 part Se. Digestion of coals was completed in 30 min. A sodium sulfide-hydroxide solution was then added and the ammonia distilled with steam. The distillate containing the ammonia was absorbed

[100] Wilfarth, H., *Z. anal. Chem.*, **24**, 454 (1885).

[101] Gunning, J. W., *Z. anal. Chem.*, **28**, 188–91 (1889).

[102] Fieldner, A. C., and Taylor, C. D., *U.S. Bur. Mines, Tech. Paper* No. 64 (1915), 22 pp. Fieldner, A. C., and Selvig, W. A., *U.S. Bur. Mines, Bull.* **492**, (1951), 51 pp. *Gr. Brit. Sci. Ind. Research, Fuel Research Paper* No. 44 (1940), 85 pp. *Inst. nacl. carbon, (Oviedo, Spain), Bol. inform.*, **5**, 226–7 (1956). *Inst. nacl. combustible, Combustibles (Zaragoza)*, **17**, 188–9 (1957).

[103] Am. Soc. Testing Materials, D271–58. Assoc. Française de Normalisation, M 03–018 (1954). Japanese Standards Assoc., JIS M 8802 (1954). Brit. Standards Inst., Brit. Standard 1016, Part 6 (1958).

[104] Crossley, H. E., *J. Soc. Chem. Ind.*, **51**, 237–8T (1932); **54**, 367–9T (1935). Carlile, J. H. G., *J. Soc. Chem. Ind.*, **52**, 306–8T (1933). Beet, A. E., *Fuel*, **13**, 343–5 (1934).

[105] Assoc. Française de Normalisation, M 03–018 (1954). Deutscher Normenauschuss, DIN 51722 (1954). Brit. Standards Inst., Brit. Standard 1016, Part 6 (1958).

[106] Beet, A. E., and Belcher, R., *Fuel*, **17**, 53–55 (1938); *Mikrochemie*, **24**, 145–8 (1938).

in a solution of boric acid, and the ammonia determined titrimetrically with a standard solution of $0.01N$ H_2SO_4. Other catalysts have been proposed. Badami and Whitaker[107] suggested either MnO_2 or $KMnO_4$; Beet,[108] $KMnO_4$; Lange and Winzen,[109] a mixture of 1.5 g of MnO_2 and 1.5 g of the mixed catalyst suggested by Beet and Belcher, and, Edwards[110] —$HgSO_4$ plus Se or V_2O_5 plus Se. Mott and Wilkinson[111] and Belcher and Bhatty[112] concluded that for coke, the use of Se as a catalyst leads to loss of nitrogen during the long digestion period required.

Dumas Method

Although the classical Dumas method has been used for many years, various modifications have been made to increase the accuracy and precision.[113]

A modification by Radmacher and Lange[114] of Unterzaucher's[115] version of the Dumas method is one of the alternative methods for nitrogen in the German standard.[116] According to the method, 40 to 50 mg of sample mixed with copper oxide is heated in a stream of moist carbon dioxide and oxygen at 950 to 1000°C. Any unburned volatile components are oxidized in contact with platinum. Excess oxygen is removed by copper, and the nitrogen is collected in a nitrometer tube over a solution of potassium hydroxide.

[107] Badami, G. N., and Whitaker, J. W., *Fuel*, **30**, 211–12 (1951).

[108] Beet, A. E., *J. Appl. Chem. (London)*, **4**, 373–9 (1954); *Nature*, **175**, 513–14 (1955).

[109] Lange, W., and Winzen, W., *Glückauf*, **89**, 324–5 (1953).

[110] Edwards, A. H., *J. Appl. Chem. (London)*, **4**, 330–40 (1954).

[111] Mott, R. A., and Wilkinson, H. C., *Fuel*, **37**, 151–8 (1958).

[112] Belcher, R., and Bhatty, M. K., *Fuel*, **37**, 159–65 (1958).

[113] Fritsche, W., *Brennstoff-Chem.*, **2**, 365–7 (1921). Baranov, A., and Mott, R. A., *Fuel*, **3**, 31–4, 49–52 (1924). Lambris, G., *Brennstoff-Chem.*, **8**, 69–73, 89–93 (1927). Terres, E., and Kronacher, H. K., *Gas u. Wasserfach*, **73**, 707–11 (1930). Banerjee, N. G., and Saha, N. C., *J. Proc. Inst. Chemists (India)*, **28**, 271–80 (1956).

[114] Radmacher, W., and Lange, W., *Glückauf*, **87**, 739–46 (1951).

[115] Unterzaucher, J., *Chem. Ingr. Tech.*, **22**, 128 (1950).

[116] Deutscher Normenauschuss, DIN 51722 (1954).

Belcher and Bhatty[112] applied Schöniger's[117] modification of the semimicro Dumas method to coal and coke. They reported that although the method was satisfactory for coal, it offered no advantages as far as speed was concerned over the Kjeldahl semimicro-method. For cokes, the method is more rapid than the Kjeldahl for single determinations, but for batch operation, a larger number of samples can be determined in a given time by the Kjeldahl method. Moreover, the Dumas-Schöniger method requires constant attention throughout the analysis, whereas the Kjeldahl does not.

Gasification Method

The gasification method based on early work by Varrentrapp and Will[118] has been brought to its highest stage of development by Mantel and Schreiber.[119] Their modification is one of the three alternative methods in the German standard.[120] Briefly, the method consists of mixing 0.5 g of coal or coke with 0.8 to 0.9 g of a mixture of 2 parts of Eschka mixture, 6 parts of soda lime, and 1 part of MoO_3. The sample then is placed in a porcelain boat, covered with platinum gauze, and inserted in a quartz tube heated to 200 to 250°C. A steam line is then connected, and the furnace raised to about 850°C for coals and 950°C for cokes. The gases pass into a solution of $0.1N$ H_2SO_4 where the ammonia is absorbed. The ammonia then is determined by one of the usual analytical techniques.

Summary

Although the Kjeldahl, Dumas, and water-gasification methods are all used for determining nitrogen in coal and coke, the Kjeldahl method is the standard procedure employed in most laboratories. However, the German school does not accept the Kjeldahl macromethod because, they claim, it is inaccurate. The Kjeldahl semimicro-method finds an important advantage in decreasing the time required for digestion, although in routine

[117] Schöniger, W., *Mikrochemie ver. Mikrochim. Acta*, **39**, 229–33 (1952).

[118] Varrentrapp, F., and Will, H., *Ann. Chem. Pharm.*, **39**, 257–96 (1841).

[119] Mantel, W., and Schreiber, W., *Glückauf*, **74**, 939–45 (1938).

[120] Deutscher Normenauschuss, DIN 51722 (1954).

laboratories where many samples are analyzed, the long digestion time required in the Kjeldahl macro-method is not serious, because the method requires little attention. Some investigators claim the Dumas method to be the most accurate, but it requires trained operators to obtain satisfactory results. It is likely that the gasification method has its best application in the analysis of cokes.

SULFUR

The origin and occurrence of sulfur in coal has been well-documented by Thiessen.[121] As is well known, sulfur occurs in coal in three forms.[122] It is present in organic combination as part of the coal substance,[123] as pyrites or marcasite, and as sulfates. The amount of organic sulfur is normally not over 3%, but in exceptional cases may be as much as 11%.[124] The sulfates, mainly calcium and iron, rarely exceed a few hundredths percent except in highly weathered or oxidized samples. Because pyrite and marcasite, the two common crystal forms of FeS_2, are difficult to distinguish from one another, they usually are designated simply as pyrite. Whelan,[125] in a study of British coals, found that the iron sulfide formed contemporaneously with the coal has the marcasite structure and exists as finely disseminated grains of $\frac{1}{2}$ to 40 μ size and as small concretions heavily impregnated with coal with a range of size between 20 and 100 μ. The iron sulfide in the cleats and other cavities is in the form of pyrites and was considered to be of secondary origin.

[121] Thiessen, G., *Chemistry of Coal Utilization*, John Wiley and Sons, New York, 1945, Vol. 1, pp. 425–30.

[122] Wandless, A. M., *J. Inst. Fuel*, **32**, 258–66 (1959).

[123] Brooks, J. D., *J. Inst. Fuel*, **29**, 82–5 (1956).

[124] Kreulen, D. J. W., *Fuel*, **31**, 462–7 (1952). Chowdhury, J. K., Datta, P. B., and Ghosh, S. R., *J. Sci. Ind. Research* (*India*), **11B**, 146–52 (1952). Hadži, D., and Novak, A., *Vestnik Sloven. kemi. društva*, **1**, 175–84 (1954). Kavčič, R., *Bull. sci., Conseil acad. RPF Yougoslavie*, **2**, 12–13 (1954). Marshall, C. E., and Draycott, A., *Univ. Sydney, Dept. Geol. Geophys. Mem.* **1954**, No. 1, 65 pp. Roy, M. M., *Naturwissenschaften*, **43**, 497 (1956). Breger, I. A., *Bull. Geol. Soc. Am.*, **67**, Part 2, 1675 (1956). Iyengar, M. S., and Lahiri, A., *Fuel*, **37**, 29–35 (1958).

[125] Whelan, P. F., *J. Inst. Fuel*, **27**, 455–8, 464 (1954).

Although the three forms of sulfur mentioned are normally the only ones considered to be present in coal, Yurovskiĭ and coworkers[126] have reported occurrences of elemental sulfur in amounts up to 0.15% in certain coals.

Determination

Because sulfur is an important consideration in utilization of coal and coke for most purposes, much work has been done in developing ways and means for improving the accuracy and precision of its determination and reducing the time required for the analysis. The three most generally accepted methods for its determination are the Eschka, bomb-combustion, and high-temperature combustion methods. The methods are based on the combustion of the sulfur compounds to form sulfate ions that can be determined either gravimetrically or volumetrically.

Eschka Method

Although Eschka[127] developed his method nearly a century ago, it continues to be one of the best known and is accepted as a national standard in several countries.[128] It has distinct advantages in that special apparatus is not required, and only simple analytical techniques are employed. Basically, the method consists of incinerating an intimate mixture of coal or coke with a mixture of 1 part sodium carbonate and 2 parts of light calcined magnesium carbonate. The sulfur compounds evolved during combustion react with the sodium carbonate and under oxidizing conditions are retained as sodium sulfate. The light magnesium carbonate reduces the bulk density and allows access of air to the mass. The sulfate in the residue is extracted in either alkaline or acid solution and determined gravimetrically as barium sulfate or by volumetric methods.

The method has been studied extensively,

[126] Yurovskiĭ, A. Z., Kaminskiĭ, V. S., and Rubinshteĭn, A. L., *Khim. i Tekhnol. Topliva i Masel*, **1957**, No. 7, 20–3.

[127] Eschka, A., *Ost. Z. Berg. u. Huttenw.*, **22**, 11–13 (1874); *Dinglers Polytech. J.*, **212**, 403–4 (1874).

[128] Assoc. Française Normalisation, M 03–008 (1944). Am. Soc. Testing Materials, D 271–58. Deutscher Normenauschuss, DIN 51724 (1955). Japanese Standards Assoc., JIS M 8802 (1954). Brit. Standards Inst., Brit. Standard 1016, Part 6 (1958).

particularly in studies of refinements, to speed up the incineration and to make the method applicable to the coals varying widely in sulfur and volatile matter contents.[129] The work has shown that, with suitable heating rates and adequately ventilated electric muffles, loss of sulfur during incineration does not occur. In addition, by supplying enough air, the sulfur is completely oxidized to sulfate, so that use of bromine to oxidize sulfites is not required.[130]

In development of an international standard, a working group of Technical Committee 27 on Solid Mineral Fuels, International Organization for Standardization, made a comprehensive investigation of the method.[131] They concluded that it is the simplest and most generally applicable for the determination of total sulfur in solid fuels. If proper precautions are taken, the method is entirely satisfactory for coals containing high amounts of sulfur, volatile matter, and barium salts. Their recommendations included the following:

1. Heat the coal in a ventilated electric-muffle furnace from room temperature to 800 ± 25°C and hold it at this temperature for $1\frac{1}{2}$ hr.
2. Extract the incinerated residue in either alkaline or acid solution.
3. Bromine water is not necessary.
4. Determine the sulfate ion gravimetrically, as barium sulfate, or volumetrically, with barium chromate.

[129] Simpkin, N., and Jones, E. H., *J. Inst. Fuel*, **12,** 168–70 (1939). Mantel, W., and Schreiber, W., *Glückauf*, **78,** 185–90, 220–23 (1942). Guérin, H., and Moreau, M., *Chim. & ind.* (*Paris*), **60,** 339 (1948). Hamaker, J., *Chem. Weekblad*, **44,** 738–41 (1948); **45,** 364–7 (1949). Teichmann, R. F. J., Bird, C. E., and van der Merwe, J. W., *S. African Ind. Chemist*, **5,** 112–14 (1951). Lissner, A., and Göbel, W., *Bergakademie*, *Freiberg. Forschungsh. A.*, *Bergbau* No. 14, 31–4 (1952). Mott, R. A., and Wilkinson, H. C., *J. Appl. Chem.* (*London*), **3,** 218–23 (1953). Rodríguez Pire, L., and Corrales Zarauza, J. A., *Inst. nacl. carbon* (*Oviedo, Spain*), *Bol. inform.*, **3,** No. 18, 13–21 (1954). Gómez Aranda, V., and Auria Arbuniés, J., *Combustibles* (*Zaragoza*), **14,** 72–9, 124–5 (1954). Mott, R. A., Ruell, D. A., and Wilkinson, H. C., *Fuel*, **34,** 78–94 (1955).

[130] Battye, T., Johnson, R., and Wilkinson, H. C., *J. Appl. Chem.* (*London*), **4,** 119–23 (1954). Marrison, F. C., and Mott, R. A., *Chem. & Ind.* (*London*), **1956,** 715–16.

[131] Jackson, P. J., *Fuel*, **35,** 212–29 (1956).

Bomb-Combustion Methods

Bomb-combustion methods have long been used for determining total sulfur in solid fuels and are specified as an alternative method in several national standards.[132] The method is favored by many analysts because of its speed and the assurance no sulfur is lost during the combustion. It is especially favored when the determination of calorific value is required as well as sulfur.

The two methods of decomposing the sample in the bomb are the Parr fusion procedure with sodium peroxide and combustion in oxygen at pressures of 20 to 30 atmospheres. The Parr method gives results in agreement with the Eschka method but requires special apparatus.

The oxygen bomb-combustion method is objected to from time to time, because it is found with certain coals that a significant proportion of the sulfur is retained in the fused cenospheres of ash.[133] To obtain correct results, the sulfur retained in the ash must be extracted and added to the bomb washings. To overcome this problem, Strambi[134] introduced a modified method in which 1 g of coal was mixed with Eschka mixture and sucrose and burned in a calorimeter bomb under a pressure of 25 atmospheres of oxygen. Sucrose was added to insure complete combustion of fuels difficult to ignite. The sulfur oxides formed were absorbed in a solution of sodium carbonate in the bomb and the sulfate ion determined gravimetrically. Strambi reported that the method gave higher results for high-sulfur coals than the Eschka—and with better precision. The method also is reported to be more rapid than the Eschka when only a small number of samples is to be analyzed. Davies and Jackson[135] found that partial fusion of the solid residue occurred, which prevented extraction of all of the sulfur; they proposed the use of acid to decompose the residue. Moszynski[136] proposed the substitution of

[132] Am. Soc. Testing Materials, D271–58. Assoc. Française de Normalisation, M 03 006.

[133] Guérin, H., and Moreau, M., *Chim. & ind.* (*Paris*), **60,** 339 (1948).

[134] Strambi, E., *Calore*, **16,** 61–5 (1943); **24,** 111–18 (1954).

[135] Davies, I., and Jackson, P. J., *J. Appl. Chem.* (*London*), **1,** S4–7 (Supplemental Issue No. 1) (1951).

[136] Moszynski, Z. K. A., *J. Appl. Chem.* (*London*), **5,** 467–70 (1955).

ammonium carbonate for the sodium carbonate in order to increase the removal of sulfur compounds retained in the gas phase. A comprehensive investigation by a working group of ISO Technical Committee 27[137] showed that with suitable modification, the Strambi method is applicable to all types of solid fuel and is a satisfactory alternative to the better-known Eschka method. The discrepancy between certain results was traced to the use of Eschka mixtures of different bulk densities. The method finally recommended is not affected by the presence of large quantities of barium in the sample. The titrimetric finish with barium chromate also can be applied.

High-Temperature Combustion

During the past twenty years, a high-temperature combustion method based largely on work by Seuthe[138] and by Beet and Belcher[139] has been adopted in many laboratories, particularly in Europe, and is an alternative German and British standard method.[140] Considerable work has been reported on refinements in the method and comparisons with other accepted methods.[141]

One of the particular advantages claimed for the method is that chlorine can be determined at the same time. Briefly, the method consists in heating a 0.5-g sample of coal in a boat placed in an electric tube furnace heated to 1250 or 1350°C. The coal is usually covered with kaolin, ferric phosphate, or aluminum oxide to assist in the decomposition of the sulfate and prevent its retention in the ash.[142] Oxygen is passed through the furnace, and the oxides of sulfur are absorbed in hydrogen peroxide, which oxidizes any SO_2 to SO_3. At the end of the test, the solution is titrated with $0.05N$ sodium borate solution, which gives the total acidity due to the hydrochloric and sulfuric acids formed. The chloride ion is then determined by adding mercuric oxycyanide solution to convert the sodium chloride to sodium hydroxide according to the equation:

$$NaCl + Hg(OH)CN \rightarrow HgClCN + NaOH$$

The liberated alkali then is titrated with $0.025N$ sulfuric acid. The calculations for obtaining the percentage of sulfur and chlorine are as follows: If a = ml of $0.05N$ sodium borate used in the determination, a_1 = ml of $0.05N$ sodium borate used in the blank, b = ml of $0.025N$ sulfuric acid used in the determination, b_1 = ml of $0.025N$ sulfuric acid used in the blank, and w = weight of coal in grams, the percentage of sulfur

$$= \frac{0.08}{w}\left[(a - a_1) - \frac{(b - b_1)}{2}\right]$$

and percentage of chlorine

$$= \frac{0.08875(b - b_1)}{w}$$

Other Methods

Reductive fusion with various metals has been employed for converting the sulfur compounds in coal finally to hydrogen sulfide and then using usual analytical techniques for estimating the sulfur from the H_2S evolved. The metals employed include lithium,[143] zinc,[144] calcium,[145] and aluminum.[146] Lange

[137] Jackson, P. J., *Fuel*, **36**, 26–38 (1957).

[138] Seuthe, A., *Arch. Eisenhüttenw.*, **11**, 343–4 (1938); *Glückauf*, **75**, 409–11, 909 (1939).

[139] Beet, A. E., and Belcher, R., *Fuel*, **19**, 42–5 (1940).

[140] Deutscher Normenauschuss: DIN 51724 (1955). Brit. Standards Inst., Brit. Standard 1016, Part 6 (1958).

[141] Hambersin, J. M., *Ind. chim. belge*, **10**, 181–7 (1939). Pieters, H. A. J., and Popelier, C., *Chem. Weekblad*, **41**, 27–33 (1945). Hamaker, J., *Chem. Weekblad*, **45**, 364–7 (1949). Barton, K., and Střižová, S., *Paliva*, **30**, 73–6 (1950). Horikawa, S., and Sato, T., *Nenryô Kyôkaishi*, **30**, 355–61 (1951). Sventy, A., and Linhart, K., *Paliva*, **32**, 166–8 (1952). Fedoseev, P. N., and Lagoshnaya, R. M., *J. Anal. Chem. U.S.S.R.*, **9**, 39–43 (1954). Pashenko, M., and Desvatova, I., *Novosti Neftyanoĭ Tekh. Neftepererabotka*, **5**, 10–14 (1954). Belcher, R., and Spooner, C. E., *Fuel*, **34**, 164–8 (1955). Mott, R. A., and Wilkinson, H. C., *Fuel*, **35**, 6–18 (1956); *Brennstoff-Chem.*, **38**, 212–5 (1957). Schuhknecht, W., and Kunz, H., *Brennstoff-Chem.*, **37**, 78–84 (1956); **38**, 313 (1957). Zörner, A., *Arch. Eisenhüttenw.*, **29**, 23 (1958).

[142] Mott, R. A., and Wilkinson, H. C., *loc. cit.* ref. 141.

[143] Radmacher, W., and Mohrhauer, P., *Z. anal. Chem.*, **141**, 419–29 (1954). Lange, W., and Mohrhauer, P., *Brennstoff-Chem.*, **39**, 55–8 (1958).

[144] Kurchatov, M. S., *Bull. inst. chim. acad. Bulgare sci.*, **2**, 221–47 (1953); *Compt. rend. acad. bulgare sci.*, **7**, No. 3, 41–4 (1954); *Zavodskaya Lab.*, **23**, 1294–6 (1957).

[145] Chernyi, A. T., and Podoinikova, K. V., *Zavodskaya Lab.*, **15**, 1002–3 (1949).

[146] Jantsch, G., Zemek, F., and Zemek, O., *Monatsh. Chem.*, **85**, 38–46 (1954).

and Mohrhauer reported that fusion with lithium gave comparable results with the Eschka and the high-temperature combustion methods.

Several investigations have suggested volumetric methods for determinations of the sulfate ion. Although they are considered to be rapid when only a few determinations are required, for mass-production work they do not show much improvement for increasing the through-put of samples. Radmacher and Mohrhauer[147] used $BaCrO_4$ to precipitate the $BaSO_4$, with the excess $BaCrO_4$ determined volumetrically. Elphick and Gerson[148] described a method for estimating sulfur in coal by conductometric titration of the washings from a bomb calorimeter. Wilkinson[149] precipitated the sulfate with 4-amino-4′-chlorodiphenyl hydrochloride and titrated a suspension of the precipitate with standard NaOH. Majumdar and Banerjee[150] suggested a method in which the barium sulfate precipitate is dissolved in an excess of a standard solution of Versenate and the excess of Versenate titrated against a standard solution of magnesium chloride. Gordiyeviskiĭ and Radnayeva[151] replaced the final gravimetric step in the Eschka method by amperometric titration with $Pb(NO_3)_2$.

Forms of Sulfur

In most schemes of coal analysis for forms of sulfur—sulfate, pyritic, and organic—the sulfate and pyritic sulfur are determined directly, and the organic sulfur is taken as the difference between the total sulfur and the sum of the sulfate and pyritic sulfur.

The methods in most general use are based on the well-known Powell and Parr procedure developed about thirty years ago. The British Standard,[152] Fuel Research Board,[153] U.S. Bureau of Mines,[154] and Spanish[155] methods are based on this work.

The method of Powell and Parr is time-consuming, with prolonged digestions with hydrochloric and nitric acids, although Powell pointed out that hot digestion with HCl for the determination of sulfate would speed up the determination. Teichmann[156] reported that prolonged extractions with dilute hydrochloric acid resulted in some oxidation and solution of the pyrite. He suggested the use of strong acid for a shorter period of 1 hr.

Mott[157] advocated that both hydrochloric and nitric acid extractions be carried out at boiling temperatures, which insures that the hydrochloric acid will not attack pyrite, and the nitric acid will not attack the organic sulfur. Also, the digestions at boiling temperatures for only 30 min result in an important saving of time.

Edwards, Daybell, and Pringle,[158] in a comprehensive investigation of methods for determining forms of sulfur, confirmed that the British Standard and the nonsimultaneous procedure of the Fuel Research Board give practically identical results, but both are liable to a slight error affecting both sulfate and pyritic sulfur. The British Coke Research Association method is not so subject to oxidation effects and is recommended where highest accuracy is required. The conclusion of Teichmann was confirmed that some oxidation of pyrite takes place by prolonged hydrochloric acid digestion for determining nonpyritic iron.

Radmacher and Mohrhauer[159] used a different approach to the determination of pyritic sulfur by basing their method on reduction of pyrite by nascent hydrogen and estimation of hydrogen sulfide produced. The pyrite was reduced to H_2S by treatment in an acid solution containing granulated zinc and chromium

[147] Radmacher, W., *Brennstoff-Chem.*, **33**, 129–35 (1952). Radmacher, W., and Mohrhauer, P., *Brennstoff-Chem.*, **34**, 33–7 (1953).

[148] Elphick, J. O., and Gerson, T., *Fuel*, **31**, 438–44 (1952).

[149] Wilkinson, H. C., *Analyst*, **81**, 9–11 (1956).

[150] Majumdar, S. K., and Banerjee, N. G., *J. Proc. Inst. Chemists (India)*, **29**, 213–15 (1957).

[151] Gordiyeviskiĭ, A. V., and Radnayeva, E. E., *Koks i Khim.*, **7**, 14–15 (1958).

[152] Brit. Standards Assoc., Brit. Standard 1016 (1942).

[153] Dept. Sci. Ind. Research, *Fuel Research Paper* No. 44 (1940), 85 pp.

[154] Fieldner, A. C., and Selvig, W. A., *U.S. Bur. Mines, Bull.* **492** (1951), 16–17.

[155] Inst. Nacl. Combustible, *Combustibles (Zaragoza)*, **14**, 181–2 (1954).

[156] Teichmann, R. F. J., *S. African Mining Eng. J.*, **56**, Part 2, 99–103 (1945).

[157] Mott, R. A., *Fuel*, **29**, 53–61 (1950); *Coke and Gas*, **12**, 369–70 (1950); *Gas World*, **132**, No. 3437, Coking Sect. 66–7 (1950).

[158] Edwards, A. H., Daybell, G. N., and Pringle, W. J. S., *Fuel*, **37**, 47–61 (1958).

[159] Radmacher, W., and Mohrhauer, P., *Glückauf*, **89**, 503–11 (1953).

powder. The method requires that the coal be crushed finely so that all the pyrite can be reached for attack by the hydrogen.

OXYGEN

Because of the lack of a simple, direct method for determining oxygen in coal, it usually is estimated by subtracting the sum of carbon, hydrogen, nitrogen, sulfur, and ash from 100. The disadvantage of such an indirect method is that all of the errors of the other determinations are placed on the oxygen, and a material balance cannot be obtained to aid in checking the accuracy of the ultimate analysis.[160]

Determination

Attempts have been made for many years to develop satisfactory direct methods, critical reviews of which have been made by several investigators.[161] Proposed methods may be divided into groups, as follows:

1. Oxidation methods
    - (a) Solid oxidizing agents
    - (b) Liquid oxidizing agents
    - (c) Gaseous oxygen
2. Reduction methods
    - (a) Gaseous hydrogen
    - (b) Carbon

Oxidation Methods

The only oxidation method that has been applied to coal is the one involving the use of gaseous oxygen. Dolch and Will[162] and Kirner[163] burned a known weight of coal with a definite quantity of oxygen from which the oxygen of the coal was determined by deducting the added quantity of oxygen from the sum of the residual oxygen and the oxygen of the oxidation products. The disadvantages of the method are the difficulties associated with the accurate measurement of the large volume of oxygen used and the amount and composition of the combustion products.

Whitaker, Chakravorty, and Ghosh,[164] in a different approach, determined oxygen by combustion of a known weight of coal in a bomb and measuring the drop in pressure due to consumption of oxygen by the "available hydrogen" and the subsequent condensation of the water so formed. The amount of oxygen is 7.94 times the difference between total hydrogen and available hydrogen. The agreement of the results by this method and that determined by difference is reported to be within 1.0 and often within 0.5 percentage points.

Reduction with Hydrogen

The determination of oxygen by reduction of the coal with hydrogen, although first suggested in 1863, was developed further by ter Meulen,[165] and his name is usually associated with the method. According to the method, the coal is pyrolyzed in a stream of hydrogen, and all of the oxygen is catalytically converted to water. Among the more recent investigations of the method are those reported by Wildenstein[166] and Guérin and Marcel,[167] who made several modifications to improve the accuracy. The criticisms of the method are that the apparatus is complicated, sulfur and chlorine may poison the catalyst, and the large factor (0.888) for computing the oxygen is not favorable.[168]

Reduction with Carbon

The most widely used procedure is what generally has become known as the Schütze-Unterzaucher method.[169] Essentially, the method consists of pyrolyzing the coal in a stream of nitrogen and passing the volatilized products over carbon at about 1100°C, which converts all of the oxygen in the volatile products to carbon monoxide. The carbon monoxide then is determined using a variety of techniques.

[160] Casamento, P., and Engles, O. G., *Australian J. Appl. Sci.*, **8,** 55–61 (1957).

[161] Spooner, C. E., *Fuel*, **26,** 15–23 (1947). Schöberl, H., *Brennstoff-Chem.*, **33,** 241–4 (1952). Siniramed, C., Manci, C., and Berti, V., *Riv. combustibili*, **9,** 607–21 (1955).

[162] Dolch, M., and Will, H., *Brennstoff-Chem.*, **12,** 141–6, 166–9 (1931).

[163] Kirner, W. R., *Ind. Eng. Chem., Anal. Ed.*, **8,** 57–61 (1936).

[164] Whitaker, J. W., Chakravorty, R. N., and Ghosh, A. K., *J. Sci. Ind. Research (India)*, **15B,** 72–7 (1956).

[165] ter Meulen, H., *Rec. trav. chim.*, **43,** 899–904 (1924); **53,** 118–25 (1934).

[166] Wildenstein, R., *Chaleur & ind.*, **32,** 293–8 (1951).

[167] Guérin, H., and Marcel, P., *Bull. soc. chim. France*, **1952,** 310–11.

[168] Spooner, C. E., *Fuel*, **26,** 16 (1947).

[169] Schütze, M., *Z. anal. Chem.*, 118, 245–58 (1939). Unterzaucher, J., *Ber. deut. chem. Ges.*, **73B,** 391–404 (1940).

Work has been done by many workers to improve the accuracy and precision of the method,[170] and it is finding increased use.

One of the main difficulties in the method is the errors caused by the minerals in the coal,[171] especially the carbonates and water of hydration, which add oxygen to the determined amount. Radmacher and Hoverath[172] have overcome the problem by first demineralizing the coal by treatment with hydrofluoric and hydrochloric acids.

CHLORINE

The determination of chlorine normally is not considered in the ultimate analysis of coal, because its amount is usually only a few tenths of a percent or less. However, its determination is becoming of increasing importance, particularly in connection with problems in coal combustion processes. Methods for liberating the chlorine in a form suitable for its analytical determination include bomb combustion, high-temperature combustion, and heating with Eschka mixture in an oxidizing atmosphere. All three methods are alternative German standards;[173] the high-temperature combustion and Eschka methods are the basis of the British standard.[174]

[170] Spooner, C. E., *Fuel*, **26**, 15–23 (1947). Georgiadis, C., *Compt. rend. congr. ind. gaz*, 66*th Congr., Lyon*, **1949**, 272–7. Deinum, H. W., and Schouten, A., *Anal. Chim. Acta*, **4**, 286–99 (1950). Bastick, M., *Bull. soc. chim. France*, **1952**, 308–9. Guérin, H., Bastick, M., and Marcel, P., *Chaleur & ind.*, **33**, 177–80 (1952). Blanzat, A., and Barbe, L., *Bull. soc. chim. France*, **1952**, 569–73. Wnękowska, L., and Lesniak, J., *Prace Glównigo Inst. Górnictwa, Komun.* No. 132 (1952), 15 pp. Bastick-Moreau, M., *Ann. mines*, **143**, 3–44 (1954). Deinum, H. W., *Brennstoff-Chem.*, **35**, 334–7 (1954). Romováčková, H., and Kessler, M. F., *Paliva*, **35**, 351–6 (1955). Guérin, H., and Bastick, M., *Compt. rend. congr. ind. gaz*, 72*d Congr. Strasbourg*, **1955**, 254–66. Drekopf, K., and Braukmann, B., *Brennstoff-Chem.*, **36**, 203–8 (1955). King, J. G., *Coke and Gas*, **18**, 31–2 (1956). Siniramed, C., and Manci, C., *Riv. combustibili*, **11**, 366–79 (1957).

[171] Schöberl, H., *Brennstoff-Chem.*, **33**, 241–4 (1952).

[172] Radmacher, W., and Hoverath, A., *Brennstoff-Chem.*, **40**, 97–104 (1959).

[173] Deutscher Normenauschuss: DIN 51727 (1955).

[174] British Standards Inst., Brit. Standard 1016, Part 8 (1959).

Determination

Bomb-Combustion Method

In the determination of calorific value of coal by the calorimetric bomb method, all of the chlorine is released and is found combined as chlorides in the filtrate from the bomb washings.[175] It then can be determined readily by usual analytical techniques for this element. If the determination of calorific value is not required, precautions may be made to minimize ejection of ash from the crucible by mixing the coal with Eschka mixture. In the German standard, 1 g of coal is mixed with 0.8 g Eschka mixture and 0.1 g benzoic acid. Use of ammonium carbonate solution instead of water in the bomb also promotes rapid removal of the acid gases from the vapor phase.

High-temperature Combustion Method

In the high-temperature combustion method, a 0.5- to 1.0-g sample of coal is burned in a stream of oxygen in a tube furnace at 1250 to 1350°C, and the acid gases are absorbed in neutral hydrogen peroxide to form sulfuric and hydrochloric acids. The solution then is neutralized with sodium borate and the chlorine determined by titration of the sodium hydroxide formed after addition of mercuric oxycyanide. The advantage of the method is that chlorine and sulfur can be determined simultaneously, as mentioned in the section on determination of sulfur.

Eschka Method

In the Eschka method, a 1- to 2-g sample of coal is mixed with about 3 g of Eschka mixture and covered with an additional 2 g. The crucible containing the coal and Eschka mixture is heated in an oxidizing atmosphere for 2 hr at 650 to 700°C to destroy the organic material and convert all of the chlorine to chlorides. The chloride in the residue then can be extracted and determined by one of the usual analytical techniques.

MINERAL MATTER

Occurrence and Origin

The so-called "mineral matter" or inorganic constituents of coal generally are considered

[175] Selvig, W. A., and Gibson, F. H., *Ind. Eng. Chem., Anal. Ed.*, **5**, 189–91 (1933).

to be the sum of all elements in coal that are not part of the organic coal substance (carbon, hydrogen, nitrogen, oxygen, and sulfur). The mineral matter includes the minerals associated with the coal and, in addition, the chemically bound elements in the organic portion of the coal other than C, H, N, O, and S.

Francis[176] grouped the inorganic constituents of coal into two classes. One class consisted of the mineral constituents derived from the coal-forming plants and not completely lost during degradation. The other consisted of the constituents added to the coal-forming deposit after the death of the plant. The two classes are sometimes known as "inherent" and "extraneous" mineral matter, respectively, although there is no general agreement on the use of the terms. Some authorities also include with the inherent mineral matter both finely-divided inseparable minerals and those ash-forming, nonvolatile elements that are organically combined, complexed, or base-exchanged. As Deul[177] has pointed out, many elements become combined with the organic matter during degradation, compaction and burial, and weathering and are properly a part of the inherent mineral matter of coal. The inclusion of both the inorganic constituents of the original plant material, the elements that become combined with the coal, and the inseparable minerals, is a useful classification when applied to practical problems of coal cleaning. It defines the lower limit to which coal can be separated physically from its mineral impurities.

Parks[178] listed the modes of occurrence of mineral matter in coal, as follows: (1) rare and trace elements; (2) disseminated mineral matter; (3) detrital mineral matter; (4) crystalline mineral matter in fractures, veinlets, petrifications, and nodular masses; and (5) banded impurities. The main source of the rare and trace elements is the living plant and the enrichment resulting from combination of these elements with the organic matter during and following decay. Deul[177] found that the ash intimately associated with the organic part of coals, compared with original coal ash and with ash of separable minerals, is always enriched in B, Be, Fe, Na, Sc, U; usually enriched in As, Ce, Ge, Ti, Zr; occasionally enriched in Co, Ni, Mn, Mo, Pb, Sn, Sr, Y, Yb; and never enriched in calcium.

Disseminated mineral crystals are commonly found in the anthraxylon and attritus of coal; their size ranges from 10 to 80 μ. The most frequent occurrence is pyrite, which seems to be of primary origin. Pyrite, calcite, and kaolinite are the mineral constituents commonly disseminated in fusain.

The detrital minerals in coal are always intermixed with attritus. They are almost wholly microscopical in size with a range of about 5 to 250 μ. Their source is weathered and eroded rock and derived soil. Although clay minerals are the chief impurity, other minerals include quartz, garnet, tourmaline, feldspar, rutile, and hornblende, to name a few.

Crystalline mineral matter also occurs in fractures, veinlets, petrifications, and nodules. They are of secondary origin, having entered the coal bed in solution followed by their precipitation and filling the fractures with crystalline mineral matter. Frequently they are of three familiar mineral species—calcite, pyrite, and kaolinite.

The banded impurities are interbedded with the coal. They may be thin or they may be thick and rock-like. Clay is the most common substance in banded impurities, consisting mainly of one or more of the three common clay minerals—kaolinite, illite, and montmorillinite.

Identification of Minerals Associated with Coal

Although for many years coal petrographers have relied on optical microscopy for determining the mineral constituents of coal, several of the more recently introduced techniques of mineralogical examination provide a means for obtaining additional information. They include X-ray diffraction,[179]

[176] Francis, W., *Coal—Its Formation and Composition*, Edward Arnold, Ltd., London, 1954, pp. 483–509.

[177] Deul, M., *Econ. Geol.*, **50**, 103–4 (1955).

[178] Parks, B. C., *2nd Conf. on Origin and Constitution of Coal*, Crystal Cliffs, Nova Scotia, 1952, pp. 272–84.

[179] Endell, J., and Endell, K., *Feuerungstech.*, **31**, 137–43 (1943). Mitra, G. B., *J. Sci. Ind. Research (India)*, **9B**, 133–5 (1950); *Fuel*, **33**, 316–30 (1954); *Brennstoff-Chem.*, **35**, 337–8 (1954). Mukherjee, B., and Lahiri, A., *J. Sci. Ind. Research (India)*, **12B**, 341–6 (1953). Schliephake, R., *Bergbau Arch.*, **19**, No. 1, 1–16 (1958).

radiography,[180] thermal and differential thermal analysis, and electron microscopy.[181] The techniques and optical methods involved are described by Gumz, Kirsch, and Mackowsky.[182]

Although many mineral constituents can be identified in the coal mass, enrichment techniques often are employed to assist in identification of the minerals. The minerals may be separated largely from the bulk of the coal by various techniques such as float-and-sink, elutriation, froth flotation, and electrostatic separation of finely crushed coal. As these methods do not entirely recover all of the minerals, other procedures may be employed. Solvent extraction of the organic coal substance would be an excellent procedure; unfortunately, no solvent has yet been found in which all the various petrographic coal components are completely soluble.

A method that is used with considerable success is the removal of the organic coal substance by incineration at low temperatures, leaving most of the mineral substantially unaltered. The technique is based on the work of Nagelschmidt and Hicks.[183] The method, as used by the British Coal Utilisation Research Association,[184] consisted of completely oxidizing the coal at 375°C in a specially designed flask through which a controlled mixture of oxygen and nitrogen was passed. Localized overheating was prevented by rotating the flask during the test. The total time required for complete oxidation was 100 to 250 hr depending on the rank of coal. The method described by Schliephake[179] is based on the same principle, except that the particles of crushed coal were distributed over the surface of a large dish that was heated to 390°C in an ordinary laboratory furnace using air to maintain combustion. The kaolin, carbonate, and chloride minerals, and many of the minor accessory minerals, are unchanged by incineration of the coal at the low temperatures employed. The shale minerals lose their oxonium ions $(H_3O)^+$ between about 50 and 300°C without affecting the basic structural arrangement. The sulfide minerals are oxidized to hematite, and the accessory minerals possessing H_2O-groups, or loosely attached hydroxyl or oxonium ions, are changed to lower hydrates or anhydrous compounds.

Mineral Composition of Coal

Table 2 shows the minerals that may occur in bituminous coals.[185,186] Nelson reported that probably more than 95% of the mineral matter associated with normal marketable coals consists of species belonging to the shale, kaolin, sulfide, and chloride groups.

Gumz, Kirsch, and Mackowsky[187] classified the minerals as to their origin, either as syngenetic or epigenetic. Syngenetic minerals are those introduced into the coal during the biochemical change in forming the coal, and epigenetic minerals are those introduced during the maturing process following the biochemical change. They reported that the carbonate and sulfide minerals and quartz may be of both origins, whereas the clay minerals are of syngenetic origin and the salt minerals of epigenetic origin.

Lissner[188] has reported that, in addition to the minerals commonly associated with bituminous coals, brown coals contain calcium combined with the organic coal substance as humates in the form of dopplerite.

Determination of Mineral Content of Coal

As some of the mineral constituents of coal are intimately bound with the coal substance, it is not possible to completely separate the mineral matter by any physical method in

[180] Nelson, J. B., *Bull. Brit. Coal Utilisation Research Assoc.*, **17,** 45–6 (1953). Macrae, J. C., and Leighton, L. H., *Trans. Inst. Mining Engrs.* (*London*), **112,** Part 6, 500–506 (*Discussion*, 506–15) (1953).

[181] Meldau, R., *Bergbau Arch.*, **11/12,** 155–64 (1950).

[182] Gumz, W., Kirsch, H., and Mackowsky, M. T., *Slackenkunde*, Springer-Verlag, Berlin, Göttingen, Heidelberg, 1958, pp. 105–23.

[183] Nagelschmidt, G., *Med. Research Council* (*Brit.*), Spec. Rept. Ser. No. 244, 95–124 (1943). Hicks, D., and Nagelschmidt, G., *Med. Research Council* (*Brit.*), Spec. Rept. Ser. No. 244, 153–86 (1943).

[184] Nelson, J. B., *BCURA Bull.*, **17,** 45 (1953); *BCURA*, 1955 Ann. Rept., 48–9 (1956).

[185] Nelson, J. B., *BCURA Bull.*, **17,** 43 (1953).

[186] Mackowsky, M. T., *Bergbau Arch.*, **4,** No. 1, 1–16 (1943).

[187] Gumz, W., Kirsch, H., and Mackowsky, M. T., *Op. cit.*, ref. 182, pp. 37–47. Mackowsky, M. T., *Proc. Intern. Comm. for Coal Petrology*, No. 2, 31–4 (1956).

[188] Lissner, A., *Sitz. ber. deut. Akad. Wiss. Berlin, Kl. Chem., Geol., u. Biol.*, **1956,** No. 1, 30 pp.

TABLE 2

MINERALS ASSOCIATED WITH BITUMINOUS COALS[185]

Group	Species	Formula
Shale	Muscovite Hydromuscovite Illite Bravaisite Montmorillonite	$(K, Na, H_3O, Ca)_2 (Al, Mg, Fe, Ti)_4 (Al, Si)_8O_{20} (OH, F)_4$
Kaolin	Kaolinite Livesite Metahalloysite	$Al_2(Si_2O_5) (OH)_4$
Sulfide	Pyrite Marcasite	FeS_2
Carbonate	Ankerite Ankeritic calcite Ankeritic dolomite Ankeritic chalybite	$(Ca, Mg, Fe, Mn) CO_3$
Chloride	Sylvine	KCl
	Halite	NaCl
Accessory Minerals	Quartz	SiO_2
	Feldspar	$(K, Na)_2O{\cdot}Al_2O_3{\cdot}6SiO_2$
	Garnet	$3CaO{\cdot}Al_2O_3{\cdot}3SiO_2$
	Hornblende	$CaO{\cdot}3FeO{\cdot}4SiO_2$
	Gypsum	$CaSO_4{\cdot}2H_2O$
	Apatite	$9CaO{\cdot}3P_2O_5{\cdot}CaF_2$
	Zircon	$ZrSiO_4$
	Epidote	$4CaO{\cdot}3Al_2O_3{\cdot}6SiO_2{\cdot}H_2O$
	Biotite	$K_2O{\cdot}MgO{\cdot}Al_2O_3{\cdot}3SiO_2{\cdot}H_2O$
	Augite	$CaO{\cdot}MgO{\cdot}2SiO_2$
	Prochlorite	$2FeO{\cdot}2MgO{\cdot}Al_2O_3{\cdot}2SiO_2{\cdot}2H_2O$
	Diaspore	$Al_2O_3{\cdot}H_2O$
	Lepidocrocite	$Fe_2O_3{\cdot}H_2O$
	Magnetite	Fe_3O_4
	Kyanite	$Al_2O_3{\cdot}SiO_2$
	Staurolite	$2FeO{\cdot}5Al_2O_3{\cdot}4SiO_2{\cdot}H_2O$
	Topaz	$(AlF)_2SiO_4$
	Tourmaline	$H_9Al_3(BOH)_2Si_4O_{19}$
	Hematite	Fe_2O_3
	Penninite	$5MgO{\cdot}Al_2O_3{\cdot}3SiO_2{\cdot}2H_2O$

order to determine it quantitatively. Therefore, other methods normally are used to estimate the mineral matter.

The method most widely used is the calculation of the mineral matter from the ash content determined by standard methods. Usually the ash is less than the mineral matter because of certain weight losses during ashing. The greatest loss of weight is usually from partial or complete loss of the water of constitution of the shale and kaolin minerals. Another weight loss results from conversion of pyrite (FeS_2) to ferric oxide during heating in an oxidizing atmosphere in which the sulfur is evolved as SO_2. Losses also occur by dissociation of the carbonate minerals into metal oxides with release of CO_2. The loss of weight is compensated for in part by reaction of the oxides formed in the decomposition of the carbonates with the sulfur oxides formed by combustion of pyrite and the organic sulfur in the coal substance. As a whole, for high-rank coals containing little calcite and pyrite minerals, the residue after ashing is about 10 to 15% less than the original weight of the mineral matter.

Several empirical formulas are used for the calculation of the mineral matter from the ash content and other analytical data. One of the best known is the Parr formula used in the ASTM system of coal classification for converting parameters of volatile matter, fixed carbon, and calorific value to a mineral-matter-free coal basis.[189] The Parr formula is as follows:

$$\text{Mineral matter} = 1.08 \times \text{ash} + 0.55 \times \text{sulfur}.$$

Variations of the Parr formula have been used either for simplification or to improve the accuracy. The ASTM classification system provides a simplified approximation formula where the highest accuracy is not required, as follows:

$$\text{Mineral matter} = 1.1 \times \text{ash} + 0.1 \times \text{sulfur}.$$

The King, Maries, and Crossley (KMC) formula[190] developed from a survey of minerals in British coals is the most comprehensive of those proposed because, in addition to the correction for water of hydration, corrections are made for the forms of sulfur, carbon dioxide in the mineral carbonates, sulfur retained in the ash, and chlorine. Their formula is:

$$\text{Mineral matter} = 1.09\ \text{Ash} + 0.5\ S_{pyr} + 0.8\ CO_2 - 1.1\ SO_{3\,\text{ash}} + SO_{3\,\text{coal}} + 0.5\ \text{Cl}.$$

Because of the large amount of analytical work required in making the corrections, Brown, Caldwell, and Fereday[191] made a statistical study that showed certain components in the formula were interrelated. A simplified formula requiring fewer analytical determinations was developed as follows:

$$\text{Mineral matter} = 1.06\ \text{Ash} + 0.53\ S + 0.74\ CO_2 - 0.32.$$

In this formula, as well as in the KMC, organic sulfur is considered as part of the coal substance. Fereday and Flint[192] devised another formula which enables a calculation to a sulfur-free ultimate analysis by including the organic sulfur as part of the mineral matter. The formula is:

$$\text{Mineral matter} = 1.06\ \text{Ash} + 0.67\ S + 0.66\ CO_2 + 0.30.$$

In a reassessment of the factors used in the KMC formula, Millot[193] obtained an average value of 11.2% for the water of constitution of the silicate minerals in 78 coals, corresponding to a correction factor of 1.13. He therefore recommended the substitution of this factor for the 1.09 factor in the original KMC formula. Pringle and Bradburn,[194] in a further study, confirmed the validity of the factor 0.8 CO_2 in the formula for making the necessary correction for the carbonate minerals.

Despite the wide usage of such empirical formulas, they often are criticized because the average factor used, particularly the factor for water of hydration of the silicate minerals, may be considerably in error when applied to a particular coal.

Another method used to some extent for determining mineral matter is a float-and-sink separation of the coal at various specific gravities and determination of the calorific value or carbon of each fraction.[195] The analytical data are plotted against the ash content and extrapolated to zero ash and, therefore, to zero mineral matter. This method is based on the assumption that the composition of the pure coal substance is uniform through

[189] *ASTM Standard Specifications for Classification of Coals by Rank*, D 388–38.

[190] King, J. G., Maries, M. B., and Crossley, H. E., *J. Soc. Chem. Ind.*, **55,** 277–81T (1936).

[191] Brown, R. L., Caldwell, R. L., and Fereday, F., *Fuel*, **31,** 261–73 (1952).

[192] Fereday, F., and Flint, D., *Fuel*, **32,** 115–16 (1953).

[193] Millot, J. O'N., *Fuel*, **37,** 71–85 (1958).

[194] Pringle, W. J. S., and Bradburn, E., *Fuel*, **37,** 166–80 (1958).

[195] Brinsmaid, W., *J. Ind. Eng. Chem.*, **1,** 65–8 (1909). Stansfield, E., and Sutherland, J. W., *Trans. Am. Inst. Mining Met. Engrs.*, **88,** 614–26 (1930). Fieldner, A. C., Selvig, W. A., and Gibson, F. H., *Trans. Am. Inst. Mining Met. Engr.*, **101,** 224–35 (1932). Radmacher, W., *Glückauf*, **87,** 1104 (1951).

the various specific gravity fractions. Unfortunately, this is not strictly so, because the various petrographic components such as vitrain, clarain, durain, and fusain are known to have different chemical compositions; the proportion of the components in the specific gravity fractions may vary from that of the whole coal.

The calculation of the mineral content from the chemical composition of coal also has been considered.[196] This method is based on the assumption that when the chemical composition of the organic coal substance is known, the mineral content is the difference between 100 and this value, as follows:

$$\text{Mineral matter} = 100 - (C + H + N + O + S)$$

all calculated on the dry coal basis. Such a method requires a direct determination of oxygen, a determination not normally made, except in a few laboratories. Also, the method is subject to error, because the water of constitution of the mineral matter is included with the hydrogen and oxygen to an unknown extent.

A method for determining the total mineral content of coal that generally overcomes objections to other methods is the one developed by Radmacher and Mohrhauer.[197] It is a so-called "direct" method by which the mineral matter is removed by digestion with hydrofluoric and hydrochloric acids, the difference between the weight of the coal before and after the acid treatment representing the mineral matter. The method requires that certain corrections be made for the pyrite not removed by the acid treatment, the small amount of residual ash, and the chlorine taken up by the coal. The method has been reported to be satisfactory for British[198] and American[199] coals, if modified slightly by a hydrochloric acid pretreatment to remove high percentages of calcium and magnesium.

[196] Flint, D., *BCURA Bull.*, **17**, 133–42 (1953). Mohrhauer, P., *Bergbau-Rundschau*, **8**, 335–47 (1956).

[197] Radmacher, W., and Mohrhauer, P., *Brennstoff-Chem.*, **36**, 236–9 (1955); **37**, 26 (1956).

[198] Bishop, M., and Ward, D. L., *Fuel*, **37**, 191–200 (1958).

[199] Tarpley, E. C., and Ode, W. H., *U.S. Bur. Mines, Rept. Invest.* 5470 (1959), 15 pp.

MINOR ELEMENTS IN COAL

Since Thiessen's review,[200] many investigations have been made of the occurrence of minor elements in coal and its ash. Most of the work has been done to determine the possibility of recovering elements of commercial value, the deleterious effect on combustion processes and equipment, and application in correlation of coal beds.[201–203]

Several surveys have provided considerable data on the occurrence of minor elements in coals of the United States. Headlee and Hunter[204] determined the quantity of 36 elements in 16 of the major coal beds of West Virginia (Table 3). Nunn, Lovell, and Wright[205] made a similar survey of Pennsylvania anthracites (Table 4), and Deul and Annell,[206] of low-rank coals from Texas, Colorado, North Dakota, and South Dakota (Table 5).

Surveys made in other countries include the following: British Columbia[207] and Nova

[200] Thiessen, G., *Chemistry of Coal Utilization*, John Wiley and Sons, New York, 1945, Vol. 1, pp. 490–3.

[201] Gibson, F. H., and Selvig, W. A., *U.S. Bur. Mines, Tech. Paper* 669 (1944), 23 pp. Crossley, H. E., *Chem. Age (London)*, **55**, 629–36 (1946); *Inst. Fuel (London), Bull.*, December, 1946, 57–60, 67; *J. Inst. Fuel*, **20**, 94–5 (1947). Goldschmidt, V. M., *Progress in Coal Science*, D. H. Bangham, Ed., Interscience Publishers, Inc., New York, 1950, pp. 238–47. Monkhouse, A. C., *Coke and Gas*, **12**, 363–8 (1950); *Colliery Eng.*, **28**, 336–41 (1951). Haught, O. L., *Missouri Univ. School Mines and Met., Tech. Ser.* No. 85, 17–23 (1954). Wandless, A. M., *J. Inst. Fuel*, **30**, 541–52 (1957). Yurovskiĭ, A. Z., *Izvest. Akad. Nauk S.S.S.R., Otdel. Tekh. Nauk*, **1958**, No. 7, 82–6. Breger, I. A., *Econ. Geol.*, **53**, 831–5 (1958).

[202] Otte, M. U., *Chem. Erde*, **16**, 237–94 (1953).

[203] Leutwein, F., and Rösler, H. J., *Freiberger Forschungsh.*, **C19** (1956), 196 pp.

[204] Headlee, A. J. W., and Hunter, R. G., *Ind. Eng. Chem.*, **45**, 548–51 (1953); *West Va. Geol. Survey, Bull.* No. 13A, 36–122 (1955).

[205] Nunn, R. C., Lovell, H. L., and Wright, C. C., *Trans. Ann. Anthracite Conf.*, Lehigh Univ., **11**, 51–65 (1953).

[206] Deul, M., and Annell, C. S., *U.S. Geol. Survey, Bull.* 1036H, 155–72 (1956).

[207] Newmarch, C. B., *Can. Mining Met. Bull.*, **43**, 141–8 (1950); *Brit. Columbia Dept. Mines, Bull.* No. 33 (1953), 107 pp.

TABLE 3

COMPOSITION OF WEST VIRGINIA COAL ASH[204]

	Column Analysis				Cube Analysis			Enrichment Ratio of Coal Ash to Earth's Crust	Earth's Crust % as Oxide
Oxide	Average %	Maximum %	Minimum %	Variance Ratio	Maximum %	Minimum %	Variance Ratio		
Li_2O	0.075	0.14	0.024	5	0.66	<0.005	>13	5	0.014
Na_2O	1.78	3.66	0.82	4	9.13	0.30	30	0.47	3.81
K_2O	1.60	2.50	0.89	3	4.86	0.36	13	0.51	3.11
Rb_2O	0.030	0.055	<0.03	>2	0.12	<0.03	>4	0.88	0.034
CaO	2.76	12.0	0.81	15	36.3	0.12	300	0.53	5.07
SrO	0.38	1.05	0.11	10	12.0	<0.01	>1200	7	0.050
BaO	0.22	0.49	0.059	8	1.80	0.02	90	5	0.043
MgO	0.98	1.71	0.58	3	3.20	0.10	32	0.28	3.48
Al_2O_3	29.9	38.7	21.8	2	53.3.	2.8	19	1.85	16.7
SiO_2	43.9	55.5	29.2	2	82.6	2.0	41	0.75	59.0
Fe_2O_3	15.9	37.8	3.87	9	91.9	1.10	94	2.19	7.26
TiO_2	1.52	2.30	1.00	2	5.73	0.17	34	1.47	1.03
Ag_2O	0.0010	0.003	<0.0005	>6	0.009	<0.0005	>18	100	0.00001
As_2O_3	<0.07	0.075	<0.07	>1	0.22	<0.07	>3	<100	0.0007
B_2O_3	0.12	0.31	0.027	11	0.66	0.017	39	133	0.0009
BeO	0.008	0.030	0.002	15	0.14	<0.0005	>280	4	0.0019
Bi_2O_3	<0.004	0.007	<0.004	>2	0.10	<0.004	>3	<200	0.00002
Cb_2O_5	0.010	0.012	<0.007	>2	0.022	<0.007	>3	3	0.0034
CoO	0.010	0.023	0.006	4	0.096	<0.005	>19	2	0.0050
Cr_2O_3	0.023	0.029	0.016	2	0.053	0.010	5	0.79	0.029
CuO	0.061	0.13	0.028	5	0.53	0.006	90	5	0.012
GaO	0.022	0.032	0.015	2	0.124	0.005	25	12	0.0019
GeO_2	0.011	0.026	<0.005	>5	0.250	<0.005	>50	11	0.0010
HgO	0.011	0.021	<0.008	>3	0.059	<0.008	>7	157	0.00007
La_2O_3	0.030	0.051	<0.035	>2	0.096	<0.035	>3	13	0.0023
MnO	0.046	0.23	0.016	15	1.16	0.006	193	0.38	0.120
MoO_3	0.016	0.040	0.005	8	0.140	<0.006	>23	6	0.0025
NiO	0.047	0.10	0.016	7	0.57	<0.005	>11	3.6	0.0127
P_2O_5	0.35	1.21	<0.20	>6	8.35	<0.2	>42	1.9	0.180
PbO	0.048	0.14	0.020	7	0.27	<0.015	>28	28	0.0017
Sb_2O_3	<0.005	<0.005	<0.005	>1	0.028	<0.005	>6	<50	0.0001
SnO_2	0.020	0.070	<0.01	>7	0.190	<0.01	>19	4.4	0.0045
V_2O_5	0.050	0.069	0.032	2	0.15	0.014	11	2.8	0.018
WO_3	<0.01	0.019	<0.01	>2	0.056	<0.01	>6	<2	0.0087
ZnO	0.053	0.11	<0.05	>2	0.24	<0.05	>5	10	0.0050
ZrO_2	0.029	0.046	0.019	2	0.089	0.009	10	12	0.0025

Scotia,[208] Canada; English coals[209] and vitrains;[210] Charleroi Basin, Belgium;[211] Germany;[212] Hungary;[213] India;[214] Japan;[215] Poland;[216] Portugal;[217] Russia;[218] Asturias, Spain;[219] and Spitzenbergen, Norway.[220]

[208] Hawley, J. E., *Can. Mining Met. Bull.* **48,** 712–26 (1955).

[209] Reynolds, F. M., *J. Soc. Chem. Ind.* (*London*), **67,** 341–5 (1948).

[210] Horton, L., and Aubrey, K. V., *J. Soc. Chem. Ind.* (*London*), **69,** Supplement No. 1, S41–8 (1950).

[211] Legraye, M., and Coheur, P., *Ann. soc. géol. Belg., Bull.* **68,** 63–7 (1944–5).

[212] Endell, J., *Braunkohle*, **4,** 446–54 (1952). Otte, M. U., *loc. cit.* Leutwein, F., and Rösler, H. J., *loc. cit.* Leutwein, F., *Freiberger Forschungsh.*, **C30,** 28–48 (1956). Lissner, A., *loc. cit.*

[213] Szádeczky-Kardoss, E., and Vogl, M., *Földtani Közlöny*, **85,** 7–43 (1955); *Hung. Tech. Abstr.*, **7,** No. 4, 10.

[214] Mukherjee, B., and Dutta, R., *Sci. and Culture* (*Calcutta*), **14,** 213–4 (1948); *Fuel*, **29,** 190–2 (1950).

[215] Uzumasa, Y., *Chem. Researches* (*Japan*), **5,** 1–17 (1949). Hashimoto, S., *Misc. Repts. Research Inst. Nat. Resources* (*Tokyo*), No. 31, 27–34 (1953).

[216] Kuhl, J., and Ziólkowski, J., *Przegląd Górniczy*, **1954,** No. 5, 180; *Polish Tech. Abstr.*, **1955,** No. 2, 3411. Kuhl, J., *Przegląd Geol.*, **5,** 248–55 (1957).

[217] de Brito, A. C., *Rep. Port., Estud., notas e trabal. serv. fomento mineiro*, **10,** 236–62 (1955).

[218] Vakhrushev, G. V., *Uchenye Zapiski, Saratov. Godsudarst. Univ. im. N. G. Chernyshevskogo*, **15,** No. 1, 124–46 (1940).

[219] Lopez de Azcona, J. M., and Camuñas Puig, A., *Bol. inst. geól. y minero España*, **60,** 391–400 (1948); *Anales fís. y quím.* (*Madrid*), **43,** 48–50 (1947).

[220] Butler, J. R., *Kgl. Ind.-Handverk-Skipsfartdept., Norsk Polarinst., Skrifter* No. 96, 1–26 (1953).

Trace elements in certain Finnish[221] and Scottish[222] peats also have been reported. Table 6 summarizes the analyses of some of these coals for their content of minor elements.

In addition to the general surveys, the occurrence of many elements in coals and coal ashes is summarized in the following paragraphs for work done principally since 1940.

Arsenic. Hertzog[223] found that the arsenic content of fourteen coals of the United States ranged from 1 to 106 ppm. The arsenic content in thirteen of the ashes ranged from 8 to 197 ppm, with one containing 1560 ppm. Duck and Himus[224] reported that the twenty-five coals they tested averaged 14 ppm arsenic with a maximum of 98 ppm. They concluded that most of the arsenic occurred in the form of arsenopyrite. Other occurrences were reported: British, less than 220 ppm, although values over 100 ppm are rare;[225] Dutch, less

221 Salmi, M., *Geol. Tutkimuslaitos, Geotek. julkaisuja*, **51** (1950), 20 pp.

222 Mitchell, R. L., *Intern. Peat Symposium, Dublin*, **1954**, Sect. B.3, 9 pp.

223 Hertzog, E. S., *Ind. Eng. Chem., Anal. Ed.*, **7**, 163–5 (1935).

224 Duck, N. W., and Himus, G. W., *Fuel*, **30**, 267–71 (1951).

225 Wandless, A. M., *J. Inst. Fuel*, **30**, 545–8 (1957).

TABLE 4

RANGE OF CONCENTRATION OF MINOR AND TRACE ELEMENTS* IN ANTHRACITE ASH AND BURNING BANK SAMPLES[205] (Percent of Ash)

Titanium oxide (TiO_2)	1.5 to 2.0
Magnesium oxide (MgO)	0.4 to 0.9
Calcium oxide (CaO)	0.2 to 0.4
Vanadium oxide (V_2O_5)	0.02 to 0.04
Germanium oxide (GeO_2)	Trace to 0.02
Manganese oxide (MnO)	0.006 to 0.008
Arsenic (As)	Trace to 0.01
Copper (Cu)	0.001 to 0.01
Chromium (Cr)	0.001 to 0.01
Lead (Pb)	0.001 to 0.01
Lithium (Li)	Trace to 0.01
Phosphorus (P)	$0.x$†
Gallium (Ga), Nickel (Ni), Tin (Sn), Zinc (Zn)	$0.0x$
Beryllium (Be), Cobalt (Co), Tungsten (W), Molybdenum (Mo)	$0.00x$

* Some trace elements (Ag, B, Bi, Cd, Hg, La, Sb, Te, Zr) reported by other investigators were not detected in these preliminary studies. More specific examination of the samples for those particular elements may show their presence but, in general, the concentration will probably not exceed $0.00x$.

† x denotes order of magnitude of the element concentration in cases where the spectrographic lines from the unknown were not within the range of the standard currently available.

TABLE 5

MAXIMUM CONCENTRATION OF MINOR ELEMENTS IN ASH OF WESTERN COALS, PERCENT[206]

Number of Samples	Locality	10–1	1.0–0.1	0.1–0.01	0.01–0.001	0.001–0.0001
151	Harding County, S. Dak.	—	B, Ba, Sr, Mn, Ti, Mo, Co, Ni, Zn, As, Pb, Zr, Be, U.	Cu, Cr, V, Li, Y, La, Ga, Ge, Sc.	Sn.	Yb, Ag.
59	Perkins County, S. Dak.	—	B, Ba, Sr, P, Ti, Mo, Zr, U.	Cu, Cr, V, Ni, Co, Pb, Mn, Ga, Ge, Zn, La.	Y, Sc, Sn.	Be, Yb, Ag.
26	Bowman County, N. Dak.	—	B, Ba, Sr, Mn, Ti, As, Pb, V, Zr.	Cu, Cr, Mo, Ni, Co, Ga, Ge, U.	Y, Sc, Sn.	Be, Yb.
4	McKenzie County, N. Dak.	—	B, Ba, Sr, Mn, Pb.	Ti, V, Cu, Li, Sn, La, Zr, Co.	Y, Mo, Ga, Ge, Ni, Cr.	Be, Yb.
35	Jefferson County, Colo.	—	B, Ba, Sr, Mn, Ti, Mo, Y.	Cu, Cr, V, Ni, Co, Zr, Pb.	Sn, Sc, Ga, U, Ge, La, Yb, Be.	
48	Milam County, Tex.	P	B, Ba, Sr, Mn, Ti, Ni, Sn.	Cu, Cr, V, Co, Pb, Zr, Zn, Y.	Mo, Sc, Ga, Ge, Be, Yb.	

TABLE 6

MINOR ELEMENTS IN CANADIAN, BRITISH, AND GERMAN COALS AND ASHES

	In Ash, ppm					In Coal, ppm				
Element	Sydney, Nova Scotia[208]		Barnsley Vitrain, Great Britain[210]	Germany[203]	Ruhr, Germany[202]	Sydney, Nova Scotia[208]		Germany[203]		Ruhr, Germany[202]
	Average	Range		Maximum	Maximum	Average	Range	Average	Maximum	Maximum
Antimony			100–200	>1000	3000			10–30	≥10	17
Arsenic	900	280–2300		10,000		100	33–270	100	500	
Barium	300	18–2200		>1000	*	35	2–257	*	100	*
Beryllium	14		50–100	4000	1000	2	1–2	13	40	20
Bismuth				2000					100	
Boron	148	52–220	200–3000	>1000		17	6–25		≥10	
Chromium	45	18–79	100–1000	*	5000	5	2–9	*	*	50
Cobalt	87	26–196	100–300	2000	2000	10	3–34	14	30	12
Copper			800–1000	≥10,000	4000			≥25	10,000	50
Gallium			80–300	>3000	1000			30	100	20
Germanium	44	9–70	300–1000	5000	1000	5	1–8	19	50	20
Lead	572		200–800	31,000	3000	66	25–120	140	3000	30
Manganese	1200	165–2200	100–2000	≥10,000	22,000	140	9–254	*	≥5000	700
Molybdenum	60	18–105	80–200	1000	6000	7	2–12	21	200	50
Nickel	131	52–645	500–3000	≥3000	16,000	15	6–74	24	≥60	30
Silver				60	*			0.3	3	*
Strontium	560	225–750		>1000	*	65	76–87	*	100	*
Tin	9	4–18	50–200	1000	6000	1		3	300	120
Titanium			3000–8000	12,000	30,000			700	1000	1500
Vanadium	120	61–244	400–5000	>1000	11,000	14	7–28	18	>100	20
Zinc	218	115–550	500–700	21,000	8000	25	13–64	170	2000	100
Zirconium			<100–500	*	7000			*	*	140

* Not significant.

than 10 ppm;[226] and South African, less than 6 ppm.[227]

Beryllium. Fleischer and Cameron[228] re-reviewed the geochemistry of beryllium and its occurrence in coal. Stadnichenko, Zubovic, and Sheffey[229] reported an average of 45 ppm in the ash of more than 1000 American coals.

Boron. The boron contents of coals and coal ashes also have been reported: English coals, 2 to 140 ppm;[225] South African coals, 11 to 60 ppm;[230] New Zealand coal ash, range of 0.1 to about 1%.[231] Roga and co-workers[232] found that vitrain in Polish coals contained the highest concentration of boron.

Chlorine. Although chlorine normally is associated with the alkali and alkaline earth metals as chlorides, investigation of British coals has shown that some of the chlorine is linked ionically with the coal substance.[233] Its occurrence and behavior during combustion and carbonization have been reviewed by Kear and Menzies.[234] Wandless[225] reported that coals containing less than 0.3% chlorine are not likely to give rise to formation of troublesome boiler deposits. Surveys of content of chlorine have been reported: United States, for twenty-one coals, 0.01 to 0.46%;[235] English coals, 0.1 to 1.0%;[236] and Upper Silesian coals in Poland, 0 to 0.025%, with the exception of a few seams where it reaches 0.19%.[237]

Fluorine. Crossley[238] discussed the geographical distribution of fluorine in British coals, the occurrence of fluorine in minerals associated with the coal, the relation between the amount of fluorine and phosphorus, the nature of the fluorine compounds, and the industrial significance of fluorine in steam raising, gas-making, and brewing. The fluorine ranged from 0 to 175 ppm, with most coals containing less than 80 ppm. Most of the

[226] Matla, W. P. M., *Geol. en Mijnbouw*, **9**, 106–11 (1947).

[227] Wyndham, J. L. P., *S. African Mining Eng. J.*, **62,** Part II, 659–61 (1951).

[228] Fleischer, M., and Cameron, E. N., *U.S. Atomic Energy Comm.*, MDDC–643, June, 1946, 27 pp.

[229] Stadnichenko, T., Zubovic, P., and Sheffey, N. B., *Bull., Geol. Soc. Am.*, **67,** Part 2, 1735 (1956).

[230] Kunstmann, F. H., and Harris, J. F., *J. Chem. Met. Mining Society South Africa*, **55,** 12–15 (1954).

[231] Rafter, T. A., *Nature*, **155,** 332 (1945).

[232] Roga, B., Ihnatowicz, A., Weclewska, M., and Ihnatowicz, M., *Prace Głównego Inst., Górnictwa, Komun.* No. 212, Series B. (1958), 7 pp.

[233] Edgcombe, L. J., *Fuel*, **35,** 38–48 (1956).

[234] Kear, R. W., and Menzies, H. M., *BCURA Bull.*, **20,** 53–65 (1956).

[235] Gibson, F. H., and Selvig, W. A., *U.S. Bur. Mines, Tech. Paper* 669, 9 (1944).

[236] Crossley, H. E., *Inst. Fuel (London), Bull.*, 60, December, 1946.

[237] Mielecki, T., *Biul. Inst. Nauk.-Badawczego Przemyslu Węglowego (Katowice), Komun.* No. 36 (1948), 5 pp.

[238] Crossley, H. E., *J. Soc. Chem. Ind. (London)*, **63,** 289–92, 342–7 (1944).

fluorine probably is combined largely with phosphate as fluoapartite. During combustion, all of the fluorine is volatilized, and during high-temperature carbonization with steaming, probably three-fourths or more is volatilized.

Little published information on other coals is available. Churchill, Rowley, and Martin[239] reported 167 ppm fluorine in a southern Illinois coal and 85 ppm in a western Pennsylvania coal. Four samples of coal from Utah and two of unknown origin contained fluorine ranging from 145 to 295 ppm. In analysis of six other Utah coals, Bradford[240] reported values of 40 to 132 ppm.

Germanium. With the post-war development of transistors in the electronic industry, intensive searches were made for commerical sources of germanium, among them coal, because earlier reports showed important concentration in certain samples. Headlee and Hunter[241] found germanium is of secondary origin since the top and bottom layers of most coals contain the highest concentration, an association that has been confirmed by other investigators.[242,243] Unusually high concentrations have been found in certain coalified logs with the ash containing up to 9% germanium.[244]

In a survey of 700 samples of American coals, Stadnichenko and co-workers[243] detected germanium in the ash of more than half of the samples. A coal from the bottom horizon of the Lower Kittanning bed in Ohio contained 0.2% germanium in the ash. From 7 to 48 ppm germanium in the coal or 36 to 680 ppm in the ash have been reported for 24 samples of Kansas coal.[245]

Analyses of British coals have shown an average of 7 ppm,[246] which is of the same order as the content of most sedimentary rocks. A Dorsetshire lignite contained an average of 1800 ppm.[247] Other analyses have been reported as concentrations in coals from several countries: Belgium, 1 to 10 ppm;[248] India, 3 to 9 ppm;[249] Italy, 30 to 50 ppm;[250] Japan, maximum of 357 ppm in a lignite from North Miyagi and South Iwate Prefectures[251] and up to 1970 ppm in a lignite from the Miyagi Prefecture;[252] Poland, 2 to 39 ppm;[253] and South Africa, 0.4 to 15.6 ppm.[254]

Concentrations that have been reported in coal ash are: Australia, 10 to 30 ppm;[255] Argentina, up to 1900 ppm;[256] Belgium, up to 60 ppm;[257] Canada, average of 40 ppm and up to 2000 ppm;[258] Czechoslovakia, 1 to 10 ppm[259]

[239] Churchill, H. V., Rowley, R. J., and Martin L. N., *Anal. Chem.*, **20**, 69–71 (1948).

[240] Bradford, H. R., *Mining Eng.* **9**, No. 1, 78–9 (1957).

[241] Headlee, A. J. W., and Hunter, R. G., *West Va. Geol. Econ. Survey, Repts. Invest.*, **8** (1951), 15 pp.; *Ind. Eng. Chem.*, **45**, 548–51 (1953). Headlee, A. J. W., *Mining Eng.*, **5**, 1011–14 (1953).

[242] Oka, Y., Kanno, T., Matsuo, S., and Haga, K., *Tôhoku Daigaku Senko Seiren Kenkyusho Ihô*, **12**, 129–40 (1956). Katchenkov, S. M., *Doklady Akad. Nauk S.S.S.R.*, **61**, 857–9 (1948). Hawley, J. E., *Can. Mining Met. Bull.*, **48**, 712–26 (1955).

[243] Stadnichenko, T., Murata, K. J., Zubovic, P., and Hufschmidt, E. L., *U.S. Geol. Survey, Circ.*, **272** (1953), 34 pp.

[244] Stadnichenko, T., Murata, K. J., and Axelrod, J. M., *Science*, **112**, 109 (1950). Breger, I. A., and Schopf, J. M., *Geochim. et Cosmochim. Acta*, **7**, 287–93 (1955).

[245] Schleicher, J. A., and Hambleton, W. W., *Kansas State Geol. Survey, Bull.* No. 109, 113–24 (1954).

[246] Aubrey, K. V., *Fuel*, **31**, 429–37 (1952).

[247] Hallam, A., and Payne, K. W., *Nature*, **181**, 1008–9 (1958).

[248] Rouir, E. V., *Ann. soc. geol. Belg.*, **77**, 284–8 (1954). Inichar, *Ann. Mines Belg.*, **53**, 39–40 (1954).

[249] Ganguly, N. C., and Dutta, D. P., *J. Sci. Ind. Research (India)*, **15B**, 327–8 (1956).

[250] Bertetti, I., *Atti accad. ligure sci. e lettere (Pavia)*, **11**, 53–8 (1954).

[251] Oka, Y., Kanno, T., Ayusawa, S., and Haga, K., *Tôhoku Daigaku Senko Seiren Kenkyusho Ihô*, **11**, 17–28 (1955).

[252] Oka, Y., Kanno, T., Haga, K., and Suzuki, S., *Nippon Kagaku Zasshi*, **77**, 1026–30 (1956).

[253] Gregorowicz, Z., *Przemysl Chem.*, **13**, 700–701 (1957).

[254] Kunstmann, F. H., and Hammersma, J. C., *J. Chem. Met. Mining Soc. South Africa*, **56**, No. 1, 11–22 (1955).

[255] Durie, R. A., and Schafer, H. N., *Australia, Commonwealth Sci. Ind. Research Organization, T. C.* **27** (1958), 55 pp.

[256] Lexow, S. G., and Maneschi, E. P. P., *Anales asoc. quim. arg.*, **38**, 225–9 (1950).

[257] *Inst. natl. ind. charbonniere, Ann. mines Belg.*, **53**, 39–40 (1954).

[258] Fortescue, J. A. C., *Am. Mineralogist*, **39**, 510–19 (1954). Hawley, J. E., *Econ. Geol.*, **50**, 517–32 (1955).

[259] Šimek, B. G., Coufalik, F., and Stadler, A., *Zprávy Ústavu Vědecký Výskum Uhlí*, **1948**, 167–74 (in English, 174–5). Šimek, B. G., *Chem. Listy*, **34**, 181–5 (1940).

and up to 1200 ppm;[260] Great Britain, average of 50 ppm;[261] India, 400[262] and 170 to 1220 ppm in ash of Assam and Hyderabad coal;[263] Japan, 100 ppm, generally,[264] and 170 to 2600 ppm in ash of coal from Hokkaido;[265] Portugal, 10 to 1000 ppm in 24 of 64 ashes tested;[217] and Russia, from 0.001 to 1.0% in coal ash.[266]

Although attempts have been made to obtain germanium in commercial quantities from coal and its products,[267] the only production has been from certain flue dusts from gas-producer plants in England. Surveys of the occurrence of germanium in flue dusts, boiler deposits, and coal ashes, and gases from power plants,[268] and tar[269] and gas liquor[270] from coal carbonization processes, have disclosed no significant concentration to warrant industrial recovery to this time.

Uranium. Appreciable quantities of uranium are present in certain low-rank coals. A bibliography[271] of more than 50 published reports of the United States Geological Survey shows that, although most of the coal beds and associated carbonaceous shales of the United States are practically nonradioactive, many assay from 0.005 to about 0.10% uranium. The results indicate that lignites contain the most uranium, coals of subbituminous B and C ranks the next largest concentration, and coals of highest rank the lowest amount, which usually is 0.001% or less. Some lignite containing more than 0.1% uranium has been found in the northern Great Plains.[272] Uranium also has been reported in coals of other

[260] Šantruček, P., *Věstnik ústředniho ústavu geol.*, **33**, 367–70 (1958).

[261] Aubrey, K. V., *Nature*, **176**, 128–9 (1955).

[262] Iyer, V. G., and Sundaram, N., *Indian Mining J.*, **3**, 182–3 (1955).

[263] Mukherjee, B., and Dutta, R., *Sci. and Culture (Calcutta)*, **14**, 538–9 (1949); *Fuel*, **29**, 190–2 (1950).

[264] Inagaki, M., *J. Coal Research Inst. (Japan)*, **3**, 261–7 (1952).

[265] *Ibid*, **2**, 379–88 (1951).

[266] Ratynskiĭ, V. M., *Compt. rend. acad. sci. U.R.S.S.*, **40**, 198–200 (1943); *Trudy Biogeokhim. Lab., Akad. Nauk S.S.S.R.*, **8**, 183–223 (1946).

[267] Reynolds, F. M., *Chem. Prod.*, **13**, 152–3 (1950). Powell, Alan R., Lever, F. M., and Walpole, R. E., *J. Appl. Chem. (London)*, **1**, 541–51 (1951). Williams, A. E., *Iron & Coal Trades Rev.*, **163**, 29–30 (July 6, 1951). Lever, F. M., *Metal Ind. (London)*, **80**, 3–4 (1952). Chirnside, R. C., and Cluley, H. J., *G. E. C. Journal*, **19**, 94–100 (1952). Powell, Alan R., *Chem. & Ind. (London)*, **1954**, 1225–6. Schreiter, W., *Chem. Tech. (Berlin)*, **6**, 141–8 (1954). Ona, K., Inada, Y., and Konno, I., *Tôhoku Daigaku Senko Seiren Konkyusho Ihô*, **11**, 159–64 (1955). Germanium Research Study Group, *Gt. Brit. Dept. Sci. Ind. Research*, April 1957, 26 pp. Losev, B. I., Mel'nikova, A. N., and El'piner, I. E., *Izvest. Akad. Nauk S.S.S.R., Otdel. Tekh. Nauk*, **1957**, No. 9, 90–5. Aubrey, K. V., *Rev. ind. minerale*, **1958**, Spec. No. July 15, 51–64. Gordon, S. A., Volkov, K. Yu., and Menkovskiĭ, M. A., *Geokhimiya*, **1958**, 384–8.

[268] Cooke, W. T., *Trans. Roy. Soc. S. Australia*, **62**, 318–19 (1938). Kakihana, H., *J. Chem. Soc. Japan*, Pure Chem. Sect., **70**, 226–9 (1949). Belugou, P., and Domoulet, P., *CERCHAR Notes Tech.*, **R520** (1953), 4 pp. Steward, K., *Gas J.*, **274**, 279–80 (1953). Inagaki, M., Oikawa, H., Moriya, S., Yamaguchi, T., and Otsu, T., *J. Coal Research Inst. (Japan)*, **5**, 359–66 (1954). Forrest, J. S., Smith, A. C., and Ward, J. M., *Nature*, **175**, 558 (1955). Howes, E. A., and Lees, B., *J. Inst. Fuel*, **28**, 298–9 (1955). Machin, J. S., and Witters, J., *Illinois State Geol. Survey, Circ.* **216** (1956), 13 pp. Inagaki, M., Moriya, S., Yamaguchi, T., Nakayama, K., and Sakai, K., *J. Coal Research Inst. (Japan)*, **7**, 193–200, 287–96 (1956). Mielecki, T., and Mroz, T., *Przegląd. Górniczy*, **13**, 145–7 (1957). Pilkington, E. S., *Australian J. Appl. Sci.*, **8**, 98–111 (1957). Corey, R. C., and Myers, J. W., *Coal Utilization*, **12**, No. 11, 33–5 (1958). Corey, R. C., Myers, J. W., Schwartz, C. H., Gibson, F. H., and Colbassani, P., *U.S. Bur. Mines, Bull.* No. 575 (1959), 68 pp.

[269] Kostrikin, V. M., *Zhur. Priklad. Khim.*, **12**, 1449–54 (1939). Yoshida, T., and Nakagawa, M., *Kooru Taaru*, **5**, 230–4 (1953). Crawley, R. H. A., *Nature*, **175**, 291–2 (1955). Diev, P., and Davydov, V. I., *Zhur. Priklad. Khim.*, **30**, 1685–7 (1957).

[270] Oikawa, H., Inagaki, M., Moriya, S., Yamaguchi, T., and Otsu, T., *J. Coal Research Inst. (Japan)*, **6**, 97–102 (1955). Inagaki, M., Oikawa, H., and Moriya, S., *J. Coal Research Inst. (Japan)*, **6**, 129–32, 161–5, 234–8 (1955). Inagaki, M., *J. Coal Research Inst. (Japan)*, **6**, 419–25 (1955). Kodama, K., *Nagoya-shi Kôgyô Kenkyûjo Kenkyû Hôkoku*, No. 14, 41–3 (1956).

[271] Kehn, T. M., *U.S. Geol. Survey, Bull.* No. 1059A (1957), 28 pp.

[272] Burton, L. R., U.S. Atomic Energy Comm., RME–1060 (1955), 13 pp. King, J. W., Foran, J. F., and Speal, A. J., U.S. Atomic Energy Comm., RME–1062 (1955), 16 pp. King, J. W., and Young, H. B., *U.S. Geol. Survey, Profess. Paper* **300**, 419–31 (1956).

countries.[273] From studies of the geochemistry of uranium in coal,[274] it probably occurs as an organo-uranium complex soluble at a pH less than 2.2, with distinct uranium minerals rarely present, and little of the element retained by ion exchange.[275] Organic material may serve as a reducing agent converting the soluble uranyl ion to an insoluble uranous form, the sulfide ion associated with the carbonaceous material may act as a reducing agent, and the uranium may be precipitated as an insoluble humate. Szalay[276] concluded that a pH of less than 3 is necessary to liberate uranium from Hungarian brown coal. Almássy[277] suggested that uranium was bound originally to humic acids. Schopf and Gray[278] found a relationship between high values of uranium and samples rich in humic matter, although no correlation was found between uranium content and any coal component.

Low-rank coals are effective materials for removing uranium from solution.[279] Probably solution of uranium by ground water from overlying volcanic rocks is followed by downward percolation through porous strata until the uranium is retained by the highest of the underlying lignite beds.[280] In the Red Desert area of Wyoming, the concentration of uranium is apparently related to the permeability of the rocks enclosing the coal beds.[281]

Miscellaneous. Other elements that have been detected include: copper,[282] iodine,[283] nickel,[284] noble metals,[285] rare earths,[286] tin,[287] and zirconium.[288]

[273] Hoffmann, J., *Chem. Erde*, **15,** 277–82 (1943). Foldvari, A., *Acta Geol. Acad. Sci. Hung.*, **1,** 37–8 (1952). Sarrot-Reynaud de Cresseneuil, J., *Trav. lab. géol. fac. sci. univ. Grenoble*, **30,** 43–54 (1953). Szalay, S., *Acta Geol. Acad. Sci. Hung.*, **2,** 299–311 (1954). Davidson, C. F., and Ponsford, D. R. A., *Mining Mag.* (*London*), **91,** 265–73 (1954); *S. African Mining Eng. J.*, **65,** Part 2, 721, 723, 725–7 (1954). Zeschke, G., *Glückauf*, **92,** 1107–8 (1956). Szalay, S., and Almássy, G., *Magyar Tudományos Akad. Kem. Tudományok Osztálvának, Közleményei*, **8,** 33–8 (1956); *Hung. Tech. Abstr.* **9,** No. 3, 21 (1957). Broda, E., Nowotny, K., Schönfeld, T., and Suschny, O., *Berg.-u. hüttenmänn. Monatsh. montan. Hochschule Leoben*, **101,** 121–4 (1956). Hecht, F., and Tomic, E., *Österr. Chemiker-Ztg.*, **58,** 221–7 (1957).

[274] Breger, I. A., *Econ. Geol.*, **53,** 823–41 (1958). Bergstrom, J. R., *N. Dakota Geol. Survey, Rept. Invest.*, No. 23 (1956), 1 p. Denson, N. M., Bachman, G. O., Gill, J. R., Hail, W. J. Jr., Love, J. D., Masursky, H., Moore, G. W., Pipiringos, G. N., Vine, J. D., and Zeller, H. D., U.S. Atomic Energy Comm., TEM-341A (1952), 44 pp. Deul, M., and Annell, C. S., *U.S. Geol. Survey, Bull.* No. 1036-H (1956), 155–72.

[275] Breger, I. A., Deul, M., and Rubinstein, S., *Econ. Geol.*, **50,** 206–26 (1955). Breger, I. A., Deul, M., and Meyrowitz, R., *Econ. Geol.*, **50,** 610–24 (1955). Breger, I. A., and Deul, M., *U.S. Geol. Survey, Profess. Paper* **300,** 505–10 (1956).

[276] Szalay, S., *Acta Geol. Acad. Sci. Hung.*, **2,** 299–311 (1954).

[277] Almássy, G., *Magyar Kém. Lapja*, **11,** 206–9 (1956).

[278] Schopf, J. M., and Gray, R. J., *U.S. Geol. Survey, Circ.* **343** (1954), 10 pp.

[279] Moore, G. W., *Econ. Geol.*, **49,** 652–8 (1954).

[280] Denson, N. M., and Gill, J. R., *U.S. Geol. Survey, Profess. Paper* **300,** 413–8 (1956).

[281] Masursky, H., *U.S. Geol. Survey, Profess. Paper* **300,** 439–44 (1956).

[282] Aubrey, K. V., *Nature*, **174,** 141–2 (1954).

[283] Stolper, W., *Bergakadamie* (*Berlin*), **8,** 587–8 (1956).

[284] Shakhov, F. N., and Efendi, M. E., *Compt. rend. acad. sci. U.R.S.S.*, **51,** 139–40 (1946).

[285] Popovic, A., *Glasnik Khem. Drushtva, Beograd*, **19,** 305–7 (1954).

[286] Mukherjee, B., *Fuel*, **29,** 264–6 (1950).

[287] Borovik, S. A., and Ratynskiĭ, V. M., *Doklady Akad. Nauk S.S.S.R.*, **45,** 128–9 (1944); *Compt. rend. acad. sci. U.R.S.S.*, **45,** 120–1 (1944).

[288] Degenhardt, H., *Geochim. et Cosmochim. Acta*, **11,** 279–309 (1957).

CHAPTER 6

CHEMICAL CONSTITUTION AND REACTIONS OF COAL

I. G. C. Dryden

British Coal Utilisation Research Association, Leatherhead, Surrey, England

Many new techniques have been developed and applied to the study of the chemical constitution and reactions of coal since the publication in 1945 of the two volumes on *Chemistry of Coal Utilization*. Three books[1,2] and many reviews[3,4] have covered the progress, but so rapid has been the pace of progress that even the latest surveys are not up-to-date on the structural side.

CHEMICAL CONSTITUTION

Ultrafine Capillary Structure

The internal capillary structure of coal particles is an important feature. It is through these very small capillaries that reagents or vapors have to pass during chemical reaction or adsorption processes, the results of which may be influenced by the restrictions imposed by narrow channels in the coal. The adsorptive properties of carbon products such as charcoal made from coal may depend critically on the nature of this pore structure. A relatively small internal surface in coals, moreover, seems to be linked more closely than any other characteristic with good coking properties—a relation yet to be explained. The absolute magnitude of the internal specific surface area has been the subject of considerable controversy.

The sizes of holes comprised in the term ultrafine structure range from gaps a few Angströms wide, caused by random packing of the large molecules, to comparatively large channels of width up to 100 A. In addition there are still larger capillaries and cracks as indicated by the mercury porosimeter, which in bituminous coals contribute from 20 to 50% of the total internal free volume.

The methods used for characterizing quantitatively the internal structure of coals involve measuring (1) helium and apparent densities, from the differences in which capillary volume distributions in relation to molecular sieve effects are determined; (2) heats of wetting, from which the corresponding internal surface area distributions are derived directly; and (3) vapor and gas sorption, from which both volume and area distributions can be derived directly.[5] Comparisons of results obtained by these three methods constitute an important way of checking the conclusions reached.

[1] Kreulen, D. J. W., *Elements of Coal Chemistry*, Nijgh & Van Dikmar N. V., Rotterdam and 's-Gravenhage, 1948, 204 pp.; Francis, W., *Coal*, Arnold, Ltd., London, 1954, 567 pp.

[2] Van Krevelen, D. W., and Schuyer, J., *Coal Science: Aspects of Coal Constitution*, Elsevier, Amsterdam, London, New York, and Princeton, 1957, 352 pp.

[3] Marshall, C. E., *Fuel*, **22,** 140–55 (1943). Horton, L., Randall, R. B., and Aubrey, K. V., *Fuel*, **23,** 65–80, 100–109 (1944). Horton, L., *J. Inst. Fuel*, **23,** 91–5 (1950); **28,** 300–305, 315 (1955). Korten, F., *Glückauf*, **85,** 271–3 (1949). Müller, K. O., *Chem. Tech.* (*Berlin*), **2,** 147 (1950). Fuchs, W., *Brennstoff-Chem.*, **32,** 12–19 (1951); *Chem.-Ztg.*, **76,** 61–6 (1952). Mackowsky, M. T., *Glückauf*, **87,** 551–7 (1951). Dryden, I. G. C., *Research* (*London*), **8,** 183–90 (1955); *Brennstoff-Chem.*, **37,** 42–6 (1956); *BCURA Monthly Bull.*, **21,** 285–90 (1957); *Fuel*, **38,** 248–9 (1959); *Proc. Conf. on Science in the Use of Coal*, Sheffield, 1957, Inst. Fuel, London, 1958, pp. A13–A24. Horton, L., *Proc. Conf. on Science in the Use of Coal*, Sheffield, 1959, Inst. Fuel, London, 1958, pp. A24–A28. Wandless, A. M., *Proc. Conf. on Science in the Use of Coal*, Sheffield, 1957, Inst. Fuel, London, 1958, pp. A1–A12; *J. Inst. Fuel*, **30,** 541–52 (1957). Van Krevelen, D. W., *Fuel*, **38,** 245–7 (1959).

[4] Dryden, I. G. C., *Brennstoff-Chem.*, **36,** 80–5 (1955).

[5] Bond, R. L., and Spencer, D. H. T., *Proc. Conf. on Industrial Carbon and Graphite*, Soc. Chem. Ind., London, 1957, 231–51.

TABLE 1

HEAT OF WETTING (cal/g, DRY BASIS) OF A CARBON BLACK* IN VARIOUS LIQUIDS[5]

Wetting Liquid	Molecular Volume (cm^3/mole)	Dipole Moment (10^{-18} e.s.u.)	Heat of Wetting (cal/g)
Methanol	40.5	1.7	21.5
Acetic acid	58	1.3	20.9
Ethanol	59	1.7	19.7
n-Propanol	75	1.7	17.1
IsoPropanol	77	1.7	17.9
Benzene	89	0	21.5
n-Butanol	92	1.7	16.0
Carbon tetrachloride	96.5	0	21.5
Toluene	107	0.4	23.6
Xylene	121	0.4	23.9
n-Hexane	131	0	21.3

Wetting Liquid	Molecular Volume (cm^3/mole)	Dipole Moment (10^{-18} e.s.u.)	Heat of Wetting (cal/g)
n-Heptanol	141	1.7	19.7
Decalin	154	0	19.9
Diethyl phthalate	207	2.7	25.6
Dibutyl phthalate	266		21.5
Tri-*o*-tolyl phosphate	312		19.9
Silicone Fluid D.C. 702	494		16.0
Silicone Fluid D.C. 703	523		13.5

* Surface = 230 m^2/g by vapor sorption at 25°C and by electron microscopy.

The heat liberated when a solid is wetted should be proportional to the area wetted, that is, to that accessible to the liquid. When an out-gassed sample is immersed in an organic liquid, a rise in temperature of the system results[6,7] proportional to the total area of surface wetted by the liquid and to the heat liberated on wetting unit area of the surface. An advantage of the method is that the results are not complicated by sorption hysteresis or by the problem of selecting the monolayer point on a continuous isotherm. But the heat per unit area is the sum of a physical heat of wetting and the heat of interaction of the wetting liquid with any functional groups on the surface, especially those containing oxygen. The heat of interaction may seriously affect the constancy of the total with porous materials like coal, because the wetting liquid must have small molecules for penetrating the smallest capillaries, good wetting properties, and must not be too volatile at room temperature; it must necessarily therefore be a *polar* organic liquid, and methanol is the best found up to the present.

The fullest use of area measurement by heat of wetting depends on the observation that the heat of wetting per *unit area* of unoxidized nonporous carbons by organic liquids is about 1 cal/10 m^2 (10.7 m^2 is used in recent work).[7] Similar observations have been made on graphite[8] and on a carbon black.[9] Table 1 shows that the value is moderately independent of the hydrogen bonding tendency, dipole moment, and size of molecule of the liquid.[5] There is a spread of some ±25% around the mean value, doubtless owing to specific effects, and liquids with very large molecules, such as silicones, give even lower values.

Bond and his colleagues have also used comparison with sorption on a carbon black as a basis for estimating areas from sorption isotherms at room temperature, using ordinary permanent gases and rare gases. This technique differs from the BET method, for which relative pressures are needed, that is, measurements must be made below the critical temperature, and from the use of the de Boer equation for permanent gases.[10] Comparison between these three methods of assessment have been made,[5] and agreement at room temperature is found to be reasonable, although differences by a factor of about 2 may be found. In a given investigation, consistency in using one of them is the important requirement. There is a possibility of error in using the direct comparison with carbon black, owing to the influence of small capillary diameters on the shape of the adsorption isotherm—for the

[6] Griffith, M., and Hirst, W., *Proc. Conf. on Ultra-fine Structure of Coals and Cokes*, BCURA, London, 1944, pp. 80–94; Maggs, F. A. P., *Proc. Conf. on Ultra-fine Structure of Coals and Cokes*, BCURA, London, 1944, pp. 95–109.

[7] Bond, R. L., and Maggs, F. A. P., *Fuel*, **28**, 169–71 (1949).

[8] Bartell, F. E., and Suggitt, R. M., *J. Phys. Chem.*, **58**, 36–40 (1954).

[9] Kiselev, V. F., *Doklady Akad. Nauk S.S.S.R*, **89**, 113–15 (1953); *Chem. Abs.*, **49**, 10038 (1955).

[10] De Boer, J. H., *The Dynamical Character of Adsorption*, The Clarendon Press, Oxford, 1953, 239 pp. Zwietering, P., and van Krevelen, D. W., *Fuel*, **33**, 331–7 (1954).

carbon black such effects are absent and the adsorbing surface is in fact strongly convex. Consequently, porous standards have more recently been used.

The equilibrium moisture contents of coals are dependent on their porous structure.[11] The variation of porosity, heat of wetting, sorption of oil, moisture content, and sorption of dyes with rank follows a curve, or band, with a minimum in the 85 to 90% range of carbon content. True density, friability, calorific value and caking power follow similar curves, but with less pronounced minima or maxima.[2]

Heat of wetting measurements have been developed for investigating molecular sieve behavior in relation to internal surface.[13] The high values of the specific internal surface area of coals found by the heat of wetting method[14] of measurement were disputed on the basis of measurements of the adsorption of nitrogen or argon at very low temperatures, from which internal areas were calculated to be smaller by a factor of about 100.[15] It has since been shown, however, that large specific internal areas are indicated by *any* method applicable at ordinary temperatures.[4,5,16,17,18] In bituminous coals these areas are associated mainly with a capillary system having passages of width 40 A or less, which passages are linked by constrictions even smaller down to 5 to 8 A in width. These lowest figures are based on the assumption that a diffusing molecule is not excluded by any constriction larger than itself. Most of the internal *surface area* in a coal particle is contained within the ultrafine region of the total porous structure, which may represent only 50 to 80% of the internal free volume.

On the smallest scale, it is difficult to envisage definite capillaries with walls, and the controversy has largely resolved itself into an argument over words: whether surface area or some other term better describes the boundaries of these "gaps between molecules." There are, however, certain features of the adsorption isotherms of coal, such as the failure of hysteresis loops to close before zero relative pressure, that suggest a parallel with imbibition rather than sorption, and have not yet been fully explained.[19]

The work of Bond[5,18] has enabled chemical effects to be distinguished from effects due to the distribution of area and volume in relation to constriction size and has also indicated that the "holes" in coal tend to be flat rather than cylindrical or spherical. This work has confirmed the view that polar interaction between coal surface and wetting liquid plays a relatively small part—although with low-rank bituminous coals not a negligible one—in determining the heat of wetting of coal-like materials.

In Fig. 1 the values for specific surface calculated from the neon sorption (Σ_N) and methanol heat of wetting (Σ_M) of some British bright coals are plotted against carbon content (Parr's basis); both curves are of similar shape, and they merge at about 94% carbon. For high-rank coals containing more than about 90% carbon and anthracites, and for carbonized coals of 90% carbon content and above prepared from coals covering a wide range of rank, no differences greater than experimental error have been found between the surface area values obtained by the two methods.[5] However, for coals of lower carbon content the values of Σ_N are always smaller than the corresponding values of Σ_M.

[11] Gauger, A. W., *Trans. Am. Inst. Mining Met. Engrs.*, **101,** 148–61 (1932).

[12] *Proc. Conf. on Ultra-fine Structure of Coals and Cokes*, BCURA, London (1944), 366 pp. King, J. G., and Wilkins, E. T., *Proc. Conf. on Ultra-fine Structure of Coals and Cokes*, BCURA, London, 1944, pp. 46–56. Bangham, D. H., *Proc. Conf. on Ultra-fine Structure of Coals and Cokes*, BCURA, London, 1944, pp. 18–34.

[13] Bond, R. L., and Maggs, F. A. P., *Fuel*, **28,** 172–5 (1949).

[14] Bangham, D. H., *loc. cit.*; *Ann. Repts. on Progr. Chem.* (*Chem. Soc. London*), **40,** 29–43 (1943). Bangham, D. H., Franklin, R. E., Hirst, W., and Maggs, F. A. P., *Fuel*, **28,** 231–7 (1949).

[15] Lecky, J. A., Hall, W. K., and Anderson, R. B., *Nature*, **168,** 124–5 (1951). Malherbe, P. le R., *Fuel*, **30,** 97–109 (1951). Malherbe, P. le R., and Carman, P. C., *Fuel*, **31,** 210–19 (1952). Zwietering, P., Oele, A. P., and van Krevelen, D. W., *Fuel*, **30,** 203–4 (1951).

[16] Zwietering, P., and van Krevelen, D. W., ref. 10.

[17] Maggs, F. A. P., *Nature*, **169,** 793–4 (1952); *Research*, **6,** 13S–14S (1953).

[18] Bond, R. L., *Nature*, **178,** 104–5 (1956); *Brennstoff-Chem.*, **37,** 233–4 (1956).

[19] Anderson, R. B., Hall, W. K., Lecky, J. A., and Stein, K. C., *J. Phys. Chem.*, **60,** 1548–58 (1956).

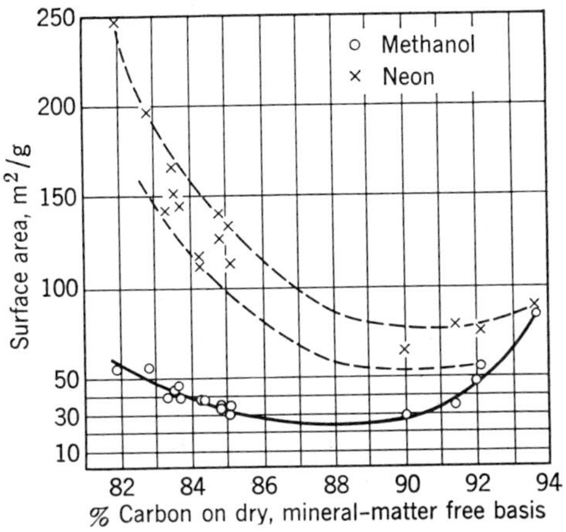

Fig. 1. Surface area of coals as estimated from neon sorption at 25°C and methanol heat of wetting.

Figure 2 shows that $(\Sigma_M - \Sigma_N)$ decreases continuously with increasing carbon content of the coals in contrast to the minimum in Fig. 1. Figure 2 also shows some results that have been obtained for carbonized coals which follow the same general trend. Clearly, the values for specific surface area obtained by the two methods are not uniquely related to one another, but addition of a function linear in carbon (or oxygen) content makes possible a reasonable correlation.

That the surface of *standard reference substances* such as carbon black is more accessible to neon than to methanol is improbable in view of the general agreement between Σ_N and Σ_M for a large number of carbonaceous adsorbents. Some possible causes of the inconsistencies between the two methods are: (1) polar interaction with release of heat may occur between the functional groups on the surface (particularly —OH) and the methanol molecules, causing unduly high estimates of Σ_M to be obtained; (2) methanol may be imbibed, that is, may prise apart molecules, also causing spuriously high values of Σ_M to be deduced; and (3) neon may be adsorbed on a smaller proportion of the sites in low-rank coals than is methanol. Several experimental observations bear on these questions.

A carbon black, containing 6.3% oxygen, of which 1% occurred as —OH, has been examined. Its heats of wetting in methanol and ethylenediamine were 21.7 and about 40 cal/g, respectively. The carbon was heated for several hours in hydrogen at a temperature of 1000°C, during which time the oxygen content fell to less than 1%, and further heat of wetting determinations were then made. The heat of wetting in methanol was unchanged, whereas that in ethylenediamine had fallen to about 20 cal/g, showing that the "cleaned" black behaved in a normal manner with respect to its heat of wetting by these organic liquids.[7] Thus, in contrast to the marked heat of interaction with ethylenediamine, it would appear that interaction between methanol and the surface groups did not contribute a significant amount of *heat* when the carbon was wetted. The values for surface area deduced by the BET method from the sorption of methanol and benzene at 25°C by this carbon black were 216 and 209 m^2/g, respectively,[5] but by nitrogen at 88°K, the value was 690 m^2/g. On the other hand, it has been reported that the amount of *sorption* of methanol by a lampblack was increased several times because of the presence of oxygen on its surface, and that the surface area values deduced from the sorption of polar and nonpolar molecules differed considerably.[20]

Mild oxidation *can* increase the methanol heat of wetting of coals by a factor of 2 or 3[19,21] for oxidized carbon blacks. A sample of

[20] Anderson, R. B., and Emmett, P. H., *J. Appl. Phys.*, **19**, 367–73 (1948).

[21] Sharma, J. N., Kini, K. A., Das Gupta, N. N., and Lahiri, A., *J. Indian Chem. Soc., Ind. & News. Ed.*, **20**, 95–8 (1957).

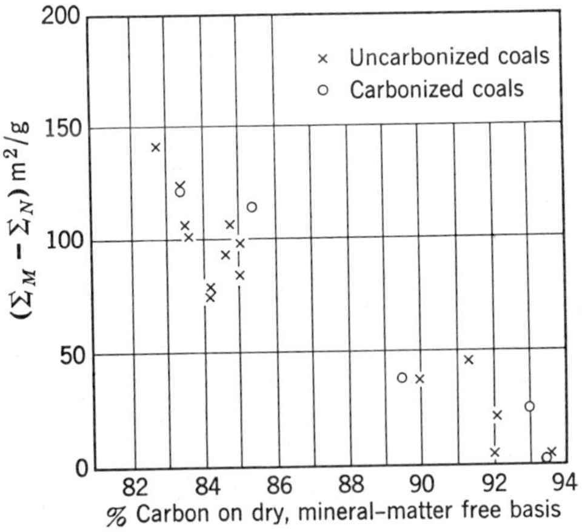

Fig. 2. Difference between estimated surface areas of a range of coals from heat of wetting in methanol (Σ_M) and neon sorption (Σ_N).

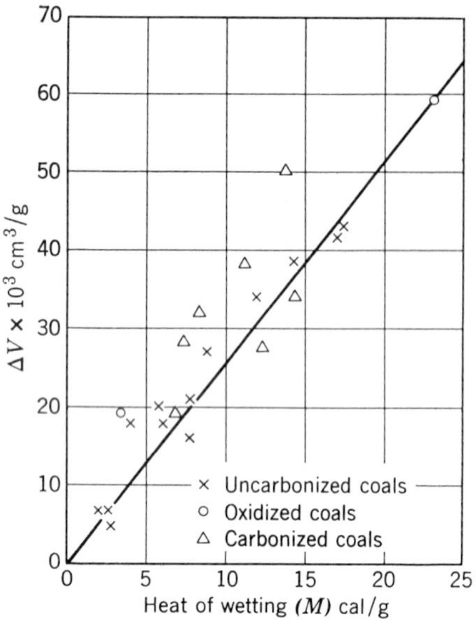

Fig. 3. Relation of the difference in specific volume determined in methanol and helium (ΔV) to the heat of wetting in methanol (M).

South Wales dry steam coal examined immediately after collection from the coal face gave identical estimates of area at room temperature (46 m^2/g) by both neon sorption and methanol heat of wetting; after oxidation the adsorption figure was 47 m^2/g but the heat of wetting figure had increased to 76 m^2/g.[22]

Franklin[23] reported that the density in methanol of high-rank coals and anthracites, as well as of coals of lower rank, always exceeded the corresponding density in helium, and showed that the difference between the specific volumes (ΔV) was linearly related to the methanol heat of wetting (M). A similar graph, Fig. 3, shows results for coals over a range of ranks, carbonized coals, and oxidized coals. The linearity and comparatively low degree of scatter suggest that there is a causal relation between these quantities regardless of the nature of the particular surface; if so, there can be no such relation (cf. Fig. 2) between ΔV and Σ_N. If ΔV were the result of a surface compression in the adsorbed layer as postulated by Franklin, the appropriate measure of the surface area would therefore seem to be Σ_M and not Σ_N.

Acetylation leads to a reduction of the methanol heat of wetting and the moisture sorption of low-rank coals.[24] A similar experiment carried out on a British low-rank coal of 83.6% carbon in which the coal was partially acetylated with ketene also gave a product whose heat of wetting in methanol was lower than that of the coal. These experiments suggest that polar interaction is partly the cause of high heats of wetting of low-rank coals, but not wholly, unless methanol interacts also with the added acetyl groups.

It may be deduced from the foregoing considerations that it is not possible at present to determine absolute values for the surface areas of low-rank coals. A set of values *between* those obtained by rare gas sorption and heat of wetting is unlikely to be greatly in error. The various contradictions may be connected with the artificiality of the concept of surface in intermolecular passages. Either a steady decrease in the adsorption of the rare gases per unit "area" of the surface with a decrease in rank, or a steady increase in the heat liberated by unit area on wetting with methanol, might be attributed to any of the physical and chemical characteristics that change monotonically with rank. Thus the extent of the internal surface area of coals might be said to be in some respects at least a matter of definition; the definition must stem from or be related to the method of measurement.

The original critics[19] of the conventional interpretation of heat of wetting measurements in methanol are now largely in agreement with an "internal surface" situated in capillaries or holes of molecular dimensions amounting to 30 to 100 m^2/g, which becomes inaccessible to gases such as nitrogen and hydrogen at characteristic temperatures in the range of 100° to 200°K, according to the molecular size of the gas.[17] Measurements made below the characteristic temperatures indicate, at least for coking coals, areas about a hundred-fold smaller, which probably results from excessively slow diffusion of gas into the porous system at low temperatures.[25]

Ingenious use of heat of wetting has been made to distinguish between molecular sieve

[22] Bond, R. L., Private communication.

[23] Franklin, R. E., *Trans. Faraday Soc.*, **45**, 274–86 (1949).

[24] Mazumdar, B. K., Bhangale, D. H., and Lahiri, A., *Fuel*, **36**, 254–6 (1957).

[25] Van der Sommen, J., Zwietering, P., Eillebrecht, B. J. M., and van Krevelen, D. W., *Fuel*, **34**, 444–8 (1955).

and chemical or polar effects.[26,27] Measurements to estimate reactive-group contents of the coal, using the differences between heats of wetting in polar and nonpolar liquids, are unsatisfactory. The differences are often within the experimental error of measurement.[26] The deduced contents of hydroxyl are too small in relation to the total oxygen in the coals, which in itself indicates the unsuitability of the method. The heat of wetting method indicated[26] only 1 to 5% as hydroxyl or 5 to 20% as hydroxyl plus carbonyl, whereas other measurements on the same coals showed[28] that more than 40% of the oxygen in fact is contained in hydroxyl groups.

The minor differences in heat of wetting according to the nature rather than the size of the molecule, although unquestionably real, are merely contributory to the factors of 1 to 4 that relate the heat of wetting or adsorption capacity of a low-rank coal, oxidized coal, or oxidized carbon to the corresponding quantity for a pure carbon surface. It must be emphasized that they have no relevance, as is still sometimes implied,[26,27,29] to the major discrepancy—up to a hundred-fold—between areas estimated from heat of wetting or adsorption at room temperature and those from the BET low-temperature adsorption method. Comparisons between permeability areas and BET areas on coals and oxidized coals[30] also have no relevance to the major discrepancy. These comparisons refer only to the surface of the particles and the macrocapillaries.

Several conclusions may be drawn from the foregoing review. Bituminous coals have specific internal surfaces in the range of 30 to 100 m^2/g situated almost entirely in the ultrafine capillaries of widths below 40 A and accessible only through constrictions of width 5 to 8 A. This surface can be measured only by methods applicable at temperatures not too far below room temperature, the exact temperature depending on the nature of the gas or vapor used; activated diffusion into the microcapillary system becomes exceedingly slow at low temperatures. Measurements at liquid air temperatures relate only to the surfaces of particles and macrocapillaries and yield values too low by a factor of up to 100. The heat evolved on wetting unit *surface* increases by a factor of up to 4, and adsorption per unit of surface in equivalent conditions also increases with the content of accessible oxygen-containing groups if the molecules of liquid or vapor are polar in nature. The major effects on the heat of wetting of unit *weight*, or extent of adsorption, are however determined by the molecular size of adsorbate or wetting liquid rather than by these factors; conversely, chemical methods for estimating reactive oxygen groups are chiefly determined by the chemical nature of the reagent and the reaction conditions employed, the molecular sieve effects here being secondary. A knowledge of total surface and molecular sieve parameters, and more exact knowledge of hydrogen-bonding energy terms and of the contribution of adsorption on nonpolar sites to the energy are necessary for estimating oxygen groups from the heat of wetting.

[26] Kini, K. A., Nandi, S. P., Sharma, J. N., Iyengar, M. S., and Lahiri, A., *Fuel*, **35**, 71–6 (1956). Kini, K. A., Nandi, S. P., Sharma, J. N., and Lahiri, A., *Fuel*, **36**, 154–8 (1957).

[27] Iyengar, M. S., and Lahiri, A., *Fuel*, **36**, 286–97 (1957).

[28] Mazumdar, B. K., Bhangale, P. H., and Lahiri, A., *J. Sci. Ind. Research* (*India*), **15B**, 44–5 (1956); *Fuel*, **36**, 307–12 (1957).

[29] Nandi, S. P., Kini, K. A., and Lahiri, A., *Fuel*, **35**, 133–4 (1956).

[30] Kini, K. A., Ganguli, N. C., and Lahiri, A., *Proc. 2nd Intern. Congr. Surface Activity*, Butterworths, London, 1957, Vol. 2, pp. 330–4.

Solvent Extraction

Extraction of coal by solvents has been for many years a method used for studying the constitution of coal and for producing products of potential industrial value. Several detailed reviews of the experimental work have been published.[31–35] Both yield and nature of the extract depend on the solvent, the conditions of extraction, and the composition of the coal.

[31] Bakes, W. E., *Gt. Brit., Dept. Sci. Ind. Research, Fuel Research Tech. Paper* No. 37 (1933), 214 pp.

[32] Kiebler, M. W., *Chemistry of Coal Utilization*, John Wiley and Sons, New York, 1945, Vol. 1, 715–60.

[33] Dryden, I. G. C., *Fuel*, **29**, 197–207, 221–8 (1950).

[34] Kröger, C., *Erdöl u. Kohle*, **9**, 441–6, 516–20, 620–4, 839–43 (1956).

[35] Roga, B., and Karabon, B., *Wiadomości Chem.*, **11**, 649–61 (1957).

Effect of Physical Properties of Solvents on Extraction

Knowledge of solubility has developed rapidly and has led increasingly into the realm of thermodynamics. The solubility of a solid in a liquid with which it forms an ideal solution, that is, one which obeys Raoult's Law, can be formulated in terms of the melting point and the heat of fusion of the solid. Such solutions are possible only when molecules differ little in size and shape and are of similar and non-polar chemical type. The heat of solution is zero, and the entropy of solution corresponds to random mixing. When the molecules differ in size, shape, and type, the situation is more complex; the entropy differs from the ideal, and the heat of solution depends on the relative cohesion of solute and solvent molecules. Furthermore, when polar molecules are involved, there must be mutual interaction between the liquid molecules themselves, the solid molecules themselves, and between the solid and the liquid molecules; in this case there may be a considerable heat of mixing, and the entropy will differ greatly from the ideal value owing to preferred orientation of the molecules, etc. In extreme cases such as this, the simple solubility equation for ideal solutions bears no relation to the observed solubilities; it is only approximately applicable even to the simplest systems.

When we consider the solubility of a non-polar polymer in liquids of chemical type similar to itself, the cohesive energy density, which is related to the latent heat of vaporization and the "internal pressure," is an important factor; swelling or solubility is a maximum when the cohesive energy densities of solid and liquid are equal. The system of rubber in hydrocarbon solvents approximately fulfills this condition. At the other extreme, we have a typical polar system, agar and water. It is possible that coal may exhibit one type of behavior with one class of liquids, another behavior with another class. Until recently, the tendency to seek empirical correlations with factors such as enter into the fundamental theory has guided research on the solubility of coal, and little beyond this is available for review.

It has been shown that solvents which cause swelling and solution of high polymers fall into zones when plotted on a diagram with coordinates μ^2/ϵ (μ = dipole moment, ϵ = dielectric constant) and surface tension; the potency increases toward the center of the zone.[36] Although this is a promising line of thought based on bulk properties alone—since μ^2/ϵ is sensitive both to chemical structure and physical condition—it has nevertheless been found to be of little help in classifying solvents for coals.[37,38]

Good solvents swell the coal substance, but so do some poor solvents.[39,40] Such a one-way connection is to be expected if the soluble portion is in some way trapped in a colloidal matrix.[41] It is highly probable that at least two, and possibly more, properties must be favorable if a substance is to be a good solvent for coal. However, up to the visible decomposition temperature, the necessary conditions grow less stringent as the temperature of extraction is increased, and much above this temperature all solvents become ineffective. The characteristics in question need not all be expressible as bulk physical properties. Furthermore, the relative solvent power of two solvents may occasionally be reversed by a change in the type of coal treated.[42]

The physical properties of solvents most usually considered are surface tension[43] and internal pressure.[32,44] These quantities are related by a length parameter related to the molecular diameter, so that for most common solvents they stand in the same relative order. The examples of glycol and water, both exceptionally poor coal solvents possessing high internal pressures owing to the hydroxyl association, show that all types of intermolecular forces contributing to internal pressure cannot be effective. The relation of yield to internal pressure at elevated temperatures is approximately linear.[32,44] The correlation of

[36] Ostwald, W., and Ortlov, H., *Kolloid Z.*, **59**, 25–32 (1932). Ostwald, W., *J. Oil & Colour Chemists' Assoc.*, **22**, 31–46 (1939).

[37] Dryden, I. G. C., *Chem. & Ind.* (*London*), **30**, 502–8 (1952).

[38] Dryden, I. G. C., *Fuel*, **30**, 39–44 (1951).

[39] Losikov, B. V., *Izvest. Vsesoyuz. Teplotekh. Inst.*, **14** (5), 26–33 (1941).

[40] Dryden, I. G. C., *Nature*, **162**, 959–60 (1948); **163**, 141 (1949).

[41] Agde, G., and Hubertus, R., *Braunkohlenarch.*, **46**, 3–30 (1936).

[42] Dryden, I. G. C., *Fuel*, **30**, 217–33 (1951).

[43] Kreulen, D. J. W., *Fuel*, **25**, 99–108 (1946).

[44] Pertierra, J. M., *Anales fis. y quim.* (*Madrid*), **37**, 58–68 (1941).

solvent power with internal pressure has been found to be inapplicable to the extraction of coal at temperatures below the normal boiling point of the solvent.[37,38]

The specific solvent power for low-rank coals of the primary aliphatic amines and related compounds has been attributed to the availability of the unshared pair of electrons on the nitrogen atom.[37,38,45] This is tantamount to saying that the amines behave as polar solvents toward coal, and therefore that coal-amine systems are to be placed in the water-agar category discussed earlier. Underlying chemical effects of this nature are reflected in the values of the thermodynamic functions such as free energy, heat, and entropy of solution; the relation of these functions to solubility behavior, being entirely based on fundamental thermodynamics, is the same for both polar and nonpolar types of system. With this in mind, an attempt has been made to deduce[46] thermodynamic functions for the coal-amine system from experimental data.[47] The values obtained show that in the expression for free energy, the heat term exceeds the entropy term, whereas the opposite is usual in the better-understood nonpolar systems.

An empirical extension of fundamental solution theory to the explanation of the variation of extraction yield with temperature[48,49] has indicated how such functions might be applied in estimating the mean molecular weight of the coal. The results of this investigation were in accordance with the hypothesis that the coal consists of an assembly of units that are large compared with ordinary molecular sizes, and with a wide size distribution. However, the data available were insufficient to enable the mean size of the unit to be estimated.

Effect of Chemical Structure of Solvent on Extraction

In any discussion of solvent extraction of coal, specific chemical effects of solvents must be considered in addition to the physical properties. It has been suggested that extraction of coal by benzene under pressure and pyridine be considered as two different processes.[40]

In extraction of coal at about 400°C, a two-ring solvent containing a hydroxyl group attached to an aromatic ring and a hydroaromatic ring (a potential hydrogenating agent) was found to be particularly effective.[50] Hydrogen under pressure was found to compensate for the lack of a hydroaromatic ring. The degree of extraction at atmospheric pressure from a bituminous coal with a number of solvents containing one or more rings was found to correlate well with the boiling point. Exceptions were observed both above and below the average, most probably owing to the specific influence of chemical structure. The correlation with boiling point was found to extend only to aromatic solvents. In the higher rank bituminous range, α- and β-naphthol were both effective solvents; with a subbituminous coal, β-naphthol produced even less extract than phenanthrene, but α-naphthol yielded about 83%. This would appear to be a specific effect. Angular ring systems such as phenanthrene appeared to be superior for extracting bituminous coals of medium rank.[51,52] In a study of extraction of Dutch bituminous coals at 200 to 400°C, the importance of simultaneous hydrogenation *via* the hydroaromatic portion of the solvents used was shown; diphenylamine was found an effective catalyst for transfer of hydrogen in this manner since it increased the yield of extract.[53]

Between 360 and 400°C, the solvent power has been found to decrease in the following order: amines, phenols, cyclic hydrocarbons, aliphatic hydrocarbons.[54] The amines were of

[45] Dryden, I. G. C., *Nature*, **166**, 561–2 (1950).

[46] Dryden, I. G. C., *Chem. & Ind.* (*London*), **30**, 193 (1952).

[47] Dresel, E. M., Pritchard, E., and Dryden, I. G. C., *Research* (*London*), **4**, 578–9 (1951). Dresel, E. M., Dryden, I. G. C., and Farenden, P. J., *Research* (*London*), **4**, 579–81 (1951).

[48] Dryden, I. G. C., *Discussions Faraday Soc.*, No. 11, 28–42 (1951).

[49] Dryden, I. G. C., *Fuel*, **31**, 176–99 (1952).

[50] Orchin, M., and Storch, H. H., *Ind. Eng. Chem.*, **40**, 1385–9 (1948).

[51] Orchin, M., Golumbic, C., Anderson, J. B., and Storch, H. H., *U.S. Bur. Mines, Bull.*, No. 505 (1951), 15 pp.

[52] Golumbic, C., Anderson, J. B., Orchin, M., and Storch, H. H., *U.S. Bur. Mines, Rept. Invest.* No. 4662 (1950), 12 pp.

[53] Oele, A. P., Waterman, H. I., Goedkoop, M. L., and van Krevelen, D. W., *Fuel*, **30**, 169–77 (1951).

[54] D'yakova, M. K., and Davtyan, N. A., *Bull. acad. sci. U.R.S.S., Classe sci. tech.*, **1945**, 203–8; *J. Appl. Chem.* (*U.S.S.R.*), **21**, 113–25 (1948).

the aniline type. Substitution of a radical for hydrogen in groups such as OH, NH, NH_2 increased the solvent power (with aliphatic amines at lower temperatures, the reverse is true). Insertion of a carbonyl group in the *para*-position caused a sharp increase in effectiveness.

Dryden[37,38,45] concluded that the important factor determining the abnormally high (specific) solvent power of amines and related solvents for lower rank bituminous coals was the availability of the unshared pair of electrons on a nitrogen or an oxygen atom. The chief specific solvents are primary aliphatic amines, with or without aromatic and hydroxylic substituents, and pyridine and certain other heterocyclic bases. The great superiority of primary over secondary and tertiary aliphatic amines as solvents for coal was explained by the absence, in the primary amines, of steric hindrance related to the presence of more than one alkyl group on the amine, and of groups surrounding the centers in the coal that interact with amines. There is evidence that the unshared electron pair in the solvent interacts, thus releasing heat, with some oxygen-containing group which is only present, or exposed to solvent action, in coals containing less than about 90% C.[55]

Many organic liquids have been suspected of exercising more than a solvent action on coal. An indication of chemical interaction is the observation that the total weight of products sometimes exceeds the original weight of coal, although up to about 5% may be the result of strong adsorption of solvents on the residue and extract. With mixed solvents, the dangers of interaction are greater.

Extraction results with dioxane may be misleading, owing to the formation of decomposition products from the solvent, and alcohol solvents may esterify when used to extract oxidized coals.[56]

Mixtures of the higher ketones and formamides have been reported to be considerably better solvents for a Northumberland coal (82% C) than either of the pure components. The best were acetophenone-monomethylformamide and methylcyclohexanone-dimethylformamide, both in equimolecular proportions.[57] Some solvent pairs show enhanced solvent power, whereas others appear to behave independently in admixture.

With regard to the successive use of different solvents at or below their normal boiling points, Kiebler considered[32] that, within a considerable experimental error, the total amount of extract obtainable had been proved to equal the yield obtained with the better solvent alone. However, it has been reported that pretreatment of coal with either retene or anthracene oil increased its solubility in pyridine.[58]

Influence of Rank of Coal on Extraction

Benzene-type solvents. The results of extracting Pakistan coals with benzene, benzene/ethanol, and pyridine have been described by Islam and Ahmed.[59] The results obtained by a number of investigators are summarized in Table 2. It is not easy to discern more than a very broad pattern in these observations; the conflict of detail suggests that differences in technique and in the petrographic composition of the coal studied influenced the results. It may be concluded that above 88 or 89% C content, or below about 25% volatile matter, the amount of extract obtainable decreases rapidly; with coals of carbon content lower than this limit, no definite trend with rank has been proved.

Pyridine-type solvents. Table 3 shows the influence of rank on extraction by pyridine-type solvents.

With ethylenediamine, a correlation between yield of extract and carbon content has been found[42] with a scatter of points about a mean line usual with any properties of a series of coals (Fig. 4). Other amine solvents showed[42]

[55] Bangham, D. H., and Dryden, I. G. C., *Fuel*, **29**, 291–2 (1950).

[56] Crick, R. G. D., *Fuel*, **30**, 187–90 (1951).

[57] Rybicka, S. M., *Fuel*, **38**, 45–54 (1959).

[58] Zabavin, V. I., and Surkova, V. L., *Bull. acad. sci. U.R.S.S., Classe sci. tech.*, **1946**, 975–88.

[59] Islam, M. S., and Ahmed, M., *Pakistan J. Sci. Research*, **5**, 45–7 (1953).

[60] Lowry, H. H., and Rose, H. J., *U.S. Bur. Mines, Inform. Circ.* No. 7420 (1947), 12 pp.

[61] Zabavin, V. I., *Bull. acad. sci. U.R.S.S., Classe sci. tech.*, **4**, 27–40 (1941). Zabavin, V. I., and Surkova, V. L., *Trudy Inst. Goryuchikh Iskopaemykh, Akad. Nauk S.S.S.R., Otdel. Tekh. Nauk*, **3**, 167–82 (1954). Novikov, V. N., *J. Appl. Chem. (U.S.S.R.)*, **24**, 314–24 (1951).

[62] Tebbett, R. W., Ward, S. G., and Weighall, J. G., *Birmingham Univ. Mining Dept., Res. Bull.* No. 1, 7–39 (1941); *Iron & Coal Trades Rev.*, **143**, 49 (1941).

TABLE 2

INFLUENCE OF RANK ON DEGREE OF EXTRACTION BY BENZENE-TYPE SOLVENTS

Yield	Solvent	Details
Increasing with decreasing rank	Naphthalene[54]	
	Hydrogenated pitch[60]	
Exhibiting a maximum	Anthracene oil, retene, pyridine[58,61]	Ten coals, 14.6 to 43.7% volatile matter, pronounced peak in 30 to 40% range with rapid decrease below about 20% volatile matter.
	Neutral tar-oil[62]	Indication of flat maximum between 25 to 40% volatile matter, but points very scattered.
	Tetralin[54]	Yield increasing with increasing rank, then decreasing.
	Phenanthrene[52]	Coals from lignite to anthracite; maximum at coal of 84.4% C.

similar but not identical behavior: for example, the yield obtained with monoethanolamine fell steeply with increase in rank even below 85% C content; on the other hand, the yield from extraction with benzylamine, piperidine, and pyridine showed much less variation with rank than that shown in Fig. 4. Although the effectiveness of all these solvents fell off rapidly above 88% C content, below this point pyridine showed a slight *increase* in extraction with increase in rank. The relative efficacies of the solvents were found to depend not only on the rank of the coal but also on the temperature of extraction.

The foregoing evidence is consistent with the view: (1) that the yield probably decreases or increases slowly with increase in carbon content from 80% to 85 or 87%, depending on the solvent; (2) that the effect in this range may be masked by petrographic differences and perhaps by the use of a solvent such as pyridine, which does not give the full possible yield without stringent precautions; (3) that above 85 to 87% C the falling off in yield accelerates; (4) that above 88 to 89% C the amount of extract is small; and (5) that beyond 92 to 93% C the yield becomes negligible. Below 80% C the yield may also possibly decrease. These conclusions agree broadly with those reached for benzene-type solvents, particularly in that the yield of extract falls to a low value above about 88% C content.

There is a correspondence between yield of extract and the saturation sorption or imbibition of solvent, independent of the rank of coal or the particular amine solvent[42,65] (Fig. 4). An adsorption isotherm for ethylenediamine vapor on a bright Northumberland coal of

[63] Blayden, H. E., Gibson, J., and Riley, H. L., *J. Chem. Soc.*, **1948,** 1693–700.

[64] Wynne-Jones, W. F. K., Blayden, H. E., and Shaw, F., *Brennstoff-Chem.*, **33,** 201–6 (1952).

[65] Dryden, I. G. C., *Nature*, **166,** 606 (1950).

TABLE 3

INFLUENCE OF RANK ON DEGREE OF EXTRACTION BY PYRIDINE-TYPE SOLVENTS

Yield	Solvent	Details
Increasing with decreasing rank	Pyridine[63]	Two coals of 90.4 and 88.5% C.
	Ethylenediamine[40,42]	British bright bituminous and anthracite coals, 78 to 94% C; particularly rapid decrease in yield between 85 and 87% C.
Exhibiting no correlation	Pyridine[64]	S. Wales coals—maximum extraction at 30% volatile matter

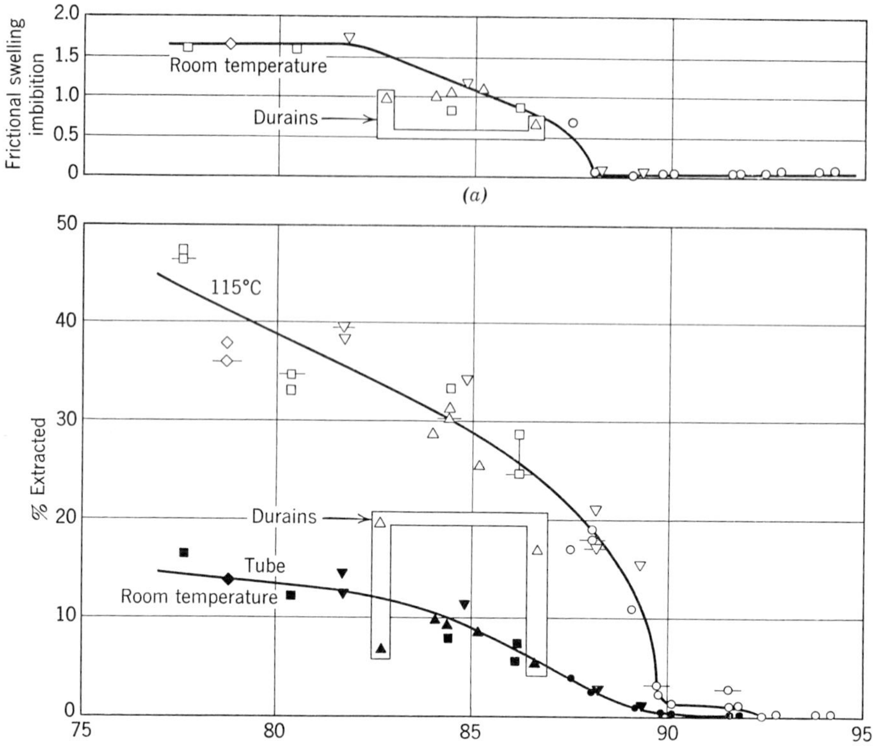

Fig. 4. Interaction of coals with ethylenediamine: (*a*) Imbibition of solvent; (*b*) extraction yield.[44]

about 82% C content showed three main features; (1) chemisorption up to 3 to 6% adsorbed, (2) a fairly normal sorption isotherm from the completion of chemisorption up to a relative pressure of at least 0.8, and (3) a steeply rising indefinite region near saturation which corresponded to observable dissolution of the coal.[47] The kinetics of sorption of pyridine has been studied for coals of different ranks.[66]

Influence of Petrological Constitution on Extraction

It is generally agreed that vitrain is the most soluble constituent of a given coal and fusain the least. The same is true of the sorptive power for solvent vapors. A report on the Pott-Broche process[60] makes it clear that, even at temperatures of about 400°C, durain is less responsive to solvents than vitrain, whereas fusain is for practical purposes insoluble. Dryden,[42] using ethylenediamine, found that the yield from a black durain was similar to those from bright coals of the same carbon content. One exinite concentrate had a lower solubility. Gray durain and fusain contained components that were inert toward the solvent. Extraction yield and imbibition were closely interrelated, as with bright coals, except for fusain and cannel coal.

Oversohl[67] claimed that treatment with benzene above approximately 280°C rendered visible a cellular structure in vitrain. He also found, as is usual, a smaller yield of extract from petrological constituents other than vitrain, and propounded a theory of extraction involving first the removal of colloidal material embedded in the cellular structure and later the bursting of the cells with solution of their contents. The relation of this hypothesis to the various micellar and molecular extraction mechanisms that have been proposed is not clear.

Coals with High Organic Sulfur Content

In a study of extraction of a high-sulfur Sardinian coal with tetrahydrofuran and also

[66] Kuznetzov, M. I., and Dybskii, V. V., *J. Appl. Chem.* (*U.S.S.R.*), **20,** 976–96 (1947). Rotinyan, A. L., *J. Appl. Chem.* (*U.S.S.R.*), **21,** 751–4 (1948).

[67] Oversohl, W., *Brennstoff-Chem.*, **31,** 103–11 (1950).

with a series of "specific" solvents, the latter gave the highest yields of extract. When the temperature of extraction was increased to 350°C, evidence of partial decomposition was noted. Observations in the infrared region from 690 to 610 cm^{-1} characteristic of aromatic C—H out-of-plane vibrations suggested that less highly substituted material was preferentially extracted at the lower temperatures. Treatment with tetralin at 350°C in an atmosphere of hydrogen yielded 87% extract.[68]

Extraction yields of the high-sulfur coal "Raša," from Istria, ranged from 0.7% in benzene at the boiling point to 24% in benzene under pressure at 250°C and 43% in pyridine.[69]

Roy[70] concluded that small amounts of diphenylamine, 2-methylnaphthalene and γ-phenylpyridine were present in the solvent extract from a sulfur-rich Assam coal.

Conditions of Extraction and Their Influence on Yield of Extract

Influence of Time

Dryden[71] has examined the effect of time of extraction on yield of extract, using bituminous coals of various ranks and primary aliphatic amines as solvents. The time effect differed according to whether (1) the coal was shaken in a tube with the solvent at room temperature, or (2) it was extracted in a Soxhlet thimble near the boiling point of the solvent; inadequate diffusion through the thimble was found to be a major limiting factor. Although the Soxhlet yield approached a limiting value asymptotically in a period of days, no evidence was found of an absolute limit over a much longer time. Comparison of yields after a standard time of extraction, for example, 48 hr in a Soxhlet apparatus at standard reflux rate, was recommended in preference to any attempt at extrapolation. When the coal and solvent were shaken together, extraction was much more rapid. With coal ground through 72 mesh, 1 hr was found to be adequate. Excessive shaking sometimes had a deleterious effect on yield. When solvent and coal are heated together, there is always a danger that the extraction yield will pass through a maximum because of aggregation or resorption of the extract, and a Soxhlet type or a countercurrent method should be used when possible.[62]

Influence of Temperature

The effect of temperature is most marked for benzene-type solvents.[72] With neutral tar-oils, a sharp rise in yield from less than 3% at about 200°C to more than 50% at between 350 and 400°C has been reported.[62] Ethylenediamine extracts at least half as much from low-rank bituminous coals at room temperature as at its boiling point (116°C).[40]

In a series of the primary aliphatic amines, the limiting yield reached in a period of days has been found to be independent of the method of extraction and depends to a first order of approximation on the temperature of extraction rather than on the particular amine used.[71] An explanation of this in terms of the known solubility behavior of high polymers has been proposed.[48] In a study of extraction of coal with tar oil at 300 to 400°C it has been reported that the logarithm of the yield of extract varied linearly with the reciprocal of the absolute temperature, a relationship also noted for amine solvents at lower temperatures.[48]

Influence of Particle Size

Yields tend to increase when the particle size of the treated coal is reduced. The greatest effect is shown when coals are milled to the order of size of 1 μ (10^{-3} mm). In extraction of bituminous coals with primary aliphatic amines, the effect of particle size over the range 72 to 240 mesh B.S.S. was always small but increased somewhat with increasing rank,[71] as would be expected from the fall in porosity.[23]

Influence of Moisture and Atmospheric Oxygen

The presence of moisture in the solvent and of oxygen in the extraction apparatus both tend to decrease the yield obtained. The efficiency

[68] Berti, V., Siniramed, C., and Manci, C., *Riv. combustibili*, **8**, 639–67 (1954). Berti, V., Casu, B., Manci, C., Padovani, C., and Siniramed, C., *Riv. combustibili*, **9**, 5–34 (1955). Berti, V., Casu, B., Manci, C., and Siniramed, C., *Riv. combustibili*, **11**, 91–104 (1957).

[69] Popović, B., and Samec, M., *Sprejeto Na Seji Razreda, Dne* **11**, April (1951). Jenčič, S., *Acad. Sci. et Art. Sloven., Classis III, Ser. A*, **3**, 77–90 (1951).

[70] Roy, M. M., *Fuel*, **36**, 344–54 (1957).

[71] Dryden, I. G. C., *Fuel*, **30**, 145–58 (1951).

[72] D'yakova, M. K., and Surovtseva, V. V., *Zhur. Priklad. Khim.*, **28**, 65–70 (1955).

of extraction of coals by alcoholic potash is reduced by the presence of water.[73] With amine solvents, the virtual absence of moisture is essential if good yields of extract are to be obtained[40] because its "specific" activity is directly affected by water.[37,38] However, the presence of moisture does not appear to be critically important with the majority of solvents that are immiscible with water. Moisture has been reported to *increase* the extraction yield with β-naphthol.[53]

It has been observed for tar-oils that the yield of extract was increased when air in the extraction apparatus was replaced either by nitrogen or by carbon dioxide.[62]

Influence of Pretreatment

The form of pretreatment most frequently studied consists in heating the coal out of contact with air. Preheating in nitrogen at temperatures up to 300°C increased the degree of extraction of a Durham coking coal (88.5% C, 27.6% volatile matter) with pyridine; a further but smaller peak was observed at 400°C. A South Wales coking coal (90.4% C, 19.2% volatile matter) exhibited a maximum at about 310°C.[63] Preheating vitrains above 300°C under water at a pressure exceeding 500 atm appears to cause them to melt and convert them to a product that can be diluted with oils.[67]

Oxidation of the coal substance, effected by pretreatment with air at atmospheric or slightly higher temperatures (100°C), also influences the amount of extract obtainable. Oxidation of a Pittsburgh Seam coal (84.4% C), with phenanthrene at its boiling point decreased the yield of extract from 95% to 30%.[51] Oxidation of a Durham coking coal (88.5% C) did not appreciably modify the yield of extract with pyridine,[63] although the corresponding X-ray *c*-dimension curve was profoundly changed.[74] The γ-fractions of both the original and the oxidized coal showed the usual *c*-dimension peak and exhibited caking properties, whereas the α and β fractions did not; oxidation depressed all the *c*-dimension values by 2 to 3 A when all the fractions had gained approximately the same absolute percentage weight of oxygen.

[73] Berkowitz, N., *Nature*, **162,** 960–1 (1948).

[74] Blayden, H. E., Gibson, J., and Riley, H. L., *Proc. Conf. Ultrafine Structure of Coals and Cokes*, BCURA, London, 1944, pp. 176–231.

On the whole, it may be said that oxidation normally decreases the yield obtainable with either benzene or pyridine. Since the effect depends upon the differential solubility of coal and its primary oxidation products, it is unlikely that a single rule can cover all cases.

Influence of Miscellaneous Factors

Berkowitz subjected the residues, the α fraction, from a 14-day extraction of various coals with pyridine in a Soxhlet apparatus to the influence of ultrasonic vibrations for 15 min, and thus obtained further yields of extract. He attributed the increase in yield to dispersion of the colloidal units constituting the residue.[75] In contrast, another study of extraction of coal with primary aliphatic amines showed that treatment with ultrasonics appeared to hasten greatly the process of solution without affecting the limiting value. Although mechanical shaking also increased the rate of solution, it was found to have an adverse effect on the yield of soluble material obtainable in a single treatment, but again the ultimate yield after several successive treatments was not affected.[65,71]

Currell and Zechmeister[76] observed that aromatic compounds, including benzene, phenol, tropolone, and pyridine, were decomposed when exposed in water to high-frequency ultrasonic radiation for 3 to 40 hr. Simple compounds such as acetylene and cyanides were recovered in substantial yields. This suggests a warning against the use of high-frequency or prolonged ultrasonic treatment to assist the extraction of coal for scientific studies.

An increase in the yield of certain solvent extract fractions from a powdered black coal has been reported when a pressure of 20 ton/cm^2 was applied for 5 min.[77]

The Extraction Products

Primary Fractionation of Extracts

Several combinations of solvents have been developed for fractionation of the extracts

[75] Berkowitz, N., *Nature*, **163,** 809–10 (1949).

[76] Currell, D. L., and Zechmeister, L., *J. Am. Chem. Soc.*, **80,** 205–8 (1958).

[77] Trifonov, I., Spasov, A., Enev, E., Torbova, M., and Petkova, Z., *Godishnik Sofiyskiya Univ., Fiz. Mat. Fak., Kniga* 2, *Khim.*, **1950,** 115–20.

from coal, many of which have been used only, or chiefly, by their originators. Among these may be mentioned morpholine and hexachlorbiphenyl,[78] anthracene oil, and retene,[58] and carbon disulfide, carbon tetrachloride, and petroleum ether.[79] It has been shown that the fractionation by pyridine and chloroform into α, β, and γ constituents does not yield the same results on a micro-scale as on a macro-scale.[63] Fractionation of extracts obtained from bituminous coal with tetralin and *m*-cresol mixtures by adding ether, dioxane and pyridine yielded fractions with different H/C ratios and molecular weights.[53] Phenanthrene extracts may be fractionated into products having different degrees of aromaticity, molecular weights, and coking properties by the use of benzene and cyclohexane.[51]

Chromatographic fractionation of coal extracts followed by spectrometric examination of the products has been studied by several investigators.[80,81] All subfractions from the γ-fraction of the pyridine extracts from two coals showed absorption maxima at 2600 and 2300 A.[81]

Fractionation of coal extracts obtained with ethylenediamine by precipitation with nonsolvents, chromatography, and electrophoresis rarely gave discrete chromatographic bands, the usual result being a continuous smudge of brown material spread out diffusely on development. Sharp differentiation either by type or by molecular weight appeared difficult.[49] In an extension of this work to partition chromatography on paper, using ethylenediamine as the stationary phase, and other solvents, such as monoethanolamine, *m*-cresol, pyridine and piperidine, all containing 10% of ethylenediamine as the mobile phase, a few discrete bands, all brown in color, were obtained.[82]

Coking Properties of Extracts and Residues

This subject has given rise to much confusion, particularly since a large proportion of the earlier work was directed toward the isolation of a hypothetical coking principle; but certain conclusions now stand out clearly.[83] There is general agreement that the swelling and agglutinating properties are found particularly in the γ (chloroform-soluble) fraction of pyridine extract, and that the residue by itself is almost or entirely inert in this respect.

It has been observed that the yield of extract obtained with benzene and cyclohexanone runs parallel to the coking power of the coal, but the yield from pyridine extraction does not.[84] The extraction yield from different types of coal in anthracene oil has been found to correlate well with the agglutinating values and with the thickness of the plastic layer.[85] Similar results using phenol, naphthalene, pyridine and anthracene oils have been reported.[86] Coking power was also found to be related to the degree of swelling in acetone, pyridine, and aniline.[86] From a study of extraction of coking coals with alcohol-benzene and phenol-benzene, it was concluded that there is a direct relation between extract yield and coking power.[87]

When remixture of extract and residue is attempted, reports of the behavior of the so-called reconstituted coals are conflicting. The great importance of the physical structure of the whole material, which is not reconstituted by mere mixing, and of the cooperation of all fractions in the coking process has been emphasized.[63,88] This can be understood

[78] Bouilloux, G., *Bull. soc. chim. France*, **9**, 779–82 (1942).

[79] Baba, A., *Bull. Resources Research Inst. Japan*, No. 3, (1952), 31 pp.

[80] Lahiri, A., and Mikolajewski, E., *Nature*, **155**, 77–8 (1945); *Fuel*, **24**, 66–73 (1945). Steinbrecher, H., and Behling, R. D., *Brennstoff-Chem.*, **39**, 146–8 (1958). Kirkby, W. A., and Sarjant, R. J., *Nature*, **170**, 597–9 (1952).

[81] Barclay, L. R. C., and Layton, T. M., *Fuel*, **35**, 31–7 (1956).

[82] Brooks, J. D., *Research (London)*, **5**, 93–4 (1952).

[83] Bakes, W. E., *Gt. Britain, Dept. Sci. Ind. Research, Fuel Research Tech. Paper* No. 37, 50, 67 (1933). Kiebler, M. W., *Chemistry of Coal Utilization*, John Wiley and Sons, New York, 1945, Vol. 1, p. 755.

[84] Funasaka, W., Yokokawa, C., Suga, S., Kito, Y., and Awaya, T., *Bull. Inst. Chem. Research, Kyoto Univ.*, **25**, 18–23 (1951).

[85] Zabavin, V. I., and Tantsyrev, I. N., *Trudy Inst. Goryuchikh Iskopaemykh, Akad. Nauk S.S.S.R., Otdel. Tekh. Nauk*, **2**, 227–35 (1950).

[86] Larina, V. A., and Galaganova, A. S., *Izvest. Fiz. Khim. nauch.-issled. Inst. Irkutsk Univ.*, **3**, 1–2, 72–84 (1953).

[87] Tuturina, V. V., and Kaigorodtseva, R. A., *Izvest. Fiz-khim. nauch.-issled. Inst. Irkutsk Univ.*, **3**, 1–2, 42–8 (1953).

[88] Kugo, M., *Mem. Fac. Eng., Hokkaido Univ.*, **8**, 116–38 (1949).

qualitatively, since the relative positions and degree of intermixture of the various structural units, and the resulting pore structure, must greatly affect the behavior of the material as a whole. But there is still need for a comprehensive theory of the coking process that will take into account all the factors governing the rates and temperatures of softening and decomposition.

Extraction at atmospheric pressure but at high temperature of several coals with polycyclic compounds removed up to 95% of a medium rank coal. Of the 95% yield obtained with phenanthrene, about 20% was benzene-soluble and appeared to embody the swelling properties, whereas the benzene-insoluble portion appeared to embody the agglomerating properties of the original coal.[51]

That the amount of material extracted by a benzene-alcohol mixture increased with preheating only in the case of coking coals,[89] may be compared with the early observation of Illingworth[90] that was followed up, using chloroform as solvent, by Dryden and coworkers.[91] The latter have examined in detail the origin and characteristics of the materials causing this enhanced yield of extract, which resemble chemically the parent coal and are in part responsible for the plasticity developed on heating. They have also studied the volatility and thermal stability of the extract obtained with chloroform, and of corresponding products of mild carbonization under high vacuum that resemble it closely.

One question has never been answered: To what extent are caking properties modified by adsorbed solvent? Even the γ fraction appears to contain adsorbed pyridine, which, since it is held on the surfaces of and at the points of contact between coal units, may possibly influence sintering and allied behavior out of proportion to its bulk concentration. Such an adsorbed solvent certainly does not confer caking properties on the residue. The same consideration renders more difficult the interpretation of figures for the volatile matter yielded by the products of extraction.

[89] Novikov, V. N., *J. Appl. Chem. (U.S.S.R.)*, **24**, 314–24 (1951).

[90] Illingworth, S. R., *Fuel*, **1**, 213–19 (1922).

[91] Dryden, I. G. C., and Pankhurst, K. S., *Fuel*, **34**, 363–6 (1955). Brown, J. K., Dryden, I. G. C., Dunevein, D. H., Joy, W. K., and Pankhurst, K. S., *J. Inst. Fuel*, **31**, 259–73 (1958).

Nature of Coal Extracts

It is well established that solutions of coal extract in solvents of the pyridine type show colloidal features. Kann[92] reported that ethylenediamine extracts prepared at atmospheric temperature consisted largely of particles a few hundred Angstrom units in diameter. A positive indication of colloidal material is likely to be correct unless, for example, much finely divided residue has passed through the extraction filter and has been wrongly classified as extract. On the other hand, optical clarity, indicating absence of colloidal particles, depends on the degree of resolution of the optical system employed. With the best ultramicroscope technique, the presence of particles as small as 100 to 200 A in diameter may sometimes be detected, but unless care is taken solutions containing particles still larger than this may appear optically clear.

A contrast with the high values of the order of 10^6 suggested by the colloidal properties of solutions of coal extracts and by the ultramicroscope observations made on them[92] is shown by the apparent molecular weights of about 300 for ethylenediamine extracts of coal found by a cryoscopic method with catechol as solvent.[49] However, in the same paper, observations on the ability of these extracts to pass through cellophane membranes were interpreted to mean that at least 90% of the material in solution was colloidal, that is, had a molecular weight not smaller than several thousands. Low values of molecular weight of extracts in pyridine,[64] in phenanthrene,[51,53] and in benzene[52] have been reported. Wynne-Jones and associates found a maximum value of about 1200 for the average molecular weight in pyridine of pyridine extracts, a maximum absolute amount of the γ fraction, and a maximum molecular weight of nearly 4000 for the β fraction, all at the same point in the coal series, namely with South Wales coals of about 30% volatile matter (88% C). The molecular weight of the γ fraction showed a steady fall from 560 to 380 with increase in rank.[64]

Attention has been drawn[93] to the correspondence between the size of particles seen in dried ethylenediamine extracts when examined under the electron microscope[94] and

[92] Kann, L., *Fuel*, **30**, 47–8 (1951).

[93] Dryden, I. G. C., *Nature*, **169**, 269 (1952).

[94] Nagelschmidt, G., and Cartwright, J., *Nature*, **169**, 270 (1952).

those seen in the ultramicroscope.[92] However, pyridine extracts dried by heating rapidly to 180 to 200°C in nitrogen before examination in the electron microscope appeared to be structureless films.[95] The units seen by the electron microscope might well have been aggregates formed during drying, whereas the structureless films might have been the results of an interaction between pyridine and the coal extract, producing a varnish-like condensate similar to that reported by Dryden.[49] The evidence of the electron microscope is therefore not conclusive in either direction.

Four factors that could contribute to a discrepancy between cryoscopic or other "osmotic" molecular weights, representing number averages, and optical estimates, which are weighted differently, are: (1) the large effect on number average produced by a small proportion of low molecular weight material; (2) the effect on osmotic pressure of the finite concentration needed for experimental measurement compared with the infinite dilution assumed in osmotic theory; (3) dissociation to a different degree in different solvents; and (4) abnormal effects of traces of impurity such as carbon dioxide and water on certain solvents, for example, amines with which they form chemical compounds, reducing the vapor pressure or freezing point out of all proportion to the molecular concentration of impurity. The impossibly low values of molecular weight deduced[47] from the vapor pressure depression in ethylenediamine must be ascribed to (4). Wynne-Jones and associates[64] varied the concentration of their pyridine extracts considerably, and may be said to have ruled out (2) as a major factor with pyridine; the high molecular weight obtained with a β fraction demonstrated the importance of (1). A combination of (1), (3), and (4) appears the most probable explanation of the discrepancy.

Given has well summarized the present position and has used measurements of ultrafiltration and diffusivity to clarify the situation.[96] The true molecular weight of an appreciable part of a low-temperature extract is in the range of 1000 to 2000, but the distribution varies with the solvent, for example, ketone or amide. Extensive aggregation can occur at between 0.01 and 0.1% concentration as is especially well brought out by osmotic pressure measurements of Hookway quoted by Given.

The translational diffusion coefficients of the β and γ fractions extracted from a South Wales coal (87.8% C) and Yubari coal (83.4% C) have been measured by the porous diaphragm method.[97] By assuming spherical molecules and using the Sutherland-Einstein relation, molecular weights of from 870 to 1170 were calculated for the γ fraction (less ether-solubles) and from 3230 to 4350 for the β fraction of the South Wales coal.

Values for the intrinsic viscosities of solutions of coal extract in ethylenediamine, although somewhat variable, have been found to be usually less than 10, which indicates that the particles in solution, *if colloidal*, must be approximately spherical in shape.[98] On the contrary, a solution of extract obtained from a medium-rank coal with anthracene oil at 350°C, shown by other methods to be colloidal, possessed a remarkably high viscosity.[53]

Relation between Extracts and Residues

When lower-rank bituminous coals are extracted with ethylenediamine below its normal boiling point, the products consist entirely of solid extract and solid residue with no appreciable amount of oily material, although the extract yields more volatile matter on heating.[49]

Work on the solution of coals in oils containing oleic acid, showed that the residues remaining after 75% of a coal had been dissolved resembled closely, in ultimate and proximate analyses, the original coals.[99] This argues a more uniform composition for coals than is sometimes supposed and is supported by a study of extraction of coals with anthracene oil at 350°C, in which it was found that the C/H ratio of the undissolved residue showed little change up to 80% extraction.[53] It has also been shown that the infrared spectra of extracts from coals in the range of carbon contents of 78 to 89% resemble closely those of the parent coals.[100] The composition,

[95] Boult, E., Thirsk, H. R., and Blayden, H. E., *Nature*, **170,** 317–18 (1952).

[96] Given, P. H., *Compt. rend. 27th congr. intern. chim. ind., Brussels,* **1954,** *Ind. chim. belge.,* **20** (Special), **2,** 102–5 (1955).

[97] Sakabe, T., *Fuel*, **35,** 310–17 (1956).

[98] Boas-Traube, S. G., and Dryden, I. G. C., *Fuel*, **29,** 260–1 (1950).

[99] Gillet, A., *Research*, **2,** 407–14 (1949).

[100] Brown, J. K., *Fuel*, **38,** 55–63 (1959).

molecular weight, X-ray diffraction characteristics, magnetic susceptibility, ultraviolet absorption, and free radical content of the fractions of a benzene extract of a Japanese coal (84% C) closely resembled the original coal.[101] When considering comparative analyses of extracts and residues, it is important to distinguish between low-temperature extractions—in which the two differ but little—and high-temperature extractions—in which the extract is richer in hydrogen.

It has been reported that the composition of the extracts obtained with phenol and tetralin from both bright and dull coals were similar.[102] In the course of a study of the interaction of coals with sulfur, it has been found that hydrogen was more readily removed by sulfur from anthracene oil or retene extracts than from the original coal. Whereas the amount of labile hydrogen estimated by this method varied according to the rank of coal, the amount in extracts of these coals was the same for coals of both low and medium rank.[103] Extraction of fifteen bright coals (82 to 93% C) and ten of the corresponding durains (84 to 90% C) with pyridine yielded extracts that were fractionated by chromatography on alumina.[104] The chloroform eluate was found to have an infrared spectrum similar to that of a Nujol suspension of the same coal.

Mechanism of Solution

An examination in considerable detail of the effect of various changes in extraction conditions on the yield of extract and on imbibition of solvent by the residue, coupled with the further conclusion that extracts and residues are fundamentally similar in type, has led to the suggestion that extraction occurs by the removal of units of colloidal size directly from the coal.[49,100] The following model was proposed. A matrix of larger and more strongly linked micelles, which can be partly and progressively dispersed in a suitable solvent by increasing the temperature, is intimately associated with a proportion of smaller less strongly bonded micelles that are normally trapped within this matrix unless it is first swollen by a solvent. The difference between these two classes of micelle was supposed to be one of degree rather than of kind. Possibly they formed a continuous series, the dividing line being determined by the temperature. The micelles, assumed to be rigid and comparatively indestructible, would be extracted as individuals by diffusion through the swollen pores of the matrix. It has been shown, however, that the large permanent micelle in the Bangham sense does not exist in coals; in the foregoing context the term may, however, be taken also to refer to single molecules or to aggregates with strong resistance to disintegration.

This conception of the mechanism of extraction has many similarities with the models proposed by Agde and Hubertus[41] and Kreulen,[43] but there are significant differences. It is supported by the close correlation observed between yield of extract and degree of imbibition of solvent;[42] for if solvent is imbibed into spaces between the larger units, which may be supposed to constitute the insoluble residue, the sizes of the spaces between these units will increase with the degree of imbibition, and the smaller units will then more easily escape from the structure. It can be seen that if a continuous distribution of sizes of unit, over a considerable range, is postulated, the observed gradual increase in degree of extraction with increase of temperature or with the use of more powerful solvents can be qualitatively explained; moreover, the extracts and residues would then be fractions in terms of size rather than of chemical type, and the observed similarity between extracts and residues would also be expected.

Quantitatively, it has been found that when the degree of imbibition is such that the average diameter of a swollen pore is of the same magnitude as the average diameter of the coal units themselves, the slope of extraction yield against temperature with a given type of amine solvent becomes the same for different ranks of coal; when the imbibition is smaller than this critical value, the slope is greater (Fig. 5).[49] This observation is also in accord with the views expressed in the last paragraph, which might be summarized in the phrase, "micellar sieve extraction mechanism."

[101] Sakabe, T., Inouye, K., Ouchi, K., and Honda, H., *J. Fuel Soc. Japan*, **32**, 610–26 (1953).

[102] Egorova, O. I., *Bull. acad. sci. U.R.S.S., Classe sci. tech.* **7–8**, 63–70 (1941).

[103] Zabavin, V. I., and Khrisanfova, A. I., *Izvest. Akad. Nauk S.S.S.R., Otdel. Tekh. Nauk*, **11**, 1629–34 (1946).

[104] Kirkby, W. A., Lakey, J. R. A., and Sarjant, R. J., *Fuel*, **33**, 480–96 (1954).

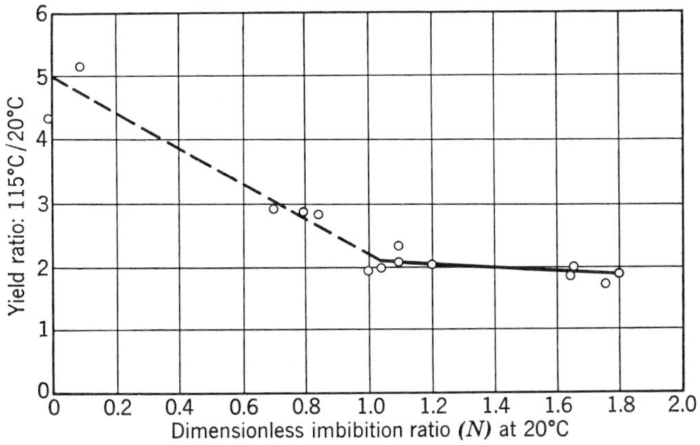

Fig. 5. Interaction of coals with ethylenediamine: relation between temperature coefficient of extraction yield and degree of imbibition.[49]

A more fundamental account of the ideas underlying this mechanism has been given in another paper.[48]

Extraction of lower-rank coals in ethylenediamine at room temperature shows several interesting features; some of these are not easily described in brief, and reference must be made to the original papers,[65,71] but a few may be mentioned here. The limiting yield of soluble material appears to depend only on the type of coal, the solvent, and the temperature of extraction; but yields less than the limiting value may be obtained if precautions are not taken, and it is essential to extract the coal with more than one successive portion of solvent. If, during extraction of a coal sample by direct shaking with the solvent, the concentration of extract in solution exceeds a certain value (at which a solution tends to form a surface skin and to throw down sludge), some of the extract that was dissolved in solvent which is itself imbibed in the pores of the residue, appears to be forced out into and increase the concentration of the free liquid phase; this is presumably an effect on the partition resulting from increased aggregation in the free solution. When a sample of coal is successively extracted with fresh portions of ethylenediamine by shaking at room temperature, the imbibition of solvent by the residue undergoes periodic fluctuations, probably owing to rearrangements of physical structure as the amount extracted increases. In a given stage of extraction, the ratio of solvent to coal exercises an important influence on the apparent yield of extract obtained, but after some four stages the total result is the same for all ratios sufficiently high to produce a mobile suspension of coal in liquid.

A rather different view of extraction mechanism has been put forward by Gillet.[105] A three-stage mechanism of extraction has also been proposed and mathematical equations derived to fit this concept.[106] In a study of the extraction of Dutch bituminous coals at 200 to 400°C with polycyclic compounds,[53] the kinetics of extraction up to 350°C were considered and activation energies between 14 and 40 kcal/mole were deduced, depending on the rank of coal and the solvent used. Special attention was paid to hydroaromatic solvents above 300°C.

The application of the Arrhenius equation to the dependence of solvent extraction *yield* (not rate) with temperature is to be deprecated since the equation is based on equilibrium thermodynamics. That the extraction yield is not an equilibrium solubility can be seen by varying the amount of solvent at a given temperature.

Applications of Solvent Extraction

The best known industrial application of solvent extraction of coal is the Pott-Broche process, closed down following war damage.[60,107] Extraction was carried out at 415

[105] Gillet, A., *Nature*, **167**, 406–7 (1951).

[106] Baba, A., *Rept. Resources Research Inst.*, No. 27 (1955), 120 pp. Baba, A., and Yuki, N., *J. Fuel Soc. Japan*, **33**, 537–44 (1954).

[107] Frese, E., Gt. Brit., Bd. of Trade, German Div., F.D. 940–49, Item 6, 10 frames (translation, 1947).

to 430°C and a pressure of 100 to 150 atm, and with a heating time of 60 to 90 min. Sixty to 70% of the coal feed was ground below 0.06 mm and one part of this coal was mixed with two parts of the solvent, which was a middle-oil obtained from the hydrogenation of pitch. The plant processed about 5 tons of coal an hour, and the yield of extract based on dry, ash-free coal averaged 78 to 84%. Recovered solvent was partly recycled, but a considerable make-up of fresh oil was essential to maintain the yield, which suggests that the process involves hydrogenation as well as extraction. The addition of hydrogen, in fact, was found to reduce the requirements of fresh oil considerably, but its use led to serious operating difficulties because of the formation of solid deposits. The extract solution was filtered at a pressure of 3 to 4 atm and a temperature of 150°C; the filter "candles" employed processed 0.6 ton of feed per square meter per hour. The residue, containing 25 to 30% ash, was normally used as works fuel. The extract, after vacuum distillation to recover solvent, melted at about 230°C and contained less than 0.05% ash. About 1% of gasoline-like oils, an unspecified proportion of resin-like oils, and 2 to 3% by weight of gas were byproducts of the process. In general, bituminous coals of more than 25% volatile matter were found to be easily extractable, but one low-rank coal (Gelsenberg) gave difficulty. Durain yielded less extract than vitrain, fusain scarcely any. The main use of the extract was in the production of a low-ash coke for electrode carbon manufacture. A possible use for the residue has also been described.[108] The residue left on the filter, which still contained a proportion of extract and solvent, was hot-pressed without addition of any kind into briquettes which were carbonized at low temperature, with solvent recovery, to form a dense strong coke.

Baba[79,109] has described developments in Japan which, although employing the same general principles as the Pott-Broche process, show considerable enterprise in extending the field of application. The solvents used were pitch, pitch-oil, tar-oil or tar, high-temperature tar being preferred because of its large content of unsubstituted polycyclic aromatic compounds. Filtration difficulties increased with increasing swelling of the coal by the solvent. The extract, with or without excess solvent, is known as "Bojuntan." Two plants for its production were begun in 1943, each to treat 50,000 tons of coal a year, but their construction was stopped in 1945. Plant on the pilot or semi-industrial scale had been set up for the following applications: (1) production of ashless coke for electrode manufacture, by carbonization of "Bojuntan"; (2) production of carbonized briquettes from good quality coal, using "Bojuntan" made from slurry as binder; (3) use of "Bojuntan" as a substitute for asphalt in road surfacing; (4) production of metallurgical coke from weakly caking coal by blending with "Bojuntan" and semicoke. Mention was also made of the use of the insoluble residue for manufacturing active carbon.

The use of coal extracts as a base for varnish manufacture[110] has been described. The use of products, obtained by extraction under pressure with hydrogenated naphthalenes, as filling agents for rubber and plastics has been discussed in a document from the files of I.G. Farbenindustrie.[111] The following applications have also been suggested: (1) addition to non-caking coals to induce caking; (2) briquette binding; (3) production of pitch containing little "free carbon"; (4) production of low-ash pulverized fuel; (5) production of colloidal fuel; (6) addition to tar; (7) production of low-ash carbon; (8) production of chemicals; and (9) production of "fillers."[62,112]

A more novel use for solvent extraction is in the removal—by ethylene dichloride—of resins from certain coals whose nonresinous constituents possess adequate coking power.[113] This process has no connection with the relation between solvent extraction and coking properties already discussed, but is concerned exclusively with the removal of an unwanted

[108] Thomas, H., Ger. Pat. Appln. (1943); Gt. Brit., Bd. of Trade, German Div., F.D. 29–46, frames 1275–7.

[109] Baba, A., Yoshida, Y., Kumai, J., and Watari, S., *Rept. Resources Research Inst. Japan*, No. 22 (1955), 62 pp.

[110] Arend, A. G., *Paint Technol.*, **10**, 139–41 (1945).

[111] I. G. Farbenindustrie, Gt. Brit., Bd. of Trade, German Div., F.D. 2615–48, frames 1–22, (1936–41).

[112] Baba, A., *J. Fuel Soc. Japan*, **19**, 115–24 (1940).

[113] Schabelitz, E. J., Brit. Pat. 654,753 (1951).

diluent. Removal of resins and hydrocarbons from coals by pyridine-chloroform, particularly from lignites, has been discussed.[56]

Hauswald has reviewed British and European patents concerned with the pressure extraction of coal.[114] A method for reducing the sulfur content of coal extracts by heating above 150°C, under pressure, with aqueous sodium hydroxide and zinc oxide, has been patented by Dreyfus.[115] In 1946, a plant was described by Crawford[115] in which coal was extracted at 250°C and 120 atm with 2.5% aqueous sodium hydroxide for 20 min, yielding a low-ash product.

Fungistatic activity has been observed in minor fractions of coal extracts.[116]

Conclusions

In view of the considerable length of this survey, it has been deemed advisable to state the broad conclusions that have emerged from it, including an appraisal from the standpoint of modern knowledge of some of the work earlier reviewed.[31,32] Exceptions to almost every statement could, however, be found in individual papers.

Of the conditions of extraction, particle size is less important than might be expected, probably owing to disintegration by the solvent; the unfavorable effect of moisture is probably confined to solvents forming azeotropes with water. The presence of gaseous oxygen during extraction decreases the yield obtained with a considerable number of solvents; oxidation of the coal substance normally tends to decrease the yield of extract, whereas partial hydrogenation increases it. Curves of adsorption of pyridine on coals that have been preheated, plotted against the temperature of preheating, appear to be characteristic of the type of coal; a large maximum in the curve, sometimes coinciding with a minimum accessible surface, is characteristic of coking coals. There is also a peak in the curve relating extraction yield to temperature of preheating; this, too, may be more marked with the coking coals and like caking properties is destroyed by oxidation. The extraction maximum, however, occurs well below the decomposition temperature of the coal, and the sorption maximum above it. The solubility in most solvents becomes negligible at about 550°C.

There are many differences in the extraction process and the nature of the extraction products depending on whether the extraction is carried out below or above about 250°C. Above this temperature, changes are probably caused in coal both as a result of the influence of temperature and of the chemical action of the solvent. A model, adapted from that proposed earlier by Agde and Hubertus,[41] is helpful for illustrating these differences. A colloidal matrix of more tightly bonded or interlocked units (molecules?) which forms the residue of low-temperature extraction, contains material, itself largely colloidal except at high dilution, extractable at the lower temperatures. Solvents that swell the matrix can release this material, whence the common observation that swelling and extraction run parallel at low temperatures; the solvent, to be effective, must also stabilize the extract in solution. At the higher temperatures the residue itself may be peptized. Thus the nature of the solvent is critical at low temperatures but much less critical at higher temperatures. Primary aliphatic and certain heterocyclic amines are the most favorable chemical types so far described in the literature. The value of correlating solvent power with internal pressure is limited. An important property at the lower extraction temperatures is the possession by the solvent molecules of a nitrogen or oxygen atom with a readily available unshared pair of electrons; and yet other features are probably also relevant.

At all temperatures, the greater part of the extracted material is probably colloidal in solution above 1% concentration, but solutions obtained above 250°C undoubtedly contain a larger fraction of volatile oils. Big discrepancies are observed between molecular weights of extracts determined cryoscopically and optically, or stoichiometrically by methods that can be interpreted alternatively in terms of sorption or chemical equivalence. The most likely explanations are: (1) the large influence on number average molecular weights of small amounts of low molecular weight material; (2) greater aggregation in some solvents; and, (3) the effect of impurities such as carbon dioxide and water. The specific

[114] Hauswald, W., *Braunkohle*, **2**, 285–91 (1950).

[115] Dreyfus, H., U.S. Pat. 2,221,866 (1940). Crawford, A., *BIOS Final Rept*. No. 522, Item 30 (1946).

[116] Kosanke, R. M., *Science*, **119**, 214 (1954). Schenk, N. C., *Science*, **119**, 213 (1954).

viscosities of high-temperature extracts in solution behave differently, with varying physical conditions, from those of their low-temperature analogues.

The value of extract fractionation as practiced at present is doubtful. Wheeler's α and β fractions appear to be similar to each other and to the parent coal, but the γ fraction differs at least in part. The application of chromatography, partition chromatography, and electrophoresis promises further advances, but there are few fractions differing in kind from the parent coal except in amounts less than 1% of the whole coal, or when pyrolysis has occurred. Coal extracts contain little mineral matter even when produced at temperatures as high as 350°C in high yield.

It is probably correct to say that in low-temperature extractions the extract, residue, and parent coal are similar to one another in composition (with the reservation that Wheeler's γ fraction contains more hydrogen than the others), and that in extractions above 350°C the initial portions of extract are closer to hydrocarbons in nature, whereas the later portions approach the composition of the coal. At low temperatures, certain solvent *mixtures* are superior to their constituents, but the results of *successive* treatment with different solvents suggest that their behavior is independent of one another; at higher temperatures neither of these statements is correct.

It therefore appears to be useful to differentiate between the low-temperature, specific solvents of the pyridine type and high-temperature solvents of the benzene type.[33,117] The latter show a greater relative variation in extraction yield with change of temperature. It may be that the temperature effect varies inversely with yield of extract. When the yield of extract obtained from coals which have been preheated is plotted against the temperature of preheating, a maximum is reached, which, with the benzene type, occurs at a considerably higher temperature than with the other. Broadly speaking, high-temperature extracts contain more simple molecular and less aggregated material than their low-temperature counterparts, unless extensive peptization has taken place.

The yield of extract decreases comparatively rapidly as the rank of coal increases from 85 to 89% C or down to about 25% volatile matter and becomes negligible at 92 to 93% C. Coals may be grouped into three categories in terms of yield of extract with ethylenediamine: (1) high volatile bituminous, (2) prime coking plus dry steam, and (3) anthracites, respectively. For some solvents, there may be a maximum yield between 87 and about 80% carbon; the direction of the change with rank in this range appears to vary with the solvent used. These statements apply broadly to all temperatures and types of solvent. Of the petrological constituents, vitrain yields the most extract and fusain the least. Pyridine sorption and extraction yield, over the bituminous range, appear to be closely related to one another regardless of rank or petrographic type, but this relation does not extend to lignites. The same is generally true of the imbibition and yield with ethylenediamine, but fusain and cannel coal appear to be exceptions.

Turning to coking properties, it is well established that the extract will swell and agglomerate on heating, whereas the residue will not. Separation of phenanthrene extract into a swelling and an agglomerating constituent has been claimed. Efforts at reconstituting the coal from its extraction products are unsatisfactory, owing probably to the overriding importance of the structure of the original material. A comprehensive theory of the coking process would greatly assist the understanding of these phenomena.

Commercial applications of the products of coal extraction have so far been remarkably few. The production of low-ash carbon for electrodes seems to have been the main outlet, although a number of other suggestions have been made, and developments in Japan have promised large scale use in the manufacture of metallurgical coke. Many direct uses for crude extracts appear to deserve further investigation. Some chemical reactions of coal extracts have been described.

The Carbon-Hydrogen Skeletal Structure

Direct Chemical Methods

It is only recently that chemical reduction of coals at temperatures below 200°C at atmospheric pressure has rendered this an instrument of some precision in the elucidation of coal structure. Previous attempts at characterizing hydrogenation products were open to

[117] Nesterenko, L. L., *Trudy S. M. Kirov Kharkov Khim.*, **1939,** No. 2, 70–90.

many of the objections that rendered oxidation an unfruitful method for studying coal constitution, although it is true that the products of hydrogenation are themselves simpler and can be specified fairly closely.

Powdered coal in suspension with lithium in ethylamine at the boiling point (17°C), and coal extracts obtained with dimethylformamide have been reduced by controlled potential electrolysis in this solvent, at −2.2 v vs. a mercury pool, using 0.1N tetraethylammonium iodide as the supporting electrolyte.[118] In known aromatic systems, a greater degree of reduction is effected by the lithium reaction than by electrolysis. The electrolysis products were acetylated *in situ* with acetic anhydride to prevent oxidation of any hydroquinones produced. The number of hydrogen atoms added per 100 carbon atoms was plotted against the ash-free carbon content of the parent coal. In both cases the greatest effect was found at about 89% C. The amount of lithium reduction reached 37 atoms per 100 at the maximum, and for the same coking coal a maximum of 75% was rendered soluble in pyridine by the treatment. It was suggested that the low-rank coals did not take up hydrogen under these conditions, as contrasted with pressure hydrogenation conditions, because of difficulty in accommodating the necessary departure from planarity when an aromatic ring is reduced, owing to rigid cross-linkage in the low-rank coals. This explanation is in accord with the higher degree of substitution of the low-rank coals, for which there is much evidence from infrared spectroscopy. A smaller degree of reduction of coals with more than 89% carbon would be expected from the larger size of their aromatic ring clusters. It is clear that reduction in these experiments was not complete and that therefore only a lower limit to the amount of aromatic carbon can be deduced from them.

It has recently been suggested that the necessity for high negative potentials, −1.95 to −2.2 v vs. a mercury pool, for the reduction of a considerable part of the aromatic systems in coal extracts by polarography in dimethylformamide indicates that nuclei such as biphenyl, naphthalene, phenanthrene, and triphenylene form an important part of the structure.[119] This work was supported by background data on hydrocarbons.[120]

Reduction was also obtained by the use of lithium in ethylenediamine at 90 to 100°C for $2\frac{1}{2}$ hr.[121] The hydrogen uptake amounted to 24 atoms per 100 carbon atoms at 83.4% C content, 20 at 87% C, and 17 for an anthracite (93.8% C). Later work not fully published at the time of writing has disclosed a peak of 40 to 50 H atoms/100 C atoms. The reduced material absorbed ultraviolet and visible light less intensely than the parent coal and this was attributed to a decrease in the number of conjugated double bonds or of free radicals. Some background work on organic compounds had previously been reported.[122] X-ray examination of the reduced products formed confirmed that the aromatic nuclei had been hydrogenated.[123]

An attempt to deduce hydrogen distribution in five coals has been made on the assumption that during oxidation in air at 170°C the aromatic skeleton is unaffected.[124] Oxidation was continued until no further change in composition took place. Correction was made for the content of quinone groups and carboxyl groups. This procedure presumably leads to an estimate of the aromatic hydrogen on a cluster after all other linkages have been eliminated. However, it is known that many of the aromatic positions in the original coal would be substituted. Moreover, reasons are given in the section on oxidation mechanism for reserving judgment on the validity of the basic assumption.[125] The correctness or otherwise of the determinations of functional groups is also of considerable importance. When calculation was made according to these

[118] Given, P. H., Lupton, V., and Peover, M. E., *Nature*, **181**, 1059–60 (1958); *Proc. Conf. Science in the Use of Coal*, Sheffield, 1958, Inst. Fuel, London, 1958, pp. A38–A42.

[119] Given, P. H., and Peover, M. E., *Fuel*, **39**, 463–73 (1960).

[120] Given, P. H., *J. Chem. Soc.*, **1958**, 2684–7.

[121] Reggel, L., Raymond, R., Friedman, S., Friedel, R. A., and Wender, I., *Fuel*, **37**, 126–8 (1958).

[122] Reggel, L., Friedel, R. A., and Wender, I., *J. Org. Chem.*, **22**, 891–4 (1957).

[123] Ergun, S., and Wender, I., *Fuel*, **37**, 503–6 (1958).

[124] Chakrabartty, S. K., Mazumdar, B. K., and Lahiri, A., *Fuel*, **37**, 498–501 (1958).

[125] Mazumdar, B. K., Chakrabartty, S. K., and Lahiri, A., *J. Sci. Ind. Research* (*India*), **16B**, 275–6 (1957). Mazumdar, B. K., and Lahiri, A., *Fuel*, **37**, 495–8 (1958).

assumptions the estimated values of aromatic hydrogen atoms per carbon atom were 25 to 100% greater than estimates[126] from infrared spectroscopy, the best agreement being obtained for the anthracite. The discrepancy may be partly the result of using a wrong basis for comparison, as already indicated, and partly due to incorrectness of the basic assumption about the mechanism of oxidation.

The reaction of sulfur with coal has also been used in order to estimate "labile" hydrogen.[103] The quantity of hydrogen removed was found to be a function of the rank of coal though solution in retene increased the absolute amount and tended to reduce the effect of rank.[127]

In pursuit of a cross-check with the statistical structural analysis of coals, a bituminous coal of 85.6% C was subjected to hydrogenolysis at 325°C and about 190 atm cold pressure, using stannous chloride and tetralin as catalyst and hydrogen carrier respectively.[128] The products were then hydrogenated in the presence of a nickel catalyst and the saturated products were fractionated by vacuum distillation. An estimate, from measurements on the fractions of the distribution by size of polynuclear ring systems in the products, was made that was assumed to give an indication of the distribution of aromatic nuclei in the original coal. From three to eleven condensed rings were indicated, and the presence of the larger pericondensed nuclei such as coronene was considered from the results to be unlikely.

Hydrogenolysis of coal under mild conditions with an iodine catalyst has been reported to split the ring structure.[129] At 350 to 400°C and 300 atm, a coal of 89% C content was converted in 30 to 35% yield to an oil containing little oxygen. Cellulose, starch, sugar, and wood were converted into somewhat similar cyclic products at 300 to 335°C, which might throw doubt on the value of the method as a means of investigating coal structure, but all these substances are known to undergo cyclization into coal-like chars when carbonized to the reaction temperature used. It is doubtful, from the rather complicated results reported, that much can be learned of the structure of the coal from a study of primary hydrogenation products from coal, even using the many techniques for fractionation and characterization.[130]

Semidirect Methods

In what follows, only a few highlights bearing on the coal molecule will be discussed. More comprehensive accounts of the results obtained by using these techniques in the study of coal are given in Chapter 2.

X-ray Diffraction. This is one of the few orthodox methods that can be used to estimate the size of the polycyclic aromatic units in the coal molecule. The X-ray diffraction patterns of coals are such that they would be difficult to interpret otherwise than in terms of a mixed aromatic/aliphatic structure.[131]

Several systematic X-ray studies of coals of a wide range in rank and of coal derivatives have been made.[74,132–134] The X-ray spectra alone cannot constitute unique evidence for the existence of special microscopic crystallites of graphite-like nature even in the highest rank bituminous coals.[133] With a complex substance like coal, no structure can be deduced from the

[126] Brown, J. K., *J. Chem. Soc.*, **1955**, 744–52.

[127] Khrisanfova, A. I., and Zabavin, V. I., *Izvest. Akad. Nauk S.S.S.R., Otdel. Tekh. Nauk*, **1948**, 367–70.

[128] Schuhmacher, J. P., van Vucht, H. A., Groenewege, M. P., and Blom, L., *Brennstoff-Chem.*, **36**, 215 (1955). Schuhmacher, J. P., van Vucht, H. A., Groenewege, M. P., Blom, L., and van Krevelen, D. W., *Fuel*, **35**, 281–90 (1956). Geerards, J. J. T. M., and Waterman, H. I., *Brennstoff-Chem.*, **37**, 267–8 (1956).

[129] Geerards, J. J. T. M., Waterman, H. I., and van Krevelen, D. W., *Brennstoff-Chem.*, **39**, 332–6 (1958). Kemkes, J. F., Meulenbeld, F., Waterman, H. I., and van Krevelen, D. W., *Brennstoff-Chem.*, **39**, 359–61 (1958). Geerards, J. J. T. M., van Krevelen, D. W., and Waterman, H. I., *Brennstoff-Chem.*, **39**, S11–14 (1958).

[130] Neuworth, M. B., *J. Am. Chem. Soc.*, **69**, 1653–7 (1947). Basu, A. N., and Glenn, R. A., *Fuel*, **27**, 96–9 (1948); **29**, 134–7 (1950). Wolfarth, J. S., and Glenn, R. A., *Fuel*, **28**, 253–7 (1949). Glenn, R. A., and DeWalt, C. W., *Fuel*, **32**, 157–68 (1953). Anon., *Chem. Eng. News*, **30**, 1954–9 (1952).

[131] Ergun, S., and Tiensuu, V. H., *Fuel*, **38**, 64–78 (1959).

[132] Blayden, H. E., Gibson, J., and Riley, H. L., *Inst. Fuel, War-time Bull.*, 117–29, February (1945). Siever, R., *Proc. 2nd Conf. on Origin and Constitution of Coal*, Nova Scotia, **1952**, Nova Scotia Research Council, Halifax, 1955, pp. 341–62.

[133] Nelson, J. B., *Fuel*, **33**, 153–75, 381–93 (1954).

[134] Hirsch, P. B., *Proc. Roy. Soc. (London)*, **A226**, 143–69 (1954).

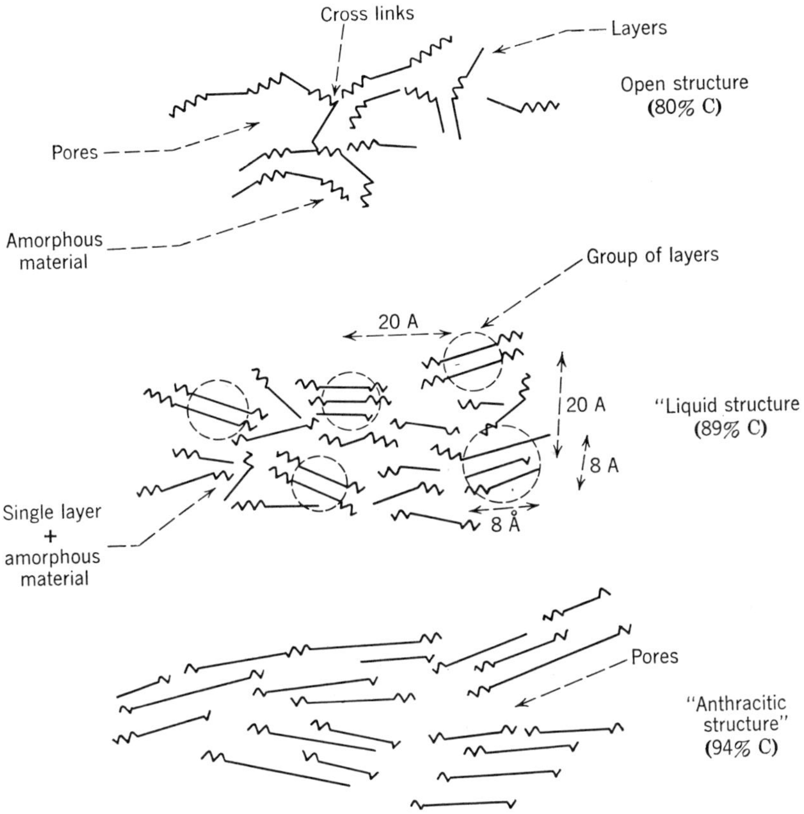

Fig. 6. Schematic model of coal structure according to Hirsch.[134]

curves. However, the curves do furnish a criterion which must be satisfied by any proposed model of the coal molecule. In vitrinites ranging from 77 to 89% C content on the Parr basis, and assuming that not less than about 75% of the carbon occurs in aromatic rings, the average size of condensed aromatic unit cannot be greater than that corresponding to four fused rings.[133] This size is smaller than that corresponding to the crystallite diameters deduced earlier.[74,134] Nelson reached this conclusion by comparing the ratio of the number of C—C distances of length 6.51 and 2.46 A respectively for coals with those observed for condensed aromatic hydrocarbons of various sizes and shapes.

In a study of a series of British vitrinites over a range of Bragg spacings from 0.4 to 5000 A, scattering at very small angles was shown to be related to the ultrafine porosity of the coal,[134] but there was no indication of the presence of close-packed micelles of the type postulated by Bangham.[14] The presence of disk-like cracks parallel to the bedding plane was indicated. Hirsch deduced a schematic structure for coal (Fig. 6) that is not easy to relate closely to that of the chemical molecule.

Improved methods of interpreting the X-ray results on coals have been developed.[135,136] Nonaromatic carbon, and oxygen, directly attached to an aromatic cluster have thus been shown to contribute to the size of the layer or cluster, and these atoms together with other nonaromatic carbon contribute somewhat to the fraction of roentgen-amorphous carbon.[137] Although the roentgen-amorphous material is qualitatively representative of the nonaromatic groupings, exact chemical equivalence must not be assumed.

The new methods of analysis gave approximate distributions, in histogram form, of the

[135] Diamond, R., *Acta Cryst.*, **10,** 359–64 (1957); **11,** 129–38 (1958).

[136] Cartz, L., and Hirsch, P. B., *Phil. Trans. Roy. Soc. London*, **A252,** 557–604 (1960).

[137] Hirsch, P. B., *Proc. Conf. Science in the Use of Coal*, Inst. Fuel, London, 1958, pp. A29–33.

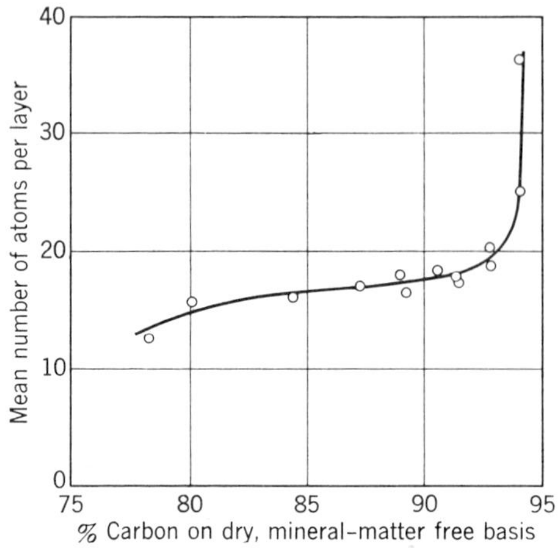

Fig. 7. Variation of the average number of planar atoms in an aromatic layer with carbon content of coal.[137]

aromatic layer sizes terminated by nonaromatic carbon or 5-membered rings in terms of arbitrary groups containing 11, 22, 32, 72, and 128 planar carbon atoms. For all the bituminous coals the lowest group predominated. The mean number of atoms per aromatic layer, including the first carbon atoms of attached nonaromatic groupings, and attached oxygen atoms, is plotted against the dry, mineral-matter-free carbon content of the vitrains in Fig. 7. The average number of carbon atoms in an aromatic cluster can be deduced from this only if the degree of aromatic substitution and the concentrations of carbonyl and hydroxyl groups are known. Hirsch concluded that the average number of aromatic rings in a cluster varied from about 2 at 78% C to about 4 at 90% C, and then rose rapidly to about 12 for the anthracites. With anthracites the distribution was much wider than for bituminous coals and the average had little significance.[137] These estimates thus conform with the limitation on the ring cluster numbers for bituminous coals to 4, earlier deduced by Nelson.[133] It appears that 70 to 90% of the layer-planes in most vitrains probably occur either singly (40 to 60%) or in pairs.[134,136] There is thus scarcely any evidence for the existence of graphitic crystallites as such.

In studies of macerals other than vitrinite, coal extracts, and carbonized coals, the "gamma band" was observed and ascribed to the presence of exinites.[138] Dried solvent extracts were very similar to the coals and appeared to contain a microporous structure like that in coals; the insoluble residues were expanded in a direction normal to the layer planes.

The method for measurement of aromatic layer size developed by Diamond[135] has been applied to a vitrinite of 84% C, extracts therefrom, and amorphous polyallylnaphthalene.[139] Measurements were also extended to other amorphous polymers, poly(9-allyl-fluorene) and a naphthalene-urea polymer. Agreement between the results predicted for the polymer structures from Diamond's theory and the experimental measurements was fairly good as regards the average parameters, particularly for the polyallylnaphthalene. However, although all these compounds contain two rings or less in the aromatic cluster, the histograms showed appreciable amounts of carbon in the group of 22 atoms and sometimes in the 32-atom group also. This was apparent even in the theoretical histograms deduced for these compounds, and appears to indicate a basic weakness in the significance of Diamond's group-size parameter; it is clearly not related simply to the size for a single cluster in these compounds, and presumably tends to overestimate the fraction of larger clusters present. On the other hand, in a coprecipitate of three large (28 C atoms) polycyclic quinones,[136] the *smaller* layer sizes appeared to be overemphasized, the average was also low ($N = 18$ carbon atoms instead of 28), and 25% of amorphous carbon was indicated. This discrepancy was tentatively attributed to straining and bending of the layers in the coprecipitate.

In a study of the X-ray scattering patterns of a "distillate" obtained by high-vacuum carbonization of a Pittsburgh vitrain in relation to those of the vitrain itself and the nonvolatile residue, it was found that both "distillate" and residue contained aromatic layers, but their scattering intensities were more diffuse than those of the original vitrain.[140] The results were consistent with the view that the aromatic layers in the "distillate" represented some of the aromatic clusters from the vitrain, set free and vaporized on heating.

There is a difficulty in the chemical interpretation of the X-ray evidence that ascribed

[138] Cartz, L., Diamond, R., and Hirsch, P. B., *Nature*, **177**, 500–502 (1956).

[139] Ruland, W., and Tschamler, H., *Brennstoff-Chem.*, **39**, 363–5 (1958).

[140] Ergun, S., *Fuel*, **37**, 365–70 (1958).

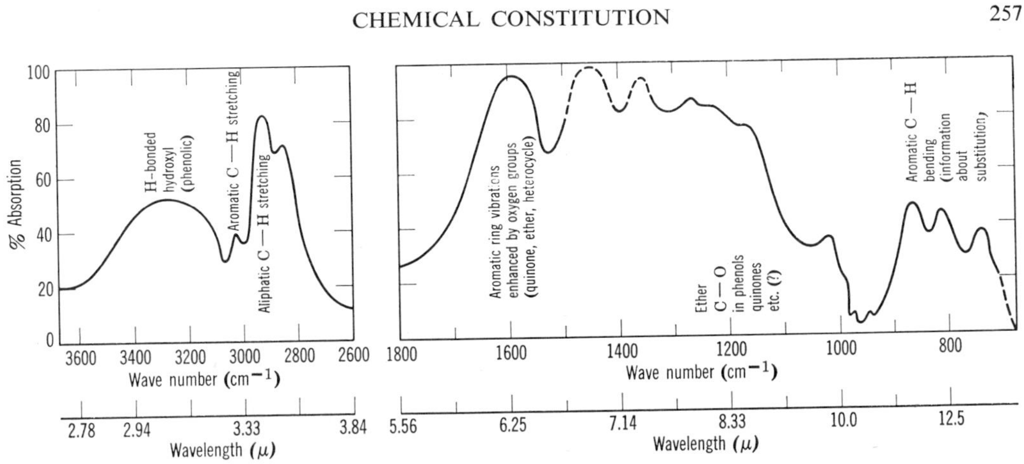

Fig. 8. Typical infrared spectrum of a medium-rank coal.

four or less rings to the aromatic cluster in bituminous coals.[141] The H/C ratio of coals is low, and infrared and other evidence indicates the presence of a considerable proportion of nonaromatic C—H for which the atomic H/C ratio must approach if not equal two. This leaves little hydrogen available to fill the edges of the aromatic clusters, and at first sight appears to require larger clusters than is suggested by the X-ray evidence. Cross-linkages either directly or through aliphatic carbon atoms would be required to eliminate the discrepancy, even after account was taken of the influence of oxygen-containing groups. These interlinkages are required in addition to the normal nonaromatic substitution of aromatic rings that is implicit in the coal structure, and which also helps to "conserve" aromatic hydrogen. This is one of the central problems in building a model of the coal molecule and severely limits the possibilities; more is said on this subject in the section "Appraisal."

Infrared absorption. Infrared spectra are helpful in providing information about the C—H, C=O, and O—H bonds in a coal, whether in aromatic or nonaromatic systems. They therefore complement the X-ray studies of aromatic carbon and provide information on the nonaromatic structures, which X-ray studies cannot elucidate.[126,142] A typical spectrum of a medium-rank bright coal is illustrated in Fig. 8. The molecular skeletons of all coals except anthracites are basically similar, and complex, and the hydroxyl groups are mainly phenolic.[143] A pyridine extract has a spectrum almost identical with that of the parent coal. The spectra of vitrains show similarities to those of chars from cellulose and sugar.[144]

Brown[126] made a study of twelve vitrains over the range of 78 to 94% C content and confirmed that the structural skeletons of all the coals studied were similar in type. The concentration of hydrogen-bonded phenolic hydroxyl groups was found to fall with increasing rank to very small values at about 89% C. Correspondingly the amount of aromatic hydrogen increased and that of aliphatic hydrogen decreased, the latter especially between 90 and 94% C. Absorption in the region from 900 to 700 cm^{-1} indicated decreasing substitution of the aromatic systems with increasing rank; the evidence on the distribution of aromatic hydrogen atoms in this region was fully discussed. The strong absorption at 1600 cm^{-1} was attributed to aromatic ring vibrations, enhanced, for example, by quinone groups. An important

[141] Dryden, I. G. C., *Fuel*, **34,** S29–35 (1955).

[142] Cannon, C. G., and Sutherland, G. B. B. M., *Trans. Faraday Soc.*, **41,** 279–88 (1945). Hadži, D., *Acad. Sci. et Art. Sloven., Classis III, Dissertationes*, **99,** 3 (1951); *Bull. sci., Conseil acad. RPF Yougoslavie*, **3,** 42–3 (1956). Gordon, R. R., Adams, W. N., and Jenkins, G. I., *Nature*, **170,** 317 (1952). Bergmann, G., Huck, G., Karweil, J., and Luther, H., *Brennstoff-Chem.*, **35,** 175–6 (1954), **38,** 193–9 (1957). Kinney, C. R., and Doucette, E. I., *Nature*, **182,** 785–6 (1958). Roy, M. M., *Fuel*, **36,** 249–50 (1957). Gordon, R. R., Adams, W. N., Pitt, G. J., and Watson, G. H., *Nature*, **174,** 1098–9 (1954). Brown, J. K., and Hirsch, P. B., *Nature*, **175,** 229–33 (1955).

[143] Cannon, C. G., *Nature*, **171,** 308 (1953).

[144] Friedel, R. A., and Pelipetz, M. G., *J. Opt. Soc. Am.*, **43,** 1051–2 (1953).

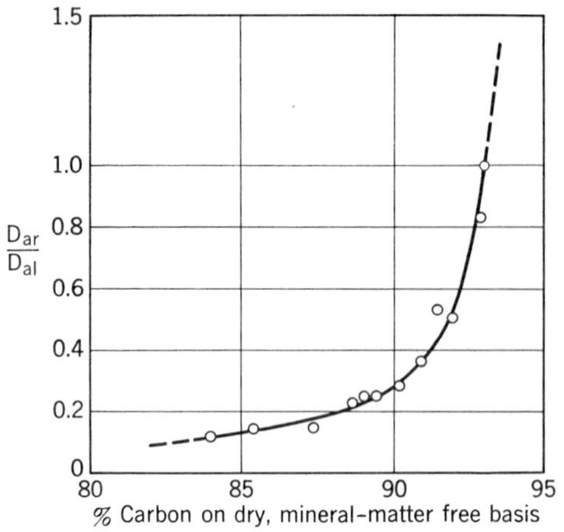

Fig. 9. Ratio of the optical densities of the 3030 cm^{-1} and 2920 cm^{-1} absorption bands plotted against carbon content for a range of vitrains. This ratio when multiplied by a factor of approximately 2 gives the ratio of aromatic to aliphatic hydrogen.[126]

feature of Brown's work was his semiquantitative estimation of the ratio of aromatic to aliphatic C—H bonds. (Fig. 9).

Spectroscopic measurements indicate that ethylenic double bonds, unless between completely substituted carbon atoms, are absent from bituminous coals.[126,145] Unconjugated carbonyl groups appear to be absent in bituminous vitrains, and the intensity of the methyl band indicates that these groups are few in relation to methylene.[146] The spectra in the region 1600 cm^{-1} have been interpreted to indicate the predominance of benzenoid rather than polyaromatic nuclei.[147] The spectra indicate that long aliphatic chains are not present in coals and that the nonaromatic hydrocarbon groupings are most likely hydroaromatic.[147] Interference due to minerals in the coal has been recognized; the absorption band at 1035 cm^{-1}, assigned by some to organic ethers, has been found to be due mainly to kaolinite.[147]

Methylation of Australian brown coals produced changes in their infrared spectra that suggested that keto-enol structures might be contributing to the intense band at 1600 cm^{-1}. The spectra of some phenol-formaldehyde polymers were found to be similar to those of brown coals.[148] By comparing the spectra of peat, fossil wood, lignite, and coal, it has been concluded that peat is related to cellulose and bituminous coal to lignite, but that lignite is not related to peat.[149]

Nuclear magnetic resonance. Richards and associates[150] were the first to apply broad-line proton magnetic resonance spectroscopy to a study of coals. Their specimens were hand-selected lumps of bright coal containing more than 50% vitrain, shaped into cylinders, and dried. The spectra were taken at 90°K. The second moments (mean square widths) of the absorption lines were found to be $21\frac{1}{3}$, $19\frac{1}{2}$, and 16 for coals containing 79, 83, and 93% carbon on the dry, mineral-matter-free basis, respectively. About one-third of the hydrogen in the highest rank coal occurred in aliphatic groups, whereas in the lowest rank coal the corresponding proportion was about two-thirds. An uncertainty arose from the fact that hydrogen in the methyl groups in the peri-position, or other situations involving bond distortion, was shown by models to behave like methylene hydrogen, whereas that in other methyl groups behaves like aromatic hydrogen.

Limits to the values of aromaticity derived from these observations were assessed in an appraisal and comparison of several methods of studying the coal molecule.[151] It was later found that aliphatic hydrogen in the form of $-\overset{|}{\underset{|}{C}}H$ also required to be bracketed with the aromatic hydrogen in this method.[152] By recalculation of Richards' results using reasonable assumptions about the relative proportions of aliphatic $-CH_3$, $-\overset{|}{C}H_2$, and $-\overset{|}{\underset{|}{C}}-H$ and assuming that no methyl was in the *peri-* or equivalent position,[153] the derived values of aromaticity could be brought more easily into

[145] Wyss, W. F., Private communication.

[146] Van Vucht, H. A., Rietveld, B. J., and van Krevelen, D. W., *Fuel*, **34**, 50–9 (1955).

[147] Friedel, R. A., and Queiser, J. A., *Anal. Chem.*, **28**, 22–30 (1956).

[148] Brooks, J. D., Durie, R. A., and Sternhell, S., *Australian J. Appl. Sci.*, **9**, 63–80 (1958).

[149] Alexanian, C., *Compt. rend.*, **246**, 1192–5 (1958).

[150] Newman, P. C., Pratt, L., and Richards, R. E., *Nature*, **175**, 645 (1955). Bell, C. L. M., Richards, R. E., and Yorke, R. W., *Brennstoff-Chem.*, **39**, S30–1 (1958).

[151] Dryden, I. G. C., *Fuel*, **37**, 444–68 (1958).

[152] Richards, R. E., Private communication.

[153] Given, P. H., *Fuel*, **39**, 147–53 (1960).

agreement with those estimated from infrared data.[126] Results of proton magnetic resonance were shown to be within the limits estimated from infrared data.

The low resolution method used by Richards can be applied to solid coals. High resolution NMR, which gives more precise information, can only be applied to liquids or solutions. This method has been used on methylene chloride solutions of several products obtained by high-vacuum carbonization at less than 400°C of a series of bright coals, and of one made by chloroform extraction of a coking coal after shock heating to 420°C.[154] The relative intensities were estimated of the peaks associated respectively with aromatic and phenolic hydrogen, hydrogen on nonaromatic carbon directly attached to an aromatic ring, and hydrogen on other saturated carbon atoms. It was proved experimentally that the spectra were representative of the materials studied; from their infrared spectra it was known that the materials closely resembled the parent coals. Another important conclusion was that methylene bridges between aromatic rings were not present in these products, and by implication probably not in the parent coals either. By a straightforward calculation involving a knowledge of the phenolic hydroxyl contents of the materials and the total hydrogen contents, and the assumption that the average atomic ratio of nonaromatic hydrogen to carbon is 2, an estimate of the proportion of carbon in aromatic form and the degree of substitution of the aromatic rings could be made (Fig. 10). Also, via H/C ratios calculated for the ring systems when stripped of their oxygen groups and nonaromatic substituents, the number of condensed rings in the average polycyclic nuclei could be estimated. Values of the ratio of aromatic to nonaromatic hydrogen were compared with those estimated by infrared measurements on the same samples, using the method earlier applied to coals.[126] Agreement was good, confirming the correctness of Brown's original choice of a value for the extinction coefficient ratio, used for converting infrared absorption intensity ratios to hydrogen ratios. The data for these carbonization products were converted to the average number of planar carbon and oxygen

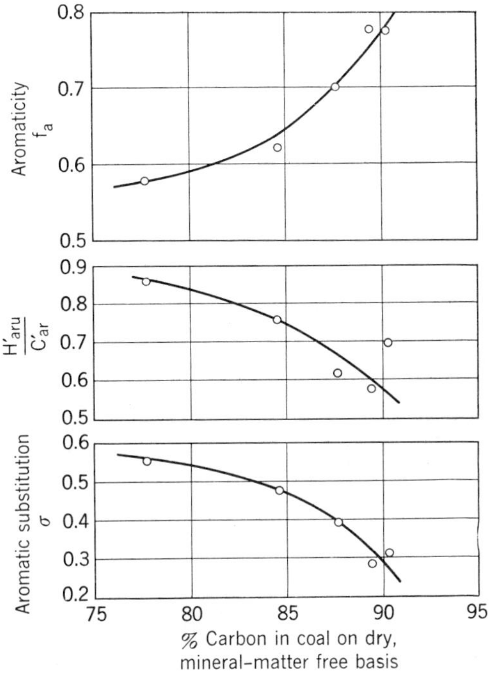

Fig. 10. Some features of the molecular structure of coals as determined by high-resolution nuclear magnetic resonance.[154]

atoms associated with an aromatic nucleus, as calculated from X-ray data,[137] and plotted against the carbon content of the parent coals. Reasonably good agreement with Hirsch's values was obtained. The meaning of this agreement is not yet clear, since the products studied were not representative of the whole coal and contained considerably more hydrogen than the parent coals. A single experiment with a chloroform extract of a coking coal that had been heated to 420°C indicated on infrared evidence that about one $—CH_3$ group was present for every four $—\overset{|}{C}H_2$ groups.

Diamagnetic susceptibility. Wooster and Wooster[155] made probably the first measurements of the magnetic susceptibilities of coals, and suggested that the method could be used to determine aromaticity. Only anthracite showed anisotropy. Honda and Ouchi[156] made a study of the diamagnetic susceptibilities of a number of organic substances, including

[154] Brown, J. K., Ladner, W. R., and Sheppard, N., *Fuel*, **39**, 79–86 (1960). Brown, J. K., and Ladner, W. R., *Fuel*, **39**, 87–96 (1960).

[155] Wooster, W. A., and Wooster, N., *Proc. Conf. Ultrafine Structure of Coals and Cokes*, BCURA, London, 1944, pp. 322–9.

[156] Honda, H., and Ouchi, K., *Rept. Resources Research Inst. Japan*, No. 1 (1952), 66 pp.

coals and a series of condensed aromatic compounds. They concluded that magnetic anisotropy in the last was proportional to the number of condensed nuclei present, but this is clearly an oversimplification. In further studies, four coals were separated into portions of different density by float-and-sink before measuring susceptibility. A relation with ash content was observed.[157]

In a final careful study of over sixty Japanese and other vitrains, special care was taken to eliminate paramagnetic mineral impurities,[158] and the diamagnetic susceptibility of the pure coal was determined from the temperature variation of the total using Curie's Law, by extrapolation to infinite temperature. Because powdered, acid-washed samples were used, the anisotropy, thought to be the best measure of aromatic characteristics, had to be calculated. It was considered to equal three times the difference between the measured true diamagnetic susceptibility and the value calculated from ultimate analysis by a modification of the well-known additive relations of Pascal.[159] This is a doubtful procedure, but it was also used in respect of a range of polynuclear aromatic hydrocarbons, so that the method was to some extent calibrated, although definitely empirical. The anisotropy divided by the number of aromatic carbon atoms was found to depend characteristically on the size and shape of the condensed ring system. The number of aromatic carbon atoms in 100 g of a coal was deduced from the values of aromaticity estimated from the density. The value of this reduced anisotropy increased gradually with rank up to about 90% C content. In the range from 70 to 90% C, the value equalled that of aromatic compounds containing between 2 and 4 condensed rings, and it then increased rapidly with further increase in rank.

Schuyer[160] developed a new empirical relation between molar diamagnetic susceptibility and the numbers of various types of atoms and condensed rings in the cluster unit—a form of statistical structural analysis. The relation was developed from measurements on forty model substances, and the number of rings in the aromatic cluster could be predicted correctly for these models within an error of, at the most, 2 rings. (The number of fused rings in the models ranged from 1 to 10.) This method gave a number of rings per aromatic cluster unit increasing from about 3 at 80% C to 9 at 90% C and over 30 at 96% C content. These figures are distinctly higher than those of Honda and Ouchi, and also higher than those derived from the latest X-ray information, especially for the coking coals and anthracites.

More rigorous methods that have not yet been applied to coals have been developed for interpretation of measurements of diamagnetic susceptibility.[161]

Indirect Methods

Statistical structural analysis. Because of the uncertainties that attended, especially in the early days, the interpretations of the results of X-ray, infrared, and other measurements on coals, and because it is fundamentally impossible to specify a molecule that is an "intrinsic mixture" by other than statistical parameters, van Krevelen and his colleagues developed methods of statistical structural analysis.[2,162] By these methods structural parameters for an aromatic/aliphatic model can be deduced from ultimate analyses and certain other properties, for example, density, refractive index, and heat of combustion. The most useful result of such developments has been a more quantitative expression of the qualitative arguments advanced by earlier workers such as Howard[163] from the consideration of ultimate analyses and other evidence. It is still not sufficiently appreciated, especially by physicists who study coal, that any structures postulated must conform to the firmly established elementary analyses of the coals

[157] Honda, H., *J. Fuel Soc. Japan*, **32,** 486–93 (1953).

[158] Honda, H., and Ouchi, K., *Fuel*, **36,** 159–75 (1957).

[159] Pascal, P., *Ann. chim. et phys.*, **19,** 5–70 (1910).

[160] Schuyer, J., *Naturwissenschaften*, **42,** 388 (1955).

[161] London, H., *J. phys. radium*, **8,** 397–409 (1937). McWeeny, R., *Proc. Phys. Soc. (London)*, **A64,** 261–75, 921–30 (1951); **A65,** 839–45 (1952); **A66,** 714–20 (1953). Rebane, T. K., *Doklady Akad. Nauk S.S.S.R.*, **114,** No. 1, 70–2 (1957); *Vestnik Leningrad. Univ., Fiz. i Khim.*, **1957,** No. 3, 19–31.

[162] Van Krevelen, D. W., *BCURA Quart. Gaz.*, No. 24 6–19 (1954).

[163] Howard, H. C., *Ind. Eng. Chem.*, **44,** 1083–8 (1952).

concerned and the valency requirements of the atoms. It is in applying such tests—crucial tests of consistency—to hypothetical models that structural analysis is likely to prove most fruitful in the future.

van Krevelen began by tracing the development band of coal rank and the course of reactions such as oxidation on a diagram with coordinates H/C and O/C.[164] To data relating mineral-free analyses to the helium densities of coals, he then applied the concept that molecular volume can be regarded as the sum of mean atomic volumes[165] and of correction factors relating to special structural features.[166] Such methods are essentially empirical; their accuracy in predicting density increases with the number of structural corrections employed, but their usefulness for deducing structural features from density decreases as more factors are taken into account. For bituminous coals, it was possible by elimination, using qualitative chemical and infrared evidence, to reduce greatly the number of structural features considered and to deduce values for the "ring condensation index" $2R/C'$, where R is the number of polycyclic rings in a ring cluster and C′ the number of carbon atoms in a cluster unit including the nonaromatic structures.

In a further paper,[167] van Krevelen and Chermin showed that the key coefficient in the basic equation should be expressed not as a constant but as a function of H'/C'. This change significantly reduced the values of the ring condensation index and indicated smaller clusters. Model polymers suitable for testing the equations were not available, and a further necessary assumption was that the coal molecules were randomly packed. In an effort to avoid these difficulties, van Krevelen and Chermin later substituted a purely graphical method for the previous algebraic one and included pitch fractions among the model compounds used to construct the master diagram.[2,168] The algebraic method gave $2R/C'$, from which the fraction of aromatic carbon f_a could be deduced, if $R \gg 1$, from the relation

$$f_a = 1 - \left[\frac{2(R-1)}{C'}\right] + \left(1 - \frac{H'}{C'}\right) - \frac{B}{C'}$$

where B is twice the average number per cluster unit of bridges linking them together to form the molecule, that is, the average number of other clusters to which any one is linked. The graphical method gave f_a directly, from which values of $2(R-1)/C'$ could then be obtained. The graphical method is clearly the sounder method, but uncertainty remained[169] because the two-dimensional parameter R was deduced from the density, a three-dimensional quantity, which implies that there is some constant feature in the packing of clusters in different coals and in the model compounds. On the other hand, the structural contributions to molecular volume are quite large compared to the basic atomic contributions.

When the foregoing equation was used to convert $2(R-1)/C'$ to f_a, or vice versa, two difficulties arose. First, the value of H′/C′ used had to be corrected to that of the equivalent hydrocarbon to allow for oxygen in the structure (not enough was known about the forms of oxygen present, and for low-rank coals estimates of the necessary correction differed significantly according to the point of view—van Krevelen assumed it to be zero). Second, values of the factor B were unknown. If R is large this is not a serious matter, but if R is small it makes for a large uncertainty.

In further papers, a more complicated relation between the molar refractions of coals and their ultimate analyses was employed to obtain directly the number of aromatic carbon atoms in a cluster;[170] the theory was elaborated and related the "structural increment" of molecular refraction to the laminar surface area of the polycyclic aromatic nucleus.[171] By using these values together with the parameters derived from density, it was possible to characterize the whole structure of the model used as working hypothesis, in terms of C′, C'_{ar}, and R. Values for R were found to be in reasonable

[164] van Krevelen, D. W., *Fuel*, **29**, 269–83 (1950).

[165] Traube, J., *Ber.* **28**, 2722–8 (1895).

[166] van Krevelen, D. W., *Brennstoff-Chem.*, **33**, 260–8 (1952).

[167] van Krevelen, D. W., and Chermin, H. A. G., *Fuel*, **33**, 79–87 (1954).

[168] van Krevelen, D. W., and Chermin, H. A. G., *Fuel*, **33**, 338–47 (1954). Van Krevelen, D. W., and Schuyer, J., *Fuel*, **36**, 313–20 (1957).

[169] Dryden, I. G. C., *Fuel*, **32**, 394–6 (1953).

[170] Schuyer, J., and van Krevelen, D. W., *Fuel*, **33**, 176–83 (1954).

[171] Schuyer, J., Blom, L., and van Krevelen, D. W., *Trans. Faraday Soc.*, **49**, 1391–401 (1953). Scholte, T. G., *Physica*, **15**, 437–49 (1949).

agreement with the corresponding aromatic layer diameters deduced from X-ray measurements,[134] but the latest refinements in X-ray interpretation and[135,136,172] statistical analysis[2,168] have both tended to suggest lower values of R, which renders previous assumptions ($R \gg 1$, $R \gg$ B) less valid. The methods have been applied to British coals.[173]

Application of "statistical structural" methods to the interpretation of calorific values of coals in chemical terms gave values of $2(R-1)/C'$ that were lower[174] than those of the original density method. The heat of combustion is related more closely than density to the chemical structure of the molecule, but lack of knowledge of the values of the intercluster linkage affects somewhat the values derived.

In most of the work so far discussed, parameters derived from more than one property of coal have been combined to build up the complete statistical picture. Starting from the fact that the ring condensation index and R are not independent variables,[175] an attempt has been made to link these quantities by means of diagrams based on intrinsic properties of the types of chemical structure considered. The values of R thus obtained were much too large, being based on van Krevelen's earliest values of the index.[166] Improvements have since been made.[151]

By van Krevelen's methods, molar refraction tends to lead to estimates of ring number larger than those from density or from the X-ray estimates. It is difficult to measure, or to find in the literature, refractivities for pure hydrocarbons in noncrystalline form that are comparable with the contributions to be expected from similar aromatic nuclei included irregularly in the coal structure. Some of the parameters in van Krevelen's equations and curves relating structure to molar refraction,[176] therefore may well be in error. Given has proceeded in an opposite manner[177] by estimating the refractive index to be expected from models such as he has proposed,[153,178] and found no serious discrepancy even though these models contain on the average only two fused aromatic rings in a cluster unit as compared with van Krevelen's estimate of 6 to 7 rings for a similar coal.

A new method of calculating molar refraction has been proposed, but it has not been applied to coals.[179] More limited application of calorific value to estimate characteristic changes in coal with carbonization has also been made.[180] Other relevant papers should be consulted.[181]

Functional Groups

The problem of determining functional groups containing oxygen in bituminous coals is simplified by the established fact that these coals contain no appreciable amounts of alcoholic hydroxyl, carboxyl, or methoxyl groups. Experimentally, therefore, phenolic hydroxyl and carbonyl have to be determined; the possibility that ester groups are present can be examined by hydrolysis, and the difference unaccounted for may be ascribed to ether and heterocyclic oxygen, which are more difficult to determine unequivocally or to recognize in the infrared spectra. It now appears that hydroxyl and carbonyl can account for all but 10 to 30% of the oxygen. Infrared spectra do not indicate unequivocally the presence of ether oxygen, but its presence may be inferred from the variation of the infrared spectra with the rank of coal.[126,182]

Reactions are more easily carried out with solvent extracts than with solid coals. Fortunately, experience has shown that extracts

172 Diamond, R., Thesis, University of Cambridge 1956.

173 Leicester, J., *J. Appl. Chem. (London)*, **6**, 96–101 (1956).

174 Dryden, I. G. C., and Griffith, M., *Fuel*, **34**, S36–47 (1955). Schuyer, J., and van Krevelen, D. W., *Fuel*, **33**, 348–54 (1954).

175 Dryden, I. G. C., and Griffith, M., *Fuel*, **32**, 199–210 (1953). Dryden, I. G. C., *Fuel*, **32**, 82–8 (1953).

176 van Krevelen, D. W., Blom, L., and Chermin, H. A. G., *Nature*, **171**, 1075–6 (1953).

177 Given, P. H., *Coal*, W. Francis, Ed., Second edition, Arnold, London, 1961, 806 pp., chapter 12.

178 Given, P. H., *Nature*, **184**, 980–1 (1959).

179 Batsanov, S. S., *Zhur. Neorg. Khim.*, **3**, 241–59 (1958).

180 Urazovskii, S. S., Voloshin, V. A., and Vysotskaya, A. I., *Doklady Akad. Nauk S.S.S.R.*, **120**, 595–8 (1958).

181 Van Krevelen, D. W., *Brennstoff-Chem.*, **34**, 167–82 (1953); *Ann. mines Belg.*, **53**, 591–8, 750–7 (1954). Huntjens, F. J., and van Krevelen, D. W., *Fuel*, **33**, 88–103 (1954).

182 Roy, M. M., *Fuel*, **36**, 249–50 (1957).

representing more than about 5% of the coal by weight are not only similar in infrared spectra to the parent coals, but contain similar amounts of the two main functional groups—possibly it would be more correct to say similar proportions of their total oxygen in these forms, although these two statements are not far from being equivalent.

Although brown coals are not dealt with in detail, two points are especially worthy of note; they contain a significant proportion of alcoholic hydroxyl groups,[183] and there is strong evidence for groupings capable of undergoing the keto-enol transition.[148]

Hydroxyl

A summary of the main determinations is given in Table 4.[184–192]

The incentive toward the development of titration methods in aqueous media is speed of determination, but the method is unsatisfactory, particularly for coals above 83% C content into which penetration by the reagent is inadequate, and the results are therefore low. In view of what is known of the constrictions within the coal structure, it is surprising that reagents with large molecules penetrate so completely as would appear from Table 4; quite high figures have been obtained by using a remarkably large reagent molecule to form trimethylsilyl ethers.[193] It seems likely that swelling of the coal by medium or reagent is responsible. Determinations of hydroxyl concentration in coal by the use of Grignard reagents,[194] which have unsatisfactory properties for use with coal, cannot be relied on.[190] Results obtained with them,[51] for example on extracts, have been omitted from the table.

An attempt has been made to calculate oxygen group concentrations from the difference between heats of wetting in polar liquids, including methanol, and heats of wetting in nonpolar liquids.[26,27] Assumptions had to be made about the energy of interaction, and the differences in heat of wetting were sometimes no greater than the experimental error. It is not surprising that the results were irregular and, for example, that the average hydroxyl concentration deduced was much smaller than most of the figures included in Table 4.[26]

An uncertainty that affects all methods of determining hydroxyl in coal is the doubt about the completeness of reaction. Infrared spectra of acetylated products, for example, usually show a residual hydroxyl band probably not the result of moisture. The ultimate slowness of the interaction, at least in the absence of a liquid medium, is shown by the fact that up to seventeen successive treatments of some coals with ketene were required before an end point was reached.[192] Moreover, there is evidence that carbonyl may exist in chelated form, and none or only part of the associated hydroxyl may be determined. In view of this evidence and of the variations shown in the investigations reported, it is probably not far wrong to assume that about 60% of the total oxygen in bituminous coal exists as hydroxyl, although this amount is not strictly constant. Above 90% C, the percentage may drop to a much lower figure, to perhaps 10%.

A graphical presentation of the results shown in Table 4 is given in Fig. 11. Comparative results from different laboratories using different methods on a single sample of vitrain

[183] Durie, R. A., and Sternhell, S., *Australian J. Appl. Sci.*, **9**, 375–9 (1958).

[184] Syskov, K. I., and Kukharenko, T. A., *Zavodskaya Lab.*, **13**, 25–8 (1947).

[185] Ihnatowicz, A., *Prace Glównego Inst. Górnictwa, Komun.* No. 125 (1952), 39 pp.

[186] Uporova, E. P., and Rafikov, S. R., *Izvest. Akad. Nauk Kazakh, S.S.R., Ser. Khim.*, **1956**, No. 9, 23–32.

[187] Mukherjee, P. N., Bhowmik, J. N., and Lahiri, A., *Chem. & Ind.* (*London*), **1957**, 633–4.

[188] Brooks, J. D., and Maher, T. P., *Research* (*London*), **7**, S30–1 (1954); *Fuel*, **36**, 51–62 (1957).

[189] Yohe, G. R., and Blodgett, E. O., *J. Am. Chem. Soc.*, **69**, 2644–8 (1947). Yohe, G. R., Wilt, M. H., Kauffman, H. F., and Blodgett, E. O., *Trans. Illinois State Acad. Sci.*, **47**, 77–80 (1955).

[190] Blom, L., Edelhausen, L., and van Krevelen, D. W., *Fuel*, **36**, 135–53 (1957).

[191] Brown, J. K., and Wyss, W. F., *Chem. & Ind.* (*London*), **1955**, 1118. Wyss, W. F., *Chem. & Ind.* (*London*), **1956**, 1095–6.

[192] *Gt. Brit., Dept. Sci. Ind. Research, Fuel Research Board, Ann. Rept.*, 1955, p. 6. Horton, L., *Congr. intern. chim. ind. Liège*, **1958**, *Ind. chim. belge*, **Suppl. Vol. 1**, 416–19 (1959).

[193] Friedman, S., Steiner, W. A., Raymond, R., and Wender, I., 132*nd Meeting of the Am. Chem. Soc.* (1957). Langer, S. H., Connell, S., and Wender, I., *J. Org. Chem.*, **23**, 50–8 (1958). Friedman, S., Kaufman, M. L., Steiner, W. A., and Wender, I., *Fuel*, **40**, 33–46 (1961).

[194] Fuchs, W., *Chem. Tech.* (*Berlin*), **6**, 378–80 (1954).

TABLE 4

Determinations of Hydroxyl Groups in Coal*

Method	Reagents	Medium	Temperature, Time	Measurement Made	Coal or Extract	Range of Rank, % C	O_{OH}, % of Coal or Extract	O_{OH}/O %
Sorption and titration[184]	$Ba(OH)_2$, $Ca(OAc)_2$	Water	Room, 4 days	Sorption of $Ba(OH)_2$ minus sorption $Ca(OAc)_2$ both by titration, meq/g	Coal residues after extraction	—	—	—
Sorption and titration[185]	$Ba(OH)_2$, $Ca(OAc)_2$	Water	Room, days	Similar to ref. 184	Coals	65 to 93	7.1 to 0.1	53 to 4
Neutralization[186]	NaOH, Na_2CO_3	40% aqueous alcohol	Room, 20 hr	meq NaOH neutralized minus meq Na_2CO_3 neutralized, by titration	Coals (including macerals)	67.5 to 85.9	2.4 to 0.0	—
Neutralization[187]	KOH, CO_2	Ethanol, 85% ethanol	Reflux, room, 1 hr	Titration of K_2CO_3 liberated by CO_2 from neutralized coal	Coals	70.5 to 89.6	4.0 to 0.9	—
Titration[188]	Na aminoethoxide	Ethylenediamine	Room, 1 to 2 hr	Na^+ equivalent to phenolic OH, etc.	Coals, Vitrains	67.1 to 94.3 71.3 to 92.1	7.8 to 0.2 7.3 to 0.4	67 to 14 64 to 17
Methylation[189]	NaOH, dimethylsulfate	Water	100°C, 4 hr	Increase in methoxyl groups in product determined	Coals	72.4 to 94.9	3.1 to 0	16 to 0
Methylation[190]	Diazomethane	Ether	−5°C, 1 week	Determination of added methoxyl groups	Vitrains	65.5 to 90.9	5.4 to 0.7	38 to 15
Acetylation[191]	Acetic anhydride	Pyridine	100°C, 5 hr	Titration of acetic acid uptake, determination of acetyl groups	Coals, Extracts	77.7 to 91.8 81.8 to 86.5	5.6 to 0.2 4.5 to 2.6	44 to 6 46 to 31
Acetylation[192]	Ketene	None	Room to 110° C, 35 to 85 hr	Determination of acetyl groups in product	Coals	81.9 to 93.0	6.4 to 0.6	100 to 42
Acetylation[28]	NaOAc	Acetic anhydride	Reflux, 3 hr	Weight increase, titration of unreacted acetic anhydride, determination of acetyl groups	Coals, Vitrains	65.5 to 91.1 79.2 to 90.0	11.4 to 0 7.2 to 0.4	49 to 0 55 to 13
Acetylation[190]	Acetic anhydride	Pyridine	90°C, days	Titration of acetic acid liberated from acetylated sample by $Ba(OH)_2$	Vitrains	65.5 to 90.9	7.8 to 0.5	71 to 17
Etherification[190]	Dinitrofluorobenzene, $NaHCO_3$	Water plus dimethylformamide	20°C, days	Determination of nitrogen added to product	Vitrains	79.5 to 90.9	3.7 to 0.3	33 to 10.3
Etherification[193]	Hexamethyldisilazane	Pyridine	Reflux, 4 hr	Determination of silicon added to product	Coals, Vitrains	66.9 to 93.3 68.5 to 91.2	7.8 to 0.2 8.8 to 0.5	56 to 14 52 to 13
Acetylation[187]	NaOAc	Acetic anhydride	Reflux, 3 hr	Same as ref. 28	Coals	70.5 to 89.6	9.5 to 0.9	—

* The ratio of functional oxygen to total is necessarily based on oxygen in *coal*, even when the former was determined on an extract. Side reactions, or reaction with groups other than those nominally determined, have been neglected.

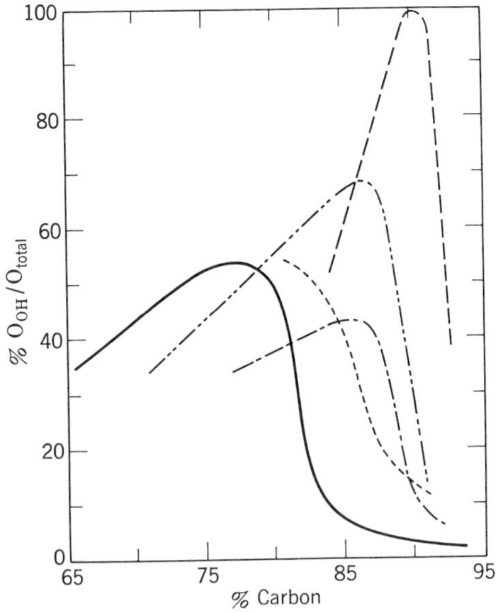

Fig. 11. Comparison of hydroxyl determinations in coal by various workers.[195] Blom[190] — · · —; Horton[192] — — — — —; Ihnatowicz[185] ———; Lahiri[28] - - - - - - - -; Brown and Wyss[191] — · — · — · —.

(86 to 87% C) gave values of the percentages of oxygen as hydroxyl from about 30 to 70.[195] In an examination of a wide range of model compounds, complete acetylation was generally observed, but examples of complete failure to react and of incomplete acetylation were both found, as well as two side reactions, one increasing and the other decreasing the apparent number of hydroxyl groups.[195]

Carbonyl

Table 5 summarizes the principal determinations of carbonyl made on coals.[196,197,198]

Polarographic reduction and related methods have been used to estimate carbonyl groups in solvent extracts of coals.[196,199] The use of dimethylformamide as medium with tetraethyl ammonium iodide or other salt as supporting electrolyte has proved most successful. The method has been refined with increasing precision and supplemented so that relative concentrations of carbonyl can be estimated with some certainty. Background work on model compounds has been carried out.[200] The identification as carbonyl depends on a comparison of the coal extract polarogram with the half-wave potentials of compounds of known structure, and the concentration is estimated from the quantity of electricity consumed during reduction over the range of potentials where carbonyl is known to react. This range, 0 to −1.3 v, is fairly well separated from the range where aromatic nuclei are reduced— −1.2 to −2.2 v against the mercury pool reference electrode. This estimation assumes that no side reactions take place. A more rigorous method[197] involves controlled-potential electrolysis of acetylated coals, followed by acetylation of the reduced product with acetic anhydride containing radioactive carbon, C^{14}. As a preparative method, this gives sufficient product for analysis. The second acetylation gives a direct measure of the hydroxyl groups produced in the reduction.

It has been known for some time from infrared evidence that unconjugated or weakly conjugated ketonic carbonyl groups do not occur in bituminous vitrains. It has therefore been assumed that carbonyl was present as quinone. It has been mentioned that the prominent band at 1600 cm^{-1} in coal spectra has been associated in part with quinone chelated to hydroxyl, and allowance has to be made in calculating the carbonyl concentrations determined by Given's methods. From a combination of polarographic and infrared evidence, together with total oxygen and hydroxyl concentrations, it has been suggested[197] that highly conjugated ketone groups chelated to hydroxyl rather than quinone groups may be the major form of carbonyl present in these vitrains. This involves a different correction to the calculation of results both for carbonyl and hydroxyl.

The method of determining carbonyl whereby coal is reacted with phenylhydrazine to form the hydrazone appears to give results of about the right magnitude.[185,198] The reaction is characteristic of ketones but may take place with at least some quinones—there is too little

[195] Halleux, A., Delavarenne, S., and Tschamler, H., *Fuel*, **38**, 283–90 (1959)

[196] Given, P. H., and Peover, M. E., *Proc. Conf. on Science in the Use of Coal*, Sheffield, Inst. Fuel, London, 1958, pp. A33–7.

[197] Given, P. H., and Peover, M. E., *J. Chem. Soc.*, **1960**, 394–400.

[198] Tadzhiev, A. T., *Doklady Akad. Nauk Uzbek. S.S.R.*, **1958**, No. 4, 29–31.

[199] Given, P. H., and Schoen, J. M., *J. Chem. Soc.*, **1958**, 2680–4.

[200] Given, P. H., Peover, M. E., and Schoen, J. M., *J. Chem. Soc.*, **1958**, 2674–9.

TABLE 5

DETERMINATIONS OF CARBONYL GROUPS IN COALS

Method	Medium	Temperature, Time	Measurement Made	Coal or Extract	Range of Rank, % C in Coal	O_{CO}, % of Coal or Extract	O_{CO}/O, %
Controlled potential electrolysis[196]	Dimethylformamide	Room, >1 hr	Quantity of electricity consumed	Extract	81.7 to 89.2	(2.1–3.3) to 2.0	(20–32) to 54
Controlled potential electrolysis + acetyl determination[197]	Dimethylformamide	Room, >1 hr	Radioactive count on product	Acetylated extract	78 to 89	(2–2.5) to 0.9	(14–18) to 20*
Formation of hydrazone[185]	Water	90°C, ½ hr	Determination of nitrogen in excess reagent	Coal	65 to 93	1.25 to 0.4	9 to 20
Formation of hydrazone[198]	Water	100°C, ½ hr	Determination of nitrogen increase in product	Coal	Brown, long flame	0.4 to 0.8	—
Formation of oxime[190]	Pyridine	90°C, 20 hr	Decrease of nitrogen content on hydrolysis	Vitrain	65.5 to 91	0.5 to 0.2	5 to 10

* Based on oxygen in *extracts*.

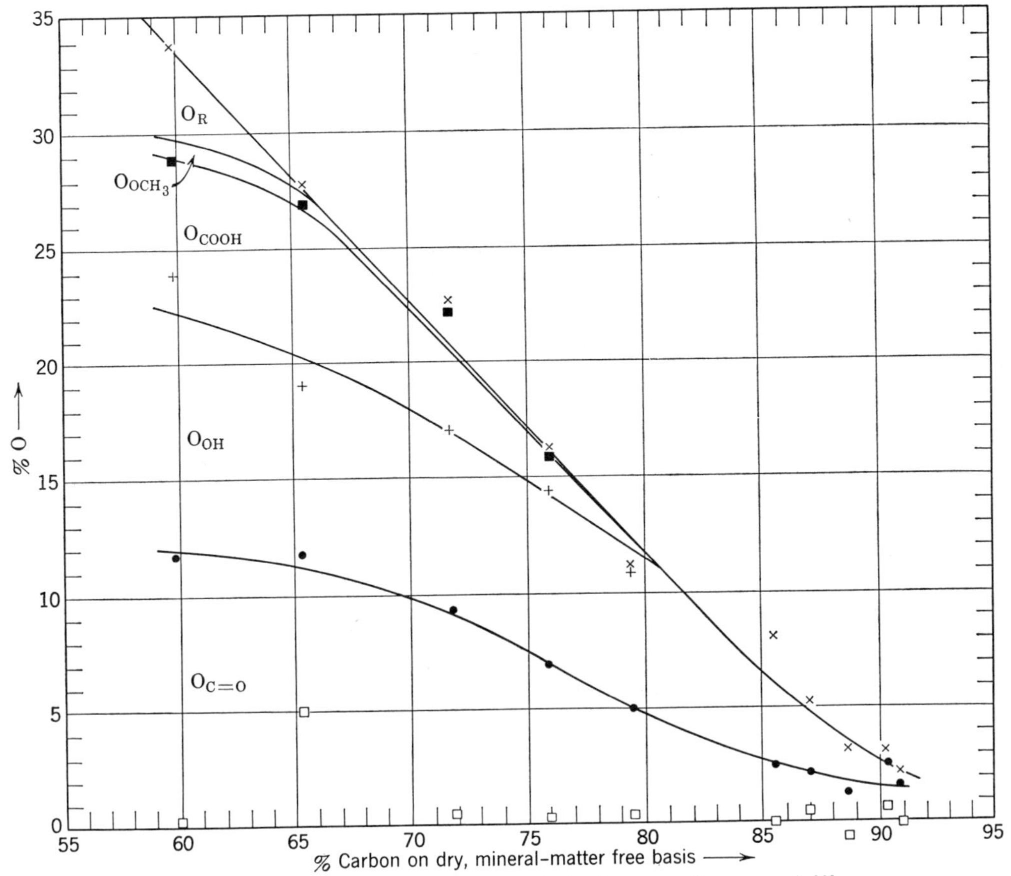

Fig. 12. Functional oxygen groups of vitrains according to rank.[202]

background information at present. With other quinones, reaction occurs but follows quite a different course. The retention of phenylhydrazine may result from a form of adsorption, and it may be wise to discount such measurements until further information is available. This method has been used by other investigators than those cited.

An unusual method, supposed by its originators to measure the carbonyl concentration in coal, was adsorption of hydrogen chloride or methylamine.[26] The irreversible adsorption of part of these vapors was used as a measure on the basis of one molecule so adsorbed per carbonyl group.

The behavior of polynuclear quinones in concentrated sulphuric acid, and their adsorption spectra, have been examined with special reference to the interpretation of the behavior of coals.[201]

[201] Durie, R. A., and Shannon, J. S., *Australian J. Chem.*, **11,** 168–88, 189–99 (1958).

Distribution of Oxygen in Bituminous Vitrains

Curves have been given that indicate the distribution of various oxgyen groups (Fig. 12) from which it may be estimated that in vitrains of 80% C content about 55% of the oxygen is in hydroxyl form and 40% is in carbonyl form; in vitrains of 90% carbon, about 50% is in each form.[202] Independent results[197] suggest the corresponding percentages of hydroxyl and carbonyl for these two levels of rank are 10% and 20%, and 60% and 25% respectively. It seems that more than 70% of the total oxygen can be accounted for in these forms.

Several investigators have found that hydrolysis with alcoholic potash increases the acidic hydroxyl concentration in some coals, and the presence of esters has been postulated.[185,203] In view of the foregoing

[202] Blom, L., Thesis, Univ. of Delft, 1960.

[203] Ahuja, L. D., Kini, K. A., and Lahiri, A., *J. Sci. Ind. Research* (*India*), **17B,** 231–2 (1958).

totals, it seems more likely that the effect of the alkali was to render the hydroxyl more completely accessible or reactive.

Coals of High Organic Sulfur Content

A number of coals, particularly from Yugoslavia, Pakistan, and parts of Australia and India, have a high content of organic sulfur and a correspondingly low oxygen content. They often, but not invariably,[204] exhibit a high degree of softening and swelling when heated, and in this respect resemble coking coals of 88 to 90% C, although their carbon content may correspond to that of a lignite. If the sulfur is added to the carbon content, the sum is roughly equal to the carbon content of a normal coal with similar behavior. Kreulen made a detailed examination of such a coal[205]—the Raša coal of Istria.[69] The nature of the sulfur linkages has also been examined by reacting the coal with methyl iodide,[206] and it was concluded that the 70% of the organic sulfur that would not react occurred in heterocyclic aromatic ring structures. The nature of organic sulfur in coals, solvent extracts, and oxidized samples of high-sulfur coals have been studied to distinguish thiophenol, sulphide, disulphide, and heterocyclic forms of sulfur.[207,208] Since more than 70% of the organic sulfur remained in the residue after oxidation, it has been concluded that this amount exists in heterocyclic linkages.[208]

The heat of wetting of Raša coal has been shown to be low for its low-carbon content—a result compatible with its coking properties.[204,209] In a study of the reactive oxygen groups in high-sulfur Indian coals, it was observed that the correlation of low heat of wetting and low concentration of hydroxyl groups with high swelling number was close.[210]

The small amount of evidence available suggests that the abnormal amount of oxygen in high-sulfur coals is distributed in much the same manner between functional groups as the oxygen in normal coals, and that the sulfur does not have the same effect on properties as would an equivalent amount of oxygen. Little is known about the carbon-hydrogen structure of these coals and its contribution to their unusual properties.

Macerals

An excellent comparative review of the properties of macerals was published early in 1959.[211] Comparative work on the chemistry of macerals has originated mainly from the laboratories of Kröger, of van Krevelen, and of Alpern and his associates.

Kröger has successfully separated exinite, vitrinite, and micrinite from five Ruhr coals by the float-and-sink technique using toluene-carbon tetrachloride mixtures.[212] According to his petrographic analyses, the concentrates obtained consisted of 99% vitrinite, 94 to 97% exinite, and 91 to 97% micrinite respectively. Van Krevelen estimated the composition and properties of the macerals from those of less pure concentrates with known petrographic analyses.[213] Alpern also used this method.[214,215] He developed an electrostatic separation method of limited application,[214] and he also succeeded in separating an exinite from its associated vitrinite by selective dissolution of the vitrinite in boiling α-naphthol. The qualitative variation in hydrogen content, volatile matter, and density was known before the complete separation of macerals was attempted. The difference between macerals becomes smaller with increasing rank of the associated vitrain; vitrinites and exinites become very similar in the region of 90% C

[204] Hadži, D., and Novak, A., *Vestnik Sloven. kemi. društva*, **1**, 175–84 (1954).

[205] Kreulen, D. J. W., *Fuel*, **31**, 462–7 (1952).

[206] Kavčič, R., *Bull. sci., Conseil acad. RPF Yougoslavie*, **2**, 12–13 (1954).

[207] Brooks, J. D., *J. Inst. Fuel*, **29**, 82–5 (1956),

[208] Roy, M. M., *Naturwissenschaften*, **43**, No. 21, 497 (1956).

[209] Novak, A., and Hadži, D., *Bull. sci., Conseil acad. RPF Yougoslavie*, **1**, 104–5 (1953).

[210] Iyengar, M. S., and Lahiri, A., *Fuel*, **37**, 29–35 (1958).

[211] Brown, J. K., *BCURA Bull.*, **23**, 1–17 (1959).

[212] Kröger, C., *Brennstoff-Chem.*, **37**, 182–6 (1956).

[213] Dormans, H. N. M., Huntjens F. J., and van Krevelen, D. W., *Fuel*, **36**, 321–39 (1957). Huntjens, F. J., Dormans, H. N. M., and van Krevelen, D. W., *Brennstoff-Chem.*, **39**, S67–70 (1958).

[214] Ladam, A., Iselin, P., and Alpern, B., *Brennstoff-Chem.*, **39**, S58–61 (1958).

[215] Alpern, B., *Compt. rend. 27th congr. intern. chim, ind., Brussels*, 1954, Ind. chim. belge., **20**, (Special), **2**, 9–12 (1955).

content, whereas micrinites differ appreciably from exinites and vitrinites until about 94% C is reached.[213]

Published data on macerals include elementary composition,[213–219] heat of wetting,[220] solubility in organic solvents,[34,42,214,221] X-ray[133,136,138,222] and infrared data,[213,223] electron spin resonance adsorption,[224] content of oxygen-containing groups,[219,223] and behavior on oxidation[214,225,226,227] and hydrogenation.[214,228]

Exinites have the most and micrinites the least hydrogen content, volatile matter, and calorific value. The order is reversed for density[213,216] and free radical content.[224] Analyses of fusinite vary little throughout the whole range of bituminous coals[228] and average 94% C, 2.8% H, and 2.3% O. Major differences between the macerals in heat of wetting are not apparent, but vitrinite exhibits the well-known minimum, exinite appears to show a maximum at 85% C, whereas that of micrinite is substantially independent of rank.[220]

The extractability of some macerals by ethylenediamine was determined from experiments on concentrates and durains of known petrological analysis.[42] It was found that exinite resembled vitrinite in content of soluble matter, whereas fusinite was completely insoluble although it swelled considerably; micrinite appeared to reduce the extractability of the associated vitrinite, as well as being itself completely inert to the solvent. One exinite concentrate appeared to contain much less soluble matter than the exinite in the black durain from which it was prepared; it may be suspected that microspores, megaspores, and other components of exinite show different behavior so that concentrates may not be representative of the rather heterogeneous "whole exinite." More recent studies suggest that exinites from low-rank coals are dispersed much less than vitrinites in pyridine[34,214] and ethylenediamine,[214] although the difference becomes small at 89% C of the vitrinite. Kröger[34] has claimed that micrinite, which he agrees has a low content of soluble matter, gives a higher yield of extract as the rank increases. Alpern found that dissolution of an exinite in boiling α-naphthol was slow, but that in α-naphthaquinoline the exinite dissolved quickly and the vitrinite slowly.[214] Dryden has confirmed the low extractability of fusinites.[42]

The only quantitative estimates available for the layer dimensions of macerals other than vitrinite are those of Kröger.[222] After abandoning the half-width method, he made the estimates by measuring the displacement of the 11 band in the observed X-ray patterns from the position expected for graphite-like layers with an average C—C distance of 1.42 A. Within experimental error, it seems that the condensed aromatic layers in different associated macerals do not differ greatly in size, and follow vitrinite in showing little variation in the range 80 to 90% C content. Although this conclusion may be qualitatively correct, some doubt is cast on its quantitative accuracy, since Kröger's estimates of the layer sizes in his vitrinites are considerably higher than the recent values obtained by other workers. X-ray analyses of gray and black durains have shown that the layers are smaller and less perfect in exinites than in vitrinites, and that they contain more amorphous material, whereas micrinites have somewhat larger and more perfect layers.[136,138] In an examination of a series of vitrains, a fusain, and several whole bands, only the layer spacing and stacking height were

[216] Alpern, B., *Proc. Intern. Comm. Coal Petrology*, No. 2, 43–6 (1956).

[217] Fenton, G. W., and Smith, A. H. V., *Proc. Conf. Science in the Use of Coal*, Sheffield, Inst. Fuel, London, 1958, pp. C20–4.

[218] Macrae, J. C., *Fuel*, **22**, 117–29 (1943).

[219] Kröger, C., Pohl, A., Kuthe, F., Hovestadt, H., and Bürger, H., *Brennstoff-Chem.*, **38**, 33–42 (1957).

[220] Kröger, C., and Bakenecker, J., *Brennstoff-Chem.*, **38**, 82–7 (1957).

[221] Kröger, C., *Ibid.*, **39**, S62–6 (1958).

[222] Kröger, C., and Ruland, W., *Brennstoff-Chem.*, **39**, 1–13 (1958).

[223] Monnot, G. A., and Ladam, A., *Compt. rend.*, **241**, 1939–41 (1955).

[224] Kröger, C., *Brennstoff-Chem.*, **39**, S29 (1958).

[225] Kröger, C., and Hillmann, D., *Brennstoff-Chem.*, **39**, 110–18, (1958).

[226] Fuchs, W., and Glaser, F., *Brennstoff-Chem.*, **39**, S56–7 (1958).

[227] Schulz, G., and Howard, H. C., *J. Am. Chem. Soc.*, **68**, 994–6 (1946). Szádeczky-Kardoss, E., *Bányász. és Kohász. Lapok*, **77**, 241–7, 253–60 (1944); *Chem. Abs.*, **42**, 2081 (1948). Shotts, R. Q., *Mining Eng.*, **187**, 889–97 (1950). Kukharenko, T. A., and Ryzhova, Z. A., *Khim. i. Tekhnol. Topliva*, **1956**, No. 4, 20–30; *Chem. Abs.*, **50**, 12440 (1956). Chandra, D., *Econ. Geol.*, **53**, 102–8 (1958).

[228] Parks, B. C., and O'Donnell, H. J., *U.S. Bur. Mines, Bull.* No. 550 (1956), 193 pp.

recorded, and these exhibited no marked systematic trends.[229] The so-called gamma band in the X-ray scattering patterns of exinite corresponds to an interlayer space of about 4.7 A as compared with 3.5 A in vitrinites and has been ascribed to structures with a high aliphatic or naphthenic content.[133,138,222]

Statistical structural analysis,[213] which gives a much smaller number of condensed rings per cluster in the exinites than in the other macerals, is qualitatively as well as quantitatively in disagreement with the X-ray work.

Statements published on the aromatic hydrogen in macerals other than vitrinite seem poorly supported by the experimental evidence[2,162,223] and in one case are highly speculative.[223] The aromatic hydrogen concentration in exinite is certainly no greater, and is probably smaller, than in vitrinite. There also is little doubt that the aromaticity, defined as the fraction of carbon in aromatic form, of low-rank exinite is smaller than that of the associated vitrinite, and that of the micrinite is probably larger.

The few data available on phenolic hydroxyl content show a decrease from micrinite through vitrinite to exinite. Ketonic carbonyl groups have been identified in a low-rank exinite and vitrinite,[223] but this particular evidence must be suspect in view of the known absence of such groups in vitrinites. Bands at 1670 to 1690 cm^{-1}, characteristic of unconjugated carbonyl, have frequently been observed in the infrared spectra of exinites and whole coals, but these bands are probably due to oxidation. It is unfortunate that chemical determinations of functional oxygen groupings in exinite and micrinite have been based[219] only on methods open to objection: hydroxyl groups by titration in aqueous medium and carbonyl by reaction with phenylhydrazine.

In a study of oxidation by chlorine dioxide[225] of exinites, vitrinites, and micrinites, vitrinites proved the most readily oxidized, as earlier work had shown to be true with other oxidizing agents.[227] Fusinite resists oxidation.[226] Vitrinites and exinites are also the most readily liquefied by hydrogenation,[214,228] and fusinite is almost completely resistant.[230]

[229] Kessler, M. F., and Vecerikova, V., *Uhli*, **6,** 28–30 (1956).

[230] Fisher, C. H., Sprunk, G. C., Eisner, A., Clarke, L., and Storch, H. H., *Ind. Eng. Chem.*, **31,** 190–5 (1939).

CHEMICAL REACTIONS OF COAL

Halogenation

With the exception of the chlorofluorocarbons, products produced by halogenation of bituminous coals have proved of singularly little interest either scientifically or commercially. For the most part they are either gases or infusible solids.[231] Chlorination of coal has been patented as a means of de-ashing coals and cokes at 1400°C.[232]

Chlorination in the Gas Phase. Chlorination of 6 to 20 mesh coals at 225°C for 24 hr has yielded products that contained about 50% chlorine.[233] A few volatile products were obtained, among them carbon tetrachloride and hexachlorobenzene. Roasting at 500°C followed by steam activation at 800°C produced an activated charcoal, of reasonable efficacy though low bulk density, from the solid residue. Chlorination of finely graded coal in a fluidized bed gave a product that was stable up to 300°C, hydrogen chloride being evolved between 300 and 600°C, and chlorine in addition above 600°C.[234] No tar was formed. Dark colored solutions were produced by extracting with aqueous alkali; concentrated ammonia and sodium carbonate removed 25 to 50% of the chlorine.

In a study of the chlorination of a medium-rank caking coal in various conditions by gaseous chlorine,[235] the most stable products were produced at the highest reaction temperatures. With chlorine alone, the maximum chlorine content of the product was 24% at 700°C; when air was added with the chlorine it was 38% at 270°C. Measurements of the hydrogen chloride formed showed that both substitution and dehydrogenation had occurred. Almost half of the chlorine introduced could be removed by refluxing with aqueous sodium hydroxide, and the whole by heating

[231] Weiler, J. F., *J. Am. Chem. Soc.*, **58,** 1112–14 (1936); *Chemistry of Coal Utilization*, John Wiley and Sons, New York, 1945, Vol. 1, pp. 337–45. Roy, M. M., *J. Sci. Ind. Research (India)*, **16B,** 46 (1957)

[232] Aluminium Laboratories, Ltd., Brit. Pat. 716,718 (1954).

[233] Boyk, S., and Hass, H. B., *Gas World*, **120,** 568 (1944); *Ind. Eng. Chem.*, **38,** 745–8 (1946).

[234] Beyers, E., and van Rooyen, D., *Fuel Research Inst., Pretoria. Bull*, No. 31 (1952), 6 pp.

[235] Pinchin, F. J., *Fuel*, **37,** 293–8 (1958).

to 550°C. The presence of chlorine during carbonization to 700 or 1100°C considerably reduced the loss of carbon from the coal, the formation of volatile matter being partially inhibited.

Chlorination in suspension in various liquids. Russian workers[236] have described the chlorination of brown coal in admixture with carbon tetrachloride or water which gave resinous products containing up to 45% chlorine that were soluble in organic solvents. The maximum chlorine content resulting from chlorination of these products in aqueous suspension at 20 and 100°C was 38% but was 60% when heated at 200 and 325°C in sealed tubes with antimony pentachloride. Only the latter products were relatively resistant to hydrolysis.[237] Ultrasonic waves have been found to accelerate greatly the rate of reaction with chlorine or bromine of a long-flame coal suspended in water.[238] During chlorination of a caking coal in suspension in water at 0 to 10°C, the chlorine content rose to 20 to 25%.[235] Ultraviolet irradiation of coal suspended in carbon tetrachloride, through which chlorine was bubbled, resulted after 72 hr in products containing about 36% chlorine; over 10% chlorine was introduced in 14 hr even in the absence of free chlorine.[239]

In chlorination with sulphuryl chloride in chloroform of ethylenediamine extracts of a low-rank bituminous coal,[49] a considerable fraction of the product containing 15 to 30% chlorine was soluble in chloroform, giving a bright red solution, part of which was volatile. It was shown that adsorbed ethylenediamine was not responsible. A gray insoluble residue containing 19% chlorine was also produced. These products might repay further investigation, although as a tool for the study of coal the method is unsatisfactory because of participation of the solvent in the reaction.[240]

The reaction of coals in suspension in carbon tetrachloride with sulphuryl chloride in the presence of benzoyl peroxide[49] is not specific for side-chain chlorination as it is with compounds such as toluene.[240] Extensive side-chain chlorination occurred in low-rank coals without added catalyst, although addition of the peroxide or of aluminum chloride did increase the total chlorine taken up by the coals. The lower-rank coal took up the least chlorine.

Chlorine attached to nonaromatic carbon atoms is usually more easily removed than chlorine attached to aromatic rings. This labile chlorine can be estimated by treatment with potassium acetate in acetic acid or with sodium hydroxide. Potassium acetate removed between 40 and 60% of the chlorine introduced by sulphuryl chloride.[240]

Treatment of a suspension of brown coal with chlorine at 140°C for 15 to 40 hr yielded a product that was soluble in common organic solvents and aqueous alkalis at room temperature, and also in boiling water[241] at higher temperatures.

Bromination and iodination. In bromination of coals and coal extracts with bromine in chloroform suspension at 0°C for 24 hr, the bromine uptake showed a steady decrease from about 18% for a coal of 77.5% C content to 6% for a coal of 90% C.[242] Aqueous sodium hydroxide removed only 30 to 40% of the bromine, suggesting that the hydrogen replaced was mainly in aromatic systems. Attempts to achieve aliphatic substitution with N-bromosuccinimide proved ineffective.

Treatment of coals with iodine solution, Wij's solution and Hübl's solution, gave an iodine uptake that depended rather complexly on the reagent and the type of coal used.[243]

Fluorination of coals. A laboratory process has been described whereby the whole of a low-rank bituminous coal can be converted by fluorination to transparent fusible solids, chlorofluorocarbon oils, and gases, without a

[236] Kretov, A. E., Savin, M. I., Shenbor, M. I., and Lev, I. E., *Ukrain. Khim. Zhur.*, **18,** 305–10 (1952).

[237] Shenbor, M. I,. Kretov, A. E., and Savin, M. I., *Ukrain. Khim. Zhur.*, **21,** 636–40 (1955).

[238] Losev, B. I., El'piner, I. E., and Mel'nikova, A. N., *Doklady Akad. Nauk S.S.S.R.*, **114,** 372–4 (1957).

[239] Bent, R., and Brown, J. K., *Proc. Conf. Science in the Use of Coal*, Sheffield, Inst. Fuel, London, 1958, pp. A48–51.

[240] Pinchin, F. J., *Fuel*, **38,** 147–54 (1959).

[241] Alekseeva, V. A., *Ukrain. Khim. Zhur.*, **20,** 319–24 (1954).

[242] Brown, J. K., Given, P. H., Lupton, V., and Wyss, W. F., *Proc. Conf. Science in the Use of Coal*, Sheffield, Inst. Fuel, London, 1958, pp. A43–7.

[243] Roy, M. M., *J. Sci. Ind. Research (India)*, **16B.** 274 (1957).

trace of tar.[244] The yields of oil were at least 100% by weight on the coal used, and of solid about 5%; the gases have been estimated to contain about 60% of the fluorine consumed, and the oil about 20%. The reagent consisted of chlorine trifluoride diluted with nitrogen, and treatment was effected in a fixed bed at 200 to 400°C. The atomic ratio of fluorine to chlorine could be controlled within the range 10 to 20. A typical analysis of the oil product was C 25%, H 0.1 to 0.4%, F 65%, and Cl 8%. The low halogen/carbon ratio suggests the presence of cyclic structures. Some typical properties of the oils after washing and fractional distillation are: boiling point at atmospheric pressure 50 to 300°C; density 2.0 g/ml; viscosity at 100°F, 20 to 800 cS; specific resistance about 10^{15} ohm-cm.

Oxidation

The oxidation of coals has continued to provide material for a large number of papers. Nonetheless, the great bulk of the work reported has merely supplemented or extended familiar lines of research, has resulted in few applications, and has contributed comparatively little to a knowledge of coal structure. The more primary oxidation products such as the humic acids, clearly more closely related to the original coal in molecular structure, have proved as difficult to study as the coals themselves. It is perhaps fair to say that owing to the complexity of coal, classical degradative methods entail immense labor and give unrewarding results. Relative conclusions regarding changes in structure with rank may, however, be usefully elucidated by quantitative oxidation studies.

General studies of the oxidation process in air, below ignition temperature, appear to have established that there are at least three stages, which result progressively in increased production of gases and changes in the nature of the solid residue. Oxidation to a substantial loss in weight appears to follow the same course up to at least 175°C. Evidence has suggested a close similarity of the basic structure of the humic acids produced to that of the parent coal, a justification in part for oxidation studies with a scientific objective, despite what has been said earlier. The humic acids obtained by the use of nitric acid are different from those obtained by oxidation in air, and there is now considerably greater knowledge of the points of difference.

There have been many investigations of oxidation mechanism. It has been well established that phenolic hydroxyl groups play a part in the mechanism of oxidation, but a study of simple phenols has shown the likelihood of dimerization, which must be borne in mind in interpreting the aromatic nuclear structure of oxidized coal. There has also been an interesting suggestion that lactone intermediates have a bearing on the increased susceptibility to oxidation of coals in alkaline media.

Measurements of the order of reaction and apparent activation energy give results covering a wide range of values, but are suspect owing to the complex nature of the coal molecule and the influence of reagent diffusion on the measured rates.

Among much work on spontaneous ignition and weathering, little that is new will be found. Coals liable to spontaneous heating can be recognized by a variety of laboratory tests, but for practical control it would seem that exclusion of air from the storage pile is the only really effective method.

Several workers have studied the oxidation of coal as a whole from various points of view.[245,246] An important thesis[247] on the subject of oxygen in coal and its effects on properties was produced in 1945. For each value of volatile matter, there was found a value of oxygen content above which swelling was zero, although fusion did not disappear until there was a higher oxygen content. Mixed coals behaved additively only in terms of their oxygen content and volatile matter, so that swelling of a blend could be predicted from general curves but not from the swelling of the individual coals. Yields of moisture and phenols on carbonization at 500°C were closely related to oxygen content. This was also true for oxidized coals. The oxygen

[244] Farenden, P. J., and Pritchard, E., *Proc. Conf. Science in the Use of Coal*, Sheffield, Inst. Fuel, London, 1958, pp. A52–5.

[245] Kramers, W. J., *BCURA Monthly Bull.*, **10**, 395–404 (1946).

[246] Yohe, G. R., *Illinois State Geol. Survey, Rept. Invest.* No. 207 (1958), 51 pp.

[247] Georgiadis, C., Thesis, Univ. of Lille (1945); *Centre document. siderurg., Circ. inform. tech.*, **8**, 837–83 (1951).

content was found to play a dominant role in determining the rate of oxidation; the high volatile matter coals showing the most swelling were the most easily oxidized.

In a study of the changes in weight and in other characteristics during oxidation of a series of bituminous coals with air, the optimum temperature for production of humic acids was found to be 200°C.[248] High yields of humic acids were obtained from all the bituminous coals but not from lignite or anthracite. The oxidation in air of younger coals has been found to lower their yield of volatile matter and tar and their solubility in anthracene oil, whereas oxidation of older coals increased their volatile matter and sorption capacity.[249] Elementary composition, sorption capacity, and the number of functional groups was at first unchanged in the younger coals, whereas the carbon content of older coals was considerably reduced. From a study of four coals treated with oxygen at 100 and 200°C, it has been concluded that the groups that form methane and tar on pyrolysis are particularly sensitive to oxidation.[250]

In an unpublished work, Farenden made a careful study of the oxidation of a bright coal of 82% C content (−72 B.S. mesh) with oxygen in nitrogen at 110° and 175°C. The coal was tumbled in a rotating glass vessel and the concentration of oxygen was kept constant by circulating a limited volume of gas with purification and make-up as necessary. Functional groups in the solid and gaseous products evolved were determined periodically. The oxidation followed a similar course at the two temperatures. Three stages were distinguished. In the first the oxygen reacting was equally divided between that found in the gaseous products and that fixed by the coal. Alkali-soluble products did not appear. The oxygen content of the solid at the end of this stage had increased from 11 to 20%. In the second stage, solubility in $N/10$ hydroxide increased rapidly and that in water slowly. The amount of oxygen in the gaseous products became greater than the amount fixed by the solid, the average ratio being 2.4. At the end of this stage the solid product contained little nonaromatic hydrogen. In the third stage the material soluble in water increased at the expense of that soluble in alkali alone, indicating further degradation of the soluble units. The total weight of oxygen in the solid product at this stage started to decrease for the first time, and humic acids showed a maximum yield at the beginning of the stage. The maximum yield of humic acids obtained at 175°C was smaller than at 110°C, possibly owing to decarboxylation or dehydration, and to faster oxidative degradation of the acids to water-soluble acids. The results were considered to justify the importance attached to the study of humic acids as a means of deducing the structure of the original coal, and to afford some evidence for the basic homogeneity of the structural units in a bright coal.

Slow oxidation of pulverized coal in air up to 225°C followed by examination of the products has shown that mainly the structures containing methylene groups are attacked.[251] The oxygen absorbed was present to 45% in carboxyl groups, most of the remainder being carbonyl and ether oxygen. Maximum alkali solubility was attained at about 225°C.

Unlike other oxidants, moist ozone converts the carbon of coal and carbon black directly to water-soluble products without any intermediate formation of humic acids.[252] The investigators suggested that coals and humic acids contain as part of their structure "stacked planes" similar to those known to be present in carbon black.

Seventy to ninety percent of a range of bituminous coals has been converted to alkali-soluble products within 24 hr at room temperature by oxidation with aqueous performic acid.[242]

The oxidation of coal with alkaline permanganate has been studied in some detail with a view to obtaining maximum yields of humic acid and seeking a humic acid which is a single chemical entity.[253] It was concluded that,

[248] Friedman, L. D., and Kinney, C. R., *Ind. Eng. Chem.*, **42,** 2525–9 (1950).

[249] Khrisanfova, A. I., *Trudy Inst. Goryuchikh Iskopaemykh, Akad. Nauk S.S.S.R.*, **2,** 179–92 (1950); *Chem. Abs.*, **49,** 4259 (1955).

[250] Bastick, M., *Bull. soc. chim. France*, **1954,** 1239–46.

[251] Grosskinsky, O., and Jüttner, B., *Brennstoff-Chem.*, **39,** S7–11 (1958).

[252] Ahmed, M. D., and Kinney, C. R., *J. Am. Chem. Soc.*, **72,** 559–61 (1950). Kinney, C. R., and Friedman, L. D., *J. Am. Chem. Soc.*, **74,** 57–61 (1952).

[253] Bailey, A. E. W., Kimberley, I. E., and Ward, S. G., *Fuel*, **33,** 209–21 (1954).

although acids of constant elementary composition could be obtained, they are not a single substance. It was found that good yields of colored, water-soluble subhumic acids could be produced by further oxidizing humic acid with alkaline permanganate.[254,255] The acids were separated from inorganic matter by extraction with organic solvents. Their properties indicated that most acids were more complex than benzenecarboxylic acids. The oxidation of humic acid with alkaline hydrogen peroxide produced smaller yields of subhumic acids than oxidation with alkaline permanganate. Nevertheless subhumic acids could be prepared with little mineral contamination by oxidizing humic acid with hydrogen peroxide in the absence of inorganic reagents, and these could be rendered soluble in ether by treatment with mineral acid or by dialysis.[256] Oxidation of 1 g of humic acid with 1.6 g of hydrogen peroxide was found to give a maximum yield, 64%, of the humic acid carbon as nonvolatile, water-soluble, subhumic acids.[257] Formic and acetic acids were identified in the volatile products. Subhumic acids were also obtained by the action of ozonized oxygen on humic acids dissolved in alkali.[258] Methods were described for controlled oxidation on a moderately large scale in a closed system. These subhumic acids were partially fractionated by solvent extraction and by stepwise acidification of the sodium salts.[259]

An investigation of the conditions governing the formation of the products obtained by oxidizing bituminous coals with hot concentrated nitric acid has been reported.[260] Cold nitric acid was used as a means for solvent fractionation of the products. Successive stages of oxidation resulted in conversion of coal to black nitric-acid-insoluble humic acids, then to brown nitric-acid-soluble humic acids of lower equivalent weight, and finally to water-soluble acids.

From the constancy of the quantities of nitrogen and nitrate produced from coals of different rank when oxidized with alkaline permanganate, it was deduced that most coals contained the same kind of nitrogen linkages.[261]

Several other papers have dealt with investigations of the foregoing general type, and the patent literature is concerned with the possible utilization of the products of oxidation in general.[255,262]

Detection of oxidation in coals. The recognition of a coal that has been oxidized, for example by outcropping of the seam, in contrast to an unweathered deep-mined coal, can be important in classification and utilization. Methods for determining the extent of the oxidation are also necessary.

A study[263] of the changes of "permanganate

[254] Bailey, A. E. W., Kimberley, I. E., Turner, L., and Ward, S. G., *Fuel*, **34,** 37–49 (1955).

[255] Roy, M. M., *Research* (*London*), **10,** 77–8 (1957).

[256] Brooks, V. T., Lawson, G. J., and Ward, S. G., *Fuel*, **35,** 385–97 (1956).

[257] Lawson, G. J., *Fuel*, **36,** 182–90 (1957).

[258] Dobinson, F., Lawson, G. J., and Ward, S. G., *Fuel*, **35,** 398–408 (1956).

[259] Ward, S. G., *Brennstoff-Chem.*, **37,** 99–100 (1956).

[260] Kinney, C. R., and Ockert, K. F., *Ind. Eng. Chem.*, **48,** 327–32 (1956).

[261] Kinney, C. R., Eckerd, J. W., Rexford, P., and Charmbury, H. B., *J. Am. Chem. Soc.*, **70,** 392–4 (1948).

[262] Roga, B., and Ihnatowicz, M., *Prace Glównego Inst. Górnictwa, Komun.* No. 82 (1951), 12 pp.; *Chem. Abs.*, **46,** 11629 (1952). Funasaka, W., Yokokawa, C., Ueno, S., Matsumiya, S., and Miyazaki, Y., *J. Chem. Soc. Japan, Ind. Chem. Sect.*, **54,** 579–81 (1951); *Chem. Abs.*, **47,** 6633 (1953). Funasaka, W., Yokokawa, C., Kajiyama, S., and Fukaura, H., *J. Chem. Soc. Japan, Ind. Chem. Sect.*, **55,** 35–6 (1952); *Chem. Abs.*, **47,** 9592 (1953). Gonzales-Sanchez, F., *Combustibles*, **15,** 273–305, 361–94 (1955). Kawana, Y., and Sato, M., *Kagaku* (*Tokyo*), **26,** 359 (1956); *Chem. Abs.*, **50,** 13671 (1956). Fujimoto, T., Takashima, S., and Watanabe, A., *Repts. Himeji Inst. Technol.*, No. 6, 72–6 (1956); *Chem. Abs.*, **50,** 17390 (1956). Higuchi, K., Arai, T., Tsuyuguchi, M., and Asakawa, K., *J. Fuel Soc. Japan*, **35,** 404–10 (1956); *Chem. Abs.*, **50,** 17383 (1956). Surkova, V. I., and Sharypkina, M. Ya., *Khim. i Teknol. Topliva*, **1956,** No. 5, 23–31; *Chem. Abs.*, **50,** 14204 (1956). Bogoyavlenskii, V. V., and Petrenko, V. G., *Koks i Khim.*, **1957,** February, 3–7; *Chem. Abs.*, **51,** 9125 (1957); Stadnikoff, G., *Brennstoff-Chem.*, **38,** 237–41 (1957). Nakagome, T., Yoshida, S., and Sugimura, H., *J. Fuel Soc. Japan*, **37,** 465–71 (1958). Kreulen, D. J. W., *Freiberger Forschungsh.*, **A119,** 5–21 (1958). Fuchs, W. M., U.S. Pat. 2,338,634 (1944). Ruhrchemie, A. G., Ger. Pat. Appl. R 117,134 (1943).

[263] Egorova, O. I., *Bull. acad. sci. U.R.S.S., Classe sci. tech.*, **1942,** No. 7–8, 107–15; *Chem. Abs.*, **38,** 4404 (1944).

number"[264] during the oxidation of various types of coal has shown that it passes through a minimum for some coals whereas for others it is constant for a time and then increases rapidly. This number can be used to determine the degree of oxidation of a coal.[263,265] Radspinner and Howard[266] examined several methods of estimating the degree of oxidation of coal, only two of which gave relatively sensitive results. Determination of carboxyl groups showed that a fairly constant fraction of the total oxygen added reacted like a carboxyl group; this fraction changed with rank. When the oxidized coals were heated in a vacuum at 350°C, a linear relation was found between the oxygen that had been added and that evolved as oxides of carbon. The heat of wetting of coals in methanol and in water changes during oxidation and may be used for determining the extent of oxidation.[267]

The discovery of an unusual characteristic of oxidized coals—that the electric charge acquired when the particles impinge on a metal surface increases with degree of oxidation—has led to the development by the British National Coal Board of a convenient instrument for determining the degree of oxidation for control purposes.[268] Other methods proposed have involved measurement of luminescence,[269] plasticity, ignition temperatures,[270] and gaseous products of low-temperature carbonization.[271]

[264] Cubic centimeters of *N* $KMnO_4$ reduced by 0.5 g (dry, ash-free) of pyridine-extracted coal in 1 hr at 100°C, Heathcoat, F., *Fuel*, **12**, 4–9 (1933).

[265] Khrisanfova, A. I., *Izvest. Akad. Nauk S.S.S.R., Otdel. Tekh. Nauk*, **1949**, 1116–21; *Chem. Abs.*, **49**, 7221 (1955).

[266] Radspinner, J. A., and Howard, H. C., *Ind. Eng. Chem., Anal. Ed.*, **15**, 566–70 (1943).

[267] Onusaitis, B. A., and Khrisanfova, A. I., *Izvest. Akad. Nauk S.S.S.R., Otdel. Tekh. Nauk*, **1947**, 895–900; *Chem. Abs.*, **43**, 3593 (1949).

[268] Thomas, D. G. A., *Brit. J. Appl. Phys.*, **4**, S55 (1953). Anon., *Iron & Coal Trades Rev.*, **166**, 941–2 (1953).

[269] Men'shchikov, F. S., and Rozmanova, Z. E., *Zavodskaya Lab.*, **21**, 1471–4 (1955); *Chem. Abs.*, **50**, 8993 (1956).

[270] Veselovskii, V. S., and Terpogosova, E. A., *Izvest. Akad. Nauk S.S.S.R., Otdel. Tekh. Nauk*, **1954**, No. 12, 140–1 *Chem. Abs.*, **50**, 11641 (1956).

[271] Zabramnyi, D. T., and Tadzhiev, A. T., *Trudy Inst. Khim., Akad. Nauk Uzbek. S.S.R.*, **1955**, No. 6, 3–14 *Chem. Abs.*, **51**, 9126 (1957).

Oxidation and coal structure. A study of the changes in analysis of the solid residue and the volatile products obtained during oxidation can throw some light on the structure of coal.

The amount of acetic acid produced by oxidation of coals with nitric acid-potassium dichromate mixture correlated well with the yield of methane on carbonizing the same coals to 500°C.[272] It was suggested that this, nearly half of the total methane evolved to 1000°C, was produced from the same groups—probably aliphatic or alicyclic material—that led to acetic acid on oxidation. In agreement with this hypothesis, acetic acid was not formed from coal carbonized above 500°C, which shows that the structural features responsible were split off by carbonization to this temperature. The concentration of these groups was found to diminish with increasing rank, in agreement with the indications about aliphatic hydrogen from the infrared spectra. However, if the yields of acetic acid and methane are plotted separately against rank of coal, it is seen that the good average agreement tends to obscure somewhat different trends in the two quantities with change of rank.

From a comparison of products formed by oxidizing coal with nitric acid with those obtained from known compounds, it has been deduced that the humic acids are derived from the aromatic ring systems and the aliphatic and benzenecarboxylic acids arise, in part at least, from the nonaromatic groups.[273] Horton has used this approach to elucidate changes in coal structure with change in rank. The oxidation, by boiling alkaline permanganate, of 7 coals (−100 B.S.S.) of different rank but from the same seam has been reported in detail.[168] With increasing rank, the yield of carbon dioxide passed through a minimum and that of nonvolatile acids other than oxalic acid through a maximum, both at about 90% C; the yield of oxalic acid and the volatile acids diminished steadily. It was deduced from these results that there are two distinct stages in coalification: as the carbon content increases from 84 to 90% there develop carbocyclic hexagons, some aromatic and some partially reduced, which may form small groups of fused rings;

[272] Kinney, C. R., *J. Am. Chem. Soc.*, **69**, 284–9 (1947).

[273] Benning, A., *Brennstoff-Chem.*, **36**, 38–43 (1955).

above 90%, these relatively simple structures fuse together to form more complex structures of lower hydrogen content.

Adams and Pitt[274] oxidized a −270 B.S.S. sample of a bituminous coal in air at 200°C for 5½ hr, and deduced from its infrared absorption spectrum that: (1) the concentrations of C—O, O—H, and hydrogen-bonded OH groups had been increased, (2) C=O groups and acid dimers had appeared, and (3) aromatic C—H bonds had been almost eliminated. They concluded that the main reaction during oxidation is the formation of carboxyl groups from the peripheral C—H groups on aromatic nuclei.

The changes in the composition of coals oxidized for 470 hr at 170°C have been interpreted on the unestablished assumption that the aromatic nuclei in the molecule are unaffected.[125] The assumption was used as basis for calculating the fraction of carbon in aromatic form from the analyses of oxidized coals, after correcting for carbon and oxygen in carboxyl groups. Pending further evidence, oxidation cannot be recommended for assessing aromaticity.

Oxygen-containing groups in relation to oxidation of coal by molecular oxygen. A phenol theory of coal oxidation,[275] now superseded by more detailed knowledge, was postulated in 1940. Yohe and Blodgett[189] were probably the first to show experimentally the importance of hydroxyl groups in coal in determining their ease and mechanism of oxidation. Suspensions of eight coals, from lignite to anthracite, in 5% sodium hydroxide solution were oxidized at 100°C with gaseous oxygen, and the rate of gas uptake measured. The coals were also subjected to methylation by dimethylsulphate. Low-rank coals were methylated to an appreciable extent, and the rate of oxidation was thereby considerably reduced. Bituminous coals of high rank and anthracite were not methylated appreciably, and the treatment did not change the rate of oxidation to any significant extent. This investigation was extended later to a study of known phenols.[276] Twelve phenols in aqueous sodium hydroxide were oxidized under conditions previously used in the study of coals. The curves of oxygen uptake against time were of similar shape. The products included oxalic and acetic acids formed by degradation and also bimolecularly condensed compounds; thus 2,6-di-*t*-butyl-4 methylphenol gives 3,5,3′,5′-tetra-*t*-butyl-4,4′-dihydroxy-1,2-diphenylethane and 3,5,3′,5′-tetra-*t*-butylstilbene-4,4′-quinone. It was pointed out that condensations of this kind may accompany the oxidation of coals to humic acids.

Kharitonov concluded that oxidation of coal is accompanied by polymerization, and that the ease with which a coal spontaneously ignites or can be oxidized by aqueous potassium permanganate depends directly on its content of functional oxygen-containing groups.[277] He had earlier shown that carboxyl groups formed in oxidation tended to reduce the oxidation associated with the phenolic groups present.[278] Thus slight oxidation with air might increase the oxidizability of a coal originally low in phenolic group content and reduce that of a coal originally high in hydroxyl content.

The lactone of 2-hydroxy-2′-diphenyl-carboxylic acid has been isolated from the water-soluble acids produced by the controlled oxidation of aqueous, alkaline suspensions of bituminous coals.[279] It is known that in acid media the lactone ring is stable and resistant to oxidation, whereas in alkaline media it is opened and yields the reactive hydroxyl and carboxyl groups. It was therefore suggested that the fact that coals are oxidized more rapidly when suspended in aqueous caustic soda than in acid or even in dilute sodium carbonate might indicate the opening of lactone rings by alkali.

It has been shown that reductive acetylation decreases the oxygen sorption of both coals and oxidized coals.[280]

[274] Adams, W. N., and Pitt, G. J., *Fuel*, **34,** 383–4 (1955).

[275] Tronov, B. V., *Zhur. Priklad. Khim.*, **13,** 1053–8 (1940); *Chem. Abs.*, **35,** 1966 (1941).

[276] Yohe, G. R., Hill, D. R., Dunbar, J. E., and Scheidt, F. M., *J. Am. Chem. Soc.*, **75,** 2688–93 (1953).

[277] Kharitonov, G. V., *Izvest. Akad. Nauk Kazakh. S.S.R., Ser. Khim.*, **1946,** No. 9, 92–100; *Chem. Abs.*, **50,** 10373 (1956).

[278] Kharitonov, G. V., *Khim. Gen. Tverd. Goryuchikh Iskopaemykh, Trudy I Vsesoyuz.*, **1953,** 176–84; *Chem. Abs.*, **49,** 2264 (1955).

[279] Entel, J., Ruof, C. H., and Howard, H. C., *J. Am. Chem. Soc.*, **75,** 3038–9 (1953).

[280] Nandi, S. P., Kini, K. A., and Lahiri, A., *J. Indian Chem. Soc., Ind. & News Ed.*, **20,** 91–4 (1957).

General Mechanism and Kinetic Equations for Oxidation of Coal by Molecular Oxygen.

In a review of the low-temperature reactions of coal with oxygen, it was pointed out that hydrogen is removed faster than carbon and a table was given illustrating their distribution in the products.[281] The reaction rate was expressed in terms of a general equation, in which some of the exponents are dependent on rank. The heats of activation were shown to increase with increasing temperature.

Oreshko[282] studied the kinetics of coal oxidation by passing oxygen through a heated insulated pile of dry coal and recording the changes in temperature and weight. He concluded that the oxidation of coal proceeded differently in three temperature ranges: up to 70 to 80°C, activation energy 3 to 4 kcal/mol; 80 to 130 or 160°C, activation energy 6 to 8 kcal/mol; and 180 to 290°C, activation energy 12 to 17 kcal/mol. He deduced that the rates of oxidation were not primarily dependent on diffusion, and developed rate equations that could be applied both to isothermal conditions and to conditions of constant rate of temperature increase.[246] Two of the oxidation periods were examined in considerable detail; the first was dominated by the disintegration of unstable peroxide complexes, the other by the formation of thermally more stable combinations of carbon and oxygen. Equations for the rate of reaction in each period were verified by tests in which coal was heated in a stream of oxygen at a rate of 1.4 to 1.6°C/min. In the first stage, the rate constants were found to be substantially independent of rank and it was supposed that the type of peroxide formed was the same for all the coals studied. In the second stage, both the entropy factor and the energy of activation were functions of rank, and an important division was made on this basis at 26% volatile matter. Below 26% volatile matter, the rate constants, even in the second stage, were independent of rank. The rate of low-temperature oxidation of certain Russian coals was found to be proportional to the total surface including the internal surface, that is, that of the capillaries.

Wood also found that from the kinetic point of view the oxidation proceeds in distinct stages, and that the kinetics of coal oxidation at 30°C can be correlated by the Elovitch equation $dq/dt = Ae^{-Bq}$, where q is the amount of oxygen sorbed, t is time and, A and B are constants.[283]

Distinct differences in the characteristics of the air oxidation of anthracite up to 350°C have been found according to whether the particle size was larger or smaller than 20 mesh.[284] Above 20 mesh, the initial rate of oxidation was directly proportional to the external particle surface and the rate passed through a maximum with increasing degree of oxidation and then decreased. With the smaller sizes, the initial rate increased more slowly with increase of external particle surface and appeared to reach a limiting value beyond which further subdivision had no effect, and the rate of oxidation decreased regularly with increasing extent of oxidation.

The amount of oxygen taken up from the gas phase in a closed system during the aging of coals at room temperature has been found to be approximately proportional to the square root of the time for many types of coals at different levels of moisture and pyrite content.[285]

Kreulen[286] oxidized coals of three different ranks in air at temperatures from 120 to 250°C, and analyzed the results by a kinetic method previously developed.[287] He found

[281] Howard, H. C., *Trans. Am. Inst. Mining Met. Engrs.*, **177**, 523–34 (1947).

[282] Oreshko, V. F., *Bull. acad. sci. U.R.S.S., Classe sci. tech.*, **1943**, No. 7, 52–61; *Chem. Abs.*, **39**, 2633 (1945); *Izvest. Akad. Nauk S.S.S.R., Otdel. Tekh. Nauk*, **1949**, 249–57, 748–59, 1642–8; **1951**, 1031–40; *Chem. Abs.*, **44**, 2200, 2201, 2202 (1950); **46**, 10579 (1952); *Ugol'*, **24**, No. 2, 10–13 (1949); *Chem. Abs.*, **45**, 5906 (1951); *Doklady Akad. Nauk S.S.S.R.*, **70**, 445–8 (1950); **71**, 331–3 (1950); **74**, 327–30 (1950); **82**, 135–8 (1952); *Chem. Abs.* **45**, 5388 (1951); **44**, 10295 (1950); **45**, 2175 (1951); **47**, 12788 (1953); *Trudy Inst. Goryuchikh Iskopaemykh, Akad. Nauk S.S.S.R., Otdel. Tekh. Nauk*, **2**, 160–78 (1950); *Chem. Abs.*, **49**, 4259 (1955); *Zhur. Priklad. Khim.*, **23**, 306–16 (1950); **25**, 373–83 (1952); *Chem. Abs.*, **45**, 2174 (1951); **46**, 7303 (1952).

[283] Wood, T., *J. Appl. Chem. (London)*, **8**, 565–71 (1958).

[284] Scott, G. S., and Jones, G. W., *U.S. Bur. Mines, Rept. Invest.* No. 3546 (1941), 15 pp.

[285] Frey, W. A., *Oel u. Kohle*, **39**, 603–15 (1943).

[286] Kreulen, D. J. W., *Rec. trav. chim.*, **62**, 215–21 (1943).

[287] Kreulen, D. J. W., and ter-Horst, D. T. J., *Rec. trav. chim.*, **59**, 1165–79 (1940).

that the formation of carbon dioxide and water was independent of that of humic acid, and he also concluded that the ratio of carbon to hydrogen in the groups that reacted was almost independent of rank.

In a study of oxidation of coals in air at 30 and 60°C at different moisture and sulfur levels,[288] moisture content *per se* and pyrite were found to have little effect on oxygen adsorption; the adsorptive capacity, however, depended on rank as reflected by the maximum inherent moisture content.

During oxidation of coals in air at temperatures between 150 and 250°C, the ratio of carbon dioxide to carbon monoxide in the products has been found to be independent of time, but the ratio of oxides of carbon to water and the apparent energy of activation changed with time.[289] The oxygen content of the solid residue reached a limiting maximum value. The rate of formation of carbon-oxygen complexes did not depend on the stage of oxidation although it varied with rank of coal; the rate of decomposition of the complexes was proportional to the amount formed but did not depend on the type of coal, and was also only slightly dependent on temperature.

Another study of the kinetics of coal oxidation showed that the apparent activation energy depends on rank, that gaseous diffusion is an important controlling factor, and that aging is primarily a surface phenomenon—caking, for example, being prevented by changes in the particle surfaces.[290] Many other studies of the kinetics and mechanism of oxidation of coal with molecular oxygen have been published.[291]

Mechanism of oxidation by electrochemical means or chemical reagents. In a study of the electrolytic oxidation of coal,[292] the main products of oxidation at a copper anode were humic acids and water-soluble acids—apparently formed independently, for the humic acids were not oxidized further at the copper anode. They could, however, be further oxidized to soluble products incapable of forming salts at lead and platinum anodes. An anthracite was found to yield much less water-soluble material at a copper anode. The β- and γ-pyridine extracts of a Yorkshire coal gave products similar to those from the coal itself. The water-soluble acids and the humic acids appeared, in general, to be similar in nature to those obtained by using chemical methods of oxidation. Small quantities of benzenecarboxylic acids, mellitic, hemimellitic, and isophthalic acids, were isolated, and oxalic acid was present in appreciable amounts, but the bulk of the material was unidentified. The humic acids contained an appreciable fraction soluble in ether that had a resinous appearance. It was suggested that coal consists of two types of molecular units, which differ considerably in size but are of a similar chemical structure. The larger units would give rise to the humic acids, and the smaller units would be capable of forming water-soluble acids as the initial breakdown product.

The effectiveness of various promoters in the

[288] Sevenster, P. G., *J. Chem. Met. Mining Soc. South Africa*, **53**, 163–76 (1952).

[289] Georgiadis, C., and Gaillard, G., *Chim. & ind. (Paris)*, **70**, 383–96 (1953).

[290] Sommers, H., and Peters, W., *Chem. Ing. Tech.*, **26**, 441–53 (1954).

[291] Grigor'ev, M. Yu., and Semenova, K. K., *Izvest. Tomsk. Ind. Inst.*, **60**, No. 3, 50–2 (1940); *Chem. Abs.*, **37**, 1029 (1943). Grigor'ev, M. Yu., and Plekhanova, A. I., *Izvest. Tomsk. Ind. Inst.*, **60**, No. 3, 53–6 (1940); *Chem. Abs.*, **37**, 1029 (1943). Constantinides, G., *Ann. triestini*, **18**, Nos. 1–2, 3–18 (1946); *Pubb. facolta sci. e ing. univ. Trieste, Ser. B*, No. 13 (1947); *Chem. Abs.*, **46**, 8345 (1952). Chereau, J., *Compt. rend.*, **234**, 1165–6 (1952). Mehta, M. M., Kini, K. A., Moitra, A. K., and Lahiri, A., *J. Sci. Ind. Research (India)*, **11B**, 418–22 (1952). Yamasaki, T., *J. Mining Inst. Japan*, **69**, 253–8 (1953); *Chem. Abs.*, **48**, 11756 (1954). Veselovskii, V. S., and Terpogosova, E. A., *Izvest. Akad. Nauk S.S.S.R., Otdel. Tekh. Nauk*, **1953**, 905–9; *Chem. Abs.*, **48**, 2343 (1954); Schroeder, H., *Brennstoff-Chem.*, **35**, 14–23 (1954). Veselovskii, V. S., and Orleanskaya, G. L., *Izvest. Akad. Nauk S.S.S.R., Otdel. Tekh. Nauk*, **1954**, No. 4, 140–6; *Chem. Abs.*, **49**, 4968 (1955). Kessler, M. F., *Uhli*, **5**, 342–7 (1955); *Chem. Abs.*, **50**, 10373 (1956). Kamiya, Y., *J. Fuel. Soc. Japan*, **33**, 412–23 (1954); *Chem. Abs.*, **48**, 14158 (1954). Higuchi, K., and Shibuya, Y., *J. Fuel Soc. Japan*, **33**, 366–73 (1954); **34**, 286–92 (1955); *Chem. Abs.*, **48**, 11029 (1954); **49**, 12816 (1955). Smirnova, A. V., and Shubnikov, A. K., *Khim. i Tekhnol. Topliva i Masel*, **1957**, No. 5, 40–8; *Chem. Abs.*, **51**, 18548 (1957) Mukherjee, P N , and Lahiri, A , *Brennstoff-Chem* , **38**, 55–8 (1957). Piercy, F. E., Murray, D. W., and Cuthbertson, A. C., *Fuel*, **37**, 129–40 (1958).

[292] Belcher, R., *J. Soc. Chem. Ind. (London)*, **67**, 213–20, 265–70 (1948).

electrolytic oxidation of coal have been compared.[293] Vanadium pentoxide was the best promoter found. Electrochemical oxidation in alkaline solution has also been reported.[294]

The oxidation with alkaline permanganate of various condensed cyclic compounds has been studied for comparison with that of a Pittsburgh Seam coal.[295] Methods were investigated for determining the carbon balance in the oxidation and for the characterization of the products. Satisfactory first-order kinetics for alkaline permanganate were found in the reactions with tetralin, pyrene, naphthacene, and triphenylene. Under the conditions used, tetralin was oxidized the most rapidly; then in descending order came bituminous coal, pyrene and naphthacene, diphenylene oxide and carbazole, decacyclene and triphenylene. Nearly saturated structures were the least reactive. A qualitative correlation was found between the rates of oxidation of some of the hydrocarbons and their "double-bond character."

In another study of the reaction of coal with boiling alkaline permanganate, a kinetic equation was derived. No support was found for the view that the reaction is unimolecular.[296] At first the reactive peripheral groups were oxidized, producing very simple products like carbon dioxide and oxalic acid, the latter being formed in the primary attack and not in subsequent reactions.

Investigations were made to determine whether satisfactory rates of conversion of coal to organic acids could be attained with combinations of dilute nitric acid and oxygen gas at elevated temperatures under pressure.[297] The reaction rate as well as the distribution of carbon among the oxidation products was markedly affected by acid concentration. At zero acid concentration and about 190°C, the reaction products were mainly CO_2 and insoluble materials including unattacked coal. With 4.5 molar acid, the fraction of the carbon of the coal appearing as the desired acids increased to about 38%, 56% appeared as CO_2 and only 6% remained insoluble. The organic acids recovered were similar to those prepared by the alkaline pressure oxidation of the same coal, but contained about 4% N compared with less than 1% for the alkaline process.

Studies of spontaneous ignition and weathering of coal. In the period under review, there have been several good general discussions of spontaneous heating.[298] There is a possibility that the heat of wetting of dry coal by rain may raise the temperature of a pile of coal above the danger point.[299] The content of phenolic hydroxyl has been shown to be an important factor in promoting spontaneous ignition of coal.[280] Several methods of estimating the tendency of coals to heat spontaneously on storage or to resist weathering have been proposed.[21,300]

The spontaneous-heating properties of a number of coals have been described and discussed by Davis,[301] summarizing the work of the United States Bureau of Mines. The tendency to spontaneous ignition with type of coal is also illustrated in three other papers.[302]

[293] Eddinger, R. T., and Demorest, D. J., *Fuel*, **26**, 157–9 (1947).

[294] Lavin, G. B., and Kuznetsov, S. I., *Zhur. Priklad. Khim.*, **26**, 1089 (1953).

[295] Ward, J. J., Kirner, W. R., and Howard, H. C., *J. Am. Chem. Soc.*, **67**, 246–53 (1945).

[296] Smith, J. W., and Mapstone, G. E., *Fuel*, **36**, 191–204 (1957).

[297] Savich, T. R., and Howard, H. C., *Ind. Eng. Chem.*, **44**, 1409–11 (1952).

[298] Blaskett, D. R., *Australia (South), Dept. Mines, Mining Rev.*, No. 86, 60–79 (1948). Van Doornum, G. A. W., *J. Inst. Fuel*, **27**, 482–95 (1954); *S. African Fuel Research Inst., Rept.* No. 2 (1958), 95 pp. Lange, W., and Winzen, W., *Brennstoff-Wärme-Kraft*, **6**, 382–5 (1954). Coward, H. F., *Ministry Fuel and Power (Gt. Brit.), Safety in Mines Research Estab., Research Rept.* No. 142 (1957), 80 pp.

[299] Berkowitz, N., and Schein, H. G., *Fuel*, **30**, 94–6 (1951).

[300] Lemaire, E., *Génie civil*, **120**, 269–70 (1943). Skochinskii, A. A., Makarov, S. Z., and Rode, T. V., *Izvest. Akad. Nauk S.S.S.R., Otdel. Tekh. Nauk*, **1944**, 29–41; *Chem. Abs.*, **39**, 1974 (1945). Virtala, V., Oksanen, S., and Frilund, F., *Valtion Tek. Tutkimuslaitos, Julkaisu*, No. 14 (1949), 57 pp.; *Chem. Abs.*, **46**, 7770 (1952). Budryk, W., *Przeglad Górniczy*, **8** (39), 443–6 (1952); *Chem. Abs.*, **48**, 14157 (1954). Chalupa, V., and Drabek, V., *Uhli*, **5**, 310–19 (1955); *Chem. Abs.*, **50**, 1289 (1956). Veselovskii, V. S., and Terpogosova, E. A., *Izvest. Akad. Nauk S.S.S.R., Otdel. Tekh. Nauk*, **1950**, 1204–11; *Chem. Abs.*, **46**, 1227 (1952).

[301] Davis, J. D., *AGA Proc.*, **27**, 426–36 (1945).

[302] Noetzold, E., *Glückauf*, **76**, 381–8, 393–7 (1940). Elder, J. L., Schmidt, L. D., Steiner, W. A.. and Davis, J. D., *U.S. Bur. Mines, Tech. Paper* No, 681 (1945), 24 pp. Marevich, N. V., and Travin, A. B., *Izvest. Akad. Nauk S.S.S.R., Otdel. Tekh. Nauk*, **1953**, 1110–17; *Chem. Abs.*, **48**, 2343 (1954).

There are many further papers dealing with the various aspects of spontaneous ignition.[303]

Several papers are concerned with the changes of coal during weathering and storage, which include increase in oxygen content, loss of coking properties, and loss of calorific value.[304] Estimates of the loss of calorific value vary widely, and it is doubtful whether sufficient valid measurements covering different types of coal and different conditions of storage are yet available, except possibly in unpublished commercial files.

A study of coal piles[305] resulted in the following recommendations for safe storage: (1) coal to be placed in thin layers about 1 ft in thickness; (2) each increment to be compacted; (3) sides gently sloping (maximum angle 14°); (4) the final surface to be smoothed; (5) coal to be removed in thin layers from the top and the new surface compacted and smoothed; and (6) alteration of the pile angle to be avoided. The storage and handling arrangements at a large gas works have been described in which the importance of sealing the bottom of the pile was pointed out.[306] In general, it is desirable to prevent aeration completely by sealing exposed surfaces or, alternatively, to ensure adequate aeration which prevents a build-up of temperature although it will not prevent weathering. Four other papers concerned with this topic may be cited.[307]

Attempts, presumably stimulated by the success of oxidation inhibitors in oil storage, have been made to find inhibitors for the weathering of coal in air. Impregnation of small samples of an Illinois coal with 5% ether solutions of antioxidants, such as hydroquinone and amines, did not make a significant difference to the rate of oxygen

[303] Tronov, B. V., *Izvest. Tomsk. Ind. Inst.*, **60**, No. 3, 11–36 (1940); *Chem. Abs.*, **37**, 1029 (1943). Tronov, B. V., and Grigor'ev, M. Yu., *Izvest. Tomsk. Ind. Inst.*, **60**, No. 3, 46–9 (1940); *Chem. Abs.*, **37**, 1029 (1943). Pogrebitskii, E. O., *Bull. acad. sci. U.R.S.S., Ser. Geol.*, **1943**, Nos. 4–5, 129–35; *Chem. Abs.*, **38**, 6517 (1944). Jessen, H. E., *Gasteknikeren*, **32**, 463–71, 481–6 (1943); *Chem. Abs.*, **39**, 1746 (1945). Scott, G. S., *U.S. Bur. Mines, Bull.* No. 455 (1944), 206 pp. Kartvelishvili, G. A., *Ugol'*, **21**, 34–5 (1946); *Chem. Abs.*, **41**, 4291 (1947). Romwalter, A., *Magyar Tech.*, **2**, 21–2 (1947); *Chem. Abs.*, **41**, 7753 (1947). Sanchez Marco, G., *Combustibles (Zaragoza)*, **7**, 124–32, 163–70 (1947); *Chem. Abs.*, **42**, 7043 (1948). Olpinski, W., *Przeglad Chem.*, **6**, 258–61 (1948); *Chem. Abs.*, **43**, 9411 (1949); *Prace Glównego Inst. Górnictwa, Komun.* No. 130 (1952), 9 pp.; *Chem. Abs.*, **48**, 14155 (1954); *Przeglad Gòrniczy*, **9**, 192 (1953); *Prace Glównego Inst. Górnictwa, Komun.* No. 139, 32–4 (1953); Olpinski, W., and Gabrys, P., *Prace Glównego Inst. Górnictwa, Komun.* No. 139, 4–13 (1953); *Chem. Abs.*, **48**, 4803 (1954). Olpinski, W., Pawlikowski, T., and Rozmus, J., *Prace Glównego Inst. Górnictwa, Komun.* No. 139, 13–20 (1953); *Chem. Abs.*, **48**, 4803 (1954). Olpinski, W., and Pawlikowski, T., *Prace Glównego Inst. Górnictwa, Komun.* No. 139, 20–31 (1953); *Chem. Abs.*, **48**, 4803 (1954). Orleanskaya, G. L., *Izvest. Akad. Nauk S.S.S.R., Otdel. Tekh. Nauk*, **1950**, 1071–9; *Chem. Abs.*, **45**, 8739 (1951). Gibinski, S., *Prace Glównego Inst. Górnictwa, Komun.* No. 97 (1951), 10 pp. Orleanskaya, G. L., *Izvest. Akad. Nauk S.S.S.R., Otdel. Tekh. Nauk*, **1952**, 877–83; *Chem. Abs.*, **47**, 8343 (1953). Mapstone, G. E., *Chem. & Ind. (London)*, **1954**, 658. Toledo, F. G. M., *Ion*, **14**, 263–8 (1954); *Chem. Abs.*, **48**, 12389 (1954). Tanaka, M., and Yamasaki, T., *J. Mining Inst. Japan*, **71**, 633–7 (1955); *Chem. Abs.*, **50**, 17380 (1956). Chakravorty, S. L., *Indian Mining J.*, **3**, No. 9, 117–21 (1955); *Chem. Abs.*, **50**, 8165 (1956). Rafikov, S. R., *Izvest. Akad. Nauk Kazakh. S.S.R., Ser. Khim.*, **1956**, No. 9, 75–91; *Chem. Abs.*, **50**, 14206 (1956). Migdalski, H., *Bergbautechnik*, **6**, 364–70 (1956); *Chem. Abs.*, **50**, 17381 (1956). Zavarzina, N. B., *Mikrobiologiya*, **25**, 200–207 (1956); *Chem. Abs.*, **50**, 12439 (1956). Kahlturin, A. I., and Azerbaev, I. N., *Izvest. Akad. Nauk Kazakh. S.S.R., Ser. Khim.*, **1958**, No. 1, 112–20; *Chem. Abs.*, **53**, 8585 (1959).

[304] Bodmer, G., *Schweiz. Ver. Gas- u. Wasserfach., Monatsbull.*, **21**, 101–10 (1941). Schmidt, L. D., *Iron Steel Engr.*, **18**, No. 3, 64–70 (1941). Weber, T., *Bergbau*, **54**, 299–303 (1941). Waard, S. de, *Brennstoff- u. Wärmewirtsch.*, **23**, 73–8 (1941). Schmidt, L. D., Elder, J. L., and Davis, J. D., *Ind. Eng. Chem.*, **35**, 150–5 (1943). Mapstone, G. E., *Chem. & Ind. (London)*, **1954**, 577–8. Teichmann, R. F. J., *S. African Mining Eng. J.*, **56**, 99–103 (1945). Jackman, H. W., Eissler, R. L., and Reed, F. H., *Illinois State Geol. Survey, Circ.* No. 227 (1957), 22 pp.

[305] Allen, R. R., and Parry, V. F., *U.S. Bur. Mines Rept. Invest.* No. 5034 (1954), 31 pp.

[306] Clark, L. J., and Walker, J. H., *Inst. Gas Engrs., Publ.* No. 386 (1951), 50 pp.; *Gas World*, **133**, 616–22 (1951).

[307] Goodman, J. B., Parry, V. F., and Landers, W. S., *U.S. Bur. Mines Rept. Invest.* No. 3915 (1946), 37 pp. Taylor, R. A. A., *J. Inst. Fuel*, **14**, 144–7 (1941). Anon., *Umschau*, **49**, 90 (1949); *Power Eng.*, **57**, 62 (1953).

adsorption.[308] However, treatment with 3% aqueous hydroquinone was reported to give protection for 24 days.[309] Petroleum products and their emulsions, soaps, fatty acids, and milk of lime have also been reported to be effective in protecting stored coal against oxidation and spontaneous combustion.[310] Spraying of lignite entering storage with a 0.165% solution of calcium bicarbonate has been claimed to control the interior temperatures of piles up to 1000 tons at around 30°C for 18 mo.[311] An untreated control pile rose to 60°C. A bitumen-clay emulsion was found to be less effective.

Active oxygen in oxidized coal. The formation, during the air oxidation of coals, of a low surface concentration of groups having the characteristics of hydroperoxides has been studied throughout the period under review. Whether these play any part in the conversion of coal to humic acids or represent an isolated phenomenon in the initial oxidation at low temperatures is still not clear. These peroxides can oxidize titanous chloride.[312] It was shown that their concentration increased to a maximum after 10 to 25 days exposure to air and then diminished gradually.[313]

The initial reaction of both coal and carbon with oxygen appears to involve chemisorption followed, in the presence of water, by the formation of a complex with oxygen and water having the oxidizing properties of peroxygen.[314] The concentration of peroxygen was determined by shaking the coal sample with a solution of ferrous thiocyanate, and the ferric thiocyanate produced estimated colorimetrically. The normal inherent water content of coals and the peroxygen content are functions of their internal surface. Up to a temperature of 70°C, the peroxygen complex was found to be relatively stable, but broke down above this temperature. The peroxygen complex on carbon was not completely destroyed until 350°C was reached. Oxidation above 70°C was found to result in an increase in both the proportion of carbon dioxide to monoxide produced and in the amount of oxygen fixed by the coal in acidic groupings. These observations were substantiated by means of gravimetric balances for the coal-oxygen reaction at 50 and 90°C. The remarkable difference in behavior above and below 70°C, a temperature long known to be critical in the spontaneous ignition of coal, is common to coal and carbon, and it was concluded that the overall oxidation was, in both cases, controlled by the rate of breakdown of the complex. The presence of water is of paramount importance.

Peroxide groups formed in coal on oxidation have been estimated by reaction with hydroquinone dissolved in water, the quinone formed being estimated iodometrically. Adsorbed oxygen did not affect the results.[315] Other methods have been described.[316]

The Russell effect—the ability of some coals to affect a photographic plate in the dark—has been attributed to the peroxides formed on oxidation.[317]

The oxygen-containing groups formed during more extensive oxidation of coals have a much smaller heat of interaction with ethylenediamine than the groups in unoxidized coals. Heats of wetting in ethylenediamine and in methanol have been compared.[55]

Oxidation and petrographic type. Schulz and Howard[227] oxidized fractions of bituminous coals, selected for their high content of various petrographic constituents, to water-soluble acids with fuming nitric acid followed

[308] Yohe, G. R., Organist, R. H., and Lansford, M. W., *Illinois State Geol. Survey Circ.* No. 201, GRY-2 (1951-2), 21 pp.

[309] Gillet, A., and Fastré, M. L., *Bull. soc. chim. Belges*, **53**, 83–96 (1944).

[310] Terpogosova, E. A., *Izvest. Akad. Nauk S.S.S.R., Otdel. Tekh. Nauk*, **1954**, No. 4, 147–51; *Chem. Abs.*, **49**, 4968 (1955).

[311] Khrisanfova, A. I., and Shubnikov, A. K., *Khim. i Tekhnol. i Masel*, **3**, 21–8 (1958); *Chem. Abs.*, **52**, 16721 (1958).

[312] Yohe, G. R., and Harman, C. A., *J. Am. Chem. Soc.*, **63**, 555–6 (1941).

[313] Yohe, G. R., and Wilt, M. H., *J. Am. Chem. Soc.*, **64**, 1809–11 (1942).

[314] Jones, R. E., and Townend, D. T. A., *Nature*, **155**, 424–5 (1945); *Inst. Gas. Engrs., Publ.* No. 271 (1945); No. 287 (1945); *Trans. Faraday Soc.*, **42**, 297–305 (1946); *J. Soc. Chem. Ind.* (*London*), **68**, 197–201 (1949).

[315] Rafikov, S. R., and Sibiryakova, N. Ya, *Izvest. Akad. Nauk Kazakh. S.S.R., Ser. Khim.*, **1956**, No. 9, 13–22; *Chem. Abs.*, 50, 8992 (1956).

[316] Kramers, W. J., and McKee, J. H., *Trans. Faraday Soc.*, **42**, 303–5 (1946). Chakravorty, S. L., Long, R., and Ward, S. G., *Fuel*, **29**, 68–9 (1950). Chalishazar, B. H., and Spooner, C. E., *Fuel*, **36**, 127–8 (1957).

[317] Yohe, G. R., *Fuel*, **29**, 163–5 (1950).

by alkaline permanganate. The yield of benzenoid acids decreased in the order fusain, opaque attritus, anthraxylon and translucent attritus, resins. The yields of mellitic acid were 25.4, 13.5, 6.2, and 1.5 g per 100 g of material on the dry, ash-free basis, respectively. It was pointed out that the yield of mellitic acid from the fusain sample was of the same order as that from a 700°C coke. There was a correlation between mellitic acid yields, carbon-hydrogen ratios, and opacity in thin sections. Esterification and fractionation of the mixed esters showed the presence of benzenepentacarboxylic acid in addition to mellitic acid in the oxidation products of fusain and in those from a splint coal that contained a high concentration of opaque attritus. The oxidation procedures used did not disclose any specific chemical differences among petrographic constituents beyond those indicated by ultimate chemical analysis.

The heat evolved and the yield of regenerated humic acids have been found to be similar for both micrinite and exinite when they were oxidized with a solution of chlorine dioxide and to be almost independent of the rank above 15% volatile matter. The values for vitrinites were much higher and dependent on rank. It was concluded from a comparison with the reaction between chlorine dioxide solution and known organic compounds that cyclic systems of the hydroaromatic type and phenolic hydroxyl groups were probably present in the coals.[225]

Several other investigations of oxidation of coal have included a study of the oxidation of the macerals separated from the coals.[227,318]

Nature and chemistry of oxidized coals and humic acids. Changes in bulk density and wettability are associated with oxidation of coal.[319] A straight-line relation has been found between the specific volume and the hydrogen content of a coal progressively oxidized in air.[320]

A "Gas Evolution Index" for methane adsorbed on coal was found to be linearly related to the degree of oxidation of a coal as measured by the consequent change in ignition temperature in the presence of a solid oxidant such as sodium nitrate.[321] It was suggested that the original coal surface was hydrophobic, and that the oxidized surface preferentially adsorbed water thus decreasing the adsorption of methane. Treatment of the oxidized coal with benzidine restored its ignition temperature to that of freshly mined coal. The ignition temperature measurement was capable of detecting a degree of oxidation too small to be recognized by chemical means.

From an investigation of the ion-exchange properties of oxidized coal, it has been concluded that the content of acidic groups and their relative strengths can be estimated from measurements with ferrous sulphate.[322]

Brown coal can be distinguished from weathered black coal by determining the "coagulation threshold" which is the minimum concentration of barium chloride to produce turbidity in an alkaline solution that is prepared in a specific manner. This value for the extracted humic acids ranged from 10.0 to 19.6 millimoles/l of barium chloride for brown coal and from 2.0 to 6.8 for weathered black coal.[323] It has been shown by oxidizing coals under standard conditions that the rate of production of humic acid depends on rank, extent of weathering, and petrological composition.[324] The properties of the humic acid measured—optical density and coagulation point—were independent of the extent of weathering and the reaction time. Thus examination of the properties of the humic acids gave a means of classifying the parent coals. The foregoing quantities were also determined for a single coal with various oxidizing agents, of which hydrogen peroxide was considered the most satisfactory for standardization.

Oxidation of bituminous and lower-rank

[318] Oreshko, V. F., *Doklady Akad. Nauk S.S.S.R.*, **81**, 663–6 (1951); *Chem. Abs.*, **46**, 9826 (1952).

[319] Zagrebel'naya, V. S., *Bull. acad. sci. U.R.S.S., Classe sci. tech.* **1946**, 157–66 (1946); *Chem. Abs.*, **40**, 4866.

[320] Farenden, P. J., and Spencer, D. H. T., *Fuel*, **34**, 131–2 (1955).

[321] Lamba, E. G., and Ettinger, I. L., *Izvest. Akad. Nauk S.S.S.R., Otdel. Tekh. Nauk*, **1955,** No. 4, 110–19; *Chem. Abs.*, **49,** 15209 (1955).

[322] Gillet, A. C., Chandelon, P., and Gillet, I. E., *Compt. rend. 27th congr. intern. chim. ind., Brussels,* **1954,** *Ind. chim. belge*, **20** (Special), **2,** 95–9 (1955).

[323] Kukharenko, T. A., *Zavodskaya Lab.*, **22,** 661–2 (1956); *Chem. Abs.*, **50,** 16077 (1956).

[324] Kukharenko, T. A., Lyubimova, S. L., and Shalyapina, A. N., *Izvest. Akad. Nauk S.S.S.R., Otdel. Tekh. Nauk*, **1956,** No. 12, 133–6.

coals in air for about 250 hr reduced their flotability with purified petroleum light oil as a collector.[325] In the initial stage, the decrease was slight, or negative, but was steady in the later stages. The extent to which the coals were affected was greatest with bituminous coals, least with anthracite and lignite.[326]

The possibility of raising the calorific value of naturally oxidized outcrop coal has been studied by stepwise thermal decomposition at atmospheric pressure.[327] Heating at 400°C increased the calorific value by 15%, yielding about 85% of the dry, ash-free coal in an improved fuel having 13,800 Btu/lb. In effect, the same amount of heat was concentrated in a smaller weight. From analyses of the gases evolved, a picture of the constitution of such coals was developed.

It has been found that the smoke yield and the caking and swelling power of coal can be greatly reduced by pretreatment with nitrogen dioxide at room temperature, and with nitric oxide or nitrous oxide at 320°C.[328] Reaction with nitrogen dioxide is highly exothermic, but is ineffective above $\frac{1}{8}$ in. particle size owing to slow diffusion. The smoke reduction was associated with a reduction in hydrogen content to about 4% and required a longer time of pretreatment than reduction in the swelling or caking properties. The heat of combustion was reduced by less than 1% when the swelling number of a coking or a caking coal was halved, and by about 2% when the smoke yield was halved. About a two-third reduction in smoke could be achieved before the loss of calorific value exceeded 5%. Loss of carbon during treatment with nitrogen dioxide was negligible, in striking contrast to the considerable loss during atmospheric oxidation of coal. Changes in analysis suggested that the reaction was equivalent to replacing two hydrogen atoms by one nitro group and one oxygen atom.

Humic acids derived from coal or pure compounds by oxidation have been considered not to differ in kind[329] from the natural variety. The products from the oxidation of a coking coal with hydrogen peroxide and a lower rank coal with nitric acid were dissolved in a 23% solution of sodium hydroxide, and it was found that the optical density of the solutions at equal carbon concentration increased with rank of coal but remained the same for a given coal whether the humic acid yield was 3.5 or 26.5%.[330] Examination of the humic acids by infrared spectroscopy and X-ray diffraction suggested that they contained the same structural units as the parent coal. X-ray observations of the solutions suggested that they were true molecular solutions and not colloidal. Coagulation thresholds with calcium chloride were recorded.

From an investigation of the nature of the acetone-soluble and -insoluble humic acids produced by the action of nitric acid on bituminous coal, it was concluded that the differences in solubility were due to relatively slight differences in hydration.[331] The use of detergents in fractionating humic acids has been proposed.[332]

The molecular weights of nitrohumic acids from bituminous coal, determined by cryoscopic and ebullioscopic methods, were found to vary with the solvent used.[333] In the highly polar solvents, the acids were more strongly associated than would be expected for carboxylic acids, and it was suggested that forces peculiar to the humic acids must play a part.

The changes in viscosity, light absorption, and electrical conductivity of alkaline solutions of humic acids during the gradual addition of sodium carbonate have been studied to elucidate the state in which the acids were present in the solution.[334]

In studies of base exchange capacities of humic acids,[335] it has been concluded that carboxyl groups alone participate in the reaction between calcium acetate and humic

[325] Sun, S. C., *Mining Eng.*, **6**, 396–401 (1954).

[326] See also, Gogitidze, T. A., and Plaksin, I, N., *Izvest. Akad. Nauk S.S.S.R.*, **1956**, No. 11, 77–81; *Chem. Abs.*, **51**, 9126 (1957).

[327] Macrae, J. C., *Fuel*, **22**, 93–102 (1943).

[328] Billington, A. H., *Fuel*, **33**, 295–301 (1954).

[329] Kukharenko, T. A., *Doklady Akad. Nauk S.S.S.R.*, **72**, 907–10 (1950); *Chem. Abs.*, **44**, 9147 (1950).

[330] Kasatochkin, V. I., and Larina, N. K., *Doklady Akad. Nauk S.S.S.R.*, **114**, 139–42 (1957); *Chem. Abs.*, **52**, 1586 (1958).

[331] Herbert, S. A., Charmbury, H. B., and Kinney, C. R., *Fuel*, **27**, 168–9 (1948).

[332] Davies, R. I., Coulson, C. B., and Luna, C., *Chem. & Ind.* (*London*), **1957**, 1544.

[333] Polansky, T. S., and Kinney, C. R., *Fuel*, **31**, 409–13 (1952).

[334] Bouilloux, M. G., *Bull. soc. chim. France*, **11**, 38 (1944).

[335] Fuchs, W., *Fuel*, **22**, 112–16 (1943).

acids from peat and from the oxidized products from bituminous coals.[336] In an investigation of the mechanism of the interaction of humic acid with bases and salts, an explanation by the classical laws of electrochemistry was found to be difficult, but a satisfactory explanation in terms of an electrical double layer was proposed.[337]

The ultimate analyses of the humic acids produced from four bituminous coals heated in an air oven at 200°C for 200 hr were almost identical.[338] The base-exchange capacities of these acids increased from 425 to about 500 meq/g with decrease in rank but fell again to 350 for a lignite. The base-exchange capacity was attributed chiefly to the presence of carboxyl groups.

Electrometric titration of humic acids prepared from coal,[339] and studies of the condition of humic acids in solution[340] and of the rheological properties of the solutions[341] indicate that the acids exhibit polymeric characteristics.

The infrared spectra of humic acids, made both by oxidizing coal and from hydroquinone or glucose, mulled in mineral oil and in hexachlorobutadiene, were similar, the differences being limited to the intensities of the bands near 1700 and 1600 cm^{-1}. The first of these bands was assigned to carboxylic carbonyl groups and the second, at least in part, to some particular types of conjugated and hydrogen bonded carbonyl groups.[342] Mulls prepared from humic acids made by oxidation of coal with 30% hydrogen peroxide and concentrated nitric acid, and water-soluble acids prepared by ozonization, showed strong absorptions for the carboxyl group, aromatic unsaturation, and for the nitro group in the humic acid prepared with nitric acid.[343] Absorption in the ultraviolet[344] and visible regions[345] have also been studied.

Humic acids made by oxidizing coals with nitric acid have been shown to give three well-defined polarographic waves at the dropping mercury electrode.[346] The waves were long and drawn-out, but otherwise normal. The first was the most persistent and was thought to be due to the reduction of nitro groups, the second was best developed in solutions in 30 to 40% alcohol, and the third in aqueous solutions of a salt such as potassium chloride. The structures responsible for these waves were not identified. From observations of the diffusion currents with the first wave, it was concluded that the molecular weights of these acids were less than 1000. Other information on molecular weights and functional group contents of humic acids from air oxidation of coal has been given.[347]

The reaction of humic acids from coal with sodium and liquid ammonia yielded products that were soluble in ether, the yield varying with the source of the humic acid, peat giving more than the mature coals. The changes in composition were reported, together with molecular weights of the products.[348] Chemical studies of the hydrogenation of humic acids with a nickel catalyst have also been made.[349]

Water-soluble acids. Oxidation of coal to water-soluble products, in relation to its

[336] Kukharenko, T. A., *Trudy Inst. Goryuchikh Iskopaemykh, Akad. Nauk S.S.S.R.*, **2**, 193–207 (1950); *Chem. Abs.*, **49**, 4263 (1955).

[337] Chatterjee, S., and Bose, S., *J. Colloid Sci.*, **7**, 414–27 (1952).

[338] Mukherjee, P. N., Bhowmik, J. N., and Lahiri, A., *J. Proc. Inst. Chemists (India)*, **28**, 4–14 (1956).

[339] Roy, M. M., *Kolloid-Z.*, **153**, 174–8 (1957).

[340] Thiele, H., and Kettner, H., *Kolloid-Z.*, **130**, 131–60 (1953).

[341] Mukherjee, P. N., and Lahiri, A., *J. Colloid Sci.*, **11**, 240–3 (1956).

[342] Čeh, M., and Hadži, D., *Fuel*, **35**, 77–83 (1956). Ziechmann, W., *Brennstoff-Chem.*, **39**, 353–9 (1958).

[343] Lovell, H. L., and Kinney, C. R., *Am. Chem. Soc., 120th Meeting, (Abstracts)*, 5 I, September (1951).

[344] Scheffer, F., and Welte, E., *Landwirtsch. Forsch.*, **3**, 1–6 (1951).

[345] Kukharenko, T. A., *Doklady Akad. Nauk S.S.S.R.*, **89**, 133–6 (1953); *Chem. Abs.*, **47**, 7757 (1953).

[346] Cody, A. F., Milliken, S. R., and Kinney, C. R., *Anal. Chem.*, **27**, 362–6 (1955).

[347] Chowdhury, J. K., and Biswas, A. B., *J. Indian Chem. Soc.*, **19**, 289–98 (1942). Higuchi, K., *J. Fuel Soc. Japan*, **29**, 104–12 (1950). Higuchi, K., Shibuya, Y., and Tsuyuguchi, M., *J. Fuel Soc. Japan*, **33**, 366–79 (1954). Higuchi, K., Tsuyuguchi, M., and Asakawa, K., *J. Fuel Soc. Japan.*, **34**, 645–53 (1955).

[348] Kukharenko, T. A., Vvedenskaya, T. E., and Bel'gova, V. A., *Doklady Akad. Nauk S.S.S.R.*, **86**, 601–4 (1952); *Chem. Abs.*, **50**, 2949 (1956).

[349] Kukharenko, T. A., and Savel'ev, A. S., *Doklady Akad. Nauk S.S.S.R.*, **76**, 77–80 (1951); *Chem. Abs.*, **45**, 8451 (1951).

constitution, has been reviewed by Gonzalez Sanchez and Uson Lacal.[350]

In a long-term program on the oxidation of coal supported at the Coal Research Laboratory of the Carnegie Institute of Technology, a process was developed for converting coal into yellow water-soluble polycarboxylic aromatic acids in a 50% yield. The components of the mixtures varied in their aromatic nuclear structure and in the number and location of carboxyl groups. Previous work had indicated the advantageous nature of the products obtained with alkaline permanganate. However, because so much of the coal was necessarily converted to CO_2, only the cheapest oxidizing agents could be considered, and gaseous oxygen in the presence of aqueous alkali was chosen. Similarly, the rank of coal used was based on a compromise, higher yields being obtained with higher-rank coals but at the expense of a lower reaction rate.

In the early experiments, 50 g samples of through 200-mesh coal were suspended in 350 ml of an aqueous solution containing 335 g of potassium hydroxide. Oxygen was supplied at a constant pressure through a reducing valve. It was found necessary to introduce a trap in the oxygen line near the autoclave to prevent plugging by the reactants. Sufficient alkali was used to fix all the carbon dioxide formed—assuming complete oxidation—thus ensuring a constant partial pressure of oxygen. Potassium hydroxide gave more complete reaction than sodium hydroxide, and only a part could be replaced with lime. In water alone, the reaction resembled a combustion and produced carbon dioxide, ash, small amounts of humic acids, and negligible amounts of the desired soluble acids. At 1.5*N*, a maximum yield of humic acids was reached which, on further increase in alkali concentration, was decreased by further oxidation to give additional quantities of the desired soluble acids. These investigations confirmed that higher yields were obtained from higher-rank coals and that catalysts were not helpful in increasing the desired product. Optimum conditions were different for different coals.

For the preparation of larger amounts of the acids for chemical characterization and commercial evaluation, an internally stirred batch autoclave was used. The necessary pilot plant equipment[351] for recovery of the acids was designed on the basis of bench-scale experiments. The stationary vertical autoclave, 19½ in. i.d. and 30 in. deep with an operating capacity of approximately 25 gal, was designed for a pressure of 1000 psi gauge at temperatures up to 300°C. In a standard run, 15 lb of Pocahontas No. 3 coal of about 90% C and 20% volatile matter pulverized to −100 mesh was stirred vigorously with 45 lb of sodium hydroxide and 15 gal of water in an external hopper. The quantity of alkali employed was just sufficient to neutralize the carbon dioxide so that sodium carbonate and only a negligible amount of bicarbonate resulted. The coal was thoroughly wetted to prevent the possibility of explosive reaction between dry coal particles and oxygen gas. The well-mixed alkaline slurry was then pumped into the autoclave and heated to about 160°C by 90 psi steam in an internal nickel coil. At this point, the steam was shut off, the electrical heaters turned on, and oxygen introduced. The oxygen flow was manually regulated to maintain the total pressure on the system at 900 psi gauge, made up of 350 psi gauge oxygen partial pressure and 550 psi gauge vapor pressure of the reaction mixture.

In the heating-up period, approximately 30% of the carbon of the coal was converted to carbon dioxide, and the remainder became soluble in alkali at the humic acid stage. The reaction conditions of 270°C and 900 psi gauge were maintained for an additional 2 hr to convert 20% more of the coal carbon to carbon dioxide. During the oxidation an approximate total of 1.63 pounds of oxygen was consumed per pound of coal charged.

After cooling and removing a small sample for analysis, the 18 gal of product were diluted with water to 22 gal and filtered through heavy cotton twill in a plate-and-frame press to remove ash, some sodium carbonate, and a small amount of insoluble carbonaceous residue. The filtrate was then poured into about 50 lb of 95% sulphuric acid in a glass-lined 50-gal tank with a cooling jacket. The acidity was adjusted to 1*N*, the optimum normality for extraction. A pitchlike material accumulated

[350] Gonzalez Sanchez, F., *Combustibles (Zaragoza)*, **14**, 3–39 (1954). Uson Lacal, R., *Combustibles (Zaragoza)*, **18**, 163–7 (1958).

[351] Franke, N. W., and Kiebler, M. W., *Chem. Inds.*, **58**, 580–1 (1946).

on the liquid surface and was removed manually. The acidified solution was then extracted twice for 2 hr at 40°C with two-thirds of its volume of methyl ethyl ketone in glass-lined extraction tanks. The tanks were equipped with stirrers and with heating coils to prevent separation of sodium sulphate. The methyl ethyl ketone was practically insoluble in the saturated solution of sodium sulphate. An efficiency of over 85% was consistently obtained in the two-stage extraction. The methyl ethyl ketone was next distilled from the extract until the residue consisted of 50 to 70% aromatic acids. This syrup was then dried in a vacuum drum dryer operated at 28-in. vacuum and 1 psi of steam on the rolls. The average nuclear size N of the product was calculated from the relation: N = mol wt − functionality × 44. This size corresponds to the nuclear size with the carboxyl-hydrogen attached to the nucleus. The mixture of acids recovered from the oxidation of coal was colored. Despite the fact that they were prepared at a temperature of 270°C, under high oxygen pressures, they were sensitive to temperatures above 150°C even in the absence of oxygen. Thus, on rapid heating up to 350°C in a molecular still, only 55 to 60% of the mixture distilled, and a coke residue and gaseous products were formed.

The acids were fractionated by distribution between ether and pentane on a molecular weight basis and between acetone and benzene on a basis of structure.[352] Their nuclear structure was investigated by hydrogenolysis of the esters.[353] The acids consisted of about one-third benzenecarboxylic acids, one-third of the molecular weight range 200 to 250 containing nuclei of two to three fused rings, and the rest of molecular weight 250 to 450 with larger polycyclic ring systems. Six acids of the benzenecarboxylic acid series, in which trimellitic acid predominated, were isolated and characterized.[353] Molecular weights were determined ebullioscopically in acetone by using a modified Menzies-Wright apparatus,[354] and equivalent weights by potentiometric titration with a glass electrode.

The products were strong organic acids and underwent the usual reactions characteristic of the carboxyl group. They reacted readily with alcohols to form esters.[355] With monohydric alcohols, the esters ranged in properties from soft resins to viscous liquids, and with polyhydric alcohols, the esters were brittle resins. The esters could be fractionated by the solvent procedure described for the acids and also by molecular distillation. Approximately 60% of the butyl esters could be distilled before decomposition began. The yield of refined esters was about 60%, and they were of improved color and viscosity. About 90% of the refined product could be molecularly distilled.[356]

In further work on the nuclear structure of the water-soluble acids, the acids were esterified and then treated with hydrogen under high pressure and at 270°C, with a copper-chromium oxide catalyst, to reduce their carboxyl groups to methyl groups.[357] The products were separated (1) into distillation fractions with different boiling ranges; (2) each of these fractions, by chromatography on bentonite, silica and alumina, into (*a*) oxygenated materials, (*b*) aromatic hydrocarbons, and (*c*) alicyclic hydrocarbons; and finally (3) into individual compounds by fractional distillation of each of the three chemical classes. Trimethylbenzenes had been isolated previously, and the investigation was therefore confined to the fractions boiling above 195°C. The compounds isolated were: phthalan, 5-methylphthalan, 4,5- and 4,7-dimethylphthalans, 2-methylcyclohexanemethanol, pentamethylbenzene, and all 3 tetramethylbenzenes. The presence of phthalan is interesting in view of the possibility that coals may contain heterocyclic oxygen.

The acids were also converted[358] into their copper salts, and these were decarboxylated by being heated to 250°C. From the products of this reaction, the following compounds were isolated: benzene, naphthalene, biphenyl, diphenylene oxide, phenanthrene, benzophenone, fluorenone, phenol, benzoic acid, pyridine,

[352] Roy, A. N., and Howard, H. C., *J. Am. Chem. Soc.*, **74,** 3239–42 (1952).

[353] Ruof, C. H., Savich, T. R., and Howard, H. C., *J. Am. Chem. Soc.*, **73,** 3873–9 (1951).

[354] Berman, N., and Howard, H. C., *Anal. Chem.*, **21,** 1200–1202 (1949).

[355] Berman, N., and Howard, H. C., *Fuel*, **29,** 109–11 (1950).

[356] Franke, N. W., Kiebler, M. W., Ruof, C. H., Savich, T. R., and Howard, H. C., *Ind. Eng. Chem.*, **44,** 2784–92 (1952).

[357] Entel, J., *J. Am. Chem. Soc.*, **76,** 3646–9 (1954).

[358] *Ibid.*, **77,** 611–15 (1955).

quinoline, α-, β-, and γ-phenylpyridine, β-naphthaquinoline and the lactone of 2-hydroxyl-2′-diphenyl-carboxylic acid. The way in which these compounds may be related to the nuclei of the aromatic acids and to the structural units of the coal itself was discussed. From the fact that phenol was found in the products, it was deduced either: (1) that the original coal structure contains phenolic groups that are protected from oxidation, or (2) that these phenolic groups are formed by the cleavage of an ether link during oxidation, but are themselves protected against further oxidation. Some of the foregoing work has been covered by patents.[359]

A number of model compounds have been prepared and examined in connection with the elucidation of the structure of these water-soluble acids.[360]

As a prelude to commercial exploitation, methods for fractionating water-soluble acids were developed. An aqueous solution of these acids can conveniently be fractionated in two ways: (1) either by extraction with a series of mixtures of a nonsolvent, carbon tetrachloride or petroleum ether, with progressively larger proportions of a solvent, methyl ethyl ketone, to yield fractions of increasing molecular weight and functionality; or (2) by acidifying the solution in steps and extracting it with this ketone between each step. Alternatively, the acids could be sublimed under 1 mm Hg pressure, yielding a mixture of phthalic and methylphthalic anhydrides and benzoic and toluic acids: the original mixture yielded only 13% of sublimate, 75% of which was phthalic anhydride, but some of the fractions obtained by stepwise acidification yielded up to 60% of sublimate which was 1.2% of the original mixture. Fifty to 90% of the sublimate consisted of monobasic acids. The presence of methyl-substituted aromatic rings was thought to indicate that methyl groups may be attached to the rings in the coal molecule itself.

In a comprehensive study of the structure of these acids, benzoic and phthalic acids were isolated by partition chromatography and tentative identifications were made, by mass spectrometry, of many more compounds than had previously been reported.[362] From the types identified and from the nuclear C/H ratio of the various fractions, it was concluded that these acids contained considerable numbers of aliphatic carbon atoms. Thirty-nine percent of the acids were identified[363] from the infrared and mass spectra of their methyl esters. The mixture contained at least 0.22% of benzene-tetracarboxylic acid, 21% of trimellitic, 16% of various phthalic acids, 0.05% of naphthoic, 0.54% of toluic, and 1.2% of benzoic acids. It was concluded that nuclei up to naphthalene or larger are present in the structure of the Pocahontas coal from which the acids were derived. The mixture of acids was decarboxylated[364] at a temperature below 265°C to avoid risk of synthesizing any of the nuclei identified by the copper-quinoline method, and the products were identified by fractionation and mass spectrometry. The benzene, naphthalene, biphenyl, indane, and phenanthrene ring systems and many of their alkyl-substituted derivatives were definitely identified.

In further work,[365] a longer time was allowed for decarboxylation; the products were separated by fractional distillation and chromatography and identified as previously. Almost all the nuclei were identified in this way and eight nuclei were found to make up 94% by weight of the total, distributed as follows: methylnaphthalene 38%, benzene 35%, biphenyl 9.0%, naphthalene 4.0%, phenanthrene 3.3%, benzene substituted with five saturated carbon atoms 1.8%, benzophenone 1.7%, and toluene 1.5%. No anthracene and no nuclei with more than three fused rings were found as would be anticipated from the conditions chosen for the preparation

[359] Carnegie Institute of Technology, Brit. Pat. 635,088 (1950), 652,935 (1951). Kiebler, M. W., U.S. Pat. 2,461,740 (1949). Howard, H. C., U.S. Pat. 2,555,410 (1951).

[360] Entel, J., Ruof, C. H., and Howard, H. C., *J. Am. Chem. Soc.*, **73,** 4152–8 (1951); **74,** 441–4 (1952); *Anal. Chem.*, **25,** 616–18, 1303–10 (1953). Entel, J., *Anal. Chem.*, **26,** 612 (1954). Ruof, C. H., Entel, J., and Howard, H. C., *Fuel*, **35,** 409–14 (1957).

[361] Montgomery, R. S., and Sienknecht, P. J., *Ind. Eng. Chem.*, **47,** 1274–6 (1955).

[362] Holly, E. D., and Montgomery, R. S., *Fuel*, **35,** 49–55 (1956).

[363] Holly, E. D., Montgomery, R. S., and Gohlke, R. S., *Fuel*, **35,** 56–59 (1956).

[364] Montgomery, R. S., Holly, E. D., and Gohlke, R. S., *Fuel*, **35,** 60–5 (1956).

[365] Montgomery, R. S., and Holly, E. D., *Fuel*, **36,** 63–75 (1957).

of the water-soluble acids. It was deduced that the part of the coal converted to carbon dioxide was similar to that converted to soluble acids, but somewhat more aliphatic in nature. The nuclei of the acids studied amounted to 23% by weight of the total coal. Biphenyl nuclei could be synthesized during the oxidation of a low-rank coal, but this seems less likely for the Pocahontas coal studied.

The copper salts of the acids were also subjected[366] to thermal decarboxylation which allowed identification of nitrogen-containing nuclei, such as phenanthridine, quinoline, pyridine, and especially phenylpyridine. Less complete decarboxylation and much greater yields of oxygenated nuclei were obtained than with the copper-quinoline method.

Other work on the production of alkali-soluble acids by oxidation in alkaline medium has been reported.[183,367,368]

Paper chromatography of water-soluble acids from Grosskinsky's method of treatment of coals[369]—using ethanol as solvent—and quantitative determinations of constituents[370] have been described. Identification by solvent fractionation of benzenecarboxylic acids made by oxidizing coal in air at 230 to 270°C has also been described.[371] The separation of isomeric phthalic acids was improved by double chromatography.[369]

Products and utilization. In a general paper, Grosskinsky reviewed the results of many years work on coal oxidation by his group at the laboratories of the Gesellschaft für Kohlentechnik.[372] The influence of reaction time, temperature, concentration, and humidity of the oxygen, and the size, rank, and layer thickness were discussed. Much of this review was concerned with carbonaceous products having properties similar to those of wood charcoal. Liquid phase studies were also reported in which nitric acid was used to oxidize: (1) coal, (2) coal preoxidized in air, and (3) coal pretreated with sulphuric acid. The resulting "primary" carboxylic acids were soluble in alkali; about half of the product was soluble in ether and the remainder in ethyl acetate. These products were nonvolatile carboxylic acids including benzenecarboxylic acids with more than two carboxyl groups in the molecule. The acids were found useful in the production of varnishes, softeners, and detergents.

Wet oxidation of coal with the object of producing useful products has also been described.[373] It was assumed that wet oxidation is necessary if acidic products of low molecular weight are to be produced, and two coals of different rank were oxidized with nitric acid under pressure. The primary products were subsequently decomposed under pressure with alkali at 420°C. The effect of rank, which was not great, and preoxidation in air on the products were investigated. The primary products contained some 20% of benzenecarboxylic acids, 2 to 3% of aliphatic acids, and about the same amount of picric acid, together with a larger amount of residue of high molecular weight. Several patents have been taken out.[374]

Oxidation of a Northumberland coal with dilute nitric acid and with sodium hydroxide and oxygen under pressure yielded humic acids close to the formula $C_{12}H_8O_4$, which were condensed with glycols and diamines to give hard, black, infusible resins which could be plasticized and filled in the usual manner.[375]

A study of the oxidation of coal in admixture with soil has shown definite acceleration in sunlight and no correlation between rate of oxidation and the bacterial count.[376] This evidence has been used to support a photochemical basis for the reaction.[376] The addition of small amounts of phosphates also

[366] Montgomery, R. S., and Holly, E. D., *Fuel*, **36**, 493–500 (1957).

[367] Durie, R. A., and Sternhell, S., *Australian J. Appl. Sci.*, **9**, 360–9, 370–4 (1958).

[368] Kamiya, Y., *Kôgyô Kakaku Zasshi*, **59**, 197–202 (1956); *Chem. Abs.*, **51**, 9124 (1957).

[369] Beck, K. G., *Brennstoff-Chem.*, **35**, 275–9 (1954).

[370] Benning, A., *Brennstoff-Chem.*, **36**, 120 (1955).

[371] Germain, J. E., and Valadon, F., *Bull. soc. chim. France*, **1958**, 1415–17.

[372] Grosskinsky, O., *Glückauf*, **86**, 988–95 (1950).

[373] Jüttner, B., *Brennstoff-Chem.*, **37**, 70–4 (1956).

[374] Jüttner, B., Ger. Pat. 743,225 (1943). Bergwerksverband zur Verwertung von Schützrechten der Kohlentechnik G.m.b.H., Brit. Pat. 705,862 (1954), 735,885 (1955), 797,726 (1958); Ger. Pat. 861,247 (1952), 870,998, 892,895 (1953).

[375] Thomson, G. H., *J. Appl. Chem. (London)*, **2**, 603–8 (1952).

[376] See e.g.: *Proc. Natl. Inst. Sci. India*, **21A**, 273–8 (1952); *Bull. Natl. Inst. Sci. India*, No. 2 (1953); *Nature*, **183**, 513–14 (1959).

affects the rate, and moreover a similar reaction proceeds reasonably quickly in the dark. There is some evidence that addition of coal with phosphates can improve the fertility of tropical soils. Much work has been done on the addition of coal-derived humic acids to soil, but evidence for a resulting increase in fertility due to the carbonaceous components is conflicting. There is no doubt that increases in crop yields have been observed with dressings of this nature, for example, air-oxidized coals, but the nitrogen content of the coal can seldom, if ever, be ruled out as the factor responsible.

Hydrolysis

A study of hydrolysis of a predominantly bright coal from Wyoming of 77% C content with 5*N* sodium hydroxide at 350°C for 24 hr has been reported.[377] Fractions were obtained by distillation and solvent extraction of the products. The solid residue contained much less oxygen than the coal. The products obtained, expressed in weight percent of the ash-free coal, were: gases, predominantly hydrogen, methane, and ethane, 2.8%; liquid phenols, molecular weight 90 to 180, 3.0%; solid phenols, molecular weight >300, 5%; fatty acids, average equivalent weight 100, 13.0%; nitrogen bases, 0.7%; ammonia, 0.5%; hydrocarbons, molecular weight 100 to 400, 15.6%; carbonates as carbon dioxide 22.0%; and insoluble residue 28.3%. Compounds identified were phenol, and acetic, propionic, butyric, and succinic acids; probably present were caproic and lauric acids, cresols and xylenols, and alkylated or partly hydrogenated cyclic hydrocarbons containing one or more condensed rings. Elemental balances showed that the overall reaction corresponded to the addition of water to the coal. It was thought that the fatty acids might have been derived from alcohols by oxidation, from ketones and aldehydes by oxidation-reduction, from anhydrides, or from esters. The products appeared to have been formed by hydrolysis, oxidation, and hydrogenation. The following possible reactions were postulated in addition to those already mentioned:

[377] Parker, F. G., Fugassi, J. P., and Howard, H. C., *Ind. Eng. Chem.*, **47**, 1586–92 (1955).

$$R'OR + H_2O \xrightarrow{NaOH} ROH + R'OH$$

$$RCH_2CH{=}CHR' + H_2O \xrightarrow{NaOH} RCH_3 + R'CH_2CHO$$

$$C + 2H_2O \xrightarrow{NaOH} CO_2 + 2H_2$$

Treatment of bituminous and other coals with a 5% solution of potassium hydroxide in alcohol-benzene solution followed by extraction with monoethyl glycolate at its boiling point gave an extract from a gas coal that consisted of a neutral portion together with a large amount of acid similar to humic acids but containing more carbon and hydrogen; coals of higher rank were comparatively inert.[378] Potash with an alcohol-pyridine mixture yielded about 40% of the coal in solution.[379] After four or six successive extractions, 80% went into solution. The proportion of neutral product to acidic product was characteristic of the type of coal.

Treatment of brown and subbituminous coals with aqueous or alcoholic alkali at about 190°C for 12 hr yielded alcohol-soluble products that contained more hydrogen and less oxygen than the original coal, although they possessed fairly similar infrared spectra.[380]

Miscellaneous

Work on the production of cation exchangers by treatment of coal with sulphuric acid has continued spasmodically.[242,381] The use of Alabama mine wastes has been investigated as a raw material for making granular cation exchangers; the coal was activated by treatment with concentrated sulphuric acid followed by washing, and autoclaving with dilute sodium hydroxide. The highest capacities realized for hydrogen-ion exchange were 7000 to 7500 grains/cu ft. The high loss of fines during activation seemed to preclude commercial development.[382] In a study of the

[378] Zabavin, V. I., Kleimenova, L. A., and Surkova, V. L., *Trudy Inst. Goryuchikh Iskopaemykh, Akad. Nauk S.S.S.R., Otdel. Tekh. Nauk*, **5**, 48–56 (1955).

[379] Zabavin, V. I., and Kleimenova, L. A., *Izvest. Akad. Nauk S.S.S.R., Otdel. Tekh. Nauk*, **8**, 72–7 (1957).

[380] Brooks, J. D., and Sternhell, S., *Fuel*, **37**, 124–6 (1958).

[381] Goedkoop, M. L., U.S. Pat. 2,735,819 (1956).

[382] Broderick, S. J., and Bogard, D., *U.S. Bur. Mines, Rept. Invest.* No. 3559 (1941), 17 pp.

interaction of various bituminous coals with sulphuric acid on a semi-industrial scale, ion exchange capacity was an optimum, 1.5 to 2 meq/g, after treatment with a concentration of 76 to 82% sulfur trioxide, and so was the mechanical strength.[383] Roy found that good results could be obtained by treating coal with sulfuric acid in the ratio of 4 cc/g and heating to 100°C in the presence of a catalyst such as mercurous oxide.[384] He determined sulfonyl and carboxyl concentrations in the products as well as the exchange capacity. The effect of 100% sulfuric acid on lignite and bituminous coal has been studied at temperatures between 150 and 250°C.[385] The ion-exchange properties of the product were found to depend on the temperature and duration of treatment.

Several extensive studies have been made of the interaction between sulfur and coal on heating.[127,386] The products of the reaction of two types of coal with abietic acid have also been investigated.[387]

APPRAISAL AND SYNTHESIS OF EVIDENCE

Several papers have been published that deal with structural models of the coal molecule or means for their development and assessment. A model has been proposed based on a polycyclic aromatic core—ovalene was taken as typical—with a "methylene system" comprising the nonaromatic constituents.[388] In addition to methylene groups, the methylene system was assumed to contain carboxyl, hydroxyl, and carbonyl constituents. This model was used in considering the changes that occur with change in rank. The coal molecule has also been pictured as a partially hydrogenated aromatic structure containing five to eight rings interlinked by carbon-oxygen bonds.[389] Although these models correspond in principle to those still favored by many investigators, details such as size and layout have been completely superseded following the acquisition of more detailed quantitative information.

The structure assigned to coals must accord with their elementary analyses. It is tempting when interpreting physical techniques to overlook this factor in the structural model developed. In fact, it is difficult for an X-ray crystallographer and an organic chemist to find common ground when discussing a substance like coal. The very concept of an organic molecule is foreign to the thought of the crystallographer, who is concerned mainly with the spatial arrangement of atoms.

The work of van Krevelen[2] brought together Gillet's chemical approach[99,390] and Waterman's physical analysis methods for oils. Although recent work has introduced many more numerical parameters to be satisfied by a coal model, the analytical one remains of fundamental importance. The most characteristic feature of the analytical requirements is that a small amount of hydrogen has to satisfy a large number of potential carbon valencies.

Among the arguments advanced by Howard[163] for the cyclic structure of bituminous coals were the following: (1) degradation products of coal, even from reactions at not more than 100°C, are predominantly cyclic; (2) the ease of removal of hydrogen is evidence for the presence of alicyclic rings; (3) the C/H ratio is high; (4) mild hydrogenation produces condensed cyclic structures that depend on the rank of coal but not on the temperature, pressure or catalyst, which suggests that they exist as such in the coal. The butyl esters of the water-soluble acids produced by oxidizing coal can be shown to contain nuclei with a C/H ratio of about 1.05. The specific refractions in relation to molecular weight of these esters show that they have nuclei more complex than benzene, and their color is also evidence in favor of condensed

[383] Jurkiewicz, J., Zielínski, H., and Laskowska, H., *Prace Glównego Inst. Górnictwa, Ser. B, Komun.* No. 151 (1954), 12 pp.

[384] Roy, M. M., *J. Appl. Chem. (London)*, **6,** 335–41 (1956).

[385] Rasajski, S. N., and Dzokic, D., *Glasnik Khem. Drushtva, Beograd*, **22,** 333–43 (1957).

[386] McCulloch, A., *Chem. & Ind. (London)*, **1945,** 374. Berkovitch, I., and McCulloch, A., *Fuel*, **25,** 36–41, 69–78 (1946). Mazumdar, B. K., Choudhury, S. S., Chakrabartty, S. K., and Lahiri, A., *J. Sci. Ind. Research (India)*, **17B,** 509–11 (1958).

[387] Zabavin, V. I., *Izvest. Akad. Nauk S.S.S.R., Otdel. Tekh. Nauk*, **1947,** 871–80.

[388] Huck, G., and Karweil, J., *Brennstoff-Chem.*, **34,** 97–102, 129–35 (1953).

[389] Orchin, M., *Ohio State Univ., Eng. Exp. Sta. News*, **25,** No. 5, 25–31 (1953).

[390] Gillet, A., *Bull. soc. chim. belges*, **57,** 298–306 (1948).

nuclei. Howard's paper represented an important step in considering all available numerical evidence together with qualitative evidence from the behavior or properties of coal. This type of approach was further extended by taking into account the newer findings of infrared spectroscopy,[391] ultrafine structure, and the heat of wetting measurements.

In 1958, Dryden[151] contributed a chemical reassessment of the quantitative evidence related to the coal molecule. It was directed to: (1) assessment of the self-consistency of structural data, (2) interconversion of structural parameters, and (3) selection of a self-consistent set of parameters lying within the experimental errors of all the available estimates. It is important to understand the limitations of this approach. The average coal molecule was assumed to have random construction, within certain limits, in respect of size of aromatic nuclei, types of internuclear linkage and nonaromatic structure, and content of functional groups. The aromatic nuclei were termed "clusters," the aromatic nuclei together with nonaromatic substituents "cluster units," and the molecule was assumed to consist of several cluster units linked together not linearly but in raft-like formation. Only 6-carbon, aromatic, rings were considered. Three independent parameters were needed to specify such a molecule: (1) an estimate of the number of fused rings in a nucleus R_a, or of the number of aromatic carbon atoms in the nucleus C'_{ar}, or of the "ring condensation index," which equals $2(R_a - 1)/C'_{ar}$; (2) a measure of the degree of aromaticity such as the fraction of carbon in aromatic structures f_a, or the ratio of aromatic to total hydrogen other than that in hydroxyl; and (3) the number of interlinkages per cluster unit B. The H/C ratio, after correction for functional groups, is an experimental parameter which is needed to interrelate the other three. Two of the three types of parameters can be determined by certain experimental techniques or can be estimated from statistical structural analysis of the physical properties; the third, B, can only be assumed, or deduced from the available evidence taken together with structural considerations such as bond angles. Whereas, for example, van Krevelen combined values of nuclear size deduced from refractrometric data with values of aromaticity from densimetric data, the new method enabled the two to be considered and compared independently, subject, however, to the assumptions made. The same treatment could be accorded separately to infrared and X-ray data.

van Krevelen's original equation relating ring condensation index with aromaticity and H/C ratio is dependent on the assumption that the number of fused rings per cluster is large compared with the value of B. By 1958, it had become clear that this was incorrect. An alternative approach was therefore adopted by Dryden. The principle was to take an average cluster unit, substitute hydrogen atoms for the intercluster linkages, the functional groups, and the nonaromatic substituents, and thus arrive at the composition of the parent aromatic hydrocarbon corresponding to an average aromatic cluster. This composition was expressed in terms of the ratio H'_{aru}/C'_{ar}.

For such a parent structure, entirely aromatic, a much simplified equation relating ring condensation index and H/C ratio holds true:

$$2(R_a - 1)/C'_{ar} + H'_{aru}/C'_{ar} = 1$$

Fortunately it was found that for the type of model assumed the relative proportions of nonaromatic hydrogen attached to chains, bridges, and alicyclic rings were of minor importance to the mathematical relations. The type of equation reached was then:

$$(H'_{aru} - B)/C'_{ar} = a - (a - H'_1/C')/f_a$$

where H'_1 = hydrogen atoms per cluster unit after replacement of oxygen-containing substituents by equivalent hydrogen. The constant "a" should range from 1.75, if all nonaromatic groups were alicyclic, to 2.0 if they were in the form of chains, and in the paper initial analyses were carried out by using a mean value of 1.875. An essential step at this stage was the use of theoretical curves for pure polycyclic hydrocarbons corresponding to two extreme types of aromatic condensation—catacondensation and pericondensation. For the raft-like structure, the limits of B were set for steric reasons at 1.5 to 2.5. Successive approximations allowed all

[391] Storch, H. H., *Proc. 2nd Conf. on The Origin and Constitution of Coal*, Nova Scotia Research Council, 1952, pp. 363–94. Storch H. H., *J. Inst. Fuel*, **28**, 154–63 (1955).

the main parameters to be deduced from whatever starting point was provided by the experimental results. Results from each experimental method required slightly different treatment based on these general principles.

Application of these methods to a few typical molecules of the type described earlier showed that the number of rings in the nucleus could be estimated fairly closely, with at most a 20% discrepancy from the true values; the calculated values were usually the lower. In such comparisons, however, errors can arise, depending on whether the H/C ratio and ring condensation index are themselves averaged for the various cluster units, or whether their components such as hydrogen content are first averaged and the ratios then calculated. Since for coals the observed quantities are averages, the parameters deduced will not necessarily be equal to the arithmetic means of the R_a values or the ring condensation indices for the individual clusters.

The general conclusion from this paper[151] was that measurements on coal by X-ray diffraction, infrared spectroscopy, and nuclear magnetic resonance could be reconciled at the cost of a moderate increase in the infrared extinction coefficient ratio ($\varepsilon_{al}/\varepsilon_{ar}$) assumed by Brown, but that those estimated by statistical structural analysis from density, refractive index, and heat of combustion could not. Of the second group, estimates from density fell closest to those in the first group. Objections can be raised to the assumed model in the light of more recent knowledge. A similar analysis could, however, be developed for a new and more appropriate set of assumptions; such a tool is of value in collating results.

The more recent work using high resolution nuclear magnetic resonance[154] has provided evidence that Brown's original estimates of H_{al}/H_{ar} and hence his choice of extinction coefficient ratio (2.0), were almost correct. The shortage of hydrogen in coal was the chief reason why the preferred values of H_{ar}/H_{al} had to be higher than Brown's in the analysis just described. This, with small numbers of rings in the nucleus and consequently a large proportion of peripheral aromatic carbon atoms, required either a high value of f_a or a higher value of B than the model could sustain. Economy of aliphatic hydrogen could also render the requirement less severe, and the nonaromatic H/C atomic ratio in coals might be less than 2 instead of slightly greater, as assumed in Dryden's analysis.

Given has found it possible to increase the value of B by developing a more regular or less random molecular model, and reduced the nonaromatic hydrogen requirements by incorporating more nonaromatic carbon in the form of —C—, and —CH groups in proportion to —CH_2 and —CH_3.[153,178] He also incorporated most of the aliphatic carbon atoms in hydroaromatic rather than aliphatic structures. The prototype of Given's structural unit is 9,10-dihydroanthracene and the average aromatic nucleus was taken to be equivalent to that of naphthalene. A more representative model would also need to show nuclei of one to three rings, and a few even larger. Such models, which conform approximately to all the known numerical properties of coals, are difficult to construct, and few appear to be possible. A typical coal molecule as envisaged by Given, for a vitrain of 82% C content, is illustrated in Fig. 13. It will be realized that a molecule of this kind is not planar, and that therefore an assembly of them would give rise to considerable entanglement, an ultrafine structure on the scale of a few Angstroms as observed, and a very limited degree of stacking between aromatic nuclei. In the course of developing this model, it was realized that structures projected on a plane could be misleading. Patient work with atomic models is necessary.

Nuclear magnetic resonance has shown no methylene bridges in the primary products obtained by vacuum distillation of coals.[154] These primary products resemble the parent coal in so many characteristics that the presence of methylene bridges in the parent coal is also unlikely. If they were present in the coal, it would be necessary to postulate their complete and selective destruction on heating to less than 400°C, and moreover a destruction that did not change the infrared spectrum appreciably. But the absence of methylene bridges invalidates Given's molecular model. If, however, the basic structure is based on dihydrophenanthrene instead of dihydroanthracene, the methylene bridges disappear and the problem is circumvented without changing the other characteristics of the model.

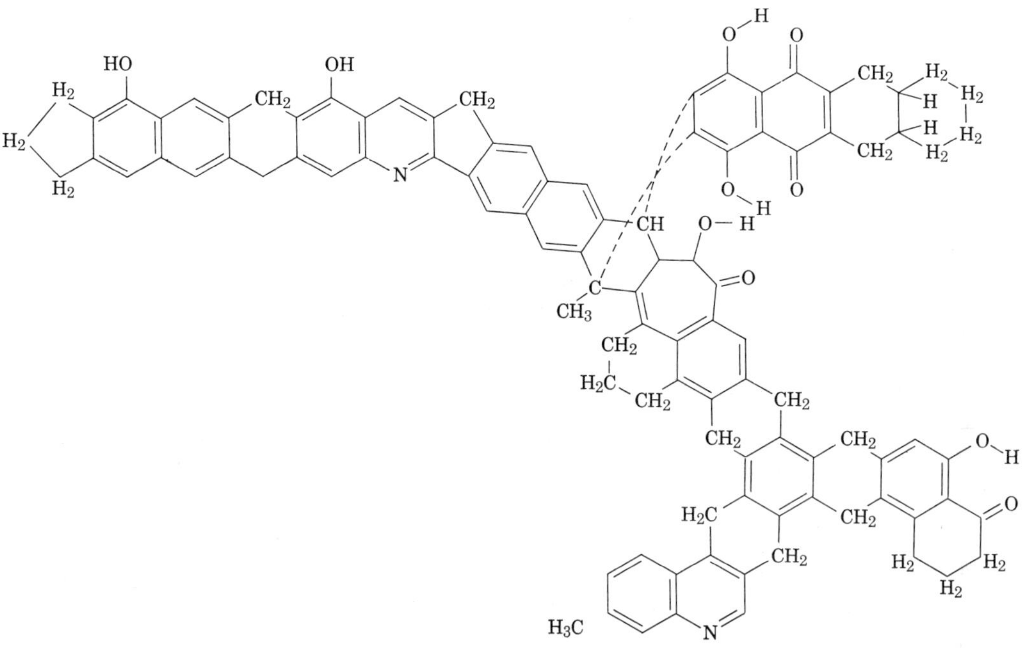

Fig. 13. Molecular structure for coal (82% C) proposed by Given.[153]

Qualitative knowledge of the carbon-hydrogen structure of the coal molecule has been given by several types of investigation. Oxidation has shown that increase in aromaticity, and formation of aromatic-hydroaromatic structures, accompanies increase in rank, and has suggested many of the types of aromatic cluster that may be present. Selective chemical reactions yield results that are in general agreement with the expected extent of substitution, etc. In the polarography of coal extracts, the frequent occurrence of half-wave potentials—for the reduction of the aromatic systems—between −1.95 and −2.2 v points to the presence of considerable proportions of biphenyl, naphthalene, phenanthrene, and triphenylene structures; the extent of reduction in this process suggests a minimum polycyclic aromaticity between 0.35 and 0.5. Methods of functional group analysis, and polarography at lower potentials (−0.5 to −1 v), have given fairly precise estimates of hydroxyl and carbonyl group concentrations that show these groups to account for 70 to 90% of the oxygen in bituminous coals. X-ray diffraction, infrared and nuclear magnetic resonance spectra, oxidation, and dehydrogenation, when combined with ultimate analysis, have given further, and quantitative, information about C—H structure. This information is summarized in Table 6 for vitrinites at two levels of rank; an indication is given, where known, of the difference to be expected in the corresponding exinite and micrinite, that is, whether the parameter for these is greater ($>$ v) or less than ($<$ v) that for the vitrinite.

The need to conform to the figures in Table 6 excludes probably all of the molecular models proposed for coal up to the present with the exception of the type put forward by Given (Fig. 13), after modification from a dihydroanthracene to a dihydrophenanthrene basis and adjustment of the number of methyl groups. One consequence is worth noting: that the aromatic hydrogen content amounts to approximately one atom per aromatic ring. This consequence goes far to explain why many reactions of aromatic substances, such as sulphonation, are observed only to a limited extent with coal. It is easy to see from a model such as Given's why coal must exhibit an irregular and amorphous packing, resist the action of solvents unless first swollen, and readily aggregate in solution even at low concentrations. The irregular packing of molecules can also satisfactorily account for the observed ultrafine structure on the scale 5 to 40 A, although a quantitative comparison would be difficult or impossible. The reason

TABLE 6

ESTIMATED AVERAGE PARAMETERS FOR COAL[392]

Parameter	Vitrinite		Exinite	Micrinite
% C, pure coal (Parr basis)	82.5	90	~v	>v
% volatile matter	39	24.5	>v	<v
H/C (atomic)	0.76	0.65	>v	<v
H_{ar}/H_{al}	0.23	0.54	<v	>v
H_{CH_3}/H_{al} (IR, oxidation)	0.21(0.1–0.3)	0.23(0.1–0.3)	—	—
$(H/C)_{al}$ (atomic)	~1.65	~1.6	—	—
CH_2 bridges: C_{ar}—CH_2—C_{ar}	Absent	Absent	—	—
% substitution on aromatic periphery	50+	30	—	—
R_a (aromatic rings per cluster)	$\ngtr 3$	~4	$\lesssim$v	$\gtrsim$v
C'_{ar} (aromatic C atoms per cluster)	$\ngtr 13$	16	~v	~v
Average stacking number of parallel aromatic clusters	1.3	1.8	—	—
O as hydroxyl, % of coal	7.0	2.0	<v	>v
O as carbonyl, % of coal	2.3	0.9	—	—
Carbon atoms per free radical	~8000	~3000	<v	>v
f_a (fraction of carbon as aromatic) derived from above	0.69–0.70	0.79–0.80	<v	>v

why very few aromatic units, on the average, line up in parallel stacking normal to their plane also becomes obvious.

The distribution of sizes of aromatic clusters or nuclei is also not shown in the present molecular diagram. The nuclei of the water-soluble acids from coal, produced by oxygen in the presence of alkali, represent 23% of the weight of original coal, are probably not dissimilar from those in the equal weight of coal converted to carbon dioxide, and have been shown to consist of single and double rings in about equal proportions by weight together with 3 to 4% of triple rings (phenanthrene, not anthracene). Ring fission during the production of the acids may have reduced the proportion of larger rings. Evidence from polarography has suggested that certain nuclei containing one, two, and three rings play an important part in coal structure. X-ray histograms suggest that about half of the carbon in the nuclei is almost equally distributed between one, two, and three rings, but these estimates may be biased toward the larger ring systems. Comparison with the figures in Table 6, based on X-ray averages, also suggests that nuclei with more than three rings do play an appreciable part in the coal structure. At this stage perhaps the best conclusion is that nuclei with one to three rings are almost equally important, and that there are appreciable proportions of larger clusters, diminishing in amount as their size increases.

The only evidence regarding the number of intercluster linkages per cluster unit (B) is that adduced by Montgomery and Holly,[365] who set B at 3.3, the average functionality of the water-soluble acids from coal. This value, however, can include any aliphatic side chains that terminate without linkage to a further nucleus.

Solvent extraction yields little information about the coal molecule, apart possibly from a molecular weight distribution. If solvent extracts contain more than 1 to 5% of the parent coal, they resemble it closely, provided they are prepared below 250°C; if prepared above 250°C, pyrolysis has clearly modified their constitution. Solutions of solvent extracts of coal may exhibit molecular association even at concentrations below 0.1%. The best specific solvents for coal contain a nitrogen atom with a readily available lone pair of electrons, and invariably swell the coal. There are other reasonably good solvents, effective at room temperature, for which this explanation

[392] Dryden, I. G. C., *Fuel*, **41**, 55–61, 301–4 (1962).

does not appear adequate; for example, dimethylformamide and methylcyclohexanone. Some mixtures, particularly of the formamides with the ketones, are much more effective solvents for coal than are their components. The trend of content of soluble matter with rank depends on the solvent; with some, such as pyridine, it exhibits a maximum in the bituminous range, and with others such as ethylenediamine it falls steadily with increasing rank through this range. Most coals containing more than 90% carbon are soluble only to a very small extent in any solvent. It seems to have been established that the molecular weights of coal extracts lie in the thousands, although aggregation can readily lead to apparently much higher values and to particles visible in the ultramicroscope.

From the point of view of variety and interest of products, hydrogenation, extensive oxidation, hydrolysis, pyrolysis, and fluorination are the most interesting reactions of coal. But these reactions furnish only limited information about the structure of the coal itself, because degradation is severe and secondary reactions undoubtedly take place.

CHAPTER 7

STORAGE, HANDLING, AND TRANSPORTATION

W. S. LANDERS and D. J. DONAVEN

Bureau of Mines, U.S. Department of Interior, Denver, Colorado.

STORAGE OF COAL

Every nation involved in World War II realized the importance of the proper storage of coal as a conservation measure to alleviate some of the burden on transportation systems and to provide some continuity of supply to consumers. The sharp rise in quantities stored in the United States during the war is shown in Fig. 1.[1] The trend from a low at the end of the war has been toward more coal in storage.

Storage Techniques

Various techniques have been developed for the storage of coal to minimize coal deterioration and prevent spontaneous combustion during storage. Considerable information about storage techniques has been summarized in a manual[2] that also discusses certain aspects of handling, freezeproofing, suppression of coal dust, and the technique of inventory of coal stored in piles.

Experimental storage of a subbituminous 2-in. slack coal containing 24.1 % moisture in concrete bins, $25\frac{1}{2}$ by 13 by 7 ft, was performed by Parry and Goodman.[3] They compared three types of storage; exposed to the weather, in a ventilated bin, and in a closed bin. Only minor changes in the physical and chemical properties were noted for the coal stored in the closed bin. After thirty-five days' storage, the reduction in average size of the coal in each sample was: coal exposed to weather, 75 %; coal in ventilated bins, 17 %; and coal in closed bin, 5 %.

Domestic storage of subbituminous lump coal and its performance in a hand-fired furnace was studied by Landers and Parry.[4] The coal was placed in two 5-ton benches in a tight storage bin and stored for 8 months. The authors reported that there was no significant physical or chemical change, and that the stored coal performed satisfactorily in a hand-fired furnace.

Storage in Stockpiles

Taylor has recommended the exclusion of air by retarding circulation within the pile, because adequate ventilation requires a number of perforated pipes projecting vertically and horizontally from the pile.[5] He stated that providing a sufficient ventilating system may be uneconomical. Ventilation may be retarded by packing coal uniformly, avoiding segregation, and blanketing the pile with fine coal. Taylor discussed temperature-time curves of the conditions within a pile which show a slow steady rise and then an accelerated heating rate to indicate an incipient fire. The "safe" temperature limits, 35 to 70°C vary with the coal.

Successful storage of subbituminous coal and lignite by compaction of large piles has been reported by Gerow.[6] The procedure with nut and slack was to prepare the pile in five 2-ft layers, each layer compacted by a bulldozer or truck before the next layer was added. Particular attention was given to avoiding loose coal around the edges of the pile. Gerow noted that the exterior of the pile

[1] *U.S. Bur. of Mines, Minerals Yearbook*, 1940 to 1960.

[2] Anon., *Power Eng.*, **60,** S–1–24 (1956).

[3] Parry, V. F., and Goodman, J. B., *U.S. Bur. Mines, Rept. Invest.*, No. 3587 (1941), 10 pp.

[4] Landers, W. S., and Parry, V. F., *U.S. Bur. Mines, Rept. Invest.*, No. 3759 (1944), 17 pp.

[5] Taylor, R. A. A., *J. Inst. Fuel*, **14,** 144–7 (1944).

[6] Gerow, C., *Modern Power and Eng.*, **38,** 50–1, 58 (1944).

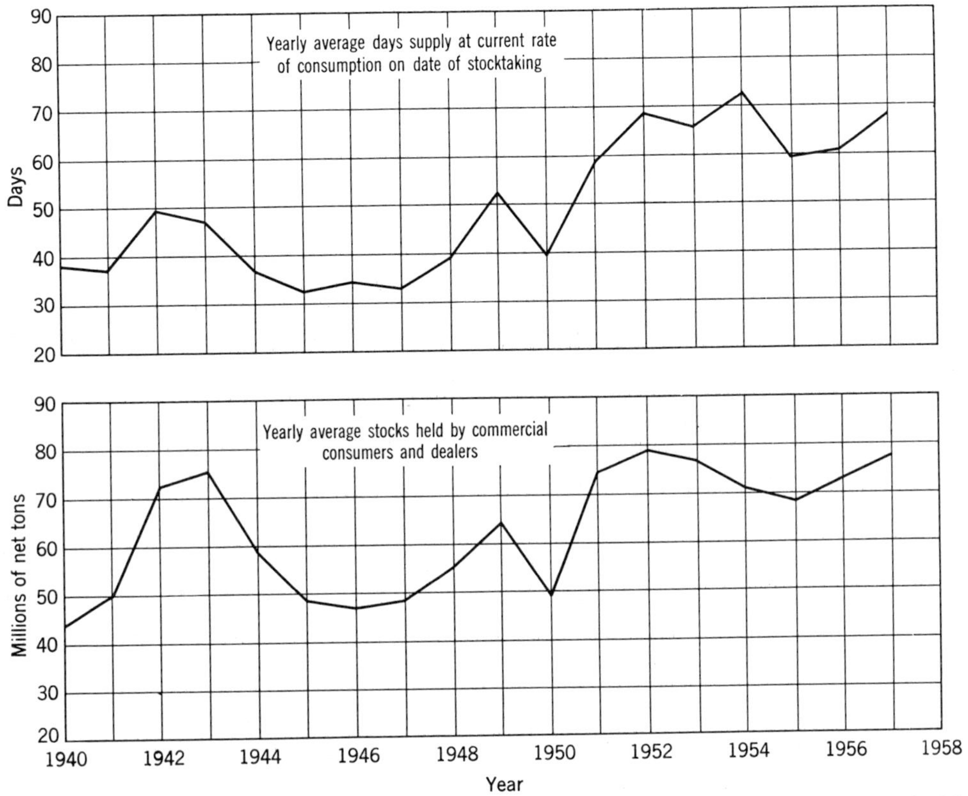

Fig. 1. Trends of stocks and equivalent consumption of bituminous coal and lignite in the United States, 1940–1957.[1]

rapidly forms a seal of disintegrated coal about 3 in. thick.

Taylor suggested that wetting a coal pile might assist in controlling the temperature rise through the critical period.[5] Hegedus stored four Hungarian brown coals from March to December in cones covered with a 10 cm layer of straw. The coals were sprayed periodically with cold water. He found that untreated coal decreased from 6 to 20% in heating value, whereas treated coal decreased 0.4 to 6%.[7]

The technique of storing coal by depositing it in thin layers and then compacting by earthmoving equipment has been successfully practiced by the Central New York Power Corporation at the Oswego Steam Station. The procedure reported by Ewart consisted of spreading the coal in layers 13 ft wide, 40 ft long and 10 in. thick by bulldozer carryall units.[8] Subsequent trips pack the layers. The coal stocked, purchased from as many as sixty mines, had a volatile content from 17 to 37%, and the sulfur content ranged from 1 to 3%. The size of the coal in the Oswego pile varied from fine slack to double-screened and run-of-mine. The pile has contained as much as 560,000 tons of coal, and some sections have been in place for as long as 3 yr; there have been no fires. Test pit results reported by Ewart indicated densities of 80 lb/cu ft.

Hurysz has suggested that hydraulic transport to the pile results in greater compaction.[9]

A study of storage procedures for low-rank coals was made by Allen and Parry.[10] The study was made on asphalt capped piles and on uncapped, compacted piles. The authors made observations on six asphalt capped subbituminous *C* slack and subbituminous *A* slack coal piles. Combustion eventually occurred in each pile. The cause of the pile failure was considered to be improper formation, resulting

[7] Hegedus, L., *Mezógazdaság és Ipar*, **3**, No. 5, 11–15 (1949).

[8] Ewart, J. N., *ASME, Trans.*, **72**, 435–7 (1950).

[9] Hurysz, Y., *Przeglad Górniczy*, **5**, 32–5 (1949).

[10] Allen, R. R., and Parry, V. F., *U.S. Bur. Mines, Rept. Invest.*, No. 5034 (1954), 31 pp.

in numerous air channels caused by segregation and lack of compaction. In most of the asphalt capped piles, compaction was neglected or confined to unidirectional traverse of the pile by a bulldozer in such a manner that the sides were not compacted. As a result of the information gained by observing these unsatisfactory storage piles, Allen and Parry recommended the following procedures: (1) build up the piles in increments not exceeding 1 ft in depth; (2) thoroughly compact each increment before placing additional coal on the pile, this compaction being applied to the entire pile top, sides, and edges; (3) maintain the side slopes at 14° or less to prevent segregation and to facilitate the above compaction; (4) smooth the pile surface to minimize differential pressures due to wind; and (5) omit the asphalt capping applied to the previous piles.

Two storage piles were formed according to these specifications, each containing about 5000 tons of subbituminous *A* slack coals from Hanna, Wyoming. The packed density of these piles was computed at 63.2 to 70.4 lb/cu ft. These two piles were in place approximately 5 yr at the time of the authors' paper, and at no time had there been any indication of abnormal heating. Temperature survey stations were established in the piles. Periodic temperature readings were taken from thirty-two stations on approximately 35-ft centers at a depth of 3 ft, and gas samples were taken from the same locations at a 4-ft depth. The results of the temperature and gas analysis are summarized in Fig. 2.

Allen and Parry served as advisers during the construction of stockpiles that eventually stored 4 to 6 million tons of lignite excavated from the site of the Garrison Dam, North Dakota, and recommended the procedures developed from their previous studies. The operation was supervised and contracted by the Corps of Engineers. The strip-mined

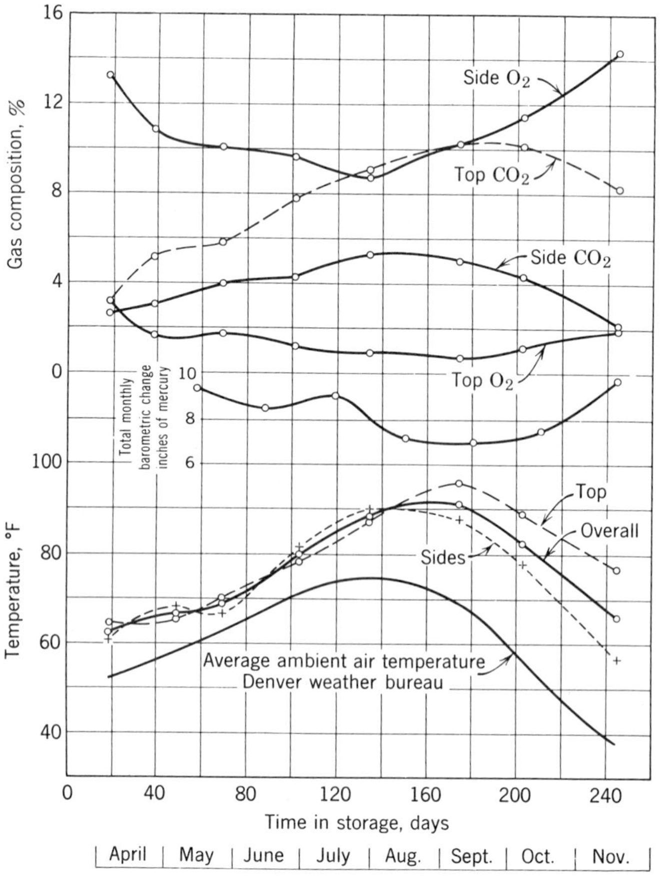

Fig. 2. Trends of average temperatures and gas analysis, Brighton, Colorado Storage pile.[10]

Fig. 3. Typical view of the recommended pile-forming techniques, as practiced at the Garrison Dam, N. Dak. Lignite dumped from Euclids is being leveled by bulldozer (right background), crushed and compacted by chopper and sheepsfoot roller (foreground), and smoothed by bulldozer (center).

lignite ranged in size from fines to lumps that had to be broken with the 1.5- and 2.5-cu yd shovels before they could be loaded into the 13-cu yd bottom dump Euclid carryalls for transportation to the pile sites.

The final procedure developed by the contractor, principally to facilitate unloading the Euclids, was to crush the lignite with a sheepsfoot roller after leveling with a bulldozer. Some six to ten passes of the roller were used after each 9-in. lift was in place. The immediately apparent advantages were that: (1) segregation and concentration of lumps were entirely eliminated; (2) void space was obviously reduced, thereby increasing the packing density of the stored lignite; (3) the reduction in voids and resultant suppression of air passages would restrict the movement of air within the pile; (4) the improved pile surface facilitated the movement of the heavy equipment over the entire pile surface, thereby permitting more rapid pile formation; (5) reclaiming of the lignite would be easier, and less crushing would be necessary when the fuel was removed for use in the local steam plant; and (6) the size reduction would be a distinct advantage if the lignite were ultimately removed for industrial use.

One additional piece of equipment was found to expedite the disintegration of large lumps. This was a "chopper," similar to a sheepsfoot roller, except that it was designed for splitting and breaking rather than compaction. Figure 3 is a view of the pile-forming techniques developed at Garrison Dam. The bulk density of the lignite in a pile as shown by churn drill tests, was 74 lb/cu ft. Observations made after 12 and 20 months of storage showed the average temperatures and gas analyses from a number of stations in the pile to be:

Location	Temp. °F	CO_2, %	O_2, %
Top, (12 mon)	65.9	10.0[11]	0.5[11]
Top, (20 mon)	61.5	12.0[11]	0.3[11]
Sides, (12 mon)	74.6	7.3	5.4
Sides, (20 mon)	65.5	11.2	2.4

Storage in Pits

Storage of coal in pits was studied by Goodman, Parry, and Landers.[12] The authors found that for safe storage, segregation should be avoided when placing the coal in layers. The layers should not exceed 3 ft in depth and

[11] 3.5-ft depth.

[12] Goodman, J. B., Parry, V. F., and Landers, W. S., *U.S. Bur. Mines, Rept. Invest.*, No. 3915 (1946), 37 pp.

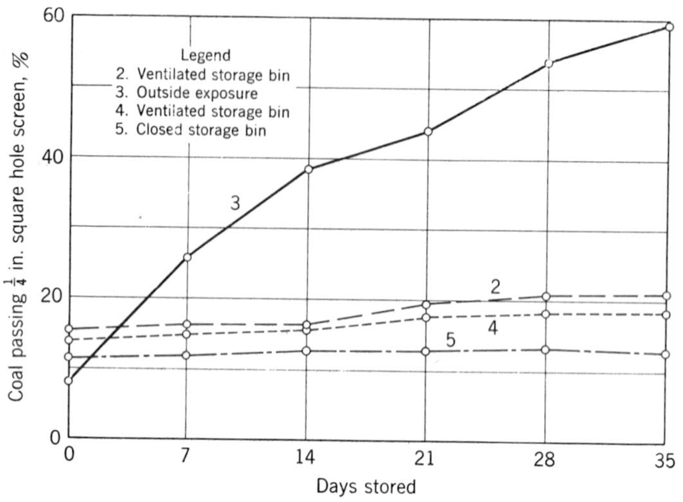

Fig. 4. Size degradation during storage as shown by passage through a $\frac{1}{4}$-in. square hole screen.

should be carefully leveled and compacted. A bulk density as high as 56 lb/cu ft was obtainable with the Colorado subbituminous *B* slack coal used.

Under-water storage of Illinois coking coal was studied by Jackman, Eissler, and Reed.[13] This method effectively prevented oxidation.

Effects of Storage on Coal Properites

Storage procedures should be applied that, in addition to preventing spontaneous combustion, will retain properties of the coal sufficient for the end use of the coal. With low-rank coals, the principal effects of storage are loss of strength, degradation, and some loss of heating value.

Parry and co-authors, in the series of papers already cited, examined some of the properties of Western low-rank coals stored by various techniques. The degradation of Colorado subbituminous *B* coal, originally minus 2-in., stored in bins, is shown by Fig. 4.[3] The storage procedures were: a ventilated bin, a closed bin, and exposure to the weather but protection from rain.

Friability tests were made on fresh and stored coals to measure the effect of storage on strength. The measured friability was 16.1% on a fresh sample. After ten days' storage in a ventilated bin, the friability increased to 21.6, and at twenty days it was 38.4%.[3] When lump subbituminous coal was stored under domestic conditions for 8 months, Landers and Parry found that the friability increased from 21% to 26%.[4] The friability of pit-stored coal was measured by Goodman, Parry, and Landers.[12] During eight months of pit storage, the Colorado subbituminous *B* coal used showed an increase in friability from 23.6% to 30.2% based on a complete sample. Samples from the top 18 in. of the pit after about five months of storage gave a friability of 30.8%, whereas the coal at a depth of 20 to 38 in. had a friability of 27.3%.

A series of tests designed to measure retention of physical properties by lignite stored under seven conditions of decreasing severity was described by Eckhardt and Yates.[14] Meteorological information related to the 160 day storage period was also reported. The data after 160 days storage are given in Table 1. The results of this series of tests indicate that favorable storage conditions were generally those that permitted control over moisture-absorption cycles. The loss of heating value of the coal was generally not great. Some values found with Western low-rank coals stored by various methods are shown in Table 2.

Effects of Storage on High-Rank Coals

Bituminous coals are generally stored to provide reserve supplies either for power

[13] Jackman, H. W., Eissler, R. L., and Reed, F. H., *Illinois State Geol. Survey, Circ.*, No. 227 (1957), 22 pp.

[14] Eckhardt, C. J., Jr., and Yates, C. W., *ASME, Trans.*, **65,** 829–36, 1943.

TABLE 1

PHYSICAL PROPERTIES OF LIGNITE STORED 160 DAYS

Class of Storage	Size Degradation, %	Size Stability, %	Reduction in Moisture Content, %	Friability, %
Atmospheric	81.1	*		64.9
Atmospheric, protected from sunshine	70.0	48.8		42.4
Atmospheric, protected from rain	28.2	59.9	21.0	36.5
Atmospheric, protected from sunshine and rain	17.1	58.8	21.0	35.5
Stored in basement	9.6	69.2	18.7	32.6
Sealed in metal cans	8.8	81.0	1.3	21.1
Submerged under water	5.2	87.5		24.7
Sample at beginning of test	3.2	88.3		15.7

* Not possible to secure an appropriate sample after 84 days.

generation or coke production. Minor changes that may occur in the physical properties of such coals during storage have little or no effect on their end uses. Oxidation of coal stored for power plant use reduces its value but does not preclude this use. Coal intended for coke production can, however, deteriorate through oxidation in storage to such an extent that it is unusable as charge material to the coke ovens. The extent and rate of oxidation of bituminous coal in storage varies widely from coal to coal and is greatly affected by the method of storage.

The weathering of coke-blending coal was studied by the Illinois State Geological Survey.[13,15] These tests utilized a 500-lb experimental slot oven. The first of the two

[15] Reed, F. H., Jackman, H. W., Rees, O. W., Yohe, G. R., and Henline, P. W., *Illinois State Geol. Survey, Bull.* No. 71 (1947), 132 pp.

TABLE 2

HEATING VALUES OF SOME WESTERN COALS BEFORE AND AFTER STORAGE

Gross Heating Value, Btu/lb, m.a.f.		Storage Time	Storage Method, Size	Coal, Data Source*
Before Storage	After Storage			
13,190	13,160	35 days	Ventilated concrete bin,	Colorado
13,190	13,100	180 days	Closed concrete bin	Subbituminous *B*;[3]
			2 × 0 in.	
13,270	13,060	8 months	Domestic bin	Colorado
			10 × 1½ in.	Subbituminous *B*;[4]
13,350	13,190	8 months (approx.)	Open pit, entire sample	Colorado Subbituminous *B*;[12]
	13,130	66 days	0–2 in. depth	
	12,640	137 days	0–2 in. depth	
	13,250	66 days	0–18 in. depth	
	13,230	137 days	0–18 in. depth	
	13,280	66 days	20–38 in. depth	
	13,300	137 days	20–38 in. depth	
			2½ × 0 in.	
12,250	11,980	2 yr	Compacted pile No. 5	Garrison Dam North Dakota, lignite[10]

* Footnotes refer to references in text.

TABLE 3

BLENDS OF 80% WEATHERED ILLINOIS NO. 6 SEAM COAL AND 20% POCAHONTAS COAL*

Pocahontas Coal, Gieseler Fluidity, Maximum Dial Div/Min	Age of Illinois Coal Since Mining—Days	Coke Properties: +2 in, Shatter, %	Tumbler Stability, % + 1 in.	Tumbler Hardness, % + ¼ in.	Average Size, inches	Breeze, — ½ in. % of coal	Apparent Specific Gravity
1224	13	61.8	48.7	66.6	2.24	3.5	0.808
	72	63.1	47.9	65.8	2.35	2.8	0.785
	132	61.5	46.0	66.2	2.34	3.0	0.798
	198	66.2	48.2	66.4	2.33	3.1	0.779
	258	69.3	42.8	62.3	2.46	4.2	0.785
13.8 average	31	63.3	49.4	67.2	2.43	3.0	0.809
	62	71.8	47.1	66.5	2.42	3.5	0.774
	94	63.5	48.4	64.8	2.51	4.2	0.756
	122	65.4	42.5	59.8	2.49	6.1	0.823
	153	69.2	37.0	51.5	2.37	9.5	0.806

* Washed, 2 by ⅜-in. Illinois No. 6 Seam was stored in conical piles of 2 to 4 tons.

publications covered some preliminary work with Illinois No. 6 Seam coal, generally considered unsuitable for stockpiling because of the gradual loss of coking properties, especially in the fine-coal sizes. The authors performed four series of tests with storage in roofed but otherwise open bins, small (2 to 4 ton) conical piles, and in a ridge-shaped pile 150 ft long and 25 ft high. The tests were generally conducted with fresh and three-month or six-month stored coals of various sizes or both. The conclusions reached were that washed, prepared sizes of Illinois coal, exclusive of fines, could be used to 25% of the total coal blend after three to six-months' storage. Satisfactory coke could be made from as much as 80% of this coal blended with a fluid, medium-volatile coal after six-months' storage. Blends of 80% Illinois No. 6 Seam stored thirty days with a low-fluidity coal gave satisfactory results. The most noticeable effect of weathering of the blending coal was an increased breeze yield and a lower hardness-factor. Table 3 contains some typical results of the 2 by ⅜-in. washed Illinois No. 6 coal stored in small conical piles.

The second Illinois State Geological survey paper[15] discussed winter and summer storage, approximately six mon each, of Illinois coals from No. 5 and No. 6 Seams. The coals were stored out-of-doors in conical piles containing about 2½ tons. The Pocahontas coal used for blending was stored for an equivalent six-mon period in covered bins, partially open on one side. The progress of weathering was studied by Gieseler fluidity, free-swelling index,

TABLE 4

COMPARISON OF SOME PHYSICAL PROPERTIES OF COMMERCIAL COKES FROM FRESH AND STOCKPILED POND-CREEK COAL BLENDED WITH POCAHONTAS NO. 3 COAL[16]

No. Ovens	Ovens Using Weathered Coal: 15		14		12		Average Results for July, 1952 with Fresh Coal, Mean
Coal Source	Pile Surface		Pile Surface		Pile Interior		
Measure of Coke Quality	Mean	Standard Deviation	Mean	Standard Deviation	Mean	Standard Deviation	
2-in. screen test	74	4.5	70	4.4	75	5.4	72
2-in. shatter test	67	5.1	61	4.5	62	7.3	66
1-in. tumbler test	46.7	1.6	46.5	1.2	44.4	1.9	48
¼-in. tumbler test	66.1	1.8	65.3	1.9	64.3	0.8	65.2

physical properties of the cokes produced, and expansion pressure. Major differences were found between winter and summer storage; Illinois No. 5 and No. 6 coal could be stocked during the winter for six mon without significant loss of coking properties, but summer storage of more than thirty days produced significant decreases in coking properties. The coal from Illinois No. 5 Seam was more resistant to weathering than No. 6.

Smith, Reynolds, and Brewer studied high volatile *A* bituminous coal from the Pond Creek Seam from Pike County, Kentucky, after eighteen months of storage in a compacted pile.[16] Samples were collected from the surface, sides, and interior of the storage pile. Three groups of tests were made with the surface and interior coal in the coke ovens of the Ford Company plant at Dearborn, Michigan. A summary of the results obtained with fresh and stored blends of 65% Pond-Creek coal and 35% low-volatile Pocahontas No. 3 coal is shown in Table 4. Storage of the high-volatile coal did not appear to produce any significant changes in coke properties.

Laboratory Studies of Storage Conditions

Schmidt, Elder, and Davis oxidized 400-lb samples of coal in air in a slowly rotating steam-jacketed drum maintained at 100°C for varying periods of time.[17] Periodically, 90-lb samples were taken for BM-AGA carbonization tests at 800°C. The authors found that the number of days of accelerated oxidation required to reduce the coke-strength index[18] by 15% (T_{15}) was a function of the fresh moisture-free and mineral-matter-free volatile-matter content (V_0) and the oxygen content on the dry, ash-free basis (O_0), as shown by the following equation:

$$T_{15} = 1.22(V_0) - (O_0)^{1.65} - 11.28.$$

Brewer and associates utilized a previously developed accelerated oxidation technique[17,19] to obtain a comparison between accelerated oxidation results and exposed storage in the laboratory. The authors gave factors for eleven coals, derived from agglutinating values and coke-strength indexes, that permit estimation of room temperature storage periods which will reduce the coking power of the coal by 15% as measured by the coke strength index. Changes in specific gravity, C, H, O, Btu content, and dry tar yield during accelerated storage were found to correlate with the oxygen consumed, expressed as percent by weight of dry, mineral-matter-free coal. The authors reported data on many of the properties of the coals and stated that the agglutinating values, tar yields, and strength indices of the cokes are the best criteria for judging extent of the oxidation of coal.

Wildenstein compared the oxygen-uptake and resulting changes in plasticity on three coals exposed to air, nitrogen, and purified nitrogen at 105 to 110°C for 2 hr.[20] The oxygen content of the air-aged coals increased and the Gieseler plasticity decreased markedly. Nitrogen drying caused less effect on the plasticity. A series of tests with room-temperature exposure to air resulted in a slight increase in oxygen content, but the plasticity changes were considerable.

Abernethy and Tarpley made a study of the changes in heating value and certain other coal properties that might occur between the time that a laboratory sample is prepared and the analysis.[21] Hand-picked lumps from four high-volatile bituminous coals were subjected to a defined collection and sampling procedure and stored in four ways. These storage conditions were: screw-top sampling cans, analyzed without 30°C air-drying after storage and prior to analysis; screw-top sampling cans analyzed after 30°C air-drying; shallow pans stored in the 30°C air-drying room; and shallow pans stored in the laboratory. The conclusions were that some loss in heating value occurred for the five-month storage period. This was true of all samples under all storage conditions. Therefore, a laboratory sample should be handled as expeditiously as possible.

Roga and Pampuch proposed to use the increase in volume after immersion in water of

[16] Smith, F. W., Reynolds, D. A., and Brewer, R. E., *AGA, Proc.*, **1953**, 835–51.

[17] Schmidt, L. D., Elder, J. L., and Davis, J. D., *Ind. Eng. Chem.*, **35**, 150–5 (1943).

[18] Schmidt, L. D., *Iron Steel Engr.*, **18**, No. 3, 67–70 (1941).

[19] Brewer, R. E., Reynolds, D. A., Steiner, W. A., and Van Gilder, R. D., *Ind. Eng. Chem.*, **40**, 1243–54 (1948).

[20] Wildenstein, R., *Chaleur & ind.*, **34**, 233–6 (1953).

[21] Abernethy, R. F., and Tarpley, E. C., *U.S. Bur. Mines, Rept. Invest.*, No. 5386 (1958), 13 pp.

TABLE 5

NUMBERS OF FISSURES FORMED IN COAL PARTICLES ON STORAGE[22]

Type of Fissure by Direction Relative to Sedimentation Plane	Storage Conditions			
	Tightly Closed Jars	Tightly Closed Jars, N_2 Atmosphere	Open Boxes Exposed to Weather	Under Water
Parallel	24	17	101	68
Right angles	81	53	36	15
Total fissures	105	70	137	83

a parallelepiped of coal, 50 by 10 by 10 mm, cut vertically and parallel to the bedding plane as a measure of the weathering of hard coals.[22] The authors classified a coal whose volume increase is greater than 2% as highly susceptible to weathering, from 0.5 to 2% as highly resistant to weathering, and a coal with less than 0.5% as nonweathering. The experimental basis for the authors' proposal was obtained from a study of the fissuring of coal particles stored in four ways, the degradation of coal stored exposed to atmospheric conditions, and observations on the swelling of coal in water. The fissuring experiments consisted of examining fifteen coal particles, each from six coals stored according to the methods in Table 5 and classifying the fissures as those forming parallel to the bedding plane or fissures forming at right angles to this plane. The fissures at right angles to the bedding plane were generally limited to the surface layer, while those parallel to the plane were extensive and deep. The conclusions reached, derived from the two-way classification shown in Table 5 were that: the fissures formed during the air-sealed storage were the same as those formed in an N_2 atmosphere; the fissures formed under water were the same as those formed in the box exposed to the weather; and the fissures formed in the sealed conditions (*A* and *B*) were essentially different than those formed under conditions of weather exposure or under water (*C* and *D*). A comparison of the weathering and swelling properties was made with twenty-nine coals ranging in moisture content from 0.6 to 16.1%. The samples, which were originally 100 to 150 mm in size, were reduced to 8 to 10 mm and stored in boxes exposed to the weather. Portions withdrawn before reducing were used for the swelling tests. The results of the comparison between the amount of plus 6 mm coal retained after ninety-day exposure and the computed increase in volume of a parallelepiped immersed in water are shown in Fig. 5.

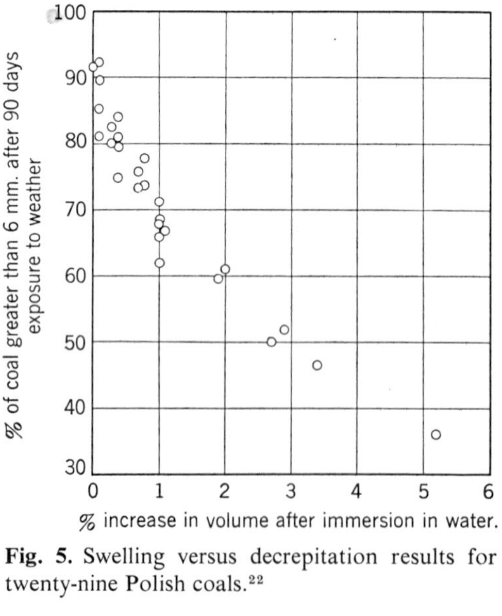

Fig. 5. Swelling versus decrepitation results for twenty-nine Polish coals.[22]

Spontaneous Combustion

Current theories of spontaneous combustion have been thoroughly reviewed by Yohe.[23] This review covers the literature on the oxidation of coal as well as storage. Scott has discussed the physics of heating and cooling caused by oxidation within a coal pile.[24] A review of spontaneous oxidation of coal by

[22] Roga, B., and Pampuch, R., *Prace Głównego Inst. Górnictwa, Komun.* No. 189 (1956), 10 pp.

[23] Yohe, G. R., *Illinois State Geol. Survey, Rept. Invest.* No. 207 (1958), 51 pp.

[24] Scott, G. S., *U.S. Bur. Mines, Bull.* No. 455 (1944), 206 pp.

Sanchez Marco covers methods of avoiding spontaneous combustion.[25]

Berkowitz and Schein proposed that the heat of wetting may be an important factor in initiating spontaneous combustion in small and medium coal piles.[26] The magnitude of the temperature rise resulting from passing water-saturated oxygen through a 40-g bed of dry lignite in a calorimeter at 25°C could be as high as 50°C. The heat released was due to wetting and not oxidation, because moist, oxygen-free nitrogen gave virtually identical results as saturated oxygen. The heat liberated by wetting could raise the temperature of the coal to the point where the oxygen sorption rate of a reactive coal would be sufficient to ignite the coal.

Mapstone has suggested that ferrous sulfide may play a role in the spontaneous combustion of pyritic coal.[27] The mechanism postulated involves production of ferrous sulfide by limited oxidation of pyrite with oxygen or a coal-oxygen complex having properties similar to organic peroxides. If finely divided ferrous sulfide were then subsequently exposed to air, the pyrophoric property of the ferrous sulfide could cause ignition of the coal.

The possibility that microbiological activity was a source of heating in coal was investigated by Zavarzina with negative results.[28]

The moisture content of coal as a factor influencing the self-heating tendencies has been examined by several investigators.[25,29] Elder and associates found that the presence of normal or a greater-than-normal moisture content in coal resulted in a decreased heating rate in an adiabatic calorimeter.[30] They studied the relative spontaneous heating tendencies of minus $\frac{1}{4}$-in. coal dried at 212°F in nitrogen prior to charging the calorimeter. They allowed the temperature of the 110-lb coal charge in an oxygen stream to rise to 300°F. The heating rates obtained were found to follow an Arrhenius type relation. Parallel tests of oxygen consumption were conducted in a rotary, steam-jacketed drum. It was found that the ratio of the heating rate of a coal in the adiabatic calorimeter to the rate of a reference coal was nearly equal to the ratio of the characteristic oxidation rate of the coal to that of the reference coal, under comparable conditions. The heat liberated by 1 lb of oxygen reacting with coal was estimated to be approximately 5000 Btu.

An expression to indicate the tendency of coal to oxidize when stored in a pile was developed by Blaskett.[31] The factor is equal to RH^2/a, where R is the initial rate of oxygen consumption, H the height of the pile, and a a constant for the coal and manner of storage. The constant a is equal to K/Ds, where K is the thermal conductivity of the pile, D the bulk density of the pile, and s the specific heat of the coal.

A method of evaluating the suitability of a coal for storage was also derived by van Doornum.[32] The author concluded from experimental calorimetric evidence that the rate of heat generation of a spontaneous heating process q over the course of the experiment could be represented by a finite series of the form:

$$q = \sum_i A_i \exp(a_i v - \alpha_i t)$$

where A_i is the rate of heat generation, cal/cm³ hr; a_i a temperature coefficient per degree centigrade; v the temperature rise above a convenient reference temperature in degrees centigrade; α_i a decay constant per hour; and t time in hours. Usually two or three terms were required. If the sum of the dimensionless parameters M_i,

$$\sum_i M_i = \Sigma_i \frac{A_i a_i}{\alpha_i \rho s}$$

evaluated at the temperature of the proposed storage pile, was less than 0.63, the coal was considered as storable without further precautions. The constant ρ represents coal density, gm/cm³ and s specific heat cal/gm per degree centigrade. Values of ΣM_i equal to or greater than 1 are considered dangerous. Intermediate

[25] Sanchez Marco, G., *Combustibles*, **7**, 124–32, 163–70 (1947).

[26] Berkowitz, N., and Schein, H. G., *Fuel*, **30**, 94–6 (1951). Schein, H. G., *Brennstoff-Chem.*, **32**, 298–301 (1951).

[27] Mapstone, G. E., *Chem. & Ind.* (*London*), **1954**, 658.

[28] Zavarzina, N. B., *Mikrobiologiya*, **25**, 200–207 (1956).

[29] Jessen, H. E., *Gasteknikeren*, **32**, 463–71, 481–6 (1943).

[30] Elder, J. L., Schmidt, L. D., Steiner, W. A., and Davis, J. D., *U.S. Bur. Mines, Tech. Paper* No. 681 (1945), 24 pp.

[31] Blaskett, D. R., *S. Australia Dept. Mines, Mining Rev.* No. 86 (1948), pp. 60–79.

[32] Van Doornum, G. A. W., *J. Inst. Fuel*, **27**, 482–95 (1954).

values of ΣM_i require secondary considerations of pile geometry and ventilation, if present.

Szadeczky-Kardoss reported the formation of oxidized zones on the surface and in the interior of the mass of brown coals, especially those subject to spontaneous ignition.[33] The author considered the formation of these particular locations by the action of oxygen as protection against further oxidation. This action is more effective with nonslacking coals.

Gibinski extracted humic acids from several coals by hot caustic soda.[34] The residues showed a greater tendency to spontaneous combustion than the raw samples.

The use of additives to prevent or inhibit oxidation of coal has been studied by Terpogosova.[35] The results were evaluated principally by the effect on ignition temperature. Effective agents were petroleum products, such as paraffin and paraffin emulsions, with NaOH. Fatty acids, soaps, and 5% slaked lime were somewhat less effective.

Khrisanfova and Shubnikov treated a storage pile of Moscow lignite with an aqueous solution of calcium bicarbonate as it was constructed.[36] This treatment kept the pile temperature near 30°C for eighteen months while the temperature of a control pile rose to 60°C. They successfully applied this technique in piles of 185 to 1000 tons.

HANDLING COAL

The handling of coal is considered here as short-range transportation. Problems in this field are perhaps more of an engineering than scientific nature but there are considerations which are peculiar to coal.

Flow of coal and granular materials, in general—from bins, chutes, and pipes—have been reviewed by Richards.[37] He pointed out that no general theory of solids flow through a bin has been established.

[33] Szadeczky-Kardoss, E., *Ungar., Z. Berg u. Hüttenwessen*, **77,** 241–7, 253–60 (1944).

[34] Gibinski, S., *Prace Glównego Inst. Górnictwa Komun.* No. 97 (1951), 10 pp.

[35] Terpogosova, E. A., *Izvest. Akad. Nauk S.S.S.R., Otdel. Tekh. Nauk*, **1954,** No. 4 pp. 147–51.

[36] Khrisanfova, A. I., and Shubnikov, A. K., *Khim. i Tekhnol. Topliva i Masel*, **1958,** No. 7, 21–8.

[37] Richards, J. C., *BCURA Bull.*, **19,** 145–58 (1955).

Significant progress has been reported by Jenike in papers describing the problems associated with flow from bins.[38] He described a measure of the compressive strength developed by a bulk solid subjected to compaction from the static pressure due to the weight of the material, impact of falling particles, and vibration of the bin. This measure, called the flow-factor, was determined experimentally. The flow-factor was used by Jenike in conjunction with other experimental factors to obtain dimensions of the bin discharge that will prevent doming and to determine the proper width of a bin to prevent funneling.

The flow characteristics of brown-coal fines and other powdered materials were evaluated by Herrmann by the ratio of the time required for the coal and for the same volume of water to flow through a standard orifice.[39]

Improvements of bin discharge performance have been studied by several groups. Brown has reported results of tests on a model bin with horizontal rods placed near the boundary between the moving central cone and the stationary material.[40] The proper placement of these rods was reported to regulate the flow for straight-through discharge.

The development of a device to eliminate the central cone and its associated problems occurring in the discharge of coal from bins has been reported by Cooper and Garvey.[41] The optimum design was capable of handling $\frac{1}{4}$ in. by 0 coal with 15% surface moisture.

Improvement of discharge by alteration of the flow characteristics of fine coal was investigated by Marks, Broadley, and Yellott by passing air at low velocity through a bed of coal.[42] At air rates less than those required for fluidization, the angle of repose of the coal was reduced from about 50 to as low as 2°. They demonstrated that a minimum angle of repose was obtained by aeration of surface-dry coal. Small amounts of surface moisture

[38] Jenike, A. W., *Univ. Utah, Eng. Exp. Sta., Bull.*, No. 64 (1954), 81 pp. *Chem. Eng.*, **61,** No. 12, 175–80 (1955). *Mech. Eng.*, **76,** 839–40 (1954).

[39] Herrmann, A., *Freiberger Forschungsh. Beihefte, A* **18,** 75–91 (1953).

[40] Brown, R. L., *Fuel*, **29,** 220 (1950).

[41] Cooper, E. D., and Garvey, J. R., *Mech. Eng.*, **80,** No. 5, 59–60 (1958).

[42] Marks, C. H., Broadley, P. R., and Yellott, J. I., *Mech. Eng.*, **78,** 272 (1956).

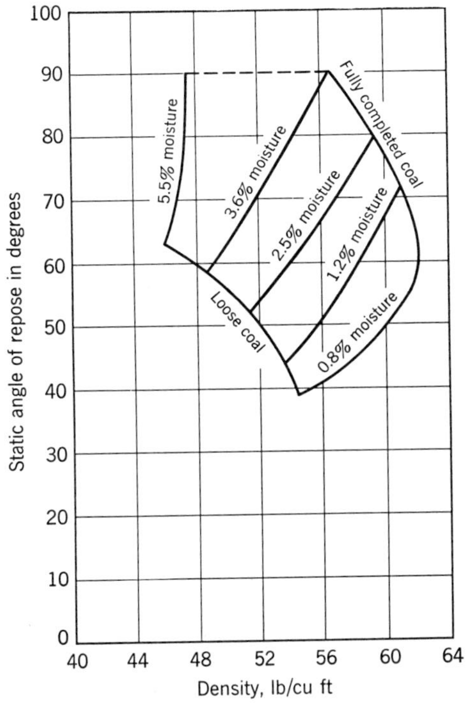

Fig. 6. Effect of moisture and degree of compacting on static angle of repose.[43]

increased the angle of repose significantly. A large fines-content of the coal was advantageous to a low angle of repose. The initial flow rate from the flat bottom bin used in these experiments increased from 10 lb/sec without aeration to 25 lb/sec with aeration.

The use of scale models to delineate problems associated with the flow of wet fine-coal from bins have been used by several investigators. Wolf and von Hohenleiten studied the arching and ratholing in bins and stoppages in chutes.[43] Plastic models were made to a linear scale of 1:10, and the particle size of the coal used was also scaled approximately 1:10. The presence of a scale factor was noted by the authors: All of the effects occurring in the prototype at a coal moisture content of 5.5 to 6% occurred in the model at moisture content of 3.5 to 3.75%. A similar scale reduction of 1:12 was used by Legget, who studied the flow of bituminous coal from bins.[44]

Wolf and von Hohenleiten determined a number of coal properties which have a bearing on the flow of coal particles. Figure 6 shows the static angle of repose obtained as a function of bulk density and moisture content of the coal. Compaction was accomplished by vibration. The high moisture coal, 5.5%, was remarkably sensitive to compaction; an increase of less than 2 lb/cu ft resulted in an increase in the angle of repose from about 63 to 90°. The influence of moisture on the pressure exerted at the base of a column of coal as a function of coal height is shown by Fig. 7. At

[43] Wolf, E. F., and von Hohenleiten, G. L., *ASME, Trans.*, **67**, 585–99 (1945); *Mech. Eng.*, **70**, 313–14 (1948).

[44] Legget, A. F., *ASME, Trans.*, **69**, 525–33 (1947).

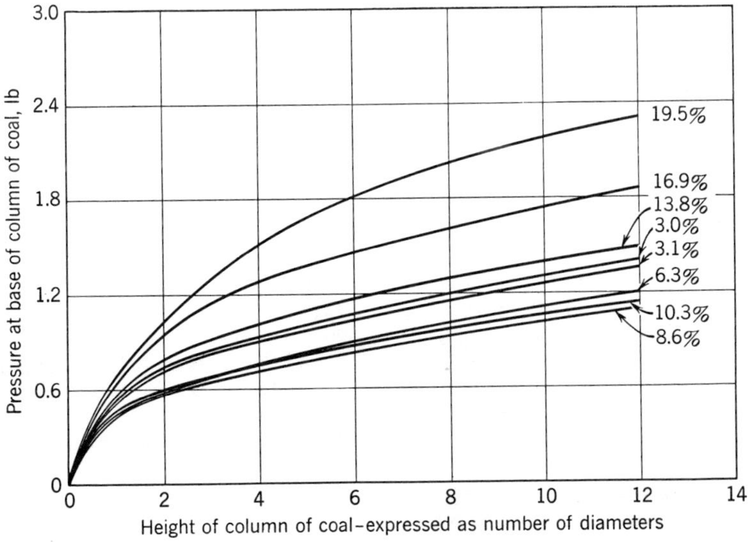

Fig. 7. Side-wall friction test, pressure at the base of a column of coal in 3-in. diameter steel tube.[43]

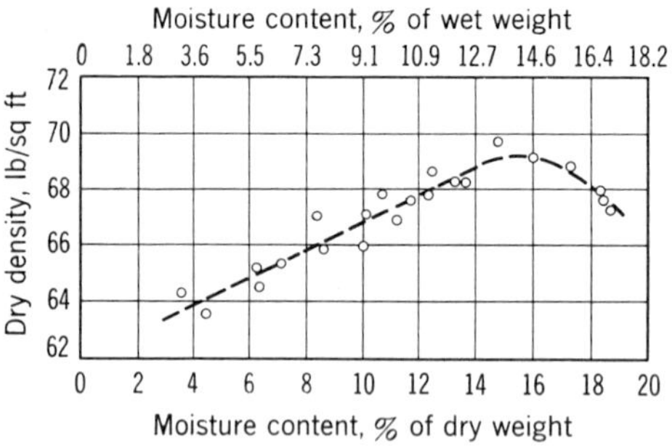

Fig. 8. Variation of dry density of coal with moisture content; compaction by Proctor test.[44]

coal heights greater than 1 diam, increasing wall friction takes a considerable portion of the load. The influence of moisture content on bulk density is shown by the arrangement of the curves; each increasing moisture content results in a decreasing pressure at the base, until a minimum appears at about 8% moisture—then the pressure begins to rise. Wolf and von Hohenleiten also determined the effect of moisture on bulk density, with the familiar concave upwards-curves resulting. Legget[44] determined the density as a function of moisture by the Proctor test.[45] The compaction produced by this test, performed by a 5.5 lb hammer falling 12 in., resulted in a density-moisture curve concave downwards, Fig. 8.

TRANSPORTATION

Pneumatic Transportation

The literature of pneumatic transportation of solids is extensive from both practical and theoretical viewpoints. The pneumatic transportation of coal is a particular case with special problems. The work of Vogt and White and a number of the papers contributed to the Fifteenth Annual Chemical Engineering Symposium of the American Chemical Society are of particular theoretical interest.[46,47]

Vogt and White found that the ratio (α) of the pressure drop of a suspension of wheat or sand to the pressure drop of the pure fluid flowing at the same velocity was best correlated by:

$$\alpha - 1 = A\left(\frac{D}{d}\right)^2\left(\frac{P}{w}\frac{r}{Re}\right)^k$$

where D = pipe diameter in feet;
d = average effective diameter of particles in feet;
P = fluid density in pounds per cubic foot;
w = solids density in pounds per cubic foot;
r = weight ratio of solid to fluid flowing;
$Re = \dfrac{DG}{v}$, the dimensionless Reynolds number;
G = fluid mass velocity in pounds per square foot;
v = viscosity of the fluid in pounds per foot per second;
A,k = empirical constants in functions of $Re\sqrt{cd}$; and
cd = drag coefficient for a spherical particle under free settling conditions.

Numerous authors have used this correlation with varying success and have proposed modifications or other correlations. In a paper describing the pressure drop resulting from the transportation of fine coal, 90% minus 200 mesh, in dense coal-air mixtures,[48] the coal-to-air weight ratios used were from 125 to 256, and some modification of Vogt and White's correlation was considered most useful. Vogt and White's solids-to-air ratios were from 0.224 to 35.5. Design of a pneumatic transport system utilizing Vogt and White's correlations has been given by Mitchell.[49]

Acceleration of the particles was not considered by Vogt and White, whereas other authors have attempted to measure the velocity of the particles by isolating a section of the system and determining the weight of the solids contained.[47,50] Alternate methods of

[45] Proctor, R. R., *Eng. News-Record*, **111**, 245 (1933).

[46] Vogt, E. G., and White, R. R., *Ind. Eng. Chem.*, **40**, 1731–8 (1948).

[47] Symposium on Dynamics of Fluid-Solid Systems, *Ind. Eng. Chem.*, **41**, 1099–1250 (1949) 21 papers.

[48] Albright, C. W., Holden, J. H., Simons, H. P., and Schmidt, L. D., *Ind. Eng. Chem.*, **43**, 1837–40 (1951).

[49] Mitchell, S. J., *Brit. Chem. Eng.*, **2**, 356–8 (1957).

[50] Clark, R. H., Charles, D. E., Richardson, J. F., and Newitt, D. M., *Trans. Inst. Chem. Engrs. (London)*, **30**, 209–24 (1952).

determining velocity utilizing radioactive tracers have been used.[51]

The weight flow-rate measurement of pulverized coal (75 to 92% minus 200 mesh) by a flow nozzle and orifice combination was determined by Carlson, Frazier, and Engdahl.[52] Coal-air-suspensions in the range 0.4 to 1.4 lb of coal per pound of air could be measured by this technique.

Hydraulic Transportation

The technique of long distance transportation of coal by pipeline has developed primarily because of post-war inflation of transportation rates. A concentrated effort by the Pittsburgh Consolidation Coal Company culminated in a 108-mile pipeline transporting 150 tons/hr.[53] The literature of hydraulic transportation has been compiled through 1950 by Dougherty[54] and through 1954 by Lammers, Allen, Donaven, Wagner, and Parry.[55]

Application of hydraulic transportation to a specific problem requires consideration of a number of variables:[55] material being transported; transport medium; pipeline length; pipeline capacity; size analysis; linear velocity; loading; pipeline diameter; and type of pump.

Theory of the flow relations occurring during the transportation of coal-water mixtures was a major topic of a colloquium held by the British National Coal Board in 1952. A paper presented there by Durand and Condolios discussed many of the practical as well as theoretical problems associated with hydraulic transport.[56] The work of Durand and Condolios has been summarized in a more accessible journal.[57] They classified solids for hydraulic transportation in three graded groups by size, as follows: (1) homogeneous mixtures, particle size less than 20 to 30 μ; (2) heterogeneous mixtures, particle size greater than 50 μ; and (3) intermediate mixtures, a transition group, 25 to 50 μ.

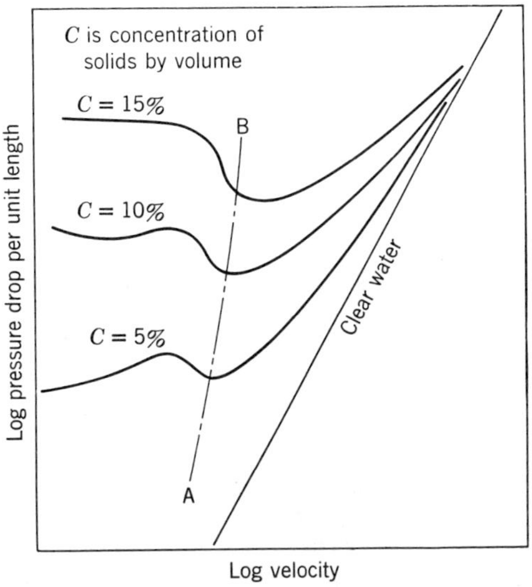

Fig. 9. Representative unit pressure drop, concentration, and velocity relations for hydraulic transportation of a heterogenous mixture. (After Durand and Gondolios.[56])

The flow characteristics of heterogeneous mixtures were found by Durand and Condolios to correspond to Bingham bodies. In the turbulent region, the solids flow without deposition but they settle when the flow is laminar. In the turbulent region, the material behaves like a true fluid, with a density equal to that of the slurry. Flow of heterogeneous mixtures as described by Durand and Condolios is characterized by some degree of gravity sedimentation. Figure 9 is a typical set of curves for a sand, adapted from Durand and Condolios. At high velocities, the pressure-drop curves approach the clear water curve asymptotically or are parallel to it. Line *A-B* separates the region of flow without deposition (to the right of *A-B*) from the region of deposition. The velocity at which deposition begins was termed the limiting velocity by the authors. It very nearly coincides with the minimum point of the pressure drop curve. Durand and Condolios found that the expression

$$\frac{U_1}{\left[\frac{2gD(\rho_s - \rho_f)}{\rho_f}\right]^{1/2}} = A$$

[51] Sauerwein, K., Hossner, R., and Rotter, W., *Atomwirtschaft*, **1**, No. 2, 71 (1956). Knitzinger, C. A. J., *Nuclear Instr.*, **1**, 66–70 (1957).

[52] Carlson, H. M., Frazier, P. M., and Engdahl, R. B., *ASME, Trans.*, **70**, No. 2, 65–79 (1948).

[53] Dauber, C. A., *Coal Age*, **62**, No. 4, 84–7, (1957).

[54] Dougherty, R. W., *U.S. Bur. Mines, Rept. Invest.* No. 4799 (1951), 22 pp.

[55] Lammers, G. C., Allen, R. R., Donaven, D. J., Wagner, E. O., and Parry, V. F., *U.S. Bur. Mines, Rept. Invest.* No. 5404 (1958), 39 pp.

[56] Durand, R., and Condolios, E., *Proc. Colloq. Hydraulic Transport of Coal, Nat. Coal Board. (London)*, November 5–6, 1952, pp. 39–51.

[57] Anon., *Colliery Eng.*, **32**, 185–90 (1955).

was a function of the particle size and volumetric concentration below a particle size of about 1 to 2 mm. Above that size the expression was nearly constant, equal to 1.34. Nomenclature of the terms used is given in Table 6. The pressure drop for a particular material

TABLE 6

NOMENCLATURE FOR HYDRAULIC TRANSPORT EQUATIONS

A = numerical value of expression 1. Constant for certain conditions, dimensionless;
B = experimentally determined constant, dimensionless;
C = solids concentration, volumetric;
D = pipe inside diameter, feet;
d = particle diameter, feet;
d_i = particle diameter of size i, (mixed sizes), feet;
g = acceleration due to gravity, ft/sec²;
H_m = head loss of slurry, feet of water per foot of pipe;
H_w = head loss of clear water, at same velocity as mixture, feet of water per foot of pipe;
k = constant, dimensionless;
ρ_f = density of fluid medium, lb/ft³;
ρ_s = density of solid, lb/ft³;
U = pipeline transport velocity, ft/sec;
U_t = terminal velocity of particle in water, ft/sec;
U_1 = limiting velocity for particle deposition in pipe, ft/sec;
W_i = weight fraction of particles of size i, (mixed sizes); and
X = drag coefficient, dimensionless.

at a velocity greater than the limiting velocity was found to be a function of the parameters:

$$\frac{H_m - H_w}{H_w C} \quad \text{and} \quad \frac{U}{(gD)^{1/2}}$$

The authors established that for any heterogeneous material the flow expression becomes:

$$\frac{H_m - H_w}{H_w C} = K\left[\frac{U^2(X)^{1/2}}{\frac{gD(\rho_s - \rho_f)}{\rho_f}}\right]^{-1.5}$$

Worster evaluated the constants in Durand's equation and obtained[58]

$$\frac{H_m - H_w}{H_w C} = 121\left\{\frac{gD(\rho_s - \rho_f)U_t}{\rho_f U^2[gd(\rho_s - \rho_f)/\rho_f]^{1/2}}\right\}^{1.5}$$

Smith compared Worster's values with the results of tests on transporting mixed-size sand-water slurrys in pipes of two sizes.[59] The two sets of data were found to be best correlated if the average size of the particle is computed as

$$d = \frac{\Sigma W_i}{\Sigma(W_i/d_i)}$$

The agreement was considered by Smith as only fair, however. The constant B, determined experimentally, for a given mixed-size slurry was found to be roughly constant over a wide range of concentration and over the normal range of velocity. The constant is evaluated as:

$$B = \frac{H_m - H_w}{H_w C}\left[\frac{U^2}{\frac{gD(\rho_s - \rho_f)}{\rho_f}}\right]^{1.5}$$

Experiments carried out[60] in a 1-in. pipe with six materials showed that Durand's equation was experimentally equivalent for heterogeneous suspensions to

$$\frac{H_m - H_w}{H_w C} = \frac{1100 gDU_t(\rho_s - \rho_f)}{U^3 \rho_f}$$

Degradation occurring during the pipeline transportation of coal has been discussed, and some experimental work has been recorded, though with inconclusive results, by various authors. Experimental determination of the degradation characteristics of a coal in the usual pilot pipeline is complicated by the usual construction of such lines as a closed loop. A long distance transportation problem, simulated by operating the pilot line for an equivalent time, results in a greatly disproportionate number of passes through the pump. The co-degradation resulting is not easily separable into pump and flow degradation. Various methods have been devised to separate these effects; for example, Worster has described the results of transporting ½-in. coal with a jet pump to reduce degradation.[58] Degradation was about ½% by weight per mile at 10 miles when velocity was 2 ft/sec. Bond, in a description of the design, testing, and operation of a 72-mile pipeline for transportation of

[58] Worster, R. C., *Proc. Colloq. Hydraulic Transport of Coal, Natl. Coal Board (London)*, November 5–6, 1952, pp. 5–20.

[59] Smith, R. A., *Trans. Inst. Chem. Engrs. London*, **33**, 85–92 (1955).

[60] Newitt, D. M., Richardson, J. F., Abbott, M. and Turtle, R. B., *Trans. Inst. Chem. Engrs. London*, **33**, 93–110 (1955).

gilsonite, mentioned both a short-circuit test of the pump alone and a test when the centrifugal pump was replaced by an air-lift.[61] The results of both tests were uncertain.

A test utilizing the pipe alone has been reported by Worster and Denny.[62] A toroidal pipe partially full of 1-in. coal-water mixture was rotated at speeds of 7.5, 9.1, and 11 ft/sec for times corresponding to distances up to 70 miles. For these conditions, and with a particular coal, the degradation was proportional to the cube of the speed and the square root of the distance traveled. The authors suggested that the damage is done by impacts between particles as well as with the side walls. The fracturing dependent on particle-wall impact may be more important than particle-particle impact.[60]

[61] Bond, R. K., *Mech. Eng.*, **79,** 1171 (1957).

[62] Worster, R. C., and Denny, D. F., *Proc. Inst. Mech. Engrs. (London)*, **169,** 563–86, 1955.

In a comparison of the degradation characteristics of two types of materials, Texas lignites and a Colorado bituminous coal,[55] the bituminous coal was found to generate less micron-size material than did the lignite.

Other problems associated with hydraulic transportation have been studied to a greater or lesser degree. Pumps are of special importance. The British have developed some remarkable pumps and feeders, especially those necessary for coal particles 3 in. or larger.[63] The Pittsburgh Consolidation Coal Company line and the gilsonite line both use reciprocating pumps for particles with top size approximately 14 and 8 mesh, respectively.[53,61]

[63] Scivier, J. B., *Colliery Guardian*, **186,** 491–4 (1953). Anon., *Colliery Guardian*, **186,** 456–8 (1953). Magill, J. A., *Colliery Guardian*, **193,** 647–52 (1956). Anon., *Colliery Guardian*, **194,** 842–3 (1957).

CHAPTER 8

CLEANING AND PREPARATION

David R. Mitchell and H. Beecher Charmbury

The Pennsylvania State University, University Park, Pa.

Even though chemistry continues to play a minor role in the cleaning and preparation of coal, the subject of coal preparation is of increasing importance to the coal industry and has a vital role in the utilization of coal. Since 1945, the percentage of total coal production mechanically cleaned in the United States increased from 26 to 66%. Changes in the methods of mining underground coal, the growth of stripping and augering operations, and the depletion of high-grade (low sulfur and ash) coals have been mainly responsible for the rapid growth in coal preparation. Mechanically cleaned coals will, in all probability, continue to be required in the future for carbonization. In certain areas, low-quality coals may be and are being used by utilities. However, with an increase in air pollution control and with a growing tendency to evaluate utility coal costs on a kilowatt basis rather than on a ton or Btu basis, the importance of preparing coal for this large market is increasing.

SAMPLING

Increasingly stringent specifications on the part of coal consumers have led to more elaborate samplings at the mine-preparation plant. This increased activity has taken two forms: (1) sampling of material processed within a plant to develop high efficiency in the various units and to obtain accurate analyses of the finished products and (2) development of automatic means for taking the gross sample and automatically reducing it to the small amount required by the analytical laboratory.

A decade ago considerable work was done in the United States to work out standards for the sizes of gross sample required to achieve reasonable accuracy. Standard procedures of sampling were established by the American Society of Testing Materials on the basis of this work. Because these standards were based on size-weight ratios, the accumulation of the gross sample could be done by taking increments either by hand or by use of a mechanical sampler. ASTM standards are also available for sample reduction, special testing, and chemical analyses. These ASTM standards are readily available in the literature and are followed closely by coal producers and consumers where high accuracy is required. At many plants, simplified procedures are used.

A common procedure in reducing the gross sample is to crush to approximately $\frac{3}{8}$ in. and to riffle out a sample of approximately 20 to 25 lb. This 25-lb minus $\frac{3}{8}$ in. sample is further crushed to minus 4 mesh, and a 10-lb sample is riffled out. The minus 4 mesh sample is then pulverized to -20 mesh and a $\frac{1}{2}$ lb sample is riffled out. The $\frac{1}{2}$ lb minus 20 sample is then pulverized to 60 mesh and riffled to a final 50-g sample for the laboratory. During all these stages, careful mixing must be done to avoid size and specific gravity segregation that could vitiate the final laboratory sample. In developing automatic procedures for sample reduction, these intermediate stages are frequently bypassed. When this is done, care must be exercised so that the size of the sample is kept to the amounts cited in the foregoing procedure at the various reduction stages.

The present procedures for automatic sampling usually involve sampling the raw feed and final products of the plant with automatic samplers and with automatic sampling reduction. With the large flows amounting to as much as 2000 tons/hr on a single belt conveyor, an automatic cutter on such a flow will cut out an increment much larger than is required in existing standards. This has caused operating companies to restudy the problems of a gross-sample accumulation and the methods for its reduction. A careful study of procedures developed at a number of plants shows that resultant sizes of the order of 4–6 in. to 0 can be automatically sampled with a high degree of accuracy, if the gross sample is of the order of 1 lb for each ton. This would mean that, for a mine with an 8000 ton/day output of clean coal, the gross sample representing a day's output should be approximately 8000 lb and be made up of increments accumulated during the day's operation. Because the individual increments are automatically reduced in size and amount to the order of 4 mesh or smaller, only about 10 lb of the sample is handled by hand by the laboratory personnel for final reduction to 60 mesh and 50 g. For top sizes less than 4 in., the gross sample can be less.

For automatic sample reduction, the increments of the gross sample generally pass to a crusher set at $\frac{3}{8}$ in., with approximately one-tenth cut out by riffling or other means after crushing of the sample and nine-tenths returned by gravity or conveyed to the coal stream. This sample then passes to a pulverizer set at about 20 mesh and about one-twentieth of the feed to the pulverizer is cut out and nineteen-twentieths are returned to the coal stream. This procedure usually gives a minus 20-mesh sample small enough to handle conveniently in the laboratory for final pulverizing to minus 60-mesh.

In addition to sampling for chemical analysis, samples must be obtained frequently for float-and-sink testing. In sampling for float-and-sink testing, an added burden is put on the procedure, in that the sample cannot be reduced prior to the float-and-sink tests. This means that considerable care must be exercised in getting the right amount of sample and in thoroughly mixing the sample prior to testing. Landry and Bailey[1] investigated this problem in some detail for a Western Pennsylvania coal. They determined the size of gross sample required statistically for various sizes of coal, and investigated the possibility of thoroughly mixing the gross sample and reducing the mixed gross sample by riffling or other means to a much smaller amount for float-and-sink testing in the laboratory. The weight of gross sample required for the coarser sizes, 4 × 2 in. ranged from 16.4 tons, for samples having a variance of 2.34 and to 387 tons for samples having a variance of 0.15. They found these samples could be reduced to one-thirtysixth by careful mixing and reduction without serious loss of accuracy. However, they stated that much more work is necessary to develop procedures that would give a high degree of accuracy for float-and-sink testing of all types and sizes of coal.

To avoid the problem of sample mixing and reduction without loss of accuracy, many companies have developed equipment where large samples can be float-and-sink tested with a minimum amount of labor.

Automatic sampling is receiving considerable attention in most of the coal producing countries in the world. Visman[2] studied the possibility of automatically sampling coking coal, $\frac{5}{16}$ in. × $\frac{1}{2}$ mm in size. As a result of these studies, he developed nomograms that can be used as a guide for determining the amount of gross sample required for any ash accuracy. He developed a constant for use in his statistical equations that shows the variance of the mean ash in the flowing coal and termed it "trend variance." He also developed an automatic sampling pipe which will take accurate increments at points within the stream of coal. Calculations and test data indicated that this sampler does a good job on the small sizes of coal tested but would be difficult to use on larger sizes.

Federwisch[3] has studied errors of sampling and has developed a sampling constant for use in statistical equations, which is the product of four parameters: the shape factor, f; the size distribution factor, g; the dissociation factor, l; and the mineralogical factor, m.

$$C = f \cdot g \cdot l \cdot m$$

These parameters are governed by the physical characteristics of the coal, size analyses, and washability data. Methods for determining these parameters and the use of the constant C

[1] Landry, B. A., and Bailey, A. L., *Proc. 1st Intern. Conf. on Coal Preparation, Paris*, **1950,** *Rev. ind. minérale*, Special Issue (1950–1), pp. 123–33.

[2] Visman, J., *1st. Proc. Intern. Conf. on Coal Preparation, Paris*, **1950,** *Rev. ind. minérale*, Special Issue (1950–1), pp. 134–44.

[3] Federwisch, J., *Proc. 3rd Intern. Coal Preparation Conf., Liège*, **1958;** *Ann. mines Belg.*, Special Issue (1958), pp. 656–661.

TABLE 1

VARIATION IN THE ACCURACY OF THE ANALYSIS OF THE GROSS SAMPLE WITH THE NUMBER OF INCREMENTS[4]

Material Sampled	Quantity Sampled, Tons	Variance in Quality Flow s_G	Error of Preparation s_V	Critical Number of Increments	Accuracy (±2 × Standard Deviation) Number of Increments $n_{critical}$	16	8	4	2	1
Washed fines	550–800	0.81	0.10	22	0.40	0.45	0.61	0.84	1.16	1.63
Coking coal	700–800	0.66	0.10	15	0.40	—	0.51	0.69	0.95	1.33
Nuts	155	0.88	0.10	13	0.52	—	0.66	0.90	1.26	1.77
Washed coal	—	0.79	0.10	17	0.43	—	0.58	0.79	1.10	1.54
Aspirated fines	160	1.06	0.15	18	0.58	0.60	0.81	1.10	1.52	2.14
Aspirated fines	700	1.34	0.15	16	0.74	0.74	0.99	1.37	1.91	2.68
Aspirated fines	—	1.20	0.15	17	0.66	0.67	0.89	1.23	1.72	2.41
Middlings:										
Heavy Medium	100	1.40	0.30	21	0.86	0.92	1.16	1.52	2.07	2.86
Jig	190	3.35	0.35	13	1.98	—	2.46	3.44	4.80	6.66
Untreated fines, *Q*	800	3.53	0.40	12	2.20	—	2.63	3.64	5.08	7.14
Untreated fines, *R*	800	9.24	0.40	21	4.12	4.68	6.60	9.30	13.10	18.50

were given. Nomograms were also given to permit easy use of the statistical data to determine the weight of the sample required, the sampling tolerance, and the sample reduction required.

Paul[4] made an exhaustive study of the whole problem of sampling, including accuracy of ash analysis, sample reduction, "trend variance," and finally, number of increments required in making up the gross sample. There is a wealth of detailed information in the results given and conclusions drawn. In general, his conclusions as to procedures are very similar to those established in the United States. The standard deviation to be expected from analysis of washed coal is of the order of 0.022 for the same laboratory and 0.035 for two laboratories. The effect of sample reduction on standard deviations of ash was studied for 0.2 to 0 mm, 3 to 0 mm, and 10 to 0 mm coal. The standard deviation for washed coal for the 0.2 to 0 mm coal averaged 0.05. For intermediate products and refuse, it ranged up to 0.13. The standard deviation for the 3 to 0 mm coal ranged from 0.06 for washed coal to 0.24 for refuse. The standard deviation for the 10 to 0 mm coal ranged from 0.06 for washed coal to 0.30 for refuse. It is apparent that the effect of variations in analyses and sample reduction is very small in regard to the accuracy of the analyses of the total sample. It was found that the chief error in sampling is in the taking of the sample. Table 1 gives the summation of the results of a series of experiments of various preparation-plant products to determine accuracy of sampling. The ash content of these various products ranged from 4 to 8% for the washed coal products, 9 to 14% for the aspirated fines, 34 to 43% for the middlings and 14 to 20% for the untreated fines.

The number of increments taken was plotted against the standard deviation for each, and a critical number was obtained beyond which increasing the number of increments did not appreciably increase the accuracy. This number is listed as well as the critical number of increments for each series of tests. As a result of this study, it is evident for most sampling problems—involving washed coal, aspirated fines, and middlings—that sixteen increments are usually sufficient in accumulating the gross sample to get a high degree of accuracy. For greatly variable unwashed fines or other variable products, it might be necessary to increase the number of increments to thirty.

In doing this work, Paul's theoretical basis was that the minimum weight in grams of the increment should not be less than 50*d*, where *d* is the diameter of the largest particle in millimeters. This theoretically gives the following:

for aspirated fines less than	1 mm—	50 g
for fine coal less than	3 mm—	150 g
for fine coal less than	10 mm—	500 g
for large coal less than	30 mm—	1500 g
for large coal less than	80 mm—	4000 g

[4] Paul, H., *Proc. 3rd Intern. Coal Preparation Conf., Liège*, **1958**, *Ann. mines Belg.*, Special Issue (1958), pp. 613–30.

In most instances, however, especially with fine and very fine coal, on practical grounds, if for no other reason, larger samples are taken. In the experiments conducted by Paul their weights were:

of aspirated fines of less than	1 mm—about	1 kg
of fine coal less than	10 mm—	1.5 to 3 kg
of large coal less than	30 mm—	2 to 3 kg
of large coal less than	80 mm—	3 to 4 kg

EFFECT OF IMPURITIES IN COAL FOR POWER-PLANT USE

High-ash coals are much more widely used on the European continent in power stations than in Great Britain and the United States. This is due to the higher percentage of true middlings in European coals compared to those of Great Britain and the United States and to higher coal costs. The recovery and use of true middlings, which have a specific gravity between 1.45 and 2.2 and range from 15 to 30% in many of the raw coals of Germany, France, Belgium, and the Netherlands, are required economically. As a consequence, 3-product separations are common and even 4-product separations are made in coal-preparation plants in these countries. Because of these conditions, coal-preparation and power-plant engineers in these countries have made detailed studies of the economics of using low-grade fuel for power generation. Their findings, available in the technical literature, can be used as a guide by engineers in other coal consuming countries faced with similar problems.

A decade ago, except in a few isolated instances, middling recovery was practically unheard of in the United States. Today, because of the necessity of mining seams containing increasing amounts of inherent impurities—with increasingly lower ash and sulfur specifications for metallurgical and other special-purpose coals—an increasing number of coal-mining companies are installing equipment to recover middlings and other off-grade products for sale to local power plants. Furthermore, many power-plant operators, particularly in utilities, are finding that the availability of high-grade fuels from accustomed areas is diminishing and that they must turn to the use of lower-grade coals in their plants. Modern power plants can and are being built to burn low-grade coals and high-ash middling products at relatively high efficiencies. Boiler efficiency is only one of many factors that must be considered.

Yeager[5] has discussed the effect of impurities in evaluating coals for steam-raising purposes and cited the results of a survey of maintenance costs of several industrial boiler plants. Plants using low-ash, high-grade coals had an average maintenance cost of $1.10/ton of coal consumed, whereas plants using high-ash low-grade coals had an average maintenance cost of $2.10/ton.

Jenkins and Booth[6] have reported trends in preparing coal for power generation in Great Britain, showing that there has been an increase in use of untreated small coal during the past decade.

Blass, Peter, and Lemke[7] made a detailed study of the relative value of low-grade coals used in Germany at power stations with boilers fired with pulverized coal. Seventeen power stations were studied, twelve having wet-bottom furnaces (liquid-slag removal) and five being normal dry-ash removal furnaces. The coals used in these plants showed great variations in kind and in inerts content:

volatile matter (dry, ash-free)	12 to 38%
ash (dry basis)	18 to 36%
moisture	5 to 15%
total inerts as fired (moisture + ash)	21.8 to 45.9%.

In order to insure comparability, all results were expressed on the basis of a "unit power station" of 100 Mw capacity rating, 25% stand-by, raising 400 tons of steam an hour at 80 atm gauge pressure, 500°C, with an annual steam production of 1,650,000 tons, having an annual heat content of 1061 × 10[9] kcal and operating 4125 hr/yr at 75 Mw output. The operating results at three power stations having wet-bottom boilers were used in deriving average costs for the reference "unit power

[5] Yeager, U. B., *Mech. Eng.*, **81**, No. 12, 104 (1959).

[6] Jenkins, G. S., and Booth, R. W., *Proc. 3rd Intern. Coal Preparation, Conf., Liège*, **1958**, *Ann. mines Belg.*, Special Issue (1958), pp. 151–9.

[7] Blass, E., Peter, G., and Lemke, K., *Proc. 3rd Intern. Coal Preparation Conf., Liège*, **1958**, *Ann. mines Belg.*, Special Issue (1958), pp. 160–79.

station." The steam cost, a quantity kept constant to make possible comparing the various coals, was found to be 14,609,600 marks/yr. The "normal coal" for the reference "unit power station" contained 15% inerts (5% ash, 10% moisture) with a heat value of 7000 kcal/kg and was assigned the price of 56.27 DM/ton which prevailed in the summer of 1957 for coal of this quality. For the reference station having 90% mean annual boiler efficiency obtained from the three power stations studied in detail, the fuel consumption as "normal coal" was 168,300 tons giving an annual fuel cost of 9,468,500 marks.

The valuation process consisted of using the reference "unit power station" consuming "normal coal" as a base and then determining the annual steam costs for the seventeen colliery power stations studied adjusted to the unit station and to the normal coal to show the effect of inerts in the fuel on the cost of operation. The following cost factors were ascertained:

K = annual steam costs in marks comprised of:
- (*a*) the quantity of fuel, B, tons used per year;
- (*b*) the price of the fuel, P, marks/ton;
- (*c*) the operating costs depending on the fuel, O, marks/yr; and
- (*d*) the capital costs depending on the fuel, C, marks/yr.

This gives:

$$K = (P \times B) + O + C \qquad \text{marks per annum} \tag{1}$$

For each of the seventeen individual power stations studied, these costs were ascertained and modified to permit comparison.

Nonat,[8] analyzed conditions in France for coal produced in the Nord and Pas de Calais districts for use in power stations, and gave prices for products of the coal-preparation plant sold to outside customers reduced to a common moisture content of 5%.

Ligny and Haumont[9] calculated the relative values of low-volatile Belgium coals of varying ash content to be used in a pulverized coal fired boiler with dry-ash removal. Their calculations were based on a unit of 115 Mw installed capacity, 127 atm (abs.) reheat cycle, temperature of preheat and reheat 540°C, thermal loading 2500 kcal/kwhr output, and operated 5500 hr/yr. The base coal contained 20% ash and 3% moisture and was 0 to 2 mm in size. However, calculations were given for a 10% ash, 3% moisture coal that permitted changing the base to this lower ash coal. This facilitated comparisons with the findings of other investigators cited herein. The value-line plotted from the Ligny-Haumont data has a slight discontinuity at 35% ash, less than 1%, owing to the reinjection of grit in coals of less than 35% ash, but not with higher-ash coals. They reasoned that reinjection of grit is not feasible with coals above 35% ash.

Dutch investigators[10] have analyzed, in considerable detail, the effect of ash on the operation of a colliery-operated pulverized-coal power station and the economics of preparing coal for such a station in the Netherlands. Van der Grinten, in an available unpublished report, analyzed the effect of inerts on mechanical grates (Martin). He concluded that, generally, the grate power station is inferior to the pulverized-fuel power station for most products of the preparation plant. One of the main factors that renders the grate power station inferior to the pulverized-fuel station is the boiler efficiency, 82% compared to 90% for a grade of coal at which the 90% figure would be reached by a pulverized-fuel power plant. This causes the sum of money to be allowed for 1 ton of coal to be approximately 6.5% lower with the grate power plant than with the pulverized-fuel plant.

Van Os and Meerman and Slotboom[11,12] analyzed the results obtained by the authors of the reports on the use of low-grade coal for boiler fuel[6–10] and concluded that "the various figures, viewpoints and conclusions . . . clearly show that low-grade fuel is far from being

[8] Nonat, L., *Proc. 3rd Intern. Coal Preparation Conf., Liège*, **1958,** *Ann. mines Belg.*, Special Issue (1958), pp. 180–91.

[9] Ligny, J., and Haumont, H., *Proc. 3rd Intern. Coal Preparation Conf., Liège*, **1958,** *Ann. mines Belg.*, Special Issue (1958), pp. 192–212.

[10] Van der Grinten, F. L. H., Van de Loo, O. F. H. H., and Beckers, J. M. H., *Proc. 3rd Intern. Coal Preparation Conf., Liège*, **1958,** *Ann. mines Belg.*, Special Issue (1958), pp. 213–31.

[11] Van Os, W. E., and Meerman, P. G., *Proc. 3rd Intern. Coal Preparation Conf., Liège*, **1958,** *Ann. mines Belg.*, Special Issue (1958), pp. 252–67.

[12] Slotboom, J. G., *Proc. 3rd Intern. Coal Preparation Conf., Liège*, **1958,** *Ann. mines Belg.*, Special Issue (1958), pp. 268–83.

TABLE 2

PERCENTAGES OF TRUE MIDDLINGS AND SHALE IN COALS FROM SEVERAL COUNTRIES

Germany[7]			Netherlands[10]			Belgium[9]				Great Britain[6]			France[8]
A	B_1	*C*	*A*	*B*	fine coal	1	2	3	4	*A*	*B*	*C*	Fine coal
True middlings, (specific gravity between 1.45 and 2.2)													
16	16	19	15	26	17	16	19	24	17	9	11	19	19
Shale, (specific gravity above 2.2)													
9	14	23	45	31	20	24	31	33	47	8	6	12	20

a semi-waste product, not deserving much attention, but that its value should be appreciated as a rather important part of the total value of the products brought to the surface."

Table 2 shows the differences in the coals studied by the authors of the various reports cited above. The relatively small amount of true middlings and the relatively small amount of shale (clean refuse) are largely responsible for the comparatively lesser interest taken in the problem of the "intermediate fraction" in Great Britain.

Slotboom[12] discussed in detail factors affecting boiler efficiencies and concluded:

(1) Under ideal conditions, that is, when the boiler is adapted to occurring variations in such a way that the crucial temperatures remain constant and the temperature in the stack remains normal (roughly 400°K), the boiler efficiency decreases only slightly with increasing incombustible content. This can be achieved when the following conditions are satisfied: ideal combustion, no ash removal from the combustion chamber, no reinjection of fly-ash, constant steam production.

(2) For every percent of infed coal which is not burnt, the boiler efficiency decreases by roughly 1%.

(3) From a heat-economical point of view, ash removal from the combustion chamber means a disadvantage.

(4) This drawback can be eliminated by reinjection of fly-ash only if the combustion is very poor.

(5) In the case of moderate to good combustion, reinjection of fly-ash means a disadvantage.

(6) Keeping the temperature along the gas path constant at increasing ash content calls for a larger heat-transmission surface in the convective part of the boiler and for a smaller transmission surface in the combustion chamber. The required size of the above surfaces depends to only a relatively low degree on the question whether or not the fly-ash is reinjected and whether or not the liquid ash is removed from the boiler.

(7) In searching for an economical boiler with a high efficiency which is suited for the firing of ash-rich fuels, it must be borne in mind that a proper aerodynamical design [self-cleaning?] and burners producing short flames are among the most important factors. A fundamental study of the combustion and the most favorable flow pattern in the vicinity of the burners is of very great importance (e.g., jet engines, gas turbines, ramjets).

(8) The firing of ash-rich fuel in a boiler for ash-poor coal, the performance of which, however, is not adversely affected by a moderate amount of ash leads to higher stack temperatures and hence to a lower efficiency. In the opposite case, the stack temperature may decrease too much so that it may be necessary to cut out part of the heating surface.

EVALUATION OF PREPARED COAL FOR METALLURGICAL PURPOSES

Coal is used in sintering, pelletizing, zinc retort-smelting, blast-furnace smelting and other metallurgical processes. For these uses, special coals prepared to rigid specifications are used to get the desired metallurgical results at lowest cost. By far the largest use is in the form of coke for the iron blast furnace. This discussion deals primarily with the interrelationship of mine, preparation plant, coke oven, and blast-furnace plant.

Conditions in the United States

The specifications of pig iron affected by coal impurities are:

Pig iron for acid Bessemer process:
- Phosphorus, maximum 0.10%, usual ranges from 0.085 to 0.10%.
- Sulfur, 0.04%, varies from 0.03 to 0.08%.
- Silicon, from 1.10 to 1.50%.

Pig iron for the basic open-hearth process:
- Phosphorus, no limit, but usually kept below 0.6%.
- Sulfur, maximum 0.05%.
- Silicon, maximum 1.25%.

Phosphorus is not usually reduced by coal washing, hence, only those coal seams low enough in P to keep the total P content of the blast-furnace burden within the range required are used. Silicon is a constituent of coal ash, all of which remains in the coke following the carbonization process. A knowledge of the amount of Si in the coke ash is necessary for proper blast-furnace burden calculations.

Because the operation of the coke plant is inextricably associated with the operation of the blast-furnace plant, this discussion of the effects of coal-preparation practices on the costs of producing a ton of molten metal considers the coke plant a part of the blast-furnace complex.

Major impurities in coal affecting blast-furnace operations that are removable or modified by preparation-plant processes are moisture, ash, and sulfur. Moisture is important only to the coke plant because it is removed in the coking process. The final quenched-coke contains approximately 3% moisture. Moisture in the feed to coke ovens ranges from 2 to 10%, depending on coal types and plant operating requirements. Ash in coal carries through the coking process to the final coke with only a slight reduction in weight due to calcining of carbonates and removal of water of constitution. The sulfur in the coke is equal to about 0.8 that of the coal carbonized.

Although the effects of high ash and high sulfur in coke for the blast furnace and high moisture in coal for the coke ovens can be evaluated rather readily for any particular situation, the effect of the variability of these constituents is more difficult. Gill,[13] in 1928, wrote:

"With the figures normally shown for blast-furnace practice, it is difficult to assess quantitatively the effect of ash variability on the operation of a furnace . . . high ash or sulfur are either of them definite evils, but if they are regularly and consistently high it is possible to so burden the furnace that the evil effects are confined to increased expense in the production of iron. On the other hand, a fluctuating ash is an unmitigated evil."

Variability of the impurity content of metallurgical coals is an important element in blast-furnace costs. Savings attributed to decreases in variability have been estimated by various investigators as being 25 to 30% of the total savings made by using prepared low-ash and low-sulfur coal over raw coal. Preparation processes, washing, and drying do greatly reduce the variability of the mined product, producing a uniform product for use.

Approximately 20% of the world's coal production is used for the manufacture of coke for the iron blast furnace. Until the early 1900's, good blast-furnace coke could be made only from coals with inherent good coking qualities. The development of the iron and steel industries in different areas of the world was facilitated or hindered by the abundance or deficiency of good coking coals.

During the first 100 yr of the iron and steel industry in the United States, coke was usually produced at or near the mines. During the past 50 yr, the manufacture of coke in the United States has largely moved to the site of the blast furnace, which may be several hundred miles from the mines supplying the coal.

As presently operated, the economics of coal preparation, transportation, carbonization, and blast-furnace operation are inextricably tied together. The type of raw coal and its washability and coking characteristics are inevitably an important factor in this complex. Because of the great number of variables from one mine to another, between coal fields, areas and countries, and in types and operation of coke plants and blast furnaces, it is practically impossible to produce figures representing savings in the production of pig iron by removing moisture, ash, and sulfur from the raw coal that could be universally applied. Figures given herein are to be considered as a guide only. Each complex must be studied in detail to arrive at valid figures.

The growth of mine mechanization and the drive to reduce the cost of producing a ton of pig iron since World War II are two forces that have caused many studies to be made of the various factors affecting the costs of producing a ton of pig iron. Mine mechanization requires practically all metallurgical coal to be washed and dewatered. Freight savings also often pay for the cost of washing and dewatering. Preparation-plant costs for cleaning approximately 4 in. top-size resultants range from about 25 cents per ton of clean coal for large capacity Baum-type jig plants for coals easily cleaned with screen and centrifuge dewatering,

[13] Gill, C. S., *J. Iron & Steel Inst.* (*London*), **117,** 91–7 (1928).

to as much as 75 cents for plants with multiple cleaning circuits for difficult coals and similar dewatering facilities. Froth flotation plants are being installed for the extreme fines with further increases in costs. Because of increases in the amount of slack sizes, many plants have installed thermal driers which further increase costs.

Because of increased rigid specifications and increased impurities in the mined product, there is a trend toward 3-product separations as in European practice, that is, a low gravity separation for metallurgical coal, a middling product for local steam plant use, and a clean refuse for disposal. The extent to which preparation can be done depends on the savings that can be effected in transportation of inert materials, savings in coke-plant and blast-furnace operation, and monetary return from middlings if recovered and sold. There have been instances where coals have been cleaned and dried lower than necessary.

Exclusive of transportation, savings at the coke plant are sometimes given in the range of 5 to 10 cents per ton for each 1% moisture reduction when considering moisture ranges of the order of 4 to 14%.

Costs for reduction of ash and sulfur content are less easily generalized. Again, exclusive of transportation costs, the reduction of 1% ash in the clean coal is given values to the blast furnace ranging from 20 to 60 cents/ton of coal with the United States average being somewhere around 35 cents. Similarly 0.10% sulfur reduction in the clean coal is given values from 10 to 50 cents/ton of coal with the average for the United States practice probably of the order of 15 cents/ton of coal.

Oil treatment may be done at the preparation plant or at the coke plant. There are advocates for each practice. If done at the preparation plant, it has the advantage of aiding drainage in transit and reducing frozen shipments.

Final pulverization is usually done at the coke plants in the United States. Often, modification of pulverization practices will improve coke yield and quality.

Bertholf[14] has given a step-by-step method for analyzing the effect of ash of coals then available at the Pueblo, Colorado plant of the Colorado Fuel and Iron Company in terms of fuel cost per ton of pig iron. For each coal, there was a point at which the fuel cost was a minimum, increasing as the ash was increased or decreased by preparation practices. Studies such as these are invaluable to management, in that moves can be made to remove high fuel-cost coals from the mix.

Price[15] analyzed this problem in a slightly different way in which all costs of the blast furnace affected by changes in the ash content of the coal were calculated per ton of pig iron produced. These included at each ash level per ton iron: washing conversion costs, coke yield and costs, coke cost per ton of iron, limestone costs per ton iron, and blast-furnace operating costs per ton iron. Not included were revenues from byproducts and coke breeze which were stated to closely match coke-oven operating costs for the plant under study and iron-ore-burden costs. The cumulative effect of these costs at various ash levels is given in Fig. 1.

King[16] has reported the results of a series of plant tests covering a 4-month period for washed and unwashed Pittsburgh Seam coal in the Pittsburgh district. A battery of ovens was charged with washed coal and other batteries with raw coal. As was to be expected, the total coke yield was greatest for the unwashed coal. However, coke from the washed coal was far superior to that from the unwashed, and the percentage of blast-furnace coke actually increased 2.2% over that from unwashed coal. For the blast furnace tests, two furnaces of identical physical dimensions and with similar accessories were used. One furnace used coke from washed coal for a 2-month period, with the companion furnace on coke from unwashed coals. For the second 2-month period this was reversed. Table 3 gives the composite data for this 4-month test period. King,[17] in a later paper commenting on these benefits derived from using coke from washed coal, stated: "I estimate that 60% of the total was directly due to the reduction of ash and sulfur. The other 40% was due to an improvement in physical properties and uniformity."

[14] Bertholf, W. M., *Steel*, 122–30, July (1946).

[15] Price, J. D., *Proc. 2nd Intern. Coal Preparation Conf., Essen*, **1954,** Steinkohlenbergbauverein, Essen, Paper B III 1 (1954), 5 pp.

[16] King, C. D., *AIME, Proc. Blast Furnace, Coke Oven & Raw Materials Comm.*, **3,** 3–23 (1943).

[17] *Ibid.*, **4,** 181 (1944).

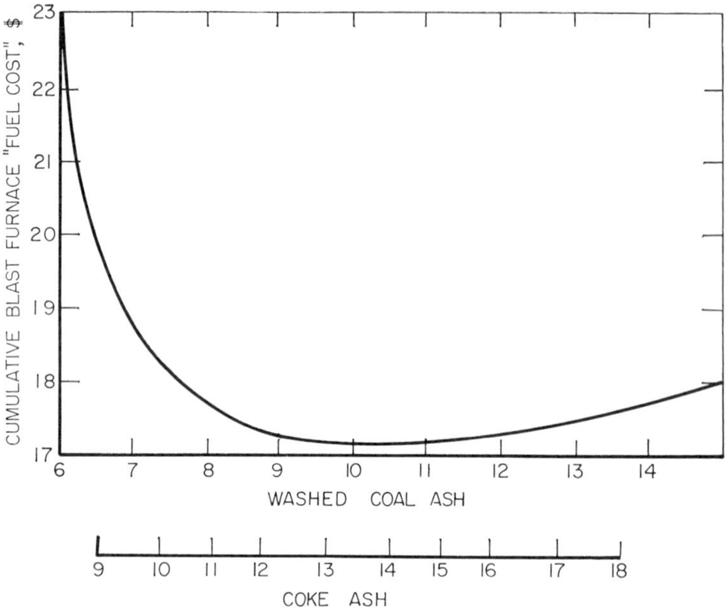

Fig. 1. Cumulative blast furnace fuel cost at various ash levels.[15]

Because of the many variables inherent in the flow of raw materials into a blast furnace complex, modern statistical methods such as multiple correlation and linear programming are being increasingly used for analyzing these problems. Flint[18] and MacFeeters[19] have given details of the multiple correlation method for studying the effect of variables on blast-furnace operation.

Several comparatively recent papers give effects of coke properties on blast-furnace operation at the Inland Steel Company, Chicago. Peters and Neustaetter[20] stated: "that physical characteristics of raw materials are at least equally as important as chemical characteristics" and Squarcy[21] discussed the use of foreign cokes. Gray and Isenberg[22] gave results of a comparative blast furnace test of coke (5.97% ash) from washed coal over coke (10.24% ash) from unwashed coal: "The iron production increased 226 tons per day or 10.42%, the coke consumption decreased 178 lbs per ton of iron or 9.34%. The furnaces operated with about the same stone consumption and there was a one percent decrease in slag volume."

[18] Flint, R. V., *AIME, Proc. Blast Furnace, Coke Oven & Raw Materials Comm.*, **11,** 49–68 (1952).

[19] MacFeeters, R. O., Joint Meeting of Eastern States Blast Furnace and Coke Oven Assoc. and The Blast Furnace and Coke Oven Assoc. of the Chicago District, Cleveland, Ohio, November, 1953.

[20] Peters, J. F., and Neustaetter, K., *AIME, Proc. Blast Furnace, Coke Oven & Raw Materials Comm.*, **13,** 12–28 (1954).

[21] Squarcy, C. M., *AIME, Proc. Blast Furnace, Coke Oven & Raw Materials Comm.*, **11,** 137–43 (1952).

[22] Gray, R. L., and Isenberg, N., *AIME, Proc. Blast Furnace, Coke Oven & Raw Materials Comm.*, **13,** 107–27 (1954).

Sulfur

In recent years most blast-furnace operators in the United States have been faced with an increasing amount of sulfur in blast-furnace coke. Various estimates of the increased cost of producing a ton of pig iron per 0.10% increase in sulfur in coal ranges from 10 to 50 cents. Various methods of combating this trend include washing coals at lower gravities and changing furnace practices to accommodate more sulfur in the coke.

Kennedy and Thornton[23] have given results of extended tests to increase the sulfur-holding ability of the blast-furnace slag. They concluded: "If uniformly low sulfur is expected, any increase in sulfur entering the blast furnace requires a corresponding increase in slag volume and/or basicity. Increased basicity is to be preferred to increased slag volume, providing the proper slag fluidity can be

[23] Kennedy, T. H., and Thornton, A. W., *Yearbook Am. Iron Steel Inst.*, **1949,** pp. 222–42.

TABLE 3

COMPARATIVE BLAST-FURNACE PERFORMANCE AND PRACTICE DATA FOR COMPOSITE 4-MONTHS TEST PERIOD[16]

	Furnace Coke from		Effect of Washing	
	Unwashed Coals	Washed Coals	Amount	%
Coke analysis (dry basis), percent:				
Ash	11.87	10.36	−1.51	
Sulfur	0.98	0.88	−0.10	
Fixed carbon	87.83	89.34	+1.51	
Day to day variations exceeding 0.30% ash in coke, percent	50.0	9.5	−40.5	
Iron production, net tons per day	781.1	844.4	+63.3	+8.1
Net coke rate, lb/ton iron	1909	1760	−149	−7.8
Flux rate, lb/ton iron	995	797	−198	−19.9
Calculated slag volume, lb/ton iron	1286	1091	−195	−15.2
Air blown, cu ft/min (carbon basis)	53,362	54,129	+767	+1.4
Blast pressure, lb/sq in.	19.3	18.6	−0.7	
Average iron analysis, percent:				
Silicon	1.13	1.09	−0.04	
Sulfur	0.029	0.028	−0.001	
Casts over 1.40% silicon, percent	8.97	4.45	−4.52	
Casts under 0.85% silicon, percent	8.63	6.17	−2.46	
Casts over 0.040% sulfur, percent	6.09	3.77	−2.32	

maintained. Up to an optimum level of somewhere between five and eight percent magnesia in the slag, its substitution for lime improves the desulfurization of the iron .... Increased magnesia beyond this range results in a reduction in sulfur absorbing power." They also gave a theoretical calculation showing that 118 additional pounds of coke and 295 additional pounds of slag would be required for a furnace producing 0.03% S iron and using 1700 lb of coke per ton of iron when the sulfur in the coke is increased from 1.00% to 1.25%. Based on these figures and with coke at $15/ton and limestone at $3/ton, the increased cost for each 0.10% S increase equals 52 cents/ton of pig iron. To this must be added the cost due to decreased iron production, because of the increased slag-volume.

No general rule can be made for increased costs due to an increase of 1.10% S. Some authorities have stated that each 0.10% S will increase costs approximately the same as 1.0% ash. This may be far from the truth, because the ability of particular furnace complexes to absorb increases vary widely. However, for a furnace operating at maximum slag basicity for absorbing sulfur, any increase will be reflected in increased costs.

Table 4 gives a summation of results given in the recent technical publications cited herein. The results are calculated to a common basis of 1% coke ash reduction and show the great variability to be expected in comparing one plant with another in the United States.

European and English Conditions

Several very valuable papers dealing with the relations of the mines, preparation, coke, and blast-furnace plants were given at the Third International Coal Preparation Congress in Liège, Belgium, in 1958, representing conditions in five countries.[24] Properties of the various coals and cokes used in these countries are given in Table 5. Prices and price ratios given must be used with a certain amount of reserve, because it is not certain that all costs are included in all prices. Conditions in

[24] Loison, R., *Proc. 3rd Intern. Coal Preparation Conf., Liège*, **1958**; *Ann. Mines Belg*. Special Issue (1958), pp. 43–65; Lemke, K., pp. 65–76; Schranz, H., pp. 77–86; Bakels, P. S., pp. 87–110; Goffart, J., and Stienon, J., pp. 111–18; Griffiths, W. H., pp. 119–29; and Reerink, W., and Lemke, K., pp. 130–45.

TABLE 4

Calculated Summation of Effect of 1% Coke Ash Reduction on Blast Furnace Operation from United States References

(Sulfur Reduction Negligible Except for Reference 16 with a 0.1% S Reduction)

Reference No.	Type of Tests or Calculation	Savings in Coke, Pounds Per Net Ton Hot Metal	Savings in Flux, Pounds Per Net Ton Hot Metal	% Increase in Hot Metal Production
15	Arithmetic	39	31	2.3
16	2-mon plant tests	105	150	4.50
16	2-mon plant tests	91	116	6.20
18	Multiple correlation studies	55		
19	Multiple correlation studies	32		5.80
22	4-mon No. 6 BF	56	7	1.16
22	2-mon No. 3 plant	42	0	2.55

these countries differ considerably from conditions in the United States, in that it is common practice to have the coke ovens at the mines, and three or more product separations are made at the preparation plants.

Table 6 gives the relations between price and quality for coking coal in those countries which have standards tied to premium and penalty provisions. No correlation between sulfur content and price of coking coal is given for any of the five countries. Also no correlation is given for moisture and ash in relation to the price of coking coal in Holland.

Table 7 gives the relation of coke quality to price in Germany, England, and France. Papers from Belgium and Holland gave no correlations.

As in the United States, changing seam qualities and mine mechanization in recent years have put an increasing burden on the preparation plant to prepare a low impurity product for coking. Because of increased ash and moisture in the fine sizes it is frequently impossible to bypass these sizes or dry clean them, so that moisture could be controlled by mixing dry slack or aspirated dust with the wet-washed coarse sizes. Almost universally, all sizes must be wet-washed and the final moisture regulated by more elaborate dewatering and thermal drying.

Moisture

The cost of reducing the moisture content to an acceptable figure of 5 to 10% depends on the type and size of the coal. For hard-structure coals with a low percentage of fines, and for those exceptional cases of low-ash and low-moisture content in the fines, it is still possible by the use of modern drainage towers, screens, and centrifuges to reach a moisture content of 8 to 10% in the final product. The cost of this type of dewatering is small.

These cases are increasingly uncommon, and eventually the only way to reduce the moisture content will be by the heat-drying of filtered slurry derived from a froth-flotation or other cleaning process. The cost of heat-drying given by Bakels, Lemke, and Loison ranged from 0.021 DM to 0.23 DM/kg of water.[24] These figures refer to drying of slurry from about 25% moisture to 5% moïsture in medium-size thermal driers running 14 hr per day. These costs, when referred to a ton of coal, are calculated to be from 0.25 DM to 0.35 DM for each 1% moisture reduction, a sizable cost-increase.

Advantages of reducing the moisture content, other than transportation costs, accrue mainly to the coke plant. These may be summarized as greater throughput, reduction of heating-gas requirements, and increased ammonia plant costs, partly compensated by increased ammonia made and decreased life of ovens. It can be said that the coke plant saves between 0.5% and 1% of the price of one ton of coal (moisture free) for each 1% reduction in moisture. Bakels[24] calculated a saving of 0.54 DM/ton of moisture-free coal for each 1% moisture reduction or 0.9% of the price of the coal. Transport and life of the ovens' costs were not included.

Various objections are raised for too-dry coal. These include excessive dust loss and fires, dust in the tar, and windage loss in

TABLE 5

CHARACTERISTICS AND PRICES OF COKING COAL AND BLAST-FURNACE COKE[24]

	Belgium	Germany (Ruhr)	England	France (Nord, Pas de Calais)	Holland
Coking coal					
Total moisture content, %	8	10	8	7	8
Ash content (moisture-free) %	8	7–8	7–8	7–8	8
Price in 1957 (DM/t)*	74.53	56.60	57.83	64.10	63.98
Blast furnace coke					
Total moisture content, %	3	5		1	1
Ash content (moisture-free) %	10	10	10	10	10
Lower size limit (mm)	60	80 (60/40)	40	60	60
Price in 1957 (DM/t)	121.28	76.60 (73.1/72.4)	87.05	91.50	86.09
Pig iron for steel production					
Price in 1957 (DM/t)	336	278	220	347	343
Price ratios in 1957					
Coking coal/blast furnace coke	1:1.6	1:1.3	1:1.5	1:1.4	1:1.3
Blast furnace coke/pig iron	1:2.8	1:3.7	1:2.5	1:3.8	1:4.0
Coking coal/pig iron	1:4.5	1:4.9	1:3.8	1:5.2	1:5.4

* Equivalent German prices are calculated using the following factors: Belgium: 1 FB = 0.084 DM; England £1 = 11.70 DM; France 1 FF = 0.012 DM; Holland 1 fl = 1.105 DM; USA $ = 4.20 DM.

TABLE 6

RELATIONS BETWEEN QUALITY AND PRICE OF COKING COAL[24]

Country	Coking Coal: Total Moisture Content	Coking Coal: Ash Content
Belgium	Basic figure: 7%. The price changes by 0.1% for a 0.1% change in moisture content.	Basic figure 10%. The price changes by 0.1% for a 0.1% change in ash content.
Germany*	For the calculation of the price the weight of a consignment is usually recalculated to a total moisture content of 10%. Premiums or penalties for lower moisture content are not usual.	Coking coal above and below 9.8% ash content (moisture free) is evaluated on the basis of coke quality (see Table 7), taking yield of coke into account.
England	The weight of the completely drained charge is calculated.	The price of coking coal and the price of coke depend on the ash content.
France (Nord, Pas de Calais)	Basic figure: 7%. The price changes by 0.1% for a 0.1% change in moisture content.	Basic figure 9%. The price changes by 0.25% for a 0.1% change in ash content.

* There are premiums for low ash content and penalties for high ash content in Germany, either for the coking coal (e.g., from or to metallurgical coke-oven plants) or for the coke (from or to blast-furnace plants). The price increments or decrements for ash content are passed back by the coke-oven to the colliery, because the latter is responsible for the value of the ash content.

TABLE 7

RELATIONS BETWEEN QUALITY AND PRICE OF COKE[24]

Country	Blast Furnace Coke: Total Moisture Content	Blast Furnace Coke: Ash Content	Total Sulfur Content
Germany	Coke + 80 mm: 3.0% + 60 mm: 4.5% + 40 mm: 5.0% Higher or lower values than the limits quoted are subject to a premium or penalty of 0.11% of the current sale price at the pit head for each 0.1%.	Limiting value for all sizes +80, +60 and +40 mm, 9.8%. Lower values than limit subject to increase of 0.3% of pit head sale price per 0.1% decrease in ash content (moisture free). Further reduction of ash content below 9% subject to premium of 0.2% of price per 0.1% ash content. Higher ash contents than 9.8% subject to deduction of 0.3% of price per 0.1% ash content (moisture free).	Tolerance 0.9–1.2 % total sulfur (moisture free). Below lower limit 0.9 subject to 0.3% premium per 0.1%. Excess over 1.2% subject to penalty of 0.3% of pithead sale price per 0.1%.
England	The price of coke depends on its moisture, ash and sulfur content. The basis for the calculation is the calorific value. The coke is also evaluated according to shatter test results at 1½ and ½ in.		
France (Nord, Pas de Calais)	Coke + 90 mm: 2% 90–60 mm: 4% 60–40 mm: 4% 0.15% of price per 0.1% moisture content for +90 mm coke.	Limit for all sizes 10.5%. 0.25% of price per 0.1% ash content.	No correlation.

transport and stocking. Each plant must evaluate the moisture problem separately.

Ash

The type of washing equipment and cost of washing were discussed in considerable detail by Lemke[24] and Bakels.[24] In most European countries, and particularly in Germany, the supply of coking coal is met by the natural yield of the 10 mm × 0 size. The examples that have been given relate to this size.

The costs used by Lemke for the various preparation processes are given in Table 8. These costs are stated to be valid for the Ruhr area for machines and equipment of medium size working at rated capacity for 14 hr a day. Overall costs for any particular coal depends on the washability characteristics of the coal, size consist, machine design, flow arrangement, and control. Where the raw dust and slurry are low in ash, they are recovered and added back without washing. However, almost always the dust and slurry (minus 0.5 mm) must be cleaned by the relatively costly cleaning stages of the froth-flotation or Convertol process. Bakels examined, in considerable detail, the effect of the ash content of coal on the prime cost of coke and pig iron for a 3-product separation plant for the tariffs and conditions existing in Holland. Washing costs for a plant equipped with heavy-medium washers for nuts, cyclones for fines, and froth flotation and thermal drying for slurry totaled 2.67 Dutch guilders per ton of feed. Increase in the value of coking coal for each 1% ash reduction was calculated to be 0.98 Dutch guilders.

It is difficult to generalize on the savings to be expected in blast-furnace operations for each 1% decrease in the ash content of the coke. Conditions vary widely in Europe and England. Generally, and exclusive of transport costs, the advantages to the blast furnace will be between 1.5 and 2.5% of the price of blast-furnace coke for each percent decrease in ash content. Using the average of 2.0% for

TABLE 8

COST OF PREPARATION[24]

Preparation Machine or Process	Cost in DM/Metric Ton of Feed (Moisture Free)
Rewashing in a dense medium (2-product separation)	1.3
Filter	1.2
Flotation (Froth)	1.2
Drying of filter cake to 5% total moisture content	6
Flotation operation	6

$20 coke, this would be equal to 40 cents/ton of coke for each 1% ash reduction. Naturally increased costs of preparation and coking must be less than this to justify producing a lower ash coke.

Sulfur

The advantage to the blast furnace in European countries for reducing the sulfur content of coke by 0.10% is of the order of 0.3 to 0.45% of the price of coke.

WASHABILITY CHARACTERISTICS

The washability characteristic of a coal is the most important tool available to determine the extent to which a coal may be cleaned. Examination of washability data for a particular coal or for a particular size of coal will reveal the quality of coal which may be obtained by mechanical cleaning as well as the quantity of coal of a particular quality. Further examination of the data will indicate the ease or difficulty with which the coal may be cleaned.

Mechanical cleaning of coal in this country was first reported by Stutz[25] in 1880. However, the first systematic study on the washability characteristics of various coals in the United States was not reported until 1924.[26] These studies were concerned with coals in the Western states. Later, in 1929, the United States Bureau of Mines[27] reported on the washability characteristics of coals from many of the important coal fields of the Eastern and Central states. Since 1950, the U.S. Bureau of Mines[28] has conducted an extensive survey on the washability characteristics of bituminous coals from Pennsylvania, Kentucky, Virginia, West Virginia, and Tennessee to determine their suitability for metallurgical use.

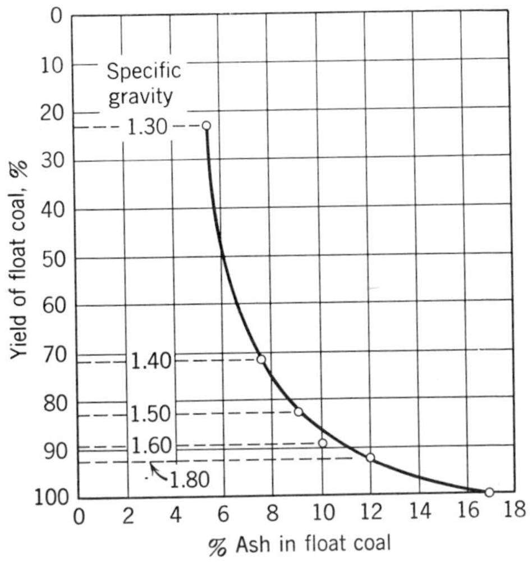

Fig. 2. Typical yield-ash curve.

The washability curves as presented in the early studies were of a yield-ash type as presented in Fig. 2. This type of curve indicated the quantity and quality of coal to be obtained by cleaning at certain specific gravities. To obtain the specific gravity to be used in order to meet ash requirement an estimate had to be made between the specific gravity levels shown in the figure. In order to simplify the selection of specific gravity, the washability curves were drawn as illustrated in Fig. 3. This method of presentation was used in this country by Bird.[29]

From Fig. 3 it may be seen that the original coal (100%) has 17.5% ash. If a clean coal of 8.0% ash were desired, the theoretical specific gravity to be used in cleaning the coal would be obtained as illustrated in the figure. Starting with the 8.0% ash value on the *B* curve, there

[25] Stutz, S., *AIME, Trans.*, **9,** 461–77 (1880).

[26] McMillan, E. R., and Bird, B. M., *Univ. Wash. Eng. Exp. Sta., Bull.* **28,** (1924), 234 pp.

[27] Yancey, H. F., and Fraser, T., *U.S. Bur. Mines, Bull.*, **300,** (1929), 250 pp.

[28] Crentz, W. L., et al., *U.S. Bur. Mines, Repts. Invest.*, 4747 (1950); 4763, 4788, 4815, 4823, 4834 (1951); 4894, 4910, 4920 (1952); 4941, 4993 (1953); 5029, 5070, 5094 (1954); 5112, 5131, 5135, 5136, 5140, 5159, 5185 (1955); 5208, 5229, 5230, 5232, 5235, 5258, 5263, 5266, 5278 (1956); 5306, 5357, 5358 (1957); 5391, 5405, 5434 (1958); 5475, 5492 (1959); 5572, 5609 (1960).

[29] Bird, B. M., *Proc. 2nd Intern. Conf. Bituminous Coal*, **2,** 82–111 (1928).

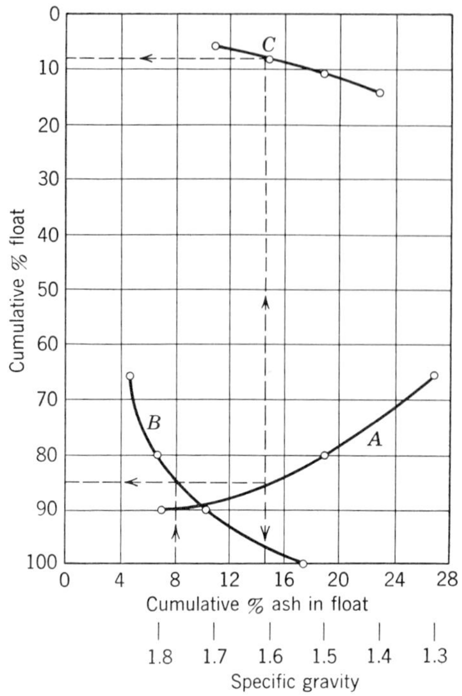

Fig. 3. Typical washability curves. *A*. Specific gravity-yield; *B*. yield-ash; *C*. ±0.1 distribution.

is indicated a yield of 85% clean coal. By taking the yield value to the *A* curve, the specific gravity may be observed to be 1.61. At this specific gravity the amount of near-gravity material present in the coal sample may be obtained from curve *C*. This amount of material is defined as the percentage of the coal sample that will float in a range within plus and minus 0.10 specific gravity of the cleaning value. In this particular example, the ±0.10 specific-gravity distribution is 8.0%. According to Mitchell,[30] the ease or difficulty to be expected in plant operation may be estimated by the amount of near gravity material at the specific gravity to be used to clean the coal. These estimates are presented in Table 9.

The data necessary to plot washability curves are obtained in the laboratory by examining the float-and-sink characteristics of the coal, or a particular size fraction of the coal. To determine these characteristics, it is necessary to have a representative sample of the coal crushed to a specific top-size. Generally the top size is determined by market specifications, by the properties of the coal, by the method of mining and by the capacity of the test equipment.

Table 10 illustrates the top size as 4 in. and the coal sized at $\frac{1}{4}$ in., so that consideration may be given to cleaning the coarse sizes, 4 × $\frac{1}{4}$ in., by one method, and the fine sizes, $\frac{1}{4}$ in. × 0, by another. The sizes are treated separately in organic liquids which have been blended to desired specific gravities. The sized sample is placed in the lowest specific gravity blend. Generally, this is 1.30 for bituminous coal and 1.40 for anthracite. The float fraction is removed, weighed, and analyzed. The weight is calculated on a percentage basis and the fraction may be analyzed for ash, sulfur, Btu, or any other particular characteristic of the coal desired for evaluation. The fraction of the coal that sinks in the lowest specific gravity fluid is then removed and placed in the next higher gravity fluid. The float fraction is again removed, weighed, and analyzed, and the sink fraction is processed in the next higher gravity. The preceding procedure is repeated until a sufficient number of float-sink fractions have been obtained for an evaluation. The

[30] Mitchell, D. R., Ed., *Coal Preparation*, AIME, New York, 1950, p. 785.

TABLE 9

SIGNIFICANCE OF AMOUNT OF NEAR-GRAVITY MATERIAL

Amount of Near-Gravity Material		Estimate of Coal-Preparation Plant Cleaning Problem
Greater than	Less than	
0%	7%	Simple
7%	10%	Moderately difficult
10%	15%	Difficult
15%	20%	Very difficult
20%	25%	Exceedingly difficult
25%		Formidable

TABLE 10

WASHABILITY CHARACTERISTICS OF A BITUMINOUS COAL

1	2	3	4	5	6	7	8	9	10	11	12	13	14	15	16	17	18	19	20	21	22
										Cumulative Float						Cumulative Sink					
Size of Coal	Size Wt %	Specific Gravity		Wt %	Size Wt %	% Ash	% S	Product Values		Wt %	Size Wt %	Ash Product	% Ash	S Product	% S	Wt %	Size Wt %	Ash Product	% Ash	S Product	% S
		Sink	Float					Ash	Sulfur												
4 × ¼ in.	60.0		1.30	70.0	42.0	5.0	1.00	210	42.0	70.0	42.0	210	5.0	42.0	1.00	100.0	60.0	1050	17.5	93.0	1.55
		1.30	1.50	10.0	6.0	20.0	1.50	120	9.0	80.0	48.0	330	6.9	51.0	1.06	30.0	18.0	840	46.6	51.0	2.84
		1.50	1.80	10.0	6.0	40.0	2.00	240	12.0	90.0	54.0	570	10.6	63.0	1.17	20.0	12.0	720	60.0	42.0	3.50
		1.80		10.0	6.0	80.0	5.00	480	30.0	100.0	60.0	1050	17.5	93.0	1.55	10.0	6.0	480	80.0	30.0	5.00
¼ in. × 0	40.0		1.30	60.0	24.0	4.0	1.00	96	24.0	60.0	24.0	96	4.0	24.0	1.00	100.0	40.0	696	17.4	53.6	1.34
		1.30	1.50	20.0	8.0	15.0	1.40	120	11.2	80.0	32.0	216	6.8	35.2	1.10	40.0	16.0	600	37.6	29.6	1.85
		1.50	1.80	10.0	4.0	35.0	1.60	140	6.4	90.0	36.0	356	9.9	41.6	1.16	20.0	8.0	480	60.0	18.4	2.30
		1.80		10.0	4.0	85.0	3.00	340	12.0	100.0	40.0	696	17.4	53.6	1.34	10.0	4.0	340	85.0	12.0	3.00
4 in. × 0	100.0		1.30		66.0			306	66.0		66.0	306	4.6	66.0	1.00		100.0	1746	17.5	146.6	1.47
		1.30	1.50		14.0			240	20.2		80.0	546	6.8	86.2	1.08		34.0	1440	42.4	80.6	2.37
		1.50	1.80		10.0			380	18.4		90.0	926	10.3	104.6	1.16		20.0	1200	60.0	60.4	3.02
		1.80			10.0			820	42.0		100.0	1746	17.5	146.6	1.47		10.0	820	82.0	42.0	4.20

data thus obtained, assuming that each float fraction and the final sink fraction have been analyzed for ash and sulfur, will provide information listed in columns 1, 2, 3, 4, 5, 7, and 8. The values listed in column 6 are obtained by multiplying columns 2 and 5. Also, the data in column 6 for the 4 in. × 0 coal is obtained as a composite of the figures above in the various float-sink fractions.

Using the values in the first 8 columns of Table 10, calculations for cumulative values may be made. The ash and sulfur product values, columns 9 and 10, are made by multiplying columns 6 and 7 for column 9, and 6 and 8 for column 10. The product values are simply unit values for the ash and sulfur. These same unit values for the entire coal, 4 in. × 0, composite may be obtained by adding the unit values in each of the specific gravity fractions for each size. Cumulative float-weight percent values, column 11, and cumulative float-size weight percent values, column 12, are obtained by cumulating columns 5 and 6, respectively, for successively higher specific gravities. The cumulative ash and sulfur products, columns 13 and 15 respectively, are obtained in the same manner from columns 9 and 10. The cumulative percent ash in the float coal is obtained by dividing the cumulative units of ash, column 13, by the cumulative size weight percent, column 12, and the cumulative percent sulfur is obtained in the same manner, that is, by dividing column 15 by column 12.

The cumulative float values indicate the quantity and quality of the clean coal to be expected by mechanically cleaning the coal at the various gravities. Thus, if the 4 × $\frac{1}{4}$ in. coal, with 17.5% ash and 1.55% sulfur, is cleaned at a specific gravity of 1.50, the float or clean coal, from column 4 (1.50 float) will amount to 80.0% of the 4 × $\frac{1}{4}$ in. size, column 11, or 48.0% of the entire 4 in. × 0 coal, column 12, and the clean coal will have 6.9% ash, column 14, and 1.06% sulfur, column 16. Also, if the $\frac{1}{4}$ in. × 0 coal, with 17.4% ash and 1.34% sulfur is cleaned at a specific gravity of 1.50, the clean coal would amount to 80.0% of the $\frac{1}{4}$ in. × 0 size, or 32.0% of the 4 in. × 0, and the clean coal would have 6.8% ash and 1.10% sulfur. The composite data also indicates that the 4 in. × 0 coal, with 17.5% ash and 1.47% sulfur, if cleaned at a gravity of 1.50 would yield 80.0% clean coal with 6.8% ash and 1.08% sulfur.

Cumulative sink values are obtained in essentially the same manner and are used to indicate the theoretical quantity and quality of refuse when the coal is mechanically cleaned at some designated specific gravity. The cumulative sink-weight percent values, column 17, and cumulative sink-size weight percent values, column 18, are obtained by cumulating columns 5 and 6, respectively, for successively lower specific gravities, that is, by taking the weight percent at the highest specific gravity and adding to it the weight percent at the next lower gravity. The cumulative ash and sulfur products, columns 19 and 21, are obtained in the same manner from columns 9 and 10, respectively. The cumulative percent ash and percent sulfur in the sink product are obtained by dividing the cumulative sink ash and sulfur products, columns 19 and 21, by the cumulative sink size weight percent, column 18. These results are presented in columns 20 and 22.

If a coal that had the characteristics given in Table 10 were cleaned at a specific gravity of 1.50, the refuse from the 4 × $\frac{1}{4}$ in. fraction would amount to 20.0% of the particular size, from column 17 for the 1.50 sink listed in column 3, or 12% of the complete size range, from column 18, and the refuse would contain 60.0% ash and 3.50% sulfur from columns 20 and 22, respectively. Similar data on the $\frac{1}{4}$ in. × 0 refuse obtained by cleaning this size at a specific gravity of 1.50 will show that the refuse would amount to 20.0% of the $\frac{1}{4}$ in. × 0 coal or 8.0% of the entire size range, and that the refuse would contain 60.0% ash and 2.30% sulfur. Also, the composite sink values indicate that if the 4 in. × 0 coal were cleaned at a specific gravity of 1.50, the refuse would amount to 20% of the original coal and contain 60.0% ash and 3.02% sulfur. Figure 4 illustrates how the washability characteristics presented in Table 10 may be used to show the possibilities of mechanically cleaning the coal. This figure presents a schematic flowsheet and material balance on the grade and recovery of all products.

Within recent years, special attention has been given to the possibility of obtaining three products from the preparation process: (1) a high-grade clean coal for metallurgical purposes, (2) a middling fuel for steam generation, and (3) a high-ash refuse. Ruell[31] has described a method of obtaining theoretical values for a middling product. The method is based upon the unit products as listed in columns 9 and 10 of Table 10. After the specific gravity is determined to obtain a high-grade low-ash or low-sulfur product or both, the number of units of ash and sulfur in this product are deducted from the total number of units. Then consideration is given to the weight percent and to the number of units to be subtracted from the total by cleaning at higher gravities. The remaining weight percent, after subtracting the weight percent of clean coal and the weight percent of clean refuse from 100%, represents the recovery of weight percent of middling product. By dividing the remaining units of ash and sulfur by the remaining weight percent, the grade of the middling will be obtained.

If, for example, it were desired to have a 3-product process for the coal described in Table 10, estimates regarding the quantity and quality of the middling product would be made as follows for the 4 in. × 0 composite: First, the coal would be cleaned at a specific gravity of 1.50 to obtain a clean coal amounting to 80.0% of the feed, and the clean coal would have 6.8% ash and 1.08% sulfur. The sink product from this process would amount to 20.0% of the original coal and would contain 60.0% ash and 3.02% sulfur. This sink product would then be cleaned at a specific gravity of 1.80 to obtain a float middling product and a sink refuse. The sink refuse would amount to 10.0% of the original coal and would contain 82.0% ash and 4.20% sulfur. Because the clean coal amounts to 80.0% of the original coal, and because the refuse amounts to 10.0%, the middling product must amount to 10.0%. Also, because the total units of ash amount to 1746, and because 546 are removed in the clean coal, column 13, and 820 are removed in the refuse, column 19, the balance or 380 remain in the middling. Thus, the 380 divided by 10.0 will yield a middling with 38.0% ash. Following the same reasoning the sulfur content of the middling would be 1.84%.

When the float-sink data on a particular coal, or a particular size of coal, contain values for a larger number of specific gravities, for example, 1.30 to 2.00 at 0.10 intervals, then a whole series of refuse and middling values can

[31] Ruell, D. A., *Fuel*, **29**, 211–15 (1950).

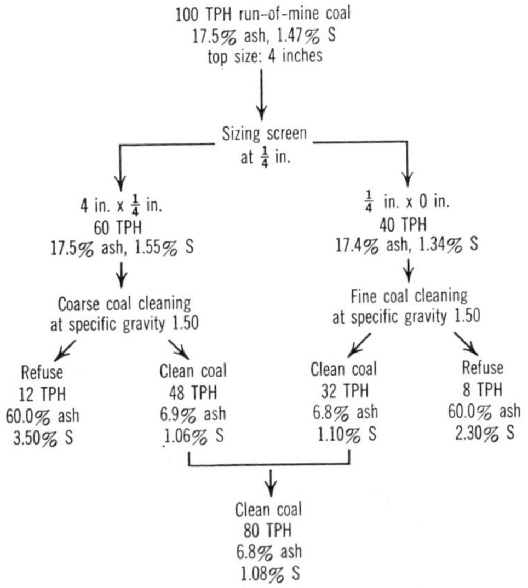

Fig. 4. Schematic flowsheet and material balance showing the theoretical possibilities for mechanically cleaning a bituminous coal based on the washability characteristics of the coal (basis—100 TPH original coal).

be obtained for a given clean coal. Thus, if a clean product is made as the 1.40 float product, the middling and refuse separation may be made at a specific gravity of 1.7, 1.8, 1.9, or 2.0, and theoretical calculations may be made for the quantity and quality of the corresponding middling and refuse products. Ruell[31] illustrated this in his paper and also showed how curves might be drawn for use in predicting the quantity and quality of middling and refuse products at intermediate specific gravity values such as 1.85. Calculations may be made and curves drawn in a similar manner from the same data based upon a clean coal product at 1.50 or some other specific gravity. Other investigators[32] have also presented a similar approach to the theoretical considerations of making a three-product separation.

After the float-sink data is obtained for a particular coal, consideration must be given to the selection of the specific gravity to be used in actual plant operation. This is necessary to obtain the optimum degree of cleaning. Because the separation of the coal does not result in a "perfect" separation of combustible and mineral matter there will be heat losses in the refuse product. Mathews[33] has pointed out

[32] Mayer, F. W., *Glückauf*, **86**, 498–509 (1950). Sommer, O., *Glückauf*, **91**, 878–89 (1955).

[33] Mathews, J. K., *J. Inst. Fuel*, **27**, 430–7 (1954).

that such items as market specifications and transportation costs are prime factors to be considered. Using these factors, along with washability data, calculations may be made and curves plotted to obtain the particular specific gravity to be used in cleaning for the optimum recovery of combustible material. Naturally, as the economic factors change, additional curves must be produced. Actual examples were presented.

Where a specific ash or sulfur content is not required to meet specifications, it is necessary to determine the point at which an increase in recovery is more than offset by a lower quality product. This has been evaluated by Dell[34] and takes into consideration not only the value of the lower grade product and transportation cost but also the cleaning cost. The basis of this method is to determine the zero-ash equivalent or the recovery of coal with zero ash which is equivalent in dollar value to the coal being considered. These figures are then related to market prices and preparation costs.

Summarizing, it may be stated that the washability characteristics, or float-sink data, for a coal may be used to determine the quantity and quality of products that theoretically might be obtained from a cleaning process. The ash, sulfur, or Btu specifications, or both, of the final product, generally dictate whether or not cleaning is necessary. After technical considerations are evaluated with the help of the washability characteristics, and if cleaning is technically sound, then economic considerations must be made. These considerations include the value of the final product and the cost of obtaining it. Cleaning may be necessary to meet specifications or to make a higher-grade product with increased financial reward.

Determination of Efficiencies

After a cleaning plant has been constructed and placed into operation based upon washability data, the coal-preparation engineer evaluates the operating efficiency. The washability data is used to determine the specific gravity of cleaning as well as the quantity and quality of all products. Thus the efficiency of the plant operation, or the operation of any particular cleaning process being used in the plant, is simply a measure of the actual results compared with the theoretical. In spite of the apparent simplicity of the problem there have been numerous methods proposed to evaluate efficiency.[35] References are listed in chronological order so that the reader may follow the developments over a 40-yr period. Because there is no method that has been generally accepted by the industry, no particular details are presented at this time.

The lack of a "standard" method of determining efficiency probably results from the large number of variables in a coal-preparation operation. The variables include the characteristics of the coal to be cleaned, the particular method of operation, the amount and nature of the impurities to be removed, the percentage of near gravity material, the friability of the coal, and the size range. The capacity and type of the operation, the selected specific gravity of cleaning, the relative amounts of the float-and-sink products, the methods of handling the feed to and the products from the operation, and the skill of the plant operator are also factors affecting efficiency.

Even though there are many methods for evaluating efficiency, the coal-preparation engineer and plant operator generally rely upon the amount of misplaced material in the products. For example, if the washability characteristics of a coal and the economics of the overall operation dictate that the coal should be cleaned at a specific gravity of 1.50, the engineer or plant operator is interested in

[34] Dell, C. C., *Colliery Eng.*, **34**, 105–9 (1957).

[35] Handcock, D., *Alabama Geol. Survey, Monograph* No. 7, (1912), pp. 234–45. Delamater, G. R., *Coal Age*, **5**, 723–9 (1914). Drakeley, T. J., *Trans. Inst. Mining Engrs. (London)*, **54**, 418–59 (1917–18); **59**, 86 (1919–20). Fraser, T., and Yancey, H. F., *Trans. Amer. Inst. Mining Met. Engrs.*, **69**, 447–69 (1923). Campbell, J. R., *Mining Congr. J.*, **14**, 555–8 (1928). Chapman, W. R., and Mott, R. A., *The Cleaning of Coal*, Chapman & Hall, London, (1928), Chap. 21. MacLaren, W., *Colliery Guardian*, **144**, 231–50 (1931). Tromp, K. F., *Glückauf*, **73**, 121–31 (1937). Reed, W., and Grounds, A., *Colliery Eng.*, **23**, 319–22 (1946). Hancock, R. T., *Colliery Eng.*, **24**, 140–2, 191–3, 237–9 (1947). Anderson, W. W., AIME, *Trans.*, **187**, 256–64 (1950). Belugou, P., and Ulmo, V., *Rev. ind. minérale*, **31**, 13–23 (1950). Raineau, —, and Belugou, P., *Rev. ind. Minérale*, **31**, 24–41 (1950). Belugou, P., and Vitoux, —, *Rev. ind. Minérale*, **31**, 209–29 (1950). Lyons, O. R., AIME, *Trans.*, **193**, 895–902 (1952).

TABLE 11

WASHABILITY CHARACTERISTICS OF AN ANTHRACITE COAL[36]

1	2	3	4	5	6	7	8
Specific Gravity				Cumulative Float		Cumulative Sink	
Sink	Float	Wt %	% Ash	Wt %	% Ash	Wt %	% Ash
	1.40	0.88	5.95	0.88	5.95	100.00	19.75
1.40	1.50	60.36	7.73	61.24	7.70	99.12	19.87
1.50	1.60	16.85	17.60	78.09	9.84	38.76	38.79
1.60	1.65	3.56	24.60	81.65	10.48	21.91	55.08
1.65	1.70	2.41	31.93	84.06	11.10	18.35	60.99
1.70	1.80	3.32	38.60	87.38	12.14	15.94	65.39
1.80	1.90	2.02	45.60	89.40	12.90	12.62	72.43
1.90	2.00	0.88	47.82	90.28	13.24	10.60	77.55
2.00	2.10	0.48	53.45	90.76	13.45	9.72	80.24
2.10		9.24	81.63	100.00	19.75	9.24	81.63

the amount of 1.50-sink material in the clean coal and the amount of 1.50-float material in the refuse. These two amounts represent material that has been misplaced during the operation. An excessive amount of misplaced material in the clean coal indicates a poorer grade than the theoretical value, and an excessive amount of misplaced material in the refuse indicates a lower recovery. Conversely, the more 1.50-sink material in the clean coal, the higher will be the recovery of the clean coal, but with a lower grade; and the more 1.50-float material in the refuse, the higher will be the grade of the refuse. The latter represents a loss of clean coal in the refuse.

In order to obtain a more satisfactory evaluation of an operation, some engineers require complete washability data on the clean coal and refuse products similar to that obtained on the original coal. This may be necessary when a particularly friable coal is being cleaned. At times, the particles of a friable coal will have a tendency to break up during the cleaning process. This may result in the liberation of pure coal from refuse and thus change the original washability characteristics. When complete washability data is available for the clean coal and refuse products, calculations may be made for the feed to the operation, and the calculated values may be compared to the original washability characteristics. Complete washability data on the products also makes it possible to determine the distribution of any specific gravity fraction in the clean coal and in the refuse.

Table 11 represents washability data on a typical anthracite coal.[36] From the data and economic conditions, it was decided to clean this coal at a specific gravity of 1.70. Examination of the data shows that the coal had 19.75% ash and that at a specific gravity of 1.70 the clean coal should amount to 84.06% of the feed and have 11.10% ash. Also, the data indicates that the refuse should amount to 15.94% of the feed and that the refuse should contain 65.39% ash. The near-gravity material amounts to 9.29%, indicating that the coal should be moderately difficult to clean at this specific gravity.

Table 12 presents the same type of washability data on the clean coal and refuse products from the plant operation. These data indicate, from column 6 or column 8 for 100% of the respective products, that the clean coal had 11.99% ash, and that the refuse had 55.57% ash. The recovery of clean coal, 82.19%, shown in Table 12 was calculated from the percentages of ash in the feed coal, clean coal, and refuse by the following equation:

$$\text{Recovery} = \frac{\%\text{ ash in refuse} - \%\text{ ash in feed}}{\%\text{ ash in refuse} - \%\text{ ash in clean coal}} \times 100.$$

The percentage recovery subtracted from 100 is the percent of refuse. Thus, the theoretical recovery of clean coal is 84.06%, and the actual recovery is 82.19%. The theoretical

[36] Decker, H. W., Jr., Thesis, Pennsylvania State Univ., 1955.

TABLE 12

WASHABILITY DATA ON THE CLEAN COAL AND REFUSE PRODUCTS[36]
(SPECIFIC GRAVITY OF CLEANING 1.70)

1	2	3	4	5	6	7	8
Specific Gravity		Wt %	% Ash	Cumulative Float		Cumulative Sink	
Sink	Float			Wt %	% Ash	Wt %	% Ash
			Clean Coal—82.19%				
	1.40	1.40	5.00	1.40	5.00	100.00	11.99
1.40	1.50	76.12	8.43	77.52	8.37	98.60	12.09
1.50	1.60	13.37	18.11	90.89	9.80	22.48	24.49
1.60	1.65	3.30	26.25	94.19	10.38	9.11	33.86
1.65	1.70	3.02	31.55	97.21	11.04	5.81	38.18
1.70	1.80	1.90	36.94	99.11	11.53	2.79	45.35
1.80	1.90	0.22	45.30	99.33	11.61	0.89	62.39
1.90	2.00	0.28	55.56	99.61	11.74	0.67	68.40
2.00	2.10	0.00	—	99.61	11.74	0.39	76.93
2.10		0.39	76.93	100.00	11.99	0.39	76.93
			Refuse—17.81%				
	1.40	0.46	4.83	0.46	4.83	100.00	55.57
1.40	1.50	8.18	9.48	8.64	9.23	99.54	55.80
1.50	1.60	6.12	18.90	14.76	13.24	91.36	59.95
1.60	1.65	8.87	26.88	23.63	18.36	85.24	62.90
1.65	1.70	7.72	31.15	31.35	21.51	76.37	67.08
1.70	1.80	9.36	40.33	40.71	25.84	68.65	71.12
1.80	1.90	6.77	49.05	47.48	29.15	59.29	75.96
1.90	2.00	6.88	56.48	54.37	32.61	52.52	79.45
2.00	2.10	2.64	70.10	57.00	34.34	45.64	82.91
2.10		43.00	83.70	100.00	55.57	43.00	83.70

grade of the clean coal is 11.10% ash and the actual grade is 11.99%. These figures give some indication as to the efficiency of the operation.

The amount of misplaced material is also given in Table 12. This material amounts to 2.79% in the clean coal and 31.35% in the refuse. The former value is obtained under the clean coal using columns 1 and 7. The 1.70 sink value in column 1 indicates 2.79% cumulative sink weight percent in column 7. The 31.35% float in the refuse is obtained from columns 2 and 5 under the refuse. The 1.70 float value in column 2 indicates 31.35% cumulative float weight percent in column 5. On the basis of 100 tons of total products there are 82.19 tons of clean coal and 17.81 tons of refuse. These values multiplied by the percentage of misplaced material in each will show 2.29 tons of misplaced material in the clean coal and 5.58 tons of misplaced material in the refuse. Therefore, the total misplaced material would be 7.87 tons or 7.87%. This may be compared to 9.29% of near gravity material in the original coal. This is another means of indicating efficiency.

Calculated results showing the weight distribution of the various specific gravity fractions in the clean coal and in the refuse are presented in columns 3 and 4, respectively, of Table 13. These values were obtained by multiplying the weight percentages listed in column 3 of Table 12 by the respective weight percent of clean coal and refuse. The total weight percent, column 5, is the sum of columns 3 and 4 in each specific gravity fraction. This may be compared to the weight percent of the original coal in each specific gravity fraction as listed in column 3 of Table 11. The values are also listed in column 7 of Table 13 for convenience. Column 6 of Table 13 is the calculated cumulative weight percent based on the washability characteristics of the products from the process. It is obtained by cumulating the weight percentages listed in column 5. Based upon the figures in column 6, the theoretical recovery of clean coal, that is, the cumulative 1.70 float, is 85.47% as compared to 84.06% indicated in column 5 of Table 11. Thus, the actual recovery of 82.19% is really worse than expected from the washability characteristics of the original coal.

The percentage distribution of the various specific gravity fractions in the clean coal and refuse are presented in columns 8 and 9 of Table 13. These values are obtained by dividing the values in columns 3 and 4, respectively, by the values in column 5. They indicate, for example, that 90.98% of the coal that sinks at a specific gravity of 1.50 and floats at a gravity of 1.60 appears in the clean coal, and that 9.02% appears in the refuse. Because all of the material with a specific gravity of 1.70 and lower should be in the clean coal and all of the material with a specific gravity higher than 1.70 should be in the refuse, the distribution of the material as shown in columns 8 and 9 is a measure of the efficiency.

Of the various methods listed above for evaluating efficiency,[35] probably the one most generally used is that proposed by Fraser and Yancey.[35] Their method takes into consideration both the quantity and quality of the preparation-plant clean-coal product and compares them with the theoretical values. The equation used to calculate efficiency is as follows:

$$\%\text{ efficiency} = \frac{\text{recovery of clean coal}}{\text{theoretical recovery}} \times \frac{\%\text{ ash in feed coal} - \%\text{ ash in clean coal}}{\%\text{ ash in feed coal} - \%\text{ ash in theoretical clean coal}}$$

Other properties of the coal, such as the percent of sulfur, Btu value etc., may be substituted for the percent ash. This makes the equation flexible enough that the efficiency may be determined based upon the most important interests of the producer, who in turn is guided by market specifications. The equation has the disadvantage, however, that a high recovery of a low grade product may result in an efficiency greater than 100%. It is also true that an exceptionally clean coal with a low recovery will produce erroneous values. Yancey and Geer[37] presented a critical review of the preceding method as well as some of the other more important methods based upon ideas developed at the First International Conference on Coal Preparation.

Using the Fraser and Yancey equation the efficiency is calculated below:

$$E = \frac{82.19}{84.06} \times \frac{19.75 - 11.99}{19.75 - 11.10} \times 100 = 87.7\%$$

Efficiency calculations, using the data in Tables 11, 12, and 13, have also been made for other methods listed above, and the results are presented in Table 14.[36] The results indicate a wide range for the efficiency values and indicate that the numerical value for a plant operation has no significance unless the method of calculation is associated with the value.

In conclusion, each plant operation is an individual case and the operation may best be

[37] Yancey, H. F., and Geer, M. R., *AIME, Trans.*, **190**, 507–17 (1951).

TABLE 13

Distribution of Specific Gravity Fractions in the Clean Coal and Refuse Products

1	2	3	4	5	6	7	8	9
Specific Gravity		Wt Distribution					% Distribution	
Sink	Float	Clean Coal	Refuse	Total	Cumulative Total	Wt % in Original Coal	Clean Coal	Refuse
	1.40	1.15	0.08	1.23	1.23	0.88	93.50	6.50
1.40	1.50	62.56	1.46	64.02	65.25	60.36	97.72	2.28
1.50	1.60	10.99	1.09	12.08	77.33	16.85	90.98	9.02
1.60	1.65	2.71	1.58	4.29	81.62	3.65	63.17	36.83
1.65	1.70	2.48	1.37	3.85	85.47	2.41	64.41	35.59
1.70	1.80	1.56	1.67	3.23	88.70	3.32	48.30	51.70
1.80	1.90	0.18	1.21	1.39	90.09	2.02	12.95	87.05
1.90	2.00	0.24	1.22	1.46	91.55	0.88	16.44	83.56
2.00	2.10	0.00	0.47	0.47	92.02	0.48	0.00	100.00
2.10		0.32	7.66	7.98	100.00	9.24	4.01	95.99
		82.19	17.81	100.00		100.00		

TABLE 14

COAL PREPARATION PLANT EFFICIENCIES[36]
(BASED UPON DATA IN TABLES 11, 12, AND 13)

Method Used[35]	% Efficiency	Remarks
Delamater's equation	95.5	Using rule 4
Drakeley's equation	77.0	At operating gravity
	80.0	At true gravity
Original Fraser-Yancey	87.7	At operating gravity
equation	91.0	At true gravity
	96.5	
Modified Drakeley's equation	80.9	At operating gravity
	80.0	At true gravity
Chapman and Mott equation	93.0	
MacLaren's equation	96.0	
Tromp's area	52.6 (cm^2)	Error area; corresponds to a fair degree of accuracy
Hancock's equation	79.1	At Hancock's critical gravity
Anderson's equation	90.7	At true gravity

evaluated by the misplaced material in the products. It is, therefore, most difficult to compare one plant operation with another or one preparation process with another. All variables must be considered and all calculations must be on the same basis.

PROCESSING DEVELOPMENTS

Three principal types of flowsheets are used in the United States for processes to clean coal mechanically.

The first, and probably the most important from a tonnage standpoint, is that in which all of the coal is cleaned in a single unit, as illustrated in Fig. 5. In this type of flowsheet, the run-of-mine coal is first screened at some specific size such as 4 in. The oversize may then be reduced directly to a top-size equivalent to the previous screening operation, or it may pass over a hand-picking table for preliminary cleaning prior to crushing or through a Bradford Breaker. After crushing, the product may be added to the original undersize; or, if sufficiently clean from hand-picking, it may bypass mechanical cleaning and go directly to the market. The Bradford Breaker undersize would be combined with the original screen

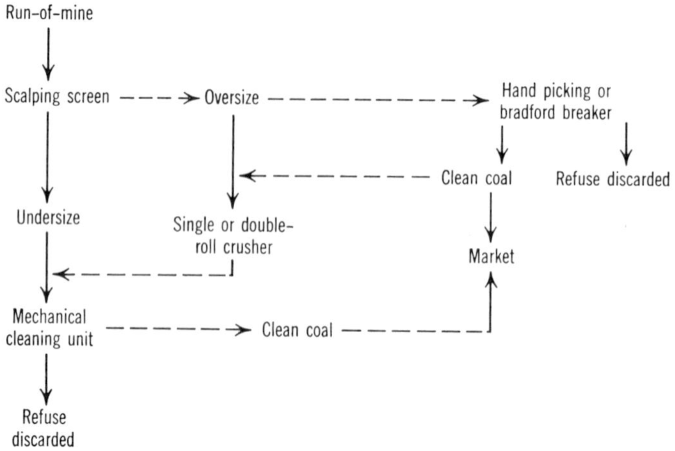

Fig. 5. Schematic flowsheet for mechanically cleaning coal in a single unit.

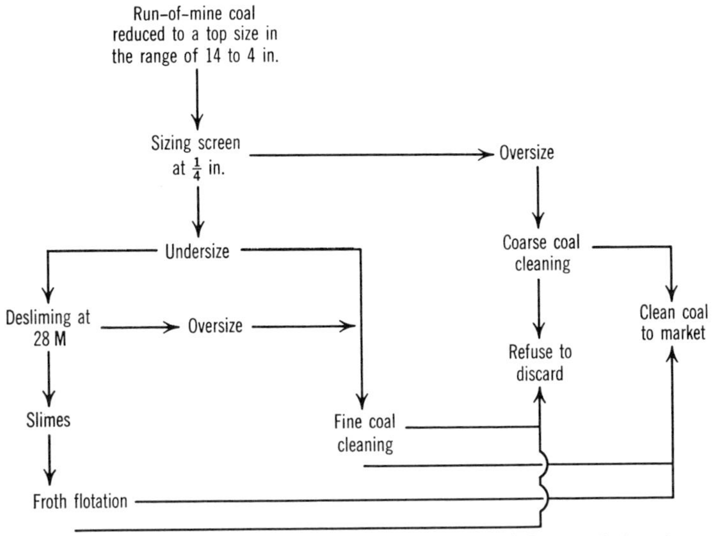

Fig. 6. Schematic flowsheet for mechanically cleaning coarse and fine coal sizes in separate units.

undersize. Usually, the top size does not exceed 7 in. and is seldom less than 2 in. The jig, to be discussed later, is the only mechanical cleaning method that can be used in this type of flowsheet. When the entire size range is cleaned in the preceding manner, the efficiency of separation decreases with a decrease in size.

The low efficiencies in cleaning the smaller sizes have been responsible in part for the development of a second type of flowsheet, in which the run-of-mine material is reduced to the same top size as shown in Fig. 5, but before being mechanically cleaned, the entire size range is divided into coarse and fine sizes. The top size in this type of flowsheet may be as large as 14 in. but seldom exceeds 4 in. The size separation is made somewhere in the range of $\frac{1}{8}$ to $\frac{3}{4}$ in. and usually at $\frac{1}{4}$ in. The coarse sizes are cleaned by one method and the fine sizes by another. The jig or a heavy media method is generally used for coarse-coal cleaning, and a tabling method is generally used for fine-coal cleaning. All of these methods will be discussed later. The jig may also be used to clean the fine sizes. More recently, a heavy media method using cyclones has been used. Regardless of the method of fine-coal cleaning, it is advantageous to deslime the fine coal and remove the extreme fine sizes before cleaning. The desliming generally occurs in the size range of 10 to 200 mesh. When it occurs in the larger part of the range, that is, 10 to 48 mesh, economic consideration is given to the possibility of cleaning the slimes or extreme fines. Cleaning of particles smaller than 10 mesh is generally performed by a froth flotation method. A schematic flowsheet showing this type of cleaning is presented in Fig. 6.

The third type of flowsheet for cleaning coal mechanically is used in the anthracite industry. The run-of-mine coal is separated into commercial sizes and each size or a combination of not more than two sizes is cleaned in individual units. This development occurred years ago as a result of rigid specifications for the anthracite sizes established by the industry and the need to recover as high a percentage as possible of domestic sizes. The cost of such an operation is high in comparison to the others and, consequently, it is gradually being replaced by coarse and fine coal cleaning.

All of the preceding methods are used to make a mechanical separation of clean coal and refuse, with the exception of froth flotation, and are based upon the difference in the specific gravities of the coal and refuse. The rate at which particles of different sizes and different specific gravities settle in water and air make it possible to obtain a separation. Thus, if particles are settling through a fluid in a cleaning vessel and the fluid is moving upward at a higher velocity than the slower settling particles and at a lower velocity than the faster settling particles a separation will occur. This principle is used in the jig. Also, heavy particles will settle faster in a flowing stream than lighter particles of the same size. This principle is used in tabling.

The sink-float principle, using organic liquids of different specific gravities for laboratory separation, has already been described. Organic liquids, however, cannot be used in commercial operations at the present time because of cost. Instead, finely divided solids are added to water in order to prepare a suspension of the specific gravity to be used for the separation. Sand and magnetite are the solids most commonly used. The suspension of finely divided solids in water is the basis for heavy media processes of coal cleaning. Calcium chloride has also been used to prepare a solution for a heavy media process. The highest specific gravity attainable with calcium chloride is, however, about 1.35. Thus, heavy media processes using calcium chloride have to rely also on the settling velocity principle as well as the sink-float principle.

Of the various methods used to clean coal, jigging continues to be the most popular. In recent years, approximately 47% of all of the coal to be mechanically cleaned in the United States (about 160 million tons per year) was cleaned by jigging. The jigging process consists of subjecting a bed of particles to the upward and downward currents of water as the particles pass over a perforated screen surface. The movement of the water causes the particles to stratify, regardless of their size, so that the light clean-coal particles are on the top of the bed at the discharge end of the screen, and the heavy refuse particles are on the bottom. The light particles are crowded and washed out of the cleaning vessel, and the heavy particles are withdrawn from the lower portion of the bed. Bird and Mitchell[38] have presented a detailed description of the different types of jigs for coal preparation and have discussed their operating characteristics. Massmann[39] has listed several faults based upon practical operation which cause erratic performance of the Baum-type jig, most commonly used in this country. Some of these faults are: an irregular and segregated feed to the jig, the improper transverse distribution, and the failure to keep the screen deck clean. Laboratory tests[40] with Pennsylvania anthracite have indicated that specification grade in all sizes from stove to No. 5 buckwheat could be met while cleaning a broad range of sizes in the Baum-type jig. These tests also indicated that treatment of the same coal in a heavy media process only provided a recovery efficiency of 1% higher than that obtained by jigging. More recent studies[41] on the theoretical and practical application of the Baum-type jig have shown the importance of air pressure, air-valve speed, and the suction phase of the jigging cycle on cleaning the finer sizes. Within recent years, the Baum-type feldspar jig has been used in this country to clean the finer sizes. This type utilizes feldspar or any material of similar gravity as an artificial bed in the jig. As of January, 1960, there were only two commercial installations in this country, but others were under construction. Plant size tests[42] with the feldspar jig have shown its ability to provide high capacity per unit area and high cleaning efficiency in sizes down to 48 mesh.

A new type of mechanical jig has been described[43] recently, and several commercial installations have been made. This is a piston-type jig designed to clean coal smaller than $\frac{1}{2}$ in. Adjustable operating features include the screen deck openings, plunger eccentric and speed, water distribution, and the refuse ejector. Good efficiencies were claimed for sizes down to 60 mesh.

Of the different methods used for coarse coal cleaning, heavy media methods using magnetite as the medium are the most popular. Within recent years the magnetite process has been growing in importance. In 1959, approximately 25% of the 245 million tons of bituminous coal that was cleaned was cleaned by the magnetite process. The process consists of circulating a suspension of magnetite in water through the cleaning vessel, adding the feed coal to the vessel, and removing the float-and-sink products. The suspension is separated from the clean coal and refuse products by drainage and rinsing screens. The magnetite is separated

[38] Bird, B. M., and Mitchell, D. R., *Coal Preparation*, D. R. Mitchell, Ed., AIME, New York, 1950, Chap. 12.

[39] Massmann, A. P., *Mechanization*, **20**, No. 5, 87–8 (1956).

[40] Yancey, H. F., and Geer, M. R., *U.S. Bur. Mines, Rept. Invest.* No. 5308 (1957), 23 pp.

[41] Singhal, S. C., and Eveson, G. F., *J. Inst. Fuel*, **32**, 210–18 (1959). Charbonnier, R. P., *Can. Mining Met. Bull.*, **52**, 385–8 (1959).

[42] Yancey, H. G., and Geer, M. R., *U.S. Bur. Mines Rept. Invest.* No. 5412 (1958), 14 pp.

[43] Citron, E. H., *Mining Eng.*, **10**, 488–9 (1958).

from fine coal, caused by degradation, by magnetic separators. There are many different types of cleaning vessels[44] used for this method. The most popular are the cone-type, the wheel-type, the drum-type, and the tank-type. In all types, the clean-coal float-product is crowded and washed out the overflow of the vessel. The methods for removing the refuse sink-product vary from a trap release to a drag mechanism scraping the refuse off the bottom and out of the vessel.

Experimental studies[45] have been made on the operating variables and theories for heavy media processes. These studies have shown the effect of turbulent motion of the medium, the viscosity of the suspension, and the tendency of the medium to flocculate, on the efficiency of the process.

The most recent development in heavy media processes has been the use of Accu-Ray[46] for control of the specific gravity of the suspension. This is a totally automatic robot based upon gamma radiation. It constantly controls the addition of water or magnetite to the system.

The heavy media process is also used in cyclones[47] for fine coal cleaning. The process was developed in Europe for cleaning sizes in the range of ½ in. to 48 mesh. Several commercial installations have been made in the United States. Pilot plant tests[48] have indicated that cyclones alone, without heavy media, may be used for doing some cleaning of the extreme fines, that is, sizes smaller than 48 mesh. Conditions for cleaning rather than sizing with cyclones alone require relatively high pulp densities and high inlet pressures.

Of the methods used for cleaning fine coal in the United States, tabling is the most widely used. Approximately 20 million tons were cleaned by this method in 1959, and this figure has been steadily increasing. The method consists of adding a suspension of the fine coal in water to a rectangular or trapezoidal table moving back and forth with a differential motion in a direction perpendicular to the flow of the pulp. The heavy refuse particles settle to the surface of the table and are trapped by riffles while the light clean coal particles are washed over the end of the table. The refuse particles are carried along the riffles to the side of the table by the differential motion.[49]

Although concentrating tables are recognized as rather efficient, except for the extreme fine sizes, they require considerable floor space. The capacity is about 10 tons/hr/table and the deck of the table requires approximately 120 sq ft of floor space. To combat this, the twin-deck suspension-type table was developed in 1956, and since that time practically all table installations have been of this type. It simply consists of placing one deck over the other and driving both with the same mechanism. Good separation efficiencies have been reported[50] with sizes down to 200 mesh.

Froth flotation is the most important method for cleaning the extremely fine sizes of coal and is growing in importance. Flotation equipment has been installed in at least nine preparation plants in 1959. This equipment has a total capacity of approximately 600 tons/hr. Flotation is used extensively in foreign countries. Its use in this country has been slow to develop because of economic reasons. The process consists of passing air through a suspension of the fine coal in water. The coal particles attach themselves to the air bubbles and are removed in the froth on top of the pulp. The refuse particles remain in the pulp. Chemicals are added to the pulp to help produce the froth; one of the most important being used is methyl isobutyl carbinol, which is used in amounts of less than 0.1 lb/ton of feed coal.

Although washability data may be used to determine the quantity and quality of coal theoretically possible from a flotation process, a more realistic method[51] has been developed. This method consists of splitting up the original

[44] Milligan, E. O., *Mining Congr. J.*, **44**, No. 6, 72–4 (1958).

[45] Whitmore, R. L., *Colliery Eng.*, **36**, 151–3 (1959). *J. Inst. Fuel*, **31**, 422–8 (1958). Geer, M. R., Sokaski, M., West, J. M., and Yancey, H. F., *U.S. Bur. Mines, Rept. Invest.* No. 5354 (1957), 64 pp.

[46] Ippoliti, J. E., Unpublished—Presented before Coal Conv. Am. Mining Congr., Pittsburgh, Pa., May, 1960.

[47] Yancey, H. F., *Coal Age*, **63**, No. 12, 118–21 (1958). de Chawloski, S., *Rev. ind. minérale*, **40**, 925–36 (1958). Krijgsman, C., *Mining Congr. J.*, **45**, No. 5, 47–52 (1959).

[48] Charmbury, H. B., and Mitchell, D. R., *Mining Eng.*, **11**, 211–14 (1959).

[49] Charmbury, H. B., and Mitchell, D. R., *Mechanization*, **20**, No. 9, 79–82 (1956).

[50] Ambrose, F. S., and Davis, D. H., *Mining Congr. J.*, **43**, No. 11, 41–4 (1957).

[51] Dell, C. C., *J. Inst. Fuel*, **30**, 532–6 (1957).

flotation froth into several fractions and refloating the fractions two or three times. Each refloated fraction is then analyzed to determine the quantity and quality of the products. In actual plant operation, the first float is referred to as a rougher operation and the subsequent flotations as cleaner and recleaner operations. The cleaning and recleaning are required, at times, to meet specification grade.

Considerable data[52] have been published on the theory and practice of coal flotation, so that the technical aspects of the problem are far in advance of the economic factors. The technical know-how is available, and the method will continue to expand at specific locations and when the economics are favorable.

At the Third International Coal Preparation Congress held in Liège, Belgium, a series of papers were presented by authors from the major coal producing countries of the world to show the trend in coal preparation in each country. A general review[53] of all of the papers was presented. This review shows many trends that are common in all of the countries. It also shows that the need for coal preparation is increasing due to the continual depletion of high quality coals and the continual increase in mechanization of mining methods. The latter is responsible for the run-of-mine product containing more foreign matter and for a change in the size consist of the mine product so that more emphasis is being placed on the preparation of fine coal. In all countries, the jig is the most important method of coal preparation, but it is gradually being replaced by the heavy media method for coarse-coal cleaning. Jigs remain the most important method of fine-coal cleaning except in the United States, where concentrating tables are used to a greater extent. Flotation is becoming of greater importance in all countries due to the increase in extreme fines and the fact that greater liberation of the pure coal from foreign material is attained in them. Finally, there is a general trend in all countries to go to larger preparation plants so that greater capacities may be attained. This has been necessary due to increasing capital investments and operating costs per ton.

Great Britain[54] produced approximately 220 million tons of mechanically cleaned coal in 1957. Based upon plant capacity, about 70% of this production was mechanically cleaned in jigs, 11% in heavy media processes, and 3½% by flotation. Installations planned for the future will increase heavy media to about 19% and flotation to 4%, again based upon plant capacity.

In Western Germany,[55] in 1957, there were 137 coal-preparation plants with a total cleaning capacity of about 43,000 tons/hr. There was a gradual increase in the average hourly capacity from 240 to 310 tons in the preceding 6 yr. Cost data showed that fine coal cleaning was about 40 to 50% higher than coarse coal cleaning, and that cleaning by flotation was about 200% higher than coarse coal cleaning.

In the Saar Area,[56] there has been a change to cleaning by heavy media processes. There has also been a decrease in the number of preparation plants, but the average capacity of the plants has increased from 280 to 390 tons/hr. There were fourteen plants in 1958 producing about 23 million tons of mechanically cleaned coal. Of this production, 63% was cleaned by jigs, 18% by heavy media, and 8% by froth flotation.

The figures presented at the Third International Conference on Coal Preparation for the United States[57] showed that there were approximately 700 coal-preparation plants producing about 250 million tons of mechanically cleaned coal.

In the Netherlands,[58] there were twelve coal-preparation plants in 1958 producing about 12 million tons of cleaned coal. About 45% of the coarse sizes and practically all of the fines were cleaned by jigs. Thirty-five percent of the coarse coal was cleaned by jigs followed by

[52] The Dow Chemical Company, "Flotation Index," Published Annually.

[53] Pozzetto, L., *Proc. 3rd Intern. Coal Preparation Conf., Liège*, **1958**; *Ann. mines Belg.*, Special Issue (1958), pp. 767–78.

[54] Grounds, A., and Reed, W., *Proc. 3rd Intern. Coal Preparation Conf., Liège*, **1958,** *Ann. mines Belg.*, Special Issue (1958), pp. 676–83.

[55] Meyer, G. A. H., *Proc. 3rd Intern. Coal Preparation Conf., Liège*, **1958,** *Ann. mines Belg.*, Special Issue (1958), pp. 689–94.

[56] Lehner, R., *Proc. 3rd Intern. Coal Preparation Conf., Liège*, **1958,** *Ann. mines Belg.*, Special Issue (1958), pp. 694–702.

[57] Charmbury, H. B., and Mitchell, D. R., *Proc. 3rd Intern. Coal Preparation Conf., Liège*, **1958,** *Ann. mines Belg.*, Special Issue (1958), pp. 703–11.

[58] Krijgsman, C., *Proc. 3rd Intern. Coal Preparation Conf., Liège*, **1958,** *Ann. mines Belg.*, Special Issue (1958), pp. 712–22.

heavy media and 20% by heavy media alone. Recently, several heavy media cyclones have been installed for fine coal cleaning. Seventy-five percent of all the coal slurries were cleaned by flotation.

Belgium[59] had eighty-two coal-preparation plants with an annual capacity of about 50 million tons. The coarse sizes were cleaned by jigs (30%) and by heavy media (61%) and the fines by jigs (60%) and heavy media cyclones (17%). Fifteen of the plants cleaned the extreme fines by flotation. Heavy media processes were gradually replacing the jigs for both coarse and fine coal cleaning.

In Japan,[60] there were about 800 coal mines with an annual production of approximately 47 million tons. Thirty-five million tons were mechanically cleaned in 287 preparation plants. Jigs were used to clean 85% of the coal, heavy media for 3%, and concentrating tables and flotation 1% each. Japan is the only other country besides the United States that reported the use of concentrating tables. Fifty of the preparation plants in Japan used flotation.

France[61] produced about 46 million tons of mechanically cleaned coal in 1957. Twenty-one of the coal-preparation plants cleaned only the large sizes, fifteen only the small sizes, and forty-six cleaned both the large and small sizes. The coarse sizes were cleaned by jigs (39%) and by heavy media (53%). The fines were cleaned by feldspar jigs (74%), other jigs (6%), and heavy media cyclones (2%). Approximately 3% of the total production was cleaned by flotation. There were twenty-five flotation plants. These twenty-five flotation plants had a total capacity of 875 tons/hr. Heavy media and flotation methods of cleaning were increasing and jigging and other methods were decreasing in France.

With the slow but gradual development of atomic energy to replace coal and other fossil fuels, future trends in coal preparation throughout the world may turn to the production of an ultra-clean coal for chemical uses. Considerable money is being spent in the United States for the discovery and development of new uses for coal, but little is being spent at the present to produce an ultra-clean low-ash organic raw material. Some work[62] has been done to prepare low-ash coal for electrode-carbon and other purposes. These methods involve the acid and alkali extraction of coal and may be referred to as coal-refining processes following coal preparation. Coal-refining processes may also be involved in the preparation of coal for carbonization. Within recent years, investigations have been made concerning the importance of petrographic constituents and their influence on carbonization. Jowett[63] has studied the possibility of separating these constituents using selective breakage, heavy media separation with cyclones, light reflection, and froth flotation. Much more is desired, however, on the benefits of petrography, the know-how of making the separations, and the economics of the problem before the so-called coal-refining processes will be made practical operations.

[59] Venter, J., and Burton, G., *Proc. 3rd Intern. Coal Preparation Conf., Liège*, **1958,** *Ann. mines Belg.*, Special Issue (1958), pp. 723–37.

[60] Imaizumi, T., *Proc. 3rd Intern. Coal Preparation Conf., Liège*, **1958,** *Ann. mines Belg.*, Special Issue (1958), pp. 738–41.

[61] Pozzetto, L., *Proc. 3rd Intern. Coal Preparation Conf. Liège*, **1958,** *Ann. mines Belg.*, Special Issue (1958), pp. 742–66.

[62] Graham, H. G., and Schmidt, L. D., *J. four élec.*, **59,** No. 3, 75–8 (1950); *J. Iron Steel Inst.* **168,** 88 (1951). Anon., *The Production of Ultra-Clean Coal, Colliery Eng.*, **28,** 25–7 (1951). Crawford, A., *Trans. Inst. Mining Engrs.*, **111,** 204–18 (1952). Inagaki, M., Japanese Pat. 7733 (1951); *J. Coal Research Inst. (Japan)*, **1,** 390–7 (1950); **2,** 72–6 (1951); **5,** 261–8 (1954). Gorman, R., Edmonds, R. G., and Doughty, E. W., Brit. Pat. 706,967 (1954).

[63] Jowett, A., *Colliery Guardian*, **197,** 693–9 (1958); *Trans. Inst. Mining Engrs.*, **118,** 466–77 (1959); Jowett, A., and Hill, N. W., *Colliery Guardian*, **200,** 177–85 (1960).

CHAPTER 9

PYROLYTIC REACTIONS OF COAL

HENRY C. HOWARD*

Bureau of Mines, U.S. Department of Interior, Pittsburgh, Pennsylvania

The chemical and physical changes that take place when coal is subjected to the action of heat are of great importance in both the science and the technology of coal. Coals, because of their mode of genesis, would be expected to be resistant to thermal change, and there is only a limited amount of experimental evidence, either direct or indirect, indicating structural changes below temperatures at which pyrolytic breakdown, as evidenced by melting and rapid evolution of volatile, organic-degradation products, occurs.

Changes below Decomposition Temperatures

A few direct methods are available for study of the changes occurring in coals below temperatures of active decomposition. Some of these are differential thermal analysis, paramagnetic resonance absorption, and diamagnetic susceptibility.

Differential Thermal Analysis

The technique of differential thermal analysis, which has been used successfully to detect phase and compositional changes resulting from the action of heat in a number of mineral systems,[1] would appear to be an excellent tool to obtain similar information concerning coal. Data on a variety of coals are now available from a number of recent investigations.[2]

Breger and Whitehead,[3] using an apparatus in which the sample was heated in vacuum, have studied the behavior of a variety of carbonaceous materials ranging in rank from peat to a Pennsylvania low-volatile coal. The curves which they obtained, representing the differential change in temperature, compared with a reference material of known behavior, as a function of temperature, they called "thermograms;" this term has been adopted generally. Their thermograms for peat give evidence of an exothermal reaction reaching a peak at around 375°C; with materials of higher rank, this peak was less pronounced and shifted to higher temperatures. In the thermogram of a high-volatile "C" coal from Sweetwater, Wyoming, there was little evidence of change of any kind below temperatures of 400°C; this was generally true of the bituminous coals. With a low-volatile bituminous coal from Somerset County, Pennsylvania, the curve deviated little from a horizontal base line until temperatures above 400°C were reached. Thermograms of the petrographic components of a coal from the Sydney area in Nova Scotia were also reported. The results with vitrain were similar to those obtained with bituminous coals; the thermogram for durain resembled that of a low-volatile bituminous coal; the clarain gave evidence of an endothermic change just below 400°C; and the fusain showed little evidence of either endothermic or exothermic change from room temperature to 800°C.

Glass[4] has reported thermograms for coals ranging in rank from peat to meta-anthracite. He used a covered cell in which the sample was heated in an atmosphere of its own decomposition products. American coals ranging in rank from peat to meta-anthracite were studied. With all of these except the meta-anthracite, an endothermic change was observed between 120 and 150°C which was ascribed to the evaporation of water. Following this endothermic change, small exothermic reactions were indicated in the temperature range 220 to 280°C

* Deceased.

[1] Grim, R. E., and Rowland, R. A., *Am. Mineralogist*, **27**, 746–61, 801–18 (1942).

[2] For a review of recent work, see Nelson, J. B., *BCURA Bull.*, **19**, 501–31 (1955).

[3] Breger, I. A., and Whitehead, W. L., *Fuel*, **30**, 247–53 (1951).

[4] Glass H. D., *Econ. Geol.*, **49**, 294–308 (1954).

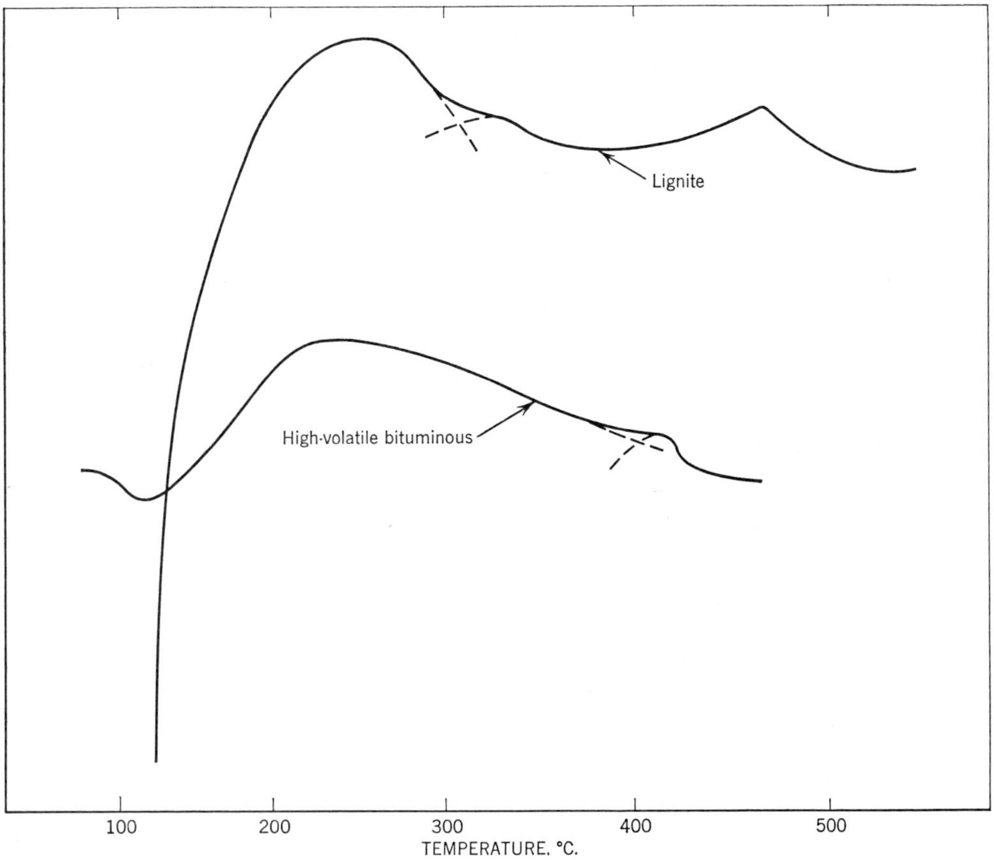

Fig. 1. Thermograms of Canadian coals by Berkowitz.[5]

for all the coals except the semianthracite, the anthracite, and the meta-anthracite.

Berkowitz,[5] using an apparatus which will be described later in which the sample was heated in a current of nitrogen, has studied a series of Canadian coals ranging, on the m.a.f. basis, from a lignite of 73.9% carbon to a low-volatile bituminous coal of 90.5% carbon. In this apparatus, all the coals gave evidence of an exothermic change in the temperature range 200 to 300°C. There was some indication of two such changes, a greater one below 300°C and a smaller one above this temperature (Fig. 1). The first exothermic change is much more marked in the lignite than in the higher rank coals, but other than this there was little effect of rank.

Kröger and Pohl[6] have reported thermograms for macerals from four samples of German coals; the carbon contents of the vitrinites of these coals ranged from 83.45 to 88.84% on the m.a.f. basis. The samples were heated in nitrogen. These macerals all showed some endothermic reaction at 100 to 120°C, presumably due to evaporation of water. Beyond this temperature range, there was no evidence for a low-temperature change, because the curves did not rise above the base line until temperatures above 400°C were reached.

As a result of the investigations just discussed, one may conclude, with reservations, that differential thermal analysis gives evidence of small exothermal changes in bituminous coals at temperatures below melting and decomposition. From this, it may be inferred that some structural changes occur in this temperature range, and refinements in this technique may produce more definite evidence for or against low-temperature changes in bituminous coals.

[5] Berkowitz, N., *Fuel*, **36**, 355–73 (1957).

[6] Kröger, C., and Pohl, A., *Brennstoff-Chem.*, **38**, 179–83 (1957).

Paramagnetic Resonance Absorption

Paramagnetic resonance absorptions have been determined[7] for the solid products resulting from heating of a bituminous coal up to 800°C, and these have been compared with that of the original coal. Marked increases in the absorption signal have been observed only at temperatures above 400°C. However, on the basis of the few data available, appearance of some unpaired electrons, "free radicals," at temperatures as low as 300°C, cannot be excluded. These may participate in reactions taking place at temperatures below those of active decomposition.

Diamagnetic Susceptibility

In measurements of the diamagnetic susceptibility of some British coals, it was suggested[8] that a study of magnetic properties combined with a knowledge of composition might furnish information concerning the content of aromatic carbon in coal and, presumably, changes in aromatic content on heating. Work along these lines has been carried out by Honda and co-workers.[9] They have investigated extensively the magnetic properties of coals and cokes and related substances. The values of magnetic susceptibility of the products from thermal treatment at 200 and 300°C were found to be almost the same as those of the parent coals; products from thermal treatment at temperatures higher than 350°C showed significant changes. The rapid decrease in diamagnetism above 400°C was interpreted to be a result of the production of free radicals.

Infrared Spectra

The infrared spectra of a bituminous coal heated in vacuum at 200°C have been reported by Adams and Pitt[10] to show only minor changes of the same nature as are presumably due to rearrangement of the oxygen-containing groups in the coal.

Preheating

Indirect evidence for structural changes below active decomposition temperature has been brought out by Warren[11] in work on the effects of heating rate in the preplastic range on the yields of products obtained in subsequent pyrolyses at higher temperatures. The effect of preheat on the solubility of coal in a variety of solvents has been recorded by a number of workers.[12] Later work on this subject has been reported by Dryden and Pankhurst.[13] Compared with the original unheated coal, these workers found up to eightfold increases in yields of chloroform-soluble material from samples of a British coking coal preheated in the range of 200 to 400°C. That the increased yield of chloroform-soluble material is to be traced to some low-temperature transformation is indicated by the following: (1) Pyrolysis up to 375°C yielded about 2% "volatile matter," whereas 8% chloroform-soluble products were formed; (2) the residue recovered after extraction of the heated coal with chloroform no longer softened or swelled on heating, whereas the swelling number of the unheated coal was not changed by chloroform extraction; and (3) the observed melting of this coal took place over a short range of 10° in the neighborhood of 400°C, whereas rapid increases in yield of chloroform-soluble material began just above 300°C.

A study[14] has also been made of the effects of a prolonged preheat treatment of a sample of a Pittsburgh-bed coal on the nature of the products obtained by subsequent hydrogenation at a higher temperature. In this work, the hydrogenations were carried out at 375°C in the presence of a copper chromium oxide catalyst at pressures of 3000 psig at reaction temperature. Sixteen experiments were made with a normal heating-up period of approximately 110 min,

[7] Ingram, D. J. E., Tapley, J. G., Jackson, R., Bond, R. L., and Murnaghan, A. R., *Nature*, **174**, 797 (1954).

[8] Wooster, W. A., and Wooster, N. A., *Proc. Conf. Ultra-fine Structure of Coals and Cokes*, Brit. Coal Utilisation Res. Assoc., 1944, p. 322.

[9] Honda, H., and Ouchi, K., *Magneto-chemical Studies of Coals and Their Related Substances*, Part I, *Repts. Resources Research Inst. (Japan)*, 1952, No. 1.

[10] Adams, W. N., and Pitt, G. J., *Fuel*, **34**, 383–4 (1955).

[11] Warren, W. B., *Ind. Eng. Chem.*, **27**, 1350–4 (1935). Davis, J. D., *Chemistry of Coal Utilization*, John Wiley and Sons, New York, 1945, Vol. 1, p. 839.

[12] Kiebler, M. W., *Chemistry of Coal Utilization*, John Wiley and Sons, New York, 1945, Vol. 1, pp. 727–8.

[13] Dryden, I. G. C., and Pankhurst, K. S., *Fuel*, **34**, 363–6 (1955).

[14] Glenn, R. A., *Fuel*, **33**, 419–35 (1954).

and the results of these were compared with three other experiments in which the temperature was held at 200°C for 24 hr before being raised to the reaction temperature of 375°C. The preheat treatment was found to increase the yield of total distillate from 55.6 to 61.2%, the yield of gas (by difference) from 6.9 to 11.9%, and to decrease the nonvolatile residue from 37.5% to 26.9%. These facts were interpreted to indicate that "rearrangements occurring during the preheat period apparently consist of aromatization through dehydrogenation, cyclo-dehydrogenation, cyclization or other similar processes." It should, however, be pointed out that, because the final hydrogenation at 375°C was carried out immediately following the 200°C preheating period without an intermediate cooling, the results observed may have been because of a very different heating rate in the temperature range where hydrogenation reactions become rapid, rather than to changes effected in the structure of the coal by the preheat at 200°C.

Other indirect methods which could be used to study the effects of low-temperature thermal treatment on coals, such as oxidation of preheated coals, have not been applied to products prepared by heating in the interesting critical temperature range.[15]

The behavior of low-rank carbonaceous materials, such as lignites, when subjected to preheat temperatures of approximately 300°C, is very different from that of bituminous coals. A typical North Dakota lignite[16] has been shown to undergo significant changes in composition when subjected to saturated steam at 314°C; on subsequent pyrolysis it gives somewhat different yields of decomposition products (Table 1). On the moisture and ash-free basis, there were definite increases in the yields of char; tar yields were unchanged; total gas yields decreased; the fractions of illuminants and of carbon monoxide in the gas increased; and those of carbon dioxide and hydrogen sulfide decreased. These changes must have resulted from loss of components other than water during the preheat treatment. The large loss of carbon dioxide, together with no increase in tar and light-oil yields, may be characteristic of the decomposition of "humic types" of compounds, where the carbon dioxide originates from carboxyl groups attached to rather large condensed cyclic units.

TABLE 1

EFFECT OF PREHEAT AT 314°C IN SATURATED STEAM ON COMPOSITION AND CARBONIZATION YIELDS OF A LIGNITE[16]

	Untreated,† %	Preheated,‡ %
Proximate Analysis*		
volatile matter	44.6	42.6
fixed carbon	55.4	57.4
Ultimate Analysis*		
Hydrogen	4.9	5.2
Carbon	75.6	76.7
Nitrogen	1.2	1.3
Oxygen (diff.)	17.2	15.7
Sulfur	1.1	1.1
Carbonization Yields*	At 500°C	
Char, %	67.4†	71.8§
Tar	5.96	6.05
Light Oil	1.47	1.47
Water	10.9	9.7
Gas	13.7	10.7

* On moisture and ash-free basis.
† Representative of charge to preheating unit.
‡ Average of two samples.
§ Average of three samples.

The loss of carboxyl groups from South Arcot, India, lignite in the temperature range 100 to 300°C has been reported,[17] with almost complete loss at 300°C (Fig. 2). Careful studies of the pyrolysis of coals below temperatures of active decomposition have furnished little evidence of gaseous and liquid products in significant yields.[18] This appears also to be generally true for petrographically enriched fractions. However, Macrae's[19] detailed investigation of the thermal decomposition of coal fractions greatly enriched in spore exines (largely exinite in maceral terminology) by prolonged heating in vacuum at temperatures from 150 to 300°C has shown that although

[15] Howard, H. C., *Chemistry of Coal Utilization*, John Wiley and Sons, New York, 1945, Vol. 1, pp. 346–7.

[16] Hoeppner, J. J., Opland, M. L., and Fowkes, W. W., *U.S. Bur. Mines, Rept. Invest.* No. 5260 (1956), 21 pp.

[17] Iyengar, M. S., and Ghosh, R. S., *Proc. Conf. on Science in the Use of Coal, Sheffield*, Inst. Fuel, London, 1958, p. C 49.

[18] Burgess, M. J., and Wheeler, R. V., *Fuel*, **5**, 65–8 (1926). Holroyd, R., and Wheeler, R. V., *J. Chem. Soc.*, **1928**, 2669–77, 3197–203.

[19] Macrae, J. C., *Fuel*, **22**, 117–29 (1943).

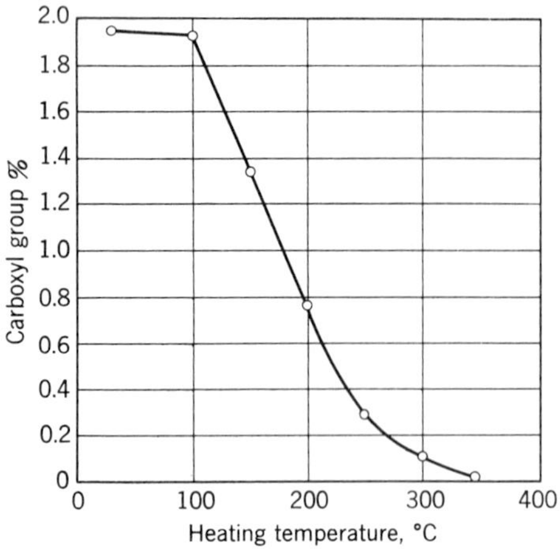

Fig. 2. Effect of temperature on elimination of carboxyl groups in South Arcot lignite.[17]

the total volume of gases evolved from both the exinite and the vitrinite sample with which it was compared was small, those from the former were about twofold those from the latter at corresponding temperatures and times of heating.

With oxidized bituminous coals, the situation is quite different. It has been shown[20] that such coals, when heated to temperatures below those at which tar formation takes place, yield significantly greater amounts of water and oxides of carbon than do unoxidized coals. This fact has been made the basis of a suggested analytical procedure for measuring the extent of oxidation of bituminous coals.[21] Data on the evolution of carbon dioxide from three oxidized American bituminous coals, oxidized under controlled conditions in the laboratory at temperatures ranging from 75 to 125°C, are shown in Fig. 3. In this work, the oxidized coals were heated at 350°C in a vacuum, the condensable volatile products were frozen out in a Dry Ice (acetone)-cooled trap, and the gases were recovered through a Toepler pump.

Reactions at and above Decomposition Temperatures

Over a wide range in rank, coals exhibit more or less definite decomposition temperatures,[18] as indicated by melting and rapid evolution of volatile products; an important part of coal science and technology has been concerned with the yields and composition of the products formed and the rates and mechanisms of the physical and chemical changes taking place during destructive distillation. Although pyrolysis is admittedly a crude tool, the results can be of the greatest significance, both with respect to the mechanisms of the changes and, by inference, the structure of the material decomposed.

For the study of coal in the range above decomposition temperatures, a great variety of tools and techniques are available. One of the most important is the simple proximate analysis, a procedure in which a weighed sample of

[20] Lefebvre, H., and Faivre, R., *Compt. rend.*, **203,** 881–3 (1936).

[21] Radspinner, J. A., and Howard, H. C., *Anal. Chem.*, **15,** 566–70 (1943).

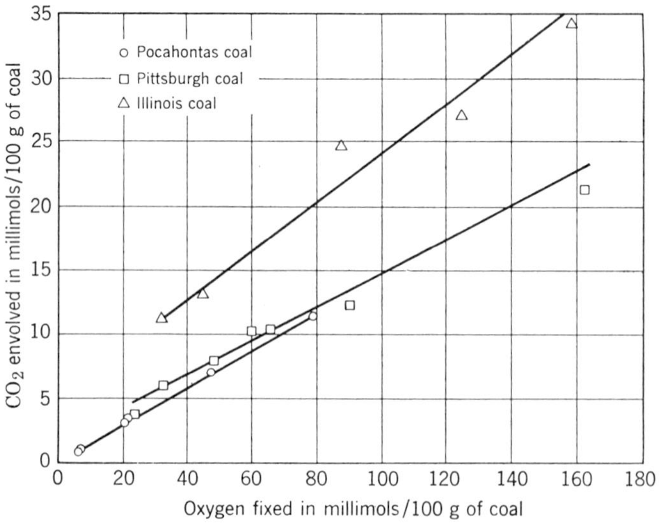

Fig. 3. Carbon dioxide from decomposition of coal as a function of degree of oxidation at 75 to 125°C.[21]

coal is subjected to destructive thermal decomposition under certain arbitrarily chosen but rigidly controlled conditions.

Proximate Analysis

Organic compounds exhibit nonvolatility prior to thermal breakdown either because of covalent bonding throughout large masses of the substance, because of strong secondary attractions between smaller covalent bonded masses, or because of combinations of these two factors. Thus, the yields of residue and volatile matter in the proximate analysis reflect the nature of the atomic bonding, and correlations would be expected to exist between fixed carbon or volatile matter and many other properties of coal. These relations are most satisfactory in the coals of high-volatile bituminous rank and upward. It is not surprising, therefore, that all coal classification systems, other than those based on ultimate composition, involve as a parameter some type of pyrolysis and fixed carbon or volatile-matter determinations.[22]

In general, chemical composition, rates of reaction, agglutinating properties, molecular size distribution, and composition of primary fragments in degradation reactions show correlations with fixed carbon. Relations between fixed carbon and content of condensed aromatic structures have been pointed out,[23,24] and the following relation has been developed:[25]

$$\text{F.C.} = 1.032 f_a \times \text{C}/100$$

where F.C. = percent fixed carbon, f_a = fraction of aromatic carbon, and C = percent carbon. As would be expected, because part of the aromatic carbon is volatile, the fraction of aromatic carbon estimated by this equation is lower than that calculated from volume-composition relations. It should be understood that equations of this type are highly empirical and are not generally applicable outside the coal series. For example, the compound dibenzanthrone, which has been used as a model for comparison in coal investigations and has been reported to yield 82% "coke" on carbonization, can be distilled without decomposition under appropriate experimental conditions.[26] Also, it is well established that the yields in the proximate analysis are a function of heating rate and apparatus design.[27]

Whether the high yield of nonvolatile residue obtained when coals are pyrolyzed is a consequence of extensive covalent bonding or of secondary attractive forces between molecules of moderate size is a matter of importance. Both types of organic structures are known. Glucose, molecular weight 180, is an extreme example of nonvolatility due to hydrogen-bonding and, as would be expected, "normal" volatility for a compound in this range of molecular weight can be obtained by destruction of the hydrogen-bonding by methylation. Another interesting example of the effect is to be found in the series of isomeric hydroxy benzoic acids in which the para isomer shows a vapor pressure at 100°C of only 1/1000 of that of the ortho, presumably because of the intermolecular hydrogen-bonding in the former.[28] When both compounds are completely methylated, the boiling points of the two isomers are only a few degrees apart.

The chemist generally associates secondary valence forces with compounds containing oxygen- or nitrogen-functional groups. Where physical properties indicate strong hydrogen-bonding as in the carbohydrates, the oxygen is present in very high atomic ratio, that is, unity with respect to the carbon atoms. Bituminous coals of about 85% carbon on the dry, ash-free basis show atomic oxygen to carbon ratios of 0.05 to 0.07, based on the total oxygen of the coal, and it has not been definitely established just what fraction of the total oxygen is of such a character as to participate in hydrogen-bonding reactions. Thus, on the basis of stoichiometry alone, the nonvolatility of coal is probably not due to hydrogen-bonding of small units. Another characteristic of organic compounds that owe their nonvolatility to hydrogen bonding is their solubility in hydrogen-bonded solvents. In this respect, bituminous

[22] Rose, H. J., *Chemistry of Coal Utilization*, John Wiley and Sons, New York, 1945, Vol. 1, pp. 50–9.

[23] Ellis, C. E., *Proc. Roy. Soc.* (*London*), **212A**, 1–12 (1952).

[24] Huck, G., and Karweil, J., *Brennstoff-Chem.*, **34**, 97–102 (1953).

[25] van Krevelen, D. W., *Brennstoff-Chem.*, **34**, 167–82 (1953).

[26] Unpublished data, Coal Research Laboratory, Carnegie Institute of Technology, Pittsburgh, Pa.

[27] Fieldner, A. C., and Selvig, W. A., *U.S. Bur. Mines, Bull.* No. 492 (1951), 51 pp.

[28] Sidgewick, N. V., *J. Chem. Soc.*, **117**, 396–406 (1920).

TABLE 2

AVERAGE YIELDS AND RANGE OF YIELDS OF FISCHER ASSAY OF VARIOUS COALS[37]

(As-received basis; maximum temperature, 500°C)

Rank of coal*	Coke, %			Tar, gal/ton			Light oil, gal/ton			Gas, cu ft/ton			Water, %		
	Number of samples	Average	Range	Number of samples	Average	Range	Number of samples	Average	Range	Number of samples	Average	Range	Number of samples	Average	Range
Sa	—	—	—	2	0.7	—	2	0.03	—	—	—	—	—	—	
Lvb	17	89.7	85.8–93.3	17	8.6	6.3–12.7	17	1.02	0.73–1.61	17	1,760	1,600–1,960	17	3.2	1.1–6.6
Mvb	30	83.3	77.4–90.4	30	18.9	9.7–25.6	30	1.67	1.07–2.30	30	1,940	1,390–2,240	30	4.1	2.8–7.0
Hvab	134	75.5	68.8–81.4	134	30.9	22.9–40.7	134	2.29	1.48–3.28	134	1.970	1,690–2,360	134	6.0	3.0–9.2
Hvbb	11	70.4	66.0–73.2	12	30.3	24.3–43.1	11	2.18	1.55–3.44	10	2,010	1,660–2,420	11	11.1	10.2–13.1
Hvcb	7	67.1	65.4–68.6	12	27.0	18.5–38.8	7	1.88	1.29–2.71	7	1,800	1,560–2,070	7	15.9	12.0–19.1
Hvch or Suba	1	59.1	—	5	20.5	18.4–24.4	5	1.65	1.36–1.93	1	2,660	—	1	23.4	—
Suba	—	—	—	2	17.8	17.6–18.0	2	1.35	1.33–1.36	—	—	—	—	—	—
Subb	6	57.6	54.8–59.9	7	15.4	13.2–16.7	7	1.33	1.14–1.55	6	2,260	1,830–2,760	6	27.8	23.3–30.4
Lignite	1	36.5	—	7	15.2	6.7–27.0	7	1.19	0.63–1.89	1	2,100	—	1	44.0	—
Cannel	7	58.8	44.1–69.0	8	73.5	53.7–108.3	8	5.06	3.66–7.42	2	1,810	1,500–2,120	7	3.7	2.0–4.8

* Sa, semianthracite; Lvb, low-volatile bituminous; Mvb, medium-volatile bituminous; Hvab, high-volatile A bituminous; Hvbb, high-volatile B bituminous; Hvcb, high-volatile C bituminous; Suba, subbituminous A; Subb, subbituminous B.

coals in the range of 85% carbon appear like many polymers[29] to occupy an intermediate position, being most soluble in solvents of high internal pressure combined with some hydrogen-bonding properties; for example, they show low solubility in the low internal pressure, aliphatic hydrocarbons on the one hand, and in such strongly hydrogen-bonded solvents as water and glycols on the other. The ease of dispersion of coals, particularly those of lower rank and, consequently, higher oxygen content in amine solvents, is evidence for the view that the forces holding large fragments of the coal structure to each other are of the hydrogen-bond type. But it seems probable that the most important factor in the nonvolatility of coals of 85% carbon and upward is the presence of covalent bonds throughout large masses of the coal substance. Recent X-ray work[30] has reduced significantly the estimated average size of the planar (aromatic) areas in bituminous coals, but it should be noted that the "out of plane," but still covalently bonded areas, may be much more extensive.

The results of the proximate analysis, because the analysis itself constitutes a small-scale carbonization test, would be expected to permit prediction of many facts observed in larger-scale carbonization experiments, and this has been found to be true. Correlating equations relating proximate analysis with such quantities as: coke yield, volatile matter, tar plus light oil, light-oil and gas yields have been published.[31]

Because, in the proximate analysis, a final constant temperature of 900°C is used, the composition of the "fixed carbon" and its calorific value is nearly constant, the latter being approximately 14,500 Btu/lb. Thus, by use of the proximate analysis, it is possible to compute the Btu in the fixed carbon, and by the difference between this quantity and the total Btu per pound in the coal, to calculate the heating units in the volatile matter. This quantity divided by the percent of volatile matter yields the "specific volatile index"[32] which has found applications in the classification of coals and in the prediction of their behavior during combustion.

The number of heat units in the volatile matter, computed as described, has been used by Parry and co-workers[33] to calculate tar yields over a wide range in rank.

Laboratory Assay Methods

Where information concerning tar yields and composition of the gaseous, liquid, and solid products resulting from the pyrolytic decomposition of coal is required, a number of small-scale laboratory assay methods are available. One of the oldest, and a still widely used procedure, was first described by Fischer and Schrader[34] and is generally known as the Fischer Assay. The U.S. Bureau of Mines adopted this procedure in 1928, at the time of the initiation of the Bureau of Mines-American Gas Association (BM-AGA) survey of the carbonizing properties of American coals. Its selection was based on a comparative study of three smallscale test procedures available at that time; the Gray-King retort,[35] a Bureau of Mines oil shale retort,[36] and the Fischer-Schrader retort. The last named was selected on the basis of its superiority in ease of manipulation and closer agreement between duplicate samples. A summary of this work, which has recently been published,[37] reports assays of some 380 samples from twenty-one states in the United States, and from British Columbia ranging in rank from lignite to semianthracite. The data are summarized in Table 2. The significant variation in yields shown for each of the "ranks" is due to variations in moisture, ash, volatile matter,

[29] Burrell, H., *Interchem. Rev.*, **14,** 3–16, 31–46 (1955).

[30] Hirsch, P. B., *Proc. Roy. Soc. (London)*, **A226,** 143–69 (1954).

[31] Lowry, H. H., Landau, H. G., and Naugle, L. L., *Trans. Am. Inst. Mining Met. Engrs.*, **149,** 297–330 (1942).

[32] Burrough, E. J., Swartzman, E., and Strong, R. A., *Can. Dept. Mines, Mines Branch, Invest. Fuels Fuel Testing*, 1930, 1931, **725,** 36–50 (1933).

[33] Parry, V. F., Landers, W. S., and Wagner, E. O., *Mining Eng.*, **8,** 54–64 (1956).

[34] Fischer, F., and Schrader, H., *Z. angew. Chem.*, **33,** 172–5 (1920).

[35] Gray, T., and King, J. G., *Gt. Brit., Dept. Sci. Ind. Research, Fuel Research, Tech. Paper* **1** (1921), 13 pp. Note: Equipment and procedures have been standardized for *International Classification of Hard Coals by Type*, United Nations Publications, 1956, II, E. 4.

[36] Stanfield, K. E. and Frost, I. C., *U.S. Bur. Mines, Rept. Invest.* No. 4477 (1949), 13 pp.

[37] Selvig, W. A., and Ode, W. H., *U.S. Bur. Mines, Bull.* No. 571 (1957), 56 pp.

Fig. 4. Sections showing typical internal structure of coke residues.[37]

and oxygen contents. In the low-rank coals, the effect of the difference in the nature of the vegetable debris which went into the coal beds is clearly evident; the wax and resin-rich lignites from Arkansas yield, on the dry basis, in some instances, fourfold as much tar as those from North Dakota. The spore-rich cannel coals gave average yields, on the dry basis, of 73.5 gal of tar per ton. Some information as to the properties of the coals with respect to swelling and coking can be obtained by examination of the residue from the retort. Typical samples are shown in Fig. 4.

The Fischer Assay is also widely used abroad, especially in the study and classification of low-rank coals where tar yields may be an important parameter. Modified forms of the Fischer Assay have been described by two groups of workers.[38–40] The first of these has developed an apparatus comprising a simple design of retort that is easily fabricated from mild steel, an accurately controlled method of heating, and a complete condensing and collection system. Hydrogen sulfide is effectively separated from the other volatile products by absorption in copper sulfate, and the assay gases are recirculated to insure correct sampling and analysis. It was stated that mass balances of 100 ± 0.3% are obtainable. With certain modifications, this apparatus has been used over the temperature range 300 to 1000°C and has been most extensively used in carbonization studies of lignites.[16] Krumin's apparatus[40] was stated to have several advantages over the earlier types, namely: (1) The heating rate can be closely controlled over a wide range, (2) conditions can be accurately reproduced, because electric heating permits better temperature control, and (3) complete mass balances can be obtained. Results of carbonization assays have been reported for test cores and for face and "run of mine" samples from underground and strip mines from two counties in Ohio.

There are available also a number of bench-scale assays especially designed for evaluation of coals for carbonization in by-product ovens. A review of these has been published.[41]

With the development of commercial interest in carbonization in fluidized beds, there has arisen a need for a laboratory method to determine yields of tar and char and particularly to determine agglutinating behavior of bituminous coals in such systems. A bench-scale apparatus has been designed for fluid-bed carbonization assay.[42] A 400 g coal sample when fluidized by a stream of dry, purified nitrogen at a velocity of 0.3 ft/sec formed a bed about 15 in. deep in a 2-in. diameter, stainless steel pipe, 25 in. long. The standard deviations for yields of char, tar, liquor, and gas were reported to be significantly less than for the Fischer assay. In addition to determining yields, this apparatus has given useful indications of the agglutinating behavior of coals in fluid beds and the effect of dilution with inerts or oxidation on this behavior.

[38] Goodman, J. B., Gomez, Manuel, Parry, V. F., and Landers, W. S., *U.S. Bur. Mines, Bull.* No. 530 (1953), 24 pp.

[39] Goodman, J. B., Gomez, Manuel, Parry, V. F., *U.S. Bur. Mines, Rept. Invest.* No. 5383 (1958), 49 pp.

[40] Krumin, P. O., *Ohio State Univ. Studies, Eng. Exp. Sta., Bull.* No. 165, 51–60 (1957).

[41] Brown, J., *Fuel*, **26**, 5–15 (1947).

[42] Curran, G. P., Kansas City Meeting, American Chemical Society, Division of Gas and Fuel Chemistry, April, 1954.

Pyrolysis at Reduced Pressures

Pyrolytic decomposition of coal at reduced pressures has thus far proved to be of little commercial value; however, for the study of the nature of the primary thermal decomposition products of coals, it has been shown to be a useful tool.[43] There is particular interest in these products in studies of the coking mechanism because it seems probable that fluidity and the agglutinating, swelling, and expansion properties of a coal are determined by their nature. One interesting technical application of vacuum pyrolysis has been described. It has been claimed[44] that alternate heating in air and heating in vacuum is an effective procedure for altering coking and swelling properties. The coal is successively dried in vacuum at 150°C, heated in air at 150°C, and vacuum pyrolyzed at the same temperature until the desired reduction in agglutinating properties has been attained. This process seems to depend on the fact that at constant temperature and oxygen partial pressure, the rate of reaction of the coal soon reaches a very low rate, and the initial rate can be restored only by some process which removes the oxygen complexes and regenerates a reactive surface. It is probable that the mechanism is analogous to that responsible for the effect of alkali in oxidative degradation of coal in a number of reactions[45] except that, in this instance, restoration of the reactive surface is effected by a pyrolytic rather than a hydrolytic reaction.

Simek, Ludmila, and Edinger[46] have studied the pyrolysis of a number of Czechoslovakian coals and their petrographic components at temperatures of 500°C and pressures of 20 to 760 mm of mercury. They concluded that the yields of tars to be expected from a given coal in technical operations can be predicted from the yields obtained in small-scale experiments at low pressures.

Guérin and co-workers[47] have pyrolyzed a number of European coals in vacuum using the Le Beau technique.[48] Coals ranging in rank from subbituminous to low-rank were pyrolyzed at 100° intervals from 200 to 1000°C. At each temperature, the compositions of the gaseous and solid products were determined. It was found that the total volume of the gas produced between 200 and 1000° increased with rank, and that the maximum gas evolution occurred at 700°C. At 400°C, the volume of gas evolved increased with decrease of rank. The ultimate composition of the residues on a H/C vs O/C type of plot (Fig. 5) shows clearly the difference between increases in rank effected by pyrolysis in vacuum as compared with natural coalification processes.

Pyrolytic decomposition in vacuum has also been used for the direct determination of oxygen in coal.[49] The sample was heated to 1000°C, and the volatile products passed through a silica cracking column at 900°. The carbon monoxide, carbon dioxide, and water formed were determined separately, and the total oxygen was calculated. For five European coals ranging in rank from subbituminous to anthracite, the results showed a high degree of reproducibility and were in good agreement with the "average value by other methods." They were 0.04 to 0.7 percentage points higher, except in one instance where the value was 0.12 percentage points lower.

Further work in "short-path" types of pyrolytic stills has been reported.[50] One of these is shown in Fig. 6. It was designed to eliminate some of the more troublesome features in an earlier type[51] and to furnish an apparatus more suitable for routine investigations. Heat transfer through a bed of coal is difficult under the most favorable conditions, and this problem is accentuated in vacuum systems.

In an attempt to insure uniform heating in the coal particles, the coal container used

[43] Howard, H. C., *Chemistry of Coal Utilization*, John Wiley and Sons, New York, 1945, Vol. 1, pp. 761–73.

[44] Wolf, C. S., U.S. Pat. 2,276,362 (1942).

[45] Entel, J., Ruof, C. H., and Howard, H. C., *J. Am. Chem. Soc.*, **75**, 3038–39 (1953). Cf. van Krevelen, D. W., and Schuyer, J., *Coal Science*, Elsevier, New York, 1957 p. 319.

[46] Simek, B. G., Ludmila, J., and Edinger, J., *Zprávy ústava vedecky vyzkum. uhli*, **4**, 145–56 (1948); *Chem. Abs.*, **43**, 1948 (1949).

[47] Guérin, H., and Margel, R., *Compt. rend.*, **231,** 1502–4 (1950). Guérin, H., and Marcel, P., *Bull. soc. chim. France*, **1956**, 1212–20.

[48] Le Beau, P., *Compt. rend.*, **177**, 319–22 (1923).

[49] Guérin, H., and Bostick, M., *Bull. soc. chim. France*, **1953**, 304–6.

[50] Sun, B., Ruof, C. H., and Howard, H. C., *Fuel*, **37**, 299–308 (1958).

[51] Juettner, B., and Howard, H. C., *Ind. Eng. Chem.*, **26**, 1115–18 (1934).

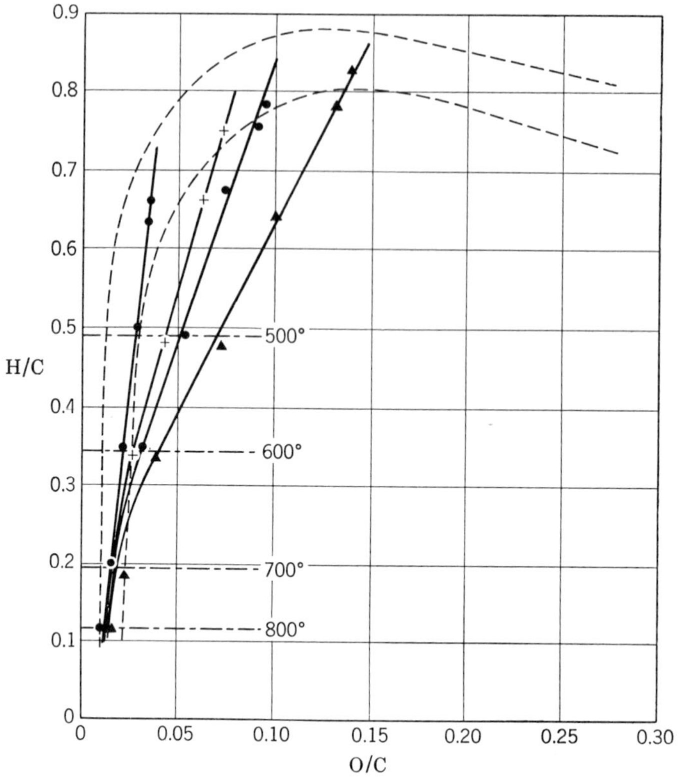

Fig. 5. Change in composition of residue.[47]

consisted of an aluminum plate in which a series of concentric square-bottomed grooves, $\frac{1}{8}$ in. by $\frac{1}{8}$ in. were machined. These grooves were filled about half full of coal particles with the purpose of heating the top particles by radiant energy from the walls of the grooves extending above them. Data for the decomposition of four American coals; a low-volatile, two bituminous coals, and a lignite were reported. Maximum temperatures used were 500 to 550°C. The most important observation was the very much higher fraction of low-volatile components recovered in the decomposition products from the coal with the strongest coking properties. These amounted to approximately 50% of the total volatile material for the Pittsburgh-bed coal, 37% for the low-volatile Pocohontas, 37% for the more weakly coking Illinois bituminous coal, and only 10% for the lignite.

An improved design[52] of this still has been used in a detailed investigation of the nature of the higher molecular-weight fraction of the volatile products from decomposition of a sample of Pittsburgh-bed coal in high vacuum. The maximum weight loss was found with this coal at about 525°C and amounted to 23%. Half of this was recovered from the condenser as a solid; the balance was made up of gaseous and liquid products not condensable at the cooling water temperature of 20 to 30°C and 10^{-3} mm pressure of mercury. As in the earlier work, the condensate was found to consist of a semitransparent, lacquer-like material, initially yellowish-orange but rapidly turning reddish-brown and then black in air and light. The properties of a typical sample are shown in Table 3. This investigator, because of the similarity of the infrared spectra of this material to that of the original coal, regarded it as derived therefrom with little decomposition. He also accepted the view[53] that unconjugated double bonds, long paraffin chains, and condensed aromatic ring structures are absent

[52] Greifer, B., *Vacuum Pyrolysis of Bituminous Coal, Thesis*, Carnegie Institute of Technology, Pittsburgh, Pennsylvania, 1957. Also see Orning, A. A., and Greifer, B., *Fuel*, **35**, 381–3 (1956).

[53] Friedel, R. A., and Queiser, J. A., *Anal. Chem.*, **28**, 22–30 (1956).

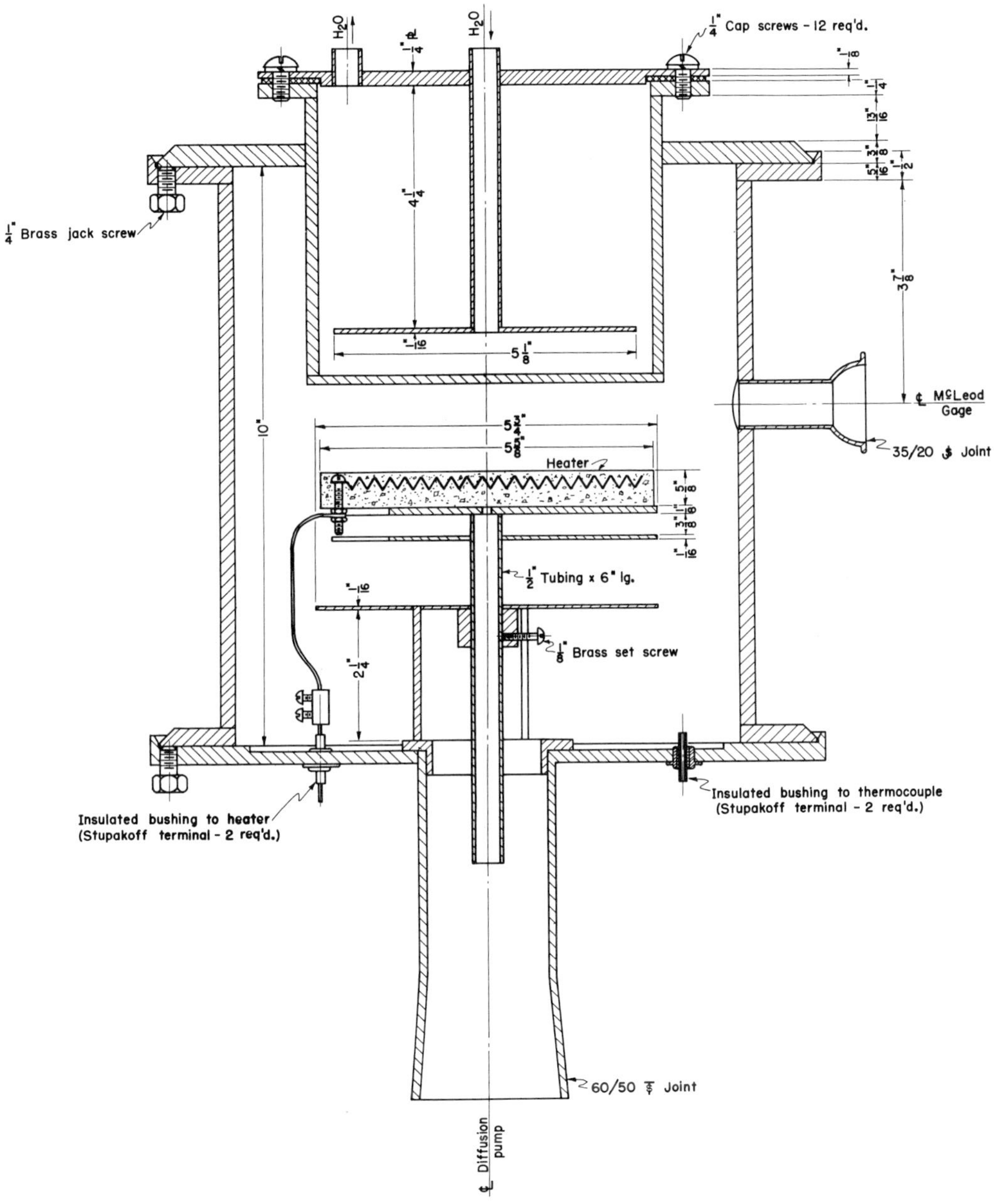

Fig. 6. Short-path vacuum still.[50]

TABLE 3

PROPERTIES OF CONDENSATE FROM VACUUM PYROLYSIS UP TO 512°C[52]

	C	H	O + N + S (Diff.)	Residue on Ignition
Ultimate Composition	84.99%	7.48%	7.06%	0.47%

Mol wt = 274 n_D = 1.668 at 25° d = 1.171 g/ml at 25°

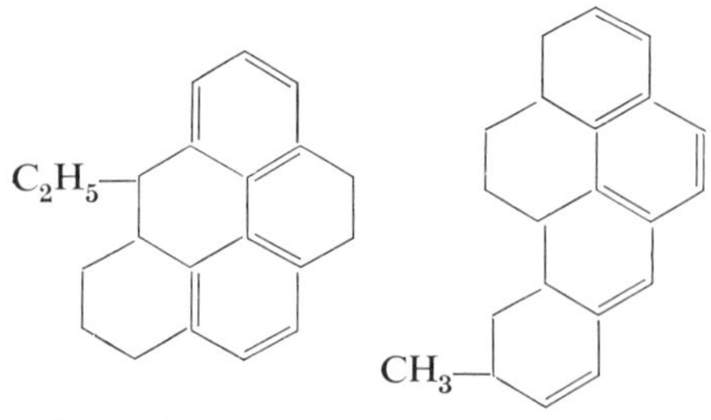

Fig. 7. Possible structures of condensate.[52]

in coals of this rank. Thus, a limited choice of structures for this decomposition product was left; a possible structure in agreement with these assumptions is shown in Fig. 7. Such a compound agrees in ultimate composition (oxygen, nitrogen, and sulfur have been neglected), hydrogen:carbon ratio, molecular weight, and molecular refraction with the values of these quantities for the material recovered. It is of interest that the fraction of aromatic carbon in the postulated structure is close to that estimated by use of van Krevelen's graphical procedure.[54]

The view expressed in this as well as in the older work that these solid asphaltene-like substances, which are characteristic of the volatile products recovered in high-vacuum pyrolytic decompositions, are primary fragments has been carefully considered by Dryden and co-workers.[55] They have stated "Since coal must have a macromolecular structure, most of the fragments released are likely to have been attached to other parts of the coal skeleton, before pyrolysis, by more than one chemical bond; in general, the larger the fragment, the greater the number of bonds. The release of large fragments involving the fission of many bonds will therefore be much less probable than the release of small fragments." As further support of the view that the high-molecular-weight volatile matter originates in small fragments, such as the smaller free radicals and unsaturated compounds broken off from larger molecules by pyrolysis, they cited, among other evidence, Wadsworth's observation that "when coal was carbonized to 400°C in an atmosphere of nitric oxide, a known inhibitor of chain reactions, a light translucent tar of low viscosity was produced which changed slowly into a more-normal type of tar."

Berkowitz and Dammeyer's work with nitric oxide leads to a somewhat different view.[56] They found two effects of nitric oxide in the low-temperature carbonization of a subbituminous coal: (1) retardation of the thermal decomposition of the coal, and (2) inhibition of secondary polymerization, but both reactions were found subject to the condition that the nitric oxide be present at the start of incipient pyrolysis, for it was found that these reactions do not occur if the nitric oxide is admitted to the retort where the temperature is above the decomposition temperature of the coal. These facts suggest that, if the high-molecular-weight products in the vacuum tars are the result of secondary reactions of small free-radicals, these reactions must occur before the volatile matter escapes from the coal particle. This view furnishes a reasonable interpretation of the high yields of high-molecular-weight products reported in early work, where the gap between evaporating and condensing surface was very short, and the condenser was cooled with liquid nitrogen. Under such conditions, secondary polymerization reactions in the gas phase or after condensation seem highly improbable. Thus, it appears that any reaction to modify the nature of tars in the sense of reducing the pitch content must be effected before the volatile matter leaves the coal particle.

In any discussion of the genesis of the asphaltene-like products recovered in vacuum pyrolysis, it should be kept in mind that similar types of products also predominate in the "extracts" recovered by treating coal with benzene at 250°C and in the products recovered from hydrogenation reactions at 300 to 350°C. In the last-named instance, it seems unlikely that, if small free-radicals or unsaturated fragments were the primary products, these would polymerize more rapidly than they would react with hydrogen.

The nonvolatile residue from high-vacuum pyrolysis shows little evidence of softening, and even the residue from highly fluid coals consists of the same discrete particles as existed in the original sample. An interesting feature is the development of a considerable porosity, 34% for the residue as compared with 5% for the

[54] van Krevelen, D. W., and Chermin, H. A. G., *Fuel*, **33**, 338–47 (1954).

[55] Billington, A. H., Dryden, I. G. C., and Napier, D. H., *BCURA Bull.*, **16**, 258–67 (1952).

[56] Berkowitz, N., and Dammeyer, W., *Fuel*, **35** 19–30 (1956).

original coal. This indicates the definite heterogeneity on a molecular scale of even those coals in which the single maceral vitrinite predominates. Such behavior is in accord with Brown's[57] picture of coal as a molecular sieve structure. Even with the strongly coking coals in which internal porosity is at a minimum, there is considerable internal surface on a molecular scale. The internal structure appears to be made up of flat cavities interconnected by small capillaries, and it is interesting that, in decomposition in high vacuum, the rupturing of bonds in this internal structure contributes a large part of the volatile matter evolved.

Dryden and co-workers,[58] in their studies of the mechanism of coke formation, especially in connection with the development of chloroform soluble products by preheat, have employed the design of the short-path vacuum carbonization apparatus shown in Fig. 8. A sample of −72 B.S. mesh coal was spread on the distilling surface in a uniform layer 1 to 2 mm thick (>5 to 10 particle diameters). Because of heat transfer problems, neither the mean coal temperature nor the gradient through the coal bed was exactly known, and temperature control was by means of a thermocouple in the heated surface on which the coal layer was spread. Because of the thermal gradient in the coal layer, it was found necessary to remove the residue remaining after a single carbonization, mix thoroughly, and recarbonize. By this two-step procedure, an additional 30% of distillate was recovered and, because it was estimated that a third stage would not increase the yield by a useful amount, the two-step procedure was adopted. Judged by infrared spectra, the products obtained in the two stages were similar.

The results for two coals, a weakly coking Northumberland and the Durham coking, were reported. They contained, respectively, 81.2 and 84.2% carbon on the moisture and ash-free basis. The mean distillation temperature was estimated to be 370°C. Under these conditions, the yield of distillate was 7.7 and 9.1%, respectively. The distillates from the two coals reflected, in a general way, the composition of the original coal. Those from the Northumberland coal, on an atomic basis, had a much larger fraction of the oxygen in the distillate, 60% compared with 24% for the Durham coal, and the atomic ratio of oxygen to carbon was significantly higher, 0.056 compared with 0.0083.

The volatility of the distillates was determined by a modification of a vapor-saturation method in which a sample was heated in a silica vaporizer in a current of oxygen-free nitrogen. The vaporizer was immersed in a salt bath maintained at a constant temperature of 350°C. The gas stream, saturated with vapors, was passed through a weighed tube of alumina adsorbent and the nitrogen was metered. The rate of vaporization of the distillate from the Northumberland coal was initially higher and remained so throughout the course of the measurement. The rates of vaporization of the distillates from both coals decreased with time, indicating either considerable molecular heterogeneity or a slow molecular growth in the product during the course of the experiment. Estimates were made of the percentages of these distillates which would be carried off in the carbonization gases at various temperatures. These are plotted in Fig. 9. It is apparent that the distillate from the Northumberland coal was more volatile over the whole temperature range, but one must agree with the authors that although "the difference in volatility is considerable it is still insufficient to account for more than a small part of the difference in the softening behavior between these coals."

The distillates from these two coals were also subjected to examination by chromatography on activated alumina. The alumina was packed into the column as a slurry in chloroform, and 10% solutions of the distillates in chloroform were added and eluted first with chloroform and, if necessary, with acetone. The distillate from the Durham coal gave four; distinct bands: (1) a dark brown band strongly adsorbed at the top of the column, 9%; (2) another brown band containing phenolic hydroxyl, but less aliphatic hydrogen than the other fractions; (3) an orange-brown band, phenolic hydroxyl absent, 80%; and (4) an orange-yellow band, phenolic hydroxyl absent, 1%. The distillate from the Northumberland coal treated in a similar fashion showed no separation into bands even when examined in ultraviolet light: The "color varied continuously from dark brown at the top of the

[57] Brown, R. L., *J. Inst. Fuel*, **29**, 218–31 (1956).

[58] Brown, J. K., Dryden, I. G. C., Dunevein, D. H., Joy, W. K., and Pankhurst, K. S., *J. Inst. Fuel*, **31**, 259–73 (1958).

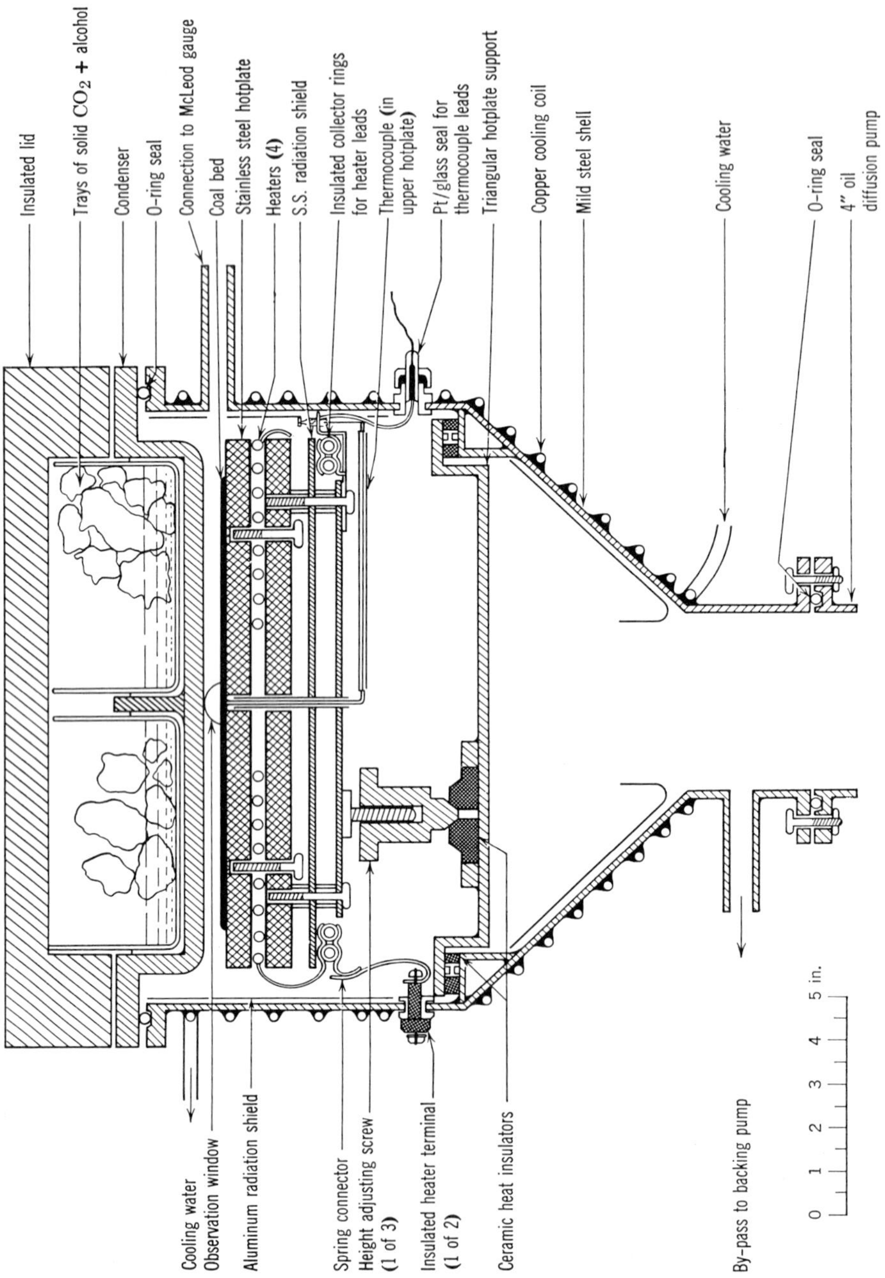

Fig. 8. Short-path carbonization apparatus.[58]

column to weak yellow brown at the bottom." These results were interpreted to indicate a definite difference in the modes of pyrolysis of the two coals.

This work represents a very careful attempt to trace the marked difference in coking behavior of these two coals to some difference in the properties of the higher-molecular-weight volatile products formed during vacuum pyrolysis. In view of the facts disclosed, it seems that some other factor or factors must be of controlling importance. Two of these which suggest themselves are (1) the rate of escape of the plasticizing material from contact with the skeletal structure, and (2) ease of plasticization of the skeletal structure itself. The importance of the first has been emphasized by Berkowitz[59] and Francis,[60] who pointed out the correlation between coking properties as measured by "swelling number," "agglutinating value," and the moisture content of a number of British Commonwealth coals. The difference in swelling observed with these coals was ascribed to "the disparity between the rates of formation and disengagement of the gaseous products of thermal decomposition of the coal." On the basis of this view, the critical factors would be the rate of formation of the volatile products of pyrolysis, their size and shape, and the pore structure of the coal. Noncoking coals have a large pore volume, and their volatile matter is characterized by a large content of water vapor and oxides of carbon. Hence, such high-molecular-weight plasticizing components as may be present are highly diluted and swept rapidly away from the surfaces of the skeletal structures. It is possible that part of the significant improvement in the physical properties of cokes from weakly coking Western American coals, effected by preheat below fusion temperatures,[61,62] is due to the separate evolution of the nonplasticizing components of the volatile matter so that they are no longer present to exert a diluting effect when fusion temperatures are reached. The importance of the rate of escape of the volatile matter as a factor in the coking phenomena is evident from the failure of even strongly coking coals to swell and fuse into coherent cokes when pyrolyzed in high vacuum, when the rate of escape of volatile products is at a maximum. The second factor is more difficult to evaluate, but it is possible that the differences in the nature of the skeletal structure between coking and noncoking coals is of as much importance as the amount and properties of the fragments evolved on pyrolysis.

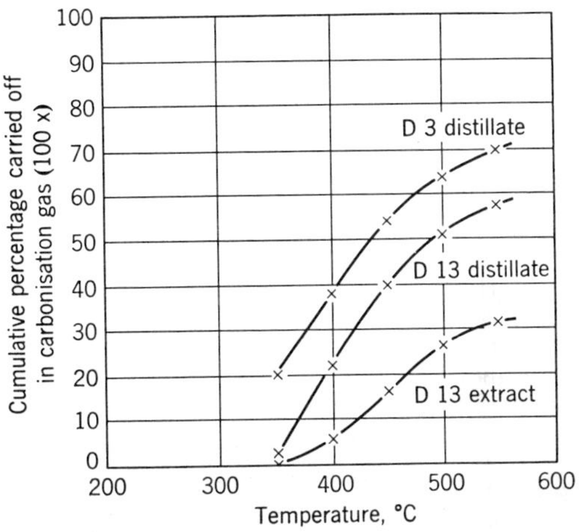

Fig. 9. Estimated proportions of the fusible pyrolysis products lost from coal in carbonizing chamber owing to vaporization.[59] D3: Northumberland weakly caking coal. D13: Durham coking coal.

Application of Mass Spectrometry

The use of mass spectrometry in the study of the primary decomposition products of a series of vitrains has been described by Holden and Robb.[63] Over the temperature range 200 to 370°C, loosely bound aliphatic compounds with a wide range of molecular weights were evolved. In addition to these, alkylbenzenes naphthalenes, and higher aromatics were also observed. Above this temperature, where decomposition was definitely indicated by the appearance of methane, series of substituted phenols and naphthols were detected. This method has also been used to characterize coal extracts and to determine distributions of molecular weights in vacuum-distillation fractions.

[59] Berkowitz, N., *Fuel*, **28**, 97–102 (1949); **29**, 138–48 (1950).

[60] Francis, Wilfred, *J. Inst. Fuel*, **25**, 15–28 (1952).

[61] Parr, S. W., and Layng, T. E., *Ind. Eng. Chem.*, **13**, 14–17 (1921).

[62] Smith, F. W., Birge, G. W., Wolfson, D. E., and Reynolds, D. A., *U.S. Bur. Mines, Rept. Invest.*, No. 5418 (1958), 26 pp.

[63] Holden, H. W. and Robb, J. C., *Nature*, **182**, 340 (1958).

Comparison of Pyrolyses at Normal and Elevated Pressures

Work in this field prior to 1939 has been reviewed by Sustman and Ziesecke;[64] the results reported in recent work are in general agreement with those of earlier investigations. When coals are carbonized in closed systems where evolution of the volatile matter of the coal takes place against an external gaseous pressure made up of either the volatile products derived from the coal or these plus an inert gas, there generally results an increase in coke and gas yields and a decrease in tar yields. In some cases, production of a "hard" coke from a "noncoking" coal has been reported. Increases in the proportion of low-boiling constituents in the tar have been found, as well as increased "aromatization," so that the tars produced at 500 to 600°C, under external pressures of 100 atmospheres, resemble, in this respect, high-temperature tars.

The effects of external pressures up to 100 atmospheres on the course of the pyrolysis of four German coals have been studied by Sustman and Ziesecke.[65] In the first, small-scale experiments, a sample of 100 g of finely ground coal was placed in an iron boat in the bottom of the pressure vessel. The internal volume of the latter was about 400 ml, and it was equipped with a central thermocouple well, a pressure gauge, and a needle valve to control the rate of escape of volatile products and, consequently, the working pressure. At the beginning of each experiment, the vessel was flushed out with nitrogen. The temperature was raised to 400°C at the rate of 6°/min and from 400 to 600°C at the rate of 4°/min. The final temperature was maintained until there was no further significant gas evolution. This usually required about 3 hr. In the pressure experiments, the valve was kept closed until the desired working pressure was reached. The results are shown in Fig. 10. These data confirm the results of earlier workers that increased pressure results in increase of coke and gas yields at the expense of the tar. The changes in composition of the gaseous products also showed a more or less constant pattern with pressure changes. Typical results are given in Table 4. The marked increase in hydrocarbons in the gas produced at the higher pressures results in significant increases in heating value per unit of volume.

The nature of the coke residues from the Saar and Upper Silesian coals appeared to be little affected by gas pressure, contrary to early work with a Ruhr coal where improvement in coke properties was noted.[66]

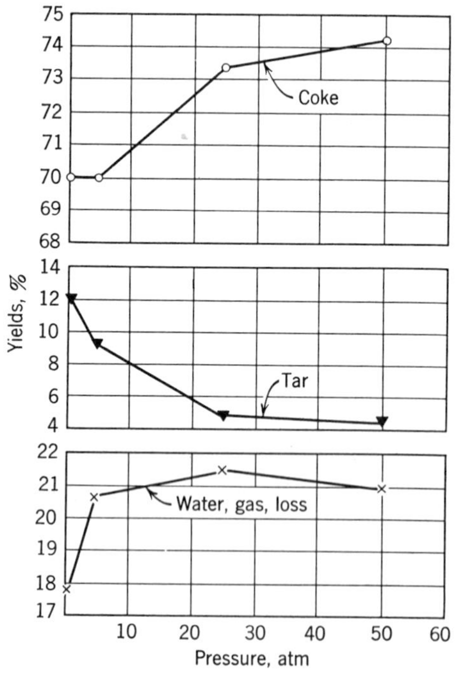

Fig. 10. Yields of products in pressure distillation to 600°C (Ruhr Gasflammkohle).[65]

TABLE 4

EFFECT OF PRESSURE ON COMPOSITION OF CARBONIZATION GASES FROM A SAAR COAL, IN LITERS AT N. T. P. /100 G[65]

	Pressure, Atmospheres			
Gas	1	5	25	100
$CO_2 + H_2S$	0.63	0.81	0.95	1.55
Unsaturates	0.36	0.33	0.23	0.15
O_2	0.05	0.02	0.06	0.03
CO	0.81	0.60	0.44	0.39
H_2	2.22	1.53	1.21	1.27
Saturated Hydrocarbons	5.68	6.74	8.25	11.47
N_2	0.75	0.57	0.56	0.64

[64] Sustman, H., and Ziesecke, K. H., *Brennstoff-Chem.*, **20,** 228–32 (1939).

[65] Sustman, H., and Ziesecke, K. H., *Brennstoff-Chem.*, **21,** 37–42, 49–56, 61–68 (1940).

[66] Fischer, F., Bahr, T., and Sustman, H., *Brennstoff-Chem.*, **11,** 1–9 (1920).

In order to obtain larger quantities of tar for more detailed characterization, carbonization under pressure was carried out in a larger vessel on 2.5-kg samples. With the Ruhr coal, there was a significant increase in the fraction of the raw tar boiling up to 360°C from 79 to 91.4% for pressures up to 50 atmospheres. The increase was largely in neutral oils, and the light-oil content of these was increased. The increase in density of the light oil was attributed to increase in aromatic content. Yields of tar acids were reported greater than those obtained in coke-oven practice. The effects of pressure on the tars from the Saxon brown coal were similar. The increase in aromatics in the light oil was shown to be largely para and meta xylenes. Experiments at a constant pressure of 50 atmospheres at temperatures of 400, 500, and 600°C showed little change in tar yields but significant increases in specific gravity. At 50 atmospheres and 400°C, this brown coal yielded a light oil containing 32% aromatic compounds.

Recently, a number of investigations on the effects of pressure on the course of coal pyrolysis have been reported in Czechoslovakian literature[67–69] but, judged on the basis of the English abstracts available, the results have not been significantly different from those reported by earlier workers.

In the most recent work,[70] the results with Estonian shale, Moscow brown coal, Donets Basin bituminous coal, and Ukranian brown coal are discussed. Samples weighing 140 g were charged to an electrically heated, stainless-steel retort heated at 15 to 20°/min to final temperatures of 550 to 630°C. Pressures of 20, 50, 100 atmospheres of nitrogen, hydrogen, or steam were used. These conditions were stated to allow the control of secondary processes and, consequently, the recovery of products of various compositions and yields. For example, it was found that the yield of gas of high calorific value was markedly increased by carbonization under pressure. It was also pointed out that the size of the equipment necessary for carbonization at elevated pressures could be significantly reduced compared with that used at normal pressures.

Decomposition of Coal in the Presence of an Added Liquid Phase

The behavior of coal when subjected to the action of heat in the presence of an added liquid phase exhibits many ramifications. The phenomena observed appear to range in character from those definitely involving chemical reaction through all gradations to what may be regarded as simple physical changes.

It is well known that when bituminous coals are subjected to the action of organic liquids even at room temperature, a certain amount of soluble extract can be recovered. With certain active solvents, significant yields of extract are obtained at temperatures no higher than 100°C, and such solvents have been designated by Dryden as "specific." The extracts recovered by the action of these specific solvents have molecular weights ranging from a few hundred to several thousand, the latter predominating.

The present discussion is concerned only with the breakdown of coal at elevated temperatures by "nonspecific" solvents. The extracts recovered under such conditions are of moderate molecular weight, ranging from a few hundred to a few thousand and are strikingly similar in general properties to the asphaltenes, recovered in short-path pyrolytic decompositions which have been discussed in a previous section.

Earlier workers in studies of the extraction of coal at a constant elevated temperature have found some correlation between yield of extract and such properties of the solvent as surface tension and internal pressure. Any correlation with one of these would be expected to exist for the other, because these two properties have been shown to be linearly related.[71] These correlations were only moderately successful. In this work it was apparently not recognized that maximum solubility occurs

[67] Šimek, B. G., Ludmila, J., and Edinger, J., *Paliva a voda*, **29**, 97–100 (1949); *Chem. Abs.*, **43**, 5569 (1949).

[68] Hrádek, V., *Paliva*, **33**, 141–4, 197–8 (1953); *Chem. Abs.*, **49**, 7835 (1955).

[69] Sisakov, N. V., *Paliva*, **34**, 8–11 (1954); *Chem. Abs.*, **49**, 7835 (1955).

[70] Altshuler, V. S., and Shafir, G. S., *Khim. Pererabotka Topliva, Akad. Nauk S.S.S.R., Inst. Goryuch. Iskopaemykh, Trudy 2-go Vsesoyuz. Soveshchaniya, Moscow*, **1954**, 76–86, published 1957; *Chem. Abs.*, **52**, 15877 (1958).

[71] Hildebrand, J. H., and Scott, R., *The Solubility of Nonelectrolytes*, Third edition, Reinhold, New York, 1949, pp. 402–3.

when solute or solvent differs least in the solubility parameter (the square-root of the internal pressure) and that in a series of solvents a maximum will be encountered in the solubility-solubility parameter relation on one side of which a *decrease* in the solubility parameter results in decrease in solubility, and on the other side of which an *increase* in the solubility parameter has the same effect. This has been shown by Gee[72] to hold even for the swelling of natural and artificial polymers.

Interesting discussions[29] of the solubility parameter as related to the solubility of high-molecular-weight compounds have emphasized the importance of consideration of hydrogen-bonding effects as well as the solubility parameter. The responsibility of hydrogen-bonding for the nonvolatility of low-molecular-weight compounds has already been pointed out as was, also, the fact that such compounds are characterized by high solubility in strongly hydrogen-bonded solvents. Because hydrogen-bonded liquids show abnormal heats of vaporization, and because this is the most important factor in the solubility parameter, it is evident that highly hydrogen-bonded solvents such as alcohols and glycols will also have large values of the solubility parameter, but that such compounds may be very poor solvents for weakly hydrogen-bonded or nonhydrogen-bonded solutes. Unfortunately, it is not possible as yet to evaluate hydrogen-bonding effects quantitatively, but it seems evident that the very small solubility of bituminous coals in alcohols and glycols, despite the large solubility parameter of these solvents, is due to their strongly hydrogen-bonded character. Thus, any simple relation between solubility and the solubility parameter alone cannot be expected.

Few studies of the effects of temperature and nature of solvents on yields of extracts from bituminous coals have been reported since the work of Kiebler.[73]

Russian workers[74] have studied the decomposition of coal in phenol under hydrogen pressure. Equal quantities of coal and phenol were heated in an autoclave at 250, 300, 350, 375, 400, and 420°C under an initial hydrogen pressure of 80 atmospheres. The heating rate was 3.3°/min. The amount of soluble material produced depended on the temperature; under suitable conditions, the coal was completely dissolved. These workers concluded that a depolymerization, rather than a decomposition, takes place at high temperatures in the presence of solvents and cited as evidence for this view the fact even at the highest temperature the gaseous products contained only small amounts of the hydrocarbon gases so characteristic of pyrolytic decompositions of coal.

The compositions of phenol and phenol-tetrahydronaphthalene extracts of bright and semidull coals have been determined and reported as nearly identical.[75]

Solution of coal in anthracene oil and in retene at 350°C has been used to study the properties of coking coals by Zabavin. Retene was stated to dissolve a smaller fraction of the coal than did anthracene oil but was more readily separated from the coal extract.[76]

Zabavin and Tantsyrev[77] studied the solubility of all grades of bituminous coal when heated in anthracene oil at 350°C and found that the solubility correlated with agglutinating value and with the thickness of the plastic layer formed on pyrolysis.[78] A comparison of the methods was reported to show that solubility in anthracene oil was more useful for characterizing "lean" coals than agglutinating value. The reverse was true for "fat" and gas coals, and both methods were found in good agreement for coking coals.

The decomposition of a Japanese coal, Yubari, in creosote oil was studied[79] over the

[72] Gee, G., *Trans. Inst. Rubber Ind.*, **18**, 266–81 (1943).

[73] Kiebler, M. W., *Chemistry of Coal Utilization*, John Wiley and Sons, New York, 1945, Vol. 1, pp. 677–760.

[74] Rappoport, I. B., and Khudyakova, A. D., *J. Appl. Chem. (U.S.S.R.)*, **12**, 759–66 (1939); *Chem. Abs.*, **34**, 3051 (1940).

[75] Egorova, O. I., *Bull. acad. sci. U.R.S.S., Classe sci. tech.*, **1941**, No. 7–8, 63–70; *Chem. Abs.*, **39**, 2190 (1945).

[76] Zabavin, V. I., *Bull. acad. sci. U.R.S.S., Classe sci. tech.*, **1941**, No. 4, 27–40; *Ibid*, **1944**, No. 7–8, 71–3; *Chem. Abs.*, **36**, 1462 (1942); *Ibid.*, **39**, 2190 (1945).

[77] Zabavin, V. I., and Tantsyrev, I. N., *Trudy Inst. Goryuch. Iskopaemykh, Akad. Nauk S.S.S.R.*, **2**, 227–35 (1950); *Chem. Abs.*, **49**, 4259 (1955).

[78] Brewer, R. E., *Chemistry of Coal Utilization*, John Wiley and Sons, New York, 1945, Vol. 1, pp. 280–98.

[79] Baba, A. J., *Soc. Chem. Ind. Japan*, **47**, 283–6 (1944); *Chem. Abs.*, **43**, 1942 (1949).

temperature (T) range 275 to 400°C. Up to 375°, the yield of soluble products (y) was given by the relation $\log y = -1637/T + 4.4157$.

Dryden[80] has made an exhaustive review up to 1949 of the action of solvents on bituminous coals. He pointed out the usefulness of differentiation between the action of low-temperature specific solvents of the amine type and high-boiling hydrocarbon types and suggested that, at temperatures above 250°C, the changes caused in coal in solvent systems are probably the result of both the influence of temperature and the chemical action of the solvent.

Oversohl[81] has studied the behavior of bituminous coals when heated in certain liquids at elevated temperatures and pressures. The investigations were carried out on a series of Ruhr coals of degrees of coalification ranging in rank from *Flamm-*, *Gas-*, and *Fett-* to a *Mager-* coal. The liquids used were water, benzene, and a tetralin-cresol mixture. About 2-cm pieces of coal enclosed in a large mesh screen cylinder were placed in an autoclave containing the liquid to be used and heated for 2 hr at 350°C. In one series of experiments, the pressure in the autoclave corresponded to the vapor pressure of the water, 150 to 160 atm, and, in another heating, was carried out with a pressure of nitrogen of as high as 700 atm above the water. In the first series, the vitrain became viscous and formed a bubbly mass due to gas evolution. At pressures above 500 atm, gas evolution was prevented and the vitrain formed a pitchlike melt. The fused vitrain could be diluted with high boiling oils and, hence, the whole coal could be converted to a product of pitch-like consistency, fusible at normal pressures. The fusibility of the vitrain was found to reach a maximum with high-volatile coals. Peat and brown coals when heated under water pressure at 350°C gave evidence of coalification; the carbon content of the peat increased from 59 to 75%; it became black and friable, and microscopic examination indicated a material resembling fusain. The carbon content of a middle-German brown coal was increased from 67 to 81% and appeared to be more like vitrain or semifusinite. With the hard brown coal, the carbon content was increased from 76 to 83%. In the experiments where the coal was heated with organic liquids to temperatures above 250°C, microscopic examination of the products indicated that the vitrain was composed to a considerable extent of small cellular structures. These experiments, describing the conversion of coal to a reversibly thermoplastic material by heating under water with an impressed inert gas pressure, are highly suggestive and indicate that the pyrolysis of bituminous coal may be made to take a very different course if escape of the primary gaseous decomposition products is inhibited.

Storch and co-workers[82] have reported the extraction of a bituminous coal with a number of organic solvents, chiefly aromatic compounds, at their normal boiling points and covering the temperature range 255 to 360°C. Two types of extraction apparatus were employed; the smaller made use of 1 g samples of 40 to 60 mesh coal wrapped in a weighed piece of glass cloth (No. 201, Filter Media Corporation) and suspended in the vapor of the boiling solvent directly below the reflux condenser. Experiments were conducted for 6 hr and generally in nitrogen. After cooling, the boiling flask containing the solvent and extracted material was replaced by another flask containing benzene, and the glass bag and residual coal were exhaustively extracted with benzene. The amount of benzene-soluble material in the extract was also determined. In larger-scale experiments conducted in a stainless steel extractor with about a 6 in. diameter, the coal sample was spread out on a circular piece of glass cloth of about 5 in. in diameter, covered with a similar piece, and these were clamped together by means of stainless steel rings in the manner in which cloth is held between embroidery rings. This sample assembly was supported just above the level of the boiling solvent by inward projecting pins fastened to the interior wall of the extractor. Experiments with this larger extractor showed that the ratio of solvent to coal could be as low as 6 to 1 without markedly affecting the yield. There are several points in this work of special interest: the low yields obtained with anthracene, compared with

[80] Dryden, I. G. C., *BCURA Bull.*, **13**, 113–38 (1949).

[81] Oversohl, W., *Brennstoff-Chem.*, **31**, 103–11 (1950).

[82] Golumbic, C., Anderson, J. B., Orchin, M., and Storch, H. H., *U.S. Bur. Mines Rept. Invest.* No. 4662, 1950, 12 pp.

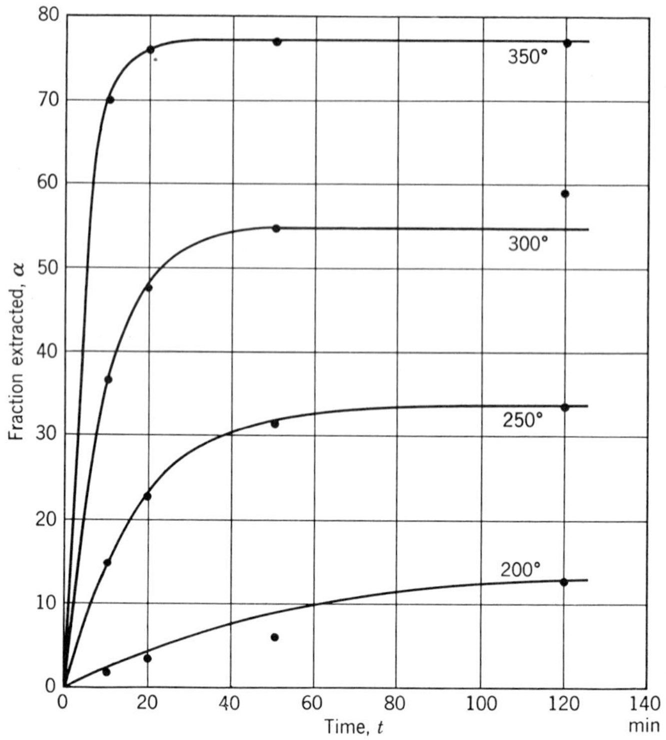

Fig. 11. Extractive disintegration of a bituminous coal with anthracene oil at different temperatures.[83]

phenanthrene, despite the higher extraction temperature used with the former; the large differences in the effectiveness of the isomers of phenylphenol; and the little difference in yield reported for the two isomeric naphthols. A limited number of comparisons were made with a subbituminous coal. With this low-rank coal, phenanthrene was shown to be a rather poor solvent compared with α-naphthol; the yield from the former was 27% compared with 83% for the latter. Also with the subbituminous coal, outstanding differences were reported with the two isomeric naphthols, the β isomer yielding only 17% extract.

Dutch workers[83] have designated as "extractive disintegration" the decomposition process involved in the "extraction" of coal with organic solvents at temperatures in the range 200 to 400°C. These workers pointed out that, in any consideration of the effect of temperature on the process, the question arises as to how far the increase in yield at higher temperatures can be attributed to purely physical factors.

Gillet and Pirlot[84] have drawn attention to the fact that in extractive disintegration of a Belgian bituminous coal with β-naphthol and with anthracene oil, an abrupt change in the course of the disintegrative process takes place at around 260°C.

They[83] have observed the same effect with anthracene oil and a Limburg coal. (See Fig. 11.) The difference between the course of the process at 200 and 300°C is very striking, and they ascribed the high temperature dependence of the rate of reaction in the early stages to the fact that, at this stage in the process, the thermal decomposition of the coal is the governing factor. They pointed out that in the initial stages of the disintegration the velocity of the process is nearly constant, but in later stages, the velocity of reverse reactions becomes appreciable and, finally, an equilibrium state is approached. They suggested that these reverse reactions consist of precipitation of dissolved coal particles on the coal surfaces.

[83] Oele, A. P., Waterman, H. I., Goedkoop, M. L., and van Krevelen, D. W., *Fuel*, **30**, 169–78 (1951).

[84] Gillet, A., and Pirlot, A., *Bull. soc. chim. Belges*, **41**, 511–20 (1932).

Thus, the forward and reverse reactions can be represented by $v_1 = k$ and $v_2 = k'\alpha$, respectively, where v_1 and v_2 are the respective forward and reverse velocities; k and k' are the corresponding rate constants, and α represents the fraction of extracted matter. The rate equation can then be written $d\alpha/dt = k - k'\alpha$, which on integration gives $\ln \frac{1 - k'\alpha}{k} = k't$. But because at equilibrium, $k = k'\alpha_{eq}$ where α_{eq} is the fraction extracted at equilibrium, then $\ln \left(1 - \frac{\alpha}{\alpha_{eq}}\right) = -k't$. Because α, at time t, and α_{eq} are available from the experimental data, it is possible to calculate k'. Knowing k', it is possible to calculate k from the relation $k = k'\alpha_{eq}$. If the rate constants, k, are calculated at a number of different temperatures, it is possible to apply the conventional Arrhenius relationship

$$\log k = \frac{E}{2.3RT} + B$$

where E is the activation energy and B the "collision factor." The resultant plots are shown in Figs. 12 and 13.

These plots bring out two important points: (1) The activation energy of the disintegration process generally increases with increase in coal rank for both of the solvents; and (2) for the higher-rank coal, 20.1% volatile matter,

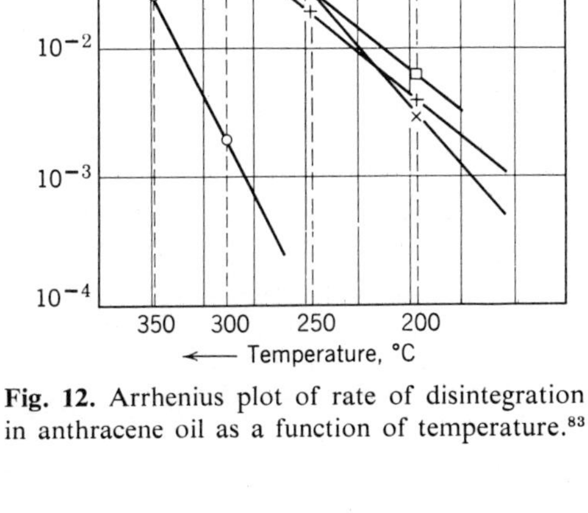

Fig. 12. Arrhenius plot of rate of disintegration in anthracene oil as a function of temperature.[83]

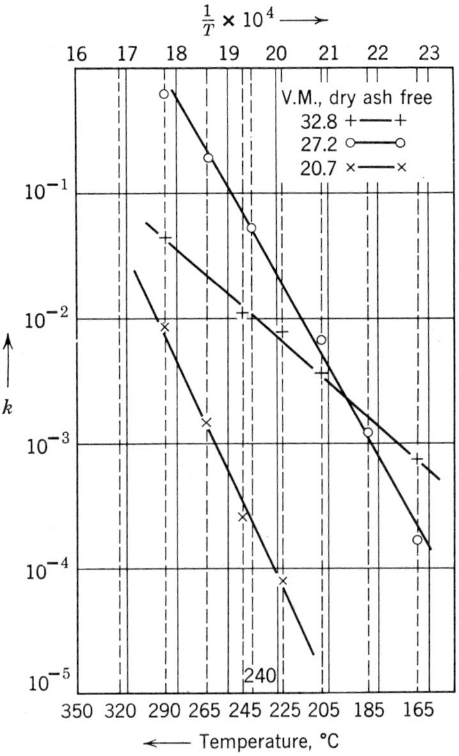

Fig. 13. Arrhenius plot of rate of disintegration in β-naphthol as a function of temperature.[83]

the activation energy of the process in β-naphthol is significantly greater than in anthracene oil. This is to be expected if hydrogen bonding is a less-important factor in the internal cohesive forces in coals of this rank than in those of lower rank, and this seems highly probable.[85]

Other investigations of these workers[85] of the action of such hydrogen-transfer solvents as tetralin and other partially hydrogenated aromatics will be discussed elsewhere (see Chapter 6). It is sufficient to mention that they pointed out that the effectiveness of such hydrogen-transfer solvents in coal disintegration may arise from the prevention of condensation reactions by inhibition of hydrogen evolution rather than by hydrogenolytic splitting of oxygen-bridge bonds. In this connection, it should also be emphasized that although such solvents as anthracene oil and β-naphthol are not usually regarded as hydrogen-transfer solvents, the possibility of their acting in this way, especially at temperatures of 300°C and

[85] Orchin, M., and Storch, H. H., *Ind. Eng. Chem.*, **40**, 1385–9 (1948).

upward, cannot be excluded. Kiebler[86] has shown that although α-naphthol gave a yield of 60% of extract at 300°C, with a 35% volatile matter coal, in the process, the solvent was very largely transformed to α-dinaphthalene oxide. He suggested that the high yield obtained with this solvent was due to hydrogen liberated by this transformation.

A number of investigations have studied the effect of organic liquids on coals when subsequently subjected to pyrolysis. Brusset and Le Rat[87] used a Valdonne lignite and a Camphausen "fat" coal. The coals were pretreated with a number of organic liquids such as benzene, acetone, meta-cresol, pyridine, and ethylenediamine. Small samples were then pyrolyzed up to 1000°C in a current of nitrogen at a heating rate of 5°/min. The curves of loss of weight as a function of temperature were stated to be reproducible to less than 1%. For the untreated coals, this relation was linear between 425 and 500°C for the lignite and between 450 and 500°C for the "fat" coal.

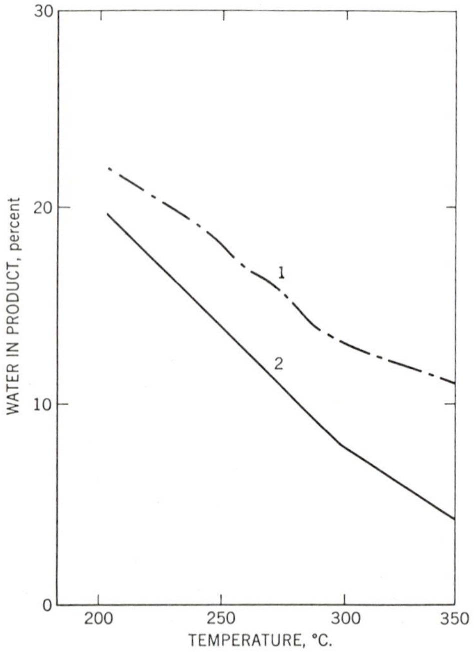

Fig. 14. Effect of heating method on water retained. 1. Dry, in absence of air. 2. In the presence of water.[90]

The effect of the treatment with the organic liquids was generally to reduce the decomposition temperature and to extend the linear portion of the weight-loss temperature curve. The effects on the total loss in weight were not marked except in one experiment where a sample of the "fat" coal was treated with ten different solvents (details of this were not given). In this instance, the total weight loss to 1000°C was 39.8% compared with 32.4% for the untreated sample. The authors suggested that the lowering of the decomposition temperature and higher yield of volatile matter were due to the solvent creating a "disengaged" state of the coal micelles, thus rendering these more mobile and permitting a more rapid escape of the products of pyrolysis from the mass.

Oranskii[88] heated mixtures of powdered coal and coal tar at 400 to 430°C in an autoclave revolving at 50 to 60 rpm. The "degree of cracking," presumably a formation of volatile matter, was determined as a function of temperature, time, and ratio of solvent to coal. In one experiment at 400°C carried out for 30 min, 35.8% of the coal and 11.6% of the solvent in the paste were "cracked," whereas when heated separately under identical conditions, the respective values were 10 and 18%. Up to 350°C, the products were primarily water of decomposition and gas. In a later paper,[89] in which work on the use of fractions of primary tar and powdered coal was described, "cracking conversions" of 63, 45, and 31% were stated to be obtained from rhabdopyssite, "ordinary" coal, and humus coal, respectively. Solvent regeneration of 80 to 90% was reported.

A comparative study of the different effects obtained when a German, woody-brown coal was heated in the presence of a water phase and dried in the absence of air has been reported by Liebnitz and co-workers.[90] Both processes had the effect of increasing the rank of the brown coal as judged by increase in carbon content, decrease in oxygen, and decrease in water holding capacity, but heating in the presence of water was significantly more effective at the same temperatures. The relations

[86] Kiebler, M. W., *Chemistry of Coal Utilization*, John Wiley and Sons, New York, 1945, Vol. 1, p. 738.

[87] Brusset, H., and Le Rat, B., *Compt. rend.*, **238**, 2533–4 (1954); *Chem. Abs.*, **49**, 692 (1955).

[88] Oranskii, I. N. I., *Zhur. Priklad Khim.*, **29**, 759–64 (1956); *Chem. Abs.*, **50**, 16074 (1956).

[89] *Ibid.*, 1086–93; *Chem. Abs.*, **50**, 17378 (1956).

[90] Liebnitz, E., Könnecke, H. G., and Schröter, M., *J. prakt. Chem.*, **6**, 18–24 (1958).

between the water retained in the product and temperatures for the two procedures are shown in Fig. 14. Another interesting feature of the pressure water heating is in the effect on the yield of bitumen, defined as alcohol-benzene soluble products. The raw coal on a dry basis yielded 6% bitumen, defined as alcohol-benzene soluble products. The yield was increased to 18.5% by pressure water heating but to only 8.7% by the dry heating in the absence of air.

Special Pyrolytic Procedures and Apparatus

Thermogravimetric Balances

Knowledge of the rate of loss of weight of a coal when heated either at constant temperature or at some controlled rate of temperature rise is of primary importance to an understanding of the nature of the thermal decomposition processes. Audibert[91] reported the results of such experiments as early as 1926. With the recent availability in the laboratory of automatic control and recording devices, there has been renewed interest in the procedure.[92] Boyer[93] has used this method to study the decomposition of French coals, and representative data are shown in Fig. 15 for two coals, coal *A* being higher in rank than coal *B*. The effect of rank on the temperature at which the maximum in the rate of decomposition curve occurs is clearly illustrated.

Waters[94] has recently described an improved type of recording thermogravimetric balance

[91] Audibert, E., *Rev. ind. minérale*, **1926,** 115–36; *Chem. Abs.*, **20,** 2239 (1926).

[92] Thévenin, R., *Bull. soc. hist. nat. Toulouse*, **85,** 383–8 (1950); *Chem. Abs.*, **48,** 4201 (1954).

[93] Boyer, A. F., *Compt. rend. congr. ind. gaz—69th Congr.* **1952,** pp. 653–58 and 668–71. Also see, Vdovenko, *Ibid.*, 70th Congr. **1953,** pp. 250–73 and 71*st Congr.* **1954,** pp. 444–60; Tanno, H., *J. Coal Research Inst. Japan*, **3,** 43–8 (1952); *Chem. Abs.*, **49,** 7219 (1955); Bracia Goyanes, C., *Inform. quim. anal.* (*Madrid*), **9,** 159–74 (1955); *Chem. Abs.*, **50,** 6247 (1956).

[94] Waters, P. L., *Nature*, **178,** 324–6 (1956); *Coke and Gas*, **20,** 252–6, 289–91, 341–43 (1958).

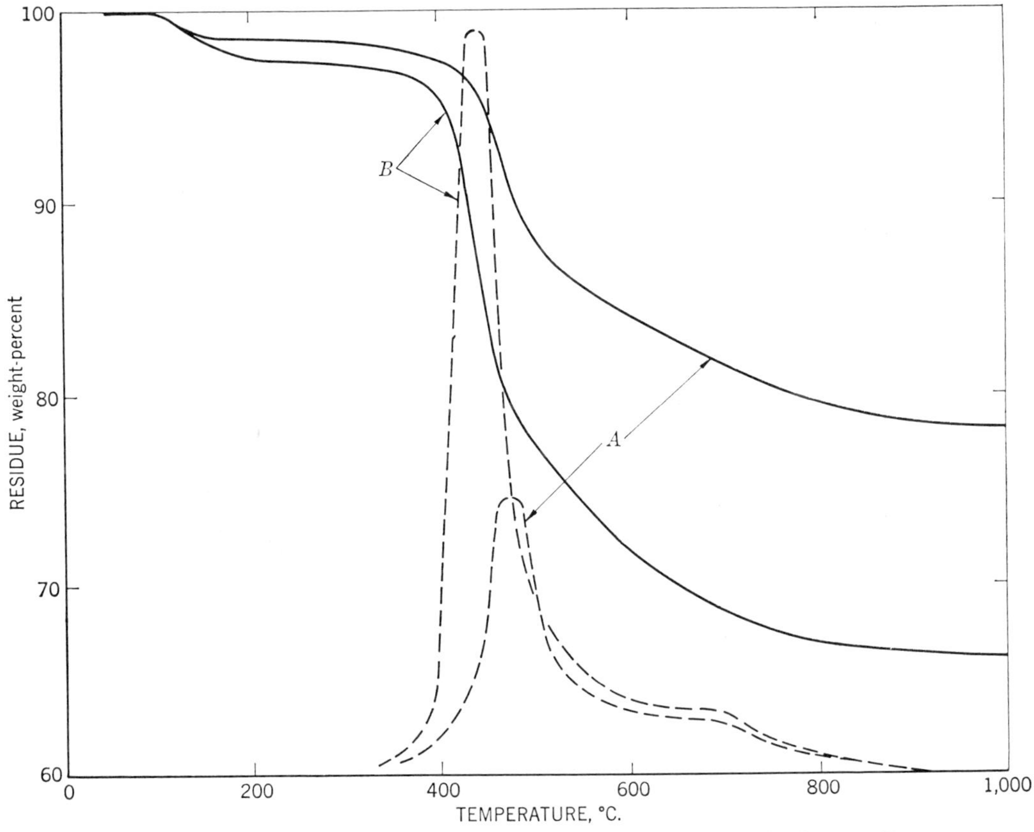

Fig. 15. Change in weight of two French coals—integral and derived curves.[93]

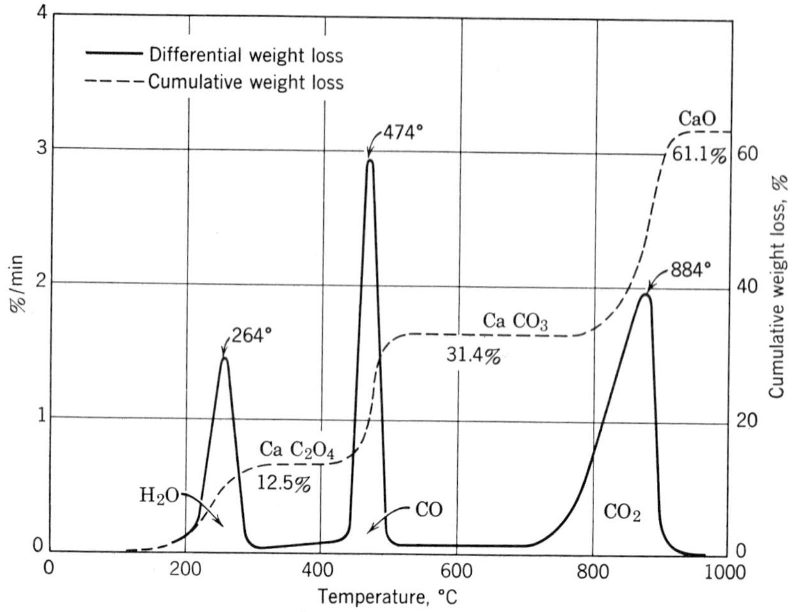
Differential weight loss
Cumulative weight loss
%/min
Cumulative weight loss, %
Temperature, °C
264°
474°
884°
CaO
61.1%
Ca CO_3
31.4%
Ca C_2O_4
12.5%
H_2O
CO
CO_2

(a)

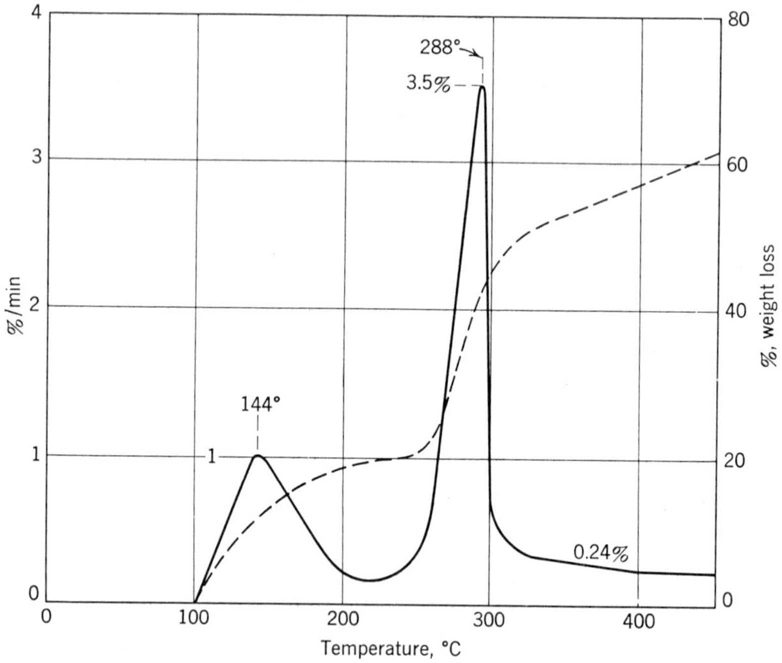
%/min
%, weight loss
Temperature, °C
288°
3.5%
144°
1
0.24%

(b)

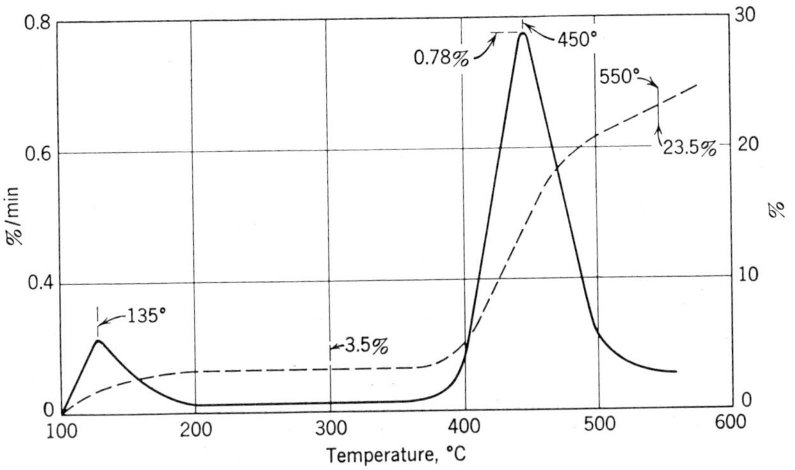

(c)

Fig. 16. Thermogravimetric curves.[94] (*a*) Calcium oxalate monohydrate (6°/min). (*b*) Potato starch (3°/min). (*c*) Victoria Tunnel Seam coal (3°/min).

by which both the cumulative weight-loss and differential curves are recorded simultaneously. With types previously available, it has been necessary to differentiate the cumulative curve graphically. With the design described, total weight changes up to 0.5 g can be recorded at a maximum rate change of 30 mg/min. The cumulative weight change is accurate to within 1%, and the differential weight change to within 0.2 mg/min. The decomposition curves for three substances with this apparatus are shown in Fig. 16. In curve "*a*," for hydrated calcium oxalate, the first peak results from the loss of water, the second, from that of carbon monoxide, and the third, from that of carbon dioxide. In curve "*b*," for starch, the first peak represents loss of water and the second, deep-seated pyrolytic breakdown. It is interesting to compare this curve with curve "*c*" for the coal from the Victoria Tunnel Seam. The very great thermal lability of the carbohydrate compared with the coal is well brought out; decomposition of the former is nearly complete at temperatures where the decomposition of the latter is scarcely started. Typical examples of the differential pyrolysis curves of Australian coals using this apparatus have been published by H. R. Brown[95] and are shown in Fig. 17. With the heating rate of 3°/min, the maximum in the rate of heating curve for the coking coals appears at temperatures above 440°C and below that temperature for the lower-rank coals. The latter are also characterized by a very much broader temperature range of decomposition. For brown coals, the start of rapid decomposition is about 270°C; for lignites, 300°C; for subbituminous, 350 to 390°C; and for weakly coking coals, about 400°C.

Rate of weight-loss curves were also obtained for several conditions of constant temperature.[95] A typical curve is shown in Fig. 18. Such curves are reminiscent of low-temperature oxidation reactions, but in these, the decreased rate at constant temperature is probably due to surface effects, whereas here the more likely explanation is, as Dryden[96] has pointed out, a range of bond strengths on the molecular scale. That this heterogeneity is on a molecular rather than on a microscopic scale is indicated by the fact that even a pure vitrain sample showed the same phenomenon.

The rate at which the coal was heated was shown to have a significant effect both on the rate of decomposition and on fluidity. The effect of heating rate on a coal from the Victoria Tunnel Seam is shown in Fig. 19. With this coal, it appears that the coal substance was relatively thermally stable up to 400°, that it first softened without thermal decomposition involving evolution of volatile products, and fuses at about 410°C. Maximum fluidity was reached during a period of rapidly increasing rate of decomposition. Thus, the softening process need not be related necessarily to the pyrolytic process resulting in evolution of volatile products. It may, as others have suggested,

[95] Brown, H. R., *J. Inst. Fuel*, **30**, 137–59 (1957).

[96] Dryden, I. G. C., *J. Inst. Fuel*, **30**, 156 (1957).

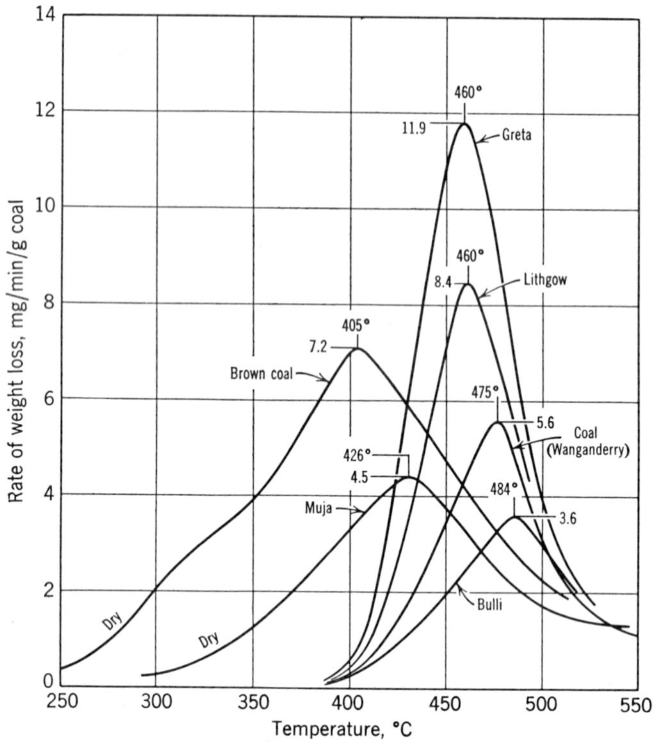

Fig. 17. Decomposition rates of various coals.[95]

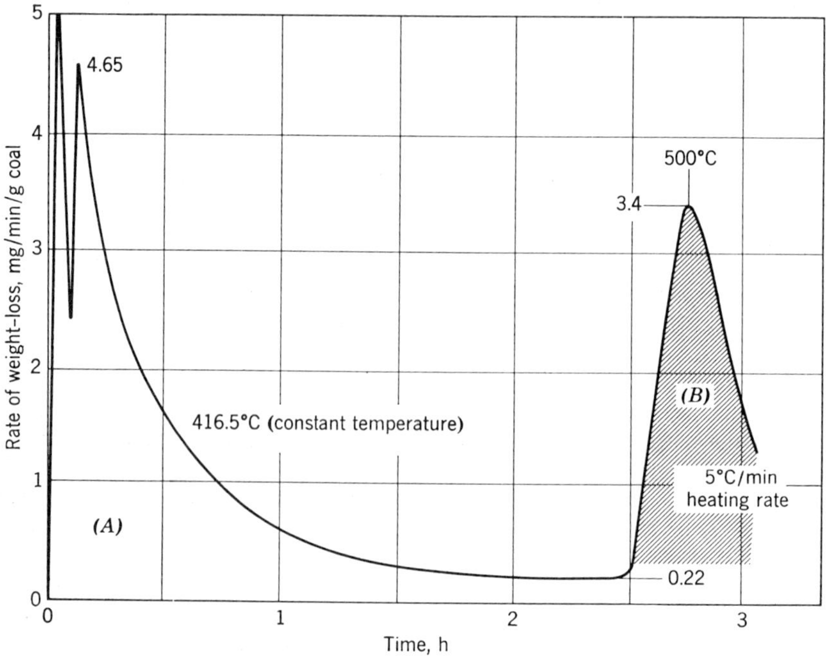

Fig. 18. Isothermal and rising temperature-weight loss.[95]

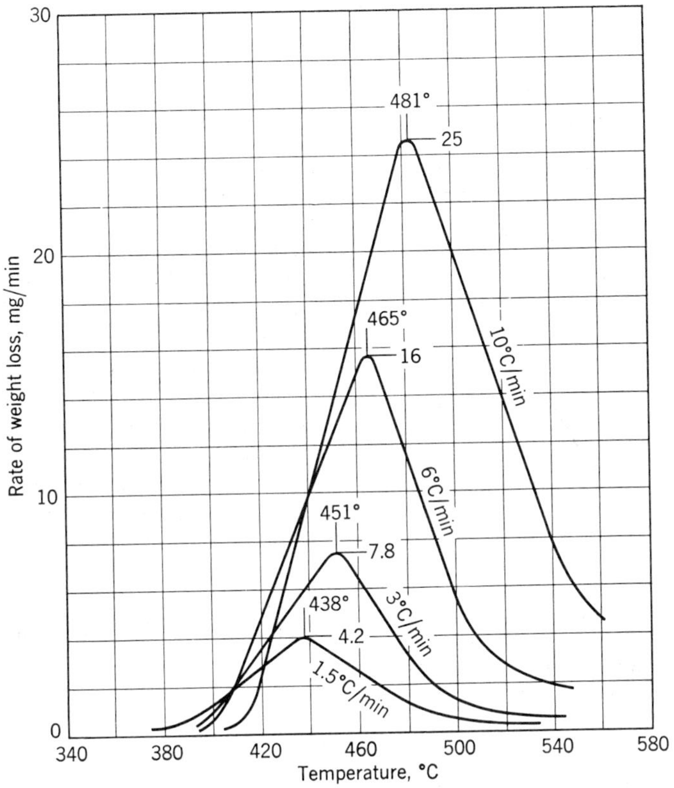

Fig. 19. Effect of heating rate on rate of decomposition.[95]

be a reversible process. Samples heated at 3°C/min up to a temperature just below fusion point and then repeatedly taken through a cycle of cooling and reheating before being allowed to reach the temperature of maximum fluidity showed variable behavior. One sample showed little reduction in maximum fluidity; others showed considerable reduction. Thus, it seems that, provided the evolution of volatile matter has not commenced, the phenomenon of softening is, in some coals, relatively reversible, and over a certain temperature range, the behavior of coal is similar to that of a thermoplastic resin.

Differential thermal analysis. This procedure, which has been mentioned briefly in earlier paragraphs, is another physical tool which has been used to obtain further insight into the processes occurring during the pyrolytic decomposition of coal. In mineral systems where the decomposition processes are simple phase changes or well-defined transitions, it has been highly successful. However, as has been indicated earlier, it is difficult to interpret the results obtained with coal in an unequivocal manner. Detailed reviews of the field have been made by Ramachandron and Bhattacharyya[97] and Nelson,[98] and a survey of methods and instruments with emphasis on application to coal has been made by Kirsch.[99]

Whitehead and Breger[100] have described a differential thermal analysis apparatus designed for operation either in a vacuum or in a controlled-gas atmosphere.[3] Typical thermograms recorded with this apparatus are shown in Fig. 20. Displacements in an upward direction from the initial point are interpreted as being the result of exothermal reactions and those in a downward direction, the result of endothermic reactions. The heating rate was 12°/min. The small peak at 575°C visible in all the curves was from the transformation

[97] Ramachandron, V. S., and Bhattacharyya, S. K., *J. Sci. Ind. Research* (*India*), **13A**, 365–72 (1954).

[98] Nelson, J. B., *BCURA Bull.*, **19**, 502–31 (1955).

[99] Kirsch, H., *Brennstoff-Chem.*, **38**, 87–92 (1957).

[100] Whitehead, W. L., and Breger, I. A., *Science*, **111**, 279–81 (1950).

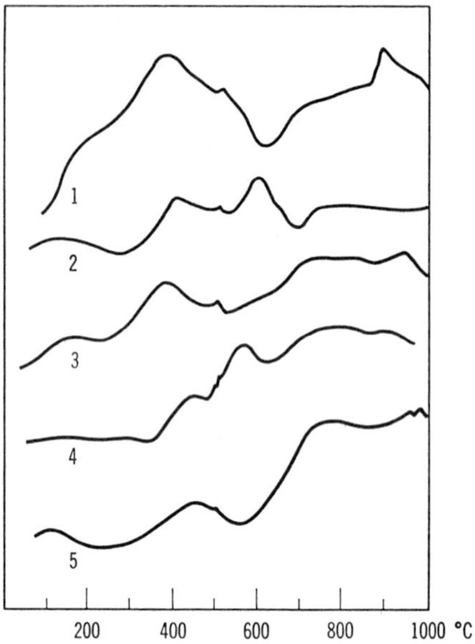

Fig. 20. Thermograms of American coals.[100] 1. North Dakota lignite; 2. Wyoming high-volatile C bituminous; 3. Colorado bituminous coal; 4. Pennsylvania low-volatile coal; 5. Semi-anthracite from Alaska.

point of the quartz which was added to the reference substance as an internal indicator. The significance of the low-temperature transformations indicated in these curves has already been discussed. Marked exothermal reactions starting below 400°C for the lignite and somewhat above for bituminous coals are shown. Another exothermic reaction is indicated for these coals above 700°C. With the anthracite, no marked change of any kind is shown until temperatures approaching 700°C are reached, reflecting its higher thermal stability. It was concluded that a "standard curve" characteristic of a given type of coal cannot be established.[101]

Yagishita[102] has subjected a number of Japanese coals to differential thermal analysis and described five stages of reaction as follows: (1) endothermic, due to evaporation of water—larger in low rank coals; (2) exothermic, with a small amount of volatilization; (3) endothermic, with considerable volatilization, but less in the low rank coals; (4) exothermic, with little volatilization, and (5) endothermic, with little volatilization. Generally, the endothermic reactions appeared to be larger with the coking coals, and this was ascribed to dehydrogenation accompanying condensation reactions.

Differential thermal analyses on a number of American coals have also been reported by Glass.[4] This work was done in conventional equipment in a nickel sample holder with a loose-fitting cover. It was stated that, because gases escaped almost continuously, no air could enter the system, and the coal was decomposed in an atmosphere of its volatile decomposition products. Half-gram samples were used, and they were packed into the well with a uniform but light pressure. A heating rate of about 10°C/min was employed. Typical curves are shown in Fig. 21. The author recognized five distinct thermal types (Table 5).

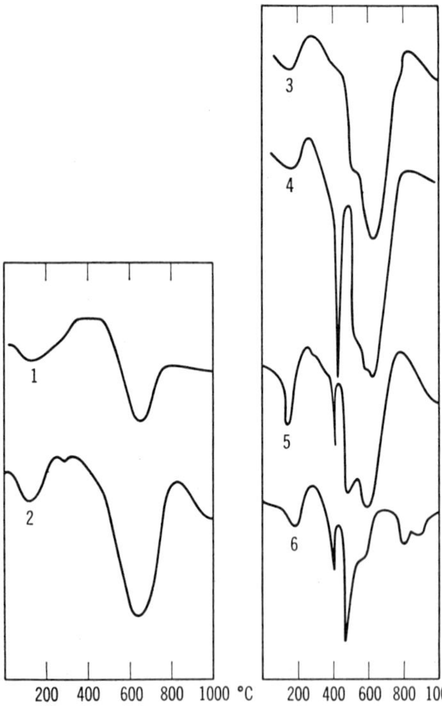

Fig. 21. Thermograms of American coals by Glass.[4] 1. Pennsylvania anthracite; 2. Arkansas semianthracite; 3. West Virginia low-volatile bituminous coal; 4. High-volatile A bituminous coal; 5. Illinois high-volatile B bituminous coal; 6. Wyoming subbituminous C coal.

[101] Whitehead, W. L., *Proc. Conf. on Origin and Constitution of Coals,* Nova Scotia Dept. of Mines, 1950, pp. 100–10.

[102] Yagishita, H., *Misc. Repts., Research Inst. Nat. Resources (Tokyo),* No. 26, 40–7 (1952); *Chem. Abs.,* **47,** 1358 (1953).

TABLE 5

TYPES OF THERMAL CURVES[4]

Type of Curve	Coal Ranks Included	Characteristic Peaks
I. Meta-anthracite	Meta-anthracite	Single endothermic 725 to 735°C
II. Anthracite	Anthracite Semianthracite	Single endothermic 630 to 680°C
III. Low Volatile	Low-volatile bituminous Medium-volatile bituminous	Two endothermic ~500°C ~620°C
IV. High Volatile	High-volatile A, B, and C bituminous	Strong exothermic reaction imposed on primary endothermic
V. Subbituminous	Subbituminous, Lignite, Peat	Large primary endothermic at 450°C; secondary endothermic may be absent or appear weakly at 550 to 600°C

A careful study of the effect of experimental factors on the nature of the curves obtained in differential thermal analysis has been made by Clegg.[103] He concluded that changes in peak location, peak height, temperature range, or interruption of devolatilization can occur when the following factors are varied: heating rate, particle size of sample, amount of inert diluent, manner of packing, manner of covering, and the geometry of the sample holder. A prime cause of variation was shown to be related to the ease and manner in which volatile decomposition products escaped and to the pressure developed inside the specimen well.

The first major endotherm is the result of loss of moisture. Exothermic reaction is thought to follow water loss. The main volatiles endotherm which usually begins to develop around 300°C reflects evolution of volatile matter. The size and shape of the endotherm, however, is controlled not only by the formation of gas but also by the ease and manner of its escape from the sample. When conditions are imposed that tend to delay or temporarily stop free escape of decomposition gases . . . the result is reflected in the thermogram, first by increased peak heights and then by interruption of the main volatiles endotherm. Careful consideration was given to the possibility that an exothermic reaction might be responsible for the interruption interval, but there is no evidence for such an interpretation.

The decomposition of cellulose, lignin, peat, xylite, and various ranks of coal has been studied by Pampuch and Roga.[104] It was reported that the release of moisture at 90 to 120°C, the release of volatile matter, the development of plasticity, 320 to 460°C, and the shrinkage and release of hydrogen, 670 to 720°C were all endothermic, whereas solidification of the plastic mass at 480 to 540°C was exothermic.

Berkowitz[105] employed an apparatus in which continuous streams of pure, dry nitrogen were passed through the test and reference cells during each run. Samples of six Canadian coals ranging in rank from a lignite of 73.9% carbon to a low-volatile bituminous coal of 90.5%, all on a dry, ash-free basis, were subjected to differential thermal analysis in this apparatus. The typical results, in contrast to those of Glass, appear to be relatively little affected by rank, and the reactions up to 550°C, the maximum temperature to which the experiments were carried, appear to be predominantly exothermic. The author concluded that the discrepancies between the results of different investigators are principally due to lack of standardization of technique.

Dielectric heating. Because of the problem of heat transfer in granular beds of coal, encountered both in the laboratory and in technology, interest has been aroused in the possible application of dielectric heating. It

[103] Clegg, K. E., *Illinois State Geol. Survey, Rept. Invest.*, No. 190 (1955), 30 pp.

[104] Pampuch, R., and Roga, B., *Prace Głównego Inst. Górnictwa, Ser. B., Komun.*, No. 183 (1956) 15 pp.

[105] Berkowitz, N., *Fuel*, **36**, 355–73 (1957).

would appear to be the ideal method for heating coal rapidly and uniformly, particularly in vacuum pyrolyses, where heat-transfer problems are especially difficult. However, in application, a number of difficulties have been reported. Because of the heterogeneous character of coal, the heating is not uniform, and because, during the decomposition process, the coal changes from a dielectric to a conducting substance, arcing may take place between the solid particles.

Sevastýanov[106] and co-workers have compared the changes in the electrical characteristics of coal brought about by dielectric heating with those effected by conventional methods. They stated that, in heating with a high-frequency field, the coal becomes conducting at 240 to 270°C, the resistance falls to a few ohm-cm, and at 400°C the resistance becomes as low as 1 ohm-cm. With conventional heating, the change in resistance was stated to be represented by the equation: $\log R = -0.011T + 11$, where R = resistance in ohms per cm and T = temperature in degrees centigrade and covers the range 350 to 450°C. With dielectric heating, the corresponding equation was stated to be: $\log R = -0.00366T + 1.60$ over the temperature range 250 to 450°C. They concluded that decomposition of coal by dielectric heating involves a different mechanism than when conventional methods of heating are used.

Patents have been issued covering the use of dielectric heating in pyrolytic processes.[107]

Karavaev[108] and co-workers have pyrolyzed a Cheremkhov coal at 1000°C. The heating was by means of a "tube generator," 7-m wavelength, and a frequency of 4.3×10^7/sec. They reported that the composition of the coke produced was little different from that obtained by conventional heating to the same temperature. The yields of coke and gas were found to be significantly lower, and the composition of the gas was different from when conventional heating methods were used.

Tyutyunnikov[109] used dielectric heating in the pyrolysis of coal up to 480°C. He stated that coal becomes a conductor at 480°C, and that conventional methods of heating had to be used above this temperature. The coal was placed between two plates of a condenser and a high-frequency current (21.4 megacycles/sec) set up. Rates of temperature rise of 33, 40, 50 and 60°/min were maintained up to 480°C. The temperature gradient from the surface to the center of the charge was less than 4° at all heating rates. With conventional heating methods, optimum coke properties were obtained with a heating rate of 1°/min; with the dielectric heating, best results were at rates of 33°/min.

Taylor[110] has described an apparatus for studying plastic changes in coal in which dielectric heating was used. The coal sample was confined in a mullite crucible between two asbestos disks, these in turn by two others of perforated alundum, and carbon electrodes were pressed against the latter. The upper electrode was movable and attached to a volume-indicating device. The movable electrode also communicated with a hydraulic system by which loads up to 62.5 lb/sq in. could be applied to the swelling coal. The radio-frequency field was applied from a valve generator set with a working frequency of 25 megacycles/sec, and the output was 200 w. The heating rate was 2.5°/min, and it was stated that "the ability of radio frequency heating to give absolutely uniform temperature distribution and rates of temperature rise permits a precise study of the individual stages in the plastic zone" (See Chapter 4).

Effects of Pretreatment and Additives on the Pyrolytic Process

Oxidation

The effect of oxidation, both of the slow, natural weathering and the accelerated type, on the course of a subsequent pyrolytic decomposition of the coal has been investigated rather thoroughly, and there is general agreement on the effects produced. These are:

[106] Sevastýanov, Y. L., Dolgopolov, N. N., Burýan, Y. L., and Margolin, V. S., *Doklady Akad. Nauk S.S.S.R.*, **74,** 755–7 (1950); *Chem. Abs.*, **45,** 2174 (1951).

[107] Hodge, V. M., U.S. Patent 2,542,028 (1951). Yoda, Y., Japanese Patent 1777 (1953).

[108] Karavaev, N. M., Sevastýanov, Y. L., Dolgopolov, N. N., and Burýan, Y. L., *Doklady Akad. Nauk S.S.S.R.*, **77,** 871–4 (1951); *Chem. Abs.*, **45,** 6819 (1951).

[109] Tyutyunnikov, Y. B., *Kok i Khim.*, **1957,** No. 12, 22–6; *Chem. Abs.*, **52,** 6758 (1958).

[110] Taylor, J. W., *Research (London)*, **10,** 108–15 (1957).

Fig. 22. One-fourth inch mesh, Pittsburgh-bed coal oxidized at 200°C in air and subsequently carbonized at 750°C.[113]

increased yields of oxides of carbon and water, reduction in the yields of tar, and changes in plastic properties. With coals of low agglutinating power, even small amounts of oxidation may result in completely non-coherent residues. With highly fluid coals, the physical properties of the cokes produced may pass through a maximum, the initial oxidation improving the physical properties, and further reaction having an adverse effect.

Macrae[111] has reported the results of pyrolyses in the Gray-King assay apparatus of two samples of a naturally weathered British coal obtained in connection with the Fuel Research Branch Coal Survey in Yorkshire. The assay was carried out at 50°C intervals over the range 200 to 600°C. One of the samples, "*A*," was taken about eight yards from the "basset edge" or perimeter of the seam (Top Beamshaw, Stanley Main); the other, "*B*," from about two yards from the edge, probably representing the extreme of alteration found in an outcrop coal. The yields of solid and liquid carbonization products on a dry, ash-free basis did not differ greatly for the two samples over the range of temperatures studied. However, the difference in weight yields of carbon dioxide was a striking 9.7% compared with 17.0%.

The marked effect of small amounts of oxidation on the agglutinating properties of coals has been well recognized for a number of years, and there now appears to be definite evidence for the view that it is principally due to a surface reaction.[112] It has been shown[113] that oxidation of the strongly agglutinating Pittsburgh-bed coal results in such a change of the character of the surface of the coal that although considerable amounts of fluid material may still be produced at fusion temperatures, this product is unable to wet neighboring oxidized surfaces, with the result that there is no general fusion and a coherent coke is not formed. (See Fig. 22.)

[111] Macrae, J. C., *Fuel*, **22,** 93–102 (1943).

[112] Sommers, H. W., and Peters, W., *Chem. Ing. Tech.*, **26,** 441–53 (1954).

[113] Carman, E. P., Geer, M. R., and Riley, H. L., *U.S. Bur. Mines, Inform. Circ.* No. 7794 (1957), p. 74.

Marshall and Draycott[114] have investigated the effect of oxidation on the carbonizing properties of Australian Coals. Twenty-gram samples were ground to pass a 60-mesh B. S. screen and spread in a thin layer in a glass tube 18 in. long and 0.5 in. in diameter. They were oxidized at room temperature in a current of dry oxygen for periods ranging from 25 to 250 hr. The treated samples were carbonized in a Gray-King low-temperature apparatus, and the products were determined by standard methods. The results for a medium, high-volatile coal from the Liddell Seam of the Durham District, New South Wales, are shown in Fig. 23. The effect of oxidation on the yields of tar, liquor, and gas is clearly shown. The tar yield, often stated to be a sensitive index of weathering, shows an almost linear decrease up to 125 hr, with little change thereafter. The changes observed are more marked than have been reported[115] for American coals.

Boyer[116] has studied, by the thermographic balance, the pyrolytic decomposition of a French *gras* coal (37.5% V.M.) and compared the results with those obtained with two samples of the same coal oxidized in the laboratory. The original coal contained 7.5% oxygen, and in the two oxidized samples the oxygen content was increased by 2.8 and 6.2%. The results are shown in Fig. 24. The heating rate was 2°/min. The curves for the oxidized samples are characterized by a much greater rate of loss of weight at the temperature of 300°C and a greatly decreased maximum in the "active decomposition" range; that is, they show a much longer temperature zone of decomposition. At corresponding temperatures, the cokes from the oxidized coals were of higher oxygen content, showed decreased anisotropy, and were graphitized with more difficulty.

The peroxide-like character of the primary-surface oxidation products of bituminous coals has been previously discussed.[117] Russian workers[118] have reported that these groups are thermally unstable and decompose with the evolution of heat. This exothermic effect was

[114] Marshall, C. E., and Draycott, A., *Fuel*, **33**, 26–41 (1954).

[115] Schmidt, L. D., *Chemistry of Coal Utilization*, John Wiley and Sons, New York, 1945, Vol. 1, p. 666.

[116] Boyer, A. F., *Ann. mines Belg.*, **1956**, 908–12.

[117] Howard, H. C., *Chemistry of Coal Utilization*, John Wiley and Sons, New York, 1945, Vol. 1, p. 348. Also see Jones, R. E., and Townend, D. T. A., *Inst. Gas Engrs., Communs.*, **271** (1943–44), 21 pp.; **287** (1944–45), 23 pp.

[118] Titov, N. G., Khrisanfova, A. I., Kanavets, P. I., and Smirnov, R. N., *Khim. i Tekhnol. Topliva*, **43,** No. 1, 9 (1956); *Chem. Abs.*, **50,** 2949, 10373 (1956).

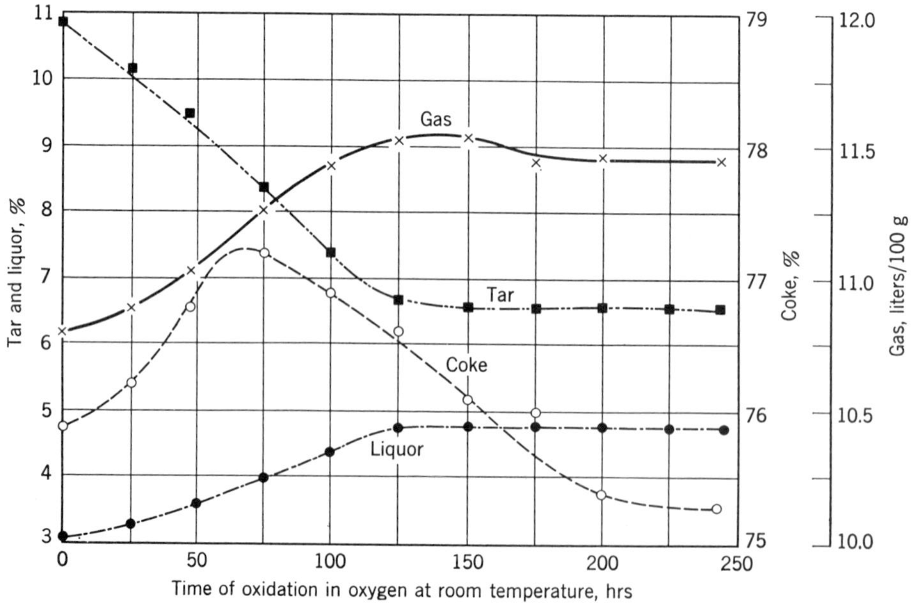

Fig. 23. Carbonization yields for ozidized Liddel Seam samples.[114]

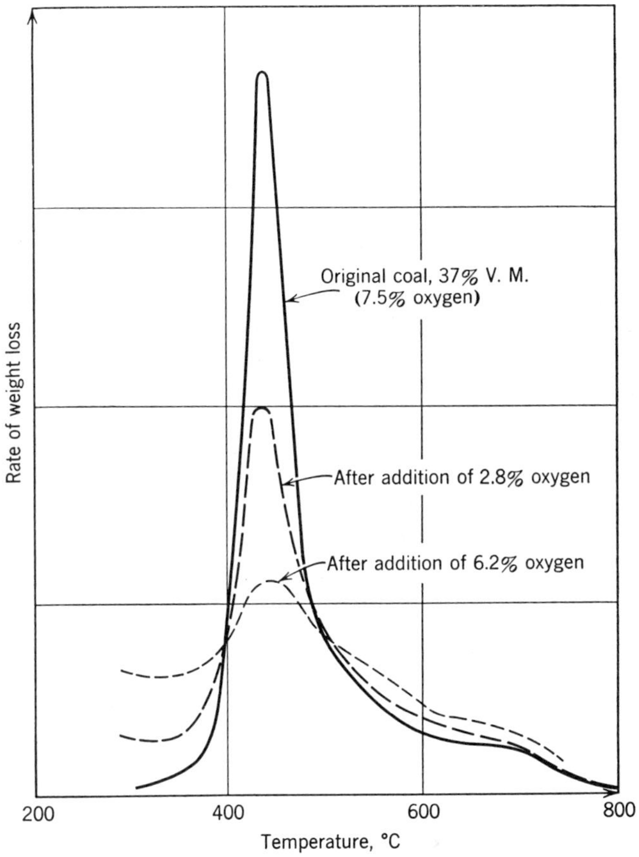

Fig. 24. Effect of oxidation on rate of weight loss.[116]

determined for sixteen samples of Russian coals using thermographic methods. The "peroxide number" determined by titration with dilute sodium hydrosulfite was determined for the same samples, and a relation was found between this number and the exothermic effects observed.

Nitric Oxide

Because of the chain inhibiting properties of nitric oxide, the pyrolysis of coal in an atmosphere of this gas has been studied by a number of workers. Dryden and co-workers[119] cited Wadsworth's observation of the production of "a light translucent tar of low viscosity, which slowly changed into a more normal tar" as evidence of its effectiveness. Billington[120] described the effect of pretreatment of bituminous coals with nitric oxide, nitrous oxide, and nitrogen tetroxide on the caking, swelling, and smoke-producing properties of these coals. The gases were passed at a rate of not more than 6 l./hr through a bed of 80 g of coal contained in a glass vessel and maintained at a specified temperature for a specified period. The ultimate composition of a Northumberland coal (type 802) was modified by treatment with nitric oxide as shown in Table 6. The changes in composition and the observed effects on caking, swelling, and tar-producing properties during subsequent

[119] Billington, A. H., Dryden, I. G. C., and Napier, D. H., *BCURA Bull.*, **16**, 263 (1952).
[120] Billington, A. H., *Fuel*, **33**, 295–301 (1954).

TABLE 6

	Untreated	Treated with NO at 320°C
	%	%
C	82.0	79.2
H	5.0	3.8
N	1.8	4.3
O (difference)	11.2	12.7
Volatile matter	36.0	27.7

pyrolysis were in accord with the oxidation indicated by the changes in ultimate composition brought about by the treatment. There is no positive evidence that the nitric oxide played a special role.

Berkowitz and Dammeyer[121] have studied the pyrolysis in nitric oxide of a subbituminous coal from the Drumheller field of Alberta, Canada. This coal is an essentially bright coal having 40.2% volatile matter on the dry, ash-free basis. Samples were carbonized at different partial pressures of nitric oxide, in a stainless steel retort of 500 g capacity at a heating rate of about 8°C/min. The maximum temperature, 580°C, was maintained for approximately 20 min. It was tentatively concluded that carbonization of this subbituminous coal in the presence of nitric oxide results in retardation of the thermal decomposition and inhibition of secondary polymerization of the tar phase. These effects were not noted if the nitric oxide was admitted to the retorts at a point where the temperature was above that of the decomposition temperature of the coal and their extent depended upon the concentration of the nitric oxide. At low partial pressures of nitric oxide, the dominant reaction was retardation of the decomposition, and significant modification of the tar was confined to high partial pressures of nitric oxide. Tars modified by nitric oxide did not appear to be more stable to light, and their infrared spectra were not significantly different from those prepared in its absence.

Reduction

As might be expected, the general effect of reducing reactions on the pyrolytic behavior of coals is the reverse of that observed with oxidation. The effects were first observed many years ago and were made the subject of a patent[122] which claimed that partial hydrogenation improved the coking properties of coals of all ranks. Kukharenko and Khrisanfova[123] have discussed, critically, the effect of mild reduction on the coking properties of noncoking coals and stated that the reduction results not only in improved coke but also in more valuable by-products. Later work by Storch[124] and co-workers has shown that the free-swelling index of an oxidized bituminous coal can be significantly increased either by catalytic hydrogenation at 400°C and 500 psi or by reduction with lithium aluminum hydride in tetrahydrofuran. Development of coking properties in a subbituminous coal was not found possible with the chemical reductant. It was also observed that restoration of coking properties, judged by the free-swelling index and nature of the coke button, took place in the catalytic hydrogenation of the oxidized bituminous coal at lower hydrogen pressures than required for the development of coking properties in the subbituminous coal, even though there was little difference in oxygen content in the two samples. Pampuch and Nadziakiewicz[125] emphasized that coals exposed to air gradually lose coking properties and that these can be restored by partial hydrogenation at 400°C and 300 to 400 atm. Plasticity, thickness of plastic layer, and shrinkage during coking were found to be greater in the partially hydrogenated coals; the amount of tars produced was two to three times greater, and these were richer in phenols and in the fractions boiling from 270 to 360°C.

Boron Trifluoride

The Research Council of Alberta has reported[126] that the presence of boron trifluoride during coal pyrolysis leads "to a complete suppression of coking properties and to drastically reduced yields of phenols." The effects noted were attributed to action of the reagent on the coal substance itself rather than on the primary tar.

Other Addition Agents

The effects of zinc chloride, aluminum chloride, sodium hydroxide, phosphoric acid, active clay, ferric oxide, and active carbon on

[121] Berkowitz, N., and Dammeyer, W., *Fuel*, **35**, 19–30 (1956).

[122] Lander, C. H., Sinnatt, F. S., King, J. G., and Crawford, A., Brit. Pat., 301,720 (1928). Crawford, A., Williams, F. A., King, J. G., and Sinnatt, F. G., *Gt. Brit. Dept. Sci. Ind. Research, Fuel Research, Tech. Paper* No. 29 (1931), 37 pp.

[123] Kukharenko, T. A., and Khrisanfova, A. I., *Izvest. Akad. Nauk S.S.S.R., Otdel. Tekh. Nauk*, **1947**, 863–70; *Chem. Abs.*, **43**, 4830 (1949).

[124] Orchin, M., Golumbic, C., Anderson, J. E., and Storch, H. H., *U.S. Bur. Mines, Bull.* No. 505 (1951), 15 pp.

[125] Pampuch, P., and Nadziakiewicz, J., *Koks Smola Gaz*, **1**, 45–52 (1956); *Chem. Abs.*, **51**, 1584 (1956).

[126] Research Council Alberta (Can.), *Ann. Rept.*, **37** (1956), 48 pp.

the pyrolytic decomposition of three French coals have been investigated by Georgiadis and Gaillard.[127] The coals ranged in volatile matter from 24.9 to 38.2%. The carbonizations were carried out to a final temperature of 525°C at a heating rate of 3°C/min in a Fischer assay retort as modified by the U.S. Bureau of Mines.[38] The amounts of additives ranged from 2 to 5%. The following effects were noted: There was a significant decrease in the yield of dry tar and a corresponding increase in coke; total gas also increased except with phosphoric acid and the activated carbon. The composition of the gas was markedly affected by the zinc chloride addition, the H_2 to CH_4 ratio being changed from approximately 0.3 to 2.0. Data obtained with the thermogravimetric balance showed, especially with zinc chloride addition, a shift in the temperature at which the maximum in the rate of weight-loss temperature curve occurs to lower temperatures and also in the maximum rate observed, as shown in Fig. 25.

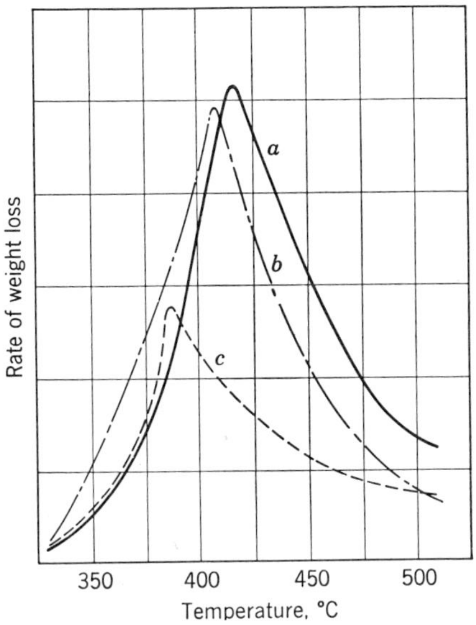

Fig. 25. Effect of addition agents on rate of weight loss.[127] *a.* Without additives; *b.* aluminum chloride; *c.* zinc chloride.

Russian workers[128] have studied the effect of the addition of three sodium compounds, hydroxide, carbonate, and chloride on the pyrolysis process. The effects decreased in the order mentioned above and consisted of lower yields of primary tar, higher yields of gas and low-boiling hydrocarbons, decrease in higher-boiling hydrocarbons, and increase in unsaturated hydrocarbons. The gas was markedly higher in hydrogen and lower in methane.

Brewer and Ghosh[129] studied the effect of carbonization in ammonia, hydrogen, and nitrogen atmospheres on the fraction of the sulfur in the coal retained in the coke. The work was done on samples of Pittsburgh-bed (high-volatile A bituminous) and Illinois No. 6 (high-volatile C bituminous) coals at temperatures of 700, 800, and 875°C, the coal sample being supported on a perforated silica plate in an electrically heated silica tube up through which the preheated gas flowed. Experiments on 15-g samples of 20 to 35 mesh, Illinois coal containing 3.55% sulfur, at a constant heating rate and a constant heating time at the carbonizing temperature of 800°C, showed that ammonia was significantly more effective in reducing the *fraction* of coal sulfur retained in the coke than was either hydrogen or nitrogen. Cooling the coke and reheating in ammonia resulted in further significant reduction in sulfur content. With the Pittsburgh-bed coal similar experiments were run at the three temperatures, comparing the fraction of the coal sulfur retained in the coke in carbonizations with and without ammonia. Significantly less sulfur was found in the coke in all instances where ammonia was used. In later experiments[130] with a high-sulfur coal from Ledo Colliery, Assam, India, containing 7.47% total sulfur, the fraction of the coal sulfur retained in the coke was reduced to 27.6% in experiments in ammonia at 800°C. The addition of aluminum oxide in amounts of 2% on the coal and carbonization in ammonia resulted in reduction of the fraction of the coal sulfur retained in the coke to 11.6% and a coke containing 1.57% sulfur.

High-sulfur coals have also been carbonized after the addition of such compounds as magnesium carbonate, calcium oxide, aluminum oxide, ammonium chloride, cuprous chloride, crude oil, thiourea, pyridine, and aniline.[131]

[127] Georgiadis, C., and Gaillard, G., *Chaleur & Ind.*, **37**, 247–58 (1956).

[128] Ravich, M. B., Lanin, V. A., Korshunov, V. I., *Zhur. Priklad. Khim.*, **24**, 970–5 (1951); *Chem. Abs.*, **48**, 8515 (1954).

[129] Brewer, R. E., and Ghosh, J. K., *Ind. Eng. Chem.*, **41**, 2044–53 (1949).

[130] *Ibid.*, **42**, 1550–58 (1950).

[131] Medvedev, K. P., Petropol'skaya, V. M., and Nikitina, K. A., *Koks i Khim.*, **1958**, No. 8, 15–8.

No reduction in the fraction of coal sulfur in the coke was found, and in some instances the sulfur in the coke was higher than in the control sample.

A study of the effect of chlorination on the thermal decomposition products of some British coals has been carried out by Pinchin.[132] He found that carbonization of chlorinated coals does not produce chlorinated tars but only hydrogen chloride along with reduced amounts of volatile matter. It thus appears that this reagent promotes condensation and molecular growth through elimination of hydrogen halide rather than degradation reactions.

Pyrolysis of Reaction Products

The pyrolysis of "humic acids" prepared by nitric acid oxidation of a high-volatile A bituminous coal from the Upper Freeport Bed in Pennsylvania has been studied by Ahmed and Kinney.[133] An approximately 4-g sample of the acids was heated in a silica tube from 150 to 950°C in 100° stages. Heating at each stage was continued until less than 0.5 ml of gas was evolved per hour. Preliminary experiments had shown that no condensable liquids or tars were evolved. The volatile products identified were water, carbon dioxide and monoxide, ammonia, hydrogen, methane, nitrogen, and oxides of nitrogen. No unsaturates, aromatics, hydrogen sulfide, sulfur dioxide, nitric oxide, or oxygen was detected. Approximately 13% of the carbon appeared as oxides of carbon and methane. The amount of carbon dioxide evolved, assumed to come from decomposition of carboxyl groups, corresponded to an equivalent weight of 242. The residue recovered at 950°C was 90.9% carbon which corresponded to approximately 70% of the original carbon of the coal, indicating that in this coal a large fraction of the carbon is in structures stable both with respect to oxidative and thermal degradation.

Popović[134] has found that a coal-like substance with coking properties can be prepared by treating lignite with a base at low pressures. The pressure, the concentration of the base, and the method of precipitation of the product from the basic solution were found to be important variables. Carbonization of the product in a laboratory retort yielded benzene and naphthalene in larger-than-usual quantities and a well-fused coke with a silvery color. The improvement observed was ascribed to decarboxylation in the pretreatment.

The effect of solution in aqueous or alcoholic alkali on the subsequent behavior on pyrolysis of Australian brown and subbituminous coals has been described by Brooks and Sternhell.[135] They found that the alkali-soluble products prepared by heating the coal with alkaline solutions at 190°C for 12 hr yielded coherent, swollen cokes on pyrolysis in a crucible test. The original coals were nonagglomerating when pyrolyzed under the same conditions. Analytical examination of the alkali-soluble products showed a definite increase in hydrogen content, some decrease in oxygen, but virtually no change in functional groups. The increase in agglutinating properties which these workers found is probably due to hydrogenation. Hydrogenation by aqueous alkali has been discussed by other investigators.[136,137]

The pyrolysis of extracts from North American lignite has been studied by Franke, Crowley, and Elder.[138] Extractions were carried out in the temperature range 370 to 440°C in a pressure vessel with the following solvents: tetralin, octahydroanthracene, octahydrophenanthrene, octahydrochrysene, and mixtures of four parts of tetralin to one part of cresol and four parts of decalin to one part of cresol. Some experiments were carried out with initial hydrogen pressures of 1000 psig. The extracts, as compared with the original lignite, when carbonized to 1000°F, showed greatly increased yields of tar and light oil and decreased gas and coke yields. Even when the weight of the residue from the extraction process was added to the weight of the coke from the pyrolysis of the extract, this figure for "total coke" was, in all instances except one, significantly

[132] Pinchin, F. J., *Fuel*, **37**, 293–8 (1958).

[133] Ahmed, M., and Kinney, C. R., *J. Am. Chem. Soc.*, **72**, 556–59 (1950).

[134] Popović, B., *Bull. sci., Conseil acad. RPF Yougoslavie*, **1**, 15 (1953); *Fuel Abs.*, **14**, No. 5, 21 (1953).

[135] Brooks, J. D., and Sternhell, S., *Fuel*, **37**, 124–6 (1958).

[136] Howard, H. C., *Chemistry of Coal Utilization*, John Wiley and Sons, New York, 1945, Vol. 1, pp. 418–24.

[137] Parker, F. G., Fugassi, J. P., and Howard, H. C., *Ind. Eng. Chem.*, **47**, 1586–92 (1955).

[138] Franke, N. W., Crowley, E. I., and Elder, H. J., *Ind. Eng. Chem.*, **49**, 1402–8 (1957).

lower than the figure for the coke from pyrolysis of the whole, dry lignite. The use of an initial hydrogen pressure of 1000 psig during the extraction was very effective in reducing the "total coke" yield. Pyrolysis of the solvent-treated but unfiltered reaction product gave higher gas yields, indicating that the residue from the extraction process pyrolyzes chiefly to gases and coke. It was also noted that the cokes from the unfiltered reaction products were significantly stronger, judged by small-scale tests than those from either the original lignite or from the extracts.

Distribution of the Elements among Pyrolysis Products

The distribution of carbon, hydrogen, and oxygen in the residues from pyrolysis of a French bituminous coal over the temperature range 200 to 900°C has been reported by Boyer.[139] The results are shown in Fig. 26. The rapid depletion of the residues in hydrogen and oxygen, starting about 400°C is clearly evident. Analysis[140] of U.S. Bureau of Mines' data for a number of American bituminous coals shows similar relations for atomic hydrogen-carbon and oxygen-carbon ratios. The nitrogen-carbon ratios show an interesting difference: they are roughly linear over the temperature range 500 to 1000°C.

The fate of the oxygen in coal during pyrolysis has been studied in some detail by Georgiadis and Gaillard.[141] Thermogravimetric studies with a Lorraine Basin (*flambant gras*) coal showed that the peak representing the maximum rate of evolution of total volatile matter occurred about 420°C, whereas that for the oxygen did not appear until about 500°C. These authors suggested that the greater difficulty in elimination of oxygen-containing compounds that was demonstrated may be responsible for the poor graphitizability of high-oxygen carbonaceous materials.

Because of its importance from the standpoint of metallurgical operations, the distribution of the sulfur between volatile products

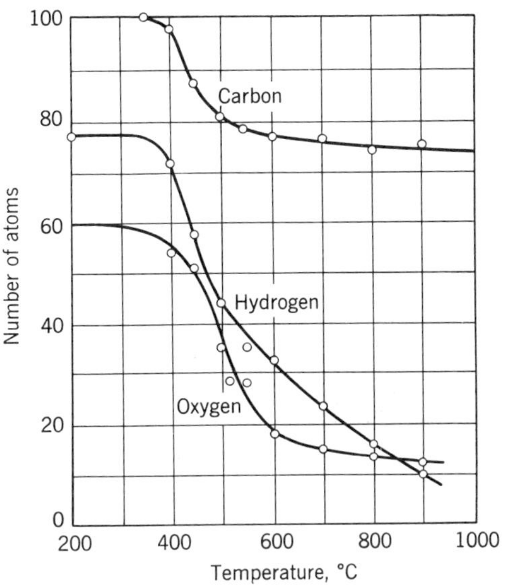

Fig. 26. Distribution of carbon, hydrogen, and oxygen in the residues from carbonization of a French bituminous coal.[139] (Numbers of oxygen atoms multiplied by 10.)

and coke has been the subject of considerable study. This problem has been discussed earlier,[142] and on the basis of coals carbonized in the U.S. Bureau of Mines-American Gas Association survey, there was a simple relation developed between the *total* sulfur in the coal carbonized and that found in the resulting coke. That this should hold irrespective of the distribution of the sulfur between organic and mineral forms in the coal is quite unexpected. These conclusions have been confirmed by an investigation[143] in which a definite proportion of radioactive, pyritic sulfur was mixed with a coal sample before carbonization. From the radioactivity of the coke and of the gaseous products resulting from the carbonization, it was shown that nearly the same proportions of pyritic and of organic sulfur are found in the coke and in the gaseous products. Of the pyritic sulfur, 66% entered the coke and 23%, the gas. For the organic sulfur, the corresponding figures were 73 and 26%.

Fuchs[144] stated that during carbonization about half of the sulfur in the coal remains in the coke, about 3% goes into the tar, and 45

[139] Boyer, A. F., *Compt. rend. congr. ind. gaz, 70th Congr.*, **1953**, pp. 242–9.

[140] Wu, W. R. K., Preprint, Division of Gas and Fuel Chemistry, American Chemical Society Meeting, May, 1958, Urbana, Illinois.

[141] Georgiadis, C., and Gaillard, G., *Compt. rend.*, **238**, 255–7 (1954).

[142] Thiessen, G., *Chemistry of Coal Utilization*, John Wiley and Sons, New York, 1945, Vol. 1, pp. 445, 447.

[143] Eaton, S. E., Hyde, R. W., and Rood, M. H., *Anal. Chem.*, **21**, 1062–6 (1949).

[144] Fuchs, W., *Brennstoff-Chem.*, **32**, 274–6 (1951).

to 50% into the gas. Russian workers[145] have studied the conversion of "mineral sulfur" to "organic sulfur" during the carbonization process by pyrolyzing coals with calcium sulfate labeled with S^{35}. Carbonizations were carried out at 900 to 1000°C on coals of various ranks. The percentage of sulfate reduced to sulfide and that converted to organic sulfur were determined, and the organic sulfur formed was found to have reached a maximum with coals of approximately 35% volatile matter. Sulfur balances during carbonization have been determined for some Irkutsk Basin coals by Okladnikova and Kalechits.[146] It was found that the relation between sulfur and the organic matter of the coal to form hydrogen sulfide is incomplete, and that ferrous sulfide formed from pyrite exerts an inhibiting effect on the decomposition of organic sulfur compounds.

The behavior of chlorine during the pyrolysis of coal has not been extensively studied. Kear and Menzies[147] reported that the fraction of the chlorine passing into the volatile products increases with increasing carbonization temperature and amounts to approximately 75% at 1000°C.

Pyrolytic Behavior of Petrographic Components

Those components of a coal bed which can be separated by hand-picking, on the basis of visual or microscopic appearance, or by gravity or flotation techniques show distinctively different behavior when subjected to pyrolytic decomposition. The observed differences are, however, generally what would be predicted on the basis of ultimate chemical composition and proximate analyses. As would be expected, the differences in composition and in pyrolytic behavior of such components are most marked in the lower-rank coals. In some Western American low-rank coals, "resins" occur in concentrations which allow ready recovery either by hand-picking or flotation techniques. Such enriched samples have atomic hydrogen to carbon ratios as high as 1.4 and yield, in the ASTM Standard proximate analysis, volatile matter as high as 95%. Tar and light-oil yields in a standard Fischer retort assay are of the order of 80%.

The thermal decomposition of enriched exinite fractions has been compared by Macrae[148] with that of vitrinite from the same bed. The exinite fractions were prepared by crushing and water elutriation of spore-rich durain bands from the Upper Beeston seam, West Yorkshire, England. The vitrinite sample was obtained by hand-selection of lumps of bright coal from the same seam. Analyses of the samples showed outstanding differences in both ultimate composition and fixed carbon. Atomic hydrogen-carbon ratios for the exinite samples averaged 1.00, and the average value for fixed carbon was 28.1%. The corresponding values for the vitrinite samples were 0.723 and 59.9%.

Pyrolyses were carried out in vacuum over the temperature range 150 to 350°C, using a Sprengel-type pump. The total volume of gases evolved from the two components over the temperature range studied did not differ greatly. However, gas evolution started at a markedly lower temperature with the exinite, 213°C, compared with 270°C for the vitrinite. Outstanding differences were noted both in the total yield and the composition of the liquid products. Typical data are shown in Table 7.

TABLE 7

LIQUID PRODUCTS FROM VACUUM PYROLYSIS OF EXINITE AND VITRINITE SAMPLES[148]

	Exinite	Vitrinite
	g/100 g	
Water	3.36	2.06
Light oil	2.79	0.99
Heavy oil	29.78	2.28
Heavy oil composition	% of Heavy oil	
Acids	0.3	2.0
Phenols	3.8	21.6
Bases	1.6	6.4
Neutral oil	90.5	70.0

The liquid products from the exinite sample were characterized by low percentages of phenolic material and high yields of neutral oils as compared with the vitrinite. Treatment

[145] Medvedev, K. P., and Petropol'skaya, V. M., *Koks i Khim.*, **1958,** No. 3, 25–9; *Chem. Abs.*, **52,** 9560 (1958).

[146] Okladnikova, Z. A., and Kalechits, I. V., *Trudy Vostochno-Sibir. Filiala, Akad. Nauk S.S.S.R., Ser. Khim.*, **1955,** No. 3, 105–26; *Chem. Abs.*, **50,** 7425 (1956).

[147] Kear, R. W., and Menzies, H. M., *BCURA Bull.*, **20,** 61–4 (1956).

[148] Macrae, J. C., *Fuel*, **22,** 117–29 (1943).

of the neutral oil with ferric chloride, which is a specific reagent for neutral oxygenated compounds, showed reaction of 28.3% for the exinite fraction compared with 14.2% for the vitrinite sample. Although the liquids from the exinite showed a much lower percentage of phenolic material, the yield of phenols per gram of solids was somewhat higher for the exinite sample.

Studies of the thermal decomposition products of samples of American bright and splint coals have led to equivocal results,[149] because of failure to recognize that comparison was being made between samples rich in one petrographic component, vitrinite, with others containing two components, micrinite and exinite, mixed in variable proportions, the *average* chemical composition of which did not differ greatly from the vitrinite. Later work in which samples of the macerals have been pyrolyzed separately has led to more definite conclusions.[150]

Halle[151] has examined the primary tar obtained by carbonizing a sample of "resinous liptobiolite" from Georgia, Russia. Oils made up 29% of the tar. They were reported to consist of condensed hydroaromatic compounds, and hexahydro and tetrahydro retenes were isolated. Resinol-type resins, with partially blocked hydroxyl groups, constituted 31.0%; resin-type acids, 8.6%; humic acids, 3.2%; high-molecular weight condensation products, 28%.

Enriched maceral fractions from four coal beds in the Ruhr field, Germany, have been prepared by Kröger[152] and thermogravimetric studies made of their behavior on pyrolysis. The beds studied were "R," "Zollverein 1," "Anna," and "Wilhelm," and the volatile-matter content of the vitrinite fractions on the dry, ash-free basis were 36.1, 31.9, 28.3, and 23.5%, respectively. In every instance, the micrinite fractions were lower in volatile matter than the vitrinite from the same bed and the exinite fractions higher in every instance except one, the Wilhelm bed, where there was little difference. The weight-loss temperature curves shown in Fig. 27 are typical for the two lower-rank seams. With those of higher rank, all the curves were closer together, and in the highest rank the curve for the exinite fraction lay between the micrinite and vitrinite curves.

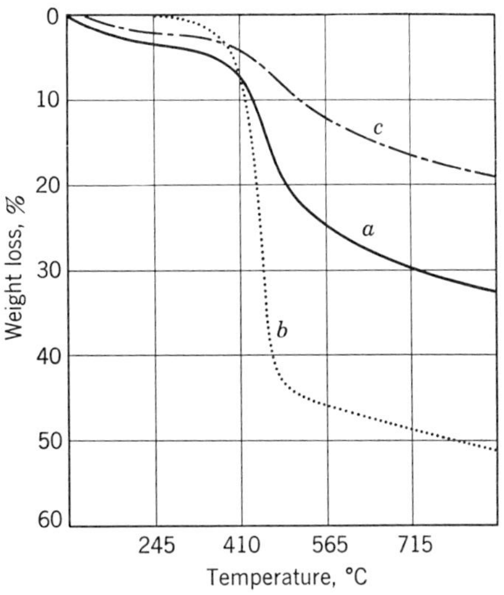

Fig. 27. Weight-loss curves of petrographic components.[152] *a*. Vitrinite; *b*. exinite; *c*. micrinite.

General Mechanisms of Thermal Decomposition

Lack of knowledge of the structure of coal means that any mechanism of its thermal decomposition must remain highly speculative. Nevertheless, even such speculative discussions appear worthwhile because of the numerous suggestions for further investigations which they engender. One common characteristic of all mild, degradative processes of coal, whether they be purely thermal or combinations of thermal and chemical changes is the production of a series of fragments differing significantly in molecular size but usually showing closely related chemical properties. To what extent the fragmentation is the result of the breaking of covalent bonds or of associative forces has not been established as yet. Coal has been pictured[153] as a "chicken wire" type of two dimensional network with the atoms held together by covalent forces over very large, atomic areas. On the basis of such

[149] Sprunk, G. C., *J. Geology*, **50,** 411–36 (1942).

[150] Parks, B. C., O'Donnell, H. J., Birge, G. W., and Gibson, F. H., *U.S. Bur. Mines, Rept. Invest.* No. 5453 (1959), 34 pp.

[151] Halle, R. R., *J. Applied Chem.* (*U.S.S.R.*), **18,** 184–90 (1945); *Chem. Abs.*, **39,** 5436 (1945).

[152] Kröger, C., and Pohl, A., *Brennstoff-Chem.*, **38,** 102–7 (1957).

[153] Lowry, H. H., *J. Inst. Fuel*, **10,** 291–301 (1937).

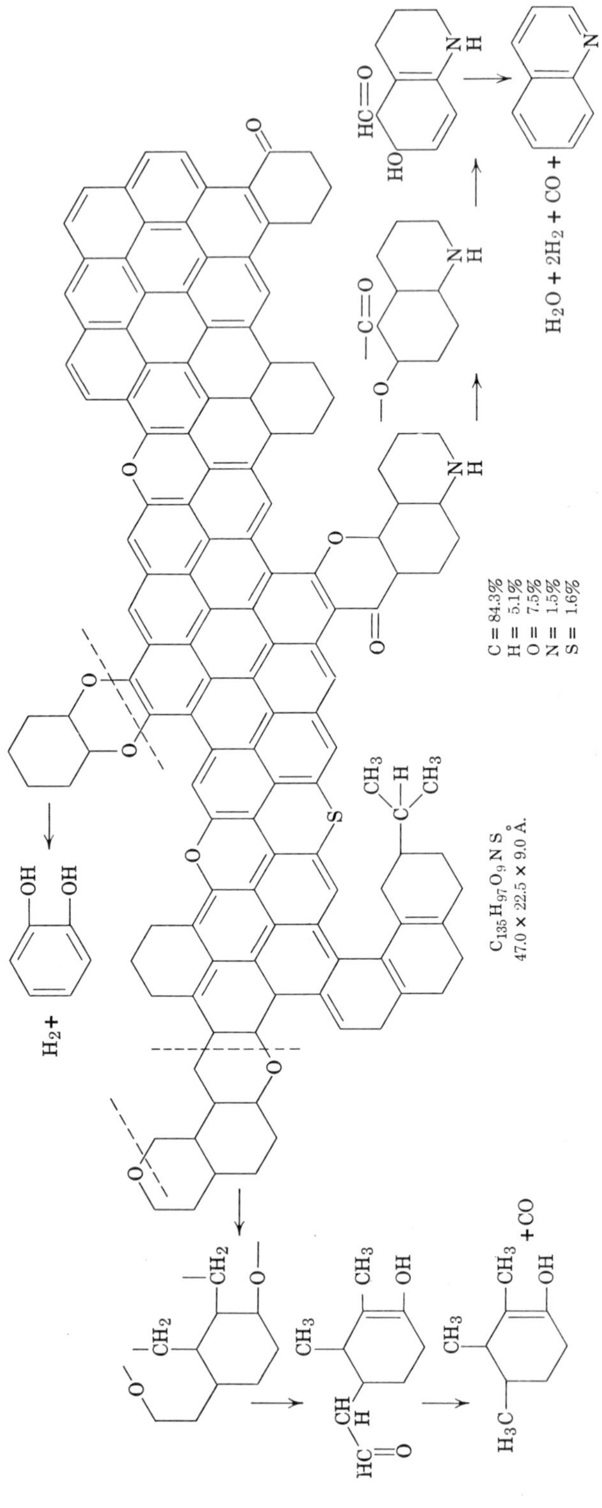

Fig. 28. Example of coal pyrolysis.[154]

a structure, Fuchs and Sandhoff[154] have pictured the pyrolytic breakdown of a bituminous coal as in Fig. 28 where rupture of the network and formation of tar and gas molecules are initiated by the splitting of rings containing hetero atoms. It was stated "it is not probable that the variety of compounds making up gas and tar originates simultaneously .... It is more probable that these compounds have originated from slightly larger units ... which within a certain temperature range will give low-temperature tar and gas and within a higher temperature range will yield high-temperature tar and gas. The solid residues will lose first their oxygen content and then their hydrogen content as the temperature rises through the low- and high-temperature carbonization ranges."

Gillet[155] and co-workers have studied the thermal breakdown of a number of coals using both an added liquid phase (anthracene oil) and inert gas atmospheres. On the basis of this work, they concluded that coals consist of condensed cyclic-hydrocarbon units of the order of 26 carbon atoms linked together by oxygen bridges. During heating either in the presence or the absence of an added liquid phase, solution or fusion of these units occurs because of rupture of the oxygen bridges.

The changes in composition during carbonization of a number of coals and two model compounds have been discussed by van Krevelen,[156] and the data has been presented on a H/C-O/C type of plot (see Fig. 29). On such a plot, all the carbonaceous materials studied approach the coal band at a value of H/C of approximately 0.28, and differences between the carbonization and the coalification processes are clearly indicated. Only with a Dutch semianthracite coal do the compositions of the carbonization products fall within the coal band.

[154] Fuchs, W., and Sandhoff, A. G., *Ind. Eng. Chem.*, **34**, 567–71 (1942).

[155] Gillet, A., *Rev. universelle mines*, **19**, 147–65 (1943); Gillet, A., and Willemme, J., Fastre, M. L., *Bull. soc. chim. Belg.*, **53**, 27–44 (1944); Gillet, A., and Danvister, G., *Bull. soc. chim. Belges*, **54**, 5–9 (1945). Gillet, A., and Willemme, J., *Bull. soc. chim. Belges*, **54**, 169–80 (1945). Gillet, A., and Genkenne, A., *Chaleur & Ind.*, **23**, 269–73 (1947).

[156] van Krevelen, D. W., *Fuel*, **29**, 269–84 (1950).

Changes in the structure and properties of bituminous coal over the temperature range 350 to 500°C have been studied by Khrisanfova.[157] The results indicated that over this temperature range the structure and properties of the residues are related to the composition of the coal substance and are characteristic for each type of coal. It was found that changes in structure of the residue can be followed by changes in solubility in "heavy solvents" and in heats of wetting.

Aronov's[158] picture of coal was of a system of hexagonal carbon rings forming condensed polycyclic structures of many sizes and shapes surrounded by a "fringe" of side groups of molecules. The nucleus was pictured as having stronger chemical bonds and greater resistance to heat than the side groups. Coking and the formation of tar and gas were thought to be determined by the quantity and quality of the side groups.

In Smith and Brown's[159] discussion of the mechanism of thermal decomposition of coal, emphasis was placed chiefly on phenomena which are believed to control coke properties. They concluded that to produce strong and coherent cokes it is necessary that an appreciable fraction of the material fuse at an elevated temperature, that its vapor pressure be quite low, and that simultaneous decomposition and repolymerization occur at or near the fusion temperature.

Chukhanov[160] has stated that there are three important stages in the carbonization reactions of coals, the first of which proceeds at very high velocities, in a hundredth or even a thousandth of a second at 450 to 550°C, with the production of a high proportion of simple oxygenated compounds such as water and oxides of carbon. The second stage of reaction consists in formation of tars and hydrocarbon gases and proceeds at a velocity hundreds of times slower than the first. The third stage, degasification of the residue, is even slower.

[157] Khrisanfova, A. I., *Trudy Inst. Goryuch. Iskopaemykh, Akad. Nauk S.S.S.R.*, **2**, 78–95 (1950); *Chem. Abs.*, **49**, 4259 (1955).

[158] Aronov, S. G., *Zhur. Priklad. Khim.*, **25**, 927–35 (1952); *Chem. Abs.*, **48**, 3008 (1954).

[159] Smith, F. W., and Brown, R. L., *AGA, Proc.*, **1952**, 719–29.

[160] Chukhanov, Z. F., *Izvest. Akad Nauk S.S.S.R., Otdel Tekh. Nauk*, **1954**, No. 8, 7–22; *Chem. Abs.*, **49**, 5808–9 (1955).

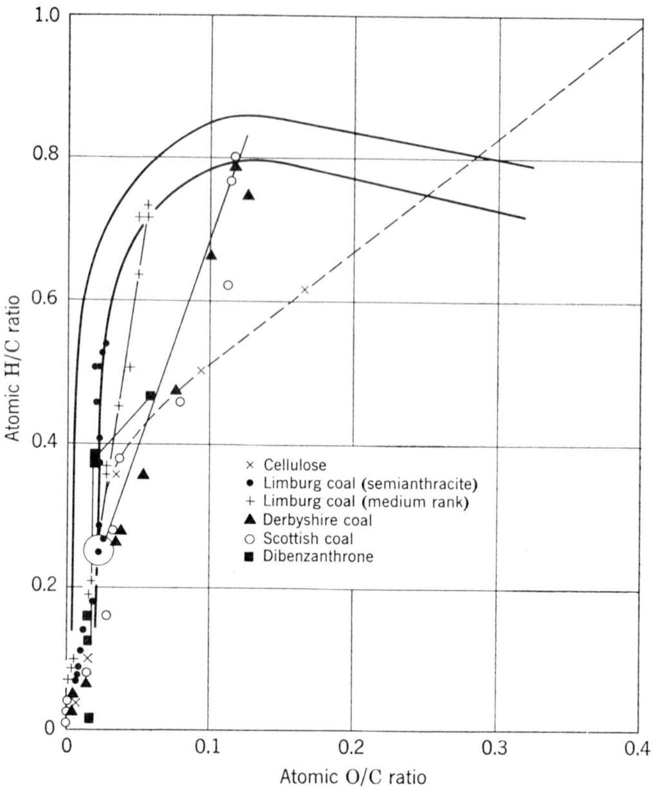

Fig. 29. Carbonization curves.[156]

Bergmann and co-workers[161] have followed the changes occurring in a coal sample heated to 300 and 400°C by infrared spectrometric measurements. Over this temperature range, no change was observed which would allow one to conclude that there had been any significant modification in the coal structure. An outstanding study of changes in the infrared spectra brought about by heating of two British coals has been made by Brown.[162] The samples used were vitrains from a weakly caking Northumberland coal and from a Durham coking coal. Samples of each were heated in nitrogen at the rate of $1\frac{1}{4}$°/min to temperatures of 300, 400, 460, 550, and 800°C and were kept at these temperatures for 1 hr. The most striking difference between the two samples of the original unheated coals was in the strong absorption and the hydrogen-bonded hydroxyl groups for the Northumberland coal compared with that for the Durham coal. Little change was noted in the spectra of the samples of either of the coals heated at 300 and 400°C. The hydrogen-bonded hydroxyl absorption band for the Northumberland coal decreased in intensity, but was still regarded as being very strong. At 460°C and upward, marked changes in the spectra of both coals were evident; aliphatic (or alicyclic) material was removed, probably by distillation and chemical decomposition, and hydrogen-bonded phenolic hydroxyl groups were also eliminated. The author suggested that the formation of C—C cross links between aromatic systems accompanied the loss of aliphatic (or alicyclic) hydrogen, and that above 550°C the process was continued by virtue of aromatic hydrogen elimination. Samples heated to 800°C were opaque to infrared radiation, over a wide range of wavelengths, presumably due to some graphitization.

Orechkin[163] pictured the transformation of

[161] Bergmann, G., Huck, G., Karweil, J., and Luther, H., *Brennstoff-Chem.*, **35,** 175–6 (1954).

[162] Brown, J. K., *J. Chem. Soc.*, **1955,** 752–7.

[163] Orechkin, D. B., *Trudy Vostochno-Sibir. Filiala, Akad. Nauk S.S.S.R., Ser. Khim.*, **1955,** No. 3, 105–26; *Chem. Abs.*, **50,** 7425 (1956).

coal to a liquid phase by pyrolysis as based on reducing depolymerization, that is, the degradation of the macromolecules of humic matter by breaking of weak C—O bonds (presumably ether bonds are meant) which join the carbon complexes composing the structural elements of coal matter. He suggested that the primary stages of the thermal decomposition can be represented by two interrelated processes: (1) reducing depolymerization, which results in the isolation of complexes, a decrease in average molecular weight, and transformation into liquid or low-melting material; and (2) a reverse process, which consists of removing oxygen in the form of water with a simultaneous condensation of the carbon complexes. A growth of the carbon skeletons and a loss of fusibility results. The predominance of either of these processes depends upon the hydrogen content of the coal and upon the medium in which the decomposition is carried out. The presence of hydrogen under pressure or of a solvent which can serve as a source of hydrogen will direct the decomposition toward the formation of liquid products, whereas oxidation of the coal or the addition of an active dehydrating agent will direct the reaction toward condensation. The rate of temperature increase may also be important in changing the direction of the process. This emphasis on competing processes during coal pyrolysis and the importance of the predominance of reducing reactions is of great interest and permits explanation of a number of phenomena such as the effect of pyrolysis under pressure and in the presence of added liquid phases such as organic liquids or water.

Tests carried out by Nadziakiewicz[164] on the devolatilization of a number of Polish coals at constant temperatures of 400, 500, and 600°C for variable time periods have confirmed the earlier conclusions of Lowry[165] that the devolatilization process is primarily a function of temperature, relatively unaffected by time and, hence, that the temperature of devolatilization of a carbonaceous residue can be estimated from its volatile content.

Brooks and Maher[166] have followed the changes with carbonization temperature of the acidic groups of an Australian coal. Samples of vitrinite from a medium-rank bituminous coal (83.1% carbon) were carbonized in nitrogen at 250, 350, 400, 500, and 600°C. The heating rate was 5°/min, and the samples were held at the designated temperature for 1 hr. The total oxygen was determined from the ultimate analysis, the acidic oxygen by titration with aminoethoxide in ethylenediamine,[167] and the nonacidic oxygen by difference. The results are shown in Fig. 30. The striking decrease in acidic oxygen and the increase in nonacidic oxygen over the 400 to 450°C range may have been due to conversion of hydroxyl groups to ethers or quinones, but there is also the possibility that the observed changes depended on changes of accessibility of the groups in the partially carbonized coals to the reagent, because it was observed that ball milling under benzene increased the apparent acidities of 450 and 500°C residues from 1.77 and 0.07 to 2.60 and 0.82 m.e./g, respectively. This work is of interest in connection with that of J. K. Brown[162] who found that the most significant changes judged by infrared spectra occurring in the early stages of coal pyrolysis involved hydroxyl groups.

Dryden[168] has emphasized the view that the volatile products obtained on heating coal are decomposition products. However, to what extent they represent fragments characteristic of the original coal structure and to what extent they are the result of secondary interactions of small fragments resulting in compounds bearing little relation to the coal structure is not certain. The older idea that coal consists of two main components, one of which yields the volatile matter on heating, is not generally accepted today. Also, the simple and attractive hypothesis, that coal consists of aromatic lamella with edgewise attachment of aliphatic and oxygen and hydrogen containing groups and that the rupture of these groups results in the volatile matter, cannot be accepted today without modification. The remarkable resemblance of the volatile products of pyrolysis obtained in short-path distillation and solvent decompositions to the original coal suggests that these are derived from parts of the coal structure which differ from the rest of

[164] Nadziakiewicz, J., *Koks Smola Gaz*, **1**, 87–9 (1956); *Fuel Abstr.*, **22**, 38 (1957).

[165] Lowry, H. H., *Soc. Chem. Ind.*, **1950**, 619–25.

[166] Brooks, J. D., and Maher, T. P., *Fuel*, **36**, 51–62 (1957).

[167] Katz, M., and Glenn, R. A., *Anal. Chem.*, **24**, 1157–63 (1952).

[168] Dryden, I. G. C., *J. Inst. Fuel*, **30**, 193–214 (1957).

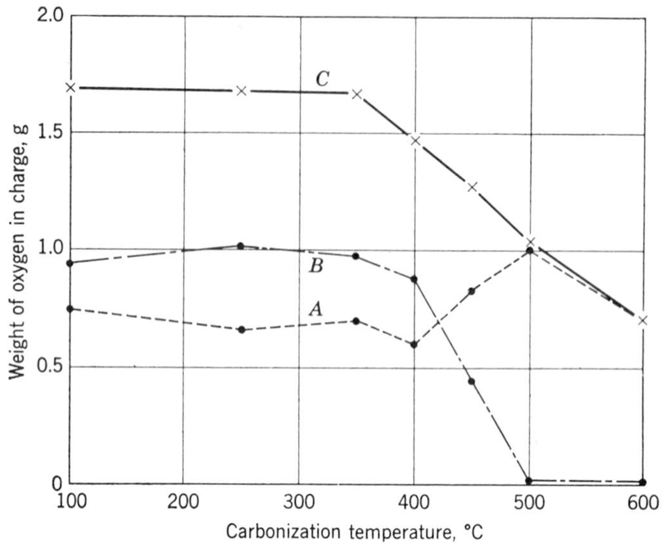

Fig. 30. Effect of pyrolysis on acidity of Lithgow vitrain: *A*. nonacidic oxygen; *B*. acidic oxygen; *C*. total oxygen.[166]

the coal only in the nature of their bonding and not in their composition. Attempts to relate carbon in the volatile matter to aliphatic carbon in the coal structure are handicapped by lack of exact knowledge concerning the fraction of aliphatic carbon in coals. Such a comparison, however, is of interest and has been made (see Fig. 31). Comparison with compounds of authentic structure, such as the vat dyes,[169] indicates the importance of the bonding involved; the completely aromatic compound 16,17-diphenoxy-dibenzanthrone gives a weight loss of 38.7% in a 900°C assay. This is presumably due to rupture of the ether links. It is also of interest to study, as Dryden[168] has done, the yields of pyrolysis products on the basis of the ultimate compositions of typical coals and of the tars produced therefrom. Because of the relative richness of the volatile matter in hydrogen, it is evident that it is the abundance of this element which limits the yield of tar. A similar analysis of the relation between oxygen content of the coal and the yield of water and of phenolic compounds indicates that simple pyrolytic decomposition of coal results in the recovery of less than 10% of the hydroxyl oxygen in coal as such compounds, for coals ranging in total oxygen content from 2.1 to 10.3%.

[169] Blayden, H. E., Gibson, J., and Riley, H. L., *Inst. Fuel (London) Wartime Bull.*, **1945,** 117–29.

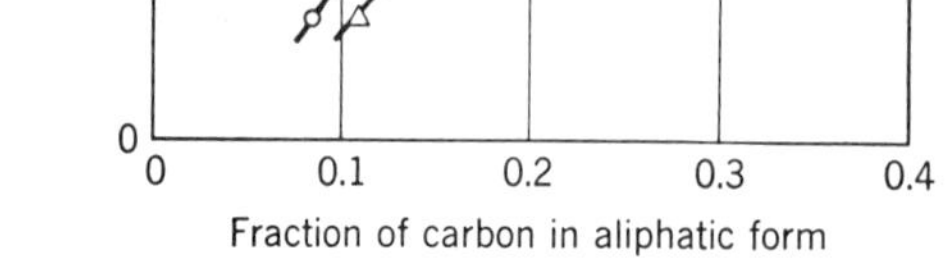

Fig. 31. Relation between aliphatic carbon and carbon in volatile products.[168] *a* From infrared spectra (Brown and Hirsch); *b* from calorific values (Dryden and Griffith).

Kinetics

During the last decade there has been growing interest in studies of the rates at which the various chemical and physical changes, which take place when coals are heated, proceed. A good part of this increased interest can be traced to the development and application of the thermogravimetric balance and other modern tools.

The rate of evolution of volatile matter for a number of Russian coals has been studied over the temperature range 300 to 600°C by Shapatina[170] and co-workers. The method consisted

[170] Shapatina, E. A., Kalyuzhnyi, V. V., and Chukhanov, Z. F., *Doklady Akad. Nauk S.S.S.R.*, **72,** 869–72 (1950); *Chem. Abs.*, **44,** 10294 (1950).

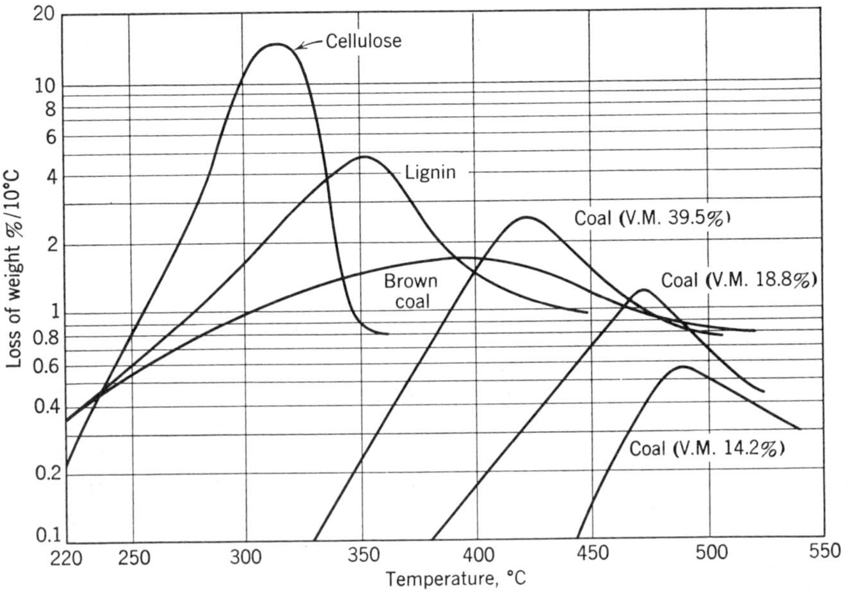

Fig. 32. Differential loss of weight curves.[171]

of dropping the powdered samples (size: 150 to 200 μ) into a furnace maintained at some constant predetermined temperature and measuring the volume of gas evolved as a function of temperature for periods of 0.8, 6.0, and 30 sec. The evolution of volatile matter at any constant temperature was reported to consist of three stages: (1) very rapid evolution of loosely bound water, carbon dioxide, and carbon monoxide; (2) a slower evolution of the main volatile matter, which was found to be first order with respect to the momentary amount of the ash-free coal; and (3) a slow residual outgassing. The first-order constants of stage (2) yielded an apparent activation energy of 5.3 kcal. This surprisingly low value has not been confirmed by kinetic studies on Western European coals.

The rates of decomposition of Dutch coals have been studied by van Krevelen, van Heerden and Huntjens[171] using the thermogravimetric balance technique in which the weight of the residue was continuously recorded as a function of temperature and where the rate of temperature rise was automatically controlled at some constant value. In such experiments, curves of the type shown in Fig. 32 are obtained. In their analysis of the curves, these authors pointed out that the rate of decomposition at any instant can logically be assumed to be some function of the fraction of undecomposed coal substance present at that instant or that $-dF/dt = kF^n$ where k is the velocity constant, F, the fraction of undecomposed coal, t, the time, and n, an exponent indicating the order. Substituting in the general relation $k = k_0e^{-A/T}$, we have $-dF/dt = k_0e^{-A/T}\,F^n$. Because these measurements, unlike usual kinetic experiments, were made at a constant rate of temperature rise, that is $dT/dt = a$, the preceding expression becomes $-\frac{dF}{F^n} = \frac{k_0}{a}\,e^{-\frac{A}{T}}\,dT$. By a process of approximations, it can be shown that, depending on the order of the reaction, the integral of $-dF/F^n$ will be $1 - F$, $-\ln F$, or $(1/F) - 1$ for zero, first- or second-order reactions, respectively. Because the data gave a linear relation for $-\ln F$ as a function of $\ln T$ and nonlinear for values of $1 - F$ and $(1/F) - 1$, it was concluded that the process is first order with respect to the quantity F, the fraction of undecomposed coal. The quantity A is related to E, the activation energy of the process by the expression $A = E/R$ where $R = 1.98$ cals. The relations between volatile matter and values of A for cellulose lignin, and a number of coals are shown in Fig. 33. The significance of the first minimum is not evident; the second may be associated with the less-extensive networks

[171] van Krevelen, D. W., van Heerden, C., and Huntjens, F. J., *Fuel*, **30,** 253–59 (1951).

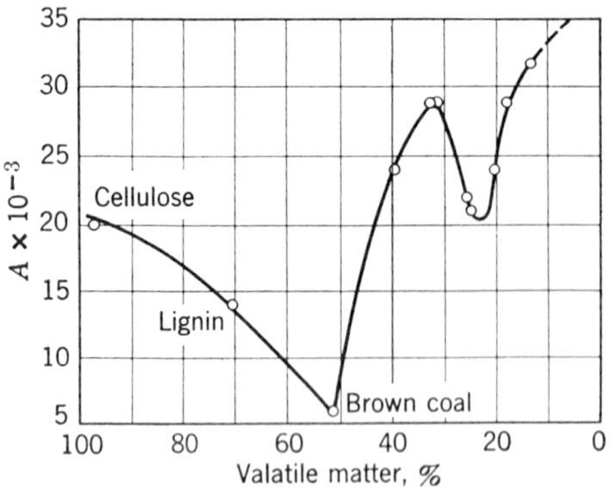

Fig. 33. Relation between activation energy and volatile matter.[171]

and, consequently, greater thermal lability of coals in this volatile-matter range.

Stone, Batchelor, and Johnstone[172] have studied rates of devolatilization of coals in a fluidized bed. Samples were withdrawn from the bed at frequent intervals and analyzed for volatile matter. The rate was assumed to be represented by the general equation $-dV/dt = k(V - V_c)^n$ where V is the volatile-matter content at any time, t is time, V_c is the constant volatile-matter content ultimately reached after heating at a constant temperature, and n is the order of the reaction. The work was done on three American bituminous coals, Illinois No. 2, Colorado Upper Mesa Verda, and Pittsburgh Seam which, on the moisture- and ash-free basis, had volatile-matter analyses of 44.9, 41.5, and 42.2%, respectively. Curves for the Pittsburgh-bed coal at three different temperatures are shown in Fig. 34. The rate data were found not to follow the rate equation given for the entire period for any constant value of n, but the results did follow a first-order process for certain periods of time. At lower temperatures, the first-order relation held over a greater time interval, but the fractional approach to completion at which the first-order reaction no longer held was roughly the same for all temperatures. For the Pittsburgh-bed coal, three stages of first-order processes in the devolatilization were discernible similar to those observed by Shapatina[170] for Russian coals. For the Pittsburgh-bed coal, the activation energies calculated for the three zones over the temperature range 770 to 950°F were 26.2, 23.4, and 6.7 kcal/g-mole, respectively. After the initial stages of devolatilization, there was a period in which the rate was practically independent of the amount of residual volatile matter and must have been controlled by some physical property such as pore structure. The decrease in activation energy with extent of devolatilization was consistent with this view. Some effect of particle size was observed; there was an increase in the initial rate of devolatilization with decrease in particle size, but the effect was not important after the first 15 sec.

Important information concerning the kinetics of the pyrolytic decomposition of coal

[172] Stone, H. N., Batchelor, J. D., and Johnstone, H. F., *Ind. Eng. Chem.*, **46**, 274–8 (1954).

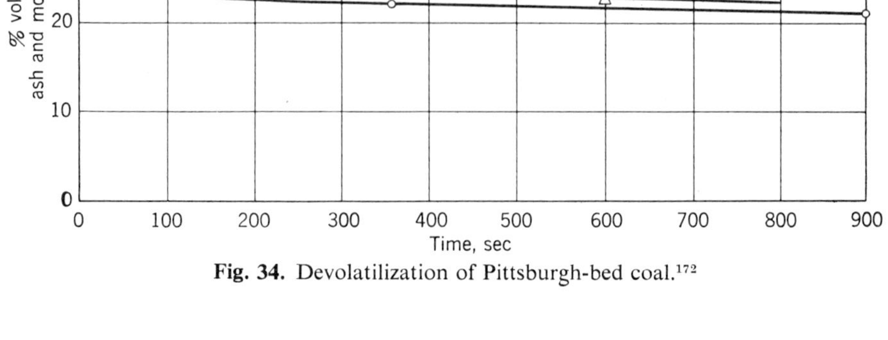

Fig. 34. Devolatilization of Pittsburgh-bed coal.[172]

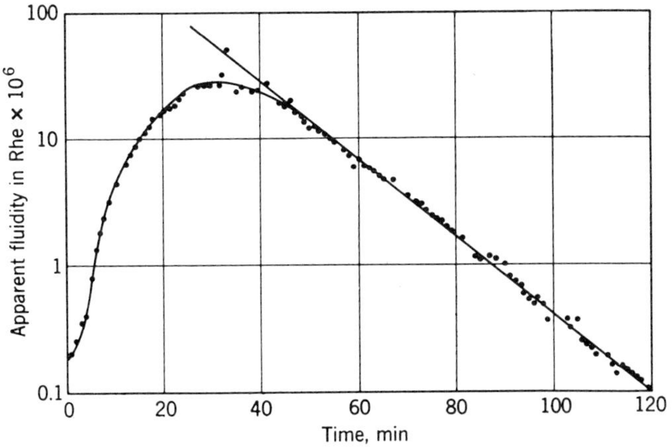

Fig. 35. Variation of apparent fluidity of Silkstone coal with time at 407°C.[173]

has resulted from Fitzgerald's studies[173] on the changes in fluidity with time of coal samples held at specified constant temperatures. A modified Gieseler plastometer was used for this work, and a typical curve illustrating the data obtained is shown in Fig. 35 where the apparent fluidity on a logarithmic scale is plotted against time. In this work, although the coal attained a constant temperature within five minutes, its fluidity continued to increase for the next 15 min, and for some coals for as long as 40 min. Because coal probably contains a large fraction of unorganized material, this period of increasing fluidity at constant temperature probably was not due to a "heat arrest" characteristic of the melting of crystalline systems. The possibility that the phenomenon is a purely physical one cannot be excluded, but Fitzgerald suggested that this initial increase in fluidity is due to a chemical reaction in which part of the solid coal is decomposed to a fluid product. The decrease in fluidity after the maximum is passed was assumed due to a second chemical reaction in which the fluid coal is decomposed into a solid (semicoke) and volatile matter. According to this view, the fluid coal is an intermediate product of two consecutive reactions, and the overall change can be represented as

$$\text{coal} \xrightarrow{k_1} \text{fluid coal} \xrightarrow{k_2} \text{semicoke}$$

where k_1 and k_2 represent the velocity constants for the reactions forming the fluid coal and its decomposition, respectively. Volatile matter is evolved in each of these reactions. Because the reactions are consecutive, the concentration of the fluid coal and, hence, the fluidity of the system pass through a maximum. To evaluate k_1 and k_2 a number of assumptions are necessary. It was assumed that each of the reactions is first order, that the amounts of volatile matter evolved are negligible, that for each reactant, mass can be substituted for molar concentration, and finally that the fluidity of the system is related to the proportion of fluid coal present in a simple way. The evaluation of k_2, the constant for the linear righthand portion of the curve, is more direct. The logarithms of the slopes for measurements at different temperatures were found to be linearly related to the reciprocal of the absolute temperatures and, hence, both the velocity constant and the activation energy could be evaluated. For ten coals, ranging in volatile content from 24 to 40%, the values of E, the activation energy, lay between 47 and 55 kcal/mole. These high values point to a chemical reaction, and their relative constancy for a variety of coals suggests that this part of the fusion process is essentially the same for all the coals. The direct evaluation of k_1, representing the rate of the process occurring before the maximum fluidity is reached, was possible with only one coal, the Silkstone, where the curves for three temperatures were said to be sufficiently regular over their initial portions to permit graphical integration. The values of k_1 so determined were three to fourfold those of k_2. Activation energies calculated from values of k_1 and k_2 over the temperature range of 397° to 415°C were the same within the limits of the experimental accuracy.

[173] Fitzgerald, D., *Trans. Faraday Soc.*, **52**, 362–69 (1956).

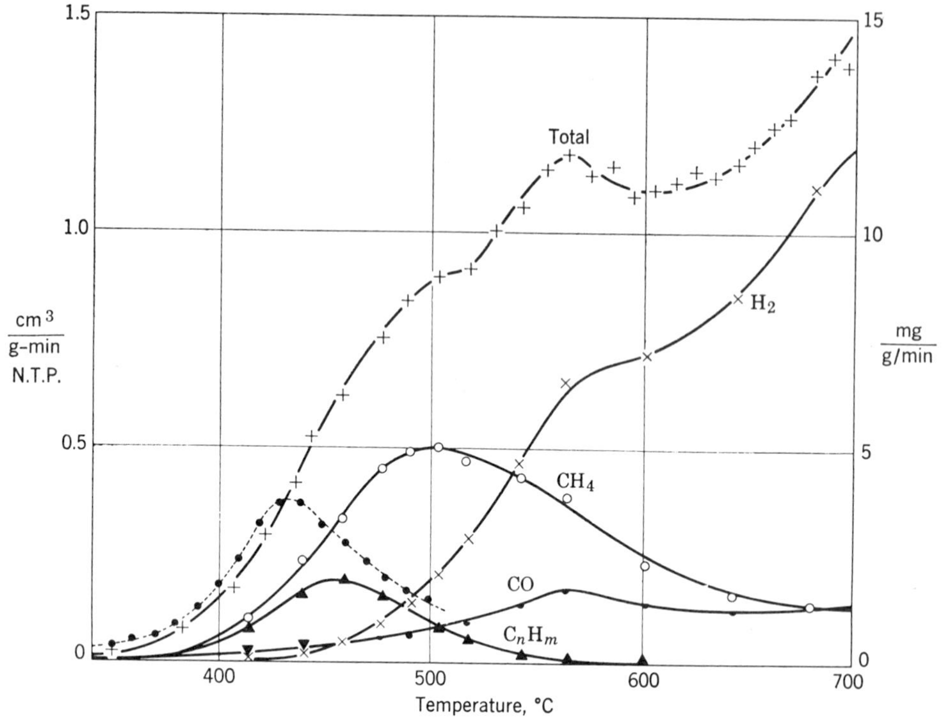

Fig. 36. Rate of gas flow against temperature for the vitrinite of high-volatile coal B at 1.8°C/min. The dashed line represents rate of weight loss against temperature.[175]

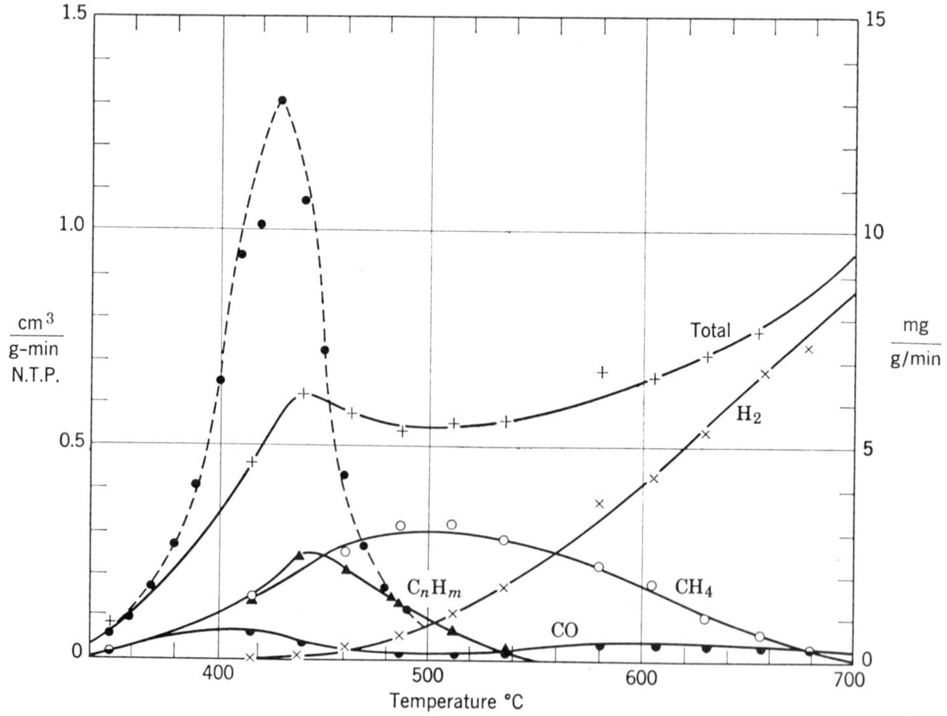

Fig. 37. Rate of gas flow against temperature for the exinite of high-volatile coal B at 1.8°C/min. The dashed line represents rate of weight loss against temperature.[175]

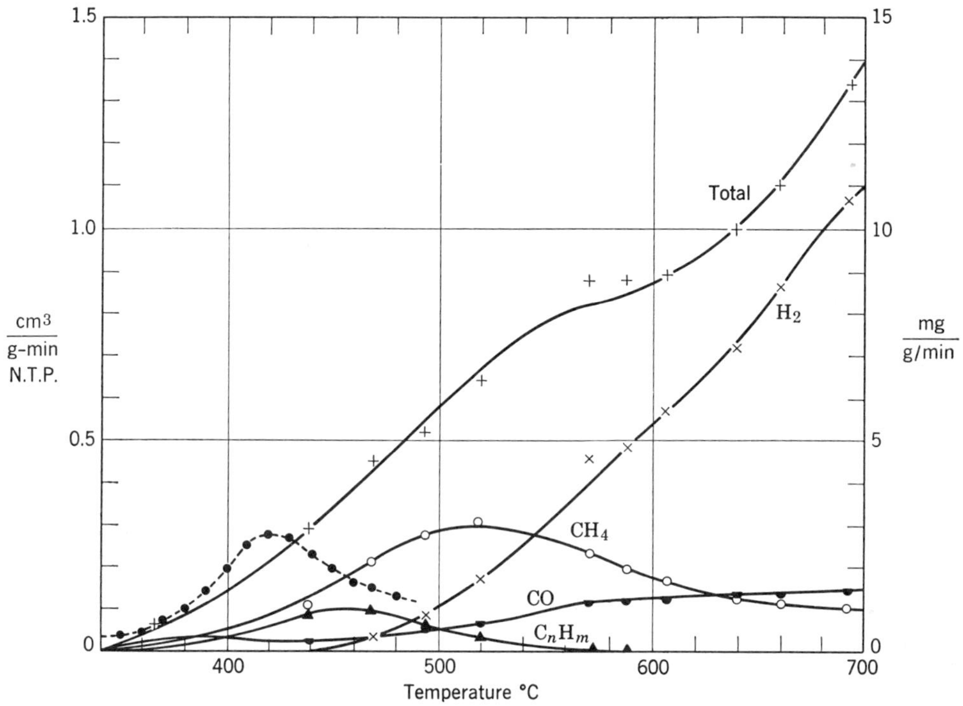

Fig. 38. Rate of gas flow against temperature for the micrinite of high-volatile coal B at 1.8°C/min. The dashed line represents rate of weight loss against temperature.[175]

Later work,[174] that attempted to relate k_2 of a coal with the $1\frac{1}{2}$ in. shatter index of the coke prepared therefrom, indicated that, as would be expected, the rate of the decomposition reaction is not the only factor determining the physical properties of the coke.

Fitzgerald[175] and Fitzgerald and van Krevelen[176] have studied the kinetics of the evolution of molecules which are gases at normal temperatures; for it is pointed out that in terms of the number of molecules these are much more important than the tars, and in kinetic studies it is the changes in molar concentrations which are of the most significance. In this work, the difficulties of interpreting the results of gravimetric experiments were avoided by measuring the flow rate of the noncondensible gases and determining their composition at frequent intervals.

The pyrolyses were carried out in a fused silica tube of 17 mm internal diameter, one end of which was placed in the furnace and contained a thermocouple well, the latter being completely within the coal sample. Approximately 3-g samples of coal were used, and helium could be introduced at the end of the tube carrying the sample, to flush the apparatus. The tube was evacuated by a "Sigma" type pump of such a pumping speed that with the size coal samples used, the pressure never rose above 20 mm Hg. A condenser filled with ice was inserted in the system ahead of the pump; following the pump there was a hydrogen-sulfide absorption tube, the sampling point, and a mercury manometer. Rate of evolution of gaseous products was determined by interrupting the flow for known periods of time and measuring the resultant pressure increase in the system, the volume of which was known. Pressure changes were recorded graphically. Heating rates of about 1.80°C were used, and six coal samples, predominantly vitrinite, ranging from 80 to 94% carbon on the dry, ash-free basis, were studied. The exinite and micrinite associated with one of the vitrinites were also studied. The results for a British high-volatile coal from the Beeston seam are shown in Figs. 36, 37, and 38. These curves were interpreted on the assumption

[174] Fitzgerald, D., *Fuel*, **35**, 178–83 (1956).

[175] Fitzgerald, D., *Inst. of Gas Engrs., Publ.* No. 516 (1957), 42 pp.

[176] Fitzgerald, D., and van Krevelen, D. W., *Fuel*, **38**, 17–37 (1959).

that the various gases appearing were the products of first-order reactions and that the rate-determining step was the breaking of the bond between the radical and the coal lamella. If V_0 is the total volume of any given gas that can be produced, and V is that produced up to a given time then $\frac{dV}{dt} = k(V_0 - V)$ where k is the velocity constant. The temperature dependence of this constant was assumed to be given by the Arrhenius relation and substituting we have: $dV/dt = k_0(V_0 - V)e^{-E/RT}$ where k_0 is the frequency factor, E, the activation energy, and R, the gas constant. By plotting $(\ln dV/dt)(V_0 - V)$ against $1/T$ a straight line having a slope equal to $-E/R$ and an intercept of $\ln k_0$ should be obtained. The plots actually obtained were far from linear. Typical plots for methane and for hydrogen are shown in Figs. 39 and 40. The Arrhenius plots for both methane and hydrogen show an initial steep part, up to about 550°C, followed by one of very much lesser slope, indicating that the velocity constants at higher temperatures are smaller than would be predicted from the values at lower temperatures. The continued slow increase in k for evolution of methane, even over the temperature range where the rate of evolution is decreasing, suggested that another reaction was responsible for the methane formation at higher temperatures, and these authors postulated "autohydrogenation," a reaction between hydrogen and carbon to form methane. The decrease in the value of k for hydrogen is, however, greater than can be accounted for by methane formation, even if *all* the methane appearing above 500°C were assumed to result from this reaction. It was suggested that owing to "the assembly of lamellae to which the hydrogen atoms are attached, surviving C—H bonds become stronger and continue to become so with increasing temperature."

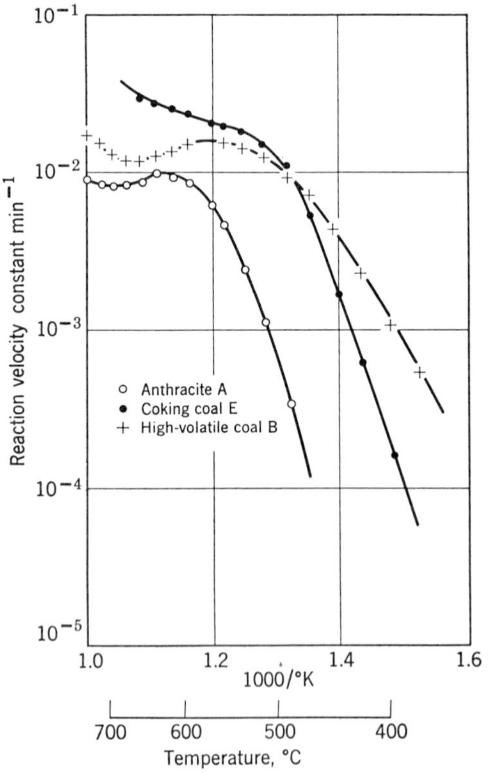

Fig. 39. Arrhenius diagram for methane for anthracite A, vitrinite of coking coal E, and vitrinite of high-volatile coal B. Other coals examined fall between these according to rank.[175]

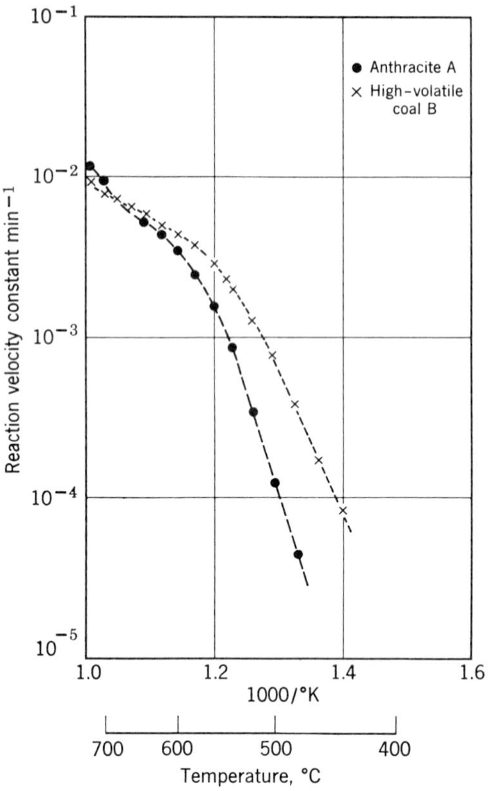

Fig. 40. Arrhenius diagram for hydrogen for anthracite A and vitrinite of high-volatile coal B. Other coals examined fall between these according to rank.[175]

van Krevelen and co-workers[177] have made use of the thermogravimetric balance to study rates of pyrolysis. Confirming earlier work in decompositions at constant temperature, the loss in weight was found to reach a limiting value, which increased with temperature. At constant heating rates, the maximum rate of loss of weight occurs early, and the lower the rank the earlier this maximum occurs. At increasing heating rates, the maximum in the rate-of-weight-loss temperature curve was shifted to higher temperatures and preheating was found to result in decrease in the amount of material which could be volatilized. The effect was most marked at 450 to 500°C, which suggests the occurrence of special reactions in this temperature range. On the basis of these facts, Chermin and van Krevelen[178] have developed a simplified mathematical picture of the pyrolysis of coal by which various quantities characteristic of the pyrolytic process can be calculated with some agreement with those from experiments. The reactions postulated were:

$$P \xrightarrow{k_1} M$$

$$M \xrightarrow{k_2} R + G_1$$

$$R \xrightarrow{k_3} S + G_2$$

where k_1, k_2, and k_3 are the velocity constants; P is the original coal, M is the metastable intermediate product, R, the semicoke, S, the coke, G_1, the primary gas, and G_2, the secondary gas. The first reaction can be regarded as a sort of depolymerization, the second, as a cracking process, and the third, as a degasification. All were assumed to be first-order processes in the mathematical development. This treatment reproduced, semiquantitatively, the course of the devolatilization of coal at constant temperature, the activation energies at constant heating rates, and the volatile matter as a function of temperature of carbonization.

Smutkina and Kasatochkin[179] have studied the kinetics of the thermal decomposition of a number of Russian coals, including the Moscow brown coal, kerogen of kukersite shale, and Olenek boghead coal. Evolution rate of volatile matter was determined by placing a 0.2-g sample in a thin-walled, quartz tube, suspended on a quartz spring and placed in a vertical, tubular furnace. The decomposition was carried out in a stream of purified nitrogen and, to protect the specimen from oxidation, a carbon tube heated to 850°C by an electric furnace was placed in the upper end of the quartz tube. Changes in the length of the quartz spring during an experiment, corresponding to changes in weight of the sample, were determined by means of a cathetometer. The integral and differential weight curves obtained were similar to those for European coals. However, changes in the velocity constant, calculated on the basis of a first-order reaction, were so great these workers concluded that the thermal decomposition of coal cannot be described by a single first-order reaction; consequently, it is not possible to delineate the process from any one value of activation energy. In contrast, the rate constants for the boghead coal and the shale had low-velocity constants which were relatively "constant." The observed increase in activation energy for coals with degree of decomposition was attributed to the selectivity of the destructive processes in which bonds with increasing thermal stability are broken in order.

The work of Dryden and Pankhurst,[180] discussed earlier in connection with the evidence for low-temperature transformations in coal, is also of interest from the standpoint of pyrolytic reactions. These workers have associated the formation of chloroform-soluble material, which is formed by "shock-heating" coking coals above 300°C with the "fluid coal" or "metaplast" postulated, as an essential intermediate in the coking process.

Pitt[181] has investigated the kinetics of the formation of this chloroform-soluble material by studying its rate of formation under as nearly isothermal conditions as possible. For this purpose, a fluid-bed apparatus based on the design of earlier workers[172] was employed. It was stated that by heating sand in a separate, fluidized bed to the desired temperature and

[177] van Krevelen, D. W., Huntjens, F. J., and Dormans, H. N. M., *Fuel*, **35**, 462–75 (1956).

[178] Chermin, H. A. G., and van Krevelen, D. W., *Fuel*, **36**, 85–104 (1957).

[179] Smutkina, Z. S., and Kasatochkin, V. I., *Khim. i Tekhnol. Topliva i Masel*, **1957**, No. 5, 27–32.

[180] Dryden, I. G. C., and Pankhurst, K. S., *Brennstoff-Chem.*, **39**, S84–6 (1958).

[181] Pitt, G. J., *Brennstoff-Chem.*, **39**, S87–9 (1958). Oxley, G. R., and Pitt, G. J., *Fuel*, **37**, 19–24 (1958).

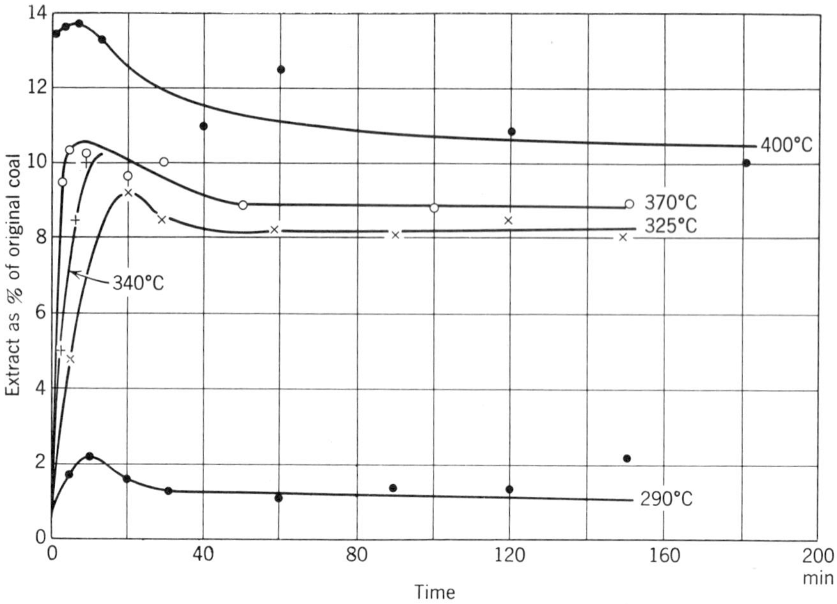

Fig. 41. Yield of extract as a function of time of carbonization.[181]

mixing the coal sample with the sand just before entrance to the reactor temperature equilibrium was attained in 15 sec. Samples were withdrawn from the reactor at the desired time intervals into an evacuated and cooled metal receiver so that the temperature of the coal was reduced below 100°C in less than 5 sec. Experiments were made with one British coking coal, Union Hill Top Drift, 30.3% volatile matter, on the dry, ash-free basis. The yield of chloroform-soluble material as a function of time at temperatures from 290 to 450°C is shown in Fig. 41. In agreement with the earlier work, it is evident that the rate of formation of chloroform-soluble material is very slow below 300°C. The initial portions of the curves for the higher temperatures rise steeply, and the slope increases with increasing temperature up to 400°C. If the rising portions are assumed to be the result of first-order reactions, the calculated activation energy is of the order of 30 kcals/mole and the frequency factor $4 \times 10^{10}\ \text{min}^{-1}$. The flatness of the curve for a temperature of 400°C indicates that the extracts are stable at this temperature for as long as 3 hr. At temperatures higher than 400°C, the yield of extract rises very rapidly for short periods and then decreases rapidly, indicating that above 400°C, molecular growth of the chloroform-soluble material to an insoluble product is occurring rapidly. These facts are of the greatest importance, because they suggest that with a coal of this type, reactions carried out below 400°C result exclusively in fragmentation and that only above this temperature do reactions involving molecular growth of these fragments become significant. Gieseler measurements were made at constant temperature using a procedure similar to that of Fitzgerald.[173] A comparison of these curves with rate curves for the yield of extract shows that at temperatures between 400 and 440°C the rate of formation of the extract is much greater than the rate of development of fluidity, and also that, while chloroform solubility remains appreciable at 425°C for periods as long as 2 hr, the fluidity becomes too small to measure under these conditions. Thus, one is forced to conclude that it is not the chloroform-soluble extract *per se* which is responsible for the fluidity.

Evidence for the presence of a more complex mechanism for the formation and destruction of chloroform-soluble material in Alberta coking coals has been presented by Den Hertog and Berkowitz.[182] They reported a double maximum in the plot of chloroform-soluble versus temperature. The minimum between the maxima occurred at almost exactly 400°C, a temperature at which the chloroform-soluble

[182] Den Hertog, W., and Berkowitz, N., *Fuel*, **37**, 358–61 (1958).

material of the British coal exhibited marked stability.

Work in this field has also been reported by Nadziakiewicz[183] who heated coal samples (presumably Polish) in an iron tube in a salt bath at a rate of 3°C/min. The samples were maintained at selected temperatures for 5 min and then rapidly cooled. After weighing, the sample was extracted with chloroform. Maximum fluidity was determined on the samples by the Gieseler procedure. A plot of the yields of chloroform-soluble material against the logarithm of the maximum fluidity, for a series of coals heated to their temperatures of maximum fluidity, indicated a linear relation between these quantities. It was concluded that it is "very probable that the plasticity of bituminous coals can be accounted for by a substance which is formed as a product of the primary decomposition of coal and which can be extracted by chloroform."

Pyrolysis of Model Compounds

In many fields, the study of the properties and behavior of so-called model compounds has resulted in significant contributions to our knowledge and to the advancement of technology. It has been the experience of many investigators in the field of coal science and technology that progress has been handicapped by lack of knowledge of the properties and reactions of substances of known composition which are believed to be similar to coal.

Because bituminous coals are resistant to hydrolytic reactions, it is probable that linkages such as ester, anhydride, acetal, or lactone are absent; and because the infrared spectra of unoxidized high-volatile bituminous coals do not show the presence of carbonyl groups, it is probable that the pyrolyzable links consist of aliphatic carbon-to-carbon or ether links.

The thermal decomposition of a "coal model" has been discussed briefly by Fuchs and Sandhoff,[154] but the most significant experimental work in this field has been carried out by Neuworth[184] and co-workers. The first models studied were synthesized by condensation of a pyrene alkanoic acid of the type $C_{16}H_9(CH_2)_nCOOH$ with 2,6-xylenol and reduction of the resulting carbonyl compounds. This led to the following type of model compound:

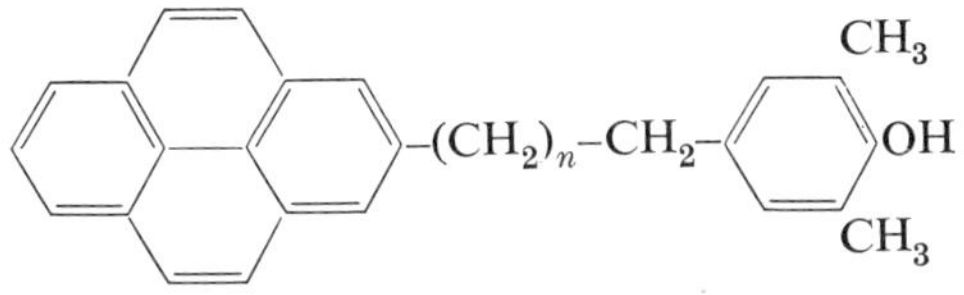

where $n = 0, 1, 3$.

In such a type of structure, rupture would be expected to occur in the aliphatic chain and, in general, the scheme of pyrolysis would be expected to be as follows: The aliphatic links would rupture producing free radicals which then would capture hydrogen from other aliphatic linkages and either distill or coke, depending on their molecular weight. The portion of the molecule which supplied the hydrogen would produce a substituted ethylene which, because of its stability, would eventually be converted to coke. Concurrently, thermal dehydrogenation could occur resulting in a substituted ethylene and hydrogen. On the average, the production of fragments of molecular weight sufficiently low to distill would require the rupture of two aliphatic linkages.

The model compounds were pyrolyzed in a modified Fischer retort to a final temperature of 425°C. Two hours were required to reach this temperature, and the time at 425° was 3 hr. The amount of compound pyrolyzed ranged from 25 to 50 g. The results of these experiments supported the mechanism outlined. For example, the "methane model" should yield two major fragments which on hydrogen capture would result in four primary products: 2,6-xylenol, 2,4,6-trimethylphenol, pyrene, and 1-methylpyrene. Actually these compounds and the secondary dealkylation products of the phenolic compounds, phenol and o-cresol, were found to comprise about 85% of the "tar." Similarly, rupture of the aliphatic linkage in the "ethane model" would produce two benzyl-type radicals which on hydrogen capture would yield 2,4,6-trimethylphenol and 1-methyl pyrene as primary pyrolysis products. These compounds were found to make up 75% of the "tar." These results furnish a mechanism for the production of monocyclic phenols from coal containing preformed phenolic nuclei as part of a repetitive polymer. Free radical rupture of the aliphatic linkage followed by

[183] Nadziakiewicz, J., *Fuel*, **37**, 361–4 (1958).

[184] Depp, E. A., Stevens, C. M., and Neuworth, M. B., *Fuel*, **35**, 437–45 (1957).

hydrogen capture results in a "primary tar" having similar ultimate composition but a higher hydrogen-carbon ratio than the parent substance. Similar relationships exist between low-temperature tar and the coal from which it is produced. Very low coke yields were obtained in these experiments which may be ascribed to the lower hydrogen requirements for rupture of monomeric models of the type used. For a repetitive polymer, rupture of two adjacent linkages is required to produce fragments of sufficient vapor pressure to distill out as "tar." Random rupture of aliphatic linkages in a polymer would produce many fragments of too-low vapor pressure to distill, and these would ultimately form coke by secondary dehydrogenation and dehydrocondensation reactions.

This work was later extended to include pyrolysis of benzyl type ethers.[185] The method of synthesis involved condensation of a pyrenyl chloride $C_{16}H_9CH_2Cl$ with the sodium salt of an alkylated phenol. With 2,4,6-methylphenol, the resultant ether had the following structure:

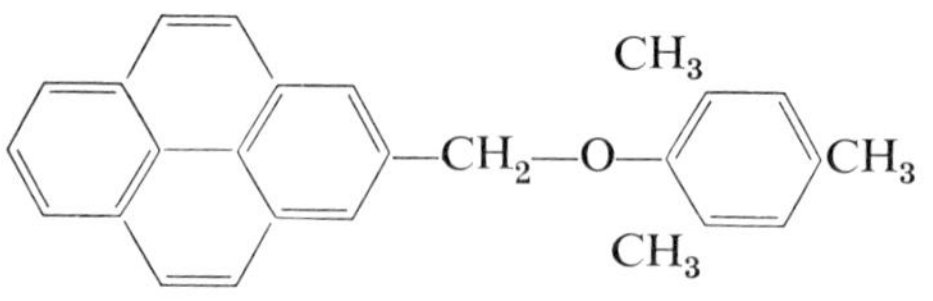

A similar ether was synthesized using 3,5-xylenol. In this compound, the methyl groups were removed from the vicinity of the ether link.

Pyrolyses of these compounds were carried out as described above. The major components of the "tar" from the first of these compounds were 2,4,6-trimethylphenol and 1-methylpyrene, which constituted 92% of the liquid pyrolyzate. These are the predicted products from the free radical rupture of the ether linkage followed by hydrogen capture. Similar results were obtained with the second ether, except that, as would be expected, the 3,5-xylenol was the predominant phenolic compound. These results point to the lability of the ether link in aromatic ethers.

In some respects, the pyrolysis products of these ethers differ markedly from those from bituminous coal. These are the high carbon monoxide content of the gases and the low ratio of oxygen in char to oxygen in the gas. The last may be ascribed to the fact that in these models all of the oxygen is in the form of ether links, whereas in bituminous coals of about this rank, at least half of the oxygen is probably present in hydroxyl groups.

[185] Depp, E. A., and Neuworth, M. B., *Brennstoff-Chem.*, **39**, S90–2 (1958).

CHAPTER 10

LOW-TEMPERATURE CARBONIZATION

PHILIP J. WILSON, JR., and J. D. CLENDENIN

United States Steel Corporation, Pittsburgh, Pennsylvania

Low-temperature carbonization of coal may be defined as carbonization in the absence of air which is carried on to a final char temperature in the range 450 to 700°C. Carbonization in the intermediate range 700°C up to a final coke temperature below 900°C is considered to be medium-temperature carbonization, and above that temperature, the carbonization process is classed as high-temperature carbonization.

No sharp line of demarcation exists between processes in the low- and medium-temperature fields, or even between many of these fields and some high-temperature processes. This will be evident from the descriptions of the processes in this chapter. Not only are strictly low-temperature processes included, but also some processes which can also be operated at higher temperatures.

Four main products are produced by low-temperature carbonization: the final semicoke or "char," a tar consisting largely of complex liquid hydrocarbons, fixed gases, and aqueous liquor. If the heating is done rapidly under conditions which prevent decomposition of the primary products, 68 to 87% of the heating value of the raw coal is ordinarily left in the char, and 10 to 23% appears in the primary tar. Approximately 7% is found in the fixed gases, and the balance, roughly 2%, is lost or dissipated in the reactions of decomposition. In the United States, the yield of tar has ranged from about 9 to 44 gal/ton, depending on the coal which is coked. Little or no ammonia is produced.[1]

Low-temperature carbonization aims generally at the production of a solid, smokeless fuel coupled with a high yield of coal tar. The solid product from low-temperature carbonization, whether finely divided or in the form of lumps, has usually been a soft, black, porous material, which still contains considerable volatile material. Because in these properties it more nearly resembles charcoal than high-temperature coke, the solid product from low-temperature carbonization has ordinarily been called "char," and this name will be used here. Another common designation is "semicoke," which is applicable because of the incomplete degree of carbonization which the coal has experienced. Strictly low-temperature tar differs materially in composition from high-temperature tar, and to distinguish it, it should be termed low-temperature tar.

The technology of low-temperature carbonization is characterized by the great number and variety of processes which have been proposed. These processes, amounting literally to hundreds, range all the way from the inventor's ideas expressed only on paper, through laboratory setups and small pilot plants, to commercial installations carbonizing hundreds of tons of coal daily. The conditions of operation of these plants vary greatly, depending on such factors as the coal to be treated, products to be made, volume of throughput, temperatures of carbonization, materials of construction, and many others.

The products from this great variety of processes necessarily vary widely in chemical and physical characteristics, their properties depending largely on the treatment to which they have been exposed. For char especially, the size has varied from a fine powder to large lumps, including, in some processes, the preparation of a shaped product or briquette. Considerable variations also necessarily occur

[1] Parry, V. F., *Combustion*, **25**, No. 9, 38–42 (1954). Selvig, W. A., and Ode, W. H., *U.S. Bur. Mines, Bull.* No. 571 (1957), 56 pp. Goodman, J. B., Gomez, M., and Parry, V. F., *U.S. Bur. Mines, Rept. Invest.* No. 5383 (1958), 49 pp.

in the compositions of the tar and in the gas.

Early in the nineteenth century, low-temperature carbonization and probably many of the early processes for producing metallurgical fuel operated at temperatures within the low-temperature range began to be developed. Late in that century the study of carbonization was greatly stimulated by two conditions, the spectacular development of organic synthesis and the increasing interest in the preservation of our natural resources. The appreciation that by heating coal to an elevated temperature in the absence of air a smokeless fuel could be produced, and that at the same time the tarry matter, which would have escaped as objectionable smoke if the coal had been burned directly, could be recovered as a useful liquid, attracted widespread attention to the procedure. These black disagreeable tars became of great interest as potential raw materials for the synthesis of such valuable products as dyes, pharmaceuticals, and other organic chemicals. The general opinion was that with the char sold at a price competitive with coal, the value of the tar for a chemical raw material—or at least as a liquid fuel—rendered either high- or low-temperature carbonization, commercially attractive. The gas also was of interest, although its relatively low yield in low-temperature carbonization made this product of minor importance.

The widespread expression of these ideas aroused tremendous interest in low-temperature carbonization, which was increased following World War I by the concern about the future supply of petroleum and natural gas.[2,3] It culminated during the late 1920's and early 1930's. Inventors and promoters on all sides turned their attention to the development of processes which were expected to be highly profitable. At that time, literally hundreds of low-temperature processes were proposed and tried, and the technical literature of the period contains many references to all kinds of processes.[4,5]

In spite of the great amount of work devoted to the problem, low-temperature carbonization was almost entirely unsuccessful in the United States. Although several commercial plants were built at that time, only one of significant size continues in operation. One authority[6] estimated that between 1920 and 1930 in the United States alone over $50,000,000 were expended on unsuccessful commercial ventures without tangible return.

A similar condition prevailed in Europe. With differing economic conditions, however, low-temperature carbonization was somewhat more successful there. Oil and natural gas were not as available as they were in the United States, so that solid fuels retained their importance, especially for domestic use. In England, much of the solid domestic fuel is burned in open grates, and several plants which furnished a satisfactory lump char for this service have continued in existence. In Germany, brown coal was carbonized by low-temperature methods on an extensive scale, and the tar products were hydrogenated to produce synthetic motor fuels. This product became especially important during World War II, and was required as an essential source of aviation fuel. Japan also obtained some needed liquid fuel by this method.

At present, low-temperature carbonization is again receiving broad study, and since World War II many processes have been described. In the work now being done, char is the prime objective, although the value of the tar is still a basic consideration in initiating a process. Successful commercial applications, however, are still to be established in the United States.

CLASSIFICATION OF LOW-TEMPERATURE CARBONIZATION PROCESSES

The great number and variety of the processes available for the low-temperature carbonization of coal make necessary some system of classification. The main factors on which classifications may be based are the general design of retorts, the character of the coal carbonized, and the character of the char or char product produced.

[2] Fieldner, A. C., *U.S. Bur. Mines, Tech. Paper* No. 396 (1926), 46 pp.

[3] MacDougall, D., and Bardgett, H., *Proc. Intern. Congr. Pure and Appl. Chem., 11th Congr., London*, **1947,** IV, pp. 451–60 (1950).

[4] Gentry, F. M., *The Technology of Low-Temperature Carbonization*, Williams & Wilkins Co., Baltimore, 1928, 399 pp.

[5] Utah Conservation and Research Foundation, *Low-Temperature Carbonization of Utah Coals*, The Quality Press, Salt Lake City, Utah, 1939, 872 pp.

[6] Soule, R. P., *Proc. Intern. Conf. Bituminous Coal, 3rd Conf.*, **1931, 1,** pp. 272–98.

Although all these factors will be considered in discussing low-temperature processes, the basis of classification used here will be the principal physical characteristics of the retorts.

VERTICAL RETORTS

Vertical retorts embrace many different carbonization devices, possibly more than any other arrangement, and they include both batch and continuous types, with some processes resembling existing types of high-temperature coke ovens. In vertical-retort processes, either batch or continuous, the coal is charged at the top, as with a high-temperature coke oven, but discharge has usually been from the bottom. Sizes of retorts are based on securing the maximum throughput per unit of capital cost. Because coal and char are very poor thermal conductors, and penetration of heat through layers of these materials into the center of the charge is slow, retorts indirectly heated by conduction of heat through the retort walls are usually narrow, resulting in relatively small capacities per retort. To increase rates of heat transfer and carbonization, some retorts have been made of steel, and in others resort has been made to internal heating of the charge. Internal heating is effected by circulating hot gases, either the products of combustion of a fuel or preheated gases, such as make-gas, through the retort. Carbonization has thereby often been accelerated, and the throughput of a retort increased materially.

Vertical retorts have been used for carbonizing a wide variety of coals, caking and noncaking, and various sizes, such as slack, lumps, and briquettes. However, all types of vertical retorts are not adapted to all coals. Thus, when caking coals are plastic they usually flow together to form large masses which adhere to the retort walls. Unless the plastic properties are reduced by a prior oxidation or other treatment, such a coal is usually best carbonized in a batch retort designed to permit ready discharge of the char. For continuous retorts, sticking of char in the retorts cannot be permitted, and only coals possessing a low caking power should be introduced.

Construction materials for vertical retorts range from steel to various refractories. Higher temperatures to speed up carbonization of the charge are, of course, applicable with refractory retorts. Similar methods of heating, indirect, direct, or a combination of both, are used with both types of materials.

Fixed-Bed Retorts

Indirectly Heated Fixed-Bed Retorts

Fixed-bed retorts consist essentially of chambers, which are heated indirectly by circulating the products of combustion of a fuel gas through flues built in the walls. Because of the lower temperatures and the consequent slower rate at which carbonization proceeds, retorts must be much narrower in width than coke ovens to obtain a satisfactory rate of production.

Krupp-Lurgi process.[7,8,9,10] The Krupp-Lurgi process was developed in Germany during the 1930's, and up to the close of World War II a number of plants, each with a daily capacity of several hundred tons, had been constructed in that country. During the war the process proved important to Germany because the low-temperature tar from the Krupp-Lurgi plants was a useful source of the fuel oils required by the German military forces. In recent years, the availability of petroleum oils appears to have greatly diminished the need for the process.

Figure 1 shows a cross section of a Krupp-Lurgi coke oven. The coal was carbonized in units, each consisting of six narrow carbonization cells contained between seven rectangular steel heating cells, through which were recirculated hot gases. A unit of six carbonization cells constituted an oven, and each cell was approximately 10.5 ft long by 6.9 ft high by 3 in. wide at the top and 4 in. wide at the bottom. Each oven had a volume of about 125 cu ft and carbonized about 2.75 metric tons/charge.

The ovens were constructed entirely of steel plate. After several years of operation, the German ovens were in good condition and a life of 10 to 12 yr was expected. The ovens were heated by recirculation of the hot gases through the heating cells. The combustion

[7] Schmidt, L. D., *U.S. Bur. Mines, Inform. Circ.* No. 7395 (1947), 18 pp.

[8] Rhodes, E. O., *U.S. Bur. Mines, Inform. Circ.* No. 7490 (1949), 84 pp.

[9] Anon., *Coke and Gas*, **8**, 77–9 (1946). Kennedy, G. L., *J. Inst. Fuel*, **33**, 598–608 (1960).

[10] Georges, P., *Ann. mines, Mém.*, **136**, No. 6, 45–83 (1947).

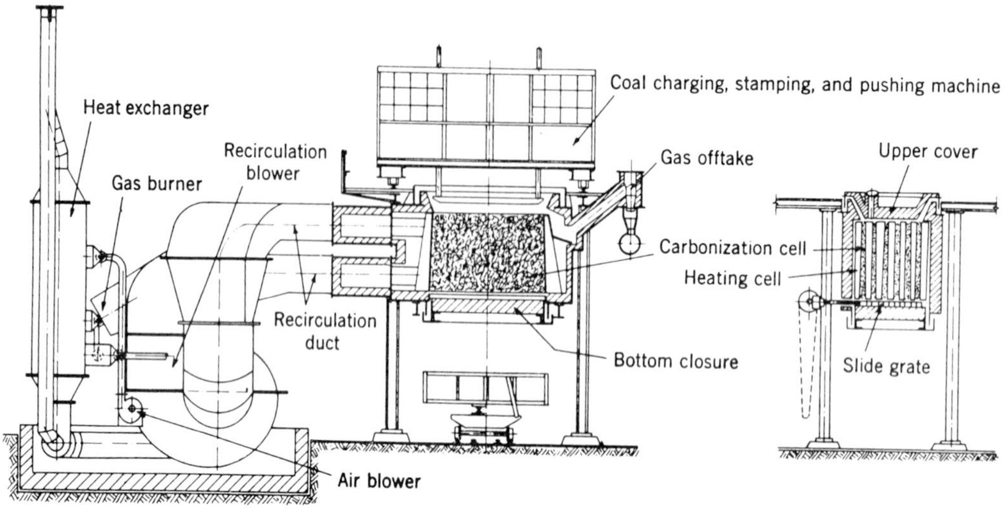

Fig. 1. Diagrammatic drawing of a Krupp-Lurgi low-temperature coke oven.[7]

gases from a common, gas-fired combustion chamber were recirculated by a fan through ducts running to and from the oven. The circulating gases flowed into the flues at the bottom of each heating cell, up between the steel walls, out at the top, and back to the furnace. At regular intervals, the direction of the combustion gases through the flues was reversed to heat uniformly the walls of the chambers. The temperature of the heating gas entering a heating cell was about 610 to 620°C and leaving the cell about 560 to 580°C. The drop in temperature of about 40°C represented sufficient heat to carbonize adequately the coal in a cell, both at the bottom where it was 4 in. thick and at the top where it was only 3 in. thick.

The flow diagram, Fig. 2, shows the manner in which the gas from the ovens was treated to recover tar, ammonia liquor, and light oils. The quantities on the flow diagram, of coal carbonized, char, and other products made, apply to 24 of the newer ovens installed during the war (1943) at the Wanne-Eickel plant of Krupp Treibstoffwerke G.m.b.H. The plant at Wanne Eickel then contained 56 ovens with an annual capacity of 204,000 metric tons of char, 1200 tons of fuel oil, and 1920 tons of motor fuel.

Normally at Wanne Eickel a blend of high-volatile Ruhr coking coals was carbonized: about 25% Hannibal coal with 22% volatile matter, and 75% Bismarck coal with 33 to 34% volatile matter. The charge mixture usually analyzed 8 to 10% moisture, 30% volatile matter, and 5.4% ash. The coal size for Krupp-Lurgi ovens was preferably 77% minus $\frac{1}{8}$ in. and about 5% minus 100-mesh. By stamping the coal during charging, the density of the charge could be increased about 8% over the normal density of 48 lb/cu ft.

A fuel gas containing 220 to 250 Btu/cu ft was burned in the combustion chambers, and the heat input amounted to 970 to 1050 Btu per pound of coal charged.

About 3350 cu ft of gas were produced per ton. The yield of dry char amounted to about 84% of the dry coal. The larger coke was consumed in a Fischer-Tropsch plant which produced water gas with a $CO:H_2$ ratio of 1.35 to 1.50:1. The smaller sizes were used in house heating and in the small mobile gas producers.[9]

German coals in the range of 27 to 38% volatile matter have been of most interest for use in the Krupp-Lurgi ovens because they have given chars of good quality plus high yields of tar and light oil. When gas coal containing 30% volatile matter, 9% moisture, and 5.8% ash was carbonized, a char was produced analyzing about 7.2% ash and 9.3% volatile matter. The tar yield was 6.4% and the light oil yield 0.67% of the weight of the dry coal.[11] Other chars from ovens in the Saar ranged in drum stability from 50 to 80%, their apparent specific gravities were 0.8, and the porosities 42 to 45%. The volatile matter ranged from about

[11] Anon., *Coke Smokeless-Fuel Age*, **5**, 10–13 (1943).

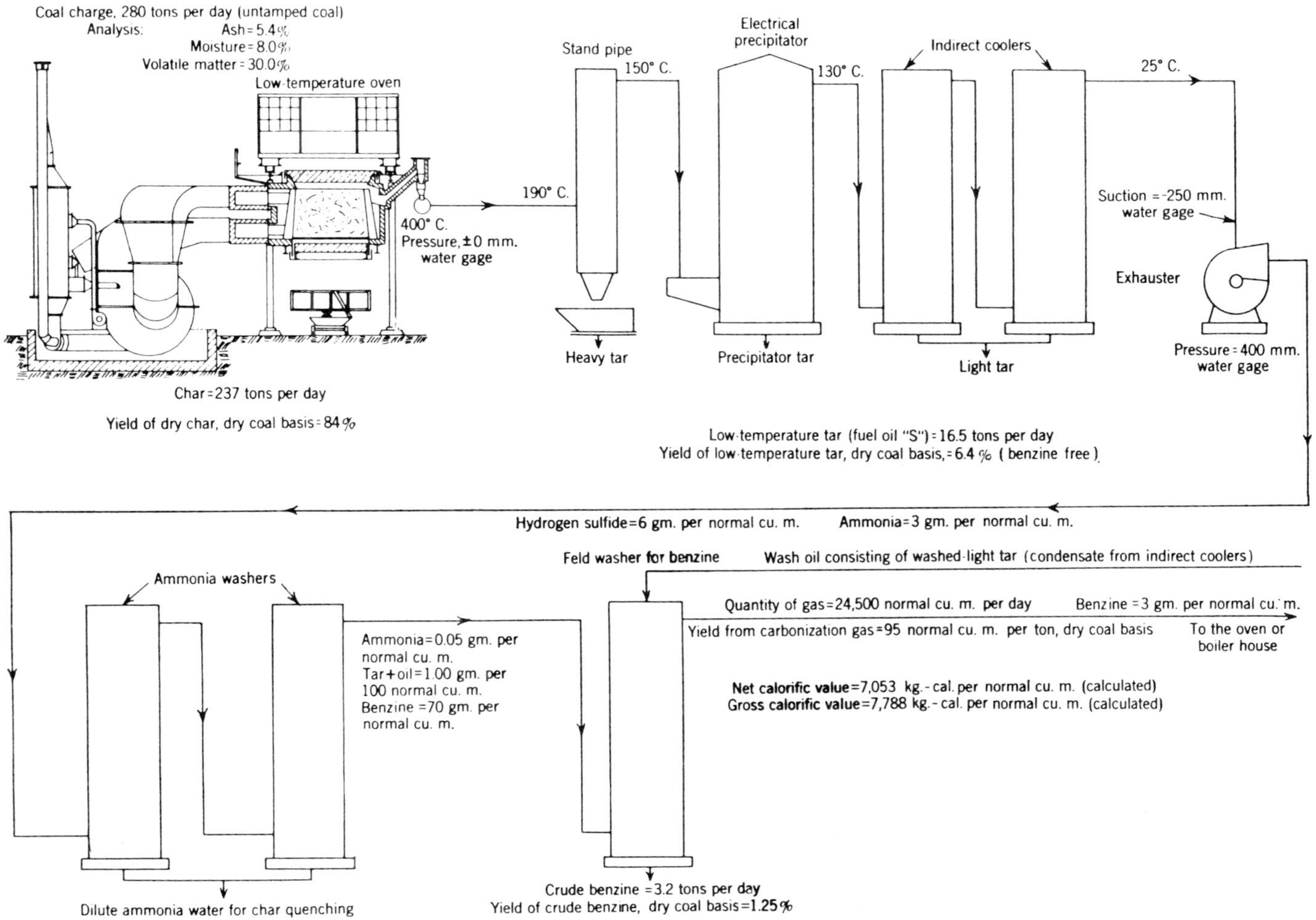

Fig. 2. Flow diagram for recovery of low-temperature char, tar, ammoniacal liquor, and benzine from Krupp-Lurgi low-temperature carbonization plant.[8]

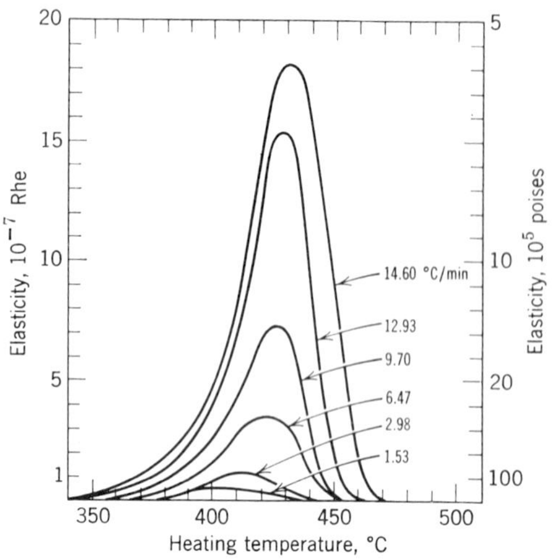

Fig. 3. Curves showing the relationship of plasticity to the rate of temperature rise on coal carbonization.[15]

12% in the large sizes to from 16 to 18% in the small. About 12% of this char was $-\frac{1}{2}$-in. breeze, which was used in the coal mix to high-temperature coke ovens. Some mixes in which it was blended with Saar coal produced high-temperature coke that was nearly as good as coke from mixes in which special Ruhr coal had been blended.

In the high-temperature coke oven, the plastic zone moves steadily inward from the walls after charging is completed until it meets in the center of the oven shortly before it is pushed. However, in the narrow Krupp-Lurgi ovens, at the lower temperatures, no definite plastic zone existed, but all the coal in the entire oven was in a plastic condition at one time. Formation of a good low-temperature char was attributed by the designers of these ovens to the rapid and simultaneous increase in temperature throughout the plastic mass (350 to 450°C). The faster the temperature increased in this range, the more complete was fusion of the coal constituents with increase in the strength of the char. Studies have shown that the temperature started to rise in the middle of the narrow, low-temperature ovens shortly after charging, even when charged with coal containing several percent of water. Special tests indicated that in $\frac{1}{2}$ hr all the water had been driven off from a coal containing 8%, in 2 hr the coal was entirely in a plastic state, and after 3 hr it had been converted to char. Unlike high-temperature coke which breaks in the center of the oven, the char normally showed no tendency to divide in the center of the oven; the pieces were uniform from wall to wall. Analyses showed virtually the same volatile content in the char from the center of an oven as from the walls. Furthermore, retention of char in the oven for an additional 2 hr after carbonization had been completed changed the properties very little.

Krupp-Lurgi retorts have been used not only for carbonization of coals or coal mixtures directly, but also for briquettes. As applied to brown coal and lignite, the Krupp-Lurgi ovens were a part of a three-stage process. The coal was first passed countercurrent through a stream of hot gases at about 1000°C. At this temperature, it disintegrated into a powder which was compressed in a ring-roll press at a pressure of 2000 to 2500 atm into briquettes. The hard and very dense ovoids were finally carbonized in the Krupp-Lurgi ovens. In Germany at the end of the last war the production of briquetted char from lignite was reported to be 3800 tons/day.[12]

Brennstoff-Technik.[13,14] The objective of the Brennstoff-Technik process was to achieve good yields of char in lump form from weakly coking coals by rapidly heating the coal in very narrow ovens. Both experience and tests have shown that the plasticity of a coal increases with the rate of temperature rise. The relationship is illustrated by Macura in Fig. 3, which shows the pronounced increase in plasticity of a coal when the rate of heating was increased from 1.43°C/min to 14.6°C/min.[13,15]

In the Brennstoff-Technik process, the desired high heating rates were secured, not by the use of high temperatures, but primarily by making the ovens narrow—usually in the range 60 to 125 mm. The narrower ovens were used for the weaker caking coals. It was claimed that these ovens might be used to process high-volatile, weakly caking coals which could not be carbonized in ordinary coke ovens.

[12] Jennings, R. J. S., *Chem. & Process Eng.*, **34,** 343–9 (1953).

[13] Barritt, D. T., and Kennaway, T., *J. Inst. Fuel*, **27,** 229–48 (1954); *Gas World*, **139,** No. 3625 Coking Sect., 23–34 (1954); *Gas J.*, **277,** 153–5, 219–21 (1954).

[14] Gollmer, W., *Glückauf*, **92,** 320–8 (1956). Lorenzen, G., *Brennstoff-Chem.*, **32,** 324–31 (1951).

[15] Macura, H., *Oel u. Kohle*, **14,** 1097–1107 (1938); **15,** 1–12 (1939).

Brennstoff-Technik ovens were constructed of steel to benefit by the much higher heat conductivity of iron as compared to silica refractories. The temperatures in the retorts seldom exceeded 650°C, a level at which the coke has not shrunk sufficiently for easy pushing. Because of this condition, and because of the complications which pushing the multiplicity of narrow ovens would entail, the heating walls, which contained the flues, were pivoted at the top so that they could be swung aside, thus allowing the coke to drop out into the quenching car.

A battery of Brennstoff-Technik ovens consisted of several blocks of ovens, each block usually containing twelve ovens. All the ovens in a block were charged simultaneously from a car running along the top of the battery. Carbonization of the oven charge, usually about a ton in weight, was complete in approximately 2 hr if the coal was briquetted and 4 hr when unbriquetted fine coal was charged. After carbonization had been completed, the walls were pushed apart by a mechanical device, and the coke char was dropped into the quenching car and was quenched with water.

TABLE 1

PRODUCT YIELDS, BRENNSTOFF-TECHNIK PROCESS[13]

	Yield Per Ton Dry Coal
Char	1500 lb
Tar	19.7 gal
Light oil	2.6 gal
Ammonia	2.2 lb
Total gas	4550 cu ft
Gross heating value of gas	680 to 793 Btu/std cu ft

The ovens were heated by hot combustion gases which were recirculated by means of a blower through the flues of the walls and then through a combustion chamber. In this chamber, fuel gas was burned with air and the products of combustion were added to the recirculating gases to maintain them at the desired temperature. The ducts which conducted the hot combustion gases in and out of the main flues of a block of ovens also served as the trunions for the movable walls. Figure 4 shows the general arrangement of a block of Brennstoff-Technik ovens.

Yields of products obtained by coking a weakly caking coal having a volatile content of about 30% are shown in Table 1. After the heating requirements for the retorts and coal dryers were met, only about 700,000 to 800,000 Btu in the surplus gas per ton of coal processed was available for distribution.

In Germany, Brennstoff-Technik ovens were employed in a number of trial plants, of which the Berlin-Neukölln Gas Works was the most important.[12] This plant, with a capacity to carbonize as much as 50 to 60 tons of coal or briquettes per day, was operated from 1938 to 1944. The tests included the use of ribbed walls which transferred the heat into the heart of the charge, and considerably increased the heating surface of the chambers so that the coking time was reduced appreciably and the production rate increased. Successful results in the Berlin-Neukölln plant led to the design of three large plants, each with an annual capacity of 500,000 tons. However, at the close of the war, work on these plants, all in various stages of completion, was suspended. In a plant erected at Marienau in the Saar, however, work on the process was continued. For a number of years, about 154 tons of coal were carbonized each day, not as slack, but in the form of briquettes, with cellulose sulfite as a binder.

Scherer[16] has reported the processing of weakly caking long-flame coals in Brennstoff-Technik ovens, each with a width of 90 mm and a carbonization time of 2.5 hr, to produce 110 tons of briquettes per day. The briquettes from the retorts were subsequently heated in a shaft retort to about 700 to 800°C by means of hot gases passed upward through the retort, thereby reducing the volatile matter to from 3 to 5%. The final coke had a drum resistance of about 90% and was suitable for both metallurgical and domestic use. The tar represented 11% by weight of the raw coal. In addition, 2470 cu ft of gas having a heating value of about 843 Btu/cu ft were produced from the ovens. In the shaft retorts, 7600 cu ft of lean gas (heating value 225 Btu/cu ft) were produced per ton of coal. Recent reports indicate that the Brennstoff-Technik ovens at Marienau have been abandoned, but a semicommercial plant may still be operating in Germany.[17]

[16] Scherer, F., *J. Mines, Metals & Fuels (India)*, Special Issue, **7**, 210–15 (1959).

[17] Private communication (1957).

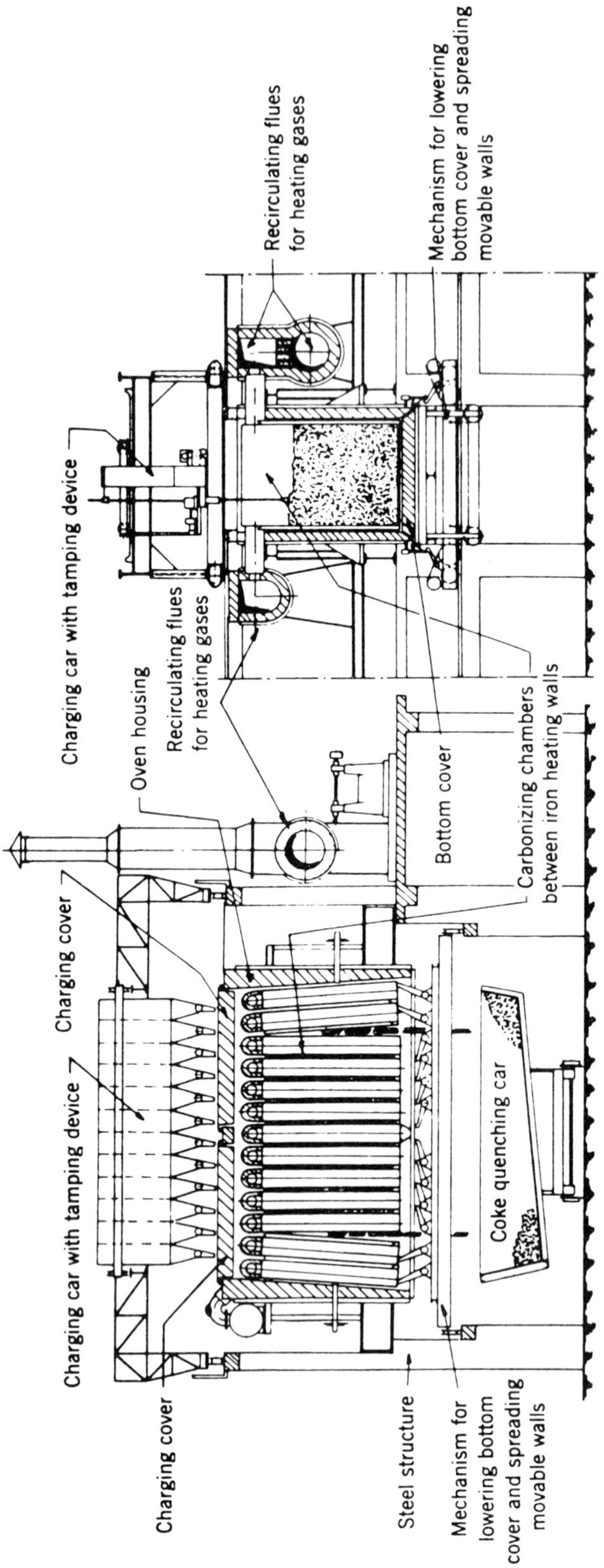

Fig. 4. Brennstoff-Technik low-temperature oven for carbonization of bituminous coal fines.[8]

Cellan-Jones ovens.[11,18] In 1941, an experimental battery containing fifty-one low-temperature coke ovens was erected in Great Britain. These ovens were built of special, high-alumina fireclay brick, and resembled conventional high-temperature coke ovens in both design and operation. Each oven was heated by burners operated on the underjet principle, with the flames burning in a horizontal combustion flue located below the level of the oven floor. The hot gases flowed from the combustion flue up through the vertical flues which were divided centrally by a thick brick partition to give a hairpin arrangement. The brickwork partition provided a reservoir of heat. The gases leaving the vertical flues were conducted through a horizontal sole flue into a vertical waste-heat flue located remote from the oven end. More heat could be supplied to the sole flue by an extra burner, if necessary. No regenerators were provided and the fuel gas was not reversed.

These British ovens, carbonizing from 5 to 6 tons of coal per charge, had a daily throughput of 200 tons with a brickwork temperature not in excess of 750°C. The Vane Tempest coal charged contained 7% moisture, 32% volatile matter, and 5% ash—with 85% of the coal passing a ½-in. screen. The yield of coke was from 72 to 73% of the coal charged, with not more than 7% of it breeze. In addition, 16 gallons of tar and 5500 to 6000 cu ft of gas, with a heating value of 680 Btu/cu ft, were produced. About 10 to 15% of the gas was surplus. The coke was sold for domestic use and as a producer fuel. It contained up to 8% volatile matter, and possessed a bulk density of 36 lb/cu ft and a Critical Air Blast reactivity value of approximately 0.03. The ovens are no longer operating.[19]

Carmaux process.[10,12,14,20] In the Carmaux process, which was operated in Southern France, coal briquettes were carbonized in tall, narrow, vertical retorts after having been subjected to a preliminary heating for about 4 hr in the presence of air at 250°C. This treatment was designed to oxidize the surface of the briquettes with formation of a thin "skin," thus preventing their softening or sticking when they were processed in the slot retorts. The preliminary heating chamber was a smaller, cylindrical vessel located above the main retort, and here the raw briquettes were heated by direct contact with a mixture of air and hot combustion gases ascending from the flues of the retort. An exhauster was used to pull the gases through the bed of briquettes.

The main retorts were constructed of steel and were heated by means of flues located in the long sides. Steam, produced in a waste-heat boiler, was introduced into a char-cooling chamber located at the lower end of the retort. Here it first cooled the char, then it carried the heat from the char up into the retort, and finally it swept the vapors evolved from the coal out of the retort and thus prevented their cracking on the retort walls.

The coal used was of 6-mm size and contained 25% volatile matter and 10% ash. About 6% pitch was added to serve as binder for the briquettes. When pretreatment of the briquettes in the upper chamber was completed, about one-third of the charge in a retort was discharged from the bottom, and the contents of the pretreating chamber were dropped into the retort. The pretreatment chamber was then recharged with fresh briquettes.

Otto retort.[12] An Otto low-temperature carbonization retort was developed from the Otto vertical chamber retort used for gas manufacture. The charge of coal briquettes was heated indirectly by gas burned in twin vertical flues, a system which was reported to provide good control of temperatures and which gave maximum tar yield and optimum char properties. Noncaking coals were desirable charge materials. An experimental retort with a capacity of 4 tons of coal per day was operated in Germany from 1938 to 1944, and from it briquettes suitable for use in generators and a dust-free low-temperature tar were obtained. Although a larger plant was designed from the results of the experimental retort, the close of the war prevented its installation.

Weber process.[8,12] Another process in which briquettes of char or of a noncaking bituminous coal were carbonized in vertical-slot, batch retorts was the Weber process. A flow sheet of the process is shown in Fig. 5. The coal was elevated to a crusher, from which it passed through a feed hopper to a mixer. Here a binder comprising concentrated sulfite pulp

[18] Cellan-Jones, G., *Coke Smokeless-Fuel Age*, **3**, 33–7 (1941).

[19] Private Communication (1960).

[20] Beaugrand, P., *Congr. intern. chauffage ind., 4th Congr., Paris*, **1952.** Group II, Sect. 21, No. 24, 8 pp.

liquor plus an aqueous suspension containing about 1% clay were added with heating and vigorous agitation. The briquettes, formed from the paste in a roll press, were dried on a slowly moving belt until most of the water had been expelled. They were then carbonized at a low temperature to produce "very reactive, dense, hard, and smokeless-burning briquettes or ovoids."

Because the binder possessed no softening tendency, as does pitch, the briquettes would not become plastic and stick together, thus obstructing the retort.

Phurnacite process.[12,21] The "Phurnacite" process carbonized coal briquettes in tall, vertical slot ovens at low or medium temperatures to produce a smokeless domestic fuel. The ovens were provided with vertical flues and recuperators, with the products of combustion of a fuel gas recycled through the flues. The briquettes were charged in at the top of an oven. After carbonizing for about 4 hr, the door at the bottom was removed and the hot ovoids discharged onto the inclined oven floor and into the quenching car. An installation erected in South Wales in 1942 consisted of two batteries, each containing forty ovens, with a daily capacity of about 1000 tons of the raw ovoids.

Parker retort.[22,23] The Parker retort for the production of a smokeless domestic fuel was an early development in British low-temperature carbonization practice. The first installation was made near Barnsley in 1927 and was followed by two further installations. A plant at Bolsover, which went into operation in 1936, was enlarged to 19 batteries with a total of 728 retorts. Another at Askern near Doncaster contained 9 batteries with 360 retorts. A retort consisted of a multiplicity of small, vertical tubes with a tapering cross section in which the coal was carbonized. These tubes were about 9 ft long, with a diameter of $4\frac{1}{2}$ in. at the top and $5\frac{3}{4}$ in. at the bottom. Twelve of these tubes contained in a single monobloc casting of special cast iron formed a retort, and forty retorts were combined into a battery. The retorts were heated only by radiant heat, which was supplied by the combustion of fuel gas in chambers arranged so that they alternated with the retorts throughout the battery. The temperature of the retorts was maintained at approximately 650°C, and approximately $4\frac{1}{2}$ hr were required for carbonization. Products of carbonization were withdrawn from an offtake near the top of the retort. A cooling chamber, large enough to hold the char from two retorts, was located below each pair of retorts. This chamber was fitted with double doors at the bottom, an inner door to support the char, and a water-sealed door below to prevent admission of air. Average yields of products per ton of coal carbonized at Bolsover were as follows:

Char	1500 lb
Crude tar	16.5 gal
Crude light oil	3.5 gal
Aqueous liquor	30.0 gal
Gas	4000 cu ft with a heating value 700 Btu/cu ft

The char, termed "Coalite," is sold for smokeless domestic fuel.

Directly Heated Fixed-Bed Retorts

In addition to the fixed-bed retorts which are heated by transmission of heat through the walls, a number of other retorts have been developed in which carbonization is effected by direct contact of the coal with hot gases.

Rexco process.[23,24] In the Rexco process, a low-temperature char in the form of lumps has been produced by carbonizing a static charge of the coal at approximately 700°C in large-capacity, vertical chambers which were heated internally by combustion of a portion of the charge. The coals used in this process were weakly coking or noncoking, and were charged in the form of lumps, 4 in. by $2\frac{1}{2}$ in., and $2\frac{1}{2}$ in. by 1 in. The retorts were vertical, cylindrical vessels, about 11 ft in diameter and 25 ft high, consisting of a mild steel shell lined with firebrick, with a capacity of about 34 tons of charge.

When a retort was filled, about 23 tons of the larger size lumps were charged into the lower portion of the retort where they were

[21] Anon., *Colliery Eng.*, **30**, 48–57 (1953); *Coke and Gas*, **14**, 5-14 (1952).

[22] Pound, G. S., *J. Inst. Fuel*, **24**, 61–8 (1951); *Coke and Gas*, **21**, 395–401 (1959). O'Connor, J. G., *Fuel Soc. J., Univ. of Sheffield*, **9**, 9–16 (1958).

[23] MacDougall, D., and Bardgett, H., *Intern. Congr. Pure and Appl. Chem., 11th Congr., London*, **1947**, 10 pp.

[24] Anon., *Coke and Gas*, **16**, 383–9 (1954); *Coke Smokeless-Fuel Age*, **2**, 124–9 (1940).

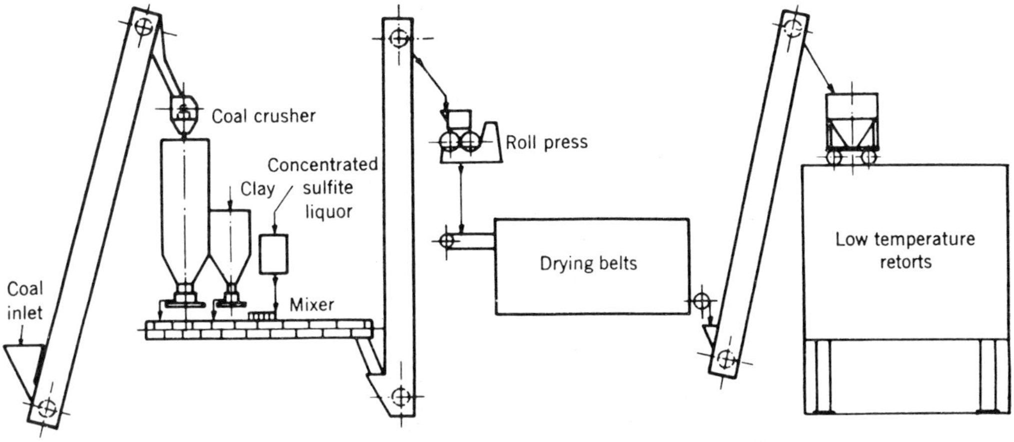

Fig. 5. Weber process for producing char briquettes.[8]

supported on a grate. The coal flowed down and was distributed uniformly by means of a telescopic chute, thus preventing the shattering of the lumps which would otherwise occur if the coal were allowed to fall freely. Above the larger lumps were laid about 11 tons of the $2\frac{1}{2}$ in. by 1 in. coal. The telescopic chute was gradually withdrawn as filling of the retort proceeded. When charging was complete, gas burners above the top layer of coal were ignited, and the hot products of combustion were drawn downward through the charge by an exhauster, thus carbonizing the coal in the retort. The volatile products of carbonization evolved from the coal joined with the hot gases, and the mixture passed downward through the grate to the tar recovery equipment and the exhauster. From the exhauster a portion of the cool, tar-free gas was recycled to the dome which formed the top of the retort. Here it entered through the two tangential burners and was burned with air to provide the heat for carbonization. Carbonization proceeded regularly for 8 or 9 hr until it was substantially completed in the upper two-thirds of the charge. At this time, sufficient sensible heat was present in the upper carbonized mass to complete carbonization of the lower portion of the charge. Accordingly, the air to the gas burners in the combustion chamber in the dome was shut off, and gas merely recirculated through the charge for about 6 hr more to complete carbonization. During the combustion stage, the amount of air to the burners was strictly controlled to reduce to a minimum the excess air which burns a portion of the charge.

As the hot carbonization zone moved downward through the retort, the volatile products which were evolved mixed with the hot gases, but the combined gases immediately entered a cool layer of coal so that secondary cracking of the tars was prevented. When carbonization was complete, the circular grate was withdrawn, and the char fell onto the inclined coke bench which was situated under the retort. The Rexco char had been substantially cooled by means of the circulated gas by the time it was discharged from a retort. It was screened into the sizes required by the domestic consumers. The entire operation required approximately $16\frac{1}{2}$ hr, made up as follows: preparing and charging the retort $1\frac{1}{2}$ hr, carbonizing the upper portion of the charge 8 hr, carbonizing the lower third of the charge and cooling 7 hr. Since about 34 tons of coal were carbonized per charge, the daily throughput of a retort was about 45 tons.

The retort was simple and sturdy in construction, without flues or heat transfer surfaces. The original Rexco plant consisted of two retorts which were installed about 1936 by the National Carbonising Co., Ltd. at Mansfield in Nottingham to supply the char, termed "Rexco," as a fuel for domestic use. Increasing demand for the product led to installation of a third retort of the same dimensions as the two original ones. The new retort was improved by redesign of the combustion chamber so that the heating gas was burned in two separate stages. With the new arrangement, better control of carbonization temperature to maintain it below 700°C was possible. In addition, the overall time of carbonizing a charge could be reduced to $13\frac{1}{2}$ hr, so that

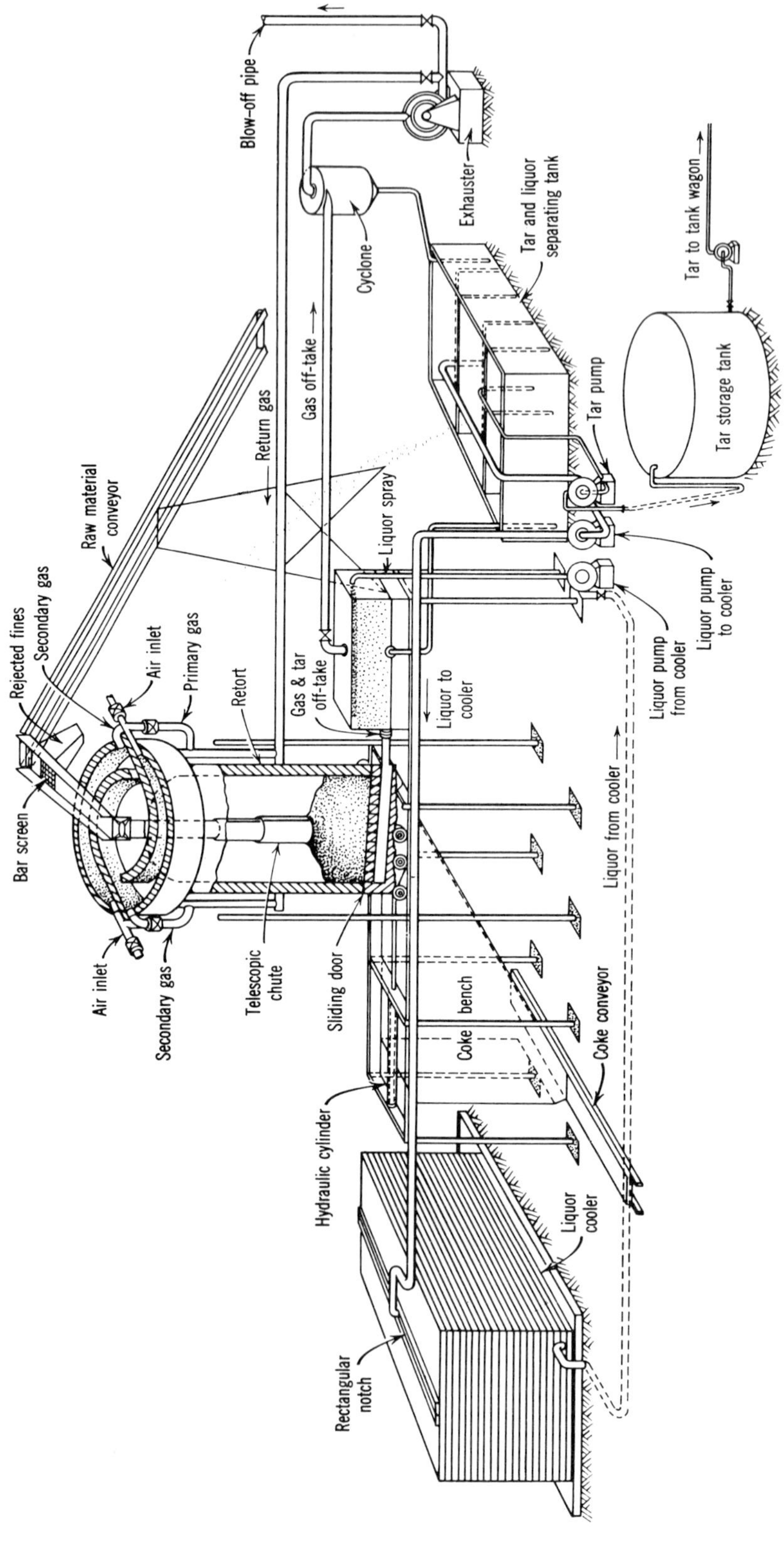

Fig. 6. Diagrammatic arrangement of a "Rexco" plant.[24]

approximately 55 tons of coal were the daily throughput. The general arrangement of a retort is shown in Fig. 6.

Average analyses when carbonizing a non-coking coal of medium quality, which were obtained during 11 years of operation, are shown in Table 2.

TABLE 2

REXCO PROCESS—ANALYSES OF COAL CHARGED AND CHAR PRODUCED[25]

	Coal	"Rexco" (Char)
Moisture, %	5.3	5.4
Volatile matter, %	34.6	6.6
Fixed Carbon, %	55.2	80.8
Ash, %	4.9	7.2

Yields of products per ton of the coal were: char, 1360 lb; tar oil, 19 gal; and gas, with a heating value of approximately 140 Btu/cu ft, 3 million Btu.[24,25]

The "Rexco" char was dense, hard, and slow-burning, with a high degree of reactivity. The results of a burning test made on Rexco by the Fuel Research Station of the Department of Scientific and Industrial Research are shown in Fig. 7. It indicates that the duration of intense fire with Rexco was more than twice that with the standard coke. Briquettes made from coal fines with a pitch binder could be carbonized in a Rexco retort with elimination of the smoke-producing quality of the briquettes.

[25] Private communication (1948).

A somewhat similar carbonization process was patented by Shaha.[26] It involved partial combustion of the coal at the bottom of a retort to furnish the hot gases for converting the upper portion of the charge into a soft coke.

The semicontinuous Horner process described by Churchill[26] bears an even closer resemblance to the Rexco process. Hot gases from a combustion chamber above the charge on a grate were drawn down through the coal while the charge was raised progressively. The finished semicoke was removed intermittently from the coking chamber by means of a horizontal slicer bar. When the entire charge was coked, the grate was lowered, a new charge added, and the cycle repeated.

Records "Coalene" retort.[5,27,28] In the Records "Coalene" retort, fine coal was coked by a batch process in an inclined retort by direct contact with steam at about 10-lb pressure and superheated to from 540 to 650°C. The coal charge which was admitted into the retort through an upper door was carbonized by direct contact with the hot steam, as well as indirectly by hot gases from a combustion chamber located below the retort. Before introduction into the retort, the steam was

[26] Shaha, A. K., Indian Pat. 36,609 (1949). Churchill, L. S., *Rept. Louisville Area Development Assoc.*, Louisville, Kentucky, September 1946, pp. 33–6.

[27] Price, J. D., Paper presented before Rocky Mountain Coal Mining Inst., June, 1956, 21 pp.

[28] Anon., *Chem. Eng. News*, **25**, 1052 (1947); *Ind. Eng. Chem.*, **39**, 12A, 14A (1947). Records, E. H., U.S. Pat. 2,385,731 (1945).

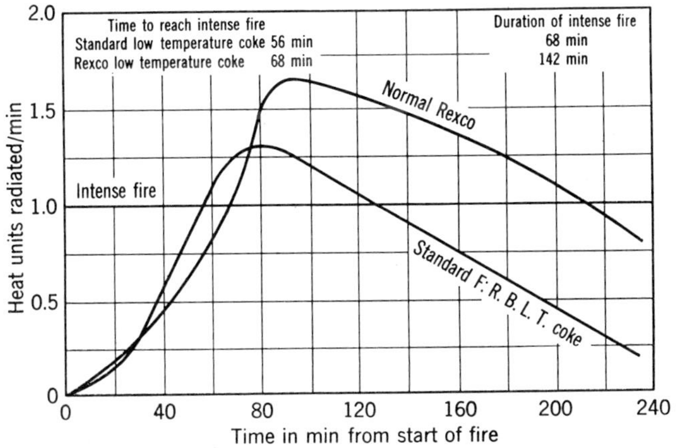

Fig. 7. Fuel Research Station tests on "Rexco" fuel. Duration of intense fire is twice as long as with a standard low-temperature char.[24]

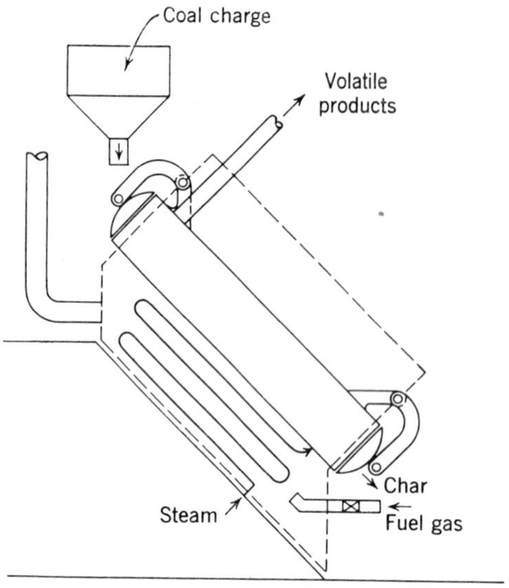

Fig. 8. Records low-temperature carbonization retort.[27]

superheated in a steam coil located in the combustion chamber. Steam and volatile products were withdrawn from the top of the retort and condensed. The char was discharged through the bottom door of the retort on completion of coking, and was quenched. Figure 8 shows a diagram of a retort.

A small, eight-retort plant, installed at Wellington, Utah during 1947, was operated for several years on Utah noncoking coals. This plant was designed to carbonize the coal charge in about 1.5 hr with a yield of from 1300 to 1600 lb of char and 40 gal of tar per ton of coal. Much of the gas was used for process heat.

Additional designs developed by the inventors included a special, vertical, cylindrical container into which the coal contained in a perforated basket was charged before the superheated steam was introduced.[29] Two of these retorts with a total capacity of from 15 to 20 tons of coal per day were installed in Tacoma, Washington. A third design was a vertical cylinder in which the coal was carbonized continuously in the same general way.[30]

Karrick retort. The Karrick process was basically much like the Records process. Superheated steam at about 650°C and 10-psi pressure was passed into the top of a vertical, cylindrical retort about 10 in. in diameter and 10 ft long. The volatile products were removed at the bottom. Both batch and continuous retorts were operated by the U.S. Bureau of Mines on a pilot scale, but the process never advanced beyond this stage. The results of the work indicated that steam is an expensive heating medium.[31]

[29] Anon., *Coke Smokeless-Fuel Age*, **4**, 248–51 (1942). Louttit, J. E., *Coal Age*, **42**, 155–7 (1937).

[30] Records, E. H., U.S. Pat. 2,397,432 (1946).

[31] Karrick, L. C., U.S. Pat. 1,894,691 (1933). Anon., *Business Week*, **38**, February 28, 1948.

Continuous Vertical Retorts

Continuous vertical retorts, as the name indicates, operate with coal charged continuously in at the top and char removed at the base. The height is sufficient so that carbonization is completed by the time the coal reaches the base. Some retorts are narrow chambers similar to the slot ovens, the majority are round or ellipsoidal shafts, usually a single shaft, but some are multiple or concentric shafts.

Continuous vertical retorts have been heated, either indirectly by hot combustion gases passing through flues in the retort walls, or directly by means of hot gases ascending through the coal charge. Narrow, slot-type ovens or thin layers of coal have been required where indirect heating alone was applied, but this method of heating has been supplemented in many retorts by introduction of hot gases or steam at the base of the retort. Introduction of a hot gas accelerates carbonization of the charge by carrying heat up uniformly through the mass of descending coal. When the coal is carbonized by direct contact with the hot gases, the cross section area of the retort may be made a great deal larger than would be possible if it were indirectly heated. The hot gases which are introduced for direct heating of the retorts include the products of combustion of fuel gas, steam, or product gas which is recirculated after preheating. Retorts using steam or product gas usually have been indirectly heated as well.

When hot gases are introduced directly, the volatile products of carbonization mix with this heating medium, and additional provision is required in the gas-processing equipment for cooling this much larger volume and treating it to separate the tar. If steam is used as the heating medium, the treatment is quite simple.

When the mixture of gases is cooled, the steam is condensed and may be separated readily from condensed tar by decantation. The presence of products of combustion in the final mixture of product gas from a retort obviously lowers its heating value materially, although its total heat content will not be altered. A lean gas is less desirable as a fuel than the higher heating-value gas which would be produced without the dilution from products of combustion.

The charge to retorts carbonized by direct contact with hot gases must be sufficiently open to permit the passage of these gases, and briquettes or a lump coal is required.

Continuous Vertical Retorts, Heated by Indirect Plus Direct Heat

Koppers continuous vertical ovens.[8] Vertical chamber ovens, heated by the combustion of fuel gas in flues, were developed by H. Koppers A.-G. in Germany for carbonizing nut coal at low temperatures by a continuous process. The coal to be carbonized was fed in at the top and cooled char removed at the base. Heat generated by burning gas in flues of silica brick was transferred to the charge by conduction through the hot brick walls. The difference from usual coke-oven practice consisted in the continuous movement of the coal downward through the retort, rather than its intermittent charging and horizontal pushing at relatively long time intervals. Furthermore, the use of a noncaking coal was necessary to avoid the provision of mechanical means for removing the char from the retort. The introduction of large quantities of steam into the continuous oven below the carbonizing zone was a third point of difference. The resulting water gas not only increased the gas yield, but, in conjunction with the undecomposed steam, served to augment the transfer of heat to the charge. The Koppers method, therefore, was a combination of external and internal heating.

Information is available on a plant that was under construction at the close of World War II at the Oheim mine of Berwerksverwaltung Oberschlesien, Katowice (Kattowitz).[8] This plant had a capacity for carbonizing 1,000,000 metric tons/yr of Upper Silesian, weakly caking, bituminous coal.

The arrangement of oven, flues, and regenerators (Figs. 9 and 10) was somewhat similar to that of conventional, Koppers high-temperature coke ovens, with the banks of flues separating the oven chambers.[32] The regenerators, however, were located beside the ovens, one above the other, instead of below them. Dimensions in feet of each oven at the Oheim plant were as follows: total chamber height, 37.7; length of chamber, 11.5; width of chamber at top, 0.8; width of chamber at a point 18.7 ft below the top, 1.1; and width of chamber at bottom, 1.1. The oven walls were lined with fire brick, but the heating flues were constructed of silica brick.

The fuel gas was burned in the flues, and the hot gases were passed alternately upward and downward through the flues and into the upper and lower regenerators. Automatic reversing equipment controlled the direction of flow of the gases. To maintain a uniform temperature over the entire height of an oven, additional heat could be obtained by use of auxiliary burners. The heat from these burners was directed into either the upper or lower regenerator as needed, thus raising its temperature and that of the air it preheated during the next reversal. Product gas from the ovens or producer gas generated from a portion of the char was used for underfiring the ovens.

Above each unit of fifteen ovens was a coal bunker from which coal was dropped into the feed hopper on the top of each oven chamber. From the feed hopper, the coal was fed continuously into the carbonizing chamber. The char was discharged into a bottom hopper, from which it was withdrawn periodically into a tilting cart for transportation to a special quenching station.

To precool the char entering the discharge chamber, a portion of the product gas, free of tar and light oils, was returned to this chamber at a point just above the char-discharge outlet. At the same time, water injected into the char was converted into steam, which mixed with the recirculated gas. In its passage through the char in the chamber, the steam formed water gas which increased the volume of low-temperature carbonization gas.

The gases flowed from each oven through two standpipes into the collector main and were conducted on through indirect coolers and then through a mechanical or electrical tar separator. The free ammonia was removed in the final cooler, from whence the gas passed to

[32] Reid, W. T., *U.S. Bur. Mines, Inform. Circ.* No. 7430 (1948), 82 pp.

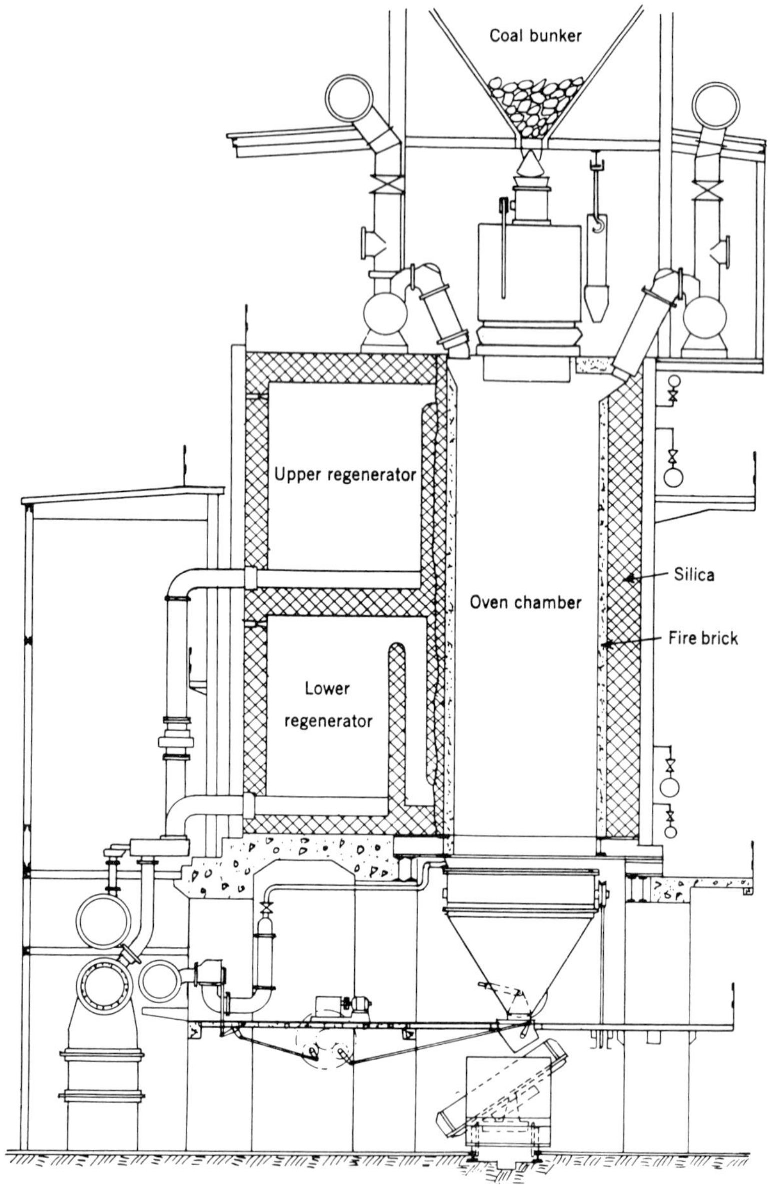

Fig. 9. Diagrammatic drawing of a Koppers continuous, vertical, steaming oven for bituminous coal.[8]

the light-oil absorption equipment, and then to a gas holder.

When carbonizing Upper Silesian, high-volatile, bituminous, nut coal containing 5% moisture and 7% ash to produce char with 3% volatile matter, results were as follows: coal throughput per oven per day, 25 metric tons; char yield (3% moisture) 64%; low-temperature tar yield, 9.2%; low-temperature light-oil yield, 0.8%; heat consumption per pound of coal, 990 Btu; gas yield per short ton (2000 lb) of coal, 10,330 standard cu ft, with a gross heating value per standard cubic foot of 470 Btu (equivalent to 2250 Btu per pound of coal). The combined tar and light-oil yield amounted to 87% of that found by the Fischer assay.

Prior to the installation at the Oheim mine, several plants of this type had been installed for other services, including calcining petroleum coke, carbonizing ovoid briquettes or weakly caking high-volatile bituminous coal at 700°C, and production of mixed gas (distillation gas plus water gas) from non-caking, high-volatile, bituminous, nut coal.

Ceramic refractories were considered to have an advantage over iron as a construction material, because of the wider range in temperatures that was possible. The ceramic ovens were not limited to a maximum temperature of about 600°C, the maximum for iron. It was also claimed that recirculation of the gases raised the throughput of the ovens 50% above that of ovens that had no gas recirculation.

Several installations of the continuous Koppers ovens were utilized in Japan during World War II, and Reid[32] has reported on their performance. The temperature in the vertical flues ranged from about 1000°C at the bottom to 850°C at the top, with 800°C as the estimated maximum temperature of the charge, which would classify the Koppers process as "medium-temperature" carbonization. The large volume of steam which was added below the carbonizing zone was the only gas introduced into the retort. Because contact with the heated walls tended to decompose tar vapors, the yield of tar in the Koppers ovens would be lower than with internally heated processes carbonizing at the same temperature, but by direct contact with the hot gases alone. These ovens produced a high yield of gas, which was usually employed as synthesis gas, the flues being heated with producer gas generated from a part of the char.

At the plant at Ube, Yamaguchi-ken, the Koppers ovens were arranged in two banks, each bank consisting of thirty-six ovens divided into six batteries of six ovens each. A battery was 26 ft long and was separated from the adjoining battery by a narrow, air-cooled chamber. The overall width of a bank was 25 ft, including the regenerators, and the height of the carbonizing zone was 31 ft. Each oven was 12 ft 9 in. long, and tapered in width from $10\frac{5}{8}$ in. at the top to $11\frac{3}{4}$ in. at the bottom, to facilitate descent of the charge. The volume of each oven was approximately 350 cu ft, so that about 0.001 ton of coal was carbonized per cubic foot of oven volume per hour. The daily rated capacity of the seventy-two ovens was 600 tons of coal, and 21 hr were required for passage of the coal through an oven. The regenerators were of conventional design and were reversed on a 30-min cycle. When rich coke-oven gas was burned in the oven flues, only the air was preheated; but with the leaner

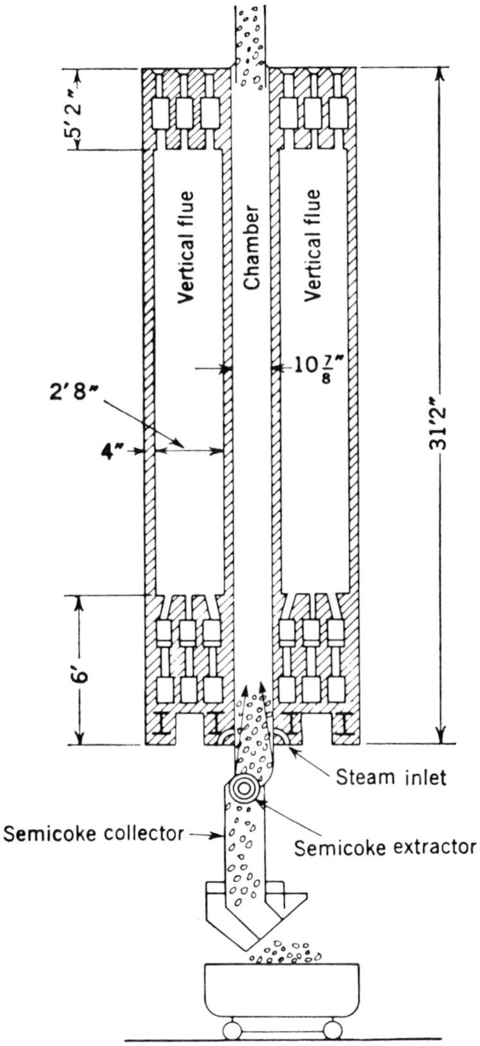

Fig. 10. Cross section through Koppers continuous, vertical, steaming coke oven.[32]

producer gas, the air and gas passed through separate passages of the regenerators.

Steam at 300°C was introduced at the bottom of the carbonizing zone. It amounted to about 360 tons/day, and about 144 tons were converted into water gas at the expense of about 175 tons of char containing an estimated 55% C. The rate of heat transfer from the hot refractory walls to the charge of coal was increased by the addition of the steam, the quantity of product gas was increased, and the retort temperature was decreased to from 700 to 850°C, although the temperature in the flues might be as high as 1000°C. About 800 to

1200 Btu per pound of coal was required for carbonization and the water-gas reaction.

The tar-recovery and byproduct-processing systems were similar to those used in coke-oven practice. Dust in the offtake gas was not particularly troublesome and was removed along with the tar by passing the gas through the usual coolers and centrifugal tar separators.

The formation of clinker in the lower sections of the ovens, and the development of cracks in the walls, permitting leakage of air into the oven chambers and the combustion of part of the char, were the two principal difficulties encountered in operating the plant. The clinker gradually accumulated on the face of the refractory, interfering with heat transfer and allowing its temperature to increase to 900 to 1000°C. The masses of clinker eventually became so thick that daily poking was required to maintain a continuous downward flow of char. Another difficulty was the clogging of the regenerators with dust, which was entrained in the dirty producer gas, so that cleaning at 3-year intervals became necessary. The presence of 660 grains of hydrogen sulfide per 100 cu ft of product gas caused severe corrosion in the gas offtake and main gas lines. In operation, it was possible to maintain the rated capacity of the retorts, approximately 600 tons of coal per day, only for short periods. At an average rate of about 480 tons of coal per day, continuous operation of the plant was possible.

Didier-Werke retorts.[8,33,34] Continuous, ceramic, vertical-chamber ovens were used for the low-temperature carbonization of noncaking coal in a commercial plant erected at Didier-Werke in Berlin. It consisted of twenty-four chambers and was designed to carbonize 300,000 metric tons of noncaking coal per year. The ovens resembled Woodall-Duckham, continuous vertical retorts, although the heating arrangement was modified to adapt them to low-temperature carbonization. Recirculated, tar-free, product gas was introduced into each oven above the char extractor. By recirculating a larger or smaller volume of this gas, the velocities of gases and vapors expelled from the coal could be regulated to control within wide limits the length of time during which the vapors and gases remained inside the chamber, and thus to adjust tar quality and yields. Sections through Didier retorts are shown in Fig. 11.

This oven was not considered satisfactory for a caking coal. To carbonize coal evenly within a reasonable time, a narrow oven chamber was required—about 8 in. in width. Because the contraction of low-temperature char is less than that of high-temperature coke, the friction between the wall faces and the surface of the char mass formed from a caking coal would be too great to be overcome by the weight of the charge. The low-temperature carbonization step was only one part of a three-stage process which yielded briquetted chars. The process comprised:

1. Carbonization of a noncaking coal of less than 2-in. size at 550 to 600°C to remove the tar in the continuous, vertical chamber ovens;
2. crushing and briquetting the crushed char in a roll press, after addition of 5 to 8% of pitch (produced by distillation of the tar) as binder, together with 10 to 15% of a coking coal; and
3. carbonization of the briquettes in either continuous or intermittent vertical chamber ovens at 900 to 1000°C. These second higher-temperature ovens were similar to those normally used in European gasworks practice, except for recirculation of clean make-gas up through the charge. Briquettes suitable for domestic or metallurgical use could be produced, depending on the conditions of operation.

In blast-furnace trials, briquettes produced from lignite chars were found to be entirely suitable as fuel.

Rochdale process.[35] The use of heated gases to carry the heat rapidly through the charge was the basis of the Rochdale process, which was developed by the British North Western Gas Board. By introduction of the gas, the coal throughput of gas-making, continuous, vertical retorts, which would otherwise be reduced, was reported to have been increased by as much as 100% above the nominal figure for which the retorts were originally designed. In

[33] Thau, A., *Coke and Gas*, **10**, 397–402 (1948). Anon., *Coke and Gas*, **18**, 347–52 (1956).

[34] Reed, F. H., *U.S. Bur. Mines, Inform. Circ.* No. 7462 (1948), 74 pp.

[35] Nicklin, T., and Redman, M., *Gas World*, **139**, 784–7, 854–62, 924–7, 998–1004 (1954). Anon., *Coke and Gas*, **18**, 375–9 (1956). Biddulph-Smith, T., *Coke Oven Managers' Year Book*, **1957**, 412–40. Nicklin, T., *Brit. Chem. Eng.*, **2**, 208–9 (1957).

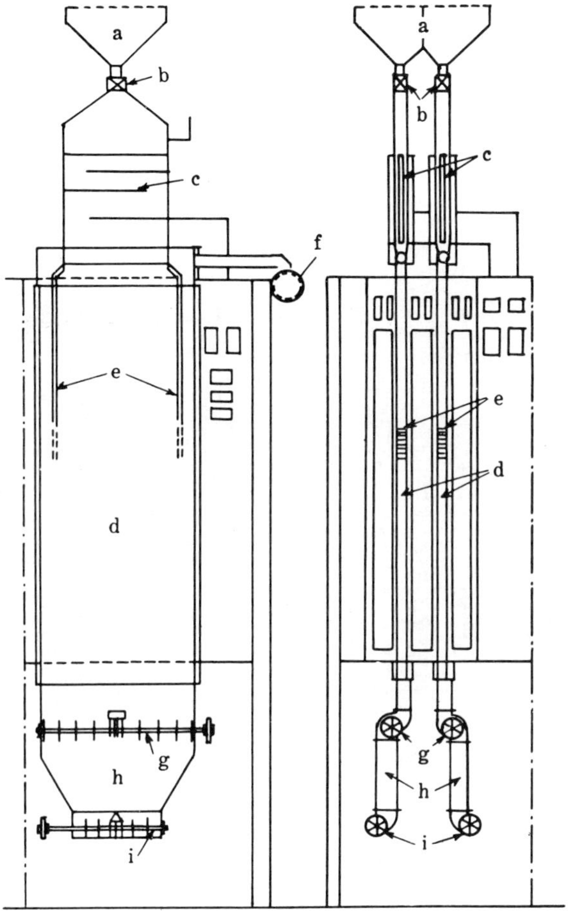

a. Coal bunker d. Carbonization chamber g. Coke extractor
b. Cell valve e. Baffle wells h. Coke–cooling chamber
c. Predrier f. Collecting main i. Coke–discharge valve

Fig. 11. Didier-Werke uniflow chamber ovens in two vertical sections.[34]

normal operation of continuous, vertical retorts, steam is admitted at the base to reclaim the heat from the hot coke before it is discharged. The steam then reacts with the coke higher up in the retort to produce carbon monoxide and hydrogen, which combines with the gases evolved from the coal and increases the volume of gas produced. In the Rochdale process, the steam was replaced by gas, either blue water-gas or producer gas, both of which have been successfully applied. Because these gases do not react with the char to any extent, all the available heat from the char is transferred to the coal charge for carbonization.

The Rochdale process was used to produce a highly reactive smokeless fuel for domestic grates. A coal of low or moderate swelling index in small nut sizes was carbonized in continuous vertical retorts at temperatures below the normal temperatures of gas-making retorts. Into the base of the retorts, blue water-gas was injected in amounts of 18,000 to 18,500 cu ft per ton of coal carbonized. Temperatures in the combustion chambers were usually maintained around 890°C, which was probably equivalent to a 700°C maximum temperature in the charge. Although carbonized at a low temperature, the volatile matter of the char was low, about 3 to 5%. The heating value was about the same as that of normal coke, but the bulk density was lower, about 17 lb/cu ft, instead of the 19 lb/cu ft with ordinary gas coke. The char ignited easily and burned completely in the open grate. In

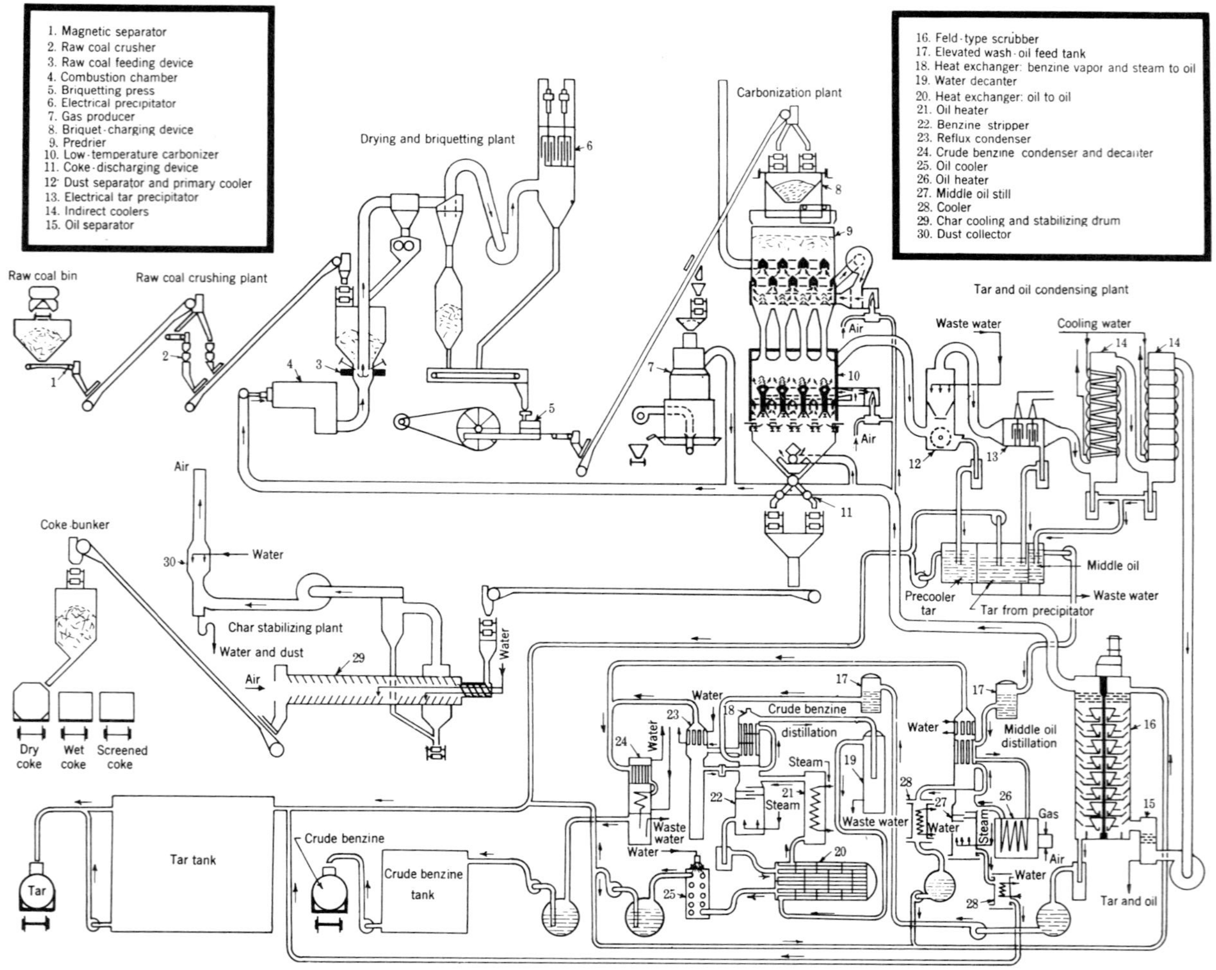

Fig. 12. Lurgi-Spülgas low-temperature carbonization plant for the production of char, tar, and benzine at Offleben, Germany.[8]

appearance it was dull, without the silvery sheen of gas coke. The price of the low-temperature char was expected to command a premium price.

Continuous Vertical Retorts; Directly Heated Only

Lurgi-Spülgas retort.[8,9,11,14] In the Lurgi-Spülgas retort, coal, usually in the form of nuts or briquettes, is dried and then carbonized as it continuously descends through the retort in direct countercurrent contact with ascending heated combustion gases. A retort is divided into two chambers, an upper chamber for drying, and a lower one in which the coal is carbonized by contact with gases at about 600 to 700°C. These two chambers are connected by several vertical pipes, through which the coal descends from the upper chamber to the lower. The hot gases used for drying and for carbonizing are produced by burning gas in two separate external combustion chambers, and the products of combustion from these chambers are introduced into the dryer and the carbonizer, respectively. The hot char is finally cooled by direct contact with cool product gas before it is withdrawn at the bottom of the carbonizer. Uniform distribution of incoming coal is necessary, and a distributing mechanism is provided for this purpose at the top of the retort. Many Lurgi-Spülgas plants were built in Germany shortly before World War II to obtain from brown coal the tar from which motor fuel could be made.

A flow sheet of a Lurgi-Spülgas plant at Offleben in Germany is shown in Fig. 12. In this plant, raw brown coal, containing about 54% moisture, was transported from the open mine to a storage and feed bunker, and from thence to a crushing plant.[8,12] There, by means of roller screens and "Titan" breakers, the coal was reduced in size to 0 by 20 mm. Before raw coal was introduced into the Lurgi-Spülgas retorts, it was dried in a separate dryer to a moisture content of about 15% by contact with combustion gases produced by burning a portion of the product gas. The dried coal was then briquetted in a press without the use of a binder, and the briquettes were elevated to the top of the low-temperature retorts, of which there were 10, each with a capacity of 300 tons of briquettes per day, and dropped into the upper drying section.

In the top zone of the Lurgi carbonizer, the briquettes were dried to 0.5% moisture or less, with the drying gases entering at about 220°C and leaving at 60 to 80°C. As drying of the briquettes progressed, they disintegrated into dry granular pieces which descended by gravity through the large down-pipes into the lower section, which was about 10 ft in diameter and 20 ft high. Here carbonization took place in the presence of oxygen-free combustion gases having a temperature of 600 to 700°C. The maximum temperature reached by the briquettes was 500°C. Granular char was discharged from the carbonizer through a lock, and cooling was completed by contact with water and air in a revolving cylinder before transporting it to the loading bunkers.

The mixture of combustion gases and volatile products leaving the carbonizer passed first through a scrubber, in which dust and heavy tar were removed, and then through an electrical precipitator to separate light tar. The tar-free gases from the electrical precipitator were cooled in indirect coolers to condense middle oil and were scrubbed with an absorbent oil to recover a light oil. The heating gases from the two chambers were kept separate and were recirculated back to the combustion chambers. In one plant, sulfur was removed from the hot gases circulated through the dryer. Yields of products expected per ton of dried briquettes prepared from a normal brown coal were: char, 45% by weight; tar, 12.5% by weight; and gas, 4600 cu ft with a heating value of 225 Btu/cu ft.

The char from the retorts contained from 10 to 12% volatile matter and up to 20% ash. It was very reactive, and at the Böhlen plant was cooled in nitrogen to avoid ignition. The size consist of the char product varied with the briquetting method used. Char from ordinary plunger presses usually contained about 20% breeze, 50% 6 to 20 mm size, and 30% over 20 mm. Char from briquettes produced in ring roll presses contained only about 5 to 8% below 6 mm in size. The char was screened into suitable sizes for industrial and domestic purposes. The fine material was used in boiler plants and in a gasification process, the middle size mainly in gas producers, and the large size as a domestic fuel and in central heating plants. This char was reported to be extremely reactive after aging, so that it required special precautions to avoid spontaneous ignition and explosions.

A Lurgi-Spülgas carbonizer was installed at Lehigh in Stark County, North Dakota, for carbonizing lignite which was mined there, and data on its operation have been published.[36] As mined, this lignite contains about 40.3% moisture, 6.7% ash, and 0.78% sulfur.

The lignite was first crushed, and the $+\frac{5}{8}$-in. size was dried in the drying section of the Lurgi unit by direct contact with the products of combustion from burning a portion of the gas produced in the carbonizer. Although the temperature of these combustion products was approximately 250°C, the temperature of the lignite was much lower—about the boiling point of water—because of its high-moisture content. Under the conditions in the dryer, the lignite disintegrated, and the material descending to the carbonizer ranged from about 0.25 to 0.75 in. in size. The dried lignite gravitated through eight pipes, each about 16 in. in diameter and 5 ft long, which were arranged to provide a seal against the upward flow of volatile products. Upward leakage was hindered by maintaining the pressure of the gas in the carbonizer at slightly less than that in the dryer.

Products of combustion at about 700°C from burning additional fuel gas were circulated through the carbonizer to heat the lignite in its descent to about 590°C and to effect carbonization. In the lower section of the carbonizer, before the char was discharged, it was cooled by recirculating cool product gas through it. Volatile products were withdrawn from the top of the carbonizing chamber and were cooled and condensed for recovery of the tar, which was distilled into creosote and a pitch binder.

The finished char was briquetted with pitch, from the tar produced, or with asphalt. About 100 tons of briquettes were obtained from 250 tons of raw lignite. The briquettes were of the pillow type, $2\frac{1}{8}$ in. square by $1\frac{5}{16}$ in. thick, with a weight of 3.5 oz. Until the briquettes had cooled and hardened, they were fragile and required gentle handling. Although the char itself was smokeless, the binder caused a little smoke when the briquettes were burned.

The Lurgi-Spülgas system has also been used successfully in several Japanese plants, particularly as a source of liquid fuels during World War II.[32] At the Wakamatsu Works of Nissan Ekitai Neuryo K.K., four carbonizers were installed, each with a design capacity for carbonizing 300 tons of Takamatsu coal per day. On the basis of the heat in the coal charged to the carbonizer, 65.6% appeared as heat of combustion in the char, 16.1% in the liquid products, and 5.1% in the excess gas, giving a total thermal efficiency of 86.8%. The heat losses—13.2%—included 0.5% in the waste gas, 0.2% as sensible heat in the char, 8.5% as radiation and other unmeasured losses, and 4% as sensible heat in the gas leaving the carbonizer.

According to Reid,[32] the plant was shut down for inspection after the first 13 months of continuous operation, and the internal condition of the carbonizer and the various units in the byproduct-recovery system were observed. Severe spalling of the brick lining in the upper part of the dryer section had occurred, probably because of attack on the mortar by sulfurous and sulfuric acids. Moisture had penetrated and caused swelling, with the result that the bricks originally 4.53 in. thick were reduced by spalling to only 2.56 in. In the lower part of the dryer, temperatures were higher and no such damage had occurred. The inside of the steel shell of the dryer was badly corroded, and the cast-iron main connecting with the blower for circulating gas also was seriously corroded. The steel duct making up the gas circulation system was perforated and required replacement. Although sulfur has usually been removed from the gas used for heating in the dryer zone, this precaution had not been taken with this unit, and the only protection for the steel work was the application of an acid-resisting paint. The paint film was inadequate and corrosion had occurred.

Since start of operation in 1941, this plant carbonized 626,240 metric tons of Takamatsu coal to produce 396,529 tons of char, and 16,872,000 gal of total liquid fuels—divided into 1,977,000 gal motor gasoline, 1,243,000 gal semidiesel oil, and 13,652,000 gal diesel fuel oil. These yields represent production by three retorts only, because one retort was kept idle as a spare. The coal may be classed as a high-volatile *C* bituminous coal. The caking tendency was slight, only 1.5 to 2.0, as measured by the Campredon-Gray method. The output of the carbonizer was determined by the caking power of the coal used, because this property affected the pressure drop through

[36] Hall, R. D., *Coal Age*, **45**, No. 8, 47–8 (1940).

the carbonization zone and consequently the mass flow of hot gases. Strongly caking coals could not be used because of the formation of lumps of char which interfered with the flow of the charge by gravity through the carbonizer and upset uniform gas flow conditions. Non-caking coals could not be used either with complete satisfaction because of size degradation, the excessive volume of fines blocking the flow of carbonizing gases. For instance, tests made with coal from the Ube field, having a caking index of zero, showed a pressure drop through the carbonizer of 3.15-in. water gauge, and the maximum input obtainable was 250 tons of coal per day. With Takamatsu coal having a caking index of 1.5 to 2.0, the pressure drop was 1.57 to 1.97 in. water gauge, and a throughput of 350 tons per day could be maintained. The char had adequate strength for the uses to which it was put, principally as a fuel for portable gas producers on automotive vehicles. Its high reactivity made it a fair substitute for charcoal in this application.

Joh[37] discussed the manufacture in Lurgi carbonizers of char for addition to the coal mix charged into coke ovens. Char containing from 18 to 20% volatile matter was the most suitable as a constituent in blends to be used for blast-furnace coke, and was produced in a Lurgi carbonizer operating at 500°C.

Production of char in two Lurgi retorts in New Zealand has been described by Jones.[38] Each of these retorts, which were erected in 1931, had a daily normal throughput of 140 tons of the fine sizes of Waikato coal, a subbituminous coal of early Tertiary age. The char was screened and the $+\frac{3}{8}$ in. size was sold as fuel for slow-combustion stoves. The smaller sizes were briquetted to form $2\frac{1}{2}$ oz ovoids, using either coal-tar pitch or bitumen as binder. These were popular for use as a domestic fuel and in bakers' ovens. A portion, to which 5 to 10% coking coal was added, was used for locomotive fuel. The briquettes possessed a high reactivity and retained their shape in the fire. After being properly ignited, they were hard and coherent. In the briquetting operation, the char, usually at about 80°C was sprayed with the binder, and the mixture was then heated in a fluxer with saturated steam and sent to the briquetting press. The briquettes from the press were cooled by water sprays for 3 min on a wire-mesh belt, and cooling was completed by the briquettes standing overnight in bins.

The time of passage of the coal through the retort was approximately 20 hr, including 5 hr in the dryer and 5 hr in the carbonizer. The char was discharged from the retort at approximately 200°C onto a rubber belt on which it was cooled with water sprays. The amount of fines ($-\frac{1}{8}$-in.) in the coal exerted an important influence on the operation of the retorts because of its effect on the flow of hot gas. The most important factor in carbonization was the volume of hot gas circulated through the carbonizer, so long as its temperature was above 600°C, because without sufficient gas the coal could not be carbonized. Moreover, too large a proportion of fines limited the cooling of the char in the lowest section of the retort. With 40% $-\frac{1}{8}$-in. coal in the feed, the maximum throughput of a retort was only 118 tons of coal per day, but this increased to 151 tons when the proportion of fines dropped to 20%.

All possible precautions were required in the design of the coal-handling and charging equipment in order to maintain a uniform mixture of fines and the coarser particles of the coal. When separation occurred, the hot gases followed the path of least resistance, and the finer sizes of coal were imperfectly carbonized. Even with a uniform mix, differences in the volatile contents of the different sizes of char occurred. In one instance an increase from 7.9% in the $+\frac{1}{4}$-in. size to 10.2% in the $-\frac{1}{16}$-in. material was observed.

Retorts required periodic shutdowns for overhauling and cleaning. With ample coal and labor, each retort was expected to average about $9\frac{1}{2}$-mon operation each year. Regular cleaning of the different parts of the retorts was especially important when carbonizing coal containing a large proportion of fines, and about 5 hr a week were lost in cleaning. After a 10-mon continuous campaign, the thorough cleaning of a shutdown retort required four men for about a month. From 9 to 10 days were consumed in shutting down a Lurgi retort, because gradual cooling by recirculating an inert gas was necessary. Starting up took 6 to 9 days. After a temporary shutdown due to mechanical troubles, the retort

[37] Joh, H., *J. Mines, Metals and Fuels* (*India*), Special Issue, **7**, 208–9 (1959).

[38] Jones, A. B., *Coke and Gas*, **12**, 287–90, 299, 331–4, 337 (1950).

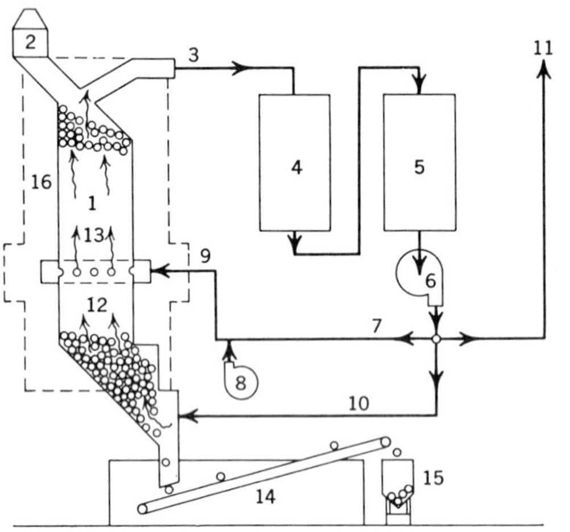

1. Retort
2. Gas-tight feeder
3. Foul gas
4. Primary cooler
5. Cottrell precipitator
6. Exhauster
7. Clean gas
8. Air
9. Combustion mixture
10. Quench or cooling gas
11. Surplus gas
12. Tuyeres
13. Heated gases, oxygen-free
14. Water-sealed discharge
15. Carbonized product
16. Insulation

Fig. 13. Flow sheet for the National Fuels Process.[41]

could be restarted immediately, if the shutdown had lasted for only about 12 hr or less. With shutdowns lasting for more than 16 hr, additional gas had to be provided to reheat the retort.

Although Jones found that a Lurgi-Spülgas retort was limited to noncaking coals, such as subbituminous or brown coals,[38] the addition of a small proportion of coking coal to the subbituminous coal was suggested as a means for increasing the size of the char. Several experiments in the New Zealand plant yielded char in larger pieces, although it was less strong than the carbonized coal briquettes.

In comparative tests on the low-temperature carbonization of some poorly coking to noncoking Indian (Raniganj) coals, the yields of products obtained by the Gray-King assay, the Fischer assay, and in a Lurgi-Spülgas plant, with a capacity of 25 tons of coal per day, were measured by Das Gupta and coworkers.[39] In the assays, the coke yield ranged from about 1250 to 1500 lb per ton of dry coal, and the tar from 16 to 31 gal. From the Lurgi-Spülgas retort, when operated at 650 to 750°C, the yield of tar was about 70 to 80% of that produced in the Fischer assay at 550°C. The char contained 5 to 7% volatile matter.

The Kollergas process resembled the Lurgi-Spülgas process. Hard coal was carbonized continuously by recirculating hot gas produced in the process up through the charge, according to Thau.[11,40] The char in the lower part of the retort was cooled before discharge by means of additional clean product gas. In a plant installed in Germany during 1938, a nut coal of 10 to 20- or 20 to 40-mm size was carbonized to produce 71 to 74% char and 6 to 9% tar. The yield of light oil was 0.9 to 1.2% of the coal. The char contained 6 to 7% volatile matter.[11]

National Fuels process.[5,13,41,42,43] The process of the National Fuels Corporation was adapted to the production of char or coke in pieces of uniform size from either coking or noncoking coals. Before carbonization a coking coal usually required an oxidation step in which the coal was brought in contact with an oxygen-containing gas at an elevated temperature to secure a uniform and acceptable coking index. The oxidized coal was then briquetted with tar or pitch, usually that produced in the process, and the briquettes were coked. The process essentially comprised the five steps: (1) crushing or grinding the coal to pass a 30-mesh or finer screen, and drying to less than 2% moisture; (2) oxidizing, where necessary; (3) briquetting in a roll press, after fluxing with a suitable binder; (4) carbonizing, by direct contact with hot product gas; and (5) recovery and treatment of the volatile products.

Figure 13 is a flowsheet of the carbonization part of the process. The carbonizing operation was so designed that the carbonized briquettes were continuously dry cooled as they descended below the maximum temperature zone in the retort, with recovery of the bulk of the

[39] Das Gupta, N. N., Moitra, A. K., Biswas, S. C., Chakravarty, H. C., and Sen Gupta, D., *J. Mines, Metals and Fuels (India)*, Special Issue, **7**, 228–31, 250 (1959).

[40] Thau, A., *Z. Ver. deut. Ing.*, **83**, 1277–82 (1939).

[41] Hamilton, J. H., and Wolf, C. S., *Ind. Eng. Chem.*, **41**, 556–63 (1949).

[42] Kruppa, W. J., U.S. Pat. 2,336,151 (1943); Canadian Pat.410,958 (1943). Zinn, R. E., *Chem. Eng. Progr.*, **50**, 3–7 (1954). Kruppa, W. J., Barclay, K. M., and Nobles, E. J., Brit. Pat. 753,084 (1956). Kruppa, W. J., and Rinckhoff, J. B., U.S. Pat. 2,815,316 (1957); Brit. Pat. 788,446 (1958).

[43] Berry, G. A., U.S. Pat. 2,131,702 (1938).

process heat. Cooling gas entered the bottom of the retort and, while ascending, simultaneously cooled the briquettes and recovered heat that was utilized higher up in the retort. Heated gas was introduced to the retort through tuyeres located at a level about one-third the height up from the bottom. This gas, heated to the desired carbonizing temperature in an external heat exchanger, supplied the process "make-up" heat. The quench gas rising from the bottom attained a temperature approximating the carbonizing temperature by the time it had reached the level of the tuyere gas inlet. From this level, the combined gases rose through the incoming charge, giving up much of their heat and leaving relatively cool from the top of the retort. As a consequence of the sweeping action of this progressively cooler gas, the volatile matter was removed with a minimum of secondary thermal decomposition. The combined gases were passed through a washer-cooler, followed by a Cottrell precipitator, and then a portion was reheated and recirculated to the retort. The balance, equivalent to the product gas, was continuously withdrawn from the circuit, subjected to further purification, and used as desired.

Preoxidation of the coal was a vital part of the process and made it possible to carbonize highly swelling coals by controlling the coking properties of the coal at a desirable level. Unless the coal was low in coking power and practically nonswelling, serious trouble from agglomeration occurred during carbonization. In the retort during carbonization, the briquettes were heated at the rate of 2 to 3°C/min while passing through the critical temperature range of approximately 350 to 500°C in which they softened. For an underoxidized coal, the particles oversoftened and swelled. The strength of such briquettes was insufficient to withstand the overhead load, and both deformation and sticking occurred.

For an overoxidized coal, on the other hand, the coking properties were almost completely destroyed. No agglomeration occurred within the briquettes and a product was obtained that was weak and impossible to handle without breakage. Although coals of a wide range of coking index from very low to very high are amenable to treatment by the N.F.C. process, in general, at least some coking property has been required. However, experience indicated that it could be so little as to make the coal wholly unacceptable in conventional coking practice.

The briquettes were formed under high pressure in the press, and the binder was required to give them sufficient strength to withstand charging into the retort and to sustain the overhead load until carbonization was complete. Dissolving of the coal by the binder at any point during passage of the briquettes through the critical temperature range in the retort was undesirable. The tar from the process itself made a good binder, although other coal tars or pitches, as well as molasses and sulfite lye, have been used instead.

The carbonizing temperature ranged from about 450°C to the temperatures of high-temperature coking, depending on the product desired. Use of the lowest temperature has produced a highly reactive fuel, attractive for domestic consumption—the highest being a coke of less than 1% volatile matter, suitable for metallurgical use. The carbonized briquettes were of uniform size, with a weight of about 42 to 47 lb/cu ft, and an apparent specific gravity of about 1.25. The briquettes were free-burning and highly reactive. Long periods of storage in the open showed no appreciable deterioration.

The N.F.C. process was developed in the 1920's by the National Fuels Corporation, then a subsidiary of Calco Chemical Co., Inc. It was tried out in a 5- to 10-ton/day pilot plant located at New Haven, Connecticut, and later in a 100-ton/day plant located in Bethlehem, Pennsylvania. In addition, study was devoted to the N.F.C. process in Utah in connection with the extensive research program sponsored by the Utah Conservation and Research Foundation.[5,44]

Wolf modified the coal tar or pitch binders for use in the N.F.C. process by addition of sulfuric acid and combining 5 to 10% of the mixture with the oxidized coal.[45] In another patent, he treated the coal with about 1 to 2% of an inorganic oxidizing acid at a temperature of 90 to 250°C for at least 5 min to reduce the swelling and caking properties of the coal.[46] Sulfuric acid was suitable and could be added in combination with oxidizing salts.

[44] Institute of Gas Technology, *Gas Making Processes*, Am. Gas Assoc., 1945, pp. 95–8.

[45] Wolf, C. S., U.S. Pat. 2,314,641 (1943).

[46] Wolf, C. S., U.S. Pat. 2,336,154 (1943); Canadian Pat. 409,133 (1942).

The North Western Gas Board proposed the preparation of briquettes by a process similar to the N.F.C. process.[35] Noncoking slack was blended approximately 90% with a higher rank coal and the mixture was then briquetted with pitch. To prevent swelling of the briquettes during carbonization with loss of shape, preoxidation of the coal was necessary, and the addition of an inorganic oxidant to the coal before briquetting was considered in an effort to reduce the cost of oxidation below that of the conventional air-oxidation step. Oxygen should be evolved by the oxidant during or just before the plastic stage is reached, about 400 to 430°C.

In carbonizing 2-in. diameter briquettes made from Yallourn brown coal in Australia, trouble was experienced because of their cracking, even at a low heating rate.[47] It was found that when a briquette was carbonized by indirect heating through the wall of the retort, the temperature of one side was raised faster than was that of the other side—a condition which increased the tendency to crack. However, by heating the retort internally by the circulation through it of gaseous products of combustion, it was possible to dry and carbonize briquettes at temperatures up to 700°C or higher—even to 900°C—without cracking, owing to the more uniform temperature conditions to which they were exposed. At higher temperatures, reaction of carbon dioxide and water vapor with the briquettes occurred.

To produce crack-free briquettes from the brown-coal char, the heating rate should be slow and carefully controlled through both the drying and distillation stages up to 600°C. Drying required not less than 5 hr, and the temperature during distillation should not be increased faster than 100°C/hr, and possibly more slowly in the case of large briquettes. Above 600°C, more rapid heating was possible. By observing these conditions, briquettes in sizes to 5 in. by $2\frac{1}{2}$ in by $1\frac{3}{4}$ in. were successfully carbonized without formation of serious major cracks and fissures.

Another treatment proposed for use in the preparation of briquettes from the Yallourn coal was to preshrink the coal particles completely before briquetting and carbonizing. This was done by precarbonizing the coal to 800°C, briquetting, and recarbonizing the briquettes at 500°C. By the preliminary carbonization to a temperature in the range 600 to 800°C, the char was adequately preshrunk, and the particles did not pull away from the pitch binder during the final carbonization. The objective in the work was the preparation of fuels suitable for foundries and other metallurgical uses.

Pintsch-Weber process.[8] A carbonization process for the brown-coal briquettes or bituminous, noncaking, lump coal (free from a large proportion of fines) being fed to the chain grate stokers of a boiler installation, was developed by Julius Pintsch Kommandit Gesellschaft in Berlin before World War II. The coal was carbonized continuously as it descended through a charging hopper located above the grate by drawing through it hot combustion gases. Hot char was delivered from the bottom of the hopper onto the grate, and the gases withdrawn from the top of the hopper were returned and burned under the boiler after the tar which they contained had been recovered. The process was of such interest that the German technical-military authorities during the war attempted to have it installed on all boilers fired with bituminous coal or brown-coal briquettes to recover tar for use as fuel oil.

In operation, a coal containing a large proportion of fines was briquetted with a binder in a roll press. The briquettes were then conveyed by a slow-moving elevator provided with broad, perforated buckets to the feed bunker of the stoker. Hot gases introduced at the bottom of the elevator housing passed up around the briquettes and hardened them as they ascended to the bunker. From the bunker, the briquettes slid down into the carbonizing shaft and were converted to char before discharge onto the chain grate.

Pieters described[48] a unit which consisted of a multiplicity of internally heated vertical retorts grouped within a casing and operated with continuous production of char.[49] The coal was first dried and then carbonized by contact with superheated steam introduced at the bottom of each retort. In the lower part of the retort with increasing temperature the coal was reported to be desulfurized by the action

[47] Anon., *Coke and Gas*, **18**, 347–52 (1956). Kennedy, G. L., and Evans, O. G., *J. Inst. Fuel*, **31**, 242–7 (1958). Megler, V. R., and Kennedy, G. L., *Fuel*, **40**, 225–74 (1961).

[48] Pieters, J., *Tech. moderne*, **38**, 254–5 (1940).

[49] Anon., *Coke and Gas*, **9**, 56 (1947).

of nascent hydrogen which was formed from steam in contact with the hot charge. Ash was removed in a dry-cleaning tower, and the resulting low-ash and low-sulfur char was briquetted and the briquettes calcined. The final briquettes were reported to form a high-grade metallurgical or domestic fuel.

Electrically heated retort.[12,50,51] In Norway, a noncaking bituminous coal, in the form of briquettes, was carbonized by the Norsk Hydro-Electrisk Kvaelstofaktieselskab on a pilot scale in a vertical shaft retort, heated by the use of electric power. The Longyear City coal was mined in Spitsbergen, and, because it was frozen as mined, it disintegrated on thawing.

The coal was crushed, and either the pulverized coal was briquetted at a very high pressure of almost 36,000 psi, or a larger size was briquetted by using a pressure of from 1400 to 2850 psi with sulfite lye as binder. The briquettes were preheated to 150 to 200°C and charged into the top of the shaft retort (Fig. 14). In their slow descent through the shaft, the briquettes were partially carbonized by the heat in the gases ascending from below, so that they became conductive. An electric furnace, containing two sets of three-phase electrodes was located in the lower end of the shaft—the briquettes forming the resistor element to provide the process heat necessary for completion of carbonization.

Below the electric furnace and above the discharge point, the briquettes were partially cooled by introduction of cool product gas. The temperature of the heated gas ascending from the cooling zone was raised further in the electric furnace to 500°C, and at that temperature some of the constituents reacted with hot carbon in the descending charge to form carbon monoxide and hydrogen. The gases continued their ascent through the shaft and provided the heat for partial carbonization of the coal above the electric furnace.

The coke product was expected to possess both the properties of high-temperature coke with 1 to 2% volatile matter and a compressive strength up to 7000 psi and, in addition, to be as reactive as low-temperature char. On the basis of the first small installation, the power consumption of the process was reported

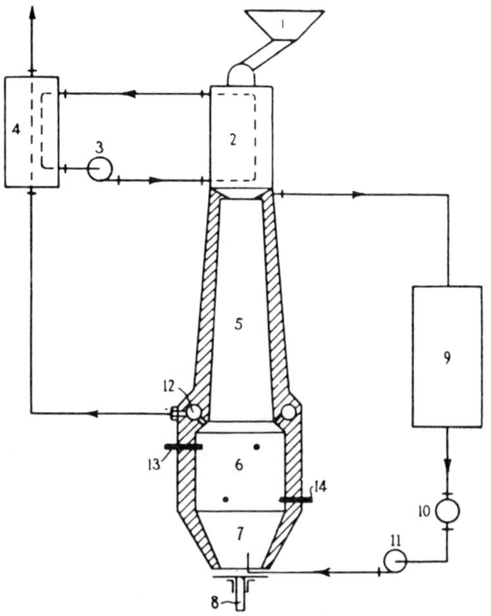

1, Briquette bin with gas-tight feeder for charging. 2, Preheater for coal briquettes. 3, Circulator for preheating gas. 4, Heat exchanger for heating the preheating gas by means of produced warm gas. 5, Coking shaft. 6, Electric furnace for heating the coke briquettes. 7, Apparatus for cooling coke briquettes by means of cold circulating gas. 8, Speed regulator for downward travel of briquettes, and gas-tight discharger. 9, Cottrell unit for tar separation. 10, Gas cooler and steam-injection unit. 11, Gas circulator. 12, Annular flue for discharge of hot gas. 13 and 14, Three-phase electrodes.

Fig. 14. Flow sheet of electric process for the carbonization of noncoking bituminous coal.[50]

to be 40 kwhr per 1000 cu ft of gas produced, which is equivalent to almost 1000 kwhr per ton of coal carbonized. A larger plant with a capacity of 25 tons of coal briquettes per day was projected; it was estimated that it could operate on 175 kwhr per ton, and that a retort with a capacity of 125 tons per day would operate on 120 kwhr per ton of coal.

Continuous Vertical Retorts Using Indirect Heat Only

Petit retort.[27,52] In the Petit process, the coal was heated by indirect heat as it descended continuously in a solid stream through a vertical retort, the steel walls of which were shaped to provide a zigzag path. By this zigzag arrangement of the walls, the coal was retarded in its descent. To insure the continuous, uniform descent of the coal and to delay buildup of carbon on the walls, the coal was agitated and kept in suspension by vibrating the walls.

[50] Jensen, O., *J. Inst. Fuel.*, **23**, 54–5 (1950).

[51] Schreiber, R., *Gas u. Wasserfach*, **91**, 222–3 (1950). Anon., *Coke and Gas*, **13**, 111–14 (1951).

[52] Petit, D., *AIME, Proc. Blast Furnace, Coke Oven & Raw Materials Comm.*, **11**, 209–13 (1952); *Blast Furnace Steel Plant*, **41**, 1107 (1953).

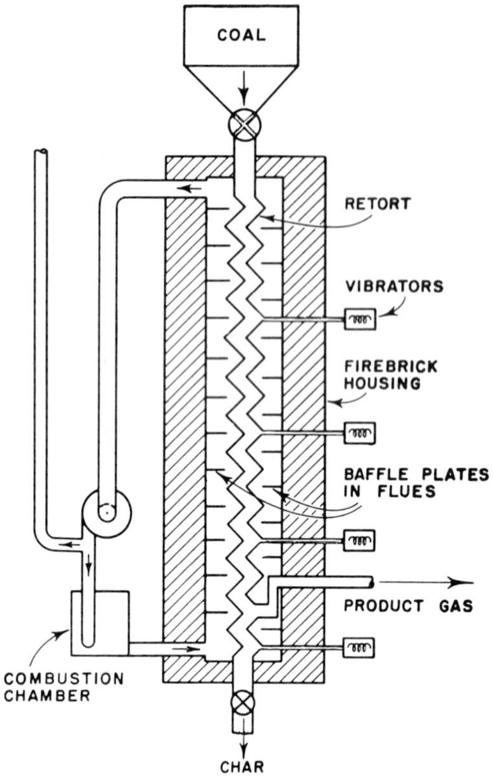

Fig. 15. Cross section through Petit retort. (After J. D. Price; Courtesy of Daniel Petit)

This was accomplished by means of several vibrators which were located at intervals up the height of, but outside, the retort, and which were connected to the wall by rods extending through the brick setting. A flow diagram of a retort is shown in Fig. 15.

The retort was mounted in a firebrick housing and was heated by gaseous products of combustion which ascended around it. Baffles located at suitable intervals in the retort setting deflected the hot gases against the walls. The volatile products were withdrawn from the retort near the bottom, and the char was removed at the bottom through a rotary valve. This retort was developed by Daniel Petit in France. A pilot plant having a rated capacity of 25 tons/day was installed by the Colorado Fuel and Iron Corporation at Pueblo, Colorado, in 1953. Other installations of similar principle are reported to have been made in Europe.

P.D.P. retort. In the effort to develop an economical source of carbon for use as a reducing agent in various processing operations, including phosphorus manufacture and nonferrous metal smelting, several pilot plants for continuous carbonization of Western coals were erected and operated in Montana by P.D.P. Processing, Inc. The process, usually known as the "Koal Krudes" process, was studied in several pilot plants, one of the last, shown in Fig. 16, being described by Berg and Atkinson.[53] It consisted essentially of four vertical, concentric cylinders of stainless steel. They made up the heated portion of the retort

[53] Berg, L., and Atkinson, D. E., *Chem. Eng. Progr.*, **56**, 73–7 (1960).

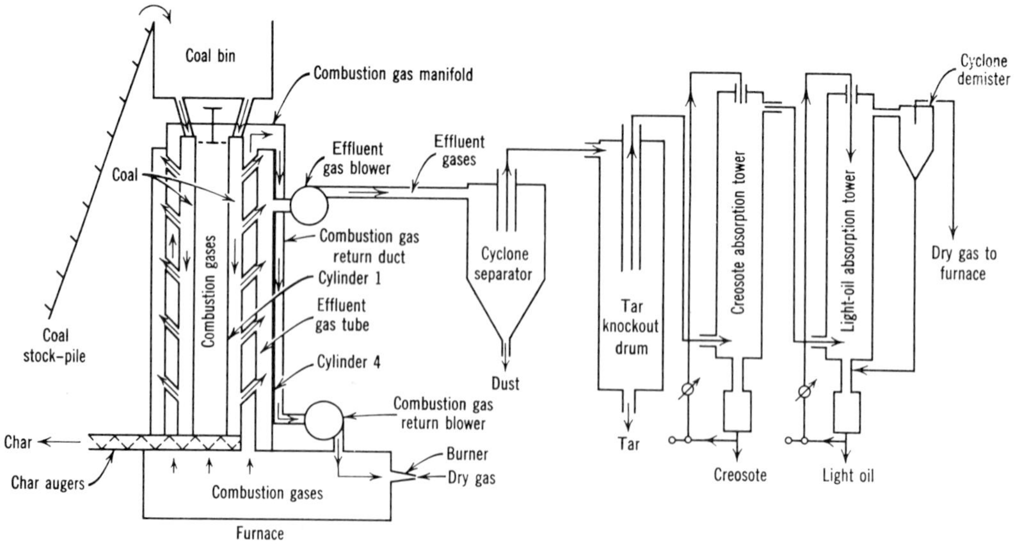

Fig. 16. Schematic diagram of pilot plant for Koal Krudes low-temperature carbonization process.[53]

and were surrounded by a mild steel shell with an outer diameter of 58 in. The inner cylinders were, respectively, 12, 20, 24, and 38 in. in diameter. The annular space between the shell and the outer stainless cylinder was packed with insulation.

The raw coal to be carbonized was fed from a hopper into the 2-in. space between the second and third stainless cylinders and descended by gravity. To assist in uniform descent of the coal without channeling, the 20-in. cylinder was slightly raised and dropped at short intervals by a mechanical lift. The coal descended through 34 in. of annular space before it passed into a funnel arrangement from which it was withdrawn by screw conveyors. Their speed could be varied to control the throughput of coal and the volatile matter of the char.

The volatile products were withdrawn from the retort through four vertical sets of louvres which bridged the outer heating annulus, and then were pulled by an exhauster through a heated manifold and a cyclone for dust removal. Tar, oils, and water were finally condensed by cooling the gas to a temperature of about 40°C.

The heat for carbonization was generated by the combustion of fuel gas in a furnace below the retort, and the combustion gases were drawn up through the annular heating channels by a blower which recirculated them to the furnace. Flow of gases between the carbonizing annulus and those through which the combustion gases passed was prevented by water seals located at the upper and lower ends of the tube. The most important temperature was that of the char leaving the retort, which was held at from 700 to 880°C.

Following pilot-plant work, a single retort of similar design but of commercial size was constructed at Red Lodge, Montana. This retort was designed to carbonize washed Red Lodge slack coal at the rate of 40 tons/24 hr. The outside dimensions of the retort were 24 ft in height and 9 ft in diameter. The char left the retort with a temperature of 590 to 650°C. About one-half ton of char and 18 to 20 gal of crude tar were produced per ton of coal. Because of financial difficulties, it is believed that work on this unit has been suspended.[27,54]

[54] Anon., *Coal Age*, **62,** No. 9, 133 (1957). *Chem. Eng.*, **65,** No. 5, 72 (1958); **67,** No. 2, 43, 45 (1960).

Stansfield retort.[55] The Stansfield retort was under development for several years by the Research Council of Alberta to upgrade Alberta noncoking, subbituminous slack coal which is produced in that Canadian province. The objective was to carbonize the coal continuously at low temperatures with production of a char which, after briquetting with a suitable binder such as petroleum asphalt, would form a solid fuel possessing a reasonably high heating value and good handling, storage, and weathering properties. The use of a continuous vertical retort employing internal heating achieved by the combustion of the volatile matter distilled from the coal was proposed for this purpose.

The slack coal contained 12 to 30% moisture and had a tendency to slake to fines. It was realized that operation with any large amount of fines or slack in a low-temperature carbonizer designed to operate with a deep bed on the principle of the gas producer could lead to irregular flow of gas and channeling. In turn, these conditions would result in uneven and incomplete carbonization of the coal, and in the production of gas of poor quality.

To avoid these difficulties, a pilot retort with a throughput of about 75 to 150 lb of coal per hour was designed on a different principle (Fig. 17). The coal passed downward between two series of closely spaced baffles or refractory louvres in such a manner that it formed a continuous, unbroken thin layer alternately changing direction, but did not descend as a cascading stream as in cascade dryers. The retort had a rectangular cross section 16.75 by 24.25 in. o.d. and was 23 ft high. The baffle plates which were 9 by 6 by 0.75 in. were made of Carborundum, and were supported in slotted Carborundum plates at an angle of 60°. The total capacity of the retort from the top of the baffles to the outlet of the char cooler was 200 lb of coal. The retort consisted of three zones: an upper drying zone, a central carbonizing zone, and a lower equalizing zone, below which was a water-cooled discharge zone. The volatile matter distilled from the coal passed between the louvres to enter the flues which were located outside of the baffle system. Here it was burned with air which entered the flues through many small ports or tuyeres located at the level of the central carbonizing zone.

[55] Gregory, J., and McCulloch, A., *Ind. Eng. Chem.*, **41,** 1003–11 (1949).

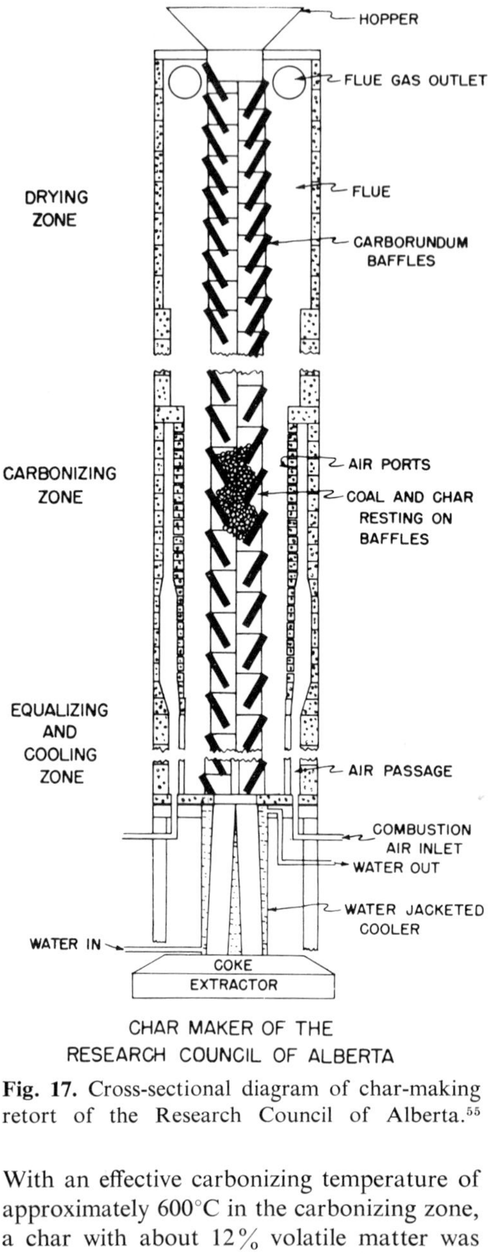

Fig. 17. Cross-sectional diagram of char-making retort of the Research Council of Alberta.[55]

With an effective carbonizing temperature of approximately 600°C in the carbonizing zone, a char with about 12% volatile matter was produced. The retort was operated at rates as high as 144 lb of coal per hour. At this rate, with an average flue-gas temperature of 500°C, the yield of char was 53%.

Archbald "Pipestem" retort.[56,57] The Archbald "Pipestem" process was reported to produce a hard, dense char containing 5 to 9% volatile matter, from a mixture of small sizes of bituminous and anthracite coals. A mixture of coking coal and inert material, such as anthracite, coke, or coal char, was wetted with tar, and the mixture was extruded at a pressure of less than 10 psi into vertical tubes installed in a furnace operating at temperatures between 650 and 760°C. Carbonization was continuous, and the time of passage of the mixture through the heated tube was reported to be between 13 and 15 min.

The tubes were originally round, but in later work hexagonal tubes with an inside diameter of 2.5 in. were used. The capacity of a tube was about 1500 lb of mixture per day.

The possible proportions of coking and inert coal varied widely with the volatile matter of coking coal. Two typical mixtures consisted of: one-third Pocahontas plus two-thirds anthracite; and 40% Clearfield coal containing 20% volatile matter plus 60% anthracite, respectively. The process thus resembles those in which an inert char is briquetted, and the briquettes carbonized. The coal was ground to pass a 10-mesh screen.

The product from the Pipestem process had a density of 45 lb/cu ft, was nonfriable and free burning.

Geissen retort.[8] In the Geissen process, brown coal was carbonized in a continuous, vertical retort containing a concentric revolving cylinder. Before carbonization, the brown coal was dried from its original moisture content of about 54% down to about 15%. The dried coal was then fed in at the top of the retort and carbonized as it descended between the inner revolving corrugated iron cylinder and an outer fixed wall consisting of a multiplicity of stationary annular collars. Revolution of the cylinder continually turned and mixed the coal.

The heat for carbonization was supplied by burning gas in a combustion chamber located inside the revolving cylinder, and then by conducting the products of combustion from the top of this cylinder down in contact with the corrugated wall. The volatile products of carbonization immediately escaped from the coal through openings between the annular collars and were withdrawn from the retort without the possibility of further cracking. The char was discharged continuously from the base of the retort.

The Geissen process was developed in Germany following World War I, and up to 1938

[56] Archbald, H., U.S. Pats. 1,828,586 (1932); 1,930,377 (1934).

[57] Private communications (1940).

about thirty-five Geissen ovens had been installed, several of which were still operated during World War II. About 150 tons of brown coal with a moisture content of 54% could be carbonized per day in each retort.

Borsig-Geissen retort.[8] The Borsig-Geissen retort, made by Borsig in 1935, was a modification of the Geissen retort. Its arrangement is illustrated in Fig. 18. In place of the corrugated, revolving cylinder consisting of heavy iron castings, the Borsig-Geissen retort employed a stainless-steel tube for the rotating cylinder. The burner was situated at the top of this rotating cylinder, and the combustion gases simply flowed downward through it and then upward through a jacket surrounding the retort without contact with the coal. Partially dried brown coal introduced at the top of the retort passed downward between the rotating cylinder and the stationary collars, as in the Geissen oven. The process required that the coal charged should not be larger than $\frac{1}{2}$ in., with not over 10% of fines below $\frac{1}{25}$-in. size. A portion of the make-gas was burned to supply the hot gases for carbonization.

Heliopore process.[58] The Heliopore process was designed to utilize the waste heat in the exhaust gas from an internal-combustion engine in carbonizing either noncoking coal or lignite. The gases distilled from the coal furnished the fuel for the internal-combustion engine, which in turn drove an electric generator. Thus the main products of a plant were expected to be electricity and char.

The principles of operation of the Heliopore process are illustrated by the flow diagram shown in Fig. 19. The primary equipment consisted of a carbonizing retort, gas cooling and scrubbing equipment, a gas storage holder, and an engine to burn the gaseous carbonization products. Coal was charged into the top of the retort while the hot exhaust gases from the engine traveled upward through special ducts in the retort walls countercurrent to the coal flow, but not in direct contact with the coal. Char was discharged at the bottom of the retort and the hot, volatile products at the top. The entire process was intended to be continuous and self-sustaining and to use coal as the only fuel.

[58] Fieldner, A. C., Davis, J. D., Parry, V. F., Schmidt, L. D., Elder, J. L., Goodman, J. B., Landers, W. S., and Goodwin, E. W., *U.S. Bur. Mines, Rept. Invest.* No. 3733 (1943), 36 pp.

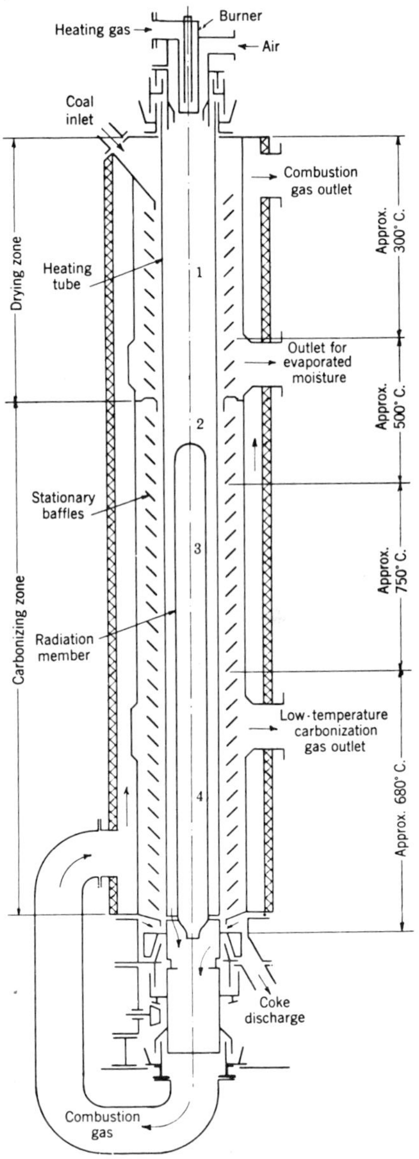

Fig. 18. Vertical cross section of Borsig-Geissen low-temperature carbonization retort.[8]

The process was tried out by the U.S. Bureau of Mines using a retort about 21 ft tall. The results showed that the engine exhaust gases were entirely inadequate as a source of carbonization heat. The tests of the Bureau of Mines showed that:

> Even with the engine running at full load on natural gas, the temperature and volume of engine exhaust gases were such that the volume of carbonization gas produced from coal satisfied about 15 percent of the fuel requirements of the

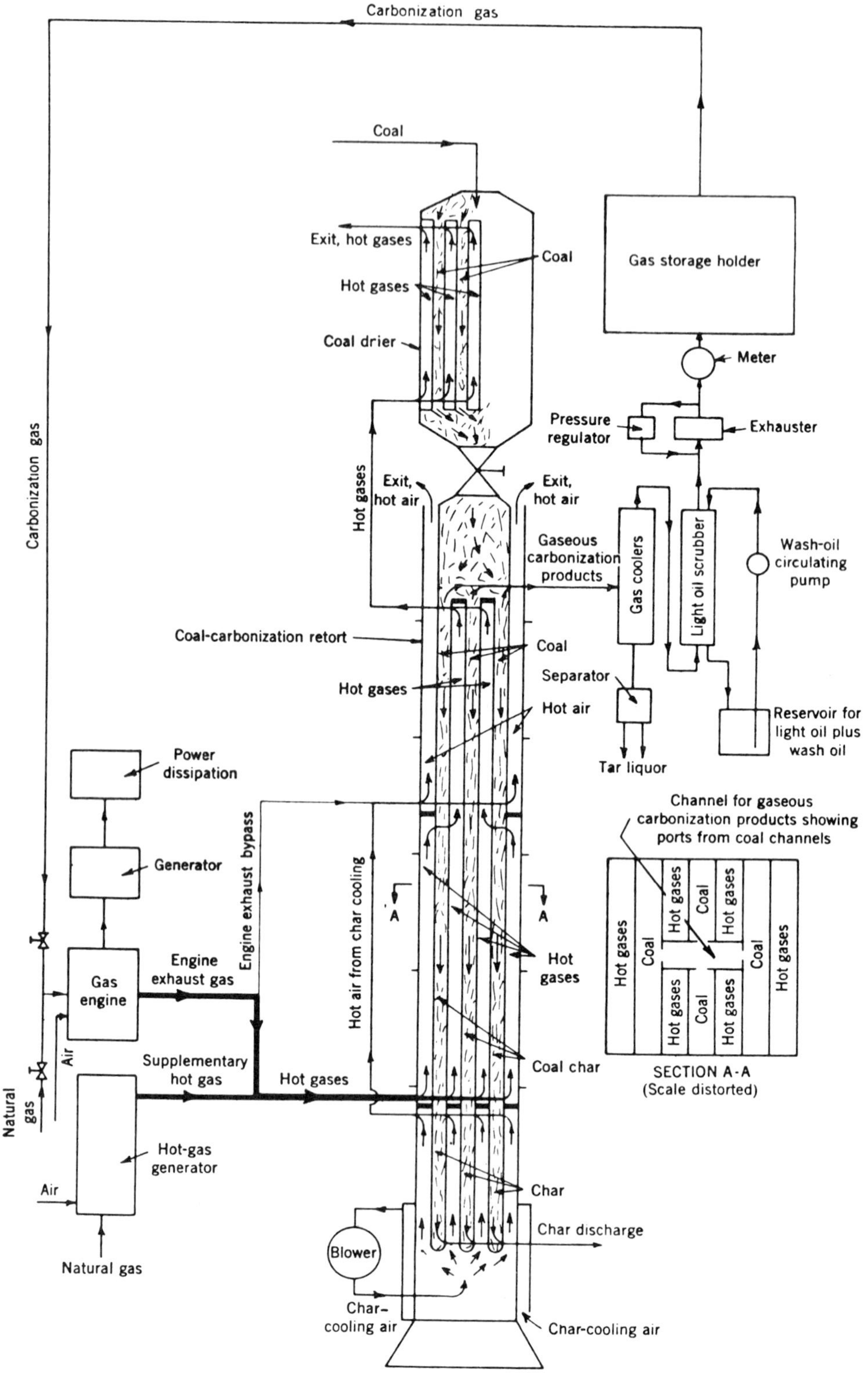

Fig. 19. Flow diagram for Heliopore coal-carbonization and power plant.[58]

engine. In subsequent tests all the supplementary hot gas that could be produced with the available equipment and at the maximum temperature attainable were mixed with the engine exhaust gas, and the volume of carbonization gas produced was slightly less than enough to satisfy the fuel requirements of the engine at full load.

The maximum value that can be assigned to the heat available in the engine exhaust gas does not exceed 6 or 7 cents per hour. Consequently, the saving of this amount of heat could not justify any very appreciable investment in special equipment. Furthermore, the low-temperature of the engine exhaust gases (1300 to 1500°F) militates against their efficient use, especially for carbonizing coal.

FLUIDIZED OR ENTRAINED CARBONIZATION

Carbonizing fine coal while fluidized by or entrained in a stream of heated gas produces a finely divided char, which may be utilized for boiler firing. In addition, such chars have found application in the preparation of metallurgical fuels—by replacement of low-volatile coal in the blends charged to coke ovens or by use in the production of briquettes which are subsequently carbonized at high temperatures.

As the name implies, a fluidized retort contains a bed of coal and char through which a gas is passed to maintain the solids in the fluidized state. Coal is added to the bed and char is withdrawn continuously. In an entrained process, on the other hand, the coal introduced into a stream of gas is carried through the retort into a separator.

Fluidized or entrained retorts are well adapted to the continuous processing of large tonnages of coal, and much useful experience on their design and operation is available to the industry from the large-scale applications of fluidized processing in other fields, such as in petroleum refining. For these reasons, fluidized processes for the carbonizing of fine coals especially have attracted considerable interest, and much research is proceeding in laboratories throughout the world.

Fluidized and entrained processes are apparently applicable to practically all types of coals, ranging from the lignites or brown coals to bituminous coking coals. With the noncoking coals the process is relatively simple, involving carbonizing the finely divided coal directly by contact with the heated gas. With coking coals, however, agglomeration occurs as the coal becomes plastic with increase in temperature, unless precautions are taken to avoid this difficulty. In most processes the coking property of such coals is destroyed by a preliminary oxidation of the coal, usually with heated air in another fluidized operation. However, other means have also been used to avoid agglomeration, including dilution of the coal mass with a large volume of inert, finished char before it is introduced into the carbonizer.

The heat for carbonization can be supplied in several ways. Although external heat has been used in several processes, internal heating is more common, either by use of preheated gases, by addition of hot char to the incoming coal, or by reaction of the oxygen in fluidizing air with the carbonaceous solids or gaseous products in the retort. This last practice tends to dilute the gas produced with the inert products of combustion, thus materially lowering its heating value.

Entrained Processes

Parry process.[1,27,59,60,61] The Parry process has been the major application of entrained carbonization in the United States. Finely divided coal entrained in air was carbonized in an externally heated retort to produce a fine char, suitable for boiler firing. The process consisted essentially of moving the dried coal with air into the base of an externally heated vertical reactor. Heat was transferred from the wall of the reactor to the suspended mixture, but the major portion of the heat required for carbonization was produced by combustion of a portion of the coal with the air. All products of reaction, including char, tar vapors, and gas, left the top of the reactor, and the separation of the solid and gaseous products was made in cyclones and the subsequent gas-processing equipment.

[59] Parry, V. F., *U.S. Bur. Mines, Rept. Invest.* No. 5123 (1955), 27 pp.

[60] Parry, V. F., Landers, W. S., Wagner, E. O., Goodman, J. B., and Lammers, G. C., *U.S. Bur. Mines, Rept. Invest.* No. 4954 (1953), 43 pp. Parry, V. F., and Landers, W. S., Sectional Meeting, World Power Conf., July, August (1954), 14 pp. Parry, V. F., U.S. Pat. 2,773,018; Brit. Pat. 756,744 (1956). Parry, V. F., Landers, W. S., and Wagner, E. O., *AIME, Trans.*, **205**, 54–64 (1956). Anon., *Chem. Eng.*, **58**, 219, 222 (1951); *Chem. Week*, **69**, No. 7, 25–6 (1951).

[61] Anon., *Chem. Eng. News*, **34**, 1318–20, 1397 (1956).

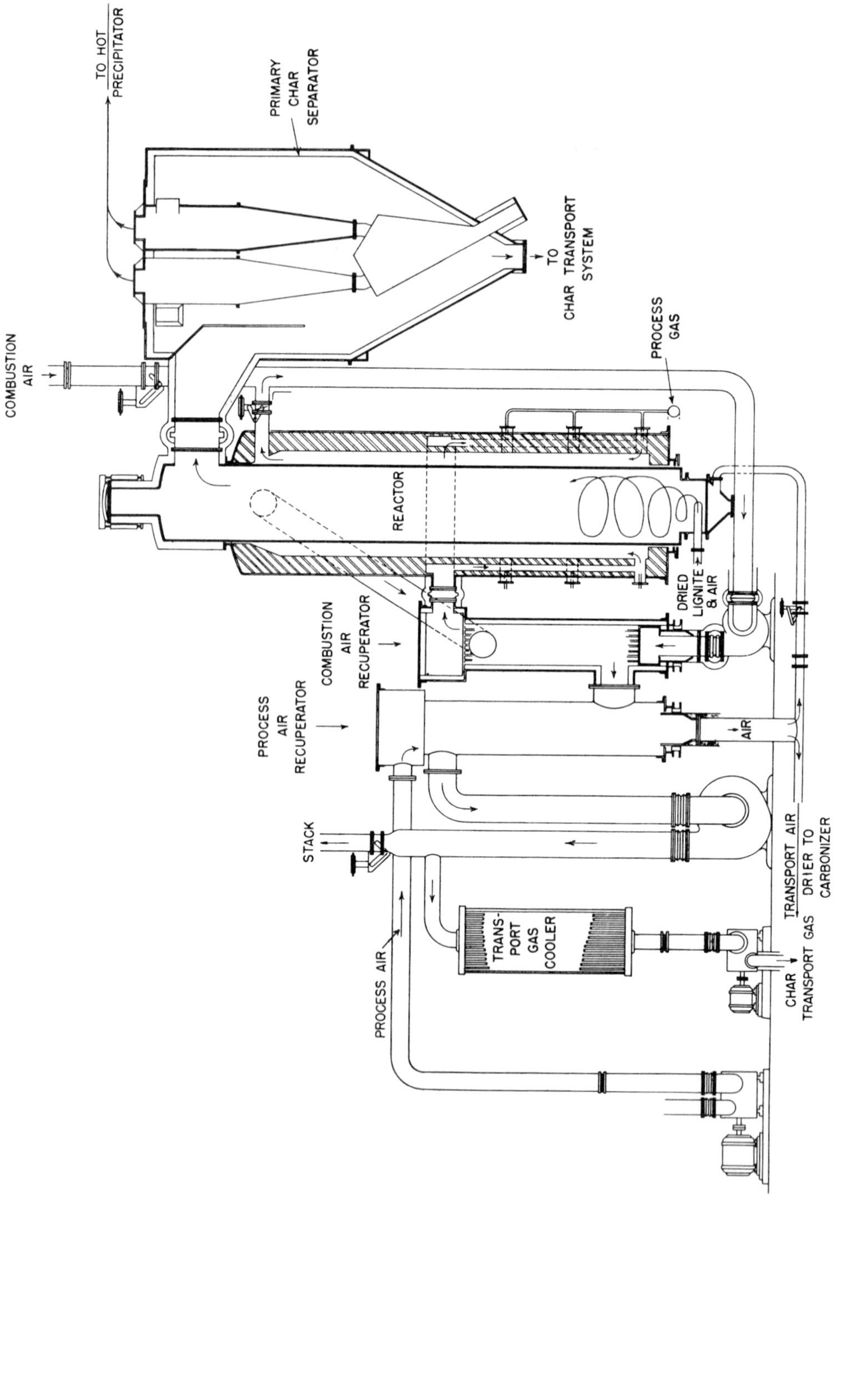

Fig. 20. Schematic diagram for Parry low-temperature process for carbonizing lignite.[59]

The Parry process was developed by the Bureau of Mines at Denver, in cooperation with the Texas Power and Light Company. As a result of their work, a prototype carbonizer for processing lignite was erected at Rockdale, Texas, in connection with the installation made by the Texas Power and Light Company to supply power to the aluminum pot lines which the Aluminum Company of America was installing at that locality. The boiler installation there was designed for combustion of either dried lignite or lignite char, the latter to be utilized after the carbonization process was perfected. The prototype carbonizer began operation during 1953 and was operated to develop and improve the process and to produce crude tar for market evaluation. No additional carbonizers have been erected as yet.

The method of drying and carbonizing the lignite has been described by Parry:[59]

> The crushed raw lignite is moved from the storage silos by a closed loop conveyor which maintains an oversupply of lignite at the small hopper above the screw feed in the drier. Products of combustion at about 816°C produced by combustion of dried lignite dust with air and recirculated gases enter the drier section at a pressure of about ¾ p.s.i.g. The wet lignite is suspended in this drying column in a turbulent state. The fuel to the furnace is regulated automatically to produce a given temperature at the exit of the drying section. Normally this temperature is about 149°C (300°F) when the stack temperature is 135°C (275°F). About 94% of the dried dust settles out in the primary separator, and the balance of the dust is separated in the secondary cyclone separator. The fine dust from the cyclones is returned to the furnace for combustion. The dried dust from the primary separator is moved with air to the boiler-plant storage silo or to the carbonizer.
>
> The prototype carbonizing unit (*shown in Fig. 20*) is designed to handle 31 tons of dried lignite per hour. This unit is designed for automatic control. The reactor is of alloy steel, 6 feet in diameter and 35 feet in height, mounted in a special furnace which is equipped with recuperators for efficient operation and preheating of process and combustion air. The process air is compressed to about 8 lbs. per sq. in. gauge and passes through a recuperator where it is preheated to about 371°C (700°F). Part of the preheated air is diverted back to the drier where it picks up the dried lignite for transport to the reactor, and part of the air goes directly to the base of the reactor. The mixture of air and coal enters the base of the reactor, and the fine coal is moved upward through the reactor entrained in the products of reaction in a turbulent condition. The solids are not in a true fluidized state because of the high superficial velocity of 6 to 8 feet per second. The temperature of the products leaving the top of the reactor is controlled automatically by regulation of the combustion around the reactor or by regulation of the air-coal ratio similar to methods employed in the smaller pilot plants.
>
> The combustion air is drawn into the system with a fan, which also circulates a portion of the products of combustion from the furnace. This mixture enters the fan at about 316°C (600°F) and is forced through the combustion air recuperator, where it is preheated to about 649°C (1200°F). From this recuperator the mixture of air and products of combustion passes to the furnace ports through a series of flues connecting the distributing bustle with each port. The air-gas ratio is controlled to provide about 5 per cent excess air, and the recirculated products of combustion are regulated to secure a maximum flame theoretical temperature in the reactor of about 1482°C (2700°F), but the developed temperature around the reactor is controlled to less than 1038°C (1900°F) to protect the heat-resisting alloy.
>
> Under these conditions of combustion the rate of heat transfer to the reactor wall is maximum because of the high heat transfer by direct radiation from the gases. Rates of heat transfer exceeding 1,000 Btu per hr. per sq. ft. have been attained in the pilot model at Denver, and it is expected that the rate will be in the range of 9,000 to 10,000 Btu per hr. per sq. ft. in the large unit when operating at rated capacity. The volume and heating surface of the reactor have been selected to permit carbonization of the coal at a rate of about 60 pounds per hr. per cu. ft. when the superficial velocity of the gases is 6 to 8 ft. per sec.

Parry selected the entrained process because the fine coal was heated in a minimum time and the capacity of a reactor was relatively high. About 12 min were required for carbonization, at the rate of 60 to 90 lb of lignite per cubic foot of reactor volume per hour, depending on the diameter and volume of the reactor. Of the heat required for carbonization, 30 to 40% was transferred into the reactor through the wall, and 60 to 70% was derived from combustion of air with carbon inside the retort. Because approximately 80 Btu were transferred to the lignite per cubic foot of air introduced, 2 to 4 cu ft of air were required per pound of dry lignite carbonized. By regulating the amount of heat transferred through the wall and by changing the air-coal ratio, the extent

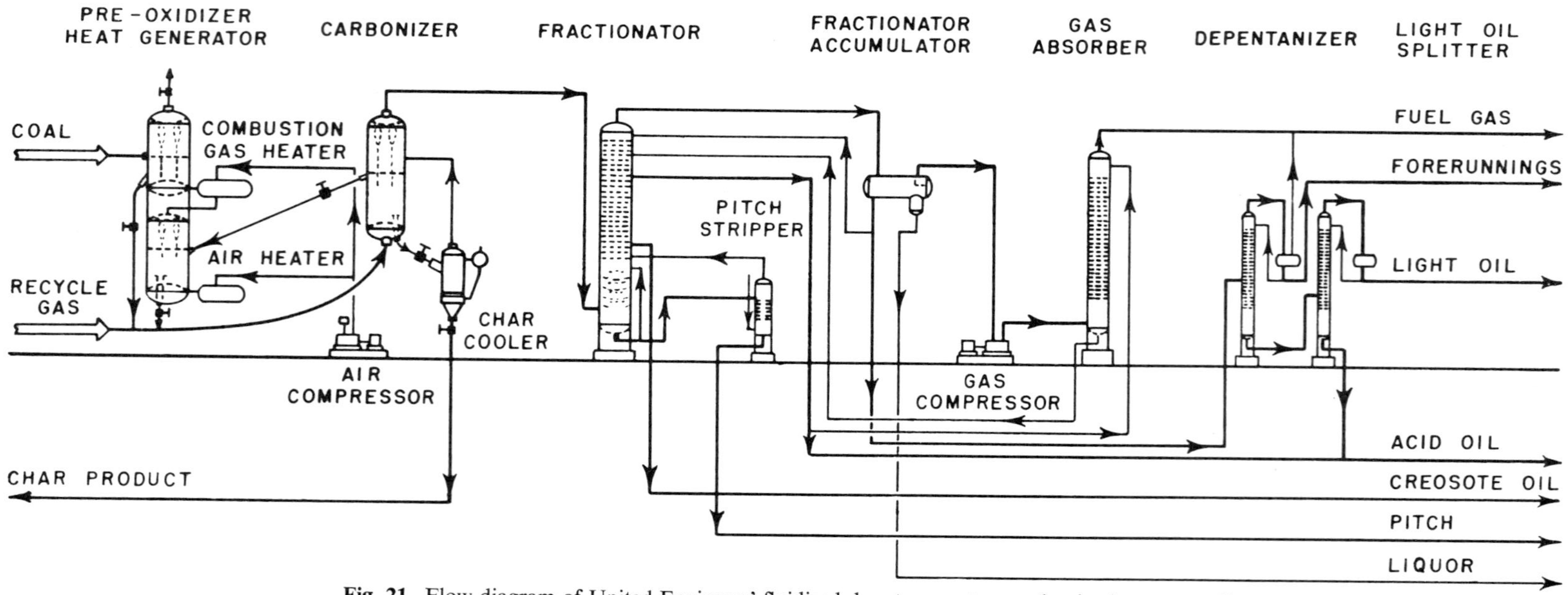

Fig. 21. Flow diagram of United Engineers' fluidized, low-temperature carbonization process.[66]

of carbonization could be controlled automatically.

Other entrained processes. An entrained carbonization process has been patented,[62] wherein finely divided coal and highly heated gas were injected into an externally heated, cyclone-shaped chamber in such a way that the mixture traveled in a helical path.[62] The differences in speeds between the slower-traveling coal and the gas increased the rate of heat transfer. By maintaining a suitable amount of oxygen in the gas mixture, the surfaces of the coal particles were rapidly oxidized, thus preventing sticking and agglomeration. An added precaution was the interposition of a smaller stream of the gas close to the wall of the cyclone. The larger particles of char were collected in a water-cooled chamber at the outlet of the cyclone and the finer sizes were removed from the gases in an electrostatic precipitator.

A circulating bed was used by Taskaev and Kozhina[63] in experiments on the low-temperature carbonization of coals. Product gas preheated to from 600 to 700°C was recirculated to entrain the coal used in sizes up to $\frac{1}{4}$ in. and to carry the particles through a vertical nozzle 1 in. in diameter and 20 in. long into a concentric tube 6 in. in diameter and 40 in. long. The coal or char particles fell through this space in a dense descending bed, and were then re-entrained by the gas and recirculated until carbonization was complete. The yield of primary tar produced at a retort temperature of 450 to 500°C was about 80% of that obtained in the laboratory in an aluminum retort. The process was recommended by the authors for the production of char to be used in industrial boilers. The product gas possessed a high heating value, about 800 Btu/cu ft because it was undiluted with gaseous products of combustion.

In a process patented by Pettyjohn,[64] a stream of 6-mesh coal granules was entrained in air at a temperature about 50°C below the softening point of the coal. Pettyjohn introduced the mixture, together with additional air at a temperature of at least 420°C, into a short, tubular vortex chamber. The temperature of the mass in the vortex was maintained at 650 to 1040°C solely by combustion of the volatile matter driven off from the coal. Devolatilization to thin, fragile granules occurred in less than 20 sec.

A free-falling process has been used by Schutte for carbonizing coal and oil together.[65] The coal in particle form, liquid oil, and preheated coke particles were introduced independently but simultaneously into the free-falling zone under turbulent conditions. The coal was rapidly heated through its plastic state and the preheated mixture descended through a reaction zone as an unagitated, gravity-packed mass. The proportions of coal and oil and of the preheated coke particles were controlled to maintain in the reactor a temperature of 540 to 760°C until carbonization had been completed, without introduction of other source of heat. The char particles and the volatile products were withdrawn at the base of the reaction zone and separated.

Fluidized Processes

United Engineers and Constructors process. The fluidized carbonization of coal for the production of a fine char suitable for boiler firing or other industrial use has been studied by United Engineers & Constructors in a pilot plant with a capacity of about 2 tons of coal per day over a period of several years. In this unit they reported that they successfully carbonized coals with a wide variety of characteristics, ranging from coking to noncoking.[27,66,67]

The pilot results led to the design of the process shown in the flow diagram, Fig. 21. This process consisted basically of three stages of coal treatment: a preoxidation, a carbonization, and a char-heating stage, each consisting of a fluidized bed. The coal, either coking or noncoking, was delivered into the plant, handled by conventional equipment, and crushed to pass a $\frac{1}{8}$-in. mesh. The fine coal from the hammer mill was continuously fed into the oxidizer, which contained a fluidized bed of coal heated to about 200°C by means of

[62] Koppers, H., GmbH, Brit. Pat. 714,925 (1950).

[63] Taskaev, N. D., and Kozhina, M. I., *Trudy Akad. Nauk Kirgiz S.S.R.*, **7**, 109–17 (1956).

[64] Pettyjohn, E. S., U.S. Pat. 2,706,706 (1955).

[65] Schutte, A. H., U.S. Pat. 2,624,696 (1953).

[66] Minet, R. G., Smith, H. B., and Trilling, C. A., *Chem. Eng. Progr.*, **50**, 342–6 (1954).

[67] Minet, R. G., *AGA*, *Proc.*, **1955**, pp. 914–19; Am. Coke & Coal Chemicals Inst. Regional Meeting, May (1955). Minet, R. G., and Mirkus, J. D., *Chem. Eng. Progr.*, **52**, 531–4 (1956). Shrikhande, K. Y., and Lahiri, A., *J. Mines, Metals and Fuels* (*India*), Special Issue, **7**, 251–9 (1959).

hot, combustion products. This fluidizing gas contained some oxygen, the concentration so regulated that by a slight degree of oxidation of the coal its coking tendency was reduced sufficiently to prevent agglomeration in the subsequent carbonizing stage.

The hot coal was transported from the oxidizer through a pipeline at high velocity by means of recycled product gas from the carbonizer into this bed. Simultaneously into the same transport line was fed a controlled quantity of char which had been heated to about 650°C in the char heater. About three parts of char were transported per volume of coal to secure the desired carbonization temperature of 480°C. Addition of this char also reduced any remaining tendency toward agglomeration of the coal in the carbonizer. From the carbonizer, char overflowed into the char heater where it was fluidized with preheated air. Approximately 3% of the char was burned, thus raising the temperature of the bed to about 650°C. The hot gases from the cyclone at the outlet of the char heater were heated further to about 980°C in a direct-fired heater and then introduced into the oxidizer. The carbonization products were withdrawn from the carbonizer, the make-char was withdrawn from the base through a cooler which reduced its temperature to about 200 to 260°C, and the volatile products—tar and gas—were withdrawn from the top, also through cyclones.

Singh process.[68–70] The Singh process also utilized three fluidized stages in series. There was an oxidation stage to reduce the agglomerating characteristics of the coal, followed by a carbonization stage. Heating of the charge in the carbonization vessel was accomplished by removal of a portion of the contents to a third bed wherein its temperature was raised by limited partial oxidation. The heated char was returned as a fluidized solid to heat the carbonization stage. A flowsheet of the process is shown in Fig. 22. A pilot plant with a capacity of 2 tons of coal per hour was erected in Chicago. The char produced consisted of granular, hollow spheres, which, however, did not possess the rib and window construction of cenospheres. It was dustless and more easily pulverized than the original coal from which it was made.

An objective in the process was to provide a suitable substitute for low-volatile coal in the coal mixes for high-temperature coke ovens. Because serious segregation of char and coal in such a mix might result from differences in the bulk densities, an effort was made to produce char with a higher-bulk density in the pilot plant. Singh indicated that the bulk densities of the chars produced by various processes apparently averaged from about 25 to 28 lb/cu ft, but in tests with Elkhorn coal he increased the density to as much as 56.6 lb by addition of blast furnace flue dust to the mixture of coal and char charged to the retort.[69] Typical process yields per ton of bituminous coal charged were given as 1320 lb char, 15 gal tar, 8200 cu ft gas (580 Btu/cu ft) and 35 lb solid sulfur. The heating value of the char, 12,180 Btu/lb, was higher than that of the raw coal, 11,460 Btu/lb.[70]

[68] Zinn, R. E., *Chem. Eng. Progr.*, **50**, 3–7 (1954).
[69] Singh, A. D., *AGA, Proc.*, **1946**, 400–409.
[70] Anon., *Chem. Eng.*, **59**, 244–6 (1952).

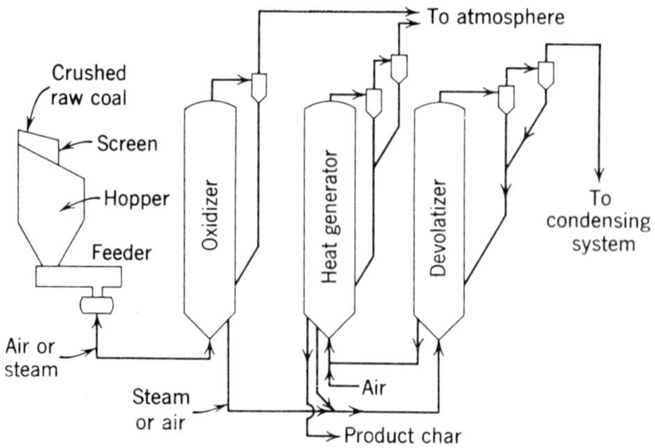

Fig. 22. Flow diagram of Singh process for the fluidized carbonization of coal.[68]

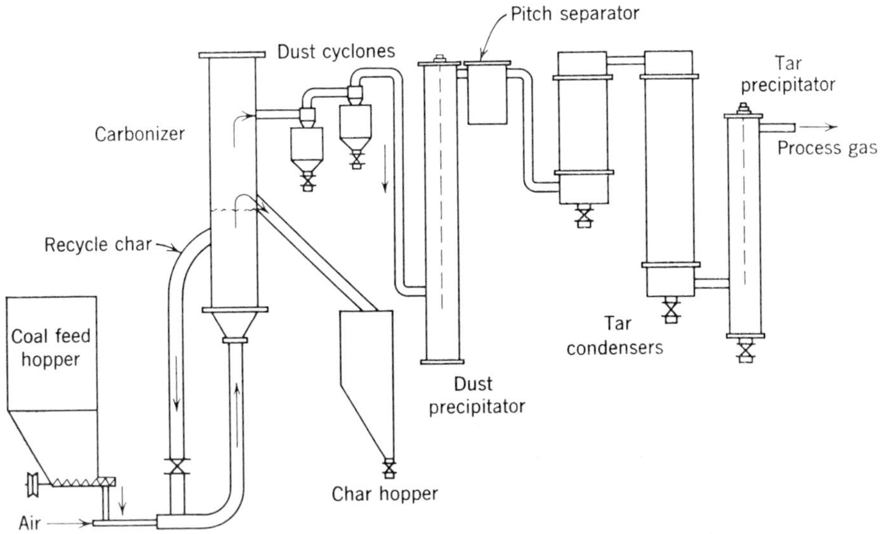

Fig. 23. Southern Research Institute fluidized carbonization process, pilot plant.[72]

Singh[71] also patented a process for producing metallurgical coke from a coking coal, which incorporated a fluidized low-temperature carbonization step. The finely-divided coal was mixed with about 0.5 to 20.0% of iron oxide, and the mixture was carbonized by direct contact with a stream of hot gases, which raised the temperature to at least 450°C. An unagglomerated char was produced, but it contained entrapped particles of oxide. Additional amounts of a coking coal were then added, and the mixture was coked at high temperature.

Other fluidized processes.[61,72,73] The Southern Research Institute has pilot planted a process for carbonizing bituminous coal using a single fluidized reactor. Their research project began in 1952 when Alabama Power Company became interested in low-temperature carbonization as a means for reducing their cost for fuel. The fluidized method that was developed reportedly handles satisfactorily coals with strongly coking or agglomerating properties.

In the SRI pilot plant (Fig. 23), agglomeration of the raw coal was avoided by mixing it, before it entered the carbonizer, with at least 10 parts of hot char which had been recirculated from the top of the carbonizer. The two materials were entrained together in a stream of high-velocity air which conveyed them upward into the carbonizer. The addition of this hot char was important because it acted as an inert diluent and prevented agglomeration.

The heat for carbonization was also produced in the same operation because the air burned a small portion of the hot char to raise the temperature of the mixture to the desired carbonization temperature—approximately 510°C. The velocity of the gases in the riser where combustion took place was about 10 ft/sec. Temperature variations in the lines and carbonizer were at a minimum due to the high rate of recycle. Combustion was very rapid, taking place in less than a second while the mixture ascended to the carbonizer. The volume of air admitted was regulated to supply all the heat for carbonization. With America Seam coal, 400 Btu were required for carbonization, so that about 5 cu ft of air were fed to the carbonizer per pound of coal.

The superficial velocity of the gases in the carbonizer was approximately 1.1 ft/sec. Although tests indicated that a retention time of less than 10 min was sufficient for carbonizing the coal, the coal was retained in the carbonizer for approximately 20 min. No significant reduction in the retention time was possible without entraining the solids in the increased gas stream from the carbonizer. The unit was operated at a feed rate of 50 lb of coal per hour per cubic foot of fluidized bed volume,

[71] Singh, A. D., U.S. Pat. 2,512,076 (1950).
[72] Lang, E. W., Smith, H. G., and Bordenca, C., *Ind. Eng. Chem.*, **49**, 355–9 (1957).
[73] Lang, E. W., and Lacey, J. C., Jr., *Ind. Eng. Chem.*, **52**, 137–40 (1960).

with maximum sustained rate of 81 lb per hour per cubic foot of volume. A stream of finished char overflowed continuously from the top of the fluidized bed to a hopper. The volatile products were withdrawn from the top of the carbonizer and passed through two dust cyclones in series to an electrostatic precipitator. A pitch separator and the tar condensers followed.

Products from the America Seam coal were approximately 80% char, 10% tar, 5.3% aqueous condensate, and 4.7% gas. Because its volatile content was over 15%, the char should form a satisfactory fuel for underfiring conventional boilers. The particles possessed a spherical, puffed shape, were friable, and the mesh size was slightly larger than that of the original coal. The gas amounted to 13,000 to 15,000 cu ft per ton of coal, with a low heating value only 120 to 140 Btu/cu ft. About 21.7 gal of tar were produced per ton of coal.

A fluidized process developed by Pittsburgh Consolidation Coal Company was disclosed in connection with the announcement of plans by Olin Mathieson to build an aluminum plant in the Ohio River valley.[61] The power was to be generated by combustion of the char and gas which were to be produced in a low-temperature carbonization plant to be built by Pitt-Consol. The process comprised crushing the local coal to −10 mesh and feeding it into a fluidized bed which was maintained at a temperature of 480 to 510°C in a reactor operated at atmospheric pressure. The gaseous products were withdrawn from the top and the char near the bottom of the reactor. The gases were cooled and the greater portion was condensed as tar, but the balance, together with the char, were to be delivered to the boiler house. About 30% of the coal was volatilized. The plant was to have a capacity for carbonizing about 5000 tons of coal a day. This low-temperature carbonization plant has not been installed. Pittsburgh Consolidation Coal Company also proposed to use the fluidized carbonization process as a primary step in a plant for the manufacture of synthetic fuels.[74] By gasifying the char, Fischer-Tropsch gas could be produced which would be converted into the synthetic oils.

The three-stage process of Welinsky for carbonizing agglomerating coals included a preheating zone, a carbonizing zone, and combustion zone.[75] The hot char from the carbonizer was partially recycled to the preheating zone to prevent agglomeration of the raw coal introduced there. Heat for the process was obtained by partial burning of another portion in the combustion zone and recycling this highly heated char to the carbonizer. The temperature in the carbonizer was held below 760°C.

In an investigation of low-temperature carbonization in a fluidized bed, which was described by Peytavy and Foch, the primary aim was to manufacture a char from Lorraine coals to replace the coke breeze, which was normally added to the coal blends charged into tamped coke ovens in the Saar-Lorraine district.[76] The retort was internally heated by partial combustion of the coal with the fluidizing air. By provision of a preheater to preheat both the coal and air, the rate of throughput of the unit was increased, the quality of the gas improved, and, in addition, the proportion of coal burned in the carbonizer was reduced to approximately 3%. The investigation was made in several pilot units with capacities ranging up to 1 ton/hr. The fluidized process with air was selected as the most promising process after other methods of carbonization were studied. The char produced from the high-volatile, weakly-caking coals was made up of swollen particles which had been subjected to fusion. The char was relatively friable, but very uniform in quality for a given set of conditions. The furnace temperature determined its volatile matter content very accurately—within about 0.5%—as shown by Fig. 24. Chars prepared at temperatures of 500 and 800°C were so closely alike in appearance as to be indistinguishable. Char produced in a rotary furnace was very different, however, from that produced in the fluidized bed. In the former, the coal particles had maintained their shape and structure, but the color of the char was brown. It appeared that the rate of heating in the different retorts was more important than temperature in producing the differences in shape and structure. Carbonization time

[74] Anon., *Petrol. Processing*, **7**, 45 (1952).

[75] Welinsky, I., U.S. Pat. 2,677,650 (1954).

[76] Peytavy, A., and Foch, P., *Chemical Engineering in the Coal Industry*, F. W. Sharpley, Ed., Pergamon Press, New York, 1957, pp. 75–99. Loison, R., *Rev. mét.*, **56**, 119–21 (1959). Foch, P., U.S. Pat. 3,011, 953 (1961).

in the rotary retort had been about 30 min as compared to about 1 min in the fluidized bed.

Nathan and co-workers patented a fluidized process in which the finely divided coal was first dried and preheated in a dense-phase, fluidized dryer.[77] The mixture passed downward as a confined stream through a carbonization zone into a pretreating zone in which it was further preheated in a dense-phase fluidized bed by partial combustion with air or oxygen until the agglomerating tendency of the coal had been reduced. The pretreated solids, together with the combustion gases, then ascended from the pretreater through a grid into the carbonization zone which contained a third dense-phase bed of char. It was maintained at a temperature above that in the pretreater by further partial combustion of the coal. Carbonization was completed in this third zone.

Coal was carbonized in a process in which hot char withdrawn from the carbonizer was added to a fluidized bed of the fresh coal to dry it, according to a patent by Nelson.[78] Not only was the coal dried, but its agglomerating tendency was eliminated by the addition of the char which entered at a temperature of 430 to 650°C. After from about 0.5 to 60 min in the fluidized bed, the mixture was transferred to the carbonizer. Nelson overcame the agglomerating tendency of certain coals during carbonization in a similar manner by injecting the particles into a turbulent suspension of hot coke or char.[79] In another patent, Nelson and Martin supplied the heat for carbonization of the coal in a lower temperature range of about 430 to 480°C by partial combustion of the char from the carbonizer.[80] The char was heated in a separate bed and was returned to the carbonizer. Practically all the oxygen in the gas used to fluidize the char in the heating bed was consumed in the process.

To render the fresh particles of a caking coal noncaking without substantial carbonization, Garbo coated the tacky surfaces of the particles with soot while fluidizing them in a dense-phase bed at a temperature of at least 320°C.[81]

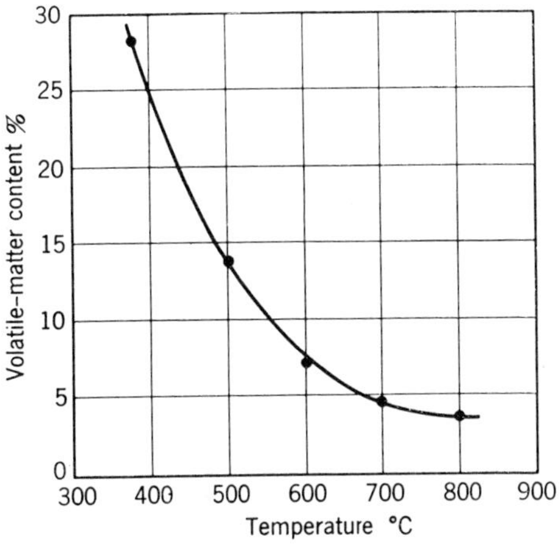

Fig. 24. Volatile-matter content of char produced by fluidization as a function of the temperature of the retort.[76]

According to Creelman, the char fines produced in fluidized carbonization may be completely burned, and the products of combustion may be returned to the fluidized bed as a source of heat for the low-temperature carbonization of the coal.[82] The fines were recovered by separating them from the volatile products with which they left the retort. Creelman also patented a process in which a portion of the char product was burned, and the hot combustion gases supplied the heat for carbonization of the coal.[83]

A low-temperature carbonization process involving what the inventor termed "thermal solution" was patented by Pevere and co-workers.[84] The coal was mixed with a hydrocarbon solvent at 290 to 370°C under sufficient pressure to keep the solvent liquid. The mixture was then atomized into an inert gas stream and passed into a carbonizing zone heated externally to 480 to 700°C where it remained until carbonization was complete.

The use of atomic energy to supply the heat needed for low-temperature carbonization has been suggested.[85] The purified gas from the

[77] Nathan, M. F., Lobo, W. E., and Skaperdas, G. T., U.S. Pat. 2,775,551 (1956). Nathan, M. F., U.S. Pats., 2,844,526, 2,841,534 (1958); Brit. Pat. 785,863 (1957).

[78] Nelson, K. J., U.S. Pat. 2,534,051 (1950).

[79] Nelson, K. J., U.S. Pat. 2,750,330 (1956).

[80] Nelson, K. J., and Martin, H. Z., U.S. Pat. 2,534,728 (1950).

[81] Garbo, P. W., U.S. Pat. 2,729,597 (1956).

[82] Creelman, G. D., U.S. Pat. 2,606,145 (1952).

[83] Creelman, G. D., U.S. Pat. 2,595,338 (1952).

[84] Pevere, E. F., Arnold, G. B., and Hess, H. V., U.S. Pat. 2,644,390 (1953).

[85] Perry, H., and McGee, J. P., *Chem. Eng.*, **66**, No. 4, 143–6 (1959).

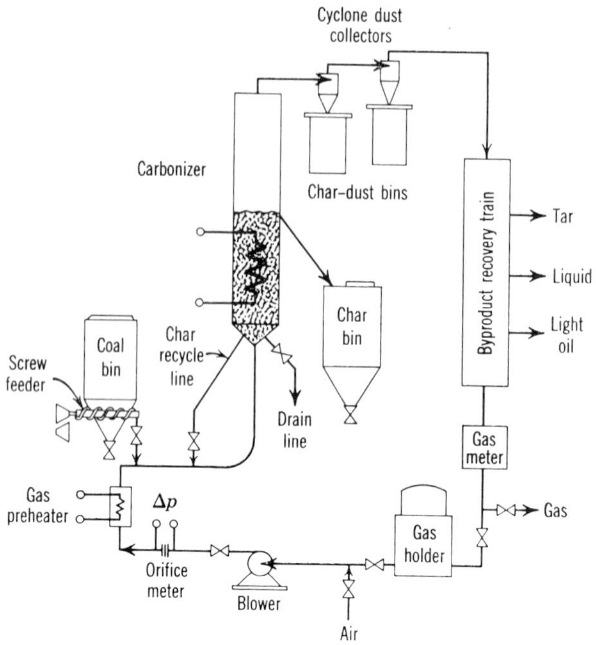

Fig. 25. Flow sheet of CSIRO pilot plant for the fluidized carbonization of coals.[86]

retorts would be preheated by recycling it through a nuclear reactor, and the heated gas would then be recirculated back through the fluidized carbonizer. Because the final product gas would be undiluted with products of combustion, its heating value would be high, about 800 Btu/cu ft.

Experimental Results from Fluidized Carbonization

The Coal Research Section of the Commonwealth Scientific and Industrial Research Organization (CSIRO) in Australia has studied fluidized carbonization in a pilot plant, with particular reference to the production of char for use in the coal mixes to coke ovens. Their reactor was 6 in. in diameter with a capacity of 50 lb of coal per hour which was carbonized usually at about 400 to 600°C, using air or gas recycled from the process as the fluidizing medium. Figure 25 is a flow sheet of the plant. When the coal was fluidized with air, the heat for carbonization was largely derived from combustion of a portion of the char product, although electrical heaters were employed when starting.[86] When reduction of the caking tendency was necessary, the coal was first oxidized at from 300 to 350°C in a similar but separate bed. The tests showed that when air was used for fluidizing, the yields of char and tar from a medium-volatile, nonagglomerating coal were lower than with recycle gas. On the other hand, the gas yield was increased although it possessed a low-heating value—60 to 70 Btu/cu ft. During carbonization, the coal particles swelled considerably, and a large proportion of the finer particles were carried over with the gases so that efficient dust-removal equipment was necessary.

In another investigation on fluidized carbonization by the Coal Research Establishment of the National Coal Board, Great Britain, preliminary measurements were made of the extent of decomposition occurring on heating high-volatile, noncoking coals for different lengths of time and at different temperatures.[87] The tests were made by subjecting static coal samples to conditions approximating those existing in a fluid bed composed of large particles of coal. The tests showed that for practical purposes decomposition of the coal was practically over after 2-min exposure at 700°C and 20 min at 450°C. Because the residence

[86] Anon., *Coal Research (CSIRO)*, No. 9, 2–8 (1960). Bowling, K. Mc G., Brown, H. R., and Waters, P. L., *J. Inst. Fuel*, **36**, 99–107 (1961). Bowling, K. Mc G., and Waters, P. L., *Brit. Chem. Eng.*, **7**, 98–101 (1962).

[87] Owen, J., *Proc. Conf. Science in Use of Coal*, Sheffield, Inst. Fuel, London, 1958, pp. C 34–9.

time for reasonably complete decomposition of the coal particles appeared to be in the range 2 to 20 min, depending on the carbonizing temperature, the fluidized unit was designed for holding times up to 30 min. The initial tests in this unit were run at temperatures in the range 400 to 650°C to produce chars of widely different volatile content for testing their smoke-producing properties. Figure 26 shows curves for the product distribution obtained on carbonizing one of the coals in the fluidized bed. As the carbonizing temperature was increased, the char yield declined rapidly, but there was a decided increase in the volume of gas produced. At about 600°C, the tar yield showed a maximum, but at higher temperatures it decreased. The yield of liquor appeared independent of the temperature.

Petit also studied the production of a char suitable for replacing low-volatile coals in the coal mixes to coke ovens.[88] In a mechanically fluidized retort, a well-fused char was produced in the form of small spheres of 1- to 3-mm diameter, which were excellent for the metallurgical-coke application. All the byproducts were recovered, including a high-heating-value gas and primary tar. Externally heated iron retorts were unsatisfactory for carbonizing coal, because a difficult-to-remove crust was rapidly formed, the life of the retorts proved short, and the quality of the char was irregular.

Material balances and the characteristics of the gases and char produced by the fluidized carbonization of four coals were studied by INICHAR.[89] The 200-g charges were fluidized with nitrogen at temperatures in the range 600 to 900°C. A high-rank coal (9.9% volatile matter), when carbonized at 865°C, yielded a gas containing 77% hydrogen with a yield of 3.65 cu ft/lb. High-volatile coals required oxidation or the admixture of an inert substance before carbonization in the fluidized state to prevent agglomeration. They yielded gas similar in composition to coke-oven gas. Preoxidation of the coal reduced the yield of tar, and the concentrations of CO and CO_2 in the gas were increased. By addition of steam to the bed, the yield of gas, including methane and its homologs, was increased.

Four Alberta subbituminous coals were

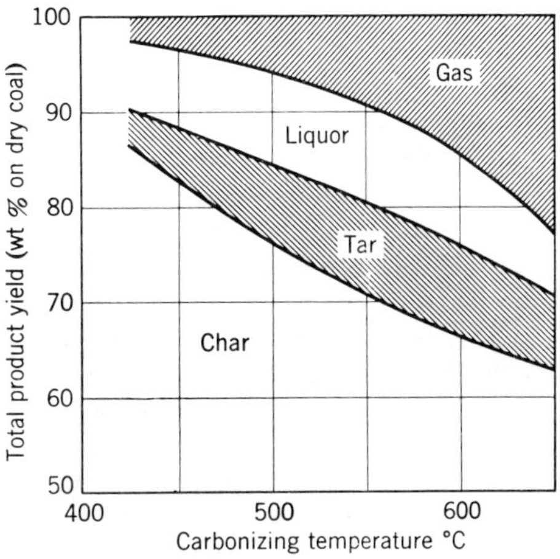

Fig. 26. Distribution of products versus the carbonizing temperature from a coal—air used for fluidization.[87]

subjected to low-temperature carbonization at 500°C with air as the fluidizing medium in tests by the Research Council of Alberta. Yields of products were: char, 57.3 to 67.4%; and tar, 2.9 to 6.7%. Volatile-matter content of the char was 15.4 to 19.5%. The gas was of low heating value because of dilution by air and combustion products.[90]

The effects of preoxidation on the yields of products resulting from low-temperature carbonization of Texas lignite have been studied by Gomez and co-workers.[91] They oxidized a lignite from Milam County in Texas with air at temperatures ranging from about 27 up to 300°C for different lengths of time, and then carbonized the coal at 500°C in a laboratory assay retort. The carbonization products were collected and analyzed. Their data suggested that, with increase in the degree of oxidation of the lignite, the tar yields dropped, and the gas yields increased in inverse ratio. The authors surmised that the structures which produce tar from a nonoxidized coal yield gas after oxidation. Structures which formed char on pyrolysis of a nonoxidized coal were changed so that on carbonization after oxidation they gave rise to additional water of decomposition. With an increasing degree of oxidation of the lignite, the proportion of carbon dioxide in the

[88] Petit, D., *AIME, Proc. Blast Furnace, Coke Oven & Raw Materials Comm.*, **11**, 209–13 (1952).

[89] INICHAR, *Bull. tech., houille dérivés*, No. 14, 449–84 (1957).

[90] Lang, W. A., *Trans. Can. Mining Inst.*, **61**, 307–13 (1958).

[91] Gomez, M., Goodman, J. B., and Parry, V. F., *U.S. Bur. Mines, Rept. Invest.* No. 5277 (1956), 17 pp.

gases evolved on carbonization increased progressively until it became almost 100% carbon dioxide. It was suggested that the decrease in tar yield with increase in the degree of oxidation might be used as a qualitative measure of the degree of oxidation. The use of transport air in fluidized or entrained processes for drying and carbonizing fine coals results in some oxidation of the coal, although the extent of oxidation is limited by the short residence time in the oxidized system.

Gas for domestic use was produced by a fluidized process in Japan, according to Hayashi.[92] A noncoking pulverized coal from the Taiheiyo coal mine was fluidized with air and steam in an alloy retort with the temperature of 800 to 850°C maintained by heating externally with producer gas. The yields when carbonizing 980 lb of coal (12,100 Btu/lb) per hour were 10,600 cu ft of gas (399 Btu/cu ft), 500 lb of char (11,240 Btu/lb), and 65.5 lb of tar. The char could be briquetted for use as smokeless fuel. The dust from the gasifier formed a low-grade of active carbon.

Terui studied the fluidized carbonization of noncaking or weakly caking fine coals in a unit where the carbonization heat was supplied by reaction of preheated air with a portion of the coal.[93] The fluidized bed was operated at 500 to 600°C with a 1- to 3-min mean residence time of the particles in the bed. The ratio of air to coal was 6.4 to 8 cu ft/lb. The gas produced had a heating value of 110 to 180 Btu/cu ft. Additional experiments were made to produce gas for domestic use from noncaking coals in an externally heated retort, with air or air plus steam used for fluidization of the 3-mm coal particles. In a retort about 4 in. in diameter, 15.45 to 44.0 lb of coal were carbonized at temperatures in the range 720 to 920°C, using 71 to 212 cu ft of air per hour. The yield of gas was improved by carbonizing at the higher temperatures, but its heating value was lower. The gas yield was also influenced by the rank of the coal, being reduced with a lower-rank coal. Gas with a heating value in the range 338 to 484 Btu/cu ft was produced.

Stone and co-workers found that the kinetics of coal carbonization in a fluidized bed did not follow a simple fixed order of reaction.[94] Instead, they showed that the order depended on the degree of devolatilization of the coal and on the temperature. At the start, oxygen and sulfur compounds are evolved. These reactions are rapid and follow approximately first-order mechanisms. At the end of the process, devolatilization approaches a zero-order reaction. The initial rate of devolatilization increases as the particle size decreases, but size is unimportant after the first 15 sec in the fluidized bed at high temperature. The banded constituents of a coal carbonize at different rates.

[92] Hayashi, G., *J. Fuel. Soc. Japan*, **38**, 599–607 (1959).

[93] Terui, A., *J. Mines, Metals & Fuels*, Special Issue 7, 260–3 (1959). *J. Fuel Soc. Japan*, **35**, 266–78 (1956); **38**, 590–8 (1959).

HORIZONTAL RETORTS

A horizontal retort consists essentially of a long, steel vessel, often cylindrical, through which the coal is moved as carbonization proceeds. Many varieties of horizontal retorts have been built and used—the types differing basically in the methods employed for transferring the heat of carbonization to the coal charge. The most common heating medium has been hot combustion gases, which have been applied indirectly by external heating of the shell; directly, by passing the hot gases through the retort in contact with the coal; and by a combination of both methods. To recover a rich product gas undiluted with combustion gases, the product gas has sometimes been reheated and recycled through the retort in direct contact with the coal. A further important method of heating is the use of preheated, solid bodies, including char, sand, or iron balls.

Cylindrical retorts may be either rotated or fixed. To keep the coal moving through the retort, the longitudinal axis of a rotated retort is frequently inclined a few degrees to the horizontal, with the coal fed in at the higher end and the char removed at the lower. Fixed horizontal retorts are provided with a screw or paddle device which by rotation conveys the coal and char from the entrance to the outlet of the retort.

Rotating horizontal retorts are operated continuously and are essentially simple in design. Addition of the drive and conveying mechanism to a fixed retort increases its complexity. The capacities of horizontal retorts are often

[94] Stone, H. N., Batchelor, J. D., and Johnstone, H. F., *Ind. Eng. Chem.*, **46**, 274–8 (1954).

relatively limited, so that a greater number of individual units may be required for a given capacity than would be needed for fluidized or many vertical retorts.

Because of the direct contact which necessarily exists between the heated coal and the walls or other parts of a retort, a caking coal will tend to stick and accumulate in large masses on these surfaces unless precautions are taken to prevent adhesion. As with other types of retorts a caking coal may be pretreated to reduce its agglomerating tendency, or the coal may be diluted with an inert medium, such as char or sand, until sticking has been prevented.

Continuous Horizontal Retorts

Indirectly Heated

Disco process.[27,92,68,95] The Disco process, the only commercially successful, low-temperature carbonization process now utilized in the United States, employs rotating horizontal retorts. The process has been operated since the early 1930's in a plant near Pittsburgh, which contains seven retorts with a total rated capacity of 800 tons of Disco char per day. The plant is located adjacent to a coal washery that belongs to the Consolidation Coal Company for the purpose of converting the fine sizes of cleaned bituminous coal produced in this washer into lump char suitable for domestic use. The cost of the present installation, which was completed in 1947, was reported to be $3,000,000.

The Disco process is unique in that it produces a lump char directly from the fine coal in the continuous rotating retorts, without the application of a briquetting step. This feature is secured by a preliminary oxidation or "roasting" of the coal in contact with air, thereby destroying a portion of its excessive caking characteristics. In the rotating retort, the oxidized coal, together with recycle char, rolls up into balls of char without adhering to the sides of the retort. The balls of char are screened and are sold as the smokeless domestic product.

In the Disco process, wet fine coal from the washery goes through five basic steps: (1) predrying, (2) "roasting," (3) carbonizing in the retorts, (4) cooling the char, and (5) screening and loading. The gas and tar are recovered as byproducts.

The drying equipment lowers the moisture of the coal to 2 to 3%, and consists of three flash dryers heated by the hot flue gas from coal-fired furnaces located below the dryers. The dried coal is carried from cyclone collectors at the top of each dryer unit by a conveyer belt to the roasting step in the carbonizer building.

In the roasting process, the coal is preheated and partially oxidized while hot by exposure to controlled quantities of air. By exposure to the air at an elevated temperature, some of the excessive agglutinating or caking property of the bituminous coal is destroyed, but enough is retained so that, under the conditions in the carbonizer, the coal will roll up into the balls, which are the characteristic form of the finished char, but will not adhere to the retort walls. The oxidizing action is closely controlled so that sticking of the coal is restricted to the extent of forming balls of char with the proper strength and reactivity but without the building up of adherent masses of coal or char on the walls of the retort. Roberts has shown that the process is governed by temperature and time factors, and that the proper conditions for a given coal are determined by experience.[96] The oxidizing operation is carried out in a roaster consisting of six, long, rectangular hearths, located one above the other, through which the coal is continuously dragged in series by rabbles. The coal is heated indirectly by hot combustion products which are circulated through flues built in the hearths. The lowest two passes differ from the four topmost ones in that some coal is recycled through them so that uniformity of quantity and quality of the feed from the roaster into the carbonizing retort may be maintained exactly. Manual air control is maintained in the ducts of the roasters. The coal from the last pass is fed to the carbonizer by a rotary-feed device at 320°C.

[95] Lesher, C. E., and Goode, J. B., U.S. Pat. 2,287,437 (1942). Denig, F., *Chemistry of Coal Utilization*, John Wiley and Sons, New York, 1945, Vol. 1, pp. 814–16. Lesher, C. E., *Proc. Smoke Prevention Assoc. Am.*, **1947,** 91–6; *Coke and Gas*, **12,** 269–72 (1950); *Mining Eng.*, **4,** 287–99 (1952). McBride, R. S., *Chem. Eng.*, **56,** No. 6, 112–14 (1949). Branin, M. L., *Utilization*, **3,** No. 5, 19–24 (1949). Pitney, A. W., *Mines Mag.*, **37,** 12–15, 23–5 (1957).

[96] Roberts, J., *Coke and Gas*, **10,** 377–80 (1948).

The carbonizing retort proper is a rotating cylindrical $\frac{5}{8}$-in. steel shell approximately 9 ft in diameter and 126 ft in length, i.d. The retort is mounted inside an insulated, stationary steel cylinder. The inner shell is mounted almost horizontally, but with a slight slope so that it rotates like a typical cement or lime kiln. Heating is external by means of flue gas circulated at high speed through the annulus between the outer stationary shell and the revolving retort. The heating gas is retained at the two ends of the heating space by rubbing strips which are attached to the outer shell and which ride on the rotary shell to give a suitable gas-tight closure. The prepared coal drops directly from the roaster, which is located just above the retort, into a screw feeder which discharges at a point close to the upper end of the rotary shell. The coal at 320°C drops into the shell and is soon heated to the plastic temperature at which the formation of lumps or balls begins. These balls roll around inside as the shell rotates, gradually working their way to the lower discharge end.

The gas generated during carbonization passes out of the retort through a takeoff at the upper (cooler) end and thence through a wash box to a direct-type primary cooler. There, tar condensing and liquor scrubbing are accomplished. The liquor for spray operation is recycled through a cooling tower. After a final scrubbing, the gas is burned in the retort furnace to provide heat for carbonization. Hot products of combustion are circulated from the furnace by a fan in a closed circuit through an annular space around the carbonizer, through ducts in the roaster, and back to the furnace again. The maximum temperature in the heating space at the coke-discharge end of the carbonizer is automatically held at 570°C.

The rate of throughput of a retort is determined by the rate of feed. After the balls are formed, they roll gradually down the slope and are discharged at the rate at which they are formed from the feed. The retort is largely free space except that, at intervals along its length, several 8-in. baffle rings dam the flow of balls and cause some piling up. The latest retorts are each designed to process 150 tons of dry coal per 24-hour day (6.25 tons per hour). The yield of char has been approximately 75% of the weight of coal, thus insuring daily 750 tons or more of marketable balls from the seven retorts of the new installation.

As normally prepared, the solid product contains 15 to 17% by weight of residual volatile matter. This percentage can be varied within limits from 14 to 20%. The solid char is screened into several sizes. The smallest size is crushed to $-\frac{1}{4}$ in., and added to the coal as the latter enters the feed hoppers above the roasters. A major control of the size and structure of the ball product is secured by varying the percentage of this undersize breeze in the feed.

Minchin described a somewhat similar process, which was operated experimentally at Marienau in the Saar, to produce a fine char suitable for use in the coal mix to coke ovens.[97] The arrangement consisted of two horizontal, rotating cylinders, one above the other. Inside the upper cylinder was a second concentric drum, and both were provided with vanes attached to the inner surfaces of the walls for movement of the coal. Inside the drums the "flambants sec" coal was subjected to a degree of oxidation by contact with a countercurrent stream of air. In the lower cylinder, carbonization was completed at 550°C. Passage of the coal through the entire unit took $1\frac{1}{2}$ hr. The drums were heated externally by hot products of combustion. The loose, friable char from the retort, when added to a mixture of local, noncoking coals, lowered the volatile content of the mixture sufficiently so that a reasonably good coke could be produced in a coke oven. The mixture charged comprised char 20%, Saar coal (Gras A) 25%, and Gras B from Lorraine 55%. The final coke was hard and good enough for a blast furnace trial.

Hayes retort.[27,42,98] The basic principles of the Hayes retort, which is shown in Fig. 27, have been described by Price:[27] A steel drum, 20 in. in diameter and approximately 25 ft long, is supported on rollers and enclosed within a brick furnace. This drum rotates at approximately $1\frac{1}{2}$ to 3 rpm. Both ends of the drum are closed by stationary end covers. At the feed end, coal is fed into the inside of the drum by a variable-speed, sealed screw feeder. The charge is moved through the retort by means of a 16-in. screw extending the full length of the retort and driven by a reversing

[97] Minchin, L. T., *Coke and Gas*, **15**, 167–71 (1953).

[98] Woody, G. V., *Ind. Eng. Chem.*, **33**, 841–6 (1941). Price, J. D., and Woody, G. V., *Am. Inst. Min. Met. Engrs., Trans.*, **157**, 317–37 (1944),

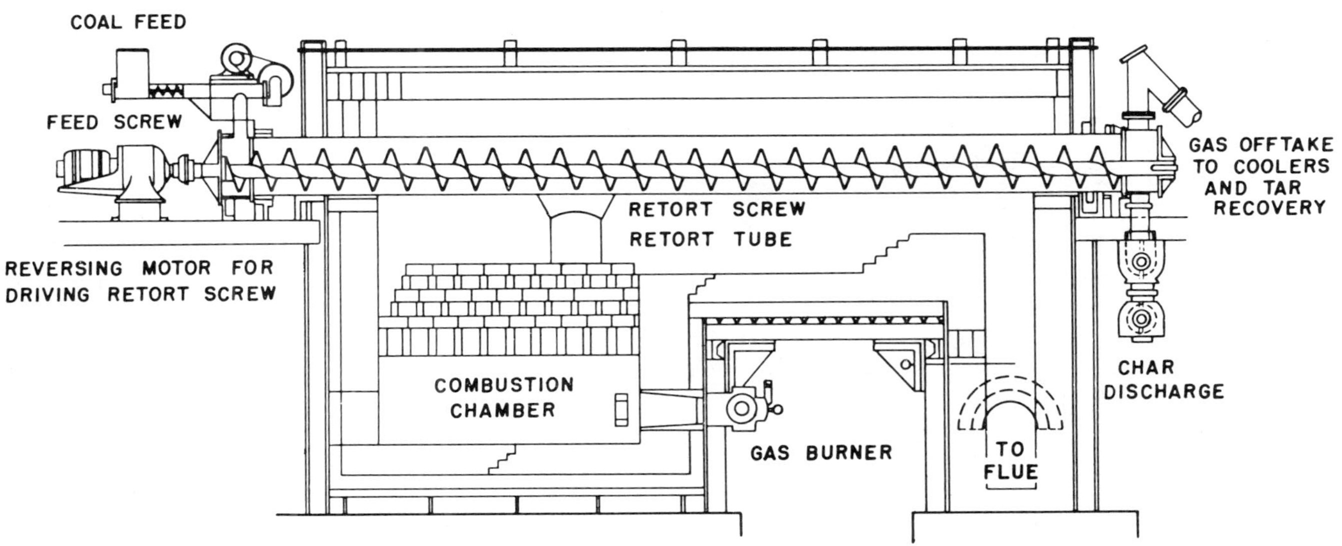

Fig. 27. Cross section through Hayes horizontal, low-temperature carbonization retort.[27]

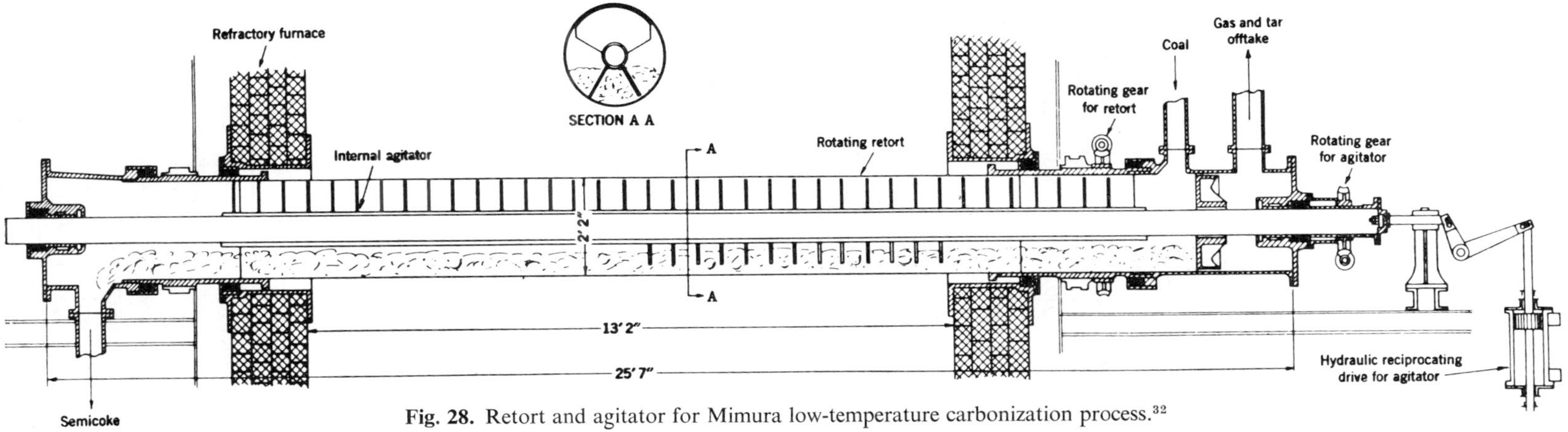

Fig. 28. Retort and agitator for Mimura low-temperature carbonization process.[32]

motor. The charge is first moved forward (toward the discharge end) and then backward for a shorter distance within the retort. This reversing screw serves several purposes. The charge is kept well agitated within the retort; the length of time the charge is held within the retort can be varied by changing the period of forward and reverse motions of the screw; and the wiping action of the screw against the walls of the retort delays the accumulation of a hard carbon deposit on the inside of the retort.

Heat is supplied to the retort through a gas burner operated on premixed air and gas and provided with automatic temperature control. At the discharge end the hot char drops into an enclosed screw type quenching conveyor, in which it is cooled with water sprays and discharged through a star discharge feeder.

The Hayes process was developed by the Allis-Chalmers Manufacturing Company. The first installation, erected at Moundsville, West Virginia, in 1928, contained five retorts and was operated until 1936 when they were damaged beyond repair by a flood. The char was rather small and suitable for boiler underfiring. In 1940, a pilot retort was installed adjacent to the coke plant of the Colorado Fuel and Iron Corporation at Pueblo, Colorado. The char was mixed with high-volatile coal to produce the charge for high-temperature coke ovens.

Mimura carbonizer.[32] The Mimura carbonizer, a Japanese development at the time of World War II, consisted of an externally heated, rotating retort, 25½ ft long and 25 in. i.d. The carbonizer was fabricated from mild steel and provided with rabble arms connected to a concentric shaft. These arms possessed two motions, a slow rotation at intervals and a horizontal motion, lengthwise of the retort. On the forward stroke of the shaft, blades on the rabble arms pushed the charge ahead of them through the retort. The agitator was then rotated until the blades were out of the coal, and on the return stroke only pins were dragged backward through the plastic mass. This agitation, together with the rotation of the retort, was intended to prevent agglomeration of large masses of material and their adhesion to the wall of the retort. A diagram of a retort and agitator assembly is shown in Fig. 28.

Operating difficulties encountered in the Nagoya plant appear typical of the possible problems which could be encountered with this type of retort. The retorts were operated at a normal temperature of 500°C to carbonize a strongly caking coal; but this procedure resulted in masses of char which adhered to the rabble arms of the internal agitator, and eventually froze the agitator in the retort. If movement of the agitator was attempted under such a condition, the arms of the agitator bent and the shaft failed. Even with only moderately caking coals, char packed between the blades of the agitator and prevented movement of the char. Attempts to eliminate the stoppages by a slight preoxidation of the coal during drying, by blending of coals, and by variations in speed of rotation of the agitator or in the carbonizing temperature were of little benefit.

Trouble was also encountered when air leaked into the combustion space surrounding the retorts and disturbed the air:fuel ratio. Local overheating occurred, with the result that temperatures of the metal parts of the retorts sometimes exceeded 700°C. Since all metal parts were of mild steel, the life of the retorts and agitators never exceeded 150 days, and failure sometimes occurred within 15 days of being placed in service. Warping of a retort induced by uneven heating caused leaking of the product gas through the end seals. In one retort, which was operated under a static pressure of 1-in. water gauge, this leaking gas usually ignited, with resulting damage to the mechanical equipment driving the retort and agitator. In another retort operated at a negative pressure of 2-in. water gauge, air leaked through the seals into the retort and burned a portion of the gas. Attempts to redesign these seals to prevent leakage were unsuccessful, the retort itself being insufficiently rigid to prevent warpage under the temperature conditions imposed.

The mechanical driving equipment was complicated; shutting down an entire bank of retorts to stop one for repairs was necessary and the hydraulic actuators for the horizontal movement of the internal agitators were inadequate. Because of jamming of the internal agitators, uncontrolled air leakage into the combustion chambers, warping of the retorts, leakage of rotating end seals, and general mechanical difficulties, not more than eighty retorts of the one hundred and twenty installed could be operated simultaneously. The retorts were designed for a daily capacity of 300 tons of coal, but under the best conditions only 82 tons/day could be carbonized in the plant.

Wanishi carbonizer.[32] The Wanishi carbonizer, designed in Japan during 1936, employed an externally heated, rotating, horizontal, cylindrical retort without internal agitation. To move the charge through the retort, a continuous internal spiral flange was provided. Each Wanishi retort was 76 ft long and 7 ft 10 in. i.d., was rotated at 0.28 rpm, and was mounted in an individual gas-fired furnace without recuperators. Four carbonizers were installed; but only three were operated at a time, one being retained as a spare. Each carbonizer was designed to handle 100 tons of coal per day, and the normal capacity of the plant was 300 tons of coal per day. The gases from the retorts passed through a hot, cyclone-type dust separator to remove tar mist and small amounts of dust. Because of the large size of a retort and its high thermal losses, the quantity of gas burned for underfiring exceeded that produced in the process, and a supplementary source of fuel gas was required. The temperature of a retort shell did not exceed 600°C and the carbonizing temperature was about 550°C. The coal was dried before carbonization in an inclined rotary drier, 73 ft long and 6 ft 4 in. i.d. which was internally heated by products of combustion leaving the main heating chamber. The drier was connected to the carbonizer by a closed hopper and a rotary feeder, which prevented loss of volatile products. The carbonizers usually were stopped annually for cleaning. The shutdown required about a month for removal of the accumulated tar and dust from the surfaces and for general repairs.

The objective of the Wanishi process was to produce a char to replace low-volatile bituminous coals in the coal blends to coke ovens. Only noncaking coals were carbonized at Wanishi, because caking coals would have built up into masses of char which clogged the retort. Because the char was required in sizes smaller than 3 mm, it was unnecessary to use a coal capable of producing lump char.

Lesher processes. In addition to the Disco process, Lesher has been responsible for the design of other retorts. In one arrangement, he proposed, first, to preheat the coal by mixing it in a screw conveyor with char heated to about 480 to 590°C, and then to devolatilize the mixture in an externally heated, horizontal rotating retort. The temperature and proportion of recycled char were regulated to raise the coal to a temperature just below its softening point. Recycling of the char was employed to ensure uniformity of the finished product. Together with control of the speed of rotation of the retort, recycling permitted the production of a granular, almost dustless char from a coking-coal feed.[27,99]

A continuous, rotating horizontal reactor which operated with a protective barrier of preheated char in contact with the walls to prevent sticking of a highly caking coal, was also patented by Lesher.[100] The char discharged from the retort was maintained at a temperature between 540 and 760°C, with a ratio of fresh coal to recycled char in the feed to the retort below 1 to 3.

Directly Heated Continuous Horizontal Retorts

University of Kentucky process.[101] Some experimental work has been done by the University of Kentucky on the production of char in a continuous rotary retort which was operated at an inclination up to 7.5° to the horizontal. The retort was heated by direct contact with gases which entered the lower end of the retort at about 540 to 760°C and flowed countercurrent to the coal. To overcome accumulation of char on the walls of the retort, a series of hammers was incorporated into the retort design to break off the accumulation by striking periodically against the shell of the retort.

Humboldt sand carbonizer.[14,102] In the Humboldt sand carbonizer, the heat for carbonization was supplied by addition of sand which had been preheated to a temperature of about 750°C. This preheated sand, together with briquettes of the coal to be carbonized, were introduced at one end of the horizontal, cylindrical drum. The drum was divided into two semicylindrical sections by means of a screen extending lengthwise through the center. Annular separating walls further divided the drum into a number of cells, and slanted

[99] Lesher, C. E., U.S. Pat. 2,723,226 (1955).

[100] Lesher, C. E., U.S. Pat. 2,622,059 (1952); Brit. Pat. 665,681 (1952).

[101] Anon., *Coal Age*, **49**, No. 7, 102 (1944). Kimberly, R. L., *Eng. Exp. Sta.*, Univ. of Kentucky, *Bull.* No. 21 (1951), 52 pp.

[102] Lameck, P., *Iron & Coal Trades Rev.*, **156,** 1173–7 (1948).

conveying ledges were attached to the partition in such a way that, as the drum revolved, the briquettes were carried upward and transferred to the next cell, and the sand dropped freely from them through the screen onto the briquettes below. In an experimental installation which was tried in Essen toward the end of World War II, the briquettes were prepared from a mixture of coal ground to less than 3 mm and iron-ore fines held together by suitable binder. The briquettes were carbonized at 550 to 600°C for about an hour in a retort 10 m long with an i.d. of 2 m, rotated at 1 rpm. The drum was surrounded with an insulating shell to retain heat. The vessel for heating the sand consisted of a refractory-lined housing fired by means of a gas burner with the sand passing countercurrent to the flame. In addition to the carbonized briquettes, about 220 lb of primary tar and 3178 cu ft of gas (heating value 843 Btu/cu ft) were recovered per ton of briquettes. In a final step of the Humboldt process, the carbonized briquettes were smelted with a proportion of lump ore in a low-shaft furnace employing cold blast. Although the combination of iron-ore fines with coke in briquettes fed to the blast furnace was not new, it was reported that the low-temperature char was advantageous because of its high reactivity. Successful results were claimed in operation of the experimental plant, but a full-scale test was needed to establish costs.

The use of hot sand in rotary retorts has been recommended by other workers because of the rapidity and ease with which heat is transmitted from the sand to the coal, a retort about 6 m long being adequate for such an operation.[103] Cooling the briquettes slowly in the hot sand prevented the development of the stresses which might occur if the briquettes were quenched rapidly in water. Jennings reported[12] that approximately two-thirds of a ton of sand per ton of coal charged was sufficient to supply the heat for carbonization, and that the conditions of carbonization could readily be adjusted to suit requirements. Further advantages were the ease of handling and safety with sand and the production by the carbonization process of an undiluted gas. No caking of briquettes containing pitch binder is possible because the pieces are separated by sand. Because of the higher temperature of the sand, any tar evolved is completely vaporized and does not adhere to the sand. Hot sand was also used by the British National Coal Board to supply the heat required for carbonization. Operation of a pilot plant indicated that carbonization with sand could be completed in about one-fourth of the time required in indirectly heated narrow ovens.[104]

Korobchanskii employed metal balls 50 mm in diameter containing a mixture of calcium chloride and sodium chloride as the medium to supply the heat needed for low-temperature carbonization of low-rank coals.[105] Aspegren has also described a horizontal vessel in which heated balls could be used.[106] Johansson has patented a process in which the coal is preheated and then carbonized in direct contact with heat-carrying balls in a rotating horizontal drum.[107] The balls removed from the carbonizer were reheated by combustion of volatile products evolved from the coal.

The use of molten metals as a means of promoting heat transfer in carbonization has also been proposed. Molten lead was used in the Piron-Caracristi process more than 30 years ago.[108] More recently, molten aluminum has been proposed.[109]

Use of a rotating horizontal kiln was proposed by Phinney for producing a char of uniform size and volatile-matter content.[110] The heat for carbonization was supplied by introducing into the kiln hot, recycled char of a size finer than that of the coal. The kiln products were discharged to a separator for gas and tar, and the solid material was then elutriated to separate the fine, recycled char from the larger lumps of freshly formed char. The finer sizes of char from the elutriator were reheated by partial combustion in a fluidized

[103] Pistorius, O., Ger. Pat. 848,644 (1950).

[104] Idris Jones, W., *Research*, **10**, 254–60 (1957). Harrison, J. A., and Madley, D. G., *Wyoming Univ. Nat. Resources Research Inst., Inform. Circ.* No. 10 (1959), pp. 70–80.

[105] Korobchanskii, V. I., *Trudy Khim. Tekhnol. Fak. Donetsk. Ind. Inst.*, **1956**, 46–56.

[106] Aspegren, O. E. A., U.S. Pat. 2,592,783 (1952).

[107] Johansson, E. A., U.S. Pat. 2,420,376 (1957).

[108] Fieldner, A. C., *U.S. Bur. Mines, Tech. Paper* No. 396 (1926), 46 pp.

[109] Farafonow, W. M., U.S. Pat. 2,787,584 (1957).

[110] Phinney, J. A., U.S. Pat. 2,654,698 (1953).

bed and recycled to the retort. Temperatures in the range 450 to 570°C were found suitable.

A flowable mass of solid particles made up of iron ore and char adhering together was produced by Lesher.[111] He first mixed preheated iron ore and the coal under conditions such that the final temperature of the mixture did not exceed the softening point of the coal. The heated mixture was then heated and agitated in a rotating retort. Under these conditions, the coal softened and became sticky in contact with the ore, and the two materials were rolled into fine solid particles. Heating was continued until the particles had reached a temperature of 450 to 540°C.

Fixed Horizontal Retorts

Rabble arms, screws, or other devices are necessary to convey the coal through fixed horizontal retorts. Fixed horizontal retorts appear to have been used to a more limited extent than have rotating retorts.

Shimomura retort.[32] The Shimomura process was used in Japan in the 1920's to produce a char suitable for use in the coal blends to coke ovens. Each stationary horizontal cylinder, constructed of cast iron, was externally heated, and was provided with rabble arms arranged on a central shaft to convey the coal through the retort. Two cylinders were combined to form a set, with one retort located above the other in the gas- or coke-fired refractory furnace. Each retort was usually 12 ft long and 2 ft in diameter, with a rated capacity per set of 8 tons of coal per day. The raw coal was fed into one end of the upper retort by a gas-tight screw conveyer. The rabble arms then propelled it along the retort to the other end where it dropped through a vertical pipe to the lower retort. The finished char left the lower retort through a screw conveyor and water seal. Volatile products were removed from the retorts at the junction pipe between them.

The rabbles were mounted on a water-cooled, concentrically located shaft, and were driven at the high speed of about 130 to 150 rpm. The functions of the rabbles were not only to propel the coal along the retort, but also to keep it from accumulating on the walls and to pulverize it. The char was produced in small, rounded particles most of which would pass a $\frac{1}{8}$-in. screen. The maximum temperature in the carbonizer was usually not over 500°C. During operation, carbon accumulated on the linings of these retorts, with a resultant decrease in the rate of heat transfer. When temperatures were increased to about 800°C to maintain output, the cylinders warped and failed. The life of the Shimomura retorts was too short to justify the expense and the process was abandoned. The Shimomura retorts were strikingly similar in principle and operating features to the primary retorts erected in the 1920's at the Clinchfield Carbocoal plant, Clinchfield, Virginia, to produce char for briquetting.[112]

Cotarco retort. In the Cotarco retort the rate of heat transfer from the metal heating surface of the wall of the horizontal retort to the coal was materially accelerated by use of a stirrer, resembling a set of ice cream mixer blades, which turned through the bed of coal at approximately 20,000 times/hr to keep the coal thoroughly mixed and to produce a finely divided char.[113] In 1955, an installation of this process was projected in Colorado.[114]

Conveyor Retorts

Processes in which the coal is carbonized continuously during its passage through a fixed horizontal duct while spread out on a conveying surface, are a third type of horizontal retort. Conveying retorts have the limitations that the moving parts of the conveyor are exposed to the temperature conditions in the retorts, and the coal must be spread out in a thin layer. Coal capacities therefore are relatively low, and a multiplicity of units would ordinarily be required for large tonnages of coal. Coking stoker units, however, are examples of conveying retorts which give relatively high capacities and, hence, are receiving much attention for producing industrial chars and cokes. In these units the coal is simultaneously carbonized to produce a char or coke, and the heat of combustion of the volatile products, together with a portion of the coal itself, is utilized in boiler firing.

111 Lesher, C. E., U.S. Pat. 2,794,728 (1957).

112 Curtis, H. A., and Daughton, E. E., *Chem. Met. Eng.*, **28,** 118–23 (1923).

113 Anon., *Petrol. Processing*, **13,** 452 (1955).

114 Anon., *Chem. Week*, **77,** No. 25, 12–13 (1955).

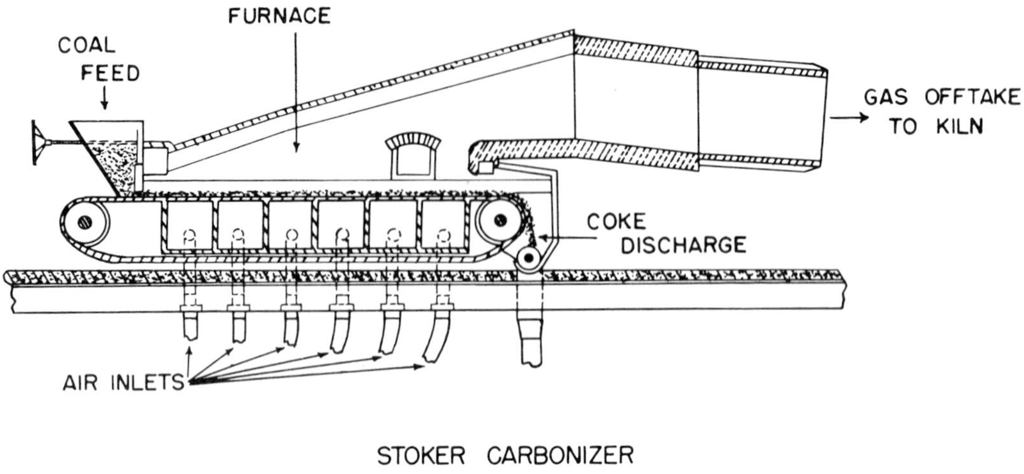

Fig. 29. Cross-sectional diagram of stoker carbonizer.[115]

Coking stoker.[14,115,116] The coking stoker process is unique in that the coal to be carbonized is fed onto the travelling grate of a boiler or other industrial heating installation. By regulating the flows of coal and combustion air, and the speed of the grate, the regular heating operation has been carried out, and at the same time a char suitable for certain uses has been discharged from the grate. The process was developed by Shawinigan Chemicals, Ltd., who applied it to horizontal lime-kiln firing at their plant in Shawinigan Falls, Quebec, in connection with calcium-carbide manufacture. Figure 29 shows the essential arrangement of the Shawinigan stoker as applied to a lime kiln. Bituminous coal from an overhead bin was spread on the travelling grate 2 to 4 in. thick. Air was blown through the grate from several compartments, so that combustion of the coal on the grate could be carefully controlled. As the grate advanced into the firebox, volatile matter was distilled from the coal and partially burned with the air above the grate to supply necessary heat for carbonization. The temperature of these gases was about 1200°C, so almost all the coal-tar products were cracked to carbon and the simpler gases, CO, H_2, CH_4, and the like. The hot volatile gases then passed through a 4-ft diameter burner pipe to the kiln, and burning was completed with secondary air in the kiln itself to supply the heat for lime burning. The presence of free carbon in the gases gave the flame in the kiln good radiant qualities. At the same time, a char containing about 3% volatile matter, which was satisfactory for manufacture of the carbide, was discharged from the grate to a conveyor. The coking operation was accomplished with a small loss (about $\frac{1}{12}$) of the original fixed carbon in the coal. The hot gases from the lime kiln were further cooled in a waste-heat boiler before leaving the stack.

C. W. & F. process.[27] A flat-bed, travelling-pan process also known as the Wallace process, was developed by the Chicago, Wilmington, and Franklin Coal Company for the preparation of a smokeless domestic fuel from the coal fines produced in the operations of this company in the Southern Illinois coal fields. The coal was dropped into a pan conveyor, levelled, and then compressed by a second lid conveyor before entering the heating chamber. Because the heat was applied from heating flues to the top surfaces of the lids and bottom surfaces of the pans, a distinct zone of separation occurred in the charge. The volatile matter was reported to be reduced uniformly from 33% in the original

[115] Hall, R. H., *Eng. J., Montreal*, **34,** 1176–82 (1951).

[116] Anon., *Chem. Eng. News*, **22,** 807 (1944). Andersen, A. H., *Trans. Can. Inst. Mining Met*,. **47,** 139–51 (1944). Andersen, A. H., and Renaud, J. E., U.S. Pat. 2,209,255 (1940). Andersen, A. H., and Rothwell, N., Brit. Pat. 563,037 (1944). Andersen, A. H., and Fasken, N. R., U.S. Pat. 2,380,930 (1945). Baum, K., *Combustion*, **26,** No. 5, 61–5 (1954). LaGrange, C. C., *J. S. African Inst. Mining Met.*, **57,** 99–114 (1956). Mansfield, V., U.S. Pat. 2,997, 426–7 (1961).

coal to 14% in the char. The final char was produced in blocks 6 by 12 by 2.5 to 3 in. in size. The coal company was reported to have installed a pilot plant with a capacity of about 10 tons of finished char per day at one of its mines in West Frankfort, Illinois.

Flat-bed retort.[27,117] A continuously operated, flat-bed, low-temperature retort consisted of a flat plate which was slightly inclined at an adjustable angle and was heated from below. Fine coal was fed onto the higher end of the plate and was moved slowly to the lower end by means of a vibrating mechanism which actuated a series of scrapers, the faces of which were shaped to resemble a snow plow. The spacing and location of the scrapers was such that the entire surface of the plate was scraped. The slope of the plate and the number of strokes of the scraper controlled the rate of flow of coal. It was claimed that by use of a metal with high heat conductivity, such as copper for the plate, and by vigorous agitation, high throughputs were secured. A pilot-scale retort was reported to have been installed at the plant of Frank Lowe and Company, in St. Louis. Figure 30 indicates the general arrangement of the device.

Storrs process.[27,118,120,124–6] The Storrs, or "Coal Logs," process produced the char in the form of cylindrical sections, or "logs," by a two-stage process, which included a flat, vibrating-bed carbonizer followed by an extrusion unit. The carbonizer consisted of a wire-mesh belt conveyor onto which the sized coal was fed to a depth of about 0.5 to 1 in. About 1 to 2 in. above the level of this belt was located a radiant heating surface. Below this surface the belt was moved at a speed generally in the range 0.75 to 1.5 ft/min and at the same time was vibrated through a vertical distance of about 0.5 in. The volatile and liquid products of carbonization escaped downward through the wire mesh and were withdrawn from collector compartments located below the belt. The coal at about 400 to 500°C and in the proper condition of plasticity was discharged over the

[117] Poindexter, F. E., and Lowe, F. W., U.S. Pat. 2,903,400 (1959).

[118] Storrs, K. L., U.S. Pats. 2,239,833 (1941), 2,349,387 (1944), 2,525,051 (1950). Anon., *Chem. Met. Eng.*, **53**, 186–88 (1946). Carter, G. W., *Chem. Eng. Progr.* **43**, 180–2 (1947); U.S. Pat. 2,809,154 (1957). Cellan-Jones, G., *Coke and Gas*, **18**, 52–5, 61 (1956).

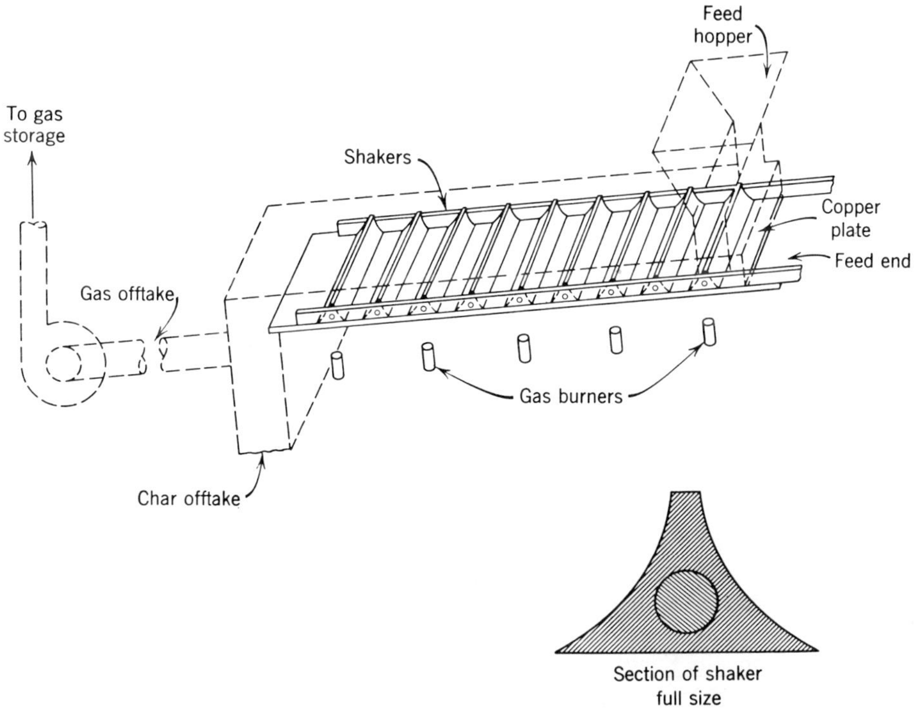

Fig. 30. Schematic diagram for Poindexter-Lowe low-temperature carbonization retort.[27]

driving pulley into the extruder. The time of heating on the belt was about 1 to 3 min. The coal was compressed in the extruder into the logs (3 to 3.5 in. in diameter) which gave the process its name.

The radiant surface of the retort was heated by gas burners located at predetermined positions to provide the desired heat gradient. In a British demonstration plant, the heating facilities consisted of twelve independently adjusted burners which were spaced along a combustion chamber 6 ft long, by 11 in. wide, and 4½ in. deep, extending the full length and width of the conveyor. This chamber was lined with refractories backed by insulating bricks. The bottom of the combustion chamber, which formed the radiant surface of the retort, was made of ⅜-in. "Staybrite" plate. The waste gases from the heating chamber could be used for heating and drying the coal or for steam generation in a waste-heat boiler.

The conveyor belt was fabricated of chrome-molybdenum wire to withstand the temperature. The extruder consisted of a continuous, variable-pitch worm rotating at about 3 rpm inside a 3½-in. diameter tube through which the log was extruded. The tube was surrounded by a jacket which was heated by gas burners to retain the plasticity of the coal during formation of the logs. As the coal was pushed through the tube, it was compacted by the pressure of the screw conveyor in combination with the resistance of the tube into a dense mass, which solidified on cooling. As a result, the specific gravity of a log was not greatly different from that of the true specific gravity of the coal from which it was made.

The carbonization step could also be operated at a higher temperature to produce merely a small size of char for direct use as a boiler fuel or for manufacture of briquettes which would then be carbonized at higher temperatures.

The Coal Logs process was first developed by the Coal Logs Company in Salt Lake City, Utah, to produce a strong, smokeless fuel from weakly caking Utah coal. Two installations, each of 100 tons/day feed capacity, were projected, one in Utah and the other at an undisclosed location. An installation was also reported at a plant belonging to the Appleby-Frodingham Company in England. A flow sheet of the Storrs process is shown in Fig. 31.

In the Utah plant, it was planned to remove the volatile matter as a series of separate fractions by withdrawing them independently from collector chambers spaced along the retort. This practice was expected to provide relatively pure materials with distinct physical and chemical properties, and to reduce the opportunity for cracking. No data on this arrangement have appeared. Pilot data from the retort indicated that a ton of Utah bituminous coal would yield approximately 1400 lb char, 23 gal oil, and 1200 cu ft fuel gas with a heating value of 1000 Btu/cu ft. This gas was sufficient to supply the energy requirements for carbonization. By further heating of the coal logs to 930 to 1090°C in a subsequent heater, a metallurgical coke was to be produced. The complete treatment was expected to furnish per ton of the Utah coal, 1200 lb coke, 29.5 gal volatile condensates, and 8000 cu ft of 600 Btu gas.

In a process resembling the Coal Log process, Reintjes coked a carbonaceous material, such as raw coal, by forcing it with or without admixing a fluid binder (depending on the thermoplastic nature of the material) under a slight or moderate pressure through a smooth-walled, externally heated metal tube, from which oxygen was substantially excluded.[119] The extent of carbonization depended on the length of the tube, temperature therein, and the time of contact. The char was extruded from the end of the tube as a substantially solid cylinder, which was, however, cracked and fissured as a result of shrinkage and the escape of vapors. As it was extruded, the cylinder broke off into lumps. The temperature of treatment was at least 430°C, preferably in excess of 590°C. Calcining of the product at higher temperatures usually proved advantageous. The process was particularly useful for preparing superior cokes for electrode-forming masses.

Tray Retorts

Retorts consisting of horizontal trays mounted in a vertical shell and resembling Nichols-Herreshoff furnaces are among the devices used for carbonizing coals.

Baumco process.[13,120] In the Baumco process for briquettes, the coal was subjected to a preliminary carbonization at about 800°C in a

[119] Reintjes, H., U.S. Pats. 2,755,234 (1956), 2,922,752 (1960).

[120] Baum, K., U.S. Pat. 2,825,679 (1958).

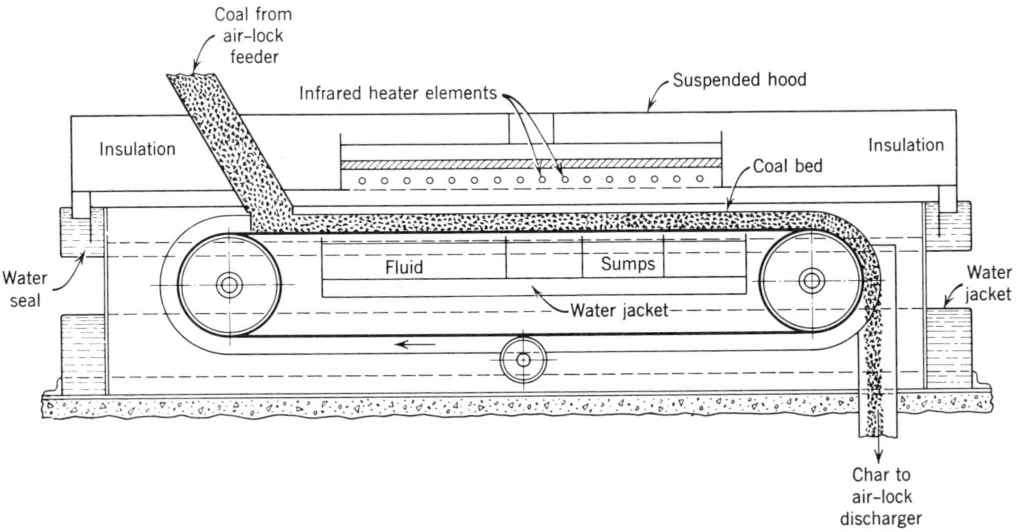

Fig. 31. Longitudinal section through Storrs low-temperature carbonization retort. (Courtesy of K. L. Storrs)

tray retort, termed a "devolatilizer," to produce a char without caking properties. This char was then mixed with a binder and with additional coal in proportions that would provide briquettes of the necessary strength. The briquettes were finally carbonized by contact with hot gases at approximately 1200°C. The process was recommended for carbonizing coals of very low coking power (probably with a swelling number not over 1), noncoking coals, and lignites.

In the primary carbonization, coal was used with a size range of up to 3 mm and containing not over 12 to 15% moisture. In the Herreshoff type of furnace, the coal was raked slowly in thin layers over the series of horizontal, circular trays, descending successively from one tray to the next. Temperatures not in excess of about 700°C were attained by direct contact with the hot gases from the second carbonization stage which were passed upward through the trays of the retort countercurrent to the flow of coal.

The char from the devolatilizer was then mixed with the binder, which consisted of up to 10% of a coking coal and up to 10% pitch. The mixture was briquetted, and the briquettes were carbonized at the elevated temperature in a continuous, vertical-shaft retort. This retort was of a type similar to ones described earlier in this chapter and was heated both directly and indirectly, with heating gases flowing through flues around the retort and upward through the shaft in direct contact with the briquettes. The lower section of the retort was used as a quenching chamber to cool the briquettes by the injection of product gas which had been cleaned and cooled in the tar-recovery system. Passage of the coal through the devolatilizer and the briquettes through the vertical retort required about an hour each. Because of dilution of the make-gas with the products of combustion, the volume was large, with a heating value of only about 140 Btu/cu ft. However, by preheating the combustion gases, it was reported that product gas with a heating value as high as 500 Btu/cu ft could be obtained. The finished briquettes were strong, dense, and uniform in size. By adjustment of conditions in the second carbonization stage, their volatile content and other properties could be varied. They appeared suitable for domestic, industrial, or metallurgical use.

Büttner retort.[8,34] The Büttner retort was developed in Germany before World War II for the low-temperature carbonization of oilshale or noncaking coal fines, such as floated coal or slimes. Figure 32 shows two cross sections of a Büttner retort. It consisted of a stationary cylindrical housing, insulated against heat losses. The housing enclosed a structure consisting of circular shelves which were revolved slowly by an electrical motor. A vertical shaft in the center of the shelf structure supported

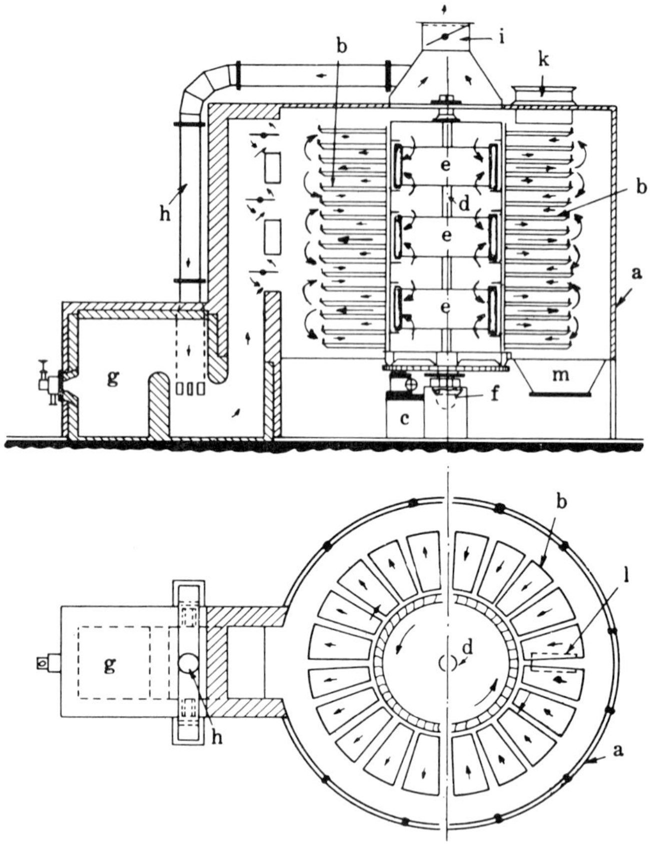

a. Housing
b. Revolving shelf with disc rings
c. Drive for shelves
d. Circulating fan shaft
e. Circulating fans
f. Gear drive for circulating fans
g. Combustion furnace
h. Recirculating pipe for combustion gases
i. Waste gas outlet
k. Coal feed
l. Coal slot
m. Char discharge

Fig. 32. Büttner coal drier adapted to low-temperature carbonization of fine coal.[8]

several circulating fans, the blades of which were sloped like those of a turbine. These fans were rotated in a direction opposite to the direction of rotation of the shelves. Hot waste gases or hot combustion gases produced by burning a fuel in a furnace were introduced into the cylindrical chamber, and then were drawn between the shelves by the turbine fans. From the top of the retort, a portion of the exit gases were recirculated back to the furnace and the remainder were released to the atmosphere. The coal to be carbonized was admitted uniformly at the top and distributed evenly on the top shelf. After the coal had completed one revolution on the top shelf, it was forced by means of a stationary adjustable scraper down through a slot onto the next lower-rotating shelf. After each revolution, the process was repeated until char was discharged at the bottom onto a belt, which transferred it to the screening plant. The coal was not agitated violently while passing through the unit, and dust formation was said to be small. No data on its performance as a drier has been found. However, it is similar in many respects to the Wyssmont turbo drier.

LOW-TEMPERATURE CHARS

Properties

Chars differ markedly from high-temperature cokes in appearance and properties. In general, pieces of low-temperature char are black to dull gray, with an open, porous structure. They are softer, more friable, and weigh less than high-temperature coke. Both the apparent and true densities are lower. The volatile content of the char usually lies in the range from about 20% down to 9 or 10%. When heated, a char gives off much hydrogen, and the gas burns with an almost nonluminous flame. Because of the high volatile content and the open structure, char ignites readily and burns freely. These qualities make a lump char an ideal fuel for the open domestic grate.

The high reactivity of chars necessitates their thorough quenching or cooling unless they are to be used very shortly. If a warm spot is retained in the center of a lump, the heat can spread, with eventual ignition of the entire mass, unless preventive action is taken. Hot, finely divided char can also ignite and cause serious trouble.

The test frequently used abroad for determining the reactivity of cokes or chars is the C.A.B. (critical air blast) method. The test is a measurement of the minimum current of air with which a given size of char will continue to burn after ignition under standardized conditions. The smaller the volume of air, the greater is the reactivity of the char, or coke, tested. Pound has published the following average values for several solid fuels, based on determinations in his laboratory:[121]

Char	0.017 C.A.B. Value
Anthracite	0.039
High-temperature coke	0.057

He further found that he could almost double the C.A.B. value of char by heating it at 900°C for 2 hr, thus decreasing its reactivity. The C.A.B. method is merely one among many such reactivity tests.

The properties of chars are greatly affected by the coals from which they have been produced and by the conditions to which they have been exposed during carbonization. With the infinite variety of coal materials, ranging all the way from peats or lignites up to the bituminous coals which have been processed, and with the hundreds of carbonization processes, each with its variety of conditions (among which temperature and time are especially important), few meaningful data correlating these variations with the properties of chars, and their relation to the different applications are available. Major problems in the commercialization of a low-temperature process have been adapting it to produce a satisfactory char from the coals available and selecting the coal giving the optimum products.

[121] Pound, G. S., *J. Inst. Fuel*, **24**, 61–8 (1951).

Utilization of Char

At the present time, low-temperature char has several potentially important applications, many of which have already been indicated in the descriptions of the processes.

Domestic or Industrial Fuel

The properties of an ideal char for domestic use were listed by Pound:[121] (1) Not friable or subject to breakage during transportation and handling; (2) suitable in size for the grate, and clean to handle; (3) readily ignitable with paper, wood, or other normal means, with the ability to catch quickly when a low fire is replenished; (4) smokeless in burning; (5) low as possible in ash; and (6) although higher in price than coal, superior to coal in heating and radiation efficiency, and consistent in quality.

The use of chars for domestic fuels finds its principal application in Europe. In the United States, gaseous and liquid fuels have largely displaced solid, smokeless fuels. Much domestic char is produced in the form of briquettes. In Great Britain, for instance, a great deal of domestic heating is still by open grate, and briquettes or lumps of char have proved attractive in this service.

Very little char appears to be used for industrial firing, this application especially being limited by the lack of good outlets for the tar. The results of preliminary combustion trials, according to Phillips,[122] indicated that standard fuel analyses were inadequate for predicting furnace performance with the chars now being produced in certain low-temperature carbonization processes. These chars, which he expected would become increasingly available as boiler fuels, were dry, fine, porous materials, usually containing less than 20% volatile matter. To aid in the selection of proper combustion equipment, Phillips studied their classification according to the ease of ignition and speed of carbon utilization by means of three test procedures: (1) an ignition temperature test; (2) a method for correlating the loss of

[122] Phillips, N. D., *Mech. Eng.*, **79**, 53 (1957).

unburned carbon in furnaces with the heating value of the volatile matter in the char; and (3) the determination of the relative combustion rates of fuels. In the ignition temperature studies, which measured the temperature at which a sample of the pulverized fuel produced a flame in air, low-temperature chars with 10 to 20% volatile matter exhibited ignition characteristics similar to those of low-volatile bituminous coals. In tests on the losses of unburned carbon made with different fuels in the same furnace, the carbon losses tended to decrease as the heating value of the volatile matter decreased. The usefulness of this method of correlation appeared to be restricted to the natural fuels, and additional information was required before such fuels as low-temperature char could be compared with these fuels of known performance. Results of the burning rate tests indicated that the low-temperature chars would burn in commercial furnaces at rates similar to fast-burning coals.

Blast Furnace Fuel

Char, even in lump form, has been unsuitable for direct use in the normal blast furnace or cupola, owing to its low strength and high reactivity. Its use in high-temperature coke mixes or in briquettes which are subsequently coked has been necessary before a char could be applied satisfactorily. In several plants, chars with volatile contents in the range of approximately 15 to 20%—near that of the low-volatile coal displaced—have been combined with the coal blend to the high-temperature ovens to produce a satisfactory metallurgical coke. The experience of operators, both in the United States and abroad, has indicated that the substitution of a slightly greater amount of char of the proper quality for the low-volatile coal has given good results.[27]

Price has listed desirable requirements for a coal to be used in the production of suitable chars:[123] (1) Capability for producing the proper quality of char from the point of view of coke improvement; (2) low ash content; (3) freedom from excessive gumminess on heating, or susceptible to blending to minimize such a tendency; (4) low contents of sulfur, phosphorus, and other objectionable impurities; (5) low moisture content, to avoid predrying; and (6) low delivered cost. Past work indicated that the following characteristics, other than the coal selected, required consideration in the production of a satisfactory coke when char is used as a component in the blend: (1) Amount of char—if more than approximately 20% is used, the resulting coke is grainy in structure and has lost strength; (2) pulverization of 100% through a $\frac{1}{8}$-in. screen has usually been adequate; (3) volatile content should preferably be at a point between 15 and 20% and variations from the selected value of more than 1% impair the effectiveness of a char, and (4) physical properties, especially surface condition. Physical characteristics have appeared largely dependent on the coal, the carbonization temperature, and the time in the retort. Better cokes have usually been prepared from chars produced as cenospheres or with a distorted surface, rather than from a char with an unaltered appearance. In general, carbonization temperatures approaching 750°C have resulted in better chars than those produced at lower temperatures of about 500°C. Although Price himself had produced chars by carbonizing Colorado noncoking subbituminous coal, he had not established to his satisfaction which low-temperature process provided the best char.

Reed and co-workers produced metallurgical coke of good quality by coking in test ovens coal mixes which included char made from Illinois coals.[124] Neither the percentage of char nor the uniformity of its volatile matter content in the range studied was found critical. However, the quality of the high-volatile coal used for blending was of great importance. Comparisons with chars produced in several types of retorts indicated that retort design and consequent operating procedure influenced the properties of the coke.

Burstlein patented a method by which durain, fusain, and clarain were separated from a coal prior to high-temperature coking.[125] These constituents were then precarbonized to char at 450 to 750°C and mixed with the balance of the coal before it was charged into the coke ovens. The constituents were separated by grinding the coal and screening.

[123] Price, J. D., Unpublished paper, Bogota, Colombia October (1952), 7 pp.

[124] Reed, F. H., Jackman, H. W., and Henline, P. W., *Illinois State Geol. Survey, Rept. Invest.* No. 187, 7–38 (1955).

[125] Burstlein, E. M., U.S. Pat. 2,752,293 (1956). Burstlein, E. M., and Société des Acieres de Longevy, Brit. Pat. 694,197 (1953).

Char was added to the coal mix in proportions of 15 to 25% by Petit for the production of metallurgical cokes.[88] Small spheres of 1- to 3-mm diameter, which were produced by a fluidized process, gave excellent results. Char from rotating horizontal retorts, however, proved unsatisfactory for the preparation of a good coke.

Japanese experiments on the use of char in the coal mixes used for the preparation of blast furnace coke have been described.[126] After World War II in Japan, it was necessary to produce coke for blast-furnace use, with at least an 85 drum index value, by using only domestic high-volatile coal. Hisada[126] also described experiments in which coke with a strength of over 90 on the drum index was obtained commercially by coking a mixture of a coal of high fluidity blended with about 30% of char containing 20% volatile matter, and ground finer than 0.3 mm in a tube mill. The optimum volatile content for the char proved to be 18 to 20%. Results of operation with the high-temperature coke produced from the blend in several blast furnaces were very good. Use of the char in the coal blends increased the proportion of carbon monoxide in the gas and the viscosity of the tar from the coke ovens.

Recent work in Australia extends earlier work on the use of char in production of metallurgical coke.[127]

When low-temperature carbonization has been used as a step in the manufacture of a high-quality metallurgical coke from a noncoking coal, precarbonizing the coal at low or medium temperatures has been the primary step. The char produced was crushed, if necessary, and the fine particles briquetted, usually using a coal-tar pitch. The briquettes were then carbonized in vertical ovens at temperatures sufficient to impart the strength required for metallurgical use, and at the same time drive off the volatile matter from the binder. The use of a coal-tar pitch of softening point 65 to 75°C, usually cutback with an equal quantity of tar oil, has been mentioned. Baum produced metallurgical coke by briquetting low-temperature char with a few percent of a swelling-type coal and a binder, and carbonizing the briquettes at high temperature in direct contact with hot gases.[120]

The successful production of coked briquettes was found to depend on securing full preshrinkage of the chars, produced on carbonization of an Australian subbituminous coal by Broadbent and co-workers.[128] They shrank approximately 20% over the temperature range, 500 to 1000°C, so it was necessary to carbonize the raw coal to at least 800°C to preshrink the char adequately. Thereafter, the char was briquetted with a suitable binder and carbonized at 500°C which gave a strong coked briquette.

Low-temperature carbonization was used as a prior stage in the production of a carbonized fuel, according to a patent of Berry, which disclosed three distinct stages of heating and carbonization, comprising: (1) heating the coal to 275 to 400°C, during which only moisture but no tarry vapors were evolved; (2) continuing heating to approximately 500°C to produce the low-temperature char; and (3) conversion of the char to high-temperature coke by heating to 900°C maximum.[129]

According to Malcor,[130] a small char (Coalite) was used directly as a fuel in a low-shaft blast furnace at Ougree, with excellent results. This char was made at Bruay from long-flame coal and had the following analysis: moisture 15% by weight, ash 12%, and volatile matter 9 to 12%. With dust-free St. Pierremont ore, the consumption of char was about the same order of magnitude as with breeze, so that the recovery on the basis of the fixed carbon was about 13% higher with the char. This indicated that the volatiles in the char played a part in reduction of the ore. The char appeared more reactive than breeze.

Reducing Agent

The reducing agent commonly used in the production of such elements as phosphorus and copper from their ores has commonly been a high-temperature coke. However, low-temperature processes, especially those which are a first step in the manufacture of briquettes, make possible the utilization of noncaking

[126] Maeda, K., *J. Fuel Soc., Japan*, **29**, 195 (1950). Kunii, S., *J. Fuel Soc., Japan*, **29**, 196 (1950). Hisada, K., Fuji Iron and Steel Co., Ltd., Tokyo, Report, October, 1958.

[127] Waters, P. L., and Hesp, W. R., *J. Inst. Fuel.*, **34**, 323–40 (1961).

[128] Broadbent, R., Donnelly, R. P., and Platell, N., *J. Inst. Fuel*, **28**, 3–16 (1955).

[129] Berry, G. A., U.S. Pat. 2,030,852 (1936).

[130] Malcor, H., *Iron & Coal Trades Rev.*, **177**, 1507–15 (1958).

coals. In this way, lower rank coals, especially the lignites and subbituminous coals in the extensive western coal fields, may be made available for wider commercial applications. The interest in this direction is indicated by the increasing attention to processes in which these western coals are employed.

Electrodes

Electrodes were produced from coal by Shea in a process which included carbonization of the coal at a temperature between about 300 and 550°C.[131] The volatile matter was reduced to between about 8 and 20%. The char was crushed, the particles briquetted under pressure without a binder, and the briquettes heated to a temperature in the same range. After a final crushing, the final char was coked at a temperature above 700°C.

Other Uses for Chars

Low-temperature carbonization has often represented a primary step in the complete gasification of coal. Noncoking coals have been charged to a carbonizing chamber located above the gas producer. Such processes have been discussed by Gumz and by Hubmann.[132] During World War II, some char was gasified in portable gas producers to drive motor vehicles, both in Japan and Germany.

Additional uses of char have been described, the service depending on some special property of the char, such as its reactivity. Munderloh pointed out that soda-treated lignite coke could be used satisfactorily as a substitute for charcoal in chemical processes, such as carbon disulfide production.[133] This finding was based both on experiments and on industrial experience.

Low-temperature carbonization was employed in laboratory tests as a step in the production of ethylene and aromatics from lignite.[134] Texas lignite proved much superior to North Dakota lignite and was carbonized at a temperature not over 550°C. The gases at the exit of the retort were immediately cracked at about 800°C to produce 6 lb of benzene and 40 lb of ethylene per ton of dried lignite.

Berg and Atkinson have pointed out that noncoking coals are distributed in great abundance throughout most of the country, and the char that would result from their carbonization is thus readily available.[135]

Because low-temperature carbonization can be applied to other carbonaceous solids, such as sawdust, coal fines, lignite, or peat, with the production of useful chars, the process may also be utilized to upgrade some of these materials, when they cannot be utilized directly with satisfactory results.[136]

PRESENT STATUS OF LOW-TEMPERATURE CARBONIZATION

Low-temperature carbonization is essentially a process for producing fuels—solid, liquid, and gaseous—from coal. The products from the various low-temperature processes, however, vary in quantities and qualities depending on the coal which is used and the nature of the treatments to which it has been exposed. The differences can be of great significance in the selection of a process for a given application, and careful study is necessary before a choice can be made. Although the recovery of chemical products from the tars and oils has frequently been proposed as a major objective, no installation in which they can be separated on a commercial scale appears in prospect.

As a producer of fuels, a low-temperature process is required to turn out products which can compete with similar types of fuels also available in liberal supply. Of the low-temperature products, the char represents the largest proportion of the original coal, and on that account its movement must be assured. Its cost is determined not only by the cost of the original coal but also by the cost of carbonization. To offset this additional processing cost, the coal must either be cheaper than the competing fuels, or the char should possess an advantage, such as better size, shape, analysis,

[131] Shea, F. L., U.S. Pat. 2,594,226 (1952).

[132] Gumz, W., *Gas-u. Wasserfach*, **88,** 192 (1957). Hubmann, O., *U.S. Bur. Mines, Special Rept.* (1957).

[133] Munderloh, H., *Brennstoff-Chem.*, **38,** 372–3 (1957).

[134] Montgomery, R. S., Decker, D. L., and Mackey, J. C., *Ind. Eng. Chem.*, **51,** 1293–6 (1959).

[135] Anon., *Chem. Week*, **83,** No. 14, 87 (1958).

[136] Frank, F., *Peat* **1,** 11–14 (1950). Christianson, B., *Acta Polytech., Chem. Met. Ser.* **3,** No. 6, 7–128 (1953). Naucke, W., *Brennstoff-Chem.*, **39,** 43–53 (1958). Oppelt, W. H., Cooney, J. P., Golob, E. F., and Kube, W. R., *U.S. Bur. Mines, Rept. Invest.* No. 5382 (1958), 55 pp.

reactivity, or other property. One example of the use of a cheaper coal would be the conversion of a low-grade coal, such as a lignite or brown coal, into an industrial or domestic fuel which is competitive with higher grades of coal. This condition will assume increasing importance as depletion of better grades of coal continues. The manufacture of formed cokes for metallurgical use, in which low-temperature carbonization is a primary step, should eventually be able to utilize such lower-grade coals, and consequently is of potential importance to the steel industry.

One desirable characteristic of char is its smokeless quality. This property, together with suitable form characteristics, has won for the lump char produced by certain processes important places in the domestic fuel market in Great Britain and in other European countries. Attention, moreover, is being given in Great Britain to conserving domestic coals of carbonizable quality for use in coking. In many localities where smoke-control regulations are in effect, smokeless fuels are being demanded, and consumers may find they do not have appliances which include devices for gas ignition. For these consumers, reactive low-temperature chars—possibly in the form of briquettes prepared from noncoking coals—which are under development, should be useful. It is expected that they will receive a premium price.

The situation of the tar is similar. Because the tar will compete with fuel oils, it should at least be comparable in characteristics with them and possess no property which will render it less desirable. The presence of excessive quantities of solid char in some tars has materially reduced their value. In Germany and Japan during World War II little petroleum was available, and the military demand for aviation fuels was acute and partially satisfied by motor fuels derived from the low-temperature tars produced in those countries. With the now adequate availability of petroleum products, the demand for the tar fractions for these uses has disappeared. Where petroleum is normally available, low-temperature tars have had difficulty in competing successfully.

The volume of the gas is small, and its heating value is often low because of dilution with combustion products. On a therm basis, the yield is about half that produced in high-temperature processes, and most of the gas is frequently consumed internally in the plant where it is made. The volume left has been insufficient to make the process of interest as a producer for city distribution.

The process descriptions have shown that many methods for the low-temperature carbonization of coal are available for use, and from them chars with a wide variety of characteristics can be obtained. The second product, also of major importance, has been the large volume of tar. It was long a common expectation that the values of these products, especially the chemicals to be derived from the tar, plus possible economies in the carbonization processes as compared to high-temperature coke ovens, would lead to attractive profits from a low-temperature plant. These opinions have been largely responsible for the great amount of work and for the development of the innumerable designs which have been proposed by engineers and inventors. Not only was much money spent in the legitimate development of new processes, but low-temperature carbonization also became a fertile field for cranks and promoters, especially during the 1920's. In 1931, Soule estimated that over $50,000,000 had been spent in the United States alone on low-temperature processes during the previous 10 years.[6] Of all the processes worked on during that period, only one—the present Disco process—survived. As a result of the widespread disillusion and the shattering losses, work on low-temperature carbonization almost dried up in the United States early in the 1930's.

The experience in Great Britain and on the Continent has been similar. However, due to the different pattern of fuel availability in these countries, especially the absence of the ample supplies of oil and gas that have been developed in the United States, work on low-temperature carbonization continued at a higher level of activity.

In spite of the years that have elapsed since the early interest in low-temperature carbonization, review of the experience of the 1920's can be of value to modern investigators. The disappointments and disillusionments resulting from low-temperature carbonization were well summarized by Soule.[6] He had just finished a long period of service with one of the largest commercial installations of a low-temperature process, which had been made in the United States and which had just ended in failure.

Low-temperature projects of the period were divided into three classes, according to their primary objectives: first, the development of low-temperature char as an improved domestic fuel; second, the use of low-temperature carbonization as a step in the complete gasification of coal to manufacture a cheap industrial gaseous fuel; and third, the replacement of bituminous coal as a boiler fuel by low-temperature char.

Justification for manufacture of low-temperature char as a domestic fuel had been based on the false premise that it could be made to sell at a cheaper price than bituminous coal. Actually, low-temperature chars at that time were unable to sell for less as a domestic fuel than did high-temperature coke. Disregarding the differences in operating techniques, there were two primary reasons for this condition: the lower byproduct credits for the low-temperature process, and the difficulties attendant on the transfer of heat into the coal charge.

The lower byproduct credits for low-temperature carbonization were due to the smaller yield of byproduct gas. This smaller yield was of vital importance, and was rarely offset by the increased yield of tar. Hence, with a lower byproduct return from the low-temperature process, it was still more essential to operate it at a lower cost in order for the low-temperature char to compete with high-temperature coke. To superficial observation, such a reduction appeared easy. Temperatures were much lower in the low-temperature processes—of the order of 540°C instead of 1100°C—and it would appear that this lower temperature level should require much less heat for carbonization of the charge. In addition, it might be expected that the capital cost of the installation could be materially reduced because of the substitution of iron and steel for the expensive refractories required in high-temperature ovens. These assumed operating economies proved illusory. Transfer of heat into coal is a very slow operation, and at low temperatures, without agitation of the charge, the rate of transfer is reduced well below that at high temperatures.[59]

Two alternatives were available, according to Soule, both of which seriously affected investment and operating costs, either coking the coal in thin layers or mechanical agitation of the charge. A thin layer construction, however, basically similar to high-temperature practice, multiplied flues and retorts and penalized the process with an excessive investment burden. To offset this, agitation of the charge was employed. Although this practice was often successful in reducing the number and cost of retorts, as a result of the agitation a weak, crumbly, light coke, much inferior in size and strength to that from the fixed retorts, was produced.

This situation left the low-temperature operator with two alternative products, a lump coke with a high investment in stationary retorts, or an inferior coke from agitated retorts of high throughput and lower investment and operating costs. Up to 1931, at the time of Soule's paper, no commercial enterprise had been able to operate profitably on either basis.

Soule considered the two other applications for low-temperature carbonization—a step in complete gasification of coal and the production of solid fuel for boiler use—more rational in conception. Complete gasification involved production of a low-heating-value gas for industrial use. His experience led to the conclusion that this should be a two-step operation, with lump char made in stationary retorts and then transferred directly, while hot if possible, to a blue-water-gas generator in which the gasification step took place. Size and properties of the char would be of lesser importance. The low-temperature tar would be a byproduct of the process, but, since, in 1931 its value only slightly exceeded that of petroleum fuel oil, the credits from its production were not enough to offset the higher costs of the lump char.

In discussing the manufacture of a boiler fuel, Soule leaned toward an operation in which finely divided coal was passed continuously through a retort, heated either externally or internally. Apparently he had in mind an approach to the present-day fluidized process. Two of the difficult operating problems he visualized were the separation of the finely divided dusts entrained with the outlet gases and the cooling of the finely divided coke. However, the char could not compete as a boiler fuel unless it could be produced for less money than the cost of the competing coal or other fuel.

Finally, great difficulty was incurred in developing applications for the fractions or components of the low-temperature tar, and this problem still appears to be unsolved.

With the advent of fluidized or entrained processes operating on finely divided coals, in units capable of handling large tonnages, some of the difficulties inherent in the earlier retorts that plagued Soule and his colleagues have apparently been reduced or even eliminated. The char from the new processes is, of course, produced in a finely divided form, suitable for industrial firing. To obtain a product in lumps, the fine char must be briquetted, and, since the binder will give off smoke on heating, a second carbonization step must be applied to the briquettes. This means that instead of a single operation—the fluidization step—three will be required, fluidization, briquetting, and final carbonization.

The alternative to producing a lump char by briquetting is to carbonize a coking coal in such a way that it forms lumps. Soule also indicated some of the difficulties which could be encountered. In a high-temperature oven, the coke shrinks and pulls away from the walls of the oven as carbonization is completed. But, in a low-temperature slot retort, carbonization may be complete when the char has actually expanded, so that discharge of the coke by pushing may be entirely impractical. To permit discharge from these low-temperature ovens, in a number of cases the walls have been considerably tapered. With other retorts, the walls have been made movable so that they can be swung aside to permit the char to drop out onto a car or conveyor. These measures can add materially to the cost of an installation.

In carbonizing a coking coal in a continuous retort, heating eventually puts it in a plastic condition where it balls up into masses that harden on the walls or screws of the retort. These accumulations may result in irregular operation or even cause production to cease. Accumulation of irregular masses on the surface causes uneven temperature conditions in the retort walls with subsequent warping and distortion. Many attempts have been made to cope with such accumulations.[112] Building heavier retorts has usually proved impractical, and periodic disassembly for cleaning has often proved to be the ultimate answer, with a resulting serious loss in production. Pretreatment of the coal can be employed, but it adds to the cost of the operation.

On the basis of his experience with low-temperature processes, Price listed the requirements he considered necessary for successfully producing a char to be used in place of low-volatile coal in the mix for blast furnace coke.[27,123] These requirements are deserving of consideration in connection with any new installation for producing a char.

> The prime requisite is that the char produced in the retort will make a satisfactory grade of coke, the satisfactory use of the coke in the blast furnaces being the criterion of its quality.
>
> Agitation of the coal within the retort appears to be desirable to prevent the coal from caking and to insure efficient and uniform heating of all coal particles.
>
> The retort should have as few interior projections as possible consistent with proper working to prevent build-up of carbonaceous material.
>
> The process of feeding coal and discharging char should be continuous.
>
> Access to the interior of the retort for inspection and cleaning should be easy.
>
> Control of heating should be adequate for handling of different grades of coal and for producing the desired quality of char therefrom, and the retort should be efficient in its consumption of fuel.
>
> It should not require that the coal be closely sized nor that all fines be removed from the coal for its satisfactory operation.
>
> It should have a relatively high rate of daily throughput in order to reduce the number of separate units required and to simplify operation practice.
>
> It must be capable of being operated at sufficiently high-temperatures to insure the production of char of the proper chemical and physical properties.
>
> It should be of sufficiently rigid construction to withstand deformation in event of emergency shutdown.
>
> It should be of simple and rugged design, making possible low repair, maintenance and operating cost.
>
> Installation cost should be as low as possible consistent with good quality of construction and with the other requirements as stated.
>
> It should be provided with adequate dust collectors in order that the tars and liquid products be not unduly contaminated by char dust.

Hansen has named five necessary conditions to be met if a low-temperature carbonization process is to be rendered technically and economically feasible: (1) universal applicability to different types of high-volatile coals; (2) production of a firm, abrasion-resistant product; (3) maintenance of desirable temperature

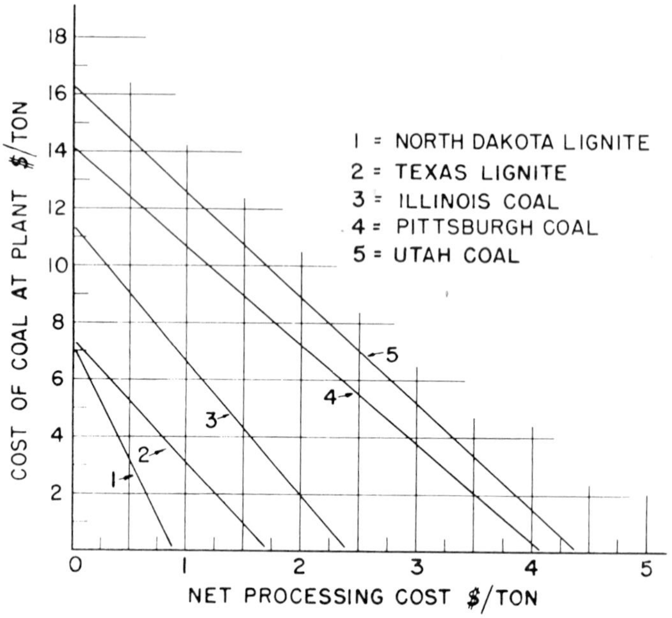

Fig. 33. Relationship of cost of coal to net processing cost when tar value is 10 cents/gal (fuel for power plants).[59]

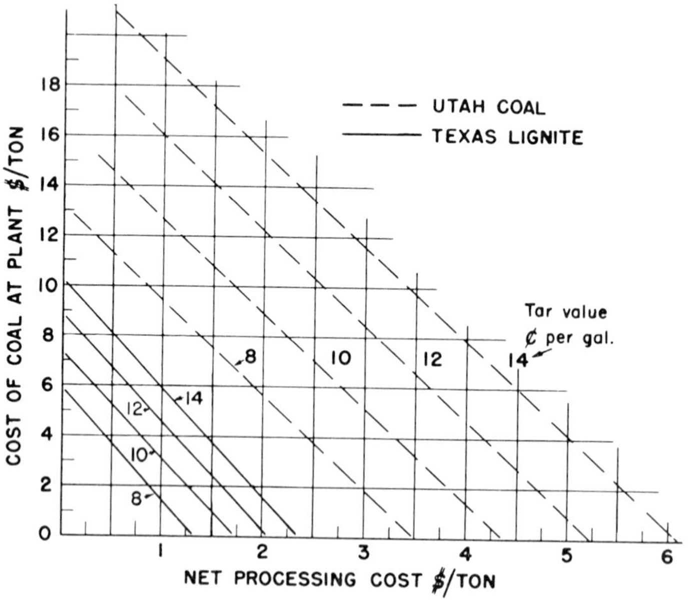

Fig. 34. Relationship of cost of tar to net processing cost for Utah coal and Texas lignite (fuel for power plants).[59]

conditions throughout the system to produce a uniform tar; (4) maintenance of an output which is economically attractive, a condition dependent on rapidity of heat transfer and involving oven design and the thickness of the coal bed; and (5) smooth, trouble-free oven operation.[137]

Low-temperature processes in the United States and Canada have operated largely on an experimental basis, with two exceptions. The two processes in regular commercial use are the Disco process of the Consolidation Coal Company, this plant being located near Pittsburgh, Pennsylvania, and the stoker carbonization process, operated by Shawinigan Chemicals Ltd. in Canada.

In addition to these plants, the Parry process has been operated experimentally on lignite by the Texas Power and Light Company at the power installation for the Aluminum Company of America at Rockdale, Texas. The Lurgi-Spülgas process also has operated in North Dakota for some time on the local lignite. A process for producing a metallurgical formed coke from Wyoming subbituminous coal was announced, and production of a low-temperature char is one of the first steps. This plant is operated by Food Machinery and Chemical Company and the United States Steel Corporation at Kemmerer, Wyoming.

In Europe, Japan, and other parts of the world, the Krupp-Lurgi, Lurgi-Spülgas, and Rochdale gas recirculation appear to be most widely in use, but work is proceeding at a number of locations on smaller-scale units of other carbonization methods.

These few plants provide a limited engineering basis for the selection or installation of a low-temperature process to meet a given need. Although many processes have been tried, they have been operated on different coals, in different ways, and for widely different objectives. Little correlation between them is available. Each process has often functioned on pretty much an experimental scale, without reaching the more critical stage of successful commercial practice. Numerous factors, such as handling the dust in the outlet gases and cooling the char, although possibly of lesser importance in a pilot installation, can become serious problems when a process becomes mature and a full-scale plant is required. The engineer entering the low-temperature field finds it almost a virgin one, which encompasses first the selection of a process for a given need, then the design and engineering with the identification and definition of the problems to be faced, and finally plant operation and meeting the unexpected developments. For these reasons commercial evaluations of processes are still very difficult.

[137] Hansen, C. J., *Brennstoff-Chem.*, **31**, 308–17 (1950).

Little significant information on the costs of low-temperature processes is available, but reliable estimates are important in considering an installation. Parry has given a useful relation between the costs of the coals employed, the cost of processing, and the value of the tar.[59] For the use of the char as a power-plant fuel he expressed this relation by an equation, which can be transposed into several forms:

$$P = VT - C(1 - R)$$

where P = net cost of processing and handling fuel to the burner, which equals the total processing cost, including crushing, drying, carbonization, tar condensing, and handling of materials, minus the total estimated cost of burning raw coal, including crushing, pulverizing, and handling; C = cost of coal delivered to the processing plant; R = net heat ratio, char/coal, obtained from assay and analysis; T = tar yield, gallons per ton, obtained from assay; and V = break-even value of the tar, cents per gallon.

Parry's formulas are illustrated graphically in Figs. 33, 34, and 35.[59]

Figure 33 compares 5 coals and shows the relationship between cost of coal and net processing cost when the crude tar is valued at 10 cents per gallon. From this graph it can be seen that, if North Dakota lignite costs $1.00 at the plant, the net processing cost must be less than $0.75 to be profitable when tar is worth 10 cents per gallon. Likewise, if Texas lignite costs $1.00 per ton, the net processing cost must be less than $1.45 per ton. On the other hand, the higher-rank coals, which produce large quantities of tar, might cost $4.00 per ton at the mine and still be profitably processed if the net processing cost is less than $3.00 per ton. This figure illustrates the wide spread in the basic economics of processing different coals because of the great variation in properties of coal.

The margin available for processing increases rapidly as the value of tar increases. Figure 34 shows this relationship when Texas lignite and a high-volatile Utah coal are considered. If the

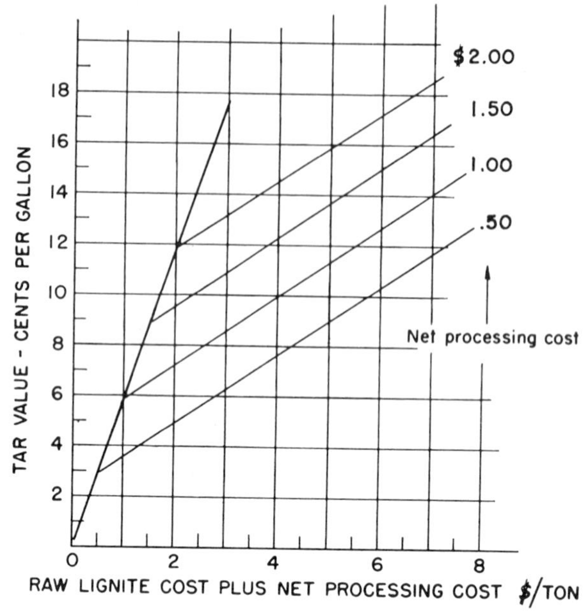

Fig. 35. Relationship of costs and tar value when processing Texas lignite for powerplant fuel.[59]

value of tar increases from 10 to 14 cents per gallon, the net processing cost of Texas lignite can increase from \$1.50 to \$2.20 per ton if the cost of lignite can be kept as low as \$1.00 per ton at the plant. When \$4.00 Utah coal is considered, the net processing cost can advance from \$3.25 to \$5.00 per ton as the value of tar increases from 10 to 14 cents per gallon. Figure 35 shows the cost relationships for Texas lignite in a different form.

In general, this study shows that processing coal for power-plant fuel will be most economical when the plant is integrated with the coal-mining operation and the coal is produced at the lowest cost. Coals having high potential tar yields might be shipped considerable distances and still be processed successfully at the plant. For example, Utah coal can be delivered on the west coast for \$11.40 per ton, but at this cost the net processing cost must be less than \$1.35 per ton to be profitable when crude tar is valued at 10 cents per gallon. But if the tar is worth 14 cents per gallon, the net processing cost may advance to about \$3.20 per ton.

There is no commercial low-volatile coal in the Western States, and the steel plants are obliged to import low-volatile coal from Oklahoma and Arkansas to blend with available high-volatile coking coals to make satisfactory coke. In some instances medium-volatile char made from the local coals can displace the imported low-volatile coal, so there is an incentive to process local coals to meet this demand. In some plants, up to 25 per cent of the coal is imported, whereas at others the amount ranges from 12 to 15 per cent.[59]

A. D. Little has estimated the gross receipts which might be derived per ton of North Dakota lignite from the char and the products potentially available from the tar.[138] The use of lignite as a fuel has been on a comparatively small scale because of its high water content—almost 50%—and its tendency to heat spontaneously in a pile, which makes storage difficult. From the \$3.03 gross receipts shown by A. D. Little should be subtracted the various costs chargeable against the operation: the cost of the lignite itself, any transportation, costs of processing and of refining the tar, the costs of transportation of the various products to the market, fixed charges on the investment, and taxes. With lignite costing about \$2.00/ton at the mine, only about \$1.00 is left to cover the other costs.

[138] Anon., *Chem. Eng. News*, **33**, 3561 (1955).

CHAPTER 11

HIGH-TEMPERATURE CARBONIZATION

W. BEIMANN, H. S. AUVIL and D. C. COLEMAN

Allied Chemical Corporation, New York, N.Y.

Coke production by high-temperature coal carbonization in horizontal chamber coke ovens has continued to parallel pig iron production in the United States, although there has been a steady rate of decrease in the coke requirement per ton of iron.[1] (Fig. 1.) Generally uneconomical, except in periods of peak demand, few beehive ovens have been constructed until 1959, when a reversal of the trend took place owing to the lower capital investment required and the increased encroachment of the petroleum industry on the markets for chemicals previously produced principally by carbonization of coal. The increased usage of natural gas had led to the decommissioning of many merchant coke plants that have lost their markets for coke-oven gas.[2]

[1] DeCarlo, J. A., Hunter, T. W., and Otero, M. M., *Minerals Yearbook* 1959, U.S. Bur. Mines, Vol. II, 187–247 (1960).

[2] President's Materials Policy Commission, *Resources for Freedom*, Supt. of Documents, Washington, D.C., 1952, vol. IV, chaps 3 and 12.

PRESENT-DAY DESIGN OF HIGH-TEMPERATURE COKE PLANTS

High-temperature horizontal-chamber coke ovens have remained essentially unchanged since the 1940's.[3] However, developments in coal preparation, coal blending, and increased knowledge of the coking behavior of coals and coal mixtures have encouraged the construction of larger ovens to meet the demand for metallurgical coke.

The 21-yr period from 1939 to 1959 witnessed a marked change in the number of coke ovens in operation, annual capacity, and number of design-constructors in the United States (Table 1).

[3] Denig, F., *Chemistry of Coal Utilization*, John Wiley and Sons, New York, 1945, Vol. 1, pp. 774–833. Grosskinsky, O., *Handbuch des Kokereiwesens*, Knapp, Düsseldorf, 1955, Vol. 1, 574 pp.; 1959, Vol. 2, 542 pp. Ress, F. M., *Geschichte der Kokereitechnik*, Glückauf, Essen, 1957, 672 pp. Farafanow, W., *Gas World*, **140**, Coking Sect. 35–44, 46 (1954).

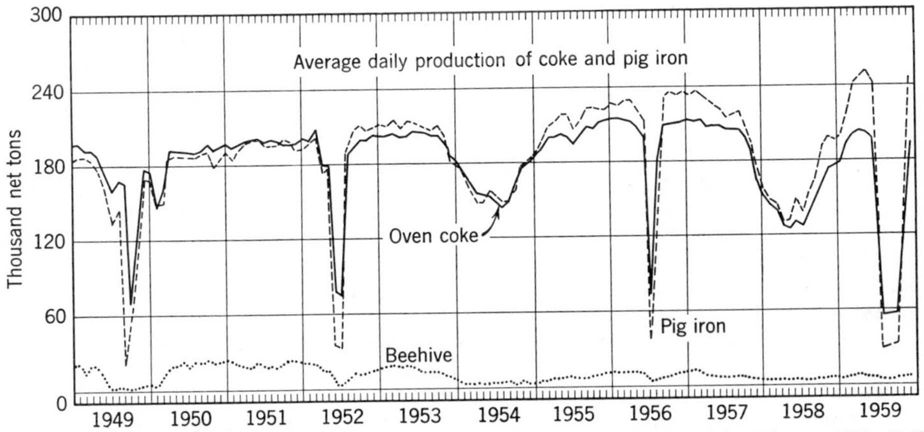

Fig. 1. Average daily production of oven and beehive coke in the United States 1949–1959.[1]

TABLE 1

NUMBER OF COKE OVENS IN THE UNITED STATES.

Make of Ovens	Number of Ovens in Operation	
	1939	1959
Koppers (including Becker)	10,061[4]	11,280[5]
Wilputte	771	3,260
Semet-Solvay	1,832	1,350
Simon-Carves	—	40
Otto	—	63
Cambria	120	—
Improved Equipment	60	—
American-Foundation	55	—
Curran-Knowles	46	—
Parker-Russell	27	—
Roberts-Morrisey	25	—
Piette	8	—
Total number of ovens	13,005	15,993
	Annual Coke Capacity in Tons	
Byproduct Coke Ovens	62,000,000[6]	81,447,700[5]
Beehive Ovens	9,850,000	4,368,800

It will be noted that the number of oven makers in the United States with ovens in operation has decreased. Simon-Carves and Otto are newcomers in the United States. The 23 – % increase in the number of ovens and the 31 – %increase in annual capacity, indicates the trend to larger-capacity ovens.[7]

Although basic designs have remained unchanged, the demand for uniformity of the coke product and labor saving have led to the increased use of mechanized coal bin gates, mechanical dischargers on coal charging car hoppers,[8] double collecting mains,[9] self-sealing doors,[10] more efficient door extractors, improved vertical heat distribution in heating flues with rich gas fuel,[11] and the use of blast furnace gas[12] for underfiring.[13]

The transfer of coal from bin to charging car through manually operated gates is not only time consuming, but erratic. Volume and density vary from hopper to hopper with the same and with different operators. Since these variations are then transmitted to the oven charge, the coke product is inconsistent from shift to shift, oven to oven, and even within a single oven. This situation has been alleviated by interconnecting all bin gates in a single row so that they are opened and closed simultaneously by a mechanism mounted on the larry car and interlocked with its drive. So operated, all charging car hoppers are filled rapidly and uniformly, and minor adjustment to compensate for variations in moisture content and bulk density is made by movable sleeves. Accurate

[4] U.S. Bur. Mines, *Directory of Byproduct-Coke Plants in the United States, July 1, 1939.*

[5] DeCarlo, J. A., and Otero, M. M., *U.S. Bur. Mines, Inform. Circ.* No. 7996, 13–20 (1959).

[6] Denig, F., *loc. cit.*

[7] von Massow, H. J., and Holt, H., *Gas World*, **141,** Coking Sect. 121–4 (1955). Kustov, B. I., *Stal'*, **15,** 391–6 (1955). Mohr, J. S., Jr., *Blast Furnace Steel Plant*, **47,** 277–80 (1959).

[8] Homan, P. J., U.S. Pats. 2,724,514 (1955), 2,865,520 (1958). Griffin, F. R., U.S. Pat. 2,827,182 (1958). Freebury, L. S., *Coke and Gas*, **18,** 353–7 (1956).

[9] Nadziakiewicz, J., Warmuziński, J., and Hoffman, H., *Prace Głównego Inst. Górnictwa, Komun.* No. 123 (1952), 19 pp.

[10] Anon., *Coke and Gas*, **14,** 107–8 (1952). Cellan-Jones, G., *Coke and Gas*, **15,** 221–2 (1953); **16,** 307–10 (1954). Homan, P. J., U.S. Pat. 2,744,858 (1956). van Ackeren, J., U.S. Pats. 2,338,675 (1944), 2,478,215 (1949).

[11] Wethly, F., U.S. Pat. 2,564,140 (1951). Litterscheidt, W., *Glückauf*, **88,** 1137–49 (1952). Seppar, A. M., and Agushevich, I. Z., *Koks i Khim.*, **1957,** No. 1, 19–23.

[12] Locke, C. R., *AGA, Proc.*, **1951,** 670–82.

[13] Beck, K. G., *Glückauf*, **95,** 65–71 (1959).

bulk density control, however, minimized this volumetric adjustment.

The next and equally important step from the point of view of both uniformity of coke product and labor saving, is the transfer of the coal charge from the charging car to the oven. This transfer has been accomplished by three means, namely: (1) rotary table feeders on each hopper, which assure positive discharge at a rapid and uniform rate; (2) the location of and (3) the shape of charging holes to permit free access for the entire charge without interruption (Fig. 2). Charge holes are spaced asymmetrically to correlate oven taper with the usual equally sized charging car hoppers and flared on the underside. So positioned and formed, the entire charge may enter the oven without resort to the leveler bar. It will be noted (Fig. 3) that the widely flared holes leave ample passage way for initial make-gas to reach the collecting mains, and since peak trimming is reduced to about 1½ strokes of the leveler bar, the charging operation is rapidly completed and the oven resealed, giving crude oven gases little opportunity to escape to atmosphere.

Further improvements are obtained by the use of two collecting mains, one on each side of the battery, operated on the Beimann cycle.[16] Dual mains and the cycle permit control of oven temperatures above the charge to prevent formation of roof carbon inherent to some coal mixtures and improve coal chemical yields. Working conditions on the battery top are also improved, and since the maximum distance gas must travel to the collecting main is reduced to less than half the battery width, there is even less opportunity for gas emission, and smokeless charging becomes a practical reality.

Practically all coke ovens constructed in the United States since 1940 have been equipped with self-sealing doors. Door luting, pug mills, and clay carriers are all obsolete. Both the oven and leveler doors are self-sealing with renewable stainless steel sealing edges (Fig. 4).

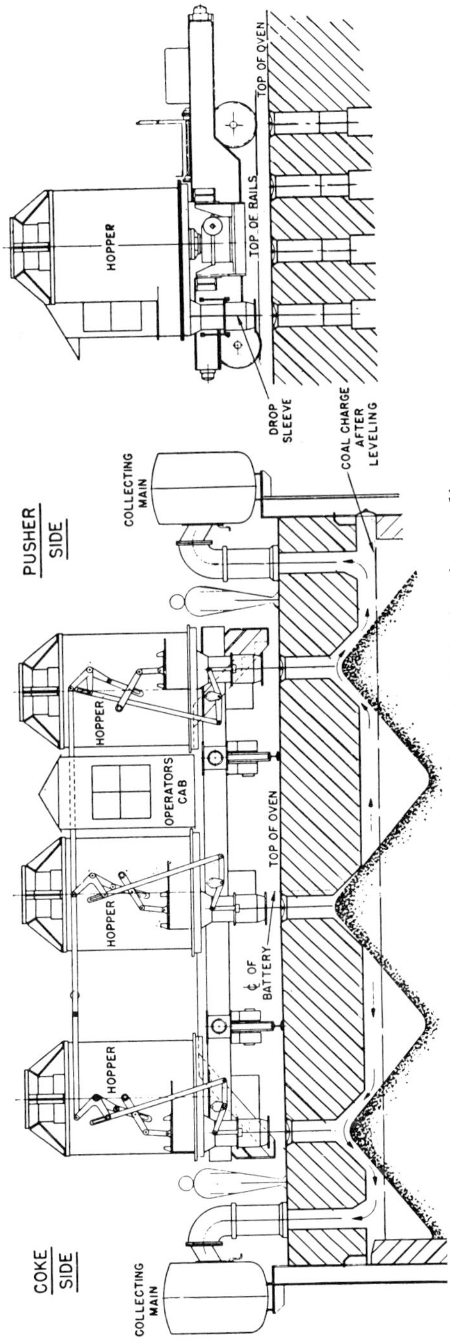

Fig. 2. Rotating table charging car.[14]

Good operation calls for removal of carbon and pitch deposits on sealing edges and surfaces each time the door is removed. So operated, these doors give many trouble-free years of service. This cleaning, however, is perhaps the most onerous task of the entire coking operation. Some progress has been

[14] Wilputte, L. N., and Wethly, F., U.S. Pat. 2,488,953 (1946).

[15] Wilputte, L. N., and Wethly, F., *Blast Furnace Steel Plant*, **34**, 355–67 (1946).

[16] Barritt, R. J., *Coke Smokeless-Fuel Age*, **7**, 25–8 (1945).

[17] Wilputte Coke Oven Division, *Allied Chemical Corporation Catalogue* (1954).

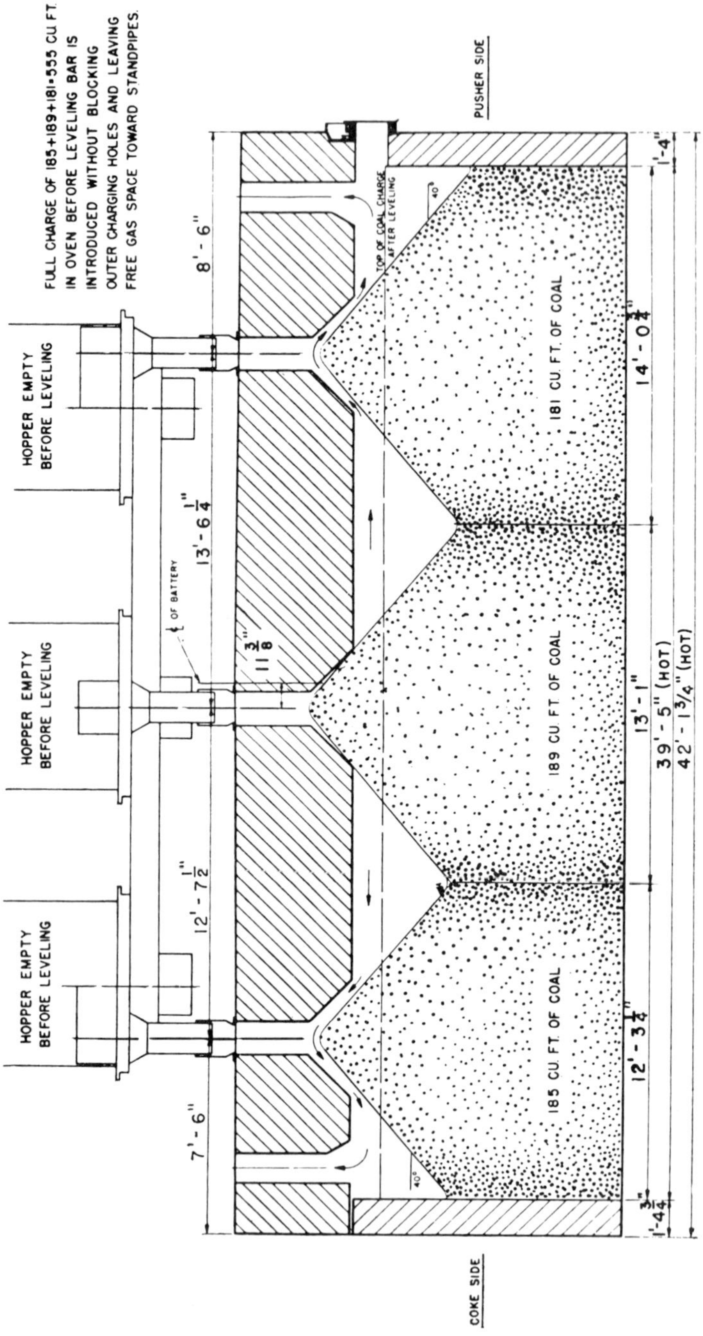

Fig. 3. Uniform oven charging.[15]

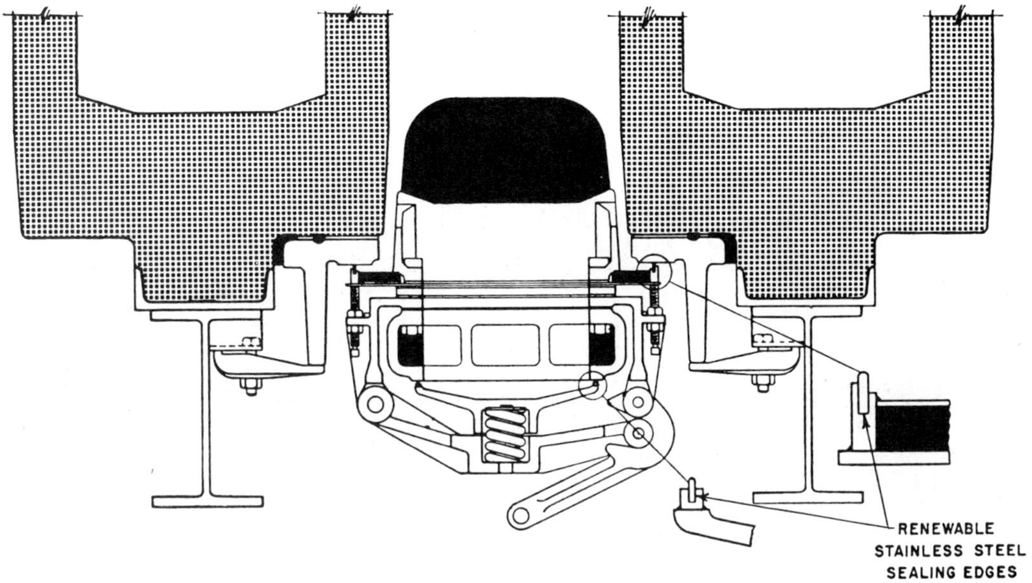

Fig. 4. Cross section through self-sealing door, door frame, and face of oven.[17]

made toward mechanical cleaning, and when a satisfactory cleaner is evolved, not only will hand cleaning be eliminated, but the hard-to-reach areas on high oven doors will not be overlooked.

Oven door handling equipment has undergone considerable development and improvement such that today the entire operation is manipulated by finger-tip controls. In both the electromechanical and electrohydraulic types commonly used in America today, stress has been placed on smoothness and continuity of the sequential movements to minimize undue stresses in the door structure. The hydraulic door extractor unit developed by Wilputte is identical for both sides of the battery. On the pusher side it is made integral with the pusher machine, and so arranged that when the extracted door is swung into cleaning position the pusher ram has direct access to the oven, the pusher machine being aligned for both operations in a single step.

Blast-furnace gas has become an increasingly popular underfiring fuel, and, where the design provided adequate regenerator capacity for the preheating of combustion air and fuel gas, the carbonizing rate has exceeded expectations.

In 1942,[18] it was predicted that straight 95 Btu blast-furnace gas was too lean to maintain a coking rate of more than 1 in./hr and that some admixture of rich gas would be required. However, several batteries have since been built and operated at $1\frac{1}{8}$ in./hr on straight blast-furnace gas analyzing only 90 Btu/cu ft.[19] The economy of blast-furnace gas underfiring for plants consisting of only a few coke-oven batteries has until recently been impaired by the requirement of a large gas holder to supply gas during interruption at the blast furnace. An automatic control system has subsequently been evolved to eliminate the holder while maintaining the same degree of safety. One installation serving a minimal combination of two blast furnaces and two coke-oven batteries has proven very successful since its completion in 1958.

The trend in coke-oven design, in keeping with similar developments in other and related fields, appears to be toward larger ovens, with more mechanization of various operations. This trend is already manifest in western Europe where large ovens up to 1160 cu ft[20] in capacity are producing acceptable metallurgical coke, and it has been reported that similar large-sized ovens are being built in Russia.[21] Inasmuch as the economy of some of the larger ovens, both as to initial costs and operating costs, is much better than that of the

[18] Rueckel, W. C., *Blast Furnace Steel Plant*, **30**, 331–8 (1942).

[19] Wilputte Coke Oven Operating Reports, unpublished.

[20] For example, at Gelsenkirchener Bergwerke, A. G., Gelsenkirchen, West Germany.

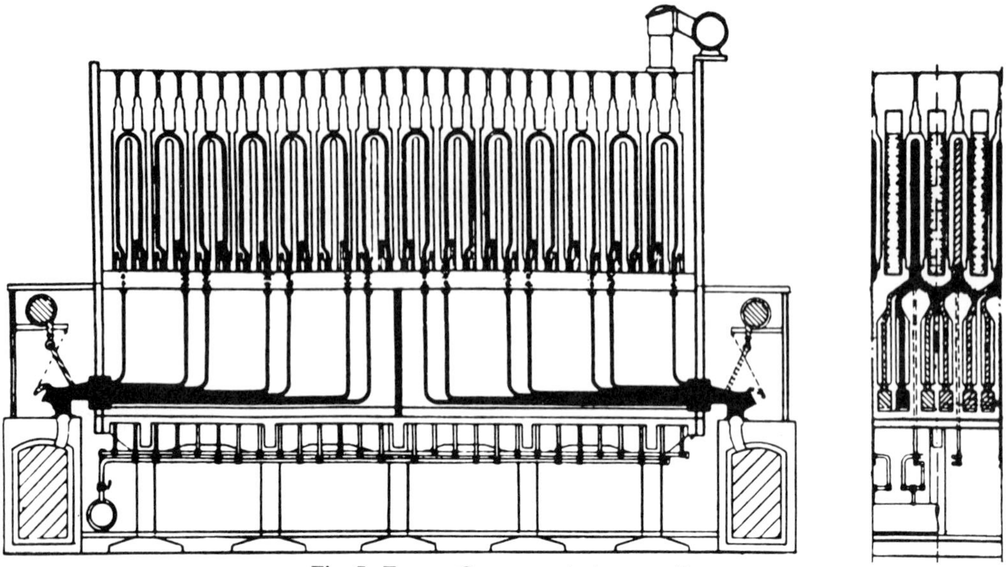

Fig. 5. Evence Coppee underjet oven.[22]

smaller American ovens, it is anticipated that they will eventually be constructed in this country.

Outside the iron curtain, nine different types of coke ovens are being built.

1. The Evence Coppee and Company, Brussels, Belgium, builds mainly underjet ovens and only occasionally gas-gun ovens, for rich- and lean-gas underfiring. The ovens are of the hairpin flue type. The hairpin flues, built in pairs, are so arranged that two adjacent flues are alternately on upflow or downflow. For vertical heat distribution, the lean gas and air meet at floor level. The rich-gas flame is raised by means of a division wall between the gas nozzle and air port, the height of the division wall being determined by the composition of the fuel gas (Fig. 5).

2. Didier-Kogag-Hinselmann G.m.b.H., Essen, Germany, builds gas-gun and underjet ovens, for rich- and lean-gas underfiring. The ovens are of the two-divided regenerator type. With gas burning in the pusher-side flues, the combustion gases pass through a horizontal flue and descend down the coke-side flues.

[21] *Am. Iron & Steel Inst., Rept., Steel in Soviet Union*, June, 1958, p. 182.

[22] Grosskinsky, O., *loc. cit.*

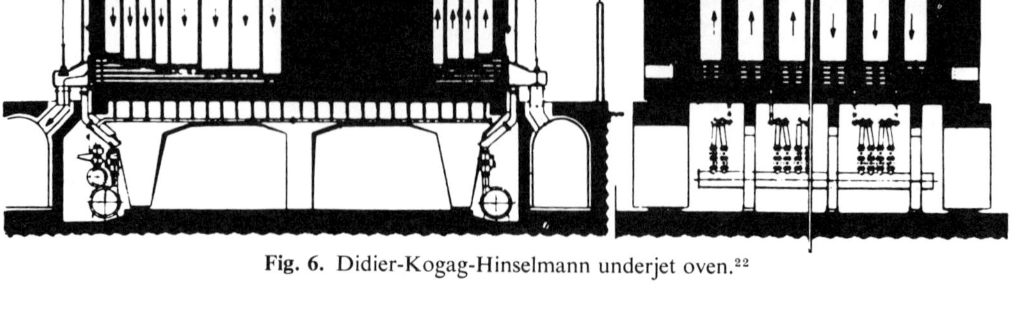

Fig. 6. Didier-Kogag-Hinselmann underjet oven.[22]

periodical reversal changes the flow in the opposite direction. The regenerators are divided into cells, each cell controlling the air or lean gas to the flues lying above. Adequate vertical heat distribution with both rich and lean gas is achieved by providing the header brick between the flues in the upper part of the oven walls with bulges. The rays of the glowing carbon particles of the short rich-gas flame are absorbed by these brick, and the heat is transmitted to the upper part of the wall. The longer lean-gas nonincandescent flame does not affect these brick. On the other hand, these brick do not impair heat distribution from long lean-gas flames (Fig. 6).

3. Les Fours Lecocq S. A., Brussels, Belgium, builds underjet ovens for rich- and lean-gas underfiring. The ovens are of the hairpin flue design with all even numbered and all uneven numbered flues alternately on upflow or downflow. The regenerators are divided into cells with one cell serving one pair of flues. The amount of air or lean gas to each cell can be regulated from the battery basement (Fig. 7).

4. The Koppers Company, Inc., Pittsburgh, Pennsylvania, builds the Koppers-Becker oven as a gas-gun or underjet oven for rich- and lean-gas underfiring. The regenerator chambers are single cells extending from pusher to coke side. The upflow lean-gas regenerator is sandwiched between two upflow air-regenerators, adjacent to which are the downflow waste-gas regenerators. All flues of every second wall burn simultaneously. The combustion products gather in a number of horizontal flues serving two or more flues, usually four, from where they are taken through ducts in the oven-roof brickwork, over the coking chambers to similar corresponding horizontal flues in the adjacent oven walls, and thence down all the heating flues of these walls. Adequate vertical heat distribution with both rich and lean gas is achieved with the Koppers-Becker underjet oven by a connection from the bottom of each heating flue to the rich-gas riser-duct to each flue of the adjacent wall. The ejector effect of the rich gas draws waste gas from the downflow heating flue, diluting the rich gas and developing a longer flame. The ejector does not function when heating with blast furnace gas, which burns with a long flame (Fig. 8).

5. Heinrich Koppers G.m.b.H., Essen, Germany, builds the so-called "Kreisstrom" oven for rich- and lean-gas underfiring. The heating flues are arranged as hairpin flues with all even numbered and all uneven numbered flues alternately on upflow or downflow. The regenerator chambers are divided along the center line of the battery. The waste gas from one-half of an oven wall passes directly to the regenerator below, the waste gas from the other half of the wall passes through a horizontal flue in the

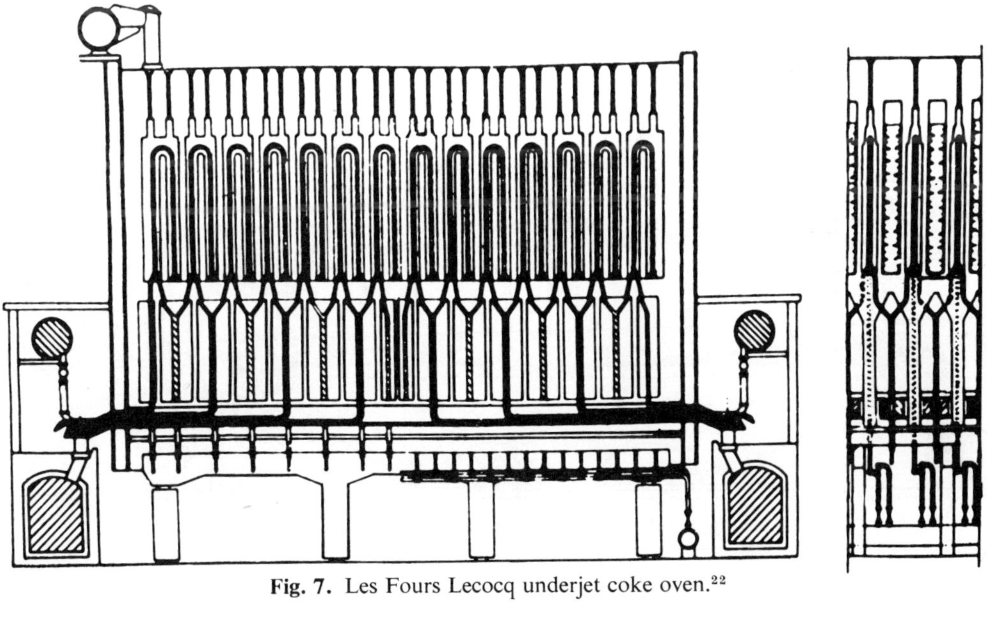

Fig. 7. Les Fours Lecocq underjet coke oven.[22]

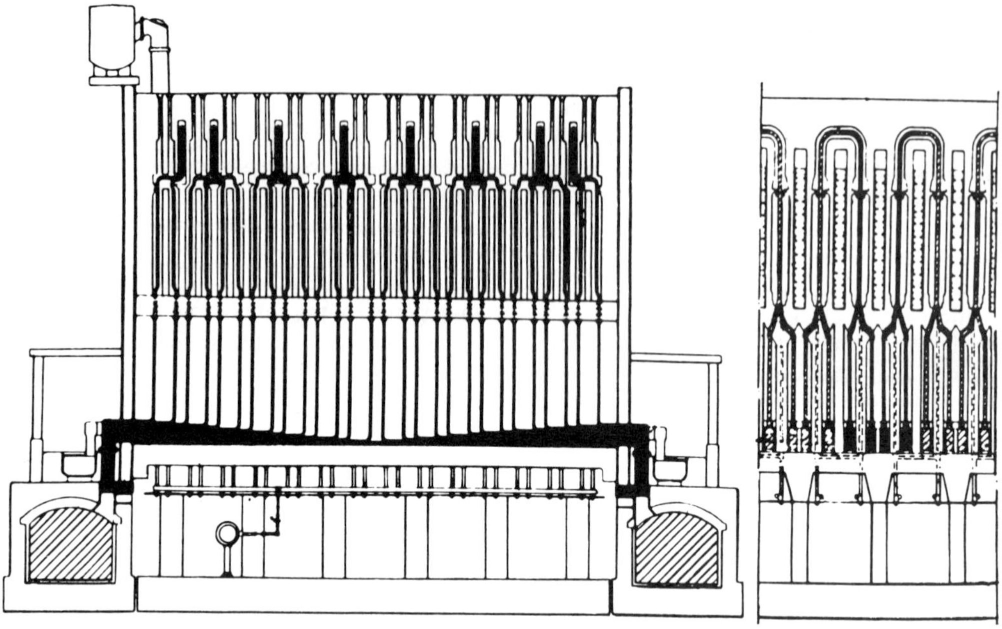

Fig. 8. Koppers-Becker underjet oven.[22]

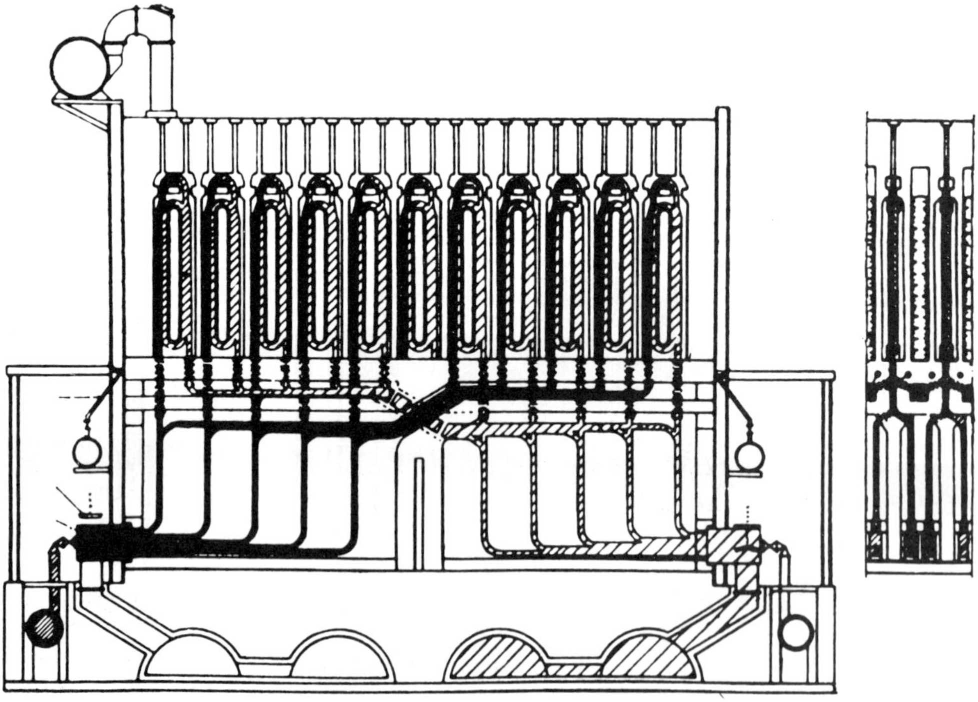

Fig. 9. Heinrich Koppers "Kreisstrom" oven.[22]

corbel and thence to the same regenerator section so that all pusher-side and all coke-side regenerator sections are alternately on upflow or downflow. To lengthen the rich gas flame, an opening is provided at the bottom of the hairpin flue pair so that waste gas dilutes the burning rich gas. The arrangement of the lean gas and air ports is such that the foregoing feature does not operate when underfiring with lean gas. The top of the hairpin flues is arranged with double crossovers, one adjustable with slide brick. In this manner, the heat in the top of the oven wall can be raised or lowered to suit coals with varying shrinkage characteristics (Fig. 9).

6. Dr. C. Otto & Company, Bochum, Germany, and Simon Carves Ltd, England, under license, build mainly underjet ovens and only occasionally gas-gun ovens, for rich- and lean-gas underfiring. The ovens are of the hairpin flue type with single cell regenerator compartments extending from pusher to coke sides. Adequate vertical heating of the oven walls with rich gas is achieved by means of high- and low-rich-gas burners in alternate flues (Fig. 10).

7. Carl Still, Recklinghausen, Germany, builds underjet and gas-gun ovens for rich- and lean-gas underfiring. The regenerators are divided along the center line of the battery. All pusher-side and all coke-side regenerators are alternately on upflow or downflow operation. This company has built the highest ovens in operation with a chamber height of approximately 20 ft. To heat these high ovens adequately on rich gas, the gas enters at the bottom of the heating flue while the air rises in a duct in the wall separating the flues. The air is then released to the flue at various levels, depending on the height of the oven, to achieve complete combustion in stages (Fig. 11).

8. Union Chimique Belge S.A., Brussels, Belgium, builds underjet and gas-gun ovens for rich- and lean-gas underfiring. The ovens are of the two-divided regenerator type. With gas burning in the pusher side flues, the combustion gases pass along a horizontal flue and descend down the coke-side flues. Periodical reversal changes the flow in the opposite direction. Rich gas, lean gas, and air enter at the bottom of the heating flues. Since this results in inadequate heating of the upper part of the walls when underfiring with rich gas, supplementary rich gas is fed into the horizontal flue. This gas burns with surplus air of the primary combustion to heat the upper part of the oven walls on downflow (Fig. 12).

9. Wilputte Coke Oven Division, Allied Chemical Corporation, New York, New York, builds underjet and gas-gun ovens for rich- and lean-gas underfiring. The larger ovens are the four-divided regenerator type, the horizontal flue being divided down the center line of the

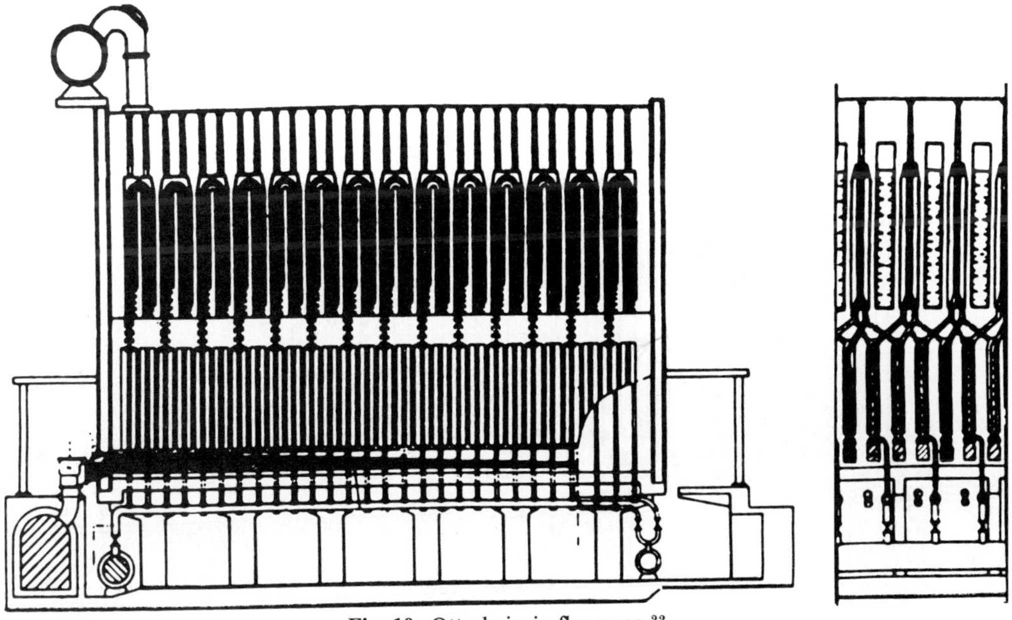

Fig. 10. Otto hairpin flue oven.[22]

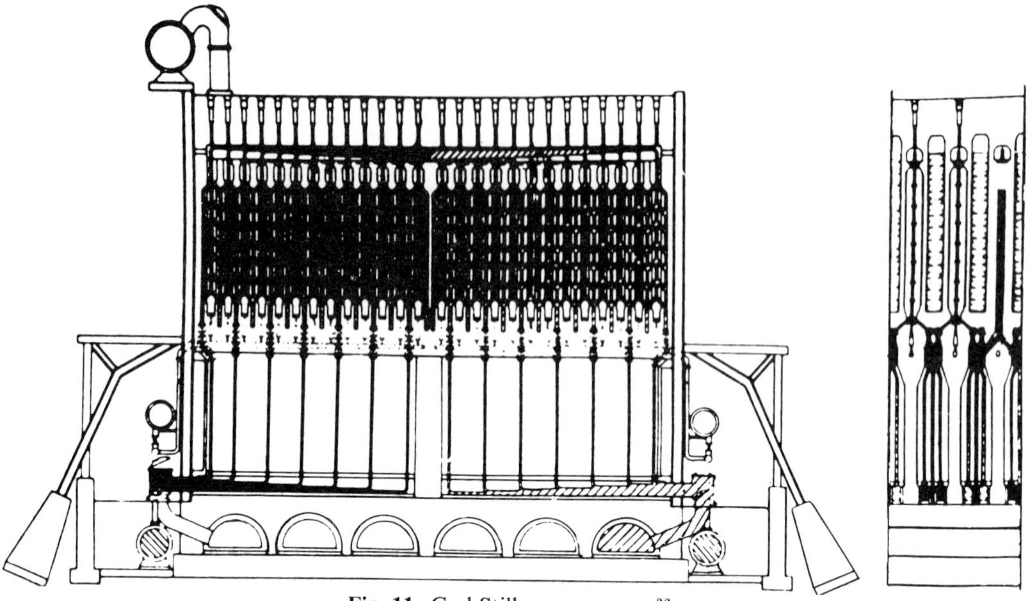

Fig. 11. Carl Still gas-gun oven.[22]

battery. Gas and air burn in the outer-zone flues—usually eight on the pusher side and six on the coke side—pass along the horizontal flue sections, and descend down the inner zone flues—normally eight for the pusher side and seven for the coke side. Periodical reversal changes the flow in the opposite direction. All pusher-side and all coke-side regenerators are alternately on upflow or downflow operation. Adequate vertical heating of the oven walls with rich gas is achieved by means of high and low burners in alternate flues. The different resistances to flow of air or air and lean gas from the bus flues to the regenerators and the waste gas from the regenerators to the bus flues are regulated by precalculated, progressively graduated, built-in Venturi-type ports.[23] (Fig. 13).

[23] Davis, G. A., U.S. Pat. 2,334,612 (1941); *Blast Furnace Steel Plant*, **33**, 568–78 (1945).

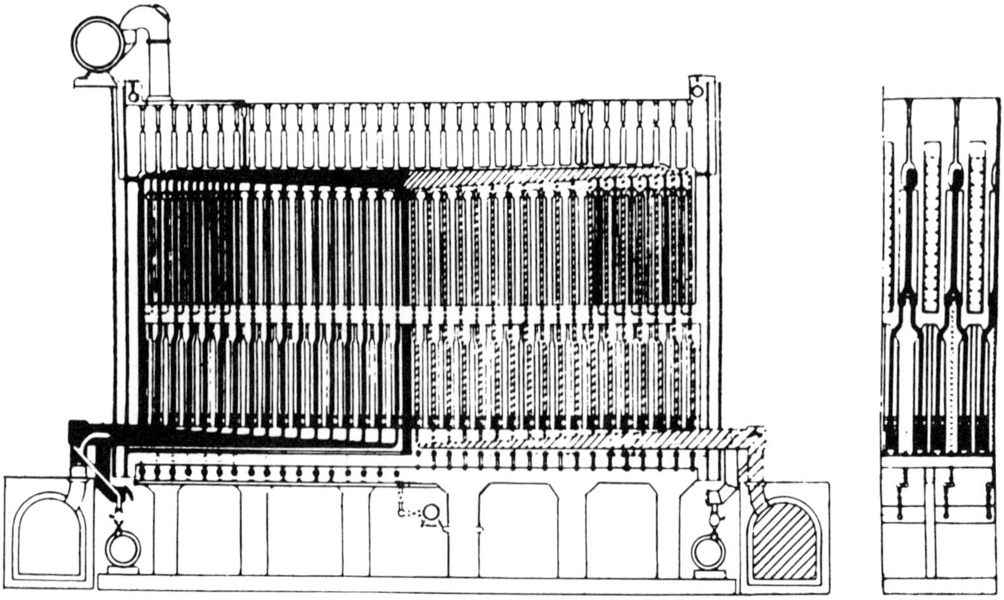

Fig. 12. Union Chimique Belge underjet oven.[22]

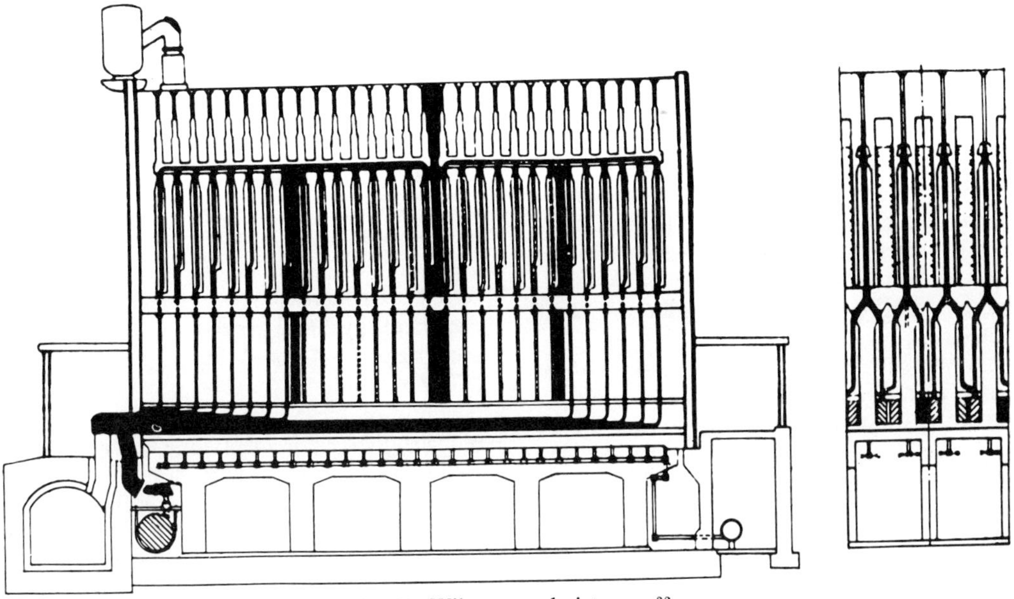

Fig. 13. Wilputte underjet oven.[22]

ECONOMIC DEMANDS FOR CHANGE IN COAL SUPPLY AND PREPARATION

One of the dominant factors that established the geographical location of many great steel-producing centers of the world was a local abundance of premium quality coking-coal, which required little if any preparation for carbonization. The increase in coke demand causing depletion of the better coal reserves, the wide dispersion of the industry throughout the world with many new plants located near poorer coking-coal supply, and mine mechanization have contributed to the upsurge in interest and importance of proper preparation techniques.

Although it has been generally understood that coke quality is affected by various preparation methods, the magnitude of this effect has until recently been underestimated. Even with high quality coking coals, a knowledge and use of the basic principles involved in their preparation for coking can lead to substantial economic gains, and in the geographic locations where poorly coking coals are the only ones locally available, proper preparation methods become a necessity.

The primary objectives of coal preparation for carbonization are to improve the physical and chemical properties of the coke, to avoid damage to the ovens during carbonization, to make use of inferior or marginal coking coals, to obtain the maximum oven efficiency consistent with costs, and to control and improve the product yield.

United States

Immense strides have been made by blast furnace operators in ore burden benefication, automatic control, and the resultant iron production increases per furnace unit. More attention is now being focused on the quality and uniformity of coke, and as a result the demand for improvement in coal preparation is increasing. The rapid advancement of mechanized mining has resulted in study and research on coal carbonization problems related to processes and techniques for the preparation of coal blends for coking. By 1959, 95.3% of the underground coal produced was cut by machine or mined by continuous machines, and 86% was mechanically loaded.[24]

Mining of the best quality coals in the first half of the twentieth century and hand loading permitted the selection of high-grade coal which did not require cleaning before use. Machine mining of complete seams generally results in higher ash and sulfur contents; and this, together with the fact that seams of higher ash and sulfur contents are often included, has necessitated the installation of numerous

[24] Young, W. H., Anderson, R. L., and Hall, E. M., *Minerals Yearbook 1959*, U.S. Bur. Mines, Vol. II, pp. 39–134 (1960).

coal cleaning plants. In 1959, 83.4% of the coal used in coke plants was washed, as compared to 31.0% in 1947 to 1949, with beneficial effects on both chemical composition and physical structure of the coke produced.[1]

The most widely used wet-washing process naturally results in a high moisture product which often presents handling difficulties, particularly after pulverization. Advanced designs of storage facilities, conveying systems and charging cars tend to minimize these problems.

Thermal systems capable of processing $1\frac{1}{2}$-in. coal at high tonnage rates have been designed and installed at several washeries to reduce the moisture content of the final product. Savings in shipping costs alone will probably justify many more such installations.

Most coal must be precrushed for efficient ash and sulfur removal by washing, and, as a result, coal shipped to coke plants now ranges in size 0 to 3 in. as compared to the 0 to 12 in. run of mine product previously considered normal. When small-sized coal is received, primary breakers are not a necessary adjunct to a coke plant preparation system.

Lorraine Area—France

Although large reserves of coal were available in this area of France, very little was used for carbonization prior to 1939 because of its poor coking quality. The high-volatile content of 38% resulted in small coke with many fractures, and the lack of bitumen content resulted in a poorly fused coke unable to withstand abrasion.[25] Since 1945, a determined, concerted, and increasingly successful effort has been made by all parties concerned, principally the French coal board and steel industry, to develop techniques for using considerable proportions of local coal. This has materially reduced the requirement for German coking coal from the Ruhr area.[26]

It was found that the mechanical properties of coke are greatly improved by homogeneity of texture and regularity of structure. A coke

[25] Burstlein, E., *Coke and Gas*, **18**, 290–5 (1956).

[26] *Ibid.*, **15**, 175–8 (1953); *Chaleur & ind.*, **35**, 351–70 (1954); **36**, 14–28 (1955). Abramski, C., *Glückauf*, **91**, 714–27 (1955). Loison, R., and Foch, P., *AIME, Proc. Blast Furnace, Coke Oven & Raw Materials Comm.*, **17**, 170–88 (1958).

Fig. 14. Preparation plant at Thionville coke ovens. Capacity, 1,400 tons/day.[25]

made from a blend of coals will be homogeneous only if the charge from which it is made is also homogeneous, both overall and within each size fraction. In practice, there are three factors that militate against this desirable end; (1) Differences in initial sizing between the coals comprising a blend; (2) differences in overall hardness between the same component coals; and (3) differences in hardness between the petrographic constituents within each coal.

A study of the coal for one Lorraine plant revealed that the average size of the coal delivered had a four-fold range in average size, and a five-fold range in overall size. Also, the Lorraine coals were twice as resistant to impact as were the Ruhr coals, and an even more marked difference appeared among the petrographic constituents. Fusain has practically no cohesive strength, durain is ten times as hard as vitrain, and shale even harder, with clarain intermediate between the durain and vitrain.[25] The result is that a normal single pass through a hammermill reduces the fusible coal to the smaller sizes, whereas the hard and relatively inert fractions will be in the larger sizes.

The Burstlein process (Sovaco) of preparing coal for coking has received much attention in recent years. The process was instituted in 1951 at the Thionville coke plant in France. Based on the principle that the petrographic constituents of individual coals used in a blend vary widely in coking properties, hardness, and grain size, a blend of homogeneous texture and uniform grain size is obtained by various combinations of open and closed circuit selective screening, individual component pulverizing, and subsequent intimate mixing. The process avoids crushing the friable or softer components, yet crushes the harder or infusible constituents to a given narrow size range. Electrically heated screens are used for screening the moist fine coal. By 1957, six coke plants in France, two in North Africa, and one each in West Germany and Italy were using the Burstlein process of coal preparation.

The Sovaco process installation at the Longwy Steel Works coke ovens, Thionville, France (Fig. 14)[25] employs the two-circuit double-effect scheme designed by Burstlein (Fig. 15). The left-hand circuit in Fig. 15, dealing with bituminous coals, has a double-decked screen. The respective mesh sizes of

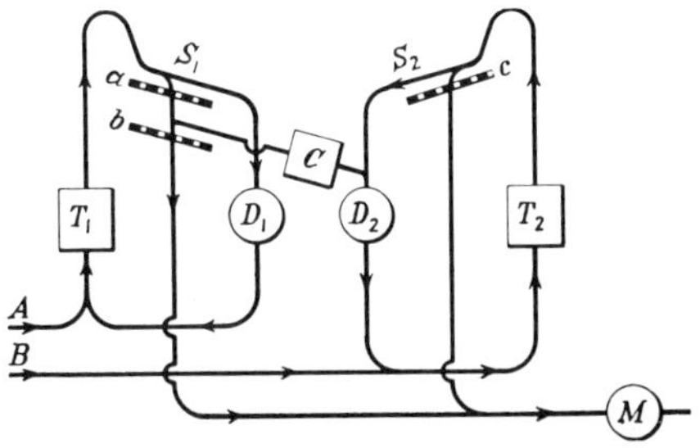

Fig. 15. Two-circuit "double-effect" scheme with dry cleaner as used at Thionville.[25]

this screen are chosen according to the coals being used so that the material oversize consists mostly of freed petrographic constituents, of which vitrain, clarain, and fusain pass *b* and go to the ovens while durain and shale remain on size *b* and go to the right-hand circuit. Here they join the predominantly inert coals, and by means of screen *c*, disintegrator D_2, and ball mill T_2 are brought within the required narrow size range. A dry-cleaning unit was installed at *C*, which removed shale equivalent to 2 points in the ash content of the coal and effected a reduction in the sulfur content of the coke. Adoption of this process at several plants of the Lorraine area has reversed the proportional requirements of Ruhr versus Lorraine coals, improved the coke quality, and realized important strategic and economic gains.

Ruhr Area

In the Ruhr, the designation "coking coal" was formerly practically synonymous with "fat" coal of 19 to 28% volatile content, for which experience has shown that coal of 22 to 25% volatile content produced the best coke. It is to this fat coal, in the final analysis, that the Ruhr district owes its overwhelming economic and political importance in Europe. A very striking example of the change in the qualitative composition of the coal output in the Ruhr area is shown in Table 2. Even in this highly productive area, it is becoming increasingly important to develop means of using the poorer coal grades to make coke, and the trend is toward improvements in coal preparation.

In addition to installations of the "Sovaco" process, impact-type pulverizers are used extensively and found advantageous in this district for crushing coal. The crushing is selective, and since the mills do not have the

TABLE 2

PERCENTAGE PRODUCTION OF TYPES OF COAL IN THE RUHR AREA.

	1936	1955	Est. 1960
Flame coal	—	2.8	12.7
Gas-flame coal	20.8	37.3	37.8
Gas coal	17.0	24.6	11.5
Fat coal	62.2	33.9	26.2
Lean coal	—	1.4	11.8

grate bars which exist in convertional hammermills, no degradation is caused by attrition. Experiments have shown that excellent quality coke can be produced from a coal blend by using all of the various grades of Ruhr coal at the estimated percentages of the 1960 production.

Great Britain

Since World War II, a large coke-oven rebuilding program has been carried out in Great Britain. The modern ovens being built are on an average 45 ft long, 13 ft high, and 16 to 18 in. wide. With the nationalization of the coal industry in 1947, a policy of centralization on larger units of 1500 to 2000 tons of coal capacity developed. Increased mechanization and larger capacity of the units resulted in a reduction of the number of men employed per ton of coke produced.[27]

This rebuilding program has set the industry's pattern for the next 20 to 25 yr and is based on the slot-type coke oven, developed over half a century. With the equipment highly standardized, any improvement in the coke quality must be sought in changing the characteristics of the coal charged or the operating methods of the ovens.[28]

As in other countries, it is recognized that selecting and preparing the coal to the ovens is an important factor influencing the quality of the coke produced. This is all the more important as there is a need to conserve the good coking coal still available. All new coking plants are being built with coal blending facilities. Blending requirements will be increased in the future when larger amounts of medium- and poor-quality coking coals will of necessity be used for the production of good quality coke.[29]

EXPERIMENTAL OVEN DEVELOPMENT IN THE UNITED STATES

The use of model ovens which simulate full-scale oven coking conditions in a number of respects has increased markedly in the United States. It is estimated that more than a hundred such ovens were in regular use in the United States during 1960. The United States Bureau of Mines, Illinois State Geological Survey, the coke-oven construction companies, and some of the coal, coke, and steel companies have each developed designs with certain desirable features. At least three companies construct experimental ovens on order, and at least one commercial laboratory provides testing service in one of the smaller ovens. Many companies producing coal and coke regularly operate several ovens. Table 3 lists ovens in use, divided into major subdivisions by type. Many of these may be considered modifications of others.

Also during this period of rapid development, each newly constructed oven of one type has more often than not been modified to some extent. Additional detail regarding the design, construction, and operation of the ovens may be found in the references given.[30–50] In the

[27] *Ministry of Labor Gazette*, London, March, 1947, September, 1954.

[28] British Coke Research Assoc., Panel No. 4 *The Study of Oven Dimensions in Relation to Coal and Coke Quality*, London (1946), 11 pp.

[29] Brown, E. G., *Trans. Inst. Mining Engrs. (London)*, **114,** 705–42 (1955).

[30] Brewer, R. E., *Chemistry of Coal Utilization*, John Wiley and Sons, New York, 1945, Vol. 1, pp. 240–78.

[31] Russell, C. C., *Trans. Am. Inst. Mining Met. Engrs.*, **139,** 313–27 (1940). Russell, C. C., Perch, M., and Farnsworth, J. F., *AIME, Proc. Blast Furnace, Coke Oven & Raw Materials Comm.*, **8,** 32–47 (1949).

[32] Fieldner, A. C., and Schmidt, L. D., *U.S. Bur. Mines, Inform. Circ.* No. 7190 (1941), 60 pp., particularly pp. 41–8.

[33] Potter, C. L., *AIME, Proc. Blast Furnace, Coke Oven & Raw Materials Comm.*, **11,** 159–73 (1952).

[34] Coleman, D. C., and Farley, P. J., *AIME, Proc. Blast Furnace, Coke Oven & Raw Materials Comm.*, **15,** 176–94 (1956).

[35] Harris, H. E., Margel, E. W., and Howell, R. D., *AIME, Proc. Blast Furnace, Coke Oven & Raw Materials Comm.*, **13,** 127–41 (1954).

[36] Brown, W. T., *Blast Furnace Steel Plant*, **27,** 172–4 (1939).

TABLE 3

EXPERIMENTAL OVENS KNOWN TO BE IN USE IN THE UNITED STATES FOR THE ASSAYING QUALITY IN COKE

Oven Type Identification	Coal Charge (lb)	Charge Area (in.)	Charge Thickness (in.)	Sides Heated 1, 2, or Sole	Heating-Wall Material	Heat Supply	Type*
A. Fuel Research, sole heated[30]	15	11 × 11	4	Sole	Si C	Elec.	CP-Exp.
B. Koppers, movable wall[30–33]	400	30 × 42	12	2	High Alumina	Gas	CV-Pres.
C. Wilputte, movable wall[34,35]	400 to 700	36 × 36	12–18	2	Si C	Elec.	CV-Pres.
D. Bethlehem-Tester[30,33,36]	52	11 × 36	4	Sole	Silica	Gas	CP-Exp.
E. Bureau of Mines, sole heated[30,37–39]	40	11 × 24	4.5–5	Sole	Si C	Elec.	CP-Exp.
F. United States Steel[40,41]	30	10 × 20	6	2	Si C	Elec.	CV-Pres.
G. Illinois Geological Survey[42,43]	500	36 × 35	14	2	Si C	Elec.	CV-No
H. Illinois modified[44]	675	36 × 35	17	2	High Alumina	Elec.	CV-Pres.
I. Bureau of Mines-AGA[45]	100 to 225	24 high	13 and 18 diameter	Circumference	Sheet Steel	Elec.	CV-No
J. Bureau of Mines, slot[32,46]	200 to 400	27 × 42	6–13	2	Sheet Steel	Elec.	CV-Pres.
K. Bureau of Mines, Tuscaloosa[47,48]	850	36 × 48	17	2	Silica	Elec.	CV-Pres.
L. Bethlehem, 18 inch[50]	700	36 × 36	18	2	Silica	Elec.	CV-Pres.

* CV = constant volume; CP = constant pressure; Exp. = expansion determined; Pres. = pressure development determined; No = no determination of pressure or expansion.

design of the larger ovens, two lines of thought have been applied in regard to heating and width of charge. In the Tuscaloosa Oven (*K*) and the Bethlehem 18-inch Oven (*L*), the heating walls simulate full-scale oven heat-transfer characteristics, and the width of ovens is within the range employed in full-scale operations. The Illinois Modified Oven (*H*) does not differ significantly from the other two just mentioned. The proponents of this type claim that better simulation of full-scale conditions is automatically obtained simply by controlling flue temperatures.

With a given flue temperature, heat flow will match full-scale ovens closely, regardless of coking rate, differences in heat demand from, for example, wet and dry charges, etc. Interpretation of results is simplified and the chances for error in interpretation are substantially lessened. The proponents of the narrower-than-full-scale-width-high-conductivity-wall ovens feel that reduced flue temperatures of time-temperature control programs may be used to compensate for the differences, and that there are no undue hazards in interpretation considering the gains from reduction in operating costs occasioned by significant reduction in weight of coal and coke samples handled, time required per test, increased heating element life, lower power requirements, etc. There presently appears to be no pronounced majority opinion. A major coal company resolved the problem in a recent (1959) installation by using a Bethlehem 18-inch Oven (*L*) with silica walls, and by mechanizing and automating the installation to a high degree. Between charging and discharging, oven operation is entirely automatic including all recording of data, and the oven is left completely unattended unless trouble develops and an attendant is called by an alarm system.

The smaller ovens holding charges of less than 60 lb are similar except for the United States Steel, 30 lb oven (*F*). The Bureau of Mines sole-heated oven (*E*) is generally viewed as a modification of the Bethlehem-Tester (*D*) and the Fuel Research oven (*A*) may be viewed as a split-chamber modification of the Bureau of Mines sole-heated (*E*) oven.

Originally, a variety of mechanical and hydraulic systems were used to measure the pressure developed in the ovens. More recently the use of load cells based on strain gauges has become the usual practice. With the newer load cell systems, interference attributed to expansion of oven parts has been noted.[40,44] Water-cooled tie-rods and maintaining surroundings at a constant temperature have been used to combat the interference. Possibly, such trouble was not previously encountered because the older systems had sufficient play in them so that minor expansions of steel or brick work were not particularly noticeable.

As indicated in the heading of Table 3, it is known that all the ovens listed are used for

[37] Auvil, H. S., and Davis, J. D., *U.S. Bur. Mines, Rept. Invest.* No. 3403 (1938), 18 pp.; *Fuel*, **17,** 335–48 (1938).

[38] Naugle, B. W., Wilson, J. E., and Smith, F. W., *U.S. Bur. Mines, Rept. Invest.* No. 5295 (1957), 15 pp.

[39] Auvil, H. S., Davis, J. D., and McCartney, J. T., *U.S. Bur. Mines Rept. Invest.* No. 3451 (1939), 21 pp.

[40] Price, J. G., Schoenberger, R. W., and Perlic, B., *AIME, Proc. Blast Furnace, Coke Oven & Raw Materials Comm.*, **17,** 212–25 (1958).

[41] Brisse, A. H., and Richards, P. L., *AIME, Proc. Blast Furnace, Coke Oven & Raw Materials Comm.*, **14,** 265–78 (1955).

[42] Reed, F. H., Jackman, H. W., and Heline, P. W., *Trans. Am. Inst. Mining Met. Engrs.*, **177,** 490–501 (1948); *Illinois State Geol. Survey, Bull.* No. 71 (1947), 132 pp.

[43] Eddinger, R. T., and Mitchell, J., *AIME, Proc. Blast Furnace, Coke Oven & Raw Materials Comm.*, **15,** 148–63 (1956).

[44] Jackman, H. W., Helfinstine, R. J., Eissler, R. L., and Reed, F. H., *AIME, Proc. Blast Furnace, Coke Oven & Raw Materials Comm.*, **14,** 204–24 (1955).

[45] Reynolds, D. A., and Holmes, C. R., *U.S. Bur. Mines, Tech. Paper* No. 685 (1946), 35 pp. Brown, R. L., Davis, J. D., Reynolds, D. A., and Smith, F. W., *U.S. Bur. Mines, Bull.* No. 536 (1954), 17 pp.

[46] McCartney, J. T., and Davis, J. D., *U.S. Bur. Mines, Rept. Invest.* No. 3644 (1942), 23 pp.

[47] Gayle, J. B., and Auvil, H. S., *U.S. Bur. Mines, Rept. Invest.* No. 4923 (1952), 16 pp. Gayle, J. B., and Gulledge, E. L., *U.S. Bur. Mines, Rept. Invest*, No. 5011 (1954), 12 pp.

[48] *Ibid.*, No. 5027 (1954), 9 pp.

[49] Gayle, J. B., Eddy, W. H., and Shotts, R. Q., *U.S. Bur. Mines, Rept. Invest.* No. 5307 (1957), 16 pp.

[50] Aikman, R. P., *AIME, Proc. Blast Furnace, Coke Oven & Raw Materials Comm.*, **15,** 163–75 (1956).

determining coke quality. Originally, such ovens were used almost exclusively for measuring expansion characteristics, but it is quite possible that the major purpose of the majority of tests presently made is the determination of the relation of the characteristics of the coals or blends under test to the quality of the coke produced. Use of the experimental ovens ranges from selecting the coal to be mined to monitoring full-scale coke-plant performance.

The literature describes use of one or more of the ovens listed in Table 3 for selection of coal to be mined or purchased;[34,36,41,49] for investigation of coal blending;[42,49,51,52] and for determining correlation between pressure measurements in test and full-scale ovens,[34,53] precision of test oven results,[54] effect of weathering of coal on coke quality,[42] effect of coking rate or flue temperatures on coke quality and expansion pressure,[40,42,43,55,56] effect of oven width,[40,44,56] and effect of charge density on coke characteristics, coking time, and pressure development.[38,48,52]

At present, there appears to be general agreement that many of the ovens will produce coke having the same screen and shatter test results as full-scale oven cokes, and that the experimental ovens produce cokes that have a somewhat lower apparent specific gravity and lower tumbler hardness indices. However, general correlations appear to hold between test and full-scale ovens for these latter indices. The test ovens do yield accurate relative indications of the likelihood of oven damage from expanding coals, but there is still considerable discussion as to the maximum pressure indication that may be considered "safe."

[51] Reynolds, D. A., *U.S. Bur. Mines, Rept. Invest.* No. 4522 (1949), 8 pp.

[52] Reynolds, D. A., Wolfson, D. E., and Smith, F. W., *AIME, Proc. Blast Furnace, Coke Oven & Raw Materials Comm.*, **14**, 224–42 (1955). Smith, F. W., and Reynolds, D. A., *U.S. Bur. Mines, Rept. Invest.* No. 5110 (1955), 24 pp.

[53] Lambert, L. L., Lancucki-Pater, W. J., and Lee, G. W., *J. Inst. Fuel*, **30**, 362–86 (1957).

[54] Naugle, B. W., Davis, J. D., and Wilson, J. E., *Ind. Eng. Chem.*, **43**, 2916–22 (1951).

[55] Gayle, J. B., and Eddy, W. H., *U.S. Bur. Mines, Rept. Invest.* No. 5125 (1955), 16 pp.

[56] Gayle, J. B., and Eddy, W. H., *AIME, Proc. Blast Furnace, Coke Oven & Raw Materials Comm.*, **17**, 193–212 (1958).

BASIC PRINCIPLES OF COAL PREPARATION FOR CARBONIZATION

Petrography

A full knowledge of the analysis and characteristics of the raw material available is certainly the primary requisite if a coal preparation system is to be designed to attain the highest coke quality possible. Standard chemical analysis should be supplemented by petrographic studies to determine the basic components present in each coal that, in turn, affect the ultimate behavior of the blended mass during carbonization. The Sovaco process, based on this concept, has been very successful in Europe, and the universal adoption of similar methods of coal preparation is inevitable.

Cleaning

Mechanized mining and the resultant higher ash and sulfur content of coal destined to be carbonized makes cleaning mandatory if the current blast furnace demands for low-ash and low-sulfur coke are to be met. Griffin[57] has shown that each 1% reduction in coke ash increases the coke value $0.45 to $0.56 per net ton and a reduction of sulfur content from 0.75 to 0.65% in 6% ash coke increases its value $0.20.

Flint[58] has shown that 1.08% C is required to smelt each percent of coke ash and 3.00% C for each percent of coke sulfur. Thus, the effective carbon in coke is reduced from a starting point of 84.5% to 77.0% if the ash is 10% and the sulfur 1%. Substantial savings result from ash and sulfur reductions in coal cleaning plants.

Pulverization

For years, one pass through a pulverizer—usually a hammermill—has been the conventional method of crushing coal for coke-plant use in the United States. Some plants pulverize the individual components of a blend separately prior to mixing; others reverse the

[57] Griffin, J., *Influence of Iron-Bearing Raw Materials on Cost Effects of Ash and Sulphur of Blast Furnace Coke*, Island Creek Coal Sales Co., Huntington, West Virginia, 1958, p. 38.

[58] Flint, R. V., *AIME, Proc. Blast Furnace, Coke Oven & Raw Materials Comm.*, **11**, 49–68 (1952).

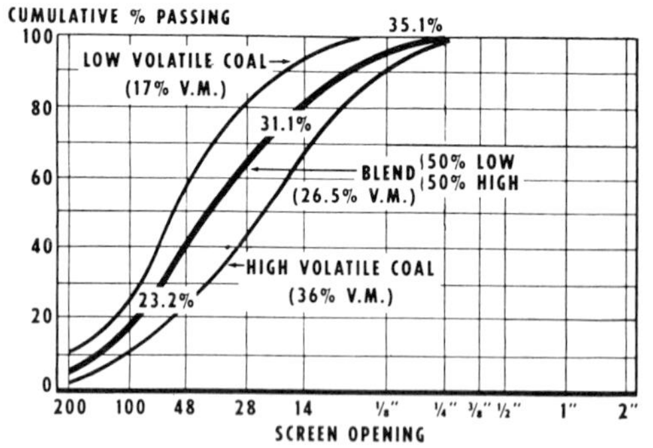

Fig. 16. Pulverization performance of hammer mills. Size distribution as related to volatile matter content of the coal feed.[59]

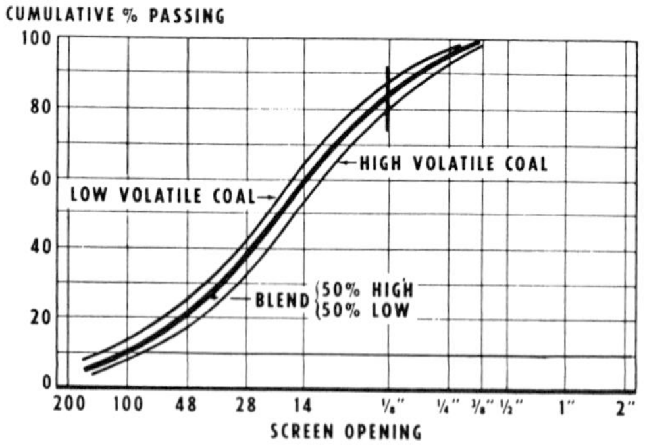

Fig. 17. Pulverization performance of hammer mills. Effect of proper mill adjustment.[59]

procedure. Very few adjustments are made to compensate for the friability of grindability indices of the various coals being processed and the resultant size consist averages 70 to 85% $-\frac{1}{8}$-in. product with up to 15% passing through a 100-mesh screen.

Brisse and Harris[59] have reported results of an investigation of the performance of hammer-mills (Fig. 16). Two coals, one low-volatile coal (17% volatile matter) and one high-volatile coal (36% volatile matter) were mixed in equal proportions and pulverized by one pass through a hammermill set to attain 90% $-\frac{1}{8}$-in. on the high-volatile fraction. The center curve shows the resultant size consist of the pulverized blend. Each coal was also pulverized separately, the low-volatile results being indicated by the curve on the left and the high volatile results by the curve on the right. The volatile content of the size fraction "$+\frac{1}{8}$ in.," "$\frac{1}{8}$ in. × 48 mesh," and "−48 mesh" were found to be 35.1%, 31.1%, and 23.2%, respectively. The curves indicate that the blend curve is approximately the arithmetical average of the two individual coal results. Several conclusions were made from these results:

1. The blend is not homogeneous in each size fraction.
2. The result would not be acceptable to the average coke-plant operator because of the excessive percentage of −100-mesh product (17%) due to overcrushing the low-volatile coal.
3. Pulverizations of +90% are not possible with one pass through a hammermill if one constituent of the blend is comparatively easy to pulverize.
4. Each coal of a blend being pulverized behaves as though it had been pulverized separately.

The results of pulverizing the same coals using mills adjusted to give similar results for each individual coal which, when subsequently mixed, have an acceptable size consist are shown in Fig. 17.[59] These two coals have widely different pulverization characteristics, and should be pulverized separately in mill with different settings. Only coals of closely related pulverization characteristics should be pulverized after blending.

Brisse and Harris[59] showed the effect of pulverization on coke strength for two basically different coal types (Fig. 18). Coal I was a two-coal blend made up of a highly banded high-volatile coal and a low-volatile coal. Coal II

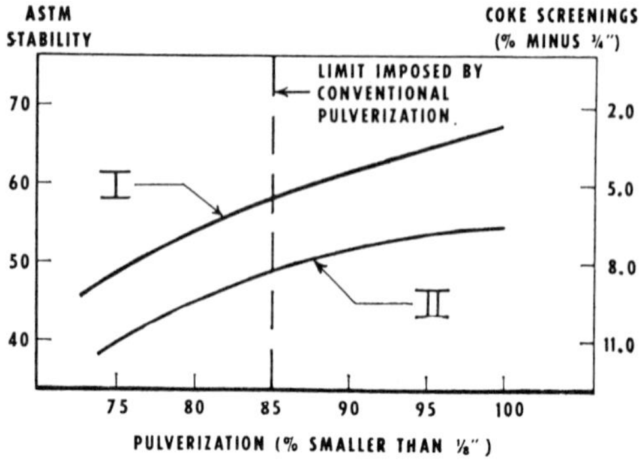

Fig. 18. Effect of coal pulverization on coke strength.[59]

[59] Brisse, A. H., and Harris, H. E., *Blast Furnace Steel Plant*, **45**, 301–7 (1957). See also Brisse, A. H., and Kuchta, B. R., *AIME, Proc. Blast Furnace, Coke Oven & Raw Materials Comm.*, **18**, 17–21 (1959).

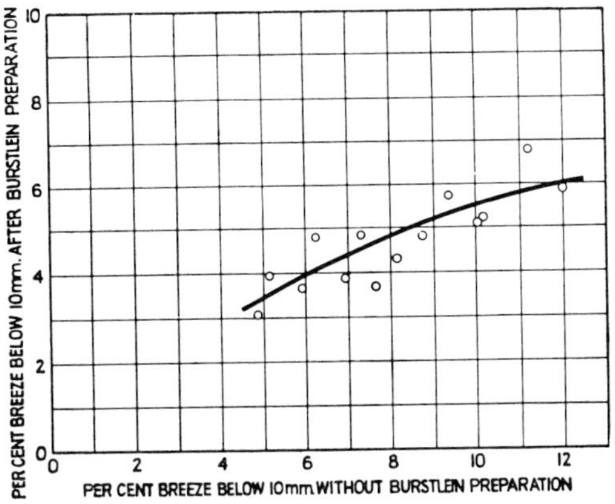

Fig. 19. Yield of breeze below 10 mm with and without Burstlein preparation.[25]

was a homogeneous coal from a petrographic point of view, with a minimum of banding, was of a high-vitrinite content and hence, was highly fluid. With Coal II, little gain in coke strength was effected at levels above 90% $-\frac{1}{8}$ in. because the coal was homogeneous petrographically.

Similar advantages in breeze yield reductions have been claimed by Burstlein[25] for the Sovaco process, which employs selective and high pulverizations (see Figs. 19–21). Additional benefits were claimed for improved coke quality indicated by the increase in Micum +40 mm values, and the decrease in Micum −10 mm values.

Unpublished tests by pulverizer manufacturers indicate that pulverizations over 90% $-\frac{1}{8}$ in. with less than 12% −100-mesh are possible by operating either hammermills or impactors in series without resorting to screening between passes. Peripheral speeds are generally slower and are varied to fit the grindability index of each individual coal. Pulverizers are now designed so they can be readily converted from hammermill to impactor, or vice versa, at relatively low cost. Although few plants have actually adopted higher pulverization practice, immense interest has developed in the United States, and a trend in that direction is indicated for the future.

Blending

Although results obtained from investigations on a laboratory scale are not necessarily the same as those found for full scale ovens, considerable work has been done to show the effects of blending on the physical or chemical properties of the coke, on expansion, and on yields of products.

Effects of Blending on the Physical and Chemical Properties of Coke

Cheradame[60] deduced two rules for the utilization of fusible coals in blends: (1) that the antifissuring "efficiency" of a coal is greater the higher the temperature of resolidification, and (2) that the richer in volatile matter and the less fusible the "flaming" coal, the more fusible

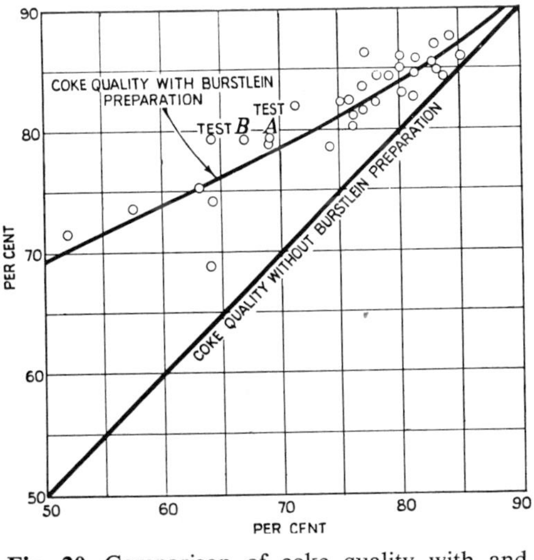

Fig. 20. Comparison of coke quality with and without Burstlein preparation. Micum +40 mm values.[25]

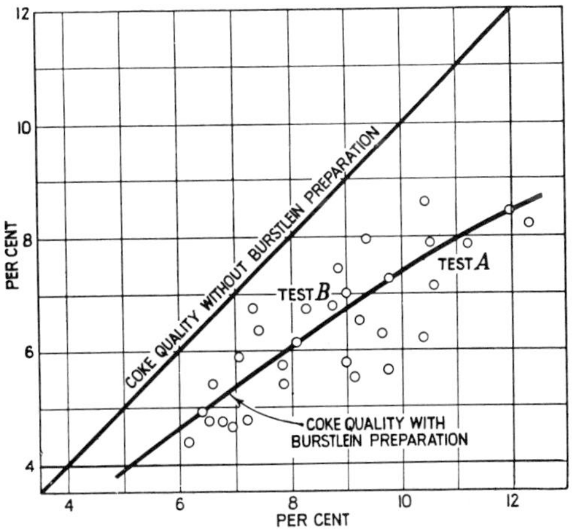

Fig. 21. Comparison of coke quality. Micum −10 mm values.[25]

[60] Cheradame, R., *BCURA Quart. Gazette*, **33**, 1–12 (1957).

and rich in volatile matter the coal mixed with it had to be.

Yoshida[61] has shown that for coals with free-swelling indices higher than $3\frac{1}{2}$ and volatile matter in the range of 18 to 42%, the drum strength in percent on a 15-mm sieve is proportional to an exponential of the percent volatile content of the coal.

Several investigators[62] have shown the effects of blending 80% of various high-volatile coals of different rank with 20% of a low-volatile coal on coke strength. It was shown that the benefits of blending were greatest for the weaker-coking high-volatile coals and decreased markedly when the strength indices for the high-volatile coals carbonized singly were high. In the blends carbonized at 900°C in the BM-AGA retort, using one low-volatile and various high-volatile coals, Reynolds[62] concluded that the coke strength indices of an appreciable proportion of the high-volatile-*A* Appalachian coals increase only slightly on raising the proportion of Pocahontas No. 3 from 20 to 30%.

On carbonizing equal parts of a high-volatile coal with several low-volatile coals, together with 5% anthracite in the Tuscaloosa oven, Gayle and co-workers[63] found that there was very little difference in coke quality especially attributable to any one of the various low-volatile coals. Reproducibility data was used to judge significant differences. In this same series of tests, a second high-volatile component and mixes of the two high-volatile coals were used with the same low-volatile coal. Results indicated only slight changes in coke quality.

The use of inerts such as anthracite fines and coke breeze has received considerable investigation. The use is widespread, especially in foundry coke blends where larger size cokes of high shatter resistance are desired. The optimum amount of an inert to use is dependent on the base coal or coals used and the coking rate at which the blend is coked.[64,65] Various other factors have been shown to be important. Cheradame[60] pointed out that inert additives must be substances that do not contract much at the temperature of resolidification, so that as a result the mean contraction rate becomes less. Gayle[66] and others[65] have shown that the size of anthracite used is important. Abradability of the coke was increased by additions of anthracite of No. 5 Buckwheat size, but it either decreased or remained substantially constant when the inert was ground to −48 mesh.

In addition to the usual increase in shatter strength and coke size with the use of anthracite, it has been shown that uniformity of the coke is also improved.[65] A survey of foundry coke characteristics[67] listed uniformity of coke size as the first in order of importance of four principal coke factors effecting commercial cupola operations.

Low-temperature chars properly utilized in blends have been shown to have effects on coke strength similar to those found by the use of low-volatile coals. In 1935, Reed, Jackman, and Henline[68] presented a review of recent developments in the production and use of char. Price[62] listed the important factors in char preparation. He concluded that the char must be in effect *tailor-made* to yield optimum results with each base coal used. Important factors were: (1) the coal from which the char is made; (2) the amount of char to be used in the final mixture; (3) the degree of pulverization of the char; (4) the average volatile content of the char and its variation from this average; and (5) the physical properties of the char, with particular reference to its surface. Reed, Jackman, and Henline[68] used Illinois char in blends with various coals, and they found that the actual volatile matter of the char within the range of 15 to 24% was not

[61] Yoshida, Y., *Nenryô Kyôkaishi*, **36**, 831–40 (1957).

[62] Reynolds, D. A., *U.S. Bur. Mines, Rept. Invest.* No. 4552 (1949), 8 pp. Smith, F. W., and Reynolds, D. A., *U.S. Bur. Mines, Rept. Invest.* No. 5110 (1955), 24 pp. Price, J. D., *Mining Eng.*, **5**, 716–20 (1953).

[63] Gayle, J. B., Eddy, W. H., and Shotts, R. Q., *U.S. Bur. Mines, Rept. Invest.* No. 5307 (1957), 16 pp.

[64] Seymour, W., and Schmidt, L. D., *U.S. Bur. Mines, Rept. Invest.*, No. 3808 (1945), 23 pp.; *Coke and Coal Research Rept.*, Am. Coke and Coal Chem. Inst., May, 1948.

[65] Clendenin, J. D., and Kohlberg, J., *Proc. 6th Ann. Anthracite Conf.*, 1948, 215–71.

[66] Gayle, J. B., Eddy, W. H., and Brown, R. L., *U.S. Bur. Mines, Rept. Invest.* No. 5287 (1956), 19 pp.

[67] Krause, D. E., and Lowrie, H. W., Jr., *Am. Foundrymen Assoc.*, **11**, No. 7, 49–58 (1947).

[68] Reed, F. H., Jackman, H. W., and Henline, P. W., *Illinois State Geol. Survey, Rept. Invest.* No. 187 (1955), 38 pp.

critical and tended to have little or no effect on coke strength.

The use of petroleum coke[69–71] in blends improves some coke qualities. Dense, low-ash, and low-sulfur cokes for foundry use have been produced from blends composed of 40 to 75% petroleum coke of 8 to 12% volatile matter, 10 to 20% coal-tar pitch of 300 to 400°F melting point, 15 to 25% Pocahontas coal of 18% volatile matter, and 5 to 15% anthracite at about 24 hr coking time.[72]

Additions of coal-tar pitch[73] to blends deficient in agglutinating constituents results in increased coke strength. Stephans[74] has shown the effects of 5 to 15% pitch in blends on Gieseler fluidity. In a blend with Bulli seam coal having a maximum fluidity of 10.4 dial divisions/min, mined in New South Wales, Australia, it was found that each 1% addition of pitch raised the fluidity by a factor of about 1.72.

Hall[70] reported that graphite additives produced coke which was superior to that made with anthrafines.

Effects of Blending on Expansion

The blending of coals with materials such as anthracite, coke breeze, coking coals, oxidized coals, semicokes, sand, and sawdust to reduce expansion has been reviewed.[30,75]

Brysch and Ball[75] reviewed data from various sources on the effects of blending coals on expansion pressures and noted that the relationship of the expansion pressure and the percentage of a given coal in a binary blend was not a straight line, and that inflection points occur at different percentages for the various blends. Russell[31] found that expansion pressures of some binary blends varied inversely with the blend fluidities.

[69] Gayle, J. B., and Eddy, W. H., *U.S. Bur. Mines, Rept. Invest.* No. 5276 (1956), 22 pp.

[70] Hall, J. A., Am. Coke & Coal Chem. Inst., Chicago, February, 1958, unpublished.

[71] Martin, S. W., U.S. Pat. 2,640,016 (1953).

[72] Bowers, T. G., U.S. Pats, 2,808,326, 2,808,370 (1957).

[73] Russell, C. C., *AIME, Proc. Blast Furnace, Coke Oven & Raw Materials Comm.*, **14**, 94–100 (1955).

[74] Stephans, J. N., *Coke and Gas*, **146**, 296–7, 302 (1958).

[75] Brysch, O. P., and Ball, W. E., *Inst. Gas Tech., Research Bull.* No. 11 (1951), 60 pp.

Previous work[30] and reviews[75,76] have indicated that volume expansion of binary blends of high- and low-volatile coals can be approximated by calculating the weighted average of the component coals. Naugle and co-workers[54] showed a consistent deviation in the expansion of the blend from the weighted average of the expansions of the component coals. The expressions:

20% low-volatile:
(30 tests)

$$E_B = 0.6(0.2E_{LV} + 0.8E_{HV}) + 21.4 \quad (r = 0.89)$$

30% high-volatile:
(23 tests)

$$E_B = 0.476(0.3E_{LV} + 0.7E_{HV}) + 28.5 \quad (r = 0.85)$$

(where E = dry, solid coal expansion, %; subscripts B, LV, and HV refer to blend, low-volatile, and high-volatile, respectively; and r = coefficient of correlation)

were found to approximate expansions of blends containing 20 and 30% low-volatile coal. Brysch and Ball,[75] using data from the Bureau of Mines sole-heated oven, found that correlation did not reveal the influence of the low-volatile coal shown by Naugle, but did show that the expansions of blends in this range of composition could be expressed as simple equations involving only the expansion of the high-volatile coal.

20% low-volatile:
(26 tests)

$$E_B = 0.728E_{HV} + 6.4 \qquad (r = 0.91)$$

30% low-volatile:
(15 tests)

$$E_B = 0.576E_{HV} + 7.1$$

(where E = percent expansion at wet bulk density of 55.5 lbs/cu ft, and subscripts B and HV refer to blend and high-volatile, respectively.)

Gayle and Eddy[77] found that additive behavior for binary blends of coking coals can be achieved either by charging the blend components in separate layers or by testing relatively coarse sizes of coal, indicating that the behavior

[76] Fieldner, A. C., and Rice, W. E., *U.S. Bur. Mines, Inform. Circ.* No. 7241 (1943), 63 pp.

[77] Gayle, J. B., and Eddy, W. H., *U.S. Bur. Mines, Rept. Invest.* No. 5464 (1959), 35 pp.

of blends may depend to a large extent on the degree of pulverization of the constituent coals.

Recent investigators[75] have shown the effects of blending inerts on expansion pressure. However, none of the studies have been as comprehensive as the findings of Gayle and co-workers[66,78] from tests in the sole-heated oven. They found that (1) substituting anthrafines for a part of a coal charge has an effect similar to that caused by an equivalent percentage reduction in dry-coal charge density insofar as expansion is concerned, (2) that the inert-free density of the coal or blend charged is the fundamental variable controlling expansion of any particular coal, and (3) that variations in inert-free charge density caused by changes in the degree of pulverization, moisture content, amount of oil added, and/or added inert all have about the same effect on expansion pressure. Previous investigations giving slot-oven expansion results were re-examined and served to confirm the conclusions based on sole-oven studies. Later,[77] it was shown that dense inerts, such as anthrafines, and iron-ore fines, markedly reduce the observed expansion of charges of a wide range of coking coals. Low-density inerts, such as vermiculite and charcoal, caused an increase in observed expansion in some instances.

Effects of Blending on the Yield of Products

When a given charge of coal is carbonized, the yields of coke and byproducts per unit weight of coal are dependent on the temperature of carbonization and independent of the dry-coal bulk density.[52]

Lowry and co-workers[79] correlated BM-AGA test data with coal analysis and found that the yields were related to the proximate analysis of the coal or blend. They concluded that, "The data presented in the Bureau of Mines assay tests do not indicate that either coal petrography or blending of coals is important in carbonization (yields) except insofar as they are reflected in the proximate analysis. This statement may be qualified by saying that it is based on the facts that the data for blends appeared generally to be additive based on the original coals, and that the extreme petrographic 'types' of coal did not in general differ significantly from the average."

Particle Size

At constant bulk density, particle size itself has been shown to exert a measureable influence on expansion.[75]

Bureau of Mines tests in the vertical slot oven have shown no noticeable change of maximum pressure with increased pulverization and is in agreement with Pontypridd tests. Volume expansion in the sole-heated oven was slightly increased by finer pulverization. Naugle[54] calculated that a 10% increase in pulverization, that is, the percent passing 4 mesh, increased the dry, solid-coal expansion 2%.

Reed and co-workers[80] showed an increase in size, shatter, and stability of coke with finer pulverization for one blend of coals. Tests on another blend[81] gave lower shatter strength, smaller size, and increased stability with increased pulverization. An optimum pulverization may exist for any given blend or coal at constant coking conditions. Burstlein[82] found that there is an optimum sizing for the individual coals used in blends. Brisse and Harris[59] agreed with the findings of Burstlein and found it advantageous to subject each coal to a pulverization treatment according to its particular characteristics. Maximum coke strength was obtained when size of particles of a given coal were held within a narrow range. Some coals should be pulverized finer than $100\% - \frac{1}{8}$ in., the most desirable size being dependent on the characteristics of the coal.

Iron Coke Blends

Previous work on the production and use of iron-coke was reviewed[83] in a report on pilot-plant and industrial scale tests wherein various iron-bearing materials were blended with various coals for coking. Addition of ores or flue

[78] *Ibid.*, No. 5176 (1955), 18 pp.

[79] Lowry, H. H., Landau, H. G., and Naugle, L. L., *Trans. Am. Inst. Mining Met. Engrs.*, **149,** 297–330 (1942).

[80] Reed, F. H., Jackman, H. W., Rees, O. W., Yohe, G. R., and Henline, P. W., *Illinois State Geol. Survey, Bull.* No. 71 (1947), 132 pp.

[81] Isenberg, N., and Jackman, H. W., *Investigations of Beckly Seam Coal Blended with Wheelwright Coal for Use in the Production of Metallurgical Coke*, Inland Steel Co., East Chicago, Indiana, 1945, p. 38.

[82] Burstlein, E. M., U.S. Pat. 2,782,147 (1957).

[83] Russell, C. C., Whitstone, P., and Liggett, R. P., *AIME, Proc. Blast Furnace, Coke Oven & Raw Materials Comm.*, **14,** 93–122 (1955); *Gas World*, **141,** 1616–22 (1955).

dusts to coal caused marked increases in size and shatter strength of the coke. Tumbler stabilities were within the range of values found for blast furnace coke used in the United States. When iron oxide was mixed with coal, approximately 80% of the sulfur in the charge remained in the iron-coke where ordinarily only about 60% was retained. Wet quenching and storage reduced the metallic iron content of the coke. Hall[70] reported similar results using 10 to 40% hematite concentrates in the blends. Barking and Eymann[84] found that the metallic iron in ferro-coke oxidized only on the surface during extended storage. In recent years, there has been much interest in production of ferro-coke.[85]

Chemical Pretreatment of Coal

The use of various compounds such as salt, oxidizing agents, sodium carbonate, and limestone to volatilize increased quantities of sulfur during the coking process or to bind the sulfur in the coke to facilitate its later removal in the slag was discussed.[30] Data by Medvedev and co-workers[86] showed that the use of various organic and inorganic additives did not reduce the amount of total sulfur remaining in the coke. However, the use of bauxite or aluminum oxide in the presence of ammonia gas during carbonization[87] has been reported to be effective. The treatment[88] of noncoking coals with lye under high pressure has been claimed to yield suitable cokes when the reaction product is blended with 10 to 60% of the same coal.

[84] Barking, H., and Eymann, C., *Stahl u. Eisen*, **75**, 386–41 (1955).

[85] Barking, H., and Eymann, C., *Glückauf*, **89**, 993–1003 (1953); *Brennstoff-Chem.*, **37**, 129–44 (1956); *J. Metals*, **10**, 274–6 (1958). Cellan-Jones, G., *Coke and Gas*, **15**, 315–20 (1953). Koppers Co., and Becker, J., Brit. Pats. 775,747–8, 777,162 (1957). Granger, G., Lach, M. J., and Mitchell, J., *AIME, Proc. Blast Furnace, Coke Oven & Raw Materials Comm.*, **18**, 152–61 (1959).

[86] Medvedev, K. P., Petropolskaya, V. P., and Nikitina, K. A., *Koks i Khim.*, **1958**, No. 8, 15–8; *Chem. Abst.*, **52**, 20998–9 (1958).

[87] Ghosh, J. K., and Brewer, R. E., *Ind. Eng. Chem.*, **42**, 1550–8 (1950). Roy, M. M., and Goswami, M. N., *J. Indian Chem. Soc., Ind. & News Ed.*, **18**, 93–9 (1955).

[88] Popovic, B., Ger. Pat. 920,063 (1954); *Chem. Abst.*, **48**, 10550 (1958).

COKING CONDITIONS

Charge Density

The effects of charge density and the factors affecting charge density have been the subject of numerous investigations. Much of the experimental work has been done on a laboratory scale, and conditions may not have approximated those occurring in full-scale practice.

The dropped-coal method was developed[89] to approximate bulk densities obtained in full-scale ovens. Auvil and co-workers[90] made a comparison between bulk densities in operating ovens at several plants and those obtained by the new method and the two previously existing methods commonly used in the United States for measuring bulk density. It was concluded that, the dropped-coal method, although less precise than the others, duplicated actual oven bulk densities with great fidelity. The two older methods, although more precise, do not give results that parallel actual oven bulk densities closely enough to allow use of correction factors. Results of comparisons are shown in Table 4.

Ramsburg and McGurl[91] and others[75] have shown that small additions of oil decrease bulk density of dry coal and increase that of wet coal, thus reducing variations in charge density due to moisture fluctuations.

Russian investigators have found that changes in the physical state of water has considerable effect on bulk density. Lowering the temperature from ambient temperature to below freezing increased bulk density from 39.6 to 48.6 lb/cu ft.[75]

Work on the effect of moisture and size consist on bulk density, using the ASTM method, indicated that the particle size distribution governs the bulk density of dry coal—wider spreads of sizes giving the higher bulk densities.[75] Figure 22, shows this relationship where $\bar{X}$ = absolute size constant, mm, and n = size distribution constant. The low-bulk densities for the finely pulverized samples show that in

[89] Landers, W. S., Schmidt, L. D., and Seymour, W., *U.S. Bur. Mines, Rept. Invest.* No. 3807 (1945), 22 pp.

[90] Auvil, H. S., Schmidt, L. D., and Graham, H. G., *U.S. Bur. Mines, Rept. Invest.* No. 3935 (1946), 21 pp.

[91] Ramsburg, C. J., and McGurl, G. V., *AGA., Proc.*, **22**, 92–104 (1940).

TABLE 4

Comparison between Test Results and Average Byproduct Oven Densities[90]
(Arranged in Order of Increasing Surface Moisture)

Coal Designation	Moisture, %		Bulk Density at Given Moisture Content, lb/cu ft				Differential between Oven and Test Values		
	Surface	Total	Oven Charge	Dropped Coal	Cone	ASTM	Dropped Coal	Cone	ASTM
C	0.7	2.2	50.2	50.9	48.1	47.1	+0.7	−2.1	−3.1
D-1	2.0	3.1	48.5	48.4	46.9	46.4	−0.1	−1.6	−2.1
D-2	2.0	3.1	47.8	48.4	46.9	46.4	+0.6	−0.9	−1.4
D-3	3.1	4.2	45.8	46.1	41.5	41.0	+0.3	−4.3	−4.8
D-4	3.7	4.8	44.8	45.2	39.7	38.8	+0.4	−5.1	−6.0
E	4.0	5.0	45.7	46.2	39.8	38.9	+0.5	−5.9	−6.8
D-5	4.2	5.3	44.1	44.5	38.5	38.0	+0.4	−5.6	−6.1
B	5.0	6.4	51.4	51.4	44.5	43.8	−0.0	−6.9	−7.6
				Oiled Coal					
G	2.6	3.9	50.7	50.5	48.2	47.4	−0.2	−2.5	−3.3
F-1	2.8	3.9	50.2	50.3	48.1	47.4	+0.1	−2.1	−2.8
F-3	2.8	3.9	51.1	50.5	48.5	47.7	−0.6	−2.6	−3.4
F-2	3.1	4.2	50.6	50.5	48.3	47.6	−0.1	−2.3	−3.0
Maximum difference between test value and byproduct oven charge							0.7	6.9	7.6
Minimum difference between test value and byproduct oven charge							0.0	0.9	1.4

this range of size, the absolute size is the dominant factor. It was found that for wet coals, the smaller the particle size, the lower the minimum bulk density, and the greater the moisture content at minimum bulk density.

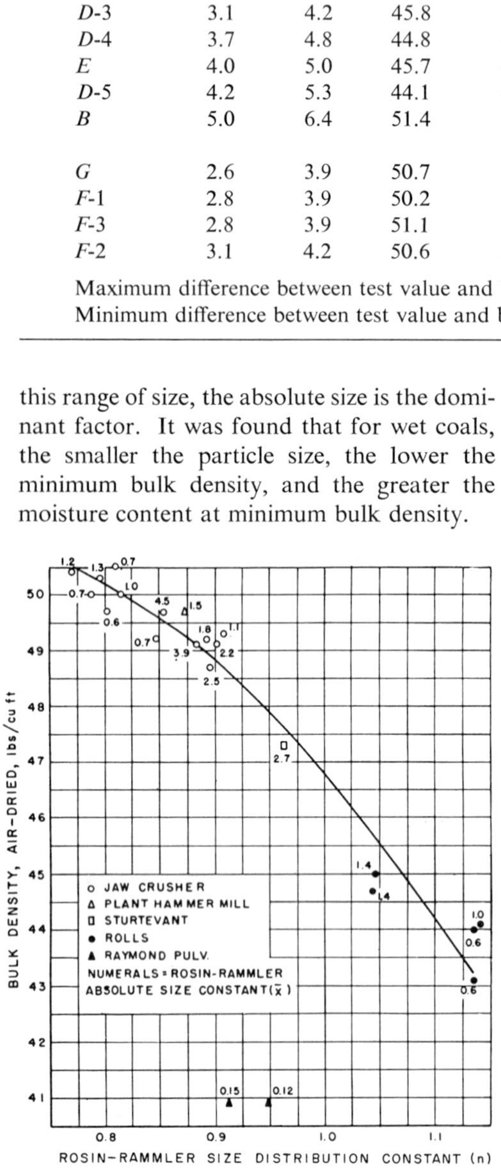

Fig. 22. Effect of size distribution on bulk density of air-dried coal.[75]

Reynolds and co-workers[52] have studied the effects of bulk density on yield and quality of products of coal carbonization. Data included tests on eleven coals and six blends in the BM-AGA retort. Changing the dry-coal bulk density had little or no effect on the yield of coke, gas, and ammonia or the heating value of the gas. It was found that: (1) The yield of tar per ton of dry coal increased and that of light oil decreased as coal bulk density was increased, the effects being relatively small, probably as a result of changes in cracking conditions occasioned by the changes in bulk density; (2) the apparent specific gravity of the coke varied directly with the coal charge density, an increase of 1 lb/cu ft in the dry coal bulk density increasing the apparent specific gravity about 1%; (3) the ASTM coke stability and hardness indices of average and strong cokes varied directly with charge density, an increase of 1 lb in bulk density increasing these indices about 0.7; and (4) the $1\frac{1}{2}$-in. shatter index decreased as the coal bulk density increased, an increase of 1 lb/cu ft decreasing this index about 1.0.

Naugle and co-workers[38] concluded that the determined apparent specific gravity of the coke carbonized under a constant load, was independent of the original coal bulk density. The effect of dry-coal bulk density on coal expansion in the sole-heated oven was also reported by Naugle and co-workers. It was shown that coal expansion in the sole-heated oven was related to dry coal bulk density by

$$E_R = \frac{BD_R(1 + E_T)}{BD_T} - 1$$

where E_R = fractional expansion at any reference bulk density,
E_T = fractional expansion at any test bulk density,
BD_R = any reference bulk density, pounds dry coal per cubic foot, and
BD_T = any test bulk density, pounds dry coal per cubic foot.

Data from various sources was assembled by personnel of the Bureau of Mines on the effect of dry-coal bulk density on maximum wall pressures in vertical slot-type ovens.[92] A new basis, relative wall force, was used to correlate the data. The relationship, shown in Fig 23., has been found useful in converting measurements at different bulk densities to a common reference for comparison, and the figure shows that an increase of 4 lb/cu ft increases the wall force 100%.[92]

Coleman and Farley[34] found that coal blends have different rates of increase in wall pressure per pound of dry bulk density above and below some critical point. Above the critical point, their curves show a 100% increase in expansion pressure as the bulk density increased from 2.5 to 3.5 lbs. For any coal, measurements at one bulk density could be approximated from results of tests at another bulk density. Any advantage to be gained by increasing bulk density to increase coke quality is limited if, at normal charge density, the coal is exerting wall forces approaching those likely to cause damage to oven walls.

Correlation of data from full-scale ovens and the British Coke Research Association's Pontypridd test oven has furnished further evidence that charges developing a maximum pressure of 2 lb/in.^{2} as measured in a movable test oven at the highest dry-coal bulk density to be encountered in full scale oven[53] are "safe" for use under usual coking conditions.

Further work reported by Gayle and Eddy[78] suggested that the inert-free density of the coal or blend charged is the fundamental variable controlling expansion of any particular coal, and that variations in inert-free charge density caused by changes in degree of pulverization, moisture content, amount of oil added, and/or added inert all have about the same effect on expansion pressure. Data from previous investigations was re-examined to confirm these conclusions.

Tests on medium- and high-volatile coals were made by the Bureau of Mines in the Tuscaloosa oven to determine the relationship between bulk density and coking time. It was found that a change in wet bulk density was accompanied by an approximately equal change in coking time on a percentage basis.[48] It was later shown that this relation held regardless of the flue temperature used in the experimental oven.[93]

Oven Width

There is evidence that expansion pressure decreases for highly expanding coals as oven widths are increased from 6 to 12 in. For widths of more than 12 in., results have been variable.[50,75]

Opinions differ on to what extent oven width influences coking time. Lowry, Landau, and Naugle,[79] working with 13-in. and 18-in. cylindrical retorts found that the coking time for a given coal was proportional to the 1.54 ± 0.12 power of the retort radius. The majority of workers use exponents in the range from 1.49 to 2.0.[94] From tests in the Tuscaloosa oven, Gayle and Eddy[56] found that coking times vary as the 1.49 power of the oven width. It was further concluded that changes in oven width over the range of 12 to 21 in. exert little or no influence on coke properties provided coking rates are maintained constant.

Coking Rate

Gayle and Eddy[56] and Veit[94] have shown that coking time is inversely related to the fourth power of the inner-wall temperature.

Auvil and co-workers,[95] in a study of the

[92] Smith, F. W., and Reynolds, D. A., *U.S. Bur. Mines, Rept. Invest.* No. 5110, (1955), 24 pp.

[93] Gayle, J. B., and Eddy, W. H., *U.S. Bur. Mines, Rept. Invest.* No. 5186 (1956), 20 pp.

[94] Veit, P., *Brennstoff-Chem.*, **31**, 14–22 (1950).

[95] Auvil, H. S., Davis, J. D., and McCartney, J. T., *U.S. Bur. Mines, Rept. Invest.* No. 3451 (1939), 21 pp.

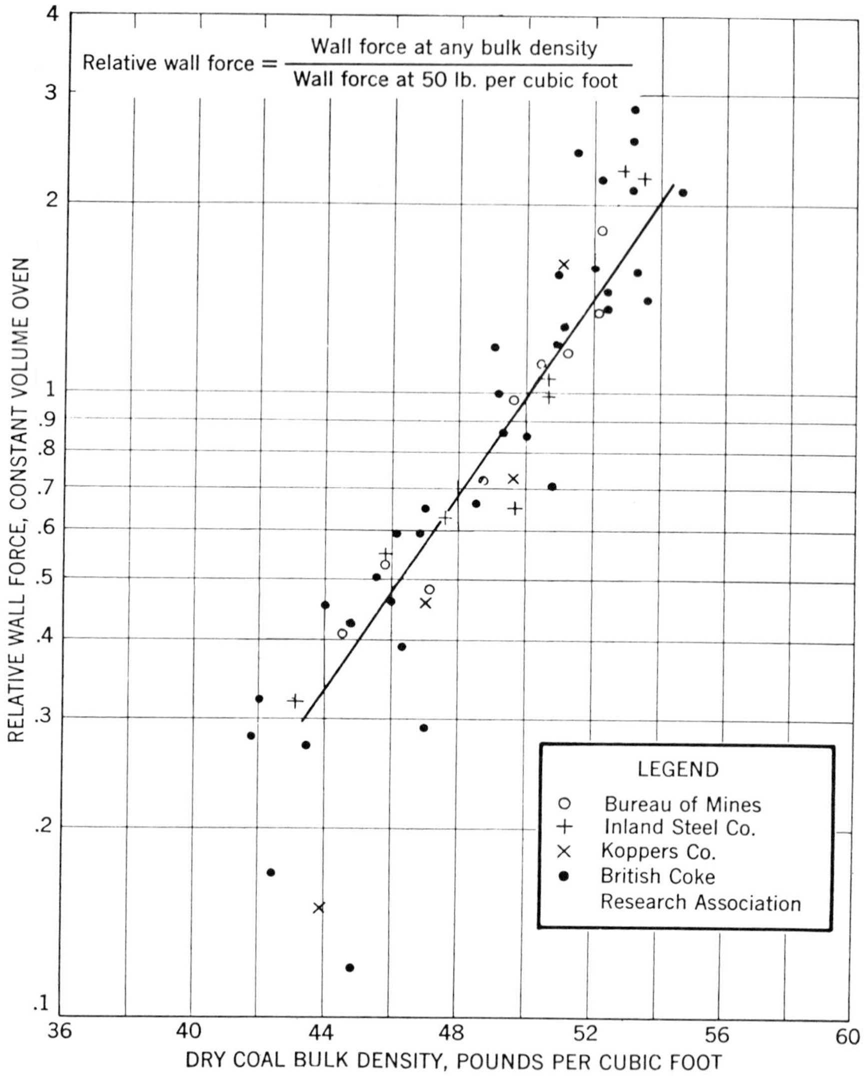

Fig. 23. Effect of dry-coal bulk density on force of coal expansion in constant-volume coke ovens.[92]

relationship between heating rate and expansion in the sole-heated oven, have reported that, with increasing heating rate, three coals showed a considerable increase, three coals showed a moderate increase, and one coal showed a decrease in expansion.

Figure 24 shows the effect of heating rate, as represented by flue temperature, on expansion pressure.[92] Results from the various sources are not in agreement, but the general trend shows increased pressures with increased temperatures. The British Coke Research Association found no general relationship between temperature and wall forces.[96]

Gayle and Eddy[56] concluded that moderate changes in flue temperature during soaking periods exerts relatively little influence on coke properties provided that coking rates are maintained constant throughout the presoaking period.

Figure 25 shows the relationship of coking time to coke quality for a blend carbonized at rates corresponding to 18, 22, 26, and 30 hr for a 19-in. oven.[97] Gayle and Auvil[47] found

[96] The Pontypridd Test Plant Committee, *British Coke Research Assoc., Tech. Paper* No. 5 (1956), 52 pp.

[97] Jackman, H. W., Henline, P. W., and Reed, F. H., *Illinois State Geol. Survey, Circ.* No. 213 (1956), 16 pp.

similar behavior in tests on a single coal. Gayle and Eddy[55] later reported results on six coals tested at flue temperatures ranging from 1500 to 2400°F. They found the trends in coke properties with carbonizing temperature were generally in qualitative agreement with the findings of previous investigators. However, the results varied enough to suggest that each coal may have its own individual sensitivity to variations in carbonizing temperature.

Charge Moisture

Reed and associates[98] found that moisture affects coke quality only insofar as it affects bulk density. Tests in the Bureau of Mines sole-heated oven indicate that there is no *per se* moisture effect on coal expansion.[52] The effects of moisture on coking time have been investigated by Gayle and Gulledge[48] and Lowry, Landau, and Naugle.[79,99] Gayle found that moisture had little effect beyond its contribution to the wet-charge density, but pointed out that the situation results in a magnification of the uncertainty with which the *net* influence of moisture content can be stated. Lowry and associates found, assuming other conditions constant, that each percent of moisture increases carbonizing time by 3.2%, but pointed out that any errors in moisture determinations

[98] Reed, F. H., Jackman, H. W., Rees, O. W., Yohe, G. R., and Henline, P. W., *Illinois State Geol. Survey, Bull.* No. 71 (1947), 132 pp.

[99] Lowry, H. H., *Chemistry of Coal Utilization*, John Wiley and Sons, New York, 1945, Vol. 1, pp. 848–62.

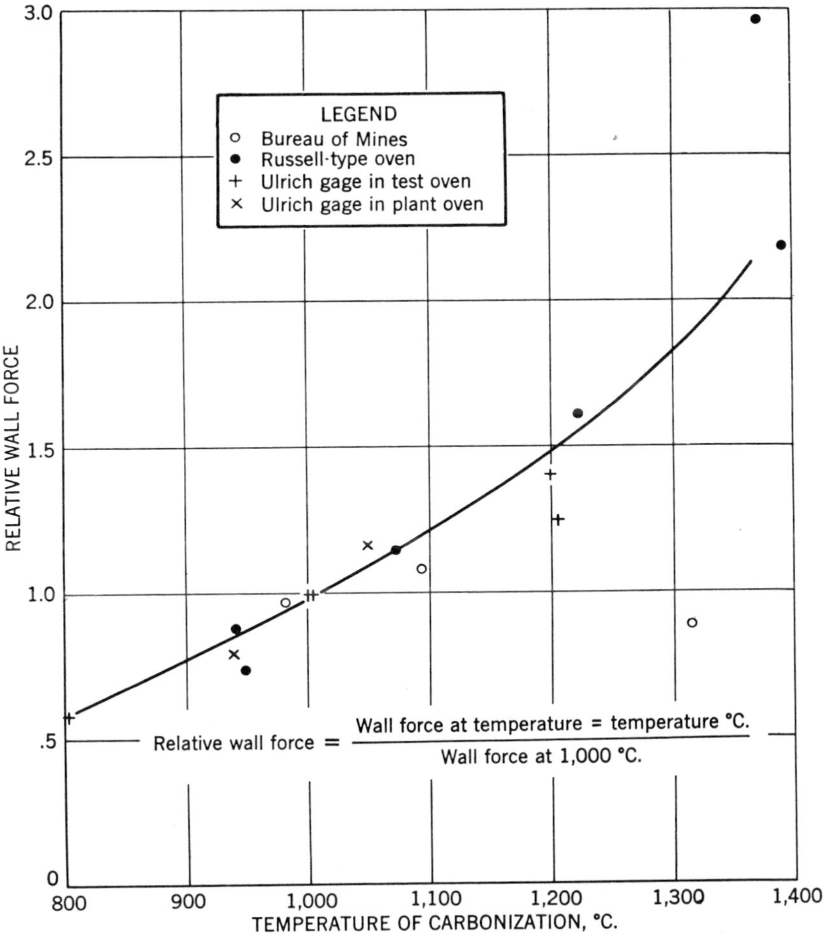

Fig. 24. Effect of temperature of carbonization on the wall force exerted in constant-volume slot oven.[92]

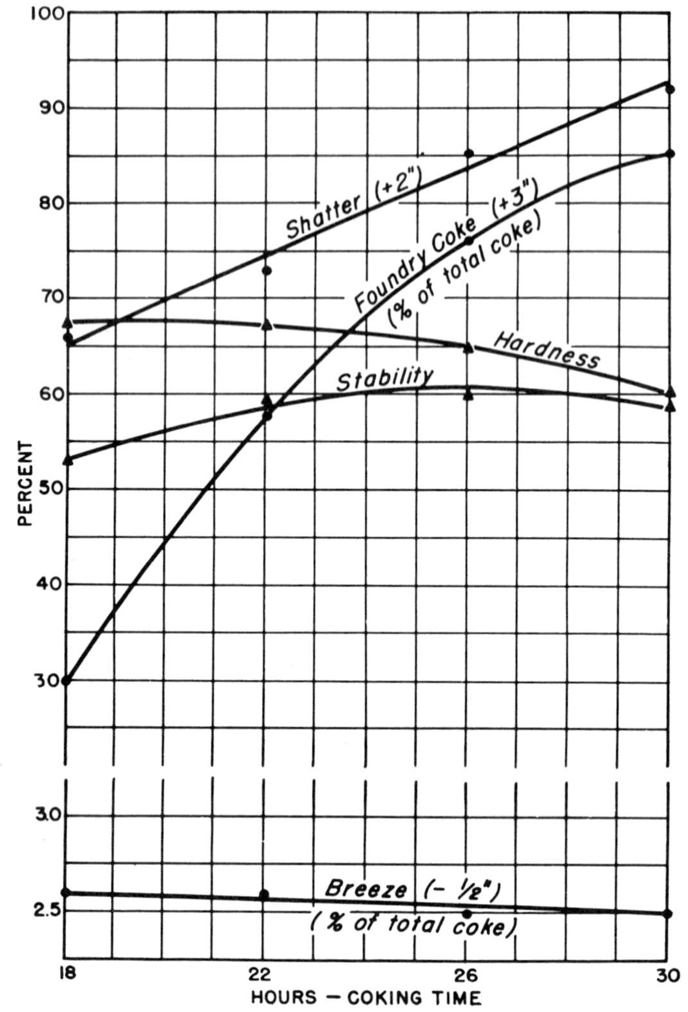

Fig. 25. Dependence of coke properties on coking time.[97]

would be magnified in the calculated carbonizing time. A review of previous work on this subject was published in 1960.[100]

Relation Between Coke Shrinkage, Coal Expansion, and Oven-Wall Damage

Differences of opinion exist on the relative importance of the effects of possible variations in coke shrinkage and coal expansion on life expectancy of coke ovens.

[100] Doherty, J. D., and Griffen, J., *AIME, Proc. Blast Furnace, Coke Oven & Raw Materials Comm.*, **19,** 117–30 (1960). Stahl, C. W., Smyser, J. M., and Aikman, R. F., *AIME, Proc. Blast Furnace, Coke Oven & Raw Materials Comm.*, **19,** 131–4 (1960).

From measurements of the shrinkage of low-temperature coke (500°C) on reheating to higher temperatures, and from measurements of the total area of fissures in coke charges from a small experimental oven, Auvil, Davis, and McCartney[95] concluded that all cokes shrink about the same extent between the formation temperature and a given pushing temperature. Data given indicated that coke from lower-rank coking coals shrank relatively more in the lower-temperature ranges (below 700°C) and relatively less in the higher-temperature ranges (700 to 950°C). Total coke shrinkage ranged from 9 to 11% on a linear basis. After further testing based largely on measurement of total area of fissures in coke made in a small oven of 11 by 24 in., these investigators reported[101] that the average linear shrinkage of cokes between formation temperature and 1000°C was about 10.5%, and the range of variations in shrinkage of cokes from various ranks of American coking coals was less than 2%. Expansion tests made at a constant pressure of 2.2 psig and coking rates approximating those in commercial ovens showed a range of more than 50% for the combined effects of coal expansion and coke shrinkage in the test charges. Since a maximum of 2% variation may be attributed to coke shrinkage effects, they concluded that any difficulty experienced in pushing should be taken as evidence of dangerous swelling unless other causes were known to exist. This conclusion was questioned by Mott[102] and Mott and Spooner,[103] but was supported editorially in *Coke and Smokeless-Fuel Age*[104] and by Foxwell[105] and Roberts.[106] Soth and Russell[107] also investigated the shrinkage of coke and reported linear shrinkages ranging from 9 to 12% between formation and pushing temperatures. More recently, Cheradame[60] has reported total shrinkage on a semicoke from a coking coal to be about 12% from coal resolidification temperature to 1000°C.

That there is still disagreement in regard to shrinkage behavior of coke, or on the definition of terms, is evidenced by the statement of Coleman and Farley[34] that, "Far more damage to oven brickwork is caused by non-contracting coals, and the resultant stickers or hard pushes, than can be caused by the internal gas pressure effect that takes place during carbonization," and the written comment by Russell and Perch[34] that, the ". . . statement deviates so far from what is believed to be generally accepted that it cannot go unchallenged. Carbonization pressure of coal has been responsible for very considerable damage to coke oven walls. This is a known fact and has been demonstrated many times." In another written discussion of the report, Brisse[34] stated that, "Considerable evidence is now available to show that carbonization pressure is not responsible for hard pushing in the coke plant. In contrast, available information appears to indicate positively that the lack of contraction of the coking charge is basically responsible for such troubles."

Review of these statements suggests that part of the disagreement is a matter of definition of terms, with both "coke shrinkage" and "lack of contraction" being used in connection with two different phenomena—(1) the shrinkage of coke per se, and (2) the apparent shrinkage of coked charges as indicated by the looseness or tightness of fit of the coke charge to the oven-chamber dimensions at the time scheduled for pushing the coke. The adjective "apparent" is used in conjunction with shrinkage because the apparent shrinkage is possibly not the real or true shrinkage of the charge. For example, it appears reasonable to postulate that charges developing little expansion pressure may not keep the ends of all pieces in contact with the oven walls. Charges are known to settle or slump, and individual coke pieces or areas of the charge may pull away from the oven wall while plastic coal is still being converted into coke. Also, Davis and co-workers[108] showed an example where the central portion of a cylindrical coking chamber was void of coke. Such phenomena could be isolated only with difficulty in an industrial oven of rectangular shape, whereas the in situ illustration of the Bureau of Mines makes it obvious that the pieces, as formed, are definitely shorter than the half-oven width dimension.

On the other hand, work by Lambert, Lancucki-Pater, and Lee[53] indicates that the coke pieces may actually be longer than half-oven width at the time the centrally located coal in any given area of the oven resolidifies to form coke. A fortuitous combination of circumstances permitted them to test normally operable ovens to destruction of the walls. Selected ovens were charged with test mixes, starting

[101] Auvil, H. E., Davis, J. D., and McCartney, J. T., *U.S. Bur. Mines, Rept. Invest.* No. 3539 (1940), 17 pp.

[102] Mott, R. A., *J. Inst. Fuel*, **13**, 189–200, 257–264 (1940); *Gas World*, **113**, 96–100 (1940); **114**, 4–6, 8–13 (1941).

[103] Mott, R. A., and Spooner, C. E., *Fuel*, **18**, 329–44, 371–9 (1939).

[104] Investigator, *Coke Smokeless-Fuel Age*, **1**, 113–6 (1939).

[105] Foxwell, G. E., *Coke Smokeless-Fuel Age*, **3**, 76–80 (1941).

[106] Roberts, J., *Coke Smokeless-Fuel Age*, **3**, 128–9, 210–13 (1941).

[107] Soth, G. C., and Russell, C. C., *Trans. Am. Inst. Mining Met. Engrs.*, **157**, 281–305 (1944).

[108] Davis, J. D., Reynolds, D. A., Sprunk, G. C., and Holmes, C. R., *U.S. Bur. Mines, Tech. Paper* No. 644 (1942), 46 pp.

TABLE 5

INFLUENCE OF BLEND COMPOSITION ON WALL MOVEMENT[53]

Test		5	2	3	6	4	7	8	9*
Bargoed coal	%	30	25	54	54	74	65	100	100
Total moisture	%	2.6	2.8	3.3	2.7	3.3	2.3	3.2	2.7
Bulk density (dry)	lb/cu ft	50.6	48.6	48.4	49.4	48.6	50.3	48.7	47.7
Maximum internal pressure	lb/in.^{2}	—	—	2.7	4.6	8.9	7.5	14.2	—
Maximum wall movement									
A: Gauge 2	in.	Nil	Nil	0.12	0.38	0.52	0.67	1.06	0.62
B: Gauge 3	in.	—	—	—	0.18	—	0.65	0.88	0.58

* Carried out in oven 3; all other tests carried out in oven 1.

with mixes expected to develop only inconsequential thrust against the walls. Mixes expected to develop successively higher thrust were used in successive tests until the wall failed, that is, until deflection during the coking period was large enough to cause visible cracking, permanent deflection of the wall, and resulted in inability to push the charge. Table 5 summarizes their results.

When discussing tests wherein wall movements of 0.12, 0.18, 0.38, and 0.52 in. were measured, Lambert and co-workers stated: "The movement of the commercial oven wall during these tests appears to have been almost entirely an elastic deflection, as the gauge readings were nearly similar after each carbonization to those observed before the test. Confirmation of this was obtained by the film record. Furthermore, there was no evidence within this pressure range of the development of any wall cracks...." Other information given in this report makes it clear that the latter statement does not apply where the wall movement was 0.52 in. In these tests, the dimension from end to end of the coke pieces would probably be larger than the normal oven width by an amount equal to the wall deflection. Apparently, as the coke shrank, walls followed until they reached their normal position, and shrinkage up to 0.38 in. was not detectable at the end of the carbonizing period. The tests seem to make it clear that coke shrinkage amounted to at least 0.38 in. plus an added amount equal to whatever shrinkage was required to provide for any "pinches," that is, decreases in width of the oven in the direction of pushing.

Auvil,[109] in an unpublished talk on "Detection and Measurement of Damage to Coke Oven Walls," showed oven-width charts taken in various operating ovens. These charts indicated width variations ranging from ovens having almost perfect taper to those having "pinches" of up to 1 in. in some areas. He reported that operators had little or no difficulty in pushing ovens with pinches of $\frac{3}{8}$ to $\frac{1}{2}$ in., provided that the oven had been completely coked out. Ovens with restrictions larger than $\frac{1}{2}$ in. were in regular use but operators classed them as "hard pushers" or cripples. An oven-width gauge[110] designed to provide a recording of actual oven-width of operating ovens was described. Since the strength of any given oven wall under existing operating conditions is unlikely to be known, and since there are many potential causes of oven damage, Auvil suggested that periodic precision measurements on a sample of the ovens in a plant was the only way they could tell whether or not conditions were under sufficient control to avoid normally undetectable and progressive damage that otherwise might not become appartent until months or years later.

Variation of Rate of Coke Shrinkage with Temperature

The rate of shrinkage of coke between coal resolidification or coke formation temperature and higher temperatures has been investigated

[109] Auvil, H. S., *Blast Furnace & Coke Oven Assoc.*, Chicago, January 30, 1953.

[110] Auvil, H. S., U.S. Pat. 2,923,062 (1960).

and the findings used in explaining the mechanism of fissure formation in coke. Cheradame,[60] for cokes from a variety of coals, showed maximum coefficients of contraction at temperatures slightly above the resolidification temperature, lower coefficients between 500 and 600°C, a general increase between 600 and 700°C, secondary maximums between 700 and 750°C, and a decreasing rate of shrinkage above the latter temperatures. The rate of shrinkage reaches a maximum between 700 and 800°C according to Nadziakiewicz.[111] He concluded that volume shrinkage on semicoke to 1000°C is about 22% of the initial volume of the coal. The rate of shrinkage of the coke existing in an oven during $\frac{1}{2}$-hr periods starting 1, 2, 4.5, 6.5, 9.5, 12, and 14 hr after charging has been given by Taits.[112] Excluding the first period, shrinkage is highest in the newly formed coke and decreases toward the walls of the oven. Maximum shrinkage in newly formed coke decreases from about 2.3% per half-hour at 2 in. from the wall to about 1.0% near the center of the coke. Coke shrinkage in the outer 2 in. is relatively high, 1.5 to 2.5%/$\frac{1}{2}$ hr, in the first 2 hr of the carbonizing period, under 0.5%/$\frac{1}{2}$ hr between 6.5 and 9.5 hr, and negligible thereafter. Taits[112] concluded that the principle factors determining the formation of fissures are the shrinkage gradient and the cohesive forces between adjoining layers of coke.

Bresler[113] has reported that semicokes shrink the same in all directions, and that cokes from various coals each have their individual shrinkage-rate behavior. The temperature at which total coke shrinkage is 3.5% and the temperature at which rate of shrinkage is the highest were reported to be coincident.

PREHEATING OF COAL FOR COKE-OVEN USE

Lowry[99] reviewed prior work on the preheating of coal and summarized existing knowledge as follows:

> If the difficulties and hazards of charging hot, dry coal to an oven can be overcome satisfactorily, preheating coal for carbonization should:
>
> 1. Increase the capacity of an oven to a greater extent than represented by the decrease in the total temperature interval traversed in the thermal pretreatment. (The decrease in coking time of the heated charge is not due to making use of exothermic heats of distillation but primarily to removal of moisture. The removal of moisture decreases the heat requirement for the charge and also increases the bulk density of the charge—a disadvantageous effect with expanding coals.)
>
> 2. Improve the economy of the carbonization process. (It is more economical, both from the standpoint of thermal efficiency and from that of fixed charges, to dry and preheat coal in separate equipment.)
>
> 3. Reduce the amount of liquor to be handled, evaporated and disposed of as a troublesome effluent.
>
> 4. Not significantly affect the yields of gas, tar, or coke. (The coke should be more uniform in cell structure and strength and might be of improved quality. The yield of ammonia would probably be decreased.)

Recent investigations on preheating of coal have provided additional information on which to base a further evaluation of its proposed inclusion in commercial practice.[114]

Effect of Preheating on Coal Composition and Properties

Parker[115] reported that preheating certain weakly coking high-volatile coals to about 400°C in the absence of air causes reduction of fluidity over the plastic range, and increases of the fusion point and loss of volatile matter, mainly oxides of carbon and water.

Smith and co-workers[116] studying Illinois coal concluded that preheating to 135°C followed by cooling did not significantly alter the dry-basis proximate or ultimate analysis. However, increasing the preheating to 320°C caused a decrease in volatile matter content from 38.1 to 34.4% and an increase of fixed carbon content from 54.6 to 57.1%. Oxygen content was

[111] Nadziakiewicz, J., *Coke and Gas*, **1**, 46–55 (1939).

[112] Taits, E. M., *Doklady Akad. Nauk SSSR*, **93**, 1077–9 (1953).

[113] Bresler, A. E., *Zavodyskaya Lab.*, **16**, 291–7 (1950).

[114] Roberts, J., *Coke and Gas*, **10**, 377–8 (1948). Puening, F., U.S. Pat. 2,552,014 (1951). Horner, H. R., U.S. Pat. 2,658,862 (1953). Makarov, G. N., Shilov, B. H., Schaschkova, T. D., Stein, I. Ya., and Giljasetdinov, L. P., *Koks i Khim.*, **1957**, No. 4, 12–7.

[115] Parker, A., *Gas. J.*, **284**, 115–18, 136 (1955).

[116] Smith, F. W., Birge, G. W., Wolfson, D. E., and Reynolds, D. A., *U.S. Bur. Mines, Rept. Invest.* No. 5418 (1958), 26 pp.

lowered from 10.2 to 8.8%. Heating value on the moist mineral-matter-free-basis was increased by about 1400 Btu/lb at all preheating temperatures, including 135°C. The agglutinating value was negligibly affected.

Effect of Preheating on Bulk Density and Wall Pressure

Perch and Russell[117] showed that coal of zero moisture content had the highest bulk density, and that the bulk density decreased with increasing temperature of preheat. Data of Smith and co-workers[116] shows that preheated coals have appreciably high bulk densities. The effect of preheated charges on oven-wall pressure requires additional investigation.

Effect of Preheating on Yields of Carbonization Products

Preheating prior to carbonization, according to Parker,[115] should reduce the volume of liquor produced but not significantly affect the yield of coke, gas, and tar. Smith and co-workers[116] agreed with Parker where preheated coal was cooled prior to carbonization. However, when hot preheated coal was carbonized, results showed an appreciable increase in tar yield, an appreciable decrease in gas yield, a slight decrease in the light-oil yield, and a small decrease in coke yield.

Effect of Preheating on Carbonizing Time

In some recent investigations,[116–118] there is general agreement that a substantial decrease in carbonizing time, probably from about 30 to 40%, can be realized through the preheating of coals prior to carbonization.

Effect of Preheating on Physical Properties of Coke

Experiments conducted with high-volatile and high-oxygen coals[116,117] reveal that preheating increases coke stability as measured by the 1-in. tumbler index. Preheating had no apparent effect, however, on apparent specific gravity. It was also concluded that preheating improved the quality of furnace coke made from high-oxygen coal about as much as did the addition of 20 to 30% low-volatile coal or 10% petroleum pitch. Improved coke cell structure[115] and uniformity of coke[118] should be expected.

Additional Effects of Preheating

Preheating to 500°F[119] could reduce the heat required for carbonization by about 25% or increase the oven capacity equivalently. Other investigations[118] have reported that preheating from about 600 to 700°F together with an increase in oven flue temperature increases the coke-producing capacity by about 50 to 75%. In addition, a wider range and variation of coal blends can be used.

Commercial Application of Preheating

Considerable advancement has been made in the application of preheating to commercial practices. Preheating, in the process described by Nathan,[120] is accomplished in a fluidized bed through indirect heat exchange between a hot fluid and a coal. This process involves the use of a conventional shell-and-tube-type heat exchanger. Another process[118] involving preheating in a fluidized bed utilizes direct solid-gas contact. In conjunction with this latter process, hot preheated coal may be pneumatically conveyed into the coke ovens. This type of system should solve some of the problems and remove some of the hazards involved in handling preheated coals.

VARIOUS PROGRAMS AT COKE PLANTS

The heterogeneous nature of coal makes manual representative sampling difficult. The old method of spot sampling with a shovel at five or more points on the top of loaded coal cars has never yielded samples which are truly representative of the large volumes of coal used for carbonization. For the precise control of product desired today, this method is obsolete.

Several types of automatic sampling devices have been on the market for years, and their use is rapidly increasing. One of the most

[117] Perch, M., and Russell, C. C., *Blast Furnace Steel Plant*, **47**, 591–7 (1959).

[118] Schmidt, L. D., and Allied Chemical Corporation, Brit. Pat. 813,791 (1959).

[119] Parry, V. F., U.S. Pat. 2,763,478 (1956).

[120] Nathan, M. F., U.S. Pat. 2,844,886 (1958).

effective designs takes continuous crosscuts of the coal stream, usually at a conveyor head, and discharges to a small pulverizer. A small portion of the pulverized coal passes to a sample container and the balance is returned to the main coal stream. Tests have proven this method to be very efficient. Some plants test each individual coal as it enters the preparation system and tests the pulverized blend as it passes to the storage bins. From the accurate data obtained, it is possible to regulate the chemical and physical properties of the coke produced within narrow limits. When coal suppliers are presented with an accurate assay of their product, they quickly realize that they must be ever vigilant if they are to supply the uniform product to the consumer.

In 1942, sixty-four of eighty-eight plants purchased coal of different volatile content. Of these, forty purchased high- and low-volatile coals; twenty purchased high-, medium-, and low-volatile coals; and four purchased high- and medium-volatile coals. The balance of the plants—twenty-four—used one coal only. In 1959, sixty-nine of seventy-five plants blended coal of different volatile content. Of these, thirty-six used high- and low-volatile coals (including six employing anthracite); twenty-eight used high-, medium-, and low-volatile coals (including thirteen employing anthracite); four used low- and medium-volatile coals; and, 1 used low- and medium-volatile coal and anthracite. Only six plants did not blend coal; one plant used straight high-volatile coals, and five used straight medium-volatile coals.[1]

It is apparent from the foregoing figures that the use of straight high-volatile coal in coke ovens, formerly normal procedure, has been virtually eliminated because of the demand for stronger coke and the depletion of premium quality high-volatile coal. In addition to the blending of different types of coal, many plants are faced with the problem of blending coal of one type from many different sources, the extreme being twenty different low-volatile coals to one plant. Efficient sampling, testing, and blending facilities are essential if any degree of coke uniformity is to be achieved under such conditions.

At coke plants, with few exceptions, coal is pulverized so that 60 to 85% will pass through a $\frac{1}{8}$-in. square mesh sieve, either individually as received or after blending. Although few plants have actually made facility changes to increase or regulate the degree of fineness of the several blend constituents, a vast amount of interest is developing in the subject. Pulverizer manufacturers have designed mills which can be converted internally from hammermill to impactor or vice versa, and manufacturers have been asked by several companies to conduct comparative tests by each method. These tests are all aimed at increasing the overall pulverization or percentage through $\frac{1}{8}$-in. mesh, without materially increasing the "bug dust" or percentage of 100-mesh coal in the product. Several installations are presently being contemplated.

The effect of oil additions to highly pulverized coal has long been known. The weight per cubic foot is increased, and fluctuations due to variations in moisture content and degree of pulverization are materially dampened. By manual sampling and regulation of the oil additions, some degree of control has been accomplished. Prellwitz[121] has stated that completely automatic units have reduced the weight variations to $\pm\frac{1}{4}$ lb/cu ft from a predetermined level with a resultant increase in production of approximately 4%.

[121] Prellwitz, S., *AIME, Proc. Blast Furnace, Coke Oven & Raw Materials Comm.*, **18**, 4–17 (1959).

CHAPTER 12

SOLID PRODUCTS OF CARBONIZATION

CH. G. THIBAUT

Association Technique de la Sidérurgie Francaise, Paris

When considering the solid products of carbonization, it should be borne in mind that semicokes or chars have very little importance in world production as compared with the production of coke. Furthermore, most of the coke is used in blast furnaces. Total coke production was more than 65,000,000 net tons in the United States in 1941, and increased during the three good years, 1955 to 1957, to 83,000,000 tons. Eighty-seven percent was produced by the steel industry and 90% was used in blast furnaces. Table 1 shows total coke production in different countries.[1]

TABLE 1

COKE PRODUCED IN THE PRINCIPAL INDUSTRIAL COUNTRIES IN 1958[1]

	Million Metric Tons	Percent of World Production*
United States	48.5	19
European Community	74.4	30
Soviet Union	51.5	21
United Kingdom	18.7	7
China	10.0	4
Japan	5.2	2
India	4.0	2
Canada	3.1	1

* World production was estimated to be 140 million metric tons in 1938 and 250 million metric tons in 1958.

The consumption of coke by industry may be slightly different in other coke- and steel-making countries, but the percentage of the total coke used in the blast furnaces is practically the same. Emphasis will therefore be on blast-furnace coke, but mention will be made of the special properties of foundry cokes, coke for gas producers, chars, and coke breeze. Coke has three functions in blast furnaces: first as a fuel; second as the provider of reducing gas, and indirectly or directly as the reducer; and last but very important, to insure in the shaft of the furnace as high a permeability as possible. Hence, as a fuel and a reducer, its calorific power and carbon content must have a high level, and as the permeability agent it must be refractory and strong enough to insure a high void volume.

CHEMICAL PROPERTIES

As a rule, methods used to determine the chemical composition of cokes and chars have been derived from those used for coals. The carbon content—the most important chemical character from the standpoint of the users—is generally calculated as 100 − (ash + moisture + volatile matter). The direct determination is more tedious. The detailed methods in use for these determinations are fully described in the ASTM Standards and in analogous standards of various countries. In the field of international standardization, the ISO Technical Committee 27, Solid Mineral Fuels, is now preparing proposals and recommendations for the analysis of coals and cokes. With regard to cokes, proposals are being studied for the determination of total moisture in the bulk sample and in the analysis sample, and of ash, volatile matter, sulfur, chlorine, phosphorus, nitrogen, carbon, and hydrogen.

The concepts of proximate analysis and of fixed carbon are not exactly the same in different countries. For instance, in the United States, it is commonly accepted that the proximate analysis includes the determinations of

[1] Ver. deut. Eisenhüttenleute, *Stahleisen-Kalendar 1960*, Stahleisen m.b.H., Düsseldorf, 1959.

TABLE 2

Moisture Content, Proximate Analysis, and Sulfur Content of Various Cokes

Countries	Plants	Moisture	Ash	Volatile Matter	Total Sulfur
		%	%	%	%
United States[2]	Clairton	7.0	17.7	1.1	1.1
	Bethlehem	4.3	11.3	1.3	1.0
	Brooklyn	6.7	8.7	0.7	0.7
	Erie	4.6	8.5	1.3	0.7
	Dearborn	7.7	10.5	0.7	0.5
	Provo	4.1	10.3	0.6	0.9
United States[3]	Everett	2.1	7.4	1.0	0.6
Germany[4]	Friedrich	2.0	9.4	1.5	1.0
	Reden	1.8	11.0	1.2	0.7
France[4]	Anzin	1.0	11.5	1.0	0.7
	Lievin	1.5	10.2	1.3	0.8
	Carling	1.5	9.7	1.3	0.8
Belgium[5]	Brabant	2.8	11.8	1.3	1.0
	Tertre	1.5	11.4	1.1	0.7
Netherlands[5]	Maurits	1.4	10.5	1.1	1.0

moisture, ash, and volatile matter. In other countries, it is considered that ash and volatile matter do not include the sulfur. In fact, the total sulfur is determined directly by the Eschka method; it contains the ash sulfur, which represents about one-tenth of total sulfur and the remaining nine-tenths are subtracted from the fixed carbon as reported in United States practice. Knowledge of these different ways of defining the fixed carbon content is useful when comparisons of carbon contents are made between American and European data. Table 2 reproduces different data from various origins. Ash, volatile matter, and sulfur are specified on a dry basis. These data cannot be compared indiscriminately because of different methods of determining ash and volatile matter. However, it appears that some American coals are cleaner than the common European ones, but, judging from the moisture content, quenching seems more regular in Europe.

Although the proximate analysis is in common use all over the world in coke-oven plants and steel plants, it must nevertheless be strongly criticized for its empirical nature and for its fundamental lack of precision. The percentage of ash is necessary for burden calculations. But the volatile matter content has no practical significance in determining the thermal characteristics of the coke. Coke-oven and blast-furnace men consider the volatile matter as indicative of the good or bad coking of the product. In practice, this is academic because screening easily eliminates the uncoked particles.

Furthermore, the differences found between laboratories determining the proximate analysis are often large. Calculated as it is from such data, the fixed carbon can in no way be considered as representative of the true calorific value and reducing power of the coke. The foregoing is no criticism of sampling or laboratory testing, but only of the principle and the concept of fixed carbon itself.

Coke users are directly interested in the main chemical features of the fuel from two points of view: (1) The calorific value and the reducing power are determining factors for the thermal balance of the furnaces, and in this respect, carbon and hydrogen contents are essential data; (2) the percent ash, sulfur, phosphorus,

[2] Am. Iron Steel Inst. and Am. Coke & Coal Chem. Inst., Contribution No. 43, Am. Iron Steel Inst., New York, 1952, 159 pp.

[3] Granger, G., Lach, M. J., and Mitchell, J., *AIME, Proc. Blast Furnace, Coke Oven & Raw Materials Comm.*, **18**, 152–61 (1959).

[4] Langlois, G., Thibaut, Ch. G., and Wildenstein, R., Publ. IRSid., Ser. B, No. 35 (1957), 116 pp.

[5] Thibaut, Ch. G., *Qualités des cokes métallurgiques*, CESSID, Metz, Second edition, 1960, pp. 1–84.

TABLE 3

COMPARISON OF PROXIMATE ANALYSIS WITH ULTIMATE ANALYSIS[5] (DRY BASIS)

Cokes	Ash	Combustible Sulfur	Volatile Matter	Hydrogen	"Fixed Carbon"	Total Carbon
	%	%	%	%	%	%
Erin	9.0	1.0	1.8	0.4	88.2	87.4
Lothringen	8.6	1.0	1.8	0.4	88.6	87.8
Viktor	9.6	1.1	1.6	0.4	87.7	86.8
Jakobi	10.0	1.2	1.7	0.6	87.1	87.7
Mont Cenis	10.2	1.1	1.0	0.8	87.7	86.4
Constantin	10.2	1.1	1.2	1.0	87.5	86.1
Anzin	11.5	0.7	1.0	0.4	86.8	85.9
Carling	9.7	0.7	1.3	0.4	88.3	87.7
Liévin	10.2	0.7	1.3	0.4	87.8	85.9
Marienau	10.1	0.8	1.4	0.4	87.7	86.9
Thionville	10.0	0.8	1.6	0.5	87.6	87.4
Boucau	8.7	0.7	0.8	0.1	89.8	89.5

chlorine, and the ash analysis have varying importance for the materials balance, including the slag basicity and the quality of the metal. Thus, it is essential that the concentrations of these constituents should be known with sufficient accuracy. This implies, of course, as perfect a sampling as possible; the manner of taking increments and reducing the analysis samples can be based on the procedure adopted by the Coke Evaluation Project.[2]

It is also of paramount importance that the conventional determination of fixed carbon should be replaced by direct determination of the true carbon content. This method requires a procedure considered until now as too difficult for common industrial laboratories. Such an objection, however, has always been advanced where an elaborate method had to be introduced into current practice. Starting from the Liebig method or from the high-temperature Sheffield method, it should be possible to develop a simplified procedure for determining the carbon and hydrogen contents of coke accurately and rapidly. A step in this direction is being made at ISO by the Testing-of-Coke Working Group. It can be reasonably hoped that in the near future everybody in industry as well as in research centers will forget the fixed carbon of our grandfathers and will speak only in terms of *ultimate analysis* for the carbonized products.

Since the late 1940's, thermal balances of the blast furnaces have been calculated in France by Irsid[5] on the basis of the percent of hydrogen and of the total carbon content in the coke. Thus, it is possible to compare a large range of values of the so-called "fixed carbon" with the exact total carbon, as in Table 3. The percent of combustible sulfur[6] is the total sulfur minus the ash sulfur. The data refer to a number of samples for each coke.[4,5,7] The juxtaposition of the columns of Table 3 is such that comparisons may easily be made between "fixed" and total carbon on the one hand, and between volatile matter and hydrogen on the other. The "fixed carbon" content C_f is calculated from:

$$C_f = 100 - \text{ash} - \text{volatile matter} - \text{combustible sulfur}$$

In eleven out of twelve cokes, the "fixed carbon" is higher than the true carbon content. It is evident that there is something fundamentally wrong with this situation. Furthermore, when the "fixed carbon" is calculated from ash and volatile matter only, as it commonly is in Britain and the United States, the values obtained for the "fixed" carbon percentage are still higher, that is, still more incorrect.

When applied to the assessment of blast-furnace performances, use of "fixed" carbon

[6] Thiessen, G., *Chemistry of Coal Utilization*, John Wiley and Sons, New York, Vol. 1, 1945, p. 435.

[7] Thibaut, Ch. G., and Astier, J., Publ. IRSid., Ser. A, No. 83 (1954), 138 pp.

leads to overestimated values of the thermal requirements, as compared principally with the quantity of heat furnished by the hot blast. This raises the question of the calorific power per unit of weight for cokes and other products of carbonization. Since the heat of combustion is different for amorphous carbon and graphite, the heat of coke depends on its structure. Some typical values are summarized in Table 4.[5]

Mott and Parker[8] have studied the difficulties and the precautions to be observed when using the bomb calorimeter to assure satisfactory precision. They showed how the carbon monoxide in the residual gases can be reduced to below 30 ppm, that is, $3 \times 10^{-3}\%$, and how the sulfur correction must be carefully applied. When the percent of total carbon has been determined, the calorific value of the coke carbon can be calculated with the values from Dewey and Harper,[9] that is, 8103 kcal/kg or 14,600 Btu/lb. But it can be noted, from the six values at the bottom of Table 4, that the calorific value of carbon in coke is also dependent on the H and S contents. These six figures are average values determined with the bomb calorimeter by three specialized laboratories, and they equal the quotient of the heat of combustion and the total carbon content in the sample. These numbers must not be compared directly with the figures preceding them in the table without taking into account the H and S percentages. But where the latter contents have been determined, the calorific value of the carbon in coke may be calculated[10] as follows:

Calorific value of C = P_c (kcal/kg)
Calorific value of coke = P_K (kcal/kg)
H, S, C = % hydrogen, combustible sulfur, and total carbon, respectively

$$P_c = \frac{1}{C}(P_K - 341.6\,H - 22.1\,S)$$

When the six values given in Table 4 are recalculated, using this equation, the values range from 8001 to 8073, with a mean value of 8040 kcal/kg, which is not far from the value of 8024 reported by Dewey and Harper.[9] The calorific value of carbon in coke is much closer to that of amorphous carbon than to that of graphite.

In reference to Table 3, the *volatile matter* percentages reported for the six cokes at the bottom were determined by eight different laboratories in France and Germany and represent

[8] Mott, R. A., and Parker, C., *Fuel*, **37**, 371–81 (1958); **38**, 189–204 (1959).

[9] Dewey, P. H., and Harper, D. R., *J. Research Natl. Bur. Standards*, **21**, 457–74 (1938).

[10] Sanna, D., Unpublished.

TABLE 4

SOME TYPICAL VALUES FOR THE HEAT OF COMBUSTION OF CARBON TO CARBON DIOXIDE[5]

Authors	Year	kcal/kg C	Forms of Carbon
Le Chatelier	1908	7900	Graphite
Roth	1937	7850	Graphite
Rossini	1938	7836	Graphite
Dewey and Harper	1938	7832	Graphite
Gumz	1950	7871	Graphite
Spiers	1950	7838	Graphite
Loison	1956	7838	Graphite
Le Chatelier	1908	8138	Amorphous Carbon
Rey	1939	8107	Amorphous Carbon
Gumz	1950	8083	Amorphous Carbon
Dewey and Harper	1938	8024	Coke carbon
Dewey and Harper	1938	8103	Coke carbon (0.3% H)
(Six cokes)[4]	1957	8180	Coke, 0.36% H, 0.71% combustible S
(Six cokes)[4]	1957	8260	Coke, 0.50% H, 0.94% combustible S
(Six cokes)[4]	1957	8260	Coke, 0.43% H, 0.74% combustible S
(Six cokes)[4]	1957	8160	Coke, 0.37% H, 0.69% combustible S
(Six cokes)[4]	1957	8250	Coke, 0.43% H, 0.80% combustible S
(Six cokes)[4]	1957	8240	Coke, 0.50% H, 0.83% combustible S

Note: Btu/lb × 0.555 = kcal/kg.

the average results from about ten samples for each coke. The original results from the different laboratories varied within a range of 0.2 to 2.8%. This demonstrates the great lack of accuracy and precision of this determination. Dahme and Echterhoff[11] have pointed out the fundamental errors in the determination of volatile matter in coke. The volatile matter can be advantageously replaced by the hydrogen content which is closely correlated with the carbonizing temperature.[12]

A rapid commercial method for the determination of the carbon and hydrogen contents in solid mineral fuels has been presented at a recent meeting of the International Standards Organization.[13] The procedure has been specially studied for cokes, but it is applicable to all solid and liquid fuels. The method is now being applied in different countries for discussions concerning its precision.

TABLE 5

SOME DATA ON THE ULTIMATE ANALYSIS OF COKES AND THE ASH COMPOSITION

Chlorine	0.15 to 0.60	% in dry coke
Phosphorus	0.01 to 0.25	% in dry coke
Total sulfur	0.70 to 1.10	% in dry coke
Ash sulfur	0.02 to 0.08	% in dry coke
Combustible sulfur	0.68 to 1.02	% in dry coke
Nitrogen	0.70 to 1.30	% in dry coke
Silica	35 to 55	% in ash
Lime	2 to 8	% in ash
Alumina	20 to 35	% in ash
Iron	8 to 15	% in ash
Sulfur	0.4 to 1	% in ash
Magnesia	1.5 to 4	% in ash
$Na_2O + K_2O$	2 to 6	% in ash

Other features of chemical analysis, concerned with the metallurgical aspects of the value of cokes, are the percentages of nitrogen, phosphorus, chlorine, etc., and the analysis of ash. To give an idea of the range of concentrations of these elements, Table 5 shows values for representative metallurgical cokes.[4,5] These values must be considered merely as a guide. Whenever a material balance is necessary, the coke and ash must be accurately sampled and analyzed.

Determination of *moisture* in coke could be the subject of a whole chapter. Cokes being hygroscopic, moisture has a great variability. Two rules should be generally adopted for analysis of the solid products of carbonization: (1) Sampling and procedures should always be in accordance with the specifications of the ASTM Standards or of ISO Recommendations; (2) It is always advisable, furthermore, to give the point of sampling, whether immediately after quenching, before or after screening, before or after handling, or just at the point of charging into the buckets or skips for feeding the blast furnace, cupola or gas producer. The accurate determination of the coke moisture is of tremendous importance for the thermal balance of blast furnaces, where the coke is weighed in its wet condition. The evaluation of the total heat accounted for by the combustion of carbon involves not only the precise knowledge of the total carbon content on a dry basis, but also of the exact moisture at the point where the wet coke is being weighed. These considerations are certainly well known, but their sound and regular application is unfortunately underestimated in current operating practice.

An estimate of the *mineral-matter* content of coke may be required for the conversion of analytical data to the dry, mineral-matter-free basis for calculating the oxygen content or assessing the calorific power of the carbon. A first approach to this problem has been described.[14] If MM, A, S, and Fe are the mineral-matter, ash, total sulfur, and iron contents in the sample, the mineral-matter content may be obtained by the following equations:

$$MM = A - \tfrac{4}{5}S - \tfrac{1}{7}Fe, \text{ when the iron content is known}$$

and

$$MM = A - \tfrac{2}{3}S, \text{ when the iron content is not known}$$

PHYSICAL PROPERTIES

Physical properties of coke may be described under three groups: the static properties considered at the scale of a lump of coke, the static

[11] Dahme, A., and Echterhoff, H., *Brennstoff-Chem.*, **40**, 337–41 (1959).

[12] Lowry, H. H., *Chem. & Ind. (London)*, **1950**, 619–25.

[13] Spiers, H. M., *Technical Data on Fuel*, 5th edition, World Power Conf., London, 1952, 517 pp. Radmacher, W., and Hoverath, A., *Brennstoff-Chem.*, **41**, 52–4 (1960).

[14] Barker, J. E., and Mott, R. A., *Fuel*, **33**, 363–5 (1960).

properties of bulk coke,[15] and the mechanical properties. In the first group, the characteristics related to the lump of coke are considered separately, even if their determination involves the consideration of a great number of lumps. For example, the microscopic examination of porosity is in no way directly concerned with coke in bulk, being related to the internal structure of one piece, or to the *average* structure of a number of lumps. The second group includes the characteristics of a very great number of grains or lumps, the size and size range, the bulk density, the void volume, and the permeability. The third group relates to the resistance of coke to breakage and abrasion, or cohesion, and to the variations of the properties classed in the first and second groups when degradations occur due to mechanical effects.

X-Ray Diffraction and Optical Anisotropy

In X-ray diffraction studies of coke, the characteristic sharp peak for graphite is replaced by a broad band which corresponds to partial organization of the carbon in a graphite lattice, so that[4] the ratio of the height of the band to its average width (x/l) can be considered as representing the degree of structural organization, proportional to the size of the crystallites. In Table 6, values of this ratio x/l observed at Cerchar are compared with the volatile matter in the coal blend and with the optical anisotropy.[16] The anisotropy in reflected light decreases with an increase in the volatile matter in the coal.

TABLE 6

OPTICAL ANISOTROPY AND X-RAY DIFFRACTION OF COKES AS A FUNCTION OF THE RANK OF THE COALS[4]

Volatile Matter of the Blend	Optical Anisotropy Lab. *C*	Optical Anisotropy Lab. *W*	X-Ray Diffraction Ratio
%			x/l
23	91	260	3.5
24	90	290	2.4
26	89	200	2.2
27	81	140	2.6
30	41	100	1.9
31	49	95	1.8

Pore Structure

A piece of coke is a spongy material with pores which have various dimensions depending on the nature of the blend of coals and on the carbonizing conditions, especially the pulverization, the method of filling the oven—top charging or cake charging, the rate of heating, and the final temperature. By heating a coal in a small steel retort at constant volume and increasing temperatures, Mott and Wheeler[17] have observed the formation of the pores. From 400 to 550°C, the total porosity increased from 3 to 51%, the latter figure being the sum of 5% for closed pores and 46% for open pores.

There is a considerable amount of discussion concerning the relative importance of open and closed pores and the part they play in the reactivity and combustibility of semicokes and cokes. It has been suggested the external surface of the lump, the internal surface area of the pores, and the "submicroscopic" surface area of the smallest pores whose diameters are only a few angstroms be considered separately.[18] The external surface for pieces of coke from 10 to 1 cm ranges from 20 to more than 500 sq m/cu m, or 6 to more than 150 sq ft/cu ft. The total area of the microscopic pores ranges from 0.5 to 3.5 sq m/g, and the surface of the submicroscopic pores, as measured by void infilling of mercury, extends to several square meters per gram.[18]

The specific surface can also be estimated with the B.E.T. method[19] using nitrogen

[15] Although M. A. Mayers, *Chem. Coal Utilization*, **1**, 906–12 (1945), employed the word "structure" for designating the properties of bulk coke considered in beds or layers or large volumes; in Europe the structural properties relate generally to the molecular and microscopic characteristics, extended up to the lump of coke regarded as a unit. The difference in use of this word must, therefore, be appreciated when dealing with the physical properties of coke.

[16] Ramdohr, P., *Arch. Eisenhüttenw.*, **1**, 669–72 (1928). Mackowsky, M. T., *Brennstoff-Chem.*, **36**, 304–14 (1955).

[17] Mott, R. A., and Wheeler, R. V., *The Quality of Coke*, Chapman & Hall, London, 1939, 464 pp.

[18] Mackowsky, M. T., *Nouvelles méthodes d'évaluation de la qualité du coke métallurgique, La Documentation Métallurgique*, No. Spécial SK 1, February 15, 1956, pp. 79–90.

[19] Brunauer, S., Emmett, P. H., and Teller, E., *J. Am. Chem. Soc.*, **60**, 309–19 (1938).

absorption at 85°K. Table 7 compares the results obtained by this method in two laboratories[4] and the specific surface as calculated from microscopic measurements of the diameters of pores by the methods of Dahme[20] and de Vries.[21] It will be noted that the values

TABLE 7

SOME EVALUATIONS OF SPECIFIC SURFACES IN COKES[4]

Cokes	B.E.T. Method		Microscopy	
	Lab. *C*	Lab. *K*	Dahme	de Vries
Carling	1.1	2.1	280	260
Friedrich	1.5	1.1	300	210
Liévin	1.2	1.2	250	190
Reden	1.7	1.8	270	230
Anzin	1.8	0.8	320	190
Thionville	—	—	290	240

show large discrepancies, so the accuracy of the determinations seems to be doubtful—the reason probably being the difficulties of sampling.

The plastic properties of coal influence[22] the formation of the pores, their average width, and the thickness of the walls which separate them. Good metallurgical cokes are obtained from coals showing a fair plasticity as measured by the dilatometer, a plastic range of about 100°C, an oxygen content not more than 4%, and a resolidification point of about 500°C. The cellular structure can be observed by use of plain photographs with enlargements of 3, for instance,[5] or more quantitatively by observing plane sections under the microscope.[23] The difficulties encountered in sampling and the need for a very large number of examinations have an important effect on the results obtained by different experiments, as shown in Table 8. The average values which would be obtained from a large number of observations are likely to be more comparable. This implies the use of costly mechanisms and statistical devices.

It is often assumed[20] that all the pores in a lump of coke intercommunicate so that there are only open pores which could be filled by a fluid, depending only on the nature of the fluid itself. Porosity can be measured directly by use of a displaced fluid which may be a liquid or a gas, or indirectly from the ratio of apparent density d_a to true density d_t; the compactness, $C = d_a/d_t$, increases with the decrease of the relative volume of pores, and inversely, the porosity, $P = 1 - (d_a/d_t)$, increases.

[20] Dahme, A., *Ing. Arch.*, **21**, 346–51 (1953).

[21] de Vries, H. L., *Mitteilungsbl. math. Statistik*, **7**, 213 (1955).

[22] Arnu, C., *Congr. assoc. tech. gaz, Paris*, **1959**, 281–314.

[23] Mackowsky, M. T., *Brennstoff-Chem.*, **36**, 304–14 (1955).

"True" and Apparent Densities

The porosity of coke presents a problem in determining the "true" density or specific gravity. The determination of true densities is generally made in a pycnometer with water or another liquid. The results are especially influenced by the grinding and size of the powdered coke. Fair reproducibility[24] is observed when toluene is used if all particles are ground to −0.2 mm. The ASTM Standard method with water and −200-mesh (0.074 mm) particles gives more satisfactory results because toluene or benzene do not fill the fine pores as completely as water containing a wetting agent, as shown in Table 9. The most accurate procedure would appear to be the method using helium as the displacing fluid.[25]

TABLE 8

WIDTH OF PORES AND THICKNESS OF WALLS (MICRONS)[4]

Cokes	Average Width of Pores		Average Thickness of Walls	
	Lab. *C*	Lab. *W*	Lab. *C*	Lab. *W*
Carling	140	130	60	80
Friedrich	150	160	90	100
Liévin	140	160	90	120
Reden	110	140	70	110
Anzin	180	180	100	110
Thionville	120	160	70	110

It appears from Table 9 that the "true" densities obtained by use of benzene and toluene are systematically lower than with water (ASTM method) and with helium. There is some correlation between the average true

[24] Thibaut, Ch. G., and Vigneron, P., *Centre document. sidérurg., Circ. inform. tech.*, No. 2, 341–4 (1956).

[25] Howard, H. C., and Hulett, G. A., *J. Phys. Chem.*, **28**, 1082–95 (1924).

TABLE 9

"TRUE" DENSITY DETERMINED WITH DIFFERENT FLUIDS[4]

Cokes	Average Diameter of Pores, μ	Results for True Density with Helium	Water	Toluene	Benzene
Anzin	180	1.99	2.00	1.88	1.95
Liévin	150	1.95	1.96	1.90	1.91
Friedrich	155	1.94	1.95	1.87	1.85
Carling	135	1.88	1.87	1.81	1.75
Reden	125	1.86	1.92	1.78	1.79

densities and the average diameter of the pores. The effective penetration of a fluid into the finest pores must be considered in finely powdered coke as well as in lumps. In fact, the porosity of the first of these five cokes is the highest and it also has the highest true density and the widest pores. Too great confidence should not be placed in this kind of determination without consideration of accurate sampling and the statistical nature of the data.

The same limitations apply to the different methods of measuring the apparent specific gravity or apparent density. The apparent density of a piece of coke or the average density of several pieces can be determined by established methods.[26] The ASTM Standard method can be used to obtain reproducible results, the results for six cokes[4] being compared in Table 10 with those obtained by another procedure,

TABLE 10

APPARENT DENSITY OF COKES[4]

Cokes	Lead-Shot Method	ASTM Method	
Thionville	0.88	0.86	0.90
Anzin	0.91	0.90	0.90
Friedrich	0.91	0.92	0.93
Carling	0.95	0.93	0.95
Liévin	0.97	0.94	0.99
Reden	1.05	1.04	1.04

in which 20 "full-length" pieces of each coke sample were placed in a steel vessel that was then filled completely with lead shot 1.5 mm in diameter.[4] The displacement lead, a non-wetting fluid, gave the value of the total porosity. This latter procedure has high sensitivity because of the great difference between the densities of coke and lead, but it requires careful sampling and numerous weighings.

The purpose of ISO Working-Group on "Testing of Coke" is to study the best methods for industrial or commercial estimations of "true" and apparent densities of coke. It appears from the results[27] of comparative experiments that the procedures using water as in the ASTM Standards can be considered the most practical and duplicable methods.

In Table 11, data collected in the United States[2] and in France,[4] using only the ASTM methods, are compared. The American cokes are characterized by the conventional letters employed by the reporters of the Coke Evaluation Project. The cokes have been listed in Table 11 according to the volatile content—dry, ash-free basis—of the average blends. The porosity is given in the last column as calculated from the apparent and "true" densities. The coke ovens of Reden and Carling were cake-charged, and the blend contained 7 to 8% finely crushed coke breeze. A tendency for the "true" density to decrease inversely with the volatile matter of the blend is indicated, but the apparent density and the porosity do not appear to be correlated with the volatile-matter content of the coal. This observation is not in accord with an assumption that coals that form much gas give lighter and more porous cokes. Operating conditions, such as grinding of the coals, oven-flue temperature, and coking time probably have a larger influence on these properties than does the analysis of the coals.

The apparent density varies with the size of the lumps,[17] the broken pieces having a smaller number of cracks, cavities, gaps, or fissures

[26] Mayers, M. A., *Chemistry of Coal Utilization*, John Wiley and Sons, New York, 1945, Vol. 1, pp. 863–95.

[27] Poiré, P., and Vigneron, P., *Comparaison de deux méthodes de mesure de la densité apparente du coke*, COMINPECC, Paris (1958), pp. 1–21.

TABLE 11

COMPARISON OF "TRUE" and APPARENT DENSITIES OF AMERICAN AND EUROPEAN COKES

Cokes	Volatile Matter % of Dry, Ash-Free Coal	Densities, ASTM		Porosity
		"True"	Apparent	%
Anzin	25	2.00	0.90	55
Friedrich	26	1.95	0.93	52
Reden	29	1.92	1.04	46
Liévin	30	1.96	0.98	50
BG	31	1.91	1.00	48
F	31	1.91	0.91	52
S	32	1.91	0.86	55
Carling	32	1.87	0.95	49
B	33	1.95	0.90	48
I	33	1.89	0.86	54
R	33	1.99	0.98	51
Thionville	33	1.91	0.89	53
C_1	34	1.99	1.01	49
K	35	1.87	0.85	55
J	36	1.92	0.90	53
C_2	37	1.94	1.02	47
G	38	1.87	0.87	53
T	40	1.83	0.84	56

than the "full-length" pieces. This is not generally so, as demonstrated by the following data[27] obtained by the two methods mentioned earlier, with samples from the same ovens:

	ASTM Method	Lead-Shot Method
Coke, +2″	0.901	0.883
Coke, $\frac{3}{8}$ × 2″	0.871	0.824

The usefulness of determinations such as dimensions of pores, "true" and apparent densities, and porosity is not established for industrial uses of coke. Even if such measurements are not directly useful for the assessment of blast-furnace performance or for other commercial uses, however, they are helpful in fundamental research for interpreting variations of other parameters.[28] A second group of physical properties is more directly related to characteristics of coke as fed into a furnace. The average size and size range, the void volume, and the bulk density have a strong influence on the behavior of a bed of coke, especially on its permeability which is essential for combustion, gasification, and reduction processes.

[28] Schenck, H., Esch, H., and Langenbeck, W., *Stahl u. Eisen*, **79,** 1933–8 (1959).

Bulk Density

Transition from studies of the properties of individual pieces of coke to those of bulk coke is found in the apparent density, which is used for calculating the porosity or void volume of individual pieces of coke. It is also used for determining the void volume v or voidage of bulk coke:

$$v = 1 - \frac{d}{d_a},$$

where d_a is the apparent density and d is the bulk density. For practical application with American units, the equation becomes:

$$v = 1 - \frac{d}{62.4g_a}$$

where g_a is the apparent specific gravity (lb/cu ft).

Much consideration has been given to the usefulness of bulk density and the best procedure for its determination.[2,4,29] Bulk density

[29] Graham, J. P., *Recent research in coking technology. Proc. 26th Blast Furnace Conf., BISRA, London,* **1959,** pp. 1–5. Syskov, K. I., *Freiberger Forschungsh.*, **A24,** 24–31 (1954).

may be defined as the weight of coke per cubic foot as actually determined either in a standard box of one or several cubic feet, or in a large container such as a railroad car. Within this range, proposals have been advanced, especially in Germany, for using a box of one cubic meter capacity. In the United Kingdom and in the United States, the bulk density is currently determined by carefully filling and weighing a box of 2 to 8 cu ft, but the determination has also been in railroad cars and in a large cylinder of 31 cu ft or 0.9 cu m.[2] In France, the determination of bulk density in a railroad car has recently been standardized.[30] A standard procedure for bulk density is being studied by a Working Party of ISO on Testing Coke.

Various comparisons between results obtained in small and large containers have been carried out in different countries. Table 12 shows the average values recorded from a number made during 2 years[4] on six European cokes. The bulk density is expressed as a density—that is, the ratio of the coke specific gravity to water specific gravity. This manner of recording the data is easy and is common with metric units because the number is the same for density and for specific gravity in metric tons per cubic meter or kilograms per liter. It is also advantageous because of the uniformity of the expressions of true, apparent, and bulk densities. Nevertheless, for the sake of American readers, the last column of Table 12 is also given in pounds per cubic foot.

Closer packing of the coke occurs in a railroad car during transportation, the smaller pieces slipping down and nesting into the voids formed between the larger lumps. Therefore, it is logical that an increase in bulk density is found between the starting point of a railroad car and the discharging point after some hundred miles on the tracks. From a commercial point of view, if utilized for a price specification, the bulk density can be determined by leveling and weighing railroad cars before dispatch. It will be noted from the data given in Table 12 that the coke on arrival at the research station was completely unloaded and screened for size analysis before being reloaded in railroad cars dispatched to the blast-furnace plants. A number of values of the bulk density were measured in containers of one hectoliter and one cubic meter. As commonly observed,[17] the wall effect decreases with increase in capacity of the container. Various studies of the wall effect and of packing,[17,31] as well as studies of the effects of size and bulk density on the permeability of a layer of coke,[32,33] have been published. When the bulk density is investigated, it is necessary to determine the size distribution at the same time in order to correlate as far as possible the void volume and the other characteristics mentioned. Joseph[31] has shown how the voids in a bed decrease from 50 to 25% when two

[30] AFNOR, NF. M. 03-022 (1959).

[31] Thibaut, Ch. G., and Vigneron, P., *68th Congr. assoc. tech. gaz, France*, June, 1951, pp. 797–827.

[32] Joseph, T. L., *Blast Furnace Steel Plant*, **45**, 489–93 (1957).

[33] Furnas, C. C., *Ind. Eng. Chem.*, **23**, 1052–8 (1931). Nixon, E. W., and Maw, F. R., *J. Iron Steel Inst. (London)*, **174**, 331–5 (1953). Ergun, S., *Ind. Eng. Chem.*, **45**, 477–85 (1953).

TABLE 12

BULK DENSITY OF DRY COKE DETERMINED IN VARIOUS CONTAINERS[4]

Cokes	In Railroad Car before Dispatch	In Railroad Car on Arrival	After Rescreening			
			In hl Box	In cu m Box	In Railroad Car	
	kg/l.	kg/l.	kg/l.	kg/l.	kg/l.	lb/cu ft
Carling	0.464	0.499	0.445	0.449	0.491	30.5
Friedrich	0.446	0.479	0.422	0.431	0.473	29.6
Liévin	0.457	0.510	0.446	0.454	0.516	32.2
Reden	0.501	0.536	0.473	0.486	0.536	33.5
Anzin	0.438	0.472	0.408	0.414	0.439	27.4
Thionville	—	—	0.404	—	0.443	27.6
Number of determinations	50	65	1195	216	37	37

sizes of a material are mixed: the maximum voidage occurs when all pieces are uniform in size and the minimum happens when the percentage of small constituents mixed with larger lumps increases up to about 35%. Of course, the decrease of the void volume does depend on the size ratio, that is, the ratio of the dimensions of the large and small particles; when the size ratio is not greater than 2.0, the decrease of the voidage is about 4%.

Thus, a study of bulk density can scarcely be separated from an investigation into mean size, size range, or size ratios. From the data recorded in Table 12, it has been possible to correlate the bulk density in hectoliters (3.5 cu ft) with the results obtained in 15-ton railroad cars, d_r:

$$d_r = (0.80 + 0.004D_{50})d_{\text{hl}}$$

This correlation is valid for mean sizes D_{50} in millimeters (as is to be defined) between 70 and 90 mm. It could be carefully extended, for instance, to American cokes such as those described in the Coke Evaluation Project[2] where the mean sizes varied from 62 to 86 mm, or 2.45 to 3.37 in. In this research, the bulk densities shown in Table 13 were determined by

TABLE 13

BULK DENSITY OF AMERICAN COKES[2]

Company	Bulk Density	Standard Deviation	Bulk Density
	lb/cu ft		kg/l
C_1	27.4	0.5	0.438
C_2	27.8	0.5	0.445
J	24.9	0.5	0.398
B	24.7	0.6	0.395
BG	27.6	0.6	0.442
I	23.8	0.5	0.381
F	25.2	0.3	0.404
S	23.6	0.3	0.378
R	26.2	0.4	0.419
G	22.9	0.3	0.366
T	21.5	0.3	0.344
K	23.1	0.5	0.370

weighing over 800 lb of top by 1.06-in. coke in a cylindrical container of 38 cu ft; twenty determinations were made for each coke with satisfactory standard deviations. These data may be compared with the numbers given in Table 12 for 1 cu m, since the capacities of the containers are similar. It will be noted that some European cokes show a greater bulk density than the American ones. This is not strongly correlated with the nature of the coals as characterized by their volatile-matter content (Table 11). It will be seen that the bulk density is largely influenced by the preparation of the blend and the coking conditions.

Further data concerning the wall effect and the capacity of the container are recorded in Table 14 from a Russian contribution[34] to this subject.

These data are of interest in completing the observations made concerning Table 12. The values given for these Russian cokes are the averages of ten determinations in each container. The increase in bulk density with increasing container capacity is as well established as in the French observations, but the size of the coke is different—the mean size varying from 61 to 63 mm (2.5 in.) for the second line of Table 14. The Russian experimenters found practically no significant difference between results obtained in cubic and cylindrical cubic-meter containers. This result is in perfect agreement with results recorded previously[5] from ninety data, which leads to the general conclusion that the shape of the container has no practical importance whenever its capacity is 1 cu m or more.

It appeared from the French research[4] that the bulk densities of six cokes (Table 12) were closely correlated with the apparent density (Table 11). This correlation may be explained by the narrow range of the size distribution of these cokes, the void volume also varying within a narrow range. In the American research,[2] bulk densities and apparent specific gravities for the twelve cokes (Tables 13 and 14) were equally correlated, despite the wider range of the mean sizes. These observations lead to the general conclusion that the determination of the apparent density is of little interest when the bulk density has been determined. Nevertheless, their determinations are important where the knowledge of porosity and void volume are concerned, especially when we are comparing in some detail the performances of two cokes which do not differ largely in other respects, or when we are dealing with two size classes of a single coke. Furthermore, the determination of bulk density—especially in railroad cars—requires little work compared

[34] ISO Tech. Comm. 27, Working Group 8, Document 37, USSR, October, 1957.

TABLE 14

RUSSIAN DATA ON BULK DENSITY OF COKE DETERMINED IN DIFFERENT CUBIC CONTAINERS[34]

	Volume of Container				Railroad Car
	0.1 cu m	0.2 cu m	0.5 cu m	1 cu m	66 cu m
Kuznetz Basin	0.427	0.446	0.463	0.462	0.516
Kuznetz Basin	0.436	0.447	0.451	0.467	0.485

with that of true and apparent densities for which careful sampling, and somewhat lengthy laboratory experiments are required.

Size Analysis

To discuss bulk density just after discussing apparent density gives a natural transition from consideration of the physical properties on the smallest scale to those on the largest scale. Size analysis can be considered the fundamental property which is so closely interrelated with bulk density that these properties can scarcely be studied separately.

Size analysis, which is carried out for the determination of the size distribution, is the series of percentages by weight of the coke pieces that pass through a screen and remain on another screen with finer apertures. The specifications for the procedure are detailed in the standards of different countries. Their common basis is found in the manner of checking the size consist of coke by using a series of sieves and weighing the oversizes retained successively on each sieve. There is also, from discussions on the subject within the ISO, a general agreement on the necessity for hand placing the pieces on the apertures of the sieves, even after mechanical shaking or careful hand shaking—at least for the pieces larger than 25 mm or 1 in., hand or mechanical shaking being sufficient only for smaller pieces. It may also be agreed that the reference method should use only hand placing on motionless sieves for all the pieces. It happens, in fact, that shaking disturbs the real size by abrasion of a not negligible number of the pieces.

On the other hand, there are two opposite points of view about the shape of the holes. In Great Britain and in the United States, the sieves are commonly perforated plates with square apertures for proper sieving purposes, as well as for checking the result of standard shatter and tumbler tests.[35] The sides of the apertures used in practical applications[2] are mentioned in Table 16. In the European countries the sieves have round holes with diameters expressed, of course, in millimeters, the series being standardized by the national organizations;[36] the usual diameters also appear in Table 16. One advantage of the square aperture is that it permits the use of wire screens. The advocates of the round hole plates point out that the circle is the most accurate and the simplest geometrical figure for defining the dimensions of differently shaped particles. It is generally admitted that for hand placing a round hole is simpler to use than a square aperture. However, since both are used commercially, it is necessary to know how to convert data obtained by one procedure to equivalent data obtained by the other method.

The ratio of the areas of a square and of a circle, the diameter of which equals the square's side, is 1.13. When a general comparison is made between experimental data[5] (Table 15) from various countries, the algebraic average value of D/L is 1.135. These data relate to different cokes, sieved on square and round perforated plates, before or after testing for strength. There is some deviation from the average value, but a detailed analysis of the data shows that the value 1.135 ± 0.04 covers 86% of the cokes.

It is not always recommended to trust the equivalence value, but it is useful when direct experimental determinations are not available—for instance, when interpreting published papers. For this reason only, it is used in

[35] ASTM Standards E11-38T, D141-23, D294-29.

[36] AFNOR, NF x 12.501 (1941). DIN, DIN 1170 (1939). Mott, R. A., ISO Tech. Comm. 27, Working Group 8, Document 70, August, 1959.

TABLE 15

SIZING OF COKE[5]
EXPERIMENTAL DATA FOR THE EQUIVALENCE COEFFICIENT D/L

Origin of Determinations	Number of Observations	Values of D/L
United Kingdom	5	1.169
United Kingdom	5	1.130
United Kingdom	3	1.146
Germany	2	1.165
Germany	6	1.091
U.S.S.R.	5	1.137
Czechoslovakia	5	1.120
Netherlands	3	1.180
Belgium	2	1.122
France	2	1.125
Algebraic mean value		1.135

Table 16 for comparison screen analyses of American and European cokes. The dimensions in square apertures of side length L in inches are converted into diameters D in millimeters of equivalent round holes by $D = L \times 25.4 \times 1.135 = 28.8L$. To facilitate the comparisons, the numbers representing the percentages of oversizes on the successive sieves are rounded off. For the same reason, the seven American sieve apertures[2] are interposed within the nine European sieve holes. The three United States cokes are chosen as representative by their various mean sizes decreasing from left to right, whereas the German (Friedrich) and French cokes (Thionville, East District, Liévin, North Basin) have increasing mean sizes.[4] The definitions of the mean size are practically equivalent in both works.[2,4] For the United States cokes, the mean size is determined by the intersection of abscissa 50% oversize or undersize with the percentages shown in Table 16 plotted on arithmetic probability paper. In the French study,[4,5] as seen in Fig. 1, the intersection of the same abscissa is taken with the straight central part of the curve obtained by plotting the percentages on logarithmic ordinates. The mean sizes in inches at the bottom of Table 16 relate to square apertures, and the mean sizes in millimeters relate to round holes.

The size spread is determined in the United States[2] by the difference between the plotted oversizes where the constructed line crosses the 25 and 50%, or 50 and 75% points—this spread in inches representing the scatter of the coke about its mean size. In France, checking the size heterogeneity has been influenced[5] by the considerations developed by Mayers.[26] When

TABLE 16

TYPICAL CUMULATIVE SCREEN ANALYSES OF AMERICAN AND EUROPEAN COKES

L in.	D mm	Koppers	Clairton	Ironton	Thionville	Friedrich	Liévin
	150				0	1	5
	125				2	4	11
4	115	29	9	6			
	100				15	22	34
3	86	59	35	22			
	80				41	54	60
	60				76	83	84
2	58	85	73	55			
	50				89	90	90
1.5	43	91	84	73			
	40				96	94	94
1.06	31	95	91	84			
0.75	24	96	94	90			
	20				99	97	97
0.53	15	96	95	93			
	10				99	98	98
0	0	100	100	100	100	100	100
Mean size (in.):		3.37	2.70	2.45	2.56	2.83	3.02
Mean size (mm):		97	78	71	74	82	87

a size distribution, as in Table 16, is plotted on a graph (Fig. 1) with arithmetic or regular coordinates for the percent oversize or undersize, and with a logarithmic scale for the round or square apertures of the sieves, the curve is linear in its central and largest length, provided that the data are obtained on a representative sample. Under these conditions, the straight part of the curve crosses the lines corresponding to 25, 50, and 75% oversize at three points in line. The corresponding ordinates D_{25}, D_{50}, and D_{75} have interesting properties: D_{50} is the mean size of the coke; the spread can be expressed by the ratio of quartiles D_{75}/D_{25}, an ideal coke having $D_{75}/D_{25} = 1$. Attempts have been made[5] to obtain a single number for the expression of mean size and spread. The slope of the line may be characterized by a number which is well known to artillery experts: $E = (D_{25} - D_{75})/2$. A study extended to a large number of European cokes has shown the possibility of adjusting the mean size D_{50} with a correction taking into account the spread or heterogeneity of the size consist, by using coefficients 2 and -4, respectively, for D_{50} and E. The corrected mean size S is then $S = 2D_{50} - 4E$ or $S = 3D_{75} - D_{25}$. The corrected mean size is expressed in millimeters or inches as is the mean size itself. It increases with the mean size D_{50} and with the size homogeneity. The mean size is equal to S when $D_{75} = 0.6D_{25}$, which corresponds to most of the commercial sized cokes used in blast furnaces. For more homogeneous cokes, $D_{75} > 0.6D_{25}$ and $S > D_{50}$; in less homogeneous cokes, $D_{75} < 0.6D_{25}$ and $S < D_{50}$.

The three curves plotted on Fig. 1 were selected as typical for different mean sizes and spreads. Their characteristics, given in Table 17, are followed by further examples for American cokes from the Coke Evaluation Project[2] and tests with ferrocoke.[3] By comparing Fig. 1 with the three first lines of Table 17, it may be seen that with the mean size D_{50} and the spread D_{75}/D_{25} or the corrected mean size S, the graph—which has been used for determining the numbers—is replaced by only two numbers which suffice for the assessment of the size consist. For example, the cokes A, B, F, where S is higher than D_{50}, or D_{75}/D_{25} higher than 0.6, are satisfactory from the point of view of the spread, but cokes C and E have poor homogeneity.

Such a rough representation is an approximation because the nonlinear parts of each curve, left and right (Fig. 1), are neglected. But this approximation can be regarded as sufficient for technical purposes since, as already mentioned,[32] the few percents of oversizes and undersizes not included in the straight part of the curve have little effect on the void volume of the coke.

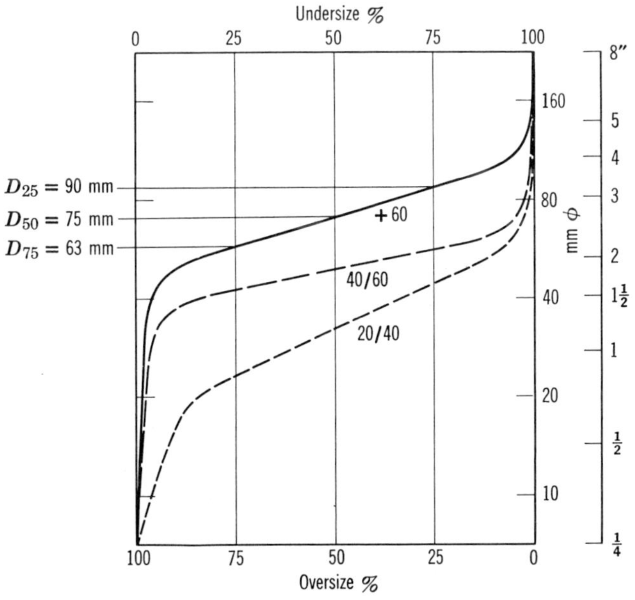

Fig. 1. Typical size distributions.

TABLE 17

EXAMPLES OF SIZE ASSESSMENT[5]
(DIMENSIONS IN MILLIMETERS)

Cokes	Mean Size D_{50}	Quartiles D_{75}	D_{25}	Ratio D_{75}/D_{25}	Size and Homogeneity S(mm)	S(in.)
Europe						
A. +60 mm[5]	75	63	90	0.70	99	3.9
B. $\frac{40}{60}$ mm[5]	50	44	57	0.77	75	3.0
C. $\frac{20}{40}$ mm[5]	37	26	47	0.55	31	1.2
United States						
D. Clairton[2]	73	56	96	0.58	72	2.9
E. Ironton[2]	65	45	95	0.47	40	1.6
F. Eastern[3]	75	60	95	0.63	85	3.3
G. Ferrocoke[3]	72	54	96	0.56	66	2.6

A proposal has been advanced[37] to replace usual size analyses by quite a different method. The author stated that "the grading of coke on sieves does not insure 'a clear definition of the size.'" He therefore suggested the use of an automatic scale for weighing the lumps of coke. Instead of linear dimensions—inches or millimeters—the distribution of a coke consignment is then obtained by the volume or weight of the lumps. The graphic definition of the distribution is reported by the numbers of lumps percent from 0 to 25 g, 25 to 50 g, and so on. Such an attempt had been made on French cokes.[38] It is doubtful whether the correlation with the permeability of coke layers or with blast furnace performance may prove better with weight distribution than with the usual size analysis.

The size consist of a coke is a dominant factor in determining the permeability of a bed of coke. For this reason, the expression of size distribution with a number such as S may prove valuable for estimating the permeability which can be obtained with a coke more or less carefully screened. With various data from Nixon and Maw[33] and from the international low-shaft furnace[5] for permeability of layers of cokes and ores differently sized, there is a very good correlation between the corrected mean size S and the drop of pressure in a layer under constant air flow.

[37] Shemeryankin, B. V., *Coke & Chem.*, **1960,** No. 4, 29–31.

[38] Thibaut, Ch. G., unpublished.

Careful permeability tests were also carried out by the test team of the Coke Evaluation Project.[2] The results correlated significantly with the mean size of the coke, but the correlations with the spread and with bulk density were poor. The negative correlation with the corrected mean size S is highly significant,[5] this factor being responsible for 78% of the variability of the permeability. In this case, partial correlations[2] showed that "the effect of spread on permeability is minor when compared with the effect of mean size," but "this low significance of the spread might be due to the small range in the value of the spread covered by the data of this investigation." From the observations reported in the preceding paragraph, it may be hoped that characteristics such as those given in Table 17 will prove useful for further studies of permeability.

The correct determination of the size of coke is necessary not only for ascertaining the quality of purchased coke, but also for observing the changes in the size consist from the coke-oven wharf to the blast-furnace top. Furthermore, partial sieve analysis is compulsory in the procedures involved in the evaluation of coke strength or abradibility. The use of a mechanical screen,[39] as shown for example in Fig. 2, may be recommended. This sieving car is used in various European coke and steel plants and in research stations. The area of

[39] Thibaut, Ch. G., *Centre document. sidérurg., Circ. inform. tech.*, **13,** 1397–9 (1956).

the perforated plates (1.5 m^2 or 16 sq ft) allows the rapid sieving of 100 kg or 220-lb samples.

Coke Strength

The estimation of coke strength was first carried out by dropping big lumps of coke on a concrete floor or on a rail. The breakage into smaller sized pieces was observed by eye and gave a rough idea of the resistance of coke to impact.

When considering the mechanical properties of coke, numerous authors[2,4,5,17,40–43] have considered the resistance to compression and the resistance to abrasion. When the coke mechanical properties are considered as a whole, the qualities that prevent the coke from collapsing or that tend to avoid the formation of smaller particles may be described by the simple word "cohesion."[41] The word "hardness" should be used only for expressing a surface property, as for metallic materials. Hardness of cokes in this restricted sense may be determined in terms of the Mohs series. It has been shown[41] that the hardness of coke, as determined on full-length pieces, is higher than the hardness of galena crystals and lower than those of baryta, and thus approaches the hardness of carbon steel. This explains why industrial steel conveyors and screens are easily destroyed by coke abrasion.

Compression tests of coke have been carried out in different laboratories,[5,40] but difficulties in sampling for sufficient duplicability always arise. The resistance to compression of coke lumps is always much higher than the actual pressure which coke must resist in cupolas or even in the highest blast furnaces.

The resistance of coke to shatter by dropping and by impact is usually determined by means of a standardized test such as the ASTM Drop Shatter Test[44] or the British method.[45] This test has been the subject of much research in comparison with other tests.[17,46,47] It has recently been recommended by ISO as a standard international method.[48] A study was carried out by Mott[46] on the effects of various factors on the results of the classical shatter test, which lead to the use of a fissure index derived from application of the degradation laws of Kick and Rittinger.[49] One significant conclusion was that "it is illusory to think that a large initial size is of any value for blast furnace practice, for after such a treatment (drops of 30 to 90 feet) the mean size of a coke may be the same for initial sizes of 2 and 6

40 Simmersbach, O., and Schneider, G., *Grundlagen der Koks-Chemie*, Springer, Berlin, 1930, 366 pp.

41 Thibaut, Ch. G., *Rev. tech. Luxembourg*, **39,** 144–66 (1947).

42 Mott, R. A., *The History of Coke-making*, Heffer, Cambridge, 1936, 139 pp. Blayden, H. E., Noble, W., and Riley, H. L., *J. Iron Steel Inst.*, **136,** No. 2, 47–76 (1937). Thibaut, Ch. G., *Rev. mét.*, **40,** 129–42 (1943); *Compt. rend. congr. intern. chim. ind., Liège*, 1958, **1,** 450–4 (1958).

43 Sapozhnikov, L. M., *Koks i Khim.*, **11,** No. 4, 1–9 (1941).

44 ASTM Standards D141–48.

45 British Standards Institution, *Standard methods of shatter test and abrasion test*, 1942.

46 Mott, R. A., *Proc. Conf. Science in the Use of Coal, Sheffield*, Inst. Fuel London, 1959, C51–5.

47 Thibaut, Ch. G., Publ. IRSid, A62, 1953, 65 pp.

48 ISO Tech. Comm. 27, Document 476 (1959), 6 pp.

49 Kick, F., *Das Gesetz der proportionaler Widerstände*, Leipzig, 1885, Chap. 1. Rittinger, P. R., von, *Lehrbuch der Aufbereitungskunde*, Berlin, 1867, Chap. 1.

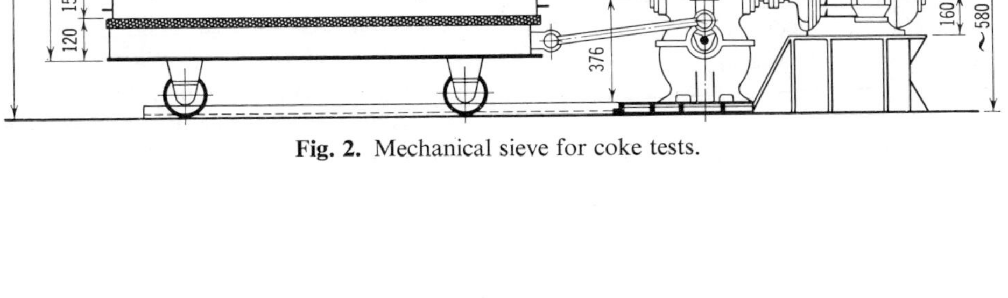

Fig. 2. Mechanical sieve for coke tests.

inches." A thorough statistical analysis of shatter-test data was made by the Coke Evaluation Project team.[2] For the twelve cokes involved in this work, the shatter indices ranged from 38 to 78 for the 2-in. index, from 58 to 90 for the 1-in. index, and from 92 to 98 for the $\frac{1}{2}$-in. index. Some typical correlation coefficients from this research are listed in Table 18, all of which are significant at the 0.01% level.

TABLE 18

SHATTER INDICES—TYPICAL CORRELATION COEFFICIENTS[2]

2-in. Screen Test Index Versus	Coefficients
2-in. shatter index	0.97
1.5-in. shatter index	0.96
0.53-in. shatter index	0.89
Screen Test Mean Size Versus	
Shatter-test mean size, +1.06-in. coke	0.87
Shatter-test mean size, all coke	0.89

It was shown[2] that the 2-in. screen test was highly correlated both with the 3-in. screen test and the mean size. Table 18 shows that the strength of coke—or more precisely, its resistance to breakage on impact—is strongly correlated with the sieve analysis or screen test. This factor suggested that "a simple screen analysis of coke on the wharf and the conventional empirical shatter indices are different measures of the same property of coke." Furthermore, the results of both tests were also highly correlated with tumbler indices, as will be seen below, so that "the choice of which test should be used for plant control would most probably be based on the simplicity of the test."

Apart from the shatter test, the most utilized devices for the determination of the mechanical resistance of coke are the drums or tumblers. The best known are the ASTM tumbler,[50] the B.S. abrasion test apparatus,[45] and the Micum drum.[51] The latter has been recommended as standard international machine by ISO.[52] There are fifteen or sixteen other drum tests. In all these empirical tests, the aim of the designers was to replace the successive drops in a shatter test by mechanical means. The first drums[5] were plain rotating cylinders whose external surface was a smooth steel sheet. It has been shown[47] that in such a drum, if the cover is perfectly adjusted, no degradation happens even after 30,000 revolutions during 20 hr. To produce abrasion or breakage of the coke pieces, it is necessary to provide the cylindrical surface of the drum with bars or angles to lift and drop the coke. Drums having cylindrical surfaces formed by steel bars have been used for 60 years. Those machines have the advantage that the steel rods leave space for the breeze produced during rotation to be continuously removed, so that it suffices to weigh the material remaining in the drum after the test. However, the rods rapidly wear out, so the spacings between them increase rapidly. Also, the dust thrown out of the drum during the tests brings about disagreeable working conditions and high maintenance costs.

For years, the standardized drum in the U.S.S.R.[53] has been a drum, the cylindrical surface of which is formed of steel rods 25 mm in diameter, spaced 25 mm apart. The drum diameter is 2 m, the length 800 mm; 410 kg (900 lb) of coke + 50 mm (+2 in.) are charged into the drum, which is rotated for 15 min at a rate of 10 rpm. The mechanical strength of the sample is characterized by the weight of coke remaining in the drum after testing and also by the weight of breeze through a 10-mm screen. Syskov[29] has given some typical values of the results obtained with the steel rods drum: 280 to 310 kg remaining in the drum, and 40 to 50 kg of −10-mm breeze. This test has now been replaced, since the beginning of 1959, by the Micum test as recommended by ISO.[52] The Heinitz drum was similarly made with steel rods 10 mm in diameter spaced 10 mm apart, into which 25 kg of coke were charged with 25 steel balls of 1 kg each.[47,54] The data obtained in the Heinitz test have been compared to other tests[47] and have proved less reproducible, the larger scatter resulting from the reasons explained earlier.

Among the tumblers provided with angles on their internal cylindrical surface, one of the first was the Cochrane drum. Many similar drums have been or are in use in different countries, most of them in the United States. Instead of describing the drums, which would

[50] ASTM Standards D294–50.

[51] Doerflinger, G., *Stahl u. Eisen*, **47**, 1867–71 (1927).

[52] ISO Tech. Comm. 27, Document 475, (1959).

[53] Russian Standard GOST, 5953–51.

[54] Sainte-Claire Deville, P., *Rev. ind. minérale*, **13**, 123–30 (1932).

TABLE 19

Main Features of Various Tumbler Tests

Types	Drum Diameter mm	Drum Diameter (in.)	Drum Length mm	Drum Length (in.)	Number of Revolutions	Rate of Rotation rpm
British Standard[45]	—	30	—	18	1000	18
ASTM Standard[50]	—	36	—	18	1400	25
Coke Evaluation Project[2]	—	36	—	18	400	25
Carnegie-Illinois Steel Corp.[2]	—	69	—	32	50	12
Interlake Iron Corp.[2]	—	69	—	32	50	12
Geneva Steel Co.[2]	—	18	—	23.5	720	24
Rice drum[4]	—	18	—	22	1240	60
Tisco Breslau drum[55]	—	36	—	36	100	25
Tisco Haven drum[55]	—	36	—	18	1500	60
German DIN drum[5]	990	—	900	—	100	25
Japanese Standard	1500	—	—	—	30	15
Micum tumbler[5]	1000	—	1000	—	100	25
Irsid tumbler[5]	1000	—	1000	—	500	25

require a large amount of space, Table 19 summarizes several of their main features. Differences occur also in the size and weight of the samples as well as in the expression of results. Table 19 does not include drums without angles, drums with rods, and also smaller machines used for laboratory tests.

It has been shown[2,17,47,55] that the size consist of the cokes after the shatter test and the tumbler test are more or less correlated. In the Coke Evaluation Project,[2,56] numerous correlations were calculated for the shatter and tumbler indices, the most typical being reported in Table 20. The degree of significance is very high, signifying that both tests measure the same physical characteristics. Furthermore,[2] "the results obtained in the early stages of the tumbler test are similar to those obtained in the shatter test .... Such high correlations indicate that determination of both indices is unnecessary."

In another study[47] where comparisons were made between the shatter test, the Micum test, and the Heinitz, Breslau, and Wolf tests, it appeared clear that only the scales were diverse. Different cokes were also tested with a Micum drum and a shatter test apparatus, the drum being stopped after every 10 revolutions to permit screen analyses, and the shatter drops being continued to 50 drops instead of the standard number of 4 drops. It was shown that practically the same results were obtained with 50 drops in the shatter test and with the normal number of 100 revolutions of the Micum tumbler test. Since the tumbler test has the advantage of being a "push-button" machine, the drop-shatter tests seem relatively impractical.

It has also been recognized[2,46] that in its classical form with 4 drops, the shatter test reproduces merely impact effects and coarse

TABLE 20

Correlations of Shatter and Tumbler Indices[2]

2-in. Shatter Index Versus	Correlation Coefficients
Screen test 2-in. index	0.97
Tumbler test stability	0.85
2-in. tumbler index (100 rev)	0.98
1.06-in. tumbler index (100 rev)	0.90
+1.06-in. Shatter Index Versus	
+1.06-in. tumbler index (100 rev)	0.99
0.53-in. Shatter Index Versus	
0.53-in. tumbler index (100 rev)	1.00

[55] Das Gupta, N. N., Rao, V. V., and Sinha, S. N., *Fuel Research Inst. News* (*India*), **4**, No. 4 (1954). Das Gupta, N. N., Das Gupta, A. R., and Roy, C. A. K., *Fuel Research Inst. News* (*India*), **5**, No. 2 (1955).

[56] Campbell, R. W., *Blast Furnace Steel Plant*, **40**, 653–60, 779–86 (1952). Mulcahy, B. P., *Blast Furnace Steel Plant*, **41**, 408–12 (1953). Rogers, W. T., *Blast Furnace Steel Plant*, **43**, 627–34 (1955).

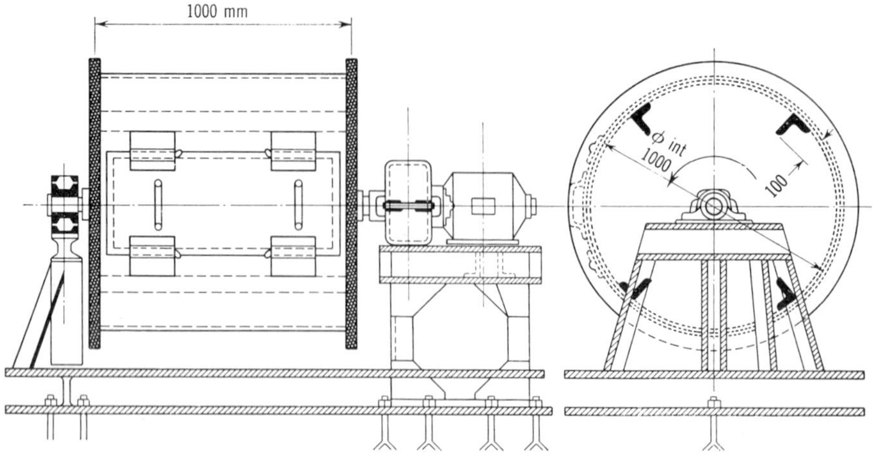

Fig. 3. Drum for Micum and Irsid tests.

breakage, the lumps of coke being divided into a limited number of smaller-sized pieces. The Coke Testing Working Group of ISO has mentioned[57] that the shatter test, compared with the Micum tumbler test, only determines the resistance to breakage of coke on dropping, whereas the Micum test determines the coke cohesion that includes resistance to breakage on impact and to abrasion.

In the United Kingdom,[58] the shatter test is still considered "the best available means of assessing the mechanical strength of coke for use in blast furnaces so far as British practice is concerned." However, the drum test in the form of the classical Micum test or of a modified Micum or half-Micum test is receiving increased attention[46] in Britain. In Germany, the DIN tumbler is now being replaced by the Micum drum, and important studies on this test are being investigated.[59]

The Micum tumbler (Fig. 3) has been used for years to obtain information on the degradation or breakage of coke. The mechanical resistance of coke to attrition, to shatter, and to breakage due to initial fissuration is reported by the three indices determined after 100 revolutions: (1) M_{10}, the undersize through 10 mm; (2) M_{20}, the oversize on 20 mm; and (3) M_{40}, the oversize on 40 mm. The "Micum cohesion" using these indices is:[41] $C_M = M_{40} + M_{20} - M_{10}$.

[57] ISO Tech. Comm. 27, Working Group 8, Rep. 4th session, January, 1959.

[58] Lee, G. W., *J. Inst. Fuel*, **32,** 154–60 (1959); *AIME, Proc. Blast Furnace, Coke Oven & Raw Materials Comm.*, **19,** 302–28 (1960).

[59] Dahme, A., *Glückauf*, **95,** 680–7 (1959).

In the past, the sample for Micum tests was taken on a fork with prongs spaced 50 mm apart. The procedure has now been standardized[52,60] with a more accurate definition of the sample by the oversize on a perforated plate with round holes of 60-mm diameter. The screen analysis after tumbling is also carried out with similar perforated plates of 40, 20, and 10-mm round holes diameters.

The different cohesion tests give similar results,[2,47] the scales of which depend on the different levels of external work. As a result, ASTM, and Micum indices are correlated with each other. Table 21 shows typical values of the indices for different cokes. Other comparisons of shatter and Micum indices have been reported.[47,55,61,62] Several comparisons of shatter and ASTM tumbler indices have also been made.[63,64] The sampling conditions and the numbers of individual tests may have been significantly different in obtaining the data reported in Table 21. Some of the British cokes were sampled at the coke-oven wharf before the screens, and others after railway transport and rescreening at the blast-furnace

[60] AFNOR, M.03.023 (1958); M.03.020 (1959).

[61] Sanderson, T., and Wattleworth, D. R., *J. Inst. Fuel*, **23,** 115–20 (1949).

[62] Thibaut, Ch. G., Publ. IRSid, Ser. B, No. 24 (1954), 26 pp.

[63] ISO Working Group 8, Document 80 (1959). Gayle, J. B., and Eddy, W. H., *AIME, Proc. Blast Furnace, Coke Oven & Raw Materials Comm.*, **17,** 193–208 (1958).

[64] Das Gupta, N. N., and Ghosh, S. R., *Indian Mining J.*, **3,** No. 10, 1–4, 6 (1955). Das Gupta, N. N., *Research & Ind.*, **2,** 283–5 (1957).

TABLE 21

SOME TYPICAL SHATTER, MICUM TUMBLER, AND ASTM TUMBLER INDICES

			ASTM Tumbler Test		Micum Test				Shatter Test	
Reference	Year	Cokes	Stability +1 in.	Hardness $+\frac{1}{4}$ in.	M_{40}	M_{20}	M_{10}	C_M	+2 in.	$+\frac{1}{2}$ in.
2	1952	Granite City	49	67	—	—	—	—	78	97.7
		Bethlehem	48	63	—	—	—	—	70	97.9
		Clairton	23	67	—	—	—	—	49	95.2
		Ironton	5	67	—	—	—	—	38	91.8
47	1953	Anzin	—	—	83	92	6.4	169	74	—
		Béthune	—	—	78	90	8.2	160	76	—
		Neunkirchen	—	—	66	83	13.0	136	73	—
		Tollcross	—	—	58	78	17.0	119	—	96.0
55	1954	Bhowrah	61	—	86	—	7.2	—	87	98.0
		Kendwadih	55	—	75	—	11.5	—	74	97.6
		Damua	45	—	78	—	16.6	—	82	95.2
		Methani	22	—	56	—	10.7	—	53	96.6
		Raniganj	19	—	48	—	8.3	—	45	96.6
64	1955	Margherita	10	—	23	—	13.0	—	59	94.0
63	1958	Woodstock Coal	26	62	—	—	—	—	67	97.9
		Mary Lee Coal	58	61	—	—	—	—	90	98.5
		Blue Creek Coal	55	60	—	—	—	—	86	98.4
63	1959	Mannesmann	53	61	81	—	7.3	—	—	—
65	1959	Everett	66	69	90	93	6.4	177	58	99.1
		Everett	62	72	80	92	6.2	166	84	98.9
		Everett ferrocoke	53	68	75	89	9.1	155	79	98.6
		Philadelphia	54	71	72	87	9.1	150	77	98.5
		Connecticut	52	71	69	88	8.0	149	69	98.2
		Eastern	53	72	68	88	9.0	147	69	98.2

bunkers. It seems likely that correlations previously established in different countries will be revised when data such as those at the end of the table[65] become available for other cokes.

The empirical character of the different methods of determination of coke cohesion has led[2,46,59,66] to fundamental and statistical work to explain the significance of the procedures used by physical theory and to increase accuracy and reproducibility of results. For instance, the ASTM tumbler test may be criticized on the empirical choice of 1400 revolutions, whereby a single rate of degradation is obtained at a level not necessarily the best for duplicability or discrimination between cokes. Therefore, observations have been made at 100, 400, 900, and 1400 revolutions, as suggested by Epstein and Lowry.[67]

[65] ISO Working Group 8, Documents 75 and 79 (1959).

[66] Ulmo, J., Thibaut, Ch. G., Vigneron, P., and Menuet, B., Publ. IRSid, Ser. A, No. 97 (1955), 28 pp.

Table 22 shows some typical results. The percentages corresponding to 1400 revolutions are the standard stability and hardness factors as recorded in Table 21. From these examples it appears that the sensitivity of the test, as appreciated by the ranges of the oversizes and undersizes, does not reach its maximum with 1400 revolutions; for instance, comparing the Granite City coke and the Ironton coke, for which the values show the largest interval, the range of the oversize on 1.06 in. is successively, for 100, 400, 900, and 1400 revolutions, 41, 50, 48, and 45, and for the undersize through 0.265 in., the corresponding values are 4, 4, 1, and 0. Thus, it is clear that the empirical choice of 1400 revolutions for this test does not lead to the best sensitivity, and it is not therefore necessary to maintain this number of 1400 revolutions. The reporters of the Coke Evaluation Project concluded quite logically that the recommended tumbler test with the ASTM

[67] Lowry, H. H., and Epstein, B., *AIME, Proc. Blast Furnace, Coke Oven & Raw Materials Comm.*, **7**, 3–23 (1948).

TABLE 22

CUMULATIVE PERCENTAGE OF COKE ON 1.06-IN. SCREEN AND ON 0.265-IN. SCREEN[2]

Cokes	Number of Revolutions	On 1.06-in. Screen	On 0.265-in. Screen
Ironton	100	44	88
	400	20	79
	900	9	72
	1400	5	67
Clairton	100	66	90
	400	45	80
	900	31	70
	1400	23	63
Bethlehem	100	85	94
	400	70	85
	900	57	75
	1400	48	67
Granite City	100	85	92
	400	70	83
	900	57	73
	1400	49	67

drum should be carried out for checking the hardness of coke with 400 revolutions. A similar research[68] was undertaken in France at about the same time. The coincidence was accidental, and American and French research workers learned of their similar conclusions much later.[69] The fact that similar studies were undertaken in the same period emphasized the real need for improvements in the field of the tumbler tests.

It has been noted that the Micum test uses coke pieces larger than 50 cm (fork) or 60 mm (round holes), the shatter tests use +2-in. pieces, and the ASTM tumbler test uses 2 × 3-in. size coke. Since more and more plants use smaller sizes of coke,[2,5,70] down to 1.5 in. or 1 in., these tests are not reliable for a large percentage of the coke that is fed into the blast furnace, or in gas producers that use small-sized coke. In the United States, the sampling of coke for chemical and physical tests[2] is generally carried out at the coke-oven plant, in most cases, either on the wharf or on the belt leading to the screens. In Europe, whenever coke qualities are related to blast-furnace operating practice,[5,71] the samples are taken at the blast-furnace bins, preferably where the skips are being filled for transit to the blast-furnace top. Both procedures may be useful, but it is important that test results under these different sampling conditions be carefully compared. The coke-oven manager wants information about coke quality, and the blast-furnace superintendent needs information on the quality of the coke being used. When comparing blast-furnace performance with coke properties, the latter should always be considered not at the coke wharf or belt, but only as charged into the furnace.

For the purpose of assessing blast-furnace performance from the point of view of coke properties, neither the shatter test nor the Micum or the ASTM tests are reliable if a large percentage of the coke is under 2 in. or 60 mm. As far as the Micum test is concerned, the indices are valuable where the coke is screened over 60 mm (round holes), which corresponds to 50 mm or 2 in. on square

[68] Thibaut, Ch. G., and Vigneron, P., *Soc. francaise de métallurgie*, October, 1952, p. 23; October, 1953, p. 35.

[69] Thibaut, Ch. G., *Centre document. sidérurg., Circ. inform. tech.*, **12,** 2233–46 (1955).

[70] Bardgett, H., *Iron & Coal Trades Rev.*, **176,** 333–7 (1958).

[71] Thibaut, Ch. G., and Sanna, D., *Centre document. sidérurg., Circ. inform. tech.*, **11,** 2109–43 (1955).

TABLE 23

AN ANALYSIS OF DEGRADATION CURVES[5]

Number of Revolutions of Micum Tumbler	% through 10-mm Screen		
	Coke 4	Coke 5	Difference
100	19	6	13
300	43	15	28
500	58	20	38
800	72	29	43
1000	76	31	45
1800	83	47	36
3800	89	66	23
8800	92	87	5

aperture screens. In fact, most of the European operators have long been accustomed[5] to using only cokes larger than 60 or 80 mm. This practice is now rapidly changing[72] so that it has been recognized as necessary[69,71] to design a new type of mechanical test for smaller cokes.

The first step was to take the oversize on 20-mm round holes, which corresponds to 0.7-in. square apertures instead of the oversize on 60-mm sieves.[60] As the second step, it was decided to use the classical Micum drum for the new test. It was not established that 100 revolutions gave best duplicability and sensitivity. Therefore, a large number of degradation tests were carried out[5] to ascertain the best conditions for the Micum test.[73] For this purpose, the drum was provided with an automatic revolutions counter, and was stopped after 100, 300, 800, 1800, 3800, and 8800 revolutions. Figure 4 shows degradation curves obtained for five typical cokes from France and Germany. It is clear that the optimum sensitivity corresponds neither to 100 nor to 8800 revolutions.

Considering cokes Nos. 4 and 5 on the graph for -10 mm, Table 23 shows the percentages of undersize and their differences[5] as a function of the number of revolutions. Thus, by ascertaining the sensitivity of the test by the differences between two different cokes, the reliability was found to be 2.9 greater after 500 revolutions than with 100 revolutions. The maximum sensitivity in this instance is obtained with 1000 revolutions, and it rapidly decreases with increase in the number of revolutions.

Considering results obtained with other cokes, and noting that 1000 revolutions take 40 min compared to 20 min for 500 revolutions, and that the increase in sensitivity between 500 and 1000 revolutions is far from proportional to the increase of time, the scientists selected 500 revolutions for further investigation.[66] Since the screening operation for the previous sample and the preparation of the next test takes about 20 min, during which time the tumbler is working by itself, the choice of 500 revolutions has added merit. The differences between the standard Micum test and the Irsid test are recorded in Table 24 and show that the disadvantages of the Micum test disappear with the Irsid test.

The Irsid test is applicable to a wider field than the Micum test, because it may be used for all solid products of carbonization down to 20 mm. Furthermore, it requires less material and work since three screens are not used and handplacing of the pieces on the screen is not required for sizes under 25 mm as previously stated.

It has been proven[66] that in spite of the large range of the coke size in the Irsid test its reproducibility and sensitivity are not worse than in the standard Micum test, but are, on the contrary,[73] higher. In the first practical use of the Irsid test, the values of I_{20} and I_{10}[5,66,68] were taken into account, I_{20} being the percent oversize on 20 mm and I_{10} the percent undersize through 10 mm. It was observed[5] that the differences between the percentages of 10- and 20-mm pieces were only 1 or 2%, so that I_{10} and I_{20} are highly correlated. It is sufficient, therefore, to measure only I_{20} with two advantages: (1) only one screen is required

TABLE 24

COMPARISON OF MICUM AND IRSID TESTS

	Micum Test	Irsid Test
Weight of sample	50 kg	50 kg
Drum	Micum	Micum
Revolutions	100	500
Screens		
For sampling	60 mm	20 mm
For product	40 mm	
	20 mm	20 mm
	10 mm	
Weighings	5	2
Handplacing	Compulsory for 3 screenings	none

[72] Thibaut, Ch. G., *Centre document. sidérurg.*, No. 9, 1863–77 (1958).

[73] Bastenaire, F., *Rev. statist. appl.*, **3**, No. 2, 83–99 (1955).

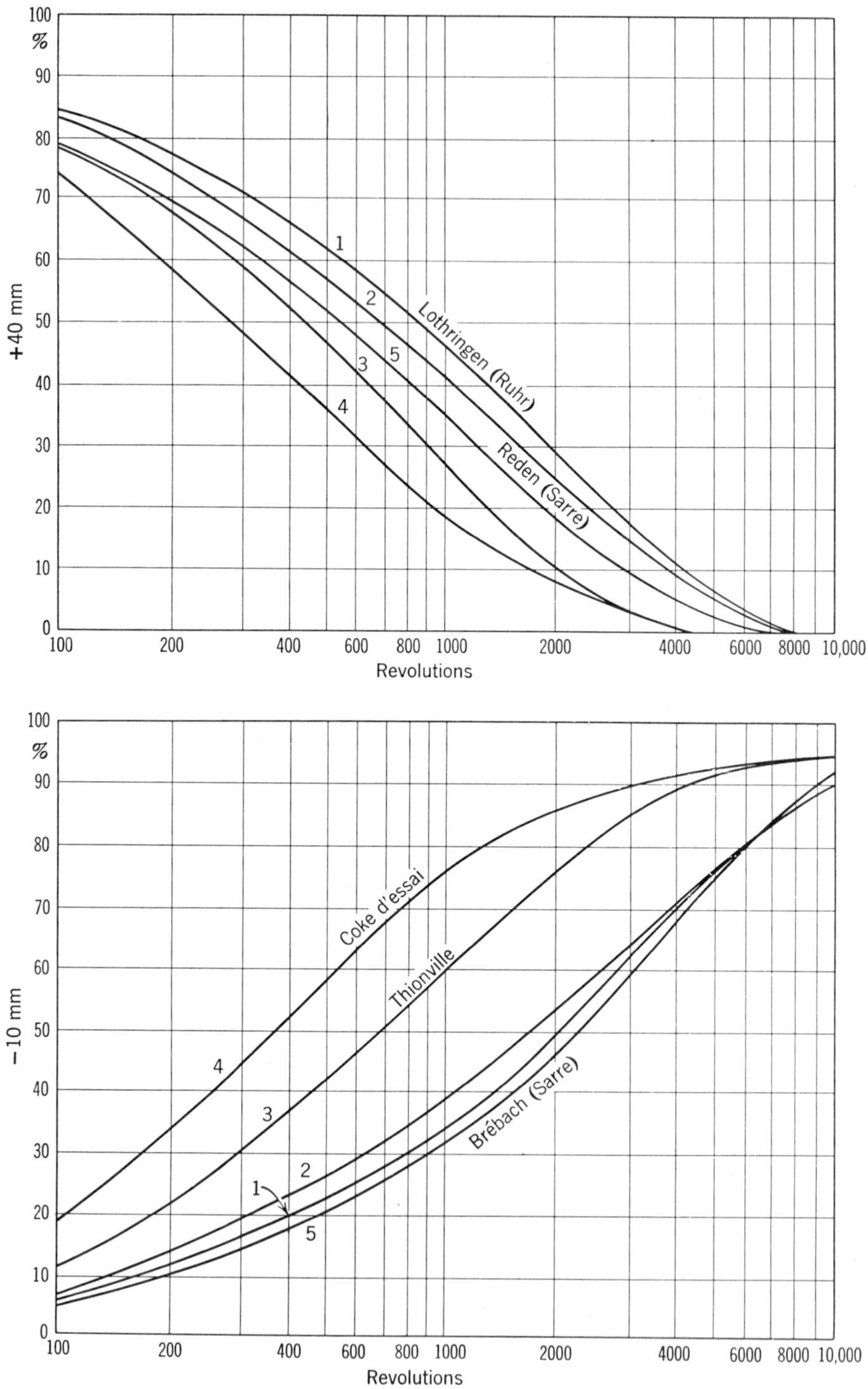

Fig. 4. Typical degradation curves of five cokes as a function of the number of revolutions of a Micum drum.[73]

for sampling and for screening after tumbling, and (2) the cohesion is given by one number that increases as the cohesion increases.

It is usual[52,60] to perform the Micum and Irsid tests, as well as the shatter test,[48] by carrying out three tests, the average values being reported. A study of the precision[66] has led to the conclusion that two cokes must be considered significantly different when the differences between the various indices are at least: 2.1 units for M_{40}, 0.85 unit for M_{20}, 0.75 unit for M_{10}, 1.3 units for I_{20}, and 3.0 units for C_M.

Comparisons between the standard Micum indices and the Irsid index are given in Table 25. These indices are highly correlated and the Micum cohesion C_M may be calculated from the Irsid index by the relation: $C_M = 2.15 I_{20}$, where C_M is equal to the sum of M_{40} and M_{20} less M_{10}. This equation was derived from a much larger number of data than those given in Table 25. The coke that showed the highest ratio of C_M/I_{20}, that is, 2.42, had a mean size D_{50} of 87 mm or 3.9 in. and showed high indices for both M_{40} and M_{20} in the standard Micum test. In this instance, the cohesion index C_M obtained with 100 revolutions is relatively high as well as the ratio C_M/I_{20} because the degradation by breakdown or attrition is partially cancelled by the initial mean size, and this coke may be considered improperly characterized by the standard Micum test. In contrast, the coke that showed a ratio of C_M/I_{20} of only 2.03 had a mean size of 64 mm or 2.5 in. and contained a large proportion of pieces which were just over 60 mm as sampled for the Micum test. The sample for the Irsid test takes a large number of smaller pieces that withstand better the tumbling up to 500 revolutions. For this smaller coke, the Micum test underestimates the cohesion, and the Irsid test gives a better evaluation of its real ability to resist degradation in transit and stock descent. The Irsid index has a large advantage compared with the standard Micum indices since it gives a more exact determination of the cohesion of the smaller pieces of coke which have a lesser tendency to be degraded. This view is supported by comparisons of smaller cokes[5] with larger cokes from the same plants, where the former generally have higher cohesion indices as measured by I_{20}.

The trend toward a modification of the standard Micum test for its application to small sizes of coke has real importance in this period where[2,46,72] emphasis is given both to the use of smaller sized cokes and to the necessity of good mechanical properties for feeding larger and larger blast furnaces.

TABLE 25

COMPARISON OF MICUM AND IRSID INDICES

Cokes	M_{40}	M_{20}	M_{10}	C_M	I_{20}	C_M/I_{20}
European cokes[5]	91	94	6	179	81	2.21
	87	93	5	175	80	2.19
	84	92	7	169	74	2.42
	79	91	7	163	77	2.11
	79	88	10	157	66	2.38
	71	89	8	152	75	2.03
	67	86	10	143	66	2.16
American cokes[65]	90	93	6	177	79	2.24
	80	92	6	166	79	2.10
	75	89	9	155	72	2.15
	72	87	9	150	73	2.06
	69	88	8	149	73	2.04
	68	88	9	147	74	1.99

Size Stabilization

Coke is submitted to continuous or discontinuous size degradation from the coke-oven wharf to the tuyères of the blast furnace or the grate of a domestic fire; it may also be submitted to screening after various stages of degradation. When considering blast-furnace, cupola, or domestic cokes as rescreened for gasification or combustion purposes, it should be noted that increasing cumulative percentages of breeze or dust are eliminated by the successive screenings. Thus, the useful coke after removal of breeze presents at each stage a higher cohesion and a lower bulk density than at the preceding stage; in other words, it has been stabilized.

Systematic checking of natural stabilization occurring with a French coke[5] has been recorded as follows (the T index is defined later):

	Micum Cohesion C_M	Bulk Density d	T Index
Before screening	150	0.445	61
After transit	151	0.435	64
At the tuyères	156	0.445	67

Another example[5] showed the Micum cohesion increasing from 149 in the bunker to 159 at the tuyère level.

TABLE 26

RESULTS OBTAINED WITH A STABILIZATION PROCESS[5]

Cokes	Treatment	Micum Indices M_{40}	M_{20}	M_{10}	C_M	Increase in % −50 mm or −2 in.
A	Usual screening	79	88	10	157	—
	Screening after drop	81	90	9	162	5.6%
B	Usual screening	82	92	6	168	—
	Screening after drop	83	93	6	170	9.1%

Table 26 summarizes[5] the results of stabilization obtained by dropping the coke, before screening, on basalt plates, the height of the drop being in each case 4 m or 13 ft.

That the increase in cohesion is more important if the initial cohesion is lower is shown by comparing coke A with coke B. The disadvantages and advantages of a deliberate stabilization must be calculated in all cases by taking into account on the one hand the increase in the percentage of small coke and breeze obtained with the operation, and on the other hand, the decrease in coke consumption which may be recorded at the blast furnaces with stabilized coke. A dominant factor may be the relative price of nuts and breeze compared with the price of blast-furnace coke. In the countries where both prices are becoming similar, the advantages of the process may easily offset the increase in breeze formation.

Stabilization has been also studied[74] by using tumbling of the coke instead of a drop before screening. After treatment by 10 revolutions in a Micum tumbler, the cohesion was increased, the bulk density remained practically constant, 3% more breeze through 10 mm was formed, and the percent coke through 60 mm increased from 33 to 53%. At the Marienau Research Station, the stabilization of a weak coke by tumbling it up to 50 Micum revolutions reduced the M_{10} indices from 18 to 11.

In this period, when there is an increasing demand for coke breeze for sintering ore fines and blast-furnace flue dust, it may prove advantageous to carbonize poorer blends of slacks in order to obtain weaker cokes and more breeze. This situation would not favor the blast-furnace coke rates, unless the quality of blast-furnace coke could be greatly improved by stabilization. In this view, the composite price of coke and breeze—and therefore the cost of pig iron—could be lowered in countries where carbonization involves poor coking coals and costly imported additional slacks.

BEHAVIOR OF COKE IN THE BLAST FURNACE

It seems likely[5,46] that the large sizes for cokes which have been requested in the past by most blast-furnace managers, expecially in Europe, are not necessary for good blast-furnace performance. The large number of observations of the tuyère zone by Wagstaff,[75] which will be considered later, have shown that good performance may be obtained with small-sized coke in the bosh of the furnace. In this respect, it is necessary to gather data on what happens in the blast furnace from the top down to the tuyère level. Investigations in the blast furnace have been carried out in the U.S.S.R. for the purpose of developing a test for measuring the dominant physical properties of coke.[43] The main conclusion[43,76] was that the mechanical test in a drum gave results that could be used to ensure correct gas permeability in the furnace.

The degradation of various cokes during travel down the stock column in the blast

[74] Sanna, D., and Poiré, M., *Assoc. tech. sidérurg.*, (1959), unpublished. Thibaut, Ch. G., *Centre document. sidérurg., Circ. inform. tech.*, No. 3, 659–67 (1961).

[75] Wagstaff, J. B., Elliot, J. F., and Buchanan, R. A., *J. Metals*, **4**, 709–17 (1952). Wagstaff, J. B., *J. Metals*, **5**, 895–902 (1953).

[76] Syskov, K. I., *Compt. rend. acad. sci., U.R.S.S.*, **39**, No. 6, 218–21 (1943).

furnace has been investigated[5,77] in different steel plants, by taking samples of the coke as charged on the bell, and other samples again 8 to 16 hr later—according to the transit time—by extraction through two tuyères on opposite sides of a blast furnace. The white-hot coke was pulled out and packed into steel containers for testing. A number of tests were carried out by different laboratories on these samples of top and hearth coke originating from twelve blast furnaces in Lorraine and Saar. After careful treatment of the samples,[78] they were submitted to size analysis, Micum tests and Irsid tests for the large coke, Irsid tests for the small coke, proximate and ultimate analyses, ash analysis, determination of specific surface, reactivity with CO_2 and H_2O, densities, and porosity. The main results of these tests were as follows: Between the bell and the tuyère zone, the mean size of coke decreased on the average by approximately 30%; the Irsid cohesion I_{20} decreased by 6%; the porosity increased by 4%; the specific surface increased; the volatile matter decreased by 30% and the hydrogen content by 50%; the variations of density and void volume were random, and the reactivity, as determined by three methods, increased in all instances, on the average by 35%, and was correlated with the increase of specific surface and porosity.

[77] Lee, G. W., *Proc. 7th Conf., Brit. Coke Research Assoc.*, (1956), 26 pp.

[78] Rieffel, E., and Danquigny, J., *Chaleur & ind.*, **30**, 302 (1954); **31**, 73–94 (1955).

In other words, the changes in thermal and physical properties that occur while the coke is passing through the blast furnace increase its rate of combustion in the tuyère zone. This increase is independent of the speed of the stock descent and the blast volume. The changes in the properties of the coke could lead to a tentative conclusion that requirements regarding the reactivity of the coke charged at the top of the furnace have no utility. This point will be discussed later in the section on thermal properties.

The degradation from the mechanical point of view, as expressed by the decrease of the Irsid cohesion I_{20}, averages 5 units of I_{20}, which corresponds to about 500 to 700 revolutions of the Micum tumbler. Table 27 gives the most typical physical data for the cokes in their initial state and after their descent through the blast furnace. Each number is the average result of three tests. In one instance—No. 11—the coke was not pulled out from the tuyères, but was pushed out by the blast through the slag notches at the shutting down of the furnace. This means of sampling is quicker than the latter, but the decrease in mean size and cohesion is higher than the average. In two instances—No. 1 and 12—the coke had been produced in cake-charged ovens. When these two cokes are considered separately, the degradation that occurred in the blast furnace, as recorded by the decrease in cohesion, was correlated with the initial quality of each coke. Another observation was that there was a tendency for the coke to be less affected when

TABLE 27

DATA ON COKE AT THE TOP AND AT THE TUYÈRES OF BLAST FURNACES[78]

Coke Number	Mean Size (mm)		Cohesion (I_{20})		Apparent Density	
	Top	Tuyères	Top	Tuyères	Top	Tuyères
4	72	51	78	78	0.86	0.88
9	71	66	81	77	0.91	0.91
3	70	48	80	76	0.89	0.83
1	73	52	80	75	0.88	0.90
8	89	57	80	76	0.86	0.85
10	71	50	83	80	0.93	0.95
2	77	57	78	71	0.88	0.93
7	82	63	81	78	0.91	0.91
11	73	48	80	72	0.90	0.88
6	85	50	81	70	0.97	0.98
5	84	65	77	74	0.95	0.94
12	64	47	75	71	—	—

the speed of the stock descent was high. However, this point requires further observation, owing to possible influences of such other factors as burden composition and method of charging.

High-Temperature Cohesion

Among recent work involved with coke physical properties, mention must also be made of tests carried out on coke at high temperatures. Some blast-furnace investigators have been doubtful about the usefulness of cohesion tests made at room temperature. This point of view implies that the breakdown of coke may happen differently at the high temperatures in cupolas, blast furnaces, and gas producers. From the data in Table 27, the stack and bosh temperatures do not appear to exert marked influences on the physical characteristics of coke. However, it is of interest to compare directly, by means of appropriate experimental devices, the behavior of cokes submitted to cohesion tests at increasing temperatures. This has been done in the United States Steel and Battelle experiments[79,80] by using a silicon carbide cylinder heated to various temperatures up to 2500°F or 1370°C. A sample of lump coke is placed in the cylinder which is rotated end-over-end on centrally placed trunions so that 10 revolutions correspond to 20 drops of a shatter test. The internal diameter and length of the tube are 10 in. and 54 in. Detailed data are not available, but Fig. 5[79] summarizes the observations recorded with different cokes at increasing temperatures. The degradation is expressed by the percent oversize on a 1-in. screen after 600 revolutions. It appears that[79] "some cokes are as resistant at high temperatures as at room temperature," whereas others "do not show the same relative resistance to impact at room temperature as at 1500°F and at 2500°F." It seems that[80] "the mechanical strength of coke at room temperature gives no indication of its mechanical strength at the higher temperatures encountered in cupola operation," but, "no attempt has yet been made to correlate the results of these tests with the performance of the coke in a cupola."

A further series of tests has been undertaken[81] to explain why certain cokes that appeared sufficiently coherent by standard tests had a bad influence on blast-furnace operation, expecially 5 to 8 hr after being charged at the top. Temperature and pressure seemed responsible for this behavior. A series of tests was made, using a Baldwin tensile-test machine, and samples that were small cubes cut from pieces of coke, the length of sides being a fraction of an inch. Table 28 reproduces some typical compression data at 1650°C compared to tumbler stability and hardness, as well as the proximate analysis. The resistance to compression was determined

[79] Lightner, M. W., *Steel*, **136**, No. 6, pp. 116, 118, 120, 122, 126, 128, 131; No. 7, pp. 92, 95–6, 101; No. 8, pp. 82, 84, 87; No. 9, pp. 96, 98, 102 (1955).

[80] Varga, J., and Lownie, H. W., Preprint No. 56-27, 60th Ann. Meeting Am. Foundrymen's Soc., Atlantic City, N.J., May 1956.

[81] Holowaty, M. O., and Squarcy, C. M., *J. Metals*, **9**, 577–81 (1957).

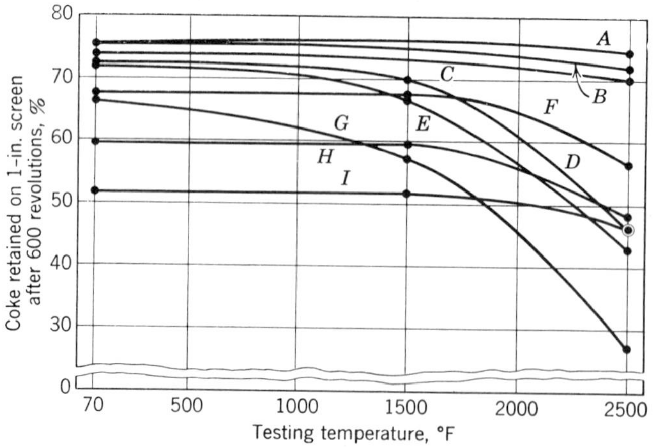

Fig. 5. Effect of testing temperature on degradation of coke.[80]

TABLE 28

COMPRESSION TESTS AT 1650°C[81]

Cokes	Volatile Matter	Ash	Fixed Carbon	ASTM Tumbler Stability	ASTM Tumbler Hardness	Resistance to Compression
	%	%	%			kg/cm²
Inland	2.3	7.1	90.6	55	69	235
Inland	0.8	6.0	93.2	54	67	372
Beehive	1.2	9.2	89.6	48	58	88
Beehive	1.3	8.1	90.6	28	56	274
Anthracite	8.1	11.5	80.4	6	—	417

as the mean result of three tests in the tensile machine in a nitrogen atmosphere. Observation of blast-furnace operation indicated that it was not correlated with the ASTM tumbler indices, whereas the pig-iron production increased as the resistance of coke to compression at high temperature. Furthermore, studies of the structure of the different cokes submitted to trials showed that the best ones had a regular distribution of the pores, with a high thickness of the walls between pores.

Reactivity

When entering into the subject of thermal properties of coke, Mott and Wheeler[17] pointed out that the word "reactivity" is "somewhat loosely used to express the ability of a coke to interact with, usually, oxygen, carbon dioxide or steam." It has been suggested[82] that the reactivities to carbon dioxide (carboxyreactivity), steam (hydroreactivity), oxygen (oxyreactivity), and sulfur (thioreactivity) be considered separately. As far as the industrial uses of coke are concerned, all have interest, and it may be suggested that the reactivity of the coke with liquid metal, as in the hearth of the blast furnace, also deserves attention. Many studies have been devoted to the theoretical and practical aspects of reactivities, to which the interested reader may be referred.[82–88]

The importance of reactivity to users of coke is not yet resolved. It is likely that the determinations of various reactivities on small particles of coke in laboratories are in no way correlated to the behavior of lump coke in gas producers, cupolas, and blast furnaces.[5,87,89] Blayden[88] has noted that "the reactivity of coke is not so much a characteristic of the fuel as an implied behavior, which is in fact dependent both on the properties of the fuel and the manner in which it is utilized in combustion or gasification processes." This opinion is confirmed both by the extensive studies of Guérin, and by field experiments carried on in cupolas and blast furnaces.

It is generally agreed[85,86,88] that the gasification reactions, such as the carbon-carbon dioxide reaction, are complex. The formation of various surface oxides makes it difficult to ascertain the detailed gasification mechanism. Different physical factors, such as specific surface, porosity, and catalytic effects, have distinct influences which are generally not separate. Laboratory determinations run on small particles can hardly be expected to correlate directly with the industrial observations. In the laboratory, the reactivity is determined by observing the changes of volumetric concentrations of gaseous mixtures before and

[82] Guérin, H., *Le Problème de la Réactivité des Combustibles Solides*, Dunod, Paris, 1945, 222 pp.

[83] Guérin, H., *Chim. & Ind. (Paris)*, **59**, 30–6, 337–38 (1948); **70**, 875–85 (1953).

[84] Guérin, H., and Fort, R., *Congr. assoc. tech. gaz, Paris*, **1958**, 803–56.

[85] Ergun, S., *J. Phys. Chem.*, **60**, 480–5 (1956).

[86] Letort, M., and Martin, J., *Nature*, **157**, 874–5 (1946). Harker, H., Marsh, H., and Wynne-Jones, W. F. K., *Conf. Ind. Carbon & Graphite*, Soc. of Chemical Industry, London, 1958, 630 pp.

[87] Boyer, A. F., and Durand, G., *Chim. & ind.*, **82**, 309–28 (1959).

[88] Blayden, H. E., *Proc. Conf. Science in the Use of Coal, Sheffield*, 1958, Inst. Fuel, London (1958), pp. E-1-6.

[89] Larsen, B. M., *Am. Inst. Min. Met. Engrs., Tech. Publ.* No. 2132 (1947), 13 pp.

TABLE 29

VOLATILE MATTER OF BLENDS VERSUS REACTIVITY* OF COKES[2]

Plants	Volatile Matter	Reactivity*	Plants	Volatile Matter	Reactivity*
	%	°C		%	°C
BG	30.9	537	K	35.4	533
B	32.7	521	J	36.0	531
R	32.8	546	C_2	37.3	545
I	33.2	542	G	37.6	551
C_1	33.7	539	T	40.0	573

* The reactivity is expressed as the temperature at which the rate of reaction of the coke and oxygen is high enough to raise the temperature of the sample under adiabatic conditions 15°/min, that is, the lower the temperature, the higher the reactivity.

after interaction with the solid mineral fuel which is concerned. The expression of results[2,4,82] may differ widely from one investigator to another. However, if different cokes are classified according to the increase in reactivity recorded by various investigators versus oxygen or carbon dioxide or steam, these classifications show little differences, provided that the sampling and sample reduction has been correct.[4,88] In other words, there is a general similarity in the relative rates of reaction, whatever be the reactant.

That "the reactivity of cokes and chars prepared at similar temperatures is influenced by the rank of the parent coals: low-rank coals tend to yield cokes or chars of high reactivity" has been clearly stated by Blayden.[88] This statement is only partially in accord with recent work carried out on typical cokes in the United States[2] and in Europe.[4] The team of the Coke Evaluation Project prepared samples for determinations of reactivity, the results of which are shown in Table 29 together with the volatile-matter content of the blends which yielded the cokes.[2] Since the lower values indicate more reactive cokes, these observations do not confirm the statement that low-rank coals yield more reactive cokes.

In the research on six cokes,[4] the reactivities were determined in five French and German research laboratories by ten different methods. Good correlations were found between reactivity determined with carbon dioxide and with steam. The correlations between reactivities and ignition temperature or critical air blast[26] were less significant, but the number of experiments for the latter tests was lower. Table 30 summarizes the results. The ten expressions of reactivity were all different so all have been compared to the highest value to get a more comparable classification, and thus the numbers in Table 30 are relative values, the lowest meaning the less-reactive coke. The dominant influence on reactivity appears to be the average rank of the parent coals as measured by their volatile matter content on a dry basis.

TABLE 30

AVERAGE REACTIVITY COMPARED WITH OTHER DATA[4]

Cokes	Mean Rank of Blend (Volatile Matter)	Average Reactivity	Ignition Temperature	Critical Air Blast	True Density	Porosity
	%		°C	l/min		%
Anzin	23	33	543	1.92	1.95	54
Friedrich	24	56	517	1.86	1.90	51
Liévin	26	63	516	1.92	1.90	48
Reden	27	72	533	1.98	1.88	45
Carling	30	93	505	1.77	1.81	47
Thionville	31	91	485	1.78	1.85	52

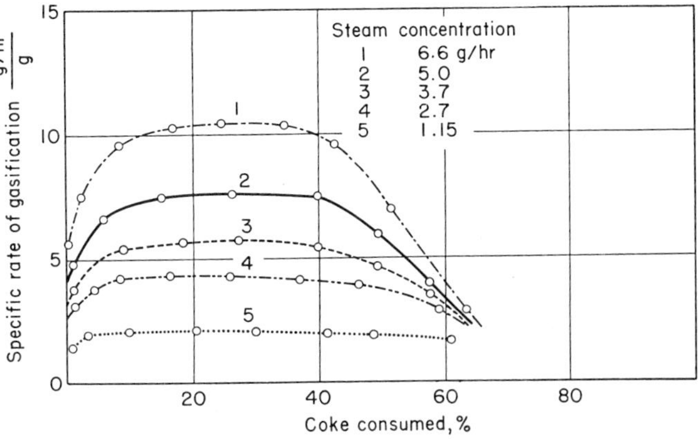

Fig. 6. Gasification rate of coke S as a function of steam concentration and weight percent consumed.[84]

It had been assumed in the past that reactivity could be determined at a given temperature by burning one or two percent of the coke in oxygen or carbon dioxide.[90] However, it is now well established that the results change during the determination itself, even at constant temperature.[83] Figure 6 shows as an example the effect of various steam concentrations in the gas brought into reaction as a function also of the percent weight of coke burned, or loss in weight.[84] These results are in accordance with the conclusions of Ergun.[85] Many investigators have recorded similar curves by changing the temperature, the size of coke, the rate of gas flow, and the thickness of the coke layer. Thus, it appears clearly that in the stack of a blast furnace, the reactivity of a piece of coke changes steadily with the composition of the neighboring gases, with the temperature, with the velocity of the gases, and with the size of the piece itself.

The data shown in Fig. 6 were obtained by use of crushed coke with mean sizes increasing from 0.36 mm (42-mesh ASTM) to 4.5 mm (about 4-mesh).[84] In another investigation,[87] prismatic samples of about a 1-in. side cut from pieces of coke were hung in a stream of carbon dioxide, and the loss in weight was obtained as a function of gas flow rate at different temperatures up to 1500°C or 2730°F. After 45 min the percent weight in loss was highly correlated with the apparent density of the cokes. Attempts were made to compare the reactivities of various cokes with their behavior in foundry cupolas,[91] where accurate heat balances were calculated, but no correlation was found between reactivity and thermal efficiency of cupolas. Yet this thermal efficiency has been well correlated with the oversize of coke on 80-mm sieve as recorded in Micum tumbler tests.[87] As a result, "it seems that foundry coke could be evaluated—apart from its chemical composition—by three tests: initial size, size after testing in the drum test, and apparent density."

Porosity is not correlated with reactivity, but the latter increases as the structural organization and the true density decrease. It seems likely that these intrinsic properties of the cokes are strongly dependent on the coal, and relatively independent on coke-plant practice.

Reactivity and Blast-Furnace Reactions

In the blast furnace, the practical aspects of reactivities are to be sought in the stack between the bell and the zone of carbon monoxide formation and in the combustion zone. At the tuyère level, the oxygen of the blast burns a quantity, C_v, of coke carbon which may be related to the weight of useful carbon, C_u, calculated as the difference of total carbon charged at the top minus the carbon in flue dust and in pig iron. It is advantageous to consider from material balances, the ratio

$$\frac{C_v}{C_u} = \frac{\text{carbon burned by the blast}}{\text{useful carbon}}$$

For accurate calculations, the blast volume or weight should not be determined from the volume delivered by the blowing engine, but

[90] Cassan, H., *Congr. assoc. techn. gaz, Paris*, **1937**, 759–89.

[91] Danis, F., *Fonderie*, **147**, 179–82 (1958).

should be calculated from a complete material balance that requires numerous gas analyses.[5,7,92,93] The difference, $C_s = C_u - C_v$, burns in the stack as defined earlier. This combustion occurs[5] in different ways and implies interactions between direct and indirect reduction of the oxides and the gasification processes involved in reactivity. If i, u, v, are respectively the indices of indirect reduction of all oxides including water, of direct reduction, and of reactivity—the latter meaning the relative weight of carbon burned by carbon dioxide in the stack—a material balance of the blast furnace yields all the numbers for which knowledge is compulsory to determine these indices. Unfortunately, it has been demonstrated that direct and indirect reductions have practically no sense, and that the system of equations leading to calculation of i, u, v, is indeterminate.[5] Simple knowledge of the initial and final states of all materials passing through the furnace do not allow an estimation of the separate effects of oxide reduction and reactivity, and may be considered as the fundamental mystery of the blast furnace.

The relative weight of carbon disappearing in the stack

$$\frac{C_s}{C_u} = \frac{\text{carbon consumed in the stack}}{\text{useful carbon}}$$

may be ascribed to a "reactivity effect"[5] or merely expressed by "solution loss," or any other complex process. The fact remains, however, that it is impossible to distinguish the separate responsibilities of the ores, the gases, and the cokes. With the mathematical tools at our disposal, it would be difficult to obtain correlations between the observations in practice and laboratory determinations of reactivities, the latter being influenced also by catalytic effects.[88,94]

The foregoing discussion implies that determination of reactivity is not significant for blast-furnace and cupola operation.[5,29,42,72] However, changes in coke reactivity may influence the efficiency and the composition of gases in gas producers.[90,95]

As far as the blast furnace is concerned, experiments carried out for weeks or for months with two different cokes[5,93] have never shown a significant increase in production or decrease in coke rate by use of a more or less reactive fuel. If it is attempted to ascertain the effect of reactivity by considering either the ratios C_v/C_u (blast combustion) or C_s/C_u (solution loss), or by comparing the top gas compositions and their CO/CO_2 ratios, no correlation[5] has been found with specific coke consumption.

Another influence of a reactivity on blast-furnace operation might be found in the combustion of coke in the tuyère zone of the furnace. Several investigations of the reactions occurring in this zone have been carried out in different countries. They started with the first observations made by van Vloten in 1893[96] and were continued by Kinney and his co-workers[26] and further German experiments.[97] In general, it was found that the linear penetration of blast was about 3 ft or 40 in.,[5] and common blast pressures and air velocities had practically no influence on this distance. The shape of the combustion zone in front of each tuyère was roughly represented as a sphere more or less distorted toward the center of the hearth, or toward the neighboring tuyères, and lengthened upward. It appeared that the lateral depth of the zone could be affected by differences in blast pressure, or shape of the tuyère nose, as well as by important changes in air velocity. It was also confirmed that a significant decrease in blast temperature tended to enlarge the combustion zone in all directions. But, no observation could relate the penetration either to the size or to other properties of coke.

All the preceding experiments were concerned with relatively small blast furnaces and low blast pressures, when the coke was not moving rapidly in front of the tuyères. Quite different information was obtained[75] with higher blast volumes, as shown by high-speed ciné films taken by Wagstaff through the tuyères. These investigations in several American blast furnaces showed a vigorous movement of the coke pieces. The shape and extent of this steady recirculation were described as a "raceway" of which diameter was roughly proportional to the tuyère diameter. Where the blast

92 Thibaut, Ch. G., Sanna, D., and Douez, F., *Rev. mét.*, **52,** Part 1, 485–507 (1955). Thibaut, Ch. G., Publ. IRSid., Ser. A, No. 61, 3–140 (1953).

93 Thibaut, Ch. G., *Rev. mét.*, **55,** 513–23 (1958).

94 Meunier, F., *Rev. mét.*, **33,** 258–64 (1936).

95 Egalon, R., Hoebeke, J., and Jarcsek, F., *Mém. soc. ingrs. civils France*, **5/6,** 249–90 (1953).

96 Wust, F., *Stahl u. Eisen*, **48,** Part 2, 1273–87 (1928).

97 Eichenberg, G., and Eilender, W., *Stahl u. Eisen*, **53,** Part 2, 997–1000 (1933).

pressure and velocity were steady, Thibaut and co-workers[98] did not find raceways. Under these conditions they found that 1% carbon monoxide appeared at about 40 cm from the tuyère nose, and the maximum percent carbon dioxide was found within 60 to 80 cm. The complete transformation into carbon monoxide, accompanied by water reduction, occurred within 90 to 120 cm or 3 to 4 ft from the tuyère nose. No relation was disclosed with any characteristic of the coke. Other observations recorded in the international low-shaft furnace[99] suggested that the combustion zone could be restricted by use of small-sized coke, but the effect of this factor was not proved because the tuyère diameters were small, as compared to those in blast furnaces. Experiments similar to those of Wagstaff have been carried out to determine, under noncombustion conditions, how the formation and shape of a raceway might be characterized. It has not been possible to define the conditions changing the nonraceway into the raceway movement. Correlations found by Wagstaff were modified and showed that linear penetration increased as the square of air velocity and the square of tuyère diameter, and also increased with the air density. Thus, an increase in blast temperature would decrease the penetration. Penetration also increased with a decrease in the apparent density of the particles (wood or semicoke).

Investigations were also undertaken with help of a small specially designed furnace by Taylor and associates.[100] Raceway measurements were recorded for noncombustion conditions, and under combustion conditions with cold and preheated blast. The penetration was higher and the raceway was more stable when the coke was being burned. These experiments indicated an increase in linear penetration as the tuyère diameter increased, and also as the airflow increased, within the range of 6 to 20 in. or 15 to 50 cm for the measured penetration with tuyères of 1- to 2-in. diameter. Direct measurements of raceway penetration were also made into six operating Scottish and Welsh blast furnaces by use of water-soaked wooden poles which measured the depth of the penetration when coming up against what felt like a wall of coke pieces. It has been confirmed that the volume of the raceway increases by about 11% when the tuyère diameter increases from 5.5 to 7 in. Like the mean size of coke, however, the tuyère diameter "has a more pronounced effect on the spread of the raceway than on the penetration." The observed penetrations changed within the range of 26 to 48 in. There was a definite trend toward an increase in raceway volume as the hearth diameter increased. In stationary beds, which are met mainly in small furnaces, the volume of the oxidizing zone decreased with smaller coke, whereas it increased under raceway conditions. This change in volume may be an indication of a beneficial effect of small mean size of coke on the stock descent in large blast furnaces. But, it would be necessary to collect other data to characterize more accurately the optimum size of coke, as far as smoothness of descent and extent of combustion zone are correlated with coke rate and iron production.

[98] Menuet, B., *Rev. mét.*, **54**, 217–39 (1957).

[99] Paschal, F., and Biver, R., *Centre document. sidérurg., Circ. inform. tech.*, No. 9, 1777–91 (1956).

[100] Taylor, J., Lonie, G., and Hay, R., *J. Iron Steel Inst.*, **187**, 330–41 (1957).

Correlation of Coke Properties and Blast Furnace Performance

The difficult subject of correlating of coke properties with its behavior in blast furnaces, cupolas, and gas producers has been treated by many authors. Generally, some of the coke properties have not been accurately described, and pertinent data have often been omitted. This omission may be explained by the fact that an operator or superintendent has noted a more or less important change in stock descent or iron production, arising at the same time as a change in coke manufacture. Such an instance draws attention only to a single aspect of the changes both in furnace performance and in coke-oven operation and leads to the consideration of only the characteristics of coke considered as major, with lack of complete information.

Neither in the work sponsored by the Coke Evaluation Project,[2,56] nor in the French study on six cokes[4] has it been possible to correlate specific qualities of cokes with blast-furnace performance. In the latter, observations of blast-furnace operations were not collected. In the former, data were recorded for hearth diameters ranging from 15 ft 6 in. to 27 ft. The coke consumptions ranged from 1500 up to 2300 lbs per net ton of iron. However, the data referred only to 11 to 15 days operation and information on the blast temperature and the top gas analysis were not given. The percent carbon burned at the tuyères was calculated from the recorded volume of wind blown, which is certainly erratic in most instances. As stated in the main report,[2] "it is unfortunate that the recommendations on sampling and testing of coke apply only to coke at point of manufacture." Thus, it is not surprising that coke qualities and furnace performances could not be correlated, for two major reasons: too short observations of blast-furnace data and too long distances between the testing and charging points of coke. It may be also suggested that, even by sampling the coke in the skips, the tumbler test designed for +2-in. coke would not have taken into account the smaller pieces representing 15 to 45% of the total weight.

In an important statistical study[101] on blast-furnace assessment, a large number of data were carefully collected in five British steel plants, and some highly interesting correlations were distinguished, involving weight of total burden, slag basicity, blast temperature, percent sinter and percent hard ore. In one instance, where some coke properties were measured, the coke rate was correlated neither with tumbler test nor with 1½-in. shatter, size, or bulk density, but this may be explained by the very small standard deviations of these values, implying regular coke. In another case, the carbon rate was found negatively correlated with the 1½-in. shatter index, the mean value of which was 78 with a standard deviation about 1.8.

A survey of British practice over 40 years[17] stated that "in times of good trade, the quality of the coke, in high demand, has deteriorated, and it has been necessary to use more of it per ton of pig iron." This is certainly true for a part of the observed data, but it is now recognized[101,102] that an increase in the driving rate of a blast furnace involves also an increase in fuel consumption. The fuel consumption increases in periods of high demand for two reasons: deterioration in coke quality and average increase in fuel consumption of the blast furnaces per unit of time. Other observations[103] have shown a relation between the cohesion of coke as measured with the Micum tumbler and the difficulty of stock descent. It has been shown also[5] that an increase of 20% in the bulk density of coke resulted in an increase of about 25% in the blast pressure.

An extensive use of multiple regression analysis[104] has been developed to derive an equation with linear regression coefficients, allowing a calculation of coke rates by taking into account characteristics of the burden, iron analysis, age of lining, blast volume, and temperature. However, the "effective carbon factor" according to Flint is actually based on ultimate carbon, ash, and sulfur contents of coke, but no effect of physical properties of coke is mentioned. The influences of sulfur and ash content are not easily calculable for general conditions because of a great diversity in the composition of burdens. For instance, a limited increase in ash content may have little importance in blast furnaces when the slag volume is high and also if the slag basicity is at a high level. But, there is no doubt that an increase of ash by 1% is definitely expensive in blast furnaces that use rich siliceous ores because of the necessity of a further addition of lime or magnesia for obtaining a fluxing slag. Various papers have been concerned with this subject of ash and sulfur contents in coke.[105]

There is no general agreement on the effects of physical properties of coke on blast-furnace operation. Mayers[26] concluded that "the properties of coke which determine the size of the pieces and the void volume in a bed of the coke as charged into the furnace are of prime importance" but that "further correlation of coke properties with the results of furnace operation by statistical methods is essential . . . ." A step in this direction was made by Thibaut in 1943[42,106] by correlating the monthly average coke rates of four blast furnaces during 8 years

[101] Ridgion, J. M., and Whitehouse, A. M., *J. Iron Steel Inst.*, **184**, 249–57, 592–5, 657 (1956).

[102] Thibaut, Ch. G., *Centre document. sidérurg., Circ. inform. tech.*, No. 7, 1541–52 (1959).

[103] Kahlhöfer, H., and Send, A., *Stahl u. Eisen*, **76**, 141–4 (1956).

[104] Flint, R. V., *AIME, Proc. Blast Furnace, Coke Oven & Raw Materials Comm.*, **11**, 49–68 (1952).

[105] Lowry, H. H., *Ind. Eng. Chem.*, **41**, 502–10 (1949). Griffen, J., *The Cost Effects in Basic Iron Blast Furnace Practice of Ash and Sulfur Levels of Coke*, 1954, 132 pp.; *Influence of Iron-bearing Raw Materials on Cost Effects of Ash and Sulfur of Blast Furnace Coke*, 1958, 71 pp., Island Creek Coal Sales Co., Huntington, West Virginia.

[106] Thibaut, Ch. G., *Publs. assoc. ingrs., fac. polytech. Mons.*, **87**, 481–512 (1944).

with the Micum tumbler indices and bulk density of different cokes. In France, the ores are constant in analysis and particular attention is paid to the properties and behavior of various cokes, especially in the Lorraine district. Coke rates have been highly correlated with a formula which takes into account the Micum cohesion C_M of the coke, as defined in formula $C_M = M_{40} + M_{20} - M_{10}$, and the bulk density d or apparent specific gravity in metric tons per cubic meter (with lb/cu ft, the term $-200d$ is changed into $-3.21d$):

$$T = C_M - 200d.$$

An essential function of coke in a furnace is to insure as perfect a permeability as possible, whatever the grade and the degree of beneficiation of iron ores. Low-bulk density favors permeability and homogeneous reaction conditions at all levels of the furnace. It is a useful guide for the assessment of coke quality since it is related to good quality of the coal blend, to adequate coking and quenching, and to high porosity, low apparent density, size homogeneity, high void volume, and low formation of smalls and breeze. High cohesion is related to other favorable properties as adequate swelling of slacks, proper crushing of the coals and adequate coking, absence of strains and shrinkage cracks, regular structure, resistance to breakage by impact, shattering, compression, or abrasion, and also high voidage with size stability.

Thus, both high cohesion and low density favorably influence the permeability in the shaft and the combustion conditions. Substitution of C_M by the Irsid cohesion I_{20} leads to the equivalent relation, $T = 2.15I_{20} - 200d$. An index, such as T, is only significant if it is correlated with blast-furnace operating data. Thus, it was necessary to compare T values to actual coke rates under various conditions. For this purpose, a method has been developed with average corrections for the adjustment of coke consumption to standard conditions: 85% C in wet coke, blast temperature, 750°C, lime to silica ratio in the slag, 1.35, hearth diameter, 20 ft, 6000 lb/ton of iron for the burden, and nominal driving rate. Such adjusted coke rates are reproduced in Table 31 in kilograms per metric ton of iron; for the sake of American readers, they are also expressed in pounds per ton of iron, corresponding to a standard burden weight of 4500 lb. Apart from the first observation mentioned in the table,[106] the following ones were carried out in different blast furnaces[71] during 1 to 6 mon for each field trial, with many samplings and other measurements by day and night. The T indices were calculated using C_M from 1942 to 1953 and using I_{20} after 1956. The case of Clairton[2] is added as an attempt to apply the method to an American blast furnace where careful tests on two cokes had been carried out; the Micum cohesion of both cokes is estimated, starting from the shatter and ASTM tumbler indices. In all these experiments, two more-or-less different cokes were compared in one blast furnace; the average standard coke rates related to both cokes are compared to the corresponding T indices. Each increase in coke physical value—as expressed by T index—produced a decrease in the coke consumption per ton of metal. An increase of one unit in T index involves an average decrease by 13 lb in the coke rate, or about 0.8%, which may be explained by the effects of cohesion and bulk density. It appears that blast-furnace operation is mainly influenced by such simple properties as densities, void volume, and cohesion of coke. It does not seem affected by more intrinsic qualities like width of pores, degree of graphitization, or "restricted reactivities" measured on powdered samples.

COKE PROPERTIES AND MANUFACTURE

Routine tests, such as tumbler tests, bulk density determinations, size analysis, and chemical composition seem sufficient for practical evaluation of cokes. However, more fundamental work is necessary for understanding the relations between the practical value of coke, the conditions of its manufacture and the properties of the parent coals. Outstanding studies[22,58,63,77] have been devoted to this important subject, and are described in other chapters on plastic properties, pyrolitic reactions, and carbonization of coal. Therefore, the relation among coal, coking conditions, and coke properties is considered in this section only in light of the numerous measurements of coke properties carried out in recent years[4,5,107]

[107] Thibaut, Ch. G., Menuet, B., and Vigneron, P., *La document. mét. No. Special, Congr. du centenaire* No. SK 2, SA 15, 149–59 (1956). Thibaut, Ch. G., *J. ind. gaz*, **1962**, No. 2, 1–10.

TABLE 31

COKE QUALITY AND BLAST FURNACE COKE RATES
(RESULTS OF EXPERIMENTS IN ELEVEN BLAST FURNACES WITH TWENTY-TWO COKES)

Plants	Years	Micum and Irsid Cohesions	Bulk Densities	*T* Index	Adjusted Coke Rates kg/metric ton *M*	Adjusted Coke Rates lb/net ton	Differences *T*	Differences *M*
La Chiers	1935	168	0.500	68	981	1512	+9	−45
	1942	174	0.485	77	936	1422		
Joeuf	1950	161	0.440	73	981	1512	+9	−59
		168	0.430	82	922	1394		
Knutange	1951	165	0.440	77	934	1418	+5	−51
		172	0.450	82	883	1316		
Thionville	1951	162	0.445	66	1129	1808	+8	−43
		168	0.425	74	1086	1722		
Uckange	1952	170	0.445	79	794	1138	+4	−23
		166	0.415	83	771	1092		
Neuves-Maisons	1953	158	0.445	69	919	1388	+7.5	−47
		172	0.480	76.5	872	1294		
Dilling	1953	145	0.410	63	992	1534	+0.5	−8
		149	0.430	63.5	984	1518		
Boucau	1956	162	0.455	70	1633	2816	+1	−9
		162	0.460	71	1624	2798		
Dilling	1957	162	0.475	67	925	1400	+7	−30
		165	0.455	74	895	1340		
Seraing	1958	165	0.480	69	927	1404	+6	−44
		163	0.440	75	883	1316		
Clairton	1949	98	0.440	10	960	1470	+13	−100
		112	0.445	23	860	1270		
Average data					995	1540	+6.4	−42

by twelve French and German research stations under leadership of the French steel industry. Complete description of the sampling and experimental procedures[4] cannot be repeated here, but the most typical results of the 80,000 determinations made have been summarized[5] and are graphically interpreted in Fig. 7 and 8. That the ordinates of individual graphs do not have the same direction is not important since only the differences are significant in showing trends.

In Fig. 7 and 8, the measured properties of coke are plotted as functions of the volatile matter on a dry, ash-free basis in the blend of slacks used for production of the cokes tested. Three groups may be distinguished: those with relatively low-, medium-, and high-volatile matter contents, the former relating to the "oldest" coals and the latter to the low-rank blends. Figure 7 indicates a general trend for the properties to stay at a definite level for the high-rank coals, and to deviate from this level as the rank decreases. Figure 8 strengthens this general impression except that the medium category behaves quite differently from the lowest and highest ones. Thus the properties listed in Fig. 7 appear to be directly affected by the blend of coals. On the other hand, it should be emphasized (see Fig. 8) that the manufacture of C and T (Carling and Thionville) cokes is especially elaborate: the crushing is finer; coke breeze addition and cake charging are carried out at Carling; special crushing and screening of the different coals are used at Thionville;[91,108,109] the coking rates are also higher in producing these cokes than for those from coals with lower-volatile contents. Figures 7 and 8 together may be considered as an attempt to separate the factors of coke quality as considered from the point of view of the coke-oven manager: on the one hand, Fig. 7

[108] Thibaut, Ch. G., *Actualités sci. et ind.* (1958).

[109] Sabatier, J. L., *Proc. 4th Conf.*, (1950), *Brit. Coke Research Assoc.* 1951, pp. 27–34.

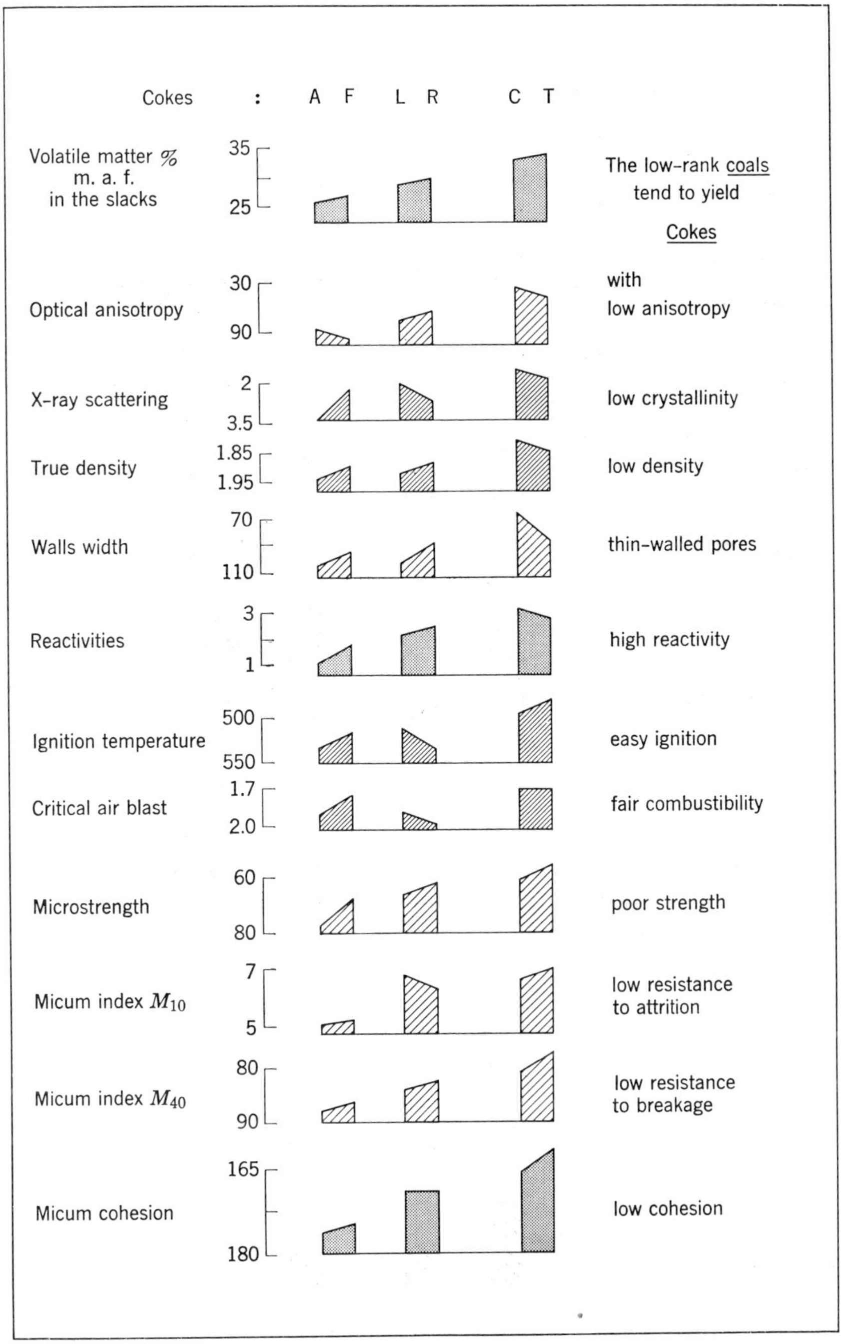

Fig. 7. Effects of the properties of coals on the properties of the cokes produced from them.

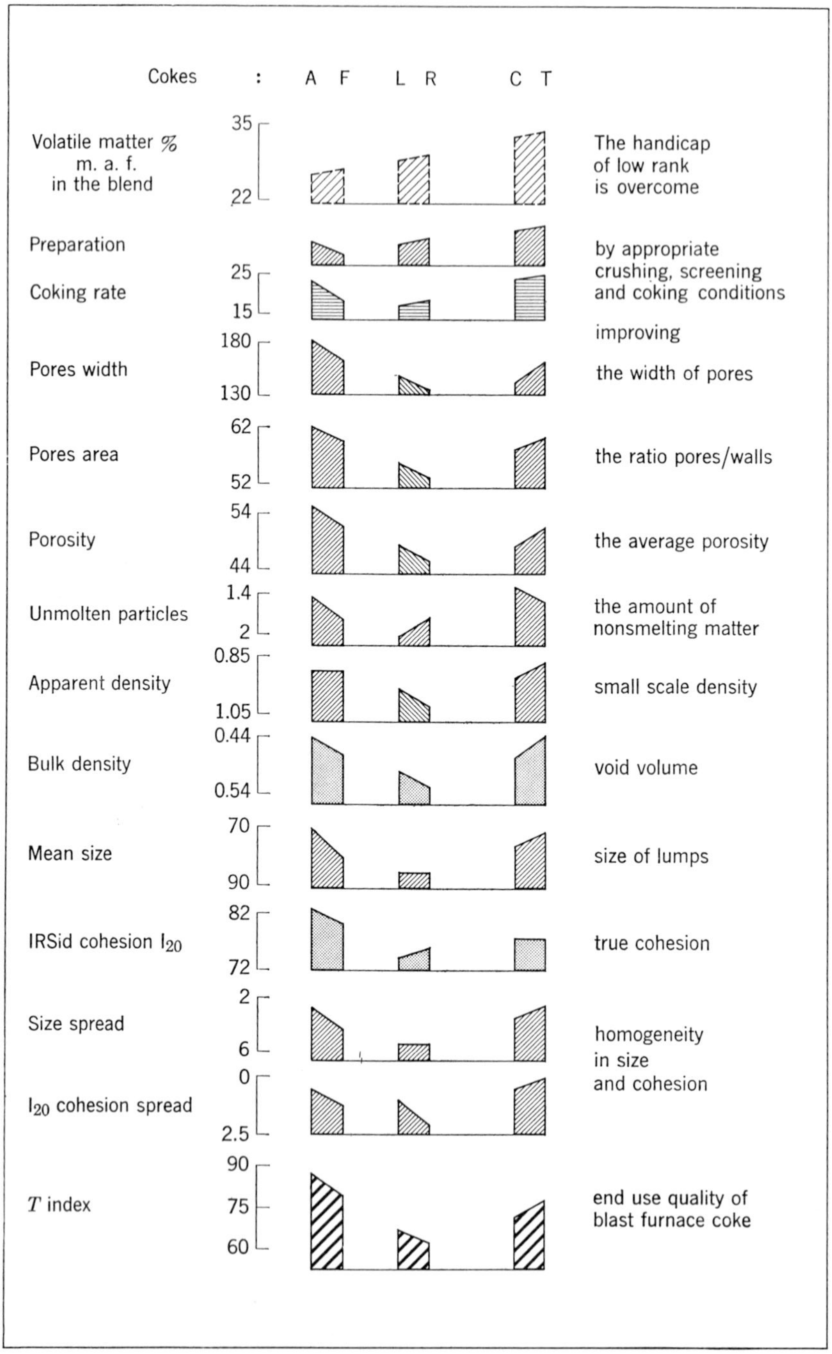

Fig. 8. Influence of coking conditions on the properties of cokes.

TABLE 32

Typical Recent Semicokes

Type of Oven	Moisture	Volatile Matter	Size Analysis −1 mm	Size Analysis $-\frac{1}{8}$ in.	True Density	Specific Surface	Microhardness	Carbonization Temperature
	%	%	%	%		m²/g	kg/mm²	°C
Chamber	12.5	10.8	—	5	1.73			600
Retort	7.1	19.0	44	80	1.54			600
Retort	—	12–20			1.47			5–600
Fluidization	—	13–14			1.58	40	50–80	500
Fluidization	—	7–8			1.62	—	—	600
Fluidization	—	3–5			1.84	45	200	700
Fluidization	—	2–3	33	78	1.79	50	220	800
Fluidization	—	14–15						600

summarizes the properties of coke which are dependent on *nature*, whereas Fig. 8 represents those on which *man* has an influence. Such a representation of facts is obviously somewhat subjective and also in accord with a definite stage in our conceptions of coal and coke. It seems, therefore, only reliable as a provisional rough estimate of how the coke manufacture is allowed to improve the quality of its product.

SPECIAL SOLID CARBONIZATION PRODUCTS

Regarding those solid products of carbonization that differ substantially from blast-furnace coke, it is perhaps unnecessary to attempt a strict classification. One of them is produced with blast-furnace or cupola coke itself, that is, breeze. Ferrocoke and similar products are manufactured under the same conditions as blast-furnace coke. Only semicoke is made under quite different conditions.

Semicokes and chars can be produced either in chamber ovens[110] or in rotating retorts or by fluidization.[111–115] Such products have been used, especially in England, but also in France, as domestic smokeless fuel as substitutes for anthracite of other coals. Their main industrial applications are either in the blends for coke manufacture, as "fillers"[110] for replacement of low-volatile coals, or as substitutes for coke breeze in the ore sintering plants. Table 32 summarizes some typical properties of different kinds of semicoke. Generally speaking, it is advisable to use these products in the form of small-sized char, which requires finer crushing before being blended with the coal or mixed with iron ore.

The moisture contents given in Table 32 refer to semicokes as delivered after transportation, and is a simple indication of the adsorption capacity. The other data relate to random observations, useless for precise calculations, and are only shown as illustrative of the range of properties. The semicokes produced by fluidization[111,112,116] have properties that depend on the coals used and on the carbonization conditions, especially the final temperature. When this temperature is increased up to 800°C, their qualities become similar to those of coke breeze. Semicokes are light, their true density is much lower than that of high-temperature coke, and their bulk density is also lower—about 0.4 or 25 lb/cuft as measured in 1-liter containers, and about 5% higher in railroad cars. The size distribution for semicokes made in chambers or retorts may allow screening resulting in various percentages of semicoke, as for example, the one (first line of Table 32) which has been used for

[110] Denig, F., *Chemistry of Coal Utilization*, John Wiley and Sons, New York, 1945, Vol. **1**, pp. 774–833.

[111] Foch, P., and Loison, R., *Rev. ind. minérale*, **38,** 1–34 (1957).

[112] Peytavy, A., and Foch, P., *Proc. Intern. Conf. Chem. Eng. in the Coal Industry*, Natl. Coal Board, 1956, pp. 75–99.

[113] Foch, P., and Geoffroy, J., *Compt. rend. congr. intern. chim. ind.*, 31*st Congr., Liège*, **1958,** 6 pp. Loison, R., and Foch, P., *Rev. ind. minérale*, **41,** 277–312 (1959).

[114] Malcor, H., *Rev. universelle mines*, **11,** No. 2, 1–23 (1955).

[115] Bouillot, A., Desfossez, P., Thibaut, Ch. G., Serruau, A., and Menuet, B., *Rev. mét.*, **52,** 63–79 (1955).

[116] Glenn, R. A., and Jacoby, J. W., *Bituminous Coal Research*, **19,** No. 3, 13–15 (1959).

field trials[114] in the international low-shaft furnace. As a product of fluidized carbonization, the size consist of the char is a function of the size of the parent coal, which is generally a high-volatile coal. The specific surface, as determined by the B.E.T. method (Table 32), is twenty to fifty times higher than that of coke. It often happens[117]—last line of Table 32—that only the volatile matter content of the semicoke is specified.

Data reported in Table 32 are mainly concerned with low-temperature carbonization in France, using low-rank coals with 35 to 40% volatile matter. Other experimental or industrial developments have also been carried out with similar raw material[118] in Italy, Poland, the United States, the United Kingdom, Germany, South Africa, and the U.S.S.R., in chamber ovens or rotating retorts. The carbonization temperatures range from 450 up to 850°C and, as a rule, the solid products are the object of much less accurate description than the gases, oils, and tars. Low-temperature carbonization is also being applied to lignites in the U.S.S.R., Eastern and Western Germany, and Czechoslovakia, mainly in byproduct—either refractory or steel—ovens. In these countries also, the description of semicokes is restricted to volatile-matter content, ranging from 2 to 6% up to 18%, according to the carbonization temperature.

The size of the solid products is usually reported as greater than 10 to 15 mm—or $\frac{1}{2}$ in.—probably after it has been screened; in some instances, finer chars are produced for use in power plants, or as replacement of coke breeze in the sintering of iron ores. In the latter application,[119] comparisons of different solid fuels have been carried out by Irsid. The composition of gases leaving the sinter charge is an interesting indication of the thermal value of various fuels mixed with the blend of ore and flue dust. When the fuel content in the mix increases, the percent oxygen in the gases decreases, but the characteristic ratio $CO/(CO + CO_2)$ remains at a constant level during the largest part of the sintering process. This ratio is substantially affected by the size and the nature of the solid fuel; for example, it increases from 0.23 up to 0.26 as the mean size of coke breeze decreases in the range of 5 down to 0.5 mm, indicating an increase in reactivity. The same tendency is observed when the volatile matter content of the fuel is increased; when coke breeze (volatile matter 4%, mean size 2.5 mm) is replaced by semicoke produced in a fluidized bed (volatile matter 6%, grain size 2 mm), the $CO/(CO + CO_2)$ ratio in the exit gases increases from 0.22 to 0.26. This apparent increase in reactivity may be related to the lower true density and higher porosity of char compared with coke breeze.

[117] Situlin, I. K., and Borisov, V. I., *Koks i Khim.*, No. 8, 61–4 (1959).

[118] Simonovitch, M., *Assoc. tech. sidérurg., Comm. Cokeries*, May, 1958.

[119] Michaud, J., *Iron & Coal Trades Rev.*, **175**, 143–5 (1957).

Ferrocoke

The manufacture of ferrocoke has been the subject of a large number of laboratory tests and field trials.[120,121] Almost all coke-oven plants in continental Europe have carried out experiments with a view of replacing low-volatile components in the blends by blast-furnace flue dust or powdered iron ores. The coals were mainly high-volatile slacks, and the ferrous raw materials have been either various kinds of flue dust, pyritic residues, or different fine iron ores, especially magnetite.[120,122] Supposed disadvantages of this special cokemaking related to coking time, yield in solid product, influence on byproduct quality, and eventual degradation oven walls have also been discussed.[121,123]

As far as the quality of the ferrocoke is concerned, it may be summarized as a mixture of coke and "ferro," the latter being itself a combination of iron oxides and reduced iron, so that[124] ferrocoke may be described, from the technical and thermal points of view, as a more or less intimate mixture of classical coke, sinter and scrap. The volatile-matter content of ferrocoke is somewhat more complex than it is in normal coke because of the interference of the volatile matter brought about by the ferrous additions, such as carbon dioxide, oxygen

[120] Barking, H., and Eymann, C., *Stahl u. Eisen*, **77**, 104–7 (1957).

[121] Poiré, M., *Assoc. tech. sidérurg., Comm. Cokeries*, March, 1959.

[122] Russell, C. C., and Whitstone, P., *Gas World*, **141**, 1616–22, 1685–7 (1955).

[123] Sanna, D., *Centre document. sidérurg., Circ. inform. tech.*, No. 12, 2583–7 (1959).

[124] Thibaut, Ch. G., *Centre document. sidérurg., Circ. inform. tech.*, No. 2, 437–50 (1960).

TABLE 33

DATA ON FERROCOKE CONSUMPTION[102]

Plants: / Fuel:		Hamborn Ferrocoke 100%	Everett Coke 100%	Everett Ferrocoke 45%
Ferrocoke consumption	lb/ton iron	4490	0	920
Total coke consumption (85% C)	lb/ton iron	2840	2060	2130
"Sinter" part of ferrocoke	lb/ton iron	1740	0	210
Total burden, including flux	lb/ton iron	4670	4470	4580
Blast temperature	°F	930	1000	1060
Top temperature	°F	420	470	450
Daily production	net tons	15	430	470
Slag volume	lb/ton iron	1680	750	820
Lime to silica ratio	—	1.68	1.36	1.39
CO/CO_2 in top gas	—	7.7	2.6	2.9
Iron temperature	°F	2620	—	over 2500
Net burden	lb/ton iron	4500	4080	4440
Standard coke rate	lb/ton iron	1720	1430	1240

of the metallic oxides, combined water and sulfur, which cannot be completely eliminated during the complex carbonization process associated with partial or total reduction of the oxides. It has been mentioned[121] that the percentage of sulfur is often higher in ferrocoke than in coke. The reactivity is also higher. The reoxidation of iron, a few days after manufacture, seems[120] limited to the external surface if the iron in the product is finely divided. Generally speaking, attempts toward industrial manufacture have been stopped in many coke-oven plants because of fear of attrition and wear of the walls. However, some important experiments have been carried out in Germany[120] and in the United States.[3,122]

The behavior of ferrocoke has been studied in both a blast furnace[120,122] and a low-shaft furnace.[125] In the latter case, the ferrocoke was made in regular coke ovens with German high-volatile coal (33.5%) and 30% of Swedish concentrates of magnetite ground to 70 μ or 200 mesh, the ferrocoke having a carbon content of 54% and an iron content of 24%. Ferrocoke was used at Hamborn in a low-shaft furnace, about 1 m in hearth diameter. From the burden composition,[125] the ferrocoke may be regarded as the sum of 61% coke with 9.5% ash, and 39% "sinter" holding 62% iron. The blast pressure was 2.8 psi and its temperature 500°C. The CO_2/CO ratio in top gas was 7.7. Some other typical data and their interpretation[102] are reported in Table 33 compared with American data.[3] Unfortunately, the operating conditions and results for the Hamborn low-shaft furnace are not given with normal coke, but the results are very similar to what is normally expected in such furnaces with ordinary driving rates and blast pressures.[114]

The most extensive manufacture and use of ferrocoke has been carried out[3] by Eastern Gas and Fuel Associates. The blend of blast-furnace flue dust with different coals was studied by carrying out many tests in a movable wall oven, and further, in a coke-quality oven, and finally in full-scale ovens. A blend of 20% blast-furnace flue dust and 80% Kopperston coal (29% volatile matter) was selected for plant trials. More than 60,000 tons of ferrocoke were produced without any noticeable degradation of the ovens. Table 34 summarizes the main characteristics of the ferrocoke compared with the normal coke manufactured in the same plant. The ferrocoke containing 10 to 12% iron may be considered as an intimate mixture of 80% regular coke (with 87% C) and 20% of "sinter" holding 50 to 60% iron. It was used in the Everett 17 ft. 6 in. blast furnace for foundry iron-making for several months. This experience is especially interesting because of the duration of the observations, and because accurate comparison is possible with current practice.[3,124]

It has been reported that ferrocoke was manufactured in North Korea on an experimental scale in 1951, using brown coal containing 55

[125] Jaeger, F., *Stahl u. Eisen*, **73**, 1526–9 (1953).

TABLE 34

COMPARISON OF FERROCOKE VERSUS REGULAR COKE[3]

		Ferrocoke	Coke
Chemical analysis:	Moisture	3.5	2.1%
	Volatile matter	1.9	1.0%
	Nonferrous ash	9.0	7.4%
	Sulfur	0.62	0.56%
	Total iron	12.4	—
Screen analysis:	Mean size	2.8	2.9 in.
ASTM tumbler:	Stability	54	55%
	Hardness	64	68%
Apparent density		1.16	0.99
Micum tumbler:	M_{40}	75	80
	M_{20}	89	92
	M_{10}	9	6
Micum cohesion:	C_M	155	166
Irsid cohesion:	I_{20}	72	79

to 65% fixed carbon.[126] Beginning in June 1958, new experiments were conducted under commercial conditions in a coke battery, with imported coals and iron ore. The ferrocoke contained up to 20% of iron. It was used in a blast furnace for a fortnight. The pig iron production increased from 488 up to 587 metric tons/day. The articles report no details about the chemical or physical characteristics of the ferrocoke. From the Everett experiments, it appears that partial replacement of coke by ferrocoke has resulted in a thermal saving of $\frac{190}{1430} = 13\%$ as expressed[102] by the decrease of standard coke or carbon rate; in other words, since the comparison is based on equal blast temperature, the replacement of 50% coke by ferrocoke results in 15% thermal saving. It is likely that such an outstanding experiment will draw attention again to ferrocoke production in various countries. The development of ore sintering in steel plants creates no obstacle indeed because even with a sintering plant on hand, it remains quite helpful to add flue dust in the coke-oven blend, for two reasons: first, the partly sintered and reduced "ferro" part of the ferrocoke seems to be very reducible and decreases the carbon rate; second, the use of the flue dust at the coke ovens leads to higher production of the sintering plant with less flue dust and more iron ore. Moreover, the ferrocoke breeze seems an ideal fuel for sintering plants.

[126] Martynov, V. V., *Coke & Chem.*, **1960**, No. 3, 53–4. Pak Syn Nok, *Métallurg.*, **4,** No. 6, 5–7 (1959).

EVALUATION AND SPECIFICATION OF COKE

This last section deals only with the metallurgical cokes, the other solid products of carbonization such as breeze, domestic coke, semicokes, and iron cokes having specifications quite different from those for blast-furnace or foundry cokes. Where specifications do exist for domestic cokes and semicokes,[17] they are practically limited to simple sizing.

Foundry coke has been discussed previously, particularly where reactivity is involved. The qualities required in cokes for cupola operation do vary widely[127,128] in different countries. In a general review of this subject, Das Gupta[128] recalled that, in the past cupolas were fed with charcoal and anthracite; nowadays, beehive cokes and byproduct cokes are used, depending on the local commercial conditions.

The chemical characteristics of foundry coke are not significantly different from those of blast-furnace cokes. They depend on the parent coals as far as ash content and reactivity are concerned. Specifications are commonly included in the following figures: moisture not over 2, 4, or 5%; ash not over 9 or 10%; and volatile matter not over 1.5 or 2%. Special specifications are generally devoted to sulfur and phosphorus contents. Because of the low-slag volume in cupolas, desulfurization has

[127] American Foundrymen's Association, *Handbook of Cupola Operation*, First edition, 1946, 459 pp.

[128] Das Gupta, N. N., and Lahiri, A., *Iron & Steel Rev. (India)*, **5,** 1–7 (1958).

little importance compared with blast-furnace practice, and practically the entire weight of phosphorus of the coke goes to the melted iron. For these reasons, specification for foundry coke look generally as follows: sulfur not over 0.7 or 1%, and phosphorus not over 0.03%.

The physical properties of foundry cokes have been investigated less than those of metallurgical cokes. The quality mainly considered is the size of the lumps. It is generally agreed[128] that pieces smaller than 3 in. are undesirable in cupolas, the reason being found in the fact that foundry coke must be unreactive; the practical reactivity is strongly influenced by the size of the lumps. The following relations between the diameter of the cupola and of the size of cokes may be considered representative:

Internal Diameter of Cupola		Coke Size	
in.	mm	in.	mm
<24	<600	2½ × 3½	60 × 90
24–36	600–900	3½ × 6	90 × 160
>36	>900	>3½	>90
<36	<900	2½ × 3½	60 × 90
36–42	900–1070	2½ × 5½	60 × 140
42–48	1070–1220	3½ × 5½	90 × 140
48–60	1220–1530	2½ × 5½	90 × 140 and +140
>60	>1530	−5½	+140

The indications, at least in principle, may be compared with some tendency at the present time to feed the blast furnaces with different sizes of cokes, the smaller sizes going to the small furnaces, and the larger lumps being screened for charging to the large furnaces. Consideration of what happens in models—and also in cupolas—supports this conception. Objectively, it must be remarked that it does not meet general agreement; some blast-furnace experts hold an opinion that the size of coke should in all cases be similar to that of the ore or sinter. It should be borne in mind, however, that the average size of sinter scarcely exceeds ½ to 1 in.[119,129]

A number of cokes have been carefully tested in a relatively small French cupola, with a hearth diameter of 27½ in. The iron temperature and the thermal yield of the cupola were the highest with cokes in which no lump was smaller than 3½ in. The thermal efficiency of the cupola was best when the cohesion of the coke, measured by the percent oversize (M_{80}) on the 80-mm sieve in this particular instance, was about 80%. From these experiments, it appears that the two main specifications for foundry coke are to be found in size—not lower than 3½ in.—and in mechanical strength expressed by M_{80}. Das Gupta has given values of M_{40} ranging from 77 to 82, and figures of shatter test as follows: percent +2 in., 80 to 93%; and percent −½ in., 2 to 4%. Generally speaking, our knowledge about the desirable properties for foundry cokes is less than in the field of blast-furnace coke. This fact is easily explained by the comparatively small tonnage of coke used in the foundry industry compared to that used in the iron and steel industry.

Even when limited to blast-furnace coke, the subject of the evaluation and optimum specification remains wide open for discussion. No definitive progress has been recorded since it was stated either that[17] "a rigid specification for metallurgical coke is not possible" or that[26] "the correct evaluation of blast-furnace coke is still an unsolved problem." In spite of this, coke has been used successfully in ironmaking since 1708.

As far as chemical analysis is concerned, it has already been mentioned that it is very difficult to establish a general rule for the percent ash and sulfur which *can* be tolerated in blast-furnace coke because this depends on the richness of the ore smelted[17] as well as on the lime silica ratio in slag. The Coke Classification Working Party (United Nations, Geneva) is considering this problem on the basis of the experience reported in various proposals by participating delegations. In Britain,[70] the ash content should be lower than 7.2% for foundry coke. In Germany,[130] an agreement between producers and users is based upon 9.8 to 10.2% ash and 0.9 to 1.2% sulfur, with appropriate penalties and bonuses for higher and lower percentages. In France,[5] the committees of the Association Technique de la Sidérurgie established a proposal based on 10.5% ash and 0.8% sulfur, lower and higher values receiving bonuses and penalties. In the United States,[2,131] the levels of sulfur and ash

[129] Millar, W. D., *Blast Furnace Steel Plant*, **48**, 933–39 (1960).

[130] Kohlenbergbau Leitung, *Qualitätsabkommen für Koks*, Essen, March, 1951, pp. 1–6.

[131] Sanna, D., *Centre document. sidérurg., Circ. inform. Tech.*, No. 9, 1967–76 (1959).

contents are often low; it seems likely that the commercial specifications[105] widely vary in the different districts and steel corporations. More generally speaking, important progress should result from replacement of "fixed carbon" by a direct measurement of "total carbon," associated with the determination of hydrogen.

The problems that arise with the sampling procedures,[2,4,17] as well as with the preparation of samples for chemical analysis[78,132] and for physical tests,[2,4,5] are commonly solved by national standards which often treat together the sampling of coke and coal. The ISO Testing and Coke Working Group is considering[57] the possibility of an international agreement on sampling of coke. It is often admitted that the coke sampling may be considered simpler than coal sampling, at least when a screened product is concerned. The choice of the point where the increments are to be taken is subject to argument. Whatever the sampling procedure is, the point of sampling should always be specified in testing reports. The coke-oven manager is mainly interested in the qualities of its coke as pushed for the control of blends and coking conditions. The merchant wishes the sampling at the dispatching or delivery point. The blast-furnace superintendent is equally interested in the latter state of purchased coke, in relation with the price he pays for it. For correlating coke properties and the blast-furnace performance,[101,104] it is important that the sampling be carried out as close as possible to where it is charged to the bell. In all the tables of this chapter, unless a special statement was made to the contrary, reference is to American cokes as sampled at the coke-oven plant, and to European cokes as charged into the blast-furnace skips.

Before dealing with the choice of characteristic physical properties of cokes, it should be noted that the dominant studies[2,4] related to coke quality in recent years support elimination for practical evaluation of laboratory tests which are undertaken on small samples, such as true density, X-ray scattering, and reactivity. These determinations are useful for ascertaining the relationships between the results of routine tests and the properties of coal, as well as for explaining what can actually be described by routine tests, but they are unable to assess the practical value of coke. An important conclusion of the French research on six cokes[4] was that the determinations carried out on small samples in laboratories are disappointing, the results being the least reproducible with the most elaborate methods. By comparison with these data, the results of routine tests—such as screen analysis, tumbler tests, and bulk density—prove much more reliable because of their better repeatability for one consignment and their better sensitivity for discarding different cokes.

Starting with data and statistical interpretations which have been commented on in the section devoted to physical properties, the report on the Coke Evaluation Project[2,56] gave definite recommendations, in accord with the program of the sponsoring Joint Committee. The sampling and testing procedures adopted by the team were accepted as adequate and recommended as "tentative standards to be followed by every plant to permit valid comparisons of data on coke quality." The conclusions related to the selection of the representative tests were presented in the following terms:[2]

> Study of the data presented with full consideration given to the high correlations found between various measures of coke quality, and numerous discussions in committee meetings, led to complete agreement on acceptance of the following properties for adequate description of coke quality—the order given has no special significance:
>
> 1. ash
> 2. sulfur
> 3. volatile matter
> 4. moisture
> 5. bulk density
> 6. 2 in. screen-index taken at the wharf
> 7. tumbler hardness index, as measured in the ASTM test after 400 revolutions
>
> .... The proposed recommendation is limited to one year of test of coke quality.

It should be noted that the recommendation related to screen analysis applied to coke taken at the oven wharf. No special point of sampling is compulsory for the measures of bulk density and for the tumbler test. The recommendation did not specify the number of determinations to be made for ascertaining

[132] Chakraborty, P., Tarafdar, T. C., Ghosal, A., and Das Gupta, N. N., *J. Sci. Ind. Research (India)*, **17B**, 408–12 (1958). Das Gupta, N. N., Tarafdar, T. C., and Sinha, S. N., *J. Iron Steel Inst.*, India, **10**, 204–9 (1958).

the valuable mean values for each characteristic. It may be noted, however, that these proposals were expressed in view of further examination after practical use in the plants during one year.

Similar suggestions have been advanced in the United Kingdom[70] with two dissimilarities: the volatile matter index was omitted, and the tumbler test was replaced by the shatter test. In Germany,[130] commercial cokes are classified according to their size range: top by 90, 60, 40 mm, or screened cokes 60 by 80 mm and 40 by 60 mm; as far as trade agreements are concerned, sulfur and ash content are taken into account but no mention is made of bulk density or a tumbler test. A proposal[18,133] has been made for use of an index for the physical value of metallurgical coke, but does not apply to cokes smaller than 60 mm.

The latter disadvantage also concerns the best known cohesion tests: the shatter test, the Micum test, and the ASTM tumbler test, where all pieces of the sample must be larger either than 2 in. on a square aperture or larger than 60 mm on round holes of perforated plates. Therefore, the French Iron and Steel Research Institute has studied[66,68] and proposed[57,60] the modified Micum test known as Irsid test, which is carried out with only one sieve of 20-mm round aperture, as well as definition of the sample for ascertaining, after tumbling, the cohesion of the coke. This test has some similarity with the modified ASTM tumbler test after 400 revolutions[2] because the number of revolutions is 500, and its field of application is wider since it applies to all cokes down to -1 in.

Apart from this difference, there is a remarkable agreement between the American[2,56] and the French[4,5] recommendations. The proposals issued by the French steel industry technical committees of coke-oven and blast-furnace engineers since 1950 relate to:[5,134]

1. Moisture
2. Volatile matter
3. Ash
4. Sulfur
5. Screen analysis
6. Micum tumbler test
7. Bulk density

It is noticeable that the French and American studies proceeded simultaneously without knowledge on the part of either team of the others activities until the conclusions were published. Thus, the agreement on the basic concepts on coke quality is more significant.

Currently, most of the work carried out on coke quality is undertaken in connection with international meetings. Two world organizations are dealing with this subject.

The Coal Committee of the Economic Commission for Europe (United Nations, Geneva) has already published[135] a special booklet on the classification of hard coals by type, with complete description of the involved procedures. Working Groups are working on the classification of cokes and their use. The Utilization Working Group chiefly considers the different end uses of cokes and chars, from the point of view of their properties, especially reactivities. Another Working Group is dealing with two types of classification of cokes, by size and by type. The classification of cokes by type will likely refer to proximate analysis and to cohesion. It has been suggested[5] that one simple classification with three parameters be considered: (1) total carbon with eventual adjustment for ash and sulfur; (2) mean size connected with size spread; and (3) bulk density associated with cohesion. Whatever advancement of these studies is made by these Working Groups in the next several years, the conclusions can be applied only if accurate rules are determined for the sampling procedures and for all chemical analysis procedures and physical tests. Formulation of these rules is the subject of programs of the International Standards Organization.

The matters related to coke quality are sponsored by Technical Committee 27: Solid Mineral Fuels. This committee created several Working Groups dealing with coal preparation, plastic properties of coal, etc. Working Group 8 studies recommendations related to all the aspects of the testing of coke. The sessions of this Working Group take place successively in the fifteen participating countries. During the first five meetings, the group has already reached agreement on detailed procedures, issued as ISO recommendations, for the shatter test and for the classical Micum

[133] Wesemann, F., Graf, U., and Warturann, R., *Stahl u. Eisen*, **76**, 133–44 (1956).

[134] Thibaut, Ch. G., *Centre document. sidérurg., Circ. inform. tech.*, No. 3, 605–21 (1958).

[135] United Nations, Secretariat of the E.C.E. *International Classification of Hard Coals by Type*, E/ECE/247, E/ECE/COAL/110, United Nations Publ. 1956, II.E.4, 52 pp.

test. Future work will be concerned with coke sampling, screen analysis, bulk density determination, comparison between ASTM and Micum tumbler tests, modification of Micum test for better sensitivity and extended application, and determination of moisture, ash, and volatile matter. It has also been suggested[5] to replace the volatile matter estimation in cokes by the hydrogen content, and to replace the calculation of fixed carbon by the accurate and rapid determination of total carbon.

The whole of these international friendly discussions is practically restricted to metallurgical coke. The question may be raised of the utility of a more extensive knowledge on coke properties and other solid products of carbonization with the eventuality of steelmaking without use of coke,[136] either by direct reduction processes or by using coal in low-shaft furnaces. It does not seem likely, however, that blast furnaces will be superseded within the present century by more economical processes for ironmaking. It has also been advanced that burden preparation and use of increasing quantities of sintered ores associated with fuel oil or natural gas injections could result in the possibility of feeding the blast furnace with "softer" cokes. This view is obviously wrong since an improved quality of ores and sinters will not permit the blast furnace to become impermeable. Moreover, all present and future progress in the field of coke utilization is strongly related to a still better knowledge of coke.

[136] Epron, P., *Rev. mét.*, **56,** No. 1, 3–12 (1959).

CHAPTER 13

LOW-TEMPERATURE TAR

CLARENCE KARR, JR.

Bureau of Mines, U.S. Department of the Interior, Morgantown, West Virginia

Low-temperature coal tar has been defined generally as that tar produced within a certain temperature range well below that used in coke ovens. There has been a tendency to call any tar produced at temperatures as high as 600 or 700°C a low-temperature tar. However, because aromatization reactions start to make a significant contribution at about 600°C, even at relatively short residence times, the upper temperature limit for a true low-temperature tar should be about 500°C. Definitions of low-temperature tar on the basis of certain specific physical or chemical properties do not appear to be feasible. There is too much variation in the nature of the tar with the type of coal used, and definitions based on the presence or absence of certain compounds or classes of compounds are presumptions that hardly seem justified in view of the lack of detailed characterization of a wide variety of tars. The one outstanding chemical characteristic that can be cited for low-temperature tars is the small amount of any individual compound, a situation in common with petroleum and shale oil, but completely different from coke-oven tars in which a single compound, like naphthalene, can account for as much as 10% of the tar.

Historically, the production of low-temperature tar in small-batch operations can be traced back at least to the seventeenth century. Low-temperature brown-coal tar production began in Germany, in 1846. It was on a commercial basis by 1855, with processing of the tar producing illuminating oil, lubricants, and paraffin wax. Full commercialization of the process as a source of oil was begun in South Wales from around 1860 to 1870, but was abruptly suspended in the latter year upon the importation of the newly discovered petroleum from the United States. To this day, the economic margin in favor of petroleum has been the major obstacle to the large-scale commercialization of low-temperature tar, except in Germany during the two World Wars. True commercialization of the low-temperature carbonization process was achieved in England, in 1906, with the formation of the Coalite Corporation. However, this first plant, as well as the modern one at Bolsover, was established for the production of a smokeless fuel for domestic heating to meet the demand for smoke control. The low-temperature tar produced was fundamentally a by-product, much as coke-oven tar. The "Disco" process of the Pittsburgh Coal Company, which culminated in a large works in 1949, was established under similar circumstances. The demand for a solid smokeless fuel in the Pittsburgh area for domestic heating has declined sharply since then due to the increased use of natural gas. Low-temperature carbonization processes specifically for the production of low-temperature tar were conducted in Germany during both World Wars. This tar was utilized almost entirely as a petroleum substitute for the production of fuels, on both occasions. These were considered uneconomic operations from a peacetime standpoint. More details about current low-temperature tar refineries in various countries will be presented later. The most important point, historically, about low-temperature tar is that it has been produced in quantities comparable to coke-oven tar only in war-time Germany, and compared to the production of petroleum, the amount has been infinitesimally small.

There are no modern texts devoted exclusively to low-temperature tar. Rhodes[1] discussed the status of the field prior to about 1943, along with material devoted mostly to

[1] Rhodes, E. O., *Chemistry of Coal Utilization*, John Wiley and Sons, New York, 1945, Vol. **2**, pp. 1287–1370.

coke-oven tar. The "Disco" process was described, and there was a discussion of the effects of temperature on coal during its carbonization that is pertinent to the formation of low-temperature tar; a table was presented which listed compounds identified from the carbonization of coal, including low-temperature carbonization.

Gentry, in his book published seventeen years earlier, devoted fifty-five pages to a discussion of low-temperature tar and covered the subject from the historical, scientific, technical, and commercial points of view.[2]

The purpose of this chapter on low-temperature tar is an exposition of modern viewpoints on the subjects of composition, processing, utilization, and economics of low-temperature tar. This is prompted by the relatively high level of current commercial interest in low-temperature tar and its potentialities as a source of speciality fuels and chemicals.

COMPOSITION

Compared to the research on coke-oven tar and petroleum, the detailed characterization of low-temperature tars has received very little attention until the last few years. Undoubtedly, this is due, in part, to the lack of large-scale commercialization of low-temperature tar; however, it is also axiomatic that the lack of understanding of the nature of low-temperature tar has been one of the obstacles to the development of suitable refining techniques and widespread utilization. The classic methods of organic chemistry by which most of the constituents of coke-oven tar were first identified are not generally applicable to low-temperature tars because, in comparison, the components present in the latter are in small amounts. Another important factor is that low-temperature tars vary widely in composition with the nature of the coal and the specific carbonizing equipment, that is, the temperature program, residence time, and maximum temperature. Coke-oven tars, in general, are comparatively uniform in composition and essentially independent of the nature of the coal, so that information from the characterization of one coke-oven tar is applicable to other coke-oven tars. There has been, therefore, sufficient justification for the compilation of data obtained at numerous laboratories over a period of many years on the composition of a variety of coke-oven tars. This cannot be done for low-temperature tars, however, because the tar from different coals can be as different as the petroleum from different oil fields.

MODERN METHODS OF CHARACTERIZATION

The methods of characterization which are truly applicable to low-temperature tars are those which will give detailed information in a relatively short period of time so that a variety of tars can be examined by the potential refiner. These are all modern methods of characterization, many of which were developed for petroleum research and give detailed information quickly on a variety of samples. Two examples are the recording infrared spectrophotometer, a product of petroleum research during the second World War, and gas-liquid chromatography, the capabilities of which were first fully demonstrated by the petroleum industry a few years ago.

Fractional Distillation and Infrared Spectrophotometry

Fair and Friedrich[3] have presented in detail a method for the quantitative infrared analysis of alkyl phenol mixtures extracted from low-temperature bituminous coal tar. The method involves distillation of the tar-acid mixture into fractions with specific boiling ranges and infrared analysis using a designated absorption band for each phenol. The fifteen lowest boiling phenols were included in this method. As an aid in determining which phenols to look for in tar-acid distillate fractions, Karr[4] compiled a list of phenols based on boiling points and tabulated physical properties on 183 different phenols boiling below 260°C. To help fill the need for spectral information on the high-boiling phenols, Yarboro and Karr[5] described a general method of synthesis

[2] Gentry, F. M., *The Technology of Low-Temperature Carbonization*, The Williams & Wilkins Co., Baltimore 1928, 399 pp., Chap. 3, *Low Temperature Coal Tar*, pp. 73–128.

[3] Fair, F. V., and Friedrich, R. J., *Anal. Chem.*, **27**, 1886–8 (1955).

[4] Karr, C., Jr., *U.S. Bur. Mines, Inform. Circ.* No. 7802, (1957), 15 pp.

[5] Yarboro, T. L., and Karr, C., Jr., *J. Org. Chem.*, **24**, 1141–3 (1959).

of high-molecular-weight tar acids and presented spectral data on several of these.

The detailed analysis of a typical low-temperature bituminous coal-tar base mixture boiling up to about 355°C has been demonstrated by Karr and Chang[6], employing the readily available techniques of fractional vacuum distillation and spectrophotometric analysis. Particular emphasis was laid on fairly detailed spectral information for each fractional boiling range. In all, fifty-one individual tar-base compounds were identified in a tar-base mixture and their quantities estimated. For the higher-boiling fractions, an ultraviolet spectrophotometric method described by Chang and Karr[7] was used to determine total pyridines and total quinolines. Karr, Estep, and Papa[8] have also published infrared spectra of ten different polymethyl quinolines and have described possible methods of correlating the infrared spectra of quinolines with their structures.

Primarily on the basis of infrared analysis of twenty-seven neutral oil distillate fractions in the range from 134 to 188°C, Jäger and Kattwinkel[9] have identified and determined the amounts of thirty-four compounds in a low-temperature coal tar, including twenty alkylbenzenes, four *n*-paraffins, six naphthenes, three isoparaffins, and an alkylthiophene.

Gas-Liquid Chromatography

Low-temperature bituminous coal-tar phenols have been identified and determined quantitatively by Karr, Brown, Estep, and Humphrey[10] using a gas-liquid chromatographic technique for low-boiling phenols first described by Karr and Brown.[11] A comparison was made with the independent technique of fractional distillation followed by infrared analysis. Low-temperature tar phenols of a somewhat higher boiling range (up to 234°C) were likewise identified and quantitatively determined by Karr, Brown, Estep, and Humphrey[12] using gas-liquid chromatography. In addition, chromatographic fractions were collected for micro infrared analysis. Comparison was again made with the technique of fractional distillation followed by infrared analysis. Phenols, the three cresols, the six xylenols, the three ethylphenols, two different ethylmethylphenols, and two different trimethylphenols were all included in the analyses. Using gas-liquid chromatography, Irvine and Mitchell[13] identified phenols of a still higher boiling range (up to 250°C). A total of twenty-eight phenolic compounds were identified, and the infrared absorption bands of authentic specimens used for identification were given; no quantitative results were presented.

Grant and Vaughan[14] used gas-liquid chromatography to determine the presence and estimate the amounts of about eight aromatic components in the total naphthas (boiling between 150° and 207°C) from two different low-temperature tars. A total of fifty-two aromatic hydrocarbons boiling up to 218°C were identified by Chang and Karr[15a] in a low-temperature bituminous coal tar by means of gas-liquid chromatography and infrared spectrophotometry. Quantitative determinations were made on forty-three compounds boiling up to 200°C. It was shown that alkylbenzenes, with equal numbers of carbon atoms in their alkyl groups, exhibited a linear relationship between the logarithm of relative retention time and normal boiling point, and that the C_8, C_9, C_{10}, C_{11}, and C_{12} alkylbenzenes lay on five parallel lines whose separation from each other is related to the logarithm of the number of carbon atoms in the alkyl group.

These workers extended their investigation to include aromatic hydrocarbons boiling

[6] Karr, C., Jr., and Chang, T. -C. L., *J. Inst. Fuel*, **31**, 522–7 (1958).

[7] Chang, T. -C. L., and Karr, C., Jr., *Anal. Chem.*, **30**, 971–2 (1958).

[8] Karr, C., Jr., Estep, P. A., and Papa, A. J., *J. Am. Chem. Soc.*, **81**, 152–6 (1959).

[9] Jäger, A., and Kattwinkel, G., *Brennstoff-Chem.*, **35**, 353–62 (1954).

[10] Karr, C., Jr., Brown, P. M., Estep, P. A., and Humphrey, G. L., *Anal. Chem.*, **30**, 1413–16 (1958).

[11] Karr, C., Jr., and Brown, P. M., *Am. Chem. Soc. 130th Meeting*, Abstracts of Papers, Atlantic City, 1956, p. 4K.

[12] Karr, C., Jr., Brown, P. M., Estep, P. A., and Humphrey, G. L., *Fuel*, **37**, 227–35 (1958).

[13] Irvine, L., and Mitchell, T. J., *J. Appl. Chem. (London)*, **8**, 425–32 (1958).

[14] Grant, D. W., and Vaughan, G. A., *Vapour Phase Chromatography*, Desty, D. H., Ed., Academic Press Inc., New York, 1957, Chap. 34.

[15] Chang, T. -C. L., and Karr, C., Jr., (*a*) *Anal. Chim. Acta*, **21**, 474–90 (1959); (*b*) *Anal. Chim. Acta*, **24**, 343–56 (1961).

through 280°C and identified thirty-nine additional compounds of which twenty were alkylnaphthalenes.[15b] Quantitative determinations were made on nearly all of the components. It was shown that alkylnaphthalenes with equal numbers of carbon atoms in their alkyl groups displayed a linear relationship between log relative retention and normal boiling point, a situation similar to that for the alkylbenzene series, and that the values for the C_{11}, C_{12}, and C_{13} alkylnaphthalenes lie on three parallel lines.

Low Ionization Voltage Mass Spectroscopy

Low ionization voltage mass spectroscopy is a modern technique which takes advantage of the fact that each class of compound has a threshold potential for fragmentation. At this ionizing voltage, only the parent mass peak is produced, so that the proportions of compounds with differing molecular weights can be determined readily from the relative intensities of the parent mass peaks. No distinction can be made, however, between isomers or compounds of identical molecular weights. Rowlands, Powell, and Kahler[16] reported that application of this technique to the high-boiling tar acids of a low-temperature lignite tar tentatively indicated the presence of indenols and acenaphthols. The application of low-ionizing voltage mass spectrometry to oils derived from coal has been outlined by Sharkey, Shultz, Wood, and Friedel.[17]

Countercurrent Distribution

Tar acids boiling between 231 and 331°C from a low-temperature bituminous coal tar have been fractionated and analyzed by Karr, Estep, and Hirst[18] using distillation and countercurrent distribution. An instrument with an automatic robot mechanism was operated so as to obtain from 100 to 105 transfers or plates, the two immiscible liquid phases being cyclohexane and an aqueous phosphate buffer. The presence of eighty-four individual compounds was demonstrated, of which twenty-four were identified with respect to specific isomers, and thirty-six were identified with respect to structural type, and the remainder were tar acids of unknown structure. Quantities of all components were determined or estimated. Classes of tar acids included polyalkylphenols, indanols, indenols, alkenylphenols, cycloalkylphenols, phenylphenols, naphthols, tetrahydronaphthols, aliphatic carboxylic acids, and alkyl indanones (impurities).

Elution Chromatography and Ultraviolet Spectrophotometry

Leibnitz, Schmidt, and Gundermann[19] have recently shown that the main groups of constituents in brown-coal tars can be separated from each other and their amounts determined by means of elution chromatography on alumina and silica gel, using a series of eluants. This isolation of classes of compounds, such as paraffins, aromatics, sulfur compounds, and asphaltic substances using a series of eluants (petroleum ether, benzene, etc.), and following the course of the fractionation by physical properties measurements, is the same as that described a few years earlier by Karr, Weatherford, and Capell[20] for high-sulfur crude petroleum oils. The separation of the crude petroleum oils into pure paraffins, naphthenes, and mononuclear aromatics, together comprising about two-thirds of the total crude oil, was demonstrated by ring analysis and ultraviolet spectrophotometry, techniques which can also be used on the fractions obtained by elution chromatography of coal tars.

As an aid in the spectrophotometric analysis of samples obtained in the fractionation of coal tars and pitches by elution chromatography, Karr[21] presented tables listing 408 polycyclic aromatic hydrocarbons and 204 polynuclear heterocyclic aromatics, all with three or more aromatic rings, according to increasing values in millimicrons of the longest wavelength absorption band. Structural formulae, the values in millimicrons of additional absorption bands,

[16] Rowlands, D. C., Powell, W. H., and Kahler, E. J., *Am. Chem. Soc. 135th Meeting*, Abstracts of Papers, Boston, 1959, p. 4J.

[17] Sharkey, A. G., Jr., Shultz, J. L., Wood, G., and Friedel, R. A., ASTM Committee E-14, Annual Meeting on Mass Spectroscopy, Los Angeles, 1959.

[18] Karr, C., Jr., Estep, P. A., and Hirst, L. L., Jr., *Anal. Chem.*, **32**, 463–75 (1960).

[19] Leibnitz, E., Schmidt, W., and Gundermann, E., *Brennstoff-Chem.*, **39**, 87–90 (1958).

[20] Karr, C., Jr., Weatherford, W. D., Jr., and Capell, R. G., *Anal. Chem.*, **26**, 252–6 (1954).

[21] Karr, C., Jr., *Appl. Spectroscopy*, **13**, 15–25, 40–5 (1959); **14**, 146–53 (1960).

solvents, and literature references were also given. In addition to conventional ring analysis methods, which can be conducted on elution chromatographic fractions, a procedure has also been given by Karr[22] for determining the arrangement of rings in condensed polynuclear aromatic compounds obtained from low-temperature coal-tar pitches by suitable fractionations. This method used the ring index (total rings per carbon atom) and atomic hydrogen to carbon ratio.

Paper chromatography has been used for the characterization of phenolic mixtures, usually with water-saturated paper and an organic eluant. Leibnitz, Behrens, and Ringpfeil[23] used paper chromatography for the analysis of the polyphenols occurring in the aqueous liquor from low-temperature carbonization of brown coals. A total of twenty-seven different compounds were shown to be present by means of color reactions and rates of movement of components. Albesmeyer and Krampitz[24] have recorded the color reactions of a number of coal-tar bases for use in chromatography. Karr[25] has recorded twenty different color reactions for sulfur compounds, including specific tests for aliphatic thiols, aromatic thiols, aliphatic disulfides, aromatic disulfides, aliphatic sulfides, cyclic sulfides, aromatic sulfides, thiophenes, and condensed thiophenes. To analyze the dilute aqueous methanol solutions of tar bases obtained by an elution chromatographic technique with cation exchange resins, Chang and Karr[26] developed a procedure based on ultraviolet spectrophotometry. Liquid-liquid chromatography was used by Barker and Hollingworth[27] to characterize the phenols in the ammoniacal liquor from low-temperature carbonization. Celite columns, impregnated with aqueous solutions of hydrogen chloride or sodium silicate, were used with cyclohexane and cyclohexane-butanol mixtures as eluants. The amounts of the compounds were determined directly in the eluates by ultraviolet spectrophotometry.

[22] Karr, C., Jr., *Fuel*, **39**, 119–23 (1960).

[23] Leibnitz, E., Behrens, U., and Ringpfeil, M., *Wasserwirtsch-Wassertech.*, **6**, 299–304 (1956).

[24] Albesmeyer, W., and Krampitz, G., *Naturwissenschaften*, **44**, 466–7 (1957).

[25] Karr, C., Jr., *Anal. Chem.*, **26**, 528–36 (1954).

[26] Chang, T. -C. L., and Karr, C., Jr., *Anal. Chem.*, **29**, 1617–19 (1957).

[27] Barker, L., and Hollingworth, N. W., *J. Appl. Chem. (London)*, **9**, 16–25 (1959).

COMPOUNDS IDENTIFIED AND AMOUNTS

Although there are many publications which list compounds identified in various low-temperature tars, most of this information, upon close examination, turns out to be of dubious value. The majority of investigators have listed only a few compounds found in a particular tar, and because the composition of low-temperature tars depends on the nature of the coal used and the carbonization temperature, combining results from different investigators can be misleading and illogical. Many identifications, especially in the older work, have been based on inconclusive evidence such as melting points of picrates. Amounts of the compounds usually have not been given, and even when given they have been based sometimes on the quantity of compound that could be crystallized from a mixture, or some such technique that does not constitute an adequate analysis. It should, therefore, be kept in mind that there are varying degrees of uncertainty about the data on identified compounds presented in the following series of tables. For recent work using modern methods of characterization, this uncertainty is largely limited to the distinction between the less common isomers.

As would be expected, there is more quantitative information available on the low-boiling phenols than on any other group of constituents. These compounds are of obvious commercial interest, they are easily isolated from the tar, and they sometimes involve the rare phenomenon of an individual constituent comprising as much as 1% of the tar. Table 1 shows the amounts of individual low-boiling phenols in various low-temperature tars as determined by Karr and associates,[12,28] Brewer and Smith,[29] Fowkes and associates,[28] and Jäger and Kattwinkel.[30]

[28] Bur. Mines Staff, *U.S. Bur. Mines, Inform. Circ.* No. 7893 (1959), 32 pp. Karr, C., Jr., Comberiati, J. R., and Estep, P. A., *Fuel*, **39**, 475–8 (1960).

[29] Brewer, J., and Smith, C. D., *Am. Chem. Soc. 135th Meeting*, Abstracts of Papers, Boston, 1959, p. 3J.

[30] Jäger, A., and Kattwinkel, G., *Brennstoff-Chem.*, **31**, 65–79 (1950).

TABLE 1

AMOUNTS OF INDIVIDUAL LOW-BOILING PHENOLS FROM VARIOUS LOW-TEMPERATURE TARS

Type of Tar	West Virginia Bituminous	Wyoming Subbituminous Stansbury	Hanna	Big Horn	Wyodak	Texas (Sandow) Lignite	North Dakota Lignite	German "Bituminous"
Investigator	Karr et al.[12]	Karr et al.[28]				Brewer et al.[29]	Fowkes et al.[28]	Jäger et al.[30]
% by Weight on Dry Tar:								
Phenol	0.07	0.27	0.63	0.75	0.81	0.80	1.5	1.4*
o-Cresol	0.18	0.19	0.37	0.57	0.36	0.39	0.7	1.6*
m-Cresol	0.25	0.31	0.43	0.81	0.72	0.50	1.2	1.55*
p-Cresol	0.17	0.36	0.60	1.1	0.74	0.42	1.2	1.55*
o-Ethylphenol	0.05	0.01	0.01	0.04	0.11	0.06	0.1	present
m-Ethylphenol	0.09	0.12	0.13	0.26	0.34	0.09	0.1	present
p-Ethylphenol	0.09	0.16	0.27	0.53	0.43	0.15	0.4	0.15
2,3-Xylenol	0.05	0.04	0.03	0.02	0.02	0.09	0.2	0.2
2,4-Xylenol	0.33	0.47	0.57	0.55	0.14	0.18	0.4	0.9
2,5-Xylenol	0.13	0.07	0.14	0.18	0.43	0.09	0.3	1.1
2,6-Xylenol	0.05	0.01	0.07	0.07	0.04	0.03	0.05	present
3,4-Xylenol	0.13	0.06	0.16	0.25	0.20	0.06	0.4	0.5
3,5-Xylenol	0.21	0.13	0.21	0.01	0.30	0.09	0.2	1.3
2,3,5-Trimethylphenol	0.12				0.06			0.28
2,4,6-Trimethylphenol	0.05				0.01			0.4
3-Ethyl-5-methylphenol	0.08				0.04			0.30
4-Ethyl-2-methylphenol	0.16		0.11	0.26	0.10			0.6

* Including amount in aqueous liquor.

The low-boiling aromatics are probably second in commercial interest to the low-boiling phenols. There is less available information and, in addition, it is difficult to compare the amounts in different tars, because the lowest-boiling aromatics (benzene, toluene, etc.) are usually collected separately at the carbonization plant as a benzole or light oil. In addition, some tars are topped at various temperatures in order to remove water. Table 2 shows the amounts of individual low-boiling aromatics from various low-temperature tars as determined by Jäger and Kattwinkel,[9] the Coal Tar Research Association,[31] Grant and Vaughan,[14] Roberts, Towler, and Holland,[32] Boyer,[33] and Chang and Karr.[15]

The low-boiling *n*-paraffins and olefins are of commercial interest, although less so than the low-boiling aromatics. There is even less available information and, here again, it is

[31] Coal Tar Research Assoc., *The Coal Tar Data Book*, Gomersal, England, 1953, 160 pp.

[32] Roberts, A. L., Towler, J. H., and Holland, B. H., *Gas Council (Gt. Brit.) Research Commun.*, GC 31 (1956).

[33] Boyer, A. F., Centre d'Études et Recherches des Charbonnages de France, Verneuil, private communication, 1958.

TABLE 2

AMOUNTS OF INDIVIDUAL LOW-BOILING AROMATICS FROM VARIOUS LOW-TEMPERATURE TARS

Type of Tar	German "Bituminous"	English, North Derbyshire†	English No. 1	English No. 2	English Derbyshire‡	French (80% C)	West Virginia Bituminous
Investigator	Jäger et al.[9]	Coal Tar Research Assoc.[31]	Grant et al.[14]		Roberts et al.[32]	Boyer[33]	Chang and Karr[15]
% by Weight on Dry Tar:							
Benzene	0.30*	0.015			3.30§	16.0‖	
Toluene	0.60	0.204			7.30	29.0	trace¶
Ethylbenzene	0.10	0.10			0.01	4.5	trace
o-Xylene	0.20	0.121			0.94	4.5	trace
m-Xylene	0.5	0.232			2.15	11.0 (m- + p-)	trace
p-Xylene	0.4	0.090			1.15		trace
Isopropylbenzene	0.05						trace
n-Propylbenzene	0.05		0.26				0.002
1,2,3-Trimethylbenzene	present		0.30	0.27	0.29		0.045
1,2,4-Trimethylbenzene	0.20		0.27	0.36	0.19	3.5	0.039
1,3,5-Trimethylbenzene	0.10		0.10	0.14	0.11	1.0	0.017
1-Methyl-2-ethylbenzene	0.03				0.12 (1-methyl-2-, 3-, 4-ethylbenzene)	1.5 (1-methyl-2-, 3-, 4-ethylbenzene)	0.001
1-Methyl-3-ethylbenzene	0.30		0.15 (1-methyl-3- + 4-ethylbenzene)	0.22 (1-methyl-3- + 4-ethylbenzene)			0.001
1-Methyl-4-ethylbenzene	0.15						0.001
Styrene		trace			0.64	4.5	0.005
Methylstyrene			0.08	0.11			0.023**
Indan			0.39	0.27			0.057
Indene					0.15		0.033
Benzofuran			0.22	0.11	0.06		0.014

* Including amount in light oil for lowest boiling members.
† Carbonization temperature 600°C.
‡ Carbonization temperature 600 to 620°C ("Coalite" process).
§ On basis of total liquid condensate at −80°C from carbonization gas.
‖ On basis of commercial benzole from 500°C carbonization.
¶ On basis of neutral oil (16.92 weight % of tar) from tar topped at 175°C; therefore, only trace amounts present in tar.
** Specifically, β-methylstyrene (trans), 0.020%, and 4-methylstyrene, 0.003%.

TABLE 3

AMOUNTS OF INDIVIDUAL LOW-BOILING NORMAL PARAFFINS AND OLEFINS FROM VARIOUS LOW-TEMPERATURE TARS

Type of Tar	German "Bituminous"	English Derbyshire*	French
Investigator	Jäger et al.[34]	Roberts et al.[32]	Boyer[33]
% by Weight on Dry Tar:			
n-Pentane	0.2	10.04†	0.5‡
n-Hexane	0.3	16.31	2.0
n-Heptane	0.15	3.54	2.5
n-Octane	0.30	4.34	1.5
n-Nonane	0.15	9.62	1.0
n-Decane	0.31	2.68	0.5
n-Undecane	present		0.2
1-Pentene	0.15	4.32 (1- and 2-Pentene)	1.0 (1- and 2-Pentene)
2-Pentene			
1-Hexene	0.15	3.03	4.0 (1-, 2- and 3-Hexene)
2-Hexene		1.42 (2- and 3-Hexene)	
3-Hexene			
1-Heptene	0.15	4.71	5.0 (1-, 2- and 3-Heptene)
2-Heptene		0.32	
3-Heptene			
1-Octenc	0.15	1.48	2.0§
1-Nonene	0.15		2.0§
1-Decene	0.15		1.0§
1-Undecene	present		0.4§

* Carbonization temperature 600 to 620°C ("Coalite" process).
† On basis of total liquid condensate at −80°C from carbonization gas.
‡ On basis of commercial benzole from 500°C carbonization.
§ Total octenes, total nonenes, total decenes, and total undecenes.

difficult to compare the amounts in different tars, because the lowest boiling paraffins and olefins are usually collected separately as a light oil. Table 3 shows the amounts of individual low-boiling *n*-paraffins and olefins from various low-temperature tars as determined by Jäger and Kattwinkel,[34] Roberts, Towler, and Holland,[32] and Boyer.[33]

The low-boiling tar bases would presumably be of some commercial interest, but the amounts present have been reported for only one tar. Table 4 gives the amounts of individual low-boiling pyridines as determined by Karr and Chang,[6] and the presence in undetermined amounts by Jäger and Kattwinkel,[35] Matsumoto and Ihara,[36] and Yeh and Kalechits.[37] Table 5 gives the amounts of individual low-boiling anilines as determined by Karr and Chang.[6] Parant[38] identified three individual anilines but did not report any amounts.

In addition to the information on the amounts of low-boiling constituents, there are some data on the quantities of individual high-boiling compounds. Table 6 gives the amounts of individual high-boiling phenols in two different tars as determined by Karr, Estep, and Hirst,[18] and by Jäger and Kattwinkel.[30] Irvine and Mitchell,[13] and Parant[38] have identified individual high-boiling phenols but have not reported the amounts. Terres and associates[39] have identified 2-isopropylphenol, as well as the common phenols, by paper chromatographic analysis. Table 7 gives the amounts of

[34] Jäger, A., and Kattwinkel, G., *Erdöl u. Kohle*, **8**, 629–36 (1955).
[35] *Ibid.*, 706–11.
[36] Matsumoto, K., and Ihara, I., *Kooru Taaru* (*Tokyo*), **3**, 224–7 (1951).
[37] Yeh, T. -H., and Kalechits, I. V., *Jan Liao Hsüeh Pao* (*Peking*), **2**, 146–55 (1957).
[38] Parant, A., *Compt. rend. congr. ind. gaz.*, **65**, 409–21 (1948).
[39] Terres, E., Gebert, F., Hülsemann, H., Petereit, H., Toepsch, H., and Ruppert, W., *Brennstoff-Chem.*, **36**, 275–80 (1955).

TABLE 4

AMOUNTS OF INDIVIDUAL LOW-BOILING PYRIDINES IN LOW-TEMPERATURE TAR

Type of Tar	West Virginia Bituminous	German "Bituminous"	Japanese "Coal"	Chinese "Coal"
Investigator	Karr et al.[6]	Jäger et al.[35]	Matsumoto et al.[36]	Yeh et al.[37]
% by Weight:				
Pyridine		present†	present	present
2-Methylpyridine		present	present	present
3-Methylpyridine		present	present	present
4-Methylpyridine		present	present	present
2,3-Dimethylpyridine	0.01*	present	present	present
2,4-Dimethylpyridine	0.45		present	present
2,5-Dimethylpyridine			present	present
2,6-Dimethylpyridine		present	present	present
3,4-Dimethylpyridine		present	present	present
3,5-Dimethylpyridine	0.04	present		present
3-Ethylpyridine				present
4-Ethylpyridine		present		present
2,3,4-Trimethylpyridine			present	
2,3,5-Trimethylpyridine	0.58	present		present
2,3,6-Trimethylpyridine			present	present
2,4,5-Trimethylpyridine		present	present	present
2,4,6-Trimethylpyridine	0.87	present	present	present
4-Isopropylpyridine	0.09			
2-Ethyl-4-methylpyridine				present
2-Ethyl-6-methylpyridine				present
3-Ethyl-4-methylpyridine	0.88			
4-Ethyl-2-methylpyridine	0.11		present	present
5-Ethyl-2-methylpyridine	0.48			
2,6-Dimethyl-4-ethylpyridine	0.45			present

* On basis of total tar bases distilling up to about 360°C (0.31 weight % of tar topped at 175°C).

† Total tar bases distilling to 300°C comprised 2.5 weight % of the tar.

TABLE 5

AMOUNTS OF INDIVIDUAL LOW-BOILING ANILINES IN LOW-TEMPERATURE TAR

Type of Tar	West Virginia Bituminous	French (Bruay)
Investigator	Karr et al.[6]	Parant[38]
% by Weight:		
Aniline	0.26*	present†
N,N-Dimethylaniline	0.06	
2-Methylaniline		present
N-Methyl-2-methylaniline	0.16	
3-Methylaniline	0.48	
4-Methylaniline		present
2,5-Dimethylaniline	0.22	
2,6-Dimethylaniline	0.60	
3,5-Dimethylaniline	0.77	
2-Ethylaniline	0.22	

* On basis of total tar bases distilling up to about 360°C (0.31 weight % of tar).

† In mixed tar; 30% produced at 475°C, 70% produced at 675°C.

individual dihydric phenols from two different tars as determined by Fowkes et al.,[28] and Barker and Hollingworth.[27] The dihydric phenols produced from the aqueous liquor by Coalite and Chemical Products, Limited have been listed by Pound[40]; Leibnitz, Behrens, and Ringpfeil[23] have reported the identification of a number of dihydric phenols in an aqueous liquor from a commercial carbonization.

Table 8 gives the amounts of individual medium-boiling aromatics in two different tars as determined by Chang and Karr[15b] and Jäger and Kattwinkel.[9] Table 9 gives the amounts of some individual high-boiling aromatics and the names of constituents identified by Chang and Karr.[15b] Jäger and Kattwinkel,[35] Parant,[38] Vahrman,[41] and the

[40] Pound, G. S., *Coke and Gas*, **14**, 355–62, 401–7 (1952).

[41] Vahrman, M., *J. Appl. Chem.* (*London*), **8**, 485–92 (1958).

TABLE 6

AMOUNTS OF INDIVIDUAL HIGH-BOILING PHENOLS FROM VARIOUS LOW-TEMPERATURE TARS

Type of Tar	West Virginia Bituminous	German "Bituminous"	English (Low-Rank)	French (Bruay)
Investigator	Karr et al.[18]	Jäger et al.[30]	Irvine et al.[13]	Parant[38]
Weight %:				
2,3,4-Trimethylphenol		0.25†		
2,3,6-Trimethylphenol	0.01*			
2,4,5-Trimethylphenol		0.75		present‡
3,4,5-Trimethylphenol	0.13	0.13	present	present
2-Ethyl-4-methylphenol		present		
2-Ethyl-5-methylphenol		present		
3-Ethyl-2-methylphenol	0.01		present	
3-Ethyl-4-methylphenol	0.01		present	
5-Ethyl-2-methylphenol	0.01	present	present	
2-Isopropylphenol§				
3-Isopropylphenol			present	
4-Isopropylphenol	0.01	present	present	present
2-*n*-Propylphenol	0.01			
3-*n*-Propylphenol	0.01	0.09	present	
4-*n*-Propylphenol	0.13	0.04	present	
2,3,4,5-Tetramethylphenol		0.18		
2,3,5,6-Tetramethylphenol	0.04	0.1		
2,3-Dimethyl-6-ethylphenol (or 3,4-Dimethyl-6-ethylphenol)		0.02		
3,5-Dimethyl-4-ethylphenol		<0.01		
2,4-Diethylphenol		0.025		
2,6-Diethylphenol		present		
3,5-Diethylphenol		0.015	present	present
2-Isopropyl-4-methylphenol		0.035		
4-Isopropyl-3-methylphenol		<0.01		
5-Isopropyl-2-methylphenol	0.01		present	
2-Methyl-4-*n*-propylphenol				present
3-Methyl-5-*n*-propylphenol	0.29			
4-Methyl-2-*n*-propylphenol		0.08		
2-*n*-Butylphenol		0.09		
4-*n*-Butylphenol		<0.01	present	
2-Ethyl-6-*n*-propylphenol	0.26			
3-Ethyl-4-*n*-propylphenol	0.12			
4-Ethyl-3-*n*-propylphenol	0.24			
4-Isopropyl-3-*n*-propylphenol	0.14			
2,4-Di-*n*-propylphenol	0.13			
2,4- or 3,4-Dialkylphenol	0.02			
3-Methyl-2-propen-1-ylphenol	0.003			
Alkenylphenol I	0.005			
Alkenylphenol II	0.001			
Alkenylphenol III	0.03			
2-Cyclohexylphenol	0.20			
2-Alkylcycloalkylphenol I	0.02			
2-Alkylcycloalkylphenol II	0.01			
Cycloalkenylphenol I	0.005			
Cycloalkenylphenol II	0.04			
4-Indanol	0.53	0.5	present	present
5-Indanol	0.49	0.4		present
3-Methyl-4-indanol	0.09			
6-Methyl-4-indanol		0.25		present

TABLE 6 (*continued*)

Type of Tar	West Virginia Bituminous	German "Bituminous"	English (Low-Rank)	French (Bruay)
Investigator	Karr et al.[18]	Jäger et al.[30]	Irvine et al.[13]	Parant[38]
Weight %:				
1-Methyl-5-indanol	0.36			
6-Methyl-5-indanol	0.005	0.3		present
7-Methyl-5-indanol	0.10			
X-Methyl-X-indanol		0.16		
4-Indenol	0.03			
5-Indenol	0.27			
1-,2-, or 3-Methyl-4-indenol	0.14			
Methylindenol	0.08			
1-,2-, or 3-Polyalkyl-4-indenol	0.13			
3,4-Dinuclearphenol	0.06			
5,6,7,8-Tetrahydro-1-naphthol	0.02			
5,6,7,8-Tetrahydro-2-naphthol	0.02			
2-Phenylphenol	0.08			
3-Phenylphenol	0.03			
4-Phenylphenol	0.37	<0.01		
1-Naphthol	0.20	0.13		present
2-Naphthol‖	1.54	0.45		present
4-Methyl-1-naphthol	0.04			
1-Methyl-2-naphthol	0.02			
3-Methyl-2-naphthol	0.008			
4-Methyl-2-naphthol	0.11			
5-Methyl-2-naphthol¶				
6-Methyl-2-naphthol	0.01			
7-Methyl-2-naphthol	0.05			present
8-Methyl-2-naphthol	0.02			
2-Ethyl-1-naphthol	0.85			
2,5- or 2,7-Dimethyl-1-naphthol	0.008			
2,6-Dimethyl-1-naphthol	0.008			
5,7-Dimethyl-1-naphthol	0.05			
Methyl-2-, 3-, or 4-fluorenol	0.06			

* On the basis of the total distillable tar acids (3.62 weight % of the tar).
† On the basis of the crude tar.
‡ In mixed tar; 30% produced at 475°C, 70% produced at 675°C.
§ Identified in a German bituminous tar by Terres, et al. (ref. 39).
‖ Quantity in North Dakota lignite tar, 0.4% of the crude tar, Fowkes et al. (ref. 28).
¶ Identified in North Dakota lignite tar by Fowkes et al. (ref. 28).

Coal Tar Research Association.[31] The data from this last source are probably not typical of true low-temperature tars, because the carbonization temperature was 600°C. Table 10 presents the quantities of individual high-boiling pyridines, anilines, and quinolines determined by Karr and Chang.[6] The details of the complete research program on tar acids and tar bases by Karr and his co-workers have been assembled in a single publication, along with the ultraviolet and infrared spectra of their authentic specimens.[42]

There is some additional qualitative information available on individual constituents of low-temperature tars. Table 11 shows several ketones which have been identified by Jäger and Kattwinkel,[43] Karr, Estep, and Hirst,[18] and Nerdel and Kayser.[44] Table 12 lists miscellaneous saturates, olefins, and thiophenes

[42] Karr, C., Jr., Estep, P. A., Chang, T. -C. L., and Comberiati, J. R., *U.S. Bur. Mines, Bull.* No. 591 (1961), 228 pp.

[43] Jäger, A., and Kattwinkel, G., *Erdöl u. Kohle*, **9**, 447–50 (1956).

[44] Nerdel, F., and Kayser, R., *Brennstoff-Chem.*, **38**, 167–70 (1957).

TABLE 7

AMOUNTS OF INDIVIDUAL DIHYDRIC PHENOLS FROM LOW-TEMPERATURE TARS

Type of Tar	North Dakota Lignite	English (Derbyshire)	English Bituminous	German (Brown Coal)
Investigator	Fowkes et al.[28]	Barker et al.[27]	Pound[40]	Leibnitz et al.[23]
Weight %:				
Catechol	0.7*	1.03†	present‡	present§
3-Methylcatechol	0.2	0.58	present	present
4-Methylcatechol	0.3	0.65	present	present
Guaiacol (2-Methoxyphenol)	0.4			
4-Methylguaiacol	0.8			
4-Ethylguaiacol	0.4			
3,5-Dimethylcatechol		}		present
3,6-Dimethylcatechol	0.03	}0.34‖		present
4,5-Dimethylcatechol		}		present
3-Ethylcatechol	0.05			
4-*n*-Propylcatechol	present			
Resorcinol		0.51	present	present
2-Methylresorcinol		}0.36	present	
5-Methylresorcinol		}	present	present
4-Methylresorcinol		0.29	present	present
2,4-Dimethylresorcinol		}	present	
4,5-Dimethylresorcinol		}0.27¶		present
4,6-Dimethylresorcinol		}		present
Hydroquinone		0.06		
2-Methylhydroquinone		0.03	present	present
2,3-Dimethylhydroquinone				present
2,5-Dimethylhydroquinone				present
2,6-Dimethylhydroquinone				present
Pyrogallol				present

* Weight % on the basis of the crude tar.
† Concentration (g/l) in the ammoniacal liquor.
‡ In the aqueous liquor from the "Coalite" process (carbonization temperature 600 to 620°C).
§ In the aqueous liquor.
‖ Concentration of "higher catechols."
¶ Concentration of "higher resorcinols."

which have been identified by Jäger and Kattwinkel,[35] Roberts, Towler, and Holland,[32] and Landa, Romováček and Romováčková.[45] Table 13 presents the names of azulenes which have been identified by Treibs and Krumbholz,[46] and Landa, Romováček and Romováčková.[47]

There are only two instances in which a significantly large number of constituents has been identified in a single low-temperature tar and the amounts of many of these determined. Jäger and Kattwinkel have identified about 140 constituents of a German tar produced at 600 to 650°C in a Krupp-Lurgi plant, and Karr and co-workers have identified over 200 constituents of a West Virginia tar produced at 500°C in a fluidized-bed pilot plant. Two points should be clear from an examination of Tables 1 through 13: (1) with a few rare exceptions, all constituents are present in quantities much less than 1% of the total tar, and among these exceptions the amounts are barely above 1%; (2) there are more qualitative and quantitative dissimilarities among the various types of tars than there are similarities. In particular, it cannot at all be assumed that each of the approximately 350 compounds named here could be detected in

[45] Landa, S., Romováček, J., and Romováčková, H., *Chem. listy*, **49**, 313–16 (1955).

[46] Treibs, W., and Krumbholz, K., *Ann. Chem. Liebigs*, **595**, 208–11 (1955).

[47] Landa, S., Romováček, J., and Romováčková, H., *Chem. listy*, **50**, 1964–8 (1956).

TABLE 8

AMOUNTS OF INDIVIDUAL MEDIUM-BOILING AROMATICS IN LOW-TEMPERATURE TARS

Type of Tar	West Virginia Bituminous	German "Bituminous"
Investigator	Chang and Karr[15]	Jäger et al.[9]
% by Weight:		
sec.-Butylbenzene		<0.005†
tert.-Butylbenzene		present
1,2-Diethylbenzene	0.012*	
1,3-Diethylbenzene	0.015	
1,4-Diethylbenzene	0.001	present
1,2-Dimethyl-3-ethylbenzene	0.085	present
1,2-Dimethyl-4-ethylbenzene	0.260	present
1,3-Dimethyl-2-ethylbenzene		0.005
1,3-Dimethyl-5-ethylbenzene	0.156	0.07
1,4-Dimethyl-2-ethylbenzene	0.110	
1,2,3,4-Tetramethylbenzene	0.153	
1,2,3,5-Tetramethylbenzene	0.253	
1,2,4,5-Tetramethylbenzene	0.159	present
1-Methyl-2-*n*-propylbenzene	0.011	
1-Methyl-3-*n*-propylbenzene	0.046	
1-Methyl-2-isopropylbenzene	0.002	
1-Methyl-3-isopropylbenzene	0.007	
1-Methyl-4-isopropylbenzene	0.009	
1,2-Dimethyl-3-*n*-propylbenzene	0.042	
1,3-Dimethyl-2-*n*-propylbenzene	0.003	
1,4-Dimethyl-2-*n*-propylbenzene	0.084	
1,4-Dimethyl-2-isopropylbenzene	0.017	
1-Methyl-3,5-diethylbenzene	0.034	
1-Methyl-2,4-diethylbenzene	0.141	
3-Phenyl-1-butene	0.017	
1-Methylindan	0.161	
4-Methylindan	0.249	
5-Methylindan	0.251	
1,6-Dimethylindan	0.340	
4,7-Dimethylindan	0.186	
3-Methylindene	0.102	
2-Ethylindene	0.186	
3-Ethylindene	0.188	
2,3-Dimethylindene	0.254	
2-Methylbenzofuran, 3-Methylbenzofuran	0.024	‡
5-Methylbenzofuran, 6-Methylbenzofuran	0.021	‡
7-Methylbenzofuran	0.011	
1,2,3,4-Tetrahydronaphthalene	0.088	
2-Methyl-1,2,3,4-tetra hydronaphthalene	0.149	
6-Methyl-1,2,3,4-tetra hydronaphthalene	0.071	

* On basis of neutral oil (16.92 weight % of tar).

† On basis of crude tar.

‡ All four isomers identified.

TABLE 9

AMOUNTS OF INDIVIDUAL HIGH-BOILING AROMATICS IN LOW-TEMPERATURE TARS

Type of Tar	West Virginia Bituminous	German "Bituminous"	French (Bruay)	English Haunchwood Peas	English N. Derbyshire
Investigator	Chang and Karr[15b]	Jäger et al.[35]	Parant[38]	Vahrman[41]	Coal Tar Research Assoc.[31]
% by Weight:					
Naphthalene	0.914*	0.25†	present‡	<0.1†	0.66§
1-Methylnaphthalene	1.728	present	present		0.24
2-Methylnaphthalene	2.315	present	present		0.20
1-Ethylnaphthalene	0.283				
2-Ethylnaphthalene	0.456				
1,2-Dimethylnaphthalene	0.510				
1,3-Dimethylnaphthalene	0.940				
1,4-Dimethylnaphthalene	0.014				
1,5-Dimethylnaphthalene	0.528				
1,6-Dimethylnaphthalene	0.926				
1,7-Dimethylnaphthalene	0.686				
2,3-Dimethylnaphthalene	0.547		present		
2,6-Dimethylnaphthalene	0.596		present		
2,7-Dimethylnaphthalene	0.690				
1-Methyl-6-ethyl naphthalene and/or 1-Methyl-7-ethyl naphthalene	0.471				
2-Methyl-6-ethyl naphthalene and 2-Methyl-7-ethyl naphthalene	0.559				
1,2,6-Trimethylnaphthalene	present				
1,2,7-Trimethylnaphthalene	present				
1,3,6-Trimethylnaphthalene	0.105				
1,3,7-Trimethylnaphthalene	0.064				
Cyclohexylbenzene	0.039				
2a,3,4,5-Tetrahydro-acenaphthene	0.167				
2,3-Dimethylbenzofuran	0.025				
Dimethylbenzofuran II	0.255				
Dimethylbenzofuran III	0.092				
Diphenyl ether	5.528				
Biphenyl	2.696		present		
3-Methylbiphenyl	0.178				
4-Methylbiphenyl	0.104				
Acenaphthene	present		present		trace
Acenaphthylene	0.099				
Fluorene			present		
Dibenzofuran	present		present		
Carbazole					1.29
Anthracene			present	0.04	0.06
Phenanthrene					1.6

* In neutral oil (16.92 weight percent of the total tar).
† In total tar.
‡ In mixed tar; 30% produced at 475°C, 70% produced at 765°C.
§ In total tar produced at 600°C.

any single tar, let alone be present in the amount indicated here. This should emphasize the necessity for rapid, detailed analysis of each tar that has a different coal source, carbonization temperature, or residence time in the carbonizer. It should be realized that the sum of the constituents listed here accounts for only a small part of total tar, because these compounds are restricted essentially to the tar distillate, and many of them are low-boiling.

THEORETICAL ASPECTS OF COMPOSITION

There are two obvious theoretical aspects of the composition of low-temperature tars. The first is the possible relationship between the structures of coals or lignites and the compounds found in their respective tars, especially tars produced at temperatures no higher than about 500°C. The second is the influence on tar composition of thermodynamic chemical equilibria and the kinetics and mechanisms of chemical reactions such as isomerization, disproportionation, dealkylation, cyclization, and aromatization which can take place under various conditions of temperature, residence time, and catalysis.

Not much is known about the dependence of tar composition on coal structure. It has been demonstrated by Orning and Greifer[48] that if coal sample is heated at 510°C under conditions of molecular distillation, the distillate has an identical infrared spectrum and, therefore, the same molecular configuration as the coal. Under these conditions, once the molecules, or radicals, are released from the pores of the coal particles, they do not enter into any collisions with each other or the walls of the vessel until they strike the condensing surface.

[48] Orning, A. A., and Greifer, B., *Fuel*, **35**, 381–3 (1956).

TABLE 10

AMOUNTS OF INDIVIDUAL HIGH-BOILING PYRIDINES, ANILINES, AND QUINOLINES IN A LOW-TEMPERATURE TAR

Type of Tar	West Virginia Bituminous
Investigator	Karr and Chang[6]
% by Weight:*	
2,3,4,6-Tetramethylpyridine	0.99
2,3,5,6-Tetramethylpyridine	0.24
3,4-Diethylpyridine	0.63
2,3-Cyclopentenopyridine	0.16
2-Phenylpyridine	1.03
4-Phenylpyridine	1.21
2-Methyl-6-phenylpyridine	0.72
N-Benzyl-2-methylaniline	0.19
N-Benzyl-4-methylaniline	0.29
2-Naphthylamine	0.63
Isoquinoline†	
Quinoline	1.86
2-Methylquinoline	2.36
4-Methylquinoline	3.21
6-Methylquinoline	0.44
7-Methylquinoline	1.02
8-Methylquinoline	0.33
2,4-Dimethylquinoline	5.92
2,6-Dimethylquinoline	1.95
2,7-Dimethylquinoline	1.36
2,8-Dimethylquinoline	1.89
3,4-Dimethylquinoline	1.33
4,6-Dimethylquinoline	1.32
2,4,6-Trimethylquinoline	3.41
2,4,7-Trimethylquinoline	2.85
2,4,8-Trimethylquinoline	2.81
2,6,8-Trimethylquinoline	2.25
Acridine (2,3-Benzoquinoline)	0.13
2,4-Dimethylbenzo[g]quinoline	present
Phenanthridine (3,4-Benzoquinoline)	0.48
Benzo[f]quinoline (5,6-Benzoquinoline)	2.13
Benzo[h]quinoline (7,8-Benzoquinoline)	1.64
2,4-Dimethylbenzo[f]quinoline	present
2,4-Dimethylbenzo[h]quinoline	present

* On basis of total tar bases distilling up to about 360°C (0.31 weight % of tar).

† Isoquinoline and quinoline were identified by Jäger et al. (ref. 35) in a German tar.

Therefore, secondary reactions are limited to those occurring before release from the coal particle and after condensation, and these must not be extensive. Under the conditions of commercial or experimental low-temperature carbonization, there are many collisions of released molecules or radicals with each other, the surface of the char particles, and the walls of the vessel. The longer the residence time, the less "primary" the tar is at the same carbonization temperature. Modern fluidized bed low-temperature carbonization can involve relatively short average-residence times (down to some seconds, only), and when conducted at 500°C or lower, presumably produces a tar with significant "primary" characteristics. Such tars change appreciably upon brief storage, as has been demonstrated by Gomez, Goodman,

TABLE 11

KETONES IDENTIFIED IN LOW-TEMPERATURE TARS

Type of Tar	German "Bituminous"	West Virginia Bituminous	German Brown Coal
Investigator	Jäger et al.[43]	Karr et al.[18]	Nerdel et al.[44]
Ketone Identified:			
Acetone	present		
2-Butanone	present		
2-Pentanone	present		
2-Hexanone	present		present†
3-Methyl-2-butanone	present		
Cyclohexanone			present
Methylcyclohexenone			present
4-Methyl-1-indanone		present*	
6-Methyl-1-indanone		present	
X-Alkyl-X-indanone		present	
Acetophenone			present

* As impurities in the high-boiling phenols.
† As impurities in the phenols from the aqueous liquor.

and Parry[49] with the ordinary, simple assay methods.

It is known that coal is heterogeneous and frequently contains organic residues which are little altered from the original plant material. These are typically water insoluble, relatively heat stable compounds, such as natural waxes. From available evidence, there seems little reason to doubt that such substances are distilled with little alteration under conditions of true low-temperature carbonization. Vahrman[50] showed that the chloroform eluates from an alumina column of both coal extracts and their respective low-temperature tars were similar ester fractions with a pine-like odor. The yields of ester fractions from the coals were paralleled by smaller, but approximately proportionate amounts of ester fractions from the respective tars. Karr, Estep, and Hirst[18] have shown that there are long-chain aliphatic carboxylic acids throughout the entire high-boiling range from about 240 to 330°C, which includes compounds like myristic acid, which are widespread in plant esters.

Vahrman[51] has also shown that a lignite contained 1.1% hydrocarbons, whereas the low-temperature tar (600°C assay) contained 25.5% on the basis of the original lignite. The remainder of the hydrocarbons had been produced from dioxane-soluble acids, esters, and alcohols (13.3%), and 10.7% from the coal matrix (dioxane-insoluble). The "resinols," namely the benzene-soluble, petroleum ether-insoluble materials of a low-temperature bituminous coal tar, were examined by Vahrman.[52] Upon pyrolysis or hydrogenation of the resinols, it was found that the main product was a mixture of lower phenols. These resinols had molecular weights of 300 to 600, were thermoplastic resins, and from their infrared spectra and other considerations were thought to be linear polymers of monocyclic rings linked by ether oxygen. This is a structure which is related, in a broad sense, to the structure commonly assigned to wood lignin.

By means of X-ray scattering intensities, Ergun[53] has shown that both the high vacuum distillate and the residue from a bituminous coal vitrinite contained aromatic layers. The vitrinite contained 36% volatile matter. It was conceived as a polymer of planar aromatic nuclei with 1 to 3 rings which depolymerize upon heating, setting some of the structural entities free as a vapor. This finding appears to be analogous to that of Vahrman with respect to the "resinols."

From the extensive studies by many investigators of the oxidation reactions and products

[49] Gomez, M., Goodman, J. B., and Parry, V. F., *U.S. Bur. Mines, Rept. Invest.*, No. 5227 (1956), 12 pp.

[50] Vahrman, M., *Chem. & Ind. (London)*, **1958,** 462–3.

[51] Vahrman, M., *Fuel*, **31,** 171–5 (1952).

[52] Vahrman, M., *Fuel*, **30,** 288–90 (1951).

[53] Ergun, S., *Fuel*, **37,** 365–70 (1958).

TABLE 12

MISCELLANEOUS SATURATES, OLEFINS, AND THIOPHENES IDENTIFIED IN LOW-TEMPERATURE TARS

Type of Tar	German "Bituminous"	English Derbyshire*	Czechoslovakian Brown Coal
Investigator	Jäger et al.[35]	Roberts et al.[32]	Landa et al.[45]
Compound Identified:			
n-Dodecane	present		present
n-Tridecane	present		present
n-Tetradecane			present
n-Pentadecane			present
2-Methylbutane	present		
2-Methylpentane	present		
2-Methylhexane	present		
2-Methylheptane	present		
3-Methylheptane	present		
1,3,3-Trimethylheptane	present		
2-Methylnonane	present		
3-Methylnonane	present		
Cyclopentane	present		
Methylcyclopentane	present		
Cyclohexane	present	present†	
Methylcyclohexane	present	present	
1,2,4-Trimethylcyclohexane	present		
1,2,2-Trimethylcyclohexane	present		
n-Propylcyclohexane	present		
tert.-Butylcyclohexane	present		
sec.-Butylcyclohexane	present		
n-Butylcyclohexane	present		
1-Dodecene	present		
1-Tridecene	present		
2-Methyl-2-butene	present		
2-Methyl-2-pentene	present	present	
3-Methyl-2-pentene	present		
2-Methyl-2-hexene	present		
2-Methyl-2-heptene	present		
1,3-Pentadiene	present		
2-Methyl-1,3-butadiene		present	
Cyclopentene	present	present	
3-Methylcyclopentene-1	present		
Cyclohexene	present	present	
1-Methylcyclohexene-1		present	
Cyclopentadiene	present	present	
Methylcyclopentadiene	present		
Thiophene		present	
2-Methylthiophene		present	
3-Methylthiophene		present	
2,5-Dimethylthiophene	present		

* Carbonization temperature 600 to 620°C ("Coalite" process).
† In total liquid condensate at −80°C from carbonization gas.

of coal, recently reviewed by Yohe,[54] the firm conclusion is drawn that coal is comprised primarily of high-molecular-weight substances containing single and condensed aromatic rings, and aliphatic structures as connecting chains between aromatic nuclei, or as parts of hydroaromatic units. It is clear that depolymerization of these high-molecular-weight substances would produce mainly alkylated and hydroxylated aromatics rather than the parent

[54] Yohe, G. R., *Illinois State Geol. Survey, Rept. Invest.* No. 207 (1958), 51 pp.

TABLE 13

AZULENES IDENTIFIED IN LOW-TEMPERATURE TAR

Type of Tar	German Brown Coal	Czechoslovakian Brown Coal
Investigator	Treibs et al.[46]	Landa et al.[47]
Azulene Identified:		
Azulene	present	present
2-Methylazulene		present
4,8-Dimethylazulene	present	present
Trimethylazulenes		present
1,4-Dimethyl-7-isopropylazulene	present	

hydrocarbons. Therefore, the lower the carbonization temperature and the shorter the residence time, the smaller will be the quantities of unalkylated aromatics, like naphthalene. Because the naphthalene nuclei would be released as alkylated derivatives, the presence of naphthalene itself would depend mainly on dealkylation reactions. Whereas coke-oven tars contain around 10% naphthalene, true low-temperature tars contain only about 0.1% naphthalene. On the other hand, the combined alkylated naphthalenes make up a significant proportion of the total low-temperature tar. Vahrman[41] found that the largest class of compounds in the pentane-soluble fraction of a low-temperature tar was alkylated naphthalenes. The data indicated a wide range of alkylation, and these naphthalenes comprised 23% of the pentane-soluble portion, or 9% of the anhydrous tar. If the tar from this coal contained a typical 9 or 10% naphthalene (itself) upon *high*-temperature carbonization, one would be tempted to conclude that it had all been produced by dealkylation of alkylated naphthalenes, especially because the yield of high-temperature tar is several times less than the yield of low-temperature tar.

Trefny[55] has examined the light oil formed at increasing carbonization temperatures in a commercial coking plant. At 400°C, with the beginning of the decomposition within the plastic zone, there were only aliphatics and no aromatics. At 450°C, toluene and xylenes were first detected (about 0.5 volume %, each, in the light oil). Benzene was not detected until the carbonization temperature reached 550°C, or somewhat above the conditions for true low-temperature carbonization. Thus, alkylated benzenes were formed in the early stages of depolymerization, but no benzene itself.

There is no question about the fact that, although low-temperature tars are rich in alkylated aromatics, there are no aromatic rings among the tar acids, tar bases, or neutral oils which have long alkyl groups attached to them. Normal-butyl groups are extremely rare; Karr and co-workers were unable to detect a single normal-butylated compound among about 200 constituents with aromatic rings, although an excellent collection of authentic specimens and spectra was available to them. It is conceivable that cyclization reactions leading to structures of the indan and tetrahydronaphthalene type have accounted for the lack of *n*-butylated aromatics and the presence of these hydroaromatic structures in the tar. However, there is every reason to believe that the hydroaromatics could have been released from the coal, and, in addition, the identification of some two dozen aromatics with *n*-propyl groups is difficult to explain if cyclization is occurring so extensively. The tentative conclusion might be reached that the connecting aliphatic chains in the coal polymers are limited essentially to three carbon atoms. Once again, this is suggestive of the polymeric structure usually assigned to wood lignin. Another obvious fact about low-temperature tar is that one can expect to find the same fundamental structures throughout all classes of aromatics, whether they are in the tar acids, tar bases, or neutral oils. This suggests a common origin in the form of repetitive units in a polymeric structure.

Working from the hypothesis that all of the aromatics found in low-temperature tar are the result of chemical synthesis, and, therefore, can be examined from the standpoint of

[55] Trefny, F., *Tech. Mitt.*, **48,** 223–33 (1955).

chemical reaction kinetics and chemical thermodynamics, Karr[56] has compared the isomeric distributions of thirty-eight alkylated aromatics from a low-temperature bituminous tar with the thermodynamic chemical equilibrium distributions at 500°C, and with the "kinetic" distributions. The latter were obtained from the free valence index, or number, which is generally accepted as a rate-determining factor in free radical substitution on an aromatic nucleus. It was found that there was no overall agreement with either thermodynamic distributions or kinetic distributions. Of particular significance was the large proportion of tar values which fell completely outside the range presented by the combined thermodynamic and kinetic systems. In addition, it was pointed out that in at least four sets of ortho-meta-para isomers there were relatively large proportions of para isomers which cannot be explained from the standpoint of either the thermodynamic distribution, which predicts a predominance of meta isomer, or the kinetic distribution, which predicts a predominance of the ortho isomer.

Particularly striking is the predominance of the para isomer among both the cresols and the ethylphenols from all four subbituminous tars listed in Table 1. The tars were obtained under very carefully controlled carbonization temperatures in a fluidized-bed carbonizer with relatively short residence time. The temperatures were as follows: Stansbury, 510°C; Hanna, 510°C; Big Horn, 499°C; and Wyodak, 482°C. Here again, the structure usually assigned to wood lignin can be called upon for a plausible explanation. The products of depolymerization of lignin are generally explained by the assumption that the main building block in lignin is something comparable to 1-*n*-propyl-3-methoxy-4-hydroxybenzene.[57] The oxygen in the hydroxy group is derived from an ether linkage between the benzene ring and one of the chain carbons, which is fundamental to the polymeric nature of the lignin, whereas the methoxy group exists as such in the polymer. In going from lignin to lignite to bituminous coal, oxygen is lost, and it is evident that one process involved could be the loss of these methoxy groups, or their equivalent, in the lignin-derived polymers.

[56] Karr, C., Jr., *J. Phys. Chem.*, **64**, 462–4 (1960).

[57] Brauns, F. E., *The Chemistry of Lignin*, Academic Press, Inc., New York, 1951, 808 pp.

This is shown by the fact that catechol derivatives are the major phenolic compounds obtained upon depolymerization of lignin and are also present in somewhat smaller, but still relatively large amounts in low-temperature lignite and brown-coal tars. However, the low-temperature tars from bituminous coals contain only small or even-trace amounts of catechols. From these considerations, it might be expected that something comparable to 1-*n*-propyl-4-hydroxybenzene would be a significant building block in the lignin-derived polymeric structures of the subbituminous coals, and that para isomers would be the most prevalent among the monoalkyl phenols released upon depolymerization.

PROCESSING

There are currently about a dozen commercial low-temperature tar refineries where significant amounts of tar are processed by standard and modern techniques. The great majority of the refining capacity is in plants constructed by Germany for wartime fuel needs that are now operated by the governments of Czechoslovakia and East Germany. These low-temperature tar production and processing plants were never associated with plants producing or processing high-temperature tar and, hence, constitute low-temperature tar refineries in the true sense of the word. During 1943, these wartime plants produced and refined 658 million gallons of low-temperature tar, as compared to 504 million gallons of coke-oven tar.[58] About 91% was produced from brown coal, and 76% was hydrogenated to fuels. A small amount went into paraffin wax and electrode coke, but the remainder also went into fuels either directly or indirectly. Japan, likewise, depended on low-temperature tar plants of German design for wartime fuel needs.[59] Total refinery capacity in 1944 was 30 million gallons, which was small compared to Germany, but significant compared to the 69 million gallons of crude petroleum available during that year. More than half of this capacity was in North Korea, Manchuria, and Karafuto (Sakhalin Island, now Russian). The economic situation,

[58] Rhodes, E. O., *U.S. Bur. Mines, Inform. Circ.* No. 7490 (1949), 84 pp.

[59] Reid, W. T., *U.S. Bur. Mines, Inform. Circ.* No. 7430 (1948), 82 pp.

as well as the policies of the occupying powers in West Germany and Japan, resulted in the closing and even dismantling of most of the refineries that were still in operating condition.

The current total world capacity of low-temperature tar refineries is conservatively estimated at about 400 million gallons, or approximately 10 million barrels, per year. Some current low-temperature tar refineries, with information made available recently (1951 through 1958), are listed in Table 14. The hydrogenation works at Most (formerly Brüx), Czechoslovakia, have been described in considerable detail.[60] Modern operations at the Böhlen[61] and Rositz[62] plants in East Germany have been mentioned in the literature. Bahmüller[63] has recently given a description of the refining processes of the Offleben works of the Braunschweigische Kohlen-Bergwerken at Helmstedt, West Germany. LeRoux[64] has given an up-to-date description of the low-temperature tar-refining units of the Houillères du Basin de Lorraine at Marienau, France. Boyer[33] has made available some current data on the "Carbolux" plant, Pas de Calais, France. The refining operations of the Bolsover works of Coalite and Chemical Products, Limited, England, have been described in detail by Pound.[65] The Velsen Saarbergwerke, Geislautern/Saar, West Germany, has been described;[66] the tar produced by this plant is the tar which was characterized by Jäger and Kattwinkel.[9,30,34,35,43] There are a number of small plants in almost every industrialized country of the world which have been conducting simple refining operations on quantities of the order of magnitude of 100,000 or 200,000 gal/yr, most of these plants being relatively old. An example for the United States is the Dakota Briquetting Company in Dickinson, North Dakota, which has been distilling lignite tar into pitch and into a liquid for wood preservation for many years. The tar produced by the "Disco" process of the Consolidation Coal Company of Pittsburgh has not been processed by that company, but has been sold to the Koppers Company of Pittsburgh, which processes coke-oven tar. In a similar manner, the tar produced by the "Rexco" process of the National Carbonising Company, Limited, Mansfield, Notts., England, is sold to coke-oven tar processers.

Pursglove[67] has described, in some detail, the low-temperature tar-refining and utilization units planned for the 675,000 kw Kammer power-station complex at Cresap, West Virginia. Consolidation Coal Company was to supply not only bituminous coal, as at present, but also low-temperature char produced at about 500°C by a fluidized-bed process. This phase of the complex has been temporarily deferred,[68] but when placed into operation, 65 million gallons of tar would be processed per year, including 40 million pounds of cresylic acids, placing it on a par with the existing refineries in eastern Europe. A plant of similar size has been proposed as part of the 360,000 kw power plant for the Aluminum Company of America, at Rockdale, Texas. Instead of using dried lignite, as is presently done, this power plant could use low-temperature char, as demonstrated on an experimental basis with the char produced at about 500°C in a fluidized-bed prototype carbonizer with an annual capacity of tar production of about 6 million gallons. If all of the lignite were carbonized, about 55 million gallons of tar would be made available per year.[69] A novel first-step refining unit for the tar was proposed, consisting of solvent extraction, as will be described later. The Institute of Heat and Power, Ukrainian Soviet Socialist Republic, published a description in 1957, that was prepared in 1954, of the power plant proposed for the city of Aleksandriya. This was to include a low-temperature carbonization plant for brown coal, and a

[60] Anon., *Most Hydrogenation Works in Czechoslovakia*, Special Rept., September, 1957; manuscript on open file, U.S. Bur. Mines, Washington, D.C.

[61] Birthler, R., and Szkibik, C., *Freiberger Forschungsh.*, **A36**, 42–57 (1955).

[62] Chilton, C. H., *Chem. Eng.*, **65**, No. 2, 53, 56 (1958).

[63] Bahmüller, *Braunkohle*, **9**, 486–93 (1957).

[64] LeRoux, P., *Goudron pour routes*, **1958**, No. 7, June, 12–25.

[65] Pound, G. S., *J. Inst. Fuel*, **24**, 61–8 (1951).

[66] Anon., *Research Report of the Economy and Trade Ministry*, *Nordrhein-Westfalen*, No. 58, Institute for Coal Technology, Dortmund (1951).

[67] Pursglove, J., Jr., *Coal Age*, **62**, 70–3 (1957).

[68] Anon., *Chem. Eng. News*, **36**, No. 21, 27 (1958).

[69] Batchelder, H. R., Filbert, R. B., Jr., and Mink, W. H., *Am. Chem. Soc. 135th Meeting*, Abstracts of Papers, Boston, 1959, p. 2J.

TABLE 14

SOME CURRENT LOW-TEMPERATURE TAR REFINERIES

Name and Location	Approximate Annual Capacity, millions gallons	Type of Tar Refined	Refining Processes	Products	Ref. No.
Most Hydrogenation Works, Czechoslovakia	90	Lignitic brown coal	Hydrogenation, distillation, and extraction	Gasoline, motor oils, phenols	60
Böhlen Plant, East Germany	70	Brown coal, Rositz raw fuels	Hydrogenation, etc.	Diesel oil, etc.	61
Rositz Plant, East Germany	60	Lignite	Cracking, distillation	Gasoline, diesel fuel	62
Offleben Works, Helmstedt, West Germany	25	Soft brown coal	Distillation, extraction	Light oils, middle oils, phenols	63
Houillères du Bassin de Lorraine, Marienau, France	9	Lorraine coal, Velsen residue	Distillation, extraction, esterification and hydrogenation of phenols	Phenols and derivatives, specialty oils	64
"Carbolux" Plant, Pas de Calais, France	9	Bruay coal	Distillation, extraction, acid treating	Phenols, bases, solvents, motor fuels	33
"Coalite" Plant, Bolsover, England	8	Bituminous coal	Distillation, extraction, chlorination, esterification, etc.	Gasoline, diesel oil, mono- and dihydric phenols and derivatives	65
Velsen Works, Saar, West Germany	2	Bituminous "fat" coal	Distillation, extraction	Phenols, naphtha, residue	66

2 million gallon per year tar refinery for producing gasoline, fuel oil, phenols, and paraffins.[70]

Selective Solvent Extraction

During the early part of World War II, the Germans completed a large low-temperature tar refinery that was based primarily on the use of selective solvent extraction. This was the 88 million gallon per year plant of A. G. Sächsischewerke at Espenhain (now in East Germany), which utilized a series of continuous extractions of brown-coal tar with sulfur dioxide, naphtha, and ethylene dichloride to yield gasoline, heating oil, diesel oil, and soft and hard waxes.[71] In certain respects, this was a unique application of selective solvents for a plant of this size and should be recognized as a full-scale demonstration of the fact that low-temperature tars are more satisfactorily refined by methods borrowed from the petroleum industry than from the coke-oven tar industry.

Of fundamental importance to the low-temperature tar refiner has been the recovery, in a pure state, of the phenols which constitute such a significant proportion of many tar distillates and also can comprise such an important part of the earnings. One of the oldest techniques, which is now performed with modern, continuous-flow equipment, is extraction with aqueous caustic soda. The extraction unit placed into service in 1954, at the Houillères du Bassin de Lorraine, has a capacity of 24 million pounds of phenols per year, with about 7 million pounds coming from low-temperature tar and the rest from coke-oven tar. A flowsheet and photograph of this unit have been published.[64] The Bolsover refinery has an extraction unit with a capacity of 8 million pounds of phenols per year. A flowsheet of this unit has also been published.[72] The Offleben plant uses extraction with water at elevated temperatures and under pressure, that is, 165°C and 240 psig, to recover phenols. A flowsheet and photograph of the unit that was expected to extract 9 million pounds of phenols in 1957 has been published.[63]

There are no commercial size low-temperature tar-refining units as such at the present time in the United States. However, the Pitt-Consol Chemical Company operates a continuous extraction plant at Newark, New Jersey, for the refining of petroleum cresylic acids that is said to be the prototype for the proposed low-temperature tar refinery at Cresap, West Virginia. The Newark plant involves refining of impure cresylic acid by continuous countercurrent extraction with a two-solvent system of light naphtha and aqueous methanol in an agitated, Scheibel-type extractor. This plant has been operating at its full-production capacity of 20 million pounds per year of refined phenols since February, 1957. A process flowsheet and photographs of the plant have been published.[73] Some of the laboratory experiments that went into the development of this extraction technique have been described.[74] The U.S. Bureau of Mines' Low-Temperature Tar Laboratory has conducted experiments on the extraction of acids from a distillate of a low-temperature subbituminous coal tar with aqueous monoethanolamine.[75]

Solvent extraction has been proposed for the first refining step of the low-temperature lignite tar plant planned for Rockdale, Texas.[69] A solvent-extraction pilot plant, with a capacity of about 350,000 gal of tar per year, has been constructed at Rockdale. This comprises a continuous countercurrent, two-solvent system of hexane and aqueous methanol in a rotating-disc contactor-type extractor. Other pilot-plant refining units include a vacuum still and a topping still. Together, these units produce five main products: (1) hexane solubles (97% neutral oils) distilling to 200°C; (2) hexane solubles (91% neutral oils) distilling between 200°C and pitch; (3) hexane-soluble pitch; (4) methanol solubles (99% caustic soluble) distilling to 235°C; and (5) high-boiling methanol solubles (57% distillable to decomposition). The fourth product is very nearly pure phenols, with essentially the

[70] Shchegolev, G. M., *Khim. Pererabotka Topliva, Akad. Nauk S.S.S.R., Inst. Goryuch. Iskopaemykh, Trudy 2-go Vsesoyuz., Soveshchaniya, Moscow*, **1954**, 45–55 (Pub. 1957).

[71] Terres, E., *Natl. Petrol. News*, **38**, No. 6, R84–8 (1946).

[72] Bristow, W. A., *J. Inst. Fuel*, **20**, 109–30 (1947).

[73] Forbath, T. P., ed., *Chem. Eng.*, **64**, No. 7, 228–31 (1957).

[74] Neuworth, M. B., Hofmann, V., and Kelly, T. E., *Ind. Eng. Chem.*, **43**, 1689–93 (1951).

[75] Chang, Y. -C., and Johannsen, R. C., *J. Inst. Fuel*, **32**, 174–8 (1959).

composition shown in Table 1 in the Texas Lignite column. It is evident that the proposed refineries at Cresap and Rockdale will have counter-current extraction units with essentially identical two-solvent systems. The first of these, however, will be used with tar distillate and the other with crude tar. It should be pointed out that this crude lignite tar is a mobile liquid, because the heavy tar condensate is recycled to the carbonizer.

Columnar Adsorptive Fractionation

According to Depp and Neuworth,[76] the low-temperature bituminous-tar phenols recovered by the aqueous methanol-hexane two-solvent system contain an amount of tar bases in excess of the highest quality petroleum cresylic acids. Also, these tar acids contain small quantities of aliphatic carboxylic acids, including butyric acid, which contaminate the phenol and *o*-cresol fractions, imparting a foul odor. The low-temperature tar phenols, as received in aqueous methanol solution, can be refined with ion-exchange resins in column operation. Tar bases are removed with a cation resin having strongly acidic sulfonic acid groups, and butyric acid is removed with an anion resin having strongly basic quaternary ammonium hydroxide groups. Brooks[77] has shown that phenols and bases can be separated from nonaqueous solutions of whole coal tar or pitch by passing the solution through columns of strongly basic and acidic ion-exchange resins, respectively.

It has been demonstrated that coal-tar phenols can be adsorbed on polyamide products and the various components recovered by elution with a solvent in a columnar process.[78] Polyesters, a vinyl polymer, keratin, or polyglucosamine comprise the various adsorbents which can be used. Tar bases can be removed from phenols by adsorption on a column of silica-alumina catalyst.[79] The bed can be regenerated, and the tar bases recovered, by elution with a methanolic solution of sulfuric acid. When refining sludge-forming feeds, the catalyst can be regenerated by a burning in air at 480 to 650°C.

Azeotropic Distillation

Azeotropic distillation has been used in the Polish coke-oven tar industry for the recovery of indole, biphenyl, benzofuran, naphthalene, phenol, and tar bases from appropriate refinery streams.[80] A large number of high-purity compounds have been made commercially available by this technique. The Gesellschaft für Teerverwertung m.b.H. in Germany has offered a wide variety of pure coal-tar chemicals from coke-oven tar on a commercial scale in recent years. Azeotropic distillation has been used by this company for the recovery, for example, of indole of 95% purity from a close-boiling coal-tar fraction, in which the azeotropes of biphenyl and methylnaphthalene with glycol distill first.[81]

Feldman and Orchin[82] have pointed out that the isomeric monomethylnaphthalenes occur together in all types of coal-tar distillates and have shown that 1-methylnaphthalene can be separated from the 2-isomer by azeotropic distillation at 260 mm with 5-ethyl-2-nonanol, with which only the 2-derivative forms an azeotrope. The alkylnaphthalenes are one of the most prevalent compound classes in low-temperature tars. Teter[83] has shown that naphthalene and its lower alkylated homologs can be separated from admixed nonaromatic hydrocarbons by azeotropic distillation with benzonitrile. The nonaromatics are removed as azeotropes with the benzonitrile, and the naphthalenes can be obtained in substantially pure form by freezing the resulting mixture.

Adduct Formation

The Midland Tar Distillers, Limited, has disclosed a method for obtaining quantitative yields of pure 2-methylquinoline from coke-oven tar bases boiling in the range 242 to 250°C by complex formation with urea or

[76] Depp, E. A., and Neuworth, M. B., *Ind. Eng. Chem.* **49,** 182–5 (1957).

[77] Brooks, J. D., *Chem. & Ind.* (*London*), **1956,** 316–17.

[78] Liquori, A. M., Mele, A., and Carelli, V., Ital. Pat. 534,674 (1955).

[79] Neuworth, M. B., U.S. Pat. 2,766,297 (1956).

[80] Gondzik, J., and Stateczny, *Przemysl Chem.*, **9,** 132–7 (1953).

[81] Moehrle, E., and Franck, H., Ger. Pat. 812,079 (1951).

[82] Feldman, J., and Orchin, M., *Ind. Eng. Chem.*, **44,** 2909–14 (1952).

[83] Teter, J. W., U.S. Pat. 2,551,912 (1951).

thiourea.[84] A 76% yield of 3-methylpyridine with 99% purity can be obtained from a tar-base fraction in the boiling range 140 to 145°C by forming an insoluble complex with cuprous or cupric chloride.[85] The Yorkshire Tar Distillers, Limited, has revealed a procedure for obtaining 95 to 98% pure 2-ethylphenol from coke-oven tar fractions boiling in the range 203 to 206°C by formation of a solid-addition complex with 2-methylquinoline.[86] Acenaphthene and its alkyl derivatives can be separated from aliphatic, naphthenic, monocyclic aromatic, heterocyclic, and naphthalene compounds by formation of a solid yellow complex with dichlorophthalic anhydride.[87] A 96% yield of carbazole with 98% purity can be obtained from mixtures with polycyclic aromatic hydrocarbons, like anthracene, by forming a molecular addition compound with N-methyl-2-pyrrolidinone.[88]

Thermal and Catalytic Cracking

Construction is being completed at the Rositz refinery of a new cracking unit which will process 23 million gallons per year of low-temperature lignite tar, along with 20 million gallons of press oils, to yield 25 million gallons of gasoline and diesel fuel, as well as gas and coke.[62] Gasoline and diesel fuel will be piped nineteen miles to the Böhlen refinery for further processing. In war-time operation, during 1942–1943, 20 million gallons of low-temperature tar were cracked at the Rositz refinery.[58] At 430 to 450°C and 588 to 822 psig the yields of products were as follows: heating oil, 47%; gasoline, 4%; diesel oil, 20%; electrode coke, 7%; fuel coke, 3%; pitch, 6%; and gas, 13%. Delayed coking of low-temperature tars to produce gasoline and coke has received considerable commercial interest in the United States.

Dell[89] has described the delayed coking of the lignite tar produced at Rockdale, Texas. The tar was topped at 235°C and the residue treated by delayed coking at 400–450°C to produce 22 to 24% green coke, 6% low-boiling phenols, liquid condensate, and gas. The green coke was calcined at 1200°C to produce a coke for use in making carbon electrodes. The high-boiling residue from the liquid condensate was recycled to the delayed coking apparatus. Lummus Company has successfully run low-temperature tars through a pilot delayed-coking unit to make gasoline and coke, demonstrating the feasibility of commercial operation. The plans for the Cresap refinery included coking of the heavy bottoms of the low-temperature bituminous tar to give green coke.[67] This would be mixed with petroleum coke and calcined to make electrode carbons.

The constituent of greatest commercial interest in the gas from thermal cracking of low-temperature tars is ethylene, and considerable attention has been given to finding the optimum conditions for producing this valuable chemical intermediate. Montgomery, Decker, and Mackey[90] found that by carbonizing Texas Sandow lignite at temperatures not exceeding 550°C, immediately followed by thermal cracking of the volatile products (tar) at 800°C and an "average" retention time of about 1.5 sec, the estimated yields per ton of dried lignite were 40 lb of ethylene, 70 lb of methane, 6 lb of benzene, 5 lb of toluene, and 3 lb of naphthalene. It is of interest to compare these reported results with the yields from the carbonization of a bituminous coal at 800°C in which the 550°C carbonization and the 800°C thermal-cracking steps might be considered to be combined in one operation. From 1 ton of coal the yields were 120 lb of methane, 6 lb of benzene, and 4 lb of toluene, with no data on ethylene.[1] Studies have been made recently in Japan on the thermal cracking of low-temperature tars to produce ethylene. Imai, Suzuki, and Nomura[91] cracked neutral oil boiling from 120 to 360°C from a low-temperature bituminous coal tar, and obtained a 22% yield of ethylene at a cracking temperature of 700 to 900°C. The same tar fraction before extraction of acids and bases yielded only 17% ethylene. Kishida and Chuto[92]

[84] Arnall, P., Brit. Pat. 736,589 (1955).

[85] Rusek, L., Austrian Pat. 172,652 (1952).

[86] Milner, D. W., and Flathers, R., Brit. Pat. 736,604 (1955).

[87] Arnold, G. B., and Hess, H. V., U.S. Pat. 2,652,438 (1953).

[88] Boente, L., Ger. Pat. 937,230 (1955).

[89] Dell, M. B., *Ind. Eng. Chem.*, **51**, 1297–8 (1959).

[90] Montgomery, R. S., Decker, D. L., and Mackey, J. D., *Ind. Eng. Chem.*, **51**, 1293–6 (1959).

[91] Imai, K., Suzuki, M., and Nomura, S., *Nenryô Kyôkaishi*, **36**, 526–34 (1957).

[92] Kishida, M., and Chuto, M., *Kooru Taaru*, (*Coal Tar, Tokyo*), **4**, 398–402 (1952).

thermally cracked heavy-oil fractions of a low-temperature coal tar in a steel vessel at 700°C at less than 10 mm Hg pressure and obtained 10.3% ethylene and 8.6% by weight propylene. Thermal cracking of low-temperature tar fractions has been studied in a Japanese continuous pilot plant.[93] These experiments were directed toward the production of solvent naphtha fractions from the paraffin-containing oil and caustic-washed middle oil and the production of phenol and cresols from the high-boiling tar acids.

An extensive examination has been made recently in the Ukrainian Soviet Socialist Republic of the catalytic cracking of fractions from low-temperature Aleksandrovsk lignite tar. The phenols were removed from the fraction boiling from 200 to 350°C, and the resulting oil was cracked on an aluminum silicate moving-bed catalyst at 450°C. A stable, liquid, rich in aromatics but very low in sulfur, was obtained, which was considered to be a good motor fuel.[94] It was found that by conducting this same catalytic cracking operation in a stream of hydrocarbon gas (mainly propane and butanes), the production of coke and gas was reduced, and the production of gasoline boiling below 200°C was increased.[95]

Dealkylation

Because low-temperature tars have a high degree of alkylation, methods of removing some or all of the alkyl groups, or decreasing their chain length, have been of considerable interest. Most of these methods consist of specific applications of thermal or catalytic cracking. Thus, Jones and Neuworth[96] have studied the thermal cracking of cresols and xylenols and compared the quantitative distributions of phenols in "Disco" low-temperature bituminous coal tar and a typical high-temperature tar. These investigators proposed a free radical mechanism for the reaction. In further work, they made a systematic study of the effect of thermal cracking variables on the yield and composition of phenols obtained by cracking "Disco" tar acids boiling from 230 to 300°C.[97] The highest yield of phenols boiling below 230°C was about 17%, based on the feed. The dealkylation of these same high-boiling tar acids was extended to include catalytic cracking in the vapor phase at 370 to 450°C, using silica-alumina catalyst with at least 5 weight % carbon deposited on it and an inert diluent.[98]

The Yorkshire Tar Distillers, Limited, has dealkylated alkylpyridines to lower alkylpyridines and pyridine by passing them with hydrogen at 400 to 500°C and atmospheric pressure over a hydrogenation catalyst consisting of nickel, cobalt, or copper molybdate or thiomolybdate.[99] The Coal Tar Research Association has studied the thermal dealkylation of coal tar naphthas at 1500 psig in the presence of hydrogen.[100] It was found that polymethylbenzenes were almost completely cracked to benzene and toluene at 650°C. This work was extended to include the thermal dealkylation of polynuclear aromatics. For example, a coal-tar naphthalene oil with 20 weight % naphthalene, 11 weight % 1-methylnaphthalene, and 17 weight % 2-methylnaphthalene was cracked at 635 to 655°C and 1500 psig in a large excess of hydrogen to give 77 weight % liquids consisting of 53 to 59 weight % naphthalene, 2 to 4 weight % 1-methylnaphthalene, and 3 to 5 weight % 2-methylnaphthalene.[101] Low-temperature tars have been effectively dealkylated by injection into a fluidized coke bed at 810°C, the carrier gas containing at least 30% hydrogen.[102] High yields of benzene and naphthalene were obtained. This process is, in a sense, a method for converting low-temperature tars into coke-oven tars.

Not all interest is directed toward the removal of alkyl groups; the reverse process of alkylation is practiced on a commercial scale in the low-temperature tar industry. The Houillères

[93] Suzuki, A., Kitahara, A., and Koga, M., *Kôgyô Kagaku Zasshi* **54,** 789–90 (1951).

[94] Dal, V. I., and Ruban, I. N., *Ukrain. Khim. Zhur.*, **23,** 411–14 (1957).

[95] Dal, V. I., and Ruban, I. N., *Ukrain. Khim. Zhur.*, **24,** 107–9 (1958).

[96] Jones, B. W., and Neuworth, M. B., *Ind. Eng. Chem.*, **44,** 2872–6 (1952).

[97] Jones, B. W., and Neuworth, M. B., *Ind. Eng. Chem.*, **45,** 2704–5 (1953).

[98] Neuworth, M. B., Brit. Pat. 743,707 (1956).

[99] Milner, D. W., and Clapham, P., Brit. Pat. 739,088 (1955).

[100] Betts, W. D., Popper, F., and Silsby, R. I., *J. Appl. Chem. (London)*, **7,** 497–503 (1957).

[101] Betts, W. D., and Popper, F., *J. Appl. Chem. (London)*, **8,** 509–13 (1958).

[102] Sawyer, E. W., Brit. Pat. 759,159 (1956).

du Bassin de Lorraine has conducted isobutylation of various phenols to obtain derivatives such as *p*-tert-butylphenol, used in the preparation of oil-soluble resins, di-tert-butyl-*p*-cresol, used as a nontoxic anti-oxidant, and di-tert-butyl-*m*-cresol, used as a stabilizing agent in fuels and in rubber.[64]

Isomerization

In dealkylation by catalytic cracking, it is not only observed that disproportionation takes place, which might be considered concurrent dealkylation and alkylation, but also, isomerization occurs. In many instances, operating conditions can be chosen that result in high yields of isomers. This is a particularly valuable process when it is desired to have larger amounts of a specific isomer than occur naturally. The natural mixture is isomerized, the desired isomer is recovered, and the other isomers are recycled to the isomerization unit. In this manner, the entire mixture can be converted into a single isomer.

The Koppers Company has shown that an approximately equal mixture of 1- and 2-ethylnaphthalene can be converted into a product containing 17% l-ethylnaphthalene and 83% 2-ethylnaphthalene by isomerization at 425°C over an aluminium fluoride or aluminum fluosilicate catalyst.[103] The cresols can be isomerized in a catalytic vapor-phase process using six moles of hydrogen per mole of cresol at atmospheric pressure and 470°C, the catalyst being alumina or bauxite which has been treated with aqueous ammonium fluoride or hydrogen fluoride.[104] At a contact time of only 0.27 sec, a 52% conversion of *m*-cresol (97% purity) was obtained, the composition of the product being: *o*-cresol 27%; *m*-cresol, 59%; and *p*-cresol, 14%. The British Coal Utilisation Research Association has studied the disproportionation and isomerization of cresols over cracking catalysts.[105] With silica-alumina and silica-zirconia-beryllia catalysts at 320–450°C and the addition of a small amount of steam to reduce phenol losses as catalyst deposit and as dehydroxylated material, a maximum conversion of cresols by disproportionation to phenols and xylenes of about 20% was observed. Indications were found that the reactions took place on the acid groups in the catalyst. These studies were extended to include xylenols, higher alkylated individual phenols, and a sample of commercial high-boiling tar acids.[106] With silica-alumina catalyst at 380 to 400°C, the isomers that predominated in the products from these feeds were *m*-cresol, 2,4- and 2,5-xylenol, and 2,3,5-trimethylphenol. When the process was conducted in the presence of benzene, there was a transfer of alkyl groups from the phenols to the benzene. Pigman, Del Bel, and Neuworth[107] studied the vapor-phase isomerization-disproportionation of cresols and xylenols over silica-alumina catalyst. The maximum conversion temperature was 344°C, the catalyst showing a very high selectivity for isomerization-disproportionation of methyl groups. Evidence was found for the step-wise shift of the methyl groups.

Hydrogenation

Without question, hydrogenation has been and still is one of the most important processes for refining low-temperature tars. During 1943, German war-time low-temperature tar refineries hydrogenated 500 million gallons of low-temperature tars, mostly from brown or lignitic coals. Holroyd[108] has assembled a complete description of these war-time hydrogenation processes, including flow diagrams for the conversion of brown-coal tar to motor gasoline and aviation-base gasoline at the Leuna Works of the I. G. Farbenindustrie A. G. at Merseburg, now in East Germany. In the pressure-hydrogenation process, 1.25 tons of brown-coal tar yielded approximately 1 ton of gasoline with an octane number of 60 to 70.[109] Gasoline obtained from low-temperature bituminous tar had an octane number of 80 to 90, due to the higher aromatic-naphthenic structure of this tar as compared to that from brown-coal tar.[110] Low-temperature (450°C) lignite tar from

[103] Nickels, J. E., U.S. Pat. 2,615,056 (1952).

[104] Good, G. M., and Holzman, G., U.S. Pat. 2,678,337 (1954).

[105] Given, P. H., *J. Appl. Chem. (London)*, **7**, 172–82 (1957).

[106] *Ibid.*, 182–93.

[107] Pigman, I., Del Bel, E., and Neuworth, M. B., *J. Am. Chem. Soc.*, **76**, 6169–71 (1954).

[108] Holroyd, R., *U.S. Bur. Mines, Inform. Circ.* No. 7370 (1946), 135 pp.

[109] Thau, A., *Brennstoff u. Wärmewirtsch.*, **24**, 7–12, 26–9, 40 (1942).

[110] Heinze, R., *Gas u. Wasserfach*, **85**, 413–25 (1942).

Bouches-du-Rhône was processed at 470°C and 6000 psig with a mixture of three parts hydrogen and one part nitrogen over molybdenum disulfide catalyst.[111] An 86 weight % yield of gasoline, boiling 55–186°C, with an octane number of 75, was obtained.

The Most refinery in Czechoslavakia has hydrogenated lignitic tar at 4500 psig and 500°C, using tungsten disulfide catalyst, to produce about 60 million gallons per year of gasoline and other liquid fuels.[60] A Czechoslovakian patent describes the catalytic hydrogenation of brown or lignitic coal tars containing monovalent and multivalent phenols.[112] The hydrogenation is carried out at 250 to 450°C and at 3900 to 4800 psig with pure tungsten disulfide catalyst, or 25% tungsten disulfide plus 3% nickel sulfide on a carrier containing 72% alumina in the presence of free aluminum, to obtain motor fuels and lower phenols. The Böhlen refinery in East Germany has hydrogenated low-temperature brown coal tar at 210 to 1000 psig and 425 to 440°C, using a catalyst of 10 to 12% molybdena on gamma-alumina.[61] The product was distilled to 350°C (approximately 60%) and recycled. A light-yellow diesel fuel oil with a cetane number of 36 was obtained. The Zeitz refinery in East Germany has hydrogenated tars and tar oils at 4200 psig to obtain gasoline, diesel oil, and a recycle stock.[113]

The hydrogenation of the low-temperature bituminous tar produced at the Cheremkhovo Plant, East Siberia, has been described recently.[114] At 420 to 460°C, 1500 to 4300 psig, and contact times of 9 to 180 min, conversions of 67 to 90% were obtained. The product contained 15 to 30 gal of phenol per metric ton of tar (about 250 gal). Recent Russian patents describe the production of high-octane gasolines from low-temperature coal tars. In one process, the whole tar is hydrocracked in the liquid phase, the hydrogenated product is freed of tar acids and bases, and then the material boiling up to 300°C is aromatized or reformed over molybdena-alumina catalyst, or chromia-molybdena-alumina catalyst, at 750 to 1120 psig to produce a high-octane gasoline.[115] In another process, the fraction boiling 300 to 400°C, obtained by vacuum distillation of a low-temperature coal tar, is hydrocracked over a tungsten catalyst, and the hydrogenation product boiling up to 300°C is aromatized or reformed over molybdena-alumina catalyst at 1500 psig to produce the desired motor fuels.[116]

The Central Fuel Research Institute, Jealgora, India, has recently reported the results of experiments directed toward the production of diesel oil by hydrogenation of a low-temperature tar.[117] The tar-fraction boiling from 250 to 300°C was treated with a high initial pressure of hydrogen at 400°C with a nickel-complex catalyst concentration of 10%. The tar acids and bases were almost completely removed and the product had 72% saturates. When the neutral oil, boiling from 250 to 300°C, was similarly treated, but with a group VI metal-sulfide promoter concentration of 4% in the catalyst, a diesel oil yield of 73%, with a saturate content of 74%, was obtained. Law[118] has emphasized the fact that low-temperature tars, by mild hydrogenation with existing petroleum refining processes, can be converted into "synthetic" petroleum crude oils which can be processed in conventional petroleum refining units. Experiments were described in which low-temperature tar from Rocky Mountain area low-rank coals was hydrogenated at pressures of 750 psig, or more, using either commercial cobalt-molybdena or metallic nickel catalysts. Hydrogen uptake averaged 1.7 weight %, equivalent to 1200 to 1500 cu ft of hydrogen per barrel (42 gal), the average volume yield being 121.2%. Conventional distillation and extraction of this "synthetic" crude oil yielded 36.4% gasoline (containing about 40% benzene and having a high octane number), 37.8% diesel fuel (with a cetane number around 40), 6.3% tar acids, 1.4% tar bases, and 18.1% recycle oil.

[111] Gossedin, A., *Rev. inst. franç. pétrole et Ann. combustibles liquides*, **1**, 145–51 (1946).

[112] Schnabel, B., and Kubička, R., Czech. Pat. 85,992 (1956).

[113] Günther, G., *Freiberger Forschungsh.*, **A23**, 5–12 (1954).

[114] Alekseeva, K. A., and Moldavskiĭ, B. L., *Khim. i Tekhnol. Topliva i Masel*, **3**, No. 10, 7–11 (1958).

[115] Lozovoĭ, A. V., Muselevich, D. L., Blonskaya A. I., Ravikovich, T. M., and Titova, T. A., U.S.S.R. Pat. 111,769 (1958).

[116] Lozovoĭ, A. V., Krichko, A. A., and Pchelina, D. P., U.S.S.R. Pat. 111,825 (1958).

[117] Anon., *Petrol. Refiner*, **38**, 336–8 (1959).

[118] Law, R. D., *Western Petroleum Refiners Association*, Rocky Mountain Regional Meeting, Casper, Wyoming, 1957, Tech. 57–29.

Lang and Lacey[119] studied the continuous hydrogenation of the heavy distillate (including tar acids), boiling from 235 to 360°C, from a low-temperature Alabama bituminous tar. At 480°C, a hydrogen pressure of 3000 psig, and with molybdena-alumina catalyst, a 92 volume % yield of product was obtained, which contained 40% aromatics and 50% boiling below 235°C. Tar acids were practically eliminated and hydrogen uptake was about 5.5 weight % of the heavy distillate. The hydrocracked oils boiling below 235°C still contained considerable amounts of naphthenes, so continuous reforming was studied. With standard molybdena-alumina catalyst, and hydrogen at 370 psig and 525°C, a 74 volume % yield of product, containing 80% aromatics, was obtained. These two operations were combined in a single operation for the neutral heavy distillate by using hydrocracking-reforming conditions of 550°C, hydrogen pressure of 500 psig, and molybdena-alumina catalyst. The yield was 74 volume %, the product containing 79% aromatics and 46% boiling below 235°C. Tar acids boiling from 235 to 300°C were hydrocracked at 460°C and 2100 psig hydrogen pressure over molybdena-alumina catalyst to give a yield of 90 volume %, the product containing 50% tar acids and 53% boiling below 235°C. Hydrocracking of the acids boiling from 300 to 360°C gave a yield of 84 volume %, the product containing 50% tar acids and 23% boiling below 235°C.

The Coal Tar Research Association has described a two-stage hydrogenation of high-sulfur coke-oven tar fractions to produce mainly naphthenic hydrocarbons, free of sulfur.[120] A coal-tar fraction boiling from 200 to 280°C, containing 0.95% sulfur, and consisting mainly of naphthalene and methylnapthalenes, was washed to remove acids and bases and hydrogenated in a continuous system at 410°C and 480 psig over cobalt-molydate catalyst. The partially hydrogenated product, which contained only 0.01% sulfur, was washed to remove hydrogen sulfide and hydrogenated in a continuous system at 300°C and 480 psig over a catalyst consisting of 65% nickel on kieselguhr, the hydrogen being in excess. The product was a water-white liquid that was 95% fully hydrogenated, consisting mainly of decalin and methyldecalins, that is, fused ring dinuclear naphthenes. It was pointed out that this material was useful as motor-, jet-, and diesel-fuels, solvents, plasticizers, and heat-exchange fluids. This same procedure gave a product of 84% fully hydrogenated material when applied to a coal-tar fraction boiling from 300 to 380°C and consisting mainly of anthracene and phenanthrene.[121] Not only dinuclear and trinuclear naphthenes, but also higher polynuclear naphthenes can be produced by complete hydrogenation of polynuclear aromatic hydrocarbons. Ruhröl G. m.b.H. has described a procedure for completely hydrogenating the polynuclear compounds in heavy coal and lignite oils.[122] Hydrogenation is conducted under high-contact concentrations with catalysts of large surface area composed of molybdenum, tungsten, chromium and vanadium oxides, sulfides, and halides supported on silica gel or alumina. At 450 to 500°C and 7500 psig, catalyst concentrations over 10% produce the completely hydrogenated products.

The Houillères du Bassin de Lorraine has conducted catalytic hydrogenation of refined xylenols to produce alkylcyclohexanols and alkylcyclohexanones.[64] The wartime operation of the Leuna Plant included the hydrogenation of large amounts of phenol to cyclohexanol by a completely continuous and automatic process.[58] Taking into account repairs and catalyst removal, the annual production capacity was 13 million pounds. The process consisted of vapor-phase hydrogenation at 150°C and 220 to 290 psig over a nickel-aluminum oxide catalyst containing 30 to 40% free nickel. Unreacted phenol was recycled and the overall conversion to cyclohexanol was about 99%. Part of this cyclohexanol was used to make cyclohexanone in a continuous catalytic vapor phase dehydrogenation plant, using zinc-iron catalyst with 8 to 12% iron and a temperature of 425 to 450°C. Conversion to cyclohexanone was 98%, and the production capacity was about 780,000 gal per year.

Besides the process just mentioned, there are other specific dehydrogenation processes of

[119] Lang, E. W., and Lacey, J. C., Jr., *Am. Chem. Soc. 135th Meeting, Boston*, Abstracts of Papers, 1959, p. 2J.

[120] McNeil, D., Waddington, W., and Popper, F. B., Brit. Pat. 769,293 (1957).

[121] McNeil, D., Waddington, W., and Popper, F. B., Brit. Pat. 761,755 (1956).

[122] Ibing, G., Ger. Pat. 867,387 (1953).

considerable value. The direct dehydrogenation of mixtures of alkylbenzenes, or alkylphenols, or alkylpyridines, containing some ethyl, isopropyl, or *n*-propyl groups to obtain vinyl derivatives is of obvious commercial interest. A C_8-aromatic distillate fraction, containing primarily ethylbenzene and *m*- and *p*-xylene, was treated at 575°C with a commercial catalyst consisting of 72.4% magnesium oxide, 18.4% iron oxide, 4.6% potassium oxide (as carbonate), and 4.6% copper oxide.[123] The yield of styrene (or vinylbenzene) based on ethylbenzene was 94%. In a similar manner, *m*- and *p*-vinyl-, and *m*- and *p*-isopropenylphenols were prepared from the corresponding alkylphenols by catalytic dehydrogenation at 600°C over chromia (15%)-alumina, with steam as a diluent.[124] A specific hydrogenation process, in addition to the production of cyclohexanol from phenol, is the conversion of catechols to the corresponding phenols by the removal of a molecule of water. This process is carried out at 360°C and a hydrogen pressure of 900 psig over a molybdena (18.5%)-alumina catalyst.[125] A catechol feedstock yielded 65.9% phenol and 19.1% water.

Oxidation and Halogenation

The processers of coke-oven tar fractions have long been familiar with the wide variety of compounds which can be obtained by oxidation of the aromatic constituents. Modern methods involve catalytic oxidation with air or oxygen, in liquid or vapor phase, in continuous processes, including those with fluidized-bed catalysts. By proper selection of catalyst, temperature, pressure, and contact time, as well as the tar fraction, the optimum yield of a particular product can be obtained. Among the products produced on a commercial scale are the following: anhydrides (maleic anhydride and phthalic anhydride), quinones (*p*-quinone, 1,4-naphthoquinone, phenanthraquinone, and anthraquinone), and aromatic carboxylic acids (*o*-, *m*-, and *p*-toluic acid, phthalic acid, isophthalic acid, terephthalic acid, and 1- and 2-naphthoic acid).

Coal-tar neutral oils, boiling from 230 to 360°C, have been oxidized with air at 430 to 450°C in a vapor-phase process with a fluidized bed of solid catalyst composed of 11 to 13% vanadium pentoxide, 2 to 3% potassium sulfate, and 84 to 87% silica gel support.[126] Yields of phthalic anhydride ranged from 23.2 to 47.0%, depending on the feed. Light and medium coal-tar distillate fractions were oxidized with various amounts of air at about 400°C in the presence of vanadium pentoxide.[127] The maximum yield of maleic anhydride from the light oil was 27.9%, and the maximum yield of phthalic anhydride from the medium oil was 47.5%. Similar oxidation of a coal tar pitch at 280 to 400°C produced 14.6 to 19.4% phthalic anhydride and 4.4 to 6.8% maleic anhydride.

Goftman and Golub[128] conducted extensive experiments in the vapor-phase catalytic oxidation of various coke-oven tar fractions in the presence of vanadium pentoxide. The light middle-oil fraction at 458°C, an air-vapor ratio of from 19 to 32:1, and a time of contact of 0.84 sec gave 25.1% phthalic anhydride and 5.5% maleic anhydride; the naphthalene fraction at 462°C, an air-vapor ratio of 13:1, and a time of contact of 0.80 sec gave 73.1% phthalic anhydride and 6.4% maleic anhydride; the phenanthrene fraction at 445°C, an air-vapor ratio of 18:1, and a time of contact of 2.05 sec gave 32.3% phthalic anhydride and 9.0% maleic anhydride; the anthracene fraction at 456°C, an air-vapor ratio of from 18 to 35:1, and a time of contact of 1.72 sec, gave 22.6% phthalic anhydride and 7.7% maleic anhydride. In addition, these investigators oxidized the pyrene and chrysene fractions at 460°C to give phthalic anhydride, maleic anhydride, *p*-quinone, and 1,4-naphthoquinone.[129] They also showed that 62.7% of the anthracene in the anthracene fraction could be converted to anthraquinone.[130]

[123] Corson, B. B., Heintzelman, W. J., and Pavlik, F. J., *Ind. Eng. Chem.*, **50**, 621–2 (1958).

[124] Corson, B. B., Heintzelman, W. J., Schwartzman, L. H., Tiefenthal, H. C., Lokken, R. J., Nickels, J. E., Atwood, G. R., and Pavlik, F. J., *J. Org. Chem.*, **23**, 544–9 (1958).

[125] Jelínek, J., Czech. Pat. 86,160 (1957).

[126] Mathur, V. K., Nair, C. S. B., Basu, A. N., Lahiri, A., and Bhatnagar, J. N., Indian Pat. 54,960 (1957).

[127] Sasayama, H., *Kôgyô Kagaku Zasshi*, **56**, 489–91, 590 (1953).

[128] Goftman, V., and Golub, A. I., *Koks i Khim.*, **1956**, No. 2, 51–5.

[129] Goftman, M. V., and Golub, A. I., U.S.S.R. Pat. 102,794 (1956).

[130] Goftman, M. V., and Golub, A. I., *Zhur. Priklad. Khim.*, **28**, 507–15 (1955).

The Bolsover refinery has a modern unit for chlorination of phenols. A flowsheet and photographs of this tar-acid chlorination plant have been published.[40] One of the chief chlorinated products made is 2,4-dichlorophenol, which is used in the manufacture of "2, 4-D," a widely used selective weed killer. Another product is a mixture of monochloroxylenols, which is used in formulating antiseptics. The British Coal Utilisation Research Association has recently described a process for the fluorination of coal to produce a mixture of liquid perfluorocarbons.[131] This process involves reaction of low-grade coal with gaseous chlorine trifluoride diluted with nitrogen. The main product is an oil boiling from 150 to 350°C with 65% fluorine and 7% chlorine. There seems to be little doubt that this process would be applicable to low-temperature tar pitch. The highly fluorinated liquids which would be produced might be useful as special lubricants, heat-transfer media, dielectric fluids, and hydraulic fluids.

UTILIZATION

Because most refining processes are conducted with some specific end uses in mind, several fields of utilization of low-temperature tar have already been indicated in the sections dealing with processing. By far the largest part of the total world production of low-temperature tar to date has been utilized as liquid fuels, that is, gasoline, diesel oil, and fuel oil. In this respect, there is no fundamental difference between utilization of low-temperature tar and utilization of petroleum. There has been a tendency throughout the whole development history of the low-temperature tar industry strongly persisting up to the present, to view the utilization of low-temperature tars as being analogous to the utilization of coke-oven tars, that is, primarily as a source of chemicals. This viewpoint is contrary to historical fact, and, in particular, contrary to the fact that low-temperature tars are very much different from coke-oven tars. There is ample evidence in the literature that this erroneous viewpoint can be attributed mainly to the use of the word "tar" in naming the product of low-temperature carbonization. It is for this reason that a number of investigators have proposed using terms like "crude oil" instead of "tar."

This is not to say that low-temperature tars have not been used as a source of chemicals. Hundreds of millions of pounds of phenols and cresylic acids have been recovered from low-temperature tars and utilized, mainly in the manufacture of plastics and resins. However, this has been the only really large-scale chemical utilization of low-temperature tars, especially in terms of the percent of total profit. Other chemicals traditionally associated with coke-oven tars, like naphthalene, are not produced from low-temperature tars. Here again, it is probably more logical to draw an analogy with the petroleum industry in which only a small proportion of the crude oil is utilized as chemicals, or so-called "petrochemicals," although the proportion of the total profit to the petroleum industry from these chemicals is fairly large.

The detailed aspects of utilization, where there is the maximum practical emphasis on chemical utilization, can best be seen in terms of two current low-temperature tar refineries. These are the 8-million-gallon-per-year Bolsover plant of the Coalite and Chemical Products, Limited, England, described in detail by Pound[40] and Bristow,[72] and the 9-million gallon per year Houillères du Bassin de Lorraine, Marienau, France, described in detail by LeRoux[64] and Sabatier.[132] These plants are small enough that they can emphasize chemical production without saturating their local markets.

The utilization of primary products from the Bolsover refinery has been reported to be as follows: (1) Gasoline, boiling from 42 to 175°C, octane number 75 (about 3% of the tar, the bulk of the gasoline being produced from 1.5 million gallons per year of benzole or "crude gas spirit"), used directly as an aviation fuel before World War II, now apparently shipped to petroleum refineries; (2) diesel oil, cetane number 25 (about 22% of the tar), used by local city buses; (3) fuel oil, specific gravity 1.032 at 60°F (about 20% of the tar, together with fluxing oil), used locally for generating steam and heat; (4) phenols and cresylics, and high-boiling tar acids (about 13% of the tar, or 8 million pounds per year) sold to

[131] Chilton, C. H., *Chem. Eng.*, **66,** No. 4, 62 (1959).

[132] Sabatier, J. L., *Chemical Engineering in the Coal Industry*, Sharpley, F. W., Ed., Pergamon Press, New York, 1956, pp. 122–8.

manufacturers of phenol-formaldehyde resins, plasticizers, pharmaceuticals, wetting agents, inhibitors, disinfectants, antiseptics, insecticides, and ore-flotation agents—approximately 49% of the total revenue of the refinery comes from the sale of these compounds; (5) catechol and homologues, and resorcinol and homologues (a total of about 400,000 lb/yr extracted from the aqueous liquor with butyl acetate), sold to manufacturers of antioxidants, inhibitors, photographic developers, azo-dyestuffs, tanning agents, pharmaceuticals, synthetic glues, and adhesives; (6) various pitch fractions (total pitch constitutes about 25% of the tar): (a) Residual pitch, used as a binder, and a high-boiling neutral oil used as a fluxing oil, for the manufacture of colored mastic flooring, similar to "asphalt tile;" (b) bituminous material, used in the manufacture of road-making emulsions, and in the linoleum industry; and (c) brown resin powder, used as a rubber extender to increase frictional wear; (7) creosote, containing about 30% tar acids, used in large quantities for timber preservation; (8) flotation oils, consisting of selected neutral oil, and acid- and base-containing oils, used in coal cleaning plants to separate coal from mineral impurities; and (9) rubber solvent (a selected gasoline fraction boiling from 90 to 150°C, containing 50% olefins, used as a particularly effective synthetic-rubber solvent, due to the high olefin content, which is a special characteristic of low-temperature tar distillates.

The utilization of secondary products synthesized at the Bolsover refinery has also been reported to be as follows[40]: (1) Trixylenyl phosphate (made from a xylenol fraction and phosphorous oxychloride, used as a plasticizer in the plastics industry; (2) chlorinated phenols (made from phenol, selected xylenols, and xylenol fractions, and chlorine gas): (a) 2,4-dichlorophenol, used in the manufacture of 2,4-dichlorophenoxyacetic acid ("2,4-D"), a selective weed killer and (b) monochloroxylenols, used in the formulation of antiseptics; (3) benzylcresols (made from a phenol-cresol mixture and benzyl chloride), used in the formulation of antiseptics.

The utilization of primary products from the Marienau refinery has been reported to be as follows[64]: (1) Phenol, cresols, xylenols, and cresylic acids (about 10% of the tar, or 7 million pounds per year), used by the chemical industry for preparation of disinfectants, insecticides, dye intermediates, pharmaceuticals, plastics and resins—about 40% of the total revenue is obtained from these compounds; (2) phenolic oil, boiling from 230 to 270°C, containing 40% phenols, such as higher alkylated phenols, indanols, naphthols, and dihydric phenols (this oil constitutes about 8% of the tar), used for wood impregnation, and as a raw material for the manufacture of disinfectants that are used as soapy emulsions; (3) catechol and methylresorcinol, isolated from the phenolic oil by selective washing with water, used by the chemical industry; (4) dephenolized oil, boiling from 185 to 230°C (about 1% of the tar), used as an additive to road tars, and as a base, or solvent, for insecticides and agricultural disinfectants; (5) denaphthalization oil, boiling from 270 to 320°C (about 5% of the tar), used to denaphthalize coke-oven gas because it will dissolve more than 5% of naphthalene, due to its mixed chemical nature (paraffinic, naphthenic, aromatic) not possessed by comparable coke-oven tar oils and can be used at temperatures as low as −5°C and will not crystallize at low temperatures like coke-oven tar oils—about 6% of the total revenue is obtained from the sale of this oil; (6) heavy oil, boiling from 320 to 400°C (about 32% of the tar), used as a fuel oil, because it mixes well with petroleum fuel oils, in contrast to coke-oven fuel oils, which do not blend well with petroleum fuel oils, and is also used as an additive to road tars—about 18% of the total revenue is obtained from this oil; and (7) pitch (about 31% of the tar), used competitively with petroleum bitumens, for water-proofing and air-tight sealing, because it has special sealing characteristics, for briquetting coal-fines, and in admixture with fairly large quantities of heavy oil and a little dephenolized oil for road surfacing—about 22% of the total revenue comes from the pitch.

The utilization of secondary products synthesized at the Marienau refinery has also been reported to be as follows[64]: (1) Triarylphosphates (made by esterification of a xylenol fraction with phosphorous oxychloride), used as a plasticizer in polyvinyl resins; (2) alkylcyclohexanols and alkylcyclohexanones (made by catalytic hydrogenation of a xylenol fraction), used as heavy solvents in the print and varnish industries, and to prepare plasticizers; and (3) isobutylation products—*p*-tert-butyl phenol, used in the preparation of oil-soluble phenolic resins, di-tert-butyl-*p*-cresol, used as a nontoxic antioxidant, and di-tert-butyl-*m*-cresol, used as a stabilizing agent in fuels and rubber.

In contrast to these two plants, the low-temperature tar refineries in East Germany and Czechoslovakia are operated with maximum emphasis on the production of liquid fuels. Each of these refineries has about ten times the capacity of either the plant in England or the one in France, which might be too large at the present time for plants producing mainly chemicals but are quite small for a refinery producing mainly liquid fuels.

Jet and Rocket Fuels

The United States Air Force has recently made public the specifications for a jet super-fuel.[133] These include thermal stability to temperatures of 700 to 800°F, with a heating value of about 18,400 Btu/lb and 135,000 Btu/gal. Research data have indicated that saturated cyclic hydrocarbons of relatively high-molecular weight, such as decalins, may be suitable. It has been shown that about 4.6% of a low-temperature bituminous tar consisted of dinuclear naphthenes, that is, polyalkyldecalins.[28] Other perhydro-condensed cyclic hydrocarbons could be prepared by catalytic hydrogenation of those low-temperature tar fractions containing alkylnaphthalenes (as much as 9% of some tars[41]), alkylphenanthrenes, and alkylanthracenes. McNeil, Waddington, and Popper[120] have demonstrated that by a two-stage hydrogenation process even high-sulfur content fractions of polynuclear aromatic hydrocarbons can be converted into their corresponding perhydro derivatives.

In a discussion of the potentialities of low-temperature tar for conversion to jet and missile fuels, Wainwright[134] pointed out that perhydrofluoranthene has a heat of combustion of 18,130 Btu/lb, and 148,450 Btu/gal. Fluoranthene and its alkyl derivatives are present in low-temperature tar. The combined decalins, tetralins, naphthalenes, phenanthrenes, anthracenes, fluoranthenes, and so forth, constitute at least 20% of a bituminous low-temperature tar, being present almost exclusively as alkylated derivatives. The perhydro derivatives already present in the tar in large quantities and the perhydro derivatives that could be prepared by hydrogenation could yield at least one barrel of super-fuel per 5 barrels of low-temperature tar. McGee[135] has also discussed the potentialities of low-temperature tar for conversion to jet and missile fuels and has presented these same conclusions. The potential military market for hydrocarbon super-fuel is estimated at 2 or 3 million barrels per year, while estimates of price range from 2.5 to 10 times as much as the price of the standard jet fuel, JP-4, which is 10 to 12 cents per gallon.[133] About 600 million gallons of low-temperature tar could conceivably be processed per year for the manufacture of super-fuel, nearly as much as the German annual wartime production of low-temperature tar and the capacity of about ten plants of the size planned for Cresap, West Virginia, or Rockdale, Texas.

Plastics and Resins

Next in importance to utilization of low-temperature tar hydrocarbons as fuels has been the utilization of the phenols for the manufacture of phenolic plastics and resins. A consideration of the modern plastics industry indicates that significant quantities of *non*phenolic resins may someday be produced from low-temperature tar fractions. Reinforced plastics are of special interest because of their potential for large-scale utilization in construction. The reinforced plastic industry expects a growth rate of at least 20% annually, mainly flat and corrugated panels for industrial and home construction but also sports car bodies and boat hulls.[136] The great majority of these reinforced plastics are polyesters, requiring, among other chemicals, phthalic anhydride, maleic anhydride, and glycols. These anhydrides could be produced by catalytic vapor-phase oxidation of appropriate low-temperature tar fractions, as already discussed, and the abundant supplies of olefins in low-temperature tars, their light oils and gases could be hydrated to glycols by conventional processes.

Another important class of plastics and resins, used for molded articles, floor tiles,

[133] Chilton, C. H., *Chem. Eng.*, **66,** No. 8, 78 (1959).

[134] Wainwright, H. W., *Missiles and Rockets*, **3,** No. 11, 43–8 (1958).

[135] McGee, J. P., *Coal Utilization*, **12,** No. 9, 31–3 (1958).

[136] Anon., *Chem. Eng. News*, **36,** No. 34, 73–7 (1958).

varnishes, and so forth, are those prepared by polymerization or copolymerization of compounds with active olefinic groups, such as styrene and indene. Much of the styrene is currently made by catalytic dehydrogenation of ethylbenzene, and alphamethylstyrene is similarly prepared from cumene (isopropylbenzene), which itself is largely synthetic. One of the earliest developments in synthetic resins, the coumarone (benzofuran)-indene resins, is still widely used in modern varnishes, floor tile, adhesives, and waterproofing materials. This has a significant bearing on low-temperature tars (and their light oils) for three reasons: (1) These important compounds (ethylbenzene, styrene, isopropylbenzene, methylstyrene, benzofuran, and indene, and their alkylated derivatives), are present in combined amounts of practical significance; (2) recent developments in catalytic dehydrogenation make it feasible to convert the ethyl and isopropyl groups in refinery distillate fractions directly to vinyl and isopropenyl groups without prior isolation of compounds, as already discussed, the content of reactive compounds being enhanced considerably, because there is a fair amount of constituents with ethyl or isopropyl groups, or both, with or without methyl groups; and (3) recent developments in the manufacture of styrene-indenecoumarone copolymers have made it feasible to prepare these resins directly from regular naphthas without prior isolation of compounds.

Greenhow and Szewczyk[137] obtained high yields of pale yellow resins of softening point from 90 to 100°C (ring-and-ball) by thermal polymerization, under pressure, of coal tar naphthas at about 200°C. The naphthas had varying contents of styrene, indene, and benzofuran, which were the constituents producing the copolymers. These investigators also obtained amber resins, similar to the regular indenecoumarone resins, by catalytic polymerization of vertical retort coal-tar naphthas with boron trifluoride catalysts at 60 to 70°C.[138] They reported that this type of copolymerization is practical on a commercial scale and can be carried out either by mixing styrene-rich naphthas (up to about 10% styrene) and indene-benzofuran-rich naphthas (up to about 20% indene plus benzofuran) and resinifying, or by polymerization of a wide boiling-range fraction, like 130 to 200°C, without closer fractionation.

The Houillères du Bassin du Nord et du Pas de Calais has described a process for preparing copolymers of styrene, indene, and benzofuran directly from a benzole or coal-tar oil fraction, boiling from 137 to 200°C, by catalytic polymerization with boron trifluoride at 60 to 70°C.[139] A typical resin had a softening point of 120°C and a molecular weight of 3500. Yasunaga, Takagi, and Ohmori[140] have described the production of resins with softening points of about 160°C directly from coal-tar solvent naphthas containing benzofuran and indene by polymerization with boron trifluoride catalyst.

The high content of alkylnaphthalenes in low-temperature tars could be used to advantage in the manufacture of thermoplastic resins, other than in oxidation to difunctional acids and anhydrides. Light-colored thermoplastic resins have been prepared from mixtures of methylnaphthalenes by formic acid catalyzed condensation with paraformaldehyde at 100°C.[141] These resins had molecular weights ranging from 350 to 530, and softening points of 70 to 115°C. The polyalkylnaphthalenes have been used as an effective plasticizer in amounts ranging from 15 to 25% in the preparation of thermoplastic resins with high softening points by the thermal polymerization of acenaphthylene at 125 to 150°C.[142]

Road Tar and Other Construction Materials

Mention has already been made of the Bolsover refinery in England supplying bituminous material for road-making emulsions, and the Marienau refinery in France supplying pitch, heavy oil, and dephenolized oil, used in admixture for road surfacing. The June, 1958, issue of *Goudron pour routes* (Road Tar) contains some discussion of the current use of

[137] Greenhow, E. J., and Szewczyk, J., *Australia, C.S.I.R.O., Coal Research Sec., Tech. Paper* No. 125 (1957), 18 pp.

[138] *Ibid.*, No. 46 (1955), 9 pp.

[139] Delassus, M., Parant, A., and Croisier, J., Fr. Pat. 966,981 (1950).

[140] Yasunaga, K., Takagi, M., and Ohmori, K., Japan Pat. 790 (1957).

[141] Gordon, L. B., May, P. D., and Lee, R. J., *Ind. Eng. Chem.*, **51**, 1275–8 (1959).

[142] Kaufman, M., and Williams, A. F., U.S. Pat. 2,632,749 (1953).

low-temperature tar products for the surfacing of highways in France.

Low-temperature tar has been used recently in Russia for a road paving process in which about 0.5% low-temperature tar and about 0.5% slaked lime are blended with the crushed stone before application.[143] Road tests have been performed in India on a road tar made of a blend of 3.5 parts low-temperature tar pitch and one part low-temperature tar anthracene oil.[144] This blend conformed to the softening point limits prescribed by Indian Standards Specifications for road tars. A road-surfacing process has been described in which low-temperature tar pitch, which contains up to 50% phenolic material, is added to the bitumen binder used for hot-process rolled asphalt.[145] It has been claimed that the high content of phenolic components, characteristic of low-temperature tar pitches, results in rapid absorption of oxygen, accelerating the hardening of the surface, and preventing the formation of a smooth surface liable to cause skidding.

As already mentioned, the Bolsover refinery of Coalite and Chemical Products, Limited, produces a mastic for flooring from low-temperature tar. A United States patent has been obtained describing this product, which includes a fluxing oil free from tar acids and tar bases, with an initial boiling point of 300°C.[146] About 40,000 tons of this flooring, which is similar to "asphalt tile," was laid in England during 1947 in a postwar housing program. Gundermann[147] has given a detailed description of the evaluation and scope of application of twenty-five different roof-coating compositions based on brown coal-tar oils. Highly viscous creosote oils were found to be high in phenolic polymerization products and could be processed directly as roof coatings without any chemical treatment. Special anticorrosion and sealing compositions have been prepared from low-temperature tars by treatment with redox catalysts in the presence of high molecular weight polymers (polystyrene and polyvinyl chloride) in an atmosphere of oxygen at 140 to 165°C.[148]

Coolant-Moderators and Similar Uses

Coolant-moderators for nuclear reactors are preferably liquids, with a high dielectric constant and stable to radiation decomposition. Good examples among organic liquids are isopropylbiphenyl and *m*-terphenyl.[149] Liquid mixtures of the lower alkylated biphenyls are recoverable from low-temperature bituminous tars.[28] It is probable that terphenyls are also present. Tests at temperatures of 400 to 450°C and the application of radiation doses approaching reactor operation conditions have shown that *o*-, *m*-, and *p*-terphenyls have sufficient stability to be used as reactor moderators and coolants.[150]

Mixtures of chlorinated biphenyl, biphenyl, *o*-terphenyl and *m*-terphenyl have been used as heat transfer and hydraulic fluids at temperatures up to 330°C.[151] The boric acid esters of mixtures of phenol, cresols, and xylenols have been used as heat exchange liquids at temperatures up to 300°C.[152]

Fungicides and Wood Preservation

The use of coal-tar creosote (a mixture rich in high-boiling phenols) for the protection of wood against fungal attack has been practiced for well over a century. The first commercial distillation of coal tar around 1850 was in response to the demand for creosote to protect the millions of railroad ties which were then being laid in the first worldwide railway-building boom. At that time, essentially nothing was known about the nature of wood decay, that is, that it was produced by fungi, and that it was the high-boiling phenolic substances that were acting as fungicides. A vast body of practical knowledge in this application of coal-tar creosote has been assembled over the years, but it has not been until recently that more refined studies have demonstrated that low-temperature tar creosote, and specific

[143] Gelfand, S. I., *Bitumen, Teere, Asphalte, Peche*, **9**, 126–7 (1958).

[144] *Ann. Rept., Regional Research Lab., Hyderabad (India)*, (1957–58), 137 pp.

[145] National Research Development Corporation, Brit. Pat. 714,153 (1954).

[146] Buist, C., U.S. Pat. 2,658, 005 (1953).

[147] Gundermann, E., *Chem. Tech. (Berlin)*, **7,** 20–6 (1955).

[148] Chemieprodukte G. m.b.H., Brit. Pat. 756,723 (1956).

[149] Parkins, W. E., *Chem. Eng.*, **64,** 253–6 (1957).

[150] Colichman, E. L., and Fish, R. F., *Nucleonics*, **15,** 72–4 (1957).

[151] Good, R. J., U.S. Pat. 2,741,598 (1956).

[152] Sully, B. T. D., Brit. Pat. 759,482 (1956).

fractions and phenolic types characteristic of low-temperature tars, may have equal or even superior fungicidal utility.

Creosotes suitable for timber preservation have been prepared from low-temperature tar in India.[144] These have been found to have good antitermite and fungicidal properties even though their specific gravities were lower than those specified for creosotes traditionally prepared from coke-oven tars. In the last few years, New Zealand and the United Kingdom have modified their standards specifications to accommodate creosotes from low-temperature tars also. It has been demonstrated that the tar acids appearing throughout the entire boiling range of a low-temperature tar creosote were very toxic to Madison 517 fungus, one of the most prevalent wood fungi.[153] Also, weathering had no adverse effects, because upon aeration these tar acids underwent no loss in toxicity.

Just as the railways have used creosote for many years, ships have an even older history of tar utilization for water-proofing and protection against mildew (fungi). The old-fashioned nickname for a sailor, "tar," comes from tarpaulin, a canvas water-proofed and protected from mildew by painting with tar. Recent research on the use of tar oils for the preservation of ropes and twine against mildew have shown that the best protection was obtained with a highly phenolic dehydrated low-temperature tar.[154]

Tar acids in the boiling range 220 to 250°C have been shown to have high efficiencies as fungicides.[155] Compounds possessing high activity, that is, 75% inhibition of growth, or more, were 2-ethyl-5-methylphenol, 3-ethyl-5-methylphenol, 4-*n*-propylphenol, 3,5-diethylphenol, 4-*n*-butylphenol, 4-indanol, 2-naphthol, and 2-phenylphenol, all of which are found in, and characteristic of, low-temperature tar. Phenylphenols, methyl- and dimethyl-naphthols have been shown to be active fungicides against *Merulius lacrymans* and *Coniophora cerebella*, which infect timber.[156]

[153] Baechler, R. H., *Proc. Am. Wood-Preservers' Assoc.*, **49**, 12–17 (1953).

[154] Anon., *Chem. Age (London)*, **71**, 723–4 (1954).

[155] Coles, G. V., Martin, J. T., Byrde, R. J. W., *Ann. Rept. Agr. Hort. Research Sta., Long Ashton, Bristol*, **1956**, 101–2.

[156] Goftman, M. V., and Kharlampovich, G. D., *Zhur. Priklad. Khim.*, **30**, 660–3 (1957).

Pharmaceuticals

The complex chemical nature of low-temperature tar offers virtually unlimited possibilities for utilization in the pharmaceutical industry. Pharmaceutical outlets make up a very small volume of the tar processed in current low-temperature tar refineries and yet amount to a sizable fraction of the economic value of the tar. Thus, as already mentioned, the refineries at Bolsover and Marienau obtain from 40 to 50% of their total revenue from the sale of phenols of various kinds, and next to utilization in the manufacture of phenolic plastics, one of the significant outlets is in pharmaceuticals, including antiseptics. Large volumes of antiseptics and disinfectants are utilized in modern agricultural practice.

An example of an individual pure compound, utilized as a pharmaceutical in large amounts, is niacin, or nicotinic acid (3-pyridinecarboxylic acid), one of the B-complex vitamins. The current demand in the United States is about 3 million pounds per year, with a growth of about 5%/yr. A major outlet for nicotinic acid is enrichment of animal feed. It is generally prepared by oxidation of 3-alkylpyridines, or quinoline and some of its carbocyclic derivatives. Bylicki,[157] in Poland, has pointed out that modern methods of separating the isomeric nicotinic acids from their mixtures make it possible to manufacture these acids by oxidation of any tar-base fractions containing the alkyl group in position 3 or 4. Tajika, Sano, and Nakano[158] have described the preparation of nicotinic and isonicotinic acids from a coal-tar base fraction, boiling 140 to 145°C, by first removing 2,6-dimethylpyridine as its urea complex and then oxidizing the mixed 3- and 4-methylpyridines into the corresponding acids.

Another type of tar-base oxidation product of pharmaceutical importance is the pyridine N-oxides. Midland Tar Distillers, Limited, has made pilot quantities of these compounds available and published examples of their application in the pharmaceutical field.[159] Reilly Tar & Chemical Corporation has described

[157] Bylicki, A., *Roczniki Chem.*, **29**, 791–802 (1955).

[158] Tajika, Y., Sano, H., and Nakano, T., *Japan. Pat.* 2628 (1955).

[159] Cox, A. J., *Mfg. Chemist*, **28**, No. 10, 463–7 (1957).

the oxidation of pyridine, 3-methylpyridine, and 3,5-dimethylpyridine with hydrogen peroxide in glacial acetic acid to obtain the corresponding 2-hydroxy-N-oxide derivatives, which are useful bactericides.[160] The high-boiling tar bases also have important pharmaceutical uses. The activity of derivatives of phenanthridine (3,4-benzoquinoline) as trypanocides and antiseptics has been outlined.[161]

Phenol itself has been in use as an antiseptic and disinfectant for almost a century, and the phenols, in general, are still widely used for these purposes. A modern example is 2-phenylphenol, which has high efficiency as a germicide, and fungicide for pathogenic fungi, yet has low toxicity to higher forms of life, so that it is popular for agricultural purposes. Derivatives of 2-naphthol have been examined for their bacteriostatic and fungistatic properties, and it was found that alkylation enhances the former property while halogenation has a beneficial effect on both types of activity.[162] It has been shown that useful bactericides can be obtained by chlorination of the indanols.[163] Alkylresorcinols have been in use for a number of years in the pharmaceutical industry. Best known among these is 4-*n*-hexylresorcinol, used in medicine as a germicide. The 4-*p*-alkylbenzylresorcinols have been shown to possess useful properties for combating parasites such as amoebae and helminths.[164]

A number of steroid hormones have been synthesized from coal-tar constituents and their derivatives. For example, the pilot plant scale production of equilenin, a steroid hormone, from 1-naphthylamine-6-sulphonic acid at the University of Wisconsin has been described.[165] Recently, the total synthesis of the steroid hormone, aldosterone, was announced by University of Wisconsin chemists, who started with 1,6-dihydroxynaphthalene, a coal-tar derivative.[166] A quantity of 2-methylresorcinol, that had been isolated from low-temperature tar acids, was announced as the starting material for a projected total synthesis of cortisone, a steroid hormone.[167] It has been shown that 2,3- and 2,4-dimethylquinoline are effective for controlling the growth of new tissue in plants, such as the inhibition of sprouting in stored potatoes.[168] 1-Methylnaphthalene has been mentioned as a possible intermediate in the preparation of the auxin, or so-called plant hormone, naphthaleneacetic acid.[169]

Fluorene derivatives prepared from 9-bromofluorene have been found to act as adrenergic blocking agents,[170] and 2-aminoalkylfluorenes have been demonstrated to be useful antispasmodic, antihistaminic, and anesthetic agents.[171]

Miscellaneous Uses

A number of miscellaneous uses of low-temperature tar fractions has been mentioned in the descriptions of processes and in the information on the refineries at Bolsover and Marienau. These include such diverse uses as electrode carbons, flotation agents, dyes, and so forth.

Studies have been made in Japan on the utilization of the olefins in low-temperature coal tar.[172] The olefins in low-temperature tar light-oil fractions boiling from 150 to 180°C, from 180 to 210°C, and from 210 to 240°C, were hydrated with peracetic acid at 70°C, most of the olefins being converted into glycols. Detergents were made from acidic oil fractions of low-temperature tars.[173] Sodium dodecylxylenolsulfonate was readily prepared from a xylenol fraction of a low-temperature tar, its aqueous solutions showing surface tensions comparable with regular commercial detergents. Nonionic detergents prepared

[160] Cislak, F. E., U.S. Pat. 2,752,356 (1956).

[161] Theobald, R. S., and Schofield, K., *Chem. Rev.*, **46,** 171–89 (1950).

[162] Baichwal, R. S., Baichwal, M. R., and Khorana, M. L., *J. Am. Pharm. Assoc., Sci. Ed.*, **46,** 603–7 (1957).

[163] Sterling Drug, Inc., Brit. Pat. 770,593 (1957).

[164] Gerzon, K., U.S. Pat. 2,719,866 (1955).

[165] Anon., *Chem. Eng. News*, **27,** No. 52, 3888 (1949); **28,** No. 6, 380 (1950).

[166] Anon., *Chem. Eng. News*, **36,** No. 25, 43 (1958).

[167] Anon., *Chem. Trade J.*, **127,** 23 (1950).

[168] Denny, F. E., U.S. Pat. 2,692,822 (1954).

[169] Coulson, E. A., *J. Soc. Chem. Ind. (London)*, **62,** 177–9 (1943).

[170] Herdegen, T. F., Heisler, R. Y., and Ullyot, G. E., *J. Am. Chem. Soc.*, **72,** 940–2 (1950).

[171] Cusic, J. W., U.S. Pat. 2,527,798 (1950).

[172] Kato, T., and Takeshita, K., *Kyûshû Daigaku Kogakubu Kogaku Shûhô*, **25,** 136–9 (1952).

[173] Ishii, Y., and Nishiyama, M., *Yushi Kagaku Kyôkaishi*, **1,** 129–32 (1952).

from dodecylated alkylphenols and polyethylene glycol showed similar surface activity.

A method for waterproofing fiber in the liberation and treatment of pulp from cotton stalks has been described in which the pulp is thoroughly mixed with a suspension of low-temperature coal tar, the emulsion subsequently being broken with a suitable coagulating agent.[174]

Goftman and Kharlampovich[175] have made extensive studies of the utilization of the high-boiling phenols in coal tar, including antioxidant properties. The antioxidant efficiencies were compared to that of a low-boiling phenol, *o*-cresol, which increased the inhibition period for oxidation of paraffin at 170°C, about sixfold, at a concentration of 0.05%. The efficiencies, relative to 1.00 for *o*-cresol, were: polyalkylphenol, 3.00; 2-naphthol, 4.50; phenylphenol, 4.67; 1-naphthol, 10.83; methylnaphthol, 13.25; and dimethylnaphthol, 15.10.

ECONOMICS

No comprehensive, realistic evaluation of the economics of processing and utilization of low-temperature tar in the United States has ever been published. This is undoubtedly partly due to the fact that as of the summer of 1959 there was no large, modern low-temperature tar refinery in operation in this country. The Dakota Briquetting Company, as mentioned before, has been distilling about 100,000 to 200,000 gal of low-temperature lignite tar a year into pitch and creosote. This has been a small-scale, intermittent operation, but it might be presumed that this plant has generally made some kind of profit over the many years it has been in existence. It has been pointed out that, in some countries at least, a low-temperature tar refinery with distillation as the sole process can obtain a satisfactory return on its investment.[132]

A number of papers mentioning the economics of low-temperature carbonization have been published, a few of which apparently present a satisfactory analysis of the economics of the carbonization itself. Almost all of these papers either present nothing on the tar, except to mention that the overall economic picture ultimately depends on the price that the tar refiner will be willing to pay, or they attempt an analysis of the economics of the tar, but base it on patently erroneous premises. As a result, it is not at all feasible, on the basis of published information, to give clear-cut answers to questions about the economics of low-temperature tar in the United States. Fortunately, it is possible to say a few things about the economics of low-temperature tar in other countries, primarily because there are low-temperature tar refineries currently in operation in these localities. There are four or five refineries in West Europe, with capacities of the order of magnitude of 10 million gallons per year, which have prospered, at least as evidenced by published accounts of periodic modernization of their plants and addition of new units. There are also at least three or four refineries in East Europe with capacities of the order of magnitude of 50 to 100 million gallons per year. These are all state-operated, so that it is difficult to determine whether they are operated at a real profit or not. However, there does not seem to be any reason for plants of this sort being operated at all unless there were some practical considerations. One obvious factor is the shortage of gasoline and diesel fuel, especially in East Germany. Mention has been made recently of modernization and expansion of these plants.

LeRoux[64] and Sabatier[132] have recently presented information on the operation of the low-temperature tar refinery at Marienau, France, which makes an elementary analysis of the economics feasible. They have presented data on the yields of the primary distillation products, the prices of products relative to the price of the tar, and the percentage of the total revenue that is obtained from each product. No actual prices have been published, but such information would be of little value anyway in terms of the market in the United States. The economic data for two processing schemes is shown in Tables 15 and 16. The first scheme, which is not actually conducted at the plant but presumably could be, is the example given by Sabatier to demonstrate a satisfactory return with distillation as the only upgrading process. A price of 7 cents/gal has been arbitrarily assigned to the tar for two obvious reasons, one being that this is the current average price of petroleum crude-oil

[174] Spencer, A. M., Jacobson, A., and Sixt, K., U.S. Pat. 2,668,110 (1954).

[175] Goftman, M. V., and Kharlampovich, G. D., *Zhur. Priklad. Khim.*, **30**, 439–46 (1957).

TABLE 15

Elementary Economics of Processing Scheme 1, (Tar Distillation, only), Marienau Low-Temperature Tar Refinery

Product	% of Total Tar	Price, cents/gal*	% of Total Revenue
Diesel fuel and gasoline	31	10	30
Denaphthalization oil	20	14	27
Heavy fuel oil	15	8	11
Pitch	34	10	32
Overall		11	

* On the basis of a cost to the refinery of 7 cents/gal for the crude tar.

in the United States, the other being that this then results in a price of 10 cent/gal for the diesel fuel and gasoline, these being current Gulf coast refinery prices for diesel oil and regular grade gasoline. The overall product value comes to 11 cents/gal, which would allow about 1.5 cents/gal for distillation costs, and about 2.5 cents/gal for revenue before any interest or taxes. It would be very difficult to translate this into the actual situation for a similar plant of about 10 million gallons per year capacity in the United States. In this processing scheme (distillation only), nearly 60% of the revenue would come from products with special properties peculiar to low-temperature tar, namely, the pitch and the denaphthalization oil. If the products were sold in their entirety as conventional petroleum refinery products, the revenue might be at least 1 cent/gal less on the overall product.

The second processing scheme, which is the one actually conducted at the Marienau refinery, includes tar distillation, extraction of phenols, and fractional distillation of the phenols. As mentioned before, 40% of the revenue is obtained from the phenols. On the basis of 7 cents/gal for the price of the tar, the overall product value is 14 cents/gal, which allows, after subtracting about 1.5 or 2 cents/gal for tar distillation, about 5 cents/gal on the tar, or 25 cents/gal on the 170 to 230°C phenolic oil (20% of the tar), for the extra costs involved in processing the phenols, plus revenue. Assuming a maximum of 9 cents/gal of phenolic oil for extraction with caustic, and 2 cents/gal for distillation of the phenols (which constitute 50% of the phenolic oil), this amounts to not more than 10 cents/gal of phenolic oil to process the phenols, leaving a minimum revenue of 3.5 cents/gal of 7-cent tar

TABLE 16

Elementary Economics of Processing Scheme 2, (Tar Distillation, Phenols Extraction, and Phenols Fractionation), Marienau Low-Temperature Tar Refinery

Product	% of Total Tar	Price, cents/gal*	% of Total Revenue
Diesel fuel and gasoline	21	10	14
Phenolic products	10	55	40
Denaphthalization oil	5	14	6
Heavy oil	32	8	18
Pitch	31	10	22
Losses	1		
Overall		14	

* On the basis of a cost to the refinery of 7 cents/gal for the crude tar.

before any interest or taxes. The overall price for the phenols obtained from 7-cent tar is 55 cents/gal, which readily could undersell petroleum cresylics, which are currently 70 cents/gal for the mixed phenols, and, of course, considerably more than this for the fractionated compounds. This processing scheme would appear to give a satisfactory return, because the overall product has twice the tar price, and Sabatier has implied that this is indeed the case. Here again, it would be difficult, although not as much as for the first processing scheme, to translate this into the situation for an American plant. The Marienau plant is very approximately similar in its processing plan to the refinery proposed for the Cresap, West Virginia, power complex. The latter, however, apparently would not include the diversity of products supplied by the Marienau plant, nor would it include the production of any liquid fuels, at least on the basis of information published to date, although its capacity would be fully as great as the typical East German liquid fuels plant.

Sabatier has pointed out that the costs of processing low-temperature tar depend on the economic conditions in each country. A crucial factor could be the policy of protecting national products like gasoline. The light oils and dephenolized oils from the tar give good gasoline and diesel oil, respectively, but the price will not be sufficient if these products must pay all of the taxes generally paid by gasoline. Another factor is the fluctuation in the price of petroleum products for combustion. The refiner could always sell the tar oils for combustion at the same price as the corresponding petroleum products. This is the minimum price, and, according to Sabatier, the refiner apparently could do this at the present time in France and make a little profit. This situation should change radically when large quantities of Sahara oil are imported. A more critical factor for plants like the one at Marienau is the price of coke-oven tar and coke-oven tar pitch. These vary a great deal from one country to another. Finally, there is that basic factor which is the hardest one to establish, namely, the price of the low-temperature tar itself. Sabatier has stated that, at the present time in France, the price should be essentially the same as that for coke-oven tar. He pointed out that some authorities think that the price of low-temperature tar in America should be set at about 20% below coke-oven tar. Because crude coke-oven tar is currently 17 cents/gal in the United States, this would price low-temperature tar at about 13 or 14 cents/gal, which is, perhaps, still too high a price to expect the refiner to pay.

Sabatier suggested that the minimum capacity of a carbonization plant that is to be an economic success is 500 metric tons of coal per day, and that a plant over 1000 metric tons per day would be quite certain to make money; the minimum capacity of a low-temperature tar refinery that is to be an economic success is about 16 million gallons per year, which is why the Marienau plant refines some coke-oven tar, using a separate tar still for it. These conclusions may possibly be true for France but are hardly applicable elsewhere; the Bolsover refinery in England has been an economic success for some years, by all accounts, and its capacity is only about 8 million gallons per year. Finally, Sabatier pointed out that in a study for a new refinery, it was found that the price for a low-temperature tar refinery was only 4% higher than for a coke-oven tar refinery; this seems plausible.

Pound[40] and Bristow[72] have presented information on the operation of the low-temperature tar refinery at Bolsover, England. This is not sufficiently detailed to make even an elementary analysis of the economics. However, it has been revealed that as much as 49% of the total revenue has been obtained from the phenolic products which constitute 13% of the tar. This is proportionately a very similar situation to that reported for the Marienau refinery, which obtains 40% of its revenue from phenols constituting 10% of the tar. The Marienau refinery obtains 14% of its revenue from diesel fuel and gasoline, which constitute 21% of the tar; 25% of the tar at Bolsover goes into diesel fuel and gasoline, but it cannot be assumed, therefore, that 16 or 17% of the revenue is derived from these products. It has been revealed that over £250,000 were spent through 1946 to construct the refinery at Bolsover. A number of new units have been added since then, so that the total investment can be estimated at very approximately $1.0 million for this refinery with about 8 million gallons per year capacity. It would be very difficult to translate this in terms of a plant of similar capacity in the United States, especially because the charges for construction labor are so widely different.

Minet, Smith, and Trilling[176] have estimated that a low-temperature tar refinery with a capacity of about 66 million gallons per year would represent a plant investment of about $6 million, mostly for distillation and extraction units. Lang and Lacey[119] have estimated that a low-temperature tar refinery with a capacity of about 42 million gallons per year would represent a plant investment of about $4 million, mostly for distillation and extraction units; a hydrocracking unit for the aromatic high boilers would bring the total refinery cost to about $6.5 million.

The refinery envisioned by Minet and associates would have the income from the tar proportioned as follows: light oil (3% of the tar), 3%; acid oil (34% of the tar), 53%; creosote oil (28% of the tar), 33%; and pitch (35% of the tar), 11%. The percentages of the income for the specified products appear to be somewhat out of proportion in this analysis. These investigators pointed out that they assumed only fuel value for the pitch, at $8.25 per ton, which would be about 4 cents per gallon. If a more realistic value of 10 cents per gallon had been assigned to the pitch (coke-oven pitch is currently about 18 cents per gallon), 22% of the income would have been obtained from the pitch, about the same as that actually experienced at the Marienau refinery. In addition, this would have increased the overall product value from 14 cents per gallon to 16 cents per gallon, after accounting for processing costs and discounts to overcome sales resistance. Their overall processing costs were about 3 cents per gallon. Minet[177] has estimated that the carbonizer could sell the tar to the refiner at about 11 cents per gallon and obtain a satisfactory return on his investment. Under these circumstances, the refinery envisioned might presumably be a financial success, at least if greater value were allowed for the pitch, because then the gross earnings would be 5 cents per gallon.

The refinery envisioned by Lang and Lacey[119] would increase the proportions of low-boiling aromatics by hydrocracking the heavy distillate. The proportions of products would, therefore, be considerably different than those obtained at either the Bolsover or Marienau refineries. The income was proportioned among the products as follows:[178] pitch and road binder, 41%; naphthalene, 19%; benzene, toluene, and xylene, 12% (total); aromatic solvent, 12%; phenol and cresylic acid, 10.5%; gasoline stock, 3%; creosote, 1.5%; and hydrogen (sent to hydrocracker), 1%. The income on the overall product was 18.5 cents per gallon, estimated processing costs were about 5 cents per gallon, and the cost of purchasing the crude tar was set at 9.5 cents per gallon. The gross earnings, before any taxes or interest, would be about 4 cents per gallon. At a tar cost of 9.5 cents per gallon, and allowing 10 cents per gallon for total pitch (Lang allowed about 11.5 cents), the refinery outlined by Minet and associates would give an overall product gross earning of 6.5 cents per gallon. The divergence in gross earnings per gallon would be even greater if Lang's product prices were assigned to Minet's products. This means that, if tar and product prices are set the same for the two proposed refineries, the one involving essentially nothing but distillation and extraction would earn about 7 cents per gallon, after deducting about 3 cents per gallon for processing, while the one adding a hydrocracking unit would earn only about 4 cents per gallon, after deducting about 5 cents per gallon for processing. The first could presumably be a financial success, while the second might be marginal. Lang pointed out that the return on the investment for the plant with the hydrocracking unit is low for a plant concentrating on chemicals. It would appear that this is because the emphasis was placed on the wrong chemical products, that is, aromatic hydrocarbons instead of phenols. One clue that this proposed plant is not as economically feasible as it could be is that 41% of the revenue depends on pitch and road binder.

The consensus, as derived from publications, is that the most likely commercialization of low-temperature carbonization in the immediate future will be as part of a large electric power-generating complex. Such complexes, without the carbonizing unit, have been in

[176] Minet, R. G., Smith, H. B., Jr., and Trilling, C. A., *Chem. Eng. Progr.*, **50**, 342–7 (1954).

[177] Minet, R. G., *Coal Utilization*, **10**, No. 2, 33–5 (1956).

[178] Lang, E. W., Southern Research Institute, Birmingham, Alabama, private communication, 1958.

existence and consist of a closely integrated system, with relatively low-cost coal or lignite from a mine on the site; the electric power plant; and a large-scale consumer of power, such as an aluminum refining plant, or a new domestic network of significant size. In discussing the power-plant complex planned for the city of Aleksandriya, Shchegolev[70] stated that the electricity would be produced more cheaply in this plant because char would be burned and the tar would be sold for a profit. Considerable emphasis was laid on the advantages of a power-plant complex for the production of liquid fuels from coal as compared to other processes used to produce synthetic liquid fuel. Shchegolev pointed out that the advantage of the power-plant method is based on the fact that the coefficient of utilization of the fuel is several times greater than in any other process used in the synthetic liquid-fuel industry. Therefore, it does not require the production of valuable byproducts to obtain good economic operating indices, and calculations for the Aleksandriya plant showed that even if the tar is allowed only value for its fuel constituents, the process is still definitely economical. Shchegolev concluded that one can produce liquid fuels from coal economically and readily in power-plant complexes, but not in plants specializing in synthetic liquid fuels.

The liquid fuel needs of the world as a whole will increase tremendously in the near future because of the world population explosion and the rapid industrialization and increase in living standards in many formerly "underdeveloped" regions. Although petroleum will be available in huge quantities for many years to come, the world reserves are not without limits, and are probably at best only 1% as great as the world reserves of coal. Perhaps of more significance is the fact that the cost of producing petroleum has been steadily increasing in recent years, while the cost of producing coal has not increased over the same period and might well decline in the future. From these considerations no one doubts that someday coal will be an important source of liquid fuels, but it is not at all possible to say when this will be. Of all the ways liquid fuels could be produced from coal, the oldest and simplest method, namely, low-temperature carbonization, appears, on the basis of published information, to be the most economically feasible. Jones[179] has pointed out that it is very unlikely that coal hydrogenation will be used in Britain to supplement liquid fuel needs. Similarly, Clough[180] has stated that economic factors have put coal hydrogenation out of the picture in Great Britain, but that it is practical to make liquid fuels by hydrogenation of coal-tar oils, including low-temperature tar. Clough pointed out that the Billingham plant of Imperial Chemical Industries, Limited, England, during 1957 was producing motor and aviation gasolines with premium grade octane numbers by catalytic hydrogenation of a coal-tar creosote fraction boiling from 200 to 340°C. This process was successfully operated in a full-scale plant from 1935 until gasoline production was discontinued in September 1958. Laughrey, Shaver, Ponchaud, and Skinner[181] have outlined in detail the advantages low-temperature carbonization has over hydrogenation of coal as a method of producing liquid fuels. As already mentioned, workers in the Soviet Union are of the opinion that the only economical method of producing liquid fuels from coal is by low-temperature carbonization in a power-plant complex, the char being used to produce electricity.

It is clear, however, that limitation of low-temperature carbonization plants to power-plant complexes would restrict the liquid fuels production to only about 5%, at best, of the total requirements. Assuming no decrease in the proportion of electricity produced from coal in the next ten years (it is certain now that the contribution from nuclear power plants will be small), the amount of coal used for this purpose will be about 320 million tons per year, because the demand for electricity will have doubled. Assuming that all of this new power was produced in complexes involving low-temperature carbonization, about 425 million tons of coal would be carbonized to produce a maximum of about 0.3 billion barrels of tar per year. The liquid fuels need for the United States ten years from now will be double the current 3 billion barrels per year. There are, of

[179] Jones, W. I., *Chem. & Ind.*, **1955,** No. 16, 436.

[180] Clough, H., *Ind. Eng. Chem.*, **49,** 673–8 (1957).

[181] Laughrey, P. W., Shaver, E. M., Ponchaud, G. C., and Skinner, L. C., *The Economics of Low-Temperature Carbonization and Hydrogenation of Pitch from Western Kentucky Coal, U.S. Bur. Mines, Internal Rept.* (1952).

course, other uses for the char, other than burning in power plants, such as in blending to make metallurgical cokes, and in producing fuel gas and synthesis gas, but it would be difficult to imagine to what extent these uses would increase the amount of tar produced. Therefore, it appears that if low-temperature tar is produced in significant commercial quantities in the United States within the next ten years it could not be of much importance as regards total liquid fuel requirements, but instead would have its greatest impact on the organic chemical industry, even though the majority of it might be processed to give liquid fuels.

CHAPTER 14

HIGH-TEMPERATURE TAR

J. F. Weiler

Allied Chemical Corporation, Edgewater, New Jersey.

High-temperature tar is the heavy-distillate fraction from the destructive distillation of coal where temperatures in excess of 800°C are used. In the United States, this means that high-temperature tar is almost exclusively the tar produced in horizontal, slot-type, byproduct metallurgical coke ovens. In the British Isles, only some 40% of the high-temperature tar comes from coke ovens, whereas in Germany about 80% comes from coke ovens. The remainder is produced in vertical and horizontal retorts used by the manufactured gas industry. In the coke ovens and in the vertical and horizontal retorts, the tar is recovered from the total distillate by partial condensation. The material escaping the initial partial condensation is "light oil" and gas.

High-temperature coal tar is essentially a high-temperature conversion product of the primary pyrolysis products of coal. The primary pyrolysis products are relatively hydrogen-rich and are comprised of naphthenic and aliphatic components of substantial oxygen, nitrogen, and sulfur contents. The conversion process is a composite of many elimination reactions. Water, ammonia, hydrogen sulfide, carbon dioxide, methane, and hydrogen are eliminated in processes resulting in ring closure, condensation, unsaturation, and aromatization. The degree of conversion is high but not complete and is variable. The high degree of conversion is evidenced by the almost 100% content of aromatic compounds found for the distillate portion of the tar. Higher conversion in some tars, as compared to others, results in an increased ratio of unsubstituted aromatics to alkyl aromatics and a decreased content of tar acids and tar bases. Roughly 40% of the tar is nondistillable. By destructive distillation to coke, some 70% of distillate can be obtained. Distillation residues have yielded small percentages of perylene and picene. These five-ring aromatics approach the size limit for distillability, and it has appeared reasonable that larger analogues of these compounds comprise the bulk of the distillation residue. One has but to reflect that the end products in the high temperature conversion of hydrocarbon materials are hydrogen and graphite, and the intermediate possibilities are evident.

GENESIS AND GENERAL PROPERTIES

Since 1940,[1] considerable data has become available to show, in a qualitative way, how high-temperature tar varies in composition with the composition of the coal used, the temperature employed, and the characteristics of the equipment used to destructively distill the coal. Mantel and Hansen[2] have published data from the operation of byproduct coke ovens of the Harpener Bergbau using Ruhr coals of a volatile-matter content from 19 to 30%. The coke ovens were the larger capacity ones installed during the late post-war period. The data on the effect of the volatile-matter content of the coals and the temperature of operation of the ovens on the properties of the resulting tars are shown in Figs. 1-6. Higher temperatures in the coke oven result in increased tar density and increased pitch content of the tar. The anthracene oil fraction diminishes with the increase in tar density, but no data are given for the actual content of anthracene. Crude phenol content decreases with increased tar density while naphthalene content increases.

[1] Rhodes, E. O., *Chemistry of Coal Utilization*, John Wiley and Sons, New York, 1945, Vol. 2, pp. 1287–370.

[2] Mantel, W., and Hansen, H., *Brennstoff-Chem.*, **34**, 138–48 (1953).

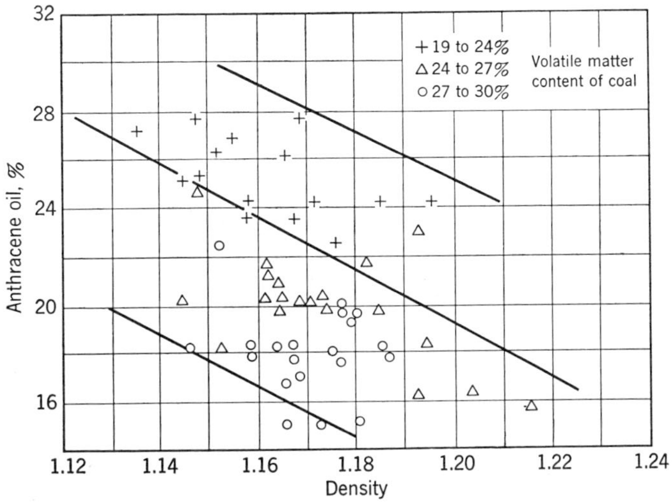

Fig. 1. Coke-oven tar density vs anthracene oil content for Ruhr coals of various volatile matter content.[2]

The benzene-insoluble content of tar increases with coke-oven temperature. The authors noted that normal coke-oven tars have benzene-insoluble contents of 1 to 7%. If the content exceeds 10%, the tar has been overheated. Decreased volatile-matter content of the coal results in tar of decreased pitch content for a given density. The ratio of toluene to phenol in tar decreases with increased coke-oven temperature. At lower coke-oven temperatures, the ratio of toluene to phenol is much higher for low-volatile coals, but the ratio decreases much faster with increased coke-oven temperature than it does for high-volatile coals.

The Coal Tar Research Association has published assays of vertical-retort tars, horizontal-retort tars, and coke-oven tars produced from coals of the British Isles.[3] With the assays of the tars, the type of coal carbonized and the type of carbonization equipment and operating conditions are given. The average properties and range of these properties for the three types of tar produced in the British Isles are shown in Table 1. The effect of higher temperature of carbonization, in the order vertical retort, horizontal retort, and coke oven is seen in the increased contents of the unsubstituted aromatics, namely, naphthalene and phenanthrene and anthracene. The decrease in total tar acids with increased temperature is also typical.

[3] The Coal Tar Research Assoc., *The Coal Tar Data Book*, Section B1, 1–3, Gomersal, England, 1957.

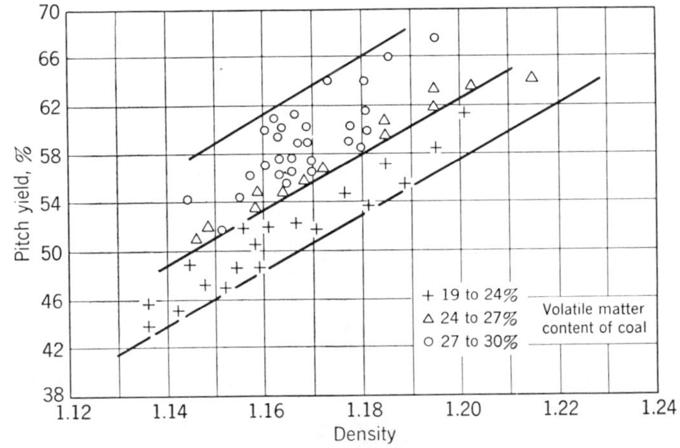

Fig. 2. Coke-oven tar density vs pitch yield for Ruhr coals of various volatile matter content.[2]

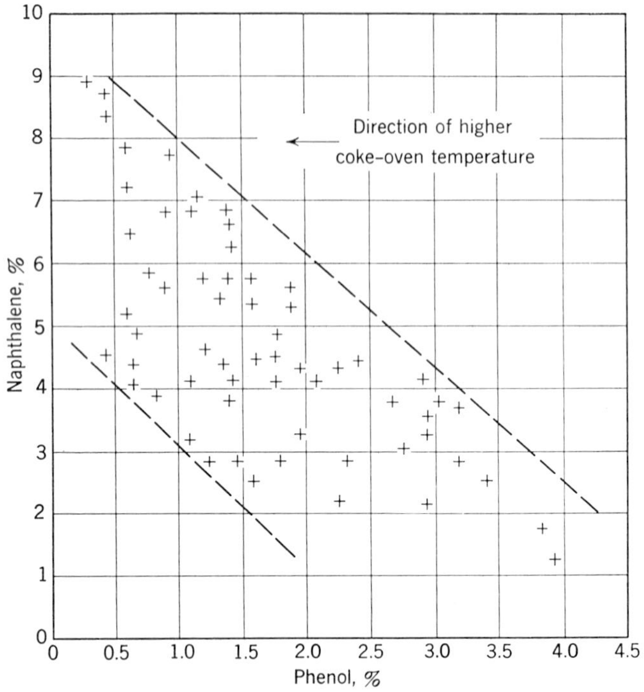

Fig. 3. Phenol content vs naphthalene content of coke-oven tars from Ruhr coals at different coke-oven temperatures.[2]

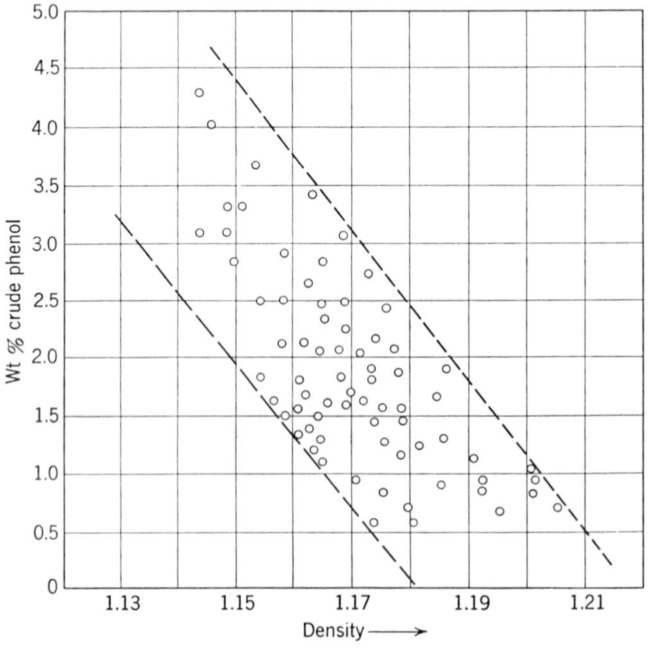

Fig. 4. Coke-oven tar density vs crude phenol content of tars from Ruhr coals.[2]

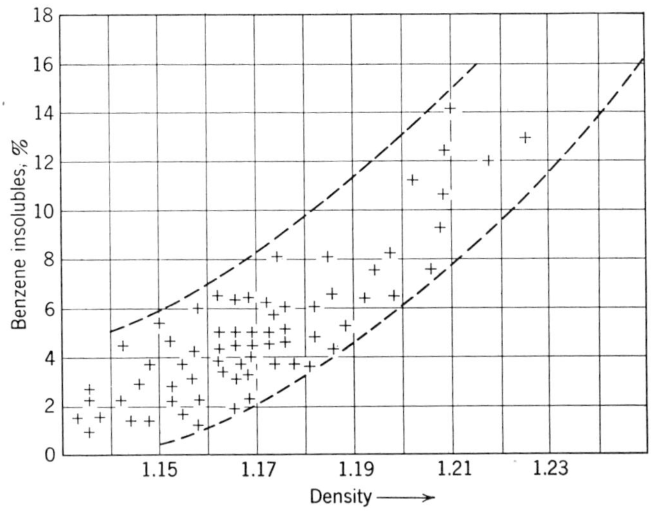

Fig. 5. Coke-oven tar density vs benzene insolubles content of tars from Ruhr coals.[2]

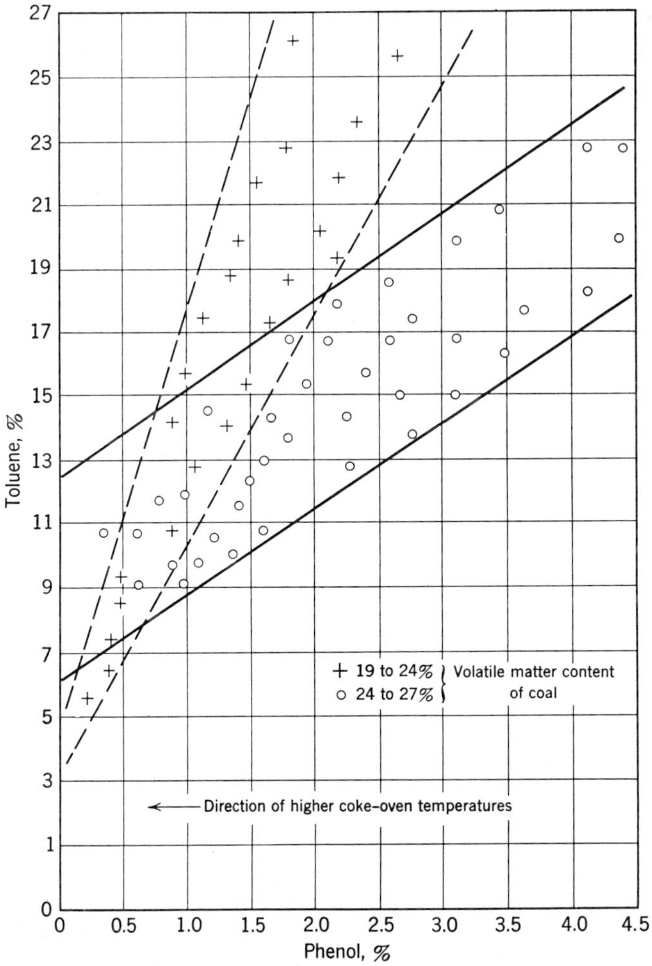

Fig. 6. Phenol content and toluene content of coke-oven tars, from Ruhr coals of various volatile matter content at different coke-oven temperatures.[2]

TABLE 1

ASSAY OF TARS FROM VERTICAL RETORTS, HORIZONTAL RETORTS AND COKE OVENS IN THE BRITISH ISLES[3]

	Type of Tar (No. of Samples)					
	Vertical-Retort (4)		Horizontal-Retort (2)		Coke-Oven (6)	
	Average	Range	Average	Range	Average	Range
Specific Gravity @ 15.5°C	1.079	0.017	1.158	0.004	1.171	0.043
Absolute Viscosity, cs @ 60°C	22.9	3.0	28.4	12.7	31.5[1]	22.4[1]
Toluene-Insoluble, %	3.7	3.9	12.5	5.1	6.6[2]	2.1[2]
Total Tar Acids, %	15.0	2.4	7.3	5.8	2.15[3]	1.92[3]
Phenol, %	0.81	0.73	1.27	0.5	0.45[4]	0.47[4]
Naphtha Fraction						
150°C/760 mm to 137°C/100 mm, %	3.12	1.02	1.93	1.1	0.71[5]	0.84[5]
Paraffins in Naphtha, wt %	18.0	14.2	5.0	1.0	1.3[6]	3.1[6]
Naphthalene, %	3.47	1.7	7.48	0.6	9.02	4.02
Phenanthrene + Anthracene, %	2.6	1.2	5.5	0.3	8.1	5.1
Pitch,[7] %	44.4	7.5	58.8	8.6	59.0	14.7

Notes: [1] One high value of 105 cs not included.
[2] One low value of 2.7% not included.
[3] One high value of 7.4% not included.
[4] One high value of 1.15% not included.
[5] One high value of 2.66% not included.
[6] One high value of 5.2% not included.
[7] Of average Softening-Point K & S 71°C, Range 5°C.

Holmes, Wilson, and Davis[4] have given data on tars from coals of Pennsylvania, Virginia, West Virginia, Kentucky, Indiana, and Washington. Seven of the coals were of low-volatile rank, four of medium-volatile and twenty of high-volatile *A* rank. The coals were carbonized by the procedure of the Bureau of Mines—American Gas Association method using an 18-in. laboratory retort. Their data have been converted to show tar composition. Figure 7 shows data for carbonization at 800, 900, and 1000°C. Tar acids decrease with temperature, while the ratio of naphthalene to anthracene increases. Both phenol content and ratio of naphthalene to anthracene are higher for higher-volatile coals.

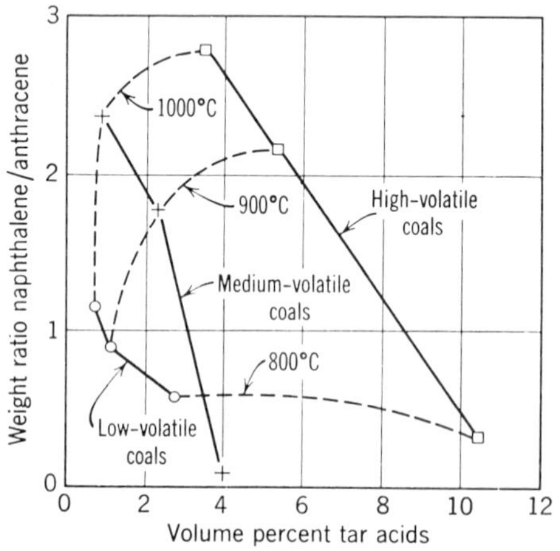

Fig. 7. Volume percent tar acids and ratios of naphthalene content to anthracene content in tars from United States coals coked at 800, 900, and 1000°C.[4]

SELECTIVE SOLVENT ACTION ON COAL TAR AND COAL-TAR PITCH

Mallison[5] has formulated an analytical test procedure based on solvent fractionation which furnishes a relative molecular-weight distribution of a pitch or whole tar. Such distributions can be useful in an empirical way for control testing in various applications to which the pitch or tar is brought. The fractions separated and the solvent used are: (1) Anthracene

[4] Holmes, C. R., Wilson, J. E., and Davis, J. D., *U.S. Bur. Mines, Rept. Invest.* No. 3695 (1943), 12 pp.

[5] Mallison, H., *Bitumen, Teere, Asphalte, Peche,* **1,** 313–17 (1950).

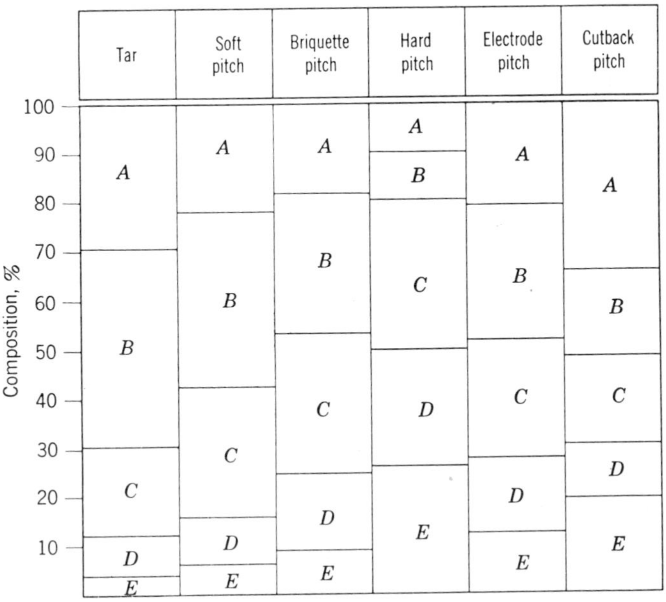

Fig. 8. Solvent fractionation of tar and pitch.[5] *A*. Low-molecular weight tar oils. *B*. Middle-molecular weight tar oils. *C*. Low-molecular weight resins. *D*. Medium-molecular weight resins. *E*. High-molecular weight resins.

oil-pyridine insoluble, high-molecular-weight resins (*H* resins); (2) benzene insoluble, high- plus medium-molecular-weight resins (*H* + *M* resins); (3) methanol insoluble, high- plus medium- plus low-molecular-weight resins (*H* + *M* + *N* resins); (4) methanol soluble-aqueous methanol insoluble, medium-molecular-weight oils (*M* oils); and (5) aqueous methanol soluble, low-molecular-weight oils (*N* oils).

Figure 8 gives the distributions found for a tar and for various types of pitches. Indicative of the nature of the three resin fractions is the equation proposed by Mallison to estimate the coke yield of a hard pitch from its contents of the various resin fractions:

$$\text{Coke yield} = 0.80(H\text{ resin}) + 0.67(M\text{ resin}) + 0.54(N\text{ resin})$$

Green and Olden,[6] working with a vertical-retort tar, have reported data on the behavior of tar-solvent systems as the ratio of solvent to tar increases. Figure 9 shows the weight percent of the insolubles precipitated from a tar by ethanol, *o*-xylene and pyridine at various solvent to tar ratios. Figure 10 shows the C/H ratio of the insolubles precipitated by ethanol and toluene at increasing solvent/tar ratios. For the poorer solvents, such as toluene and ethanol, Green and Olden considered that the initial tar dilution causes flocculation of colloidal material with co-precipitation of the less soluble of the tar-soluble material. The larger aromatic structures, and especially those with the more condensed ring structures, precipitate first. These structures have the higher C/H ratio, so that this ratio must decrease with added solvent as long as more material is precipitated. At some point, increased dilution causes a leaching of compounds from the precipitated material. The smaller structures are dissolved first, so that C/H ratio of the remaining insoluble material increases now with further dilution.

Green and Thakur,[7] also working with pitch from vertical-retort tar, studied the alkaline permanganate oxidation of C_1 and C_2 resins. The C_1 and C_2 resins are the equivalents of Mallison's *H* resin and *M* resin, respectively. C_2 resin was more easily oxidized than C_1 and gave less carbon dioxide and more steam-volatile acids and oxalic acid. C_2 behaved as a less-condensed and less-aromatic structure than C_1, a behavior supported by elementary analysis. C_1 was compared to coke or matured

[6] Green, S. J., and Olden, M. J. F., *J. Appl. Chem.* (*London*), **1**, 433–43 (1951).

[7] Green, S. J., and Thakur, B., *J. Soc. Chem. Ind.* (*London*), **69**, 238–44 (1950).

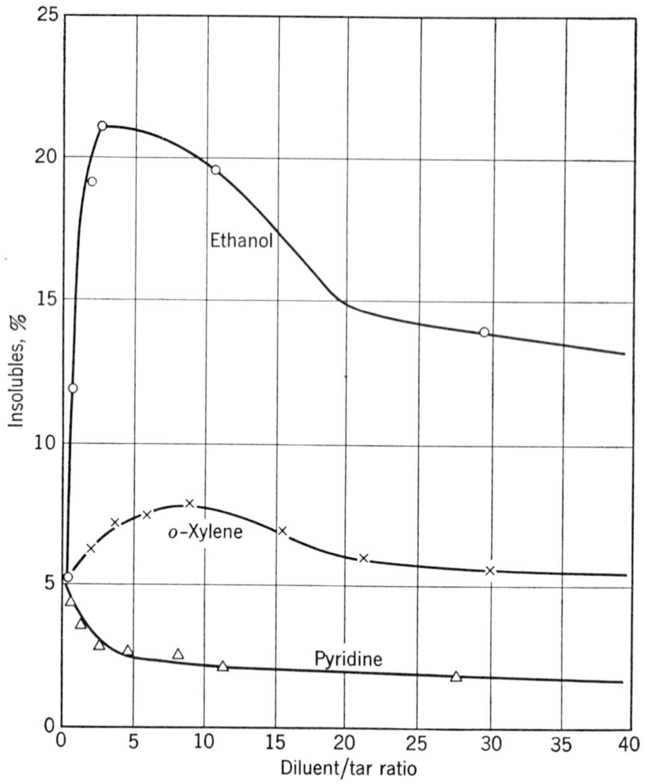

Fig. 9. Percent of insoluble matter separated from tar on addition of the solvents, ethanol, *o*-xylene, and pyridine, to tar at various solvent tar ratios.[6]

coal, while C_2 was likened to the tar resins, that is, the higher-molecular-weight material of the benzene-soluble portion of the tar.

Green and Mukherji[8] examined vertical-retort tar and pitch by solvent fractionation. The tar and pitch were digested in hot benzene. After cooling and filtering the digested mix, the benzene solution was diluted many-fold

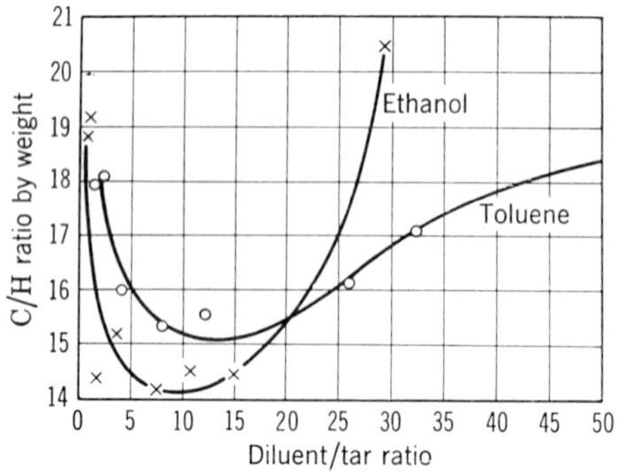

Fig. 10. C/H ratio of insoluble matter separated from tar on addition of the solvents, ethanol, *o*-oxylene, and pyridine, to tar at various solvent tar ratios.[6]

with petroleum ether to precipitate resinous matter. The yield of resinous matter was 23% on pitch and 11% on tar. The resinous matter was purified by solution in chloroform and reprecipitation by petroleum ether. The purified resins were treated in solution successively with Na_2CO_3, NaOH, and HCl to separate acidic, phenolic, and basic components. The pitch resins were 99% neutral component. The neutral portion had a molecular weight in chloroform of 815. On fractional precipitation from chloroform, the precipitate became progressively lighter in color and decreased in molecular weight through seven fractions from 934 to 304. Compared to this are the properties of the resins recovered from the tar before distillation. The resinous material from the tar was 36% phenolic, 34% neutral, and 15% precipitated by HCl.

The neutral resinous matter from the pitch was subjected to various chemical treatments

[8] Green, S. J., and Mukherji, S. M., *J. Soc. Chem. Ind. (London)*, **67**, 438–41 (1948); **68**, 16–19 (1949).

to reveal its structure. Reaction with bromine was largely by substitution. Benzene carboxylic acids were among the products of alkaline permanganate oxidation. Hydrolysis with 20% NaOH in glycol monomethylether, thermal decomposition at 1 mm Hg pressure at 140 to 200°C, and hydrogenation at 220 to 240°C under 175 atm pressure with Raney Ni in alcohol all produced alkyl OH and phenolic OH material, suggesting the presence of alkylaryl ether linkages. Nitrogen of the resins was not removed, but H_2S was split out. Dehydrogenation with Se gave anthracene.

CHEMICAL CHARACTERIZATION OF COAL TAR AND COAL-TAR PITCH

In 1940, Kruber[9] reviewed the results of work to that time on isolation and identification of the components of coal tar. A list of compounds identified as components of high-temperature tar up to 1940 was also presented by Rhodes.[1]

Since 1940, most contributions to the list of compounds identified in coal tar have been compounds that occur in the pitch fraction. The Coal Tar Research Association has published a list, complete to 1956, of compounds found in coal tar.[10] Both Franck[11] and Schäfer[12] have published lists of compounds found up to 1955 in the higher-distilling fractions of coal tar. Since these listings, Lang and associates[13] have added six more compounds. To appreciate the significance of such lists, one must bear in mind the following rough estimate of the material balance between total material and that accounted for by identified components of tar fractions:

Boiling-Point Range, °C	Fraction of Tar, %	Fraction Accounted for by Identified Components, %
to 250	20	90
250 to 360	15	70
360 to 450	15	35
>450	50	5

In the higher distilling portion of coke-oven tar represented by the pitch fraction, where 5000 compounds have been estimated to be present, only 75 have been separated. Schäfer[12] quoted the literature in saying that about two-thirds of these are isocyclic, the other third heterocyclic. All are condensed polynuclear ring compounds. Approximately one-half of these compounds are substituted, and only a few are partially hydrogenated. About 60% of the substituted compounds are methyl substituted, another 30% are phenolic. The phenyl group, nitrile group, and the carbonyl group each appear once as substituent groups. No compounds with methyl groups distill at over 410°C, and none with the phenolic hydroxyl distill at over 438°C. The highest boiling substance is fulminene, which boils at 555°C. In general, the heterocyclics have higher boiling points than the corresponding isocyclic hydrocarbons. Schäfer emphasized the stability of aromatic compounds at high temperature due to the resonance energy associated with the aromatic conjugated double-bond system. Their stability accounts for their exclusive presence in coal tar with its history of exposure to high temperature. The various polynuclear compounds vary widely in extent of aromatic conjugation and, therefore, in stability, or conversely, in reactivity. The reactive structures are those which cause tar and pitch to increase in viscosity and softening-point on heating.

After this background material, experimental data are given on the characterization of fractions separated from pitch by solvent fractionation. In making a choice of solvents to effect fractionation, the use of polar compounds is avoided. The high affinity of aromatic hydrocarbons for molecules containing dipoles, as shown by molecular-compound formation, makes it difficult to free the pitch fractions from solvent. Coal-tar pitch was first separated into alpha-bitumen (insoluble in benzene), into beta-bitumen (soluble in benzene—insoluble in petroleum ether), and into gamma-bitumen (soluble in petroleum ether). The

[9] Kruber, O., *Angew. Chem.*, **53**, 69–74 (1940).

[10] The Coal Tar Research Assoc., ref. 3, Section A4.

[11] Franck, H. G., *Brennstoff-Chem.*, **36**, 12–20 (1955).

[12] Schäfer, H. G., *Freiberger Forschungsh.*, **A51**, 35–66 (1956).

[13] Lang, K. F., Buffleb, H., and Schweym, E., *Brennstoff-Chem.*, **40**, 369–70 (1959).

TABLE 2

SEPARATION OF COAL-TAR PITCH (SUPPLIED BY THE FIRM COLAS) FROM ZWICKAU INTO α-, β-, AND γ-BITUMEN, AND PROPERTIES OF THESE FRACTIONS[12]

Designation	Yield from the Initial Pitch	Appearance	Softening Point, or Solidification Point	Elementary Composition C	H	O + N + S	Coke Residue	Molecular Weight of the Bitumen
			°C	%	%	%	%	
Coal-tar pitch from Zwickau, supplied by the firm Colas, Dresden	—	Black, shiny, shell-like fraction; ash 0.28 %; total S 0.86 %	According to Kraemer-Sarnow 69°	91.7	4.22	4.08	38.5	—
Bitumen obtained from this pitch								
α-Bitumen (insoluble in benzol)	27.4	Black powder	No softening up to 300°	90.14	3.32	6.54	87.7	800
β-Bitumen (soluble in benzol, insoluble in petroleum ether	40.7	Solid black-brown mass	According to Kraemer-Sarnow 92.9°	90.75	4.38	4.87	35.3	320
γ-Bitumen (soluble in petroleum ether)	31.9	Brown, viscous	Solidifying Point 8°	90.30	5.17	4.53	3.43	210

properties of these fractions are shown in Table 2. The molecular weights were determined cyroscopically in phenanthrene. Both phenanthrene and pyrene were found to be outstanding materials for this purpose. An attempt to further fractionate the beta-bitumen by solution in chloroform and precipitation by successive additions of petroleum ether gave precipitates which contained chlorine even after vacuum drying. No correlation could be shown between chlorine content and molecular size. Fractionation of beta-bitumen was obtained by solution in benzene and reprecipitation by successive additions of petroleum ether. Gamma-bitumen was refractionated into the following fractions: (1) Insoluble from 100 g bitumen in 480 cc warm heptane; (2) soluble in warm heptane, insoluble in cold heptane; (3) insoluble upon addition of 300 cc of methanol to the cold heptane solution (two layers formed); (4) soluble in heptane layer; and (5) soluble in methanol layer. Table 3

TABLE 3

FRACTIONS PROGRESSIVELY PRECIPITATED FROM SOLUTIONS OF BETA- AND GAMMA-BITUMEN PREVIOUSLY SEPARATED FROM ZWICKAU COAL TAR[12]

Beta-Bitumen	Properties	Yield	Molecular Weight	Elementary Composition C	H	N	S	O	Coke Residue
		%		%	%	%	%	%	%
1	Powder, black	32.2	600	91.23	4.46	1.27	0.83	2.21	50.2
2	Powder, black	5.6	530	91.31	4.60	1.27	0.80	2.02	53.2
3	Powder, dark brown	5.6	400	91.75	4.53	1.24	0.81	1.67	35.3
4	Powder, dark brown	7.9	360	91.53	4.62	1.35	0.81	1.69	33.4
5	Powder, dark brown	4.7	340	91.88	4.72	1.08	0.81	1.51	30.5
6	Powder, brown	3.6	320	91.50	4.86	0.67	0.69	2.28	17.9
7	Powder, red brown	10.5	310	91.89	4.94	1.53	0.66	1.98	20.6
8	Sticky mass, red brown	8.7	240	91.58	4.97	0.93	0.60	1.92	7.48
9	Salve-like, red brown	21.0	230	92.67	5.11	0.63	0.42	1.17	4.37
Gamma-Bitumen									
1	—	—	270	91.59	5.04	0.62	1.00	1.75	—
2	—	—	206	91.60	5.23	0.49	1.00	1.68	—
3	—	—	269	89.67	5.14	0.65	0.87	3.67	—
4	—	—	200	92.09	5.28	0.25	0.78	1.60	—
5	—	—	245	92.64	5.40	0.43	0.97	0.56	—

shows the properties of fractions obtained from refractionation of beta- and gamma-bitumen.

Attempts were also made to refractionate the bitumen fractions by column chromatography. Separation into various colored portions was obtained, but it was found that photo-chemical reactions took place with gamma-bitumen and, to a lesser degree, with beta-bitumen on activated alumina. Paper chromatography by the ring technique was then tried. In this manner, using mixtures of benzene, petroleum ether, and propanol, whole bitumen (benzene soluble) was divided into three parts, a dark brown center spot, considered α resin; an ochre-colored ring surrounding the center, considered β resin; and a yellow ring outside the ochre ring, the γ resin. Beta-bitumen was completely soluble in a mixture of benzene and petroleum ether. By adding a lyophobic component, such as propanol, the solubility was gradually decreased, and the rate of elution by the solvent became different for the various components. Separation resulted in the three characteristic zones described above. From this work, it was concluded that solvent precipitation, or solvent extraction techniques, did not yield pure resin fractions but fractions enriched in α, β, and γ resins, respectively. Mutual solubilization prevented a better separation.

Estimates were made of the number of rings in the polynuclear structures of the substances in the three classes of bitumen by means of the additive properties, density, and molar refraction. From analytical data and molecular weight, the average formula for pure beta-bitumen was $C_{23.78}H_{13.78}O_{0.46}N_{0.20}S_{0.08}$ and for gamma-bitumen $C_{20.07}H_{13.86}O_{0.41}N_{0.12}S_{0.06}$. From the number of carbon atoms per molecule and the number of rings per carbon atom, R/C (also called ring index), the number of rings per molecule was calculated. (The ring index is a function of density.) Molar refraction, calculated as an additive property of atomic refractions and identified by symbol $R'm$, was compared with measured molar refraction, Rm. The difference, the molar refraction increment, Im, is a function of the number of condensed rings in the molecule.

The structures of the two bitumens, estimated from ultimate analyses, molecular weights, and densities on the one hand, and from molar refractions on the other, are given immediately below:

Bitumen	R/C	Carbons per Molecule	Rings per Molecule
Beta	0.219	23.78	5.21
Gamma	0.209	20.07	4.19

	Rm	*R'm*	*Im*	Rings per Molecule
Beta	100.96	77.21	33.75	5 to 7
Gamma	91.84	67.17	24.67	4 to 5

Solid fractions from beta-bitumen showed strong diamagnetic properties when pulverized, but the information led to no quantitative conclusions. Diamagnetic anisotropy in polycyclic compounds increases with an increase in the number of rings, but the increment does not follow a simple law. It is dependent on the extent of aromatic conjugation in the molecule.

Work on the volatilizable oils from a pitch from vertical-retort tar[14] affords us a detailed look at the high-boiling material of a tar which has undergone less thermal conversion than the highly aromatic coke-oven tars. The separation techniques of this work are typical. A soft pitch, produced by pot-still distillation of a British vertical-retort tar, was steam distilled to a hard pitch. The distillate obtained in the steam distillation was examined in this work. It was diluted with an equal volume of benzene and extracted with 10% sodium hydroxide and 20% sulfuric acid. The oil contained 30% by weight of alkali-soluble material and 6% of acid solubles. The neutral oil was fractionally distilled under vacuum. Five fractions boiling between 205 and 270°C at 10 mm pressure were collected for study. Lower-boiling material was discarded as anthracene-oil components of no interest in this study. Each of the fractions was treated with light petroleum to separate small amounts of insolubles for separate study. The light-petroleum solutions were fed to a chromatographic column and the same solvent

[14] Harris, A. S., White, E. N., and McNeil, D., *J. Appl. Chem. (London)*, **3**, 443–51 (1953).

used for development and initial elution. The chromatographic column consisted of a silica gel layer between activated alumina layers. Further elution was done successively with benzene, chloroform, and ethanol. The chromatographic separation produced four distinct types of materials which were eluted in the following order: (1) Soft, colorless aliphatic waxes with no fluorescence in ultraviolet light; (2) pale-yellow oils exhibiting an intense purple fluorescence in U.V.; (3) yellow, oily or waxy solids; and (4) dark brown viscous oils. The percent of insoluble material and of each of the chromatographic fractions in the distillate fractions were:

Fraction: boiling range at 10 mm Hg	Relative Amounts	Percentage Composition of each Fraction: Insoluble	Waxes	Fluorescent Oils	Yellow Solids	Brown Oils	Loss
°C	%						
205–220	30.3	4.6	8.6	32.2	35.7	17.1	1.8
220–235	31.3	0	13.0	39.8	20.4	25.4	1.4
235–245	19.5	8.4	15.2	30.8	24.6	20.0	1.0
245–255	10.5	2.6	13.6	30.5	24.2	25.2	3.9
255–270	8.4	3.8	19.2	37.2	15.1	23.0	1.7

The yellow, oily or waxy solids received principal attention. They were fractionally distilled and the fractions examined by fractional crystallization, chromatography, and picrate formation. Isolated and identified were fluoranthene, pyrene, brazan, 2,3-benzfluorene, 1,2-benzanthracene, 4-methylpyrene, 1,2,8-trimethylphenanthrene, naphthacene, 5,12-dihydronaphthacene, alkyl brazan, triphenylene, and chrysene.

The fluorescent oils contained about 25% of picrate-forming oils which were fully aromatic polycyclic compounds. The nonpicrate-forming oils had molecular weights of about 250 and refractive indices (n_D^{25}) from 1.54 to 1.62. Absorption spectra corresponded to those of substituted naphthalenes, namely, 2-phenylnaphthalene, but with a bathochromic shift of about 100A. Dehydrogenation techniques produced no change in the character of the material. It is of interest that neutral pitch-oils derived from coke-oven tar, produced under conditions of higher temperature, contain less than 5% of fluorescent oils.

The waxes and brown oils were examined only briefly. The refractive indices of the waxes (n_D^{25}) were in the range 1.46 to 1.49 and the melting points in the range of 50 to 60°C. Approximately 52% formed a clathrate with urea. A lighter fraction of the wax analyzed 84.7% C and 14% H. The brown oils contained a much higher proportion of oxygen and nitrogen than the other chromatographic fractions. A typical analysis was C, 82.5; H, 7.8; N, 1.6; S, 0.5, and O by difference, 7.6%. The nitrogen heterocycle, 1,2-benzcarbazole, was recovered from this fraction.

In a further attempt[15] to study still more of the pitch, a blend of soft pitch, 60% of vertical-retort tar origin and 40% of horizontal-retort tar origin was distilled to a coke residue. The distillate itself distilled 5% by volume to 300°, 15% to 380°, and 25% to 415°C at atmospheric pressure. It was considerably less volatile than the pitch-steam distillate studied previously. This distillate was fractionated much as was the steam distillate. The proportions by weight of waxes, fluorescent oils, solids, and brown oils was 4, 5, 69, and 22, respectively, compared to 13, 35, 30, and 22 for the steam distillate. Components of the coking distillate which were not found in the steam distillate were 1,12-benzperylene, picene, naphthalfluorene, perylene, 3,4-benzpyrene, 1,2-benzpyrene, 11,12-benzfluoranthene, 3-methyl chrysene, 2,3-benzcarbazole, phenanthridone, 1,9-benzxanthene, propionamide, and 2,3-dimethylanthracene.

Sack[16] endeavored to go still further in volatizing pitch. He sought to "depolymerize" pitch by heating to a high temperature in a short-path still with a cold finger to collect

[15] *Ibid.*, **6,** 293–7 (1956).

[16] Sack, H. A., *J. recherches centre natl. recherche sci., Labs. Bellevue (Paris)*, **16,** 21–30 (1951).

volatilizing "monomer" units. The pitch appears to have been a pitch from a vertical-retort tar (supplied by Gaz de Paris, it had a density of 1.275 g/cc at 15°C, softening point, Kraemer and Sarnow, of 68°C, benzene insolubles of 20.4%, and nitrobenzene insolubles of 1.2%). Operating at 15 mm Hg pressure, three fractions were obtained: (1) Between 300 and 380°C, 15% of the pitch, having a clear yellow color; (2) between 380 and 450°C, 30% of the pitch, consisting of bright red resins of varying hardness; and (3) between 450 and 550°C, 25% of the pitch in the form of an orange-yellow powder. Fractions 2 and 3, 1.5 g each, required 2 and 4 hr, respectively, to collect. The factor of thermal decomposition must have played a considerable part in the process. The following properties of the fractions were noted: (1) Fractions 2 and 3, corresponding to 55% of the pitch, were completely soluble in absolute alcohol immediately on removal from the condenser while only 3% of the original pitch dissolved; (2) on exposure to sunlight for several days, the products gradually lost their solubility in alcohol; (3) heating fractions 2 and 3 at 120°C for 24 hr with benzoyl peroxide converted them into a pitch and concentrated mineral acid added to a benzene solution of these fractions also caused pitch formation; (4) the "repolymerized" pitch could be again "depolymerized" to give fractions 2 and 3; and (5) the benzene solutions instantaneously decolorized a dilute solution of $KMnO_4$.

Ultimate analyses of the fractions and the original pitch were:

Element	Fraction 2	Fraction 3	Original Pitch
C	90.81%	91.94%	90.8%
H	5.27	5.58	5.1
O (Direct)	2.42	2.58	—
O (By difference)	—	—	2.3
S	0.74	—	1.2
N	—	—	0.6
C/H (Atomic)	1.45	1.39	1.49

The molecular weights, determined cryoscopically in benzene, were 250 and 170 for fractions 2 and 3, respectively. These properties were said to indicate ethylenic monomers having a ring system of 3 or 4 fused rings, one ring being heterocyclic or alicyclic and containing the ethylenic linkage. Polymerization and depolymerization through the ethylenic bond were proposed to account for the evidence of a reversible association–dissociation of relatively small stable structures of fused-ring aromatic character.

Franck[11] questioned that the few isolated compounds were typical of the composition of pitch. He noted the plastic character of pitch, and even of fractions of pitch, obtained by selective fractionation processes and compared this with the well-defined crystalline character of the few isolated components. It may be that the properties of adhesiveness and plasticity derive from macromolecular components which do not crystallize. If the macromolecular material is a condensation product of known tar constituents, its molecular weight must be at least over 350, which would correspond to a trimer of a two-ring aromatic monomer. Data on molecular weights of pitch fractions were presented to resolve this question. By vacuum fractional distillation, a series of distillate fractions totaling 50% of the pitch was obtained. The molecular weight of each fraction was determined ebullioscopically in benzene and chlorobenzene. Figure 11 shows the fractional distillation results. The boiling-point curve shows concentration of components at certain boiling points. The crystallization-point curve shows more effectively the concentration of these components, concentration being proportional to the breadth of a maximum or minimum. Components are identified with the various maxima or minima by comparison with known boiling-points and melting-points. Molecular-weight determinations were consistent with the preceding selections.

By exhaustive extraction, a benzene-soluble fraction of the pitch was obtained in 80% yield. The molecular weight of this benzene-soluble fraction was 276 by ebullioscopic measurement in benzene. Furthermore, the benzene-insoluble portion of the pitch was extracted exhaustively with chlorobenzene. The additional 3% of pitch taken into solution showed an average molecular weight of 329 by ebullioscopic determination in chlorobenzene. A molecular weight distribution for 83% of the pitch is shown in Fig. 12. The mean molecular weight for the 30% of the benzene solubles not obtained as distillate is calculated from the molecular weights of the total solubles and of the distillate fractions. Because no molecular weight over about 330 was found, it was concluded that macromolecules were not present. Hence, some other explanation for the plastic

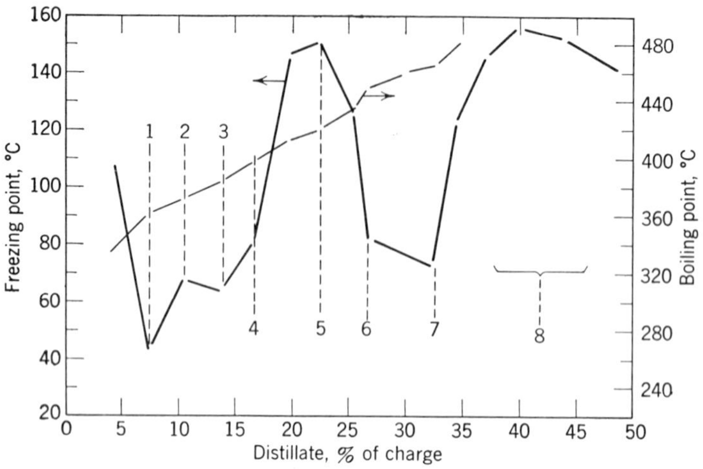

Fig. 11. Fractional distillation of coal-tar pitch at 5 mm Hg pressure, 15:1 reflux ratio, in a column of eighteen theoretical plates. Boiling points corrected to atmospheric pressure.[11] 1. Phenanthrene; 2. fluoranthene; 3. pyrene; 4. benzofluorene; 5. chrysene; 6. benzofluoranthene; 7. benzopyrene; 8. picene.

properties of pitch was required. For the pitch itself, the concept of an eutectic of many components was acceptable. For a fraction of the pitch obtained by selective solvent extraction, it might be argued that a multiplicity of components would not be available. Solvent-separated fractions were prepared from 20 kg of pitch and were analyzed, in detail, to show what number of components might be expected. The solvent fractionation was carried out in the following manner: Powdered pitch was treated with an equal weight of warm chlorobenzene, and this partial solution was cooled and mixed with eighteen times its weight of decalin to precipitate decalin insolubles. The insolubles were filtered off, dried in vacuum and treated with an equal weight of chlorobenzene. This partial solution was then mixed with eighteen times its weight of xylene to precipitate xylene insolubles. In a final repetition of this procedure, pyridine insolubles were prepared. The corresponding filtered solutions were freed of solvent by vacuum distillation to give the soluble fractions.

The yields and properties of the fractions so obtained from 20 kg of pitch are shown in Table 4. An analytical distillation of the decalin solubles yielded 70% distillate, 30% boiling below chrysene, the other 40% comprising chrysene and picene. Of the xylene solubles,

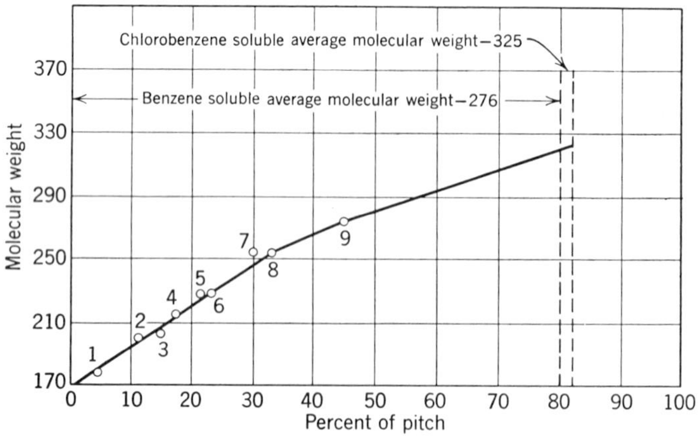

Fig. 12. Molecular weight distribution of fractions obtained from coal-tar pitch by fractional distillation and solvent extraction.[11] 1. Phenanthrene; 2. fluoranthene; 3. pyrene; 4. benzofluorene; 5. 1,2-benzanthracene; 6. chrysene; 7. benzofluoranthene; 8. benzopyrene; 9. picene.

30% was taken as distillate, the first 5% boiling below chrysene. The conclusion was drawn that macromolecules were not present in the soluble fraction of pitch, and that macromolecules were not necessary to explain the properties of pitch. The higher-molecular weight, soluble components of pitch were comprised of condensed-ring aromatic compounds of up to seven rings. The most plentiful of these components was the four-ring compound, fluoranthene, present in tar at 2.2% concentration, which is exceeded only by naphthalene and phenanthrene.

Wood and Phillips[17] were concerned that little or no data were available to define the nature and quantity of the dispersed phase implicit in the explanation of tar properties as those of a colloidal system. They sought to provide the missing data through ultracentrifuge measurements and molecular-weight measurements by an osmotic pressure technique. They advanced certain objections to cyroscopic and ebullioscopic methods for the molecular-weight determination of the higher-pitch fractions. In order to get sufficiently large temperature differences, one must use high concentrations, often approaching saturation, where large departures from Raoult's law may occur. Also, as both methods determine number-average molecular weight and are responsive to retained solvent molecules, traces of retained solvent will considerably depress the results. The high concentrations needed restrict the use of these methods to fractions that are more than sparingly soluble in the appropriate solvents. Osmotic pressure methods require only very low concentrations, and retained solvent has a negligible effect, because the solvent passes through the membrane.

Five commercial pitches of softening-point 75 to 80°C (ring-and-ball) were examined in detail: (1) Pitch obtained from tar from high-temperature coke ovens (CO1) by distillation in a pot still; (2) pitch from a blend of tars produced in both low- and high-temperature coke ovens and distilled in a Wilton pipe still (CO2); (3) pitch from a blend of 80% horizontal retort and 20% coke-oven tars, distilled in a Lennard pipe still (HR); (4) pitch from a blend of vertical-retort tars distilled in a Foster-Wheeler pipe still (VR); (5) pitch from low-temperature tar distilled in a pipe still (LT); and two other materials (6) a hard pitch, 170°C (ring-and-ball) produced by blowing air through pitch CO2 at 325°C (AB); and (7) a 200-penetration Venezuelan straight-run bitumen with 10.8% asphaltene content. A preliminary separation of each pitch was made by exhaustive extraction by boiling under reflux with light petroleum (100 to 120°C boiling range), benzene, nitrobenzene, and quinoline, in succession.

For molecular-weight determinations, the extracts were concentrated by distillation to give a volume of 50 ml/g of original pitch. The benzene extract was then fractionated by stepwise addition of petroleum ether, each addition followed by filtration. The nitrobenzene extract was similarly processed with petroleum

TABLE 4

PROPERTIES OF A COAL-TAR PITCH AND OF FRACTIONS FROM THE SOLVENT FRACTIONATION OF 20 KG OF THE PITCH: (I) DECALIN SOLUBLE, (II) XYLENE SOLUBLE, (III) PYRIDINE SOLUBLE, (IV) PYRIDINE INSOLUBLE, AND (V) WHOLE PITCH[11]

Material	I	II	III	IV	V
Yield, %	46	20	23	11	100
Softening Point (Kraemer-Sarnow), °C	15	70	240	Nonmelting	68.5
Brittle Point, °C	−6	43.7	—	—	38.0
Number-average molecular weight	240	280	360	—	—
C, %	92.5	91.9	90.4	92.3	92.1
H, %	5.1	4.7	4.1	3.0	4.5
N, %	1.1	1.2	1.5	0.8	1.3
S, %	0.8	0.7	0.6	1.0	0.8
O, % (by difference)	0.5	1.5	3.4	2.9	1.3

[17] Wood, L. J., and Phillips, G., *J. Appl. Chem. (London)*, **5**, 326–38 (1955).

TABLE 5

SEPARATION OF VARIOUS COAL-TAR PITCHES INTO FRACTIONS BY SOLVENT EXTRACTION: YIELDS AND NUMBER-AVERAGE MOLECULAR WEIGHT ($\overline{M}_n$) OF FRACTIONS[17]

Extraction Solvent	Coke Oven Tar, 1		Coke Oven Tar, 2		Horizontal Retort Tar	
	Yield	$\overline{M}_n$	Yield	$\overline{M}_n$	Yield	$\overline{M}_n$
	%		%		%	
Petroleum Ether	46.1	250	46.1	213	41.1	221
	6.3	356	5.5	314	8.4	314
	2.2	444	2.3	399	12.8	411
	9.3	482	11.0	399	11.4	512
Benzene	1.0	421	1.0	534	0.4	684
	2.0	646	6.2	613	1.7	1173
	2.3	771	1.6	832	2.1	1409
	2.7	973	3.9	1083	0.6	1461
			2.4	1180	1.5	1549
	5.2	1085	5.0	1414	3.5	1810
Nitrobenzene	0.9	1175	1.1	1719	2.0	1929
	1.6	1405	5.3	1834	3.7	2300
	7.7	1623	4.6	2031		
Quinoline	6.5	1794	0.3	—	3.6	2615
(Insoluble)	6.2	—	3.7	—	7.2	—

Extraction Solvent	Vertical-Retort Tar		Low-Temperature Tar		Air-Blown Pitch		Bitumen	
	Yield	$\overline{M}_n$	Yield	$\overline{M}_n$	Yield	$\overline{M}_n$	Yield	$\overline{M}_n$
	%		%		%		%	
	41.1	223	55.6	248	12.2	309	78.8	377
Petroleum Ether	23.1	341	7.8	334	0.9	303	4.8	556
	5.2	486	11.8	464	1.7	566	16.4	3632
	7.0	512	8.5	346	30.7	861	8.5	2196
Benzene	0.7	620	0.8	334	1.3	973	7.1	4797
	1.3	777	2.8	364	2.5	1071		
	1.8	901	2.3	438	1.7	1272		
	1.5	914	1.1	447	0.7	1223		
	3.1	976	2.9	447	0.1	—		
	2.7	1001	1.8	664	6.1	1667		
Nitrobenzene	1.3	1107	1.2	827	4.0	2683		
	2.0	1268	2.8	1326	2.9	4460		
	1.3	1428			2.5	5089		
Quinoline	0.0		0.0	—	12.3	5568	0.8	5115
(Insoluble)	7.9		0.6	—	20.4	—	Trace	—

ether. The preliminary petroleum-ether extract was subdivided by successive exhaustive extractions. The quinoline extract was not further subdivided. In the molecular-weight determinations, static measurements of osmotic pressure were made in the apparatus of a modified Zimm-Myerson[18] design by using polyvinylalcohol membranes.[19] The equilibrium or the

[18] Zimm, B. H., and Myerson, I., *J. Am. Chem. Soc.*, **68,** 911–12 (1946).

[19] Hookway, H. T., and Townsend, R., *J. Chem. Soc.*, **1952,** 3190–5.

half-sum methods were used, and after each measurement a semimicro concentration determination was made on the sample removed from the osmometer. No appreciable loss of solute through the membrane occurred with samples of average molecular weight exceeding 350. Values as low as 183 were obtained by the half-sum method, in close agreement with cryoscopic measurements. Measurements were made at three concentrations to permit extrapolation to zero concentration. The reduced osmotic pressure nearly always decreased with increased concentration, but so slightly that the departure from van't Hoff's law was negligible. Pure aromatic compounds, ranging in molecular weight from 252 to 516, were correctly measured by the method. The results were independent of choice of solvent between benzene, nitrobenzene, and quinoline. Results are shown in Table 5.

Figure 13 shows the molecular-weight distribution curves for the five commercial pitches, the air-blown pitch, and the bitumen. The curve for the pitch from coke-oven tar (CO2) includes the cryoscopic molecular weights of six cuts obtained from a high efficiency fractional distillation and comprising 20% of the pitch. The distillate fractions form a smooth continuous curve with the solvent separated fractions.

The curves of the five commercial pitches are almost identical over the low-molecular-weight range, 50% of each comprising molecules of molecular weight between 150 and 350. There are substantial differences between the remaining 50% of the components of the pitches. The upper limit of molecular weights extends from 1400 for the low-temperature-tar pitch to 2800 for horizontal-retort-tar pitch. The molecular weights of the heavier half of these pitches

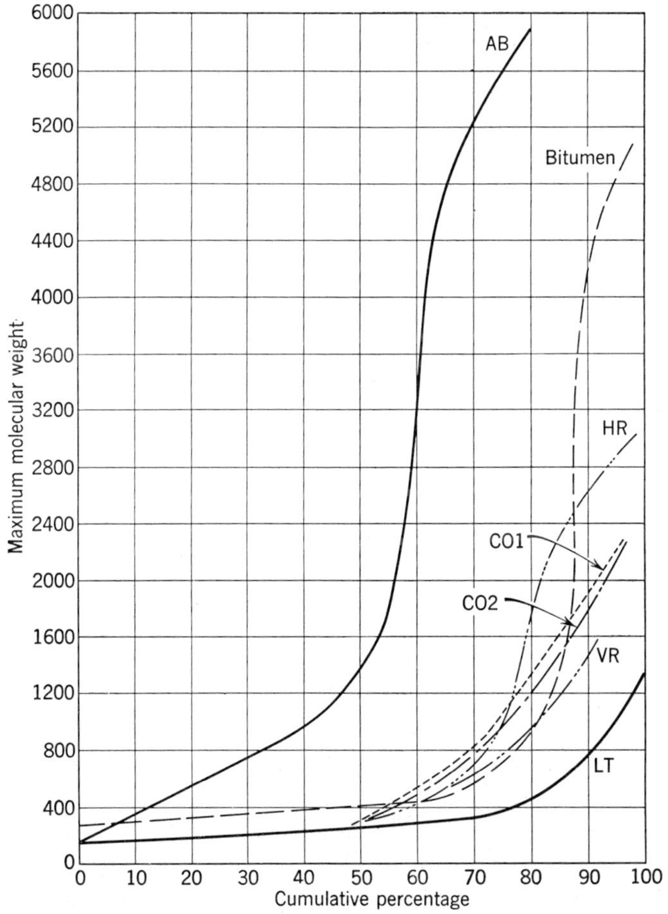

Fig. 13. Molecular weight distributions for pitches from coal tars of various types and for a typical bitumen.[17]

increases regularly with the carbonization temperature producing the respective tars.

The curve for the air-blown pitch is quite different. It starts at the same low-molecular weight but climbs more steeply to a higher order of upper-limit molecular weight. A comparison of molecular-weight distribution before and after air-blowing indicates that polymerization or condensation of molecular species in all molecular-weight fractions has occurred. The formation of water in the air-blowing process suggests that condensation is the predominant reaction. This is a significant point in respect to the chemical constitution of tar. It calls for the presence in tar of a substantial content of components containing reactive hydrogen groups such as the methylene group. If the coke-oven-tar pitch no. 2 and the air-blown pitch are adjusted to the same hardness as the bitumen (by needle-penetration test) by addition of tar oil, all three molecular-weight-distribution curves are alike except at the high-molecular-weight range where the bitumen and air-blown pitch curves extend to much higher molecular weights. It is suggested that the much lower temperature susceptibility of viscosity of the bitumen and air-blown pitch, compared to the normal coke-oven-tar pitch, is due to the extreme range of molecular weight and not to chemical differences.

The preceding molecular-weight study was augmented by elementary analysis. Further samples of the five commercial pitches were separated into primary fractions as before by successive extraction with solvents of increasing power. Elementary analyses, including direct determination of oxygen, were made for each fraction. Because these primary fractions were obtained in nearly the same yield as before, their number-average molecular weights

TABLE 6

AVERAGE COMPOSITION OF SOLVENT SEPARATED PITCH FRACTIONS AS SHOWN BY EMPIRICAL FORMULAS[17]

Extract	Pitch	Ash	$\overline{M}_n$[1] (Ash Free)	Atoms/Molecule C	H	O	S	N	Atomic C/H
		%							
	CO1	0.7	258	19.9	13.3	0.10	0.05	0.19	1.50
	CO2	1.0	222	17.0	10.9	0.22	0.05	0.10	1.56
Light Petroleum	HR	0.2	251	18.9	14.8	0.14	0.07	0.28	1.28
	VR	0.8	264	19.2	18.8	0.54	0.05	0.27	1.02
	LT	0.2	270	19.3	19.6	0.68	0.06	0.43	0.98
	CO1	0.3	559	42.4	27.1	0.59	0.13	0.61	1.56
	CO2	0.3	532	40.6	25.1	0.43	0.08	0.65	1.62
Benzene	HR	0.0	643	48.5	33.1	0.53	0.12	1.03	1.47
	VR	0.4	621	44.4	31.2	2.07	0.16	1.30	1.42
	LT	0.3	395	27.5	23.6	1.79	0.11	0.65	1.17
	CO1	1.8	1476	109.5	57.8	3.8	0.50	1.78	1.90
	CO2	0.7	1697	122.3	66.8	6.1	0.89	2.41	1.83
Nitrobenzene	HR	0.8	2130	141.9	86.4	12.9	1.93	4.90	1.64
	VR	3.6	1209	83.6	48.1	7.4	0.38	1.86	1.74
	LT	0.8	988	61.1	45.8	8.3	0.90	3.25	1.33
Quinoline	CO1	2.4	1660	120.8	67.2	5.6	0.37	2.75	1.80
				C, %	Atoms/Atom of Carbon				
	CO1	3.6	—	85.5	0.43	0.047	0.003	0.014	2 32
	CO2	3.2	—	87.2	0.38	0.041	0.003	0.013	2.68
(Insoluble)	HR	2.3	—	87.6	0.40	0.037	0.006	0.015	2.50
	VR	7.9	—	79.4	0.35	0.078	0.003	0.011	2.86
	LT	10.6	—	67.8	0.71	—	—	—	1.41

Note: [1] Number-average molecular weight.

were taken to be the same as before, the values of which were calculated from the molecular weights of the subfractions shown in Table 5. Table 6 shows the molecular weights and calculated empirical formulas. On the basis of these empirical formulas, structural limits with respect to number of rings and manner of attachment of the rings can be defined. The empirical formulas of the light petroleum extracts of pitch from coke-oven tar require a highly condensed ring structure, namely, perylene for coke-oven tar 1 and pyrene for coke-oven tar 2. Because molecular weight and composition are average values, one might expect a spread in structures sufficient to allow the presence of methyl or methylene groups which are indicated to explain the condensation effected by the air-blowing of pitch. A 6-, 7-, or 8-ring highly condensed structure presents the necessary deficiency in hydrogen compared to the indicated average of four or five rings, to compensate for the excess hydrogen required by smaller structures containing methylene groups. It may, therefore, be deduced that these fractions, which account for just over 50% of the pitch, consist, predominantly, of condensed aromatic hydrocarbons containing three to eight rings and exhibiting a small amount of methyl substitution or hydrogenation of the ring.

The benzene extracts also have high C/H ratios requiring that at least six out of seven carbon atoms be aromatic. The average structure must be about ten rings. By the same arguments used in the case of the light-petroleum extracts, constituents up to twenty-five rings may be included and, therefore, also a significant amount of methylene groups or their equivalent. The structural formulas (1), (2), and (3) of Fig. 14 are isomeric. In structure (1), the high C/H ratio of the highly compact ring structure balances the low C/H ratio of the added methyl and methylene groups. Because of the allowable molecular-weight spread, these methyl groups could be absent from (1) and added to structure (2) and (3).

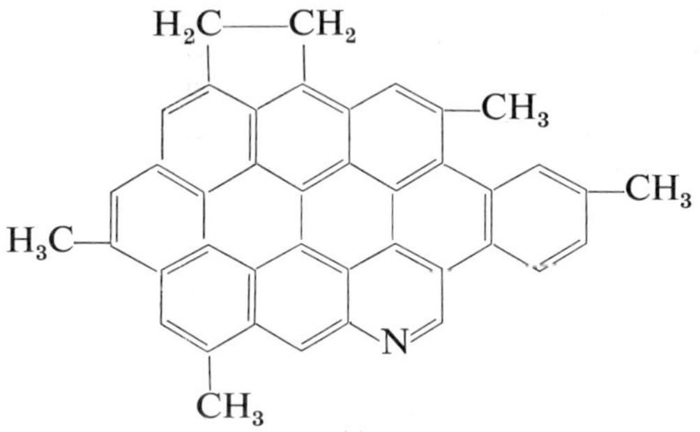

(I)

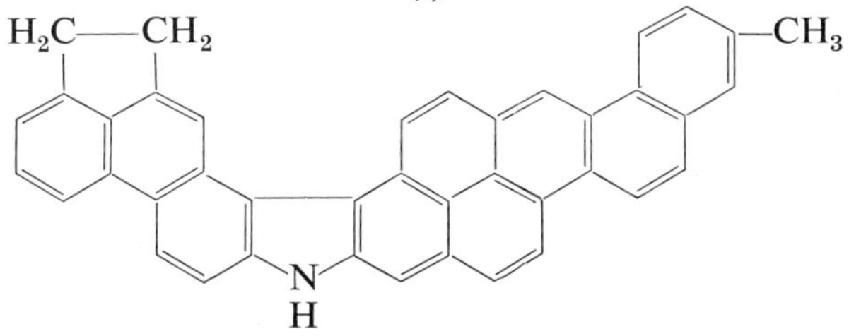

(II)

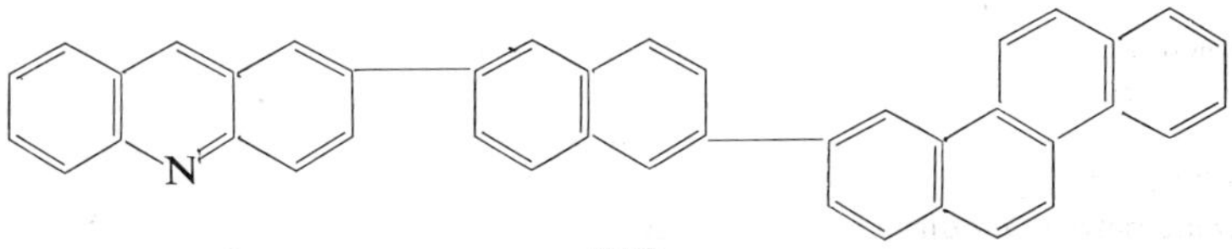

(III)

Fig. 14. Isomeric condensed-ring structures conforming to the formula $C_{41}H_{24}N$ and showing possible variation in methyl and methylene group content. Formula approximates that of the benzene extract of a coke-oven-tar pitch.[17]

The nitrobenzene and quinoline extracts of the coke-oven pitches have higher C/H ratios, which require a still more compact packing of aromatic rings. Even so, approximately one carbon in eight must be aliphatic. The insoluble fractions of the pitches from coke-oven tar have even higher C/H ratios, about 2.5, but even this does not allow for an unsubstituted aromatic structure of a very high degree of compactness and high molecular weight. An unsubstituted aromatic of closest packing, a honey-comb structure of fourteen rings has a C/H ratio of 2.6 and a molecular weight of only 520. To reach molecular weights of 1500 requires the linking of these smaller compact structures and the presence of aliphatic carbons to compensate for the hydrogen which linking eliminates.

The soluble fractions of pitch from horizontal-retort tar have slightly lower C/H ratios than the corresponding fractions from coke-oven-tar pitch. The insoluble fraction is very similar to the insoluble fraction from the pitch from coke-oven tar.

The petroleum extract of pitch from vertical-retort tar is less aromatic than that from coke-oven or horizontal-retort tar. As indicated by the work of Harris, White, and McNeil,[14] this can be attributed to the presence of paraffin waxes and partly saturated aromatic compounds. The highly condensed aromatic compounds such as pyrene are major components, as was found for pitch from coke-oven tar. The less-soluble fractions have the same aromaticity as do the corresponding fractions of pitch from horizontal-retort and coke-oven tar. Their oxygen content is generally higher.

Wood and Wilman[20] extended this analytical study by use of infrared spectroscopic analysis, especially in the region of 3000 cm^{-1}. Their main object was to estimate the content of aromatic C—H groups and, thereby, fix more exactly the degree of compactness of the aromatic structures in question. They worked with the same pitch sample as Wood and Phillips,[17] made similar solvent fractionations, and obtained much the same molecular weights and C/H ratios. The lower-molecular-weight fractions were analyzed in carbon disulfide solution, the higher fractions as dispersions in pressed potassium chloride disks. The problem in using the aromatic C—H band for quantitative purposes is that its strength varies from compound to compound. An average value was sought from measurement of twenty-two compounds soluble in carbon disulfide and twenty-nine other compounds, many of low solubility, all known to have been found in pitch. The twenty-two compounds were measured in carbon disulfide at 3052 cm^{-1} frequency, the twenty-nine compounds in potassium chloride disks at 3047 cm^{-1}. These frequencies were chosen because the pitch fractions absorbed here for the two techniques. The data on known compounds showed nothing to suggest that the absorptivity changed in any regular way with the increase of

[20] Wood, L. J., and Wilman, W. G., *Proc. Conf. on Science in the Use of Coal*, Sheffield, 1958, Inst. of Fuel, London, 1958, pp. C10–14.

TABLE 7

AVERAGE FORMULA AND NUMBER OF AROMATIC C—H GROUPS/MOLECULE IN COKE-OVEN-TAR PITCH FRACTIONS[20]

Fraction No.	Soluble in	Yield	Molecular weight Range	Average Chemical Formula	Average Number C—H Group	Number of Rings/Molecule
		%				
1	Petroleum ether (40 to 60°C)	41.4	180–260	$C_{17}H_{13}$	10–11	3–4
2	Petroleum ether (100 to 120°C)	9.9	260–350	$C_{22}H_{15}$	11–12	5–6
3	Benzene/petroleum ether	21.2	350–800	$C_{42}H_{27}N$	22	7–15
4	Benzene	5.4	800–1100	$C_{72}H_{42}NO$	29–30	16–20
5	Pyridine	10.6	1100–1800	$C_{96}H_{64}N_2O_4S$	43	21–35
6	Quinoline	6.0	1800–2100	$C_{144}H_{88}N_3O_7$	44–45	30–40

molecular size and the number of rings in the molecule. The assumption was therefore made that the same absorptivity applied up to the 2000 molecular-weight maximum being considered. Table 7 summarizes the analysis of data on fractions from a pitch from coke-oven tar.

The first four fractions were also examined in a carbon disulfide solution over the wider range of wavelengths obtained with the rock-salt prism. In all these four spectra, the C—H out-of-plane deformation vibrations[21] show an outstanding feature. The ortho-substitution bands at 750 cm^{-1} are much stronger than bands due to other types of substitution. And further, there is no absorption due to unsubstituted phenyl groups. It is hard to explain such strong ortho-substitution bands except by a fair number of unsubstituted end-rings, that is, rings fused to only one other ring. It was concluded that these molecules are largely branched, rather than a continuous line of fused benzene rings. Structures (1) and (2) of Fig. 15 were regarded as typical for these pitch fractions. Three molecules like (2) joined together by coumarone rings would represent the largest pitch molecules.

Analysis of the data on vertical-retort-tar pitch is shown in Table 8. The average composition of the lowest molecular-weight fraction corresponds closely to that of tetramethylphenanthrene. From the work of Harris[14,15] and associates, it is certain that this fraction is a mixture of unsubstituted condensed aromatic, less-aromatic, and even paraffinic compounds. As regards the high-molecular-weight fractions, the large proportion of hydrogen not present in the form of aromatic C—H still leaves the composition very much a matter of speculation.

Hügel[22] has recently reported additional data on the distribution of nitrogen, oxygen, and sulfur compounds in a pitch from a vertical-retort tar. The pitch was separated into a fraction, 50% by weight, insoluble in cold toluene and a fraction soluble in cold toluene. Both fractions had the same percentage content of nitrogen, oxygen, and sulfur. The toluene-soluble fraction was again separated into

[21] Bellamy, L. J., *The Infra-Red Spectra of Complex Molecules*, Second edition, John Wiley and Sons, New York, 1958, p. 75.

[22] Hügel, G., *Brennstoff-Chem.*, **39**, 213–17 (1958).

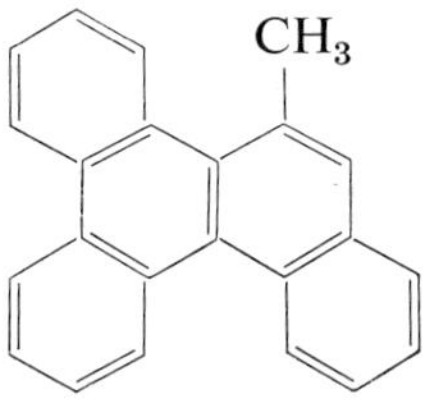

(I) $C_{23}H_{16}$

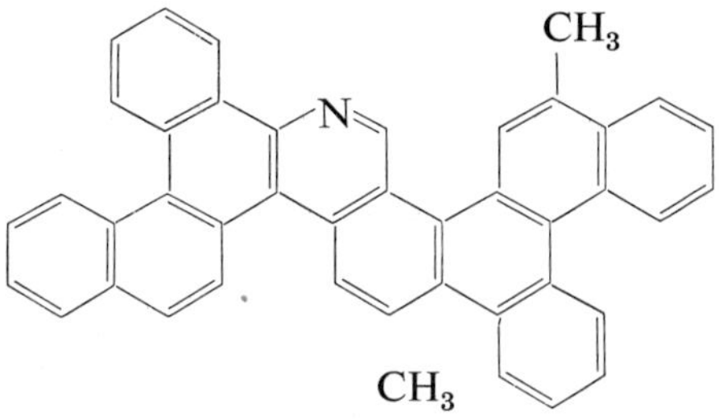

(II) $C_{43}H_{27}N$

Fig. 15. Structural formulas showing large number of unsubstituted end-phenyl-rings required by infrared spectra of solvent separated fractions of a coke-oven-tar pitch.[20]

a part precipitated as a thiourea adduct and a part not precipitated. The two fractions so obtained were each treated with a 60% solution of perchloric acid producing in each case a basic and a nonbasic fraction. Molecular weights, cryoscopic in acetophenone, were given for the toluene soluble whole and the four fractions into which it was separated. No mention was made of reasons for an apparent

Fraction	Molecular Weight
Toluene-soluble whole	240
Ppt'd by thiourea	
(A) Basic constituents	480
(B) Nonbasic	333
Not ppt'd by thiourea	
(C) Basic constituents	241
(D) Nonbasic	282

discrepancy between the molecular weight of the whole and those of the component fractions. The ratio of basic to nonbasic component was 0.3 for fractions resulting from thiourea treatment. Not all the nitrogen content of the pitch appeared in the basic fractions. The basic fractions did contain two to four times the concentration of nitrogen that the nonbasic fractions contained. Of the total nitrogen contents of the fractions forming and not-forming thiourea adducts, basic nitrogen accounted for 44% and 63%, respectively. Oxygen concentrated in the nonbasic fractions, while sulfur appeared uniformly distributed.

GRAPHITE-LIKE ORDERED STRUCTURES IN COAL TAR AND COAL-TAR PITCH

Wide-angle X-ray analysis can measure the extent of a structure comprising pairs of atoms separated by a frequently encountered vector distance or of systems of atoms arranged in a periodic fashion (lattice structure). Volkmann[23] has reported the results of X-ray examination of a coal tar, of several coal-tar pitches of different softening points, of samples of the suspended phase (C_1 resin), of coal tar and, finally, of coal-tar-pitch coke. In Fig. 16, the relative intensity of X-ray diffraction is plotted against the diffraction angle to give the (002) band profile. A curve for graphite is included; its intensity and that of the pitch coke are reduced by half to show the peaks. The system of atoms producing the diffraction peaks shown is that of structures similar to the layers of carbon atoms in graphite. The angle at which the diffraction intensity is maximum measures the most prevalent distance between the layers according to the relation, Distance, A $= \frac{1}{2} \sin \theta$. The larger the angle the smaller the distance, which for graphite is 3.35 A. The distances for the other materials are 5.5 A for tar, 4.5 A for 95° softening point pitch, 4.3 A for C_1 resin, and 4.25 A for pitch coke. The interpretation of these data is that the removal of lower-boiling tar components permits the residual lamellar molecules to be held closer together by their attractive forces.

Blayden, Gibson, and Riley[24] studied by X-ray diffraction the extent of a graphite-like structure developed in a pitch and in fractions separated from it by solvents on heating to various temperatures. A hard coal-tar pitch was extracted with carbon tetrachloride. A portion of the carbon tetrachloride insoluble material was further extracted with benzene, carbon bisulfide, and acetone. The whole pitch, the carbon tetrachloride extract and insoluble residue, and the residue after extraction with acetone were carbonized in a nitrogen atmosphere at 5°C/min rise to various final

[23] Volkmann, E. W., *Fuel*, **38**, 445–68 (1959).

[24] Blayden, H. E., Gibson, J., and Riley, H. L., *Inst. Fuel Wartime Bull.* 117–29 (1945).

TABLE 8

AVERAGE FORMULAS AND ESTIMATES OF NUMBER OF AROMATIC C—H GROUPS IN VERTICAL-RETORT-TAR PITCH FRACTIONS[20]

Fraction No.	Soluble in	Yield	Aromatic C—H Wt. Fraction	Molecular Weight Range	Average Chemical Formula	Average Number Aromatic C—H Groups	Number of Rings/Molecule
		%					
1	Petroleum Ether (40 to 60°C)	50.5	0.355	180–350	$C_{18}H_{19}$	7	3–5
2	Petroleum Ether (100 to 120°C)	22.2	0.253	350–500	$C_{29}H_{25}$	8	6–7
3	Benzene/petroleum ether	3.4	0.255	450–550	$C_{35}H_{33}NO_2$	10	8–9
4	Benzene	5.6	0.249	550–900	$C_{51}H_{40}NO_3$	14	10–14
5	Pyridine	13.1	0.207	900–1500	$C_{79}H_{60}N_2O_8S$	19	15–25
6	Quinoline	1.2	—	—			

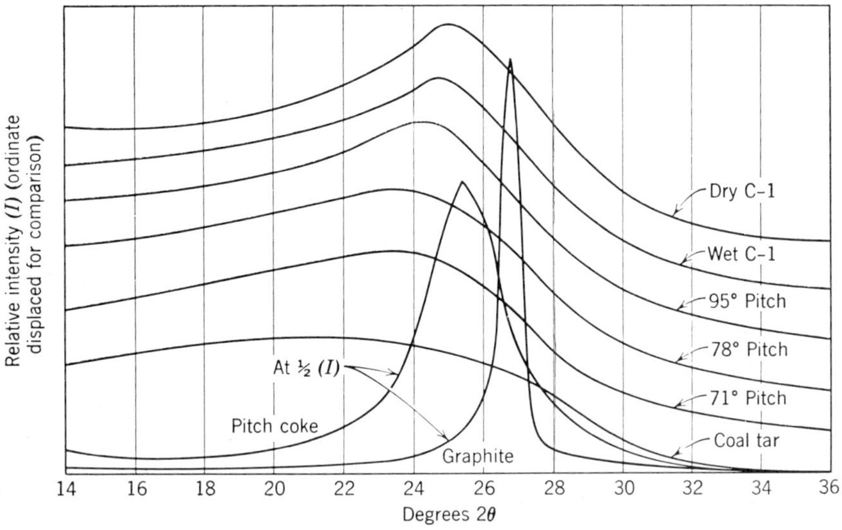

Fig. 16. X-ray diffraction by coal tar and coal-tar pitches of various softening points, by C-1 fraction from coal-tar pitch and by pitch coke, compared with diffraction by graphite.[23]

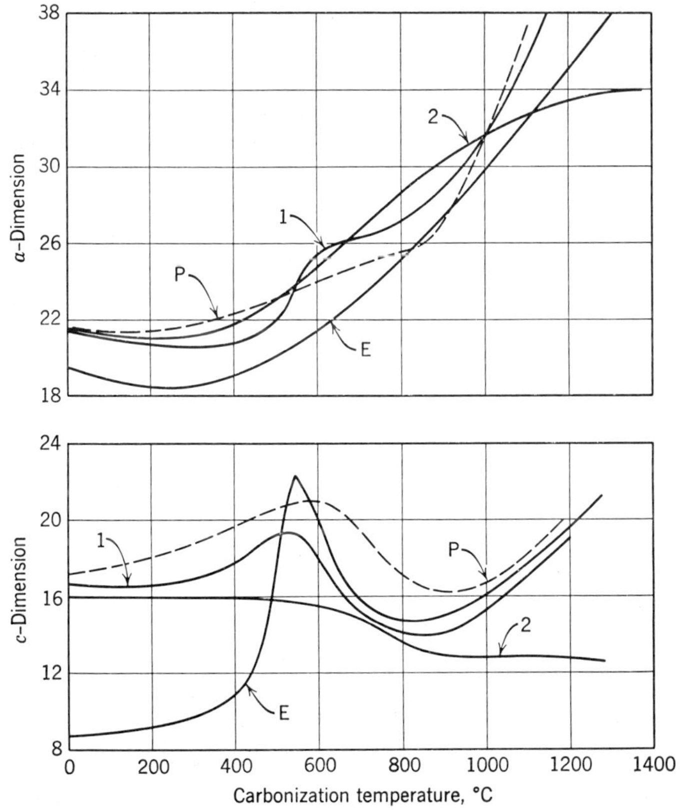

Fig. 17. Dimensions of graphite-like crystallites found in chars from coal-tar pitch and its fractions carbonized at various temperatures (in Angstrom units normal to plane), *c*-dimension, and parallel to plane of carbons in the hexagonal structure, *a*-dimension.[24] P. Whole pitch; E. carbon tetrachloride extract of the pitch; 1. residue insoluble in CCl_4; 2. residue insoluble in CCl_4, benzene, and acetone.

TABLE 9

PROPERTIES OF PITCH AND ITS SOLVENT-SEPARATED FRACTIONS FOR A PITCH BEFORE AND AFTER OXIDATION BY SPARGING WITH AIR AT 270°C UNTIL SOFTENING POINT ROSE FROM 79.1 TO 103°C[25]

	Normal				Oxidized			
	Yield, %	% C	% H	% O,N,S	Yield, %	% C	% H	% O,N,S
Original Pitch		92.0	4.7	3.3		92.2	4.6	—
α Fraction	30	90.2	3.8	6.0	28	91.5	4.0	4.5
β Fraction	15	89.6	4.9	5.5	27	91.2	5.1	3.7
γ Fraction	55	91.4	5.3	3.3	45	92.5	5.3	2.2

temperatures with a 10-min soak at the final temperature. Figure 17 shows the results of the study. The pronounced maximum in the *c*-dimension curve of the char from the carbon tetrachloride extract was attributed to the presence of thermally stable, lamellar molecules which undergo a unidimensional crystallization under the stimulus of increased thermal vibration. At higher temperatures, thermal instability wrecks this structure, and normal development of graphite lattice proceeds.

Ghosh and Riley[25] reported a similar study comparing a coal-tar pitch before and after oxidation by blowing air through the pitch at 270°C until the softening point, ring-and-ball, increased from 79.1 to 103°C. The pitches were separated into benzene insoluble (α pitch), benzene soluble-petroleum ether insoluble (β pitch), and petroleum ether soluble (γ pitch). Yields of the fractions and elementary analyses are shown in Table 9. Oxidation increased the β pitch content at the expense of the γ pitch. These fractions were carbonized as in the previous work and the chars examined by X-ray diffraction. The curves for the crystallite dimensions in the normal and oxidized pitch chars followed each other, but the curves for oxidized pitch were depressed about 2 A.

PHYSICAL PROPERTIES OF TAR AND PITCH

Temperature Coefficient of Specific Gravity

Mapstone[26] has published data on the temperature coefficient of specific gravity of coal tars and coal-tar distillates and has correlated this property with specific gravity and with boiling point. Many types of tars were evaluated, and the correlations hold equally well for all. The data is therefore of interest in the instance of high-temperature coal tar. Figure 18 shows a plot of specific gravity at 15.56°C versus temperature coefficient of specific gravity for tars. Figure 19 shows the temperature coefficient of specific gravity plotted against the 50% point of distillation in degrees centigrade for coal-tar distillates.

Berneis and Wood[27] have reported on the temperature coefficient of specific gravity of coal-tar pitch in the temperature range 20 to 70°C. They confirmed older work showing a transition-point temperature below which the temperature coefficient decreases abruptly. The transition temperature is a function of the softening point of the pitch; see Fig. 20. It was noted that all glasses, for example, glycerine and thermoplastic resins show discontinuities in values of physical constants when cooled to the temperature where their viscosity is 10^{13} poises. By extrapolation of viscosity values obtained to 10^{11} poises on a medium-hard pitch, 10^{13} poises was estimated for the pitch at 10 to 15°C.

Heat Capacity and Heat Content

Hyman and Kay[28] offered the following equations for calculating the heat capacity and heat content of a tar or pitch:

$$C = 1/D(0.302 + 0.000656T),$$

or

$$C = S(0.321 + 0.000694T)$$

[25] Ghosh, S. P., and Riley, H. L., *Fuel*, **27**, 56–60 (1948).

[26] Mapstone, G. E., *J. Appl. Chem. (London)*, **5**, 582–8 (1955).

[27] Berneis, K., and Wood, L. J., *Brennstoff-Chem.*, **36**, 301–4 (1955).

[28] Hyman, D., and Kay, W. B., *Ind. Eng. Chem.*, **41**, 1764–8 (1949).

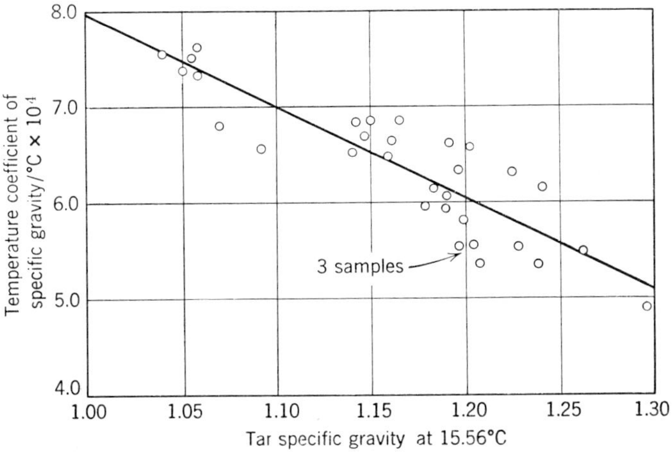

Fig. 18. A plot of specific gravity of coal tars against their temperature coefficient of specific gravity.[26]

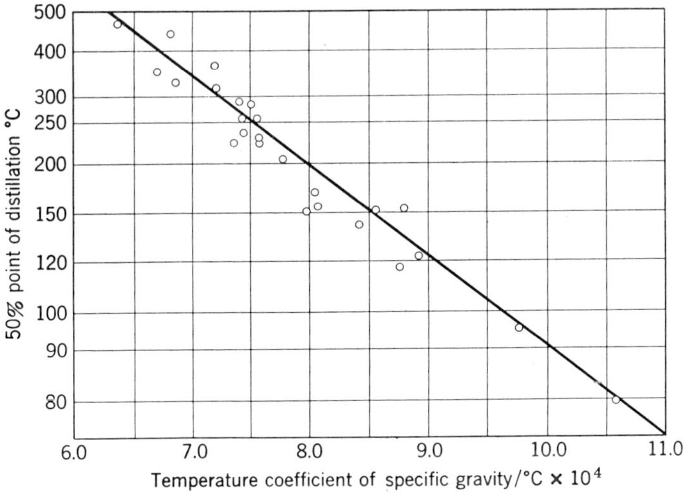

Fig. 19. A plot of the temperature coefficient of specific gravity of coal-tar distillates against the 50% point distillation temperature.[26]

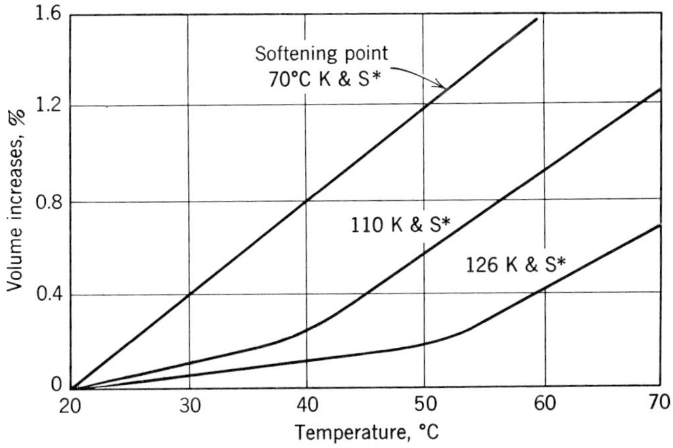

Fig. 20. Thermal expansion of coal-tar pitches. Curves for hard, medium, and soft pitch.[27]
* K & S = Kraemer and Sarnow.

where C = heat capacity in Btu/°F/lb,
D = specific gravity at 60°/60°F,
T = Temperature, °F, and
S = the benzene-soluble fraction of the material;

and $H = 1/D(0.302T + 0.000328T^2 - 19.3)$,
or $H = S(0.321T + 0.000347T^2 - 20.5)$,
where H is heat content in Btu/lb.

Latent Heat of Vaporization

Briggs and Popper[29] gave equilibrium flash-vaporization data on tars in the form of Cox charts with naphthalene as the reference substance. Charts for horizontal- and vertical-retort tars, coke-oven tar, and carburetted water-gas tar were shown. Briggs and Popper[30] have also determined the latent heat of vaporization of various coal-tar distillate fractions of the preceding types of tar. The distillates ranged from light naphthas through heavy anthracene oils. Molar latent heats of vaporization ranged from 8 kcal for light naphthas to 15 kcal for anthracene oils. The corresponding specific latent heats ranged from 83 cal/g to 65 cal/g. A simple relationship was developed, namely,

$$L_e = d_{20}(116.1 - 0.143T_b),$$

where L_e = specific latent heat of vaporization at the atmospheric boiling point (cal/g), and
T_b = average boiling point (°K).

Thermal Conductivity

Briggs and Popper[31] have measured the thermal conductivity of pitches of different softening point. The thermal conductivity is independent of temperature over the range 25 to 105°C. Pitches of higher toluene-insoluble content had higher conductivities.

Surface Tension

From a report of work on insoluble matter in coal tar by Green and Olden,[6] the surface tension of a vertical-retort tar is 36 dynes/cm.

Gogoleva and Stepanenko[32] found that the surface tension of a pitch of K&S softening-point of 65°C was 45 to 26 dynes/cm in the temperature range, 185 to 225°C. The values for a pitch of softening point of 145°C were 62.4 to 30.5 dynes/cm in the temperature range of 270 to 335°C.

Viscosity

In common with other bituminous substances, high-temperature tar and pitch are most easily characterized by their viscosities and the change of their viscosity with temperature. Lee, Warren, and Waters[33] and Höppler[34] have reported that coal tar and pitch are quite like other bitumens in their rheological behavior. High viscosity, a high temperature coefficient of viscosity, and failure to obey Newton's law of viscous flow are characteristic of bituminous materials. Lee and associates found that the change in viscosity with temperature followed the relation $\log \eta = a - n \log T$ where η is viscosity and n is the temperature coefficient and equals $(\log \eta_1 - \log \eta_2)/(\log T_2 - \log T_1)$. This relation is not exact but serves to indicate comparative temperature susceptibilities of bitumens. The deviations from Newtonian behavior grow less and disappear at higher temperatures where the bitumens are quite fluid.

Fair and Volkmann[35] used a falling, coaxial-cylinder viscometer in a study of coal-tar pitches. The instrument permitted measurement of shear stress and shear rate in absolute units. A number of soft pitches (ring-and-ball softening points from 52 to 57°C) were studied. Shear rates of 2×10^{-5} to 45×10^{-5} sec^{-1} were measured at shear stresses of 1400 to 7100 dynes cm^{-2} in determining the viscosities of these pitches at 25°C. All of five pitches studied showed pure Newtonian behavior at viscosities from 1.6×10^7 to 6×10^7 poises; that there was no evidence of departure from Newtonian behavior could be due to the rather low shear stresses employed. To estimate temperature coefficient of viscosity, a kinematic

[29] Briggs, D. K. H., and Popper, F., *Trans. Inst. Chem. Engrs. (London)*, **35**, 369–73 (1957).
[30] *Ibid.*, 374–9.
[31] Briggs, D. K. H., and Popper, F., *Fuel*, **33**, 222–9 (1954).

[32] Gogoleva, T. Ya., and Stepanenko, M. A., *Koks i Khim.*, **1959**, No. 4, 42–5.
[33] Lee, A. R., Warren, J. B., and Waters, D. B., *J. Inst. Petrol.*, **26**, 101–28 (1940).
[34] Höppler, F., *Mitt. Dachpappen-Ind.*, **14**, 179–87 (1941).
[35] Fair, W. F., Jr., and Volkmann, E. W., *Ind. Eng. Chem., Anal. Ed.*, **15**, 235–9 (1943).

viscosity of 1×10^6 centistokes was assumed for the various pitches at their ring-and-ball softening-point temperature, and the viscosity values at 25°C were converted to kinematic viscosity values. The temperature coefficients so obtained were greater than values previously determined at a much higher temperature level, so the present studies were extended to cover the range of temperatures from 25 to 85°C and checked against measurements made by a capillary method. As shown in Fig. 21, the temperature susceptibilities of the pitches were less at temperatures below the ring-and-ball softening point than at temperatures above this point.

The work of Sack[16] shows the nature of the deviation from Newtonian behavior exhibited by coal-tar pitch. Newtonian flow requires that under constant load, deformation occur in equal increments for equal time intervals, that is, a plot of deformation versus time is a straight line. Also, when the load is removed, all deformation stops, that is, no further deformation occurs, and all deformation to the point of load removal is retained. Figure 22 shows the deformation of a pitch specimen under constant load and the elastic recovery that takes place when the load is removed. Note the high, almost instantaneous deformation with initial application of load. The rate of

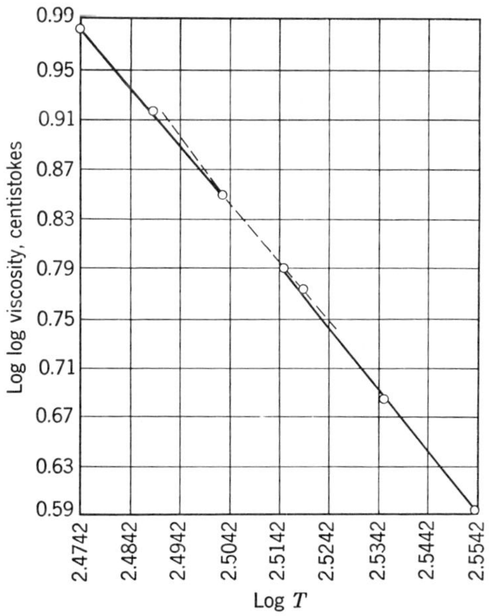

Fig. 21. Temperature coefficient of viscosity of coal-tar pitch as shown by viscosity-temperature plot over the range 25 to 85°C.[35]

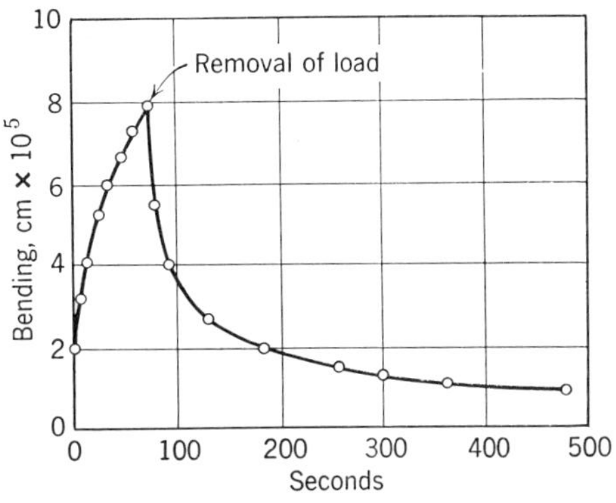

Fig. 22. Deformation and elastic recovery curve of a coal-tar pitch specimen under constant load, followed by instantaneous load removal.[16]

deformation falls off to a constant value with increased time. Upon removal of the load, there occurs a high, almost instantaneous recovery. The rate of recovery then falls off and becomes zero before the specimen returns to its original unloaded condition. The amount by which the specimen fails to return to its original condition is a measure of the viscous flow which has occurred as opposed to the elastic deformation occurring simultaneously. Such behavior is called viscoelastic flow. Figure 23 shows the deformation and elastic-recovery curve of a pitch plasticized with 10% nitrobenzene. Curves for four different loads are shown. For the lowest load, no elastic recovery is evident. The entire deformation has been by viscous flow. This behavior is characteristic of viscoelastic materials. There is a minimum load below which the flow is Newtonian, that is, no elastic flow can be detected. Viscoelastic behavior is shown by the heaviest oils that can be distilled from coal tar. The elastic effect is noted only under very high loads. After heating this oil with 10% of aluminum chloride for 8 hr at 280°C, a dark substance is obtained which shows an elastic effect of the same magnitude as shown by pitches under comparable load.

Sack concluded that marked anisotropic points and positions of minimum energy therefore exist. Molecules, or aggregates of molecules, tend to reform or regroup into these preferred orientations after deformation. A spheri-colloidal system, being isotropic, has no such positions of minimum energy, and the internal structure of pitch is not that of a spheri-colloid. Sack ascribed to tar what he

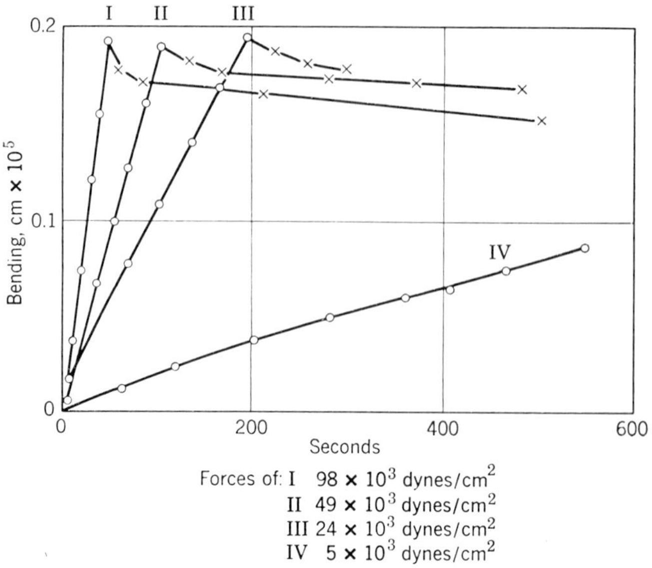

Fig. 23. Deformation and elastic recovery curve of a specimen of coal-tar pitch, plasticized with 10% nitrobenzene, under constant load followed by instantaneous load removal.[16]

called an elastic limit, see Fig. 24 which compares tar with concentrated solutions of pitch in nitrobenzene and in aniline. The tar curve (1) is extrapolated linearly to zero rate of shear where it intersects the force axis at about 18,000 dynes cm^{-2}, the elastic limit. This corresponds to the Bingham yield point for plastic flow. In view of the behavior of other viscoelastic materials, the linear extrapolation is probably not valid. A more probable extrapolation is shown. A comparison of curves 1, 2, and 3 shows the effect of different solvents on molecular aggregates in pitch. Tar itself can be considered a solution of pitch in tar-distillate oils. The structure set up by molecular aggregates in pitch is shown to persist even in relatively dilute solutions by the fact that the solutions show hysteresis cycles, as in Fig. 25, typical of thixotropy.

Franck and Wegener[36] measured the viscosity of the more important coal-tar products over a range of temperatures in absolute units. The temperature coefficients of viscosity were calculated and equations developed to relate

[36] Franck, H. G., and Wegener, O., *Brennstoff-Chem.*, **39**, 195–207 (1958).

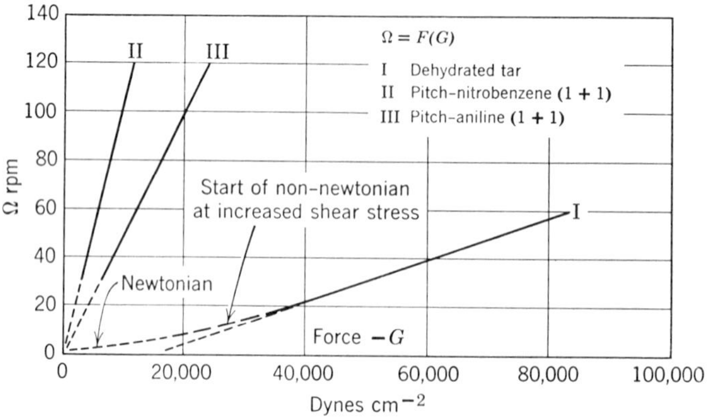

Fig. 24. Curves showing shear rate (revolutions per minute of viscometer shear bob) as a function of applied force for tar and for solvent-pitch systems.[16] Linear extrapolation of curve I would indicate a plastic yield point of 18,000 dynes cm^{-2}.

these data in fundamental units to the values for various empirical flow tests used in the trade to characterize these products. The tests considered were: *The softening-point, Kraemer-Sarnow*, which measures the temperature at which a 5 mm deep column of pitch in a tube of 8 mm diameter becomes soft enough that an overlying layer of 5 g of mercury breaks through the pitch; *the softening-point, ring-and-ball*, which measures the temperature at which a steel ball, supported on pitch molded within a metal ring, passes through the horizontally positioned ring; *road-tar-viscometer-seconds*, which is the efflux time, at a stated temperature, of 50 cc of tar product from standard orifices of either 10 mm or 4 mm diameter as designated; *equiviscous-temperature*, which is the temperature at which a tar product has a standard efflux time from a standard orifice; and *Engler-degrees*, which is a specific viscosity compared to water at 20°C. Viscosity measurements in fundamental units were made with the Höppler viscometer and the viscobalance which are based on Stokes' law (falling-ball principle). Because the preceding are suitable only for Newtonian flow, their use was supplemented where necessary by measurements on a rotational viscometer, the Epprecht-viscometer, which permits measurements under known shear stresses of widely different values.

Figure 26 shows temperature-kinematic viscosity curves for four briquette pitches of different softening-points produced by continuous distillation of coke-oven tar in the Ruhr district, Germany. A straight-line relation is obtained by plotting the double logarithm of the viscosity against the logarithm of the absolute temperature. The temperature coefficient of viscosity is the same for the four pitches (parallel lines) and the softening point (Kraemer and Sarnow) can be related to viscosity at a temperature by the expression,

$$\log \log (V + 0.8) = 0.85 + 5.3 \log (SP + 273) - \log T,$$

where V = Viscosity of pitch centistokes,
SP = Softening-point, °C, Kraemer-Sarnow, and
T = Temperature of pitch, °K.

This equation corresponds to that developed by Walther for mineral-oil greases and holds for briquette pitches made by continuous distillation of the coke-oven tar, wide departures being noted for pitches produced from coke-oven tar by batch distillation (pitches from low-temperature tars by continuous or batch distillation show wide departures). For the coke-oven-tar pitches, the departures are related to the content of toluene insolubles. Pitches which have been subjected to the high temperature of tar-distillation practice for a long time as in batch distillation, have relatively larger contents of toluene insolubles, that is, higher-molecular-weight materials. The viscosity-temperature data for eight pitches with softening-points between 62.4 and 76.5°C and toluene-insoluble content

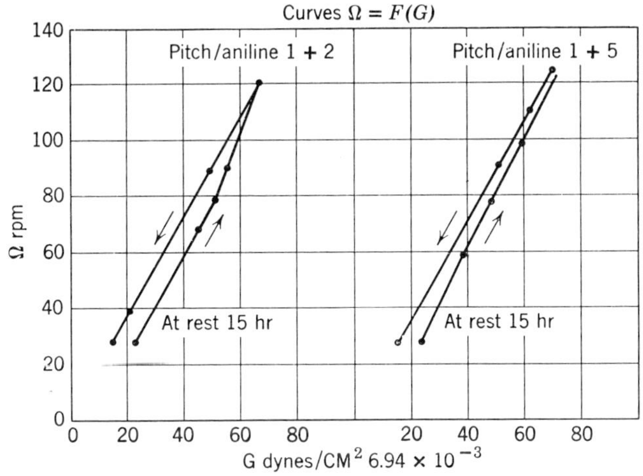

Fig. 25. Curves showing shear rate (revolutions per minute of viscometer shear bob) as a function of shear stress at increasing then decreasing shear stress to show thixotropy.[16]

from 4.29 to 30.24% are shown in Fig. 27. From these data, the above equation was modified by introduction of a toluene-insoluble (TI) term, namely,

$$\log \log (V + 0.8) = 0.80 + 0.04 \log TI + (5.8 - 0.03TI) \log (SP + 273) - \log T,$$

in which the multiplier $(5.8 - 0.03TI)$ is obviously the temperature coefficient of viscosity which decreases linearly with increased toluene-insoluble content, see Fig. 28.

The softening point, whether Kraemer-Sarnow or ring-and-ball, is nominally an equiviscous temperature. If we substitute $T = SP + 273$ in the equation we obtain the relation

$$\log \log (V + 0.8) = 0.80 + 0.04 \log TI$$

where V is the viscosity at the softening point. The viscosity at the softening point is not a constant, and values are calculated that range from 4.8×10^6 cst, for a pitch having 4.29% toluene-insoluble content, to 17.6×10^6 cst, for a pitch of 30.24% toluene-insoluble content.

Ubbelohde, Ullrich, and Walther[37] have noted that viscosity of a pitch at its softening point increases with increased softening point. This observation is consistent with the experience that higher softening point pitches have a higher toluene-insoluble content. The ring-and-ball softening points, were determined for a number of pitches prepared by continuous distillation and having 17% toluene-insoluble content. The viscosity at the ring-and-ball softening point, was calculated to be 750,000 cst. A relation between the ring-and-ball, SP_1, and the Kraemer-Sarnow, SP_2, softening points is given by

$$SP_1 = 1.035SP_2 + 9.6$$

which applies only to pitches of 17% toluene-insoluble content. Viscosity data were obtained from eight road tars, and it was found, as with pitches, that the double log of the viscosity was linear with the log of the absolute temperature. All tars showed the same temperature coefficient of viscosity, the viscosity-temperature relation being

$$\log \log (V + 0.8) = 13.8 - 5.3 \log T$$

An equation for converting efflux times to centistokes was given. When the efflux time was determined for the same tars with a 4 mm

[37] Ubbelohde, L., Ullrich, C., and Walther, C., *Oel, Kohle, Erdoel, Teer*, **11**, 684 (1935).

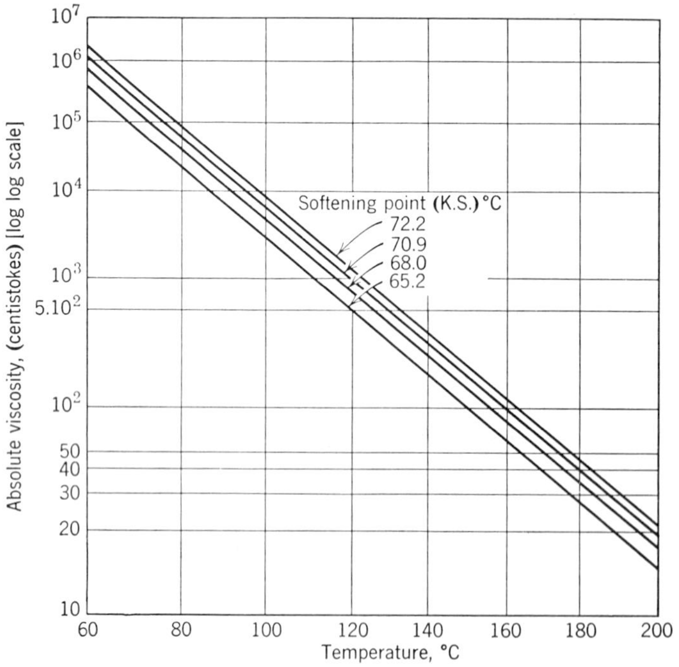

Fig. 26. Viscosity of four coal-tar briquette pitches, produced by continuous distillation, as a function of temperature.[36]

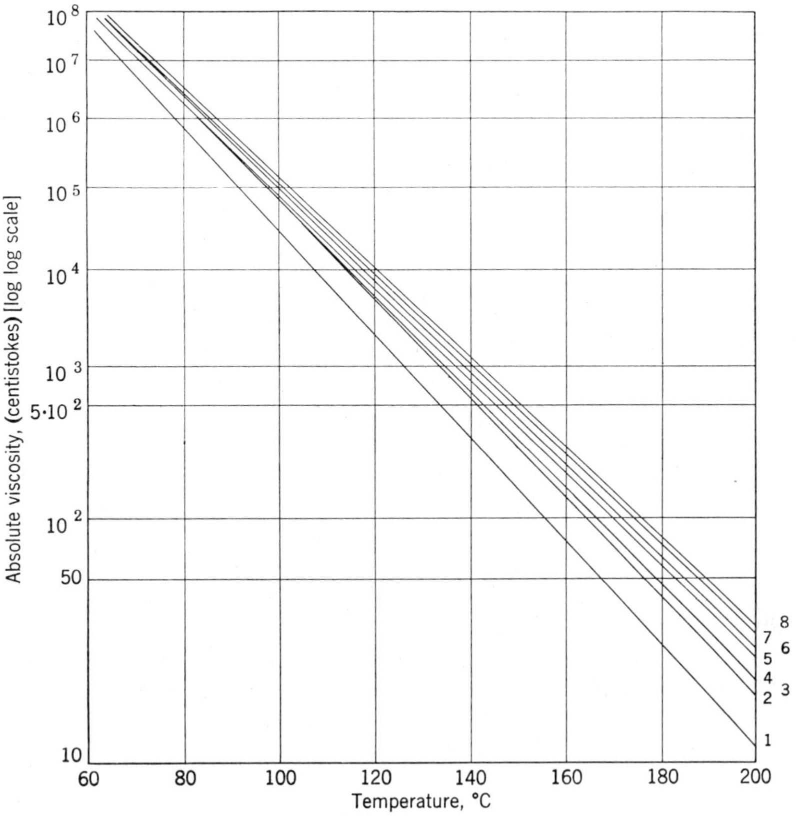

Fig. 27. Viscosity as a function of temperature for eight coal-tar pitches of toluene insolubles content increasing from 1 to 8.[36]

orifice, all tars had equal-temperature coefficients of viscosity, but the value was markedly greater than that obtained using a 10 mm orifice. A different equation with different constants converts efflux times with a 4 mm orifice to centistokes. Subtracting the two equations permits conversions between efflux times with a 10 mm orifice and efflux times with a 4 mm orifice.

Intrinsic Viscosity

The determination of intrinsic viscosity for estimation of molecular weight has been applied to coal tar. Rao, Basu, and Lahiri[38] studied fractions from a neutral tar-oil, several solvent-separated fractions of pitch from a coke-oven tar, and benzene-alcohol extracts from a sample of Assam coal. Viscosities in benzene solution, molecular weights, and specific gravities were determined. The applicability to their data of the equation of Kraemer, $[\eta] = K_1M$, and a modification of the same by Mark and Houwink, $[\eta] = KM^a$ was studied.

A coke-oven tar (Bhowra) was freed of acids and bases by extraction with 20% sulfuric acid and 10% sodium hydroxide. The neutral tar was fractionally distilled under vacuum. The pitch, when dissolved in twenty-eight parts of benzene gave 4.5% residue on filtration. To

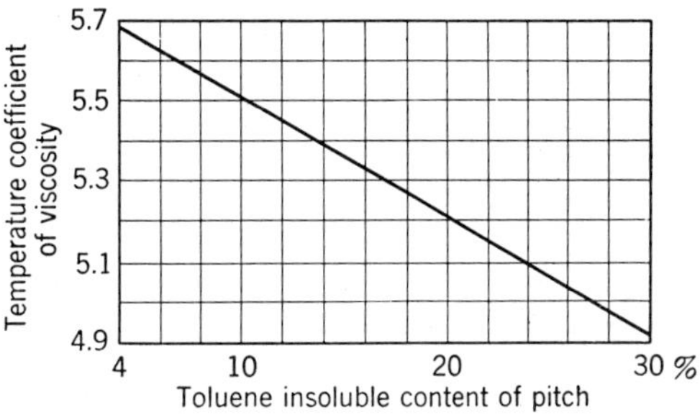

Fig. 28. Temperature coefficient of viscosity of coal-tar pitches as a function of their toluene insolubles content.[36]

[38] Rao, H. S., Basu, A. N., and Lahiri, A., *J. Sci. Ind. Research* (*India*), **B11**, 234–9 (1952).

TABLE 10

FRACTIONATION OF A COAL-TAR PITCH BY PRECIPITATION FROM A BENZENE SOLUTION* BY PETROLEUM ETHER[38]

Fraction Designation	Percent of Petroleum Ether in Equilibration Mix	Fraction of Pitch Precipitated	Cumulative Fraction	Molecular Weight
O	20	0.069	0.069	
P	29	0.034	0.103	
Q	31	0.002	0.105	
J	40	0.047	0.152	1022
R	41	0.001	0.153	
I	60	0.462	0.615	538.5
G	80	0.206	0.821	351.0
E	—	0.15	0.971	315.0

* 1 part pitch in 28 parts benzene.

fractionate the pitch, the benzene solution was cooled to 5°C and petroleum ether added stepwise with stirring. After a few hours of equilibration, the precipitated material was filtered off and washed with a benzene-petroleum ether mixture having 10% more petroleum ether than the equilibration solvent. The results of a typical pitch fractionation are shown in Table 10. In determining the intrinsic viscosities, concentration of solute was varied from 12 g/100 of solution, down to 4 g/100 for oil fractions, and from 6 down to 2 g/100 for pitch fractions. No significant change in intrinsic viscosity with concentration was noted. The data are shown in Table 11. For molecular weights below 350, $[\eta]$ versus M gave a straight line, making the equation, $[\eta] = K_1M$, applicable. For molecular weights above 350, the constant K_1 decreased. The use of equation $[\eta] = KM^a$ also failed to give constant values of K for molecular weights above 350. The values of K_1 for naphthalene and phenanthrene were found to be 6.76×10^{-5} and 6.97×10^{-5}, respectively. It was argued that the low K_1 values for pitch fractions does not allow them to be polynuclear like phenanthrene.

Badenhorst and Perold[39] studied the viscosity of dispersions made by redispersing tar insolubles in the remainder of the tar from which the insolubles had been removed. Three groups of insolubles were removed from a tar, namely, the quinoline insolubles (C_1 resins), the quinoline soluble-toluene insolubles (C_2 resins), and the toluene soluble-light petroleum

[39] Badenhorst, R. P., and Perold, G. W., *J. Appl. Chem. (London)*, **7**, 32–8 (1957).

TABLE 11

INTRINSIC VISCOSITIES IN BENZENE, MOLECULAR WEIGHTS, AND VALUES OF $K_i = [\eta]/M$ FOR COAL-TAR FRACTIONS[38]

Fraction	Specific Gravity	Molecular Weight	$[\eta]$	$K_i = [\eta]/M$
Oil #1	1.019	187.5	0.0100	5.334×10^{-5}
#2	1.028	211.4	0.0112	5.298
#3	1.034	224.1	0.0125	5.578
#4	1.042	237.2	0.0126	5.313
Pitch *E*	—	315.0	0.0174	5.508
*F**	—	323.3	0.0171	5.290
G	—	351.0	0.0186	5.284
*H**	—	477.4		4.610
I	—	538.5		3.825
*K**	—	629.5		2.938
J†	—	1022		2.770

* From a different pitch than described in Table 10.
† At concentrations of 0.08 to 0.03 g/100-g solution.

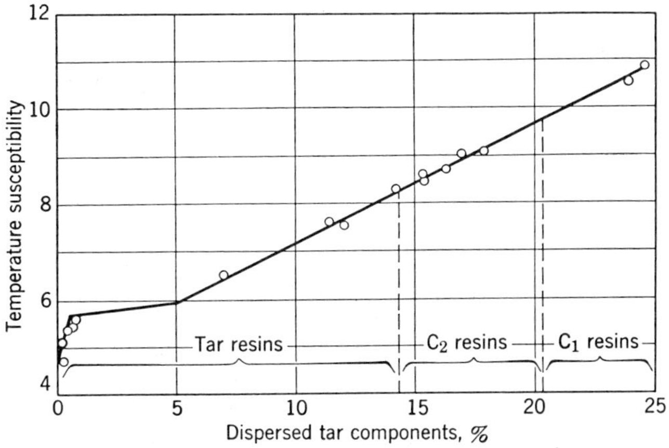

Fig. 29. The temperature susceptibility of viscosity of dispersions made by adding previously solvent precipitated components of coal tar back to the solvent-freed soluble tar.[39] The components were added back in the order of decreasing solubility.

insolubles (tar resins). Starting with the tar resins and proceeding with the C_2 resins and finally with the C_1 resins, these materials were added in successive small increments to the light-petroleum-soluble portion of the tar and the flow properties noted after each addition. Figure 29 shows the temperature susceptibility of viscosity of the dispersions. Figure 30 is a plot of the viscosities with the region below 5% added insolubles enlarged to show irregularities not apparent on the main plot. Figure 31 is a plot of Einstein's hydrodynamic constant for the viscosity of dilute suspensions as applied to the dispersion of the tar components back into tar, again with enlargement of the plot from 0 to 6% added insolubles. In all these plots, deviations from linearity occur at two concentration levels, just below 1% and again at 15% of dispersed resin material. This behavior was claimed to be explicable only on the basis of a colloidal structure for the dispersions as prepared in this study. At the lowest concentration of tar resins, maximum adsorption of dispersing medium on the dispersed tar resin units may be assumed. The individual unit may therefore be quite large. On slightly increasing the concentration of tar resins, near maximum size should be maintained while the increased concentration leads to a shortened distance between dispersed units and the development of an incipient brush-heap structure. The brush-heap structure is responsible for increased viscosity of the dispersion. This is the condition up to and at 0.63% concentration of added resins. At this point, the concentration of C_2 resins has reached 0.2% as the result of the fact that the tar resins could not be recovered completely free of C_2 resins. The C_2 component may be regarded as a very active nucleus for micelle formation. A stable micelle structure, therefore, now develops on the pattern of a C_2 nucleus surrounded by layers of the lighter tar resins with peripheral layers of dispersing medium. The growth of such micelles will, therefore, proceed by way of consumption of tar resins and dispersing medium and will, thus, lead to a diminution in the size and numbers of the initially formed and less-stable tar-resin micelles. This involves a gradual breakdown of the first-formed brush-heap structure and, hence, leads to a lesser increase in viscosity and smaller deviation from the behavior demanded by the Einstein relation. With increasing concentration of the C_2 component, the new micellar structure will in its turn develop to a stage where brush-heap formation due to interaction once again assumes significant proportions and, thus, causes an ever-increasing development of greater viscosity and larger deviation from the Einstein relation in accordance with the results. The final brush-heap structure is, therefore, based on a C_2 resin-tar oils micelle, which may, therefore, be regarded as the truetar micelle of the colloid structure of coal-tar pitch.

Dickinson[40] has shown that the different insoluble fractions obtained from coal tar by

[40] Dickinson, E. J., *J. Soc. Chem. Ind. (London)*, **64**, 121–30 (1945).

solvent action behave differently in their effect on viscosity of the mixture when dispersed in tar-distillate oils. Also, the same insoluble fractions from different tars behave differently, and the type of oil influences the behavior. Data are given for C_2 resin (benzene insol.-pyridine sol.) and resin **B** (n-hexane insol.-benzene sol.) from a vertical-retort tar (low-aromatic tar) and a horizontal-retort tar (high-aromatic tar). Oils from the same tars were used as examples of high- and low-aromatic distillate oils. Departures from Einstein's law were greater for C_2 resin than for resin **B**; greater for low-aromatic C_2 than for high-aromatic C_2 in high-aromatic distillate oil, and were the same for the two resin-**B** samples; the same for the two C_2 resin and the high-aromatic resin **B** and less for low-aromatic resin **B** in low-aromatic distillate oil; greater for low-aromatic C_2 in high-aromatic distillate oil than in low-aromatic oil; and the same for high-aromatic C_2 in high-aromatic distillate oil as in low-aromatic oil.

The microscopic appearance of various of these dispersions was noted. In particular, C_2 resins were dispersed in tar-distillate oil from the tar yielding the resin and in tar-distillate oil from a tar of opposite aromaticity character. Low-aromatic C_2 would not form a colloidal (homogeneous) dispersion in low-aromatic distillate-oil but would in high-aromatic oil.

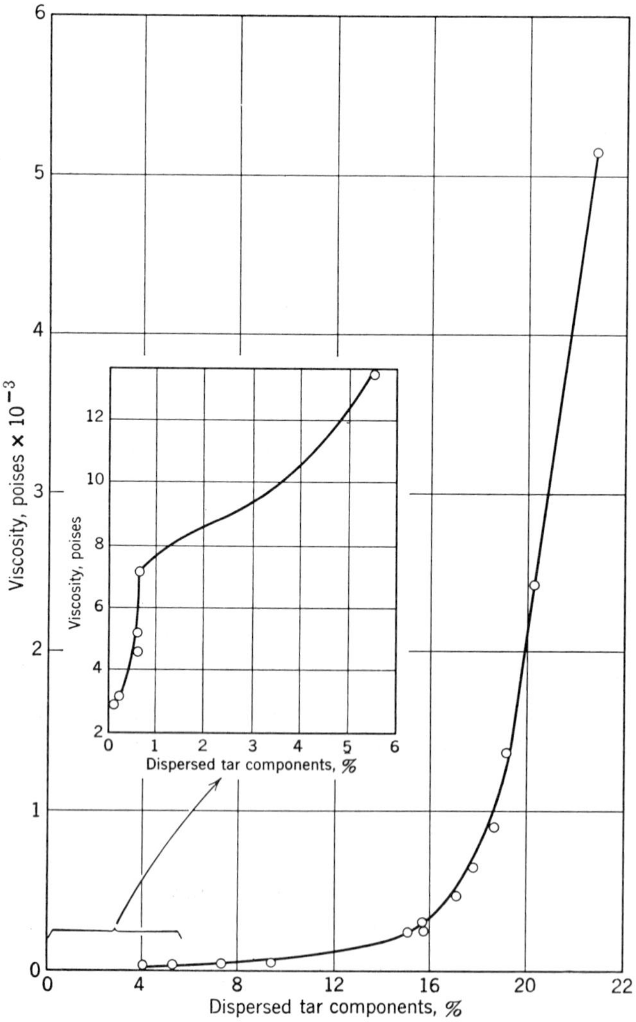

Fig. 30. Viscosity as a function of percentage of dispersed tar components in coal tar.[39] Previously solvent precipitated components were added back to the solvent-freed soluble tar.

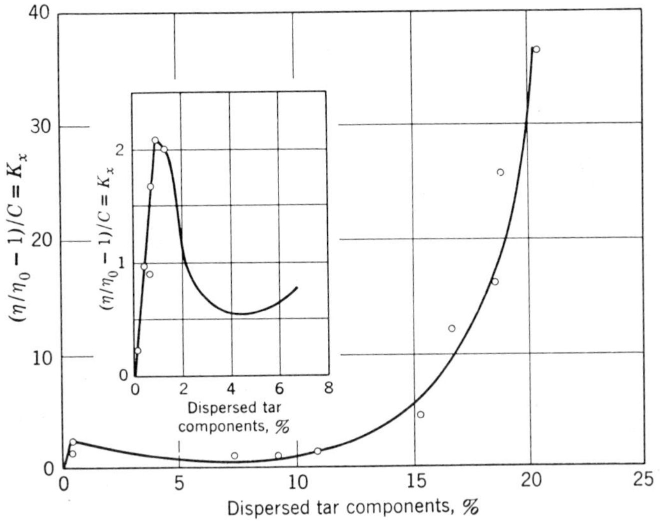

Fig. 31. Plot of the Einstein constant, $K_x = (\eta/\eta_0 - 1)/C$ against percentage of dispersed insoluble coal-tar component in tar.[39] Dispersions were made by adding previously solvent precipitated components back to the solvent-freed soluble tar.

High-aromatic C_2 resin formed a colloidal dispersion in both high-aromatic distillate oil and low-aromatic oil. The chemical nature of both the dispersion medium and the disperse phase has an effect on the degree of dispersion. Further behavior of low-aromatic C_2 resin was noted. Where all products were from the same low-aromatic tar, C_2 resin would not disperse colloidally in tar-distillate oil plus resin A (*n*-hexane soluble) but would disperse in tar distillate oil plus resin B. The stabilizing effect of resin B is most reasonably explained in terms of a colloid system where it provides a solvation layer for micelles of C_2 resin. All of the above described behavior is consistent with a colloidal system where varying chemical structure causes varying ease and degree of solvation of micelles of the more associated high-molecular-weight resins. The existence of such a structure explains the marked dependence of viscosity on temperature as being due to the change of degree of dispersion (micelle size) with temperature and also accounts for the gradual transition in physical properties from fluid tars to hard pitches.

Dielectric Constant and Dielectric Dispersion

Walther[41] has used dielectric constant measurements to afford an insight into the physical-chemical constitution of coal-tar pitch. In particular, he has measured the dielectric constant, at a constant frequency, of coal-tar pitches during their solidification as the temperature is reduced. The particular range of temperature is that within which anomalous dispersion occurs, that is, the temperature range within which a relatively large decrease in dielectric constant occurs with decreasing temperature. Above and below this temperature range, a relatively small increase in dielectric constant occurs with decrease of temperature. The curve of the plot of dielectric constant against temperature, $\epsilon = f(T)$, in the anomalous dispersion range has the *S* shape, Fig. 32, of a probability summation curve. The rate of change of the dielectric constant with temperature, $\Delta\epsilon/T = f'T$, has the bell shape of the frequency-distribution curve, the Gaussian error curve. Because of the similarity of the dispersion curves to the probability curves, the characteristics of the dispersion curves are analyzed by the techniques applied to probability curves. A comparison was made of the dielectric behavior of three soft pitches of the same softening point but of different "plasticity," all made from the same tar. One pitch, pitch *A*, was made by distillation of the tar to a final pitch softening-point, Kraemer-Sarnow, of 19°C. Pitches *B* and *C* were prepared by distillation to successively higher softening points and then cutback with flux

[41] Walther, H., *Kolloid-Z.*, **117**, 75–96 (1950).

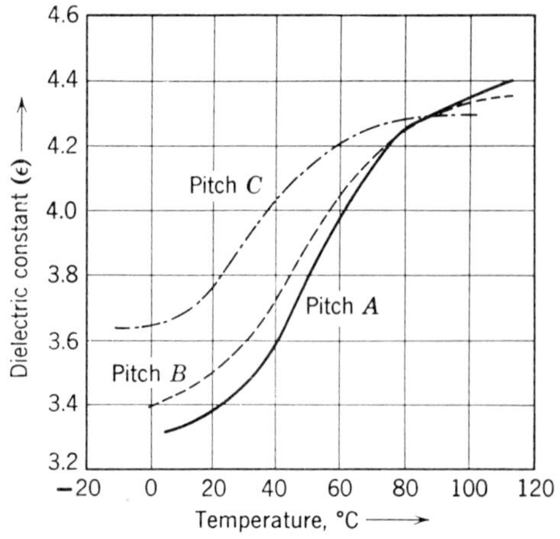

Fig. 32. The dielectric constant of coal-tar pitches *A*, *B*, and *C* as a function of temperature in the anomalous dispersion range.[41]

oil to a final softening point the same as pitch *A*. Data on the preparation conditions and the properties of the pitches are given in Table 12. Additional rheological properties of these pitches are shown in Figs. 33 and 34.

Figure 32 shows the *S*-shaped curves for the dielectric constant as a function of temperature for the three pitches. The analysis of such curves comprised determining whether the data represent a single normal distribution or a summation of two or more normal distributions. A single normal distribution can be recognized in that a plot of the cumulative frequency percentage (cumulative sum of the $\Delta\varepsilon/\Delta T$ values to a given temperature as a percentage of the total change) versus temperature on probability paper gives a straight line. Figure 35 shows, on probability paper, the cumulative frequency percentage of the dielectric constants as measured and of the two separate normal distributions, T_1 and T_2, into which the original measurements were divided. The original data failed to yield a straight line, the deviation being greater in the order *A*, *B*, *C*. For pitch *A* there appears to be a primary normal distribution extending over the full temperature range of the anomalous dispersion plus a secondary normal distribution covering only the lower part of the temperature range. For pitches *B* and *C*, the secondary distribution becomes increasingly large.

Figures 36, 37, and 38 show for each of the pitches, *A*, *B*, and *C*, the two normal distributions plotted as cumulative frequency-percentage curves on probability paper. The use of probability paper in place of Cartesian coordinates converts the *S* shape of the lips of the bell into straight lines for easier plotting and extrapolation. Before discussing the significance of the preceding phenomena, data on a series of pitches from different tars are presented. Including pitch *A*, five pitches were prepared by distilling batchwise at atmospheric pressure five different tars to an overhead temperature of 240°C. Properties of the pitches

TABLE 12

PREPARATION, PROPERTIES AND DIELECTRIC BEHAVIOR OF THREE COAL-TAR PITCHES OF SAME SOFTENING POINT BUT DIFFERENT "PLASTICITY" PREPARED FROM SAME TAR BY CUTBACK AFTER DISTILLATION[41]

Preparation/Properties	Pitch *A*	Pitch *B*	Pitch *C*
Initial Distillation of Tar:			
Final Temperature, °C	250	360	400
Softening point, Kraemer-Sarnow, °C of distillation residue	19	87	135
Oil for cutback, %	—	33	48
Final Pitch:			
Softening point, Kraemer-Sarnow, °C	19	18.0	18.5
Break point, °C	−10.0	−15.5	−18.5
Specific gravity at 20°C	1.207	1.217	1.237
Benzene Insoluble, %	9.0	19.5	35.7
Falling ball, height to break at 0°C, m	1.6	2.4	14.0
Viscosity at 20°C, cP × 10^{-7}	7.00	7.08	7.57
Dielectric Constant at 20°C	3.44	3.490	3.775
$\epsilon_1 - \epsilon_0$	1.250	0.940	0.655

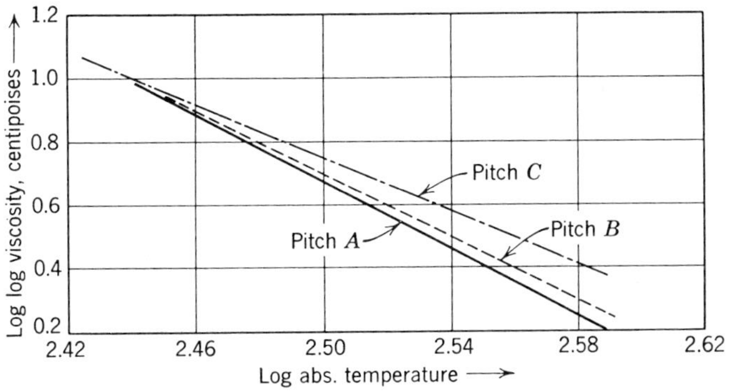

Fig. 33. Viscosity as a function of temperature for pitches *A*, *B*, and *C* used in study of anomalous dispersion of dielectric constant.[41]

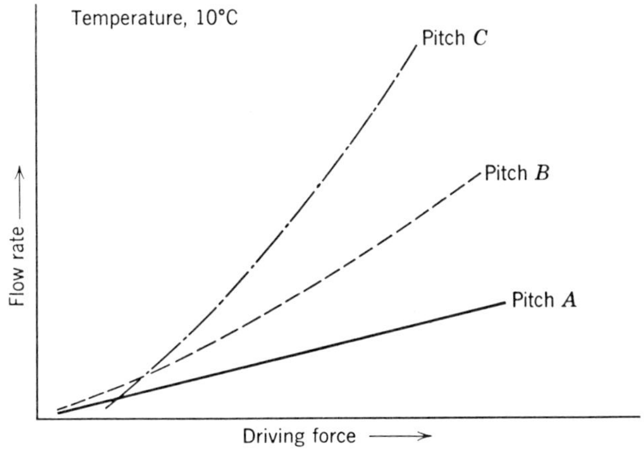

Fig. 34. Flow properties showing departure from Newtonian behavior for pitches *A*, *B*, and *C* used in study of anomalous dispersion of dielectric constant.[41]

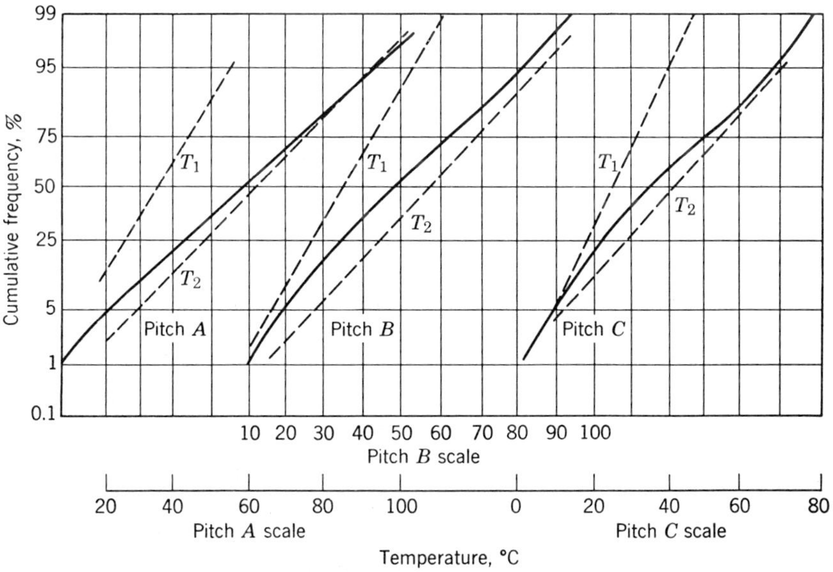

Fig. 35. The cumulative frequency percent of the dielectric constant as a function of temperature for pitches *A*, *B*, and *C*. Original measurements shown in solid line. The dotted lines T_1 and T_2 are the two normal distributions which combined give the measured values.[41]

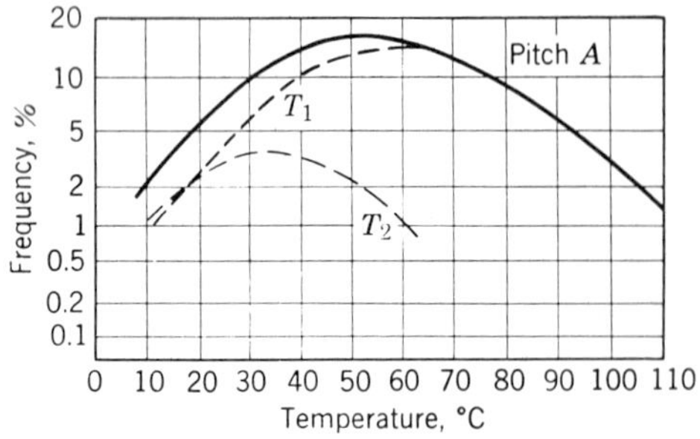

Fig. 36. Pitch *A*: Frequency curves for the two normal distributions shown in Fig. 35.[41]

are shown in Table 13. The percentage values for T_1 and T_2 refer to the relative contribution that each makes to the total. This is measured by the area under the distribution curve when plotted as a cumulative frequency percentage. It will be noted that the sum of the percentages for T_1 and T_2 do not add to 100. This results from the fact that only 98% of the area under the original data curve was used in the analysis. One percent at each end of the curve was discarded as subject to too much error.

In interpreting these anomalous dispersion data, the following mechanistic description of the dispersion process was offered. From higher temperatures down to that where anomalous dispersion begins, the dielectric constant of a pitch rises slowly, as dipoles are better able to maintain orientation because of diminished Brownian motion. At some point, the increase in viscosity of the pitch starts to interfere with free orientation of the dipole with the high frequency field, and the dielectric constant starts falling. The rate of decrease follows a specific probability curve which is associated with the properties of a molecularly

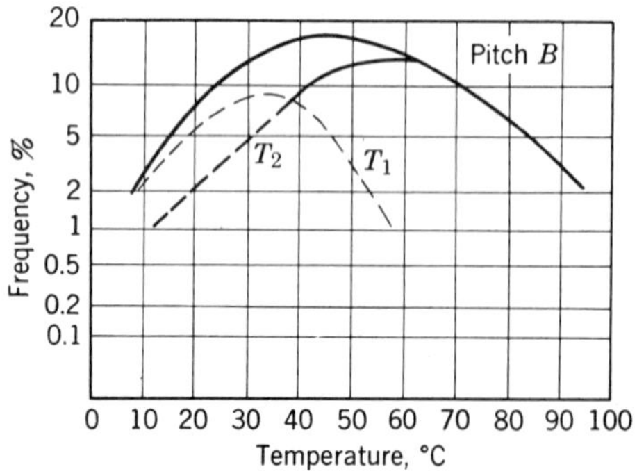

Fig. 37. Pitch *B*: Frequency curves for the two normal distributions shown in Fig. 35.[41]

dispersed solution. At some lower temperature, the rate of fall of the dielectric constant departs from the initial specific probability curve. At this temperature, aggregate formation such as molecular association or micelle formation is presumed to start and contributes an additional, probability-controlled mechanism of interfering with free orientation of dipoles with the high frequency field. This continues with a drop in temperature until the dipoles are completely "frozen," from which point on the dielectric constant changes only slightly, corresponding to the volume contraction of the pitch.

The preceding picture assumes essentially iso-gel structure for the pitch, that is, the liquid and solid components are chemically similar materials of different molecular weight with intermediate materials affording a smooth

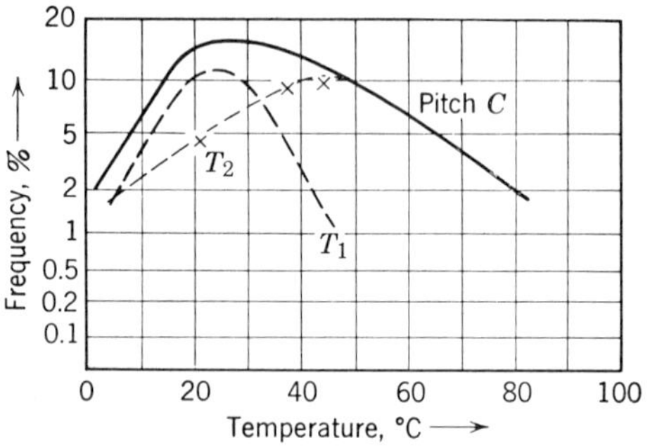

Fig. 38. Pitch *C*: Frequency curves for the two normal distributions shown in Fig. 35 for Pitch *C*.[41]

transition between the two. High values of the dielectric constant and of $\epsilon_1 - \epsilon_0$ usually indicate essentially molecularly dispersed materials. Low dielectric constant and low $\epsilon_1 - \epsilon_0$ are usually accompanied by large deviations from a single normal probability distribution. Large deviations from a single normal-probability curve are reflected in the flow properties of the pitch. Structural viscosity is marked, the temperature coefficient of viscosity is low, and plastic rather than brittle properties are evident. Some materials, showing unusual amounts of structural viscosity such as air-blown asphalts, give $\Delta\epsilon/\Delta T = f'T$ curves which appear to be the sum of three or even more normal curves. Such materials are probably not iso-gels.

The anomalous dispersion curves of the three pitches *A*, *B*, and *C*, from the same tar, can be satisfactorily resolved by a maximum of

TABLE 13

PROPERTIES, INCLUDING DIELECTRIC CONSTANT, OF A NUMBER OF SOFT COAL-TAR PITCHES FROM TARS OF VARIOUS SOURCES[41]

	(Leipzig) Pitch A	Upper Silesian	Waldenburg	Ruhr	Saar
Softening point, Kraemer-Sarnow, °C	19.0	22.5	18.0	19.0	24.5
Specific gravity, 20°C	1.207	1.243	1.224	1.220	1.246
$\epsilon_{20°}$	3.44	4.25	3.705	3.56	4.33
$\epsilon_1 - \epsilon_0$	1.250	1.145	1.100	1.050	1.005
Benzene insolubles, %	9.0	15.7	11.8	10.0	17.2
T_1, %	7.1	35.8	23.8	12.2	44.1
T_2, %	90.9	62.2	74.2	85.8	53.9

two normal distributions. The pitches behave as iso-gels. The second, or low-temperature distribution, becomes successively larger in the order *A*, *B*, *C*. This trend is in agreement with the rheological difference between the pitches, as shown in Figs. 33 and 34, which require increased micelle or gel structure in the order *A*, *B*, *C*.

The anomalous dispersion data on the five pitches from five different tars, Table 13, indicate that structural viscosity should increase in the order *A*, Ruhr, Waldenberg, Upper Silesia, and Saar. This same order might be expected for extent of structural viscosity from the benzene-insoluble contents of the pitches. Actually, the behavior shown in Fig. 39 suggests that other factors must be present. The pitch *A* and the Saar pitch are correctly placed as Newtonian and highly structural, respectively. Ruhr pitch appears to have more structural viscosity than Upper Silesian pitch on the basis that departure from linearity indicates structural viscosity.

Physical Constitution of Coal Tar

Although it has been possible to define, within limits, the range of molecular sizes and types found in coal tar and coal-tar pitches, attempts to resolve their physical structure and its relationship to physical properties have been less successful. That coal tar is a heterogeneous system was concluded by Nellensteyn[42] from examination of coal-tar solutions with an optical microscope. Horizontal-retort tar and coke-oven tar showed a relative profusion of

[42] Nellensteyn, F. J., *Z. angew. Chem.*, **43**, 402 (1930).

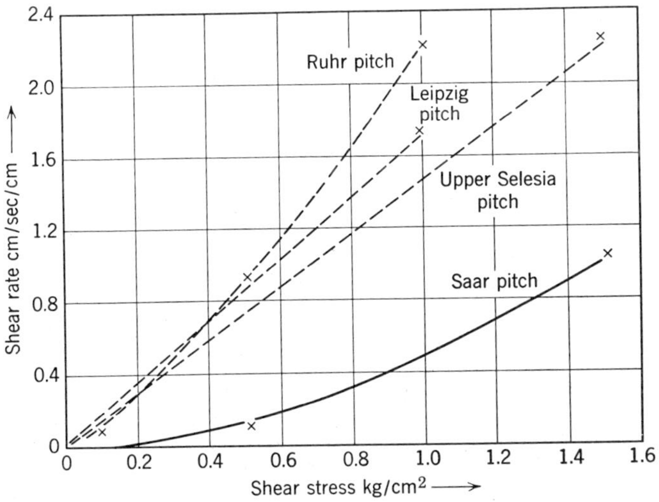

Fig. 39. Shear rate-shear stress plot of four coal-tar pitches to illustrate extent of structural viscosity. Departure from linearity indicates structural viscosity.[41]

particles ranging from 10-μ size down to the limits of visibility. Vertical-retort tar showed only a small fraction of the particles evident in the former two tars. Sarkar,[43] using an electron microscope, noted that in horizontal-retort and coke-oven tars the particles are comparatively large, 0.5 to 1 μ, and infrequent, whereas in vertical-retort tar the particles are numerous but smaller. The particles in horizontal-retort tar and coke-oven tar appear to be clusters of small particles similar to the small dispersed particles of vertical-retort tar.

Dickinson[40] has examined, under the electron microscope, filtered vertical-retort tar and filtered pyridine solutions of horizontal-retort and coke-oven tar. The vertical-retort tar showed a well-defined and regularly dispersed particle structure, the particles having diffuse perimeters and a visible diameter of about 300 *A*. In the other tars, the colloidal dispersion was less regular.

Wood and Phillips[17] sought to assess the amount of colloidal suspended particles as well as of micelles formed by aggregation of relatively small molecules. They ultracentrifuged filtered crude tars and tar solutions. Centrifuging of two vertical-retort tars alone and mixed with pyridine, and of horizontal-retort tar alone and mixed with various proportions of anthracene oil, led to few positive results. The only sedimentation detected was of 1 to 2% of a toluene-insoluble fraction in the undiluted vertical-retort tars which passed from the inner to the outer section of the spinning top of the ultra-centrifuge. The authors stated their realization that molecules tending to form micelles would be much less associated at the temperatures giving viscosities at which ultracentrifuging was possible than in pitch at ordinary temperatures.

The possible nature of the disperse phase and the dispersion medium have been subjects of speculation. Frydlender[44] suggested that the disperse phase is comprised of organographites, presumably a descriptive name for polynuclear condensed-ring hydrocarbons of a high degree of condensation and high molecular weight. The continuous phase is composed of simple aromatics, presumably the well-known one-, two- and three-ring compounds. A resinous fraction which peptizes the disperse phase is constituted of intermediate materials between the organographites and the simple aromatics.

Sack[16] has described pitch as a plastic material which is a resin-plasticizer-filler system. The resin is polymerized heterocyclic and aliphatic monomers. The plasticizer is monomeric material or a polymer of small-molecular weight. The filler is the organographite material mentioned earlier. Active centers of the filler were described as polarizing neighboring hydrocarbon molecules and forming a type of chemical bond with the hydrocarbon. Sack[16] rejected a sphericolloidal system, because of the visco-elastic properties of coal-tar pitch which requires anisotropy not possible in a spherical system.

Wood and Phillips[17] and Franck[12] supported the view of Krenkler and Wagner[45] that bituminous materials are supercooled liquids. Coal tar and coal-tar pitch are regarded as supercooled liquids with some 5% of finely dispersed inert filler which is of little consequence in determining properties. The failure of the ultracentrifuge to achieve any appreciable separation of a disperse phase is held to preclude a colloidal system. Concerning the indirect evidence for a colloidal system, it was argued that the solubility of compounds in nonelectrolytes is controlled largely by internal pressure, on which surface tension is dependent, and that flocculation may be considered a normal solution phenomenon and does not throw any light on the state of aggregation in the original solution. It was also argued that all viscous liquids (inorganic and organic), thermoplastic resins, and supercooled liquids have high temperature coefficients of viscosity; it is not possible at present to relate the values of these to a knowledge of the intermolecular forces involved.

TAR PROCESSING

To further the utilization of coal tar, much work has been done to recover and refine chemicals from the tar and to improve and evaluate oil and pitch fractions for various uses. The scheme of preliminary fractionation of the tar by distillation has not changed, although the efficiency of the fractionation has increased. The tar is still fractionally distilled into one or

[43] Sarkar, A. K., *Kolloid-Z.*, **130,** 116–20 (1953).

[44] Frydlender, J. H., *Rev. prod. chim.*, **52,** 209–15 (1949).

[45] Krenkler, K., and Wagner, R., *Erdöl u. Kohle*, **1,** 280–5 (1948).

more fractions containing the tar acids, tar bases, and naphthalene and into a creosote-oil fraction leaving "pitch" as a residue. The tar acids and tar bases are recovered by extraction with aqueous caustic and sulfuric acid, respectively, leaving a neutral naphthalene oil. The naphthalene is separated by fractional distillation or by crystallization or by a combination of the two procedures.

Tar Acids

The separation of the tar-acid fraction from the distillate fraction containing them has been the object of much work. A patent[46] claims the separation of phenols from hydrocarbons without the use of caustic soda, by a method based on distillation in the presence of steam, to give a residue of high-phenol content. The separation with methanol-water mixtures has been studied,[47] and equilibrium data and diagrams are given for twenty-one ternary and two quaternary systems. The most effective solvent appears to be 70% methanol. The extract of phenols is not hydrocarbon free where the phenols have been extracted from aromatic hydrocarbons but can be freed of hydrocarbon by extraction of the methanol extract with a paraffinic solvent. The more highly alkylated phenols are more soluble in aromatic hydrocarbons, which renders extraction more difficult. This trend is not evident for paraffinic hydrocarbons.

Neuworth[48] and associates described a two-solvent fractional extraction process using aqueous methanol and a paraffinic solvent for the treatment of low-temperature tar-acid oils. Under certain conditions, this process may have application to high-temperature tar-acid oils. The extracted phenols contain considerably greater amounts of bases than do caustic extracted phenols. Depp and Neuworth[49] also described a process, based on ion-exchange resin action, to remove tar bases from the extracted phenols.

[46] Milner, D. W., U.S. Pat. 2,437,649 (1948). Yorkshire Tar Distillers Ltd. and Milner, D. W., Brit. Pat. 595,406 (1947).

[47] Prutton, C. F., Walsh, T. J., and Desai, A. M., *Ind. Eng. Chem.*, **42,** 1210–17 (1950).

[48] Neuworth, M. B., Hoffmann, V., and Kelly, T. E., *Ind. Eng. Chem.*, **43,** 1689–94 (1951).

[49] Depp, E. A., and Neuworth, M. B., *Ind. Eng. Chem.*, **49,** 182–5 (1957).

Cumming and Morton[50] have studied glycerol, triethylene glycol, and their aqueous solutions for the extraction of phenol from coal-tar hydrocarbons. Glycerol was the best extractant for phenol itself, but triethylene glycol was more efficient for the homologues.

A process for the separation of tar acids from tar oils by washing with concentrated aqueous sodium phenolate solution has been patented.[51] The rich phenolate solution, after being washed with a paraffinic solvent to take out entrained hydrocarbons, is extracted with an ether, ketone, or ester. The phenols are recovered from the oxygenated solvent by distillation. The use of hot water under pressure, as an extraction medium for phenols of high temperature tar, is referred to in an abstract of a Japanese paper.[52] The solubility of alkyl phenols in water has been reported,[53] and a countercurrent extraction using water at 200 to 250°C has been patented.[54] Some Japanese work using liquid ammonia and water has been described.[55] Two papers by Japanese workers[56] describe the extraction of phenols with aqueous solutions of sodium salts of sulfonic acids. In the first of these, solubility data at 25 or 60°C were given for phenol, the cresols, xylenols, benzene, toluene, xylenes, ethylbenzene, and some paraffinic and olefinic hydrocarbons in a 30% aqueous solution of sodium xylene sulfonate. The solubility of the phenols was over 1300 cc/100 cc of solvent, whereas that of the hydrocarbons was little higher than in water itself. The second paper gave data on 1 to 36% aqueous solutions of sodium or calcium 1- or 2-naphthalene sulfonates as solvents.

[50] Cumming, A. P. C., and Morton, F., *J. Appl. Chem. (London)*, **2,** 314–23 (1952).

[51] Doughty, E. W., Murray, J. V., Jr., and Fales, J. D., U.S. Pat. 2,595,516 (1952).

[52] Fuchi, I., Iio, K., Yamuzaki, Y., and Ishibashi, T., *Kooru Taaru*, **1,** 202–8, 242–4, 277–81 (1949); *Chem. Abs.*, **46,** 7740 (1952).

[53] Erichsen, L. V., and Dobbert, E., *Brennstoff-Chem.*, **36,** 338–45 (1955).

[54] Metallgesellschaft A. -G., Brit. Pat. 710,137 (1954).

[55] Ishida, K., *Sci. Repts. Research Insts. Tohoku Univ.*, Ser. A, **5,** 377–94 (1953); *Chem. Abs.*, **48,** 9045 (1945).

[56] Kozaki, S., Kido, T., Matsuzaua, S., and Kiyama, M., *J. Chem. Soc. Japan*, Ind. Chem. Sect., **58,** 307–12 (1955); *Chem. Abs.*, **49,** 14297 (1953).

The recovery of the individual phenols from the tar-acid fraction has been the subject of many studies. The recovery of pure metacresol and paracresol from the mixture of the isomers has been and still is a big problem. Stevens[57] has reviewed methods proposed to 1942 and has given details on the separation by alkylation of the mixture to di-tertiarybutyl derivatives, separation of the derivatives by fractional distillation, and dealkylation again to the respective cresols.

Cislak and Otto[58] used a selective solvent, such as 95% ethanol, to separate the reaction products of the two cresols with hexamethylenetetramine and then decomposed the separated reaction products with ether or dioxane. The same men used 2,6-lutidine to separate an addition compound with *m*-cresol at -5 to 15°C. Distillation of the addition compound yielded *m*-cresol of 80% purity. By using 4-picoline, an addition compound with *p*-cresol may be separated.[59] Engel[60] added phenol and a paraffinic solvent to a *m*- and *p*-cresol mixture to crystallize out the phenol-*m*-cresol addition compound. Engel also was able to crystallize *p*-cresol from a hydrocarbon solution of the mixed cresols at 0°C. Othmer[61] and associates showed that separation of mixtures of *m*- and *p*-cresol could not be effected by any of fourteen entrainers in either azeotropic or extractive distillation. Savitt and Othmer[62] modified the method of Bentley[63] using benzidine to form an addition compound with *p*-cresol. Freezing-point data were given on the metacresol-paracresol binary and binary systems having metacresol and paracresol as one component and various amine compounds as the second component.

The formation of the addition compound with urea has been used to increase the *m*-cresol content of a mixture from 48 to 78%. Similarly, a 98 to 100% *m*-cresol was obtained from an 85% starting material.[64] The successive use of benzidine and sodium acetate to form addition compounds with *p*-cresol can enrich a 60% *p*-cresol mixture to 97% *p*-cresol in two stages.[65] White and Grant[66] described the complete separation of the three isomeric cresols by liquid-liquid partition chromatography using a buffered stationary phase. The separation of *m*- and *p*-cresol by extractive crystallization with acetic acid as solvent also has been described.[67] The distribution of the two cresols between aqueous caustic and an organic solvent is the basis for a separation method.[68]

Cislak and Otto[69] claimed the separation of 2:4-xylenol from its mixture with 2:5-xylenol by the formation of the metacresol complex with the 2:4-isomer. These same isomers were separated by alkylation with excess di-isobutylene, the 2:5-isomer remaining unchanged. After separation by distillation, the 2:4-isomer alkylation product was debutylated.[70] Separation of these isomers was claimed through the formation of a complex between the 2:4-isomer and one of several bases such as the methyl quinolines.[71] It was reported that 3:5-xylenol does not react with isobutylene; this behavior was used to separate it from other phenols boiling in the range 202 to 230°C.[72]

Additional separations of close boiling tar-acids have been reported. The mixture, *o*-ethylphenol, *o*-cresol, and 2:6-xylenol was treated with quinaldine which produces a

[57] Stevens, D. R., *Ind. Eng. Chem.*, **35,** 655–60 (1943).

[58] Cislak, F. E., and Otto, M. M., U.S. Pat. 2,310,616 (1943).

[59] Cislak, F. E., and Otto, M. M., U.S. Pats. 2,432,062–3 (1947).

[60] Engel, K. H., U.S. Pats. 2,339,388 (1944), 2,382,142 (1945).

[61] Othmer, D. H., Savitt, S. A., Krasner, A., Goldberg, A. M., and Markowitz, D., *Ind. Eng. Chem.*, **41,** 572–4 (1949).

[62] Savitt, S. A., and Othmer, D. H., *Ind. Eng. Chem.*, **44,** 2428–31 (1952).

[63] Bentley, W. H., and Catlow, B., U.S. Pat. 1,980,901 (1934).

[64] Anazawa, I., *J. Chem. Soc. Japan*, Ind. Chem. Sect., **58,** 231–3 (1955); *Chem. Abs.*, **49,** 13625 (1955).

[65] Zajaczkowska, B., *Acta Polon. Pharm.*, **11,** 45–7 (1955); *Chem. Abs.*, **50,** 10379 (1956).

[66] White, D., and Grant, D. W., *Nature*, **175,** 513 (1955); *Chem. Abs.*, **49,** 10607 (1955).

[67] Chivate, M. R., and Shah, S. M., *Chem. Eng. Sci.*, **5,** 232–41 (1956).

[68] Walker, C. A., *Ind. Eng. Chem.*, **42,** 1226–30 (1950).

[69] Cislak, F. E., and Otto, M. M., U.S. Pat. 2,456,582 (1948).

[70] Roberts, E., and Rose, M. J., Brit. Pat. 645,446 (1950).

[71] Cislak, F. E., and Otto, M. M., U.S. Pat. 2,526,807 (1950).

[72] Kel'tseva, O. B., and Moshkin, P. A., *Khim. i Tekhnol. Topliva i Masel*, **1957,** No. 2, 11–17; *Chem. Abs.*, **51,** 11275 (1957).

crystalline complex with both monoalkyl phenols. The separated monoalkyl phenols could be separated by fractional distillation.[73] Terres and associates[74] have reported on the separation of complex tar-acid mixtures by paper chromatography. Previous work was reviewed.

Tar Bases

Only the higher boiling bases are considered here. The separation of isoquinoline from quinoline and quinaldine can be accomplished by treating with an aqueous solution of H_3PO_4. Isoquinoline forms an extremely water-soluble phosphate in contrast to the limited solubility of the phosphates of the other components.[75] The use of complexes with phenol and 2-naphthol for the purification of isoquinoline[76] and with phenols having at least one free meta position for the purification of quinaldine has been claimed.[77] The isolation of quinoline as its monophosphate has been described.[78] Quinoline from vertical-retort tars is improved by treatment at 160°C with 1 to 5% sulfur followed by distillation.[79] The use of phthalic acid for purifying quinoline and isoquinoline has been described. The salts are crystallized from water or alcohol and the base regenerated by alkali.[80] Purification of quinoline may be accomplished by the formation of an insoluble quinoline acid tartrate in water. The acid tartrate may then be recrystallized from water before treatment with alkali to regenerate the purified base.[81]

Isoquinoline is purified by solution in glacial acetic acid and treatment with 25% hydrogen peroxide at 80 to 90°C for 12 hr. The acetic acid is removed in vacuum, and the residue is made alkaline and extracted with chloroform. Both quinoline N-oxide and isoquinoline N-oxide so recovered are treated by known methods to regenerate the bases.[82] The separation of quinoline and isoquinoline by chromatography on special cupric-chloride-treated paper is described. A melting point diagram has been given for mixtures of quinoline and isoquinoline sulfates.[83]

It has been claimed that countercurrent extraction by the two solvent system 2N hydrochloric acid-benzene can recover from a 53:47 mixture of quinoline and isoquinoline a benzene extract of quinoline of 99.5% purity.[84] Pure quinoline can be readily prepared from a quinoline fraction by addition of 10 to 15% water, followed by centrifuging and dehydrating the resulting crystalline hydrate.[85]

Naphthalene

Woolhouse[86] has recently reviewed various methods for the production of naphthalene of 78°C freezing point and investigations concerning its characteristics. McNeil[87] has also recently discussed the various factors in the production of naphthalene. The importance of the removal of acids and bases before further processing was stressed.

The desulfurization of naphthalene, by the use of 3 to 15% acid clay at 200°C or by activated clay at 150°C, followed by distillation of the naphthalene, has been claimed.[88] To prevent color bodies in naphthalene, refined naphthalene, obtained by treatment with sulfuric acid followed by sublimination, may be passed molten through a bed of silica or ferric-oxide gel at 80 to 100°C.[89] The use of metallic sodium as a means of purifying naphthalene

[73] Milner, D. W., and Flathers, R., Brit. Pat. 736,604 (1955).

[74] Terres, E., Gebert, F., Hülsemann, H., Petereit, H., Toepsch, H., and Ruppert, W., *Brennstoff-Chem.*, **36**, 162–9, 228–36 (1955).

[75] Engel, K. H., U.S. Pat. 2,456,774 (1948).

[76] Cislak, F. E., and Otto, M. M., U.S. Pat. 2,432,065 (1947).

[77] Cislak, F. E., and Otto, M. M., U.S. Pat. 2,432,064 (1947).

[78] Soc. Chem. Ind. in Basle, Brit. Pat. 588,277 (1947).

[79] Yorkshire Tar Distillers Ltd. and Milner, D. W., Brit. Pat. 661,482 (1951).

[80] Shannan, W. V., Brit. Pat. 701,976 (1954).

[81] Arnall, P., Brit. Pat. 744,235 (1956).

[82] Ochiai, E., and Ikebara, M., Jap. Pat. 3,326 (1954).

[83] Profft, E., and Buchmann, G., *Chem. Tech. Berlin*, **8**, 30–5 (1956); *Chem. Abs.*, **51**, 11348 (1957).

[84] Yamamoto, A., Arakawa, H., Higucki, H., and Yoshimura, K., Jap. Pat. 5,720 (1957).

[85] Oberkobusch, R., *Brennstoff-Chem.*, **39**, 349 (1958).

[86] Woolhouse, T. G., *J. Appl. Chem.* (*London*), **7**, 573–83 (1960).

[87] McNeil, D., *Gas World*, **149**, Coking Sect., 43–47, 52 (1959).

[88] Kajima, T., Yamamuro, T., and Sahan, T., Jap. Pat. 175,175 (1947).

[89] Armstrong, W. E., Densham, A. B., and Griffith, R. H., Brit. Pat. 651,154 (1951).

containing 0.13 to 0.16% sulfur, prior to its hydrogenation, is mentioned in an abstract of a Polish paper.[90] Patents have been issued covering the use of sodium metal.[91] Patents have also been granted covering a process for purification of naphthalene by recrystallization from methanol.[92] A patent specifies the recovery of naphthalene from tar oils by precipitation with liquid ammonia.[93] Oxidizable impurities may be removed from hot-pressed naphthalene by passing the vapor and a limited amount of air through a fluidized bed of mild oxidation catalyst at 350 to 500°C. The preferred catalyst, on account of its cheapness, is iron oxide.[94] Naphthalene of 94 to 95% purity, from which tar acids and tar bases have been removed, may be further purified by treatment in the molten state with anhydrous aluminum chloride. Impurities are polymerized and the product is obtained by fractional distillation.[95] A process has been described which eliminates sulfuric acid washing as a stage in the refining of naphthalene. The crude naphthalene was treated with less than 1% of isobutylene or diisobutylene in the presence of sulfuric acid as an alkylation catalyst. The impurities are more readily alkylated than the naphthalene, which can be easily separated by distillation. Butylation and sublimation of naphthalene, m-pt 78°C, gives white naphthalene of 79.6 to 79.8 m-pt.[96] The thionaphthene content of crude naphthalene can be reduced to less than 1.3% by fractional distillation after reaction with a limited amount of formaldehyde in the presence of sulfuric acid. The formaldehyde reacts preferentially with thionaphthene.[97]

When naphthalene is purified by azeotropic distillation with cresols and by partial sulfonation, the product still contains 0.04% sulfur in the form of thionaphthene and cannot be used for hydrogenation over nickel catalyst. Subsequent partial chlorination followed by rectification reduces the sulfur to 0.02%, but chlorine content of 0.06 to 0.16% is difficult to remove.[98] A study of the azeotropes formed by ternary systems of naphthalene, a mixture of *m*- and *p*-cresol, and each of three pyridine fractions has been made. The isobaric boiling points of the three ternary mixtures are approximately the same and slightly below the maximum azeotropic boiling point of the respective cresol-pyridine fraction mixture.[99]

A number of patents have been issued describing naphthalene purification by washing naphthalene cake on a centrifuge with aqueous solutions of surface active agents.[100] Patents also describe the crystallization of naphthalene from emulsified naphthalene-bearing oils containing surfactants.[101] The purification of naphthalene, as well as benzene, toluene, and xylene by catalytic hydrogenation over a platinic oxide-alumina catalyst at 350 to 550°C has been patented. Hot-pressed naphthalene, crystallizing point 79.19°C, after hydrogenation and fractional distillation, crystallized at 80.1°C.[102] It has been reported that the system naphthalene:2-methylnaphthalene forms a eutectic mixture at 30.1 mole percent naphthalene. The vapor-liquid equilibrium of the system has been determined and expressed by the equation

$$y/(1 - y) = Kx/(1 - x)$$

where x and y are the mole fractions of the more volatile component in the liquid and vapor, respectively, and K is the relative volatility and equals 1.5.[103] The same authors gave K values from 1.6 to 2.66 for naphthalene and for tar fractions boiling between 137 and

[90] Treszezanowiez, E., and St. Ciborowski, *Przemsyl Chem.*, **6,** 132–4 (1950); *Chem. Abs.*, **46,** 3989 (1952).

[91] Tatterson, B. F., U.S. Pats. 2,626,236 (1953), 2,927,074 (1960).

[92] Scott, R., and Joscelyne, E. H., U.S. Pat. 2,598,449 (1952).

[93] Jarry, R. M., Fr. Pat. 973,567 (1951).

[94] Armstrong, W. E., Densham, A. B., Griffith, R. H., and Marsh, J. D. F., Brit. Pat. 692,098 (1953).

[95] Yamamoto, A., Jap. Pat. 4,484 (1952).

[96] W. Butler & Co., and Poynton, N. H., Brit. Pat. 701,097 (1953).

[97] Schulze, W., Ger. Pat. 840,393 (1952).

[98] Zieborak, K., *Przemsyl. Chem.*, **9,** 339–41 (1953); *Chem. Abs.*, **49,** 5813 (1955).

[99] Zieborak, K., and Markowaka-Majewaka, M., *Bull. acad. polon. sci.*, Classe III, **2,** 341–4 (1954); *Chem. Abs.*, **49,** 5910 (1955).

[100] Ogilvie, C. B., U.S. Pat. 2,678,955 (1954). Apgar, F. A., U.S. Pat. 2,704,779 (1955).

[101] Krzesz, Z., U.S. Pat. 2,711,432 (1955). Detrick, R. S., and Laver, G. G., U.S. Pat. 2,890,254 (1959).

[102] Densham, A. B., Gough, G., and Griffith, R. H., Brit. Pat. 735,706 (1955).

[103] Karavaev, N. M., Zykov, D. D., and Garber, Yu. N., *Ukrain. Khim. Zhur.*, **21,** 178–81 (1955); *Chem. Abs.*, **50,** 1439 (1956).

211°C, 219.9 and 283.6°C, 219.9 and 250°C, and 250 and 283.6°C, respectively. The last experimental point on each of the equilibrium curves corresponded to a 80 to 93 mole percent naphthalene in the vapor phase. Because these curves should not be extrapolated, they cannot be used where a pure naphthalene distillate is desired.[104]

The boiling point isobars of naphthalene and monocyclic aromatic fractions boiling at 210.48°C and 212.35°C have been studied; an azeotrope was not formed with the first fraction, but was with the second, with a very small boiling-point depression.[105]

Distillation of naphthalene oil has been studied. It was found that naphthalene of above 78°C crystallizing point could be obtained from acid- and base-free oils (presumably of coke-oven origin), with recoveries of about 63%. Better recoveries resulted, averaging 73%, by washing the feed with 75% sulfuric acid before distillation. The naphthalene produced could be upgraded to a crystallizing point of 79.3 to 79.55°C by conventional methods.[106]

Thionaphthene may be removed from naphthalene by treatment with an aliphatic peracid followed by extraction with water and redistillation.[107] Mastrangelo and Dornte have used a solid-solution treatment for investigation of the thionaphthene impurity in naphthalene, using an adiabatic calorimeter; the phase diagram and heats of fusion for systems with various proportions of naphthalene were given.[108]

Creosote Oil

Rhodes,[109] in 1951, reviewed the history of creosote production and pointed out the effect of the type of coal tar, the type of tar still, and of the tar-chemicals market on creosote composition. In the early days of creosote production, both low- and high-temperature tar were used, but since 1940, mainly high-temperature tar has been used in creosote production. Before substantial markets for high softening-point pitches developed, the creosote produced in tar distillation was so-called "low residue" creosote, that is, having less than 20% residue on distillation to 355°C vapor temperature. Such creosote contained the higher-boiling portion of the tar acid-naphthalene fraction, the methylnaphthalenes, and the lower-boiling portion of the anthracene fraction. When high softening-point pitch is being produced, that is, pitch for production of carbon electrodes, foundry cores, clay "target" pigeons, and pulverized fuel, creosote of as high as 50% residue is produced. Before World War II, when relatively small amounts of phenol, naphthalene, and anthracene were recovered, most of the coal-tar chemicals were still contained in American creosotes. During and after World War II, a major effort was made by the American tar-distilling industry to recover substantial quantities of tar acids, tar bases, and naphthalene. Phenol and cresols were removed rather completely, but no attempt was made to recover high-boiling tar acids. Creosote continued to contain an appreciable amount of naphthalene, perhaps half of that normally contained in creosote. While the removal of substantial amounts of these salable low-boiling constituents from creosote is not considered to vitally decrease the effectiveness of creosote, it is noted that the distillation-range requirements of creosote specifications are designed to standardize permanence-toxicity balance by requiring sufficient oil distilling below 270°C to insure adequate toxicity and sufficient oil above 355°C to insure permanence.

Stasse,[110] in 1954, using data obtained from the fractional distillation of eight domestic creosotes and two English creosotes, showed the major components of coke-oven creosotes to be naphthalene (10 to 30%) and phenanthrene (9 to 14%), with smaller percentages of methyl naphthalenes (1 to 6%), acenaphthene (2 to 5%), fluorene (2 to 4%), fluoranthrene (2 to 5%), pyrene (2 to 3%), carbazole (2 to 3%), anthracene (1 to 2%), and diphenylene oxide (1%). No attempt was made to

[104] Karavaev, N. M., Zykov, D. D., and Garber, Yu. N., *Ukrain. Khim. Zhur.*, **21**, 410–15 (1955); *Chem. Abs.*, **50**, 3737 (1956).

[105] Lisecki, Z., *Roczniki Chem.*, **29**, 763–7 (1955); *Chem. Abs.*, **50**, 6119 (1956).

[106] Potashnikov, M. M., *Koks i Khim.*, **1956**, No. 6, 46–9.

[107] Murray, J. V., Jr., and Anderson, J. R., U.S. Pat. 2,779,722 (1957).

[108] Mastrangelo, S. V. R., and Dornte, R. W., *Anal. Chem.*, **29**, 794–7 (1957).

[109] Rhodes, E. O., *Proc. Am. Wood-Preservers' Assoc.*, **47**, 40–61 (1951).

[110] Stasse, H. L., *Proc. Am. Wood-Preservers' Assoc.*, **50**, 13–40 (1954).

estimate "minor" components and tar acids and bases.

Mayfield,[111] in 1951, reviewed the literature regarding the toxic properties of high-temperature coke-oven tar creosote. It was his conclusion that all the fractions of the different boiling ranges of creosote are toxic, the toxicity varying somewhat with the organism. The value of creosote as a wood preservative depends largely on whether or not it remains in the wood under the conditions and throughout the period of service.

Chemicals from Creosote Oil

The recovery of the more plentiful constituents in creosote oil as refined chemicals has been the subject of considerable work. The principal constituents studied have been phenanthrene, anthracene, and carbazole. The solid crystalline material that separates from the creosote-oil fraction upon cooling is comprised chiefly of these three substances and is known as anthracene salts or crude anthracene cake.

A method of separating phenanthrene from fluorene and carbazole has been described, using azeotropic distillation with diethylene glycol. Phenanthrene and anthracene form azeotropes, each containing about 4.5% of the hydrocarbon at 20 mm pressure. Fluorene forms a lower-boiling azeotrope rich in hydrocarbon, while carbazole does not form an azeotrope.[112]

The separation of anthracene and phenanthrene from carbazole in a semirefined anthracene has been claimed by distillation in the presence of a monohydric alcohol as solvent and carrier. Carbazole is left behind, while anthracene and phenanthrene pass over with the carrier.[113]

The continuous fractional distillation of tar to separate a fraction containing a mixture of phenanthrene and anthracene and a fraction enriched in carbazole is also covered by a patent.[114]

For the separation of carbazole from crude anthracene pastes, which crystallize from anthracene oils, sulfonation with chlorosulfonic acid at room temperature has been described. Carbazole is converted to carbazole-N-sulfonic acid which is water soluble; other material remains unchanged. Carbazole is regenerated from the sulfonic acid by heating with dilute acid.[115] Another method describes the precipitation of carbazole sulfate from a benzene solution of the anthracene paste by an excess of 80 to 90% sulfuric acid.[116]

A Dutch patent covers the isolation of pure anthracene and carbazole from crude anthracene. Crude anthracene is extracted with toluene and then with acetone. The acetone extract is solvent freed to yield crude carbazole. Extraction of the crude carbazole with toluene leaves refined carbazole. The toluene extract of the carbazole is concentrated to permit a relatively pure anthracene to crystallize out. The mother liquor is used to extract the crude anthracene starting material.[117]

The use of acetone or methylethylketone containing 5 to 20% water as a selective solvent for separating anthracene from carbazole has been claimed. The water in the solvents decreases anthracene solubility without decreasing carbazole solubility.[118]

Fluorsulfonic acid has been used as a selective solvent to separate anthracene from phenanthrene.[119] Anhydrous HF has been used for the same purpose.[120] Crystallization from furfural also has been used to recover a refined anthracene from its mixture with phenanthrene and carbazole.[121] The purification of anthracene by recrystallization from a monoalkyl ether of diethylene glycol, containing up to four carbon atoms in the alkyl group, has been described.[122]

[111] Mayfield, P. B., *Proc. Am. Wood Preservers' Assoc.*, **47,** 62–84 (1951).

[112] Feldman, J., and Orchin, M., U.S. Pat. 2,590,096 (1952).

[113] Andrews, J. W., U.S. Pat. 2,675,345 (1954).

[114] Ratte, H., and Geller, J., U.S. Pat. 2,795,538 (1957).

[115] Borodkin, V. F., *J. Appl. Chem.* (*U.S.S.R.*), **23,** 763–6 (1950).

[116] De Bruijn, H., U.S. Pat. 2,575,314 (1951).

[117] Directie van de Staatsmijnen in Limburg, Dutch Pat. 74,130 (1954). DeJong, F. P. K., U.S. Pat. 2,764,595 (1956).

[118] Nickolls, K. R., and Williams, V. C., U.S. Pat. 2,783,287 (1957).

[119] N. V. DeBataafsche Petroleum Mastschoppy, Ger. Pat. 939,572 (1956).

[120] Lien, A. P., U.S. Pat. 2,426,624 (1947).

[121] Quaker Oats Co., *Chem. Dept. Tech. Bull.* No. 118A (1957), 3 pp.

[122] Kleiss, L. D., and Marwil, S. J., U.S. Pat. 2,715,647 (1955).

The simultaneous purification of anthracene and production of N-ethyl carbazole from a mixture of the two has been described. The mixture is reacted with diethyl sulfate and caustic potash in aqueous acetone to give the N-ethyl carbazole. The anthracene, unchanged, is filtered off and washed with methanol to yield anthracene of 93% purity.[123] Refined anthracene is separated from crude fractions by formation of its Diels-Alder adduct with maleic anhydride. The separated adduct is dissociated by heat.[124]

The selective hydrogenation of anthracene in a mixture of it with carbazole and phenanthrene permits distillation of phenanthrene and dihydroanthracene away from the carbazole, which can be recovered from the residue by recrystallization from xylene. Dehydrogenation of dihydroanthracene permits recovery of anthracene.[125]

A process has been described for the recovery of anthracene from anthracene oil by treatment of a fraction with air at 300°C for less than 30 min and subsequent treatment of the product with a benzene: alkali mixture to give 60 to 70% of 98% anthracene.[126] The redistillation of crude anthracene before recrystallization from solvent has been suggested as a means of obtaining pure anthracene and pure carbazole. A single recrystallization of redistilled material from a mixture of toluene and light pyridine bases, 7.5 to 8 parts to one part of carbazole, yields 93 to 95% anthracene and 95 to 97% carbazole.[127]

Carbazole can be separated from anthracene and phenanthrene by treating with potassium alcoholate to convert the carbazole to a metal-carbazole derivative.[128] Sodium phenolate has been used to form the metal-carbazole derivative.[129] Precipitation of carbazole sulfate from a benzene solution of carbazole has been described.[130]

A process has been claimed for the separation of α- and β-methyl naphthalenes by fractional azeotropic distillation with ethylene glycol. The method is applicable to mixtures in diphenyl and also to tar-oil fractions.[131] The same separation has been described using 5-ethyl-nonane-2-ol, with which only the β derivative forms an azeotrope.[132]

Separation of acenaphthene from neutral acenaphthene oil by several methods has been reported. Deep refrigeration was most effective. Dibenzofuran was the impurity most difficult to remove because it forms a eutectic with acenaphthene.[133]

Patents cover the use of sulfonation to remove impurities from a chrysene fraction obtained in the fractional distillation of coal tar. Sulfur-free chrysene is recovered.[134]

The separation of 1- and 2-methyl naphthalenes by alkylation with a tertiary olefin has been described. The two alkylated methyl naphthalenes are separated by fractional distillation and are then dealkylated to methyl naphthalenes.[135] A method of separating acenaphthene, diphenylene oxide, and fluorene from coal-tar oil has been described. A coal-tar oil, of boiling range 260 to 303°C, plus added recycle oil is first separated into fractions boiling at 265 to 279°C, 279 to 288°C, and 288 to 299°C, by fractional distillation; the crystals formed on cooling are separated from the oils to give acenaphthene, diphenylene oxide, and fluorene, each substantially free of the others. The mother liquor oil is recycle oil. It was stated that the original oil should be at least three times the weight of the three final products, and this was accomplished by adding the proper amount of recycle oil.[136]

[123] Buc, S. R., Randall, D. I., Smith, C. M., and Jackson, R. D., U.S. Pat. 2,783,244 (1957).

[124] Insginer, T. H., U.S. Pat. 2,440,688 (1948).

[125] Corson, B. B., and Detrick, R. S., U.S. Pat. 2,438,148 (1948).

[126] Rutgerswerke A. -G., Ger. Pat. 914,004 (1954).

[127] Gluzman, L. D., *Koks i Khim.*, **1958,** No. 4 35–42.

[128] Insinger, T. H., U.S. Pat. 2,464,810 (1949).

[129] Swietowslawki, W., U.S. Pat. 2,464,833 (1949).

[130] Imperial Chemical Industries, Ltd. and DuPont de Nemours and Co., Brit. Pat. 548,877 (1944).

[131] Milner, D. W., and Jowett, J., Brit. Pat. 668,853 (1952).

[132] Feldman, J., and Orchin, M., *Ind. Eng. Chem.*, **44,** 2902–14 (1954).

[133] Fumakubo, E., and Matsumoto, Y., *Kooru Taaru,* **1,** 134–7, 161–3 (1949); *Chem. Abs.*, **46,** 7751 (1952). Fumakubo, E., and Yoshida, S., *Kooru Taaru,* **2,** 384–90 (1950); *Chem. Abs.*, **46,** 7751 (1952). Fumakubo, E., and Tokura, J., *Kooru Taaru,* **3,** 4–5 (1951); *Chem. Abs.*, **46,** 7751 (1952).

[134] Kruber, O., and Franck, H., Ger. Pat. 807,682 (1951).

[135] Nickels, J. E., U.S. Pat. 2,598,715 (1952).

[136] U.S. Steel Corp., Brit. Pat. 768,267 (1957).

Refined Tars

Refined tars are dehydrated, partially distilled, and sometimes fluxed, coal tars that are used in the manufacture of various coal-tar paints and coatings, as saturants for felts and fabrics and as preservatives for fish nets and ropes. It is principally their water-resistant properties that make them valuable. Usually, refined tars are made to customers' particular requirements.[137]

Road Tar

Coal-tar road materials have been used in the U.S. for the construction of highways since about 1870. Special methods have been developed for the use of road tars, including the stabilization of soil bases, prime coating of subgrades, hot and cold surface treatments, drag leveling courses, mulch treatment, road mixes, penetration construction, tar-concrete construction, joint- and crack-filling and hot and cold patching. This large variety of maintenance and construction methods requires a wide range of consistencies of tar. Within this range, twelve grades of road tar designated RT-1 to RT-12 and two grades, RTCB-5 and RTCB-6 are recommended by the American Society for Testing Materials and the American Association of State Highway Officials.[137]

The characteristics which make road tars especially suitable for their job are deep penetration into soils, good wetting of aggregate particles, superior adhesion to aggregate surfaces, resistance to water, insolubility in lubricating oils, and petroleum fuels. Service life is limited by weathering and ageing, which is held to be a combination of evaporation of lighter constituents, crystallization, and oxidation.[138] Increased service life is obtained by increasing the amount of high-boiling oils preferably boiling over 350°C.[139] Factors influencing the adhesion of road tars are surface tension, viscosity, and molecular constitution, especially dipole moment. It is shown that a tar, free of acids and bases, is quickly stripped from aggregate by the action of water.[140]

The rapid change of the viscosity of coal tars with temperature makes it difficult to achieve a balance between too great a fluidity in summer and brittleness in winter. The addition of materials like natural rubber and synthetic elastomers in amounts under 5% has been proposed as a means of lowering the temperature susceptibility of road tar.

Roofing and Waterproofing Pitches

Because of its extreme resistance to penetration by water and also deterioration by water action and because of self-healing characteristics resulting from cold flow, coal tar pitch is a superior water-proofing material for flat roofs on buildings and for reservoirs, bridge decks, foundations and the like. The pitch is usually applied to such structures between alternate plies of coal-tar saturated felt or fabric. In the case of roof decks, bridge decks, and other horizontal surfaces, the multiply membrane formed in this manner is covered with slag or gravel. This covering affords protection of the tar against the action of sunlight. Roofing and waterproofing pitches and tar-saturated felts and fabrics usually are made to comply with the requirements of standard specifications promulgated by the American Society for Testing Materials and the Federal Specifications Board.[137]

Fiber Pipe and Conduit Pitch

Coal tar pitch is used for the impregnation under vacuum of fiber pipes or tubes which are used as electrical conduits and for drainage purposes. Coal-tar pitch is superior for such purposes because of its resistance to water, fungi, tree roots, and acid and alkaline soil conditions. A description[141] of the manufacture of these pipes which are 25% cellulosic fiber and 75% pitch has been published.[137]

[137] Rhodes, E. O., *Encylopedia of Chemical Technology*, Eds. Kirk, R. E., and Othmer, D. F., The Interscience Encyclopedia Inc., New York, 1954, Vol. 13, pp. 626–31.

[138] Franck, H. G., *Bitumen, Teere, Asphalte, Peche*, **6,** 185–7 (1955). Lee, A. R., and Dickinson, E. J., *Gt. Brit., DSIR, Road Research Tech. Paper* No. 31 (1954), 48 pp.

[139] Dickinson, E. J., *Road Tar*, **9,** 31–4 (1955). Krenkler, K., *Forsch. ges. Strassenwesen, Forschungsarb. aus dem Strassenwesen, Neue Folge*, **14** (1954), 51 pp.

[140] Franck, H. G., *Bitumen, Terre, Asphalte, Peche*, **4,** 32–4 (1953).

[141] Anon., *Pipes and Pipelines*, **4,** 36–7 (1959), October.

Carbon Electrode Pitch

Coal-tar pitch is the preferred binder for use in the manufacture of amorphous and graphitized carbon electrodes, blocks and other shaped forms. The pitch is hot-mixed with sized petroleum coke, anthracite coal, carbon black and scrap carbon, and the hot mix is molded or extruded into desired shapes. These shapes are heated to 1000°C to form amorphous carbon articles or to 3000°C to give graphitized carbon forms. The pitch yields a coke during this heating which cements the carbonaceous aggregates together in a dense product of high electrical conductivity.[137]

Darney,[142] in an account of practices in Europe and England, has reported on the superiority of pitch from high-temperature tar, especially coke-oven tar, for this purpose. Pitches from low-temperature tar are unsuitable. In general, a high carbonizing temperature in tar production correlates with a high coking value and high carbon/hydrogen ratio in the pitch therefrom. These high values correlate, in turn, with high electrode strength.[143] The behavior of the binder during its transition from a liquid to a solid coke is important. Coke-oven tar pitches give a less-porous coke than does vertical-retort tar pitch. Also, coke-oven tar pitches shrink less on transition from liquid to coke and, thus, leave less free space between the particles of aggregate, giving a denser product. Shrinkage is influenced by the decomposition temperature of the pitch. The lower the temperature at which decomposition occurs, the greater the temperature range of shrinkage on baking. In testing pitches, it is usual to perform a distillation test up to 360°C and then determine the softening-point of the residue to ensure that decomposition does not take place at this relatively low temperature.

The viscosity of the pitch is an important processing variable. Usually, the critical adhesion range of viscosity corresponds with a temperature range of 30 to 40°C above the Kraemer-Sarnow softening-point or 20 to 30° above the ring-and-ball value. The actual softening-point required depends on the mixing and forming facilities but should always be as high as permissible because the harder pitches have a higher coke value.

Darney also discussed the role of the various insolubles fractions in pitch. It is generally agreed among European electrode manufacturers that a content of 30 to 35% free carbon (C_1 + C_2 resins) in the pitch binder is desirable, especially for Soderberg (continuous anode) paste. Continental European manufacturers specify a minimum of 20% beta-resin (C_2 resins). Free carbon has no wetting properties; the compounds which are benzene soluble contribute to this property. To ensure adequate wetting and binding properties, it is usually specified that the coking value percent minus free carbon percent should not be less than 20.

McNeil and Wood[144] reviewed the methods of characterizing pitches for electrode binder usage, especially the use of solvent fractionation. Charette and Bischofberger's[143] characterization factors were discussed, namely, (1) the coking value times the C/H ratio of the whole pitch and (2) the percentage of benzene-insoluble material times the C/H ratio of the benzene insoluble fraction times C/H ratio of the benzene-soluble fraction. These factors gave a straight-line relationship with the compressive strength of test electrodes made from fourteen coal-tar pitches and ten petroleum pitches. McNeil and Wood noted that results are available to show that vertical-retort tar pitches give compressive strengths higher than calculated by Charette and Bischofberger's criteria. They concluded that there is no unanimity in views put forward for the properties of the ideal binder pitch. Both the coking value and the ratio of solvent fractions appear to be important. Chemical condensation, air-blowing, and thermal treatment can alter both values, but the results differ with different pitches and must be evaluated for each individual case.

More recently, Dell[145] has compared the aromaticities of a number of coal-tar pitches with characterization by the usual determinations such as coking value, C/H ratio, and solvent fractionation. He concluded that the quality of carbon anodes is related directly to

[142] Darney, A., *Proc. Conf. on Industrial Carbon and Graphite, London*, **1957**, Soc. Chem. Ind., London, 1958, pp. 152–61.

[143] Charette, L. P., and Bischofberger, G. T., *Ind. Eng. Chem.*, **47**, 1412–15 (1955).

[144] McNeil, D., and Wood, L. J., *Proc. Conf. on Industrial Carbon and Graphite, London*, **1957**, Soc. Chem. Ind., London, 1958, pp. 162–72.

[145] Dell, M. B., *Fuel*, **38**, 183–7 (1959).

the aromaticity of the binder pitch. The usual tests for pitches such as quinoline-insoluble and benzene-insoluble contents and coking value did not correlate with aromaticity. The more useful characterizations for aromaticity include C/H of the carbon bisulfide-soluble fraction, sulfonation index, and infrared index[146] of the whole pitch and the distillate fraction. The coke-oven tar pitches ranked highest in aromaticity of the pitches tested.

Target Pitch

The brittleness of hard coal-tar pitch makes it especially suitable for the production of clay pigeons used as targets in trap-shooting and skeet-shooting. The targets are made by mixing the pitch and suitable filler such as limestone dust, heating the mixture and forming it into saucer-shaped disks. The brittle target disintegrates when struck by one or more pellets from a shot-gun shell.[137]

Enamels and Coatings

Hot-application enamels made from coal-tar pitch and suitable fillers are used for the protection of iron and steel surfaces against corrosion. Internal and external surfaces of steel ships, dry docks, lock gates, pen stocks, water lines, oil lines, gas lines, underground tanks, and piling are structures protected with coal-tar enamels. Their extreme resistance to penetration by water and to destructive effects of soil conditions has been largely responsible for their outstanding success for such purposes.[137]

Fair[147] has published an account of the properties, specifications, tests, and recommendations for use of coal-tar coatings, both hot-applied and cold-applied. Data were given showing that water absorption by coal-tar coatings is less than 20% of that of other bituminous coatings. This is particularly important in view of the recommended practice of using cathodic protection in conjunction with bituminous coatings. McComb[148] has stated that water absorption decreases the electrical resistivity of the coating, and cathodic protection becomes uneconomical.

Core-Binder Pitch

Core compounds containing high-softening-point pitches are used extensively in the preparation of foundry molds and cores. Sand, clay, and core compounds are intimately mixed, moistened with water, and rammed into or around a pattern having the shape and dimensions of the desired casting or cavity. The pattern is then removed from the mold or core, and the latter is baked to remove moisture and to melt and carbonize the core compound sufficiently to cement the sand and clay particles firmly together.[137]

After the molds and cores have been used in the foundry, they are disintegrated and conditioned for re-use by the addition of more core compound. The suitability of coal-tar pitch for this purpose is due to its inherent binding capacity, which imparts strength to the molds and cores prior to and during the casting of the metal.[137]

Pitch Coke

Pitch coke is used in the manufacture of prebaked and Soderberg-type electrodes. It also is used in foundries, in open hearth furnaces, and in the manufacture of silicon carbide and other carbon compounds. It is especially suitable for such purposes because of its low ash and sulfur contents.[137]

Coal-Tar Fuels

Coal tars and tar residuals are used for fuel purposes both in this country and abroad. Foreign practice usually consists in atomizing the molten pitch alone or in combination with creosote into the combustion chamber of the still, boiler, or other type of unit to be heated. The largest users of coal-tar fuels in the U.S. are steel plants. Coal-tar fuels are especially desirable for certain purposes where low ash and sulfur contents and highly luminous flames are desirable.[137]

[146] Montgomery, D. S., and Goodspeed, F. E., Presented at the New York Meeting of the Am. Chem. Soc., September, 1957.

[147] Fair, W. F., Jr., *Corrosion*, **12,** 579T–87T, 605T–10T (1956).

[148] McComb, G. B., *Petrol. Engr.*, **29** (7), D38–44 (1957).

CHAPTER 15

LIGHT OIL AND OTHER PRODUCTS OF COAL CARBONIZATION

R. E. MUDER

Koppers Company, Pittsburgh, Pa.

The title of this chapter suggests the changes that have taken place since the original two volumes of *Chemistry of Coal Utilization* were planned. In 1939, it was thought worthwhile to prepare separate chapters on the recovery of ammonia and other nitrogen compounds, on light oil, on the removal of miscellaneous constituents from gas, and on ammonia liquor. In 1958, one brief dissertation was considered enough to bring these subjects up to date. The decision of the committee preparing this work reflects the economic changes in the byproduct coking industry, at least in the United States. Some protest the use of the adjective "byproduct" and like to speak of "the coke and coal chemicals industry." This is unrealistic. Coke ovens are now seldom built for any purpose other than the production of metallurgical coke. And even in those countries where carbonization in vertical retorts, etc. for the production of city gas is still much practiced, the recovery of ammonia, light oil, etc. must be looked on as a necessary purification of the gas which may yield a profit.

The markets for coal-carbonization byproducts have not declined but have, rather, expanded vigorously. The coking industry continues to move its products into these markets, but they have grown so large that coal carbonization supplies only a small portion of the demand as shown by Fig. 1–4. Also, the presence of supplies from other sources places an effective ceiling on prices.

Figure 1 shows the recent trends in ammonia production where the coke oven material, most of it sold as ammonium sulfate containing

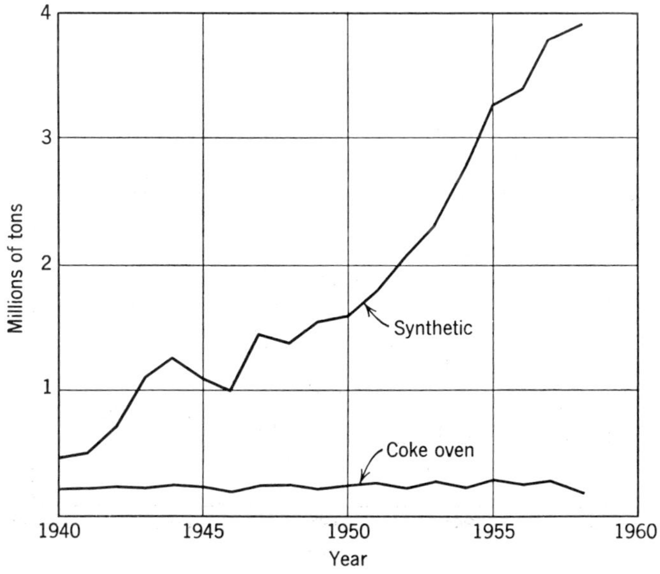

Fig. 1. Production of ammonia in the United States. (From *Minerals Year Books*, U.S. Department of the Interior and reports of U.S. Tariff Commission.)

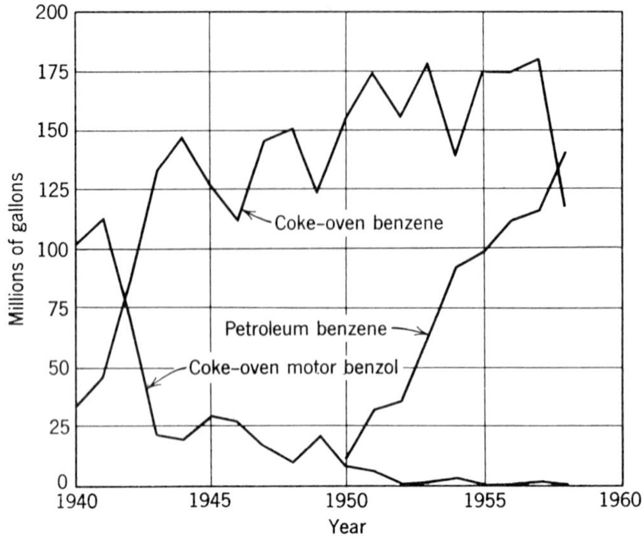

Fig. 2. Production of benzene in the United States. (From *Minerals Year Books*, U.S. Department of the Interior and reports of U.S. Tariff Commission.)

only 25% of ammonia, has fallen far behind the synthetic product, which has the advantages of being used as such and of being easily convertible to any desired salt or to urea. The rate of production of coke-oven ammonia is at present almost completely dependant on the rate of production of coke. Although it is possible to increase the yield of ammonia by various means, they are not such that any coke-plant operator today would consider. Nor is another byproduct coke plant likely to be built for the primary purpose of recovering ammonia as was the first one built in the United States in 1896. Rather, some plants

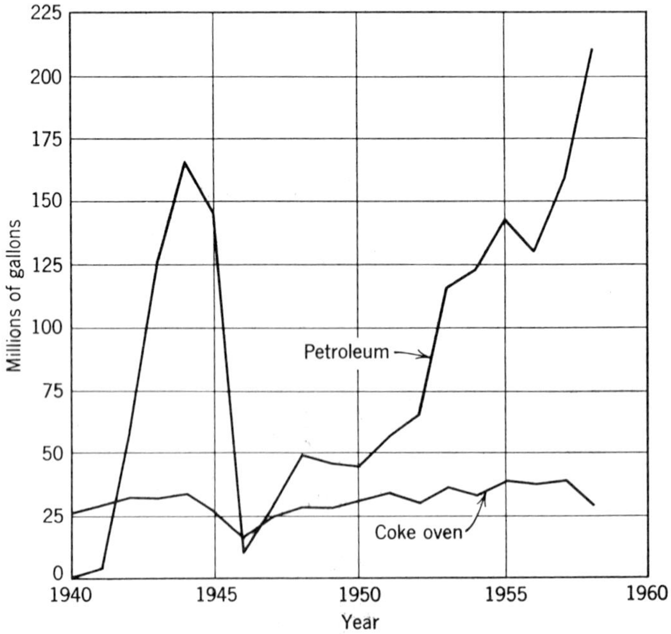

Fig. 3. Production of toluene in the United States. (From *Minerals Year Books*, U.S. Department of the Interior and reports of U.S. Tariff Commisssion.)

are beginning to consider ammonia a noxious waste product, and it is recovered only because it would be even more expensive not to.

The situation in light oil products is shown in Fig. 2, 3, and 4. Benzene, sold mainly for motor fuel prior to World War II, enjoys a steadily increasing demand which is being met mainly by increases in production from petroleum. This industry showed its ability to produce toluene for TNT during the war and has since supplied the growing needs, as it has done also for xylene. These aromatics can by no means be considered waste products at the coke plant, and their recovery should remain profitable as long as they can be sold at prices close to those now prevailing. When left in the gas, they can be valued only as contributors to its heating value, and so they will continue to be recovered if the income derived is sufficient to cover all the expenses incurred. The minimum prices at which these compounds are profitable to recover at any coke plant are not difficult to determine. The petroleum industry, on the other hand, does not think so much in terms of the costs or prices of individual products because the object is the maximizing of the return on processing of their crude stock. Thus the cost of producing benzene and toluene is a function of such variables as the costs of various crudes at the refinery and the relative demand for "regular" and "premium" gasoline.

The greatest problem the petroleum producers present to the coke plants is the very high quality of their products. The aromatic hydrocarbons are produced by processes of catalytic reforming and contain less than 1 ppm of sulfur. The reformates contain only 20 to 50% of aromatic hydrocarbons, and so a process of purification, usually solvent extraction, must be used to produce materials salable such as benzene, etc. It costs very little more to make almost pure products than to make the traditional "nitration grade" ones. Thus, the petroleum products are generally better than those produced by the customary acid-washing and distillation procedures of the coke plants. The latter have at their disposal methods of making comparable products, and these are

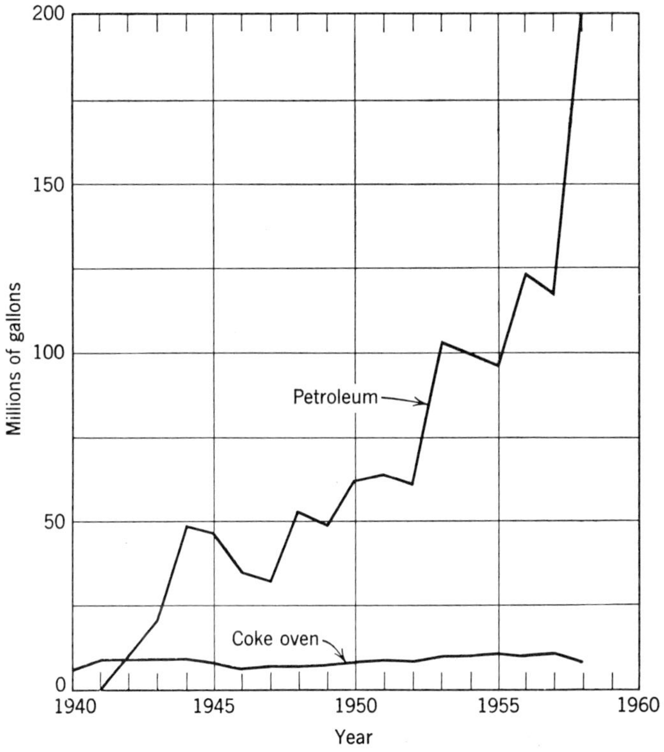

Fig. 4. Production of xylene in the United States. (From *Minerals Year Books*, U.S. Department of the Interior and reports of U.S. Tariff Commission.)

TABLE 1

COMPOSITION OF CONDENSABLE COMPOUNDS FROM COAL GAS[1,2]
(PERCENTAGES BY VOLUME)

	Horizontal Retort	Intermittent Vertical Chamber Oven	Coke Oven	Continuous Vertical Retort	Low-Temperature Retort
Monoolefins	1.61	1.89	1.64	2.33	16.26
Dienes	3.03	4.39	2.48	2.58	1.36
Cycloolefins	1.64	4.69	5.37	1.16	9.55
Paraffins	5.84	7.17	0.34	22.37	46.53
Naphthenes	1.85	1.72	0.21	3.62	8.00
Aromatics	79.62	74.31	85.26	63.04	15.56
Indene	0.79	0.33	1.13	0.72	0.15
Carbon disulfide	0.39	0.38	0.40	0.66	0.06
Thiophenes	0.52	0.62	0.67	0.33	0.66
Other compounds	4.71	4.49	2.50	3.19	1.87

[1] Claxton, G., *Benzoles: Production and Uses*, The National Benzole and Allied Products Assoc., London, 1961, 979 pp.

[2] Roberts, A. L., Towler, J. H., and Holland, B. H., *Gas Council (Gt. Brit.) Research Commun.* **GC 31** (1956), 25 pp.

being used. Such further refining is economically justified, however, only at installations that handle the production from a very large coke plant or from several smaller ones.

LIGHT OIL

Light oil derived from the carbonization of coal is a foul-smelling amber liquid composed mainly of mononuclear aromatic hydrocarbons. Although some use is made of highly unsaturated fractions containing indene, coumarone, and dicyclopentadiene for resin manufacture, light oil is recovered primarily for the sake of the benzene, toluene, and xylene it contains, and refining processes are directed toward bringing these compounds to usable levels of purity. The composition of light oil depends only slightly on the kind of coal carbonized. It depends much more on the temperature to which the primary products of carbonization are subjected. As Table 1 shows, the oil from low-temperature carbonization is highly paraffinic, whereas that derived from coke oven gas is predominantly aromatic.

The identity of the minor constituents of light oil has been established by means of gas chromatography, which allows the complete explication of the composition of complex mixtures. The hydrocarbons found in four light oils resulting from the carbonization of Durham coal, as determined in the laboratory of Benzole Producers, Ltd., are shown in Table 2. French, of the same organization, investigated the distribution of sulfur in crude light oils as shown in Table 3.[3] Carbon disulfide and thiophenes account for most of the sulfur found in crude light oil. The former is removed by fractional distillation; most of the latter is removed by the usual washing with sulfuric acid, as are the sulfides and disulfides. These form azeotropes with paraffins and naphthenes and are found in fractions distilling over at temperatures below their normal boiling points. Mercaptans are not usually found in crude light oils, but they may occur in refined light oils as a result of the action of sulfuric acid in the presence of hydrogen sulfide.

The composition of refined light oil has also been elucidated in recent years. Analyses by gas chromatography of an American rectified light oil before and after refining by hydrogenation are presented in Table 4. A similar comparison of an English crude and the hydrogenated product prepared from it is shown in Table 5. Table 6 consists of a comparison of the nonaromatic hydrocarbons in hydrogen- and acid-refined light oils. Hydrogen-refined light oils contain more cycloparaffins resulting from saturation of cyclopentadiene, benzene,

[3] French, K. H. V., *The National Benzole Assoc., Research Paper* 1–1953, 24 pp.

TABLE 2

HYDROCARBONS FOUND IN CRUDE LIGHT OIL[1,4]

Hydrocarbon	% by Weight: Intermittent Vertical Chambers	Horizontal Retorts	Coke Oven (*a*)	Coke Oven (*b*)
Aromatic				
Benzene	45.5	68.4	72.4	66.1
Toluene	18.9	11.6	13.5	16.0
Ethylbenzene	0.47	0.17	0.18	0.21
m- and *p*-Xylene	6.51	2.04	2.24	4.02
o-Xylene	1.20	0.49	0.68	0.87
Prehnitene	1.48	<0.01	<0.01	0.07
Naphthalene	0.98	0.72	1.22	5.00
Unsaturated				
Styrene	1.23	0.61	1.11	1.43
Indene	1.62	0.69	1.57	1.49
Cyclopentadiene and dicyclopentadiene	0.49	1.79	0.14	0.14
Paraffinic				
Isobutane	<0.01	0.13	N.D.	N.D.
n-Butane	0.05	0.04	N.D.	N.D.
Isopentane	0.09	0.06	N.D.	0.01
n-Pentane	0.20	0.09	<0.01	<0.01
2,2-Dimethylbutane	<0.01	<0.01	<0.01	0.02
2-Methylpentane	0.01	0.08	<0.01	<0.01
3-Methylpentane	0.16	0.02	>0.01	<0.01
n-Hexane	0.52	0.25	0.05	0.02
n-Heptane	0.71	0.07	0.09	0.09
2-Methylheptane	0.18	<0.01	<0.01	<0.01
3-Methylheptane	<0.01	<0.01	<0.01	0.04
2,2,4-Trimethylpentane	<0.01	<0.01	<0.01	0.02
n-Octane	0.69	0.13	0.07	<0.01
3-Methyloctane	<0.01	<0.01	<0.01	0.11
2,2,5-Trimethylhexane	<0.01	0.04	<0.01	<0.01
n-Nonane	0.15	0.05	0.05	0.06
n-Decane	0.11	0.03	<0.01	<0.01
n-Undecane	0.33	0.04	<0.01	<0.01
n-Dodecane	<0.01	<0.01	0.05	<0.01
Naphthenic				
Cyclopentane and/or 2,4-dimethylpentane	0.23	0.10	<0.01	<0.01
Methylcyclopentane	0.39	0.21	0.03	0.16
Cyclohexane	1.12	0.35	0.15	0.11
Methycyclohexane	0.64	0.22	<0.01	0.48
Cycloheptane and/or *trans*-1,2-dimethyl cyclohexane	0.02	0.04	<0.01	0.04
Cis-1,3-dimethylcyclohexane	>0.01	0.04	<0.01	0.14
Ethylcyclohexane	0.33	0.02	0.01	<0.01

N.D. = No determination

[4] Benzole Producers, Ltd., *Half-Yearly Report of the Technical Dept.*, September, 1958, 14 pp.

TABLE 3

SULFUR COMPOUNDS OCCURRING IN CRUDE LIGHT OIL FRACTIONS BELOW 160°C.[3]

Type of Plant	Continuous Vertical	Intermittent Vertical Chamber	Horizontal	Coke Oven	Coke Oven	Coke Oven	Coke Oven
Source of coal	Mainly Yorkshire	Durham and Northumberland	Yorkshire	Glamorganshire	Monmouthshire	South Yorkshire	South Yorkshire
Total sulfur in coal, % by weight	1.7	1.6	1.4	0.7	1.1	1.6	1.4
Carbonization temperature, °C	1080 Top 1330 Bottom	970 Top 1330 Bottom	1350–1360	1000	1230–1240	1210	1325
Specific gravity of crude (15.5°/15.5°C)	0.837	0.860	0.876	0.878	0.884	0.885	0.898
Sulfur occurring as:	% by weight	% by weight	% by weight	% by weight	% by weight	% by weight	% by weight
Carbon disulfide	0.308	0.451	0.621	0.035	0.279	0.372	0.474
Thiophene	0.196	0.159	0.393	0.040	0.109	0.350	0.364
Methylthiophenes	0.158	0.088	0.182	0.019	0.025	0.181	0.077
Ethyl and dimethylthiophenes	0.075	0.032	0.056	0.006	0.004	0.084	0.038
Propylthiophenes	0.026	0.002	0.008	0.000	0.001	0.010	0.003
Sulfides	0.115	0.015	0.009	0.011	0.000	0.017	0.014
Disulfides	0.032	0.035	0.053	0.016	0.010	0.072	0.016
Total	1.910	0.782	1.322	0.127	0.428	1.086	0.986

TABLE 4

ANALYSIS OF CRUDE AND HYDROGENATED LIGHT OILS[5]

Compound	% by Weight	
	Crude	Hydrogenated
i-Pentane	0.0	0.7
n-Pentane	0.0	0.1
2,2-Dimethylbutane	0.0	0.25
Cyclopentene + 2-methylpentane	0.0	0.1
Cyclopentadiene	0.15	0.0
3-Methylpentane + 2,3-dimethylbutane	0.0	0.25
n-Hexene	0.1	0.0
n-Hexane	0.0	0.2
Methylcyclopentane + methylcyclopentene	0.1	0.3
2-Methylhexane + 2-methylhexene	0.05	0.05
3-Methylhexane	0.1	0.1
Cyclohexane	0.3	1.2
n-Heptene + *n*-heptane	0.35	0.35
Benzene	61.35	59.4
Thiophene + methylcyclohexane	1.0	0.9
3- and 4-Methylheptane, -heptene	0.1	0.1
n-Octane + *n*-octene	0.15	0.2
Toluene	23.2	24.0
Ethylcyclohexane	0.3	0.25
n-Nonane + *n*-nonene	0.15	0.15
Ethylbenzene	0.6	2.3
m-Xylene + *p*-xylene	7.25	7.0
o-Xylene	1.8	1.5
Styrene	2.3	0.0
Dicyclopentadiene	0.15	0.0
Unknown	0.50	0.60

[5] Reitz, O., Badische Anilin- & Soda-Fabrik, A. -G., private communication.

toluene, etc. and also more aliphatic substances from hydrogenation of olefins and destruction of phenols and heterocyclic compounds.

The impurities present in a number of British commercial benzenes are listed in Table 7. An American sample was very thoroughly examined by Glick, Miskalis, and Kessler.[6] This sample was an acid-washed material with a boiling range of 0.6°C, a freezing point of 5.07° (equivalent to a purity of 99.30 mole %), a thiophene content of 444 ppm, and a sulfur content of 177 ppm. Progressive freezing concentrated the impurities which were examined by means of gas chromatography and mass spectrometry. The analysis in Table 8 accounts for all but 4% of the nonbenzenoid material originally present.

[6] Glick, C. F., Miskalis, A. J., and Kessler, T., paper presented at meeting of Am. Chem. Soc., Atlantic City, N.J., September 17, 1959.

Lists of compounds found in the light-oil fraction of coal tar have been published.[7]

Formation of Light Oil

Several mechanisms have been proposed for the formation of the mononuclear aromatic hydrocarbons that make up most of coke-oven light oil. The changes in composition and quantities of the liquid byproducts of coal carbonization at different temperatures are well known and are shown schematically in Fig. 5.[8] It is generally agreed that the

[7] Jäger, A., and Kattwinkel, G., *Brennstoff-Chem.*, **35**, 353–62 (1954). Kruber, O., Raeithel, A., and Grigoleit, G., *Erdöl u. Kohle*, **8**, 637–43 (1955). Jurkiewicz, J., and Rosinski, S., *Koks, Smola, Gaz*, **2**, 22–8, 62–5, 98–107, 144–9, 180–8 (1957); **3**, 20–5 (1958).

[8] *Koppers Handbuch der Brennstofftechnik*, H. Koppers, Essen, 1953, Third edition, p. 204.

TABLE 5

COMPOSITION OF DURHAM COKE-OVEN LIGHT OIL AND HYDROFINED LIGHT OIL PREPARED AFTER RECTIFICATION FOR REMOVAL OF HEAVY SOLVENT[9]

Compound	% by weight	
	Crude	Rectified, Hydrogenated
Hydrocarbons, boiling point below 35°C	0.3	
Propane		0.06
Isobutane		0.04
n-Butane		0.26
Isopentane		0.17
n-Pentane		0.26
Cyclopentadiene	0.8	
Carbon disulfide	0.5	
2,3-Dimethylbutane, cyclopentane, 2-methylpentane		1.08
n-Hexane		0.19
Methylcyclopentane + dimethylpentenes	0.2	0.30
Cyclohexane	<0.05	0.2–0.4
C_6-C_7 unsaturates	1.5	
Benzene	65.0	74.0
Thiophene	1.0	0.002
n-Heptane	0.1	0.1–0.2
Methylcyclohexane	0.1	0.3–0.6
C_7-C_8 unsaturates	1.0	
Toluene	14.0	16.0
Methylthiophenes	0.2	0.003
C_8-C_9 paraffins	0.2	
Ethylbenzene and xylenes	3.0	4.0
Indene, coumarone, and higher alkylated benzenes	3.0	
Naphthalene	3.0	
Residue	6.1	<3
Total	100.0	approx. 100

[9] Spencer, D., *Ind. Chemist*, **34,** 287–93 (1958).

maximum yield of aromatic hydrocarbons is achieved when the primary distillation products are heated to a temperature of 700 to 900°C.[10] As the temperature is raised within this range or above it, the ratio of benzene to toluene increases.[11]

Fuchs and Sandoff[12] expanded and modified the conclusions of Morgan and Soule[13] which were in accord with those of most earlier workers:

The initial decomposition of coal giving semicoke and low-temperature tar is brought about by (*a*) splitting off of radicals containing heterocyclic and homocyclic rings, (*b*) opening of some heterocycles to intermediate aldehydes which give rise to phenols, amines, and naphthenic compounds, (*c*) progressive dehydrogenation and splitting off of side chains by the action of the developing hydrogen.

Final decompositions are at a maximum between 600° and 800°C and are marked by (*a*) loss of hydrogen and other simple gases from semicoke and hydroaromatic volatile compounds, (*b*) hydrogenation of phenols to aromatic hydrocarbons with the formation of methane, ether, and water, and (*c*) formation of higher aromatics from semicoke proper and by secondary pyrogenetic syntheses.

[10] Korobchanskii, I. E., *Trudy Donets. Ind. Inst. Yubileinyi Sbornik*, **1941,** 52–9; *Khim. Referat. Zhur.*, **4,** No. 9, 117–18 (1941). Warmuzinski, J., *Przegląd Górniczy*, **7,** 413–18 (1951). Trefny, F., *Brennstoff-Chem.*, **33,** 257–60 (1952). van Ahlen, A., *Glückauf*, **78,** 633–9 (1942).

[11] Mantel, W., and Hansen, H., *Brennstoff-Chem.*, **35,** 97–104 (1954).

[12] Fuchs, W., and Sandhoff, A. G., *Ind. Eng. Chem.*, **34,** 567–71 (1942).

[13] Morgan, J. J., and Soule, R. P., *Chem. & Met. Eng.*, **26,** 1025–30 (1922).

TABLE 6

NONAROMATIC HYDROCARBONS IN HYDROGEN-REFINED AND ACID-WASHED LIGHT OILS[14]

Hydrocarbon	Volume % of Original Light Oil	
	Hydrogen-Refined	Acid-Washed
Pentane + traces of C_3, C_4	0.18	0.02
Cyclopentane	0.41	—
2-Methylpentane, 3-Methylpentane	0.06	—
n-Hexane	0.11	0.02
Methylcyclopentane	0.08	0.01
Cyclohexane	0.11	0.01
Dimethylpentane, Methylhexane, Dimethylcyclopentane	0.09	0.02
n-Heptane	0.16	0.04
Methylcyclohexane	0.14	0.03
Methylheptane, Dimethylhexane, *n*-Octane, Dimethylcyclohexane	0.27	0.09
Distilling between 128 and 150°C	0.33	0.09
Distilling above 150°C	0.76	0.12
Total	2.70	0.45

[14] Grothe, W., *Erdöl u. Kohle*, **6,** 450–4 (1953).

On the basis of thermodynamic analyses and experimental studies, Gruzdeva[15] proposed these mechanisms: (1) Phenolic products of low-temperature tar are transformed into polycyclic aromatic hydrocarbons, such as naphthalene and anthracene, rather than into monocyclic ones. (2) Reduction of xylenols and cresols transforms them directly into benzene rather than through intermediate products such as xylene and toluene. (3) Toluene is transformed into naphthalene by a condensation reaction rather than into benzene by demethylation.

Nikol'skii[16] questions the participation of

[15] Gruzdeva, N. A., *Zhur. Priklad. Khim.*, **25,** 980–93, 1045–56 (1952).

[16] Nikol'skii, N. A., *Trudy Donets. Ind. Inst. Yubileinyi Sbornik*, **1941,** 236–50; *Khim. Referat. Zhur.*, **4,** No. 9, 117 (1941).

TABLE 7

IMPURITIES IN BRITISH COMMERCIAL BENZENES[1]

	Freezing Point, °C						
	5.10	5.18	4.99	5.18	5.38	5.33	5.50
	% by Weight						
2,4-Dimethylpentane	0.02	0.00	0.01	0.00	0.00	0.00	0.00
Methylcyclopentane	0.11	0.09	0.11	0.14	0.09	0.09	0.09
3-Methylhexane	0.05	0.00	0.04	0.05	0.00	0.00	0.00
Cyclohexane	0.09	0.02	0.14	0.12	0.05	0.04	0.00
n-Heptane	0.15	0.11	0.19	0.17	0.07	0.07	0.00
Methylcyclohexane	0.09	0.05	0.36	0.10	0.00	0.00	0.00
Toluene	0.00	0.00	0.00	0.00	0.00	0.00	0.00
S as carbon disulfide	<0.02	<0.02	<0.12	<0.010	0.020	0.015	0.003
S as thiophene	0.12	0.13	0.11	<0.005	0.006	0.006	0.001

TABLE 8

IMPURITIES IN COKE-OVEN BENZENE[6]

Normal Boiling Point of Pure Component, °C	Component	Mole %
79.205	2,2-Dimethylpentane	0.00012
80.51	2,4-Dimethylpentane	0.00060
90.05	2-Methylhexane	0.0230
86.071	3,3-Dimethylpentane	0.00085
71.812	Methylcyclopentane	0.00656
89.79	2,3-Dimethylpentane	0.0145
91.95	3-Methylhexane	0.024
98.428	*n*-Heptane	0.0913
87.84	1,1-Dimethylcyclopentane	0.0069
93.468	3-Ethylpentane	0.0047
80.738	Cyclohexane	0.063
91.87	1-*trans*-2-dimethylcyclopentane	0.050
90.77	1-*trans*-3-dimethylcyclopentane	0.047
	A trimethylcyclopentane (?)	0.0082
	An octane (?)	0.0010
100.934	Methylcyclohexane	0.222
99.53	1-*cis*-2-dimethylcyclopentane	0.019
109.28	1-*trans*-2-*cis*-4-trimethylcyclopentane	0.0071
84.1	Thiophene	0.0466
110.623	Toluene	0.0334
Total		0.670

phenols in the formation of low-boiling aromatic hydrocarbons.

Recovery of Light Oil from Gas

Condensation

Of the three physical processes used to separate light oil from coke oven-gas—simple condensation, adsorption, and absorption—the first remains simplest in theory. However, the relatively high costs of the necessary compression and refrigeration have limited its application to those plants where coke-oven gas is compressed for other purposes such as the recovery of relatively pure hydrogen or distribution over long distances. The application of pressure and the removal of heat are

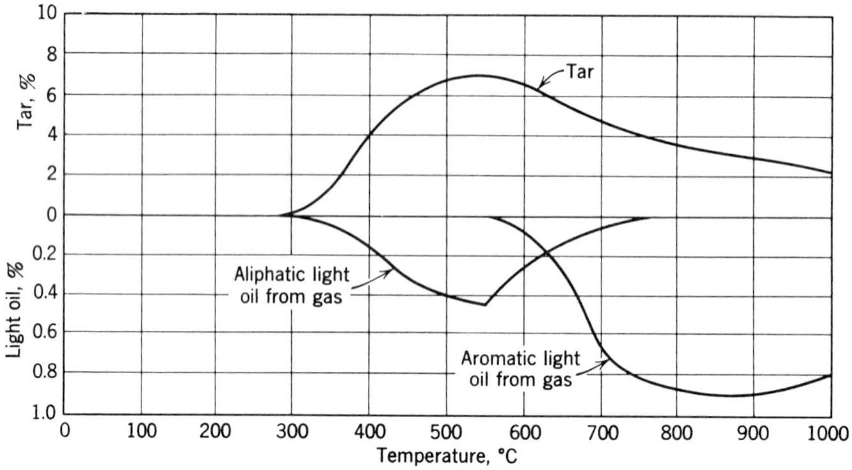

Fig. 5. Distribution of the liquid products of coal carbonization.[8]

accomplished in diverse ways.[17] One new development in Germany is the cooling of the compressed gas by scrubbing with calcium chloride brine and crude heavy solvent. The increasing use of oxygen in the steel industry suggests the possibility that somewhere in the air-separation plants lie sources of cheap refrigeration, which could be used for this purpose, just as there are large quantities of cheap heat in flushing-liquor available for those applications that can use energy at such a temperature level.

Adsorption

European interest in recovery of light oil by adsorption on porous solids has continued although it has not challenged the lead of liquid absorbents. Attention has been paid to the mechanism of adsorption of benzene on active carbon,[18] and descriptions of the application of this mechanism are available.[19]

Absorption

Most light oil is still recovered by countercurrent washing of the gas with a liquid absorbent. As it is not the purpose of this text to discuss developments in equipment, no detailed descriptions of the newer types of light-oil scrubbers will be given. However, some notice must be given to the growing use of spray scrubbers, one example of which is pictured in Fig. 6. The arrangements in scrubbers built by various vendors vary somewhat, but all spray scrubbers consist of series of large empty chambers in which the gas is in contact with sprays of oil. Such scrubbers have the great advantage that no packing is present to serve as a trap for sludge in the wash oil, and so they exhibit a very low-pressure drop throughout their service life. A disadvantage is that the spray chamber provides fewer transfer units than do equal heights of the better tower packings. To realize the pressure-drop advantage of the spray tower requires, for equal light-oil recovery, either a somewhat taller tower or a greater circulation of oil.

According to the two-film theory of mass transfer, the overall mass transfer coefficient K_G may be defined by the equation

$$1/K_G = 1/k_G + 1/Kk_L$$

[17] Société des usines Quini et Cie., and Herenc, A., Fr. Pat. 847,846 (1939). Kravchenko, V. M., *Bull. acad. sci. U.R.S.S., Classe sci. tech.*, **1940,** No. 3, 17–26, 69–84. Guinot, H. M., Brit. Pat. 711,067 (1954). Schmalenbach, A., Pummer, H., and Brandes, H., U.S. Pat. 2,794,515 (1957). H. Koppers G.m.b.H., Brit. Pats. 765,469 (1957); 790,987 (1958). L'Air Liquide, Brit. Pat. 683,008 (1953). Engel, E., and Harpener Bergbau A. -G., Brit. Pat. 727,028 (1955). Grunberg, J., U.S. Pat. 2,743,590 (1956). Haber, F., *Brennstoff-Chem.*, **38,** 366–71 (1957).

[18] Alekseevskii, E. V., and Gorchakov, N. D., *J. Appl. Chem. (U.S.S.R.)*, **17,** 289–98, 394–408, 487–94 (1944). Amphoux, B., *Ann. chim.*, **19,** 1–57 (1944). Alekseevskii, E. V., and Vanyushina, Z. S., *J. Appl. Chem. (U.S.S.R.)*, **18,** 193–207, 658–65 (1945). Gorchakov, N. D., and Pogodin, I. I., *J. Appl. Chem. (U.S.S.R.)*, **18,** 538–47 (1945). Vreedenberg, H. A., and van Nouhuys, H. L., *Rec. trav. chim.*, **65,** 235–45 (1946). Sastri, M. V. C., and Krishnaswami, K. R., *Quart. J. Indian Inst. Sci.*, **5,** No. 2, 42 pp. (1942). Poggi, A. R., *Ann. chim. applicata*, **33,** 191–3 (1943). Dubinin, M. M., and Timofeev, D. P., *Compt. rend. acad. sci. (U.R.S.S.)*, **54,** 701–4 (1946); *J. Phys. Chem. (U.S.S.R.)*, **21,** 1213–22 (1947); **22,** 133–43 (1948). Hougen, O. A., and Marshall, W. R., *Chem. Eng. Progr.*, **43,** No. 4, 197–208 (1947). Dubinin, M. M., Zaverina, E. D., and Raduskevich, L. V., *J. Phys. Chem. (U.S.S.R.)*, **21,** 1351–62 (1947). Josefowitz, S., and Othmer, D. F., *Ind. Eng. Chem.*, **40,** 739–43 (1948). Mennesier, A., and Boucher, R., *Compt. rend.*, **226,** 914–16 (1948). Davies, R. G., *Chem. & Ind. (London)*, **1952,** 160–5. Dubinin, M. M., and Timofeev, D. P., *Doklady Akad. Nauk S.S.S.R.*, **76,** 555–8 (1951). Kadaner, D. G., Luk'yanovitch, V. M., and Radushkevich, L. V., *Doklady Akad. Nauk S.S.S.R.*, **87,** 1001–4 (1952). Avgul, N. N., Dzhigit, O. M., and Kiselev, A. V., *Doklady Akad. Nauk S.S.S.R.*, **86,** 95–8 (1952); **89,** 97–9 (1953). Puri, B. R., Lakhanpal. M. L., and Varma, B., *J. Indian Chem. Soc.*, **29,** 841–5 (1952). Everett, D. H., *Ind. chim. belge*, **20,** 257–66 (1955). Timofeev, D. P., *Zhur. Fiz. Khim.*, **27,** 1642–9 (1953). Polyakova, M. M., and Tesner, P. A., *Doklady Akad. Nauk S.S.S.R.*, **93,** 855–8 (1953).

[19] Dowding, G. F., and Ward, S. G., *Eng. J. (Montreal)*, **29,** 719–21 (1946). Deringer, H., *Trans. Fuel Economy Conf., World Power Conf.*, Sect. A4, Paper No. 4 (1947), 7 pp.; *Am. Gas J.*, **168,** No. 1, 22–3 (1948). Rosendahl, F., *Österr. Chemiker-Ztg.*, **53,** 237–41, 258–62 (1952); *Chem. Ztg.*, **65,** 245–8 (1941). Activkohle-Union Verwaltungs G.m.b.H., Belg. Pat. 446,771 (1942). Schunck, R., *Magyar Mérnök-Épitészegylet Közlönye*, **76,** 161–6 (1942). Robson, G. W., *Gas World*, **123,** 399–400 (1945); *Gas. J.*, **246,** 687, 689 (1945). Chemical Engineering and Wilton's Patent Furnace Co., Ltd. and Wilton, N., Brit. Pat. 570,270 (1945). Fielder, H. J., *Gas Times*, **62,** 185–6, 191, 194, 197 (1950). Meslans, M., *Compt. rend. congr. ind. gaz, Paris*, **67,** 396–408 (1950).

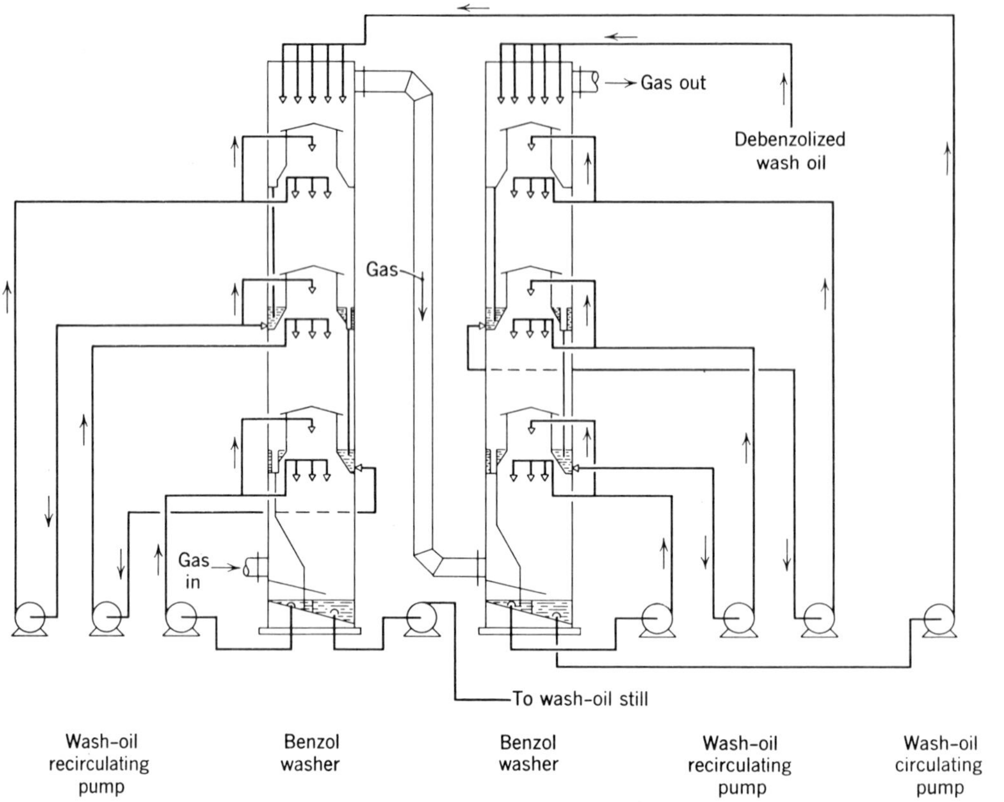

Fig. 6. Flow diagram of Koppers spray light-oil scrubber.

where k_G is the gas film coefficient, k_L is the liquid film coefficient, and K is the partition coefficient or the ratio of the concentration of a substance in the liquid to its concentration in the gas at equilibrium, both expressed in the same units.

Silver and Hopton[20] have determined these partition factors for a number of light-oil constituents in wash oil (Table 9). The wash oil described in Table 9 is typical of those used for washing coke-oven gas, and, for purposes of scrubber design, many engineers would assign it an apparent molecular weight of 260. Silver and Hopton calculated the apparent molecular weight M by means of the expression

$$M = \frac{62{,}300d(273 + \theta)}{PK}$$

where d is the specific gravity of the solution, θ is the temperature in °C, and P is the vapor pressure of the pure liquid solute (Table 10).

[20] Silver, L., and Hopton, G. U., *J. Soc. Chem. Ind.*, **61**, 37–45 (1942).

Thus, the assumed molecular weight of 260 when used in calculations of the absorption of benzene, toluene, and xylene at ordinary absorption temperatures is not an unrealistic figure.

In countercurrent absorption, the minimum flow ratio required for complete absorption is given by

$$qK = 1$$

where q is the ratio of the volume of oil to the volume of gas measured in the same units. In any real scrubbing process, equilibrium is not reached, and an excess of oil such that $qK = 1.3$ may be provided. The calculated flow ratios for absorption of light-oil constituents at various temperatures were calculated as in Table 11.

Similar calculations for the stripping still utilizing 30% more than the minimum amount of steam are shown in Table 12.

A typical light-oil scrubber is operated at 25 to 30°C with an oil circulation rate of 15 gal/1000 cu ft. The wash-oil still operates usually between 120 and 130°C and with a

TABLE 9

PARTITION FACTORS FOR LIGHT-OIL CONSTITUENTS IN WASH-OIL[20]

Solute	Boiling Point	Partition Factor, K 25°C	80°C	130°C
Carbon disulfide	46°C	235	55.9	22.3
Benzene	80°C	650	114	36.4
Toluene	111°C	2030	280	78.8
Xylene	138°C	7570	716	170
Naphthalene	218°C	171,000	8620	1320

The petroleum wash oil used had the following properties:

Specific gravity at 20°C	0.871
Viscosity at 20°C	12.4 centipoises
Viscosity at 100°C	1.6 centipoises
Boiling point:	
First drop	270°C
50%	331°C
75%	380°C

TABLE 10

APPARENT MOLECULAR WEIGHT OF WASH OIL DETERMINED FROM THE DENSITY OF SOLUTIONS OF LIGHT-OIL CONSTITUENTS AT THREE TEMPERATURES[20]

Solute	Apparent Molecular Weight of Wash Oil 25°C	80°C	130°C
Carbon disulfide	187	161	143
Benzene	260	215	199
Toluene	271.5	228	193
Xylene	247.5	222	200
Naphthalene	321	276	251

TABLE 11

CALCULATED FLOW RATIOS FOR ABSORPTION OF LIGHT-OIL CONSTITUENTS AT FOUR TEMPERATURES[20]

Substance to be Absorbed	Flow Ratio, gal/1000 cu ft ($qK = 1.3$) 15°C	20°C	25°C	30°C
Carbon disulfide	24.1	28.7	34.5	40.5
Benzene	8.2	10.2	12.5	15.1
Toluene	2.42	3.11	4.01	4.99
Xylene	0.623	0.832	1.08	1.40
Naphthalene	0.0247	0.0353	0.0475	0.0667

TABLE 12

STEAM RATIOS IN STRIPPING STILL FOR RECOVERY OF LIGHT-OIL CONSTITUENTS[20]

	Steam Ratio in lb/gal of Oil				
Pressure	10 in. Hg Absolute		1 lb/sq in. Gauge		
Temperature	80°C	100°C	110°C	120°C	130°C
Carbon disulfide	0.157	0.100	0.264	0.222	0.181
Benzene	0.321	0.187	0.464	0.375	0.295
Toluene	0.791	0.434	1.07	0.843	0.639
Xylene	2.01	1.02	2.40	1.86	1.36
Naphthalene	24.3	10.4	22.6	16.5	10.7

steam rate of 0.65 lb/gal. These figures, along with those of Silver and Hopton, explain why the proportion of toluene recovered is generally higher than that of compounds boiling at both higher and lower temperatures.

Further discussions of the principles of scrubbing light oil from gas, which can be found in the listed references,[21] exhibit quite various opinions on what constitutes a good wash oil.

Sludge Formation in Wash Oil

The absorbent oil used for light-oil recovery is generally used over and over again, with fresh oil added periodically to make up for losses by leaks and distillation. The oil is repeatedly brought into contact with gas containing a very large variety of compounds—heated, steamed, and cooled—and, in many plants, it comes into contact with air in storage tanks. It is not surprising, therefore, that there often is a noticeable deterioration of the oil. In creosote oils, the most apparent change is an increase in viscosity. Petroleum wash oils, on the other hand, develop an insoluble sludge or "muck," which is very troublesome, for it is deposited in the wash-oil heaters and in the packing of the scrubbers. Although the wash oil itself may take part in some of the reactions which result in these phenomena, it mainly acts as a menstruum in which constituents derived from the gas react. The differences in behavior of the two types of wash oil result from the fact that the aromatic creosote is a good solvent for the reaction products

[21] Meier-Grolman, F. W., Oppelt, W., and Schmitt, H., *Oel u. Kohle*, **36,** 457–69 (1940). Koeppel, C., *Gas-u. Wasserfach*, **83,** 73–9 (1940). Mantel, W., and Hansen, H., *Brennstoff-Chem.*, **36,** 328 (1955); **37,** 19 (1956). Gvozdev, V. D., Zykov, D. D., and Karavaev, N. M., *Izvest. Akad. Nauk S.S.S.R., Otdel Tekh. Nauk*, November, 1956, 64–76. Scheer, W., *Oel u. Kohle*, **38,** 1262–3 (1942). Weiler, W., *Gas-u. Wasserfach*, **91,** No. 19, 253–4 (1950). Fitz, W., *Feuerungstech.*, **30,** 73–84, 113–17 (1942). Pluckthun, J., *Oel u. Kohle*, **39,** 91–5 (1943). Alfs, M., *Oel u. Kohle*, **39,** 20–9 (1943). Greenfield, G. J., *Gas World*, **117,** No. 3026, Coking Sect., 59–69, 72 (1942). Wunderlich, G., *Brennstoff-Chem.*, **34,** 367–73 (1953). Kalinowski, B., Grodon, A., and Gregor, A., *Koks, Smola, Gas*, **2,** 153–6 (1957). Chlosta, J., *Oel u. Kohle*, **37,** 961–2 (1941). Demann, W., *Reichsamt Wirtschaftsausbau, Pruf-Nr.* 37 (PB52005), 59–66 (1940). Montgomery, C. R., *Iron Steel Engr.*, **27,** No. 12, 74–8 (1950). Cowley, L. R., *Gas J.*, **230,** 627–8 (1940). Revyakin, A. A., and Zakharov, M. L., *Coke and Chem.* (*U.S.S.R.*), **1939,** No. 4–5, 55–8. Kopelevich, G. V., *Coke and Chem.* (*U.S.S.R.*), **1940,** No. 1, 33–4. Levich, I. A., and Sud'ya, P. A., *Coke and Chem.* (*U.S.S.R.*), **10,** No. 1, 32 (1940). Benes, V., *Chem. obzor.*, **24,** 25–9 (1949). Krasnodebski, K., and Mlynarski, A., *Przegląd. Górniczy*, **5,** No. 1, 21–7 (1949). Eckert, F., *Gas-u. Wasserfach*, **92,** No. 19, 279–80 (1951). Figg, L. J., Jr., and Shaulis, E. E., Brit. Pat. 548,907 (1942); U.S. Pat. 2,294,972 (1943). Landa, S., Brit. Pat. 792,919 (1958); *Paliva*, **35,** 315–18 (1956). Avdeeva, A. V., and Pitelina, N. P., *Khim. Prom.*, **1947,** No. 2, 19–21. Scheer, W., *Oel u. Kohle*, **38,** 1262–3 (1942). Petingell, N. W., *Gas Rev.*, **1,** 526, 528–32, 535–6 (1947); *Natural Gas Bull., Australia*, **11,** No. 4, 18–20 (1947). Illingworth, H. S., *Gas World*, **131,** No. 3424, Coking Sect., 37–41 (1950); Rosendahl, F., *Gas-u. Wasserfach*, **93,** 369–73 (1952). Otto, H., *AIME, Proc. Blast Furnace, Coke Oven & Raw Materials Comm.*, **8,** 50–60 (1949). Ganguly, A. K., Basak, N. G., and Lahiri, A., *Chem. Age India*, **7,** No. 2, 94–6 (1956). Kotlik, B. E., *Koks i Khim.*, **1956,** No. 5, 36–40. Kafarov, V. V., Dytnerskii, Y. I., and Kulik, I. I., *Koks i Khim.*, **1956,** No. 3, 47–9. Shevlyagin, V. N., *Koks i Khim.*, **1958,** No. 9, 32–36.

whereas the aliphatic petroleum oil is a poor one.

The reactions that cause thickening and sludging are many and complicated. However, an extensive investigation of the Benzole Technical Committee set up by the British Ministry of Fuel and Power has elucidated them considerably[22] by summarizing the work of earlier investigators, by studying plant conditions and their effect on sludging in a large number of coke-oven and gas works, and by laboratory investigations.

The compositions of sludges vary greatly.[23] Ultimate analyses show:

Carbon	36 to 75%
Hydrogen	4 to 7%
Oxygen	5 to 37%
Nitrogen	2 to 5%
Sulfur	1 to 14%
Iron	0 to 8%

Tar fog in the gas entering the light-oil scrubbers is a factor in sludging in plants that do not have effective electrostatic tar precipitators, but in most plants this is a minor cause of trouble.

The role of sulfur is much more important. Sludging troubles in Great Britain were more prevalent at coke plants where no purification from H_2S was practiced than at plants making city gas, which are required to remove H_2S. This has also been found to be true in the United States where the only two coke plants in which light-oil scrubbing is practiced after dry box purifiers are singularly free from sludge troubles. In the presence of iron and oxygen, H_2S is converted to elementary sulfur, which reacts with unsaturated compounds such as indene and dicyclopentadiene to form compounds containing a thiophene ring:

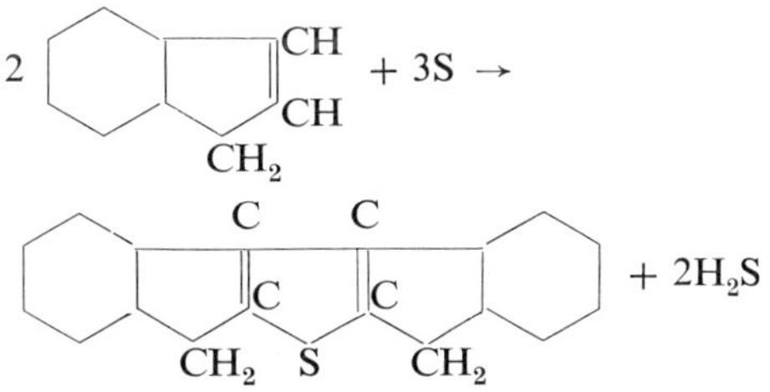

[22] Ministry of Fuel and Power, *Report on Investigations of the Benzole Technical Committee* 1942–1946 *on Sludging and Corrosion in Benzole-Absorption Plants*, H.M.S.O., London, 1947, 110 pp.

[23] Dick, F. A., Claxton, G., and Hoffert, W. H., *J. Soc. Chem. Ind.* (*London*), **66,** 238–48 (1947).

Compounds of this type are soluble, but when heated they can react with more of the unsaturated compound and polymerize, or they can polymerize directly to form substances of the type $(X_3S)_n$ or $(X_2S)_n$ (X = unsaturated compound). Such compounds are insoluble in the oil. Kattwinkel[24] agreed that H_2S is chiefly responsible for the thickening of tar oils and proposed a mechanism involving interaction between organic mercaptans, disulfides, heterocyclic sulfur compounds, and unsaturated hydrocarbons.

Sludge is also formed by oxidation of the constituents of light oil. When indene is present, sludge is formed in much greater quantity since this compound is easily attacked by oxygen at the methylene group. A hydroperoxide is formed and undergoes polymerization or intermolecular condensation.[25] The presence of oxides of nitrogen, particularly NO, promotes the formation of sludge by what is probably also an oxidation reaction. When present in low concentrations (about 0.2%), high-boiling tar acids have an inhibiting effect on oxidation-sludge formation; at higher concentrations they have a powerful promoting effect. The lower-boiling acids, such as phenol and the cresols, are stronger inhibitors. Phenols also contribute to the thickening of absorbing oils by condensing with HCN to form hydroxyaldehydes which further react to form resinous materials.[26] Primary amines behave as inhibitors, but pyridine and quinoline are powerful catalysts in the oxidation and condensation reactions that lead to sludge formation.

Sludge is removed from petroleum wash-oil systems by periodic pumping from the bottoms of decanters, scrubbers, and tanks where it accumulates. It has a tendency to collect at the interface between oil and water and to induce emulsification. The emulsions may be broken by heating them until the water boils.

Wash oils may be regenerated by distillation. Pipe-stills either at ordinary or subatmospheric pressures have been used for this purpose. A common method of continuous regeneration

[24] Kattwinkel, R., *Brennstoff-Chem.*, **22,** 37–41 (1941).

[25] Barash, M., *Coke and Gas*, **10,** 176–9 (1948).

[26] Wilson, O. B., *Coke Smokeless-Fuel Age*, **7,** 34–8, 53–6, 58, 73–7, 93–9, 231–5 (1945); **8,** 12–14, 18, 38–41, 64–6 (1946).

is to treat a side stream of the oil from one of the trays of the wash-oil still in a separate vessel with the steam used for debenzolization. The resulting mixture of vapors enter the base of wash-oil still and are then used to strip light oil from the main flow of wash oil. The residue in the wash-oil purifier is allowed to accumulate and is removed periodically.

Sludging in wash oil and corrosion resulting from its use have been the subjects of many papers.[27] Corrosion is generally ascribed to NH_4CNS formed from ammonia and carbon disulfide.

$$CS_2 + 4NH_3 \rightarrow NH_4CNS + (NH_4)_2S$$

Refining of Light Oil

Acid Washing

Although other methods of refining, particularly hydrogenation, are becoming more popular, the greater portion of light oil is still refined by means of washing with sulfuric acid, usually in a batch agitator. The process combines simplicity with effectiveness. In one vessel it accomplishes the following, as listed by Glowacki:[28] (1) Sulfur compounds are dissolved or oxidized. (2) Nitrogen bases and some unsaturates combine with the acid. (3) Unsaturates are polymerized. (4) Aromatic compounds are sulfonated. (5) Some aromatics condense with unsaturates. (6) Some unstable hydrocarbons are oxidized. In addition, any phenols that resist the action of sulfuric acid are removed by a neutralizing wash with sodium hydroxide. Improvements of the process are directed toward accomplishing the removal of unsaturates—pyridine, thiophene, and phenol—with a minimum loss of aromatic hydrocarbons, a minimum use of acid, and a minimum evolution of corrosive SO_2 upon subsequent distillation.

As customarily practiced, a light-oil fraction containing compounds boiling up to approximately 150°C is subjected to long periods of contact with the acid in a cylindrical agitator with a conical bottom. Long periods of agitation are necessary to produce products meeting stringent acid-wash test specifications, although most of the desirable action is completed in a few minutes. The removal of most of the sulfur present as thiophene is accomplished very quickly. Longer agitation decreases the yield of products and increases the formation of substances that result in evolution of corrosive SO_2 upon distillation[29] (Table 13).

TABLE 13

EFFECT OF TIME OF AGITATION ON WASHING OF CRUDE LIGHT OIL[29]

Time of Agitation min	Distillate to 160°C		
	Yield %	Total S %	Dissolved SO_2 lb/10,000 gal
0	90.8	0.55	7
5	86.4	0.25	32
10	85.0	0.21	40
20	85.1	0.20	55
30	84.2	0.20	57
60	83.1	0.19	56

To reduce the contact time, reduce corrosion, and increase the yield, Lawton[30] devised a process in which the oil is washed for a short period of time with strong H_2SO_4; this is then diluted with water in the agitator. The decanted oil is steam-distilled to remove the resulting resins. This modification of the Ufer process has been made continuous by the

[27] Clark, A. A., and Sowden, W., *Gas World*, **124,** No. 3213, Coking Sect., 23–30 (1946). Newall, H. E., Ackroyd, G. C., and Cowley, C. M., *Fuel*, **27,** 73–81 (1948). Middleton, A. C., *Coke and Gas*, **11,** 309–12, 325 (1949). Lorenzen, G., *Brennstoff-Chem.*, **30,** 430–1 (1949). Amemiya, F., *Kooru Taru*, **2,** 328–36 (1950). Ambler, C. M., U.S. Pat. 2,380,139 (1945). Larke, R. H., *Coke Oven Managers' Assoc. Yearbook*, **1941,** 110–21. Zhadanoz, V. V., and Zil'bermann, D. E., *Coke and Chem.* (*U.S.S.R.*), **9,** No. 3, 42–4 (1939). Cawley, C. M., *Indian Eng.*, **121,** 79 (1947). Segur, D. K., U.S. Pat. 2,418,988 (1947). Wells, J. H., and Wilson, P. J., Jr., U.S. Pat. 2,389,570 (1945). Cawley, C. M., and Newall, H. E., *J. Soc. Chem. Ind.* (*London*), **64,** 285–90 (1945). Claffey, J. B., U.S. Pat. 2,344,969 (1944). Koeppel, C., *Glückauf*, **78,** 533–6 (1942); *Gas World*, **117,** 538–9 (1942). Thau, A., *Teer u. Bitumen*, **40,** 47–52 (1942). Jifuku, S., Hosoi, F., and Ito, H., *Kooru Taru*, **8,** 452–5 (1956). Carabasse, J., *Chim. & ind.* (*Paris*), **58,** 228–33 (1947).

[28] Glowacki, W. L., *Chemistry of Coal Utilization*, John Wiley & Sons, New York, 1945, vol. 2, p. 1206.

[29] Dummett, G. A., *Gas World*, **134,** No. 3490, Coking Sect., 7–16 (1951).

[30] Lawton, J. J., U.S. Pat. 2,720,486 (1955).

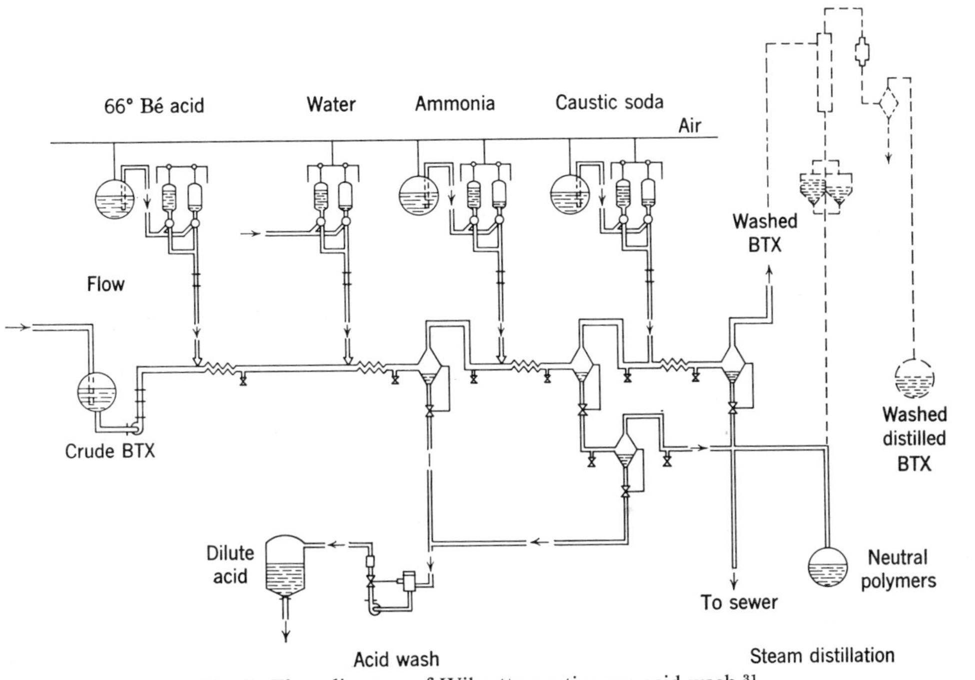

Fig. 7. Flow diagram of Wilputte continuous-acid wash.[31]

Wilputte Coke Oven Division[31] (Fig. 7). Crude light oil and 66° Bé sulfuric acid are pumped through pipe coils designed to obtain turbulent flow and intimate mixing. Water is then injected. The mixture goes through another mixing coil where the acid is diluted; most of the resinous material formed in the first step dissolves in the light oil. The two phases are separated in a decanter. The oil phase is partially neutralized in a third mixing coil with ammonia liquor. The mixture flows to a decanter where three phases—oil, ammonium sulfate solution, and polymeric material—are separated. Completion of neutralization of the oil with sodium hydroxide is accomplished in a fourth mixing coil. The oil is steam distilled to remove dissolved reaction products which are responsible for evolution of sulfur dioxide during later fractional distillation.

In the continuous A. B. Separator-Nobel process,[32] light oil is first treated with sodium hydroxide solution, then countercurrently in two steps with sulfuric acid, again with caustic soda, and finally with water. The various phases are first mixed in pumps and fed through mixing towers to centrifugal separators. Lower acid consumption, smaller washing losses, and better sulfur removal are claimed. The importance of proper mixing and control of reaction time have also been stressed by Kolyandr.[33] Great reductions in washing losses by the adoption of a continuous process have been reported in Japan.[34]

Washing the entire light oil has the advantage of simplicity. There are serious disadvantages. The benzene fraction, although it has the lowest concentration of unsaturated compounds, is the most difficult to wash; large quantities of strong acid are necessary. Sulfonation of toluene and, especially, xylene results. The presence of indene, styrene, etc., leads to the formation of condensation products through reactions such as

$$C_6H_5{-}CH{=}CH_2 + C_6H_6 \rightarrow C_6H_5{-}CH(CH_3){-}C_6H_5$$

[31] Hartman, L. F., Lawton, J. J., and Carbone, W. E., paper presented at the joint meeting of the Blast Furnace and Coke Oven Associations of the Eastern States and the Chicago District, Chicago, Ill., October 24, 1958.

[32] Sager, F., and Palmquist, F., *Erdöl u. Kohle*, **6**, 783–7 (1953).

[33] Kolyandr, L. Y., *Koks i Khim.*, **1957**, No. 9, 47–51.

[34] Yoshida, T., Hibino, O., Kurokawa, K., Ikenaga, M., and Murayama, M., *Coal Tar (Tokyo)*, **8**, 107–12 (1956).

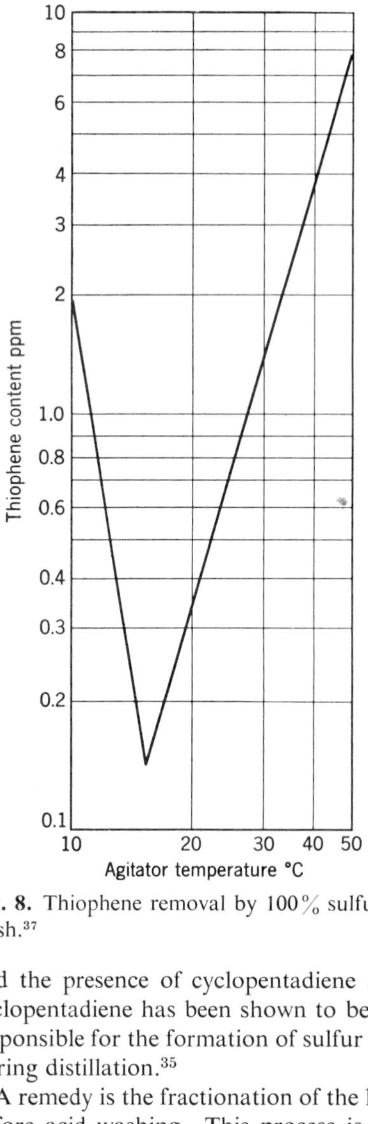

Fig. 8. Thiophene removal by 100% sulfuric-acid wash.[37]

and the presence of cyclopentadiene and dicyclopentadiene has been shown to be chiefly responsible for the formation of sulfur dioxide during distillation.[35]

A remedy is the fractionation of the light oil before acid washing. This process is carried out extensively at a plant in France.[36] Here, benzene and toluene fractions are produced with boiling ranges no greater than 0.5°C. Benzene is continuously washed in a series of three agitators water-jacketed to hold the temperature at 35°C. The benzene is first washed with acid sludge from the other agitators. In these vessels, the washing agent is fresh 100% sulfuric acid. Toluene is washed similarly with 96% sulfuric acid.

The chief sulfur compound removed from benzene by acid-washing is thiophene which occurs in crude benzene in concentrations as high as 1%. Acid-washing, as ordinarily practiced, reduces the thiophene content to 0.02 to 0.08%. More drastic action is necessary to produce "thiophene-free" benzene, a term which may be applied to material containing less than 1 ppm of thiophene. Washing with oleum, the most common method, results in high losses of benzene through sulfonation unless the process is closely controlled. Hartmann and associates[31] included a two-stage oleum wash in their scheme when thiophene-free benzene was desired.

Losses can be reduced by the use of nonfuming acid, that is, 100% H_2SO_4. Carabasse and associates[36] claimed that the benzene resulting from their method of washing gave a negative test for thiophene. Kovacic, Bretz, and Muder[37] applied 100% H_2SO_4 to nitration benzene and showed that the degree of thiophene removal is dependent on the temperature (Fig. 8). Best results are obtained at 15°C. They also found that reducing the strength of acid, even to 99%, resulted in much poorer thiophene removal. Increasing the strength of the acid, that is, having free SO_3 present, resulted only in greater losses. No sludge is produced, but, rather a clear acid useful for washing crude light oil or for the production of $(NH_4)_2SO_4$.

Washing with 66° Bé sulfuric acid in the presence of starch, sucrose, or cellulose for complete removal of thiophene has been patented.[38] This process has the disadvantage of producing a useless sludge containing much carbonaceous matter.

An interesting method of acid washing petroleum products that might be applied to light oil has been described by Phillips and Napier.[39] Oil and acid go through a vessel with turbine-bladed mixers whose function is to reduce the acid particle size to an optimum diameter. The mixture passes to a reactor where an electrical potential is applied in such a manner that violent motion is imparted to the electrically

[35] Claxton, G., and Hoffert, W. H., *Gas World*, **113,** No. 2927, Coking Sect., 70–6 (1940); *Coke Oven Managers' Assoc. Yearbook*, **1941,** 146–64.

[36] Carabasse, J., Delassus, M., and Mennessier, A., U.S. Pat. 2,680,707 (1954).

[37] Kovacic, J. E., Bretz, G. M., and Muder, R. E., Can. Pat. 563,919 (1958).

[38] Lauer, H. V., and Ponchaud, G. C., U.S. Pat. 2,346,524 (1944).

[39] Phillips, R. J., and Napier, H. G., *Petrol. Engr.*, **31,** No. 3, C18–22 (1959); *Petrol. Refiner*, **38,** No. 2, 151–4 (1959).

conductive acid particles. Neutralization and washing are conducted with similar electrostatic mixing.

Hydrogenation

In 1925, Badische Anilin- & Soda-Fabrik A. G. found it possible to treat crude light oil with hydrogen so that all the olefinic material was saturated without appreciable hydrogenation of the aromatic hydrocarbons. At the same time, compounds containing oxygen, nitrogen, and sulfur were reduced.

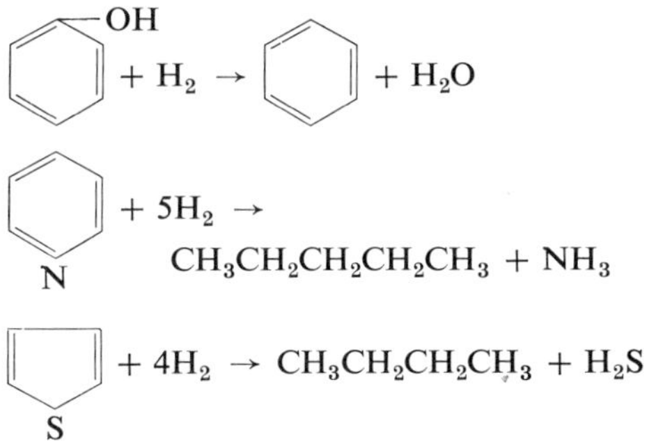

Various catalysts were found useful, especially oxides and sulfides of metals of the sixth and eighth groups of the periodic system such as tungsten-nickel and, most particularly, molybdenum-cobalt. During the 1930's, this method was demonstrated on a large scale, but no commercial application was made until 1950 when Scholven-Chemie A. G. adapted equipment used for coal hydrogenation for this purpose. A raffinate was produced from which both motor fuel and aromatics for chemical purposes were prepared. The success of the plant led to the erection of other plants listed by Jäckh[40] in Table 14.

TABLE 14

LIGHT-OIL HYDROGENATION PLANTS[40]

Plant Location	Capacity tons/mon
Gelsenkirchen-Buer, Germany	22,000
Dortmund, Germany	6000
Stuttgart, Germany	500
Frankfurt a. M., Germany	1100
Hervest-Dorsten, Germany	500
Brambauer/Westf., Germany	1500
Essen, Germany	1500
Dortmund, Germany	4500
Carlingen/Lorraine, France	6000
Saarbrücken, Germany	6000
Duisberg-Hamborn, Germany	4100

In addition, a plant has been constructed in Italy of 750 tons/mon capacity, one in Belgium of 2000, one in India of 1500, and two in Japan of 3600 tons each. In the United States, a plant of this type has been constructed at the Clairton, Pennsylvania, plant of the United States Steel Corporation with a capacity of 14,000 tons/mon.[41] Jones and Laughlin Steel Corporation at Aliquippa, Pennsylvania have installed a plant with a capacity of 6000 tons/mon based on an adaptation of the Hydrofining process.[42] In 1955, 65% of Russian benzene was purified in this way.[43]

A flow diagram of a plant based on the BASF–Scholven process is shown in Fig. 9. Crude light oil is heated by heat exchange, pumped to a tower where it is retained long enough to allow gum-forming constituents to polymerize, and then vaporized at 20 to 30 atm in a stream of hydrogen. The vapor stream is further heated to 300 to 400°C and is passed over the catalyst where the aforementioned reactions take place. These reactions are exothermic. The vapors are cooled and condensed. The condensate is sent to a stabilizing column where H_2S and light hydrocarbons are removed. The hydrogen is recycled, although a portion is bled off and replaced by fresh hydrogen to maintain the proper hydrogen partial pressure. The light oil not vaporized in the evaporator is distilled to remove resins and is blended with the fresh light-oil feed.

The hydrogen used may be obtained by the catalytic reforming of natural gas with steam, by the reaction of iron with steam, by the dissociation of ammonia, or by the distillation of coke-oven gas. Coke-oven gas itself may be used, but the operating pressure of the plant must be increased so that the partial pressure of hydrogen is the same as when relatively pure hydrogen is used. The concentration of carbon monoxide must always be kept low; otherwise, the consumption of hydrogen will be excessive. Also, CO is a temporary catalyst poison. When

[40] Jäckh, W., *Erdöl u. Kohle*, **11**, 625–9 (1958).

[41] Smeaton, R. M., and Joseph, J. S., paper presented at meeting of Eastern States Blast Furnace and Coke Oven Association, Pittsburgh, Pa., November 13, 1959.

[42] Anon., *Chem. Eng.*, **65**, 110–13 (1958).

[43] Kondukov, N. B., *Koks i Khim.*, **1956**, No. 6, 42–5.

coke-oven gas is used, removal of unsaturated compounds, oxygen, sulfur compounds, and nitrogen oxides is advisable to avoid difficulties from polymerization and condensation reactions.

One advantage of hydrogenation over the conventional acid-wash process is the conservation of the olefins present in the light oil; the resulting paraffins are useful in those fractions used as motor fuel or solvents. Also, the considerable amount of styrene in crude light oil is recovered as ethyl benzene. Other advantages are the small losses of aromatics and the total removal of sulfur compounds. The earlier German plants reduced the total sulfur content of the raffinate to 20 ppm and the thiophene content of the benzene fraction to a level somewhat below 10 ppm. When the American plants were built, a thiophene content of less than 1 ppm was desired. Slight modifications in the process accomplished this easily. The disadvantages of the hydrogenation process are the high capital cost of the plant, making it uneconomical if less than 5000 tons a month of light oil are to be treated, and the increased difficulty in preparing nitration grade products by subsequent fractional distillation because of the greater concentration of paraffins.

Fuller descriptions of this process may be found in the listed references.[44]

A combination of cracking and noncatalytic hydrogenation promises a one-step purification for light oil. Sawyer[45] has claimed that treatment with hydrogen at 650 to 900°C will produce a product substantially free from nonaromatic hydrocarbons.

Thiophene Removal

Although sulfur occurs in light oil in the form of carbon disulfide, alkyl sulfides, etc., the customary refining methods, if followed by efficient fractional distillation, remove all sulfur compounds from the benzene fraction except thiophene. Therefore, special efforts must be exercised to remove it if a sulfur-free benzene is desired.

Physical methods of separation are not very efficacious. Thiophene's normal boiling point is 4°C above that of benzene. The two compounds do not form an azeotrope. The relative volatility in the range 0 to 1% of thiophene is 1.1045.[46] Thus, it is possible to separate these substances by fractional distillation. Practically, however, it is extremely difficult to prepare benzene containing less than 1 ppm of thiophene by this means. It has been shown that using a column containing seventy theoretical plates at a very high reflux ratio would permit the recovery of 90 parts of benzene containing 0.001% of thiophene from 100

[44] Urban, W., *Erdöl u. Kohle*, **4**, 279–82 (1951). Nonnemacher, H., Reitz, O., and Schmidt, P., *Erdöl u. Kohle*, **8**, 407–11 (1955). Helm, J., and Vodicka, J., *Paliva*, **33**, 79–83 (1953). Thomas, A., *J. usines gaz*, **81**, 203–5 (1957). Strzeszewka, M., and Wiszniowski, K., *Koks, Smola, Gaz*, **2**, 114–20 (1957). Novak, H., and Liebich, H. G., *Brennstoff-Chem.*, **35**, 308–10 (1954). Hayaski, S., *Coal Tar (Tokyo)*, **6**, 466–72 (1954). Fitz, W., *Coke and Gas*, **21**, 321–5 (1959). Sansam, W., and Gigiel, J., *Gas World*, **141**, Coking Sect. **46**, 91–9 (1955). Eisenberg, A., Peter, K., Pernecker, M., Ritter, H., Koster, L., Brodkorb, F., and Wiegner, F., Brit. Pat. 776,145 (1957). Scholven-Chemie A.-G., Brit. Pat. 789,986 (1958). Yorkshire Tar Distillers Ltd., Brit. Pat. 525,814 (1940). Badger, E. H. M., Griffith, R. H., and Newling, W. B. S., *Proc. Roy. Soc. (London)*, **A197**, 184–201 (1949). Plant, J. H. G., and Newling, W. B. S., *Gas World*, **129**, 807–8, 922–7 (1948); *Gas Journal*, **256**, 654, 657, 659 (1948); Inst. Gas Engrs., Copyright Publ., No. 345/158,(1949), 26 pp.; Inst. Gas Engrs., Commun. No. 345. Grothe, W., *Erdöl u. Kohle*, **6**, 450–4 (1953). Boye, E., *Chemiker Ztg.*, **80**, 279–81 (1956). Trefny, F., *Erdöl u. Kohle*, **8**, 874–9 (1955). Densham, A. B., Gough, G., and Griffith, R. H., Brit. Pat. 735,706 (1955). Ando, S., and Ushiba, F., *Coal Tar (Tokyo)*, **7**, 563–6 (1955). Niewiadomski, F., and Wiszniowski, K., *Chemik (Gliwice)*, **9**, No. 6, 167–70 (1956). Ritter, H., *Erdöl u. Kohle*, **10**, 433–5 (1957). Vykoukal, J., *Paliva*, **37**, 120–6 (1957). Müller, K., *Gas-u. Wasserfach*, **98**, 656–8 (1957). Ivanovskii, F. P., Kal'varskaya, R. S., Beskova, G. S., and Sakolova, N. P., *Zhur. Fiz. Khim.*, **30**, 1860–6 (1956). D'yakova, M. K., Vol'Epshtein, A. B., Aleksi, E. A., and Vasil'chikova, E. I., *Khim. i Tekhnol. Topliva*, **1956**, No. 9, 44–51. Zykov, D. D., Karavaev, N. M., and Kondukov, N. B., *Koks i Khim.*, **1957**, No. 12, 33–7. Gelsenkirchener Bergwerks A.-G., Brit. Pat. 805,916 (1958). British Petroleum Co., Ltd., Brit. Pats. 805,304, 805,050 (1950). Metallgesellschaft, A.-G., Brit. Pat. 766,625 (1957); Ger. Pats. 930,225, 930,223, 931,616 (1955). Scholven-Chemie A.-G., Brit. Pat. 789,986 (1958). Schmalenbach, A., U.S. Pat. 2,827,504 (1958).

[45] Sawyer, E. W., Brit. Pat. 757,345 (1956).

[46] Coulson, E. A., Hales, J. L., and Herington, E. F. G., *Trans. Faraday Soc.*, **44**, 636–44 (1948).

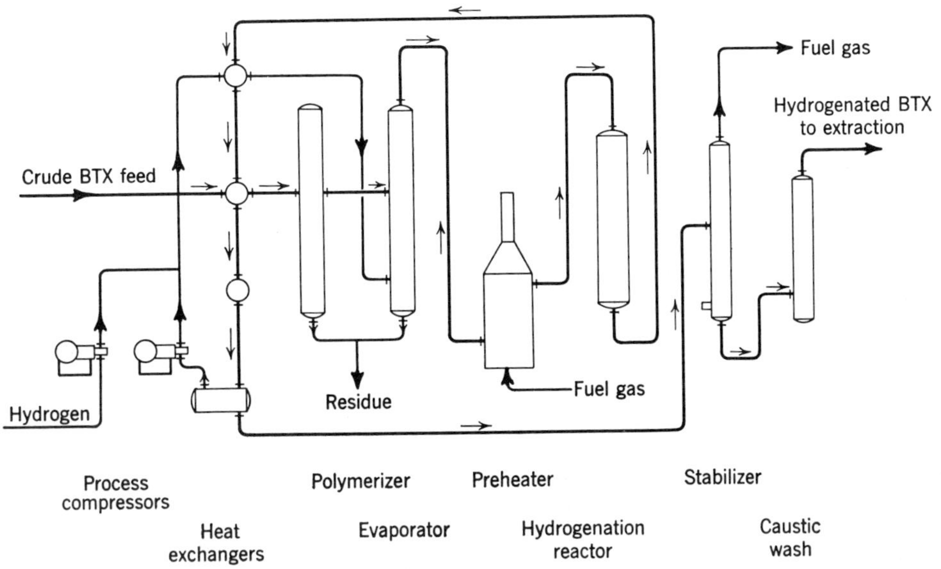

Fig. 9. Flow diagram of light-oil hydrogenation process.

parts of benzene containing 0.1 %. Crystallization is even less useful as the freezing point-composition diagram shows (Fig. 10).[47] Other observers have confirmed that the two compounds form a continuous series of solid solutions[48] although Kitaigorodskii has denied this because of the dimensional incompatibility of the molecules.[49]

Thiophene has been called a superaromatic. It undergoes the characteristic reactions of benzene, such as sulfonation and nitration, at higher rates, at lower temperatures, and at lesser concentrations of reagents. Its ring, however, is not so stable as that of benzene. The presence of the sulfur atom leads to reactions which are not shared by hydrocarbons. Thus, many chemical methods of thiophene removal have been devised.

It was shown previously that washing with sulfuric acid, as normally practiced, removes a great deal of the thiophene present in crude benzene, and that complete removal can be achieved by more vigorous washing. Hydrogenation also reduces thiophene to any desired level.

Aluminum chloride is used commercially to prepare thiophene-free benzene. E. I. duPont de Nemours and Co.[50] continuously add benzene and anhydrous $AlCl_3$ to a boiling, saturated solution and distill off thiophene-free benzene. A British patent[51] describes the introduction of the benzene into a reaction vessel

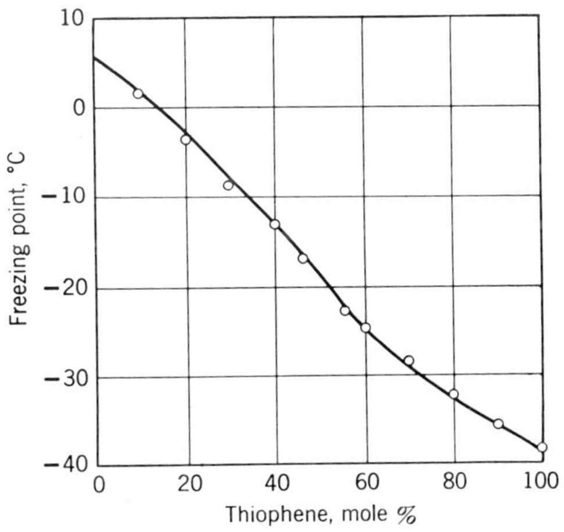

Fig. 10. Freezing point diagram of system: benzene-thiophane.[47]

[47] Fawcett, F. S., and Rasmussen, H. E., *J. Am. Chem. Soc.*, **67**, 1705–9 (1945).

[48] Kofler, W., Kofler, A., and Kofler, L., *Mikrochimie ver. Mikrochim. Acta*, **38**, 218–31 (1951). Kravchenko, V. M., *J. Applied Chem. (U.S.S.R.)*, **23**, 301–11 (1950); *Zhur. Priklad. Khim.*, **23**, 288–98 (1950); *Doklady Akad. Nauk S.S.S.R.*, **79**, 443–6 (1951).

[49] Kitaigorodskii, A. I., *Doklady Akad. Nauk S.S.S.R.*, **113**, 604–6 (1957).

[50] Peterson, M. D., U.S. Pat. 2,311,189 (1943).

[51] Refiners Ltd. and Scott, T., Brit. Pat. 571,256 (1945).

containing an $AlCl_3$-benzene complex, whereupon thiophene reacts with the salt, and pure benzene overflows.

Raney nickel has been claimed to bring about complete removal of thiophene by boiling the benzene in its presence for a short time.[52]

Either NaOH or KOH, water, and a low-boiling aliphatic alcohol added to benzene in a still allow the removal of the benzene overhead as an azeotrope containing no thiophene.[53]

Catalytic treatment in the gas phase has been claimed to remove thiophene. Suitable catalysts are the oxides and salts of calcium, magnesium, aluminum, lead, zinc, copper, and molybdenum and metallic selenides, sulfides, and tellurides.[54] Claims have also been made for alkaline earth, silver, nickel, cobalt and iron vanadates, chromates, molybdates, and tungstates.[55]

Other reagents proposed for thiophene removal are bleaching earths,[56] anhydrides of fatty acids,[57] hydrofluoric acid,[58] olefins in the presence of phosphoric acid,[59] alkaline iron hydroxide with fuller's earth and aluminum-magnesium silicate,[60] nitric and acetic acids,[61] and chlorine in the dark in the absence of catalysts.[62]

Removal of Paraffins

In discussions of coke-oven light oil, the term, "paraffins," is used to denote all saturated hydrocarbons that appear in the refined products. It includes aliphatic hydrocarbons, such as *n*-heptane and 2-methylhexane, and alicyclic compounds, such as cyclohexane. Acid washing does not remove them; hydrogenation engenders them. Their removal is difficult since they are generally less reactive than aromatics. Hence, chemical means of removing them are limited in number. Physical methods, such as distillation, crystallization, and extraction, have usually been used.[63]

If acid-washed light oil contained only aromatic hydrocarbons, the preparation of very pure benzene and toluene by modern distillation techniques would be a simple matter. Table 8 showed that one benzene sample contained only 0.03% of toluene, yet the total impurities amounted to 0.67%. This circumstance must be ascribed to the carbonization conditions rather than to poor refining techniques. Actually, this was a very well-refined benzene produced in a modern continuous distillation plant,[64] as attested by its boiling range of 0.6°C. Why, then, did this material contain so many impurities whose normal boiling points are 10 or 20° higher than that of benzene and so little material boiling near 80°C? The vapor-liquid equilibrium diagrams for the systems benzene-*n*-heptane[65] and benzene-methyl-cyclohexane[66] (Fig. 11 and 12) show that, in the regions of high benzene concentration, the vapor-liquid equilibrium curves are almost tangent to the 45° line. This means that vapor and liquid have almost the same composition, and that a very large number of theoretical plates is necessary to effect any great change in composition. The relative sparseness of materials whose boiling points are near 80°C is a consequence of the formation of azeotropes[67] which are removed with

[52] Bougault, J., Cattelain, E., and Chabrier, P., *Bull. soc. chim.*, **7,** 780–9 (1940). Graul, R. J., and Karabinos, J. V., *Science*, **104,** 557 (1946).

[53] Wilson, P. J., Jr., Wells, J. H., and Sommerfield, P. M., U.S. Pat. 2,427,988 (1947).

[54] Fohlen, J. L., Fr. Pat. 853,404 (1940).

[55] Plant, J. H. C., and Newling, A. B. S., *Inst. Gas Engrs., Commun.* No. 344; *Gas World*, **129,** 807–8, 922–7 (1948); *Gas J.*, **256** 654, 657, 659 (1948); Inst. Gas Engrs., Copyright Publ. No. 345/158 (1949), 26 pp.

[56] Litachrom, Societa anon. italiana per l'industria delle terre decoloranti ed. affini, Ital. Pat. 445,059, 445,060 (1949).

[57] Hartough, H. D., Kosak, A. I., and Reid, E. E., U.S. Pat. 2,462,391 (1949).

[58] Lien, A. P., and Evering, B. L., U.S. Pat. 2,464,520 (1949).

[59] Vesely, J. A., U.S. Pat. 2,563,087 (1951).

[60] Schwenke, H., Brit. Pat. 669,949 (1952).

[61] Imoto, E., and Motoyama, R., *J. Chem. Soc. Japan*, **55,** 305–7 (1952).

[62] Limido, G. E., *Compt. rend.*, 27 *congr. intern. chim. ind., Brussels*, **1954,** 2; *Ind. chim. belge*, **20,** Special No., 217–21 (1955).

[63] A very good general discussion of physical methods of hydrocarbon separation is *Hydrocarbons from Petroleum* by F. D. Rossini, B. J. Mair, and A. J. Streiff, Reinhold, New York, 1953, 556 pp.

[64] Seyler, H. W., *World Petroleum*, **11,** No. 11, 88–95 (1940).

[65] Sieg, L., *Chem. Ing. Tech.*, **22,** 322–6 (1950). Clark, A. M., *Trans. Faraday Soc.*, **41,** 718–52 (1943). Myers, H. S., *Ind. Eng. Chem.*, **47,** 2215–19 (1955).

[66] Sieg, L., *Loc. cit.* Myers, H. S., *Ind. Eng. Chem.*, **48,** 1104–8 (1956).

[67] Marschner, R. F., and Cropper, W. P., *Ind. Eng. Chem.*, **38,** 262–8 (1946); **41,** 1357–62 (1949).

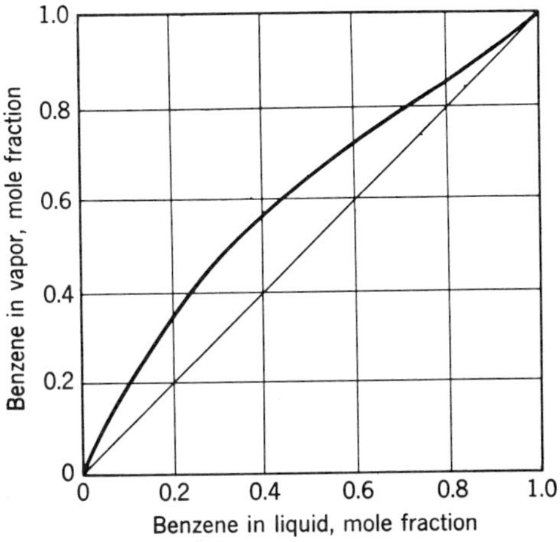

Fig. 11. Vapor-liquid equilibrium diagram of the system: benzene-*n*-heptane.[65]

the forerunnings. Cyclohexane, 2,4-dimethylpentane, and 2,2,3-trimethylbutane boil at 80.74, 80.51, and 80.8°C, respectively. The corresponding azeotropes boil at 77.5, 75.2, and 76.6°C. These azeotropes contain about equal quantities of each constituent, and so some benzene is lost. This loss is trifling, however. The situation is very much the same in the case of toluene. The problem, then, is the removal of paraffins whose normal boiling points are substantially above those of the aromatic compounds.

Azeotropic and extractive distillation. Azeotropic distillation has often been suggested as a means of removing paraffins from aromatics. Mair, Glasgow, and Rossini have discussed the principles involved in an excellent paper[68] which antedates much of the further work that has been reported. An understanding of this paper would have prevented the writing of some claims that a single azeotropic distillation is capable of removing completely the paraffinic impurities from coke-oven benzene or toluene. The following is quoted from this paper.

> Aromatic hydrocarbons may be separated from naphthalenes and paraffins with relative ease by one appropriate azeotropic distillation in an efficient rectifying column, in mixtures in which the aromatic hydrocarbons normally have boiling points *not significantly lower* than the boiling points of the naphthenes and paraffins in the given mixture.

[68] Mair, B. J., Glasgow, A. R., Jr., and Rossini, F. D., *J. Research Natl. Bur. Standards*, **27**, 39–63 (1941).

Unfortunately, in the mixtures dealt with here, the aromatic hydrocarbons do have boiling points significantly lower than those of the naphthenes and paraffins. Table 15 lists the azeotropes of certain hydrocarbons with methanol, ethanol, and acetone, which have frequently been suggested as entrainers for paraffins in the azeotropic purification of benzene.[69,70] The alcohols and the common impurities in benzene and toulene form azeotropes which have higher boiling points than the corresponding aromatic azeotropes. Acetone is more selective and can be used for purifying benzene, especially if water is also present.[71]

Tables 16 and 17 present comparisons of the effects of various azeotropic agents on the purification of a single benzene and a single toluene.[72] The benzene used was from the same source as that whose composition has been presented in Table 4. The toluene had a boiling range of 0.6°C and a paraffin content of 0.95%. Mixtures of benzene or toluene and the azeotropizing agents were distilled at a reflux ratio of 20 through a batch column having thirty-five theoretical plates until all of the added material had been brought over. The still residue was then analyzed. No really

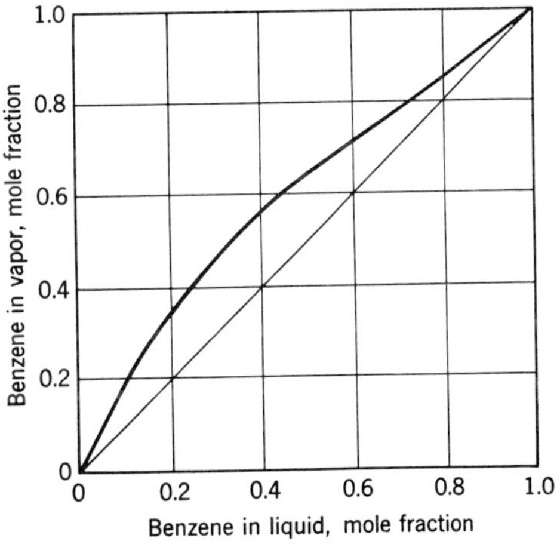

Fig. 12. Vapor-liquid equilibrium diagram of the system: benzene-methylcyclohexane.[66]

[69] Horsley, L. H., *Anal. Chem.*, **19**, 508–600 (1947).

[70] Horsley, L. H., *Azeotropic Data*, Am. Chem. Soc., Washington, D. C., 1952, 328 pp.

[71] Griswold, J., and Bowden, R. H., *Ind. Eng. Chem.*, **38**, 509–12 (1946).

[72] Bridgewater, H. L., Koppers Co., Inc., Unpublished data.

TABLE 15

HYDROCARBON AZEOTROPES[70]

Hydrocarbon	Boiling Point of Hydrocarbon °C	Boiling Point of Azeotrope, °C		
		Methanol	Ethanol	Acetone
n-Hexane	68.7	50	58.7	49.7
Benzene	80.1	57.5	68.2	nonazeotrope
Cyclohexane	80.8	54	64.9	53.0
n-Heptane	98.4	59.1	70.9	55.8
Methylcyclohexane	100.9	59.2	72.0	nonazeotrope
Toluene	110.6	63.8	76.7	nonazeotrope
n-Octane	125.6	63.0	77	nonazeotrope

TABLE 16

AZEOTROPIC DISTILLATION OF BENZENE

Azeotropic Agent	% by Volume	Freezing Point of Residue, °C
Methanol	10	5.34
Ethanol	10	5.25
Acetone	2.5	5.19
Acetone	10	5.31
Acetone	22.5	5.33
Acetone	45	5.36
Acetonitrile	2.5	5.26
Acetonitrile	10	5.40
Acetonitrile	22.5	5.40

TABLE 17

AZEOTROPIC DISTILLATION OF TOLUENE

Azeotropic Agent	% by Volume	% Paraffins in Residue
Methanol	10	0.45
Methanol	50	0.16
Ethanol	10	0.35
Methyl ethyl ketone	10	0.72
Acetonitrile	2.5	0.75
Acetonitrile	5	0.50
Acetonitrile	10	0.46
Dioxane	10	0.55
Ethylene diamine	2.5	0.80
Ethylene diamine	5	0.45
Ethylene diamine	10	0.40
Ethylene glycol monomethyl ether	10	0.42
Diethylamine	10	0.80
Piperidine	10	0.85

pure aromatic compounds were obtained even though large quantities of the polar compounds were used.

The formation and behavior of azeotropes has been discussed fully by Swietoslawski.[73] He overstated the case when he claimed that azeotropic distillation is incapable of yielding completely pure aromatics.[74] The difficulties, however, have been underestimated by many others. Anderson, a former co-worker of Swietoslawski, has expressed the opinion[75] that many of the results claimed for azeotropic

[73] Swietoslawski, W., *Azeotropia i Poliazeotropia*, Warsaw, 1957, 272 pp.

[74] Swietoslawski, W., *Przemysl Chem.*, **26**, 33–6 (1947).

[75] Anderson, J. R., Mellon Institute, Pittsburgh, Pa., private communication.

distillation could be achieved by ordinary distillation if conducted just as carefully. It was Anderson's understanding of the azeotropism of benzene that led to a process in which an equal quantity of a nonaromatic hydrocarbon boiling between 68 and 90.5°C, such as cyclohexane, is added to form an azeotrope that can be separated from the high-boiling paraffins. The added hydrocarbon can be separated from the benzene by a second distillation with one of the polar azeotropic agents.[76]

There are many publications on this subject.[77] The reader should be wary of data obtained with materials of wide boiling range and of claims based on synthetic mixtures of aromatics with paraffins of similar boiling points. Azeotropic and extractive distillation are similar in that a substance is added to the system whose components are to be separated. The added substance changes the relative vapor pressures so that the separation becomes easier. In azeotropic distillation, the added agent forms an azeotrope with at least one of the components of the mixture and is recovered as a distillate. In extractive distillation, the separating agent is one of relatively low-vapor pressure. The agent is added near the top of the column, scrubs the ascending vapors, and is recovered from the bottom of the column. A simplified flow diagram of an extractive distillation process is shown in Fig. 13.

Extractive distillation has been used extensively for the preparation of nitration-grade toluene from petroleum. It is not believed to have been applied to coke-oven light-oil distillates because the relatively small volumes of toluene produced in most plants have precluded its use for economic reasons. However,

[76] Anderson, J. R., U.S. Pat. 2,581,344 (1952).

[77] Benedict, M., Johnson, C. A., Solomon, E., and Rubin, L. C., *Trans. Am. Inst. Chem. Engrs.*, **41,** 353–92 (1945). Clark, C. R., U.S. Pats. 2,407,364 (1946), 2,388,040 (1945). Colburn, A. P., and Schoenborn, E. M., *Trans. Am. Inst. Chem. Engrs.*, **41,** 421–43 (1945). Deanesly, R. M., U.S. Pat. 2,360,655 (1944). Engel, K. H., U.S. Pat. 2,376,870 (1945). Greenburg, R. B., U.S. Pat. 2,352,534 (1944). Lake, G. R., U.S. Pats. 2,406,695 (1946), 2,432,771 (1947). Geckler, R. D., and Fragen, N., U.S. Pat. 2,316,126 (1943). Scarth, V., U.S. Pat. Reissue 22,811 (1946). Coulson, E. A., and Holt, E. C., *J. Soc. Chem. Ind.* (*London*), **63,** 329–36 (1944). Lake, G. R., *Trans. Am. Inst. Chem. Engrs.*, **41,** 327–52 (1945). Keyes, D. B., *Ind. Eng. Chem.*, **33,** 1019–21 (1941). Woodall-Duckham, Ltd., Underwood, A. J. V., and Spiers, H. M., Brit. Pat. 580,108 (1946). Berg, L., Carpenter, H. C., and Daly, J. B., *Chem. Eng. Progr.*, **46,** No. 6, 283–9 (1950). Hachmuth, K. H., *Chem. Eng. Progr.*, **48,** 617–26 (1952). Kortüm, G., *Chem.-Ztg.*, **74,** 151-4 (1950). Berti, V., and Bottiglia, F., *Riv. combustibili*, **5,** 329–60 (1951). Harrison, J. M., U.S. Pat. 2,561,624 (1951). Keeling, W. O., U.S. Pat. 2,484,834 (1949). Yoshida, H., Hibino, O., Hara, K., and Tamura, N., *Coal Tar* (*Tokyo*), **2,** 5–9 (1950). Chadder, W. J., U.S. Pat. 2,416,377 (1947). Davidson, M. W., U.S. Pat. 2,506,858 (1950). Dummett, G. A., and Ralph, S. J., U.S. Pat. 2,417,355 (1947). Lake, G. R., and Stribley, J. M., U.S. Pat. 2,477,303 (1949). Ralph, S. J., and Dummett, G. A., U.S. Pat. 2,450,381 (1948). Standard Oil Development Co., Brit. Pat. 593,463 (1947). Waldron, J. W., U.S. Pat. 2,419,521 (1947). Woodall-Duckham (1920) Ltd. and Spiers, H. M., Brit. Pat. 569,405 (1945). Woodall-Duckham (1920) Ltd., Spiers, H. M., and Chadder, W. J., Brit. Pat. 569,416 (1945). Denyer, R. L., Fidler, F. A., and Lowry, R. A., *Ind. Eng. Chem.*, **41,** 2727–37 (1949). Leffert, C. B., and Berg, C. H. O., U.S. Pat. 2,563,344 (1951). Berg, C. H. O., U.S. Pat. 2,428,611 (1947). Berg, L., and Harrison, J. M., *Chem. Eng. Progr.*, **43,** No. 9, 487–96 (1947). Cubicciotti, R., U.S. Pat. 2,483,625 (1949). D'Arcangelo, F., *Bol. inform. petrol.* (*Buenos Aires*), **24,** 387–99 (1947). Hartley, F. L., *Petrol. Refiner*, **24,** 519–24 (1945). Latchum, J. W., Jr., and Connors, J. S., U.S. Pat. 2,434,322 (1948). Petry, F. A., Lee, R., Biles, J. L., and Betts, W. V., U.S. Pat. 2,428,467 (1947). Reed, H. C., and Holt, B. M., U.S. Pat. 2,413,245 (1946). Spiers, H. M., U.S. Pat. 2,386,755 (1945). Foxon, G. H., and Imperial Chemical Industries, Ltd., Brit. Pat. 646,116 (1950). Miller, S. P., Can. Pat. 451,985 (1948). Koch, H., and van Raay, H. G., *Brennstoff-Chem.*, **35,** 105–12 (1954). Swietowslawski, W., and Zieborak, K., *Bull. intern. acad. polon. sci., Classe sci. math. nat.*, **Ser. A.,** 9–33 (1950). Zieborak, K., *Roczniki Chem.*, **25,** 388–91 (1951); *Prace Glównego Inst. Chem. Przemysl*, No. 1, 1–44 (1951). Rius, A., and Otero de la Sandara, J. L., *Anales real. soc. espan. fis. y quim.* (*Madrid*), **48B,** 569–88 (1952). Weck, H. I., and Hunt, H., *Ind. Eng. Chem.*, **46,** 2521–3 (1954). Williams, G. C., Rosenberg, S., and Rothenberg, H. A., *Ind. Eng. Chem.*, **40,** 1273–6 (1948). Chadder, W. J., and Spiers, H. M., *Proc. XIth Intern. Congr. Pure and Appl. Chem.* (*London*), **5,** 1003–14 (1947). Anderson, J. R., U.S. Pat. 2,618,591 (1952). Foxon, G. H., U.S. Pat. 2,672,436 (1954). Redlich, O., and Gable, C. M., U.S. Pat. 2,604,935 (1954). Petry, F. A., and Holm, H. K., U.S. Pat. 2,400,654 (1946).

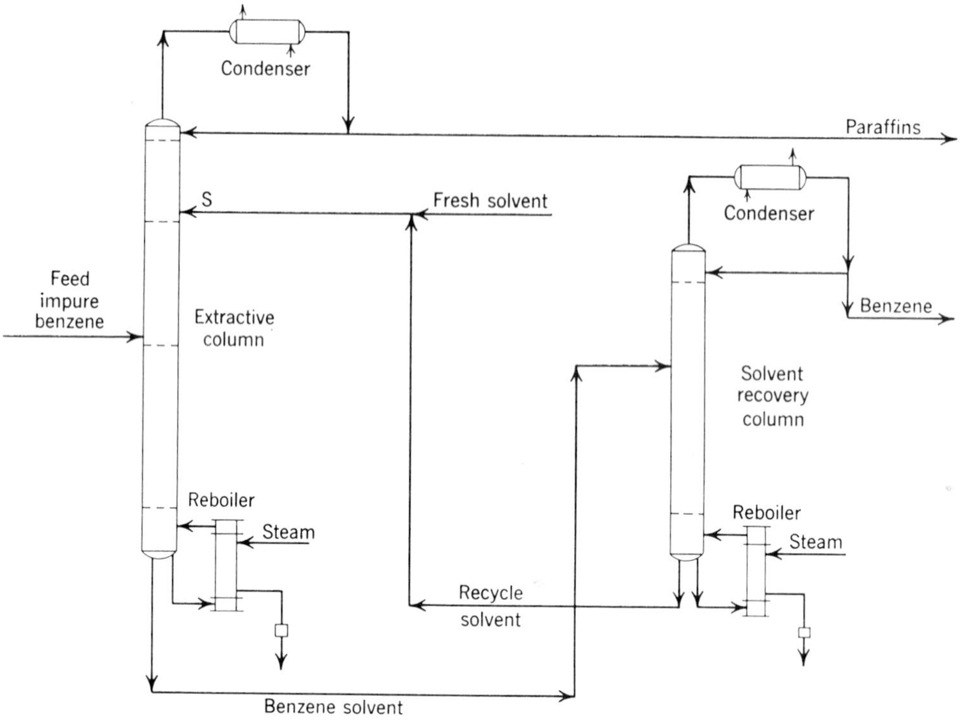

Fig. 13. Flow diagram of extractive distillation of benzene.

TABLE 19

SOLVENTS FOR THE PREPARATION OF TOLUENE BY EXTRACTIVE DISTILLATION[77a]

Solvent	Boiling Point, °C	Relative Volatility
Furfural	163	2.30
Acetonylacetone	188	2.20
Nitrobenzene	211	2.16
Nitrotoluene	222	2.16
Phenol	182	2.10
Aniline	183	2.08
Chlorex (dichloroethyl ether)	178	2.09
Phenyl cellosolve	240	2.01
Phenol-cresol (60–40)	192	1.98
Acetophenone	202	1.85
1-Methoxy-2-hydroxy-3 phenoxypropane	278	1.92
Ethyl carbitol	202	1.85
m-Cresol or *p*-cresol	202	1.85
1-Ethoxy-2,3-dihydroxypropane	222	1.80
1-Isopropoxy-2,3-dihydroxypropane	226	1.70
Diacetone glycol	190	1.64
1-Methoxy-2,3-dihydroxypropane	220	1.46

[77a] Dunn, C. L., Millar, R. W., Pierotti, G. U., Shiras, R. N., and Souders, M., Jr., *AIChE, Trans.*, **41,** 631–44 (1945).

there is no technical reason why it should not be just as effective for coke-oven toluene or benzene.

Table 19 lists some of solvents proposed for separating toluene from other hydrocarbons. The relative volatilities represent y_1x_2/y_2x_1 where y and x refer to the solvent-free concentrations in the vapor and liquid phases at equilibrium. The subscripts 1 and 2 refer to "non-toluene," a mixture of saturated hydrocarbons boiling between 99 and 113°C, and to toluene, respectively. The data were obtained with mixtures containing 25% of toluene, 25% of "non-toluene," and 50% of solvent. Of the solvents listed, phenol was considered to exhibit the best combination of selectivity, stability, and ease of recovery.

Most of the literature references[78] pertinent to the purification of the lower-boiling mononuclear aromatic hydrocarbons by extractive distillation are derived from the petroleum industry where the problem is the recovery of benzene, toluene, or xylene from a naphtha in which it is present in a concentration of less than 50%. With the coke-oven products, the problem is the removal of small amounts of higher-boiling paraffins and naphthenes.

Solvent extraction. All liquid hydrocarbons are miscible with each other. However, they display differences in solubility in nonhydrocarbon solvents, and so they can be separated by extraction. The differences in solubility are generally not very great, and multistage contact is necessary. Lists of two-phase liquid systems containing hydrocarbons have been compiled by Francis[79] and by Gladel and Durandet.[80]

With the exception of some perfluorinated compounds, solvents exhibit preferential solubility for aromatic hydrocarbons. This circumstance is unfortunate since it requires, for coke-oven light-oil products, that 98 to 99% of the material be extracted leaving only a small portion of paraffin-rich raffinate.

Solvent extraction methods of purifying aromatic hydrocarbons have the advantage over those previously discussed in that they can be applied to fractions of a wide boiling range. Thus, it is possible to treat a petroleum naphtha or a coke-oven light-oil so as to obtain paraffin-free extracts from which pure benzene and toluene can be separated by fractional distillation.

The most widely used extraction process is the Udex process described by a number of writers.[81] This process was originated by the Dow Chemical Company and fully developed

[78] Pierotti, G., and Dunn, C. L., U.S. Pat. 2,341,812 (1944). Shiras, R. N., and Souders, M., U.S. Pat. 2,350,256 (1944). Shiras, R. N., and Johnson, A. J., U.S. Pat. 2,351,028 (1944). Sweeney, W. J., U.S. Pat. 2,378,808 (1945). Dunn, C. L., and Liedholm, G. E., *Oil Gas J.*, **51,** No. 5, 68–70, 96 (1952); *Proc. Am. Petroleum Inst., 7th Mid-Year Meeting*, **32M,** Sect. 3, 208–11, 235–7 (1952); *Petrol. Engr.*, **24,** No. 8, C7–10 (1952). Drickamer, H. G., and Hummel, H. H., *Trans. Am. Inst. Chem. Engrs.*, **41,** 607–29 (1945). Fenske, M. R., Carlson, C. S., and Quiggle, D., *Ind. Eng. Chem.*, **39,** 1322–8 (1947). Berti, V., and Leidi, G., *Riv. combustibili*, **3,** 255–69 (1949). Durrum, E. L., U.S. Pat. 2,407,820 (1946). Griswold, J., Andres, D., Van Berg, C. F., and Kasch, J. E., *Ind. Eng. Chem.*, **38,** 65–70 (1946). Griswold, J., and Van Berg, C. F., *Ind. Eng. Chem.*, **38,** 170–7 (1946). Griswold, J., and Morris, J. W., *Ind. Eng. Chem.*, **41,** 331–6 (1949). Griswold, J., U.S. Pat. 2,537,459 (1951). Morrell, C. E., Carlson, C. J., and Nelson, J. F., U.S. Pat. 2,567,228 (1951). Morrow, M. R., and Smalling, J. N., U.S. Pat. 2,570,066 (1951). N. V. de Bataafsche Petroleum Maatschappij, Dutch Pats. 64,693 (1949); 66,635 (1950); 66,636 (1950); Brit. Pat. 681,434 (1952). Standard Oil Development Co., Brit. Pat. 571,913 (1945). Woerner, R. C., U.S. Pat. 2,494,274 (1950). Cooper, H. B. H., and Medcalf, E. C., U.S. Pat. 2,655,467 (1953). Herbst, W. A., U.S. Pat. 2,665,315 (1954). Matsumura, H., *J. Chem. Soc. Japan*, Ind. Chem. Sect. **54,** 730–2 (1951); *Kumamoto J. Sci.*, Ser. A., *Math. Phys. Chem.*, **1,** No. 1, 41–9 (1952). Lebeis, E. H., and Bondor, F. S., U.S. Pat. 2,768,131 (1956).

[79] Francis, A. W., *Physical Chemistry of Hydrocarbons*, A. Farkas, Ed., Academic Press, Inc., New York, 1950, pp. 241–313; *Solubilities of Inorganic and Organic Compounds*, Supplement to 3rd edition, A. Seidell and W. F. Linke, Eds., Van Nostrand, New York, 1952, pp. 1077–1115.

[80] Gladel, Y. L., and Durandet, J., *Rev. inst. franc. petrole*, **10,** 258–61 (1955); **11,** 488–9 (1956).

[81] Read, D., *Oil Gas J.*, **51,** No. 7, 82–9 (1952); *Petrol. Refiner*, **31,** No. 5, 97–103 (1952); *Proc. Am. Petroleum Inst., 17th Mid-Year Meeting*, **32M,** III, 258–73 (1952). Reidel, J. C., *Oil Gas J.*, **52,** No. 35, 72–5, 91 (1954). Resen, F. L., *Oil Gas J.*, **50,** No. 43, 55–6 (1952). Thornton, D. P., Jr., *Petrol. Processing*, **7,** 498–501 (1952), **8,** 384–7 (1953). Bloch, H. S., and Wackher, R. C., *Petrol. Refiner*, **34,** No. 2, 145–9 (1955). Grote, H. N., *Chem. Eng. Progr.*, **54,** No. 8, 43–8 (1958).

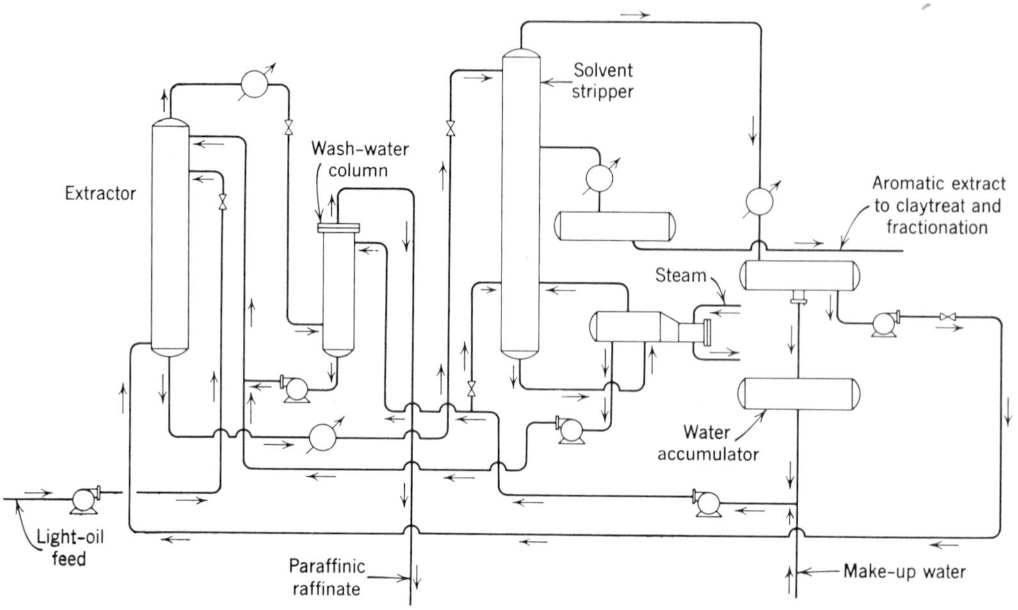

Fig. 14. Flow diagram of Udex extraction of hydrogenated light oil.

by Universal Oil Products Company. Although first employed for the extraction of aromatics from catalytically reformed gasolines, it has recently been installed for treatment of the products from two American coke-oven light-oil hydrogenation plants. The solvents employed are mixtures of various polyethylene glycols and water. The particular solvent used in any application depends on the nature of the feedstock and the completeness of recovery and the purity of the aromatics desired. Usually, mixtures of diethylene glycol and water are used. A study of the liquid system, benzene-heptane-diethylene glycol-water, showed that although water reduces the solubility of benzene in the solvent, selectivity is improved.[82] Figure 14 presents a flow diagram of the process. The feed stock to the unit is fed to an appropriate point in the extraction column. The solvent-aromatic stream goes to a distillation column where the aromatics are stripped from the solvent by means of steam. The paraffinic raffinate is washed with water to remove traces of the solvent.

Udex plants can be designed to recover aromatics to almost any degree of completeness and of almost any degree of purity. Since benzene is the most soluble, its recovery is generally very close to complete, whereas toluene and xylene are recovered to a lesser extent. The aromatic extracts are usually given a clay treatment to destroy traces of olefins and then are fractionally distilled.

For many years, the Edeleanu[83] process has been used to remove aromatics from kerosene and Diesel fuels. The solvent used is liquid sulfur dioxide, which is not selective enough for the preparation of pure benzene and toluene. During World War II, the process was modified to prepare a large proportion of the toluene derived from petroleum.[84] The modification consisted in washing the SO_2 extract with a high-boiling kerosene which displaced the paraffinic hydrocarbons with others more easily separable by distillation. This process is giving way to the Udex process.

Many other solvents have been proposed, among them furfural,[85] water at 1700 lb/in.^{2} and 300°C,[86] mixtures of methanol, ethylene

[82] Johnson, G. C., and Francis, A. W., *Ind. Eng. Chem.*, **46,** 1662–8 (1954).

[83] Edeleanu, L., Brit. Pat. 11,140 (1908).

[84] Moy, J. A. E., *Ind. Chemist*, **24,** 433–40, 505–8 (1948). Hightower, J. V., *Chem. Eng.*, **56,** No. 5, 139–41 (1949). Ratliff, R. A., and Strobel, W. B., *Oil Gas J.*, **53,** No. 4, 87–9 (1954); *Petrol. Refiner*, **33,** No. 5, 151–5 (1954); *Petrol. Engr.*, **26,** No. 12, C26–34 (1954).

[85] Kemp, L. C., Jr., Hamilton, G. B., and Gross, H. H., *Ind. Eng. Chem.*, **40,** 220–7 (1948).

[86] Arnold, G. B., and Coghlan, C. A., *Ind. Eng. Chem.*, **42,** 177–82 (1950).

glycol, and dipropylene glycol,[87] ethylene carbonate with glycerol and glycol,[88] dimethyl sulfate and oleum,[89] beta, beta′ thio-, imino-, and oxydipropionitriles,[90] and liquid ammonia.[91]

Perfluoro compounds, that is, organic compounds in which all of the hydrogen has been replaced by fluorine, are unusual in that they selectively extract paraffins from aromatics. This fact has been the basis for several patents.[92] Commercialization of these patents has been prevented by the high cost of the solvents.

Crystallization. Crystallization can be used to purify the major constituents of light oil. There is no evidence for the existence of solid solutions or molecular compounds of benzene or toluene and their paraffinic impurities at the concentrations found. It is possible, therefore, by chilling the appropriate fraction, to obtain crystals of benzene and toluene free of all paraffinic impurities. The problem, then, is the mechanical separation of the crystals from the mother liquor containing the impurities.

Keeling and Glowacki[93] patented a process for the continuous refining of benzene by crystallization in which the impure benzene is chilled to form a slurry. After centrifuging, the crystals are washed with pure benzene. The mother liquor and wash liquor are recycled and mixed with fresh benzene. The purity is controlled by the amount of mother liquor continuously discarded. Starting with a material having a freezing point of 4.78°C, they found that a side stream amounting to 5% of the feed resulted in a product with a freezing point of 5.32°C. For 10 and 20% sidestreams, the corresponding freezing points were 5.41 and 5.50°C. Keeling[94] later mixed the impure benzene with propane and isopropyl alcohol under pressure, released the pressure to evaporate the propane and thus chill the mixture, and separated the crystals from the mixture. Zimkin[95] refined crude benzene containing unsaturates by crystallizing from methanol. Little or no commercial application has been made of these processes because of the high cost compared to the cost of distillation and extraction. The crystallization of toluene is less attractive because of its low freezing point— −95°C. Crystallization is used to separate *m*- and *p*-xylene.

Other methods. Adsorption on porous solids, particularly silica gel, is a powerful tool for separating hydrocarbons and has been used extensively by Rossini and his colleagues[96] and discussed by Juhola.[97]

Two quotations from Rossini serve to explain this technique.

> The results of the work of the API Research Project 6 with regard to the influence of molecular size and structure on adsorbability, for adsorption from equivalent concentrations in the liquid phase, with silica gel as the adsorbent, may be summarized as follows:
>
> In order of decreasing adsorbability, the several structural components may be listed as (*a*) dinuclear or polynuclear aromatic; (*b*) mononuclear aromatic; (*c*) cycloparaffin and paraffin, with (*a*) and (*b*) being as a class very much more strongly adsorbed than (*c*).
>
> It is to be emphasized, however, that with mixtures in which the concentration of one of the two components becomes very small, the order of adsorbability will not necessarily remain as listed above for equivolume mixtures.

The separation of benzene and toluene from the products of catalytic reforming of petroleum fractions has been commercially practiced.

[87] Smith, A. S., and Funk, J. E., *Trans. Am. Inst. Chem. Engrs.*, **40**, 211–25 (1944).

[88] Badertscher, D. E., Francis, A. W., and Johnson, G. C., U.S. Pat. 2,688,645 (1954).

[89] Francis, A. W., U.S. Pat. 2,776,327 (1957).

[90] McKennis, A., U.S. Pat. 2,441,827 (1948). Wilkes, B. G., Jr., U.S. Pat. 2,439,534 (1948). Skinner, D. A., *Ind. Eng. Chem.*, **47**, 222–9 (1955).

[91] Fenske, M. R., *Chem. Eng. Progr.*, **53**, No. 1, 124 (1957). Fenske, M. R., McCormick, R. H., Lawroski, H., and Geier, R. G., *AIChE Journal*, **1**, 335–41 (1955).

[92] Egan, C. J., U.S. Pat. 2,582,197 (1952). Braae, B., U.S. Pat. 2,649,467 (1953); Swed. Pat. 136,843 (1952). Francis, A. W., and Johnson, G. C., U.S. Pat. 2,663,670 (1953). Ruehlen, F. N., U.S. Pat. 2,614,965 (1952).

[93] Keeling, W. O., and Glowacki, W. L., U.S. Pat. 2,400,883 (1946).

[94] Keeling, W. O., U.S. Pats. 2,438,368 (1948), 2,984,834 (1949).

[95] Zimkin, E., Brit. Pat. 586,790 (1947).

[96] Rossini, F. D., Mair, B. J., and Streiff, A. J., *Hydrocarbons from Petroleum*, Reinhold, New York, 1953, pp. 131–71.

[97] Juhola, A. J., *Physical Chemistry of Hydrocarbons*, Vol. II, A. Farkas, Ed., Academic Press, Inc., New York, 1953, pp. 317–89.

The Arosorb process[98] consists of adsorption on silica gel, displacement of saturates with butane or pentane, and elution of the aromatics with xylene. Light paraffin hydrocarbons can serve as elutriants if large quantities are used.[99]

"Molecular sieves" is the term applied to a series of synthetic zeolites, crystalline sodium, and calcium alumina-silicates, which have been heated to remove their water of hydration. The resulting crystals contain pores, 4 or 5 A in diameter. Substances with smaller molecular diameters are adsorbed; larger ones are not. Thus, normal paraffins can be removed from aromatics, but branched-chain compounds and cycloparaffins cannot be.[100]

In 1939, Bengen[101] discovered that linear aliphatic compounds formed crystalline adducts with urea. These are not molecular compounds with definite compositions. Rather, urea uses the straight chain as a mandrel and wraps itself into a helix. Later it was discovered that thiourea forms similar adducts with cycloparaffins and isoparaffins.[102] These facts have been made the basis of a method of separation which has been dubbed "extractive crystallization."[103] Molony has concluded that it was not suitable for purifying benzene.[104] Separation of rather pure benzene as a clathrate with nickel ammonium cyanide has been claimed.[105]

Paraffins can be completely removed by treatment with chlorine in the presence of ultraviolet light and the absence of iron.[106] Thiophene is chlorinated at the same time. The benzene may be distilled from the chlorinated products. Commercialization of this process has been retarded by the high cost of the stoneware and glass construction necessary. Chlorination may also be applied to alkylbenzenes.[107]

Passage of benzene vapor over chrome-alumina catalysts will dehydrogenate paraffins and naphthenes so that they can be removed by subsequent washing with H_2SO_4.[108] Pyrolysis at 650 to 900°C in the presence of hydrogen has been claimed.[41]

AMMONIA

The subject of ammonia recovery from coal gases bulks very large in the history of coal carbonization technology, since coal for many years was the only commercial source of this important chemical. However, the great growth of the synthetic ammonia industry in the past twenty years has reduced the coal carbonization industry's share in the ammonia market to a small portion of the total, although the total production from coal has changed comparatively little. The situation at present is that some coke plants, with little trouble, can sell ammonia in various forms at a profit whereas others cannot. The problems of the future of coke oven ammonia have been considered by Schulte-Mattler,[109] by Wilson and Wells,[110] and by Black.[111]

[98] Harper, J. I., Olsen, J. L., and Schuman, F. R., Jr., *Chem. Eng. Progr.*, **48**, 276–80 (1952). Davis, W. H., Harper, J. I., Weatherly, E. R., *Oil Gas J.*, **51**, No. 2, 112–4, 116, 119–20 (1952).

[99] Eagle, S., and Scott, J. W., *Petrol. Processing*, **4**, 881–4 (1949). Lipkin, M. R., U.S. Pat. 2,398,101 (1946). Hirschler, A. E., and Lipkin, M. R., U.S. Pat. 2,441,572 (1948).

[100] Lautensack, H., and Suval, M. L., Linde Air Products Co., private communication.

[101] Bengen, F. M., Ger. Pat. Appl. OZ123438 (1940). Bengen, F. M., and Schlenk, W., Jr., *Experentia*, **5**, 200 (1949).

[102] Angla, B., *Compt. rend.*, **224**, 402–4 (1947); *Ann. chim.* **4**, 639–98 (1949). Fetterly, L. C., U.S. Pats. 2,499,820 (1950), 2,520,715 (1950), 2,520,716 (1950), 2,569,984 (1951), 2,569,986 (1951).

[103] Zimmerschied, W. J., Dinerstein, R. A., Weitkamp, A. W., and Marshner, R. F., *Ind. Eng. Chem.*, **42**, 1300–6 (1950); *J. Am. Chem. Soc.*, **71**, 2947 (1949). Redlich, O., Gable, C. M., Beason, L. R., and Millar, R. W., *J. Am. Chem. Soc.*, **72**, 4153–62 (1950). Domask, W. G., and Kobe, K. A., *Petrol. Refiner*, **34**, No. 4, 128–33 (1955). Swern, D., *Ind. Eng. Chem.*, **47**, 216–21 (1955). Fetterley, L. C., *Petrol. Refiner*, **34**, No. 4, 134–7 (1955).

[104] Molony, O. W., National Benzole Co. Research Rept. 55–5.

[105] Jones, A. L., and Fay, P. S., U.S. Pat. 2,732,413 (1956).

[106] United States Steel Corporation, Brit. Pat. 785,659 (1957).

[107] Reiff, O. M., and Moore, A. M., U.S. Pat. 2,464,769 (1949).

[108] Wilson, P. J., Jr., Wells, J. H., and Liebel, C. R., U.S. Pat. 2,455,634 (1948). Concordia Bergbau A.-G., Brit. Pat. 722,180 (1955).

[109] Schulte-Mattler, W., *Brennstoff-Chem.*, **23**, 91–6, 103–7, 115–19 (1942).

[110] Wilson, P. J., Jr., and Wells, J. H., *Blast Furnace Steel Plant*, **36**, 806–12, 961–4 (1948).

[111] Black, J. H., *AIME, Proc. Blast Furnace, Coke Oven and Raw Materials Comm.*, **18**, 259–90 (1959).

Since recovery of ammonia is becoming less attractive economically, consideration must be given to destroying it or simply not recovering it. The latter course has been forced on some plants during emergencies and has been tried designedly. Conclusions were that the greater part of the ammonia must be removed to prevent stoppages from ammonium bicarbonate in gas lines, to prevent the formation of resinous materials, and to prevent widespread corrosion, particularly in the light-oil plant. Destroying as much as 95% of the ammonia formed in the oven by adding iron oxide to the upper portion of the coal has been claimed in a recent patent.[112] This claim is not substantiated by experience in the production of "iron coke." The addition of as much as 40% of iron ore to coal before carbonization resulted in no noticeable change in the yield of ammonia.[113] According to Schulte, removing the ammonia from the gas by water washing, distilling, and burning the vapors result in such a lowering of capital costs that it may be advantageous to install the necessary equipment in a new plant rather than using the conventional saturator.[114]

Yields of Ammonia

The subject of the yield of ammonia from coal carbonization has been discussed in Chapter 27 of the original edition of this work.[115] The diversity of observers, coals, and methods there presented make it impossible to draw any satisfying conclusions. More recently, Reynolds and Wolfson,[116] in the admirably systematic way characteristic of the Bureau of Mines, have studied the yields of ammonia obtained from fifty-eight American coals carbonized by the BM-AGA[117] method at 800, 900, and 1000°C. It was found very roughly that ammonia yields varied inversely with geological age and directly with moisture and nitrogen content, but no close correlations could be made. The yields were high for Washington, Alabama, and Utah coals, about average for Kentucky and Oklahoma coals, and low for Pennsylvania and West Virginia coals.

Recovery

As Ammonia Liquor

The recovery of ammonia by scrubbing with water and distilling the resulting solution to prepare a concentrated ammonia liquor is still practiced to some extent. The costs of shipment of the liquor limit its economical production to especially situated plants.

The partial pressures of ammonia over aqueous solutions and its rate of absorption have been determined by several observers.[118]

As Ammonium Sulfate

Most recent publications about ammonium sulfate have been concerned with improvements in apparatus for the purposes of lowering

112 Bergwerksgellschaft Hibernia A.-G., Brit. Pat. 746,697 (1956).

113 Russell, C. C., and Whitstone, P., *AIME, Proc. Blast Furnace, Coke Oven & Raw Materials Comm.*, **14**, 93–116 (1955).

114 Schulte, E. V., *AIME, Proc. Blast Furnace, Coke Oven & Raw Materials Comm.*, **17**, 237–46 (1958).

115 Hill, W. H., *Chemistry of Coal Utilization*, John Wiley & Sons, New York, 1945, vol. 2, pp. 1017–31.

116 Reynolds, D. A., and Wolfson, D. E., *U.S. Bur. Mines, Rept. Invest.* No. 4526 (1949), 15 pp.

117 Fieldner, A. C., and Davis, J. D., *U.S. Bur. Mines, Monograph* No. 5 (1934), 164 pp. Reynolds, D. A., and Holmes, C. R., *U.S. Bur. Mines, Tech. Paper* **685** (1946), 35 pp.

118 Kharbanda, P., *Coke and Gas*, **18**, 139–41 (1956). Pexton, S., and Badger, E. H. M., *J. Soc. Chem. Ind. (London)*, **57**, 106T (1938). Badger, E. H. M., and Wilson, D. S., *J. Soc. Chem. Ind. (London)*, **66**, 84–6T (1947). van Krevelen, D. W., and Huntjens, F. J., *Rec. trav. chim.*, **68**, 191–216 (1949). Gibadlo, F., *Gas Age*, **105**, No. 5, 28–30 (1950). Peakin, F. H., *J. Soc. Chem. Ind. (London)*, **59**, 81–7T (1940). Kronacher, H., *Gas- u. Wasserfach*, **86**, 193–7 (1943). Terres, E., Attig, W., and Tscherter, F., *Gas- u. Wasserfach*, **98**, 512–16, 577–82 (1957). Eymann, C., *Gas- u. Wasserfach*, **84**, 573–9 (1943); **89**, 10–13 (1948); **90**, 505–12, 534–8, 568–70, 577–8 (1949). Koopsmans, H., and Debey, J. J. H., *Gas C's Gravenhagel*, **63**, 159–63 (1943). Brodovich, A. I., and Krasnovskaya, R. V., *Koks i Khim.*, **1941**, No. 1, 27–30. Molstad, M. C., McKinney, J. F., and Abbey, R. G., *AIChE, Trans.*, **39**, 605–62 (1943). Dodge, B. F., and Dwyer, O. E., *Ind. Eng. Chem.*, **33**, 485–92 (1941). Williams, G. C., Akell, R. B., and Talbott, C. P., *Chem. Eng. Progr.*, **43**, 585–96 (1947). Bond, J., and Donald, M. B., *Chem. Eng. Sci.*, **6**, 237–44 (1957). Dunn, B. D., *Koks i Khim.*, **1937**, No. 2, 44–50.

pressure drops and increasing the size of the crystals.[119]

A very large part of the cost of production of ammonium sulfate is the purchase of sulfuric acid. Cheaper sources of the sulfate ion have been sought with disappointing results. Gypsum—$CaSO_4 \cdot 2H_2O$—is a cheap and easily obtainable material which can react with ammonia and carbon dioxide to form insoluble calcium carbonate and ammonium sulfate.

$$2NH_3 + CO_2 + H_2O + CaSO_4 \rightarrow CaCO_3 + (NH_4)_2SO_4.$$

Klempt[120] described laboratory experiments in which ammonia mixed with CO_2 and water vapor was bubbled through aqueous slurries of $CaSO_4$. Conversion of 90 to 95% of the ammonia was achieved. The presence of H_2S was not harmful. However, when coke-oven gas was used, poor recoveries were obtained because of the low concentration of ammonia. It was concluded that the process would be practical only if the ammonia were first concentrated and additional carbon dioxide were supplied.

[119] Otto, C., U.S. Pats. 2,435,898 (1948), 2,509,520 (1950), 2,759,796 (1956); Brit. Pat. 647,742 (1950). Otto, H., *AIME, Proc. Blast Furnace, Coke Oven & Raw Materials Comm.*, **8,** 50–60 (1949); Brit. Pat. 768,345 (1957). Mantel, W., and Hansen, H., *Brennstoff-Chem.*, **33,** 69–75 (1952). Schulte, E. V., U.S. Pat. 2,595,104 (1952). Tiddy, W., U.S. Pat. 2,383,171 (1945); Can. Pat. 421,160 (1944). Wethly, F., *Blast Furnace Steel Plant*, **33,** 976–80 (1945). Thau, A., *Oel u. Kohle*, **37,** 217–29, 302–15 (1941); **39,** 240–56 (1943). Appleby-Frodingham Steel Co., Ltd., and Stephenson, H. P., Brit. Pat. 542,493 (1942). Shevlyagin, V. I., *Coke and Chem. (U.S.S.R.)*, **8,** No. 10, 44–5 (1938). Weber, T., *Glückauf*, **75,** 890–1 (1939). Koppers Co., Inc., Brit. Pats. 653,325 (1951), 773,147 (1957). Maleson, L., U.S. Pat. 2,324,222 (1944). Biddulph-Smith, T., *Coke Oven Managers' Assoc. Year Book*, **1942,** 152–67. Deshalit, G. I., Soldatchko, I. S., and Romanii, E. V., *Koks i Khim.*, **1958,** No. 8, 33–4. Kennaway, T., Wood, C. W., and Box, P. L., *Gas World*, **143,** No. 3733, Coking Sect., **48,** No. 517, 49–58 (1956). Otto, A., *Gas- u. Wasserfach*, **97,** 369–72 (1956). Sorkin, M. M., Gavrilova, G. E., Mezhueva, E. A., and Kogan, M. G., *Koks i Khim.*, **1956,** No. 1, 55–6. N. V. Otto-Simon-Carves, Dutch Pat. 79,457 (1955).

[120] Klempt, W., *Ber. Ges. Kohlentech.*, **5,** 401–13 (1950).

Ferrous sulfate in the form of pickle liquor seems attractive for this purpose since it is a waste product whose disposal is a problem. Hoak[121] discussed various processes which have been proposed and the difficulties which are encountered in applying them. The desired reaction is

$$2NH_4OH + FeSO_4 \rightarrow (NH_4)_2SO_4 + Fe(OH)_2$$

Unfortunately, the hydrogen sulfide present reacts to form finely divided, difficultly filterable ferrous sulfide, and hydrogen cyanide forms complex ferrocyanides. In the Elzi[122] process, pickle liquor is neutralized with scrap iron and is used to scrub coke-oven gas countercurrently so that the pH is raised to 7.0 to 7.5. In a second tower, the pH is raised to 8.0 to 8.5 with vapors from the ammonia still. Aeration in a third tower oxidizes the iron compounds to ferric oxide, which is filtered off. The filtrate may be evaporated or sent to the saturators. Tiddy[123] described a modification of this process in which most of the precipitated iron compounds are removed before aeration. In a third process,[124] purified ammonia and pickle liquor are fed to a special reactor where the iron is precipitated completely as ferrosoferric oxide. When applied to coke-oven gas, the process is modified to oxidize the ferrous sulfide formed to ferric oxide and sulfur.[125] Other modifications have been described.[126]

A means of recycling sulfuric acid has been described by Delaplaine and McCullough.[127] Ammonium sulfate is heated in the presence of inert particles impregnated with zinc oxide. Ammonia and water vapor are driven off. The zinc sulfate formed is decomposed at a higher temperature to yield sulfur trioxide and zinc

[121] Hoak, R. D., *Disposal of Spent Sulfate Pickling Solutions*, Ohio River Valley Water Sanitation Comm., Cincinnati, 1952, pp. 55–8.

[122] Elzi, F. A., U.S. Pat. 2,427,555 (1947).

[123] Tiddy, W., U.S. Pat. 2,511,306 (1950).

[124] Hoak, R. D., U.S. Pat. 2,529,874 (1950).

[125] Hoak, R. D., U.S. Pat. 2,712,980 (1955).

[126] Meissner, H. P., and Hyde, R. W., U.S. Pat. 2,584,280 (1952). Burden, F. A., and Robinson, C. H., Brit. Pat. 704,690 (1954). Prchlik, J., Hlinak, L., and Kartakova, C. D., *Paliva*, **35,** 157–62 (1955). Dixon, T. E., *Iron Age*, **175,** No. 12, 91–3 (1955). Jäppelt, A., Doerges, A., and Schade, H., U.S. Pat. 2,848,302 (1958).

[127] Delaplaine, J. W., and McCullough, R. F., *Chem. Eng. Progr.*, **51,** 499–503 (1955).

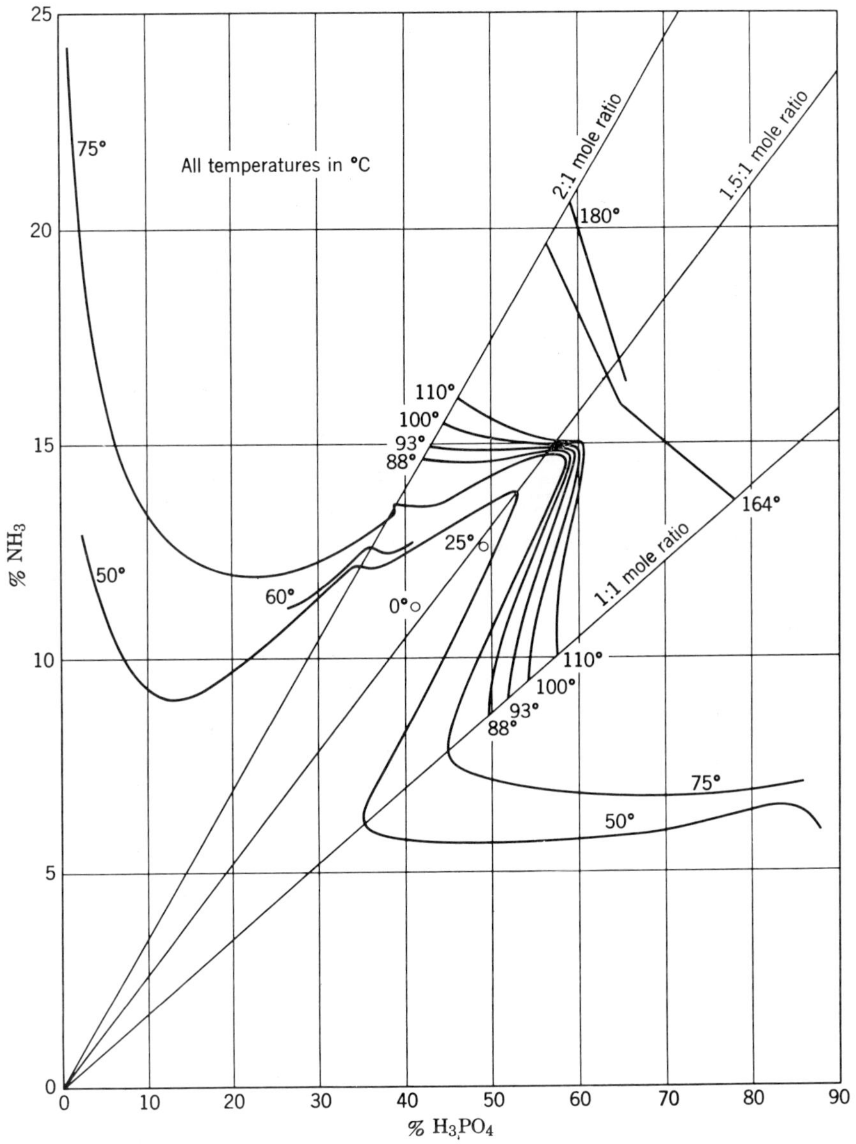

Fig. 15. Solubility diagram of the system: NH_3—H_3PO_4—H_2O.

oxide. This process eliminates the cost of sulfuric acid, but adds additional equipment to the usual ammonia recovery system. Extra energy is also required to decompose ammonium sulfate. This is essentially a method of preparing high-grade concentrated ammonia liquor.

As Ammonium Phosphates

The advantages of ammonium phosphates over ammonium sulfate as fertilizers are obvious. Three-quarters of the weight of $(NH_4)_2SO_4$ is a sulfate ion whose only function is to fix the ammonia in solid form. The ammonium phosphates contain no such inert material and supply two of the three elements needed in a "complete" fertilizer. The phosphates are particularly useful in sugar beet fields that are naturally rich in potassium, and for this reason the production of phosphates has been most practiced at coke plants in the western United States. Early attempts at preparing these salts in an ammonia saturator were unsuccessful because the wet-process phosphoric acid used contained relatively large quantities of alumina which resulted in unfilterable gels. With the

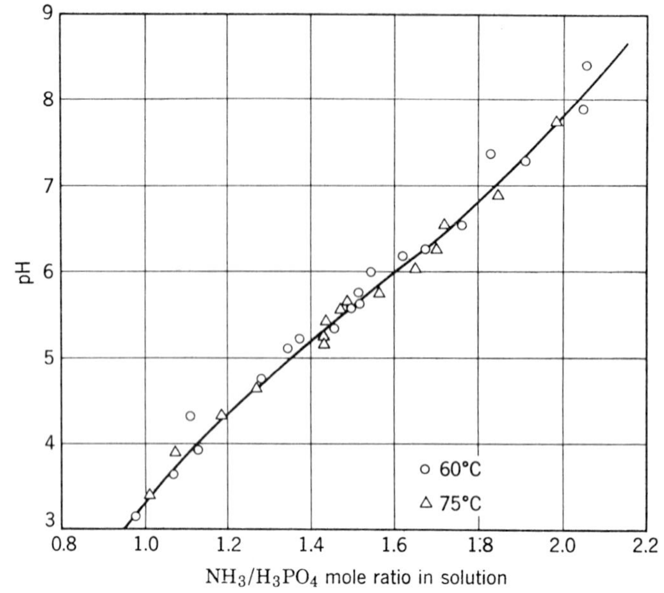

Fig. 16. pH of saturated solutions in system: $NH_3—H_3PO_4—H_2O$.[129]

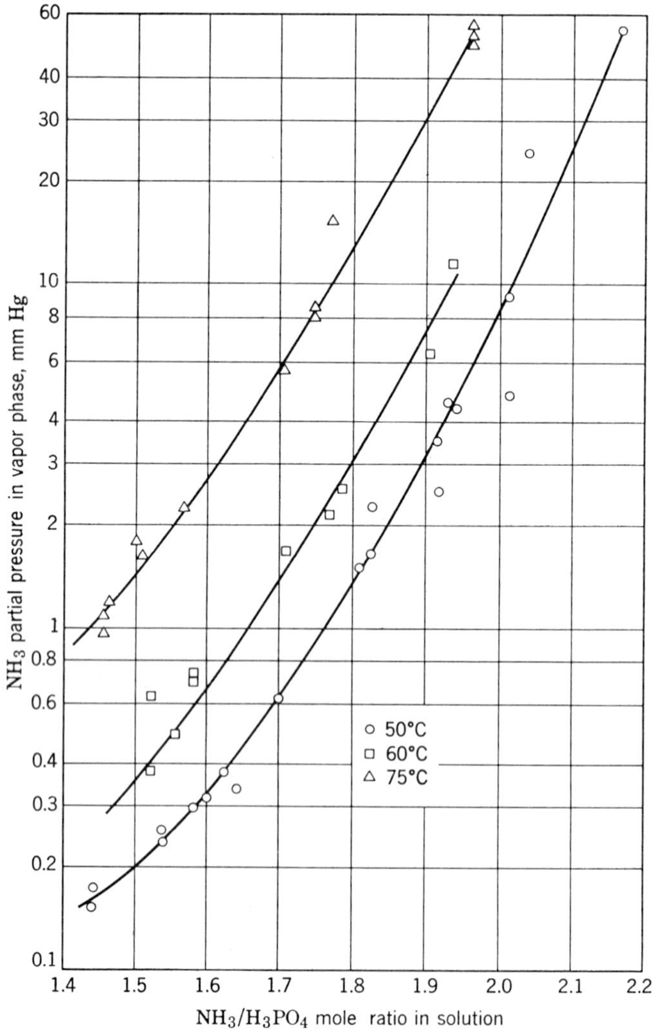

Fig. 17. Partial pressure of NH_3 over saturated solutions in system: $NH_3—H_3PO_4—H_2O$.

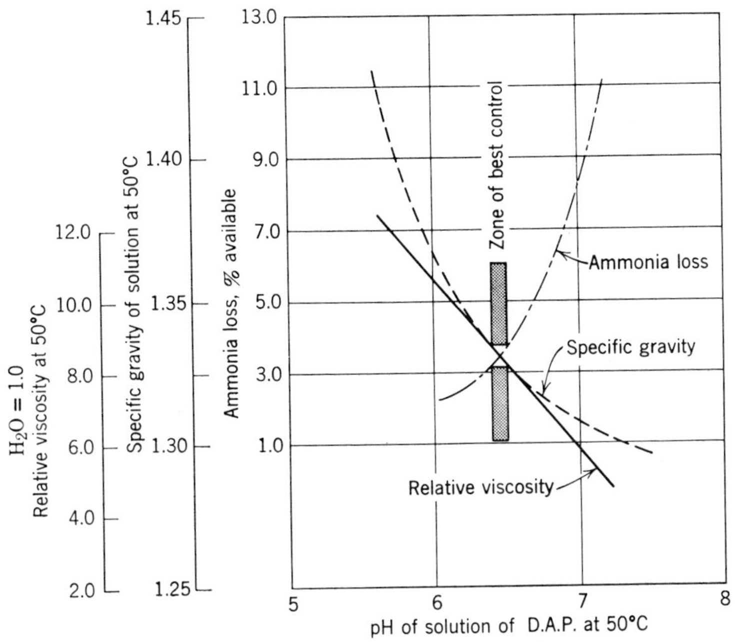

Fig. 18. Effect of pH in system: NH_3—H_3PO_4—H_2O.[130]

advent of the purer electric-furnace acid, it became possible to prepare rather pure mono- and diammonium phosphate.

The solubility relationships in the system NH_3—H_3PO_4—H_2O, presented in Fig. 15 are compiled from a number of sources.[128] The locations of the points of inflection depend somewhat on temperature, but generally at ammonia: phosphoric acid mole ratios between 1.5 and 2.0 the solid phase is diammonium phosphate. At mole ratios below 1.5, monoammonium phosphate crystallizes; above ratios of 2.0, the solid is the relatively unstable salt, $(NH_4)_3PO_4 \cdot 2(NH_4)_2HPO_4$. The dependence of pH on NH_3:H_3PO_4 mole ratio is shown in Fig. 16.[129] Experiments of these authors showed that good crystals of pure $(NH_4)_2HPO_4$ were obtained in the pH range of 5.8 to 6.6. Below pH 5.8, the product contained monoammonium phosphate. Above pH 6.6, considerable amounts of ammonia remained in the exit gas as is shown in Fig. 17.

Pearson[130] has concluded that diammonium phosphate is best crystallized from a saturator bath held at a pH of 6.4 to 6.5. As shown in Fig. 18, the ammonia losses become high above this range. When the solution has a lower pH, its higher viscosity and specific gravity hinder the settling of the crystals.

[128] Apfel, O., Dissertation, Tech. Hochschule, Darmstadt, 1911. Bel'chev, F. V., and Bergman, A. G., *J. Applied Chem.* (*U.S.S.R.*), **17,** 520–6 (1944). Bergman, A. G., and Bachkareff, P. F., *Bull. acad. sci. U.R.S.S., Classe math. nat., Ser. chim.*, **1938,** 237–66; *J. Applied Chem.* (*U.S.S.R.*), **10,** 1531–6 (1937). Brosheer, J. C., and Anderson, J. F., Jr., *J. Am. Chem. Soc.*, **68,** 902–4 (1946). Polosin, V. A., and Ozolin, R. K., *Kalii* (*U.S.S.R.*), No. 10, 31–4 (1937). Buchanan, G. H., and Winner, G. B., *J. Ind. Eng. Chem.*, **12,** 448–51 (1920). D'Ans, J., and Schreiner, O., *Z. physik. Chem.*, **75,** 95–107 (1910). Jänecke, E., *Z. physik. Chem.*, **127,** 71–92 (1927); **A177,** 7–16 (1936). Muromtsev, B. A., *Kalii* (*U.S.S.R.*), No. 1, 36–9 (1937). Muromtsev, B. A., and Nazarova, L. A., *Bull. acad. sci. U.R.S.S., Classe sci. math. nat., Ser. chim.*, **1938,** No. 1, 177–84. Polosin, V. A., *J. Phys. Chem.* (*U.S.S.R.*), **20,** 1471–4 (1946). Polosin, V. A., and Shakhparonov, M. I., *J. Phys. Chem.* (*U.S.S.R.*), **21,** 119–23 (1947). Ross, W. H., Merz, A. R., and Jacob, K. D., *Ind. Eng. Chem.*, **21,** 286–9 (1929). Shpunt, S. Y., *J. Applied Chem.* (*U.S.S.R.*), **13,** 9–18 (1940). Vol'fkovich, S. I., Berlin, L. E., and Mantsev, B. M., *Trans. Sci. Inst. Fertilizers Insectofungicides U.S.S.R.*, No. 153, 228–41 (1940).

[129] Thompson, H. L., Miller, P., Johnson, R. M., McCamy, I. W., and Hoffmeister, G., Jr., *Ind. Eng. Chem.*, **42,** 2176–82 (1950).

[130] Pearson, C. K., Paper presented at the meeting of the Blast Furnace and Coke Association of the Chicago District, Chicago, January 27, 1955.

If complete ammonia removal is desired while diammonium phosphate is being produced, two saturators can be run in series. The second is kept at a lower pH, and its contents are transferred to the first for conversion to the diammonium salt.[131]

Other Ammonium Salts

At first glance, the production of ammonium nitrate would appear to be attractive since this material, too, contains no inert material. The nitric acid needed, furthermore, can be manufactured on the site by oxidation of half the ammonia.[132] The hygroscopic nature of NH_4NO_3, its instability in the presence of certain organic matter, and the corrosiveness of nitric acid, however, have made its manufacture in the coke plant uneconomical. Its production in a saturator nevertheless has been proposed.[133] It is claimed that the reduction of nitric acid by hydrogen sulfide can be prevented by the addition of $(NH_4)_2HPO_4$.[134]

Ammonium chloride makes up the greater portion of the "fixed" ammonia in ammonia liquor. Its recovery in a multiple effect evaporator has been patented.[135] Ammonium chloride darkens when exposed to air because of the presence of tar acids.[136] These acids can be removed by heating the dried salt to 325°C and filtering off the char thus formed from a solution of the salt.[137] Traces of salts of organic acids that are harmful in galvanizing may be washed out with alcohol.[138]

Ammonium bicarbonate, NH_4HCO_3, can be prepared by adding CO_2 to concentrated ammonia liquor.[139] By heating under pressure, ammonium carbamate, $NH_4CO_2NH_2$, is obtained.[140]

Shaw[141] proposed adding an equal volume of ethanol to ammonia liquor concentrated by evaporation. Two liquid phases result. Ammonium thiosulfate may be separated from one, and ammonium thiocyanate from the other.

Anhydrous Ammonia

A method of producing anhydrous ammonia has been patented by Holowaty.[142] Coke-oven gas is treated in a conventional saturator with ammonium bisulfate solution. The resulting ammonium sulfate is dried and then heated to 650°F to produce gaseous ammonia and molten ammonium bisulfate.

PYRIDINE BASES

Although a large part of the organic bases resulting from coal carbonization are found in the tar, the relatively high prices they command have encouraged the recovery of the quantity that remains in the gas after cooling. The sulfuric acid in the ammonia saturator will remove most of the tar bases in the gas if the pyridine in the saturator liquor is continuously or intermittently removed. This removal is generally accomplished by neutralizing a portion of the liquor with the vapors from the ammonia still, by steam distilling the liberated bases, and by separating the bases from the aqueous condensate. This separation may be made more complete by the addition of salts, with ammonium sulfate being most often used. Although pyridine is completely miscible with pure water, its solubility in a saturated ammonium sulfate solution at 25°C is only 0.37%.[143] Details of the variations introduced into this procedure are given in the papers listed.[144] Pyridine bases can also be removed

[131] Helm, E. J., U.S. Pat. 2,921,837 (1960).

[132] Umbach, H., *Chem. Fabrik*, **1940,** 253–6.

[133] Dohse, H., Klempt, W., and Umbach, H., Ger. Pat. 715,379 (1941). Klempt, W., Schroter, J., Umbach, H., and Wiebeck, E., *Ber. Ges. Kohlentech.*, **5,** 1–97 (1939).

[134] Dohse, H., *Ber. Ges. Kohlentech.*, **5,** 158–66 (1946).

[135] Heimbach, C. H., and Thompson, N. B., U.S. Pat. 2,179,183 (1940).

[136] Walker, R., and Applebee, H. C., *Gas World*, **115,** 167–9 (1941).

[137] Bone, C. H., *Blast Furnace Steel Plant*, **33,** 1507–10 (1945).

[138] Infante Arias, J., *Rev. ind. y fabril* (*Madrid*), **7,** 92–7 (1952).

[139] Klempt, W., *Ber. Ges. Kohlentech.*, **5,** 414–20 (1950); Ger. Pat. 809,192 (1951). Gras, W., Brit. Pat. 710,617 (1954).

[140] Koeppel, C., Ger. Pat. 734,076 (1943).

[141] Shaw, J. A., U.S. Pat. 2,386 985 (1945).

[142] Holowaty, M. O., U.S. Pat. 2,899,277 (1959).

[143] Peakin, F. H., *J. Soc. Chem. Ind.*, **59,** 81–7T (1940).

[144] Gluzman, L. D., and Grigor'ev, S. M., *Coke and Chem.* (*U.S.S.R.*), **11,** No. 2, 29–35 (1941). Kvasha, A. S., Glaser, I. G., and Podolskii, N. S., *Coke and Chem.* (*U.S.S.R.*), **11,** No. 2, 35–6 (1941). Borts, A. G., and Farafonov, V. M., *Coke and Chem.* (*U.S.S.R.*), **11,** No. 2, 36–8 (1941). Ustyugov, P. A., *Coke and Chem.*, (*U.S.S.R.*), **11,** No. 2, 39–40 (1941). Wald, M. D., and Karnatz, F. A.,

from the gas leaving the saturator by washing with concentrated sulfuric acid.[145] When recovery of diammonium phosphate is practiced, pyridine can be recovered by passing the gas into a second absorber containing a phosphoric acid solution kept below pH 4.[146] The pyridine can be recovered from the solution by vaporization.

The small amount of pyridine bases in crude light oil can be recovered by washing with dilute sulfuric acid.[147]

HYDROGEN CYANIDE

Hydrogen cyanide is not usually recovered from coke-oven gas since most plants have no compelling reason to do so. Allowed to remain in the gas, it corrodes the distribution system. In Europe, where coal gases are still widely distributed, there is more interest in its removal than in the United States where most coke-oven gas is consumed in steel mills. Hydrogen cyanide is recovered at only one United States coke plant, although many plants remove most of it from the gas coincidental to the removal of hydrogen sulfide. The very common system of effecting the final cooling of gas by direct contact with water removes most of the HCN. The Seaboard process of H_2S removal does the same and converts most of it to thiocyanate. Any other process which involves contact of the gas with alkaline media removes HCN from the gas, but its final disposal varies. In at least one plant that utilizes the vacuum carbonate process, HCN in fairly concentrated form is separated from the H_2S before it is burnt to make sulfuric acid and is then evaporated back into the fuel gas stream.[148]

The removal of as much as 96% of HCN from coal gas by water washing requires the use of 20 to 30 gallons per thousand cubic feet.[149] The necessary rate is dependent on the temperature since the partial pressure of HCN in aqueous solution varies greatly in the range of ordinary scrubbing temperatures (Fig. 19).[150,151] The aqueous solution can be freed of HCN by using it to cool producer gas, or it can be distilled to produce highly concentrated HCN.[152]

If HCN is recovered concomitant to H_2S removal by scrubbing with alkaline solutions, there is a problem concerning the location of the scrubber. It is probably best placed after the light-oil scrubber, which requires that open cooling towers for the final cooler water be replaced by indirect cooling coils so that the recirculated water after a short time builds up such a concentration of HCN that it removes no more from the gas.[153] If the scrubber is placed between the ammonia absorber and the final cooler, the alkaline absorbent will pick up a small amount of naphthalene, which is enough if the vacuum carbonate process is used to very quickly stop the condenser that cools the actifier vapors. Washing these with a sidestream of wash oil from the light-oil recovery system has been tried.[154] Deposition of naphthalene in the condenser was prevented, but the wash oil soon became polluted with large quantities

U.S. Pat. 2,279,585 (1942). Schutt, J. W., U.S. Pat. 2,311,134 (1943). Boardman, H. S., *Blast Furnace Steel Plant*, **31**, 332–6 (1943). Van der Hoeven, B. J. C., U.S. Pat. 2,393,666 (1946). Wells, J. H., and Wilson, P. J., Jr., U.S. Pat. 2,435,663 (1948). Ishibashi, F., and Emoki, R., *Coal Tar (Tokyo)*, **2**, 169–74 (1950). Takeishi, M., and Watabe, F., *Coal Tar (Tokyo)*, **2**, 10–2 (1950). Kurata, J., *Coal Tar (Tokyo)*, **2**, 398–401 (1950). Yamamoto, A., Kawashima, K., and Ishiwatari, F., *Coal Tar (Tokyo)*, **2**, 163–8 (1950). Hoek, T. J. J., U.S. Pat. 2,564,433 (1951). Hisamura, M., Okazaki, F., and Kato, I., *Coal Tar (Tokyo)*, **4**, 282–8 (1952). Tanemura, T., Higuchi, S., Tamakim, K., and Nagao, M., *Coal Tar (Tokyo)*, **6**, 61–4 (1954). Meredith, H. J., *Am. Gas Assoc., Proc.*, **23**, 656–67 (1941); *Gas J.*, **237**, 166, 169 (1942). Sweeney, M. P., U.S. Pat. 2,799,678 (1951). Papov, K. P., and Karvatskaya, M. A., *Koks i Khim.*, **1958**, No. 2, 41–3. Shimazu, T., and Hosoi, F., *Coal Tar (Tokyo)*, **4**, 9–18 (1952).

145 Klempt, W., and Dohse, H., Ger, Pat. 725,390 (1942).

146 Langguth, R. P., U.S. Pat. 2,849,288 (1958).

147 Nettlenbusch, L., *Brennstoff-Chem.*, **30**. 185–6 (1949). Thomas, A., *Compt. rend. congr. ind. gaz, 70th Congr., Paris*, **1953**, 984–91. Wiszniowski, K., *Koks Smola Gaz*, **1**, 68–71 (1956).

148 Kurtz, J. K., *Air Pollution*, F. S. Mallette, Ed., Reinhold, New York, 1955, pp. 215–21.

149 Eymann, C., *Gas-u. Wasserfach*, **88**, 86–9 (1947).

150 Grosskinsky, O., *Handbuch des Kokereiwesens*, Knapp, Düsseldorf, 1958, Vol. 2, p. 327.

151 Horiuchi, J., and Tanabe, K., *J. Research Inst. Catalysis, Hokkaido Univ.*, **1**, 117–31 (1951); *Shokubai*, **7**, No. 8, 47–51 (1951).

152 Mitchell, J., Hill, W. H., and Gollmar, H. A., U.S. Pat. 2,419,225 (1947). Klempt, W., and Umbach, H., Ger. Pat. 744,971 (1943).

153 Schreiber, F. D., U.S. Pat. 2,386,359 (1945).

154 Denig, F., U.S. Pat. 2,458,505 (1949).

under pressure of coke-oven gas with potassium carbonate. He concluded that selective adsorption on active carbon or silica gel was not so effective in separating HCN and H_2S as a wash with water followed by distillation.

Various processes involving the formation of ferrocyanides have continued to be used in Europe. In an investigation of the old Bueb process, in which gas containing both NH_3 and HCN is washed with ferrous sulfate, the resulting mud was found to contain $Fe(NH_4)_2$-$Fe(CN)_6$, FeS, $(NH_4)_4Fe(CN)_6$, $(NH_4)_2SO_4$, $(NH_4)_2S$, NH_4CNS, $(NH_4)_2CO_3$, and NH_4Cl.[159] Treatment of the mud with lime results in $Ca_2Fe(CN)_6$, which when it reacts with KCl yields $CaK_2Fe(CN)_6$. This product can be converted into $K_4Fe(CN)_6$ by means of K_2CO_3. $Na_4Fe(CN)_6$ can be prepared from the first-mentioned calcium salt by treatment with Na_2CO_3.[160] Sodium ferrocyanide is prepared directly by scrubbing coke-oven gas with sodium carbonate and iron carbonate.[161] Contact with iron filings in the presence of K_2CO_3, Na_2CO_3 or ammonia liquor also results in the formation of alkali ferrocyanides.[162] The reactions involved are:

$$2K_2CO_3 + Fe + 6HCN \rightarrow K_4Fe(CN)_6 + H_2 + 2CO_2 + 2H_2O$$

$$4NH_4OH + Fe + 6HCN \rightarrow (NH_4)_4Fe(CN)_6 + H_2 + 4H_2O.$$

The ammonium ferrocyanide reacts with the simultaneously formed ferrous sulfide to yield $(NH_4)_2Fe[Fe(CN)_6]$. When oxidized with air, this forms $NH_4Fe[Fe(CN)_6]$. This substance can be used to scrub H_2S from the gas.

The preparation of ammonium thiocyanate by washing gas containing HCN with ammoniacal suspensions of sulfur is a very old practice. Modifications that have been described include the purification of crude thiocyanate solution by blowing with air at elevated temperatures.[163] It has been made the basis of an unusual, if marginally economical, way of preparing ammonium sulfate[164]:

$$NH_4CNS + H_2SO_4 + H_2O \rightarrow (NH_4)_2SO_4 + COS$$

The solution resulting from treating coke-oven gas with ammonia water is oxidized with a mixture of air and COS at 160°C and 25 atm to form additional $(NH_4)_2SO_4$ and CO_2.

Recovery of HCN from the final cooler water by precipitation as zinc cyanide has been proposed.[165] The resulting sludge also contains zinc sulfide, calcium carbonate, and calcium hydroxide. Acidification to such an extent that HCN is liberated, but not H_2S, allows the recovery of the zinc sulfide.

THE DISPOSAL OF AMMONIACAL LIQUOR

The cooling of coke-oven gas generates a large quantity of ammonia liquor which has little usefulness aside from its ammonia content. With the increase in population in industrial areas and the growing realization that streams and rivers have some utility beyond being convenient dumping places for household sewage and industrial wastes, there has been increased interest in the rendering of coke plant wastes less objectionable.

The quantities of contaminated wastes from seventeen coke plants have been listed by the Ohio River Valley Water Sanitation Commission as shown in Table 20.[166] Table 21 gives analyses of these wastes. These tables show that the ammonia still waste is the real offender both as to quality and quantity.

Crude ammonia liquor and ammonia still waste are objectionable because they contain specific substances that are toxic to aquatic life and that deplete the streams' dissolved oxygen. Marsden[167] has listed in the relative toxicity and oxygen absorption of different constituents in coke plant ammonia liquors (see Table 22). Free ammonia, sulfide, and cyanide

[159] Barlet, C., *Compt. rend. congr. ind. gaz, 65th Congr. Paris*, **1948**, 325–32.

[160] Michel, J., *Compt. rend. congr. ind. gaz, 66th Congr. Lyon*, **1949**, 647–54.

[161] Gollmar, H. A., and Meredith, H. J., U.S. Pat. 2,378,403 (1945).

[162] Leith, F., *Brennstoff-Chem.*, **22**, 27–34 (1941).

[163] Beier, H., Ger. Pat. 708,621 (1941). Saeki, O., Kagami, H., and Fujita, K., *Coal Tar (Tokyo)*, **7**, 309–13 (1955). Saeki, O., and Fujita, K., *Coal Tar (Tokyo)*, **7**, 361–4 (1955). Hashimoto, S., Kiriya, Y., and Saeki, O., Japan. Pat. 4784 (1954).

[164] Bergwerksband zur Verwertung von Schutzrechten der Kohlentechnik G.m.b.H., Ger. Pat. 909,138 (1954).

[165] Washburne, R. N., and Spence, L. U., U.S. Pat. 2,267,293 (1942).

[166] Ohio River Valley Water Sanitation Commission, *Reducing Wastes from Coke Plants*, Cincinnati, 1953, 36 pp.

[167] Marsden, A., *Gas World*, **133**, 524–7, 653–6, 672–7 (1951).

TABLE 20

VOLUMES OF CONTAMINATED WASTES FROM COKE PLANTS[166]

Source of Waste	Gallons/Ton of Coal Carbonized		
	Min	Max	Avg
Ammonia still with dephenolizer	18.5	41.9	31.4
Ammonia still without dephenolizer	23.0	75.0	36.8
Light-oil decanter	3.0	29.0	13.0
Miscellaneous	0.12	0.25	0.2

TABLE 21

ANALYSES OF COKE PLANT WASTES[166]

	Concentration in ppm								
	Ammonia Still Wastes						Light-Oil Decanter Waste		
	With Dephenolizer			Without Dephenolizer					
Determination	Min	Max	Avg	Min	Max	Avg	Min	Max	Avg
Phenols	40	400	158	350	2820	1320	28	164	77
Ammonia	20	255	110	21	470	155	2	34	17
Chlorides	2350	7500	5400	1150	5500	4350	7	107	51
Sulfates	215	1330	744	145	940	468	5	135	37
pH	9.5	11.5	10.0	8.8	12.8	11.3	5.2	8.5	6.2

TABLE 22

TOXICITY AND OXYGEN ABSORPTION OF AMMONIA LIQUOR CONSTITUENTS[167]

Constituent	Toxicity	Oxygen Absorption*	Rate of Biochemical Oxidation
Free ammonia	High	Nil	Very slow
Fixed ammonia	Nil	Nil	Very slow
Sulfide	High	1345	Fast
Cyanide	High	600	Very slow
Thiosulfate	Low	856	Fast
Phenol	Medium	1780	Fast
Higher tar acids	Low	High	Very slow

* Grams oxygen consumed per 100,000 parts.

are toxic but are removed by distillation. Thiosulfate and phenols show less toxicity but have a high-oxygen demand.

Greatest attention has been paid to the removal of phenols from waste liquors, because the presence of only traces of these compounds in drinking water is noted by the general public. As little as 0.1 ppm of phenol can be detected by gustatory experts. Some treatments of water with chlorine make the situation even worse, for they result in the formation of chlorophenols which can be detected by sensitive palates in concentrations as low as 3 ppb.

The most commonly used method of phenol removal is countercurrent extraction of the ammonia liquor with good solvents, such as crude or refined light-oil fractions. These solvents are available at the coke plant and can be regenerated, if necessary, in existing equipment. Some of the hydrocarbons, particularly benzene, dissolve in the ammonia liquor whence they are recovered by injection into the gas as part of the vapors from the ammonia

the removal of H_2S and other acidic impurities which eventually consume alkali.

Steam stripping of ammonia liquor followed by absorption of the volatilized phenols in caustic soda remains a practical method of dephenolization. Thompson[180] reported on operating data from a plant which started operation in 1950. The results were somewhat erratic, but they showed that with close control the phenol concentration in the liquor leaving the stripping tower can be reduced to 10 ppm or less. European interest in this method has continued.[181]

Adsorption on active carbon is capable of completely removing phenol from dephenolizer effluents since these contain less tarry materials than the crude liquors. Regeneration of the carbon is the most difficult problem encountered and is accomplished by treatment in a special furnace.[166] The use of superheated steam or solvents such as benzene has not been effective.[182]

A variety of oxidation methods have been investigated. A cooperative investigation of the use of chlorine, ozone, and chlorine dioxide was organized by the Ohio River Valley Water Sanitation Commission.[183] The rates of removal are shown by the reaction curves in Figs. 23, 24, and 25. The oxygen consumed and the biochemical oxygen demand values were reduced more than 60% by each of the oxidants. No correlation could be found between oxidant demand and either phenol content or oxygen consumed value. Partial treatment with ozone or chlorine dioxide

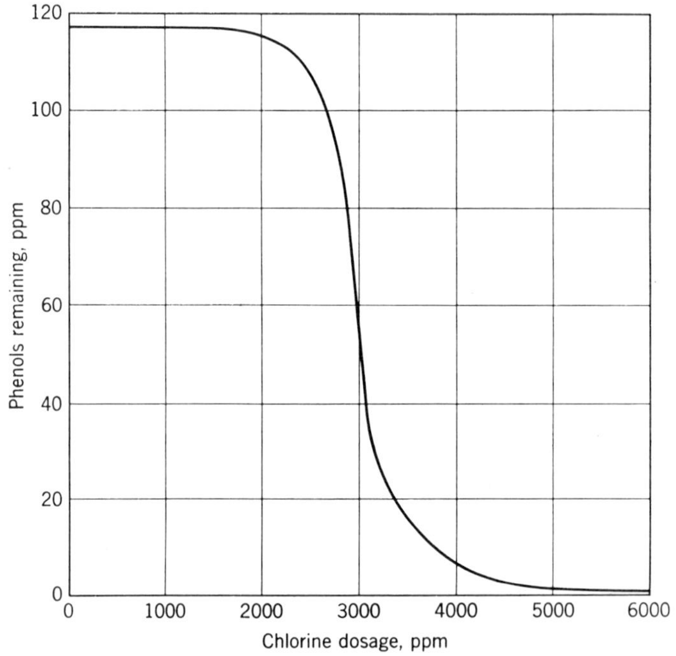

Fig. 23. Chlorine treatment of dephenolized ammonia still waste.[183]

Since both the extraction and steam-stripping processes have been refined to the point that they are capable of removing 99% of the phenols in ammonia liquor, there should be a lessening interest in what have been termed "post-dephenolization" methods; that is, methods of removing the residual phenol from the effluents from the previously described processes.

[180] Thompson, C. V., quoted in ref. 166.

[181] Grinberg, A. M., *Koks i Khim.*, **1957,** No. 3, 34–7. Kazachkov, L. I., *Sanit. Karakteristika Prom. Stochnykh Vod*, **1940,** 37–67. Wilks, F., *Sewage and Ind. Wastes*, **22,** 196–200 (1950). North Thames Gas Board, Shannon, W. V., and Golding, W. E., Brit. Pat. 699,453 (1953).

[182] Duchemin, M., and Lecomte, H., *Compt. rend. congr. ind. gaz, 69th Congr.*, **1952,** 920–34.

[183] Cleary, E. J., and Kinney, J. E., *Proc. 6th Ind. Waste Conf., Purdue Univ. Eng. Bull., Ext. Ser.* No. 76, 1951, pp. 158–70. Ohio River Valley Water Sanitation Commission, *Phenol Wastes: Treatment by Chemical Oxidation*, Cincinnati, 1951.

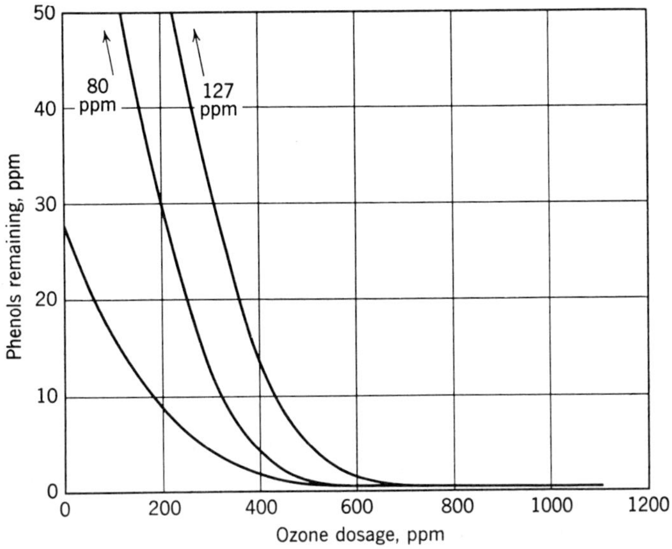

Fig. 24. Ozone treatment of dephenolized ammonia still waste.[183]

was possible, but complete treatment with chlorine was necessary to prevent formation of chlorophenols. These pilot plant studies indicated that such treatment would be very expensive. The use of ozone was also discussed by Hall.[184] Chlorine has been used to destroy phenols in the effluent from a rubber reclaiming plant.[185] Aeration in the presence of pyrolusite[186] or lime[187] has also been used.

Since the days of Lister, carbolic acid has in many minds been synonymous with antisepsis and has contributed its aroma to the happily disappearing "hospital smell." It is used as a

[184] Hall, D. A., *Gas World*, **147**, No. 3829, Coking Sect., **52**, No. 539, 7–13 (1958).

[185] Sechrist, W. D., and Chamberlin, N. S., *Proc. 6th Ind. Waste Conf., Purdue Univ. Eng. Bull., Ext. Ser.*, No. 76, 1951, pp. 396–412. Chamberlin, N. S., and Griffin, A. E., *Sewage and Ind. Wastes*, **24**, No. 6, 750–60 (1952).

[186] Ruff, V. F., and Martynova, O. I., *Vodosnabzhenie i Sanit. Tekh.*, **15**, No. 2-3, 98–102 (1940).

[187] McIntire, F. L., and Manka, D. P., U.S. Pat. 2,635,073 (1953).

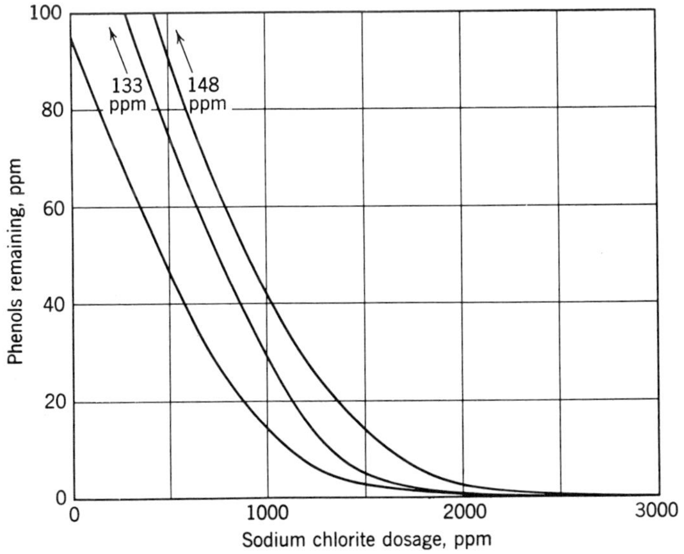

Fig. 25. Chlorine dioxide treatment of dephenolized ammonia still waste.[183]

Physically, the change from brown coal to bituminous coal is gradual; the nature of the forces of adhesion involved in briquetting coals of different rank is the same, but the pressure required for their agglomeration differs. For this reason, brown coal and bituminous coal are not briquetted by the same process; in general brown coal is briquetted by the application of pressure alone, and bituminous coal is briquetted after admixture with adhesive binders. Considerable research effort has been devoted in Germany, Great Britain, and the Netherlands to briquetting bituminous coal without a binder.

BRIQUETTING BITUMINOUS COAL WITH BINDERS

One of the earliest records of briquetting bituminous coals with a binder occurs in the eighteenth century.[3] The first factory for manufacturing briquettes was built in Saint Etienne in France in 1842; a few years later a plant using hard pitch as binder was opened in Newcastle in England.[2,3] In Germany, a plant was built at Wiesche colliery in 1861. As may be expected in such an old industry, many of the techniques used are based on tradition. Little major change has taken place since the end of the last century, and the position of the industry and its techniques at that time are dealt with admirably by Franke.[3] No major change in method has been made, but improvements of importance have been introduced as a result of systematic experimentation. The most recent comprehensive review of the subject is that by Haacke and Meyer.[4]

The Briquetting Process

The briquetting process, in essence, is one in which coal particles, covered with an adhesive, are pressed into molds. This method entails the operations of drying and grinding the coal, mixing it with the binder and heating the mixture, pressing the mixture into molds, and cooling and loading the briquettes. Pitch is used almost exclusively as a binder in Europe, whereas bitumen is used in America; the following observations relate to pitch except where it is stated otherwise. The quality of the briquettes is determined by the type of coal, the nature of the binder, and the precise way in which the various operations are carried out. The layout of an average coal briquetting plant is shown schematically in Fig. 1.

The wet clean coals from the bunkers *a* are blended from the feed tables into the conveyors leading to the dryer *b*. The coal is usually less than $\frac{1}{2}$ in. in size and may contain more than 15% moisture. The dryer is heated by means of flue gas generated in the furnace *c* which burns pulverized fuel *d*. The dried coal is transferred to the dry-coal storage hoppers *e*. Hard pitch, broken to about $\frac{1}{2}$ in. in size is fed from the hopper *f* through a crusher into a service hopper *g*. Coal from *e* and pitch from *g* are proportioned and fed into a cage disintegrator *h*. The mixture is crushed and further mixed in this mill from which it is transferred to a "pug" or steam-kneading cylinder *i*. In this cylinder, the mixture is heated to the required temperature, after which it passes via a conveyor, sometimes called a tempering conveyor, to a double-roll press *j*. The ovoid briquettes produced fall onto a screen which removes the "flash" or fines, and then onto a wire mesh conveyor *k* on which they cool before being loaded into wagons.

Drying

Many different types of dryer are in common use in the industry. The vertical cascade dryer is much used and the horizontal rotating-drum dryer has long been in favor. Latterly suspension drying has been introduced.[5]

The coal entering the dryers may have moisture contents in excess of 15%, which must be reduced to 2% to 4%. Moisture in excess of this leads to a higher pitch requirement and larger steam usage in the steam-kneading stage.

Grinding

The dried coal is ground, usually in cage disintegrators or swing-hammer crushers, to a size where approximately 93% is less than 0.88 mm and 80% is less than 0.5 mm. If the coal is larger than 2 to 3 in., a preliminary jaw crusher or cone mill may be used. For large rectangular briquettes, a larger size of crushed

[3] Franke, G., *Handbuch der Brikettierung*, Enke, Stuttgart, 1930, Vol. 1, pp. 2–4.

[4] Haacke, A., and Meyer, A., *Handbuch der Brikettierung*, Enke, Stuttgart, 1930, Vol. 1, pp. 57–62.

[5] Shaw, H. P., and Forster, C. E., *Steinkohle Briquetting in Germany*, Control Commission, Villa Hugel, Essen, July, 1947.

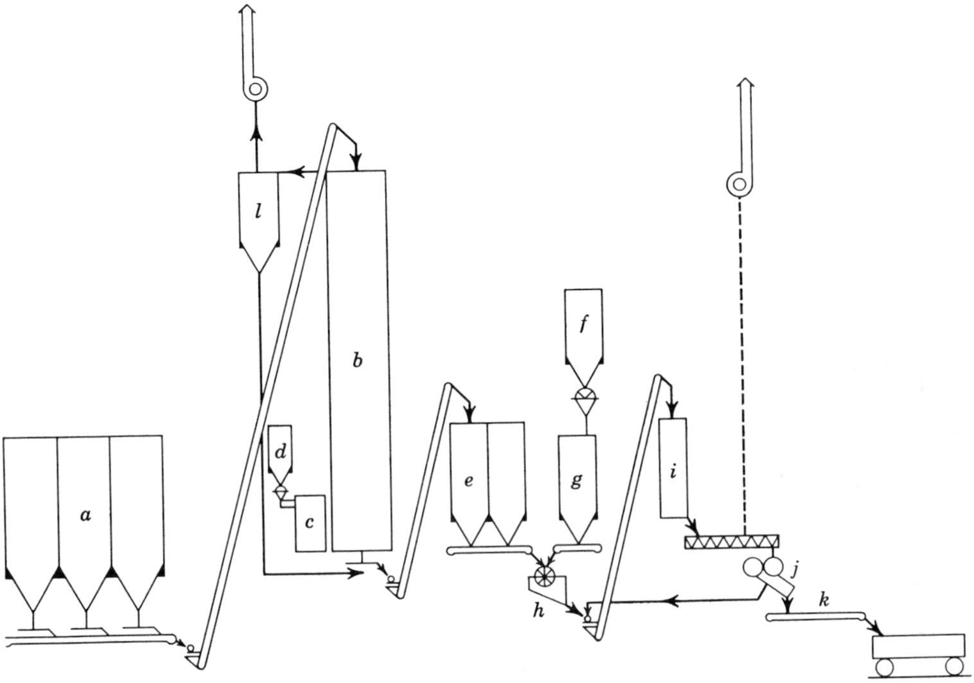

Fig. 1. Schematic drawing of a briquetting plant. *a.* Clean coal bunkers; *b.* vertical cascade dryer; *c.* furnace; *d.* fuel hopper and disintegrator; *e.* dry-coal hoppers; *f.* raw-pitch hopper and disintegrator; *g.* crushed-pitch hopper; *h.* cage disintegrator; *i.* pug; *j.* double-roll press; *k.* cooler conveyor; *l.* dust extractor.

coal can be used than for ovoids—the surface area of coal used for large briquettes is about 80 mm²/g, whereas that needed for ovoid briquettes is about 100 mm²/g.[5] The pitch is ground so that between 60 and 90% is less than 0.5 mm in size.

Mixing and Heating

The mixing and heating of the coal with the binder is the most critical step in the manufacture of briquettes. The binder may be used in liquid or solid form. If solid pitch is used, and this is the general practice in Europe, the pitch itself is ground, preferably finely, before it is added to the coal. It is proportioned by either volume or weight, and in many cases it is added to the coarse coal, the subsequent grinding assisting the mixing; in some cases mechanical mixers are also used.

Pitch is sometimes melted and sprayed onto the coal as in the Fohr-Kleinschmidt method.[6] This procedure has the advantage of obviating the difficulties associated with grinding the pitch and of eliminating the health hazard resulting from the pitch dust formed in the process. Although sprayed pitch has been used since 1912 at the Engelsburg colliery in Germany, this method has not been widely adopted. Bitumen is always used in liquid form.

The pitch and coal mixture is passed into a heater or "pug" which consists of a vertical cylindrical vessel possessing a central shaft on which paddles are mounted. The paddles are so arranged that when the shaft is rotated they agitate and mix the charge by lifting it. Superheated steam is injected through a number of nozzles directly into the mixture, thereby heating it to between 95 and 100°C. The mixture enters and leaves the pug continuously. (Some of the conditions found in practice are set out in Table 2.)

The design of this vessel has hardly changed in more than 100 years, and a full explanation has not yet been produced of why this method of heating has been employed rather than any other. It has been postulated that pitch spreads in a different manner over coal in the presence of water, and it has been established that a

[6] Meyer, G. A. H., *Der deutsche Steinkohlenbergbau, vol. 3, Brikettierung der Steinkohle*, Glückauf, Essen, 1958, pp. 90–2.

Fig. 4. A double-roll press. (Courtesy of Yeadon & Co., Leeds.)

addition to a falling stream of coal rather than mixing in worm conveyors gives a better distribution in the mixture. The pitch requirement may also be reduced by treating the coal, particularly low-volatile coal or anthracite, with a small quantity of oil (4 pt/ton).[10]

There has been a trend to use liquid bitumen in briquetting plants in Europe. Kegel[11] has discussed the advantages and disadvantages of the solid and liquid methods; the advantage of the former is the better mixing achieved, and of the latter the elimination of the health hazard.

The use of very hard pitch[12] which may be handled entirely mechanically, has also been suggested to reduce the health risk.

[10] Rhys Jones, D. C., Gregory, D. H., and Idris Jones, W., Brit. Pat. 510,095 (1939).

[11] Kegel, K., *Bergbau u. Energiewirtsch.*, **2,** 7/8, 213 (1949).

[12] Gregory, D. H., Rhys Jones, D. C., and Idris Jones, W., Brit. Pat. 606,064 (1948).

Types of Briquettes

Fuel briquettes are used for firing locomotives, for domestic heating, and industrially in small hand-fired boiler installations.

Large briquettes or fuel blocks weighing up to 28 lb are made for locomotive and industrial use. They are rectangular in shape and convenient to pack and store, but are unsuitable for mechanical handling either at the briquetting factory or at the point of use. They must be broken before use. Since 1940, the trend has been to produce cobble briquettes of about $1\frac{1}{2}$ to 2 lb in weight for industrial and locomotive use, and the most recent development in South Wales—the main briquette manufacturing region in Great Britain—is to manufacture these in double-roll presses of large diameter.[13]

Many ingenious devices have been proposed from time to time to give briquettes of different shapes. In France,[14] a press has been produced in which one of the rolls has a series of hollows of trapezoidal or rounded shape, and the other roll has a series of crowns conforming to the form of the hollows, the crowns being cut with molds of the appropriate shape. The agglomerates assume approximately a tetrahedral or "quarter-orange" shape.

Smaller briquettes, both for domestic purposes and more recently for the industrial field, are also made in double-roll presses. The cups cut in the rolls of double-roll presses must be shaped in such a way that the briquettes made in them fall away freely from the underside of the rolls. This requirement limits the shape of the briquette, and such presses generally produce shapes which are variously described as cushion, ovoid, or boulet. The cross-sectional shape is further limited by the requirement that the pattern of cups cut into the roll surface should leave as little flat or "land" surface as possible between the cups; this land surface usually ranges between 15% and 25% (Figs. 5 and 6).

The Quality of Briquettes

Standard tests for briquettes have been formulated in Germany, DIN 23081,[15] but so far no such tests have been specified by the American Society for Testing Materials or the British Standards Institute, although commercial standards exist.

TABLE 3

BRIQUETTE QUALITY

	British	German
Calorific Value (Btu/lb)	14,200	—
Moisture (%)	2	1
Ash (%)	8–10	9
Ash fusion (Ox) (°C)	1350	1300
Sulfur (%)	1	1
Pitch content (%)	7–9	6.8–7.5
Volatile matter (%)	17–19	17–21

The levels of the various parameters used in assessing briquette quality in Britain and Germany are shown in Table 3.

A further important quality is the strength of the briquette. This can be measured in various ways; the methods used in different countries are similar in principle[16] but different in detail. One test is concerned with the ability of the briquette to withstand the attrition arising in transport, and for large block briquettes, which are manually loaded, this is of primary importance. The abrasion test, known as a cohesion test in South Wales, consists in rotating briquettes in an iron drum and measuring the degradation. One hundred

[13] Gregory, D. H., *Design of Experimental Cobble-briquette Press*, *Natl. Coal Board, Coal Research Estab., Rept. No.* 32 (1950), 5 pp.

[14] Kleinberg, G., *Colloque sur l'Agglomeration des Fines de Charbon, Charbonnage de France, Paris,* **1959,** 15 pp.

[15] *Richtlinien für Abnahme und Oberwachung von Steinkohlen Brikettfabriken,* Entwurf 1, DIN 23081, 1950, p. 56.

[16] Charbonnier, R. P., and Lusinchi, J., *Wyoming Univ., Nat. Resources Research Inst., Inform. Circ.* No. 8, 44–53 (1955).

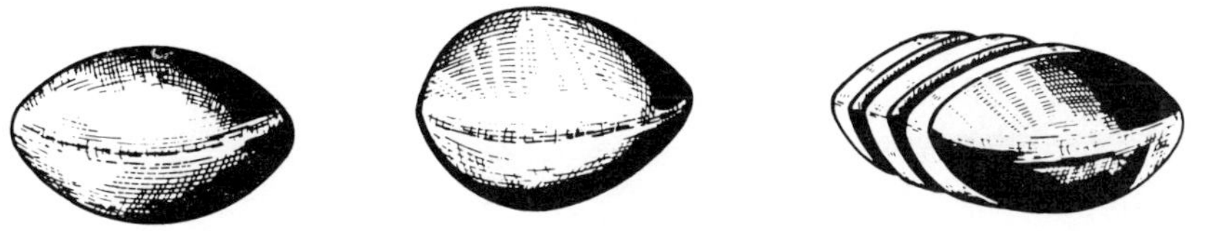

(*a*) OVOID (*b*) RATIONAL (*c*) "PHURNACITE"

Fig. 5. Briquette shapes.

TABLE 4

RELATIONSHIP BETWEEN COAL RANK, HARDNESS, AND HEAT OF WETTING

Coal	Rank	Hardgrove Index (%)	Heat of Wetting (cal/g)	Volatile Matter dry, mineral-matter free (%)	B.S. Swelling
Anthracite	100	34A	6.8	4.3	1NA
	100	57A	5.0	6.0	1NA
	100	69A	5.1	8.4	1NA
Semianthracite	201	93	3.7	12.6	1½
Steam coal	202	94	3.4	13.7	2½
	203	95	—	15.2	4½
	204	101	3.4	17.0	8
Coking coal	301	103	—	19.4	8½
	301	108	3.0	22.8	9
	201	102	3.6	28.4	7½
Gas coal	501	101	6.6	36.5	7
	501	74	5.1	31.0	7½
	501	68	4.5	34.3	7
	502	58	6.0	34.2	6½
	601	75	6.0	32.9	5
Flame coal	801	57	9.2	35.7	2½
	802	57	13.2	37.2	1½
	902	56	16.7	37.6	1½
	902	55	16.3	38.0	1A

NA = Nonagglomerating.
A = Agglomerating.

coarse coal the application of pressures in excess of 1 ton/sq in. produced weaker briquettes, may be similarly explained.

The petrographic composition of the coal also plays a part; some of the macerals are less strong than others, and when crushed they contribute a higher proportion of the fine component. It has been suggested that fusain absorbs pitch into its pores, but there is insufficient evidence available to confirm that different coal types absorb different quantities of pitch.[20]

The ease with which coals can be ground depends on their hardness or elasticity. The hardness of British coals of different rank has been reported by Hawksley[23] who used the Hardgrove test for grindability.[24] This test consists in grinding 50 g of closely sized coal (16 to 30-U.S. mesh) for 3 min, and the index of grindability is calculated from the amount of coal in the ground sample passing a 200-U.S. mesh screen. A relationship has been established between the elastic properties of coal and its "heat of wetting"[25] (Table 4).

Little is known of the differences in wettability of the surfaces of different coals by different pitches, but it would be reasonable to expect that those coals with a greater affinity for water than oil would provide a less strong bond, which would account for the lower strengths of briquettes with a high-shale content.[20] It would also explain the higher pitch requirements for low-rank coals and lignites, but would not account for the higher pitch requirements for anthracite.[26]

The pitch requirement depends not only on the size of the coal, but also on its size distribution and the magnitude of its external surface. A number of attempts have been made to provide an ideal size distribution which would give the densest packing for any particular surface area.

[23] Hawksley, P. G. W., *BCURA, Inform. Circ.* No. 36 (1950), 12 pp.

[24] Selvig, W. A., *ASTM, Proc.*, **37** (1), 815 (1937).

[25] Bangham, D. H., and Maggs, F. A. P., *Proc. Conf. on Ultra-fine Structure of Coals and Cokes, BCURA, London*, **1944**, pp. 118–30.

[26] Launder, J. L. M., *Natl. Coal Board, Coal Research Estab., Rept.* No. 103, 1951, 10 pp.

Although the strength of a briquette is dependent on the extent to which the surface of the coal particle is covered by the pitch, the amount of pitch required is not only dependent on the area of that surface, but also, to some extent, on the volume of the interstices between the particles. Much attention has been paid to the influence of size distribution on briquette quality, and to finding the size distribution, which, for a given top size, possesses both the smallest surface area and the smallest voidage.

The problem is one of general interest, and has been studied by Horsfield[27] in connection with road surfaces, and by Furnas[28] for concrete aggregates. Calculations made using a formula derived for the close-packing of spheres yielded results which were of little value because of the difference in shape, and, of course, the impossibility of attaining anything other than a random geometric distribution of particles.[29] Hudson[30] has shown that in a regular close-packed system of spheres, the misplacement of only a few of the smaller spheres throws the whole system out.

Kegel[31] has used the Fuller-Graf curve as representing the most favorable size-distribution for close packing, but many other curves, such as the Fuller-Thompson curve,[32,33] have been produced. Rogers and Homer[34] have described an experimental method of arriving at a dense packing of anthracite. Five different size-fractions were taken from anthracite $\frac{1}{4}$- to 0-in. size, the top size being $\frac{1}{4}$ to $\frac{1}{8}$ in., and the bottom $\frac{1}{120}$ to 0 in. To the top size were added increments of the next smallest size until the densest mixture was obtained. To this dense mixture was added the next smallest fraction and so on (Fig. 8). Petersen[35] recommended using sizes given by the Fuller curve, but commented on the technical difficulty of attaining this in practice. Wise[36] has stated that dense random packing is impossible unless there is a considerable spread in the distribution radii. Vollmaier[37] has shown that for Rhine coal the density decreases with sizes below 0.5 mm or above 1 mm. He referred to the larger pressures required to briquette the finer sizes. Spilker and Born[38] have also experimentally shown the significance of particle size on briquette strength.

Broche and Nedelmann[20] have made briquettes from coals of different size gradings using 7% of pitch, and have concluded that the presence of some fine material is not harmful—indeed, it is beneficial in limited quantities. They also showed that a reduction in the top size of particle was beneficial.

Much still remains to be explained. No theory propounded can explain the extraordinary increase in briquette strength achieved by comparatively small additions of very fine coal to the size gradings used normally.[39] Experiments have shown a marked rise in strength by the addition of less than 1% of coal ground to less than 5 μ in size.

Haacke and Meyer[4] have commented on the interest shown in size grading during the last half century and cited the Fuller curve, but they pointed out that the size grading used in practice differs greatly from this. Meyer[40] quoted the actual size grading used in nineteen different plants in 1956 (Fig. 9).

Broche and Nedelmann found[41] that the residence time of the briquetting mixture in the pug also affected the strength of the briquettes. In a series of briquetting trials, it was noticed that after the plant had run for about 12 min, the moisture content of the briquettes had fallen sharply and the bulk density had risen.

[27] Horsfield, H. T., *J. Soc. Chem. Ind.*, **53**, 107–15T (1934).

[28] Furnas, C. C., *Ind. Eng. Chem.*, **23**, 1052–8 (1931).

[29] Dallavalle, J. M., *Micromeritics*, Second edition, Pitman, New York, 1948, 555 pp.

[30] Hudson, D. R., *J. Appl. Phys.*, **20**, 154–62 (1949).

[31] Kegel, K., *Braunkohlenarch.*, **51**, 375 (1938).

[32] Fuller, W. D., and Thompson, S. E., *ASCE, Proc.*, **33**, 222 (1907).

[33] Dammann, K., Brit. Pat. 179,166 (1923).

[34] Rogers, J., and Homer, J. R., *Trans. Inst. Mining Engrs.*, **76**, 403–5 (1928).

[35] Petersen, W., *Der deutsche Steinkohlenbergbau vol. 3, Brikettierung der Steinkohle*, Glückauf, Essen, 1958, p. 40.

[36] Wise, M. E., *Philips Research Repts.*, **7**, 321–43 (1952).

[37] Vollmaier, A., Dissertation, Univ. Freiberg (Sa), 44, 53 (1937).

[38] Spilker, A., and Born, G., *Brennstoff-Chem.*, **11**, 307–18 (1930).

[39] Powell-Duffryn Coal Co., Private communication.

[40] Meyer, G. A. H., *2nd Symposium on Coal Preparation, 1959, Dept. Mining, Leeds Univ., Leeds*, **1959**, p. 139.

[41] Broche, H., and Nedelmann, H., *Glückauf*, **69**, 233–40, 257–67 (1933).

TABLE 6

VARIATION OF BRIQUETTE STRENGTH WITH ALTERATION OF THE PROPERTIES OF THE PITCH USED AS BINDER

Pitch sample number		172	173	174	175	176	177	178	179	180	181
Source		CO	VR	CO	LT	HR	CO	?	HR	VR	VR
Crystalloids	(%)	50.4	69.6	60.5	78.5	57.3	47.5	50.2	44.4	59.3	68.5
Resins	(%)	21.0	14.1	24.2	15.5	22.8	24.0	18.4	16.1	17.2	14.5
C_2 Content (Soluble in pyridine, insoluble in benzene)	(%)	14.7	14.2	8.9	6.8	11.8	12.8	12.9	13.0	15.0	12.8
C_1 Content (Insoluble in pyridine)	(%)	14.9	4.2	7.8	0.4	10.7	16.2	20.4	28.4	10.0	4.9
Insoluble in toluene	(%)	27.4	12.9	12.2	3.9	18 2	24.2	28.8	38.5	22.0	13.1
Volatile matter	(%)	66.6	79.5	74.5	91.2	70.9	65.5	63.5	57.0	72 0	80.0
Carbon: hydrogen atomic ratio		1.54	1.21	1.56	1.05	1.52	1.71	1.62	1.76	1.41	1.16
Logarathmic temperature Coefficient		20.8	21.3	23.2	23.9	21.9	20.4	21.1	19.1	20.3	21.9
Viscosity at 65°C	($p \times 10^{-5}$)	2.45	4.53	4.10	3.47	2.55	3.14	3.92	1.83	1.59	2.36
Ductability at 75°C	(cm)	20.5	29.1	30.7	41.1	40.0	28.0	24.0	19.4	21.4	25.5
Grindability index	(%)	20.6	22.8	30.4	29.6	23.2	—	—	23.6	21.0	16.8
Order of hardness of briquettes:											
Coal 201 24 h		8	5	2	1	4	7	9	10	6	3
Coal 800 24 h		9	3	1	2	5	7	8	10	6	3
Coal 201 and 800 1 h		3	7	2	9	1	4	5	6	—	8
Works grading		1	8	2	9	3	4	6	5	—	7

CO = Coke-oven source.
HR = Horizontal-retort source.
VR = Vertical-retort source.
LT = Low-temperature tar source.

hot strength is of importance to the manufacturers and the cold strength to the distributor and user. After 24 hr, the hardest briquettes were those made from pitches with low-insoluble matter and low aromaticity, but immediately after manufacture the strongest briquettes were those made from pitch with the highest-resin content. These conclusions were based on correlations that may, of course, be fortuitous, and some are contrary to conclusions reached by other workers. For example, it has been claimed that pitches of high aromaticity are preferable for briquetting.[46,47]

The effect of the pitch properties on the strength of the briquette is shown in Tables 6 and 7. This is by no means the only attempt that has been made to produce an adequate

[46] Audibert, E., *Comite d'organisation de l'Industrie des Combustibles Mineraux Solides, Note Tech.* E 7/6, July, 1941. *Fabrication des Agglomeres, Note No.* 5.

[47] Broadbent, R., *J. Inst. Fuel*, **28**, 3–16 (1955).

TABLE 7

COEFFICIENTS OF RANKED CORRELATIONS OF COLD-BRIQUETTE OR HOT-BRIQUETTE STRENGTHS WITH PITCH PROPERTIES

Pitch Property	Coefficient	
	Cold	Hot
Crystalloid content, γ	0.81	−0.43
Resins content, β	−0.04	0.83
C_2 Content (soluble in pyridine, insoluble in benzene)	−0.73	0.02
C_1 Content (insoluble in pyridine)	−0.80	0.43
Insoluble in toluene	−0.94	0.27
Volatile matter	0.83	−0.45
Carbon: hydrogen atomic ratio	−0.67	0.45
Logarithmic temperature coefficient	−0.89	−0.13
Viscosity at 22°C	0.67	0.00
65°C	—	0.05
85°C	—	0.13
Ductability at 75°C	0.90	0.00
Grindability	0.48	0.18
Correlation with washing and grinding	—	0.90

specification for pitch. Spilker and Born[38] attacked the problem in 1930, and Broche and Nedelmann[41] in 1933. The former produced a specification for making briquettes in the laboratory, and measured the quality of the pitch in terms of the strength of the briquettes; the latter aimed at a more exact method involving less time, and they have put forward levels of grindability, ductility, and CS_2-insoluble material. After testing fifteen pitches, Broche and Nedelman concluded that a good pitch required a high grindability index, 60% or more, a ductility of more than 80 cm, and a CS_2-insoluble content of less than 20%. They found, however, that some pitches which complied with this specification were of poor quality in practice.

The function of the binder is undoubtedly that of an adhesive, and it must wet and cover the surfaces to be stuck together. Reerink and Goecke[48] defined a good briquetting pitch as that "which can, at the operating temperature, completely wet the greatest number of coal particles and provide these with a covering film." They suggested that the most important characteristic of a pitch is its viscosity at the operating temperature, and that this must be determined in the presence of steam.

The behavior of an adhesive also depends on the nature of the surfaces involved. The surface of coal is strongly lyophilic, is amenable to wetting by pitch, and is porous. It is possible that the porosity, by causing some absorption of the pitch, contributes to the strength of the bond. The degree of coverage of the surface depends on the amount of pitch present, the amount absorbed into the coal, and the degree of fluidity of the pitch at a temperature of 95 to 100°C. The surface tension of the molten pitch might also be expected to play a part. Horizontal-retort pitches are slow to spread over the coal surface, but have a good setting property.

Since the fineness of grinding of the pitch has a considerable influence on the strength of the briquette,[41] it is important that the pitches should be hard and brittle enough, even in the summer months, to be crushed without difficulty.

Pitch is a mixture of a vast number of different substances. These may be grouped into three broad categories; those soluble in light petroleum and that are crystalline, those soluble in solvents such as benzene and that are resinous, and those soluble, if at all, in pyridine. These last appear to be compounds of great complexity with a high degree of polymerization. It is possible that the resinous fractions might be derived from the polymerization of indene, coumarone, and such compounds which are known to occur in coal tar.[49]

Different workers have used different solvents, and confusion arises because of this and the use of different terms for the various fractions. Wood[45] drew attention to this confusion and defined his terms. Although different opinions[50,51] are expressed, there appears to be general agreement that the resinous component plays a considerable part in the mechanism of adhesion.

Carbon disulfide has been used for some time to determine the solubility of pitch, but other solvents, such as benzene, toluene, pyridine, and tetralin-cresol mixtures, are in common use. Wood[45] used the letters C_1 and C_2 to define those portions soluble in pyridine, but insoluble in benzene. Resins were defined as being soluble in benzene or toluene, but not in petroleum. The crystalloids are soluble in petroleum. He found a correlation between the strength of cold briquettes and the low insoluble-matter content in the pitch on the one hand, and between the hot strength and the high content of fractions insoluble in petroleum, but soluble in benzene or pyridine, on the other. Broche and Nedelmann[41] used the symbol α (C_1 and C_2) to denote the fraction insoluble in petroleum and benzene, β (resins), the fraction soluble in benzene but not in petroleum, and γ (crystalloids), the fraction soluble in petroleum and benzene. For a good pitch the α fraction should not exceed 19 to 26%, and β and γ should be present in equal proportions. Demann[52] preferred this ratio to be 1 : 2.

Different solvents of similar type, for example, benzene and toluene, produce different

[48] Reerink, W., and Goecke, E., *Glückauf*, **71**, 77–85, 105–14 (1935).

[49] Krämer, G., and Spilker, A., *Ber.*, **23**, 78–83, 3276–83 (1890).

[50] Broche, H., and Nedelmann, H., *Glückauf*, **69**, 257–67 (1933).

[51] Rick, A. W., *Bitumen, Teare, Asphalte, Peche*, **3**, 316–8 (1952).

[52] Demann, W., *Brennstoff-Chem.*, **14**, 121–3 (1933).

only three provide information on which their quality as a binder may be judged. These tests are the "softening point," the "penetration," and "ductility."

The limits proposed for softening point vary; a range of 125 to 150°F is quoted by one user, and 150 to 165°F by another. The choice of bitumen is related to the mode of usage and the substance briquetted. For the briquetting of chars, for example, a softening point of 150 to 200°F is advocated. The bitumen is used in the liquid state and requires to be heated to between 300 and 400°F in order to be pumped.

A penetration of 10 to 15 at 77°C has been specified, and a definite but usually unspecified amount of ductility is required.

As for pitch, solvent analysis has been used as a means of assessing bitumen, and the solvents used differ only slightly from those used for the former. Insolubility in petroleum naphtha (68°C) is an indication of the hardness and degree of polymerization of the bitumen, whereas the part insoluble in naphtha but soluble in carbon disulfide, the "asphaltenes," is considered to be responsible for its plastic properties.[67]

Swartzman[67] and Corbett[68] have reported extensive investigations carried out to relate the quality of bitumen as a binder to some of its specific characteristics. Swartzman concluded that the present tests are inadequate and that some property of asphalt may assess its properties more accurately. This conclusion is similar to that reached by Wood[45] in his studies of pitch.

Other Binders

Although pitch and bitumen are the binders used in the major briquetting plants, a very wide variety of other substances has been used or suggested. Some of these substances are inorganic, such as cement or sodium silicate, but the use of cement is not seriously considered by manufacturers of good quality briquettes because it increases the ash content. Cement and sodium silicate both possess the virtue of being smokeless, and the former is also resistant to weathering. Suggestions have been made for making silicate-bound briquettes[69] impervious to water by selecting the correct pH value, or by a baking treatment.

The other binders are organic in nature and all, with the exception of the recently proposed synthetic resins, suffer in that they are not weatherproof. Strong and associates[70] have listed them in a bibliography drawn up to 1936, and since then the proposals have been legion.

Only four of these other kinds of binder have received serious consideration. Sulfite lye or sulfite pitch, a concentrated liquor neutralized with lime, has been used for binding many substances, and is widely used in foundries for cores. As a binder for coal[71] it produces briquettes of high strength, but the binder has two drawbacks: it has a high-sulfur content and it has a low resistance to water. This latter failing may be overcome by baking the briquettes at 600°F, but in spite of this the process has not been an important one in fuel-briquette manufacture.[72]

In the 1940's, experimental plants were erected in South Wales to examine a briquetting process[73] in which peat, treated with soda ash, was used in quite small amounts to bind coal particles together. The peat was treated with sufficient soda ash to make it almost neutral, and then was kneaded into the coal which had been crushed to the usual briquetting size. About 2% of peat was used, and the soda ash could in part be replaced by sodium silicate. The mixture was briquetted in a double-roll press at pressures in the region of 2 ton/in.^{2} The briquettes were then dried at a temperature of about 120°C. This drying or baking treatment was required to produce stable briquettes resistant to weathering. The briquettes could be carbonized. No commercial development followed these experiments in South Wales.

Cellulose has long been suggested as a binder for the manufacture of coal briquettes. The binder is used in the finely divided state or after treatment with alkali; the briquettes are not weatherproof unless they are subsequently

[67] Swartzman, E., *Wyoming Univ., Natl. Resources Inst., Inform. Circ.* No. 3, 70–96 (1949).

[68] Corbett, L. W., *Wyoming Univ., Natl. Resources Inst., Inform. Circ.* No. 3, 55–65 (1949).

[69] Anon., *Chem. Trade J.*, **117**, 609 (1945).

[70] Strong, R. A., Swartzman, E., and Burrough, E. J., *Fuel Briquetting*, Bur. Mines, Ottawa, Canada, 81–9, (1937).

[71] *Ibid.*, pp. 28–9.

[72] See ref. 5, Appendix 14 and 15.

[73] Squire, L. R. L., Brit. Pat. 502,619 (1939).

specially treated. Such a process has been fully investigated in Germany.

Jäppelt's initial work in 1943 on the briquetting of low-temperature coke from brown coal was followed up later and applied to coal. Petersen has recently described[74] the application of this method for the manufacture of briquettes made from anthracite and steam coals ranging in volatile content from 9.2 to 13.4%. The coal, of briquetting size, was mixed with a viscose binder and briquetted at a pressure of about 225 kg/cm^2 to make briquettes weighing 80 g. The binder was prepared by Müshenborn's method[75] by converting cellulose into a xanthate and was used in the form of a viscous 4% solution (based on dry cellulose) or sodium cellulose xanthate. Between 10 and 15% of this solution was added to the coal as a binder, amounting in practice to between 0.3 and 0.5% of solid binder on the weight of finished briquettes. [The binder may also be added directly as the solid xanthate in amounts of 0.3 to 0.5% (based on dry α cellulose).] This binder is unstable and coagulates within about 24 hr, even when maintained at a temperature of 15°C. This coagulation confers strength upon the briquettes, and may be accelerated by drying the briquettes initially at a temperature of about 50°C before heating them to 100°C to dry and harden the briquettes further. Higher temperatures are harmful. If the binder is added in solid form,[76] the coal need not be specially dried. The briquettes soften on contact with water, but they may be protected by coating with a water-repellent substance, such as paraffin wax. The briquettes ignite well, but on combustion the cellulose binder burns away from the coal particles and excessive amounts of combustible matter are lost through the grate bars. This rather serious fault may be overcome by introducing finely ground coking coal into the blend; when burned, the coking coal cements the briquette, which then not only burns satisfactorily, but also burns relatively smokelessly. The sensitivity of the binder, its tendency to decompose, and the need to introduce a coking coal led to a reluctance to develop the process commercially.

One of the most recent ventures is that of manufacturing fuel briquettes using a binder of starch mixed with a synthetic resin. A small pilot plant has been erected in South Wales to manufacture smokeless briquettes from anthracite fines and steam coals. The binder is a solution containing equal parts of phenol-formaldehyde resin mixed with starch; between 2 and 4% of this mixed binder is mixed with the coal, and the mixture is pressed under normal conditions in a double-roll press making briquettes of approximately $1\frac{1}{2}$ oz. The briquettes are then heated to about 150°C for approximately $\frac{1}{2}$ hr to polymerize the resin and render the briquette impervious to water. The briquettes ignite readily and produce a very satisfactory smokeless fire in the open grate.[77,78]

BRIQUETTING BROWN AND BITUMINOUS COAL WITHOUT BINDERS

Brown Coal

Brown coal, as mined, has a high-moisture content (up to 60%) and a low-calorific value. Thus it is of low commercial value, especially if it is also small in size, until it has been dried and briquetted. Millions of tons are briquetted annually and the briquettes find a ready sale on the domestic market; 80% of the brown coal briquettes produced in Germany, the largest producer, is sold to domestic consumers, and the remaining 20% is disposed of to industry.

Brown coal is converted into briquettes without the aid of a binder; pressure alone is used to compact the coal, and the success of the process depends on the careful preparation of the coal and the proper design of the presses. Full and detailed descriptions of the process are given in Franke's classical textbook[3] on the subject, and an up-to-date description of Australian practice has been given by Barangwanath[79] and Herman.[80]

[74] Petersen, W., *Der deutsche Steinkohlenbergbau, vol. 3, Brikettierung der Steinkohle*, Glückauf, Essen, 1958, p. 140.

[75] Müschenborn, W., German Pat. 810,153 (1948).

[76] Müschenborn, W., German Pat. 803,597 (1948).

[77] Fuge, E. T. J., French Pat. 1,177, 293 (1959).

[78] Cunningworth, D. E., Brit. Pat. 27,852 (1959).

[79] Baragwanath, G. E., *Brown Coal, Its Mining and Utilization*, P. L. Henderson, Ed., Melbourne University Press, 1953, pp. 150–231.

[80] Herman, H., *Brown Coal*, State Electricity Commission of Victoria, 1952, pp. 192–471.

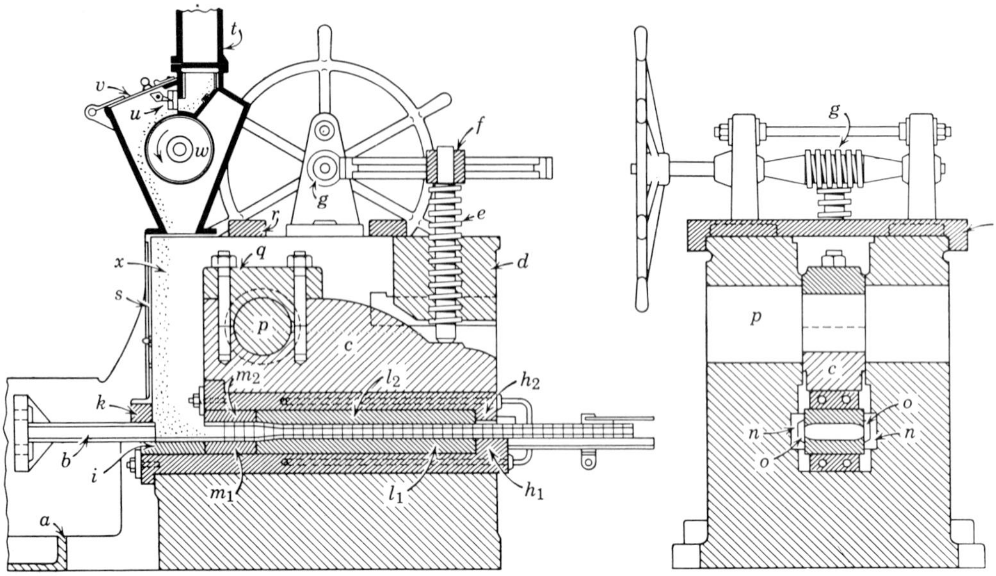

Fig. 11. The Exter press.[80] Press block, press feed, etc., of an old type single press. *a*. Press frame; *b*. stamp; *c*. tongue; *d*. press block part with spindle female thread; *e*. tongue-adjusting spindle; *f*. gear wheel; *g*. worm drive; h_1. and h_2. form hooks; *i*. and *k*. lower and upper guide pieces; l_1. and l_2. lower and upper long molds; m_1. and m_2. working pieces; *n*. side wedges; *o*. packing strips; *p*. tongue bolt; *q*. bolt cover; *r*. ties; *s*. door; *t*. feed pipe; *u*. damper; *v*. cleaning door; *w*. feed roll; *x*. coal inlet shaft.

buckel, and that under certain conditions they could be of satisfactory quality. He presented the same picture of the briquetting mechanism as did Agde (see p. 697) and stated that the final ordered structure is effected at the buckel. The buckel is considered to play an important part in counteracting deficiencies in drying the coal, and only when the moisture content of the finest and coarsest grains are equal does it lose its prime importance.

Theories of Briquetting Brown Coal

Theories of briquetting have been based on theories of the physical structure of brown coal. Copious literature exists on these subjects, and two admirable reviews of the theory of briquetting have recently appeared, one by Werner[87] and the other by Baragwanath.[88] The earliest theory which postulated the existence of "bitumen" in the coal—this "bitumen" acting as a binder—has long been abandoned, and has been followed by many others.

In brown-coal briquetting, the outstanding importance of the moisture content of the coal has led to much thought and research concerning both the nature of the coal-water relationship and the nature of the adhesion of coal particles to one another. Coal is honeycombed with capillaries of various sizes, and the moisture, up to about 12%, is held in the smaller capillaries of micron or submicron diameter; moisture in excess of this is held in the larger pores. Vapor-pressure curves have led to the view that brown coal is a gel, in which, according to some, the water is held in the disperse phase. According to others, the colloidal particle consists of a coal nucleus surrounded by an envelope of water.

The "capillary theory" propounds that when the brown-coal powder is subjected to pressure, a point is reached when the water from the capillaries is exuded and coats the surface of the particles, thus improving the intimacy of contact. When the pressure is removed, not all the water is reabsorbed by the capillaries, and a surface-adsorbed layer remains which maintains the close contact between the particles. This theory requires an optimum moisture content; sufficient moisture must be held in the capillaries for some to be exuded under the pressures used, and there must not be so much that an excessive amount is left between

[87] Werner, O., *Leitfaden der Brennstoff-Brikettierung*, Enke, Stuttgart, 1953, pp. 152 ff.

[88] Baragwanath, G. E., ref. 79.

the particles after the pressure is released. This optimum lies between 12 and 20%. In support of this theory, Fritzsche[89] has mentioned that the strength of an "optimum" briquette cannot be increased by increasing the pressure further, and that an overdried coal cannot give a strong briquette at the normal Exter-press pressure of 1400 kg/cm². According to the "capillary" theory, a pressure of 2000 kg/cm² is required for a moisture content of 5%. Ring-roll presses working within the range of 1600 to 2000 kg/cm² require the moisture content of the coal to be between 6 and 10%. This theory is useful and helps in the day-to-day practice of brown-coal briquetting.

The "colloidal theory" brings perhaps a greater refinement to the theory. It also recognizes the importance of the coal-water structure, and emphasizes the difference between the part played by the water held in capillaries, and by the water envelope or lyosorbed water of the colloidal particles. In medium-pressure briquettes, both forms of water are involved, but in high-pressure briquettes, only the lyosorbed water is needed.[90] Agde was also concerned with the chemical structure of the colloidal particles, their plate-like shape, and their mode of packing under pressure in layers parallel to the major axis. From X-ray work, it was concluded that the centers of the cohesive forces must occur on the borders of the plate-like humic acid molecules. Agde also recognized different stages of order introduced

[89] Fritzsche, A., *Braunkohle*, **40**, 509–26 (1941).

[90] Agde, G., et al., *Braunkohle*, **32**, 386–90 (1933); **43**, 269–78 (1944).

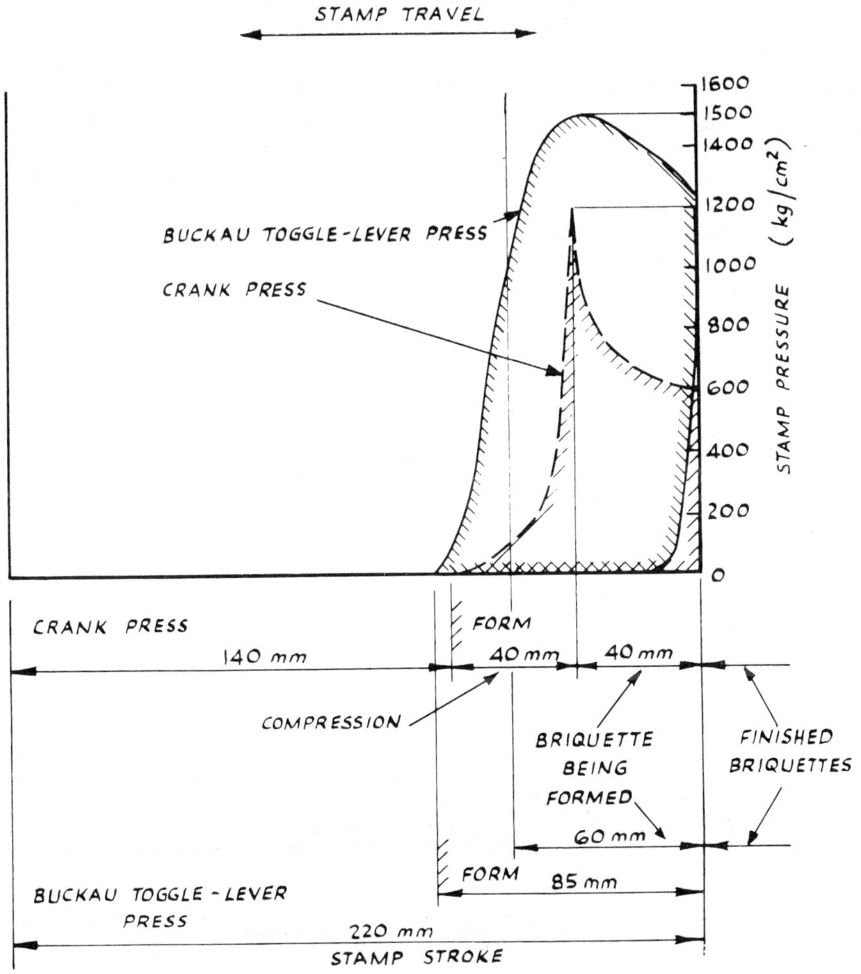

Fig. 12. Comparison of the pressure development in crank and toggle presses.[80]

into the structure of the briquette by the applied pressure. According to the magnitude of the pressure, the random structure changes first into one in which the plate-like colloidal particles arranged themselves in randomly dispersed layers, and then at still higher pressures these layers orientate themselves so that they lie parallel to the major axis.

Hock's[91] theory considered the size distribution of the particles, the closeness of packing and the hardness of the coal. He has studied the physical properties of brown coals and has shown that there is no difference, from the briquetting point of view, between brown coal and other organic substances of high molecular weight. He discarded, therefore, the capillary theory and explanations involving a dependence on the presence of water films. The adhesions between particles are considered to be due to forces of molecular origin, and, as the magnitude of these adhesion forces increases with the degree of proximity, the aim in briquetting must be to bring the particles so close together that the natural adhesion forces become effective.

Many other theories have been advanced, but most of these are excessively complex. Broadly speaking, the gel structure of brown coal is accepted, and its complexity in terms of coal types is appreciated; the forces of adhesion are the simple cohesion forces, the van der Waals forces, which are effective over distances of 0.1 to 0.05 μ and decrease accordingly to the sixth power of the distance between the surfaces. The part that hardness plays in preventing the close proximity of surfaces is accepted, and methods of testing coals to determine their hardness have been proposed.

The harder the coal, the more difficult it is to briquette. Remesnikov[92] has examined the relationship between the elastic and plastic properties of coal, and has shown that a high degree of elasticity and a low degree of plasticity are related to an inferior briquetting capacity. To describe the briquettability of a coal he has devised a "plasticity index" which is the ratio of the elastic energy E_1 to the plastic energy E involved in compressing it. This information may be derived from compression diagrams of the form shown in Fig. 14 in which E_1 is represented by the hatched area and E by the cross-hatched area under the curves. A comparison between the plasticity index and the quality of the briquette made from a series of different coals is shown in Table 10.

[91] Hock, H., *Braunkhole, Wärme u. Energie*, **3**, 48–58 (1951).

[92] Remesnikov, I. D., *Ugol'*, **28**, No. 10, 27–30 (1953).

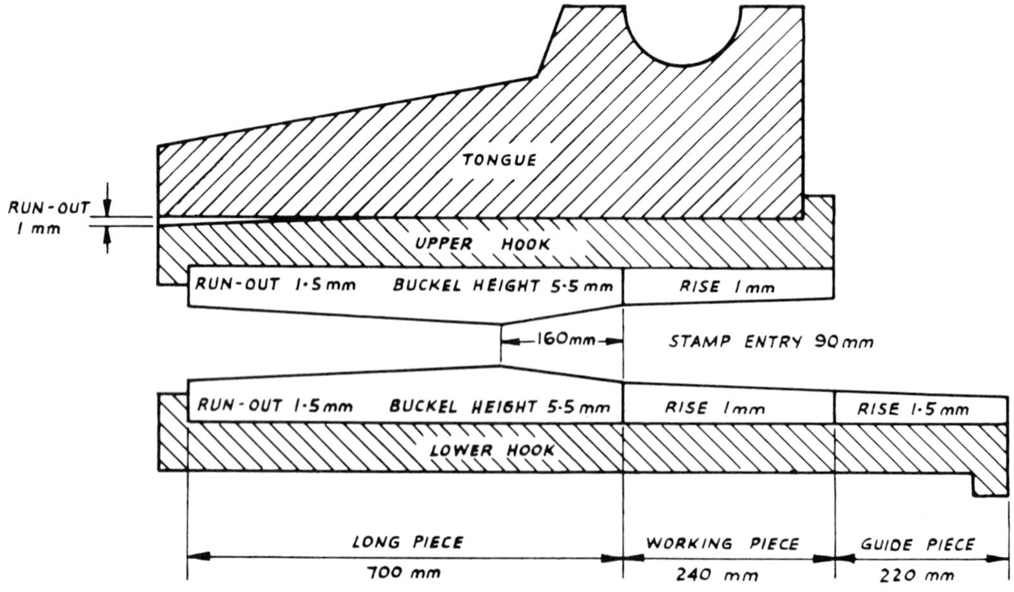

Fig. 13. The buckel.[80]

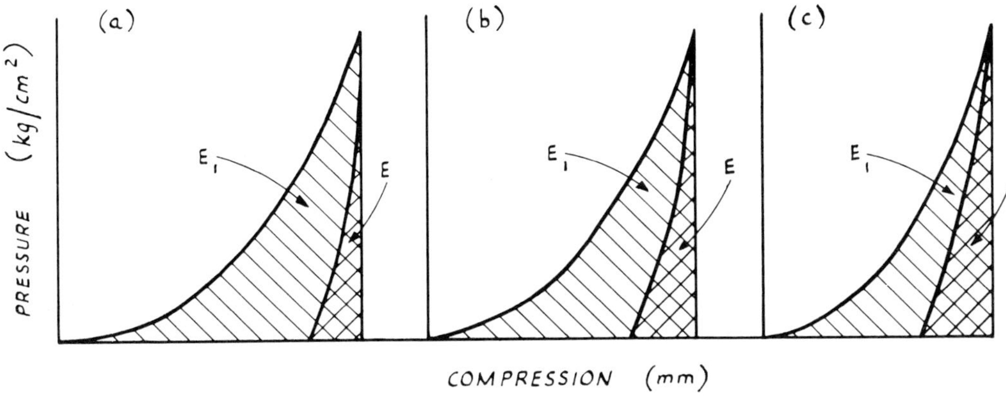

Fig. 14. Relationship between elastic and plastic deformation on compression.[92]

Briquette Quality

Brown-coal briquettes are judged according to their heating value and their physical strength. The actual standards and methods of test are different in different countries, but a committee was set up in 1957 to formulate international test methods.[93]

A high calorific value is desirable and figures of between 4000 and 5600 cal/g are usual.[94] The moisture content is determined by the requirements of the briquetting method used and is between 10.5 and 18%. The ash content is usually between 4 and 10%.

Compression strengths are of the order of 100 to 200 kg (using 3-cm diameter faces) and the bending strength 12 to 15 kg/cm^2.

[93] *Wyoming Univ., Nat. Resources Research Inst., Inform. Circ.* No. 9 (1957), viii–xiii.

[94] Franke, G., *A Handbook of Briquetting*, Enke, Stuttgart, 1930, pp. 556 ff.

Hard Brown Coal

As coal becomes more mature, it becomes harder and more difficult to briquette without a binder. The change in briquettability is gradual; the more recalcitrant harder brown coals may be briquetted by increasing the pressure, for example by using a toggle press or a ring-roll press,[95] but still higher pressures have to be used for bituminous coal.

Recent opinion[96] tends to the view that, physically, the change from brown coal to bituminous coal is a gradual one, and that the theories of adhesion in brown coal briquetting would also be applicable to bituminous coal briquettes.

The properties of brown coal that have been considered to play a part in the mechanism of

[95] Pakhalok, I. F., *Ugol'*, **30,** No. 9, 35–7 (1955).

[96] Mertins, A. J., *Braunkohle*, **4,** 19–20, 353 (1952).

TABLE 10

RELATIONSHIP BETWEEN PLASTICITY INDEX AND BRIQUETTE QUALITY

Coal	Plasticity Index	Crushing Resistance (kg/cm^2)	Residue after Grinding (%)
Far Eastern	15.6	410	92
Babayevsk	15.0	360	91
Angrensk	9.2	188	90
Moscow Region	8.9	177	89
Western Siberian	8.6	170	87
Kzyl-Kiy	4.1	12	38

TABLE 11

VARIATION IN COAL PROPERTIES WITH RANK

Coal	Carbon Content	Moisture[97] Content	Reflectance[98] in Cedar Oil		Density[99]
	(%)	(%)	max	min	(g/cm³)
Peat	58.0	20	0.26	0.26	1.49
Lignite	70.5	17	0.35	0.35	1.425
	75.5	15	0.51	0.51	1.385
	81.5	9.0	0.67	0.67	1.320
	85.0	4.0	(0.92)	(0.90)	1.283
Bituminous	89.0	1.0	(1.26)	(1.18)	1.296
Semibituminous	91.2	1.0	1.78	1.55	1.352
	92.5	1.0	2.37	1.84	1.400
	93.4	1.0	3.25	2.06	1.452
Anthracite	94.2	2.0	4.17	2.22	1.511
	95.0	3.0	5.2	2.64	1.587
	96.0	4.0	6.6	3.45	1.698
	100.0	—	11.0	—	2.25

[97] Rumsby, W., Private communication.
[98] van Krevelen, D. W., and Schuyer, J., *Coal Science*, Elsevier, Amsterdam, 1957, p. 168.
[99] van Krevelen, D. W., and Chermin, H. A. G., *Fuel*, **33**, 85 (1954).

briquetting are its moisture content, hardness, elasticity, and plasticity, and the chemical and physical nature of its surface. There is no sudden break in any of these properties as the rank of coal changes from brown coal through its various more mature stages to bituminous coal and on to anthracite. The inherent moisture gradually decreases, the elasticity increases, plasticity decreases, and the surface gets less hydrophilic. The variation in moisture content, density, and reflectance with increasing degree of coalification is shown for a series of vitrinites in Table 11.

Binderless Briquetting of Bituminous Coal

There have been numerous attempts to briquette bituminous coal without a binder, and in general these have used much higher pressures than are required by brown coal. Pressure is not enough, however; the coal must be ground to a much finer size, and both the fine size of the coal and the high pressure bring their own additional problems and increase the difficulty of briquetting.

Most of the processes attempted are described in the patent literature, and theoretical papers are few and meager; however, a general pattern of thought can be discovered, and a theory of adhesion emerges which is similar to that now generally accepted for brown coal—as, of course, it should.

The Sutcliffe Process

The earliest important attempt to briquette bituminous coal directly is that made by Sutcliffe based on patents by Phillips[100] in 1911 and Sutcliffe[101] in 1914. It is noteworthy that this process is still in use today for the manufacture of "pure coal briquettes" which are converted into active carbon at the works of Sutcliffe Speakman Ltd, Leigh, England.

In his patent, Phillips stated that dry, finely comminuted bituminous coal can be compressed into lump coal, but the objection to this on the commercial scale is the difficulty of ridding the powder of its large volume of air. Phillips overcame this problem by precompressing a stiff magma of wet coal, heating it to about 100°C, and repressing the warm coal at 15 tons pressure. This solution has been used more recently.[102]

Sutcliffe identified a second problem which

[100] Phillips, H. J., and Phillips, A., Brit. Pat. 14,841 (1911).

[101] Sutcliffe, E. R., Phillips, H. J., and Thomas, D. A., Brit. Pat. 20,679 (1914).

[102] Idris Jones, W., and Rhys Jones, D. C., Brit. Pat. 616,857 (1949).

arises from the high pressures used. When high pressures are used, the high residual elasticity in the briquette makes it difficult to discharge it from the mold, because the uneven and sudden release of pressure causes the briquette to disrupt. Sutcliffe's solution controlled the rate of discharge, and the molds were so shaped that a lateral as well as a linear expansion was permitted during discharge. In later patents,[103] he laid more and more emphasis on this aspect of the process, and also adopted means to prevent air entrainment in the briquette. To release the air, the molds were provided with slots, and the briquette was subjected to repeated pressures.[104] To relieve the internal stresses, the briquette was allowed to expand during the intervals between the compressions.

By this means a divided mold was used in making large blocks—10 × 7 × 4 in.—so that the sides might be opened to allow lateral expansion of the block when the pressure was released from the ram. For smaller briquettes, a tapered mold was used; for example, a briquette of 3 in. in diameter and 2 in. deep would have a taper of $\frac{1}{16}$th in. In Fig. 15, the taper is exaggerated. Finely divided coal *a*, minus 200 mesh, is placed in a mold *b* which is then raised until the plunger *c* covers and enters the top of the mold. Pressure is applied to the movable plunger *d* and the coal is compressed. The mold *a* is then lowered to allow the briquette to be discharged while a certain residual pressure is maintained between the plungers *c* and *d*. It was claimed that, by this means, briquettes can be made in one operation from good coking coals, but with higher-rank coals it is better to effect compaction with two or more compressions. It is not possible to briquette anthracite.

The briquettes are excellent in strength and appearance and it has been stated that they are less smoky when burned than the original coal is. The fine grinding of the coal and the high pressures required in manufacturing briquettes by this process make them expensive, and therefore they are not sold as fuel but are processed further for conversion into active carbon.[105]

[103] Sutcliffe, E. R., Brit. Pat. 102,918 (1916); U.S. Pats. 1,267,711 (1918), 1,335,303 (1920).

[104] Sutcliffe, E. R., Brit. Pat. 5,019 (1915). Bolk, W., U.S. Pat. 2,078,832 (1937).

[105] Brownlie, D., *Proc. S. Wales Inst. Engrs.*, **42**, 273–6 (1926).

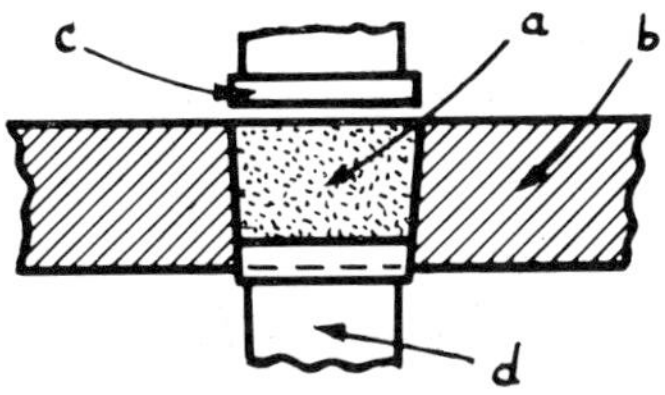

Fig. 15. The Sutcliffe mold.[103]

The Ten Bosch Process

In the early 1930's, a series of patents in the name of Ten Bosch Octrovien appeared. They described a process for briquetting bituminous coal without a binder, and adopted the same procedure for the formation of the briquettes as was put forward by Sutcliffe. Coal, reduced to a condition of "extreme fineness" was briquetted at high pressure, 3000 kg/cm^2. The difficulties caused by entrained air and by residual elastic stress were recognized in the design of the press.

It was stated that briquettes made from bituminous coal at these high pressures break up when the pressure is released as they are discharged from the mold. A number of the Ten Bosch patents[106] described molds with movable side walls which allow the briquette to expand in a "direction transverse to the line of compression." To obtain an uncracked briquette, the walls of the mold are allowed to move outward while the pressure is being removed from the ram, so that the briquette can expand in all directions and is free from constraint when it is discharged from the mold. Many versions of such molds were mentioned, but no mold with movable sides appears to have been successful in practice, although theoretically the solution is an ideal one. Later patents by Ten Bosch[107] described another solution, and one was eventually adopted on a large press. In this solution, the mold is shaped in such a way that, as the briquette is removed, the elastic strain is accommodated by the changing profile of the mold. It is not possible to provide a perfect solution in this way because a compacted powder does not possess uniform elastic stresses throughout its form. The elastic recovery in the direction of pressing is far greater than transverse to this direction, and

[106] Ten Bosch Octrovien, N. Y., Brit. Pats. 406,492 (1934), 427,969 (1933); U. S. Pat. 2,133,675 (1938).

[107] Ten Bosch Octrovien, N. Y., Brit. Pat. 450,633 (1936).

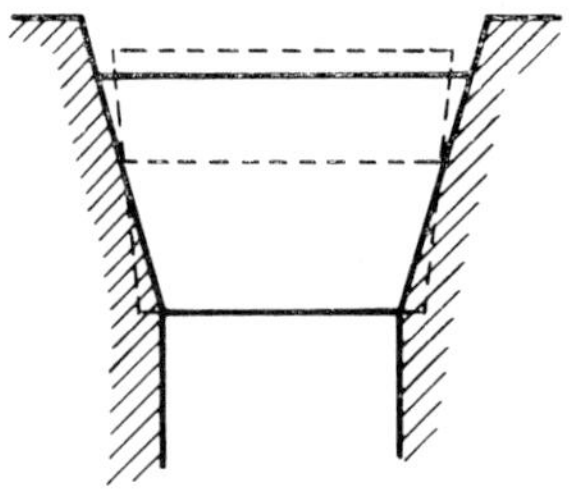

Fig. 16. Diagrammatic representation of a briquette under pressure and free in a mold.[107]

the transverse recovery is not uniform. In Fig. 16, the bold line depicts the briquette under compression in the mold, and the dotted line depicts its shape when free from constraint. In practice, when the ram pressure is removed, the top portion is free, and the bottom portion is locked under constraint; this condition causes radial cracks in the top face and circumferential cracks in the bottom part. Such cracking may be prevented[108] by using a mold which, during the pressure relief, allows the pressure to be reasonably uniform over the entire surface of the briquette. Such a condition may be achieved with a mold having a surface of rotation with a profile

$$\frac{x}{x_0} \log (1 + \epsilon_a) = \frac{y}{y_0} \log (1 + \epsilon_r)$$

where ϵ_a is the axial expansion and ϵ_r the radial expansion.

The coal is compressed into this mold by a ram which moves into the mold until its leading edge just touches the walls of the mold; at this point compression should be complete. Such a system, of course, calls for a precise filling of the mold, and this is not easy. The difficulty is partly overcome by using a stepped ram head, and by pressing in two stages, the first at low pressure, and the second at the required operating pressure.[109] The removal of entrained air is effected[110] while the mold is chamber in which a paddle is rotated rapidly, being filled; the coal powder is fed into a and this has the effect of throwing the coal particles into the mold or premold cavity. A further development of this idea appears to have been incorporated in the press manufactured to carry out the Ten Bosch process.[111]

An experimental plant of 120 kg/h capacity was operated at Gutehöffnungshütte, Osterfeld[112] between 1940 and 1944, and produced briquettes weighing 15 to 18 g. A larger press was built subsequently and produced 2000 kg/h of briquettes weighing 80 g. The press is shown diagrammatically in Fig. 17. Clean coal from an electrostatic separator was reduced in a pneumatic impact pulveriser to a size of 40% minus 10 μ. This coal, heated to 70 to 80°C, was fed from a hopper *a* into a feed chamber containing a rapidly rotating paddle *b* which fed the coal at some velocity into the feed chamber *c* of the press when the low-pressure ram retracted. The press consisted of a robust table *d* rotating in a vertical plane; the table contained three identical molds shaped to allow easy release of the briquettes. The three rams entered the molds simultaneously; the first delivered the coal from the feed chamber into the top mold *e* and compressed it to 100 atm pressure, the second ram with a specially shaped profile entered the second mold *f* containing a precompressed briquette and pressed it at 4000 to 5000 atm, and the third ram moving in the opposite direction ejected the finished briquette from the mold *g*. The cylinder rotated through 120°C after each stroke of the ram. The briquettes were hard, dense, and of good appearance.

It has been stated that the profile of the high-pressure ram and the shape of the mold were intended to effect a uniform development and release of pressure, but that different coals possessing different elastic characteristics required different profiles.[113] Petersen has commented[74] on the performance of these presses: Briquetting of anthracite and semi-anthracites without binders is not possible even when high pressures are used. Satisfactory briquettes can be made from coals of 14% volatile matter and over, that is, from steam coals and coking coals. Only poor briquettes are made from gas coals and gas flame coals. Petersen concluded that although binderless briquetting of steam coal is technically possible by the Ten

[108] Ten Bosch Octrovien, N. Y., Brit. Pat. 440,811 (1936).

[109] Ten Bosch Octrovien, N. Y., Brit. Pat. 456,114 (1936).

[110] Ten Bosch Octrovien, N. Y., Brit. Pat. 453,010 (1936).

[111] Ten Bosch Octrovien, N. Y., Brit. Pat. 505,464, 512,374 (1940).

[112] Hendricks, T. A., *JIOA*, *Final Rept.*, No. 35, Exhibit 39, 1–3 (1945).

[113] Lilpins, W., *BIOS*, *Rept.* (Trip 3265).

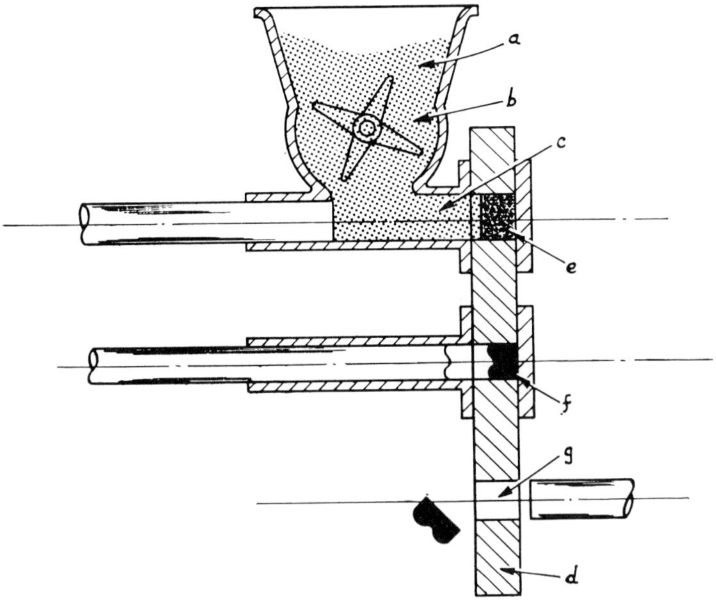

Fig. 17. The Ten Bosch press.

Bosch method, the higher costs for drying and grinding, and the very high capital cost of the plant make the process uneconomic at the present time. Kegel[11] stated that the Ten Bosch process can produce bituminous coal briquettes without a binder at pressures of 5000 to 6000 kg/cm^2, but the output of the press is too small and the process too costly. There appears to have been no further commercial development of this method.

The Komarek-Greaves Press

Double-roll presses have been used for compacting bituminous coal without binders, and Komarek-Greaves supplied three plants for Canada in 1941.[114] These presses were intended for use with a binder, but were operated for a short time to make binderless briquettes. The binderless briquettes did not stand up to handling and were very dusty,[115] but they appeared to improve when stored.[116]

Haigh[114] has also referred to the high pressures needed to briquette coal, 25,000 to 30,000 lb/in.^{2}, and has drawn attention to the difficulties experienced in finding a steel which will stand up to the heavy duty involved by developing these pressures. For small rolls, a heat-treatment process has been found satisfactory, but for large rolls, of the size required for coal, the problem was not as easily solved. Midrole—M.17 steel, a stainless steel—is used in the Duluth high-pressure briquetting unit.

The pressure developed by double rolls depends to a large extent on the nature of the material to be briquetted. Dry, finely powdered coal is not one which, unaided, allows the pressure to build up. High-pressure briquetting may be achieved by force feeding the rolls. The earliest of these feeders was a combined worm feeding from both sides to a central paddle. A single paddle has proved to be satisfactory. This paddle may consist of shoes which press the material to be briquetted into the vee between the rolls (Fig. 18).

In this form of briquetting, as in others, attention must be given to the preparation of the coal. Haigh cited a size of 100% through 100 mesh—96% actual, and 60% through 200 mesh. It is undesirable to have much coal below 325 mesh since this causes difficulties in feeding. The Komarek-Greaves process uses coal at temperatures of 400 to 180°F, temperatures sufficient to permit the coal to flow plastically, but much below that necessary to cause

[114] Haigh, P. D., *Wyoming Univ., Nat. Resources Research Inst., Inform. Circ.* No. 3, 16 (1949).

[115] Miller, J. G., *Wyoming Univ., Nat. Resources Research Inst., Inform. Circ.* No. 5, 14 (1951).

[116] Morrison, R. L., *Wyoming Univ., Nat. Resources Research Inst., Inform. Circ.* No. 5, 19 (1951).

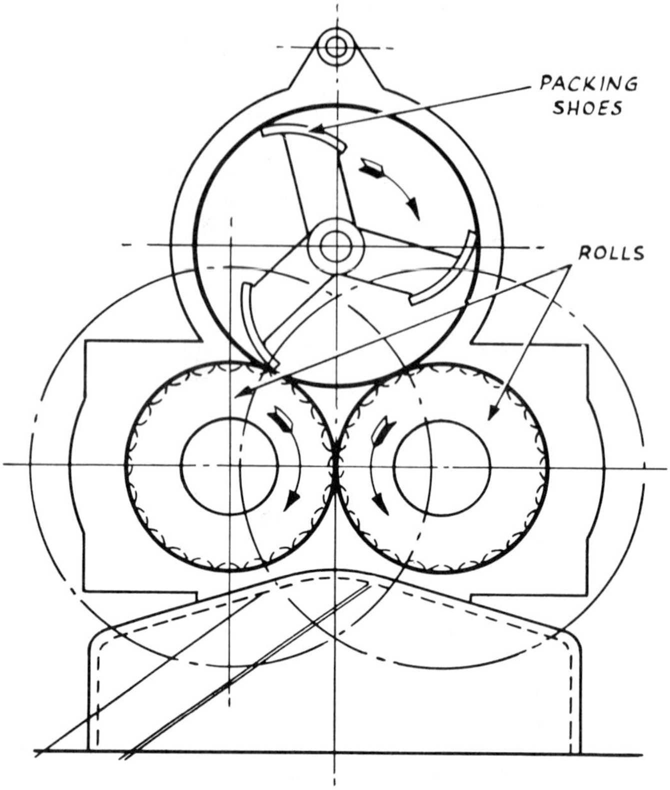

Fig. 18. Komareck-Greaves press.

fusion. Meyer[117] has reported that successful experiments were carried out with double-roll presses at Gutehöffnungshütte at Oberhausen. The pressure had been increased from the normal 2100 to 3500 lb/in.^{2} to 70,000 lb/in.^{2}, but under these conditions the process did not prove economical. An extension of the Gutehöffnungshütte process heated the coal to a higher temperature—to its softening temperature, but below its carbonizing temperature—and the softened coal was pressed into briquettes. These tests were abandoned because it was too difficult to maintain the narrow temperature range on which the process depended.

Ring-Roll Presses

An important alternative to the double-roll press has been developed for high-pressure work. The powdered coal is pressed between a roll and the inner surface of a ring; thus a very narrow angle of entry is achieved, and with it, of course, considerable drag, which obviates forcible feeding of the powdered coal. Such a system has many advantages, but also some disadvantages that have not yet been completely overcome. Ring-roll presses are used for making briquettes from brown coal, and are said to be specially suitable for the harder brown coals.

Spooner[118] referred to the earliest press as having been made in 1880. The Allen-Hutchinson briquetting press was put into operation in 1907 at the Allen plant in Pittsburgh.[119] This press consisted of a ring and roll lying horizontally on a table; both the ring and the roll were profiled so that the briquettes were approximately cylindrical in shape, and the ring and the roll were positively driven to ensure registration. The briquettes made were of 8 to 10 oz each and had a specific gravity 1.14; the capacity of the unit of the Allen plant was 5 ton/hr.

[117] Meyer, G. A. H., *Wyoming Univ., Nat. Resources Research Inst., Inform. Circ.* No. 6, 71 (1953).

[118] Spooner, E. C. R., *BIOS, Final Rept.* **1830** (1947), 21 pp.

[119] Hutchinson, E. C., U.S. Pat. 851,007 (1907).

A second type of press, very similar in design to the Allen-Hutchinson one, was developed by Jenkins[120] for the manufacture of peat fuel for the New Era Development Co. This press differed from the Hutchinson press in that the ring was smooth. Both the roll and ring were rotated by a geared drive, and the briquettes formed were semicylindrical, rather like pig-metal ingots.

The Apfelbeck Press

In 1925, Apfelbeck[121] invented a version of the ring-roll press. In 1930, one of these presses was installed at the Rtanj mine in Yugoslavia, followed by a second one in 1935. One of these presses is still in operation briquetting Rtanj coal.[122]

Such a press, shown in Fig. 19, consists of a ring *a* of about 180 cm diameter, and a roll *b* of 90 cm diameter. The ring is supported upon and driven by two rollers *c* at a peripheral speed of 75 cm/sec. The roll is mounted upon a pair of steel arms pivoted at *d* and held by means of a tie-rod *e* with a heavy spring. This arrangement permits the roll to exert a pressure of 1600 kg/cm^2 on the coal held in the ring. The ring consists of two separate parts which are held together at the bottom by means of two pairs of tapered rollers *f* mounted on the steel frame of the press. Coal is fed in at *g*, the overspill collects at *h*, and the briquettes are discharged at *i*. The ring is formed in two parts to permit easy discharge; the separate parts are held firmly together where the briquette is formed at the bottom of the press, where they are subjected to the maximum pressure. In a press of these dimensions, the pressure becomes apparent 11 cm from the bottom of the roll at *b* shown in Fig. 20.[123]

The two parts of the ring are the widest apart at the top, the gap amounts to about 5 mm, and there is, of course, a gradual release of the lateral pressure, permitting expansion in the briquette both in the direction of the applied pressure and transverse to this. It has been claimed that this gradual release is necessary for the discharge of unfissured briquettes. This claim is supported by the quality of the briquettes made at Rtanj.[124] The operation of the ring-roll press allows the pressure to build up smoothly as indicated in the time-pressure diagram. The smooth development of pressure makes the escape of air easy, but it is this escape that limits the rate of operation to peripheral speeds of about 100 cm/sec.

Ronge[125] has compared the operation of this

[120] Jenkins, C. D., U.S. Pat. 1,376,096 (1921).

[121] Apfelbeck, H., Czechoslov. Pat. 23,198 (1927).

[122] Ronge, W., *Berg-u. hüttenmänn. Monatsh. montan. Hochschule Leoben*, **8,** 112 (1955).

[123] Illiwitzki, J., *Braunkohle*, **35,** 597–600, 618–21 (1936).

[124] Holik, J., *Glückauf*, **70,** 385–93 (1934).

[125] Ronge, W., *Berg-u. hüttenmänn. Monatsh. montan. Hochschule Leoben*, **3,** 59–62 (1950).

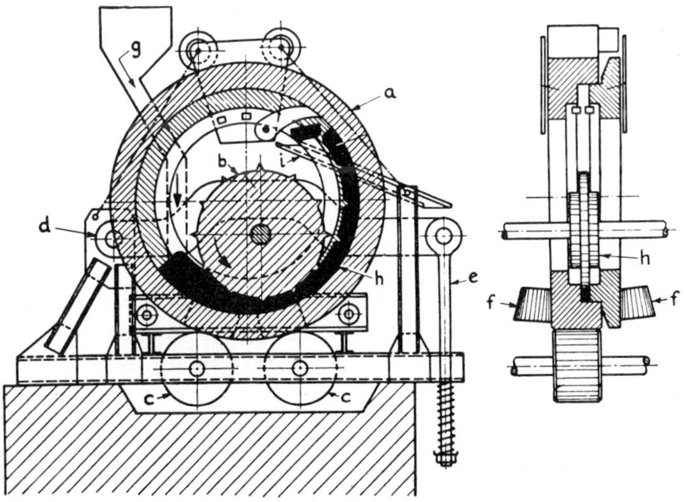

Fig. 19. The Apfelbeck press.

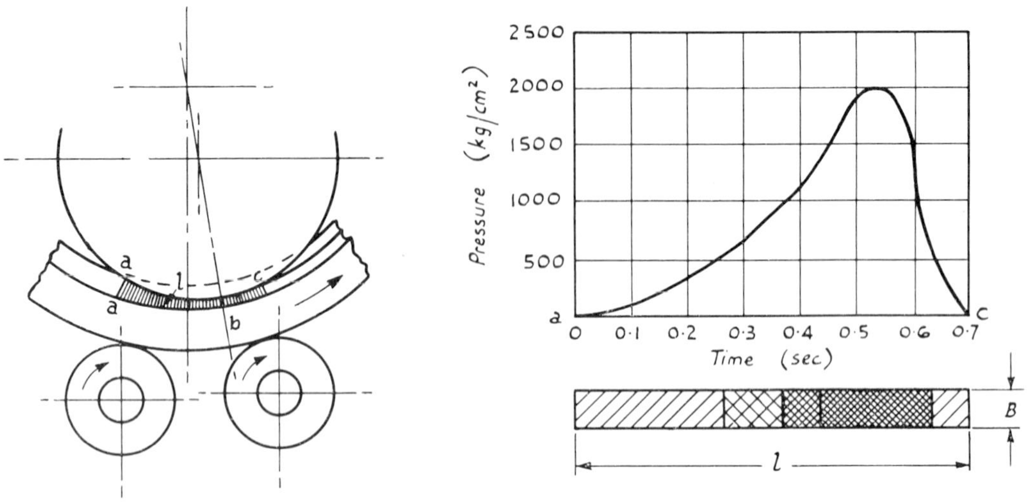

Fig. 20. Pressure-time diagram for the Apfelbeck ring-roll press.[123]

ring-roll press with that of an Exter extrusion press (see Table 12).

TABLE 12

COMPARISON OF OPERATING CONDITIONS OF EXTER AND RING-ROLL PRESSES

Press	Exter	Ring-Roll
Pressure (kg/cm²)	1000–1200	1600
Output (ton/hr)	20	9.6
Power consumption (kwhr/ton)	19–35	7.5
Production before renewal of mold forms (ton)	1000	45,000

The coal mined at Rtanj is bituminous but is of an especially plastic nature. Its analysis,[122] on the dry, ash-free basis, is carbon 85.1, hydrogen 4.73, sulfur 7.3, volatile matter 23.6%, and calorific value 15,400 Btu/lb. The coal yields a silver grey, swollen coke. Kegel[11] described this coal as being a relatively soft Jurassic bituminous coal, and stated that all attempts made in Germany to briquette hard coal in a ring-roll press were unsuccessful. Piersol[126] stated that the Apfelbeck design of ring-roll press was not satisfactory for briquetting Illinois coal without a binder.

The Krupp-Herglotz Press

Spooner[118] has described the Krupp-Herglotz press which was developed at approximately the same time as the Apfelbeck press. The chief difference between these presses lies in the means adopted for discharging the briquettes. A solid ring is obviously preferable to a split ring, and in the Krupp-Herglotz press the briquettes are formed in the ring groove in pairs separated by a projecting circumferential rib on the periphery of the roll (see Fig. 21).

Powdered coal is fed into the ring by means of a screw conveyor, and the surplus coal is recirculated. The coal is drawn into the narrowing gap between the roll and ring, the rib separating the compacting coal into two briquettes. After passing the narrowest point, the rib withdraws from the coal and finally frees the two briquettes which are eventually scooped out of the ring groove. Further improvements have been made[127] in which the release of the briquettes becomes still easier, and in one of the latest forms the protruding rib lies on the inside of the ring and the groove in the roll. The released briquettes fall away from the ring. There is apparently still room for improvement, particularly in respect to the uneven filling of the coal into the press. The development of flats on the surface of the ring when driven by friction rollers has been eliminated by driving the roll.

These presses, although capable of developing high pressures smoothly are now said to be unsuitable for the briquetting of bituminous

[126] Piersol, R. J., *Illinois State Geol. Survey, Bull.* No. 72 (1948), 198 pp.

[127] Streppel, A., German Pat. 713,849 (1941).

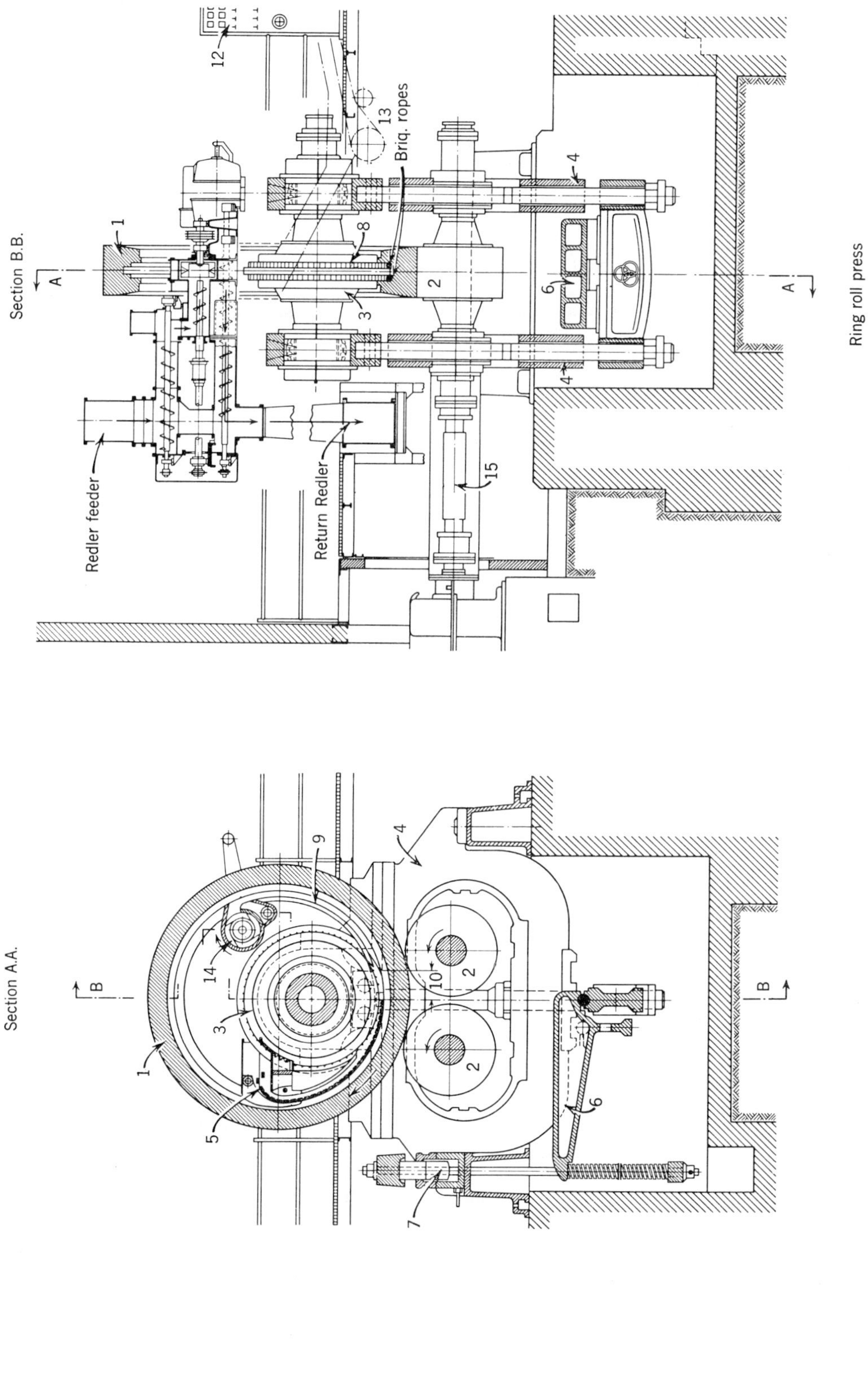

Fig. 21. The Krupp-Herglotz ring-roll press.[118] 1. Press ring; 2. driving rollers; 3. press roll; 4. frame; 5. pressure lever.

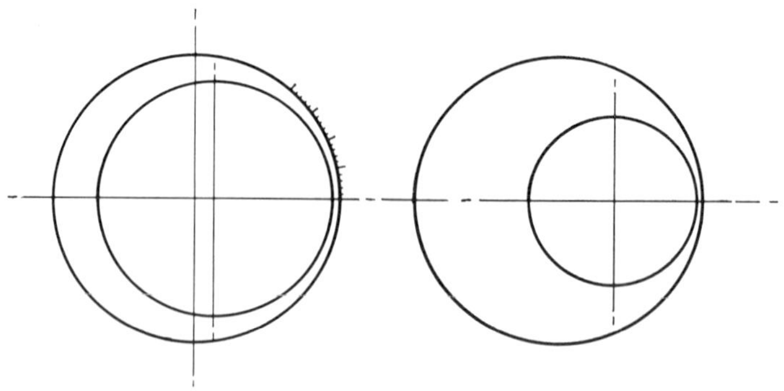

Fig. 22. Diagram of eccentricity of inner and outer dies in Piersol and Apfelbeck briquetting presses.[126]

coal, and even with the addition of a small quantity of pitch to the coal, the product is not considered satisfactory.

Piersol Press

In a program of research initiated in 1931, Piersol[128] studied the manufacture of smokeless briquettes from Illinois coals, and found that such briquettes could be made without the addition of binders by impacting powdered coal. A search for means to implement this mode of compaction in a commercial press led to a consideration of the existing presses used for briquetting bituminous coal or lignite without binders, and to the conclusion that the ring-roll of press might be developed into a satisfactory machine. Samples of Illinois coals were sent for test to Germany during the 1930's, but Apfelbeck reported that he found it necessary to add 2% of asphalt as a binder. The Apfelbeck design[126] does not work on Illinois coals because the diameter of the inner roll is too small compared with the ring, and because the cavities prevent the backward and forward extrusion of the individual briquettes (see Fig. 22). Piersol made this ability to extrude individual briquettes backward and forward the essential feature of his press[129]—a ring diameter of 45 in., a roll diameter of 37 in. with a difference in centers of 3 in. The ratio of the diameters of the roll and ring is thus 1 : 1.2, and the ratio for the Apfelbeck press is 1 : 1.7. The thickness of the briquette at the point of maximum pressure was 1 in. and the distance between the ring and roll was 2 in. at an angle of 55° from the roll center, and 3 in. at 77.5°. The maximum distance was 7 in. The dimensions of the two presses are compared in Table 13.

[128] Piersol, R. J., *Illinois State Geol. Survey, Rept. Invest.* No. 31 (1933), 70 pp.; No. 37 (1935), 75 pp.

[129] Piersol, R. J., U.S. Pats. 2,021,020 (1934); 2,119,243 (1938); 2,321,238 (1941).

TABLE 13

Comparison of Press Dimensions for the Piersol and Apfelbeck Presses

Press	Piersol	Apfelbeck
Ratio–ring:roll	1.216:1	1.71:1
Minimum gap x (in.)	1	x
Angle—roll center, $2x$ (°)	55	53.5
Angle—roll center, $3x$ (°)	77.5	75
Maximum gap (in.)	7	$7.75x$

The inner roll and packing ring in the Piersol press are rigidly connected, whereas in other versions of the ring-roll press, movement is permitted between ring and roll, and the pressure builds up against springs.

Although excellent briquettes were produced, some difficulty was experienced with the experimental press owing to the high pressures used. Various materials were used for the ring and roll dies; soft iron was no good, and chromium-steel castings eventually broke, because of excessive brittleness developing as a result of work hardening. Manganese steel was better, but eventually showed cracks; "Evansteel" proved satisfactory. The only other difficulty recorded concerning the operation of the press was that of feeding the coal. The auger feeder was unsatisfactory and was eventually replaced by an electro-magnetic vibrator feed. Piersol's object was to produce smokeless briquettes, and although he has recorded, as have others (Sutcliffe), that the smokiness of the briquette was less than that of the lump coal, he found it beneficial to subject the briquettes to a heat treatment to reduce the smoke still further. He extended his experiments into the field of briquetting hot coal or char.

The Coal Research Establishment Direct Briquetting Process

In 1953, a new concept was introduced into the technique of briquetting.[130] It was found that the density of a powder compact could be materially increased for a given applied pressure if, while still under load, it was subjected to shear strain. The strength was increased and the elastic recovery was reduced. The gain in strength with the application of additional shear strain under load was substantial, but the full benefit was obtained only if the shear strain was introduced under maximum load. Compacts made in this way possessed greater strength and higher density than the limiting values obtained by simple pressing, and, indeed, in many cases the density of the compact approached that of the particles themselves.

The failure to achieve complete compaction in conventional, simple pressing systems may be explained in this way. First, as the briquetting pressure is applied, it is opposed by forces set up in the powder and by frictional forces on the walls of the mold which resist the movement of the particles and prevent intimate contact between the surfaces of the particles. Second, when the external pressure is removed, the deformed particles recover their shape elastically, in part at least, and the compact expands; the voids within the briquette increase and residual strains remain.

If high shear strains are caused to occur when the compact is under load, internal arching systems are disrupted, internal frictional forces are reduced, and hydrostatic-strain conditions within the compact favor the plastic deformation of otherwise brittle material. Plastic deformation in turn causes the production of new clean surfaces which favor the development of strong adhesive forces.

The technique described, when applied to bituminous coals, resulted in the formation of very strong briquettes. Tensile strengths in excess of 1000 lb/in.^{2} have been obtained in the laboratory.

[130] Rhys Jones, D. C., and Gregory, H. R., Brit. Pat. 807,433 (1953).

The relationship between shear strain and briquette strength is given in Fig. 23. Excessive shear strain can cause the formation of slip planes and weak briquettes. The limiting shear stresses are given by equation

$$S = \alpha \tan \phi + C$$

where S is the ultimate shearing stress, α is the normal principal stress, ϕ is the angle of internal friction, and C is the cohesion of the material.

Experimental work,[131] using a simple shear box apparatus (see Fig. 24), has not been developed into a commercial machine, but the principle of compacting by introducing shear strain under load has been applied in two other ways; the first uses a hydraulic press with an indexing table, and the second uses an extrusion press powered either hydraulically or mechanically.

The indexing table press (see Fig. 25) consists of a horizontal rotary table and two "duplex" plungers. Twelve molds are located in the table, and the operations of filling, compacting, and ejecting take place simultaneously in the different molds. The "duplex" plunger (see Fig. 26) consists of a central core and an outer annulus; initially both parts move together to compact the powder at a pressure of 6 ton/in.^{2}, and then the pressure of oil in the locking cavity lifts a relief valve to allow the core to move forward relative to the annulus. This relative movement sets up high-angular shear strain in the briquette while it is under conditions of maximum load. Emphasis is placed on uniform release of residual strain in order to achieve a crack-free product.

In the extrusion press, the high angular shear strain is introduced by moving the briquette under pressure through an extrusion channel of a changing cross section.[132] The system is unstable, and it is necessary to use an automatic control which alters the briquetting pressure by changing the frictional resistance of the briquettes as they move through the nozzle.[131]

[131] Gregory, H. R., *J. Inst. Fuel*, **33**, 447–61 (1960).

[132] Coal Industries Patents, Ltd., Rhys Jones, D. C., and Gregory, H. R., India Pat. 65,320 (1957). Hock, H., and Roenneke, H., German Pat. 823,442 (1949).

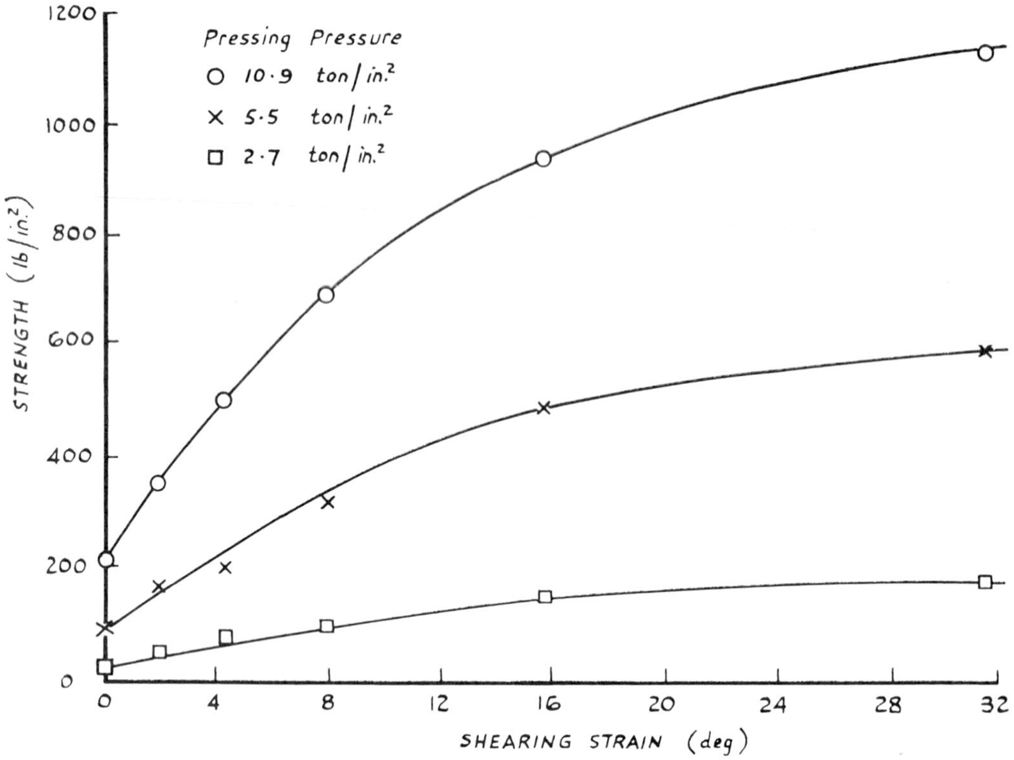

Fig. 23. Effect of shearing strain on briquetting strength.

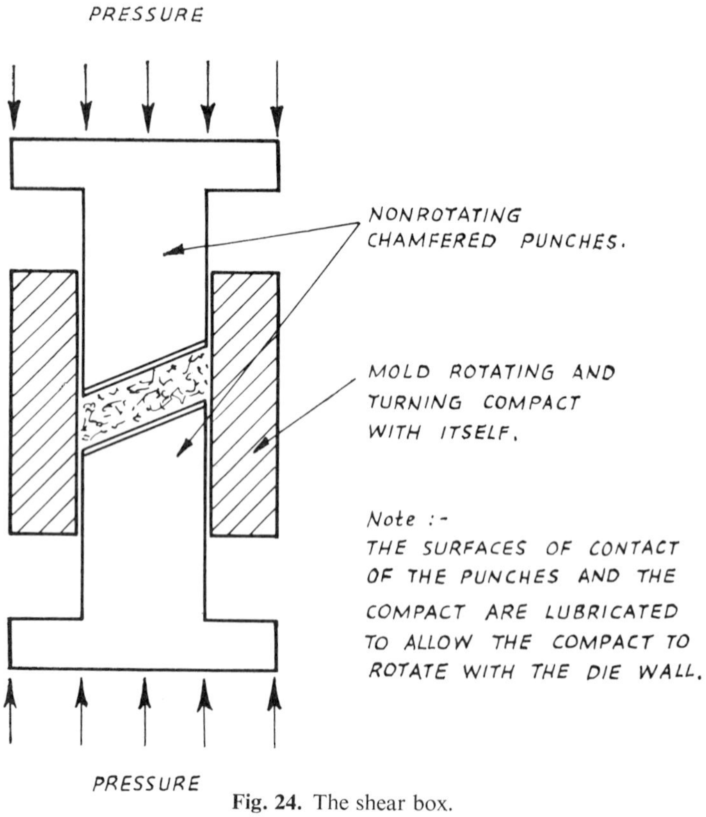

Fig. 24. The shear box.

Fig. 25. The horizontal turntable press.

The pressure travel diagrams for the duplex-plunger press and the extrusion press are shown in Fig. 27.

A pilot plant has been built in which one duplex-plunger press and two extrusion presses, one hydraulically operated and one mechanically operated, are installed. There has as yet been no commercial development. Various modifications to this system have been put forward.[133] All ranks of coal can be made into hard briquettes by this process, and the technique appears to be applicable to all substances which undergo some plastic deformation at the compaction pressures used. Low-volatile anthracites are completely elastic at pressures in the region of 6 ton/in.^{2} unless they are very finely ground.

The important process variables are moisture content, temperature, and particle size. The optimum moisture content is approximately 75% of the inherent moisture content, and the optimum temperature is between 90 and 120°C. Low-rank coals are more sensitive to temperature than high-rank coals; and little is gained by briquetting above these temperatures until the fusion range 400 to 500°C is approached. The speed of pressing has little effect provided due care is given to the problems of air entrainment; these can be severe in the "duplex"-plunger press. High speed also increases the mechanical difficulties associated with the uniform release of residual strain. The briquette size is not a limiting factor. The technique has been applied with equal success to briquettes ranging from 0.5 to 4.5 in. in diameter.

Briquettes made from bituminous coal are of high strength and good appearance and burn freely, but, as with all binderless briquettes made from fine coal, the broken surface is

[133] Gregory, H. R., Brit. Pat. 839,764 (1960).

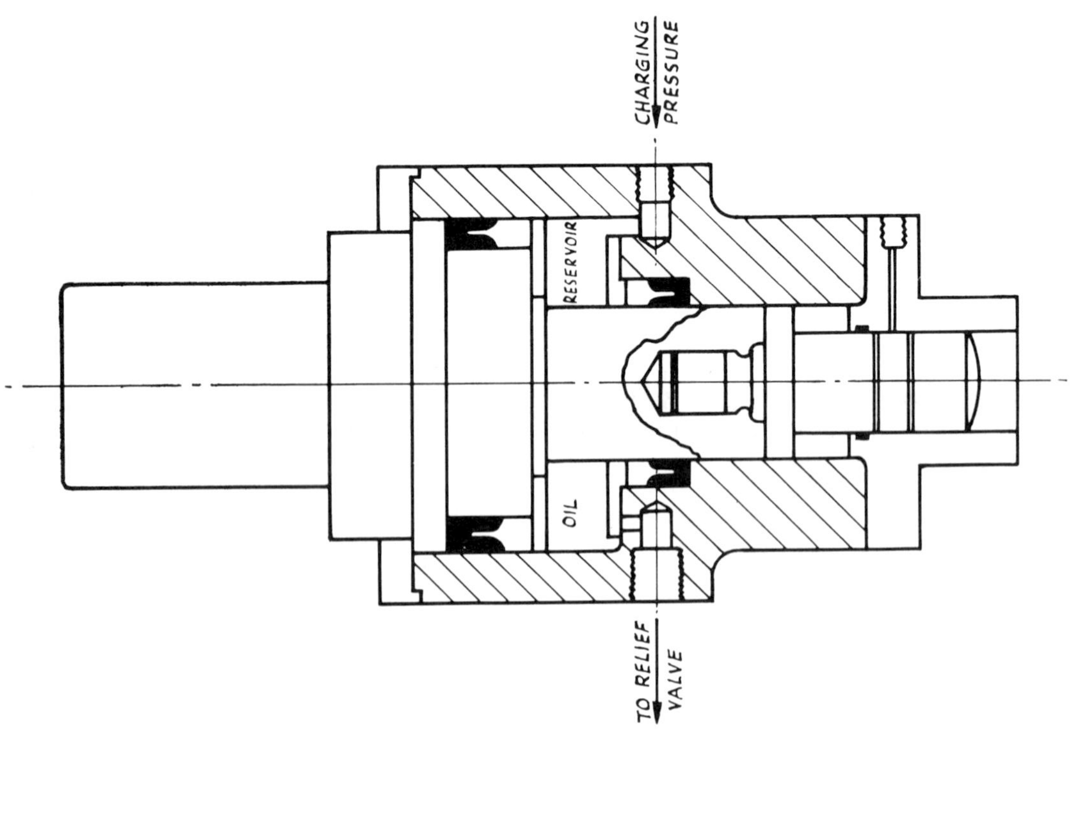

Fig. 26. Simple "duplex" plunger.

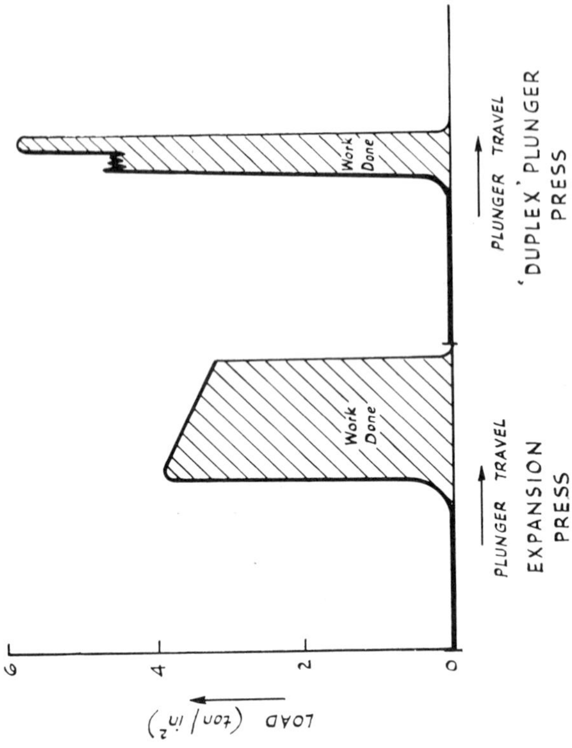

Fig. 27. Comparison of pressure-travel diagrams for "duplex" and expansion presses.

dirty. Briquettes made from high-volatile coals weaken when exposed to weather, but briquettes with a volatile content below 30% are not affected.

Binderless briquettes from noncoking coals have been carbonized to yield undistorted hard coke briquettes of very high density and strength. Briquettes made from coking coals distort and stick together unless the coking property is destroyed by oxidation or other chemical means.

The Mechanism of Briquetting

To summarize, brown coal and bituminous coal are members of a series, and the transition in properties from one to the other is a gradual one. It is unlikely that the nature of the cohesion between brown-coal particles is different from that between bituminous-coal particles.

The methods which succeed in making hard briquettes are those which aim to bring the particles close together. With bituminous coal, which is harder than brown coal, this is achieved by the adoption of one or more of the following conditions: (1) the use of high pressures; (2) the introduction of sheer strain under load; and (3) the preparation of the coal, by fine grinding, by controlling the moisture content, and by briquetting at the appropriate temperature so that it is in its optimum state of plasticity.

These conditions lead to the formation of a hard briquette in the mold, but other conditions have to be observed in order to extract the briquette without damaging it. These conditions are: (1) the uniform release of the elastic stress; and (2) the reduction in the volume of entrained air.

There have been numerous theories of binderless briquetting, but with bituminous coal it is possible to dismiss the "capillary" theory of adhesions and also perhaps the "bitumen" theory. Three mechanisms then are left, all of which may play a part.

Borchert[134] has suggested that adhesion is caused by the interlocking of irregularities in the surfaces of particles when they are pressed together, and Phillips[135] has shown how a number of "Perspex" rods can be rigidly interlocked by making use of the elastic properties. (Fig. 28.)

Photomicrographs of briquette sections have shown interlocking of particles by deformation, but the interpretation of a planar section of what is essentially a three-dimensional arrangement is difficult. Some welding of particles by actual fusion of the coal particles appears to be quite unavoidable under the conditions of briquetting, which involve the rubbing of particles against each other while under pressure. The development of hot spots by friction[136] could cause the coal to melt at these points. As the average temperature rise on briquetting is so small, this form of adhesion cannot happen to any very large extent. It is difficult to avoid the conclusion that the major cause of attachment between particles is the natural cohesion force of the substances itself. To allow these forces to come into play, the surfaces must be brought very close together, and the strength of the attachment will depend on the area of surface in contact. It is also probable that the cleanliness of the surface is important; any contamination of the surface would tend to reduce its cohesive powers. Deformation of the particles not only allows larger surface areas to come closer together, but also creates fresh uncontaminated surfaces. Plastic deformation will allow the surfaces to remain together, whereas elastic deformation will tend to re-establish the original particulate state.

CARBONIZATION OF BRIQUETTES

There are some industrial areas where good quality coking coals are becoming scarce[137] but where noncoking coals are still available. There are also areas in some parts of the world which, although provided with noncoking coal, have not developed as fully as they might because of the lack of coking coal. In some countries there is a demand for smokeless fuels for domestic use where lack of coking coal restricts the availability of low-temperature coke. If cokes possessing the necessary hardness, density, and reactivity could be provided in pieces of uniform size and shape, it is likely that efficiency of both blast-furnace and cupola operation would be increased. Future

[134] Borchert, H., *Braunkohle*, **11**, 62–4 (1959).

[135] Phillips, J. W., National Coal Board, Private communication.

[136] Bowden, F. P., and Thomas, P. H., *Proc. Roy. Soc. (London)*, **223A**, 29–40 (1954).

[137] Bardgett, H., *J. Inst. Fuel*, **27**, 274–91 (1954).

Fig. 28. Adhesion of "Perspex rods" by elastic deformation.[135]

specifications for coke are likely to recognize this.[138]

The carbonization of briquettes makes it possible to manufacture coke from noncoking coals for both the industrial and the domestic market, and to manufacture this coke to a specification which includes uniformity of shape and size. The uniformity of shape and size in briquettes and the fact that they are discrete pieces opens up interesting possibilities for their carbonization. It is impossible to carbonize these briquettes in a conventional metallurgical coke oven, but inclined ovens or vertical retorts would be suitable. It is well to consider what advantage could be taken of the particulate nature of the charge and to what new methods of rapid carbonization this could lead. The manufacture of carbonized briquettes invokes an additional stage over and above that required for making coke by conventional means, and this factor will naturally affect the economics adversely. On the credit side, however, may be the availability of cheaper coal and the higher throughput in the ovens.

For these reasons, much research has been devoted to this subject. Among the earliest workers was Parr;[139] indeed, he may be truly described as the research pioneer in this field, and his scientific studies are of importance. Sutcliffe and Evans[140] were also early in the field, but as inventors and practical engineers. They were followed by many others. The work of these men has led to the establishment of a

[138] Lee, G. W., *Coke and Gas*, **20,** 526–9 (1958).

[139] Parr, S. W., and Layng, T. E., *AIME, Trans.*, **63,** 953–9 (1920).

[140] Sutcliffe, E. R., and Evans, E. C., *Iron & Coal Trades Rev.*, **104,** 607 (1922).

number of factories, but the proportion of briquettes carbonized is small in comparison with the total briquette production.

Two processes which started during the interwar years, 1920 to 1940, are active and expanding in output. The "Phurnacite" process,[141] which uses a low-volatile steam coal, was established in South Wales in 1938 and has increased its output steadily from about 150,000 tons per annum in 1939 to 650,000 tons in 1959. The second process is the Lurgi process,[142] which was developed in Germany for the upgrading of brown coal. This process has spread to other countries.

The quality of the briquettes will obviously depend on the nature of the coal and the nature of the binder, if one is used. The quality will also depend on the way in which the briquettes are carbonized. Research has elucidated some of the problems, and experiments in the laboratory, on the pilot plant and even on a large scale, have been carried out. As in most commercial activities not as much information has been published as exists in the archives of industrial laboratories.

The economics of the process of briquette carbonization are affected by the nature and yield of the byproducts obtained, which, of course, are not the same as those obtained by the present conventional methods of manufacturing coke. The difference in yield and composition results from the following factors:

1. The coals carbonized are different, being either high-volatile or low-volatile noncoking coals.
2. The time and temperature of carbonization are different.
3. The charge is particulate and permits the easy removal of gas from the hot zone.
4. The coals may have been pretreated.
5. Two stages of carbonization may have been involved. Examples of actual results are given under the sections dealing with the "Carbocoal," "N.F.C.," and "Lurgi" processes.

Theoretical Considerations

Unless special precautions are taken, when coal briquettes are carbonized, they may fuse together, they may swell and become misshapen, or they may crack or even fall to pieces. The difficulties inherent in the manufacture of carbonized briquettes may be appreciated by a study of what happens when compacts of coal are heated, and how this behavior is influenced by the presence of a bituminous binder.

Gregory[143] has summarized the available information in two graphs, Fig. 29 and 30.

Idris Jones[144] has referred to his experiments leading to the development of the "Phurnacite" process for the Powell Duffryn Co. From these experiments, the optimum conditions for briquette composition and carbonization were determined. High rates of heating were essential, but the temperature and time of carbonizing were determined by the product required (see Tables 14 and 15).

The coals used in the process were "dry steam coals" of 9.6 to 14% volatile content, but even these possessed some slight caking powers which could be an embarrassment, and it was necessary to know the effect of different levels of this. The effect of the agglutinating index of the coal is shown in Table 16.

It is evident that there is little hope of making sound carbonized briquettes from coals lying in the coking range, and the safe approach would be to use coals possessing no marked softening properties.

The quantities of naturally suitable coals are limited, and the need to convert the more abundant coals into suitable material was appreciated at an early stage in the history of briquette carbonization. It was first achieved[145] by low-temperature carbonization, which converted coal with excessive caking powers into a semicoke possessing little or none. The caking properties of a coal may be destroyed in other ways, chlorine and other halogens being very effective.[146] Sulfur is also effective,[147] and the effect of oxygen on coal has been known for a very long time. Oxidation reduces its

[141] Anon., *Coll. Eng.*, **30,** 48–57 (1953).

[142] Anon., *Coke & Smokeless-Fuel Age*, **8,** 77–9 (1946).

[143] Gregory, D. H., *Chemical Engineering in the Coal Industry*, F. W. Sharpley, Ed., Pergamon Press, London, 1956, pp. 58–9.

[144] Idris Jones, W., *J. Imp. Coll. Chem. Eng. Soc.*, **11,** 23–8 (1957).

[145] Smith, C. H., U.S. Pat. 1,177,727 (1916).

[146] Eccles, A., and McCulloch, A., *Fuel*, **10,** 308–19 (1931).

[147] Postovski, I. Y., and Harlampovich, A. B., *Fuel*, **15,** 229–32 (1936).

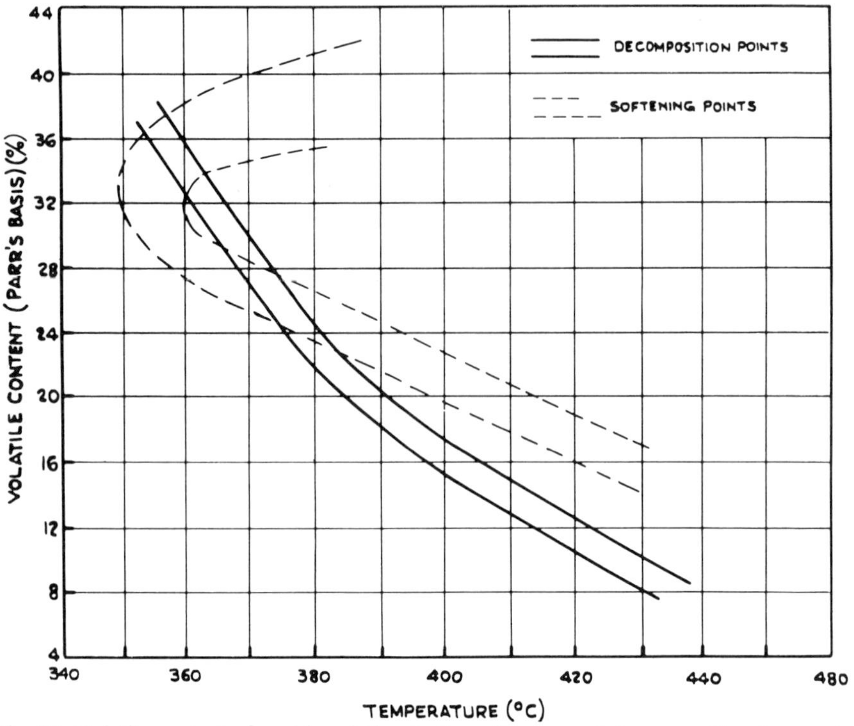

Fig. 29. Volatile content of coal in relation to the softening and decomposition temperatures.[143]

"stickiness"[148] and its fluidity[149] as well as its swelling power.

Jenkins has studied the mechanism of oxidation.[150] He has stated that by the addition of small amounts of oxygen to coal it is possible to make pitch-bound briquettes which are nonsticking and nonswelling when carbonized. The amount of oxygen required is small, and the required degree of oxidation can be obtained in between 3 and 20 min when heating coal in air to temperatures within the range 250 to 450°C; the reaction is exothermic (see Fig. 31). The time required for the necessary degree of oxidation to be effected is dependent on the temperature of the reaction (see Fig. 32). The rate of oxidation is also dependent on the size of the coal—the smaller the size the more rapid the oxidation (see Fig. 33). Williams[151] has reported that size is important for high-rank coals, but that for low-rank coals the rate of oxidation is independent of particle size for particles less than 500 μ in size.

Bond[152] has suggested that other chemicals such as boric acid and nitric oxide, which destroy the coking power, may be effective when incorporated in briquettes, and Alpern[153] has described the reduction in fluidity effected by ferric oxide.

Partington and Sidebottom[154] have investigated the effect of 5% sodium hydroxide and 5% ferric oxide on the distortion of briquettes during carbonization. The briquettes were made from a weakly coking Yorkshire coal of rank 702, and they were carbonized at atmospheric pressure and reduced pressure (1 mm of mercury). Briquettes containing either sodium hydroxide or ferric hydroxide were also carbonized. The briquettes carbonized at normal pressure were distorted, but the others were

[148] Lesher, C. E., and Zimmerman, R. E., *Trans. Am. Inst. Mining Met. Engrs.*, **139**, 328–63 (1940).

[149] Jäppelt, A., and Steinmann, A., *Oel u. Kohle*, **13**, 1027–30 (1937).

[150] Jenkins, G. I., *Chemical Engineering in the Coal Industry*, F. W. Sharpley, Ed., Pergamon Press, London, 1956, pp. 25–31.

[151] Williams, E. H., *Chemical Engineering in the Coal Industry*, F. W. Sharpley, Ed., Pergamon Press, London, 1956, p. 38.

[152] Bond, R. L., *Chemical Engineering in the Coal Industry*, F. W. Sharpley, Ed., Pergamon Press, London, 1956, p. 37.

[153] Alpern, B., *Rev. ind. minérale*, **38**, 168 (1956).

[154] Partington, R. G., and Sidebottom, R., *J. Inst. Fuel*, **32**, 417–21 (1959).

not (see Fig. 34). Thus, both the reduced pressure and the chemical reagent had the same effect. At reduced pressure, however, the loss in weight during carbonization was greater than when a chemical reagent was used (see Fig. 35). It would appear that when the coal alone was carbonized at atmospheric pressure, some of the liquid products of pyrolysis plasticized the mass and conferred a fluidity on the system; the mass might then be distorted by such forces as external loads and internal gas pressure. Subsequently, this liquid partly solidified and partly volatilized.

Gillmore and associates[56] have studied the effect of the binder on the strength after carbonization of briquettes made from anthracite fines (Table 17). They concluded that the strength is related primarily to the amount of coke deposited by the binder in the interstices between the anthracite particles. They showed, however, that there was a limit to the amount of pitch that could be incorporated, and that this limit lay below that required to fill the interstices. If more pitch was used, the carbonized briquettes were swollen or cracked, and it was considered that the swelling was

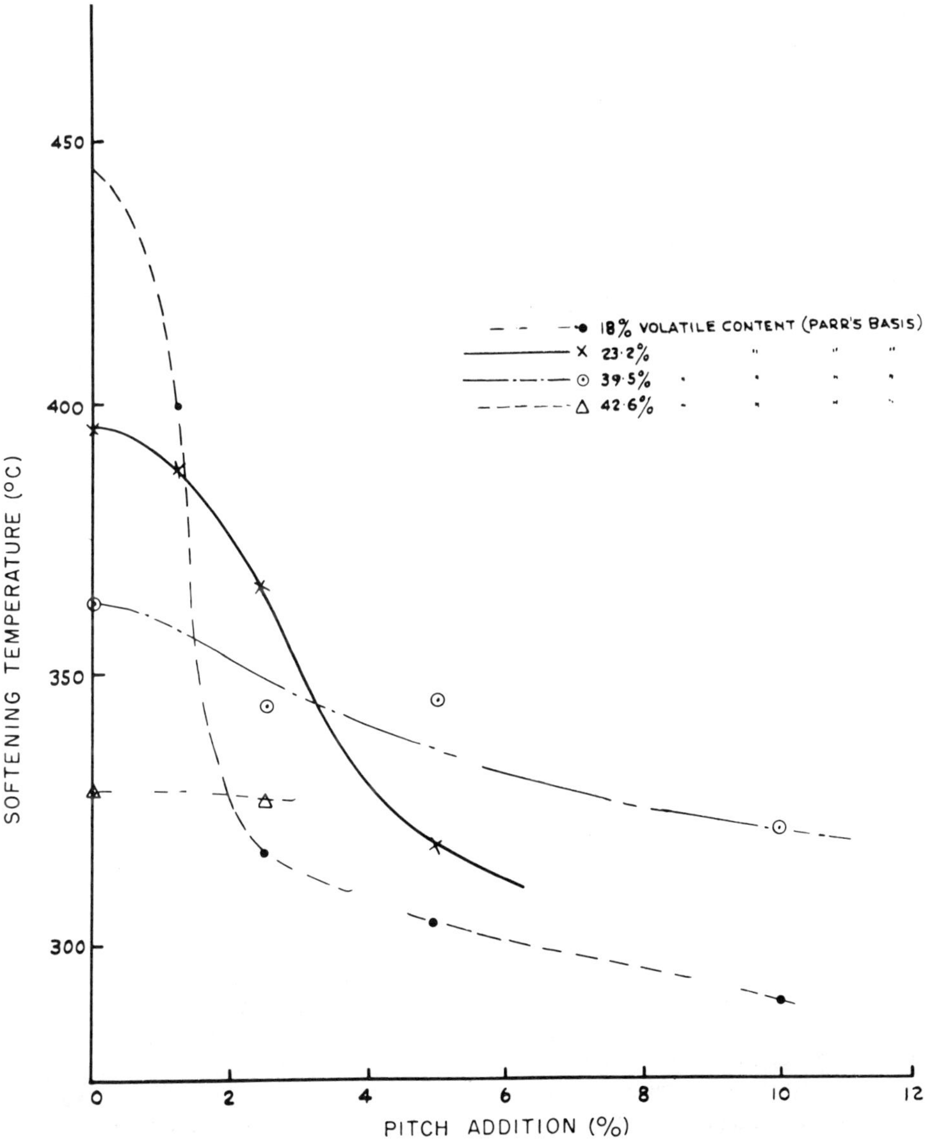

Fig. 30. Effect of pitch addition on softening temperature of pitch and coal mixture—Gieseler method.[143]

TABLE 14

EFFECT OF CARBONIZING TEMPERATURE

Flue Temperature	Yield of Residue	Proximate Analysis of Residue (Dry) V.M.	Fixed C	Ash	Bulk Density	Crushing Strength	Porosity	Loss in Weight	Loss in Volume
(°C)	(%)	(%)	(%)	(%)	(lb/cu ft)	(lb)	(%)	(%)	(%)
600	88.3	7.5	87.5	5.0	41.8	350	17.0	11.7	5.3
700	87.4	6.2	88.5	5.3	42.5	510	19.0	12.6	7.3
800	86.5	4.8	89.3	5.9	43.5	660	20.5	13.5	10.4
900	84.4	3.5	90.3	6.2	45.5	820	22.5	15.6	16.4
1000	83.0	1.6	91.5	6.9	50.0	1080	24.0	17.0	22.7

TABLE 15

EFFECT OF CARBONIZING TIME

Flue Temperature	Carbonizing Time	Yield on Charge 1sts	2nds	Breeze	Proximate Analysis Ash	V.M.	Fixed C	Crushing Strength	Bulk Density
(°C)	(hr)	(%)	(%)	(%)	(%)	(%)	(%)	(lb)	(lb/cu ft)
720	2	63.8	18.1	11.0	5.4	12.0	82.6	80	41.5
720	4	69.4	16.1	1.0	5.5	6.0	88.5	470	43.0
920	4	65.2	15.6	5.6	5.6	4.4	90.0	600	43.8
920	6.5	66.7	12.9	1.1	6.3	3.2	90.5	900	47.4
920	8	68.0	11.8	2.1	6.3	1.9	91.8	1150	48.3

TABLE 16

EFFECTS OF AGGLUTINATING INDEX* OF COAL

Agglutinating Index of Coal	Yield of Grade 1	2	3	Clusters	Proximate Analysis of Grade 1 H_2O	Ash	V.M.	Crushing Strength	Bulk Density
	(%)	(%)	(%)	(%)	(%)	(%)	(%)	(lb)	(lb/cu ft)
6	86.4	8.8	4.8	Nil	3.9	6.5	4.0	1410	47.0
7	86.1	9.0	4.8	0.1	3.4	6.6	4.3	1290	45.8
8	85.2	6.7	5.4	2.7	3.9	6.6	4.1	1100	44.5
9	84.9	5.5	5.6	4.0	3.8	6.5	4.1	900	43.5

* As determined by Brit. Standards Inst., No. 1016 (1942).

caused by the expansion of the entrapped gas.

Dainton and Kaye[155] have studied the swelling and contraction of briquettes during carbonization, and have put forward rather a complicated picture, but one which recognizes a briquette as a porous structure which can set up a resistance to the escape of gas from its interior. The magnitude of this resistance depends on the size of briquette and the fineness of the coal. This resistance may be increased or decreased according to the physical property of the coal. If the coal fuses, it sets up a greater resistance to the escape of gas and the briquette swells. This may be increased by

[155] Dainton, A. D., and Kaye, W. G., *J. Inst. Fuel*, **32,** 571–8 (1959).

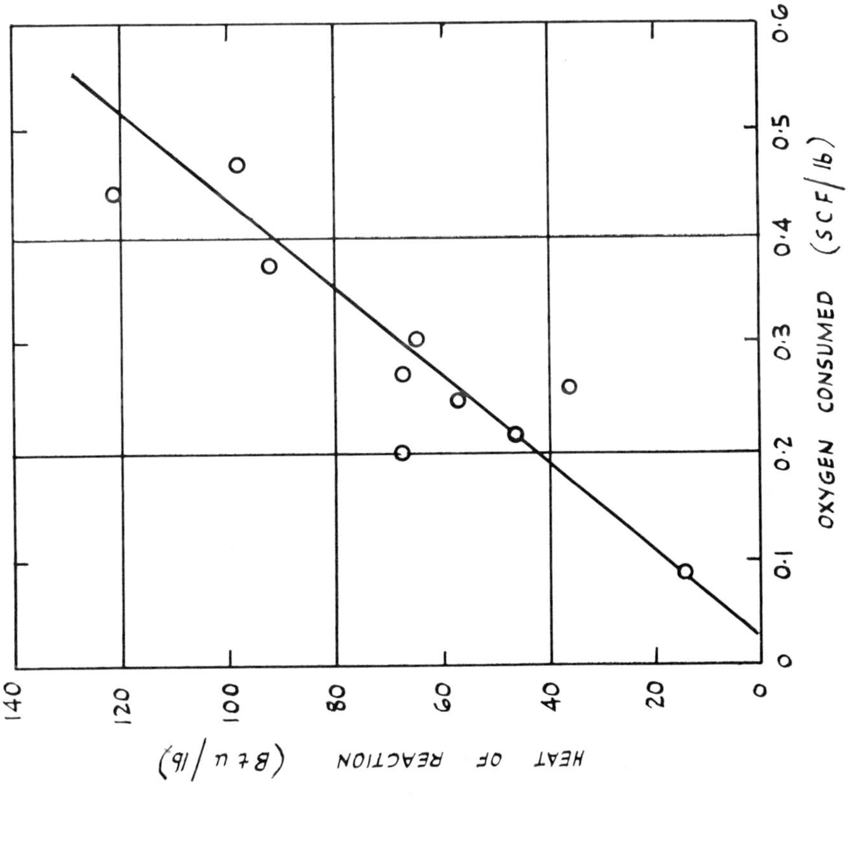

Fig. 31. Relation between oxidation and heat evolution.[150]

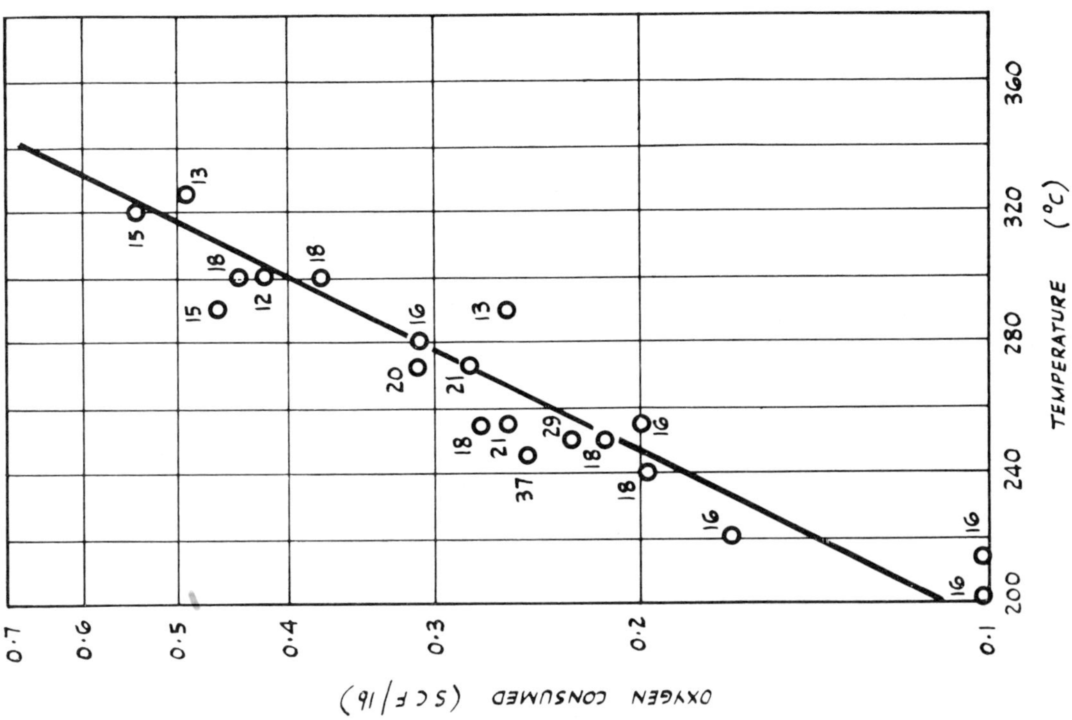

Fig. 32. Effect of temperature on oxidation.[150]

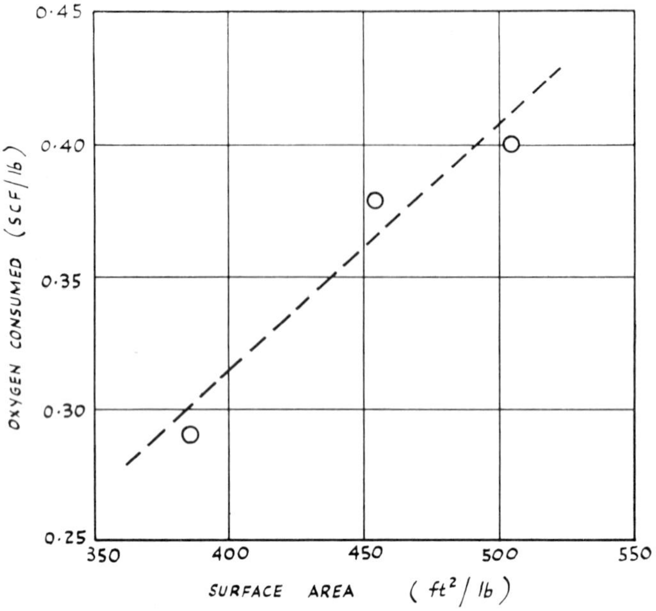

Fig. 33. Effect of particle size on oxidation rate.[150]

the use of a pitch binder, and also by changing the petrographic composition of the coal. It is not clear whether exinite is to be identified with the plasticizing material mentioned by Partington.[154]

Dainton and Phillips[156] studied the rapid carbonization of briquettes and found that small briquettes may be carbonized at higher rates than large briquettes (see Fig. 36). There is a critical rate for each size, and as this rate is exceeded, the briquettes become weaker because of the development of cracks in them.

The briquettes used in these experiments were made without a binder and from a low-rank coal. The coal softened slightly when heated and permitted some further adhesion of the particles. During carbonization, the

[156] Dainton, A. D., and Phillips, J. W., *Proc. Residential Conf. on Science in the Use of Coal*, Inst. Fuel, London, 1958, pp. C40–3.

Fig. 34. Carbonized briquettes.[154]
Sample A. Coal carbonized at atmospheric pressure.
Sample B. Coal carbonized at reduced pressure.
Sample C. Coal and ferric oxide carbonized at atmospheric pressure.
Sample D. Coal and ferric oxide carbonized at reduced pressure.

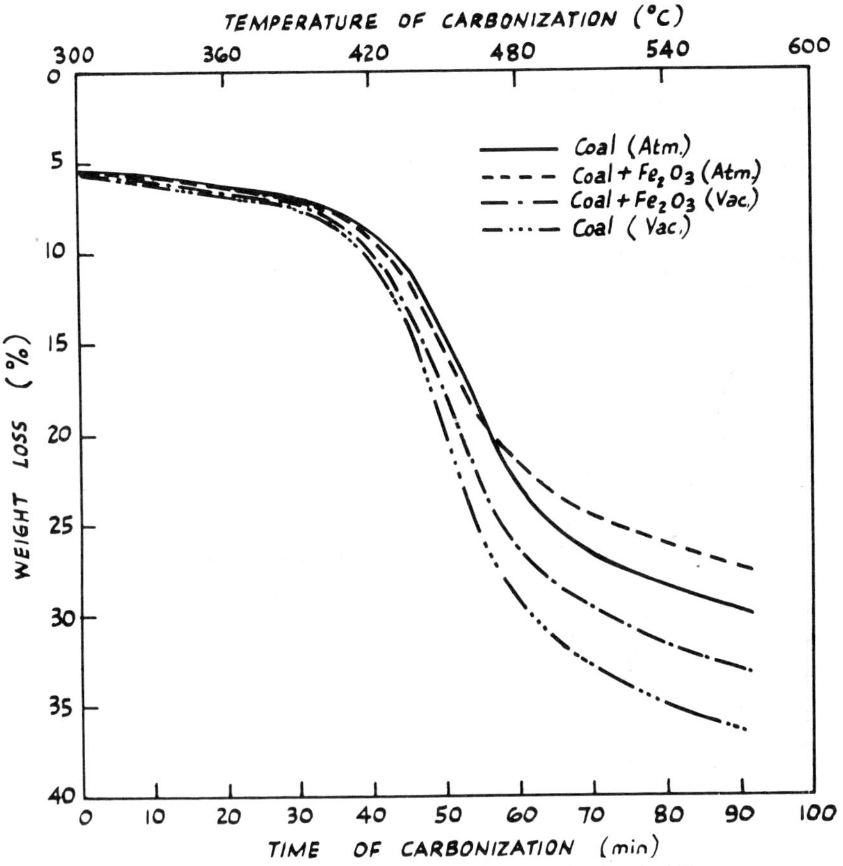

Fig. 35. Effect of ferric oxide on carbonization.[154]

TABLE 17

EFFECT OF PERCENTAGE BINDER* ON THE STRENGTH OF CARBONIZED BRIQUETTES[56]

(Briquetting Pressure: 3920 lb/in.^{2}; carbonized at 10°C/min to 750°C; 1 hr at 750°C)

Pitch Binder	Raw Briquettes			Carbonized Briquettes		
	Density†	Cell Space‡	Void Volume§	Density†	Volume‖ Shrinkage	Breaking Strength
(%)	(g/cm^3)	(%)	(%)	(g/cm^3)	(%)	(lb/in.^{2})
0	1.208	31.4	31.4	—	—	—
5	1.271	31.4	26.2	1.261	3.0	c. 385
7.5	1.312	31.0	22.9	1.307	4.5	915 ± 32
10	1.351	31.0	19.9	1.329	4.1	1270 ± 37
12.5	1.390	31.0	16.7	1.340	3.3	1440 ± 45
15	1.442	30.4	12.6	(**)	(**)	1370 ± 58
20	1.459	33.8	9.8	(**)	(**)	1030 ± 116

* Pitch 18 in Table 8.
† Calculated from weight and measured dimensions of briquette.
‡ Cell space is the calculated percentage of the briquette volume that is not coal.
§ Void volume is the calculated percentage of the briquette volume that is not coal plus binder.
‖ Calculated from measured dimensions of briquette.
** Carbonized briquettes swollen or cracked.

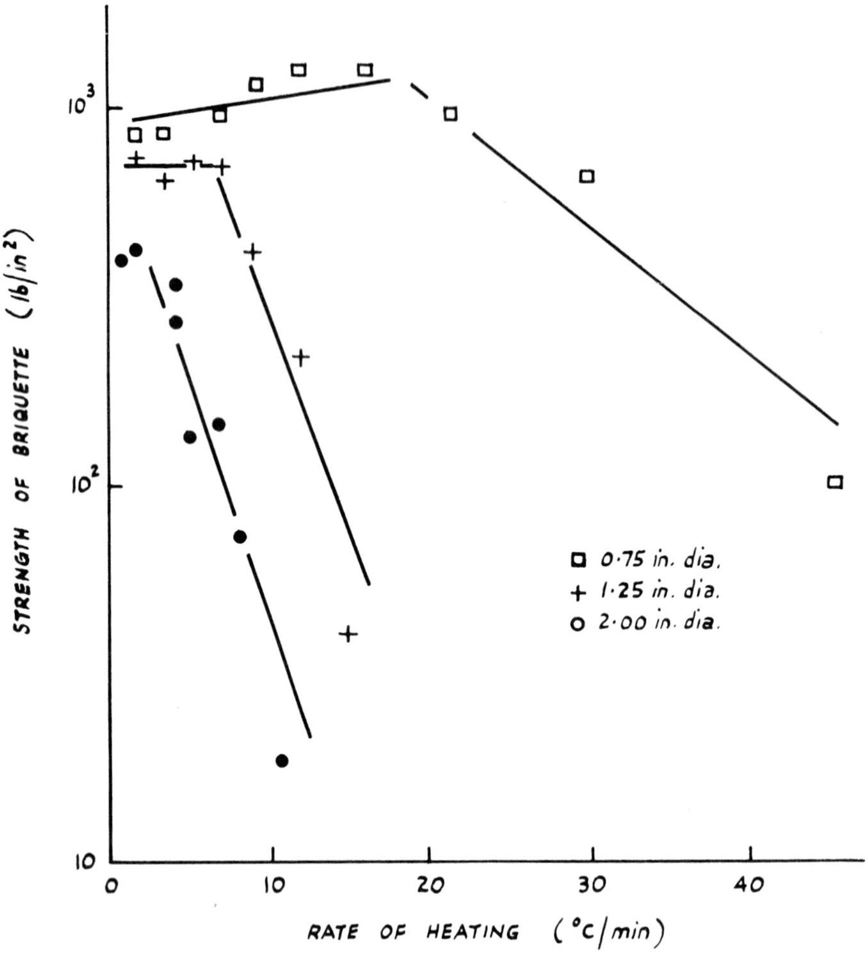

Fig. 36. Crushing strength of carbonized briquettes of three sizes plotted as a function of the rate of heating.[156]

Fig. 37. Internal fissures in briquettes.[156]
A. Heating rate 2.2°C/min.
B. Heating rate 3.8°C/min.

Fig. 38. External fissures in a briquette heated at 100°C/min.[156]

briquette lost about 30% of its weight, and this loss was accompanied by a shrinkage in linear dimensions of about 10%.

The cracks, which occurred in these briquettes as a result of excessive rates of heating, developed internally (see Fig. 37) at moderately high rates, and externally at very high rates, (see Fig. 38). An explanation of the development of these cracks was provided based on a model similar to that proposed by Boyer and associates[157] to explain the behavior of coke during bulk carbonization. The cracks are caused by the tensile stresses set up by differences in the degree of contraction in different parts of the briquette. The low thermal conductivity of the briquettes leads to large temperature differences, and thus large differences in contraction. The degree of contraction is not uniform over the whole temperature range (see Fig. 39), and it follows from Dainton and Phillips' explanation that the critical rates will be different in different temperature ranges. The thermal conductivity increases in the higher-temperature ranges, and this, too, would tend to permit higher rates of heating. The most rapid process of carbonization would therefore be one in which different rates of heating would be used in different temperature ranges (see Fig. 40).

[157] Boyer, A. F., Ladam, A., and Soulè, J. L., *Rev. ind. minérale*, **34**, 592–616 (1953).

Gage[158] has shown that the speed of quenching has an effect on briquette strength, particularly in the low-temperature carbonization range; this is to be expected from the foregoing explanation. It was observed that the strength of briquettes which had been quenched with water showed a pronounced minimum at a carbonization temperature of about 500°C, and the sectional briquettes showed cracks not present in briquettes which had been carried to a higher temperature before quenching.

All coals soften to some extent when heated and reharden when heated further,[159,160] and the difficulties in briquette carbonization arise from the time or temperature dependence of this behavior. The presence of pitch increases the softening. If the softening is too great, the briquettes may swell or collapse under the load, and in extreme cases they will stick together and agglomerate. Also, if the briquettes are heated too rapidly, or if they are too large, their low-thermal conductivity will cause them to crack. These difficulties may be overcome in two ways: first, the composition

[158] Gage, L. E., *Natl. Coal Board, Coal Research Estab., Rept.* No. S8 (1955), 13 pp.

[159] Dryden, I. G. C., and Pankhurst, K. S., *Fuel*, **34**, 363–6 (1955).

[160] Finlayson, P. C., and Macrae, J. C., *Proc. Residential Conf. on Science in the Use of Coal*, Inst. Fuel, London, 1958, pp. 15–16.

of the briquettes may be controlled by selecting coals of the right quality, by adding chemicals to modify the behavior of the coal or by using a suitable quantity of binder; second, by selecting the correct heating rate for the size and quality of the briquette, or by selecting a form of retort which will prevent excessive loads being applied to the briquettes while they pass through the critical temperature range.

Briquette Carbonization Processes

Examples of carbonization processes have been selected mainly from those ventures that have reached the pilot-plant or commercial scale, and they have been grouped in Table 18 according to the particular method adopted to overcome the difficulties inherent in briquette carbonization.

"Carbocoal" Process

The first commercial plant to manufacture carbonized briquettes from semicoke was built at Irvington, Virginia between 1918 and 1920. In 1922, it was closed down owing to technical difficulties.[161] The process, invented by Smith,[145,162] comprised two carbonization stages; in the first the coal was converted into semicoke to destroy its coking properties and in the second the briquetted semicoke was carbonized in an oven designed to minimize the load on the briquettes. In the first stage, bituminous coal, 32 to 36% volatile content, was pulverized and charged continuously into a horizontal retort, where it was heated to 440 to 650°C for 2½ to 3 hr. The semicoke contained 8 to 10% volatile matter. The gaseous and volatile materials were collected separately.

The long horizontal retorts were heart-shaped in cross section, had carborundum-lined inner walls, and possessed double intermeshing paddles to move the charge continuously through the oven. The oven was heated externally, and the shell temperature was 650 to 760°C. The pulverized coal was fed in at one end through a hopper, and the semicoke was discharged in a granular form at the other end, the coal having sintered on the way through. The daily throughput at Irvington was 400 tons, twenty-four ovens being used, each having a throughput of about 1 ton/h.

[161] Brownlie, D., *Proc. S. Wales Inst. Engrs.*, **42**, 322 (1926).

[162] Smith, C. H., Brit. Pat. 120,585 (1917); U.S. Pats. 1,276,427–9 (1918), 1,334,180 (1920).

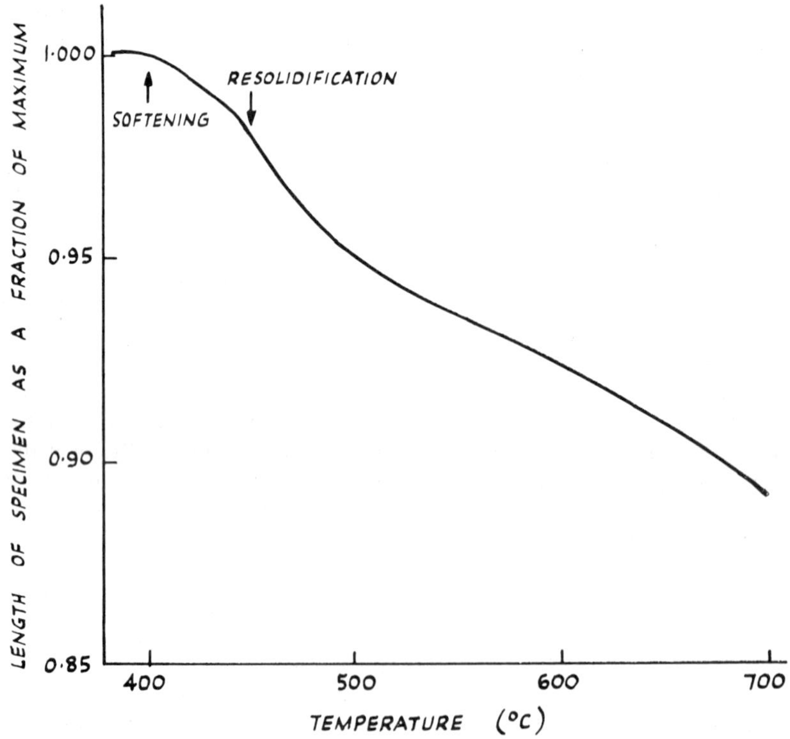

Fig. 39. Variation in length of a briquette as a function of temperature.[156]

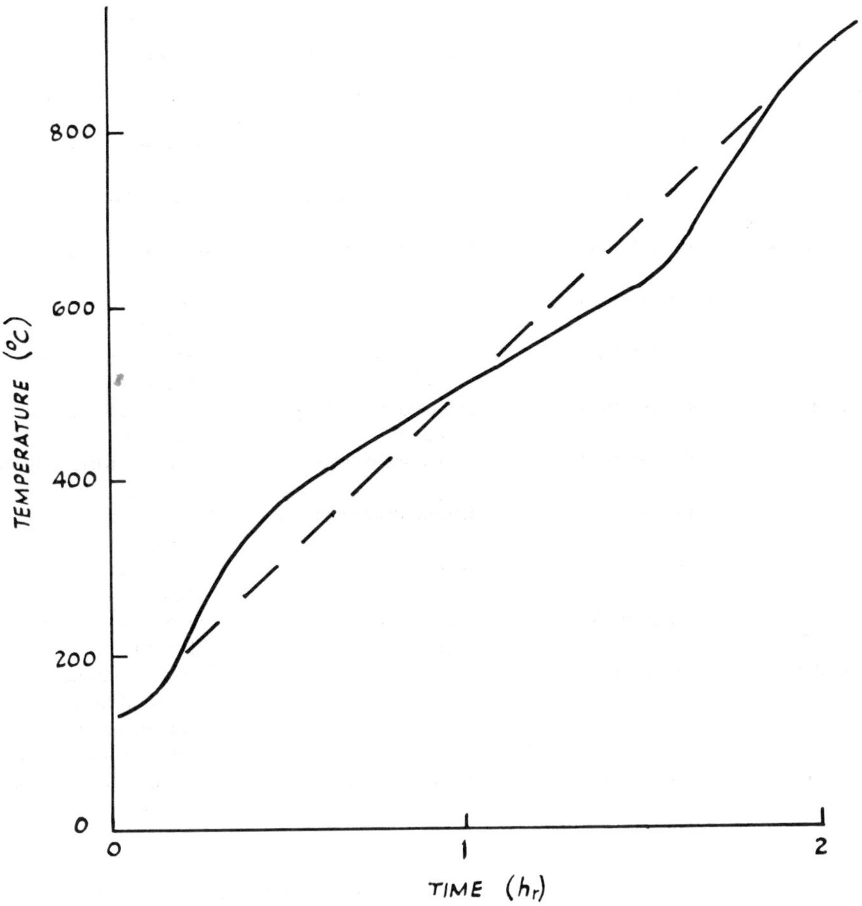

Fig. 40. The full curve represents a safe nonsteady program for 3 × 2 in. low-rank briquettes. The equivalent steady program shown as a dashed line, results in the fissuring of the briquettes.[156]

The granular semicoke was crushed to pass a 20-mesh screen in a disintegrator and was briquetted in conventional equipment with 10 to 12% of pitch which was derived from the tar produced. The briquettes were cooled on a long conveyor belt to harden them, and were then charged into an inclined oven, where they were carbonized. The oven, with an inclined sole, was rectangular in cross section, and, to minimize the load in the oven, it was provided with a "sole" built down the middle of the oven, and by two vertical internal supporting walls. The oven was thus divided into six chambers, each of which had a capacity of 2.5 tons. The briquettes were carbonized at about 980 to 1000°C for between 5 and 8 hr, and were quenched with water when discharged. The product, "Carbocoal," contained 3.6% volatile matter, and the yield per ton of ingoing coal was 1575 lb. It was a hard, reactive smokeless fuel. Analyses for the coal, semicoke, and product and the yields of the products obtained are given in Tables 19 and 20.

One of the disadvantages of the Virginia plant was the high-pitch requirement, which was in excess of the pitch produced by the process; this, together with major technical difficulties in maintaining the operation of the paddles in the pretreatment retort contributed to the early closure of the plant. Thau[163] stated that this process was modified and applied commercially in Upper Silesia.

Many other suggestions have been made. Carter,[164] in 1910, described a process for carbonizing gas-coke breeze-briquettes bound with 4% petroleum pitch by heating them in a

[163] Thau, A., *Brennstoffschwelung*, Knapp, Halle, 1949, Vol. 1, p. 157.

[164] Carter, R. A., U.S. Pats. 968,123, 972,333 (1910).

TABLE 18

PROCESSES FOR THE MANUFACTURE OF CARBONIZED BRIQUETTES

Processes	Briquetting	Special Treatment	Carbonization
Carbocoal	Pitch-binder	Coal precarbonized	Inclined hearth, narrow chamber ovens
Suncole	Pitch-binder	Coal precarbonized	Shallow layers on rotary hearth
Ocean Coal Company "OCO"	Pitch-binder	Selected noncoking coal	Shallow layers on moving hearth
Shawinigan	Pitch-binder		Shallow beds on moving hearth
Sutcliffe	No binder	Selected noncoking coal	Vertical retort; internal heating
Humphreys & Glasgow	Pitch-binder	Selected noncoking coal	Vertical retort; internal heating
NFC Process	Pitch-binder	Coals preoxidized	Vertical retort; internal heating
Carmaux	Pitch-binder	Briquettes preheated	Vertical retort; internal heating
Lurgi Spülgas	No binder	Brown coal	Vertical retort; internal heating
Phurnacite	Pitch-binder	Selected noncoking coals	Narrow ovens; inclined hearth
Brennstoff Technik	Pitch-binder		Narrow chamber ovens; external heating
Weber	Pitch-binder		Briquettes embedded in sand; rotary retort
National Coal Board	Pitch-binder	Selected noncoking coal	Vertical retort; hot sand as heat transfer medium

patented "volatilizer" in the form of an inclined retort. This process, however, has not found commercial application. A recent process making use of the precarbonization of coal is the "Baumco" process.[165] The coal is first

TABLE 19

ANALYSES OF COAL, SEMICOKE AND "CARBOCOAL"

	Coal	Semicoke	"Carbocoal"
Fixed carbon (%)	56	66.6	84
Volatile matter (%)	32	14	3.6
Ash (%)	8	8.4	9.2
Moisture (%)	4	11	3.2
Calorific value (Btu/lb)		13,900	11,000–13,400

heated in a devolatilizer to 800°C to produce a char which has no coking properties. It is then blended with pitch and a little coking coal. The former acts as a binder for the briquette when it is cold, and the coking coal acts as a binder for the carbonized briquette.

[165] Baum, K., U.S. Pat. 2,825,679 (1958). Barritt, D. T., and Kennaway, T., *J. Inst. Fuel*, **27**, 240–1 (1954).

Simpson[166] has described a process in which subbituminous or lignitous coals are precarbonized at 800°C before they are briquetted and carbonized. The carbonized briquettes were claimed to be suitable for metallurgical

TABLE 20

YIELDS OF PRODUCTS FROM 1 LONG TON OF COAL

		Primary Retort	Briquette Carbonization	Combined Yield
Gas yield	(cu ft)	3850	5500	9400
Calorific value	(Btu/cu ft)	750–800	325–350	535
Light oil or motor spirit	(gal)	1.10	0.35	2.0 1.47
Tar (less motor spirit)	(gal)	18.50	6.50	25.0
Ammonium sulfate	(lb)	6.5	12.8	19.3
Solid fuel	(lb)	1450	1575	1575

use. In this process, the precarbonization stage was used essentially to shrink the coal and thus prevent excessive shrinkage in the briquettes during carbonization.

[166] Simpson, C. H., Australian Pat. 153,801 (1953).

"Suncole" Process

In this process also the coal is precarbonized, but the briquettes are protected from excessive load during their subsequent treatment by being carbonized in shallow layers on a moving hearth. A large-scale plant was operated successfully by "Suncole," (Nottingham) Ltd. during the 1930's, but it was eventually obliged to close down because of the war.[167] It has not been reopened since. In this process, coal was precarbonized in an "L and N" retort[168] at temperatures between 630 and 700°C, and the clean semicoke breeze made was crushed and briquetted with pitch.

Earlier attempts to carbonize these briquettes in conventional ovens had resulted in a heavy loss through breakage, but success was achieved when they were carbonized in a rotary hearth oven. This oven, adopted from steelworks practice, was robust in design; it was 70 ft in diameter, and the rotary hearth, 12 ft wide, was carried on single rollers and brackets mounted on concrete columns. The hearth made one revolution in about 40 min; special charging and discharging devices were used to ensure perfect seals and to prevent the entry of air. The combustion chamber, in which the multiflame method of combustion was used, radiated heat through the muffle arch on to the briquettes, and heated them to 550 to 650°C.

The carbonized briquettes, ovoid in shape and possessing the approximate dimensions $3\frac{1}{2}$ in. × $2\frac{1}{2}$ in. × $2\frac{1}{2}$ in., were obtained in a yield of over 94% and the amount of breakage was negligible, being below 1%. The briquettes were smokeless and highly reactive, having an ignition temperature of 245°C, a volatile content of 15%, ash 5%, and moisture about 7%.

A schematic layout of the Suncole plant is shown in Fig. 41.

"OCO" Process (Ocean Coal Company)

This process[169] was used commercially by the Ocean Coal Company in South Wales for a number of years until it was closed down in 1947.[170] The briquettes were carbonized in shallow layers, but, on a moving chain grate. Anthracite or semianthracite was briquetted with pitch in a conventional manner in a double-roll press, and the briquettes were delivered to charging hoppers which fed them into a slowly moving chain grate, which passed into a retort or combustion chamber, as illustrated in Fig. 42. The briquettes were discharged and quenched by passing through a water seal.

The speed of the grate and the depth of the bed of briquettes on the grate could be raised, but the speed was eventually set at about 8 in./min, and the depth between 4 and 6 in. The retort was heated by burning the volatile matter from the briquettes, and the temperature could be controlled by controlling the rate at which air was supplied, or by controlling the speed of the grate. Each grate produced 10 tons of product daily. Both the temperature and duration of carbonization depended on the composition and size of the briquettes, and the desired residual volatile matter in the briquettes. Thus, with high-rank coal (20% volatile matter), briquettes with a residual volatile matter of 7% were obtained by carbonizing at 600°C for 45 min; by carbonizing at 500°C and under otherwise similar conditions, a volatile content of more than 10% could be obtained. These examples refer to ovoid briquettes in which the major axis was 2 in. and the minor was $1\frac{1}{4}$ in. The time required for carbonizing briquettes of different sizes changes roughly as the square of the minimum dimension. The process was simple and effective, but had two drawbacks; it was wasteful in that it consumed volatile matter to provide the heat, and it was not easy to maintain a uniform temperature in the bed—the briquettes occasionally showed signs of burning.

Shawinigan Process

A similar carbonization process has been described more recently by Andersen and Fasken of Shawinigan Chemicals, Ltd.;[171] in this process coal was used. An experimental plant using this process has been built by the Steinmuller Company in Germany (see Fig. 43),

[167] Anon., *Petrol. Technol.*, **6,** 143 (1940).

[168] Brownlie, D., *Ind. Eng. Chem.*, **19,** 43 (1927). Anon., *Coal Carbonisation*, **1,** 38–40 (1935). Utah Conservation & Research Foundation, *The Low-temperature Carbonization of Utah Coals*, Salt Lake City, Utah, 1939, pp. 303–16.

[169] Anon., *The Suncole Distillation Process*, British Coal Distillation Ltd., London, 1938, p. 11.

[170] Illingworth, S. R., and Hughes, M. E., Brit. Pat. 532,945 (1941). Illingworth, S. R., U.S. Pat. 2,071,082 (1937).

[171] Andersen, A. H., and Fasken, N. R., U.S. Pat. 2,380,930 (1942). Andersen, A. H., *Trans. Can. Inst. Mining Met.*, **47,** 139–51 (1944).

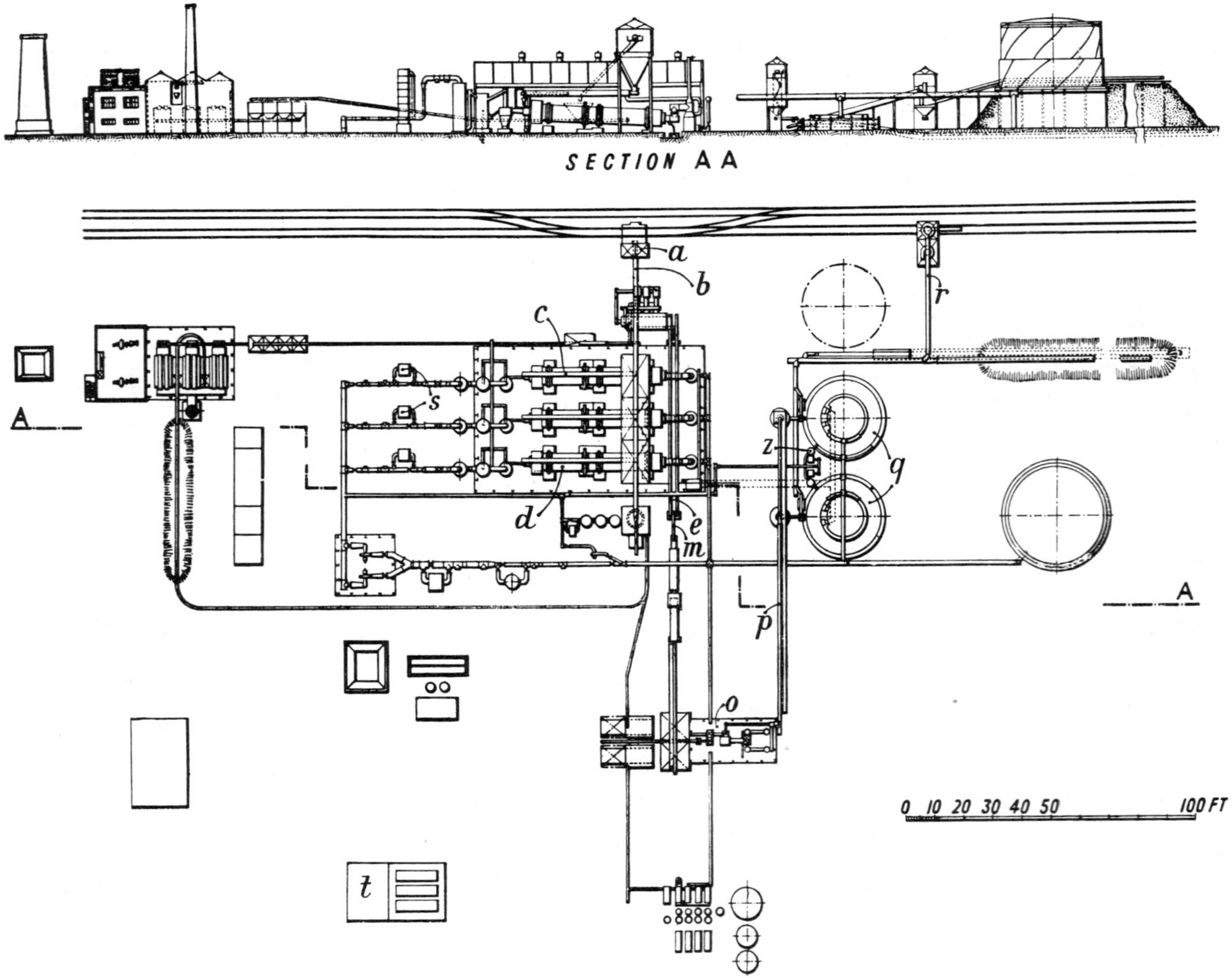

Fig. 41. Schematic layout of Suncole plant. *a.* Coal-loading pit; *b.* elevator; *c.* dryers; *d.* retorts; *e.* conveyor; *m.* conveyor; *o.* briquetting plant; *p.* cooling elevator; *q.* storing ovens; *r.* elevator; *z.* condensers for oil recovery; *s.* cooling condensers; *t.* spirit plant.

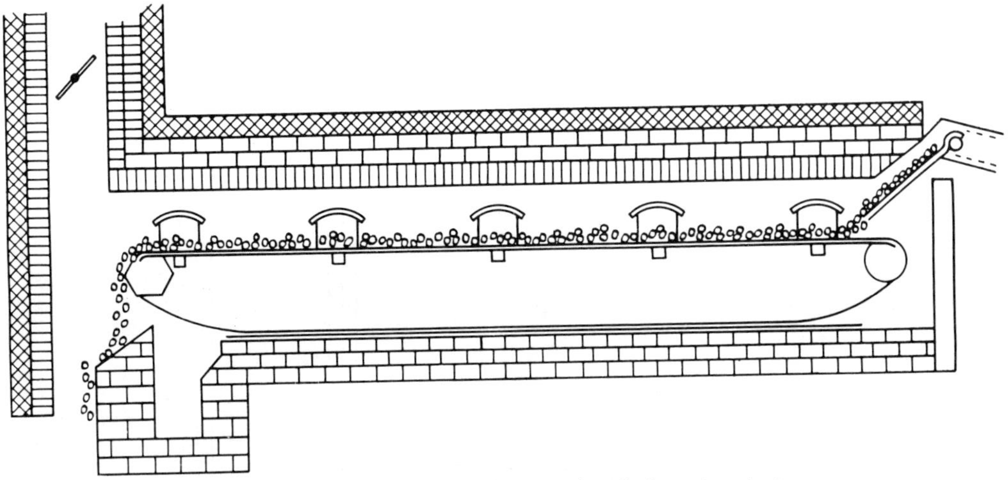

Fig. 42. Ocean Coal Company moving chain-grate retort.

and Reerink[172] has described its use for carbonizing briquettes.

Greater control is attainable in this moving-grate retort than in the previous ones. The air-cooled grate moves slowly at speeds adjustable between $1\frac{3}{4}$ and $13\frac{1}{2}$ in./min, and the bed thickness may be adjusted up to $8\frac{1}{2}$ in. The heating area is 48 sq ft. Much greater control of the air supply to the bed is also possible because of the inclusion of a number of air boxes capable of adjustment, and because the grate is sealed along its sides against a water-cooled wall, thereby preventing undue air leakage.

This process, like the preceding one, has the

[172] Reerink, W., *Chemical Engineering in the Coal Industry*, F. W. Sharpley, Ed., Pergamon Press, London, 1956, pp. 46–8.

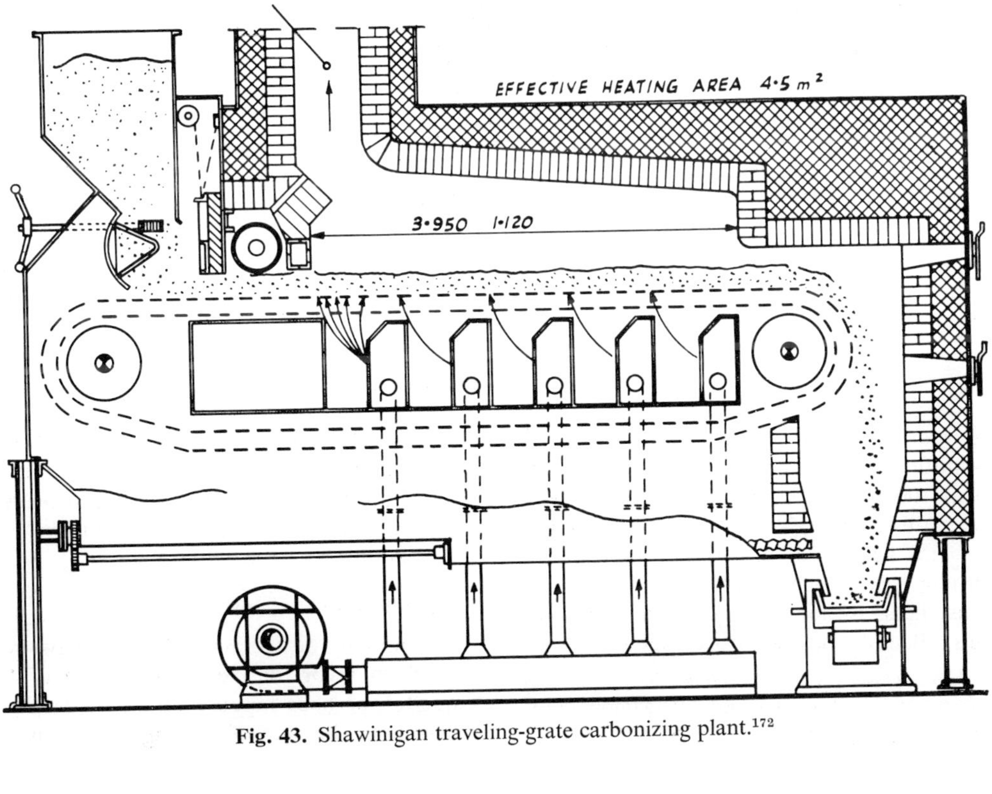

Fig. 43. Shawinigan traveling-grate carbonizing plant.[172]

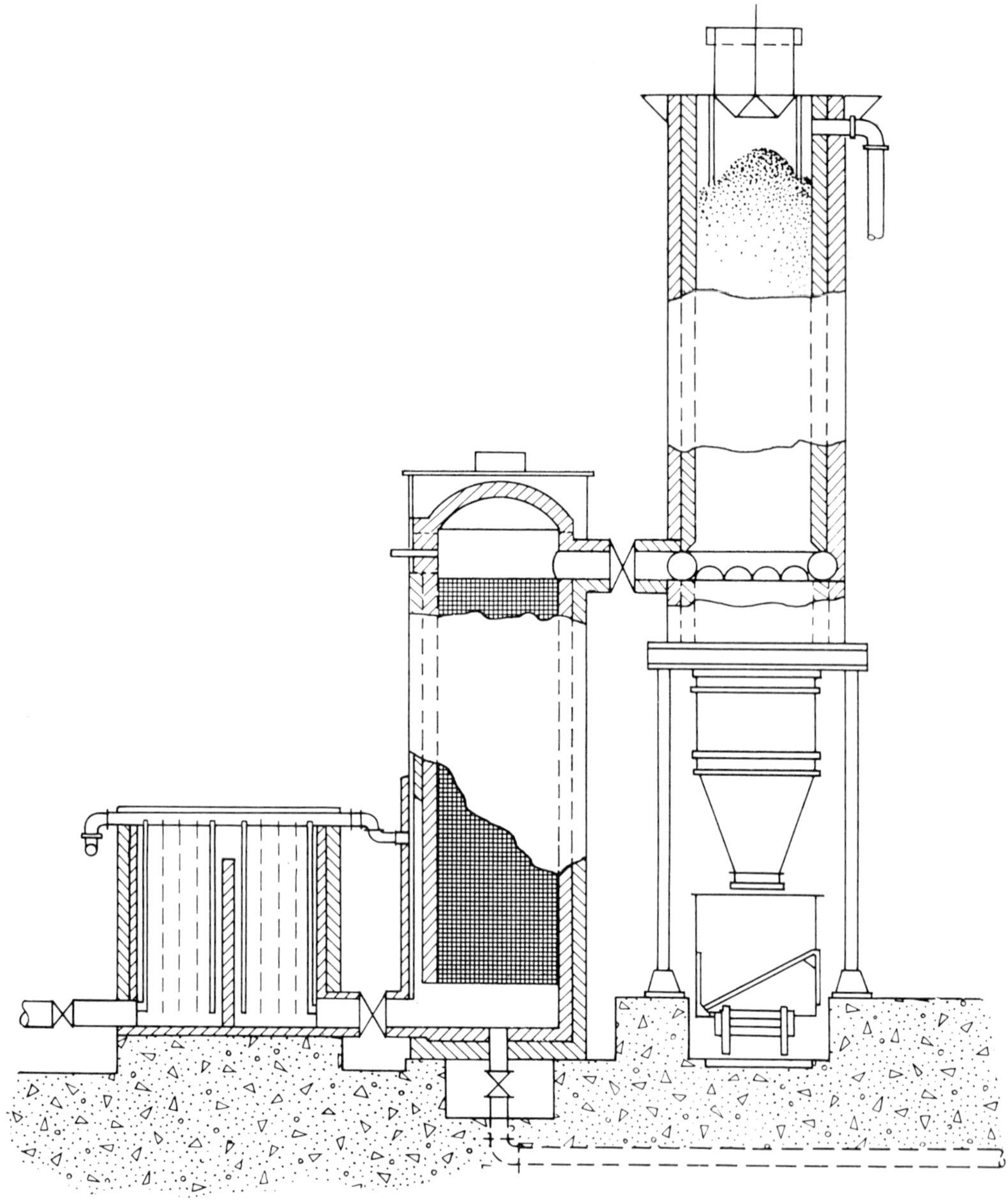

Fig. 44. Sutcliffe-Evans retort.[174]

advantage of having a low-capital cost, and of permitting very rapid carbonization. Times of 20 to 100 min have been quoted. However, there still remains the disadvantage of generating the heat required for carbonization by burning volatile matter, which, in normal circumstances, is not the most economical of fuels. Baum[173] has described methods of air control which permit a more economical combustion of volatile matter, but even so, there is much room for improvement both in thermal efficiency and in achieving a uniform bed temperature.

[173] Baum, K., *Chemical Engineering in the Coal Industry*, F. W. Sharpley, Ed., Pergamon Press, London, 1956, pp. 52–4.

Sutcliffe's Pure Coal Briquettes

Internally heated vertical retorts have been much in favor for the carbonization of briquettes, perhaps because of the control they permit in the rate of heating. The Sutcliffe and Evans retort shown in Fig. 44, was erected by Pure Coal Briquettes, Ltd., in 1926.[101,174,175] It consisted of a vertical steel cylinder lined with refractory, and was heated by means of an intensely superheated mixture of rich gas and steam which entered the retort through inlet ports situated about one-third of the way up the retort. The gas and steam mixture was heated in two vertical checkerwork regenerators; a portion of the rich gas from the retort (440 Btu/cu ft) was burned in the combustion chamber of one regenerator, the hot flue gas being drawn through the brickwork and then through a steam superheater. Meanwhile, the superheated steam mixed with another portion of rich gas was passed through the other hot regenerator and so heated. As soon as one regenerator had delivered up its heat and cooled down, the heating gas was directed to the other, which had meanwhile been reheated. Briquettes were charged into the top of the retort through a hopper and were partially cooled in the bottom part of the retort before being discharged into the quenching car.

The briquettes were made from various non-coking coals by the Sutcliffe-Speakman process, which does not require the use of a binder. Originally, the carbonized briquettes were used as a fuel, but at the present time the process is used for the manufacture of reactive carbon.

Humphreys & Glasgow Retort

Originally, in 1938, "Phurnacite" was made in an internally heated vertical retort constructed by Humphreys and Glasgow.[176] This plant operated for 2 years at Aberaman in Wales, but was later displaced by externally heated "Disticoke" retorts. The Humphreys and Glasgow retort was similar to that described below in the National Fuels Corporation process, hot rich gas being used for internally heating the retort. The rich gas was heated externally to the retort in a recuperative regenerator, and was introduced through tuyères in the lower third of the retort. The hot briquettes were quenched first with cold gas, and then by being discharged continuously through a water seal. The retort was charged periodically with a 3-ft depth of raw ovoids when the top gas temperature reached 600°C. The briquettes were made from noncoking low-volatile steam coals and the quality of the combined product was equivalent to the present day "Phurnacite."

National Fuels Corporation (N.F.C.) Process

In 1938, a small plant producing 5 to 10 ton/day of carbonized briquettes was erected at New Haven, Conn., and this was followed by a 100 ton/day plant at Bethlehem, Pennsylvania. Both plants were closed down by 1951, and a new plant of 10 to 15 ton/day was planned to be built at Linden, New Jersey.

The process is of particular interest for two reasons; it provided for the oxidation of the coal prior to forming it into briquettes, and it used an internally heated vertical retort for carbonizing the briquettes to a closely controlled temperature program. It was claimed[177] that this process can produce carbonized briquettes of good quality from any coal that has a B.S. swelling number of 1 or more, but whether the process is quite as comprehensive as was claimed can only be determined by experience.

After the coal was dried, it was ground to −0.5 mm in size, and left the collecting cyclone to enter the oxidizer (see Fig. 45) at about 83°C, and with less than 1% of surface moisture.[178] The coal was introduced with air into the oxidizer by means of a Fuller-Kinyon pump; the oxidizer consisted of a 3-in. tube of heat-resisting steel externally heated by means of Dowtherm vapor. The oxidation was carried out at carefully controlled temperatures up to 475°C, but usually nearer 250°C. Although about 5 lb of air were used for each pound of coal, the air consumption could rise as high as 15 lb; the contact time in the oxidizer was less than 4 sec. The terms "fluid bed" and "fluidized bed" have been used to describe the process, but the oxidation was

[174] Sutcliffe, E. R., and Evans, E. C., *Proc. S. Wales Inst. Engrs.*, **38**, 341–80 (1922). Sutcliffe, E. R., *Fuel*, **7**, 326–7 (1928); U.S. Pat. 1,767,231 (1930).

[175] Gentry, F. M., *The Technology of Low-temperature Carbonisation*, Bailliere, Tindall & Cox, London, 1929, pp. 249–52.

[176] Humphreys & Glasgow, Ltd., Brit. Pat. 469,175 (1937).

[177] Berg, G. A., and Beardsley, A. P., Brit. Pat. 340,717 (1928).

[178] Barritt, D. T., and Kennaway, T., *J. Inst. Fuel*, **27**, 232–40 (1954). Utah Conversation & Research Foundation, *The Low-temperature Carbonization of Utah Coals*, Salt Lake City, Utah, 1939, 872 pp.

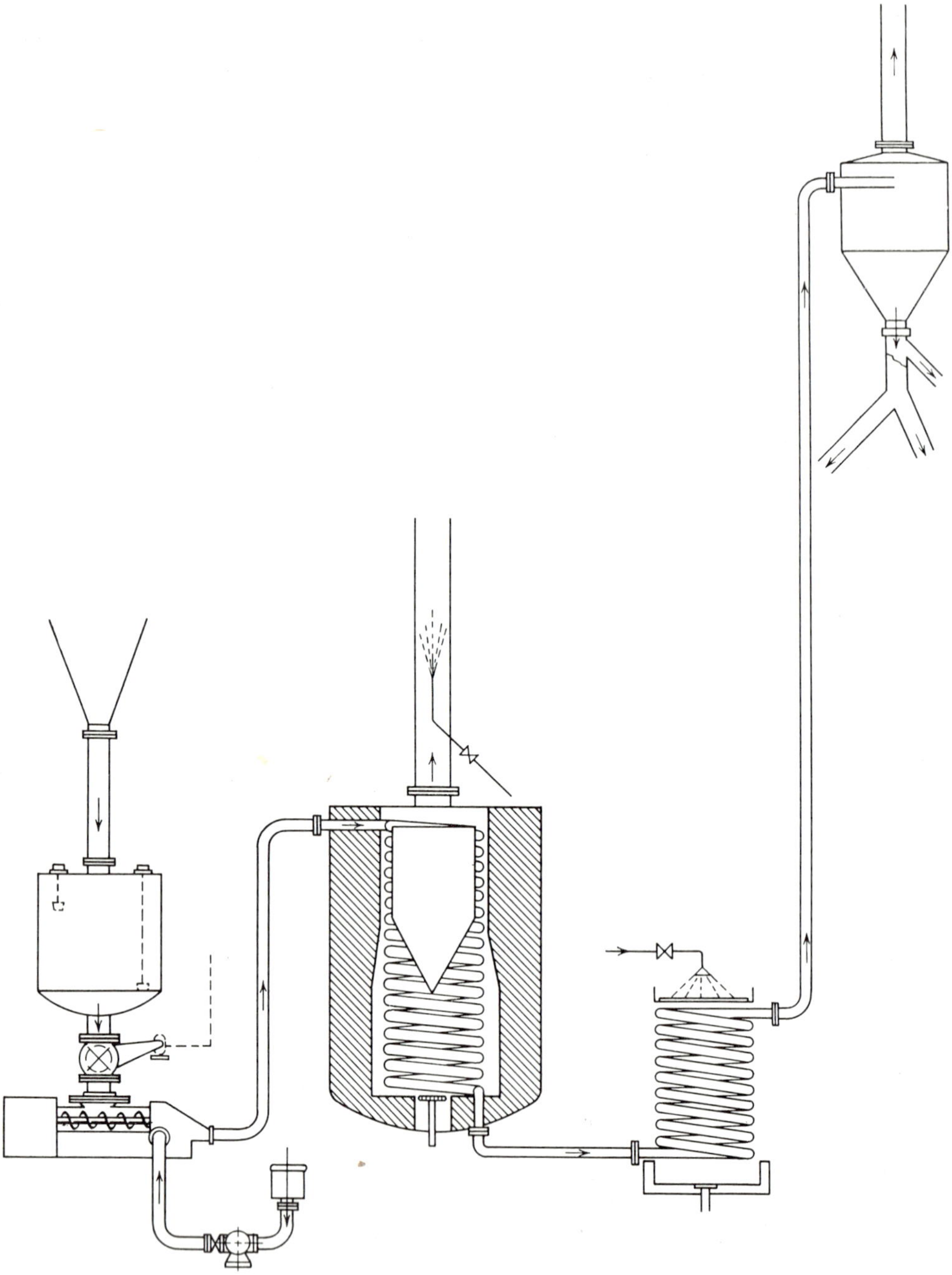

Fig. 45. Oxidizer used in National Fuels Corporation process.[178]

carried out essentially in a dilute disperse-phase of coal in air.

The hot coal was passed through a cooler from which it emerged at about 60°C and was passed to the briquetting plant where, mixed with about 8% of pitch, it was pressed by conventional presses into ovoids. The ovoids were cooled on a conveyor belt until sufficiently hard to be introduced into the retort.

The designers of the retort recognized the need to control the rate of heating strictly in a certain critical temperature range if nonswollen, unfused coke briquettes were to be produced. This range was between 350 and 500°C.

The retort, Fig. 46, was a vertical cylindrical steel vessel lined with refractory material. It was about 8 ft in diameter and 20 ft high, and had a capacity of 100 ton/day. The raw briquettes were fed into the top continuously and discharged through a quenching section.

The retort was heated internally by partially burnt gas introduced at about one-third of the way up the retort. The briquettes entering the retort were preheated to some temperature below 150°C by means of a hot flue gas, and the hot briquettes were quenched in the base of the retort by means of cold gas. On their way through the retort, the briquettes passed through the preheating zone where their temperature was raised to 350°C, and then through the next section where their temperature was raised gradually to 500°C. Below this zone the temperature was raised more rapidly to 900°C, or any other desired temperature, at the point of entry of the heating gases. After this step, the briquettes entered the quenching zone. The rate of heating at any point was determined by the cross-sectional area of the retort, and the rate could be changed by raising or lowering the positions of the 350 and 500°C-temperature regions.

The gas from the retort passed to a conventional byproduct plant where tar, liquor, and ammonia were separated. The gas was then recirculated in three streams to the retort for purging, heating, and quenching. The gas

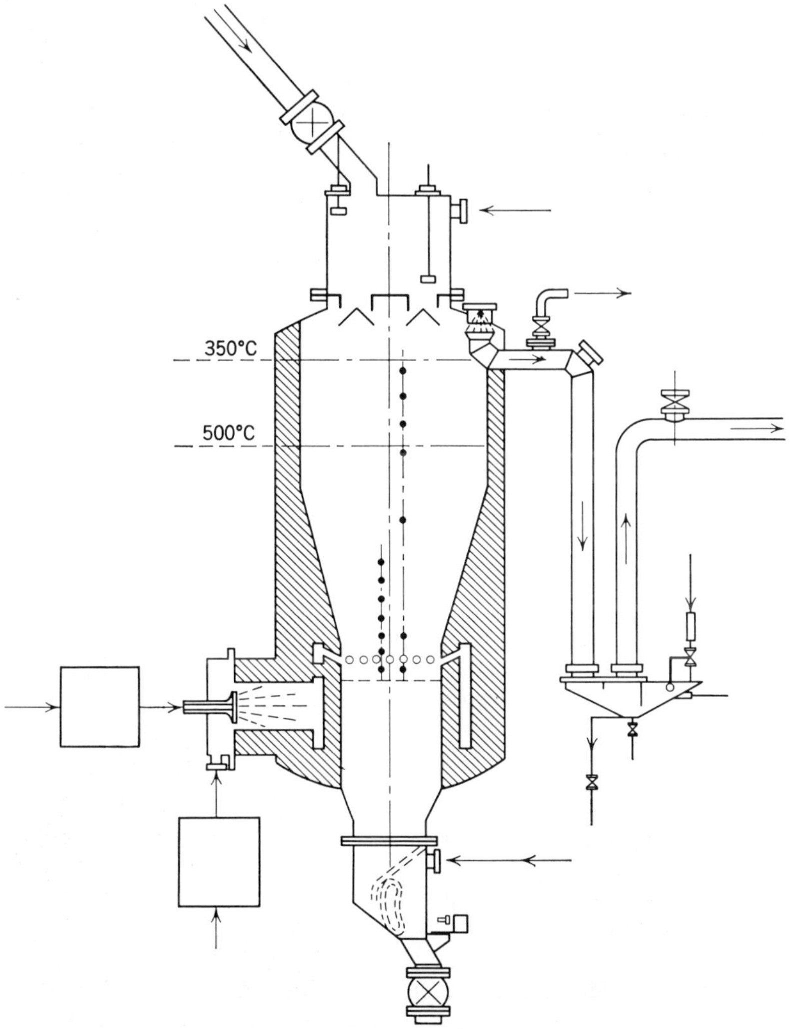

Fig. 46. National Fuels Corporation retort.[178]

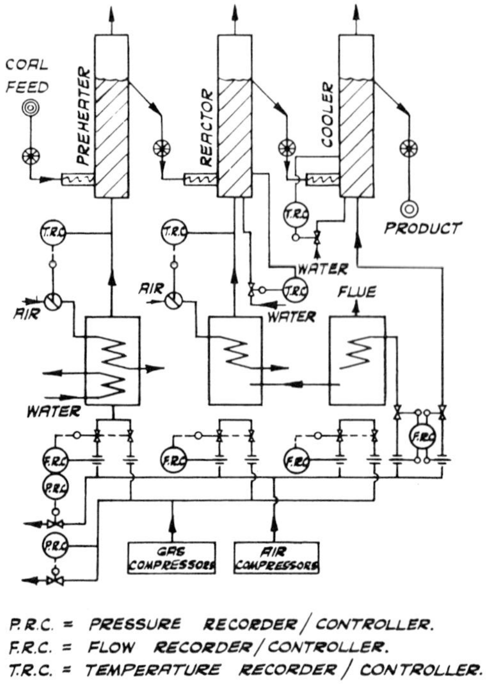

Fig. 47. Flow diagram of National Coal Board process pilot plant.[150]

yield varied little with high- or low-volatile coal—between 7500 and 8500 cu ft/ton—and had a calorific value of 525 to 550 Btu/cu ft.

The tar yield, however, varied between about 63 gal/ton for low-volatile coals and 237 gal/ton for high-volatile coals. The briquettes had about the same specific gravity as the original coal, about 1.25, and were dense in structure and strong. They were free-burning even when the volatile content was below 1%, and they stood up to weathering.

National Coal Board Oxidation Process

A pilot plant of 1 ton/hr capacity was operated by the N.C.B. in 1954 for the oxidation of coal in a fluidized bed.[150,179] The purpose of the plant was to destroy the coking properties of certain high-rank coals to make them suitable for conversion into "Phurnacite" carbonized ovoids (see Fig. 47).

The plant consisted of three vessels, each of which was a vertical steel cylinder with means of introducing air at the bottom and means of conveying powdered coal from one to the other. The first of the three vessels was a preheater, the second an oxidizer, and the third a cooling vessel. At the conclusion of the work on the pilot plant, it was shown that for a commercial plant only the oxidizer would be required, and that oxidation could be effected by fluidizing the coal with cold air. In it a high-rank coal of 17% volatile matter could be oxidized sufficiently in less than 15 min at a temperature of 400°C for conversion into carbonized briquettes.

[179] Idris Jones, W., Rhys Jones, D. C., and Launder, J. L. M., Brit. Pat. 731,944 (1955).

Carmaux Process

This process[180] introduced a novel procedure. The briquettes themselves were pretreated. They were heated in the presence of air in a vertical steel cylinder situated above the carbonizing retort, and this pretreatment provided the briquettes with a thin outer skin of hard coke which prevented them from sticking together in the carbonizing stage (see Fig. 48).

The briquettes used were made from coking-coal fines and pitch, and in the pretreatment stage they were heated for 4 hr at 250°C in flue gas to which about 60% of air had been added. The treated briquettes were carbonized for 12 hr in the carbonizing retort which was an internally heated vertical retort.

[180] Jennings, R. J. S., *Chem. & Process Eng.*, **34**, 343–9 (1953).

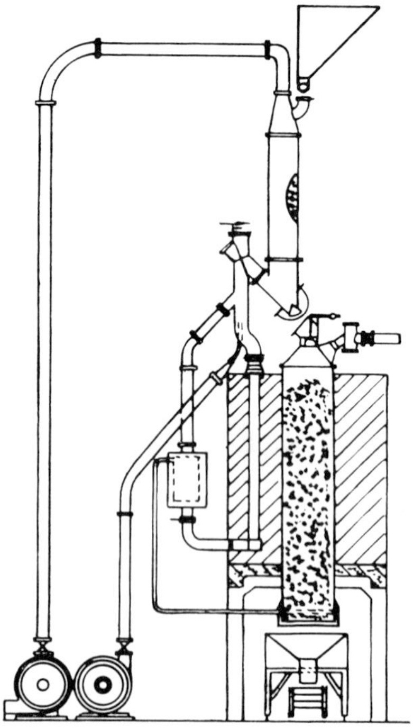

Fig. 48. Section through a Carmaux vertical-chamber furnace.[180]

Lurgi Spülgas Process

The Lurgi Spülgas process[181] was developed to produce low-temperature tars and oils by the low-temperature carbonization of brown coal. The technical success of the process may be assessed by the large number of plants built to date. This process employs a shaft furnace internally heated by Spülgas (oxygen-free heating gas).

A second and later process[182] was developed in collaboration with F. Krupp A-G. which used narrow steel-chamber ovens which were externally heated; the process resembles to some extent the Brennstoff-Technik process. When a particulate charge was used—nut coal or briquettes—Spülgas could be internally circulated to speed the carbonization. This process seems to have little advantage over the former when dealing with briquettes unless a richer gas is required for sale to the domestic market, when internal heating would not be permissible. One plant has been built at Velsen in the Saar, not far from Marienau.[183]

The "Spülgas" process produced a granular low-temperature friable breeze or hard-lump low-temperature smokeless coke, according to the quality of the brown-coal briquettes carbonized. It also produced a low-temperature tar particularly amenable to hydrogenation, and a lean gas of low-calorific value. The liquid products were the most important, and it was said that the chief purpose of the plants built in Germany was to produce a raw material for the production of motor spirit.[181,182]

In most of the plants in operation today, the brown-coal briquettes charged to the carbonizer are manufactured in normal extrusion presses, and during carbonization these briquettes disintegrate and break down into small sizes. At Regis, for example, the solid product contains 20% below 6 mm in size, and 50% between 6 and 20 mm.[181] If the process is to be used primarily for the manufacture of carbonized briquettes, the raw brown-coal briquettes must be of the superior quality produced in the ring-roll press.[184] The following stages are involved: (1) handling, drying, crushing and briquetting brown coal; (2) carbonizing and cooling the briquettes; and (3) recovering the byproducts.

The Lurgi Spülgas carbonizing unit consists of a vertical-shaft furnace of rectangular cross section divided into three parts. In the top part, the briquettes are dried to a moisture content of about 0.5% or less, the heat for doing this being provided by the recirculation of combustion gases obtained by burning some of the gas from the second part in which the briquettes are carbonized. These gases enter the drying section at about 220°C and leave at 60 to 80°C. The briquettes are preheated for 5 hr to about 150°C before they enter the carbonization zone, where they remain for another 5 hr. The total time to pass through the retort is 18 hr.[185] The plant is shown diagrammatically in Figs. 49, 50, and 51.

The briquettes are carbonized in the middle zone to a temperature reported variously to be as low as 500°C in some cases, and in others as high as 850°C. In the middle zone, the heating is effected by recirculating hot carbonization gas, part of which has been burnt to raise the temperature to the necessary level. The two gas cycles are entirely independent, and a delicate balance of pressure is maintained between the zones. In one case at least, the gas used in the drying zone has been purified from sulfur, although crude gas was used in the middle zone.

In the bottom zone, the carbonized briquettes are quenched by passing the carbonization gas from the byproduct recovery plant through them. The briquettes are cooled, discharged from this zone, and in some cases they are quenched further by water sprays. The carbonizing gas recuperates heat from the briquettes before being fed to the carbonization zone. The recycled gas is, of course, diluted in the second stage by admixture with combustion gas, and its calorific value ranges from 1700 to 2700 kcal/m^3 (180 to 280 Btu/cu ft) according to the fuel being carbonized. Between 35 and 48% of the gas made is required for the drying and heating stages, depending on the moisture content of the briquettes and the nature of the fuel. The surplus gas is generally used for steam raising.

[181] Anon., *Coke & Smokeless-Fuel Age*, **8**, 77–9 (1946). CIOS File XXX–13, Item 30; File XXVIII–23, Item 30; File XXXII–91, Item 30, London, H.M.S.O. (1948).

[182] *BIOS*, *Final Rept.* No. 616, Item 30, London, H.M.S.O., (1948).

[183] Georges, P., *Ann. mines*, **136,** 76–9 (1947).

[184] Anon., *The Low-temperature Carbonisation of Brown Coal and Lignite*, Frankfurt-am-Main, Lurgi Gesellschaft für Wärmetechnik mbH, 1955, 25 pp.

[185] Lurgi Gesellschaft für Wärmetechnik, Private communication.

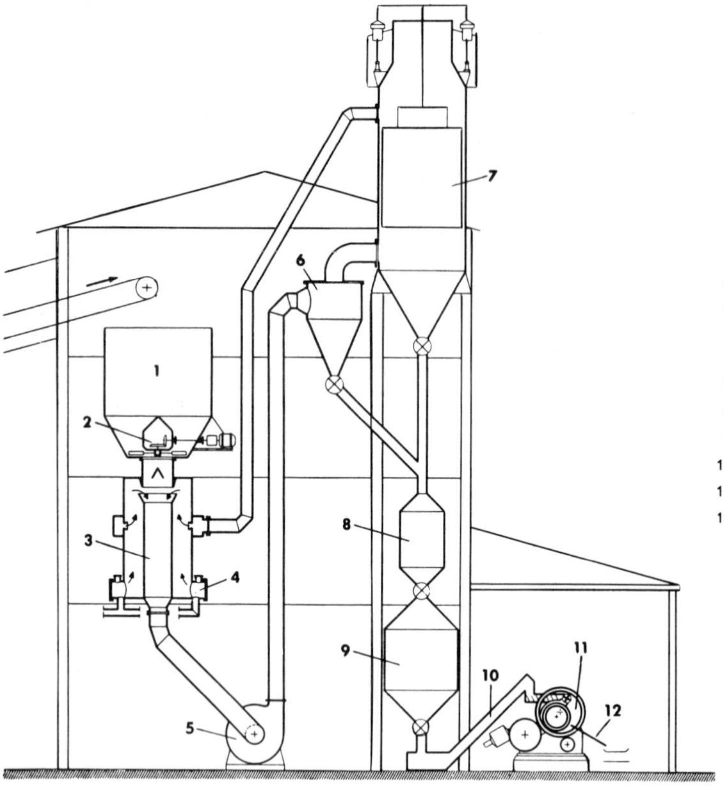

1 — raw coal bunker
2 — coal feed mechanism
3 — shaft furnace for drying coal
4 — gas burners of drier
5 — pulverizing fan
6 — cyclone
7 — electrostatic dust precipitator
8 — cooling vessel for pulverised coal
9 — pulverized coal hopper
10 — pulverized coal feeder
11 — ring-roll briquetting press
12 — briquefte conveyor

Fig. 49. Diagram of a Lurgi flash-drying, pulverizing, and ring-roll briquetting plant for brown coal.

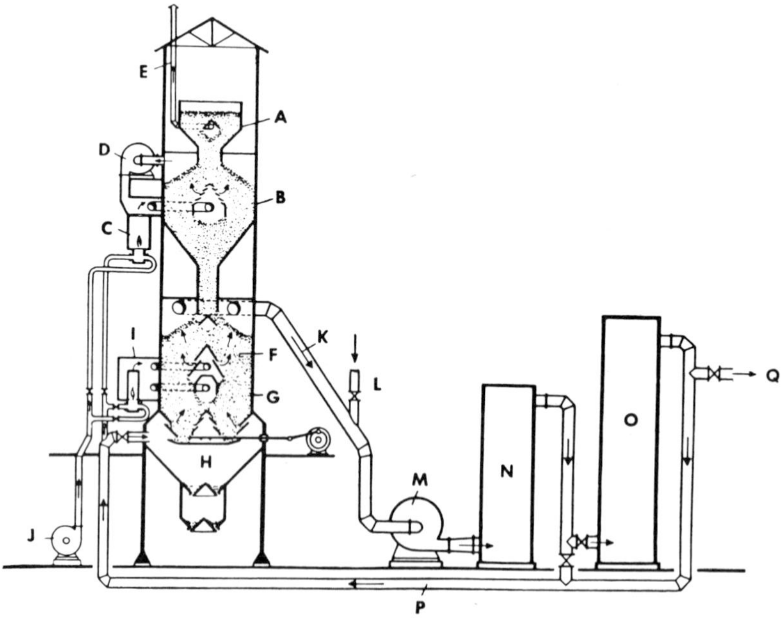

Fig. 50. Lurgi Spülgas carbonization plant.

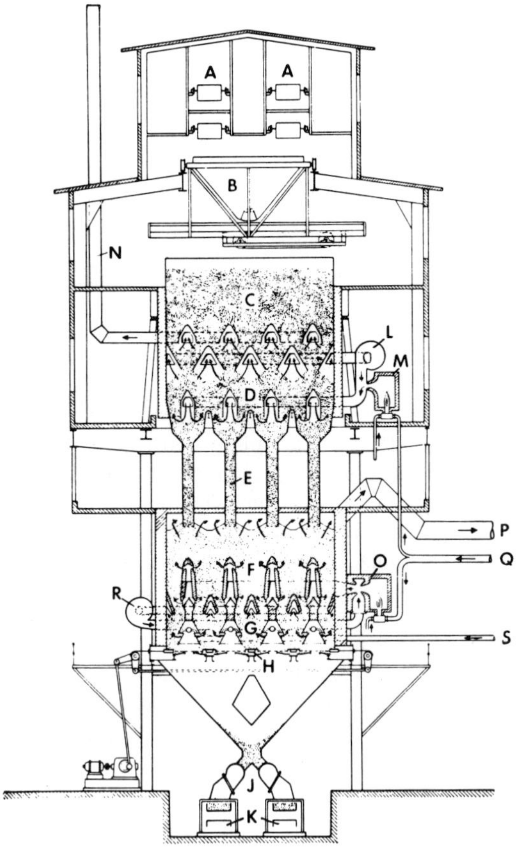

Fig. 51. Diagram of large Lurgi "Spülgas" carbonization plant.[184]

When hard, dense briquettes, made in ring-roll presses, are used, the yield of breeze below 6 mm in size falls to between 5 and 8% and that above 30 mm lies between 30 and 34%. The raw briquettes are made in the form of long cushions 7.3 cm long × 3.32 cm deep × 4.6 cm wide; these briquettes shrink when carbonized, but they retain their shape. These briquettes are denser than those made from extruded briquettes, take up less water when quenched, and are less liable to ignite spontaneously. They have a volatile content of 10 to 12%, and they contain up to 20% ash. The calorific value is about 20% higher than that of the briquettes charged to the retorts, that is about 6000 kcal/kg (11,000 Btu/lb), and they burn smokelessly. The gas leaving the central zone after dust removal is cooled to 110 to 130°C by an injection of liquor through sprays. The tar fog is separated by electrostatic precipitation and, when the gas is cooled further, a middle oil and liquor condensate are thrown down. The gas then passes to the oil washer.

The tars are paraffinic in nature,[184] are amenable to hydrogenation by the Bergius process, and yield a motor spirit. They may be subjected to modern extraction and refining methods yielding paraffin, fuel oil, diesel oil, and electrode coke, and phenols may be recovered from the middle oil and from the liquor by the "Phenosolvan" process. At Espenhain, 150,000 tons of light oil were sent for hydrogenation, and 205,000 tons of tar yielded

Diesel oil	64,000 tons
Fuel oil	71,000 tons
Hard-paraffinic wax	13,000 tons
Soft-paraffinic wax	13,000 tons
Electrode wax	12,000 tons

Lurgi Spülgas plants are built in units having a throughput of up to 500 tons of raw briquettes a day, and factories possessing a number of units have been in operation for many years.

"Phurnacite" process. In this process,[144,186] low-volatile bituminous coal is briquetted with pitch, and the resulting ovoids are carbonized. The first large-scale plant was built near Aberdare in South Wales in 1942.[187] The briquetting section was conventional, and the carbonizing plant consisted of inclined chamber ovens similar to a plant built by the Disticoke Company at Noeux les Mines in the Pas de Calais. The plant has been quadrupled in size and now produces 650,000 tons of carbonized ovoids per annum. The coal supplies are drawn from a number of different collieries in the Aberdare Valley, a region that produces dry steam coals of volatile contents ranging from 10 to 12%. These coals are washed to below 6% ash, and the washed duffs and fines are used. These range from $\frac{3}{8}$ × 0 in. and $\frac{1}{2}$ mm × 0 mm in size. Their analyses and properties are given in Table 21.

The product, sold under the name of "Phurnacite," consists of carbonized briquettes which are uniform in size, extremely hard, and free from dust. They are intended specifically for use in closed stoves, heat-storage cookers, domestic hot-water boilers, central-heating boilers, and are included in the anthracite class of fuels. Over 90% of whole ovoids are produced together with a small quantity of breeze which is recirculated. The remainder is sold as "splits" or broken ovoids.

[186] Smith, J. H., Cohen, H. T., and Humphreys & Glasgow, Ltd., Brit. Pat. 469,175 (1937). Anon., *Colliery Eng.*, **30,** 48–57 (1953).

[187] Anon., *Coke & Gas*, **14,** 5–14 (1952). Evans, S. I., *Colliery Eng.*, **35,** 523 (1958).

TABLE 21

ANALYSES AND PROPERTIES OF WELSH DRY STEAM COAL USED IN THE MANUFACTURE OF PHURNACITE[188]

Moisture	(%)	0.7–4.0
Proximate Analysis (dry basis)		
Volatile matter	(%)	9.5–13.8
Fixed carbon	(%)	78.0–87.0
Ash	(%)	3.0–8.0
Ultimate Analysis (dry basis)		
C	(%)	85.0–89.0
H	(%)	3.8–4.2
O	(%)	1.4–3.0
N	(%)	1.2–1.4
S	(%)	0.6–2.0
Ash	(%)	3.0–8.0
Calorific Value (dry, ash-free)	(Btu/lb)	
Gross		15,600–15,760
Net	(Btu/lb)	15,240–15,350

The "Phurnacite" plant consists of the following sections (1) coal recovery, storage and drying units; (2) pitch-handling, coal- and pitch-blending, grinding and briquetting units; (3) carbonizing plant; (4) screening and loading plant; and (5) byproduct recovery plant.

The sequence of operations is shown as a flow diagram in Fig. 52.

The coal, containing, on average, 15% of water, is delivered into a number of bunkers, which may be used to control the blend of coals and to recirculate breeze into the mixture. From these, the coal is discharged onto a belt which delivers the mixture to a set of vertical "Rheol" dryers. The dryers are supplied with hot gases at between 500 and 600°C from pulverized-fuel furnaces operating at 1000°C, the furnace gases being let down by air to the required temperature. The gas effluent from the dryers contains fire-dust, but this is removed by electrostatic precipitators.

[188] *Solid Smokeless Fuels*, Solid Smokeless Fuels Federation, London, 1956, 4 pp.

a Dry coal bunker
b Coal measuring belt
c Pitch
d Mixture disintegrator
e Mixture elevator
f Pug
g Cooling worm
g Roll press
i Shaker screen
j Rejects and fins belt
k Chain conveyor
l Trommel screen
m Fins bunker
n Raw ovoids belt to ovens
o Fins conveyor return to process

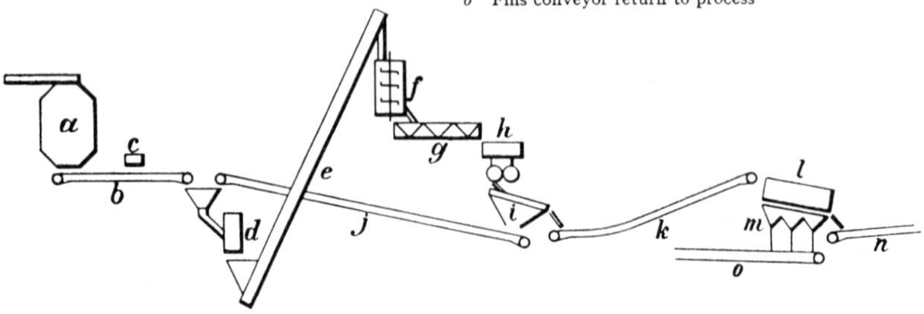

Fig. 52. Flow diagram of briquetting section of Phurnacite plant.

The briquetting plant is laid out in independent units each of 25 ton/hr capacity. The dry coal, in its original size, and the appropriate amount of pitch ground to below $\frac{1}{8}$ in. is mixed, and the mixture is passed through disintegrators where it is crushed to a fine size. These disintegrators are of the Carr type in which the coal is broken by impact against the hard steel bars of two concentric "squirrel cages" rotating in opposite directions. The mixture is heated in a "pug" supplied with superheated steam at about 230 to nearly 100°C and discharged from the pug at between 95 and 98°C. When the mixture reaches the press, the temperature is about 78°C and the mixture is pressed at from $1\frac{1}{2}$ to 2 ton/in^2. Each set of rolls produces about 80,000 tons of briquettes before it has to be renewed. The ovoids are delivered onto a chain belt where they cool, and from there they pass to a trommel screen which removes the fines from them. These rejects are returned via the disintegrators to the briquetting plant. From the trommel screen, the briquettes are delivered on a belt to the individual service hoppers that feed the carbonizing ovens.

The carbonization plant consists of batteries of forty ovens each built to the Disticoke Company's design (see Fig. 53). The ovens are of the recuperative type, heated with recycled oven gas. The ovens are 8 in. wide and each has a capacity of 2 tons of ovoids. The ovens are heated by gas burning upwards in vertical flues, the waste gas being passed downwards to recuperators and then to the chimney flue. The incoming air makes a vertical pass upward and is then distributed by an enclosed channel to the base of the vertical flues where it meets the gas.

The raw ovoids are stored in compartments in the bunkers above the ovens; each compartment holds about $2\frac{1}{2}$ tons of ovoids and charges into its particular oven. The ovens are operated in blocks of eight; each block requires a charge of 18 tons of raw ovoids and the throughput per battery is about 500 tons daily. The briquettes are carbonized at 800°C for 8 hours. The ovens are discharged in blocks of eight by a unique combination of door extractor, quenching car, and water tank. The hot ovoids are run into water held in the traveling car, and when quenched the water is run off and the ovoids discharged onto a sloping bench; from there they pass to the screening and loading plant. The ovoids are 35 × 38 × 28 mm, their bulk density is 48 to 50 lb/cu ft and each weighs 33 to 35 g.

The fluctuations in gas volume are severe owing to eight ovens being charged simultaneously. The gases are saturated with tar and liquor which are collected in a main tank, from which they are run to tar and liquor separating tanks, while the gases, still hot, pass from the main tank through a Simon-Carves electro-detarrer to a direct gas cooler and then to the exhausters. From the exhausters, which are automatically controlled to give constant suction, the gases pass through an ammonia and naphthalene washer. The gas, which has a calorific value of 500 Btu/cu ft, is returned to the oven via a balancing gas holder of 100,000 cu ft capacity. A standby gas-producer plant is maintained in a state of readiness to guard against a failure in the gas supply.

Synthracite process. This process is similar to the "Phurnacite" process and is operated by the Dutch State Mines in Limburg. The product, produced at the rate of 20 ton/hr is a hard, smokeless coke briquette of high quality.

Brennstoff-Technik process. An all-iron retort is used in the Brennstoff-Technik process.[178,183,189] These retorts are narrow chamber ovens 100 mm wide and are externally heated; they are built in blocks of twelve chambers, each chamber holding about 1 ton of briquettes. Carbonization takes 2 hr, and the temperatures do not exceed 650°C. Although originally designed for the low-temperature carbonization of coal, their dimensions and rapid rate of heating make them ideally suitable for briquette carbonization. A detailed description of the plant at Marienau in France is given by Georges.[183]

Processes Using Sand, etc.

Various processes have been developed from time to time in which briquettes have been embedded in sand or coke breeze to protect them during carbonization. Pieters[190] designed a vertical externally heated retort for such a process. One retort was erected at Le Boucau (1919) and another at Liege (1925). In the

[189] Thau, A., *Brennstoffschwelung*, Knapp, Halle, 1949, Vol. 1, pp. 75–83.

[190] Pieters, J., Brit. Pat. 287,934 (1927); Ger. Pat. 474,877 (1928); *Proc. 2nd Intern. Conf. Bit. Coal*, **1928**, Carnegie Inst. Technology, Pittsburgh, Pa., 1929, Vol. 2, pp. 707–25.

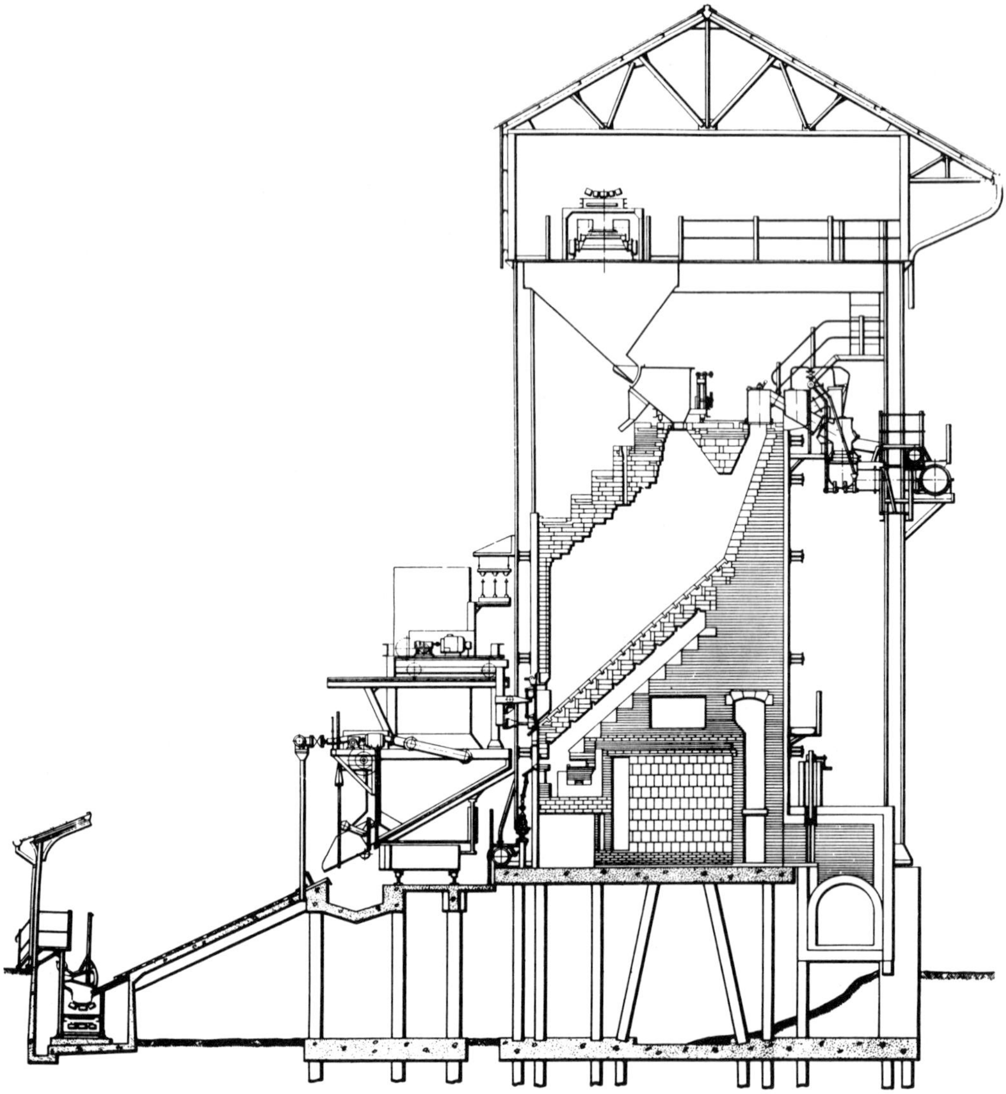

Fig. 53. Cross section through Disticoke ovens.

Pieters process, coal was briquetted with an oxidized pitch, and the briquettes were carbonized together with the inert material which had been suitably pulverized. The process was continuous and slow, the full cycle requiring about 20 hr for a metallurgical coke and 18 hr for an artificial anthracite. The temperatures of carbonization were 950 and 8000°C respectively.

Weber Process

Inert particulate material has recently been used as a means of transmitting heat rapidly to the briquettes. Weber[180,191] has advocated such a process, and a rotary retort[192] of special design has been selected as a convenient means of adaption. Briquettes are fed in at one end of the retort and hot sand at the other, the briquettes move down the slightly inclined retort while the sand moves countercurrent to the briquettes by means of helical grooves in the refractory lining of the retort. These grooves also raise the hot sand and cause it to spill over the briquettes (see Fig. 54).

[191] Thau, A., *Brennstoffschwelung*, Knapp, Halle, 1949, Vol. 1, pp. 136–44.

[192] Pistorius, O., and Demag Co., Ger. Pat. 848,644 (1950).

National Coal Board Hot Sand Process

A process[193] similar in principle to that proposed by Weber has been developed by the National Coal Board in Britain. Hot sand is used to transfer heat to the briquettes, but the sand is allowed to percolate over and through a mass of briquettes held in a vertical retort. It is found that the hot sand flows freely, and no holdup occurs. A static vertical retort is generally to be preferred to a rotary retort, but it has the disadvantage of not permitting counterflow of the heating medium and avoidance of high thermal shock to the green briquettes.

Both the briquettes and the hot sand enter the retort at the top and move downward, the sand much more rapidly than the briquettes. At the bottom of the retort the briquettes pass over a static grid and are discharged through a water seal while the sand passes through the grid into a hopper from which it is fed into a transport line to a heater and from there through a cyclone back into the top of the retort (see Fig. 55).

The pilot plant has demonstrated the feasibility of using sand as a means of heat transfer and of reducing the carbonizing time of briquettes to less than 1 hr. Two-thirds of a ton of sand are said to be required per ton of briquettes carbonized, and this sand is heated in simple equipment. The temperature of carbonization is limited to the sintering point of the heat-transfer material. In operation, none of the troubles feared happen—the sand does not stick to the briquettes, or abrade their surfaces, or soak up the volatile matter produced.

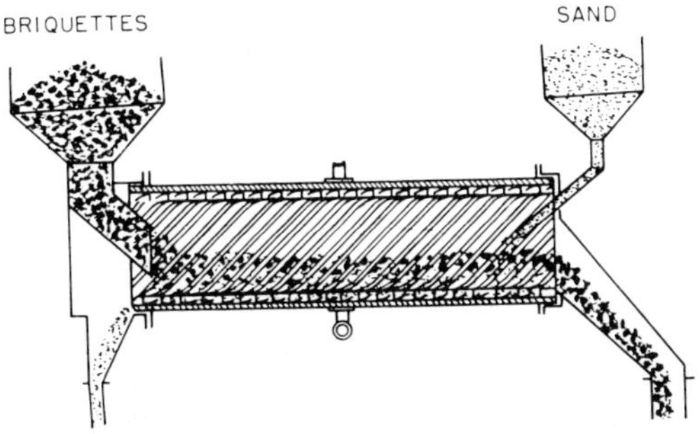

Fig. 54. Apparatus for carbonization of briquettes with hot sand as the heat transfer medium.[192]

HOT BRIQUETTING AND MISCELLANEOUS PROCESSES

Bituminous coals soften when heated, and many attempts have been made to convert them to a lump fuel in this state. The simplest way of doing this is to pass hot powdered coal through a slowly rotating drum[194] or over a vibrating plate,[195] so that the soft and sticky coal agglomerates and rolls itself into balls. Although this method is simple in concept, technical success has been limited, and this method is less attractive than those which make the agglomeration a deliberate and controlled act by pressing the soft coal in molds.

The earliest proposals to briquette hot coal by the application of pressure to the heated coal are those of Dowson in 1856,[196] and Crampton in 1872.[197] In the former process, solid blocks of fuel were manufactured by mixing coal dust, small coal, and coke, heating the mixture to 300 to 900°F for 1 to 6 hr in an oven, and then compressing the residual pasty mass into solid blocks by means of an hydraulic press. In Crampton's process, coal powder was allowed to fall down the interior of a very long vertical cast-iron pipe of narrow diameter, the pipe being heated externally for a definite length in the middle. The coal reaching the bottom had lost some of its volatile matter, and was passed while still hot into a briquetting press of conventional design and formed into blocks. In 1908, a press was invented in which hot coal was pressed into 11-lb blocks,[198] and, in 1912, a process was described for briquetting blocks of coal at 400°C and 10 ton/in.^{2} pressure.[199]

Proposals have been made from time to time to heat coal while it is subjected to pressure, and experiments[200] have shown this to be an admirable way of consolidating coal powders.

[193] Launder, J. L. M., Rich, P. D., and Smith, R., Brit. Pat. 741,679 (1955). Harrison, J. S., and Madley, D. G., Private communication.

[194] Davidson, T. M., Brit. Pat. 331,906 (1929); U.S. Pat. 1,831,704 (1931); Wisner, C. B., Brit. Pat. 217,554 (1925); U.S. Pat. 1,748,815 (1930).

[195] Allan, G. W. C., and Sparham, G. A., and C.U.R.A. Patents Ltd., Brit. Pat. 719,999 (1954).

[196] Dowson, A., Brit. Pat. 148 (1856).

[197] Crampton, T. R., Brit. Pat. 2,262 (1872).

[198] Galloway, W., *Proc. S. Wales Inst. Engrs.*, **26**, 424 (1908–10).

[199] Phillips, H. J., and Phillips, A., Brit. Pat. 14,406 (1912).

[200] Parr, S. W., and Olin, H. L., *Univ. Illinois Bull.*, No. 60 (1912), 24 pp.

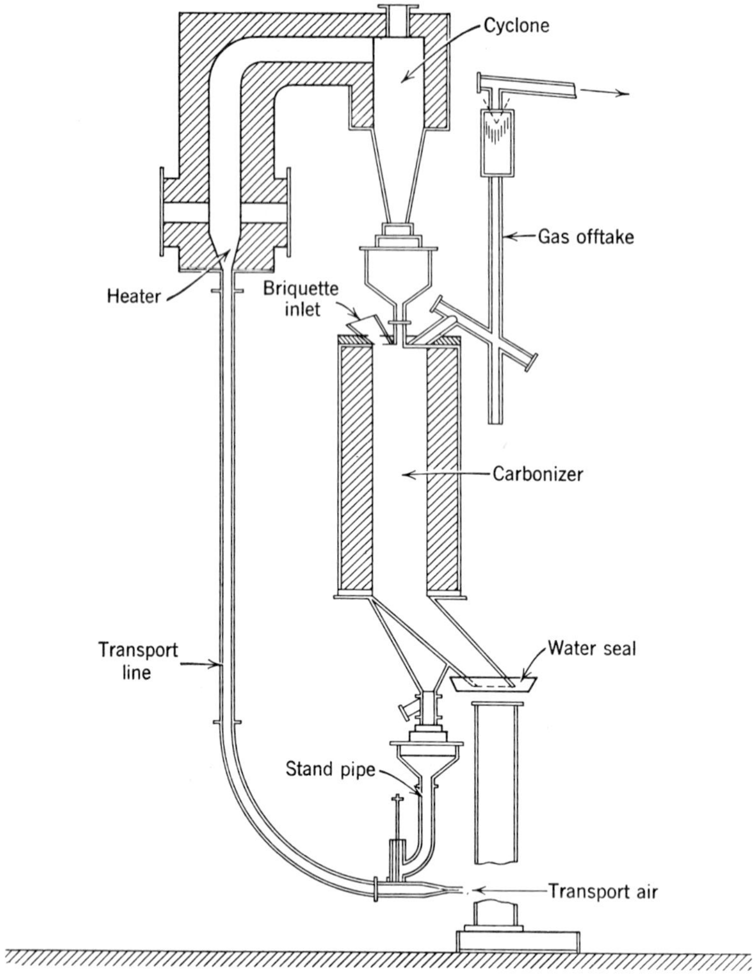

Fig. 55. National Coal Board hot sand process.[193]

Processes have been described, in which mixtures of powdered coking and noncoking components are heated while under pressure in the molds, to a temperature at which the coal softens but does not carbonize.[201] The apparatus required to do this, however, is rather complicated,[202] and the method does not appear to have been used commercially.

Commercial-scale tests have been carried out with a much simpler apparatus. In this, the coal is heated in a slowly rotating retort, and inside the retort are heavy rollers which ride on and compress the heated coal.[203]

In the processes already described, two alternative schemes are used. In the earlier, the coal is first heated and then briquetted; in the latter, presumably in an attempt to obtain greater control, the coal is first pressed and then heated while under pressure. Neither method has yielded a process which is satisfactory commercially. It has been said[204] that the essential difficulty is that of achieving sufficient control in the small-temperature range between the onset of plasticity and the commencement of the evolution of gas and tar.

Although the temperature to which the coal

[201] Manson, L., and Smith & Coventry Ltd., Brit. Pat. 1,118 (1914). Beaudequin, J., Brit. Pat. 225,849 (1926).

[202] Vloeberghs, A., Brit. Pats. 641,166 (1950); 655,314 (1951).

[203] Dobbelstein, O., Brit. Pats. 148,115 (1920); 164,751 (1921).

[204] Petersen, W., *Der deutsche Steinkohlenbergbau, vol. 3, Brikettierung der Steinkohle, Glückauf*, Essen, 1958, p. 145.

is raised in this form of briquetting lies in or near its decomposition range (and this may give rise to difficulties in briquetting owing to the development of gaseous pressures within the briquette), such a process has the virtue of producing a briquette which can be smokeless without any substantial portion of volatile matter being lost.[205] In the most recent developments in this field, this aspect has been by no means a secondary consideration.

Although no hot-briquetting process has, as yet, been established commercially, considerable effort is being devoted to the problem in Great Britain, Holland, and Germany,[206] and developments are likely in the near future.

Theory of Hot Briquetting

With increase in temperature, coal becomes progressively less elastic and more plastic, but this change is not regular. At a temperature of about 400°C, most coals exhibit a very marked change in plasticity, and it is in this region that hot briquetting becomes practicable. As the temperature approaches about 400°C, the coal behaves as though it were composed of a nonfusible and a fusible component, the latter decomposing when further heated into a solid. The coals that appear to contain a large proportion of the fusible component are coking coals, and those with small quantities are noncoking. This picture, given by Jäppelt,[207] is a simple one, and to a large extent it is adequate as a basis for understanding the mechanism of the various hot-briquetting processes devised.

Van Krevelen and associates[208] have described the phenomenon of carbonization in mathematical terms, and have referred to the rate of loss of volatile matter from the coal when heated in this temperature region. A more complicated picture has been put forward[209] that postulates a coal component which, when heated, turns into a material which melts and decomposes into a solid and vapor. The relationship between temperature, time, and fluidity is shown in Fig. 56, and the difference in behavior of coals of different rank is illustrated in Fig. 57.

The simple picture given previously must be considered in the light of how heat passes from the surface of a coal particle to its center. Coal is an extremely poor thermal conductor,[210] and when a piece is heated the surface will, in most cases, reach the fusion temperature before the center does; under conditions of rapid heating, it is quite possible that within a mass of particles, the surfaces of the individual particles will fuse and become sticky, and the center particles remain hard. In addition, coal heated up to within this temperature region decomposes and loses volatile matter, some of which is condensible and may well condense, under certain conditions of heating, on the surface of other cooler particles and thereby provide a sticky surface.[211]

Parr[212] at the University of Illinois was probably the first to study phenomenologically the thermal decomposition of coal and to relate the physical structure of the coked fuels to their free-burning properties. Beilby[213] later made a similar study using British coals.

On the Continent, Dunkel[211] examined various Upper Silesian coals, both coking and low-rank noncoking bituminous coals. His studies were intended to lead to commercial application, and a number of patents were filed in conjunction with Hoffmann.[214] In these studies, Dunkel brought out the relationship between the extent of the softening of the coal

[205] Gregory, D. H., and Adams, W. N., *Intern. J. Air Pollution*, **1**, 263–5 (1959). Gaines, A. F., and Pitt, G. J., *Intern. J. Air Pollution*, **2**, 266–75 (1959).

[206] Williams, F. A., Brit. Pat. 788,068 (1957). Stamicarbon, N. V., Brit. Pat. 768,910 (1957). Collin, F. J., A.G., Brit. Pat. 787,147 (1957). Repetzki, K., and Puff, W., Brit. Pat. 790,024 (1958). Rhys Jones, D. C., *New Scientist*, **5**, 1008–10 (1959).

[207] Jäppelt, A., *Erdöl u. Kohle*, **6**, 613–6 (1953).

[208] Chermin, H. A. G., and van Krevelen, D. W., *Fuel*, **36**, 85–104 (1957).

[209] Bronowski, J., Fitzgerald, D., Gillings, D. W., and Rhys Jones, D. C., *Nature*, **171**, 389–90 (1953).

[210] Clendenin, J. D., Barclay, K. M., Donald, H. J., and Gillmore, D. W., *Penn. State Coll., Tech. Paper*, **160** (1949), 178 pp.

[211] Dunkel, M., *Z. oberschles. berg-u. hüttenmänn. Ver. Katowice*, **65**, 360–5 (1926).

[212] Parr, S. W., *Illinois State Geol. Survey, Bull.*, **60** (1931), pp. 111–15.

[213] Beilby, G. T., *Trans. Soc. Chem. Ind. (London)*, **41**, 341–4T (1922); *Rept. Fuel Research Board*, London, 1920–1, London, H.M.S.O. (1922), 127 pp.

[214] Hoffmann, F., et al., Ger. Pats. 431,502 (1926); 454,520, 455,015, 458,247, 460,131 (1928); 495,624, 500,503, 500,599, 501,378 (1930); 539,637 (1931). Hoffmann, F., French Pat. 632,019 (1927); Brit. Pat. 293,109 (1928).

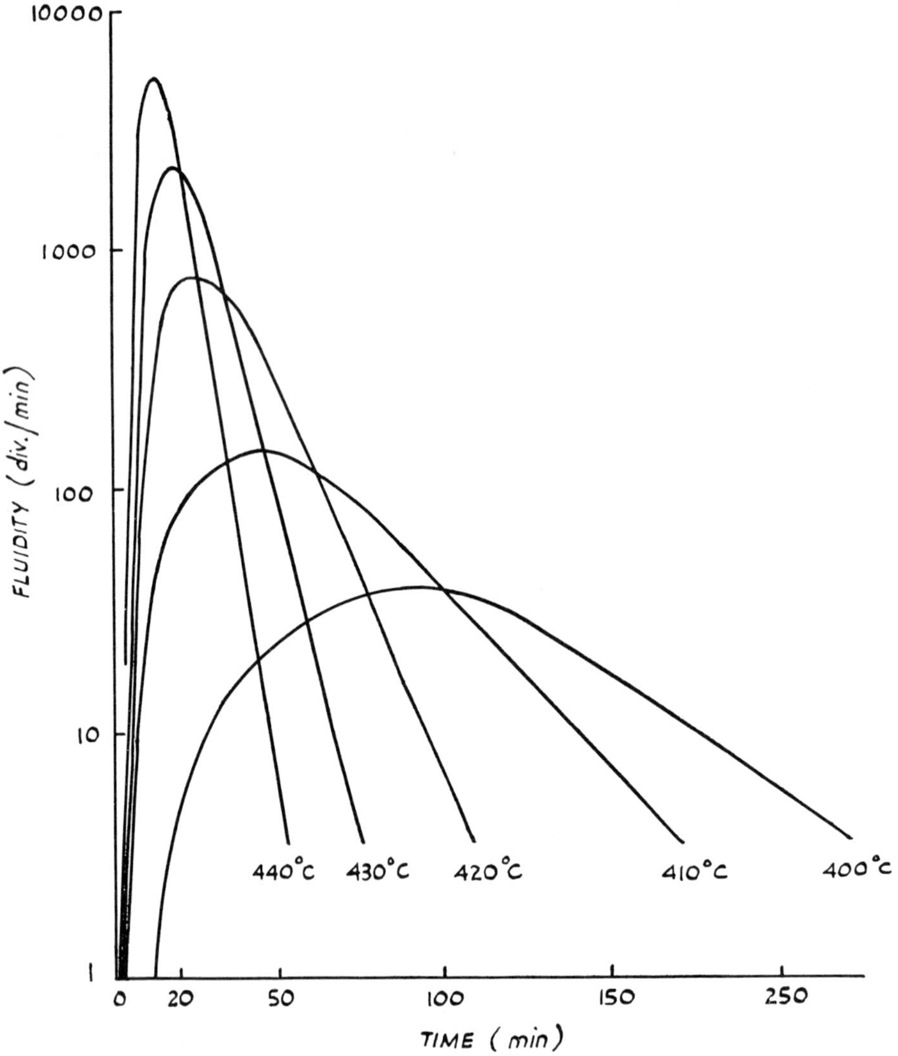

Fig. 56. Effect of temperature of heating on width of plastic range and the maximum fluidity.[209]

and its temperature. He measured the pressure required to produce an acceptable briquette at different temperatures (see Table 22), and he concluded that below 320°C the pressure required would be too high for practical application, and that above 375°C the coal would be undesirably soft. He also showed that the plasticity of the coal could be modified by the introduction of nonsoftening materials such as coke breeze.

Still later, in 1929 and 1930, Swietoslawski, Roga, and Chorazy[215] made an important contribution to knowledge in this field of briquetting. They studied both coking and noncoking

[215] Swietoslawski, W., Roga, B., and Chorazy, M., *Fuel*, **9**, 421–39 (1930).

TABLE 22

PRESSURE REQUIRED TO BRIQUETTE COAL AT DIFFERENT TEMPERATURES[211]

Temperature	Pressure
°C	kg/cm²
20	5000
240	2800
280	2100
320	1400
350	350
375	150
400	50

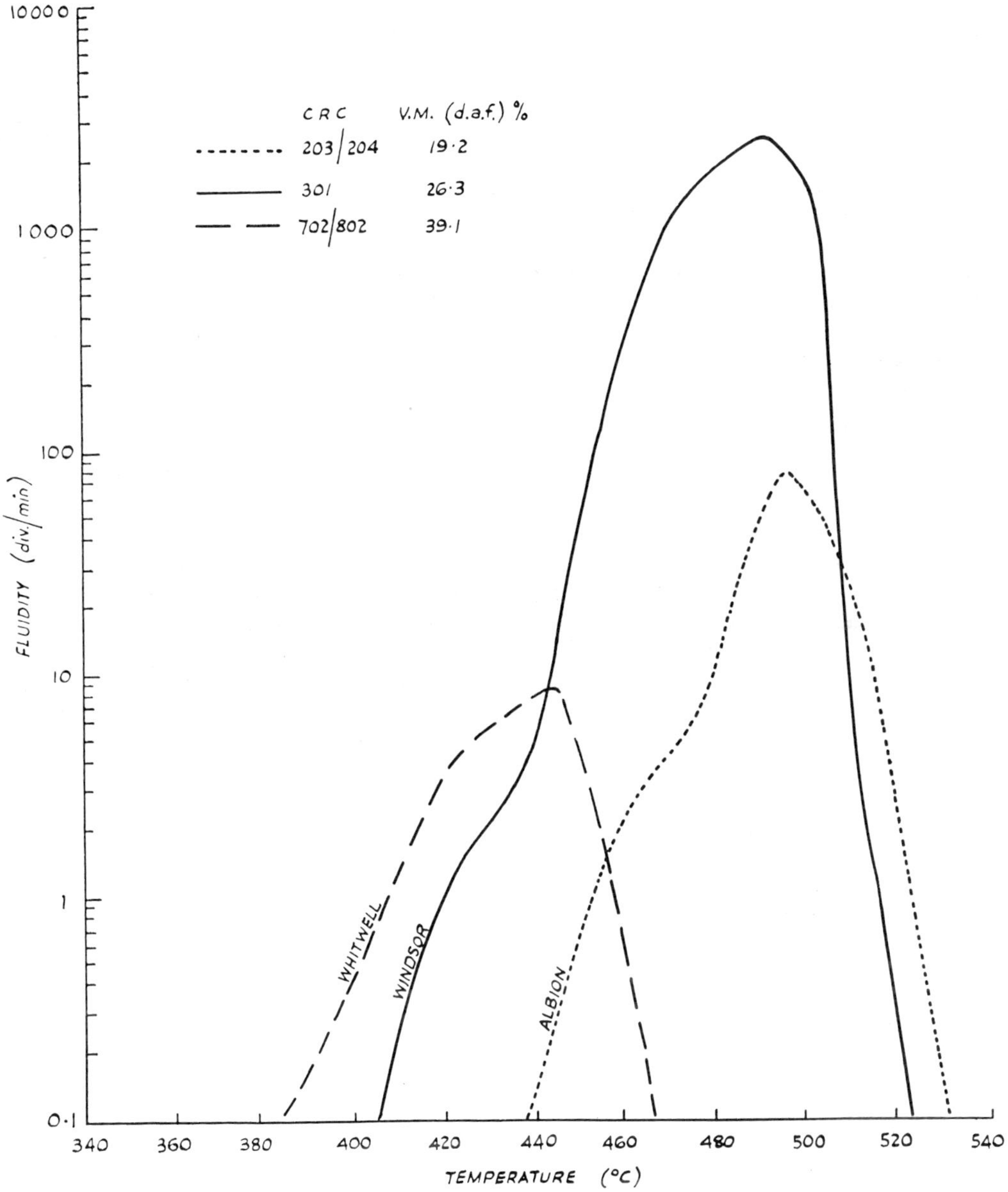

Fig. 57. Effect of rank on the fluidity and width of the plastic range.[209]

Upper Silesian coals and the noncoking coals from a number of different areas. These studies brought out the overriding importance of the plastic properties of coal in briquetting. They also showed that the plasticity of a coal can be influenced by the rate at which the coal is heated and the temperature to which it is heated.

Plasticity is essential for the binderless briquetting of coal. Anthracite and other coals which do not become plastic when heated cannot be briquetted without a binder, even at pressures as high as 3000 kg/cm^2. Coking coals and coals that soften when heated may be briquetted at pressures of 200 to 400 kg/cm^2.

The optimum conditions for briquetting vary according to the nature of the coal, its plasticity, caking power, and volatile content. Strongly coking coals may be too fluid at their point of maximum plasticity, and thus it is necessary to briquette at a temperature at which the plasticity has not yet reached its maximum or passed it. Briquettes may be made by pressing the coal when its temperature is just below the temperature of incipient plasticity, or after the coal has been held for a long period at a temperature within the fusion range to destroy the excess plasticity. It may save time to heat the coal to a higher temperature where the excess of bituminous substance is destroyed more rapidly. Coals with less bituminous substance, such as the low-rank noncoking coals or subbituminous noncoking coals, have a narrow briquetting-temperature range; for some coals it is only 20°C in which case it is necessary that the rate of heating should be rapid and the duration of heating short. Under optimum conditions of temperature and duration of heating, these coals may be briquetted at a pressure of 400 kg/cm^2.

The briquettes so made possess considerable mechanical strength and a calorific value greater than that of the original coal; they are weather resistant, ignite easily, burn well, and are less smoky than the coal from which they are made. The briquettes made at temperatures below the point of incipient plasticity have a granular structure, whereas those made at higher temperatures from coking coals have a fused structure and are more dense. The greater the potential plasticity of the coal, the greater the latitude in design of the briquetting process—at least under the rigidly controlled conditions of the laboratory.

Piersol[128] conducted an extensive series of experiments on the subbituminous coals of western United States, and he carried out trials on coals from Illinois, West Virginia, and Utah in an experimental 1 ton/hr-plant.[126] Piersol was searching for a method of briquetting these coals without a binder, and he was impressed by the "plasticity of coal developed under pressure at comparatively low temperatures of about 300°C." The relationship between temperature and pressure of briquetting for a Washington-County coal is shown in Fig. 58.

The importance of the plastic character of the coal implies that the duration of compression might be a significant variable in the manufacture of briquettes. Piersol examined the extremes—impact and sustained pressure—and in the latter extreme he also examined the effect of duration of heating on the coal. His results are not obviously in accord with the findings of other workers, which suggests that the nature of the coal plays a greater part in hot briquetting than even Swietoslawski suggested. The coals used by Piersol were strongly coalified, low-rank coals of the Cretaceous formation, and were similar to the Upper Bavarian bright coals.[204]

Good briquettes were made from these coals when they were heated to a temperature of 300°C and briquetted under a pressure of 15,000 lb/in.^{2} sustained for 15 min, or at a

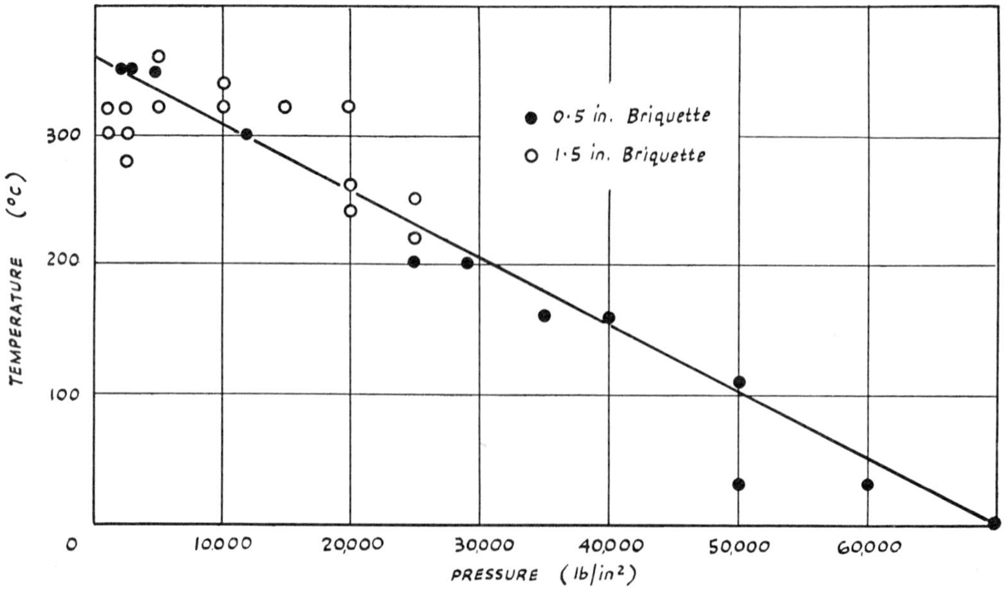

Fig. 58. Relationship between the temperature and the pressure of briquetting.[128]

pressure of 20,000 lb/in.^{2} sustained for at least 3 min. These times are too long for the commercial application of such a press. At temperatures above 350°C, excessive melting occurred, and it became difficult to confine the fluid coal to the die.

Impact compaction, in which the impact was obtained by dropping a 500-lb weight 4½ ft on to a 1½-in. diameter briquette showed that good briquettes could be obtained from coal preheated to 300°C. The application of sudden pressure to coal preheated to 300°C always resulted in a rise of internal temperature; in the six experiments quoted, the rise measured was 35°C, but since the temperature measured was that of the mold, the rise in temperature of the coal must have been considerably greater than this. This rise in temperature on rapid compression has been referred to more recently by Repetzki and Puff.[206]

In a series of experiments, the coals were preheated to temperatures between 100 and 400°C for periods of 5 to 60 min and briquetted by impact with loads of from 1500 to 3000 ft-lb for a briquette weighing 0.1 lb.[129] Piersol[129] built a pilot-scale plant in 1941 and described a method of making briquettes in which coal, preheated to about 350°C, is pressed at about 30,000 lb/in.^{2}, applied for a period of about 0.1 sec. This was done in the Piersol ring-roll[129] press. The method of heating the coal, although convenient for the pilot plant operated, was not suitable for commercial development. None of the processes described by Piersol appears to have been exploited commercially.

Jäppelt[207] has investigated the effect of the temperature of ejection from the mold on briquette quality. Since the temperature of briquetting in most cases lies near or in the plastic range, some decomposition takes place —part of the volatile matter lost occurs as CO_2 or methane—and internal pressures may build up within the briquette. If the temperature is held constant, the evolution of volatile matter, which is rapid initially, virtually ceases in a very short time. Its cessation is even more rapid if the temperature is allowed to fall. These considerations lead to the rejection of a roll press as a machine suitable for briquetting hot coal and to favoring an extrusion press for this purpose.

Petersen[204] has reported on the experimental work carried out in the experimental establishment of Klöckner-Humbolt-Deutz A.G. during 1934 to 1936 and the briquetting trials carried out at Bottrop by the Arenberg Bergbau GmbH since 1955. He summarized the work in the former establishment and stressed the importance of the speed of pressing and the need for the compact to be cooled in the press itself. In double-roll presses, the residence time is too short even at slow rates of rotation. Trials with a Couffinhal press gave satisfactory briquettes when operated slowly, and it was evident from this work that an extrusion press would be the most suitable. At Bottrop, both semianthracite and steam coals were briquetted successfully.

Processes Devised for Agglomerating Hot Coal

The Disco Process

The Disco plant[149,216] operating at the Pittsburgh Coal Carbonization Company is one example of successful operation of a process in which coal is carbonized in a revolving steel retort to produce a fuel in the form of irregular but homogeneous balls. The agglomeration of coal takes place by virtue of the surface of the coal particles becoming fluid and sticky, adhering to each other, and building up into balls—as snowballs are formed—on their passage through the rotating retort. Each particle becomes sticky on the outside at the melting point, the temperature does not exceed the melting temperature, and the product is a truly homogeneous low-temperature coke.

The property that allows this agglomeration to take place is one which causes difficulty in the retort. The sticky particles adhere not only to themselves, but to the shell of the retort and cause the growth of a thermally insulating scale. This adherence may be prevented by mixing the coal with a noncoking coal or coke breeze, as in the Davidson process,[217] or by preoxidizing the coal and thermalizing it, as in the method developed by Wisner.[218] It is

[216] Lesher, C. E., and Zimmermann, R. E., *Coal Age*, **44**, (3), 45–9 (1939). Roberts, J., and Jenkner, A., *International Coal Carbonisation*, Pitman, London, 1934, pp. 233–4.

[217] Utah Conservation & Research Foundation, *The Low-temperature Carbonization of Utah Coals*, Salt Lake City, Utah, 1939, pp. 284–91.

[218] *Ibid.*, pp. 299–303.

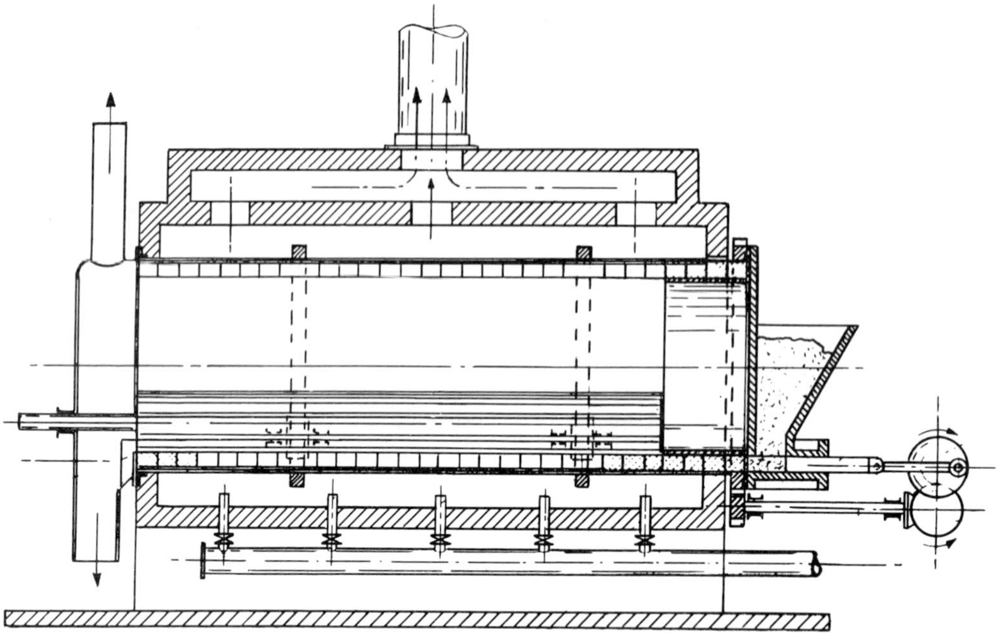

Fig. 59. Longitudinal section on the "Raffloer" retort for the continuous production of hard, free-burning, smokeless fuel, involving compression of the viscous charge.[220]

this latter process, sometimes referred to as the "Carbocite" process that is used to produce the semicoke balls sold as "Disco."

Heating under Pressure

The Dobbelstein–Raffloer Process

As may well be imagined, the "balling" process is one of considerable sensitivity, since so many factors determine the degree of stickiness of the coal, and it is not surprising that attempts have been made to reduce its sensitivity by relying less on the stickiness of the coal for agglomeration and more on its general increase in plasticity.

At Mülheim, Fischer was reported[219] to have experimented with an externally heated rotary retort in which he placed a heavy roller which extended the full length of the retort. This intermittent process was not suitable for commercial use, and its development was undertaken by Dobbelstein,[203] who built a small plant of 10 to 15 ton/day throughput at Prosper Colliery, near Essen, and by Raffloer,[220] who designed a plant for gas works in New York. These retorts (see Fig. 59), which were of the externally heated cylindrical rotary type, the inner surfaces of which were divided into compartments, were provided with one or more rollers[219] and were capable of being operated continuously. No process using these methods is known to be in commercial operation at present, although improved methods are still being suggested.[202,221]

Heating and then Pressing

Most effort and ingenuity have been put into developing processes in which the coal is first heated to a prescribed temperature, and then briquetted in a more or less conventional way. Indeed, the earliest attempt at hot briquetting was made in this way in 1856, and many attempts have been made since.

The Hardy Process

The Hardy[222] process was developed between 1929 and 1936, but it never became established on the large commercial scale. In this process, the coal, in granular form, was heated to comparatively low temperatures in a

[219] Brownlie, D., *Proc. S. Wales Inst. Engrs.*, **42,** 297–310 (1926).

[220] Brownlie, D., *Colliery Guardian*, **131,** 1023–4, 1162–4 (1926).

[221] Mory, B., Brit. Pat. 372,282 (1920).

[222] Hardy, H., Brit. Pats. 356,236, 357,330 (1931); 445,207, 445,208 (1936); 461,394 (1937).

retort in which it was agitated to ensure uniformity of heating, and to prevent agglomeration of the particles. The hot coal was then passed into a double-roll press where it was briquetted. Hardy was at pains to describe in detail the degree of softening to be reached by the coal particles before they were pressed. Unlike earlier processes, the heating was stopped before the coal had reached the state of pasty fusion, but after the particles had reached the globular or spheroidal form. The temperature at which this softening takes place varies from coal to coal. The stage is reached, for example, with an English coking coal of 27.5% volatile matter at 350°C, whereas a Belgian coal containing 14.5% volatile matter required to be heated to 500°C.

These examples appear to be the extreme conditions. If bituminous coals alone are used, they give hard, dense briquettes, and such briquettes show a tendency to cracking when cooled, which appears to be caused by differential thermal contraction and may be prevented by introducing a nonfusing component such as coke. If the temperatures of globulation are exceeded and fusion of the particles is permitted, the results are disastrous; the tendency for this to happen is less, the less marked the coking properties of the coal. The danger may also be reduced by using larger grain sizes as well as introducing a nonfusible component. Darmont[223] has described the operation of a small plant and reported that the quality of the briquette was determined very largely by the temperature, by the proportion of coke introduced into the mixture, and by the pressure of briquetting. An example of the relationship is set out in Table 23.

TABLE 23

RELATIONSHIP BETWEEN TEMPERATURE, COKE CONTENT, BRIQUETTING PRESSURE, AND BRIQUETTE STRENGTH[223]

Coke Content	Temperature	Pressure (kg/cm²)			
		300	500	800	1200
(%)	(°C)	Crushing Strength (kg)			
0	375	—	Cracked	—	—
10	344	68	88	98	Cracked
50	360	80	121	183	196
50	400	256	367	380	400
75	350	20	32	48	58
75	400	75	76	120	150

The presses used in the Hardy process were of the double-roll type, and these have difficulty in dealing with a mix of varying density. These presses are designed to work at a constant volume input and have very little tolerance. Darmont has given compression curves for heated coal and has stated that in practice it is impossible to maintain the density of the hot mix to even 1%. If 500 to 800 kg/cm² are the extreme limits between which the working pressure may vary, it can be shown, by calculation from the curves, that the corresponding weights per unit volume of material are 0.98 and 1.01. Since the variation in practice is wider than 1%, it is concluded that double-roll presses may not be used for a process of the Hardy type, and the ring-roll type of press is preferred.

Recent Developments

The satisfactory briquetting of hot coal is so dependent on the degree of softening, and the softening of coal is dependent on so many factors, that as experience has shown, the success of this process is determined by how readily the heating of the coal may be controlled.

Hardy[224] aimed at keeping the temperature of the coal just below its fusion point, and he relied on the pressure to produce a satisfactory briquette. Piersol worked at temperatures within the same region, and emphasized the rapid application of pressure.[129] Repetzki[206,225] claimed that the work done by the application of pressure to the coal powder would generate sufficient heat to raise the temperature to that at which the coal agglomerates. In an extrusion press, the external friction also contributes to the temperature rise. It is conjectural, however, to what degree this rise in surface temperature contributes to the formation of the briquette, since coal has such a low thermal conductivity.

Although the work done in overcoming both internal and external friction causes the temperature to rise, calculation shows that the work done by external friction will, under suitable circumstances, be 2.36 cal/g for each 100 kg/cm² of applied pressure. The internal work done will be less than one-quarter of the external frictional work, and a total of 3 cal/g per 100 kg/cm² would be a generous estimate of

[223] Darmont, G., *Fuel*, **31**, 75–90 (1952).

[224] Hardy, H., Brit. Pat. 460,394 (1937).

[225] Repetzki, K., Brit. Pat. 520,728 (1940).

the total work done in an extrusion press operating on hot coal. The amount of work done varies with the complexity of the coal; for a coal of rank 900, the calculated heat rise generated by internal friction is given in Table 24.

TABLE 24

HEAT GENERATED BY INTERNAL FRICTION IN EXTRUSION BRIQUETTING

Carbonizing temperature (°C)	375	400	425	450
Work done (Btu/lb)	5.98	4.36	2.75	5.80
Temperature Rise in Extrusion Nozzle (°C)	6.6	4.8	3.1	6.4

The difficulty of heating coal to the point where it becomes soft but not fluid cannot be overemphasized. Frequent reference has been made to the use of blends of coking coal and noncoking coal, and an ingenious device has been suggested in a patent by the Chemische-Technische Gesellschaft.[226] The noncoking component may be heated to a temperature (above 600°C) so far in excess of that required at the press, that when mixed with the cold coking-coal component (below 300°C) immediately prior to pressing, the mixture will at once attain the desired temperature. This system is being used successfully in Holland[227] where anthracite is heated by means of hot gases in a fluid bed to a temperature of 550°C. A diagram of the Stamicarbon process is shown in Fig. 60. Noncoking coal is transported by the screwconveyor *a* from the feed hopper *b* to the fluid bed *c* where it is heated to the required temperature. The hot coal is discharged into the tube *d* and along which it is transported in an inert gas. Coking coal is fed into *d* from the hopper *e*. The mixture of coals is separated from the carrier gas in *f* and passed to the briquetting rolls *g*.

Another version of this technique has been described by Gregory[228] in which a hot semi-coke powder is mixed with a comparatively

[226] Chemische-Technische Gesellschaft, Brit. Pat. 300,195 (1927).

[227] Stamicarbon, N. V., Brit. Pat. 787,147 (1957).

[228] Gregory, D. H., *Chemical Engineering in the Coal Industry*, F. W. Sharpley, Ed., Pergamon Press, London, 1956, pp. 56–71.

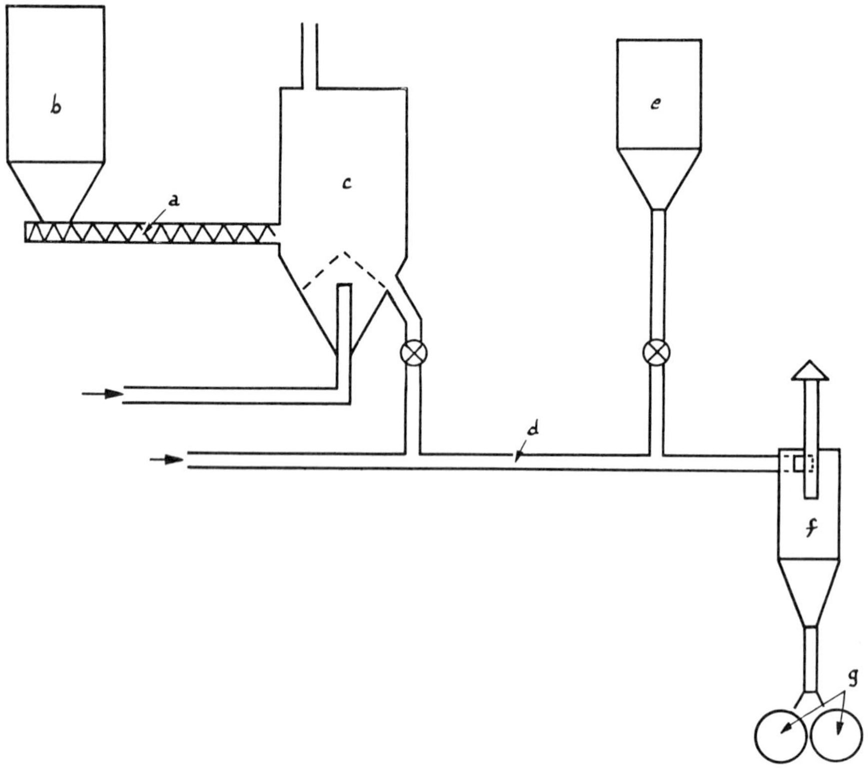

Fig. 60. Schematic drawing of the Stamicarbon process.[227]

small quantity of a fusible component which contains a proportion of coking coal and pitch. The mixture is pressed at a temperature in the region of 350 to 400°C. The product is smokeless and highly reactive.

The methods of heating coal for briquetting have been varied and numerous,[229] but the more recent attempts have been concerned with heating coals in a dense-phase fluid bed or in a dispersed phase. The Standard Oil Development Co.[230] has described a process in which the heating in the fluid bed carbonizes the coal to render it smokeless, and the hot devolatilized coal is then mixed with a binder and heated further (but to a lower temperature) in a rotary retort to agglomerate the material to suitably sized balls.

A method devised by the National Coal Board at its Coal Research Department[231] uses the fluid-bed technique for heating the coal to a temperature of about 400°C, the heat being generated by partial combustion of the coal by the fluidizing gas. The partially devolatilized coal is briquetted without a binder in an extrusion press. The temperature may be controlled to within 4°C, and difficulties of over-softening are avoided by using noncoking coal. The briquettes so made are smokeless, reactive, and dense. A pilot plant of 250 ton/day capacity has been built at Birch Coppice, England, to use this process.

Pettyjohn[232] has described a process in which a smoky coal containing 25 to 38% volatile matter is suspended in a current of hot air, and is passed into vortex chamber from which it is withdrawn after a residence time of 0.25 to 20 sec. The fluid bed provides for residence times of any duration, but the fully entrained disperse-phase system of heating is suitable for only very short contact times. A number of cyclones in sequence may be used.[233]

Sapozhnikov has described a process for the manufacture of metallurgical coke from noncoking coals in which the pulverized coal in suspension was heated in a matter of seconds to the plastic state, maintained at this temperature for $\frac{1}{2}$ to 2 min, and finally pressed at 2 kg/cm^2 to 5 kg/cm^2. The briquettes so made were carbonized at 750 to 800°C.[234]

The Coal Logs Process

A process for extruding plastic coal into cylinders or logs has been invented by Storrs,[235] and a pilot plant of 35 to 40 ton/day was operated in Salt Lake City, Utah, by the Coal Logs Co. Inc.[236] Coal was fed onto a wire belt conveyor inside a horizontal retort and the belt was rotated both horizontally and vertically as it took the coal through. The retort was heated to between 400 and 500°C, and the hot coal, in its plastic state, was fed into the screw conveyor of an extrusion machine. The screw applied a pressure of 40 lb/in.^{2} and fuel logs 3 in. to $3\frac{1}{2}$ in. in diameter were produced. The pressure required was determined by the plastic nature of the coal, and could be changed by altering the length of the pipe. In the pilot plant, the belt was 37 in. wide × 13 ft long, the depth of coal was $\frac{1}{2}$ in. to $1\frac{1}{2}$ in., and the time of carbonization 1 to 3 min. Weakly caking Utah coals were used, and it was claimed that the process is suitable for noncoking coals also. The product was a free-burning fuel containing 18 to 20% volatile matter and was smokeless. The coal logs could be subjected to a second carbonization at a higher temperature of 800 to 1000°C. Here a carbonization time of 10 to 15 min was required to give a product with 2% volatile matter and was said to be suitable for metallurgical purposes. A demonstration plant has been reported to have operated in Durham,[237] but it was said that for this the coals should possess sufficient coking properties to show a clearly defined coking stage.

De-smoking of Briquettes

The manufacture of briquettes from low-volatile coals without a binder, the carbonization of briquettes, and the making of briquettes by compressing hot coal may all result in the

229 Honigmann, L., U.S. Pat. 1,704,796 (1929). Hodges, F. A. J., Brit. Pat. 704,745 (1954).

230 Standard Oil Development Co., Brit. Pat. 631,355 (1949).

231 Rhys Jones, D. C., *New Scientist*, **5**, 1008–10 (1959).

232 Pettyjohn, E. S., U.S. Pat. 2,706,706 (1955).

233 Collin, F. J., A.G., Brit. Pat. 787,147 (1957).

234 Sapozhnikov, L. M., *Koks i Khim.*, **1959**, No. 3, 22–7.

235 Storrs, K. L., Brit. Pat. 693,214 (1953).

236 Carter, C. W., *Chem. Eng. Progr.*, **43**, 180–2 (1947).

237 Cellan-Jones, G., *Coke & Gas*, **18**, 52–5 (1956).

production of smokeless briquettes. This is a matter of some importance in Great Britain and in some Continental countries since there are circumstances in which, on economic grounds, briquettes can be manufactured only in existing plants.

Comparatively simple methods have been devised in which pitch-bound briquettes are rendered virtually smokeless. Essentially, these processes involve heating the briquettes and subjecting them to an atmosphere containing some oxygen. During this treatment, some volatile matter is lost, but the yield of briquettes by weight can be 95% or more. Some oxygen is taken up by the briquettes, and the calorific value may be reduced by 900 Btu/lb.

It is suggested that the first necessary stage in the operation is the expulsion of some of the light volatile oils by heating, followed by the exposure of the briquettes to an oxygen-containing atmosphere at a lower temperature to permit the oxygen to enter into the briquette. The action is held to be that of oxidation of the pitch and polymerization of the remaining volatile matter so that it burns smokelessly. The briquettes, however, are described as having a gray lustrous shell and black interior. It is certainly possible to make high-volatile coal briquettes smokeless by this means.[238] In all cases described, the briquettes are of small size and weigh about 50 g.

In 1932, the Maurel Investment Corporation[239] of the United States was interested in a process in which briquettes were heated first to about 500°F and then gradually cooled to 300°F in the presence of a very small percentage of oxygen or air. It was considered advisable to heat the briquettes initially in the absence of air, and it was found necessary to cool the briquettes slowly. The briquettes used were made from anthracite or a smokeless char derived by carbonizing a bituminous coal. The product was a hard, smokeless briquette.

In 1954, Couture[240] described the operation of a plant making smokeless briquettes by exposing pitch-bound briquettes to air at a temperature of 300°C. Patents in the name of Houillers de Basin Nord et du Pas de Calais[241] describe a process and plant. The plant at Douai,[241] owned by Charbon Richesse National, has operated successfully for a number of years and has recently been extended. It consists of a number of tunnel kilns, each of which is subdivided into a number of cells, which are grouped into three main sections. The first section is for heating, the second for oxidation, and the third for the initial cooling. The briquettes are carried through the kiln in trucks with perforated bottoms designed to allow the free passage of hot gas through them. The trucks are also provided with end baffle plates shaped to the profile of the kiln channel to ensure the proper circulation of the heating gases. The trucks are moved forward intermittently by a hydraulic ram from cell to cell and remain for 15 min in each position. The heating is effected by the circulation of substantial volumes of hot gas.

In the first stage, the briquettes are raised to a temperature of 300 to 350°C, and in the second stage the temperature is raised to about 400°C, the circulating gas containing some 10% of oxygen. Both the heating and oxidation are completed in less than 3 hr. The briquettes are then cooled to below about 200°C in the kiln, and the final cooling takes place in air outside the kiln. Each kiln has a capacity of 14 ton/hr. The briquettes used are manufactured from coal possessing 10 to 12% volatile matter, and 7% pitch is used as a binder. The products are hard, gray briquettes with a dark core, and they burn smokelessly.

More recently at Rouen, Les Fils Charvet[242] have built and operated a 1 ton/hr pilot plant for the same purpose, but here a circular kiln possessing a rotary table is used that carries the briquettes through the kiln. Other investigators[243] have claimed that the process is capable of extension to briquettes made from high-volatile coals, the coal itself being smoky.

Attempts have been made to decrease the noxious fumes from briquettes by adding chemicals of different kinds. Lime has been added to decrease the evolution of sulfur

[238] Coal Industry (Patents) Ltd., Gordon, R. R., and Gregory, D. H., Brit. Pat. 791,807 (1958).

[239] Maurel Investment Corporation, Brit. Pat. 407,797 (1932).

[240] Couture, J., *J. Inst. Fuel*, **27**, 3–15 (1954).

[241] Houillers de Basin du Nord et du Pas de Calais, Brit. Pats. 716,537 (1954); 722,495 (1955).

[242] Les Fils Charvet, French Pat. 1,197,956 (1958).

[243] Gregory, D. H., and Griffiths, D. M. L., Brit. Pat. Appl. 17821/58.

dioxide, and although the process has been reasonably successful, the quantities required increase the ash content of the fuel to a point which makes it unacceptable as a "premium" fuel.[244] There is little information available regarding the use of materials such as sodium phosphate and volatile iron compounds such as "ferrocene," the latter being so successful with oils.

[244] Sutcliffe, E. R., Brit. Pat. 288,811 (1940).

CHAPTER 17

THERMODYNAMICS AND KINETICS OF SOLID COMBUSTION

M. W. THRING and R. H. ESSENHIGH

University of Sheffield, England

The combustion of coal, either in a bed or as pulverized coal in a suspension, is primarily a matter of combustion of carbon with sequential or parallel combustion of volatile matter. The prime concern of this section, is with the thermodynamics and kinetics of the carbon reactions with only brief consideration of the reactions of the volatile matter.

THERMODYNAMICS

Since the early 1940's there has been no change in the knowledge of the thermodynamics of the carbon reactions so far as this concerns heats of reaction and equilibria; for reference purposes, however, the significant data compiled by Walker, Rusinko, and Austin[1] are set out below in summary.

Heats of Reaction

These data are reduced to 18°C and 1 atm and refer to β graphite; if the heat of formation of β graphite is taken as zero, then by comparison amorphous carbons are reported[2] to have positive heats of formation ($+\Delta H$) ranging from 1.7 to 2.6 kcal/mole.

Equilibria

By the law of mass action, the rate of any reaction of the form

$$A^a + B^b + C^c \cdots \underset{k_b}{\overset{k_f}{\rightleftharpoons}} X^x + Y^y + Z^z \cdots$$

[1] Walker, P. L., Rusinko, F., and Austin, L. G., *Advances in Catalysis*, Eley, O. D., Selwood, P. W., and Weisz, P. B., Eds., Academic Press, Inc., New York and London, 1959, Vol. 11, pp. 134–221.

[2] Hougen, O. A., and Watson, K. M., *Chemical Process Principles*, Wiley and Sons, New York, 1950, Part I, p. 254.

is determined by the products of the powers of the concentrations and the velocity constants (k_f and k_b); thus

$$k_f = (A)^a(B)^b(C)^c \cdots; \quad k_b = (X)^x(Y)^y(Z)^z \cdots \tag{1}$$

and at equilibrium these balance so that the equilibrium constant K_p is given by

$$K_p = \frac{k_f}{k_b} = \frac{[X]^x[Y]^y[Z]^z \cdots}{[A]^a[B]^b[C]^c \cdots} \tag{2}$$

For the reactions listed in Table 1, the equilibrium constants have been calculated and are given in Table 2 in logarithmic form; the concentrations are partial pressures.

The chief significance of the data in Table 2 is the indication that, on equilibrium considerations up to 4000°K, there is no significant restriction of the oxidation of carbon to CO or

TABLE 1

HEATS OF REACTION[1]

C (Solid) Reacting with O_2	ΔH (kcal)
(1) $C + O_2 = CO_2$	−94.0
(2) $C + \frac{1}{2}O_2 = CO$	−26.6
(3) $CO + \frac{1}{2}O_2 = CO_2$	−67.4
C (Solid) Reacting with CO_2	
(4) $C + CO_2 = 2CO$	+40.8
C (Solid) Reacting with H_2O (gas)	
(5) $C + H_2O = CO + H_2$	+31.1
(6) $CO + H_2O = CO_2 + H_2$	−9.7
(4) $C + CO_2 = 2CO$	+40.8
(7) $C + 2H_2 = CH_4$	−17.9
C (Solid) Reacting with H_2	
(7) $C + 2H_2 = CH_4$	−17.9

TABLE 2

EQUILIBRIUM CONSTANTS FOR GAS-CARBON AND ASSOCIATED REACTIONS*

Temperature °K.	$\log_{10} K_p$						
	Reaction 1	Reaction 2	Reaction 3	Reaction 4	Reaction 5	Reaction 6	Reaction 7
300	+68.67	+23.93	+44.74	−20.81	−15.86	+4.95	+8.82
400	+51.54	+19.13	+32.41	−13.28	−10.11	+3.17	+5.49
500	+41.26	+16.26	+25.00	−8.74	−6.63	+2.11	+3.43
600	+34.40	+14.34	+20.06	−5.72	−4.29	+1.43	+2.00
700	+29.50	+12.96	+16.54	−3.58	−2.62	+0.96	+0.95
800	+25.83	+11.93	+13.89	−1.97	−1.36	+0.61	+0.15
900	+22.97	+11.13	+11.84	−0.71	−0.37	+0.34	−0.49
1000	+20.68	+10.48	+10.20	+0.28	+0.42	+0.14	−1.01
1100	+18.80	+9.94	+8.86	+1.08	+1.06	−0.02	−1.43
1200	+17.24	+9.50	+7.74	+1.76	+1.60	−0.16	−1.79
1300	+15.92	+9.12	+6.80	+2.32	+2.06	−0.26	−2.10
1400	+14.78	+8.79	+5.99	+2.80	+2.44	−0.36	−2.36
4000	+5.14	+5.84	−0.70	—	—	—	—

* The reactions are numbered to conform to Table 1.

CO_2; that is, the gasification of carbon goes to completion in the presence of an excess of oxygen.

CO/CO_2 Equilibrium

One particular equilibrium that requires special consideration is the ratio between the production of CO and of CO_2 by the first stage in the reaction of solid carbon with oxygen. The earliest work[3,4] indicated that both gases were produced together in a temperature-dependent ratio. Rhead and Wheeler,[4] particularly, showed that the CO/CO_2 ratio increased with temperature, and suggested that CO was the important product at flame temperatures (above 1000°C). Langmuir and others,[5] working mostly at low pressures, challenged this, and the view that CO_2 was the main product at flame temperatures was in vogue for about 40 years. The difficulty in settling the matter is due to the alteration of the initial equilibrium between CO and CO_2 by the secondary gas-phase reaction of CO to CO_2; for that reason the low-pressure experiments, which were assumed to eliminate gas-phase reactions, were thought to have greater weight. However, Strickland-Constable[6] has now demonstrated that irregularly shaped carbon filaments at low pressures can generate thermions that upset the initial equilibrium of the primary reaction so that the low-pressure results using electrically heated filaments are no longer valid. It has also been demonstrated[7] that, if gas-phase inhibitors such as $POCl_3$ are introduced to check the gas-phase reaction, a CO/CO_2 ratio is obtained that is largely in agreement with the earlier results. In particular, Arthur[7] showed that the ratio was controlled by the following relation:

$$\text{ratio}\ \frac{CO}{CO_2} = F e^{-\Delta E/RT} \qquad (3)$$

and he quoted values of $10^{3.4}$ for the frequency factor F, and 12.4 kcal for the difference in activation energies, that for CO formation therefore being the greater. The ratio seemed to be independent of carbon reactivity, and agreed well with Rhead and Wheeler's data.[4]

[3] Baker, C. J., *J. Chem. Soc.*, **51**, 249–58 (1887). Baker, H. B., *Phil. Trans. Roy. Soc. London*, **179A**, 571–91 (1888).

[4] Rhead, T. F. E., and Wheeler, R. V., *J. Chem. Soc.*, **101**, 846–56 (1912).

[5] Langmuir, I., *J. Am. Chem. Soc.*, **37**, 1139–66 (1915); **38**, 1145–56, 2221–95 (1916). Meyer, L., *Z. phys. Chem.*, **B17**, 385–404 (1932); *Trans. Faraday Soc.*, **34**, 1056–61 (1938). Sihvonen, V., *Trans. Faraday Soc.*, **34**, 1062–74 (1938).

[6] Strickland-Constable, R. F., *Trans. Faraday Soc.*, **34**, 1074–80 (1938); **40**, 333–43 (1944).

[7] Bridger, G. W., *Nature*, **158**, 236 (1946). Arthur, J. R., *Nature*, **157**, 732–3 (1946); *Trans. Faraday Soc.*, **47**, 164–78 (1951). van Loon, W., and Smeets, H. H., *Fuel*, **29**, 119–21 (1950).

The equation predicts that the chief product of primary surface reaction at flame temperatures is CO. This last result has now also been confirmed by high-velocity experiments without inhibitors.[8]

GENERAL HETEROGENEOUS KINETICS

In contrast to the thermodynamics, there has been one major factor introduced in the last 15 years in the analysis of the reaction system, namely the importance of internal surface. There has also been an improvement in the systematics of the reaction system so that an increasing quantity of experimental data can be discussed and explained on the basis of an integrated theory. In particular there is an increased appreciation of the probable importance of the chemistry of the system, which in some quarters has been neglected in preference to a study of the physics of the system, although both should be considered together.[9]

Chemical Kinetics

The basis of the chemical kinetics of combustion is the Langmuir adsorption isotherm which was in fact originally developed[5] to explain the results of the carbon : oxygen reaction, although its subsequent applications have been mainly to heterogeneous catalysis. It is assumed (1) that there are active adsorption sites filled by chemisorption of oxygen at a rate proportional to the oxygen concentration (p_s), the constant of proportionality being the velocity constant k_1, and (2) that, the sites are made available again by the desorption of the oxygen, which carries with it the underlying carbon atoms to form CO or CO_2, the rate of desorption being the velocity constant k_2. If a dynamic balance between the two is achieved, with the surface partly covered, the reaction rate per unit area of surface R_s is given by

$$R_s = \frac{k_1 k_2 p_s}{k_1 p_s + k_2} \tag{4}$$

This is the fundamental reaction equation, but it has two limiting cases of special importance. In the normal analysis of the Langmuir isotherm, there are two extreme cases: (1) at high pressures the reaction is zero order, and (2) at low pressures it is first order with respect to oxygen. The same result can be achieved by alteration of temperature if k_1 and k_2 have different dependencies on temperature. The two limiting cases are

$$\text{When } k_1 p_s \gg k_2; \qquad R_s = k_2 \tag{5}$$

$$\text{When } k_2 \gg k_1 p_s; \qquad R_s = k_1 p_s \tag{6}$$

Influence of Resorption

Simple extension of the Langmuir theory predicts that, if one of the gaseous combustion products is resorbed, the simple isotherm equation becomes

$$R_s = \frac{k_1 k_2 p_s}{k_1 p_s + k_2 + k_3 p_3} \tag{7}$$

where k_3 and p_3 are the velocity constant and partial pressure respectively for the resorption. In effect, the resorption "poisons" the surface by occupying adsorption sites that would otherwise be available for the continuance of the forward reaction.

Equation 7 has been tested only on the CO_2[10] and steam (H_2O) reactions,[11] not on the oxidation reaction, and on these tested reactions it has been found to hold. There is still argument, however, over the detailed sequence of events that on completion is represented by the chemical equations (4) and (5) (Table 1). Because this argument is not yet resolved and does not affect the ultimate and known consequences of the reaction, it need not concern us here; details may, however, be found in the original papers or in a recent review.[1]

Reaction Steps

Of greater importance are the activation energies E and frequency factors A involved in

[8] Grodzovskii, M. K., and Chukhanov, Z. F., *Fuel*, **15**, 321–8 (1936).

[9] Thring, M. W., *Proc. Conf. on Science in the Use of Coal*, Sheffield, Inst. Fuel, London, 1958, p. D–58.

[10] Semechkova, A. D., and Frank-Kamenetskiĭ, D. A., *Acta phys.-chim. U.S.S.R.*, **12**, 879–98 (1940). Gadsby, J., Long, F. J., Sleightholm, P., and Sykes, K. W., *Proc. Roy. Soc.*, **A193**, 357–76 (1948). Lewis, W. K., Gilliland, E. R., and McBride, G. T., *Ind. Eng. Chem.*, **41**, 1213–26 (1949). Reif, A. E., *J. Phys. Chem.*, **56**, 785–8 (1952).

[11] Gadsby, J., Hinshelwood, C. N., and Sykes, K. W., *Proc. Roy. Soc.*, **A187**, 129–51 (1946). Long, F. J., and Sykes, K. W., *Proc. Roy. Soc.*, **A193**, 377–99 (1948). Johnstone, H. F., Chen, C. Y., and Scott, D. S., *Ind. Eng. Chem.*, **44**, 1564–9 (1952). Jolley, L. J., and Pohl, A., *J. Inst. Fuel*, **26**, 33–44 (1953).

the velocity constants k. In general, these may be written

$$k = A\,e^{-E/RT} \qquad (8)$$

Where such an equation summarizes the net effect of a whole sequence of intermediate steps, the values of E and A are sometimes termed "global" values and have practical, but not fundamental, significance. In the carbon reaction system, Gulbransen[12] has listed the following steps, neglecting diffusion, that may possibly be involved:

1. Immobile adsorption:
   (*a*) rate control by adsorption
   (*b*) rate control by dissociation,
2. Mobile adsorption:
   (*a*) unactivated (Hertz-Knudsen equation)
   (*b*) activated
3. Chemical surface reaction:
   (*a*) first-order kinetics
   (*b*) zero-order kinetics
4. Desorption

Distinction has already been made between adsorption and desorption in formulating equation 4, but the activation energy of either adsorption or desorption could be a "global" value incorporating the effect of a suggested surface reaction. However, although the surface reaction may in fact exist, no proof of its existence has yet been given. Its possible existence may therefore be reasonably disregarded, in which case the activation energies for both adsorption or desorption are likely to be single-step energies with precise meaning. Gulbransen[12] has used the Eyring Absolute Rate theory to predict reaction rates, and on the basis of comparison with experiment[13] he concluded that, under the particular conditions of experiment considered, the rate control was likely to be either immobile adsorption with dissociation (1*b*) or activated mobile adsorption (2*b*). On the basis of the picture given by Sherman and Eyring[14] of a molecule dissociating as it moved in to be adsorbed, the former step was thought to be the more probable.

[12] Gulbransen, E. A., *Ind. Eng. Chem.*, **44**, 1045–7 (1952).

[13] Gulbransen, E. A., and Andrew, K. F., *Ind. Eng. Chem.*, **44**, 1034–44, 1048–51 (1952).

[14] Sherman, A., and Eyring, H., *J. Am. Chem. Soc.*, **54**, 2661–75 (1932).

Diffusional Kinetics

Before the oxygen can react, it has to reach the solid surface by diffusion through a boundary layer of nitrogen and combustion products. There are two stages in diffusion: first, from the main stream to the nominal surface where part of the oxygen then reacts; and second, if the solid is fissured or porous, there is diffusion down the pores into the interior of the particle.

Boundary Layer Diffusion

If the velocity constant for crossing the boundary layer is k_0, by definition the net quantity of oxygen crossing the boundary layer is $k_0(p_0 - p_s)$ where p_0 is the oxygen partial pressure in the mainstream. This net quantity must then be proportional to the reaction rate and, if k_0 is defined in appropriate units, we may write

$$R_t = k_0(p_0 - p_s) \qquad (9)$$

where R_t is the total reaction rate per unit area of nominal surface. Because k_0 is also known as the mass transfer coefficient, where the physical picture of the system is taken to be a quiescent boundary layer of thickness δ, bounded by the solid on one side and a turbulent mainstream on the other, k_0 may be written

$$k_0 = \frac{D}{\delta} \qquad (10)$$

where D is the diffusion coefficient. The reasons for specifying the boundary layer as quiescent and the mainstream as turbulent are: (1) diffusion in the boundary layer may then be regarded as molecular; and (2) turbulence in the mainstream flattens out the concentration profiles so that the mainstream oxygen concentration (p_0) may be assumed constant right up to the edge of the boundary layer. It is thus clear, even with this approximate picture, that increased intensity of microturbulence in the mainstream will reduce δ since turbulent diffusion will carry the mainstream oxygen concentration closer to the solid wall. In the traditional terminology, which originates from rates of solution studies,[15] the quiescent boundary layer is referred to as being stagnant.

[15] Noyes, A. A., and Whitney, W. R., *Z. phys. Chem.*, **23**, 689–92 (1897).

Increased velocity, which increases the microturbulence intensity, is regarded as "scrubbing away" the stagnant film. However, whatever the terminology or the physical model invoked, the consequence of increased velocity is nearly always to increase the reaction rates.

Pore Diffusion

When the oxygen has reached the exterior surface, it may still have to diffuse down pores into the interior of the solid before it is adsorbed.[16,17] The usual model is a cylindrical pore with reaction taking place all the way down its walls. Excluding any reaction at the exterior surface, it has been shown[1,17] that the specific reaction rate with respect to the exterior surface (R_v) is given by

$$R_v = \sqrt{k_s SD} \cdot p_s^{(m+1)\frac{1}{2}} \tag{11}$$

where k_s has been written for the surface velocity constant and may take the value k_1 or k_2, S is the interior surface area per unit volume, and m is the order of reaction. The extreme values of m, as predicted by equations 5 and 6 are zero and unity.

Zero reaction order. According to equations 5 and 8, we may write for either the exterior or interior surface reaction

$$R_s = k_2 = A_2 e^{-E_2/RT}, \tag{12}$$

which is independent of oxygen concentration. In equation 11, we then have that $k_s = k_2$ and $m = 0$, hence

$$R_v = \sqrt{k_2 SDp_s} \tag{13}$$

This holds under such temperature conditions that the limiting equation 5 is operative. For the oxygen reaction, this range is generally less than 800°C.

First-order reaction. In the high-temperature region (>1200°C), the surface reaction is given by equation 6; hence

$$R_s = k_1 p_s = A_1 p_s e^{-E_1/RT} \tag{14}$$

which is first order with respect to oxygen concentration. In equation 11, we therefore have $k_s = k_1$, and $m = 1$ so that

$$R_v = \sqrt{k_1 SD}\, p_s \tag{15}$$

Combined Kinetics

It will be clear from the foregoing treatment that a single comprehensive equation can be formulated combining: (1) the boundary diffusion; (2) pore diffusion; (3) chemisorption kinetics; and (4) desorption kinetics. This comprehensive equation is cumbersome, but it may, however, be reduced to special cases for examination as already indicated in the preceding section. The most convenient basis for simplification is on a temperature basis.

Low Temperature ($T < 800$°C, Ignition Range)

Reaction, in theory, is taking place both at the surface and in the interior, so that the total reaction rate per unit area of nominal external surface R_t is given by

$$\begin{aligned} R_t &= R_s + R_v = R_s(1 + R_v/R_s) \\ &= A_2 e^{-E_2/RT} + \sqrt{A_2 SDp_s e^{-E_2/RT}} \\ &= A_2 e^{-E_2/RT} [1 + \sqrt{SDp_s/A_2 e^{-E_2/RT}}] \end{aligned} \tag{16}$$

The ratio R_v/R_s is clearly a significant factor. If this is small, the reaction is zero order as found by several investigators;[18] if the reaction is large, the reaction is half-order at low total pressures, as found by Blyholder and Eyring.[19] There is also a change in the activation energy. The energy E_2 is that required for decomposition of the oxide surface layer. By decomposing a surface oxide layer over a range of temperatures in a vacuum,[13] this energy was found to have a value of about 40 kcal. Since this was the only reaction that occurred, identification was certain. Blyholder and Eyring[19] also obtained this value—but when the reaction was half-order, so that pore diffusion was operative ($R_v > R_s$). On very thin coatings of very high purity spectroscopic

[16] Predvoditelev, A. S., and Khitrin, L. N., *Combustion of Carbon*, Moscow, U.S.S.R. (1949), Chap. 3. Weisz, P. B., and Prater, C. D., *Advances in Catalysis*, Frankenburg, W. G., Kamarewsky, V. I., and Rideal, E. K., Eds., Academic Press, New York, 1954, Vol. 6, 143–96.

[17] Wheeler, A., *Advances in Catalysis*, Frankenburg, W. G., Kamarewsky, V. I., and Rideal, E. K., Eds., Academic Press, New York, 1951, Vol. 3, pp. 249–327.

[18] Tu, C. M., Davis, H., and Hottel, H. C., *Ind. Eng. Chem.*, **26**, 749–57 (1934). Arthur, J. R., and Bleach, J. A., *Ind. Eng. Chem.*, **44**, 1028–34 (1952).

[19] Blyholder, G. D., and Eyring, H., *J. Phys. Chem.*, **61**, 682–8 (1957).

TABLE 3

VALUES OF E_2 FOR OXIDATION REACTION[21]

Investigator	E_2	Temperature Range, °C	Type of Carbon
Sihvonen	29	800–1200	Carbon filaments
Smith and Gudmundsen	22		Electrode carbon
Meyer	20–30	900–1300	Electrode carbon graphite coated
Lambert	44		Diamond
	49		Graphite
	31		Anthracite
	30		Iron-treated charcoal
	29		Coconut charcoal
Tu, Davis, and Hottel	30	750–850	Carbon spheres
Parker and Hottel	44	700–1100	Carbon plug
Tsukhanova	29	500–1000	Electrode graphite tubes
Baetke	28	900–1100	Spectroscopic carbon tubes
Patai, Hoffman, and Rajbenbach	25		Carbon disks
Kuchta, Kant, and Damon	20–25	1050–1300	Spectroscopic carbon cylinders
Gulbransen and Andrew	36		(During reaction)
	40		(By film decomposition alone)
Wicke	29	600–700	(With diffusion)
	43	1000–1100	(With diffusion)
	58	600–700	(Without diffusion)
	85	1000–1100	(Without diffusion)
Vulis*	17.5		Electrode carbon
	26.5		channel
Khitrin*	20		Electrode carbon particles
Nikolaev*	30		Electrode carbon particles
	33		Coke particles
Golovina*	31.5		Coke particles
	35.5		Electrode carbon particles
Blinov and Golovina*	37		Electrode carbon particles

* Quoted by Frank-Kamenetskiĭ.[23]

graphite where pore diffusion could not operate, the reaction order was zero ($R_v < R_s$) and the activation energy was 80 kcal or twice the previous value, as is to be expected from equation 16. Most other results show lower values than these because of the effect of impurities. This impurity effect was studied by Patai, Hoffmann, and Rajbenbach[20] who examined the effects of impurities on carbon black by introducing in intimate admixture various metal chlorides and salts of potassium. There are therefore two reasons for the range of activation energies reported in the literature[21] (see Table 3)—the influence of impurities and the influence of pore diffusion. With regard to the latter, Vulis[22] has been reported[23] as having found a log-linear relation between the frequency factor A_2 (see Table 4) and the activation energy E_2. This relation can be explained by reference to equation 16. If we assume that, under the conditions of measurement, the ratio R_v/R_s ranges between its two possible extremes the variation of A_2 changes as $A_2{}^n$, where $1 > n > \frac{1}{2}$; but E_2 changes as nE_2 and hence follows the relation observed by Vulis.

This treatment has so far neglected the boundary-layer resistance. This is a valid omission with respect to R_s since from equation 12 we see that the oxygen partial pressure is not involved; p_s then equates to p_0 and there is no diffusional boundary layer. The omission is also valid with respect to R_v when the pore resistance is large compared with the boundary-layer resistance, since reaction then occurs only in the pore mouth, and the system is equivalent to external surface reaction. If the reverse is true, the effect on p_s is to reduce it appreciably below p_0; then, as p_s in the Langmuir equation 4 drops, the limiting equation 5 no longer applies. It is therefore a valid omission as a first approximation to neglect the influence of the boundary layer in this temperature range; or else, if the boundary-layer resistance is important, the relevant equations in this section are different from those in the next.

High Temperature ($T > 1200$°C, Flame Temperature Range)

In this region, the boundary-layer resistance cannot be neglected, at least in setting up the

TABLE 4

FREQUENCY FACTORS R_2 FOR $C + O_2$
(Compiled by Frank-Kamenetskiĭ[23])

Carbon	R_2, cm/sec	Investigator
Electrode carbon	5×10^4	Vulis (channel)
Electrode carbon	1×10^3	Khitrin (particles)
Electrode carbon	2×10^5	Smith and Gudmundsen (particle)
Electrode carbon	1.1×10^3	Tsukhanova (channel)
Electrode carbon	2.5×10^6	Vulis (channel)
Electrode carbon	3.5×10^6	Nikolaev (particle)
Coke	1.6×10^7	Golovina (particle)
Coke	1.8×10^7	Nikolaev (particle)
Electrode carbon	2.2×10^7	Tu, Davis, and Hottel (spheres)
Electrode carbon	5.6×10^7	Golovina (particle)
Electrode carbon	4×10^7	Blinov and Golovina (particle)
Electrode carbon	2.9×10^9	Wicke (channel)
Electrode carbon	2.6×10^9	Wicke (channel)

[20] Patai, S., Hoffmann, E., and Rajbenback, L., *J. Appl. Chem. London*, **2**, 306–10 (1952).

[21] Essenhigh, R. H., and Perry, M. G., *Proc. Conf. on Science in the Use of Coal*, Sheffield, Inst. Fuel, London, 1958, p. D-3.

[22] Vulis, L. A., *Zhur. Tekh. Fiz.*, **16**, 83–8 (1946).

[23] Frank-Kamenetskiĭ, D. A., *Diffusion and Heat Exchange in Chemical Kinetics*, Princeton Univ. Press, 1955, pp. 46–127.

operative equations. From equations 14 and 15 we have

$$R_t = R_s + R_v = (k_1 + \sqrt{k_1 SD})p_s \quad (17)$$

Eliminating p_s by equation 9, we obtain

$$\frac{1}{R_t} = \frac{1}{k_0 p_0} + \frac{1}{(k_1 + \sqrt{k_1 SD})p_0} \quad (18)$$

Therefore, whatever the relative importance of the three terms, the reaction is first order with respect to the oxygen partial pressure as confirmed by experiment.[18,20]

Equation 18 has another interesting facet concerned with the temperature coefficient of the total reaction rate. It is generally supposed that chemical velocity constants as expressed by equation 8 have very marked temperature coefficients, but this is only true if the activation energy is greater than 10 kcal. For values lower than 10 kcal, it is found that over the flame temperature range (1000 to 2000°C) the change of the Arrhenius factor with temperature can be represented by a T^2 or even a lower-power law. Blyholder and Eyring[19] measured E_1 (the activation energy involved in k_1) at low absolute pressures to eliminate the boundary layer complications. They found that on thin films, where there was no pore diffusion, E_1 had a value of 8 kcal, and on thick ones, E_1 dropped to 4 kcal. These values are in agreement with those previously assigned[24] to this parameter after a reanalysis of existing data, and they are also in agreement with values obtained by Bannerjee and Sarjant[25] in a coke bed and by Day[26] at such high relative velocities that the boundary-layer resistance must have been small, although this is not the interpretation given by Day. If these values are accepted, all three terms in equation 18 have roughly the same temperature coefficient. Their relative importance will therefore be determined by such factors as boundary layer thickness and porosity under the exact experimental conditions prevailing. To determine their relative importance, the experimental conditions must be known.

[24] Essenhigh, R. H., *Fuel Soc. J., Univ. Sheffield*, **6,** 15–24 (1955).

[25] Bannerjee, S., and Sarjant, R. J., *Fuel*, **30,** 130–9 (1951).

[26] Day, R. J., Thesis, Pennsylvania State Univ., 1949.

KINETICS OF A SPHERICAL PARTICLE

Single Particle

The single spherical particle of carbon has been studied both theoretically[27] and experimentally[28,29] as a microcosm of a pulverized coal flame. The theories particularize the general treatment outlined in the previous section. The system is idealized to the following extent: the particles are assumed to be pure carbon; to react at the exterior surface only; and to be burning in the flame temperature region (1000 to 2000°C). The specific reaction rate R_t is given by equation 18 in the reduced form that neglects internal reaction:

$$\frac{1}{R_t} = \frac{1}{k_0 p_0} + \frac{1}{k_1 p_0} \quad (19)$$

In this equation, k_0 is a function of radius but k_1 is not. The radius function, in a simplified analysis, is established as follows. Suppose a particle of radius r is surrounded by a quiescent mixture of oxygen and nitrogen, neglecting the combustion products. Because of surface reaction, a concentration gradient is established in the oxygen (see Fig. 1) so that it diffuses in toward the solid surface. If the diffusion flux is G—molecules per unit area

[27] Nusselt, W., *Z. Ver. deut. Ing.*, **60,** 102–7 (1916); **68,** 124–8 (1924). Burke, S. P., and Schumann, T. E. W., *Ind. Eng. Chem.*, **23,** 406–12 (1931); **24,** 451–3 (1932). Gumz, W., *Theorie und Berechnung der Kohlenstaubfeuerungen*, Springer, Berlin, 1939, 92 pp. Hottel, H. C., and Stewart, I. McC., *Ind. Eng. Chem.*, **32,** 719–30 (1940). Traustell, S., *Feuerungstech.*, **29,** 1–6, 25–31, 49–60 (1941). Ledinegg, M., *Dampferzeugung*, Springer, Wein, 1952, p. 238. Godsave, G. A. E., Natl. Gas Turbine Estab., Rept. 126 (1952), 33 pp. Spalding, D. B., *Proc. Inst. Mech. Engrs. (London)*, **168,** 545 (1954).

[28] Griffin, H. K., Adams, S. R., and Smith, D. F., *Ind. Eng. Chem.*, **21,** 808–15 (1929). Godbert, A. L., *Fuel*, **9,** 57–75 (1930). Godbert, A. L., and Wheeler, R. V., Safety in Mines Research Board (London), Paper 73 (1932), 21 pp. Smith, D. F., and Gudmundsen, A., *Ind. Eng. Chem.*, **23,** 277–85 (1931). Orning, A. A., *ASME, Trans.*, **64,** 497–506 (1942). Omori, T. T., and Orning, A. A., *ASME, Trans.*, **72,** 591–7 (1950). Essenhigh, R. H., and Thring, M. W., *Proc. Conf. on Science in the Use of Coal*, Sheffield, Inst. Fuel, London, 1958, pp. D21–7.

[29] Essenhigh, R. H., Thesis, Univ. of Sheffield, 1959.

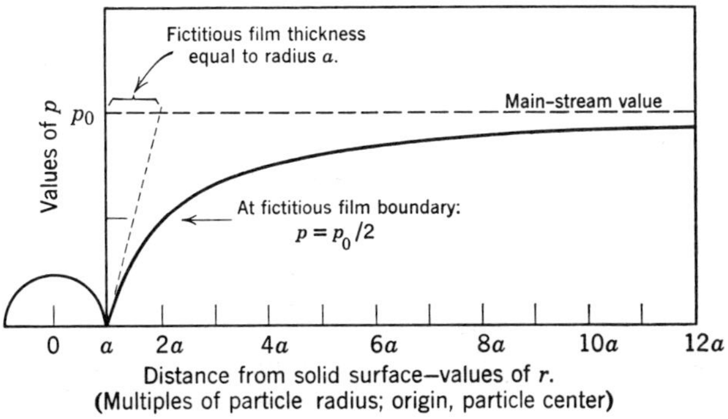

Fig. 1. Variation of oxygen partial pressure p with distance from the surface of a solid spherical particle (of radius a); the curve obeys the equation $p = p_0(1 - a/r)$. This diagram illustrates the relationship between the physical thickness and the fictitious thickness of the mass-transfer boundary layer, the fictitious film thickness being given by the intercept on the line at $p = p_0$, generated by the slope to the curve at the solid surface.[29]

per second—at a point x from the particle center where the oxygen concentration is N—molecules per unit volume—then by Fick's law

$$G = -D\frac{dN}{dx} \tag{20}$$

Because of the radial nature of the flow, we have the continuity equation

$$r^2 G_s = x^2 G$$

where G_s is the diffusion flux at the solid surface. Hence,

$$G_s = -D\frac{x^2}{r^2}\frac{dN}{dx}$$

and since G_s is constant under steady-state conditions, this integrates for boundary conditions $N = N_s$ at $x = r$ and $N = N_0$ at $x = \delta$ to

$$-G_s = \frac{D(N_0 - N_s)}{r(1 - r/\delta)} \tag{21}$$

where δ is the boundary layer thickness. G_s is inherently negative as it is diffusing in toward the surface. If the ambient gas is quiescent, δ is large and the term r/δ can be neglected. Equation 21 then has the same form as equation 9 but with

$$k_0 = \frac{D}{r} \tag{22}$$

Compared with equation 10, we see that for a *plane* surface near a *turbulent* mainstream, the *true* boundary layer thickness δ is used; for a *sphere* in a *quiescent* stream, the boundary layer thickness reduces to a value equal to the particle radius, which is Nusselt's result when the Nusselt number equals 2, and this is known as the "effective" or "fictitious" boundary layer thickness.

For the whole particle of mass m we have $m = (\frac{4}{3}\pi r^3)\sigma$; so $dm/dt = (4\pi r^2)\sigma\,(dr/dt)$ where σ is the particle density. Since

$$R_t = \frac{-dm/dt}{4\pi r^2}$$

it follows that

$$R_t = -\sigma\frac{dr}{dt} = \frac{p_0}{(r/D + 1/k_1)}. \tag{23}$$

With limits $t = 0$ at $r = a$, and $t = t_b$ at $r = 0$, equation 23 gives on integration

$$t_b = \frac{\sigma}{p_0}\left(\frac{a}{k_1} + \frac{a^2}{2D}\right) \tag{24}$$

In most, if not all, traditional analyses,[27] it is assumed, following Nusselt,[27] that the important factor is diffusion (i.e., $k_0 \ll k_1$) so that the chemical factors can be ignored. This point, however, has never yet been proved, and the chemical factors are certainly taken into account in the contemporary Russian analyses.[30] Equation 24 will, therefore, have two limiting conditions—diffusional control and chemical control—which may conveniently be considered separately.

Diffusional Control ($k_0 \ll k_1$)

If the chemistry can be neglected, equation 24 reduces to a square law that is more

[30] Khitrin, L. N., *Physics of Combustion and Detonation*, Moscow University Press, 1957, 143 pp.

TABLE 5

Predicted and Experimental Values of the Burning Constant K_D[29]

Coal	% C (dry, mineral-matter free)	n (square-law index)	K_D (experimental)	K_D (calculation)	Ratio
1. Stanllyd	93.0	2.02	2125	2720	1.28
2. Five Foot	91.8	1.94	1290	2620	2.03
3. Two Foot Nine	91.2	2.25	1470	2070	1.41
4. Red Vein	89.7	2.09	1475	2275	1.54
5. Garw	88.9	2.01	1410	2030	1.44
6. Silkstone	86.9	2.25	1110	1655	1.49
7. Winter	84.0	2.18	1125	1710	1.52
8. Cowpen	82.7	1.94	1060	1775	1.68
9. High-Hazel	81.9	2.14	1450	1725	1.19
10. Lorraine	79.3	2.20	992	1890	1.91
				Mean ratio	1.55

generally written in terms of diameter; thus

$$t_b = K_D\, d_0^2 \tag{25}$$

where d_0 is the initial diameter, and K_D is the diffusion-control burning constant given by[29]

$$K_D = \frac{\sigma}{3\rho_0 D_0 p_0 (T/T_0)^{0.75}} \tag{26}$$

where ρ is the air density and the zero subscripts refer to normal temperature and pressure. In this formulation, the conversion factor of grams of oxygen arriving to grams of carbon removed (i.e., $\frac{12}{16}$) has been included. By calculation, K_D is found to have a value of the order of 1000, and Table 5 lists data[29] for the carbon residues of single coal particles. The 50% discrepancy between prediction and experiment is accounted for by swelling, and a mean swelling factor determined photographically was found to have a value of about 1.5. The index n was also found to be close to 2. Other experimental data, with liquid fuels for comparison, are illustrated graphically in Figs. 2,[28,29,31–7] the burning values being given in Table 6.

In this connection, it is worth noting that a square-law equation is predicted similarly for burning of liquid fuels, with a value for the burning constant of about 100 sec/cm^2. It would seem, then, that volatile components of coal behave as liquid drops. Measurements have confirmed that their burning times obey a square-law equation, and the burning constants have values of the order of 100.

Chemical Control ($k_1 \ll k_0$)

With this alternative condition, equation 24 reduces to a linear law that may be written

$$t_b = K_c d_0 \tag{27}$$

and, from kinetic theory, K_c is given by[29]

$$K_c = \frac{(4\sigma/3)(2\pi RT/M_0)^{1/2}}{p_0\, e^{-E_1/RT}} \tag{28}$$

where M_0 is the mean molecular weight of the gas mixture. For a value of 4 kcal for E_1,[19] M_0 has a value, at 1500°C, of about unity. This confirms that, for large particles of the order of 1 mm, the chemistry of the system can be neglected, which is not true, however,

[31] Burgoyne, T. H., and Cohen, L., *Proc. Roy. Soc.*, **A225**, 375–92 (1954).

[32] Hall, A. R., and Diederichsen, J., *Symposium on Combustion*, *4th Symposium*, **1952,** Cambridge, Massachusetts, (1953), pp. 837–46.

[33] Spalding D. B., *Symposium on Combustion*, *4th Symposium*, **1952,** Cambridge, Massachusetts, (1953), pp. 847–64.

[34] Godsave, G. A. E., *Symposium on Combustion*, *4th Symposium*, **1952,** Cambridge, Massachusetts, (1953), pp. 818–30.

[35] Simpson, H. C., Thesis, Massachusetts Institute of Technology, 1954.

[36] Rosin, P. O., *Proc. Intern. Conf. Bituminous Coal*, **1,** 838–64 (1926). *Braunkohle*, **24,** 241–59 (1925).

[37] Essenhigh, R. H., *Proc. 2nd Pulverised Fuel Conf.*, Inst. Fuel, London, 1957, pp. B1–7.

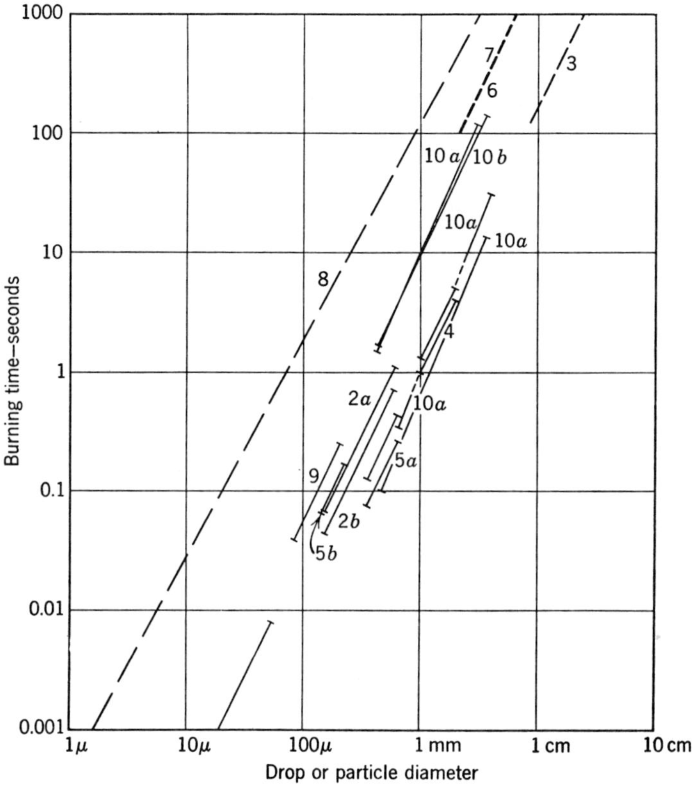

Fig. 2. Dependence of burning time of various liquid and solid fuels on drop or particle diameter.[37]

A. Liquid fuels
B. Solid carbon
C. Pulverised coal

	Investigator	Fuel
A	1. Burgoyne and Cohen[31]	Tetralin drops in flame
	2*a*. Hall and Diederichsen[32]	Kerosine
	2*b*. Hall and Diederichsen[32]	Tetralin
	3. Spalding[33]	Kerosene on 1-in. sphere
	4. Godsave[34]	Various fuels (16)
	5*a*. Simpson[35]	Various fuels (5)
B	5*b*. Simpson[35]	Coke residue from production oil
	6. Spalding[33]	(Predicted)
	7. Godsave[34]	(Predicted)
C	8. Rosin[36]	(Equation) particles in flame
	9. Orning; Griffin, Adams, and Smith[28]	Single particles
	10*a*. Essenhigh[29]	Captive particles (volatiles)
	10*b*. Essenhigh[29]	Captive particles (residue) (2 coals: High-Hazel and Lorraine)

TABLE 6

VALUES OF THE BURNING CONSTANT K^{21}
(Units: t_b measured in seconds and d_0 in centimeters)

Investigator	Fuel	K	Range of Drop or Particle Diameter Microns
	Liquid Fuels		
1. Burgoyne and Cohen[31]	Tetralin drops in flame	312.5	8 to 50
2a. Hall and Diederichsen[32]	Kerosine	300	150 to 600
2b. Hall and Diederichsen[32]	Tetralin	195	150 to 600
3. Spalding[33]	Kerosine on 1-in. spheres	150	(Equation tested indirectly)
4. Godsave[34]	Various fuels	100–130	1000 to 2000
5a. Simpson[35]	Various fuels	60–100	350 to 650
	Solid carbon		
5b. Simpson[35]	Coke residue from production oil	330	135 to 220
6. Spalding[33]	(Predicted)	2000	—
7. Godsave[34]	(Predicted)	2060	—
	Pulverized coal		
8. Rosin[36]	Equation—particles in flame	4360	—
9. Orning; Griffin, Adams, and Smith[28]	Single particles (free in furnace)	625	80 to 200
10a. Essenhigh[29]	Single (captive) particles (volatiles)	100 (L)* 131 (H.H.)*	650 to 4060
10b. Essenhigh[29]	Single (captive) particles—residue	988 (L)* 1424 (H.H.)*	325 to 4060

* L: Lorraine; H.H.: High-Hazel.

for small particles, $<10\ \mu$ in diameter. Data on the significance of this factor are not easy to find since most of the controllable experiments were carried out under conditions where it was unimportant, and the interpretations of the data generally neglected this factor. This factor appears to have been given prominence in Russia,[30] but its relative importance is not substantiated by reported experiments.

Two additional factors have to be considered. (1) The affect of internal burning will be to increase the effective value of k_1 to a value k'_1 given, from equation 18 as

$$k'_1 = k_1(1 + \sqrt{SD/k_1}) \tag{29}$$

Under conditions such that the external boundary layer is unimportant, it will be important to determine the ratio SD/k_1. If k_1 increases, the burning constant K_c will decrease, but by how much cannot yet be stated without correct measurement in an experiment. It should be invariant with temperature. (2) The other factor concerns the influence of the boundary layer in determining the nature of the reaction control. Although this factor was subsequently eliminated, it appears in equation 21. If we may suppose to a first approximation that the quantity δ is determined by the degree of turbulence in the ambient gas stream, and that the ratio r/δ is reasonably constant, equation 24 may be rewritten

$$\frac{p_0}{\sigma} t_b = \frac{a}{k_1} + \frac{a^2}{2D}(1 - a/\delta_0) \tag{30}$$

where δ_0 is the true boundary layer thickness when $r = a$. If δ_0 is reduced sufficiently by a high-stream velocity—up to 1000 ft/sec—as was done in some of the experiments carried out by Tu, Davis, and Hottel[18] and by Day,[26] the diffusional resistance can easily drop by a factor of 10 (when δ_0 is 11% greater than a), and can conceivably drop by a factor of 100 (when δ_0 is reduced to 1% greater than a). Again, exact determination requires examination by experiment.

Polyparticulate Assembly

The justification for considering the single particle in detail is the assumption already stated that it is representative of the flame in general, formed by combustion of a multitude of particles. In a flame, each particle may be thought of as existing in a closed volume which, in a stoichiometric cloud, can be treated as a sphere of radius twenty-five times the initial particle radius. The effect of being in a cloud in a closed system such as a furnace is therefore that the oxygen concentration is no longer constant, in contrast to that for the individual particle in an infinite atmosphere, and the consequence is that the burning times are increased. As before, there are two extreme conditions to be considered: (1) when $k_0 \ll k_1$; and (2) when $k_1 \ll k_0$. In either condition, however, the oxygen concentration appears to the first power, as in equation 23, so that the correction term for variable oxygen concentration is the same in both instances.

For this finite system where the oxygen concentration is varying, equation 23 may be written as

$$\frac{dr}{dt} = \frac{p/\sigma}{r/D + 1/k_1} \tag{31}$$

The oxygen partial pressure, which is proportional to the number of molecules present, is related to the dust concentration D_c, this being the concentration in the furnace at any point in the flame or moment in time. If D_0 is the initial concentration and requires a number of molecules N_0 for complete combustion, and N is the number required to consume a weight of dust D, then for a fractional excess air E we have

$$\frac{p}{p_0} = \frac{N + N_0E}{N_0(1 + E)}$$

but clearly $N/N_0 = D_c/D_0$,

$$\text{so} \quad p = p_0\frac{(E + D_c/D_0)}{1 + E} = \frac{p_0(E + r^3/a^3)}{1 + E}$$

where r is the radius at any time and

$$dt = \frac{\sigma(r/D + 1/k_1)(1 + E)}{E + r^3/a^3}\,dr \tag{32}$$

Diffusional Control ($k_0 \ll k_1$)

If we neglect the chemical terms, equation 32 integrates to give[37]

$$t_b = F_D K_D\, d_0^2 \tag{33}$$

$$\text{where} \quad F_D = \frac{1 + E}{3E^{1/3}} f(E), \tag{33a}$$

$$\text{and} \quad f(E) = 2\sqrt{3}\tan^{-1}\left(\frac{\sqrt{3}}{2E^{1/3} - 1}\right) - \ln\left[\frac{(E^{1/3} + 1)^3}{E + 1}\right] \tag{33b}$$

Equation 33 differs from equation 25 by the addition of the multiplying factor F_D. The equation for F_D was first given by Nusselt[27] whose equation differs only slightly from this.

Chemical Control ($k_1 \ll k_0$)

Similarly, if the diffusional terms are neglected, equation 32 integrates[29] to give

$$t_b = F_c K_c\, d_0 \tag{34}$$

$$\text{where} \quad F_c = \frac{(1 + E)}{6E^{2/3}} g(E) \tag{34a}$$

$$\text{and} \quad g(E) = 2\sqrt{3}\tan^{-1}\left(\frac{\sqrt{3}}{2E^{1/3} - 1}\right) + \ln\left[\frac{(E^{1/3} + 1)^3}{E + 1}\right] \tag{34b}$$

This equation also differs from equation 27 only by the addition of a multiplying factor: in the respective factors, relation 33b differs from 34b only by the changed sign between the two terms. Equation 34 is not known to have been previously derived.

Theoretical Check

To compare the possible relative importance between the factors mentioned, burning times have been calculated[29] from equations 33 and 34, taking K_D and K_c as being 1000 and unity, respectively. Three excess air levels were taken: 10%, 100%, and infinity (single isolated particle); and the calculated values are compared in Fig. 3 on a double log plot. This shows quite clearly that, if the assumptions on which the plot is based are valid, diffusion is dominant in the large particle range; but in the pulverized fuel range there may well be a reversal of roles. This possibility must be checked by experimental test.

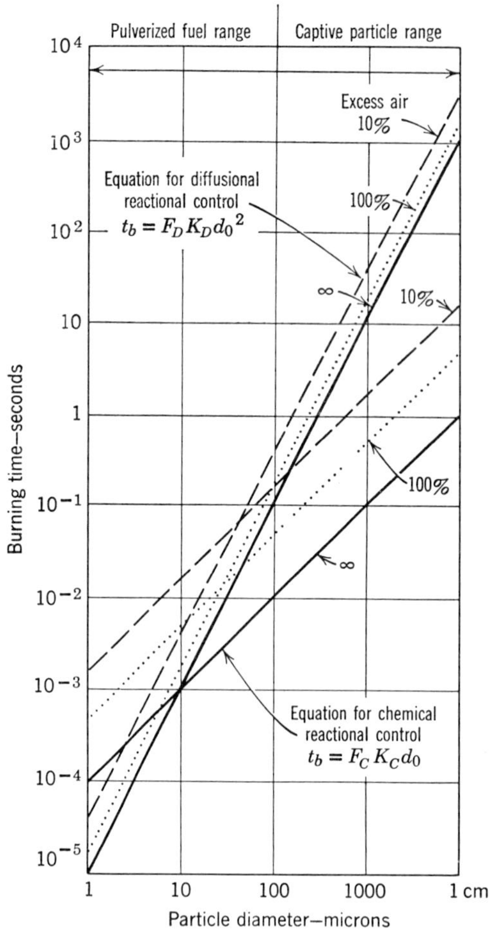

Fig. 3. Comparison of calculated burning times in a flame, at 1500°C, on respective hypotheses of diffusional and chemical reaction control.[29]

Experimental Check

Choice between equations 34 and 35 for predicting the behavior in real systems has not yet been made, but those data that are available favor equation 34. Notably, there is the check provided by Hottel and Stewart[27] who wrote equation 23 in the form

$$R_t = \frac{\text{constant}}{d^m}$$

where they expected m to take the value unity if control was diffusional, and the value zero if control was chemical, and an intermediate fractional value if both factors were of similar importance. They then determined the best fit of this equation to a variety of experimental data and found, contrary to their expectations, that the best fit was achieved for a zero value of m. This result has never been regarded as conclusive, since the data were obtained from recirculating flames and there was doubt about the validity of the interpretation.

KINETICS OF A CARBON CHANNEL

Single Channel

The simple carbon channel plays the same relative part with respect to a solid fuel bed that the single particle plays with respect to a pulverized fuel flame. The single channel has been studied much less either theoretically or experimentally, than the single particle. The most important experimental study, is that by Chukhanova, whose published results are summarized by the graph of Fig. 4, taken from Frank-Kamanetskiĭ.[23] The break between the two curves is at the transition (Re = 2000) between laminar flow, at low velocities, and turbulent flow at the higher velocities. In interpreting these data, Frank-Kamanetskiĭ did not explicitly assign the laminar region to chemical control and the turbulent region to diffusional control, but he stated, "If the rate of reaction depends on the velocity of flow, this is evidence that the process takes place in the diffusional range." This widespread belief, and its converse that chemically controlled systems must a priori be independent of flow velocity, should be treated with reserve, as will be shown. In fact, there is sound theoretical reason, apart from common sense, to show that the reverse interpretation is the correct one, and that diffusional control operates in the laminar region with chemical in the turbulent.

Laminar Flow

The problem here is illustrated in Fig. 5. The limiting condition is that reaction products originating at the upstream end of the carbon test section travel downstream by virtue of the stream velocity (and diffusion) and radially inward by diffusion. This means that dilution of the oxygen on the axis starts at some distance downstream from the mouth of the tube test section, the distance downstream depending on the stream velocity. Consequently, the mainstream or initial oxygen concentration is maintained some distance inside the tube, this "entry distance" decreasing

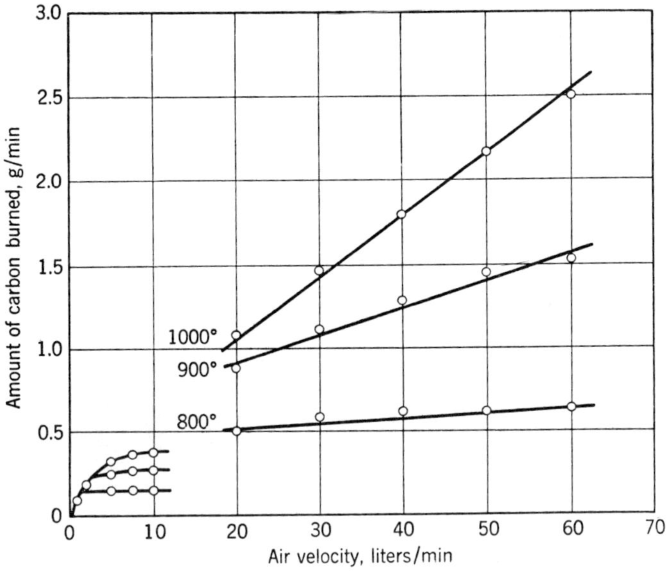

Fig. 4. Dependence of combustion rate of a carbon channel on air velocity; After O. A. Chukhanova, quoted by Frank-Kamenetskiĭ.[23]

toward the walls. Figure 5 thus illustrates the probable variation of the mainstream profile with increasing velocity. Simple theory indicates that if the velocity distribution across the radius is parabolic, so also is the concentration profile. The influence of this on reaction rate will now be self-evident. The profiles as drawn represent the gas-stream limit of the diffusional boundary layer. The nose of this profile will of course progress down the axis of the tube in equal increments as the velocity increases in equal increments; that is, the axial entry distance increases linearly with distance. In contrast, the boundary-layer thickness at any cross section ($X - X$) varies nonlinearly, decreasing rapidly to begin with as the entry distance increases to the end of the test section, but after that it becomes less

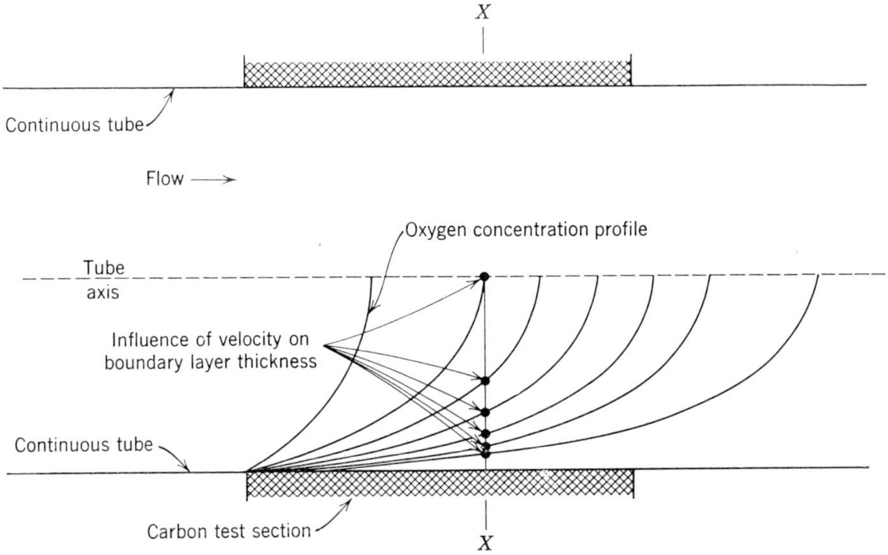

Fig. 5. Diagram of a burning carbon tube under laminar flow conditions illustrating effect of velocity on profile of mainstream oxygen concentration.

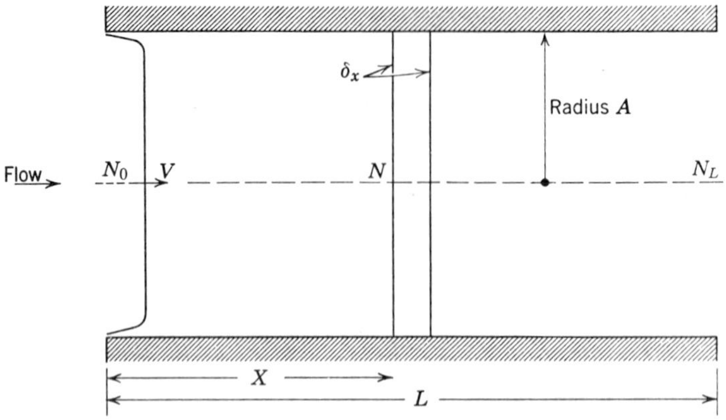

Fig. 6. Carbon tube in turbulent flow. Uniform velocity profile across tube; oxygen concentration varies from N_0 at tube mouth to N_L at tube exit.

and less sensitive to velocity. We may therefore reasonably expect the reaction rate to become decreasingly sensitive to velocity as found in practice by Chukhanova (Fig. 4). Preliminary calculations have substantiated this view.

Turbulent Flow

As the velocity increases, the laminar flow eventually breaks up, and the turbulence generated breaks down the diffusion layer, reducing it to a sublaminar layer of unimportant magnitude. This system is now amenable to a simple mathematical analysis for which the following approximations may be introduced: (1) that the velocity profile across the radius is constant with value v; (2) that turbulence is so intense that radial mixing of reaction products is instantaneous; (3) that the reaction at the surface is first order with respect to the oxygen concentration as in equation 6.

From Fig. 6, consider a volume element of width δx assumed stationary. The solid surface area bounding this element is $2\pi A\ \delta x$. If the oxygen concentration is N molecules per unit volume, the specific reaction rate is proportional to this, so in a time δt, the mass of carbon removed from the surface element is m, where m is proportional to k_1, N, $2\pi A$, δx, and δ_t. If in this period of time the number of molecules removed from the volume element to react with the carbon is δN per unit volume (removed equally from each element of volume because of the rapid turbulent exchange), the total number of molecules involved is $\pi A^2\ \delta x\ \delta N$; and this is also proportional to m. Hence

$$2\pi A\ \delta x k_1 N\ \delta t \propto \pi A^2\ \delta x\ \delta N. \tag{35}$$

If at the end of the time element this volume element moves discontinuously into the adjacent position downstream, we have $v\ \delta t = \delta x$, and in the limit,

$$\frac{dN}{N} = -\frac{ck_1}{\pi A v}\,dx \tag{36}$$

where c is the constant of proportionality involved. If the volume flow is V, then $V/A = \pi A v$, and by integrating between limits 0 to x,

$$N = N_0\, e^{-Ack_1x/V} \tag{37}$$

Since the reaction rate R, at the element surface per unit length is $m/\delta t$, then $R = B(2\pi A)k_1N_0\ e^{-Ack_1x/V}$ where B is also a constant of proportionality, and total reaction rate down the tube is the integral of this over its length; hence

$$\int_0^L R\,dx = \frac{2\pi N_0 Bv}{ck_1}\,(1 - e^{-k_1AcL/V}) \tag{38}$$

and the average specific reaction rate $\bar{R}_t$ is

$$\bar{R}_t = \frac{N_0 B}{AcL}\left(1 - \frac{N_L}{N_0}\right)V \tag{39}$$

To a first approximation, therefore, the mean reaction rate is proportional to the flow velocity or volume, as found experimentally by Chukhanova (see Fig. 5). To a second approximation, the exponential term can be expanded, giving a velocity term that may be written $V(1 + AcL/v)$. This accounts for the slight curvature found by Chukhanova and also demonstrates that, for reasons of *dilution* not *diffusion*, the reaction rate can be velocity dependent, although the reaction may be in the chemical control region and not in the diffusional control region.

Broken Bed

Although the study of the simple channel provides much useful and interesting data on the kinetics of the reactions, it is generally regarded as having less relevance to a broken bed formed from lumps of coal or coke than has the single particle to the dust flames. The channels found in broken beds are tortuous and of variable diameter and have sides that are difficult for any aerodynamic definition. This does mean that there will be no delay in the onset of turbulence, with the consequent breakup of the diffusion layers. The results of the single channel may then be introduced in the following way. If it is possible to imagine the fuel bed as an array of tubes varying only slowly in diameter and direction, the additional spaces not included in the tube array can be regarded as large diameter pores. In the exponential term of equation 37, the velocity constant k_1 can then be replaced by the k_1^1 of equation 29:

$$k_1^1 = k_1(1 + \sqrt{SD/k_1})$$

If we bear these points in mind, equation 39 provides an interesting contrast with experimental results on fuel beds of varying depth.

Relative Carbon Saturation

As one result of the experiments on the CO/CO_2 ratio of the primary reaction, the old "two-zone" theory of fuel bed combustion has now been displaced by the "three-zone" theory. The former two-zone theory was that the primary reaction, in the first zone adjacent to the entry ports, was a reaction mainly to CO_2 even at high temperatures; in the second zone, if the fuel bed was deep enough, this CO_2 was reduced to CO. With the knowledge that the primary reaction gave mainly CO at high temperatures, the two-zone theory was discarded. In the alternative three-zone theory, the following sequence is assumed to occur: (1) Oxygen in the fuel channels reacts at any available solid surface to generate CO; (2) this CO migrates back to the channel main-

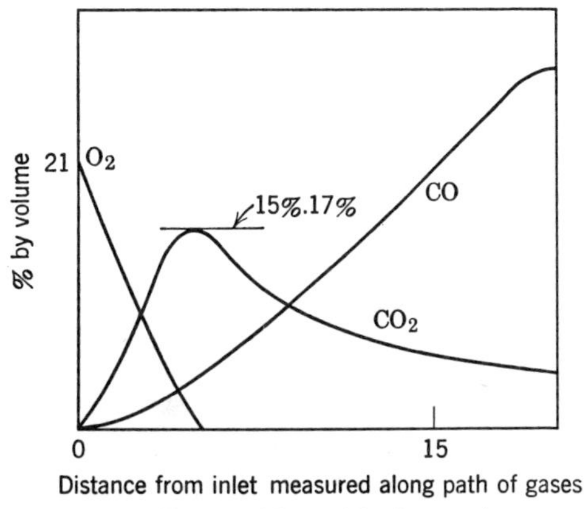

Fig. 7. Typical gas analysis in fuel bed.

streams and is oxidized with any further available oxygen to CO_2; and (3) this CO_2 is then reduced, as in the old theory, to CO. This three-stage process did not occur in Chukhanova's experiments simply because the tubes used were not long enough or were too wide. Its effect in a fuel bed is illustrated by Fig. 7, which shows the concentrations of CO, CO_2, and oxygen at different places through a fuel bed.

Because of the incomplete oxidation of the carbon removed from the fuel bed, measurements of weight loss alone give no indication of the oxygen used; and conversely the measurement of the oxygen used gives no indication of the carbon removed; for example, one molecule of oxygen can remove one atom of carbon as CO_2, or two as CO. A reaction rate index that takes incomplete oxidation into account automatically is the Relative Carbon Saturation (R.C.S.),[38,39] and is defined for pure carbon combustion by the relation

$$\text{R.C.S.} = \frac{1 - 0.019\,CO_2 - 0.048\,O_2}{1 + 0.010\,(CO_2 + O_2)} \qquad (40)$$

If there is no reaction, the R.C.S. is zero, for a value of 0.5 there is the maximum possible content of CO_2 (assumed 21%), and for a value of unity there is the maximum possible CO (34.6%) corresponding to producer gas.

Influence of Bed Depth

The R.C.S. has been used as the basis for some semiempirical equations relating mean

[38] Thring, M. W., *Fuel*, **31**, 355–64 (1952).

[39] Thring, M. W., *Coal Research* (*BCURA*), September 1944, pp. 70–2.

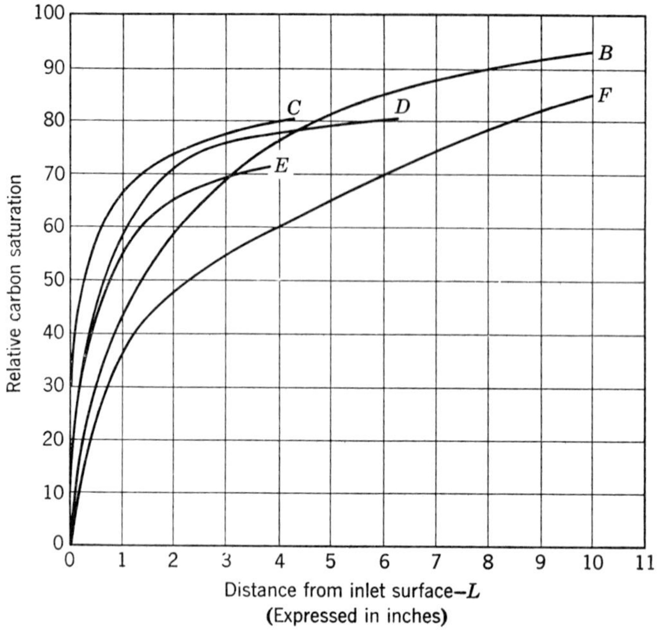

Fig. 8*a*. Relative carbon saturation plotted against distance from inlet surface. Fuel size $\frac{1}{4}$ to $\frac{3}{8}$ in.[40]

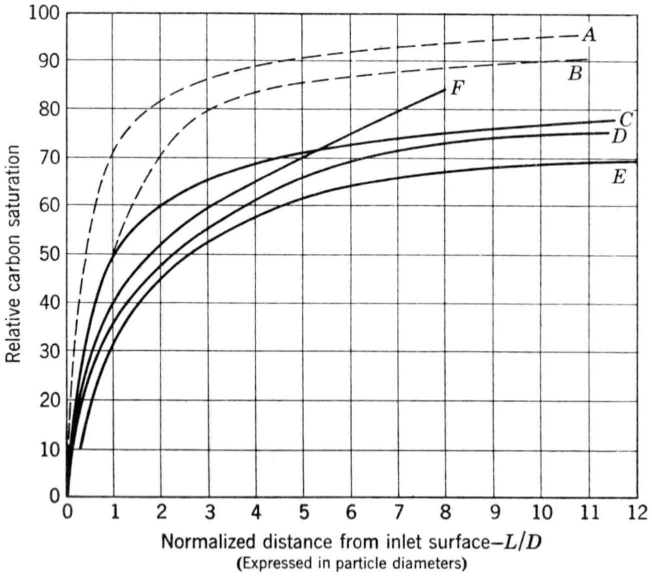

Fig. 8*b*. Relative carbon saturation plotted against ratio of distance L from inlet surface to average particle size D.[41]

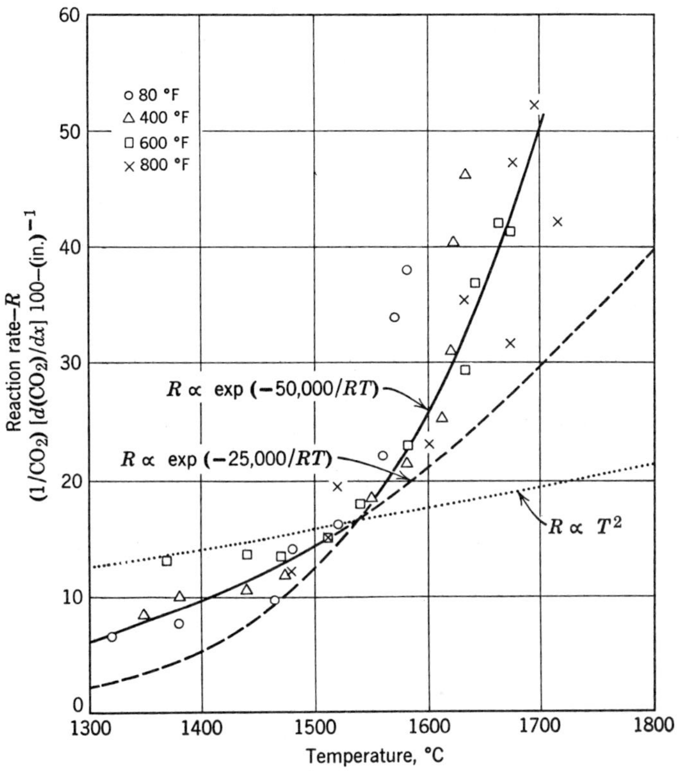

Fig. 9. Activation energies of carbon: CO_2 reaction at high temperatures.[40,43]

reaction rate with bed depth L. Figure 8a[40] and 8b[41] show R.C.S. as a function of L and L/d respectively, where d is the diameter of the particle forming the bed. Kreisinger and coworkers[42] obtained somewhat similarly shaped curves, and they were found to fit the empirical equation

$$\text{R.C.S.} = K(1 - e^{-\alpha L/d}) \tag{41}$$

where K and α were arbitrary constants with the following values:

1. Cold air: $K = 0.7$ to 0.9 and $\alpha = 0.3$ to 0.6. K was high for reactive fuels and for low wall loses, possibly because of the higher temperature, and α increased with reactivity and air rate.

2. Preheated air at 300°C: $K = 0.97$, and $\alpha \doteqdot 1.0$.

[40] Nichols, L. H. F., *J. Inst. Fuel*, **14**, 71–3 (1940).

[41] Mott, R. A., and Wheeler, R. V., *Midland Coke Research Comm., Tech. Rept.*, No. 1 (1930). Nicholls, P., *U. S. Bur. Mines, Bull.*, No. 378 (1934), 76 pp.

[42] Kreisinger, H., Ovitz, F. K., and Augustine, C. E., *U.S. Bur. Mines, Tech. Paper* No. 137 (1916), 76 pp.

Equation 41 has the same form as equation 39 which may be rewritten (with velocity v in place of volume V)

$$\bar{R}_t = \frac{N_0 B V}{AcL}\left(1 - \exp\frac{-ck_1}{\pi v}\frac{L}{A}\right) \tag{42}$$

In the exponential term, the channel radius A replaces the particle diameter d. They are directly related so that equations 41 and 42 have the L/d term in common. For the α term, equation 42 is in agreement with the empirical result that this should increase with reactivity (represented by k_1), but is at variance with the conclusions on the influence of air rate. In the tube, however, there was no opportunity for the reduction reaction to influence the reaction rate. As equation 37 shows, the effect of velocity is to increase N at any point in the bed. The effect of velocity will therefore be to increase the distance from entry before the reduction starts, and this presumably will increase the R.C.S. as found experimentally.

For the K term, there was less agreement altogether, since its equivalent in equation 42 carries no reactivity term at all. There is also disagreement over the effect of velocity, which may be a result of the reduction reaction. The similarity of equation 41 and 42 carries the implication that the flow in the channels in the bed is turbulent, and that the single-channel experiments show a closer relation to the broken bed than might have been expected. Support for this implication is provided by data[40,43] illustrated in Fig. 9 on the reduction of CO_2 by a coke bed. The temperature coefficient is far too high for any diffusional process to be operative, so it follows that since the reaction is chemically controlled, the flow is very likely to be turbulent. What is not established is the order of the reaction. In view of this and other points, it would seem therefore that a more detailed study of single channels using longer tubes might be profitable.

[43] Thring, M. W., *J. Inst. Fuel*, **14**, 47–62 (1940).

CHAPTER 18

COMBUSTION PROCESSES

Ralph A. Sherman and Bertrand A. Landry

Battelle Memorial Institute, Columbus, Ohio

Since the early 1940's, the burning of coal has continued to present challenges to improve equipment, to select coals better suited technically and economically to the existing equipment, and to reach a better understanding of combustion reactions. The challenges have been magnified by two important factors. One is the intensified competition from natural gas and fuel oils, which necessitates the development of coal-burning equipment having improved automatic features and flexibility, and large decrease in smoke and dust emission. The other has been the need to provide for higher heat releases and higher thermal efficiencies in larger power-plant furnaces, and to minimize investment and labor costs.

Although notable improvements have been made in the design of heating equipment for residential use and for small industry, coal has nearly lost this market owing to the natural advantages of the fluid fuels. The steam locomotive which represented a large market for coal has all but disappeared. Electric power plants have, however, greatly increased their total consumption of coal so that, in the United States, they, together with medium- and large-size industrial plants and certain other industrial uses, consume more than 80% of the coal produced which is burned directly. By and large, the challenges still found in this sector are not strictly problems in combustion but are problems associated with the behavior of the ash in coal.

BASIC INVESTIGATIONS

It is agreed that the burning of coal involves three stages: (1) the release of the volatile matter resulting from the heating of the coal, (2) the burning of the volatile matter in the gas phase, and (3) the burning of the solid residue of carbonaceous matter. Agreement has not been reached entirely, however, on the mechanism of combustion of the solid residue of carbonaceous matter, despite the considerable amount of experimental and interpretive work performed.

Classically, the burning of carbon in deep fuel beds has been considered to take place in two zones that are not sharply divided. The first zone is the "oxidation" zone where the main product of combustion is CO_2. The second zone is the "reduction" zone where the main product is CO. In the "twilight" zone between these two zones, the three gases CO_2, CO, and O_2 exist together; there is little or no free oxygen in the reduction zone and little or no CO in the initial portion of the oxidation zone. The CO found in the reducing zone is the result of the endothermic reaction $CO_2 + C \rightarrow 2CO$. With deep fuel beds, the CO/CO_2 ratio approaches a constant value, which is larger when preheated air is used than when preheat is not provided.[1] The extent of the conversion of CO_2 to CO depends, on the heat losses to the outside and the endothermicity of the reaction.

Hiles and Mott[2] have reviewed fully the experimental work carried out in the early 1930's.[3] This work had raised doubts regarding the validity of the oxidation zone concept; they were led to postulate that CO was the initial or primary product of combustion, and that the presence of CO_2 in the so-called oxidation zone was due to the oxidation of the CO

[1] Nicholls, P., *U.S. Bur. Mines*, No. 378 (1934), 76 pp.

[2] Hiles, J., and Mott, R. A., *Fuel*, **23**, 154–71 (1944).

[3] Bangham, D. H., and Bennett, J. G., *Fuel*, **19**, 95–101 (1940). Grodzovskii, M. K., and Chukhanov, Z. F., *Fuel*, **15**, 321–8 (1936).

to CO_2 in the void space surrounding the carbon pieces. The main facts brought out by these investigators were that the percentage of CO_2 decreased and the percentage of CO increased markedly with increase in the rate of air supply or, in other words, with increase in the rate of combustion. Thus it appeared that with increased air velocity the oxidation of "primary" CO in the void space decreased because of the shorter times available. The Russian experimental work was performed, as a rule, in small furnaces, and the fuel was composed of very small pieces.

Kreisinger and associates[4] had obtained a similar result earlier by using larger initial sizes of fuel. They had explained the effect of air velocity on the grounds that with higher rates of combustion more heat was available for the endothermic reaction $CO_2 + C \rightarrow 2CO$, in view of the fact that heat losses to the surroundings would remain about constant and be independent of rate of combustion. Chukanov's results[3] where higher percentages of CO were obtained than had been obtained by Kreisinger[4] could be explained on the basis that the heat relations given earlier were magnified because of the scale of their experiments. Hiles and Mott concluded (in part): "With such a complex series of reactions, and overlapping zones, the consideration of predominating products is preferable to barren discussion of 'primary' and 'secondary' reactions."

Arthur and others, in the early 1940's, were led to reopen the question of primary formation of CO partly on the basis of Nicholl's observation that attempted energy balances in the oxidation zone led to an excess of energy over that accountable from observed temperatures and gas compositions. It seemed possible that part of the CO_2 obtained in the sample resulted from the oxidation of "primary" CO within the sampler in spite of the fact that the latter was water cooled. To investigate this, a highly turbulent air stream was passed through channels in graphite blocks externally heated to 850 to 900°C. A blue glow appeared near the wall. Sampling the gases near the carbon wall showed, however, a very low CO/CO_2 ratio of about $\frac{1}{20}$; even drastic cooling of the probe caused only a slight increase in CO. This showed either that CO_2 was the main primary product, or that CO was perhaps the sole primary product, but that the diffusion layer where the CO burned to CO_2 was exceedingly thin.[5]

Attacking the problem in a different way, Arthur[6] reasoned that if CO_2 results mainly from the burning of "primary" CO, then inhibition of this homogeneous reaction by the introduction of "chain breakers," such as chlorine or chlorine compounds, in the combustion air, would result in an increase of the CO/CO_2 ratio in the gases. This increase was found to a marked degree, and where there had been a glow the glow disappeared. Arthur raised the question, however, of whether the chlorine might not so modify the carbon surface that CO might be produced in greater relative amounts than CO_2. This possibility does not seem to have been pursued further by Arthur, although it will be seen later that such an effect has been observed.

In September 1949, an international conference on the mechanism of the combustion of carbon took place in Nancy. The complete set of papers has been published[7] and summarized by Bangham and Crone.[8] At this meeting, Arthur and Bangham presented a rather complete account on the work done with chlorinated additives.[9] Figure 1 shows gas compositions obtained as a function of percentage of $POCl_3$ added to dry air passing through hollow electrode carbon cylinders externally heated to 850°C, the diameter of the bore being 1 cm and the rate of air flow 300 cc/hr. It is seen that, with a concentration of about 0.2% $POCl_3$, that of CO is about 19% and remains essentially constant with further increase in concentration of $POCl_3$; the CO_2 remains about 3% and the residual oxygen ranges from 5 to 8%.

Figure 2 gives the gas compositions, with 0.28% CCl_4 as additive, as a function of percentage of added water vapor. It will be noted that with zero moisture added, and actually up

[4] Kreisinger, H., Augustine, C. E., and Ovitz, F. K., *U.S. Bur. Mines, Tech. Paper* No. 137 (1916), 76 pp.; see pp. 31–4.

[5] Arthur, J. R., Bangham, D. H., and Crone, H. G., *Ind. Eng. Chem.*, **43**, 525–33 (1951).

[6] Arthur, J. R., *Nature*, **157**, 732 (1946). See also: Bridger, G. W., *Nature*, **158**, 236 (1946).

[7] *J. chim. phys.*, **47**, 313–78, 525–85 (1950).

[8] Bangham, D. H., and Crone, H. G., *BCURA Bull.*, **13**, 413–19 (1949).

[9] Arthur, J. R., and Bangham, D. H., *J. chim. phys.*, **47**, 559–62 (1950). See also ref. 5.

to 0.2% added moisture, the gas compositions are essentially as reported in Fig. 1; that is, around 19% CO, 5% CO_2, and 3% O_2, which shows that CCl_4 behaves like $POCl_3$. With the addition of 0.4% water vapor, however, the situation changes completely. The percentage of CO_2 is now 18 to 19%, the percentage of CO is about 2 to 3%, and the oxygen remains at about 1%. Further increase in the concentration of water vapor has essentially no effect.

Figure 3 shows the relations between the total amount of carbon in the gases as a function of distance traveled along the carbon tube for (1) where $POCl_3$ is added (upper curve) and (2) where no additive is used (lower curve) the air rate being the same for both conditions. The amount of carbon gasified is about twice as high when $POCl_3$ is added. Since the velocity of the air streams was the same, it is clear that in the first instance the gases were composed mostly of CO and in the second, instance, mostly of CO_2.

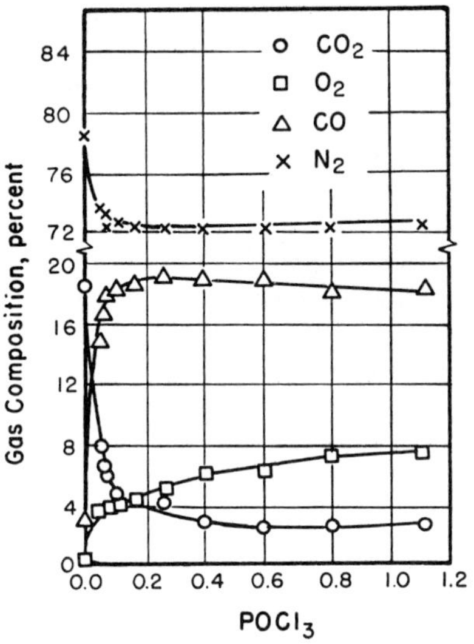

Fig. 1. Relation of gas composition to percentage of $POCl_3$ added to combustion air.[9]

From these data, Arthur and Bangham concluded that (1) chlorinated compounds inhibit the oxidation of CO to CO_2 in a dry atmosphere, (2) moisture destroys the inhibiting property of these compounds, and (3) inhibition makes oxygen available over a longer distance to burn carbon.

Wicke[10] has reported experiments involving the passing of dry and moist air through beds of small particles of carbonaceous material externally heated to temperatures ranging from 647 to 890°C. At the lowest temperature, there was no appreciable difference in the CO/CO_2 ratio for wet and dry air, and the addition of $POCl_3$ had no noticeable effect. At about 830°C, however, and with no additive, the CO/CO_2 ratio was approximately six times larger $\left(\frac{3.58}{0.59}\right)$ for dry air than for wet air, there being almost no residual oxygen in both instances. The effects increased to the highest temperature studied (890°C).

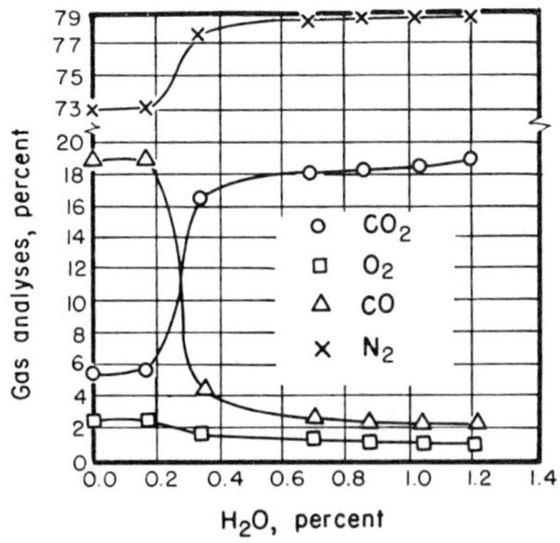

Fig. 2. Relation of gas composition to percentage of H_2O vapor added to combustion air containing 0.28% CCl_4.[9]

Van Loon and Smeets[11] introduced a glowing particle of "graphite" in a vortex of oxygen and observed that the ratio $CO/(CO + CO_2)$ increased with the velocity of the oxygen stream. The inference was that with higher

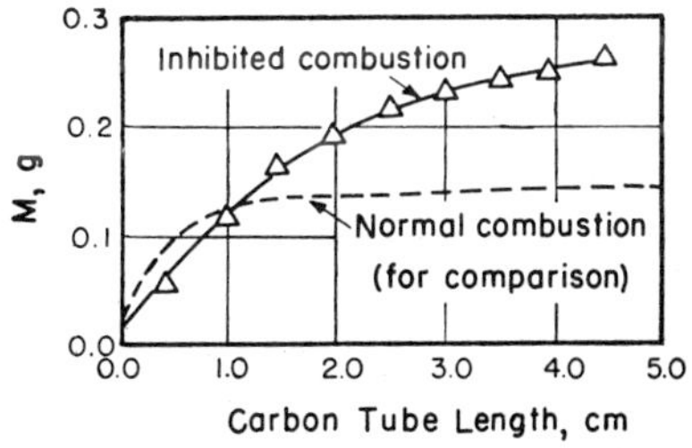

Fig. 3. Relation of mass of carbon gasified (M) as a function of length of carbon tube.[9]

[10] Wicke, E., *5th Symposium on Combustion, Pittsburgh*, **1954**, Reinhold, New York, 1955, pp. 250–1.

[11] Van Loon, W., and Smeets, H. H., *Fuel*, **29**, 119–21 (1950).

velocity there was a simultaneous decrease in the thickness of the diffusion layer surrounding the particle and, hence, insufficient time to burn the CO to CO_2. Extrapolation of their data to infinite velocity pointed to a final ratio of unity, that is, zero CO_2. That ozone appeared in the gases was carefully verified by appropriate chemical indicators.

Bridger and Appleton[11] confirmed the increase found in the CO/CO_2 ratio when chlorinated compounds are added to the combustion air supplied to a bed of coke particles, and extended their experiments to the gas phase combustion of CO in the absence of carbon. A combustible gas (undried) having the composition 11% CO, 10% O_2, and 79% N_2 was passed through a vertical, externally heated silica tube which, in most instances, was filled to a depth of 50 cm with silica chips. Measured quantities of CCl_4, HCl, or Cl_2 were added to the combustible gas and preheated together in the silica column. The rate of heating was adjusted so that the temperature of the mixture reached 1000°C in the first 35 cm of the column. Under these experimental conditions, the CO mixture burned completely to CO_2 when no additive was present, but when the chlorine compounds were added in sufficient quantity, little burning took place. Thus, with 2.5% added Cl_2, the gas composition above the bed was 0.4% CO_2, 10% O_2, and 11.4% CO. When the flow of Cl_2 was stopped, the combustion to CO_2 was complete. The concentration of additive required to give a $CO/(CO + CO_2)$ ratio, say, of 0.9, was found to be different for the three types of additives, least for CCl_4, intermediate for Cl_2, and highest for HCl.

In subsequent series of experiments, the additive was limited to Cl_2, and the effects were determined, in part, of varying the concentration of Cl_2 and of the temperature of the mixture, and of the separate preheating of the CO component, by introducing it through a central tube having its top opening at the 35 cm level of the column. Bridger and Appleton concluded from this work that (1) CO is the primary product of the reaction between carbon and oxygen, (2) the effect of inhibitors on the oxidation of CO has been demonstrated, and (3) the mechanism of inhibition cannot be determined without further investigation.

[12] Bridger, G. W., and Appleton, H., *J. Soc. Chem. Ind.*, **67,** 445–9 (1948).

Discussion

One conclusion of the experimental work described is that certain chlorinated compounds do interfere with the reaction $2CO + O_2 \rightarrow 2CO_2$. Another is that water vapor catalyses this reaction, and if present in sufficient concentration it can neutralize the "inhibitive" action of chlorine compounds. All the chlorinated compounds used react with water vapor to form HCl, and this acid, as well as phosphoric acid, has a very strong affinity for water vapor. Thus, insofar as moist gases are concerned, the effect of chlorine is to dry the mixture and prevent the oxidation of CO by removing water vapor, a known catalyst for this reaction. An increase in temperature will require additional HCl; an increase in moisture content of the gases also requires additional HCl. With a constant supply of chlorine and increasing moisture of the gases, a concentration of moisture is reached whereby further saturation of HCl is impossible and the burning of CO is resumed.

To explain completely the action of chlorine when carbon is present and the combustion air is dry, however, requires another mechanism. Wicke[10] found that with dry air and added $POCl_3$ the CO/CO_2 ratio increased to 16.9 from 3.6 without additive, and he also noted noted that the structure of the carbon was markedly changed. It may be concluded that chlorine can produce a change in the structure of burning carbon so that its reactivity increases with a resulting increase in the CO/CO_2 ratio.

Fundamentals of Fuel Beds

At one time, the fuel bed was the only known method of successfully burning coal; in fact, coal is said to have first appeared on the industrial scene when English blacksmiths began using it for heating iron in their smithies.[13] The coal so used was traditionally "sea coal." The meaning of the word is obscure, but it implies that a selection was involved to obtain best results. The fuel bed is particularly sensitive to coal properties, which is partly the reason why other methods of burning, less sensitive to fuel characteristics, have been developed. The fuel bed is still an important method of burning coal, however,

[13] Guernsey, L., *Focus* (*Am. Geograph. Soc.*), **10,** No. 3, November (1959).

and it has been the object of a great deal of research.

The classification of elementary fuel beds into three types by Nicholls[14] has made it possible to understand the effects of the relative directions of movement of the fuel and of the air supplied on the rate of ignition, the rate of burning, and the rate of ash release in the various types of beds, including the shallow beds of hand-fired or spreader-stoker-fired boiler furnaces, the variable bed depths of steam locomotives, the deep beds of gas producers, the beds of underfeed single- or multiple-retort stokers, and the beds of traveling-grate stokers. For a discussion of fuel characteristics affecting fuel beds and a description of the elementary types of fuel beds and how these serve to explain the industrial beds, the reader is referred to a review by Landry.[15]

Analytical Approach to Fuel-Bed Theory

The work of Nicholls, his predecessors, and successors was almost entirely empirical. Although it has provided an excellent basis for understanding and sometimes for preselecting beds having known or desired characteristics, the data obtained fail to explain the mechanism of burning, the rate of heat transfer required for burning stability, or the diffusion or mass transfer rates required to bring the reactant gases to the surface of the burning solid and to remove the products of combustion from the active surface. The existence of the empirical data did tempt various workers to try to construct theoretical models of fuel beds to which could be applied various chemical and physical laws regarding rate of reaction and rates of heat and mass transfer.[16] However, no completely satisfying theory has evolved from these studies.

Hougen and Watson,[17] in approaching this problem, considered that "since mass-transfer rates are little affected by temperature, it is possible for the overall rate of a reaction to be controlled entirely by chemical steps at low temperatures and by mass transfer at high temperatures where the chemical rates are relatively high . . . and where mass-transfer resistances are dominant, temperature has little effect on the (overall) rate of reaction."

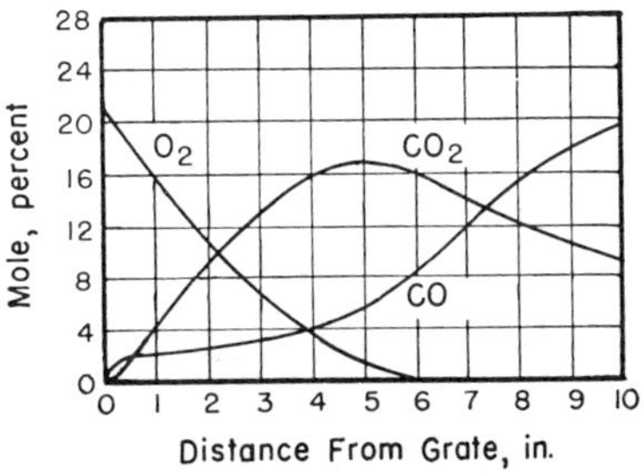

Fig. 4. Relation of gas composition to distance from grate.[17]

The basic working assumptions used by these authors were as follows: (1) The rate of reaction of O_2 with the carbon surface to form CO and CO_2, and the rate of reaction of CO_2 with carbon to form CO are both so rapid that equilibrium is reached at the various bed temperatures; (2) the reaction between outwardly diffusing CO and inwardly diffusing O_2 to form CO_2 is a homogeneous reaction whose rate approximates a third-order mechanism; (3) the size of the fuel pieces remains constant, inasmuch as the ash coating on each piece offers resistance to mass transfer; (4) mass velocity, viscosity, density, and diffusion coefficients remain constant throughout the bed, the reference temperature being 2600°F; (5) O_2 and CO_2 each diffuse independently in an atmosphere of nitrogen only, and the diffusion coefficients may be calculated by means of an empirical relation; and (6) the bed is composed of uniform-sized particles packed to yield a void space of 50%.

By using these simplifying assumptions, a set of three difference equations for O_2, CO_2, and CO were obtained in terms of the determining rates and of successive increments of bed height. These equations were integrated step by step the partial pressures being adjusted at each step to correspond to the calculated compositions. Figure 4 shows the gas compositions, as a function of distance above the grate, obtained by the foregoing process.

[14] Nicholls, P., *Trans. Am. Inst. Mining Met. Engrs., Coal Division*, **119**, 183–97 (1936).

[15] Landry, B. A., *Encyclopedia of Chemical Technology*, Kirk, R. E., and Othmer, D. F., Eds., **6**, 903–13 (1951). See also: Carman, E. P., Graf, E. G., and Corey, R. C., *U.S. Bur. Mines, Bull.* No. 563 (1957), 92 pp.

[16] Mayers, M. A., *Chemistry of Coal Utilization*, John Wiley and Sons, New York, 1945, Vol. 2, pp. 1501–14.

[17] Hougen, O. A., and Watson, K. M., *Chemical Process Principles*, John Wiley and Sons, New York, 1947, Vol. 3, pp. 1068–74.

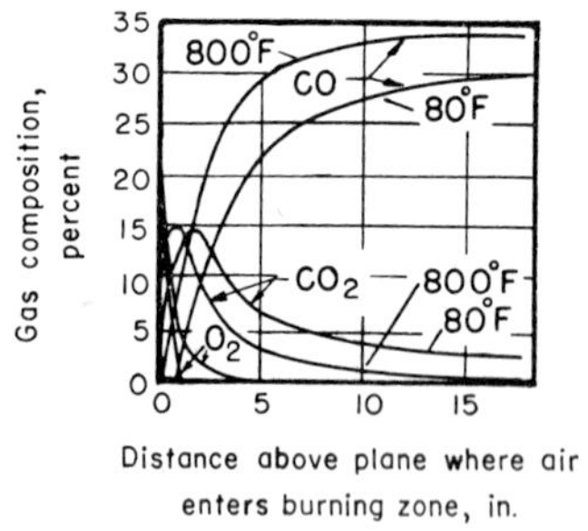

Fig. 5. Relation of gas composition to distance from ignition zone.[15]

Figure 5 shows a typical plot of Nicholls' results.[15] It is of interest to note that there is good agreement in the O_2 curves but not in the CO and CO_2 relations, Nicholl's results showing a higher CO/CO_2 ratio than that calculated by Hougen and Watson.[17]

Thring's[18] approach to this problem involved a slightly different set of assumptions: (1) All oxygen reaching the carbon surface reacts to form CO; (2) all CO meeting with O_2 forms CO_2; (3) all CO_2 reaching carbon is reduced to CO; (4) all of the foregoing reactions occur at an infinite rate; (5) the partial pressures of O_2 and CO_2 are zero at the carbon surface, and the partial pressure of CO in the void space is zero wherever the partial pressure of O_2 is not zero; (6) the rate of mass transfer is calculated by analogy with heat transfer in a bed of broken solids; and (7) all partial pressures are small compared with that of nitrogen, the terms involving the second power of partial pressure ratios being neglected, therefore, and all species diffuse independently of the others. Using a modified form of Stefan's Equation, Thring set up three differential equations for the three components CO_2, O_2, and CO which when integrated yield partial pressures which can be plotted against depth of fuel bed expressed as a nondimensional ratio. One result is that the CO_2 peak occurs precisely at the same distance where all O_2 has disappeared, which agrees well with many empirical data; another result is that CO starts to appear only when oxygen has disappeared. Thus, the three gases are never simultaneously present, a result opposite to some findings.[2]

Thring also calculated mass-transfer rates from empirical expressions to yield finally the ratio of carbon in gases W to the maximum amount that would combine with O_2 to form CO, W_0, as a function of the ratio of path length x to (original?) particle size d. Figure 6 shows a comparison of Thring's results with various experimental results recomputed by Thring. Curve A shows results with 800°F preheat, and curve B shows those with 80°F air. Curves C, D, and E, are for rates of air supply of 60, 40, and 20 cu ft/min. Curve F is from the work of Mott and Wheeler cited by Thring. Curves G, H, and I, are Thring's based on various expressions for mass-transfer rates and bed voidage. Figure 6 shows that the agreement is good for the early part of the abscissae, but that it deteriorates later. Thring concluded, in part, that the rate of the reaction $CO_2 + C \rightarrow 2CO$, which occurs in the latter stages of the fuel bed, is governed by both chemical and physical factors; in other words, the rate of this reaction is somewhere between the infinite rate assumed by Thring and the third-order mechanism rate envisaged by Hougen and Watson.

Silver,[19] in his analytical approach to the problem of fuel-bed combustion, used the following assumptions:

(1) The flow of air and of combustion products takes place on the inside of a tube having a carbon surface rather than through a bed.

(2) The rate of reaction is controlled only by the rate of molecular migration to the locus of the reaction which includes a region near the tube as well as the surface of the tube.

(3) The fraction of the total number of molecules which reaches the surface of the wall is proportional to the length of the tube.

(4) The number of any species striking the

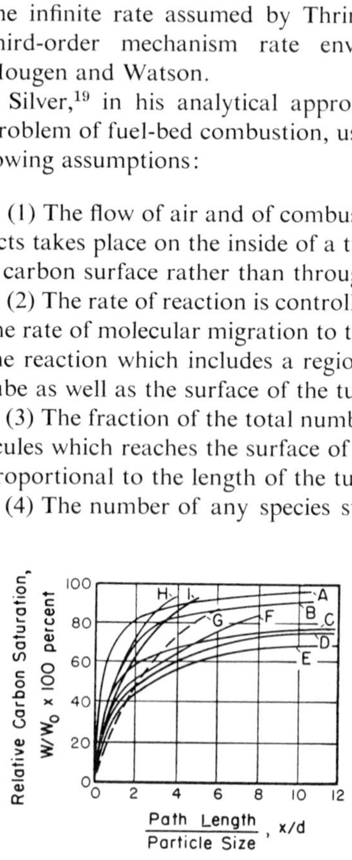

Fig. 6. Relation of carbon saturation to the ratio path length/particle size.[18]

[18] Thring, M. W., *Fuel*, **31,** 355–64 (1952).

[19] Silver, R. S., *Fuel*, **32,** 121–50 (1953).

wall is the product of the total number striking the wall and the concentration of the species.

(5) Every O_2 molecule striking the wall reacts to give a CO to CO_2 ratio of $x/(1-x)$.

(6) Every CO_2 molecule striking the wall reacts to form 2CO.

(7) O_2 and CO react in the gas phase to give CO_2, but this reaction is limited to a region near the wall, the width of the region depending on the rate of molecular transport.

From these assumptions, Silver set up three differential equations in terms of the concentrations of O_2, CO, and CO_2 and of a composite variable $\beta l = u$, which is the product of the proportion of oxygen molecules moving to the wall, β, and the length of path, l. These equations were integrated with the simplification that the increase in volume due to carbon monoxide formation may be neglected in view of the relatively high concentration of nitrogen. In view of the fifth assumption, it is evident that the concentrations obtained as a function of u will depend on the value given to x. Figure 7 is a plot of the gas concentrations when $x = 0$, that is, when there is no carbon monoxide formed by the reaction of O_2 with C. Silver did not give the corresponding plot when $x = 1$, that is, when all O_2 reacting with C forms only CO; he explained, however, that the O_2 curve remains the same, the CO curve rising faster and the CO_2 maximum being much lower.

Since Fig. 7 does not agree well with experiment, Silver suggested that experimental results may be affected by the reaction of CO with O_2 in the sampling tube to form CO_2. Figure 8 shows the results that might be obtained from the composition of Fig. 7 by allowing the CO to burn to CO_2 to the extent of the available O_2. The curves of Fig. 8 are, of

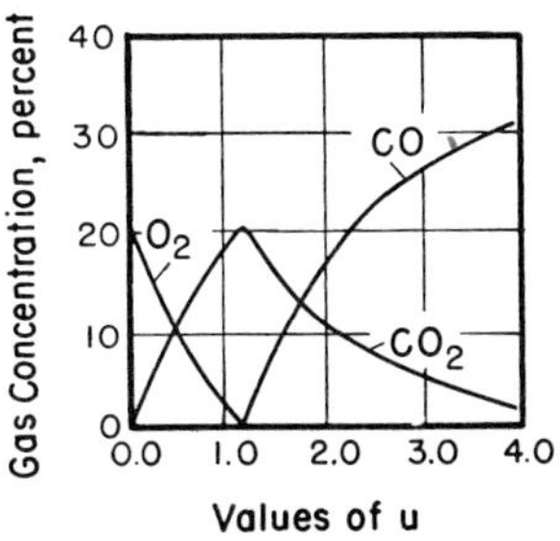

Fig. 8. Relation of gas composition to adjusted path length (no simultaneous CO and O_2).[19]

course, almost indistinguishable from those calculated by Thring,[18] inasmuch as the assumptions now coincide. Silver concluded, in part, that the value chosen for x should be closer to 1 than 0, thus allowing for the primary formation of both CO and CO_2.

In an important paper on the calculation of mass-transfer rates in absorption, vaporization, condensation, and combustion processes, Spalding[20] developed a fuel-bed theory using his concept of mass-transfer potential, with the further assumptions that only CO is produced at the surface but that one-half of the CO produced burns to CO_2. Therefore as long as O_2 is available in the void space, the concentrations of CO and CO_2 are equal. These assumptions appear to have been introduced to make the results of his computations agree with the experiments of Kolodtsev,[21] who found that up to a certain height of fuel bed the CO and CO_2 were equal in percentages, approximately. A similar result had been obtained by Hiles and Mott[2] with wood charcoal, but no investigator seems to have reported this type of result with other fuels.

Apart from various assumptions as to the "primary" and "secondary" reactions of O_2 with carbon and CO, and of CO_2 with carbon, theoretical investigators have also used various mass-transfer expressions. Silver and Mackay[22] have compared the mass-transfer expressions used by Thring, by Silver, and by Spalding after reducing them to common symbols. All of them are shown to define a parameter α which, under certain conditions, is

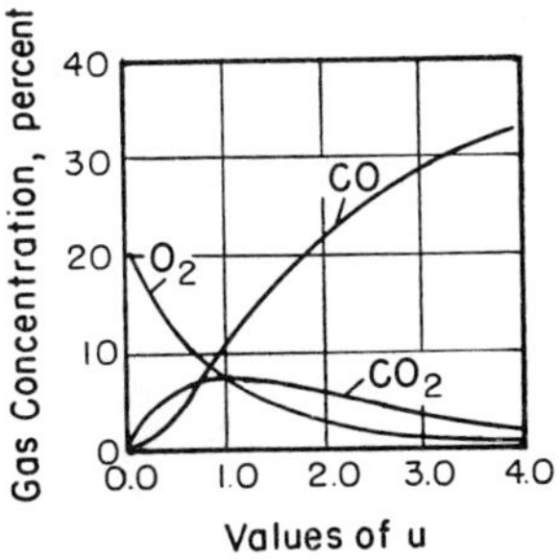

Fig. 7. Relation of gas composition to adjusted path length for $x = 0$.[19]

[20] Spalding, D. B., *Proc. Inst. Mech. Engrs. (London)*, **168**, 545–70 (1954).

[21] Kolodtsev, K. I., *J. Phys. Chem. (U.S.S.R.)*, **19**, 417–28 (1945).

[22] Silver, R. S., and Mackay, R. W., *Brit. J. Appl. Phys.*, **6**, 267–71 (1955).

Silver's β, and under other conditions is equal to 2.08 β in terms of fractional voidage, diameter of particle, and an equivalent Reynolds number. Thring and Spalding also included the Schmidt number. The expressions are different and yield widely different values of α, which indicates in part that the experiments by which different mass-transfer expressions were obtained were themselves different with none corresponding to the conditions that exist in an active fuel bed. Silver and Mackay suggested that the procedure be reversed, and that α be determined from the experimental results. They did not suggest, however, what set of differential equations should be used in terms of $\alpha l = u$.

Progress has been made in the theoretical approach to fuel-bed compositions and temperatures as a function of distance from the grate. One difficulty is that the various sets of experimental results available depend principally on the type of fuel used, on the rate of burning, or rate of air supply, and on the temperature of the air entering the grates. The equations set up to solve the problem must therefore have enough flexibility to account for differences among fuels as well as for differences in the flow rate and temperature of the primary air. Under these conditions, the mass transfer parameter may well become an empirical value whose relation to such variable fuel-bed conditions as local size of fuel pieces, local temperature, density, viscosity, and packing ratio may only be determined by future research.

Fundamentals of Pulverized-Coal Combustion

The burning of pulverized coal occurs under conditions so vastly different from those existing in a fuel bed that, in retrospect, it seems improbable that the pioneers in the art would have been successful. The cloud of fine coal particles leaving the burner is ignited mainly by radiation from the adjacent flame, and the rate of flame propagation to maintain a steady flame is strongly dependent on the air/fuel ratio and on the percentages of volatile matter and of ash in the coal.[23] No limitation of this type exists in overfeed beds, in which convection and conduction of heat play important parts, and the air/fuel ratio can vary widely without loss of ignition. Again, in contrast to the fuel bed, the burning of pulverized coal with the transporting air requires that the volatile matter be burned first, and the carbon residue burned last. These two steps occur within, or very nearly within, the volume of air supporting and immediately surrounding the burning pulverized coal, so that combustion of the residue is in an atmosphere in which the concentration of oxygen is continuously decreasing. Hence, to obtain a high combustion efficiency, that is, low carbon loss and low excess air, the time of residence in the primary furnace to allow the remaining oxygen to reach the particle must be longer by several magnitudes than for air-flow through a fuel bed.

Data obtained by Sherman[24] showed that the two main factors affecting residence time required for the same carbon loss were volatile-matter content and fineness of pulverization. Residence time decreased with increase in volatile matter and with increase in fineness of the coal. An increase in secondary air was also favorable but to a lesser degree.

Attempts to arrive at theoretical burning times from general principles have been summarized by Orning[25] for the period preceding this review. The general method of approach, since then, has remained about the same: (1) To derive, by various means, at expressions for the rate of burning of single particles; and (2) to assume that the rate of burning of the particles in the cloud is governed essentially by the same factors. Allowance must be made for differences in the gradient of oxygen concentration between the atmosphere surrounding the burning cloud of particles and that surrounding a single particle within the cloud. For a single particle, the supply of oxygen is not affected by the proximity of other burning particles.

For a typical, but straight-forward approach to the theoretical problem, the reader is referred to Yagi and Kunii[26] whose treatment

[23] De Lorenzi, O., *Combustion Engineering*, Combustion Engineering Co., Inc., New York, 1952, Sect. 8, p. 3.

[24] Sherman, R. A., *ASME, Trans.*, **56**, 401–9 (1934).

[25] Orning, A. A., *Chemistry of Coal Utilization*, John Wiley and Sons, New York, 1945, Vol. 2, pp. 1555–81.

[26] Yagi, S., and Kunii, D., *5th Symposium on Combustion*, Reinhold Publishing Co., New York, 1955, pp. 231–44.

covered the range from single-carbon particles, through clouds of them, to fluidized beds. The authors arrived at an overall rate coefficient of combustion for particles in the pulverized-coal flame, such that the reciprocal of this coefficient is equal, by analogy to heat transfer, to the sum of the reciprocals of (1) the rate coefficient of the chemical reaction, (2) the mass transfer coefficient through the solid layer of ash or of the cenosphere, (3) the mass transfer coefficient through the gas film multiplied by 3, and (4) the mass transfer coefficient of turbulent diffusion. All of these coefficients refer to the instantaneous size of the burning particle. Therefore, to obtain total burning time, these rates must be integrated over the range of sizes from the initial to the final size of the particle, and the initial size has to encompass all the sizes initially present properly weighted for their relative amounts. Unfortunately, the integration described is carried out by the authors only for the case of fluidized beds in which further complications result from carry over. Saji[27] approached the problem in a manner in part similar to that of Yagi and Kunii, but the detail of the rate expressions is different from theirs, except for the eddy-diffusion term. A main difference is that Saji concerns himself with the swelling index of the coal as affecting combustion rate, but does not seem to be concerned with ash or cenosphere formation. Saji carried out the integrations required to take into account the weighted size distribution of the burning particles and followed a method analogous to that used by Hottel and Stewart[28] which was based on the Rosin and Rammler expression for size consist.[29]

A paper by Khitrin[30] on combustion of coal in the fuel bed and in the pulverized form includes a long list of U.S.S.R. references. Unfortunately, most of the nomenclature used is not defined, although it may be in the references. Khitrin's conclusions, supported by his equations, were that: (1) relative velocity between the fuel and the gases is the most important factor for "intensifying" combustion in fuel beds, (2) residence time takes the place of relative velocity in pulverized-coal combustion, and (3) increased pressure is favorable for higher rates of heat release per unit volume.

In his approach to the problem, Essenhigh[31] considered that the time required to burn a particle of coal is proportional to the square of its initial diameter, the factor of proportionality depending on the nature of the material burned. Thus, Essenhigh started with the "integrated form" rather than with the differential form of the rate of burning, and he implied that the "diameter" of the particle during combustion is a "fictitious" diameter which becomes the true diameter only when the reaction rate is of first order. For a single particle, the partial pressure of the oxygen atmosphere surrounding the particle is, of course, constant. By generalization of the size and partial pressure parameters and differentiation of the original equation, the author arrived at an expression for the reaction rate in the pulverized-coal flame in terms of oxygen partial pressure. The variable partial pressure of oxygen was then computed on the basis of (1) perfect mixing with preassigned values of excess air and of coal dust density, and (2) the same average size for all burning particles. Integration is then possible, thus bypassing the complicating effect of size consist, which is relegated to the status of an experimental parameter. Because of the form of the equations developed, a particle of coal, enough time being available, will always burn out.

In summary, we may question the value of obtaining by analytical methods results which can only confirm what is generally known about the important factors that govern particle residence time in industrial furnaces.

The Burning of Coal under Pressure

Following the successful application in the late 1930's of hot catalytic refinery waste gases to operate gas-turbine air blowers and particularly the application near the end of World War II, of gas turbines for jet propulsion of aircraft, attention was directed to the possibilities of burning coal under pressure to supply energy to gas turbines delivering shaft power. First

[27] Saji, K., *5th Symposium on Combustion*, Reinhold, New York, 1955, pp. 252–60.

[28] Hottel, H. C., and Stewart, I. McC., *Ind. Eng. Chem.*, **32,** 719–30 (1940). See also: Stewart, I. McC., *J. Inst. Fuel*, **32,** 45 (1959).

[29] Rosin, P., and Rammler, E., *J. Inst. Fuel*, **7,** 29–36 (1933).

[30] Khitrin, L. N., *6th Symposium on Combustion*, Rheinhold, New York, 1957, pp. 565–73.

[31] Essenhigh, R. H., *2nd Conf. on Pulverized Fuel*, Inst. Fuel, London, Paper B-1 (1957).

attempts were made with pulverized coal,[32] but these attempts were soon followed by experiments with pressurized gas producers[33] and experiments with pressurized cyclone furnaces.[34] A review of these and other attempts to use coal for gas turbines, with an excellent analysis of the imposed conditions, has been prepared by Hazard.[35] Important conclusions reached were as follows:

(1) The time required for combustion of pulverized coal is independent of pressure so that the fuel can be fired at a rate proportional to pressure at constant air velocity through the combustor, the length of the flame remaining constant for any pressure.

(2) Similar relations hold for the fuel-bed of gas producers with the added advantages that carry over does not increase with burning rate so long as the air velocity remains constant and only pressure is varied.

(3) Similarly, operation of the cyclone furnace under pressure increases its turn-down flexibility, without loss of ignition, as compared to operation at atmospheric pressure.

In brief, it can be said that the burning of coal under pressure offers advantages for application to gas turbines, but there is difficulty in pressurizing reliably and simply the coarse or pulverized fuel to bring it to the operating pressure, and the removal of all or nearly all of the entrained solids to prevent excessive wear or coating of the fixed or rotating blading of the gas turbine, is an unsolved problem. Inasmuch as it is only under unusual circumstances that gas turbines fired with relatively low-cost residual oil can compete with diesel engines, it is not expected that the coal-fired gas turbines will overcome its technical shortcomings under present competitive fuel conditions.

APPLICATION OF PRINCIPLES

Smokeless, Hand-Fired Heating Equipment

Among the technical advances that were made in the last two decades, the achievement of "smokeless," hand-fired house heating equipment, burning bituminous coal, deserves mention because the problem had earlier been believed to be hopelessly insoluble. During the period from 1940 to 1950, new design principles based on the newly recognized elementary types of fuel beds, led to the development of practical operating heaters, warm-air furnaces, and low-pressure heating boilers that were essentially smokeless, had improved combustion efficiency, and permitted much longer periods between firings. Two different and independent approaches were made: one was first conceived in relation to a warm-air furnace and subsequently extended to a heater, and the other was limited initially to heaters but was later extended to warm-air furnaces and boilers. Field trials of a number of these units, from a few to several hundred, confirmed the claims made for them, but commercial exploitation has not followed because the convenience of the fluid fuels, which had become available throughout almost all of the United States in 1950 when the field tests were completed, could not be matched.

Figure 9 shows a vertical cross-sectional view of the unit developed directly for the warm-air furnace.[36] The figure is largely explanatory. A coking chamber is shown into which a mildly caking coal is fired and is gradually devolatilized by heat radiated from the hot baffle wall and from the burning fuel below. Part of the fuel burns by the cross-feed action of the "primary coking" air and part as a result of the introduction of undergrate air. Secondary air, admitted in a common stream with undergrate air, finds its way to the refractory-lined combustion chamber where it mixes with and burns the volatiles. Fellows[36] reported that operation of the furnace on heat-demand cycles similar to those of a residential stoker brought about a momentary appearance of smoke with change in damper position, and these smoke occurrences were mild by comparison with stoker operation at high rates of burning, but there was little difference in the amount of smoke produced by the two devices at moderate burning rates.

Figure 10 shows a cross-sectional view of the

[32] Hazard, H. R., and Buckley, F. C., *ASME, Trans.*, **70,** 729–37 (1948).

[33] Buckland, B. O., Hillman, A. Y., and Nelson, H. W., *Gen. Elec. Rev.*, **54,** No. 12, 12–26 (1951).

[34] Roxbee Cox, H., *Proc. Inst. Mech. Engrs.*, **164,** 407–24 (1951).

[35] Hazard, H. R., *Joint Conf. on Combustion*, Sect. 5, Inst. Mech. Engrs., (London), and Am. Soc. Mech. Engrs., (1955), pp. 10–16.

[36] Fellows, J. R., *Univ. Illinois, Eng. Exp. Stat., Circ. Ser.* No. 58 (1948) 38–58.

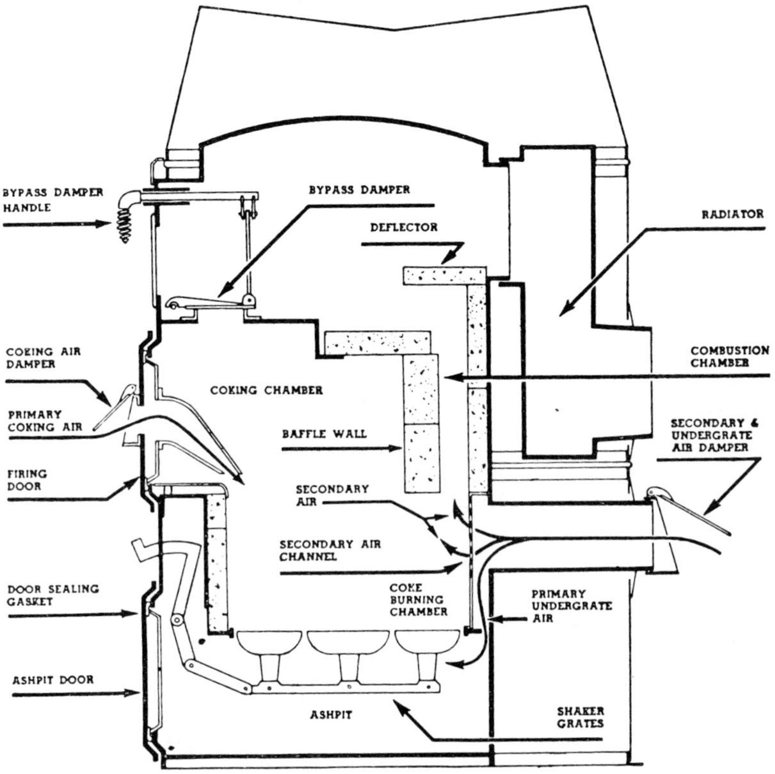

Fig. 9. Cross-sectional view of a smokeless furnace.[36]

design of smokeless heater[37] that was also applied to a warm-air furnace and to a heating boiler. The arrows represent air flow. The upper part of the cross-hatched area represents the "magazine" into which either strongly caking or mildly caking coals could be fired. Automatic gravity feed of the fuel down to the burning zone is accomplished, indirectly, as a result of the admission of air near the top of the fuel charge. This air flow induces the upward travel of ignition with consequent oxidation of the tars that would otherwise have led to the consolidation of the cake. Thus, a free-burning coke is produced to replenish the fuel which is burned, in the lower portion of the hearth, by the cross-feed flow of air; no air is admitted through the grate. Secondary air, admitted from both sides of the heater through the slot shown in the refractory arch, mixes with the hot gases as both flows hug the bottom of the arch on their way to the refractory-lined combustion chamber where combustion is completed. Although the emission of smoke cannot be avoided during the start of a new fire, in this type of heater, continued operation at high, medium, or banking rates yields no visible smoke except at refiring times, when about 5 min above No. 2 Ringelmann and a total of about 10 min above No. 1 can be expected as a maximum. Calorimeter room tests[38] showed a thermal efficiency of 63%, as compared to 51 to 54% for conventional heaters, with an unaccounted for loss of 3% as compared with 12 to 17% for conventional heaters.

Residential Stokers

Automatic stokers designed to use bituminous coal or anthracite for firing boilers and warm-air furnaces had a steadily increasing acceptance in the United States prior to 1941, when they reached their peak in sales of 176,000 units. In 1946, the sales were 172,000 units; in the years 1954 to 1958, inclusive, the average yearly sales have been 11,600 units. These stokers are, with few exceptions, of the

[37] Landry, B. A., and Sherman, R. A., *ASME, Trans.*, **72**, 9–17 (1950).

[38] Tieman, J. W., and Bagby, F. L., *Heating, Piping, Air Conditioning*, **23**, 117–28 (1951).

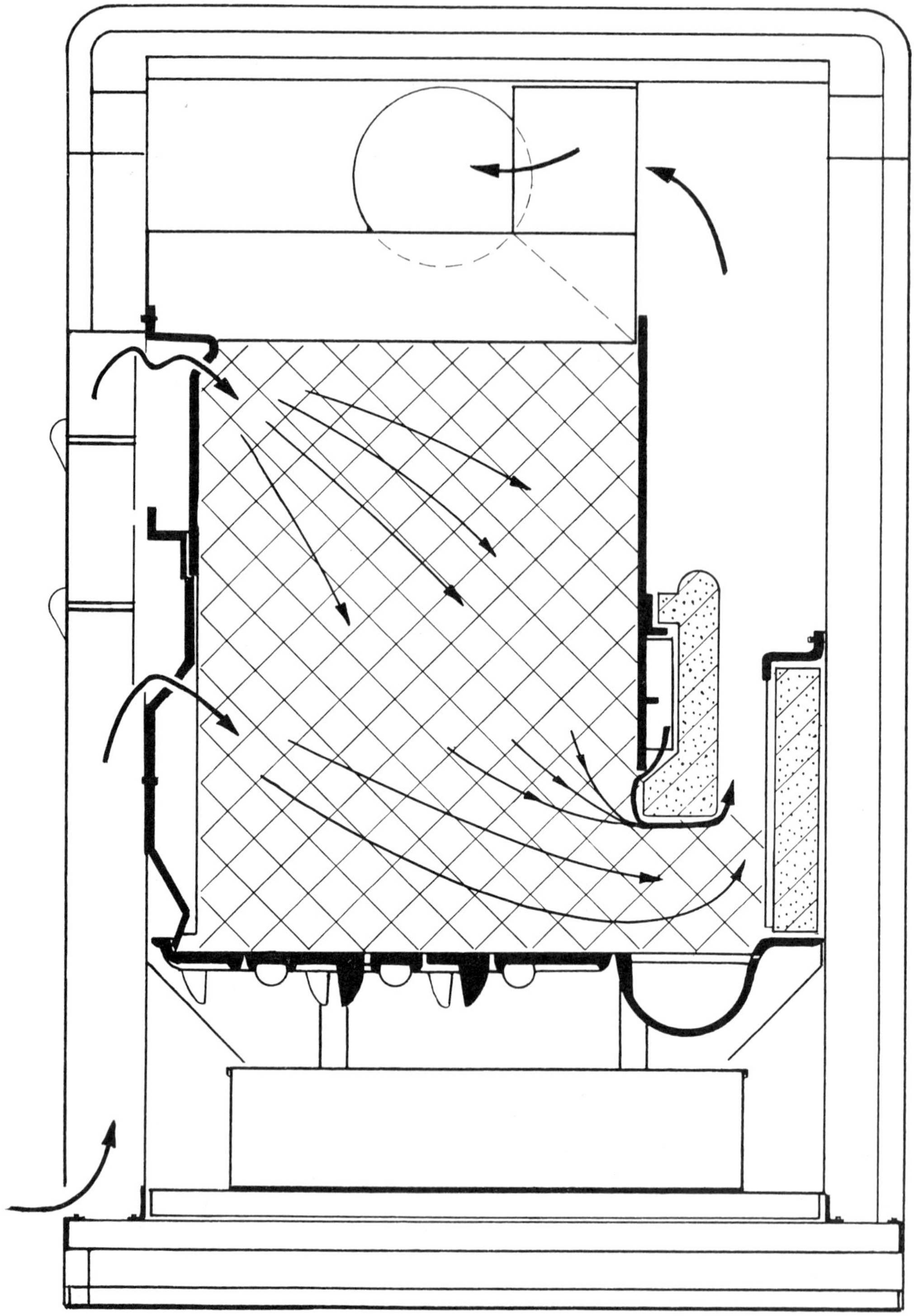

Fig. 10. Cross-sectional view of a smokeless heater.[37]

so-called underfeed type. That is, the coal is fed from a bin or hopper by a feed screw into the bottom of a conical retort, through the inner and outer walls of which air from a motor-driven fan is supplied for combustion. Air is usually supplied through a series of tuyeres on the outside of the top of the retort. This air is available for burning the coke that falls onto the hearth that surrounds the retort.

On many stokers designed for burning anthracite, a grate surrounding the retort is moved mechanically to separate the ash from the fuel. The ash is often times removed by mechanically driven conveyors for storage in cans for removal from the premises. With such a conveyor for the ash and with provision for feeding the anthracite from the bin, completely automatic heating is obtained.

The possibility of a fully automatic stoker with anthracite results from the fact that anthracite is completely noncaking. With no agglomeration, the burning bed is uniform, and a uniform response in burning rate to a supply of air is obtained. Because the ash-softening temperatures of anthracite ashes are generally in the high range—2600°F and above—clinkering is not a problem at the burning rates used, and automatic removal of ash is feasible. Bituminous coals, on the other hand, cake, and the fuel bed of a small underfeed stoker contains masses of coke, the size and strength of which depend on the caking tendency of the coal. As a result, the combustion of the coke is often limited to the periphery of large pieces of coke. To obtain the overall rates of heat release required, extremely high rates of burning and, consequently, high temperatures develop locally. The ash, even those with softening temperatures of 2600°F or higher, fuses into a clinker. Because of the clinkering of the ash from bituminous coals, residential stokers marketed for these coals, with essentially only one exception, have been designed for manual removal of the ash as a clinker. Hence, they are not completely automatic and defer rather than eliminate the work of ash removal as compared with hand firing. However, the excellent performance in maintaining uniform comfort under the command of a thermostat has given them an excellent public acceptance.

The one exception to the general rule that residential stokers for bituminous coal operated with manual removal of clinker was a stoker manufactured and marketed successfully for a period of years by a producer and seller of an eastern low-volatile coal. This stoker differs from others for bituminous coal principally in that the coal leaving the central retort falls onto an outer rotating ring grate through which the ashes are shaken for automatic removal to a container. A certain amount of the ash does agglomerate, but, by rotating action of the grate, it is formed into balls resembling potatoes. Dependent on the coal used and the method of operation, manual removal of these small clinkers is necessary only at intervals of a few weeks or even not through an entire heating season. The success of this stoker resulted from careful selection of the coal used on it, from a conservative rating to avoid high rates of burning, and from an excellent mechanical design.

Characteristics of Bituminous Coals Suitable for Residential Stokers

Because the operator of a residential stoker is not expected to make adjustments in operation to suit variable characteristics of coals, it is important, if satisfaction is to be obtained, that coals marketed for these stokers be suitable for them. Hence, a considerable amount of research has been done in an attempt to relate performance of coals on stokers to the characteristics of coal as measured by usual laboratory tests.

Helfenstine and Cady[39] have reported the results of an extended series of burning tests of a large number of coals on a typical residential underfeed stoker in a hot-water boiler. An analysis was made of the useful heat output, the uniformity of the rate of heat release, the responsiveness to heat demands, and the clinkering characteristics in relation to such factors as the heat value of the coal, the ash content, the caking characteristics of the coal as measured by the Gieseler plastometer and the free-swelling index, the vitrain content of the coal, the softening temperature of the ash, and

[39] Helfinstine, R. J., and Cady, G. H., *ASME, Trans.*, **71**, 471–5 (1949). Summarizes four detailed reports: (a) McCabe, L. C., Konzo, S., and Rees, O. W., *Illinois State Geol. Survey, Rept. Invest.* No. 78 (1942), 20 pp. (b) Helfinstine, R. J., and Boley, C. C., *Illinois State Geol. Survey, Rept. Invest.*, No. 120 (1946), 20 pp. (c) Helfinstine, R. J., *Illinois State Geol. Survey, Rept. Invest.*, No. 133 (1948), 47 pp, No. 151 (1951), 46 pp.

the shatter index of the clinker. Most of the coals studied came from the Illinois fields, but some out-of-state coals were used. The authors' conclusions were that the only good correlation obtained was that of the heat output to the heating value of the coal. The correlation of the uniformity of burning with the ash content, with the fluidity by the Gieseler test, and with the vitrain content was fair. Also fair was the correlation of the ratio of ash to clinker formed during the test with the shattering index of the clinker, but that with the ash-softening temperature was poor.

Purdy and Nelson[40] have given the results of similar studies on thirty-three coals, mainly from the eastern coals of high rank. The principal objective of the experimental work was not to correlate the results with the measurable properties of the coal, but to obtain a measure of the suitability of the coal from its performance in the burning test of 5 days' duration with a definite operating schedule. The judgment was based on the ability of the ash to form a suitable clinker, the rate of initial ignition of the coal, the rate of response to heat demands during intermittent operation, and the smoke emission. Of the thirty-three coals reported on, five were determined to be unsatisfactory, eight were marginal, and twenty were satisfactory. Poor correlations of the performance with the characteristics of coal as measured by usual laboratory tests were found in agreement with the results reported by Helfinstine and Cady.[39] The conclusions from the two investigations were that only by a burning test of a coal on a residential underfeed stoker under controlled conditions can the suitability of a coal for this use be established. Fortunately, a burning test of this type is not excessively time-consuming or expensive.

[40] Purdy, J. B., and Nelson, H. W., *ASME, Trans.*, **71**, 477–84 (1949).

The Control of Coke-Tree Formation

Because of the difficulties in obtaining uniform combustion resulting from the large and strong coke formation in small underfeed stokers, numerous attempts have been made to control the formation of coke trees. Various arrangements of tuyeres in the retort, provision of gooseneck arrangements to direct air downward onto the bed, and equipping the stoker with a mechanically operated poker that rises periodically from the bottom of the retort to break up the coke have been tried. The last measure had a certain degree of success, and one manufacturer marketed a stoker with this device.

A method for the control of coking based on a mild oxidation of the coal as it rose in the retort has been described.[41] Figure 11 shows a section of a stoker that illustrates the principle. The coal was fed upward in the retort by a vertical screw. At the top of the retort were the usual main tuyeres, and about half way down the vertical screw were tuyeres for the admission of preoxidizing air. A damper in the duct leading from the blower was adjustable to permit the proportioning of the air to the two sets of tuyeres. The vertical screw agitated the coal in the retort to a certain extent, but not enough to prevent the formation of strong masses of coke. When 15 to 30% of the total air was admitted through the lower tuyeres, the formation of large masses of coke was prevented. When a $\frac{1}{2} \times 0$-in. Central Pennsylvania coal which had

[41] Wright, C. C., and Spicer, T. S., *Trans. Am. Inst. Mining Engrs.*, **149**, 270–82 (1942). Wright, C. C., *Pennsylvania State Coll., Mineral Ind. Exp. Sta., Bull.* No. 42 (1945), pp. 68–78.

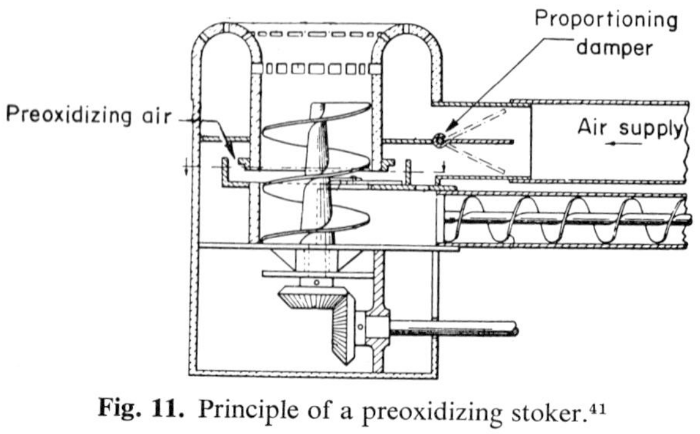

Fig. 11. Principle of a preoxidizing stoker.[41]

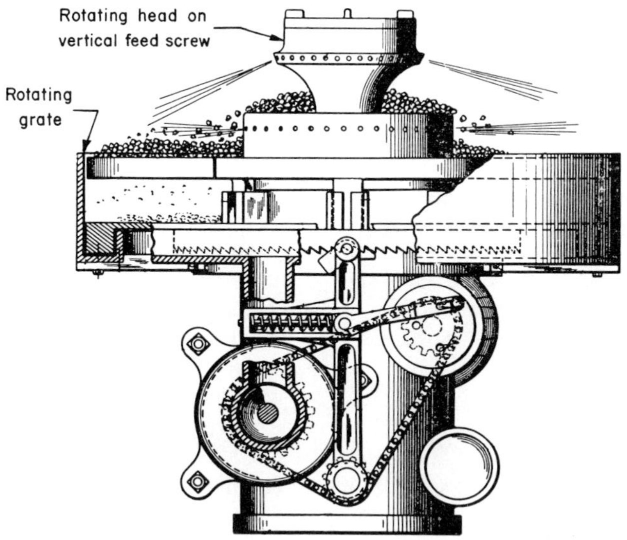

Fig. 12. Principle of a stoker developed for minimizing coke-tree formation.[42]

a free-swelling index of 9 was burned, samples taken at various levels in the retort without admission of peroxidizing air all had an index of 9. However, with the admission of air near the base of the retort, the index of samples had decreased to 1 by the time the coal had risen 2 in. above the level of admission of the air.

Figure 12 shows a section of a design of stoker developed by Sherman[42] in which the size of the pieces of coke was controlled by subjecting the coal, which was in a plastic state as it rose to the top of the vertical retort, to a shearing force of a rotating mushroom-shaped head mounted on the vertical feed screw. Air fed through the hollow vertical screw to tuyeres in the lower edge of the head served a dual purpose of cooling the head and of providing secondary air. This air was highly useful in decreasing smoke produced from combustion of volatile matter that evolved for several minutes when the stoker stopped following a period of heat demand.

Cross-Feed Stokers for Bituminous Coal

Two cross-feed stokers that were originally designed and successfully used with anthracite, one for residential use, the other for commercial use with burning rates of 60 to 1300 lb coal/hr, have been adapted to bituminous coal by Bituminous Coal Research, Inc. On both units, the coal is fed by a screw feeder transversely across the front of a grate with a slight downward incline. The fuel bed is moved across the grate by periodic movement of the grate. The ash and refuse fall off the end of the grate. As would be expected, the caking characteristics of the coal strongly influence the performance of such stokers, and to insure satisfactory performance research has been conducted to develop a laboratory method for the selection of suitable coals. Preliminary results were obtained with a fuel bed about 4-in. thickness and 1-sq ft area,[43] in an arrangement similar in principle but simpler than that described by other investigators.[44]

The coals were ignited at the top of the bed with air supplied from below. Thus, the ignition and initial burning were "underfeed" in principle. Data taken included the rate of ignition, the rates of burning during the underfeed burning period and the subsequent overfeed period, and the maximum swelling of the bed. Coals which performed well on cross-feed stokers had low maximum fluidity in the Gieseler test, a low free-swelling index, and high rates of ignition and of underfeed burning. As was expected coals which were strongly caking and gave difficulties on cross-feed stokers had a high maximum fluidity in the

[42] Sherman, R. A., U.S. Pat. 2,371,191 (1945).

[43] Zawadzki, E. A., and DuBreuil, F., *Mech. Eng.*, **81,** No. 12, 104 (1959).

[44] Carman, E. P., and Reid, W. T., *ASME, Trans.*, **67,** 425–36 (1945). Carman, E. P., Graf, E. G., and Corey, R. C., *U.S. Bur. Mines*, Bull. No. 563 (1957), 92 pp.

Gieseler test, a high free-swelling index, and low rates of ignition in underfeed burning. It is to be hoped that a continuation of the research will make it possible to classify coals in the borderline areas.

Industrial and Utility Utilization of Coal

Table 1 presents data from two annual surveys of industrial and utility power plants[45] which illustrate the changes that have occurred in the methods of burning coal from 1949 to 1959. In the utility plants, the percentage of units fired with pulverized coal decreased somewhat and the cyclone furnace, which was relatively new in 1949, made important gains, accounting for 15% of the installations in 1959 as compared with 2% in 1949. The percentage of spreader stokers remained constant, and the traveling grate and underfeed stoker essentially no longer have a place in this category as the steam generators have continued to increase in size.

TABLE 1

SURVEY OF TYPES OF FIRING COAL[45]

	1949 Percent	1959 Percent
Utility Plants		
Pulverized coal	84	77
Cyclone	2	15
Spreader stoker	8	7
Traveling grate	4	1
Underfeed stoker	2	—
Total	100	100
Industrial Over 200 psi		
Pulverized coal	41	42
Cyclone	—	4
Spreader stoker	43	31
Traveling grate	11	19
Underfeed stoker	5	4
Total	100	100
Industrial Under 200 psi		
Pulverized coal	5	4
Cyclone	—	—
Vibrating grate	—	11
Spreader stoker	41	52
Traveling grate	12	7
Underfeed stoker	42	26
Total	100	100

Among the industrial boilers of over 200 psi having a wide range of capacities, pulverized coal retained an almost constant percentage of installations. The cyclone furnace entered this field among some of the largest boilers. The spreader stoker apparently lost some ground in this category, whereas the traveling grate gained and the underfeed stoker held constant. This change in the position of the spreader stoker may have been only temporary because of an unusually large number of large boilers which this stoker could not serve well.

In the group of units operating at pressures less than 200 psi, large boilers are not numerous, and the percentage of pulverized-coal installations was sharply lower than among the larger installations both in 1949 and 1959. A new type of unit, the vibrating grate appeared in this class. The spreader stoker continued to gain in percentage at the expense of both the traveling grate and underfeed stoker, but the latter still accounts for a quarter of the units installed in 1959.

Down-Jet Combustion

The analytical and experimental research that led to the development of the system of down-jet combustion was carried out at the laboratories of the British Coal Utilisation Research Association during World War II. Since it was closely related to research aimed at the development of a gas producer for road vehicles, the research was "classified" and reports were published only during 1945 and later. Thring[46] has presented the basis for the conception of the down-jet burner. When air is blown into a fuel-bed of pieces of carbon of uniform size, experimental work has shown that, if the bed could be maintained at a thickness of three particles with uniform packing, the gases leaving the top of the bed would consist of CO and O_2 in the proper proportion to give only CO_2 and N_2 and no excess O_2. It is impossible in actual fuel beds, particularly

[45] Skrotzki, B. G., *Power*, **93,** No. 9, 82–96 (1949). Follansbee, J. W., *Power*, **103,** No. 10, 114–18 (1959).

[46] Thring, M. W., *Trans. Faraday Soc.*, **42,** 366–75 (1945).

with thin beds, to maintain such uniform conditions so the gases leaving the top of the fuel bed have either a deficiency of O_2 or an excess. Both conditions may be present in various parts of the bed, and provisions for mixing must be made to avoid losses in unburned CO.

Figure 13 shows the principle of the arrangement proposed by Thring[46] to obtain an approach to the ideal conditions of complete combustion of the CO with no excess O_2. The air was supplied through a jet that impinged on the surface of the bed of fuel. The sides and bottom of the bed of fuel were enclosed, with the result that the gaseous products of reaction of the oxygen of the air with the fuel were forced to leave the bed through its upper surface. The desired depth of penetration of the air into the bed was dependent on the size of the pieces of fuel used, the angle of the jet with the surface of the bed, the area of the nozzle opening, and the distance of the jet above the bed. A variable-aperture type of nozzle was developed; with this nozzle the depth of penetration could be maintained constant over a range of total air delivery.

Thring reported that in an early experimental model using coke of $\frac{1}{2}$ by 1 in. in a refractory-lined combustion chamber with an air rate of 60 cfm, temperatures of about 3000°F were measured at 3 to 4 in. below the surface of the bed on the axis of the jet of air. At such temperatures, the ash of the coke was quite fluid and flowed down among the pieces of fuel. Because the temperature decreased quite sharply in the lower part of the bed, the fluid slag formed a clinker.

Down-jet burning of coke. Figure 14 is a sectional view of the fuel hopper and combustion chamber of a down-jet coke burner.[47]

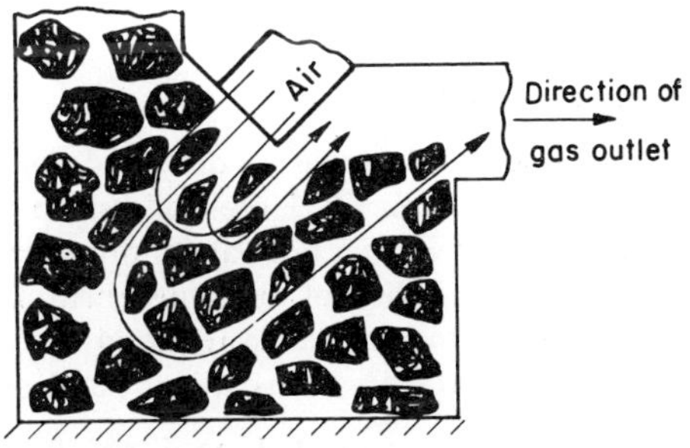

Fig. 13. Principle of a down-jet firing proposed by Thring.[46]

[47] Ross, F. F., and Sharp, G. C. H., *J. Inst. Fuel*, **23,** 24–6 (1950).

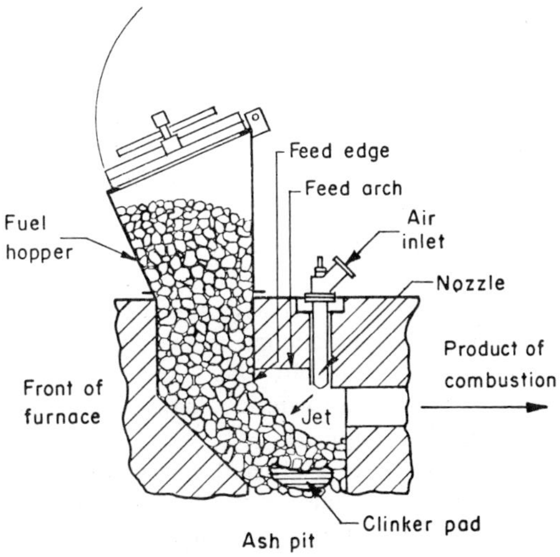

Fig. 14. Cross-section of a downjet coke burner.[47]

The pad of clinker formed is shown in the lower part of the bed. Some of the design parameters are:

1. Maximum burning rate, 50 lb/hr-ft of effective width-in. of particle diameter.
2. Air pressure, 3 to 14 in. water gauge.
3. Jet velocity, 75 to 225 ft/sec dependent on exit gas composition desired.
4. Maximum burning rate, 75 lb/hr-sq ft of active bed surface.
5. Size of coke, a mixture of $1\frac{1}{2}$-in. coke with as much as 33% of coke breeze was burned satisfactorily.
6. Size of exit throat, such as to give a gas velocity of not less than 50 ft/sec.

The down-jet burner was not considered suitable for domestic appliances since the minimum width of burner for $\frac{3}{4}$-in. coke would be 8 in., and the burning rate of 15 lb/hr would require an air pressure of 12-in. water gauge. Suggested uses were for refractory firing, the crucible melting of steel, and other metallurgical applications.

Sharpe[48] has described the application of the down-jet principle to various types of industrial uses involving both high and low temperatures and a small package stream generator in which coke was fired by the down-jet system has also been described.[49] A considerable number of boilers designed for coke

[48] Sharpe, G. C. H., *J. Inst. Fuel*, **23,** 27–31 (1950).

[49] Karthauser, F. B., and Sharpe, G. C. H., *J. Inst. Fuel*, **23,** 24–6 (1950).

burning were installed in the United Kingdom but a recent communication[50] states that they were unsuccessful because of the behavior of the ash. Mechanical clinker removal was unsuccessful because of the extremely high temperatures in the fuel bed. Manual removal was satisfactory when the boiler was operating at about half capacity and, if the boiler was operating at full capacity, it was necessary to reduce the load for a half hour to allow the clinker to harden enough to permit removal. Furthermore, difficulty was had with fireside corrosion of the water-cooled walls of the combustion chamber.

Down-jet burning of coal. Successful burning of bituminous coal by the down-jet principle with smokeless combustion has been reported when lower air pressures were used.[46] Difficulty with regular feeding of the coal on a sloping hearth required a poking mechanism.[51] Some success was had with feeding coal upward toward the jet nozzle through the retort of an underfeed stoker.

Frazier and Engdahl[52] reported the results of a rather comprehensive investigation of the application of the down-jet principle to burning bituminous coal. Trials were made of various methods of feeding the coal to the jet by means of a screw, onto a downward sloping hearth, an upward sloping hearth, into a cone having a horizontal axis, and into a cone having a vertical axis. A relatively weakly caking coal of the Indiana No. 3 seam, $\frac{3}{4}$-in. by 10 mesh was used. The results, in brief, were that no truly satisfactory performance was obtained, largely because of the caking characteristics of the coal. If conditions of furnace temperature and distribution of air were such as to avoid caking, excessive amounts of fuel were blown from the bed. With strongly caking coals, the jet of air did not penetrate the bed. These disadvantages outweighed the advantages of smokeless burning and the flexibility in the rate of burning.

Bagby and Erickson[53] have reported better success in the application of the down-jet principle to the use of bituminous coal on a much smaller scale, that of a domestic cooking range. Firing the coal by hand in small increments at intervals of about 30 min afforded an opportunity to break up a caked fuel bed. Jets directed into the issuing stream of gases to provide secondary air were required for smokeless operation. Using an Illinois No. 6 Seam coal of $1\frac{1}{2} \times 2$-in. size, excellent operation was obtained with favorably low time for boiling water on the stove and with ease of removal of the ash as a clinker which was manually removed through the firing door.

Down-jet burning of peat. Miller[54] has successfully applied the down-jet burner to the combustion of Irish peat for firing small commercial and industrial boilers. The peat used was that known as sod peat, that is, a peat mechanically harvested from the bog but dried by sun and wind to a moisture content of 30 to 40%. No difficulty was encountered in feeding the completely noncoking fuel by gravity from a hopper over an inclined hearth. The sods were broken to give pieces of which 60% passed a 2-in. screen and 15% passed a $\frac{1}{2}$-in. screen. The angle of repose of the fuel was 40 to 50°. The arrangement of hearth, air jet, and exit throat was similar to that shown by Thring.[46] Burners have been built and are in use which have outputs of 100,000 to 9,000,000 Btu/hr. For this range of sizes, the fuel-bed area was 0.44 to 6.25 sq ft, the air velocity at the jet was 3350 to 7200 ft/min, and the rate of heat release was 220,000 to 1,440,000 Btu/sq ft-hr. Ash was removed manually at intervals as pads of clinker through a side door on the burner. Figure 15 shows the percentage CO_2 maintained in the products of combustion over a two-hour period. Some 160 of these units have been applied to boilers in Ireland with continued successful operation.

Underfeed Stokers

The underfeed stoker is not used for the firing of the extremely large boilers, up to 1 million lb of steam per hour and more, because of the impossibility of building them large enough to burn coal at the required rate. In the intermediate sizes, this stoker has lost its

[50] Private communication from Joshua Bigwood and Son, Ltd., Wolverhampton, September 30, 1959.

[51] Mayorcas, R., and Thring, M. W., *Coal Research, BCURA*, 131–8, December (1945).

[52] Frazier, P. M., and Engdahl, R. B., Battelle Memorial Institute Report No. 274 to Bituminous Coal Research, Inc., December, 1949, 28 pp.

[53] Bagby, F. L., and Erickson, E. A., Battelle Memorial Institute Report No. 200 to Bituminous Coal Research, Inc., July, 1947, 24 pp.

[54] Miller, H. M. S., *International Peat Symposium, Dublin*, July, (1954).

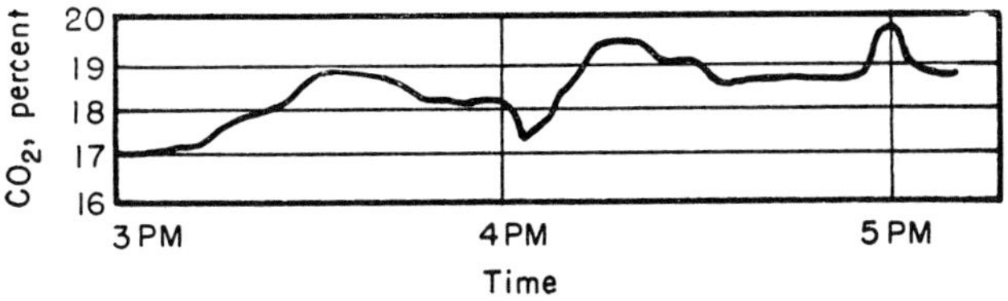

Fig. 15. CO_2 content of fuel gases over 2-hr period when burning Irish peat on a down-jet stoker.[54]

formerly favored position because of its sensitivity to the caking and ash fusion characteristics of coal. Where an assumed long-time supply of coal of the proper characteristics is available, the underfeed stoker is still frequently chosen for these intermediate-sized units and, as the data show for the smaller boilers, the single-retort type of underfeed stoker has a wide acceptance which will probably continue.

Neither research nor development on the underfeed stoker has been extensive in recent years. Investigations have been made of the crushing and segregation[55] in feeding coal to the fuel bed and to methods for making the stoker move adaptable to a wider range of coals. At the Pennsylvania State University,[56] an experimental investigation in the laboratory, and trials of several years on an operating stoker of a capacity of 500 to 1000 lb/hr, proved that a refractory arch suspended on water-cooled tubes at 1 to 2 ft above the grate improved the performance of the unit on a variety of coals. Beneficial effects included less manual attention to the fuel bed, higher boiler efficiency, and a lesser emission of smoke and fly ash.

Chain and Traveling-Grate Stokers

In the majority of furnaces recently installed with chain or traveling grates, the coal is being fired by a spreader. However, on a considerable number of these moving grates the coal is fed from a hopper under an adjustable guillotine-type gate, which controls the thickness of the fuel bed. This latter type of equipment continues to be eminently suitable for those plants which can be assured of a long-time supply of a suitable coal and which do not require rapid changes of load. The problems of smoke at low loads and that of carry-over of fly ash are much less acute than they are with spreader feeding of coal.

Although the furnaces in which traveling-grate stokers are installed may be of an open type with neither front nor rear arch over the stoker, at least a rudimentary front arch and often a long rear arch are still frequently used. Over-fire air jets to insure mixing of the gases are almost universally installed. By the use of properly installed jets at the front wall, a mass of burning and recirculating gas at high temperature can be maintained above the incoming fuel equivalent in maintaining ignition to the front arch formerly used.[57]

Free-burning fuels such as coke, anthracite, mid-west bituminous coal, and lignites are considered to be the most suitable for the traveling-grate stoker. MacLachlan,[58] from a survey of plants using Pittsburgh Bed coal on traveling grates, found that the plants that used coal from strip mines with shallow cover had the best results, although some plants were successfully using coal from deep mines. Solari[59] found that a reduction of spacing between links of a stoker gave such improved distribution of the air to the fuel bed that coals having a free-swelling index of 9 could be successfully burned.

Tempering of Coal. Tempering of coal, that is, the addition of water to coal, has long been advocated and practiced with traveling grates. Reduction in difficulty from coke formation with strongly coking coals,[60] reduction in ash-pit loss, and other advantages have been claimed, but numerical data relating the addition of moisture to the effects have been

[55] Baumann, H. A., Spicer, T. S., and Wright, C. C., *Trans. Am. Inst. Mining Engrs.*, **187**, 266–74 (1950).

[56] Spicer, T. S., Grace, R. J., and Wright, C. C., *Combustion*, **23**, No. 9, 48–54 (1952).

[57] De Lorenzi, O., *Combustion Engineering*, Combustion Engineering Co., Inc., New York, (1952), Sect. 6, p. 21. Young, N. W., *Mech. Eng.*, **80**, No. 3, 109 (1958).

[58] MacLachlan, J. M., *Mech. Eng.*, **77**, No. 1, 47 (1955).

[59] Solari, J. J., *J. Inst. Fuel*, **26**, 243–9 (1953).

[60] Carman, E. P., and Reid, W. T., *U.S. Bur. Mines, Bull.* No. 563 (1957), 92 pp.

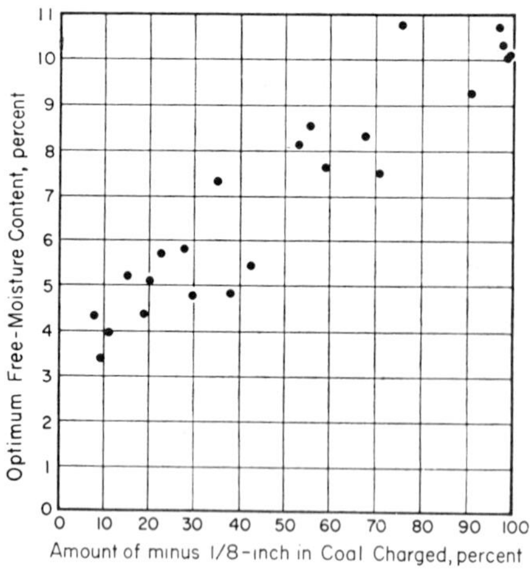

Fig. 16. Relation of optimum free-moisture content of three British coals to amount of fines, $-\frac{1}{8}$ in., in coal when burned on a traveling-grate stoker.[61]

lacking. McDonald and Murray[61] have obtained data on the effect of the free-moisture content of three British coals when burned on a traveling grate in a shell boiler. They related the amount of free moisture to the optimum CO_2, to the loss in combustible to the ash pit, and to the boiler efficiency. They concluded that there was an optimum free moisture dependent on the amount of $-\frac{1}{8}$-in. size in the coal burned. Figure 16 shows the relation that was established. A considerable scatter is evident, but this is to be expected in view of other factors that can affect the performance of the coal. McDonald and Murray concluded, as have many others, that the effect of moisture additions is physical rather than chemical. The free moisture permits the agglomeration of the fines with the larger particles and results in more uniform distribution of the air to the incoming fuel and more uniform ignition.

[61] McDonald, E. J., and Murray, M. V., *J. Inst. Fuel*, **28**, 479–98 (1955).

Vibrating-Grate Stoker

Figure 17 is a photograph of a stoker which can be considered as a variation of the traveling-grate type. The metallic belt of the latter is here replaced by a flat inclined grate.

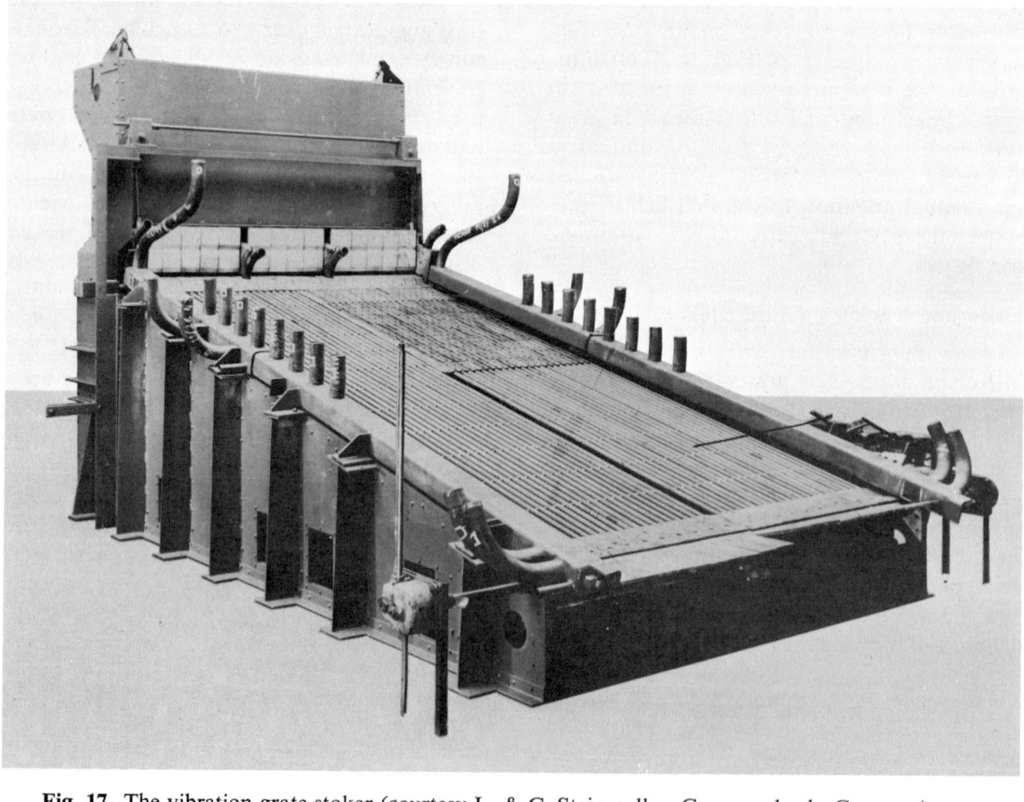

Fig. 17. The vibration-grate stoker (courtesy L. & C. Steinmuller, Gummersbach, Germany).

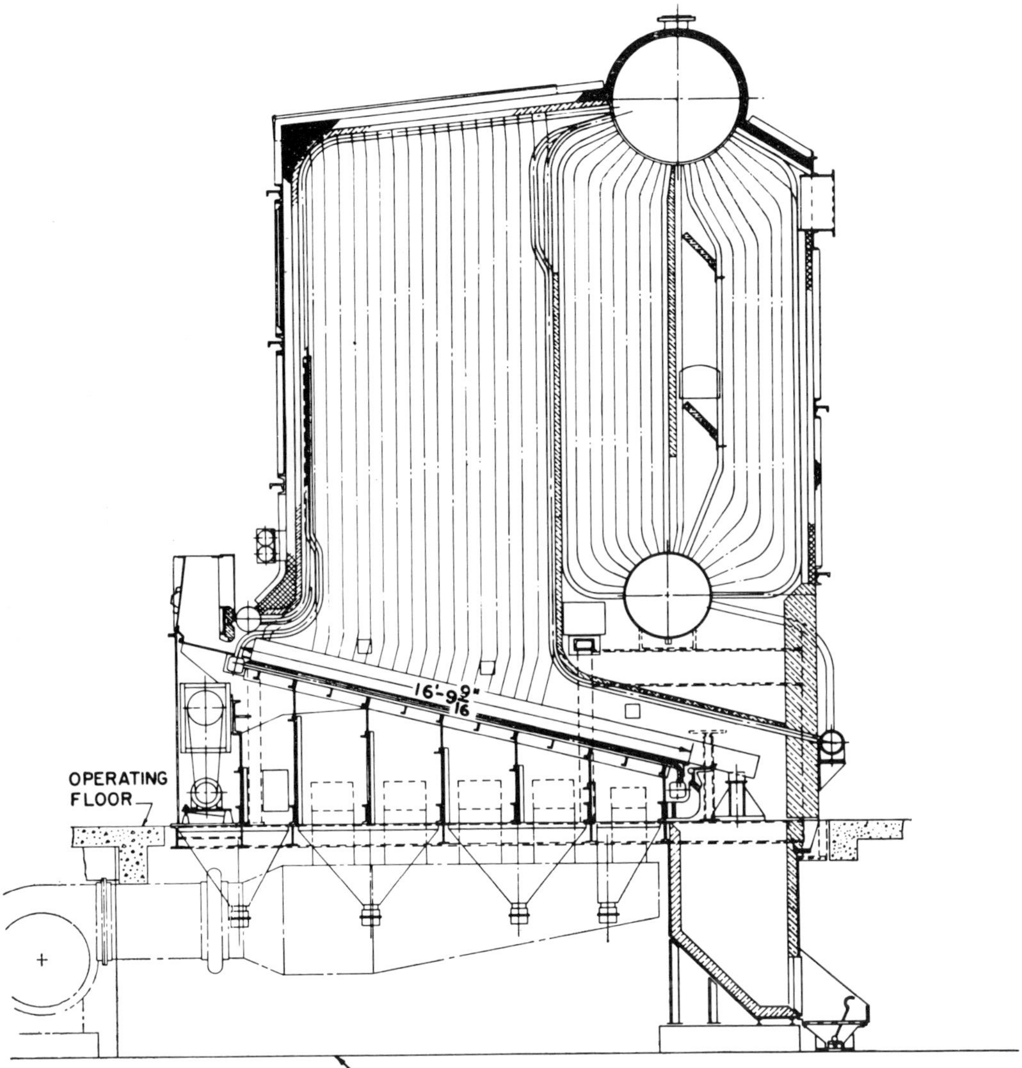

Fig. 18. A United States-built vibrating-grate stoker in a boiler having 60,000 lb/hr steam capacity. (American Engineering Co.)

The feeding of the coal onto the grate from the hopper, the movement of the bed down the grate, and the discharge of the refuse are effected by a periodic vibration of the entire grate forward and backward in an essentially horizontal plane. The grate is made up of a series of parallel boiler tubes welded to header boxes at either end which, in turn, are connected into the boiler circulation. The surface that supports the fuel bed is made up of a series of iron castings firmly wedged between the tubes with a cement between the castings and the tubes to improve the thermal conductivity. Air enters the bed through narrow slots between adjacent castings. The entire grate assembly is supported on several vertical steel plates that divide the air plenum chamber into zones. These vertical supports are flexible to permit the vibration of the grate assembly by means of an arrangement of unbalanced weights rotated by a motor. The frequency of vibration is about 600 cycles/min, thereby avoiding operation at the resonant frequency of any part of the installation. The normal method of operation is to start the vibration of the stoker every 2 min and to operate it for a period of a few seconds; the length of the period of operation is governed automatically by the load on the boiler.

The vibration not only feeds coal out of the hopper onto the grate and discharges refuse to the ash pit but, by the lifting action on the bed and subsequent compacting, is said to maintain a uniform density of the bed and thus uniform burning. Overfire air is usually provided through jets in the front wall. In the event that carry-over of ash and cinders is high, because of the agitation of the bed during vibration, provision can be made to throttle the undergrate-air supply during vibration. The German firm that developed the stoker has installed a considerable number of them in Germany and other countries. Coals of 19 to 30% volatile matter are burned in the German installations at rates of up to 40 lb/sq ft-hr. The capacity of the boilers have ranged from 20,000 to 50,000 lb steam/hr.[62]

Figure 18 is a sectional elevation of a United States version of the vibrating stoker in a boiler having a capacity of 60,000 lb of steam per hour. Within the past several years, a number of these stokers have been installed in this country under boilers of capacities of 30,000 to 60,000 lb steam per hour. They have successfully burned coals of 17 to 40% volatile content, of 3 to 14% ash content, with free-swelling indexes of 1.5 to 9, and with ash-softening temperatures of 2040 to 2700°F. The rate of firing is usually limited to 30 to 35 lb of coal/sq ft-hr.

Figure 19 presents data[63] on the fly-ash carry-over at the boiler outlet on an installation burning a low-volatile coal, of which 80% was less than $\frac{1}{8}$-in. in size. Only at the maximum firing rate did the carry-over of solids attain the value of 0.85 lb per 1000 lb of gas, the maximum allowable under many air-pollution-control ordinances.

Figure 20 shows the construction of a vibrating grate designed for application to a spreader stoker as described by Mignacca.[64] Its function is thus only to move the bed and discharge the ash, but not to feed the coal. The grate is flatter than those previously described and is not water cooled. It is also driven for a few seconds at invervals of up to 10 min by weights eccentrically mounted on a shaft supported by bearings on the stoker frame.

Figure 21 shows a side sectional elevation of a packaged steam generator of a capacity of 2500 lb steam/hr which Bituminous Coal

[62] Personal communication from Eduard Weissgerber, Director, L. & C. Steinmüller, Gummersbach, September 30, 1959.

[63] Personal communication from F. C. Messaros, American Engineering Co., October 9, 1959.

[64] Mignacca, J. K. L., *Power Eng.*, **62**, No. 6, 76–8 (1958).

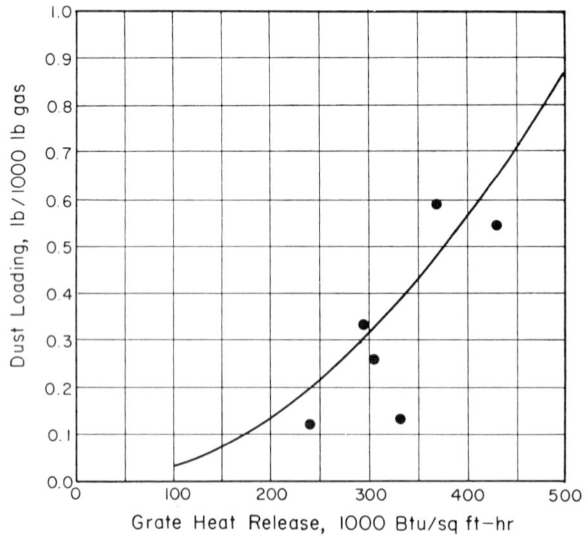

Fig. 19. Relation between fly-ash carry-over and hourly rate of heat release per square foot of grate on a vibrating stoker.[63]

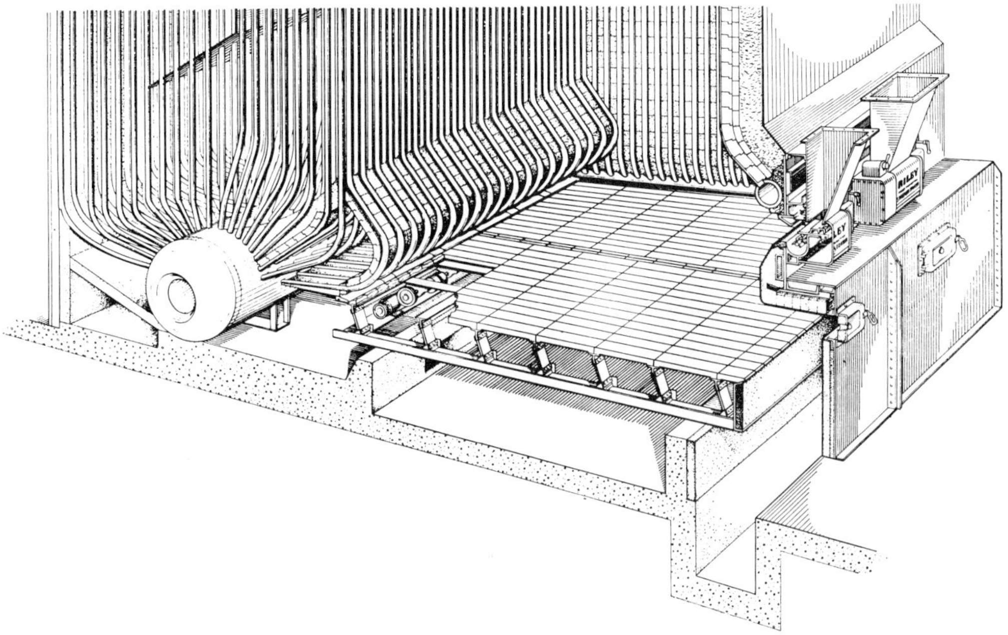

Fig. 20. Oscillating-grate stoker for spreader-stoker firing.[64] (Riley Stoker Company.)

Research, Inc.,[65] has designed to be fired with a vibrating stoker. The drive of this stoker is by a cam and follower to give an amplitude of $\frac{5}{32}$-in. at 396 vibrations per minute. The rate of travel of the fuel bed is 1 in./2 sec of operation of the stoker. The maximum burning rate used is 24 lb/sq ft-hr. Coals with free-swelling indices of 3 to 9 and with ash of softening temperatures of 2000 to 2700°F have been burned successfully. Strongly caking coals gave no difficulty under continuous operation, but caking under long periods of banking retarded the rate of pickup when there was call for steam.

[65] Kock, P. O., *Mech. Eng.*, **80,** No. 5, 108 (1958).

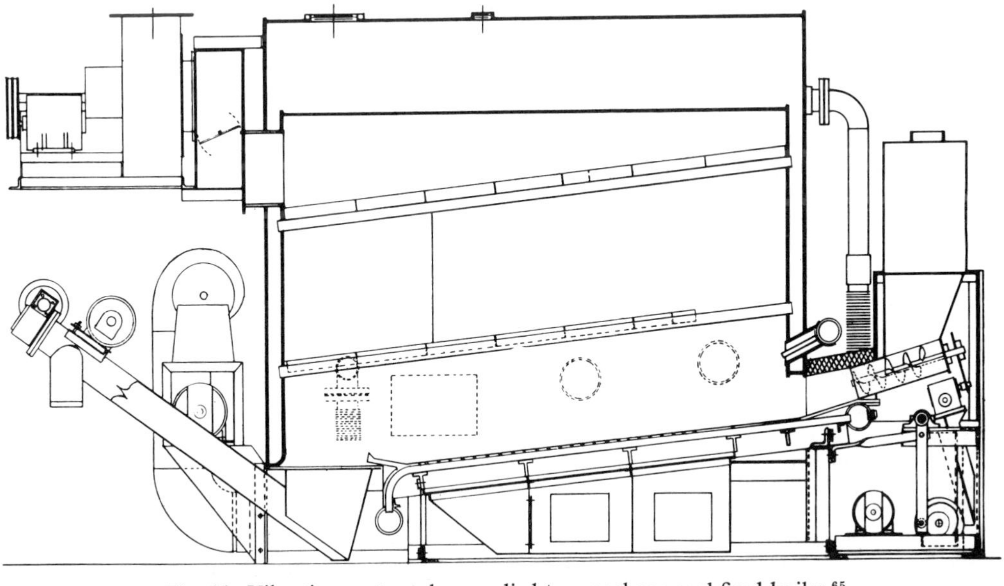

Fig. 21. Vibrating-grate stoker applied to a package coal-fired boiler.[65]

Spreader Stokers

The advantages of the spreader stoker—adaptability to a wide range of types and sizes of coals, ability to respond quickly to change of load, and relative freedom from slag and deposit problems in the furnaces or on heating surfaces—have contributed to its wide acceptance for firing boilers up to an output capacity of 150,000 to 200,000 lb steam/hr. The disadvantages of this type of stoker, the tendency to excessive smoke emission at low loads, and a high carry-over of fly ash and cinder at high loads, have been minimized by overfire jets for increased turbulence in the furnace to reduce smoke and by the use of dust collectors to reduce the emission of fly ash and cinder from the stack. A further advantage of the spreader stoker is its comparatively low auxiliary power requirements. Miller and Hawley[66] have cited an example of a boiler with a capacity of 150,000 lb steam/hr which would require 49 hp for operation of a spreader stoker and 426 hp for pulverized-coal firing.

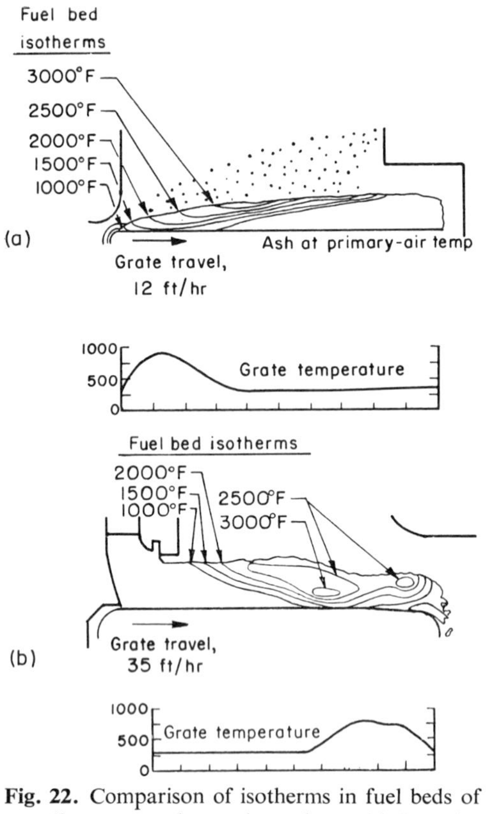

Fig. 22. Comparison of isotherms in fuel beds of traveling-grate and spreader stokers. (*a*) Spreader stoker, 70 lb/sq ft-hr; (*b*) traveling-grate stoker, 65 lb/sq ft-hr.[67]

Figure 22 shows data for coal burned at similar rates on spreader and traveling-grate stokers.[67] The fact that the areas enclosed by the 2500 and 3000°F isotherms are smaller in the spreader bed than in the traveling-grate bed explains why the tendency to slag and clinker is less with the spreader stoker than it is with the traveling-grate stoker. The authors stated that the traveling-grate stoker invariably had higher maximum fuel-bed temperatures than did the spreader. In the traveling-grate bed, the maximum temperatures were 2750 to 3150°F and in the spreader bed, 2400 to 2900°F. Despite the adaptability of the spreader stoker to a wide range of coals, a method was devised for pilot-scale testing of fuels for the spreader stoker.[68] It consisted of spreading the fuel by a rotating nozzle onto an ignited layer of charcoal on a grate in a small "pot-type" combustion chamber. By measurement of the temperatures within the bed of fuel as it built up, the depth of the equilibrium bed and the time required to burn-out the bed were determined. Thus, the design of the grate and its speed of movement could be predicted for various coals. An estimate of the amount of carry-over could also be made. The work reported covered a wide range of rank of fuels from peat to bituminous coals with free-swelling indexes of 9. Of the latter, only one was found to give difficulty because of caking.

Meissner and Funk,[69] in an attempt to measure the fraction of coal burned in suspension in spreader-stoker firing, weighed the carry-over from the furnace that was collected in the boiler passes or stack, and assumed that all the ash in this carry-over came from coal burned in suspension. They presented curves, without data points, that indicated that the amount of coal burned in suspension was 12 to 20% of that fired when burned at a rate of 15 lb/sq ft-hr and 25 to 45% when burned at 40 lb/sq ft-hr. For coal with the same top size, the amount burned in suspension was greater for high-volatile Indiana or Ohio coal than for

[66] Miller, E. C., and Hawley, C. F., *Mech. Eng.*, **79**, 23–26 (1957).

[67] Marskell, W. G., and Pratt, C. W., *J. Inst. Fuel*, **25**, 212–21 (1952).

[68] Marskell, W. G., Miller, J. M., and Joyce, W. I., *Fuel*, **31**, 91–109 (1952).

[69] Meissner, H. G., and Funk, M. O., *Combustion*, **16**, No. 3, 42–6 (1944).

a low-volatile Pocohantas coal. For coal from the same bed, the amount burned in suspension was greater the smaller the top size of the coal.

Major[70] quoted the American Boiler and Affiliated Industries Association specification of the preferred size of coal for spreaders as follows: Eastern bituminous, friable $1\frac{1}{4}$-in. nut and slack, less than 50% $-\frac{1}{4}$ in.; Eastern bituminous, nonfriable, and Midwest bituminous, $\frac{3}{4}$-in. nut and slack, less than 50% $-\frac{1}{4}$ in.

Major also stated that maximum rates of burning on the grate should be as shown in Table 2.

Turbulence with overfire air or steam jets, preheated overfire air, and good distribution of coal have been proposed as remedies for low-load smoke.[71] Holton and Engdahl[72] agreed, but they found experimentally that,

TABLE 2

RECOMMENDED BURNING RATES ON DIFFERENT TYPES OF GRATES[70]

Type Grate	Btu/hr-sq ft Max. Continuous	2-Hr Peak
Stationary	400,000	450,000
Dumping	475,000	525,000
Traveling	575,000 to 700,000	650,000 to 700,000

even with jets, smoke often was undesirably high at light loads, and they also found no relation between the size of the coal and the smoke density.

Miller[73] has presented the following data on the amount of fly ash caught in a collector of a spreader-stoker-fired boiler and its combustible content for coals of different rank (Table 3).

Because of the loss of heat in the unburned combustible, reinjection into the furnace of the fly ash collected in boiler hoppers and in mechanical collectors has become increasingly prevalent. Gausmann[74] obtained a reduction of 2.7% in carbon loss by reinjection of the fly ash on a spreader stoker burning Indiana coal. Holton and Engdhal[72] obtained a decrease of 4 to 6 percentage points, that is, from losses of 6 to 8% to those of 2%, by reinjection of the dust from the boiler hoppers and from the dust collector on a stoker burning Pittsburgh Bed coal at rates up to 600,000 Btu/sq ft-hr. Unfortunately, the reinjection of fly ash.

TABLE 3

FLY ASH FROM A SPREADER-STOKER-FIRED BOILER[73]

Coal	Burning Rate Btu/sq ft/hr	Fly Ash, Fraction of Coal Fired, %	Combustible in Fly Ash, %
Eastern low-volatile	420,000	3.6	62
Eastern high-volatile	405,000	1.8	46
Southern Illinois	438,000	3.6	37
North Dakota Lignite	386,000	1.5	21

although highly desirable from the standpoint of reduction of loss in combustible, has the disadvantage that the loading of dust in the gases is greatly increased. Holton and Engdahl reported an increase, by reinjection, of up to 160% in the dust loading at the stack after the collector.

Results have also been reported for an experimental investigation on a boiler of 100,000 lb steam/hr fired by a spreader stoker with front discharge of ash and reinjection of fly ash which proved that, with a high-performance collector, such a stoker can meet air-pollution-control ordinances.[75] When firing a 0 × 2-in. washed nut and slack with 17% excess air, the loss of combustible in the fly ash was decreased from 5.1% without reinjection to 0.2% with reinjection. Reinjection increased the dust loading at the boiler outlet by 75% but the dust loading after the collector was only 0.3 lb/1000 lb of gas.

Pulverized Coal

Pulverized-coal firing attained its prominence in the utility and industrial field because of its adaptability to boilers of high steaming

[70] Major, W. S., *Combustion*, **21**, No. 1, 41–2 (1949).

[71] Messaros, F. C., Wagner, H. L., Miller, E. C., Mosshart, D. J., and Funk, M. D., *Mech. Eng*., **74**, 453–64 (1952).

[72] Holton, W. C., and Engdahl, R. B., *ASME, Trans*., **74**, 207–15 (1952).

[73] Miller, E. C., *Proc. 27th Appalachian Coals Fuel Eng. Conf.*, 1949, pp. 34–5.

[74] Gausmann, R. W., *Power*, **95**, No. 1, 78–81 (1951).

[75] Morrow, C. H., Holton, W. C., and Wagner, H. L., *ASME, Trans*., **75**, 1863–70 (1953).

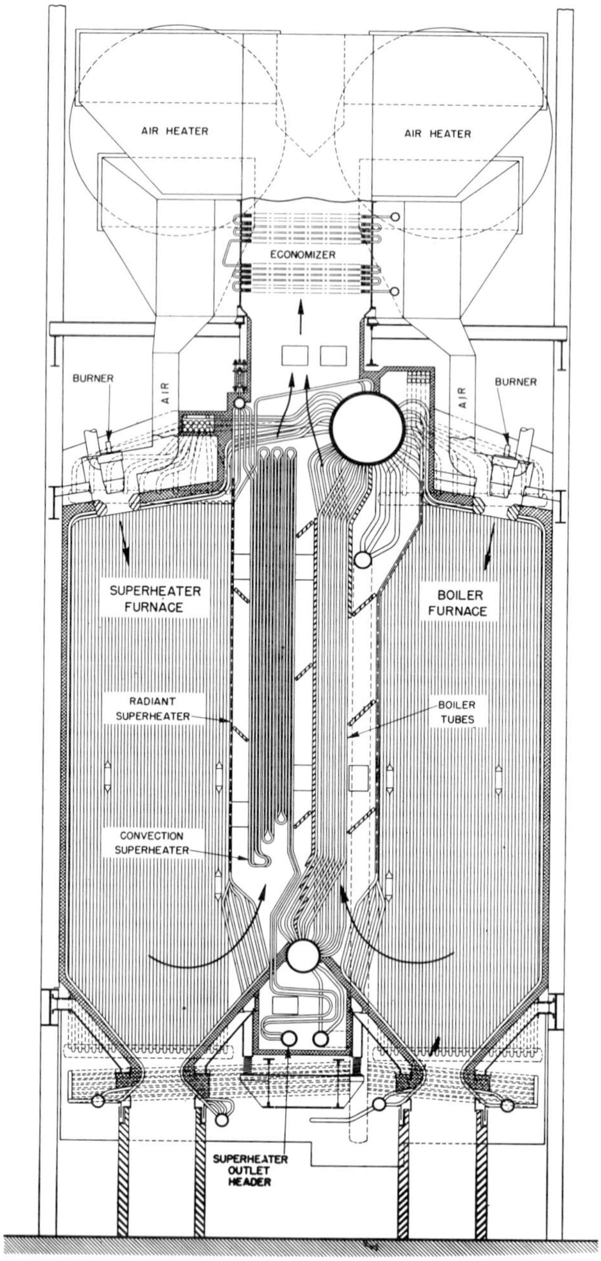

Fig. 23. Pulverized-coal burners mounted in roofs of twin-furnace boiler.[77] (Foster-Wheeler Co.)

capacity that were required by the increase in the size of turbogenerators. Boilers with an output of 1 to 2 million pounds of steam per hour are now numerous, and one with a capacity of 5 million pounds/hour is being installed. The limit, insofar as the burning of pulverized coal is concerned, may not as yet have been reached. So long as the combustion space can be increased and space can be provided for pulverizers and burners, satisfactory operation and efficient combustion appear to be assured.

The rate of heat release in modern pulverized-coal-fired steam generators is usually of the order of 20,000 Btu per hour and cubic foot of furnace volume. This rate may be higher, of course, but is often lower. With a firing rate of 20,000 Btu/hr-cu-ft, an excess of air of 20%, and a mean furnace temperature of

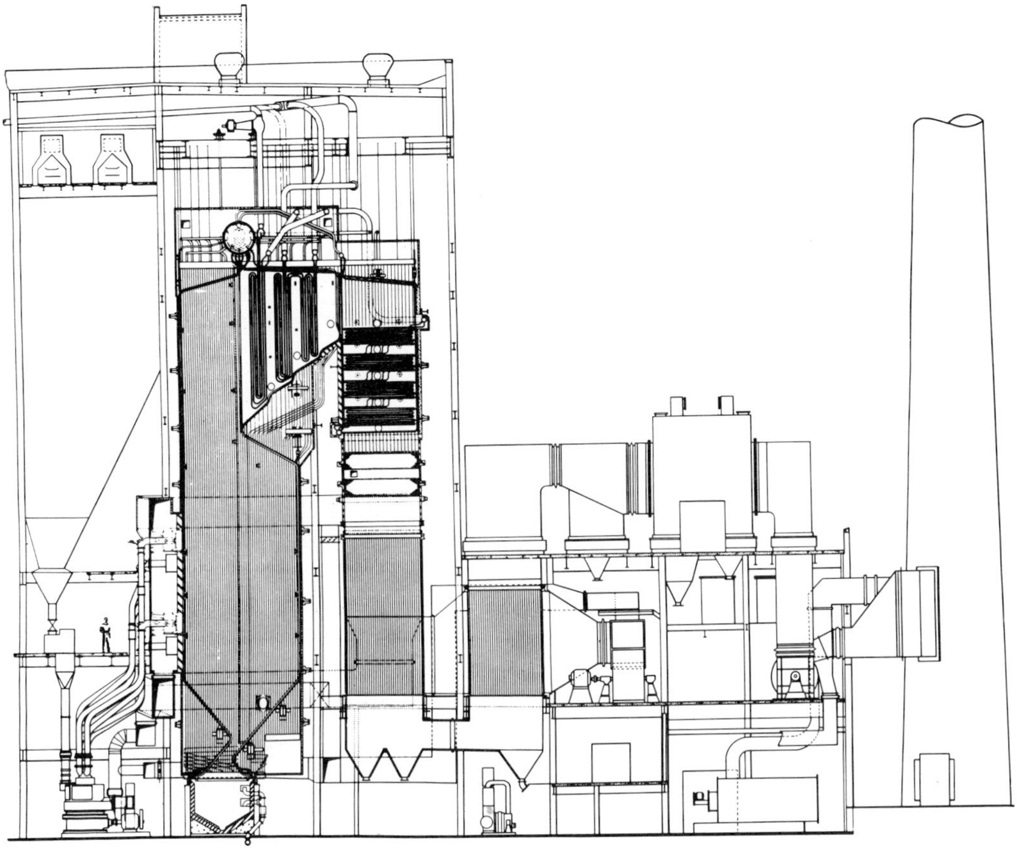

Fig. 24. Circular pulverized-coal burners firing horizontally.[78] (Babcock and Wilcox Co.)

3000°F, the residence time of a particle of coal in the furnace is about 2 sec. Inasmuch as the work of Sherman[76] and others has shown that the burning time required for bituminous coals pulverized to the usual degree of fineness, 70 to 80% through 200 mesh, is $\frac{1}{4}$ to $\frac{1}{3}$ sec, to reduce the unburned carbon to less than 1%, it is obvious that there should be no difficulty in obtaining efficient combustion in present practice.

Furnace and Burner Design. The relative ease in obtaining efficient combustion of coal with pulverized firing permits a considerable degree of flexibility in arrangement of burners in the furnace and in the design of the burners. Figures 23 to 27 show some of the arrangements that are in current use. Figure 23[77] is a sectional elevation of a twin furnace boiler in which the steam generating surface is in the right furnace and steam reheating surface is in the left furnace. The coal is fired vertically downward through burners arranged between the tubes forming the roof of the furnace. Figure 24[78] shows the application of circular burners firing horizontally across the furnace of a large reheat boiler. Figure 25[78] illustrates the firing of the coal through a series of intertube burners in the slightly inclined arch downward onto the hearth of the primary furnace from which the ash is removed continuously as molten slag. The studded wall tubes of the primary furnace are covered with refractory to reduce the rate of heat absorption and thus to insure temperatures favorable for melting the ash.

Figure 26[79] is a sectional elevation of another boiler with a slagging bottom in which three circular burners are placed in each of the front

[76] Sherman, R. A., *ASME, Trans.*, **53,** 401–10 (1934).

[77] Hemenway, H. H., *ASME, Trans.*, **74,** 537 (1952).

[78] Loughin, P. R., and Poor, H. H., *ASME, Trans.*, **74,** 529 (1952).

[79] Feeley, F. G., *ASME, Trans.*, **78,** 1747–55 (1956).

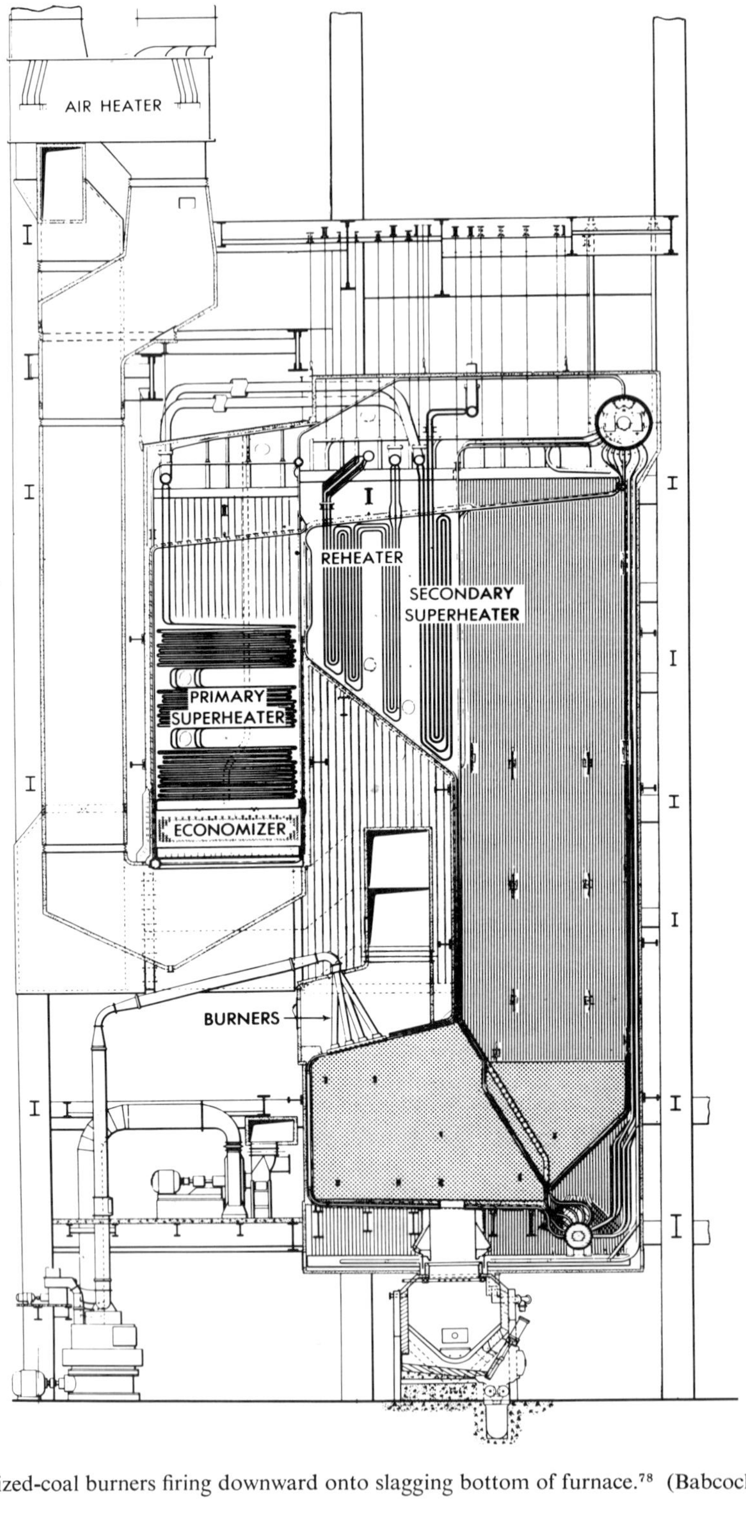

Fig. 25. Pulverized-coal burners firing downward onto slagging bottom of furnace.[78] (Babcock and Wilcox Co.)

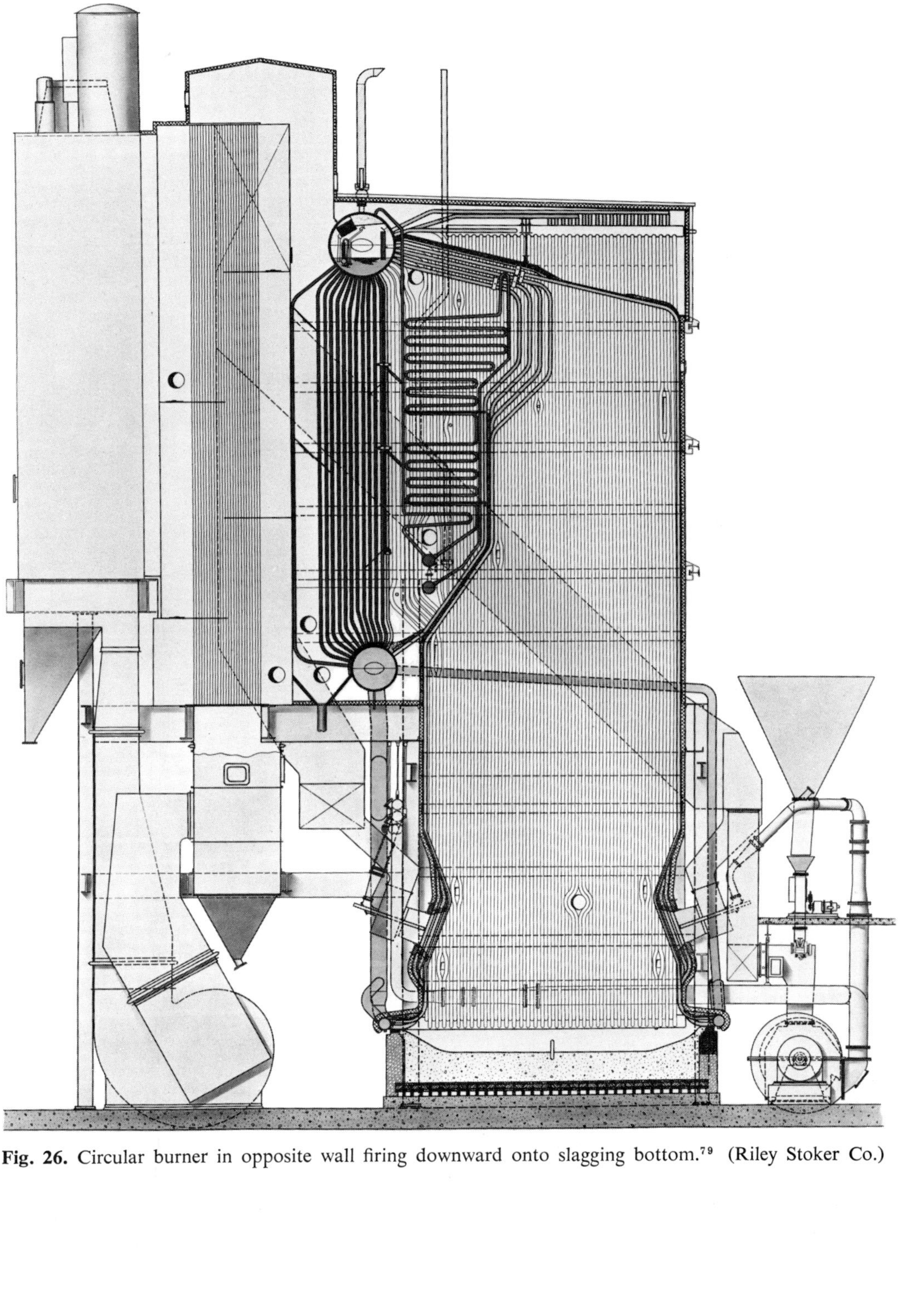

Fig. 26. Circular burner in opposite wall firing downward onto slagging bottom.[79] (Riley Stoker Co.)

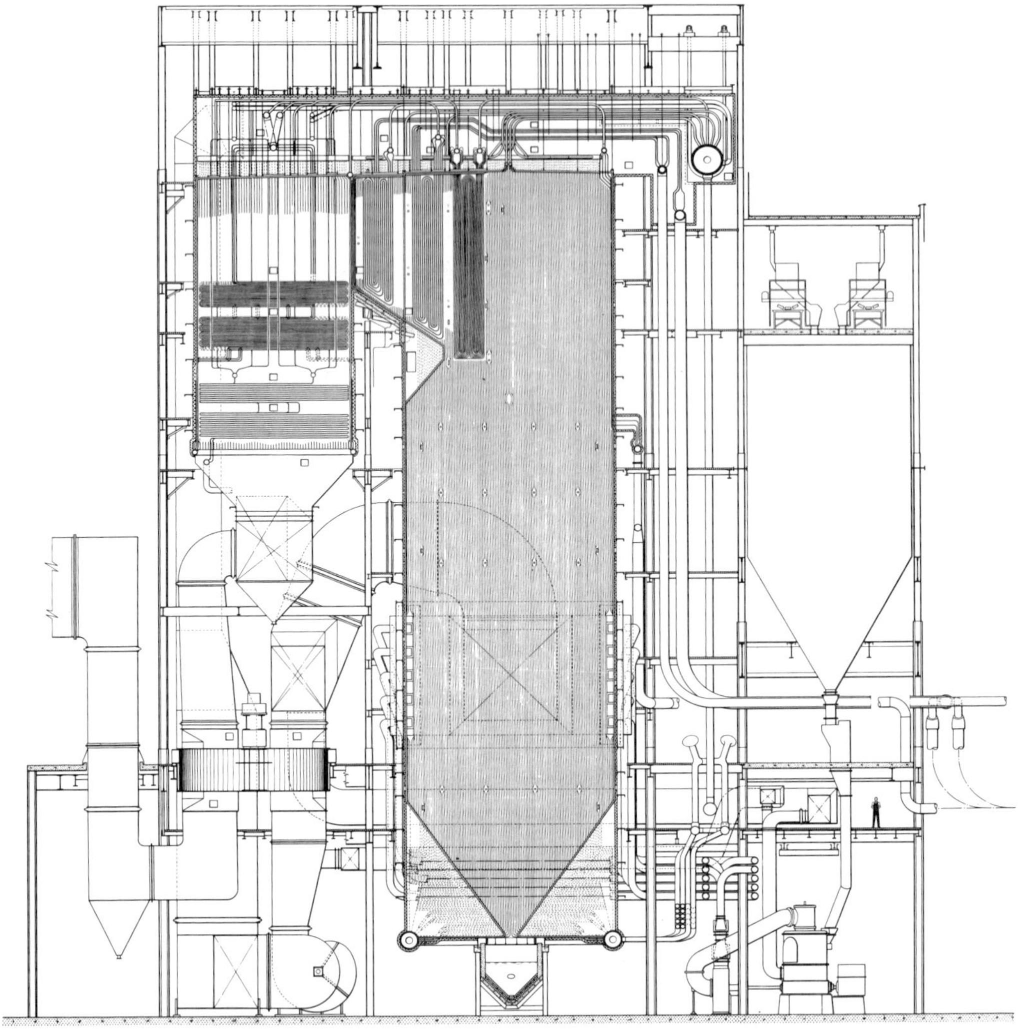

Fig. 27. Pulverized-coal burners firing tangentially from four corners of furnace. (Combustion Engineering, Inc.)

and back walls that are inclined slightly inward. The burners are not directly opposite in the two walls but are slightly offset.

Figure 27 shows the application of burners in each of the four corners of a furnace so arranged that they fire tangentially to an imaginary circle within the furnace. The burners can be tilted up or down, primarily for the purpose of control of the superheated steam temperature. The tilting feature is also of value when these burners are applied to slag-tap furnaces. Special note should be taken that the design capacity of this boiler is 3,850,000 lb of steam/hour. The overall inside height of the furnace is 154 ft.

Figures 28, 29, and 30 show that burners remain simple. In the multitip intertube burner of Figure 28, the circular burner of Figure 29, and in the corner burner of Figure 30, the principles are much the same. The coal and conveying or primary air, amounting to about 15 to 20% of the air for combustion, are discharged into the furnace at velocities of 2000 to 4000 ft/min and the remainder of the air is admitted around the rich mixture of fuel and air. Mixing is promoted by swirl vanes in the burner air passages such as Figure 29, by high velocity of the secondary air, and by impinging the coal-air stream on the protected arch tubes as in the left of Figure 28, and by the interaction of streams from the burners as in Figure 30 and Figure 26.

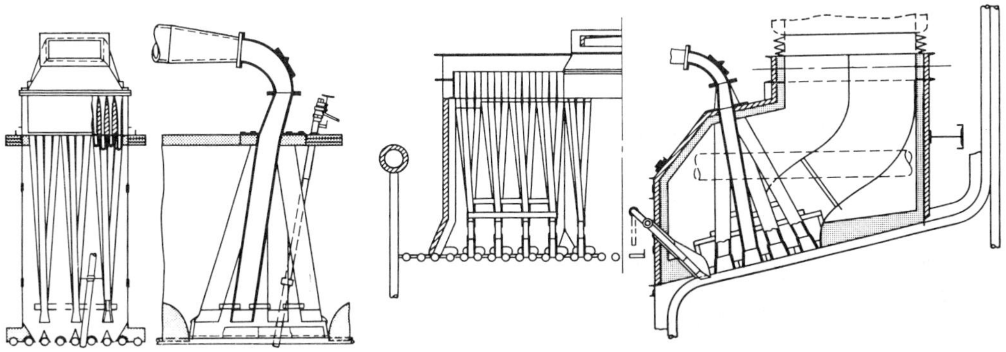

Fig. 28. Arrangement of pulverized-coal burners to fire between furnace tubes. (Babcock and Wilcox Co.)

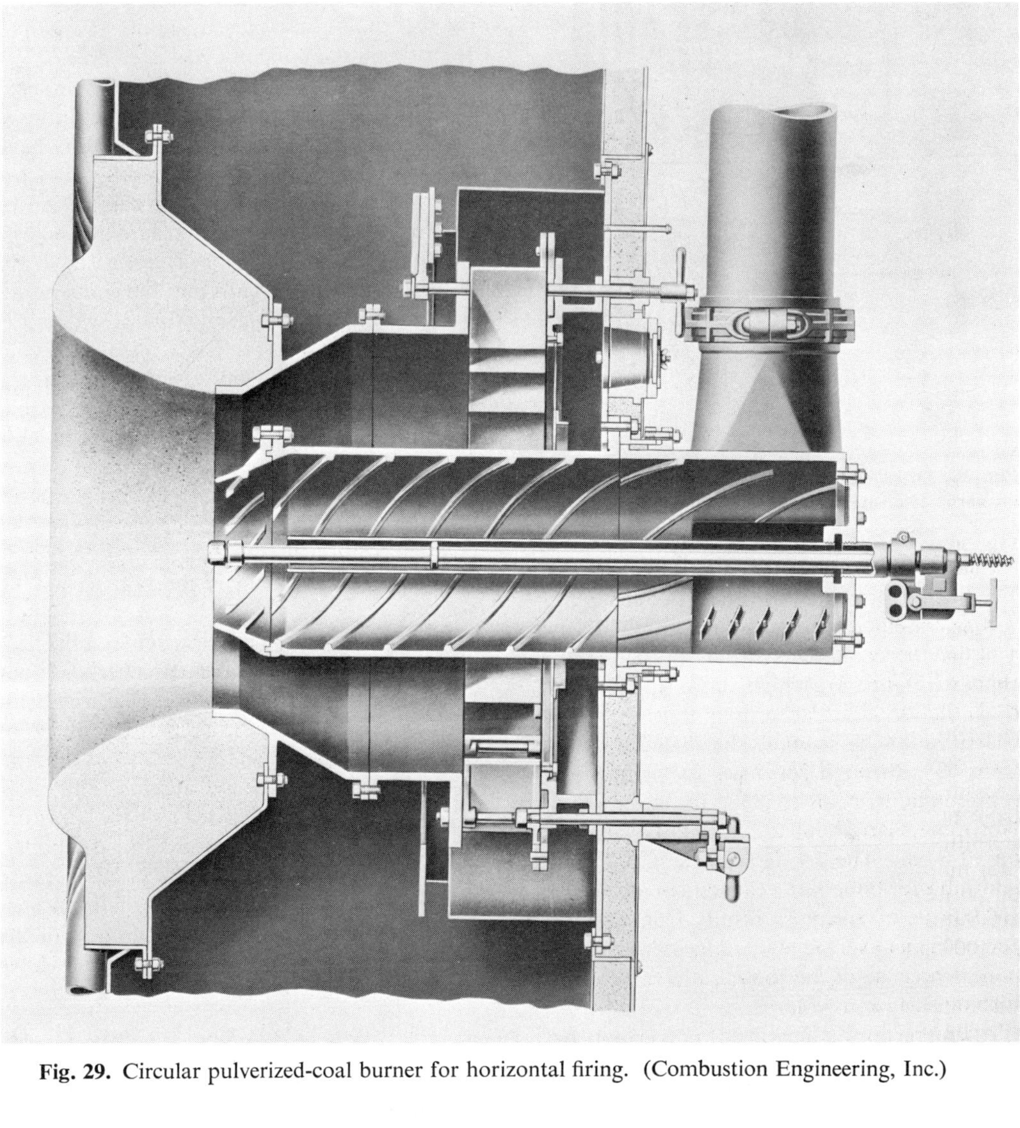

Fig. 29. Circular pulverized-coal burner for horizontal firing. (Combustion Engineering, Inc.)

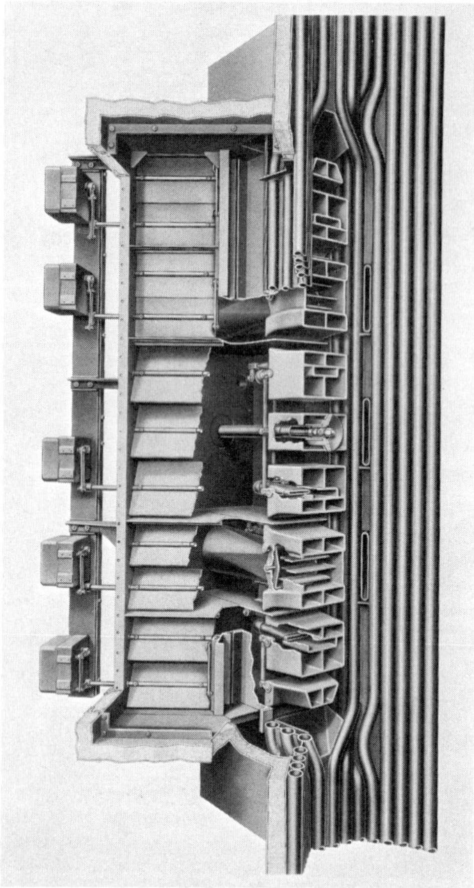

Fig. 30. Details of corner-tiltable pulverized-coal burners. (Combustion Engineering, Inc.)

Fineness of Pulverization and Rank of Coal. The fineness to which the coal should be pulverized for efficient combustion varies with the rank of the coal. Duzy[80] has presented the chart of Figure 31, which shows the fineness required for coals according to the ASTM Classification by Rank. The fineness ranges from 60% through 200 mesh for lignites and subbituminous coals to 85% for anthracites. Not more than about 2% should be retained on 50 mesh. The 5% less through 200 mesh allowable for boilers of a capacity of more than 100,000 lb of steam/hr results from the expected higher average temperatures in the large boilers because of the lower ratio of heat absorbing surface to volume.

An interesting application of pulverized-coal burners and slagging furnace without pulverization of the fuel is found in the 800,000 lb/hr boiler of the Texas Power and Light Company at Rockdale, Texas. The fuel used is either lignite char or dried lignite of which 36% and 18%, respectively, will pass 200 mesh. The reactivity of the fuels is such that Caracristi and Mumper[81] have reported that the loss in heat in unburned combustible is only 0.3 to 0.5% when fired through tangential burners without further pulverization. When the fuel contained several percent of particles larger than 8 mesh, many fell from the flame into the slag in which iron was reduced. This molten iron caused several failures of water tubes under the slag floor. By control of the size of the fuel to limit the percentage of +8 mesh to 1% or less, the difficulty was overcome.

Rates of Heat Release in Boiler Furnaces. The tremendous volume and the cathedral-like proportions of modern steam generators have developed, as it has been pointed out, not because the space is required for completion of combustion. Rather, it is to insure that a large enough fraction of the heat released is absorbed by the water walls that the temperature of the ash is reduced below that at which it is "sticky." This absorption is necessary to avoid accumulation of ash on superheater and reheater tubes, which impair heat transfer and cause excessive pressure drops in the gas passages. For this reason, the parameter of rate of heat release per unit area of furnace heat-absorbing surface is of greater significance for characterization of the design of pulverized-coal-fired furnace than is the rate per unit of volume.

Figure 32 gives the relation between the heat available and the heat absorbed per sq ft per hour for dry-ash and slag-tap pulverized-coal-fired furnaces according to the experience of one manufacturer.[82] Heat available is the sum of the heat input in the fuel and in the air less a correction for the latent heat in the moisture in the coal for the radiation from the boiler. The heat absorption rates lie in bands because of differences in characteristics of the ash of the coals, and differences in cleanliness of the furnace-wall tubes. Other manufacturers have curves of similar nature.

[80] Duzy, A. F., *Mech. Eng.*, **82,** No. 4, 109 (1960).

[81] Caracristi, V. Z., and Mumper, H. D., *Combustion*, **31,** No. 1, 34–44 (1959).

[82] Babcock and Wilcox Co., *Steam, Its Generation and Use* (1955), pp. 7–21.

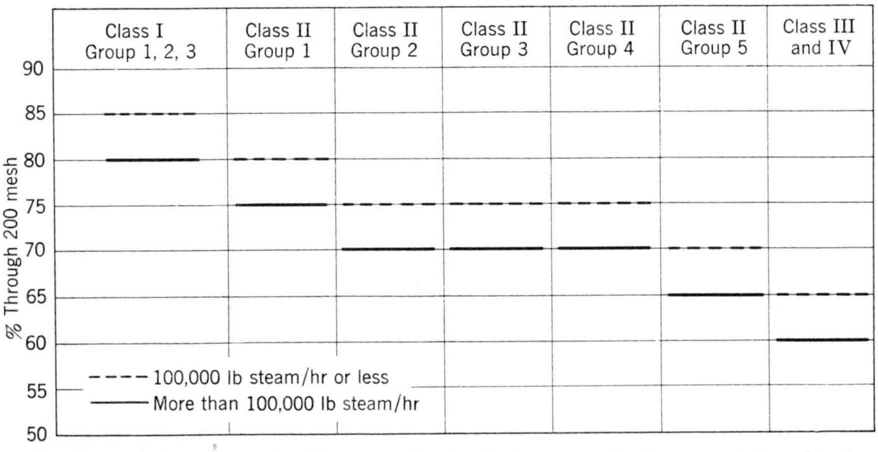

Fig. 31. Relation of required fineness of pulverization to rank of coal and size of boiler.[80]

The rate of heat input for slag-tap furnaces may go well beyond that for dry-ash furnaces, but the two overlap in the 75,000 to 150,000 Btu/cu ft-hr range. Although a much higher rate of heat release per unit area of radiant heating can be allowed when the furnace is designed for slag-tap than for dry-bottom operation, these high rates result in such high temperatures at the exit of the primary furnace that fouling of super-heaters and secondary evaporating surfaces would result. Hence, if such rates are used in the primary furnace, a secondary, open-pass furnace with radiant heat-absorbing surfaces must be provided to bring the exit temperature from the furnace below the "sticky" temperature of the ash.

This necessity to provide large furnaces with radiant heat-absorbing surfaces even with removal of ash as molten slag has led, in the United States particularly, to a deceleration in the rate of adoption of the slag-tap furnace. Whereas the 1941 *Power* survey[83] of high-pressure utility and industrial boilers showed that 27% of the pulverized-coal installations covered in the survey operated with removal

[83] Anon., *Power*, **85**, 654–5 (1941).

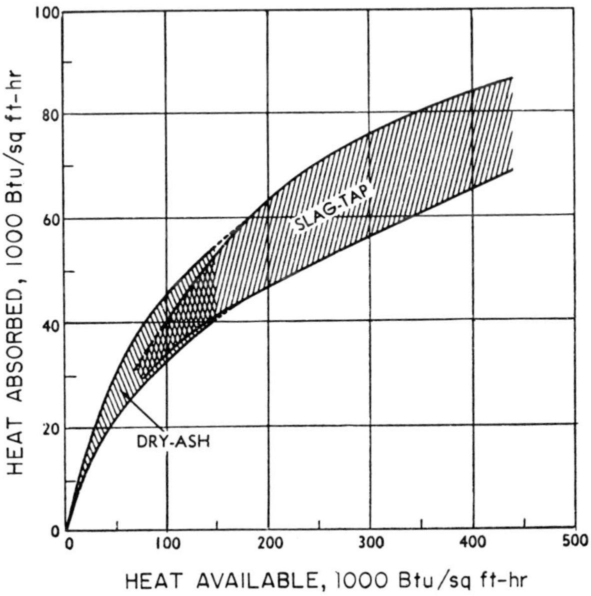

Fig. 32. Range of furnace heat absorption in relation to rate of heat release for dry-bottom and slag-tap furnace.[82]

of ash as molten slag, the similar survey for 1959[84] showed that only 5% had slag-tap furnaces.

Other factors also have contributed to the decrease in the number of pulverized-coal-fired slag-tap units. One is the fact, as discussed in more detail later, that slagging operation does not, to the degree hoped and expected, decrease the percentage of the ash leaving the furnace as fly ash. Another is the fact that coal with an ash fluid-temperature of above about 2600°F is not too well suited to slag-tap firing. If there should be a possibility that such a coal might become economically available, its use would not be possible. Inasmuch as ashes having fluid temperatures above 2600°F and as low as 2200°F can be handled in dry-bottom furnaces, this type has been found to be more flexible than the slag-tap furnace.

Cyclone Firing

Coals from various sources have wide variations in their caking properties and in the quantity and fusion characteristics of their ash. By comparison with the then existing types of fuel beds, which were highly sensitive both to caking and to ash behavior, the introduction of pulverized-coal firing, in the early 1920's, represented a notable improvement in that caking properties were found not to affect combustion performance. Moreover, experience was to show that pulverized-coal flames could be made to respond to load fluctuations with far less inertia than fuel beds. The ash problem still remained present, however. For those coals whose ash fusion temperature is high, it was found possible, eventually, to arrive at burner arrangements and, particularly, at dry-bottom furnace dimensions such that ash fusion and clinker formation could be largely avoided under wide variations of load.

For a while, it was thought that the slag-tap furnace, in which most of the ash is caused to melt at the bottom of the furnace, could solve the problems associated with the utilization of coals having low-ash fusion. As discussed earlier, however, experience was to show that this could be accomplished only with large losses of flexibility owing to the large area of slag that had to be kept molten at all times. The development of cyclone firing can be said to have removed this limitation by localizing the molten-slag area to the relatively minor dimension of the cyclone itself. Although the cyclone burner has a limited turn-down ratio, large furnaces can be provided with a number of such burners whose independent operation provides the needed flexibility along with slagging operation.

Figure 33 illustrates the present design of the cyclone burner.[85] This burner consists of a water-cooled horizontal cylinder at one end of which crushed coal is introduced where it is picked up by a tangential stream of air and brought into the cyclone proper where tangential admission of the remainder of the air gives it a rapid rotary motion. Instead of being entirely pulverized, the coal is crushed only sufficiently so that about 95% is smaller than $\frac{1}{4}$ in. The presence of the fines normally produced is necessary to maintain stable ignition.

Once the operation has been started, it is found that a layer of slag is formed on the water-cooled wall; with increasing thickness of the slag layer, the surface becomes viscous enough to hold the larger pieces of coal projected thereon by centrifugal action, and these pieces burn *in situ* because of the high relative velocity of the air stream. It is probable that most of the coal fired burns in suspension within the rotating mass while the latter spirals to the outlet because of net pressure differences. Fused ash, because of its density, is projected outwardly so that molten slag, kept hot by the high rate of heat release (about 500,000 Btu/hr-cu ft) and high temperatures (about 3000°F), can flow down the sides and out through the tap hole on the furnace end.

A historical and technical account of the first installations has been published.[86] The initial commercial burner installation was 8 ft in diameter, 11 ft long, and the burner was inclined 5° from the horizontal for ease of slag removal. The design of this unit, installed in 1944, had been evolved from successive pilot-scale units 3 and 5 feet in diameter. Subsequently, vertical cyclones were tried but were not successful. In the present designs, the cyclone burner is horizontal which has greatly simplified the design and joining of the water-cooled tubing to that of the furnace and facilitated the sealing of the burner to the furnace walls.[85]

[84] Anon., *Power*, **103,** 182–9 (1959).

[85] Rowand, W. H., and Kispert, E. G., *Mech. Eng.*, **82,** No. 4, 103 (1960).

[86] Grunert, A. E., Skog, L., and Wilcoxson, L. S., *ASME, Trans.*, **69,** 613–34 (1947).

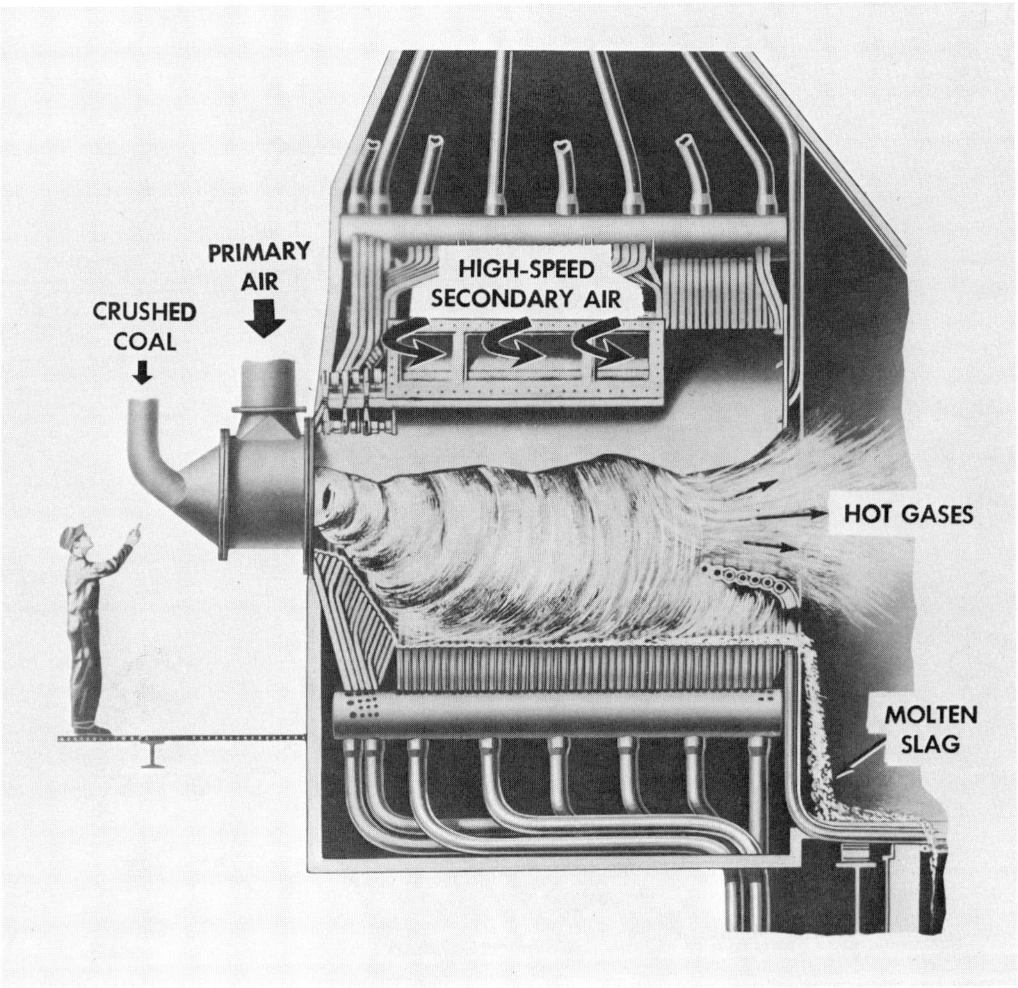

Fig. 33. View of cyclone burner.[85]

Figure 34 shows a typical installation of cyclone burners to a utility boiler furnace having a capacity of 1,200,000 lb of steam/hr; this unit is fired by five cyclone burners. During the period from 1944 to date, more than eighty boilers fired with cyclone furnaces have been installed or are now under construction.[85] Acceptance of this method of burning coal has been rapid, not only in the United States but in Europe, where a large number of installations exist, particularly in Germany.

Ledinegg[87] has written a paper that presents theoretical methods for calculating pressure drop through the cyclone furnace from the aerodynamic conditions that have to be imposed, in terms of the size consist of the coal, to obtain the desired slagging action during the resulting residence time and as a function of rate of heat release.

An extensive experimental study has been made on a small cyclone combustor[88] and the main result was the determination of the upper limit of the silica ratio of the ash,

$$\frac{SiO_2}{SiO_2 + \text{equiv. } Fe_2O_3 + MgO + CaO}$$

in the coal fired, that would allow satisfactory slag discharge performance. It was found that this ratio should not be greater than 72 for the

[87] Ledinegg, M., *Z. Ver. deut. Ing.*, **94,** 921–7 (1952).

[88] Hoy, H. R., Roberts, A. G., and Wilkins, D. M., *J. Inst. Fuel*, **21,** 429–57 (1958).

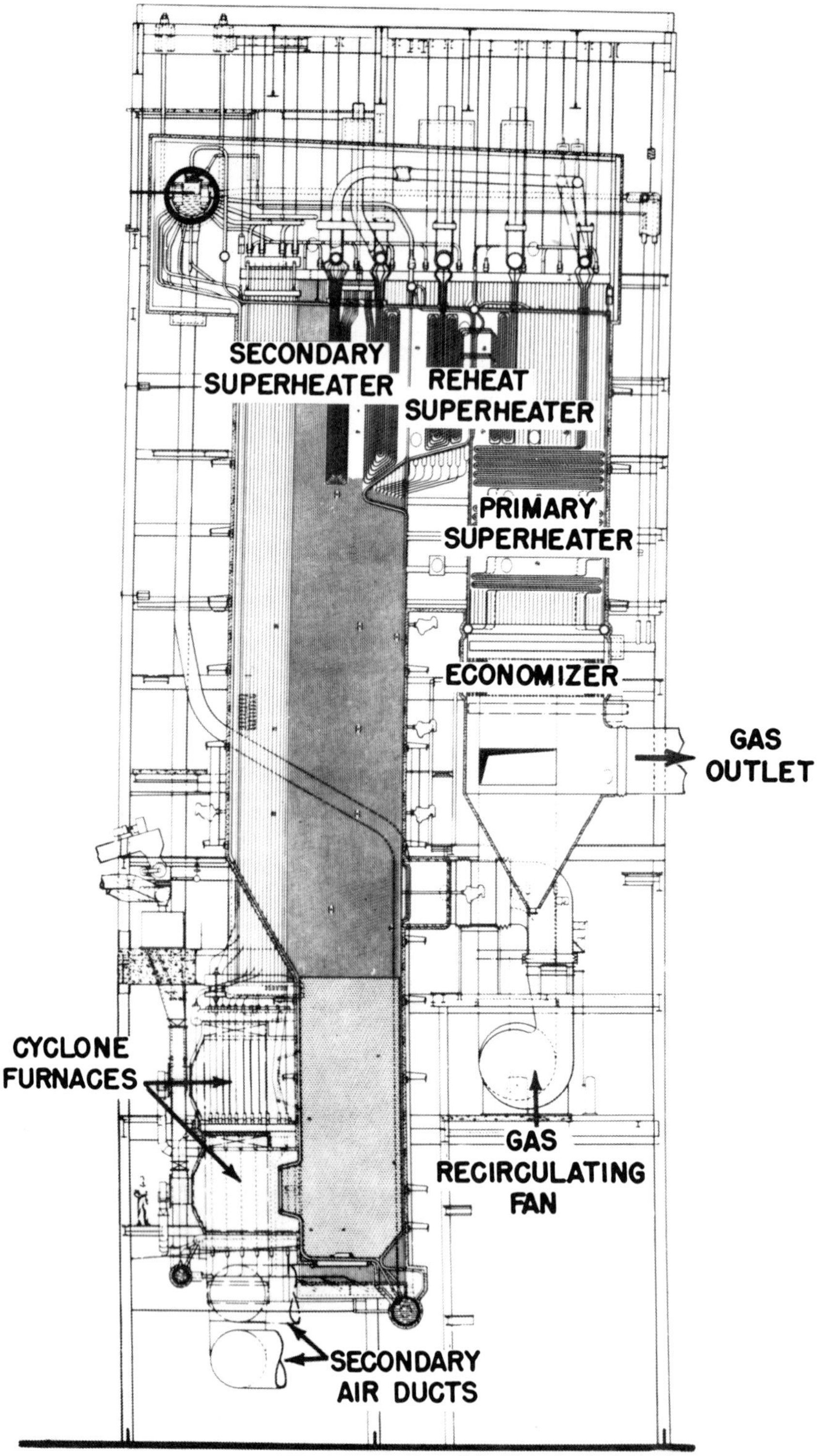

Fig. 34. Application of cyclone burners to a large boiler.[85]

(British) coals tested; otherwise, the temperature-viscosity characteristics of the slag affected the operation of the cyclone adversely. Other findings, which may reflect in part the relatively small size of the cyclone burner investigated were: (1) an indicated air/fuel ratio of 105 to 110% of stoichiometric, (2) a desirable fuel size-consist such that from 25 to 30% of the coal was smaller than 200 mesh (B.S.S.), (3) an increase in the silica ratio required an increase in the secondary air temperature or an increase in the rate of heat input, and (4) a readjustment of swirl velocity was necessary with variation in coal-ash characteristics and rate of heat input, and significant changes in the area of the swirl vane assembly and in the pressure drop were required to achieve a turn-down ratio of 2 to 1. It was also found that the ash content of the coal had to be above a certain undetermined minimum to maintain a sticky condition in the slag, and that control of the moisture of the coal was required for proper feeding.

These findings generally agree with American experience. For American coals, it is recommended[85] that, with a preheated air temperature of 600°F, the volatile matter should exceed 15% on the dry basis, the viscosity of the slag should be less than 250 poises at 2600°F, and that the ash-softening temperature in a reducing atmosphere should be no higher than 2500°F. Operational savings due to decreased power requirements and lesser wear in the crushers are significant as compared with pulverized-coal firing. However, this economic advantage with coals of high calorific value is offset by the power required to induce the swirling air motion in the cyclone.

Control of Air Pollution from Combustion Processes

The pollutants of the atmosphere that may be emitted from equipment in which coal is burned can be classified under three heads: (1) smoke, (2) cinder and fly ash, and (3) the oxides of sulfur. In the early days of municipal or other governmental control of air pollution, the principal attack was on the problem of smoke emission. Now, with the problem of the emission of smoke having become less serious, both because of improved equipment and operation and the replacement of equipment using coal by fuels of a lesser smoke-producing potential, attention is directed more toward the reduction in the emission of cinder and fly ash and toward the problem of emission of oxides of sulfur.

Smoke

Smoke is here to be understood as the finely-divided carbon, less than one micron, which results from the decomposition of hydrocarbons evolved during the process of combustion. It is well recognized that the logical method for the prevention of the emission of smoke is that of prevention of its formation rather than of attempts to burn the smoke after it has been formed. The key to success in the prevention of the formation of smoke is the admission of air in proper amounts and with sufficient turbulence for mixing with the hydrocarbons as they are released from the coal, together with adequate temperature and residence time.

Earlier in this chapter, the methods developed for the elimination or reduction of smoke emission from hand-fired heating stoves, furnaces, and boilers have been discussed. On commercial and industrial equipment, the use of overfire jets of air or steam or both have long been used to bring air and turbulence into the combustion zone. The design of such jets was an empirical, cut-and-try process until Engdahl and Holton[89] published the results of an analytical and experimental investigation. These data and the results of further research have been assembled by Bituminous Coal Research, Inc., in a bulletin[90] that presents in an authoritative manner information on how to design blower jets, steam-air jets, and plain steam jets for application to various types of stationary coal-burning equipment. Although jets are not a cure-all, as previously brought out in the discussion of low-load smoke from spreader stokers, intelligent application of overfire jets can do much on many installations to reduce smoke emission to below the nuisance level.

Fly Ash and Cinder

The emission of fly ash and cinder from static beds of fuel such as used in hand firing and underfeed or traveling-grate stokers occurs when the velocity of the gases is great enough

[89] Engdahl, R. B., and Holton, W. C., *ASME, Trans.*, **65**, 741–54 (1943).

[90] BCR Aid to Industry 500–300, Bituminous Coal Research, Inc., Pittsburgh, Pennsylvania, 1951, 24 pp.

TABLE 4

TYPICAL AMOUNT AND SIZE OF CARRY-OVER OF FLY ASH AND CINDER

		Underfeed Stoker	Spreader Stoker	Pulverized Coal: Dry Bottom	Pulverized Coal: Slag Tap	Cyclone Furnace
Carry-over, fraction of ash in coal, %*		20–30	30–80	60–85	45–55	10–15
Combustible content, %		10–50	20–60	5–10	5–10	5
Dust loading, lb/1000 lb gas†		0.5–4	1.5–10	6–9	4–6	0.6–1.5
Typical size consist,[91] cumulative percent						
Microns	Mesh					
<10		7	11	40		86
<20		15	23	70		91
<44	325	30	42	80		95
<74	200	38	56	97		98
<149	100	57	73	100		100

* Includes combustible.

† Based on coal of 15% ash content. Dust in gas leaving furnace before any collectors. Corrected to 50% excess air.

to lift particles of ash and carbon from the bed. The probability of the emission of excessive amounts of solids from such static beds is less than from combustion processes in which a part or all of the fuel is burned in suspension as with spreader stokers or pulverized coal.

Table 4 presents a brief summary of the range of emission of fly ash and cinder from several types of industrial fuel-burning equipment in terms of the fraction of the coal fired, the combustible content of the material, the dust loading in the gases, and the size consist to be expected. A wide range of carry-over of combustible and of dust loading is shown for each type of equipment. The amounts depend on a number of factors including the rate of burning, the size of the fuel fired, the caking characteristics of the fuel, and the fusion characteristics of the ash. A discussion of the effect of each is beyond the scope of this book.

A frequently used bench mark for judging the emission of solids is the value of 0.85 lb/1000 lb flue gas, when adjusted to 50% excess of air. This is used because it is the maximum allowable dust loading under many air-pollution-control ordinances of the United States. With a knowledge of the dust loading and of the size consist of the particles, we can judge the necessity of collectors on the equipment before the gases pass to the stack, and the type of collector required to reduce the emission to below the allowable limit.

Table 4 shows that the emission of solids from stoker-fired furnaces may fall within the customarily allowable limit. Also, it shows that the size of the particles is relatively large. For example, if the dust loading of the gases from such a stoker were 2 lb/1000 lb gas, a collector would be required, but its overall efficiency would have to be only $[(2 - 0.85) \times 100]/2 = 57.5\%$. With 70% of the particles larger than 44 μ, a medium- or low-draft-loss collector should suffice to effect the removal.

The emission from spreader stokers is not only greater than from those with static fuel beds, but the combustible content may be higher and the size consist somewhat smaller. Hence, more efficient collectors must be used on equipment fired by this type of stoker. If, for example, the dust loading should be as high as 8 lb/1000 lb gas, the collector should have a minimum efficiency of $[(8 - 0.85) \times 100]/8 = 89.4\%$.

When coal is burned in pulverized form with a dry-bottom furnace, all of the ash is originally in suspension and only that which is agglomerated into large particles or accumulates on

[91] Hummell, J. D., *A Review of Available Information on Corrosion and Deposits in Boilers and Gas Turbines*, ASME, New York, 1959, p. 185.

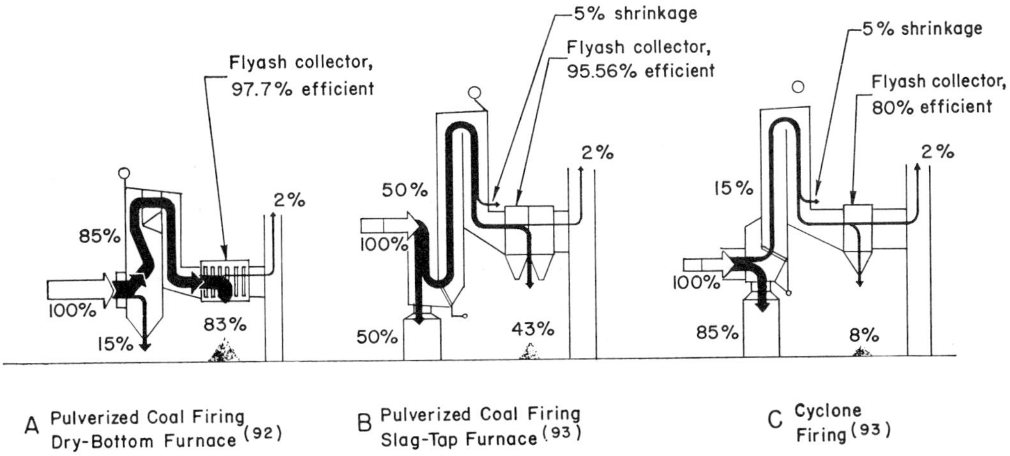

Fig. 35. Distribution of fly ash in dry-bottom, slag-tap pulverized coal and cyclone furnaces.

walls, later to fall into the ash pit, does not leave the furnace with the gases. As much as 85% of the ash in the coal may leave the furnace. The combustible content will, however, normally be low.

With removal of ash as molten slag, there is a greater chance for agglomeration of particles because of the higher furnace temperatures. The burners are frequently directed to impinge on the molten slag bath, which gives an opportunity for better collection. Nevertheless, the efficiency of the furnace for collection of ash is not great enough to avoid the necessity for the installation of efficient ash collectors. Because of the small size—70% may be smaller than 20 μ—collectors of high efficiency for small-sized particles are required.

Owing to the centrifugal forces and to the presence of molten slag on the walls in a cyclone furnace, a comparatively small fraction of the ash is discharged with the combustion gases from the cyclone. With coals of low-ash content, the emission from the stack may be below the legally allowed limit. However, because of the extremely small size—90% may be below 20 μ—the visibility of the plume is high, and the installation of collectors may be necessary and is a commonly accepted practice.

Figure 35 shows graphically the distribution of the fly ash from dry-bottom and slag-tap pulverized-coal furnaces and cyclone furnaces as reported by Grossman[92] and Schroeder and Strasser.[93] Comparison is made of the amounts caught by the collector for a desired emission of 2% of the ash of the coal from the stack. With a dry-bottom furnace, on the assumption that only 15% is caught in the furnace, the collector must be 97.7% efficient. With the slag-tap furnace, assuming 50% remains in the furnace, the collector need be of somewhat lower efficiency—95.6%. With cyclone firing, where 85% is caught in the cyclone, the efficiency of the collector need be only 80%, but it must have this efficiency for extremely small particles. Notable in Figure 35 is the fact that with both the slag-tap and cyclone furnace, a shrinkage of 5% in the ash is assumed. This shrinkage results from the high temperatures in these furnaces, which cause a volatilization of part of the ash which then escapes as a fine fume that cannot be collected.

Types of Collectors

Three general principles are used in the design of collectors for fly ash and cinder: gravity, inertia, and electrostatic. The use of gravity is typified by settling chambers in which by enlargement of the gas duct the velocity is reduced, thus allowing some of the particles to fall out. This type is effective in removing only the very large particles, and its overall efficiency is low. The inertia principle is most commonly seen in cyclone separators in which, by tangential introduction of the dust-laden gas, a free vortex is established and the particles are thrown to the wall by centrifugal force. The force acting to throw particles out of gas is inversely proportional to the radius of

[92] Grossman, P., *J. Air Pollution Control Assoc.*, **7,** 222–6 (1957).

[93] Schroeder, H. C., and Strasser, R. J., *ASME, Trans.*, **74,** 573–8 (1952).

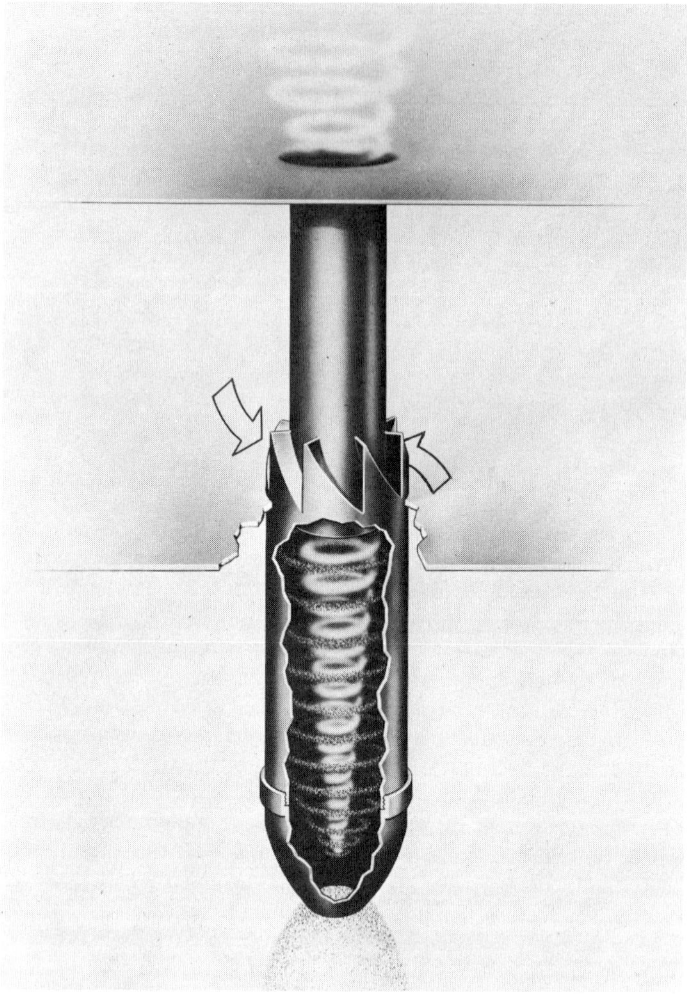

Detail of single tube

Fig. 36a. A small-diameter cyclone. Aerotec Industries, Inc.

curvature of the gas stream; hence, the efficiency of a cyclone increases with a decrease in its diameter. Present practice favors the use of a multiplicity of small cyclones; the diameters may be as small as 3 to 6 in. Figure 36 shows the construction of one type of the small collectors into which there is a multiplicity of entrances. The assembly of many of these small tubes into a unit to handle the flue gas and fly ash from a large boiler is also shown.

Figure 37 shows the principle of the van Tongeren collector in which a shave-off or skimmer plate inside the vertical wall forms a channel through which a portion of the gas with a heavy loading of dust is bled off. It reenters the cyclone below the exit of the main gas stream and thus lessens the possibility of re-entrainment.

The inertia principle is also incorporated in the louver collector[94] of which an assembly is shown in Figure 38. The dust-laden gas enters the large end of the cone in the walls of which are stamped louvers whose opening on the outside is downward. By their inertia, the particles continue toward the small end of the cone whereas the clean gas passes to the stack. The dust with only a few percent of the original gas passes from the small end of the cone to secondary cyclone collectors. The outlet of these

[94] Linderroth, E. T., U.S. Pat. 2,506,273 (1950).

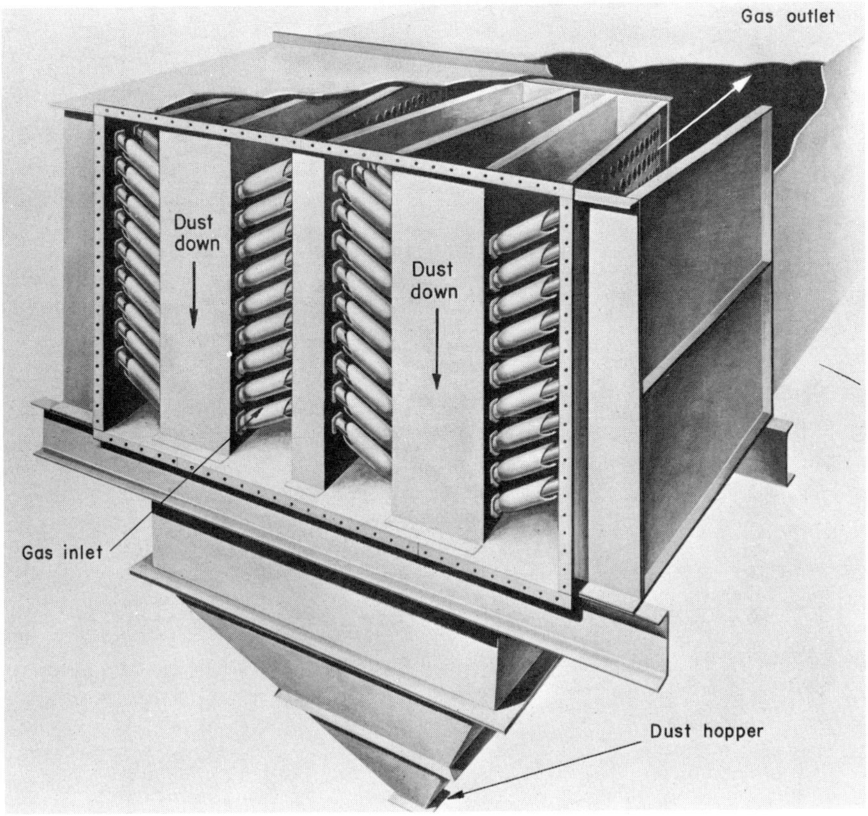

Multiple assembly

Fig. 36b. Application of small-diameter cyclones in a multiple assembly. (Aerotec Industries, Inc.)

cyclones may be recycled through the louver collectors.

Electrostatic collectors consist essentially of a series of tubes or plates among which are suspended wires or sharp-edged rods. A potential difference of 40,000 to 60,000 v is applied between the wires and the tubes or plates. The ionization of the gases by the high potential charges the particles which then migrate to the plates or tubes and are deposited. Periodically or continuously the plates or tubes are rapped to discharge the dust which falls into the collector hopper.

Electrostatic collectors are particularly effective on the smallest particles, whereas mechanical collectors are more effective on the larger particles. Furthermore, the efficiency of collection of a given dust on electrostatic collector decreases with increasing velocity through the collector, whereas the reverse is true in a mechanical collector. Thus, the efficiency of a mechanical collector will decrease when operated below the load for which it was designed, and that of an electrostatic collector will increase.

Figure 39[95] shows this relation of efficiency of collection of fly ash from a boiler to the percentage of full load at which the boiler operates. A third curve shows that a combination of a mechanical collector followed by an electrostatic collector will give an almost constant efficiency at all loads.

Figure 40 illustrates a combined mechanical collector of the type shown in Fig. 36 with a plate-type electrostatic collector. To insure a minimum of air pollution from pulverized-coal-fired boilers, such combined collectors are frequently used.

Fly-ash refiring. The discussion of spreader stokers pointed out the prevalence of the practice of reinjection into the furnace of fly ash and cinder collected in boiler hoppers and by

[95] O'Mara, R., *Combustion*, **21,** No. 10, 38–43 (1950).

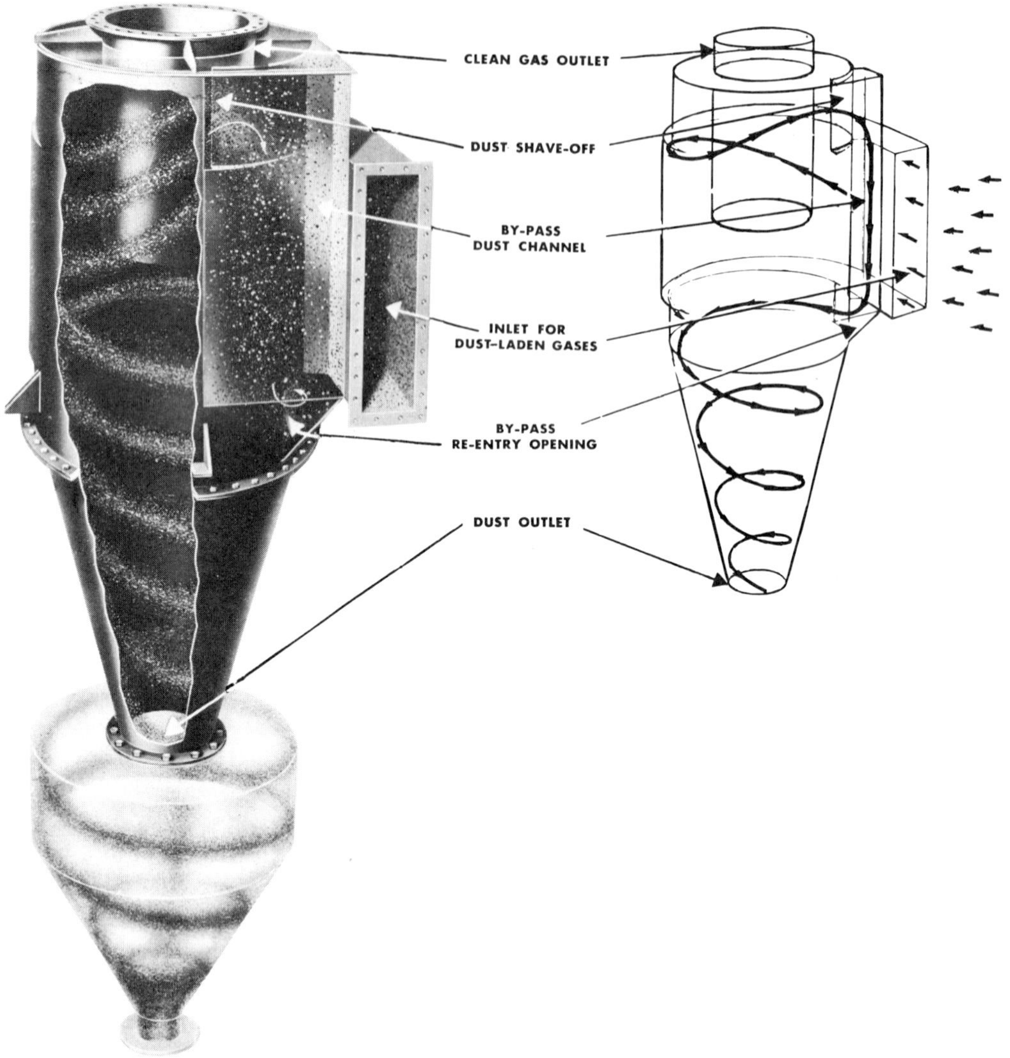

Fig. 37. The van Tongeren cyclone collector with "shave-off" plate. (Buell Engineering Co.)

the fly-ash collectors. The primary purpose of such refiring is to increase the efficiency of operation of the boiler by recovery of the heat from the combustible matter in the fly ash. The savings are often worth the difficulties inherent in the much larger circulating load of fly ash in the system.

Feeley[96] has recounted experiences in the refiring of fly ash collected by a mechanical collector and an electrostatic precipitator into a pulverized-coal-fired boiler with a slag-tap furnace. The purpose, which was successfully accomplished, was to recover as much of the ash as possible as a slag which was more readily disposable than was the fly ash. Since most of the reinjected ash was melted into slag at the first reinjection, the circulating load of ash through the boiler was not substantially increased and the rate of emission from the stack was less than 0.5 lb/1000 lb of gas when the boiler designed for an output of 350,000 lb of steam/hr was operated at an output of 300,000 lb/hr.

The practice of refiring fly ash has become common in Germany. In addition to the recovery of the ash as more readily handled slag, a gain in efficiency is claimed because of the recovery of heat in the combustible in the

[96] Feeley, F. G., Jr., *ASME, Trans.*, **78,** 1747–51 (1956).

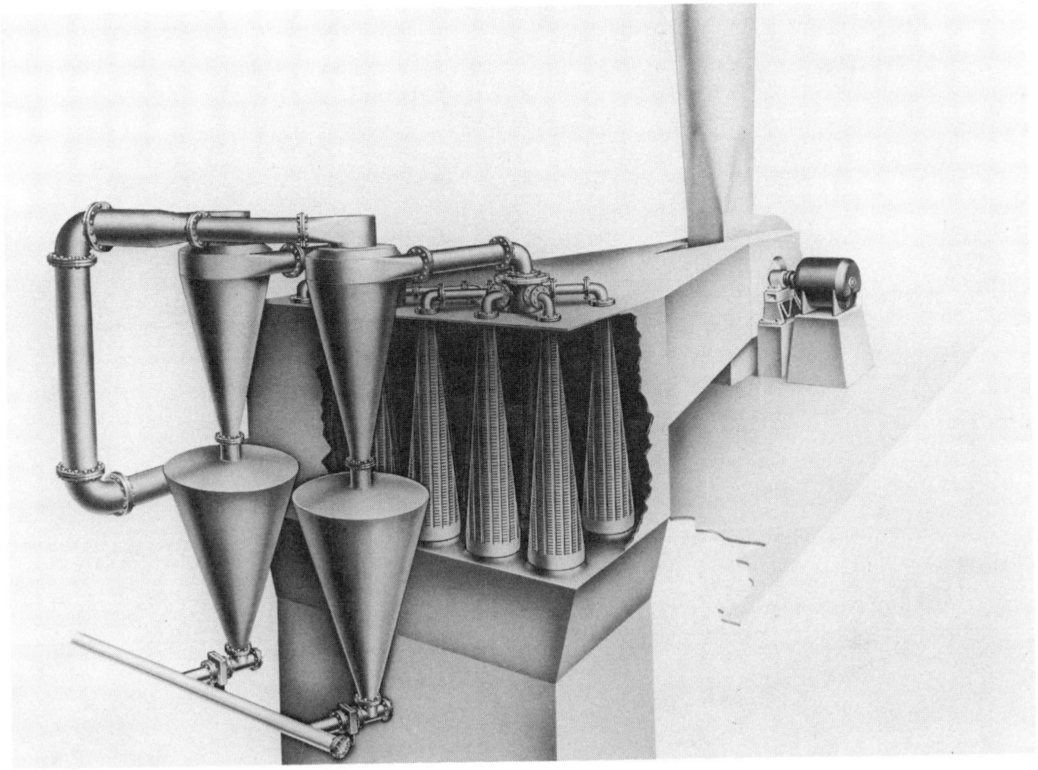

Fig. 38. The Louver collector. (The Green Fuel Economizer Co., Inc.)

fly ash. It has been pointed out, however, that this may be fallacious.[97] Assuming the total heat required to melt the ash to a fluid slag is of the same order as that of blast furnace slags, each pound would require 730 Btu. Thus, if the fly ash contains 5% combustible and 95% ash, the 0.9 lb would require 684 Btu. However, the recovery from the 0.05 lb of carbon would be $0.5 \times 14,600 = 730$ Btu. Essentially then, the heat requirements and heat recovery balance. Because all the heat in the slag leaves the furnace, no gain is realized. If the ratio of 5 combustible to 95 ash in the fly ash occurred with a coal containing 15% ash, the heat loss in the combustible would be about 1 percent. Hence, it may be concluded that unless the heat loss in combustible in the fly ash is greater than 1%, no gain in efficiency would be realized by fly-ash refiring. However, the heat in the combustible would be enough to place the ash in the form of slag if this was desirable.

At stations of the Public Service Electric and Gas Company,[98] experience has shown that the injection of the ash collected in the electrostatic precipitators into the slag-tap furnace increased the fraction of the ash of the coal removed as slag from 27 to 78% with no increase in the emission from the stack. When the combustible in the precipitator ash was as

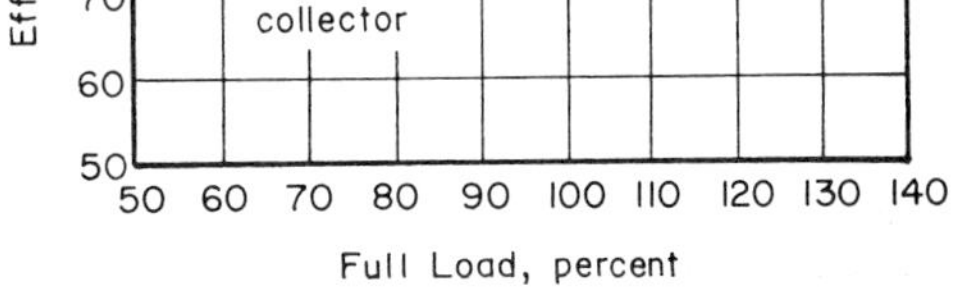

Fig. 39. Relation of efficiency of collection of mechanical and electrostatic collectors and of their combination to the load on the boiler.[95]

[97] Gumz, W., Kirsch, H., and Mackowsky, M. T., *Schlackenkunde*, Springer-Verlag, Berlin-Göttingen-Heidelberg, 1958, 522 pp.

[98] Maull, W. M., and McMahon, E. C., *Combustion*, **28**, No. 5, 40–47 (1956).

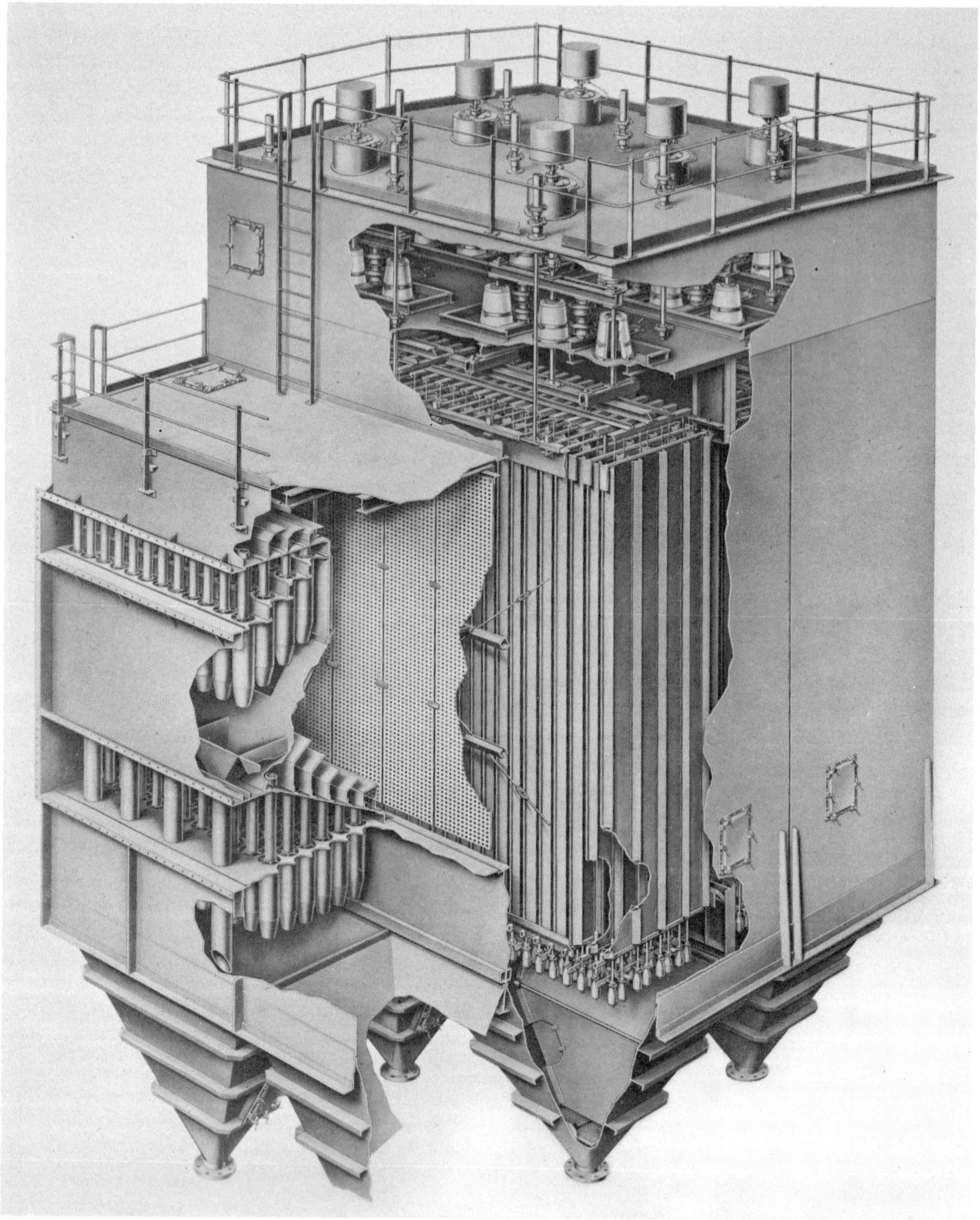

Fig. 40. Combined mechanical and electrostatic collector. (Research-Cottrell Corp.)

much as 9%, the boiler efficiency could be increased by 0.5 to 1.0% by reinjection. When the combustible content was as low as 3%, there was no gain in boiler efficiency, but reinjection was considered desirable because of the better form of the ash for removal and handling.

The Emission of Oxides of Sulfur

All coals contain a certain amount of sulfur. In the process of combustion, some of the sulfur is combined with the ash or slag. This amount will vary with the composition of the ash and the method of burning, but, in any event, it is not a major part.

Johnstone[99] found, as a result of experimental surveys, that 70% of the sulfur was found in the flue gases as oxides of sulfur when the coal was burned on stokers and about 90% when burned in pulverized form.

Of the SO_2 formed, 0 to 7% is oxidized to SO_3 during the passage of the combustion gases through the boiler or other equipment. The percentage of SO_2 oxidized to SO_3 is reported to be less the greater sulfur content of the fuel and is dependent on the degree of catalytic action of materials over which the gases pass, on the excess air, and on the temperature of the combustion process. The formation of SO_3 is of much concern in the operation of boilers since a small amount raises the dew point of the flue gases and leads to corrosion of steel in air heaters and economizers. Thus, the presence of SO_3 causes an economic loss either from damage to equipment by corrosion, or from loss of efficiency of the unit if it must be operated at high enough flue-gas temperature to avoid corrosion. The mechanism of the oxidation of SO_2 to SO_3 in combustion is by no means completely understood, but the available knowledge has been summarized by Krause.[100]

Although the percentages by volume of SO_2 and SO_3 in the products of combustion are small, the total amounts discharged to the atmosphere are large. For example, from a boiler plant that burns in a day 1000 tons of coal containing 2% sulfur, there will be discharged 40 tons of SO_2 and 1 ton SO_3 per day, assuming that 2% of the SO_2 is oxidized to SO_3. The relation of SO_2 to human and animal health has not been definitely established, although there is no question of its irritating effect on the eyes, nose, throat, and respiratory passages. In the London disaster of 1952[101] in which 4000 more deaths than normal occurred during a period of a few days of heavy smog, unusually large concentrations of SO_2 in the atmosphere were measured. Although the maximum concentration of 1.3 ppm was ten times normal, it was far less than the 10 ppm considered safe for an eight-hour exposure. SO_2 is deleterious to certain plant life and undoubtedly is responsible, either as such or as SO_3 or as H_2SO_4 that results from its oxidation in the atmosphere, for much damage to fabrics, and to metal, stone, and mortar of structures. For these reasons, efforts to decrease the emission of sulfur oxides have been increasing.

The decrease in the emission of sulfur oxides from the combustion of coal could be approached in two ways. The first would be the removal of the sulfur from the coal. Rose[102] has reviewed this problem and has established the difficulties in effecting a major reduction in the sulfur content of coal by processes that would be economically feasible for electric generating stations. The second way to abate sulfur emission is to remove the sulfur oxides from the products of combustion. The Bureau of Mines[103] has recently published a comprehensive review of the chemistry of sulfur dioxide and its removal from waste gases. The processes for removal are classified as absorption, both by liquids and solids, adsorption, reaction with metals and metallic oxides, reduction by gases or by carbon, and the oxidation by air or ozone. Of the various processes, the only ones that have had large-scale trials on flue gases from coal-burning plants are those based on absorption by liquid scrubbing. Others have had only bench-scale or small pilot-plant trials.

The use of large quantities of naturally alkaline water, to which some $CaCO_3$ is added, has been practiced at the Battersea generating station in London since 1935. Some $MnSO_4$ is added to the effluent from the scrubber to promote the oxidation of sulfites. After passing through a settling tank, 35 tons of water per ton of coal of 1% sulfur is returned to the river. The efficiency of removal of SO_2 is 92 to 97%.

The Howden-ICI process[104] is similar in the use of $CaCO_3$ but, after settling, the water is recirculated through the system. Plants in Wales and at the Fulham Station, London, operated for several years but closed down during the war and were not reopened. The process can have an efficiency of removal of SO_2 up to 99%.

[99] Johnstone, H. F., *Univ. Illinois, Eng. Exp. Stat. Bull.*, No. 228 (1931), 122 pp.

[100] Krause, H. H., *A Review of Available Information on Corrosion and Deposits in Boilers and Gas Turbines*, ASME, New York, 1959, Chap. 2.

[101] Wilkins, E. T., *Mech. Eng.*, **76**, 426–8 (1954).

[102] Rose, H. J., ASME, Paper No. 58-A (1958).

[103] Bienstock, D., Brunn, L. W., Murphy, E. M. and Benson, H. E., *U.S. Bur. Mines Inform. Circ.*, No. 7386 (1958), 96 pp.

[104] Pearson, J. L., Nonhebel, G., and Ulander, P. H. N., *J. Inst. Fuel*, **8**, 119–56 (1935).

TABLE 5

Summary of Costs per Ton of Coal of Scrubbing SO_2 from Flue Gas* of a Power Plant of 120,000 kw Capacity[108]

	SO_2 Content of Flue Gas Before Scrubbing, Vol. %					
	0.083		0.30			
	90% removal		90% removal		70% removal	
Total Investment†						
Limestone process	$1,646,750		$1,992,200		$1,747,200	
Ammonia process	3,221,100		4,945,400		4,055,400	
Sodium sulfite process	2,433,000		3,105,800		2,633,100	
Annual operating costs‡	$/ton coal consumed					
	No Credit for Products	Credit for‡ Products	No Credit for Products	Credit for § Products	No Credit for Products	Credit for § Products
Limestone process	1.24	—	1.93	—	1.69	—
Ammonia process‖	2.99	1.62	6.54	1.43	5.38	1.41
Sodium sulfite process	2.10	1.97	3.21	2.17	2.75	1.91

* 475,000 tons/year coal consumed; sulfur content of 1.5 and 5%, respectively.

† Including working capital at 10% of fixed capital.

‡ No provision for return or interest on investment, but provides for depreciation at 5% and taxes and insurance at 2% of fixed capital.

§ Ammonium sulfate at $32.00/ton; sulfur at $28.00/ton; SO_2 at $14.00/ton (Sulfuric acid equivalent price.)

‖ Anhydrous ammonia charged at $100.00/ton delivered.

Processes that have had extensive pilot-scale trial include the Fulham-Simon-Carves[105] ammoniacal liquor process in which, by autoclaving after the addition of sulfuric acid to the solution after absorption, ammonium sulfate and free sulfur are produced. The process can have an efficiency approaching 100%. Another process uses a sodium sulfite solution[106] as the absorption liquid to give sodium bisulfite. The addition of zinc oxide and heating with steam leads to precipitation of zinc sulfite crystals. When these crystals are calcined, SO_2 is driven off and can be recovered for the market. The efficiency of the process can be 90% or greater.

Research has recently been carried on by Bituminous Coal Research, Inc., on the possibilities of the oxidation of the SO_2 to flue gases to SO_3 by the use of ozone. The objective was to take advantage of the greater solubility of SO_3 in water as compared with that of SO_2. It has been reported[107] that, although technically feasible, the cost of producing the ozone was prohibitive for commercial application of the process.

The Bureau of Mines has presented an analysis[108] of the costs of removing SO_2 from flue gas by the Howden-ICI limestone process, by the Fulham-Simon-Carves ammonia process, and by the Johnstone-Singh sodium-sulfite process for coals of 1.5 and 5.0% sulfur content burned in a plant of relatively small, 120,000 kw, capacity.

Table 5 presents a summary of these cost analyses. The costs range from $1.24 to $2.17 per ton of coal or approximately one-third of the cost of the coal, estimated to be $4 to $6

[105] Newall, H. E., *Proc. 1st International Congress on Air Pollution; Mech. Eng.*, **77**, No. 5, 449 (1955).

[106] Johnstone, H. F., and Singh, A. D., *Illinois, Eng. Exp. Stat. Bull.* No. 324 (1940), 137 pp.

[107] Grohse, E. S., and Saline, L. E., *J. Air Pollution Control Assoc.*, **8**, 255–66 (1958).

[108] Field, J. H., Brunn, L. W., Haynes, W. P., and Benson, H. E., *U.S. Bur. Mines, Rept. Invest.* No. 5469 (1959), 58 pp.

per ton. The Bureau of Mines estimate of the cost of the limestone process checks closely with that made by Rees[109] of 10s ($1.40) per ton of coal. Rees made the point that this cost, as well as that of the Battersea process which was 7s ($0.98) per ton, was actually greater than the estimate made in the Egerton Committee Report of 1945 for the damage to the community from the emission of sulfur dioxide. When the estimate of 1945 was corrected for the increased prices and inflation to 1953, the estimated damage was calculated to be 6s 8d ($0.92) per ton of coal.

Although the fluid-scrubbing processes discussed are the lowest in cost of the processes yet conceived and proved feasible, their costs are still excessive in that they would require heavy capital investment and would materially increase the cost of the power or other product of the combustion of fuels containing sulfur. In addition, these fluid-scrubbing processes so lower the temperature of the flue gases that they do not rise and disperse as they would when hot. Thus, even with greatly reduced sulfur content, they may be of more nuisance and of more potential damage than the gases before scrubbing. A process that will remove the sulfur without major reduction in temperature is badly needed.

[109] Rees, R. L., *J. Inst. Fuel*, **25**, 351–7 (1953).

CHAPTER 19

COAL ASH—ITS EFFECT ON BOILER AVAILABILITY

F. G. ELY and D. H. BARNHART

The Babcock & Wilcox Co. Alliance, Ohio

Coal is an abundant natural resource in many countries of the world. Despite its value as a source of organic chemicals, its predominant use is as a fuel for the production of heat and power. In both respects, it is competitive with petroleum and natural gas. It is grossly distinguished from them, however, in being a solid instead of a fluid material, and in containing substantial proportions of nonfuel impurities or "ash," which form a residue when the fuel is burned. These properties, particularly the latter, impose difficulties upon its utilization.

Elementary steps in the use of coal for power generation[1] comprise: (1) conversion of its stored chemical energy to the form of thermal energy at high temperature by combustion with air in a furnace; (2) transfer of heat from the products of combustion to an intermediate working fluid; (3) expansion of the working fluid through a prime mover to produce mechanical energy; and (4) conversion of mechanical energy to electrical energy for distribution and application. The usual working fluid is steam, generated at high pressure and superheated to higher temperatures before delivery to an engine or turbine. The spent products of combustion, principally gases, are discharged at lower temperature to the atmosphere. Many arrangements are possible in the assembly of components forming the boiler unit.[2,3] One such arrangement is shown for reference in Fig. 1.

[1] Solberg, H. L., Cromer, D. C., and Spalding, A. R., *Thermal Engineering*, John Wiley and Sons, New York, New York, 1960, 649 pp.

[2] Combustion Engineering Co., *Combustion Engineering*, Riverside Press, Cambridge, Massachusetts, 1947, Chap. 15.

[3] Babcock and Wilcox Co., *Steam*, Thirty-seventh edition, McKibbin, New York, 1955.

The purpose of this chapter is to review significant literature of the past two decades pertaining to the effects of coal impurities on the performance and availability of steam-generating equipment.

At the beginning of the period, war-time demands accented the need for sustained operation and aggravated the causes of difficulty by requiring excessive loading of equipment and curtailment of cleaning and maintenance procedures.[4] General relief was afforded at the end of the war by return to normal operation. Subsequent economic growth has necessitated more than three-fold increase in generating capability, which has been met by installation of new, and in many cases, advanced types of equipment.

Post-war developments[5] are marked by the great increase in size and capacity of individual units; complete water cooling and omission of brickwork from the furnace-wall structure; transition from fuel-bed to pulverized-coal or cyclone firing for the larger coal-burning units; more widespread use of oil or gas as principal or alternate fuels; increase in design pressure

[4] Bailey, E. G., *Mech. Eng.*, **65,** 770–2 (1943). Christie, A. G., *Mech. Eng.*, **71,** 715–8 (1949). Crossley, H. E., and Marskell, W. G., *Proc. Joint Conf. on Combustion, Inst. Mech. Engrs. (London)*, **1955,** pp. 72–5.

[5] McClure, J. B., and Mellor, A. G., *Mech. Eng.*, **78,** 521–8 (1956). Gumz, W., *Proc. 2nd Conf. on Pulverized Fuel, Inst. Fuel (London)*, **1957,** pp. D11–25. Powell, E. M., and Argersinger, J. I., *Mech. Eng.*, **80,** No. 11, 72–5 (1958); Wilson, D. R., *Mech. Eng.*, **80,** No. 11, 75–8 (1958); Stewart, D. B., *Mech. Eng.*, **80,** No. 11, 78–81 (1958); Hawley, C. F., *Mech. Eng.*, **80,** No. 11, 81–3 (1958). Rowand, W. H., and Kispert, E. G., *Mech. Eng.*, **82,** No. 4, 103–4 (1960).

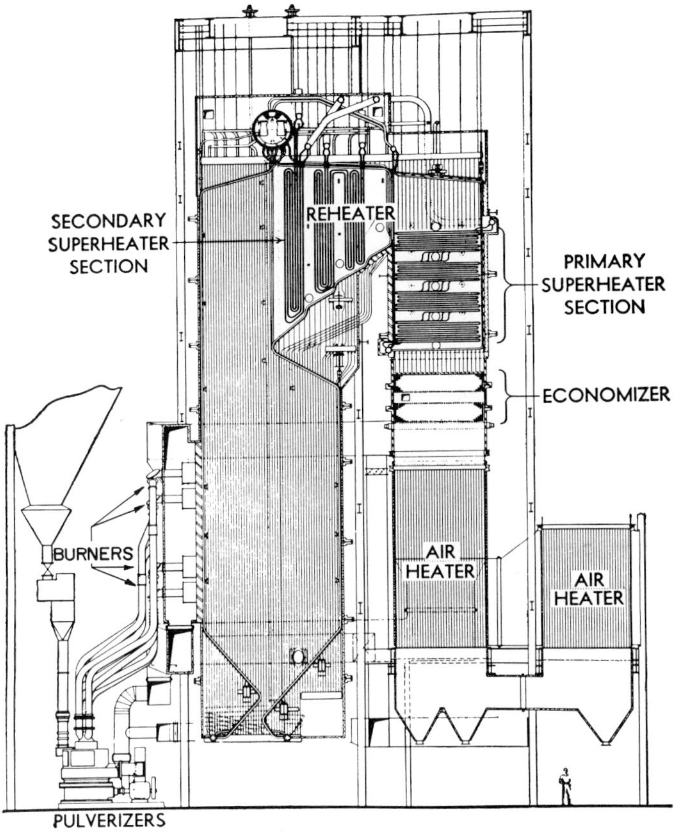

Fig. 1. Steam generating unit. Pulverized-coal fired, dry-ash removal.[3]

and temperature of steam; extensive application of automatic controls; adoption of single-boiler-turbine installations; and use of steam-reheat cycles.

These trends, in conjunction with improved technology in boiler design and operation, have been accompanied by changes in direction and emphasis of investigative effort.[6] Clinkering in fuel beds is no longer an urgent problem. Obstructive fouling of convection heating surface has diminished in prominence and is now rarely a cause of forced outage. Closer margins are imposed, however, on control of steam and metal temperatures,[7] which are affected by the extent of ash or slag deposits on the heat-absorbing surfaces, and which are severely limited by corrosive characteristics of some types of ash and reaction products. A considerably increased share of attention has been given to problems of related nature in the fields of oil firing and gas-turbine development, notably with respect to the reactions of sulfur and of ash components involved in fouling, corrosion, erosion, and atmospheric pollution.

Since the efficiency of energy conversion is dependent on attaining high initial steam temperature and low exit-gas temperature, continuing efforts are being made to extend the practical limits of these values. High specific rates of output from equipment are desirable for economic reasons. Reliable service and availability are of even greater economic importance. It is obvious that many factors unrelated to the fuel, such as quality of feed-water and the adequacy of auxiliary equipment, are vitally involved in such performance. Nevertheless, the fuel is by far the greatest source of the contaminating materials that are responsible for some of the most complex, obscure, and variable problems encountered in design and operation of steam-generating equipment.

[6] Lawton, F. W., *J. Inst. Fuel*, **32**, 120–43 (1959).

[7] Rowand, W. H., and Kessler, G. W., *Trans. 5th World Power Conf.*, **10**, 3221–35 (1956).

TABLE 1

CHARACTERISTIC MINERALS IN COAL ASH[11]

Group	Typical Species and Approximate Formula	
Shale	Muscovite (common potash mica), $K_2O{\cdot}3Al_2O_3{\cdot}6SiO_2{\cdot}2H_2O$	
	Illite (soda mica), $Na_2O{\cdot}3Al_2O_3{\cdot}6SiO_2{\cdot}2H_2O$	
	Montmorillonite (clay), $(Mg,Ca)O{\cdot}Al_2O_3{\cdot}5SiO_2{\cdot}nH_2O$	
Kaolin	Kaolinite (clay), $Al_2O_3{\cdot}2SiO_2{\cdot}2H_2O$	
	Metahalloysite, $Al_2O_3{\cdot}2SiO_2{\cdot}4H_2O$	
Sulfide	Pyrite and Marcasite, FeS_2	
Carbonate	Calcite (limestone), $CaCO_3$	
	Dolomite, $CaCO_3{\cdot}MgCO_3$	
	Ankerite, $2CaCO_3{\cdot}MgCO_3{\cdot}FeCO_3$	
Chloride	Halite (common salt), NaCl	
	Sylvite, KCl	
Accessory Minerals	Occurring in minor quantities associated with the Shale group	
	Quartz, SiO_2	Apatite, $9CaO{\cdot}3P_2O_5{\cdot}CaF_2$
	Felspar, $(K,Na)_2O{\cdot}Al_2O_3{\cdot}6SiO_2$	Zircon, $ZrSiO_4$
	Garnet, $3CaCO{\cdot}Al_2O_3{\cdot}3SiO_2$	Diaspore, $Al_2O_3{\cdot}H_2O$
	Hornblende, $CaO{\cdot}3FeO{\cdot}4SiO_2$	Magnetite, Fe_3O_4
	Gypsum, $CaSO_4{\cdot}2H_2O$	Hematite, Fe_2O_3

IMPORTANCE OF ASH

All types of coal contain substances other than the useful combustible matter.[8] In general, these comprise heterogeneous mixtures of inorganic compounds, collectively designated as "ash" or as "mineral matter." The ash content, ranging from 5% to 40% by weight of the coal, represents a mass of material which must be carried through the combustion process and ultimately removed for disposal. Although some clinker is used as light-weight aggregate and some fly-ash in concrete, in general the ash has little economic value.

The presence of ash is accounted for by minerals associated with initial vegetal growth or which entered the coal seam from external sources during or after the period of coal formation. Appreciable quantities of inorganic materials may be contributed to the commercial fuel by partial inclusion of adjacent rock strata in the mining process.[9]

[8] Selvig, W. A., and Gibson, F. H., *U.S. Bur. Mines, Bull.* No. 567 (1956), 33 pp.

[9] Grounds, A., and Wandless, A. M., *J. Inst. Fuel*, **25,** 170–7 (1952).

Identification, origins, and occurrence of mineral species have been reported by many investigators and are discussed in several excellent reviews.[10–13] Nelson has classified the principal forms into six characteristic groups[11] as shown in Table 1. The first four groups usually account for more than 95% of the ash-forming material. Their relative proportions are significant, and trace quantities of other elements in the remainder may have important effects. Many of these constituents undergo physical and chemical changes at the high temperatures encountered in combustion, by decomposition, melting, volatilization, and accelerated chemical activity. The behavior of specific minerals and of significant combinations at high temperature has also been

[10] Thiessen, G., *Chem. of Coal Utilization*, John Wiley and Sons, New York, 1945, Vol. 1, pp. 485–95.

[11] Nelson, J. B., *BCURA Bull.*, **17,** 41–55 (1953).

[12] Flint, D., *BCURA Bull.*, **17,** 133–42 (1953).

[13] Battelle Memorial Institute, *Corrosion and Deposits in Coal- and Oil-fired Boilers and Gas Turbines*, ASME, New York, 1959, H. W. Nelson, pp. 7–29.

studied and is discussed in several reviews.[13,14]

Although the character, quantity, and distribution of minerals may vary considerably within the coal seam, much of this diversity is masked in the as-fired fuel as a result of cutting, transporting, size reduction, washing, classifying, storing, and pulverizing—operations which tend to distribute the mineral matter content more uniformly throughout the coal. Large differences do exist, however, in the ash characteristics of coals from different mines or regions. To some extent they can be identified with the coal and with methods of mining and preparation, thus providing a rough basis for selection of fuel or choice of burning equipment.

Since quantitative evaluation of mineral forms is extremely difficult, the composition of coal ash is customarily determined by chemical analysis of the residue produced by burning a sample of coal at a slow rate and at moderate temperature (1350°F) under oxidizing conditions in a laboratory furnace. It is thus found to be comprised chiefly of compounds of silicon, aluminum, iron, and calcium, with smaller amounts of magnesium, titanium, sodium, and potassium.[8] Although reported as oxides of the principal or significant elements, the constituents are actually present as silicates, oxides, sulfates, or other compounds, and may differ appreciably from the original forms of mineral matter, as well as from the eventual compounds that may develop in operating furnaces. Alumina-silica compounds of the kaolin group are primarily refractory. Ferrous iron, calcium, and magnesium act as fluxes and tend to form complex silicates of lower melting temperatures than their oxides. The alkali metals also act as fluxes, and even in small percentages they play an important role in tube-bank fouling because of volatilization and later concentration in certain sections of the boiler unit. Sulfur, which is present in the fuel both in organic and in mineral combination (notably iron sulfide), has received a great deal of attention because of the reactions of SO_2 and SO_3 in fouling and corrosion.

The formation of massive "clinkers" in fuel beds, particularly if they become attached to furnace wall or grate surfaces, disturbs the normal flow and distribution of fuel and combustion air and may also cause damage to equipment. Accumulation of ash or slag on steam generating and superheating surfaces retards heat transfer. Obstruction of gas flow passages through tube banks increases the draft loss, and may lead to maldistribution of flow and temperature, reduction of capacity, or forced shutdown. Corrosive compounds formed during or after combustion may attack pressure parts or supporting structures. Severe abrasion of the heat-transfer surfaces by particles carried in the gas stream is sometimes experienced. Atmospheric pollution by effluent solids and gases extends the field of concern beyond the immediate environs of plant and equipment.

Although coal is purchased primarily for its calorific value, and allowance can readily be made for the diluent effect of ash content, increasing discrimination is being made with respect to identifiable characteristics of the ash that cause difficulty in boiler operation.[15,16] Plant-scale test and experience provide the most reliable basis for evaluation[17] but they do not usually permit direct application of technical information to other conditions because of sheer size of equipment and multiplicity of factors involved in operation. The more flexible methods of laboratory and pilot-scale investigation, under closer control and analysis, yield more fundamental information but cannot duplicate the influence of size, continuous running, and statistical effects of incompletely recognized or unmeasurable factors existing in full-scale operation.

In consequence, many hypotheses are proposed for the solution of operating problems which are supported by evidence from test and observation, which eventually may be confirmed, disqualified, or outdistanced by advances made in the engineering and practical art of boiler design and operation. Significant gains in technical knowledge become accepted and applied in this advance. With successive

[14] Whittingham, G., *BCURA Bull.*, **18**, 581–90 (1954). Watt, J. D., *BCURA Bull.*, **23**, 49–69 (1959). Gumz, W., Kirsch, H., and Mackowsky, M. T., *Schlackenkunde*, Springer, Berlin, 1958, Chap. 6.

[15] Simonson, W. F., *J. Inst. Fuel*, **25**, 226–34 (1952). Wolin, R. H., *Mining Congr. J.*, **44**, 39–42 (1958). Hubert, D. G., *Combustion*, **30**, No. 10, 38–45 (1959).

[16] Gumz, W., *Combustion*, **27**, No. 10, 47–54, No. 11, 53–4 (1956).

[17] Marskell, W. G., and Miller, J. M., *J. Inst. Fuel*, **29**, 380–7 (1956).

changes in form and size of equipment or required performance, new problems emerge which in turn demand investigation. The quality of investigation is correspondingly enhanced by contribution from scientific concepts and advancement in techniques and measuring devices.

It seems certain that ash will never cease to be an objectionable factor to contend with in the use of coal as fuel. Although separation and removal of some ash fractions is practiced to an increasing extent in commercial production[18] economic as well as practical limits are encountered and the trend toward dependence on lower quality reserves is widely recognized.[9,19] More competent understanding of the nature and behavior of ash is, therefore, important to further progress in the use of coal for power generation.[16,20]

ASH REMOVAL FROM BOILER FURNACES

There are two principal methods of burning coal, (1) by means of a fuel bed[21] in which the fuel is nominally at rest upon a grate while the combustion air is passed through it and (2) by suspension burning[22,23] in which the coal after being pulverized is blown into the furnace cavity by the air stream. Intermediate forms are represented by the spreader stoker[24] and by the cyclone method of firing.[25] Limitations on burning rate per square foot of grate area restrict the fuel bed to units of approximately 200,000 lb steam/hr maximum capacity. Above this, to units of 5,000,000 lb steam/hr capacity—equivalent to 750 megawatts electrical, the field is pre-empted by pulverized-coal and cyclone-furnace firing.

For continuous operation, removal of ash is essential to all methods. With the fuel bed, it is accomplished by intermittent shaking or dumping of grate sections, or by continuous slow movement of residual material toward a point of discharge brought about by motion of the grate members.[26] In suspension firing, the ash particles are either carried out of the furnace by the gas stream, or retained by settling or by adherence to boundary surfaces. Retained material of solid form is removed by periodic cleaning. If temperatures are sufficiently high, the retained ash is molten and may be drained continuously through a tap hole. Slag thus discharged is chilled and disintegrated to granular solid form by quenching with water in a disposal pit. For either firing method, cinders or fly-ash particles carried through the boiler are collected by settling in hoppers, by inertial, electrostatic, or water-scrubbing separators, or are discharged with the gases to the atmosphere. Characteristic distribution of ash for different methods of firing is shown in Fig. 2 and is discussed by Grossman,[27] and others.[28]

Fuel Bed Burning

The temperatures developed in fuel beds are influenced primarily by rate of burning and, to some extent, by rank and sizing of coal, mechanics of firing and thickness of bed.[29] Values as high as 2800 to 3100°F may be attained.

Concentration of ash occurs by progressive burning of combustible matter. Coalescence and mixing of fused ash particles and comparatively long retention time at high temperature promote reactions that frequently produce mineral species or glasses melting at

[18] Young, W. H., Anderson, R. L., and Hall, E. M., *U.S. Bur. Mines, Minerals Yearbook, 1959*, **2,** 93–7.

[19] Daniel, G. H., *Proc. 2nd Conf., on Pulverized Fuel, Inst. Fuel, London*, **1957,** pp. A1–9. Pracht, P., Hennecke, H., and Redottee, G., *Mech. Eng.*, **82,** No. 4, 109–10 (1960).

[20] Landry, B. A., *Mech. Eng.*, **81,** No. 6, 70–3 (1959).

[21] Mayers, M. A., *Chem. of Coal Utilization*, John Wiley and Sons, New York, 1945, Vol. 2, pp. 1482–521.

[22] Orning, A. A., *Chem. of Coal Utilization*, John Wiley and Sons, New York, 1945, Vol. 2, pp. 1522–67.

[23] Rosin, P. O., *J. Inst. Fuel*, **31,** 346–55 (1958).

[24] Beers, R. L., *Mech. Eng.*, **64,** 867–71 (1942).

[25] Grunert, A. E., Skog, L., and Wilcoxson, L. S., *ASME, Trans.*, **69,** 613–34 (1947).

[26] Murray, M. V., *BCURA Bull.*, **13,** 333–47 (1949).

[27] Grossman, P. R., *J. Air Pollution Control Assoc.*, **7,** 226–36 (1957).

[28] Kaiser, E. R., *Combustion*, **22,** No. 11, 53–7 (1951). Seidl, H., *Proc. Joint Conf. on Combustion, Inst. Mech. Engrs. (London)*, **1955,** pp. 92–103. Grasme, P., *Brennstoff-Wärme-Kraft*, **8,** 278–84 (1956); *Combustion*, **28,** No. 8, 34–41 (1957).

[29] Marskell, W. G., and Pratt, C. W., *J. Inst. Fuel*, **25,** 212–21 (1952).

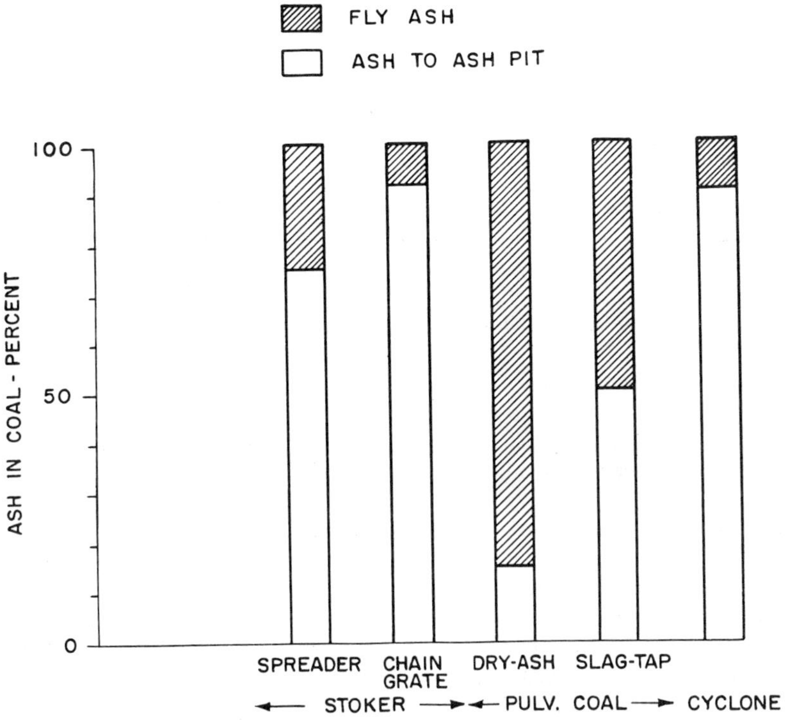

Fig. 2. Typical ash balance for principal firing methods.[27]

lower temperature than the ash does. Eventual cooling results in solidification of fused conglomerates and entrapment of incompletely burned pieces of fuel which are carried to waste. The processes and mechanisms involved in producing "clinker" are extremely varied and complex, and the degree of clinkering or its effects on operation are not amenable to precise definition.[30] Small or moderate-sized formations are advantageous in some respects because of improved character of refuse for handling and decreased amount of dust.[29,31] Large growths in the form of sheet or massive clinkers are difficult to remove and may cause loss of capacity or forced shutdown.

The tendency to form massive clinkers is most pronounced with underfeed stokers, because of greater agitation and thickness of bed normal to this method of firing. It presented a major problem with multiple-retort stokers,[32] which were widely used for power generation prior to World War II. This type is no longer being manufactured in the United States. The single-retort underfeed stoker continues to be popular for small industrial units burning caking coals. Clinker formation is usually moderate in the thinner and less-agitated beds of chain or traveling grate stokers, and on dump-grate or traveling grates used with spreader firing.

Slag deposits, comparable to clinker and often more troublesome, may form above the fuel bed as a result of air-blast entrainment of ash and fuel particles which strike and adhere to the furnace wall.[29,33] In extreme cases, these formations grow to large size and protrude into the bed, interfering with movement of the fuel and promoting the formation of clinker.

Ash-Fusing Temperature

The earliest technical investigations of ash behavior were concerned with evaluation of clinkering tendencies of different coals. Extensive studies in the United States, England,

[30] Nicholls, P., and Selvig, W. A., *U.S. Bur. Mines, Bull.* No. 364, 71 pp.

[31] Harrington, J., *J. Inst. Fuel*, **14**, 230–5 (1941).

[32] Ricketts, E. B., *U.S. Bur. Mines, Bull.* No. 364 (1932), 71 pp. Tobey, J. E., *ASME, Trans.*, **60**, 17–24 (1938).

[33] Mumford, A. R., *Combustion*, **14**, No. 10, 43–7 (1942).

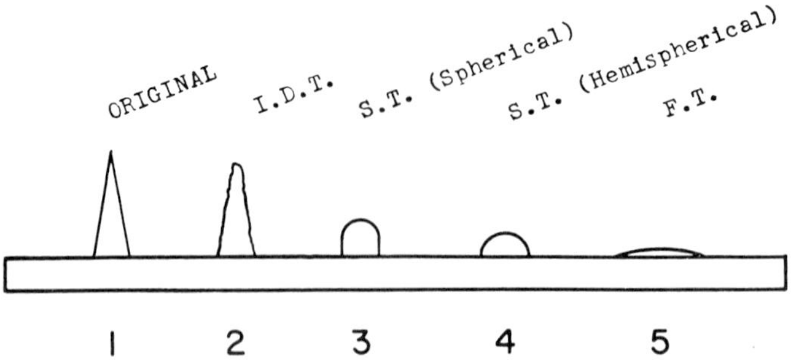

Fig. 3. Ash fusion test—ASTM method.[34]

and Germany have culminated in three standardized test methods[34–36] that are currently widely used. Although they differ in apparatus and measuring techniques, all are based on observing the temperature at which successive characteristic stages of fusion occur in a specimen of ash when heated in a laboratory furnace under specified conditions of atmosphere and rate of temperature rise. The subject has been reviewed by Barrett.[37] Although empirical in nature, the results are reasonably comparative and reproducible.

In the ASTM method,[34] the sample of finely ground ash is pressed into a standard mold to form a slender triangular pyramid $\frac{3}{4}$-in. high (Fig. 3) which is set vertically upon a ceramic base within a gas-fired or electrically heated furnace. The atmosphere is maintained in a mildly reducing condition by excess fuel gas or by using prepared gas mixtures. Temperatures are measured by optical pyrometer or thermocouple. The Initial Deformation Temperature (I.D.T.) is noted when the sharp apex or edges of the specimen first become rounded. With slow increase of temperature, melting proceeds and the Softening Temperature (S.T.) is recorded when the specimen has deformed to approximately spherical shape. A final measurement of Fluid Temperature (F.T.) is made when the ash has become completely fused and has spread to a nearly flat liquid film upon the base. Recent practice has recognized the S.T.–hemispherical (height equal to half-width) as a more identifiable and more commonly attained condition than the spherical shape. It occurs at slightly higher temperature than the S.T.–spherical. It is expected to be adopted as the preferred reference standard.

[34] ASTM, D-271-48 (1948).

[35] Brit. Standards, BS 1016 (1942).

[36] DIN Standards, 51730 (1954).

[37] Barrett, E. P., *Chem. of Coal Utilization*, John Wiley and Sons, New York, 1945, Vol. 1, pp. 496–572.

The S.T. value alone is frequently reported as the index of ash fusibility. It serves to indicate broad distinctions, but it is only partly descriptive of melting behavior. Inasmuch as coal ash is a mixture rather than a compound with a well-defined melting point, significance is attached to all reference temperatures and to the intervals between them, which vary appreciably for different ashes. The I.D.T. is broadly identified with a condition of surface stickiness, the S.T. with plastic distortion or sluggish flow, and the F.T. with liquid mobility. These properties are related not only to the mechanics of clinker formation, but also the the nature of clinker or slag produced. The so-called "long" slags tend to develop tough dense formations; those of "short" fusion interval, which freeze abruptly, are more likely to form porous or friable structures.[16]

With oxidizing conditions in the test furnace, many ashes show higher values of fusion temperature than are observed in reducing atmosphere (Fig. 4).[3] The differences are greater for those of high iron content because of the the difference in fluxing action of ferrous iron from that of ferric iron.[38] Although the reducing atmosphere provides a more dependable basis for standardization of test measurements and is regarded as characteristic of the fuel bed, oxidizing conditions in local areas or in other parts of the furnace may alter the behavior of the ash, independently of actual change in temperature. This effect is significant in the fouling of heating surfaces as discussed in later sections.

[38] Bailey, E. G., and Ely, F. G., *ASME, Trans.*, **63,** 465–71 (1941).

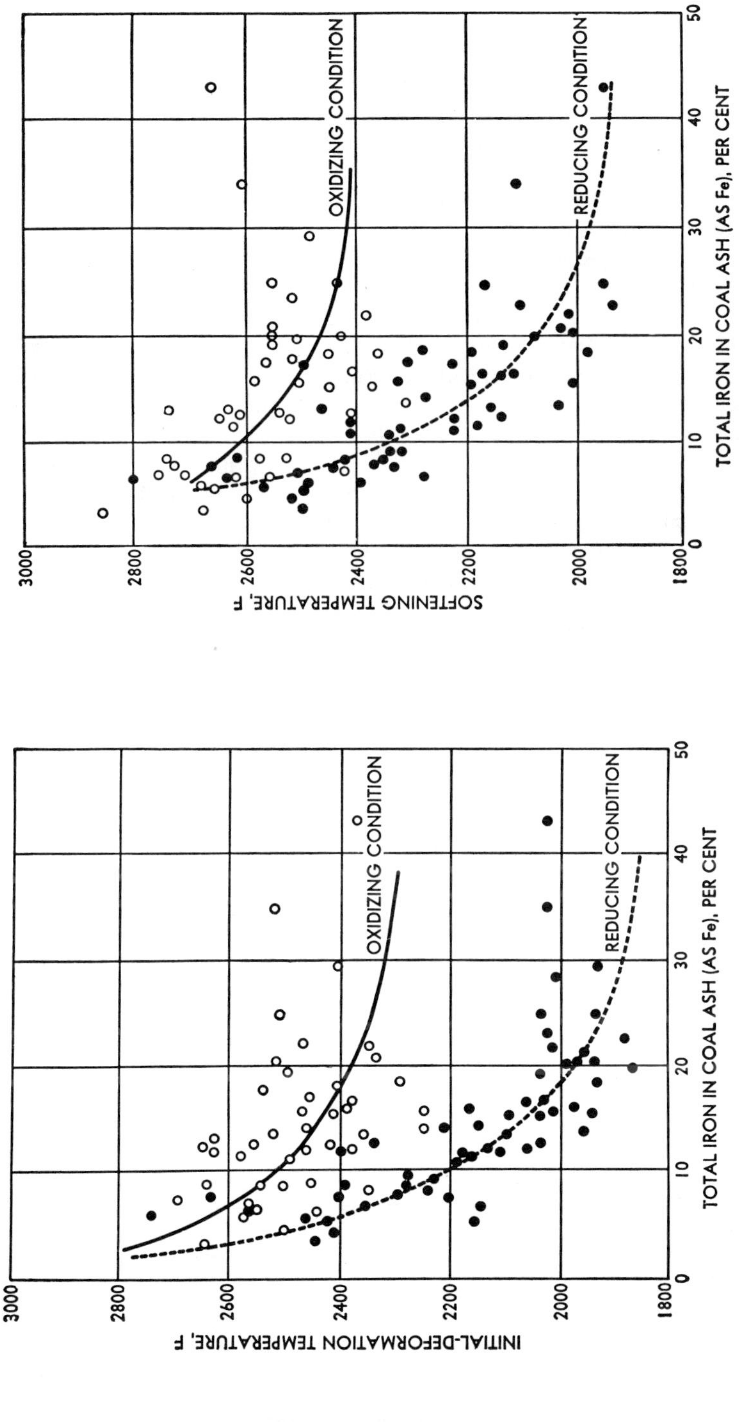

Fig. 4. Influence of iron on coal-ash fusion temperatures.[3]

Many writers have called attention to inconsistencies between the indications of fusion test data and observed clinkering behavior of ash from different coals in operating furnaces.[29,30,37,39] These inconsistencies arise in part from sampling discrepancies and also from disparity of conditions between the finely ground and well-mixed test specimen, and the indeterminate effects of segregation, movement, temperature, and reaction history of ash constituents in the fuel bed. Nevertheless, the test continues to be reported as a principal reference, possibly because of simplicity of concept and measurement, and for lack of superior methods.[40] Continuing efforts are being made to improve interpretation and to attain uniformity of techniques and agreement of results from different laboratories.

Cinder Carry-Over

Some particles of fuel are inevitably lifted from the bed by air or gas entrainment and are carried out of the furnace as "cinders" in a partially burned condition.[28,29] This action is aggravated by high rates of operation, fine sizing of coal, and irregular distribution of airflow through the bed. It is more pronounced with spreader-stoker firing,[41] in which method all particles of fuel are thrown into the furnace space, and a high percentage of fines ignite and burn in suspension without falling to the grate. Return of cinders to the furnace is practiced in many cases for recovery of combustible as well as for disposal of residue to the ash pit. It is desirable to introduce them at a quiescent portion of the grate and under a layer of fresh coal if possible, to prevent re-entrainment and recycling of an excessive dust burden through the boiler passes.

The "sand-blast" action of cinder carry-over from spreader stokers is partially beneficial in that it tends to prevent the formation of ash deposits on boiler and superheater surfaces. Under extreme conditions, this action causes wear. Problems of erosion are discussed in a later section.

Suspension Burning

Furnaces for pulverized-coal firing are specifically designed for dry or liquid ash removal.[3,5]

[39] Dunningham, A. C., and Grumell, E. S., *J. Inst. Fuel*, **14**, 221–5 (1941). Cave, G. A., *BCURA Bull.*, **11**, 141–9 (1947).

[40] Edgecombe, L. J., and Manning, A. B., *J. Inst. Fuel*, **25**, 166–71 (1952).

[41] Powell, E. M., *Mech. Eng.*, **71**, No. 11, 947 (1949).

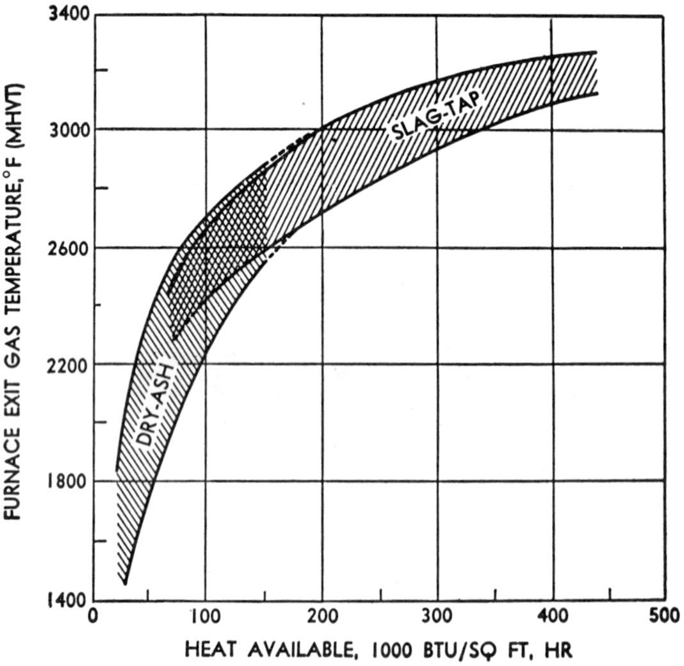

Fig. 5. General range of furnace exit temperature for dry-ash and slag-tap pulverized-coal firing.[3]

The condition of the ash depends on the temperature attained in relation to its composition or fusion characteristics, and this in turn is affected by the amount of cooling or loss of heat that occurs during and after burning of the combustible matter. All of the ash associated with the finely ground fuel particles is initially dispersed in the gas stream. The time of transit through the furnace, nominally 1 to 3 sec, is extremely short. Cooling takes place chiefly by radiation to surrounding water-cooled surfaces.

Coals having an ash of high fusion temperature (S.T. above 2600°F) are particularly suited to dry-ash operation; ashes of low fusion temperature (S.T. 2200°F and below) are suited to liquid ash removal. Most coals are of intermediate characteristics, but virtually all can be used with either method in units having exposed cooling surface or other means of temperature control correctly proportioned to the rate of heat release. Since temperatures vary with rate of firing, other conditions being the same, it is apparent that limits will be encountered, insofar as ash condition is concerned, at high rates for the dry-ash furnace and at low rates for the slag-tap furnace.

Temperatures within the furnace are affected by location and type of burners, by gas-flow pattern, and most significantly by arrangement and cleanliness of cooling surfaces. Because of the complexity of conditions influencing heat transfer,[42] estimation of heat absorption and temperature of gases at the furnace outlet must depend on empirical design factors established by experience with different furnace arrangements, knowledge of ash characteristics, and data acquired from plant-scale tests, as exemplified in Fig. 5. Gas temperature at the furnace outlet is of primary importance to the design and performance of the superheater. It also provides a reference for estimating the tendency of ash to deposit on the superheater tubes. Temperatures are shown on the basis of the multiple-shield high-velocity thermocouple (MHVT), which minimizes radiation error and approximates true gas temperature.[43]

[42] Thurlow, G. G., *BCURA Bull.*, **17**, 173–99 (1953).

[43] Mullikin, H. F., *Temperature*, Reinhold, New York, N.Y., 1941, pp. 775–804. Cohen, P., Corey, R. C., and Myers, J. W., *ASME, Trans.*, **71**, 965–76 (1949).

Slag Viscosity

The conventional measurement of fusion temperature gives a rough indication of ash behavior. Being relatively easy to obtain, this information is usually available and is taken into consideration. With respect to liquid slag removal, a more dependable guide is afforded by viscosity measurements, or calculation of viscosity from chemical analysis.[44,45] Measurement of viscosity is used chiefly for appraising the suitability of coals for slag-tap operation.

These studies showed that, at sufficiently high temperatures, all slag compositions became molten and had the characteristics of Newtonian liquids. Although widely different values of viscosity were observed, a consistent relationship was noted between viscosity and its rate of change with temperature which permitted correlation in terms of Equivalent Silica Content, as shown in Fig. 6 for 2600°F. The basis of correlation (referred to by others as Silica Ratio) is defined as:

$$\text{Equivalent Silica Content} = \frac{SiO_2}{(SiO_2 + \text{Total Iron as } Fe_2O_3 + CaO + MgO)}$$

Important differences appeared as the slags were slowly cooled below the liquid range (Fig. 7). A few having glassy characteristics showed uniform increase to higher values of viscosity, reaching the condition of supercooled liquids without phase change. Others showed abrupt freezing or crystallization in a narrow range of temperature. Between these types many slags developed plastic characteristics, as solid phases formed within the melt causing increase of internal shear stress, through a wide temperature range before solidification. As shown in Fig. 7, the apparent viscosity of the plastic condition is appreciably higher than that of the liquid and is more sensitive to temperature change.

The point of transition, corresponding to the nominal liquidus temperature of the system, is termed the Temperature of Critical Viscosity (T_{CV}). It varies with specific composition of the slag. Since calcium and ferrous iron are the principal fluxing constituents, charts were

[44] Nicholls, P., and Reid, W. T., *ASME, Trans.*, **62**, 141–53 (1940).

[45] Reid, W. T., and Cohen, P., *ASME, Trans.*, Special Section, **66**, 83–97 (1944).

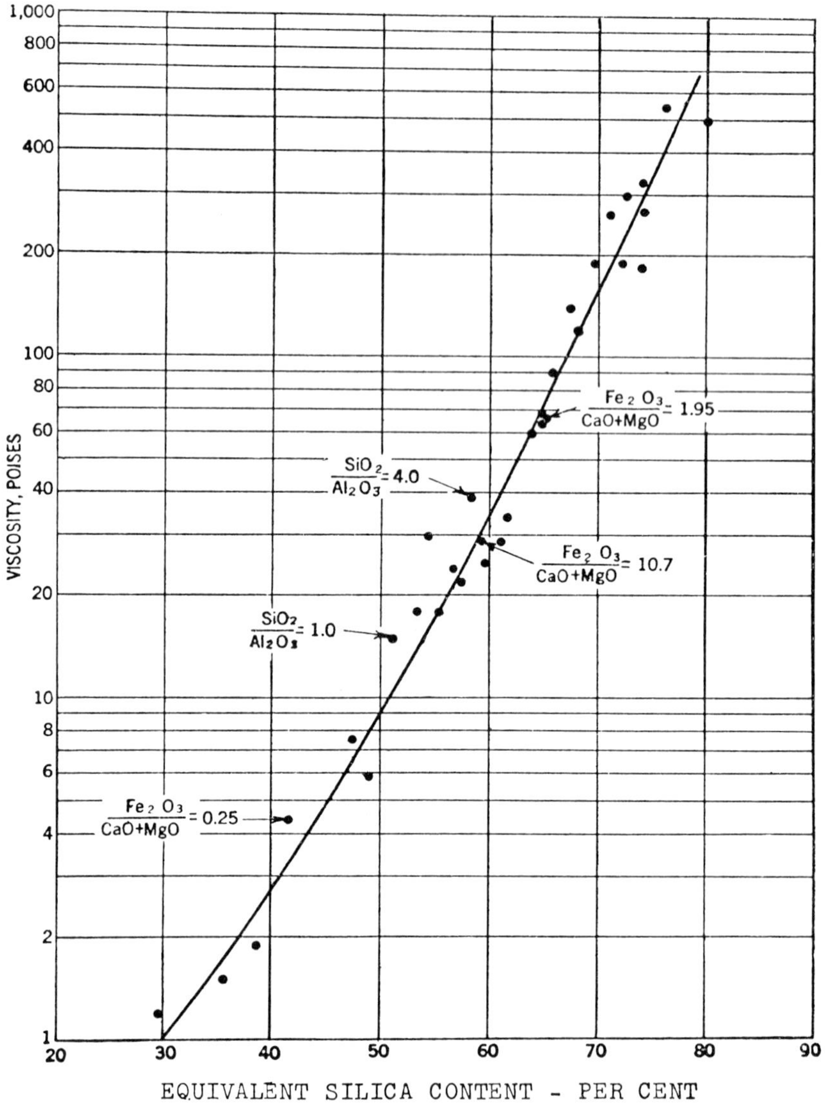

Fig. 6. Viscosity of coal-ash slags at 2600°F in air.[44] (Notations on individual points indicate wide range of compositions included in the correlation.)

compiled from the experimental data for separate evaluation of their effects on T_{CV}, Fig. 8. The state of oxidation of the iron is expressed as:

Ferric Percentage

$$= \frac{Fe_2O_3}{Fe_2O_3 + 1.11FeO + 1.43Fe}.$$

Actual measurement of viscosity is time consuming and requires specialized equipment not commonly available. Effort was therefore made to coordinate flow characteristics with the more readily obtained data on fusion temperature. Comparisons indicated that no unique value of viscosity could be identified with the I.D.T., S.T., or F.T. points, but fair correlation was found between the S.T. and T_{CV}. Work in other laboratories[46,47,48] has substantially corroborated the Reid and Cohen data and in some degree has extended the range

[46] Sage, W. L., and McIlroy, J. B., *ASME, Trans.*, **82,** 145–55 (1960).

[47] Heldt, K., *Brennstoff-Wärme-Kraft*, **7,** 321–4 (1955).

[48] Hoy, H. R., Roberts, A. G., and Wilkins, D. M., *J. Inst. Fuel*, **31,** 429–57 (1958).

of investigation. Shaw[49] has presented a comprehensive review of this subject, including a discussion of measuring techniques.

In boiler practice, tapping of slag in the liquid state is readily accomplished at viscosities of 150 poises or lower, and may be feasible for some compositions up to 250 poises at operating temperature. Tapping is more difficult and may become impossible for slags approaching the plastic condition. It is, therefore, important to recognize the approximate transition point in relation to factors in furnace operation which will provide a margin of temperature to maintain fluidity.

Sage and McIlroy[46] described a method of appraisal, resulting from laboratory and field investigation of United States coals, which employs the Reid and Cohen procedure for calculating viscosity of the fluid range and assigns an estimated value of T_{CV} at a temperature 200° above the S.T. determined by the conventional fusing test. As shown in Fig. 9,

[49] Shaw, J. T., *BCURA Bull.*, **23,** 170–90 (1959).

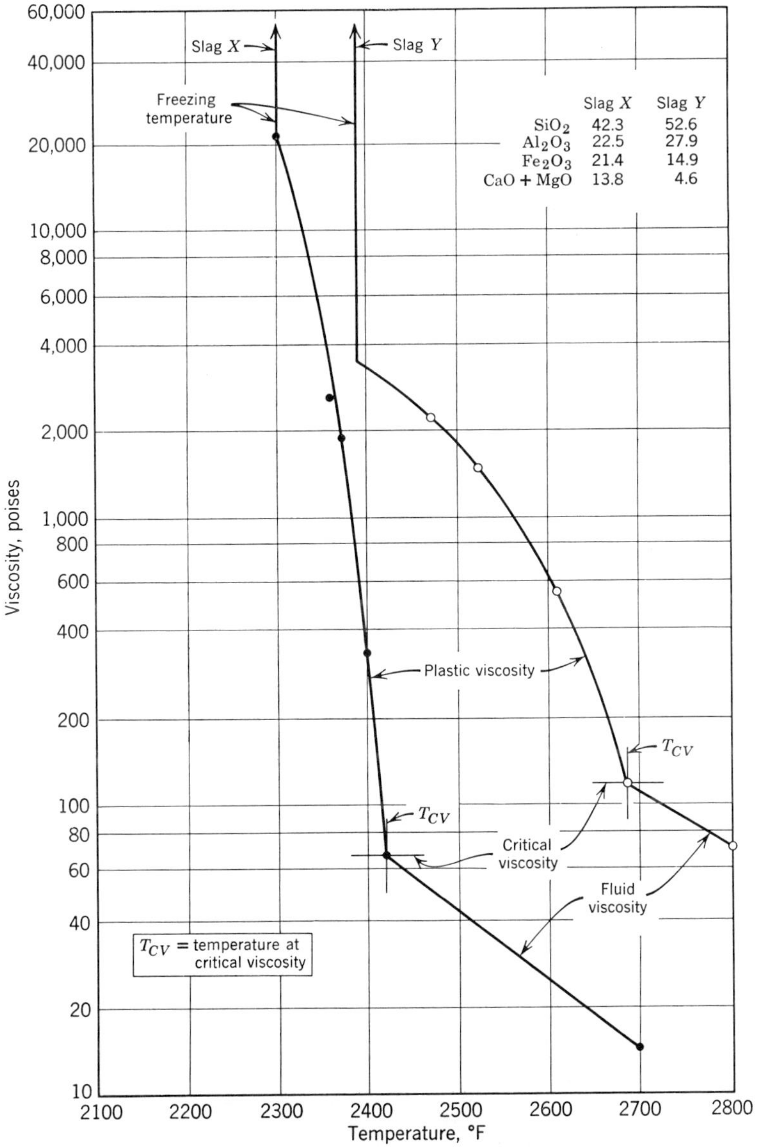

Fig. 7. Effect of cooling on solidification of two typical coal-ash slags, showing critical viscosity and region of plastic viscosity, for equilibrium conditions in air at each indicated temperature.[45]

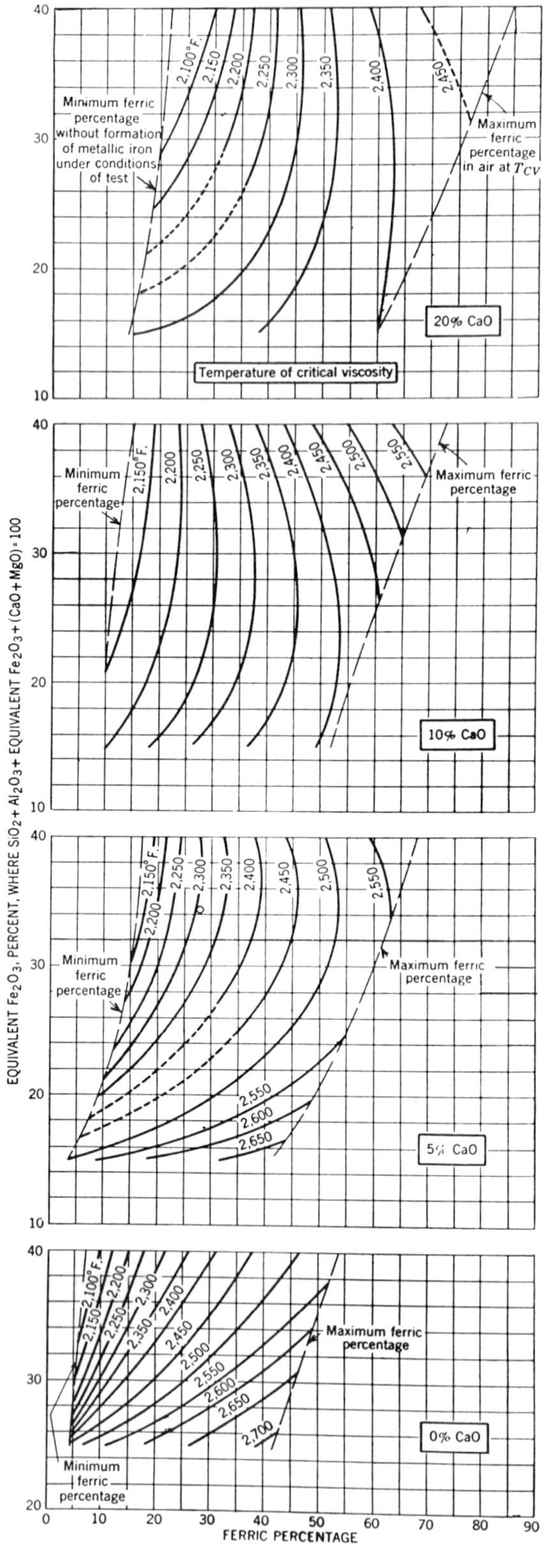

Fig. 8. Temperature of critical viscosity of coal-ash slags as a function of composition and ferric percentage for a silica:alumina ratio of 2.[45]

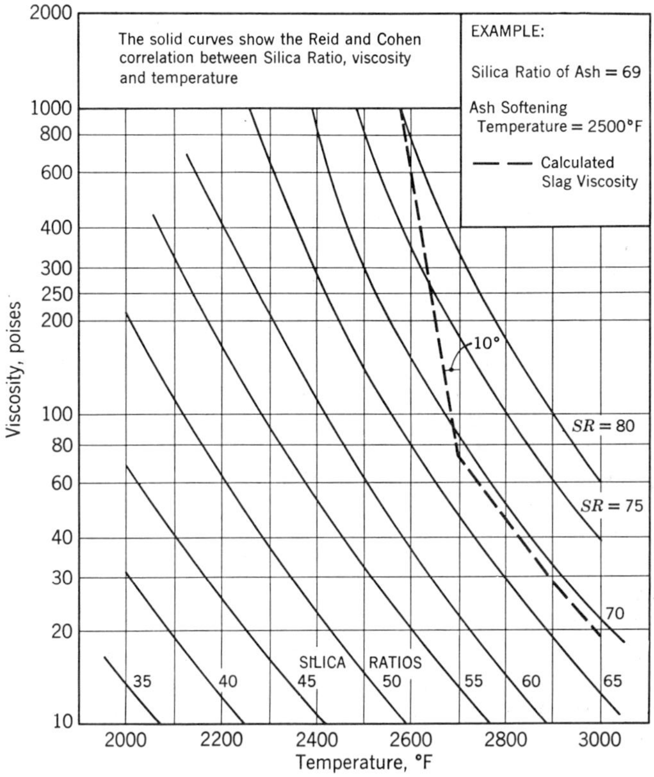

Fig. 9. Calculation of slag viscosity.[46]

viscosity for the plastic range is approximated by a line drawn from this point at an angle of 10° from the ordinate. In a more abbreviated manner, which omits the chemical analysis, Heldt[47] proposed a method of estimating fluid viscosity, Fig. 10, based on a general correlation between the ash-softening temperature of German coals and their measured viscosity at 1450°C. Variation of viscosity with temperature, as confirmed by Heldt's measurements, follows the characteristic Reid and Cohen curve. This method indicates the relative ease of tapping different slags when in the fluid condition, but does not identify critical temperatures of solidification.

Slag-Tap Furnaces

To develop high temperatures for tapping, burners are arranged for concentrating heat release in a small portion of the furnace, with the flame directed close to the floor and tap hole. In some designs, the zone of primary combustion is shielded from the main furnace by division walls and screen tubes to minimize the extent of exposed cooling surface. Adjacent walls become coated by an insulating layer of slag which tends to reach equilibrium thickness by flow from the molten surface.[50]

Tapping of viscous slag can be facilitated by blending coals[19,46] of suitable characteristics or by using fluxes added with the fuel or introduced to the slag pool on the furnace floor. Compounds of calcium, iron or alkali metals have sometimes been found effective.[46,51] With slags having appreciable iron content, an increase in the fluid range may be obtained by lowering the excess air supplied to the furnace or to individual burners. Under such conditions, however, incomplete oxidation of pyrite may result in an accumulation of molten iron sulfide or metallic iron at the bottom of an inadequately drained slag pool,[38,52] this

[50] Reid, W. T., and Cohen, P., *ASME, Trans.*, **66,** 685–90 (1944).

[51] Nicholls, P., and Reid, W. T., *ASME, Trans.*, **54,** 167–90 (1932).

[52] Caracristi, V. Z., and Mumper, H. D., *Mech. Eng.*, **81,** No. 8, 95 (1959).

accumulation can cause severe damage to the floor structure or solidify in massive formations that are difficult to remove.

Beyond the slagging zone it is desirable to prevent deposition on the furnace walls to obtain effective cooling of suspended ash particles and gases before they reach the superheater. Periodic removal by in-service cleaning equipment is often necessary in the intermediate zone.

Dry-Ash Furnaces

In a dry-ash furnace, larger space is provided for the combustion zone, with a relatively greater amount of exposed cooling surface. Fused deposits may form near the burners where molten particles are carried to the wall. At greater distance, they tend to be more scattered and of less-dense sintered structure, diminishing to a nominal dust-like coating in the cooler regions. When there is no molten drainage, the thickness of deposit is limited only by strength of its attachment to the wall. Dislodgement by gravity or by cleaning operations may occur in small or large masses which fall into a disposal hopper at the bottom of the furnace.

Retained Ash

As shown in Fig. 2, retention of ash is appreciably greater in the slag-tap furnace than in the dry-ash furnace because of the molten condition of particles and intercepting surfaces. A certain amount of initial segregation occurs due to more probable adherence of readily fused components. The time of retention in a high-temperature zone, for materials deposited on the walls or in a slag pool, is of course much longer than for particles carried by the gas stream, and opportunity is thus presented for mixing, chemical reactions, or other processes to take place in this portion of the total ash, with possible development of different materials, forms, and reaction products. Migration, selective flow, and sintering processes add strength to deposit formations. Evolution of gases may contribute to corrosion mechanisms.

Entrained Ash

Particles which are not arrested in the furnace are carried as "fly-ash" through the cooler regions of the steam-generating unit. Some physical deposition occurs on heat-absorbing surfaces, complicated by condensation of vapors as discussed in later sections. To reduce the amount of ash discharged to the atmosphere, principally for nuisance control, mechanical or electrostatic separators are employed almost universally on large pulverized-coal-fired installations.[53] Fly-ash thus recovered presents a difficult disposal problem because of its bulk and fine particle size. It is often returned to a slagging type of furnace for

[53] Fitton, A., *J. Inst. Fuel*, **25**, 178–81 (1952).

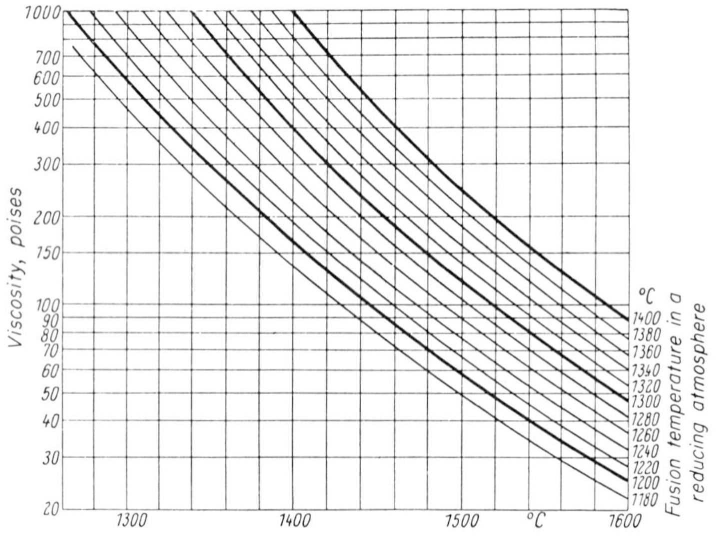

Fig. 10. Slag viscosity as related to fusion temperature.[47]

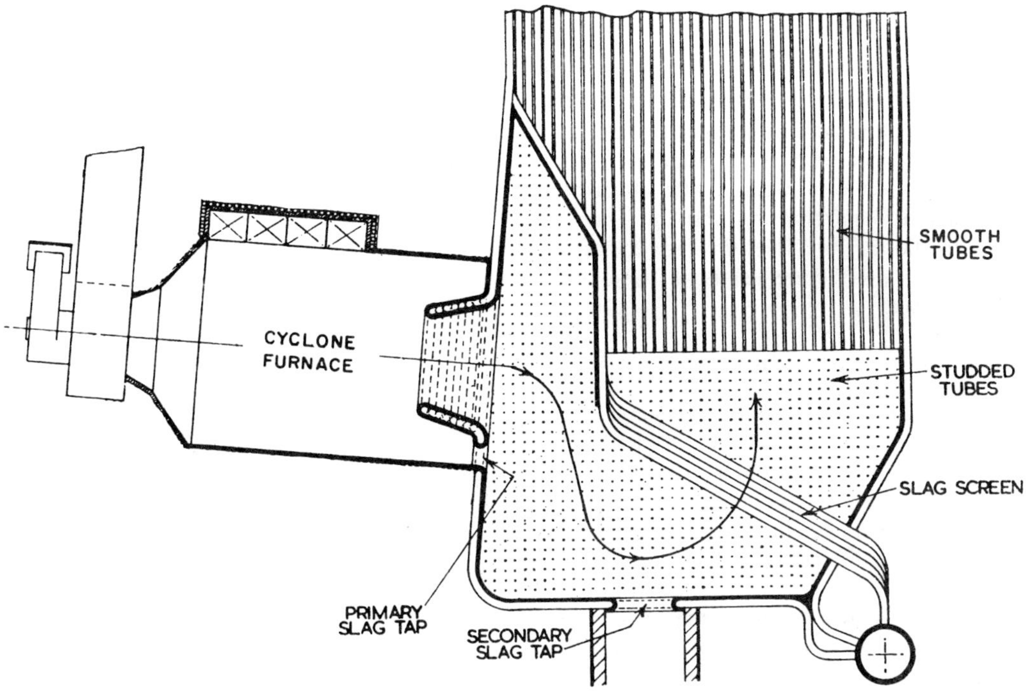

Fig. 11. Cyclone furnace.[57]

conversion to liquid form.[54] By virtue of its pozzolanic properties, fly-ash is of value for addition to concrete mixtures and can be economically marketed for this purpose in some localities.[55]

Cyclone Firing

The cyclone method of firing, Fig. 11, is of comparatively recent origin, but it has found wide application in the United States and Germany, and has been described in many articles.[25,56,57] This method is particularly adaptable to high-capacity units and to the use of low-grade coals. It is inherently restricted to slag-tap operation.

Coal crushed to $\frac{1}{4}$-inch maximum size is fed in such a manner as to be distributed to the wall surface of the cylindrical water-cooled combustion chamber. Tangential admission of air at high velocity creates centrifugal forces and accelerated rates of burning that result in high retention of particles and rapid melting of ash constituents. Gases are discharged through a re-entrant throat connected to a secondary furnace. The molten slag, which forms as a coating over the cyclone enclosure, flows toward a tap-hole for discharge into the secondary furnace.

The presence of a slag film is essential to proper functioning because it entraps incoming particles of fuel and holds them against the velocity of the air stream and thus promotes high burning rates and development of high temperatures. In this respect, the ash is beneficial to the combustion process, and can be tolerated in large percentages. Cyclone firing provides a ready means for disposal of fly-ash collected from other units and for the use of high-ash coals. It is not suitable, however, for ash of high-fusion temperature which cannot be handled in other forms of slag-tap operation.

Some of the finer fuel particles burn in suspension, and their ash is carried by the gas stream that issues from the furnace throat. Minor quantities of coarse slag drops and fuel

[54] Feeley, F. G., *ASME, Trans.*, **78,** 1747–55 (1956). Maull, W. M., and McMahon, E. C., *Mech. Eng.*, **78,** No. 1, 53 (1957).

[55] Dawson, S. H., *J. Inst. Fuel*, **31,** 401–5 (1958).

[56] Kessler, G. W., *Proc. Am. Power Conf.*, **16,** 78–90 (1954).

[57] Roberts, A. G., *BCURA Bull.*, **24,** 206–54 (1960).

particles, entrained at the throat and taphole, contribute to the formation of deposits on secondary furnace walls and screen tubes, with effects that are comparable to pulverized-coal firing, although slower in rate of accumulation. Slag discharged from the cyclone, in addition to that collected in the secondary furnace, is drained through a taphole in the secondary furnace floor. As shown in Fig. 2, approximately 90% of the fuel ash is recovered in the form of slag.

ASH FOULING

With all methods of coal-firing, some portion of the ash and its derivatives is carried by the combustion gases through the unit. The amount and type of these contaminants vary, somewhat characteristically, with firing method and with the nature of the fuel. All contaminants, in varying degrees, include a wide range of "particle" sizes,[58] from truly molecular gas and vapor, through aerosol and microscopic suspensions, to macroscopic and relatively coarse particle entrainments.

Both physical and chemical actions are operative as the heterogeneous mass of gases and entrained material sweep past a stationary surface.[59] Large quantities and long periods are involved, and "optimum" conditions for any of these actions do not necessarily exist. The smallest particles tend to be deflected with the gases. If contact is made, a strong attachment may develop through surface energy and chemical bonds. Larger particles tend to strike obstructing surfaces and adhere if sticky or rebound if solid. Energy of impact of the larger and harder particles may remove previously deposited material or actually erode the surface.

The physical aspects of particle movement and the nature of forces acting on them to cause coalescence, fracture, or separation from the gas stream have been discussed by Ungar and Putnam.[60] Diffusion of gases and vapors is primarily a function of molecular movement and vapor pressure gradients and is involved in the selective condensation of volatile constituents in some sections of the boiler unit, hence contributing to the formation of bonded deposits. For discrete solid or liquid particles smaller than 1μ in diameter, the Brownian motion and the related phenomenon of thermal precipitation[61] resulting from unbalanced molecular collisions become significant and may account for agglomeration and for traverse of the boundary layer on gas-swept surfaces. Electrical charges produced by thermal action or friction may also be significant[62] and are, in fact, imposed as a principal separating force in electrostatic dust-recovery equipment. The movement of larger particles is governed by the resultants of gravitational, drag and inertial forces, which cause separation in zones of direct impingement, turbulent vortices, and major changes in flow direction.

Fouling is here considered to be any form of ash deposit that retards heat transfer or obstructs the flow of gases through the unit. For convenience of discussion, Crossley and Marskell[4] have classified the forms of fouling as:

1. Fused-slag deposits occurring principally on surfaces exposed to radiant heat transfer, such as on the furnace walls and the first few rows of boiler or superheater tubes.
2. High-temperature bonded deposits occurring on convection-heating surfaces in regions of moderately high gas-temperature, principally in the superheater.
3. Low-temperature deposits occurring on convection heating surfaces of economizer and air heater.

Fused Slag Deposits

The formation of a slag deposit is caused mainly by physical transport of fused or partially fused particles entrained by the gas stream. When the particles strike the wall or tube surface, they are chilled and solidified. Strength of attachment is influenced by the temperature and physical contour of the

[58] Gumz, W., *Combustion*, **30**, No. 8, 51–5 (1959). Hodson, P., *ASME, Trans.*, **77**, 279–86 (1955).

[59] Barkley, J. F., Karlsson, H., Berk, A. A., Stark, C. F., and Burdick, L. R., *U.S. Bur. Mines, Rept. Invest.* No. 4996 (1953), 23 pp.

[60] Battelle Memorial Institute, *Corrosion and Deposits in Coal- and Oil-fired Boilers and Gas Turbines*, ASME, New York, N.Y., 1959, E. W. Ungar and A. A. Putnam, pp. 78–98.

[61] Bowden, A. T., Draper, P., and Rowling, H., *Proc. Inst. Mech. Engrs. (London)*, **167**, 291–300 (1953).

[62] Gumz, W., Kirsch, H., and Mackowsky, M. T., *Schlackenkunde*, Springer, Berlin, 1958, Chap. 9.

surface, and by the direction, force of impact, and wetting characteristics of the slag. Insulating effects are roughly proportional to thickness of deposit and extent of area covered.[63]

In slag-tap furnaces, firm attachment and complete coverage normally exist on the refractory-surfaced stud-tube wall construction frequently used in the combustion zone. Less secure attachment occurs on tangent smooth tube wall constructions,[64] where the slag tends to form a latticework of separate or interlaced clusters or streamers, solidified at the interface of contact with the tubes, but molten and dripping at the exposed surface. As the slag continues to accumulate, the spaces fill and stronger anchorage develops against the corrugated contour of the wall. Coverage of this type may shed from large areas if cooled abruptly by change of load or separated by other causes.[65] Beyond the slagging zone, the tendency to form fused deposits is diminished by progressive cooling of the ash. Conditions comparable to those of the dry-ash furnace prevail as gases approach the furnace outlet.

As distinguished from slag deposits, a related type, known as "bird-nest" or "re-fused" deposit has been described,[66] which may develop at the tube-bank entrance, where temperatures are in the transition range between liquid and solid condition of the ash. The mechanism of formation is attributed to a surface stickiness of particles that causes them to adhere to one another without complete fusion. The formation increases in strength by sintering processes.

Accumulation of wall deposit delays cooling with a resultant increase in the temperature of the gases delivered to the superheater. This tends to raise final steam temperature and also causes deposits to advance into the cooler portions of the furnace. Because of the necessity of maintaining superheat within a close range, it is important to limit the extent of ash coverage and to provide separate means of steam temperature control[67] to compensate for variations. Difficulties from fouling by fused slag deposits, although inherent with many coals, can usually be avoided by suitable design or kept in check by judicious operation and by use of "on-line" cleaning equipment.

High-temperature Bonded Deposits

A great deal of attention has been given to a class of fouling descriptively known as high-temperature bonded deposit, which occurs under some conditions in the superheater or reheater at a critical range of temperature appreciably below the fusion temperature of the ash. Such deposits are particularly troublesome because they obstruct gas flow and are difficult to remove from the closely spaced tubes. They may also cause corrosion of tubes operating at metal temperature above 1150°F.

Characteristic formations[68–71] comprise a relatively thin inner layer of distinctive composition and appearance and an outer layer having the composition of normal fly ash, Fig. 12. The inner layer, which is tightly attached to the tube surface, frequently shows evidence that a liquid phase existed at operating temperatures. The outer layer of semifused or sintered fly ash forms the main bulk of the deposit and may become quite massive. Materials of the inner layer often extend as a bonding matrix into the outer layer.

Several types of bonded deposit have been identified and are referred to as the alkali, calcium, phosphate, or silica types according to the principal bonding agent, although they may exist in combination. All appear to involve volatilization of elements from the ash and selective deposition upon or in the vicinity of the tube surface, thereby producing local concentrations several times greater than their normal proportion in the ash. The process is accentuated and compounded by chemical reactions that may occur during combustion, in

[63] Mumford, A. R., and Bice, G. W., *ASME, Trans.*, **70**, 601–14 (1948).

[64] Keppler, P. W., *Combustion*, **11**, No. 8, 35–9 (1940).

[65] de Lorenzi, O., *Proc. Joint Conf. on Combustion, Inst. Mech. Engrs. (London)*, **1955**, pp. 59–67.

[66] Boiler Availability Committee, *Bull. MC*/161 (1947).

[67] Kessler, G. W., *ASME, Trans.*, **77**, 919–26 (1955).

[68] Michel, J. R., and Wilcoxson, L. S., *Mech. Eng.*, **78**, 464 (1956).

[69] Crossley, H. E., *J. Inst. Fuel*, **25**, 221–5 (1952).

[70] Weber, F., *Brennstoff-Wärme-Kraft*, **3**, No. 6, 177 (1951). Gumz, W., Kirsch, H., and Mackowsky, M. T., *Schlackenkunde*, Springer, Berlin, 1958, Chap. 7.

[71] Anderson, C. H., and Diehl, E. K., *Mech. Eng.*, **78**, No. 3, 271–2 (1956).

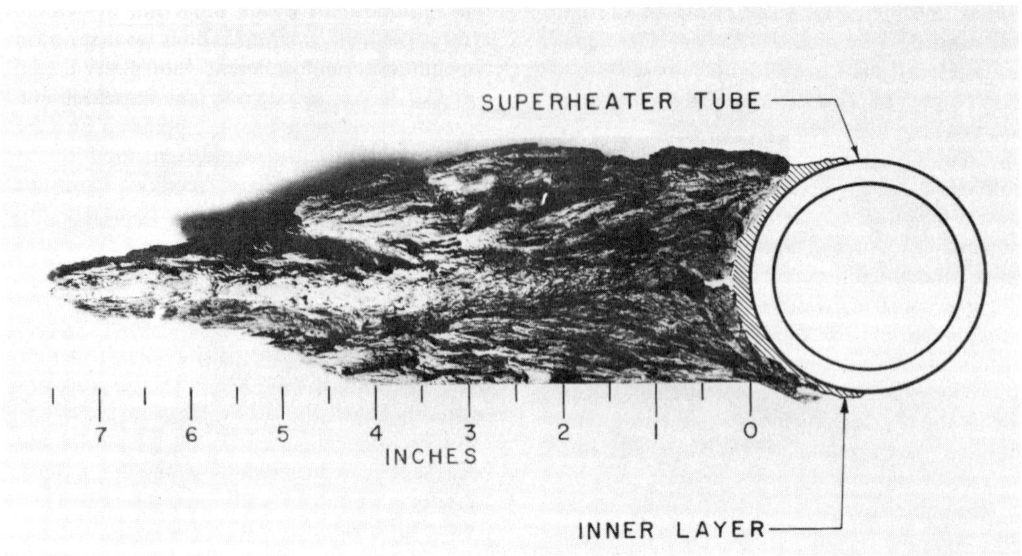

Fig. 12. Ash deposit on superheater tube.[71]

transit, or after deposition on the tubes. The occurrence and severity of fouling by these kinds of deposit depend chiefly on the composition of the coal ash, but are influenced by the method of firing, the design of equipment, and the conditions of operation.

Alkali-bonded Deposits

Alkali-bonded deposits are probably the most prevalent and widely recognized type. The nominally white inner layer, which is largely water soluble, is found to be rich in sodium and potassium and usually contains sulfate in excess of the alkali normal equivalent—a factor attributed to pyrosulfate or to complex sulfate forms involving iron and aluminum.

One mechanism of formation[72] considers that the condensation of low-melting sodium and potassium compounds upon the tube metal provides an initial sticky surface to which fly ash particles can adhere. As the composite grows in thickness, its outer surface attains higher temperature and reaches and eventually exceeds the limit of condensation or stability of the inner layer material, but promotes accretion of ash particles. Because the inner materials are shielded by outer layers, they cool and solidify. Condensation and bonding progress into the intermediate zone. Using an experimental fuel bed, Crone found that during a certain stage of combustion, corresponding to the burning of fixed carbon, characteristic spectra of Na, K, and Li appeared in the secondary flame, and a white deposit with adherent particles of ash collected on the surface of an air-cooled tube located above the bed.

[72] Crone, H. G., *Coal Research, BCURA*, 86–95, March (1945).

This explanation was found to be inconsistent with the observation that formation of the inner layer took place as a delayed development resulting from reactions between sulfur oxides and dry dust collected on the tubes.[71] From periodic inspection of an air-cooled probe, which was operated for an 11-week period in a cyclone-fired unit and regulated to maintain metal temperature at 1050°F comparable to the superheater tubes, the only deposit collected in the first week was a coating of loosely held dust that had no indication of white bonding material. At the end of the second week, a thin fused white layer was detected at the tube surface beneath the brown fly ash material. Its thickness increased in successive weeks, accompanied by growth of the outer deposit as shown in Fig. 13.

Sodium and potassium compounds represent a minor constituent of most coal ashes, but they are brought sharply into prominence by the difficulties occasioned by alkali-bonded deposits. Volatile transport has been demonstrated by heating Na_2SO_4 and also Na_2CO_3 at

2500°F in a gas-fired furnace arranged to prevent mechanical entrainment.[73] These salts were identified as the deposits that collected on a water-cooled section of the furnace roof held at 600 to 800°F. The presence of a yellow flame indicated some degree of dissociation. The carbonate deposit was partially converted to sulfate when SO_3 was added to the furnace atmosphere. Measurements of the volatility of NaCl and of Na_2SO_4 heated in a stream of dry nitrogen, Fig. 14, have been reported.[74] A rapid increase with temperature is indicated, even below the melting point of NaCl (1470°F), and above the melting point of Na_2SO_4 (1625°F). In measurements of the loss of sodium and chlorine from heated carbon pellets containing small amounts of NaCl (0.25 to 2.54%), it was found that chlorine was released at a faster rate than sodium, thus indicating that processes more complex than simple volatilization may be involved.[75] The addition of kaolin to the pellets decreased the sodium loss, attributed to formation of silicates, but did not affect the release of chlorine.

The alkali metals are present in British coals as water-soluble chlorides, or in combination with silica and alumina as insoluble and stable silicates.[69] At combustion temperatures, the former are readily volatilized. Reaction with SO_3 or with SO_2 and oxygen from the flue gases may occur in the vapor phase or after condensation has occurred on the tubes or on the surfaces of fly ash particles. Since analytical methods for determining these metals were then comparatively difficult and did not differentiate between volatile and stable forms, Crossley recommended using the chlorine content of the coal, ranging from 0.01 to 1.0%, as an index of the readily volatilized alkalis. Industry-wide surveys[69] indicated that little fouling was experienced in stoker-fired boilers when coals having less than 0.3% chlorine were burned, but that the difficulty increased at higher percentages and became acute above 0.5%. Pulverized-coal-fired units were at first thought to be immune to alkali-bonded deposits, but were later found to have difficulty with chlorine content of

[73] Corey, R. C., Grabowski, H. A., and Cross, B. J., *ASME, Trans.*, **71**, 951–63 (1949).

[74] Wickert, K., *Energie*, **10**, 267–77 (1958).

[75] Brinsmead, K. H., and Kear, R. W., *Fuel*, **35**, 84–93 (1956).

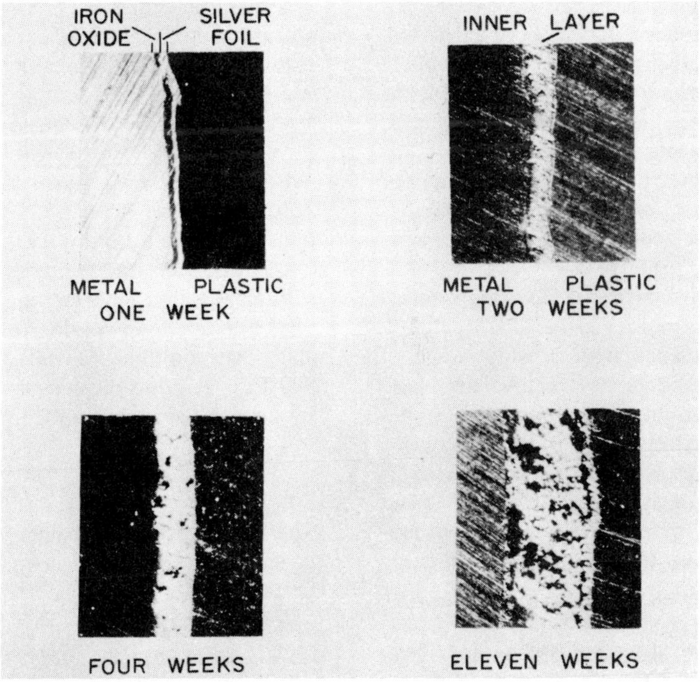

Fig. 13. Cross section of innerlayer deposit on test probe.[71]

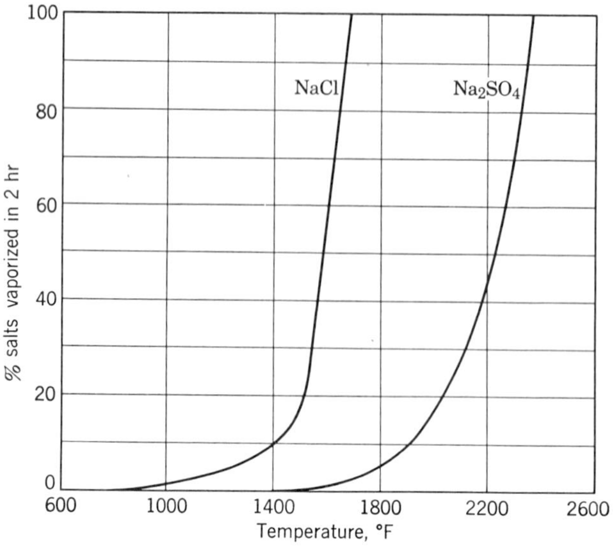

Fig. 14. Volatility of NaCl and Na_2SO_4 in stream of dry nitrogen.[74]

0.6% and higher.[76] The high-chlorine coals could be used satisfactorily when blended with coals of low-chlorine index.

Subsequent investigations[77] showed that the chlorine–alkali ratio was not stoichiometric and that a major portion of the total chlorine was present in the form of chloride ions attached to the coal substance by a linkage with ion-exchange properties. In view of general experience, it was concluded that the chlorine value continued to provide a reasonable working guide for appraising British coals, since high chlorine was generally associated with high-alkali content. It has also been suggested that chlorine itself may be instrumental in fouling[17] through reactions occurring in the combustion zone forming other volatile chlorides.

For American coals, the chlorine content appears to be less indicative than alkali content. Operating experience with a wide range of coals having 0.4% or less equivalent Na_2O showed little trouble from bonded-type deposits in superheaters; fouling increased abruptly, however, when alkali exceeded 0.5%, irrespective of method of firing.[68] Field investigations[71] were supplemented by laboratory studies of the reactions and melting behavior of alkali-sulfate mixtures to which iron and aluminum were added in proportions found in the inner layer of superheater deposits. When heated at temperatures of 1000 to 1300°F in a flue-gas atmosphere containing SO_2, alkali-iron and alkali-aluminum trisulfates were formed, similar to components identified in tube deposits, and having a liquid phase in this temperature range corresponding to tube metal temperatures. Local oxidation of SO_2 by the catalytic action of ferric oxide was sufficient to cause and maintain sulfation. It was held that pyrosulfates would decompose at these temperatures, whereas the normal sulfates, which melt above 1500°F, could not account for the observed fused condition of material at the tube surface.

Identification of the complex alkali–iron sulfates was first reported as a result of studies of the enamel-like coating found beneath the slag layer on furnace wall tubes at 600 to 800°F.[78] It was shown that formation occurs by the following reaction:

$$3(\text{K or Na})_2SO_4 + Fe_2O_3 + 3SO_3 \rightleftharpoons 2(\text{K or Na})_3Fe(SO_4)_3$$

The presence of these materials in the inner layer of superheater deposits has been confirmed.[79] Migration of components within a

[76] Jackson, P. J., and Ward, J. M., *J. Inst. Fuel*, **29**, 154–64 (1956).

[77] Edgcombe, L. J., *Fuel*, **35**, 38–48 (1956). Daybell, G. N., and Pringle, W. J. S., *Fuel*, **37**, 283–92 (1958).

[78] Corey, R. C., Cross, B. J., and Reid, W. T., *ASME, Trans.*, **67**, 289–302 (1945).

[79] Nelson, W., and Cain, C., Jr., *ASME, Trans.*, **82**, 194–200 (1960).

deposit where temperature gradients exist that account for layer-type formations on heat absorbing surfaces, but not on adjacent uncooled spacers or structural members, have been described.[62] Migration toward the cooler region of a temperature gradient has been demonstrated in experimental apparatus.[79] The complex sulfates decompose thermally into their reactants, depending on local vapor pressure of SO_3, and are unstable above 1300°F. Growth of the inner layer is thereby limited in thickness but, within such limits, is increased by longer periods of formation. At the higher temperature of the outer deposit, normal sulfates contribute to the bonding and sintering of ash particles and may form a glassy alkali–silicate bond in the extreme outer layers.

Calcium-Bonded Deposits

Superheater deposits, similar in appearance to the alkali type, but having a high percentage of calcium sulfate in the matrix, have been reported for a pulverized-coal-fired unit burning coal of high chlorine content.[80] On further investigation, it was proposed that a separate designation "calcium-bonded" deposit be recognized.[81] A simple test for its identification was described based on the appearance of a large quantity of needle-shaped crystals of $CaSO_4$ when a solution leached by boiling water from the crushed deposit sample was evaporated. In contrast, the alkali-bonded type from stoker-fired units showed few, if any, such crystals. Because of the high melting-point of calcium sulfate (2370°F), direct volatilization was thought to be of little consequence. Calcium chloride, however, (m.p. 1400°F) is relatively volatile, and if present it could be deposited as such, with subsequent conversion to the sulfate by reaction with sulfur oxides from the flue gases. An analysis of British coals indicated that calcium chloride in the fuel was a possible source. Although the chloride was not found in tube deposits, a significant amount was detected in samples of flue dust obtained from a unit burning coal of particularly high chlorine content (0.94%).

To explore the possibility of calcium chloride being formed in the combustion gases, which might take place by the reaction of HCl with calcium carbonate or oxide from the ash, experiments were carried out in a laboratory furnace burning kerosene doped with soluble calcium naphthenate, chlorobenzene, and carbon disulfide. Fly ash was formed containing both calcium sulfate and calcium chloride; the relative quantities varied with the ratio of sulfur to chlorine in the fuel but favored the formation of sulfate as predicted by thermodynamic studies. It was postulated that calcium-bonded deposits could be formed by this mechanism, and would be more likely to occur in dry-ash pulverized-coal-fired units because of the greater amount of lime particles in the gases and the lesser vaporization of alkali-metal chlorides as compared to stoker-fired units.

Alkaline earth compounds represent one of the principal groups of ash-forming materials and are present to some degree in all deposits. Troublesome superheater fouling in which calcium sulfate is the major constituent has been experienced with some lignites and brown coals having high percentages of calcium in the ash. In an examination of such deposits from the reheater of a slag-tap unit burning Texas lignite with ash containing up to 50 percent of $CaSO_4$, the underlying deposit next to the tubes had an enamel appearance, and contained a high proportion of calcium sulfate.[52] The thick outer portion consisted of fly-ash particles sintered together by unidentified chemical bonds. Although alkali in the fuel was quite low, it was thought that its volatile products contributed to the bonding action. In a 1:1 mixture of Na_2SO_4 and $CaSO_4$, sintering can occur at temperatures as low as 900°F—well below the melting point of 1680°F—which may be an important factor in the formation of inner-layer deposits on superheater tubes.[74]

Phosphate-Bonded Deposits

Superheater deposits rich in phosphorus have been identified in certain stoker-fired units, although their occurrence and the conditions of formation are considered to be rare. Phosphorus is present as a minor constituent in coal ash,[69] principally as the mineral fluorapatite [$Ca_{10}F_2(PO_4)_6$], which readily loses fluorine as HF when heated in moist air. Under reducing conditions and at temperatures above 2900°F, such as may exist in deep fuel

[80] Carlile, J. H., *J. Inst. Fuel*, **25**, 256–60 (1952).
[81] Crumley, P. H., Fletcher, A. W., and Wilson, D. S., *J. Inst. Fuel*, **28**, 117–20 (1955).

beds, elemental phosphorus is evolved. Oxidation by over-fire air and reaction with water vapor produce pyrophosphoric acid ($H_4P_2O_7$), which may condense on boiler, superheater, or economizer surfaces and react there with fly-ash to form hard, insoluble crystalline phosphates.

Silica-Bonded Deposits

Deposits rich in silica have been reported for stoker-fired units, although they have not been conspicuous with other methods of firing. The mechanism of bonding is somewhat obscure, and does not appear to involve a liquid phase. Normal vaporization of SiO_2 is negligible below 3200°F. Within the fuel bed, however, formation of silicon monoxide (SiO) or silicon sulfides (SiS and SiS_2) may occur through reduction by carbon or reaction with metallic sulfides.[82] These volatile compounds are oxidized readily by encounter with excess air or water vapor. If they persist beyond the furnace, condensation may take place on tube surfaces with subsequent oxidation to SiO_2. If oxidation occurs in transit, silica "smoke" is formed, which would be expected to travel freely with the gases. Gumz and collaborators[14] called attention to electron microscope photographs of such particles showing their extremely small size—0.1 to 0.2 μ, but with a form of concentration occurring through agglomeration in which bonding by sublimation of alkalis may be a factor. There is also the possibility of physical entrapment at the tube surface by minute whiskers of iron oxide,[62] which are shown under similar magnification to be large in relation to the particles of silica fume.

Sintered Deposits

A large percentage of the fly ash carried by gases from all types of firing, but particularly from pulverized-coal firing, is in the form of small discrete spheres with a glassy appearance, which indicates that fusion has occurred at some stage of the combustion process. The individual particles range from colorless or white beads of varying size, through shades of red and brown to black, which indicates diverse composition and to some degree the type and condition of constituents. Under the petrographic microscope, most of the beads appear to be in amorphous or glassy form but many show the presence of crystalline material[58] condensed on or adhering to their surfaces, with irregularities of shape suggesting reaction of these materials with the body of the particle.

The cementing action of matrix material is recognized as a factor that contributes strength to the inner layer of some bonded deposits. Disintegration may occur when the deposits are leached with water or dilute acid, leaving a residue of separate insoluble beads.[69] External layers, however, appear to derive their principal strength from a sintering action at temperatures above the stable range of matrix material, but several hundred degrees below the fusion point of the ash as defined by its Initial Deformation Temperature. This part of the structure is resistant to leaching and is usually described as having the same composition as the fuel ash.

Physical factors in sintering have been studied, largely in relation to ceramic and metallurgical processes where it is desirable to produce solid bodies from powders heated below the melting temperature.[83] Sizing, shape, and compaction are important in providing contact and promoting surface and lattice diffusion among particles of homogeneous material. In ash formations, it is probable that pure sintering is overshadowed by the effects of surface deposits, even in minute quantity, which impart stickiness and viscous interchange or chemical bonding between particles.[76]

The difficulty of removing such deposits from the superheater was studied by measuring the crushing strength of cylindrical pellets molded of fly ash from operating boilers and heated at various temperatures for a period of 15 hr.[84] As shown in Fig. 15, large differences were found between the fouling and nonfouling types of ash, and an approximate correlation was indicated between sintered strength and alkali content of the coal. Supplementary tests showed that strength increased with longer periods of heating. Leaching the ash with water resulted in a marked decrease of sintered strength, as did the

[82] Lange, W., *Glückauf*, **79**, 410–13 (1940). Zinzen, A., *Z. VDI*, **88**, 171–8 (1944). Whittingham, G., *J. Soc. Chem. Ind.*, **67**, 411–14 (1948).

[83] Crone, H. G., and McKee, J. H., *BCURA Bull.*, **14**, 329–40 (1950).

[84] Barnhart, D. H., and Williams, P. C., *ASME, Trans.*, **78**, 1229–36 (1956).

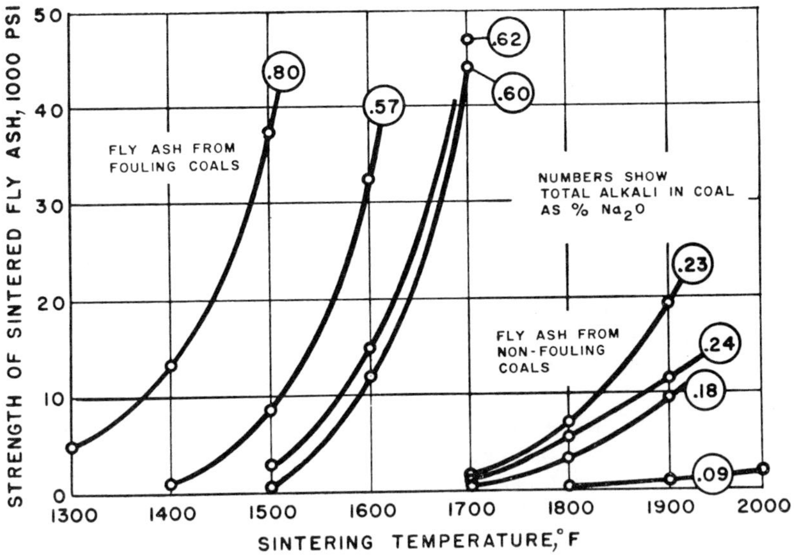

Fig. 15. Sintered strength of fly ash from fouling and nonfouling coals.[84]

admixture of materials such as kaolin and dolomite to the original fly ash. These findings support the practical observations that boiler cleaning is facilitated by prompt removal of deposits, by reducing the amount of soluble-alkali constituents, and by the use of additives to weaken the structure of the deposit.

Other methods have been used to measure sintering characteristics. One method detects the onset of sintering by abrupt decrease in rate of expansion when a lightly compacted sample of ash is heated in a refractory crucible.[85] Measurement is made by a sensitive dilatometer resting on the sample surface. Sintering temperature is defined by the limit of linearity in the temperature-dilation curve This method is somewhat analogous to the Bunte-Baum[86] observation of change in the height of a cubical specimen under the slight load of an indicating pointer. In a procedure developed at the Fuel Research Station, the first shrinkage due to sintering is detected by sudden increase in the flow of nitrogen through a heated tube containing a carefully packed plug of the ash sample.

Measurements of the sintering temperature and particularly of the sintered strength provide criteria of fouling difficulty that are of practical use in evaluating coals and equipment designs when fly ash from operating units is available for testing. The preparation of ash in pilot-scale combustors has not as yet successfully duplicated these characteristics, presumably because of differences in surface condition. The artificially produced fly ash usually shows lower sintered strength.

Low-Temperature Deposits

Deposits in the cooler sections of the unit are caused primarily by condensation of aqueous vapors and entrapment of fly ash. Surface temperatures below the acid or water dewpoint of the gases may exist in portions of the economizer and air heater where cooling fluids are first admitted, or on uninsulated areas of ducts and hoppers, particularly during the winter when air temperatures are low. These conditions may occur temporarily at low rates of operation or during periods of abnormally low feedwater temperature when feedwater heaters are out of service. They may also exist in localized sections as a result of maldistribution of air flow and gas flow through the equipment. Condensation is aggravated by the presence of deposits, since they impose a barrier to heat transfer but permit diffusion of vapors to the underlying cooled surface.

[85] Smith, E. J. D., *J. Inst. Fuel*, **29**, 253–60 (1956).

[86] Bunte, K., and Baum, K., *Gas-u. Wasserfach*, **71**, 97–101 (1928).

Sorption, solution, drainage, re-evaporation, and chemical reaction tend to consolidate the material into pasty masses or strong encrustations that are difficult to remove from the narrow gas-flow spaces and may eventually form complete obstruction. Depending on ash composition, the deposits may contain a great variety of elements and compounds but are commonly found to be high in water-soluble sulfates of iron and aluminum.[59,87] Such deposits are frequently associated with attack of the metal surfaces because of the acidic nature of the condensate and are discussed later in relation to corrosion.

Deposits can be avoided by maintaining all temperatures at a safe margin above the dew-point range,[88] but the desirability of obtaining higher thermal efficiencies by recovering more heat from the gases[88,89] has led to extensive study of sulfur and chlorine reactions in the combustion process, to investigations of corrosion resistant materials of construction,[90] and to improved design of equipment and facilities for deposit removal.

METHODS OF COMBATTING FOULING

Two aspects of dealing with the fouling problem may be considered: (1) prevention or alleviation, and (2) removal of deposits. Apart from selecting fuels of minimum fouling characteristics, which is impractical or uneconomical in many instances, the features of design and proportioning of steam-generating equipment are probably of greatest consequence,[76] and indeed are instrumental in both categories.

Equipment Design

Furnace volume and geometrical arrangement should initially be adapted to the method of firing for burnout of combustible and discharge of ash. It is equally as important that a sufficient area of radiant heat absorbing surface or other means of temperature control be provided to cool entrained ash particles below the sticky condition before they are admitted to convection banks of closely spaced tubes. Relatively large furnace height, and the addition of water-cooled screens, division walls, divided furnaces, or open-pass arrangements are examples of methods used to extend the area. Maximum cooling is obtained with smooth tubes free of ash deposits. Studding the tubes for ash retention in local zones provides a functional advantage for liquid slag removal, for tube protection, and for stability of ignition at burners. Spacing the burners to prevent flame impingement minimizes ash deposits. Arrangement of burners for tilting to different positions[65] permits changing location of the combustion zone and adjusting the amount of heat absorbed in the furnace. Recirculating cooled flue gas from the economizer pass to mix with combustion products in the furnace is used to limit outlet temperatures as well as to control superheat.[91]

To produce high final steam temperatures, required for the most efficient power cycles (1050 to 1200°F) and made feasible by the development of high-strength but costly alloy tube materials, the superheater must be placed in a region of gas temperature sufficiently high (2000 to 2300°F) to attain specified performance with an economical amount of surface and yet maintain metal temperatures within safe limits determined by operating stress and resistance to oxidation or corrosion. The high temperature may result in the delivery of ash in a molten or sticky condition to the first section of the superheater. Tubes are therefore placed on a wide spacing—from 3 to 24 in., depending on the nature of the ash—to prevent bridging by fused or bonded formations and to facilitate deposit removal.[65,92] In some designs, the high-gas-temperature portion of the superheater is arranged in the form of widely separated platens projecting into the furnace cavity, or as a part of the wall surface, thus being exposed to radiant heat transfer, but offering minimum obstruction to gas flow.

In regions of cooler gas temperature—below 1600°F—close spacing of tubes to a minimum of approximately 1-in. clearance is

[87] Rylands, J. R., Jenkinson, J. R., *Proc. Inst. Mech. Engrs. (London)*, **151**, 291–3 (1944).

[88] Koch, P. H., *Babcock & Wilcox Sales Bull.* 3-509, March (1951).

[89] Karlsson, H., and Hammond, W. E., *ASME, Trans.*, **75**, 711–8 (1953).

[90] Piper, J. D., and Van Vliet, H., *ASME, Trans.*, **80**, 1251–63 (1958).

[91] Andrew, J. D., Frendberg, A. M., and Koch, P. H., *Proc. Am. Power Conf.*, **21**, 228–35 (1959).

[92] Crossley, H. E., and Hamilton, E. M., *Proc. 2nd Conf. on Pulverized Fuel, Inst. Fuel, London,* **1957**, B 37–41.

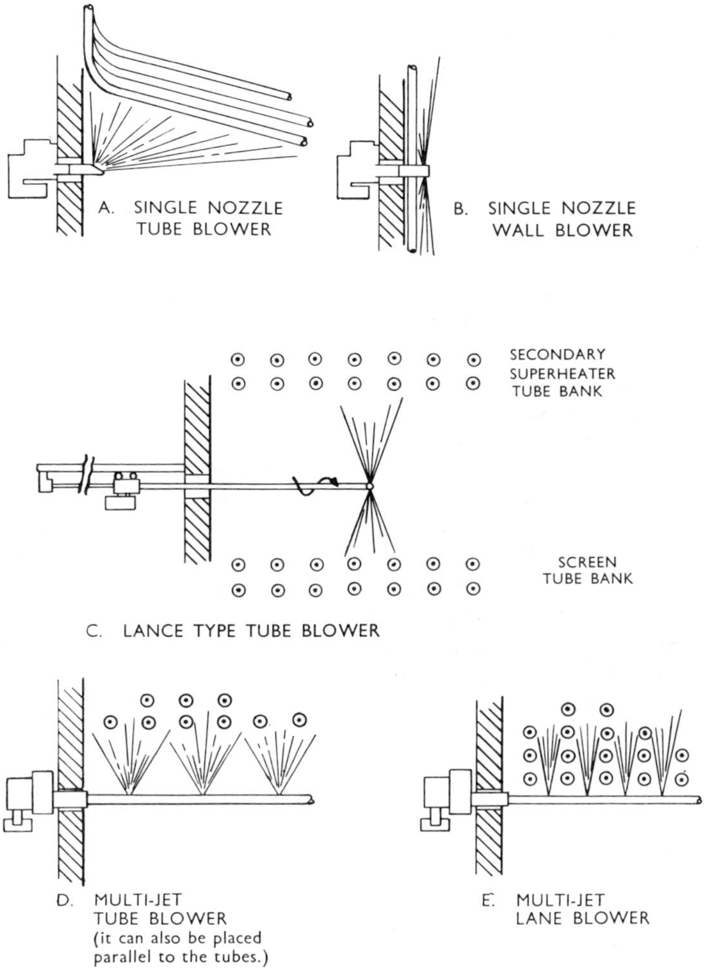

Fig. 16. Soot blower types (schematic).[94]

advantageous for improving convection transfer. Deposits of dry dust reduce thermal efficiency but are usually not difficult to remove. As temperatures approach dewpoint, however, deposits bonded by condensation of liquids may cause serious restriction because of the conventionally narrower flow areas employed in the economizer and air heater. Provisions in design for uniform distribution of gas and air flow, and auxiliary heating of inlet air or other means for maintaining adequate metal temperatures at low load should be considered.[88,93] The low-temperature portion of the air-heater is many times arranged as a separate section to facilitate cleaning and replacement of elements or for the economical use of corrosion-resistant materials.

[93] Rothemich, E. F., and Parmakian, G., *ASME, Trans.*, **75,** 723–7 (1953).

Removal of Deposits

During operation, deposits are dislodged by periodic blowing with steam or compressed-air jets applied in various ways. In the simplest form, a manually operated lance tube is inserted through openings in the setting to prod or blow deposits from the tubes. Mechanized "soot-blowers," permanently installed at selected locations are used extensively and are essential for large units. Several types are shown schematically in Fig. 16.[94] Modern installations employ programmed control for automatic sequential or functional operation of blowing elements throughout the unit.[95]

Slag is removed from the furnace wall

[94] Boiler Availability Committee, *Bull.* MC/286 August (1957).

[95] Schueler, L. B., *Mech. Eng.*, **78,** 1149 (1956).

surfaces by short retractable blowers having jets directed essentially parallel to the plane of the wall. The elements are rotated to sweep the surrounding area and to undercut deposits. Long retractable elements which rotate as they progress into the bank are used for tube banks. The blowing medium cools the lance tube while it is exposed to hot gases, and issues from nozzles at the end of the element that are directed toward the bank. The effective cleaning radius is a function of nozzle size, blowing pressure, tube arrangement, and character of deposit; consequently, the number and location of elements varies for different installations. For dust removal in the cooler passes, fixed or rotating multiple-nozzle blowers are frequently employed. The elements remain in place within the setting, and nozzles are aligned with the tube spacing. Water is sometimes used as a jetting or washing medium in air heaters.[96] The greater impact force is advantageous in clearing the small passages of regenerative-type heaters, and solvent action weakens the bonding of crystalline materials.

Friable or dust-like deposits can be removed or prevented from forming in convection-tube banks and recuperative-type air heaters by "shot-cleaning" methods used extensively in Sweden and Germany, instead of conventional blowers.[97] Cleaning is performed by the impact of steel or cast iron pellets, $\frac{1}{4}$-in. mean diameter, that are introduced through distributing devices at the top of the unit and fall by gravity through one or more tube banks to a collection hopper at the bottom of the pass. Dust is separated by elutriation and re-entrainment in the gas stream. The pellets are returned by a pneumatic or mechanical elevator to the point of introduction for periodic or continuous recycling through the system. The convection pass should be arranged to prevent segregation or undue concentration of pellets as they cascade through the bank, and to prevent their rebounding into adjacent passes or ducts from which they cannot be recovered.

[96] Hupfer, H. J., Stanley, R. J., and Van Sickle, A. H., *ASME, Trans.*, **80,** 217–22 (1958). Canaday, B. L., *Combustion*, **29,** No. 8, 55–8 (1958).

[97] Theobalt, H., *Brennstoff-Wärme-Kraft*, **3,** 194 (1951). Chappell, R. E., and Mayer, R. D., *Mech. Eng.*, **81,** No. 4, 107 (1959). Humphrey, D., *J. Inst. Fuel*, **33,** 173–9 (1960).

In regions of high temperature where sintered or bonded deposits tend to form, excessive growth has been prevented in experimental installations by applying vibration to the tubes.[68] The effectiveness of this method is enhanced by design for flexibility and adaption to resonant frequencies of the elements.

Cleaning the heating surfaces while the unit is out-of-service is a normal and important procedure for insuring subsequent availability.[94] During brief outage periods or even under low-load conditions, thermal contraction causes deposits to loosen and facilitates removal by blower equipment. During longer outages, at the time of annual inspection, thorough cleaning is strongly advised. Accumulations which form beyond the range of soot blowers, or fragments which become lodged among the tubes may require manual removal. Bonded deposits can be separated by water-washing or steam-soaking because of the solubility of inner layer materials which account for primary attachment to the tube surface. Cleaning of economizers and air heaters is similarly benefitted by washing. Surveys made by the Boiler Availability Committee have demonstrated[94] that complete cleaning to bare metal substantially prolongs the time of operation as compared to superficial cleaning (Fig. 17).

Prevention of Fouling

Many methods for preventing or reducing fouling tendencies have been explored and, to a varying degree, are useful and effective. Selection and blending of coals is well known and widely practiced. Commercial cleaning of coal by gravity separation reduces ash quantity, and selectively decreases iron and sulfur content by removal of pyrite. It has been observed and tests have confirmed that fouling is reduced for high-alkali coals by weathering or leaching.[68,98] The benefits are marginal and processing is not considered economically justified. In an opposite sense, fouling has been aggravated by the addition of salt to coal for prevention of freezing in transport and handling.

Humidification of combustion air was found to decrease superheater deposits in an underfeed stoker-fired unit burning United States

[98] Gillham, E. W. F., *J. Inst. Fuel*, **33,** 193–8 (1960).

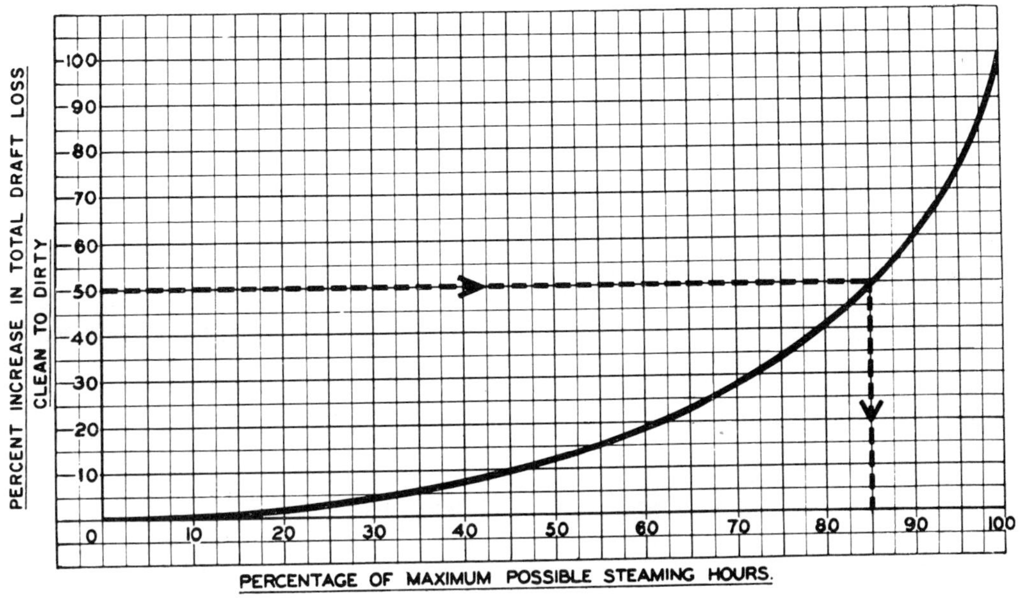

Fig. 17. Typical rate of fouling as indicated by draft loss.[94]

bituminous coal of low alkali content.[99] Based on the observation that fouling occurred more rapidly in winter, low-pressure steam was mixed with the preheated air in controlled amount to maintain normal summer conditions. Systematic study of deposits from test probes and boiler unit showed a pronounced decrease in silicon compounds, although causes and mechanisms of formation were not clearly defined. The principle of humidification was explored in Germany[100] and in England[101] by full-scale tests on fuel-bed and pulverized-coal-fired units with indications that benefits were minor except for underfeed firing.

Reduced fouling has been observed when flue gases were mixed with the combustion air supplied to underfeed and traveling-grate stokers.[102] The effects were attributed to probable lowering of fuel-bed temperature and decreased volatilization of ash constituents. Limits were encountered because of combustion difficulties and increased carbon loss. Compound firing of pulverized coal over stoker fuel beds for the purpose of improving operating flexibility was also found to reduce tube deposits.[102] This result was attributed in part to lower burning rates in the bed and to absorption of sulfur compounds by the fly ash, decreasing the potential for formation of bonded deposit. Similar effects have been noted in laboratory and field experiments in which zinc oxide, hydrocarbon, and silica smokes or other kinds of additives were introduced in the gas stream. The use of additives for control of deposits with coal-firing has in general been confined to experimental work because of economic factors and alternative means of correction. Additives are effectively used with oil-firing where ash quantities are extremely low but of more troublesome characteristics, and they may become more significant, as discussed in relation to corrosion problems.

For a given type of fuel and steam-generating equipment, important means of minimizing fouling and corrosion are available through adherence to good operating and cleaning practices. These means can be more clearly recognized from the negative point of view that difficulty is aggravated by sustained and excessive loading beyond design capacity of the unit, by departure from the proper range of excess air with unbalanced distribution to

[99] Murphy, P., Piper, J. D., and Schmansky, C. R., *ASME, Trans.*, **73**, 821–43 (1951).

[100] Schwarz, K., *Mitt. Ver. Grosskesselbesitzer*, **31**, 297–308 (1954).

[101] Freedman, A. M., *J. Inst. Fuel*, **29**, 165–70 (1956).

[102] Boiler Availability Committee, *Bull* MC/234, February (1953).

the fuel bed or to the burners, and by neglect of cleaning schedules or breakdown of cleaning devices.

Erosion

Erosion of boiler surfaces by solid particles in the gas stream does not occur in all units but is a matter which receives continual attention in inspection and maintenance of equipment.[41,103] It is concerned with all types of particles including fuel and cinder carry-over as well as ash.

Erosive tendencies are more pronounced if coarse material becomes concentrated at sharp turns of the gas stream, at intercepting baffle surfaces, and at points of extreme turbulence. The effects are aggravated by an increase in quantity, abrasiveness, and velocity of the coarser particles. Various degrees of erosion, not uncommon in operating experience, range from innocuous "polishing" of upstream tube surfaces and anticipated gradual wear of induced-draft fan blades and housings to rapid localized cutting by high-velocity re-entrainment which may occur at the discharge edge of baffles, or in the path of soot-blower jets.

In studies related to gas-turbine development, erosion experiments were conducted using fly ash from pulverized-coal-fired boilers in an air-jet directed against a steel surface.[104] When particles larger than 10 μ were removed from the ash, protective deposits formed on the target. Stronger adherence was noted for material having greater surface-to-mass ratio. Erosion was predominant in fly ash containing the coarser particles, although both deposition and erosion occurred in some cases. Stoker[105] reported experiments of similar nature using silica sand and catalyst particles against iron and gypsum targets. Erosion rates were strongly increased by higher velocity and varied with direction of the jet, being greatest at an impingement angle of approximately 30°. Hummell[106] classified erosion mechanisms as friction wear by which metal fragments are pulled or cut from the surface, and impaction by which momentum causes local stress failure. He listed the major factors as (1) physical properties of impinging particles (size, shape, weight, hardness and strength); (2) velocity of particle before impact; (3) angle of impact; (4) hardness, strength, and elasticity of the eroded material; and (5) heterogeneous nature of the metal.

TABLE 2

RELATIVE EROSION LIMITS (STEVENS[103])

Fuel and Firing Method	Apparent Allowable Velocity Ratio	
	Note (1)	Note (2)
Pulverized coal	0.75	1.00
Spreader stoker Anthracite, chain-grate Coke breeze, chain-grate	0.60	0.75
Cyclone furnace	No erosion limit	
Underfeed stoker Blast-furnace gas	0.75	1.00
Bituminous, chain-grate (Arbitrary standard—No erosion experienced)	1.00	1.20
Wood or other waste fuels containing sand	0.50	0.60
Cement dust	—	0.45

NOTES:

1. Limit at, or anywhere downstream of, a baffle which tends to concentrate ash and then permits the concentrated stream of ash to impinge on the boiler tubes.

If the concentrating type baffle is equipped with a "cinder catcher" or slot arranged to prevent the concentrated ash stream from striking the tubes, the limits in column (2) may be used.

2. Single-pass boilers or those with baffle arrangements which do not concentrate the fly ash.

Although specific factors cannot be evaluated in boiler practice, experience has led to the recognition and correction of adverse design features and the establishment of practical limits of gas velocity to reduce the probability of erosion. Stevens[103] proposed relative velocity ratios for different fuels and firing methods as shown in Table 2.

[103] Weismantle, A. R., and Artsay, N., *Mech. Eng.*, **71,** 946 (1949). Andrews, L. V., and Hawley, C. F., *Mech. Eng.*, **71,** 948 (1949). Stevens, W. D., *Mech. Eng.*, **71,** 947 (1949).

[104] Fisher, M. A., and Davis, E. F., *Mech. Eng.*, **71,** 481–7 (1949).

[105] Stoker, R. L., *Ind. Eng. Chem.*, **41,** 1196–9 (1949).

[106] Battelle Memorial Institute, *Corrosion and Deposits in Coal- and Oil-fired Boilers and Gas Turbines*, ASME, New York, 1959, J. D. Hummell, pp. 185–98.

CORROSION

External corrosion of heat absorbing surfaces may be divided into two general categories: (1) *low-temperature corrosion* of economizers, air heaters, and duct work by sulfuric acid from the flue gases; and (2) *high-temperature corrosion* of furnace walls, superheater, and reheater tubes by molten ash constituents. In either instance, corrosion or loss of metal necessitates periodic repairs or replacement of these parts, which is generally a time-consuming and costly procedure. Perhaps even more important is the fact that corrosion, or corrosion with attendant deposits, frequently limits the capacity and efficiency at which the boiler can be operated and in some instances causes unscheduled outages of equipment. The economic consequences of such outages can be very serious, especially in the electric utility industry, and consequently a considerable amount of investigative work has been done in an effort to find satisfactory solutions to these problems.

Low-Temperature Corrosion

The basic causes of low-temperature corrosion are well known, although the precise mechanism of the reactions taking place is still controversial. Briefly, the majority of the sulfur in the fuel is oxidized to SO_2 during combustion, and a very small part of the SO_2 so formed is further oxidized to SO_3. As the flue gases are cooled in passing through the boiler, the SO_3 reacts with water vapor to form sulfuric acid vapor. Although the amount of acid vapor present in the flue gases is quite low, 10 to 50 ppm, it is sufficient to raise the dewpoint temperature of the flue gases from the water dewpoint, usually around 100 to 115°F, to an acid dewpoint that may range between 250 to 350°F. If the temperature of the metal is below the acid dewpoint, a film of relatively strong sulfuric acid (70 to 90 percent) condenses on these surfaces and corrosion results. In the steam generating units of today, the weight of flue gas is normally so great that the total amount of sulfuric acid formed in this way is quite capable of seriously corroding the air heater and economizer in a short time. Fortunately, most of the acid passes through the boiler and is carried away in the gas stream.

The formation of a film of sulfuric acid on the metal depends directly on the temperature of the metal, and only indirectly on the temperature of the flue gases. The temperature of the gases may be far above the dewpoint, whereas that of the metal may still be below the condensation temperature, owing to the cooling effect of water or air in the economizer or air heater, respectively.

If the temperature of the metal falls below the water dewpoint, as may happen during periods of low-load or shut-down, corrosion is apt to occur through the action of a dilute solution of acids. The main constituents are sulfurous acid and sulfuric acid, although it has recently been shown that the presence of hydrochloric acid from the combustion of fuels containing sodium chloride can greatly accelerate corrosion of this type.

Corrosion is usually accompanied by the gas passages being plugged by fly ash adhering to the wetted metal surfaces. The acid not only attacks the heating surfaces, but may also react with fly ash in the cooler parts of the boiler to produce hygroscopic substances such as ferric sulfate, aluminum sulfate, and the bisulfates of sodium and potassium, which adhere to the metal surfaces, block the gas passages, and promote corrosion even at temperatures above the acid dewpoint.

It has been the practice to maintain economizers and air heaters at temperatures well above the acid dewpoint to prevent condensation and subsequent corrosion and deposits. The result, however, is an undesirable discard of heat still available in the gases, and other means of controlling corrosion are continually sought to permit operation at lower stack temperatures with improvement in thermal efficiency. Corrosion of coal-fired units has been the most pronounced in stoker installations and in underfeed stoker boilers in particular; pulverized-coal-fired boilers and cyclone-furnace boilers have been comparatively free from these troubles even though many of these installations utilize high-sulfur coals and operate with stack gas temperatures of 300°F and below. Low-temperature corrosion has been a fairly serious problem in boilers burning high-sulfur oils, and as a consequence a considerable amount of significant experimental work has been done in this area. Some of the findings have an important bearing on the problem in coal-fired boilers and therefore will be cited in this review.

Investigative work has been concerned with a number of aspects including thermodynamic analyses of the corrosion system, theories of SO_3 formation and the factors affecting it, methods of assessing corrosion, the mechanism of acid condensation and the factors influencing its rate of attack on heat absorbing surfaces, and the development of remedial measures. Significant findings in these areas are included in this review.

The literature on low-temperature corrosion in boilers and other combustion equipment has been reviewed extensively.[107] Two other reviews discuss the literature on the behavior of sulfur oxides in boilers and gas-turbine systems.[108] Annotated bibliographies of papers pertaining to boiler-plant corrosion and fouling have been published.[109]

Thermodynamic Considerations

To better understand the conditions under which SO_3 is formed, converted into sulfuric acid, and finally condensed on the boiler heat absorbing surfaces, it will be helpful to consider the thermodynamics of the overall process. Although this approach does not in itself lead to an accurate determination of the SO_3 content of the flue gases, it is valuable, nevertheless, in establishing the trend of the reactions involved and in determining the limiting values under equilibrium conditions. In attempting to calculate the actual values for the SO_3 content, the difficulty lies in estimating the extent to which equilibrium is reached in the boiler. This is complicated by the gas-temperature gradient in the boiler, the effect of catalytic surfaces, such as ash, tube metal oxides, and refractory, and the general lack of data on the kinetics of the pertinent reactions. Consequently, it has been necessary to measure the various properties of the flue gases where quantitative information is desired.

Three aspects will be considered; (1) the oxidation of SO_2, (2) the formation of sulfuric acid, and (3) the relationship between the dew-point and SO_3 content of the flue gases.

Oxidation of SO_2. Only a small portion of the sulfur in fuels is converted to SO_3 during combustion. Johnstone[110] found that only 2 to 3% of the sulfur in coal is present as SO_3 in flue gases in stoker-fired boilers and less than 1% in pulverized-coal-fired boilers. In cyclone-furnace boilers, this value is usually between 0.25 and 0.50%.

The oxidation of SO_2 can be written as

$$2SO_2 + O_2 \rightleftharpoons 2SO_3 \tag{1}$$

Using the data of Bodenstein and Pohl,[111] Gumz[112] derived the following expression for the equilibrium constant of this reaction:

$$\log K_p = \frac{10,373}{T} + 1.222 \log T - 13.5 \tag{2}$$

The equilibrium relationship between SO_2 and SO_3 is shown in Fig. 18 for various conditions of temperature and oxygen availability. It is evident that SO_3 formation is favored by low temperatures and high excess air. The rate of the reaction, however, greatly decreases as the flue gases cool, so that the total yield at the lower temperatures is extremely small compared to the equilibrium value.

When the excess air is increased, the excess oxygen for the reaction of equation 1 likewise increases tremendously. Thus, for example, when the excess air is 10%, a 100-fold excess of oxygen results, whereas when excess air is 50%, the oxygen excess for the reaction in equation 1 amounts to some 250 times. As a result of this rapid rise in the excess oxygen, the formation of SO_3 increases sharply, although this trend is opposed by the dilution with nitrogen, so that the absolute level of SO_3 decreases with excess air.

[107] Pray, H. A., Fink, F. W., and Peoples, R. S., Rept. No. 1 to Am. Gas Assoc., Battelle Memorial Inst., (1947), 29 pp. Whittingham, G., *Corrosion Technol.*, **1**, 182–7, 193 (1954). Hare, F. B. E., *Modern Power and Eng.*, **49**, 73–6 (1955). Kear, R. W., *Corrosion Technol.*, **3**, 125–7 (1956). Moskovits, P. D., *Ind. Eng. Chem.*, **51**, 1305–12 (1959).

[108] Lyons, C. J., *A Review and Analysis of Information on External Deposits and Corrosion Phenomena in Coal-Fired Boilers and Gas Turbines*, Battelle Memorial Inst., Columbus, Ohio, 1956, pp. 71–91. Battelle Memorial Institute, *Corrosion and Deposits in Coal- and Oil-fired Boilers and Gas Turbines*, ASME, New York, N.Y., 1959, Krause, H. H., pp. 44–77.

[109] Boiler Availability Committee, *Bull.* MC/235, February (1953); *Bull.* MC/236, February (1953); *Bull.* MC/299, March (1959).

[110] Johnstone, H. F., *Univ. Illinois Eng. Expt. Sta., Bull.* No. 228, (1931), 122 pp.

[111] Bodenstein, M., and Pohl, W., *Z. Elektrochem.*, **11**, 373–84 (1905).

[112] Gumz, W., *Brennstoff-Wärme-Kraft*, **5**, 264–9 (1953).

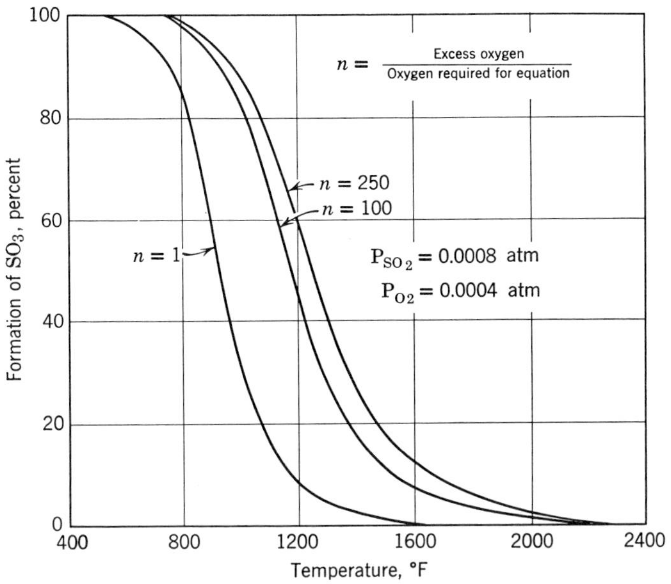

Fig. 18. Equilibrium yields of SO_3 for the Equation $2SO_2 + O_2 \rightleftharpoons 2SO_3$.[112]

The sum of these two opposed effects, as shown in Fig. 19, results in a maximum SO_3 level at around 50% excess air, which is in the range where stoker-fired boilers are generally operated. Gumz claimed that this readily explains Johnstone's[110] observations that stoker-fired boilers are troubled more with low-temperature corrosion than are pulverized-coal-fired boilers burning the same coal.

Formation of H_2SO_4. As the flue gases cool, the SO_3 reacts with the water vapor, usually present in great excess, to form sulfuric acid. Thus,

$$SO_3 + H_2O \rightleftharpoons H_2SO_4 \tag{3}$$

The dissociation of sulfuric acid has been investigated in the temperature range of 320 to 500°C, using 85, 94, and 100.5% acids.[113] It is necessary, therefore, to do considerable extrapolation of the data to reach the lower

[113] Bodenstein, M., and Katayama, M., *Z. Elektrochem.*, **15**, 244–9 (1909).

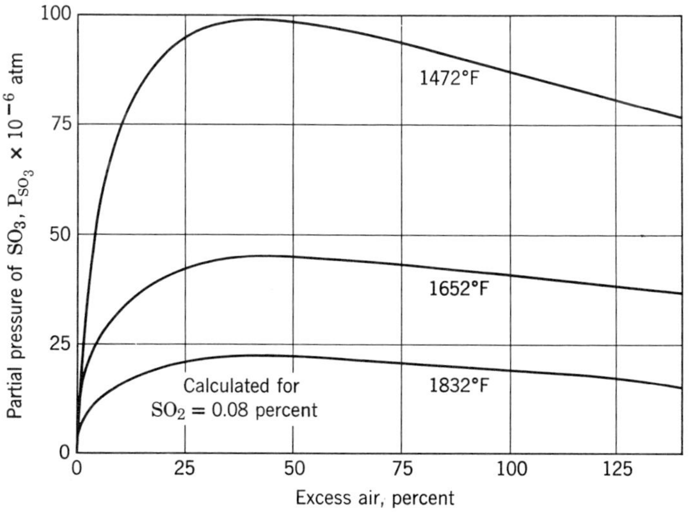

Fig. 19. Effect of excess air on SO_3 content of flue gases.[112]

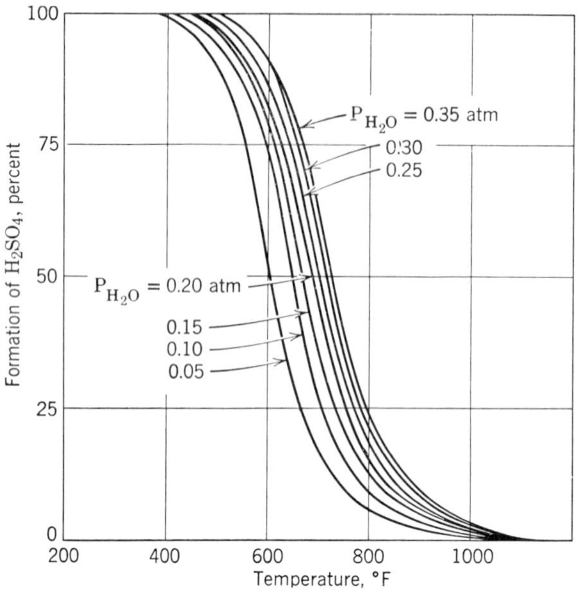

Fig. 20. Formation of H_2SO_4 according to the equation $SO_3 + H_2O \rightleftharpoons H_2SO_4$.[112]

temperatures and acid concentrations encountered in boiler practice. In his treatment of these data, Gumz[112] expressed the equilibrium constant in the following terms:

$$\log K_p = \frac{5000}{T} - 1.75 \log T + 0.00057T - 3.00. \quad (4)$$

The formation of sulfuric acid vapor is shown in Fig. 20 as a function of temperature and the partial pressure of the water vapor. At atmospheric pressure, sulfuric acid vapor begins to form between 900 and 1000°F, rising rapidly with further reductions in temperature and reaching completion between 300 and 500°F.

The reaction rate is high in the temperature range between 600 and 800°F, but becomes steadily slower as the temperature falls.[113] The reaction rate may, therefore, exert a slight influence on the attainment of equilibrium and

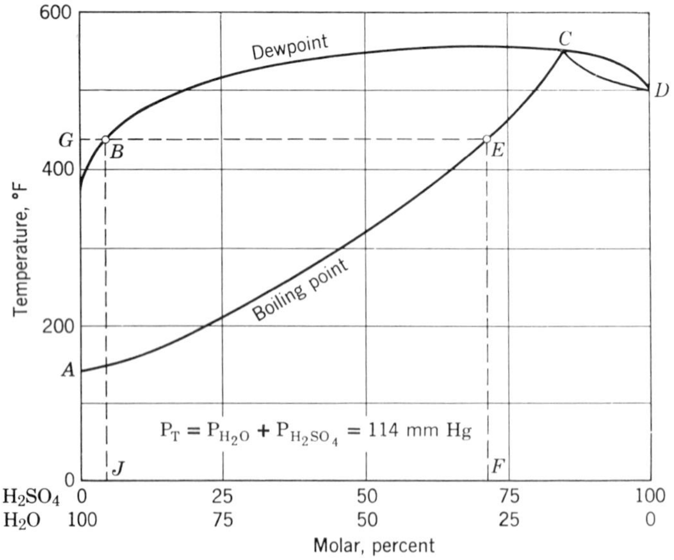

Fig. 21. Dewpoint and boiling point curves for the sulfuric acid–water system.[114]

hence on the dewpoint, since H_2SO_4 and SO_3 condense simultaneously.

The H_2SO_4–H_2O system. Since it is likely that all or most of the SO_3 has reacted with the water vapor to form H_2SO_4 vapor by the time flue-gas temperatures fall below 300°F, the dewpoint of this binary mixture, rather than the dewpoint of water vapor must be considered. Because of the high boiling point, that is, the extremely low vapor pressure, of sulfuric acid, a trace of this vapor in the gas phase is sufficient to cause a high dewpoint as shown in Fig. 21.[114] Curve *ABCD* shows the temperatures at which mixtures of water and sulfuric acid vapor are in equilibrium with the liquid mixture at a given pressure. Curve *AECD* shows the concentration of liquid that attains a total pressure *P* at any given temperature, that is, the boiling point concentration curve for the mixture at a total pressure *P*. Thus, a vapor represented by concentration *J* has a dewpoint *G*, and the liquid in equilibrium at the dewpoint is given by point *F*. In other words, a mixture of water vapor and sulfuric acid vapor containing very little of the latter is condensed at relatively high temperatures and produces a strong acid condensate.

The importance of the relationship between the dewpoint and the sulfuric acid concentration in the flue gases was recognized many years ago. One of the earliest investigators in this field was Johnstone,[115] who devised an instrument based on the electrical conductivity principle to measure the dewpoint of the flue gases directly. The essential feature of this dewpoint meter is a glass element carrying two platinum electrodes across which a known potential is applied. The element is inserted into hot gases containing H_2SO_4 vapor, and progressively cooled by air. When a film of acid forms on the element, there is a sudden increase in conductance. The temperature at which this occurs is designated as the acid dewpoint. Using the dewpoint meter, he demonstrated that sulfuric acid could condense on surfaces at a temperature appreciably higher than the dewpoint of water vapor alone.

This work was extended to show that for laboratory-prepared gases of controlled water-vapor and SO_3 content, the experimental relationship between dewpoint temperature and SO_3 content is simple and well defined.[116] The dewpoint curve devised by Taylor (see Fig. 22) was confirmed and accepted by a

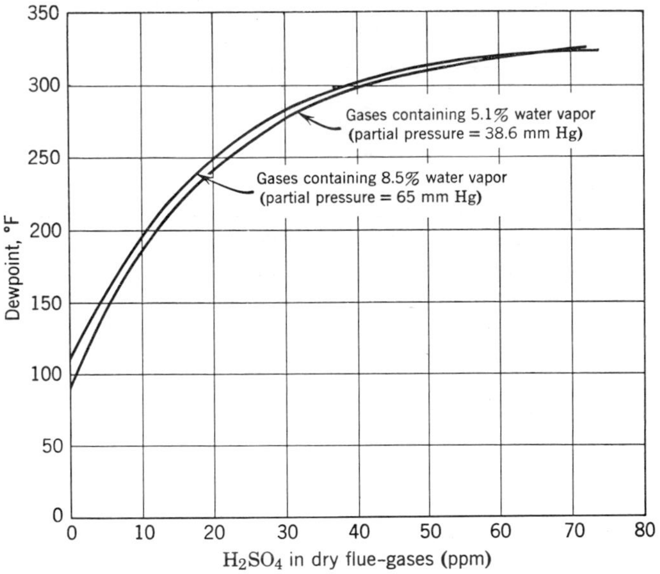

Fig. 22. Relationship between dewpoint and H_2SO_4 content of clean gases.[116]

[114] Francis, W. E., *Gas Research Board Commun.*, GRB 64 (1952), 37 pp.

[115] Johnstone, H. F., *Univ. Illinois Eng. Expt. Sta. Circ.* No. 20 (1929), 22 pp.

[116] Taylor, A. A., *J. Inst. Fuel*, **16**, 25–8 (1942).

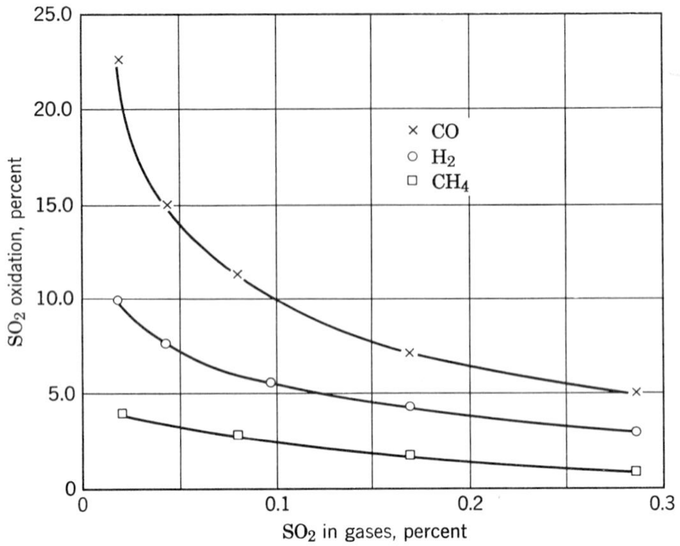

Fig. 23. Oxidation of SO_2 in various flames.[117]

number of investigators[117] as an accurate and rapid method of determining the SO_3 content of clean gases. As will be shown later, the problem of evaluating the SO_3 or H_2SO_4 content of dust-laden boiler flue gases from dewpoint measurements is difficult and complicated. Others[118] found, also, that the rate of increase in electrical conductance at temperatures below the dewpoint, frequently called the rate of acid build-up, could be regarded as the rate of acid condensing and thereby correlated with the severity of corrosion in plant practice. They viewed the dewpoint as the temperature at which corrosion would begin and rate of build-up as giving some indication of the rate of corrosive attack.

Theories of SO_3 Formation

Since sulfur trioxide in flue gases plays such an important role in ash deposition and in the corrosion in boilers, the mechanism of its formation has been the subject of a great deal of investigation. Three principal theories of SO_3 formation have emerged from this work:

1. The decomposition of sulfates in coal, mainly $CaSO_4$.[110]

2. The homogeneous oxidation of SO_2 by atomic oxygen during the burning-out stages of flames.[117]

3. The heterogeneous oxidation of SO_2 catalyzed by the Fe_2O_3 scale on high-temperature superheater surfaces.[119]

Sulfate decomposition. In 1929, Johnstone[115] presented data from which he concluded that the small amount of SO_3 in boiler flue gases could be produced entirely by the decomposition of the mineral sulfates in coal and also that it was not necessary to assume that any of the SO_2 is oxidized to SO_3. Some weight was given to this theory by the constancy of the SO_3 values throughout the boiler and the fact that the temperatures at which the sulfates decompose are several hundred degrees below furnace temperatures.

Oxidation in flames. In an investigation into the oxidation of SO_2 during combustion processes, Dooley and Whittingham[117] studied the effect of additions of SO_2 to town-gas flames and to flames of some of the separate constituents of town gas. In comparing the oxidation produced in flames of carbon monoxide, hydrogen, and methane, they noted marked differences, as shown in Fig. 23. For example, for a 0.1% addition of SO_2, the oxidation was of the following order: CO, 10%; H_2, 6%; and CH_4, 2.5%. Since the

[117] Dooley, A., and Whittingham, G., *Trans. Faraday Soc.*, **42,** 354–66 (1946). Whittingham, G., *Trans. Faraday Soc.*, **44,** 141–50 (1948).

[118] Flint, D., *J. Inst. Fuel*, **21,** 248–53 (1948). Corbett, P. F., Flint, D., and Littlejohn, R. F., *Fuel*, **25,** 246–51 (1952).

[119] Harlow, W. F., *Proc. Inst. Mech. Engrs.* (*London*), **151,** 293–8 (1944); **160,** 359–68 (1949).

order of oxidation of SO_2 is in line with the atomic oxygen level, as revealed by flame spectra, they concluded that atomic oxygen present in the combustion flames is responsible for the oxidation of SO_2,

$$SO_2 + O \rightarrow SO_3 \quad (5)$$

Furthermore, they suggested that the oxygen atoms could come from the following reactions:

$$CO + O_2 \rightarrow CO_2 + O \quad (6)$$

$$H + O_2 \rightarrow OH + O \quad (7)$$

If this atomic oxygen theory is valid, it should be possible to reduce the formation of SO_3 by introducing atoms having a greater affinity for oxygen atoms than does sulfur. To examine this theory, Whittingham[117,120] studied the oxidation of SO_2 during the slow combustion of carbon monoxide in a heated quartz tube. In these experiments, SO_2 was introduced into gases drawn from a Bunsen-burner flame, and the amount of SO_3 was measured. A number of compounds capable of preferentially removing the oxygen atoms were introduced, including ethanol, benzene, pyridine, and carbon tetrachloride. It was found that the amount of SO_3 could be lowered 50% when only 3% of some of these inhibitors was present.

Another test of this theory consisted of exposing the atoms in the flame to the presence of solid bodies capable of absorbing the energy of recombination of the atoms, permitting the removal of the oxygen atoms. Various silica and metallic "smokes" were tried, but the results were not conclusive because of secondary reactions between SO_3 and the smokes.

Catalytic oxidation. It has long been known that iron compounds are capable of catalyzing the oxidation of SO_2 to SO_3. The old Mannheim process for the manufacture of sulfuric acid used ferric oxides as the catalyst. Harlow[119,121] has expressed the view that since the temperature in the furnace is usually greater than 2500°F, any SO_3 that might be present will dissociate into SO_2 and oxygen. He pointed out that low-temperature corrosion had not been troublesome in boilers operating at relatively low superheat temperatures, but that it had become more general in boilers having superheater tubes approaching the temperature of maximum catalytic activity of approximately 1100°F. Consequently, Harlow contended that the formation of SO_3 is caused entirely by the heterogeneous oxidation of SO_2 on high-temperature superheater surfaces, with Fe_2O_3 as the principal catalyst. To demonstrate this, he conducted experiments in which flue gases containing SO_2 from a small underfeed stoker-fired boiler were passed over two series of clean steel specimens of superheater tubing held at temperatures comparable to the metal temperatures existing in a boiler from superheater to air heater. Measurements of the amount of sulfuric acid condensed on the specimens for various conditions, as shown in Fig. 24, led him to the conclusion that SO_3 was produced at the higher metal temperatures, and that it could be greatly reduced by lowering the temperature of the superheater tubes below that of maximum catalytic activity.

Tolley[122] presented supporting evidence by showing that the formation of Fe_2O_3 on a mild steel surface can induce oxidation of SO_2, with the optimum temperature for maximum yields occurring around 650°C. However, the presence of water vapor greatly reduced the speed of this reaction.

The kinetics of the contact oxidation of SO_2 in the presence of Fe_2O_3 have been studied over the range 640 to 670°C,[123] and the reactions of SO_2 with CO and with H_2 on Al_2O_3 catalysts have been investigated.[124] A small pulverized-coal-fired furnace has been used to determine the effects of rusty steel on the oxidation of SO_2 to SO_3.[125] By means of the dew-point meter in addition to chemical analysis for measuring SO_3, it was found that rusty steel initially had a pronounced catalytic effect, but that this soon subsided. This change in behavior was attributed to the conversion of Fe_2O_3 to inactive Fe_3O_4 by the accumulation of fly-ash on the steel surfaces. It was concluded that the catalytic effect of steel surfaces on the oxidation of SO_2 is not of major significance.

120 Whittingham, G., *3rd Symposium on Combustion*, **1948**, Williams & Wilkins, Baltimore, 1949, pp. 453–9.

121 Harlow, W. F., *J. Inst. Fuel*, **32**, 126–9 (1959).

122 Tolley, G., *J. Soc. Chem. Ind.*, **67**, 369–73, 401–4 (1948).

123 Boreskov, G. K., and Sokolova, T. L., *J. Phys. Chem. U.S.S.R.*, **18**, 87–101 (1944).

124 Bogoyaulenskaya, M., and Kovalskiĭ, A., *J. Phys. Chem. U.S.S.R.*, **20**, 1325–31 (1946).

125 How, M. E., Kear, R. W., and Whittingham, G., *J. Inst. Fuel*, **31**, 71–9 (1958).

A similar conclusion was reached from an investigation of the effects of superheater metal temperature on the acid dewpoint and SO_3 content of flue gases from a gas-fired pilot unit.[126] With tube-metal temperatures in the range of 1100 to 1200°F, the average SO_3 concentration increased from 50 ppm ahead of the superheater bank to 70 ppm beyond it as shown in Fig. 25. The change in measured dewpoint was 10 to 15°F. Since experiments on both pilot-scale and full-scale boilers have shown that much of the SO_3 present in the flue gases already exists ahead of the superheater, these investigators felt that catalytic oxidation was not a significant factor. In discussing this work, Harlow stated that the SO_3 in the flue gases ahead of the superheater probably came from the catalytic action of the refractory surfaces in the furnace of the test unit. Moreover, he contended that the chrome-molybdenum steel used in these experiments would not exhibit a catalytic effect for several hundred hours, and then would behave like mild steel.

Thermodynamic and kinetic considerations have shown that both catalytic oxidation on hot surfaces and homogeneous oxidation in flames probably do take place in service.[127] The conclusion has then been that the major portion of the SO_3 is formed in the flame and a minor portion is formed by catalysis.

From a consideration of reaction kinetics, it can be shown that the catalytic formation of SO_3 would be expected to increase linearly with the sulfur content of the fuel.[128] According to the atomic oxygen theory, a less rapid increase would occur. Since all literature tends to show less than a linear increase in SO_3 with increasing sulfur content, most of the SO_3 would therefore seem to originate in flame.

Still another mechanism of SO_2 oxidation occurring in the cooler sections of the boiler was suggested by Rylands and Jenkinson.[87] According to this theory, a thin film of sulfuric acid forms on the economizer tubes over a period of several months and attacks the metal, forming ferrous sulfate, which is then oxidized to ferric sulfate by oxygen from the flue gases. Because of the strong catalytic action of ferric sulfate, the formation of additional sulfuric acid from the SO_2 and moisture in the flue gases proceeds rapidly. Data indicate that the acid may be produced by a secondary reaction of surface oxidation of SO_2, with iron sulfates acting as the catalysts.[129] It has been found[130] that in an initially clean oil-fired boiler,

[126] Burnside, W., Marskell, W. G., and Miller, J. M., *J. Inst. Fuel*, **29**, 261–9 (1956).

[127] Widell, T., *Tek. Tidskr.*, **82**, 1095–6 (1952); *Combustion*, **24**, No. 12, 53–5 (1953).

[128] Whittingham, G., *J. Inst. Fuel*, **32**, 135 (1959).

[129] Whittingham, G., *J. Appl. Chem.* (*London*), **5**, 316–22 (1955).

[130] Anderson, D. R., and Manlik, F. P., *ASME Trans.*, **80**, 1231–7 (1958).

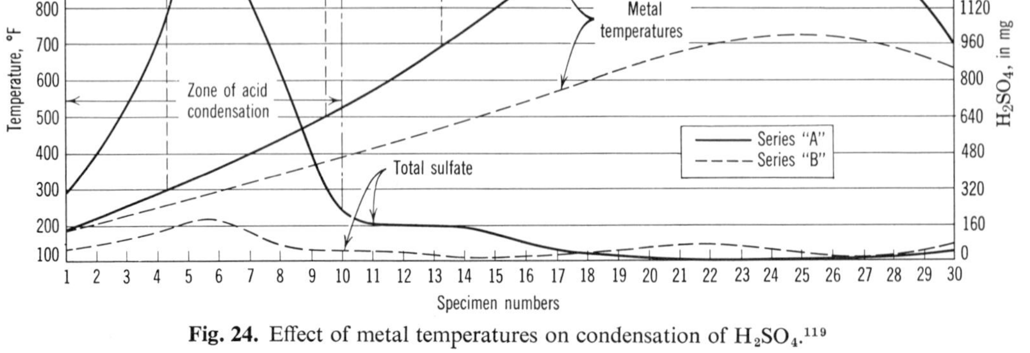

Fig. 24. Effect of metal temperatures on condensation of H_2SO_4.[119]

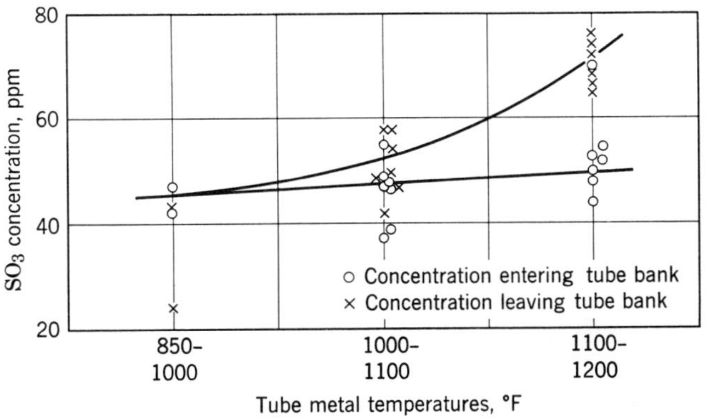

Fig. 25. Conversion of SO_2 to SO_3 across tube bank.[126] Test conditions: Flue-gas temperature, 1300 to 1400°F; CO_2, 7 to 8%; excess air, 60%; SO_2 concentration, 0.11 to 0.14%.

nearly equal amounts of SO_3 are produced in the flame, the furnace, and the convection sections, and less SO_3 is formed in the economizer and air-heater sections.

It is probable that all three mechanisms play a part in the production of SO_3 and that the relative importance of each is governed by the actual conditions prevailing in the specific installation. It is desirable in practice to minimize SO_3 formation, and various means of accomplishing this will be discussed later in this chapter.

Factors Affecting SO_3 Formation

Data on the SO_3 content of flue gases from numerous coal-fired and oil-fired boilers make it clear that SO_3 formation depends on a large number of variables which have been classified by Gumz[131] into three groups. The first group comprises the properties of the fuels inasmuch as the fuels contribute the corrosion agents, that is, sulfur and inorganic components. The nature of the compounds formed by the combustion process and during passage through the boiler is controlled by the second group of factors, the combustion conditions. The third group comprises temperatures affecting the formation, type, and nature of the condensates, that is, gas temperatures, and especially the temperatures of the heating surfaces involved. The overall course of corrosion is determined by the combined action of these three groups of factors.

Fuel properties. All coal contains sulfur which is present in three main forms: pyritic or sulfide; organic, that is, combined or otherwise held in the coal structure; and sulfatic.[132] Some of the inorganic sulfur is removed in the coal preparation plant, but the remainder, together with the combined sulfur, gives a sulfur content that may be as low as 0.8% or as high as 5.0%. Most coal prepared for use in power boilers, however, contains between 1.5 and 2.0% sulfur.

During combustion, most of the sulfur is oxidized to SO_2, of which only 3% is further oxidized to SO_3; the balance of the sulfur remains in the coal ash and slag. When coal is fired on a stoker, approximately 70% of its sulfur content goes off with the flue gases as SO_2 and SO_3.[110] This figure is 85 to 90% for pulverized-coal firing and 95% or more for cyclone-furnace firing.

In laboratory experiments with gas and oil firing, it has been shown that a parabolic relationship exists between the amount of SO_3 formed and the sulfur content of the gas, with the rate of increase of SO_3 decreasing with a rise in the sulfur content.[117,120,133] This relationship is not as simple or as well defined in the combustion of coal. Tests of a number of British coals,[134,135] using the combustion pot technique, showed that the dewpoint and hence the SO_3 content of the flue gases behave in a very complex manner. During the ignition

[131] Gumz, W., *Mitt. Ver. Grosskesselbesitzer*, **56,** 307–19 (1958).

[132] Corbett, P. F., *BCURA Bull.*, **15,** 169–81 (1951).

[133] Crumley, P. H., and Fletcher, A. W., *J. Inst. Fuel*, **29,** 322–7 (1956).

[134] Whittingham, G., *Corrosion Technol.*, **1,** 182–7, 193 (1954).

[135] Arthur, J. R., *Combustion Eng.*, **5,** 5–13 (1951).

stage of combustion when the flame consisted of burning volatiles, only the water dewpoint was recorded. However, acid dewpoints as high as 340°F were obtained during the burnout stage, when the coke was consumed. Acid dewpoints were not detected even then with the low-sulfur coals. In addition, the ratio of chlorine to sulfur in the coal had a pronounced effect on the dewpoint temperature and the rate of acid build-up. Measurements on stoker-fired boilers have shown that the rate of acid build-up increases with the sulfur-to-chlorine ratio of coals burned, and that the addition of coke breeze to high-sulfur coals reduces the rate of build-up.[134]

A considerable amount of research work on low-temperature corrosion has been sponsored by the Boiler Availability Committee. From a number of field and laboratory investigations,[118,136] a wide knowledge of the SO_3 content of the flue gases from coal-fired boilers is available.

In reviewing experience in British power stations, Crossley[92] stated that for pulverized-coal-fired boilers the apparent immunity to air-heater corrosion has been due mainly to the relatively low sulfur content of British coals. Until the sulfur content is higher than about 3%, there is little chance that the flue gases will contain SO_3 in significant amounts. However, corrosion in stoker-fired boilers generally becomes significant when sulfur content is greater than 1.5% and severe when coals contain 2.5 to 3.0% sulfur. Then, dewpoints range from 270 to 290°F[118] with correspondingly high SO_3 levels.

Others[137] have investigated the effect of a number of factors on SO_3 formation in a full-scale, oil-fired boiler. They did not find a relationship between SO_3 determined chemically and the sulfur content of the fuel, although there was a relationship between the dewpoint and sulfur content. Test data[138] collected from numerous boilers fired by oil, gas, and coal containing from 1 to 5% sulfur, showed no correlation between the SO_3 content of flue gases, as measured by the dewpoint meter, and the fuel sulfur.

Using fuel oil, which was doped with carbon disulfide and petroleum disulfides, close agreement was found in the amount of SO_3 formed during combustion of the two fuels.[139] Since the carbon–sulfur bond is of a different type than that of petroleum disulfides, the form in which the sulfur is present in the oil is considered unimportant. This conclusion is not valid for coal firing, especially for those coals containing an appreciable amount of sulfate sulfur.[110]

It would appear, therefore, that although the amount and type of sulfur in the fuel have a bearing on the amount of SO_3 produced in boilers, they are generally overshadowed by other factors that have a greater influence on SO_3 formation, that is, the combustion temperature and the presence of reactive dusts in the flue gases.[140] At any rate, as long as there is a small amount of SO_3 in the flue gases, corrosion is apt to occur if metal temperatures fall below the dewpoint of the flue gas. It has been stated that it would be necessary to reduce the sulfur content of oil below 0.5% by weight to reduce the dewpoint appreciably.[139] The sulfur content of the fuel has little effect on the dewpoint of boiler flue gases until a certain threshold value is reached—possibly 0.9 to 1.0% with fuel oil and 1.5 to 2.0% with coal.[112]

From experimental work on a small laboratory-scale, pulverized-coal-fired furnace, it was found that progressively increasing the ash content of coal from 16.1 to 35.0% had little effect on the dewpoint characteristics.[125] However, since the removal of about 90% of the dust burden increased corrosion five-fold, it was concluded that variations in ash content probably influence corrosion and dewpoint phenomena to a much greater extent at the lower-ash levels than at the higher-ash levels.

The water-vapor content of flue gases from the combustion of the primary fuels, coal, oil, and gas, generally amounts to between 2 and 10% by volume. This amount is many times that needed to combine with all of the SO_3 in the flue gases to form sulfuric acid. Some of the effects of adding moisture to coal or to the combustion air have been reviewed.[141]

[136] Corbett, P. F., and Flint, D., *J. Inst. Fuel*, **25**, 410–17 (1953).

[137] Corbett, P. F., and Fereday, F., *J. Inst. Fuel*, **26**, 92–106 (1953).

[138] Rogener, H., *Brennstoff-Wärme-Kraft*, **9**, 126–8 (1957).

[139] Rendle, L. K., and Wilsdon, R. D., *J. Inst. Fuel*, **29**, 372–80 (1956).

[140] Stratmann, H., *Mitt. Ver. Grosskesselbesitzer*, **52**, 23–30 (1958).

[141] Whittingham, G., *BCURA Bull.*, **17**, 229–36 (1953).

It has been found that increasing the amount of water vapor increased the amount of SO_3 produced from the gas phase oxidation of SO_2, but in other experiments it had a poisoning effect on the surface oxidation of SO_2 carried out in a packed quartz tube.[120] Harlow[119] claimed that increasing the moisture content of flue gases by adding steam greatly increased the amount of sulfuric acid formed when the iron oxide catalyst was present. When the catalyst was absent, the steam has only a small effect. Another investigator[122] found that the presence of water vapor leads to a considerable decrease in the catalytic oxidation on mild steel surfaces, but the inhibiting effect of water is not as marked at the 0.2% SO_2 concentration as it is at the 4.0% level.

The overall effect of water vapor on SO_3 formation in boilers, therefore, depends largely on the relative amount of SO_3 produced by atomic oxygen in the flame and by catalysis of hot metal and refractory surfaces.

Combustion conditions. The effect of the method of firing on SO_3 formation has been investigated by a number of workers. In 1931, Johnstone[110] reported that the SO_3 concentrations in the flue gases from pulverized-coal firing are 70% lower than those from the same coal when fired in a stoker. This finding has since been confirmed by a number of investigators.[92,118,119,134,142] The high-dust burden of the flue gas has been accepted as the reason for the low SO_3 concentrations and dewpoints of pulverized-coal-fired boilers and for the low incidence of corrosion in these units. Consequently, a great deal of work has been carried out in both the laboratory and the field to determine the effect on SO_3 formation of fly ash, carbon, and other dusts likely to be present in the flue gases of pulverized-coal-fired boilers.

As early as 1944, Harlow[119] showed that coating specimens of rusty mild steel with fly ash from a pulverized-coal-fired boiler greatly reduced SO_3 formation. He suggested that injecting pulverized-fuel fly ash into stoker-fired boilers as a means of coating superheater tubes might reduce SO_3 formation and thereby alleviate corrosion of the air heaters. BCURA workers[136,143] found that the acid dewpoint of the flue gases from a stoker-fired boiler dropped to the water dewpoint when pulverized-coal fly ash was injected into the system. One investigation showed that the magnetite particles present in pulverized-fuel fly ash were capable of reducing SO_3 to SO_2 at temperatures higher than 980°F.[144] The coke-like particles of the same fly ash had a similar effect at temperatures even below 400°F. This behavior of coke-like matter received further attention[145] when it was found that the reactivity of cokes to SO_3 varied considerably with different types of coke. The beneficial effect of adding coke breeze to coal has been reported.

Under some conditions, a small amount of smoke or aerosols of volatilized minerals may also be present in the flue gases and because of their large surfaces, their effect on SO_3 formation may be pronounced. The utilization of smokes formed from the fuel during combustion offers a possible means of reducing the SO_3 content of flue gases. The feasibility of using hydrocarbon smokes to diminish the formation of bonded deposits on an experimental boiler has been described.[146] It was suggested that the smoke might interfere with the formation of SO_3.

Laboratory studies[147] have shown that the presence of reactive carbon smokes beyond a certain minimum concentration, depending on the nature of the carbon, can considerably reduce the SO_3 concentration in the flue gas. Tests on a full-size boiler using auxiliary flames from solid and liquid fuels above the fuel bed of the stoker have shown that the quantity and nature of suspended carbon particles have a decided effect on the dewpoint characteristics.[148]

Introduction of silica smokes into coal-gas flames containing sulfur oxides greatly reduced

[142] Boiler Availability Committee, *Bull.* MC/160, April (1947). Corbett, P. F., *J. Inst. Fuel*, **24**, 247–51 (1951). Henning, F., and Rogener, H., *Brennstoff-Wärme-Kraft*, **5**, 269–72 (1953).

[143] Clinch, W. N. C., *Proc. Inst. Mech. Engrs. (London)*, **161**, 129–42 (1949).

[144] Crossley, H. E., Poll, A., and Sweett, F., *J. Inst. Fuel*, **21**, 207–9, 213 (1948).

[145] Kear, R. W., and Whittingham, G., *Fuel*, **32**, 265–78 (1953).

[146] Marskell, W. G., *Fuel and The Future*, HMSO, London, 1948, **1**, 33–6.

[147] Whittingham, G., *J. Appl. Chem. (London)*, **1**, 382–8 (1951). Kear, R. W., *J. Appl. Chem. (London)*, **1**, 393–9 (1951).

[148] Barker, K., and Corbett, P. F., *J. Inst. Fuel*, **27**, 495–502 (1954).

the sulfur trioxide content, as measured by the dewpoint meter.[149] This reduction was attributed to both interference with the oxidation of SO_2 and to physical adsorption of SO_3 on the large surface area of the smoke particles.

Published data[150] that summarize German experience on solid and liquid fuels employing various methods of firing are shown graphically in Fig. 26.

In general, the dewpoint and hence the SO_3 content of the flue gases are highest for oil-fired boilers and decrease in the following order for stoker, pulverized-coal, and cyclone-furnace boilers, with those for cyclones corresponding to the water dewpoint region. The differences in dewpoint levels for various firing methods appear to be due to the reactivity of dusts as well as to the quantity present in the flue gases, which in turn depend on the firing method and the amount and nature of the ash in the fuel.

It appears that very little work has been done on the effect of furnace temperatures on SO_3 formation in coal-fired boilers, probably because the influence of temperature is difficult to isolate from the greater combined effect of other variables in the system. Experimental work on a pilot-scale pulverized-coal furnace indicated that SO_3 formation increased with furnace temperatures, that is, the dewpoint and rate of build-up increased with increased heat release within the combustion zone.[125] There are also a number of investigations of oil-firing systems[133,151,152] which have shown that as temperature rises there is an increase in the SO_3 content.

In work with three small oil-fired furnaces, it was found that the dewpoint varied from furnace to furnace.[151] Production of SO_3 seemed to be the greatest in the refractory-lined furnace where flame temperatures were highest, probably because of a shorter residence time of the gases in the high-temperature zone of the furnace. In similar experiments with a small bricklined furnace, it was found that combustion-chamber temperature played a large part in the amount of SO_3 formed,[133] as shown in Fig. 27. The amount of SO_3 formed at first increased with flame temperature, which was attributed to the greater abundance of atomic oxygen at higher temperatures, although the leveling off at 3200°F was not explained.

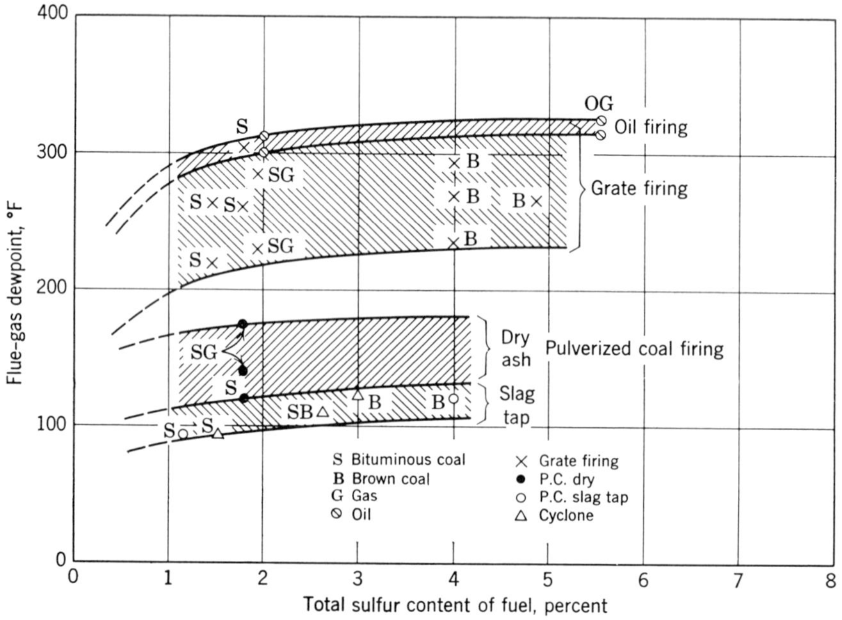

Fig. 26. Relation of flue-gas dewpoint to fuels and firing methods.[150]

[149] Whittingham, G., *Proc. Intern. Congr. Pure and Appl. Chem.*, 11*th Congr.*, *London*, **1947, 4,** 591–9.

[150] Rosahl, O., *Mitt. Ver. Ind. Kraftwirtsch.*, **4,** 53–65 (1956).

[151] Flint, D., Lindsay, A. W., and Littlejohn, R. F., *J. Inst. Fuel*, **26,** 122–7 (1953).

[152] Huge, E. C., and Piotter, E. C., *ASME Trans.*, **77,** 267–74 (1955).

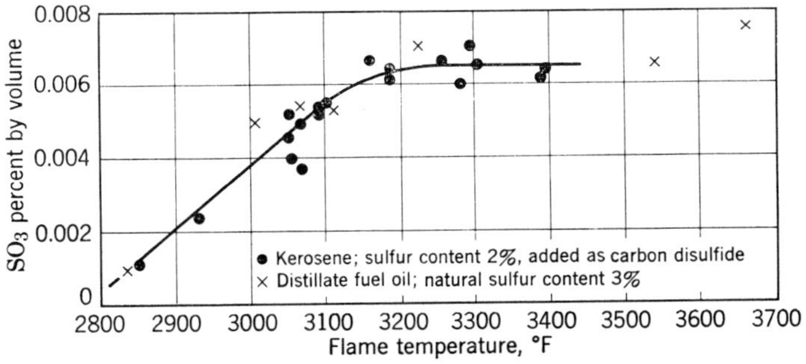

Fig. 27. Variation in SO_3 content of flue gases containing 12% CO_2 with flame temperature.[133]

The effect of temperature has also been investigated on a full-scale oil-fired boiler,[152] and it was found that reducing the load on the boiler decreased the amount of SO_3 in the flue gases. Thus, at full load with presumably maximum flame temperatures, the SO_3 content and dewpoint were 39 ppm and 340°F, respectively, and at half load with lower flame temperatures, the values were 27 ppm and 267°F. These data have been confirmed by Geissler,[153] who found that operation at lowered combustion chamber loads lessened the oxidation of SO_2.

A thermodynamic analysis of the oxidation of SO_2 referred to earlier in this chapter showed the importance of excess air in SO_3 formation.[112] This analysis has been confirmed by experimental work on laboratory combustion equipment[125,133,151] and boilers in the field.[154]

[153] Geissler, T., *Energie*, **9**, 57–9 (1957).

Crumley and Fletcher[133] determined the SO_3 content of flue gases from kerosene containing 2% sulfur at excess air levels of 9, 28, and 70%, while the furnace temperature was held at 2400°F. The results, given in Table 3, show that the overall amount of SO_3 formed and therefore the percentage oxidation of SO_2 to SO_3 is greater as the amount of excess air is increased. Similar tests on a laboratory-scale, pulverized-coal-fired furnace[125] showed that the dewpoint and rate of build-up of the flue gases reached a maximum at 45% excess air, as shown in Fig. 28. The authors suggested that changes in flame temperature can perhaps explain the effects of excess air in producing a maximum rate of acid build-up, but other factors become operative as the amount of excess combustion air is reduced. When the excess air is high, little SO_3 is formed because of the relatively cool flame. As the quantity of

[154] Lees, B., *J. Inst. Fuel*, **29**, 171–5 (1956).

TABLE 3

EFFECT OF EXCESS AIR ON THE SO_3 CONTENT OF FLUE GASES[133]

Excess Air, %	CO_2, %	Furnace Temperature, °F	SO_3 in Flue Gases, %		Oxidation of SO_2 to SO_3, %
			Found	Calculated*	
70	9	2399	0.0057	0.0076	6.9
70	9	2399	0.0061	0.0081	7.4
70	9	2383	0.0053	0.0071	6.5
28	12	2390	0.0053	0.0053	4.8
28	12	2395	0.0054	0.0054	4.9
9	14	2480	0.0047	0.0040	3.6
9	14	2444	0.0049	0.0042	3.8
9	14	2426	0.0041	0.0035	3.2

* Corrected to comparable basis of 12% CO_2.

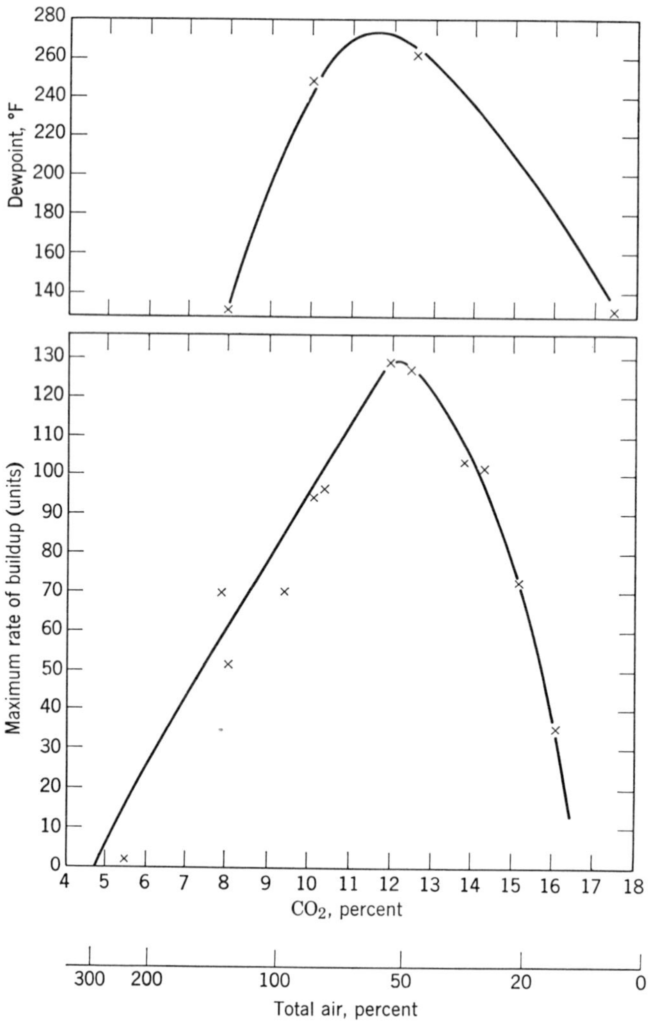

Fig. 28. Variation of dewpoint and rate of buildup with total air.[125]

excess air progressively decreases, not only is the primary production of SO_3 increased but the dilution of flue gas is decreased. However, as the total air approaches the stoichiometric value, small amounts of smoke, carbon particles, and CO are likely to be produced, whose presence in the flue gases greatly lowers SO_3 formation.

Similar results were obtained by use of a small refractory furnace burning a number of oils ranging from heavy fuel oil with 3.5% sulfur to a gas-oil with 0.75% sulfur.[151] Tests on all of the oils showed an increase in dewpoint when the excess air was increased, but this was more pronounced with some oils than with others and did not seem to be related to their sulfur content. Furthermore, there did not appear to be any clear-cut relationship between excess air and the rate of acid build-up.

Tests on a 300,000 lb/hr, oil-fired boiler[154] showed that the dewpoint and rate of acid build-up decreased as the excess air on the boiler was reduced. A decrease in the excess air from 95 to 45% caused the dewpoint to drop from 345°F to below 300°F as shown in Fig. 29. Even more marked was the decrease in rate of build-up, as shown in Fig. 30. Thus, at 260°F the rate of acid build-up was reduced from 750 units at 95% excess air to less than 150 units when the excess air was 45%. Operation at the low excess air level has increased the life expectancy of the air-heater tubes more than ten times. All of this has been attributed to the increased concentration of solid combustible content of the flue gases as the excess air was decreased.

Burner construction and operation are important factors in SO_3 formation that have been only partially explored. A 25% reduction in the SO_3 content of the flue gases could be obtained by coarser atomization of the oil brought about by changing burner heads or by reducing the amount of primary air to the burner.[133] A blue flame favored SO_3, and slow-burning, luminous flames gave less SO_3.

Some oil-fired boilers equipped with steam atomizers have experienced less cold-end corrosion than those equipped with mechanical atomizers,[155] possibly because of better atomization by steam or because steam modifies the combustion reaction to form less SO_3.

A statistical analysis of dewpoint and SO_3 data obtained from a 30,000 lb/hr oil-fired boiler showed that the percentage oxidation of SO_2 to SO_3 was dependent on the concentration of SO_2 and the sampling point.[137] Other investigators[152] have observed that the SO_3 content of the flue gases is greater at the air-heater inlet than it is at the outlet, which they have attributed to sulfuric acid condensing on the air-heater tubes.

From these considerations, it is apparent that a variety of factors affect the production of SO_3 in boiler flue gases. Probably the most important is the amount and nature of the fine fly ash, carbon, and volatilized minerals present in the flue gas. It has been shown that fine reactive dusts affect not only the dewpoint but also the rate of acid condensation on surfaces below the dewpoint. This suggests that one approach for controlling corrosion is to generate such dusts in the combustion process or inject them into the boiler setting as additives—a subject that will be dealt with later in this chapter.

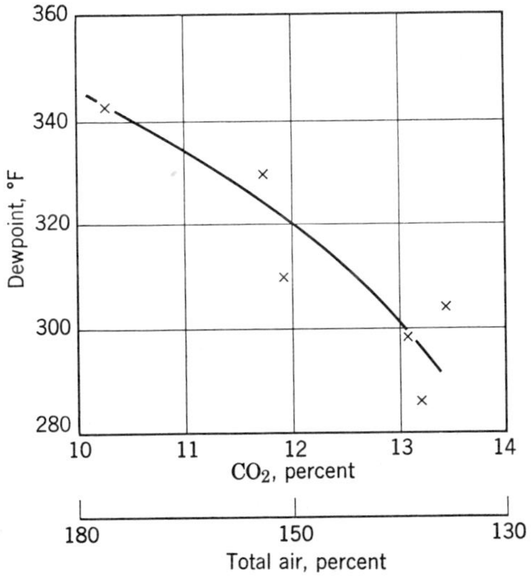

Fig. 29. Variation of dewpoint with percentage CO_2 in the flue gas.[154]

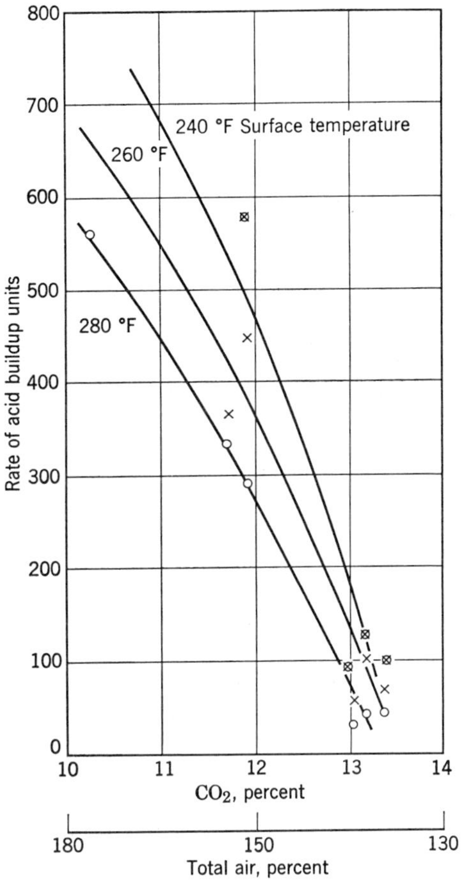

Fig. 30. Variation of rate of acid buildup with percentage CO_2 in the flue gas.[154]

Methods of Assessing Corrosion

One of the major problems in boiler corrosion research has been the difficulty of devising a reliable test for rapidly determining the corrosion rate on boiler surfaces and the effect of remedial measures on this rate. A number of methods have been used as guides for evaluating the corrosiveness of boiler flue gases, mainly (1) the determination of the SO_3 content by chemical analysis, (2) the dewpoint measurement, and (3) the determination of

[155] Marskell, W. G., *J. Inst. Fuel*, **32**, 130–2 (1959).

corrosion rates by means of test probes or the boiler surfaces themselves.

Determination of SO_3. There have been many attempts to estimate the SO_3 and SO_2 content of flue gases by absorption in aqueous reagents and chemical analysis. The work reported before 1930 has been summarized by Johnstone.[110] A later survey was made by Corbett and Crane.[156]

In one method, a sample of flue gas is drawn, and the SO_2 and SO_3 is absorbed in a solution. The amount of each is then determined analytically.[157] In practice, this proves to be difficult, since relatively large quantities of SO_2 are present that readily oxidize to SO_3. Furthermore, the SO_3 is really present as H_2SO_4, having combined with the water in the flue gases,[112] and will condense on the walls of the sampling equipment ahead of the absorber. These difficulties seem to have been overcome by passing the gases through a heated filter and a sampling probe assembly to prevent condensation and entry of metal oxides, and then into a bubbler containing 80% isopropyl alcohol and 20% water.[157] This solution acted as an inhibitor of oxidation and absorbed part of the SO_2 and SO_3. The remaining SO_3 was trapped as sulfuric acid by means of a sintered glass filter of fine porosity (G4 filter). Final absorption of SO_2 was obtained by oxidation in a hydrogen peroxide solution. Aliquot analysis of both solutions provided a measurement of total sulfur. In this method, the SO_2 retained in the isopropyl alcohol solution was removed by nitrogen gas before determining sulfate by a gravimetric procedure, but the method has been modified so that it is now possible to determine the amount of SO_3 in the isopropyl alcohol-water mixture even when SO_2 is present.[158] The technique involves the formation of a barium sulfate suspension under controlled conditions and the optical measurement of its intensity by turbidimeter. The method has the advantage of speed, but it needs care in maintaining standard conditions of precipitation. It was the first practical method for measuring the SO_3 content of the flue gases in power station boilers and the one most widely used for this purpose. A diagram of the apparatus is shown in Fig. 31.

Francis,[114] working with laboratory gases containing between 2 and 22 ppm SO_3, found Flint's method unsatisfactory because of the difficulty of trapping all of the sulfuric acid mist that formed. Others have stated that the apparatus used by Francis was inadequate because the conditioning chamber used in his experiments was too small. A somewhat more convenient apparatus for field testing,[159] and simpler absorption apparatus for laboratory use[160] have been developed. Improvements in the reproducibility of the Corbett method in the low SO_3 range (1 to 3 ppm) have been claimed by further refinement of the precipitation procedure and by using a 20-cm instead of 12-cm turbidimeter cell.[161]

An instrument for automatically determining SO_3 and sulfuric acid vapor in the 10 to 100 ppm range, based on the Corbett method, has been described.[162] The equipment, however, has the disadvantage of being large, heavy, and complicated. A compact portable apparatus has been developed that can be used in the laboratory and field to make an SO_3 determination every 15 to 20 min.[163] It was claimed that this instrument is more sensitive than other equipment described in the literature and has proven itself in power-plant testing.

Dewpoint measurement. The acid dewpoint, as stated previously, is the temperature at which sulfuric acid condensation begins when the flue gases are cooled at constant pressure. The water dewpoint is lower than and independent of the acid dewpoint and can be calculated because it is the saturation temperature corresponding to the partial pressure of the water vapor in the flue gases. The acid dewpoint, on the other hand, cannot be calculated with any degree of accuracy because of the influence of the many factors of boiler design and operation; hence, the need for determining the dewpoint of the flue gases directly.

[156] Corbett, P. F., and Crane, W. M., *BCURA Bull.*, **16,** 1–11 (1952).

[157] Flint, D., *J. Soc. Chem. Ind.* (*London*), **67,** 2–5 (1948).

[158] Corbett, P. F., *J. Soc. Chem. Ind.* (*London*), **67,** 227–30 (1948).

[159] Matty, R. E., and Diehl, E. K., *Power Eng.*, **57,** 87, 118, 121 (1953).

[160] Fletcher, A. W., *Chem. & Ind.* (*London*), **16,** 303–4 (1956).

[161] Mott, R. A., and Parramore, K., *J. Inst. Fuel*, **30,** 123–6 (1957).

[162] Crumley, P. H., Howe, H., and Wilson, D. S., *J. Inst. Fuel*, **31,** 378–82 (1958).

[163] Fielder, R. S., Jackson, P. J., and Raask, E., *J. Inst. Fuel*, **33,** 84–9 (1960).

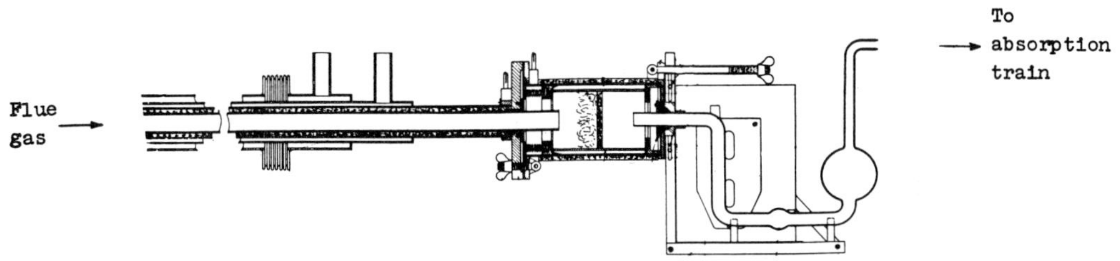

The sampling tube, dust filter, and bubbler

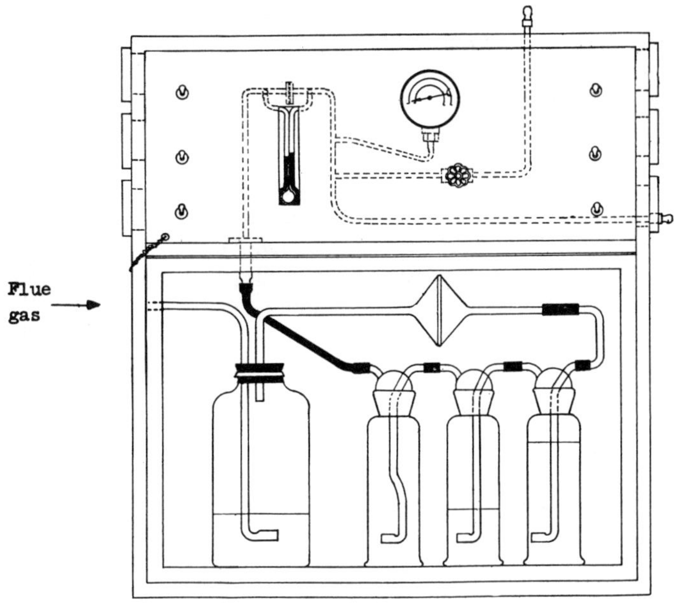

The absorption train and control box

Fig. 31. Corbett's apparatus for determining the SO_3 content of flue-gases.[158]

Although psychrometric methods and optical measurement of surface clouding are useful in other fields, they are not suitable for use under boiler conditions because of difficulty in application, lack of fundamental data, and slow formation of acid film on the detector element.

Most of the flue-gas dewpoint instruments in use today have been patterned after the one developed by Johnstone[115] in 1929. (Fig. 32). Its operating principle is based on detecting the presence of a condensed film of acid by measuring change in electrical resistance between two electrodes set in a smooth glass surface. The surface may be cooled progressively and its temperature measured. The temperature at which the resistance begins to break down is taken as the acid dewpoint. Difficulties with this and other types of recorders are caused by deposition of flue dust, soot, or conducting salts. Even when filters are used, small particles or solids condensing from the vapor phase always require periodic cleaning of the detector element.

Refinements and variations of the apparatus and electrical circuits have been reported by many researchers attempting to develop an instrument for the rapid assessment of potentially corrosive flue gases. A 1952 review[156] cited some of these modifications, and others have

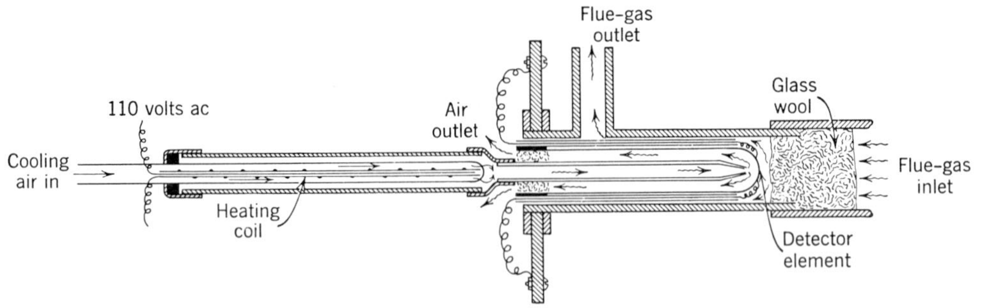

Fig. 32. Apparatus for determining dewpoint of gases.[115]

since been described.[114,118] Fundamentally, they are concerned with rearranging the electrodes and improving the accuracy of the temperature measurement. In England, the British Coal Utilisation Research Association has concentrated on the development of the dewpoint meter. Flint[118] described the "internal" and "external" dewpoint meters shown in Fig. 33. With the internal instrument, the detector surface is inserted into the boiler passes and brought into direct contact with the main gas stream. With the external type, the flue gas is withdrawn through a sampling tube and passes over the detector element which is mounted outside the boiler. Gases from high-temperature sections of the boiler may be studied with this device when the sampling tube is water-cooled, as shown in the illustration. From initial experiences in the use of these instruments it was concluded that it was possible to differentiate between the effects produced by condensed films of (1) water, (2) sulfuric acid, and (3) alkali metal salts, which are mainly mixtures of acid sulfates and are partially molten over well-defined temperature ranges.

In extending this work, a survey was made of the dewpoints and condensation characteristics of the flue gases from a number of coal-fired boilers in the British Electric Power Industry.[118] The main conclusion was that the rate of "build-up" of acid films on the detector at temperatures below the dewpoint is a very important factor in assessing potential air-heater trouble. Furthermore, although laboratory work with clean gases indicated that the acid dewpoint could be used as a measure of the SO_3 content, the results of the survey suggested that this is probably not true under actual boiler conditions. Consequently, the emphasis on further work by the BCURA was directed toward defining standard conditions for measuring the rates of acid deposition, studying the factors affecting this property of the flue gases, and correlating the rates of acid deposition with corrosion.

Modifications were made accordingly, in the design and operation of the BCURA instrument to study the rate of acid condensation in a quantitative manner. This was achieved by changing the electrode arrangement on the sensing element, using alternating current to prevent polarization, and applying a potential above a certain minimum voltage. After these changes, the rate of buildup, as shown by the meter, seemed to be a reliable index to the corrosive potential of flue gases.

Data on the dewpoint-SO_3 relationship for clean laboratory gases were obtained by a number of investigators[114,116,117,164] who considered the effect of water-vapor content. Kear[107] made a study of these data and found considerable divergence in the results. Burnside, Marskell, and Miller[126] plotted their own data along with that of others,[114–116,164] as shown in Fig. 34, and found little agreement. This lack of agreement cannot be attributed to condensable components other than sulfuric acid and water, as was demonstrated by showing that nitrogen, oxygen, carbon dioxide, sulfur dioxide, and the oxides of nitrogen had no effect on the acid dewpoint of the flue gases.[114] Likewise, hydrochloric acid had no effect on the acid dewpoint.[164,165]

Criticisms of the dewpoint meter have been raised by various investigators, notably Rylands and Jenkinson[164] who studied the condensation of acid films in the laboratory. They claimed (1) that acid existing in the form of mist will not deposit in true proportion on the

[164] Rylands, J. R., and Jenkinson, J. R., *J. Inst. Fuel*, **27**, 299–318 (1954).

[165] Kear, R. W., *Fuel*, **33**, 119–20 (1954).

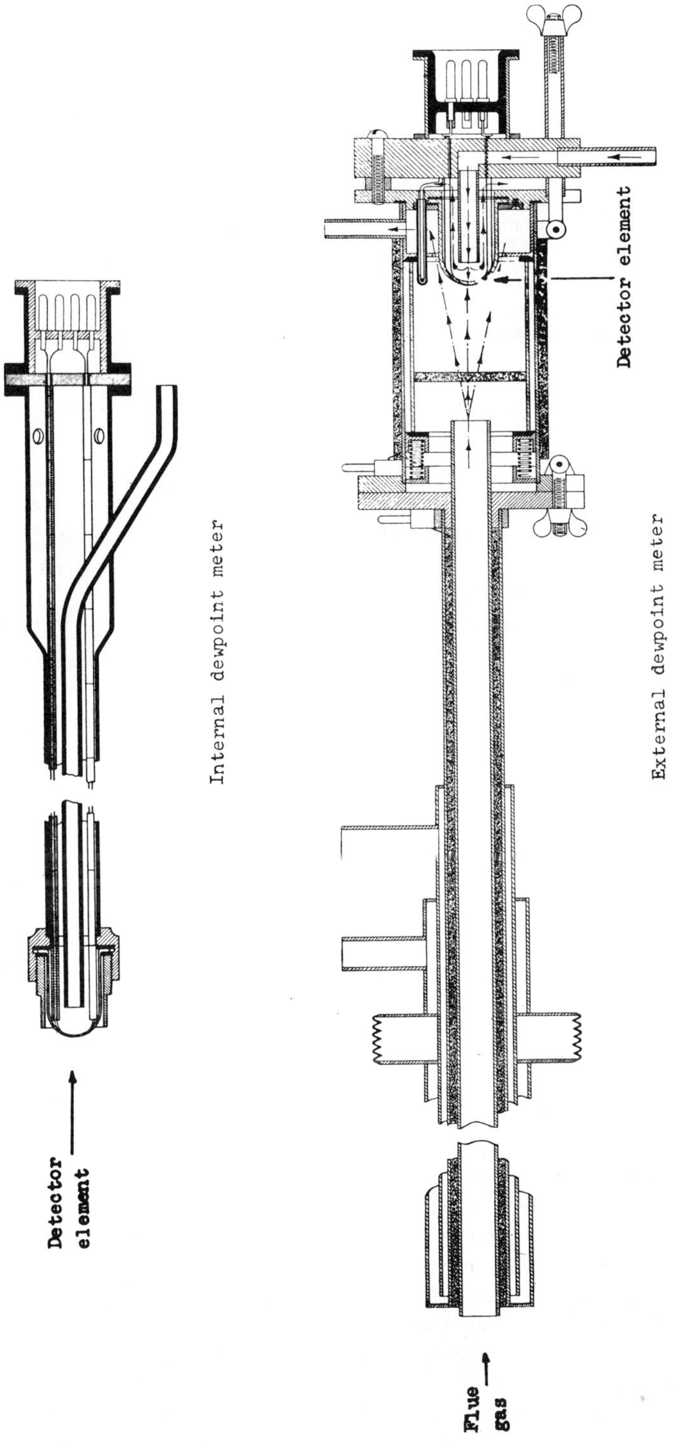

Fig. 33. Dewpoint meters for boiler testing by BCURA.[118]

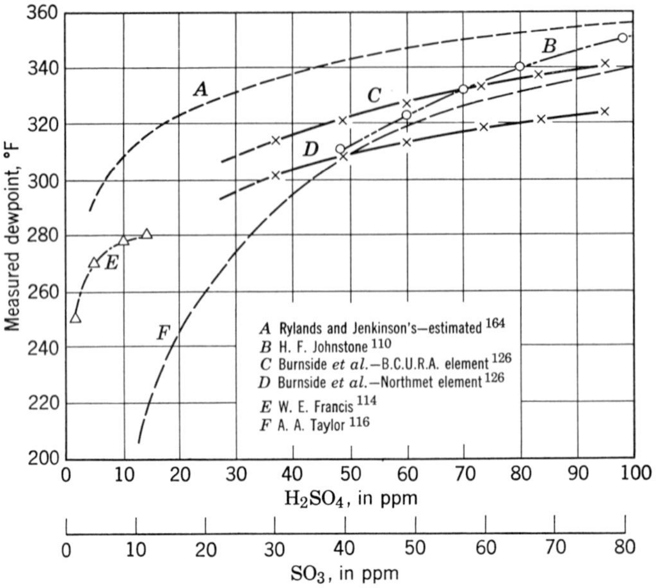

Fig. 34. Relationship between dewpoint and SO_3 concentration in gases according to various sources.[107]

dewpoint element and will thus escape detection and (2) that above the acid dewpoint there is acid deposition resulting from surface adsorption, a condition that dewpoint measurements cannot detect. Furthermore, they stated that the dewpoint changes very little with acid content of the flue gas, that is, a gas containing 1 ppm acid has a dewpoint nearer 250°F than the 120°F instrument reading. If the concentration of the acid in the gas is raised to 220 ppm, an abnormally high value for boilers, the dewpoint temperature increases only 70°F. It had been observed earlier that the extent of dewpoint rise with fuel sulfur content is smaller than expected.[166] Increasing the sulfur content of the fuel from 1 to 4% raised the dewpoint only from 239 to 288°F. In another investigation, it was shown that in the range of 1 to 5% sulfur content, the dewpoint rose only 7.2°F for each percentage.[139] It has also been claimed that the dewpoint is not a reliable measure of the acid content of flue gas, and that this in turn is not necessarily a measure of the corrosive potential of the flue gases.[164] Once acid and corrosion products are present on boiler surfaces, corrosion continues even after the metal temperature is raised to or above the dewpoint,[167] another condition that dewpoint measurements cannot detect. Furthermore, a virulent corrosive attack—caused by hydrochloric acid—has been observed[164,168] below the water dewpoint, a third condition not shown by dewpoint measurements.

Corrosion probes. Many workers prefer to use the air-cooled probe technique[169] because it provides a direct assessment of corrosion. In its present simple form, the air-cooled probe can be used for only a few hours of testing, but it is well known that corrosion rates reach a maximum in the early stages when the test surface is clean and when the corrosion mechanism is not impeded by reaction products. Kear pointed out[170] that the short duration of the tests means that the results, in many instances, are only relative and that the necessity of carrying out an analysis of the probe washings to determine the amount of metal corroded makes the method time consuming.

Laboratory results are available on the effect of progressively longer tests, up to 24 hr, but no similar information is available for prolonged tests on boilers. In the absence of such knowledge, the interpretation of short corrosion tests in terms of corrosion in a boiler is

[166] Taylor, R. P., and Lewis, A., *Proc. 4th Intern. Congr. on Ind. Heating, Paris*, **1952,** Group 2, Sect. 24, No. 154, 8 pp.

[167] Kear, R. W., *J. Appl. Chem.* (*London*), **5,** 260–6 (1955).

[168] *Ibid.*, 237–42.

[169] Flint, D., and Kear, R. W., *J. Appl. Chem.* (*London*), **1,** 388–93 (1951). Thurlow, G. G., *J. Inst. Fuel*, **25,** 252–55, 260 (1952).

[170] Kear, R. W., *J. Inst. Fuel*, **32,** 267–73 (1959).

full of uncertainties. Any attempt to extend the period of testing would mean sacrificing the simplicity of the air-cooled probe, with the detachable head for assessing corrosion by weight loss, and requiring a complicated system for controlling the cooling air and test surface temperature.

Two new probe techniques have been devised to overcome some of these difficulties. Anson[171] described a probe consisting of a metal tube that has a removable radioactive mildsteel test surface at one end. The temperature of the radioactive element was measured by spring loaded contact thermocouples and regulated by a jet of cooling air. The probe was inserted into the flue gases for 10 min. After the probe was removed from the gas stream, the condensed H_2SO_4, which had formed $FeSO_4$ with the radioactive iron of the test specimen, was washed off with distilled water. The iron content of the wash water was then measured by radioactive counting methods. The author claimed that results are quickly obtained and are reproducible. However, since the method requires careful and skilled operation and expensive counting equipment, its use would appear to be limited to the laboratory.

Kear[170] described a probe that was designed to measure corrosion over a relatively long period, several hundred hours or more. It was simple to use and the results were obtainable without the work of analysts. The probe assembly consisted of a water-cooled reflux condenser mounted on top of a removable cylindrical test piece that was closed at the bottom. The test piece was filled with a liquid that refluxes when immersed in the hot gases. The test piece was thus maintained at a constant temperature corresponding to the boiling point of the liquid it contained. Obviously, the chosen liquid must boil within the limits defined by the temperature of the gases being examined. In this way, the test piece could readily be maintained at a fixed temperature and the conditions for any of the cool metal parts of the combustion system reproduced. After exposure to the flue gases, the probe was withdrawn, the test piece was removed when cool, and the corrosion products removed by washing. The test piece was dried and the iron loss determined directly by weighing.

Techniques have been developed for estimating the rate of corrosion in the boiler installation itself by measuring the metal components under examination. This approach gives valuable information, but represents cumulative effects and cannot identify specific conditions or causes of corrosion. This criticism can be made of the scraper technique,[154] in which a procedure for sampling the corrosion products was devised and the analysis of the scrapings was related to an equivalent amount of iron corroded from the metal under test. Representative sampling is the main difficulty associated with this method. This limitation led to the development of another method employing the sampling and analysis of water used for washing corroded equipment. A possible error in that method is that there may be just as much corrosion caused by diluted acid during water washing as occurs during operation when only relatively inert concentrated acid is present.

The relative merits of the various tools used in the study of corrosion—the chemical determination of SO_3, dewpoint measurements, and corrosion probes—is still a controversial issue. However, the situation has been summarized by Crossley,[172] who concluded that all three techniques are needed simultaneously when investigating acid troubles in a boiler plant. He pointed out that there is still much to be learned about all three, and that it is by no means unusual for one of the three to give indications that are definitely contradictory to those of the other two. No single technique can be accepted as entirely trustworthy, and it is for this reason that all three still have to be employed simultaneously with consequent heavy calls on scientific manpower.

Extent of Corrosion

The amount of corrosion that occurs in boilers depends on the rate at which sulfuric acid condenses on the heat-absorbing surfaces, the strength of the acid, the presence of ash deposits and corrosion products, the type and amount of surfaces involved, and finally, the exposure time and temperature of these surfaces under various operating regimes including start-up, onload, low-load, banking, and shutdown. These, in turn, are influenced by other conditions. For convenience, the various factors involved will be grouped into three

[171] Anson, D., *J. Inst. Fuel*, **32**, 10–14 (1959).

[172] Crossley, H. E., *J. Inst. Fuel*, **32**, 121–3 (1959).

categories: those affecting the rate of sulfuric acid condensation; those controlling the concentration of the acid; and those influencing the rate of attack of the acid film on the metal surfaces.

Sulfuric acid condensation. The rate at which sulfuric acid condenses at temperatures below the acid dewpoint is one of the more important factors in low-temperature corrosion.[118,169,173,174] Experience has indicated that the rate of acid deposition is more significant than the actual dewpoint temperature. The increase in conductivity at about 50 to 70°F below the dewpoint. Moreover, rates of acid buildup are independent of gas temperature and velocity in the ranges of 600 to 1000°F and 5 to 50 ft/sec, respectively. Variations in rates of buildup are due primarily to changes in the chemical composition of the flue gases and its entrained solids.

Fereday's statistical analysis of Corbett's[137] data on an oil-fired boiler indicated that the rate of buildup increases with increased load and smoke number, but decreases with in-

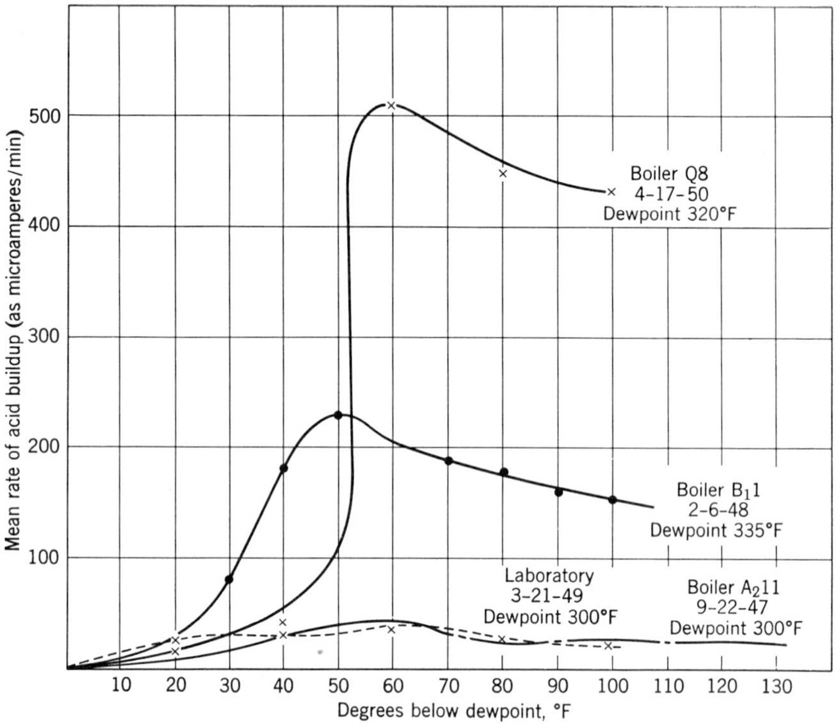

Fig. 35. The effect of element temperature on the rate of acid buildup.[118]

dewpoint temperature identifies the range of metal temperature over which corrosion is likely to occur, whereas the rate of acid deposition gives some indication of the severity of corrosion.[102]

For a given flue-gas composition, the rate of acid buildup as measured by the dewpoint meter is determined primarily by the temperature of the sensing element.[118] Typical curves are shown in Fig. 35. Their most prominent feature is the presence of a peak in the rate of creased CO_2 content. Furthermore, the class of fuel oils burned and the sampling point on the boiler also have an effect on the rate of acid buildup.

One investigator[138] correlated data from a number of boilers and suggested that a logarithmic relationship exists between the rate of buildup and the dewpoint, as shown in Fig. 36. The rate of buildup may be expected also to be proportional to the acid concentration in the flue gas. In practice, however, its value may vary by a factor of 1 to 100 from that expected.

From Taylor's[116] data, it can be shown that in a static system the weight of acid that will

[173] Corbett, P. F., *J. Inst. Fuel*, **26**, 92–106 (1953).

[174] Thurlow, G. G., *Proc. Inst. Mech. Engrs. (London)*, **168**, 571–5 (1954).

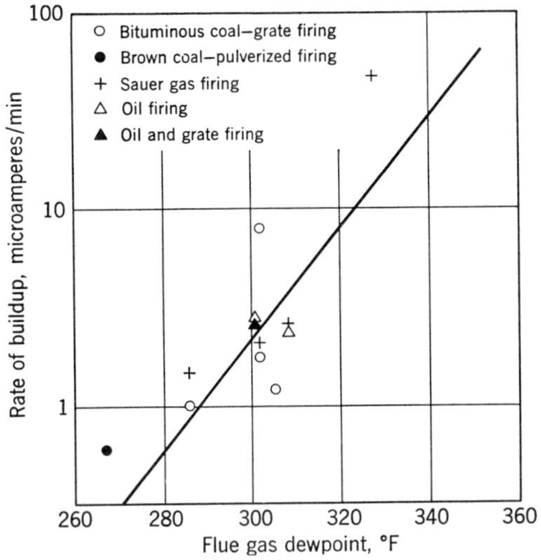

Fig. 36. Relationship between rate of buildup and dewpoint.[138]

condense at a given temperature is directly proportional to the difference between the partial pressure of H_2SO_4 at that temperature and its partial pressure at the dewpoint.[118] In a system of flowing gases, the quantity of acid available for condensation at a given temperature also increases in proportion to the difference between the dewpoint of the gas and the surface temperature, but under these conditions not all the available acid reaches the surface.[174] A peak in the rate of acid deposition occurs, furthermore, at a surface temperature between 80 and 90°F below the dewpoint, and the magnitude of the peak increases with dewpoint temperature, as shown in Fig. 37. The rate of acid deposition also decreases as the water vapor of the flue gases increases. A mechanism was suggested for this phenomenon in which the first stage in the transfer to the surface is the formation at the dewpoint of mist particles containing sulfuric acid and water.[174] For surface temperatures just below the dewpoint, only a small quantity of acid will diffuse to the wall under the influence of a slight temperature gradient. As the surface temperature is lowered, the mist particles in the gas grow at a rapid rate and produce the condensation peak rate. Below a certain critical surface temperature, the particles are of such a size that there will be a slow rate of diffusion to the surface and hence a low rate of condensation.

In a more detailed study of the mechanism of mist formation and acid deposition, it was concluded[164] that an increase in water vapor content produces an increase in the rate of condensation contrary to Taylor's findings.[175] In discussing Taylor's theory, it was pointed out that the moisture content of the gases would be expected to produce a more dilute surface condensate and, hence, for a given SO_3 content, a greater diffusion rate.[164] This was later confirmed in a study of the corrosion of iron by flue gases containing SO_2/SO_3 and relatively small amounts of water vapor (0.5 to 1.2%).[129] The Boiler Availability Committee[102] recognized that there is a peak in acid condensation rate at a surface temperature 110 to 120°F below the dewpoint, and that this peak increases as the dewpoint is raised, but is independent of the water vapor content.

Tests on a model rotary air heater[174] under conditions similar to those in industrial practice showed that both the rate of acid deposition and the rate of corrosion increase to a maximum value as the surface temperature falls below the dewpoint, as shown in Fig. 38. These peak rates occur at surface temperatures in the range of 215 to 245°F for gases having dewpoints of 240 to 320°.F

From these investigations, it is readily apparent that the rate of acid buildup varies with temperature gradients, SO_3 and water vapor in the flue gases, excess air, boiler load, smoke number, amount and nature of entrained fly

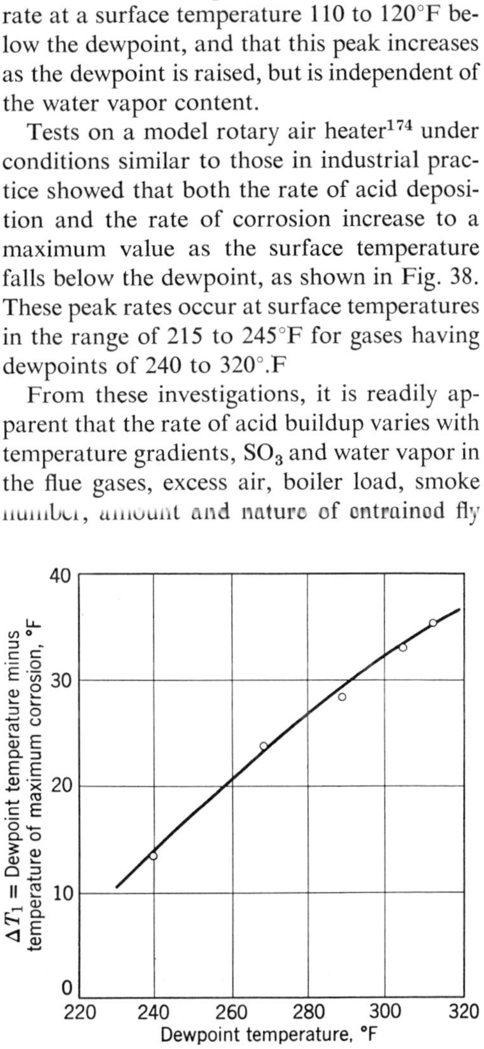

Fig. 37. Location of the temperature of maximum corrosion as a function of the dewpoint.[175]

[175] Taylor, H. D., *Trans. Faraday Soc.*, **47**, 1114–20 (1951).

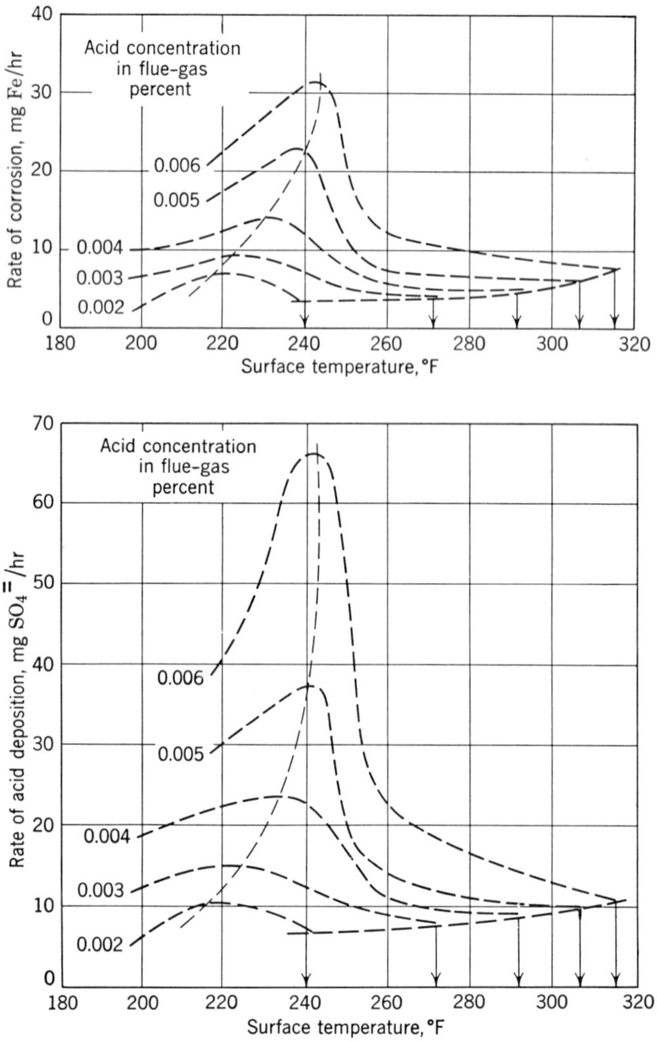

Fig. 38. Variation of corrosion and condensation with surface temperature of various acid concentrations in flue-gas.[174] Arrows denote dewpoint temperatures of various acid concentrations.

ash, and many other factors. The rate of acid build-up and the rate of corrosion generally increase together, and their maxima may coincide under some conditions.

Concentration of condensed acid. As stated previously, the concentration of acid condensing is one of the more important factors governing the extent of corrosion. In 1942, Taylor[116] found that the concentration is a function of the surface temperature and the water vapor content of the flue gases, as shown in Fig. 39. Since the moisture content of the flue gases in a coal-fired boiler seldom exceeds 10%, the acid condensing out has a concentration ranging between 67 and 87% when the surface temperatures vary between 200 and 300°F, respectively. Flint[118] found that the concentration of the acid from flue gases is primarily dependent on the temperature of the surface and the sum of the partial pressures of water vapor and SO_3.

Later, it was found that the concentration of the acid condensing is primarily influenced by the temperature of the surface condenser and secondarily by the water-vapor content of the flue gases.[175] At a given surface temperature, a change of H_2SO_4 content of the gases from 0.007 to 0.07% by volume resulted in no change in the concentration of acid in the condensate. The Boiler Availability Committee[102] has concluded that the concentration of the condensate is not affected by the proportion of SO_3 present in the flue gases, but is a function only of the surface temperature and of the water vapor content of the flue gases The effect of the water vapor throughout the range of surface temperatures is to fix the concentration of the condensate.[164]

All the foregoing considerations deal with the strength of the pure acid as it first forms, but the final strength of the acid is not known because of surface tension effects and the influence of salts in solution, which alter the partial pressure of constituents in the gases immediately above the acid film.[176]

Rate of attack. The reactions between metals and gaseous combustion products, like all other reactions between gases and solids, are relatively slow.[177] It is evident, therefore, that corrosion by flue gases must occur through reactions other than gaseous to reach serious proportions. It is now generally accepted that corrosion begins after a condensed liquid phase is formed.

It has already been shown that when sulfuric acid condenses on the heating surfaces of the boiler, the strength of the acid is governed by the temperature of these surfaces. The corrosive attack on the metal is influenced by the strength and temperature of the acid and by the composition and structure of the metal.

[176] Jenkinson, J. R., *Mech. Eng.*, **79**, 481 (1957).

[177] Kear, R. W., *Corrosion Technol.*, **3**, 59–64 (1956).

Johnstone,[110] one of the early workers in this field, exposed mild-steel specimens to an atmosphere containing sulfuric acid vapor and water vapor at 200°F. He found that the corrosion caused by the concentrated acid condensing upon the surface was very slight, the penetration amounting to only 0.0045 in./yr. In another set of experiments, he immersed mild steel specimens in sulfuric acid of varying concentrations containing ferric sulfate, and found that the presence of ferric sulfate accelerated the rate of attack, especially in the acid concentration range of 20 to 50%, as shown in Fig. 40. For comparison, the rate of corrosion of iron by sulfuric acid solution is also shown. As the concentration of acid increases, the rate of attack first increases and then decreases until, at a concentration of about 50%, the attack is very slow. Johnstone attributed this behavior to the presence of corrosion products that reduce the rate at which acid can diffuse to the corroding surface. More recently, laboratory tests of various metals in sulfuric acid have shown that the presence of air-heater deposits may change the rate of attack as much as 100-fold, sometimes preventing it and at other times causing it to take place.[59]

The action of sulfuric and sulfurous acids and their products on the corrosion of cast iron and steel has been investigated.[178] With boiling sulfuric acid, a major peak in rate of attack occurs with a 20% acid and a minor peak with

[178] Rylands, J. R., and Jenkinson, J. R., *Proc. Inst. Mech. Engrs. (London)*, **168**, 105 18 (1940).

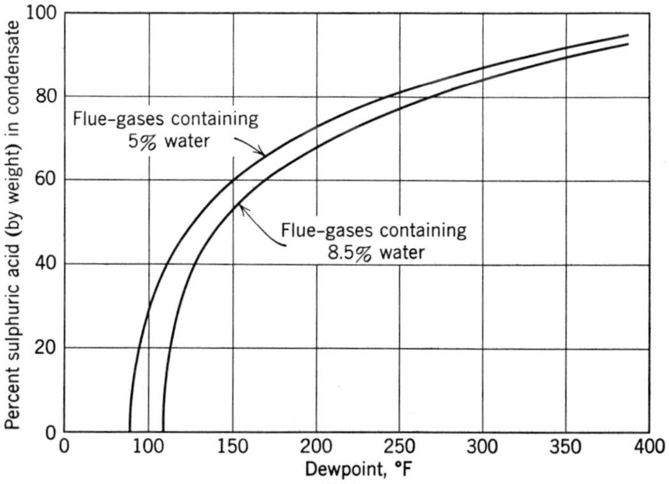

Fig. 39. Relation between the dewpoint of clean flue gas and the strength of the acid deposited.[116]

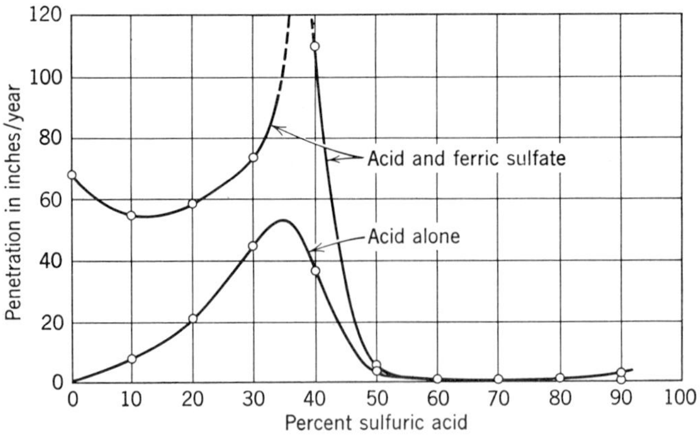

Fig. 40. Corrosion of pure iron at 200°F by acid solutions.[110]

85 to 90% acid, the latter affecting steel specimens but having little effect on cast iron. The limiting factor that determines the extent of attack with acid solutions above 20% is the diminishing solubility of ferrous sulfate as the acid concentration increases. It was found that corrosion was minimum with 70% acid. The superiority of cast iron over steel in resisting attack was attributed to the silicon content of the cast iron. A protective skin of a silicon compound quickly formed on the cast iron, and the attack was greatly reduced. This action, however, did not take place in very dilute acids in which silicon is relatively soluble, its solubility decreasing with increased acid concentration.

Flint and Kear[169] have drawn attention to the limitations of tests where samples are immersed in solutions of sulfuric acid at particular concentrations and temperatures, and have allowed sulfuric acid to condense from a synthetic gas of known SO_3 content on a metal surface at a controlled temperature. They claimed that this arrangement more nearly approaches actual service conditions. Using a small, mild-steel, hemispherical probe, they studied the effect on corrosion of dewpoint, surface temperature, and time (up to 1 hr). The results show that the rate of corrosion increased with dewpoint temperature, a peak value in acid deposition and corrosion being obtained at a surface temperature of 40 to 80°F below the dewpoint. In further test work, the time of testing was extended to 24 hr and it was shown that after 8 hr the rate of corrosion decreased and, furthermore, that the accumulation of corrosion products caused corrosion to continue even when the metal temperature was raised to or above the dewpoint.[167] This confirms similar observations made by other investigators.[87] Recent tests of 100-hr duration, using a special constant-temperature probe, gave corrosion rates that were only 10 to 25% of those obtained during the 1-hr tests.[170] The characteristic peak in corrosion at 40 to 60°F below the dewpoint also occurs in tests lasting 175 hr.

Thurlow[169] found a general relationship between the weight of acid condensed and the weight of iron corroded, and in a later investigation he showed that the ratio of acid condensed to the iron corroded was about 1.72, indicating that all of the acid was converted to $FeSO_4$. In a subsequent study, Whittingham[129] showed that this ratio varied between 6.2 and 8.0. He concluded that there was a considerable excess of acid over that needed to form ferric sulfate.

The corrosion of boiler-heating surfaces by sulfurous acid has been assumed to be negligible compared to that of sulfuric acid, but laboratory experiments[178] have shown that a 5% sulfurous acid solution corroded twenty-five times as much steel and four times as much cast iron as did a 5% sulfuric acid solution. Whittingham has called attention[107] to the rapid rate of corrosion by sulfurous acid during periods of start-up, banking, and low-loads.

Another acid that may accelerate corrosion at temperatures below the water dewpoint is hydrochloric acid. In spite of the relatively large quantity of chlorides found in many coals, very little chloride is found in any boiler deposits, and it is probable that any chlorides carried in the gas stream are changed to sulfates with the liberation of hydrochloric acid, which is carried in the gas stream toward the cooler parts of the boiler.[176] The acid, however, readily dissolves in water and is thereby trapped on surfaces below the water dewpoint. The occurrence of hydrochloric acid in boiler flue-gas condensate has been reported by Harlow,[119] and he suggested that this might account for the exceptional difficulty experienced in resisting corrosion by special metals, even though they have withstood the action of sulfuric acid during controlled tests. Others[164] have shown in laboratory experiments that 85% of the hydrochloric acid in the flue gases appears in the condensate at temperatures around 80°F. Studies on the corrosive nature of flue gases containing both sulfuric and hydrochloric acids,[165,168] have shown that they are much more corrosive than flue gases containing only the sulfur oxides, especially when the metal surfaces are below the water dewpoint, as shown in Fig. 41.

Methods of Controlling Corrosion

The problem of plugging and corrosion in air heaters and economizers is important, but it should be realized that the great majority of installations is being operated continuously between annual inspection periods with relative freedom from difficulties such as would cause forced or unscheduled outages.[89] A considerable amount of research work has been carried out since 1945, however, because of acute trouble on a comparatively small

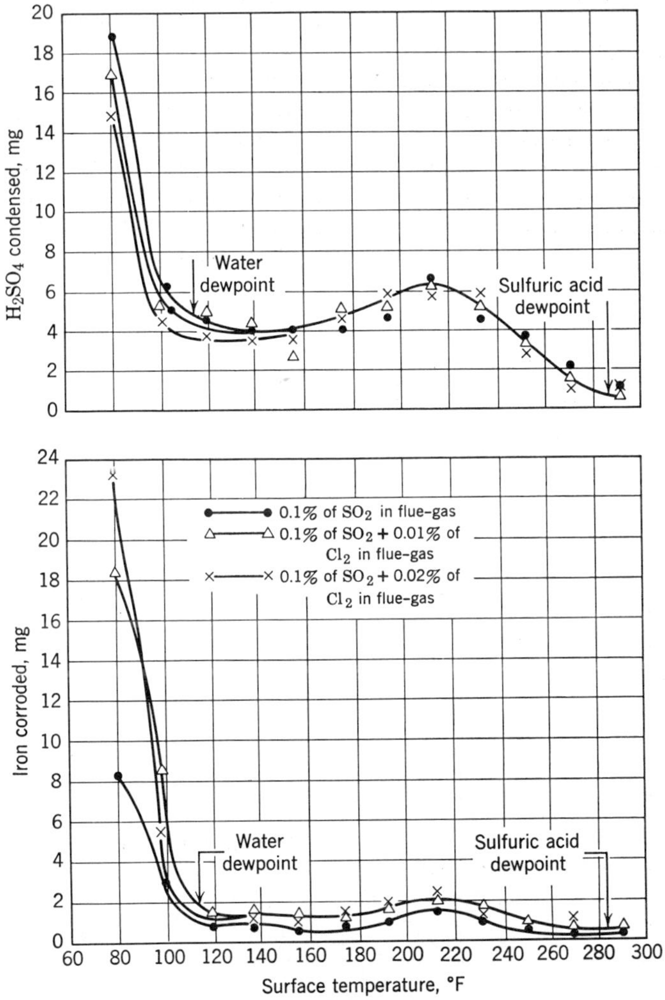

Fig. 41. The effect of chlorine addition on corrosion of mild steel by combustion gases containing oxides of sulfur.[168]

number of stoker-fired and oil-fired boilers, and also for the purpose of obtaining information to permit designing for a lower exit-gas temperature.[90]

In general, methods of controlling corrosion involve some change in fuel supply, boiler design or operation, or the use of additives in the fuel or the flue gases.

Fuel supply. The earlier consideration of the causes of corrosion would indicate that care in selecting the fuel would be a natural starting point. This has been recommended by a number of investigators.[102,110,131] Generally, low-sulfur fuels will mean a reduction in corrosion.[59,179] Unfortunately, however, distribution problems do not always allow a free choice of fuel, and it is often necessary to burn high-sulfur fuels that are known to give corrosion troubles.

The elimination of sulfur from solid or liquid fuels is neither simple nor economical. It can be accomplished by hydrogenation which requires expensive hydrogen and makes the final product more costly. In addition, adequate processing equipment would have to be installed first. Mixing of fuels from different mines makes it possible to avoid too unfavorable extremes.[131]

The presence of chlorine in coal has been discussed earlier in this chapter, and there is some indication that coals with a high chlorine-to-sulfur ratio may cause less low-temperature corrosion, but may, on the other hand, increase the tendency to form high-temperature deposits.[81]

Equipment design. Manufacturers of boilers and related equipment recognize that corrosion depends primarily on the type of fuel, the sulfur content of the fuel, the metal temperature, and the method of firing.[88,89] Drawing on a large body of experience, these organizations have drawn up design curves (Fig. 42) giving the acceptable minimum metal temperatures for the various fuels and firing methods normally employed.[3,89] If the design of a unit is such that these minimum temperatures are approached at full-load operation, some means of controlling the metal temperatures at reduced loads must be provided. Furthermore, in those installations where the ambient temperature can drop below the designed ambient temperature, it is necessary to

[179] Kear, R. W., *Corrosion Technol.*, **3**, 125–7 (1956).

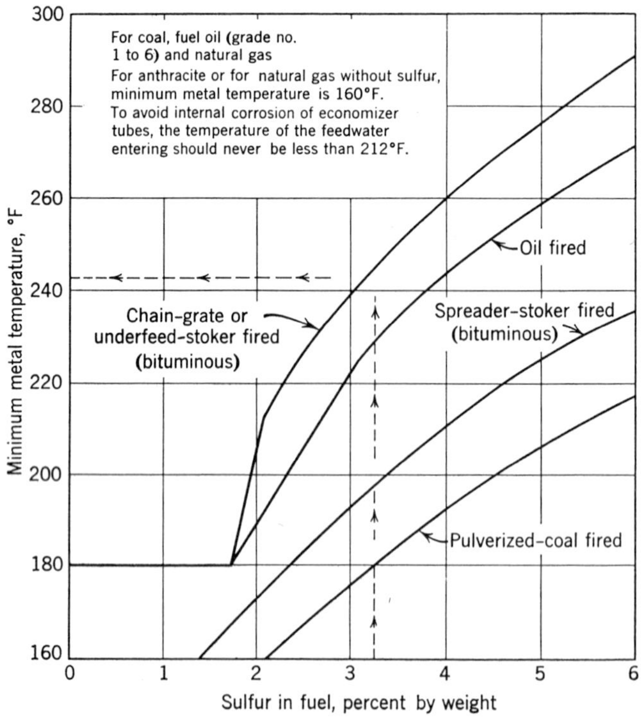

Fig. 42. Minimum allowable air-heater tube metal temperature.[88]

provide temperature control at full-load operation.

The various methods of controlling metal temperatures in air heaters include cold-air bypass, hot-air recirculation, steam air heaters, economizer by-pass, flue-gas recirculation, parallel counter-flow combination, and reduction of surface. Some of these methods have been evaluated for tubular air-heaters[88] and regenerative air-heaters.[89,180]

Particular importance is attached to suitable air distribution, especially at the air-heater inlet to avoid corrosion caused by excessively low metal temperatures.[88,89,181]

Various corrosion resistant materials for air-heater construction have been evaluated in the laboratory[110,179] and in the field.[59,90] The use of expensive high-alloy steels for this purpose is not economical, as a rule, since service life does not increase in proportion to the cost of the material.[131] This does not exclude the use of relatively cheap alloys, such as copper-bearing steel, to secure better protection when metal temperatures occasionally drop below the dewpoint or when the boiler is out of service. In these instances, low-alloy steels are technologically and economically superior to high-alloy steels.[59,90]

The Boiler Availability Committee[102] has stated that cast iron resists corrosion better than steel, and the experimental work of Rylands and Jenkinson[178] shows why this is so.

The application of various coatings, such as plastics, paints, or baked lacquers, is always difficult when large bulky assemblies like air-heater sections are involved.[182] Such equipment is hard to handle for thermal treatment or curing processes. Ceramic coatings, enamels, or glasses on tubes or flat parts require special attachment techniques to avoid damage, and exceptional care is needed in handling during transportation and erection. Experience to date on coatings has been unfavorable, and service life has been limited because the coatings become porous or break off or char under service conditions.[183]

Boiler operation. As mentioned earlier in this chapter, it has been known for many years that air heaters and economizers of pulverized-coal-fired boilers are far less susceptible to corrosion than those of stoker-fired units.[110] In fact, auxiliary pulverized-coal firing has been used on chain-grate-fired and retort-fired boilers and the corrosive nature of the flue gases greatly reduced.[136,184] For this reason, the Boiler Availability Committee[102] and others[119] have long advocated the adoption of pulverized-coal-firing as a means of avoiding this problem. The cyclone-furnace method of firing offers even greater freedom from low-temperature deposits and corrosion by virtue of the low dewpoint of the flue gases, which seldom exceeds the water dewpoint.[150]

Regardless of firing method or fuel, combustion control is the primary approach to the solution of the corrosion problem inasmuch as boiler operation with low excess air has minimized corrosion difficulties in a wide variety of industrial plants and power stations.[128,154] Zaczek and Grindley[183] have stated that operation with reduced excess air lowers SO_3 formation, corrosion, and fouling. Adjustment is critical, however, and if overdone, soot and smoke emission result. Soot collects in the cooler parts of the system and concentrates acid from the flue gases. The acid particles so formed may be discharged intermittently from the stack and deposited in the neighborhood.

A number of investigators[89,185] have stated that high air temperatures as well as low-excess air will reduce corrosion because of the lower SO_3 concentration in the flue gases brought about by the higher combustion temperatures. Others[135,186] have reduced the corrosive nature of flue gases by recirculating relatively cold flue gases through the fuel bed. The reduction in SO_3 content was attributed to lower fuel-bed temperatures.

The Boiler Availability Committee[187] and other investigators[179] have emphasized the fact that thorough cleaning of heating surfaces

[180] Widell, T. A., and Juhasz, S. I., *Trans. Roy. Inst. Technol. Stockholm*, No. 54 (1952), 50 pp.

[181] Lees, B., *J. Inst. Fuel*, **28,** 433–40, 450 (1955).

[182] Gumz, W., *Mitt. Ver. Grosskesselbesitzer*, **44,** 325–42 (1956).

[183] Coit, R. L., *ASME, Trans.*, **78,** 89–94 (1956). Zaczek, B. J., and Grindley, R., *Engineering*, **184,** 825 (1957).

[184] Henning, F., and Rogener, H., *Brennstoff-Wärme-Kraft*, **5,** 269–72 (1953).

[185] Robertson, R. J., and Gurney, W. B., *Mech. Eng.*, **72,** 1006 (1950).

[186] Marskell, W. G., *Combustion Eng.*, **5,** 13–16 (1951).

[187] Boiler Availability Committee, *Bull.* MC/131, September (1945).

at regular intervals is one of the most effective methods of increasing boiler availability by reducing deposition and corrosion. The corrosion products should be completely removed when cleaning the surfaces because they can cause corrosion even when the metal temperatures are above the dewpoint.[119,167,178]

The washing of economizers and air heaters while the boiler is operating was first applied in the early 1920's and later abandoned, but in the last few years this technique of cleaning has become popular again. In a recent paper, Jenkinson described the method now employed on cast iron economizers in England.[188] Operating experience with in-service washing of air heaters has shown that the method is sound, provided the system has been designed for this purpose and adequate control of the washing procedure is followed.[89,96,189]

Additives. In general, the use of additives in industrial and electric utility boilers is aimed at modifying the ash deposits that form on superheater and reheater tube banks to render them more easily removable by the cleaning equipment and to reduce their corrosive tendencies. Additives are also used to alleviate corrosion and pluggage of economizers and air heaters resulting from the deposition of sulfuric acid.

Many types of materials that differ greatly in cost and availability are described in technical and patent literature. These materials can be classified into three main groups, depending on the mechanism by which the SO_3 level in the flue gases is lowered.[190]

1. Materials that physically adsorb SO_3, such as silica or carbon.[147] For efficient adsorption, these materials must be introduced in activated, finely divided forms such as dusts or smokes so that a maximum specific surface area is offered to the gases.
2. Materials that combine preferentially with the atomic oxygen in the flame zone or which promote recombination of the oxygen atoms with consequent inhibition of the reaction:

$$SO_2 + O \rightarrow SO_3 \quad (5)$$

Some relatively inert materials, such as carbon and silica, may also belong to this group, since it is known that the recombination of atoms is facilitated by the presence of solid surfaces capable of absorbing the energy of recombination. Carbon tetrachloride, pyridine, ethyl nitrate, ethyl alcohol, and others have been studied as a means of reducing flame speed.[117] Iron pentacarbonyl and tetraethyl lead added to flames containing SO_2 eliminated SO_3 formation entirely.[120] Coal-tar bases sprayed into a town's gas flame reduced the rate of acid build-up and corrosion at all temperatures below the dewpoint of 280°F.[191]

3. Materials that combine with SO_3 to form noncorrosive compounds, which are carried away in the stack gases. Compounds of zinc[136,151,192] and magnesium fall into this category. For efficient contacting with the SO_3, a high, active surface-to-volume ratio is again necessary.

Insofar as low-temperature corrosion is concerned, smokes of silica, carbon, zinc oxide, and flame-speed inhibitors have not been applied on a commercial scale despite the fact that laboratory and full-scale tests showed many of them to be technically feasible. There are many other factors that are important in evaluating the overall performance of additive materials, including their effect on stack emission, its appearance and toxicity, boiler efficiency, and ash deposition. An otherwise promising material is often ruled out because the cost and expense of handling it in the boiler plant exceed the savings through its use. One electric utility found that the cost of kaolin additive,[68] 10¢/ton of coal or the equivalent of 2.0% in boiler efficiency, was excessive for ash deposit control on a large coal-fired boiler. Another utility plant using dolomite for combating ash deposition in the superheater as well as corrosion and plugging of the air heater in an oil-fired boiler found it profitable to use an amount of additive costing $\frac{1}{4}$ cent per barrel of oil. This is equivalent to 0.1% in boiler efficiency.[152] Still another plant, using magnesium hydroxide for minimizing high-temperature, oil-ash corrosion on boilers delivering 1100°F steam, spends $\frac{1}{2}$ cent per barrel of oil, which is equivalent to 0.7% in boiler efficiency.

At present, additives are used on coal-fired boilers for experimental purposes. They are used in few installations, if any, as a regular operating procedure. Because acute fouling

[188] Jenkinson, J. R., *Mech. Eng.*, **79,** 481 (1957).

[189] Bice, G. W., and Baran, W. K., *Electric Light and Power*, **37,** 51–3 (1959).

[190] Rendle, L. K., and Wilsdon, R. D., *Combustion*, **29,** No. 1, 39–46 (1957).

[191] Kear, R. W., *J. Appl. Chem. (London)*, **4,** 674–9 (1954).

[192] Littlejohn, R. F., *J. Appl. Chem. (London)*, **2,** 289–94 (1952).

and corrosion problems are more frequently encountered on oil-fired boilers, most of the literature on the use of additives concerns oil-fired applications. An excellent summary of this literature has been prepared by Jarvis.[193] Since much of the material discussed is also applicable to coal-fired boilers, some of this experimental work will be reviewed here.

The particular attention that dolomite has received as an additive material stems from the original work on an oil-fired boiler at Inglis Station of the Florida Power Corporation.[194] This material was used primarily to control ash deposition on the superheater, but it was found to have beneficial effects on air-heater pluggage and corrosion as well. In extending these tests, it was found that a 0.1% addition of dolomite based on the fuel oil could eliminate corrosion.[152] The dewpoint at the air-heater inlet was lowered from 342 to 302°F, and from 309 to 284°F at the air-heater outlet. Furthermore, SO_3 measurements indicated that without dolomite a large amount of sulfuric acid was condensing in the air heater, as shown in Table 4, but when dolomite was used, little or no acid was deposited. Also, with dolomite, the deposits on the air-heater tubes became thin, dusty, and light colored. Prior to the use of additives, the deposits were dark, moist, and thicker.

TABLE 4

TABULATION OF SO_3 IN AN AIR HEATER OF OIL-FIRED BOILER[152]

Boiler Load	Additive Rate lb/hr	SO_3 Entering Air Heater lb/hr	SO_3 Leaving Air Heater lb/hr	SO_3 Condensed lb/hr
Full	None	50.0	38.0	12.0
Half	None	23	16.7	6.3
Full	30	33.4	36.2	−2.8
Full	60	25.5	22.8	2.7

After these tests, dolomite was used at Kearny Station of the Public Service Electric and Gas Company, Newark, New Jersey. The results were negative at first,[195] but when the quantity of additive was doubled, a reduction in air-heater corrosion occurred in spite of the fact that no clear-cut changes in the dewpoint of the flue gases could be observed.[196]

In tests on a cyclone-furnace boiler, it was found that dolomite injected in the lower secondary furnace for controlling high-temperature deposits also reduced the SO_3 content of the flue gases from 12 to 4 ppm, and the dewpoint from 275 to 140°F.[68] Since this boiler was not experiencing corrosion or plugging, it was not possible to evaluate the effect of dolomite on these cold-end problems directly.

To reduce the catalytic oxidation of SO_2, Harlow[197] recommended that superheater tubes be coated with a cadmium-aluminum alloy and treated with a lime slurry to which a number of bonding agents could be added. This treatment was used successfully by Walsh,[198] who showed that washing and neutralizing the boiler surfaces and coating with lime slurry prevented corrosion, but that feeding lime to the furnace appeared to be of no value.

Laboratory experiments and field tests have also been described where small quantities of lime and soda were sprayed on superheater materials to inhibit the catalytic oxidation of SO_2 to SO_3.[199] It was found that with lime the inhibiting effect failed after 24 hr and after 556 hours with soda. Tests on operating boilers using soda-ash sprayed through soot blowers were promising but not conclusive.

Experience of intermittent spraying of additives on superheater and reheater tubes through the soot blowers has been described as a means of controlling ash deposition and corrosion in more than forty boilers in the Florida Power and Light system.[200,201] In addition to weakening the high-temperature ash deposits so that they were easily removed by the soot blowers, the spraying of magnesium oxide and calcium oxide in this way reduced the SO_3 content of the flue gases (Fig. 43).

[193] Jarvis, W. D., *J. Inst. Fuel*, **3**, 480–91 (1958).

[194] McIlroy, J. B., Holler, E. J., Jr., and Lee, R. B., *ASME, Trans.*, **76**, 31–44 (1954).

[195] Weisberg, H., *ASME, Trans.*, **76**, 44–5 (1954).

[196] Zawadowicz, S. J., *ASME, Trans.*, **77**, 277–8 (1955).

[197] Harlow, W. F., Brit. Pat. 614,120 (1946).

[198] Walsh, E. F., *Mech. Eng.*, **71**, 516 (1949).

[199] Harlow, W. F., *ASME, Trans.*, **80**, 225–32 (1958).

[200] Keck, J. W., *Southern Power and Ind.*, **78**, 30–3 (1960).

[201] Cantieri, W. F., and Chappell, R. E., *Mech. Eng.*, **83**, No. 5, 101 (1961).

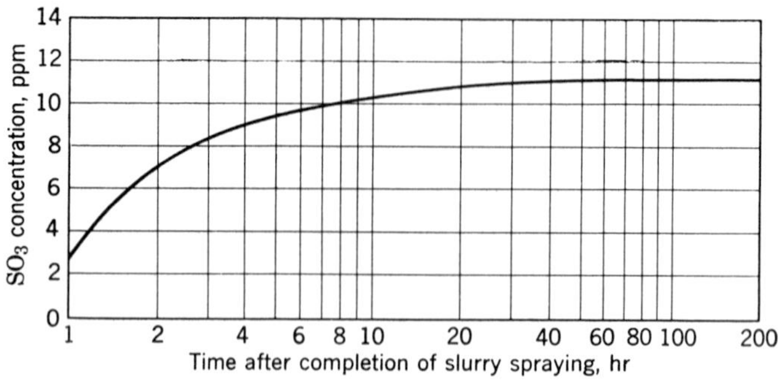

Fig. 43. Effect of lime spraying on the SO_3 content of flue gases in an oil-fired boiler.[201]

Ammonia gas was first used as an additive in 1944 to neutralize the acid flue gases of a stoker-fired boiler that was experiencing serious corrosion in a regenerative air heater.[202] Deposits of ammonium salts caused rapid plugging of the air heater, so the use of ammonia was terminated.

Recently, the use of ammonia has been recommended over all other additive types, including dolomite, magnesium and zinc naphthenates, magnesium soaps, magnesium oxide, powdered zinc, silica, carbon, and nitrogenous materials.[139] The method involves injecting ammonia into the flue gases. The reactions that occur in the NH_3-SO_3-H_2O system can be summarized as follows:

$$2NH_3 \xrightarrow{1132^\circ F} N_2 + 3H_2, \qquad (8)$$

$$\underset{(gas)}{SO_3} + \underset{(gas)}{NH_3} + H_2O \xrightarrow{662^\circ F} \underset{(liquid)}{NH_4HSO_4}, \qquad (9)$$

$$\underset{(liquid)}{NH_4HSO_4} + \underset{(gas)}{NH_3} \overset{662^\circ F}{\rightleftharpoons} \underset{(solid)}{(NH_4)_2SO_4} \text{ (slow reaction).} \qquad (10)$$

From equation 8, it can be seen that ammonia decomposes at 1132°F, so that the point of injection is restricted to a location where the flue gases are considerably below this temperature. In practice, this point usually occurs before the gases reach the air heater or economizer.

Equation 9 represents the reaction most likely to occur in the flue gases when ammonia is injected. Since the prevailing temperatures at which the acid sulfate is formed are higher than its melting point, 297°F, a liquid phase will be present. In the presence of excess ammonia, the solid sulfate will be formed as in equation 10. It will be noted that this reaction is reversible, and also that it is a comparatively slow liquid-gas reaction. For this reason, there will be considerable quantities of liquid ammonium bisulfate present in the system, which may lead to plugging. These substances are also somewhat corrosive. Under normal operating conditions, however, the authors claimed that the deposits will remain dry and corrosion should be negligible. Trials on oil-fired boilers indicated that periodic outages are required to water-wash the air heaters, and that corrosion is not always arrested.[193]

Ammonia has a more favorable result with coal-firing, since the acid ammonium sulfate is not formed in such large amounts. This is perhaps because of a reaction between the bisulfate and the fly ash to form neutral alums, or possibly because ash particles act as centers of condensation for the bisulfate mist, which is then carried out in the stack gases.

Trials of ammonia on stoker-fired units in England[193] and America[203] have been reported with favorable results. The successful operation on a number of large boiler installations in Germany was recently reported.[204] In an oil-fired boiler, ammonia formed a strong aerosol that discharged from the stack as a detached dense white plume.[205]

For many years, coal-tar bases have been used to inhibit the acid in baths used for pickling iron. Acid containing basic inhibitors

[202] Wootton, W. R., *Elec. Rev.*, **136,** 861–4 (1945).

[203] Kato, R., and Paris, B. E., *Mech. Eng.*, **83,** No. 5, 101 (1961).

[204] Hoffman, W., *Mitt. Ver. Grosskesselbesitzer*, **46,** 47–52 (1957).

[205] Austin, H. C., and Chadwick, W. L., *Mech. Eng.*, **82,** No. 4, 63–6 (1960).

dissolves the scale, but does not attack the metal. The Midland Tar Distillers Ltd. of England have a patented process that applies this principle to the control of cold-end corrosion. Coal-tar bases, which are mainly substituted pyridines, are injected into the flue gases in much the same way as ammonia. They immediately vaporize and diffuse throughout the flue gases, and thereafter, whenever any H_2SO_4 is deposited, it contains up to 10% of these bases, since their boiling range corresponds to the dewpoint range of the acid. The attack of the acid on the metal surfaces is thus inhibited. Laboratory studies indicate that corrosion is reduced by about 80%.[191] Trials on a number of boiler installations in England have been reported to be successful.[206]

Other proposals. It has been suggested that much of the SO_3 in the flue gases reacts with water vapor and forms a very persistent acid mist which is carried through the boilers without causing damage.[164] The introduction of a super-cooled surface into the boiler was proposed to intiate mist formation and reduce the degree of supersaturation so that more mist would form and less acid would condense on boiler surfaces. This can also be accomplished by spraying large quantities of cold water into the gas passes. Recent experiments in Belgium using this principle have given satisfactory results.[207]

Summary

Practically all of the sulfur in fuel is converted to SO_2 during combustion, but a small percentage, usually less than 3%, is converted to SO_3. Flue gases containing even traces of SO_3 have high-acid dewpoints, up to about 350°F, and the liquid that condenses at these temperatures contains high concentrations of sulfuric acid and causes rapid corrosion and pluggage of air heaters and economizers.

The amount of corrosion that occurs in a boiler depends on exposure time, types and quantities of corrosive agents present, the design of the heat exchanger—that is, the arrangement and temperature of surface, the material employed, and, of course, the corrosion rate. The corrosion rate in turn depends on the rate at which acid condenses on the metal surfaces, the concentration of the acid, and the metal temperature. Furthermore, the rate of acid condensation depends on temperature gradients, concentration of SO_3, water vapor in the flue gases, boiler load, excess air, nature and amount of smoke and dust in the flue gases, mist formation, and possibly other factors.

The acid concentration in the condensate film is controlled by the metal temperature and the water-vapor content of the flue gases and not by the SO_3 concentration. This is perhaps one of the reasons why corrosion rates do not correlate with the SO_3 level in the flue gases. Other reasons may be the presence of other corrosive agents including hydrochloric acid, sulfurous acid, and ferric sulfate.

The dewpoint meter has been the main tool used in studying low-temperature corrosion, but it now appears advisable to use other techniques as well, that is, apparatus for chemically determining the SO_3 in the flue gases and corrosion test probes to measure corrosion rates. This is especially true in testing boiler installations having dust-laden gases.

The prevention of low-temperature corrosion begins in the design of the boiler and its components. Provisions for controlling metal temperature are usually incorporated in the design of the air heaters if the metal temperatures fall below the minimum during lower boiler loads or if ambient air temperatures are lower than design. Particular attention should be given to the arrangement of air and flue-gas ducts to ensure uniform distribution of air and flue gases through the air heater so that all heat-transfer elements are always above the minimum metal temperature. This is a very important rule to follow in the design of tubular and plate-type recuperative air heaters, but less so for regenerative air heaters.

Boiler operation is important in avoiding air-heater corrosion. Excess air should be low enough to inhibit the formation of SO_3 but high enough to avoid carbon deposits in the air heater. Ash deposits should not be allowed to accumulate to any great extent, even though the capacity of the boiler is not affected, because metal temperatures will decrease and corrosion beneath the deposit is apt to occur. Effective in-service and out-of-service cleaning methods are available to keep these surfaces relatively clean.

[206] Davies, E. B., and Alexander, B. J., *J. Inst. Fuel*, **33**, 163–73 (1960).

[207] Jenkinson, J. R., *J. Inst. Fuel*, **32**, 129–30 (1959).

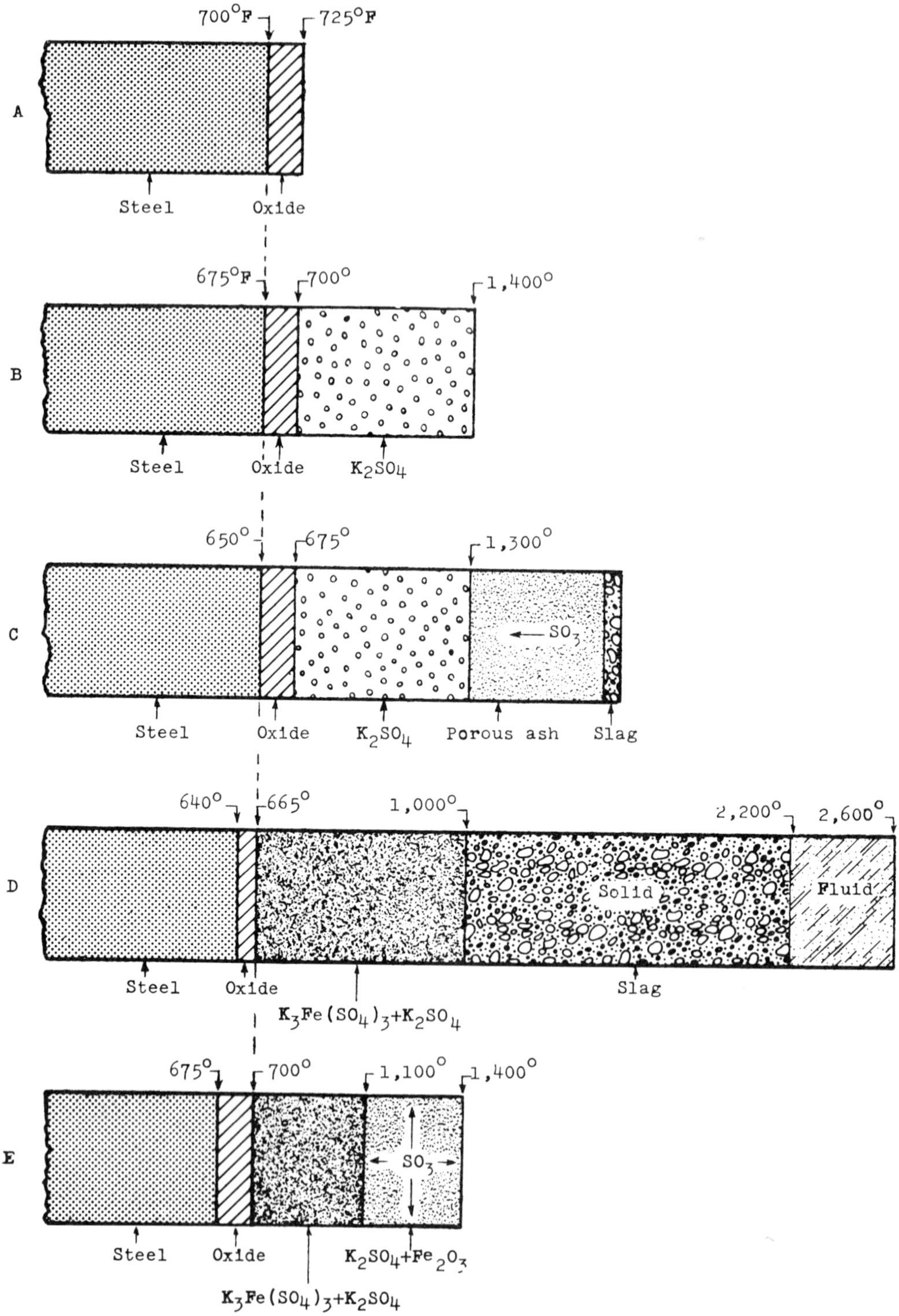

Fig. 44. Mechanism of corrosion of wall tubes by sulfate deposits.[78]

If design and operational approaches to the problem are not adequate, basic additives such as dolomite, magnesia, and lime may be used to combat corrosion. The selection of the additive and the method of applying it depend on the circumstances, which must be fully evaluated in each instance.

High-Temperature Corrosion

As shown earlier, some of the slag and ash released during the combustion of coal deposits on boiler heat absorbing surfaces in a selective manner. These deposits are a potential source of corrosion, since at high temperatures they completely destroy the protective oxides on the metal and permit accelerated oxidation to occur. A well-known example of this corrosion is the attack by oil ashes containing vanadium oxides.[208,209] A similar phenomenon occurs in coal-fired units, but here the attack is caused by alkali compounds that concentrate on the tube surfaces under the slag or ash deposit.

High-temperature corrosion in coal-fired boilers may be classified into two categories according to the type of heating surfaces involved: (1) furnace-wall tube wastage where metal temperatures generally range between 550 and 800°F with the surrounding flue gas between 2500 and 3200°F; and (2) superheater and reheater tube bank corrosion where the metal temperatures may range between 1100 and 1300°F with flue gas temperatures between 1500 and 2300°F.

The occurrence and nature of both types of corrosion will be reviewed in this section with some emphasis given to the mechanisms of attack that have been proposed and the remedial measures that have been adopted by the designers and operators of boiler equipment.

Furnace-Tube Wastage

Probably the first serious corrosion of high-temperature, heat absorbing surfaces in coal-fired boilers was the external wastage of furnace-wall tubes that occurred during the early 1940's. This trouble was confined largely to the pulverized-coal-fired slag-tap boilers, although it was occasionally observed in dry-ash furnaces in relatively small areas around the burners. Tube wastage was recognized as being due to corrosion, and it generally occurred under a lacy network of slag in an area where flames swept the wall and carbon monoxide was present in the flue gases.

In an extensive research program,[73,78,210] two types of deposits next to the furnace-wall tubes were recognized, sulfate or "enamel" deposits and sulfide deposits. In a survey of sixteen furnaces, the enamel deposit proved to be the more prevalent and most of the laboratory work was conducted on this material (see Table 5 for analyses). As a result of studies on the chemical and physical properties of the enamel, a mechanism of corrosion was proposed[73,78] in which the sequence of events is shown schematically in Fig. 44 and described as follows:

TABLE 5

ANALYSES OF TYPICAL ENAMEL DEPOSITS FROM FURNACE-WALL TUBES[210]

Sample Designation	C-64	C-70	C-3	C-8	C-26
Analysis, %					
SiO_2	5.3	7.2	2.6	7.3	8.6
Al_2O_3	4.1	4.2	5.3	0.4	3.7
Fe_2O_3	4.9	15.9	1.5	27.7	38.3
CaO	0.6	0.2	1.8	5.1	1.6
MgO	0.1	0.0	0.0	0.8	0.3
Na_2O	35.5*	22.7†	14.7	25.0*	17.8*
K_2O	—	—	20.0	—	—
$SO_3^=$	49.2	47.5	54.4	33.7	29.7
$S^=$	0.0	0.0	0.0	0.0	—
Sol. in H_2O	—	87.7	96.0	62.2	44.1
pH	3.3	4.0	3.5	4.9	3.7

* Total alkalies by difference.
† Total alkalies computed as Na_2O.

Step A. As a result of normal oxidation between 600 and 800°F, a thin layer of oxide forms on the tube metal. The rate at which this oxide grows decreases with its thickness, and in the absence of adverse conditions, a more or less limited thickness is approached.

Step B. A layer of alkali-metal sulfates forms on the oxide. This is believed to result from the deposition on the tubes of alkali-metal oxides, which originate from the flame and from the molten slag on the hearth by volatilization, and the subsequent reaction with SO_3 in the surrounding atmosphere forming alkali-metal sulfates.

Step C. As the layer of alkali sulfates thickens,

[208] Sachs, K., *Metallurgia*, **57**, 123–36, 167–73, 224–32 (1958).

[209] Natl. Assoc. Corrosion Engrs., Task Group, Rept. T-58-3; *Corrosion*, **14**, 369–72t (1958).

[210] Reid, W. T., Corey, R. C., and Cross, B. J., *ASME, Trans.*, **67**, 279–88 (1945).

the surface temperature increases to a point where it becomes somewhat sticky, and ash particles begin to adhere to it. As a consequence of its porous structure, the initial deposit of ash has a low thermal conductivity. Therefore, as it thickens, the outside temperature increases to the point where the ash begins to form a layer of slag. During this process, SO_3 is evolved as the result of reactions in the ash during the melting process.

Step D. The SO_3 diffuses through the alkali-metal sulfates and reaction occurs at the oxide-sulfate interface, forming

$$K_3Fe(SO_4)_3 \quad \text{or} \quad Na_3Fe(SO_4)_3,$$

thus reducing the thickness of the oxide:

$$3K_2SO_4 + Fe_2O_3 + SO_3 \rightarrow 2K_3Fe(SO_4)_3 \quad (11)$$

The metal then will oxidize further and renew the oxide layer. At the same time, the thickness of the slag increases and reaches equilibrium and the temperature at the surface of the enamel deposit decreases because of the low thermal conductivity of the slag.

Step E. Cycles of slag build-up and shedding normally occur on furnace walls. Deslagging increases the temperature of the enamel, causing a portion of it to exceed the decomposition temperature of the $K_3Fe(SO_4)_3$ and $Na_3Fe(SO_4)_3$, and SO_3 is released, part of which diffuses to the oxide-sulfate interface and reacts with the oxide and the excess K_2SO_4 and Na_2SO_4. Meanwhile more alkali-metal sulfate is depositing on the enamel until another layer of slag forms and the cycle is repeated.

This corrosion mechanism is a solid state reaction, but under certain conditions, mainly at higher SO_3 concentrations, liquid pyrosulfates are formed next to the oxide layer and attack the oxide at a much higher rate.[73]

In a few cases, severe tube corrosion was found where, after the slag was detached from the walls, the tubes were covered mainly with a dull black iron sulfide scale, as shown in the analyses in Table 6, although a thin layer of enamel deposit was also present. It was suggested[73] that partially-burned coal containing relatively large amounts of pyrite adhered to the alkali-sulfate deposit and subsequent oxidation of the pyrites liberated SO_2 and SO_3 which then reacted with the enamel to cause corrosion to proceed by the mechanism described earlier. Direct attack on the tube metal by the formation of Fe-FeS or FeO-FeS low-melting eutectics was not considered possible because the temperature of the tubes is several hundred degrees below that required to form these eutectics.

TABLE 6

Analyses of Typical Iron Sulfide Deposits from Furnace-Wall Tubes[73]

Sample Designation	A	B	C	D
Analysis, %				
SiO_2	2.1	4.8	2.5	2.5
Al_2O_3	2.7	4.7	3.3	3.7
Fe_2O_3	79.0	68.0	85.6	81.6
CaO	1.5	2.9	0.1	0.2
MgO	0.3	0.4	0.2	0.0
$SO_3^{=}$	3.36	2.47	0.2	0.1
$S^{=}$	23.7	21.2	26.3	32.3
Equiv. FeS	65.5	58.4	72.3	88.8
Carbon	0.5	2.0	6.8	1.0

The Vereinigung der Grosskesselbesitzer conducted a survey of the extent of tube wastage in slag-tap furnaces in Germany.[211] They found that one-third of all slag-tap boilers reported tube wastage, and that the total number showing some form of wastage is probably twice as great.

Research work on this problem in Germany appears to have been concerned mainly with sulfide attack. Gas samples taken from the corrosion zones indicated the presence of H_2S, CO, and H_2, and elemental sulfur,[212] and the corrosion products had high iron sulfide content with evidence of intercrystalline attack at the boundary layer. Wastage was attributed to attack by H_2S or elemental sulfur in the combustion gases, and it was confirmed experimentally that H_2S could cause excessive wastage of steel specimens even at temperatures as low as 550 to 750°F,[212] as shown in Fig. 45. Similar observations have been made in several Russian slag-tap boiler furnaces that use anthracite culm.[213]

Two other types of tube wastage have occurred in slag-tap furnaces. The formation of pools of iron by operation under reducing conditions can lead to overheating and thinning of the tubes when iron is tapped, the most likely region of attack being in the secondary furnace near the slag tap. It can be avoided by proper attention to combustion conditions and careful design to avoid the collection of

[211] Bachmair, A., *Mitt. Ver. Grosskesselbesitzer*, **54**, 146–53 (1958); **64**, 1–12 (1960); *Intern. Combustion Monthly Bull.*, **14**, 198 (1960).

[212] Pracht, O. P., *Energie*, **7**, 323–7 (1955).

[213] Volkov, V. M., Kaluzhskaya, T. M., and Klyachko, B. I., *Elekt. Stantsii*, **29**, 27–30 (1958).

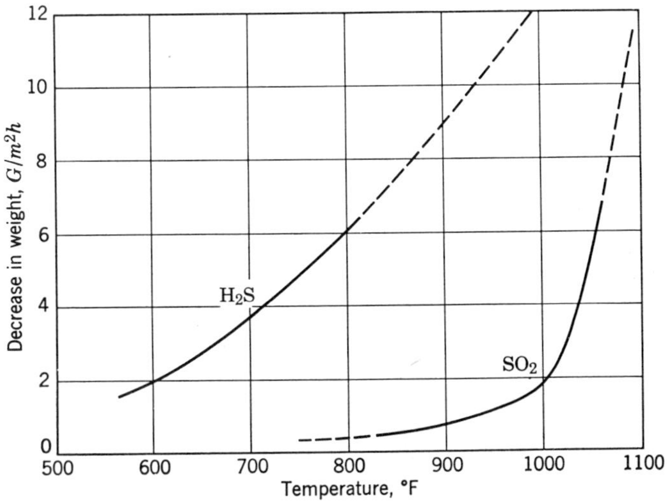

Fig. 45. Rate of attack of H_2S and SO_2 on carbon steel.[212]

pools of iron or slag. If the heat transfer to the tubes is excessive, that is, if there is no slag layer, or the circulation in the tubes is poor, or both, film boiling can occur with dissociation of the steam leading to internal corrosion of the tubes and perhaps failure. This has occurred in a number of slag-tap units in Germany.[211]

In a comprehensive review of the cyclone-firing method, Roberts[57] pointed out that although instances of tube wastage have occurred in the United States,[214] the general problem of furnace-tube corrosion does not seem to be as serious as in Europe. He attributed this to some intrinsic advantage in the axially fired design and to the generally better appreciation of the value of maintaining the studded sections in the cyclone.

The primary cause of external furnace-tube corrosion is always a local reducing atmosphere near the affected tubes. Such wastage has been largely overcome by properly positioning the burners in the design stage to avoid flame impingement and uniformly proportioning the coal and air to the burners. In some installations, the use of tertiary air or air-blanketing has been successful.

Superheater, Reheater-Tube Corrosion

Deposits rich in alkali-metal sulfates commonly occur on the heat-absorbing surfaces of boilers, irrespective of firing method or the coal burned. Chemical analyses of the deposits from stoker-fired,[142,215] pulverized-coal-fired,[142] and cyclone-fired boilers[71] have shown them to be strikingly similar in composition. The part played by deposits of this general type in furnace-wall corrosion was reported in 1945,[210] and the first mention of superheater-tube corrosion was noted at this time. Later, Crossley[216] warned that the oxidizing action of the acid sulfates is a potential danger to superheaters of boilers designed to attain a final steam temperature of 1050°F, if these boilers are subjected to deposits containing pyrosulfates.

In 1955, severe corrosion of the secondary superheater platens was discovered in several cyclone-furnace boilers.[217] At about the same time, extensive wastage of reheater and superheater tubes was observed in a number of dry-ash pulverized-coal-fired boilers.[218] Werner[219] also reported unexpected corrosion observed

[214] Newkirk, M., *ASME, Trans.*, **73,** 215–23 (1951). Moran, L. L., *Proc. Am. Power Conf.*, **16,** 125–9 (1954). Hoffman, L. W., *Elec. World*, **151,** 48–9, 90 (1959).

[215] Barkley, J. F., Burdick, L. R., and Berk, A. A., *ASME, Trans.*, **70,** 81–9 (1948).

[216] Crossley, H. E., *Combustion*, **24,** No. 4, 61–3 (1952).

[217] Sedor, P., Diehl, E. K., and Barnhart, D. H., *ASME, Trans.*, **82,** 181–90 (1960).

[218] Koopman, J. G., Marselli, E. M., Jonakin, J., and Ulmer, R. C., *Proc. Am. Power Conf. 21st Annual Meeting*, 236–45, March 31, April 1–2 (1959). Jonakin, J., Rice, G. A., and Reese, J. T., *Mech. Eng.*, **81,** No. 12, 103 (1959).

[219] Werner, M., *Werkstoffe u. Korrosion*, **8,** 587–96 (1957).

on the secondary superheater of a new high-pressure coal-fired boiler in a German power station.

Since there were strong indications that corrosion of high-temperature surfaces might become a widespread problem on the superheaters and reheaters of boilers designed to deliver 1050°F steam, a number of research investigations were undertaken to study the nature and causes of this corrosion and to develop corrective measures. Separate investigations were made of the corrosion phenomena in pulverized-coal-fired[79,220] and cyclone-furnace[217] boilers by the laboratories of the two major boiler manufacturers in the United States. There have been striking similarities in the findings of the two research groups, but there have also been important differences. For the sake of clarity, the corrosion phenomena for the two methods of firing will be treated separately.

Pulverized-coal-firing. Several years ago, extensive corrosion was discovered on the reheaters and secondary superheaters of six 1050°F, 1005°F, 1,200,000 lb/hr capacity boilers at the Joppa Station of Energy, Inc., Joppa, Illinois.[218] Although these pulverized-coal-fired, dry-ash furnace boilers had been in service only two years, some of the 18 Cr-8 Ni (Type 321) stainless steel reheater tubes had lost about 50% of their 0.125-in. wall thickness. There was very little wastage of the chrome-ferritic tubes, however. Corrosion was confined, for the most part, to the front-row elements in the reheater and secondary superheater banks, as shown in Fig. 46, and to the side of the tube facing the flow of flue gases. These tubes had an appreciable layer of ash deposits during operation, but these shed upon cooling. Screen and water-wall tubes

[220] Cain, C., Jr., and Nelson, W., *Mech. Eng.*, **83,** No. 5, 101 (1961).

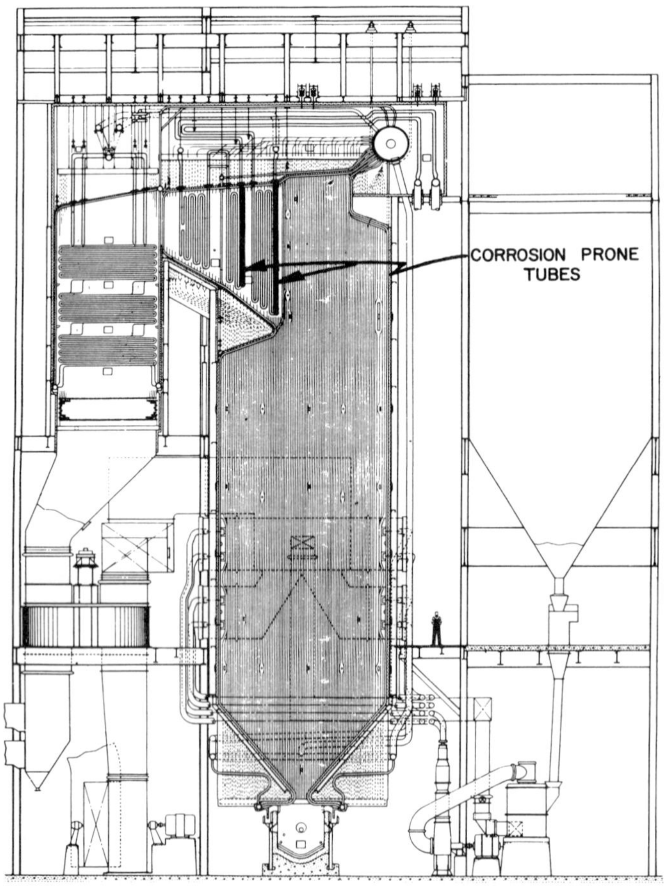

Fig. 46. Location of corrosion on a pulverized-coal-fired unit.[79]

often had ash deposits, but never pronounced metal loss.

Ash deposits from the corroded tubes were found to be multilayered with a white-to-yellow layer next to the tube (Fig. 47). The white-layer ash shown in Table 7 was rich in water-soluble, alkali sulfates that hydrolyzed to a strong acid solution and generally contained an excess of sulfate over that required to form the normal sulfates of potassium, sodium, calcium, and magnesium that were present in the deposit. X-ray diffraction confirmed the presence of potassium iron trisulfate and potassium aluminum trisulfate in these deposits which Corey had found earlier as end-products of furnace-wall tube wastage, and Anderson and Diehl had found in bonded deposits on the secondary superheaters of cyclone-furnace boilers.[71]

Laboratory studies[79] showed that the complex sulfates found in the ash deposits on the superheater became corrosive when molten. This occurred at approximately 1050°F, the metal temperature corresponding to 950°F steam. The molten compounds migrate through the ash deposit to the tube-metal surface as a result of the thermal gradient existing in the deposit and then react with the metal according to the following equation:

$$\underset{\text{(tube metal)}}{3Fe} + \underset{\text{(molten complex sulfate)}}{SO_4^{=}} \rightarrow \underset{\text{(iron oxide)}}{Fe_3O_4} + \underset{\text{(sulfide)}}{S^{=}} \qquad (12)$$

The metal temperature range for this rapid liquid-phase attack is bracketed by (1) the melting temperature of the mixture of complex alkali sulfates present, and (2) their thermal-stability limit. Corrosion by the complex alkali sulfates generally covers a band of temperatures between 1000 and 1300°F.

Further studies[220] were made of factors influencing the temperature range and the rate of corrosion by molten complex alkali sulfates. Figure 48 shows the corrosion curve for Type 321 alloy in contact with complex sulfate as a function of temperature. The tests were made for 100 hr with metal coupons in contact with sodium sulfate, potassium sulfate, and iron oxide (molar ratio—1.5:5.1:1.0), and synthetic flue gas containing SO_2. Reaction of the corrosive mixture with SO_3, produced by the catalytic action of Fe_2O_3 on SO_2 and oxygen in the flue gases, resulted in the formation of

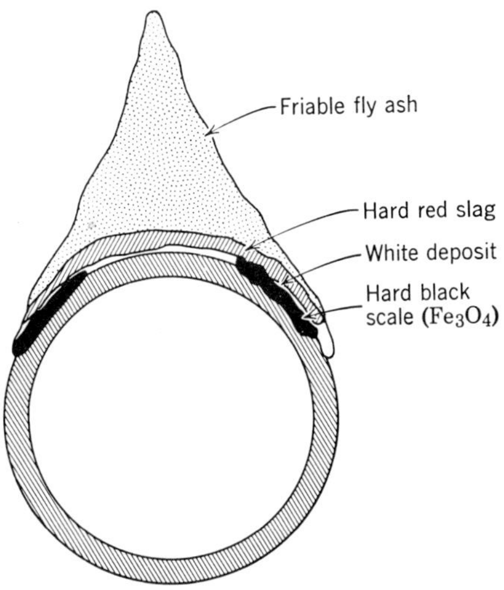

Fig. 47. Typical ash deposits on a reheater tube from a pulverized-coal-fired boiler.[79]

complex alkali sulfates. The dotted curve shows weight gain by the contents of the corrosion boat as a function of temperature. At lower temperatures, the equilibrium favors production of the complex sulfates. As the temperature increases above 1100°F, the equilibrium concentration of the complex decreases. Even so, the corrosion rate continues, owing to the greater activity of complex sulfate with metal at higher temperatures. At about 1300°F, the concentration of the complex

TABLE 7

TYPICAL COMPOSITION OF REHEATER DEPOSIT LAYERS FROM PULVERIZED-COAL-FIRING OF 1050°F BOILER[79]

Sample Designation	Outside (Friable)	Middle (Hard, Red)	Inner (White)
Analysis, %			
SiO_2	45.9	15.6	52.4
Al_2O_3	18.7	15.0	7.2
Fe_2O_3	23.1	54.8	8.6
CaO	2.5	3.2	2.4
MgO	1.9	2.2	2.7
Na_2O	1.1	0.3	0.6
K_2O	2.2	0.6	2.5
SO_3	0.8	7.5	15.6
Moisture	0.0	0.0	5.4
Other	3.8	0.8	2.6
Total	100.0	100.0	100.0
Excess SO_3, %	0.0	0.0	3.9
pH of 1% water solution	7.1	4.4	2.5
Water soluble, %	2.5	9.5	25.2

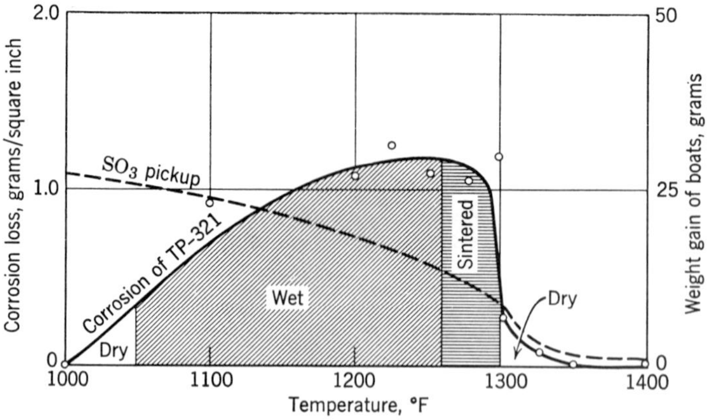

Fig. 48. Liquid-phase corrosion of type-321 alloy by complex alkali sulfates.[220]

becomes so low that the corrosion rate drops abruptly to that of a flue-gas attack. On stainless steel, the gas-phase corrosion rate is negligible compared to that of the molten complex sulfates.

The corrosion rate is also affected by changes in the sodium-to-potassium ratio.[220] Figure 49 shows the corrosion curve for Type 321 alloy in a 1:1 molar sodium-to-potassium corrosion mixture and in a 3:1 mixture. The peak corrosion temperature for the 3:1 mixture occurs about 150°F lower than that observed for the 1:1 mixture, and the corrosion rate is considerably slower. Laboratory experiments have shown that sulfides alone are quite injurious to metals, particularly the low-chrome ferritic steels around 1100°F. It was suggested that rapid corrosion may occur as a result of large build-up of sulfide in the scale, which is itself a product of the corrosion reaction.

Another investigator,[221] studying the corrosion of ferritic and austenitic steels by K_2SO_4 and gases containing SO_3, found that corrosion at 1200°F is linear with respect to time, as shown in Fig. 50. It was pointed out that this is to be expected when the corrosion product, $K_3Fe(SO_4)_3$, is liquid. When it is liquid, the liquid-metal reaction rate is the controlling rate, and diffusion processes inside the scale film are relatively unimportant.

Various methods of combatting corrosion of superheater and reheater tubes have been used or suggested, including the following:

1. The use of stainless steel shields[218] as a temporary means of protecting the tubes.
2. Coal selectivity and the grinding of coals to a finer size.[218]
3. Improved combustion conditions,[131] that is, fast ignition, good mixing, adequate excess air.

[221] Von Rahmel, A., *Arch. Eisenhüttenw.*, **31,** 59–65 (1960).

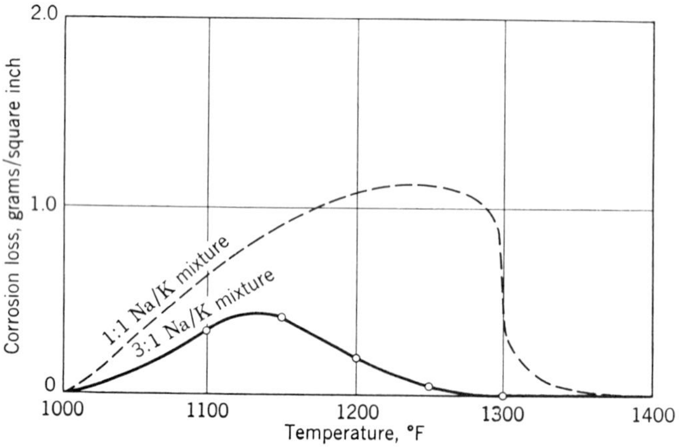

Fig. 49. Variation of corrosion rate with sodium-potassium ratio.[220]

4. Possibly the use of chromium, silicon, and other elements to improve the resistance of high-temperature alloys, as illustrated in Fig. 51.

None of these measures can be considered as a solution to the high-temperature corrosion problem at the present time. Considerably more research work is necessary before this is possible. The most effective countermeasure will, of course, be one that eliminates the basic causes of corrosion instead of treating the symptoms.

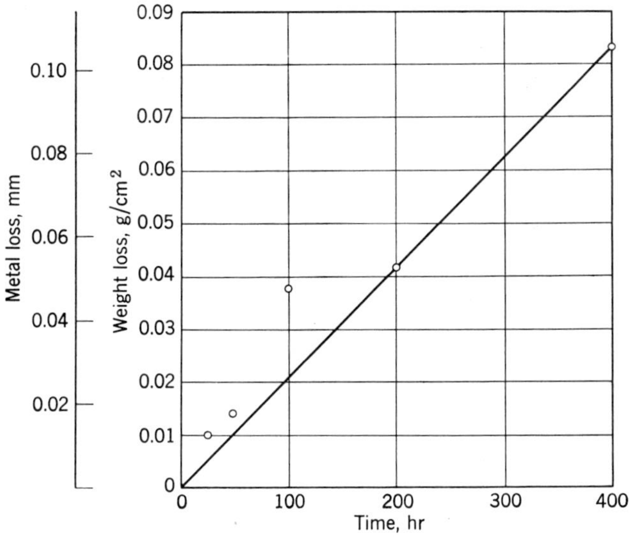

Fig. 50. Influence of time on the weight loss of specimens inserted in pure potassium sulfate by heating in an air-water vapor mixture containing sulfur (VI)—oxide at 1200°F (dewpoint: 330°F).[221]

Cyclone-furnace firing. The deposits produced when burning high-sulfur, high-alkali coals can lead to serious corrosion when gas and metal temperatures are very high. This type of corrosion has been experienced on the more recent cyclone-furnace boilers at Ridgeland and Will County Stations[217] when Central Illinois coals were burned. These coals produced ash deposits similar to those described in the section on bonded-ash deposits. The boilers operate with 1050°F main steam and 1000°F reheat steam, and the measured maximum metal temperatures have been 1175 to 1225°F, with gas temperatures 2100 to 2300°F. The superheater tubes that were corroding were made of 18-8 stainless steel (Type 321). Corrosion had not been experienced in boilers with steam temperatures lower than 1050°F, nor with the earlier cyclone-furnace units at Ridgeland Station burning the same coals but designed for 1050°F steam without reheaters.

Corrosion, which was associated with the thin white-layer deposit, was of the pitting type accompanied by surface carburization and subsurface sulfide formation. No intergranular corrosion took place. Laboratory tests on metal specimens immersed in samples of these inner deposits showed that metal loss was negligible at 1100°F, but increased rapidly in the 1200 to 1300°F range. "Catastrophic" corrosion occurred when either a reducing or an alternately reducing and oxidizing atmosphere was used. The use of 25-chromium, 20-nickel (Type 310) specimens significantly reduced the corrosion rate, as did the addition of magnesia to the deposit material. The exact corrosion mechanism has not yet been identified, but it is possibly associated with sulfur absorption from the sulfates in the deposits.

A number of modifications were carried out more or less simultaneously on the boilers. Additional slag-screen surface was installed to reduce the gas temperatures, the excess air was increased from 8 to 12%, some of the soot blowers were repositioned, and the relevant parts of the Type 321 tubes were clad with a $\frac{1}{8}$-in. thickness of 25-20, Type 310, steel. As a result, the corrosion rate in local areas was reduced from a maximum of 50%/yr in the

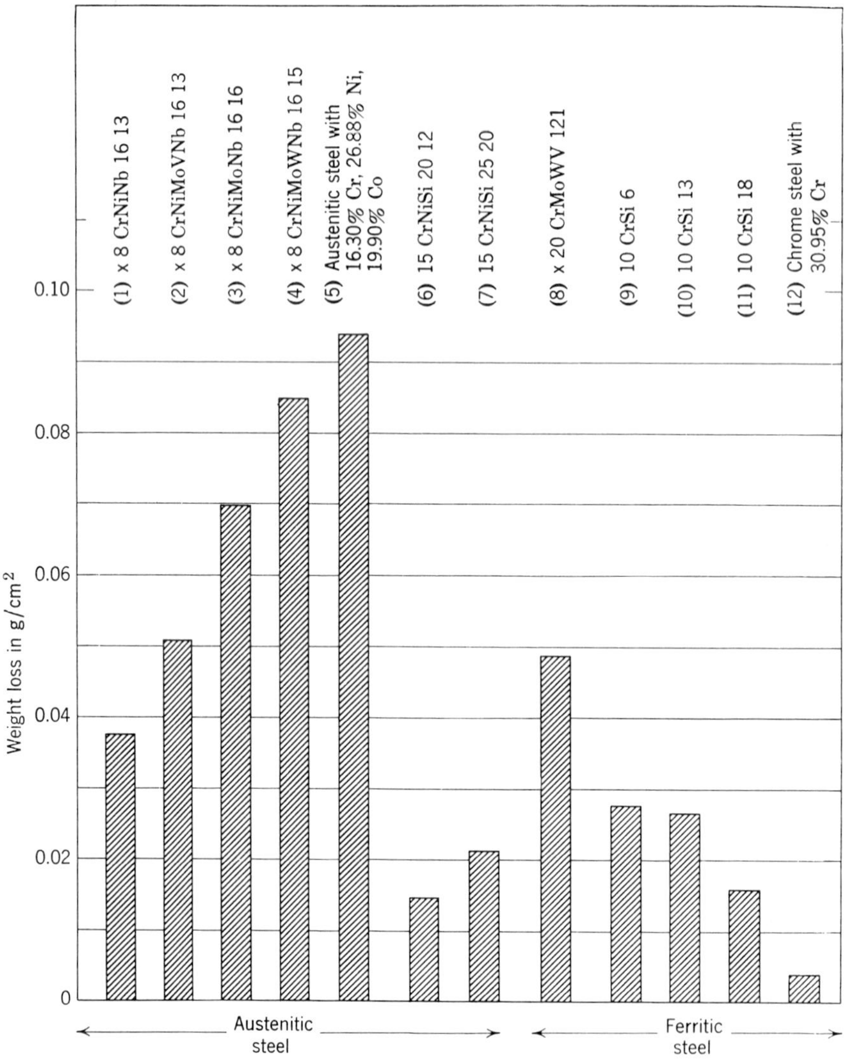

Fig. 51. Weight loss of various heat and scale resistant steels inserted in pure potassium sulfate after a 100-hr annealing at 1200°F in an air-water-vapor mixture containing sulfur (VI)—oxide (dewpoint: 330°F).[221]

period of 1953 to 1955 to 6%/yr from 1957 to 1959. In addition, the area of corrosion was greatly reduced, and it is suggested that corrosion can be kept under control by pad-welding the wasted areas every three or four years.

Summary

Furnace-wall tube wastage is still a serious problem on boilers with slag-tap furnaces. It rarely occurs on boilers having dry-ash furnaces. For reasons unknown, furnace wastage is much more prevalent on slag-tap furnaces in Germany than on those operating in the United States.

Various investigations of this problem have shown that tube wastage may be caused by the chemical reaction between the tube metal oxides and (1) deposits of alkali sulfates, pyrosulfates or sulfides or (2) H_2S and elemental sulfur. This type of corrosion is always associated with a local reducing atmosphere near the affected tubes, so that the first step in preventing corrosion is to improve combustion conditions by changes in operation or design of the burner-furnace system. Blanketing the tubes with air in the affected area has proved to be effective in pulverized-coal-fired boilers.

Superheater and reheater corrosion has become a serious problem since 1955 on both pulverized-coal-fired boilers and cyclone-furnace boilers. Corrosion occurs mainly on the high-temperature elements of the superheater and reheater made of 18-8 stainless steel, Type 321. The destruction of the normal protective oxide scale on the surface of the tubes by the complex alkali sulfates in the ash deposits appears to be the primary cause of corrosion, although other corrosion mechanisms may also be involved. Observations of corroding surfaces and laboratory experiments on the corrosive ash deposits themselves have shown that corrosion occurs in a narrow band of metal temperatures, 1100 to 1300°F, where the complex alkali sulfates are molten. The rate of attack rises with temperature, passes through a maximum, and then falls off to a low value in accordance with the thermal stability of these sulfates.

Corrective measures include improvement of combustion conditions as a primary step, that is, balanced fuel and air inputs to individual burners, proper excess-air level, fine grinding of coal, and rapid mixing of fuel and air at the burners. Protective shields and the use of higher alloy steels in the affected zones have been effective in reducing the rate of attack. Lowering the metal and gas temperatures is effective where this can be done. The use of additives and selectivity in the use of fuels offer attractive possibilities as means of controlling corrosion, but they have not been investigated to any great extent up to the present time.

The problem of superheater and reheater corrosion appears to be increasing in prominence and is likely to receive considerably more investigative effort for developing means of prevention and to permit further advance to higher-steam temperatures.

CHAPTER 20

COAL GASIFICATION

C. G. von Fredersdorff and Martin A. Elliott
Institute of Gas Technology, Chicago, Illinois

The term *gasification* as used here signifies the reaction of solid fuels with air, oxygen, steam, carbon dioxide, or mixtures of these, to yield a gaseous product that is suitable for use either as a source of energy or as a raw material for the synthesis of chemicals, liquid fuels, or other gaseous fuels. Thus, gasification yields a product that can be handled with maximum convenience and minimum cost, and in addition, greatly extends the uses to which solid fuels may be put. For both technical and economic reasons, most gasification processes for synthesis gas production or for the production of energy as gaseous fuel strive for total gasification of the solid fuel.

The science and technology of the complete gasification of coal and other solid fuels have advanced significantly. This progress has been a result of the considerable expansion in research and development work on the fundamental chemistry and physics of gasification reactions, and on gasification and purification processes. This expansion in effort stems in part from increased interest in the use of gas as a raw material for the synthesis of chemicals and liquid fuels, and in part from the need to develop methods for gasifying coal and other solid fossil fuels to ensure the long-range supply of energy in the form of gaseous fuel.

Many advances in the field of coal gasification have been made possible by advances in other fields. Fundamental research on the mechanism and kinetics of chemical reactions, on mass and heat transfer, and on fluid dynamics, has contributed to a better understanding of the phenomena occurring in gasification processes. The development of tonnage oxygen plants, improved materials, and new methods for handling and contacting solids with gaseous reactions, have opened new avenues for process development. Finally, new concepts in gas purification have led to commercial development of improved processes.

For many years air-blown, fixed-fuel-bed processes were used almost exclusively for the complete gasification of coal or coke. The basic technology and industrial applications of such processes are represented by the continuous gas producer and the cyclic water-gas generator.

Several new concepts in gasification have been introduced since 1940; the use of oxygen, operation at elevated pressure, and operation with a fluidized bed. Extensive work has been done on the gasification of coal in suspension with oxygen and steam at atmospheric and elevated pressures. In addition, hydrogasification processes, in which hydrogen reacts directly with coal or other carbonaceous material, are being developed to produce intermediate and high calorific value gases.

THERMODYNAMICS OF GASIFICATION REACTIONS

Thermodynamic analyses of gasification systems are of considerable value in that the results may be used for (1) comparison with actual operating data to indicate areas of possible improvement of the process, or to indicate approximate agreement between theory and practice, (2) evaluation of process design and prediction of reactor performance from known input conditions, (3) estimation of process requirements and yields at preset operating conditions, and (4) examination of the probable effects of varying any of several interrelated variables.

A thermodynamic analysis must necessarily be based on numerous assumptions concerning the range of variables or parameters which may be encountered in actual operation of a gasification reactor. Since thermodynamics cannot predict rates of reactions, a tacit assumption is that one or more of the chemical reactions is sufficiently rapid to approach thermodynamic equilibrium at the temperature existing in the fuel bed where such reactions occur. Reactions at incomplete equilibrium may be treated by means of an estimated or known equilibrium approach factor. In general, only processes that are capable of attaining approximately steady-state operating conditions can be examined in this way—specifically, continuous gasification processes employing fixed or fluidized beds with steam or oxygen, or steam and/or hydrogen, as gasifying media, and pulverized coal suspensions with air or oxygen, and steam.

The information available for thermodynamic analyses includes the material balances for each of the constituents, the applicable equilibrium relationships modified, if necessary, by an "approach factor," and the heat balance, modified as necessary for estimated heat losses. Gumz[1] applied these procedures for estimating gas compositions from fixed fuel beds. Calculations and graphical results for estimating gas yields in oxygen-steam gasification of pulverized coal have been published.[2,3]

In continuous gasification of solid fuels with oxygen and steam and/or hydrogen, the principal gaseous products are CO, H_2, CO_2, undecomposed steam, small to moderate amounts of methane, some nitrogen, and traces of sulfur-bearing gases. The principal components of the product gas are formed by a combination of the following overall reactions:

1. Sequential oxidation of carbon to CO and then to CO_2,

$$C + O_2 \rightarrow CO_2 \qquad \text{(I)}$$

2. In pulverized coal gasification, where the coal is suspended in and introduced concurrently with the gasifying medium, partial oxidation of the hydrogen in the volatile matter occurs,

$$H_2 + \tfrac{1}{2}O_2 \rightarrow H_2O \qquad \text{(II)}$$

3. Gasification of carbon with steam and carbon dioxide,

$$C + H_2O \rightarrow CO + H_2 \qquad \text{(III)}$$

$$C + CO_2 \rightarrow 2CO \qquad \text{(IV)}$$

4. Carbon monoxide shift,

$$CO + H_2O \rightarrow CO_2 + H_2, \qquad \text{(V)}$$

5. Where the coal feed is introduced countercurrent to the gas flow, as in a fixed-bed gasifier, the preheat zone evolves among other gases, CH_4, and higher hydrocarbons, through volatile matter decomposition,

$$C_nH_m \rightarrow \frac{m}{4} CH_4 + \frac{n-m}{4} C \qquad \text{(VI)}$$

6. Depending on the temperature, and at pressures above atmospheric, hydrogenation of volatile matter and "active" carbon to methane and higher hydrocarbons occurs,

$$C_nH_m + \frac{2n-m}{2} H_2 \rightarrow nCH_4 \qquad \text{(VII)}$$

$$C + 2H_2 \rightarrow CH_4 \qquad \text{(VIII)}$$

At gasification conditions, the oxygen-consuming reactions I and II are so rapid that they proceed essentially to completion with respect to oxygen disappearance. The carbon gasification reactions III and IV are never at equilibrium at exit conditions. Equilibrium in these reactions requires practically 100% steam decomposition, and negligible CO_2 content, at temperatures above 2000°F and pressures from 1 to 20 atm.[4] Reaction IV is generally slower than III at equal reactant concentrations.

The water-gas shift reaction V is considered to be primarily a heterogeneous phenomenon which occurs on the fuel surface, with very little reaction in the gas phase. It approaches thermodynamic equilibrium generally as a function of steam decomposition or fuel-bed depth at a given temperature, although other important factors, such as fuel reactivity and possible catalytic activity of fuel ash constituents, must be considered. At low steam

[1] Gumz, W., *Gas Producers and Blast Furnaces*, John Wiley and Sons, New York, 1950, 316 pp.

[2] Batchelder, H. R., and Sternberg, J. C., *Ind. Eng. Chem.*, **42,** 877–82 (1950).

[3] Edmister, W. C., Perry, H., Corey, R. C., and Elliott, M. A., *ASME, Trans.*, **74,** 621–36 (1952).

[4] Parent, J. D., and Katz, S., *Inst. Gas Technol. Research Bull.* No. 2 (1948), 110 pp.

decompositions, this reaction is apparently never at equilibrium, and its rapid approach to equilibrium thereafter is thought to occur primarily by virtue of steam disappearance through reaction III, which, in effect, increases the partial pressure product of $(CO_2)(H_2)/(CO)(H_2O)$.

At sufficiently high steam decomposition, and with moderate amounts of CO_2 present, it is possible to exceed water-gas shift equilibrium, and the reaction tends to operate in reverse. The rapidity of approach to equilibrium with steam decomposition is, of course, temperature dependent. At temperatures above 2000°F, thermodynamic equilibrium may be assumed without serious error in most cases over the range of steam decompositions encountered in practice. This is particularly true in pulverized coal gasification with oxygen and steam at ash-slagging conditions.[5-8]

Methane-forming reactions are of importance primarily in gasification processes operated at fuel-bed temperatures below 2000°, at or above atmospheric pressure, with coals or chars containing reactive volatile components. High-temperature cokes or fuels containing carbon substantially as fixed carbon yield negligible or no methane through direct hydrogenation. Methane yields from volatile matter hydrogenolysis vary with the rate of heating, the method of contacting, the temperature level, and the hydrogen and steam partial pressure, as the important variables. Rapid heating of the fuel in the presence of hydrogen reduces methane formation as compared to slow heating. This may be the result of surface annealing and graphitization, which makes the fuel less reactive toward hydrogenation.

Slow heating in the presence of hydrogen and steam leads to more favorable conditions for methane formation. Reaction of hydrocarbonaceous material with steam causes surface activation resulting in more rapid hydrogenation of the "activated areas" to methane.[9] Furthermore, the yields of methane in continuous feed fluid-bed processes operated below 2000°F are substantially higher than in fixed-bed gasification. This may be due to the continuous mixing and fuel circulation in fluid beds which brings "activated" fuel particles from regions of high-steam partial pressure in the lower portion of the bed to regions of high-hydrogen partial pressure in the upper portion of the fuel bed. In fixed-bed gasification, on the other hand, the upper portions of the fuel bed are not reactivated by steam at a rate sufficient to offset surface annealing, because the steam-carbon reactions are depressed by the retarding effect of hydrogen,[10] and the steam partial pressure is low. This results in substantially reduced rates of methane formation, with effluent concentrations considerably below that predicted from the β-graphite-hydrogen equilibrium.

Reactive fuels in fluid beds yield methane concentrations substantially in excess of the β-graphite-hydrogen equilibrium. Figure 1 presents an Arrhenius-type plot of the methane "equilibrium ratio" for a variety of fuels gasified by steam and oxygen-steam mixtures in fluid beds (adapted from Squires[7]). These data are based on effluent concentrations, and as such do not represent true equilibrium; however, a "quasi-equilibrium" state is suggested which exceeds the equilibrium ratio over β graphite by about an order of magnitude in the extreme cases. Interpreted as activity of the fuel, the excess free energy which may be attributed to active fuel carbon ranges from 1.8 to about 5.0 kcal/gram-mole above β graphite. In Fig. 1, the dashed line with numbers refers to distances above the grate for small-scale Lurgi fixed-bed gasification of high-volatile bituminous coal.[11]

Data of the type given in Fig. 1 may be used for estimating equilibrium approach factors for methane formation in thermodynamic calculations for similar fuels. With untested fuels, such estimates must be viewed with caution. It is presently not possible to predict reactivity properties of solid fuel constituents,

[5] von Fredersdorff, C. G., Pyrcioch, E. J., and Pettyjohn, E. S., *Inst. Gas Technol., Research Bull.* No. 7 (1957), 71 pp.

[6] von Fredersdorff, C. G., *Inst. Gas Technol., Research Bull.* No. 19 (1955), 75 pp.

[7] Squires, A. M., Paper presented at Inst. Chem. Engrs. Meeting, London, February 2, 1960.

[8] May, W. G., Mueller, R. H., and Sweetser, S. B., *Ind. Eng. Chem.*, **50**, 1289–96 (1958).

[9] Zielke, C. W., and Gorin, E., *Ind. Eng. Chem.*, **49**, 396–403 (1957).

[10] Gadsby, J., Hinshelwood, C. N., and Sykes, K. W., *Proc. Roy. Soc.* (*London*) **A 187**, 129–51 (1946).

[11] Hebden, D., Edge, R. F., and Foley, K. W., *Gas Council* (*Gt. Brit.*), *Research Commun.* GC 14 (1954), 32 pp.

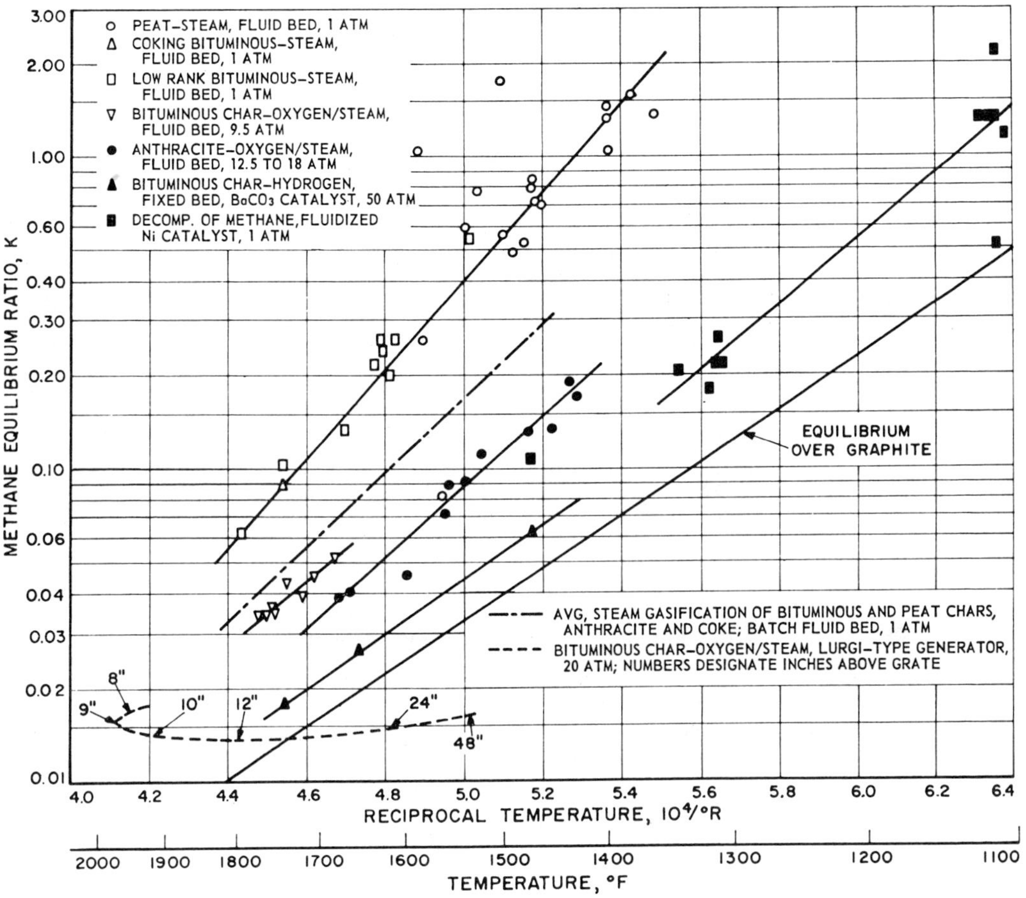

Fig. 1. Methane-hydrogen equilibrium ratios for various solid fuels in different gasification processes.[7]

TABLE 1

EQUILIBRIUM CONSTANTS OF COMMON GASIFICATION REACTIONS[12]
(All compounds as gases; carbon as β-graphite)

Temp., °K	Equilibrium Constants, atmosphere partial pressure units*						
	$C + \frac{1}{2}O_2 \rightarrow CO$	$C + O_2 \rightarrow CO_2$	$H_2 + \frac{1}{2}O_2 \rightarrow H_2O$	$C + CO_2 \rightarrow 2CO$	$C + H_2O \rightarrow CO + H_2$	$CO + H_2O \rightarrow CO_2 + H_2$	$C + 2H_2 \rightarrow CH_4$
298.16	1.1163(24)	1.2337(69)	1.1149(40)	1.0100(−21)	1.0013(−16)	9.9126(4)	7.916(8)
600	2.1676(14)	2.5167(34)	4.2922(18)	1.8669(−6)	5.0500(−5)	2.7050(1)	1.000(2)
700	9.2232(12)	3.1826(29)	3.8309(15)	2.6729(−4)	2.4076(−3)	9.0072	8.972
800	8.5483(11)	6.7080(25)	1.9433(13)	1.0893(−2)	4.3988(−2)	4.0380	1.413
900	1.3352(11)	9.2570(22)	3.1429(11)	1.925(−1)	4.2483(−1)	2.2059	3.250(−1)
1000	3.0034(10)	4.7511(20)	1.1474(10)	1.8985	2.6176	1.3787	9.829(−2)
1100	8.8006(9)	6.3451(18)	7.6012(8)	1.2206(1)	1.1578(1)	0.94850	3.677(−2)
1200	3.1462(9)	1.7376(17)	7.8762(7)	5.6966(1)	3.9945(1)	0.70120	1.608(−2)
1300	1.3111(9)	8.2517(15)	1.1497(7)	2.0831(2)	1.1404(2)	0.54742	7.932(−3)
1400	6.1659(8)	6.0484(14)	2.2057(6)	6.2856(2)	2.7954(2)	0.44473	4.327(−3)
1500	3.1946(8)	6.2903(13)	5.2538(5)	1.6224(3)	6.0805(2)	0.37478	2.557(−3)
1750	8.441(7)	6.744(11)	2.9604(4)	1.0565(4)	2.851(3)	0.2699	—
2000	3.064(7)	2.241(10)	3.3934(3)	4.189(4)	9.029(3)	0.2155	—
2250	1.375(7)	1.575(9)	6.2664(2)	1.200(5)	2.194(4)	0.1828	—
2500	7.153(6)	1.892(8)	1.6127(2)	2.704(5)	4.435(4)	0.1640	—

* Numbers in parentheses designate multiplication by that power of 10.

TABLE 2

SENSIBLE ENTHALPIES OF SYSTEM COMPONENTS[12]
(All compounds as gases; carbon as β-graphite)

Temp., °K	Enthalpy, cal/g-mole									
	O_2	H_2	H_2O	N_2	C	CO	CO_2	CH_4	C_2H_4	C_2H_6
0	0	0	0	0	0	0	0	0	0	0
298.16	2,069.8	2,023.8	2,367.7	2,072.3	251.56	2,072.6	2,238.1	2,397	2,525	2,856
600	4 279.2	4,129.5	4,882.2	4,198.0	1,198.2	4,209.6	5,322.6	5,549	6,732	8,016
700	5,057.5	4,831.5	5,771.5	4,925.3	1,621.9	4,945.5	6,481.3	6,871	8,727	10,280
800	5,856.0	5,537.4	6,689.6	5,668.6	2,081.6	5,700.0	7,689.6	8,321	10,480	12,760
900	6,669.9	6,248.0	7,634.7	6,428.0	2,569.5	6,471.0	8,938.8	9,887	12,560	15,440
1,000	7,497.0	6,965.8	8,608.0	7,202.5	3,075.0	7,257.0	10,222.0	11,560	14,760	18,280
1,100	8,335.8	7,692.0	9,606.3	7,991.5	3,596	8,056.4	11,535	13,320	17,070	21,290
1,200	9,183.6	8,427.5	10,630	8,792.8	4,130	8,868.0	12,874	15,170	19,470	24,420
1,300	10,041.2	9,173.2	11,679	9,604.7	4,680	9,690.2	14,234	17,100	21,950	27,680
1,400	10,904.6	9,928.7	12,753	10,425.4	5,242	10,521.0	15,611	19,090	24,490	31,040
1,500	11,776.5	10,694.2	13,848	11,253.6	5,814	11,358.0	17,004	21,130	27,100	34,500
1,750	13,981	12,654.1	16,683	13,357.4	7,280	13,482	20,545	—	—	—
2,000	16,218	14,671.6	19,630	15,494.8	8,780	15,636	24,140	—	—	—
2,250	18,493	16,738.9	22,666	17,659.9	10,305	17,813	27,788	—	—	—
2,500	20,800	18,850.5	25,770	19,839.0	11,850	20,008	31,480	—	—	—

TABLE 3

HEAT OF REACTION OF COMMON GASIFICATION REACTIONS[12]
(All compounds as gases; carbon as β-graphite)

Temp., °K	Heat of Reaction, kcal/g-mole*						
	$C + \frac{1}{2}O_2 \rightarrow CO$	$C + O_2 \rightarrow CO_2$	$H_2 + \frac{1}{2}O_2 \rightarrow H_2O$	$C + CO_2 \rightarrow 2CO$	$C + H_2O \rightarrow CO + H_2$	$CO + H_2O \rightarrow CO_2 + H_2$	$C + 2H_2 \rightarrow CH_4$
0	−27.2019	−93.9686	−57.107	39.5648	29.905	−9.660	−15.987
298.16	−26.4157	−94.0518	−57.7979	41.2204	31.382	−9.838	−17.889
600	−26.330	−94.123	−58.493	41.463	32.163	−9.300	−19.893
700	−26.407	−94.167	−58.696	41.353	32.289	−9.064	−20.401
800	−26.511	−94.215	−58.882	41.193	32.371	−8.822	−20.823
900	−26.635	−94.268	−59.055	40.988	32.420	−8.578	−21.166
1000	−26.768	−94.318	−59.214	40.782	32.446	−8.336	−21.43
1100	−26.909	−94.364	−59.360	40.546	32.451	−8.095	−21.65
1200	−27.056	−94.410	−59.497	40.298	32.441	−7.857	−21.79
1300	−27.212	−94.456	−59.622	40.032	32.410	−7.622	−21.92
1400	−27.376	−94.505	−59.735	39.753	32.359	−7.394	−22.00
1500	−27.545	−94.555	−59.841	39.465	32.296	−7.169	−22.06
1750	−27.99	−94.68	−60.067	38.70	32.08	−6.62	—
2000	−28.46	−94.83	−60.258	37.91	31.80	−6.11	—
2250	−28.94	−94.98	−60.426	37.10	31.49	−5.61	—
2500	−29.45	−95.14	−60.588	36.24	31.14	−5.10	—

* Negative sign designates exothermicity.

because these may be affected by previous thermal history, by thermal contact in the gasification unit, and by contact with reactants such as steam and hydrogen.

Equilibrium Compositions and Heat Effects

Tables 1 to 3 present data on equilibrium constants, enthalpies, and heats of reaction of the carbon-oxygen-steam system reactions, adapted from published data.[12] In Figs. 2 to 7 the computed equilibrium gas compositions of the carbon-steam and carbon-oxygen-steam

[12] Rossini, F. D., Pitzer, K. S., Arnett, R. L., Braun, R. M., and Pimental, G. C., *Selected Values of Physical and Thermodynamic Properties of Hydrocarbons and Related Compounds*, Carnegie Press, Pittsburgh, 1953, 1050 pp.

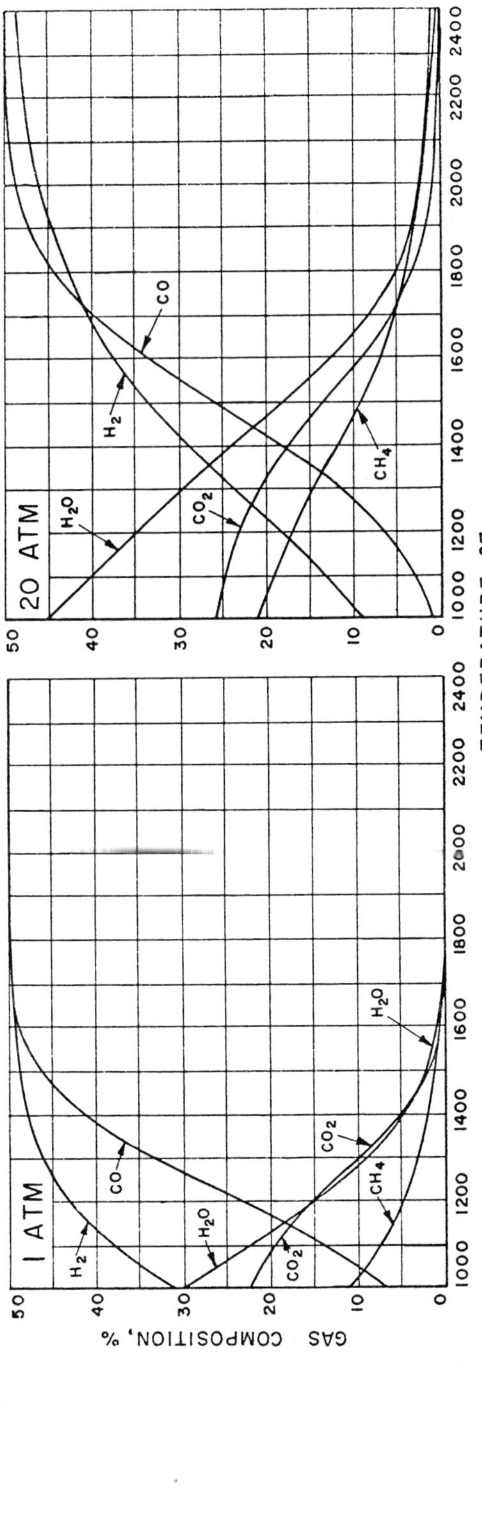

Fig. 2. Equilibrium gas compositions of the carbon-steam system.[4]

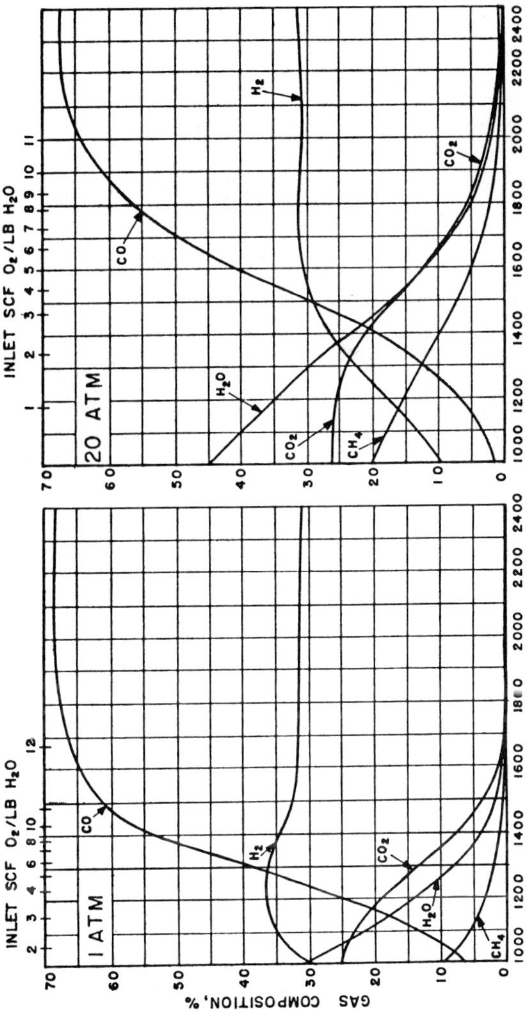

Fig. 3. Equilibrium gas compositions of the carbon-oxygen-steam system at adiabatic conditions.[4]

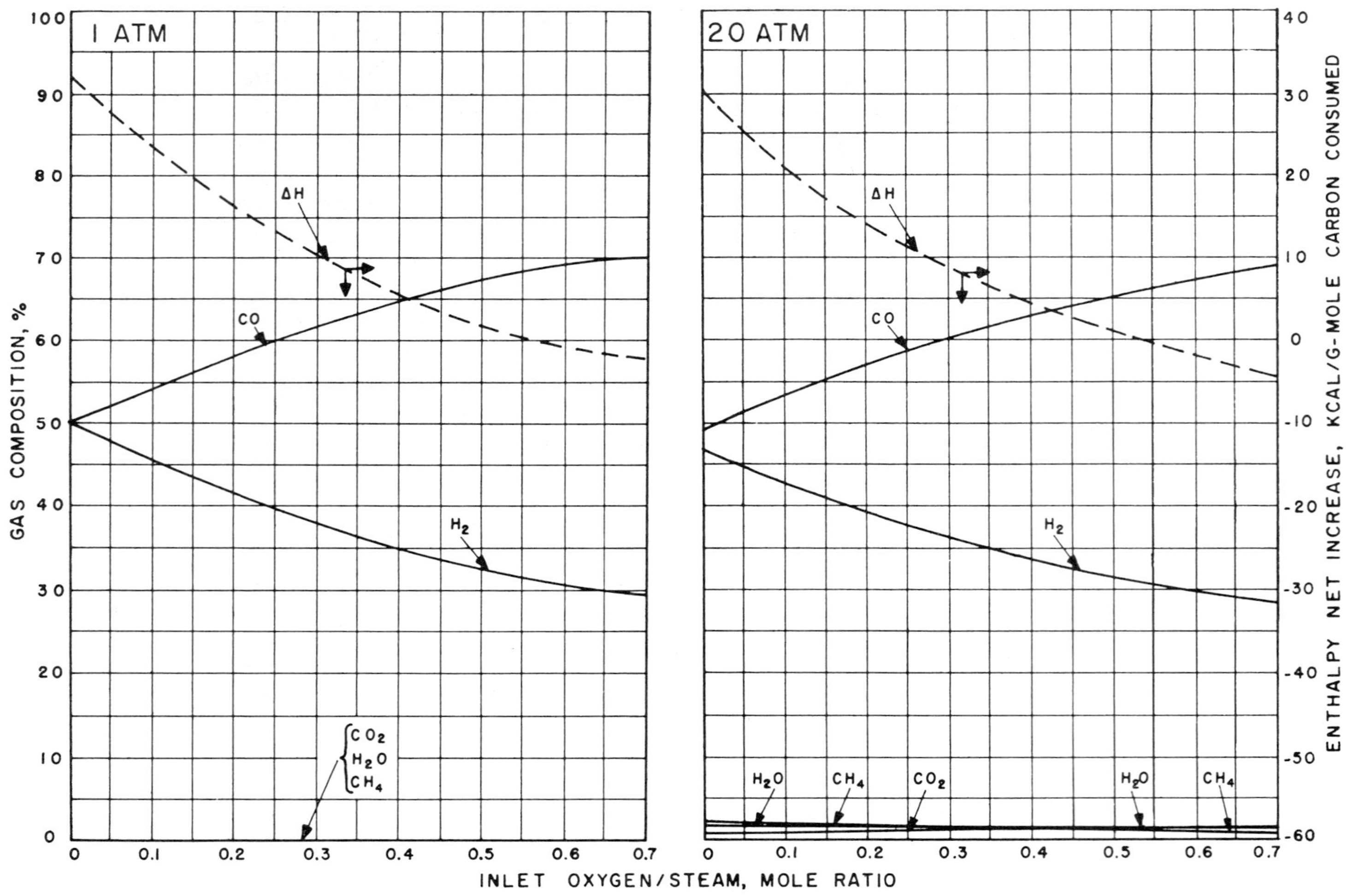

Fig. 4. Equilibrium gas composition and enthalpy change of the carbon-oxygen-steam system at 1400°K.[4]

systems are shown for a selected range of conditions.

Figure 2 is for the carbon-steam system at 1 and 20 atm and 1000 to 2400°F.[4] The overall enthalpy change for the system is always endothermic, and methane formation is thermodynamically favored at temperatures below 1600 or 1700°F. In comparison, Fig. 3 shows the equilibrium compositions for the carbon-oxygen-steam system under zero enthalpy change and the same pressure-temperature conditions as in Fig. 2, in which the theoretical oxygen/steam ratio is also shown. The principal difference between the two systems lies in the product gas H_2/CO ratio.

The effect of the oxygen/steam ratio on equilibrium gas composition and enthalpy change for the carbon-oxygen-steam system at 1400°K and 1 and 20 atm is given in Fig. 4. The system becomes thermally balanced at approximately 0.55 oxygen/steam mole ratio. These plots are based on the system defined by reaction I to completion, and equilibrium in reactions III or IV, V, and VIII. However, an actual gasification process does not achieve equilibrium steam decomposition as defined by reactions III and V, as already discussed. Depending on the type of process, the usual steam decomposition may range from 25 to 85%.

A more realistic thermodynamic analysis would, therefore, consider reaction I to completion, reaction V in equilibrium, and arbitrary assignment of steam decomposition values. With this approach, a series of plots of gas composition and system enthalpy change versus steam decomposition are obtained for particular values of input oxygen/steam mole ratio, pressure, and temperature, as illustrated in Fig. 5 for 1 atm and 2000°F. These plots apply equally well to higher pressures, and the effect of temperature is small. In Fig. 5, at $0.15 O_2/H_2O$ mole ratio the system is theoretically adiabatic at about 65% steam decomposition, with reactants and products at 2000°F. At lower steam decompositions, the process becomes exothermic to compensate for heat losses and reactant preheat requirements. It has been found that gas compositions of actual gasification processes, where small or negligible quantities of methane are produced, generally agree satisfactorily with Fig. 5, provided that the effect of coal moisture, oxygen, and hydrogen content is taken into account in computing the steam decomposition and the "effective" or equivalent oxygen/steam input mole ratio.

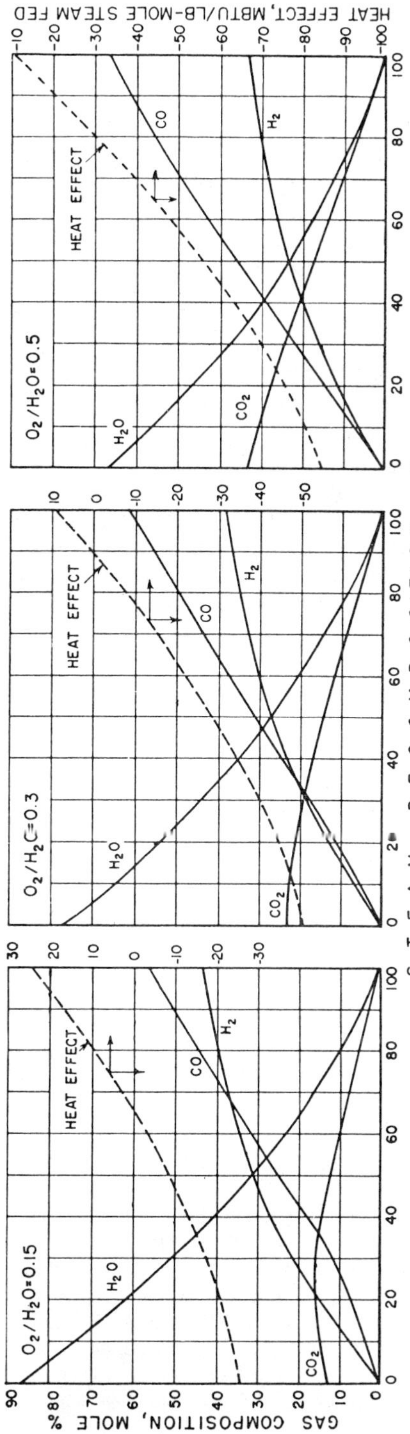

Fig. 5. Carbon-oxygen-steam water-gas shift equilibrium at several oxygen-steam ratios, 2000°F, and atmospheric pressure.[5]

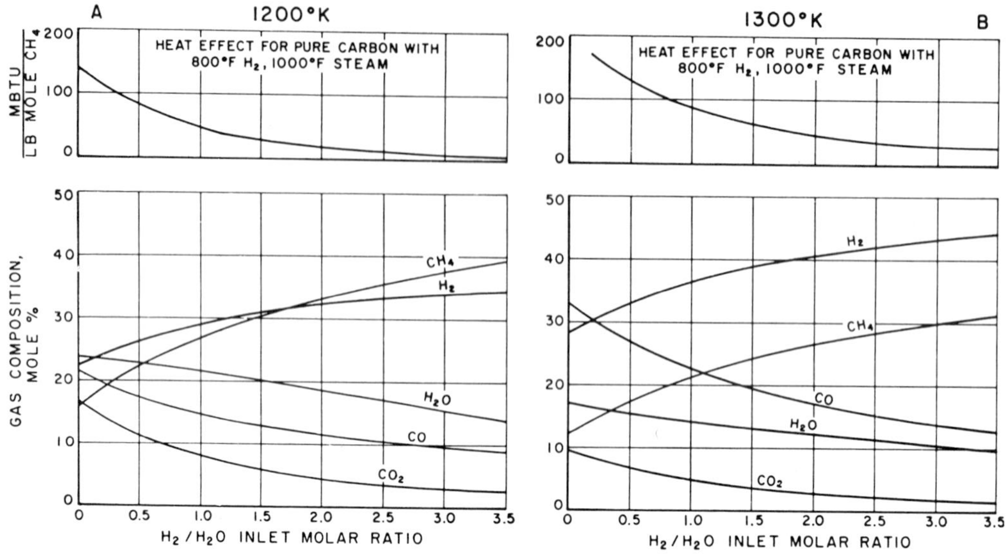

Fig. 6. Carbon-hydrogen-steam equilibrium compositions and heat effects at 200 atm pressure and two temperatures.

There is current and increasing interest in the development of pressure gasification processes which will produce gases of high methane content. These developments are being directed along the lines: (1) hydrogasification of coal with hydrogen,[13–15] and (2) simultaneous pressure gasification and hydrogenation with steam-hydrogen or steam-synthesis gas mixtures.[16,17] The latter is illustrated in the reactions which are believed to occur in the Lurgi pressure-generator fuel bed. Equilibrium compositions for the carbon-hydrogen-steam system at 200 atm as affected by H_2/H_2O inlet mole ratio and temperature, at 1200°K and 1300°K, are shown in Fig. 6. The heat effect with 800°F hydrogen and 1000°F steam is always endothermic, as shown, but it approaches small values, or thermal balance, at high H_2/H_2O inlet mole ratios. These plots are based on equilibrium in reactions III or IV, and V and VIII.

A different version of the carbon-hydrogen-steam system equilibrium plots is given in Fig. 7, where oxygen admission is employed to fulfill the heat requirements, and arbitrary steam decompositions of 30 and 40% are assumed. Here the heat effect becomes exothermic with 800°F hydrogen and 1000°F steam. These results are based on reaction I to completion, and reactions V and VIII in equilibrium.

[13] Channabasappa, K. C., and Linden, H. R., *Ind. Eng. Chem.*, **48,** 900–905 (1956).

[14] Hiteshue R. W. Anderson, R. B., and Schlesinger, M. D., *Ind. Eng. Chem.*, **49,** 2008–10 (1957).

[15] Newman, L. L., and Pipilen, A. P., *Gas Age*, **119,** No. 10, 16–21; No. 11, 18–20, 44 (1957).

[16] Dent, F. J., *Gas J.*, **244,** 502–7 (1944).

[17] von Fredersdorff, C. G., *Ind. Eng. Chem.*, **52,** 595–8 (1960).

Application of Thermodynamic and Stoichiometric Calculations to Gasification Processes

For continuous gasification processes, particularly in suspension gasification of pulverized coal with oxygen and steam, the relationships between gas composition, oxygen/steam input mole ratio, and the appropriate decomposition variable can be developed in readily usable form from the stoichiometric quantities involved.

In the suspension process, the pulverized coal flows concurrently with the gasifying medium, and rapid oxidation and gasification occurs, usually at temperatures above the ash-fusion point. A reasonable assumption is that all the coal moisture, and practically all of the coal oxygen and hydrogen, are available as their equivalent in oxygen and steam for gasification. Denoting the weight fraction, on a dry basis, of coal carbon, oxygen, hydrogen, moisture and sulfur contents by y_C, y_O, y_H, y_{H_2O} and y_S, respectively, the equivalent

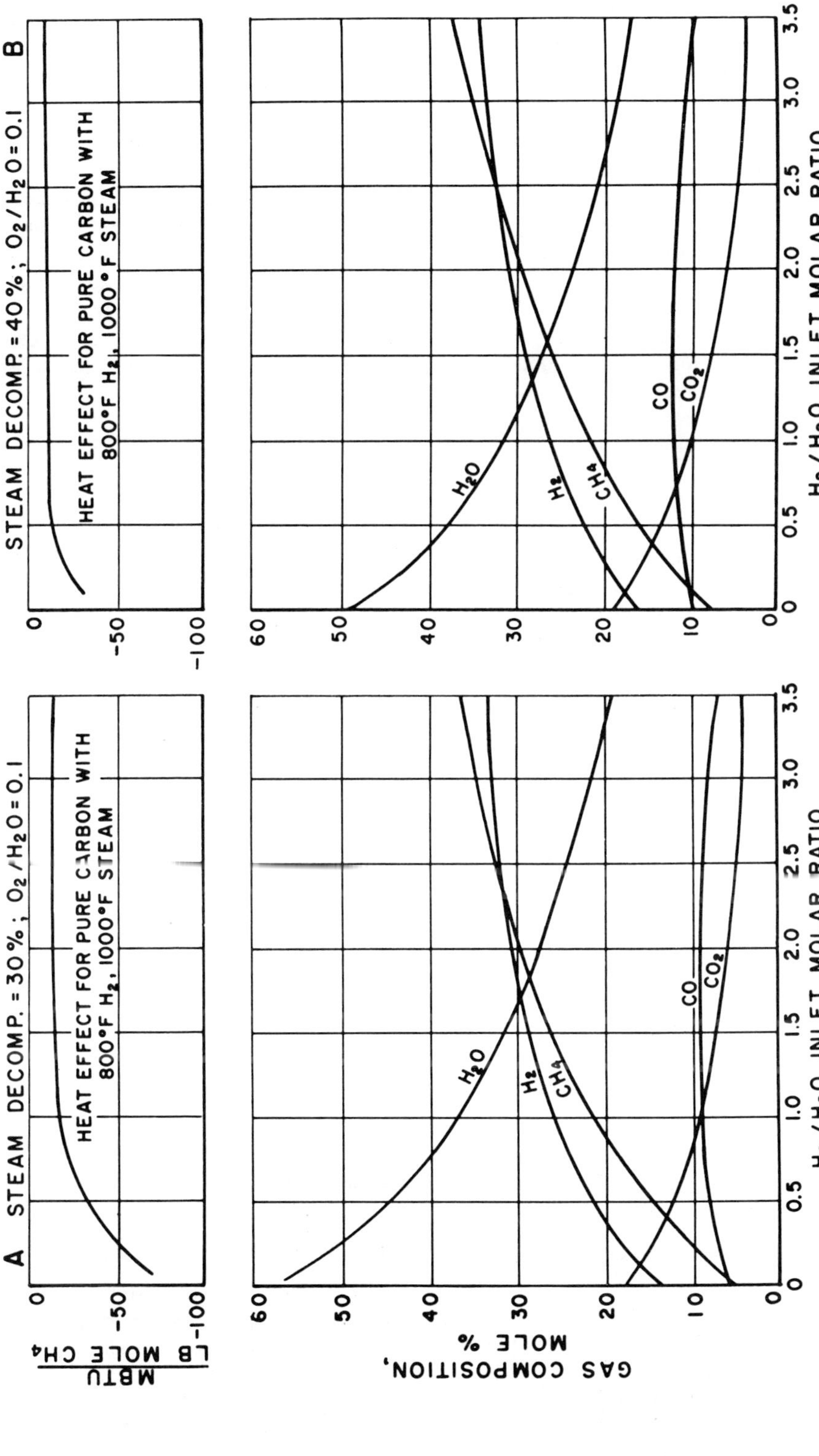

Fig. 7. Carbon-oxygen-hydrogen-steam equilibrium compositions and heat effects at 200 atm pressure and arbitrary steam decompositions.

oxygen and steam input ratios, M_{O_2} and M_{H_2O}, respectively, in moles/mole carbon input, are expressed as:

$$M_{O_2} = \frac{3[R_{O_2} + y_O - y_S - 8y_H]}{3y_C} \quad (1)$$

where R_{O_2} = external oxygen input, lb/lb dry coal.

$$M_{H_2O} = \frac{2[R_{H_2O} + y_{H_2O} + 9y_H]}{3y_C} \quad (2)$$

where R_{H_2O} = external steam input, lb/lb dry coal. These equations and the relations given subsequently also apply as an approximation for fixed-bed and fluid-bed gasification, where the gasifying medium and fuel enter countercurrently, provided corrections are made for the estimated coal moisture evolution and volatile matter decomposition occurring during preheating. The latter are factors which reduce the equivalent steam and increase the equivalent oxygen available for gasification.

In writing stoichiometric relationships for gas yield and composition, one convenient method is to express all components in terms of steam decomposition or the equivalent thereof. A second convenient way is to express all gas quantities in terms of fractional input carbon gasification. Both methods are illustrated here.

By considering reactions I, II, IV, V, and VIII, or any equivalent combination thereof, we see that the mole ratio, N, of components, in moles/mole carbon input, are:

$$N_{CO} = M_{H_2O}[(2\omega - 1)(X_{H_2} + X_{O_2}) + X_{O_2}] - 2M_{CH_4} \quad (3)$$

$$N_{H_2} = M_{H_2O}X_{H_2} = M_{H_2O}(\mathrm{SD}) - 2M_{CH_4} \quad (4)$$

$$N_{CO_2} = M_{H_2O}(1 - \omega)(X_{H_2} + X_{O_2}) + 2M_{CH_4} \quad (5)$$

$$N_{H_2O} = M_{H_2O}(1 - X_{H_2}) - 2M_{CH_4} \quad (6)$$

$$N_{\text{total}} = M_{N_2} + M_{H_2O} + M_{O_2} - M_{CH_4} + M_{H_2O}\omega(X_{H_2} + X_{O_2}) \quad (7)$$

where M_{N_2} = nitrogen in gasifying medium and evolved from fuel, M_{CH_4} = estimated methane produced in gasification zone. In pulverized coal gasification where exit temperatures exceed 2000°F, M_{CH_4} is in the range 0.0 to 0.02 moles/mole input carbon, depending on the pressure. In atmospheric pressure fixed-bed and fluid-bed processes, M_{CH_4} is usually small, 0.03 to 0.05 moles/mole input carbon, depending on coal volatile content. In fixed-bed and fluid-bed pressure gasification with oxygen and steam, M_{CH_4} may be as high as 0.1 to 0.23 moles/mole input carbon. X_{H_2} = moles hydrogen/mole equivalent steam, which is related to fractional steam decomposition, SD, as in equation 4; $X_{O_2} = M_{O_2}/M_{H_2O}$ = equivalent oxygen/equivalent steam input mole ratio, and ω = expansion factor, which varies with steam decomposition. Based on exit compositions, the expansion factor as defined here is in the 0.5–0.7 range for most gasification processes utilizing oxygen and steam. In terms of exit composition x, it is defined as:

$$\omega = \frac{x_{CO} + x_{H_2} + 2x_{CH_4}}{2(x_{H_2} + M_{O_2}/N_{\text{total}})} \quad (8)$$

To calculate exit compositions in terms of steam decomposition, the expansion factor is estimated the most conveniently from assumption of approach to water-gas shift equilibrium at the estimated exit temperature of the gasification zone. This gives:

$$a_5K_5 = \frac{N_{CO_2}N_{H_2}}{N_{CO}N_{H_2O}} \quad (9)$$

where a_5 = approach factor, which for most instances may be taken as unity. Solving for the equilibrium expansion factor from equations 3 to 6 and 9, we obtain:

$$\omega(\text{equilibrium}) = \frac{X_{H_2}(X_{H_2} + X_{O_2} + 2X_{CH_4}) + a_5K_5(X_{H_2} + 2X_{CH_4})(1 - X_{H_2} - 2X_{CH_4})}{X_{H_2}(X_{H_2} + X_{O_2}) + 2a_5K_5(X_{H_2} + X_{O_2})(1 - X_{H_2} - 2X_{CH_4})} \quad (10)$$

where $$X_{CH_4} = \frac{M_{CH_4}}{M_{H_2O}}$$

If $X_{CH_4} = 0$, equation 10 simplifies to:

$$\omega(\text{equilibrium}) = \frac{X_{H_2}[1 + a_5K_5(1 - X_{H_2})/(X_{H_2} + X_{O_2})]}{X_{H_2}(1 - 2a_5K_5) + 2a_5K_5} \quad (11)$$

If X_{CH_4} is otherwise not estimable, it may be computed approximately by trial solution of the following relationship with equation 10:

$$[1 + X_{N_2} + X_{O_2} + \omega(X_{H_2} + X_{O_2}) - X_{CH_4}] \times X_{CH_4} = PX_{H_2}^2a_8K_8) \quad (12)$$

where a_8 = approach factor for equilibrium in reaction VIII, P = total pressure, atm, and K_8 = equilibrium constant of Reaction VIII. A restriction on the use of these relationships is that the total carbon gasified, C_g, in moles/mole input carbon, cannot exceed unity. This means that:

$$C_g = M_{H_2O}\omega(X_{H_2} + X_{O_2}) + M_{CH_4} \leqq 1.0$$

or
$$X_{H_2} + X_{O_2} \leqq \frac{1.0 - M_{CH_4}}{\omega M_{H_2O}} \qquad (13)$$

Figure 8 illustrates the agreement between computed gas composition and experimental data on pulverized coal pressure gasification in a particular case where about 42% decomposition of equivalent steam was achieved at approximately 2200°F gasifier exit temperature.[5]

The alternate method of writing gas compositions in terms of carbon gasification C_g yields the following relationships from the material balances:

$$N_{CO} + N_{H_2} = 2(C_g - M_{O_2} - 2M_{CH_4}) = A \qquad (14)$$

$$N_{CO} + N_{CO_2} = C_g - M_{CH_4} = B \qquad (15)$$

$$N_{H_2O} + N_{CO_2} = 2M_{O_2} + M_{H_2O} - C_g + M_{CH_4} = D \qquad (16)$$

Here the quantities A, B, and D are fixed through assumption of input ratios M_{O_2} and M_{H_2O}, and through assumption of C_g and M_{CH_4}. All other gas quantities may be expressed in terms of N_{CO}, and N_{CO} is computed on the basis of approach to equilibrium in the water-gas-shift reaction. Thus:

$$a_5K_5 = \frac{(A - N_{CO})(B - N_{CO})}{N_{CO}(D - B + N_{CO})} \qquad (17)$$

which yields:

$$N_{CO} = \alpha - \sqrt{\alpha^2 - \beta} \qquad (18)$$

where
$$\alpha = \frac{a_5K_5(D - B) + A + B}{2(1 - a_5K_5)}$$

and
$$\beta = \frac{AB}{1 - a_5K_5}$$

This method, together with heat balances, has been used for estimating product yields from pulverized coal gasification with oxygen and steam.[3]

The relationships presented here are also readily applied to continuous gasification of fuel with steam and use of external heating,

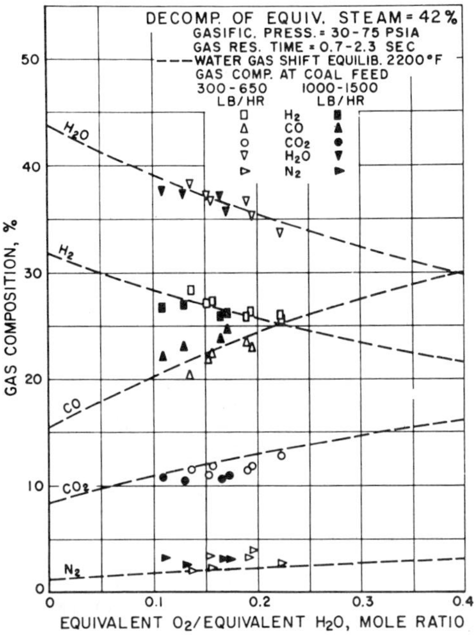

Fig. 8. Actual and computed gas compositions in pulverized coal pressure gasification.[5]

and to the steam uprun in a water-gas machine. If negligible methane is formed, and with no external oxygen:

$$N_{CO} = M_{H_2O}(2\omega - 1)X_{H_2} \qquad (3a)$$

$$N_{H_2} = M_{H_2O}X_{H_2} = M_{H_2O}(\text{SD}) \qquad (4a)$$

$$N_{CO_2} = M_{H_2O}(1 - \omega)X_{H_2} \qquad (5a)$$

$$N_{H_2O} = M_{H_2O}(1 - X_{H_2}) \qquad (6a)$$

$$N_{\text{total}} = M_{N_2} + M_{O_2} + M_{H_2O}(1 + \omega X_{H_2}) \qquad (7a)$$

If the oxygen contribution of the solid fuel is negligible, then $M_{O_2} = 0$, and the expansion factor is defined as (since $x_{H_2} = x_{CO} + 2x_{CO_2}$):

$$\omega = \frac{1 - x_{CO_2}}{x_{H_2}} \qquad (8a)$$

Some times it is possible to relate ω empirically by plotting calculated values from available data against steam decomposition. The normal scattering of points usually justifies nothing other than a straight line, or

$$\omega = aX_{H_2} + b \qquad (8b)$$

where a and b are constants dependent on temperature and fuel characteristics.

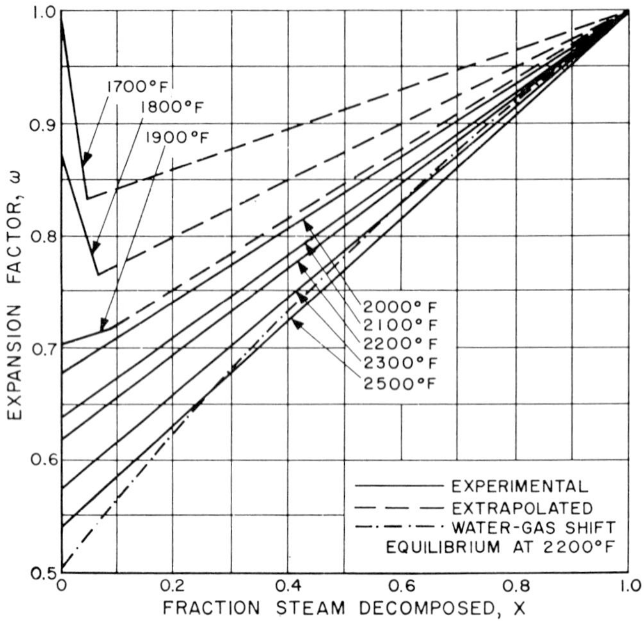

Fig. 9. Dependence of expansion factor on steam decomposition in the steam gasification of pulverized coke.[6]

Alternatively, ω may be computed from assumption of approach to water-gas shift equilibrium, which here is:

$$\omega(\text{equilibrium}) = \frac{X_{H_2}(1 - a_5K_5) - a_5K_5}{X_{H_2}(1 - 2a_5K_5) - 2a_5H_5} \tag{11a}$$

Figure 9 illustrates the ω dependence on steam decomposition and temperature, which was found in an experimental study of steam gasification of pulverized pitch coke.[6]

PHYSICS OF GASIFICATION SYSTEM

Physical phenomena play just as important a role in gasification as do the chemical reactions that bring about the conversion of carbonaceous materials of fuel gases. Important considerations are the pressure required to maintain the flow of reactants and products through the fuel bed, and the fluid and particle dynamics which directly influence the extent of longitudinal and radial mixing and the rates of heat and mass transfer at the contacting surfaces, on which factors the chemical rates largely depend. Rates of pressure loss, heat transfer, and mass transfer are recognized to be interrelated in a complex way with the temperature, gas velocity, fluid physical and transport properties, density, viscosity, thermal conductivity, diffusivity and solids properties, bulk density, size consist and specific surface area which is aerodynamically effective, fractional void volume, and thermal diffusivity, among other variables.

Interpretation of transfer processes in large-scale gasification reactors is generally difficult, even on a gross or overall basis, because of the varying conditions of temperature, gas composition, fuel size, etc., imposed, and this variation is not known as a function of distance through the fuel bed. We resort to the results of laboratory-scale or model experiments with packed beds in simplified systems where the influence of one or more variables at a time is studied under controlled or known conditions. It is assumed that this information in the form of empirical correlations and equations or verified theoretical relationships will apply, as an approximation, at actual gasification conditions.

A frequently studied aspect of the fluid dynamics problems is the pressure drop and flow distribution incurred in randomly packed and fluidized beds of uniform or closely sized solids. The results have been expressed as semiempirical relations which yield values of the friction factor in terms of fractional void volume and modified Reynolds number based on average particle size. Relationships for heat and mass transfer in packed beds have been derived on the basis of the analogy between heat, mass, and momentum transfer, and the resulting equations have been experimentally tested.

Fluid Flow Through Particle Beds

Static Bed

When a fluid is passed through a fixed bed of solids, there exists a pressure drop ΔP per unit of length L which is proportional to the rate of fluid flow, fluid viscosity and density, and the size, shape, and packing of the solid particles. Qualitatively, a packed bed may be considered as representing to fluid flow a series of sinuous paths comprising converging, diverging, and intersecting flow channels. At low flow rates, or more precisely at low values of the modified Reynolds number based on particle size $D_p\bar{u}_s\rho/\mu$, the energy dissipation or pressure loss is due primarily to viscous effects or particle surface shear forces of a fluid in laminar flow. These losses are directly proportional to the viscous forces or to the product of the viscosity and velocity of the gas in the flow channels. At high-flow rates, corresponding to fully developed turbulent flow, energy dissipation results primarily from kinetic energy losses. These losses, which occur many times per unit length of flow channel in a packed bed, are directly proportional to the inertial forces or to the product of fluid density, and square of the channel velocity. At intermediate flow rates, both viscous and kinetic effects are of approximately equal importance.

A heretofore generally accepted equation for pressure drop per unit length is due to Kozeny and Carman.[18] This relationship applies for the viscous flow range, that is, $\phi D_p\bar{u}_s\rho/\mu(1 - \epsilon) < 20$:

$$\frac{-dP}{dL} = \frac{\Delta P}{L} = \frac{C(1 - \epsilon)^2\mu\bar{u}_s}{g_c\phi^2 D_p{}^2\epsilon^3} \quad (19)$$

where C = constant = 180, dP/dL = pressure gradient in pounds per square feet per foot, ϵ = fractional void volume or bed voidage, μ = fluid viscosity in pounds per foot per second, $\bar{u}_s$ = mean superficial velocity, based on empty cross section in feet per second, ϕ = shape factor for nonspherical particles or the ratio of the specific surface of spheres to the specific surface of particles, and D_p = particle diameter, in feet. Representative shape factors, recommended by Carman,[18] are:

Material	Remarks	ϕ
Pulverized coal		0.73
Natural coal dust	Minus $\frac{3}{8}$ in.	0.65
Fusain		0.38
Flue dust	Fused, aggregates	0.55
Flue dust	Fused, spherical	0.89
Mica	Flakes	0.28
Sand	Average	0.75
Flint sand	Jagged	0.65
Flint sand	Jagged flakes	0.43
Sand	Rounded	0.83
Sand	Nearly spherical	0.95

Leva and associates[19] presented an identical equation for the viscous flow range, except that the constant C = 200. From the usual definition of a friction factor f

$$f = -\frac{Dg_c\,dP/dL}{2\rho u^2}$$

the modified friction factor for a packed bed f_b with viscous flow becomes, from equation 19,

$$f_b = -\frac{D_p\epsilon^3 g_c\,dP/dL}{2(1 - \epsilon)^2\rho\bar{u}_s{}^2} = \frac{C}{2\phi^2(Re)_p} \quad (20)$$

where $(Re)_p = D_p\bar{u}_s\rho/\mu$ = modified Reynolds number based on particle size, and ρ = fluid density in pounds per cubic foot. Figure 10 illustrates the behavior of f_b with $(Re)_p$.[19] In the viscous range, the log-log plot closely follows equation 20 with C = 200. Recommended values of the shape factor are ϕ = 1.0 for smooth spheres, ϕ = 0.862 for round sand or smooth coal particles, ϕ = 0.667 for particles such as coarse sand or coke, and ϕ = 0.578 for a Fischer-Tropsch iron catalyst.[20,21] At high modified Reynolds numbers, f_b becomes less dependent on the Reynolds

[18] Carman, T. C., *Trans. Inst. Chem. Engrs. (London)*, **15,** 150–66 (1937). Kozeny, I. J., *Sitzber. Akad. Wiss. Wien., Math.-naturw. Kl., Abt. IIa.*, **136,** 271–306 (1927).

[19] Leva, M., Grummer, M., Weintraub, M., and Pollchik, M., *Chem. Eng. Progr.*, **44,** 511–20 (1948).

[20] *Ibid.*, pp. 619–26.

[21] Leva, M., Grummer, M., Weintraub, M., and Storch, H. H., *Chem. Eng. Progr.*, **44,** 707–16 (1948).

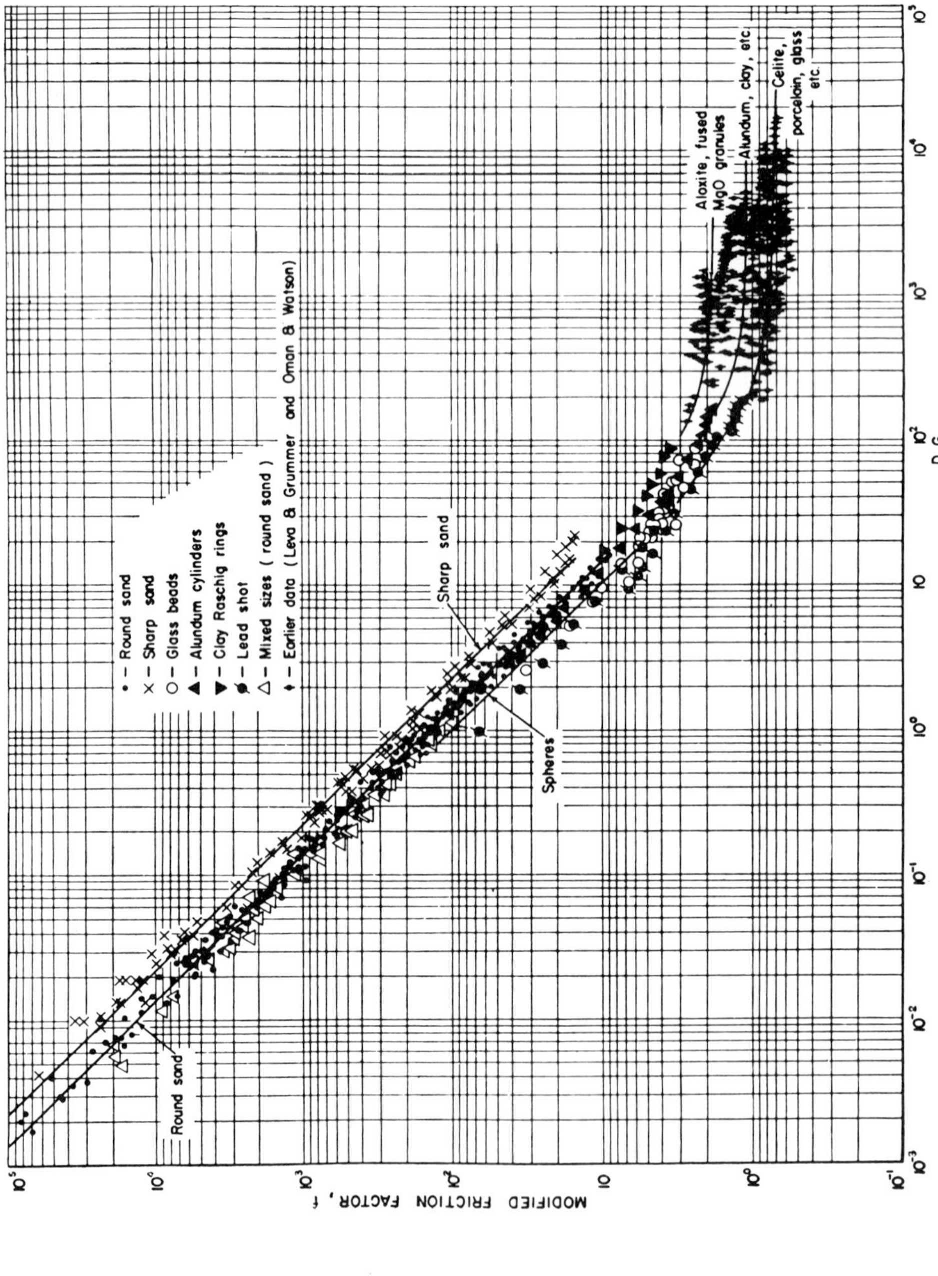

Fig. 10. Effect of Reynolds number on friction factor in flow-through packed beds.[19]

number, approaching nearly constant values for particular materials in the turbulent flow regime.

Equation 19 is generally applicable for particles less than 1.0 in. in diameter. For larger particles, the constant C would have to be modified. The quantity ϵ refers to the fractional void volume or "bed porosity" which is effective for fluid flow. It varies with distance longitudinally and radially through the bed, with the ratio of particle diameter to vessel diameter, particle size distribution, shape, and roughness, and with the method of charging the particles to the vessel. Graphical correlations for predicting porosity for dumped solids, covering a wide range of smooth and rough materials, have been presented.[22] The porosity was found to increase approximately linearly with the ratio of particle diameter to vessel diameter D_p/D_t. However, an average value, sufficiently accurate in most applications, may be estimated from the apparent density of solid particles ρ_p, the weight of solids in the bed, W_s, and the volume of the bed V_b, according to:

$$1 - \epsilon = \frac{W_s}{\rho_p V_b} \tag{21}$$

Values of ϵ for fixed beds in practice may range from a minimum of about 0.30, which corresponds roughly to minimum packing volume for spheres in rhombohedral configuration,[23] to a maximum of about 0.50, which corresponds roughly to the loosest configuration for spheres in cubic orientation.

Another form of the pressure-drop equation, applicable for the range in which flow is predominantly turbulent, that is, $200 < D_p\bar{u}_s\rho/\mu < 20{,}000$, has been given.[24] This relation was developed from analogy of fluid flow in empty pipes, and the necessary constants were determined experimentally:

$$-\frac{dP}{dL} = \frac{3.5(1 - \epsilon)\mu^{0.1}\rho^{0.9}\bar{u}_s^{1.9}}{g_c\phi^{1.1}\epsilon^3 D_p^{1.1}}, \tag{22}$$

where $\phi = 4.85 v_p^{2/3}/a_p$ = shape factor, a_p is the area of particle in square feet, and v_p is the volume of particle, in cubic feet. Carman gave a similar relationship, where $\dfrac{\phi D_p \bar{u}_s}{1 - \epsilon} > 100$:

$$\frac{-dP}{dL} = \frac{2.88(1 - \epsilon)^{1.1}\mu^{0.1}\rho^{0.9}\bar{u}_s^{1.9}}{g_c\phi^{1.1}\epsilon^3 D_p^{1.1}}. \tag{22a}$$

The relationships 19 and 22 suggest a dependence of energy loss on ϵ in proportion to $(1 - \epsilon)^2/\epsilon^3$ in the viscous region, and $(1 - \epsilon)/\epsilon^3$ in the turbulent region. Theoretical considerations suggest further that viscous and turbulent losses for fluid flow in fixed beds are independent and additive. From dimensional analyses in some early work, Burke and Plummer[25] recommended the form:

$$\frac{-dP}{dL} = \left(\frac{C_1 \rho \bar{u}_s^2 S_v}{g_c \epsilon^3}\right)\left(\frac{\mu S_v}{\rho \bar{u}_s}\right)^{2-n} \tag{23}$$

where C_1 and n = constants, and S_v = particle surface area per unit volume of bed. Here the "state of flow" exponent n varies between 1 and 2; for viscous flow, $n = 1$, and for turbulent flow, $n = 2$. Furthermore, if the area of a single particle S_p is defined in terms of the diameter of an equivalent sphere, D_s,

$$S_p = \frac{6}{D_s} \tag{24}$$

then the particle surface area per unit volume of bed becomes

$$S_v = (1 - \epsilon)\frac{6}{D_s} \tag{25}$$

With $n = 1$, dP/dL for viscous flow may be written:

$$\frac{-dP}{dL}\text{ (viscous)} = \frac{36 C_1 (1 - \epsilon)^2 \mu \bar{u}_s}{g_c D_s^2 \epsilon^3} \tag{26}$$

and with $n = 2$, dP/dL for turbulent flow takes the form:

$$\frac{-dP}{dL}\text{ (turbulent)} = \frac{6 C_2 (1 - \epsilon)\rho \bar{u}_s^2}{g_c D_s \epsilon^3} \tag{27}$$

where the constants C_1 and C_2 are not necessarily the same.

By use of the concept of the additive nature of viscous and turbulent flow energy losses, Ergun and Orning[26] and Ergun,[27] in a comprehensive treatment, developed a generalized

[22] Leva, M., and Grummer, M., *Chem. Eng. Progr.*, **43,** 713–18 (1947).

[23] Dallavalle, J. M., *Micromeritics,* Pitman Publishing Corp., New York, 1943, 423 pp.

[24] Leva, M., *Chem. Eng. Progr.*, **43,** 549–54 (1947).

[25] Burke, S. P., and Plummer, W. B., *Ind. Eng. Chem.*, **20,** 1196–200 (1928).

[26] Ergun, S., and Orning, A. A., *Ind. Eng. Chem.*, **41,** 1179–84 (1949).

[27] Ergun, S., *Chem. Eng. Progr.*, **48,** 89–94 (1952).

pressure-drop equation applicable over the complete flow range with a variety of smooth and rough particles up to 0.5 in. in diameter. The basic premises are as follows: (1) The viscous losses in a flow channel of diameter D_c may be approximated by the Poiseuille equation appropriately modified by a constant k_1,

$$\frac{-dP}{dL}\text{(viscous)} = \frac{32k_1\mu\bar{u}}{g_c D_c^2}$$

where $\bar{u}$ = mean fluid velocity in the channel; and (2) the kinetic energy losses per unit length of path, which are normally expressible as

$$\frac{-dP}{dL}\text{(kinetic)} = \frac{\rho\, d(\alpha\bar{u}^2)}{2g_c dL}$$

where α = velocity distribution correction factor, may be approximated by,

$$-dP/dL\text{(kinetic)} = k_2\rho\bar{u}^2/2g_c D_c$$

where k_2 = a constant. The total pressure drop is then

$$\frac{-dP}{dL} = \frac{32k_1\mu\bar{u}}{g_c D_c^2} + \frac{k_2\rho\bar{u}^2}{2g_c D_c} \tag{28}$$

A more useful form of this relationship results from eliminating $\bar{u}$ (mean channel velocity), and D_c (channel diameter), in terms of ϵ (void fraction), $\bar{u}_s$ (mean superficial velocity), and S_p (specific surface per unit volume of particles). From consideration of the geometry of a packed bed, $\bar{u}_s = \epsilon\bar{u}$, and $D_c = 4\epsilon/(1-\epsilon)S_p$. Equation 28 then becomes:

$$\frac{-dP}{dL} = \frac{2k_1(1-\epsilon)^2\mu\bar{u}_s S_p^2}{g_c\epsilon^3} + \frac{k_2(1-\epsilon)\rho\bar{u}_s^2 S_p}{8g_c\epsilon^3} \tag{29}$$

An alternate form may be written from the consideration that the aerodynamic surface exhibited by a rough particle may be approximated by the geometric area of an equivalent sphere, as in equation 24. Then

$$\frac{-dP}{dL} = \frac{72k_1(1-\epsilon)^2\mu\bar{u}_s}{g_c\epsilon^3 D_s^2} + \frac{3k_2(1-\epsilon)\rho\bar{u}_s^2}{4g_c\epsilon^3 D_s} \tag{30}$$

For small pressure drops and otherwise constant conditions through the length of the bed, the integrated result may be written in terms of average superficial velocity $\bar{u}_m$, bed friction factor f_b, and Reynolds number.

$$\frac{\Delta P g_c\epsilon^3 D_s}{L\rho\bar{u}_m^2(1-\epsilon)} = f_b = \frac{72k_1(1-\epsilon)}{(Re)_p} + \frac{3k_2}{4} \tag{31}$$

The left-hand member represents the ratio of total energy loss to kinetic energy loss. Experimentally, $k_1 = 2.08$, and $k_2 = 2.34$. A log-log plot of equation 31 and experimental data are given in Fig. 11. From a fundamental fluid dynamics analysis, equation 31 is equivalent to:

$$\frac{6\tau}{\rho\bar{u}_m^2} = \frac{72k_1\mu}{6r_h\bar{u}_m\rho} + \frac{3k_2}{4} \tag{31a}$$

where $\tau = \Delta P g_c\epsilon D_s/6L(1-\epsilon)$ = shear stress per unit area of solids exposed aerodynamically, and $r_h = \epsilon/S_v = D_s\epsilon/6(1-\epsilon)$ = hydraulic radius or ratio of free volume to area of solids.

Other forms of integrated results are as follows : (1) For isothermal expansion of an ideal gas entering a static bed at superficial velocity $\bar{u}_1$, measured at entering conditions P_1 and T_1, and where ϵ is constant, but the change in gas density with distance is not negligible,

$$\frac{P_1^2 - P_2^2}{2L} = \frac{150(1-\epsilon)^2\mu P_1\bar{u}_1}{g_c\epsilon^3 D_s^2} + \frac{1.75(1-\epsilon)P_1^2\bar{u}_1^2}{g_c\epsilon^3 D_s R T_1} \tag{32}$$

where R = gas constant, and values of $k_1 = 2.08$, and $k_2 = 2.34$, have been adopted. (2) For isothermal expansion of an ideal gas entering a static bed at superficial velocity $\bar{u}_1$, measured at entering conditions P_1 and T_1, where ϵ is substantially constant, and a chemical reaction $C_{\text{solid}} + A_{\text{gas}} \rightarrow \nu B_{\text{gas}}$ occurs at a rate such that the fractional conversion X of reactant A is linear with distance through the bed:

$$\frac{P_1^2 - P_2^2}{2L} = \left[\frac{150(1-\epsilon)^2\mu P_1\bar{u}_1}{g_c\epsilon^3 D_s^2}\right] \times \left[1 + \frac{\omega X_2}{2}\right] + \left[\frac{1.75(1-\epsilon)P_1^2\bar{u}_1^2}{g_c\epsilon^3 D_s R_A T_1}\right] \times \left[1 + \frac{(\nu r + \omega - 1)X_2}{2} + \frac{\omega(\nu r - 1)X_2^2}{3}\right] \tag{33}$$

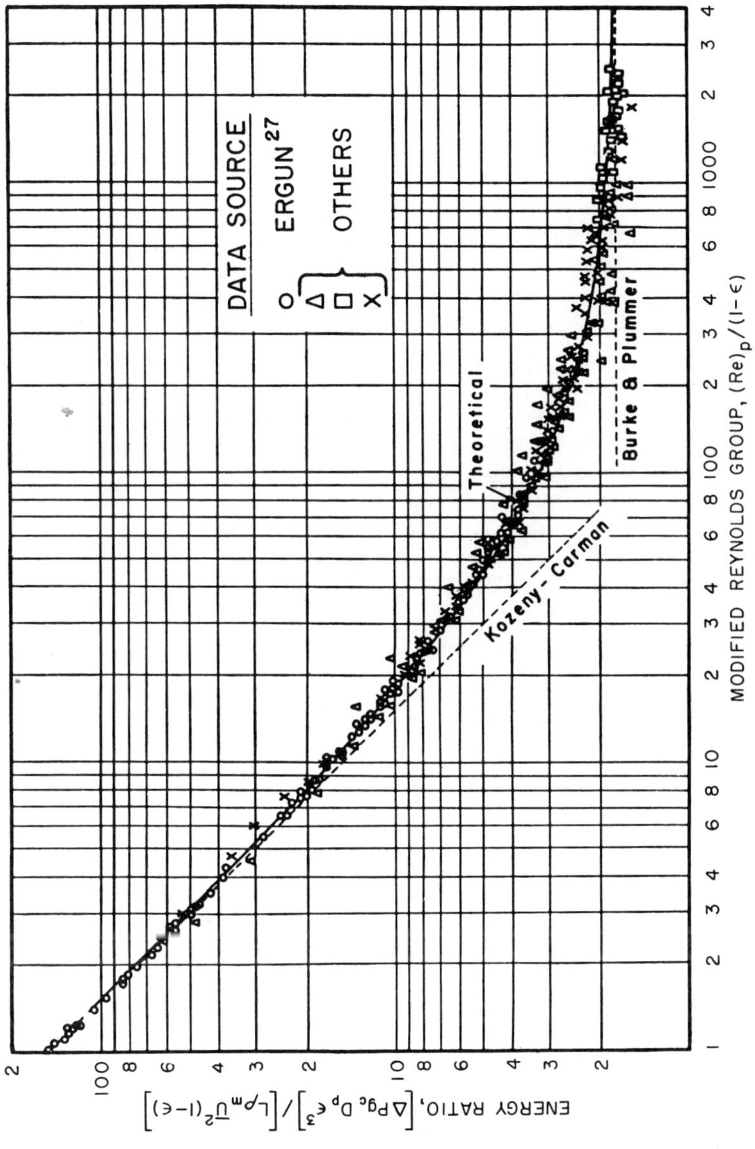

Fig. 11. Comprehensive representation of the pressure drop in packed columns.[27]

where $\omega = \nu - 1$ = expansion factor, R_A = gas constant for reactant A, X_2 = outlet fractional conversion, and r = ratio of molecular weight of B to A. (3) For flow with substantially large kinetic losses through a bed of varying cross section, and temperature varying with distance, where ϵ and D_s are substantially constant, and the volumetric flow rate Q (referred to standard conditions) at any point in the bed is given in terms of the inlet blast rate Q_1 (measured at standard conditions P_1 and T_1) by the relation: $Q = \omega Q_1$, then,

$$\frac{P_1^2 - P_2^2}{2} = \frac{1.75(1-\epsilon)\rho_1 P_1 \omega^2 Q_1^2}{g_c \epsilon^3 D_s T_1} \int_0^L \frac{T}{A^2} dL \tag{34}$$

where T and A are, respectively, the temperature and cross-sectional area at any distance L through the bed, and the integral usually is to be evaluated graphically. Equation 34 has been used to predict pressure drops in blast furnaces within 10 to 15% of measured values, provided no channeling occurred.[28]

The relationships presented here adequately describe the gross effects of the principal variables in respect to viscous and inertial forces, which largely determine the pressure gradient in flow through fixed beds. As such, predictions or computations can be made concerning the magnitude of any one of the variables, provided all other variables are accurately known. A further use of these equations is to estimate whether channeling or bypassing of flow occurs, through comparison of measured and computed pressure drops. One of the more difficult values to estimate is the fractional void volume ϵ. This value may be measured experimentally by the ratio of flow residence times of the bed, determined by use of tracers, to the computed residence time based on superficial velocity. For fixed beds having a range of particle sizes, as is normally encountered in practice, an effective average particle diameter $\bar{D}_p$ is defined by weighting the size distribution by either of two methods:

$$\bar{D}_p = \frac{\Sigma y_i}{\Sigma(y_i/D_i)}$$

or

$$\bar{D}_p = \frac{\Sigma y_i D_i}{\Sigma y_i}$$

where y_i is the weight fraction of average diameter D_i.

[28] Ergun, S., *Ind. Eng. Chem.*, **45**, 477–85 (1953).

Fluidized Bed

Figure 12 demonstrates typical pressure drop-velocity curves for upflow through beds of closely sized particles (curve I), and beds having a wide particle size distribution range (curve II). The approximately linear portions of the curves, AB and $A'B'$, correspond to the fixed bed, where negligible bed expansion occurs. With increasing gas velocity, in the absence of channeling, particle readjustment to 5 to 15% greater fractional void space causes the nonlinear portion BC of curve I. Here the pressure drop approaches the net effective, or buoyant weight, of the solids per unit area. At point C, the loosest fixed-bed packing configuration of the particle is realized, such that with slight further increase in gas velocity some of the particles become continuously agitated, exhibiting the onset of "quiescent" fluidization. At point D, fluidization is complete, and the entire mass may form a dense phase or quiescent bed, with a well-defined upper surface exhibiting little or no mixing of solid particles—analogous to a liquid at rest.

With many gas-solid systems, the bed expands to only a limited extent, and portions of the gas tend to accumulate as bubbles passing through the bed, termed aggregate or two-phase fluidization. The fluidization curve DE exhibits substantially constant pressure drop per unit of initial bed depth, nearly equivalent to the hydrostatic head of the particles, provided smooth fluidization occurs. As the velocity is increased along DE, increased turbulence in fluidization characteristics may occur as a result of gas-bubble formation, solids appear to be rapidly mixed, and the upper surface of the bed is diffuse and not well

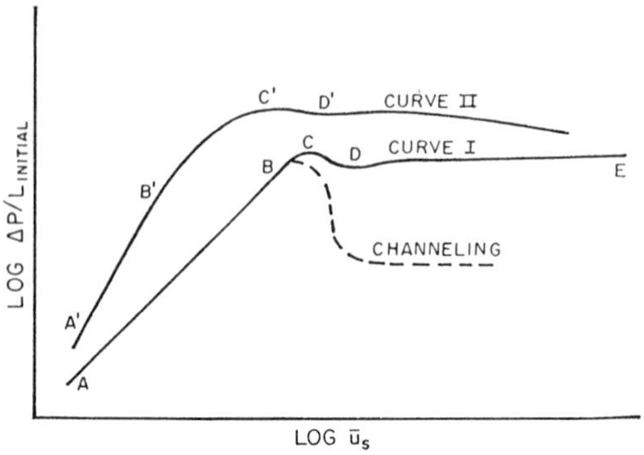

Fig. 12. Characteristic pressure drop through fixed and fluidized beds.

defined. Point E is the terminus of fluidization, where the superficial velocity equals the free-fall velocity of particles, and the bed becomes a disperse suspension.

There are three basic differences between curves I and II. For a bed having a wide particle size distribution for which curve II applies, the initial slope of $A'B'$ is higher due to smaller fraction voids; particle readjustment to greater void fraction occurs at lower velocity B', some of the finer particles are blown out of the bed and may fluidize above a static bed of coarse particles, curve $B'C'$, until complete fluidization is achieved at point D', and with further velocity increases, the terminal velocity of progressively larger particles is exceeded, these leave the bed, and the pressure drop decreases.

Phenomena which prevent smooth fluidization are channeling or bypassing of gases, giving a lower-than-normal pressure drop, and bubble formation, which in small-diameter vessels gives rise to "bumping" and "slugging," or violent fluctuations in pressure drop. The causes of unsatisfactory fluidization are believed to be related with a high length-diameter ratio of the bed, particle roughness, agglomeration tendency, high particle density, and closeness of particle size distribution. Fluidized beds have an inherent tendency toward segregation or rapid settlement of particles in localized zones. This tendency is self-accelerating, and is more pronounced at the bottom of the bed, particularly with high-density large particles. An excellent evaluation is given by Morse.[29] Beds of all-coarse particles tend toward slugging, all-fine particles tend toward channeling.[30] Good fluidization characterized by an even agitation of all particles in the bed exhibits rapid mixing of particles through top-to-bottom circulation of the solids.

Contrary to common supposition, the velocity distribution of gases through a packed bed may be far from uniform. Experimental evidence[31] for nonchanneling flow through small-diameter fixed beds indicates a flat average-point velocity profile around a central core, with the velocity increasing to a maximum at a distance about one particle diameter from the vessel wall. At the wall, the velocity decreases to zero, because of wall friction. This effect is explained qualitatively on the basis that the void fraction approaches unity close to the wall, since particles are in point contact at the wall and the flow path is least sinuous, resulting in the least resistance to flow. In the central core, the void fraction is at an effective minimum, offering maximum resistance to flow. In large-diameter beds, the effect of high velocities near the wall is of lesser significance in relation to the total mass of gas flowing. Schwartz and Smith[31] have presented theoretical velocity distribution relationships which agree closely with experimental results. The effect of backmixing may be estimated by methods of Danckwerts[32] and Levenspiel.[33]

Fluidized beds, on the other hand, exhibit a velocity distribution pattern which may be considered a compromise between piston flow and complete mixing. Qualitatively, the degree of approach to piston flow depends on the ratio of bed height to diameter; at ratios above 2, the effect of gas backmixing appears small.[34] However, large concentration differences can exist over short distances within the bed, as indicated by tracer studies,[35] seemingly because of local bypassing and transient gas bubbles which permeate the bed.

A mathematical estimate of fluidization velocity, or of fractional void volume corresponding to a selected fluidization velocity, is obtained by equating the pressure gradient due to the buoyant weight of the bed,

$$\frac{-dP}{dL} = (1 - \epsilon)(\rho_s - \rho) \qquad (35)$$

—where ρ_s is the apparent density of solids, and ρ is the density of fluid—to the pressure gradient given by an acceptable pressure drop relationship. With equation 30, for example, under the condition of a small-pressure drop such that mean fluid density $\bar{\rho}$ and mean

[29] Morse, R. D., *Ind. Eng. Chem.*, **41**, 1117–24 (1949).

[30] Parent, J. D., Yagol, N., and Steiner, C. S., *Chem. Eng. Progr.*, **43**, 429–36 (1947).

[31] Schwartz, C. E., and Smith, J. M., *Ind. Eng. Chem.*, **45**, 1209–18 (1953).

[32] Danckwerts, P. V., *Chem. Eng. Sci.*, **2**, No. 2, 1–13 (1953).

[33] Levenspiel, O., and Bischoff, K. B., *Ind. Eng. Chem.*, **51**, 1431–4 (1959).

[34] Gilliland, E. R., and Mason, E. A., *Ind. Eng. Chem.*, **41**, 1191–6 (1949).

[35] Gilliland, E. R., Mason, E. A., and Oliver, R. C., *Ind. Eng. Chem.*, **45**, 1177–85 (1953).

superficial velocity $\bar{u}_m$ along the column length may be used, we obtain the fluidization equation:[26]

$$\frac{\epsilon^3}{1-\epsilon} = \frac{72k_1\mu\bar{u}_m}{g_c D_s^2(\rho_s - \bar{\rho})} + \frac{3k_2\bar{\rho}\bar{u}_m^2}{4g_c D_s(\rho_s - \bar{\rho})(1-\epsilon)} \tag{36}$$

For viscous fluidization (modified Reynolds number less than 10.0), the term containing $\bar{\rho}\bar{u}_m^2$ may be neglected. For turbulent fluidization, the term containing $\mu\bar{u}_m$ is negligible. This indicates that the constancy of fluidization characteristics of systems employing different gases, but otherwise equal conditions, may be judged on the basis of the product $\mu\bar{u}_m$ at viscous flow, or the product $\bar{\rho}\bar{u}_m^2$ at turbulent flow.

However, no fluidization equation which appears universally applicable has thus far been proposed. A number of relationships reported in the literature are similar in form, but they differ in the values of the constants or the functional dependence of the variables.[20,21,36,37] The work of Leva and associates[19–21] suggests a fluidization equation for viscous flow of the form:

$$\frac{\epsilon^3}{1-\epsilon} = \frac{200\mu\bar{u}_m}{g_c\phi^2 D_s^2(\rho_s - \bar{\rho})} \tag{37}$$

and this relationship has been employed for correlation of data on fluidization of anthracite particles[38]—size range 0.0066 to about 0.04 in., for which the shape factor $\phi = 0.625$ was recommended. Steinour[39] presented a similar equation with the value of the constant as 147, based on identifying fluidization with the process of hindered settling. Other forms of fluidization equations are given by the relationships for terminal velocity of falling particles, appropriately corrected for particle shape and effect of other particles in the bed. At modified Reynolds numbers in the range $0.0001 < (Re)_p < 2.0$, Stokes' law may be applied:

$$\bar{u}_m = \frac{g_c D_p^2(\rho_s - \bar{\rho}_b)}{18\phi^2\mu_b} \tag{38}$$

[36] Thompson, B. H., and Garside, J. E., *Inst. Gas Engrs., Rept.* No. 437 (1953), 32 pp.

[37] Wilhelm, R. H., and Kwauk, M., *Chem. Eng. Progr.*, **44,** 201–18 (1948).

[38] Leva, M., Weintraub, M., Grummer, M., and Pollchik, M., *Ind. Eng. Chem.*, **41,** 1206–12 (1949).

[39] Steinour, H. H., *Ind. Eng. Chem.*, **36,** 618–24 (1944).

where $\bar{\rho}_p = \epsilon\bar{\rho} + (1-\epsilon)\rho_s$ = average bulk density of expanded bed, $\mu_b = \mu\psi$ = "effective viscosity" of fluidized bed, where ψ is a function of void fraction ϵ given variously as: (1) $\psi = 1 + k(1-\epsilon)$, where k is a value between 1.0 and 4.5, (2) $\psi = 1/[1-(1-\epsilon)^{1/3}]$, and (3) $\psi = [1 + 0.5(1-\epsilon)]/\epsilon^4$. Here the suggested dependence on ϵ is such that equation 38 reduces to the terminal free-fall velocity at $\epsilon = 1.0$. At modified Reynolds numbers $2.0 < (Re)_p < 500$, the intermediate law for falling particles may be applied to fluidization:

$$\bar{u}_m = \frac{0.153 g_c^{0.71} D_p^{1.14}(\bar{\rho}_s - \bar{\rho})^{0.71}\epsilon^{0.71}}{\phi^{1.14}\bar{\rho}_b^{0.29}\mu^{0.43}\psi^{0.43}} \tag{39}$$

Newton's Law for $500 < (Re)_p < 200{,}000$ takes the form:

$$\bar{u}_m = 1.74\left[\frac{g_c D_p(\rho_s - \bar{\rho})\epsilon}{\phi[\epsilon\bar{\rho} + (1-\epsilon)\rho_s]}\right]^{1/2} \tag{40}$$

Other empirical relationships have been proposed of the form: (1) $\bar{u}_m/u_t = \epsilon^n$ applicable at low Reynolds numbers, where u_t is the terminal velocity of particles in free-fall, and n is a constant between 4.5 and 6.0, and (2) $\bar{u}_{min}/u_t = k\epsilon_{max}^6$, where $k = 3.28$, $\bar{u}_{min}$ is the minimum fluidization superficial velocity, and ϵ_{max} is the maximum static bed void fraction corresponding to the point of incipient fluidization.

Heat Transfer

Transmission of heat at the 1500 to 3000°F level in gasification reactors proceeds by the combination of conduction, convection, and radiation between fuel particles, gases, and reactor wall surfaces. Interparticle transfer occurs primarily by radiation, and particle-to-gas transfer primarily by conduction and convection.

In one approach to heat transfer analyses of fixed beds, the total effect of different modes of transfer are referred to an estimated overall coefficient or effective conductivity of particles-plus-gases. This requires a numericial solution of appropriate differential equations which take into account the chemical reaction rates.[40] To establish the validity of the assumptions involved in this complex method, the computed

[40] Smith, J. M., *Chemical Engineering Kinetics*, McGraw-Hill, New York, 1956, 402 pp. Singer, E., and Wilhelm, R. H., *Chem. Eng. Progr.*, **46,** 343–57 (1950). Baron, T., *Chem. Eng. Progr.*, **48,** 118–24 (1952).

temperature distributions at given positions in the reactor must match with measured distributions—actual data for systems of interest here are not available.

As an alternate approach, both heat and mass transfer coefficients may be estimated from analogy to momentum transfer. Transfer process analogies have been proposed and evaluated by several investigators.[41]

Finally, heat-transfer studies have been made with externally and internally heated laboratory scale fixed-bed and fluid-bed vessels operated with nonreacting gases and inert solids, as well as with coal and coke. As these experiments were conducted at low temperatures, usually below 400°F, the results are indicative of the magnitude of convection-heat transfer coefficients which may be expected in actual systems, exclusive of the effects of radiation.

Fixed Bed

The presence of solid particles increases the wall heat transfer coefficient by several times over that of an empty vessel at the same gas flow rate. This increase varies with the ratio of particle diameter to vessel diameter, D_p/D_t, as shown by typical values:[42]

$\frac{D_p}{D_t}$	0.05	0.10	0.15	0.20	0.25	0.30
$\frac{h_w}{h_{empty}}$	5.5	7.0	7.8	7.5	7.0	6.6

A maximum h_w/h_{empty} at D_p/D_t of about 0.15 seems indicated. Turbulence at the wall caused by the presence of solid particles may be responsible, in part, for the increased convection coefficient. This turbulence limits the buildup of the laminar sublayer and also increases radial heat transfer rates in the vessel.

[41] Colburn, A. P., *Ind. Eng. Chem.*, **22,** 967–70 (1930). Chilton, T. H., and Colburn, A. P., *Ind. Eng. Chem.*, **26,** 1183–7 (1934). Bar-Ilan, M., and Resnick, W., *Ind. Eng. Chem.*, **49,** 313–20 (1957). Wilhelm, R. H., Johnson, W. C., Wynkoop, R., and Collier, D. W., *Chem. Eng. Progr.*, **44,** 105–16 (1948). Gamson, B. W., Thodos, G., and Hougen, O. A., *AIChE, Trans.*, **39,** 1–35 (1943). Ergun, S., *Chem. Engr. Progr.*, **48,** 227–36 (1952). Glaser, M. B., and Thodos, G., *AIChE J.*, **4,** No. 1, 63–68 (1958).

[42] Smith, J. M., see ref. 40.

Wall convection coefficients in the range 25 to 40 Btu/sq ft-hr-°F have been found[43] for heating and cooling of fixed beds of various smooth and rough particles through which air or carbon dioxide was introduced. The data for heating were correlated empirically:

$$h_w \text{ (heating)} = \frac{0.813 f_k k}{D_t} \epsilon^{-6D_p/D_t} \left(\frac{D_p \rho \bar{u}_s}{\mu}\right)^{0.90} \tag{41}$$

where k = gas thermal conductivity, Btu-ft/sq ft-hr-°F, D_p = particle diameter in feet, D_t = vessel diameter in feet and f_k = correction factor for conductivity of solids as given below:

Solids conductivity, Btu ft/sq ft-hr-°F	<0.05	25	50	75
Factor f_k	1	1.2	1.25	1.3

For carbonaceous fuels, $f_k \simeq 1.0$. Equation 41 is satisfactory for D_p/D_t in the range 0.05 to 0.3, but serves only as an approximation for vessel diameters larger than 3 in. It predicts a maximum h_w at $D_p/D_t = 0.15$. The equation may be used for most common gases or gas mixtures having a Prandtl number, $C_p\mu/k$, in the range 0.74–0.84. For mixed particle sizes, D_p is taken as the arithmetic average of the size distribution. No coordinated effect of void fraction ϵ on the h_w values was found. Particle surface roughness increases h_w slightly; for very rough particles, h_w is about 10% higher than computed from equation 41. Wall coefficients for cooling were higher than for heating, as given by the empirical correlation:

$$h_w \text{ (cooling)} = \frac{3.50k}{D_t} e^{-4.6D_p/D_t} \left(\frac{D_p \rho \bar{u}_s}{\mu}\right)^{0.7} \tag{42}$$

This relation is based on smooth, low-conductivity, spherical particles and vessel diameters up to 2 in. It predicts a maximum cooling coefficient at $D_p/D_t = 0.153$, which was verified by experiment. Application of equations 41 and 42 at modified Reynolds numbers below about 250 to 300 is not recommended.

[43] Leva, M., *Ind. Eng. Chem.*, **39,** 857–62 (1947). Leva, M., and Grummer, M., *Ind. Eng. Chem.*, **40,** 415–9 (1948). Leva, M., Weintraub, M., Grummer, M., and Clark, E. L., *Ind. Eng. Chem.*, **40,** 747–52 (1948).

Consideration of fluid dynamics of packed beds would seem to indicate the turbulence level near the vessel wall and between particles should be of the same order of magnitude. The average shear stress per unit wall area, however, is greater than that per unit exposed surface area of particles, and since the mass velocity near the wall approaches a maximum, it is to be expected from heat transfer and fluid friction analogy that the convection coefficient for particles to gas is appreciably smaller than the wall coefficient. This smaller convection coefficient is also expected in fluidized systems, where the effect of "laminar film erosion" at the wall, and the lower relative velocity between particles and gas compared to that at the wall, should cause a relatively higher-wall coefficient, which has been verified experimentally.

Eichorn and White,[44] by means of dielectric heating of solids at temperature levels somewhat above 100°F, determined fixed-bed particle-to-gas coefficients in the range 1.5 to 13 Btu/sq ft-hr-°F (about an order of magnitude lower than Leva's wall coefficients), at low modified Reynolds numbers corresponding to viscous flow, the coefficients generally increaseing with mass velocity and with particle size. Particle-to-gas coefficients increased slightly with temperature, owing to some radiation absorption; typical coefficients for a pebble-bed heater[45] operated at 2000°F are 6 to 15 Btu/sq ft-hr-°F. Baumeister and Bennett[46] correlated particle-to-gas convection coefficients in terms of a heat transfer factor j_h, void fraction, and modified Reynolds number. Their expression is about ±50% accurate, in view of considerable scattering of the experimental data:

$$j_h = \frac{h_p}{C_p \rho \bar{u}_s}\left(\frac{C_p \mu}{k}\right)^{2/3} = \frac{1.33\epsilon^3 f}{3(1-\epsilon)(Re)_p^{0.12}} \tag{43}$$

where f = friction factor $= 30.7/(Re)_p^{0.12}$, C_p = specific heat of gas, Btu/lb-°F, h_p = particle-to-gas convection coefficient, Btu/sq ft-hr-°F.

Ergun[41] developed a modified analogy between heat transfer factor j_h and bed friction factor f_b (see equation 31):

$$j_h = \frac{6\epsilon h_p}{C_p \rho \bar{u}_s}\frac{C_p \mu}{k} = f_b = 1.75 + 150\frac{1-\epsilon}{Re_p} \tag{44}$$

His corresponding analogy for mass transfer factor J_d and fluid friction factor is:

$$J_d = \frac{6\epsilon k_g M_m p_{BM}}{\rho \bar{u}_s}\frac{\mu}{\rho D} = f_b \tag{45}$$

where k_g = mass transfer coefficient, lb-moles/sq ft-hr-atm, M_m = mean molecular weight of gas, D = fluid diffusivity in square feet per hour, and p_{BM} = log mean inerts partial pressure in atmospheres.

These analogies appear satisfactory for liquids, but they give only a very rough approximation for the transfer coefficients in gas-solids systems. In testing this analogy for gases and spherical solids, Glaser and Thodos[41] showed that f_b is not independent of particle size, as stated in equation 44, and the computed h_p exceeds by ten to fifty times the experimental values. They pointed out that, although heat transfer and momentum transfer are directly related for fluid flow in circular conduits, the analogy is not as satisfactory for packed beds because momentum transfer involves surfaces, corners, and edges, whereas heat transfer is restricted to surface areas. This same criticism appears valid for mass-transfer and momentum transfer analogies, according to the data of Bar-Ilan and Resnick.[41]

The contribution of interparticle radiation to the total heat flux in fixed beds appears to be significant at temperatures as low as 300 to 400°F for particle sizes down to 0.2 in. The radiation portion of the effective bed conductivity increases with particle size and with the cube of the average absolute temperature at a particular position in the bed. This relationship based on the Stefan-Boltzmann equation has been given[47,48] for two radiating surfaces of equal emissivities:

$$K_R = (4)(1.73)(10^{-9})\epsilon\frac{D_p T^3 \epsilon_H}{2-\epsilon_H} \tag{46}$$

[44] Eichhorn, J., and White, R. R., *Chem. Eng. Progr., Symposium Ser.*, No. 4, 11–18 (1952).

[45] Lancashire, R. B., Lezberg, E. A., and Morris, J. F., *Ind. Eng. Chem.*, **52**, 433–4 (1960).

[46] Baumeister, E. B., and Bennett, C. O., *AIChE J.*, **4**, No. 1, 69–74 (1958).

[47] McAdams, W. H., *Heat Transmission*, Third edition, McGraw-Hill, New York, 1954, 490 pp.

[48] Argo, W. B., and Smith, J. M., *Chem. Eng. Progr.*, **49**, 443–51 (1953).

where ϵ = void fraction, ϵ_H = emissivity, which for carbonaceous materials[47] ranges from 0.75 to 0.98, T = absolute temperature, °R, and K_R = radiation component of bed conductivity, Btu-ft/sq ft-hr-°F. The total equivalent bed conductivity k_e has been estimated[48] via the contributions assignable to the parallel mechanisms of conductivity of gas in the void spaces, ϵk_c, conductivity due to eddy (turbulent) diffusion in the void spaces, ϵk_{td}, radiation, K_R, and conductivity of the solids fraction $(1 - e)K_s$:

$$k_e = \epsilon k_c + \epsilon k_{td} + K_R + (1 - \epsilon)K_s \quad (47)$$

where k_{td} is given in terms of particle size, fluid properties, and mass transfer Peclet number, $(Pe)_m$, or reciprocal dispersion coefficient, by $k_{td} = C_p D_p \rho \bar{u}_s / \epsilon (Pe)_m$, with values of $(Pe)_m = D_p \bar{u}_s / D$ fairly constant at 2.0, according to Levenspiel,[33] or $(Pe)_m/\epsilon$ ranging from 8 to 16 with Reynolds number.[49] K_s is given in terms of particle size, particle-to-gas heat transfer coefficient, h_p, and solid particle conductivity, k_s, by $K_s = h_p k_s D_p / (2k_s + h_p D_p)$.

Fluidized Bed

In the absence of detrimental fluidization characteristics, such as slugging and channeling, extremely high internal heat transfer rates are possible in fluidized beds, primarily as a result of rapid convective or eddy transport of heat by the moving solid particles.[50] In addition, extremely rapid exchange between gas and solids occurs by virtue of the large surface area per unit volume of bed, and the high particle-to-gas heat-transfer coefficients attainable with small particles. The latter increases with decreasing particle size. This means that with moderate rates of heat addition, rapid equalization of gas and particle temperature occurs, and the particle temperature tends to be uniform throughout the fluidized bed. In an externally heated system for conducting an endothermic chemical reaction, the system capacity is limited by heat-transfer rates through the wall surfaces. Even with high rates of external heat transfer to the bed, the temperature gradients within the fluidized solids are small, relative to the temperature drop at the heat transfer surface.[47]

Heat-transfer studies of fluidized beds reported in the literature indicate that a diversity of correlation formulas have been used, and when applied, these formulas do not yield the same numerical value for the heat transfer coefficients. This occurrence is undoubtedly due to the empirical approach employed in most of the studies, wherein transfer rates are arbitrarily related to a select few of the apparently large number of variables which influence the performance of fluidized beds. Besides gas velocity and gas physical and transport properties, this list of variables could include particle geometry, particle physical and transport properties, bed density, and vessel geometry.

A measure of internal heat circulation by solids movement is given by the equivalent or eddy conductivity, $k_e = \rho_b C_s D_e$, in which ρ_b = bed density, C_s = solids specific heat, and D_e = eddy diffusion coefficient of solid particles. The value of k_e is of the order 160 Btu-ft/sq ft-hr-°F, if ρ_b = 40 lb/cu ft, C_s = 0.2 Btu/lb-°F, and $D_e = (5.4)(10^{-3})$ sq ft/sec.[50] That this mode of heat transfer within the bed outweighs all other modes is readily seen, since gas molecular conductivity is of the order of 0.015 to 0.15 Btu-ft/sq ft-hr-°F, carbonaceous solids molecular conductivity is about 0.5 to 1.0 Btu-ft/sq ft-hr-°F, and equivalent conductivity of radiation at 2400°R with 0.05-in. particles is about 0.15 Btu-ft/sq ft-hr-°F (equation 46), when $\epsilon_h = 0.85$, and $\epsilon = 0.5$.

The rates of heat transfer between particles and gas at steady state are estimable from q Btu/cu ft-hr $= h_p A_e (t_g - t_p)$, in which A_e = effective area per unit bed volume, t_g = gas temperature, and t_p = particle temperature. On the other hand, if we know q as the heat of reaction at an average reaction rate, we can compute $t_g - t_p$ from values of $A_e h_p$, in which A_e may be taken as the area of a unit volume of equivalent spheres multiplied by a shape factor. The particle-to-gas coefficient h_p is available from theoretical relationships[47] and from empirical correlations. The theoretical basis for h_p is the steady-state solution of heat conduction through a stationary film of gas surrounding an assumed spherical particle. This gives values of the Nusselt number $h_p D_p / k = (1 + 2F)/F$, in which F is the film thickness expressed as a fraction of particle

[49] Bernard, R. A., and Wilhelm, R. H., *Chem. Eng. Progr.*, **46**, 233–44 (1950).

[50] van Heerden, C., Nobel, A. P. P., and van Krevelen, D. W., *Ind. Eng. Chem.*, **45**, 1237–42 (1953).

diameter. The minimum h_p would result for infinite film thickness, $F = \infty$ in which case $h_p = 2k/D_p$. Typically, with $k = 0.02$ and $D_p = 0.05$ in., h_p (minimum) = 9.6 Btu/sq ft-hr-°F. In a practical case with small particles, F is about 0.25, which would give $h_p = 6k/D_p$, or three times the minimum value. The value F is primarily dependent on gas flow rate past the particle, and it has been correlated with modified Reynolds number in an equivalent expression:[47]

$$\frac{h_p D_p}{k} = 0.37(Re)_p^{0.6} \tag{48}$$

which is valid for $(Re)_p > 17$. Experimental verification of h_p values for fluidized beds is difficult, owing to the inability to know the effective bed area exactly, and the small temperature differences encountered and probable inaccuracies in gas and solids temperature measurement. In one study, h_p values of 6 to 40 were determined[51] for fluidized beds of 20- to 48-mesh coal particles (0.028 to 0.014 in.). The results were correlated empirically, with about ±40% maximum deviation from the curve:

$$\frac{h_p D_p}{k} = 0.0028\left(\frac{D_t}{D_p}\right)^{0.2}(Re)_p^{1.7} \tag{49}$$

Here the Nusselt group falls below the minimum value of 2.0 suggested from theory, which probably indicates that not all of the computed surface area of the fluidized bed was effective for heat transfer. In another study,[52] h_p values of 3 to 10 were found by water evaporation from 14- to 48-mesh (0.039 to 0.014 in.) silica gel and alumina particles. Their correlation, differing from Walton's,[51] gave

$$\frac{h_p D_p}{k} = 0.0135(Re)_p^{1.30} \tag{50}$$

Smaller-than-anticipated h_p values are likely to result from bed segregation and gas bubble formation in fluidized beds.

Heat-transfer rates between wall surface and fluidized bed are estimated by the usual relation, $q = h_w A_w(t_w - t_b)$. It is normally assumed that the major resistance to heat flow occurs at a laminar film of δ thickness at the wall. Then, if molecular film conduction predominates, h_w is approximated by k_f/δ, where k_f is the film conductivity, including the effect of the additional conductivity provided by the solids at the wall, or $k_f \simeq \epsilon k_{\text{gas}} + (1 - \epsilon)k_{\text{solids}}$. Typically, with $\delta = 2$ millimeters (0.00656 ft), $\epsilon = 0.5$, $k_{\text{gas}} = 0.02$, and $k_{\text{solids}} = 0.5$, we have $h_w = 40$ Btu/sq ft-hr-°F, which is in agreement with the range of reported values. With radiant transfer as significant at high temperatures, the wall coefficient is increased by the additive factor,

$$h_R = (4)(1.73)(10^{-9})\frac{\epsilon_H T^3}{2 - \epsilon_H} \tag{51}$$

which typically at $T = 2400$°R, $\epsilon_h = 0.85$, gives $h_R = 70$ Btu/sq ft-hr-°F. However, this is an admittedly oversimplified analysis of the wall-convection coefficient.

It is not possible to assign, a priori, the film thickness δ, since it certainly depends on gas viscosity, mass velocity, and manner in which solids are circulating at the wall. When there is good fluidization, solids are moving continuously downward along the wall, and particle exchange between bed wall and interior occurs, indicating that the solids' specific heat will affect h_w. With finer particles, a greater concentration of solids in the so-called laminar film gives rise to a higher film conductivity or increased h_w. The effect of increasing gas mass velocity appears to be two-fold. Its normal effect is to cause an increased h_w by virtue of greater shearing at the wall. With decreasing bed density or decreasing solids concentration at the wall, however, h_w decreases through loss of effective film conductivity, which suggests that h_w increases to a maximum, and then declines with gas flow rate, an effect which has been confirmed[53] in studies of lean-phase fluidization.

The factors described here, as well as other effects on h_w, are indicated by several empirical correlations: fluidized spherical beads, 0.002 to 0.018 in. in diameter, in 1- and 4-in. columns with internal bayonet heating, and where the bed density ρ_b was varied over the range 4 to 80 lb/cu ft, gave h_w values of 10 to 120 Btu/sq ft-hr-°F as a function only of ρ_b and

[51] Walton, J. S., Olsen, R. L., and Levenspiel, O., *Ind. Eng. Chem.*, **44,** 1474–80 (1952).

[52] Kettenring, K. N., Manderfield, E. L., and Smith, J. M., *Chem. Eng. Progr.*, **46,** 139–45 (1950).

[53] Mickley, H. S., and Trilling, C. A., *Ind. Eng. Chem.*, **41,** 1135–47 (1949).

D_p (physical properties were not varied).[53] Equation 52 represents the data within $\pm 25\%$.

$$h_w \text{ (internal heating)} = 0.0433 \frac{\rho_b^{\,2}}{D_p^{\,3}} \qquad (52)$$

where D_p is in feet. For external heating, an effect of gas mass velocity was found that caused a maximum in the wall coefficient. Here h_w ranged from 10 to 120 Btu/sq ft-hr-°F with bed densities of 3 to 70 lb/cu ft, and gas velocities of 0.8 to 15 ft/sec as related within $\pm 25\%$ accuracy by:

$$h_w \text{ (external heating)} = 0.0118 \left(\frac{\rho_b G}{D_p^{\,3}} \right)^{0.263} \qquad (53)$$

where G = gas mass velocity in pounds per square foot per hour.

A more generalized correlation[54] was based on external heating of dense-phase fluidized Aerocat catalyst, finely divided coke, and powdered iron, 0.0027- to 0.0067-in. size range, in 2- and 3-in. columns, where h_w ranged from 20 to 174. The effects of the ratio of vessel diameter to bed height $\frac{D_t}{L}$, vessel to particle diameter, and physical properties were found in accordance with the following expression which represents the data within $\pm 20\%$:

$$\frac{h_w D_t}{k_{\text{gas}}} = 0.50 \left(\frac{D_t}{L} \right)^{0.65} \left(\frac{D_t}{D_p} \right)^{0.17} \times \left(\frac{\rho_s C_s}{\rho_g C_g} \right)^{0.25} \left(\frac{D_t G}{\mu_g} \right)^{0.80} \qquad (54)$$

Equation 54 has been criticized on the basis that the length term, L should be the heated length, and not the height of the fluidized bed.[50]

An important effect on h_w of solids feed rate in fluidized systems is found where solids are continuously discharged upwardly.[55] The wall coefficient for pulverized coal particles ranged 40 to 90 Btu/sq ft-hr-°F, increasing with coal feed rates, 400 to 800 lb/sq ft-hr, and with air flow rates, 2000 to 14,000 lb/sq ft-hr. Other solids gave similar results.

Wender and Cooper[56] collected and analyzed empirically a large mass of fluidization heat transfer data reported in the literature, including some results from commercial scale units. This source appears to be the most reliable correlation to date from the standpoint of the number of data points and ranges of the variables. For externally heated beds, the correlation is a graph of a quantity X versus $(Re)_p = D_p G/\mu$, where

$$X = \frac{10^4 h_w D_p / [k_{\text{gas}}(1 - \epsilon)(C_s \rho_p / C_g \rho_g)]}{1 + 7.5 e^{-(0.44 L_H C_g / D_t C_s)}}$$

and L_H = heated length, ρ_p = apparent density of particles. Specific values are:

$(Re)_p$	X	$(Re)_p$	X
0.04	0.275	4.0	19.9
0.1	0.83	10.0	28.1
0.4	3.80	40.0	36.8
1.0	8.0		

The relation between X and $(Re)_p$ is given to a good approximation by:

$$\ln X = 4.47 \ln (Re)_p + 76.25 - 74.1 (Re)_p^{0.05} \qquad (55)$$

For internal-tube-heated fluidized beds, these authors gave

$$\frac{h_w D_p}{k_{\text{gas}}(1 - \epsilon)} \left(\frac{k_{\text{gas}}}{C_g \rho_g} \right)^{0.43} = 0.033 C_R \left(\frac{C_s}{C_g} \right)^{0.80} \left(\frac{\rho_p}{\rho_g} \right)^{0.66} (Re)_p^{0.23} \qquad (56)$$

where C_R = correlation factor for nonaxial heater tube location, which is related with the ratio of the radial distance of the heater axis from the axis of the fluidization vessel to the radius of the vessel R_h/R_t as follows:

R_h/R_t	C_R	R_h/R_t	C_R
0	1.0	0.4	1.77
0.1	1.48	0.5	1.72
0.2	1.65	0.6	1.66
0.3	1.75	0.7	1.58

The ranges of variables covered in these correlations are summarized in Table 4. The wall coefficient correlation for external heating indicates that h_w increases with Reynolds number, gas conductivity, solids heat capacity, particle density, and vessel diameter. It decreases with increasing particle size, increasing void fraction, gas heat capacity, and heated length.

54 Dow, W. M., and Jakob, M., *Chem. Eng. Progr.*, **47**, 637–48 (1951).

55 Schmid, W. E., Bartkus, E. P., and Corrigan, T. E., *Chem. Eng.*, **59**, No. 10, 172–3 (1952).

56 Wender, L., and Cooper, G. T., *AIChE J.*, **4**, No. 1, 15–23 (1958).

TABLE 4

Ranges of Variables in Fluid-Bed Wall Coefficient Correlation

Variable	External Heating	Internal Heating
D_p, ft	0.000162–0.00278	0.000133–0.00288
C_s, Btu/lb-°F (solids)	0.117–0.276	0.20–0.316
ρ_p, lb/cu ft (particles)	52–328	49–179
ϵ, void fraction	0.436–0.95	0.397–0.945
C_g, Btu/lb-°F (gas)	0.235–1.24	0.125–1.24
ρ_g, lb/cu ft (gas)	0.0069–0.185	0.011–0.485
μ, centipoise	0.0193–0.0290	0.0105–0.0382
D_t, vessel, ft	0.083–0.394	0.24–6.33
L_H, heater length, ft	0.142–3.33	0.33–17.5
L_B, bed height, ft	0.142–8.33	0.83–26.6
h_w, Btu/sq ft-hr-°F	3.3–174	10.0–181

For internal heating, according to equation 56, h_w increases with nonaxial heater location up to a ratio $R_h/R_t = 0.4$, with Reynolds number, solids heat capacity, particle density, and gas conductivity. It decreases with increasing particle size, increasing void fraction, and gas heat capacity and density.

KINETICS OF GASIFICATION REACTIONS

With the exception of gas-phase volatilization reactions, the reactions occurring in oxidation, steam and carbon dioxide decomposition, or in hydrogenation of solid carbonaceous materials are, in the main, heterogeneous, since at least two phases participate in the interactions. Heterogeneous reactions occur by way of a series of diffusional and chemical steps, commonly described as follows:

1. Mass transfer and diffusion of reactants from a fluid stream through a relatively stagnant film to the solid surface and to available capillary areas or pore structure.
2. Adsorption, physical and/or chemical, of reactants on the solid.
3. Surface chemical reaction, which may be preceded by reactant dissociation.
4. Release of surface reaction products, and
5. Diffusion and mass transfer of products to the bulk gas stream.

Thus, in a broad sense, the heterogeneous reaction mechanism comprises a diffusional part, the rapidity of which depends on concentration differences and molecular velocities of gases near the surface and in the porous structure; and a chemical kinetic part, the rate of which depends on external and pore surface interactions.

A study of gas-carbon reaction kinetics must necessarily consider both diffusional and chemical kinetic effects. For relatively nonporous and small-particle-size materials, the ratio of external surface to available pore surface is large, and the effect of internal diffusion on the overall rate process is small. Hence the major effect of diffusion is the resistance offered by external diffusion to and from the surface, and this may be treated independently of the surface chemical kinetics. With large, porous particles, the reaction rate is significantly influenced by internal diffusion. Even though the effects of external diffusion still may be isolated, the chemical kinetics of the rate process, based on bulk gas properties, is inseparably related to the diffusion process. A measured reaction rate will always include some effect of the resistance because of internal diffusion.

The relative importance of diffusion increases with the reactivity of the carbon surface, particle diameter, and small pore size. When these factors result in appreciable diffusion resistances, the form of the deduced rate equation, order of reaction, measured activation energy, and other kinetic phenomena become dependent on particle and pore geometries or the degree of penetration of the reacting gases into the solid. In a strict sense, the true chemical kinetic effects may be studied if all diffusional resistances are negligible.

With the exception of internal diffusion and internal reaction, the five heterogeneous reaction steps previously listed occur in series. Under "steady-state" conditions, the total external diffusion rate equals the total reaction rate, and the internal diffusion rate is essentially equivalent to the reaction rate in the pore structure. The rate of reaction at the external surface would be equal to the difference between external and internal diffusion rates. A convenient, though not rigorous approach in evaluation of overall reaction rates, is the concept of additive reaction resistances.

If the resistance attributable to diffusion comprises nearly the total resistance, the reaction is considered "diffusion controlled." Analogously, the reaction is "kinetically controlled" at negligible diffusion resistance. If, however, internal diffusion resistance is appreciable, the concept of additive resistances is not valid. Conventionally, the procedure then is to correct the specific reaction rate based on unit amount of solid by an "effectiveness factor" to take into account the additional diffusion resistance for the portion of the reaction occurring internally. This is admittedly an approximation because chemical kinetics of pore reactions may not be the same as those of external-surface reactions. Further details are given in a subsequent portion of this section.

Transition of one type of rate control to another is usually indicated when a significant change of apparent activation energy with temperature occurs as obtained from an Arrhenius-type plot of the initial reaction rates. Kinetically controlled rates exhibit true activation energies on the order of 50 to 100 kcal, but the particular value found may not remain constant, since either the rate-controlling step of the kinetic mechanism or the mechanism itself may change with temperature. Apparent activation energies of substantially diffusion-controlled reactions are less than 5 kcal. Between these extremes, all intermediate values of apparent activation energy are possible, dependent on the resistances of the diffusion processes relative to those of the chemical kinetic processes. In particular, if internal reaction in the pore structure predominates, and if the external diffusion resistance is small, the measured activation energy of the reaction will be approximately one-half of the true activation energy of the kinetic process. On the other hand, an indication of the presence of appreciable external diffusion resistance is given by the effect that increasing gas flow rates will increase the measured reaction rates. This flow-rate effect usually occurs at temperatures above 2000°F in the gas-carbon reactions because kinetic rates are slow at lower temperatures, and these rates increase much more rapidly with temperature than do the external diffusion rates. At atmospheric or higher pressures and normal combustion temperatures, the carbon-oxygen reaction is substantially diffusion-rate controlled.

Considerable experimental work directed toward a basic understanding of gas-carbon reaction kinetics and the application of theoretical approaches to these problems has been reported in the literature. Certain of these results are available in articles by Walker and associates,[57] who give an excellent general review. The literature includes papers on carbon oxidation,[58] on the carbon-carbon dioxide reactions,[59] on the carbon-steam reactions,[10,60] on the reactions of carbon and hydrogen,[61] and on the kinetics of the carbon-hydrogen-steam system.[62] These and other investigations indicate there is general agreement that gas-carbon reactions follow a Langmuir-type rate equation over limited temperature ranges, that certain of the reactants and reaction products inhibit or slow down the rate as conversion increases, and the order of the reaction may vary between zero and unity, depending on the temperature and reactant partial pressure.

[57] Walker, P. L., Jr., Rusinko, F., Jr., and Austin, L. G., *Advances in Catalysis*, Eley, O. D., Selwood, P. W., and Weisz, P. B., Eds., Academic Press, Inc., New York, 1959, vol. XI, pp.133–221.

[58] Gulbransen, E. A., and Andrew, K. F., *Ind. Eng. Chem.*, **44,** 1034–51 (1952). Day, R. J., Thesis, The Pennsylvania State University, 1949. Day, R. J., Walker, P. L., Jr., and Wright, C. C., *Industrial Carbon and Graphite*, Soc. of Chem. Ind. London, 1958, pp. 348–70.

[59] Gadsby, J., Long, F. J., Sleightholm, P., and Sykes, K. W., *Proc. Roy. Soc. (London)*, **A 193,** 357–76 (1948). Reif, A. E., *J. Phys. Chem.*, **56,** 778–88 (1952). Ergun, S., *J. Phys. Chem.*, **60,** 480–5 (1956).

[60] Johnstone, H. F., Chen, C. Y., and Scott, D. S., *Ind. Eng. Chem.*, **44,** 1564–9 (1952).

[61] Zielke, C. W., and Gorin, E., *Ind. Eng. Chem.*, **47,** 820–5 (1955). Blackwood, J. D., *Australian J. Chem.*, **12,** 14–28 (1959).

[62] Goring, G. E., Curran, G. P., Tarbox, R. P., and Gorin, E., *Ind. Eng. Chem.*, **44,** 1051–65 (1952). Goring, G. E., Curran, G. P., Zielke, C. W., and Gorin, E., *Ind. Eng. Chem.*, **45,** 2586–91 (1953). Blackwood, J. D., and McGrory, F., *Australian J. Chem.*, **11,** 16–33 (1958).

Reaction mechanisms comprising elementary kinetic steps have been proposed which lead to the correct form of the rate equations; however, this does not necessarily validate a particular mechanism, since invariably more than one mechanism can be postulated for the same rate equation. A basic assumption common to all explanations of gas-carbon reaction rates is the existence of active carbon-free-sites distributed in small concentration throughout the carbon structure. These sites are capable of chemisorption of reacting gas constituents, in whole or in part, forming either adsorbed molecules or oxygenated-surface-complexes.

The reaction rate and order of the reaction depend on the formation and removal rates of the surface complexes and on the number and extent of coverage of the active free-sites. For chemisorption to occur, surface carbon atoms must have free valence electrons to form chemical bonds with the reacting gas constituents, and these bonds must be stronger than the carbon–carbon bonds in the lattice so that carbon atoms will be liberated as product-gas molecules. The active free-sites are considered to constitute unpaired electrons at the edges and at imperfections of the carbon lattice. The presence of such unpaired electrons in various carbon types has been confirmed through electron paramagnetic-resonance adsorption techniques.[63] The differences in reactivities of various carbons may by definition be assumed to be the result of differences in the number density of active free-sites.

Carbon is a heterogeneous structure comprised of crystallites in more or less random orientation. Physically, the reactivity appears to depend on degree of surface heterogeneity, which, in turn, may be related to the size and orientation of crystallographic planes, the number of crystal lattice discontinuities, and impurity atoms in the lattice. Additional factors that influence reactivity are chemical composition, particularly the amount of hydrogen, and sulfur which will affect the carbon structure; ash constituents which may provide catalytic effects; and temperature level and previous temperature history, which largely determine the extent of annealing, graphitization, or order and size of the structure crystallites.

An additional environmental factor also complicates the interpretation of gas-carbon reaction rates. Sometimes, depending on the reactor design, temperature, and carbon sample geometry, the reaction is at unsteady state because of a change in reacting surface area until a particular burnoff is achieved. When this period of constant rate—in terms of weight of carbon reacting per unit time—is attained,[57] it is kinetically more significant because it reflects a near-equilibrium condition with respect to surface area, and the results are more reproducible.

Reaction Rates and Mechanism

Carbon-Oxygen Reaction

Much information, including discussion of numerous experimental studies, concerning the carbon-oxygen reaction was presented in *Chemistry of Coal Utilization.*[64] A more recent review of kinetics of combustion and gasification, in general has been published.[65] One of the relatively few investigations of pressurized combustion of pulverized coal has also been summarized.[66]

Because the combustion of carbon with oxygen is highly exothermic, experimental investigation of the reaction at high-oxygen concentrations and high temperatures is difficult without complicating the picture in respect to diffusional effects. As a consequence, chemical kinetic studies of the reaction have employed, in the main, high-purity, relatively low-porosity carbons which were reacted at one, or a combination of low pressures, low temperatures, or high-streaming velocities of the gases. These conditions minimize bulk diffusion effects and limit the occurrence of gas-phase secondary reactions when the nature of the primary reaction products is being studied.

[63] Hennig, G. R., and Smaller, B., *Proc., 2nd Conf. on Carbon*, Univ. of Buffalo, Buffalo, New York, 1956, pp. 113–13. Winslow, D. H., Baker, W. O., and Yager, W. A., *Proc., 2nd Conf. on Carbon*, Univ. of Buffalo, Buffalo, New York, 1956, pp. 93–102.

[64] Mayers, M. A., *Chemistry of Coal Utilization*, John Wiley and Sons, New York, 1945, Vol. 2, pp. 1482–521. Orning, A. A., *Ibid.*, pp. 1522–567. van der Hoeven, B. J. C., *Ibid.*, 1586–672.

[65] Batchelder, H. R., Busche, R. M., and Armstrong, W. P., *Ind. Eng. Chem.*, **45,** 1856–62 (1953).

[66] Hazard, H. R., and Buckley, F. D., *ASME, Trans.*, **70,** 729–37 (1948).

The general conclusion is that the mechanism of oxidation must involve preferential attack of the carbon lattice at specific areas or active sites, also termed reaction centers. This elementary act is thought to result in the formation of various surface oxides or complexes of different stability. Each may decompose at a characteristic rate, increasing with temperature. At least two forms of surface oxides[67] are postulated in connection with attempted explanations of carbon oxidation rates. One form is supposed to be quite stable, the other is not. The normal reaction between carbon and oxygen is associated with the least stable of the surface oxides, whereas the stable form[68] effectively removes active centers from further participation in the reaction, thus decreasing the rate.

As the surface oxides decompose, new active sites are generated. At low temperature, most of the active sites may be covered with relatively stable oxides, and the reaction order approaches zero, that is, independence of oxygen partial pressure. At high temperatures, when only a small coverage of active sites is believed to occur, the reaction order approaches unity, since the rate should be dependent on the product of available active sites and frequency of molecular collisions, the latter being proportional to oxygen partial pressure.

Experimental studies[69] have given strong evidence that the so-called new reaction centers or active sites arising from lattice attack of the reacting molecules disappear spontaneously by thermal healing or annealing of the surface, and this annealing process increases rapidly with temperature. Thus, a new active site is either thermally stable, remaining active, or is capable of two modes of disappearance: (1) reaction with an impinging molecule or (2) thermal neutralization. The ratio of one mode to the other depends on the average time between molecular collisions on the surface and the average lifetime of an active site. This hypothesis is consistent with the observed phenomenon that the carbon oxidation rate attains a maximum at a particular temperature.

Duval's hypothesis is based on discovery that "hysteresis" or unsteady-state effects occurred in the rates of oxidation of graphitized and ungraphitized carbon filaments when the temperature or pressure was suddenly varied within certain ranges. The experimental apparatus comprised a vacuum flow system operated at about 10^{-3} mm Hg and 1100 to 2300°K. Similar effects were noted in studies of the reactions of carbon with carbon dioxide and with steam.[70] Thus, when the temperature of the carbon filament is suddenly increased, the reaction rate, initially at steady state at the lower temperatures, increased instantaneously to a higher, time-dependent value which levelled off to a new stationary rate. Conversely, when the filament temperature was decreased, the initial rate was lower than the stationary rate at the new temperature. Analogous effects occurred with increases or decreases in oxygen partial pressure.

The temperature-hysteresis effects, which were observed in the 1200 to 1800°K range, were explained on the basis of independent variations of the elementary collision yield f and fraction of surface covered with active sites x_{active}. The product of these two factors, or fx_{active}, is the reaction-rate constant. The elementary collision yield is a fraction defined as the ratio of the number of collisions leading to reaction per unit active area to the total number of collisions occurring per unit active area.

It was also found that the stationary reaction rate constant for ungraphitized filaments increases to a maximum at about 1400°K and decreases thereafter (Fig. 13). Graphitized filaments yielded similar results. The reaction rate constant F, as defined here, is the total collision yield or ratio of the measured number of collisions r, causing reaction per unit of geometric surface area to the total number of collisions N, occurring per unit geometric area, the latter being given from kinetic theory (Hertz-Knudsen Equation):

$$N = \frac{p_{O_2}}{(2\pi mkT)^{1/2}} \tag{57}$$

[67] Frey, H. M., *Proc. Roy. Soc. (London)*, **A 228**, 510–18 (1955). Sihvonen, V., *Z. Elektrochem.*, **40**, 456–60 (1934); *Trans. Faraday Soc.*, **34**, 1062–74 (1938).

[68] Langmuir, I., *J. Am. Chem. Soc.*, **37**, 1139–67 (1915).

[69] Duval, X., *Ann. chim. (Paris)*, **10**, 903–67 (1955).

[70] Boulangier, F., Duval, X., and Letort, M., *Proc. 3rd Conf. on Carbon*, Pergamon Press, New York, 1959, pp. 257–64.

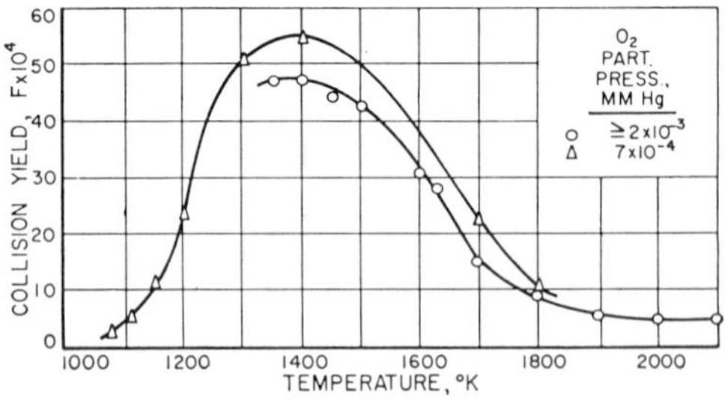

Fig. 13. Collision yield of the oxygen reaction on nongraphitic carbon filaments at low pressures.[69]

where m is the mass per molecule, and k is the Boltzmann constant. In symbolic form:

$$F = \frac{r}{N} = fx_{\text{active}} \tag{58}$$

Referring to Fig. 13, below 1400°K the collision yield or rate constant F increases rapidly with temperature, and the disappearance of active sites through thermal annealing is relatively slow. Above this temperature, thermal annealing increases rapidly. At 1800°K and above, where hysteresis effects were not detected, the elementary collision yield f approaches unity, becoming independent of temperature. Thermal annealing is extremely rapid, and the rate constant approaches a temperature-independent value.

The reaction order approached unity at pressures above 1.5×10^{-3} mm Hg at all temperatures studied; it was exactly unity at 1900 to 2000°K. In the range of 1100 to 1200°K, the reaction-rate constant increased exponentially with an Arrhenius temperature coefficient approximating 40 kcal. At 1700 to 1800°K, the rate constant decreased exponentially, at approximately −40 kcal temperature coefficient. Such variation in the measured heat effect or "activation" energy is consistent with equation 58, wherein f increases with temperature, perhaps in accordance with the Arrhenius law over small temperature ranges. But the quantity x_{active} cannot have a true temperature coefficient, since it is a concentration term which is temperature as well as time dependent. The steady-state value is the dynamic balance between thermal annealing and average time between effective molecular collisions. Thus, under conditions where x_{active} changes significantly with temperature, measured activation energy has little fundamental meaning, and accordingly, is interpreted simply as a temperature coefficient.

If this hypothesis is correct, unless both the elementary collision yield and the active site concentrations always vary with temperature in the same way for different carbons, no two carbons of different reactivity can yield fundamentally the same temperature coefficient.

Whether the primary reaction products in carbon oxidation are carbon monoxide or carbon monoxide and carbon dioxide, has not been absolutely extablished. It is plausible that the primary CO/CO_2 product ratio is solely a function of temperature and surface properties of the carbon. Duval found mostly carbon monoxide at 1100 to 2300°K. Low-temperature chemisorption studies[71] indicate that carbons release adsorbed oxygen primarily as carbon dioxide below 600°C, but at higher temperatures, greater proportions of monoxide are released.

A mechanistic explanation indicated that the relative decomposition rates of different forms of surface carbon oxides determine the primary product CO/CO_2 ratio.[72] Substantiation of the formation of both oxides as primary products has been given by Arthur,[73] who employed a flow system at atmospheric pressure with $POCl_3$ injection to inhibit gas-phase secondary oxidation of carbon monoxide. At 460 to 900°C, the CO/CO_2 ratio for two carbons of widely different reactivity increased exponentially with temperature:

$$\frac{CO}{CO_2} = 10^{3.4}e^{-12,400/RT}$$

There is general confirming evidence from other investigators that the CO/CO_2 formation ratio increases sharply with temperature, but that it is not necessarily independent of carbon type. Day,[58] in studies at atmospheric pressure and at gas velocities up to 1000 ft/sec, impinging on a carbon rod, concluded also that both CO and CO_2 are primary products. However, Blackwood and McTaggart[74] have contended that carbon dioxide is always formed as a

[71] Anderson, R. B., and Emmett, P. H., *J. Phys. Chem.*, **56,** 753–5 (1952). Norton, F. J., and Marshall, A. L., *ASME, Trans.*, **156,** 351–71 (1944).

[72] Lowry, H. H., and Hulett, G. A., *J. Am. Chem. Soc.*, **42,** 1408–19 (1920). Meyer, L., *Z. Physik. Chem.*, **B 17,** 385–404 (1932).

[73] Arthur, J. R., *Trans. Faraday Soc.*, **47,** 164–78 (1951).

[74] Blackwood, J. D., and McTaggart, F. K., *Australian J. Chem.*, **12,** 114–21 (1959).

secondary product through collisions of the primary carbon monoxide molecules with the carbon surface; the transfer of oxygen is effected at, or near, the surface. Their work disclosed that atomic or dissociated oxygen, prepared by passage through a high-frequency electromagnetic field, reacts rapidly, and stoichiometrically, with carbon at as low as ambient room temperatures. Such indications may show that the normal oxidation process is related to the production of atomic oxygen or free radicals at the surface. In partial agreement with this view, it has been suggested[75] that oxygen at the carbon surface may be chemisorbed in atomic form, and readily released as the primary product, carbon monoxide. A decrease in the oxygen partial pressure favored the production of monoxide over dioxide.[76] This effect was tentatively explained, on the grounds that as pressure is decreased, the mean free path of the molecules increases, and the probability of carbon monoxide molecules returning to the surface to be oxidized decreases.

Carbon dioxide formation has been attributed to oxidation of the primary carbon monoxide in the carbon body interstices where unreacted oxygen is diffusing toward the internal surface.[77] It appears that this mechanism should be operative when partial pressures or oxygen concentrations are sufficiently high that the mean free path is small compared to pore diameters, permitting molecular diffusion in the ordinary sense. A related view of the mechanism of carbon combustion,[78] which may be applicable for high oxygen partial pressures is that the process is always diffusion controlled. The oxidation of the primary carbon monoxide to dioxide is involved. Carbon dioxide reforms to the monoxide by reacting with the carbon surface, where the overall rate is determined by the rate of oxygen diffusion through the carbon dioxide layer, not by the speed of the chemical reaction between oxygen and carbon.

[75] Gauger, A. W., and Zelinski, J. J., paper presented at the Division of Gas and Fuel Chemistry, Amer. Chem. Soc., New York, September, 1951.

[76] Meyer, L., *Trans. Faraday Soc.*, **34**, 1056–61 (1938).

[77] Arthur, J. R., Bangham, D. H., and Thring, M. W., *J. Soc. Chem. Ind.*, **68**, 1–6 (1949).

[78] Mayers, M. A., *Chem. Rev.*, **14**, 31–53 (1934).

Based on the foregoing discussion, the simplest, but unproved, kinetic mechanism for carbon oxidation which would yield carbon monoxide as the primary product at high temperatures may be written:

$$\text{Step 1.} \quad C_f + O_2 \xrightarrow{k_1} C(O) + O$$

$$\text{Step 2.} \quad C_f + O \xrightarrow{k_2} C(O)$$

$$\text{Step 3.} \quad C(O) \xrightarrow{k_3} CO + nC_f$$

$$\text{Step 4.} \quad C_f \xrightarrow{k_4} C(\text{inactive})$$

where C_f denotes an active carbon free-site capable of reaction. This mechanism is chosen, not on the basis of preference, but rather to illustrate the type of reasoning that may be applied from basic theory. In Step 1, the dissociation of oxygen is proposed where C(O) represents chemisorbed atomic oxygen, and the free radical (O) is consumed in Step 2. If the concept of thermal annealing is accepted (step 4), then to achieve a steady-state reaction rate the liberation of carbon monoxide, step 3, must involve the production of more than one, or n, active sites per carbon atom gasified. The statistical average n may be a value other than an integer.

Assuming that the law of mass action applies for the elementary reactions as written, then at the steady state:

$$r_4 = k_4 C_f = (n - 1)k_3 C(O)$$

or, the ratio of occupied to unoccupied sites would be,

$$\frac{C(O)}{C_f} = \frac{k_4}{(n-1)k_3}$$

Since $r_3 = r_1 + r_2$ and $r_1 = r_2$, or $r_3 = 2r_1$, we also obtain

$$\frac{C(O)}{C_f} = \frac{2k_1 p_{O_2}}{k_3}$$

The number density of total sites C_t is $C(O) + C_f$; this yields the expression for the fraction of total sites occupied by oxygen $\theta(O)$:

$$\theta(O) = \frac{C(O)}{C_t} = \frac{2k_1 p_{O_2}/k_3}{1 + 2k_1 p_{O_2}/k_3}$$

This is a Langmuir-type adsorption isotherm, which at low values of oxygen partial pressure, such that $k_1 p_{O_2}/k_3 \ll 1$, yields

$$\theta(O) = \frac{2k_1 p_{O_2}}{k_3} \simeq 0.$$

The reaction rate $r_3 = k_3C(O) = k_3C_t\theta(O) = 2k_1C_tp_{O_2}$ is of the first order with respect to oxygen, provided C_t does not vary with pressure. Reaction orders above or below unity are possible if C_t increases or decreases with pressure. At low pressures, $C_t \simeq C_f$ and the rate constant, the product of two factors, becomes $2k_1C_f$. Here the rate is controlled by rate of oxygen adsorption.

Since by equations 57 and 58, $r = FN = fx_{\text{active}}p_{O_2}/(2\pi mkT)^{1/2}$, the fundamental relation between the constant k_1 collision number, and collision yield for this mechanism is $2k_1(2\pi mkT)^{1/2}/f = x_{\text{active}}/C_f$. At sufficiently high oxygen partial pressures and suitable temperatures that $k_1p_{O_2}/k_3 \gg 1$, $\theta(O) = 1$, and the rate of carbon monoxide formation, $r_3 = k_3C(O) = k_3C_t$, approaches apparent zero order, provided C_t is pressure independent. Here the rate is controlled by oxygen complex desorption.

The type of mechanism proposed suggests that the carbon-oxygen reaction will be retarded by the surface oxide-complex if this complex is sufficiently stable to cover a large fraction of the active sites or reaction centers. Artificial removal of stable oxides through a reversible reaction with excess CO; that is:

$$C(O) + CO \rightleftharpoons CO_2 + C_f$$

has been suggested[57] as explaining the observation that the carbon-oxygen reaction rate was twice as fast with CO dilution as with nitrogen dilution at 450°C and 1.1 atm total pressure.[79]

Returning to the mechanism discussed earlier, we find that the measured "activation energy" for low or high pressures would correspond to the "exponential temperature coefficients" of the quantities k_1C_f in the one case, and $k_3C(O)$ in the other. Since k_1 and k_3 should have fundamentally different temperature characteristics, this finding emphasizes that the measured activation energy may be pressure dependent. Furthermore, the quantity C_f, being a concentration term, has no true temperature coefficient. Its value is time dependent and a function of temperature, related in a theoretically complex way with molecular collision frequency and rate of thermal annealing. If $[C_f]$ remains relatively constant or increases only slightly over the low-temperature range, but then decreases at higher temperatures, the empirical exponential temperature coefficient E_f of this quantity would vary from zero or small positive values to negative values. If we accept this, we may expect variations with temperature and with carbon type of the measured temperature coefficient $E = E_1 + E_f$. Similar deductions may be derived from other postulated heterogeneous reaction mechanisms.[80] The specific arguments given here are justifiable as possible explanations for the widely varying reactivities, reaction orders, and "activation energies" reported in the literature for oxidation of different types of carbons at kinetically controlled conditions.

Activation energies ranging from 46 to 58 kcal have been reported for three graphitized carbon blacks, and two graphite dust samples reacted in 0.1 atm of oxygen at 550 to 600°C.[81] The reaction was nearly first order with respect to oxygen partial pressure. Numerous other investigators have reported activation energies in the range 17 to 100 ± 30 kcal.[57] The reaction approximated first order, invariably, except at higher temperatures—around 1800°K—where it approached zero order. The reported activation energies as low as 17 kcal probably resulted from diffusional effects present in the experimental determinations. Some instances of the reaction with spectroscopic graphite were zero order at 1.5 mm Hg, 450 to 500°C, but progressed toward first order at 10 cm Hg at the same temperature.[58,82] Day,[58] in studies of combustion of carbon electrodes at atmospheric pressure, 1550 to 2200°K carbon surface temperatures, and high-streaming velocities of gases containing 37 to 99 mole % oxygen, found a first order reaction. This result occurred because at these conditions the external mass-transport resistance predominated because the temperature coefficient of the reaction was less than 8 kcal. Certain of the experimental results reported in the literature where high temperatures were achieved by passage of electric current through

[79] Paxton, R. R., Thesis, Massachusetts Institute of Technology, Cambridge, Massachusetts, 1949.

[80] Glasstone, S., Laidler, K. J., and Eyring, H., *The Theory of Rate Processes*, 339–99, McGraw-Hill, New York, 1941, 611 pp.

[81] Armington, A. F., Thesis, The Pennsylvania State University, University Park, Pennsylvania, 1960.

[82] Blyholder, G., and Eyring, H., *J. Phys. Chem.*, **61**, 682–8 (1957).

the carbon samples at low pressures may contain inherent errors owing to electric discharge or thermionic emission effects. Duval[69] showed that this effect begins at about 1800 to 1900°K, giving spurious increases in reaction rate and erroneous "apparent activation energies" in the range of 70 to 90 kcal.

Carbon-Oxygen-Steam Reaction

This reaction is of some practical interest in gasification systems employing oxygen-steam blast, because a small portion of the fuel bed which corresponds to the oxygen consumption distance, undergoes simultaneous combustion and gasification by steam. No fundamental chemical kinetic work has been reported in the literature. Because of the complex system which results from these simultaneous reactions, the reactions are studied separately. About the only practical information available on the carbon-oxygen-steam reaction is gas composition data taken at various distances through an operating fuel-bed. These data show CO_2 to be the major product, attaining a maximum just prior to essentially complete consumption of oxygen. Batchelder and associates[65] calculated this variation of gas composition with reaction time in the oxygen-steam gasification of pulverized coal: oxygen is essentially completely consumed at reaction times of the order of 10 msec,—CO_2 attains a maximum at this time—and significant concentrations of CO and H_2 already appear at 3 msec reaction time. A large fraction of the primary CO arising from this reaction, as well as portions of the CO and H_2 products of the simultaneous carbon-steam reaction, are oxidized by the counterdiffusing oxygen. These reactions are extremely rapid in the gas phase; CO combustion is accelerated in the presence of water vapor, and a fraction of the reaction probably occurs at the carbon surface or within molecular distances of it.

Under conditions of external diffusion-rate control, the carbon-oxygen reaction is decelerated in the presence of steam because oxygen partial pressure is lowered. However, the carbon-steam reaction approaches the rate of the carbon-oxygen reaction at combustion temperatures,[57] and produces gases very rapidly combustible in oxygen. The latter effect accelerates the oxygen consumption rate by decreasing the distance over which oxygen must diffuse before it reacts. Thus, despite lower partial pressure, the total oxygen-disappearance rate in the presence of steam should not be appreciably altered.

The chemical kinetics of the carbon-oxygen, carbon-steam, and the concomitant carbon-carbon dioxide reaction are similar in that the gaseous reactants, being oxidizers, deposit oxygen as chemically bound to the carbon surface. It is not certain, however, that the nature of the oxide complexes would be the same for each reaction. These three reactions are similar in that the reactant and one or more of the reaction products, particularly in the carbon-steam and carbon-carbon dioxide system as discussed subsequently, behave as a retardant or "poison." When all these reactions are combined, as in the carbon-oxygen-steam system, how the kinetics of one will be affected by the others is not clear. It seems certain, however, that retardation effects of one will interact with the others; in particular, the hydrogen retardation of the carbon-steam reaction (an experimental effect explained as chemisorption of hydrogen on the carbon lattice), may well decrease through removal of chemisorbed hydrogen by collision with oxygen molecules, an event which is characterized by a relatively high accomodation coefficient.

Carbon-Carbon Dioxide Reaction

The reaction of CO_2 and carbon can occur to a considerable extent immediately beyond the oxygen-consumption zone of autothermic reactors. This reaction always proceeds as a part of the carbon-steam reaction. Compared with the other gas reactions of carbon, the rate is relatively slow; only the reaction between carbon and hydrogen is slower. It has been estimated that at 800°C and 0.1 atm partial pressure the rate is about one-third as fast as the carbon-steam reaction, and about 10^{-5} as fast as carbon combustion.[57] At higher temperatures, these ratios will increase because the temperature coefficient of the carbon-oxygen reaction (about 50 kcal) is lower than that of either the carbon-carbon dioxide (60 to 90 kcal), or the carbon-steam reaction (about 70 to 90 kcal).

The carbon-carbon dioxide reaction has been extensively studied to gain some understanding of the possible mechanisms. Adsorption and degassing experiments show that an initially clean carbon surface reduces CO_2 at temperatures as low as 500 to 600°C. The oxygen removed from the dioxide becomes chemically bonded to the carbon surface, and when heated carbon oxides are liberated. Only a small fraction of the carbon surface is capable of reacting in this way, a phenomenon ascribed to existence of active centers. The carbon surface oxides can be removed by a stream of CO or H_2,[83] yielding CO_2 or water vapor. Thus, an indicated mechanism (designated as mechanism *A*) for $C + CO_2 \rightarrow 2CO$ is the reversible oxygen exchange surface reaction:

$$\text{Step 5.}\quad C_f + CO_2 \underset{k'_5}{\overset{k_5}{\rightleftharpoons}} C(O) + CO$$

and the unidirectional decomposition or gasification step:

$$\text{Step 3.}\quad C(O) \xrightarrow{k_3} CO + nC_f,$$

where the *k*'s are the elementary reaction rate constants, and nC_f denotes that an average of *n* active centers must be generated for each carbon atom gasified. Based on work on graphite filaments[70] at low pressures, the CO_2 reaction exhibits hysteresis effects with respect to temperature in the range from 1400 to 1900°K. The collision yield or rate constant increases approximately exponentially according to Arrhenius temperature dependence in the range from 1300 to about 1450°K; the yield attains a maximum at about 1550°K, and decreases to small asymptotic values at 1900°K. Such a behavior is quite similar to that discussed in the carbon-oxygen reaction; the ascribed explanation is that thermal annealing of active centers occurs, or

$$\text{Step 4.}\quad C_f \xrightarrow{k_4} C(\text{inactive})$$

Thus, Duval's work gives good evidence that the concentration of active centers is determined by competition between the processes of collision by reacting molecules and thermal annealing. The concentration decreases rapidly at temperatures above about 1600°K.

Although mechanism *A* (Steps 5 and 3) is not the only one hypothesized by various investigators, it has received considerable attention in the literature. A number of studies with a radioactive tracer $C^{14}O_2$ indicate oxygen transfer between CO_2 and CO in the presence of carbon (Step 5) occurs at measurable rates at temperatures below those required for gasification (Step 3). In experiments with a flow system employing $C^{14}O_2$ at low temperatures, 650 to 900°C the rate of transfer of radioactivity in the gas phase was found to be dependent on the sum of two first-order terms containing CO_2 and CO partial pressures,[84] or, rate of radioactivity transfer $= -k_5 p_{C^{14}O_2} + k'_5 p_{C^{14}O}$, which indicates that this oxygen exchange reaction is reversible. At these temperatures, where negligible carbon gasification occurred, substantially no radioactivity was transferred to the solid phase, which suggests negligible chemisorption, and that carbon initially in the gas phase maintains its identity as monoxide plus dioxide. The rapidity of the oxygen exchange reaction was dependent as much on the physical state of the carbon as on the gas composition and the presence of catalytic agents, activated charcoal and coke, with a 6% K_2CO_3 catalyst being the most reactive in this respect. The temperature coefficient for r_5 (forward) $= k_5\,[C_f]\,p_{CO_2}$ was found to be 43 kcal for high-temperature coke, and 58.9 kcal for graphite.

If charcoal samples in a circulating gas system with continuous monitoring of C^{14} activity are used, substantially no gas phase–solid-phase carbon-exchanges were found at 500°C, but at higher temperatures, 700 to 850°C some carbon interchange apparently occurred during the course of decomposition of carbon surface oxides which were formed in the initial step of the reaction.[85] It was found that (1) a reaction of the type given by the forward step 5 of mechanism *A* occurred very rapidly at 700 to 850°C on the charcoal, and (2) the subsequent slower decomposition of the surface oxides followed a rate law comprising two first-order terms, suggesting that at least two types of oxygenated charcoal sites, possibly associated with edges and faces of the crystallites, must have been participating in the reaction. The decomposition-rate temperature coefficient for $r_3 = k_3[C(O)]$ was 8 kcal for one type of surface oxide, and 21 kcal for the other type.

In another radioactive tracer study, where the C^{14} activity imparted by carbon dioxide

[83] Marsh, J. D. F., *Inst. Gas Engrs., Commun.* No. 393 (1951), pp. 167–86.

[84] Orning, A. A., and Sterling, E., *J. Phys. Chem.*, **58**, 1044–7 (1954).

[85] Bonner, F., and Turkevich, J., *J. Am. Chem. Soc.*, **73**, 561–4 (1951).

to the solid phase was measured,[86] it was shown that carbon-carbon transfer does not occur at or below 700°C with graphite and pure sugar-carbons. However, at about 900°C a small fraction of the CO_2 molecules, 1 to 2% for sugar-carbon, strike the surface in such a way that the carbon atoms enter the lattice, leaving the oxygen atoms to depart with two new carbon atoms derived from the surface.

Mechanism *A* is consistent with a Langmuir rate equation, which states that the rate of carbon gasification by CO_2, r_{CO_2}, is a quotient of terms involving CO_2 and CO partial pressures:

$$r_{CO_2} = \frac{kp_{CO_2}}{1 + ap_{CO} + bp_{CO_2}} \tag{59}$$

where k, a, and b are functions of one or more of the rate constants characterizing the elementary reactions; here, $k = k_5(C_t)$, $a = k'_5/k_3$ and $b = k_5/k_3$. There is general agreement,[59,87] that experimental rate data at low to near atmospheric pressures fit the form of equation 59 over at least 100 to 200°C temperature ranges, at CO_2 conversions substantially removed from thermodynamic equilibrium, and that the constants a and b decrease sharply with temperature. Values of k, a, and b expressed in terms of constants of the Arrhenius equation $Ae^{-E/RT}$ are summarized in Table 5 for several types of carbons reported in the literature.

Equation 59 states that the kinetics of the reaction will be first order with respect to CO_2 partial pressure if $ap_{CO} \ll 1$ and $bp_{CO_2} \ll 1$. These inequalities prevail at low CO_2 concentrations and low temperatures, at which small amounts of CO are formed. The inequalities are also satisfied at higher partial pressures and higher temperatures, above about 1800°K, where, the constants a and b generally decrease to small or negligible values. Zero-order kinetics is indicated if $ap_{CO} \ll 1$ and $bp_{CO_2} \gg 1$, which conditions are fulfilled at low temperatures where p_{CO} is small and at high CO_2 partial pressures. Under conditions where all terms in the denominator are significant, equation 59 is interpreted as stating that the order is unity with respect to CO_2, with rate hindrance by CO and CO_2, the measured order with respect to CO_2 being a fractional value which varies continuously according to percent CO_2 conversion. As expected, reported reaction orders[81,88] vary from zero to unity depending on temperature, pressure, gas composition, characteristics of the carbon, sample dimensions, and type of experimental system.

The work with recycling radioactive CO_2 indicated zero-order kinetics at or below 850°C,[85] because the rate was controlled by slow decomposition of surface oxides. Essentially first-order kinetics at CO_2 partial pressures 50 to 200 mm Hg, and temperatures of 770 to 900°C with ash-free sugar carbons were found.[89] Vulis and Vitman[88] indicated that the reaction in graphite tubes approaches unity order at 900 to 1060°C. Strickland-Constable[88,90] found mixed-order kinetics of the reaction on carbon filaments; zero-order at 700°C, approaching 0.7-order at 1100°C, and first-order at 1800°C. In work with graphitized and ungraphitized carbon filaments, exactly unity order was found at CO_2 partial pressures 2 to 11 × 10^{-3} mm Hg and 1400 to 1600°K.[70]

Because mechanism *A* is the most widely accepted explanation of the course of the CO_2 reaction below or near atmospheric pressure, some consideration must be given to the significance of the temperature coefficients of the terms comprising its rate equation. The measured activation energy E_k, corresponding to the rate constant k in equation 59, does not always need to refer to the same rate controlling step, since this step determines reaction order. At conditions $ap_{CO} = k'_5p_{CO}/k_3 \ll 1$ and $bp_{CO_2} = k_5p_{CO_2}/k_3 \ll 1$ where first-order kinetics is expected, the fraction coverage by surface oxides $\theta(O) = k_5p_{CO_2}/k_3 \cong 0$, $C_t \cong C_f$, and the rate of gasification becomes

[86] Brown, F., *Trans. Faraday Soc.*, **48,** 1005–14 (1952).

[87] Lewis, W. K., Gilliland, E. R., and McBride, G. T., Jr., *Ind. Eng. Chem.*, **41,** 1213–26 (1949). Wu, P. C., Thesis, Massachusetts Institute of Technology, Cambridge, Massachusetts, 1950.

[88] Strickland-Constable, R. F., *Proc. Roy. Soc. (London)*, **189A,** 1–10 (1947). Vulis, L. A., and Vitman, L. A., *J. Tech. Phys. (U.S.S.R.)*, **11,** 509–18 (1941). Duval, X., *J. chim. phys.*, **44,** 296–301 (1947). Key, A., *Gas Research Board, Commun.* No. 40, (1948), 36 pp.

[89] Semechkova, A. F., and Frank-Kamenetskiĭ, D. A., *Acta Physicochim., U.R.S.S.*, **12,** 879–98 (1940).

[90] Strickland-Constable, R. F., *Trans. Faraday Soc.*, **43,** 769–78 (1947).

TABLE 5

ARRHENIUS CONSTANTS FOR THE CARBON-CARBON DIOXIDE REACTION

Carbon Type	Temperature Range, °C	Constant k: A_k g-mole/g-min-atm	Constant k: E_k kcal	Constant a: A_a 1/atm	Constant a: E_a kcal	Constant b: A_b 1/atm	Constant b: E_b kcal
Coconut-shell charcoal[59]	734–830	6.3×10^{8}	58.8	1.26×10^{-8}	−45.5	3.16×10^{6}	30.1
New England coke[87]	800–1090	6.9×10^{5}*	47.6*	1.4×10^{-2}*	−15.0*	0.21*	−6.3*
New England coke[87]	—	3.16×10^{7}	61.7	4.0×10^{-6}	−40.3	3.16×10^{-2}	−6.1
Electrode carbon[87]	—	1.0×10^{6}	50.1	3.16×10^{-9}	−60.6	0.16	−6.6
Pitch coke[6]	926–1150	1.05×10^{7}	40.1	2.0×10^{-9}	−55.1	—	—
Anthracite[87]	800–1090	2.2×10^{3}*	32.5*	4.6×10^{-3}*	−16.9*	1.8×10^{-3}*	−16.6*

* Computed from author's graphical results.

$r_{CO_2} = r_3 = k_5(C_f)p_{CO_2}$, indicating $E_k = E_5 + E_f$. At zero-order kinetics, that is, $ap_{CO} = k'_5p_{CO}/k_3 \ll 1$ and $bp_{CO_2} = k_5p_{CO_2}/k_3 \gg 1$, we find $\theta(O) \simeq 1.0$ and $r_{CO_2} = k_3\,C(O) \simeq k_3(C_t)$, indicating $E_k = E_3 + E_t$. Thus E_k is dependent on reaction order or the rate-determining step. If E_k is evaluated from temperature dependence of k in the complete rate equation, it corresponds to $E_k = E_5 + E_t$.

Table 5 shows E_k values ranging from 32.5 to 61.7 kcal. If C_t does not change appreciably over the temperature ranges indicated, that is, $E_t \simeq 0$, these E_k values may be considered as the activation energy of the forward step of the reversible oxygen exchange reaction, step 5, mechanism A. The physical meaning of the other activation energy terms, E_a and E_b, in Table 5 is interpreted on the basis of sums and differences between the activation energies of the elementary reactions: $E_a = E'_5 - E_3$, $E_b = E_5 - E_3$, and letting $E_5 = E_k$, we have $E_3 = E_k - E_b$, and $E'_5 = E_a + E_k - E_b$. Except in one case, the data in Table 5 fulfill the conditions $E_k > 0$, $E_a < 0$, $E_b < 0$, and $E_a - E_b = E'_5 - E_5 < 0$. The computed results in Table 6 include determinations of several other investigators, and indicate E_3 ranges from 28.7 to 93 kcal, and E_5 (only positive values have significance) from 27.5 to 38.9 kcal. The difference $E_5 - E'_5$ is in all cases positive and indicates, as expected, that it is energetically more difficult to deposit oxygen from CO_2 upon the carbon surface than to remove it by reaction with CO. Granted that small experimental errors and the presence of diffusional effects can lead to wide variations in measured activation energy, and that these errors may be cumulative in calculation of

TABLE 6

TEMPERATURE COEFFICIENTS OF ELEMENTARY REACTIONS OF MECHANISM A

Carbon type	$E_5 \simeq E_k$ kcal	E'_5 kcal	$\Delta H_5 = E_5 - E'_5$ kcal	E_3 kcal
Coconut-shell charcoal[59]	58.8	−16.8	75.6	28.7
New England coke[87]	47.6	38.9	8.7	53.9
New England coke[87]	61.7	27.5	34.2	67.8
Electrode carbon[87]	50.1	−3.9	54.0	56.7
Anthracite[87]	32.5	32.2	0.3	49.1
High-temperature coke[84]	43	—	—	—
Graphite[84]	58.9	—	—	—
Three different carbons[59]	—	—	23	59
Charcoal[85]	—	—	—	8 & 21
Electrode carbon, activated charcoal[91]	—	—	—	86
Graphitized carbon-black and graphite wear dust[81]	—	—	—	73–93

[91] Wicke, E., *Symposium on Combustion, 5th Symposium, Pittsburgh*, 1954, Reinhold, New York, 1955, pp. 245–52.

E'_5 and E_3, the data shown here give little support to independence of temperature coefficients of elementary rate constants on the characteristics of the carbon. However, it has been found[92] that $E_3 = 86$ kcal is the same for a high-purity electrode carbon and a medium-purity activated charcoal.[91] Ergun[59] also gave positive evidence in this respect, in that $E_3 = 59$ kcal (27 to 28 kcal less than Wicke[91]), and $E_5 - E'_5 = 23$ kcal was the same for three types of carbon of considerably different physical characteristics—activated carbon, activated graphite, and Ceylon graphite. The constancy of E_3 was found to be possible only under two conditions: (1) that the number of active sites is independent of temperature, or (2) that the number of sites varies exponentially with temperature with the same Arrhenius coefficient. In view of the results of Duval already discussed, it would appear that condition 2 is the more plausible.

Mechanism *A* states that the experimentally determined rate hindrance of CO (the term ap_{CO} in equation 59) is the result of the removal of a fraction of the surface oxide by the reverse reaction of step 5. An alternate explanation of the hindrance is blockage of active sites through chemisorption of CO, in this way indirectly reducing the fractional coverage by gasifiable surface oxides. In this view, an alternate mechanism, designated "mechanism *B*," comprises unidirectional oxygen exchange at the surface:

$$\text{Step 5. forward.}\quad C_f + CO_2 \xrightarrow{k_5} C(O) + CO$$

reversible chemisorption upon free sites.

$$\text{Step 6.}\quad C_f + CO \underset{k'_6}{\overset{k_6}{\rightleftharpoons}} CO \text{ adsorbed}$$

and unidirectional gasification of surface oxides,

$$\text{Step 3.}\quad C(O) \xrightarrow{k_3} CO + nC_f$$

in the presence of thermal annealing,

$$\text{Step 4.}\quad C_f \xrightarrow{k_4} C\ (\text{inactive})$$

where the condition of steady state in steps 5 forward and 3 with equilibrium in step 6 yields rate equation 59, the elementary rate constants being related through $k = k_5[C_t]$, $a = k_6/k'_6$, and $b = k_5/k_3$. Reif[59] in considering various mechanisms, showed that only one type of reaction can control rate hindrance at one time. Hindrance, for example, by a combination of steps 5 reverse and 6 forward, yields an unacceptable rate equation containing an additional denominator term in $p^2_{CO_2}$, contrary to experimental findings.

Gadsby and associates[59] supported mechanism *B* on the basis of charcoal chemisorption experiments at 750°C and about 300 to 400 mm Hg total pressure, which indicated that wide variation in CO_2 partial pressures corresponded more closely to variations in CO partial pressure than to changes in amount of oxygen complex absorbed. In addition, the quantity of oxygen complex remained relatively constant with variation in CO partial pressure at constant CO_2 partial pressure. This was taken as evidence that rate inhibition must have been by CO chemisorption, not by the reverse oxygen exchange of step 5, mechanism *A*. Assuming that mechanism *A* is valid, these authors computed $E'_5 = -16.8$ kcal (Table 6, coconut-shell charcoal), which cannot be correct because of the negative sign.

Finally, two of these authors[93] found that CO and hydrogen both inhibit the C—CO_2 reaction, whereas only hydrogen inhibits the carbon-steam reaction. They concluded that both reactions are, therefore, inhibited via the mechanism of chemisorption, reasoning as follows: (1) The carbon-steam reaction may take place on edge carbon atoms having only one unshared electron, and consequently, this reaction cannot be inhibited by CO since this gas would chemisorb with surface atoms containing two unshared electrons. (2) The C—CO_2 reaction may occur on sites containing two unshared electrons, and since CO and hydrogen are capable of attaching to dual-electron sites, both gases can inhibit this reaction.

Blackwood and Ingeme,[94] however, found no retarding effect due to hydrogen with the same type of carbon at 10 to 20 atm pressure levels. This anomaly may be due to the difference in pressure levels employed in these two investigations. Reif,[59] on the other hand, supports mechanism *A* on the basis of the disagreement shown between his value of

[92] Rossberg, M., *Z. Elektrochem.*, **60,** 952–6 (1956).

[93] Long, F. J., and Sykes, K. W., *Proc. Roy. Soc. (London)*, **A193,** 377–99 (1948).

[94] Blackwood, J. D., and Ingeme, A. J., *Australian J. Chem.*, **13,** 194–209 (1960).

$E_6 - E'_6 = 8$ kcal, derived from high-temperature-coke chemisorption studies and the values of E_a of other investigators (large negative in Table 5) to which it should correspond if mechanism B were correct. He pointed out that Gadsby and associates[59] misinterpreted their own chemisorption results in favor of mechanism B, stating further that Gadsby's $E'_5 = -16.8$ is not in accord with the positive values (Table 6) of other studies which support the thermodyanmic feasibility of step 1 reverse, mechanism A.

Reif[59] concluded that the chemisorption reactions between carbon and CO and CO_2 are too small in magnitude and slow in rate at 700 to 1000°C to account for the observed inhibition, but the possibility of having two inhibiting mechanisms operative at different relative rates for different carbon types at different pressure-temperature conditions is not necessarily precluded.

Ergun[59] has strongly favored mechanism A. His treatment of fluidized-bed rate data for the CO_2 reaction with three different carbons at 700 to 1400°C and atmospheric pressure required the plausible assumptions $k'_5 p_{CO} \gg k_3$ and $k_5 p_{CO_2} \gg k_3$, yielding the simplified form of Equation 59:

$$r_{CO_2} = \frac{k_3 C_t}{1 + p_{CO}/K_5 p_{CO_2}} \qquad (59a)$$

where $K_5 = k_5/k'_5 = [C(O)]p_{CO}/[C_f]p_{CO_2}$, the equilibrium constant of the oxygen-exchange surface-reaction, step 1, mechanism A. A mathematical integration of equation $59a$ and plotting the data gave K_5 values (dimensionless) independent of carbon type, and k_3C_t values (sites/g-sec) which exhibited Arrhenius dependence at about 59 kcal temperature coefficient, as already discussed:

K_5 (800 to 1400°C) $= 4.15 \times 10^3 e^{-22,700/RT}$

k_3C_t (Ceylon graphite, 1000 to 1400°C)

$= 1.30 \times 10^{28} e^{-59,300/RT}$

k_3C_t (activated graphite, 900 to 1150°C)

$6.9 \times 10^{28} e^{-59,000/RT}$

k_3C_t (activated carbon, 700 to 1025°C)

$= 71.5 \times 10^{28} e^{-57,800/RT}$

Here the pre-exponential factors of the Arrhenius forms representing $k_3[C_t]$ are interpreted as being proportional to the reactivities of the particular carbons, whereas if k_3 were fundamentally the same for all carbons, the concentration of active sites C_t would be an absolute measure of reactivity.

Because of the high-temperature coefficient of K_5, about 23 kcal, the overall rate of gasification is seen to be strongly affected by the equilibrium of the oxygen-exchange surface reaction. Since the gasification rate $r_{CO_2} = r_3 = k_3\,[C(O)] = k_3\theta(O)C_t$, this influence on the rate is through the effect of K_5 on the fraction of occupied sites θ_0, or

$$\theta_0 = [CO)]/[C_t] = \frac{1}{1 + p_{CO}/K_5 p_{CO_2}}$$

where with $p_{CO}/p_{CO_2} = 1.0$, for example, θ_0 varies 0.0215 to 0.81 in going from 700 to 1400°C. In passing, if Ergun's treatment were applied to mechanism B, we are forced to reconcile his value $\Delta H_1 = E_5 - E'_5 = 23$ kcal with the quantity $E_5 - E_3 - (E_6 - E'_6) = E_b - \Delta H_6 = -7$ plus -8 (Table 5 and Reif[59]) or -15 kcal, to which it would correspond if mechanism B were correct.

There is meager experimental evidence concerning the behavior of the CO_2 reaction at pressures substantially above atmospheric. According to recent studies[94] the rate of charcoal reaction with CO_2 at 790 to 870°C increases more rapidly with CO_2 partial pressure, up to 40 atm, than is predicted by equation 59. This seems to indicate that CO has a less retarding effect at high pressures, and/or the surface coverage by adsorbed oxygen is increased. Based on an analogy with the carbon-steam system at 1 to 50 atm,[62] a pressure mechanism involving oxygen deposition and CO adsorption is suggested which accounts for reaction acceleration via step 7, and reversibility via step 8, of the following sequence:

Step 5. $C_f + CO_2 \xrightarrow{k_5} C(O) + CO$

Step 3. $C(O) \xrightarrow{k_3} CO + nC_f$

Step 6. $C_f + CO \underset{k'_6}{\overset{k_6}{\rightleftharpoons}} [CO]$ adsorbed

Step 7. $C_f + [CO] + CO_2 \xrightarrow{k_7} 2CO + C(O)$

Step 8. $[CO] + CO \xrightarrow{k_8} CO_2 + C$

This leads to the rate equation:

$$r_{CO_2} = \frac{kp_{CO_2} + cp_{CO}p_{CO_2} + dp^2_{CO_2} - ep^2_{CO}}{1 + ap_{CO} + bp_{CO_2} + fp_{CO}p_{CO_2} + gp^2_{CO_2}}, \qquad (60)$$

where $k = k_5C_t$, $a = k_6/k'_6 + k_8/k'_6$,
$b = k_5/k_3 + k_7/k'_6$,
$c = (k_6k_7/k'_6 + k_5k_8/k'_6)C_t$,
$d = k_5k_7C_t/k'_6$, $e = k_6k_8C_t/k'_6$,
and $f = k_6/k'_6 + k_8/k'_6$, $g = k_5k_8/k_3k'_6$

A comparison between computed and observed CO_2 reaction rates with charcoal at 790 to 870°C, 4.5 to 36 atm CO_2 partial pressure, is given in Fig. 14, based on the terms $cp_{CO}p_{CO_2}$, $fp_{CO}p_{CO_2}$, and $gp^2_{CO_2}$ in equation 60 being negligible, and on the following experimental constants:

Temp, °C	k g-mole/g-min-atm	a 1/atm	b 1/atm	d g-mole/g-min-atm
790	0.52×10^{-5}	13.7	0.23	0.004
830	2.3×10^{-5}	17.1	0.17	0.02
870	6.5×10^{-5}	12.7	0.31	0.07

As a final consideration of the CO_2 reaction mechanism, it must be pointed out that although equation 59 is an acceptable rate expression for moderate CO_2 conversions, it does not permit a zero rate at thermodynamic equilibrium. Key[88] and Strickland-Constable[95] favor mechanism *A* and provide for this condition by introducing step 3 as a reversible reaction:

$$C(O) \underset{k'_3}{\overset{k_3}{\rightleftharpoons}} CO + nC_f$$

and, as a consequence of equilibrium, Duval's thermal annealing, step 4, must also be reversible:

$$C_f \underset{k'_4}{\overset{k_4}{\rightleftharpoons}} C \text{ (inactive)}.$$

The more complex rate expression then becomes

$$r_{CO_2} = \frac{[(k_5p_{CO_2} - k'_5k'_3p_{CO})/k_3]C_t}{1 + [(k_5 + k'_3)/k_3]p_{CO} + (k_5/k_3)p_{CO_2}}, \quad (59b)$$

where k'_3 is usually small in accordance with the large values of the equilibrium constant $K = k_5k_3/k'_5k'_3$ at gasification temperatures.

Carbon-Steam Reaction

The kinetics of the carbon-steam reaction is in many respects analogous to the CO_2 reaction. A unified theory of the course of gas reactions of carbon, where the reactants are oxidizing, holds that their mechanisms or elementary steps should be similar. However, the C-CO_2 system comprises only two constituents—carbon and oxygen—allowing less uncertainty about the choice of its limited number of possible mechanisms. The carbon-steam system, on the other hand, poses the added complexity contributed through possible roles of hydrogen and/or other intermediate dissociation products of water vapor—factors which are not elucidated without ambiguity at this writing. Furthermore, the CO_2 reaction with carbon is occurring simultaneously, but usually only to a small extent—sometimes negligible—in the carbon-steam system.

There seems to be general accord that the primary products of the steam reaction with relatively pure carbons at atmospheric or lower pressures are CO and hydrogen,[10,60] whereas CO_2 is a secondary product arising through the water-gas shift reaction which is known to be "catalyzed" by carbon surfaces.[59,96] As with CO_2, water vapor deposits oxygen upon a carbon surface at temperatures below those for gasification, and this oxygen exchange is reversible through addition of CO or hydrogen. Surface oxides decompose at higher temperatures, liberating carbon into the gas phase. Key[88] and Strickland-Constable[95] considered a mechanism consistent with these facts comprising three elementary steps, two of which are the same as for the CO_2 reaction. Designating this as carbon-steam mechanism *A*, in which previous steps given in the C-CO_2 reaction retain the same nomenclature, we have: reversible oxygen exchange between active sites and steam

Step 9. $C_f + H_2O \underset{k'_9}{\overset{k_9}{\rightleftharpoons}} C(O) + H_2$,

decomposition of surface oxides

Step 3. $C(O) \underset{k'_3}{\overset{k_3}{\rightleftharpoons}} CO + nC_f$

[95] Strickland-Constable, R. F., *J. chim. phys.*, **47**, 356–60 (1950).

[96] Behrens, H., *Z. physik. Chem.*, **195**, 24–36 (1950). Dolch, P., *Brennstoff-Chem.*, **14**, 261–3 (1933).

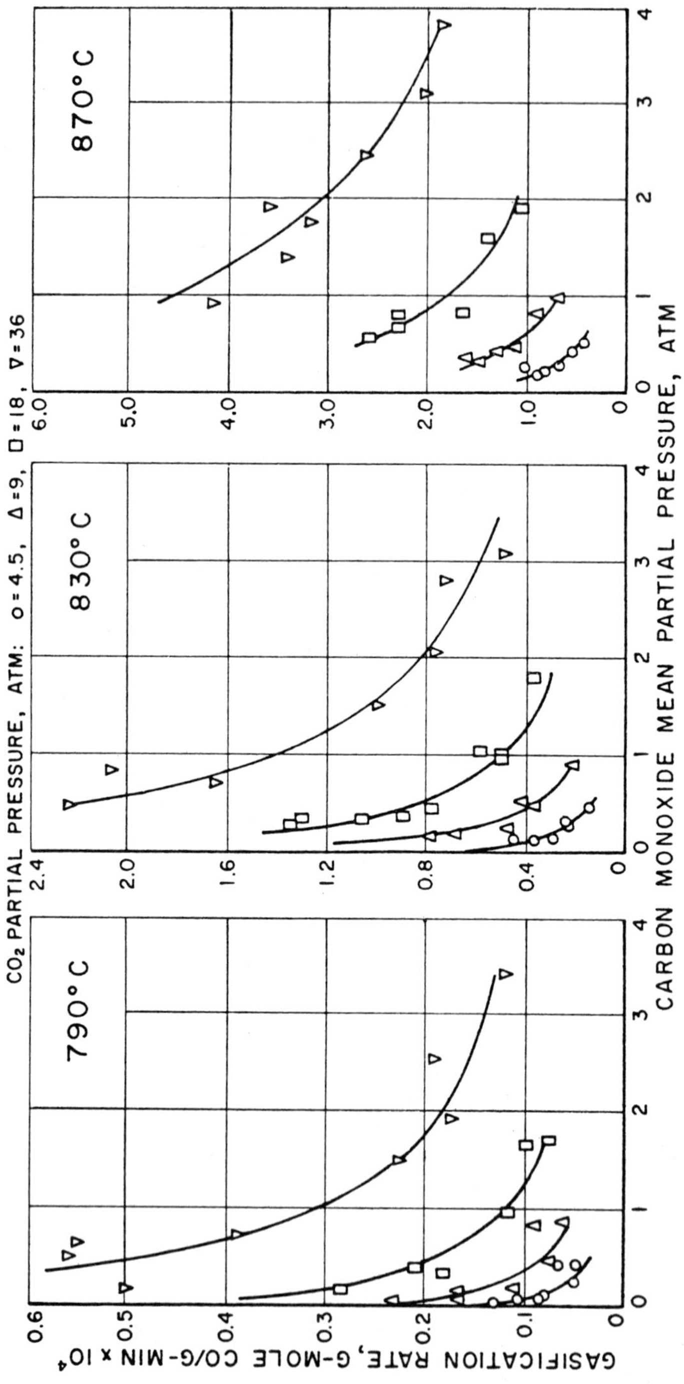

Fig. 14. Comparison of computed and observed rates of gasification by carbon dioxide as a function of the partial pressure of carbon monoxide.[94]

and reversible oxygen exchange between surface oxides and CO

Step 5. $C(O) + CO \underset{k_5}{\overset{k'_5}{\rightleftharpoons}} CO_2 + C_f$

Steps 5 and 9 provide for CO_2 formation by combining as the water-gas shift reaction, although it is recognized that steam and CO will react at other than carbon surfaces or in the gas phase,[97] yielding additional CO_2. The carbon-steam reaction also exhibits hysteresis effects with respect to pressure at vacuum conditions, and with respect to temperature in the range 1450 to 1900°K.[70]

The collision yield, or rate constant, increases in accordance with Arrhenius temperature dependence at 1350 to about 1550°K, the yield at 3×10^{-3} mm Hg partial pressure attaining a maximum at about 1650°K and decreasing rapidly above 1700°K to some asymptotic value at above 1900°K. This, again, is indicative of thermal annealing of active centers, the annealing process with graphitized and ungraphitized carbon filaments accelerating in rapidity at above 1700°K:

Step 4. $C_f \xrightarrow{k_4} C$ (inactive)

Duval's pressure-hysteresis effect, which was observed at 1700 to 1800°K between 0.29 to 4×10^{-3} mm Hg, may be described as follows: instantaneous increase of water-vapor partial-pressure gives an initial rate lower than the stationary rate at the new pressure; conversely, decrease of pressure yields an initial rate higher than the new stationary rate. This pressure phenomenon was rationalized on the basis of the decrease in average collision time with increasing pressure, and the smaller amount of thermal annealing occurring at these decreased collision times. The pressure-hysteresis effect decreased with pressure level, and tended to disappear or become undetectable at pressures approaching atmospheric, at which molecular collision frequencies are extremely high.

Experimental rate data[10,60,88,93] on steam gasification of carbon are in general accord with a Langmuir-type equation over 100 to 200°C ranges at 700 to 1100°C temperature levels and at steam decomposition substantially removed from thermodynamic equilibrium of

$C + H_2O \rightarrow CO + H_2$:

$$r_c = \frac{k p_{H_2O}}{1 + a p_{H_2} + b p_{H_2O}} \qquad (61)$$

Here the rate of carbon gasification r_c is equal to the rate of steam decomposition r_{H_2O} via carbon-steam reaction only; that is, exclusive of water-gas shift. Equation 61 states that the rate process is slowed, or inhibited, by hydrogen and water vapor. The constants k, a, and b are functions of one or more of the rate constants characterizing the elementary reactions, with a and b generally decreasing with temperature. Some values of these constants for particular carbons at specified conditions are summarized in Table 7. For porous graphite tubes at 862 to 938°C, the constants k, a, and b tend to increase with carbon burnoff in the zero to 7.5% range investigated.[60] In terms of the Arrhenius constants A and E, this trend is illustrated in Table 8.

Goring and associates,[62] in a study of the effect of pretreatment on the reactivity of low-temperature (Disco) char to steam and CO_2 in

[97] Campbell, W. M., and Johnstone, H. F., *Ind. Eng. Chem.*, **44**, 1570–5 (1952). Graven, W. M., and Long, F. J., *J. Am. Chem. Soc.*, **76**, 2602–7 (1954). Bortolini, P., *Chem. Eng. Sci.*, **9**, 135–44 (1958).

TABLE 7

RATE CONSTANTS FOR THE CARBON-STEAM REACTION

Carbon Type	Temp °C	Constant, k $\frac{\text{g-mole}}{\text{g-min-atm}}$	Constant, a 1/atm	Constant, b 1/atm
Coconut-shell charcoal[10]	700	3.6×10^{-4}	40	1.0
Coal charcoal[10]	800	9.0×10^{-4}	150	2.3
Pitch coke[6]	982	0.37×10^{-4}	10.0	—
Pitch coke[6]	1038	0.68×10^{-4}	4.25	—
Pitch coke[6]	1093	4.0×10^{-4}	3.28	—
Pitch coke[6]	1149	6.1×10^{-4}	0	—

TABLE 8

ARRHENIUS CONSTANTS FOR THE CARBON-STEAM REACTION

Carbon Type and % Burnoff	Temperature Range, °C	Constant k: A_k $\frac{\text{g-mole}}{\text{g-min-atm}}$	Constant k: E_k kcal	Constant a: A_a 1/atm	Constant a: E_a kcal	Constant b: A_b 1/atm	Constant b: E_b kcal
Graphite tube, 0%[60]	862–938	29.6	32.7	9.42×10^{-11}	−60.8	7.07×10^{-16}	−79.3
Graphite tube, 1%[60]	862–938	2.24	26.0	6.48×10^{-11}	−62.0	6.19×10^{-16}	−79.7
Graphite tube, 2%[60]	862–938	0.357	21.2	4.21×10^{-11}	−63.1	4.80×10^{-16}	−80.2
Graphite tube, 5%[60]	862–938	0.0615	16.0	1.00×10^{-11}	−66.7	2.13×10^{-16}	−82.5
Graphite tube, 7.5%[60]	862–938	0.0235	13.1	0.377×10^{-11}	−69.2	1.96×10^{-16}	−82.9
Coconut-shell charcoal[93]	684–770	1.58×10^{10}	62.3	33	0.0	3.16×10^{4}	20.1
Graphite[98]	850–1000	—	49.7	—	—	—	—
Graphite[98]	1000–1160	—	35.1	—	—	—	—
Coke[99]	900–1000	—	26.8	—	—	—	—
Electrode graphite[100]	900–1300	—	35.6	—	−75	—	—
Graphitized coke[101]	1000–1100	—	40.7	—	—	—	—
Pitch coke[103]	980–1150	—	45.0	—	—	—	—
Pitch coke[6]	980–1150	7.75×10^{5}*	59.1*	1.71×10^{-3}	−20.5	—	—
Three coal cokes[102]	800–950	—	40	—	—	—	—

* Revised values based on recalculation of data.

a fluidized bed, verified that reaction rates at 1600°F may either decrease or increase with 0 to 28% carbon burnoff, depending on short or long pretreatment times. Data of Jolley and Pohl[102] (Table 9), in terms of specific rates, gram carbon/gram-hr, indicate acceleration of gasification with carbon burnoff for coal carbonized at 850 and 1000°C.

TABLE 9

SPECIFIC GASIFICATION RATES AT 850°C OF CARBONIZED BADDESLEY COAL[102]

Carbon Burnoff, %	Carbonization Temperature, C°: 850	Carbonization Temperature, C°: 1000
	Rate, g C/g C-hr	
0	0.246	0.085
10	0.273	0.109
20	0.310	0.133
30	0.355	0.156
40	0.400	0.180
50	0.450	0.203
60	0.520	0.225

[98] Mayers, M. A., *J. Am. Chem. Soc.*, **56**, 70–6 (1934).

[99] Fleer, A. W., and White, A. H., *Ind. Eng. Chem.*, **28**, 1301–9 (1936).

[100] Blyholder, G., and Eyring, H., *J. Phys. Chem.*, **63**, 693–6 (1959).

[101] Pilcher, J. M., Walker, P. L., Jr., and Wright, C. C., *Ind. Eng. Chem.*, **47**, 1742–9 (1955).

[102] Jolley, L. J., and Pohl, A., *J. Inst. Fuel*, **26**, 33–44 (1953).

[103] Hunt, B. E., Mori, S., Katz, S., and Peck, R. E., *Ind. Eng. Chem.*, **45**, 677–80 (1953).

Interpretation of the order of the carbon-steam reaction based on equation 61 follows the same reasoning as was applied for the C-CO_2 reaction, equation 59. Orders ranging from zero to unity are expected, depending on pressure-temperature conditions, extent of

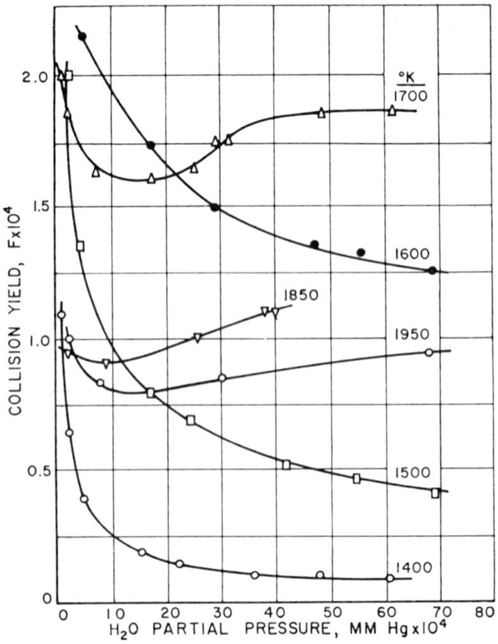

Fig. 15. Collision yield of the water vapor reaction on carbon filaments at low pressures.[70]

steam decomposition, and other factors. Because the constants a and b decrease approximately exponentially with temperature, the reaction order approaches unity at sufficiently high temperatures. This has been shown to be the case at 1370°C and above.[65] Many other investigators have reported orders between zero and unity.[6,76,88,93,101,103,104] The results of Duval and associates,[70] Fig. 15, are interpreted as indicating an approach to unity order at water vapor partial pressures above 7×10^{-3} mm Hg at all temperatures investigated, 1400 to 1950°K.

A comparison between the rate equations derivable from mechanism A with the experimental equation 61 shows that at steady state, the rate of total steam decomposition, including the water-gas shift contribution is:

$$r_{H_2O}\text{ (total)} = \frac{\left(k_9 p_{H_2O} - \frac{k'_3 k'_9}{k_3} p_{CO} p_{H_2} + \frac{k'_5 k_9}{k_3} p_{CO} p_{H_2O} - \frac{k_5 k'_9}{k_3} p_{H_2} p_{CO_2}\right) C_t}{1 + \frac{k'_9}{k_3} p_{H_2} + \frac{k_9}{k_3} p_{H_2O} + \left(\frac{k'_5 + k'_3}{k_3}\right) p_{CO} + \frac{k_5}{k_3} p_{CO_2}} \tag{62}$$

which incorporates the thermodynamic equilibrium conditions, K (carbon-steam) $= k_3k_9/k'_3k'_9$, and K (water-gas shift) $= k'_5k_9/k_5k'_9$.

Equation 62 gives a unified picture of the two modes of steam decomposition. The first two terms in the numerator refer to the carbon-steam reaction: the latter two terms refer to the water-gas shift reaction (however, this relation[95] has not been experimentally verified). Usually k'_3 is negligible because of the back reaction of step 3 being considered small at gasification conditions; the second numerator term then drops out. At small-steam decompositions, the denominator terms in p_{CO} and p_{CO_2} are comparatively small. The form of the remaining terms has been verified experimentally[10,105] to the extent that these investigators found the rate of the forward and reverse water-gas shift reaction to be proportional to the products $p_{CO}p_{H_2O}$ and $p_{H_2}p_{CO_2}$, with rate hindrance attributed to a combination of p_{H_2O}, p_{H_2}, p_{CO}, and p_{CO_2}.

Specifically, it has been stated that the reaction $CO + H_2O \rightleftharpoons CO_2 + H_2$ takes place predominantly on the carbon surface.[10] The forward reaction is nearly first order with respect to CO and fractional with respect to H_2O, hindered by H_2 and to a lesser extent by CO_2; the reverse reaction is of a fractional order with respect to both CO_2 and H_2, hindered by H_2O and unaffected by CO. The rate equation.

$$r = \frac{k[(p_{H_2}p_{CO_2} - p_{CO}p_{H_2O})/K]}{1 + K_A p_{CO_2} + K_R p_{CO}} \tag{63}$$

was shown to be applicable for the catalytic reverse water-gas shift.[105]

Mechanism A also yields the steady-state rate of carbon gasification—equivalent to steam-decomposition rate via $C + H_2O \rightarrow CO + H_2$. If none of the elementary steps are assumed in equilibrium, we obtain:

$$r_c = r_3 = k_3[C(O)] - k'_3(C_f)p_{CO}$$

$$= \frac{\left(k_9 p_{H_2O} - \frac{k'_3 k_9}{k_3} p_{CO} p_{H_2} + k_5 p_{CO_2} - \frac{k'_5 k'_3}{k_3} p_{CO}^2\right) C_t}{1 + \frac{k'_9}{k_3} p_{H_2} + \frac{k_9}{k_3} p_{H_2O} + \frac{k'_5 + k'_3}{k_3} p_{CO} + \frac{k_5}{k_3} p_{CO_2}} \tag{64}$$

where the thermodynamic equilibrium constants, K (carbon-steam) $= k_3k_9/k'_3k'_9$, and $K(C\text{-}CO_2) = k_5k_3/k'_5k'_3$. Equation 64 is simplified by neglecting the C—CO_2 reaction (third and fourth numerator terms), and by removing the second numerator term, since usually $k'_3 \simeq 0$; however, with this simplication, it is still not in accord with equation 61 because carbon-steam reaction hindrance by CO and CO_2 has no basis in fact. It was found the addition of CO does not hinder the steam reaction other than its resulting in the production of more hydrogen,[10] which does retard the reaction. Mechanism A would not be contradictory with respect to CO hindrance, and the form of equation 61 would be obtained directly from it if it were assumed that $k'_3 \simeq 0$, and that the oxygen exchange between CO and

[104] Key, A., and Cobb, J. W., *J. Soc. Chem. Ind.*, **49**, 439–44T (1930). Warner, B. R., *J. Am. Chem. Soc.*, **65**, 1447–51 (1943); **66**, 1306–9 (1944).

[105] Barkley, L. W., Corrigan, T. E., Wainwright, H. W., and Sands, A. E., *Ind. Eng. Chem.*, **44**, 1066–71 (1952).

surface oxides (step 5) was reversibly rapid to be always in thermodynamic equilibrium. Then, in terms of elementary rate constants, $k = k_9(C_t)$, $a = k'_9/k_3$, and $b = k_9/k_3$.

Mechanism *A* defines the experimentally observed rate hindrance of hydrogen as arising from surface oxide removal through the back reaction of step 9. Rate hindrance by water vapor follows directly from the algebra of this mechanism. However, there is evidence[93,99,106] that the hindrance may result from chemisorption at gasification temperatures. The alternate, mechanism *B*, involving water vapor and hydrogen sorption upon active sites has been proposed,[10,60]

Step 10. $C_f + H_2O \underset{k'_{10}}{\overset{k_{10}}{\rightleftharpoons}} C(H_2O),$

Step 11. $C(H_2O) \xrightarrow{k_{11}} CO + H_2 + nC_f$

Step 12. $C_f + H_2 \underset{k'_{12}}{\overset{k_{12}}{\rightleftharpoons}} C(H_2)$

where, in view of the concept of thermal annealing, the previous step 4 is added. This yields equation 61 directly, with $k = k_{10}k_{11}[C_t]/(k'_{10} + k_{11})$, $a = k_{12}/k'_{12}$ and $b = k_{10}/(k'_{10} + k_{11})$. If the evaporation rate, step 10 reverse, is negligible or $k'_{10} = 0$, then $k = k_{10}[C_t]$ and $b = k_{10}/k_{11}$.

Mechanism *B* can be written more in accord with the concept of oxygen deposition by steam; substituting for step 11 the previous steps 9, and 3 forward, of mechanism *A*. Long and Sykes[93] have also discussed in more detail steps 10 and 11 of mechanism *B*. They proposed steam dissociation at the carbon surface into hydroxyl radicals and hydrogen atoms, both rapidly chemisorbed at adjacent active sites. This is followed by a surface reaction involving hydrogen interchange. In detail, their mechanism is:

Step 13.

$$C_f \text{ (dual site)} + H_2O \underset{k'_{13}}{\overset{k_{13}}{\rightleftharpoons}} C(H)(OH)$$

Step 14. $C(H)(OH) \xrightarrow{k_{14}} C(O)(H_2)$

Step 3. $C(O) \xrightarrow{k_3} CO + nC_f$

Step 12. $C_f + H_2 \underset{k'_{12}}{\overset{k_{12}}{\rightleftharpoons}} C(H_2)$

[106] Muller, S., and Cobb, J. W., *J. Chem. Soc.*, **1940**, 177–83. Sherman, A., and Eyring, H., *J. Am. Chem. Soc.*, **54**, 2661–75 (1932). Barrer, R. M., *Proc. Roy. Soc.* (*London*), **A149**, 253–69 (1935).

which yields equation 61, where $k = k_{13}k_{14}[C_t]/(k'_{13} + k_{14})$, $a = k_{12}/k'_{12}$, and $b = (1/k_3 + 1/k'_{12} + 1/k_{14})[k_{13}k_{14}/(k'_{13} + k_{14})]$.

Blackwood and McTaggart[107] have studied reactions of carbons with atomic and radical species H, O, and OH produced in high-frequency electromagnetic fields, and supported step 13 of Long and Sykes' mechanism. They concluded, however, that hydrogen rate-hindrance is due to its causing a reduction in concentration of atomic species at the surface, rather than to the occupation of active sites by the atomic or molecular form of this gas. Their results indicate that these atomic species can react completely at the carbon surface, but the reaction rate is not influenced by the nature or number of oxygen-containing groups. Since they found that with molecular species the reaction rate was controlled by the oxygen groups, and with atomic species the rate was independent of them, it appears that the effect of such oxygen groups in molecular reactants lies in the production of atomic reactants.

The Blackwood and McTaggart mechanism, designated "mechanism *C*," comprises:

Step 13. $C_f + H_2O \xrightarrow{k_{13}} C(H)(OH)$

Step 15. $C(H)(OH) \underset{k'_{15}}{\overset{k_{15}}{\rightleftharpoons}} C(O) + H_2$

Step 3. $C(O) \xrightarrow{k_3} CO + nC_f$

which, if $C(O) \simeq C_f$, also leads to the rate equation 61, where $k = k_{13}[C_t]$, $a = k'_{15}/k_{15}$ and $b = k_{13}/k_3 + k_{13}/k_{15}$. A further amplification of their mechanism was suggested: at vacuum to atmospheric pressure levels, hydrogen may react with an hydroxyl radical at, or near, the surface to re-form water and atomic hydrogen, the latter recombining near the surface;

Step 16. $H_2 + C(OH) \rightleftharpoons H_2O + H + C_f$

Step 17. $H + H \rightarrow H_2$

At higher pressures, the mechanism must be further altered to account for the production of CH_4, which will be discussed subsequently.

Another investigator, favored the concept of reaction by atomic species, and concluded that the carbon surface accelerates the water-gas

[107] Blackwood, J. D., and McTaggart, F. K., *Australian J. Chem.*, **12**, 533–42 (1959).

TABLE 10

TEMPERATURE COEFFICIENTS OF ELEMENTARY REACTIONS OF CARBON-STEAM MECHANISMS

Temperature Coefficient, kcal	E_k	E_a	$E_k - E_b$	$E_b - E_a$	$E_a + E_k - E_b$
Corresponding Elementary Reaction Temperature Coefficients					
1. Mechanism *A*	E_4	—	E_2	ΔH_4	E_4'
2. Mechanism *B*[10]	E_5	ΔH_7	E_6	—	—
3. Mechanism *B*[93]	E_8	ΔH_7	E_2	—	—
4. Mechanism *C*	E_8	$-\Delta H_{10}$	E_2	—	—
Carbon Type, and % Burnoff					
Graphite tube, 0%[60]	32.7	−60.8	112.0	−18.5	51.2
Graphite tube, 1%[60]	26.0	−62.0	105.7	−17.7	43.7
Graphite tube, 2%[60]	21.2	−63.1	101.4	−17.1	38.3
Graphite tube, 5%[60]	16.0	−66.7	98.5	−15.8	31.8
Graphite tube, 7.5%[60]	13.1	−69.2	96.0	−13.7	27.8
Coconut-shell charcoal[93]	62.3	0.0	82.4	20.1	82.4
Graphite[98]	49.7	—	—	—	—
Graphite[98]	35.1	—	—	—	—
Coke[99]	26.8	—	—	—	—
Electrode graphite[100]	35.6	−75	—	—	—
Graphitized coke[101]	40.7	—	—	—	—
Pitch coke[103]	45.0	—	—	—	—
Pitch coke[6]	59.1	−20.5	—	—	—
Three coal cokes[98]	40	—	—	—	—
Graphite[110]	—	—	69	—	—
Graphite[110]	—	—	60	—	—

shift reaction through initiation of a chain sequence near the surface:[108]

Step 18. $C(H) \rightarrow H + C_f$

Step 19. $H + H_2O \rightleftharpoons OH + H_2$

Step 20. $OH + CO \rightleftharpoons CO_2 + H$

Graven and Long[97] supported a similar mechanism[109] for forward and reverse gas-phase water-gas shift. If this thesis represents actual fact, this means that "catalysis" of water-gas shift by carbon and the observed lack of CO hindrance to the carbon-steam reaction need not be contradictory phenomena.

The foregoing discussion points up the lack of sufficient confirmatory evidence to set forth carbon-steam reaction mechanisms with certainty. Indeed, as has been described, four or more postulated mechanisms lead to the same rate equations, the respective forms differing only in the functional relation of the elementary constants which combine as the rate constants. Computed activation energies of the elementary chemical steps based on measured temperature coefficients of the rate constants may in certain instances confirm or negate the thermodynamic consistency of a particular mechanism. Some of these interpretations are given here. Following the same procedure as in the C-CO_2 reaction, Table 10 was prepared from Table 8 for each of the carbon-steam mechanisms through the following relations:

1. Mechanism *A*: $E_9 \cong E_k$; $E_9' = E_a + E_k - E_b$; $E_3 = E_k - E_b$; $\Delta H_9 = E_9 - E_9' = E_b - E_a$;

2. Mechanism *B*:[10] $E_{10} \cong E_k$; $E_{11} = E_k - E_b$; $\Delta H_{12} = E_{12} - E_{12}' = E_a$

3. Mechanism *B*:[93] $E_{13} \cong E_k$; $E_3 = E_k - E_b$; $\Delta H_{12} = E_{12} - E_{12}' = E_a$

4. Mechanism *C*: $E_{13} \cong E_k$; $\Delta H_{15} = E_{15} - E_{15}' = -E_a$; $E_3 = E_k - E_b$ (if $k_{13}/k_{15} \ll k_{13}/k_3$)

As is evident, the quantities E_k and $E_k - E_b$ are common to all four mechanisms. In one

[108] Ingles, O. G., *Trans. Faraday Soc.*, **48**, 706–12 (1952).

[109] Bradford, B. W., *J. Chem. Soc.*, **136**, 1557–63 (1933).

instance, E_k refers to activation energy of surface oxygen-exchange, and, in three instances, E_k is the activation energy of chemisorption of water vapor or its dissociation products. In all mechanisms, the quantity $E_k - E_b$ refers to activation energy of a surface complex decomposition. Activation energies of 69 kcal and 60 kcal have been reported[110] for graphite rods under conditions producing a zero-order reaction. This would correspond to $E_k - E_b$, their values being somewhat lower than the average in Table 10. In mechanism *A*, the data of Table 10 indicate the heat effect of the oxygen exchange surface reaction (step 9) to be about 17 kcal exothermic with graphite, and 20 kcal endothermic with coconut-shell charcoal. Summation of bond dissociation energies[111] for this reaction (assuming no effect upon carbon-carbon bond strengths), gives 31 kcal endothermic if surface oxygen forms a C—O bond, and 67 kcal exothermic if the oxygen forms a C=O bond. Although the numerical values are not expected to agree, this does point up that reaction energetics are primarily dependent on bond strengths or electronic nature of the carbon surface. In mechanism *B*, hydrogen chemisorption (step 12) was estimated to be 108 kcal exothermic[93] from bond energies; Table 10 indicates a range of 20 to 75 kcal exothermic, and Barrer[106] found an average 50 kcal exothermic for hydrogen sorption on charcoal at 772 to 950°C. In mechanism *C*, the decomposition of adsorbed dissociated water vapor (step 15) is about 104 kcal endothermic according to the selected bond energies; this value should be compared with $-E_a$ in Table 10.

As a final consideration of the carbon-steam system, Blackwood and McGrory[62,113] demonstrated that the rate of reaction with purified coconut charcoal does not follow the Langmuir-type equation 61 over their range of conditions investigated, 750 to 830°C, 1 to 50 atm total pressure. At constant hydrogen partial pressure, the rate of CO production was greater than could be accounted by this equation; at constant steam partial pressure, the retarding effect of hydrogen was not as great as expected. These deviations were greater at high-steam partial pressures and lower temperatures, 750 to 790°C, at which a large fraction of the carbon gasified appeared

[110] James, V. E., Thesis, West Virginia University, 1957. Binford, J. S., Jr., and Eyring, H., *J. Phys. Chem.*, **60**, 486–91 (1956).

[111] Bond dissociation energies at 25°C (kcal) adopted from references 93 and 112: H—H, 104; HO—H, 119; O—H, 102; C—H (trigonal), 106; C—O, 86; C=O, 184.

[112] Steacie, E. W. R., *Atomic and Free Radical Reactions*, Second edition, Reinhold, New York, 1954, vol. I, 485 pp. Hottel, H. C., Williams, G. C., and Satterfield, C. N., *Thermodynamic Charts for Combustion Processes*, John Wiley and Sons, New York, 1949, Part I, 75 pp.

[113] Blackwood, J. D., *Coke and Gas*, **22**, 190–4, 218 (1960).

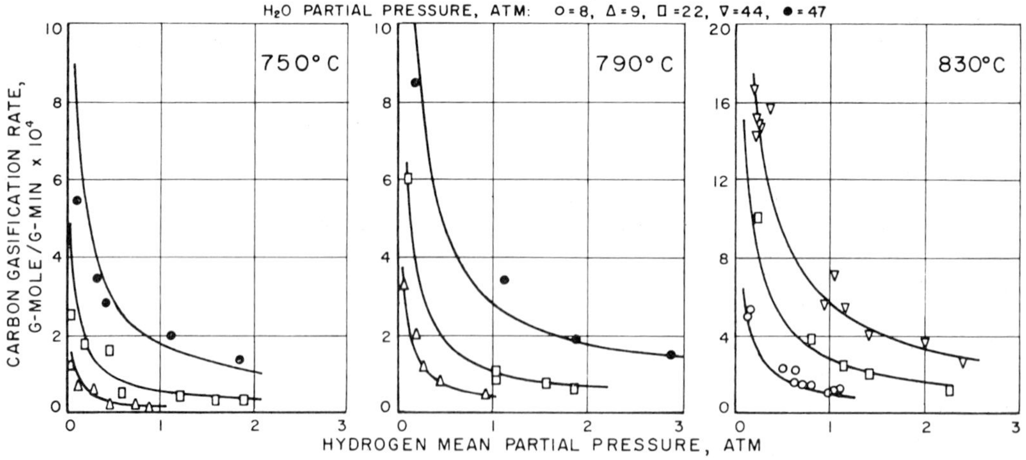

Fig. 16. Effect of hydrogen and steam partial pressures on the rate of the carbon-steam reaction at high temperatures (Blackwood and McGrory[62]).

as methane. Methane-formation rate increased directly with steam partial pressure, but was substantially independent of hydrogen in the range 0.1 to 2.5 atm partial pressure.

Since the reaction order with respect to steam was greater than unity, these investigators attributed the acceleration of gasification to a mechanism of methane formation. For example, a reaction step involving direct attack of steam on a carbon atom with at least two hydrogen atoms already attached, forming methane, and accompanied by deposition of oxygen which is liberated as CO, would account for the observed results. The proposed mechanism comprises Long and Sykes' mechanism *B* modified for methane formation:

Step 13.

$$C_f \text{ (dual site)} + H_2O \underset{k'_{13}}{\overset{k_{13}}{\rightleftharpoons}} C(H)(OH)$$

Step 14. $C(H)(OH) \xrightarrow{k_{14}} C(O)(H_2)$

Step 3. $C(O) \xrightarrow{k_3} CO + nC_f$

Step 12. $C_f + H_2 \underset{k'_{12}}{\overset{k_{12}}{\rightleftharpoons}} C(H_2)$

Step 21.

$$C_f + C(H_2) + H_2O \xrightarrow{k_{21}} CH_4 + C(O)$$

In this mechanism, it was suggested that the active sites for oxidation are positions in the carbon lattice where oxygen is already attached. CO liberation is accompanied by deposition of another oxygen group adjacent to each one destroyed, and a carbon atom suitable for hydrogen adsorption is exposed. The mechanism leads to the following rate equations:

$$r_C = r_3 = \frac{kp_{H_2O} + cp_{H_2}p_{H_2O} + dp^2_{H_2O}}{1 + ap_{H_2} + bp_{H_2O} + ep_{H_2}p_{H_2O} + fp^2_{H_2O}} \quad (65)$$

$$r_{CH_4} = r_{21} = \frac{(cp_{H_2}p_{H_2O} + dp^2_{H_2O})/2}{1 + ap_{H_2} + bp_{H_2O} + ep_{H_2}p_{H_2O} + fp^2_{H_2O}} \quad (65a)$$

where $k = [k_{13}k_{14}/(k'_{13} + k_{14})]C_t$, $a = k_{12}/k'_{12}$

$b = k_{21}/k'_{12} + k_{13}(k'_{12} + k_{14})/k'_{12}(k'_{13} + k_{14}) + k_{13}k_{14}/k_3(k'_{13} + k_{14})$,

$c = k_{12}k_{21}C_t/k'_{12}$, $d = 2k_{13}k_{14}k_{21}C_t/k'_{12}(k'_{13} + k_{14})$,

$e = k_{12}k_{21}/k_3k'_{12}$, and

$f = k_{13}k_{21}(k_3 + 2k_{14})/k_3k'_{12}(k'_{13} + k_{14})$.

A comparison between computed and observed carbon gasification rates is shown in Fig. 16, based on the terms $ep_{H_2}p_{H_2O}$ and $fp^2_{H_2O}$ in equation 65 being negligible, and the following experimental constants for charcoal:

Temp., °C	*k* g-mole/ g-min-atm	*a* 1/atm	*b* 1/atm	*c* g-mole/ g-min-atm	*d* g-mole/ g-min-atm
750	0.36×10^{-4}	35	0.06	0.3×10^{-4}	1.5×10^{-6}
790	1.25×10^{-4}	35	0.09	0.5×10^{-4}	1.5×10^{-6}
830	3.7×10^{-4}	35	0.14	1.05×10^{-4}	1.5×10^{-6}

The authors pointed out that although methane rates can be accounted for by step 21 of the proposed mechanism, this is not intended to overlook methane formation by hydrogenation without the action of steam.

Carbon-Hydrogen Reaction

The generation of methane and small amounts of higher hydrocarbons from pressure hydrogenation of carbonaceous solids is receiving increased attention by gas and fuel technologists. Experimental studies have demonstrated technical feasibility, but existing knowledge of both physical and chemical mechanisms of this process is very incomplete.

Gases containing as much as 85% methane—about 900 Btu/scf heating value—can be produced by direct dry-hydrogenation of the volatile matter of untreated pulverized coal[114] Effective hydrogenation requires pressures in the range of 30 to 200 atm at 900 to 1700°F and coal reaction times approaching ½-hr or more with 40- to 60-mesh particles. The rates of the hydrogenation and hydrogenolysis reactions increase proportionately with hydrogen partial pressure,[61,115] accelerate with temperature,[61] and are dependent on carbon

[114] Bray, J. L., and Howard, R. E., *Purdue Univ., Eng. Bull., Research Ser.*, No. 90, (1943), 55 pp.

[115] Bray, J. L., and Morgal, P. W., *Purdue Univ., Eng. Exp. Sta. Research Ser.*, No. 93, (1944), 33 pp.

properties roughly related to coal rank and heat pretreatment. Ash constituents may catalyze methane formation and certain side reactions, such as hydrogenation of CO or CO_2.

The solid-fuel oxygen content appears to bear a major and direct influence on the rapidity of the hydrogenation reaction (Blackwood and McGrory[61]). The so-called active centers for these reactions, according to some investigators are associated with at least two types of oxygen-containing structures in the carbon. One type, considered a lactone structure,[116] is responsible for an initial rapid evolution of methane plus water. This structure is destroyed during the first several minutes of hydrogenation. The other type, a chromene or benzpyran structure,[117] which is not destroyed during reaction, appears to be responsible for the subsequent slow, steady, evolution of methane. The desensitization of active centers by adsorption of electrophylic mineral acids—a result observed by Frey[67]—may be attributed to the presence of such chromene groups.

Bituminous coals, lignites, and low-temperature chars are much more rapidly hydrogenated than graphitic structures such as high-temperature coke.[118] Pure natural graphite, containing no chemically combined oxygen, is inert to hydrogen. In coals and low-temperature chars, up to 80% of their carbon content, corresponding to much more than just the volatile matter, can be hydrogenated to gaseous hydrocarbons. The reaction progresses in two stages.[118] The first stage, where hydrocarbon production is more rapid and the oxygen content of the fuel decreases most rapidly, involves pyrolysis and hydrogenation of the "coal molecules," yielding initially some fractions of ethane, ethylene, higher homologs, aromatics, and other cyclic structures. Successive hydrogenolysis reactions finally yield chiefly methane.[119] The second stage involves the much slower reactions between hydrogen and the residue from pyrolysis and hydrogenolysis.

Since the properties of the carbon are continuously changing with reaction, the reaction rates generally decrease with carbon burnoff, the latter being also related to residual oxygen content, which seems to be a good indicator of fuel reactivity. In the analysis of results of hydrogenation of other than the purest forms of carbons, it is usually difficult to define the quantities of methane formed by the carbon-hydrogen reaction as distinct from that formed through hydrogenolysis of unsaturated materials, or other possible reaction mechanisms.

In a study by Gorin and associates[61,62] with fluidized low-temperature char at 1500 to 1700°F and up to 30 atm total pressure, differential methane formation rates r_{CH_4} corresponding to initial rates at 0% hydrogen utilization, were correlated satisfactorily for 1700°F but not for lower temperatures, by an equation of the form:

$$r_{CH_4} = \frac{kp_{H_2}^2}{1 + ap_{H_2}} \tag{66}$$

The rates at 30 atm followed an Arrhenius-type plot (Fig. 17) for 0 to 60% carbon gasified, and the composite temperature coefficients, based on these rates, increased from about 17 kcal to about 50 kcal in this range of carbon burnoff. The increase in temperature coefficient was attributed to the carbon structure becoming more graphitic with extent of reaction. Equation 66 states that the reaction should be second order with respect to hydrogen at such low pressures that $ap_{H_2} \ll 1$, although there is some evidence[120] of the order being 0.5 at these conditions. The reaction is retarded by hydrogen, but apparently not by methane other than through equilibrium considerations. At pressures where $ap_{H_2} \ll 1$, the reaction approaches first order, a result also confirmed by Blackwood[61] in studies with coconut char and other types of carbons. Zielke and Gorin's empirical correlation[61] indicates that the apparent order of the reaction is a function also of carbon burnoff; at 1700°F, 20 atm, and 40% burnoff, the apparent order is 1.39, which decreases to 1.20 at 30 atm and 50% burnoff.

Equation 66 is consistent with a tentative

[116] Garten, V. A., Weiss, D. E., and Willis, J. B., *Australian J. Chem.*, **10**, 295–308 (1957).

[117] Garten, V. A., and Weiss, D. E., *Australian J. Chem.*, **10**, 309–28 (1957).

[118] Stockman, C. H., and Bray, J. L., *Purdue Univ., Eng. Bull., Research Ser.* No. 111 (1950), 44 pp.

[119] Channabasappa, K. C., and Linden, H. R., *Ind. Eng. Chem.*, **50**, 637–44 (1958).

[120] Gilliland, E. R., and Harriott, P., *Ind. Eng. Chem.*, **46**, 2195–202 (1954).

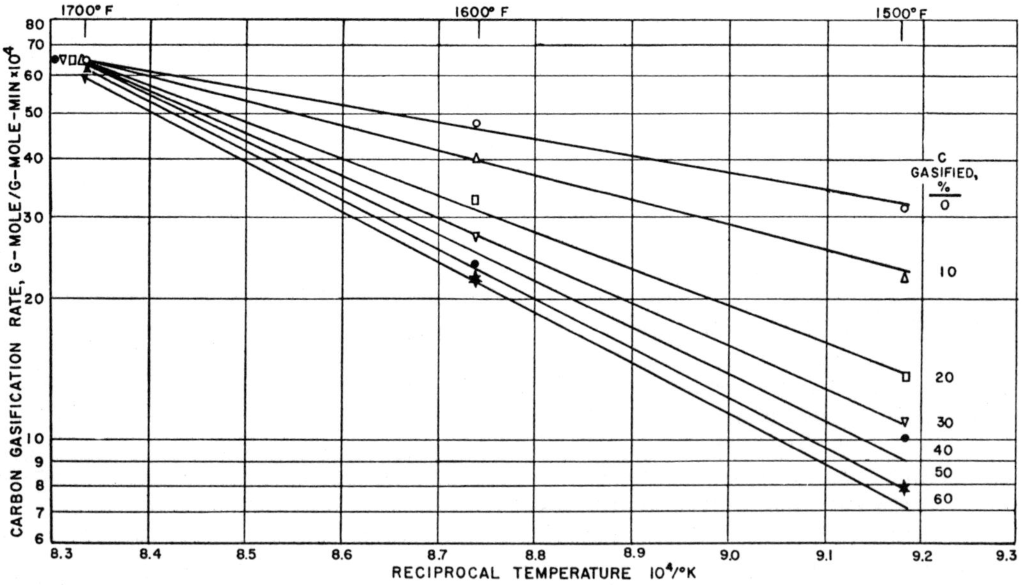

Fig. 17. Arrhenius plot of rate of gasification of a low-temperature char as a function of carbon gasified (Zielke and Gorin[61]).

mechanism (Zielke and Gorin[61]) based on hydrogenation of exposed edges of the crystallite lattice represented as aromatic linkages —HC═CH—. Successive additions of hydrogen molecules break the double bond, form methyl groups, and finally methane, with regeneration of a new —HC═CH— group for each one destroyed:

At the steady state, with step 24 rapid relative to 22 and 23, the appropriate form is obtained where $k = k_{22}k_{23}C_f/k'_{22}$, $a = k_{23}/k'_{22}$, and C_f represents the number of active —HC═CH— groups per unit amount of carbon.

Blackwood and McGrory[61] demonstrated that the methane formation rate from hydrogenation of 7- to 14-mesh coconut char at 650

Step 22.

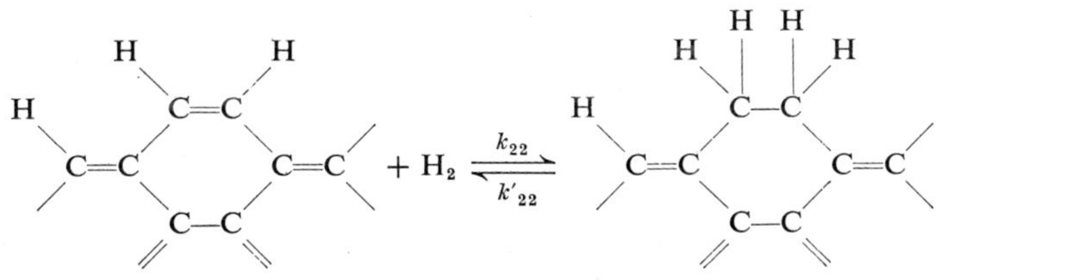

Step 23.

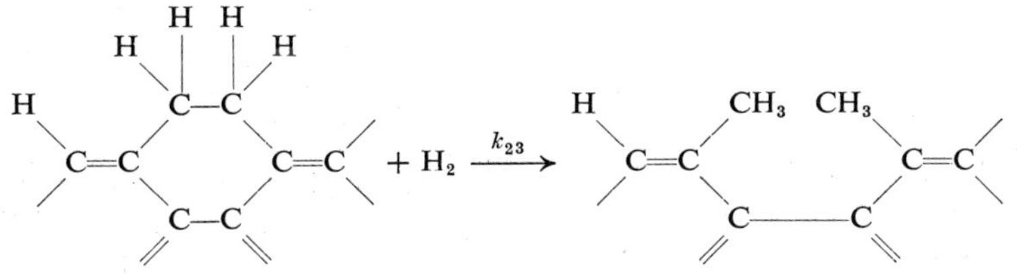

Step 24.

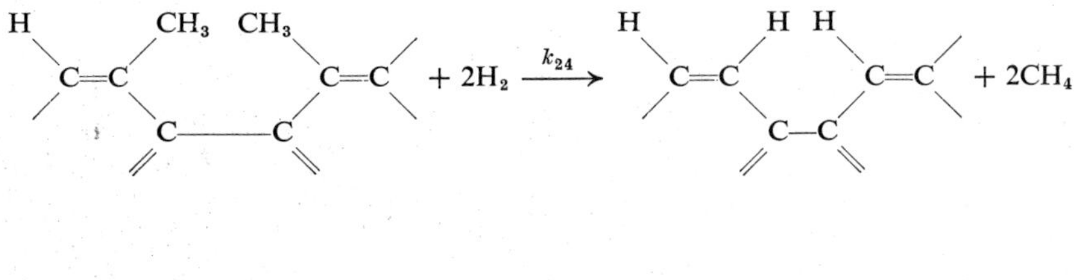

to 870°C, 5 to 40 atm, follows a first-order rate law:

$$r_{CH_4} = kp_{H_2} \qquad (67)$$

at low methane partial pressures, and under conditions where external diffusion to the carbon surface is not a controlling factor. The rate constant k, moles CH_4/g-min-atm, can be expressed as:

$$k = 17.7e^{-35{,}600/RT}$$

which indicates a 35 kcal temperature coefficient, a result in close agreement with 36 ± 6 kcal determined in atmospheric-pressure reactivity studies of carbons deposited from hydrocarbons upon porous carriers.[120] If the hydrogenation mechanism discussed earlier is correct, under the condition that the reaction is first order, these activation energies correspond to the initial hydrogen chemisorption in the forward reaction of step 22. Barrer,[106] on the other hand, found the chemisorption of hydrogen on charcoal at 695 to 864°K a variable-activation-energy process, which exhibited 10 kcal at zero saturation of the surface, and increased continuously and asymptotically to 35 kcal at complete saturation. Methane formation occurs only when chemisorption is virtually completed; the activation energy, 35 kcal, corresponds to the point of surface saturation.

The rate of methane formation from chars prepared at lower or higher than hydrogenation temperatures generally tends to increase, attain a maximum, and then decrease with the temperature of char preparation,[119] a factor which seems to be related to the residual volatile matter and oxygen content. Control of the atmosphere, such as air or steam, used in preparation of the char to improve initial hydrogenation reactivity, has been demonstrated.[119]

There is some conflicting evidence concerning the effect on the methane formation rate of a steam addition during the course of hydrogenation. Methane formation during the reaction of steam with carbon has been attributed to direct combination of the hydrogen generated with the carbon surface, particularly a surface "activated" by steam attack (Key[88]). This activating effect has also been suggested[11] and demonstrated[9] by others. Methane from the steam-charcoal reaction can be accounted for by (1) hydrogenation of C—H groups in the carbon,[121] or (2) direct steam attack on sites where hydrogen is already adsorbed (Blackwood and McGrory[61]) These investigators,[61] however, found no activating effect on methane formation of oxygen carriers such as steam or CO_2. This was explained on the basis of nonregenerability of lactone groups which are destroyed during the early periods of hydrogenation, when methane formation rates are the most rapid.

On the other hand, in studies with 65- to 100-mesh Disco char reacted in hydrogen-steam mixtures at 1500 to 1700°F, a significant accelerating effect of steam on the methane rate was found.[9] This effect diminishes with pressure and with carbon burnoff; typically the methane rate at 1700°F and 30% burnoff with hydrogen-steam mixtures ranges from 100 to 2 times greater at 1 to 30 atm than the rate in pure hydrogen at comparable partial pressures (Table 11). From the authors' tabular data, Figs. 18 and 19 (extrapolated to 100 atm) for 1700°F and 1.0 H_2/H_2O reactant mole ratio were prepared showing that (1) the total carbon gasification rate (the sum of carbon converted to methane and gasified by steam) decreases with burnoff (Fig. 19) and is an exponential function of total pressure of

TABLE 11

COMPARISON OF METHANE RATES IN PURE HYDROGEN AND IN HYDROGEN-STEAM AT SAME PARTIAL PRESSURE OF HYDROGEN (Temp = 1700°F; burnoff = 30%)

Reactants, %				Methane Rate*	
H_2	H_2O	P_{H_2O}, atm	P_{H_2}, atm	H_2/H_2O	Pure H_2†
25	75	0.75	0.25	9.8	0.076
25	75	4.5	1.5	34	0.99
25	75	22.5	7.5	95	10.0
50	50	0.50	0.50	6.7	0.20
50	50	3.0	3.0	24.8	2.67
50	50	15.0	15.0	83	24.6
67	33	0.33	0.67	5.7	0.31
67	33	10.0	20.0	75	37

* g C/g C-min × 10^4.

† Calculated from correlation for rate of direct hydrogenation (Zielke and Gorin[61]).

[121] Lewis, W. K., Gilliland, E. R., and Hipkin, H., *Ind. Eng. Chem.*, **45**, 1697–703 (1953).

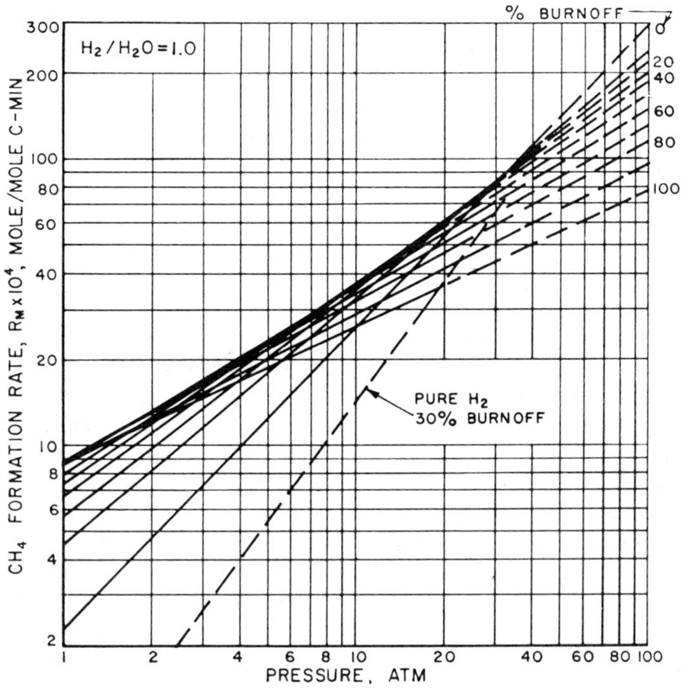

Fig. 18. Rates of methane formation from Disco char in hydrogen and hydrogen-steam at 1700°F.[9]

the form aP^b, (2) the methane rate, also an exponential function of total pressure, and a more complex function of carbon burnoff, approaches the total carbon gasification rate at the higher pressures, and (3) the methane rate in pure hydrogen at 30% carbon burnoff is always less than the corresponding rate in the 1:1 hydrogen-steam mixture up to 30 atm total pressure. This accelerating effect at high-steam concentrations appears to be too great to be explained on the basis of steam activation of the carbon surface, or of steam causing a rapid development of surface area. A form of activated diffusion or surface migration of adsorbed species through distorted lattice structure is suggested, making surface areas available that are not ordinarily penetrated by gaseous molecules.

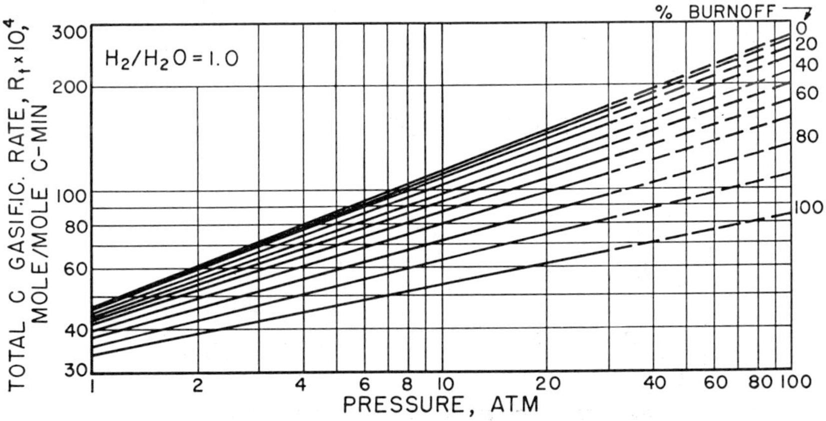

Fig. 19. Gasification rates of Disco char in hydrogen-steam at 1700°F.[9]

INTERPRETATION OF DIFFUSIONAL EFFECTS IN GAS-CARBON REACTIONS

Bulk diffusion of reactants and products to the exterior reacting surface, and internal diffusion through the pore structure, constitute the two important transport effects in heterogeneous reactions. These factors have a direct bearing on the type of interpretation that is to be made of experimental rate studies, particularly of the apparent reaction order and apparent activation energy or temperature coefficient.

As discussed by many investigators,[57,91,92,122–124] the variation of reaction rate with temperature of gas-carbon reactions can be divided theoretically into three principal and two transition zones. This is shown in Fig. 20 in the Arrhenius-type plot, and in the corresponding gas concentration profiles of the carbon samples designated as rectangles.

In zone 1, the low-temperature zone, the reaction rate is so slow that it is controlled by chemical reactivity of the solid only. Diffusion of reactants is sufficiently rapid that substantially no concentration gradient exists from the bulk gas stream, c_g, through the film thickness, δ, and through the carbon interior. Since the reaction rate is nearly nonvariant in the interior, the quantity η, defined as the ratio of the actual rate to the rate that would occur if the gas concentration within the solid were constant and equal to the bulk concentration c_g approaches unity. The measured temperature coefficient E_m is practically identical with the "true activation energy," E, only if other energy effects related to temperature and reactant partial pressure, such as those of thermal annealing, are negligible. When the temperature increases, a transition from zone 1 to zone 2 occurs, characterized by a decrease in reactant concentration with distance into the carbon particle up to the point where the concentration approaches zero at the center of the particle, zone 2.

[122] Hedden, K., *Chem. Ing. Tech.*, **30**, 125–32 (1958).

[123] Rossberg, M., and Wicke, E., *Chem. Ing. Tech.*, **28**, 181–9 (1956).

[124] Hedden, K., *Ullmann's Encyklopadie der technischen Chemie*, Third edition, Urban and Schwarzenberg, Munich, 1958, Vol. 10, pp. 362–76.

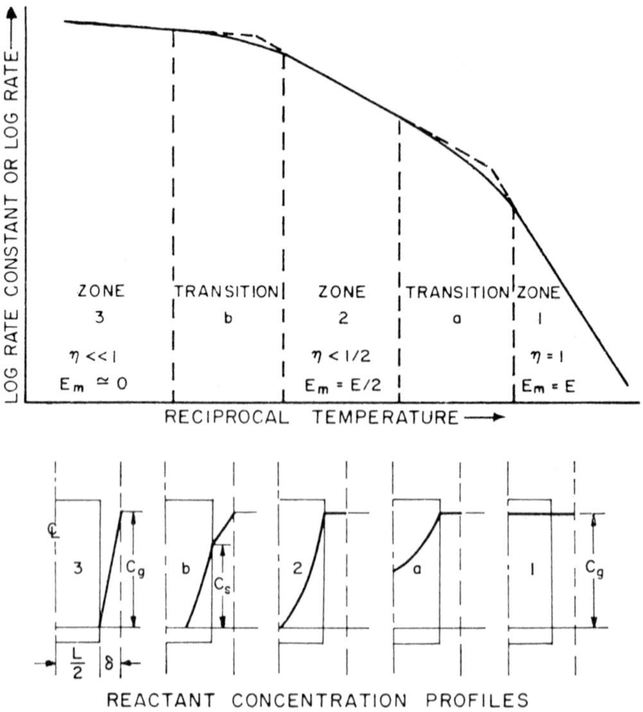

Fig. 20. Characteristic temperature zones of gas-carbon reactions.

In zone 2, the ratio η would be exactly 0.5 for a rectangular carbon slab, if the concentration profile were linear, the reaction first order, and the internal surface equally distributed. In a practical case where internal surface is large compared to external surface, η is generally less than 0.5 because the concentration profile is curved, and the gradient through the film, which is dependent on particle diameter and mass flow rate, may not be negligible. Furthermore as already discussed, reaction products retard the gas-carbon reactions. Thus, in zone 2, the reaction retardation increases as a function of distance into the carbon particle in accordance with increased reaction product concentration buildup toward its center.

The measured order of the reaction with respect to reactant partial pressure in the bulk gas stream represents an integrated average of the apparent order taken at all reacting surfaces. As shown by Wheeler[125] and Weisz and Prater,[126] the measured temperature coefficient E_m is one-half of the true activation energy E in this temperature range, provided other temperature-dependent energy effects are again negligible. Because the type of rate-controlling step in the kinetic mechanism may change with temperature in going from zone 1 to zone 2, twice the measured temperature coefficient in zone 2 may not necessarily equal the activation energy corresponding to zone 1. With further temperature increase, the reactant concentration profile becomes steeper, and transition to the high-temperature zone 3 occurs.

In zone 3, the reactant concentration decreases practically to zero at, or immediately below, the exterior surface. The reaction rate is substantially controlled by mass transport from the bulk stream—a first-order process (Fick's law) with a small temperature coefficient. The ratio η must be much less than unity here, since at most a small fraction of internal surface is participating in the reaction. A true zone 3 gas-carbon reaction requires that the rate be independent of carbon reactivity. Combustion and high-temperature gasification approach this condition, but there is usually some small reactivity effect which would indicate that these reactions are occurring in the transition region b rather than in zone 3.

A few of the mathematical details for rationalizing reaction rates in zones 1 to 3 are shown. The treatment is limited to first-order reactions occurring on surfaces and in pores. The solutions are based on Cartesian geometry. Extensive details for reactions other than unity order, and geometries other than Cartesian, are available elsewhere.[57,125,127]

Reactions below the external surface of a fuel particle are visualized as occurring in pores of average radius $\bar{r}$ and average length $2\bar{L}$. This pore model has been applied in the field of catalysis for computing kinetic factors,[125] and is just as applicable for gasification, provided pore dimensions do not change drastically with reaction time. Assuming no radial concentration gradients in the pore, the differential equation for the steady-state concentration of reactant $c_A = p_A/\mathrm{RT}$ is:

$$\frac{d^2c_A}{dl^2} = \frac{2kc_A}{\bar{r}D} \tag{68}$$

where k = rate constant based on surface area, l = distance into pore measured from pore opening, and D = diffusivity, area/unit time. With the boundary conditions:

$$c_A = c_S \text{ at } l = 0 \text{ (particle surface)}$$

$$\frac{dc_A}{dl} = 0 \text{ at } l = \bar{L} \text{ (particle center)}$$

the solution is

$$c_A = c_S \cosh \frac{q(1 - l/\bar{L})}{\cosh q} \tag{69}$$

where $q = \bar{L}\sqrt{2k/\bar{r}D}$.

The reaction rate in the pore is proportional to the diffusivity, pore cross section, and concentration gradient at the surface, or

$$r_{\text{pore}} = -D\pi\bar{r}^2 \left|\frac{dc_A}{dl}\right|_{l=0} = -D\pi\bar{r}^2 c_S q \tanh\frac{q}{L} \tag{70}$$

An effectiveness factor, $\eta = r_{\text{pore}}/r_{\text{max}}$ = actual rate/theoretical rate at zero concentration

[125] Wheeler, A., *Advances in Catalysis*, Frankenberg, W. G., Komarewsky, V. I., and Rideal, E. K., Eds., Academic Press Inc., New York, 1951, Vol. III, pp. 249–327.

[126] Weisz, P. B., and Prater, C. D., *Advances in Catalysis*, Frankenberg, W. G., Komarewsky, V. I., and Rideal, E. K., Eds., Academic Press Inc., New York, 1954, Vol. VI, pp. 143–96.

[127] Thiel, E. W., *Ind. Eng. Chem.*, **31**, 916–20 (1939).

gradient where $r_{max} = 2\pi k c_S \bar{L}$ is now found to be

$$\eta = \tanh q/q \tag{71}$$

which is utilized to express the pore rate in terms of reactant concentration at the surface, or

$$r_{pore} = 2\pi \bar{r} k c_S \bar{L} \eta \tag{72}$$

By letting m_p = particle mass, V_p = particle volume, V_g = pore volume per unit weight, S_p = external surface area per particle, S = internal surface area per unit weight, and ϵ_p = particle void fraction, the pore dimensions and the number of pores per particle n may be expressed (Smith[40]) as

$$\bar{r} = \frac{2V_g}{S}$$

$$\bar{L} = \frac{\sqrt{2}\,V_p}{S_p} = \frac{\sqrt{2}\,m_p V_g}{S_p \epsilon_p}$$

$$n = \frac{\epsilon_p S_p}{\sqrt{2}\pi \bar{r}^2}$$

The reaction rate per unit mass of particles then becomes simply:

$$r\text{ (per unit mass)} = k c_s S \eta \tag{73}$$

In equation 71, $\tanh q \simeq 1.0$ if $q > 2$, so $\eta \simeq 1/q$. Under this condition, equation 73 becomes:

$$r\text{ (per unit mass)} = \frac{\sqrt{kDV_g S}\, S_p c_s}{\sqrt{2}\,V_p} \tag{73a}$$

which indicates that the rate is proportional to the square root of the rate constant, or that the measured temperature coefficient in zone 2 is one-half the "true activation energy." This has also been shown to be generally true for reaction orders other than unity.[57,125] If the true order of the reaction is m, it turns out that the measured order of a reaction in zone 2 is $(m + 1)/2$. In particular, for zero order, equation 73a becomes:

$$r\text{ (per unit mass)} = \frac{\sqrt{kDV_g S c_s} S_p}{V_p} \tag{73b}$$

In zone 1, the measured order is generally the true order.

For spherical geometry, Wheeler[125] has shown the effectiveness factor:

$$\eta = \frac{3}{q_s} \frac{1}{\tanh q_s} + \frac{1}{q_s} \tag{71a}$$

where $q_s = D_p \sqrt{S_v k/D}/2$, D_p = particle diameter, and S_v = surface area per unit volume of particles, which at $q_s \geq 9$ becomes $\eta = 3/q_s$, and the reaction rate per unit mass simplifies to:

$$r\text{ (per unit mass of spherical particles)} = \frac{6c_s\sqrt{kDS}}{D_p\sqrt{\rho_p}}$$

where ρ_p = particle density, and $S = S_v/\rho_p$.

The overall rate of pore reactions cannot be greater than the maximum quantitative rate of reactant diffusion into the pores. Two types of diffusion are recognized as possible:

1. If the mean free path,

$$\lambda = 97(10^{-24})\frac{T}{\sigma^2 P}$$

where σ = effective molecular diameter in centimeters, T in degrees Kelvin, P in atmospheres, and λ in centimeters, is small in relation to pore diameter, collisions between molecules or ordinary molecular diffusion predominate. The molecular diffusivity may be estimated by the methods of analogy discussed under *Heat Transfer*, or it may be estimated from experimental data or from kinetic theory.[128,129]

2. If the mean free path is large relative to pore size, molecule collisions with the pore walls control the diffusion rate. This type, Knudsen flow, is much slower than ordinary molecular diffusion; the rate is proportional to pore radius $\bar{r}$ and the Knudsen diffusivity D_k is independent of pressure,

$$D_k\,(\text{cm}^2/\text{sec}) = \frac{4\bar{r}}{3}\sqrt{2k_B T/m}$$

$$= 9.7(10^3)\bar{r}\sqrt{T/M} \tag{74}$$

where k_B = Boltzmann constant, m = molecule mass, and M = molecular weight. In addition, forced-flow effects in pores influence the diffusion rate;[125] the net rate of reactant influx will be greater or smaller than the quantity obtained by direct application of Fick's law, depending on whether a volume contraction or expansion occurs in the chemical reactions.

As a closing consideration, an attempt is made to define a general rate expression for

[128] Hougen, O. A., and Watson, K. M., *Chemical Process Principles*, John Wiley and Sons, New York, 1947, Part 3, 303 pp.

[129] Present, R. D., *Kinetic Theory of Gases*, McGraw-Hill, New York, 1958, 228 pp.

first-order reactions in the three temperature zones (Fig. 20) based on some of the results just mentioned. If c_g represents the reactant concentration in the bulk gas stream, c_s the concentration at the surface, and A_e the external surface per unit mass, the reaction rate per unit mass is given by the identity:

$$r \text{ (per unit mass)} = \frac{D_f A_e(c_g - c_s)}{\delta} = kSc_s\eta + kA_ec_s \quad (75)$$

where D_f = average diffusivity of fluid film, and the right hand members represent the internal and external rates respectively. By eliminating c_s, we obtain

$$r \text{ (per unit area)} = \frac{c_g}{\delta/D_fA_e + 1/(kS\eta + kA_e)} \quad (76)$$

In zone 1, the term D_fA_e is small as compared with the second denominator term, and $c_s = c_g$. When reaction occurs at appreciable penetration into the solid as in zone 2, the quantity kA_e is negligible relative to kS. For fine particles reacted at high temperatures (large external surface), kA_e is large with respect to kS. In zone 3, where k is large, the only term remaining is $D_fA_ec_g/\delta$, where δ may be estimated from equation 48, or

$$(\text{Nu})_p = 2 + \frac{D_p}{\delta} = 0.37\,(\text{Re})_p^{0.6}$$

yielding

$$\delta = \frac{D_p}{0.37\,(\text{Re})_p^{0.6} - 2} \quad (48a)$$

or at high Reynolds numbers, simply

$$\delta = 2.7D_p^{0.4}\left[\frac{\mu}{\rho\bar{u}_s}\right]^{0.6} \quad (48b)$$

GASIFICATION PROCESSES

General Considerations in Complete Gasification

In the complete or total gasification of coal or other solid fuels, the major portion of the energy in the solid reactant is converted to energy in useful gaseous products. This may be done in either a single- or multistage process, yielding gas as the principal product. Thus, a complete gasification process may be regarded as a self-contained unit operation which may be used as a part of any overall processing scheme. This is an important characteristic when the gaseous product is used for synthesizing chemicals, liquid fuels, or other gaseous fuels, and is in marked contrast to most partial gasification processes, which are dependent for their economic success on the marketability of coproducts or byproducts.

The ideal complete gasification process would be a single-stage, continuous process which uses air as the oxidizing medium and which converts any type of coal into a combustible or synthesis gas low in inerts. Commercial processes, and processes under development, embody many of the characteristics of the ideal process, but as yet no single commercial process embodies all of these characteristics. The various methods used for the complete gasification of coal are outlined below.

Carbonization in Coke Ovens Followed by Gasification of Coke

In this method, caking coals are converted into coke which is gasified with air and steam in a cyclic water-gas set. This method was used in Germany for the production of synthesis gas for chemicals and liquid fuels, and is being used in Britain in the synthesis of ammonia and methanol. Although extensively used in the United States prior to the end of World War II, very little coke is currently gasified to produce synthesis gas because of the widespread availability of low-cost natural gas which can be converted to synthesis gas by partial combustion with oxygen or by catalytic steam-reforming. The coal carbonization-coke gasification sequence may be considered broadly as a method for complete gasification of coal, since gas is the principal product. However, substantial yields of tar are obtained in the carbonization step, which not only reduces the yield of gas from the coal, but also, in certain instances, introduces the problem of byproduct marketing. The greatest disadvantage of this method is the excessive handling of solid materials, first in the carbonization step and again in the gasification step. This is costly not only in investment and labor charges, but also in the loss of energy contained in the hot coke. These obvious deficiencies, along with the desire to use a wider variety of coals, have prompted the development of more suitable methods for complete gasification of coal.

Two-Zone Gasification

The disadvantages of the separate carbonization-gasification sequence discussed previously can be overcome by providing two reaction zones as integral parts of a single process. Noncaking or weakly caking coals are dried and carbonized in the first zone. The hot, solid product from this carbonization zone passes to a second zone, in which it is gasified with steam. As in the case of the separate carbonization-gasification sequence, two-zone processes are not complete gasification processes in the strictest sense, because significant quantities of tar are produced.

Gasification of Coal in a Conventional Water-Gas Set

Bituminous coal may be gasified completely in a conventional or modified cyclic water-gas set to yield a low-inert gas containing high concentrations of carbon monoxide and hydrogen, and low but significant concentrations of hydrocarbons. Even with closely-sized weakly caking coal, the gas output obtained from a standard water-gas set is only 70 to 80% of that obtained when coke is used.[130] This process was used at one time in the United States, and is used to a very limited extent in other countries today.

Autothermic Gasification

In an autothermic process, the energy required by the process is liberated continuously within the vessel in which the reactions occur. In autothermic gasification, the heat absorbed in the reaction of the material being gasified with steam or carbon dioxide is supplied by the oxidation of some of this material by oxygen in air, oxygen-enriched air, or high-purity oxygen. In addition, the oxidation reactions must liberate enough energy to heat all the reactants to reaction temperature.

The fixed-bed, air-blown producer was the first autothermic gasifier. The product gas has a relatively low calorific value and a high nitrogen content, which does not impair its usefulness as industrial fuel, but makes it unsuitable for synthesis and for distribution by public utilities. The development of processes for the commercial production of oxygen has made possible the development of a variety of autothermic gasification processes employing fixed beds, fluidized beds, and suspension systems.

Continuous Gasification Using Process Fluids and Solids as Heat Carriers

The heat absorbed by gasification reactions may be supplied by heating some of the reactants or some other heat transfer medium outside the reaction vessel. When these preheated streams are brought into the reaction zone they carry with them the heat required by the gasification process. This heat is transferred directly to the reacting mixture. This type of process approaches the ideal because it can be made to operate continuously without the use of high-purity oxygen. A variety of energy carriers have been used, including steam, a moving fuel bed, a moving pebble bed, and a circulating slag bath.

Continuous Gasification in Externally Heated Vessels

The heat absorbed by the gasification reactions may also be supplied by a transfer of heat to the reacting mixture through the walls of the reaction vessel. Although such processes can be operated continuously without using oxygen, they have the disadvantage of low throughput, which is limited by the rate at which heat can be transferred into the gasifier. External heating has been applied to fixed beds and fluidized beds.

Reviews of Gasification Processes

Several comprehensive treatises[1,131–133] and reviews[134–138] of gasification processes have

[130] Embling, P. M. K., *Intern. Conf. Complete Gasification Mined Coal, Liège*, 1954, pp. 65–79.

[131] Hoy, H. R., and Wilkins, D. M., *BCURA Bull.* **22,** 57–110 (1958).

[132] von Gratkowski, H. W., *Ullmanns Encyklopadie der technischen Chemie*, 3rd Ed., Urban & Schwarzenberg, München, 1958, Vol. 10, pp. 360–458.

[133] Meunier, J., *Gasification and Oxidation of Combustibles*, Masson et Cie, Paris, 1958, 550 pp.

[134] Shires, G. L., *Chem. Eng. Mining Rev.*, **50,** No. 11, 43–50; No. 12, 41–7 (1958).

[135] Wood, F. C., *J. Inst. Fuel*, **31,** 55–62 (1958).

[136] Newman, L. L., *Intern. Conf. Complete Gasification Mined Coal, Liège*, **1954,** pp. 48–64.

[137] King, J. G., *J. Inst. Fuel*, **24,** 147–51 (1951).

[138] Demann, W., *Brennstoff Chem.*, **30,** 187–204 (1949).

been published. Summary tables and other pertinent material from these publications are included in the present review. Certain of these recent publications contain information on gas producers and cyclic water-gas generators that also appeared in the previous edition of this work. Since this information is not included in the present review, the forementioned publications should be consulted.

Classification of Gasification Processes

Gasification processes may be classified in a variety of ways:

A. By the method of supplying the heat required by the gasification reactions
  1. Internal heating
    a. Autothermic
    b. Cyclic
    c. Heat-carrying fluids or solids
  2. External heating
    a. Heat transferred through walls of reaction vessel
B. By the method of contacting reactants
  1. Fixed bed
  2. Fluidized bed
  3. Suspension of particles in gasifying medium
C. By the flow of reactants
  1. Cocurrent
  2. Countercurrent
D. By the gasifying medium
  1. Steam with oxygen or air or oxygen-enriched air
  2. Hydrogen
E. By the condition of the residue removed
  1. Dry ash in nonslagging operation
  2. Slag in slagging operation

For the purposes of this review, processes will be grouped on the basis of the gasifying medium which relates to the chemistry of the process. The main subgroup under each gasifying medium is the method of contacting the reactants, which is one of the physical aspects of the process. Von Gratkowski[132] has summarized the major characteristics of typical gasification processes as shown in Table 12.

GASIFICATION WITH OXYGEN AND STEAM OR AIR AND STEAM

The extensive use of industrial oxygen in gasification processes has been one of the major technological advances of the past 25 years. The principal advantages of using oxygen are as follows: (1) The gasification process can be operated continuously because the heat required by the gasification reactions is supplied cocurrently in the same vessel by the reaction of oxygen with a portion of the combustible; (2) With high-purity oxygen the product gas contains only minor concentrations of nitrogen; (3) continuous operation and elimination of heat losses from large quantities of inerts increases efficiency; (4) operation at elevated pressures is feasible; and (5) the rate of gas production per unit of reactor volume is increased as a result of continuous operation. The early work on the use of oxygen in gasification is summarized in volume II of *Chemistry of Coal Utilization.*[139]

The development of the technology of gasification with oxygen in fixed and fluidized beds has brought forth important principles that are basic to this type of operation.[140] The mixture of oxygen with steam or carbon dioxide must be homogeneous as it enters the tuyeres, or as it is fed under the grate. In practice, a mixing distance of 10 to 15 m has been found adequate, particularly if bends or restrictions occur in the mixing section. Homogeneity of the gasifying medium is essential to avoid damage by high local concentrations of oxygen, and becomes increasingly important as the concentration of oxygen in the gasification medium increases. Gasifiers operating on oxygen must be carefully supervised and instrumented to prevent oxygen breakthrough into the space above the fuel bed. The usual precautions must be observed in compressing and handling the oxygen fed to the gasifiers.

When oxygen is used, it represents about 25 to 35% of the cost of the gas. This factor has stimulated development work on continuous gasification processes using air but producing a gas containing low concentrations of nitrogen. These developments have not as yet led to commercial processes. The only continuous commercial processes using air are those producing low calorific value gas.

Fixed-Bed Processes

The term *fixed bed* is used in gasification to signify a fuel bed supported by a grate or by

[139] van der Hoeven, B. J. C., *Chemistry of Coal Utilization*, John Wiley and Sons, New York, 1945, Vol. 2, pp. 1629–35.

[140] Sabel, O., BIOS Final Rept. No. 199, Item No. 30, London, H.M.S.O., 1945, 32 pp.

TABLE 12

GASIFICATION PROCESS CLASSIFIED ACCORDING TO KIND AND SIZE OF FUEL AND METHOD OF HEAT ADDITION[132]

(Processes shown in parentheses have been applied in semicommercial or small-scale studies)

Fuel			
Permissible caking characteristics	Non- to weakly caking	Weakly to moderate caking	Strongly caking
Size, mm diam	Lumps (8–50)	Granular (1–8)	Dust (0–1)
Condition during gasification	Static bed	Turbulent bed (fluidized)	Suspension (entrained dust cloud)
Reactants			
Direction of motion	Countercurrent	—	Cocurrent
Heat addition			
By partial combustion of the generator fuel in the gasifier			
Continuous air	Generators with grates, generators with slag tapping	Winkler, Winkler-Flesch,* BASF-Flesch-Demag*	Ruhrgas vortex chamber, (Ruhrgas resonant tube)
Continuous oxygen	Generators with slag tapping, generators with grates	Winkler, Winkler-Flesch,* BASF-Flesch-Demag*	Koppers-Totzek, Babcock, and Wilcox, Du Pont, (Bureau of Mines)
Continuous oxygen under pressure	Lurgi-Pressure Generator	(Hydrocarbon Research)	Texaco, (Inst. Gas Tech., Chicago), (Bureau of Mines)
Cyclic with air, and heat storage in bed	Coke-coal water-gas generator	—	—
By partial combustion of the generator fuel or other combustible outside the gasification space			
Externally heated	Didier-Bubiag	(Fuel Research, London), (Bureau of Mines)	—
Heat stored in moving			
Solid heat carrier (heated pebbles)	—	(ICI)†	(Lurgi-Ruhrgas)‡
Gaseous heat carrier	Pintsch-Hillebrand, Koppers, (Schwarzheide)	—	Schmalfeld, (Lutzkendorf)§ Panindco-Rouen
Liquid heat carrier	—	—	(Rummel-Otto Process)

* Preparation of the fuel bed (charging, slag elimination) in turbulent bed, gasification in fixed bed from top to bottom.
† Heat added by partial combustion in suspension, gasification in fluidized bed with circulating coke as the heat carrier.
‡ Countercurrent heat exchange.
§ Combined heat addition part added directly by partial combustion (O_2).

other means and maintained at a constant depth above the support. Thus, the upper and lower extremities of the bed are fixed in space, but within the bed fuel moves slowly from the top down through the gasification zone, and residue is discharged at the bottom. In the strictest sense, this is a moving bed with fixed extremities, but *fixed bed* is most commonly used to describe this system in gasification technology, and will be used here.

Fixed beds have several inherent characteristics that are advantageous in gasification processes. Flow of fuel and residue countercurrent to gasification medium and products of gasification leads to maximum heat economy. Relatively long residence time of the fuel in the reaction vessel permits high-carbon conversion. The product gas is not contaminated with solids to any great extent, and plug flow of solids minimizes loss of ungasified fuel in the residue.

There are several principal disadvantages of fixed beds. Caking coals cannot be used without pretreatment to render them nonagglomerating or without modifying the mechanical design or operating conditions of the gasifier.[141] Sized fuel must be used for maximum output; however, small sizes may be

[141] Nelson, H. W., *Intern. Conf. Complete Gasification Mined Coal, Liège*, **1954,** pp. 86–95.

used in processes operating at elevated pressure. The tar produced in the lower temperature zones near the top of the beds must be removed from the gas. The production of hydrocarbons in these zones is undesirable when the gas is to be used for synthesis, but desirable when the gas is produced for distribution by public utilities.

The fixed fuel bed is generally the most efficient method for converting carbonaceous fuels into gas. In such processes, the descending movement of solids in the bed is countercurrent to the reacting gases. The fuel bed is characterized by zones of different temperature which may be designated grossly as: (1) ash zone at the bottom of the generator; (2) oxidation zone or region of heat supply; (3) reduction zone where steam is decomposed by reaction with carbon and with carbon monoxide, and where carbon dioxide reacts with carbon, and (4) preheat zone where the incoming fuel charge to the top of the fuel bed is dried and heated by the hot gases flowing upward. Fuels containing distillable volatile matter are partially devolatilized in the preheat zone, the volatile products appearing as methane, higher hydrocarbons, and tars in the product gas.

Fixed-bed gasification processes have been developed to handle a variety of solid fuels. The more important characteristics of the fuel affecting its use in fixed beds are: particle size and particle size distribution, agglomerating tendency, ash fusion temperature, and reactivity. Uniformly sized nonagglomerating fuels containing a minimum of fines and having reasonable mechanical strength are the most suitable. With such fuels, carryover losses and resistance to flow of gaseous reactants are minimized. Flow of gas is distributed uniformly, which results in stable reaction zones and a uniform pattern of temperature distribution in the bed. In nonslagging operations, the ash-fusion temperature imposes an upper limit on the bed temperature. Under all operating conditions, the reactivity of the fuel determines the minimum practical temperature in the gasification zone.

Processes Using Air, Operating at Atmospheric Pressure

Continuous Gas Producers. The conventional gas producer gasifies coke, anthracite coal, and noncaking or weakly-caking coal, with air and steam continuously under nonslagging conditions. The fundamentals of producer gas generation and the operation and mechanical features of conventional gas producers have been discussed by van der Hoeven[142] and also by Meunier[133] and von Gratkowski.[132] An extensive thermodynamic analysis of gas producers has been published by Gumz.[1]

Steam added to the blast of a producer cools the fire zone by endothermic reaction with carbon, and prevents fusion of ash. The carbon monoxide and hydrogen produced in the steam-carbon reaction increase the calorific value of the gas. The steam used is generally in the range of 0.4 to 0.7 lb/lb of fuel, and approximately 60 to 90% of the steam fed is decomposed. The effect of the steam content of the blast on the composition and calorific value of the gas, and on the gasification efficiency as determined by thermodynamic calculations for carbon, is shown in Fig. 21.[1]

Although modern gas producers can use a variety of fuels, the size range $\frac{1}{2}$ to $1\frac{1}{2}$ in.[130] or 2 in.[134] is preferred. The sizes should be uniformly distributed within the range, and the fuel should be free of fines. An ash-fusion temperature greater than 1150°C and a British standard swelling index less than 2, are desirable.[130] The size consist of the fuel has a significant effect on the specific rate of gasification (lb/sq ft-hr). For example, a maximum rate of 41.5 lb/sq ft-hr was obtained with anthracite peas, $\frac{3}{8}$ to $\frac{5}{8}$ in., as compared with a maximum rate of 55 lb/sq ft-hr with anthracite nuts, $1\frac{1}{4}$ to $2\frac{1}{4}$ in.[143] In tests with coke the following data were reported:[144]

Fuel	Fines Below $\frac{1}{8}$ in. %	Maximum Specific Gasification Rate, lb/sq ft-hr
Graded coke, $\frac{3}{4}$ by $1\frac{1}{2}$ in.	—	57
3 parts graded coke, 1 part coke breeze (0 to $\frac{3}{8}$ in.)	19	21.5
2 parts graded coke, 1 part coke breeze (0 to $\frac{3}{8}$ in.)	24	15.2

[142] van der Hoeven, B. J. C., *Chemistry of Coal Utilization*, John Wiley and Sons, New York, 1945, Vol. 2, pp. 1586–672.

[143] Forman, H. R., *J. Inst. Fuel*, **17**, 109–15 (1944).

[144] Grainger, J. A., *J. Inst. Fuel*, **28**, 594–600 (1955).

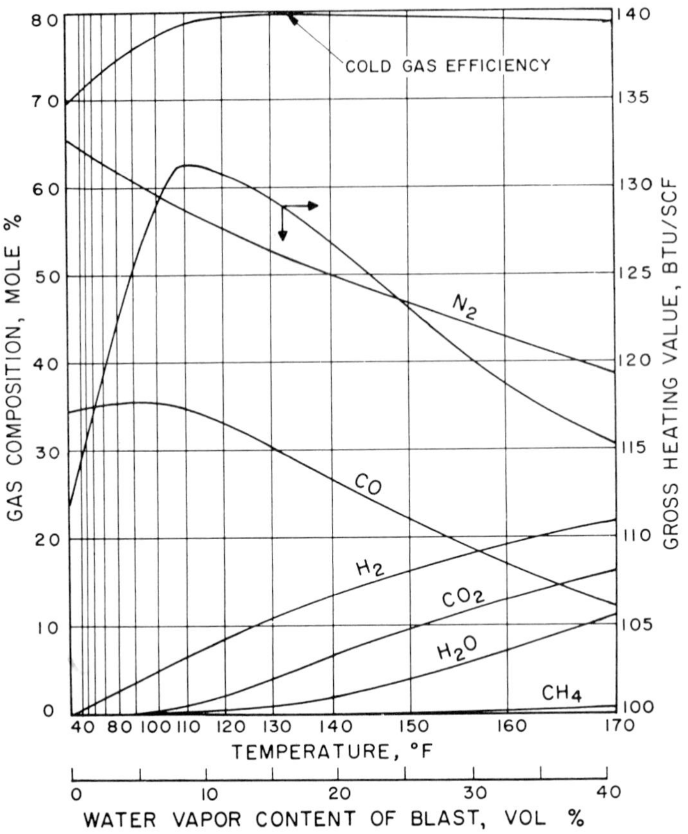

Fig. 21. Effect of steam content of blast on composition and heating value.[1]

Although these results indicate an effect of the quantity of fines, it was pointed out that the distribution of sizes is also important in determining the maximum gasification rate.

Performance data on modern producers when gasifying a variety of coals are summarized in Table 13. Data on the economics of producer gas generation have been published.[145,146]

Continuous and Cyclic Two-Zone Processes. Two-zone gasification processes have been developed to utilize coals unsuitable for carbonizing,[130] to simplify the removal of tar from the gas stream, and to improve the quality of the tar. The two-zone gasifier, which is also referred to as a double gas generator, a carbonizing generator, or a two-stage generator, consists essentially of a low-temperature retort superimposed on a high-temperature gas generator. The fuel passes down through the retorting zone, where it is heated either directly by purge gases from the high-temperature generator, or indirectly by hot gases passing through flues in the walls and in vertical, radial partitions dividing the retorting zone into four compartments.

Two-zone gasifiers designed to operate continuously by blasting with saturated air yield a gas similar to producer gas from coal, but having a slightly greater gross heating value, ranging from about 165 to 185 Btu/scf.[147] The cold gas efficiency ranges from 70 to 78%.[148] Approximately 40 to 50% of the gas from the high-temperature gasification zone passes through the retorting zone and leaves at a

[145] Hendryx, D. B., *ASME, Trans.*, **68,** 877–82 (1946).

[146] Bass, G. O., and Milner, G., *Gas World*, **138,** *Coking Sect.*, 43–54 (1953).

[147] British thermal unit per standard cubic foot of gas measured at 60°F and 30 in. of mercury, dry. Also, throughout this chapter, 1 Mcf = 1000 scf, 1 MMcf = 1000 Mcf; 1 MBtu = 1000 Btu, 1 MMBtu = 1000 MBtu; 1 therm = 100,000 Btu.

[148] Schussl, F., *Trans., World Power Conf., 5th Conf., Vienna,* **1956,** Vol. 7–8, 1681–716 (1959).

temperature of 100 to 150°C, depending on the moisture content of the fuel. Tar can be removed from this gas stream before it is recombined with the remainder of the gas from the high-temperature zone. Coal in the size range of approximately 10 to 50 mm is suitable for use.

Two-zone gasifiers operated cyclically with alternate blasting and steaming of the fuel bed produce a gas low in inerts. The blast gases leaving the high-temperature zone pass through flues and give up heat to the retorting zone. The blast gas is then burned with secondary air to heat the checkerbrick in a carburetor and superheater. In plants designed for making uncarbureted water gas, steam is admitted at the top of the superheater after a short steam purge in the direction of the blast. Preheated steam from the superheater (no carburetor is required for straight water-gas production) reacts in the high-temperature zone, and the products give up heat directly to the fuel in the retorting or predistillation zone. This type of operation preserves the gaseous hydrocarbons from the devolatilization of the coal and yields a gas sometimes referred to as coal-water-gas, having a heating value somewhat greater than that of conventional blue water gas. Results obtained in the gasification of coal in a Gas Integral (formerly VIAG) generator are:[148]

Average daily gas production corrected to 15°C and 760 mm Hg	3.56×10^6 scf
Higher heating value of gas at 0°C and 760 mm Hg	354 Btu/scf
Cold gas efficiency, percent	62.14
Heat recovered in tar and tar oils, percent	13.17
Heat in excess steam generated, percent	2.40
Total thermal efficiency, percent	77.7

The coal used was a mixture of Ruhr coal, two thirds "Furst Leopold" and one-third "Brassert," having a gross heating value of 13,400 Btu/lb and the following screen analysis:

Over 1.58 in., percent	40.5
0.79 to 1.58 in., percent	47.1
0.39 to 0.79 in., percent	7.0
0 to 0.39 in., percent	5.4

A typical analysis of the gas produced was: CO_2, 8.9; C_nH_m, 0.7; CO, 30.1; H_2, 47.9; CH , 5.2; N_2, 7.2%. The gross heating value was 357 Btu/scf.

The specific gasification rate of G.I. (VIAG) generators ranges from 1020 scf/sq ft-hr for generators 5.26 ft i.d., to 1610 scf/sq ft-hr for generators 11.8 ft i.d.

Detailed descriptions and operating results have been published for the G.I. (VIAG) process,[148,149] the Tully,[150] and the Power-Gas,[130,151] for the production of carbureted water gas; and for the Industriebedorfges (IBEG) generator[152] and the Heurtey[153] for producer gas.

Processes Using Oxygen, Operating at Atmospheric Pressure

Nonslagging processes. Anthracite coal and coke have been gasified completely with an oxygen-steam blast in fixed-bed gasifiers operated nonslagging. The operating results obtained with a Wellman-Galusha producer,[154] a converted UGI water-gas generator,[155] a Kerpely producer[156] and a Pintsch-Brassert generator[140] are summarized in Table 14. The heating value of the gas and the fuel required per Mcf of gas are within a relatively narrow range for all operating conditions. However, the oxygen used per MMBtu of gas ranged from 742 to 936 scf. Oxygen consumption is affected by exit gas temperature, and heat lost to the jackets. Newman[136] pointed out that heat lost to the jacket is principally a function

[149] Spaggiari, M. A., *Intern. Conf. Complete Gasification Mined Coal, Liège*, **1954**, 150–60.

[150] Tully, J. E., *Gas J.*, **279**, 782–7 (1954); *Gas Times*, **81**, No. 20, 28, 33–8 (1954). Harthill, I. J., *Gas J.*, **288**, 441–4 (1956).

[151] Dyde, J. H., *Gas World*, **143**, 580–5; 634–9, 688–91 (1956).

[152] Vacek, A., *Trans. World Power Conf.*, 5th *Conf.*, *Vienna*, **1956**, Vol. 7–8, 1717–32 (1959).

[153] Gagnaire, L., *Intern. Conf. Complete Gasification Mined Coal, Liège*, **1954**, 162–8.

[154] Wright, C. C., and Newman, L. L., *AGA, Proc.*, **1947**, 701–12. Wright, C. C., Barclay, K. M., and Mitchel, R. F., *Ind. Eng. Chem.*, **40**, 592–600 (1948).

[155] Blatchford, J. W., *AGA, Proc.*, **1950**, 652–8.

[156] Batchelder, H. R., Dressler, R. G., Tenney, R. F., Kruger, R. E. and Segur, R. D., *AGA, Proc.*, **1950**, 348–54. Willmott, L. F., Batchelder, H. R., and Tenney, R. F., *U.S. Bur. Mines, Rept. Invest.* No. 5108, (1955), 16 pp.

TABLE 13

OPERATING RESULTS IN THE GASIFICATION OF COAL IN MODERN MECHANICAL PRODUCERS[130]

Fuel Used	Baddesley Coal	Langwith Coal	Belgium Steam Coal	Pakistan Coal	Anthracite Briquettes
Data on Fuel					
Nominal size, in.	$\frac{1}{2}$–$1\frac{1}{2}$	$\frac{1}{2}$–$1\frac{1}{2}$	$\frac{3}{4}$–$3\frac{1}{8}$	$\frac{1}{2}$–3	$2 \times 1\frac{1}{2} \times \frac{7}{8}$
Below $\frac{1}{4}$ in., %	5.0	7.0	3.8	7.3	2.1
Average analysis					
Moisture (as fired), wt %	5.5	13.5	2.6	5.8	5.1
Ash (dry basis), wt %	9.0	5.9	20.0	12.3	21.9
Volatile matter (dry basis), wt %	40.0	38.5	12.5	43.3	8.6
Fixed carbon (dry basis), wt %	51.0	55.6	67.5	44.4	69.5
British standard swelling index	No. 1 Agglommerate		No. 1 Nonagglommerate		
Gross heating value (as fired), Btu/lb	12,080	12,060	11,980	11,340	10,800
Data on producer and average operating conditions					
Producer diameter, ft	10.5	9	8.5	8.5	8.5
Specific gasification rate, lb sq ft-hr	52	37	30	23	25
Approximate load factor, % rated capacity	105	81	—	—	—
Producer outlet temperature, °C	343	380	340	419	405
Blast saturation temperature, °C	62.8	55	66.0	58.1	59.9
Gas composition, vol. %					
CO_2	6.7	2.8	7.5	6.0	5.5
C_nH_m	—	—	—	0.25	0.2
O_2	—	—	0.2	0.25	0.1
CO	25.3	29.2	22.7	27.1	26.8
H_2	21.0	13.1	18.4	14.7	13.6
CH_4	1.8	3.3	1.9	2.0	0.5
N_2	45.2	51.6	49.3	49.7	53.3
Gross heating value, Btu/scf	165.6	168	150	158	138
Yields/ton of coal as fired					
Gas, scf, Therms (gross)	116,100	115,000*	120,900	101,800	123,900
	192.4	193.2	181.4	161	170.8
Dry tar, gal	21	20*	1.0	16.3	9.5
Efficiency, %					
Cold gas $= \frac{\text{gross heating value of gas produced}}{\text{gross heating value of coal fired}}$	71	71.5	67.6	63.4	70.6
Tar $= \frac{\text{gross heating value of tar produced}}{\text{gross heating value of coal fired}}$	13.9	13.1	0.7	11.6	7.5
Gas + tar	84.9	84.6	68.3	75.0	78.1

* Approximate.

of the depth of the reaction zone, and that, in general, with different size fuels the depth of the reaction zone is constant when expressed as the number of particles of average diameter it contains. On this basis, the depth of the reaction zone increases as average particle size increases. This may account for the increase in oxygen requirement with particle size in the results with anthracite shown in Table 14.

Operations with oxygen under nonslagging conditions have presented no unusual difficulties. In the tests with anthracite, the gasification rates with oxygen were twice those obtained with air. Operation with small sizes of anthracite required manually poking the fire to break up ash crust formations resulting from intermittent motion of the grate.[136] This difficulty would not be encountered with a grate designed for continuous motion with the size of fuel used in the test.

The effect of nitrogen in the blast on the results obtained in a Kerpely producer operating on pea coke is shown in Figs. 22 and 23.[156] Coke and oxygen requirements increase, cold gas efficiency and heating value decrease, as the nitrogen in the blast increases. The steam-coke ratio was increased with decreasing nitrogen to maintain approximately the same quality of the ash removed. The percent of steam decomposition decreased with decreasing nitrogen in the blast, but the absolute quantity of steam decomposed per pound of coke increased.

Slagging processes. Operation of gasifiers under slagging conditions removes limitations on attaining optimum performance imposed by the fusibility of the ash. For example, under equilibrium conditions the maximum gasification efficiency results when the oxygen concentration in the oxygen-steam mixture fed to

TABLE 14

OPERATING RESULTS WITH FIXED-BED NONSLAGGING OXYGEN-BLOWN PRODUCERS

Gasifier							
Type	Wellman-Galusha Producer[154]			Converted UGI Water-Gas Set[155]		Kerpely Producer[156]	Pintsch-Brassert Generator[140]
Internal diameter, ft	10			9		7	10.5
Fuel							
Type	Rice Anthracite	Barley Anthracite	Coke	Coke	Coke	Coke	Coke
Size, in.*	$\frac{5}{16} \times \frac{3}{16}$	$\frac{3}{16} \times \frac{3}{32}$	$2\frac{5}{8} \times \frac{5}{16}$	Run of oven	Run of oven	Nut and pea	
Gross heating value (as received), Btu/lb	12,250	11,590	12,000	—	—	12,515	—
Moisture (as received), wt %	6.4	7.6	5.0	—	—	2.2	3.9
Ash (as received), wt %	10.2	12.8	11.9	10.5	10.5	9.4	7.9
Ash fusion temp, °F	2810	2810	2200	2760	2760	2690	—
Input, lb/hr	2771	1335	2490	8196	11450	1715	—
Gas composition, vol %							
CO_2	16.50	17.20	11.3	16.9	16.6	19.6	22.7
O_2	0.15	0.20	0.6	0.1	0.0	—	—
CO	40.00	42.40	54.1	46.6	47.8	39.9	34.7
H_2	41.00	37.65	31.0	34.0	33.6	38.7	40.3
CH_4	0.85	0.75	0.4	0.2	0.2	—	1.4
N_2	1.50	1.80	2.6	2.2	1.8	1.8	0.9
Gross heating value, Btu/scf	271	266	258	262	265	254	257
Gasifier outlet temp, °F	879	730	508	1209	1211	—	932
Output, Mcf/hr	119.5	51.5	102.4	364	501	78.9	—
Material requirements							
Per Mcf of CO + H_2							
Fuel, lb	28.7	32.3	28.7	27.9	28.4	27.6	30.7
Oxygen (100%), scf	249	275	274	302	308	264	267
Steam, lb	40.2	44.5	28.9	38.2	38.5	47.8	58.3
Per therm of gas							
Fuel, lb	8.56	9.69	9.46	8.59	8.64	8.54	8.96
Fuel, therms	1.049	1.123	1.135	—	—	1.069	—
Oxygen (100%), scf	74.2	82.7	90.3	92.8	93.6	80.3	77.9
Steam, lb	12.0	13.4	9.5	11.8	11.7	14.8	17.0
Performance							
Specific gasification rate, therms/sq ft-hr	4.13	1.74	3.36	14.96	20.83	5.21	6.13
Specific gasification rate, therms/cu ft-hr	0.472†	0.201	0.383	—	—	0.744†	—
Steam decomposition, %	51.2	42.0	61.4	53.5	51.0	46.2	43
Carbon conversion	—	—	—	—	—	—	93

* Round-hole screen size.
† Estimated.

the gasifier is 40 to 48%,[157] corresponding respectively to maximum temperatures in the reaction zone of 1500 to 1800°C, which are well into the slagging region for most residues. Under nonslagging conditions, the oxygen concentration in the oxygen-steam blast is generally in the range of 22 to 30%, corresponding respectively to maximum reaction zone temperatures of 1200 and 1300°C.

The effect of oxygen concentration in the oxygen-steam mixture fed to a gasifier on the steam required, and the percent steam decomposition, is shown in Fig. 24.[140] For the fuels used, operation in fixed beds was impossible in the range of 25 to 35% oxygen in the gasifying medium. The marked effect of operating under slagging conditions on the reduction of steam requirements, and on the increase in steam decomposition, is evident from this figure.

Operation under slagging conditions leads to simplicity in the design of the gasifier by eliminating grates and mechanical devices for removing ash. Instead, arrangements must be provided for tapping slag periodically, and operating conditions must be such that the slag has the proper fluidity for free flow. Excellent comprehensive reviews of the methods for assessing the flow properties, formation, and mineralogy of coal-ash slags have been published.[158] For practical purposes, the flow properties of coal-ash slags may be characterized by the relation between viscosity of the liquid slag and its temperature.[159] This relation can be predicted from the silica ratio of the ash ($SiO_2 \div SiO_2$ plus equivalent Fe_2O_3 plus MgO plus CaO). In a cyclone combustor, it has been shown[160] that operability under

157 van Loon, W., *Intern. Conf. Complete Gasification Mined Coal, Liège*, **1954,** 136–41.

158 Shaw, J. T., *BCURA Bull.*, **23,** 169–90 (1959). Watt, J. D., *BCURA Bull.*, **23,** 49–69 (1959).

159 Reid, W. T., and Cohen, P., *ASME, Trans.*, **66,** 685–90 (1944). Cohen, P., and Reid, W. T., *U.S. Bur. Mines, Tech. Paper* No. 663 (1944), 22 pp.

160 Hoy, H. R., Roberts, A. G., and Wilkins, D. M., *J. Inst. Fuel*, **31,** 429–57 (1958).

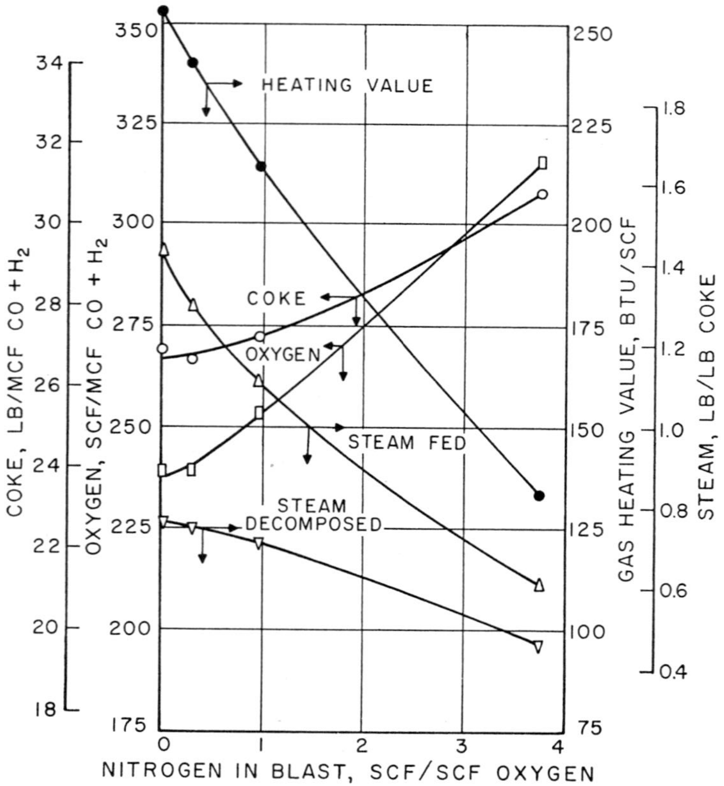

Fig. 22. Effect of nitrogen in blast on the performance of a Kerpely producer.[156]

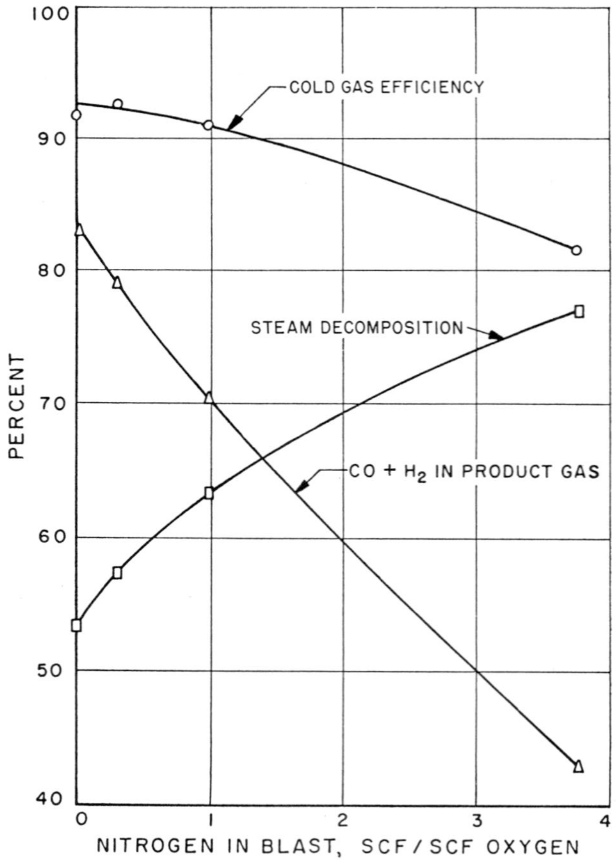

Fig. 23. Effect of nitrogen in blast on operating results with a Kerpely producer.[156]

slagging conditions depends on the intensity of heat liberation, the silica ratio of the ash, and the rate of ash-input. The same factors are important in gasification, but definitive studies similar to those made for the cyclone combustor have not been published. The flow characteristics of slags may be improved by the addition of fluxing agents such as lime, dolomite, or blast furnace slag.[160,161] The quantity and type of flux added depends on the properties of the ash and on economic considerations. In one series of tests,[161] 16 to 18% of open hearth slag and coke slag were added.

Operating results obtained in slagging gasifiers with oxygen-steam and oxygen-carbon dioxide as the gasifying medium are shown in Table 15. The Thyssen-Galoczy generator has been described in the literature.[132,136] The construction of the Leuna slagging generator is shown in Fig. 25;[162] substantially the same generator was used in Kalmykov's tests.[161]

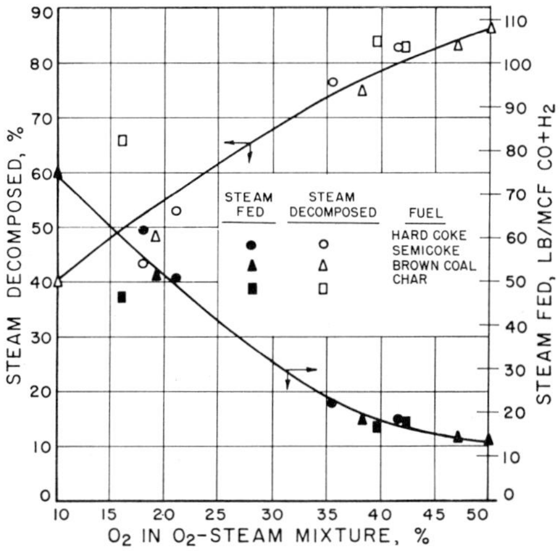

Fig. 24. Effect of oxygen concentration in blast on steam usage and decomposition.[140]

Processes Using Oxygen or Air, Operating at Elevated Pressure

Nonslagging processes. The Lurgi process for gasifying coal in a fixed bed with oxygen and steam at pressures up to 30 atm was first operated commercially in 1936.[162,163] After World War II, the process was further developed, and is now operated on a large scale at several locations to produce synthesis gas[164,165] and gas for public consumption.[166,167]

Gasification with oxygen and steam at elevated pressure has the following advantages[165] in all gasification processes: (1) high gasification rates per unit of reactor volume, with attendant reduction in the percentage of heat lost from the reactor; (2) minimum cost of compression, since it is necessary to compress only the oxygen (which amounts to about 20 to 35% of the volume of gas produced), instead of compressing all of the product gas; (3) gas purification and synthesis reactions can be accomplished more economically because of the lower investment cost in smaller vessels; (4) pressure facilitates heat exchange in waste-heat recovery and cooling of gas streams; and (5) pressure shifts the equilibrium of hydrogenation reactions toward increased concentrations of methane, with an attendant increase in the heating value of the gas. Advantages of pressure gasification specific to fixed-bed processes are:[165] (1) small sizes of fuel may be gasified with minimum carryover because of the low gas velocities resulting from the higher gas density; (2) less reactive fuels, and fuels rich in ash, can be utilized because of the higher gas residence time resulting from lower gas velocities; (3) with countercurrent flow and gasification in the presence of excess carbon, hydrogenation reactions occur and the exothermal heat of these reactions is recovered, resulting in reduced oxygen requirements per therm of gas, and in addition, carbon gasified as methane reduces the quantity of carbon that must be gasified with steam;[11] and (4) the yield and quality of byproducts is improved. The disadvantages of using pressure in fixed-bed processes are:[165] (1) methane must be removed when the gas is used for the ammonia synthesis; (2) the cost of treating the gas liquor must be recovered by the sale of phenol and ammonia;

[161] Kalmykov, N. V., *Gazovaya Prom.*, **1**, 8–14 (1956).

[162] Newman, L. L., *Ind. Eng. Chem.*, **40**, 559–82 (1948).

[163] Weir, H. M., *Ind. Eng. Chem.*, **39**, 48–54 (1947).

[164] Linton, J. A., and Tisdall, G. C., *Coke and Gas*, **19**, 402–7, 442–7 (1957); **20**, 148–52, 159 (1958).

[165] Hubmann, O., *Brennstoff-Chem.*, **40**, 65–71 (1959).

[166] Bieger, F., *Ges. Ber. Betrieb u. Forschung, Ruhrgas A.-G.*, No. 7, 1958.

[167] Andrews, R. S., *Inst. Gas Engrs., Commun.* No. 503, 1957, 31 pp.; *Erdöl u. Kohle*, **8**, 703–5 (1955).

TABLE 15

Operating Results with Slagging Gasifiers

Gasifier									
Type	Thyssen-Galoczy[162]	Leuna-Würth[140]						Slagging Producer[161]	
Internal diameter, ft	10	6.89 at bottom, 11.15 at top						5.56 at bottom, 10.75 at top	
Fuel									
Type	Coke	Brassert Refuse	Deuben Char	Metallurgical Coke				Low-Temperature Coke	
Size, in.	$2\frac{1}{2} \times 1\frac{1}{2}$	—	—	$1\frac{1}{2}$ and larger				3 × 40 mm**	3×40mm§§
Gross heating value (as received), Btu/lb	12,400	—	—	—	—	—		11,750	11,800
Moisture (as received), wt %	5.0	—	18.0	1.8	—	1.0		11.70	11.23
Ash (as received), wt %	8.5	50 to 55†	17.4	9.1	—	11.8		8.04	8.00
Ash fusion temp	2372 °F	—	—	—	—	—		2012	2012
Input, lb/hr	3600	—	—	—	—	—		12,400††	19,100‖‖
Gas composition, vol %									
CO_2	2.8	9.7	5.4	6.8	5.7	3.0		5.57	0.99
O_2	0.1	—	—	—	—	—		—	—
CO	70.4	66.5	62.4	61.4	64.6	92.5		83.64	66.12
H_2	23.1	22.9	31.2	31.0	28.7	3.0		8.76	31.01
CH_4	0.2	0.0	0.0	0.0	0.1	0.0		0.61	0.68
N_2	3.4*	0.9	1.0	0.8	0.9	1.5		1.42‡‡	1.20¶¶
Gross heating value, Btu/scf	299	288	302	298	301	307		304	320
Gasifier outlet temp, °F	—	392	464	752	734	500		986	1022
Output, Mcf/hr	134	304‡	257‡	254‡	580‡	257‡		401	629
Material requirements									
Per Mcf of CO + H_2									
Fuel, lb	28.4	—	47.4	29.0	29.4	31.4		30.9	30.5
Oxygen (100%), scf	310	347§	276§	262§	260§	310§		240	213
Steam, lb	13.1	12.3	15 8	20.5	17.4	nil		nil	11.4
Carbon dioxide, scf	—	—	—	—	—	258		218	—
Per therm of gas									
Fuel, lb	8.87	—	14.7	8.99	9.11	9.76		9.49	9.27
Fuel, therms	1.100	—	—	—	—	—		1.175	1.094
Oxygen (100%), scf	96.9	107.6§	85.5§	81.2§	80.6§	96.4§		72.9	64.8
Steam, lb	4.09	3.82	4.90	6.36	5.39	—		—	3.47
Carbon dioxide, scf	—	—	—	—	—	80.2		66.2	—
Performance									
Specific gasification rate, therms/sq ft-hr	5.11	10.95‖	9.71‖	9.48‖	21.85‖	9.87‖		13.42	22.18
Specific gasification rate, therms/cu ft-hr	—	1.33¶	1.17¶	1.15¶	2.55¶	1.20¶		—	—
Steam or carbon dioxide decomposition, %	90.1	98	84	76.5	83	86		74.2	80.6
Carbon conversion	—	—	80	93	93	87		94.0	94.2

* Oxygen purity, 90%.
† Fuel contained 45 to 50% C.
‡ Based on an estimate of 80 sq ft cross section at top of fluid bed.
§ Includes 1 to 2% N_2.
‖ Based on cross-section at top of fuel bed.
¶ Approximate value based on 13 ft, height of fuel bed and average cross section.
** 6.5% over 40 mm, 3.3% under 3 mm.
†† Exclusive of flux (17.2% of fuel input).
‡‡ Includes 0.04% H_2S.
§§ 11.0% over 40 mm, 3.0% under 3 mm.
‖‖ Exclusive of flux (18.11% of fuel input).
¶¶ Includes 0.09% H_2S.

(3) direct production of carbon monoxide-rich gases is not possible except in slagging processes—in nonslagging processes, such as the Lurgi, CO/H_2 ratios of 1.2 to 1.5 may be obtained without additional cost by replacing part of the gasification steam with carbon dioxide; and (4) some coals that are noncaking at atmospheric pressure develop caking tendencies when gasified under pressure. One complication common to all pressure gasification processes is the feeding of a solid fuel into a pressure vessel.

Basic studies of the reactions occurring in the fuel bed of a pressure gasifier 2 ft in diameter with a 7-ft fuel bed have been made by Hebden and co-workers.[11] High-reactivity, high-volatile-matter fuel gave the lowest consumption of steam and oxygen, and the highest heating value gas. Volatile matter content determines the quantity of fixed carbon to be gasified, and, consequently, the consumption of oxygen and steam. Reactivity affects the extent of steam decomposition. The consumption of fuel corrected for tar and oil produced was not affected significantly by the nature of the fuel. Increasing pressure increased the proportion of the carbon appearing as methane (14.4% at 10 atm; 22.2% at 40 atm) and, hence, increased the heating value. The reduction in oxygen consumption below that observed at one atmosphere resulted from increased methane formation at the higher pressures, and amounted to 13.0% at 10 atm, and 21.6% at 40 atm. For the three fuels studied (N.C.B. Rank Code No. 702, anthracite, and oven coke) the authors concluded that methane must have been produced by the hydrogenation of carbon. Figure 26 shows

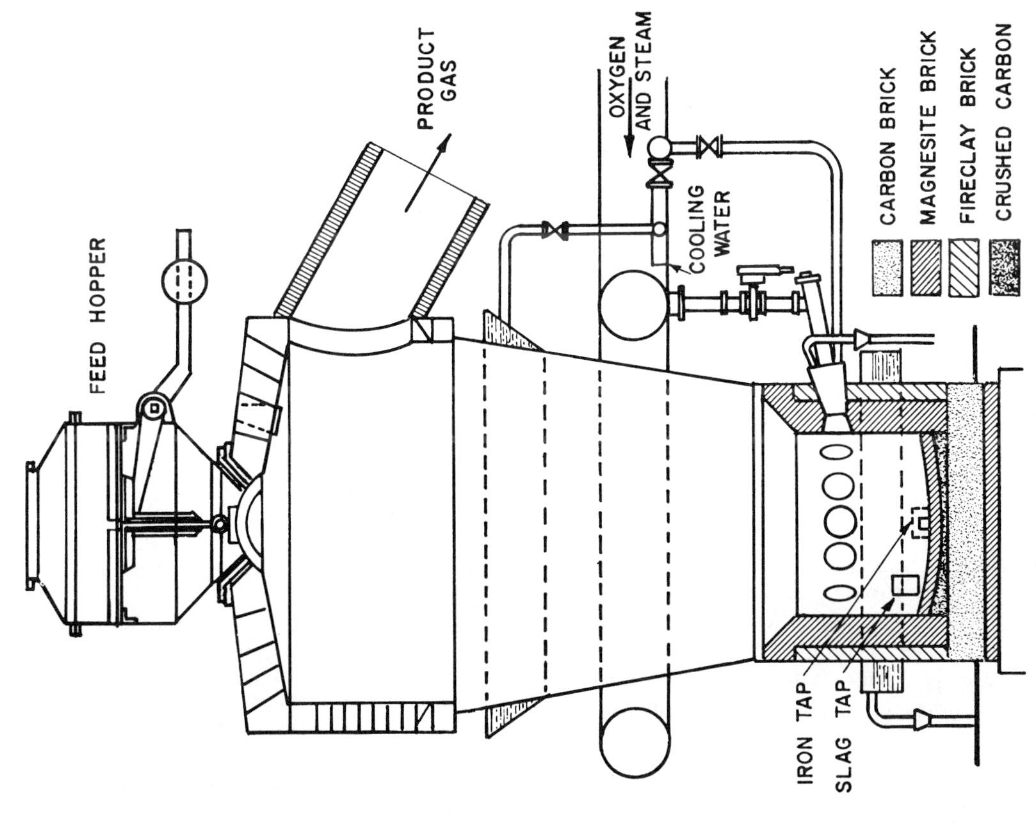

Fig. 25. Leuna slagging generator.[162]

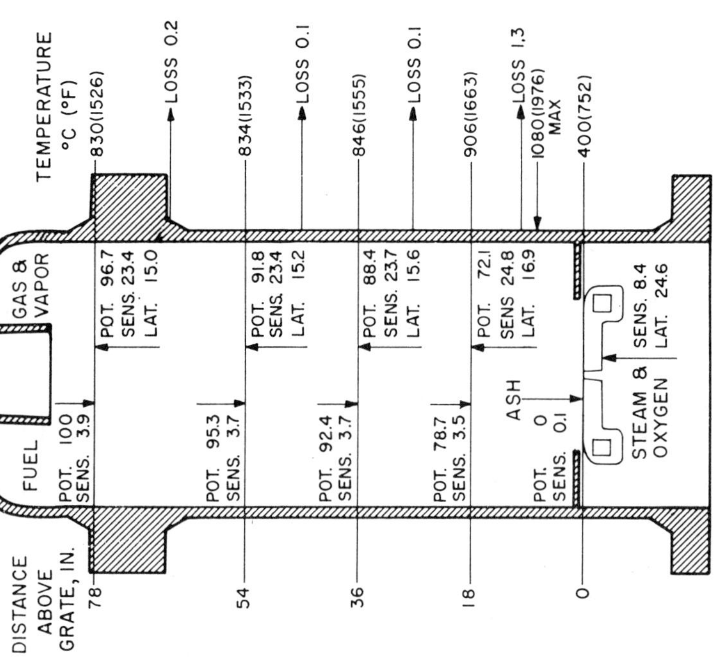

Fig. 26. Heat balances at various levels in the fuel bed when gasifying type 702 coal at 20 atm pressure and a steam-oxygen volumetric ratio of 8.75.[11] (Basis: 100 therms of fuel entering at the 78-in. level.)

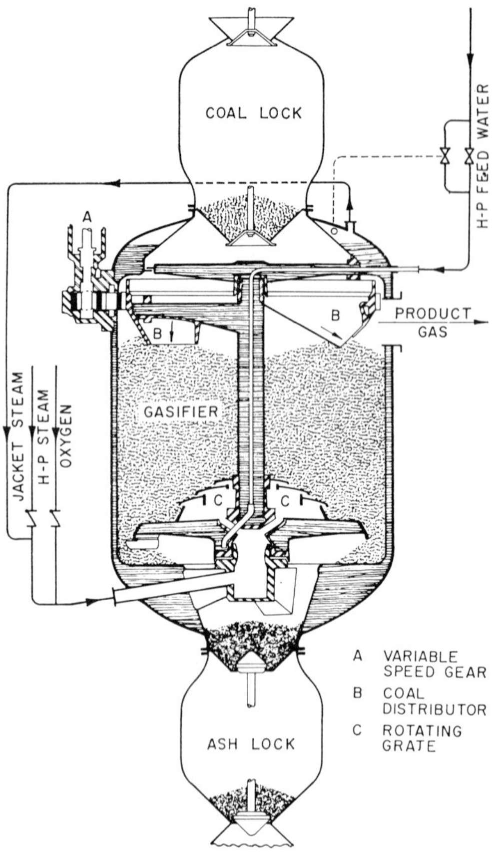

Fig. 27. Lurgi pressure gasifier.[164,165]

the temperatures at various levels in the bed, and the heat balance at these levels.

The gasification of pretreated American coals has been studied in a pressure gasifier having a cross-sectional area of 1 sq ft.[168]

Figure 27 shows a cross section of a modern Lurgi generator operating at SASOL.[164,165] The gasifier is a water-jacketed vessel, 13 ft o.d., 12 ft i.d., and 26 ft high. Coal is fed to the gasifier through a lock hopper at the top. The grate and coal distributor are mounted on a hollow, water-cooled shaft, and are driven from a ring gear at the top. The water jacket is maintained at a pressure slightly higher than gasification pressure so the inner vessel must withstand only a small differential pressure. The inner vessel is unlined, and unalloyed steels have given satisfactory service against normal causes of corrosion.[165] However, difficulties have been experienced[165] when in operation with coals having an abnormally high chlorine[169] or alkali[170] content.

To extend the range of coals suitable for use in the Lurgi generator, an experimental unit having a cross-sectional area of 8.6 sq ft and a working pressure of 440 psi was constructed and operated with various coals. Experience in this plant led to the development of a stirring device near the top of the bed which permitted utilization of fuels having a Damm agglutinating index up to 12[171] and a British Standard caking index of 10. The results of tests with various fuels are given in Table 16.[171,172]

The operating results in full-scale plants are summarized in Table 17.[165] The results of full-scale tests of Pennsylvania anthracite in one of the Lurgi generators at the Dorsten plant of Steinkohlengas, A.G. and in other fixed bed generators are compared in Table 18.[172]

The gasification of bituminous coals with air at 6 atm in an overfeed updraft experimental producer having a grate area of 1 sq ft has been studied.[173] With air preheated to 600°F, and with a range of process variables and types of coal, the heating value of the gas ranged from 104 to 164 Btu/scf, and the specific gasification rate from 42 to 224 lb/sq ft-hr.

Slagging gasifiers. Gasification at elevated pressures under slagging conditions is still in developmental stages.[174,175] The results of tests at 75 lb/sq in., in a 3 ft 9 in. i.d. gasifier having a length of 19 ft 9 in., on the effect of steam-oxygen in the gasification of petroleum coke are shown in Fig. 28.[174] The reduction in

[168] Cooperman, J., Davis, J. D., Seymour, W., and Ruckes, W. L., *U.S. Bur. Mines, Bull.* No. 498 (1951), 38 pp.

[169] Bieger, F., *Ges. Ber. Betrieb u. Forschung, Ruhrgas A.-G.*, No. 8, 1959, pp. 13-9.

[170] Bate, J. G., paper presented at the Ann. Conf. Australasian Inst. Mining and Met., August, 1959.

[171] Hubmann, O., *Intern. Conf. Complete Gasification Mined Coal, Liège*, **1954**, 169–77.

[172] Morgan, R. E., Eckerd, J. W., Ratway, J., and Baker, A. F., *U. S. Bur. Mines, Rept. Invest.* No. 5420 (1958), 22 pp.

[173] Nelson, H. W., Kouns, H. H., and Buckland, B. O., *Gasification and Liquefaction of Coal*, AIME, New York, 1953, pp. 109–21.

[174] Hebden, D., and Edge, R. F., *Gas Council (Gt. Brit.), Research Commun.* GC 50 (1958), 29 pp.; GC 50/44 (1959), 8 pp.

[175] Kolodtsov, K. I., and Babiĭ, V. I., *Teploenergetika*, **5**, No. 5, 25–31 (1958).

TABLE 16

RESULTS OF TESTS WITH VARIOUS FUELS IN LURGI GASIFIER AT HOLTEN PLANT[171]

Fuel	Rhenish Brown Coal		Ruhr High-Volatile Coal		English W. Midlands Coal		Scottish Bogside Coal		Australian Bituminous Coal	
Fuel										
Type										
Size, mm	1–5		10–30		10–30		10–30		10–30	
Gross heating value, Btu/lb	4743		6039		—		4632		6670	
Moisture, wt %	20.20		2.52		7.5		8.4		2.5	
Ash, wt %	4.65		22.41		7.5		28.5		15.67	
Fixed carbon, wt %	35.40		45.19		49.6		38.4		52.36	
Volatile matter, wt %	39.75		29.88		35.4		24.7		29.47	
Tar content (Fischer assay), wt %	10.00		14.50		12.26		5.7		15.66	
Sulfur, combustible, wt %	0.28		1.33		1.10		1 3		0.88	
Ash fusion temperature (Leitz blobbling temperature), °C	1380		—		1495		>1500		>1500	
Gasification pressure, psig	367.5		333.5		301		294		372	
Steam/oxygen ratio, lb/100 scf	45.4		42.7		37.5		40.7		40.7	
Gas composition, vol %	Crude	Purified	Crude	Purified	Crude	Purified	Crude	Purified	Crude	Purified
CO_2	32.2	2.0	28.8	2.0 }	28.9	2.0	{ 30.5	2.0	27.3	2.0
H_2S	0.1	—	0.4	— }			{ 0.5	—	0.2	—
C_nH_m	0.5	0.5	0.5	0.5	0.3	0.3	0.2	0.2	0.5	0.5
CO	17.4	25.2	18.5	25.6	19.5	26.9	17.4	24.7	21.2	28.7
H_2	37.2	54.0	41.1	57.0	40.7	56.2	42.5	60.3	39.3	53.2
CH_4	12.1	17.6	10.0	13.9	10.1	13.9	8.4	12.1	11.0	14.9
N_2	0.5	0.7	0.7	1.0	0.5	0.7	0.5	0.7	0.5	0.7
Gross heating value, Btu/scf	300	435	300	410	—	410	—	395	—	415
Throughput (raw coal), lb/sq ft-hr	323		242		178.5		229		178	
Crude gas yield, scf/lb raw coal	22.8		27.8		30.9		22.3		29.7	
Purified gas yield, scf/lb raw coal	15.7		20.1		22.5		15.4		22.0	
Oxygen used, scf/scf crude gas	0.0924		0.1499		0.1340		0.151		0.1482	
Chemical energy in gas per scf oxygen used, Btu	3520		2130		2370		1920		2230	
Steam consumption, lb/100 scf crude gas	4.19		6.39		5.02		6.20		5.92	
Gasification efficiency (coal basis), %	96.2		90.6		91.3		87.9		91.2	
Tar yield, % of assay yield	83.8		68.5		62.2		74.7		72.6	
Tar oil (sp gr = 1.0*), gal/ton coal	12.54		14.56		14.11		6.61		13.26	
Light oil (sp gr = 0.88*), gal/ton coal	7.13		2.80		4.84		2.80		8.58	

Fuel	South African Bituminous Coal		Pakistani Bituminous Coal		Belgian Low-Volatile Coal		Ruhr Low-Volatile Coal		Ruhr Coke Pearls	
Fuel										
Type										
Size, mm	15–50		3–20		3–15		5–20		5–15	
Gross heating value, Btu/lb	4875		6180		7551		8004		5650	
Moisture, wt %	10.50		4.5		0.99		2.80		20.5	
Ash, wt %	21.76		12.9		9.93		4.67		9.7	
Fixed carbon, wt %	44.45		42.6		77.37		83.58		68.5	
Volatile matter, wt %	23.29		40.0		11.71		8.95		1.3	
Tar content (Fischer assay)	4.69		14.6		1.08		2.31		—	
Sulfur, combustible, wt %	0.75		5.45		1.88		0.93		—	
Ash fusion temperature (Leitz blobbling temperature), °C	1325		1224		1420		—		1420	
Gasification pressure, psig	360		151		301		294		294	
Steam/oxygen ratio, lb/100 scf	42.2		43.6		44.3		35.4		46.1	
Gas composition, vol %	Crude	Purified	Crude	Purified	Crude	Purified	Crude	Purified	Crude	Purified
CO_2	30.2	1.0	30.2	2.0	27.7	2.0	26.5	2.0 }	37.0	2.0
H_2S	0.3	—	1.3	—	0.4	—	0.2	— }		
C_nH_m	0 2	—	0.6	0.6	0.2	0.2	0.1	0.1	—	—
CO	17.4	24.9	16.5	23.7	19.6	26.7	21.3	28.5	17.0	26.4
H_2	42.0	60.0	41.8	60.0	43.7	59.6	43.4	58.0	42.3	65.8
CH_4	9.4	13.4	8.8	12.6	7.9	10.8	8.0	10.7	3.2	5.0
N_2	0.5	0.7	0.8	1.1	0.5	0.7	0.5	0.7	0.5	0.8
Gross heating value, Btu/scf	290	405	—	405	—	385	290	385	—	345
Throughput (raw coal), lb/sq ft-hr	304		149.5		136		134		80.6	
Crude gas yield, scf/lb raw coal	26.3		27.9		43.1		46.1		38.6	
Purified gas yield, scf/lb raw coal	18.3		18.0		30.8		34.5		24.9	
Oxygen used, scf/scf crude gas	0.139		0.132		0.146		0.145		0.2118	
Chemical energy in gas per scf oxygen used, Btu	2200		2200		2010		2140		1125	
Steam consumption, lb/100 scf crude gas	5.88		5.75		6.69		5.13		9.76	
Gasification efficiency (coal basis), %	92.9		84.6		94.2		96.4		84.4	
Tar yield, % of assay yield	57.0		73.0		—		—			
Tar oil (sp gr = 1.0*), gal/ton coal	3.07		17.55		—		—		—	
Light oil (sp gr = 0.88*), gal/ton coal	3.34		7.59		—		—		—	

* Specific gravities assumed for conversion to B.I. units—Translator.

process steam required, and the increase in steam decomposition and temperature at the outlet of the gasification zone under slagging conditions, are clearly indicated.

The minimum in oxygen consumption per therm of gas at intermediate steam-oxygen ratios occurs primarily because of two opposing factors affecting the sensible heat losses from the gasification zone: (1) a decrease in sensible heat lost in the undecomposed steam

TABLE 17

Operating Results of Large Lurgi Plants[165]

Plant	Morewell-Australia Pipeline Gas 6 Generators 2.6 m		Sasol-South Africa Fischer-Tropsch-Synthesis 10 Generators 3.7 m				Ruhrgas Dorsten Pipeline Gas 6 Generators 2.6 m			
Fuel										
Type	Brown Coal		Bituminous Coal				Gas Coal			
Size	Briquettes		7–20 mm or 20–60 mm				6–30 mm			
Moisture, wt %	15.6		9.7				5.0			
Ash, wt %	1.6		26.4				22.0			
Fixed carbon, wt %	42.2		42.8				42.0			
Volatile matter, wt %	40.6		21.1				31.0			
Tar content (Fischer-Hempel), wt %	5.6		2.1				10.5			
Gas composition, vol %	Crude	Purified	Crude	Purified	Crude	Purified	Crude	Purified	Crude	Purified
$CO_2 + H_2S$	34.0	4.5	28.0	0.8	32.3	0.9	27.0	2.0	30.5	2.0
O_2	0.1	0.1	0.0	—	0.01	—	0.0	—	0.0	—
C_nH_m	0.4	0.5	0.2	0.2	0.2	0.2	0.4	0.5	0.4	0.5
CO	14.5	20.7	22.4	31.5	16.4	24.8	21.0	28.5	17.0	24.1
H_2	35.8	52.6	38.0	53.3	39.4	59.3	40.5	54.1	40.5	57.3
CH_4	13.5	19.4	10.9	13.4	11.3	14.0	10.1	13.6	10.7	14.9
N_2	1.7	2.2	0.5	0.8	0.4	0.8	1.0	1.3	0.9	1.2
Gross heating value, purified gas, Btu/scf		457		421		424		415		428
Raw gas per generator, Mcf/hr	260–372		930–1190				559–596			
Oxygen used, scf/scf	0.121	0.178	0.139	0.213	—	—	0.17	0.23	0.16	—
Steam used for gasification from outside sources, lb/Mcf ranges	51	—	35.5	—	—	—	41.5	—	56.3	—
Tar and light oil yield, wt %	3.5		1.6		—	—	7.5			
Refined light oil yield, wt %	0.75		0.4		—	—	1.67			
Ammonium sulfate yield, wt %	—	—	5.2		—	—	2.98			
Low-pressure steam produced per scf raw gas, lb/Mcf	62.1		33.2		—	—	35.5			
Efficiency referred to raw gas, %	Gross heating value (gas + byproducts)/Gross heating value of Coal = 88–92%									

as the steam to oxygen ratio decreases; and (2) an increase in the sensible heat lost in the gases as the steam to oxygen ratio decreases. The first of these effects predominates at steam-oxygen ratios above 4, and the second at lower ratios. Although the oxygen consumption is higher under slagging conditions, it was suggested[174] that savings in capital charges on a properly designed gasification and purification plant would result in an appreciable reduction in overall cost of gas. This saving occurs because of the reduction

TABLE 18

Test Results in Gasification of Anthracite in Fixed-Bed Generators[172]

Place	Dorsten Plant		Hirschfelde Gas Works[161]	Holten Plant*		Nechells Gas Works[11]	USBM Lurgi Gasifier[168]	Wellman-Galusha Trail Tests[154]
Coal	Chestnut Anthracite	Rice-Buckwheat Anthracite	Wiesche Lean Coal	Korean Anthracite	Korean Anthracite	British Anthracite	Buckwheat No. 4 Anthracite	Rice Anthracite
Moisture, wt %	—	—	—	—	—	—	—	6.4
Ash, wt %	26.4	18.7	5.4	15.5	24.8	2.5	11.6	10.9
Fixed carbon, wt %	67.7	75.6	86.9	82.2	71.9	92.7	80.4	84.3
Volatile matter, wt %	5.9	5.7	7.7	2.3	3.3	4.8	8.0	4.8
Coal (dry), lb/hr	15,547	15,147	525	879	1029	587	10.3	2594
Coal (dry, ash-free), lb/hr	11,329	12,236	494	743	774	572	9.1	2311
Oxygen (100%), scf/hr	79,510	83,612	3230	7483	7435	2952	88.5	23997
High-pressure steam, lb/hr	23,627	26,970	1325	3134	3161	1209	25.3	3879
Gas produced, scf/hr	430,629	500,387	22,115	36,597	33,097	19,006	405	119,430
Gas (CO_2-free), scf/hr	316,607	338,003	16,185	25,325	21,679	13,247	283	99,724
Gas composition, vol %								
CO_2	26.0	24.9	27.0	30.8	34.5	30.3	30.1†	16.5
C_nH_m	0.2	0.2	0.2	0.1	0.2	—	0.3	0.15
CO	26.3	26.9	20.3	24.0	21.3	19.8	24.0	41.00
H_2	40.6	40.6	38.5	41.2	40.7	43.3	35.1	40.00
CH_4	5.4	5.8	12.3	3.1	2.8	6.2	6.9	0.85
N_2	1.5	1.6	1.7	0.8	0.5	0.4	3.1	1.5
Steam temperature, °F	591	596	914	871	882	—	—	—
Generator pressure, atm	23.0	23.0	21.0	20	20	20.0	30.0	1
Grate area, sq ft	61.4	61.4	11.2	10.86	10.86	3.1	0.196	78.5
Gasification rate, scf/sq ft-hr	7026	8164	1450	3370	3048	6053	2066	1521
Coal (dry) gasified, lb/sq ft-hr	254	247	46.8	80.9	94.8	187	52.5	33.0
Ratios per Mcf gas								
Coal (dry), lb	36.1	30.3	23.7	24.0	31.1	30.9	25.4	21.7
Coal (dry, ash-free), lb	26.2	24.4	22.3	20.3	23.4	30.1	22.5	19.4
Oxygen (100%), scf	185	167	146	204	225	155	219	201
Steam, lb	54.9	53.9	59.9	85.6	95.5	63.6	62.5	32.5

* Pressure gasification trials on Korean Anthracite; prepared for Chemical Construction Corp. as part of Gesellschaft für Wärmetechnik, m.b.h., contract with United Nations Korean Reconstruction Agency.

† Gas contains 0.5% oxygen.

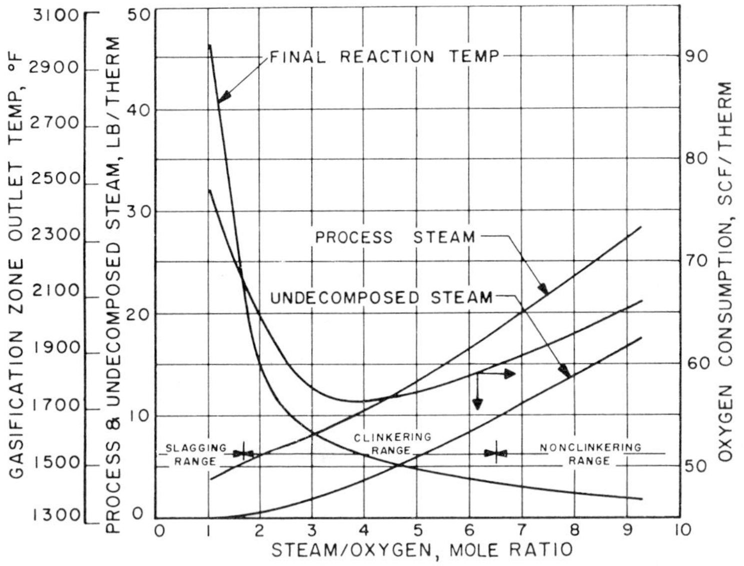

Fig. 28. Effect of steam-oxygen ratio on oxygen-steam consumption, and temperature.[174]

in steam required and the attendant reduction in liquor to be handled.

Operation under slagging conditions with coke and added slag (about 30% ash-plus-slag in the mixture) was eventually brought under control by intermittent slag tapping, with the hot gases from an auxiliary burner being used to heat the slag-tap tube and control the flow of slag. The authors pointed out that continuous slag tapping probably would be feasible in larger units. Nevertheless, independent control of slag temperature and the ability to interrupt slag flow are features of the intermittent system worthy of further examination.

In estimating the performance on coal, the authors[174] pointed out that the outlet temperatures at the top of a slagging gasifier operating on high-moisture, high-ash fuels may be so low that heavy tar fractions will accumulate and tend to agglomerate the coal. For example, a gas outlet temperature of 155°C is estimated for a coal containing 12.5% moisture and 25% ash as charged, compared with a temperature of 400°C which probably would be required at high outputs in a slagging gasifier. One method of achieving higher gas outlet temperatures is by injection of pulverized fuel through the tuyeres. With the foregoing coal, approximately 23% of the total fuel would have to be supplied as pulverized coal through the tuyeres to maintain an outlet gas temperature of 400°C.

A discussion[174] of the use of pulverized fuel in connection with fixed-bed processes directed attention to the importance of this technique in using fine sizes of fuel and recycling carry-over. An extensive study of the use of pulverized coal in a fixed-bed slagging gasifier operating at elevated pressure was undertaken at the Leatherhead Station of the British Coal Utilisation Research Association.[176] Work was also being done[177] at the same location on a slagging gasifier with injection of fines operating at a pressure 2 to 3 psig. With the present design of gasifier, an efficiency (potential heat in gas ÷ potential heat in coal) ranging from 88 to 91% has been obtained in purely fixed-bed operation with coke, as compared with 80 to 83%, when up to 47% of the total fuel was fed in the pulverized form.

Pulverized coal has been gasified with air at 3.5 to 7 atm in a fixed fuel bed operated under slagging conditions.[175] Lump anthracite (10 to 50 mm) having a gross heating value of 12,300 Btu/lb was used in the bed, and pulverized anthracite or low-volatile coal was fed through the tuyeres. The experimental unit was operated at a constant air rate of 2940 lb/hr and at a fuel loading of 359 lb/sq ft-hr, or 14.1 therms/cu ft-hr. The consumption of

[176] Ministry of Power (Gt. Brit.), *Rept. Comm. Coal Derivatives*, 1960, 79 pp.

[177] Anon., *BCURA Quart. Gaz.* No. 37/38, 18–22 (1959).

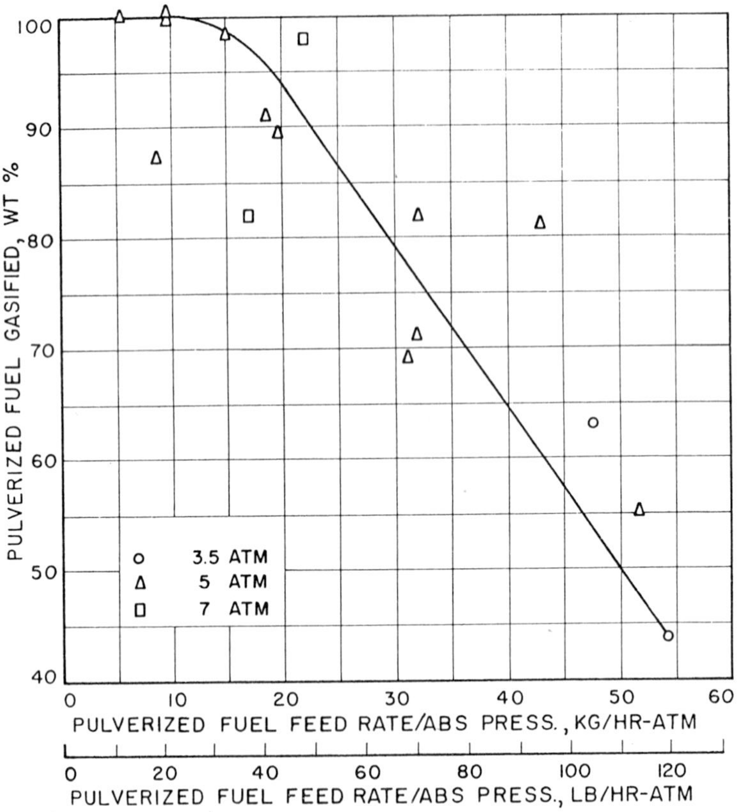

Fig. 29. Effect of feed rate and pressure on percentage of pulverized fuel gasified.[175]

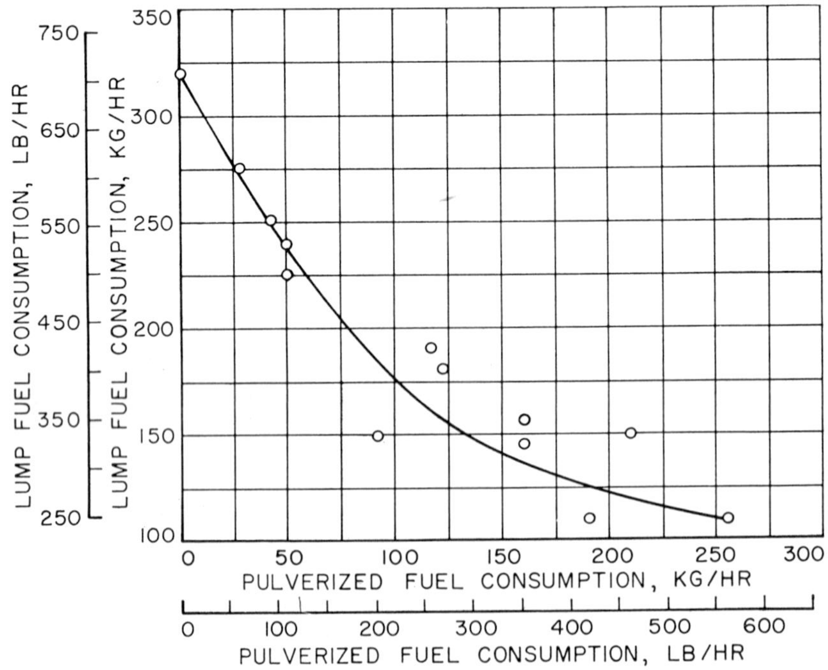

Fig. 30. Relationship between consumption of lump and pulverized fuel in tests at 5 atm pressure.[175]

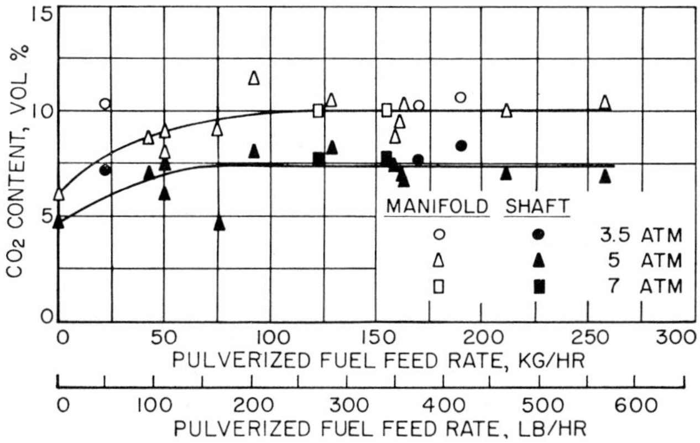

Fig. 31. Carbon dioxide content of the gas at various pulverized fuel feed rates.[175]

lump fuel ranged from 695 to 230 lb/hr, and of pulverized fuel from 0 to 565 lb/hr. From the data given, the diameter of the gasifier appears to be about 1.6 ft. The depth of the fuel bed was also about 1.6 ft.

The proportion of pulverized fuel gasified as a function of the ratio of pulverized fuel rate to absolute pressure is shown in Fig. 29. Since the bed volume is constant, this latter parameter is proportional to the pounds of pulverized fuel fed per cubic foot of reaction volume per hour per atmosphere. When the value of this parameter exceeds about 33 lb/hr atm, the percentage of pulverized fuel gasified decreases. Although not stated by the authors, it would appear that the specific gasification rate per cubic foot of reaction volume was approximately proportional to the first power of the absolute pressure, since data at 3.5 and 7 atm show no greater dispersion than data at 5 atm, when plotted against the feed rate divided by the absolute pressure.

The consumption of lump fuel as a function of pulverized fuel consumption is shown in Fig. 30, and the carbon dioxide content of the gas leaving the top of the bed and in the manifold is shown in Fig. 31. Some gas, higher in carbon dioxide, left the reactor with the slag and was recombined with the major gas stream in the offtake pipe so that the carbon dioxide content of the gas in the manifold was higher. Over the range of pulverized fuel rates, the gas leaving the top of the bed had the following approximate composition: CO_2, 7%; CO, 22 to 23%; H_2, 3 to 4%; CH_4 0.4 to 0.8%; gross heating value 96 Btu/scf. After mixing with gas from the slag tap, the heating value was 84 to 96 Btu/scf, approximately 17 Btu/scf lower than when gasifying lump fuel alone.

From the data presented in Figs. 29 and 30, the relation between the fraction of pulverized fuel gasified and the fraction of pulverized fuel in the total fuel fed may be derived, and is shown in Fig. 32. More than 90% of the pulverized fuel is gasified, until the pulverized fuel fed exceeds 40% of the total. Above this, the percentage gasification decreases rapidly, and would probably reach some low value at the limiting condition corresponding to the injection of pulverized fuel into an inert bed.

Fluidized-bed Processes

Fluidization techniques were first applied to gasification on a commercial scale in the Winkler process in the 1920's. The principal advantages claimed[178] are: (1) the ability to

[178] Flesch, W., *Intern. Conf. Complete Gasification Mined Coal, Liège*, **1954**, 235–44.

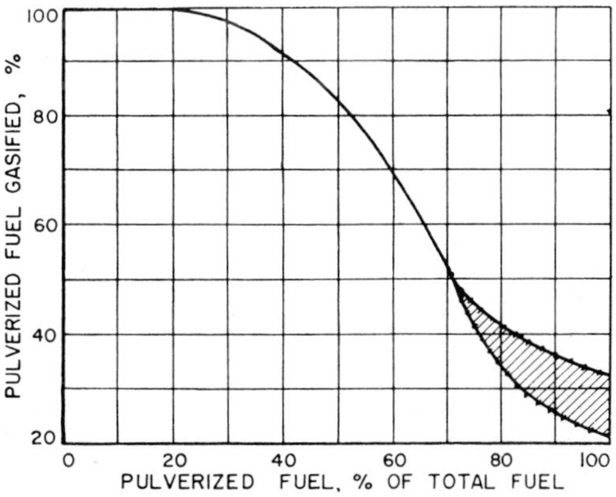

Fig. 32. Relationship between percentage of pulverized fuel in feed and percent gasified.

tolerate wide variations in the quality of the fuel—high-ash fuels having particle sizes in the range 0 to 8 mm for average fuels, and 0 to 20 mm for friable fuels, may be used; (2) minimum ground area is required because of high specific gasification rate; (3) operability over a wide range of outputs without significant loss in efficiency; (4) high degree of reliability and safety because the high inventory of fuel in the reactor minimizes the possibility of oxygen breakthrough if the fresh fuel feed is stopped; and (5) local overheating, with the possibility of clinker formation, is avoided by the mobility of particles and high rates of heat transfer in fluidized beds. Disadvantages of fluid-bed gasification are: (1) the high loss of sensible heat in the exit gases as a result of substantially thermal equilibrium between these gases and the uniformly-mixed solids in the fluidized bed at gasification temperature, this sensible heat being capable of recovery in a waste heat boiler but not available directly for the gasification process as it is in counter-current flow in fixed beds; (2) the loss of ungasified fuel as entrained dust that increases with the friability of the fuel and is greater with unreactive fuels, the energy being recoverable from the dust either by burning under boilers or gasification under more severe conditions; (3) complete mixing of solids in the fluidized bed leads to a high steady-state ash content in the bed to avoid loss of combustible with the residue leaving the gasifier; (4) slugging (discussed elsewhere in this chapter); and (5) fluidization characteristics of the fuel place restrictions on the range of possible operating conditions.

Processes Using Oxygen and Steam and Air and Steam

The Winkler process has been used commercially for the gasification of brown coal, brown-coal char, semicoke, caking coal, and high-rank brown coal.[178] The more recent generators are about 13 to 15 ft in diameter and 65 ft high.[162] The process operates at atmospheric pressure with either air, oxygen-enriched air, or oxygen and steam as the gasifying medium. Depending on the reactivity of the fuel, the temperature in the fluidized bed ranges from 800 to 1000°C. The depth of the bed in the settled condition is 3 m (9.8 ft). Some operating results are summarized subsequently in Table 24. Data on installations of Winkler generators,[178] their development,[179] mechanical features, and operating results,[162] are given in the literature.

Low-reactivity fuels, and fuels with a low ash-fusion temperature, are not suitable for use in a fluidized-bed process such as the Winkler.[178] Such fuels may be utilized in the Flesch-Winkler, or in the BASF-Flesch Demag process, which is a subsequent development. The latter process employs twin generators, and is operated as a combination fixed-bed, fluidized-bed process, as shown in Fig. 33. The two generators are operated in parallel, with the gasifying medium passing down through the two fixed beds as shown in Fig. 33*a*. The high-temperature gasification zone is at the top of the fuel bed, and as gasification proceeds, clinker is formed at the top of the bed. After 10 min of downrun operation, the generators are operated in series with the upstream generator (I) fluidized, as shown in Fig. 33*b*. Fluidization breaks up the clinker, which falls to the bottom of generator I, where it is removed. During fluidization, coal is introduced into generator I, and any fines from the coal or the fluidized bed are filtered out by the fixed bed in generator II. The fluidization part of the cycle lasts for about 1 min, and is followed by another 10-min downrun with the generators in parallel. The cycle is completed by fluidizing generator II upstream from generator I. The gasifying medium may be either air and steam or oxygen and steam. Results obtained in manufacturing producer gas are shown in Table 19,[178] and results in making synthesis gas are shown subsequently in Table 24. Gumz also has given data on both types of operation.[180]

The Flesch-Demag process normally uses fuel in the size range 0 to 8 mm. Excessive loss of carbon with the clinkers occurs if the feed contains a significant fraction of particles greater than 8 mm.[178] To ensure high-gasification rates and minimize carryover losses during fluidization, the −1 mm fraction of the feed should not exceed 30% with caking coal, and 50% with noncaking fuel.

A semicommercial scale fluidized-bed gasifier has been operated at elevated pressures by Hydrocarbon Research, Inc.[7] The gasifier was of 26.5 in. i.d., and was operated with a bed

[179] Portatius, B. von, *Freiberger Forschungsh.*, **A69**, 5–25 (1957).

[180] Gumz, W., *J. Inst. Fuel*, **28**, 393–415 (1955).

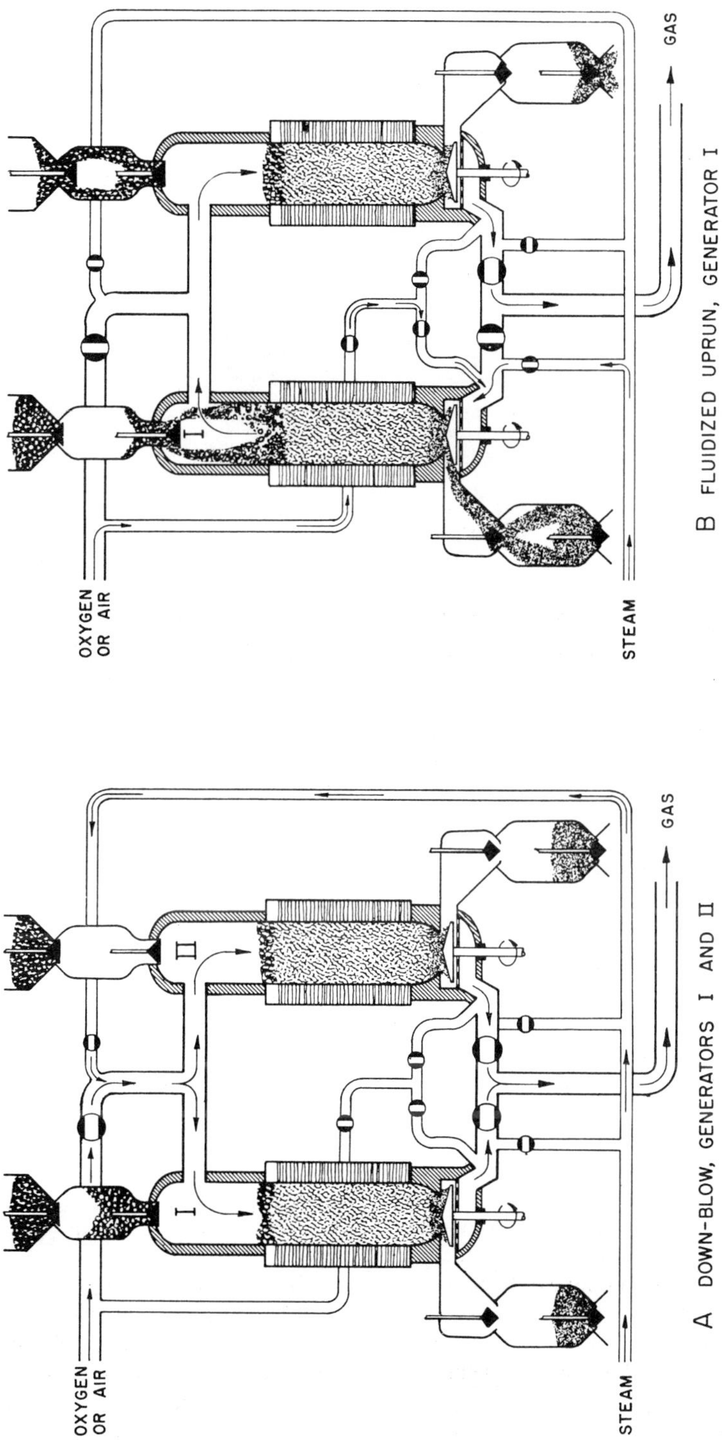

Fig. 33. BASF-Flesch-Demag process for gasifying coal fines.[178]

TABLE 19

OPERATING RESULTS IN THE MANUFACTURE OF PRODUCER GAS IN A FLESCH-DEMAG GENERATOR[178]

Fuel	Coke Fines	Brown Coal	Low-Volatile Bituminous Coal Fines
Moisture, wt %	0	17	1
Ash, wt %	22	7.5	12
Gross heating value, Btu/lb	11,340	8640	13,140
Output			
Gas composition, vol %			
CO_2	3.6	7.0	4.0
C_nH_m	0.0	0.4	0.1
O_2	0.0	0.2	0.0
CO	29.4	26.7	26.5
H_2	7.4	16.5	14.0
CH_4	0.4	1.4	1.3
N_2	59.2	47.8	54.1
Gross heating value, Btu/scf	123	161	146
Steam/Mcf gas, lb	11.8	5.9	8.9
Consumption			
Fuel/Mcf gas, lb	13.6	23.7	Not measured because mixed with dry lignite when gasified.
Steam/Mcf gas, lb	12.4	10.1	

depth of 25 ft. Anthracite fines were gasified with steam and oxygen at pressures ranging from 170 to 245 psig, and temperatures from 1430 to 1676°F. The unit produced up to 650,000 scf per day of hydrogen, carbon monoxide and methane. Kinetic data obtained in this unit are discussed elsewhere in this chapter.

Processes Employing Direct or Indirect Heat Transfer

The I.C.I. moving burden process[181] was designed to produce synthesis gas continuously in a fluidized bed by using solid particles heated by combustion with air in a separate vessel to supply the endothermic heat of reaction with steam. The original concept envisioned a three-vessel system consisting of a carbonizer fluidized with carbonization gases, a gasifier fluidized with steam, and a combustion tube.

From studies in a semitechnical plant with coke in a two-vessel system, and from preliminary economic evaluations, it was concluded that a two-vessel system would be more desirable. Tests with this system showed that caking coal could be fed directly to the fluidized gasifier without difficulty from agglomeration or adhesion to the walls. The plant was stable, and gasification temperatures could be controlled within ±9°F at 1832°F. However, attrition of the particles in the circulating burden resulted in a loss of carbon as dust (88 % as particles less than 10 μ) which amounted to from 20 to 25 % of the energy in the fuel and which could not be utilized directly in the process. It was concluded that the process would be suitable for the production of water gas where there was a use for a considerable amount of byproduct steam. The process has not as yet been developed for full-scale operation.[131]

The kinetics of fluidized-bed gasification of low-rank bituminous coal with steam at atmospheric pressure discussed elsewhere in this chapter were studied in a pilot-scale unit in which the endothermic heat of the steam-carbon reaction was supplied by heat transferred through the walls of the vessel.[182]

Fully-entrained Suspension or Dilute-Phase Gasification

The most important inherent characteristic of fully entrained suspension or dilute-phase gasification processes is their ability to utilize any grade or class of coal. Size-consist or friability of the fuel have no significance

[181] Rayner, J. W. R., *J. Inst. Fuel*, **25**, 50–9 (1952)

[182] Jolley, L. J., Pohl, A., Noaks, C. C., and Stantan, J. E., *Intern. Conf. Complete Gasification Mined Coal, Liège*, **1954**, 246–54.

because the fuel must be pulverized. Swelling and caking characteristics of the fuel do not affect the operability of the process, since each particle is discrete and separated from other particles in the flowing gas stream. Ash-fusion temperature is not important insofar as interaction between particles is concerned, but, as in other processes, it is important in determining the conditions of nonslagging and slagging operations. Although fully-entrained processes can use any type of coal, the volume of the reactor will be least for the higher-volatile fuels and for fuels having more reactive residues. Fully-entrained systems inherently produce a gas containing no tar and very little methane, because the volatile matter of the fuel is almost completely oxidized in the early stages of gasification. This characteristic is important if the gas is to be used for synthesis of ammonia or liquid fuels.

The disadvantages of fully entrained processes result from the low concentration of fuel in the gasifying medium and from the cocurrent flow of reactants. The inventory of fuel in the reaction zone of a fully entrained gasifier is low because of the low concentration of fuel in suspension. Furthermore, the concentration of carbon in the reacting mixture decreases progressively from inlet to outlet. Since temperature also decreases from inlet to outlet, rates of gasification are lowest at the outlet, where both the concentration and reactivity of the fuel residue are least. Thus, relatively large increases in the volume of the reaction zone, with attendant increases in the residence time of the reactants, will have comparatively little effect on the extent of carbon conversion. For this reason, it is not economically feasible to gasify more than 85 to 90% of the carbon in a single pass. Higher carbon conversions may be obtained by recycling the solid residue. Cocurrent flow in fully entrained processes necessitates heat exchange to improve thermal efficiency. The heat-recovery problem is complicated by the presence of solids and by need for avoiding the possibility of contact between suspended molten ash particles and heat-transfer surfaces.

In spite of some of the disadvantages mentioned earlier, overall energy production rates per unit volume of reaction space in fully entrained gasifiers are considerably greater than those of fixed-bed and fluid-bed gasifiers operating under comparable conditions of ash removal. Thus, under practical operating conditions the adverse effect of the low carbon inventory is more than offset by the high-reaction rates resulting from the large surface area of the pulverized fuel particles. The adaptability of fully entrained processes to operation under slagging conditions and at elevated pressure is also conducive to maximum specific gasification rates.

Inter-Relation of Process Variables in Suspension Gasification

Suspension gasification processes may be analyzed stoichiometrically to develop the basic interrelations between process variables. Such an analysis is possible if it is assumed that all of the hydrogen in the fuel reacts with oxygen to form water, and that no methane is present in the product gas. Both assumptions closely approximate observations in suspension gasifiers.

The reaction between fuel, oxygen and steam may be represented by the equation:

$$wCH_aO_b + yO_2 + zH_2O \rightarrow CO + H_2 + \left[y - \frac{w(a-2b)}{4}\right]CO_2 + \left[\frac{wa}{2} + z - 1\right]H_2O + xC \qquad (77)$$

in which w = moles of fuel per mole of CO, x = moles of unreacted carbon per mole of CO, y = moles of O_2 per mole of CO, z = moles of H_2O per mole of CO, a = atoms of hydrogen per atom of carbon, and b = atoms of oxygen per atom of carbon. From a carbon balance of (77),

$$w = \frac{1 + x + y}{1 + a/4 - b/2} \qquad (78)$$

Letting r = moles O_2/mole C in fuel, R = scf O_2/lb fuel, g = moles CO + H_2/mole C in fuel, G = scf CO + H_2/lb fuel, $f = 1/g$ = mole C in fuel/mole CO + H_2, $F = 1/G$ = lb fuel/scf CO + H_2, c = fraction C gasified, M = lb C, H and O in fuel/mole C in fuel, m = lb C, H, and O/lb fuel, and $\bar{y} = y/2$ = mole O_2/mole CO + H_2, we obtain from equations 77 and 78 for the yield of CO + H_2 per unit of fuel:

$$g = 2\left(c + \frac{a}{4} - \frac{b}{2} - r\right) = \frac{1}{f} \qquad (79)$$

and

$$G = \frac{2(378.5)m}{M}\left(c + \frac{a}{4} - \frac{b}{2}\right) - 2R = \frac{1}{F} \quad (80)$$

and for the oxygen required per unit of $CO + H_2$,

$$\bar{y} = \frac{0.5}{\dfrac{c + a/4 - b/2}{r} - 1} = \frac{r}{g} = rf \quad (81)$$

and

$$\bar{y} = \frac{0.5}{\dfrac{378.5m\,(c + a/4 - b/2)}{MR} - 1} = \frac{R}{G} = RF \quad (82)$$

An analysis of equation 79 shows that g must reach a maximum at some value of r, since

$$\frac{dg}{dr} = 2\left(\frac{dc}{dr} - 1\right) \quad (83)$$

and since dc/dr is always positive. The maximum theoretical value of dc/dr is 4, as indicated by the following equation based on no net heat of reaction with products and reactants at 60°F;

$$1.333C + 0.333O_2 + H_2O(g) \rightarrow 0.333CO_2 + CO + H_2 \qquad \Delta H^0_{60°F} = 0 \quad (84)$$

The maximum value of g occurs at a value of r at which $dc/dr = 1$. The cold gas efficiency varies with r in a manner identical to that of g, since it equals g multiplied by a constant. Since $f = 1/g$, it follows that the fuel consumed per unit of $CO + H_2$ passes through a minimum at some value of r. This minimum is usually above the normal operating range of tests of suspension gasifiers (6 to 12 scf O_2/lb fuel). However, statistical analysis of data from an atmospheric pressure[183] and a pressure[184] gasifier has indicated a minimum (see Figs. 42 and 51).

An analysis of equation 81 shows that $\bar{y}$ passes through a minimum at some value of r, since

$$\frac{d\bar{y}}{dr} = f\left(1 - \bar{y}\frac{dg}{dr}\right) \quad (85)$$

and depending on whether $\bar{y}(dg/dr) > 1$ when dg/dr is positive. In most gasifiers, $\bar{y}$ increases with r, and the minimum occurs at a value of r below the normal operating range. However, one set of tests analyzed statistically[183] shows a minimum value of $\bar{y}$ in the normal operating range of oxygen-coal ratios (see Fig. 41). The possibility that $\bar{y}$ will have a minimum at some value of r in the normal range is favored by high values of $\bar{y}$. The minimum theoretical value of $\bar{y}$ is 0.167 from equation 84, and, depending on process conditions and heat loss, $\bar{y}$ may be as high as 0.4 to 0.5. The maximum possible value of dg/dr is 6 from equation 84, and values of 2 to 3 are observed at normal values of r. Thus, if $\bar{y} = 0.4$, and $dg/dr = 3$, at the beginning of the operating range, $\bar{y}$ could pass through a minimum at normal values of r, since dg/dr is decreasing as r increases.

The effect of the steam-fuel ratio on material requirements cannot be predicted from stoichiometry alone. To do this, it is necessary to make an enthalpy balance, assuming water-gas shift equilibrium in the exit gases. From stoichiometry and the assumption of water-gas shift equilibrium, it is possible to calculate the effect of the steam-coal ratio on steam decomposition and on the H_2/CO ratio in the product gas. Results of tests of suspension gasifiers have shown that increasing the steam-coal ratio increases both the fuel and oxygen required per scf of $CO + H_2$.[183] Carbon conversion decreased with increasing steam-coal ratio in tests at atmospheric pressure,[183] but increased in tests at elevated pressures.[185] In the latter tests, the decomposition of total steam, that is, the steam fed plus steam from the H_2 in coal, decreased with the steam-coal ratio, and at any given ratio was slightly higher at the higher oxygen-coal ratios. The H_2/CO ratio decreased as the steam-coal ratio decreased and as the oxygen-coal ratio increased. As mentioned previously, the H_2/CO ratio is determined in suspension gasifiers by the water-gas shift equilibrium in the exit gases.

In autothermic gasification processes using oxygen, heat lost as radiation and convection increase both oxygen and fuel requirements. In smaller-scale gasifiers with relatively high heat losses, the material requirements and operating parameters will be significantly

[183] Strimbeck, G. R., Holden, J. H., Bonar, F., Plants, K. D., Pears, C. D., and Hirst, L. L., *U.S. Bur. Mines, Rept. Invest.* No. 5559 (1960), 68 pp.

[184] Strimbeck, G. R., Holden, J. H., McGee, J. P., and Hirst, L. L., *AGA, Proc.*, 1955, 1006–37.

[185] von Fredersdorff, C. G., and Pyrcioch, E. J., *AGA, Proc.*, 1953, 968–1001.,

different from those that would be attained in an adiabatic reactor. Operating results can be corrected for heat losses if kinetic expressions are known for the reaction system. When such expressions are not available, the correction may be approximated by assuming that the moles of unreacted carbon per mole of $CO + H_2$ is the same in the adiabatic reactor. This is somewhat simpler, but almost equivalent to assuming a constant value for the fraction of the available carbon that reacts with steam, as proposed for making more precise corrections for heat loss.[186] With the simplified assumption, the correction of operating results for variation in heat loss may be made as follows:

L = heat lost by radiation and conduction under test conditions, Btu/mole $CO + H_2$;

L' = heat lost by radiation and conduction under corrected condition, Btu/mole $CO + H_2$ = zero for an adiabatic reactor;

L_F = heat lost by radiation and conduction under test conditions, Btu/lb coal;

L'_F = heat lost by radiation and conduction under corrected conditions, Btu/lb coal;

H_{fi} = enthalpy of C, H, and O in fuel at inlet conditions, Btu/mole C in fuel;

H_{O_2i} = enthalpy of O_2 at inlet conditions, Btu/mole; and

H_{H_2Oe}, H_{CO_2e} = enthalpy of H_2O and CO_2 at exit conditions, Btu/mole.

All enthalpies are above the elements at 0^0R and 0 psia. Using the previous nomenclature and denoting by a prime the value of a quantity at the corrected heat loss, we obtain from an enthalpy balance based on equation 77:

$$\frac{\bar{y} - \bar{y}'}{L - L'} = \frac{1 + a/4 - b/2}{H_{fi} + (1 + a/4 + b/2)H_{O_2i} - (a/2)H_{H_2Oe} - H_{CO_2e}} = D \tag{86}$$

In this balance it is assumed that $z = z'$, $x = x'$, and that the inlet conditions and exit gas temperature are the same for both values of the heat loss. If the coal is not preheated, $H_{fi} \doteq 0$.[3] With this approximation, Fig. 34 shows the value of D as a function of exit gas

[186] Elliott, M. A., Perry, H., Jonakin, J., Corey, R. C., and Khullar, M. L., *Ind. Eng. Chem.*, **44**, 1074–82 (1952).

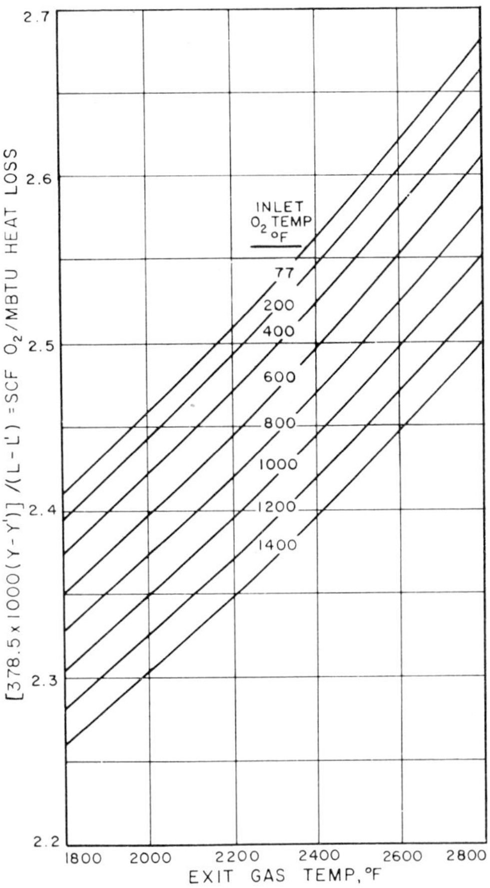

Fig. 34. Correction for heat loss.

temperature and inlet oxygen temperature for a coal having the composition $CH_{0.837}O_{0.085}$ (Sewickley coal).[183] It will be observed that there is comparatively little variation in D over a wide range of inlet and exit conditions.

The operating results at any value of heat loss L_F may be corrected to results at some other value of heat loss, L'_F, by the following relations:

$$F' = \frac{F(1 - MDL_F/378.5m)}{1 - MDL'_F/378.5m} \tag{87}$$

$$F'_{\text{adiabatic}} = F(1 - MDL_F/378.5m) \tag{88}$$

$$\bar{y}' = \bar{y} - \frac{DL_F F(1 - L'_F/L_F)}{1 - MDL'_F/378.5m} \tag{89}$$

$$\bar{y}'_{\text{adiabatic}} = \bar{y} - DL_F F \tag{90}$$

$$c' = 1 - \frac{F}{F'}(1 - c) \tag{91}$$

$$c'_{\text{adiabatic}} = 1 - \frac{1 - c}{1 - MDL_F/378.5m} \tag{92}$$

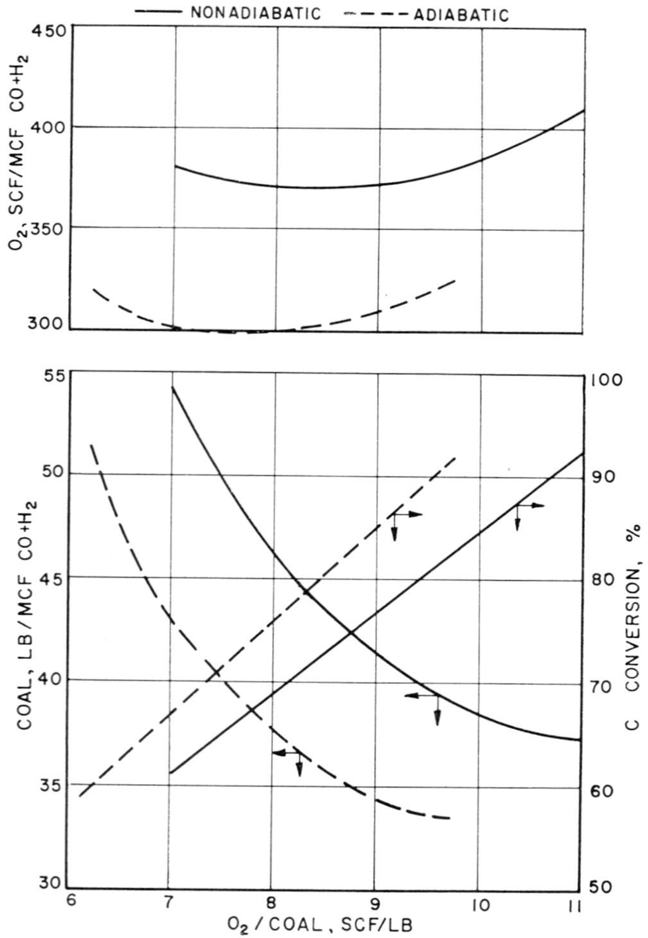

Fig. 35. Comparison of gasification results under adiabatic and nonadiabatic conditions.[183]

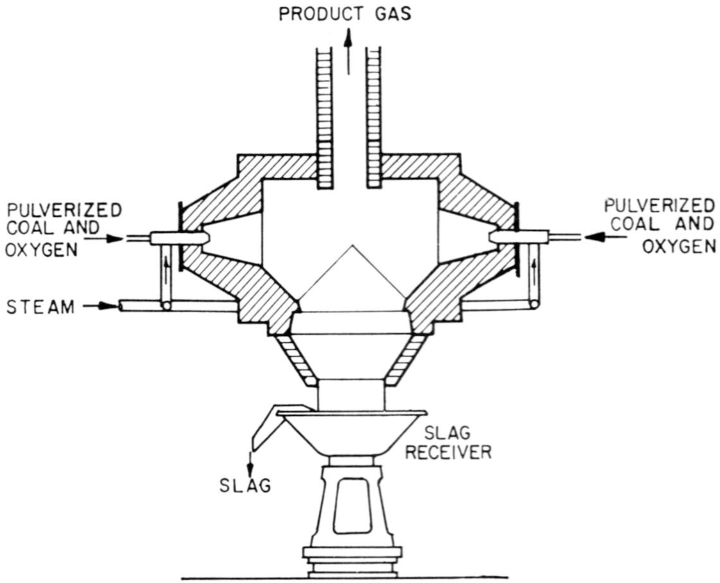

Fig. 36. A Koppers-Totzek gasifier.[190]

Figure 35 shows the effect of correcting operating results obtained in a suspension gasifier[183] to results that would be obtained in an adiabatic reactor. In these tests, the heat loss ranged from 500 to 1000 Btu/lb of coal. The results shown emphasize the importance of correcting for heat loss, particularly if the heat loss varies significantly throughout the range of process variables studied. Under such conditions the real effects of process variables may be obscured by variations in heat loss. The problem of correcting for heat losses has not received sufficient attention in experimental studies of the performance of reactors for gasifying coal in suspension with oxygen and steam, although a method has been proposed in one investigation[184] and is subsequently outlined (see p. 986). The effect of heat loss to the walls of a gasifier on the cost of gas has been computed.[187]

Processes Operating at Atmospheric Pressure

The Koppers-Totzek process. This process for gasifying various types of coal with ash contents up to 30% in suspension with oxygen and steam is being used to produce synthesis gas in several commercial synthetic ammonia and chemical plants.[188] The early history of this development, which began in 1938, has been summarized.[189,190] The first large-scale unit (1 ton of coal per hour) was operated at the Louisiana, Missouri, Gas Synthesis Demonstration Plant of the U.S. Bureau of Mines.[191] Extensive tests of this unit furnished data on the effect of process variables and design modifications on performance. These tests also disclosed the need for improving carbon conversion, refractory performance, and methods for removing the comparatively small quantities of slag produced. The experience gained in this operation was used in designing the first large industrial plant, which was put into operation in Finland in 1952.

The Koppers-Totzek gasifier, shown schematically in Fig. 36, is a refractory-lined, horizontal, cylindrical vessel with conical ends. A homogeneous mixture of oxygen and pulverized coal is introduced through coaxial burners at each end. Steam is introduced around these burners to shroud the high-temperature reaction zone and protect the burners and refractories from excessive temperatures. It has been stated[192] that the gasification reactions are completed within the coned ends of the gasifier. The products leave through a water-cooled central vertical outlet. In the gasifier at Oulu, Finland, 95% of the ash leaves with the product gas, and 5% is removed from the bottom of the gasifier, whereas in the gasifier at Ouahama, Japan, 40% of the ash is removed as slag from the bottom. Depending on the reactivity of the fuel, the exit gas temperature may range from 1000 to 1300°C. In the Oulu gasifiers, approximately 26% of the sensible heat in the exit gases is used for generating steam.[190] Residues removed from the exit gases may be recycled or used as fuel for drying the coal. The cold gas efficiency is 67%, and the overall efficiency, including waste heat recovery, is 75.5%. Performance data for these gasifiers are given in Table 24. Performance data for gasifiers in Japan,[193] France,[194] and Belgium[195] have been given in the literature. The Koppers-Totzek process has also been adapted for the production of synthesis gas from heavy oils.[196]

The Panindco process. In this process, which is an outgrowth of the Wintershall-Schmalfeldt process,[197] pulverized coal is gasified with highly preheated oxygen or air and steam in a laminar-flow system. An experimental plant which was operated at coal-feed rates up to 1600 lb/hr was installed near Rouen, France, in 1950. The gasifier is a vertical, cylindrical, refractory-lined vessel. Coal suspended in

[187] Grossman, P. R., and Curtis, R. W., *ASME, Trans.*, **76,** 689–95 (1954).

[188] Fitz, W., *Tech. Mitt.*, **47,** 561–3 (1954).

[189] Totzek, F., *Brennstoff-Chem.*, **34,** 361–7 (1953).

[190] Osthaus, K. H., *Intern. Conf. Complete Gasification Mined Coal, Liège*, **1954,** 255–64.

[191] Dressler, R. G., Batchelder, H. R., Tenney, R. F., Wenzell, L. P., Jr., and Hirst, L. L., *U.S. Bur. Mines, Rept. Invest.* No. 5038 (1954), 35 pp.

[192] Koppers, H. H., *J. Inst. Fuel*, **30,** 673–80 (1957).

[193] Terui, S., XI Sectional Meeting World Power Conf. Belgrad, **1957,** Paper 26, Sect. B.2.

[194] Charbonnages de France, Centre d'Etudes et Recherches, Ann. Rept., **1956,** p. 81.

[195] Macq, A., *Rev. gen. gaz.*, **81,** 89–97 May–June (1959).

[196] Fitz, W., *Trans. World Power Conf., 5th Conf., Vienna*, **1956,** Vol. 7–8, 1761–6 (1959).

[197] Morley, R. J., *BIOS Final Rept.* No. 1142, Item No. 30, H.M.S.O., London, 1946, 33 pp.

carrier gas (cold air in the tests) is fed axially at the center of the domed top of the gasifier. Oxygen, or a mixture of oxygen and steam, preheated in regenerators, is fed in an annular opening surrounding the coal feed. Low-temperature (nonpreheated) steam is admitted through several nozzles arranged symmetrically around the two central streams. Additional steam is admitted into a lower section of the gasifier to insure nonslagging operation (the maximum exit-gas temperature is 900°C). The use of steam for supplying heat to the gasifier, and for cooling to prevent slagging, results in a high-steam consumption. Performance data for the gasification of high-volatile coal from Provence with oxygen at two preheat temperatures[198] are as follows:

Preheat temperature, °C	750	480
Coal feed, lb/hr	758	739
O_2/coal ratio, scf/lb	7.91	8.19
Steam for gasification (exclusive of cooling steam), lb/lb	2.03	1.64
Coal, lb/Mcf $CO + H_2$	40.0	46.8
Steam, lb/Mcf $CO + H_2$	81.5	76.7
Oxygen, scf/Mcf $CO + H_2$	315	382
Preheat, MBtu/Mcf $CO + H_2$	64.8	36.1
Carbon conversion, %	96.4	85.0

The reduction in oxygen requirement of 2.33 scf per MBtu preheat is roughly comparable to the reduction in oxygen per MBtu heat loss (see Fig. 34).

Apparently all work on coal gasification has been suspended by Charbonnages de France[199] including experimental work on the Panindco process. However, it has been stated[133] that a unit for gasifying about 55 tons of dried peat per day has been constructed at Feuchy, France.

B & W—du Pont gasifier. Development work on this gasifier was done initially by the Babcock and Wilcox Company in cooperation with the U.S. Bureau of Mines.[187,200] Subsequently, a semiplant-scale unit, and eventually a full-scale unit (see Fig. 37), were installed at the Belle, West Virginia Works of the du Pont Company. The full-scale plant had a capacity

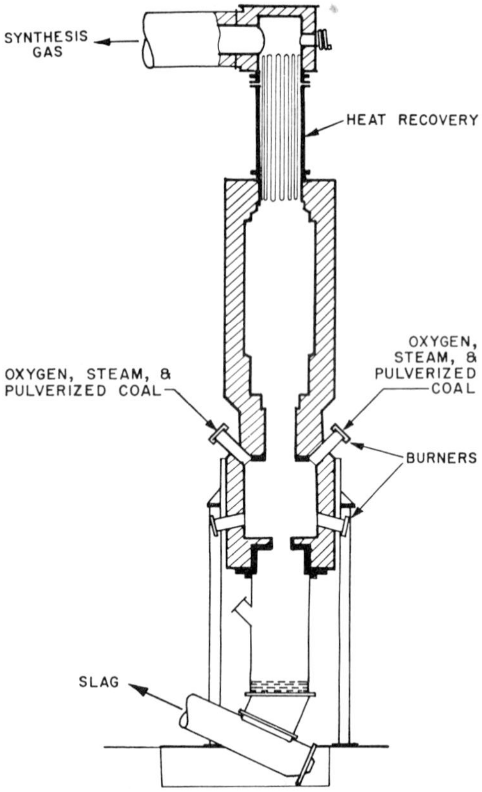

Fig. 37. A B & W—du Pont gasifier at Belle, West Virginia.[187]

of about 17 tons of coal per hour and could produce 25 MMcf of $CO + H_2$ per 24-hour day. The gasifier was a refractory-lined vessel having a circular cross-section and a primary and a secondary reaction zone. Temperatures in the primary zone were high enough to melt the coal ash and the molten slag was continuously tapped from the bottom of the chamber. A portion of the refractory surfaces in the primary reaction zone was subjected to attack by slag, and periodic replacement was necessary. Coal was conveyed into this unit by steam at a temperature of 325°F. Reactants were injected downward with flow reversal occurring in the primary zone (see Fig. 37). Operating data on the full-scale unit have not been published. Detailed data on the Bureau of Mines' pilot-plant work have been published,[200] and typical operating results of the semiplant-scale unit[187] are presented in Table 24.

Rummel single-shaft slag-bath gasifier. This gasifier was developed by the Union Rheinische Braunkohlen Kraftstoff, A.-G. in

[198] Foch, P., and Loison, R., *Intern. Conf. Complete Gasification Mined Coal, Liège*, **1954**, 224–34.

[199] Charbonnages de France, Centre d'Etudes et Recherches, Ann. Rept., **1958**, p. 96.

[200] Strimbeck, G. R., Cordiner, J. B., Jr., Baker, N. L., Holden, J. H., Plants, K. D., and Schmidt, L. D., *U.S. Bur. Mines, Rept. Invest.* No. 5030 (1954), 37 pp.

Wesseling, West Germany.[201,202] A cross-section of the large-scale unit with associated heat recovery equipment is shown in Fig. 38. The gasifier is a cylindrical, water-cooled vessel, 5 ft 11 in. i.d. An annular slag bath approximately 20 in. deep, and with a central outlet, is maintained at the bottom of the gasifier. The reactants are injected into the slag bath through twelve tangential water-cooled nozzles. Nozzles for fuel and carrier gas are placed between nozzles for steam and oxygen around the periphery. The kinetic energy in the entering reactants imparts a rotary motion to the slag bath, which has a peripheral velocity of 10 to 20 ft/sec. Thus, the reactants are brought into intimate contact with each other and with the slag. The slag acts chiefly as a heat storage and transfer medium and holds larger particles at the surface for reaction with the gasifying medium. Depending on the reactivity of the fuel, sizes in the range of −1 mm to −5 mm may be gasified. The product gas flows up through the gasifier and heat-recovery system. Entrained solids are removed in a cyclone separator and recycled as feed to two of the fuel nozzles. Under such conditions 99% of the carbon fed is gasified.

[201] Rummel, R., *Chem. Ing. Tech.*, **28,** 25–30 (1956).

[202] Rummel, R., *Coke and Gas*, **21,** 493–501, 520 (1959).

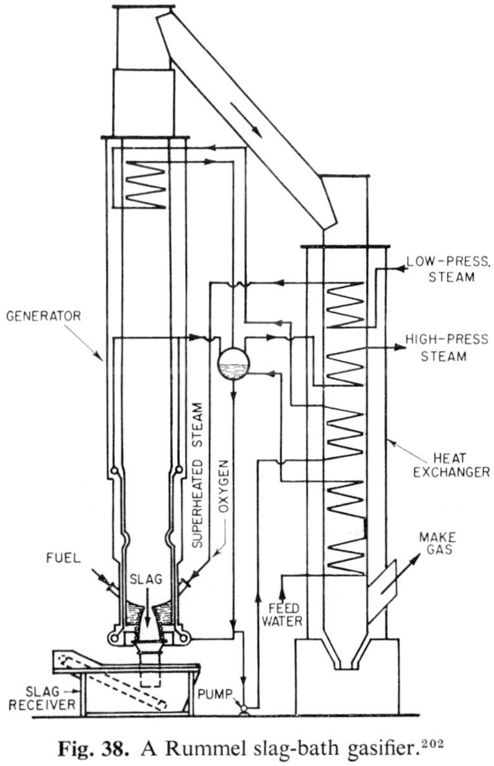

Fig. 38. A Rummel slag-bath gasifier.[202]

An equilibrium layer of slag builds up on the water-cooled sections and is held in place by studs. The slag continuously overflows and is quenched in a water bath. The viscosity of the slag preferably should be less than 10 poises. With slag viscosities greater than 100 poises, the addition of a flux is desirable. It is stated that iron oxide in the slag acts as an oxygen transfer agent through alternate reduction with carbon and oxidation by oxygen.

The full-scale plant has produced, from brown coal, up to 560,000 scf of synthesis gas per hour containing 84.1% of $CO + H_2$ by volume. The capacity of the unit was limited by the heat-recovery system, and it is claimed that the generator is capable of producing 680,000 scf of gas per hour. Low-heating value gas may be produced from air and steam. In addition, liquid fuels may be gasified by the use of suitable nozzles. Extraneous slag must be added with low-ash liquid fuels. Performance data for both the large-scale and pilot unit in gasifying coal with oxygen and steam are shown in Table 24.

Rummel double-shaft slag-bath gasifier.[202] This is an interesting variant of the development of slag-bath gasifiers. Air and fuel are injected tangentially in one shaft (see Fig. 39), and steam and fuel are injected into the other shaft. Vertical partitions slightly submerged in the slag bath separate the two shafts so that products of combustion and synthesis gas leave in separate streams. The endothermic heat of the gasification reactions is supplied by the heat generated in the combustion shaft and transferred by the rotating slag bath to the gasification shaft. This gasification system has been studied experimentally in a unit having an i.d. of approximately 3 ft, but performance data have not been published. If this development is successful, a continuous process for producing synthesis gas with use of air instead of oxygen will be available.

Ruhrgas vortex (Wirbelkammer) gas producer. This unit is a refractory-lined cylindrical vessel with water-cooled walls in the cyclone section at the base.[203,204] The cyclone chamber

[203] Nistler, F., *Coke and Gas*, **19,** 54–7 (1957).

[204] Traenckner, K., *ASME, Trans.*, **75,** 1095–101 (1953).

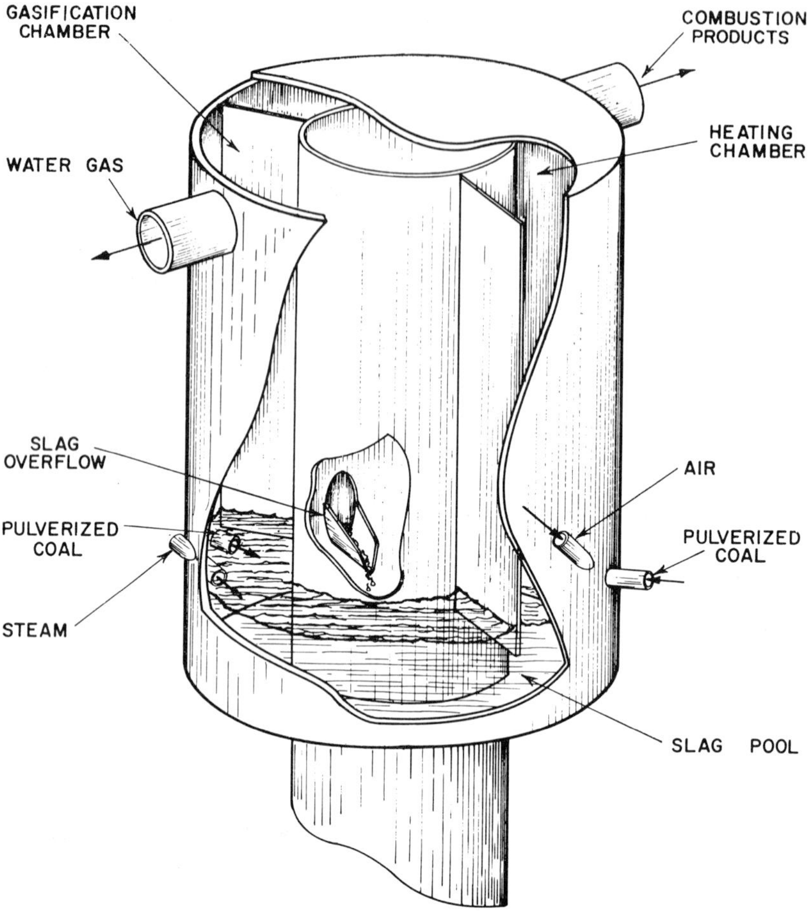

Fig. 39. A Rummel double-shaft slag bath gasifier.[202]

has an internal diameter of $6\frac{1}{2}$ ft and is $7\frac{1}{2}$ ft high. Above the cyclone chamber the diameter of the shaft increases to $9\frac{1}{2}$ ft and the shaft is 75 ft high. Fuel is injected with preheated air (1300°F) into the cyclone chamber through six tangential nozzles. Two of these nozzles are fed with recycle residue dust. When gasifying 100 tons/day of fresh bituminous coal dust containing 17% ash, 30 tons of dust containing 30% ash are removed in cyclone separators and recycled. Thus, the total feed to the cyclone is 130 tons/day of solid fuel having an ash content of 20%. Slag flows through a tap at the bottom of the cyclone chamber and is granulated by quenching in a water seal. Operating data in the generation of producer gas from bituminous coal, and synthesis gas from lignite, using oxygen instead of air, are presented in Table 20.

Intermediate and small-scale gasifiers. A number of experimental studies of the gasification of pulverized coal with oxygen and steam under slagging conditions have been made by the U.S. Bureau of Mines and others. These studies have developed basic information on the effect of process variables, and have provided operating experience useful in designing large-scale units.

The first pilot plant at the Morgantown Station of the Bureau of Mines[205] was designed to yield information on the effect of steam-preheat temperature on the coal and oxygen required for synthesis gas production. The gasifier was a refractory-lined cylindrical vessel having an internal diameter of 3 ft $1\frac{1}{4}$ in. and an effective internal volume of 86 cu ft.

[205] Strimbeck, G. R., Holden, J. H., Rockenbach, L. P., Cordiner, J. B., Jr., and Schmidt, L. D., *U.S. Bur. Mines, Rept. Invest.* No. 4733 (1950), 41 pp.

TABLE 20

PERFORMANCE DATA FOR RUHRGAS CYCLONE GASIFIER[203]

	Producer Gas from Bituminous Coal	Synthesis Gas from Lignite
Gas composition, vol %		
CO_2	5.1	16.8
CO	22.8	48.5
H_2	8.0	33.2
CH_4	0.0	0.5
N_2	64.1	1.0
Gross heating value, Btu/scf	98	265
Make-gas yield scf/ton fuel	141,500	—
Steam produced/ton fuel, ton	2.15	—
Potential heat in cold make gas, %*	61	75
Enthalpy of steam generated, %*	23	12

* Percentage of potential heat in the fuel.

The reactants were fed near the bottom tangential to a 2-ft diameter circle. Steam was preheated to temperatures ranging from 2600 to 3300°F in a pebble stove fired with natural gas, and when necessary, oxygen-enriched air. Strongly caking coal was gasified at feed rates of 200 to 450 lb/hr. Preheating steam reduced the coal and oxygen requirements as shown in Table 21. This work demonstrated the feasibility of operating with high-steam preheat temperatures, but did not consider in detail the overall economics of steam preheat in large-scale gasifiers. It was subsequently reported[200] that preheating steam to 2500 to 3000°F "involved mechanical difficulties that largely offset the savings in oxygen cost, at least under some conditions."

The pilot plant designed and operated at the Morgantown Station in cooperation with the B & W Company[200] was a refractory-lined cylindrical vessel having an internal diameter of 30 in. The gasifier was separated into two reaction zones, with a constriction to prevent slagging over the reactant injectors and to prevent recirculation of gases between the secondary and primary zones (see Fig. 40). The effective volume of these zones was 19 and 14 cu ft, respectively.

This unit was operated over a wide range of feed conditions to determine the effect of process variables and coal rank on performance and material requirements.[183] The last runs made in this unit were analyzed statistically by standard regression calculations assuming a second-order equation, and interaction between any two of the three independent process variables—coal feed rate, oxygen-coal ratio, and steam-coal ratio. Data were obtained on the gasification of anthracite, high-volatile bituminous A and subbituminous C coal at feed rates ranging from about 400 to 1000 lb/hr. The effects of oxygen-coal and steam-coal ratios at a given coal feed rate on oxygen requirement, coal requirement, carbon conversion, exit gas temperature, and heat loss in the gasification of bituminous coal, are shown in Figs. 41 to 45. The trends shown in these figures are typical of those generally observed in suspension gasifiers. As mentioned previously, a minimum in oxygen requirement is theoretically possible, but is not frequently observed in the range of oxygen-coal ratio shown.

Increasing the coal feed rate twofold reduced the fuel and oxygen requirement as a result of the reduction in the fraction of heat lost as radiation and convection. Carbon conversion was always the greatest at the highest fuel rate. This apparent anomaly may be explained when it is realized that reaction temperatures are much lower at the lower coal-feed rates as a result of the high fraction of heat lost. The data indicate that the decrease in residence time at the higher coal-feed rates was more than offset by the increase in reaction rate at the higher-gasification temperature.

The results obtained in the gasification of the three types of coal are summarized in Table 22. On the basis of oxygen and carbon requirement, the order of preference of rank for gasification under the conditions studied

TABLE 21

EFFECT OF STEAM PREHEAT TEMPERATURE ON AVERAGE REQUIREMENTS PER MCF CO + H_2 PRODUCED[205]

Steam Temperature, °F	Coal, lb	Oxygen, scf	Steam, lb
2904	33.3	171	81.1
1899	37.3	326	42.5
238	42.3	394	29.1

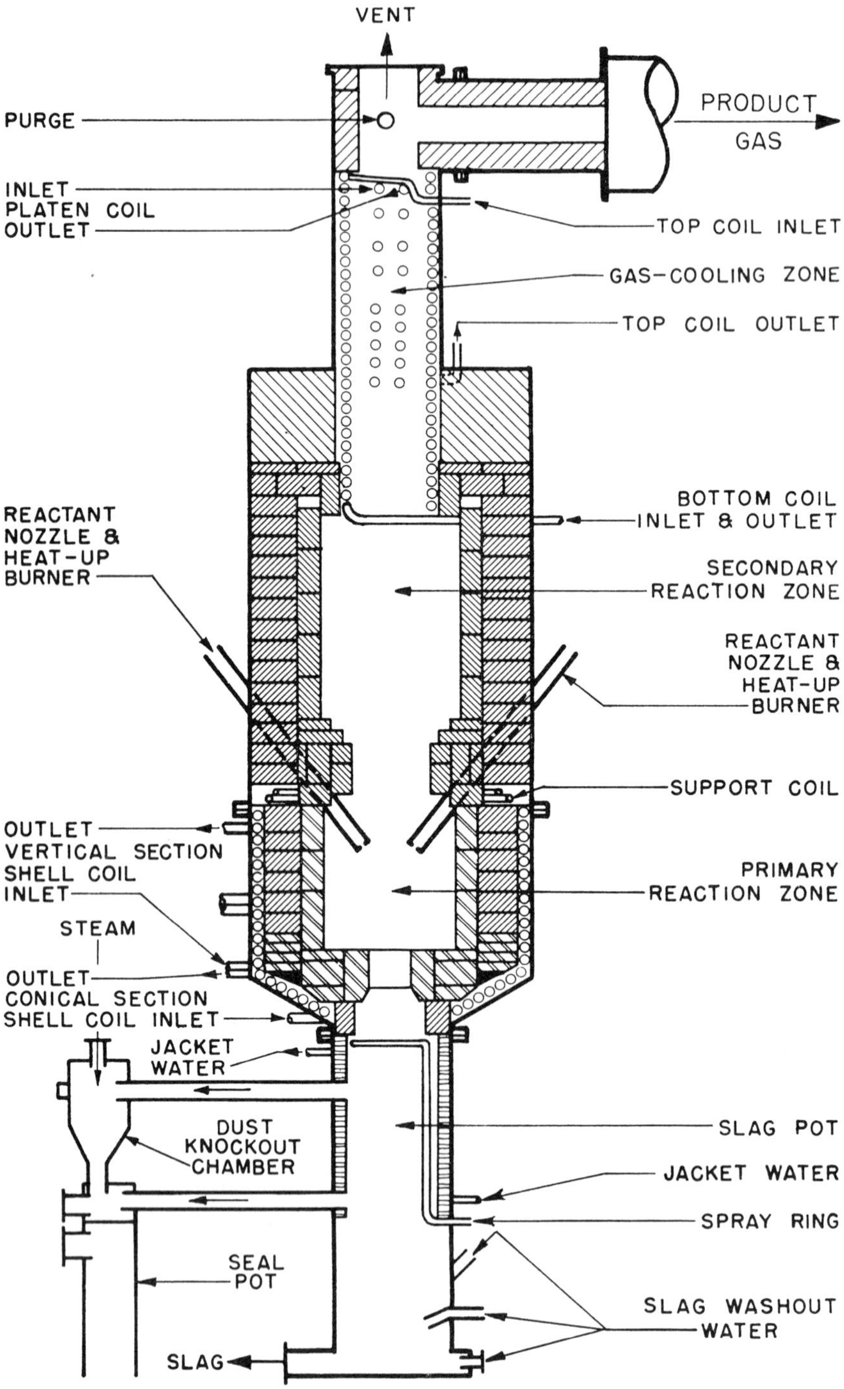

Fig. 40. Bureau of Mines atmospheric pressure gasifier No. 4, design 3.[200]

TABLE 22

COMPARISON OF BITUMINOUS A, ANTHRACITE, AND SUBBITUMINOUS C AT EQUIVALENT CARBON INPUT AND STEAM-CARBON RATIOS[183]

Coal Rank	Oxygen-Carbon Ratio scf/lb	Material Requirements per Mcf CO + H_2 Oxygen, scf	Coal, lb	Carbon Gasified, %	Calculated Exit-gas Temperature, °F	Heat Loss, Btu/lb Coal
High-volatile bituminous A	10.5	360	48	65	2100	535
	15.0	410	38	90	2300	950
Anthracite	10.5	465	53	61	2000	925
	15.0	540	42	80	2100	925
Subbituminous C	10.5	310	53	79	1800	475
	15.0	400	47	92	2100	475

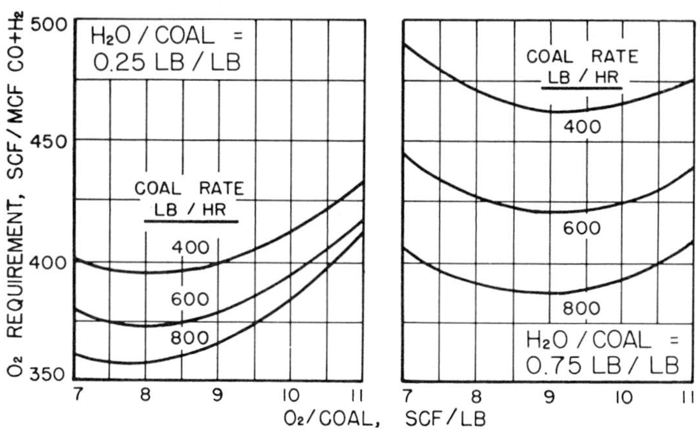

Fig. 41. Effect of oxygen-coal ratio on oxygen requirement with high-volatile bituminous A coal.[183]

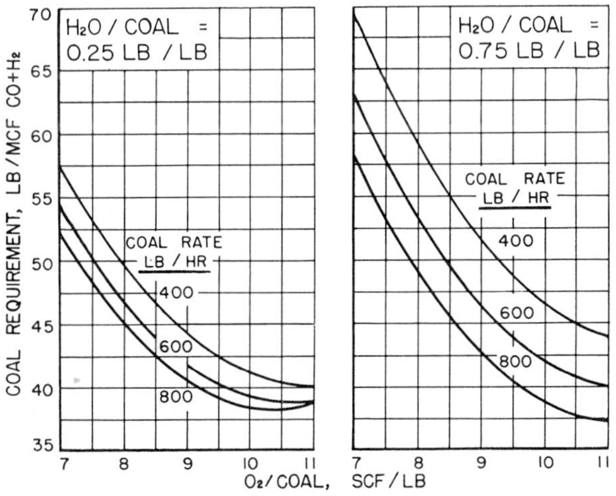

Fig. 42. Effect of oxygen-coal ratio on coal requirement with high-volatile bituminous A coal.[183]

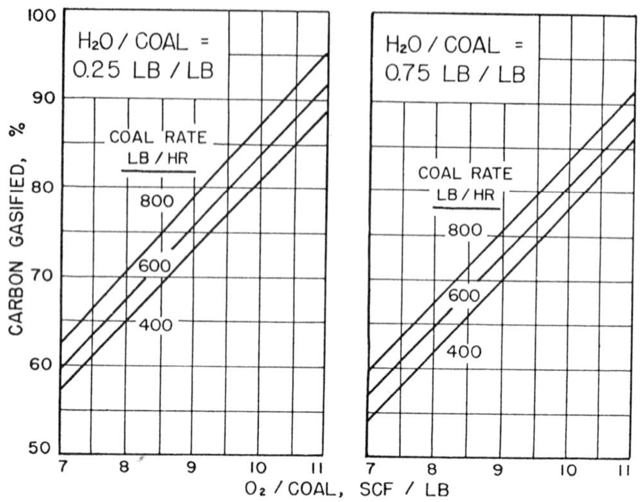

Fig. 43. Effect of oxygen-coal ratio on percentage of carbon gasified with high-volatile bituminous A coal.[183]

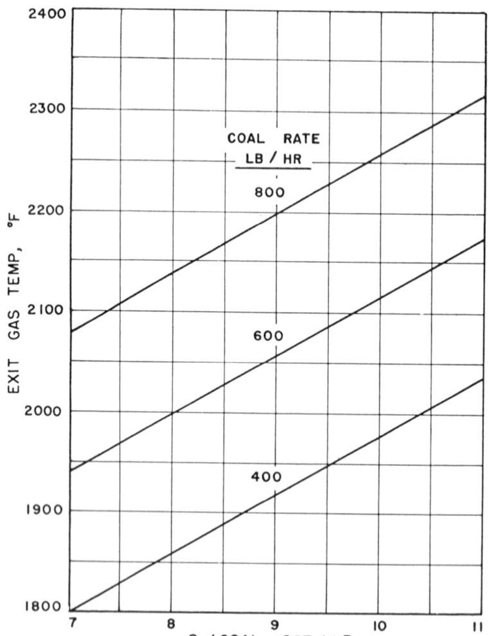

Fig. 44. Effect of oxygen-coal ratio on exit gas temperature with high-volatile bituminous A coal.[183]

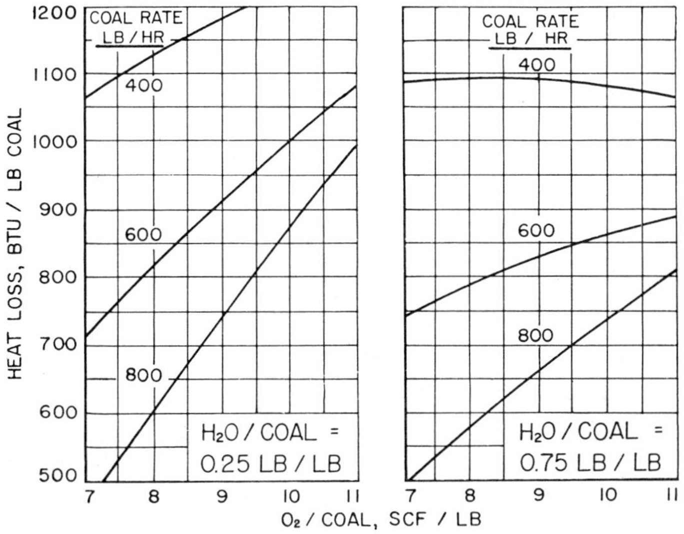

Fig. 45. Effect of oxygen-coal ratio on heat loss with high-volatile bituminous A coal.[183]

would be (1) subbituminous C, (2) high-volatile bituminous A, and (3) anthracite. This same order is indicated for reactivity as measured by carbon conversion and as shown in Fig. 46.

A cylindrical upflow slagging gasifier lined with a "ram mix" of high-purity alumina and having an internal diameter of 4 ft 6 in. and a height of 20 ft 4 in. was tested at the Louisiana, Missouri, Synthetic Fuels Demonstration Plant of the U.S. Bureau of Mines.[206] Considerable difficulty experienced with erosion

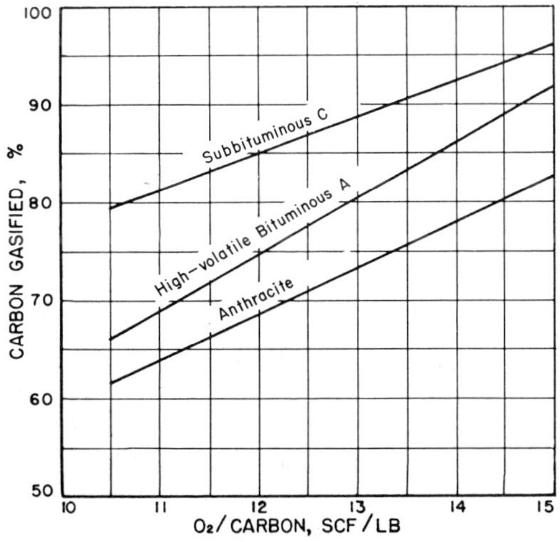

Fig. 46. Effect of oxygen-carbon ratio on percentage of carbon gasified in tests with three coals.[183] (Carbon input, 570 lb/hr; steam-carbon ratio, 0.42 lb/lb.)

of the lining was alleviated to some extent by placement of burners and the use of superheated steam instead of oxygen as a carrier for coal. The project was terminated before the development work was completed and extensive performance data obtained.

Comparatively small-scale pilot plant studies of the gasification of pulverized coal with oxygen and steam have been made in a tubular reactor under laminar flow conditions,[207] in a cyclone gasifier,[5] and in a vortex reactor.[186,208] The latter investigation presented information on some of the basic characteristics of gasification reactions in a vortex, and proposed a method for extrapolating experimental results to those that would be obtained in an adiabatic reactor. An experimental cyclone generator (2 ft in diameter × 2 ft deep) for gasifying pulverized coal with air has been studied by the Fuel Research Board.[209]

Laboratory-[210] and intermediate-scale investigations[204,211] of resonant-tube reactors for

[206] Batchelder, H. R., and Hirst, L. L., *Ind. Eng. Chem.*, **47**, 1522–8 (1955).

[207] Sebastian, J. J. S., *Ind. Eng. Chem.*, **44**, 1175–84 (1952).

[208] Perry, H., Corey, R. C., and Elliott, M. A., *ASME, Trans.*, **72**, 599–610 (1950).

[209] Fuel Research Bd. (Gt. Brit.), *Rept. for the Year 1955*, 15–16.

[210] Secretary of the Interior, *U.S. Bur. Mines, Rept. Invest.* No. 4865 (1952), 86 pp.

[211] Sommers, H., *Tech. Mitt. Krupp*, **16**, 258–62 (1958).

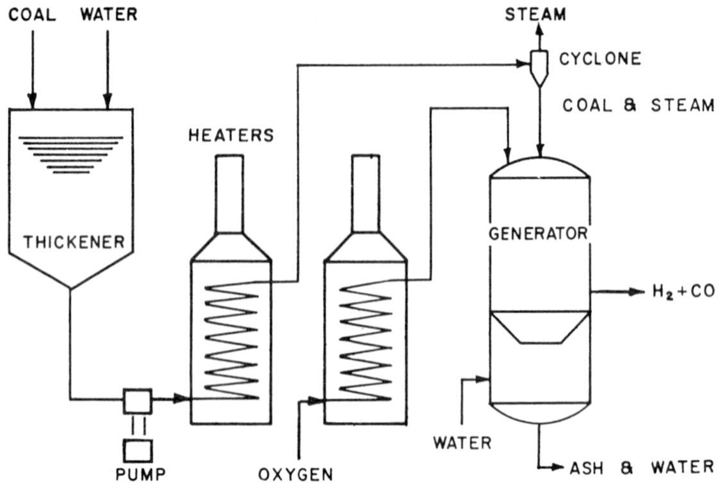

Fig. 47. Simplified flow diagram of the Texaco gasifier.[213]

gasifying bituminous coal with air have demonstrated high specific throughputs as a consequence of turbulence and high relative motion between suspended particles and the gasifying medium. This work directed attention to the many mechanical, material and noise problems that would have to be solved if this system were to be developed further.

Superatmospheric Pressure Processes

The advantages and disadvantages of operating gasification processes at elevated pressures have been discussed in connection with fixed-bed operations. This discussion is pertinent also to high-pressure suspension processes. However, operation of these processes involves the additional problem of feeding coal suspended in a carrier gas into a pressure vessel. Furthermore, all of the high-pressure processes developed or studied in large pilot plants are carried out under slagging conditions, with the attendant problem of handling slag, and, in industrial gasifiers, of removing it continuously from a pressure vessel.

The Texaco gasifier. This is the only high-pressure suspension gasification process that has ever been operated on an industrial scale. One unit operating at 400 to 500 psig was installed at the plant of the Olin Matheson Chemical Corporation in Morgantown, West Virginia[212] for production of gas for the synthesis of ammonia. The reported capacity of this unit was 5,840,000 scf/day, but no data on its performance have been published.

Some information has been published on the Texaco pilot plant,[213] a schematic diagram of which is shown in Fig. 47. A coal-water slurry containing 40 to 60 wt. % coal is pumped through a preheater in which the water is vaporized and the mixture heated to temperatures of 1000°F or more. The steam-coal ratio is adjusted by separating some of the steam from the mixture in a cyclone ahead of the gasifier. This steam-coal mixture enters at the top, either axially or tangentially. Preheated oxygen is fed through a separate water-cooled nozzle oriented to prevent impingement of oxygen on the wall of the reactor. Slag flows through a constricted opening at the bottom, is granulated by quenching in water, and is removed as a finely divided suspension in water. The gas leaves at the bottom, and passes through a waste heat boiler. The pilot unit operates at 200 psig, and has a reaction volume of 2 cu ft.[214] The temperature in the reaction zone may range from 2000 to 2500°F.

Throughputs of 300 lb of coal per cu ft of reaction volume per hour have been attained, although it is reported that throughputs were limited by the capacity of the feed system. The only performance data reported[214] indicated carbon conversion ranging from 90 to 98 %, thermal efficiency ranging from 75 to

[212] Anon., *Chem. Week*, **78,** No. 26, 76, 78, 80 (1956).

[213] Eastman, du B., *Gasification and Liquefaction of Coal*, AIME, New York, 1953, pp. 73-9.

[214] MacFarlane, W. A., *J. Inst. Fuel*, **28,** 109–13 (1955).

80%, and oxygen requirements of 350 to 375 scf/Mcf of make gas. The gas contained about 8% CO_2, 1% CH_4, and had a H_2/CO ratio approximating unity. It has also been reported[213] that, "The oxygen consumption rates ultimately expected are 15,000 to 20,000 scf per ton of dry coal with 80 to 90% of the heating value of the coal appearing in the form of hydrogen and carbon monoxide in the synthesis gas."

U.S. Bureau of Mines pressure gasifier. The high-pressure (450 psig) gasification pilot plant operated at the Morgantown Station of the U.S. Bureau of Mines has yielded extensive information on operability of pressure gasification reactors with different methods of feed introduction and reactor construction.[184,215,216] As originally designed, the reactor was lined with refractory with cooling coils next to the wall of the pressure vessel. Reactants were first introduced tangentially at the top. However, rapid erosion and fusion of the refractories in the high-temperature zone necessitated a design modification with axial down-jet introduction of reactants through a combination coal-steam-oxygen nozzle. This system was more satisfactory, but did not prevent continued erosion and dissolution of the confining refractory walls by molten slag. To stabilize the dimensions of the walls of the high-temperature reaction zone, a studded water-wall was installed with a castable refractory surface held in place by the studs.

The latest design, which has a high-temperature reaction zone 10 in. in diameter and 37 in. long, is shown in Fig. 48.[217] Coal is metered and conveyed into the reactor by means of a pressurized fluidized feeder.[218] Performance data[219] for this reactor when operated at 300 psi are given in Table 24. The use of water-cooled walls stabilized the dimensions of the high-temperature reaction zone, but increased the heat lost from this zone in comparison with that observed in the refractory-lined chamber (see Table 23).[131,184] Gasification temperature and consequently reaction rates were lower in the water-cooled chamber under comparable feed conditions. Accordingly, the unconverted carbon was higher, and the proportion of the heat input appearing as potential heat in the gas was lower.

A statistical analysis has been made of the results of tests in the water-cooled pressure gasifier showing the effect of pressure, coal rate and oxygen-coal ratio on the performance of the unit.[217] The results are presented in Figs. 49 to 53 for operating pressures of 75and 300 psig and a constant steam-coal ratio of 0.3 lb/lb. In all instances, the oxygen-coal ratio is an important process variable. In interpreting the results at the two pressure levels, the effect of residence time and heat loss must be considered. For a given coal-feed rate, residence time is least at the lowest pressure and heat losses (see Fig. 53) are greatest. Heat loss per pound of coal decreases as the coal feed rate increases (see Fig. 53). Thus, the reversal in the effect of coal-feed rate on carbon conversion, and oxygen and coal requirements, occurs because residence time is

TABLE 23

Heat Losses in Two Designs of USBM Pressure Gasifier[131]

	Refractory Chamber	Water-Cooled Chamber
Operating conditions		
Coal feed, lb/hr	434	434
Oxygen-coal ratio, lb/lb	9.30	9.33
Steam-coal ratio, lb/lb	0.43	0.40
Operating pressure, psig	100	101
Reaction chamber volume, cu ft	1.46	2.25
Heat input		
Potential heat in coal, %	96.1	97.2
Total heat input, %	100	100
Heat out		
Potential heat of gases, %	64.6	61.5
Heat loss through chamber walls, %	1.0	5.2
Potential heat of unreacted carbon, %	9.3	14.8
All other losses	25.1	18.5

[215] Strimbeck, G. R., Cordiner, J. B., Jr., Taylor, H. G., Plants, K. D., and Schmidt, L. D., *AGA, Proc.* **1952,** 778–817.

[216] Strimbeck, G. R., Cordiner, J. B., Jr., Taylor, H. G., Plants, K. D., and Schmidt, L. D., *U.S. Bur. Mines, Rept. Invest.* No. 4971 (1953), 27 pp.

[217] Holden, J. H., Strimbeck, G. R., McGee, J. P., Willmott, L. F. and Hirst, L. L., *U.S. Bur. Mines, Rept. Invest.* No. 5573, (1960), 56 pp.

[218] Barker, K. R., Sebastian, J. J. S., Schmidt, L. D., and Simons, H. P., *Ind. Eng. Chem.*, **43,** 1204–09 (1951).

[219] Secretary of the Interior, *U.S. Bur. Mines, Rept. Invest.* No. 5043 (1954), 66 pp.

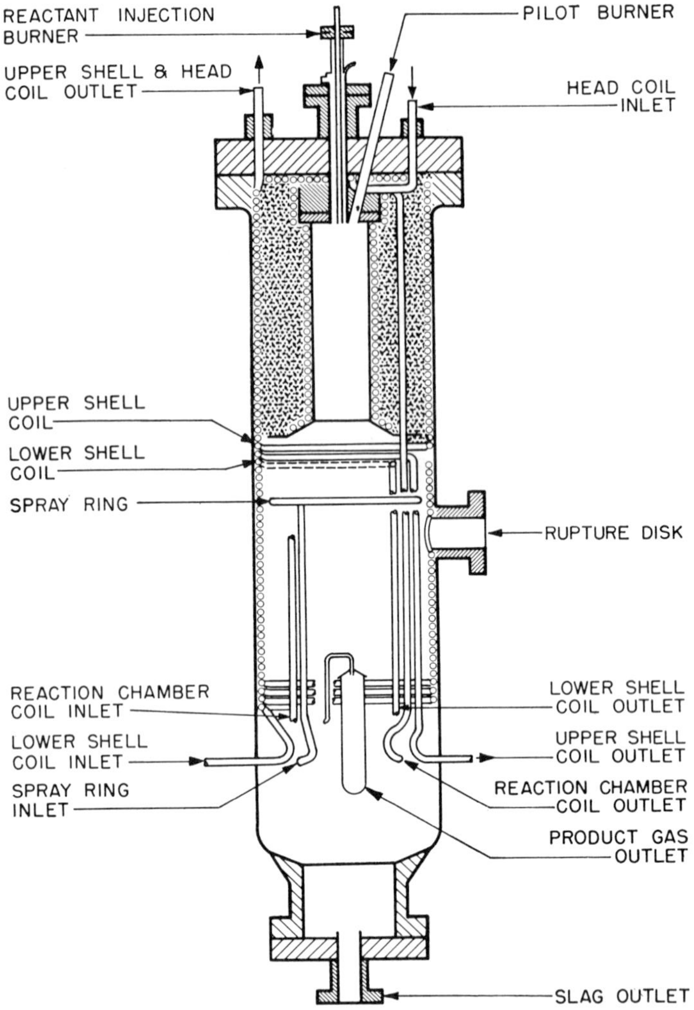

Fig. 48. Bureau of Mines pressure gasifier No. 3.[217]

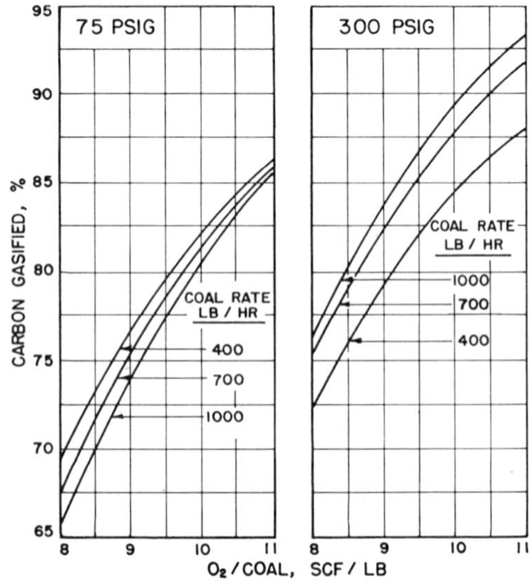

Fig. 49. Effect of oxygen-coal ratio on percentage of carbon gasified at two test pressures (bituminous high volatile A coal).[217]

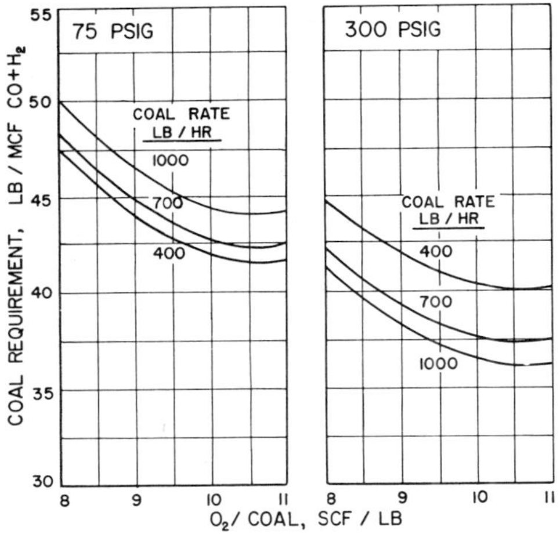

Fig. 50. Effect of oxygen-coal ratio on oxygen requirement at two test pressures (high-volatile bituminous A coal).[217]

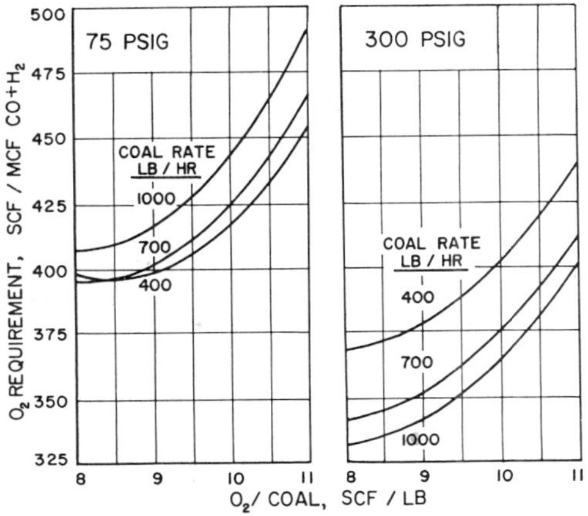

Fig. 51. Effect of oxygen-coal ratio on coal requirement at two test pressures (high-volatile bituminous A coal).[217]

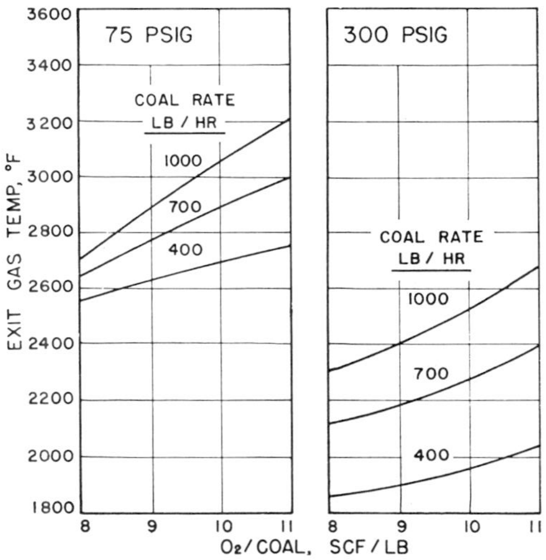

Fig. 52. Effect of oxygen-coal ratio on exit gas temperature at two test pressures (high-volatile bituminous A coal).[217]

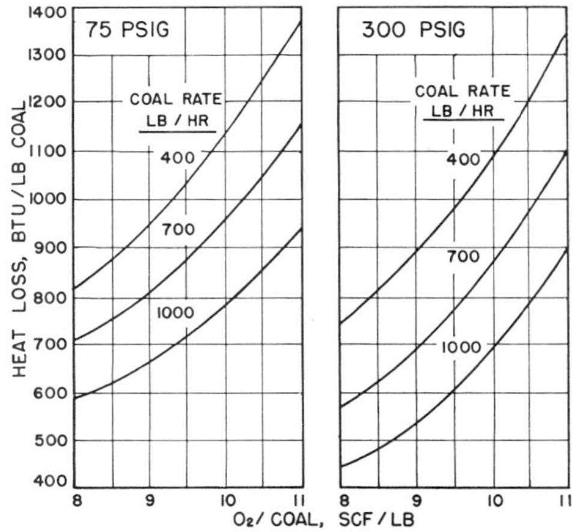

Fig. 53. Effect of oxygen-coal ratio on heat loss at two test pressures (high-volatile bituminous A coal.[217]

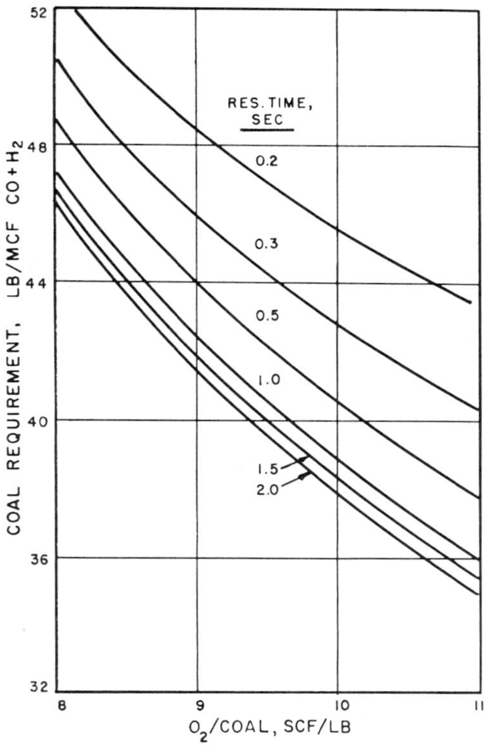

Fig. 54. Effect of oxygen-coal ratio and residence time on coal requirement with high-volatile bituminous A coal.[217]

the controlling factor at the low pressure, whereas heat loss is controlling at the high pressure.

To eliminate the effect of heat loss, results from the water-cooled reactor were corrected to a constant heat loss of 36 Btu per scf of $CO + H_2$.[184] This correction was based on the observed relation between scf of oxygen per Mcf of $CO + H_2$, and heat loss per Mcf of $CO + H_2$, which had a slope of 3.0 scf of

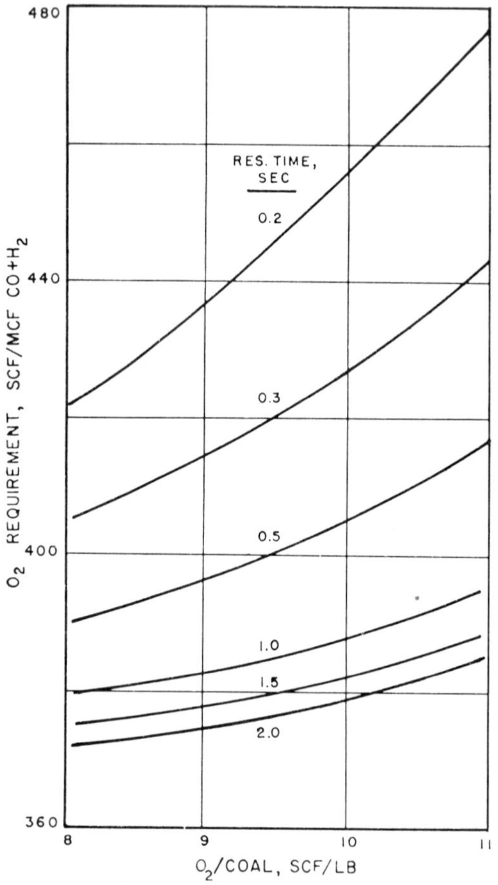

Fig. 55. Effect of oxygen-coal ratio and residence time on oxygen requirement with high-volatile bituminous A coal.[217]

O_2 per MBtu heat loss (compare Fig. 34). With effects due to heat losses eliminated, the effect of residence time (volume of reaction zone divided by volume of gas at reaction conditions) may be shown (see Figs. 54 to 56). In this analysis, it is assumed that pressure has no effect on the kinetics of the reaction, but merely affects residence time at a given

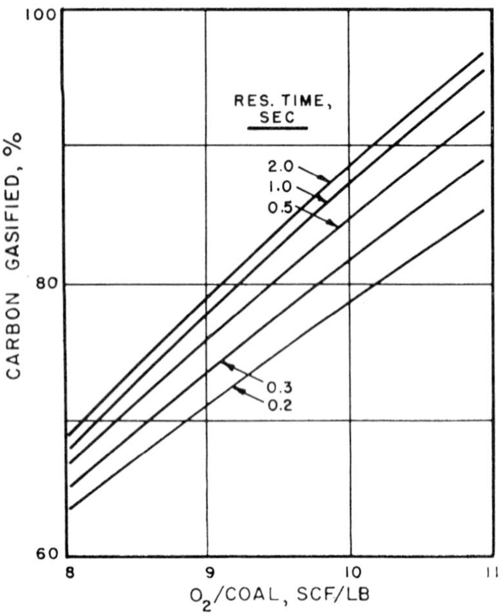

Fig. 56. Effect of oxygen-coal ratio and residence time on percentage of carbon gasified (high-volatile bituminous A coal).[217]

throughput. It was stated that coal and oxygen requirements are approximately linear functions of the reciprocal of the residence time. Figure 56 indicates an extremely rapid initial rate of carbon conversion as a result of the rapid oxidation reactions, and the gasification of the more reactive carbon. The rate of carbon conversion decreases markedly as

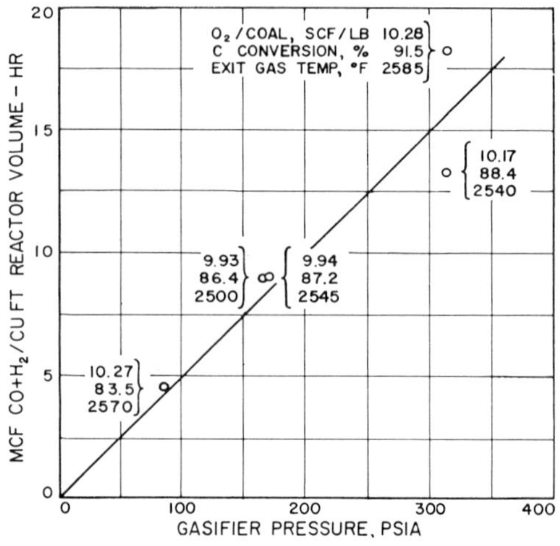

Fig. 57. Proportionality of gasification rate and absolute pressure.

the carbon conversion increases. For example, increasing the residence time from 1 to 2 secs increases the carbon conversion by about 1%.

If the only effect of pressure is to change residence time, the specific gasification rate at a given reaction zone temperature should be directly proportional to the absolute pressure. Figure 57 shows that this direct proportionality applies for data on the water-cooled reactor.[217] The data were selected so that exit-gas temperature was in the range of 2540 to 2585°F, and the oxygen-coal ratio was in the range 9.93 to 10.28 scf/lb. Carbon conversion ranged from 83.5 to 91.5%. At a given pressure, reaction rates were a function of only feed composition and temperature. Consequently, variations in heat loss should not affect any conclusions drawn from Fig. 57 if the exit-gas temperature is a proper indication of reaction-zone temperature.

Part of the Bureau of Mines' gasification program has been concerned with feeding coal-carrier gas suspensions into pressure vessels.[220] Basic heat transfer and operating data are being obtained on the slurry system, and studies are being made of a steam-pickup system. Development work is also being done on a feeder fluidized in two annular zones to improve the performance of the pneumatic feeder developed at Morgantown.

Institute of Gas Technology gasifier. Pilot plant studies of pulverized coal gasification at pressures up to 7 atm have been made by IGT[5] in a downflow slagging system operated at coal-feed rates of up to 1450 lb/hr. The gasifier consisted of a refractory-lined noncooled reactor (11.65 cu ft effective reaction zone volume) connected to an insulated slag catchpot (see Fig. 58). Sensible heat was recovered from the product gas in a refractory-lined heat exchanger. Pulverized coal and superheated steam were introduced tangentially into the upper cylindrical section or "cyclonizer" chamber of the reactor. Oxygen was introduced axially through multiple jets immediately below the outlet of the cyclonizer chamber. Hot gasification products passed through the slag catchpot, deposited liquid coal ash which solidified, and then tangentially entered the heat exchanger, where fly ash and unreacted carbon were deposited in the insulated receptacle at the bottom. One feature of the system was the use of a flow nozzle discharging tangentially into a cylindrical attrition chamber in the coal-transfer line to the gasifier. This feature accomplishes some pulverization of the coal and obviates fine grinding.

The IGT pilot plant was operated on caking and noncaking coal at pressures up to 5 atm. Coarsely pulverized (−20 mesh) coal was used, with additional fineness imparted by the nozzle system. The effect of pressure, oxygen concentration, and oxygen-coal ratio, on gas composition, material requirements, and gasifier performance was investigated extensively. The effect of the oxygen-coal ratio on material requirements and carbon conversion with noncaking high-volatile "B" bituminous coal is shown in Fig. 59. Optimum requirements corresponding to 90% carbon conversion are: 31 lb of coal, 33 lb of steam, and 312 scf of total oxygen at 115 psia. Typical operating results are given in Table 24.

Operating experience with this reactor showed very little erosion of refractory in the cyclonizer section. Uncooled high-temperature refractories in the reaction zone, such as silicon carbide, zirconium silicate, and alumina brick, could not be maintained over reasonable periods without spalling and slag penetration at ash-slagging temperature. Slag removal by a catchpot was satisfactory for intermittent pilot plant tests, but obviously would not be suitable for continous operation.

Bianchi process. This process for gasifying Provencal lignite under nonslagging conditions at pressures up to 25 atm was studied by Gaz de France at the Marseilles Gas Works.[221] Steam and oxygen preheated to about 1000°F were injected tangentially into a cylindrical chamber with conical ends. The reacting mixture left this chamber at the top and flowed upward through a water-cooled shaft. The temperature in the reaction zone was 1600°F. Additional steam was injected at the outlet of the reaction chamber to promote conversion of CO into H_2. Only limited data have been published because of difficulties with the coal-feed system. At 14 atm, the pilot unit produced gas having a CO_2-free heating value ranging from about 320 to 390 Btu/scf, and containing up to 5% CH_4 in the raw gas.

[220] Secretary of the Interior, *U.S. Bur. Mines, Rept. Invest.* No. 5118 (1955), 73 pp.

[221] O.E.E.C., Rept. PRA/CR/WP3(53)1 Paris. The Council, 1954.

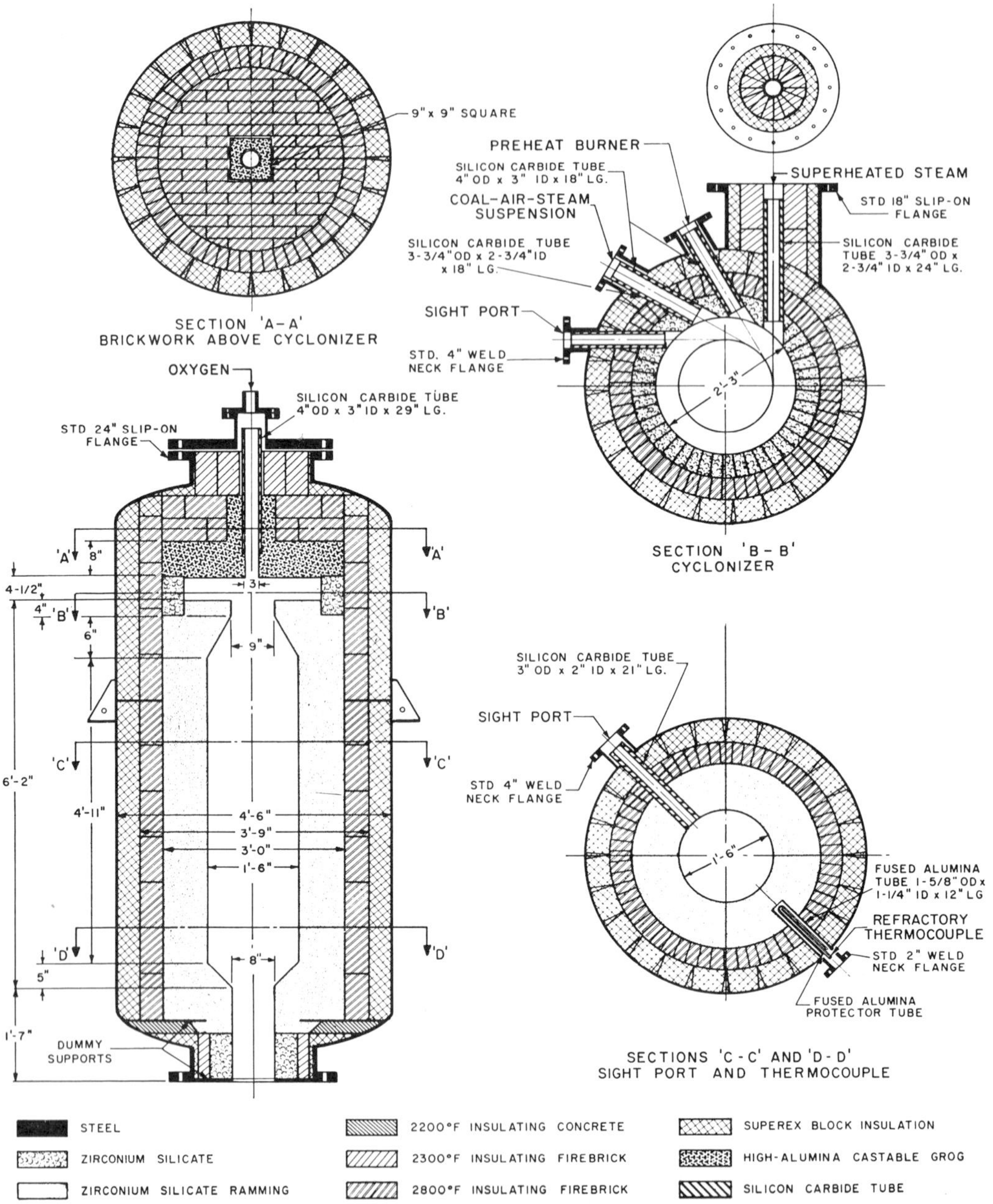

Fig. 58. Details of the Institute of Gas Technology pressure gasifier.[5]

Bases for Comparing Oxygen-steam Gasification Processes

In the final analysis, gasification processes must be compared on the basis of the total cost of the finished product. However, since material costs comprise a significant fraction of the total cost of gas, it is frequently helpful to use such criteria as carbon conversion, steam decomposition, fuel, oxygen, and steam requirements, and thermal efficiency in comparing different processes. An excellent and extensive discussion of the comparison of gasification processes has been published by Hoy and Wilkins,[131] and much of the following discussion of this subject is based on their presentation, including the performance data for a number of gasification systems shown in Table 21.

Carbon Conversion

Carbon conversion is defined as the fraction of the carbon in the solid feed converted to carbon-containing constituents in the gas. By this definition, all of the carbon dioxide in the gas is considered as carbon converted, whereas only that portion originating from the water-gas shift reaction:

$$CO + H_2O \rightarrow CO_2 + H_2 \qquad (93)$$

measures the conversion of carbon into useful energy in gaseous products. The remainder of the carbon dioxide is produced in exothermic reactions which supply the endothermic heat of the gasification reactions and other heat requirements of the process.

Since any carbon converted to carbon dioxide is not available for gasification, it would seem that a more critical and meaningful criterion of carbon gasification per se would be the fraction of the available carbon converted to energy in the gaseous form. Such a criterion has been used in analyzing the results of tests in a vortex gasifier.[186] With minor modifications to account for carbon appearing in the gas as methane and in any liquid products, this concept can be applied to any continuous autothermic process.

The carbon available for gasification is defined:

$$C_a = C_u + C_s + C_h + C_l \qquad (94)$$

in which, C_a = available carbon, moles per pound of fuel, C_u = unreacted carbon, moles per pound of fuel, C_s = carbon converted to synthesis gas ($CO + H_2$), moles per pound of fuel, C_h = carbon converted to gaseous hydrocarbons, moles per pound of fuel, and C_l = carbon converted to liquid products, moles per pound of fuel. The conversion of available carbon to energy in the gaseous form is given by:

$$x = \frac{C_s + C_h}{C_u + C_s + C_h + C_l} \qquad (95)$$

From the stoichiometry of the gasification reactions:

$$C + H_2O \rightarrow CO + H_2 \qquad (96)$$

$$C + 2H_2O \rightarrow CO_2 + 2H_2 \qquad (97)$$

$$C + CO_2 \rightarrow 2CO \qquad (98)$$

it is evident that all of these reactions produce two moles of synthesis gas (either CO, H_2, or

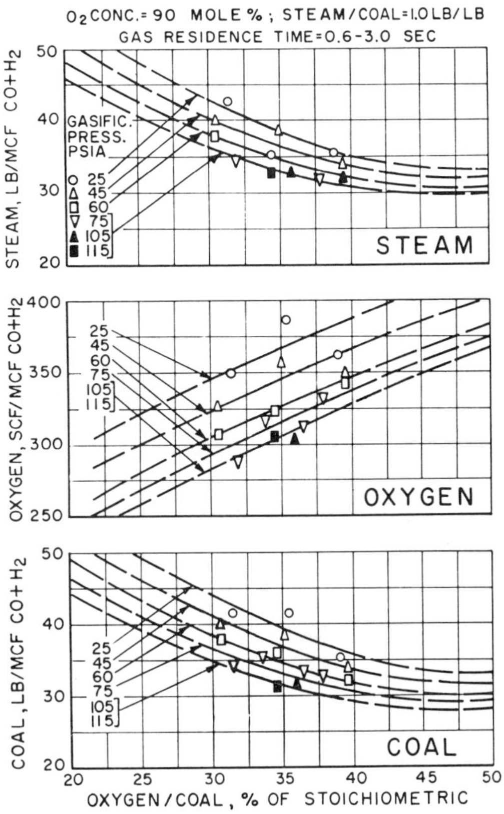

Fig. 59. Reactant requirements per Mcf $CO + H_2$ in IGT tests with noncoking coals at feed rates of 300 to 600 lb/hr.[5]

mixtures of these) per mole of carbon reacted. Accordingly,

$$C_s = \frac{S}{2} \qquad (99)$$

in which S = moles of $CO + H_2$ produced per pound of fuel, and,

$$x = \frac{1}{1 + \dfrac{2(C_u + C_l)}{S + 2C_h}} \qquad (100)$$

In fixed-bed gasifiers, in which some carbon leaves the reactor as tar before reaching the high-temperature gasification zone, and in most suspension gasifiers, in which no liquid carbon-containing products are produced, a more appropriate expression for conversion of available carbon is:

$$x = \frac{1}{1 + \dfrac{2C_u}{S + 2C_h}} \qquad (101)$$

TABLE 24

Performance Data for Some Gasification Systems[131]

	Industrial					Experimental				
Gasifier	Koppers Totzek	B&W du Pont	Winkler	Würth	Rummel	Lurgi	Flesch Demag	Rummel	IGT	USBM
Site	Oulu	West Virginia	Leuna	Leuna	Wesse-ling	Holten	Ludwigs-hafen	Wesse-ling	Chicago	Morgan-town
Fuel	Bitu-minous Coal	Bitu-minous Coal	Brown Coal	Coke	Brown Coal	Bitu-minous Coal	Bitu-minous Coal	Ruhr Coal	Bitu-minous Coal	Bitu-minous Coal
Operating pressure, atm	1	1	1	1	1	20	1	1	5	20
Input										
Fuel, lb/hr	4280	2620	64,800	12,900	20,280	2270	3430	3300	438	1287
Fuel (based on gross heating value), therms/hr	517	380	6330	1600¶	2025	250¶	460	450	53	178
Oxygen, lb/hr	3352	2204	31,000	10,000	12,015	795	2885	2690	380	1040
Steam, lb/hr	2315	2290	18,380	10,830	2645	4440	4100	2360	657	390
Steam temperature, °C	128	163	—	—	750	500	—	710	680	480
Gas composition, vol %										
CO_2	12.6	16.2	15.7	3.5	13.6**	30.6	10.7	13.6	21.6	11.1
C_nH_m	0	0.4	0	0	—	0.3	0.6	0	0	0.5
O_2	0.1	0.2	0	0	—	0	0.2	0	0	0
CO	51.1	40.2	44.4	68.0	52.0	16.2	46.1	42.5	32.6	52.5
H_2	34.0	39.4	36.0	28.0	32.1	42.9	37.2	42.5	40.9	32.4
CH_4	0.1	0.7	1.6	0	0.4	9.2	1.8	0.2	0	0.3
N_2	1.9	2.9	0.8	0.5	1.9	0.5	3.4	1.2	4.9	3.2
Hydrogen/carbon monoxide ratio	0.67	0.98	0.81	0.41	0.62	2.62	0.81	1.0	1.25	0.64
Gross heating value, Btu/scf	271	268	267	306	272	287	290	273	234	282
Gasifier outlet temperature, °C*	—	1200	800–900	380	1000	570	300–400	1200	1480	—
Materials requirements/Mcf CO + H_2										
Fuel, lb	39.5	33.9	55.2	27.8	40.6	54.0	30.6	26.8	31.8	38.6
Fuel, therms	4.77	4.91	5.39	3.44	4.07	6.80	4.10	3.62	3.86	5.00
Oxygen (97%), lb	31.4	28.5	26.6	21.5	24.0	20.2	21.7	21.8	26.7	30.4
Steam, lb	21.3	29.6	16.0	23.3	5.3	112.6	36.6	19.1	48.0	11.7
Steam generated/Mcf CO + H_2, lb	42.4	>59.2	0	0	41.9	22.5	15.1	28.2	0	29.7††
Performance										
Specific gasific. rate, therms/sq ft-hr†	—	19.4	70.6	16¶	69.3	22.6	7.1	83.3	32.0	174
Specific gasific. rate, therms/cu ft-hr‡	1.1	2.1¶	0.38	1.33	2.35	9.0	0.50	2.8	3.8	60
Steam decomposition, %§	29.7	40.0	40.2	85	62.7	35.7	35.2	75.0	31.7	52.0
Carbon conversion, %‖	88.6	84.1	80.0	—	98.6	84¶	—	99.0	100	88

* Temperature at entry to heat exchanger or spray-cooling system.
† Potential heat of make-gas/cross-section area of shaft. In fixed-bed gasifiers, this relates to cross-section area at top of shaft.
‡ Height of shaft below heat exchangers only used in calculating the volume.
§ Unless marked otherwise, steam decomposition has been calculated:

$$\text{Fixed-bed gasifiers } \frac{H_2 \text{ in gas} - H_2 \text{ in coal} + (O_2 \text{ in coal}/8)}{H_2 \text{ in steam blast}}$$

$$\text{Suspension gasifiers } \frac{H_2 \text{ in gas} - H_2 \text{ in coal} + (O_2 \text{ in coal}/8)}{H_2 \text{ in (steam blast + moisture from coal)}}$$

‖ Carbon in gases/carbon in input fuel.
¶ Approximate value.
** Includes CO_2 for fuel conveying.
†† Equivalent steam generated (water-cooled gasifier).

The concept of conversion of available carbon makes it possible to separate for practical purposes the reaction of carbon in gasification reactions from its reaction to form carbon dioxide. Conversion of available carbon is, therefore, a useful and rational parameter for analyzing data on rates of gasification reactions in both pilot-plant and large-scale gasifiers, although it has not been used to any great extent.

Steam Decomposition

Maximum steam decomposition is desired in any gasification process because both sensible and latent heat are lost from the gasifier in the undecomposed steam. As mentioned previously, undecomposed steam also increases the problem of liquor treatment and disposal.

Steam decomposition may be expressed in the following ways:

Decomposition of added process steam

$$= \frac{\text{Free } H_2 \text{ in product gas} - \text{net } H_2 \text{ in coal}}{H_2 \text{ in added steam}}, \tag{102}$$

and

Decomposition of total equivalent steam

$$= \frac{\text{Free } H_2 \text{ in product gas}}{H_2 \text{ in added steam} + \text{total } H_2 \text{ in coal}}. \tag{103}$$

Equation 102 applies to fixed-bed gasifiers, in which most of the hydrogen in the fuel is evolved before the fuel is oxidized. In entrained, suspension gasifiers, equation 103 is more appropriate, because the moisture in the coal as well as the hydrogen in the volatile matter are rapidly converted to steam in the oxidation reactions. This steam is then just as available as a reactant as added process steam, and should be considered in computing steam decomposition. This is not always done in computing steam decomposition in suspension gasifiers.

When steam is the only gasifying medium, as it is in the make period of the cyclic water-gas process, steam decomposition is a direct criterion of the conversion of carbon to gaseous products by reaction with steam. However, in autothermic processes in which both steam and oxygen comprise the gasifying medium, steam decomposition is not necessarily a measure of the extent of the direct reaction of steam with carbon. Under such conditions, steam may react with carbon monoxide formed in the primary oxidation reactions.

Steam decomposition is the highest in fixed-bed slagging gasifiers because of the high-carbon inventory, which insures an excess of carbon, and because of favorable time-temperature conditions. Steam decomposition is lower in slagging suspension gasifiers operating at comparable temperatures (see Table 24), because in a flowing system, carbon is fed to the reacting system as it is used, and there is comparatively little excess carbon inventory in the reactor.

Oxygen Requirements

The theoretical oxygen required for combustion to supply the endothermic heat of the steam-carbon reactions with reactants and products at 60°F is 167 scf/Mcf of $CO + H_2$ (equation 84). In practice, combustion of a portion of the fuel with oxygen must supply the difference between the enthalpy of the reactants entering and the products leaving, and also any heat losses to water-cooled reactor walls, and by radiation and convection from the reactor. Accordingly, oxygen requirements decrease with increasing preheat of reactants, with decreasing exit-gas temperature, and with heat losses. Furthermore, exothermic gasification reactions yielding combustible gases, such as the reaction of hydrogen with carbon to form methane, also reduce oxygen requirements. The production of methane by this reaction is favored by increased pressure and lower temperatures. These conditions exist in fixed-bed countercurrent processes operated at elevated pressures, such as the Lurgi. Consequently, this is one of the factors responsible for the low oxygen requirements attained in that process (see Table 24).

The cost of oxygen represents about 25 to 35% of the total cost of synthesis gas, depending on the process and the unit cost of oxygen. This emphasizes the importance of minimizing oxygen requirements. As shown in Table 24, oxygen requirements range from about 240 to 370 scf/Mcf of $CO + H_2$—about 40 to 120% greater than the theoretical value.

Fuel Requirements

The theoretical carbon required per Mcf of $CO + H_2$ is 21.1 lb, which corresponds to an energy input of 2.97 therms according to equation 84. This theoretical requirement is exceeded in practice, as shown in Table 24, by fuel requirements which range from 3.44 to 5.39 therms/Mcf of $CO + H_2$ for processes producing only minor quantities of methane. As for oxygen, fuel is utilized to supply the heat requirements of the process. In addition, fuel requirements increase with decreasing carbon conversion.

TABLE 25

ESTIMATED TOTAL FUEL REQUIREMENTS OF SELECTED GASIFICATION PROCESSES[131]

Process	Koppers-Totzek Typpi Oy	USBM Morgan-town	Flesch-Demag Ludwigs-hafen	Würth Leuna
Operating pressure, atm	1	20	1	1
Total heat input				
To gasifier, %	86.3	93.1	79.1	74.1
To boiler, for steam and power, %	13.7	6.9	20.9	25.9
Total heat input recovered				
Potential heat in gas, %	58.7	60.1	67.1	68.6
Potential heat of tar and spirits, etc., %	0	0	0	0
Total fuel consumption				
Therms/therm make gas	1.70	1.66	1.49	1.46
Therms/Mcf $CO + H_2$	5.52	5.72	5.20	4.65

Assumptions: Boiler efficiency, 84%. Energy for oxygen production 375 kwh/ton. Efficiency of power generation, 35%. Energy for gas-oxygen compression 3.75 kwh/ton.

Since fuel is the only reactant containing potential chemical energy, it is common practice to use the "cold gas efficiency" (potential heat in gas ÷ potential heat in coal), or so-called "thermal efficiency," as a measure of the fuel requirements of the process. Hoy and Wilkins[131] pointed out that, to be valid as a criterion for fuel requirements, thermal efficiencies must be based on a "heat input," which includes the thermal equivalent of all power and steam required for oxygen manufacture, compression of reactants, fuel preparation and charging, etc. These authors have attempted to assess the overall energy requirements of some gasification processes, as shown in Table 25, on the assumption that in each instance, gas is required at a pressure of 20 atm, and that all energy is produced from solid fuel fed either to the gasifier or to a steam boiler to generate steam required in addition to that generated by waste heat recovery.

Heat Losses and Heat Recovery

Hoy and Wilkins[131] have discussed heat losses and heat recovery under three main categories: (1) gaseous losses, (2) solid losses, and (3) heat losses by radiation and convection from the external surfaces of the gasifier, and the like.

Gaseous losses. A comparison between the heat content of the gases before and after waste-heat recovery in a cocurrent and a countercurrent gasifier is shown in Table 26.[131] Before waste-heat recovery, the enthalpy of the exit gases from the cocurrent gasifier is twice that from the countercurrent unit. After waste-heat recovery, there is comparatively little difference. However, it must be recognized that heat recovered in the countercurrent

TABLE 26

GASEOUS LOSSES IN COCURRENT-FLOW AND FIXED-BED GASIFIERS WITH AND WITHOUT WASTE-HEAT RECOVERY[131]

Gasifier	Koppers-Totzek	Leuna-Würth	
Flow pattern in gasifier	Cocurrent	Countercurrent	
Before waste-heat recovery			
Sensible heat in gases, % heat input	9.7	5.5	
Enthalpy of undecomposed steam, % heat input	5.7	1.4	
Enthalpy of moisture from fuel, % heat input	0	0.8	
Total	15.4	7.7	
After waste-heat recovery			
Temperature, °C	280	380	(280)*
Sensible heat in gases, % heat input	3.0	5.5	(4.0)
Latent heat in steam, % heat input	3.43	1.83	(1.83)
Sensible heat in steam, % heat input	0.47	0.37	(0.26)
Total	6.9	7.7	(6.09)
Waste-heat recovery			
Steam production, % heat input	8.5	0	(1.61)

* Figures in parentheses indicate the effect of making outlet temperatures equal.

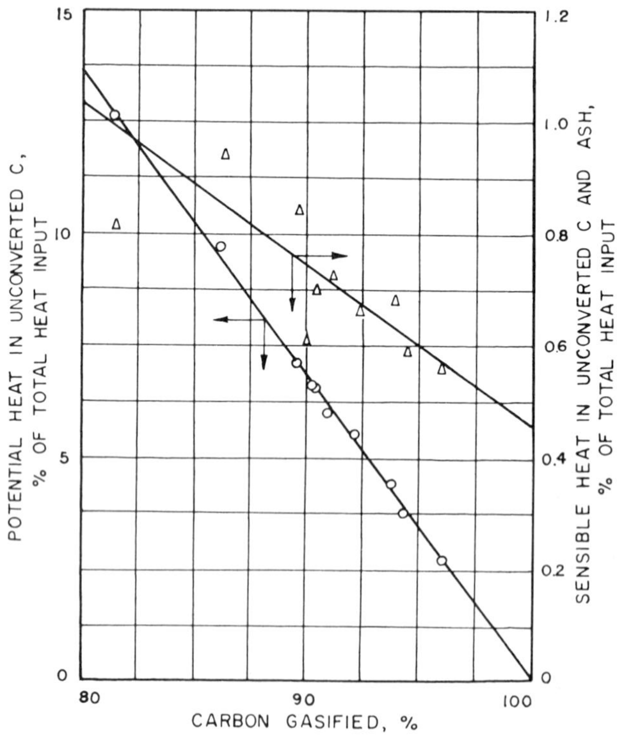

Fig. 60. IGT pressure gasifier test results showing heat loss in unconverted carbon.[5]

gasifier is directly available to the process, and thus reduces oxygen consumption, whereas heat recovered as steam improves thermal efficiency but does not reduce oxygen consumption. To reduce oxygen consumption in cocurrent process with high exit-gas temperatures, recuperation is essential. For example, steam in the Rummel generator is heated by recuperation so that the entering oxygen-steam mixture is at a temperature of 500°C.[202] However, recuperation of heat from gases at high temperatures is limited by available materials.[131]

Solid losses. The carbon conversion in fixed-bed gasifiers approaches 100% and, therefore, the loss of carbon in the ash is negligible. In cocurrent gasifiers, carbon conversion may be considerably less than 100% (see Table 24), depending on the time-temperature environment, the method of contacting reactants, and the reactivity of the fuel. Figure 60 presents data obtained in the IGT pressure gasifier[5] which shows that the potential heat in the unconverted carbon is about 7% of the heat input, even at 90% carbon conversion. In these tests, the sensible heat in the ash and unconverted carbon ranged from about 1.0 to 0.5% of the total heat input in the range of carbon conversion from 80 to 96%. In cocurrent gasifiers, the residue from the gasifier is frequently recycled to improve carbon conversion.[189,203]

External losses.[131] Heat losses by radiation and convection from the gasifier and its associated heat recovery equipment depend on the size of the unit and the specific throughput. Test of the Ruhrgas-Vortex process[204] showed external heat losses of 17% in a pilot plant having a capacity of 14 tons of coal per day, as compared with 6% in a 100-ton/day commercial plant. Increases in specific throughput by operation under pressure and by use of oxygen reduce percentage heat losses.

In gasifiers with water-cooled walls, heat is lost from the gasification zone, although it is recovered as heat in steam. However, it should be recognized in oxygen gasification processes that increases in heat flowing to water-cooled walls results in increased oxygen consumption.

Economic Considerations

An accurate economic analysis of any gasification process requires a detailed estimate of the investment cost for the gasifier and auxiliary equipment, and of the cost of materials, labor, and supporting services. It is not the purpose of this section to discuss such detailed estimates, but rather to direct attention to the magnitude of some of the major items of cost.

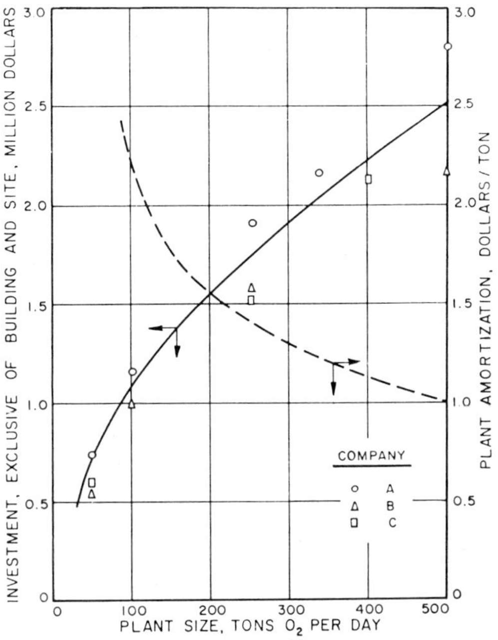

Fig. 61. Oxygen plant investment as a function of capacity.[222]

Oxygen Production Costs

Industrial oxygen having a purity of 95 to 98% is produced by liquefying air and separating the oxygen by distillation or rectification of the liquid, The average investment cost (in 1955 dollars) of a turnkey plant, exclusive of building and site, is shown as a function of the size of the plant in Fig. 61.[222] Amortization charges based on a 15-yr life, and operation 352 days/yr at 95% of capacity, are also shown. The operating cost of a 500-ton/day oxygen plant, exclusive of return on investment, is shown in Table 27. If the net return on investment were 6% (12% gross, before income taxes) the cost of oxygen would be increased by \$2.16/ton, making the total cost \$7.67/ton. If the oxygen is compressed to 500 lb/sq in., the

[222] Lyons, C. S., and Batchelder, H. R., *AGA, Proc.*, **1956,** 606–21.

TABLE 27

ANNUAL OPERATING COST OF A 500-TON/DAY OXYGEN PLANT[222]

	Cost, dollars‡	%
Operating payroll		
Labor, 2 men/shift at \$2.00/hr	35,050	
Supervision, 10% of labor	3,505	
	38,555	4.2
Maintenance		
2% of total investment	60,480	6.6
Operating supplies		
10% of operating payroll	3,856	0.4
Utilities		
Power, \$0.01/kwhr, 400 hp-hr/ton O_2	501,000	54.4
Water, \$0.01/1000 gal, 10,000 gal/ton O_2	16,768	1.8
General overhead		
100% of operating payroll	38,555	4.2
Depreciation		
6.67% total investment*	201,701	21.8
Taxes and insurance		
2% of total investment	60,480	6.6
Total	922,395	100.0
Cost of oxygen, 167,675 tons/year, \$5.50/ton†		

* Manufacturer's cost \$2,520,000
Site preparation, 20% investment 504,000

Total investment cost \$3,024,000

† Operating 352 days/year at 95% capacity.
‡ Computed in 1955 dollars.

additional energy required is 4.55[223] to 5.6[224] kwhr/Mcf (108 to 133 kwhr/ton). At \$0.01 per kwhr this would add an additional \$1.08 to \$1.33/ton to the cost of oxygen. For the same size plant, the cost of oxygen compressed to 450 psig, with 12% gross return, has been estimated at \$9.97 per ton (1959 dollars) by Katell and Faber.[225]

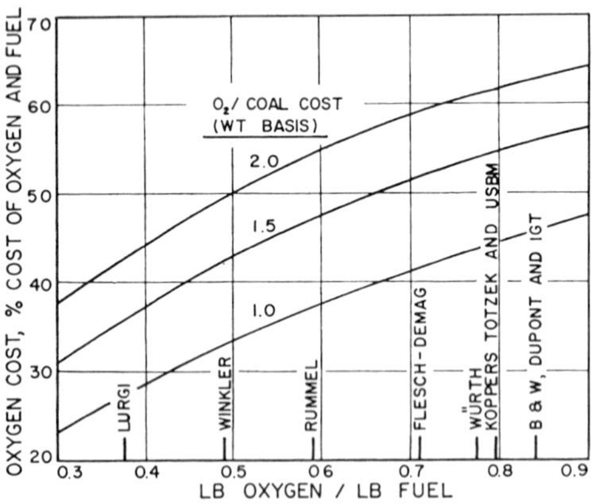

Fig. 62. Relative contribution of the cost of oxygen and coal to the cost of gas produced by various processes.

[223] McGee, J. P., Schmidt, L. D., Danko, J. A., and Pears, C. D., *Gasification and Liquefaction of Coal*, AIME, New Yoık, 1952, pp. 80–108.

[224] Anon., *Coke and Gas*, **20,** 158–9 (1958).

Relative Raw Material Costs

Considering the two primary gasification raw materials, oxygen and coal, the relative importance of the contribution of each to the cost of gas is shown in Fig. 62 for a range of oxygen-coal ratios, and of ratios of cost of oxygen per ton to cost of coal per ton. At oxygen-coal cost ratios between 1.5 and 2, the cost of oxygen represents about 40% of the raw material cost for the Lurgi process, and about 60% for most of the suspension gasification processes. Since the cost of oxygen is such an important factor, it is apparent that production of oxygen in large plants at minimum cost is essential. It should be recognized that the cost of oxygen represents a smaller percentage of the total cost of gas than is shown by Fig. 62. In addition to raw materials, this total cost includes, (1) the capital charges on the investment in the gasifier and auxilliary equipment, (2) operating expenses, and (3) the fuel required for steam generation. This latter factor may range from about 7 to 26% of the coal fed to the gasifier (see Table 25).

In suspension gasifiers, in which carbon conversion varies with the oxygen-coal ratio, the material cost reaches a minimum at some particular oxygen-coal ratio.[185]

GASIFICATION WITH HYDROGEN

The gasification of coal by reaction with hydrogen will be referred to as "hydrogasification," in contradistinction to general coal hydrogenation reactions in which both gaseous and liquid products are formed. The importance of hydrogasification as a processing technique was not recognized until 1937, when Dent and his coworkers[226] reported their results on the synthesis of gaseous hydrocarbons at high pressure. Since that time, the production of high-heating-value gases by the direct hydrogasification of coals and chars at elevated temperatures where essentially no liquids are produced has been investigated in the laboratory and pilot plant by these and other workers, and several reviews of this

[225] Katell, S., and Faber, J. H., *Chem. Eng.*, **66,** No. 6, 107–10 (1959).

[226] Dent, F. J., Blackburn, W. H., and Millett, H. C., *Inst. Gas Engrs. Commun.* No. 167 (1937), 58 pp. *Trans. Inst. Gas Engrs.*, **87,** 231–87 (1937–38).

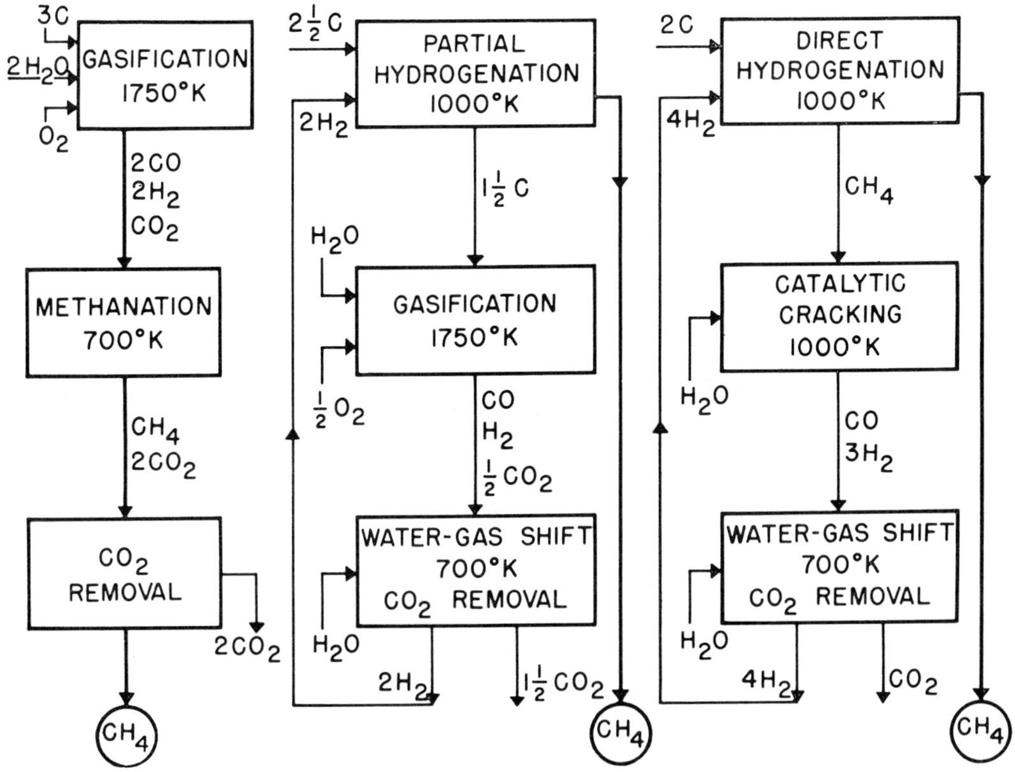

Fig. 63. Theoretical process requirements for conversion of carbon to methane.[13]

TABLE 28

HEAT BALANCES AND EFFICIENCIES OF IDEAL PROCESSES FOR CONVERSION OF CARBON TO METHANE[13]

(Based on enthalpies and standard heats of formation of β-graphite and ideal gases H_2O, O_2, H_2, CO, CO_2, and CH_4, and reference conditions of 25°C, 1 atm and liquid water)

	Gasification and Methanation (3 Moles C/Mole CH_4)	Hydrogenation: H_2 from Carbon (2½ Moles C/Mole CH_4)	Hydrogenation: H_2 from Product CH_4 (2 Moles C/Mole CH_4)
	Btu/lb-mole of Product Methane		
Heat of combustion of reactant carbon	507,879	423,233	338,587
Heat requirements			
Sensible heats	131,380	89,635	84,769
Endothermic heats of reaction	—	—	96,977
Latent heats	37,870	37,870	37,870
Total input	677,129	550,738	558,203
Heat of combustion of product methane	383,040	383,040	383,040
Heat available from sensible heat content of final and intermediate products	127,985	85,334	81,697
Exothermic heats of reaction	166,104	82,364	93,466
Total output	677,129	550,738	558,203
	Thermal Efficiency, %		
Without heat recovery*	56.6	69.6	68.6
With recovery of sensible and latent heat requirements†	75.4	90.5	87.9‡

* Heat of combustion of product CH_4/(heat of combustion of reactant carbon + total heat requirements).
† Heat of combustion of product CH_4/(heat of combustion of reactant carbon + net heat requirements).
‡ Endothermic heat of reaction assumed to be supplied from external source at 1000°K.

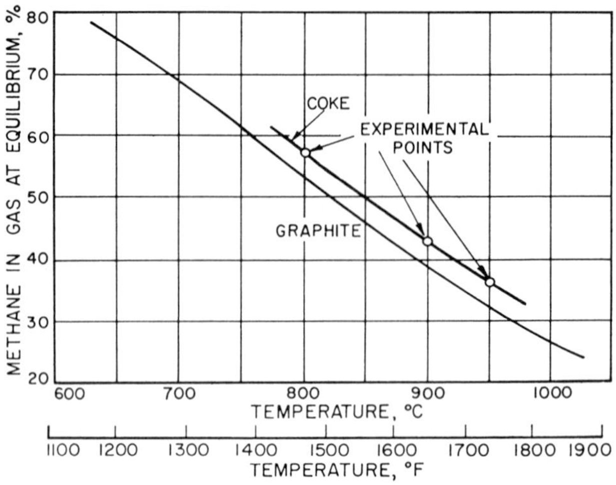

Fig. 64. Percentage of methane in a mixture of methane and hydrogen in equilibrium with graphite and coke at various temperatures and 50 atm pressure.[226]

field have been made.[15,227,228] Although beyond the scope of this review, it should be mentioned that oil shale may also be hydrogasified directly.[229]

The hydrogasification of coal to produce methane has an inherent thermochemical advantage, in comparison with the catalytic methanation of synthesis gas produced by gasifying coal with oxygen and steam. This is shown in Table 28[13] by the significantly higher-ideal thermal efficiencies of two variants of the hydrogasification process. The theoretical material requirements for the three ideal processes compared in Table 28 are shown in Fig. 63. Partial hydrogenation appears to be the most promising alternative, because it has the highest-potential efficiency, and is more generally applicable than complete hydrogenation, which probably cannot be attained in practice except possibly with lignite as a feed.

The maximum concentration of methane, and hence the maximum heating value of the gas which can be produced by the reaction of hydrogen with either graphite or coke at various temperatures and pressures, is shown in Fig. 64.[226] High pressure and low temperature favor the production of methane in accordance with equilibrium considerations in the reaction:

$$C + 2H_2 \rightarrow 2CH_4 \qquad (104)$$

The effect of pressure on the yield of gaseous hydrocarbons and on the absorbtion of hydrogen during the hydrogenation of semicoke at 800°C is shown in Table 29.[227]

Laboratory Studies

Laboratory studies of coal hydrogasification in fixed beds[226,230] and in batch autoclaves[13,114,115,118,119] have shown that the rapid reaction of hydrogen with coal or char begins at approximately 500 to 550°C. The rate of production of hydrocarbons is greatest in the first stage of conversion of coal into gas, which is characterized by pyrolysis and hydrogenolysis of the more reactive coal constituents. As the conversion of coal increases, the residue becomes progressively less reactive as the carbon structure becomes more graphitic in nature. In one study,[230] an increase in the rate of methane produced per unit of carbon ungasified was observed at the higher carbon conversions when the temperature was 900 to 950°C. This study was made in a fixed bed,

[227] Dent, F. J., *Intern. Conf. Complete Gasification Mined Coal, Liège*, **1954**, 113–24.

[228] Birch, T. J., Hall, K. R., and Urie, R. W., *Gasification of Brown Coal with Hydrogen in a Continuous Fluidized Bed Reactor*, CSIRO, Chem. Research Laboratories, Chem. Eng. Sect., April, 1959.

[229] Shultz, E. B., Jr., and Linden, H. R., *Ind. Eng. Chem.*, **51**, 573–6 (1959).

[230] Dent, F. J., Blackburn, W. H. and Millett, H. C., *Inst. Gas Engrs. Commun.* No. 190 (1938), 69 pp. *Trans. Inst. Gas Engrs.*, **88**, 150–217 (1938–39).

TABLE 29

RELATIONSHIP BETWEEN POTENTIAL HEAT IN GASEOUS HYDROCARBONS PRODUCED AND HYDROGEN ABSORBED DURING HYDROGENATION OF SEMICOKE AT 800°C[227]

Pressure of Hydrogen atm	Gaseous Hydrocarbons Produced therms/ton of coke*	Hydrogen Absorbed, therms/ton of coke*	Therms in Hydrogen Absorbed / Therms in Gaseous Hydrocarbons Produced, therm/therm
1	33.8	—	0.0
5	92.1	32.9	0.357
10	132.9	60.8	0.458
25	202.5	107.7	0.532
50	276.0	153.5	0.556
100	311.2	193.5	0.622

* 1 ton = 2240 lb.

with the temperature measured outside the wall of the reactor. Inasmuch as the forementioned observation is not borne out by studies of the same investigators in batch fluidized beds,[231] or by studies of the kinetics of the hydrogen-carbon reaction discussed earlier in this chapter it is possible that the observed increase in reaction rate at higher-carbon conversion was due to an increase in the bed temperature above that at the outside wall. Such an increase could occur under certain conditions of heat flux and reaction rate as a result of the exothermicity of the carbon-hydrogen reaction.

The extremely rapid rate of the initial hydrogasification reactions of coal and char is clearly shown by Figs. 65 and 66, which present the results of tests[14,232] in which coal was heated rapidly to 800°C in a stream of hydrogen flowing at a linear velocity of 1.2 ft/sec. In these tests, "zero" residence time was defined as the time at which the coal or char attained reaction temperature (approximately 2 min). Depending on the pressure, between 4 and 42% of the maf coal (Pittsburgh seam containing 40% volatile matter), was converted to gaseous hydrocarbons at "zero" residence time (see Table 30). Comparatively little difference was observed between the gasification of coal and of char prepared from the coal by heating for 2 hr in helium at 600°C, as

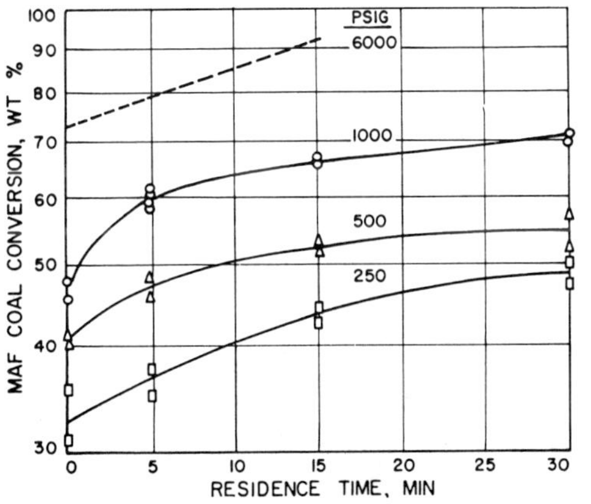

Fig. 65. Effect of residence time on the hydrogasification of coal at 800°C.[14,232]

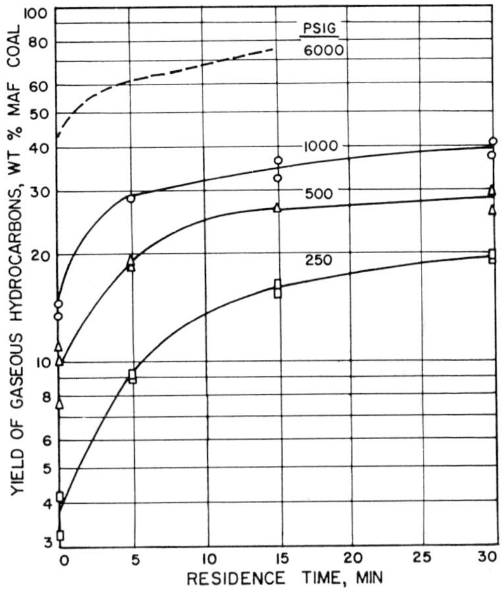

Fig. 66. Effect of residence time on the yield of gaseous hydrocarbons in hydrogasification of coal at 800°C.[14,232]

TABLE 30

YIELD OF GASEOUS HYDROCARBONS, IN WEIGHT PERCENT MAF COAL, OBTAINED BY HYDROGENATING COAL AT 800°C[232]

Residence Time, min	Pressure, psig 250	500	1000	6000
0	4	10	14	42
15	16	27	35	76
30	20	29	40	—

[231] Gas Research Bd., (Gt. Brit.), *Rept. of the Director for the Year 1948*, *Commun.* GRB 45 (1949), pp. 6–13.

[232] Hiteshue, R. W., Lewis, P. S., and Friedman, S., Am. Gas. Assoc. Production Conf. Paper CEP-60-9, 1960, 5 pp.

TABLE 31

Yield of Gaseous Hydrocarbons, in Weight Percent MAF Char, Obtained by Hydrogenating Char at 800°C[232]

Residence Time, min	Pressure, psig 250	500	1000
0	1	5	12
5	8	18	31
30	18	28	46

shown in Table 31. The gasification rates observed in these tests apply to a large excess of hydrogen, since the concentration of hydrocarbons in the effluent gas was only 2 to 5% by volume.

At all stages of gasification, the addition of sodium carbonate increases the rate of evolution of gaseous hydrocarbons from coals and cokes heated in a stream of hydrogen at elevated pressures in the temperature range 800 to 900°C.[230] For example, with the addition of 10% sodium carbonate by weight, 98% of the carbon in coke was gasified before the heating value of the outlet gas decreased below 500 Btu/scf.[132] At one stage of the gasification with alkalies present, it appeared that the rate of reaction was high enough for equilibrium to be reached between the coke, hydrogen, and methane. Under these conditions, the observed methane concentrations were greater than would be calculated for equilibrium with graphite, which was interpreted as indicating the higher energy level of the coke carbon. Dent[227] suggested that the acceleration of reaction rates by alkalies is due mainly to their adsorption by the carbon, preventing graphitization of the surface exposed to the gases.

Dent[227] observed, as others also have, that coals ranging from subbituminous to semianthracites gave coherent cokes when heated under pressure. He suggested that this is not the result of the chemical influence of hydrogen, but rather an effect of pressure, which reduces the capacity of the flowing gas for removing heavy tar vapors from the bed, and therefore creates conditions favorable for binding the particles in the bed together. To minimize experimental difficulties with caking coals, samples were pretreated by heating to various temperatures in the range 350 to 500°C, and then cooled immediately. Even with this pretreatment, coals heated to temperatures below 400°C still showed signs of fusion when treated with hydrogen at 50 atm.

Recognizing the importance of avoiding agglomeration when coal is hydrogenated continuously in either a fixed or fluidized bed, scientists have made studies[119] of the effect of conditions of pretreatment on hydrogasification results, and on agglomerating tendencies of the pretreated feed. Lignite, subbituminous,

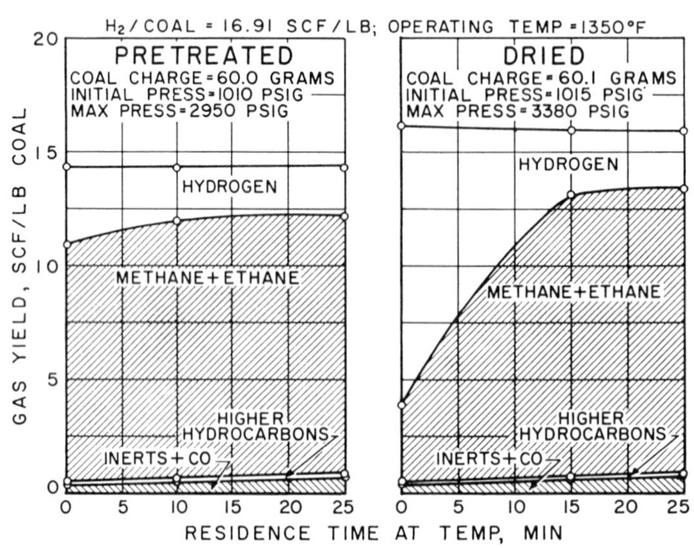

Fig. 67. Effect of residence time on hydrogasification yields from pretreated (in air at 705°F) and dried bituminous coals.[119]

TABLE 32

HYDROGASIFICATION YIELDS WITH AND WITHOUT PRETREATMENT OF COALS[119]

Pretreatment	Steam, 620°F (Run PR-19)	None
Products		
Char (moisture-, ash-free)	90.4	—
Net water and other condensates	4.4	—
CO, CO_2	1.8	
Hydrogen	Trace	—
Gaseous hydrocarbons	3.4	—
	Minutes at 1350°F	
Hydrogasification	0 (Run 112)	25 (Run 67)
Feeds		
Char or coal (moisture-, ash-free)	90.4	100.0
Hydrogen	8.9	9.4
Products		
Residue (moisture-, ash-free)	43.1 (43.1)*	39.8
Condensates	1.9 (6.3)	5.0
CO, CO_2, N_2	3.4 (5.2)	4.7
Hydrogen	1.5 (1.5)	1.4
Gaseous hydrocarbons	49.4 (52.8)	58.5

* Values in parentheses represent total pretreatment and hydrogasification yields.

and high-volatile B bituminous coal were pretreated for 50 to 75 min with either nitrogen, air, carbon dioxide, or steam in a laboratory-scale fluidized bed at temperatures ranging from 400 to 720°F. Under all pretreating conditions, the quantity of dried coal converted to gases was small, ranging from 2.5 to 9.9 wt %. The heating value of the total product gases from pretreatment ranged from 200 to 1000 Btu/scf when steam was used, but was only 25 to 190 Btu/scf with other pretreating agents. Hydrogasification of the pretreated coal in a batch autoclave, at nominal conditions of 1350°F and a maximum pressure of 3000 psig, showed that under optimum pretreatment conditions chars more reactive than the parent coal could be prepared (see Fig. 67). Optimum pretreatment temperatures were 600°F for the bituminous and subbituminous coals, and 500°F for the lignite. The existence of an optimum pretreatment condition for maximizing hydrogasification yields results from two factors: (1) the increase in reactivity of the char to a maximum value as pretreatment temperature increases, followed by a decrease in reactivity as low-temperature carbonization temperatures are approached; and (2) the continuous decrease in the content of relatively easily hydrogenated materials as pretreatment temperature is increased.

Although hydrogasification yields from pretreated coals can be maximized, such yields, even when adjusted for yields of products from pretreatment, are less than hydrogasification yields from the original coal (see Table 32). Even with pretreatment, a slight tendency toward agglomeration under hydrogasification conditions was observed, and was measured by the quantity of +40-mesh material in the residue. The feed for hydrogasification was −60 mesh. With this criterion, tendency toward agglomeration increased with increasing pretreatment temperature in the range 400 to 720°F, and increased with pretreatment atmospheres in the order: air, nitrogen, steam, and carbon dioxide. Pretreatment of lignite is not required to control agglomeration, but carbon dioxide removed in pretreatment makes possible about a 10% increase in the heating value of the gas from hydrogasification.

Pilot Plant Studies

The first pilot-scale tests on hydrogasification were made by Dent and coworkers, first in a small batch reactor,[230] and later in a batch reactor capable of operating at pressures up to 50 atm with a fixed fuel bed 12 in. in diameter and 8 ft deep.[233] Hydrogen was preheated at the top of the reactor, and flowed down through the fuel bed. Under these conditions, the process is autothermic, and the high-temperature (1000 to 1200°C) reaction zone moves down through the fuel bed until it reaches the bottom. To initiate the reaction with semicoke, hydrogen must be preheated to 500 to 550°C, and with coal to 700 to 750°C. Results obtained at 50 atm with semicoke and noncaking coal are summarized in Table 33.[231] The coke

TABLE 33

RESULTS OF PRESSURE HYDROGENATION TESTS WITH SEMICOKE AND COAL[231]

	Semicoke	Coal
Pressure, atm	50	50
Depth of charge, ft	8.3	8.3
Weight of dry charge, lb	149	253
Size of fuel, in.	$\frac{1}{4}$–$\frac{1}{8}$	$1\frac{1}{4}$
Reaction temp, °C	1270	1175
Solid residue, dry, lb	108.3	155.1
Solid volatilized, %	27.3	38.8
Hydrogen used in reaction period, scf	2675	3760
Gas produced in reaction period, scf	2803	4517
Gas heating value, Btu/scf	456	493
Methane in gas made, scf	652	1256
Hydrogen absorbed, scf	827	459
Quantities per ton of dry material		
Hydrogen supplied, scf	40,400	33,300
Gas produced, scf	42,200	40,000
Gas produced, therms	192	197
CH_4 produced, therms	97.5	111
Gas, CH_4 content, %	21.4	26.75
Tar, gal	1.3	11.4
Equivalent ammonium sulfate, lb	80	90

[233] Gas Research Bd., (Gt. Brit.), *6th Ann. Rept. Council, 1944–45, Commun.* GRB 18 (1945), pp. 8–18.

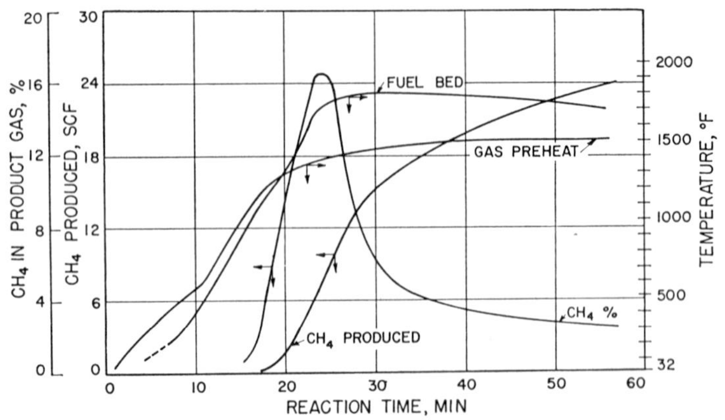

Fig. 68. Results from hydrogasification of char in a fluidized bed.[231]

residue (about 61 to 72% of the charge) exhibited a high degree of reactivity toward steam, carbon dioxide, and oxygen,[234] and was low in sulfur (0.15 wt% being typical). This residue could be completely gasified, and for coal, 1370 lb of residue per ton (2240 lb) would produce at least the 33,300 scf of hydrogen required for hydrogasification, so an overall yield of 197 therms of town gas would be obtained in the complete gasification of 2240 lb of dry coal.

The yield of methane in the large-scale fixed-bed tests did not vary widely from 100 therms/ton. It consisted of the methane produced in raising the reactants to an equilibrium temperature above 1000°C, plus the methane evolved in the devolatilization of the coal in the heating-up process. Dent[227] pointed out that hydrogenation has several important inherent characteristics as a method for distillation of coal by internal heating: (1) heat is generated when the transition from coal to coke is occurring, and therefore distillation is extremely rapid; for example, when the reaction zone traversed the charge at a rate of 10 ft/hr, it moved at more than 100 times the rate of heat penetration in conventional carbonization; and (2) the thermal efficiency is high because heat is supplied by the exothermic heat of the hydrogasification reactions instead of by combustion processes. Furthermore, in a large-scale operation, products would leave at a comparatively low temperature (about 400°C).

Hydrogasification in a fluidized bed gives considerably more latitude in methods of contacting reactants, and in methods for altering and controlling the conditions under which the reaction occurs. Part of this flexibility stems from the uniformity of temperature in completely back-mixed fluidized beds. Dent's first experiments[231] were made with semicoke to avoid agglomerating problems. A 3-lb charge of semicoke was used, and formed a static bed 4 in. in diameter and 9 in. deep in a batch reactor. The actual linear velocity of the

[234] Gas Research Bd. (Gt. Brit.), *Rept. for Year 1951, Commun.* GRB 67, (1952), pp. 6–10.

TABLE 34

Results of the Hydrogenation of Fluidized Semicoke at 20 atm Pressure in a Pilot Plant[227]

Test No.	Reaction Temperature, °C	Hydrogenating Gas	Average Residence of Fuel in Bed, min	Contact Time of Gases, sec	Methane Yield therms/ton	Heating Value of Gas Produced, Btu/scf
10	820 to 850	Hydrogen	45	20	125	435
11	800 to 830	$H_2/CO = 13/1$	30	18	90	425
12	800	$H_2/CO = 13/1$	17	14	60	455
13	800 to 840	(recycle)	30	18	65	462
14	825	Blue water-gas	42	18	63	336
15	830	Blue water-gas	24	25	29	340

fluidizing preheated hydrogen was 0.3 ft/sec. Results obtained at 25 atm are shown in Fig. 68. The rate of methane production was greatest at the onset of reaction and decreased progressively as the gasification of the charge increased. The total yield of methane was 220 therms/ton (2240 lb) of semicoke, most of which was produced in an interval of 15 to 20 min (see Fig. 68). The yield in the fluidized bed tests was significantly greater than that obtained in the fixed-bed tests as a consequence of the ability to control reaction temperature.

Two pilot-plant studies[228,235] have been made to determine the effect of process variables in the continuous hydrogasification of coal and char in a fluidized bed. Both pilot plants were operated with cocurrent upflow and with product gas and char removed through a standpipe in the center of the reactor.

The essential features of the unit operated by the Institute of Gas Technology are shown in Fig. 69. The inside diameter of the reactor was 2 in. and the inside length 113 in. With a 1-in. diameter standpipe at a height of 4.7 ft, the bed volume was 0.074 cu ft. In most tests, the superficial velocity of the fluidized gas at reaction conditions was 0.06 ft/sec, the level required to maintain free flow of solids (particle size, −60 by +325 mesh) through the reactor.

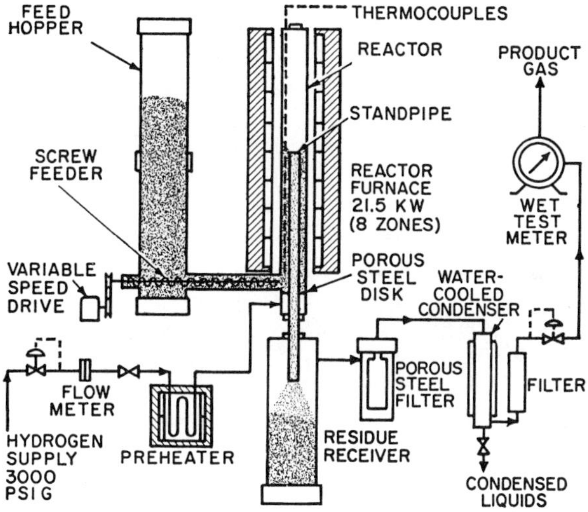

Fig. 69. Schematic diagram of the Institute of Gas Technology pilot unit for continuous gasification of coal and char in a fluidized bed.[235]

The first studies of the continuous hydrogasification of solid fuel in a fluidized bed at elevated temperatures and pressures were made by Dent and co-workers.[227] The fluidized bed was contained in a reaction space 12 in. in diameter and 9 ft deep, with baffles at various levels to restrict bubble formation. A mixture of coal or semicoke and hydrogen preheated to 500°C was fed in at the bottom, and solid residue was removed with the product gas through a dip tube at the top of the bed. Preliminary results obtained from this unit are presented in Table 34. The yield of methane per ton was significantly lower than that obtained in the small batch reactor. The difference (about 95 therms/ton) could not be explained by the lower temperatures and pressures. However, Dent suggested that difficulties in maintaining thorough contact between gas and solids when fluidized beds are increased in size may be a contributing factor.

The results of tests using the equipment just described on the hydrogasification of char at various feed rates, at 1400°F and pressures ranging from 500 to 2000 psig, are shown in Fig. 70. The gas residence time was 1.3 to 1.5 min in most tests, and 0.7 min in one series of tests at 500 psig. The reciprocal of the char space velocity was proportional to char residence time, which can be computed by assuming a typical bulk density of the expanded bed of 25 lb/cu ft. Increased char residence time and increased pressure increased the

[235] Pyrcioch, E. J., and Linden, H. R., *Ind. Eng. Chem.*, **52**, 590–4 (1960).

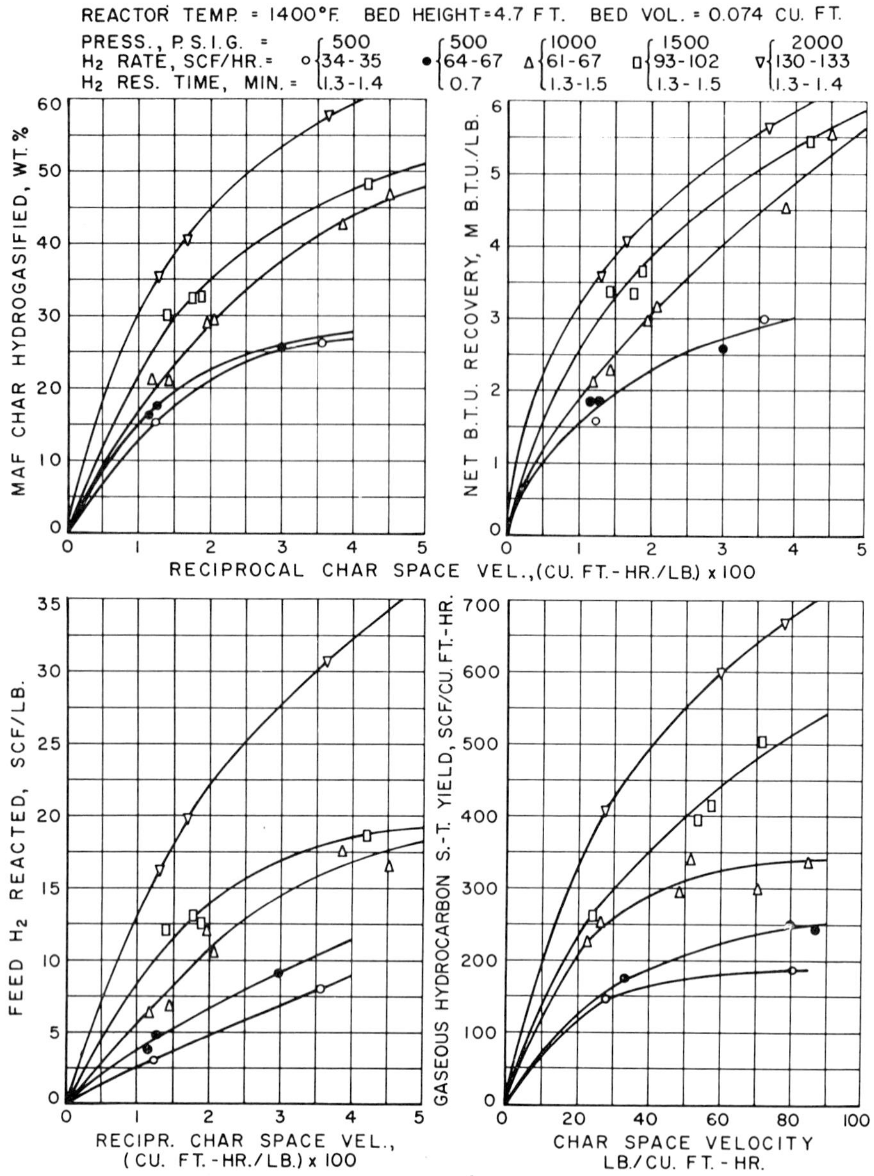

Fig. 70. Effects of pressure and hydrogen feed rate in the hydrogasification of char.[235]

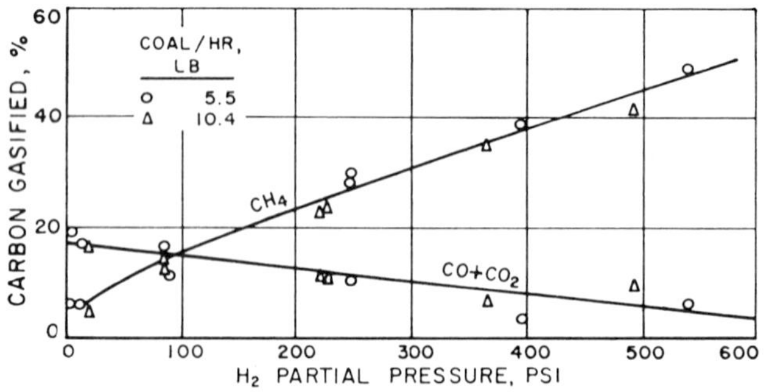

Fig. 71. Effect of hydrogen partial pressure on carbon conversion in hydrogasification of brown coal at 850°C (1562°F).[228]

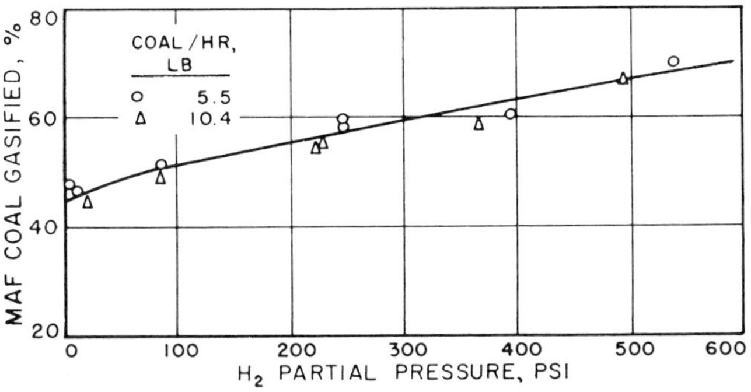

Fig. 72. Effect of hydrogen partial pressure on coal utilization in hydrogasification of brown coal at 850°C.[228]

percentage of char hydrogasified, the net Btu recovery, and the feed hydrogen reacted. The space-time yield of gaseous hydrocarbons increased with pressure and with char space velocity. Because there was no clearly defined tendency for a shift in the source of hydrogen contained in the gaseous hydrocarbons, the effect of char space velocity on hydrocarbon space-time yield was primarily the result of an increase in the reactivity of the char bed. At 500 psig, an increase in hydrogen rate, to double the level required to maintain minimum superficial velocity, increased char conversion gaseous hydrocarbon space-time yield and quantity of feed hydrogen reacting.

Birch and co-workers[228] studied the hydrogasification of brown coal in a pilot plant similar to the one shown in Fig. 69, except that coal was fed into the preheated feed gas stream before entering the reactor. The reactor was 3 in., i.d. with a standpipe 3 ft from the bottom. The free cross-sectional area was 0.047 sq ft, giving an expanded bed volume of 0.141 cu ft. Brown coal was hydrogasified in this unit at pressures ranging from 0.8 to 606 psig, at temperatures from about 500 to 950°C, superficial gas velocities generally in the range of 0.4 to 0.7 ft/sec. The results of tests at 850°C and two different feed rates are shown in Figs. 71 to 73 as a function of partial pressure of hydrogen.

In the results shown, variations in partial pressure of hydrogen were obtained either by varying the concentration of hydrogen in a hydrogen-nitrogen mixture while maintaining the total pressure constant at about 600 psig, or by changing the total pressure with substantially pure hydrogen. Conversion of carbon to methane, and hydrogen consumption, increased with hydrogen partial pressure. Coal feed-rate had comparatively little effect in the range studied. In this connection it should be recalled that the results of Pyrcioch and Linden[235] showed that the effect of coal-feed rate was least at the lowest pressure (see Fig. 70).

Certain comparisons of the results of Birch and co-workers with those of Pyrcioch and Linden are of interest. Figure 74 shows the relation between methane produced and hydrogen consumed. The relations are very similar in nature, except that at zero hydrogen consumption the brown coal produces more methane than the char. At low hydrogen consumptions, the slope of the relation for brown coal is approximately $\frac{1}{3}$, indicating that three volumes of hydrogen are consumed for each volume of methane. If the carbon-hydrogen reaction was the only one involved, the slope would be 0.5. Thus, in the early stage of hydrogasification, hydrogen is reacting with oxygen in the coal, or with carbon dioxide from the

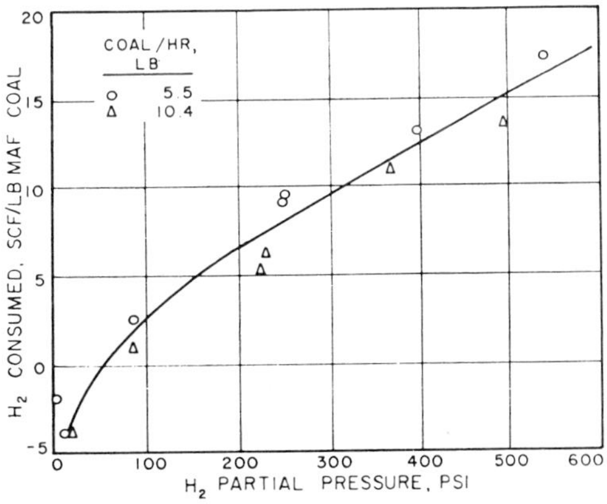

Fig. 73. Effect of hydrogen partial pressure on hydrogen consumption in hydrogasification of brown coal at 850°C.[228]

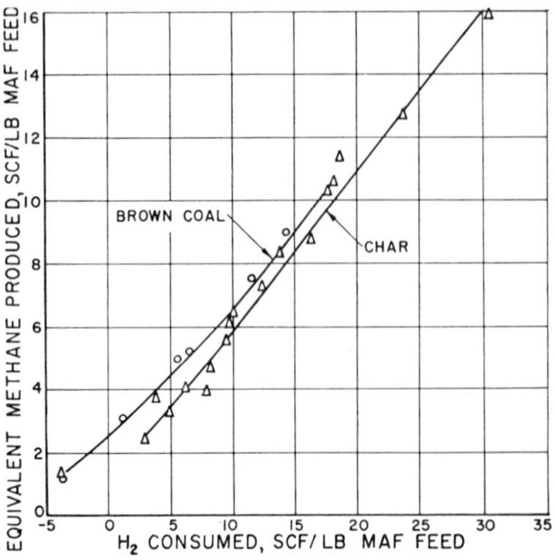

Fig. 74. Relationship between methane produced and hydrogen consumed in hydrogasification of brown coal and char.[228,235]

coal, in a reverse water-gas shift reaction. Birch and co-workers found that the composition of the wet-product gases corresponded closely to equilibrium for the water-gas shift reaction.

A second comparison of the results of these two pilot plant studies is made in Fig. 75, which shows the average reaction rate expressed as pounds of carbon appearing as methane per hour per pound of carbon in the bed, divided by the average hydrogen partial pressure in atmospheres. Analysis of the rate data on the basis of this ratio (first-order rate-constant) is justified by Blackwood's results, discussed previously in this chapter, which showed that the reaction rate was directly proportional to the hydrogen partial pressure. Although there is some scatter, the results clearly show the progressive decrease in first-order rate constant as the carbon gasified increases, and less reactive material remains behind. The comparison also shows the higher reactivity of brown coal as compared with char. Although the data are somewhat limited, the effect of temperature in increasing the rate constant is indicated. For comparative purposes, Blackwood's results with lower reactivity coconut char are shown.

At present (1960) there are no full-scale hydrogasification processes in operation, although there are several process concepts under

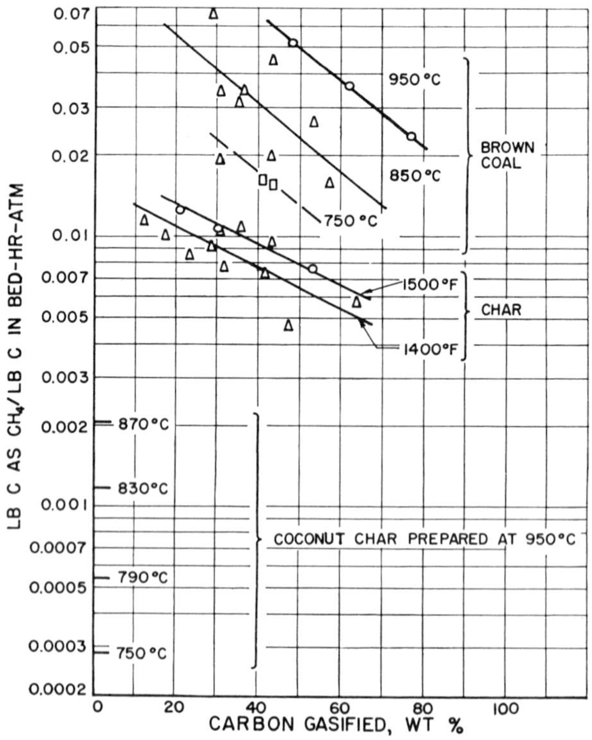

Fig. 75. Comparison of rate constants in hydrogasification of various materials.[228,235]

consideration. Dent's original work on hydrogasification evolved from studies of the reactions occurring in the upper section of a Lurgi generator, with a view toward obtaining a higher degree of enrichment. One concept for accomplishing this, with an extended hydrogenation zone above the steam-oxygen gasification zone, is shown in Fig. 76.[236] A high partial pressure of hydrogen is obtained by removing carbon dioxide and undecomposed steam from the gases leaving the high-temperature gasification zone. A variant of this concept for producing a substitute for natural gas involves the separation of methane by liquefaction, with recycle of the hydrogen. This proposed hydrogasification process gives theoretical fuel and oxygen requirements per Mcf of methane considerably less than would be obtained in the methanation of synthesis gas[17] (compare Tables 35 and 36).

A hydrogasification process using a combination of fluidized beds and suspension gasification (Fig. 77) has also been conceived by Dent,[236] who has estimated the overall

[236] Dent, F. J., *Gas Council, (Gt. Brit.), Research Commun.* GCI (1952), 41 pp.

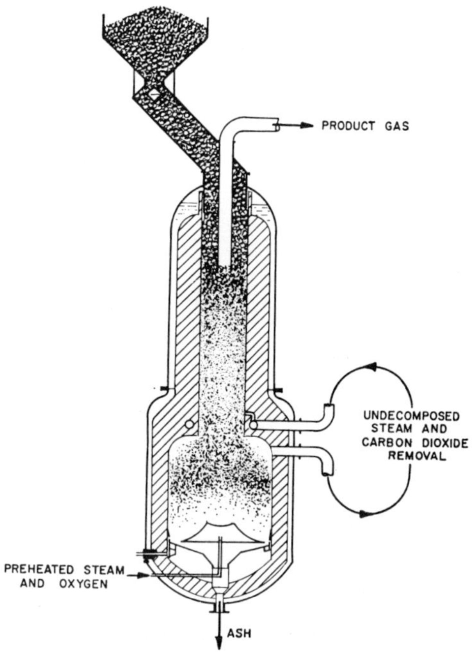

Fig. 76. Modified Lurgi generator with upper hydrogenation chamber.[236]

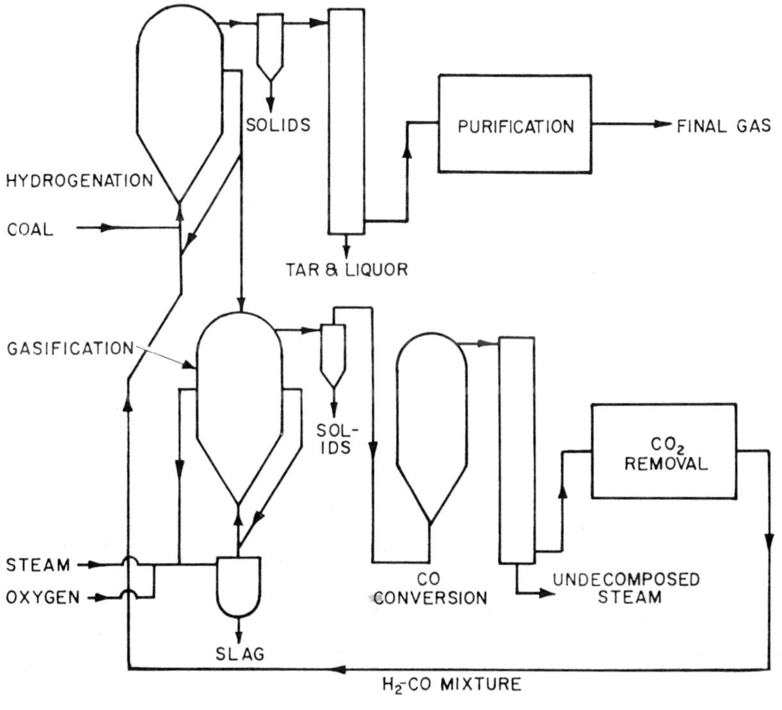

Fig. 77. Proposed hydrogasification process employing a combination of fluidized-bed and suspension gasification.[236]

TABLE 35

THEORETICAL MATERIAL REQUIREMENTS (REACTOR ONLY) FOR PRODUCTION OF METHANE FROM SYNTHESIS GAS[17]

(Inlet temperature: steam, 1000°F; hydrogen, 800°F; oxygen and coal, 60°F. Outlet temperature: 1700°F)

	Pressure, atm			
	50		200	
Oxygen-water inlet mole ratio	0.1	0.15	0.1	0.1
Hydrogen-water inlet mole ratio	1.8	2.4	0.75	1.0
Inlet steam decomposition, %	30	40	30	40
Requirements/Mcf CH_4				
Pure C, lb	67.5*	68*	64.5	63
Coal, at 95% gasification, lb	95*	96*	90.5	88
Steam, lb	111	79	151	110
Oxygen (100%), scf	256*	273*	318	230
Theoretical C appearing as CH_4, %	43	43	49	51
Hydrogen recycle, moles/mole CH_4	3.5	3.4	1.8	1.7

* Increases by 10% over theoretical values.

thermal efficiency, including oxygen manufacture, as 80% in the production of gas with a heating value up to 600 Btu/scf. Birch and co-workers[228] suggested a three-section countercurrent reactor for hydrogasification. Coal would be fed into the top section, which would be fluidized, and in which the reactive constituents of the coal would be hydrogasified at 750 to 800°C. The char, still fluidized would pass through an intermediate baffled section to maintain a temperature gradient of 800 to 950°C through the section. The lower section would be fluidized to make the heat of reaction available for heating incoming gas to reaction temperature.

Hydrogasification processes may be broadly grouped into several basic processing schemes. The following outline applies to the production of high-Btu gas as a substitute for natural gas. However, each of the basic methods can be

TABLE 36

MATERIAL REQUIREMENTS (COAL GASIFIER ONLY) FOR PRODUCTION OF METHANE FROM SYNTHESIS GAS BY VARIOUS PROCESSES[17]

	Process*						
	USBM	PAN	K-T	IGT	L	WGS	PROD
Coal type	Bituminous	Bituminous	Bituminous	Bituminous	Char	Coke	Barley Anthracite
Pressure, atm	21	1	1	7.5	21	1	1
Outlet temp, °F	2500	2350	2300	2500	1100	1200	880
Oxygen-water inlet mole ratio	1.5	0.65	0.81	0.45	0.77	0.38	0.29
Inlet steam temperature, °F	625	930	250	1500	500	210	285
Inlet oxygen temperature, °F	625	930	100	100	500	210	285
Inlet steam decomposition, %	25	25	25	31	34	54	52
C gasified, %	90	84	85	92	97	99.5	99.4
Requirements Mcf CH_4							
C, lb	96.5	119	112	93.5	84	100	87
Coal (or coke), lb	137	185	168	125	112	111	110
Steam, lb	42	77	85	131	360	152	154
Oxygen (100%), scf	1332	1060	1444	1248	584	1144	951
C appearing as CH_4 by methanation, %	33	26.7	28	34	38	32	36.6

* USBM = U.S. Bureau of Mines; PAN = Panindco process; K-T = Koppers-Totzek gasifier; IGT = Institute of Gas Technology; L = Lurgi unit; WGS = water-gas set; PROD = producer.

appropriately modified to produce town gas or intermediate calorific value gas.

1. Hydrogasification of coal or char with an external hydrogen supply.

a. Isothermal operation through indirect heat exchange to produce a high heating value gas directly, which requires operation below the stoichiometric hydrogen feed ratio and at relatively low temperatures, and will normally result in a substantial ungasified residue, preferably in an amount allowing balanced operation of a combined hydrogasification-hydrogen production plant.

b. Adiabatic operation with an excess amount of hydrogen, and preheat of the reactants to somewhat above the 1000°F threshold temperature. (By the use of approximately three times the stoichiometric hydrogen requirements, essentially complete conversion to methane should be feasible without exceeding temperatures of 1800 to 2100°F. Because of the low methane content of the product gas, this system would require a hydrogen separation and recycle step.)

2. Gasification of coal or char with steam-hydrogen mixtures.

a. Adiabatic operation with preheat to threshold temperature only, and use of just enough steam to balance the exothermic hydrogenation reactions with endothermic steam decomposition reactions, which will not eliminate the need for an external hydrogen supply, and may or may not result in essentially complete conversion, depending on the hydrogen feed ratio and temperature level.

b. Operation with sufficient external heat supply to satisfy the hydrogen requirements entirely by steam decomposition, and attain essentially complete gasification.

c. With each of the foregoing two reaction systems, a high heating-value gas can be obtained by complete carbon monoxide shift, purification and hydrogen separation, and recycle, or by partial carbon monoxide shift to adjust H_2/CO ratio, purification and methanation.

3. Complete adiabatic gasification with steam and oxygen, and without an external source of hydrogen.

a. Single-zone gasification with oxygen and steam at high pressure and relatively low temperature, followed by partial carbon monoxide shift, purification and methanation.

b. Two-zone gasification at high pressure, with a high-temperature steam-oxygen zone preceding a lower-temperature zone to which recycle hydrogen is introduced, followed by complete carbon monoxide shift, purification, and methane separation.

Development work has been done or is in progress on all of the basic variants listed previously. The eventual processing schemes selected will depend on the results of this work, along with the type of fuel to be processed and the desired characteristics of the product gas.

GASIFICATION AS A PROCESSING STEP

Synthesis gas, a mixture of hydrogen and carbon monoxide in various proportions, is the starting material for the production of liquid and gaseous hydrocarbons and other chemicals, and for the production of hydrogen. The various uses of synthesis gas in catalytic processes have been summarized by Hoy and Wilkins[131] and include, for example, conversion to (1) liquid hydrocarbons by Fischer-Tropsch synthesis,[237] (2) methane or synthetic high-Btu pipeline gas by methanation,[238] (3) methanol and higher alcohols directly,[237] (4) higher alcohols by reaction with olefins in oxo-synthesis,[237] (5) hydrogen for use in the hydrogenation of coal to produce either liquid fuels and byproducts or gaseous hydrocarbons, and for use in the synthesis of ammonia,[239] and (6) hydrogen or hydrogen-rich mixtures in reduction reactions in the steel industry.[240,241]

Economic considerations and availability of raw materials determine the feed used for the production of synthesis gas. The results of an economic analysis for synthetic ammonia production[239] are summarized in Table 37, with operating requirements for various raw

[237] Storch, H. H., Golumbic, N., and Anderson, R. B., *The Fischer-Tropsch and Related Syntheses*, John Wiley and Sons, New York, 1951, 610 pp.

[238] Pyrcioch, E. J., Dirksen, H. A., von Fredersdorff, C. G., and Pettyjohn, E. S., *AGA, Proc.*, **1954,** 813–36.

[239] Duff, B. S., *Petrol Processing*, **10,** 223–8 (1955).

[240] Wright, E. C., *Metal Progr.*, **72,** No. 10, 99–104 (1957).

[241] Staratt, F. W., *J. Metals*, **11,** 315–18 (1959).

TABLE 37

Operating Requirements for Ammonia Manufacture (Per Ton of Ammonia)[239]

	Natural Gas	Fuel Oil	Coal	Coke-Oven Gas	Reformed Gas
Gas feed, Mcf	26.0	—	—	127	75.7
Oil feed, bbl	—	4.27	—	—	—
Coal feed, ton	—	—	1.28	—	—
Shift catalyst, lb	0.30	0.35	0.35	—	—
Synthesis catalyst, lb	0.50	0.50	0.50	0.50	0.50
Caustic, lb	8.0	8.0	8.0	6.0	5.5
Monoethanolamine, lb	0.30	0.40	0.40	—	—
Lube oil, gal	0.5	0.5	0.5	0.6	0.5
Fuel gas, MM Btu*	22.0	12.0	14.2	−36.7†	−4.2†
Fuel oil, bbl	—	2.54	—	—	—
Fuel coal, ton	—	—	0.56	—	—
Electricity, kwhr	108	120	125	100	80
Treated water, M gal	1.07	1.36	1.50	1.16	—
Raw water, M gal	4.5	5.0	5.0	3.2	2.2
Operators per shift	7	7	7	6	5
Capital Investment, thousand $					
100 ton/day plant	3950	4098	4248	3620	2980
200 ton/day plant	6843	7084	7344	6231	5106
Manufacturing costs, $/ton					
100 ton/day plant	47.82	50.93	53.38	40.96	34.34
200 ton/day plant	41.25	44.16	46.51	35.01	29.34

* Based on gas-driven compressors.

† Includes credit for recovered hydrocarbon gases.

materials. The costs include depreciation at 10%, taxes, and interest and insurance at 6%, but do not include cost of land, administration and shop buildings, roads and rail sidings. All raw materials are charged at 30 cents per MMBtu. In this analysis, both capital investment and operating costs are lowest for reformed gas, and highest for coal. However, the author emphasized that local variations in raw material cost could influence the manufacturing costs by as much as 7 or 8 dollars per ton in some instances. Under conditions existing in the United States, direct gasification of coal is not competitive with natural gas, which is the predominant raw material for the production of synthesis gas.[242] In fact, at the end of 1958, only one synthetic ammonia plant in the United States was using coke as a raw material.[243] In other parts of the world where economic conditions are favorable, synthesis gas is produced by the direct gasification of coal.[164,165,178,190]

The largest plant for producing liquid fuels and petrochemicals from coal is situated in the Orange Free State, South Africa, and operated by South African Coal, Oil and Gas Corporation, Ltd. A simplified process-flow diagram showing the designed production of the variety of products from this plant is shown in Fig. 78.[244] Nine Lurgi pressure gasifiers, each having a design capacity of 837 Mcf/hr of raw gas, and one of which is standby, are used for producing synthesis gas.

In connection with the U.S. Bureau of Mines synthetic fuels research program, material requirements and cost estimates[245,246] were

[243] DeCarlo, J. A., Hunter, T. W., and Otero, M. M., *U.S. Bur. Mines, Minerals Yearbook*, 1958, Govt. Print. Office, Washington, D.C. 1959, Vol. II, p. 199.

[242] Bateman, R. L., *Oil Gas J.*, **56**, No. 35, 126–49 (1958).

[244] Johnson, W. B., *Petrol. Refiner*, **35**, No. 12, 222–8 (1956).

[245] U.S. Bur. Mines, *Cost Estimate for Coal Hydrogenation*, Ms. Rept. January 11, 1952.

[246] National Petroleum Council Committee on Synthetic Liquid Fuels, *Supplementary Report on Synthetic Liquid Fuels Production Cost*, February 1, 1953, 270 pp.

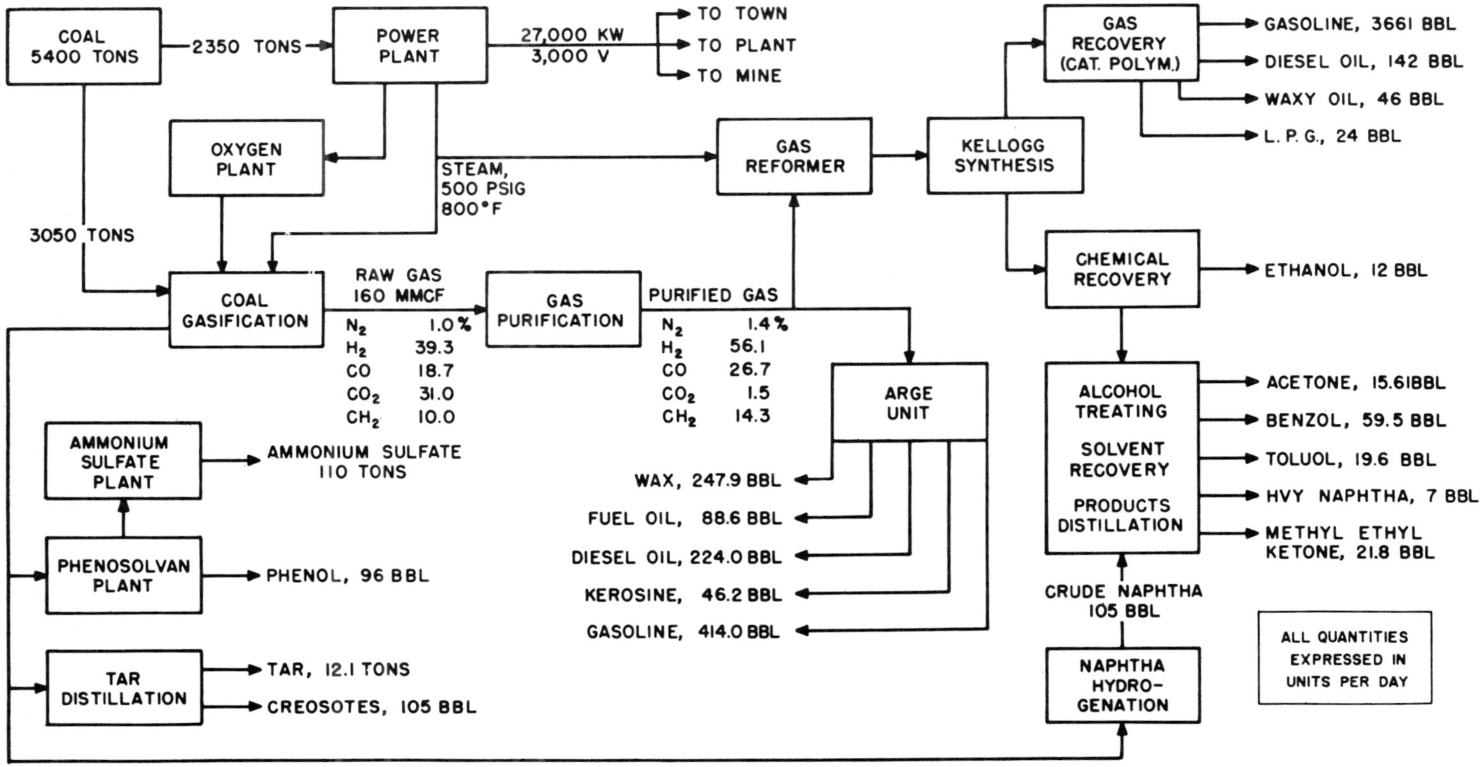

Fig. 78. Simplified process flow diagram of the Sasol plant for production of liquid fuels and petrochemicals from coal.[244]

TABLE 38

ESTIMATED OPERATING PERSONNEL AND MATERIAL REQUIREMENTS FOR A 30,000 BARREL DAY COAL HYDROGENATION PLANT[247]

	Bureau of Mines data[245]	National Petroleum Council data[246]
Coal (as recorded), tons/day	11,820†	12,990
Hydrogenation	6,142	6,422
Fuel	3,330	3,800
Gasification	2,348	2,220
Hydrogen production, MMscf/day	242	239
Low-temperature separation	39	38
Off-gas conversion*	117	80
Coal gasification	86	121
Hydrogen consumption, MMscf/day	242	239
Liquid phase	165	—
Vapor phase	77	—
Oxygen (95%), tons/day	1,952	2,615
Electric power, kw	194,710	213,880
Steam, M lb/hr	—	5,214
High-pressure	2,700	2,634
Process and heating	556‡	2,580
Cooling water, gal/min	250,000	227,555
Fresh water, gal/min	18,263	19,616
Operating labor and maintenance (plant)	2,400	3,349
Total (including supervisory)	3,020	5,250**
Construction metals		
Tons	173,267§	188,600
Tons/bbl-day	5.8	6.3
Total investment cost, $/bbl-day	12,500‖	14,560‖
Cost of equivalent gasoline at 6% return on total investment after 50% income tax, cents/gal	19.1¶	34.8

* Bureau of Mines by steam cracking, National Petroleum Council by partial combustion with oxygen.
† No loss included.
‡ Does not include extracted steam.
§ Steel only.
‖ Computed in 1952 dollars.
¶ The lower gasoline cost results principally from higher returns assumed for byproduct chemicals.
** Includes 605 miners.

made for the production of liquid fuel by the hydrogenation of a high-volatile bituminous C coal (Table 38),[247] and by Fischer-Tropsch synthesis with a high-volatile bituminous A coal as the raw material for synthesis gas production (Table 39).[247] In both process estimates, it was assumed that synthesis gas would be produced by pressure gasification of coal in suspension with oxygen and steam. In the hydrogenation of coal, approximately 20% of the coal fed to the plant is required for gasification, as compared with about 85% for the Fischer-Tropsch plant. The cost of compressed hydrogen is approximately 35 to 40% of the cost of liquid fuel produced in the hydrogenation of coal. In the Fischer-Tropsch synthesis, the cost of synthesis gas is approximately 50 to 60% of the cost of the finished product.

The production of high-Btu pipeline gas from coal is of interest principally in the United States as a means for eventually supplementing natural gas.[248] Methanation of synthesis gas and hydrogasification of coal and oil shale[229] are under consideration. Various possible processes for hydrogasification have been discussed, but definitive economic analysis of any of these have not as yet been published.

Economic evaluations have been made of the methanation process with synthesis gas produced in Lurgi generators[249,250] and in suspension gasifiers operating at elevated pressure.[251] A process flowsheet for the latter case is shown in Fig. 79. The investment cost (in 1958 dollars) for a plant capable of producing 90 MMcf of 930 Btu/scf gas/day is $1055/daily Mcf with diethanolamine purification, and $900/daily Mcf with hot carbonate purification.[251] With a 330-day operating year, and coal at $4 per ton, the selling price of gas with

[247] Bituminous Coal Staff, *U.S. Bur. Mines, Rept. Invest.* No. 5506 (1959), 306 pp.

[248] Elliott, M. A., *Proc. Pacific Gas Assoc.*, **50**, 26–32 (1959).

[249] Alberts, L. W., Bardin, J. S., Beery, D. W., Jones, H. R., and Vidt, E. J., *Chem. Eng. Progr.*, **48**, 486–93 (1952).

[250] Katell, S., Faber, J. H., and Constantine, M. T., *U.S. Bur. Mines, Rept. Invest.* No. 5547 (1959), 11 pp.

[251] Katell, S., *Coal Age*, **63**, No. 9, 116–18 (1958).

TABLE 39

ESTIMATED OPERATING PERSONNEL AND MATERIAL REQUIREMENTS FOR A 30,000 BARREL DAY FISCHER-TROPSCH PLANT[247]

Coal (as received) tons/day	18,900
Fuel	3,025
Gasification	15,875
Oxygen (95%), tons/day	11,700
Electric power, kw	96,800
Steam, M lb/hr	2,900
Cooling water, gal/min	555,750
Fresh water, gal/min*	3,240
Personnel†	6,442
Construction metals	
Tons	117,000
Tons/bbl-day	6.55
Total investment cost, $/bbl-day	12,660‡
Cost of equivalent gasoline at 6% return on total investment after 50% income tax, cents/gal	29.2

* Boiler-plant makeup only
† Includes 2720 miners
‡ Computed in 1952 dollars

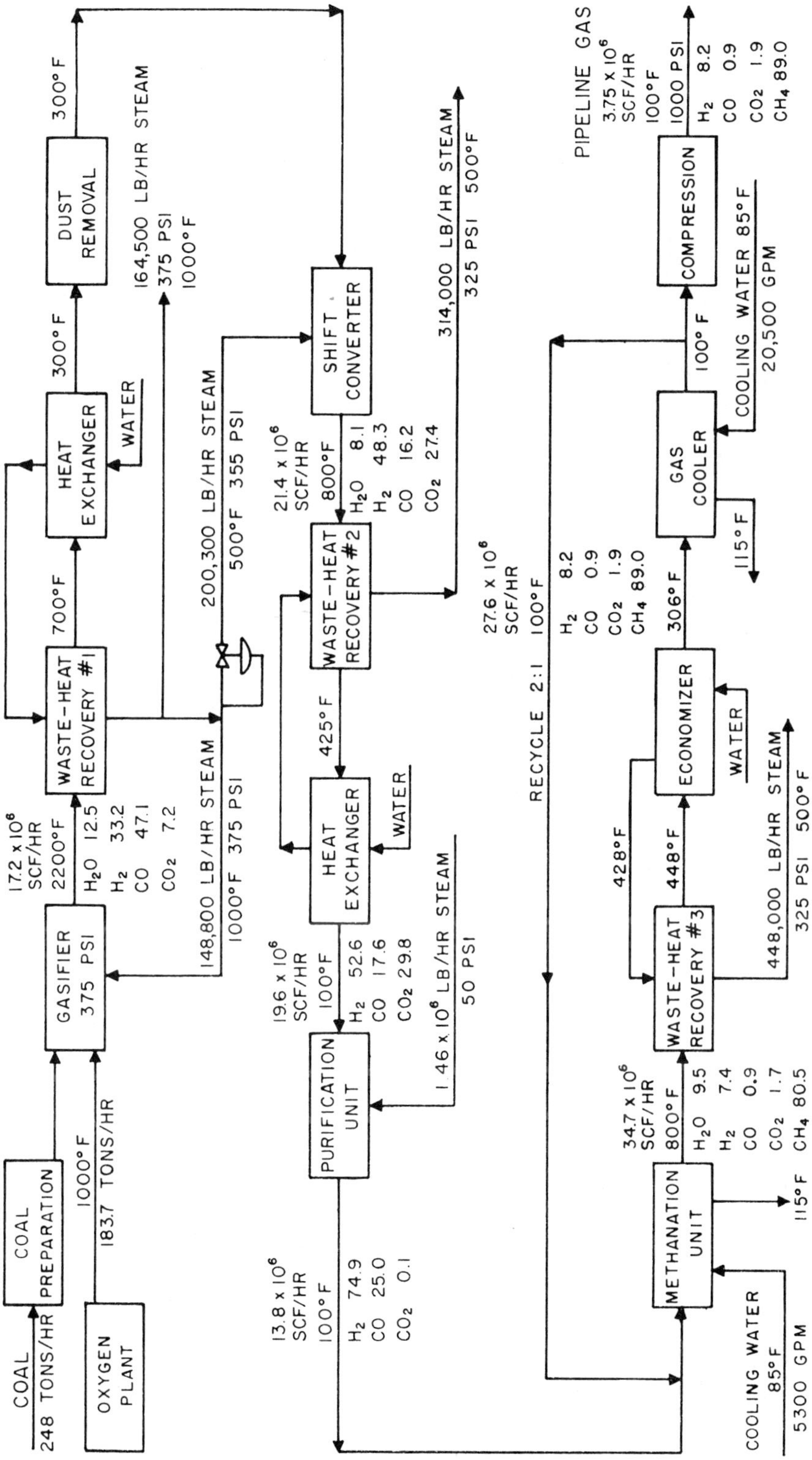

Fig. 79. Process flowsheet for production of high-Btu gas by oxygen-coal pressure gasification followed by methanation.[251]

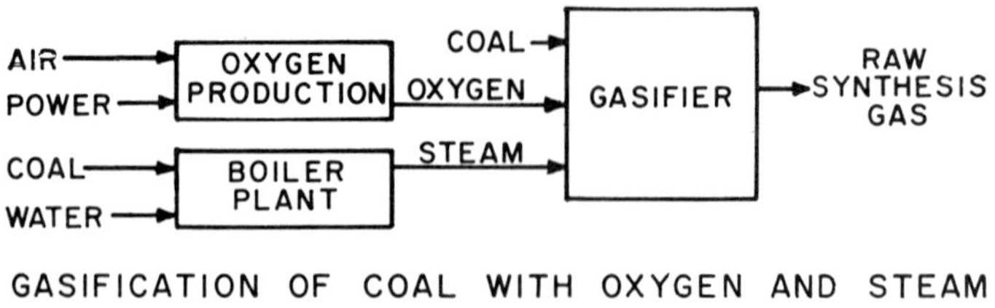

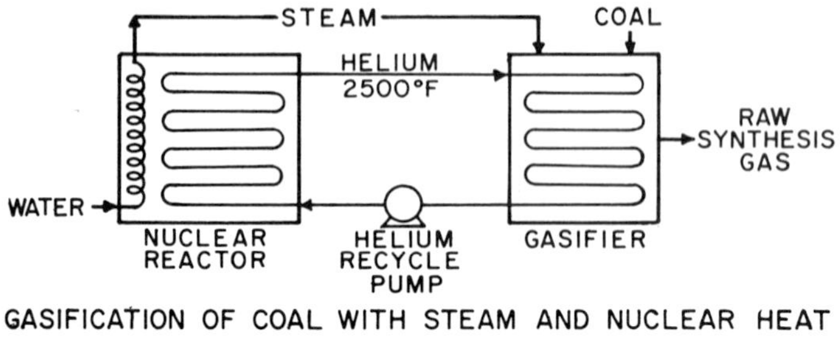

Fig. 80. Comparison of coal gasification by conventional and nuclear processes.[253]

12% gross return (defined as return on investment before income taxes) is estimated at $1.13 and $1.04 for the first and second cases, respectively. A deviation of $1/ton in coal price will vary the cost of gas by $0.08/Mcf.

The investment cost for a plant using Lurgi generators and producing 90 MMcf of 940 Btu/scf gas/day is estimated at $850/daily Mcf. With a 330-day operating year, with char at $5.00 and coal at $4.00 per ton, the selling price of gas with 12% gross return on investment is estimated at $0.92 per Mcf. In this process, the hourly feed to the Lurgi gasifiers would be 184 tons of coal, 97.4 tons of oxygen, and 695,500 lbs of steam. Coal used for other purposes is estimated at 37 tons/hr. Approximately 50,000 horsepower is required for the production of oxygen and is supplied by steam turbines. Carbon dioxide is removed by the hot carbonate process. A breakdown of investment costs is shown in Table 40, and of operating costs, in Table 41.

TABLE 40

ESTIMATED CAPITAL REQUIREMENTS FOR PRODUCTION OF HIGH-BTU GAS BY METHANATION OF LURGI GAS[250]

Unit	Cost, dollars*	%
Gasification	18,533,400	24.2
Oxygen plant	15,000,000	19.6
Shift	796,100	1.0
Primary waste-heat recovery	875,000	1.2
Purification	10,881,000	14.2
Methanation	683,600	0.9
Secondary heat recovery	3,189,600	4.2
Pipeline compression	1,561,900	2.0
Sulfur recovery	515,000	0.7
Catalyst preparation	1,800,000	2.4
Steam and powerplant	4,320,000	5.6
Plant facilities	5,815,600	7.6
Plant utilities	7,676,500	10.0
Total construction	71,647,700	93.6
Initial catalyst requirement	607,000	0.8
Total plant cost	72,254,700	94.4
Interest (during construction)	1,445,100	1.9
Subtotal for depreciation	73,699,800	96.3
Working capital	2,800,000	3.7
Total investment	76,499,800	100.0

* Computed in 1959 dollars.

The use of coal gasification as a processing step in the chemical, steel, gas, and petroleum industries will become increasingly important in the future as fluid fossil fuels become more difficult to find and more expensive to produce. Much of the work on the newer techniques for gasifying coal is still in the developmental stages. Until full-scale plants are built and operated, a definitive economic analysis of the various gasification processes cannot be made.

NUCLEAR ENERGY IN GASIFICATION

The use of nuclear energy to supply the heat required for gasifying coal is being studied cooperatively by the U.S. Bureau of Mines and the Atomic Energy Commission.[252] The present concept of the use of nuclear heat in gasifying coal with steam is shown diagrammatically in Fig. 80, in comparison with conventional processes using oxygen. In the

[252] Strimbeck, G. R., McGee, J. P., and Katell, S., *AGA Monthly*, **38**, No. 10, 35–6, 50–2 (1956).

nuclear plant, helium would be used as a heat-transfer medium in a closed cycle to prevent the gas and residue from becoming radioactive. The nuclear plant would eliminate the oxygen-production plant, the coal used to supply the endothermic heat of the gasification reactions, and the coal used for generation of power and process steam. According to speculative estimates,[251] high-Btu gas could be produced in a nuclear plant either by steam-coal or hydrogen-coal gasification for about 20% less than in the conventional methanation process. These estimates are predicated on the development of economical and effective methods for transferring the heat from helium at 2500°F to the reacting mixture of coal and steam.

TABLE 41

ESTIMATED ANNUAL OPERATING COST FOR PRODUCTION OF HIGH-BTU GAS BY METHANATION OF LURGI GAS[250]

	Cost, Dollars†	%
Direct cost		
Raw materials and utilities		
Char, 184 tons/hr × 24 × 330 × $5/ton	7,286,400	
Coal, 37 tons/hr × 24 × 330 × $4/ton*	1,172,200	
Water, 8000 gpm × 60 × 24 × 330 × $0.10/M gal	380,200	
Catalyst and chemicals	652,500	
Less sulfur produced, 48 tons/day × 330 × $25/ton	396,000	
Total raw materials and utilities	9,095,300	50.0
Direct labor		
1,200 man-hours, $2.25/hr × 365	985,500	
Supervision, 15%	147,800	
Total direct labor	1,133,300	6.2
Plant maintenance		
125 men, $5,000/year	625,000	
Supervision, 20%	125,000	
Material, 50% of labor	312,500	
Total plant maintenance	1,062,500	5.8
Payroll overhead, 18.5% of payroll	348,400	1.9
Operating supplies, 20% of plant maintenance	212,500	1.2
Total direct cost	11,852,000	65.1
Indirect cost, 50% labor, maintenance, and supplies	1,204,200	6.6
Fixed cost		
Taxes and insurance, 2%	1,445,100	8.0
Depreciation, 5%	3,685,000	20.3
Total operating cost	18,186,300	100.0

Annual production = 90 million × 330 = 29.7 billion scf

$$\text{Cost/Mcf} = \frac{\$18,186,300}{29.7 \times 10^6} = \$0.61$$

With char at $4/ton, cost/Mcf = $0.56

With char at $6/ton, cost/Mcf = $0.66

* Coal for steam-plant.

† Computed in 1959 dollars.

Studies of materials of construction and of high-temperature gas technology are underway in simulated nuclear reactors.[247]

Other possibilities for applying developments in nuclear technology to gasification reactions involve the use of gamma or neutron radiation for promoting or accelerating a chemical reaction.[253] This effect is important not only as a means of increasing the output of a given reactor, but also as a means for reducing the temperature at which a reaction proceeds at an acceptable rate. This is particularly important in those instances in which reactions must be carried out at elevated pressures. Table 42 summarizes possible areas of application of radiation to processes of interest to the gas industry. In every instance, improvement factors ranging from 1000 to 100,000 will be required before any of the concepts could be considered for active development. The first two concepts would be of considerable importance in the production of gaseous fuel in a nonfossil-fuel era.

TABLE 42

CONCEPTUAL APPLICATIONS OF RADIATION TO PROCESSES FOR FUEL GAS PRODUCTION[253]

Conceptual Application	Remarks
Decomposition of water by radiolysis by gamma or neutron radiation.	Improvement factor of 1200 to 2000 to obtain H_2 at 30 to 45 cents/Mcf considering only radiation costs.
Regeneration of carbon dioxide to carbon monoxide by nuclear reactor radiation.	Improvement factor of 1000 to 2000 to obtain carbon monoxide at radiation cost of 5 cents/Mcf.
Coal degradation by gamma radiation as an aid to coal hydrogenation.	Improvement factor of 10^4 to 10^5 required to attain radiation cost of 6 cents/Mcf of methane.
Hydrogenation of char in presence of gamma radiation.	Negligible effects observed to date.
Activation of methanation catalysts with gamma and neutron radiation.	Negligible effects observed to date.

PURIFICATION OF SYNTHESIS GAS

Raw synthesis gas may contain particulates such as tar and dust, and hydrogen sulfide, organic sulfur, and carbon dioxide, which must be reduced to acceptable concentrations determined by subsequent processing steps. Sulfur compounds and carbon dioxide are always present in raw synthesis gas. Tar is present in the gas produced in fixed-bed coal gasifiers, but not in that produced in fully entrained gasifiers. In contrast, the dust content of the gas from fixed-bed gasifiers is considerably below that of the gas from fluidized-bed or fully entrained gasifiers. The large number and variety of processes that are available for reducing the concentration of all of these constituents cannot be reviewed in detail. Instead, some of the more recent developments will be discussed, and, where available, reference will be made to appropriate reviews or texts.

The concentration of contaminants depends on the coal used and on the gasification process. The range of concentrations observed in suspension gasification processes is shown in Table 43[5,190,254] along with approximate specifications for purified synthesis gas.[254,255,256] The sequence of processes for removing the various constituents and the type of process selected are determined by such factors as temperature level desired in various processing steps, temperature levels feasible for heat recovery, tolerance of catalysts to impurities, and allowable level of inert diluents in the product gas. Evaluation of all of these factors involves both technical and economic considerations. For example, a relationship exists between the sulfur content of the feed gas and catalyst life at an acceptable level of activity.[257] Therefore, the optimum concentration of sulfur in the feed to a catalytic process is determined by minimization of the sum of the cost of sulfur removal and the cost of catalyst replacement.

Removal of Particulates

Particulates in synthesis gas must be removed to prevent deposition and subsequent plugging

[253] Elliott, M. A., and von Fredersdorff, C. G., paper presented before the Financial Forum, Boston, Massachussetts, May 11–12, 1959.

[254] Wainwright, H. W., Kane, L. J., Wilson, M. W., Shale, C. C., and Ratway, J., *Ind. Eng. Chem.*, **48**, 1123–33 (1956).

[255] Dirksen, H. A., and Linden, H. R., *Ind. Eng. Chem.*, **52**, 584–89 (1960).

[256] Dent, F. J., and Hebden, D., *Gas Research Bd., Comm.* GRB 51 (1949), 52 pp.

[257] Wainwright, H. W., Egelson, G. C. and Brock, C. M., *U.S. Bur. Mines, Rept. Invest.* No. 5046 (1954), 10 pp.

TABLE 43

RANGE OF IMPURITIES IN RAW AND PURIFIED SYNTHESIS GAS FROM SUSPENSION GASIFIERS

Gasifier	Raw Synthesis Gas			Purified Synthesis Gas (Approximate Specification)
	U.S. Bureau of Mines (Morgantown)[254]	Institute of Gas Technology Pressure[5]	Koppers-Totzek[190]	
Dust, grains/Ccf	3000+	—	—	<0.1[254]
Carbon dioxide, vol %	7 to 26	—	—	2 to 3[254]
Oxygen, vol %	0.2 and less	—	0.04 to 0.10	<0.2[254]
Nitric oxide, vol %	—	—	0.0005 to 0.0006	—
Organic sulfur, grains/Ccf	4 to 40	17 to 20	17.5 to 18.4	—
Hydrogen Sulfide, grains/Ccf	80 to 400	78 to 189	130 to 175	—
Sulfur dioxide, grains/Ccf	—	—	0.65 to 0.87	—
Total sulfur, grains/Ccf	—	—	—	<0.1[254] <0.01[255] <0.002[256]

of process equipment, as, for example, by unreacted fuel or solid residues. In some processes, partially reacted fuel must be removed and recovered for recycling and, in others, tar is recovered not only to prevent operating difficulties but also to realize byproduct credits.

The removal of particulates from fluid streams involves the application of the basic principles of fine-particle technology[23] and fluid and particle mechanics[258] or physics.[259] Dust collection or removal from a gas stream is accomplished by use of one or more of the following forces or mechanisms:[258] (1) gravitational, (2) inertial, (3) filtration, (4) electrostatic, (5) physicochemical, (6) thermal, and (7) sonic. More than one of these effects may be used in dust-collection equipment, which is, therefore, more conveniently classified by type than by operating principle. The various types of dust and mist collection equipment are listed in Table 44 along with their general characteristics and approximate cost.

Basic principles of operation and performance with descriptions of each type, have been presented by Lapple.[258,260] The more important types of equipment for removing particulates from gases produced from coal are cyclones, scrubbers, electrostatic precipitators, and packed towers. The performance, cost, and application of cyclones for dust removal have been presented,[261] and the design and performance of cyclones for separating coarse and fine dust at temperatures from 1562 to 1652°F from gas produced in Winkler generators have also been discussed by von Portatius.[179]

Electrostatic precipitation is widely used for removing tar from gases.[262,263,264] The history of the development of electrostatic precipitation has been presented,[264] and basic principles, design, performance, economics and applications have been treated extensively.[263] The performance of electrostatic precipitators, scrubbers, and coke bed filters on a laboratory and pilot-plant scale has also been studied.[257]

[258] Lapple, C. E., and Alves, G. E., *Fluid and Particle Mechanics*, University of Delaware, Newark, Delaware, 1951, 353 pp.

[259] Ehrenfeld, J. R., Smith, W. J., Surprenant, N. F., and Wiederhorn, N. M., *Am. Gas J.*, **186,** No. 1, 27, 30–33; No. 3, 32–34, 36; No. 4, 28–32; No. 5, 57–59 (1959).

[260] Lapple, C. E., *Chemical Engineers' Handbook*, Perry, J. H., Ed., McGraw-Hill, New York, 1950, pp. 1013–50.

[261] Heinrich, R. F., and Anderson, J. R., *Chemical Engineering Practice*, Cremer, H. W., Ed., Academic Press, New York, 1957, Vol. 3, pp. 464–83.

[262] Thompson, R. J. S., and Cosby, W. T., *Gas World*, **148,** Coking Sect., 23–26, 28, 30 (1958).

[263] Heinrich, R. F., and Anderson, J. R., *Chemical Engineering Practice*, Cremer, H. W., Ed., Academic Press, New York, 1957, Vol. 3, pp. 484–534.

[264] White, H. J., *J. Air Pollution Control Assoc.*, **7,** 166–77 (1957).

TABLE 44

Dust and Mist Collection Equipment[258]

Classification of Types and High-Spot Performance Characteristics

Type Collector	Approximate Minimum Particle Diameter Collectable, μ	Approximate* Purchase Cost $ (cu ft/min)	Outstanding Characteristics or Application
Gravity settling chamber	40	—	Simple, large, low Δp; precleaner; natural draft exhausts
Howard type	10	—	Difficult to clean
Impingement separators	10	0.10 to 0.20	Adaptable to existing flues
Calder Fox	2	—	Sulfuric acid mist removal
Cyclone separators	5	0.04 to 0.08	Simple, inexpensive most widely used
Multiple units	2	0.15 to 0.50	
Mechanical centrifugal separators	5	0.10 to 0.30	Compact: ventilation booths
Miscellaneous inertial separators			May be adaptable to existing flues
Packed beds			
Coke box	0.5	2.00 to 5.00 (special)	Sulfuric acid mist removal
Cloth collectors	0.1		High-collection efficiency; temperature and moisture limitations
Baghouse		0.50 to 2.50	Simple, large
Unit filter		0.15 to 1.50	For small capacities (< 1000 cu ft/min)
Mechanical filter		0.10 to 0.70	For high capacities
Automatic filter		0.50 to 2.00	For high-dust concentrations (> 20 grains/cu ft)
Scrubbers	0.1–1		Dust wetted; sludge disposal and corrosion problems
Chamber type		0.15 to 1.50	Simple
Cyclone type		0.30 to 1.00	
Inertial type		0.30 to 1.00	
Mechanical type		1.00	
Packed type			
Film type			
Miscellaneous			
Electrical Precipitators	0.01	0.30 to 1.50	For process gases; heavy duty; expensive on first cost and operation
Single-stage (Cottrell)		2.00 to 10.00 (special)	
Plate type			
Pipe type			
Two-stage		0.10 to 0.30	Small, compact; air conditioning applications
Alternating current			Laboratory instrument
Air filters			High capacity, low resistance, low-dust concentration; air conditioning
Viscous	5	0.005 to 0.030	
Dry	2	0.005 to 0.050	
Automatic		0.05 to 0.10	
Miscellaneous			
Wetted cells			
Electrostatic		0.10 to 0.30	
Pipe-line			
Activated carbon		0.10 to 0.20	Odor removal
Miscellaneous			
Sonic			Only recently of industrial importance
Thermal			Not developed industrially

* For steel construction except where otherwise noted; includes only directly chargeable and necessary auxiliaries; not installed. Based on prices in period 1945 to 1946.

Note: The above are order-of-magnitude values only applying to the majority of units in each category and do not necessarily apply to all the commercially available types of equipment in each category. In cases where values are very diverse for the specific types or applications in each category, values are not listed.

Removal of Sulfur Compounds and Carbon Dioxide

In the complete gasification of coal, the major portion of the sulfur in the coal appears as hydrogen sulfide and organic sulfur in the gas. This is shown in Table 45 by sulfur balances for suspension gasifiers, in which the sulfur appearing as hydrogen sulfide ranges from 65 to 75%, and as organic sulfur, from 1 to 4%. The distribution of sulfur among the various sulfur compounds in the gas produced in the gasification of coal with oxygen and steam at elevated pressure in the Institute of Gas Technology suspension gasifier is shown in Table 46. The bulk of the sulfur is present as hydrogen sulfide (see also Table 45), but in suspension gasifiers, carbonyl sulfide is the predominant organic sulfur compound. The concentration of mercaptans and thiophene is negligible (less than 0.1 grain/Ccf). In processes where there is an opportunity for volatile

TABLE 45

SULFUR BALANCES IN SUSPENSION GASIFICATION[254]

Run No.	Sulfur Fed to Gasifier, lb/hr	Sulfur in Gas, %		Other Sulfur, %			
		H_2S	Organic	In Residue	In Scrubber Water		Unaccounted For
	Atmospheric Pressure Gasifier No. 4						
7	11.9	71.0	3.0	21.2	4.7		0.1
15	3.6	68.0	2.8	13.7	7.8		7.7
36	5.9	69.1	4.0	20.2		6.7	
43A	3.1	66.0	2.9	25.4		5.7	
	Pressure Gasifier No. 3						
25	20.6	69.4	3.6	15.7	8.8		2.5
26	23.1	65.9	3.1	21.8	7.5		1.7
27	20.8	74.7	1.3	16.2		7.8	
32B	13.7	74.8	3.2	18.7		3.3	

TABLE 46

DISTRIBUTION OF SULFUR IN SYNTHESIS GAS FROM A PRESSURE GASIFIER[5]

Run No.	18C	19D	20B	27C
	Grains S/Ccf			
Hydrogen sulfide	117.0	116.0	115.0	107.0
Mercaptans	0	0	0	—
Thiophenes	0	0	0	—
Carbon disulfide	2.4	2.0	1.2	0
Carbonyl sulfide	12.6	17.5	18.0	16.2
Unidentified	2.0	0	0.4	0.9
Total	134.0	135.5	134.6	124.1

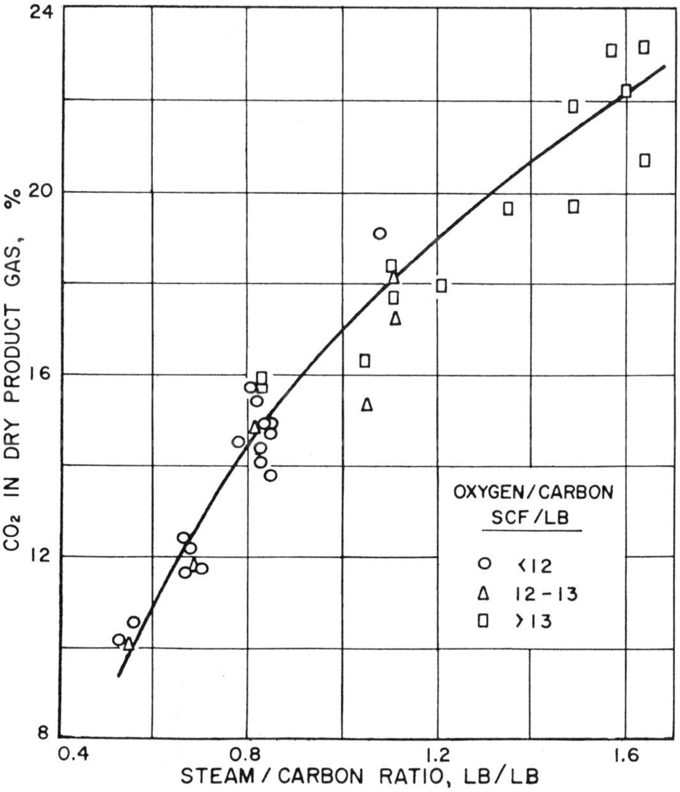

Fig. 81. Effect of steam-carbon ratio on CO_2 content of product gas in suspension gasification.[257]

matter to be distilled at relatively low temperatures and in the absence of substantial concentrations of steam, carbon disulfide may predominate, and thiophene and mercaptans may be present.[265]

The carbon dioxide content of synthesis gas depends on the type of process and the operating conditions. In suspension gasifiers, the steam-carbon ratio generally is the determining factor (see Fig. 81).[257] Carbon dioxide is also produced when synthesis gas is processed by the water-gas shift reaction,[266] either to adjust the carbon monoxide-hydrogen ratio, or to produce hydrogen. Carbon dioxide must be removed because it is an undesirable constituent in some processing, or because it reduces heating value and increases transmission costs in high-Btu gas production processes.

Processes for removing hydrogen sulfide, carbon dioxide, and organic sulfur have been described in considerable detail in the literature.[265–268] Dry oxidation processes for removal of hydrogen sulfide, and in some instances organic sulfur, are listed in Table 47. Liquid processes for removing hydrogen sulfide by oxidation are listed in Table 48. Processes for removing carbon dioxide, hydrogen sulfide, and in some instances, other constituents, are listed in Table 49. A summary of processes particularly applicable for purifying synthesis gas used in the production of synthetic liquid fuels is presented in Table 50.[269]

Oxidation of Sulfur Compounds

Dry methods for removing sulfur compounds from gases have been reviewed by several investigators.[267,270] Fluidization techniques have been applied to the removal of hydrogen sulfide and organic sulfur from coke-oven gas.[271] In this process, sulfur is removed by the sulfiding of iron oxide in a fluidized bed at approximately 350°C. The fouled oxide enters a regenerating chamber, where it is

[265] Gollmar, H. A., *Chemistry of Coal Utilization*, John Wiley and Sons, New York, 1945, Vol. 2, pp. 947–1007. See Table III, p. 950.

[266] Kohl, A. L., and Reisenfeld, F. C., *Gas Purification*, McGraw-Hill, New York, 1960, 556 pp.

[267] Kohl, A. L. and Reisenfeld, F. C., *Chem. Eng.*, **66,** No. 6, 126–78 (1959).

[268] Katell, S., and Fan, L. T., *U.S. Bur. Mines, Inform. Circ.* No. 7886 (1959), 43 pp.

[269] Field, J. H., Benson, H. E., and Anderson, R. B., *Chem. Eng. Progr.*, **56,** No. 4, 44–48 (1960).

[270] Griffith, R. H., *Removal and Recovery of Sulfur from Fuels*, Inst. Fuel, London, 1954, Vol. 1, pp. 48–58. Smith, B., *Trans. Chalmers Univ. Technol.*, Gothenburg, No. 184 (1957), 64 pp.

[271] Reeve, L., *J. Inst. Fuel*, **31,** 319–24 (1958).

TABLE 47

DRY OXIDATION PROCESSES FOR REMOVAL OF SULFUR COMPOUNDS

Designation or Developer of Process	Constituents Removed	Basic Characteristics	Reference, Page No. 265	Reference, Page No. 266
		Ordinary-Temperature Operation		
Conventional dry boxes	H_2S	Reaction of H_2S with hydrated Fe_2O_3: $Fe_2O_3 + 3H_2S \rightarrow Fe_2S_3 + 6H_2O$; $Fe_2S_3 + \frac{3}{2}O_2 \rightarrow Fe_2O_3 + 3S$ (applies to all five processes bracketed)	960–6	241–6
Tower purifiers (Thyssen-Lenze)	H_2S		966–8	246–7
Continuous (North Thames Gas Board)	H_2S			247
Continuous (Gas Technik)	H_2S			248–51
High pressure	H_2S		968	251–2
IG Farbenindustrie	H_2S	Oxidation of H_2S to S in presence of activated carbon	969, 995–7	260–4
		Elevated-Temperature Operation		
Katasulf	H_2S, Organic S	Oxidation to SO_2 in presence of catalyst.	969	266–71
North Thames Gas Board	Organic S	Oxidation to chiefly SO_2 in presence of nickel subsulfide.	1816	271–5
Soda-iron	Organic S	Oxidation to chiefly SO_2 in presence of hydrated iron oxide and sodium carbonate.		275–8
Appleby-Frodingham	H_2S, Organic S	Adsorption on iron oxide in continuous fluidized system at 660°F. Regeneration of fouled oxide with air at 1110 to 1470°F		278–9

TABLE 48

Liquid Processes for Removal of Hydrogen Sulfide by Oxidation

Designation or Developer of Process	Constituents Removed	Basic Characteristics	Reference, Page No. 265	Reference, Page No. 266
Ferrox sulfur recovery (soda-ash)	H_2S	Oxidation of H_2S to S after absorption and reaction in weakly alkaline solution of soda ash or ammonia containing iron oxide in suspension.	971–3	286–90
Gluud (ammonia)	H_2S		972	290–1
Manchester (soda-ash)	H_2S			291–5
Burkheiser (ammonia)	H_2S		970	285–6
Thylox	H_2S, HCN	Oxidation of H_2S to S after absorption and reaction in essentially neutral solution containing a thioarsenate salt.	973–81	295–303
Giammarco-Vetrocoke	H_2S*	Oxidation of H_2S to S after absorption and reaction in solution containing mixture of alkaline arsenites and arsenates.		303
Gesellschaft für Kohlentechnik	H_2S, HCN	Oxidation of H_2S to S after absorption and reaction in solution containing iron-cyanide complex.		304
Fischer	H_2S, HCN			304–5
Staatsmijnen-Otto and Autopurification	H_2S, HCN			305–8
Perox	H_2S	Oxidation of H_2S to S after absorption in aqueous ammonia solution containing organic oxidation catalyst.		318–9

* CO_2 may also be removed by proper adjustment of solution pH and arsenite-arsenate concentration.

TABLE 49

Process for Removing Carbon Dioxide, Hydrogen Sulfide, and Other Constituents by Absorption and Stripping

Designation or Developer of Process	Constituents Removed	Basic Characteristics	Reference, Page No. 265	Reference, Page No. 266
Seaboard soda liquid	H_2S, HCN	Absorption of H_2S in dilute Na_2CO_3 solution and regeneration with air.	984–5	120–2
Vacuum carbonate	H_2S, HCN	Absorption of H_2S in alkaline carbonate solution and regeneration by vacuum distillation.	985–8	122–30
Hot potassium carbonate	CO_2, H_2S	Absorption in hot potassium carbonate solution and regeneration by steam flashed on pressure reduction of solution		137–50
Ethanolamine	Acidic gases	Absorption in monoethanolamine and diethanolamine solutions; regenerated by heat supplied by steam.	990–1	18–86
Tripotassium phosphate	H_2S	Absorption in tripotassium phosphate solution; regenerated by heat supplied by steam.	991	130–5
Alkacid process	CO_2, H_2S	Absorption in solution of sodium salts of amino acids; regenerated by heat supplied by steam.	991–2	135–7
Giammarco-Vetrocoke	CO_2	Absorption in solution of alkali-carbonate activated by organic and inorganic (arsenious trioxide, selenious and tellurous acid) additives; regeneration by heat supplied by steam.		151
Water washing	CO_2, H_2S	Absorption in water at elevated pressure; desorption by flashing at reduced pressure.		153–71
Rectisol	CO_2, H_2S, organic S, HCN, benzene, and gum-formers.	Absorption in methanol at low temperature; regenerated by flashing at reduced pressure and by heat supplied by steam.		513–8

TABLE 50

PURIFICATION PROCESSES APPLICABLE TO SYNTHESIS GAS[269]

Process	Constituents Removed	Basis	Operating Characteristics
Water	CO_2, H_2S, part of organic S	Solubility	Elevated pressure, 3 to 5% loss of synthesis gas by solubility.
Rectisol	CO_2, all S compounds	Solubility in methanol at low temperatures	Elevated pressure and refrigeration to about −70°F does complete purification.
Ethanolamines	CO_2, H_2S	Chemical reaction	Elevated pressure desirable, but can operate at atmospheric pressure.
Hot carbonate	CO_2, H_2S, COS	Chemical reaction for CO_2 and H_2S. Hydrolysis of COS.	Elevated pressure; operation at about 200 to 250°F.
Iron oxide boxes$_3$	H_2S	Chemical reaction with Fe_2O_3	Used for final H_2S purification.
Activated charcoal	All S compounds	Adsorption	Used for final H_2S and organic S removal.

oxidized with air in a fluidized bed at 600 to 800°C for recycle to the purification stage. Approximately 80% of the hydrogen sulfide and organic sulfur have been removed in a single oxide bed, and 98% with two beds in the absorbing vessel. These results were obtained in a large continuous pilot plant capable of treating 2.5 million scf/day of gas containing 580 to 675 grains of H_2S/Ccf, and 13 to 39 grains of organic sulfur/Ccf.

The alkalized-iron or soda-iron process was used extensively for the purification of synthesis gas in the German synthetic liquid-fuel industry.[266,272] In this process, organic sulfur and low concentrations of hydrogen sulfide are oxidized by the addition of low concentrations of air at temperatures ranging from 300 to 500°F, in the presence of natural iron ore containing about 30% (by weight) of sodium carbonate. Organic sulfur content can be reduced from about 20 grains to 0.05 grains/Ccf.[272] Carbonyl sulfide, carbon disulfide, and mercaptans are oxidized in the soda-iron process, but the conversion of thiophene is far from complete.

Other catalytic processes have been developed for oxidizing organic sulfur. In the North Thames Gas Board Process,[273] oxidation occurs with small concentration of added air in the presence of nickel sulfide at a temperature of 380°C. The sulfur dioxide formed is removed by washing with a weak solution of sodium hydroxide, and the low concentrations of hydrogen sulfide formed are removed by conventional iron oxide purifiers. A plant using this process and having a capacity of 1.5 million scf of town gas per day was operated from 1937 to 1953.

A cyclic catalytic process has been developed for simultaneously removing hydrogen sulfide and organic sulfur from carbureted water-gas,[274] and the application of this process to the purification of synthesis gas has been studied.[254,275,276] This process uses a copper-chromium-vanadium catalyst, which becomes sulfided during conversion and removal of organic sulfur and hydrogen sulfide at temperatures ranging from 300 to 450°C. The sulfided catalyst is regenerated with air and reduced periodically at 450°C.

Wet processes, which have been applied principally for removal of hydrogen sulfide from coke-oven gas, would also be adaptable to removing hydrogen sulfide from synthesis gas. The basic characteristics of wet oxidation processes are given in Table 48. A general review of wet processes for removing hydrogen sulfide from coal gas has been published.[277]

[272] Sands, A. E., Wainwright, H. W., and Schmidt, L. D., *Ind. Eng. Chem.*, **40**, 607–20 (1948).

[273] Plant, J. H. G., and Newling, W. B. S., *Inst. Gas Engrs., Commun.* No. 344; *Trans. Inst. Gas Engrs.*, **93**, 308–33 (1948). King, J. G., *Gas Times*, **92**, No. 922, 25–6 (1958).

[274] Huff, W. J., and Logan, L., *AGA, Proc.*, **1936**, 724–59.

[275] Sands, A. E., Wainwright, H. W., and Egleson, G. C., *AGA*, *Proc.* **1950**, 564–602.

[276] Storch, H. H., *Chemistry of Coal Utilization*, John Wiley and Sons, New York, 1945, Vol. 2, pp. 1816–7.

[277] Craxford, B. A., and Parker, A., *Removal and Recovery of Sulfur from Fuels*, Inst. Fuel, London, 1954, Vol. 1, pp. 59–66, 1954.

Processes Employing Absorption or Adsorption and Stripping

Ethanolamine processes are used extensively in the removal of carbon dioxide and hydrogen sulfide from gases, particularly natural gas. Acidic gases are absorbed in the amine solution, which is regenerated by heat supplied indirectly by steam. There are various ethanolamine process flow systems, and the design and operation of ethanolamine plants differ.[266]

The hot potassium carbonate process was originally developed for removing carbon dioxide and hydrogen sulfide from synthesis gas,[278,279] but has also been applied for the purification of natural gas and other gases.[280,281] The absorber operates at elevated pressure (300 psig for synthesis gas purification) and at temperatures ranging from 225 to 240°F. The spent (rich) carbonate solution is depressurized to 5 psig, stripped by heat supplied by steam, and then pumped back to the absorber. Since the absorber and stripper operate at substantially the same temperature, steam requirements are minimized, and heat-exchange equipment between these units is eliminated. Data on the equilibrium pressures of hydrogen sulfide and carbon dioxide over solutions of potassium carbonate are available,[282] and design and operation of various flow systems have been discussed.[266,280]

Other processes for removing carbon dioxide and hydrogen sulfide include: (1) The Giammarco-Vetrocoke processes,[283] which may be used for selective or simultaneous removal of carbon dioxide and hydrogen sulfide, depending upon the solution used (see Tables 48 and 49); (2) the Alkazid process,[284] which is a further development of the Alkacid process (see Table 48, and ref. 265, pp. 991–2), and which also may be used for the simultaneous or selective removal of carbon dioxide and hydrogen sulfide; and (3) the Fluor solvent carbon dioxide removal process[285] for removing high concentrations of this gas from gases at high pressure, such as natural gas. Solvents employed may be propylene carbonate, glycerol triacetate, butoxy diethylene glycol acetate, and methoxy triethylene glycol acetate.

Selection of the most economical process for purifying a gas stream depends on such factors[279] as operating pressure, volume of gas to be treated, concentration of impurities, degree of purification required, presence of interfering impurities or constituents, availability and cost of utilities, and the ease of integration of the purification process into the overall processing scheme. Each case must therefore be estimated individually to determine the most economical process.

For carbon dioxide removal, several methods[283,286] have been compared for removing carbon dioxide (34% by volume) from 12 MMcf/day of ammonia synthesis gas at a pressure of 350 psig. The treated gas contained 10 to 20 ppm of carbon dioxide. The results of this comparison are presented in Table 51.

The Rectisol process[287] removes carbon dioxide, hydrogen sulfide, organic sulfur, hydrogen cyanide, benzene, and gum-forming constituents by physical absorption in methanol at temperatures ranging from −70 to −20°F, and at a pressure of 325 psig. This process is used at the SASOL plant[287,288] for the complete purification of synthesis gas prior to catalytic conversion to liquid fuel; the three Rectisol units have a total raw gas capacity of 165 MMcf/day. The absorption is accomplished in three steps: prewash, main wash, and fine wash. Raw gas is cooled by heat exchange, and enters the prewash first and then main-wash sections of a single absorption vessel. The last traces of aromatics, most of the carbon dioxide, and nearly all of the hydrogen sulfide are removed in this operation by countercurrent scrubbing with

[278] Benson, H. E., Field, J. H., and Jimeson, R. M., *Chem. Eng. Progr.*, **50**, 356–64 (1954).

[279] Benson, H. E., Field, J. H., and Haynes, W. P., *Chem. Eng. Progr.*, **52**, 433–38 (1956).

[280] Benson, H. E., and Field, J. H., *Petrol. Refiner*, **39**, 127–32 (1960).

[281] Benson, H. E., Paper CEP-61-6, Operating Sect. Am. Gas Assoc., 1961.

[282] Tosh, J. S., Field, J. H., Benson, H. E., and Anderson, R. B., *U.S. Bur. Mines, Rept. Invest.* No. 5622 (1960), 25 pp.

[283] Reisenfeld, F. C., and Mullowney, J. F., *Petrol. Refiner*, **38**, 161–8 (1959).

[284] Leuhddemann, R., Noddes, G., and Schwarz, H. G., *Oil Gas J.*, **57**, No. 32, 100, 102, 104 (1959).

[285] Kohl, A. L. and Buckingham, P. A., *Oil Gas J.*, **58**, No. 19, 146–7, 150–3, 156 (1960).

[286] Mullowney, J. F., *Oil Gas J.*, **56**, No. 6, 93–8 (1958).

[287] Herbert, W., *Erdöl u. Kohle*, **9**, 77–81 (1956).

[288] Linton, J. A., and Tisdall, G. C., *Coke and Gas*, **20**, 148–52, 159 (1958).

TABLE 51

ECONOMICS OF CARBON DIOXIDE REMOVAL BY VARIOUS PROCESSES[283]

Process Used	Giammarco-Vetrocoke-Monoethanolamine	Hot Potassium Carbonate-Monoethanolamine	Monoethanolamine	Hot Potassium Carbonate-NH_4OH-NaOH	Water-Monoethanolamine
Plant investment	$372,000	$545,000	$607,000	$619,000	$778,000
Direct operating costs					
Utilities, $/year					
Steam at $0.55/1000 lb	23,000	93,000	262,000	96,000	10,000
Electricity at $0.01/kwhr	25,000	19,000	19,000	21,000	41,000
Cooling water at $0.02/1000 gal	4,000	18,000	32,000	19,000	7,000
Process water at $0.10/1000 gal	—	—	—	1,000	—
Gas losses at $0.50/Mcf	6,000	7,000	4,000	5,000	37,000
Chemical losses at cost	7,000	11,000	2,000	26,000	2,000
Subtotal	65,000	148,000	319,000	168,000	97,000
Operating labor	47,000	47,000	47,000	66,000	47,000
Maintenance at 4% plant cost	15,000	22,000	24,000	25,000	31,000
Plant general	18,000	20,000	21,000	27,000	22,000
Subtotal	80,000	89,000	92,000	118,000	100,000
Total direct costs	145,000	237,000	411,000	286,000	197,000
Indirect operating costs, $/year					
Depreciation at 10% plant cost	37,000	55,000	61,000	62,000	78,000
Taxes and insurance	9,000	14,000	15,000	15,000	19,000
Total indirect costs	46,000	69,000	76,000	77,000	97,000
Total operating costs	191,000	306,000	487,000	363,000	294,000
Cost of profit (before income tax) at 25% of plant cost	93,000	136,000	152,000	155,000	195,000
Grand total treating cost $/year	284,000	442,000	639,000	518,000	489,000
Treating cost-dollars/Mcf gas treated	$0.07	$0.11	$0.15	$0.12	$0.12

methanol, which enters at −70° and leaves at −20°F. The methanol from these sections is regenerated by flashing and vaporization on reduction of pressure to subatmospheric, which cools the regenerated methanol to −80 to −90°F. The gas is finally purified (of trace concentrations of sulfur compounds, and carbon dioxide concentration, 1% by volume), by countercurrent scrubbing with methanol at −50°F in a second absorption vessel. The methanol from the final wash is regenerated by reduction in pressure and distillation, and is then cooled by heat exchange with the regenerated methanol from the main wash. Auxiliary cooling for the entire process is supplied by an ammonia refrigeration system.

Water has been used extensively to remove acid gas impurities.[266] It is applicable at low pressures with reasonably soluble gases, and may be used with the less soluble gases if their partial pressures are increased by operation at elevated total pressure. Under such conditions, primary gases may also be dissolved, which may adversely affect the economics. The use of water as an absorbent has been discussed by Kohl and Reisenfeld,[266] and the design and economics of a water-washing process for removing carbon dioxide and hydrogen sulfide from natural gas has been presented.[289]

Organic sulfur may be removed directly from a gas stream by adsorption on activated carbon, the spent carbon being reactivated by steam. Data have been presented on the removal of thiophene and carbonyl sulfide by various carbons[275] and on the effect of process variables on the adsorption of carbonyl sulfide by activated carbon.[254]

[289] Burnham, J. G., *Petrol. Eng.*, **31**, No. 2, C15-16, 18, 21 (1959).

CHAPTER 21

THE UNDERGROUND GASIFICATION OF COAL

JAMES L. ELDER

Bureau of Mines, U.S. Department of Interior, Grand Forks, North Dakota

The gasification of coal in place has as its objective the recovery of the energy or the chemical values of the coal without mining. The products are obtained in a gaseous form and might be utilized for the production of electric power, the manufacture of organic chemicals, and the synthesis of liquid or gaseous fuels. An advantage of the gasification of coal in place is the elimination of underground mining operations and possibly the utilization of coal in beds that are not profitable to mine or which remain in areas where mining operations have been completed.

The idea of underground gasification of coal is not new. Siemens suggested the gasification of slack and waste coal in the mine as early as 1868.[1] In 1888, Mendeleev in Russia suggested the true underground gasification of coal.[2] In 1909, a British patent was granted to Betts who proposed a method of gasifying coal in the bed by igniting the base of one or more shafts or boreholes, supplying it with air and steam, and withdrawing the gas formed through either the same or other shafts or boreholes.[3] Sir William Ramsay made at least two references to the underground gasification of coal in the London Times of 1912. In one, he suggested that coal be retorted within the earth for the production of gas. A borehole could be put down to the coal strata, the coal ignited, and air and water passed down, producing a gas containing hydrogen and carbon monoxide. The gases produced might be used in gas engines at the pit mouth for the production of electricity. Although there is no record of publications on the subject by Ramsay, it is known that he initiated work on an underground gasification in Durham. Before completion of the installation, war broke out and the experiment was abandoned and not taken up again because of Ramsay's death in 1916.

In Russia, Lenin recommended experimentation on underground gasification in his country. Little work was done until the government subsidized preliminary experiments in 1931. Two years later they formed a state trust known as "Podzemgaz," which was to carry out experimental work with the objective of establishing industrial plants. Five experimental stations were established in the Lower Moscow, Donetz, Kuznetzk, and North Caucaucus coal fields. From 1933 to 1939, three distinct methods of gasification of coal in place were developed. Considerable success was claimed, and at least one method was tried on an industrial scale. Concurrent with the work of the experimental stations, theoretical and laboratory studies of chemical and physical processes involved in underground gasification were conducted at the Moscow Academy of Sciences, the Donetz Institute of Coal Chemistry, and elsewhere. Nusinov has published an account of the work during this period.[4] An extensive bibliography of the early work in underground gasification of coal covering available literature through 1944 has been published.[5]

Since 1957, a Moscow publishing house, Ugletechizdat, has published a quarterly magazine, *Podzemnaya Gazifikatsiya Ugleĭ.*

[1] Siemens, W., *Trans. Chem. Soc.*, **21**, 279 (1868).

[2] Mendeleev, D. I., *Severny Vestnik* (*St. Petersburg*), **1888**, No. 8, Sect. 2, 27–73; No. 9, Sect. 2, 1–35; No. 10, Sect. 2, 1–21; No. 11, Sect. 2, 1–16; No. 12, Sect. 2, 1–26.

[3] Betts, A. G., Brit. Pat. 21674 (1909).

[4] Nusinov, G. O., *Underground Gasification of Coal*, Moscow-Leningrad Gosstoptekhnisdat, Second edition, 1941, 140 pp.

[5] Jolley, L. J., and Booth, N., Fuel, **24**, 31–7, 73–9 (1945).

The magazine is devoted to underground gasification and contains many articles dealing with geology, chemistry, and the engineering phases of the work. Status of much of the Russian work has been summarized in a recent Belgian article.[6]

THE CHAMBER METHOD

The first field-scale tests of underground gasification in Russia used the chamber method with two variations. In the first, coal was isolated from the rest of the seam by brickwork walls and panels constructed, much as is shown in Fig. 1. The air blast was applied from the gallery at the previously ignited face of one side of the panel, and the gas produced was removed by the gallery at the opposite side of the panel. In the initial tests, the gas flow through the panel depended on the natural fissures and porosity of the coal bed. The rates of gasification and combustion were low, and the product gas was of variable composition and at times contained unconsumed oxygen. A criticism of these experiments is that only carbonization of the coal took place, and a great deal of coke was left undergound. The second application of the chamber method involved breaking up the coal within the panels (see Fig. 1). After the coal was broken, it was fired at one side of the panel and the gases removed from the opposite side. Combustible gases were produced, and it was estimated in Russia that gas produced by this method would cost about one-fifth that of ordinary producer gas. The chamber method is also known as the warehouse method.[7]

A modification of the chamber method was developed to avoid much of the underground work. A number of holes were drilled into the coal seam and charged with dynamite. Gasification was started at a borehole, and it was hoped the coal would be crushed by a succession of explosions in advance of the reaction zone. In practice, it was found that the explosions and the breakage of the coal were irregular, that gas was obtained only intermittently, and that the gas contained unconsumed oxygen, large quantities of carbon dioxide, and little carbon monoxide and hydrogen. Bypassing of air around the reaction zone, and remixing of combustible gas with unreacted air beyond the reaction zone resulted in almost complete combustion of the product before it could be withdrawn from the system.

THE BOREHOLE-PRODUCER METHOD

The experiments with the chamber or warehouse methods yielded gas generally of poor quality, although there were times when a gas with a heating value of 145 to 200 Btu/cu ft could be obtained. Experiments showed that, where intimate contact between coal and all reactants was achieved, combustible gas was produced. Further studies led to development

[6] Crombrugghe, O. de, *Ann. mines Belg.*, **1959**, 478–534.

[7] Kirichenko, I. P., *Ugol'*, **1934**, No. 101, 90–98.

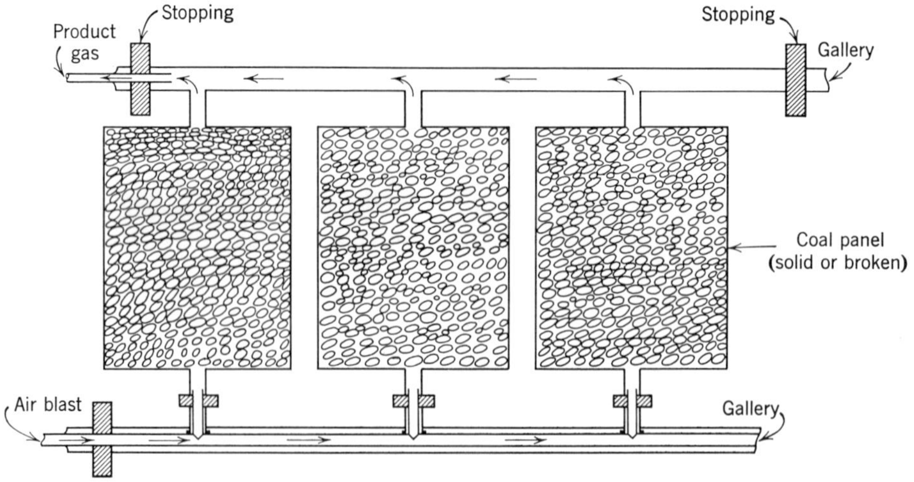

Fig. 1. The chamber method.[5]

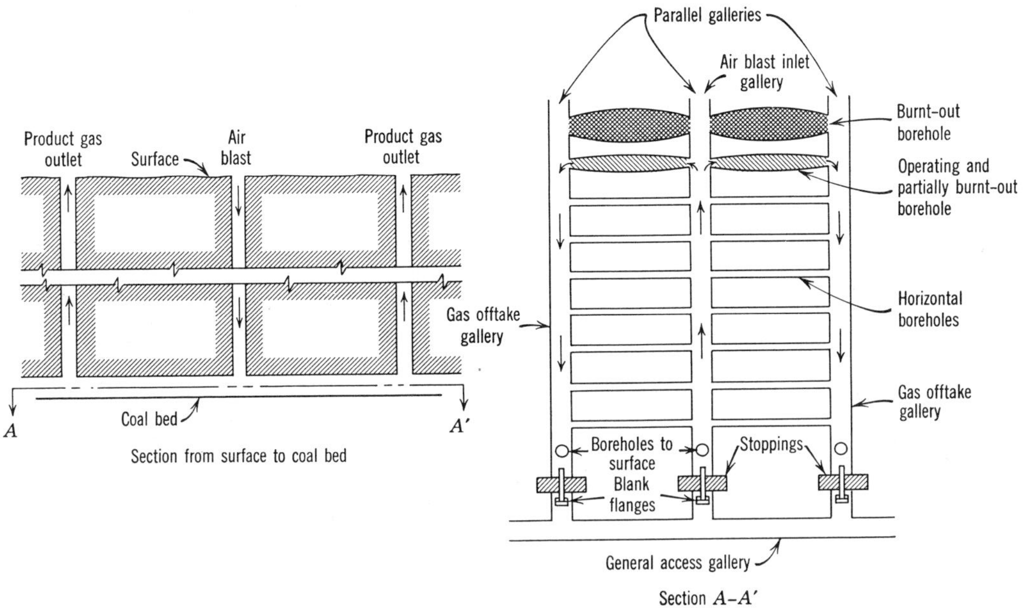

Fig. 2. The borehole producer method.[5]

of the borehole-producer method illustrated in Fig. 2. Parallel galleries within the coal bed were constructed, which served for blast inlet and for product gas offtake. Horizontal boreholes approximately 4 in. in diameter were drilled from one gallery to the other, the length being limited only by the drilling resources which in these tests was estimated at 100 to 120 yards. The boreholes were about 5 yards apart and were fitted with valves at their inlet ends and with iron seatings at their outlet ends. Provision was made for opening the inlet valves and for electric ignition of the coal in each by remote control. Boreholes were ignited either in succession or in small groups, and the process managed so that the maximum possible amount of coal was burned from each hole while maintaining an approximately constant rate of overall gas production.

The borehole-producer method is applicable to relatively horizontal coal beds. Because of the large amount of underground labor required in the construction of the galleries and horizontal boreholes, however, the method is limited in application. By 1935, it was reported that the borehole-producer method was being successfully operated at the Lower Moscow station.[8–11]

THE STREAM METHOD

At the beginning of 1934, another method was devised by Skafa, Matveev, and Filippov at the Donetz Institute of Coal Chemistry. This process, called the stream method, has been used at Gorlovka on an industrial scale with some success. A typical installation is illustrated in Fig. 3 and has been reproduced from a description given in 1939.[12] It is described in two Russian patents[13,14] and in numerous publications.

Application of the stream method applies generally to steeply pitched coal beds. Galleries are constructed following the dip of the seam and are joined at the lower ends by a horizontal gallery. Connections with the surface are made through vertical shafts. The galleries following the dip of the seam are generally about 60 yards long and 100 yards apart, which forms a rectangular panel of coal generally containing about 11 to 12 thousand tons. The coal is ignited by lighting a fire in

[8] Kirichenko, I. P., and Ton, V. S., *Gornyĭ Zhur.* 111, No. 7, 10–19; No. 8, 10–21; No. 9, 5–14 (1935).

[9] Buyalov, S. I., *Ugol'*, **1938,** No. 11, 17–24.

[10] Sazonov, N. I., *Khim. Tverdogo Topliva*, **6,** 17–28 (1935).

[11] *Ibid.*, pp. 861–9.

[12] Efremenko, I. E., and Talyzin, N., *Soviet Sci.*, **5,** 169–83 (1939).

[13] Korobchevskii, I. E., Russ. Pat. 44294 (1935).

[14] Korobchevskii, I. E., Russ. Pat. 53992 (1938).

the horizontal gallery; air is supplied by one access gallery, and the product gases are withdrawn at the other. As the coal above the horizontal gallery burns away, the fire zone advances upward along the bed. The ash, together with any of the roof which may fall, collects in the space below the fire zone. This process is illustrated in the portion of the drawing labeled "Coal panel in process of gasification." In this manner, the burning coal face is kept free of accumulations of ash, and the airflow is maintained in contact with the burning coal. Roof falls do not readily block the passage or engulf unburned coal. The stream method applied as illustrated in Fig. 3 was the most successful of the methods developed to that time and continues to be one of the important systems for the underground gasification of coal. Combustible gas has been produced from coal in place by installations following this general layout and procedure.

PERCOLATION OR FILTRATION METHOD

The percolation method was suggested in 1939 by the Power Institute of the Moscow Academy of Science.[15–17] The method depends on the presence or the formation of shrinkage cracks and fissures in the coal which render it permeable to gas. These fissures and cracks may be formed by heating or by some other method. The procedure employs the principles suggested in Betts' British patent.[3] A general layout for the percolation method is given in Fig. 4.

Several methods of connecting boreholes 1 and 2 of the percolation method by a passage through the coal bed have been attempted. Much experimentation and investigation have been directed toward the development of this passage. The percolation method and variants are attractive because of the elimination of underground work. The original method suggested for operating the system was to drill and case the boreholes as shown in Fig. 4. The coal was ignited at the base of the boreholes and the product gases were brought up the annulus formed by the air injection pipe and the wall of the borehole. After a time, sufficient coal would be burned away from the base of boreholes 1 and 2 so that a connection was made between them. Then the coal lying between the boreholes was gasified by passing air down borehole 1 and removing the product gases at borehole 2. It was suggested that a jet of oxygen could be used to burn a path between the boreholes. Other methods, including electrical-path preparation procedures and increase of coal-bed permeability by

[15] Chukhanov, Z. F., and Sagaidak, M. Ya., *Bull. Acad. Sci. U.R.S.S., Classe sci. tech.*, **8**, 3–18 (1939).

[16] Kulebakin, V. S., *Avtomat. i Telemekh.*, **4**, 3–6 (1939).

[17] Gumz, W., *Glückauf*, **76**, 670–2 (1940); *Feuerungstech.*, **28**, 273–5 (1940).

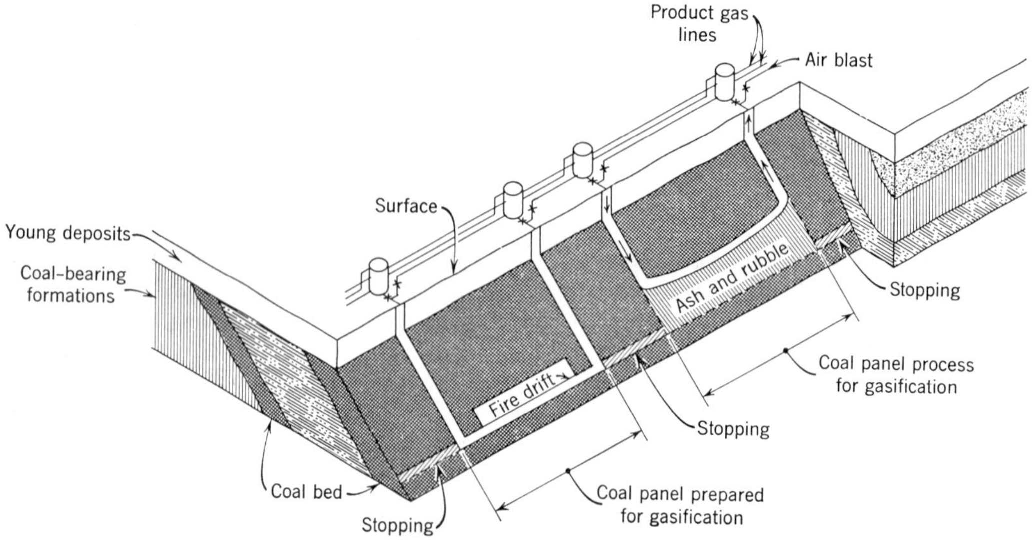

Fig. 3. The stream method.[5]

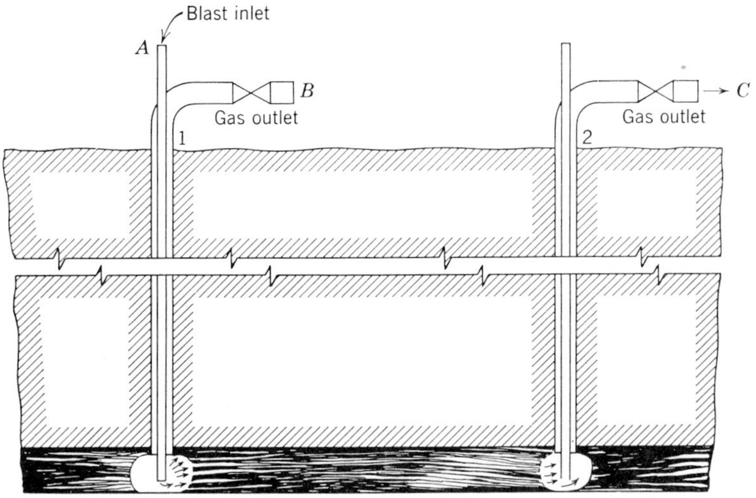

Fig. 4. The percolation method, section through boreholes.[5]

means of coal-bed fracturing techniques, were suggested and will be discussed later.

By 1944, progress had been made toward the solution of some problems confronting underground gasification, and various process requirements had been identified. These included the development of a remote-control drill capable of following a coal bed for 250 yards or more when the bed is inclined at angles ranging from 0 to 90°. This equipment is needed for path preparation in stream-method installations of the type illustrated in Fig. 3.

In unidirectional flow systems, the outlet borehole or passage from the coal bed to the surface requires a lining to eliminate gas leakage and to prevent excessive damage to the strata through which the opening passes. The inlet borehole in a unidirectional flow process does not require a lining. In a process where the flow is reversed, both inlet and outlet boreholes require linings because they are alternately subjected to high temperatures. The design and installation of such linings involve problems in selection of material and in construction methods. Borehole-producer installations require the development of borehole inlet valves capable of remote control and of rugged construction, eliminating all manual maintenance.

In all underground gasification systems, the need was recognized for developing methods of locating the reaction zone without recourse to thermocouples and test drillings. Methods of process control also needed development. The high temperature of the reaction zone and the resultant heat transfer causes disintegration and fusion of the adjacent strata which results in subsidence of the surroundings, blockage of paths, and bypassing of the reaction zone. The characteristics of the ash have an influence on the gasification reactions. The heat loss to adjoining strata decreases gasification efficiencies and may prevent the production of combustible gas. Water in the adjacent strata is very important in the heat loss considerations of underground gasification systems, owing to the magnitude of the latent heat of vaporization.

FIRST EXPERIMENTATION OUTSIDE RUSSIA

Until 1944, Russia alone was actively interested in the development of underground gasification of coal. The work in Russia aroused interest in other countries, and reviews covering much of this experimental work in the period 1944 to 1952 have been published.[18,19] In 1944, interest in Belgium led to the formation of a government subsidized cooperative society, Socogaz. Shortly after, plans were initiated in the United States to proceed with a field-scale experiment. During 1947, possible application of underground

[18] Fies, M. H., *Gasification and Liquefaction of Coal*, AIME, New York, 1953, pp. 122–72.

[19] Organization for European Economic Cooperation, *The Gasification of Coal*, 2 Rue André Pascal, Paris, France, TAR/99(52)1, 1953, 11–45.

gasification was studied in France, and in the following year Charbonnages de France entered into an agreement with Socogaz whereby the subsequent experiments at Bois-la-Dame were made jointly. Another agreement was drawn up with Charbonnages de Pologne, an organization representing coal-mining interests in Poland. The supervision of the experiments under these agreements was entrusted to a Tri-Partite committee, the executive side to Socogaz. In 1949, France began an additional experiment at Djerada in French Morocco. Shortly after the war, a committee to study the process was appointed in Great Britain by the Ministry of Fuel and Power. This committee recommended undertaking experimental work, and by the latter part of 1949, experiments had been started near Chesterfield in Derbyshire. In 1947, Socogaz collaborated with the Valdarno Company and carried out an experimental field test in underground gasification at Banco-Casino in Italy. Later, a second experiment was attempted at Terni.

Figure 5 shows the plan of the first underground gasification experiment at Gorgas, Alabama.[20] The plan of the second underground gasification experiment at Gorgas, Alabama,[20] was quite similar except that a straight line passage was used, the reaction zone was connected to the surface by boreholes, and the undertaking was generally more extensive. The installation plan of the Valdarno experiment at Banco-Casino, Italy,[22] was similar to the first Gorgas experiment.

These field-scale experiments are illustrative of stream-type experiments in a relatively flat-lying bed of coal. In the American experiments, the coal beds were of high-volatile A bituminous rank and 34 to 46 in. thick. The Valdarno experiment was conducted in a lenticular deposit of brown coal which had a thickness varying from 15 to 60 feet. The first

[20] Dowd, J. J., Elder, J. L., Capp, J. P., and Cohen, P., *U.S. Bur. Mines, Rept. Invest.* No. 4164 (1947), 62 pp.

[21] Elder, J. L., Fies, M. H., Graham, H. G., Montgomery, R. C., Schmidt, L. D., and Wilkins, E. T., *U.S. Bur. Mines Rept. Invest.* No. 4808 (1951), 72 pp.

[22] Loison, R., and Venter, J., *Underground Gasification, Russian, Italian, Belgian, and French Experiments*; Centre d'etudes et recherches des charbonnages de France (Cerchar), Paris, France, and Institut National de l'Industrie Charbonnière (Inichar), Liège, Belgium, February, 1952, 37 pp.

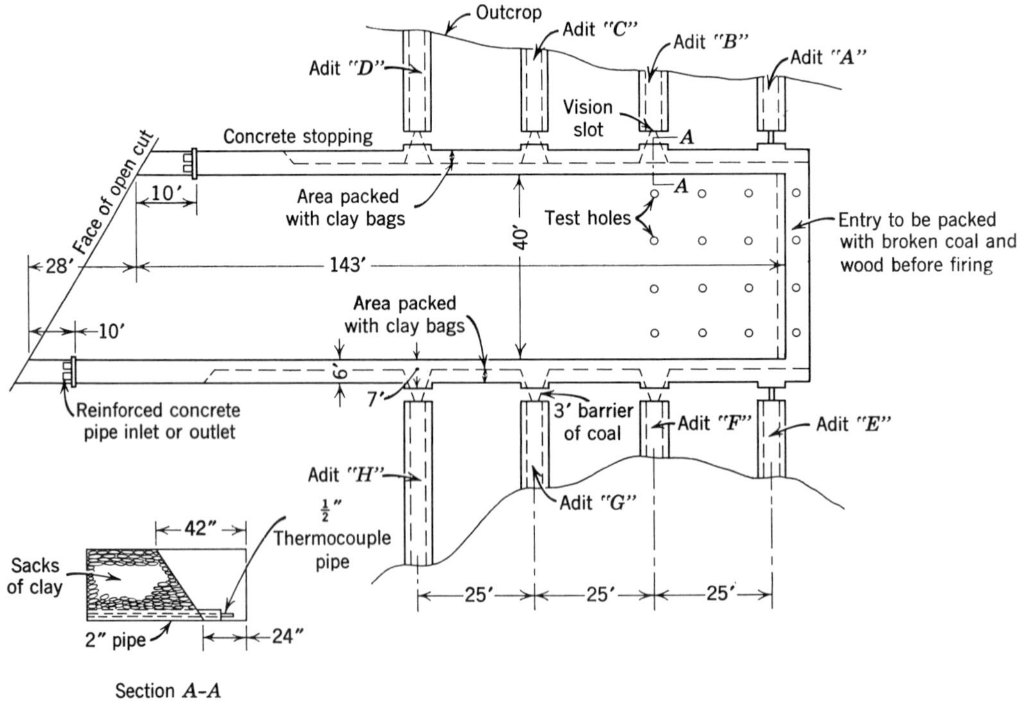

Fig. 5. Plan view of a gasification mine—first Gorgas experiment.[20]

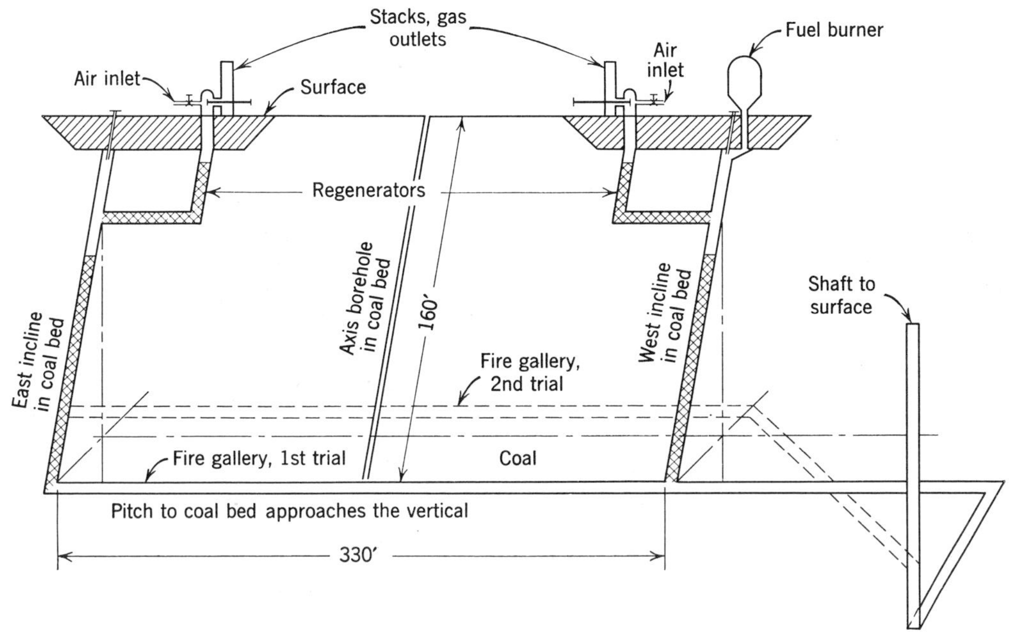

Fig. 6. Pictorial sectional view of stream method installation at Djerada.[22]

Gorgas experiment and the Valdarno experiment used roughly U-shaped passages within the coal bed, and the second Gorgas experiment used a straight-line passage. One or two additional experiments were conducted in the United States by private industry in which approximately straight-line passages were tried. All installations using the stream method in a flat-lying coal bed gave similar results. At times, combustible gases were produced, but in general there was excessive bypassing of the reaction face by the air blast, subsequent mixing of unreacted air with combustible gases, and combustion of the mixture prior to leaving the system. These several installations were operated both on a reversing direction of flow and on a unidirectional flow basis. Detailed results of each may be found in the references given.

In the French experiment at Djerada in Morocco, Fig. 6, the intention was to ignite the coal panel at the bottom and, in much the same manner as the Russians at Gorlovka, to burn the panel from the bottom upward. As shown, the fire gallery is connected to the surface by inlet and outlet shafts, the shafts in turn are filled with refractory rings, and the passages are fitted with regenerators so that in using reversing flow operation some of the sensible heat in the product can be saved and used for preheating the inlet air. In this experiment the operation was continuous for 5 months. A combustible gas was produced, the oxygen content was nil and remained so throughout the experiment, but the percentage of combustible components varied widely. Operational difficulties, such as leakage at the surface and high resistance to flow within the passages were encountered.[23,24]

Figure 7 is illustrative of the Belgian experiment by Socogaz at Bois-la-Dame.[25] The coal is a semianthracite containing about 13% volatile matter. The bed is approximately 3 ft thick, dips at an angle of 87°, and contains approximately 20% ash. In the experimental installation, the fire drift was constructed at an angle of 37°. The objective was to burn the coal panel in a generally sidewise direction instead of upward as in the Russian work at Gorlovka. A combustible gas was produced

[23] Loison, R., *Charbonages de France, Note Technique* **14/53** (1953), 14 pp.

[24] Loison, R., *Underground Gasification Experiment in Djerada, 1st Intern. Congr. Underground Gasification*, Birmingham, Ala., Centre d'études et recherches des charbonnages de France (Cerchar), Paris, France, (1952), 32 pp.

[25] *Underground Gasification, Belgian Experiments at Bois-la-Dame, Liège*, Institut National de l'Industrie Charbonnière, (1952), 81 pp.

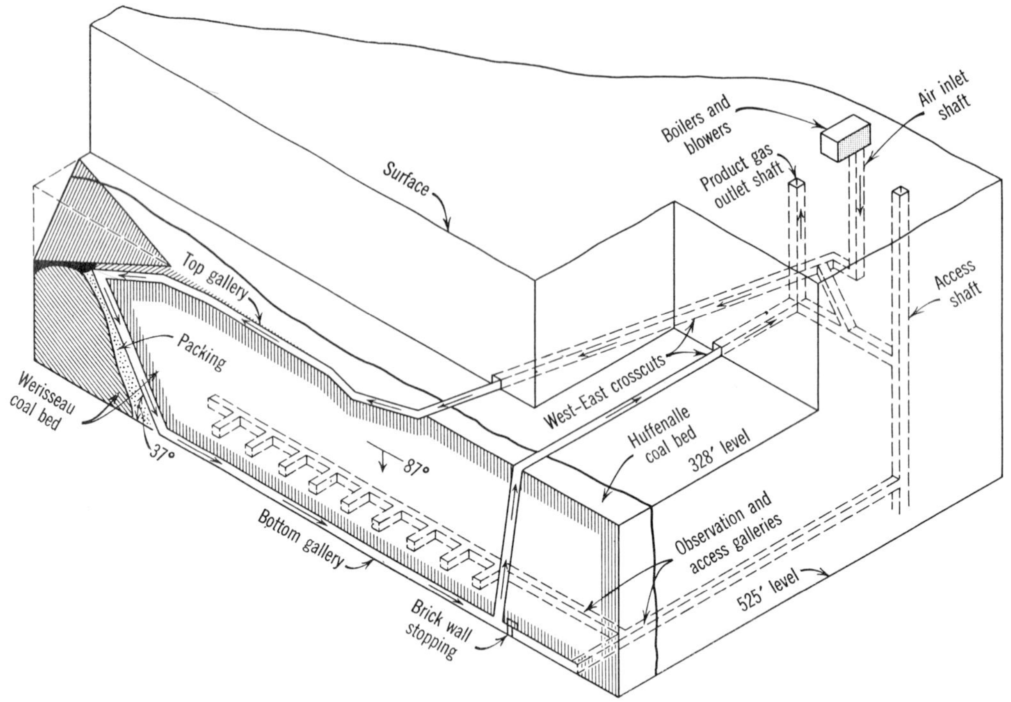

Fig. 7. Isometric sectional view of panel No. 1 at Bois-la-Dame.[25]

at Bois-la-Dame on several occasions. The sidewise burning of the coal panel in this experiment was not successfully achieved with good gasification results.

The British Ministry of Fuel and Power began extensive underground gasification tests in 1949.[26] These tests were conducted in flat-lying coal beds and have been concerned either with modification of the borehole producer method, utilizing horizontal-directed drilling within a coal bed, or with modifications of the percolation method wherein high-pressure fracturing of the coal or electrolinking of boreholes in the coal have been used to form a passage through the bed.

In Fig. 8 the first British experiment on underground gasification at Chesterfield in Derbyshire is illustrated. A 4-inch-diameter hole was drilled from an exposed coal bed face through the coal bed and was connected to the surface by vertical boreholes. This system was

[26] Ministry of Fuel and Power, *British Trials in Underground Gasification, 1949–1955*, H.M.S.O., London (1956), 90 pp.

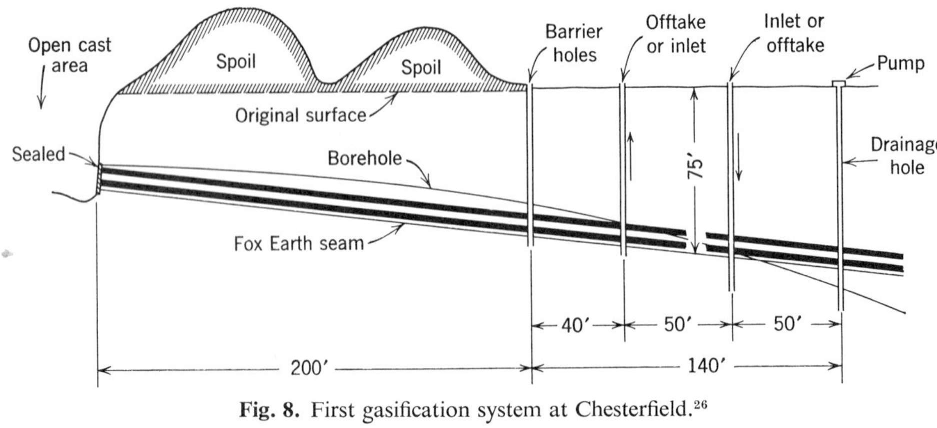

Fig. 8. First gasification system at Chesterfield.[26]

operated using a reversing flow system, and during part of the testing period a product gas having a heating value varying between 60 and 100 Btu per cubic foot was produced.

Laboratory work has been conducted during various phases of underground gasification experimentation. One test demonstrated the action of roof rock during gasification and showed what happens at the coal-roof interface when coal is gasified in place.[20]

Experiments simulating the gasification of coal in place using stream-method gasification in a laboratory-scale model have also been described.[27]

Experimentation with laboratory-scale models was conducted in Belgium by the Socogaz group following the initial experiments at Bois-la-Dame and in Italy. The purpose of the model experiments was to obtain fundamental knowledge of the physical-chemical process of integral gasification of a solid block of fuel. The solid block of fuel consisted of a mixture of coal, cement, and water poured into the model much like ordinary concrete. A passage was provided for the flow of the gas-making fluids. The scale models were 32 in. in diameter and 66 ft long.[28] Similar experiments were conducted in Poland.[29] Laboratory investigations were also conducted in England on the fracturing of coal in place by the application of hydraulic or pneumatic pressures followed by subsequent gasification through the cracks.

UNDERGROUND GASIFICATION SINCE 1951

Underground gasification processes may be classified either as shaftless systems or shaft systems. In shaftless systems, no underground work is required and the systems are essentially modifications or a development of the percolation method. In shaft systems, extensive underground work is usually required, and the systems are generally applications of the stream method in pitching coal beds, as exemplified by the Gorlovka experiment, and modifications of the borehole producer method. In Russia, both the shaft and shaftless systems are being applied, with the emphasis being placed to a larger and larger extent on the shaftless systems. Commercial installations using both methods have been built. In shaft systems, all Russian applications appear to be of the Gorlovka type, that is, application of the stream method to steeply-pitching beds. In Great Britain, experimentation has been conducted on the shaftless systems and on the development of a modification of the borehole-producer method, a shaft system. In the United States, after the second experiment at Gorgas, all experimentation has been conducted on the shaftless systems. In neither Great Britain nor the United States have there been any commercial installations of underground gasification processes.

In the shaftless systems, access openings to a coal bed are prepared by drilling boreholes to the bed. The major problem is the preparation of satisfactory gasification paths through the coal bed connecting the inlet and outlet of the system. In England, Russia, and the United States, two methods of gasification path preparation have been tried: one utilizes the electrolinking of boreholes within the coal bed, followed by electrical carbonization of the path between them and subsequent gasification of the path prepared; the other involves the formation of fractures within the coal bed by the application of hydraulic or pneumatic pressure, then the preparation of the gasification path by a system known as "backwards burning," and finally gasification of the coal adjacent to the path. Russian investigators have stated that approximately similar results were obtained in the U.S.S.R. and in the United States at the Gorgas experimentation site when the operations were conducted in comparable coal deposits.

Several methods of preparing paths within the coal bed were suggested during the early work of the Russian experimentators.[11,30,31]

[27] Fies, M. H., and Elder, J. L., *Laboratory and Field-Scale Experimentation on the Underground Gasification of Coal, Proc. United Nations Sci. Conf. on the Conservation and Utilization of Resources*, August 17–September 6, 1949, United Nations Dept. of Econ. Affairs, **3**, Fuels and Energy Resources (1951), 144–150, 153–4; *Mech. Eng.*, **71**, 850–1 (1949),

[28] DeSmaele, A., *Contribution Towards Underground Gasification of Coal With Oxygen and Steam*, Socogaz, Brussels, Belgium, 1952, 70 pp.; *Ann. Mines Belg.*, **51**, 544–73 (1952).

[29] Dziunikowski, K., *Wiadomości Gornicze*, **1957**, 284–9.

[30] Kirichenko, I. P., *Ugol'*, **101**, 90–8 (1934).

[31] Kirichenko, I. P., *Vestnik Inzhenerov i Tekh.*, **5**, 280–4 (1940).

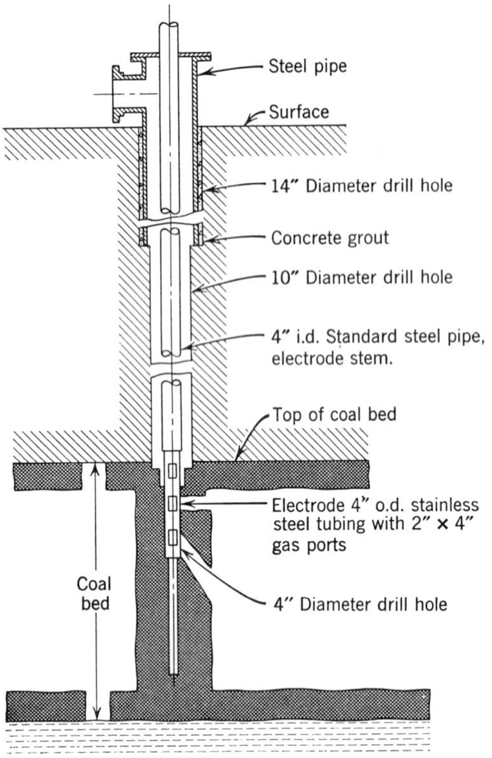

Fig. 9. Installation of electrodes and well construction of boreholes.[33]

Among them was the electrolinking method.

The process of electrolinking, electrocarbonization of coal, was first tried in the United States in the laboratory at the University of Missouri; later, field tests were made under a cooperative contract between the University of Missouri and the Sinclair Coal Company of Kansas City.[32] Subsequently, in cooperation with the Sinclair Coal Company, the Bureau of Mines conducted an additional investigation on the electrolinking and electrocarbonization of coal in place.[33] In England, study and investigation was also directed to the application of the technique to underground gasification processes.[26] In later Russian work, electrolinking-carbonization was successfully applied in commercial installations.

In the process of electrolinking, electrodes are seated within the coal bed at a given spacing, electric current is passed between the electrodes, the coal is carbonized, and by carbonization a path is formed through which gases may be pumped. Figure 9 illustrates the construction of a borehole and the seating of an electrode within the coal bed, as was done in the Gorgas, Alabama, experiments. It should be noted that this was an unlined hole, and the electrode and electrode stem, which in this installation was approximately 150 feet long, was kept in tension by the use of counterweights so that it would not buckle and strike the side of the hole. Counterweighting was done to about 80% of the gross weight of the electrode. In the Gorgas tests, electrode spacings of approximately 150 feet were used successfully. Figure 10 is a plot of resistance versus time during three tests. The general shape of each of these curves is similar, and the initial resistance for each curve is approximately 14 ohms or approximately 0.1 ohm per linear foot between the electrodes. The resistance decreased gradually to 3.5 to 4.0 ohms and then almost instantaneously to a relatively steady state of approximately 1 ohm. The initial rapid decrease from the 14-ohm range was primarily a decrease in contact resistance between the electrodes and the coal bed. The slow decrease in resistance to 3 to 4 ohms should be considered to be the electrolinking period. This is the time when a satisfactory electrical path is established between the two electrodes. The very sharp decrease in resistance from the 3- to 4-ohm value to approximately 1 ohm was attributed to electrical breakthrough, and the final very slow reduction in resistance with time was considered to be the electrocarbonization period of the test. The gases evolved are typical of carbonization products, and their analyses change as the temperatures within the bed increase.

Referring to the plot of resistance versus time (Fig. 10), the curves are quite similar, as stated, except that the period of time required to attain the electrical breakthrough for the three systems is very different. Figure 11 showing electrolinking-carbonization path, illustrates one of the difficulties with electrolinking-carbonization. These paths were determined by test-drilling after gasification of the site had been completed. The path from BH 8 to BH 12 was electrolinked and gasified first. The path from BH 14 to BH 15 was

[32] Forrester, J. D., and Sarapuu, E., *Univ. Missouri, School Mines and Met. Bull.*, No. 78 (1952), 84 pp.

[33] Elder, J. L., Fies, M. H., Graham, H. G., Capp, J. P., and Sarapuu, E., *U.S. Bur. Mines, Rept. Invest.* No. 5367 (1957), 101 pp.

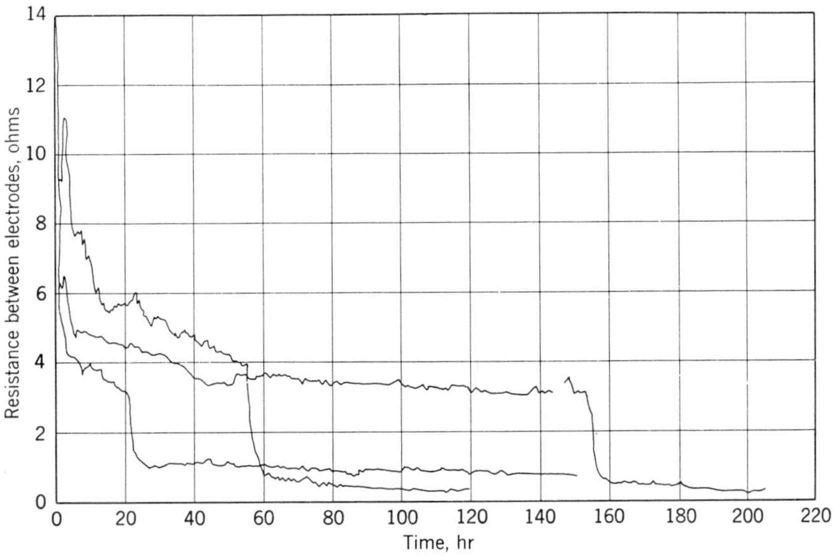

Fig. 10. Resistance time curves for electrolinking-carbonization tests.[33]

electrolinked and gasified second. It will be noted that the actual reaction path was deflected to the left and that BH 15 never actively entered the reaction zone. A path from BH 16 to BH 17 was tried next. Here the actual path formed was from BH 16 to BH 14 at right angles to the path expected. The results show conclusively that the path of the current between two electrodes may be markedly affected by changes in the coal bed resulting from prior experiments.[26,33] Although the unpredictability of electrolinking in the establishment of single gasification paths appears to offer a major difficulty, if multiple paths are established simultaneously, as they would be in commercial applications of area gasification, the path between two individual electrodes may be unimportant.

The nominal linear distance between electrodes on the electrolinking-carbonization tests shown in Fig. 10 and 11 was 150 ft; however, the actual path lengths were considerably different. Estimates made during the tests indicated that about 80% of the current passed through the coal bed during the electrolinking phase, and about 97% during the electrocarbonization phase. The extent of electrocarbonization required to achieve gasification conditions is not well established. It is quite possible that extended electrocarbonization could result in decreased pressure drops through an underground system during the

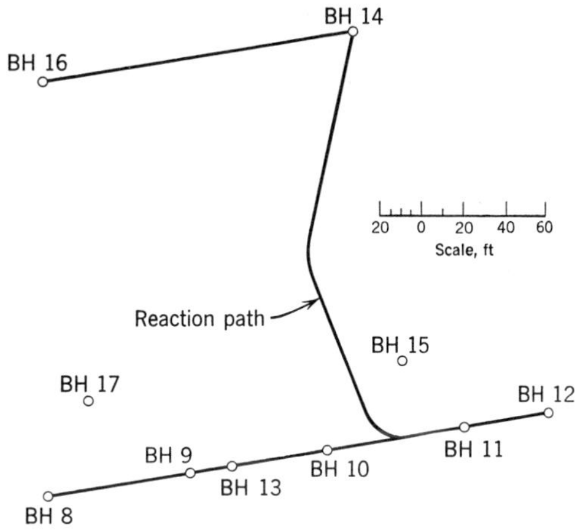

Fig. 11. Plan view of reaction paths initiated by electrolinking.[33]

actual gasification which follows. A great advantage of electrolinking-carbonization, which is not common to any other method of path development, is the fact that the reaction path formed is at a temperature high enough to permit immediate gasification of the carbonaceous material and no further path preparation is necessary.

Coal in place has a natural permeability. If an inlet and an outlet to and from the coal bed are provided, and air under pressure is pumped into the inlet, some air will be recovered at the outlet. The pressure drop and the volumetric recovery are a measure of the permeability of the coal bed. This phenomenon is the basis for another method of path preparation when using a shaftless process for underground gasification of coal. In Russia, in England, and in the United States, attempts have been made to use the normal permeability of coal beds to develop gasification paths within the seam. In England, high-pressure air has been used to effect path development. In Russia, high-pressure air as well as other procedures have been used. In the United States, high-viscosity liquids have been injected into the coal bed at relatively high rates of flow to increase the permeability of the coal bed.

Investigation has shown that the permeability of coal varies with its rank.[34] Brown coals have permeabilities from 20 to 3000 times as great as bituminous coals and probably average about 1000 times as great. This is a factor in applying pneumatic or hydraulic techniques to increase the permeability of a coal bed.

There are advantages and disadvantages in the use of electrolinking-carbonization or methods for increasing coal-bed permeability. The relatively long-path lengths and the fact that the path is already at reaction temperature are advantageous. In the United States, experimentation has shown that the quantity of coal which can be gasified from the electrolinked path is much greater than can be obtained from a path developed by backward burning following hydraulic fracturing. Experimentation in England has shown the reverse. In Russia, electrolinking has been applied industrially, the paths being established quickly and cheaper than by other methods. Backward burning techniques following hydraulic fracturing have one advantage over electrolinking technique: the path can be predicted and controlled more easily. In electrolinking, it has been shown that there is a lack of control in establishing a path between two points, whereas in backward burning the path is directly between the inlet and outlet boreholes. This factor, of course, is extremely important in establishing single reaction paths underground, which have been the nature of the experimentations in the United States and, to a great extent, in England. The factor may be of less importance in a commercial installation where a pattern of multiple paths may be the most economical arrangement.

[34] Fedorov, N. A., *Podzemnaya Gazifikatsiya Ugleĭ*, **1947,** No. 1, 16–21.

In England, the tendency has been to concentrate on the application of horizontal drillings and the development of borehole producers for the gasification of coal in place.[26] In one series of trials, approximately 5000 ft of 4-inch-diameter horizontal hole was drilled at an average rate of 3 ft/hr, varying from a maximum of 8.4 ft/hr to a minimum of 0.8 ft/hr. In these tests, some of the 4-in.-diameter drillholes were reamed to 8 in. in diameter, the reaming rate averaging approximately 2 ft/hr.

In developing a borehole producer-type system, the English experimented with two types of boreholes, (Fig. 12). Borehole (*a*), is open at each end, and the airflow and the reaction zone are shown. Such a borehole would be used between galleries and would be typical of the borehole producer method as previously described (see Fig. 2). Borehole (*b*) is known as a "blind borehole" in which the air is admitted through a metal pipe in the

TABLE 1

SUMMARY OF AVERAGE OPERATING RESULTS, GASIFICATION WITH AIR, UNIDIRECTIONAL FLOW[33]

	Peak Operation	Entire Operation
Nominal length of system, ft	155	155
Operating life of unit, days	21	59.1
Air input, cfm*	1890	1160
Input pressure, psig	27	30
Leakage of input air from system, percent	5.2	13.8
Gas production, cfm*	2160	1170
Composition of dry product gas, %:		
Carbon dioxide	9.7	10.5
Illuminants	0.3	0.3
Oxygen	0.5	0.9
Hydrogen	9.2	8.4
Carbon monoxide	13.1	10.7
Methane	1.5	1.8
Nitrogen	65.7	67.4
Specific gravity	0.94	0.95
Heating value, Btu/cu ft*	93	86
Moisture content, Mols H_2O per mol dry gas	0.11	0.15
Rate of coal utilization,† tons of moisture- and ash-free coal per day:		
Completely gasified	11	6
Carbonized only	16	11
Total affected	27	17
Gas produced per ton of moisture- and ash-free coal completely gasified, cu ft × 10^3*,†	280	290

* 60°F, 30 in. Hg, dry.
† Calculated from material balances.

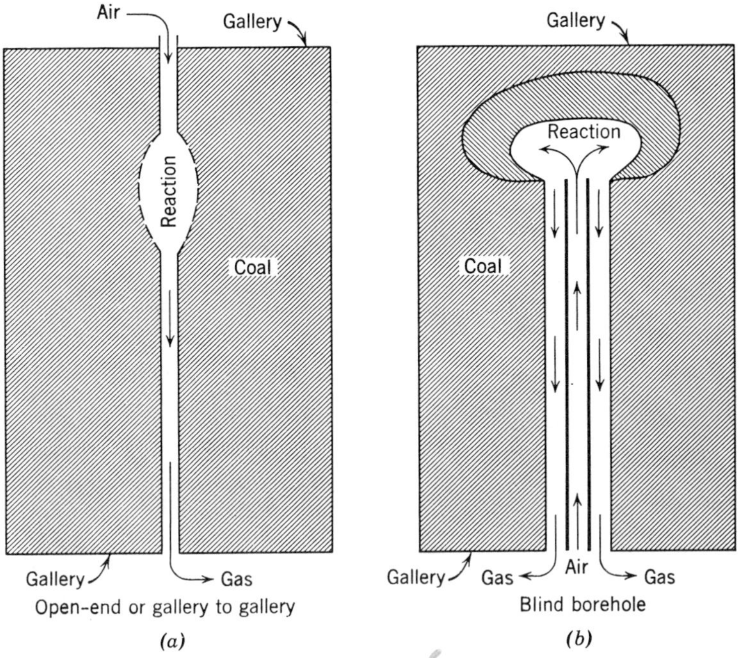

Fig. 12. Types of horizontal borehole installations—borehole producer.[26]

center of the borehole, combustion and gasification occur at the inby end, and the gases are brought out the annulus around the pipe. It is intended that the metal pipe burn off as the coal is consumed and reaction zone approaches the gallery.

The underground gasification of coal is essentially a batch operation. As operation is continued, and more and more coal is removed from the vicinity of the original reaction zone, difficulty is encountered in maintaining contact between the reactants, finally resulting in deterioration in product quality and making it necessary to move to a newly established reaction path. Air or oxygen can be used continuously for gasification, or intermittently with steam for water-gas operation. Generally, any gasification process now in use above ground can be made to work using coal in place.

The quality of the gas obtained from underground gasification depends on many factors. When coal in place is gasified with air, using unidirectional flow, the operating results in Table 1 are typical. These results could be improved by preheating the air and developing methods of controlling the water inflow to the reaction zone to reduce the heat losses from the system.

Oxygen can be used to produce a product having a higher content of carbon monoxide and hydrogen that might be suitable for synthesis of liquid fuels or organic chemicals. The results given in Table 2 are for unidirectional flow, using oxygen for gasification. This table is illustrative of the general results that may be expected. The data presented are from a short run as no data are available from longer operating periods.

Results obtained from a series of water-gas-type runs with intermittent air-blow and steam-run are given in Table 3. These results are qualitative since they do not represent average data performed for a large number of tests. A heat-balance problem is encountered in using this type of operation for at the present time it is not known how long the air-blow should be conducted to meet the heat requirements of the steam run.

In Tables 1, 2, and 3, coal utilization in tons per day is expressed as that completely gasified, carbonized only, and total affected. These values were calcuated using simultaneous equations developed from the material balances.[33] Other workers in the field of underground gasification have also developed methods whereby the quantities in question can be evaluated.[22,24,25,28]

TABLE 2

GASIFICATION WITH OXYGEN[33]

Operating time, hr	36
Oxygen input rate, cfm*	604
Composition of product gas, %:	
Carbon dioxide	47.8
Illuminants	0.3
Oxygen	0.2
Hydrogen	24.5
Carbon monoxide	21.2
Methane	4.1
Nitrogen	1.9
Heating value, Btu/cu ft*	195
Specific gravity	1.0
Moisture content, mols H_2O/mol dry gas	0.34
Product gas flow, cfm*	1335
Oxygen usage:*	
Cu ft/M of make gas	452
Cu ft/M of ($CO + H_2$)	990
Cu ft/10^6 Btu in product gas	2320
Coal utilization, tons of moisture- and ash-free coal per day:†	
Completely gasified	33.1
Carbonized only	0

* 60°F, 30 in. Hg, dry.
† Calculated from material balances.

Figure 13 is a plot showing the coal utilization for an underground system wherein the gasification was conducted using unidirectional air flow. Curves 1, 2, and 3 show the relationship between the coal gasified, the coal carbonized, and the total coal affected. Curve 4 is a calculation based on carbon balance only, which is the value sometimes used to estimate the quantity of coal gasified. In all underground gasification systems, a certain amount of coal is completely utilized and some is devolatilized only. The effect is important in determining the quantity of coal that can be obtained from a given deposit. The determination of these quantities was based on the operation of the single-path system. It is not known how these quantities would be affected when multiple-path and commercial-type operations are used. The curves as given in Fig. 13 are of a qualitative nature. They are probably not of general application but they do point out the possibility of a significant fraction of the coal being devolatilized only. It seems likely that operating conditions can be achieved that will minimize this effect.

TABLE 3

WATER-GAS OPERATION, AIR-BLOW AND STEAM-RUN[33]

	Air-Blow	Steam-Run	Air-Blow	Steam-Run	Air-Blow	Steam-Run
Individual cycle time, hr	18.8	4.2	6	6	6	6
Input rates:						
Air, cfm*	1240		2520		2500	
Steam, lb/hr		1480		1150		1200
Product-gas output rate, cfm*	1850	585	2990	250	2580	190
Composition of product gas, %:						
Carbon dioxide	13.0	20.8	16.2	35.8	16.0	40.8
Illuminants	0.3	0.4	0.4	0.6	0.3	0.4
Oxygen	0.3	0.4	0.4	0.3	0.6	0.6
Hydrogen	13.4	47.9	3.5	24.1	2.2	21.0
Carbon monoxide	10.1	12.0	2.1	8.9	1.3	6.9
Methane	3.0	7.6	2.0	7.5	1.2	7.2
Nitrogen	59.9	10.9	75.4	22.8	78.4	23.1
Heating value, Btu/cu ft*	112	279	47	195	30	172
Specific gravity	0.91	0.62	1.02	0.92	1.04	0.98
Moisture content, mols H_2O/mol dry gas	0.19	0.56	0.22	1.23	0.24	1.78
Coal utilization, tons/day†						
Completely gasified	10.3	6.3	12.3	4.7	9.3	3.7
Carbonized only	14.4	1.4	22.5	—	19.1	—
Total affected	24.7	7.7	34.8	4.7	28.3	3.7

* 60°F, 30 in. Hg, dry.
† Basis, moisture- and ash-free coal; calculated from material balances.

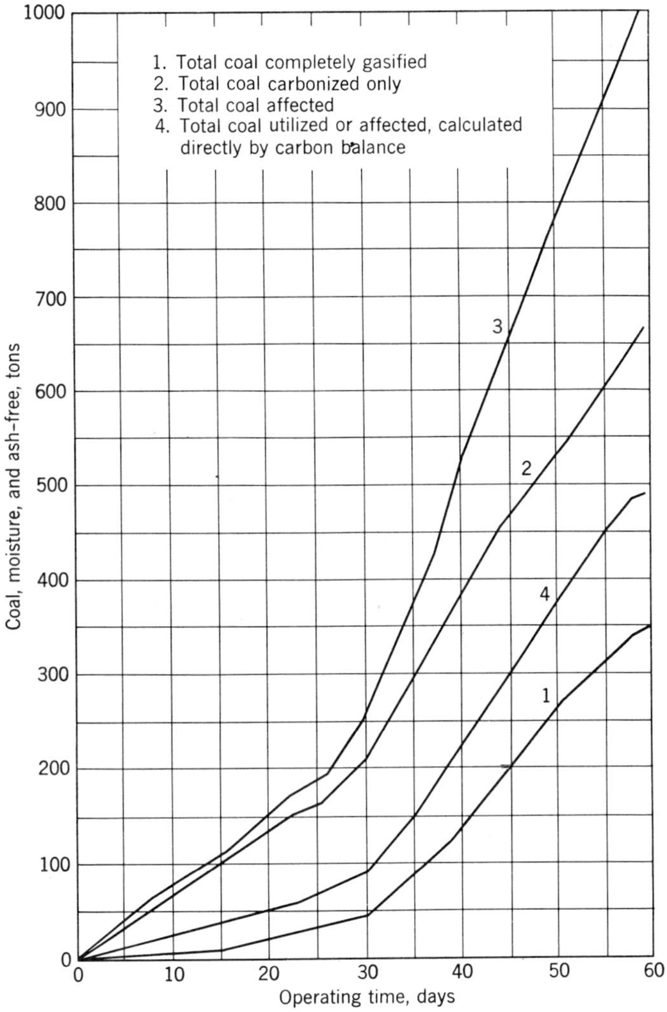

Fig. 13. Coal utilization.[33]

Heat losses occur because of both water inflow into the reaction zone and water in the surrounding strata. The latter can cause heat loss because of its vaporization, subsequent condensation, and re-evaporization. Water which enters the reaction zone not only causes heat losses but, by the water-gas reaction, results in increased hydrogen content in the effluent gas.

The problem of gas leakage from a site is unique to each individual installation of underground gasification. Steps must be taken at each installation to minimize leakage effects. From what is known, it seems probable that underground gasification will have to be conducted at operating pressures of 50 psig or less. Underground gasification has never been studied where gas leakage was not a factor, and it is not known what might be accomplished at elevated pressures.

COMMERCIAL APPLICATION OF UNDERGROUND GASIFICATION OF COAL

Industrial plants for the underground gasification of coal have been installed only in the U.S.S.R. Table 4 lists the installations where underground gasification is being conducted or planned.[35] Two of these plants have been in operation for a period of years, the Moscow

[35] Bakulev, G. D., *An Economic Analysis of Underground Gasification of Coal*, Acad. Sci. U.S.S.R., Moscow, 1957, 133 pp.

TABLE 4

PLANTS FOR UNDERGROUND GASIFICATION OF COAL[35]

Plant	Location	Type of Operation	Production, billion cu ft/yr*	Heating Value Btu/cu ft*	Production Cost Ratio: Moscow Region Plant = 1†
Plants in commercial operation					
Moscow region	Moscow basin	Shaftless	15.6	85	1
Lisichansk	Donbas	Shaft	4.3	85	2.2
Plants under construction					
Shatskaya‡	Moscow basin	Shaftless	25.1	94	0.55
Yuzhno-Abinskaya‡	Kuzbas	Shaft and shaftless‖	22.1	115	0.33
Agrenskaya§	Ukbekistan		87.9	105	0.40
Kamenskaya	Donbas	Shaftless	13.8	94	0.52
Plants planned for the period 1956–1960					
Yuzhno-Kuzbaskaya	Kuzbas	‖	303	126	0.11
Stalinskaya	Kuzbas	‖	151	126	0.12
Yuzhno-Abinskaya (new)	Kuzbas	‖	173	126	0.11
Gryzlovskaya	Moscow basin	‖	119	94	0.26
Severo-Tul'skaya	Moscow basin	‖	18.7	94	0.62

* Measured at 60°F, 30 in. Hg, saturated.
† Thermal basis, delivered to user.
‡ Experimental panels operating since 1955.
§ Reported in commercial operation, December 1959.
‖ Type of operation not stated, probably most are shaftless.

region plant since 1947 and the Lisichansk plant since 1950. Two of the plants under construction have operated experimental panels since 1955, and in effect they have been in limited commercial production for the product has been distributed to users. One plant, Agrenskaya, the largest of those groups listed as operating or under construction, was reported in operation in December 1959.

The production, heating-value, and relative-cost columns are average operating figures or design estimates. The heating value of 115 Btu/cu ft estimated in the design for the Yuzhno-Abinskaya plant can be compared to a 10-mon 1956 average of 119 Btu/cu ft for gas produced by the experimental panel; the panel production for this year was 4 billion cu ft.

The cost of production of gas is expected to decrease as new plants are installed. In the largest of these planned installations, costs are expected to be about one-ninth of those now realized at the Moscow region plant. It is expected that the gasification costs will be less, possibly one-half, than the cost of mining coal in areas adjacent to these plants. Much of the gas produced will be used in electric power plants.

The overall gas production of the Moscow region and Lisichansk plants is given in Table 5. The average heating value for the product gas from the two plants is also included. The gas produced at the Lisichansk plant has a slightly higher heating value than that of the Moscow region plant, but the values are so close that the average is essentially correct.

The Moscow region plant[36] operates on coal of the Novo-Basovsk bed, which is a brown coal. It contains approximately 30% moisture, 37% ash, and has a heating value of 4900 Btu/lb. The bed ranges in thickness from $1\frac{1}{2}$ to 15 ft, but coal less than 3 ft thick is not gasified. The plant uses a shaftless method of underground gasification, using the natural permeability of the coal bed which may be increased by use of high-pressure air or electrolinking-carbonization. When pneumatic methods of

[36] Kulish, F. D., *Podzemnaya Gazifikatsiya Ugleĭ*, **1957**, No. 2, 27–31.

TABLE 5

Commercial Production of Gas by Underground Gasification of Coal in the U.S.S.R.[35]

Year	Production, Billions of cu ft/yr* Moscow-region Plant	Lisichansk Plant	Total	Average Heating Value, Btu/cu ft*
1947	1.9		1.9	80
1948	4.6		4.6	83
1949	6.4		6.4	88
1950	7.8	3.7	11.5	93
1951	8.3	1.0	9.3	87
1952	9.0	2.8	11.8	83
1953	11.0	4.1	15.1	85
1954	13.4	3.4	16.8	83
1955	15.6	4.1	19.7	85
1956	15.2	4.3	19.5	85
Total	93.2	23.4	116.6	85

* Measured at 60°F, 30 in. Hg pressure, saturated.

increasing the permeability are used, backward burning is applied to develop the reaction zone.

Gas production at the Moscow region plant had reached approximately 90% of designed capacity in 1955–6. In 1955, the annual gas production of the plant was 15.6 billion cu ft of gas (see Table 5). This gas may be estimated to be adequate to operate a 15,000-kw electric generating station on the basis that 10,000 Btu is required to generate 1 kwhr of electric energy.

The Moscow region plant in operation uses a system of boreholes spaced approximately 75 ft apart and established on a quadratic pattern. The production of 45 million cu ft of gas per day requires at least four generators, each composed of not less than eight boreholes operating simultaneously. Each generator is prepared for regular operation in two stages: (1) four boreholes in a row are ignited and linked to each other, forming a 225-ft firefront; and (2) four boreholes in a second row, 75 ft distant, are ignited and linked to the corresponding borehole in the firefront of the first row (see Fig. 14). The result is a generator composed of four parallel gasification passages, each 75 ft long, terminating at right angles in a firefront 225 ft long. Gasification proceeds by blowing air at low pressure. The air enters the boreholes in the second row and the gas is removed from the boreholes in the first row.

While gasification is being conducted, linkage is being established between a third row of boreholes and the second row. When the gasification and establishment of this linkage are completed, gasification begins between the third and second rows. Additional rows of boreholes are linked and gasified during the continuing operation of the generator.

The underground gasification plant at Lisichansk operates in the Donbas,[37] some 4 to 5 km from the Donetz river. Until recently, this plant used the shaft method of gasification which is similar to the experimental installation previously described at Gorlovka (see Fig. 3). The coal is bituminous and has a heating value of 11,600 Btu/lb. As shown in Table 5, the annual production of gas from the plant in 1956 was 4.3 billion cu ft. This gas production is adequate to operate a 4300-kw capacity electric generating station, again using a factor of 10,000 Btu required to produce 1 kwhr of electricity.

Table 6 compares the equivalent cost of gas produced from the Moscow region plant and the Lisichansk plant in terms of standard fuel.[35] The cost of coal in terms of standard fuel is also included for comparative purposes. It

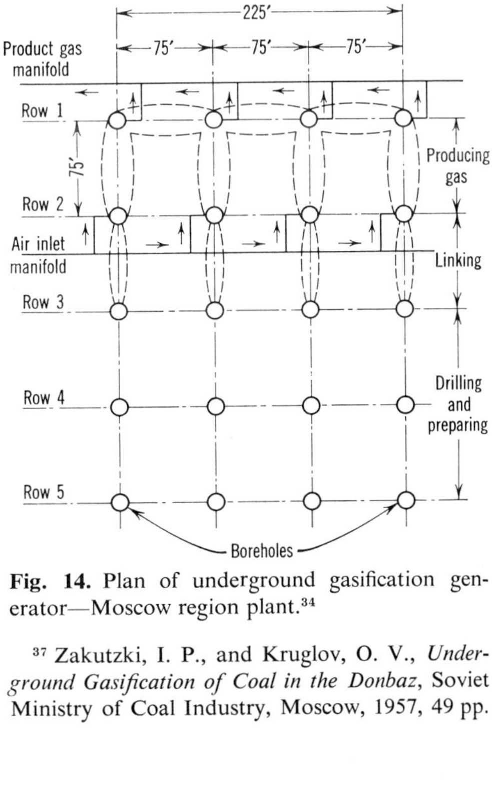

Fig. 14. Plan of underground gasification generator—Moscow region plant.[34]

[37] Zakutzki, I. P., and Kruglov, O. V., *Underground Gasification of Coal in the Donbaz*, Soviet Ministry of Coal Industry, Moscow, 1957, 49 pp.

TABLE 6

COMPARISON OF PRODUCTION COSTS AND COST DISTRIBUTION[35]

	Gas from Underground Gasification of Coal		
	Moscow-Region Plant	Lisichansk Plant	Coal*
Equivalent cost per ton of standard fuel,† rubles	252	545	180
Ratio to coal	1.40	3.03	1
Production cost distribution, %			
Wages and salaries	15.0	10.0	60.7
Materials	8.0	2.0	18.7
Fuel	—	—	0.8
Electric power	40.0	24.0	3.4
Amortization	20.0	39.0	6.7
All other	17.0	25.0	9.7
Total	100.0	100.0	100.0

* Coal mined by Tulaugol combine, Moscow basin lignite.

† One ton of standard fuel is defined as 1 metric ton having a calorific value of 7×10^6 kcal.

will be noted that the gas produced from the Moscow region plant is approximately 40% more expensive than the cost of coal mined in the area. The Lisichansk plant produces gas which is about three times as expensive as coal, and gas from this plant is about 2.2 times as costly as that from the Moscow region plant. Table 6 includes a cost distribution, which shows that amortization costs at the Lisichansk plant are large.

In the same area as the Moscow basin plant, gas production by surface generators (using 1 ton of standard fuel for comparative purposes) costs about 50% more than coal. Thus, gas produced by underground gasification of coal is somewhat cheaper in the area.

It is difficult to compare the costs for industrial operations in one section of the world with similar operations in another. Table 6 compares the cost of gas produced by underground gasification with the cost of coal as mined. The productivity of labor goes part way toward establishing a basis for comparing such costs. In underground gasification in the Moscow region plant, the productivity of labor is about 30 tons/man-month in terms of this coal. At a nearby mine having an equivalent capacity, the 1954 coal production was 33.4 tons/man-month.[36] Although this is not in agreement with the wage and salary data of Table 5, Kulish estimated that increasing the Moscow region plant capacity by a factor of 2.5 would increase the labor productivity to 100 tons/man-month.

Processes of underground gasification are still in their infancy as far as commercial application is concerned. Workable processes have been developed and are currently in limited application. Further development can be expected when the economic environment is suitable. Such cases may occur where geological conditions make the cost of conventional mining excessive and fuel from other sources either too expensive or unattainable. In certain limited cases where the use of gas is mandatory and where the cost of producing gas is higher than the cost of producing an equivalent quantity of heat by conventional mining but the overall cost of mining plus gasification is higher than for underground gasification, such application development may be expected. At present, it seems likely that the underground gasification of lower-rank coals is preferable to the underground gasification of bituminous coals.

CHAPTER 22

HYDROGENATION OF COAL AND TAR

E. E. DONATH

Koppers Company, Inc., Pittsburgh, Pennsylvania

At the time of the writing of Volume 2 of *The Chemistry of Coal Utilization* in 1939, the annual production of liquid fuels from coal and tar amounted to 1,150,000 tons in seven plants in Germany and to 150,000 tons in Billingham in Great Britain. The total number of hydrogenation plants in Germany increased during the war to twelve, and the total capacity to about 4 million metric tons/year, corresponding to approximately 100,000 barrels/day of finished products, mostly aviation and motor gasoline. None of the hydrogenation plants that remained in the Western world are at present being used for the hydrogenation of coal or tar. The plant in Billingham was shut down in 1958. Three plants in West Germany are used for the hydrogenation and refining of petroleum; one is shut down. Only in Eastern Europe and Siberia are coal and tar hydrogenation plants still in operation.

This is not only a consequence of World War II but even more an indication of the change of the economic conditions. Instead of the expected shortages of petroleum and the exhaustion of petroleum reserves, there are huge new oil fields and an abundance of petroleum. Simultaneously, coal prices in many countries have increased drastically. Thus the question arises "can coal hydrogenation become economically attractive in the future?" In view of the present prices of coal and hydrogen produced from coal in countries that possess or have free access to crude oil there is very little prospect of economic production of *gasoline* or *liquid fuel* from coal, unless a drastic change in the relationship of availability and price of crude oil and coal occurs. The status of coal and tar hydrogenation plants in countries of Eastern Europe will illustrate this. Three plants in the Soviet zone of Germany and one in Czechoslovakia are in operation. Equipment from dismantled German plants has been used to erect a large coal hydrogenation plant for the production of aviation fuels in Siberia. The local scarcity of petroleum made the operation of these plants attractive. However, it appears that, more and more, crude oil will be used in these plants following the western trend.

There remain at least two areas in which the coal hydrogenation process may have promise in the not too distant future in the western world.

The first possibility is the production of commodities that command a higher price than fuels, such as aromatics, tar acids, special fuels, and other chemicals. The other possibility is the production of natural gas substitutes. It seems that transportation costs and the ratio of consumption to reserves of natural gas will lead to price increases of natural gas, especially in certain regions. On the other hand, simplified techniques for the hydrogenation of coal to a natural gas substitute are being developed and may lead to reduced cost of such a gas.

Commercial Plant Developments

The first commercial coal-hydrogenation plant was built at the Leuna works of the I.G. Farbenindustrie A.G. The research and development[1] work for the process was done at the BASF[2] plant in Ludwigshafen. This

[1] Pier, M., *Z. Elektrochem.*, **57**, 456–60 (1953).

[2] Abbreviations:

ANIC—Italian Government owned synthetic liquid fuels company.

BASF—Badische Anilin und Sodafabrik. A.G., in Ludwigshafen, Rh, Germany; until 1945, part of I.G.

I.C.I.—Imperial Chemical Industries, England; special reference is to the Billingham plant.

I.G.—Farbenindustrie A.G., Frankfurt a.M. Germany; liquidated in 1945.

I.H.P.—International Hydrogenation Patents Company with offices in The Hague.

TABLE 1

GERMAN COAL AND TAR HYDROGENATION PLANTS[3]

Start-Up Date	Plant	Present Location	Raw Material	Pressure atm	Annual Capacity Products 1943/44 Metric Tons	Present Status
1927	Leuna	Central Germany	Brown coal Brown-coal tar	250	650,000	In operation
1936	Böhlen	Central Germany	Brown-coal tar	300	250,000	In operation
1936	Magdeburg	Central Germany	Brown-coal tar	300	220,000	Dismantled
1936	Scholven	West Germany	Bituminous coal	300	280,000	Coke-oven light-oil refining Crude-oil hydrogenation
1937	Welheim	West Germany	Pitch	700	130,000	No hydrogenation
1939	Gelsenberg	West Germany	Bituminous coal	700	400,000	Crude-oil hydrogenation
1939	Zeitz	Central Germany	Brown-coal tar	300	280,000	In operation
1940	Lützkendorf	Central Germany	Tar, oil	500	50,000	No hydrogenation
1940	Pölitz	Polish administration	Bituminous coal, oil	700	700,000	Dismantled
1941	Wesseling	West Germany	Brown coal	700	250,000	Crude-oil hydrogenation
1942	Brüx (Most)	Czechoslovakia	Brown-coal tar	300	600,000	In operation
1943	Blechhammer	Polish administration	Bituminous coal, tar	700	420,000	Dismantled

laboratory was the center of an international exchange of know-how, catalyst and process data between I.G., Standard Oil of New Jersey, I.C.I., Shell Oil, and later ANIC; it was effective until World War II. This exchange was formalized by the founding of the Dutch I.H.P. and contributed materially to process development and erection of commercial plants.

The German coal and tar hydrogenation plants are listed in Table 1. A brief characterization of these plants follows.[3]

At first, the Leuna plant produced synthetic ammonia and methanol at high pressure, and the production of synthesis gas for these processes was easily adapted to the production of hydrogen for the coal hydrogenation process. Originally built for a capacity of 100,000 metric tons of gasoline per year at 200 atm pressure, the plant subsequently was enlarged to a capacity of 650,000 metric tons[4] gasoline/yr at 250 atm[4] pressure. The raw materials for this plant were Central German brown coal and brown-coal tar. The main products of the plant were aviation and motor gasoline.

In 1936, operation of tar-hydrogenation plants at Böhlen and Magdeburg was begun. Tar from low-temperature carbonization of Central German brown coal was the raw material. These plants were integrated with power plants that used the char as fuel. Aviation and motor gasoline were produced at 300 atm pressure by liquid phase and two-stage vapor-phase hydrogenation.

A third brown-coal tar-hydrogenation plant was started in 1939 in Zeitz. The process was different in that a fixed-bed catalyst was used for the hydrogenation of brown-coal tar to yield Diesel fuel, lubricating oil and paraffin wax. This so-called TTH process can be regarded as a forerunner of the hydrorefining process for petroleum fractions. The same process is used in Spain for the refining of shale oil in a plant in Puertallano that began operation in 1956.[5]

Another plant for the hydrogenation of brown-coal tar was built in Czechoslovakia, again in combination with a power plant. The tar from the Bohemian coal used in this plant is considerably lower in hydrogen content than was Central German brown-coal tar.

[3] Hoering, M., and Donath, E. E., *Ullmanns Encyklopädie der Technischen Chemie*, Third edition, Urban and Schwarzenberg, München, 1958, Vol. 10, pp. 483–569, see p. 488.

[4] 650,000 metric tons/yr = 16,000 bbl/d; 100,000 metric tons gasoline/yr = about 850,000 barrels/yr or 2500 barrels/day; the pressure for the German hydrogenation plants is given in "technical atmospheres = atm = 1 kg/cm^2 = 14.3 psi."

[5] Reitz, O., and Kohrt, H., *Erdöl u. Kohle*, **11**, 18–22 (1958); Sanchez, R., and Rodriguez-Jurado, J., *World Petrol. Congr., Proc. 4th Congr., Rome*, **1955,** Sect. III, pp. 445–56.

TABLE 2

SYNTHETIC FUELS PRODUCTION IN EAST GERMANY[7]

	Gasoline 1000 metric tons/year	Diesel Fuel 1000 metric tons/year	Paraffin Wax 1000 metric tons/year
1949	460.0	200–240	—
1950	491.7	437.1	—
1951	603.2	521.8	—
1952	672.1	562.4	132.8*
1953	762.0	630.0	164.2*
1954	—	—	—
1955	830.0	698.0	—
1956	883.0†	830.0†	—

* Includes crude paraffin wax.
† Estimated.

Three plants in the Soviet zone of Germany and in Czechoslovakia (Table 1) are in operation. Two were dismantled after World War II, and the equipment was used in a coal hydrogenation plant of 1,000,000 metric tons/yr capacity of aviation gasoline in Siberia,[6] as mentioned before.

In the Leuna plant after World War II, hydrogenation of brown coal and brown-coal tar was resumed and the hydrogenation of petroleum oil added. In Böhlen and Zeitz, brown-coal tar is being hydrogenated. The production figures are shown in Table 2.[7]

In 1936, operation of the Scholven plant was begun for the hydrogenation of bituminous coal from the Ruhr district at 300 atm pressure. Since experiments in the BASF laboratories had indicated that hydrogenation of aromatic raw materials like bituminous coal and tar can be more readily achieved at higher pressure than 300 atm, subsequent plants for these raw materials were built in Germany with a liquid-phase unit at 700 atm pressure.

The first such plant was built at Welheim for the hydrogenation of pitch and the production of gasoline and heavy fuel oil. It was unique in one respect; a pressure of 700 atm was used for the hydrogenation in a single-stage vapor phase operation, to produce a highly aromatic gasoline. This plant also contained a demonstration unit for the extraction of coal according to the Pott-Broche process.[8] The plant was not reactivated after the war.

A plant for the hydrogenation of tars from brown coal and of petroleum oils of small capacity at high pressure was erected in Lützkendorf in Central Germany. This plant ceased operation after the war. The first coal-hydrogenation plant for the conversion of bituminous coal to gasoline at 700 atm pressure was built at Gelsenberg for the hydrogenation of Ruhr coal. The 700-atm plant at Pölitz was originally designed for the hydrogenation of cracking residues. However, during the war, the raw materials for this plant were bituminous coal, shale oil, and crude oil.

The hydrogenation plant at Wesseling used brown coal from the Cologne district as raw material. This brown coal of lower hydrogen and tar content than Central German brown coal required 700 atm pressure in the liquid-phase hydrogenation stage.

A plant for the hydrogenation of Upper Silesian coal at 700 atm was erected in Blechhammer. In this plant, coal was converted to gasoline and heavy fuel oil used by the Navy. Although in all other coal hydrogenation plants heavy oil was used as pasting oil, this plant used a blend of heavy oil and middle oil.

Analyses of the coals used as raw materials in these hydrogenation plants are shown in Table 3. The brown coals from the two districts shown were quite different in hydrogen content. The Central German brown coal with its higher-hydrogen content as well as greater-tar yield and montan wax content was very suitable for the hydrogenation process. It was readily converted to liquid fuels at 250 atm with a lower hydrogen consumption than required for other coals. The bituminous

TABLE 3

COALS USED IN HYDROGENATION PLANTS[3]

	Brown Coal		Bituminous Coal		
	Central Germany	Cologne District	Ruhr	Ruhr	Upper Silesia
Ash (% in dry coal)	12.8	5.9	4.0	3.0	6.6
Volatile matter (% maf coal)	57.9	53.5	38.5	37.5	37.8
Carbon (% maf coal)	71.9	68.7	83.8	83.8	81.9
Net H g/100g C	4.1	2.4	4.5	4.6	3.9

[6] Anon., *Oil Gas J.* **50,** No. 51, 168 (1952).

[7] Anon., *Chem. Tech.* (*Berlin*), **11,** No. 9, 503 (1959) and private communication.

[8] Broche, H., and Reinmuth, W., *Ullmanns Encyklopädie der technischen Chemie,* Third edition, Urban and Schwarzenberg, München, 1958, Vol. 10, pp. 570–5.

TABLE 4

TARS AND OILS USED IN HYDROGENATION PLANTS[3]

	Brown-Coal Tar		Bituminous-Coal Tar			
	Central Germany	Czecho-slovakia	Low-Temperature Carboni-zation	Coking	Shale Oil (asphaltic)	Crude Oil Mixed Base
Specific gravity at 50°C	0.929	1.020	1.048	1.156	0.974	0.877
Asphaltene, %	1.3	11.3	5.3	16.7	6.5	0.4
Boiling range, wt. %						
−200°C	1.8	—	5.2	—	1.8	9.4
−325°C	40.0	28.8	38.3	22.4	27.7	21.8
Residue above 325°C	58.1	71.1	57.4	77.6	70.4	68.7
Carbon, wt. %	83.4	81.8	85.7	91.2	82.3	85.1
Net H g/100g C	12.9	10.5	9.2	6.2	11.7	14.6
Residue above 325°C						
Specific gravity at 100°C	0.913	1.000	1.086	1.173	0.988	0.884
% boiling above 325°C in vacuo	30.1	23.4	28.2	42.6	31.1	63.1

coals were all of low rank. Again the difference in hydrogen content between coals from the Ruhr and Upper Silesia was reflected in the hydrogenation behavior as will be shown later.

Properties of the tars and oils used as raw materials are shown in Table 4. A wide range of raw materials was used. In general, the suitability for the hydrogenation process improves with increasing hydrogen content and decreasing content of asphaltenes and vacuum residue.

Three coal-hydrogenation plants in West Germany were converted after the war to the processing of other materials under high pressure. Some equipment was used for the production of ammonia and methanol. In the Scholven plant the catalytic hydrofining of coke-oven light-oil began in 1950. In the Gelsenberg, Scholven, and Wesseling plants the production of fuels was subsequently resumed, and the plants were equipped for the hydrogenation and the refining of crude oil. Recent capacities of these plants,[9] supplemented by new distilling, cracking, and reforming units, are shown in Table 5.

The first commercial plant for the hydrogenation of bituminous coal erected in Great Britain in Billingham,[10] started operation in 1935, and led to important process and catalyst developments. The plant was built by the Imperial Chemical Industries, Ltd., in connection with a synthetic ammonia plant. The operating pressure of 250 atm was the same as that of the ammonia plant. It was designed to produce 150,000 tons/yr of gasoline, of which two-thirds was to come from coal and the remainder from creosote oil. Raw material for this plant was gradually shifted to more and more creosote oil, and during the war only the vapor-phase section of this plant was in operation. The production of the plant was 200,000 tons/yr of aviation gasoline. The development of improved vapor-phase catalysts made it possible to improve the quality of the gasoline, as shown in Table 6. Based on the experience of Billingham, a plant for hydrogenation of gas oil to aviation gasoline was erected during the war at Heysham/Cheshire.

TABLE 5

REFINING CAPACITY OF WEST GERMAN HYDROGENATION PLANTS IN 1000 METRIC TONS/YR IN 1959[9]

	Distillation	Hydrogenation	Cracking	Reforming
Gelsenberg	4,500	720	450	500
Scholven	2,500	450	240	350
Wesseling	3,000	450	550	485
	10,000	1,620	1,240	1,335

[9] *Erdölinformationsdienst* (*Hamburg*), **14**, No. 6, (1960).

[10] Cockram, C., and Sawyer, E. W., *Ind. Chemist*, **35**, 221–9 (1959). Gordon, K., *J. Inst. Fuel*, **20**, 42–58 (1956).

TABLE 6

COMPOSITION OF GASOLINES MADE WITH VARIOUS CATALYSTS AND PROCESSING TECHNIQUES
CREOSOTE OIL HYDROGENATION, BILLINGHAM[10]

Process		Single Stage	Single Stage	Two Stage	Two Stage	Two Stage	Two Stage
Catalyst		64*	5058†	5058† and 6434‡	5058† and 231 §	Co Mo Al¶ and 231 §	Co Mo Al¶ and 7679‖
Aromatics	Vol %	43	3	7	15	18	23
Naphthenes	Vol %	33	50	54	48		
Branched-chain paraffins	Vol %	7	31	33	32		
Straight-chain paraffins	Vol %	14	16	6	5		
Octane Number C.F.R. motor method		80	68	74	76		
Octane Number Research method 1.25 ml/TEL gal				87	89	91.5	94.5

* Mo-catalyst.
† WS_2.
‡ 10% WS_2 on HF activated Fullers earth.
§ Fe on HF activated Fullers earth.
‖ Ni on silica-alumina.
¶ Cobalt molybdate on activated alumina.

Small-scale coal hydrogenation experiments were made at the British Fuel Research Station. The chemistry[11] of the liquid-phase reactions was studied.

The same basic hydrogenation process developed for the liquid- and vapor-phase hydrogenation of tar was applied for the hydrogenation of heavy petroleum residues in two plants in Italy. In 1936, plants in Bari and Livorno were built by the ANIC,[12] each with a capacity of 120,000 metric tons/yr of gasoline.

Development work in France and Japan lead to the erection of liquid-phase coal-hydrogenation[13] plants used mainly for the production of fuel oil.

Development Work

The U.S. Bureau of Mines[14] interest in the production of liquid fuels from coal and lignite began in 1924. Experimental work on the liquefaction of coal started in 1936, when an experimental coal-hydrogenation plant capable of hydrogenating about 100 lb of coal in 24 hr was designed and constructed. The passage of Public Law 290 in April 1944 permitted expansion of the Bureau's program in this field. New laboratories and pilot plant facilities were erected at Bruceton, Pennsylvania. A demonstration plant for the production of 200 barrels of oil day by hydrogenation of about 50 tons of coal/day was erected at Louisiana, Missouri and began operation in 1949.

[11] Cawley, C. M., *Research* (*London*), **1,** 553–61 (1948).

[12] Fauser, G., *Chim. e Ind.* (*Milan*), **19,** 113–22, 599–600 (1937). Kaftal, G., *Chim. e Ind.* (*Milan*), **24,** 53–62 (1942).

[13] Audibert, E., *Génie civil*, **108,** 514–16 (1936). Shibata, K., *Chem. Eng.*, **60,** 214–18 (1953).

[14] Fieldner, A. C., Storch, H. H., and Hirst, L. L., *U.S. Bur. Mines, Tech. Paper* No. 666 (1944), 69 pp. Summaries and further references are given in *U.S. Bur. Mines, Inform. Circ.* No. 7322 (1945), 79 pp.; No. 7352 (1946), 103 pp.; No. 7417 (1947), 142 pp.; No. 7446 (1948), 133 pp.; No. 7518 (1949), 87 pp.; No. 7565 (1950), 97 pp.; No. 7618 (1951), 80 pp.; No. 7647 (1952), 81 pp.; No. 7663 (1953), 67 pp.; No. 7699 (1953), 102 pp.; No. 7756 (1954), 80 pp.; No. 7794 (1955), 103 pp.; No. 7904 (1959) 112 pp.; No. 7905 (1957), 85 pp.; and in the *Synthetic Liquid Fuels Annual Report of the Secretary of the Interior:* 1947 report has no number; then, *U.S. Bur. Mines, Rept. Invest.*, No. 4456 (1948), 62 pp.; No. 4651 (1949), 62 pp.; No. 4770 (1950), 74 pp.; No. 4865 (1951), 83 pp.; No. 4942 (1952), 85 pp.; No. 5043 (1953), 66 pp.; No. 5118 (1954), 73 pp.; No. 5236 (1955), 42 pp.

The laboratory and pilot plant work originally was concerned with a variety of problems. Coal-assay work and study of petrographic constituents was combined and operating variables were investigated, such as temperature, reaction time, pressure, nature of pasting-oil and catalysts, as they affected production of various fuels, gasoline, Diesel oil, and fuel oil. The work established the suitability of most lignites and bituminous coals in the United States for the hydrogenation process. The characterization of coal-hydrogenation products was a further large activity. The kinetics of coal liquefaction and of the further conversion steps of asphaltenes were investigated in autoclave experiments.

A bibliography of pressure hydrogenation up to the year 1948 was assembled.[15] The preparation characteristics of some coals available for synthetic liquid fuels plants were obtained.[16]

Operation of the demonstration plant[17] was begun in 1949. During 4 years of operation, six American coals including high-volatile bituminous, subbituminous coal, and lignite were processed into gasoline and other fuels. American-made equipment was tested, and engineering data were obtained during operation of the plants.

Cost studies[18] showed that the coal-hydrogenation process for the production of gasoline from coal was not competitive with gasoline produced from petroleum. Operations of the demonstration plant at Louisiana were therefore terminated in June 1953. However, work on a laboratory and pilot-plant scale continued to obtain fundamental information on the structure and reactivity of coal, and to develop economically practicable processes for production of liquid fuels.

Pilot-plant work was centered around variations of the coal hydrogenation process. In the high-temperature coal-hydrogenation process, active catalysts and high temperatures and pressures were used to convert coal injected as coal-paste in one stage into hydrocarbon gases and gasoline or low-boiling aromatic hydrocarbons. The hydrocarbonization process converted dry coal at lower pressure (250 to 1000 psi) into hydrocarbon gases.[19] The plastic state of coal poses a problem in the operation of such processes. Another problem is finding operating conditions that lead to a product gas of suitable heating value without having to separate hydrogen from hydrocarbon gases. This work is a continuation and extension of laboratory experiments and larger-scale development in Great Britain.[20] Later laboratory work at Purdue University[21] was followed by development work at the Institute of Gas Technology[22] in Chicago. The hydrogenation of coal at high temperature and comparatively low pressure differs in many respects from that in the conventional liquid-phase hydrogenation. The fact that anthracite and chars react quite rapidly indicates that it is to a great extent a reaction of carbon with hydrogen forming methane rather than a hydrogenolysis of the coal structure. The production of gaseous hydrocarbons by coal hydrogenation at low pressure is treated in the chapter on "Gasification".

Until 1954, basic research in the hydrogenation of coal was conducted at the Carnegie Institute of Technology Coal Research Laboratory in Pittsburgh. It had as its purpose the elucidation of the coal structure rather than the development of a coal-hydrogenation process. Hydrogenation was performed below 400°C, that is at temperatures below those normally used in coal hydrogenation. Large amounts of Adkins' catalyst (Ca-Cu-Cr-oxides) were used in a "cold head" autoclave which permitted separation of distillable fractions. Chromatography and other methods were

[15] Wiley, J. L., and Anderson, H. C., *U.S. Bur. Mines, Bull.* No. 485, I (1950), 306 pp., II (1951), 186 pp., III (1952), 54 pp. See also Sternberg, W. M., *Classified Index of TOM Reels, U.S. Bur. Mines, Administrative Publication*, **I** (1949), 197 pp., **II** (1949), 221 pp., **III** (1949), 195 pp., **IV** (1949) 271 pp., *Supplement* (1951), 97 pp.

[16] Fraser, T., Crentz, W. L., and Barrett, O. T., *U.S. Bur. Mines, Bull.* No. 495 (1950), 132 pp.

[17] Kastens, M. L., Hirst, L. L., and Chaffee, C. C., *Ind. Eng. Chem.*, **41,** 870–85 (1949).

[18] Bituminous Coal Staff, *U.S. Bur. Mines, Rept. Invest.* No. 5506 (1959) 306 pp.

[19] Clark, E. L., Pelipetz, M. G., Storch, H. H., Weller, S., and Schreiber, S., *Ind. Eng. Chem.*, **42,** 861–5 (1950). Newman, L. L., and Pipilen, A. P., *Gas Age*, **119,** No. 10, 16–21, No. 11, 18 (1957).

[20] Dent, F. J., *Gas J.*, **244,** 502–7, 612, 639–40, 802–4 (1944); *Riv. combustibili*, **8,** 611–38 (1954).

[21] Stockman, C. H., and Bray, J. L., *Purdue Univ., Research Ser.* No. 111 (1950), 44 pp.

[22] Chanabasappa, K. C., and Linden, H. R., *Ind. Eng. Chem.*, **48,** 900–5 (1956).

used for the resolution of the product. Stepwise hydrogenation[23] gave indications that the coal substance may be similar in nature to a polymer consisting of repetitive units. Low-temperature coal hydrogenation in the laboratories of the Dutch State Mines[24] indicated that the units were relatively small and contained only a few condensed rings.

A novel technique related to coal hydrogenation—the electrochemical reduction of coal[25]—has been applied to the study of the constitution of coal. Coals with 70 to 75% C gave a 45 to 50% yield of soluble products; preoxidation of the coal (e.g., with H_2O_2) increased the solubilization. UV-spectra indicated that the fractions of the electrochemical reduction product of coal consisted to a great extent of benzenoid materials. In addition, structures based on naphthalene, phenanthrene, and benzphenanthrene nuclei were identified.

In 1952, the Carbide and Carbon Chemicals Company announced[26] a 300 tons/day coal-hydrogenation plant designed specifically for the production of chemicals. Carbide's interest in coal hydrogenation began in 1935. Laboratory scale bench work led to a small pilot plant and eventually to the 300 tons/day unit. According to publications, coal was hydrogenated at 3000 to 6000 psi pressure and 450 to 550°C whereby 3 to 6% by weight of hydrogen was consumed. In commercial operation in the past, a temperature of 500°C apparently has not been exceeded. The use of such higher temperatures could account for the higher content of ring structures in the liquid product obtained by this process. For the process, no new major discoveries in catalyst, reactor design, or paste preparation have been claimed. However, it has been stated that the residence time was reduced drastically to $4\frac{1}{2}$ minutes. Apparently, coarser coal was used than heretofore, and improvements[27] were made in preheating, temperature control, and product separation.

The simplified flow sheet[26] published for this process suggests that emphasis has been in improvements in the separation of products. Continuous processing involving extraction separation, reforming, and distillation columns for the production of tar acids and bases and aromatics is indicated. New separation procedures[28] to obtain clean hydrocarbon, nitrogen-base, and phenolic portions are described. The phenols were separated into conventional phenol, cresol, and xylenol fractions, an intermediate fraction called Metaphenol 220 boiling between 207 and 230°C and a high-boiling (240 to 270°C) fraction of phenols. In 1955, the larger-scale separation of aromatic hydrocarbons and nitrogen bases was started. For a commercial plant, a size of 1500 to 6000 tons of coal/day was indicated. The effect of such a commercial plant on the chemical raw material economy was described[28] as shown in Table 7 in qualitative terms since the optimum mode of operation was not fixed.

The operation of the 300 tons per day plant ceased[29] in 1956 and a 12-ton/day unit has been used to improve techniques developed during Carbide's five years of operating the large pilot plant at the same site at Institute, West Virginia.

The production of chemicals by the coal hydrogenation process has been evaluated by other investigators. German correlations for 700 atm pressure and 480°C and Bureau of Mines results with Illinois No. 6 coal have been used to estimate the attainable yields of chemicals by known techniques[30] including hydroforming of light oil and higher fractions.

[23] Glenn, R. A., *Fuel*, **33**, 419–35 (1954); **34**, 201–11 (1955). Falkum, E. and Glenn R. A., *Fuel*, **32**, 133–52 (1952).

[24] Schumacher, J. P., van Vucht, H. A., Groenewege, M. P., and Blom, L., *Brennstoff-Chem.*, **36**, 215, (1955); see also van Krevelen, D. W., and Schuyer, J., *Coal Science*, Elsevier, Amsterdam, 1957, 352 pp.

[25] Fuchs, W., Schmidt, I., and Veiser, O., *Erdöl u. Kohle*, **12**, 542–7 (1959).

[26] Anon., *Chem. Eng. News*, **30**, 1953–9 (1952). Callaham, J. R., *Chem. Eng.*, **59**, No. 6, 152–61 (1952). Anon., *Chem. Eng.*, **60**, No. 3, 195–6, No. 9, 122, 124, No. 12, 180–4 (1953). Guthrie, V. B., *Petrol Processing*, **1**, 833–7 (1952). Anon., *Coal Age*, **57**, 70–5 (1952).

[27] Edmonds, R. G., and Doughty, E. W., U.S. Pat. 2,784,148 (1957). Doughty, E. W., Howell, J. H., and Eccles, M. A., Brit. Pat. 727,274 (1955). Alspaugh, P. L. Brit. Pat. 759,298 (1956).

[28] Woodcock, W. A. and Tenney, A. H., *Joint Symposium on Future of Aromatic Hydrocarbons*, Am. Chem. Soc., April 4–7, 1955.

[29] Anon., *Chem. Eng. News*, **34**, 4009 (1956).

[30] Donath, E. E., *AIME, Trans.*, **193**, 381–5 (1952); *Ind. Eng. Chem.*, **46**, 2032–5 (1954).

TABLE 7

PRODUCTION OF COMMON CHEMICALS IN A SINGLE COMMERCIAL COAL-HYDROGENATION PLANT[28]

Benzene	Insignificant
Toluene	Insignificant
Aromatic solvent	Appreciable
Naphthalene	Significant
Substituted naphthalenes	Significant
Phenol	Appreciable
Higher phenols (Many are phenol equivalents)	Significant
Aniline	Insignificant
Toluidines (mixed)	Excessive
Xylidines (mixed)	Excessive
Quinoline	Excessive
Indole	Excessive
Hydrocarbon gases (for petrochemicals)	Appreciable

Explanation of terms

Insignificant	Less than 5% of U.S. 1953 Use
Appreciable	Order of 10–20% U.S. 1953 Use
Significant	20 to 40% of U.S. 1953 Use
Excessive	Greater than U.S. 1953 Use

Data have been obtained in a pilot plant of the Koppers Company, Inc.[31] on hydrogenation conditions that produce fractions similar to products obtained from coal tar. The data shown as an example in Table 8 were obtained by liquid-phase hydrogenation of Ohio No. 6 seam coal with 1% $FeSO_4$ as catalyst at 5000 psi pressure and a maximum temperature of 488°C. The products of Table 8 were obtained with a yield of from 75 to 80% of the maf coal. Varying ratios of light to heavy fractions can be obtained by changes in operating conditions.

The tar acids were lower in sulfur content than those from normal coke-oven tar; however, cresols and xylenols were present in greater amounts. The neutral oil and pitch had a higher hydrogen content than similar coke-oven tar fractions. In many respects, the pitch resembled coal-tar pitch. Other process studies indicated that advantages exist in combinations of the coal-hydrogenation process with the carbonization[32] of the hydrogenation residue.

[31] Private communication.

[32] Frese, E. L. Schappert H. M., and Simmat, W. E., U.S. Pat. 2,738,311 (1956). Gilbert, W. I., and Montgomery, C. W., U.S. Pat. 2,654,695 (1953).

The hydrofining of coke-oven light-oils and of petroleum fractions is an outgrowth of the vapor-phase coal-hydrogenation process. Laboratory work for the hydrofining of coke-oven light-oil began in 1925, and led to semicommercial work at the Ludwigshafen plant of the BASF (then I.G.) in 1931. The first commercial plant was built by Scholven-Chemie, Gelsenkirchen-Buer,[33] in 1948. Since then several more plants have been built in Europe (Germany, Great Britain, France). In the United States, two plants[34] for the refining of coke-oven light-oil have been built and started operation in 1957 at Aliquippa, Pennsylvania and in 1958 at Clairton, Pennsylvania. These plants have a throughput of 55,000 and 137,000 gallons/day respectively of the benzene, toluene, and xylene (BTX) fraction of the coke-oven light-oil. Both plants are equipped with selective extraction units for the separation of nonaromatics and thus produce all the benzene, toluene, and xylenes contained in the light oil in pure form.

TABLE 8

PRODUCT COMPOSITION HYDROGENATION OF OHIO NO. 6 SEAM COAL

Salable tar acids (C_6-C_8)	7.0 Wt. %
High-boiling tar acids	2.0 Wt. %
Tar bases	3.5 Wt. %
Neutral oil	39.4 Wt. %
Pitch 100° softening point	36.9 Wt. %
Heavy oil not recovered from centrifuge cake	11.2 Wt. %
	100.0

COAL AND TAR HYDROGENATION IN LIQUID PHASE

General Observations

The production of liquid fuels such as gasoline or Diesel fuel from coal is a two-stage

[33] Nonnenmacher, H., Reitz, O., and Schmidt, P., *Erdöl u. Kohle*, **8**, 407–11 (1955).

[34] Vollmer, R. W., and Simmat, W. E., Eastern States Blast Furnace and Coke Oven Assn., February 17, 1956. Cronan, C. S., *Chem. Eng.*, **66**, No. 1, 110–13 (1958). Smeaton, R. M., and Joseph, J. S., Eastern States Blast Furnace and Coke Oven Assn., November, 12–14, 1959.

process. In the first stage—the liquid phase hydrogenation—coal or tar or oil residues are converted into a middle oil.[35] This in turn is hydrogenated with a fixed bed catalyst in vapor phase into gasoline or other fuels. After World War II, data from German commercial and continuous pilot-plant operation became available.[36]

[35] Definition of several terms in this chapter:

Asphaltene	Benzene-soluble, hexane-insoluble material.
Bayermasse	Luxmasse, Lautamasse; iron oxide residue from alumina recovery from bauxite.
Brown coal	In general, lignite; however, this term is narrower than the European usage of brown coal.
Coal liquefaction	% of carbon or coal substance converted into liquid and gaseous products.
Gasification, gas	Portion of the product consisting of gases especially CH_4 to C_4H_{10}
Heavy-oil	Oils boiling above middle-oil.
HOLD	Heavy-oil let down—the bottoms product from the hot catchpot containing residual coal and ash.
Middle-oil	This term is used for oils that are suitable for hydrogenation in vapor phase into gasoline. In general the boiling range is 180 to 325°C.
Solids	Benzene insoluble material.

[36] Holroyd, R., *BIOS Overall Rept.* No. 1, *Report on the Petroleum and Synthetic Oil Industry of Germany* (1947), 134 pp. Farragher, W. F., and Horn, W. A., *FIAT Final Rept.* No. 422, *Manufacture and Regeneration of Catalysts at I. G. Ludwigshafen-Oppau* (1945), 14 pp. Holroyd, R., Ed., CIOS XXXII–107, *I. G. Leuna* (1945), 209 pp. Oriel, J. A., Jones, I. H., and Weir, H. H., CIOS XXVIII–40, *The High Pressure Hydrogenation Plant, Brown Coal, Wesseling* (1945), 154 pp. Cockram, C., CIOS XXX–102, *Scholven* (1945). 41 pp. Holroyd, R., CIOS XXX–103, *I. G. Werke Ludwigshafen-Oppau* (1945), 133 pp. Cockram, C., CIOS XXX–104, *Welheim* (1945), 32 pp. Cockram, C., CIOS XXX–105, *Gelsenberg* (1945), 39 pp. Ellis, J. F., and Morley, R. G., CIOS XXXIII–24, *Zeitz* (1945), 26 pp. Hollings, H., CIOS XXXII–90, *Wintershall A. G. Lützkendorf* (1945), 37 pp. Ellis, J. F., CIOS XXXII–92, *Brabag Böhlen* (1945), 41 pp. Sherwood, P. W., *FIAT Final Rept.* No. 952 (1947), 244 pp. Krönig, W., *Katalytische Druckhydrierung*, Springer, Berlin, 1950, 266 pp. Hoering, M., and Donath, E. E., ref. 3, pp. 483–569.

Initially for commercial coal hydrogenation, 200 to 300 atm pressure was used and considered satisfactory. However, it was found for high rank bituminous coals, some low hydrogen content brown coals and, oils and tars of high asphaltene content that higher pressures[10,37] were required to obtain satisfactory conversion in commercial plants. At the higher pressure not only were better yields obtained, but also higher temperatures, and therefore higher throughputs could be used without danger of coking. More recent Russian autoclave experiments have shown that this trend continues into the range of higher pressures,[38] up to 1700 atm. The results of continuous experiments with a bituminous coal from the Ruhr district at 300 and 700 atm and the maximum temperature and throughput consistent with continuous operation and conversion of coal into a product boiling below 325°C are shown in Table 9.[39]

TABLE 9

PRESSURE AND TEMPERATURE, HYDROGENATION OF A RUHR COAL[39]

Pressure, atm	300	700	
Catalyst	SnC_2O_4 + NH_4Cl	Iron Oxide	SnC_2O_4 + NH_4Cl
Reaction temp, °C, average	465	475–480	475
Coal liquefaction, %	93	96	97
Oil make −325°C pounds/(cu ft reactor vol hr)	11.3	17	20
% Gas/(oil make + gas)	25	21	18
% Asphaltene in HOLD*	18	11	9

* The asphaltene content is an indication of the viscosity of the oil and thus the separability of the solids from the oil and the yield of recovered oil.

The table shows the influence of pressure on coal liquefaction. Even more important is the decrease in formation of normally gaseous hydrocarbons (and therefore hydrogen consumption) and asphaltene content, and the increase in the oil space-time-yield. Equilibrium considerations indicate that not only the absolute pressure but also the ratio of the pressure of hydrogen to oil vapor would influence the reaction. However, in practice the oil partial pressure is usually below 10%, since the minimum hydrogen-recycle rate is

[37] Pier, M., Trans. Faraday Soc., **35**, 967–79 (1939).

[38] Kazanskii, B. A., Gonikberg, M. G., Lozovoi, A. V., Gavrilova, A. E., and Blonskaya, A. I., *Trudy Inst. Goryuchikh Iskopaemykh, Akad. Nauk S.S.S.R., Otdel. Tekh. Nauk*, **6**, 3–15 (1955).

[39] Sherwood, P. W., *FIAT Final Rept.* No. 952 (1947) 244 pp.

determined by heat-transfer considerations and cooling-gas requirements. The Sn-Cl catalyst combination produces better results than the iron catalyst. However, it was not used commercially at the higher pressure because of the problem of corrosion control at 700 atm and because of wartime shortages of tin.

For the hydrogenation of German brown coal, iron catalysts were used. They replaced the small amounts of molybdenum oxide initially used and did not require neutralization of the calcium humates in the brown coal with sulfuric acid. In general, it was found that tin compounds are a better catalyst for bituminous coals and molybdenum compounds for lower-rank subbituminous and brown coals. The activity of the tin catalyst is greatly enhanced by addition of chlorine compounds that form hydrochloric acid under reaction conditions.

Although the tin oxalate (0.06% on coal) and NH_4Cl (1.15%) were added to the coal paste and only mechanically mixed, the use of other catalysts required greater care. Molybdena (0.02 to 0.06%) was impregnated on brown coal from ammoniacal solution to attain good distribution and activity. Iron catalysts such as red mud (1.5 to 2%) were added to the paste, but the coal was usually impregnated in addition with an iron sulfate (1.3% $FeSO_4 \cdot 7H_2O$) solution.

The mode of catalyst addition has been investigated more recently in autoclave[40] experiments for the hydrogenation of coal without pasting oil. The activity of a number of catalysts was found to be much greater when impregnated on the coal from aqueous solution than when added as powder to the coal. Ballmilling of coal and catalyst was of intermediate effectiveness. The fine distribution of the catalyst becomes more important as the catalyst concentration decreases.

The catalysts in liquid-phase coal hydrogenation are used on a once-through basis. Catalyst recovery schemes have been suggested, but so far none has been found practical for the recovery of catalysts such as tin or molybdenum. From a practical point of view, the catalysts should then be compared on an equal cost basis: Fe catalyst in amounts of 1% is then comparable to Mo or Sn quantities that are of the order of 0.01%.

Based on autoclave experiments with various catalysts, especially the high activity of tin and the promoting action of HCl added as NH_4Cl, a reaction scheme[41] for the coal liquefaction has been suggested. Because the asphaltene fraction[42] is an important intermediate product in coal hydrogenation, the study of its hydrogenation is helpful for a quantitative evaluation. The conversion of asphaltene to *n*-hexane-soluble oil is first order with respect to the asphaltene remaining and also with respect to hydrogen pressure. The average activation energy in the range 430 to 440°C is 36 kcal per mole. Of the asphaltene converted, about 92% goes into the production of oil and 8% into hydrocarbon gas at a process temperature range of 400 to 440°C. This gas formation figure may appear low in comparison to figures given elsewhere in this chapter. It must be remembered that in these experiments the hexane-soluble oil produced is higher boiling than the middle-oil of commercial coal hydrogenation.

The new technique of mass spectrometry[43] with low ionizing voltage has been applied to the analysis of liquid products from coal hydrogenation. Resolution of the product into classes of composition and molecular weight as well as identification of individual compounds is possible.

For the hydrogenation of tar and oil residues, activated carbon-supported catalysts[36] were used in amounts of 0.3 to 1.5% of the feed. Char such as that from Winkler gasifiers using brown coal in a fluidized bed was suitable and available at low cost. It was first impregnated with 5% $FeSO_4$ and then with an equivalent amount of caustic. The catalyst was then ground in an asphaltene-free heavy-oil to a 40% slurry of less than 60 μ particle size. For low-sulfur feed stocks, Na_2S was also added. The Fe-containing liquid-phase catalyst replaced molybdenum satisfactorily and the somewhat higher oxygen,

[40] Weller, S., Pelipetz, M. G., Friedman, S., and Storch, H. H., *Ind. Eng. Chem.*, **42**, 330–4 (1950). Weller, S., and Pelipetz, M. G., *Ind. Eng. Chem.*, **43**, 1243–6 (1951).

[41] Weller, S., Clark, E. L., and Pelipetz, M. G., *Ind. Eng. Chem.*, **42**, 334–6 (1950).

[42] Weller, S., Pelipetz, M. G., and Friedman, S., *Ind. Eng. Chem.*, **43**, 1572–5 (1951).

[43] Sharkey, A. G., Jr., Wood, G., Schultz, J. L., Wender, I., and Friedel, R. A., *Fuel*, **38**, 315–28 (1959); Sharkey, A. G., Jr., *Chem & Ind.* (*London*), **1958**, 833–4.

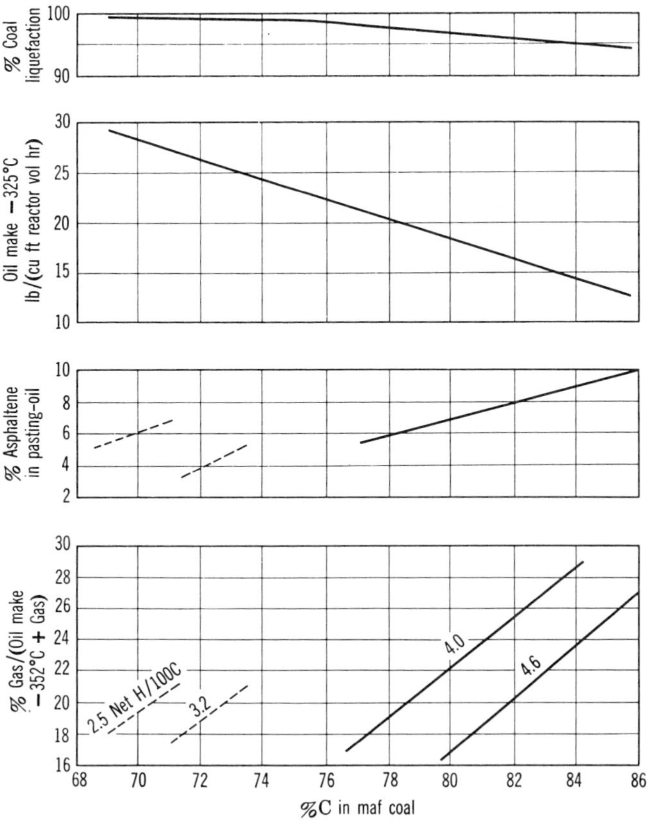

Fig. 1. Coal analysis and hydrogenation correlation.[44]

nitrogen, sulfur, and olefin content of the liquid-phase products was not harmful if these fractions were used as vapor-phase feed stocks.

Pilot plant experiments at 9000 psi hydrogen pressure with iron catalyst according to I.G. reports[44] from Ludwigshafen with a number of brown and bituminous coals for the production of middle-oil (325°C end point) gave the correlation shown in Fig. 1. The significance of this correlation lies in the fact that pilot-plant operating conditions were used that produced results closely resembling those attained in German commercial plants. Coal liquefaction and middle-oil production rate decreased with increasing carbon content of the coal. The asphaltene content of the recycle pasting-oil and the yield of gas expressed as % of the liquefied coal increased in general with the carbon content of the coal. However, different groupings of coal of different origin must be made. It seems that the net-hydrogen[45] content of coal permits a satisfactory grouping. Thus, a brown coal of the same carbon content but a higher net-hydrogen content than another coal will produce a recycle pasting oil of a lower-asphaltene content. Less gaseous hydrocarbons will be formed and, thus, this coal will be more suitable for liquid-phase hydrogenation. For bituminous coals of different hydrogen contents, different levels of gas formation are observed, but the experimental material does not indicate differences in the asphaltene content of the pasting oil. The differences of coals of similar carbon but different net-hydrogen content could probably also be expressed by indicating petrographic differences. However, petrographic data for these coals were only partly available.

Individual coals may show significant deviations from these correlations. Furthermore, the parameters shown, like asphaltene content of the pasting oil, gas loss, and oil make rate

[44] TOM Reel 143, Frames 62–98; See Sternberg, W. M., ref. 15.

[45] g hydrogen per 100 g carbon after deduction of the H-equivalent of the O, N, and S in the coal.

TABLE 10

COAL ANALYSES, BUREAU OF MINES DEMONSTRATION PLANT[46]

Coal	Wyoming, Rock Springs	West Kentucky	Illinois No. 6	Wyoming, Lake DeSmet
% Ash, dry basis	5.2	7.2	8.3	13.0
V.M., maf basis	46.9	42.5	46.7	49.8
Ultimate, maf basis				
H	5.3	5.6	5.6	5.4
C	78.6	80.0	78.7	71.1
N	1.7	1.5	1.5	1.4
O	13.3	9.9	10.8	21.3
S	1.1	3.0	3.4	0.8
Net H/100 C	4.3	4.8	4.7	3.4

TABLE 11

LIQUID-PHASE HYDROGENATION, BUREAU OF MINES DEMONSTRATION PLANT[46]

Coal	Wyoming, Rock Springs	West Kentucky	Illinois No. 6	Wyoming, Lake DeSmet
Pressure, psi	7,900	8,300	7,900	8,100
Mean temperature, °F/°C	875/468	865/463	863/462	851/455
Coal throughput, maf basis				
tons/day	52.4	54.9	57.1	57.5
lb/cu ft reactor vol hr	16.8	25.4	26.4	26.6
Coal liquefaction, %	94.5	95.3	94.2	96.6
Oil make, −325°C*				
bbl/ton maf coal	3.15	3.44	3.52	3.00
wt. % maf coal	51.0	55.5	57.0	48.5
Heavy-oil make,				
bbl/ton maf coal	0.55	0.89	0.70	0.77
Gas, wt, % of maf coal	26.8	13.7	12.1	14.9
Gas, wt. % of (oil make −325° + gas)	34.4	19.8	17.5	23.5
Reaction H_2,				
wt. %/maf coal	7.2	6.5	6.1	6.8
Carbon balance, %	105.4	96.5	96.0	98.0
Pasting-oil, 10% pt, °F/°C	595/312	604/318	596/313	613/323
Pasting-oil, % Asphaltene	25.3	8.4	7.1	10.7

* Vapor-phase charging-stock.

depend greatly on the amount of asphaltene purged in the HOLD oil-recovery system and blending components of the recycle pasting-oil. The data shown are based on purge amounts that were used and found possible in German plants. They will be shown later in diagrams. For example, an increase of the purge rate would result in a greater-oil make and lower-gas formation. For a determination of the economic optimum of the purge rate, a rather extensive experimental study would be required. The pasting oil used was substantially free of fractions boiling below 325°C.

After these remarks, it is of interest to compare data obtained in the Bureau of Mines demonstration plant[46] with various American coals in Tables 10, 11, and 12. For comparison with the data of Fig. 1, these coal hydrogenation data were extrapolated to the pressure of 9000 psi using data for two pressures shown in Table 9, and the results are shown in Table 13. It is evident that the data for the first coal obtained during initial operation are, as stated by Chaffee and Hirst,[46] out of line because of operating problems and unsatisfactory asphaltene- and solids-purging methods. The results obtained for the Kentucky and Illinois coal conform to Fig. 1 reasonably well. The oil make −325° is somewhat lower as well as the mean operating temperature. This is a result of throughput limitations in the coal preparation units of the Bureau of Mines demonstration plant. The data for the fourth coal (DeSmet, Wyo.) deviate considerably from the curves shown in Fig. 1. This may by the result of differences in operating conditions or coal composition. The first point of view is supported by the fact that only a limited amount of work with this coal was done, and only one catalyst was tested, and the coal

TABLE 12

LIQUID-PHASE HYDROGENATION, BUREAU OF MINES DEMONSTRATION PLANT[46]

[Yield of Low-boiling (below 225°C) tar-acids, Weight % maf coal]

Coal	Wyoming, Rock Springs	West Kentucky	Illinois No. 6	Wyoming, Lake DeSmet
Phenol	1.1	1.2	1.2	0.9
o-Cresol	0.5	0.4	0.5	0.3
m-Cresol	1.1	1.2	1.2	0.9
p-Cresol	0.3	0.6	0.6	0.7
Sub total	3.0	3.4	3.5	2.8
Mixed xylenols	2.0	1.8	1.8	1.9
Total	5.0	5.2	5.3	4.7

TABLE 13

COAL HYDROGENATION, BUREAU OF MINES DEMONSTRATION PLANT

(Results extrapolated to 9000 psi according to Table 9 for comparison with Fig. 1)

Coal	Wyoming, Rock Springs	West Kentucky	Illinois No. 6	Wyoming, Lake DeSmet
Coal, % C maf	78.6	80.0	78.7	71.1
net H/100 C	4.3	4.8	4.7	3.4
Coal liquefaction, %	95.5	96.3	95.2	97.6
Oil make −325°C, lb/cu ft reactor vol hr	10.5	16	17	15
% gas/(oil make −325° + gas)	33	18.4	16.1	22
Asphaltene in pasting oil, %	23	7.0	5.7	9.2

[46] Chaffee, C. C., and Hirst, L. L., *Ind. Eng. Chem.*, **45**, 822–38 (1953).

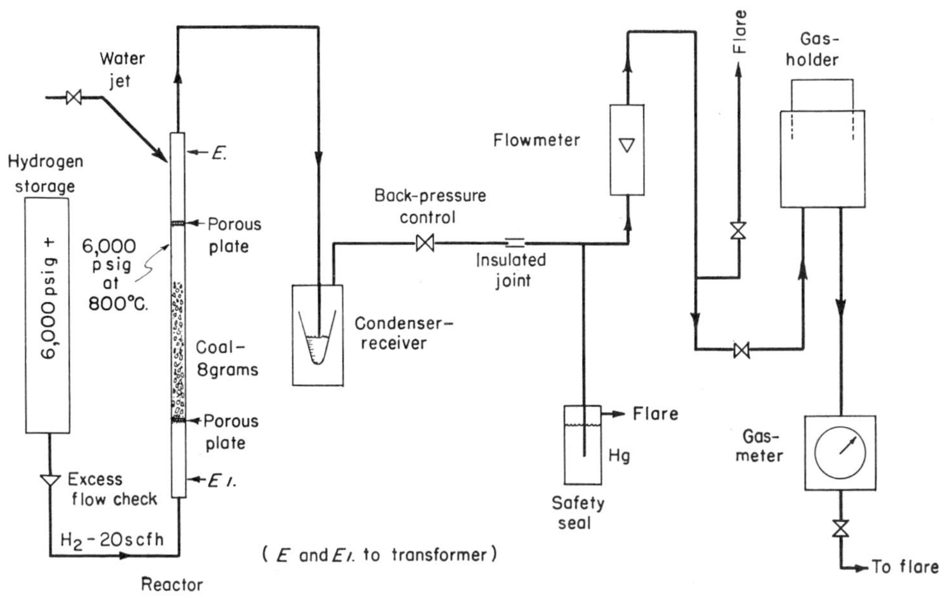

Fig. 2. Flowsheet for hydrogenating coal at 800°C and 6000 psi.[47]

throughput was small because of limitations of the coal-preparation—especially drying—plant.

The experiments described have been made at temperatures around 480°C. This is in agreement with the upper limit for the temperatures used in commercial liquid-phase coal-hydrogenation plants of about 500°C. Above this temperature range, the yield of oil declines with a corresponding increase in hydrocarbon gas formation and hydrogen consumption. Furthermore, the exothermic reaction makes temperature control increasingly difficult and vaporization of pasting-oil causes operating problems.

To study the hydrogenation of coal at 800°C and 6000 psi pressure[10,47] required design and development of apparatus to avoid the long heating and cooling periods inherent in tests made in autoclaves that do not permit accurate measurement of the effect of residence time. A simplified flowsheet and details of the reactor are shown in Fig. 2 and 3. A bed of 30- to 60-mesh coal was supported and retained by porous plates in a tubular reactor. After establishing a hydrogen flow of 20 std. c.f.h., the charge was heated to 800°C in about $1\frac{1}{2}$ min by passing an electric current directly through the reactor. At the end of the desired reaction time (1 to 15 min), the temperature of the reactor and its contents was rapidly decreased to room temperature by quenching it with water. High-boiling oils and unreacted solids were washed from the reactor with benzene. The amount of remaining benzene-insoluble solids served to calculate the extent of conversion of coal. Overhead oils were weighed after separation from water and analyzed by a mass spectrometer, as were the hydrocarbon gases. The initial experiments were made with bituminous C coal from Rock Springs, Wyoming, impregnated with an aqueous solution of ammonium molybdate to provide a catalyst concentration of 1% molybdenum, based upon maf coal.

The product yields are shown in Fig. 4. The maximum oil yield was achieved after about 2 min. However, the bulk of this oil at first remained in the reactor and was split and distilled out of the reactor within 8.5 min. This overhead oil contained about one-third each of benzene and naphthalene, the remainder being other aromatic hydrocarbons. The hydrocarbon gases contained 80 to 90% methane, the rest being ethane and propane. It is interesting to note that about one-half of the oil and gas were formed during the $1\frac{1}{2}$ min preheating time needed to reach the temperature of 800°C.

[47] Bur. Mines Staff, *U.S. Bur. Mines, Inform. Circ.* No. 7904 (1959), p. 100; No. 7905 (1959), p. 80; Pelipetz, M. G., U.S. Pat. 2,860,101 (1958).

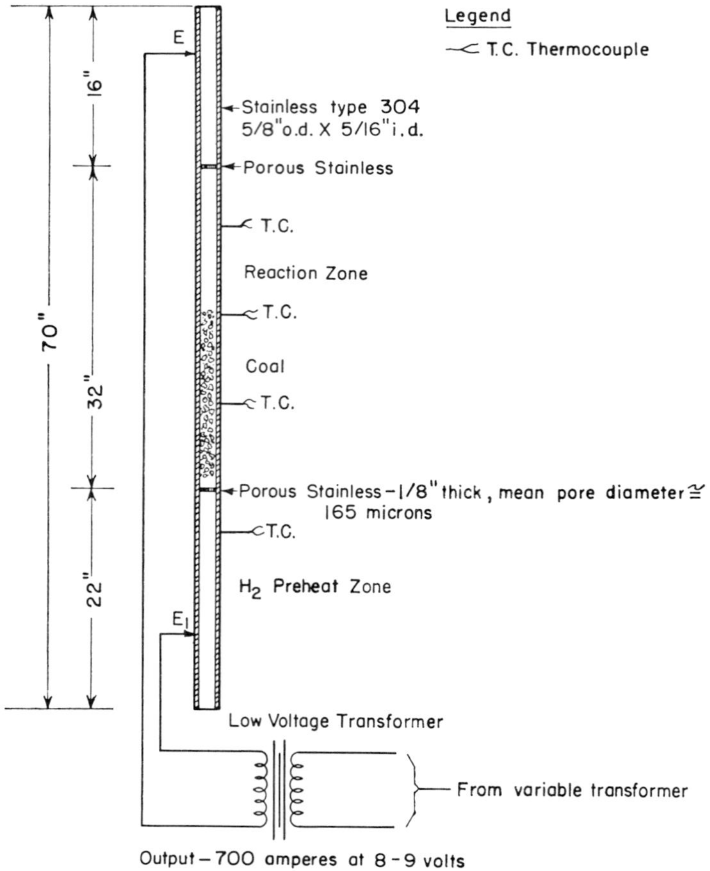

Fig. 3. Reactor for hydrogenating coal at 800°C and 6000 psi.[47]

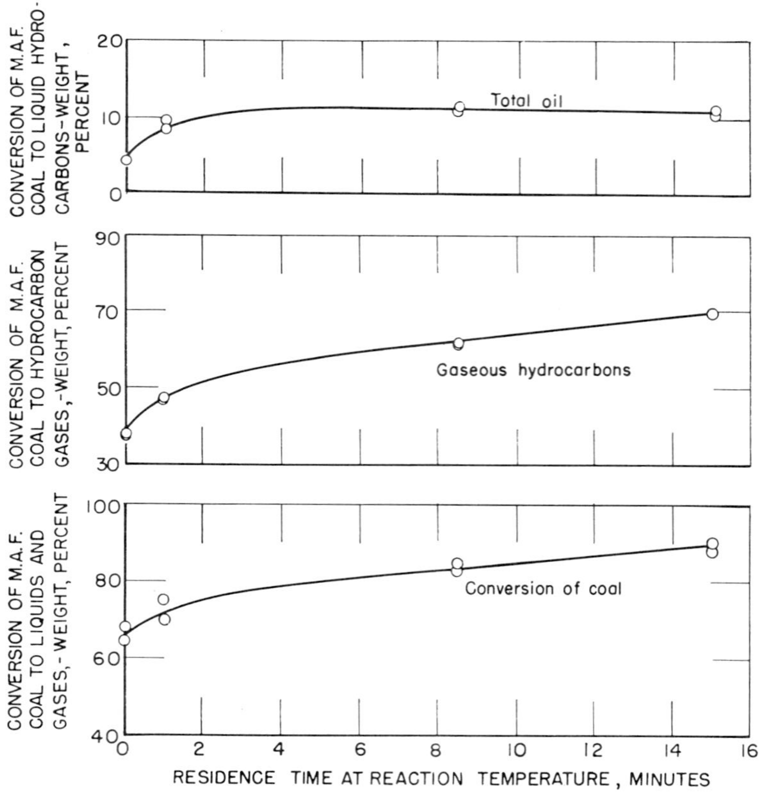

Fig. 4. Residence time vs coal conversion showing liquid and gaseous hydrocarbon yields.[47] Rock Springs, Wyoming, coal, 6000 lb/in.^{2} at 800°C, catalyzed with 1% molybdenum as ammonium molybdate.

TABLE 14

ULTIMATE ANALYSES OF KEROGEN AND OIL-SHALE PRODUCTS ON AN ASH-FREE BASIS[50]

	Kerogen	Extracted Bitumen*	N-T-U- Oil†	Hydrogenated Products	
				Benzene-soluble‡	Trichloro-ethylene-soluble §
C	78.28	82.59	84.53	83.93	81.03
H	10.55	11.04	11.32	13.00	12.28
S	1.34	1.19	0.75	0.12	0.18
N	2.59	0.81	1.77	0.36	0.57
O (by difference)	7.24	4.37	1.63	2.59	5.94
H/C (Atomic ratio)	1.62	1.60	1.61	1.86	1.82

* Benzene-soluble bitumen extracted from raw oil shale; not considered as kerogen in this study.
† Shale oil from an N-T-U retort comprising approximately 60% of the kerogen.
‡ Benzene-soluble material comprising 80% of carbon in the kerogen.
§ Trichloroethylene-soluble material comprising 6% of the carbon in the kerogen.

In contrast to these results at 800°C, experiments have shown that dicobalt octacarbonyl[48] in the presence of carbon monoxide and hydrogen, functions at much lower temperature than even the 400 to 475°C range of conventional liquid-phase hydrogenation as a selective homogeneous hydrogenation catalyst for polynuclear aromatic hydrocarbons at about 3000 psi pressure. Isolated benzene rings are stable in this system. Naphthalenes are slowly reduced to tetralins. Linearly condensed compounds such as anthracene are readily hydrogenated at the meso-positions. Phenanthrene-type compounds are reduced very slowly at 200° to dihydro derivatives. More highly condensed systems are reduced to yield phenanthrene derivatives. Such reactions are similar to some of those occurring in liquid-phase coal or tar hydrogenation. It would be of great impact on the coal-hydrogenation process if the use of some catalyst would permit such a drastic reduction of the operating temperature. However, the necessary splitting of the coal structure to yield products of lower molecular weight appears to necessitate temperatures in the range 300 to 450°C.

Halogens, including iodine, have been used commercially as catalysts for the hydrogenation of coal. Iodine has been used at 350 to 400°C to elucidate the structure of coal.[49] In addition to coal, materials such as wood, carbohydrates, and aromatic model substances were investigated.

The research effort of the Bureau of Mines in the field of synthetic liquid fuels has encompassed oil shale as well as coal. Retorting of the oil shale to obtain oil for further refining is at present the accepted method. However, the direct hydrogenation of oil shale or of kerogen concentrates from such shale will produce higher oil yields. Such a process is probably not economically attractive. Nevertheless, recent experiments[50] may be of interest since they show that such kerogen has a much higher hydrogen content than coal and undergoes hydrogenolysis readily. At 355°C, with catalysts such as $NiCl_2$ or $SnCl_2$, a hydrogen pressure of 4200 psi, and 4 hr residence time, conversion of up to 98% of the kerogen into liquid and gaseous products was achieved. About 86% of the carbon was obtained as oil, 8% as gas, and 6% remained in the residue. By comparison, retorting of oil shale produces an oil yield of 60%. The gas is rich in CO_2; its main hydrocarbon component is ethane, the methane content is small. The ultimate analysis of kerogen and its hydrogenation products are shown in Table 14 in comparison with shale oil.

Commercial Hydrogenation of Coal in Liquid Phase

For the commercial hydrogenation of brown coal from Central Germany, a pressure of 250

[48] Friedman, S., Metlin, S., Svedi, A., and Wender, I., *J. Org. Chem.*, **24**, 1287–9 (1959).

[49] Geerards, J. J., van Krevelen, D. W., and Waterman, H. I., *Research (London)*, **10**, 164–5 (1957); *Brennstoff-Chem.*, **39**, 332–6, 511–4 (1958).

[50] Hubbard, A. B., and Fester, J. I., *U.S. Bur. Mines, Rept. Invest.* No. 5458 (1959), 26 pp.

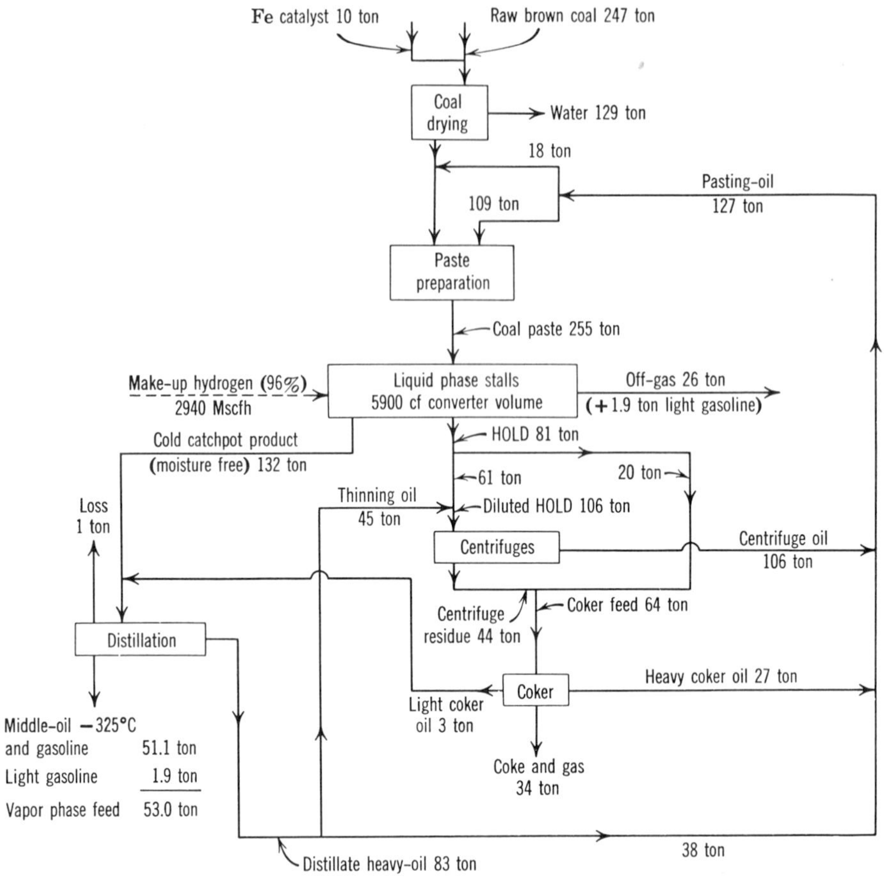

Fig. 5. Liquid-phase hydrogenation of central German brown coal to vapor-phase feed stock. Feed rate 100 ton/hr maf coal.[51]

atm corresponding to about 200 atm hydrogen partial pressure was used. A simplified flow sheet[51] is shown in Fig. 5. A liquid-phase hydrogenation stall with four converters produced 70,000 metric tons/yr of a product suitable as vapor-phase feed stock or about 1500 barrels/day. A vapor-phase feed stock yield of 53% based on maf coal was obtained with a hydrogen consumption of 29,400 scf/ton maf coal or about 9000 scf/bbl product. The isolation and identification[52] of the chemical constituents in high-boiling residues from the high-pressure hydrogenation of Central German brown coal has been the object of investigations. Many aromatics have been identified, but only pyrene has been found in significant quantities.

Hydrogenation of the Rhenish brown coal of much lower hydrogen content required 700 atm pressure. Similar equipment was used, although details such as heating of the coal paste, etc., required variations. For the preheating of the Rhenish brown coal, a top temperature of 425°C was the limit, since, at higher temperatures, $CaCO_3$ deposits were formed in the heater tubes. Small amounts of converter burden were withdrawn from the bottom of the first converter to remove sand and $CaCO_3$ agglomerates of about 2- to 4-mm size called "caviar," during hydrogenation of both brown coals.

A brief comparison[53] of the hydrogenation

[51] Krönig, W., *Katalytische Druckhydrierung*, Springer, Berlin, 1950, 266 pp.

[52] Runge, F., and Meckelburg, A., *Die Bergakademie, Freiberg. Forschungsh.*, **A17**, 54–8 (1953).

[53] Holroyd, R., Ed., CIOS Rept. No. XXXII–107 *I. G. Leuna*, (1945), 209 pp.; Oriel, J. A., Jones, I. H., and Weir, H. H., CIOS XXVIII–40 *The High Pressure Hydrogenation Plant, Brown Coal, Wesseling* (1945), 154 pp.

results obtained with these two brown coals is shown in Table 15.

The hydrogenation of bituminous coal became commercial several years after that of brown coal. In general, the equipment was similar to that used for brown-coal hydrogenation. The preheating of bituminous coal in shell and tube-type heat exchangers required precaution because of the high viscosity of coal-paste in the plastic range of the coal. At first, coal-paste was preheated with some hydrogen to 280°C, below the plastic range, then the remainder of the recycle gas was added after being heated by heat exchange and gas-fired heater. Heating while in the plastic range was thereby practically avoided. The converter inlet temperature was reached after another passage of paste and hydrogen through the gas-fired preheater. A better utilization of heat by heat exchange was achieved when it was found that coal-paste with not more than 45% coal could be successfully heated in shell and tube heat exchangers. This thin paste was then preheated in heat exchangers, and a smaller amount of thick paste with 55% coal was directly admixed in the gas fired heater.

TABLE 15

BROWN-COAL HYDROGENATION[53]

	Central German	Rhenish
	Brown Coal	
Pressure, atm	250	650
Catalyst, % maf coal	Up to 10% Bayermasse (red mud)	6% Bog iron ore +1.2% S
Temperature, °C	480–490	475–480
Throughput:		
coal maf, lb/cu ft reactor vol hr	29	33
recycle gas, scf/lb coal	42	47
coal/pasting oil	$\frac{50}{50}$	$\frac{40}{60}$
Coal liquefaction, %	98.5	99.6
Oil make −350°, lb/cu ft reactor vol hr	15	17
Gas, %	26	24–25
% Asphalt/heavy oil	5	8

As mentioned before, for the hydrogenation of bituminous coal at 4500 psi pressure, tin oxalate and NH_4Cl were used as catalysts. In the temperature range of the condensation system of the coal-hydrogenation stall, the volatile chlorine compounds were highly corrosive. To avoid this corrosion, the vapors leaving the last converter or the hot catchpot were neutralized[54] by scrubbing with a 20% suspension of Na_2CO_3 in oil at the reaction temperature. In the German 700-atm plant, less active iron catalysts were used and thus the Cl-corrosion problem was avoided. The main advantage of the use of this higher pressure was a better yield due to higher coal liquefaction and decreased gas formation. A brief comparison of commercial results is shown in Table 16.[55] These results differ somewhat from those shown in Table 9 because the coals are different.

[54] Holroyd, R., and Cockram, C., Brit. Pat. 432,958 (1934). Pier, M., Urban, W., and Donath, E. E., Ger. Pat. 692,631 (1934).

[55] Cockram, C., CIOS Rept. No. XXX–102, *Scholven*, (1945) 41 pp.; Cockram, C., CIOS Rept. No. XXX–105, *Gelsenberg* (1945), 39 pp.

TABLE 16

BITUMINOUS COAL HYDROGENATION AT DIFFERENT PRESSURES[55]

% C in maf coal (Ruhr)	82.8	83.9
Pressure atm	300	700
Catalyst/maf coal %	0.06% SnC_2O_3 + 1.15% NH_4Cl	1.2% $FeSO_4 \cdot 7H_2O$ + 2% red mud + 0.3% Na_2S
Temperature, °C max.	476°	485°
Throughput maf coal, lb/cu ft reactor vol hr	19	26
Recycle gas, scf/lb coal	68	76
Coal liquefaction, %	92–94	97
Oil make −325°C, lb/cu ft reactor vol hr	12	16
Gas, %	26	25
% Asphalt/heavy-oil	11	9

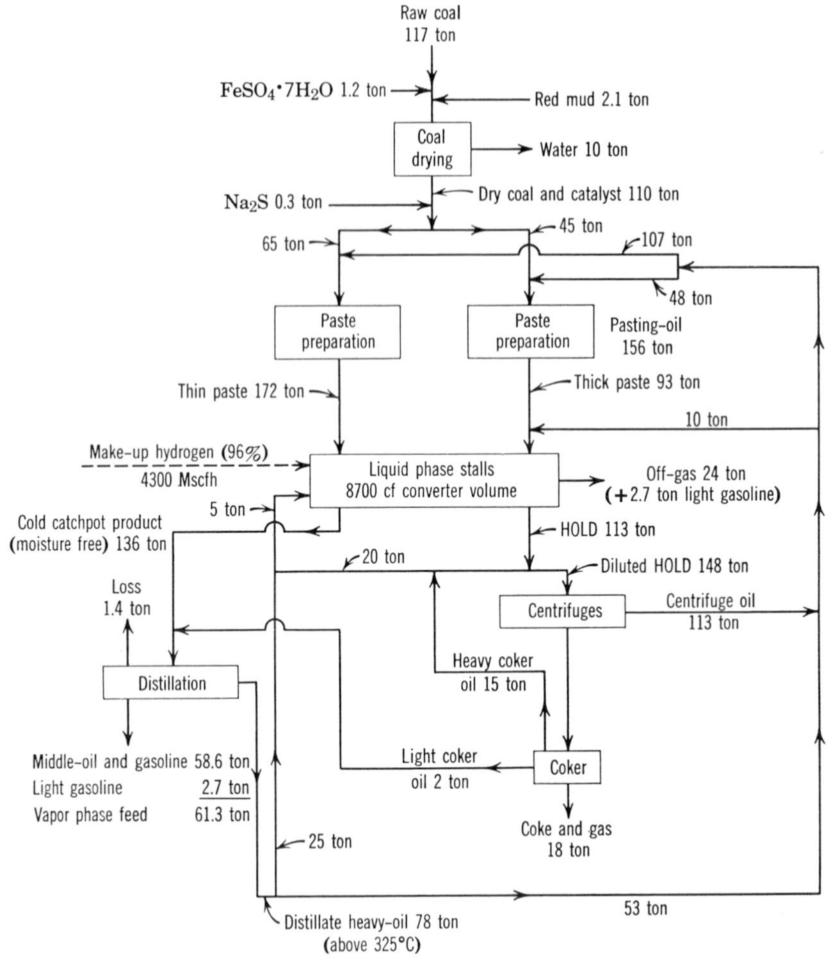

Fig. 6. Liquid-phase hydrogenation of bituminous coal at 10,500 psi to vapor-phase feed stock. Feed rate 100 ton/hr maf coal.[56]

A simplified flow sheet[56] of the hydrogenation of bituminous coal to a product suitable for vapor-phase hydrogenation (boiling below 325°C) is shown in Fig. 6. The yield of product was 61% of the maf coal. The total hydrogen consumption was 43,000 scf/ton maf coal or about 11,000 scf/bbl of product. The distillate heavy-oil (cold catchpot residue) obtained from bituminous coal was a premium Navy fuel-oil of high specific gravity and low pour-point which was miscible with fuel oils of every origin. Production of such heavy-oil was studied with Upper Silesian coal. A comparison of the production of heavy-oil with that of middle-oil is shown in Table 17. A slight increase in temperature was needed to obtain the additional distillate heavy-oil make. However, a further increase of the throughput was not possible. It would have increased the asphaltene content and thereby the viscosity of the HOLD to a point where its stability during preheating and reaction became insufficient and the centrifuging and coking equipment could not handle it any more because of high viscosity and deposit formation.

[56] Krönig, W., *Katalytische Druckhydrierung*, Springer, Berlin, 1950, 266 pp.

To maintain operability of the liquid-phase units for the heavy-oil production shown in Table 17, the addition of middle-oil to the pasting-oil was needed. This and other measures involved in the production of distillate heavy-oil can be seen in the flow sheet Fig. 7.[56] The hydrogen consumption is 36,400 scf/ton maf coal or about 10,000 scf/bbl of liquid-phase product. While these data pertain to the making of distillate heavy-oil, the production of a residual heavy-oil containing asphaltene can be achieved under milder conditions, but the separation of the oil from residual coal and ash then becomes a problem.

To illustrate this, the results of pilot-plant experiments[57] for the production of heavy fuel-oil by hydrogenation of coal at 1500 psi hydrogen pressure will be reported. Rock Springs, Wyoming, high-volatile C bituminous coal of 77.8% C on a maf basis, with 6.3% ash and 2.2% H_2O, a very reactive coal, was the raw material. The pilot plant used a paste containing 35% coal. The hydrogenation product was centrifuged to obtain recycle pasting-oil as well as product oil. The reaction volume of the two converters of the pilot plant was 0.232 cu ft. Results obtained with different catalysts and operating conditions are summarized in Table 18. Only with 0.5% molybdenum as ammonium molybdate was a fuel oil containing less than 1% ash and 10% solids obtained. The use of such an amount of molybdenum as a catalyst is not economically feasible without catalyst recovery. With the less active catalysts, similar oils could probably be obtained at lower throughputs. The hydrocarbon gas formation was comparatively high and therefore the hydrogen consumption would be higher than for the production of a comparable product at higher pressure. Hydrogenation at low pressure thus would appear to be suitable for the production of residual oils that may have significant ash contents. It seems that the use of such pressures in commercial operation would require further information as to whether commercial heat-exchange equipment could be used without the formation of deposits.

HOLD Solids Removal

For the separation of the cold catchpot product and the gases into fractions, conventional processes such as distillation are available. The heavy-oil letdown (HOLD), which contains the solid residue from coal hydrogenation, required the development of new equipment. The purpose of the process is the separation of the solids from the oil and asphaltene and the recovery of a solids-free oil with a low asphaltene content and with as little oil loss as possible. The quantity of the solid material depends on the coal liquefaction, the ash content of the coal, and the catalyst. The quantity ranges from just a few percent for low-ash coals to almost 25% of the total dry coal in the case of some high-ash content brown coals. The solids content of the HOLD ranged between 20 and 40%. It was attempted to attain a high-solids content but solid concentrations above 40% usually involved the danger of coke deposits in the hot catchpot. The solids in the HOLD are very fine, their particle size being usually from 40 μ to submicron sizes.

The separation of the solids from the oil is difficult because of the presence of asphaltenes which cause low filtration rates and, if subjected to coking, produce a comparatively low yield of oil. Therefore, a combination process consisting of centrifuging in the presence of a diluting oil and coking of the centrifuge concentrate was used. The direct coking of an adjusted part of the HOLD was used in the case of some coals to increase the asphaltene purge from recycle pasting-oil (Fig. 5). In some plants the coking step for the HOLD was replaced by a distillation, which led to a high-melting point pitch instead of coke as residue.

[57] Clark, E. L., Hiteshue, R. W., and Kandiner, H. J., *Chem. Eng. Progr.*, **48**, 15–20 (1952).

TABLE 17

BITUMINOUS COAL HYDROGENATION: MIDDLE AND HEAVY OIL PRODUCTION[56]

Percent in Product:			
Gasoline and Middle-oil	100	65	50
Distillate Heavy-oil	0	35	50
Temperature, °C	476	478	480
Throughput maf coal, lb/cu ft reactor vol hr	27	39	40
Pasting-oil composition			
% Heavy-oil	100	85	73
% Middle-oil	0	15	27
Coal liquefaction, %	96	96	96
Oil make			
Middle-oil —325°C, lb/cu ft reactor vol hr	16	16	14
Heavy-oil, lb/cu ft reactor vol hr	0	9	12
Total	16	25	26
Gas, %/(oil make + gas)	23	20	18.5
% Asphaltene in HOLD heavy-oil	14.7	18.0	19.3

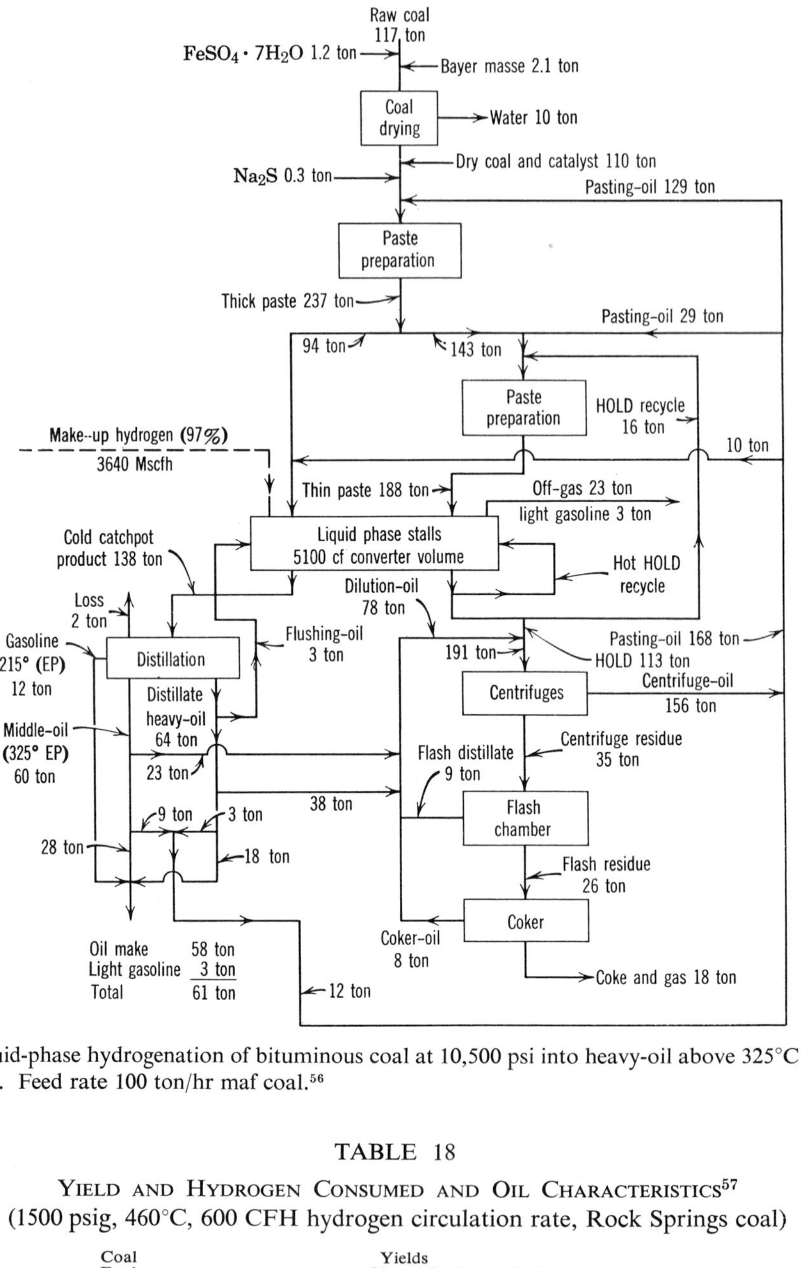

Fig. 7. Liquid-phase hydrogenation of bituminous coal at 10,500 psi into heavy-oil above 325°C and vapor-phase stock. Feed rate 100 ton/hr maf coal.[56]

TABLE 18

Yield and Hydrogen Consumed and Oil Characteristics[57]
(1500 psig, 460°C, 600 CFH hydrogen circulation rate, Rock Springs coal)

Catalyst	Paste Feed Rate lb/hr	Coal Feed Rate lb/cu ft Reactor vol hr	H_2-Absorbed wt % Coal	Coal Lique-faction %	Yields wt % maf Coal: H.C. Gas	Yields wt % maf Coal: H_2O	Yields wt % maf Coal: Soluble Oil	Oil Characteristics: Viscosity at 180°F	Oil Characteristics: % Total Insoluble	% Ash	% Asphaltene	Boiling −300°C %
0.5% Ni*	10.4	15.7	6.2	87.8	22.8	13.9	54.8	125.7 SSU	13.0	2.0	10.9	11.1
0.5% Ni*	16.5	25.0	4.6	86.2	15.5	11.0	61.5	139.5 SSF	17.3	2.9	19.4	11.8
0.5% Mo†	10.1	15.3	6.6	87.3	24.2	15.0	52.0	56.6 SSU	8.5	0.4	7.1	15.4
0.5% Mo†	16.7	25.2	5.5	90.7	18.0	13.1	62.7	82.4 SSU	7.4	0.5	13.7	13.8
1.0% Fe‡	9.9	15.0	4.7	87.2	17.9	9.5	60.6	250.5 SSU	14.3	1.9	17.0	7.6
1.0% Fe‡	16.8	25.4	3.9	77.0	11.2	12.6	53.1	203.7 SSF	20.4	3.2	21.9	7.3
0.5% Sn§	10.1	15.3	6.1	91.7	20.8	12.1	61.5	116.9 SSU	10.7	1.3	13.1	11.2
0.5% Sn§	16.6	25.1	4.7	89.6	15.0	9.6	66.5	159.4 SSU	10.8	1.5	15.4	15.1

* 0.5% Ni on coal as charged as $NiCl_2$.
† 0.5% Mo on coal as charged as ammonium molybdate; coal neutralized with H_2SO_4.
‡ 1.0% Fe on coal as charged as $FeSO_4$.
§ 0.5% Sn on coal as charged as $SnCl_2$.

After tests with semicontinuous centrifuging, the deLaval centrifuge with 3500 rpm and about 3-ton/hr throughput was finally adopted. The residue from the centrifuge was coked in two types of ovens depending on the raw material used.[58] The main problem in coking was heat transfer to a semiliquid, viscous material that was being converted into a sticky paste and coke. It has been suggested that fluidized beds be used for this purpose but experimental results are not available.

Commercial Hydrogenation of Tar in the Liquid Phase

The hydrogenation of tar and oil residues is similar in many respects to that of coal, the main difference is that the amount of HOLD is considerably smaller. Tars and oils boiling above 300°C are suitable for liquid-phase hydrogenation. Tars and shale oils usually were distilled to obtain a residue above 325°C for liquid-phase hydrogenation and the lower-boiling material for direct vapor-phase hydrogenation. With unstable materials that decompose during distillation, the distillation of the tar was usually possible after dilution with the liquid-phase hydrogenation product. Liquid-phase hydrogenation of residues becomes easier with increasing hydrogen content and decreasing asphaltene content as well as a decreasing content of non-distillable materials or residues from vacuum distillation. Hydrogenation of mixtures of various raw materials is possible as long as they are miscible and flocculation of asphaltenes does not occur during the hydrogenation reaction. The properties of some tars and oils used in hydrogenation plants have been given in Table 3.

[58] Sherwood, P. W., *FIAT Final Rept.* No. 952 (1957), 244 pp.

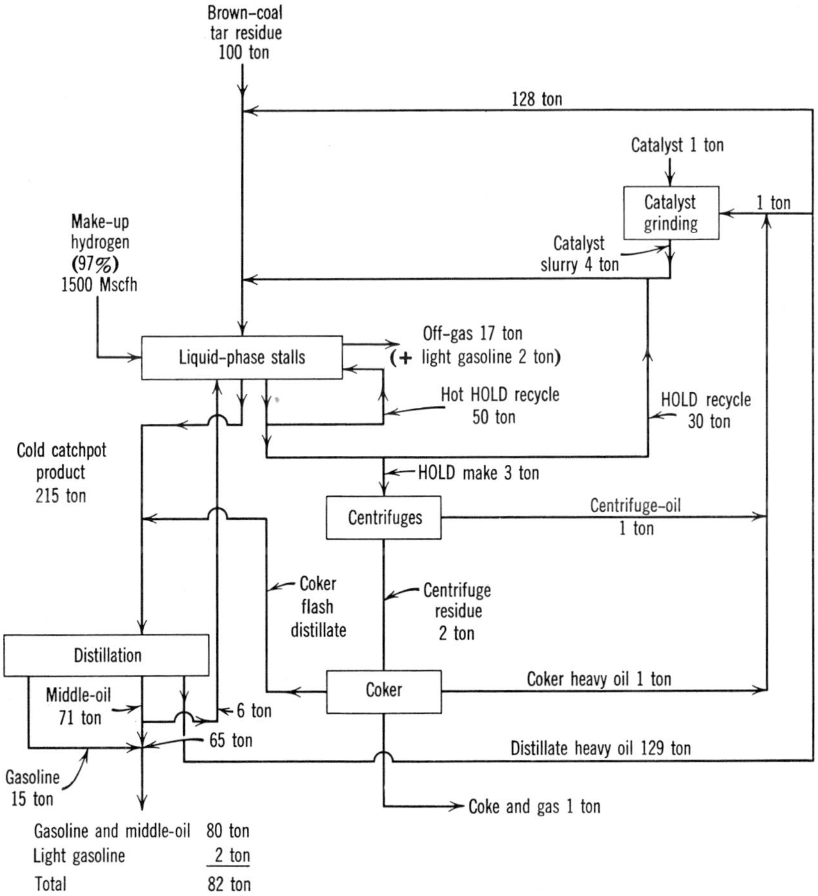

Fig. 8. Liquid-phase hydrogenation of brown-coal tar at 3000 psi. Feed rate 100 ton/hr, tar residue above 325°C.[59]

The residues were hydrogenated as such or together with recycle heavy-oil in the presence of 0.3 to 1.5 wt.% of a finely ground catalyst. The catalyst concentration in the reactors is usually increased by the recycle of part of the HOLD. Raw materials that contain dust, for instance brown-coal tar, were centrifuged before hydrogenation to avoid dilution of the catalyst with solids and settling of coarse particles. The catalyst was ground in asphalt-free oil to a 40% paste and a particle size of below 60 μ. To avoid settling of the catalyst, a light support such as charcoal (e.g. brown-coal char) was used.

A flow sheet[59] for the hydrogenation of brown-coal tar from central Germany in Leuna at 200 atm pressure is given in Fig. 8. One hundred tons of the tar residue above 325°C produced 82 tons of liquid-phase middle-oil and gasoline. In addition, the original carbonization products contained about 100 tons of middle-oil and gasoline which was used as vapor feed together with the liquid-phase product.

Tar from Bohemian lignite was similar but not as favorable in its hydrogenation properties. It was lower in hydrogen content and higher in oxygen and asphaltenes. Even more difficult was the hydrogenation of bituminous coal tars. Tars obtained in Lurgi low-temperature carbonizers could be hydrogenated at 300 as well as at 700 atm. Coke-oven tars with higher aromatic and asphaltene content required higher pressure, such as 700 atm. A comparison of the hydrogenation results[60] obtained with different residues is shown in Table 19. The products of the liquid-phase hydrogenation in their general character are similar to corresponding fractions obtained by low-temperature carbonization.[61,62] The hydrogen consumption amounts to 15,000 scf/t of tar residue corresponding to about 3000 scf/bbl of liquid-phase middle-oil from tar residue or about 1300 scf/bbl of total middle-oil and light-oil contained in the tar.

TABLE 19

LIQUID-PHASE HYDROGENATION OF RESIDUAL OIL[60]

Residue above 325°C from	Crude Oil	Brown-Coal Tar	Coke-Oven Tar
Pressure, atm	700	300	700
Temperature, °C	474	485	482
Feed, lb residue/cu ft reactor vol hr	40	24	14.5
Fresh feed/recycle feed	40:60	30:70	45:55
Oil make −325°C, lb/cu ft reactor vol hr	34	19	11.2
Gas, %	12	17	21.5
Hydrogen consumption, Mscf/ton oil make, −325°C	14.5	18.7	55.5

Hydrogenation at Low Temperature

Commercial operation of the Pott-Broche process from 1938 to 1944 to produce a coal extract at the Welheim plant in the Ruhr district has been described.[63] The extract was used for a short time as feed-stock for hydrogenation at 700 atm pressure. The hydrogenation behavior was similar to that of the coal from which the extract was obtained. The normal feed-stock of the Welheim plant was coal-tar pitch, which was converted into aromatic gasoline and fuel-oil.

The combination of a coal-extraction plant with pitch hydrogenation permitted an important saving in the extraction step. Initially, the solvent for the coal extraction was a 80/20 tetralin-cresol mixture. In commercial operation, this mixture was replaced by the middle-oil obtained by liquid-phase hydrogenation of coal-tar pitch. The spent, dehydrogenated middle-oil from the extraction process was used as vapor-phase feed stock. It was replaced by fresh liquid-phase middle-oil. Thus, the costly steps of naphthalene rehydrogenation and the use of cresol were avoided.

Hydrogen-donor solvents such as tetralin have found application in the hydrogenation of petroleum residues to more valuable lower-boiling products with little degradation to coke

[59] Hoering, M., and Donath, E. E., *Ullmanns Encyklopädie der Technischen Chemie*, third edition, Urban and Schwarzenberg, München, 1958, Vol. 10, p. 521.

[60] Holroyd, R., *BIOS Overall Report* No. 1, *Report on the Petroleum and Synthetic Oil Industry of Germany* (1947), 134 pp.

[61] Storch, H. H., Hirst, L. L., Fisher, C. H., and Sprunk, G. C., *U.S. Bur. Mines, Tech. Paper* No. 622 (1941), 110 pp.

[62] Basu, A. N., and Glenn, R. A., *Fuel*, **27**, 96–9 (1948); **29**, 134–7 (1950). Wolfarth, J. S. and Glenn, R. A., *Fuel*, **28**, 253–7 (1949).

[63] Kröger, C., *Erdöl u. Kohle*, **9**, 441–6, 516–20, 620–4, 839–43 (1956). See also Ref. 8.

and gas.[64] Various petroleum fractions containing polycyclic aromatics have been hydrogenated and used as hydrogen donors. The fraction boiling from 700 to 900°F was more effective as a hydrogen donor than the lower-boiling fractions in the conversion of petroleum residues into gasoline and gas oil.

In the Pott-Broche process, the coal used for the extraction was dried to 1% H_2O and ground to from 60 to 70% below 60μ. The coal was mixed with 2 parts middle-oil from coal-tar pitch hydrogenation in the liquid phase. The coal paste was pressurized to 1500 psi. It was then heated to 415 to 435°C and entered the tubular reactor in which the reaction took place with a residence time of 1 hr. After cooling to 150°C, the product was depressurized and entered the filtration plant. Semiautomatic ceramic cartridge filters were used at 150°C and a pressure of 45 to 60 psi. The filtration rate was about 125 to 145 lb of total product per square foot of filter area per hour. It was necessary to wash and separate the filter residue from the extract solution as fast as possible without reheating to avoid repolymerization or oxidation of the extract which would lead to the formation of insoluble extract fractions and deterioration of the filtration rate. The filter residue with a solvent content of about 35% was freed from solvent in rotary kilns at 360°C. The filtrate was distilled in vacuo to separate spent solvent from the extract for use as vapor-phase feed stock.

The coal liquefaction was from 80 to 82% and the extract yield 66 to 69% on a dry coal basis. The plant capacity at Welheim was 80 to 90 metric tons of extract/day. Properties of the extract in comparison to the coal are shown in Table 20. The extract had a higher hydrogen and lower oxygen and sulfur content than the coal.

Mixed solvents were used for the industrial extraction of coal. It is interesting to note that the properties of a good coal-extraction, solvent normally containing condensed aromatic-naphthenic rings with phenolic groups, can be combined in one[65] molecule such as *o*-cyclohexylphenol or 1,2,3,4-tetrahydro-5-hydroxynaphthalene.

TABLE 20

PROPERTIES OF BITUMINOUS COAL AND ITS EXTRACT[63]

	Extract	Coal
Softening point, °C	200–240	
Analysis, H_2O free		
Ash, %	0.04–0.06	5–6
V.M.	40–42	29–31
C	88.5	85.2
H	5.4	5.2
S	0.6	0.9
Solubility in		
Benzene, %	35	
Trichlorethylene, %	44	
Tetralin, %	69	
Cresol, %	100	

The aforementioned solvents react with coal as hydrogen donors, however a considerable solubilization of coal can be achieved with polycyclic aromatics which would not be expected to act as hydrogen donors. It appears possible, however, that hydrogen-transfer reaction could occur within the coal substance at the extraction temperature. Investigations with phenanthrene[66] and other high-boiling condensed aromatics have shown that 95% of the organic matter of Pittsburgh bed coal can be dispersed in boiling phenanthrene. Solvent fractionation by means of benzene and cyclohexane separates the extract into fractions of different degree of aromaticity and different coking properties. Dispersions of coal[67] in coke-oven tar pitch and heavy oil have been used as roofing pitch and sheet steel and pipeline coatings.

The limited hydrogenation of coal by means of hydrogen-donating solvents leads to solubilized coal that, while still in solution, can be separated from the coal ash by filtration. A similar product can be obtained as was first

[64] Langer, A. W., Am. Chem. Soc. Meeting, April 13–18, 1958, Petroleum Div. Preprints, p. 35. Varga, J., Rabo, G., and Székely, A., *Acta Chim. Acad. Sci. Hung.*, **5**, 443–51 (1955). Varga, J., *Magyar Tudományos Akad., Kem. Tudományok Osztályának, Közleményei*, **5**, 167–82 (1954). Birthler, R., Deutloff, E., Karolyi, J., Steingaszner, P., and Zalai, A., *Brennstoff-Chem.*, **39**, 348 (1958).

[65] Orchin, M., and Storch, H. H., *Ind. Eng. Chem.*, **40**, 1385–9 (1948).

[66] Orchin, M., Golumbic, C., Anderson, J. E., and Storch, H. H., *U.S. Bur. Mines, Bull.* No. 505 (1951), 15 pp.

[67] Rose, H. J., and Hill, W. H., U.S. Pat. 1,925,025 (1933). Fair, W. F., Jr., U.S. Pat. 2,395,041 (1946).

shown by Uhde and Pfirrmann[68] by hydrogenation with gaseous hydrogen at pressures up to 4500 psi pressure, about 410°C, and a coal throughput of 30 lb/cu ft reactor vol/hr. The softening point of the resulting extract, 60 to 120°C, was lower than that obtained with solvents. The extract usually contained one-fourth more hydrogen than the coal from which it was made, whereas the O, N, and S content were reduced by 20 to 35%. The amount of recycle hydrogen was smaller than in the previously described coal hydrogenation processes, namely about 35 Mscf/ton of coal.[69] This amount results in a high-hydrogen partial pressure at the converter inlet which decreases considerably to the converter outlet as the hydrogen consuming reaction proceeds. Thus high liquefaction is achieved whereas the formation of products of higher hydrogen content that reduce the filtration rate is apparently prevented. An understanding of the factors that influence the filtration rate of coal-hydrogenation products is rather limited, however.

Experimentally, coal extracts were used in coke ovens[8] for the production of electrode carbon or as fuel for coal dust engines or in dissolved or molten form as fuel for stationary Diesel engines. In spite of a hydrogen content of less than 7%, ignition properties similar to those of normal Diesel fuels were observed.

Another possible use[69] of coal extracts, especially of extracts prepared with gaseous hydrogen, is in the production of metallurgical coke. For this use, the ash content is not a hindrance and the costly filtration step can be omitted. The extract in amounts of from 5 to 12% was added to slightly coking or noncoking coals before they were used in coke ovens. The preferred procedure was the preparation of briquettes which then were carbonized. It was claimed that mixtures of coal with coal extract produced a stronger coke than mixtures with coke-oven tar pitch.

The preparation of such ash-containing extracts is not limited to operation at the low temperatures around 400°C. Use of the higher temperatures of the normal coal hydrogenation process is feasible with correspondingly shorter residence time or higher throughput. The extracts obtained under such conditions normally gave much lower filtration rates. It is evident that decrease of the throughput in such an operation will lead to extracts with decreasing softening points and finally to oils with decreasing asphaltene content and boiling range.

VAPOR PHASE HYDROGENATION

General

In the second step of the coal-hydrogenation process, the vapor-phase hydrogenation, fixed-bed catalysts are used. Further details about this subject can be found in recent summary articles.[3,70] At first the primary objective was the conversion of middle-oil, vaporized in a stream of hydrogen under pressure, into gasoline. This two-stage process, originally conceived by Pier for the hydrogenation of coal, also proved useful for the hydrogenation of tars and crude oils.

A flow sheet[71] of a vapor-phase hydrogenation unit is shown in Fig. 9. The middle oil from liquid-phase hydrogenation or the distillation of tar or crude oil is brought to reaction pressure and preheated together with part of the recycle hydrogen in shell-and-tube-type heat exchangers by the reaction products. Feed and hydrogen then enter the gas-fired or electrically heated preheater. The mixture leaving the heat exchangers passes on to the converters. The remainder of the recycle gas is added as cooling gas directly into the high pressure converters to control the reaction temperature. Normally two to four converters of 300 cu ft vol are used in series in one stall. The residence time of the reactant in the converters at 300 atm is from 2 to 4 min,

[68] Uhde, F., French Pat. 800,920 (1936); Pfirrmann, T. W., U.S. Pat. 2,167,250 (1939).

[69] Waldmann, H., *Ullmanns Encyklopädie der technischen Chemie*, Third edition, Urban and Schwarzenberg, München, 1958, Vol. 10, pp. 576–81.

[70] Donath, E. E., *Advances in Catalysis*, Frankenburg, W. G., Komarewsky, V. I., and Rideal, E. K., Eds., Academic Press, New York, 1956, Vol. 8, pp. 239–92. Arnold, M.R., *Ind. Eng. Chem.*, **48,** 1629–42 (1956). Bradbury, J. T., *Ind. Eng. Chem.*, **49,** 1523–31 (1957). Arnold, M. R., Bradbury, J. T., Keely, W. M., and O'Hara, F. J., *Ind. Eng. Chem.*, **50,** 1370–9 (1958). Bradbury, J. T., Keely, W. M., O'Hara, F. J., and Vance, R. F., *Ind. Eng. Chem.*, **51,** 1111–20 (1959).

[71] Clough, H., *Ind. Eng. Chem.*, **49,** 673–8 (1957), flow sheet page 674 modified.

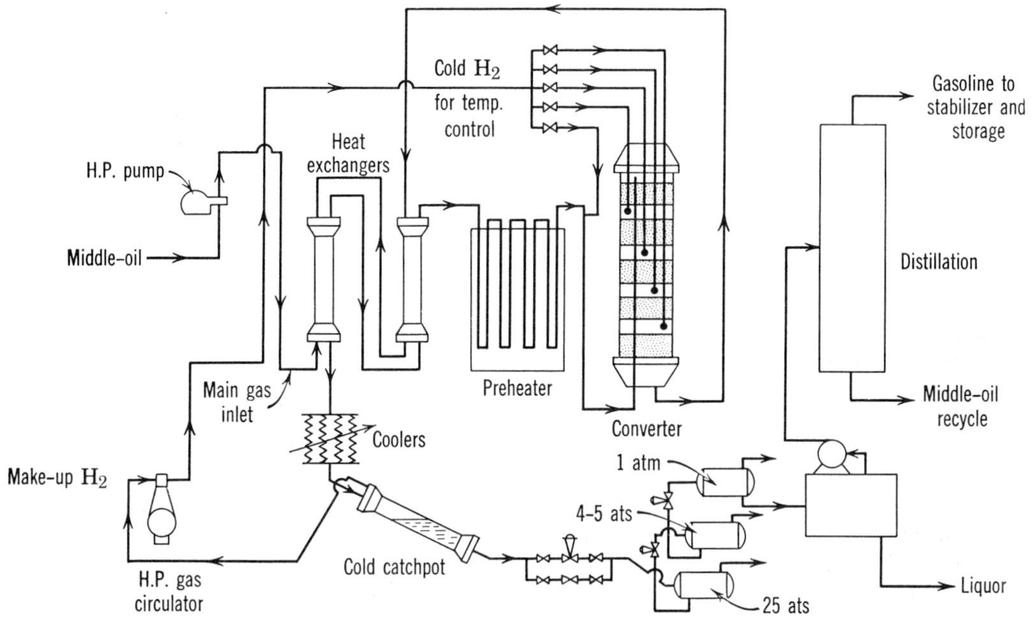

Fig. 9. Flow sheet vapor-phase hydrogenation.[71]

assuming empty reactors. Down-flow is normally used in the reactors since otherwise there is the danger of incipient fluidization which would lead to rapid abrasion of the catalyst followed by dust deposits in the equipment, especially the heat exchangers.

Vapor-phase converters contain several beds of catalyst each in a separate basket, and the spaces between catalyst beds are used for the thorough mixing of the cooling gas with the reactants. This separation is necessary to avoid thermal shock and cold spots on the catalyst and partial product condensation. The control of the reaction temperature thus requires an adjustment of the preheater outlet temperature and the amounts of cooling gas added to the individual catalyst beds. The use of external cooling agents has been repeatedly suggested[72] with utilization of the heat of reaction for the generation of steam. However, so far externally cooled tubular converters have not been used commercially in the coal or oil hydrogenation process at high pressure.

After the reaction products leave the reactors, they enter the inside of the tubes in the heat exchangers and are there cooled to a temperature of 120 to 150°C. In a final cooler, a temperature of 30°C is reached. Lower temperatures are avoided because of the danger of the formation of solid gas hydrates, and at higher temperatures the condensation of light ends would be unsatisfactory. The cooler is flushed with water to avoid formation of deposits and plugging with ammonium carbonate or sulfide. In the cold catchpot, recycle hydrogen is separated from the liquid products.

The recycle hydrogen in the vapor-phase hydrogenation stage returns directly to the recycle compressors. Byproduct hydrocarbons and nitrogen, introduced with the make-up hydrogen, are sufficiently soluble in the liquid product to maintain the required hydrogen concentration. In vapor-phase hydrogenation, it is not necessary to scrub the recycle gas for the removal of gaseous hydrocarbons. By comparison, in the liquid-phase hydrogenation of coal and tar residues, the formation of methane is higher and scrubbing of the recycle gas with oil is normally needed to avoid too high a concentration of gaseous hydrocarbons and nitrogen. It has been suggested that the scrubbing of the recycle gas be avoided by increasing the total pressure and using a recycle gas with a hydrogen concentration below the normal 70 to 90%. Indications were obtained that this may introduce the danger of product foaming and unsatisfactory liquid-gas separation.

[72] Skinner, L. C., Dressler, R. G., Chaffee, C. C., Miller, S. G., and Hirst, L. L., *Ind. Eng. Chem.*, **41**, 87–95 (1949).

The liquid product is depressured in stages to obtain rich and lean gas in separate fractions. Sometimes—for example, with oils that require little hydrogen or with make-up hydrogen of low nitrogen content—it is advantageous to utilize part of the hydrogen contained in the lean gas by adding it to the make-up gas.

After separation of the liquor—the aqueous phase—the hydrocarbon phase is distilled into gasoline and recycle middle oil. The gasoline after stabilization is the finished product. The middle-oil can be used as low-sulfur Diesel or jet fuel or as fuel oil.

Vapor-phase hydrogenation without a catalyst is completely impractical. Whereas coal liquefaction in the liquid phase occurs to some extent without a catalyst, in most vapor-phase hydrogenation processes virtually no reaction occurs without a catalyst. The influence of the catalyst on rate and direction of the reaction is far greater than in liquid-phase hydrogenation.

Catalytic cracking and hydrorefining encompass reactions closely related to those in the vapor-phase step of coal hydrogenation. The development of hydroreforming reaction and of the catalysts used there can be traced to the vapor-phase hydrogenation.[73]

The first catalysts developed for the hydrogenation of middle oils into gasoline were based on molybdenum and tungsten. The first catalyst used commercially at Leuna, about 1926, had the following composition: MoO_3, 53.5%; ZnO, 30%; and MgO, 16.5%. The catalyst was prepared by mixing and reacting an aqueous paste of its components followed by shaping and drying. The activity depended on the purity of the components and the particle size of the zinc and magnesium oxide. Brown-coal middle oil at 450°C and 200 atm pressure gave about 30% conversion to gasoline at an hourly space-time yield of gasoline of 6 to 12 lb/cu ft/hr. The formation of gaseous hydrocarbons was about 20% of the converted feed. In commercial operations, this catalyst showed decreased activity with time. This lessened activity was caused by high-boiling condensation products which were recycled and which formed in increasing amounts as the operating temperature increased. The catalyst remained active longer when feed stocks of lower end point, or with a smaller content of heavy ends, were used. It was also found that the catalyst was completely sulfided during its use. Tests showed that sulfide formation was not responsible for the decrease in activity. Therefore, various methods of preparing sulfided and sulfur-containing catalysts were tested. In 1930, a very active tungsten sulfide was prepared from ammonium sulfo-tungstate. With this catalyst, the splitting of middle oil proceeded at about 400°C with a hourly space time yield of about 30 lb/cu ft of catalyst. The conversion to gasoline was 50% or more and the gas formation based on converted feed was only around 8%. With the increased gasoline yield, greatly increased plant capacity and improved catalyst life were achieved. This catalyst also gave a very satisfactory conversion of highly aromatic bituminous coal or tar middle-oils. The Mo—Zn—Mg catalyst was not satisfactory for the splitting of such oils, the space time yield was lower and catalyst life was shorter than with brown-coal middle oils.

The tungsten disulfide catalyst, however, had one disadvantage. It produced gasolines of high-hydrogen content and a very low content of aromatics and therefore a low octane-number. Although the catalyst produced a higher proportion of branched chain paraffins than the Mo—Mg—Zn catalyst, the increase was not sufficient to compensate for the loss of knock resistance caused by the lower content of atomatics unless gasolines of lower end-point were made. The splitting reactions with this catalyst do not start until the hydrogenation of aromatic rings is nearly complete.

The high-splitting activity of the catalyst and the high content of branched chain paraffins in the product is probably because a carbonium ion mechanism governs the splitting reactions on this catalyst. The products obtained by splitting with the Mo—Zn—Mg catalyst and in the liquid phase in the presence of hydrogen, on the other hand, resemble, except for the low olefin content, the products obtained by thermal cracking of oil which follows a free radical mechanism. These differences

[73] Voge, H. H., *Catalysis*, Emmett, P. H., Ed., Reinhold, New York, 1958, Vol. 6, pp. 407–93. Ciapetta, F. G., Dobres, R. M., and Baker, R. W., *Catalysis*, Emmett, P. H., Ed., Reinhold, New York, 1958, Vol. 6, pp. 495–692. McKinley, J. B., *Catalysis*, Emmett, P. H., Ed., Reinhold, New York, 1957, Vol. 5, pp. 405–526.

TABLE 21

COMPOSITION OF GASEOUS HYDROCARBONS FROM HYDROGENATION OF BITUMINOUS COAL

	Liquid Phase	Mo—Zn—Mg Catalyst	WS_2 Catalyst	Fullers Earth-HF-WS_2 Catalyst
CH_4, Wt. %	26	19	14	1
C_2H_6, Wt. %	24	23	10	1
C_3H_8, Wt. %	32	37	29	15
C_4H_{10}, Wt. %	18	21	47	83
% *i*-C_4/total C_4	25	18	50	75

in the splitting mechanism are clearly reflected in the composition of the products of reaction and are most readily recognizable in the composition of the gaseous hydrocarbons formed as byproducts of the various types of coal hydrogenation processes. Table 21 shows the composition of the gaseous hydrocarbons—methane to butane—and the isobutane content in the total butane.

It may be noted that the isobutane concentration with the WS_2 and supported WS_2 catalyst is above the equilibrium value of about 35% at the operating temperature. This indicates that longer chains undergo isomerization before the splitting off isobutane.[74] Such a splitting mechanism is even more pronounced on the Fullers-earth or aluminum-silicate supported catalysts described later.

Since the octane number of gasoline obtained with the WS_2 catalyst was low, the search for catalysts was continued in two directions. The first was the development of a catalyst of the WS_2 type with a lower hydrogenation activity and a higher activity for the production of branched-chain hydrocarbons. The second was the development of a catalyst for the production of highly aromatic gasolines. This meant finding catalysts that gave higher conversions and lower formations of gaseous hydrocarbons than the Mo—Zn—Mg catalyst.

In 1934, an improved WS_2-type catalyst was obtained by the use of activated Fullers earth. This was, after treatment with hydrofluoric acid, suitable as support for a catalyst containing 10% WS_2. This diluted catalyst gave the same conversion with petroleum oils as WS_2 and in addition produced gasolines with a 5- to 10-point higher octane number. However, if used with brown- or bituminous-coal middle-oils, it rapidly lost its activity. It was found that nitrogen compounds and, to some extent, oxygen compounds were the catalyst poisons. Extraction of tar acids and bases with caustic and acid was possible. However, the loss caused by such treatment made the removal of oxygen and nitrogen from middle-oil by selective hydrogenation to water and ammonia more attractive.

Vapor phase hydrogenation results led to the conclusion that catalysts on "basic" supports are suitable for nonsplitting prehydrogenation-type reactions, and that "acidic" supports are the best used for splitting catalysts. Activated alumina was found to be the best support for prehydrogenation catalysts because of the rapid reduction of tar acids. Alumina precipitated from aluminum salts at a constant pH was especially satisfactory and produced catalysts that could be formed into pellets of high-mechanical strength.

Coal and tar middle-oils, therefore, had to be pretreated by vapor-phase hydrogenation (saturation or prehydrogenation) to remove oxygen and nitrogen compounds. Tungsten sulfide as a saturation catalyst gave satisfactory removal of these compounds; however, it produced some gasoline of comparatively low octane number. It was found that the splitting activity of tungsten sulfide can be virtually suppressed by adding 15% nickel sulfide. This catalyst found commercial application for some time, especially for the hydrogenation of diisobutene to isooctane. It may be mentioned that a catalyst with a higher nickel sulfide content was suitable for dehydrogenation reactions. A prehydrogenation catalyst of similar properties and activity, but lower cost, was obtained with activated alumina as a carrier. This catalyst contained 70% activated alumina, 27% tungsten sulfide, and 3% nickel sulfide and was used commercially as a saturation catalyst.

Improved Fullers earth-WS_2 splitting catalysts were also developed. Hydrofluoric acid-treated Fullers earth alone is a splitting catalyst that produces high-octane gasoline. It required higher operating temperatures and therefore, to avoid loss of activity, had to be used at pressures above 300 atm. Iron compounds on HF-treated Fullers earth also gave gasolines of improved octane number. Catalysts of this type were used commercially by the I.C.I. in Billingham, England for the

[74] Donath, E. E., *Advances in Catalysis*, Frankenburg, W. G., Komarewsky, V. I., and Rideal, E. K., Eds., Academic Press, New York, 1956, Vol. 8, pp. 254–7.

hydrogenation of creosote oil and by ANIC[75] in Italy for the hydrogenation of petroleum oil.

Further octane-number improvement, in comparison with the WS_2-Fullers earth catalyst, was obtained with synthetic aluminum silicates as carrier. Such a catalyst with 0.5% MoO_3 as activator produced gasolines with an aromatics content of about 50% from N_2-free aromatic feedstocks at 400°C.[76] Thus, a catalyst type was found which permitted splitting of aromatic rings, probably by a carbonium ion mechanism, at temperatures around 400°C simultaneously with the formation of paraffins with predominantly branched chains. This catalyst type is basically different from catalysts used for the manufacture of highly aromatic gasolines at 500°C. Catalysts of this type are now being used commercially[77] for the high-pressure hydrogenation of petroleum oils in Germany and were used for the hydrogenation of creosote oil in England.[10]

Apparently, substantial progress in the development of splitting or hydrocracking catalysts has been achieved.[78] This "isocracking" process operates at unusually low temperatures (400 to 700°F or 204 to 371°C) and pressures (500 to 2000 psi or 33 to 135 atm). Low-boiling feed stocks with an end point of 450 to 570°F and 40 to 50% aromatics content such as light coker-distillate or recycle-oil are used as feed. A conversion of 50 to 70% is obtained in recycle operation and gasolines with highly isoparaffinic, low-boiling fractions and 40% aromatics in the high-boiling fraction are produced. The off-gas is rich in propane and butane and its isobutane content is above the equilibrium concentration for the reaction temperature as with some of the aforementioned catalysts used at high pressure. The C_1 to C_3 formation is only 2 to 5 wt.%, and the total volumetric yield of butane and higher boiling material is 114 to 116%. Hydrorefining of the feed stock may be required. The influence of this development on the economics of the coal-hydrogenation process can not be evaluated at this time. However, if these new catalysts could be used in liquid-phase hydrogenation at lower temperatures and pressures, the influence could be great.

Parallel with this development of low-temperature vapor-phase catalysts was the development of catalysts which at higher temperatures (about 450 to 500°C) permitted conversion of aromatic middle-oils into highly aromatic gasoline in one stage. Two such commercially used high-temperature catalysts may be mentioned here. For operation at a pressure of 300 atm, a catalyst containing 80% activated carbon, 15% chromium oxide, and 5% vanadium oxide was used. For operation at a pressure of 700 atm at the Welheim plant of Ruhröl GmbH, a catalyst using Fullers earth as the carrier was developed. This catalyst contained 0.6% molybdenum, 2% chromium, 3% zinc, and 5% sulfur on hydrofluoric acid-treated Fullers earth. It gave higher conversion and longer catalyst life than the 300 atm catalyst. Both catalysts gave gasoline containing about 50% aromatics from bituminous-coal middle-oil.

The catalysts used for the prehydrogenation of middle-oil were suitable for the hydrorefining of petroleum fractions. In view of the smaller amount of impurities contained in most petroleum oils, the use of lower pressures became commercially possible. Most of the petroleum hydrorefining catalysts are alumina-molybdena-type catalysts which may contain additional activating elements such as cobalt, nickel, or others. Even spent noble-metal (e.g. platforming) catalysts have been found useful for the hydrorefining of crude oil fractions. To indicate the importance of hydrorefining reactions in the petroleum industry, it may be worth noting that today 16%[79] of all the crude oil processed in the United States is subjected to hydrogen treatment at some point during its passage through the refinery. This wide application of hydrorefining processes became economically possible because of the availability of cheap byproduct hydrogen from catalytic reforming reactions. A comprehensive treatise of this subject mainly from the point of view of catalysis[80] has been published.

Reactions and Catalysts

Reactions observed in vapor-phase hydrogenation are of two general types: (1) reactions

[75] TOM, Reel 173, Frames 418, 529; Reel 254, Frames 418–23. See Sternberg, W. M., ref. 15.

[76] TOM, Reel 253, Frames 958–84. See Sternberg, W. M., ref. 15.

[77] Oettinger, W., *Erdöl u. Kohle*, **6**, 693–6 (1953).

[78] Stormont, D. H., *Oil Gas J.*, **57**, No. 44, 48–9 (1959).

[79] Anon., *Petrol. Week*, **9**, No. 10, 36–9 (1959).

[80] McKinley, J. B., ref. 73.

without change of the carbon skeleton of the feed; and (2) reactions that involve change of the carbon skeleton of the feed. The first group of reactions includes hydrogenation of olefins and aromatic rings and reduction of organic sulfur, oxygen, and nitrogen compounds to the corresponding hydrocarbons and H_2S, H_2O, and NH_3. Temperatures of 200°C or more are used. The reactions of the second group, like isomerization and splitting, require a higher temperature of about 400°C. The catalyst activity is enhanced by sulfur and decreased by nitrogen and oxygen compounds, and the presence of these elements influences the reaction temperature.

Reactions of pure compounds have been studied with tungsten sulfide in continuous systems, that is, while a stream of hydrogen and oil is fed at reaction temperature and pressure over the catalyst. Reactions without change of the carbon skeleton are shown in Table 22. Hydrogenation of olefinic double bonds becomes rapid at temperatures around 200°C, and for the hydrogenation of aromatic rings to hydroaromatics, temperatures of 300 to 350°C are suitable. For complete reduction of oxygen, nitrogen, and sulfur compounds as contained in coal-hydrogenation middle-oils, a temperature of 380°C is needed.

Tungsten sulfide catalyst should accelerate not only the hydrogenation reactions, but also dehydrogenation of hydroaromatics to aromatics. Calculations show that such reactions are thermodynamically possible at about 500°C and a hydrogen pressure of about 50 atm. Use of the tungsten sulfide catalyst under such conditions results in partial dehydrogenation accompanied by extensive splitting of the hydroaromatics with rather rapid loss of catalyst activity. However, as mentioned before, the splitting activity of tungsten sulfide can be suppressed by addition of nickel sulfide. With a nickel sulfide-tungsten sulfide catalyst at 485°C a total pressure of 50 atm and a 6-atm partial pressure of methylcyclohexane or cyclohexane, the corresponding aromatics can be obtained with a conversion of 90 and 80%, respectively. A liquid hourly space velocity of 0.6 vol/vol can be used, and side reactions, like splitting into gaseous hydrocarbons, occur to only a minor extent. This precursor of other hydroforming catalysts loses activity with time and cannot be regenerated by simple burn-off of the deposited hydrocarbons. The loss of catalyst activity appeared to be more rapid when the content of nonhydroaromatic hydrocarbons in the feedstock was high.

To effect changes in the carbon skeleton of the feed with a tungsten sulfide catalyst, temperatures above 380°C are needed, and for commercial operation about 400°C and higher are used. Examples of such reactions obtained with pure compounds are shown in Table 23. Isomerization reactions with tungsten sulfide as catalyst are possible. Normal butane can be converted to isobutane. For this reaction, a high partial pressure of the butane can be used. The conversion to isobutane reaches the

TABLE 22

HYDROGENATION REACTIONS WITHOUT CHANGE OF THE CARBON SKELETON WITH TUNGSTEN DISULFIDE AS CATALYST[81]

Reaction Type	Feed	Product	Reaction Conditions: Temperature °C	Pressure atm	Feed Partial Pressure atm	Feed Rate vol/vol/hr	Conversion, %
Hydrogenation of olefins	Diisobutylene	Isooctane	216	250	20	2.5	99
Hydrogenation of aromatics	Benzene	Cyclohexane	322	200	3	0.1	99
	Naphthalene	Decalin	336	200	3	1.0	90
Reduction of organic compounds of	Lignite middle oil containing:	Prehydrogenated oil	380	200	8	1.1	
Oxygen	18% Phenols						99.5
Nitrogen	4% Nitrogen bases						99.5
Sulfur	2.5% sulfur compounds						99

[81] Donath, E. E., ref. 44, pp. 251–2.

TABLE 23

HYDROGENATION REACTIONS WITH CHANGE OF THE CARBON SKELETON, TUNGSTEN DISULFIDE AS CATALYST[81]

Reaction Type	Feed	Product	Temperature °C	Pressure atm	Feed Partial Pressure atm	Feed Rate vol/vol/hr	Conversion, %
Isomerization	*n*-Butane	Isobutane	400	200	160	0.6	35
	Cyclohexane	Methylcyclopentane	408	250	27	1.2	90
	Benzene	Methylcyclopentane	408	250	27	1.1	90
Splitting of paraffins	Isooctane	Lower-boiling hydrocarbons	400	250	18	1.7	40
	Paraffins 260°–320°C	Gasoline 180°C end point	408	220	5	1.1	90
	n-Heptane	Lower-boiling hydrocarbons	435	250	20	1.7	15
Splitting of naphthenes	Decahydronaphthalene	Lower-boiling hydrocarbons	408	220	5	1.2	90

thermodynamic equilibrium. Practically no splitting to lower molecular-weight hydrocarbons was observed under these reaction conditions with butane. Isomerization of higher paraffins was obtained together with splitting to lower molecular-weight hydrocarbons. The conversion of cyclohexane to methylcyclopentane is nearly complete at 408°C. Under identical conditions, benzene is also converted to methylcyclopentane. This shows that the hydrogenation of benzene to cyclohexane, which can be considered the first step in this direction, is very rapid.

Under approximately the same reaction conditions, isooctane can be split, with isobutane as the main product. Similarly, a paraffinic oil boiling between 260 and 320°C and decahydronaphthalene are converted almost completely into lower-boiling hydrocarbons. Splitting of normal heptane, however, is slower. Even at 435°C, the conversion to lower-boiling hydrocarbons is only 15%. Decahydronaphthalene is split almost completely at 408°C. Similar reaction products and rates are obtained with naphthalene in pilot-plant experiments. In commercial plants, removal of the heat of reaction is often the controlling factor for the throughput. Hydrogenation of naphthalene, because of the greater heat of reaction, would then require a lower throughput than hydrogenation of decahydronaphthalene.

The activity of vapor-phase hydrogenation catalysts is impaired by the presence of highly condensed aromatic rings such as may occur in asphaltenes or may form by condensation reactions in the absence of a sufficient pressure of hydrogen. This has been demonstrated in some detail in experiments with additions of coronene.[82] Probably the dehydrogenation, condensation, and irreversible adsorption of highly condensed aromatic hydrocarbons is a major cause of the loss of activity of WS_2 as well as of other coal-hydrogenation catalysts. As an example,[83] experiments with a paraffin-base petroleum-oil fraction boiling between 200 and 325°C were made at pressures well below 200 atm. The results in Table 24 were obtained. The catalyst activity, measured by the conversion to gasoline, dropped immediately after the pressure was decreased from 200 to 100 atm and dropped further very rapidly during 6 days of operation at 100 atm. After returning to 200 atm pressure, conversion to gasoline increased slowly, but the catalyst

TABLE 24

WS_2 CATALYST ACTIVITY AT 200 AND 100 ATM[83]

Hydrogen Pressure, atm	Conversion to Gasoline, %
200	90 (Constant value)
100	
First Day	45
Sixth Day	22
Return to 200	
First Day	27
Final value after several days	42

[82] *Ibid.*, pp. 257–8.

[83] TOM, Reel 205, Frames 792–9. See Sternberg, W. M., ref. 15.

activity remained well below the original value. Thus, poisoning of the catalyst by operation at 100 atm pressure is partly reversible and partly irreversible. The reversible part of the catalyst deterioration changes gradually into irreversible damage. For instance, after a short period of operation at 150 atm, the catalyst activity was recovered completely by operating again at 200 or 250 atm, whereas prolonged operation (e.g. for 1 week) at 150 atm caused permanent decrease of activity. The lowest pressure at which high catalyst activity could be maintained for months depended on the operating temperature, throughput, hydrogen-oil ratio, and boiling range, and aromaticity of the feedstock. For the prehydrogenation of coal middle-oil and for the splitting hydrogenation, this pressure seems to be just slightly below 200 atm at 400°C, for the hydrorefining of coke-oven light-oil, petroleum fractions, etc. at lower temperature, pressures of 20 atm are sufficient.

The reduction of the activity of hydrogenation catalysts by nitrogen and oxygen compounds has been mentioned. Experimental studies of the increase of the cetane number of Diesel fuels by hydrogenation with WS_2 catalyst indicate that ammonia or nitrogen compounds may even be beneficial, since they reduce the splitting rate to a greater extent than the pure hydrogenation activity. The splitting activity of a WS_2 catalyst for petroleum middle-oils is apparently not noticeably affected by the addition of aniline up to a nitrogen content of 0.02% based on the feed. An increase of the organic nitrogen content to 0.05% of the feed decreased the splitting rate considerably. A temperature increase of around 20°C was necessary to maintain the same conversion to gasoline compared to no addition of aniline. The activity of nitrogen bases as catalyst poisons depends on their structure. It appears that high-boiling bases with nitrogen in the nucleus are the most deleterious. The WS_2 catalyst is less sensitive to nitrogen bases than the strongly splitting catalysts on Fullers earth as support which have weaker hydrogenating activity. It is more sensitive, however, than the Mo-Zn-Mg catalyst. The carbonium-ion-splitting mechanism appears especially sensitive to ammonia or organic nitrogen compounds. Similarly, cracking catalysts lose their activity in the presence of nitrogen compounds.

On the other hand, the activity of WS_2 is enhanced by the presence of hydrogen sulfide. The presence of sulfur is necessary to maintain the activity of the catalyst, especially when oxygen-containing oils are treated. Usually, a sulfur content of the feedstock above 0.2% is sufficient to prevent a decrease in activity of the WS_2. Higher-sulfur contents up to 2% increase the catalyst activity by 10 to 20%, but a further increase has little influence. Increased oxygen content of the feed seems to increase the required amount of sulfur. In commercial use of WS_2 as a prehydrogenation catalyst, sulfur addition was not necessary with most commercial sulfur-containing oils. Sulfur as such, or as hydrogen sulfide, was added to prehydrogenated, practically sulfur-free feedstocks of splitting catalysts. Because of the corrosion caused by hydrogen sulfide, its addition was kept as low as permitted by catalyst activity.

The mechanism of the catalyst activation by sulfur is not understood. The amount of sulfur compounds necessary to maintain or increase the catalyst activity depends in some cases on the stability of the heavy-metal sulfide component of the catalyst. Thus, molybdenum sulfide seems to require a higher hydrogen sulfide concentration than tungsten sulfide. However, some catalysts that do not contain elements that can form sulfides under reaction conditions also showed an increased activity when sulfur compounds were added to the feed. Hydrogen sulfide in many cases decreases the catalyst sensitivity to nitrogen compounds and thus causes an activity increase. Sufficient data for pure compounds are not available to permit segregation of these effects.

Nonsplitting Catalysts

The tungsten sulfide catalyst permitted conversion of middle-oils to gasolines of any desired boiling range at high-conversion rates and with high yields. The disadvantage of the catalyst was the comparatively low octane number of the gasoline. Catalysts that were developed to produce gasolines of higher octane number using WS_2 on supports were poisoned by nitrogen bases and to some extent by oxygen compounds. It was therefore necessary to remove nitrogen and oxygen from coal middle-oils. A separate hydrogenation step, the prehydrogenation, was used for this purpose. It preceded the second vapor-phase stage, in which middle-oil underwent splitting into gasoline. The prehydrogenation thus is a combined refining and hydrogen addition step for middle-oil.

The WS_2 catalyst was used at first as a prehydrogenation catalyst. The cost of this catalyst and the properties shown in Table 25 were the reasons for the development of other prehydrogenation catalysts. The data in Table 25 were derived from experiments for the prehydrogenation of bituminous-coal middle-oil at 250 atm hydrogen pressure and a throughput of 50 lbs/cu ft/hr (about 0.9 vol/vol/hr of catalyst).

TABLE 25

TEMPERATURE OF PREHYDROGENATION REACTIONS WITH WS_2 CATALYST[84]

Reaction	Begin	Complete
Tar acid reduction	310	380
Tar base reduction	310	380
Hydrogenation	310	Maximum 410–460°C
Splitting	350	*

* At 380°C 12% gasoline boiling below 150°C formed by splitting hydrogenation.

Practically complete removal of nitrogen compounds is needed for the splitting catalyst. Therefore, a prehydrogenation temperature above that at which splitting begins must be used. Since the gasoline formed by the prehydrogenation catalyst has a lower octane number than that obtained from the splitting stage, a prehydrogenation catalyst with minimum splitting activity is desirable.

[84] Donath, E. E., ref. 74, p. 265

The first suitable activated-alumina catalyst contained 10% MoO_3 activated by addition of 3% nickel oxide. This catalyst was superseded by a more active tungsten catalyst containing 70% activated alumina, 27% tungsten sulfide, and 3% nickel sulfide. This catalyst was used commercially for the prehydrogenation of middle-oils and also for the direct hydrogenation of shale oil and lignite tar (TTH process).

The Al_2O_3-W-Ni catalyst was used for the prehydrogenation of coal middle-oils at a pressure of 300 atm, a temperature around 420°C and a throughput of 40 to 65 lbs feed/cu ft/hr. Higher throughputs were possible with the more paraffinic feed stocks. The hydrogen consumption amounted to 3.5 to 5% by wt. of the feed. Recycle gas rates as high as 70 scf/lb feed were used to obtain safe removal of the heat of reaction and a catalyst life of 1 yr or more.

The removal of oxygen and nitrogen compounds from various middle-oils gave products with yields from 93 to 99% by weight and a hydrocarbon off-gas formation of 2 to 3% by weight. Product yield and gasoline content of the product are shown in Table 26.

TABLE 26

YIELD AND PRODUCT COMPOSITION FROM MIDDLE-OIL PREHYDROGENATION[85]

	Prehydrogenation Product, wt. % of		In Prehydrogenation Product	
	Liquid-Phase Middle-Oil and Gasoline	Raw Material	Wt. % Motor Gasoline	Wt. % Aviation Gasoline (Octane Number, 87 Leaded)
Raw Material				
Bituminous coal	97	65	38	27
Brown coal	95	52	45	25
Coke-oven tar	99	79–84	22	15
Low-temperature tar from				
Bituminous coal	96	84	38	27
Brown coal	93	84	45	20
Crude oil	98	92	50	(30)

[85] Hoering, M., and Donath, E. E., *Ullmanns Encyclopädie der Technischen Chemie*, Third edition, Urban and Schwarzenberg, München, 1958, Vol. 10, p. 528.

The oil boiling above gasoline is suitable for the conversion into gasoline or for use as Diesel or jet fuel.

The comparatively high gasoline content is due to the inclusion of the liquid-phase gasoline in the feed-stock. Its octane number was not very high. To improve its octane number in the I.C.I. plant in England,[86] the liquid-phase gasoline was refined separately at high throughput and low temperature.

The degree of refining required for the splitting hydrogenation step varied with the character of the middle-oil. The analytical criterion used was the content of bases. It was expressed as a NH_3-equivalent and determined by titration in an anhydrous medium. The maximum NH_3-equivalent content of prehydrogenated middle-oils from coke-oven tar and bituminous and brown coal was 5 to 10 mg NH_3/l, for petroleum middle-oils it was 20 to 30 and for middle-oil from Central Germany lignite tar it was 70 mg/l. This reduction in content of bases was always accompanied by a reduction of the tar acid content to below 0.5%. At the same time, the end-point of the liquid-phase middle-oil was reduced from 325 to about 290°C.

For the prehydrogenation of creosote oil, Al_2O_3-Mo catalysts[71] have been used commercially to reduce the nitrogen content of about 1% sufficiently for further splitting hydrogenation without the strong saturation of aromatics obtained with WS_2. Table 27 gives a comparison of results obtained under experimental conditions that produce a product with a nitrogen content of 10 ppm. The high-aromatics content with the alumina-cobalt-molybdate catalyst is evident. Such oils

TABLE 27

TUNGSTEN SULFIDE COMPARED WITH COBALT-MOLYBDATE-TYPE CATALYST[71] (3000 lb/sq in. gauge; 97% hydrogen)

Catalyst	Wt. % Co	Wt. % Mo	Average Temperature for 10 ppm Nitrogen	Aromatics Vol. %
Tungsten sulfide			365°C	28
7424	1.6	6.3	390	40
7457	0.8	6.5	403	50
7493	8.4	6.5	425	53

[86] Gordon, K., *Chem. Age* (*London*), **55**, 795–804 (1946).

produced gasolines of considerably higher octane number in splitting hydrogenation with suitable catalysts, as will be seen later.

The more pronounced refining activity of the Al_2O_3-W-Ni catalyst in comparison with WS_2 has also been observed in the TTH process, for the low temperature hydrogenation of lignite-tar for the preparation of Diesel fuel, lubricating oil, and paraffin wax. Data[87] obtained over a range of temperatures, throughputs, and hydrogen-feed ratios have been published and are shown in Table 28.[87] With the alumina tungsten-nickel-sulfide catalyst, a product with a tar-acid content of 0.05 to 0.1% was achieved at a lower temperature than with WS_2. At the same time, the product had a smaller content of fractions boiling below 250°C.

Changes of the hydrogen rate over a wide range do not have a pronounced effect on the results. The hydrogen rate used in commercial operation is about 45 cu ft/lb of feed. Essentially the same result has been obtained in the prehydrogenation of middle-oils. In splitting-hydrogenation reactions, however, especially at temperatures above 450°C, the hydrogen-feed ratio has a noticeable effect with a flat optimum at 45 cu ft/lb.

In the prehydrogenation of middle oils and in the TTH process, tar acids and bases and sulfur compounds are converted into hydrocarbons and the hydrogen content of the hydrocarbons is increased by the saturation of olefins and aromatics as shown, for example, by the aniline-points of the middle-oils in Table 28. The former reaction occurs in the vapor phase, whereas in the TTH process part of the tar remains a liquid and reacts in this form. For both reactions, pressures of 200 atm or higher are used. The hydrogen consumption in the TTH treatment of brown-coal tar is about 4.5 wt.%.

Similar to these reactions and a transition to the hydrorefining processes are the processes used for the improvement of lubricating oils, Diesel fuels, and kerosenes. They require less hydrogen than the TTH treatment of tar. Similar[88] operating conditions and catalysts are suitable. The main result and objective of the increase in hydrogen content in the case of

[87] Guenther, G., *Freiberger Forschungsh.*, **A17,** 38–43 (1953).

[88] Hoering, M., and Donath, E. E., ref. 3, pp. 546–8.

TABLE 28

INFLUENCE OF THE H_2-FEED RATIO IN THE HYDROGENATION OF BROWN-COAL TAR WITH WS_2 AND Al_2O_3-WS_2-NiS[87]

Pressure: 280 atm (80% H_2, 19% $CH_4 + N_2$, 1% H_2S)
Throughput: 23 lbs tar/cu ft catalyst and hour
Feed: Specific gravity, about 0.92; tar acids, 10%; middle-oil aniline point, 32°C.

	Product			
Cu ft gas/lb feed	Specific Gravity	% Tar Acids	Middle-Oil Aniline-Point, °C	% −250°
WS_2 catalyst, 331°C				
46	0.837	0.05	66	32
170	0.840	0.16	64	30
WS_2 catalyst, 314°C				
46	0.889	3.4	43	22
170	0.888	2.5	41	22
Al_2O_3-W-Ni catalyst, 327°C				
12	0.861	0.10	51	27
22	0.864	0.05	52	26
46	0.860	0.10	53	27
56	0.866	0.05	51	26
85	0.866	0.08	52	27
170	0.861	0.09	53	24

lubricating oils is a flatter viscosity-temperature curve. In Diesel fuels, the tendency to engine-knock is decreased and in kerosenes the burning properties are improved.

These processes led to the pure hydrorefining processes of the petroleum industry that use pressures below 70 atm (1000 psi). The purpose of these hydrorefining processes can be: improvement of odor, color, and color stability; removal of impurities for catalytic refining and cracking feed stocks; removal of S, N, and O impurities and olefins; and improvement of thermal and storage stability and lead susceptibility. The raw materials are distillation and cracking products ranging from gasoline to high-boiling vacuum distillates. The most widely applied catalyst is cobalt-molybdate on activated alumina as support. Similarly coke-oven light-oil or other fractions such as naphthalene can be purified by hydrorefining (see Chapter 15, pp. 647–8).

Splitting Catalysts

In the commercial plants, for the splitting-hydrogenation of middle-oils into gasoline, normally a pressure of 300 atm and a temperature of about 400°C were used. In the German hydrogenation plants, a catalyst (I.G. No. 6434) containing 10% WS_2 on activated Fullers earth treated with hydrofluoric acid was used. During the war in England, an iron-containing catalyst (Table 6) that gave gasolines of higher octane number was introduced. After the war, further progress was achieved by the use of specially prepared synthetic aluminum silicate as the catalyst support.

The WS_2-Fullers earth catalyst had a life of 1 to 2 yr, depending on the feed stock. The conversion of prehydrogenated middle-oil as feed stock was 60 to 70% when motor gasoline was produced and 40 to 50% when aviation gasoline was produced. The fractions boiling above gasoline were normally recycled. The recycle gas rate was about 25 to 35 scf/lb total feed and the throughput 60 to 125 lb total feed per cubic foot of catalyst per hour.

The yield based on liquid-phase middle-oil and gasoline was 90 to 94 wt.% of motor gasoline and 80 to 84 wt.% of aviation gasoline which, upon addition of 4 cc/gal lead tetraethyl, had an octane number of 87.[89] The volumetric yield was higher and normally

[89] *Ibid.*, p. 529.

TABLE 29

MOTOR GASOLINE AND AVIATION GASOLINE (OCTANE NUMBER 87)[89]
WEIGHT % YIELD

	Yield Wt. %			
	On Liquid-Phase Middle-Oil and Gasoline		On Raw Material maf Basis	
	Motor Gasoline	Aviation Gasoline	Motor Gasoline	Aviation Gasoline
Raw material:				
Bituminous coal	92	83	62	55
Brown coal	92	81	50	44
Coke-oven tar	93	84	74–78	67–71
Low-temperature tar from				
Bituminous coal	90	81	79	71
Brown coal	90	80	82	73
Crude oil	94	84	88	79

exceeded 100%. The yield of byproduct gaseous hydrocarbons was 6 to 20 wt.%. The sensitivity of this catalyst to nitrogen bases is shown in Fig. 10 with aniline as model substance.[90] The reversibility of the decrease in activity is readily visible.

Prehydrogenated middle-oils are very suitable raw materials for catalytic cracking.[91] The prehydrogenation gasolines derived from coal can be readily used as catalytic reforming feed stocks. Because of high purity and the high ring-compound content very favorable yields of aromatics are obtained.

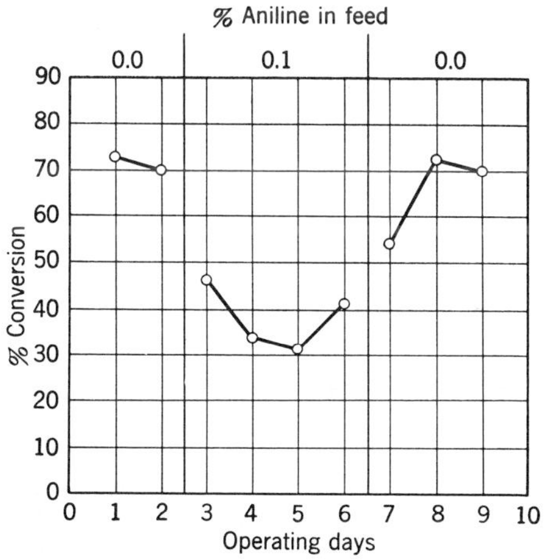

Fig. 10. Nitrogen bases and activity of Fullers earth–WS_2 catalyst.[90]

[90] Donath, E. E., ref. 74, p. 279.

[91] Free, G., *Erdöl u. Kohle*, **3**, 162–7 (1950).

With the catalyst containing 10% WS_2 on HF activated Fullers earth,[92] the gasolines obtained had higher octane numbers than those made over WS_2 catalyst alone. A comparison of this catalyst with WS_2 is given in Table 30. The experiments were made with a

TABLE 30

COMPARISON OF CATALYSTS

Catalyst:	WS_2	HF-Fullers earth + 10% WS_2
Conversion	65%	65%
Hydrocarbon gas as % of converted feed	8	8
Gasoline		
Specific gravity	0.716	0.722
%, 100°C	40	41
End point, °C	188	202
Octane number	57	64
% Paraffins	78	72
% Naphthenes	20	19.5
% Aromatics	1.5	8
% Olefins	0.5	0.5

petroleum distillate boiling between 209 and 327°C, of 0.85 sp. gr., at 200 atm pressure, 407°C, and an hourly throughput of 120 lb of feed/cu ft of catalyst. The data show that conversion and gasoline yield are alike. The gasoline obtained with the Fullers earth catalyst has a higher-octane number. Only about one-half of this increase in octane number can be ascribed to the higher content of aromatics.

[92] Pier, M., Simon, W., and Jacob, P., U.S. Pats. 2,154,527 (1939), 2,194,186 (1940).

This indicates a higher content of more knock-resistant branched-chain paraffins. Similar data for a coke-oven tar middle-oil and various catalysts have been shown in Table 6.

The influence of feed fractions of various boiling ranges on the overall conversion was investigated with a bituminous-coal middle-oil. The oil was prehydrogenated with the Al_2O_3-W-Ni catalyst and the middle oil obtained was distilled into six equal fractions. The properties of the fractions and the results of splitting hydrogenation with the Fullers earth-WS_2 catalyst at 250 atm and an hourly throughput of 93 lb of feed/cu ft of catalyst are shown in Table 31. The gasolines were cut to an end-point of 150°C. The lowest and highest fractions show deviations from the four middle fractions. The high conversion and low gasoline volatility obtained with the first fraction are probably interrelated. This fraction had a high content of single-ring naphthenes. A single opening of a naphthenic ring of this low-boiling fraction would decrease its boiling point sufficiently to convert it into gasoline without formation of fractions boiling below 100°C. Fractions 2 to 5 produce gasolines of similar nature; the gasoline volatility (% at 100°C) decreases, as expected, with increasing molecular weight. The highest fraction requires somewhat higher temperatures for a splitting rate comparable with that obtained with the other fractions. The content of tar acids and bases of this fraction was not higher than in the lower fractions and cannot be the cause. Either the need for a greater number of splitting steps to convert this fraction, which consists of molecules with an average of seventeen carbon atoms, into gasoline or the presence of hydrocarbons of a condensed nature might be the cause of the refractory behavior.

The Fullers earth catalyst produces more branched-chain paraffins and does not saturate aromatics to the same extent as WS_2 does alone. It even permits splitting of part of the aromatics of the feed without saturation. Further development of catalysts, which permit preservation of a higher percentage of the aromatics in the feed and production of more branched-chain paraffins, was successful in England and Germany.

Specially prepared synthetic alumina-silica catalysts containing heavy metals were used for the hydrogenation of petroleum oils. Apparently, precipitation[94] of the silicate at a low pH or coprecipitation with metals is essential. A comparison of gasolines produced from petroleum gas-oils with the Fullers earth-WS_2 and with the silicate catalyst is shown in

[94] Oettinger, W., and Göhre, O., Ger. Pat. 821,684 (1951). Nonnenmacher, H., Oettinger, W., and Reitz, O., Ger. Pat. 854,346 (1952). von Füner, W., Oettinger, W., Nonnenmacher, H., and Reitz, O., Ger. Pat. 869,200 (1953). Oettinger, W., *Erdöl u. Kohle*, **6**, 693–6 (1953).

TABLE 31

INFLUENCE OF FEED FRACTIONS ON CONVERSION

Feed Fraction:	1	2	3	4	5	6
Specific gravity	0.825	0.841	0.856	0.875	0.887	0.894
50% point, °C	164	192	214	243	263	290
Aniline-point	43.5	45.5	48.5	51.5	54.0	58.5
Nitrogen bases (mg NH_3/l.)*	1.0	1.7	3.4	4.8	4.8	1 0
Splitting hydrogenation						
Temperature, °C	374	374	374	374	374	391
Gasoline						
Space time yield, lb/(cu ft)(hr)	55	40	51	51	48	55
Specific gravity	0.735	0.735	0.725	0.730	0.736	0.730
% boiling below 100°C	38	62	58	51	48	45
Paraffins	48	39	44	43	42	42.3
Naphthenes	49	59	54	56	53	48.5
Aromatics	3	2	2	1	5	3
Octane number (motor method)	71	75	76	75	74	71.5

* The content of nitrogen bases was determined by anhydrous titration with perchloric acid in glacial acetic acid and expressed as mg NH_3/liter oil.[93]

[93] Wittman, G., *Angew. Chem.*, **A60**, 330–3 (1948).

TABLE 32

FULLERS EARTH AND SILICATE CATALYSTS

Catalyst:	HF-Fullers earth + WS_2	Silicate
Gasoline		
Specific gravity	0.715	0.723
% Boiling to 100°C	47	45
End-point, °C	190	190
% Aromatics	8	17.5
Octane number		
Motor method, clear	63	72
Motor method + 1.5 ml TEL/gal	76	83
Research method, clear	64	74
Research method + 1.5 ml TEL/gal	77	85

Table 32. With the silicate catalyst, gasolines of higher aromatics and isoparaffin content were obtained. The yield was said to be somewhat lower. The additional aromatics were concentrated in the high-boiling gasoline fractions.

The development in England was concerned with the improvement of catalysts for the hydrogenation of coke-oven tar middle-oil (creosote oil).[71] At first, in the Fullers earth catalyst, the WS_2 was replaced by iron compounds. This brought an octane number improvement of 2 points. Further improvement was attained by using a prehydrogenation catalyst that gave satisfactory nitrogen removal with comparatively little hydrogenation of aromatics (see Table 27). For the splitting-hydrogenation of such oils, Ni-containing silica-alumina catalysts were developed. The improvement in octane rating is shown in Table 33, a brief summary of the catalyst development since about 1930 (see Table 6).

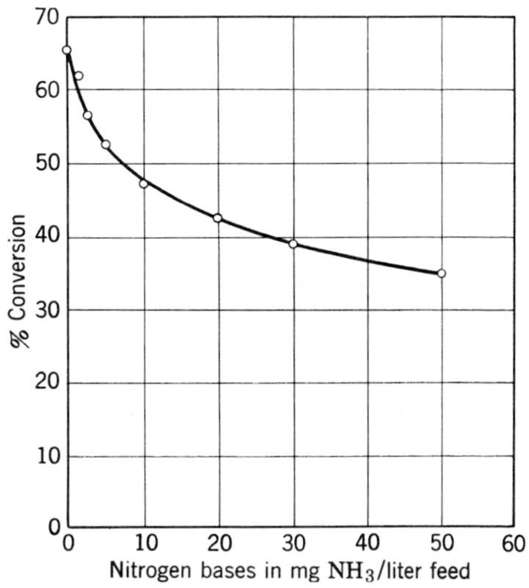

Fig. 11. Influence of increasing amounts of nitrogen bases on the activity of the Fullers earth-WS_2 catalyst.[90]

The activity of the WS_2 catalyst is decreased by nitrogen bases. The nitrogen sensitivity of the Fullers earth-supported catalyst, however, is much greater. Figure 11 shows the change in conversion of an acid-washed, nitrogen-free gas-oil with the addition of nitrogen bases.[90] Addition of the equivalent of 50 mg NH_3/liter decreased conversion by about 50%. Figure 11 indicates that increasing amounts of bases become proportionally less effective. However, different feed stocks seem to have different sensitivity to N-bases; the catalyst is practically inactive at 400°C when bituminous-coal middle-oil containing tar bases equivalent to 5g NH_3/liter is used as feed. However,

TABLE 33

COMPARISON OF CATALYST 231 AND 7679

(Pressure 3000 lb/in.² 97% hydrogen; throughput, 1 kg/l./hr of cobalt-molybdate-refined creosote; aromatics 46 vol %)

			6 psi Reid Vapor Pressure Gasoline Properties				
						Octane Number 1.25 Ml. TEL/US Gal.	
Catalyst	Temperature, °C	Conversion, %	Distillate and loss at 100°C, %	End-Point, °C	Aromatics	Motor Method	Research Method
231*	375	68	50	163	18	85.5	91.5
7679†	380	75	51	146	23	87.0	94.5

* Fe on HF-activated montmorillonite.

† Ni on silica-alumina.

Fullers earth-supported catalysts show remarkable splitting activity with nitrogen-containing feed stocks at temperatures higher than 450°C.

The catalyst is not permanently damaged by tar bases. Figure 10 shows the decrease in conversion observed with petroleum gas-oil when 0.1% of aniline is added. Operating conditions similar to those shown in Table 28 were used. After aniline addition, conversion drops to one-half of the original value. When the aniline addition is discontinued, the conversion rapidly returns to its original value.

Single-Stage Splitting Catalysts

The first two commercially used catalysts (Mo-Mg-Zn and WS_2) belong in this category. The WS_2-Fullers earth catalyst developed later can be regarded also as a single-stage splitting catalyst for N-free oils such as most petroleum gas-oils. The high content of aromatics in all coal-derived oils invited the development of catalysts that permit splitting of these oils into highly aromatic gasolines. Two such catalysts were used commercially, one for operation at 300 atm, the other for 700 atm pressure.

The 300 atm catalyst was used commercially in the Scholven plant for the production of aviation gasoline blending-stock with an aromatics content of 50 to 55% from bituminous-coal middle-oil. The catalyst contained 15% Cr_2O_3 and 5% V_2O_5 on activated carbon. The space-time yield was 30 lbs/cu ft/hr of gasoline in recycle operation with a conversion of 30% at a temperature of 500°C. The recycle gas rate was 35 scf/lb total feed and the formation of gaseous hydrocarbons 20 to 25 wt.% of the fresh feed depending on gasoline end-point.

For use at 700 atm, a Fullers earth-supported catalyst was developed at the Welheim plant. It contained, in addition to Fullers earth, HF, Cr_2O_3, ZnO, S, and 0.5% MoO_3. From coke-oven tar middle-oil, gasolines with about 50% aromatics content were obtained at about 450°C. Throughput and gasoline space-time yield in recycle operation thereby were about 38 and 22 lb/cu ft/hr respectively and the gas formation 12 to 15% of the fresh feed. The hydrocarbon off-gas contained only about 30% total butane with an isobutane content of 25%. With the Fullers earth catalysts, which were used at 300 atm, the corresponding figures were 75 and 75%. This difference is not caused by the pressure but mainly by the difference in temperature,[95] which was only around 400°C for the 300 atm catalyst using prehydrogenated middle-oil.

The 700-atm vapor-phase catalyst is of interest in plants that use 700 atm in the liquid-phase stage and produce only highly aromatic gasoline. The Bureau of Mines used this catalyst in experiments at the Louisiana, Missouri demonstration Plant,[17,96] and considered its use for the manufacture of gasoline from coal. The catalyst was reproduced and tested in experimental equipment at the Bruceton Station of the Bureau of Mines.[97]

The splitting-hydrogenation of middle-oils with an end-point of 325 to 350°C is a well-developed process. Attempts have been made to use splitting-hydrogenation processes with fixed-bed catalysts for high-boiling oils. The first successful attempt was connected with the TTH hydrogenation of brown-coal tar with WS_2 or Al_2O_3-WS_2-NiS catalysts. The normal operating temperature was 360°C. It was found possible to operate without catalyst deterioration at higher temperatures and thereby increase gasoline yield and decrease the production of high-boiling oils. Table 34 gives yields and product properties for the MTH process[98] that were practiced at the Zeitz plant. This process would, because of the high yield, appear very attractive for the production of middle-oil as feed for the splitting hydrogenation into gasoline. However, the octane number of the gasoline obtained from this highly saturated middle-oil was poor.

Splitting of Colorado shale oil has been attempted[99] with a hydrorefining cobalt-molybdate catalyst. The authors stated that the

[95] Pier, M., *Brennstoff-Chem.*, **32**, 129–33 (1951).

[96] Clarke, E. A., Chaffee, C. C., and Hirst, L. L., *U.S. Bur. Mines, Rept. Invest.* No. 4676 (1950), 27 pp.; No. 4944 (1953), 80 pp.; No. 5043 (1953), 66 pp.

[97] Wolfson, M. L., Pelipetz, M. G., Damick, A. D., and Clark, E. L., *Ind. Eng. Chem.*, **43**, 536–40 (1951). Pelipetz, M. G., Frank, L. V., Ginsberg, H. H., Wolfson, M. L., and Clark, E. L., *Chem. Eng. Progr.*, **50**, 626–8 (1954).

[98] Hoering, M., *Die Chemie der Braunkohle*, W. Knapp, Halle, 1953, Vol. II, pp. 283–93.

[99] Cottingham, P. L., White, E. R., and Frost, C. M., *Ind. Eng. Chem.*, **49**, 679–84 (1957).

catalyst would require periodic regeneration in view of the indicated catalyst deposit. This would be a major difference between this operation and the aforementioned TTH and MTH processes. Inspection data for the shale oil are given in Table 35 and selected yield and product inspection data in Table 36.

TABLE 34

PRODUCTS FROM TTH AND MTH BROWN-COAL TAR HYDROGENATION PROCESSES[98]

Operating temperature		TTH 360°C	MTH 390°C
Throughput lb/cu ft/hr		25	25
	Feed	Products	
Specific gravity 50°C	0.918	0.828	0.805
% −200°C	0	11	29
% −200 to 350°C	51	53	65
% >325°C	48	36	6
Tar acids, %	7	0.2	<0.1
Cetane No., 200 to 350°C fraction	20	>50	>50
Hydrocarbon off-gas, wt. %	—	3	6
H_2 consumption, wt. % of tar	—	4.4	5.6

Cobalt-molybdate on alumina is a hydrorefining or prehydrogenation catalyst and produces at the lowest used temperature of 759°F a substantially sulfur-free product, which still retains about 20% of the original nitrogen content. At increasing temperatures, the gasoline content of the product increases. The octane number of the gasoline remains substantially unchanged and low until, at temperatures above 900°F, the formation of gaseous hydrocarbons and the aromatics content of the gasoline increase. This and the low propane-butane content of the hydrocarbons off-gas indicate that the splitting is not of the type obtained with the Fullers earth or aluminum-silicate supported catalysts. The product with a boiling range above gasoline decreases in specific gravity up to a reaction temperature of 900°F, owing to a decrease in molecular weight and an increase in hydrogen content. Above 900°F, an increasingly higher specific gravity due to a higher aromatics content is observed. Such a process could be of interest for a combined production of gasoline and natural gas substitutes.

The TTH and MTH processes are suitable for the treatment of shale oil. A commercial plant using this process is in operation[100] in Puertollano, Spain. More readily hydrogenated than shale oil are distillates obtained by the coking of shale oil. Lower pressures have been used for the hydrogenation of such distillates than for shale oil itself. Catalysts normally used for refining reactions have been used for this purpose at temperatures at which splitting is pronounced. Results[101] obtained with a cobalt-molybdate catalyst at a pressure of 1500 psi, 835°F, and liquid hourly-space-velocities of 0.9 to 1.1 show that Colorado shale-oil coker distillate is converted with 96 to 98 vol % yield into 60% jet fuel and 40% Diesel fuel. Runs in excess of 300 hours were possible before catalyst regeneration became necessary. Processes of this type are envisaged at present as one step in the conversion of shale oil into products normally obtained from petroleum.

TABLE 35

PROPERTIES OF COLORADO SHALE OIL PREPARED IN GAS COMBUSTION RETORT[99]

Specific gravity 60°/60°F	0.9393
Pour point, °F	79
Viscosity S.S.U. at 100°F	311
Ultimate Analysis, wt. %	
Sulfur	0.67
Nitrogen	2.10
Carbon	84.90
Hydrogen	11.32
ASTM distillation corrected to 760 mm, °F	
Initial boiling point	273
Recovery, %	
5	369
10	427
20	548
50	695
End-point	733
Total recovery, vol %	70
Loss, vol %	2

Splitting-hydrogenation of heavy oil is applied in Scholven using crude oil residues as feed stock. The liquid-phase hydrogenation product, after passing through hot catchpots for the separation of heavy asphaltic oil and

[100] Sanchez, R., and Rodriguez-Jurado, J., *World Petrol. Congr., Proc. 4th Congr., Rome*, **1955**, Sect. III, pp. 445–56.

[101] Clark, E. L., Hiteshue, R. W., Kandiner, H. J., and Morris, B., *Ind. Eng. Chem.*, **43**, 2173–8 (1951).

TABLE 36

MATERIAL BALANCES AND LIQUID PRODUCT PROPERTIES[99]

(Single-pass hydrogenation of raw gas-combustion shale oil over cobalt-molybdate catalyst; 3000 psi, 1.0 vol/vol/hr; 6000 cu ft H_2/bbl feed)

Average catalyst temperature, °F/°C	759/403	842/450	890/477	933/491	979/526	1010/543
Product Yield, wt. %						
Gasoline	15.7	26.2	38.3	47.9	49.4	38.6
Recycle-oil	79.7	66.8	46.6	28.4	13.3	6.4
Total product oil	95.4	93.0	84.9	76.3	62.7	45.0
Gas	3.0	4.0	12.2	20.9	34.1	52.3
Catalyst deposit	0.7	1.0	1.1	0.9	1.1	1.2
Product specific gravity 60/60°F	0.8768	0.8287	0.7955	0.7782	0.7623	0.7696
Gasoline						
Specific Gravity 60/60°F	0.7667	0.7560	0.7432	0.7465	0.7354	0.7465
% S	0.02	0.01	0.02	0.01	0.01	0.02
% N	0.13	0.03	0.01	0.005	0.005	0.005
% Aromatics	15	18	14	22	25	37
50% Point °F	315	289	276	257	222	205
Octane number Research	49.8	50	50	55	70	79
Recycle oil						
Specific gravity 60/60°F	0.9026	0.8643	0.8530	0.8582	0.8922	0.9804
% S	0.02	0.01	0.03	0.02	0.07	0.07
% N	0.45	0.16	0.03	0.06	0.15	0.13

liquid-phase catalyst, enters directly[102] into converters containing a catalyst in a fixed bed which reduces the heavy-oil content from about 50 to 25%. Stable and low-sulfur gasoline and Diesel fuel are obtained as the main products. For the splitting and refining of a heavy oil–middle-oil–gasoline mixture, catalysts using MoO_3 or WS_2 on aluminum-silicates[77] were suitable. The catalyst can be reactivated by careful burning-off of carbonaceous deposits. This type of combined operation could be useful for the hydrogenation of coal or tar if recovery of tar acids is not intended since the fixed-bed catalyst used to date will reduce the tar acids and bases drastically. A drawback of this combination, especially in hydrogenation of low-rank coals and lignite, is that reduction of the CO and CO_2 formed in the liquid-phase unit into CH_4 would occur in the vapor phase. This would cause additional consumption of hydrogen. In normal liquid-phase operation, CO and especially CO_2 are removed by being dissolved in product oil and recycle-gas scrubbing-oil.

[102] Urban, W., *Erdöl u. Kohle*, **8**, 780–2 (1955). A similar process has been proposed by Skinner, L. C., Donath, E. E., Schappert, H., and Frese, E., *Petrol. Refiner*, **29**, No. 7, 83–7 (1950).

CHAPTER 23

NONFUEL USES OF COAL

RICHARD A. GLENN

Bituminous Coal Research, Inc, Monroeville, Pennsylvania

The nonfuel uses of coal are considered to be those in which: (1) coal is converted into other useful products by such processes as carbonization, gasification, hydrogenation, oxidation, extraction, combustion, hydrolysis, sulfonation, halogenation, and amination; (2) coal or coke is used to furnish carbon for reduction of ores to free metals, for the reduction of nonmetals and their salts, and for the conversion of oxides of metals to chlorides and the conversion of oxides of both metals and nonmetals to carbides; (3) coal is used primarily for some particular physical property, for example, for the manufacture of activated carbon, artificial graphite, protective coatings, and molding materials; and, (4) other products from the preparation and utilization of coal are used as raw materials, as sources of chemicals, or as materials of construction.

In this review of the nonfuel uses of coal, information is presented primarily on those processes which have been developed to or beyond the pilot-plant stage with emphasis being placed on the unit requirements of raw materials or unit yields of primary products and the present state of development of the process. For many purposes coal and coke are used to supply both heat for the process and carbon for reduction; such uses are being included as nonfuel uses.

About 100 million tons of coal are consumed annually in the United States for nonfuel purposes. The consumption of bituminous coal in the United States during 1959 is shown in Fig. 1 according to the principal uses.[1] The nonfuel uses are included in the two categories designated as "carbonization" and "other

[1] U.S. Bur. Mines, *Weekly Coal Rept.* No. 2216 (1960), 15 pp.

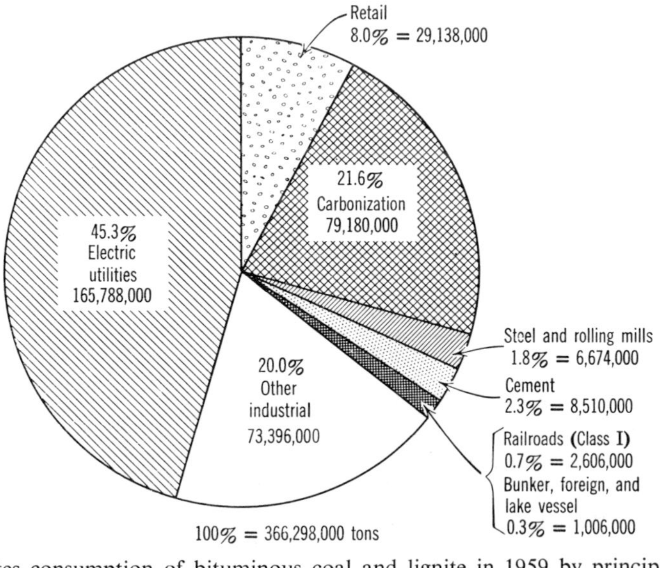

Fig. 1. United States consumption of bituminous coal and lignite in 1959 by principal uses.[1] (Exports excluded.)

industrial," with an undesignated portion of the latter being coal for fuel purposes.

PROCESSES FOR CONVERSION OF COAL TO OTHER PRODUCTS

The most important nonfuel use of coal is as a raw material for conversion into chemicals. The processes for converting coal are as follows, arranged in approximate order of their importance: carbonization, gasification, hydrogenation, oxidation, extraction, combustion, hydrolysis, sulfonation, halogenation, and amination.

Individual chemical reactions taking place during the conversion of coal into chemicals by these processes are complex—a natural consequence of the complex nature of the coal substance itself.

Carbonization

By far, the largest nonfuel use of coal is for carbonization, that is, for destructive distillation with the resultant production of coke, tar, gas, and other products. Detailed data on the consumption of anthracite, bituminous coal, and lignite are reported monthly by the U.S. Bureau of Mines and summarized annually in the "Minerals Yearbook." During the 10-year period, 1950–1959, the consumption of bituminous coal and lignite in slot-type ovens varied from a low of 75,563,000 tons in 1958 to a high of 104,648,000 tons in 1953.[2] On December 31, 1959, there were seventy-four slot-type oven coke plants in the United States together with forty-five beehive oven plants with 5148 beehive ovens in operating condition.[3]

The estimated world production of metallurgical coke in 1958 was 289 million tons; this was some 5% lower than the previous year in which almost 294 million tons of oven and beehive coke were produced. As a result of the marked drop in coke production in the United States during 1958, the U.S.S.R. led the world in the production of metallurgical coke by producing 56,100,000 tons.[4]

Altogether, 366,298,000 tons of bituminous coal and lignite were consumed in the United States during 1959. Of this amount, 79,180,000 tons or 21.7% were used in the production of coke—77,340,000 tons in slot-type ovens and 1,840,000 tons in beehive ovens. The relation of the amount of coal used for carbonization and that used for other purposes, on a percentage basis, is presented in Fig. 1.[1] The steel industry uses coke in large quantities as a source of carbon for both reduction and fuel. Tar and gas are processed for their chemical values and then used for fuel and other purposes. Figure 2 shows graphically the distribution of coke by use in the United States for selected years.[5] Use of coke in the blast furnace continues to be the major market for coke.

Carbonization in Slot-Type Ovens

The yields of the various principal products from the carbonization of coal in slot-type ovens with chemical recovery vary with the carbonizing conditions and with the type of coal charged.[6] On the average, the yields of the principal products may be stated as being: 1520 lb coke and breeze, 78 lb tar, 20 lb light oil, 20 lb ammonium sulfate, and 280 lb (10,500 cu ft) coal gas.[7] The average yield of coke in slot-type ovens in the U.S. for the six-year period 1953–1958 was about 1400 lb/ton coal carbonized.[8]

The yields of selected individual chemicals recovered as byproducts from carbonization of coal in slot-type ovens are given in Table 1.[7] The yield of benzene, 11.8 lb/ton coal, is the largest for any individual liquid chemical, and the yield of methane, 132 lb/ton coal, is largest of any of the gaseous products. The combined yield of the individual chemicals listed represents only about one-third the total yield of the crude tar and light oil. The complexity of the condensible materials is also shown by the

[2] Young, W. H., Anderson, R. L., and Hall, E. M., *U.S. Bur. Mines*, "*1958 Minerals Yearbook*," Vol. 2, 126 (1959).

[3] Bureau of Mines Staff, *U.S. Bur. Mines, Inform. Circ.* No. 7996 (1960), 20 pp.

[4] Young, W. H., Anderson, R. L., and Hall, E. M., *op. cit.*, p. 242.

[5] Sheridan, E. T., Ryan, E. E., and DeCarlo, J. A., *U.S. Bur. Mines, Mineral Rept.* MMS No. 2950 (1959), 41 pp.

[6] Denig, F., *Chemistry Coal Utilization*, John Wiley and Sons, New York, 1945, Vol. 1, pp. 774–833. Davis, J. D., *Chemistry of Coal Utilization*, John Wiley and Sons, New York, 1945, Vol. 1, pp. 834–47.

[7] Rhodes, E. O., *Encyclopedia of Chemical Technology*, Kirk, R. E., and Othmer, D. F., Eds, John Wiley and Sons, New York, 1954, **13**, p. 625.

[8] U.S. Bur. Mines, *1953–1958 Minerals Yearbooks.*

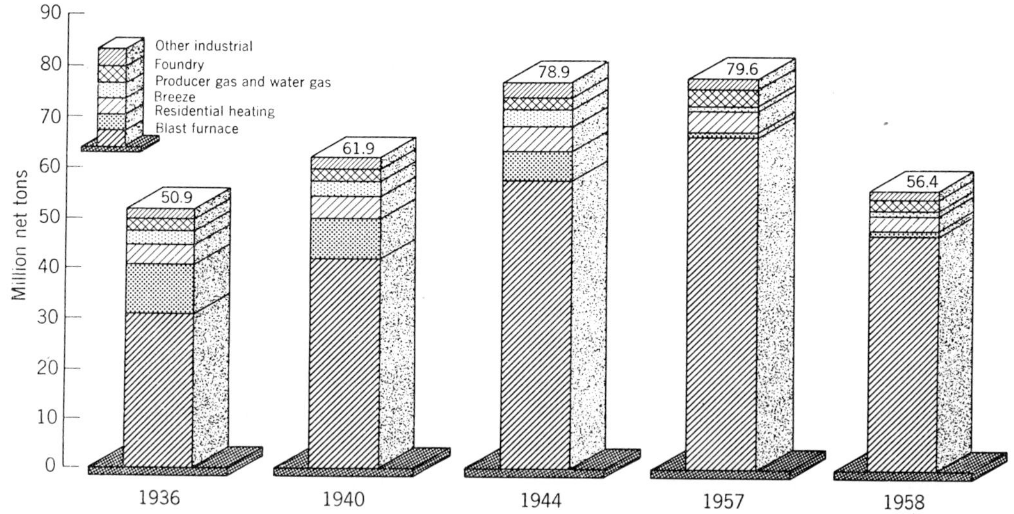

Fig. 2. Distribution by principal uses of oven and beehive coke in the United States in 1958.[5]

fact that with one exception, each individual chemical represents less than 0.5% of the coal charged. Even so, as a result of the large tonnage of coal carbonized each year, there is a large total production of the various coal chemicals. Data on the amounts of those chemicals recovered by coke-oven operators by distillation of only one-third (34%) of the total crude tar produced in the United States in 1958 are listed in Table 2.[9] The production of these chemicals by tar distillers from the processing of more than half the tar is not included in the table; about 15% of the tar was burned as fuel without any prior processing.[10]

TABLE 1

YIELDS OF SELECTED CHEMICALS FROM HIGH-TEMPERATURE CARBONIZATION OF COAL[7]

Coal Tar and Light Oil	Lb/Ton	Coal Gas	Lb/Ton
Benzene	11.80	Carbon monoxide	43.2
Toluene	2.72	Hydrogen	30.4
Xylene	1.33	Methane	132.0
Naphthalene	6.48	Hydrogen sulfide	6.7
Phenanthrene	2.26	Hydrogen cyanide	1.7
Anthracene	0.64	Ethylene	19.6
Pyridine	0.16	Propylene	3.4
Quinoline	0.13		
Phenol	0.95	Liquor	
		Ammonium sulfate	20.0

[9] Young, W. H., Anderson, R. L., and Hall, E. M., *op. cit.*, p. 247.

[10] *Ibid.*, p. 253.

Carbonization in Beehive Ovens

The carbonization of coal in beehive ovens yields only coke; the byproducts, such as tar and gas, are not recovered. The average yield of beehive coke for the 6 yr period 1953–1958 was about 1200 lb/ton coal charged.[8]

Carbonization in Mixtures with Iron Ore

The possibility of adding iron ore to the coke-oven charge to produce ferrocoke has been investigated widely in recent years, particularly in Germany and the United States.[11,12] Production of ferrocoke on an industrial scale for almost two years was reported to have been entirely satisfactory; 20% blast-furnace flue dust was used in the coke-oven charge.[13]

In addition to the production and use of ferrocoke in blast furnaces, the use of ferrocoke in the Thyssen gas producer has been reported and discussed by Paschke.[14]

[11] Barking, H., and Eymann, C., *J. Metals*, **10,** 274–6 (1958); *Brennstoff-Chem.*, **37,**129–44(1956); *Stahl u. Eisen*, **75,** 386–91 (1955); *Glückauf*, **88,** 1090–4 (1952); **89,** 993–1003 (1953).

[12] Russell, C. C., Whitstone, P., and Liggett, R. P., *AIME, Proc. Blast Furnace, Coke Oven, & Raw Materials Comm.*, **14,** 93–113 (1955).

[13] Granger, G., Lach, M. J., and Mitchell, J., *AIME, Proc. Blast Furnace, Coke Oven, & Raw Materials Comm.*, **18,** 152–161 (1959).

[14] Paschke, M., *Proc. Intern. Conf. on Complete Gasification of Mined Coal, Liège*, **1954,** Paper C6; *Ann. mines Belg.*, **1954,** Special No. 211–8.

TABLE 2

UNITED STATES PRODUCTION OF COAL-CHEMICAL MATERIALS AT COKE-OVEN INSTALLATIONS IN 1958[9]

Crude tar	669,316,299 gal
Crude chemical oil	22,058,867 gal
Pitch of tar	824,470 tons
Ammonia (NH_3 equivalent all forms)	381,151,834 lb
Gas	789,828,396 Mcf
Crude light oil	218,229,276 gal
Benzene	119,669,528 gal
Toluene	28,072,423 gal
Xylene	8,408,431 gal
Solvent naphtha	4,419,272 gal
Sulfur	2,581,780 lb

Note: Production by tar distillers not included.

Carbonization on Traveling Grates

Traveling-grate coking stokers are being used in some areas for the continuous production of coke for use as a reactant in certain chemical and metallurgical processes. In Canada, coke for carbide manufacture has been produced commercially for several years in stokers adapted to rotary lime-burning kilns. About 385 tons of coke are produced daily on traveling grates at 1100 to 1200°C from coal with about 38% volatile matter.[15] Of the heat in the coal as fuel, 46.6% is recovered in the coke, 47.6% goes to the kilns in the gas, and 5.8% is lost. The theory of the reactions taking place and the calculations needed to establish the basic laws governing carbonization on traveling grates together with various applications have been discussed.[16]

The operation of a traveling-grate coking stoker in South Africa has been investigated by the Fuel Research Institute and reported to be operating successfully with a fixed carbon recovery of 91 to 97%.[17] In the United States, the continuous production of coke on a water-cooled vibrating-grate stoker has been the subject of an investigation in which the heat generated was employed for raising steam in a boiler. The overall energy recovery of the stoker-boiler combination in the form of steam and coke was found to approach 89% of that represented by the total energy in the starting coal.[18] No commercial operations of traveling-grate coking stokers in the United States have been reported.

Carbonization in Coal-Gas Retorts

City-owned gas companies in the United States have replaced coke-oven gas with natural gas where available. At the end of 1958, only three city-owned plants were in operation, and one of these had announced plans to close in 1959. During 1958, the amount of coal carbonized in city-owned gas plants amounted to approximately 870,000 tons or only two-thirds of that carbonized the previous year.[19]

Carbonization by Medium- and Low-Temperature Processes

The large yield of tar from medium- and low-temperature carbonization of bituminous coal and lignite continues to stimulate research and development on a significant scale in the United States.[20] Low-temperature tar is quite different in character from high-temperature tar, however, and much research and development work will be required to develop markets for such tar in amounts and at prices which will justify low-temperature carbonization of coal in the United States. Markets for low-temperature char for power generation may be developed sooner than markets for the tar. World production of gashouse, low-temperature, and medium-temperature coke (excluding breeze) amounts to approximately 52 million tons annually; of this, more than 43 million tons are produced in Europe including about 13 million tons from Great Britain alone.[4]

In the United States, the economics of medium- and low-temperature carbonization on a commercial scale continues to be unattractive, except in certain areas. In 1956, the latest year for which data are available, only 182,000 tons of gashouse, low-temperature, and medium-temperature coke combined were produced.[21]

[15] Andersen, A. H., *Trans. Can. Inst. Mining Met.*, **47**, 139–51 (1944).

[16] Baum, K., *Combustion*, **26**, No. 5, 61–5 (1954); *Brennstoff-Wärme-Kraft*, **6**, 244–9 (1954).

[17] LaGrange, C. C., *J. S. African Inst. Mining Met.*, **57**, No. 3, 99–114 (1956).

[18] Grace, R. J., and Doherty, J. D., Paper 5819P6, Ann. Meeting AIME, February, 1958.

[19] Young, W. H., Anderson, R. L., and Hall, E. M., *U.S. Bur. Mines, 1958 Minerals Yearbook*, Vol. 2, 260–1 (1959).

[20] Price, J. D., *Proc. 52nd Meeting, Rocky Mountain Coal Mining Inst.*, **1956**, 30–48.

[21] Young, W. H., Anderson, R. L., and Hall, E. M., *op. cit.*, p. 243.

Rotary kilns. In a plant with a design capacity for processing 300,000 tons/year, bituminous coal is subjected to low-temperature carbonization in rotary kilns after a prior, oxidation treatment. By this process, 1 ton of high-volatile bituminous coal yields: 1440 lb semicoke; 150 lb (15.8 gal) tar; 16 lb (2.5 gal) light oil; 250 lb (3720 cu ft) gas; and, 110 lb (6.6 gal) liquor. The semicoke is sold commercially as a smokeless fuel.[22]

Fluidized beds. Processes based on fluidized-suspension carbonization have received particular attention in recent years in the United States. A 50-ton/day pilot-plant for carbonizing bituminous coal by a fluidized technique has been reported to have been operated experimentally more than one year. By using this technique, 150 to 400 lb of tar are produced from a ton of coal—as much as four times that by carbonization in high-temperature coke ovens. In addition, 1000 to 1400 lb char and 700,000 to 4,000,000 Btu in gas are obtained.[23] Plans for a plant to supply char to a large electric power plant have been discussed.[24]

Continuous, fluidized-bed, low-temperature carbonization of Alabama coking coals have been investigated in a pilot plant.[25] Internal blending of raw coal with at least 10 parts of hot char was found to be the best procedure for overcoming agglomeration. In the carbonization of America Seam coal by this process, approximately 80% of the coal was converted to char, 10% to tar, 5.3% to water, and 4.7% to gas.

Fluidized-bed carbonization of lignite has been investigated extensively on a pilot-plant scale.[26] Commercial-scale operation of the process using Texas lignite was studied at Rockdale, Texas, in a prototype unit.[27] The properties and processing of the low-temperature tar from Texas lignite[28] and from America Seam coal[29] have been investigated in an effort to find markets for these materials.

One of the major potential markets for low-temperature char is for fuel for power generation; however, for low-temperature carbonization to be attractive to power generating companies, the processing and fixed costs of carbonization, together with the loss in fuel value, must be offset by the credit received from the sale of the tar.[30]

Vertical retorts. Low-temperature carbonization of North Dakota lignite briquettes in vertical retorts by the Lurgi-Spülgas process is being conducted in a 200 to 300 ton/day full-scale commercial plant producing 750 lb char and 50 lb tar per ton lignite in addition to gas.[31]

Continuous carbonization of subbituminous coal at about 1200°F in vertical retorts has been under investigation on a pilot-plant scale for several years.[32] Char for electric furnace production of phosphorus and for roasting and smelting of nonferrous metals is produced together with roofing tar, light oils, and creosote. Construction of a 100,000-ton/yr plant has been reported[33] but has not been commercially operated.[34]

Gasification

Hydrogen and/or carbon monoxide are used in large quantities as the starting materials in various synthetic processes. The manufacture of ammonia and methanol, the manufacture of synthetic liquid fuels, the hydrogenation of vegetable oils and other miscellaneous organic chemicals, the manufacture of formic and

[22] Denig, F., *Chemistry of Coal Utilization*, John Wiley and Sons, New York, 1945, Vol. 1, pp. 774–883. Lesher, C. E., *AIME, Trans.*, **139,** 328–63 (1940); *Mining Eng.*, **4,** 287–99 (1952).

[23] Anon., *Coal Age*, **57,** No. 10, 70–5 (1952).

[24] Pursglove, J., Jr., *Coal Age*, **62,** No. 1, 70–3 (1957).

[25] Lang, E. W., Smith, H. G., and Bordenca, C., *Ind. Eng. Chem.*, **49,** 355–9 (1957).

[26] Parry, V. F., Landers, W. S., Wagner, E. O., Goodman, J. B., and Lammers, G. D., *U.S. Bur. Mines, Rept. Invest.* No. 4954 (1953), 43 pp. Parry, V. F., Landers, W. S., and Wagner, E. O., *Mining Eng.*, **8,** 54–64 (1956).

[27] Parry, V. F., *U.S. Bur. Mines, Rept. Invest.* No. 5123 (1955), 27 pp.

[28] Batchelder, H. R., Filbert, R. B., Jr., and Mink, W. H., *Ind. Eng. Chem.*, **52,** 131–6 (1960).

[29] Lang, E. W., and Lacey, J. C., Jr., *Ind. Eng. Chem.*, **52,** 136–40 (1960).

[30] Minet, R. G., Smith, H. B., Jr., and Trilling, C. A., *Chem. Eng. Prog.*, **50,** 342–7 (1954). Landers, W. S., *Proc. Amer. Power Conf.*, **19,** 344–51 (1957).

[31] Bur. Mines Staff, *U.S. Bur. Mines, Inform. Circ.* No. 7692, Part II, 45–7 (1954).

[32] Anon., *Chem. Eng.*, **63,** No. 7, 138, 140 (1956).

[33] Anon., *Mechanization*, **22,** No. 7, 98 (1958); *Chem. Eng.*, **67,** No. 9, 105 (1960).

[34] Berg, L., and Atkinson, D. E., *Chem. Eng. Prog.*, **56,** No. 1, 73–7 (1960).

TABLE 3

Coal and Coke Requirements for the Major Types of Coal Gasification Processes[37]

Process	Raw Materials Per 1000 cu ft	Tons Coal and Coke Consumed
Lump Coal or Coke		
Blue water gas	32 lb coke	(1953) 300,000 coke for methanol
(290 Btu/cu ft)	55 lb steam	545,000 coke for ammonia
		(1954) 1,030,569 coke for all purposes
		(1975) 1,000,000 coke
Producer gas, hot raw	16.7 lb bituminous coal	(1948) 4,500,000 bituminous coal
(167 Btu/cu ft)	6.7 lb steam	(1953) 421,272 coke
		(1954) 352,152 coke
Producer gas, cold clean	13.5 lb coke	(1975) 2,000,000 coke
	7.2 lb steam	
Pulverized Coal for Synthesis Gas		
Atmospheric pressure	39 lb coal	Experimental and semicommercial
	20 lb steam	plants in operation in the United
	350 cu ft oxygen	States; large scale unit reported
		under construction.
Elevated pressure	36 lb coal	Pilot plants only
	11 lb steam	
	340 cu ft oxygen	
(1975) 150 to 175 million tons of coal will be gasified directly, primarily for synthetic liquid fuels.		

oxalic acids, and the manufacture of oxygenated chemicals by the Oxo process are examples.

Commercially, lump coal or coke is used in water-gas sets and gas producers to obtain a product suitable for fuel and as a source of hydrogen and/or carbon monoxide. In the past decade, producers of methanol and ammonia in the United States have switched from coal to natural gas as a source of synthesis gas and heat. Thus, in 1958, only 625,000 tons coke were used for gas-making for all purposes; this is only one-fifth of that used for this purpose in 1947–1949.[35] The consumption of coal in gas producers in the United States by the steel industry has also declined. In 1959, only 67,667 tons coal were used for this purpose.[36] Data on consumption of coal for producer gas by other industries, for example glass, brick, lime, are not available.

The basic raw material requirements for the major types of coal gasification processes are summarized in Table 3 together with data on the current and future consumption of coal and coke.[37] The world's first pulverized coal gasification plant[38] began operating on a commercial scale in Finland in 1952. The original plant, based on a process investigated by the U.S. Bureau of Mines at Louisiana, Missouri,[39] had three gasifier units, one of which functioned as a reserve. Construction of other Koppers-Totzek plants in Japan, Spain, Belgium, and France have been reported.[40]

The world's first commercial peace-time plant to produce synthetic liquid fuels from coal has been operating for several years near Coalbrook, Union of South Africa.[41] Pulverized coal is gasified at about 380 lb/sq in. by the Lurgi process. The development of the Texaco process of gasifying coal by partial oxidation was studied recently at Morgantown, West Virginia, using an 80 ton/day

[35] Young, W. H., Anderson, R. L., and Hall, E. M., *op. cit.*, p. 230.

[36] American Iron and Steel Inst., *Annual Statistical Report for 1959*, New York, 1960, p. 36.

[37] Rose, H. J., and Glenn, R. A., *The Chemical and Other Process Uses of Coal*, Paper presented before Division of Gas and Fuel Chemistry, Am. Chem. Soc. Cincinnati, Ohio, April, 1955.

[38] Totzek, F., *Brennstoff-Chem.*, **34**, 361–7 (1953); *Chem. Eng. Prog.*, **50**, 182–7 (1954).

[39] Willmott, L. F., Batchelder, H. R., and Tenney, R. F., *U.S. Bur. Mines, Rept. Invest.* No. 5038 (1954), 35 pp.

[40] Hoy, H. R., and Wilkins, D. M., *BCURA Bull.* **22,** 57–110 (1958).

[41] Hoogendoorn, J. C., and Salomon, J. M., *Brit. Chem. Eng.*, **2,** 238–54 (1957).

unit.[42] Continuous gasification of high-temperature coke and low-temperature char from the Disco process has been demonstrated in a full-scale Kerpely producer using oxygen or oxygen-enriched air.[43]

The gasification of pulverized coal at atmospheric pressure with steam and oxygen was studied by Strimbeck and co-workers.[44] Full-scale operation of the process for synthesis gas production was conducted at Belle, West Virginia,[45] for a number of years before being discontinued. The basic raw material requirements for the pressure gasification of pulverized coal with steam and oxygen were reported in 1953.[46] More recently a statistical analysis of the effects of process variables has been reported.[47]

Additional studies on a pilot-scale unit have been conducted at the Institute of Gas Technology and reported in detail.[48]

The economics of fuel gas manufactured from coal have been outlined in relation to the application of technically feasible gasification processes.[49] The various unit operations involved in the manufacture of high Btu gas from coal have been reviewed and discussed.[50] An excellent review of the developments in the total gasification of coal was reported in 1958 by Hoy and Wilkins,[40] and the status of the various gasification processes has been summarized (Table 4).[51] The economics of coal gasification have not yet justified commercial operation of any of these processes in the United States.

Hydrogenation

The high-pressure hydrogenation of coal in the liquid phase at elevated temperatures is another outstanding example of the conversion of coal into products useful as chemicals and intermediates. The chemical reactions occurring during the hydrogenolysis of coal are as complex as the reactions occurring during carbonization. Consequently, they are described simply as the reaction of coal with

[42] Anon., *Chem. Week*, **82,** No. 12, 62 (1958); *Chem. Eng. News*, **34,** 2855–6 (1956).

[43] Batchelder, H. R., Dressler, R. G., Tenney, R. F., Kruger R. E., and Segur, R. D., *AGA, Proc.*, **1950,** 348–54. Willmott, L. F., Batchelder, H. R., and Tenney, R. F., *U.S. Bur. Mines, Rept Invest.* No. 5108 (1955), 16 pp.

[44] Strimbeck, G. R., Cordiner, J. B., Jr., Baker, N. L., Holden, J. H., Plants, K. D., and Schmidt, L. D., *U.S. Bur. Mines, Rept. Invest.* No. 5030 (1954), 37 pp.

[45] Grossman, P. R., and Curtis, R. W., *ASME, Trans.*, **76,** 689–95 (1954); *Mech. Eng.*, **76,** 47 (1954).

[46] McGee, J. P., Schmidt, L. D., Denko, J. A., and Pears, C. D., *Gasification and Liquefaction of Coal*, AIME, New York, 1953, pp. 80–108. Strimbeck, G. R., Cordiner, J. B., Jr., Taylor, H. G., Plants, K. D., and Schmidt, L. D., *U.S. Bur. Mines, Rept. Invest.* No. 4971 (1953), 27 pp.

[47] Holden, J. H., Strimbeck, G. R., McGee, J. P., Willmott, L. F., and Hirst, L. L., *U.S. Bur. Mines Rept. Invest.* No. 5573 (1960), 56 pp.

[48] von Fredersdorff, C. G., Pyrcioch, E. J., and Pettyjohn, E. S., *Inst. Gas Tech. Research Bull.* No. 7 (1957), 71 pp.

[49] Waring, B. W., and Foster, J. F., *Economics of Fuel Gas from Coal*, Foster, J. F., and Lund, R. J., Eds., McGraw-Hill, New York, 1950, Chap. 5.

[50] von Fredersdorff, C. G., and Vandaveer, F. E., *AGA, Proc. Operating Sect.*, **1959,** CEP-59-8, pp. 41–64.

[51] Field, J. H., Benson, H. E., and Anderson, R. B., *Chem. Eng. Prog.*, **56,** No. 4, 44–8 (1960).

TABLE 4

Present Status of Coal Gasification Processes[51]

Process	Lurgi	Entrained		Winkler	Flesch-Demag
Type of operation	Fixed-bed, nonslagging	Vertical or cyclone gasifier, slagging	Koppers horizontal gasifier, slagging	Fluidized, nonslagging	Cyclic, fixed-bed gasification
Status	Commercial	Mainly pilot plant	Commercial	Commercial	Large pilot plant
Capacity, tons coal gasified/hr	15 to 20	0.2 to 4	5	45	4
Pressure	Elevated	Elevated	Atmospheric	Elevated	Atmospheric
Oxygen consumption, cu ft/1000 cu ft $H_2 + CO$	230	300 to 350	300 to 350	300 to 350	250
Plant Location or research groups	Germany Czechoslovakia South Africa Australia Pakistan	Texas Co. Institute of Gas Technology Bureau of Mines Fuel Research Station BCURA, Ruhrgas	Finland Japan	Leuna, Germany Zeitz, Germany	Badische Anilin-Soda-Fabrik, Germany

hydrogen to yield motor fuel, oil, pitch, and gases.

Production of Liquid Fuels

Hydrogenation of coal by the Bergius process was used widely in Germany during World War II for the production of motor fuels.[52] It has been investigated extensively in the United States both in the laboratory and in a demonstration plant.[53] The data and results obtained in these studies have been used by a number of workers to estimate the output of proposed full-scale commercial plants.[54] Even after allowing credit for the various byproduct chemicals such as phenol, cresols, xylenols, benzene, toluene, xylenes, naphthalene, and L.P.G., the economics of the process have not led to commercialization of the process in the United States.

Production of Chemicals

The production of chemicals rather than liquid fuels by liquid-phase hydrogenation of coal has been investigated in a "pilot plant" processing 300 to 500 tons of coal per day. By reducing the contact time to a few minutes instead of a few hours, increased yields of valuable chemicals were claimed to be possible[55] (see Fig. 3). A 12-ton/day unit was announced in 1956 for use in improving techniques developed during 5 yr of operation of the larger pilot plant.[56]

[52] Parker, A., *J. Inst. Fuel (London)*, **20**, 128–9 (1946). Cockram, C., CIOS Rept. Nos. 30–105 (1945), 39 pp. Sherwood P. W., FIAT Rept. No. 952 (1947), 244 pp. Gordon, K., BIOS Rept. No. 1 (1947), 134 pp.

[53] Kastens, M. L., Hirst, L. L., and Chaffee, C. C., *Ind. Eng. Chem.*, **41**, 870–85 (1949). Chaffee, C. C., and Hirst, L. L., *Ind. Eng. Chem.*, **45**, 822–38 (1953). Clarke, E. A., Chaffee, C. C., and Hirst, L. L., *U.S. Bur. Mines, Rept. Invest.* No. 4944 (1953), 80 pp.

[54] Ford, Bacon, and Davis, Inc., *The Synthetic Liquid Fuel Potential of the U.S.*, New York, 1952, pp. 14–15. Donath, E. E., *Ind. Eng. Chem.*, **46**, 2032–5 (1954). President's Materials Policy Commission, *Resources for Freedom*, 1952, Vol. IV, p. 177. Glenn, R. A., and Rose, H. J., *The Metallurgical, Chemical, and Other Process Uses of Coal*, Bituminous Coal Research, Inc., Pittsburgh, Pennsylvania, 1958, p. 41.

[55] Anon., *Chem. Eng. News*, **30**, 1954–9 (1952); *Coal Age*, **57**, No. 10, 70–75 (1952). Kirkpatrick, S. D., *Chem. Eng.*, **60**, No. 12, 180–4, 250–3 (1953).

[56] Anon., *Chem. Eng. News*, **34**, 4009 (1956).

Liquid-phase coal hydrogenation is widely recognized for its potential as a process for the volume production of aromatic chemicals. The realization of this potential depends, however, on the advances made in the technology of the process and of the recovery and utilization of the resultant products (see Chapter 22).

Oxidation

The conversion of coal into chemicals by oxidation processes has been studied extensively. Many different kinds of oxidation reagents have been used—nitric acid, potassium permanganate, ozone and oxygen.

Alkali-Oxygen Oxidation

The oxidation of alkaline slurries of coal with oxygen under pressure at elevated temperatures has been studied on both a laboratory and a pilot-plant scale. About half the carbon in the coal could be recovered in the form of carboxylic acids following a 2-hr oxidation at 270°C and an oxygen pressure of 900 psi. In addition to 1000 lb of aromatic acids, 40 lb of acetic acid, 170 lb of oxalic acid, and about 3000 lb of carbon dioxide were produced per ton of coal[57] (see Fig. 4).

Thirty-five individual components have been identified in the reaction mixture including the mono-, di-, tri-, tetra, penta-benzene-carboxylic acids.[58] By investigation of the material from the decarboxylation of the oxidation product, it was also shown[59] that the mixture contains a large number of compounds with different nuclei in their structure.

The acids obtained by this process are of potential value as drilling-mud additives, and as intermediates for thermosetting resins, and water-soluble films, and similar materials.[60] The future of the process will depend on progress in developing economical methods to process the complex mixture of acids produced and on the availability of cheap oxygen.

[57] Franke, N. W., Kiebler, M. W., Ruof, C. H., Savich, T. R., and Howard, H. C., *Ind. Eng. Chem.*, **44**, 2784–92 (1952).

[58] Roy, A. N., and Howard, H. C., *J. Am. Chem. Soc.*, **74**, 3239–42 (1952). Holly, E. D., and Montgomery, R. S., *Fuel*, **35**, 49–55 (1956).

[59] Montgomery, R. S., and Holly, E. D., *Fuel*, **36**, 63–75, 493–500 (1957), **37**, 181–6 (1958).

[60] Anon., *Chem. Eng. News*, **37**, No. 39, 96, 98 (1959).

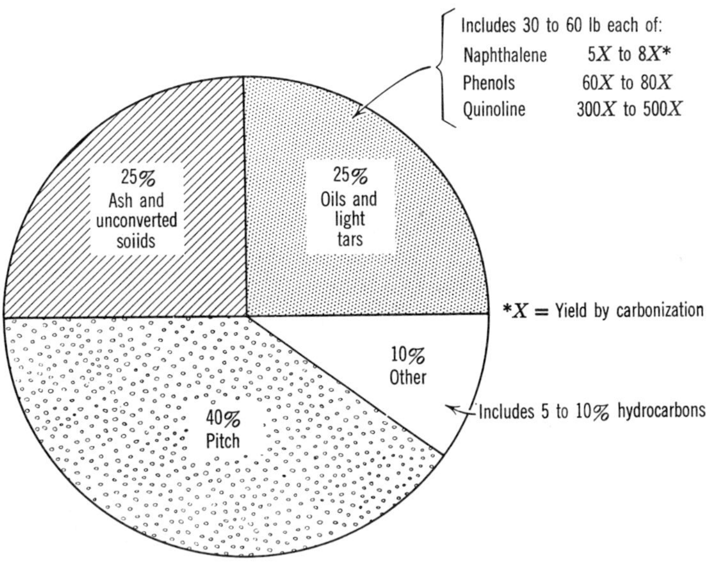

Fig. 3. Distribution of products from hydrogenation of coal to produce chemicals.

Nitric Acid Oxidation

Hot, concentrated nitric acid rapidly attacks bituminous coals and converts the major portion of the carbon of the coal to alkali-soluble, humic acid-like, oxidation products containing several percent nitrogen.[61] Field tests using ammonium and potassium salts of these humic acid-like materials have shown that such derivatives of bituminous coal may be as satisfactory as some commercial types of plant food.[62] However, as far as is known, nitric acid oxidation on a commercial scale has never been attempted in the United States.

[61] Savich, T. R., and Howard, H. C., *Ind. Eng. Chem.*, **44**, 1409–11 (1952). Kinney, C. R., and Ockert, K. F., *Ind. Eng. Chem.*, **48**, 327–32 (1956).

Extraction

The extraction of coal with solvents has been studied perhaps more extensively in the laboratory than any other process for the resolution

[62] Charmbury, H. B., and Polansky, T. S., *Mechanization*, **24**, No. 1, 65–7 (1960).

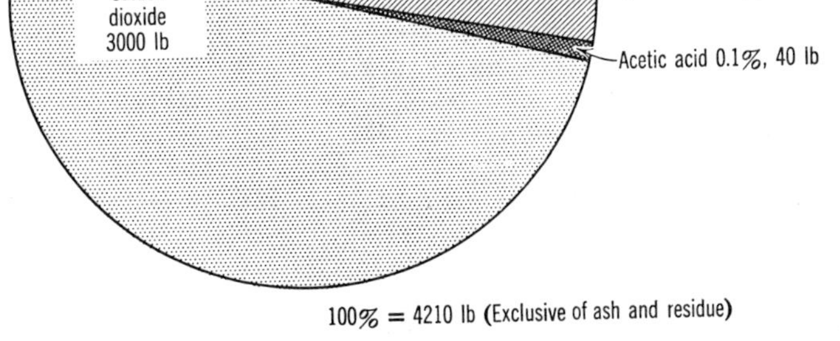

Fig. 4. Distribution of products from pressure oxidation of 1 ton of coal with alkali and oxygen.

and conversion of coal into simpler materials. All types of solvents have been tried on all types of coal at various temperatures. The processes tried may be readily grouped into two classes according to the temperatures employed—namely, processes above and below a temperature of 350°C.

The extraction of coal at temperatures above 350°C was conducted on a commercial scale during World War II in Germany to obtain a very low-ash product which was valuable as a raw material for the manufacture of electrode carbon.[63] About 5.3 tons of coal were processed per hour with a yield of 3.7 tons of nearly ash-free extract. This extract was processed further by carbonization in regular coke ovens. Although there is a huge market for electrode carbon in the United States today, the high-temperature extraction of coal to separate it from the ash has not been attempted on a commercial scale.

Low-temperature extraction of certain low-rank coals has been commercially successful on a limited scale.[64] Certain Utah coals contain from 5 to 7% resin and waxes which have been extracted commercially and processed into dielectric enamels and similar materials.[65]

At least one 25-ton/day plant has extracted California lignite with solvents to recover 280 lb of crude montan wax per ton. The lignite was extracted further to recover a quantity of humate material, and finally the spent lignite was used as fertilizer filler.[66]

Another plant has extracted North Dakota leonardite, a naturally oxidized lignite, to recover alkali-soluble materials valuable for their wood-staining properties. Leonardite, however, is used mainly as a drilling mud additive.[67]

[63] Lowry, H. H., and Rose, H. J., *U.S. Bur. Mines, Inform. Circ.* No. 7420 (1947), 12 pp.

[64] Kleptko, E., *Eng. Mining J.*, **150**, No. 1, 81–3 (1949); *AIME, Trans.*, **177**, 433–48 (1948).

[65] Anon., *Coal Age*, **62**, No. 1, 126 (1957).

[66] Anon., *Chem. Processing*, **16**, 20–21 (1953). Bur. Mines Staff, *U.S. Bur. Mines, Inform. Circ.* No. 7692, Part II, 114–16 (1954). McMurtrie, R., and Oppelt, W. H., *Ind. Eng. Chem.*, **48**, 1996–2000 (1956).

[67] Budge, C. E., *N. Dakota Research Found. Bull.*, No. 8, 32, 39 (1959). Bur. Mines Staff, *U.S. Bur. Mines, Inform. Circ.* No. 7692, Part 2, 113 (1959). Fowkes, W. W., and Frost, C. M., *U.S. Bur. Mines, Rept. Invest.* No. 5611 (1960), 12 pp.

Combustion

Ordinarily combustion is not considered as a process for the conversion of coal into chemicals. It is used however, as a means of converting coal into carbon dioxide both as the main product and as a byproduct. Both "dry ice" and liquid carbon dioxide are produced commercially from coal—the combustion of coal producing both the carbon dioxide to be recovered and the heat and power needed in the process. One ton of dry ice from 18% flue gases requires 0.85 ton of coke or 1 ton of coal plus 286 kwhr electric energy.[68] The production of 1 ton of carbon dioxide from limestone and coke requires an estimated 0.14 ton of coke.

Coal combustion products are also processed along with the gases from the decomposition of carbonate in many processes where carbon dioxide is required in quantity. The soda ash, beet sugar, and the basic magnesia industries are examples. The quantities of coal so used are not available. The production of 1 ton of 58% soda-ash by the ammonia-soda process requires 190 to 220 lb of coke for calcining limestone plus 820 to 1500 lb of coal for heat and power.[69]

One ton of refined beet sugar requires 0.05 ton of coke for calcination of limestone and carbon dioxide production plus 0.66 ton of coal for heat and power.[70]

Hydrolysis

Low-rank coals may be converted in good yield to soluble products by hydrolysis. The alkaline hydrolysis of a Wyoming subbituminous coal has been investigated on a laboratory scale at temperatures up to 425°C, and the composition of the products from hydrolysis at 350°C with 5N alkali for 24 hr has been studied in detail.[71] The yields of the various classes of compounds present in these products are given in chart form in Fig. 5. Altogether, three-fourths of the coal was converted

[68] Shreve, R. N., *The Chemical Process Industries*, Second edition, McGraw-Hill, New York, 1956, p. 127.

[69] *Ibid.*, pp. 289–95. Faith, W. L., Keyes, D. B., and Clark, R. L., *Industrial Chemicals*, Second edition, John Wiley and Sons, New York, 1957, p. 659.

[70] Shreve, R. N., *op. cit.*, p. 651.

[71] Parker, F. G., Fugassi, J. P., and Howard, H. C., *Ind. Eng. Chem.*, **47**, 1586–92 (1955).

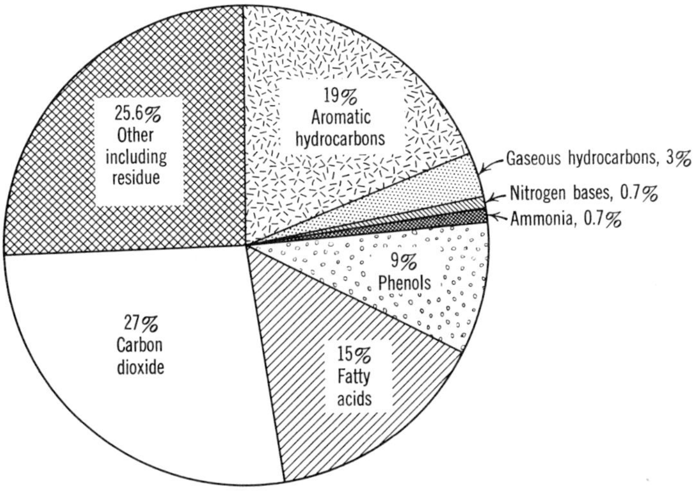

Fig. 5. Distribution of products from alkaline hydrolysis of Wyoming subbituminous coal.

into soluble or gaseous products. The phenols positively identified in the products included phenol, 3,5-dimethylphenol, *o*-ethylphenol, and *m*-ethylphenol, and the carboxylic acids included acetic, propionic, *n*-butyric, and succinic acids. The relative high yields of low-boiling phenols, 3.7%, and of fatty acids, 15.7%, indicate that hydrolysis is potentially significant as a process for the conversion of low-rank coal into chemicals.

Sulfonation

Bituminous coal may be sulfonated to produce material having a sufficient cation-exchange capacity to be of commercial value. A ton of this material requires 1.4 tons of coal refuse beside that required for heat and power.[72] Only one manufacturer is known to be making such materials from coal in the United States.[73] Considerable effort has been devoted in various countries, however, particularly in India, to a study of the preparation and properties of sulfonated products from coal.[74] An efficient drying agent has been prepared with a bulk density of 25 lb/cu ft and reported to have promise as an industrial desiccant.[75]

Sulfonation of air-dried, heated lignite has been reported to yield a cation resin suitable for hard-water treatment that compares favorably with available commercial products.[76]

Halogenation

Liquid products from the fluorination coal have been reported as having properties similar to those of commercial perfluorocarbons. These materials are produced by treating low-grade coal with gaseous chlorine trifluoride diluted with nitrogen with an efficiency of carbon utilization in oil production of 50%. The products contain from 60 to 72% fluorine, have thermal stability, and are nonflammable and hydrophobic; they have been reported to be useful as high-temperature lubricants, heat-transfer media, dielectric materials, and hydraulic fluids.[77] Production of these

[72] Broderick, J. J., and Bogard, D., *U.S. Bur. Mines, Rept. Invest.* No. 3559 (1941), 17 pp.; *Ind. Eng. Chem.*, **33,** 1291–6 (1941). Weast, R. C., and Buswell, A. M., *Chem. Met. Eng.*, **49,** No. 11, 169–70 (1942).

[73] Zimmerman, O. T., and Lavine, I., *Handbook of Material Trade Names*, Industrial Research Service, Dover, New Hampshire, 1953, p. 632.

[74] Datta, D., and Mukherji, B. K., *Sci. and Culture (Calcutta)*, **21,** 217 (1955). Roy, M., *J. Appl. Chem.*, **6,** 335–41 (1956). Kosaka, Y., *J. Fuel Soc. (Japan)*, **30,** 179–89 (1951). Stach, H., and Teichmüller, M., *Brennstoff-Chem.*, **34,** 275–81 (1953).

[75] Sarin, B. K., and Puri, R. P., *J. Sci. Ind. Research (India)*, **14B,** 587–91 (1955).

[76] Seitz, D. S., Minnear, F. L., and Dunbar, R. E., *Ind. Eng. Chem.*, **52,** 313–4 (1960).

[77] Farenden, P. J., and Pritchard, E., *Proc. Conf. on Science in Use of Coal*, Inst. Fuel, London, 1958, A52–5. Anon., *Chem. Eng.*, **66,** No. 4, 62 (1959).

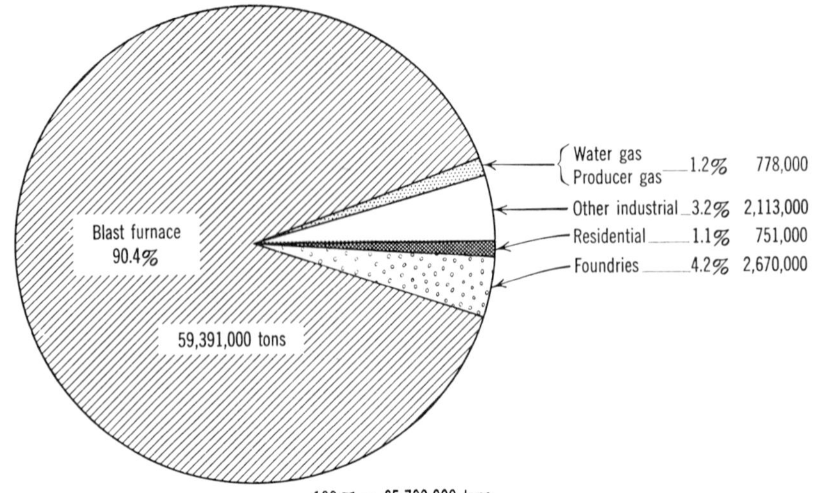

Fig. 6. Consumption of coke in the United States by principal uses—average for 5-yr period 1955–1959.[79] (Breeze and exports excluded.)

materials on a commercial scale has not been reported.

Amination

Preparation of ion-exchange materials from coal by amination has been reported to be feasible by research workers in India using two different chemical processes: in one, a cation-exchange material with a capacity of 160 to 300 milliequivalents per 100 g was produced from coal by nitration and reduction, in the other, anion-exchange materials with a capacity of 75 to 80 units were obtained from low-rank coals by treatment with ethylene dichloride and ammonia.[78] It was not reported, however, whether commercial production of such materials has been attempted.

PROCESSES REQUIRING COKE OR COAL TO SUPPLY CARBON FOR REDUCTION

The largest chemical use of coke or coal by far is to supply carbon to the metallurgical industries for use in the reduction of ores. For example, the reduction of iron oxides in the blast furnace alone consumed some 90.4% of all the coke consumed during the 5-yr period 1955–1959; consequently, the demand for coke is largely dependent on the demand for steel (see Fig. 6).[79]

[78] Central Fuel Research Inst., *Report of Director for the Year* 1957–58, Jealgora, India, April, 1958, p. 83.

Reduction of Oxides with Production of Free Metals

The production of free metals from ores by reduction with carbon at high temperatures is based on the conversion of the ore, if necessary, to the oxide form, usually by roasting and then heating the oxide with carbon in the form of coke or coal in a suitable furnace. The heat required may be supplied by the combustion of an added amount of coal or coke, or it may be supplied separately from electric power or by the combustion of either solid or gaseous fuels.

Pig Iron

The reduction of iron ore in the blast furnace to produce pig iron is by far the greatest user of coke for any purpose. In 1959, there were over 47 million tons of coke consumed by blast furnaces in the manufacture of pig iron alone; for each ton of pig iron there was used 0.79 ton of coke. This ratio of coke used to pig iron produced has been decreasing steadily.[80] The consumption of coke for pig iron manufacture has been estimated to amount to 90 million tons/year by 1975.[81]

[79] U.S. Bur. Mines, *1953–1958 Minerals Yearbooks. U.S. Bur. Mines, Monthly Coke Rept.* No. MCR 383, February 11, 1960, plus Supplement.

[80] American Iron and Steel Inst. *Annual Statistical Report for* 1959, New York, 1960, pp. 44–5.

[81] President's Materials Policy Commission, *Resources for Freedom*, 1952, Vol. II, p. 118.

Sponge Iron

The direct reduction of iron ore at temperatures below the melting point of iron is economical for at least one use. One company has operated a 45 ton/day Bruckner furnace to produce sponge iron suitable for use in the recovery of copper from solution. For each ton of sponge iron produced, 890 lb of slack coal were required.[82]

A general description of the methods that have been proposed or tried for the production of sponge iron and direct iron was reported by Barrett in 1954[83] and a brief account of thirteen important processes was reported by Raleigh in 1959.[84] Commercial applications of these processes in the United States may occur in the near future.

Ferroalloys

At present, ferroalloys are produced in blast furnaces and in electric furnaces. The blast-furnace production of ferroalloys in the United States for the 5-yr period 1955–1959 averaged 4,028,600 tons annually and required 1.63 tons of coke per ton of alloy.[85] The production of ferromanganese requires 2.25 tons of coke per ton of ferromanganese.[86] In 1959, the blast-furnace production of 635,337 tons of ferroalloys required 1,018,702 tons of coke. The amount of coke for 1,349,048 tons of ferroalloys made in electric furnaces has not been reported.[87]

Conversion of Oxides of Metals to Chlorides

Coal or coke is used in the production of certain chlorides of metals to supply carbon for the reduction of the oxide to the free metal as well as to supply heat for the process.

Aluminum Chloride

One ton of aluminum chloride produced from bauxite requires 275 lb of coke and 80 lb of asphalt binder in addition to the fuel needed for the calcination of the ore and for the baking of briquettes of bauxite and coke.[88] In 1958, ten or more plants produced 49,635 tons of aluminum chloride of all grades,[89] which is equivalent to the use of 6775 tons of coke.

Titanium Tetrachloride

For the production of one ton of titanium tetrachloride in a shaft furnace, 0.86 ton of coal and 0.46 ton of titanium dioxide are required.[90] In 1953, an estimated 11,000 tons or more titanium tetrachloride were produced using an estimated 9500 tons or more of coal. The production of titanium tetrachloride is expected to increase rapidly according to the increasing demand for titanium metal.

Magnesium Chloride

One ton of magnesium chloride produced from magnesite requires about 0.5 ton of coal as reducing agent and 0.4 ton of coal as fuel for calcination in addition to electric power.[91] This method of producing magnesium chloride has not been economical in peacetime.

Reduction of Nonmetals

Coal and coke are also used as reducing agents for nonmetals existing as salts of metals. The reduction of phosphates and sulfates are examples.

Phosphorus

In the electric furnace process for making elemental phosphorus by the reduction of calcium phosphate, 1.33 to 1.50 tons of coke or carbon are required as the reducing agent for each ton of phosphorus produced in addition to 12,000–13,200 kwhr and 34 to 50 lb of

[82] Frick, F., *Mining Eng.*, **5**, 83–4 (1953).

[83] Barrett, E. P., *U.S. Bur. Mines, Bull.* No. 519 (1954), 143 pp.

[84] Raleigh, W. A., Jr., *Coal Age*, **64**, No. 3, 66–71 (1959).

[85] American Iron and Steel Inst., *Annual Statistical Reports*, 1955–1959.

[86] National Security Resources Bd., *Materials Survey—Manganese*, Oct. 1952, p. 10, Chap. VI.

[87] American Iron and Steel Inst., *Annual Statistical Report for* 1959, New York, 1960, pp. 42, 45.

[88] Shreve, R. N., *The Chemical Process Industries*, Second edition, McGraw-Hill, New York, 1956, p. 127.

[89] Young, W. H., Anderson, R. L., and Hall, E. M., *U.S. Bur. Mines, 1958 Minerals Yearbook*, Vol. 1, 214 (1959).

[90] McBerty, F. H., *U.S. Office Publication Board*, PB-22,626 (1946), 32 pp. (*FIAT Final Rept.*. No. 774).

[91] Riegel, E. R., *Industrial Chemistry*, Fifth edition, Reinhold, New York, 1949, pp. 352–5. Shreve, R. N., *The Chemical Process Industries*, First edition, McGraw-Hill, New York, 1945, pp. 309–10.

electrode carbon.[92] One ton of 85% phosphoric acid by the electric furnace process requires only 0.38 to 0.44 ton of coke breeze in addition to 3460 to 4070 kwhr electric power.[93] In the blast furnace, 3.5 tons of coke per ton 100% phosphoric acid are required for reducing agent and fuel. In 1957, 383,000 tons of elemental phosphorus were produced in the United States in thirty-two plants,[94] and required an estimated 574,500 tons of coke and 5 billion kwhr of electrical energy.

Black Ash from Salt Cake

Sodium sulfide is produced primarily by the reduction of sodium sulfate with powdered coal in a reverberatory furnace. In addition to the coal needed for fuel, 0.5 ton of coal per ton of salt cake is used in the charge as a reducing agent.[95] For the 5-yr period 1954–1958, the average production of sodium sulfide (60 to 62% Na_2S) was 38,375 tons,[96] and required about 42,000 tons per year of coal for fuel and as a reducing agent.

Sodium Sulfite from Salt Cake

In the manufacture of glass, 10 lb of coal per ton of glass sand is added in the charge to reduce the salt cake to sodium sulfite.[97] The consumption of coal for this purpose was estimated to be 26,000 tons and that for 1975 to be 40,000 tons. In 1954, the glass industry consumed 1.07 ton bituminous-coal equivalent as fuel per ton of glass sand used, 0.14 ton being coal and coke; in addition 0.18 tons of bituminous-coal equivalent was used as purchased electricity.[98]

Barium Sulfide from Baryte

The process for the production of barium sulfide from baryte is similar to that for production of black ash from salt cake. About 0.33 ton of coal is required per ton of baryte ore for reducing agent in addition to 0.35 ton of coal for fuel.[99]

The average yearly production of black ash, including lithopone, decreased 12% during the last 5 years from that of the previous 5-year period. The average annual production for the period 1954–1958 was only 117,659 tons in comparison to 128,299 tons for the period 1949–1953.[100] Thus, the estimated amount of coal consumed has decreased during this same period from 87,000 to 80,000 tons. The production of one ton of lithopone requires 1.1 tons of coal (half for steam) in addition to 100 kwh electrical energy.[101] In 1956, the latest year for which data are available, consumption of coal for production of 38,434 tons lithopone[102] was reported to have been 42,300 tons.

Sulfuric Acid from Waste Pickle Liquor

The sulfuric acid present in the free and combined states in the waste pickle liquor of the steel industry and in the waste liquor from the titanium dioxide industry may be recovered by roasting the recovered salts with coal in a

[92] Van Wazer, J. R., *Encyclopedia of Chemical Technology*, Kirk, R. E., and Othmer, D. F., Eds., John Wiley and Sons, New York, 1953, Vol. 10, p. 447. Faith, W. L., Keyes, D. B., and Clark, R. L., *Industrial Chemicals*, Second edition, John Wiley and Sons, New York, 1957, pp. 602–6. Waggaman, W. H., *Phosphoric Acid, Phosphates, and Phosphatic Fertilizers*, Reinhold, New York, 1952, p. 151. Waggaman, W. H., and Ruhlman, E. R., *U.S. Bur. Mines, Inform. Circ.* No. 7951 (1960), 36 pp. Barber, J. C., Marks, E. C., and Megar, G. H., *Tenn. Valley Authority, Chem. Eng. Bull.* No. 4 (1960), 53 pp.

[93] Shreve, R. N., *op. cit*, pp. 352–3. Faith, W. L., Keyes, D. B., and Clark, R. L., *op. cit.*, pp. 592–4.

[94] Waggaman, W. H., and Ruhlman, E. R. *loc. cit.*

[95] Kobe, K. A., *Inorganic Process Industries*, MacMillan, New York, 1948, p. 88–9. Riegel, E. R., *op. cit.*, pp. 80–3.

[96] *Chemical Statistics Handbook*, Fifth edition, Mfg. Chemists' Assoc., Inc., Washington, D.C., 1960, p. 11.

[97] Scholes, S. R., *Modern Glass Practice*, rev. ed., Industrial Publications, Chicago, Illinois, 1952, pp. 65, 96. Shreve, R. N., *The Chemical Process Industries*, Second edition, McGraw-Hill, New York, 1956, p. 237.

[98] U.S. Bureau of Census, *Census of Manufactures:* 1954, Vol. II, Industry Statistics, Part 2, p. MC-32A (1957). U.S. Bur. Mines, *1954 Minerals Yearbook* Vol. I, 980 (1958).

[99] Shreve, R. N. *op. cit.*, pp. 507–8. Preisman, L., and Davis, L. W., *Encyclopedia of Chemical Technology*, Kirk, R. E., and Othmer, D. F., Eds., John Wiley and Sons, New York, 1948, Vol. 2, pp. 320–3.

[100] Young, W. H., Anderson, R. L., and Hall, E. M., *U.S. Bur. Mines, 1958 Minerals Yearbook*, Vol. 2, 126 (1959).

[101] Shreve, R. N., *op. cit.*, p. 507.

[102] Lansche, A. M., and Miller, E. B., *U.S. Bur. Mines, Mineral Market Rep.*, MMS No. 2648, 1957, 8 pp.

rotary kiln[103] or on a Dwight-Lloyd sintering machine and recovering the resultant sulfur dioxide for subsequent conversion to sulfuric acid. The reduction of one ton of dry ferrous sulfate on the sintering machine requires 0.1 ton of coal in addition to ignition fuel. A 50-ton/day semicommercial plant was reported to be operating by the sinter process in 1950.[104]

Conversion of Oxides to Carbides

The oxides of both metals and nonmetals are converted into the corresponding carbides by reaction with carbon at high temperatures. The carbides of nonmetals, for example, silicon carbide and boron carbide, are valuable as abrasives because of their hardness. The carbide of calcium is valuable as a source of acetylene and as an intermediate in the manufacture of calcium cyanamide.

Calcium Carbide

One ton of calcium carbide requires from 0.65 to 0.68 ton of coke as a source of carbon in addition to 33 to 40 lb of electrode paste and 2900 to 3000 kwhr electric power.[105] The 1,018,184 tons of calcium carbide produced in 1959[106] required about 660,000 tons of coke and 3,020,000,000 kwhr electric power besides that required for manufacturing the necessary quicklime. The consumption of coke for carbide manufacture in 1975 has been estimated to be 1,000,000 tons.[107]

Calcium Cyanamide

Calcium carbide reacts with nitrogen gas to form calcium cyanamide. The manufacture of 1 ton of fixed nitrogen as calcium cyanamide from limestone requires from 1.75 to 1.86 tons of coke in addition to 1.0 ton of coal and 9000 kwhr electric energy for heat and power.[108] Thus, 1 ton of calcium cyanamide (35% N) requires 0.65 ton of coke, 0.35 ton of coal, and 3150 kwhr electrical energy. The annual Canadian production of calcium cyanamide is approximately 200,000 tons/yr and consumes an estimated 500,000 to 600,000 tons of coal and coke plus 1,800,000,000 kwhr electric power.

Silicon Carbide

The manufacture of 1 ton of silicon carbide from silica requires 1.4 tons of coke and 8000 to 10,000 kwhr of electric power.[109] In 1958, the United States and Canada produced 110,456 tons of silicon carbide[110] which required about 157,500 tons of coke and 980,000,000 kwhr electric power.

Conversion of Ores to Oxides

The production of certain metallic oxides from ores requires coal or coke to supply carbon for reduction and for fuel.

Zinc Oxide

By the American (Wetherill) process, franklinite, an oxidic ore, is charged together with coal to a special furnace; the zinc oxide is reduced, distilled, and reoxidized outside the furnace. One ton of zinc oxide requires 2 to 5 tons of anthracite, low-volatile bituminous coal, or coke as reducing agent and as fuel in addition to 360 kwhr electric power.[111] By the French process, zinc spelter is vaporized in a retort with indirect heat and in the presence of carbon monoxide; the vapors are burned in a special chamber to produce zinc oxide and carbon dioxide. Coal is used to produce the carbon monoxide and to furnish fuel for the reheat furnace. One ton of zinc oxide requires 0.7 ton of coal for fuel and carbon monoxide plus 35 kwhr electricity.

[103] Shreve, R. N., *The Chemical Process Industries*, First edition, McGraw-Hill, New York, 1945, p. 357.

[104] Bartholomew, F. J., *Chem. Eng.*, **57,** No. 8, 118–20 (1950).

[105] See Shreve, R. N., *op. cit.*, p. 336, Abbott, A. J., *Encyclopedia of Chemical Technology*, Kirk, R. E., and Othmer, D. F., Eds., John Wiley, and Sons New York, 1948, Vol. 2, pp. 838–46. Faith, W. L., Keyes, D. B., and Clark, R. L., *Industrial Chemicals*, Second edition, John Wiley and Sons, New York, 1957, pp. 34–5.

[106] U.S. Dept. Commerce, *Chemical and Rubber Ind. Rept.*, Vol. 7, No. 3, p. 10 (1960).

[107] President's Materials Policy Commission, *Resources for Freedom*, 1952, Vol. IV, pp. 162, 191.

[108] Shreve, R. N., *op. cit.*, p. 398. Kastens, M. L., and McBurney, W. G., *Ind. Eng. Chem.*, **43,** 1020–33 (1951).

[109] Mantell, C. L., *Industrial Electrochemistry*, Third edition, McGraw-Hill, New York, 1950, pp. 634–8. Shreve, R. N., *op. cit.*, p. 332.

[110] Young, W. H., Anderson, R. L., and Hall, E. M., *U.S. Bur. Mines, 1958 Minerals Yearbook*, Vol. 1, 141 (1959).

[111] Pierce, W. M., *Mining and Met.*, **29,** 286–8 (1948). Shreve, R. N., *op. cit.*, p. 505.

Fused Alumina

Bauxite is used for the manufacture of aluminous abrasives. One ton of fused alumina requires 60 lb of petroleum coke or 180 lb of coke or anthracite, and 35 to 60 lb of electrode carbon in addition to 2000 to 2600 kwhr electric power.[112]

Titanium Dioxide

Ilmenite and titaniferous ores may be selectively reduced in an electric furnace to produce a high-grade pig iron and a titanium dioxide-rich slag, the latter being valuable as an intermediate for titania manufacture. One ton of titanium dioxide concentrate (65%) produced by an electric furnace requires 0.25 to 0.30 ton of coke, 2.4 to 2.6 tons of ilmenite ore, and 7100 to 8300 kwhr electric power[113] One plant has been operating commercially in Canada since about 1952 processing a titaniferous ore in an electric furnace with low-ash coal. One ton of titanium slag (70%) requires 0.33 ton of coal, 2.28 tons of ore, and an unspecified amount of electricity.[114]

PROCESSES USING COAL PRIMARILY FOR ITS PHYSICAL PROPERTIES

In addition to the uses of coal and coke based on the chemical reaction between carbon from coal or coke with some element, there are a number of uses based on the physical properties of either processed or unprocessed coal.

Activated Carbon

Bituminous coal is being used as the raw material for the manufacture of granular activated carbon. Texas lignite is being used in the manufacture of a powdered activated carbon.[115] An estimated 4 to 5 tons of bituminous coal are required for heat and power and as the raw material for the manufacture of 1 ton of granular activated carbon. The unit processes involved include a low-temperature carbonization followed by a high-temperature steam activation. The other carbonaceous materials which may be used include coal refuse[116] and anthracite.[117] In 1954, there were 46,200 tons of activated carbon produced from all types of raw materials.[118]

Electric Furnace Graphite

Artificial electric furnace graphite was produced by five companies operating nine plants in 1958 by the graphitization of carbonaceous materials at high temperatures.[119] A ton of synthetic graphite requires 1.15 tons of anthracite, petroleum coke, or retort coke plus 6000 to 10,000 kwhr electric power.[120]

Graphite Electrodes

The production of 1 ton of high-calcined graphite electrodes requires 0.4 ton of coal-tar pitch and 1.25 ton of petroleum coke in addition to 0.3–1.0 ton of coal for producer gas for baking the green electrodes and 11,600 to 17,000 kwhr electric power for calcination. The 1953 production of graphite electrodes has been reported as 170,000 tons and that for 1963 has been estimated as 230,000 tons.[121]

Amorphous Electrode Carbon

Certain electric furnace processes employ an electrode carbon calcined at a low temperature. The electrodes are prepared on the site and consequently no data are available on the total produced. The electrolytic production of aluminum is a good example of such a process.

[112] Edwards, J. D., and Abbott, A. J., *Encyclopedia of Chemical Technology*, Kirk, R. E., and Othmer, D. F., Eds., John Wiley and Sons, New York, 1947, Vol. 1, pp. 646–53. U.S. Bur. Mines, *Materials Survey on Bauxite* (August, 1953), p. 19, Chap. 6.

[113] Stoddard, C. K., Kerby, C., Cole, S. S., Eck, L. T., and Davis, C. W., *U.S. Bur. Mines, Rept. Invest.* No. 4750 (1950), 15 pp.

[114] Knoerr, A. W., *Eng. Mining J.*, **153,** No. 3, 72–9 (1952).

[115] Bur. Mines Staff, *U.S. Bur. Mines., Inform. Circ.*, No. 7692, Part II, 110–11 (1954).

[116] Broderick, S. J., and Hertzog, E. S., *U.S. Bur. Mines, Rept. Invest.* No. 3548 (1941), 27 pp.

[117] Anon., *Heating and Ventilating*, **49,** No. 7, 98 (1952). Hassler, J. W., *Pennsylvania State Univ., Mineral Inds. Exp. Stat., Bull.* No. 70 (1957), 1–5.

[118] McCann, P. W., *Chem. Eng. News*, **34,** 5671 (1956).

[119] Young, W. H., Anderson, R. L., and Hall, E. M., *U.S. Bur. Mines, 1958 Minerals Yearbook*, Vol. 1, 497 (1959).

[120] Abbott, H. W., *Encyclopedia of Chemical Technology*, Kirk, R. E., and Othmer, D. F., Eds. John Wiley and Sons, New York, 1949, Vol. 3, pp. 1–34. Mantell, C. L., *op. cit.*, pp. 644–8.

[121] Hader, R. N., Gamson, B. W., and Bailey, B. L., *Ind. Eng. Chem.*, **46,** 2–11 (1954).

Foundry-Mold-Facing Materials

Little statistical information is available on the manufacture or foundry-mold-facing materials from bituminous coal, but this is an established use. For example, one plant has a 140 ton/day capacity.[122] About 1.0 ton of coal is required per ton of product besides that needed for heat and power.

Protective Coatings

Coal-tar pitch is used in the manufacture of quality enamels and protective contings which, when properly applied over a primed surface, meet many of the requirements of an ideal hot organic coating. The addition of finely divided fillers and fluxes improves the properties of coal-tar pitch considerably. Some of the more widely used fillers are slate flour, bituminous coal dust, asbestos, limestone, silica, and carbon. Plasticizers and other organic liquids may also be added.[123] In some preparations, bituminous coal of selected quality is blended with creosote and light oil in amounts up to 25%, or with other appropriate coal-tar products in various percentages to give products of desired physical properties.[124] The production of pipeline enamel and other protective coatings is known to be large, but the consumption of coal for this purpose has not been disclosed.

Paving Binder

An improved paving binder has been reported using high- or low-temperature coal tar with digested coal for hot-mix, hot-lay bituminous surfaces. Field tests comprising twelve $\frac{1}{2}$-mile test sections of highway were made in Kentucky.[125]

USES FOR RESIDUAL PRODUCTS FROM COMBUSTION OF COAL

The utilization of the residual products from the combustion of coal is becoming of greater and greater importance, particularly in areas having air-pollution and waste-disposal regulations.

Fly-Ash and Flue-Dust

In power plants burning pulverized coal, about 90% of the ash present in the coal is discharged to the atmosphere as fly ash unless steps are taken to collect it. The amount of fly ash potentially available in the fourteen Appalachian states has been estimated at 5,000,000 tons/yr. Eleven utility companies now sell fly-ash. Two companies sell approximately 100,000 tons a year each.[126] The largest use of fly ash is in the building industry. About 20 to 30% of the portland cement in concrete may be replaced by fly ash for many purposes at a considerable saving in costs and improvement in quality of concrete. This is particularly true of concrete dams.[127] Fly-ash is also used in asphalt highway construction, in the manufacture of lightweight aggregates for concrete, and in the manufacture of building stones and of cinder blocks.[128]

There have been extensive investigations of the germanium content and distribution in United States coals and in power-plant and coke-oven products. The potential supply is large, but the concentrations are in general too low to be economically attractive.[129] Flue dust is the commercial source of germanium in England. About 9 lb of germanium are obtained from 1 ton of flue dust.[130]

Coal Ash and Cinders

The manufacture of cinder blocks for the

[122] Anon., *Mining Eng*., **6,** 777 (1954). Axelson, F., *Coal Utilization*, **11,** No. 5, 35–7 (1957).

[123] Seymour, R. B., *Organic Coatings*, Reinhold, New York, 1959, p. 38.

[124] Glenn, R. A., and Rose, H. J., *The Metallurgical, Chemical, and Other Processes of Coal*, Bituminous Coal Research, Inc,. Pittsburgh, Pennsylvania, 1958, p. 54.

[125] Anon., *Black Diamond*, **144,** No. 3, 22 (1960).

[126] Anon., *Bituminous Coal Research*, **14,** No. 3, 3–6, 16 (1955).

[127] Anon., *Bituminous Coal Research*, **13,** No. 4, 4–5 (1954).

[128] Anon., *Bituminous Coal Research*, **13,** No. 2, 4–5, 12 (1953). Littlejohn, C. E., *Clemson A & M Coll. Eng. Expt. Stat., Bull.* No. 6 (1954), 42 pp.

[129] Stadnichenko, T., Murata, K. J., Zubovic, P., and Hufschmidt, E. J., *U.S. Geol. Survey, Circ.* No. 272 (1953), 34 pp. Headlee, A. J. W., and Hunter, R. G., *Ind. Eng. Chem*., **45,** 548–51 (1953). Headlee, A. J. W., *Mining Eng*., **5,** 1011–14 (1953). Corey, R. C., Myers, J. W., Schwartz, C. H., Gibson F. H., and Colbassani, P. J., *U.S. Bur. Mines, Bull.* No. 575 (1959), 68 pp.

[130] Johnstone, S. J., *Minerals for the Chemical Industry*, John Wiley and Sons, New York, 1954, pp. 191–6.

building industry consumes large quantities of coal cinders yearly.

Coal ash is a potential source for several of the uncommon elements such as vanadium, lithium, germanium and even uranium.[131] Lignitic uranium ores of commercial importance have been discovered in several areas of the United States.[132] Coal ash has also been used during wartime as a source of alumina; 1 ton of alumina is obtained from 10 to 12 tons of ash.[133]

Carbon Dioxide

The combustion of coal produces carbon dioxide in large quantities, since coal itself is about 85% carbon. The "dry ice" and liquid carbon dioxide industry in certain areas of the country are based on the recovery of the carbon dioxide from stack gases from the combustion of coal. Similarly, the gases from the combustion of coal to supply heat for the calcination of limestone and dolomite are processed in many plants for the carbon dioxide they contain.

Sulfur Dioxide

The major portion of the sulfur in coal is converted to sulfur dioxide during combustion and is discharged into the atmosphere with the other stack gases. The removal and/or recovery of the sulfur dioxide from stack gases is becoming of considerable economic importance as a result of air-pollution control. Several processes have been proposed and at least one is under investigation for the recovery of the sulfur as ammonium sulfate for sale as a fertilizer in England.[134] Recovery or removal of sulfur dioxide from power-plant stack gases on a commercial scale has not been attempted in this country. Each percent of sulfur in a ton of coal represents 20 lb of sulfur or 40 lb of sulfur dioxide. The total sulfur in fuels consumed annually in the United States exceeds the nation's total sulfur requirements considerably.

[131] Davidson, C. F., and Ponsford, D. R. A., *Mining Mag. (London)*, **91,** 265–73 (1954). Simons, H., *Science News Letter*, **67,** No. 6, 90–1 (1955).

[132] Anon., *Mining Cong. J.*, **41,** No. 2, 159 (1955). Waylett, W. J., *Ibid.*, 134–9. Vine, J. D., *Proc. Intern. Conf. Peaceful Uses Atomic Energy, Geneva*, **1956,** Vol. 6, pp. 405–11.

[133] Lea, F. M., *Chem. Met. Eng.*, **53,** No. 6, 210 (1946). Anon., *Chem. Age (London)*, **54,** 573–6 (1946).

[134] Rees. R. L., *Problems and Control of Air Pollution*, Mallette, F. S., Ed., Reinhold, New York, 1955, pp. 149–51.

USES OF MATERIALS ASSOCIATED WITH COAL

The materials associated with coal in the ground vary considerably in composition from one region to another. In certain areas, these materials are of economic importance and may be used in considerable quantity for various purposes. The annual consumption of these materials is seldom reported.

Shales

Lightweight aggregate for concrete is made from certain shales. In 1955, a West Virginia plant was reported in operation producing 120 tons/day of aggregate, and additional plants were reported operating in Ohio and Tennessee.[135] In 1956, plants reporting the use of lightweight aggregate in building block manufacture in the United States obtained 15.9% of such aggregate from expanded shale (partly from coal mining) or clay.[136]

The construction of a 40,000 ton/year plant in the United States has been announced for the extraction of aluminum sulfate from coal-shale wastes and other materials such as high-iron clays and bauxites.[137]

Fireclay

Building bricks are made from the fireclay and burned-out mine rejects found in certain areas. In England in 1955, the seventy-six brickworks of the National Coal Board produced 528 million bricks.[138] In the United States, selected shale from Pennsylvania anthracite reject piles has been used for the commercial production of buff face-brick.[139]

[135] Anon., *Coal Age*, **60,** No. 11, 78–81 (1955). Carman, E. P., and Caldwell, D. L., *U.S. Bur. Mines, Inform. Circ.* No. 7756, 56 (1956).

[136] Peck, R. L., *Pit and Quarry*, **49,** 336–40 (1957)

[137] Anon., *Chem. Eng.*, **67,** No. 13, 58 (1960).

[138] National Coal Bd., *Ann. Rept. and Statement of Accounts for* 1955, London, 1956.

[139] Campbell, H. L., Jr., *Trans. Ann. Anthracite Conf. Lehigh Univ.*, **4,** 181–6 (1941).

The burned-out shales from gob piles which are commonly called "red dog" are used as a roadbuilding material in many areas. Sandstone, limestone, and similar stones encountered in stripping operations are saleable in certain localities.

Pyrite

Iron sulfide or pyrite occurs in considerable quantities in certain coal seams, and has been recovered commercially at various times. In the United States, the commercial production of pyrite from coal is small and variable, being dependent on the market demand and prices offered. During World War II, a plant was operated in Kansas;[140] however, the most consistent production has been at an Indiana mine.[141]

In England during the period 1940–1945, a total of 115,000 tons of pyrite of 42.3% sulfur content were recovered at British coal mines. In 1954, the recovery rate was 11,000 tons/yr.[142]

In 1943, in the Ruhr area of Germany, ten jigplants recovered 12,000 tons of pyrite.[143]

[140] Spencer, K. A., *Mining Cong. J.*, **26,** No. 5, 28 (1940).

[141] Josephson, G. W., and Barsigian, F. M., *U.S. Bur. Mines, 1950 Minerals Yearbook*, 1187–91 (1953).

[142] Parker, A., *Removal and Recovery of Sulphur from Fuels*, Inst. Fuel, London, 1954, Vol. 1, pp. 5–10.

[143] Lowry, H. H., and Rose, H. J., *U.S. Bur. Mines, Inform. Circ.* No. 7422, (1947), 27 pp.

Methane (Firedamp)

Methane occurs in sufficient concentration in certain mines for it to be recovered commercially for use as a fuel. Although it is recovered from numerous mines in Europe, it is not being recovered in the United States. The total amount of methane recovered in firedamp in Western Europe in 1954 was 14 billion cu ft, equivalent to more than 500,000 tons coal. The recovery of methane in 1960 has been predicted as attaining a level of 32 billion cu ft per year.[144] In England in 1956, one or two individual pits were reported producing methane commercially.[145]

CONCLUSION

About one-fourth of the bituminous coal consumed annually in the United States is used for nonfuel purposes. As a result of the large demand for metallurgical coke, the carbonization industry is by far the largest user of bituminous coal as a raw material. The use of bituminous coal and coke as a raw material by other industries amounts to only about one-tenth that used by the carbonization industry. As newer and improved processes are developed, the direct use of coal as an industrial raw material will increase in importance.

[144] The European Productivity Agency of the Organization for European Economic Cooperation, *Drainage and Use of Methane from Coal-Fields (Fire-Damp)*, Pub. TAR163(55), **1,** Paris, 1956, p. 12.

[145] Anon., *Coke and Gas*, **18,** 228–30 (1956).

AUTHOR INDEX

SUBJECT INDEX

* United States.

* United States.

* United States.